Matthews' Plant Virology

'The farther backwards you can look, the farther forward you are likely to see.'

Winston Churchill

Matthews' Plant Virology

Fourth Edition

Roger Hull

Emeritus Research Fellow
John Innes Center
Norwich Research Park
Colney, Norwich

ELSEVIER
ACADEMIC
PRESS

AMSTERDAM • BOSTON • HEIDELBERG • LONDON • NEW YORK • OXFORD
PARIS • SAN DIEGO • SAN FRANCISCO • SINGAPORE • SYDNEY • TOKYO

This book is printed on acid-free paper

Elsevier Academic Press
525 B Street, Suite 1900, San Diego, California 92101-4495, USA
http://www.elsevier.com

Elsevier Academic Press
84 Theobald's Road, London WC1X 8RR, UK
http://www.elsevier.com

Library of Congress Catalog Number: 2001089791

British Library Cataloguing in Publication Data
A catalogue record for this book is available from the British Library

ISBN 0-12-361160-1

Typeset by J & L Composition Filey, North Yorkshire
Printed and bound by Krips, The Netherlands

02 03 04 05 06 07 08 9 8 7 6 5 4 3 2

About the Author

Roger Hull graduated in Botany from Cambridge University in 1960, and subsequently studied plant virus epidemiology at London University's Wye College, gaining a PhD in 1964. He lectured on agricultural botany there between 1960 and 1965.

He was seconded to Makerere University in Kampala, Uganda in 1964 where he taught, and learnt, tropical agricultural botany and studied the epidemiology of groundnut rosette disease. By watching aphids land on groundnut plants he gained an understanding of the edge effect of spread of virus into the field. In 1965 Roger Hull joined Roy Markham at the ARC Virus Research Unit in Cambridge where he worked on biophysical and biochemical characterization of a range of viruses, especially *alfalfa* mosaic virus. This work continued when he moved to the John Innes Institute, Norwich with Roy Markham in 1968. There Dr Hull became a project leader and

deputy head of the Virus Research Department. In 1974 he spent a sabbatical year with Bob Shepherd in the University of California, Davis where he worked on the characterization of cauliflower mosaic virus. There he was introduced to the early stages of molecular biology which changed the direction of his research. On returning to the John Innes Institute he applied a molecular biological approach to the study of cauliflower mosaic virus elucidating that it replicated by reverse transcription, the first plant virus being shown to do so. Involvement with the Rockefeller Rice Biotechnology Program reawakened his interest in tropical agricultural problems and he led a large group studying the viruses of the rice tungro disease complex. He also promoted the use of transgenic technology to the control of virus diseases and was in the forefront in discussing biosafety issues associated with this approach. Moving from rice to bananas (plantains) his group was among those who discovered that the genome of banana streak badnavirus was integrated into the host genome and in certain cultivars was activated to give episomal infection – another first for plant viruses. He retired at the statutory age in 1997.

Dr Hull is an Honorary Professor at East Anglia University and Peking University, a Doctoris Honoris Causa at the University of Perpignan, France, and a Fellow of the American Phytopathological Society. Today Dr Hull is an Emeritus Research Fellow at the John Innes Centre and still continues research on banana streak virus. He is involved in promoting the uptake of transgenic technology by developing countries as one approach to alleviating food insecurity. His other interests are gardening, bird watching, travelling and his children and grandchildren.

Contents

Plate section appears between pages 74 and 75.

Preface

Plant Virology is synonymous with the name of R.E.F. Matthews, who wrote the first three editions of this standard text. It was a great loss, not only to the plant virology community but also to the scientific community as a whole, that Dick Matthews died in 1995. Obituaries to Dick Matthews published at that time by Bellamy (*Virology* 1995; **209**, 287; *Virology* 1995; **211**, 598; *Arch. Virol.* 1995; **140**, 1885) and Harrison (*Biographical Memoires of Fellows of the Royal Society* 1999; **45**, 297–313) describe his contribution to plant virology. This edition is dedicated to his memory.

New editions of *Plant Virology* have been published at 10-year intervals, and the fourth edition follows this timing. As was noted in the prefaces for previous editions, each has chronicled ever-increasing major advances in the subject. The last decade has been no exception—if anything, the rate of progress has increased almost exponentially. This is illustrated in the graph of annual numbers of publications shown in Fig. 1.3. The advances have been due to several technologies, including the ability to clone and manipulate plant viral genomes, be they RNA or DNA, the ability to express viral (and other) sequences integrated (transformed) into the plant genome, and non-destructive techniques for observing the behavior of the virus within the plant cell.

Over the last 10 years, the classification of plant viruses has been rationalized with the general acceptance of taxa such as genera and species. With the increase of taxonomic information, this has led to, and is continuing to raise, difficulties of definition, which the International Committee on the Taxonomy of Viruses will have to resolve. For instance, it is becoming increasingly apparent that there is considerable nucleotide variation in isolates of certain viruses that cause similar symptoms in a specific (crop) plant. Should these be considered as one or several species and, if the latter, are there common criteria for viruses from different genera? The wealth of data used in virus taxonomy has now allowed 977 species in 70 genera to be recognized, compared with the 590 species (viruses) and 35 genera (groups) of 10 years ago. The genomes of representatives of all but one of the genera have now been fully sequenced.

From the sequence data has come a greater understanding of the genes that viruses encode and how these genes are expressed in a controlled manner within the plant. The sequence data have also given a clearer understanding into virus evolution and especially the role played by recombination.

The expression of viral sequences integrated in plant genomes by transformation techniques has broadened the understanding of viral gene function and opened up the field on using viral sequences to confer protection against target viruses. This, in turn, has revealed a previously unknown generic resistance system in plants and other organisms, and is opening up the way to capitalize on this system in areas as different as disease resistance and genomics.

These rapid developments over the past decade have necessitated a substantial rewriting and reorganization of this edition. However, I have considered it important to retain material from the third edition giving description of phenomena studied over the years as these can, and do, form the basis of understanding newly recognized mechanisms. I hope to have expressed the dynamism of the subject and I have tried in various places to point to future directions that may prove to be scientifically profitable. The chapters are now arranged to lead the reader through the subject logically, building on the information in previous chapters.

I have started with an introduction to the subject and a description of each group of viruses, the principles of the architecture of their particles and their genome organizations. This lays the ground for the molecular information given in subsequent chapters, such as the mechanisms by which viral genomes are expressed and replicated, and how the genomes interact with host genomes.

The descriptions of how viruses move from host to host is followed by a chapter that brings together the various interactions involved in the full functioning of a virus. After a chapter on virus-like agents such as viroids and satellites, virus detection, control and evolution are discussed.

Over 60% of the more than 360 illustrations are new, including several in color, as in the first edition. The reference list has been expanded from about 3000 in the previous edition to about 4500 in this. In the spirit of remembering important contributions to the subject, many of the important 'older' references are retained. In other places, references to reviews are used to limit the overall number. References are also given to some non-virological subjects that are important in understanding the interactions of viruses with their hosts and vectors.

I am greatly indebted to a large number of colleagues for their helpful discussion on various topics and for access to pre-publication material. My eternal gratitude goes to my wife, Jennifer, who has tolerated the 'piles of papers' all over the house and who has given me continuous encouragement.

Roger Hull
May 2001

Richard Ellis Ford Matthews
(1921–1995)

CHAPTER 1

Introduction

I. HISTORICAL BACKGROUND

The scientific investigation of plant diseases now known to be caused by viruses did not begin until the late nineteenth century. However, there are much earlier written and pictorial records of such diseases. The earliest known written record describing what was almost certainly a virus disease is a poem in Japanese written by the Empress Koken in 752 AD and translated by T. Inouye as follows:

> In this village
> It looks as if frosting continuously
> For, the plant I saw
> In the field of summer
> The color of the leaves were yellowing

The plant, identified as *Eupatorium lindleyanum*, has been found to be susceptible to TLCV*, which causes a yellowing disease (Osaki *et al.*, 1985).

In Western Europe in the period from about 1600 to 1660, many paintings and drawings were made of tulips that demonstrate flower symptoms of virus disease. These are recorded in the Herbals of the time (e.g. Parkinson, 1656) and some of the earliest in the still-life paintings of artists such as Ambrosius Bosschaert. During this period, blooms featuring such striped patterns were prized as special varieties leading to the phenomenon of 'tulipomania' (see Blunt, 1950; Pavord, 1999). The trade in infected tulip bulbs resulted in hyperinflation with bulbs exchanging hands for large amounts of money or goods (Table 1.1).

* Acronyms of virus names are shown in Appendix 1.

TABLE 1.1 Tulipomania: the goods exchanged for one bulb of Viceroy tulip

4 tons of wheat	4 barrels of beer
8 tons of rye	2 barrels of butter
4 fat oxen	1000 lb cheese
8 fat pigs	1 bed with accessories
12 fat sheep	1 full dress suit
2 hogsheads of wine	1 silver goblet

One of the earliest written accounts of an unwitting experimental transmission of a virus is that of Lawrence (1714). He described in detail the transmission of a virus disease of jasmine by grafting. This description was incidental to the main purpose of his experiment, which was to prove that sap must flow within plants. The following quotation from Blair (1719) describes the procedure and demonstrates, rather sadly, that even at this protoscientific stage experimenters were already indulging in arguments about priorities of discovery.

> The inoculating of a strip'd Bud *into a plain stock and the consequence that the Stripe or Variegation shall be seen in a few years after, all over the shrub above and below the graft, is a full demonstration of this Circulation of the Sap. This was first observed by Mr. Wats at Kensington, about 18 years ago: Mr. Fairchild performed it 9 years ago; Mr. Bradly says he observ'd it several years since; though Mr. Lawrence would insinuate as if he had first discovered it.* (Lawrence, 1714)

In the latter part of the nineteenth century, the idea that infectious disease was caused by microbes was well established, and filters were available that would not allow the known bacterial pathogens to pass. Mayer (1886) (Fig. 1.1) described a disease of tobacco that he called *Mosaikkrankheit*. He showed that the disease could be transmitted to healthy plants by inoculation with extracts from diseased plants. Iwanowski (1892) demonstrated that sap from tobacco plants displaying the disease described by Mayer was still infective after it had been passed through a bacteria-proof filter candle.

This work did not attract much attention until it was repeated by Beijerinck (1898) (Fig. 1.1) who described the infectious agent as 'contagium vivum fluidum' (Latin for contagious living fluid) to distinguish it from contagious corpuscular agents (Fig. 1.2). The centenary of Bejerinck's discovery, which was considered to be the birth of virology, was marked by several publications and celebratory meetings (see Zaitlin, 1998; Bos, 1999a, 2000a; Harrison and Wilson, 1999; Scholthof *et al.*, 1999a; van Kammen, 1999).

Baur (1904) showed that the infectious variegation of *Abutilon* could be transmitted by grafting, but not by mechanical inoculation. Beijerinck and Baur used the term 'virus' in describing the causative agents of these diseases, to contrast them with bacteria. The term 'virus' had been used as more or less synonymous with bacteria by earlier workers. As more diseases of this sort were discovered, the unknown causative agents came to be called 'filterable viruses'. The papers by Mayer, Iwanowski, Beijerinck and Baur have been translated into English by Johnson (1942).

Between 1900 and 1935, many plant diseases thought to be due to filterable viruses were described, but considerable confusion arose because adequate methods for distinguishing one virus from another had not yet been developed. The original criterion of a virus was an infectious entity that could pass through a filter with a pore size small enough to hold back all known cellular agents of disease. However, diseases were soon found that had virus-like symptoms not associated with any pathogen visible in the light microscope, but that could not be transmitted by mechanical inoculation. With such diseases, the criterion of filterability could not be applied. Their infectious nature was established by graft transmission and sometimes by insect vectors. Thus, it came about that certain diseases of the yellows and witches' broom type, such as aster yellows, came to be attributed to viruses on quite inadequate grounds. Many such diseases are now known to be caused by mycoplasmas (phytoplasma and spiroplasmas), and a few, such as ratoon stunting of sugarcane, by bacteria.

An important practical step forward was the recognition that some viruses could be transmitted from plant to plant by insects. Fukushi

Fig. 1.1 Left: Martinus Willem Beijerinck (1851–1931). Right: Adolf Eduard Mayer (1843–1942). Photographs courtesy of the historical collection, Agricultural University, Wageningen, Netherlands.

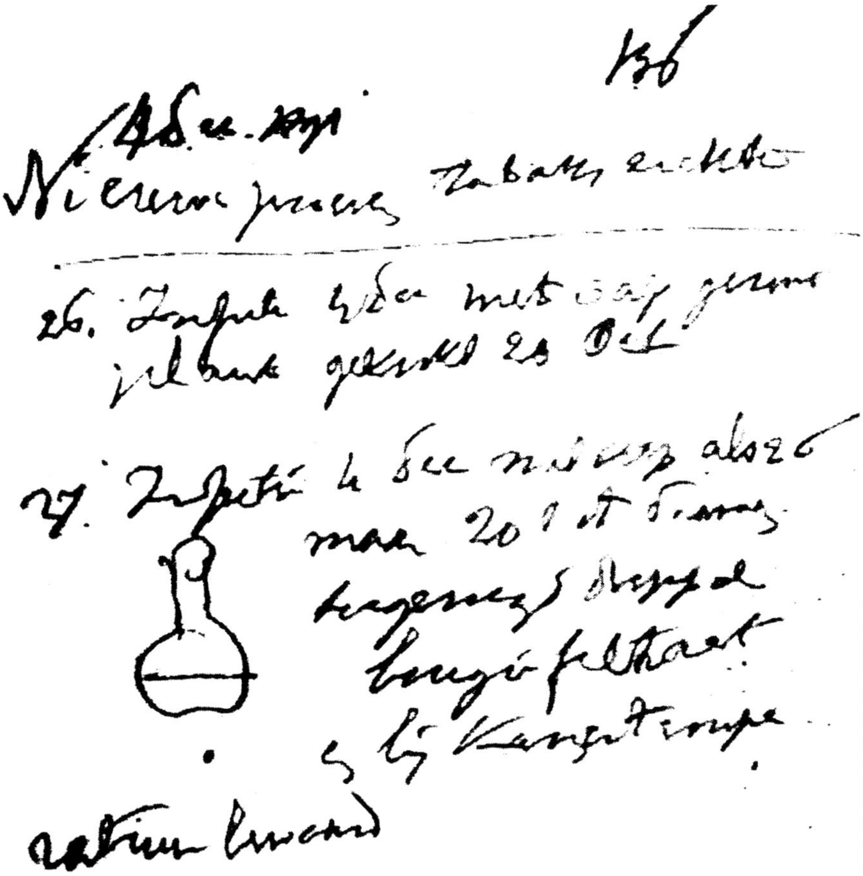

Fig. 1.2 Page from Lab Journal of W. M. Beijerinck from 1898 relating to TMV. (© Kluyver Institute). Courtesy of Lesley Robertson, Curator, Kluyver Laboratory Collection, Delft University of Technology.

(1969) records the fact that in 1883 a Japanese rice grower transmitted what is now known to be RDV by the leafhopper *Recelia dorsalis*. However, this work was not published in any available form and so had little influence. Kunkel (1922) first reported the transmission of a virus by a planthopper; within a decade, many insects were reported to be virus vectors.

During most of the period between 1900 and 1935, attention was focused on the description of diseases, both macroscopic symptoms and cytological abnormalities as revealed by light microscopy, and on the host ranges and methods of transmission of the disease agents. Rather ineffective attempts were made to refine filtration methods in order to define the size of viruses more closely. These were almost the only aspects of virus disease that could be studied with the techniques that were available. The influence of various physical and chemical agents on virus infectivity was investigated, but methods for the assay of infective material were primitive. Holmes (1929) showed that the local lesions produced in some hosts following mechanical inoculation could be used for the rapid quantitative assay of infective virus. This technique enabled properties of viruses to be studied much more readily and paved the way for the isolation and purification of viruses a few years later.

Until about 1930, there was serious confusion by most workers regarding the diseases produced by viruses and the viruses themselves. This was not surprising, since virtually nothing was known about the viruses except that they were very small. Smith (1931) made an important contribution that helped to clarify this situation. Working with virus diseases in potato, he realized the necessity of using plant indicators—plant species other than potato, which would react differently to different viruses present in potatoes. Using several different and novel biological methods to separate the viruses, he was able to show that many potato virus diseases were caused by a combination of two viruses with different properties, which he named X and Y. Virus X was not transmitted by the aphid *Myzus persicae*, whereas virus Y was. In this way, he obtained virus Y free of virus X. Both viruses could be transmitted by needle inoculation, but Smith found that certain solanaceous plants were resistant to virus Y. For example, by needle inoculation of the mixture to *Datura stramonium*, he was able to obtain virus X free of virus Y. Furthermore, Smith observed that virus X from different sources fluctuated markedly in the severity of symptoms it produced in various hosts. To quote from Smith (1931):

> *There are two factors, therefore, which have given rise to the confusion which exists at the present time with regards to potato mosaic diseases. The first is the dual nature, hitherto unsuspected, of so many of the potato virus diseases of the mosaic group, and the second is the fluctuation in virulence exhibited by one constituent, i.e., X, of these diseases.*

Another discovery that was to become important was Beale's (1928) recognition that plants infected with tobacco mosaic contained a specific antigen. Gratia (1933) showed that plants infected with different viruses contained different specific antigens. Chester (1935, 1936) demonstrated that different strains of TMV and PVX could be distinguished serologically. He also showed that serological methods could be used to obtain a rough estimate of virus concentration.

Since Fukushi (1940) first showed that RDV could be passed through the egg of a leafhopper vector for many generations, there has been great interest in the possibility that some viruses may be able to replicate in both plants and insects. It is now well established that plant viruses in the families *Rhabdoviridae* and *Reoviridae* and the *Tenuivirus*, *Tospovirus* and *Marafivirus* genera multiply in their insect vectors as well as in their plant hosts.

The high concentration at which certain viruses occur in infected plants and their relative stability turned out to be of crucial importance in the first isolation and chemical characterization of viruses, because methods for extracting and purifying proteins were not highly developed. In 1926, the first enzyme, urease, was isolated, crystallized, and identified as a protein (Sumner, 1926). The isolation of others soon followed. In the early 1930s, workers in various countries began attempting to isolate and purify plant viruses using methods similar to those that had been used for enzymes. Following detailed chemical studies suggesting that the infectious agent of TMV might be a protein, Stanley (1935) announced the isolation of this virus in an apparently crystalline state. At first Stanley (1935, 1936) considered that the virus was a globulin containing no phosphorus. Bawden *et al.* (1936) described the isolation from TMV-infected plants of a liquid crystalline nucleoprotein containing nucleic acid of the pentose type. They showed that the particles were rod-shaped, thus confirming the earlier suggestion of Takahashi and Rawlins (1932) based on the observation that solutions containing TMV showed anisotropy of flow. Best (1936) noted that a globulin-like protein having virus activity was precipitated from infected leaf extracts when they were acidified, and in 1937 he independently confirmed the nucleoprotein nature of TMV (Best, 1937b).

Electron microscopy and X-ray crystallography were the major techniques used in early work to explore virus structure, and the importance of these methods has continued to the present day. Bernal and Fankuchen (1937) applied X-ray analysis to purified preparations of TMV. They obtained accurate estimates of the width of the rods and showed that the needle-shaped bodies produced by precipitating the virus with salt were regularly arrayed in only two dimensions and, therefore, were better described as paracrystals than as true crystals. The isolation of other rod-shaped viruses, and spherical viruses that formed true crystals, soon followed. All were shown to consist of protein and pentose nucleic acid.

Early electron micrographs (Kausche *et al.*, 1939) confirmed that TMV was rod-shaped and provided approximate dimensions, but they were not particularly revealing because of the lack of contrast between the virus particles and the supporting membrane. The development of shadow-casting with heavy metals (Müller, 1942; Williams and Wycoff, 1944) greatly increased the usefulness of the method for determining the overall size and shape of virus particles. However, the coating of metal more or less obscured structural detail. With the development of high-resolution microscopes and of negative staining in the 1950s, electron microscopy became an important tool for studying virus substructure.

From a comparative study of the physico-chemical properties of the virus nucleoprotein and the empty viral protein shell found in TYMV preparations, Markham (1951) concluded that the RNA of the virus must be held inside a shell of protein, a view that has since been amply confirmed for this and other viruses by X-ray crystallography. Crick and Watson (1956) suggested that the protein coats of small viruses are made up of numerous identical subunits arrayed either as helical rods or as a spherical shell with cubic symmetry. Subsequent X-ray crystallographic and chemical work has confirmed this view. Caspar and Klug (1962) formulated a general theory that delimited the possible numbers and arrangements of the protein subunits forming the shells of the smaller isodiametric viruses. Our recent knowledge of the larger viruses with more complex symmetries and structures has come from electron microscopy using negative-staining and ultrathin-sectioning methods.

Until about 1948, most attention was focused on the protein part of the viruses. Quantitatively, the protein made up the larger part of virus preparations. Enzymes that carried out important functions in cells were known to be proteins, and knowledge of pentose nucleic acids was rudimentary. No function was known for them in cells, and they generally were thought to be small molecules. This was because it was not recognized that RNA is very susceptible to hydrolysis by acid, by alkali, and by enzymes that commonly contaminate virus preparations. Markham and Smith (1949) isolated TYMV and showed that purified preparations contained two classes of particle, one an infectious nucleoprotein with about 35% of RNA, and the other an apparently identical protein particle that contained no RNA and that was not infectious. This result clearly indicated that the RNA of the virus was important for biological activity. Analytical studies (e.g. Markham and Smith, 1951) showed that the RNAs of different viruses have characteristically different base compositions while those of related viruses are similar. About this time, it came to be realized that viral RNAs might be considerably larger than had been thought.

The experiments of Hershey and Chase (1952), which showed that when *Escherichia coli* was infected by a bacterial virus, the viral DNA entered the host cell while most of the protein remained outside, emphasized the importance of the nucleic acids in viral replication. Harris and Knight (1952) showed that 7% of the threonine could be removed enzymatically from TMV without altering the biological activity of the virus, and that inoculation with such dethreonized virus gave rise to normal virus with a full complement of threonine. A synthetic analog of the normal base guanine, 8-azaguanine, when supplied to infected plants was incorporated into the RNA of TMV and TYMV, replacing some of the guanine. The fact that virus preparations containing the analog were less infectious than normal virus (Matthews, 1953c) gave further experimental support to the idea that viral RNAs were important for infectivity. However, it was the classic experiments of Gierer and Schramm (1956), Fraenkel-Conrat and Williams (1955) and Fraenkel-Conrat (1956) that demonstrated the infectivity of naked TMV RNA and the protective role of the protein coat. These discoveries ushered in the era of modern plant virology. The remainder of this section summarizes the major developments of the past 45 years. Subsequent chapters in this book deal with many of the developments in more detail.

The first amino acid sequence of a protein (insulin) was established in 1953. Not long after this event, the full sequence of 158 amino acids in the coat protein of TMV became known (Anderer *et al.*, 1960; Tsugita *et al.*, 1960; Wittmann and Wittmann-Liebold, 1966). The sequence of many naturally occurring strains and artificially induced mutants was also determined at about the same time. This work made an important contribution to establishing the universal nature of the genetic code and to our understanding of the chemical basis of mutation.

Brakke (1951, 1953) developed density gradient centrifugation as a method for purifying viruses. This has been an influential technical development in virology and molecular biology. Together with a better understanding of the chemical factors affecting the stability of viruses in extracts, this procedure has allowed the isolation and characterization of many viruses. The use of sucrose density gradient fractionation enabled Lister (1966, 1968) to discover the bipartite nature of the TRV genome. Since that time, density gradient and polyacrylamide gel fractionation techniques have allowed many viruses with multipartite genomes to be characterized. Their discovery, in turn, opened up the possibility of carrying out genetic reassortment experiments with plant viruses (Lister, 1968; van Vloten-Doting et al., 1968).

Density gradient fractionation of purified preparations of some other viruses has revealed non-infectious nucleoprotein particles containing subgenomic RNAs. Other viruses have been found to have associated with them satellite viruses or satellite RNAs that depend on the 'helper' virus for some function required during replication. With all of these various possibilities, it is in fact rather uncommon to find a purified virus preparation that contains only one class of particle.

The 1960s can be regarded as the decade in which electron microscopy was a dominant technique in advancing our knowledge about virus structure and replication. Improvements in methods for preparing thin sections for electron microscopy allowed completed virus particles to be visualized directly within cells. The development and location of virus-induced structures within infected cells could also be studied. It became apparent that many of the different groups and families of viruses induce characteristic structures, or viroplasms, in which the replication of virus components and the assembly of virus particles take place. Improved techniques for extracting structural information from electron microscope images of negatively stained virus particles revealed some unexpected and interesting variations on the original icosahedral theme for the structure of 'spherical' viruses.

There were further developments in the 1970s. Improved techniques related to X-ray crystallographic analysis and a growing knowledge of the amino acid sequences of the coat proteins allowed the three-dimensional structure of the protein shells of several plant viruses to be determined in molecular detail.

For some decades, the study of plant virus replication had lagged far behind that of bacterial and vertebrate viruses. This was mainly because there was no plant system in which all the cells could be infected simultaneously to provide the basis for synchronous 'one-step growth' experiments. However, following the initial experiments of Cocking (1966), Takebe and colleagues developed protoplast systems for the study of plant virus replication (reviewed by Takebe, 1977). Although these systems had significant limitations, they greatly increased our understanding of the processes involved in plant virus replication. Another important technical development has been the use of in vitro protein-synthesizing systems such as that from wheatgerm, in which many plant viral RNAs act as efficient messengers. Their use allowed the mapping of plant viral genomes by biochemical means to begin.

During the 1980s, major advances were made on improved methods of diagnosis for virus diseases, centering on serological procedures and on methods based on nucleic acid hybridization. Since the work of Clark and Adams (1977), the ELISA technique has been developed with many variants for the sensitive assay and detection of plant viruses. Monoclonal antibodies against TMV were described by Dietzen and Sander (1982) and Briand et al. (1982). Since that time, there has been a very rapid growth in the use of monoclonal antibodies for many kinds of plant virus research and for diagnostic purposes.

The late 1970s and the 1980s also saw the

start of application of the powerful portfolio of molecular biological techniques to developing other approaches to virus diagnosis, to a great increase in our understanding of the organization and strategy of viral genomes, and to the development of techniques that promise novel methods for the control of some viral diseases.

The use of nucleic acid hybridization procedures for sensitive assays of large numbers of samples has been greatly facilitated by two techniques: (1) the ability to prepare double-stranded cDNA from a viral genomic RNA and to replicate this in a plasmid grown in its bacterial host, with the batches of cDNA labeled radioactively or with non-radioactive reporter molecules to provide a sensitive probe; and (2) the dot blot procedure, in which a small sample of a crude plant extract containing virus is hybridized with labeled probe as a spot on a sheet of nitrocellulose or other material. The polymerase chain reaction, also dependent on detailed knowledge of genome sequences, is being increasingly used in virus diagnosis.

Our understanding on the genome organization and functioning of viruses has come from the development of procedures whereby the complete nucleotide sequence of viruses with RNA genomes can be determined. Of special importance has been the ability to prepare *in vitro* infectious transcripts of RNA viruses derived from cloned viral cDNA (Ahlquist *et al.*, 1984b). This has allowed techniques such as site-directed mutagenesis to be applied to the study of genome function. Nucleotide sequence information has had, and continues to have, a profound effect on our understanding of many aspects of plant virology, including the following: (1) the location, number, and size of the genes in a viral genome; (2) the amino acid sequence of the known or putative gene products; (3) the molecular mechanisms whereby the gene products are transcribed; (4) the putative functions of a gene product, which can frequently be inferred from amino acid sequence similarities to products of known function coded for by other viruses; (5) the control and recognition sequences in the genome that modulate expression of viral genes and genome replication; (6) the understanding of the structure and replication of viroids and of the satellite RNAs found

associated with some viruses; (7) the molecular basis for variability and evolution in viruses, including the recognition that recombination is a widespread phenomenon among RNA viruses and that viruses can acquire host nucleotide sequences as well as genes from other viruses; and (8) the beginning of a taxonomy for viruses that is based on evolutionary relationships.

In the early 1980s, it seemed possible that some plant viruses, when suitably modified by the techniques of gene manipulation, might make useful vectors for the introduction of foreign genes into plants. Although this has been achieved for several genes in model experiments, the concept has only been demonstrated to any practical significance in a few cases. However, some plant viruses have been found to contain regulatory sequences that can be very useful in other gene vector systems.

In the early decades of this century, attempts to control virus diseases in the field were often ineffective. They were mainly limited to attempts at general crop hygiene, roguing of obviously infected plants, and searches for genetically resistant lines. Developments since this period have improved the possibilities for control of some virus diseases. The discovery of two kinds of soil-borne virus vectors (nematodes, Hewitt *et al.*, 1958; fungi, Grogan *et al.*, 1958) opened the way to possible control of a series of important diseases. Increasing success has been achieved with a range of crop plants in finding effective resistance or tolerance to viruses.

Heat treatments and meristem tip culture methods have been applied to an increasing range of vegetatively propagated plants to provide a nucleus of virus-free material that then can be multiplied under conditions that minimize reinfection. Such developments frequently have involved the introduction of certification schemes. Systemic insecticides, sometimes applied in pelleted form at the time of planting, provide significant protection against some viruses transmitted in a persistent manner by aphid vectors. Diseases transmitted in a non-persistent manner in the foregut or on the stylets of aphids have proved more difficult to control. It has become increasingly apparent that effective control of virus disease in a particular crop in a given area usually requires an integrated and continuing program involving more than one kind of control measure. However, such integrated programes are not yet in widespread use.

Cross-protection (or mild-strain protection) is a phenomenon in which infection of a plant with a mild strain of a virus prevents or delays infection with a severe strain. The phenomenon has been used with varying success for the control of certain virus diseases, but the method has various difficulties and dangers. Powell-Abel *et al.* (1986) considered that some of these problems might be overcome if plants could be given protection by expression of a single viral gene. Using recombinant DNA technology, they showed that transgenic tobacco plants expressing the TMV coat protein gene either escaped infection following inoculation or showed a substantial delay in the development of systemic disease. These transgenic plants expressed TMV coat protein mRNA as a nuclear event. Seedlings from self-fertilized transformed plants that expressed the coat protein showed delayed symptom development when inoculated with TMV. Thus, a new approach to the control of virus diseases emerged. Since these experiments, the phenomenon has been shown to be widespread and two basic types of protection have been recognized—one based on the expression of the gene product and the other being RNA-based. Both of these are leading to economically useful protection against specific viruses in several crops but are raising various non-scientific and ethical questions about the acceptability of this approach.

The late 1980s and 1990s was a period where molecular biological techniques were applied to a wide range of aspects of plant virology. As well as those areas described above, reverse genetics is being used to elucidate the functions of viral genes and control sequences. This approach, together with others such as yeast systems for identifying interacting molecules and to transform plants to express viral genes, and coupled with the ability to label viral genomes in such a manner that their sites of function within the cell are known, is revealing the complexities of the interactions between viruses and their hosts. The advances in plant

genome sequencing are identifying plant genes that confer resistance to viruses. A major advance in the late 1990s arising from the work on transformation of plants with viral sequences was the recognition that plants have a generic defense system against 'foreign' nucleic acids. Coupled with this is the identification of viral genes that suppress this defense system.

In recent years, considerable progress has been made in the development of a stable and internationally accepted system for the classification and nomenclature of viruses. Nine hundred and seventy seven plant viruses have been placed in 14 families and 70 genera. The 14 virus families and most (but not all) of the genera are very distinctive entities. They possess clusters of physical and biological properties that often make it quite easy to allocate a newly isolated virus to a particular family or genus. The rapidly expanding information on nucleotide sequences of viruses infecting plants, invertebrates, vertebrates and micro-organisms is emphasizing, even more strongly than in the past, the essential unity of virology. The time is therefore ripe for virologists to consider more grouping into higher taxa.

These advances in our understanding of plant viruses, how they function and how this knowledge can be applied to their control has resulted in a burgeoning of papers on the subject. This is illustrated in Fig. 1.3 which shows the numbers of papers that have *virus* + *mosaic* and *virus* + *mottle* in their titles, abstracts and key words. Obviously this survey does not include all the plant virus literature, as papers devoted to viruses such as those with 'streak' or 'stripe' in their names would not be included.

In spite of our greatly increased understanding of the structure, function and replication of viral genomes, there is still a major deficiency. We have little molecular understanding of how an infecting virus induces disease in the host plant. The processes almost certainly involve highly specific interactions between viral macromolecules (proteins and nucleic acids) and host macromolecular structures. At present, we appear to lack the appropriate techniques to make further progress. Perhaps an understanding of disease processes will be the exciting area of virology in the first decade of the twenty-first century. More details of the historical development of plant virology are given by Zaitlin and Palukaitis (2000); and a collection on seminal papers on TMV, which have led many of the conceptual advances, has been published by Scholthof *et al.* (1999a). Hull *et al.* (1989) provide a useful directory and dictionary of viruses and terms relating to virology.

II. DEFINITION OF A VIRUS

In defining a virus, we have to consider those of all organisms. The size of viral nucleic acids ranges from a monocistronic mRNA in the satellite virus of tobacco necrosis virus (STNV) to a genome larger than that of the smallest cells (Fig. 1.4). Before attempting to define what viruses are, we must consider briefly how they differ from cellular parasites on the one hand and transposable genetic elements on the other. The three simplest kinds of parasitic cells are the Mycoplasmas, the *Rickettsiae* and the *Chlamydiae*.

Mycoplasmas and related organisms are not visible by light microscopy. Cells are 150–300 nm in diameter with a bilayer membrane, but no cell wall. They contain ribosomes and DNA. They replicate by binary fission, and

some that infect vertebrates can be grown *in vitro*. Their growth is inhibited by certain antibiotics.

The *Rickettsiae*, for example the agent of typhus fever, are small, non-motile bacteria, usually about 300 nm in diameter. They have a cell wall, plasma membrane, and cytoplasm with ribosomes and DNA strands. They are obligate parasites and were once thought to be related to viruses, but they are definitely cells because (1) they multiply by binary fission, and (2) they contain enzymes for ATP production.

The *Chlamydiae*, for example the agent causing psittacosis, include the simplest known type of cell. They are obligate parasites and lack an energy-generating system. They have two phases in their life cycle. Outside the host cell they exist as infectious *elementary bodies* about 300 nm in diameter. These bodies have dense contents, no cell wall, and are specialized for extracellular survival. The elementary body enters the host cell by phagocytosis. Within 8 hours it is converted into a much larger non-infectious *reticulate body*. This is bounded by a bilayer membrane derived from the host. The reticulate body divides by binary fission within this membrane, giving thousands of progeny within 40–60 hours. The reticulate bodies are converted to elementary bodies, which are released when the host cell lyses.

No. of publications

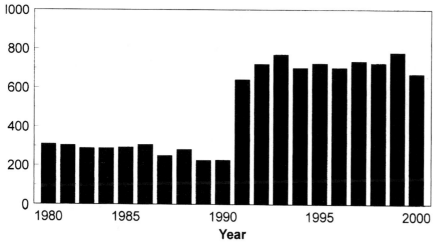

Fig. 1.3 Publications on 'mottle + virus' and 'mosaic + virus' from 1980 to 2000.

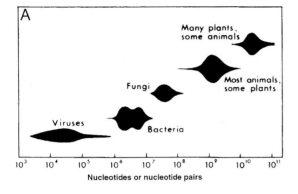

A

Many plants, some animals

Fungi

Most animals, some plants

Viruses

Bacteria

10^3 10^4 10^5 10^6 10^7 10^8 10^9 10^{10} 10^{11}

Nucleotides or nucleotide pairs

B

Escherichia coli

Vaccinia Virus

Bacteriophage T2 — DNA of T2

Tobacco Mosaic Virus

Bacteriophage M13

Adenovirus

Polyoma virus — DNA of polyoma virus

Poliovirus

Bacteriophage f2 — RNA of bacteriophage f2

Fig. 1.4 Size comparisons of different organisms.
(a) Organisms classified according to genome size. The vertical axis gives an approximate indication of relative numbers of species (or viruses) within the size range of each group. Modified from Hinegardner (1976).
(b) Size comparisons between a bacterium, several viruses and the viroid. From Diener (1999), with kind permission of the copyright holder, © Springer-Verlag.

There are several criteria that do not distinguish all viruses from all cells:

1. Some pox viruses are bigger than the elementary bodies of *Chlamydiae*.
2. The presence of DNA and RNA is not a distinguishing feature. Many viruses have double-stranded (ds) DNA like that of cells, and in some the DNA is bigger than in the *Chlamydiae*.
3. A rigid cell envelope is absent in viruses and mycoplasmas.
4. Growth outside a living host cell does not occur with viruses or with many groups of obligate cellular parasites, for example, *Chlamydiae*.

5. An energy-yielding system is absent in viruses and *Chlamydiae*.
6. Complete dependence on the host cell for amino acids, etc., is found with viruses and some bacteria.

There are three related criteria that do appear to distinguish all viruses from all cells:

1. There is no continuous membrane separating viral parasite and host during intracellular replication. Cellular parasites that replicate inside a host cell appear always to be separated from host-cell cytoplasm by a continuous bilayer membrane (see Fig. 11.24).

2. There is no protein-synthesizing system in viruses.
3. Replication of viruses is by synthesis of a pool of components, followed by assembly of many virus particles from the pool. Even the simplest cells replicate by binary fission.

Plasmids are autonomous extrachromosomal genetic elements found in many kinds of bacteria. They consist of closed circular DNA. Some can become integrated into the host chromosome and replicate with it. Some viruses infecting prokaryotes have properties like those of plasmids and, in particular, an ability to integrate into the host cell chromosome. However, viruses differ from plasmids in the following ways:

1. Normal viruses have a particle with a structure designed to protect the genetic material in the extracellular environment and to facilitate entry into a new host cell.
2. Virus genomes are highly organized for specific virus functions of no known value to the host cell, whereas plasmids consist of genetic material often useful for survival of the cell.
3. Viruses can cause death of cells or disease in the host organism but plasmids do not.

We can now define a virus as follows: *A virus is a set of one or more nucleic acid template molecules, normally encased in a protective coat or coats of protein or lipoprotein, that is able to organize its own replication only within suitable host cells. It can usually be horizontally transmitted between hosts. Within such cells, virus replication is (1) dependent on the host's protein-synthesizing machinery, (2) organized from pools of the required materials rather than by binary fission, (3) located at sites that are not separated from the host cell contents by a lipoprotein bilayer membrane, and (4) continually giving rise to variants through various kinds of change in the viral nucleic acid.*

To be identified positively as a virus, an agent must normally be shown to be transmissible and to cause disease in at least one host. However, the *Cryptovirus* group of plant viruses is an exception. Viruses in this group rarely cause detectable disease and are not transmissible by any mechanism except through the seed or pollen.

The structure and replication of viruses have the following features:

1. The nucleic acid may be DNA or RNA and single- or double-stranded. If the nucleic acid is single-stranded it may be of positive or negative sense. (Positive sense has the sequence that would be used in an mRNA for translation to give a viral-coded protein.)
2. The mature virus particle may contain polynucleotides other than the genomic nucleic acid.
3. Where the genetic material consists of more than one nucleic acid molecule, each may be housed in a separate particle or all may be located in one particle.
4. The genomes of viruses vary widely in size, encoding between 1 and about 250 proteins. Plant viral genomes are at the small end of this range, encoding between 1 and 12 proteins. The viral-coded proteins may have functions in virus replication, in virus movement from cell to cell, in virus structure, in transmission by invertebrates or fungi, and in suppression of host defense systems.
5. Viruses undergo genetic change. Point mutations occur with high frequency as a result of nucleotide changes brought about by errors in the copying process during genome replication. Other kinds of genetic change may be due to recombination, reassortment of genome pieces, loss of genetic material, or acquisition of nucleotide sequences from unrelated viruses or the host genome.
6. Enzymes specified by the viral genome may be present in the virus particle. Most of these enzymes are concerned with nucleic acid synthesis.
7. Replication of many viruses takes place in distinctive virus-induced regions of the cell, known as viroplasms.
8. Some viruses share with certain non-viral nucleic acid molecules the property of integration into host-cell genomes and translocation from one integration site to another.
9. A few viruses require the presence of another virus for their replication.

III. ABOUT THIS EDITION

This edition follows many of the features of previous editions but has been reorganized to take account of the greater understanding of how viruses function. The first chapters describe the basic features of viruses, their classification, the symptoms they cause, how they are purified, what they are made of and the structure of their particles. The next two chapters recount how viruses express their genetic information and replicate themselves. This is followed by several chapters discussing the interactions between viruses and their hosts and vectors in disease manifestation and transmission, culminating in a chapter on how these functions are integrated and how some of them have been put to other uses. The last set of chapters deal with other virus-like sequences, the detection, control and evolution of plant viruses. It is hoped that this will form a logical sequence and will reveal the breadth and dynamism of the subject.

In such a dynamic subject, there has been an avalanche of publications over the last 10 years since the previous edition (see Fig. 1.3). In many cases, I have referred to review papers on specific topics where the original papers on that topic can be found. I have retained many of the older references from the previous edition as these describe phenomena and results that can assist in the interpretation of the new phenomena that are being unveiled. The older references also put a perspective on the subject which can be swamped by the new 'in vogue' topics.

CHAPTER 2

Nomenclature and Classification of Plant Viruses

I. NOMENCLATURE

A. Historical aspects

In all studies of natural objects, humans have an innate desire to name and to classify. Virologists are no exception. Virus classification, as with all other classifications, arranges objects showing similar properties into groups and, even though this may be a totally artificial and human-driven activity without any natural base, it does have certain properties:

- It gives a structured arrangement of the organisms so that the human mind can comprehend them more easily.
- It helps with communication between virologists.
- It enables properties of new viruses to be predicted.
- It could reveal possible evolutionary relationships.

In theory, it is possible to consider the problems of naming and classifying viruses as separate issues. In practice, however, naming soon comes to involve classification.

Early workers generally gave a virus a name derived from the host plant in which it was found together with the most conspicuous disease symptom, for example, tobacco mosaic virus. Viruses were at first thought of as stable entities, and each disease condition in a particular host species was considered to be due to a different virus. However, by the early 1930s three important facts began to be recognized:

Virus acronyms are given in Appendix 1.

1. Viruses can exist as different strains, which may cause very different symptoms in the same host plant.
2. Different viruses may cause very similar symptoms on the same host plant.
3. Some diseases may be caused by a mixture of two unrelated viruses.

J. Johnson in 1927 and in subsequent work stressed the need for using some criteria other than disease symptoms and host plants for identifying viruses. He suggested that a virus should be named by adding the word virus and a number to the common name for the host in which it was first found; for example, tobacco virus 1 for TMV. Johnson and Hoggan (1935) compiled a descriptive key based on five characters: modes of transmission, natural or differential hosts, longevity *in vitro*, thermal death point, and distinctive or specific symptoms. About 50 viruses were identified and placed in groups.

K. M. Smith (1937) outlined a scheme in which the known viruses or virus diseases were divided into 51 groups. Viruses were named and grouped according to the generic name of the host in which they were first found. Successive members in a group were given a number. For example, TMV was *Nicotiana* virus 1, and there were 15 viruses in the *Nicotiana* virus group. Viruses that were quite unrelated in their basic properties were put in the same group. Although Smith's list served for a time as a useful catalog of the known viruses, it could not be regarded as a classification.

Holmes (1939) published a classification based primarily in host reactions and methods of transmission. He used a Latin binomial–trinomial system of naming. For example, TMV became *Marmor tabaci*, Holmes (*Marmor* meaning marble in Latin). His classification was based on diseases rather than the viruses, and thus 53 of the 89 plant viruses considered by Holmes fell in the genus *Marmor*, which contained viruses known even at that time to differ widely in their properties.

Between 1940 and 1966, various schemes were proposed either for plant viruses only or for all viruses. None of these schemes was adopted by any significant number of virologists. It became increasingly apparent that a generally acceptable system of nomenclature and classification could be developed only through international co-operation and agreement, with the opinions of a majority of working virologists being taken into account.

At the International Congress for Microbiology held in Moscow in 1966, the first meeting of the International Committee for the Nomenclature of Viruses was held, consisting of 43 people representing microbiological societies of many countries. An organization was set up for developing an internationally agreed taxonomy and nomenclature for all viruses. Rules for the nomenclature of viruses were laid down. The subsequent development of the organization, now known as the International Committee for Taxonomy of Viruses (ICTV), has been summarized (Matthews, 1983a; 1985a,b,); the ICTV has presented seven reports (Wildy, 1971; Fenner, 1976; Matthews, 1979, 1982; Francki *et al.*, 1991; Murphy *et al.*, 1995; van Regenmortel *et al.*, 2000) and published intermediate reports in the *Archives of Virology*. The main features of the agreed nomenclature and taxonomy as they apply to plant viruses are considered in the following sections.

B. Systems for classification

Organisms may be classified in two general ways. One is the classic monothetic hierarchical system applied by Linnaeus to plants and animals. This is a logical system in which decisions are made as to the relative importance of different properties, that are then used to place a taxon in a particular phylum, order, family, genus, and so on. Maurin *et al.* (1984) proposed a classification system of this sort that embraces viruses infecting all kinds of hosts. While such systems are convenient to set up and use, there is, as yet, no sound basis for them as far as viruses are concerned. The major problem with any such system is that we have no scientific basis for rating the relative importance of all the different characters involved. For example, is the kind of nucleic acid (DNA or RNA) more important than the presence or absence of a bounding lipoprotein membrane? Is the particle symmetry of a small RNA virus (helical rod or icosahedral shell) more important than some aspect of genome strategy during virus replication?

An alternative to the hierarchical system was proposed by Adanson (1763). He considered that taxa were best derived by considering all available characters. He made a series of separate classifications, each based on a single character, and then examined how many of these characters divided the species in the same way. This gave divisions based on the largest number of correlated characters. The method is laborious and has not been much used until recently because at least about 60 equally weighted independent qualitative characters are needed to give satisfactory division (Sneath, 1962; Harrison *et al.*, 1971) (Table 2.1). Although the availability of computers has renewed interest in this kind of classification, so far they have not really been used for this to any great extent.

In practice, some weighting of characters is inevitably involved even if it is limited to decisions as to which characters to leave out of consideration and which to include. Its main advantages at present may be to confirm groupings arrived at in other ways and to suggest possibly unsuspected relationships that can then be checked by further experimental work.

The rapid accumulation of complete nucleotide sequence information for many viruses in a range of different families and groups is having a profound influence on virus taxonomy in at least three ways:

1. Most of the virus families delineated by the ICTV, mainly on morphological grounds, can now be seen to represent clusters of viruses with a relatively close evolutionary origin.

TABLE 2.1 Descriptors used in virus taxonomy

I. Virion properties

A. Morphology properties of virions
1. Size
2. Shape
3. Presence or absence of an envelope or peplomers
4. Capsomeric symmetry and structure

B. Physical properties of virions
1. Molecular mass
2. Buoyant density
3. Sedimentation coefficient
4. pH stability
5. Thermal stability
6. Cation (Mg^{2+}, Mn^{2+}, Ca^{2+}) stability
7. Solvent stability
8. Detergent stability
9. Radiation stability

C. Properties of the genome
1. Type of nucleic acid, DNA or RNA
2. Strandedness: single-stranded or double-stranded
3. Linear or circular
4. Sense: positive, negative or ambisense
5. Number of segments
6. Size of genome or genome segments
7. Presence or absence and type of 5′ terminal cap
8. Presence or absence of 5′ terminal covalently-linked polypeptide
9. Presence or absence of 3′ terminal poly(A) tract (or other specific tract)
10. Nucleotide sequence comparisons

D. Properties of proteins
1. Number
2. Size
3. Functional activities (especially virion transcriptase, virion reverse transcriptase, virion hemagglutinin, virion neuraminidase, virion fusion protein)
4. Amino acid sequence comparisons

E. Lipids
1. Presence or absence
2. Nature

F. Carbohydrates
1. Presence or absence
2. Nature

II. Genome organization and replication
1. Genome organization
2. Strategy of replication of nucleic acid
3. Characteristics of transcription
4. Characteristics of translation and post-translational processing
5. Sites of accumulation of virion proteins, site of assembly, site of maturation and release
6. Cytopathology, inclusion body formation

III. Antigenic properties
1. Serological relationships
2. Mapping epitopes

IV. Biological properties
1. Host range, natural and experimental
2. Pathogenicity, association with disease
3. Tissue tropisms, pathology, histopathology
4. Mode of transmission in nature
5. Vector relationships
6. Geographic distribution

From Fauquet (1999), with permission.

2. With the discovery of previously unsuspected genetic similarities between viruses infecting different host groups, the unity of virology is now quite apparent. The stalling approach of some plant virologists to the application of families, genera and species to plant viruses is no longer tenable.
3. Genotypic information is now more important for many aspects of virus taxonomy than phenotypic characters.

However, there are some limitations to the use of genotype (sequence) data alone. An important one is that with the present state of knowledge it is very difficult, or impossible, to predict phenotypic properties of a virus based on sequence data alone. For example, if we have two viral nucleotide sequences differing in a nucleotide at a single site, we could not, in most cases, deduce from this information alone that one led to mosaic disease and the other to lethal necrosis in the same host. However, with increasing understanding of gene function, we could now decide from the nucleotide sequences alone which of two rhabdoviruses replicated in plants and insects and which in vertebrates and insects.

Any classification of viruses should be based not only on evolutionary history, as far as this can be determined from the genotype, but should also be useful in a practical sense. Most of the phenotypic characters used today in virus classification will remain important even when the nucleotide sequences of most viral genomes have been determined.

C. Families, genera, species and groups

At a meeting in Mexico City in 1970, the ICTV (then called the ICNV) approved the first taxa for viruses (Matthews, 1983a). There were two families and type genera for these, plus 22 genera not placed in families for viruses infecting vertebrates, invertebrates, or bacteria. Sixteen taxa designated as groups were approved for viruses infecting plants. While other virologists subsequently moved quite rapidly to develop a system of families and genera for the viruses, plant virologists clung to the notion of groups. Some plant virologists had especial difficulty with the species concept, believing that it should not be applied to viruses. Their main reason is that because viruses reproduce asexually the criterion of reproductive isolation cannot be used as a basis for defining virus species (e.g. Milne, 1988).

The main building block of a biological classification is the species. The pragmatic view of Davis and Heywood (1963) in discussing angiosperm taxonomy was that:

> There is no universally correct definition (of a species) and progress in understanding the species problem will only be reached if we concentrate on the problem of what we shall treat as a species for any particular purpose.

The species concept applied to plants is also discussed by Wagner (1984). In day-to-day practice, virologists use the concept of a 'virus' as being a group of fairly closely related strains, variants or pathovars. A virus defined in this way is essentially a species in the sense suggested by Davis and Heywood, and defined by the ICTV. In 1991, the ICTV accepted the concept that viruses exist as species, adopting the following definition (van Regenmortel, 1990):

> A viral species is a polythetic class of viruses that constitutes a replicating lineage and occupies a particular ecological niche.

The species has formed the basis of modern virus classification being firmed up in subsequent ICTV reports, especially the seventh in which a 'list of species-demarcating criteria' is provided for each genus. This enables viruses to be differentiated as species and tentative species. Guidelines to the demarcation of a virus species are given in van Regenmortel *et al.* (1997).

With the species forming the basis of the classification system, they can be grouped into other taxa on various criteria. To date, the taxonomic levels of Order, Family and Genus have been defined by the ICTV (Table 2.2) and it is likely that there will be pressure for further higher and intermediate taxa. For example, Hull (1999b) argued that the development of an overall classification for viruses and elements that replicate by reverse transcription necessitates creating the taxa of Class and Sub-order.

A detailed discussion of virus classification, the currently accepted taxa and how the ICTV operates is given in Fauquet (1999) and by van Regenmortel *et al.* (2000).

1. Delineation of viruses (species)

The delineation of the kinds of virus that exist in nature, that is to say virus species, is a practical necessity, especially for diagnostic purposes relating to the control of virus diseases.

TABLE 2.2 Criteria demarcating different virus taxa

I. Order
Common properties between several families including:
Biochemical composition
Virus replication strategy
Particle structure (to some extent)
General genome organization

II. Family
Common properties between several genera including:
Biochemical composition
Virus replication strategy
Nature of particle structure
Genome organization

III. Genus
Common properties with a genus including:
Virus replication strategy
Genome size, organization and/or number of segments
Sequence homologies (hybridization properties)
Vector transmission

IV. Species
Common properties within a species including:
Genome rearrangement
Sequence homologies (hybridization properties)
Serological relationships
Vector transmission
Host range
Pathogenicity
Tissue tropism
Geographical distribution

From Fauquet (1999), with permission.

For some 20 years, B. D. Harrison and A. F. Murant acted as editors for the Association of Applied Biologists (AAB) to produce a series of some 354 descriptions of plant viruses and plant virus groups and families. Each description was written by a recognized expert for the virus or group (genus) in question. The two editors used common-sense guidelines devised by themselves to decide whether a virus described in the literature is a new virus, or merely a strain of a virus that has already been described. When they published a new virus description, they were, in effect, delineating a new species of virus. The AAB descriptions of plant viruses are widely used by plant virologists and accepted as a practical and effective taxonomic contribution. The descriptions now cover an increasing proportion of the known plant viruses and are being updated and extended as a CD-ROM produced by the AAB (see Appendix 3).

Viruses are now recognized as 'species' or 'tentative species' which are viruses that have not yet been sufficiently characterized to ensure that they are distinct and not strains of an existing virus or do not have the full characteristics of the genus to which they have been assigned. Of the 977 plant viruses listed in the ICTV seventh report, 701 are true species and 276 tentative species.

2. Virus strains and isolates

A common problem is to determine whether a new virus is truly a new species or a strain of an existing species. Conversely, what was considered to be a strain may, on further investigation, turn out to be a distinct species. This is because of the population structure of viruses which, because of continuous production of errors in replication, can be considered a collection of quasispecies. The concept of quasispecies is discussed in more detail in Chapter 17 (Section I.A).

Various characters are considered in designating a strain. These are described in more detail in relation to virus variation and evolution in Chapter 17.

3. Naming of viruses (species)

Questions of virus nomenclature have generated more heat over the years than the much more practically important problems of how to delineate distinct virus species. At an early stage in the development of modern viral taxonomy, a proposal was made to use cryptograms to add precision to vernacular names of viruses (Gibbs et al., 1966). The proposal was not widely adopted except, for a time, among some plant virologists.

When a family or genus is approved by the ICTV, a type species or type virus is designated. Some virologists favor using the English vernacular name as the official species name. Using part of a widely known vernacular name as the official species name may frequently be a very suitable solution, but it could not always apply (e.g. with newly discovered viruses). Other virologists favor serial numbering for viruses (species). The experience of other groups of microbiologists is that, while numbering or lettering systems are easy to set up in the first instance, they lead to chaos as time passes and changes have to be made in taxonomic groupings. The idea of Latinized binomial names for viruses was supported by the ICTV for many years but never implemented for any viruses. The proposal for Latinization has now been withdrawn.

In successive editions of the reports of the ICTV (e.g. Matthews, 1982), virus names in the index have been listed by the vernacular name (usually English) followed by the family or genus name; for example, tobacco mosaic *Tobamovirus*, Fiji disease *Fijivirus*, and lettuce necrotic yellows *Rhabdovirus*. This method for naming a plant virus is becoming increasingly used in the literature.

4. Acronyms or abbreviations

Abbreviations of virus names have been used for many years to make the literature more easy to read and more succinct to present. The abbreviation is usually in the form of an acronym using the initial letters of each word in the virus name. As the designation of the acronym was by the author of the paper, it was leading to much overlap and confusion. For instance, AMV was used to designate alfalfa mosaic virus and arabis mosaic virus and could also justifiably be used for abutilon mosaic virus, agropyron mosaic virus, alpina mosaic virus, Alstromeria mosaic virus, Alternantha mosaic virus, Aneilema mosaic virus or Anthoxanthum mosaic virus. Therefore, in 1991 the Plant Virus Section of the ICTV initiated a rationalization of plant virus acronyms (Hull *et al.*, 1991) and has subsequently updated the list regularly (Fauquet and Martelli, 1995; Fauquet and Mayo, 1999). The designation of the abbreviations is based on the following principles:

- Abbreviations should be as simple as possible.
- An abbreviation must not duplicate any other previously coined and still in current usage.
- The word 'virus' in a name is abbreviated as 'V'.
- The word 'viroid' in a name is abbreviated as 'Vd'.

A set of guidelines is laid out in Fauquet and Mayo (1999). Although these, and the acronyms derived from them, are not officially sanctioned by the ICTV, the acronyms are used in the seventh report (van Regenmortel *et al.*, 2000) and it is hoped the plant virology community will adopt them. The guidelines are:

1. When similar virus names contain the terms 'mosaic' and 'mottle', 'M' is chosen for 'mosaic' and 'Mo' for 'mottle' except where Guidelines 7 and/or 8 apply. For example, *Cowpea mosaic virus* and *Cowpea mottle virus* are abbreviated as CPMV and CPMoV respectively. However, there are still a few abbreviations in which 'Mo' is not used for 'mottle', as there was no need to do so. These will remain unchanged unless change becomes necessary.
2. The word 'ringspot' is abbreviated as 'RS' in many, but not all, instances even if 'R' could have sufficed.
3. The word 'symptomless' is abbreviated as 'SL' in many, but not all, instances even if 'S' could have sufficed.
4. The second or third letter, or sometimes the second consonant or last letter, of the host plant name, in lower case, can be used to differentiate certain conflicting abbreviations; e.g. CdMV for *Cardamom mosaic virus*.
5. Abbreviations that use the same letters, but differ only by the case used (upper or lower), should be avoided.
6. Abbreviations for single words should not normally exceed two letters.
7. Secondary letters in abbreviations are omitted when their use would make the abbreviation excessively long, generally more than five letters. For example, CGMMV is preferred to the longer CGMoMV, even though that complies with Guideline 1.
8. Abbreviations in current and widespread use are retained except where their use could cause confusion. However, abbreviations can and have been changed in common usage. The purpose of the advisory list is to suggest targets for harmonization of abbreviations. It is accepted that some abbreviations are unlikely ever to be changed.
9. When a particular combination of letters has been adopted for a particular plant species, new abbreviations for virus names containing the same host will normally use the same combination.
10. When several viruses have the same name and are differentiated by a number, the abbreviations will have a hyphen between the letters and numbers, e.g. *Plantain virus 6* and *Plantain virus 8* are abbreviated as 'PlV-6' and 'PlV-8'.
11. When viruses are distinguished by a letter, this letter is added at the end of the abbreviation without a hyphen, e.g. *Plantain virus X* is abbreviated PlVX, in agreement with most of the cases in common usage in plant virology (PVY, PVX, PVM, PVS, etc.).
12. When viruses are distinguished by their geographic origin or any other combination of letters, a minimum number of letters (two or three) is added to the virus abbreviation and a hyphen is used between the two sets of letters when the are added after the word 'virus'; e.g. *Tomato yellow leafcurl virus* from Thailand is abbreviated 'TYLCV-Th'. (Abbreviations for country names are given in Appendix 1.)
13. When a virus name comprises a disease name and the words 'associated virus', these are abbreviated as 'aV'. For example, *Grapevine leafroll associated virus 2* is abbreviated to 'GLRVaV-2'.

14. In some instances, where a plant name is abbreviated as two capital (upper case) letters, this usage is retained as an exception; e.g. 'CPRMV' for *Cowpea rugose mosaic virus*.

The current preferred list of abbreviations or acronyms is given in Appendix 1.

D. Plant virus families, genera and orders

The current classification of plant viruses is given in Appendix 2 and shown diagrammatically in Fig. 2.1.

There is no formal definition for a genus, but it is usually considered as 'a population of virus species that share common characteristics and are different from other populations of species'. The criteria for demarcating a genus are listed in Table 2.2. Currently there are 70 genera of plant viruses recognized. In some cases (e.g. the *Rhabdoviridae*), there are numerous viruses which obviously belong to that family but for which there is insufficient information to place them either in existing genera or for creating new genera; these are listed as 'unassigned'. Genera are named either after the type species (e.g. *Caulimovirus* after cauliflower mosaic virus) or are given a descriptive name, often from a Greek or Latin word, for a major feature of the genus (e.g. *Closterovirus* from the Greek κλωστηρ (kloster) which is spindle or thread, descriptive of the virus particle shape; and *Geminivirus* from the Latin *geminus* meaning twins to describe the particles).

Similarly, genera are grouped together into families on common characteristics (Table 2.2). There are 14 families recognized for plant viruses (Appendix 2); some, such as *Reoviridae* and *Rhabdoviridae*, are in common with animal virus families. Twenty two of the genera have not yet been assigned to families and are termed 'floating genera'. The acquisition of further data on these 'floating genera', together with changing attitudes on virus classification, will no doubt lead to the designation of further plant virus families. The family is either named after the type member genus (e.g. *Caulimoviridae* named after the genus *Caulimovirus*), or given a descriptive name, as with the genus, for a major feature of the family (e.g. *Geminiviridae*, descriptive of the virus particles).

Only three orders have been accepted thus far by the ICTV. The *Mononegavirales* contains, amongst other families, the *Rhabdoviridae* in which there are two plant virus families.

E. Use of virus names

The ICTV sets rules, which are regularly revised, on virus nomenclature and the orthography of taxonomic names (see Pringle, 1998; van Regenmortel *et al.*, 2000). The last word of a species is 'virus' and suffix (ending) for a genus is '...virus', for a subfamily is '... virinae', for a family is '... viridae', and for an order is '... virales'. In formal taxonomic usage the virus order, family, subfamily, genus and species names are printed in italics (or underlined) with the first letter being capitalized; other words in species names are not capitalized unless they are proper nouns or parts of proper nouns. Also, in formal use, the name of the taxon should precede the name being used, e.g. 'the family *Caulimoviridae*', 'the genus *Mastrevirus*', and 'the species *Potato virus Y*'. In informal use, the family, subfamily, genus and species names are written in lower case Roman script, the taxon does not include the formal suffix and the taxonomic unit follows the name being used; e.g. 'the caulimovirus family', 'the mastrevirus genus', and the 'potato virus Y species'. In even less formal circumstances, but widely used, the taxonomic unit is omitted and the taxon for higher taxa can be in the plural; e.g. 'caulimoviruses', 'mastreviruses', 'potato virus Y'.

Informal usage arises from practicalities and can lead to the adoption of more formal use. For instance, the genus *Badnavirus* was not adopted in 1991 but was used widely in the literature and was adopted in the 1995 ICTV report. However, the year 2000 report limited its use to certain DNA viruses with bacilliform particles excluding RTBV. As will be apparent in this book, it is necessary to distinguish the reverse transcribing DNA viruses that have isometric particles from those that have bacilliform particles; the informal usage with be 'caulimoviruses' for the former and 'badnaviruses' for the latter.

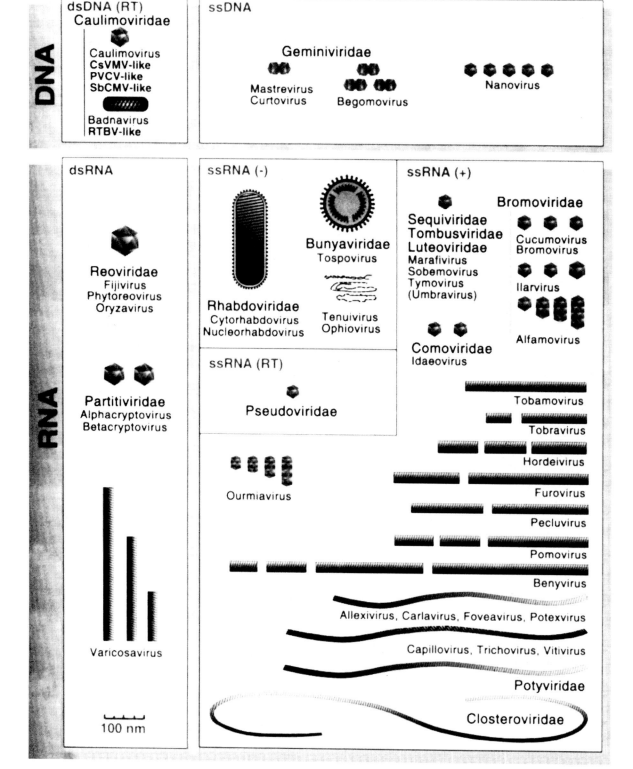

Fig. 2.1 The families and genera of viruses infecting plants. Outline diagrams are drawn approximately to scale. From van Regenmortel *et al.* (2000), with permission.

II. CRITERIA USED FOR CLASSIFYING VIRUSES

There are two interconnected problems involved in attempting to classify viruses. First, related viruses must be placed in genera and higher taxa. For this purpose, the more stable properties of the virus—such as amount and kind of nucleic acid, particle morphology, and genome strategy—are most useful. The second problem is to be able to distinguish between related viruses and give some assessment of degrees of relationship within a group. Properties of the virus for which there are many variants are more useful for this purpose. These include symptoms, host range, and amino acid composition of the coat protein. Certain properties, such as serological specificity and amino acid sequences, may be useful both for defining groups and for distinguishing viruses within groups.

Two general problems must be borne in mind when considering the criteria to be used for virus classification. One is the problem of weighting—the relative importance of one character as compared to another. Some weighting of characters is inevitable, whatever system is used to place viruses in taxa. The second problem concerns the extent to which a difference in one character depends on a difference in another. For example, a single-base change in the nucleic acid may result in an amino acid replacement in the coat protein, which in turn alters serological specificity and electrophoretic mobility of the virus. The same base change in another position in a codon may not affect the gene product at all.

The preceding discussion relates to phenotypic characters. As the complete nucleotide sequences of more and more viruses are being determined, virus genotype is becoming increasingly important. What weight is to be given to sequence data in virus classification? A geneticist may say that the evolutionary history of a virus and therefore its relationships with other viruses are completely defined, as far as we will be able to know it, by its nucleotide sequence. This is certainly true in a theoretical sense, but there are also practical aspects to be considered as discussed in Section I.B.

The overall properties used in classifying a virus are listed in Table 2.1. Properties that are useful for characterizing strains of a virus are discussed in Chapter 17 (Section II).

A. Structure of the virus particle

The importance of virus structure in the classification of viruses is summarized in Fig. 2.1. With isometric viruses, particle morphology, as revealed by electron microscopy, has not proved as generally useful as with the rod-shaped viruses. This is mainly because many isometric particles lie in the same size range (about 25–30 nm in diameter) and are of similar appearance unless preparations and photographs of high quality are obtained (Hatta and Francki, 1984). Where detailed knowledge of symmetry and arrangement of subunits has been obtained by X-ray analysis or high-resolution electron micro- scopy, these properties give an important basis for grouping isometric viruses. For large viruses with a complex morphology, the structure of the particle as revealed by electron microscopy gives valuable information indicating possible relationships.

B. Physicochemical properties of virus particles

Physicochemical properties and stability of the virus particle have sometimes played a part in classification. These properties are discussed in Chapter 15 (Section III). Non-infectious empty viral protein shells and minor noninfectious nucleoproteins are characteristic of certain groups (e.g. tymoviruses). The existence of these components reflects the stability of the protein shell in the absence of the full-length viral RNA.

C. Properties of viral nucleic acids

The organization and strategy of viral genomes, as discussed in Chapter 6, is now of prime importance for the placing of viruses into families, and genera, or for the establishment of a new family or genus. The luteo-viruses provide a good example (Mayo and D'Arcy, 1999). They were first formally recognized as a group comprising BWYV, SbDV and the RPV, RMV and SGV strains of BYDV in the second ICTV report (Fenner, 1976). Over the next few years groupings of these viruses increased in a variable manner, depending upon the criteria used. By the sixth report (Murphy *et al.*, 1995) the previous 'luteovirus group' was named as the genus *Luteovirus* and the species divided into two subgroups typified by BYDV-PAV and PLRV. By this time, many of the luteoviral genomes had been sequenced and it was becoming very apparent that there needed to be a rethink on the classification of this group (D'Arcy and Mayo, 1997). Not only were there differences in genome organizations but there were different polymerase enzymes in different species, some having a 'carmovirus-type' polymerase and others a 'sobemovirus-type' polymerase (see Chapter 8, Section IV, for polymerases). Based mainly on these sequence data the group was divided into three genera (see Section III.P) that were contained in the family *Luteoviridae* (Mayo and D'Arcy, 1999; van Regenmortel *et al.*, 2000).

Another example is the delimitation of the genera within the family *Caulimoviridae*. Previously, four of the genera (*Caulimovirus*, 'Soybean chlorotic virus-like', 'Cassava vein mosaic virus-like' and 'Petunia vein clearing virus-like') were classed in the genus *Caulimovirus*, and two (*Badnavirus* and 'Rice tungro bacilliform virus-like') in the genus *Badnavirus* (Murphy *et al.*, 1995). However, because of differences in genome organization these have now been separated into different genera (see Appendix 2) in the family *Caulimoviridae* in recognition of some basic similarities in genome organization and their replication involving reverse transcription (van Regenmortel *et al.*, 2000).

D. Viral proteins

The properties of viral proteins, and in particular the amino acid sequences, are of great importance in virus classification at all levels: for delineating strains, as discussed in Chapter 17 (Section II.A.2); for viruses, as discussed in this section; and for indicating evolutionary relationships between families and groups of viruses.

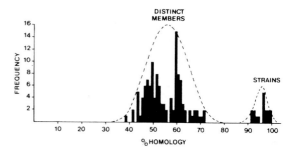

Fig. 2.2 Demarcation between the extent of amino acid sequence homologies in coat proteins amongst distinct individual potyviruses (left-hand distribution) and between strains of the same virus (right-hand peak). The 136 possible pairings between 17 strains of eight distinct viruses were analyzed. The homologies between distinct viruses had a mean value of 54.1% and a standard deviation of 7.29%, while the homologies between strains of individual viruses showed a mean of 94.5% and standard deviation of 2.56%. The dashed curves show that all values for distinct viruses and strains fall within ±3 standard deviations from their respective mean values. From Shukla and Ward (1988), with permission.

Coat protein amino acid sequence homologies have been used to distinguish between distinct potyviruses and strains of these viruses (Fig. 2.2) and to estimate degrees of relationship within the group (Fig. 2.3). Dendrograms indicating relationships within a virus group have also been based on amino acid composition of the coat proteins, for example, with the tobamoviruses (Gibbs, 1986). A dendrogram based on peptide patterns obtained from *in vitro* translated 126-kDa proteins of eight tobamoviruses was very similar to that obtained with coat proteins (Fraile and García-Arenal, 1990). Once a set of viruses such as the tobamoviruses has been delineated based on coat protein amino acid composition, new isolates may sometimes be readily placed as a strain of an existing virus (e.g. Creaser *et al.*, 1987).

Using a statistical procedure known as principal component analysis, Fauquet *et al.* (1986) showed that groupings of 134 plant viruses and strains obtained using the amino acid composition of their coat proteins (Fig. 2.4) correlated well with the groups established by the ICTV (Matthews, 1982).

However, phylogenetic relationships based on one protein may not be similar to that on

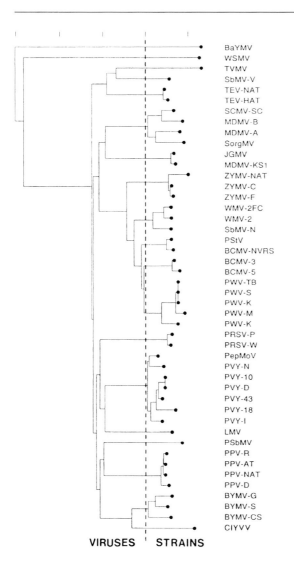

BaYMV
WSMV
TVMV
SbMV-V
TEV-NAT
TEV-HAT
SCMV-SC
MDMV-B
MDMV-A
SorgMV
JGMV
MDMV-KS1
ZYMV-NAT
ZYMV-C
ZYMV-F
WMV-2FC
WMV-2
SbMV-N
PStV
BCMV-NVRS
BCMV-3
BCMV-5
PWV-TB
PWV-S
PWV-K
PWV-M
PWV-K
PRSV-P
PRSV-W
PepMoV
PVY-N
PVY-10
PVY-D
PVY-43
PVY-18
PVY-I
LMV
PSbMV
PPV-R
PPV-AT
PPV-NAT
PPV-D
BYMV-G
BYMV-S
BYMV-CS
CIYVV

VIRUSES ¦ **STRAINS**

Fig. 2.3 Relationship between some potyviruses and their strains. The dendrogram is based on the amino acid sequence of the core of the viral coat proteins, omitting the more variable terminal sequences. The core regions were aligned, and the percentages of different amino acids in each sequence (that is, their distance) were calculated. Gaps were counted as a 21st amino acid. The distances were then converted into the dendrogram using the neighbor-joining method of Saitou and Nei (1987). The divisions in the horizontal scale represent 10% differences. The broken vertical line indicates a possible boundary between virus species and strains. From Matthews (1991), with permission.

another protein. This is suggestive of recombination between viruses (see Chapter 8, Section IX.B).

In recent years, it has proved easier to obtain the sequence of nucleotides in genomic nucleic acids than that of the amino acids in the gene products. Thus, it is possible to identify amino acid sequence homologies between potential gene products, as has been done for some geminiviruses (Mullineaux et al., 1985) and potyviruses (Domier et al., 1987). Such studies have shown that there may be amino acid sequence similarities between different taxonomic groups of viruses in their non-structural proteins. Coat proteins, on the other hand, have an amino acid composition that is characteristic of a virus group (Fig. 2.4), and there are usually no significant amino acid sequence similarities between groups. Thus, in spite of the fact that coat proteins represent only a small fraction of the information in most viral genomes, they appear to be the most useful gene products for delineating many distinct viruses and virus groups.

However, functional equivalence of viral coat proteins may not always be associated with amino acid sequence similarities. Thus, the coat proteins of AMV and TSV are required to activate the genomic RNAs to initiate infection (see Chapter 8, Section IV.G). The coat proteins of these two viruses are able to activate each other's genome, recognizing the same sequence of nucleotides in the RNAs, but there is no obvious amino acid sequence similarity between them (Cornelissen and Bol, 1984).

An important goal for using properties of viral proteins in classification is to know the three-dimensional structure of the protein of at least one member of a group. This then allows amino acid substitutions in different viruses in a group to be correlated with biological function. This was achieved for TMV and six viruses related to it (Altschuh et al., 1987). The amino acid sequence homologies of these seven tobamoviruses ranged from about 28% to 82%. Twenty-five residues are conserved in all seven sequences (Fig. 2.5).

Twenty of these conserved residues are concentrated in two locations in the molecule: at low radius in the TMV rod near the RNA binding site (residues 36–41, 88–94 and 113–120) and at high radius forming a hydrophobic core (residues 61–63 and 144–145). Where viruses within this set of seven differ in sequence, the

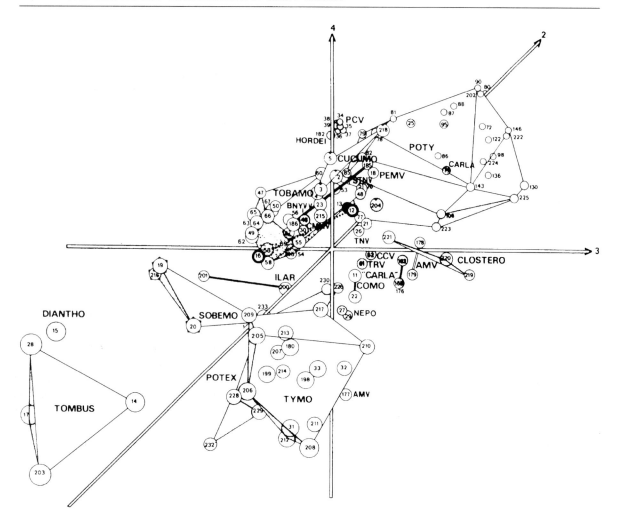

Fig. 2.4 Three-dimensional diagram illustrating factors 2, 3 and 4 of a principal-components analysis of 122 data sets of plant virus coat proteins compared by their amino acid composition. The three axes contain 62% of the information. The positions of the viruses on axis 2 are indicated by the sizes of the circles. The numbers within the circles are codes for individual viruses. From Fauquet *et al.* (1986), with kind permission of the copyright holder, © Karger, Basel.

differences are often complementary. For example, among buried residues, a change to a large side-chain in one position may be compensated by a second change to a smaller one in a neighboring amino acid. This study with tobamoviruses, while clearly delineating functionally critical regions of the molecule, did not lead to any clear evidence of particular evolutionary relationships within the group.

Other viral gene products have been used in taxonomy. As described in Chapter 8 (Section IV.B), there are major groups of RNA-dependent RNA polymerases. However, a re-evaluation of these suggested that this criterion was insufficient to support evolutionary groupings of RNA viruses (Zanotti *et al.*, 1996).

E. Serological relationships

Serological methods for determining relationships and their limitations are discussed in Chapter 17 (Section II.B). In the past, these methods have been the most important single criterion for placing viruses in related groups. Members of some groups may be all serologically related (e.g. tymoviruses), whereas, in

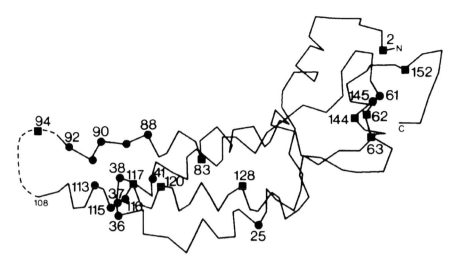

Fig. 2.5 Conserved amino acid residues in the coat proteins of seven tobamoviruses. The α-carbon chain tracing of one subunit is shown viewed down the disk axis. The positions are marked for hydrophobic (■) and hydrophilic (●) residues, which are invariant in all seven viruses From Altschuh *et al.* (1987), with permission.

others, only a few may show any serological relationship (e.g. nepoviruses). In the groups of rod-shaped viruses defined primarily on particle dimensions, many viruses within groups are serologically related, and little or no serological relationship has been established between groups. It seems probable that serological tests will remain an important criterion upon which virus groups are based. Serological tests may also be used to estimate degrees of relationship between viruses in a group (e.g. Fig. 2.6).

For both tombusviruses and tymoviruses there is no correlation between SDI values as illustrated in Fig. 2.6 and estimates of genome homology. For tobamoviruses, however, there are clear correlations between SDI values, amino acid sequences, and estimates of genome homology (Koenig and Gibbs, 1986). Serological methods can be used to designate a set of virus strains as constituting a new virus within an established group; for example, PMMoV virus in the genus *Tobamovirus* (Pares, 1988). When it comes to defining degrees of relationship within a group, borderline situations will sometimes be found. In that circumstance, additional criteria will be needed.

Using a broadly cross-reactive antiserum against the viral coat protein core, Shukla *et al.* (1989d) showed that potyviruses transmitted by mites or whiteflies were serologically related to a definitive potyvirus with aphid vectors. Distinct potyviruses have often been difficult to define because serological relationships have been found to be complex and often inconsistent. Shukla *et al.* (1989c) applied a systematic immunochemical analysis to some members of this group. This method involved the use of overlapping peptide fragments to define the parts of the protein combining with antibodies in particular antisera and to make it possible to develop both virus-specific and group-specific antisera.

F. Activities in the plant

The use of biological properties such as host range and symptoms in particular host plants as major criteria led to considerable confusion in the identification and classification of viruses. As knowledge of physical and chemical properties of viruses increased, the biological properties were rated as much less important. However, biological properties must still be given some importance in classifying viruses.

Macroscopic symptom differences will often reveal the existence of a different strain of a virus where no other criterion, except a full nucleotide sequence comparison, will do so. The detailed study of the cytology of infected cells made possible by electron microscopy has shown that many groups of viruses cause characteristic inclusion bodies or other cytological abnormalities in the cells they infect (see

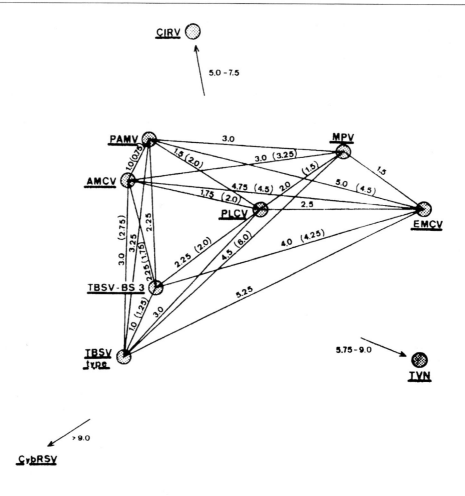

Fig. 2.6 The use of serology to estimate degrees of relationship between viruses in a group. The diagram illustrates a classification of 10 tombusviruses with distances representing the mean serological differentiation index of reciprocal tests (RT-SDI). RT-SDI values have been rounded to the nearest 0.25 and, when in order to represent the relationships in two dimensions, the 'observed' and 'diagrammatic' (in parentheses) RT-SDI values differ, the two values are shown. CIRV, TVN and CybRSV are only distantly related to one another and to the other tombusviruses, which form a central cluster. In a multidimensional system, these three viruses would have to be arranged in planes above and below that of the central cluster. The arrows indicate the average distance of these viruses from the central cluster as a whole, but not from individual viruses. TVN has been renamed NRV. From Koenig and Gibbs (1986), with permission.

Chapter 3, Section IV). Ultrastructural changes in cells usually appear to be much more stable characteristics of virus groups than macroscopic symptoms. Certain tombusviruses can be distinguished from other members of the group by their cytopathological effects, especially in *Chenopodium quinoa* (Russo *et al.*, 1987).

Cross-protection in the plant has been used as a useful indicator of relationship between viruses (see Chapter 17, Section II.C.4).

G. Methods of transmission

As discussed in Chapter 11, most viruses have only one type of vector, and usually all the viruses within a group have the same type of vector. Over the last few years, this character has been considered important in defining new genera within, for example, the original *Potyvirus* and *Geminivirus* groups. In general, the type of vector appears to be a stable character that is useful in delineating major groups of viruses. However, within a group of viruses or virus strains some may be transmitted efficiently by a vector species, some inefficiently, and some not at all. Details of the way in which a virus is transmitted by a vector (e.g. non-persistent or persistent aphid-transmitted viruses, or on the surface or within the spores of a fungus) may provide further criteria for the grouping of viruses. However, under certain conditions a virus culture may lose the ability to be transmitted by a vector; for example, WTV (see Chapter 11, Section IV.D.2).

III. FAMILIES AND GENERA OF PLANT VIRUSES

Below are given 'thumb-nail' sketches of salient features of members of plant virus families and genera. More details of various features are given in subsequent chapters, in reviews and books referred to here and in the seventh ICTV report (van Regenmortel *et al.*, 2000). Properties of these virus taxa are summarized in Appendix 2.

A. Family *Caulimoviridae* (reviewed by Hohn, 1999; Schoelz and Bourque, 1999)

This family contains all the plant viruses that replicate by reverse transcription. The viruses all have circular double-stranded DNA genomes with gaps or discontinuities at specific sites, one in one strand and between one and three in the other strand; these discontinuities represent priming sites for DNA synthesis during replication. Transcription is asymmetric to give a more-than-genome length RNA which is both the template for reverse transcription and the mRNA for at least some of the gene products. Other mRNAs may also be transcribed. See Chapter 8 (Section VII) for details of replication and Chapter 7 (Section V.H.1) for details of genome expression.

There are six genera within this family, falling into two groups: the caulimoviruses that have isometric particles and the badnaviruses that have bacilliform particles. The genera are distinguished on their genome organization.

1. Genus *Caulimovirus* (type species: *Cauliflower mosaic virus*)

Caulimoviruses have isometric particles of about 50 nm diameter with a T = 7 multilayered structure found within cytoplasmic inclusion bodies. The genome contains six open reading frames (ORFs), the 5' ones being expressed from the more-than-genome length RNA and the 3' ORF 6 from a separate mRNA. There are two other possible ORFs but no products have yet been found from them. Most species are transmitted by aphids in the semi-persistent manner, transmission requiring a virus-coded helper protein. See Chapter 5 (Section VI.B.7) for details of particle structure, Chapter 6 (Section IV.A.1) for details of genome organization, Chapter 7 (Section V.H.1) for details of genome expression, and Chapter 11 (Section III.F) for details of transmission.

2. Genus 'Soybean chlorotic mottle virus-like' (type species: *Soybean chlorotic mottle virus*) (reviewed by Reddy and Richins, 1999)

This resembles the genus *Caulimovirus* but with differences in genome organization (see Chapter 6, Section IV.A.2).

3. Genus 'Cassava vein mosaic virus-like' (type species: *Cassava vein mosaic virus*) (reviewed by de Kocho, 1999)

This resembles the genus *Caulimovirus* but with differences in genome organization (see Chapter 6, Section IV.A.3).

4. Genus 'Petunia vein clearing virus-like' (type species: *Petunia vein clearing virus*)

This resembles the genus *Caulimovirus* but with differences in genome organization (see Chapter 6, Section IV.A.4). Recent sequencing indicates that there is only one large ORF comprising ORFs 1 and 2 illustrated in Fig. 6.1 (K.R. Richert-Pöggeler and T. Hohn, personal communication).

5. Genus *Badnavirus* (type species: *Commelina yellow mottle virus*) (reviewed by Lockhart and Olszewski, 1999)

The genus contains several viruses of importance to tropical agriculture. Badnaviruses have bacilliform particles usually about 130×30 nm though there are some instances of longer particles (up to 900 nm long) not associated with cytoplasmic inclusion bodies. The DNA genome encodes three ORFs thought to be translated from the more-than-genome length RNA transcript. ORF III expresses a polyprotein processed to give, amongst other products, the virus coat protein, an aspartate protease, reverse transcriptase and RNaseH. Many members are transmitted in a semi-persistent manner by mealybugs. See Chapter 6 (Section IV.A.5) for details of genome organization.

6. Genus 'Rice tungro bacilliform virus-like' (type species: *Rice tungro bacilliform virus*) (reviewed by Hull, 1996, 1999a)

Together with RTSV, this genus causes rice tungro disease; it depends on RTSV for its leafhopper transmission. Particles are very similar to those of the genus *Badnavirus*, but the genome has four ORFs, the 3' one being translated from an mRNA spliced from the more-than-genome length RNA. See Chapter 5 (Section VI.B.5) for details of virion structure, Chapter 6 (Section IV.A.6) for details of genome organization, and Chapter 7 (Section V.H.1) for details of genome expression.

B. Family *Geminiviridae* (reviewed by Buck, 1999a)

This is one of the two families of plant viruses that have circular ssDNA genomes. The main distinguishing feature is that these genomes are contained in geminate virus particles consisting of two incomplete icosahedra (see Chapter 5, Section VI.B.3). The family comprises four genera that are distinguished from one another by their genome organization and their vectors.

1. Genus *Mastrevirus* (type species: *Maize streak virus*) (reviewed by Palmer and Rybicki, 1998)

This genus contains the economically important MSV. All members have narrow host ranges and, with the exception of TYDV and BeYDV which infect dicotyledenous species, their host ranges are limited to species in the *Poaceae*. Mastreviruses are transmitted by leafhoppers in a circulative, non-propagative

manner (see Chapter 11, Section IV.D.1). They have genomes comprising a single component of ssDNA (see Chapter 6, Section V.A.2).

2. Genus *Curtovirus* (type species: *Beet curly top virus*)

The type member of this genus used to be an economically important virus of sugarbeet and has a wide host range. Curtoviruses are transmitted by leafhoppers in the circulative non-propagative manner (see Chapter 11, Section IV.D.1). They have a monopartite ssDNA genome (see Chapter 6, Section V.A.3).

3. Genus *Begomovirus* (type species: *Bean golden mosaic virus*)

This is the largest geminivirus family, members of which have narrow host ranges among dicotyledonous species. They are transmitted by whiteflies (see Chapter 11, Section V.B). Most members of this genus have their genomes divided between two DNA molecules (see Chapter 6, Section V.A.4), though the genomes of some members are monopartite.

4. Genus *Topocuvirus* (type species: *Tomato pseudo-curly top virus*)

This is a very recently designated genus (Pringle, 1999a) that has been split off from the *Curtovirus* genus. It has a similar genome organization to the curtoviruses but it is transmitted by the treehopper, *Micrutalis malleifera*.

C. Family *Circoviridae*

This family contains two genera, one of viruses that infect animals and the other of viruses that infect plants. They have small virus particles, 17–22 nm in diameter, with icosahedral symmetry that contain small circular ssDNA genomes.

1. Genus *Nanovirus* (type species: *Subterranean clover stunt virus*)

This genus contains the economically important virus BBTV. The viruses have a genome of at least six circular ssDNA molecules of approximately 1 kb in size encapsidated in an icosahedral virion about 20 nm in diameter. See Chapter 6 (Section V.B.1) for genome organization and Chapter 8 (Section VIII.E) for the replication mechanism.

D. Family *Reoviridae*

The *Reoviridae* comprises genera of viruses that infect vertebrates, invertebrates and plants. Some members are restricted to vertebrates but others infect both vertebrates and invertebrates. Those that infect plants also infect their invertebrate vectors. The virus particles are complex, made up of one, two or three distinct shells each with icosahedral symmetry. The particles of most of the genera have surface spikes. The genomes comprise 10, 11 or 12 segments of linear dsRNA depending upon the genus. There are three genera that infect plants which are distinguished on their structure and the number of segments into which their genome is divided.

1. Genus *Fijivirus* (type species: *Fiji disease virus*)

Fijiviruses have 10 dsRNA segments (see Chapter 6, Section VI.A.1, for genome organization) encapsidated in a double-shelled particle 65–70 nm in diameter (see Chapter V, Section VII.2, for structure). These viruses cause hypertrophy of the phloem leading to vein swelling and sometimes enations. They are transmitted in delphacid planthoppers in a circulative propagative manner and infect monocotyledonous plants of the *Graminae* and *Liliaceae* families.

2. Genus *Oryzavirus* (type species: *Rice ragged stunt virus*)

Members of this genus have genomes comprising 10 dsRNA segments (see Chapter 6, Section VI.A.2, for genome organization) encapsidated in a double-shelled particle 75–80 nm in diameter (see Chapter 5, Section VII.3). They infect monocotyledonous plants of the family *Graminae* and are transmitted by delphacid planthoppers in the circulative propagative manner. Oryzaviruses resemble fijiviruses in having 10 dsRNA segments and being transmitted by delphacid planthoppers. The two genera are distinguished by particle size and by nucleic acid and serological similarities within, but not between, the genera.

3. Genus *Phytoreovirus* (type species: *Wound tumor virus*)

Phytoreoviruses have 12 dsRNA genome segments (see Chapter 6, Section VI.A.3, for genome organization) encapsidated in a double-shelled particle about 70 nm in diameter (see Chapter 5, Section VII.1). They infect both dicotyledonous and monocotyledonous species and are transmitted by leafhoppers in the circulative propagative manner.

E. Family *Partitiviridae* (reviewed by Milne and Marzachi, 1999)

Members of the *Partitiviridae* have small isometric non-enveloped virions 30–40 nm in diameter with a dsRNA genome divided into two segments, the smaller encoding the coat protein and the larger the RNA-dependent RNA polymerase.

Two of the genera of this family infect fungi and two infect plants. The plant-infecting viruses are known as 'cryptic viruses' as they cause no, or very few, symptoms. They are not graft-transmitted and they have no biological vector, but are highly seed-transmitted. Presumably, this leads to full systemic infection as they move through the plant as cells divide.

1. Genus *Alphacryptovirus* (type species: *White clover cryptic virus 1*)

The 30-nm diameter isometric particles of members of this genus appear smooth in outline, lacking fine structural detail. The genome segments are 1.7 and 2 kbp in size and the capsid is made up of molecules of a single protein species of 55 kDa.

2. Genus *Betacryptovirus* (type species: *White clover cryptic virus 2*)

The isometric particles are 38 nm in diameter, show prominent subunits, and contain genome segments of about 2.1 and 2.25 kbp. The protein subunit weight has not yet been determined.

F. No family

1. Genus *Varicosavirus* (type species: *Lettuce big-vein virus*)

This genus has rod-shaped virions of 320–360 nm length and 18 nm diameter that contain two dsRNA molecules encapsidated in multiple copies of a single protein species of about 48 kDa. They are transmitted by chytrid fungi (*Olpidium*).

G. Family *Rhabdoviridae* (reviewed by Jackson *et al.*, 1999)

Rhabdoviruses have characteristic bullet-shaped or bacilliform membrane-enveloped particles. The outer surface has glycoprotein spikes that pass through the membrane. On the inner surface of the membrane is a layer of protein (the matrix protein) that contains the helically arranged nucleocapsid. The nucleocapsid is made up of (–)-strand RNA and nucleoprotein together with small amounts of other virus-coded proteins. Their structure is discussed in Chapter 5 (Section VIII.A). Rhabdoviruses infect vertebrates, invertebrates and plants.

There are two genera of plant-infecting rhabdoviruses that resemble the animal-infecting viruses except in having an extra gene (in the few viruses that have been sequenced), which is thought to encode a protein for cell-to-cell movement (for genome arrange see Chapter 6, Section VII.A). The plant-infecting rhabdoviruses are transmitted by insects in a circulative, propagative manner. The most common vectors are leafhoppers, planthoppers and aphids, though mite- and lacebug-transmitted viruses (one each) have been identified.

1. Genus *Cytorhabdovirus* (type species: *Lettuce necrotic yellows virus*)

Members of this genus are thought to replicate in the perinuclear space or in the endoplasmic reticulum and to mature by budding into the cytoplasm. Many of the members have relatively narrow (about 60 nm) diameter particles.

2. Genus *Nucleorhabdovirus* (type species: *Potato yellow dwarf virus*)

Members of this genus are thought to replicate in the nucleus and to mature by budding through the inner nuclear membrane into the perinuclear space. Many of the members have relatively broad (about 90 nm) diameter particles.

3. Unassigned

Many plant rhabdoviruses have not been assigned to a genus, as they have not been sufficiently characterized. Some of these appear to lack the outer membrane and essentially be just nucleoprotein 'cores' (Francki *et al.*, 1985a).

H. Family *Bunyaviridae*

The *Bunyaviridae* is a large family of viruses, most of which infect both vertebrates and invertebrates. There is one genus that infects plants and also its invertebrate vectors. The virions are spherical or pleomorphic with surface glycoprotein spikes embedded in a lipid bilayer envelope. Within this envelope the genome, comprising three RNA species, is associated with a nucleoprotein forming a nucleocapsid. The 5' and 3' terminal nucleotides of each segment are complementary, leading to the formation of 'panhandle' structures. The RNAs are either (–)-sense or have their coding regions in an ambisense arrangement (see Chapter 7, Section V.B.12).

1. Genus *Tospovirus* (type species: *Tomato spotted wilt virus*) (reviewed by Moyer, 1999)

Tospoviruses have three ssRNA species, two of which have an ambisense strategy (see Chapter 6, Section VII.B.1), the largest being of (–)-sense. These are contained within membrane-bound particles (see Chapter 5, Section VIII.B, for structure). One of the gene products is involved in cell-to-cell spread of the virus (see Chapter 9, Section II.D.1.j) which distinguishes members of this genus from other Bunyaviruses.

Tospoviruses are transmitted by thrips in a circulative, propagative manner (see Chapter 11, Section VI.B). TSWV has a host range of more than 925 species belonging to 70 botanical families.

I. No family

There are two genera that have affinities to the *Bunyaviridae* but for which there is not sufficient information to formally place them in that family

1. Genus *Tenuivirus* (type species: *Rice stripe virus*) (reviewed by Falk and Tsai, 1998; Falk, 1999)

The virions of tenuiviruses so far characterized comprise very long filamentous nucleoproteins, 3–10 nm in diameter, which often form a circular outline. Although these nucleoproteins have various features of the genome organization that indicate a relationship to bunyaviruses, particularly the *Phlebovirus* genus, no membrane-bound particles have yet been detected. The genome comprises four or more segments, the largest having negative polarity and three or more species with an ambisense arrangement (Chapter 6, Section VII.C).

Tenuiviruses infect members of the *Graminae* and are transmitted by planthoppers in a circulative, propagative manner.

2. Genus *Ophiovirus* (type species: *Citrus psorosis virus*)

Ophioviruses have genomes similar to those of tenuiviruses, being filamentous nucleocapsids about 3 nm in diameter and often taking on a circular form. The genome is divided into three segments forming nucleoproteins with a protein species of 43–50 kDa.

No natural vectors have been identified for ophioviruses. Members of this genus infect both dicotyledonous and monocotyledonous plant species.

J. Family *Bromoviridae*

Members of this family have isometric particles with T = 3 icosahedral symmetry, 26–35 nm in diameter or bacilliform particles whose symmetry is based upon the icosahedron. The genomes of linear positive-sense ssRNA are divided between three molecules with genome organizations as shown in Chapter 6 (Section VIII.A). The subgenomic RNA for coat protein is often also encapsidated; several members also encapsidate satellite or defective interfering RNAs.

1. Genus *Bromovirus* (type species: *Brome mosaic virus*) (reviewed by Ahlquist, 1999)

The virions of bromoviruses are isometric, about 27 nm in diameter and are stabilized by a pH-dependent protein:protein interaction and also a protein:RNA interaction. Above pH 7 the particles swell and become salt labile. The capsids are made up of 180 copies of a single protein species of 20 kDa.

The natural host range is narrow. All species are thought to be transmitted by beetles, and some species are possibly naturally mechanically transmitted (see Chapter 12, Section II.F).

2. Genus *Cucumovirus* (type species: *Cucumber mosaic virus*) (reviewed by Roossinck, 1999a)

Cucumoviruses have isometric particles, about 30 nm in diameter, that are stabilized by protein:RNA interactions and thus are salt labile. The capsid comprises 180 copies of a single protein species of about 24 kDa.

CMV has an extensive host range infecting over 1000 species in more than 85 plant families. Other cucumoviruses have a narrower host range. All are transmitted by aphids in a nonpersistent manner.

CMV is divided into two subgroups (I and II) based on serology and nucleotide sequence identity (Table 2.3). PSV probably also has several subgroups.

TABLE 2.3 Nucleotide sequence similarities between CMV subgroups and *Cucumovirus* species for each RNA. From Roossinck (1999a), with permission.

Virus	CMV-I	CMV-II	PSV	TAV
RNA1				
CMV-I	100	78	68	70
CMV-II		100	70	69
PSV			100	68
RNA2				
CMV-I	100	74	65	64
CMV-II		100	66	66
PSV			100	64
RNA3				
CMV-I	100	78	64	65
CMV-II		100	65	66
PSV			100	72

3. Genus *Alfamovirus* (type species: *Alfalfa mosaic virus*) (reviewed by Bol, 1999a)

The virions of this monospecific genus are bacilliform, 18 nm in diameter and 30–57 nm long depending on the RNA species, encapsidated in a protein of 24 kDa (see Chapter 5, Section VI.B.2, for details of structure). They are stabilized primarily by protein:RNA interactions and thus are salt labile. The presence of coat protein or the subgenomic RNA encoding it is required for virus replication (see Chapter 8, Section IV.G). Because of sequence similarities, it has been suggested that AMV is included in the genus *Ilarvirus* (Scott *et al.*, 1998).

AMV has a wide host range and is transmitted by aphids in a non-persistent manner.

4. Genus *Ilarvirus* (type species: *Tobacco streak virus*) (reviewed by Bol, 1999a)

Ilarvirus particles have quasi-isometric shapes varying from being roughly spherical to being bacilliform. They are stabilized by protein:RNA bonds and thus are salt labile. As with AMV, the coat protein (24–26 kDa) or its subgenomic mRNA is required for replication; AMV coat protein can be substituted for ilarvirus coat protein.

Viruses in this genus infect mainly woody plants and are transmitted by pollen and seeds.

Ilarviruses are divided into seven or eight subgroups or clusters based on serological relationships (Bol, 1999a).

5. Genus *Oleavirus* (type species: *Olive latent virus 2*)

Virions of this monospecific genus have different shapes and sizes ranging from quasi-spherical with a diameter of 26 nm to bacilliform of diameter 18 nm and lengths of 55, 48, 43 and 37 nm. They encapsidate the three genomic RNAs and a subgenomic RNA that is not the coat protein mRNA. The coat protein is 20 kDa.

No natural vector is known for OLV-2.

K. Family *Comoviridae*

Virions are isometric with T = 1 (pseudo T = 3) icosahedral symmetry (see Chapter 5, Section VI.B.6.a, for structure). The capsids are made up of one or two coat protein species and encapsidate the genome comprising two positive-sense ssRNA species. The RNAs are expressed as polyproteins (for genome organization, see Chapter 6, Section VIII.B) that are processed to give the functional proteins (see Chapter 7, Section V.E.8).

1. Genus *Comovirus* (type species: *Cowpea mosaic virus*) (reviewed by Lomonossoff and Shanks, 1999)

The capsids are constructed from two polypeptide species (M_r 40–45 kDa and 21–27 kDa) expressed from the smaller RNA species (see Chapter 6, Section VIII.B.1, for genome organization).

Comoviruses are transmitted by Chrysomelid beetles.

2. Genus *Fabavirus* (type species: *Broad bean wilt virus 1*) (reviewed by Cooper, 1999)

As with comoviruses, the capsids of fabaviruses are constructed from two protein species of similar sizes to those of comoviruses and also expressed from the smaller RNA species (see Chapter 6, Section VIII.B.2, for genome organization).

Fabaviruses are transmitted by aphids in a non-persistent manner.

3. Genus *Nepovirus* (type species: *Tobacco ringspot virus*) (reviewed by Murant *et al.*, 1996; Mayo and Jones 1999a)

Capsids of TRSV are composed of a single polypeptide species (52–60 kDa) while those of other nepoviruses may be made up of two or three smaller proteins. The genome organization (see Chapter 6, Section VIII.B.3) is similar to that of comoviruses.

Nepoviruses often cause ringspot symptoms. Many are transmitted by longidorid nematodes (see Chapter 11, Section XI.E), some by pollen, one (BRAV) by mites and others have no known vector.

The species in this genus are clustered into three subgroups 'a', 'b' and 'c', based on the length and packaging of RNA2 and on serological relationships (Murant, 1981).

L. Family *Potyviridae* (reviewed by Shukla *et al.*, 1994; López-Moya and García, 1999)

Members of the *Potyviridae* have flexuous particles 650–900 nm long (genus *Bymovirus* with two particle lengths) and about 11–15 nm in diameter. The genome is positive-sense ssRNA with a VPg at the 5′ end and a 3′ poly(A) tract; the bymovirus genome is divided between two segments. The genome is expressed as a polyprotein (for genome organization, see Chapter 6, Section VIII.C) that is cleaved to functional proteins (see Chapter 7, Section V.B.1.b).

All members of the *Potyviridae* form cylindrical inclusion bodies in infected cells (see Chapter 3, Section IV.C). They are transmitted by a variety of vectors—one of the characters that define the genus.

1. Genus *Potyvirus* (type species: *Potato virus Y*)

Virus particles are 680–900 nm long and 11–13 nm in diameter encapsidating a genome of about 9.7 kb with multiple copies of a single protein species of 30–47 kDa.

Potyviruses are transmitted by aphids in the non-persistent manner using a helper component (see Chapter 11, Section III.E). Some of the members of this genus are seed transmitted (see Chapter 12, Section III.A). This is the largest of the genera of plant viruses (91 species and 88 tentative species) and contains some economically important virus such as PVY, BYMV, PPV and PRSV.

2. Genus *Ipomovirus* (type species: *Sweet potato mild mottle virus*)

Virus particles are 800–950 nm long containing a genome of 10.8 kb encapsidated in multiple copies of a single protein species of about 38 kDa. The natural vector is the whitefly *Bemisia tabaci* with which the virus has a non-persistent relationship.

3. Genus *Macluravirus* (type species: *Maclura mosaic virus*)

Macluravirus particles are 650–675 nm long and contain an RNA of about 8 kb encapsidated in multiple copies of a single protein species of 33–34 kDa. They are transmitted by aphids in a non-persistent manner.

4. Genus *Rymovirus* (type species: *Ryegrass mosaic virus*)

Members of this genus have particles of 690–720 nm length and 11–15 nm diameter that contain an RNA genome of 9–10 kb with the capsid comprising many copies of a single protein species of about 29 kDa. They are transmitted by eriophyid mites (see Chapter 11, Section IX.A).

5. Genus *Tritimovirus* (type species: *Wheat streak mosaic virus*)

Tritimovirus particles are 690–700 nm long and contain an RNA of about 8.5–9.6 kb in size, the capsid comprising multiple copies of a protein of about 32 kDa. They are restricted to the *Graminae* and are transmitted by eriophyid mites possibly in a persistent manner (see Chapter 11, Section IX.A).

6. Genus *Bymovirus* (type species: *Barley yellow mosaic virus*)

The virions are flexuous rods of 13 nm width and two modal lengths, 250–300 and 500–600 nm. The genome is divided between two RNA species, the longer particle containing RNA of 7.5–8.0 kb in size and the shorter particle RNA of 3.5–4.0 kb in size. The capsid is made up of multiple copies of a single protein species of 28–33 kDa.

Members of this genus are restricted to the *Graminae* and are transmitted by the plasmodiophora fungus *Polymyxa graminis*.

M. Family *Tombusviridae*

Viruses in this family have isometric icosahedral (T = 3) particles, 32–35 nm in diameter, with well-defined structure (see Chapter 5, Section VI.B.4.d). The particles contain a single species of positive-sense ssRNA with a size ranging from 3.7 to 4.7 kb. The genome organization differs between genera in this family (see Chapter 6, Section VIII.D). The unifying feature of the family is that each member has a highly conserved RNA-dependent RNA polymerase that is interrupted by an in-frame termination codon that is periodically suppressed (see Chapter 7, Section V.B.9).

Individual members have relatively narrow host ranges and can infect either mono- or di-cotyledonous species, but not both. These viruses are relatively stable and many are found in natural environments such as surface waters and soils from which in many cases they can be acquired without the need for a biological vector.

1. Genus *Tombusvirus* (type species: *Tomato bushy stunt virus*) (reviewed by Rochon, 1999)

Members of this genus have genomes of about 4.7 kb encapsidated in particles of 32–35 nm, the detailed structure of which has been well characterized (see Chapter 6, Section VIII.D.1). They are composed of 180 subunits of a single protein species of 41 kDa.

The transmission of most viruses in this genus is soil-borne with no obvious biological vector. CNV is transmitted by the fungus *Olpidium bornovanus*.

2. Genus *Aureusvirus* (type species: *Pothos latent virus*)

The single species in this genus, PoLV, has a genome of 4.4 kb contained in particles of 30 nm diameter that have a rounded outline and knobby surface made up of 180 copies of a single 40-kDa protein species.

PoLV has only been reported from southern Italy. Its natural transmission is through the soil or circulating solution in hydroponics.

3. Genus *Avenavirus* (type species: *Oat chlorotic stunt virus*)

The single species in this genus, OCSV, has a coat protein significantly larger (48.2 kDa) than those of most other members of the family *Tombusviridae* with a protruding domain. The isometric particles of 35 nm diameter contain a genome of 4114 nt with a genome organization described in Chapter 6 (Section VIII.D.3).

OCSV has only been described in oats (*Avena sativa*). Its natural transmission is soil-borne possibly by zoosporic fungi.

4. Genus *Carmovirus* (type species: *Carnation mottle virus*) (reviewed by Qu and Morris, 1999)

The 3.88–4.45 kb RNA genome of members of this genus is contained in isometric particles of 32–35 nm diameter. The particles have a regular surface structure under the electron microscope, giving a granular appearance but not showing subunit arrangement of its T = 3 icosahedral symmetry comprising 180 copies of a single protein species of 36–41 kDa.

5. Genus *Machlomovirus* (type species: *Maize chlorotic mottle virus*) (reviewed by Lommel, 1999b)

This is a monospecific genus. MCMV particles are approximately 30 nm in diameter and comprise a genomic RNA of 4.4 kb and capsid protein subunits of 25 kDa. The genome organization is discussed in Chapter 6 (Section VIII.D.5).

The host range of MCMV is restricted to members of the family *Graminae*. The virus is seed-transmitted and has been reported to be transmitted by chrysomelid beetles and thrips.

6. Genus *Necrovirus* (type species: *Tobacco necrosis virus* A) (reviewed by Meulewaeter, 1999)

Necrovirus virions are about 28 nm in diameter and are made up of 180 subunits of a 30-kDa capsid protein. The genome is linear ssRNA of 3.7 kb, the organization of which is shown in Chapter 6 (Section VIII.D.6).

Necroviruses have wide host ranges including both monocotyledonous and dicotyledonous species. In nature the infection is usually restricted to the roots. These viruses are transmitted by the chytrid fungus, *Olpidium brassicae* (see Chapter 11, Section XII).

7. Genus *Panicovirus* (type species: *Panicum mosaic virus*)

The 4.3 kb genomic RNA is contained in isometric particles of approximately 30 nm diameter comprising 180 copies of a single protein species of 26 kDa. As with the machlomoviruses, the panicovirus RNA-dependent RNA polymerase has an N-terminal extension fused to the rest of the polymerase that is conserved in the family *Tombusviridae* (see Chapter 6, Section VIII.D.7, for genome organization).

The host range is restricted to the *Graminae*. Natural transmission is most likely mechanical.

8. Genus *Dianthovirus* (type species: *Carnation ringspot virus*) (reviewed by Lommel, 1999a)

Dianthoviruses differ from the rest of the *Tombusviridae* in that their genomes are divided between two species of ssRNA; the genome organization is shown in Chapter 6 (Section VIII.D.8). These two RNAs are contained in virions of 32–35 nm diameter that have icosahedral symmetry being made up of 180 subunits of a 37–38-kDa coat protein.

The moderately broad host range of dianthoviruses is restricted to dicotyledonous plant species. The virus appears to be readily transmitted through the soil but no biological vector is known.

N. Family *Sequiviridae*

Members of this family have isometric particles, 30 nm in diameter, that contain positive-sense ssRNA 9–12 kb in size. The virion is made up of three protein species of approximately 32, 26 and 23 kDa; these proteins are in equimolar amounts. The RNA encodes a polyprotein that is cleaved to give the functional proteins.

1. Genus *Sequivirus* (type species: *Parsnip yellow fleck virus*) (reviewed by Mayo and Murant, 1999)

The RNA is about 10 kb and is not polyadenylated (see Chapter 6, Section VIII.E.1, for genome organization). The viruses are transmitted by aphids in a semi-persistent noncirculative manner; transmission of PYFV depends on a helper waikavirus, AYV.

2. Genus *Waikavirus* (type species: *Rice tungro spherical virus*) (reviewed by Gordon, 1999)

The RNA is more than 11 kb and has a 3′ poly(A) tail. One or more small ORFs near the 3′ end of the RNA have been reported but there is uncertainty as to whether these are expressed (see Chapter 6, Section VIII.E.2, for genome organization).

Waikaviruses are transmitted either by leafhoppers or aphids in a semi-persistent noncirculative manner and transmission is thought to involve one or more virus-encoded helper proteins.

O. Family *Closteroviridae* (reviewed by German-Retana *et al.*, 1999)

Members of the *Closteroviridae* have very flexuous filamentous particles about 12 nm in diameter, the length being characteristic of the genus. The genomic nucleic acid is linear, positive-sense ssRNA that may be mono- or bipartite, dependent on genus.

The host ranges of individual species are usually narrow. These viruses are usually phloem limited and cause yellowing-type symptoms or pitting or grooving of woody stems. Transmission is a characteristic of the genus but they may be reclassified into aphid-, whitefly- and mealybug-transmitted genera.

1. Genus *Closterovirus* (type species: *Beet yellows virus*) (reviewed by Bar-Joseph *et al.*, 1997; German-Retana *et al.*, 1999)

The flexuous virions are 1250–2000 nm long, with one end coated with an anomalous coat protein giving a 'rattlesnake' structure (see Fig. 5.12). For most species, there is a single characteristic particle length, but CTV also has shorter particles containing subgenomic or defective RNAs. Full-length particles contain a single RNA species of 15.5–19.3 kb in size, the genome organization of which is described in Chapter 6 (Section VIII.F.1).

Natural transmission is by aphids, mealybugs or whiteflies (*Trialeuroides*), dependent on species. Both CTV and BYV cause important diseases.

2. Genus *Crinivirus* (type species: *Lettuce infectious yellows virus*)

Crinivirus particles have two modal lengths, 700–900 nm and 650–850 nm and, as with Closteroviruses, has an anomalous protein at one end giving a 'rattlesnake' structure; the major coat protein is 28–33 kDa. The genome is in two segments that are separately encapsidated.

The natural vectors of criniviruses are whiteflies (*Bemisia* and *Trialeuroides*) which transmit in a semi-persistent manner.

P. Family *Luteoviridae* (reviewed by Smith and Barker, 1999; Mayo and Miller, 1999; Miller, 1999)

The viruses in this family have isometric icosahedral (T = 3) particles 25–30 nm in diameter that are hexagonal in outline and comprise 180 subunits of a single protein species of 21–23 kDa. The particles contain a single molecule of positive-sense ssRNA 5.7–5.9 kb in size. Genera are distinguished on genome organization (see Chapter 6, Section VIII.G).

Many members of the *Luteoviridae* are phloem-limited, causing 'yellows'-type diseases—hence the name luteovirus from the Latin 'luteus' meaning yellow. Transmission is by aphids in a specific, circulative, non-propagative manner (see Chapter 11, Section III.H.1.a). These viruses often assist the transmission of other viruses (see Chapter 11, Section III.H.1.a).

The sequencing of genomes of members of this family has led to considerable reorganization of the classification.

1. Genus *Luteovirus* (type species: *Barley yellow dwarf virus*-PAV)

Members of this genus are restricted to *Graminae* where they cause serious disease.

2. Genus *Polerovirus* (type species: *Potato leafroll virus*)

This genus has members that infect dicotyledonous species and others that are restricted to monocotyledonous species. They also cause serious diseases.

3. Genus *Enamovirus* (type species: *Pea enation mosaic virus 1*)

Pea enation mosaic disease is caused by a complex of two viruses, PEMV-1 which is a member of the *Luteoviridae*, and PEMV-2 which is an umbravirus (Demler *et al.*, 1993; 1994a). Two sizes of isometric particles are found, that of PEMV-1 (28 nm diameter) having icosahedral symmetry and that of PEMV-2 (25 nm diameter) having quasi-icosahedral symmetry (see Chapter 5, Section VI.B.4.f). Both types of particle are composed of coat protein encoded by PEMV-1. PEMV-1 can infect protoplasts but does not spread in plants unless PEMV-2 is present. Unlike other members of the *Luteoviridae*, pea enation mosaic is transmissible mechanically, as well as by aphids, and spreads to mesophyll tissues; these two properties are conferred by PEMV-2. Virions of some strains of PEMV-1 + PEMV-2 also contain a satellite RNA.

Aphid transmission of PEMV-1 + PEMV-2 is conferred by PEMV-1. Aphid transmissibility can be lost after multiple passages of mechanical transmission. Differences have been found in the electrophoretic profiles of the particles of aphid-transmissible and non-transmissible isolates (Hull, 1977b) (Fig. 2.7) but the cause of these differences has not been fully resolved.

4. Unassigned

Eleven viruses that have properties suggestive of the family *Luteoviridae* have not been assigned to genera because of lack of full characterization.

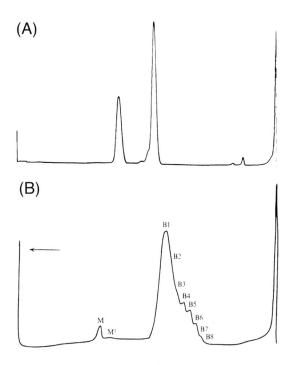

Fig. 2.7 Densitometer traces of the nucleoproteins of **(a)** a PEMV isolate not aphid transmissible, and **(b)** an aphid-transmissible PEMV isolate, electrophoresed in 3.4% polyacrylamide gels at 16 V/cm for 16 h using the buffer system described by Hull and Lane (1973). From Hull (1977b), with permission.

Q. Floating genera

There are currently 20 genera of (+)-strand ssRNA plant viruses that have not been placed in families. I will describe these in an order that brings together viruses with similar properties — firstly rod-shaped viruses and then viruses with isometric particles. The first two genera have rigid rod-shaped particles and several similarities in genome organization. Genera 3–9 have flexuous rod-shaped particles with a single RNA genomic species; several of these have been reclassified relatively recently from larger groupings. Genera 10–14 have rod-shaped particles encapsidating divided genomes; several of these also have recently been reclassified. Genera 15–18 have isometric particles, genus 19 has bacilliform particles, and genus 20 does not self-encapsidate. It is likely that some of the arbitrary juxtapositions will be grouped into families as and when more information becomes available.

1. Genus *Tobamovirus* (type species: *Tobacco mosaic virus*) (reviewed by Lewandowski and Dawson, 1999; Knapp and Lewandowski, 2001)

As described in Chapter 1, TMV was the first virus to be recognized as a pathogenic entity that differed from bacteria, and this virus has been in the forefront of many of the advances in plant virology. The centenary of its recognition was celebrated in 1999 with various meetings and publications (see Scholthof *et al.*, 1999a; a special issue of *Proc. Roy. Soc. Lond. B*). In fact, this has led to the problem that TMV has the image of the archetypical plant virus.

Tobamoviruses have rod-shaped particles 300–310 nm long and 18 nm in diameter; the structure of these particles has been studied in detail (see Chapter 5, Section III.B). The particles contain a single positive-strand ssRNA of 6.3–6.6 kb in size, the genome organization of which is described in Chapter 6 (Section VIII.H.1). The capsid comprises multiple copies of a single polypeptide species of 17–18 kDa.

Tobamoviruses are very stable with purified preparations and leaf material retaining infectivity for more than 50 years. They are naturally transmitted mechanically (see Chapter 12, Section II.F).

2. Genus *Tobravirus* (type species: *Tobacco rattle virus*) (reviewed by Visser *et al.*, 1999)

Tobraviruses have rod-shaped particles of 20–22 nm diameter and with two predominant lengths, 180–215 nm and 46–115 nm. The genome is positive-sense ssRNA and is divided between two species, the size of that in the long particles being about 6.8 kb and in the short particles ranging from 1.8 kb to about 4.5 kb (see Chapter 6, Section VIII.H.2, for genome organization). The capsid is made up of many copies of a single protein species of 22–24 kDa.

Tobraviruses have wide host ranges including both mono- and dicotyledonous species. The natural vectors are Trichodorid nematodes (see Chapter 11, Section XI.E). PEBV has been reviewed by Boulton (1996).

3. Genus *Potexvirus* (type species: *Potato virus X*) (reviewed by AbouHaidar and Gellatly, 1999)

Potexviruses have flexuous, filamentous particles 470–580 nm long and 13 nm in diameter. These virions contain a single species of linear, positive-sense ssRNA of about 5.9–7.0 kb encapsidated in multiple copies of a single species of coat protein of 18–27 kDa. The RNA is capped at the 5′ end and is 3′ polyadenylated. The genome organization is described in Chapter 6 (Section VIII.H.3).

Viruses in this genus are transmitted by mechanical contact.

4. Genus *Carlavirus* (type species: *Carnation latent virus*) (reviewed by Zavriev, 1999)

Members of this genus have slightly flexuous filamentous particles, 610–700 nm long and 12–15 nm in diameter. These virions contain a single molecule of linear, positive-sense ssRNA of 7.4–8.5 nm in size that has a 3′ poly(A) tract; some species encapsidate subgenomic RNAs in shorter particles. The capsid is composed of a single polypeptide species of 31–36 kDa. The genome organization is described in Chapter 6 (Section VIII.H.4).

Most carlavirus species are transmitted by aphids in the non-persistent manner. However, CPMMV is transmitted by the whitefly *Bemisia tabaci*. CPMMV is also seed-transmitted, as are PeSV and RCVMV. Some species (e.g. CLV and PVS) are naturally mechanically transmitted.

5. Genus *Allexivirus* (type species: *Shallot virus X*) (see Kanyuka *et al.*, 1992)

The virions of allexiviruses are highly flexuous and filamentous, about 800 nm long and 12 nm in diameter. They contain a single species of linear, positive-sense ssRNA about 9.0 kb in size which is encapsidated in a single species of coat protein of 28–36 kDa. The genome organization is described in Chapter 6 (Section VIII.H.5).

Allexiviruses have very narrow host ranges and are naturally transmitted by eriophyid mites.

6. Genus *Capillovirus* (type species: *Apple stem grooving virus*) (reviewed by Salazar, 1999)

Capilloviruses have flexuous, filamentous rod-shaped particles 640–700 nm long and 12 nm in diameter containing a linear, positive-sense ssRNA genome of 6.5–7.4 kb in size. The RNA is polyadenylated at the 3' end and is encapsidated in a single protein species of 24–27 kDa. The genome organization is described in Chapter 6 (Section VIII.H.6).

No natural vectors are known for viruses in this genus; ASGV is seed-transmitted.

7. Genus *Foveavirus* (type species: *Apple stem pitting virus*) (see Jelkmann, 1994; Martelli and Jelkmann, 1998)

Members of this genus have flexuous filamentous virions about 800 nm long and 12 nm in diameter. The virions comprise a single molecule of linear, positive-sense ssRNA, 8.4–9.3 kb in size and a single species of coat protein (28–44 kDa). The genomic RNA is 3' polyadenylated and its organization is described in Chapter 6 (Section VIII.H.7).

The natural host range of species in this genus is restricted to one or a few plant species. There is no known natural vector for any of the species.

8. Genus *Trichovirus* (type species: *Apple chlorotic leaf spot virus*) (reviewed by German-Retana and Candresse, 1999)

Members of this genus have very flexuous rod-shaped particles ranging from 640 nm to 760 nm in length and having a diameter of 12 nm. The virions contain a single species of linear positive-strand ssRNA of about 7.5 kb in size with a polyadenylated 3' terminus which is encapsidated in many copies of a single polypeptide species of 22–27 kDa. The genome organization is described in Chapter 6 (Section VIII.H.8).

The host ranges of trichoviruses are relatively narrow. No natural vector has yet been identified, although there is evidence for field spread of GINV. PVT is seed-transmitted in several hosts.

9. Genus *Vitivirus* (type species: *Grapevine virus A*) (reviewed by German-Retana and Candresse, 1999)

Viruses in this genus have flexuous filamentous particles 725–825 nm long and 12 nm in diameter. The particles contain a single species of linear, positive-stranded ssRNA of about 7.6 kb in size, capped at the 5' terminus and 3' polyadenylated. The capsid comprises multiple copies of a single protein species of 18–22 kDa. The genome organization is described in Chapter 6 (Section VIII.H.9).

The natural host range of individual virus species is restricted to a single plant species, though the experimental host range may be large. GVA and GVB are transmitted in the semi-persistent manner by mealybugs (genera *Pseudococcus* and *Planococcus*) and GVA is also transmitted by the scale insect *Neopulvinaria innumerabilis*. HLV is transmitted semi-persistently by aphids in association with a helper virus.

10. Genus *Furovirus* (type species: *Soil-borne wheat mosaic virus*) (reviewed by Shirako and Wilson, 1999)

The virions of furoviruses are rigid rods of about 20 nm diameter and two predominant lengths of 260–300 nm and 140–160 nm. The genome is bipartite, linear positive-sense ssRNA that in the longer particles is about 6–7 kb in size, and in the shorter particles 3.5–3.6 kb; shorter particles of SBWMV may contain populations of deletion mutants. The capsid comprises multiple copies of a single polypeptide of 19–21 kDa. The genome organization is shown in Chapter 6 (Section VIII.H.10).

As the name of the type species and the sigla for the generic name (furo- = *fu*ngus-borne, *ro*d-shaped) suggest, viruses in this genus are transmitted by fungi. SBWMV is transmitted by the plasmodiophora fungus *Polymyxa graminis*.

11. Genus *Pecluvirus* (type species: *Peanut clump virus*) (reviewed by Reddy *et al.*, 1999; Shirako and Wilson, 1999)

Viruses in this genus have rod-shaped virions of about 21 nm diameter and two predominant lengths of 245 and 190 nm. The particles contain linear positive-sense ssRNA of two sizes, that from the long particles being about 5.9 kb and that from the short particles about 4.5 kb encapsidated in multiple copies of a single protein species of 23 kDa. The genome organization is described in Chapter 6 (Section VIII.H.11).

Pecluroviruses are transmitted by the plasmodiophorid fungus *Polymyxa graminis* and by seed (in groundnuts).

12. Genus *Pomovirus* (type species: *Potato mop-top virus*) (reviewed by Torrance, 1999; Shirako and Wilson, 1999)

Pomoviruses have rod-shaped particles 18–20 nm in diameter and of three predominant lengths, 290–310 nm, 150–160 nm and 65–80 nm. These contain a genome of linear, positive-sense ssRNA divided into three molecules of about 6, 3–3.5 and 2.5–3.0 kb in size encapsidated in multiple copies of a major 20-kDa coat protein. The genome organization is described in Chapter 6 (Section VIII.H.12).

Viruses in this genus have narrow host ranges, so far limited to dicotyledonous species. They are transmitted in the soil by fungi; *Spongospora subterranea* and *Polymyxa betae* have been identified as vectors of PMTV and BSBV respectively.

13. Genus *Benyvirus* (type species: *Beet necrotic yellow vein virus*) (reviewed by Tamada, 1999)

Benyviruses have rod-shaped particles of 20 nm diameter and four predominant lengths, about 390, 265, 100 and 85 nm. These contain the genome of linear, positive-sense ssRNA which is divided into four molecules of about 6.7, 4.6, 1.8 and 1.4 kb in size; some sources of BNYVV have a fifth RNA molecule of 1.3 kb in size. The largest two make up the infectious genome, the other RNAs being ancillary and influencing transmission and symptomatology (see Chapter 14, Section II.B.3.c). These RNAs are 5′ capped and, unlike the RNAs genomes of other rod-shaped viruses, are 3′ polyadenylated. The capsid is made up of a major protein species of 21–23 kDa. The genome organization is described in Chapter 6 (Section VIII.H.13).

The type species of this genus, BNYVV, is economically very important in many sugarbeet growing areas. It and the other species in this genus, BSBMV, are transmitted by the fungus *Polymyxa betae*.

14. Genus *Hordeivirus* (type species: *Barley stripe mosaic virus*) (reviewed by Lawrence and Jackson, 1999)

Hordeiviruses have rigid rod-shaped particles of about 20 nm diameter and ranging in length from 110 to 150 nm. The genome is divided between three molecules of linear, positive-sense ssRNA designated α, β and γ; α ranges in size between species from 3.7 to 3.9 kb, β from 3.1 to 3.6 kb and γ from 2.6 to 3.2 kb; between strains of a given species the sizes of α and β are relatively similar but that of γ can vary (see Chapter 6, Section VIII.H.14, for genome organizations). The capsid is composed of multiple copies of a single coat protein species of 17–18 kDa.

There are no natural vectors known for any member of this genus with field spread thought to be by leaf contact. BSMV and LRSV are efficiently seed-transmitted, this providing primary foci from which secondary spread occurs.

15. Genus *Sobemovirus* (reviewed by Sehgal, 1999; Tamm and Truve, 2000)

Members of this genus have isometric virions (T = 3) of about 30 nm made up of 180 subunits of a single capsid protein species of about 26–30 kDa. Each particle contains a single molecule of positive sense ssRNA of about 4.1–4.5 kb which has a VPg at the 5′ end. The genome organization is described in Chapter 6 (Section VIII.H.15).

Several sobemoviruses are seed-transmitted, most are beetle-transmitted and one is transmitted by myrids (see Chapter 11, Section VII.B).

16. Genus *Marafivirus* (type species: *Maize rayado fino virus*)

Purified preparations of marafiviruses sediment as two components, B component that contains the viral genome and T component that contains no RNA. The icosahedral particles have a diameter of 28–32 nm and most contain a major protein of 21–22 kDa and a minor protein of about 24–28 kDa. The 6.5 kb genome in the B component is capped at the 5′ end, polyadenylated at the 3′ end and has a high content of cytidine. The genome organization is shown in Chapter 6 (Section VIII.H.16).

Marafiviruses have narrow host ranges restricted to the family *Gramineae*. They are transmitted by leafhoppers and possibly replicate in their vectors.

17. Genus *Tymovirus* (type species: *Turnip yellow mosaic virus*) (reviewed by Gibbs, 1999a)

The virions of tymoviruses are isometric, with diameter about 30 nm, and have icosahedral T = 3 symmetry with 180 subunits of a protein of 20 kDa. Both full (B component) and empty (T component) particles can be distinguished by electron microscopy and sedimentation. The full particles contain a single species of linear, positive-sense ssRNA of about 6.3 kb in size which has a high cytidine content, resembling marafiviruses. The genome organization of this RNA is described in Chapter 6 (Section VIII.H.17). The replication of the viral RNA is associated with vesicles in the periphery of chloroplasts (see Chapter 8, Section IV.K.2).

Tymoviruses appear to be restricted to dicotyledonous hosts. They are transmitted by beetles of the families *Chrysomelidae* and *Curculionidae*.

18. Genus *Idaeovirus* (type species: *Raspberry bushy dwarf virus*) (reviewed by Mayo and Jones, 1999b)

This monospecific genus has isometric virions about 33 nm in diameter that contain three species of linear, positive-sense ssRNA of about 5.5, 2.2 and 1 kb in size. The larger two make up the viral genome (see Chapter 6, Section VIII.H.18, for genome organization); the smallest RNA is a subgenomic mRNA for the 30-kDa coat protein.

The natural host range of RBDV is restricted to *Rubus* species. The virus is transmitted both vertically through seed and horizontally through pollen.

19. Genus *Ourmiavirus* (type species: *Ourmia melon virus*) (see Accotto *et al.*, 1997)

Viruses in this genus have bacilliform particles of 18 nm diameter and 62, 46, 37 and 30 nm length (see Fig. 5.26); the ends of the particles are conical (hemi-icosahedral). The genome is linear, positive-sense ssRNA divided into three segments of approximately 2.9, 1.1 and 1.0 kb (estimated from M_r measured by gel electrophoresis). The capsid comprises copies of a single protein species of 25 kDa.

No natural vectors have been identified.

20. Genus *Umbravirus* (type species: *Carrot mottle virus*) (reviewed by Robinson and Murant, 1999)

Umbraviruses do not encode conventional coat proteins and their naturally transmitted particles are formed by association with a helper virus, usually from the family *Luteoviridae*, which contributes its coat protein. The unencapsidated umbravirus RNA is surprisingly stable, being mechanically transmissible, though sensitivity to organic solvents suggests that it might be associated with lipid membranes.

Umbravirus genomes are linear positive-stranded ssRNA of about 4 kb in size; the genome organization is shown in Chapter 6 (Section VIII.H.20).

Members of this genus have very limited host ranges and, as noted above, their natural transmission is by aphids in association with a helper virus.

IV. RETROELEMENTS

Retroelements have been grouped into viral retroelements, eukaryotic chromosomal non-viral retroelements, and bacterial chromosomal retroelements (see Hull and Covey, 1996). Plant viral retroelements, classified in the family *Caulimoviridae,* are pararetroviruses which do not include an integration phase in their replication cycle. Plant genomes (and those of organisms from other kingdoms) contain a variety of retroelements which resemble integrating animal retroviruses in genome organization and replication cycle but are not known to cause infections. These elements are known as retrotransposons. Structurally, they have long terminal repeat (LTR) sequences, which are involved in the integration into the host genome. As well as the basic RNA → DNA → RNA replication, these elements have several other features in common. The enzyme complex of active retroelements comprises reverse transcriptase (RT) and ribonuclease H (RNaseH) and an open reading frame encoding a nucleic acid-binding protein termed *gag* for retroviruses and coat protein for pararetroviruses (see Mason *et al.,* 1987). These elements also encode an aspartate proteinase and an integrase (*int*) (except pararetroviruses). Some of these retrotransposons form virus-like particles for which the ICTV has recently proposed two new families, *Pseudoviridae* and *Metaviridae* (van Regenmortel *et al.,* 2000).

A. Family *Pseudoviridae*

This family comprises retrotransposons that form isometric virus-like particles containing linear positive-strand ssRNA capable of undergoing a retrovirus-like replication cycle. They form an intrinsic, and often significant, part of the genomes of many eukaryotic species, especially plants. In many cases the plant genome contains numerous integrated copies of apparently defective forms of these elements that mutated and were unable to replicate. The virion is thought to play an essential role in the replication cycle but is unlikely to be infectious and capable of horizontal spread.

1. Genus *Pseudovirus* (type species: *Saccharomyces cerevisiae Ty1 virus*)

The particles of members of this genus are 40–60 nm in diameter and are heterodisperse

structurally. Some particles with icosahedral symmetry have been recognized. The particles contain an RNA of 5.6 kb in size that has two ORFs. The 5′ ORF encodes the coat protein (*gag*) and the 3′ ORF (pol) has aspartate protease, integrase and reverse transcriptase activities. Both ORFs are expressed from the genomic RNA, the 3′ ORF by frameshift from the C terminus of the 5′ ORF (see Chapter 7, Section V.B.10, for frameshift).

There are numerous pseudoviruses found in higher plants and some in lower plants such as yeasts.

2. Genus *Hemivirus* (type species: *Drosophila melanogaster copia virus*)

Members of this genus have isometric particles but little is known of their structure.

The 5.1 kb genome has one ORF that encodes the *gag* and *pol* functions. The expression is controlled by splicing which produces an mRNA for *gag*.

Although hemiviruses resemble pseudoviruses and members of the *Caulimoviridae* in using initiator methionine tRNA as a primer for synthesis of (−)-strand DNA from the genomic RNA, they differ from these other viruses/elements in that they use only half the tRNA generated by cleavage in the anticodon loop.

There are no species of this genus known yet in higher plants and only one in yeast.

B. Family *Metaviridae*

The particles formed by members of this family are poorly characterized and often heterogeneous; they are frequently referred to as virus-like particles (VLPs).

1. Genus *Metavirus* (type species: *Saccharomyces cerevisiae Ty3 virus*)

The heterogeneous VLPs of metaviruses are generally spherical, about 50 nm in diameter. They contain a polyadenylated RNA of about 5 kb. The genome of *Saccharomyces cerevisiae Ty3 virus* resembles that of pseudoviruses in having two ORFs but differ in the order of functional domains (protease, reverse transcriptase, integrase) in the *pol* ORF.

This genus contains species from higher plants (*Lilium henryi del1 virus*) and from yeasts and other fungi.

2. Genus *Errantivirus* (type species: *Drosophila melanogaster gypsy virus*)

Members of this genus have enveloped irregular particles of about 100 nm diameter containing an RNA of about 7.5 kb in size. The genome encodes three ORFs, a 5′ ORF encoding the *gag* gene, a *pol* ORF with the functional domains (protease, reverse transcriptase, integrase) in a different order from those of *Pseudoviridae*, and an *env* (envelope) ORF. The *pol* ORF is expressed by frameshift from the *gag* ORF and the *env* ORF is expressed from a spliced mRNA.

There are no errantiviruses known yet in higher or lower plants.

V. VIRUSES OF LOWER PLANTS

A. Viruses of algae

Two groups of viruses have been found infecting algae, large viruses placed in the family *Phycodnaviridae* and a virus that morphologically resembles TMV. There are also one or more uncharacterized viruses.

1. Large algal viruses (reviewed by van Etten, 1999; van Etten and Meints, 1999)

Virus-like particles (VLPs) have been observed in thin sections of many eukaryotic algal species belonging to the *Chlorophyceae*, *Rhodophyceae* and the *Phaeophyceae*. The particles are polygonal in outline and vary in diameter from about 22 to 390 nm. Some have tails reminiscent of bacteriophage. The most studied viruses are those infecting *Chlorella*-like green algae. Members of this group are very diverse biochemically (van Etten *et al.*, 1988). An important technical advance was made when van Etten *et al.* (1983) developed a plaque assay for a virus called Paramecium bursaria chlorella virus 1 (PBCV-1) infecting a culturable *Chlorella*-like alga. As a consequence, most is now known about the properties of this virus.

The particles of PBCV-1 are large icosahedra with multilaminate shells surrounding an electron-dense core. The outer capsid comprises 1692 capsomeres arranged in a $T = 169$ skew icosahedral lattice. Hair-like fibres extend from some of the particle vertices. The genome of PBCV-1 is dsDNA of 330 kb with covalently closed hairpin termini. It contains 701 potential coding ORFs that are arranged in 376, mostly overlapping, ORFs that are believed to encode proteins, and 325 minor ORFs that may or may not encode proteins. The encoded proteins include replication enzymes (e.g. DNA polymerase, DNA ligase and endonuclease), nucleotide metabolism enzymes (e.g. ATPase, thioredoxin and ribonuclease reductase), transcription factors (e.g. RNA transcription factors TFIIB and TFIIS and RNase III), enzymes involved in protein synthesis, modification and degradation (e.g. ubiquitin C-terminal hydrolase, translation elongation factor 3 and 26S protease subunit), phosphorylating enzymes, cell-wall degrading enzymes, DNA restriction and modification enzymes, sugar and lipid manipulation enzymes, and various structural proteins.

Similar large DNA viruses have been found in marine algae (see van Etten and Meints, 1999).

2. Small algal viruses

Skotnicki *et al.* (1976) described a virus infecting the eukaryotic alga *Chara australis*. The virus (CAV) has rod-shaped particles and some other properties like those of tobamoviruses. However, its genome is much larger (11 kb, rather than 6.4 kb for TMV), and about 7 kb of the genome has been sequenced, revealing other relationships (Matthews, 1991): (1) the coat protein of CAV has a composition closer to BNYVV, and to TRV than to TMV; (2) the GDD-polymerase motif of CAV is closest to that of BNYVV; and (3) the two GKT nucleotide-binding motifs found in CAV are arranged in a manner similar to that found in potexviruses.

Thus, CAV appears to share features of genome organization and sequences found in several groups of rod-shaped viruses infecting angiosperms. It appears to have strongest affinity with the *Furovirus* genus and has no known angiosperm host. For these reasons it is most unlikely that CAV originated in a recent transfer of some rod-shaped virus from an angiosperm host to *Chara*.

In its morphology, *Chara* is one of the most complex types of *Charophyceae*, which is a well-defined group with a very long geographical history (Round, 1984). Based on cytological and chemical similarities, land plants (embryophytes) are considered to have evolved from a charophycean green alga. *Coleochaete*, another of the more complex types among the *Charophyceae*, has been shown to contain lignin, a substance thought to be absent from green algae (Delwiche *et al.*, 1989). Molecular genetic evidence supports a charophycean origin for land plants. Group II introns have been found in the tRNA^{ala} and tRNA^{ile} genes of all land plant chloroplast DNAs examined. All the algae and eubacteria examined have uninterrupted genes. Manhart and Palmer (1990) have shown that introns are present in three members of the *Charophyceae* in the same arrangement as in *Marchantia*, giving strong support to the view that they are related to the lineage that gave rise to land plants. Tree construction suggests that the *Charophyceae* may have acquired the introns 400 to 500 million years ago. Thus, the virus described in *Chara* is probably the oldest recorded virus infecting a plant on or near the lineage that ultimately gave rise to the angiosperms.

Straight (*c.* 280 nm) and flexuous (*c.* 700–900 nm) virus particles have been found associated with dieback symptoms in the brown alga *Eklonia radiata* in New Zealand (Easton *et al.*, 1997).

B. Viruses of fungi

Representatives of five viral families infect fungi, if one excludes yeast. Isometric particles of 60 nm diameter and containing a 16.8 kbp dsDNA are found in the aquatic fungus *Rhizidomyces* (Dawe and Kuhn, 1983). This virus belongs to the genus *Rhizidovirus* and appears to be transmitted in a latent form in the fungal zoospores being activated under stress conditions.

As shown in Table 4.1, the majority of fungal viruses have dsRNA genomes. These viruses can influence the biology of plant pathogenic fungi (McCabe and van Alfen, 1999). In many cases, they are cryptic but some induce the fungus (e.g. *Ustilago maydis*) to produce killer toxins and others reduce the virulence of a pathogenic fungus.

DsRNA viruses of fungi are found in three families. The members of the *Hypoviridae* have pleomorphic vesicle-like particles 50–80 nm in diameter that contain a single molecule of dsRNA of about 9–13 kbp (reviewed by Nuss, 1999). The coding region may be divided into two ORFs or expressed as a single ORF depending upon species. Hypoviruses infect the chestnut blight fungus *Cryphonectria parasitica*, reducing its virulence on chestnut trees. The family *Partitiviridae* has two genera that infect fungi (reviewed by Ghabrial and Hillman, 1999). As with the plant-infecting partitiviruses (Section III.E), the isometric particles, 30–40 nm in diameter, contain two segments of dsRNA, the larger encoding the viral polymerase and the smaller the virus coat protein. These viruses are associated with latent infections of fungi. Members of the genus *Totivirus*, family *Totiviridae*, encapsidate a single molecule of dsRNA (4.6–6.7 kbp) in isometric particles, 40 nm in diameter (reviewed by Ghabrial and Patterson, 1999). These viruses are associated with latent infections of their fungal hosts.

The one family of (+)-strand ssRNA genome viruses that infects fungi, the *Barnaviridae*, has bacilliform particles 18–20 nm wide and 48–53 nm long (reviewed by Romaine, 1999). The virion contain a single RNA molecule of 4 kb that has a VPg at the 5′ end (Revill *et al.*, 1998) and lacks a poly(A) tail at the 3′ end. The genome contains four major ORFs (Revill *et al.*, 1994). The genome arrangement is similar to that of poleroviruses (Section III.P.2). The function of ORF 1 is unknown. ORFs 2 and 3 make up the viral polymerase containing a protease, VPg and RdRp domain. ORF 4 encodes the 22-kDa coat protein that is expressed from a subgenomic RNA (Revill *et al.*, 1999). The one species in this genus, *Mushroom bacilliform virus*, infect *Agaricus bisporus* and *A. campestris*. It is possibly associated with La France disease of cultivated mushrooms.

C. Viruses of ferns

No viruses have been reported from bryophytes. A virus with particles like those of a *Tobravirus* was found in hart's tongue fern (*Phyllitis scolopendrium*) by Hull (1968).

D. Viruses of gymnosperms

A disease of *Cycas revoluta* has been shown to be due to a virus with a bipartite genome and other properties that place it in the *Nepovirus* group (Hanada *et al.*, 1986). The virus was readily transmitted by mechanical inoculation to various *Chenopodium* species. It was also transmitted through the seed of these species.

There have been a few reports of pines being infected experimentally with viruses from angiosperms (see Fulton, 1969; Jacobi *et al.*, 1992). There are a few reports of naturally occurring virus-like diseases in other gymnosperms but the viral nature of the diseases has not been demonstrated (e.g. Schmelzer *et al.*, 1966). However, because of the presence of substances such as tannins, there are technical difficulties in attempting to isolate viruses from gymnosperms.

E. Summary

The existence of a (+)-sense ssRNA virus infecting the genus *Chara* suggests an ancient origin for this type of virus. Other than this example, the meager information about viruses infecting photosynthetic eukaryotes below the angiosperms can tell us very little about the age and course of evolution among the plant viruses. The cycads are regarded as living fossils, being in the record since early Mesozoic times. However, the *Nepovirus* found in *Cycas revoluta* is quite likely to have originated in a modern angiosperm, since it readily infects *Chenopodium* spp. The *Phycodnavirus* PBCV-1 infecting a *Chlorella*-like alga is much more likely to be of ancient origin. However, based on structure they do not appear to be primitive viruses. They are much larger and more complex than any known viruses infecting angiosperms, with a genome of about 300 kbp and at least 50 structural proteins (Meints *et al.*, 1986).

VI. DISCUSSION

As noted in the introduction to this chapter, the classification system for viruses, though fully justified, is without a natural base. This is primarily because there is no time-related information on the evolution of, and relationships between, virus species and genera. Thus, one cannot distinguish with certainty between convergent and divergent evolution. There is further discussion on virus evolution in Chapter 17.

Notwithstanding these limitations, an effective system for classifying plant viruses has been developed over the last 20 years or so. The development of this system has given rise to controversies and, no doubt, there will be others in the future (see Bos, 1999b, 2000b; Pringle, 1999b; van Regenmortel, 1999, 2000). This system is fulfilling most of the four criteria that I gave in Section I.A but, as pointed out above, there are some reservations on evolutionary relationships. The system is dynamic and is being modified and refined to take account of new research findings. New virus genera are being created and genera grouped together on common features into families. For instance, the grouping of luteoviruses into three genera in one family (described in Section III.P) has helped significantly with the understanding of these viruses. Other virus groups are being, or are likely to be, merged. For instance, it is becoming increasingly apparent that AMV shares many properties with the ilarviruses (Section III.J.3) and there is beginning to be a strong case for merging the *Ilarvirus* and *Alfamovirus* genera. However, if this occurs recognition will have to be given to the differences in particle structure between the two groups. One large grouping that needs further consideration is that of the 20 floating genera of (+)-strand ssRNA viruses. In Section Q, I have brought together genera that appear to have common properties to highlight the possibilities for creation of new taxa.

With the increasing masses of information, there is beginning to be pressure for the creation of higher taxa. The example of viruses (and other elements) that involve reverse transcription in their replication is noted in Section I.C. This would bring together viruses from different kingdoms, especially eukaryotic. There are, of course, virus groups such as the *Reoviridae* and *Rhabdoviridae* that span the plant and animal kingdoms. The real need now is to develop systems that reflect commonalities in viruses with (+)-strand ssRNA genomes. As is discussed further in Chapters 7 and 8, viruses are faced with various problems to overcome in infecting their host, and common solutions have been found for plant- and animal-infecting viruses. This should be reflected in the classification.

The current state of plant virus taxonomy is reviewed by Mayo and Brunt (2001).

Disease Symptoms and Host Range

Viruses are economically important only when they cause some significant deviation from normal in the growth of a plant. For experimental studies, we are usually dependent on the production of disease in some form to demonstrate biological activity. Symptomatology was particularly important in the early days of virus research, before any of the viruses themselves had been isolated and characterized. Dependence on disease symptoms for identification and classification led to much confusion, because it was not generally recognized that many factors can have a marked effect on the disease produced by a given virus (see Chapter 10).

Most virus names in common use include terms that describe an important symptom in a major host or the host from which the virus was first described. There is a vast literature describing diseases produced by viruses. This has been summarized by Smith (1937, 1972), Bos (1978), Holmes (1964), and in the CMI/AAB *Descriptions of Plant Viruses* issued from 1970 onwards and edited by B. D. Harrison and A. F. Murant—now available on CD-ROM from the Association of Applied Biologists (see Appendix 3). Some viruses under appropriate conditions may infect a plant without producing any obvious signs of disease. Others may lead to rapid death of the whole plant. Between these extremes, a wide variety of diseases can be produced.

Virus infection does not necessarily cause disease at all times in all parts of an infected plant. We can distinguish six situations in which obvious disease may be absent: (1) infection with a very mild strain of the virus; (2) a

tolerant host; (3) nonsterile 'recovery' from disease symptoms in newly formed leaves; (4) leaves that escape infection because of their age and position on the plant; (5) dark green areas in a mosaic pattern (discussed in Chapter 10, Section III.O.4); and (6) plants infected with cryptic viruses (see Chapter 2, Section III.E).

I. ECONOMIC LOSSES DUE TO PLANT VIRUSES (reviewed by Waterworth and Hadidi, 1998)

One of the main driving forces for the detailed studies on plant viruses described in this book is the impact that the diseases they cause have on crop productivity worldwide; yet it is difficult to obtain firm data on the actual losses themselves. The losses due to fungal and bacterial pathogens are well documented and it is common to see lists of loss estimates attributable to specific named fungi or bacteria. In these compendia, the losses due to viruses are lumped together in categories such as 'virus diseases' and 'all other' or 'miscellaneous' diseases. However, viruses are responsible for far greater economic losses than is generally recognized. This lack of recognition is due to several factors, especially their insidious nature. Virus diseases are frequently less conspicuous than those caused by other plant pathogens and last for much longer. This is especially true for perennial crops and those that are vegetatively propagated.

One further problem with attempting to assess losses due to virus diseases on a global basis is that most of the data are from small comparative trials rather than widescale comprehensive surveys. Even the small trials do not necessarily give data that can be used for more global estimates of losses. This is for several reasons, including: (1) variation in losses by a particular virus in a particular crop from year to year; (2) variation from region to region and climatic zone to climatic zone; (3) differences in loss assessment methodologies; (4) identification of the viral etiology of the disease; (6) variation in the definition of the term 'losses'; and (6) complications with other loss factors.

Several publications give guidance on approaches to overcome these problems. These include one published by FAO on loss assessment methods (Chiarappa, 1971) and with chapters on the rationale and concepts of crop loss assessments and modeling of crop growth and yield for loss assessment (Teng, 1987). Others have discussed the nature of crop losses (Main, 1983), identification and assessment of losses (McKenzie, 1983), crop destruction and classification of crop losses (Main, 1977), assessments of plant diseases and losses (James, 1974), methods for determining losses (Bos, 1982), and terms and concepts of crop losses (Nutter *et al.*, 1993).

In addition to the obvious detrimental effects such as reduced yields and visual product quality, virus infections often do not induce noticeable disease but influence their effects on plants in a variety of more subtle ways. Table 3.1 identifies some of the ways in which viruses can damage crop plants. From this it can be seen that the effects of virus infection extend into areas far beyond the actual reduction in yield and quality. Loss estimates do not take account of these indirect factors.

In spite of all these limitations, there have been various collections of loss data (e.g. Hull and Davies, 1992; Waterworth and Hadidi, 1998). Some examples are given in Table 3.2.

II. MACROSCOPIC SYMPTOMS

A. Local symptoms

Localized lesions that develop near the site of entry on leaves are not usually of any economic significance but are important for biological assay (see Chapter 12). Infected cells may lose

TABLE 3.1 Some types of direct and indirect damage associated with plant virus infections

Reduction in growth
  Yield reduction (including symptomless infection)
  Crop failure

Reduction in vigour
  Increased sensitivity to frost and drought
  Increased predisposition to attack by other pathogens and pests

Reduction in quality or market value
  Defects of visual attraction: size shape, color
  Reduced keeping quality
  Reduced consumer appeal: grading, taste, texture, composition
  Reduced fitness for propagation

Cost of attempting to maintain crop health
  Cultural hygiene on farm including vector control
  Production of virus-free propagation materials
  Checking propagules and commodities on export/import (quarantine programs)
  Eradication programs
  Breeding for resistance
  Research, extension and education

Data from Waterworth and Hadidi (1998), with permission.

chlorophyll and other pigments, giving rise to chlorotic local lesions (Plate. 3.1). The lesion may be almost white or merely a slightly paler shade of green than the rest of the leaf. In a few diseases, for example in older leaves of tomato inoculated with TBSV, the lesions retain more chlorophyll than the surrounding tissue. For many host–virus combinations, the infected cells die, giving rise to necrotic lesions. These vary from small pinpoint areas to large irregular spreading necrotic patches (Plate 3.2). In a third type, ring spot lesions appear. Typically, these consist of a central group of dead cells. Beyond this, there develop one or more superficial concentric rings of dead cells with normal green tissue between them (Plate 3.3). Some ring spot local lesions consist of chlorotic rings rather than necrotic ones. Some viruses in certain hosts show no visible local lesions in the intact leaf, but when the leaf is cleared in ethanol and stained with iodine, 'starch lesions' may become apparent.

Viruses that produce local lesions when inoculated mechanically onto leaves may not do so when introduced by other means. For example,

TABLE 3.2 Some examples of crop losses due to viruses

Crop	Virus	Countries	Loss/year
Rice	Tungro	SE Asia	1.5×10^9
	Ragged stunt	SE Asia	1.4×10^8
	Hoja blanca	S. and C. America	9.0×10^6
Barley	Barley yellow dwarf	UK	£6 $\times 10^6$
Wheat	Barley yellow dwarf	UK	£5 $\times 10^6$
Potato	Potato leafroll	UK	£3–5 $\times 10^7$
	Potato virus Y		
	Potato virus X		
Sugarbeet	Beet yellows	UK	£5–50 $\times 10^6$
	Beet mild yellows		
Citrus	Citrus tristeza	Worldwide	£9–24 $\times 10^6$
Cassava	African cassava mosaic	Africa	2×10^9
Many crops	Tomato spotted wilt[a]	Worldwide	1×10^9
Cocao	Cocoa swollen shoot	Ghana	1.9×10^8 trees[b]

[a] Data from Prins and Goldbach (1998), with permission; references to other data given in Hull and Davies (1992), with permission.
[b] Number of trees eradicated over about 40 years.

BYV produces necrotic local lesions on *Chenopodium capitatum*, but does not do so when the virus is introduced by the aphid *Myzus persicae* feeding on parenchyma cells (Bennett, 1960). However, AMV does produce local lesions following aphid transmission.

B. Systemic symptoms

The following sections summarize the major kinds of effects produced by systemic virus invasion. It should be borne in mind that these various symptoms often appear in combination in particular diseases, and that the pattern of disease development for a particular host–virus combination often involves a sequential development of different kinds of symptoms.

1. Effects on plant size

Reduction in plant size is the most general symptom induced by virus infection (Plate 3.4). There is probably some slight general stunting of growth even with 'masked' or 'latent' infections where the systemically infected plant shows no obvious sign of disease. For example, mild strains of PVX infecting potatoes in the field may cause no obvious symptoms, and carefully designed experiments were necessary to show that such infection reduced tuber yield by about 7–15% (Matthews, 1949d). The degree of stunting is generally correlated with the severity of other symptoms, particularly where loss of chlorophyll from the leaves is concerned. Stunting is usually almost entirely due to reduction in leaf size and internode length. Leaf number may be little affected.

In perennial deciduous plants such as grapes, there may be a delayed initiation of growth in the spring (e.g. Gilmer *et al.*, 1970). Root initiation in cuttings from virus-infected plants may be reduced, as in chrysanthemums (Horst *et al.*, 1977).

In vegetatively propagated plants, stunting is often a progressive process. For example, virus-infected strawberry plants and tulip bulbs may become smaller in each successive year.

Stunting may affect all parts of the plant more or less equally, involving a reduction in size of leaves, flowers, fruits, and roots and shortening of petioles and internodes. Alternatively, some parts may be considerably more stunted than others. For example, in little cherry disease, fruits remain small owing to reduced cell division, in spite of apparently ample leaf growth. A reduction in total yield of fruit is a common feature and an important economic aspect of virus disease. The lower yield may sometimes be due to a reduction in both size and number of fruits (e.g. Hampton, 1975). In a few diseases, for example prune dwarf, fruits may be greatly reduced in number but of larger size than normal. Healthy cherry trees pollinated with pollen from trees infected with this virus (Way and Gilmer, 1963) or necrotic ring spot virus (Vértesy and Nyéki, 1974) had a reduced fruit set. Seed from infected plants may be smaller than normal, germination may be impaired, and the proportion of aborted seed may be increased (Walkey et al., 1985).

2. Mosaic patterns and related symptoms

One of the most common obvious effects of virus infection is the development of a pattern of light and dark green areas giving a mosaic effect in infected leaves. I will describe the phenotypes in this section, and in Chapter 10 (Section III.P) I discuss the current molecular understanding of mosaic pattern formation.

The detailed nature of the pattern varies widely for different host–virus combinations. In dicotyledons, the areas making up the mosaic are generally irregular in outline. There may be only two shades of color involved—dark green and a pale or yellow–green, for example. This is often so with TMV in tobacco (Plate 3.5), or there may be many different shades of green and yellow, as with TYMV in Chinese cabbage (Plate 3.6). The junctions between areas of different color may be sharp and such diseases resemble quite closely the mosaics produced by inherited genetic defects in the chloroplasts. AbMV is a good example of this type (Plate 3.7). TYMV in Chinese cabbage may approach genetic variegation in the sharpness of the mosaic pattern it produces.

The borders between darker and lighter areas may be diffuse (Plate 3.8). If the lighter areas differ only slightly from the darker green, the mottling may be difficult to observe as with some of the milder strains of PVX in potato. In mosaic diseases infecting herbaceous plants, there is usually a fairly well-defined sequence in the development of systemic symptoms. The virus moves up from the inoculated leaf into the growing shoot and into partly expanded leaves. In these leaves, the first symptoms are a 'clearing' or yellowing of the veins; as described in Chapter 10 (Section III.O.2), it is suggested that this symptom is an optical illusion. However, chlorotic vein-banding is a true symptom and may be very faint or may give striking emphasis to the pattern of veins (Plate 3.9). Vein-banding may persist as a major feature of the disease.

In leaves that are past the cell division stage of leaf expansion when they become infected (about 4–6 cm long for leaves such as tobacco and Chinese cabbage) no mosaic pattern develops. The leaves become uniformly paler than normal. In the oldest leaves to show mosaic, a large number of small islands of dark green tissue usually appear against a background of paler color. The mosaic areas may be confined to the youngest part of the leaf blade, that is, the basal and central region. Although there may be considerable variation in different plants, successively younger systemically-infected leaves show, on the average, mosaics consisting of fewer and larger areas. The mosaic pattern is laid down at a very early stage of leaf development and may remain unchanged, except for general enlargement, for most of the life of the leaf. In some mosaic diseases the dark green areas are associated mainly with the veins to give a dark green vein-banding pattern.

In monocotyledons, a common result of virus infection is the production of stripes or streaks of tissue lighter in color than the rest of the leaf. The shades of color vary from pale green to yellow or white, and the more or less angular streaks or stripes run parallel to the length of the leaf (Plate 3.10).

The development of the stripe diseases found in monocotyledons follows a similar general pattern to that found for mosaic diseases in dicotyledons. One or a few leaves above the inoculated leaf that were expanded at time of inoculation show no stripe pattern. In the first leaf to show striping, the pattern is relatively fine and may occur only in the basal (younger) portion of the leaf blade. In younger leaves, stripes tend to be larger and occur throughout the leaf. The patterns of striping are laid down at an early stage and tend to remain unchanged for most of the life of the leaf. Yellowed areas may become necrotic as the leaf ages.

A variegation or 'breaking' in the color of petals commonly accompanies mosaic or streak symptoms in leaves. The breaking usually consists of flecks, streaks or sectors of tissue with a color different from normal (Plates 3.11 and

3.12). The breaking of petal color is frequently due to loss of anthocyanin pigments, which reveals any underlying coloration due to plastid pigments. In a few instances, for example in tulip color-adding virus, infection results in increased pigmentation in some areas of the petals. Nectar guides in petals are often invisible to humans (but visible to bees) because they involve pigments that absorb strongly only in the ultraviolet region (Thompson *et al.*, 1972). The effects of virus infection on these nectar guides and on the behavior of honeybees do not appear to have been studied.

Infected flowers are frequently smaller than normal and may drop prematurely. Flower breaking may sometimes be confused with genetic variegation, but it is usually a good diagnostic feature for infection by viruses producing mosaic symptoms. In a few plants, virus-induced variegation has been valued commercially. Thus, as described in Chapter 1 (Section I), at one time virus-infected tulips were prized as distinct varieties. As with the development of mosaic patterns in leaves, color-breaking in the petals may develop only in flowers that are smaller than a certain size when infected. Thus, in tulips inoculated with TBV less than 11 days before blooming no break symptoms developed even though virus

was present in the petals (Yamaguchi and Hirai, 1967). Virus infection may reduce pollen production and decrease seed set, seed size and germination (e.g. Hemmati and McLean, 1977) (Fig. 3.1).

Fruits formed on plants showing mosaic disease in the leaves may show a mottling, for example, zucchinis infected with CMV (Plate 3.13). In other diseases, severe stunting and distortion of fruit may occur. Seed coats of infected seed may be mottled.

3. Yellow diseases

Viruses that cause a general yellowing of the leaves are not as numerous as those causing mosaic diseases, but some, such as the viruses causing yellows in sugarbeet, are of considerable economic importance. The first sign of infection is usually a clearing or yellowing of the veins (Plate 3.14) in the younger leaves followed by a general yellowing of the leaves (Plate 3.15). This yellowing may be slight or severe. No mosaic is produced, but in some leaves there may be sectors of yellowed and normal tissue. In strawberry yellow edge disease, yellowing is largely confined to the margins of the leaf. When severe, a yellows disease may lead to a total loss of the crop (e.g. Weidemann *et al.*, 1975).

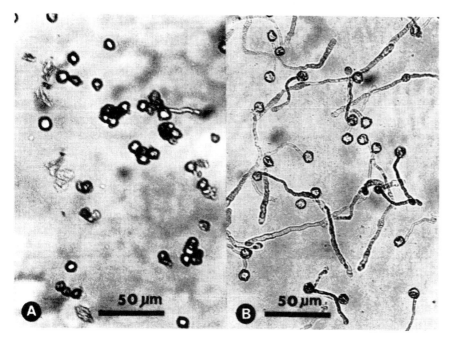

Fig. 3.1 Effect of TRSV infection on germination and germ tube growth of soybean pollen. The pollen grains were germinated overnight in 30% sucrose. **(A)** Infected; **(B)** healthy. From Yang and Hamilton (1974), with permission.

4. Leaf rolling

Virus infection can result in leaf rolling, which is usually upwards (Plate 3.16) or occasionally downwards. Pronounced epinasty of leaf petioles may sometimes be a prominent feature.

5. Ring spot diseases

A marked symptom in many virus diseases is a pattern of concentric rings and irregular lines on the leaves (Plate 3.17) and sometimes also on the fruit (Plates 3.18 and 3.19). The lines may consist of yellowed tissue or may be due to death of superficial layers of cells, giving an etched appearance. In severe diseases, complete necrosis through the full thickness of the leaf lamina may occur. With the ring spot viruses, such as TRSV, there is a strong tendency for plants to recover from the disease after an initial shock period. Leaves that have developed symptoms do not lose these, but younger growth may show no obvious symptoms in spite of the fact that they contain virus. Ring spot patterns may also occur on other organs, for example bulbs (Asjes et al., 1973), and the tuber necrotic ring spot strain of PVY causes rings on potato tubers, often around the eyes, that become sunken and necrotic (see Beczner et al., 1984; Weidemann and Maiss, 1996).

6. Necrotic diseases

Death of tissues, organs or the whole plant is the main feature of some diseases. Necrotic patterns may follow the veins as the virus moves into the leaf (Plate 3.20). In some diseases, the whole leaf is killed. Necrosis may extend fairly rapidly throughout the plant. For example, with PVX and PVY in some varieties of potatoes, necrotic streaks appear in the stem. Necrosis spreads rapidly to the growing point, which is killed, and subsequently all leaves may collapse and die. Wilting of the parts that are about to become necrotic often precedes such systemic necrotic disease.

7. Developmental abnormalities

Besides being generally smaller than normal, virus-infected plants may show a wide range of developmental abnormalities. Such changes may be the major feature of the disease or may accompany other symptoms. For example, uneven growth of the leaf lamina is often found in mosaic diseases. Dark green areas may be raised to give a blistering effect, and the margin of the leaf may be irregular and twisted (Plate 3.21). In some diseases, the leaf blade may be more or less completely suppressed, such as in tomatoes infected with CMV and/or TMV (Plate 3.22) (Francki et al., 1980a).

Some viruses cause swellings in the stem, which may be substantial in woody plants, such as in cocoa swollen shoot disease. Another group of growth abnormalities is known as enations. These are outgrowths from the upper or lower surface of the leaf usually associated with veins (Plate 3.23). They may be small ridges of tissue, or larger, irregularly shaped leaflike structures, or long filiform outgrowths. Conversely, normal outgrowths may be suppressed. For example, a potyvirus infection in Datura metel causes the production of fruits lacking the normal spines (Rao and Yaraguntiah, 1976).

Viruses may cause a variety of tumor-like growths. The tumor tissue is less organized than with enations. Some consist of wart-like outgrowths on stems or fruits. The most studied tumors are those produced by WTV which are characteristic of this disease (Plate 3.24). In a systemically infected plant, external tumors appear on leaves or stems where wounds are made. In infected roots they appear spontaneously, beginning development close to cells in the pericycle that are wounded when developing side-roots break through the cortex. The virus may also cause many small internal tumors in the phloem of the leaf, stem and root (Lee and Black, 1955).

Stem deformation such as stem splitting and scar-formation is caused by some viruses in some woody plants.

Virus infection of either the rootstock or scion can cause necrosis and/or failure of the graft union (Plate 3.25).

One of the unusual symptoms of BSV in some Musa cultivars is that the fruit bunch emerges from the side of the pseudostem instead of from the top of it (Plate 3.26). This is due to necrosis of the cigar leaf.

8. Wilting

Wilting of the aerial parts frequently followed by death of the whole plant may be an important feature (e.g. in virus diseases of chickpea; Kaiser and Danesh, 1971).

9. Recovery from disease

Not uncommonly, a plant shows disease symptoms for a period and then new growth appears in which symptoms are milder or absent, although virus may be still present. This commonly occurs with *Nepovirus* infections. Many factors influence this recovery phenomenon. The stage of development at which a plant is infected may have a marked effect on the extent to which symptoms are produced. For example, tobacco plants inoculated with BCTV develop disease symptoms. This stage is frequently followed by a recovery period. If very young seedling plants are inoculated, a proportion of these might never show clear symptoms, even though they can be shown to contain virus (Benda and Bennett, 1964). The environment can also affect recovery from disease as can host species or variety and virus strain.

The molecular aspects of disease recovery are discussed in Chapter 10 (Section III.R).

10. Reduced nodulation

Various workers have described a reduction in the number, size and fresh weight of nitrogen-fixing *Rhizobium* nodules induced by virus infection in legumes (e.g. AMV in alfalfa: Ohki *et al.*, 1986; CMV in pea: Rao *et al.*, 1987). In general, overall nitrogen fixation is reduced by virus infection. With AMV infection in *Medicago*, nitrogen fixation per unit of nodule fresh weight is the same as in healthy plants, but infected plants produce less nodule tissue. Hence, nitrogen fixation per plant is reduced (Dall *et al.*, 1989).

11. Genetic effects

Infection with BSMV induces an increase in mutation rate in *Zea mays* and also a genetic abnormality known as an aberrant ratio (AR) (Sprague *et al.*, 1963; Sprague and McKinney, 1966, 1971). For example, when a virus-infected pollen parent with the genetic constitution A_1A_1, *PrPr, SuSu* was crossed with a homozygous recessive line (a_1a_1, *prpr, susu*) resistant to the virus, a low frequency of the progeny lines gave significant distortion from the expected ratios for one or more of the genetic markers. (A_1-a_1 = presence or absence of aleurone color; *Pr–pr* = purple or red aleurone colour; *Su–su* = starchy or sugary seed.) This AR effect was observed only when the original pollen parent was infected and showing virus symptoms on the upper leaves. The AR effect is inherited in a stable manner, with a low frequency of reversion to normal ratios. It is inherited in plants where virus can no longer be detected. WSMV also induces the AR effect (Brakke, 1984). The genetics of the AR effect are complex, and probably more than one phenomenon is involved. There is no evidence that a cDNA copy of part or all of the viral genome is incorporated into the host genome. The AR effect has been reviewed by Brakke (1984). Further work may show that similar phenomena occur in other virus-infected plants.

C. Agents inducing virus-like symptoms

Disease symptoms, similar to those produced by viruses, can be caused by a range of physical, chemical and biological agents. Such diseases may have interesting factors in common with virus-induced disease. The activities of such agents have sometimes led to the erroneous conclusion that a virus was the cause of a particular disease in the field. Further confusion may arise when a disease is caused by the combined effects of a virus and some other agent.

1. Small cellular parasites

A group of diseases characterized by symptoms such as general yellowing of the leaves, stunting, witches-broom growth of axillary shoots, and a change from floral to leaf-type structures in the flowers (phyllody) was for many years thought to be caused by viruses. Diseases of this type are not transmissible by mechanical means. They were considered to be caused by viruses because (1) no bacteria, rickettsiae, fungi or protozoa could be implicated; (2) they were graft transmissible; (3) they were transmitted by leafhoppers; and (4) some, at least, could be transmitted by dodder.

Mycoplasmas and spiroplasmas belong to the *Mycoplasmatales*. They are characterized by a bounding unit membrane, pleomorphic form, the absence of a cell wall, and complete resistance to penicillin.

Rickettsia-like organisms are distinguished from the *Mycoplasmatales* by the possession of a cell wall and by their *in vivo* susceptibility to penicillin. They have been found in the phloem of various species.

From the pioneering experiments of Doi *et al.* (1967) and Ishiie *et al.* (1967) a new branch of plant pathology was opened up by the demonstration that mycoplasmas (now termed phytoplasmas), spiroplasmas and rickettsia-like organisms can cause disease in plants (e.g. Windsor and Black, 1973). These agents are generally confined to the phloem or xylem of diseased plants. They are too small to be readily identified by light microscopy. The widespread availability of electron microscopy provided the key technique that has allowed their importance in plant pathology to be recognized. Phytoplasma diseases of plants are described in Maramorosch and Raychaudhuri (1988) and recent advances in their study by Davis and Sinclair (1998) and Lee *et al.*, (1998a). They are classified on molecular properties such as 16S ribosomal RNA (see Marcone *et al.*, 1999; Webb *et al.*, 1999).

A simultaneous virus and phytoplasma infection may give rise to a more severe disease than either agent may alone. For example, OBDV and the phytoplasm agent of aster yellows are both confined to the phloem, and both can be transmitted by the leafhopper *Macrosteles fascifrons* (Stål.). In a mixed infection, they cause more severe stunting than either agent alone (Banttari and Zeyen, 1972).

2. Bacteria

Some bacteria can cause virus-like symptoms. For instance, mechanical inoculation of cowpea leaves can give rise to necrotic local lesions caused by unknown, adventitious bacteria.

3. Toxins produced by arthropods

Insects and other arthropods feeding on plants may secrete very potent toxins, which move systemically through the plant and produce virus-like symptoms. *Calligypona pellucida* (F.) (Homoptera) produces salivary toxins that cause general retardation of growth and inhibition of tillering. Only females produce the toxins (Nuorteva, 1962).

Eryophyid mites feeding on clover may induce a mosaic-like mottle in the younger leaves. One mite is sufficient to induce such symptoms. These mites are small and may be overlooked since they burrow within the leaf. A virus-like mosaic disease of wax myrtle (*Myrica cerifera*) was found to be caused by a new species of eryophyid mite (Elliot *et al.*, 1987).

Maize plants showing wallaby ear disease are stunted and develop galls on the underside of the leaves, which are sometimes darker green than normal. This virus-like condition has been shown to be due to a toxin produced by the leafhopper *Cicadulina bimaculata* (Ofori and Francki, 1983).

4. Genetic abnormalities

Numerous cultivated varieties of ornamental plants have been selected by horticulturists because they possess heritable leaf variegations or mosaics. These are often due to maternally inherited plastid defects. The variegated patterns produced sometimes resemble virus-induced mosaics quite closely. However, in mosaics due to plastid mutation, the demarcation between blocks of tissue of different colors tends to be sharper than with many virus-induced mosaics.

Other virus symptoms may be mimicked by genetic abnormalities. For example, Edwardson and Corbett (1962) described a 'wiry' mutant in Marglobe tomatoes that gave an appearance similar to that of plants infected with strains of CMV and TMV. All the leaves of the mutant resembled the symptoms of virus infection in that upper leaves lacked a lamina as shown in Plate 3.21. However, the disease could not be transmitted by grafting, and genetic experiments suggested that a pair of recessive genes control the mutant phenotype.

5. Transposons

Transposons are mobile elements that can move about the plant (or animal) genome.

There are two types of transposons, what may be termed true transposons such as the *Ac/Ds* and *MuDR/Mu* maize transposons (reviewed by Walbot, 2000) and retrotransposons, which are described in more detail in Chapter 2 (Section IV). If transposons move into a gene or genes that control leaf or flower color during the development of that organ, they can cause flecks or streaks of different colors that resemble virus symptoms. This can be especially noticeable in flowers where one can find apparent flower color break most likely caused by the movement of a transposon.

6. Nutritional deficiencies

Plants may suffer from a wide range of nutritional deficiencies that cause abnormal coloration, discoloration or death of leaf tissue. Some of these conditions can be fairly easily confused with symptoms due to virus infection. For example, magnesium and iron deficiency in soybeans leads to a green banding of the veins with chlorotic interveinal areas. The yellowing, however, is more diffuse than is usual in virus infection and is usually in the older leaves. In sugarbeet, yellowing and necrosis due to magnesium deficiency may be similar to the disease produced by BYV. Potassium and magnesium deficiency in potatoes produces marginal and interveinal necrosis similar to that found in certain virus diseases.

7. High temperatures

Growing plants at substantially higher temperatures than normal may induce virus-like symptoms. When *N. glutinosa* plants were held at 37.8° for 4–8 days and then returned to 22°, new leaves displayed a pattern of mosaic, vein-clearing, chlorosis, and other abnormalities with a resemblance to virus infection (see Fig. 3.2). These symptoms gradually disappeared in newer leaves, but could be induced in the same plants again by a second treatment at high temperature (John and Weintraub, 1966). Mechanical inoculation and grafting tests to various hosts and electron microscopy failed to reveal the presence of a virus in heated plants.

8. Hormone damage

Commercially used hormone weedkillers may produce virus-like symptoms in some plants. Tomatoes and grapes are particularly susceptible to 2,4-dichlorophenoxyacetic acid (2,4-D). Growth abnormalities in the leaves caused by 2,4-D bear some resemblance to certain virus infections in these and other hosts. Compounds related to 2,4-D can cause almost complete suppression of mesophyll development, giving

Fig. 3.2 Virus-like symptoms induced in *Nicotiana glauca* following a period of growth at high temperature (37.8°C). From John and Weintraub (1966), with permission.

a plant with a 'shoestring' appearance. Alternatively, vein growth may be retarded more than the mesophyll. Mesophyll may then bulge out between the veins to give an appearance not unlike leaf curl diseases.

9. Insecticides

Certain insecticides have been reported to produce leaf symptoms that mimic virus infection (e.g. Woodford and Gordon, 1978).

10. Air pollutants

Many air pollutants inhibit plant growth and give rise to symptoms that could be confused with a virus disease. For example, chimney gases from a cement factory caused *Zea mays* seedlings to become stunted and yellowed, with necrotic areas and curled leaf margins (Cireli, 1976).

D. The cryptoviruses

Cryptoviruses escaped detection for many years because most of them cause no visible symptoms or, in a few situations, very mild symptoms. They are not transmissible mechanically or by vectors, but are transmitted efficiently in pollen and seed. They occur in very low concentrations in infected plants (reviewed by Boccardo *et al.*, 1987). Nevertheless, they have molecular characteristics that might be expected of disease-producing viruses.

As described in Chapter 6 (Section VI.B), the genome consists of two dsRNA segments and these viruses share some properties with the reoviruses. There is no indication, other than the low concentration at which they occur, as to why they cause symptomless infection.

III. HISTOLOGICAL CHANGES

The macroscopic symptoms induced by viruses frequently reflect histological changes within the plant. These changes are of three main types—necrosis, hypoplasia and hyperplasia—that may occur singly or together in any particular disease. For example, all three are closely linked in the citrus exocortis disease (Fudl-Allah *et al.*, 1971).

A. Necrosis

Necrosis as the major feature of disease was discussed in Section II.B.6. In other diseases, necrosis may be confined to particular organs and tissues and may be very localized. It commonly occurs in combination with other histological changes. It may be the first visible effect or may occur as the last stage in a sequence. For example, necrosis of epidermal cells or of midrib parenchyma may be caused by lettuce mosaic virus in lettuce (Coakley *et al.*, 1973). Necrosis caused by TNV is usually confined to localized areas of the roots (e.g. Lange, 1975). Late infection of virus-free tomato plants with TMV may give rise to internal necrosis in the immature fruits (e.g. Taylor *et al.*, 1969).

In the potato leaf roll disease, the phloem develops normally but is killed by the infection. Necrosis may spread in phloem throughout the plant, but is limited to this tissue (Shepardson *et al.*, 1980).

In *Pelargonium* infected with TRSV, histological effects seen by light microscopy were confined to reproductive tissues (Murdock *et al.*, 1976). Pollen grain abortion and abnormal and aborted ovules were common. The symptoms could be confused with genetic male sterility.

B. Hypoplasia

Leaves with mosaic symptoms frequently show hypoplasia in the yellow areas. The lamina is thinner than in the dark green areas, and the mesophyll cells are less differentiated with fewer chloroplasts and fewer or no intercellular spaces (Fig. 3.3).

In stem-pitting disease of apples, pitting is shown on the surface of the wood when the bark is lifted. The pitting is due to the failure of some cambial initials to differentiate cells normally, and a wedge of phloem tissue is formed that becomes embedded in newly formed xylem tissue (Hilborn *et al.*, 1965). The affected phloem becomes necrotic.

The major anatomical effect of ASGV in apple stems is the disappearance of the

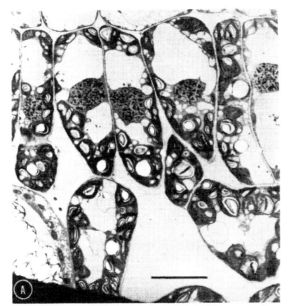

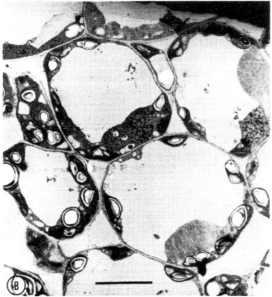

Fig. 3.3 Histological and cytological effects of TMV in tobacco. **(A)** Section through palisade cells of a dark green area of a leaf showing mosaic. Cells are essentially normal. **(B)** Section through a nearby yellow-green area. Cells are large and undifferentiated in shape. Nuclei are not centrally located as in the dark green cells. Bar = 20 μm. (Courtesy of P. H. Atkinson.)

cambium in the region of the groove. Normal phloem and xylem elements are replaced by a largely undifferentiated parenchyma (Pleše *et al.*, 1975).

Reduced size of pollen grain and reduced growth of pollen tubes from virus-infected pollen may be regarded as hypoplastic effects (Fig. 3.1). A variety of other effects on pollen grains has been described (e.g. Haight and Gibbs, 1983).

C. Hyperplasia

1. Cells are larger than normal

Vein-clearing symptoms are due, with some viruses at least, to enlargement of cells near the veins (Esau, 1956). The intercellular spaces are obliterated, and since there is little chlorophyll present the tissue may become abnormally translucent.

2. Cell division in differentiated cells

Some viruses such as PVX may produce islands of necrotic cells in potato tubers. The tuber may respond with a typical wound reaction in a zone of cells around the necrotic area. Starch grains disappear and an active cambial layer develops (Fig. 3.4).

Similarly, in a white halo zone surrounding necrotic local lesions induced by TMV in *N. glutinosa* leaves, cell division occurred in mature palisade cells (Wu, 1973).

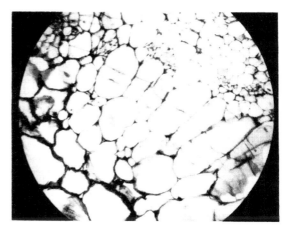

Fig. 3.4 Section through a potato tuber infected with PVX, showing a cork cambial layer being developed near a group of necrotic cells (bottom left).

3. Abnormal division of cambial cells

The vascular tissues appear to be particularly prone to virus-induced hyperplasia. In the diseased shoots found in swollen shoot disease of cocoa, abnormal amounts of xylem tissue are produced but the cells appear structurally normal (Posnette, 1947). In plants infected by BCTV, a large number of abnormal sieve elements develop, sometimes associated with companion cells. The arrangement of the cells is disorderly and they subsequently die (Esau, 1956; Esau and Hoefert, 1978). OBDV causes abnormalities in the development of phloem in oats, involving hyperplasia and limited hypertrophy of the phloem procambium (Zeyen and Banttari, 1972).

In crimson clover infected by WTV, there is abnormal development of phloem cambium cells. Phloem parenchyma forms meristematic tumor cells in the phloem of leaf, stem and root (Lee and Black, 1955).

Galls on sugarcane leaves arise from *Fijivirus*-induced cell proliferation. This gives rise in the mature leaf to a region in the vein where the vascular bundle is grossly enlarged (Fig. 3.5). Two main types of abnormal cell are present: lignified gall xylem cells and non-lignified gall phloem (Hatta and Francki, 1976). Hyperplastic growth of phloem was marked in plum infected with PPV (Buchter *et al.*, 1987).

IV. CYTOLOGICAL EFFECTS

The cytological effects of viruses have been a subject of interest ever since the early searches with light microscopes for causative organisms in diseased tissues. About the beginning of the twentieth century, these studies led to the discovery of two types of virus inclusion: amorphous bodies or 'X bodies' and crystalline inclusions. The X bodies resemble certain microorganisms. Some workers erroneously considered that they were in fact the parasite or a stage in the life cycle of the parasite causing the disease. These early conclusions were not entirely wrong since many of the X bodies are in fact virus-induced structures in the cell where the components of viruses are synthesized and assembled.

A. Methods

Light microscopy is still important in the study of cytological abnormalities for several reasons:

1. Much greater areas of tissue can be scanned, thus ensuring that samples taken for electron microscopy are representative.
2. It may allow electron microscopic observations to be correlated with earlier detailed work on the same material using light microscopy.
3. Observations on living material can be made using both phase and bright-field illumination.

Improvements in procedures for the fixing, staining and sectioning of plant tissues over the past 40 years and the widespread availability of high-resolution electron microscopes has led to a substantial growth in our knowledge about the cytological effects of viruses in cells. As in other fields of biology, electron microscopy is providing a link between macroscopic and light microscope observations on the one hand and molecular biological and biochemical studies on the other.

Examination of stained thin sections remains the standard procedure, but freeze-fracturing can give useful information on virus-induced membrane changes (Hatta *et al.*, 1973). Scanning electron microscopy is of less value in the study of virus diseases but is sometimes useful (Hatta and Francki, 1976). It must be remembered that small differences in conditions under which plants are grown before sampling and in the procedure used to prepare tissue for electron microscopy can have a marked effect on the appearance and stability of organelles and virus-induced structures (e.g. Langenberg, 1982). Chilling of tissue before fixation may improve the preservation of very fragile virus-induced structures (Langenberg, 1979).

To relate any observed cytological effects to virus replication, it is very useful to be able to follow the time course of events in infected cells. In principle, protoplasts infected *in vitro* should provide excellent material for studying the time course of events. However, various limitations are becoming apparent:

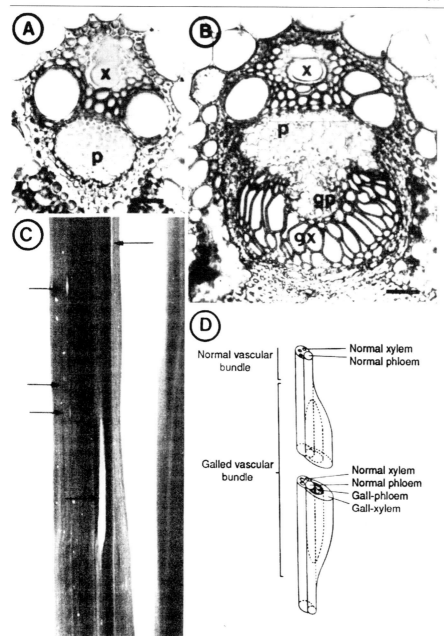

Fig. 3.5 Structure of leaf galls on sugarcane infected with FDV. **(A)** Transverse section of vascular tissue in a leaf vein from a healthy sugarcane plant, showing the xylem (x) and phloem (p) tissues. **(B)** A transverse section of vascular tissues of a vein on a galled leaf of an FDV-infected sugarcane plant, showing the gall phloem (gp) and gall xylem (gx), in addition to normal phloem (p) and xylem (x) tissues. (Bars = 0.5 μm). **(C)** Part of a sugarcane leaf infected with FDV, showing small and large gall (arrows). **(D)** A diagram of the tissue distribution in the vein of an FDV-infected sugarcane leaf showing normal and gall tissues. From Egan *et al.* (1989), with permission.

1. Some ultrastructural features seen in TMV-infected tobacco leaf cells were not observed during TMV replication in protoplasts (Otsuki *et al.*, 1972a).
2. Crystalline inclusion bodies in TMV-infected leaf cells were degraded in protoplasts made from such leaves (Föglein *et al.*, 1976).
3. Development of cytological changes in protoplasts infected *in vitro* may be much less synchronous than indicated by growth curves.

The uses of fluorescent dyes and confocal microscopy are described in Chapter 9 (Section II.B).

B. Effects on cell structures

1. Nuclei

Cytopathological effects of virus infection have been well illustrated by Francki *et al.* (1985a,b). Many viruses have no detectable cytological effects on nuclei. Others give rise to intranuclear inclusions of various sorts and may affect the nucleolus or the size and shape of the nucleus, even though they appear not to replicate in this organelle.

Shikata and Maramorosch (1966) found that in pea leaves and pods infected with PEMV, particles accumulate first in the nucleus. During the course of the disease, the nucleolus disintegrates. Masses of virus particles accumulate in the nucleus and also in the cytoplasm. PEMV also causes vesiculation in the perinuclear space (De Zoeten *et al.*, 1972). Virus particles of several small isometric viruses accumulate in the nucleus (as well as the cytoplasm). They may exist as scattered particles or in crystalline arrays (e.g. SBMV: Weintraub and Ragetli, 1970; TBSV: Russo and Martelli, 1972). Masses of viral protein or empty viral protein shells have been observed in nuclei of cell infected with several tymoviruses (Hatta and Matthews, 1976) (Fig. 3.6).

Crystalline plate-like inclusions were seen by light microscopy in cells infected with severe etch in tobacco (Kassanis, 1939). The plates were birefringent when viewed sideways and were a very regular feature of severe etch infection. Intranuclear inclusions have been described for some other potyviruses.

The virus-coded proteins involved in potyvirus nuclear and cytoplasmic inclusions are discussed in Chapter 6 (Section VIII.C.1.b).

Electron-lucent lacunae appeared in the nucleolus of *Nicotiana* cells infected with PVA (Edwardson and Christie, 1983). For some rhabdoviruses, viral cores appear in the nucleus and accumulate in the perinuclear space (see Fig. 8.19).

Geminiviruses cause marked hypertrophy of the nucleolus, which may come to occupy three-quarters of the nuclear volume. Fibrillar rings of deoxyribonucleoprotein appear, and masses of virus particles accumulate in the nucleus (e.g. Rushing *et al.*, 1987). An isolate of CaMV has been described in which nuclei become filled with virus particles and greatly enlarged (Gracia and Shepherd, 1985). The virus particles were not embedded in the matrix protein found in cytoplasmic viroplasms.

2. Mitochondria

The long rods of TRV may be associated with the mitochondria in infected cells (Harrison and Roberts, 1968) (see Fig. 8.17) as are the isometric particles of BBWV-1 (see Fig. 3.9C) (Hull and Plaskitt, 1974). The mitochondria in cells of a range of host species, and various tissues, infected with CGMMV develop small vesicles bounded by a membrane and lying within the peri-mitochondrial space and in the cristae (Hatta *et al.*, 1971).

Aggregated mitochondria have been observed in *Datura* cells infected by the potyvirus HMV (Kitajima and Lovisolo, 1972),

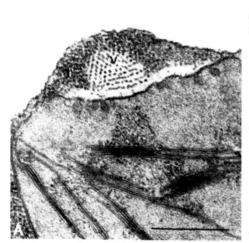
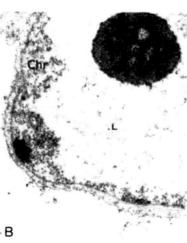

Fig. 3.6 The intracellular sites of TYMV and TYMV protein accumulation. **(A)** Small crystalline array of TYMV particles associated with electron-lucent area in the cytoplasm of a stage C cell subjected to plasmolysis. Bar = 500 nm; r, ribosomes; v, TYMV particles. From Hatta and Matthews (1974), with permission. **(B)** Accumulation of TYMV coat protein (electron-lucent material, L) in the nucleus of a stage D cell not subjected to plasmolysis. Bar = 500 nm. Modified from Hatta and Matthews (1976).

but there was no indication that these aggregates were involved in virus synthesis. The development of abnormal membrane systems within mitochondria has been described for several virus infections (Francki, 1987). They have no established relation to virus replication and are probably degenerative effects. For example, in some tombusvirus infections multivesiculate bodies appear in the cytoplasm. These have been shown to develop from greatly modified mitochondria (Di Franco *et al.*, 1984; Di Franco and Martelli, 1987). In infections with other tombusviruses, multivesiculate bodies originate from modified peroxisomes (Martelli *et al.*, 1984). In *Sonchus* infected with BYSV, virus particles are found in phloem cells. The flexuous rod-shaped particles are frequently inserted into the cristae of the mitochondria (Esau, 1979).

3. Chloroplasts

The small peripheral vesicles and other changes in and near the chloroplasts closely related to TYMV replication are discussed in Chapter 8 (Section IV.K.2). TYMV infection can cause many other cytological changes in the chloroplasts, most of which appear to constitute a structural and biochemical degeneration of the organelles. The exact course of events in any mesophyll cell depends on: (1) the developmental stage at which it was infected, (2) the strain of virus infecting, (3) the time after infection, and (4) the environmental conditions (Matthews, 1973; Hatta and Matthews, 1974).

In inoculated leaves, the chloroplasts become rounded and clumped together in the cell. There is little effect on grana or stroma lamellae. The chloroplasts become cup-shaped, with the opening of the cup generally facing the cell wall. Starch grains accumulate. 'White' strains of the virus cause degeneration of the grana in inoculated leaves. In expanded leaves above the inoculated leaf that become fully infected without the appearance of mosaic symptoms, the effects of infection on chloroplasts are similar to those seen in inoculated leaves.

In leaves that are small at time of infection and that develop the typical mosaic, a variety of different pathological states in the chloroplasts can readily be distinguished by light microscopy in fresh leaf sections. Islands of tissue in the mosaic showing various shades of green, yellow and white contain different strains of the virus, which affect the chloroplasts in recognizably distinct ways (see Fig. 9.18). In dark green islands of tissue, which contain very little virus, chloroplasts appear normal.

The most important changes in the chloroplasts seen in tissue types other than dark green are: (1) color, ranging from almost normal green to colorless; (2) clumping to a variable extent; (3) presence of large vesicles; (4) fragmentation of chloroplasts; (5) reduction in granal stack height; (6) presence of osmiophilic globules; and (7) arrays of phytoferritin molecules. Some of these abnormalities are illustrated in Fig. 9.18.

Different strains of TYMV produce particular combinations of abnormalities in the chloroplasts. In blocks of tissue of one type, almost all cells show the same abnormalities and these persist at least for a time as the predominating tissue type when inoculations are made to fresh plants.

In contrast to the small peripheral vesicles, which appear to be induced by all tymoviruses in the chloroplasts of infected cells, none of the changes noted is an essential consequence of tymovirus infection; nor can they be regarded as diagnostic for the group. For example, no clumping of the chloroplasts occurs in cucumbers infected with OkMV. On the other hand, clumping of chloroplasts is induced by TuMV in *Chenopodium* (Kitajima and Costa, 1973). Several viruses outside the *Tymovirus* genus induce small vesicles near the periphery of the chloroplasts. These vesicles differ from the *Tymovirus* type in that they do not appear to have necks connecting them to the cytoplasm. In *Datura* leaves infected with TBSV, the thylakoid membranes undergo varied and marked rearrangements (Bassi *et al.* 1985). For most of these viruses, the vesicles or other changes appear to be degenerative consequences of infection. However, for BSMV, the vesicles appear to be associated with virus replication (Lin and Langenberg, 1984a).

In many infections, the size and number of starch grains seen in leaf cells are abnormal. In mosaic diseases, there is generally speaking less starch than normal, but in some diseases (e.g. sugarbeet curly top and potato leaf roll) excessive amounts of starch may accumulate. Similarly in local lesions induced by TMV in cucumber cotyledons, chloroplasts become greatly enlarged and filled with starch grains (Cohen and Loebenstein, 1975.) (See also Section IV.D.)

4. Cell walls

The plant cell wall tends to be regarded mainly as a physical supporting and barrier structure. In fact, it is a distinct biochemical and physiological compartment containing a substantial proportion of the total activity of certain enzymes in the leaf (Yung and Northcote, 1975). Three kinds of abnormality have been observed in or near the walls of virus-diseased cells:

1. Abnormal thickening, owing to the deposition of callose, may occur in cells near the edge of virus-induced lesions (e.g. Hiruki and Tu, 1972). Chemical change in the walls may be complex and difficult to study (Faulkner and Kimmins, 1975).
2. Cell wall protrusions involving the plasmodesmata have been reported for several unrelated viruses. The protrusions from the plasmodesmata into cells may have one or more canals. They may be quite short or of considerable length. They appear to be due to deposition of new wall material induced by the virus, and they may be lined inside and out with plasma membrane (e.g. Bassi *et al.*, 1974).
3. Depositions of electron-dense material between the cell wall and the plasma membrane may extend over substantial areas of the cell wall (as with ONMV: Gill, 1974) or may be limited in extent and occur in association with plasmodesmata (as with BSMV: McMullen *et al.*, 1977). They have been called *paramural bodies*.

The major cytopathic effect of CEVd is the induction of numerous small membrane-bound bodies near the cell wall with an electron density similar to that of the plasma membrane (Semancik and Vanderwoude, 1976). They are found in all cell types.

5. Bacteroidal cells

The first phase in the infection of soybean root cells by *Rhizobium* (i.e. development of an infection thread and release of rhizobia into the cytoplasm) appears not to be affected by infection of the plant with SMV. In the second stage a membrane envelope forms around the bacterial cell to form a bacteroid. Structural differences such as decreased vesiculation of this membrane envelope were observed in virus-infected roots (Tu, 1977).

6. Myelin-like bodies

Myelin-like bodies consisting of densely staining layers that may be close-packed in a concentric or irregular fashion have been described for several plant virus infections (e.g. Kim *et al.*, 1974). They probably reflect degenerative changes in one or more of the cell's membrane systems. They may be associated with osmiophilic globules, which are thought to consist of the lipid component of cell membranes. Kim *et al.* suggested that myelin-like bodies in bean leaf cells infected with comoviruses may be formed from the osmiophilic globules.

7. Cell death

Drastic cytological changes occur in cells as they approach death. These changes have been studied by both light and electron microscopy, but they do not tell us how virus infection actually kills the cell.

C. Virus-induced structures in the cytoplasm

The specialized virus-induced regions in the cytoplasm that are, or appear likely to be, the sites of virus synthesis and assembly (viroplasms) are discussed in Chapter 8. Here other types of inclusions will be described. These are usually either crystalline inclusions consisting mainly of virus, or the pinwheel inclusions characteristic of the potyviruses. Light and electron microscopy of these inclusions is reviewed in Christie and Edwardson (1977).

1. Crystalline inclusions

Virus particles may accumulate in an infected cell in sufficient numbers and exist under suitable conditions to form three-dimensional crystalline arrays. These may grow into crystals large enough to be seen with the light microscope, or they may remain as small arrays that can be detected only by electron microscopy.

The ability to form crystals within the host cell depends on properties of the virus itself and is not related to the overall concentration reached in the tissue or to the ability of the purified virus to form crystals. For example, TYMV can readily crystallize *in vitro*. It reaches high concentration in infected tissue but does not normally form crystals there. By contrast, sTNV occurring in much lower concentrations frequently forms intracellular crystals.

a. TMV

In tobacco leaves showing typical mosaic symptoms caused by TMV, leaf-hair and epidermal cells over yellow–green areas may almost all contain crystalline inclusions, while those in fully dark green areas contain none. The junction between yellow–green and dark green tissue may be quite sharp, so that there is a zone where neighboring leaf hairs have either no crystals, or almost every cell has crystals. Warmke and Edwardson (1966) followed the development of crystals in leaf-hair cells of tobacco. Virus particles were first seen free in the cytoplasm as small aggregates of parallel rods with ends aligned. These aggregates increase in size. The growing crystals are not bounded by a membrane, and as they become multilayered they may sometimes incorporate endoplasmic reticulum, mitochondria, and even chloroplasts between the layers.

The plate-like crystalline inclusions are very unstable and are disrupted by pricking or otherwise damaging living cells. They are birefringent when viewed edge-on, but not when seen on the flat face. They contain about 60% water and otherwise consist mainly of successive layers of closely packed parallel rods oriented not quite perpendicularly to the plane of the layers. Rods in successive layers are tilted with respect to one another. This herringbone effect can be visualized in freeze-fractured preparations for some strains of TMV (Fig. 3.7). Sometimes long, curved, fibrous inclusions, or

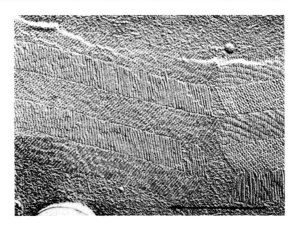

Fig. 3.7 Crystal of TMV rods in a freeze-etched preparation. Part of TMV crystal lying within a mesophyll cell that has been penetrated by glycerol. The crystal has retained its herringbone structure and the lattice spacing of about 24 nm, despite the fact that the tonoplast was ruptured. Bar = 1 μm. From Willison (1976), with permission.

spike-like or spindle-shaped inclusions made up largely of virus particles, can be seen by light microscopy. Different strains of the virus may form different kinds of paracrystalline arrays. Most crystalline inclusions have been found only in the cytoplasm but some have been detected in nuclei (e.g. Esau and Cronshaw, 1967).

b. BYV

The inclusions found in plants infected with BYV occur in phloem cells and also appear in other tissues, for example, the mesophyll. By light microscopy, the inclusions are frequently spindle-shaped and may show banding (Fig. 3.8A). Electron microscopy reveals layers of flexuous virus rods (Fig. 3.8B). Most of the viral inclusions occur in the cytoplasm. Smaller aggregates of virus-like particles were seen in nuclei and chloroplasts (Cronshaw *et al.*, 1966).

c. Other helical viruses

Rod-shaped viruses belonging to other groups may aggregate in the cell into more or less ordered arrays. These can frequently be observed only by electron microscopy, and other material besides virus rods may be present in the arrays. As well as such aggregates of

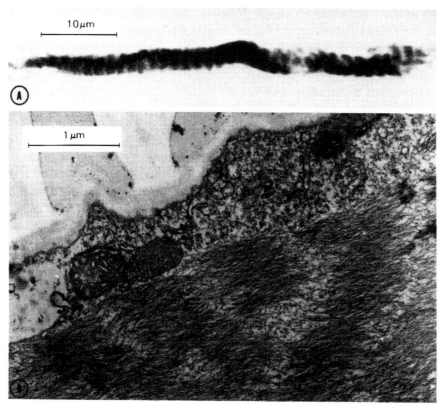

Fig. 3.8 Inclusion bodies caused by BYV in parenchyma cells of small veins of *Beta vulgaris*. The banded form of these inclusions is shown **(A)** by light microscopy and **(B)** by electron microscopy. Bands are made up of flexuous virus particles in more or less orderly array. From Esau *et al.* (1966), with permission.

virus rods, RCVMV induces the appearance of large crystals in the cytoplasm (Khan *et al.*, 1977). These contain RNA and protein and consist of a crystalline array of polyhedral particles about 10 nm in diameter. No virus rods were present. The significance of these unusual crystals in unknown.

d. Small icosahedral viruses

Many small icosahedral viruses form crystalline arrays in infected cells (see Fig. 3.9D). Sometimes these are large enough to be seen by light microscopy (e.g. TNV: Kassanis *et al.*, 1970). Icosahedral viruses that do not normally form regular arrays may be induced to do so by heating or plasmolyzing the tissue to remove some of the water (Milne, 1967; Hatta, 1976).

Many strains of BBWV induce cylindrical tubules of virus particles (Fig. 3.9A,B). Russo *et al.* (1979) described a strain that formed unusual hollow tubules that are square or rectangular in section. The walls are made up of two parallel rows of virus particles.

e. Reoviruses and rhabdoviruses

Plant cells infected with viruses belonging to the *Reoviridae* or *Rhabdoviridae* frequently contain masses of virus particles in regular arrays in the cytoplasm. With some rhabdoviruses the bullet-shaped particles accumulate in more or less regular arrays in the perinuclear space (see Fig. 8.19) (see also Francki *et al.*, 1985c).

2. Pinwheel inclusions

Potyviruses induce the formation of characteristic cylindrical inclusions in the cytoplasm of infected cells (e.g. Hiebert and McDonald, 1973). The most striking feature of these inclusions viewed in thin cross-section is the presence of a central tubule from which radiate curved 'arms' to give a pinwheel effect. Reconstruction from serial sections shows that the inclusions consist of a series of plates and curved scrolls with a finely striated substructure with a periodicity of about 5 nm. The bundles, cylinders, tubes and pinwheels seen in section are aspects of geometrically complex

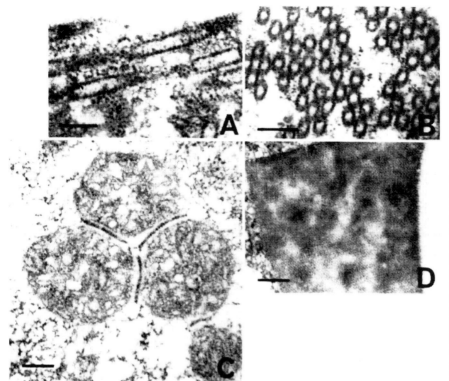

Fig. 3.9 Electron micrographs of thin sections of *Nicotiana clevelandii* leaves infected with BBWV-1 (then named petunia ringspot virus). **(A,B)** Longitudinal and transverse sections of tubes composed of virus particles. **(C)** Virus particles arranged between mitochondria. **(D)** Crystals of virus particles. Markers = 250 nm. From Hull and Plaskitt (1974), with kind permission of the copyright holder, © Karger, Basel.

structures. The general structure has been confirmed by examination of freeze-etched preparations (McDonald and Hiebert, 1974) and by the use of a tilting stage together with computer-assisted analytical geometry (Mernaugh *et al.*, 1980) (Fig. 3.10). For some potyviruses, the pinwheels are tightly curved. For others they are more open. The inclusions induced by members of the group may differ consistently in various details.

Studies using confocal laser scanning microscopy show that the three-dimensional structure of ZYMV cytoplasmic inclusions is a linear filamentous structure of varying thickness and length (Lim *et al.*, 1996). The average length varied from 9.4 ± 0.3 to 20.1 ± 0.4 µm and the average width from 2.1 ± 0.1 to 3.7 ± 0.1 µm.

Pinwheel inclusions originate and develop in association with the plasma membrane at sites lying over plasmodesmata (Lawson *et al.*, 1971; Andrews and Shalla, 1974). The central tubule of the pinwheel is located directly over the plasmodesmata and it is possible that the membranes may be continuous from one cell to the next. The core and the sheets extend out into the cytoplasm as the inclusion grows. Later in infection, they may become dissociated from the plasmodesmata and come to lie free in the cytoplasm. Virus particles may be intimately associated with the pinwheel arms at all times and particularly at early stages of infection (Andrews and Shalla, 1974). The virus-coded protein that is found in these inclusions is discussed in Chapter 6 (Section VIII.C).

3. Caulimovirus inclusions

Two forms of inclusion bodies (also termed viroplasms) have been recognized in the cytoplasms of plants infected with CaMV and other 'caulimoviruses' (see Fig. 8.23); they are not seen in cells of plants infected by the closely-related 'badnaviruses'. Both forms contain virus particles. Electron-dense inclusions are made up of ORF VI product and are considered to be the sites of virus synthesis and assembly (see Chapter 8, Section VII.B.2). Electron-lucent inclusion bodies are made up of ORF II product, one of the proteins involved in aphid transmission (see Chapter 11, Section III.F).

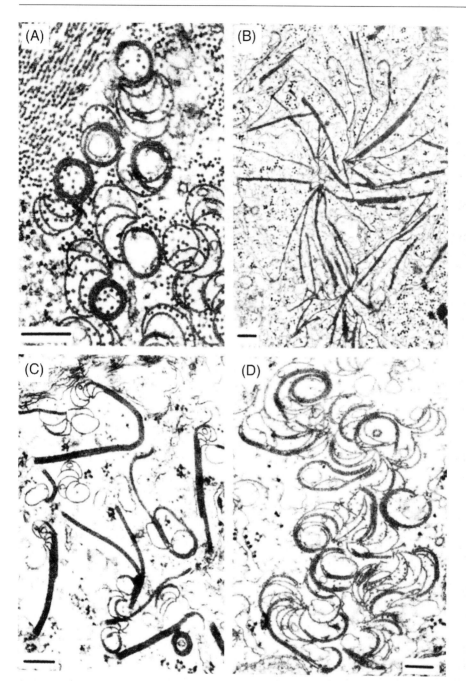

Fig. 3.10 Morphological modifications of cytoplasmic inclusions according to the subdivisions of Edwardson (1974a) and Edwardson *et al.* (1984). **(A)** Scrolls induced in *Nicotiana clevelandii* by an unidentified isolate from eggplant in Nigeria. **(B)** Laminated aggregates induced by statice virus Y in *Chenopodium quinoa*. **(C)** Scrolls and laminated aggregates induced by strain O of TuMV in *N. clevelandii*. **(D)** Scrolls and short curved laminated aggregates in *N. tabacum* infected by PVY. Bars = 200 nm. From Lesemann (1988), with permission.

D. Cytological structures resembling those induced by viruses

Some normal structures in cells could be mistaken for virus-induced effects—for example, crystalline or membrane-bound inclusions in plastids (e.g. Newcomb, 1967). Prolamellar bodies in chloroplasts give the appearance in section of a regular array of tubes. These bodies may be induced by certain chemical treatments (Wrischer, 1973).

Phosphorus-deficient bean leaves (Thompson *et al.*, 1964) showed degenerative changes in the chloroplasts like those seen in some virus infections. Similarly, in sulfur-deficient *Zea mays*, chloroplasts contained many osmiophilic granules and small vesicles (Hall *et al.*, 1972).

Bundle sheath chloroplasts of C4 plants contain numerous small vacuoles near their periphery and vacuoles also have been described for chloroplasts in certain tissues of C3 plants (e.g. Marinos, 1967). In a spontaneous plastid mutant of *Epilobium hirsutum*, the degenerate grana and vacuolation of the mutant plastids seen by electron microscopy (Anton-Lamprecht, 1966) bear some resemblance to the pathological changes induced by TYMV infection in Chinese cabbage.

Nuclei of healthy cells sometimes contain crystalline structures that might be mistaken for viral inclusions (e.g. Lawson *et al.*, 1971).

Such virus-induced effects as disorganization of membrane systems, presence of numerous osmiophilic granules, and disintegration of organelles are similar to normal degenerative processes associated with aging or degeneration induced by other agents.

E. Discussion

There are three main cytological effects of virus infection. Firstly, cellular components are frequently affected. Sometimes the impact of infection is major, inducing marked changes in the structures of organelles such as chloroplasts or even breakdown of internal components leading to death of the cell. These frequently play a role in the symptoms that infection produces. There are also many minor and transitory effects that virus infection has on cells. I will describe in Chapter 13 some of the dynamism involved in virus replication and expression.

Secondly, there are often aggregates of virus particles. As will be described in subsequent chapters, viruses can go through numerous rounds of replication, which if allowed to happen in an uncontrolled manner could result in cell death. Thus, it is possible that aggregation of a virus is a mechanism for removing it from the replication pool.

Thirdly, there are aggregates of virus gene products. The ORF VI inclusion bodies of CaMV are the sites of virus expression and replication. However, the transactivation properties of the ORF VI product, described in Chapter 7 (Section V.B.7), could be deleterious to the normal function of the cell. Thus, it is possible that the aggregation of the ORF VI protein is a mechanism of sequestering it away from the cell components. Similarly, the potyviral genome is expressed from a polyprotein which gives rise to as many copies of protease and replicase molecules as of coat protein molecules. It is likely that the proteases and replicases could be deleterious to the cell and hence they are sequestered into inclusion bodies.

V. THE HOST RANGE OF VIRUSES

Since the early years of the twentieth century, plant virologists have used host range as a criterion for attempting to identify and classify viruses. In a typical experiment, the virus under study would be inoculated by mechanical means to a range of plant species. These would then be observed for the development of virus-like disease symptoms. Back-inoculation to a host known to develop disease might be used to check for symptomless infections. In retrospect, it can be seen that reliance on such a procedure gives an oversimplified view of the problem of virus host ranges. Over the past few years, our ideas of what we might mean by 'infection' have been considerably refined, and some possible molecular mechanisms that might make a plant a host or a non-host for a particular virus have emerged.

The term 'host' is sometimes used rather loosely. Technically, it is defined as 'an organism or cell culture in which a given virus can replicate'. This would mean that a plant species in which the virus can replicate in the initially infected cell (subliminal infection) is a host. However, this is impractical and for this book I will use the terms *local host* for a species in which the virus is restrained to the inoculated leaf and *systemic host* for a species in which the virus spreads from the inoculated leaf to other, but not necessarily all, parts of the plant. More detailed aspects of the interaction of viruses with plants are discussed in Chapter 10.

A. Limitations in host range studies

1. General

Almost all the plant viruses so far described have been found infecting species among the angiosperms. Only a minute proportion of the possible host–virus combinations has been tested experimentally. The following arithmetic indicates the scale of our ignorance. Horváth (1983) tested the host range of 24 viruses on 456 angiosperm species. He found 1312 new host–virus combinations, that is, 12% of those he tested. There may be about 250 000 species of angiosperms (Heywood, 1978) and over 900 plant viruses have been recorded. If the 12% rate applied on average to all these plants and viruses, then there may be more than 25×10^6 new compatible host–virus combinations awaiting discovery. In relation to this figure, the number of combinations already tested must be almost negligible.

Our present knowledge of the occurrence and distribution of viruses among the various groups of plants is both fragmentary and biased. There are four probable reasons for this. First, plant virologists working on diseases as they occur in the field have been primarily concerned with viruses causing economic losses in cultivated plants. They have usually been interested in other plant species only to the extent that they might be acting as reservoirs of a virus or its vector affecting a cultivated species. Thus, until fairly recently all the known plant viruses were confined to the angiosperms. Within this group, most of the known virus hosts are plants used in agriculture or horticulture or are weed species that grow in cultivated areas.

Most of the world's population is fed by 12 species of plant (three cereals: rice, wheat and corn; two sugar plants: beet and cane; three root crops: potato, sweet potato and cassava; two legumes: bean and soybean; and two tree crops: ·coconut and banana). Eighty-seven of the 977 viruses recognized by the ICTV have names derived from these 12 host plants. The number of angiosperm species is commonly reported to be about 250 000. Thus, about 9% of the viruses have been described from about 0.005% of the known species. Tobacco and tomato provide another example. Both these annual crops are of high value commercially on a per-hectare basis. Fifty-five viruses are listed with tobacco and tomato as part of the vernacular name. Not many of the species in lower phyla are used commercially. It is not surprising, therefore, that the first virus found in these groups was one associated with cultivated mushrooms.

The second reason for our incomplete knowledge is a more speculative one. It seems likely that widespread and severe disease in plants due to virus infection is largely a consequence of human agricultural manipulations. Under natural conditions, viruses are probably closely adapted to their hosts and cause very little in the way of obvious disease. Thus, casual inspection of plants growing in their natural habitat may give little indication of the viruses that might be present. Adequate testing of a significant number of such species by means of inoculation tests, both by mechanical transmission and with possible invertebrate vectors, would be laborious and time-consuming. Very little systematic testing of this sort has been carried out.

Thirdly, the selection of 'standard' test plants for viruses is to a great extent governed by those species that are easy to grow in glasshouses and to handle for mechanical and insect vector inoculation.

Fourthly, the genera and species chosen for a host range study may not form a taxonomically balanced selection. Watson and Gibbs (1974) pointed out that most virologists, working in the north temperate zone, use mainly festucoid grasses in host range studies; whereas in other parts of the world non-festucoid groups predominate in the flora and in agricultural importance.

2. Technical limitations

There are a number of technical difficulties and pitfalls in attempting to establish the host range of a virus. These will be discussed in relation to diagnosis in Chapter 15 (Section II). Another kind of potential limitation in the use of mechanical inoculation is illustrated by experiments with the viroid PSTVd. Based on mechanical inoculation tests, a particular accession of *Solanum acaule* has been considered immune to PSTVd. However, *Agrobacterium*-mediated inoculation of this plant led to the systemic replication of the viroid, as did grafting with viroid-infected tomato (Salazar *et al.*, 1988b).

B. Patterns of host range

In spite of the limitations alluded to above, some general points can be made. Different viruses may vary widely in their host range. At one extreme BSMV is virtually confined to barley as a host in nature (Timian, 1974). At the other extreme, CMV, AMV, TSWV, TMV and TRSV have very wide host ranges. For instance, TSWV has a host range of over 925 species belonging to 70 botanical families, and CMV can infect more than 1000 species in more than 85 botanical families.

The host range of one virus may sometimes fall completely within the host range of another apparently unrelated virus. Thus, Holmes (1946) found that among 310 species tested, 83 were susceptible to both TMV and TEV, 116 were susceptible to TMV but not TEV, none to infection by TEV but not TMV, and 111 were not susceptible to either virus. There were also great differences in symptom response (Table 3.3).

Bald and Tinsley (1970) showed that eight cocoa virus isolates could be placed in a host range containment and divergence series. The most virulent isolate infected 21 of the 26 species examined, and this host range contained the host ranges of the other seven isolates.

Bald and Tinsley (1967) re-examined data of earlier workers by statistical procedures. There was a tendency for phylogenetically more advanced groups of plants to contain a higher proportion of species susceptible to the viruses examined than the lower phylogenetic groups. Generally speaking, viruses can infect a higher proportion of species in the family of the common field host and closely related families than in distantly related families.

C. The determinants of host range

Biological and statistical studies of the sort described in the previous section cannot lead us to an understanding of reasons why a virus infects one plant species and not another. For this we need biochemical, molecular biological and genetic information of the sort that is beginning to become available. There is no doubt that very small changes in the viral genome can affect host range. For example, Evans (1985) described a nitrous acid mutant of CPMV that was unable to grow in cowpea but that could grow in *Phaseolus vulgaris*. The mutation was in the B-RNA.

On the basis of present knowledge there are four possible stages where a virus might be blocked from infecting a plant and causing systemic disease: (1) during initial events—the uncoating stage; (2) during attempted replication in the initially infected cell; (3) during movement from the first cell in which the virus replicated; and (4) by stimulation of the host's cellular defenses in the region of the initial infection. These stages will be considered in turn. Molecular aspects of host range determination are discussed in Chapters 9 and 10.

1. Initial events
a. Recognition of a suitable host cell or organelle

Bacterial viruses and most of those infecting vertebrates have specific proteins on their surface that act to recognize a protein receptor on the surface of a susceptible host cell. The surface proteins on plant rhabdo-, reo- and tospo-viruses may have such a cell recognition function. Such a function is unlikely to be of use in plants. For instance, virus particles of LNYV, whose outer membranes have been removed with detergent, are infectious to plants (Randles and Francki, 1972). However, the G

TABLE 3.3 Host ranges of, and symptom responses to, TMV and TEV

TMV	Localization	Symptoms?	TEV
111[a]	Immune/subliminal	No symptoms	227
100	Local	No symptoms	15
15	Systemic	No symptoms	8
27	Local	Symptoms	7
57	Systemic	Symptoms	53

[a] Number of species.
Data from Holmes (1946), with permission.

protein spikes or projections (see Fig. 5.40) are likely to be important in recognizing membranes of insect vectors. Surface proteins on plant reoviruses almost certainly have a recognition role in their insect vectors. There is no evidence for plant cell recognition receptors on the surface of any of the ssRNA plant viruses; however, there are receptors on the surface of RNA viruses that have a circulative interaction with their biological vector (see Chapter 11, Section III.H.1.a). The evidence available for these small viruses suggests that host range is usually a property of the RNA rather than the protein coat. When it has been tested, the host range of a plant virus is the same whether intact virus or the RNA is used as inoculum. Attempts to extend host ranges by using infectious RNA rather than virus have generally been unsuccessful. Hiebert *et al.* (1968) showed that artificial hybrid *Bromovirus* particles, consisting of CCMV RNA in a coat of BMV protein, could still infect cowpea, which is immune to BMV. The hybrid could not infect barley, the normal host of BMV. These and similar tests showed that the host range of a viral RNA cannot be extended by coating the RNA in the protein of a virus that can attack the host. Atabekov and colleagues (Atabekov, 1975) have studied the host ranges of many 'hybrid' viruses and found examples where a heterologous coat protein limited host range. Thus, BMV RNA in a coat of TMV was unable to infect its normal hosts (Atabekov *et al.*, 1970a). On balance, it appears that viral coat proteins play little if any positive part in cell recognition. This view is supported by the fact that viral uncoating following inoculation appears not to be host-specific (see below). Surface recognition proteins may be of little use to a virus in the process of infecting a plant because of the requirement that they enter cells through wounds on the plant surface.

Leaf-hair cells have been infected with various viruses by introducing the virus directly into the cell with a microneedle (see Chapter 9, Section II.B), thus presumably bypassing any virus–cell surface interaction. Similarly, intact virus particles or other infectious materials are able to pass from cell to cell through the plasmodesmata and cause infection while remaining within the plasma membrane (see Chapter 9, Section II).

From experiments using model membranes and plant viral coat proteins in various aggregation states, Datema *et al.* (1987) concluded that hydrophobic lipid-coat protein interactions do not occur. Experiments on the binding and uptake of CCMV by cowpea protoplasts gave no evidence to support an endocytic uptake of virus mediated by specific receptors (Roenhorst *et al.*, 1988).

As discussed in Chapter 8, stages in the replication of many viruses take place in association with particular cell organelles. Recognition of a particular organelle or site within the cell by a virus (or by some subviral component or product) must be a frequent occurrence. Plant viruses may have evolved a recognition system basically different from that of viruses that normally encounter and recognize their host cells in a liquid medium or at a plasma membrane surface.

b. Lack of specificity in the uncoating process

Various lines of evidence suggest that there is little or no host specificity in the uncoating process. Thus, for TMV (Kiho *et al.*, 1972) and TYMV (Matthews and Witz, 1985), virus was uncoated as readily in non-hosts as in host species. Gallie *et al.* (1987c) showed that, when mRNA coding for the enzyme chloramphenicol acetyltransferase was packaged into rods with TMV coat protein and inoculated to protoplasts or plant leaves, the RNA became uncoated. Francki *et al.* (1986b) showed that, when VTMoV and the viroid PSTVd were inoculated together on to a susceptible host, the viroid was incorporated into virus particles. When such virus was inoculated to a species that was a host only for the viroid, viroid infection occurred, indicating that the nucleic acids had been uncoated. However, these experiments did not eliminate the possibility that VTMoV replicated only in the initially infected cells of the presumed non-host species.

2. Replication

Following inoculation of TMV to plant species considered not to be hosts for the virus, viral

RNA has been found in polyribosomes (Kiho *et al.*, 1972). Furthermore, TMV particles uncoat and express their RNA in *Xenopus* oocytes (P.C. Turner *et al.*, 1987b). However, there is some evidence that specific viral genes involved in replication may also be involved as host range determinants. Various host proteins have been found in replication complexes of several viruses (see Chapter 8, Sections IV.E.5 and IV.H.5). Mouches *et al.* (1984) and Candresse *et al.* (1986) obtained evidence that the replicase of TYMV consists of a 115-kDa viral-coded polypeptide and possibly a 45-kDa subunit of plant origin. The 115-kDa polypeptide from different tymoviruses showed great serological variability. They suggested that this variability might be a consequence of the need for the viral-coded polypeptide specifically to recognize the host subunit in different host species in order to form a functional replicase. Thus, the virus-coded peptide might be directly involved in defining host specificity.

Most strains of CaMV infect only *Brassicaceae*. The few strains that also infect some solanaceous species can be divided into three types according to which species they infect. To determine which CaMV genes determine this host range, Schoelz *et al.* (1986) made recombinant viruses by exchanging DNA segments between the cloned strains. The resulting hybrids were then tested on the relevant solanaceous species. These experiments indicated that the first half of gene VI, which codes for the inclusion body protein (see Chapter 6, Section IV.A.1.b), determines whether the virus can systemically infect *Datura stramonium* and *Nicotiana bigelovii*.

Squash leaf curl disease in the United States is caused by two distinct but highly homologous bipartite geminiviruses. The host range of one virus is a subset of the other (Lazarowitz, 1991). Analysis of agroinfected leaf disks indicated that virus replication was involved in the host restriction of one of the viruses. Replication of the restricted virus was rescued in *trans* by coinfection with the non-restricted virus. Sequence analysis revealed that the restricted virus had a 13-nucleotide deletion in the common region. In other respects the sequences of the two common regions were almost identical. Lazarowitz (1991) suggested that this deletion may have been involved in the host range restriction.

As discussed in Chapter 7 (Section V.B.9), some viruses synthesize 'read-through' proteins. Successful read through depends on the presence of an appropriate suppressor tRNA in the host plant. The presence of such a tRNA may, in principle at least, be a factor determining host range for viruses depending on the read-through process.

3. Cell-to-cell movement

Two lines of evidence strongly support the view that possession of a compatible and functional cell-to-cell movement protein is one of the factors determining whether a particular virus can give rise to readily detectable virus replication in a given host species or cultivar.

First, the experiments summarized in Chapter 9 show that many viruses contain a gene coding for a cell-to-cell movement protein. Viruses cannot usually infect leaves through the cut end of petioles or stems. However, Taliansky *et al.* (1982a,b,c) showed that, if the upper leaves were already infected with a helper virus, a virus in solution could pass into the upper leaves and infect them. This happened with mutants defective in the transport function, and also with viruses that were not normally able to infect the species of plant involved. For example, they showed that BMV can be transported from conducting tissues in tomato plants preinfected with TMV. Tomato is not normally a host for BMV. The same occurred with TMV introduced into wheat plants already infected with BSMV. A similar situation appears to exist in field infections in sweet potato, where prior infection with SPFMV allows CMV to infect this host (Cohen *et al.*, 1988). Thus, when the resistance of an apparent non-host species is due to a block in the transport function, this block can be overcome by a preinfection with a virus that has a transport protein compatible with the particular host. This appears to be a fairly general phenomenon. However, in interpreting these phenomena involving the joint infection with two or more viruses, the possibility that one of the viruses can suppress the plant defense system must be borne in mind (see Chapter 10, Section IV.H, for suppression of host defense systems).

The second line of evidence concerning the importance of cell-to-cell movement in host range is that a number of viruses have been shown to be able to infect and replicate in protoplasts derived from species in which they show no macroscopically detectable sign of infection following mechanical inoculation of intact leaves. For example, Beier *et al.* (1977) attempted to infect 1031 lines of cowpea by mechanical inoculation with CPMV. Sixty-five lines were defined as operationally immune because no disease developed and no virus could be recovered. Protoplasts could be prepared from 55 of the 'immune' lines. Fifty-four of these could be infected with CPMV. Similar results have been obtained with other hosts and viruses.

Sulzinski and Zaitlin (1982) mechanically inoculated cowpea and cotton plants with TMV. They isolated protoplasts at intervals and determined the proportion of infected cells at various times using fluorescent antibody. Only about 1 in $5-15 \times 10^4$ protoplasts were infected and this number remained unchanged for at least 11 days. These results show that subliminally infected plants can support virus replication in individual cells, but that the virus cannot move out of the initially infected cells. Viruses that can be induced to invade and replicate systemically by a helper virus, as described in the first part of this section, can presumably infect the host concerned on their own in a similar subliminal fashion. This idea is supported by the observation that the *Geminivirus* BGMV, normally confined to phloem tissue, moved to cells of many types in bean leaves doubly infected with BGMV and TMV (Carr and Kim, 1983).

However, not all sequence differences in the movement proteins can be involved in defining host range. Solis and García-Arenal (1990) showed that there were major differences in the C-terminal region of the movement proteins of TMV, another *Tobamovirus*, TMGMV, and TRV, and yet all of these viruses share the same range of natural hosts in the *Solanaceae*.

A functional cell-to-cell movement protein may not be the only requirement for systemic movement of a virus. CCMV infects dicotyledons (cowpea) and the related BMV replicates in monocotyledons (barley). Both viruses replicate in protoplasts of both hosts. If the movement protein was the only factor controlling systemic movement, transferring RNA3 between the two viruses should switch the host range. This does not occur. Therefore, systemic infection must also involve factors coded for by RNAs 1 or 2 or both (Allison *et al.*, 1988). Since BMV must be encapsulated for systemic movement, it is probable that all four genes are involved in systemic movement and host specificity.

4. Stimulation of host-cell defenses

In some host–virus combinations, the virus stimulates host-cell defense mechanisms that may be a factor limiting virus replication and movement to cells near the first infected cell, giving rise to local lesions, without subsequent systemic spread. This phenomenon, which makes the plant resistant to the virus from a practical point of view, is discussed in Chapter 10 (Section III). Other aspects of host defense systems are discussed in other sections in Chapter 10.

5. Host genes affecting host range

Host genes must be involved in all interactions between a virus and the plant cell following inoculation, whatever the outcome may be. The effects of certain host genes that limit virus infection to local lesions have been studied in some detail. These are discussed in Chapter 10 (Section III).

6. Summary

Concerning the molecular basis for viral host ranges, present evidence suggests the following.

1. There is usually no receptor recognition system for the host plasma membrane and little host specificity in the initial uncoating process.
2. In some plant–virus combinations, molecular mechanisms may exist for blocking virus at some stage in the replication process in the cells where virus first gained entry.

This forms the basis for true immunity to infection.

3. In many host–virus combinations, replication can occur in the first infected cells but movement out of these cells is not possible because of a mismatch between a viral movement protein and some host cell structure. This situation gives rise to subliminal infection and, for practical purposes, a resistant species or cultivar.

4. In other virus–host combinations, movement may be limited by host responses to cells in the vicinity of the initially infected cell, giving rise to a 'local lesion host', which from a practical point of view is 'field resistant'. Using these ideas, the different kinds of host response to inoculation with a virus are defined in Table 10.1.

VI. DISCUSSION AND SUMMARY

The ability to be transferred to healthy host plant individuals is crucial for the survival of all plant viruses. Viruses cannot, on their own, penetrate the undamaged plant surface. For this reason, each kind of virus has evolved ways to bypass or overcome this barrier. Viruses that are transmitted from generation to generation with high frequency via pollen and seed or in plant parts in vegetatively reproducing species can avoid the need to penetrate the plant surface.

Many groups of plant viruses are transmitted by either invertebrate or fungal vectors, which penetrate the plant surface during the process of feeding or infection, carrying infectious virus into the plant cells at the same time. These are described in Chapter 11.

Mechanical transmission is a process whereby small wounds are made on the plant surface in the presence of infectious virus. For many groups of viruses, if conditions are favorable, infection follows. Mechanical inoculation is a very important procedure in experimental virology. In the field, it is not considered to be of great importance except for a few groups such as tobamoviruses and potexviruses, which appear to have no other means of transmission. However, there are several viruses for which a biological vector is not known and which may also be transmitted in this manner. These viruses are, however, well adapted to this form of transmission as they are relatively stable and occur in high concentration in infected leaves. These features make transmission possible when leaves of neighboring plants abrade one another or roots come in contact.

It is probable that successful entry of a single virus particle into a cell is sufficient to infect a plant but, in the normal course of events, many may enter a single cell. Following initial replication in the first infected cell, virus moves to neighboring cells via the plasmodesmata, this being a relatively slow process. Virus then enters the vascular tissue, usually the phloem, movement in this tissue being much more rapid. Most viruses code for specific gene products that are necessary for the virus to move out of the initially infected cell and through the plant.

The final distribution of virus through the plant may be quite uneven. In some host–virus combinations virus movement is limited to local lesions. In others, some leaves may escape infection, while in mosaic diseases dark green islands of tissue may contain little or no virus. Some viruses are confined to certain tissues. For example, luteoviruses are confined mainly to the phloem. Some viruses penetrate to the dividing cells of apical meristems. Others appear not to do so.

There is clear evidence that both virus particles and infectious viral RNA can move from cell to cell, but this may not be the form in which all viruses move. Evidence suggests that TMV may move in the form of nucleoprotein complexes made up from viral RNA, virus-coded proteins, and some host components. Virus-coded movement proteins may exert their effect by altering the properties of the plasmodesmata to facilitate viral passage, but this has not been proven (virus movement is discussed in Chapter 9).

The molecular basis for the host range of viruses is not fully understood. Most plant viruses do not have a mechanism by which intact particles can recognize a host cell that is suitable for replication. Uncoating of the viral nucleic acid appears to occur in hosts and non-hosts alike. No particular step in the replication of a virus has yet been implicated in limiting host range. However, there is good evidence that in 'non-hosts' of some viruses the virus can replicate in the first infected cell but cannot move to neighboring cells. This is because the viral-coded cell-to-cell movement protein does not function in the particular plant. For practical purposes, such a plant is resistant to the virus in question. As described in Chapter 10 (Section IV), plants have a general defense system against 'foreign' nucleic acids that can be suppressed by viruses that infect that host. It is likely that the interaction between the host defense system and the viral defense suppression system is the major determinant of both host range and symptom expression.

Plate 3.1 Chlorotic local lesions in *Chenopodium quinoa*. **(A)** Coarse chlorotic lesions turning necrotic caused by BYMV. Note the hole in the tip of the leaf to indicate an inoculated leaf. Courtesy A. Brunt. **(B)** Fine chlorotic lesions induced by GRV. (Courtesy Association of Applied Biologists.)

Plate 3.2 Necrotic local lesions in *Nicotiana tabacum* NN induced by TMV. (Courtesy Association of Applied Biologists.)

Plate 3.3 Necrotic ringspots in inoculated leaf of *Nicotiana tobacum* caused by ToRSV.

Plate 3.4 Effect of groundnut rosette virus disease on size of groundnut (*Arachis hypogea*) plant. Healthy plant above and infected plant below. (Courtesy Association of Applied Biologists.)

Plate 3.5 Systemic mosaic symptoms on TMV in *Nicotiana tabacum* cv. Samsun three weeks post-inoculation. (Courtesy Association of Applied Biologists.)

Plate 3.6 Chlorotic mosaic induced by TYMV in Chinese cabbage.

Plate 3.7 Mosaic of defined areas of chlorosis and dark green symptomatic of a West Indian isolate of AbMV in *Abutilon sellovianum var marmorata*. (Courtesy Association of Applied Biologists.)

Plate 3.16 Leafrolling induced by BLRV in broad bean (*Vicia faba*). Note the upcurling of the leaflets.

Plate 3.17 Systemic ringspotting caused by Ullman's strain of TMV in *Nicotiana clevelandii*. Note that the ringspots are spreading and fusing with one another. (Courtesy A. Brunt.)

Plate 3.18 Symptoms of PRSV on the fruits of papaya. Some of the dark green ringspots are turning necrotic.

Plate 3.19 Necrotic ringspot on tomato caused by tomato streak disease (ToMV + PVX). This fruit was from the plant shown in Plate 3.20.

Plate 3.21 Mosaic and leaf deformation caused by the Uganda variant of ACMV in field-infected cassava plant.

Plate 3.20 Tomato streak disease (ToMV + PVX) causing severe stem necrosis in a tomato plant. The plant would subsequently collapse and die.

Plate 3.22 Leaf deformation resulting from the infection of a tomato plant with CMV. Note the complete reduction of the lamina of several leaves.

Plate 3.23 Enations induced by PEMV on the underside of a pea (*Pisum sativum*) leaf. Note that the outgrowths are on the veins. (Courtesy Association of Applied Biologists.)

Plate 3.24 Tumor caused by WTV on *Melilotus alba*. Note that the tumor is often in the hypercotyl region at the soil surface.

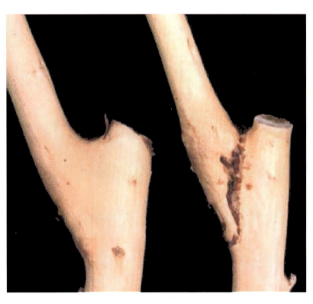

Plate 3.25 Necrosis of a graft union of Virginia Crab and a ASGV-infected rootstock (right) compared with a healthy union (left). (Courtesy Association of Applied Biologists.)

Plate 3.26 Infection of a plantain by BSV causing the fruit bunch to emerge from the side of the pseudostem instead of the top. The emerging unrolled leaf (cigar leaf) is necrotic.

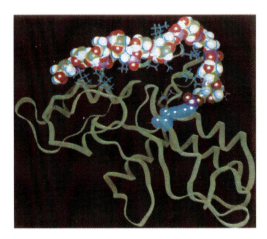

Plate 4.1 Schematic ribbon drawing illustrating interactions of PVY VPg with RNA fragment. The RNA backbone and Tyr-64 of the VPg are shown in space-filling representation. The side-chain residues that are in close contact with RNA are shown in blue. From Plochocka *et al.* (1996), with kind permission of the copyright holder, © National Academy of Sciences, USA.

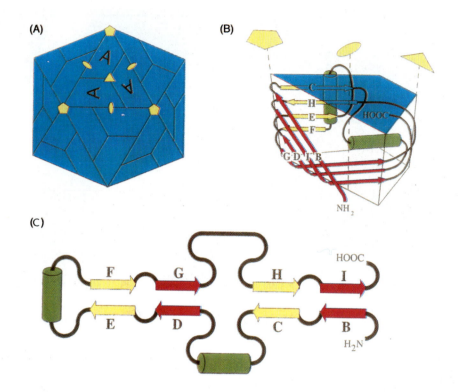

Plate 5.1 (A) The icosahedral capsid contains 60 indentical copies of the protein subunit — in blue, labeled A. These are related by 5-fold symmetry elements (pentagons at vertices), 3-fold symmetry elements (triangles in faces) and 2-fold symmetry elements (ellipses at edges) — shown in yellow. For a given-sized subunit, this point group symmetry generates the largest possible assembly (60 subunits) in which every protein lies in an identical environment. **(B)** Schematic of the subunit building block found in many RNA, and some DNA, viral structures. Such subunits have complementary interfacial surfaces which, when they repeatedly interact, lead to the symmetry of the icosahedron. The tertiary structure of the subunit is an eight-standard β-barrel with the topology of the jellyroll — see part (C). In this diagram the β-strand and helix coding are identical to that in (C). Subunit sizes generally range between 20 and 40 kDa with variation between different viruses occurring at the N and C termini and in the size of insertions between strands of β-sheet. These insertions generally do not occur at the narrow end of the wedge (B–C, H–I, D–E and F–G turns). **(C)** The topology of viral β-barrel showing the connections between strands of the sheets (represented by fat arrows) and positions of the insertions between strands. The green cylinders represent helices that are usually conserved. The C–D, E–F and G–H loops often contain large insertions. From Johnson and Spier (1999), with permission.

Plate 5.2 Models of MSV-N coat protein and complete capsid. Ribbon drawings of polypeptide chain of **(A)** STNV coat protein and **(B)** MSV-N coat protein. In these drawings β-strands are labelled according to convention. **(C)** Fit of 110 copies of the MSV-N pseudo-atomic model, shown as Cα tracings (in yellow), the two apical capsomeres, 10 peripentonal capsomeres, and 10 equatorial capsomeres. The rectangular boxed region indicates the position of close-up views of the models shown in (D) and (E) and the double-ended arrows the cross-sectional views shown in (F) and (G). **(D)** Close-up view of the 2- and 3-fold icosahedral and **(E)** equatorial capsomere interactions. Cross-sectional slices of the MSV-N particle **(F)** through the equatorial region and **(G)** peripentonal capsomeres of the particle. The icosahedral 2-, 3- and 5-fold axes are indicated as a black ellipsoid, triangle and pentamer respectively. From Zhang *et al.* (2001), with permission.

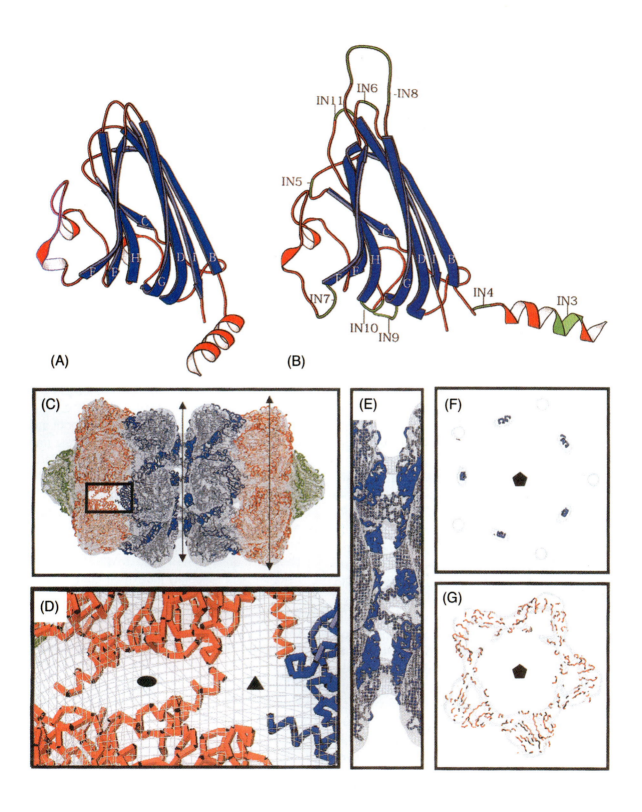

(A)

(B)

IN11 IN6 -IN8
IN5 -
C
H D   B
F F G
IN7 -
IN10 IN9
IN4 IN3

(C)

(D)

(E)

(F)

(G)

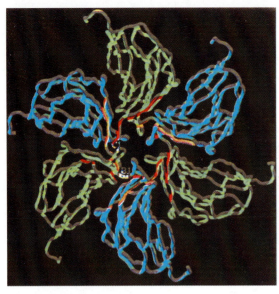

Plate 5.3 Image of the annuli made by the N termini of the B- (blue) and C- (green) subunits interior to the loops between β-strands F and G (FG loops), viewed along a quasi-sixfold axis. Residues 1–10 of each subunit are colored orange (B) and red (C). The N-terminal peptide, which is not seen in the A-subunit, crosses under each subunit and contacts a neighbour near residue 5. From Canady *et al*. (1996), with permission.

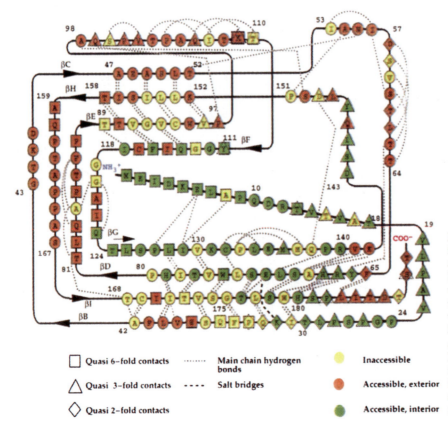

Plate 5.4 Hydrogen bonding, inter-subunit contacts, and accessibility diagram for the B-subunit of TYMV. Atoms were assumed in contact if they were 4.11 Å apart for van der Waals interactions, 3.3 Å for hydrogen bonds, and 3.8 Å for salt bridges. Residues were considered inaccessible if they had less than 5 Å² accessible surface area. From Canady *et al*. (1996), with permission.

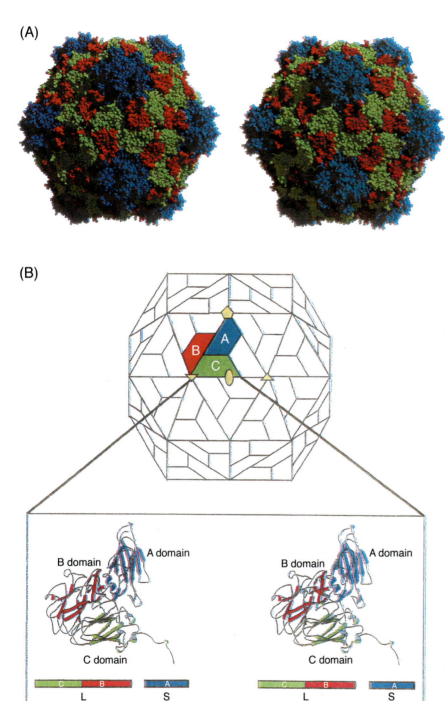

Plate 5.5 The structures of viral capsid and icosahedral asymmetric unit of RCMV. Two proteins (S- and L-subunits) are in the RCMV capsid. The S-subunit forms the A domain (in blue), while the L-subunit forms the B (red) and C (green) domains. **(A)** Stereoview of a space-filling drawing of the RCMV capsid. All atoms are shown as spheres corresponding to a diameter of 1.8 Å. The pentameric S-subunits form the protrusion. **(B)** At the top, the icosahedral asymmetric unit of the capsid is color-coded in the schematic presentation of the CPMV capsid. The S-subunit occupies the A position, forming the A domain around the 5-fold axis; the two domains of the L-subunit occupy the B and C positions. Positions A, B and C are quasi-equivalent positions of identical gene products on a *T* = 3 surface lattice. At the bottom, a stereoview of a ribbon diagram of the icosahedral asymmetric unit is shown. All three domains are variants of the jellyroll β-sandwich structure. The schematic presentation of composite proteins, L- and S-subunits, is also shown. From Lin *et al.* (2000), with permission.

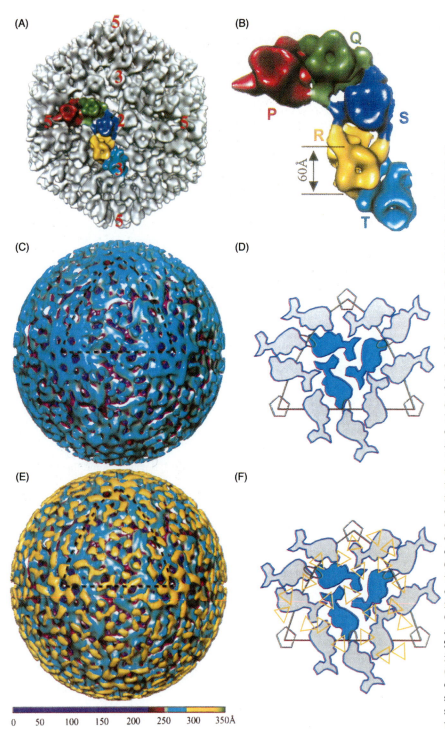

Plate 5.6 Structure of RDV. **(A)** Shaded surface view of the reconstruction of the full RDV as viewed along the icosahedral 2-fold axis. The numbers 5, 3 and 2 designate the icosahedral 5-, 3- and 2-fold axes. Highlighted in colour are a contiguous group of five trimers found in each asymmetric unit. **(B)** Blown-up view of the group of five trimers computationally extracted from panel (A). These trimers at the distinct quasi-equivalent positions are designated P, Q, S, R and T, using the convention set out for BTV (Grimes *et al.*, 1997). **(C)** Inner shell computationally extracted at 590 Å diameter, showing a *T* = 1 lattice. The dashed triangle designates one triangular face of the icosahedron. **(D)** Schematic of fish-shaped density distribution within a triangle of a *T* = 1 lattice. **(E)** Capsid computationally extracted at 604 Å diameter, showing the interface between the outer shell (yellow) and inner shell (blue) proteins. Color bar shows the coding as a function of radius. **(F)** Schematic illustrating the interaction pattern of trimers (yellow triangles) of the outer shell with the fish-shaped densities of the inner shell. From Lu *et al.* (1998b). with permission.

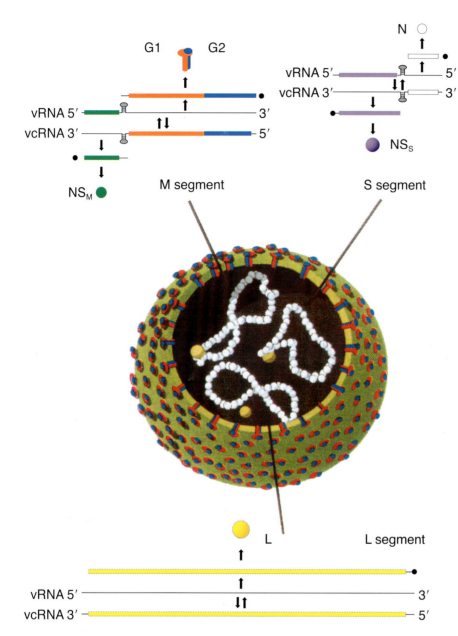

Plate 5.7 Morphology and genome expression of TSWV. The core of the TSWV particle consists of three viral segments (L, M and S) tightly associated with nucleoprotein (N, white), to which the viral RNA-dependent RNA polymerase (L, yellow) attaches. The virus particle is enveloped by a membrane in which two types of glycoprotein (G1 and G2, red and blue respectively) are present. The ambisense segments M and S each encode two proteins on the viral RNA (vRNA) and viral complementary RNA (vcRNA) strands respectively, whereas the L RNA encodes only the large L ORF. All viral proteins are translated from mRNAs whose transcription is initiated by 'cap snatching' from a host mRNA (indicated by a black dot). The glycoproteins, G1 and G2, result from the cleavage of the precursor protein; NS_s (purple) and NS_m (green) are non-structural proteins. Double arrows indicate replication, whereas single arrows indicate transcription and translation. From Prins and Goldbach (1998), with permission.

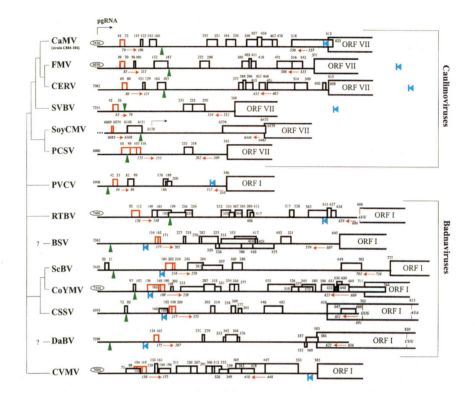

Plate 7.1 Comparison of the primary structures of the pgRNA leader of plant pararetroviruses. On the left, a phylogenetic tree shows the relationship of their reverse transcriptases (according to Richert-Pöggeler and Shepherd, 1997; note that BSV and DaBV were not included in that comparison). Caulimoviruses and badnaviruses are grouped (as indicated on the right). The leader sequence preceding the first long ORF (ORF VII or ORF I) is depicted as a thick line; the sORFs are indicated by boxes, with internal start codons shown by vertical lines. The numbered genome position of the pgRNA 5′-end is enclosed within an ellipse if mapped by primer extension or not enclosed if putative. The numbering within the leader is from the 5′-end (except for SoyCMV where the latter is unclear). Red arrows under the leader define the complementary sequences that form the base of the large stem-loop structures. The conserved sORFs preceding the structures are also in red. A green triangle indicated a putative or, in the case of CaMV, FMV and RTBV (Rothnie, 1996), mapped poly (A) signal. An arrowhead adjacent to a vertical line (in blue) shows the location of the primer binding site for reverse transcription. From Pooggin *et al.* (1999), with permission.

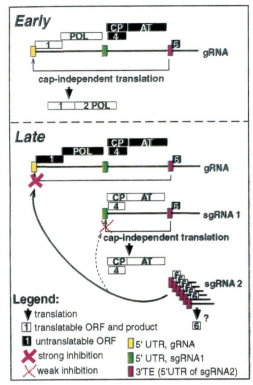

Plate 7.2 Translational switch model for *tran*-regulation of BYDV gene expression by sgRNA2. Open boxes indicate translational ORFs and their translation products (below large arrows). Black boxes indicate ORFs that are not translated. **Early:** polymerase is translated from the genomic (g)RNA (the only viral RNA at this stage) via the 3′TE (red box) *in cis*. **Late:** As abundant sgRNA2 accumulates, it specifically inhibits gRNA (bold cross) in preference to sgRNA1 (dashed cross), via the 3′TE *in trans*. This allows almost exclusive translation of late genes from sgRNA1. The different 5′ UTRs of gRNA (gold box) and sgRNA1 (green box) contribute to differential inhibition. The role of ORF 6 encoded by sgRNA2 is unknown (?) but it is not necessary for *trans*-inhibition. From Wang *et al.* (1999c, where the model is discussed in detail), with permission.

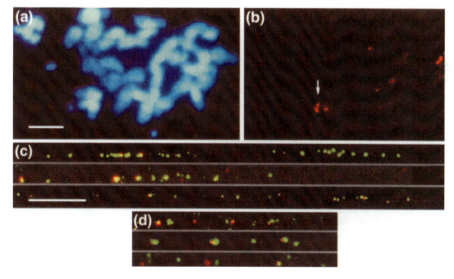

Plate 8.1 *In situ* hybridization of *Musa* Obino l'Ewai chromosomes. **(a,b)** Hybridization of metaphase chromosomes. **(a)** Chromosomes stained blue with the DNA stain DAPI (4,6-diamidino-2-phenylindole). Bar = 5 μm. **(b)** Hybridization sites of BSV (red) showing one major site in each metaphase (arrowhead) and at least one minor site near the limits of hybridization sensitivity (arrow). **(c,d)** Hybridization of extended chromosomal DNA fibres. Two different-length hybridization patterns, with chains of dots representing probe hybridization sites, were detected with BSV (green) and *Musa* flanking sequence (red probes). **(c)** Three independent aligned long fibres. Both BSV and *Musa* flanking sequence are present in multiple copies in a 150-kb structure. Bar = 5 μm. **(d)** Three independent aligned short fibres in a 50-kb structure. From Harper *et al.* (1999b), with permission.

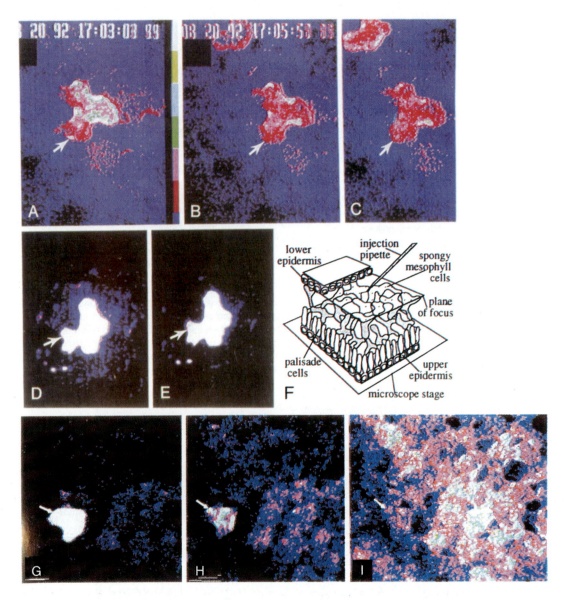

Plate 9.1 Microinjection of fluorescently labeled 10-kDa Lucifer yellow dextran **(A–E)** or 20-kDa fluorescein isothiocyanate-labelled dextran **(G–I)**, together with TMV MP **(A–C and G–I)** and without TMV MP **(D,E)** into spongy mesophyll cells of mature tobacco leaves. Magnifications: A–C, ×230; D and E, ×130. The time course of 10-kDa dextran movement is shown. Elapsed time after injection: (A) 1 min; (B) 4 min; (C) 5 min; (D) 1 min; (E) 30 min; (G) 3 min; (H) 14 min; (I) 34 min. Arrows point to the injected cell. Pictures are *averaged*: colours are derived by false color imaging and represent intensity of fluorescent signal ranging from blue (lowest intensity) to white (highest intensity) with black and blue representing the background; see color bar at right of fluorescent image in (A). In the top part of the picture an automatic timer gives date and time of experiment. The timer has been turned off in image (C) to allow a better view of the fluorescent cells. **(F)** Schematic view through a tobacco leaf. Spongy mesophyll cells form a three-dimensional interconnected network. To inject into spongy mesophyll cells, a small portion (5 mm²) of the lower epidermis of the leaf is peeled off. The leaf is then placed, abaxial side up, on the microscope stage. Only surface cells in the plane of focus can be injected and subsequently monitored. These cells are indicated in a lighter pattern. From Waigmann *et al.* (1994), with kind permission of the copyright holder, © The National Academy of Sciences, USA.

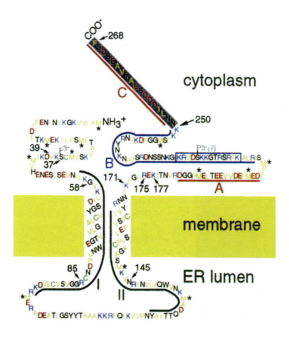

Plate 9.2 Topological model of the TMV MP. The amino acid sequence was deduced from the ORF encoding the MP. Hydrophobic amino acids are in yellow, basic residues are blue, acidic residues are red and Cys residues are green. The trypsin-resistant core domain contains the first 249 or 250 residues. The C-terminus (grey) was rapidly removed by tryspin. Domains I (residues 56–96) and II (residues 125–164) are regions that are conserved among tobamovirus MPs and are outlined in black; domains A (residues 183–200) and C (residues 252–268) are acidic and are outlined in red; and domain B (residues 206–250) is basic and is outlined in blue. Cytoplasmic, transmembrane, ER luminal and core domains were inferred from western blots of proteinase K-treated microsomes, hydropathy analysis, fluorescent microscopy, trypsin susceptibility, and mass spectrometry. Transmembrane domains are presumed to be α-helical. Hydrophobic peptides (38–85 vs 58–85) differed in length between the monomer and core, suggesting that the C-terminus of the monomer may interact with its N-terminus. Serine-37 is phosphorylated (p³-) and S S 218 may be phosphorylated (p³?) based on Prosite prediction and sequence conservation among tobamoviruses. Met residues (*) have been reported to generally be involved in protein-protein interactions. From Brill *et al.* (2000), with kind permission of the copyright holder, © The National Academy of Sciences, USA.

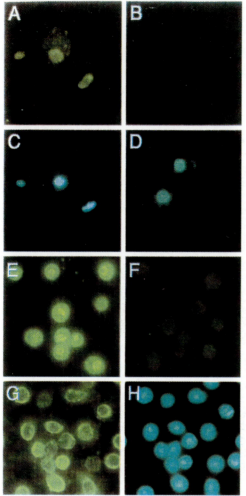

Plate 9.3 Subcellular localization of MSV coat protein in **(A–D)** tobacco protoplasts and **(E–H)** insect cells determined by immunofluorescent staining. BY2 tobacco cells were transfected either (A and C) with pJITMSVCP (the MSV coat protein fragment cloned into a pUC-based transient expression vector, pJIT163), or (B and D) with pJIT163, and sampled after 48 hours' incubation. **(A,B)** Fluorescein isothiocyanate (FITC) fluorescence is identified by excitation at 488 nm, and indicates location of coat protein. **(C,D)** Images of the same cells, photographed after excitation at 366 nm, to show 4,6-diamidine-2-phenylindole dihydrochloride (DAPI) fluorescence indicating the location of nuclei. **(E–H)** Sf21 insect cells were infected with recombinant virus containing (E) the wild-type (wt) coat protein sequence, or (G and H) mutant CP201, which lacks the 20 N-terminal amino acids (Liu *et al.*, 1997), or (F) with wt baculovirus. Cells were sampled 30 hours post-inoculation. Presence of wt or mutant coat protein is identified by fluorescence at 488 nm (E and G). For wt coat protein, most of the FITC signal was present in the nuclei, whereas for mutant CP201, comparison of the images produced at 488 and 366 nm (G and H, which depict the same cells) showed that the mutant protein did not accumulate in the nuclei. Nuclei were identified by DAPI staining and excitation at 366 nm (H), the nuclei staining intensely, but less intensive staining was always seem in surrounding insect cell cytoplasm. From Liu *et al.* (1999), with kind permission of the copyright holder, © The American Phytopathological Society.

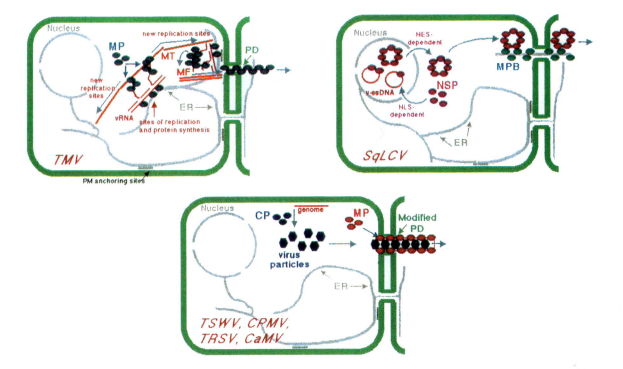

Plate 9.4 Models for plant virus intracellular movement. **(Top left)** TMV. MP complexed with viral genomic RNA is proposed to move along microtubules from ER-associated sites of viral replication and protein synthesis ('viral factories') to establish additional viral factories at other ER sites. From viral factories associated with cortical ER, actin microfilaments may deliver the MP-genome complexes to putative cell-wall adhesion sites and plasmodesmata. **(Top right)** The geminivirus SqLCV. NSP is a nuclear shuttle protein that moves newly replicated viral ssDNA genomes from the nucleus to the cytoplasm. MPB, associated with ER-derived tubules, traps the NSP-ssDNA complexes in the cytoplasm and is proposed to guide these along the tubules and through the cell wall into adjacent cells. In these adjacent cells, NSP-ssDNA complexes would be released and target the viral DNA to the nucleus to initiate a new round of replication and infection. **(Bottom)** Viruses that form MP tubules: TSWV, CPMV, TRSV and CaMV. These viruses require coat protein or equivalent as well as MP for cell-to-cell movement. The viral genome, encapsidated in virus particles, moves through MP-containing tubules that appear to emerge from highly modified plasmodesmata. CP, coat protein; ER, endoplasmic reticulum; MF, microfilament; MP, movement protein; MT microtubule; NES, nuclear export signal; NLS, nuclear localization signal; PD, plasmodesma; vRNA, viral RNA genome; v-ssDNA, viral ssDNA genome. From Lazarowitz and Beachy (1999), with permission.

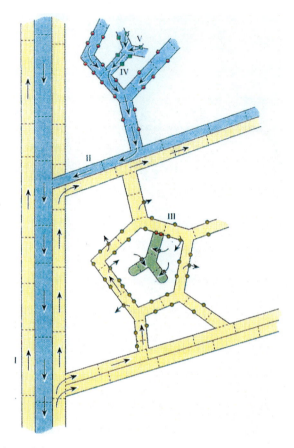

Plate 9.5 Schematic model of the sink/source transition in a developing *Nicotiana benthamiana* leaf. At the top, the diagram depicts the exporting (source; blue) region of the leaf, and the lower part depicts the importing (sink; yellow) region of the leaf. Symplastic phloem unloading of solute and virus occurs from the class III vein network (yellow circles). In the sink portion of the leaf, the minor veins are immature (or non-functional; green shading) and receive assimilate (and virus) directly from the mesophyll by cell-to-cell movement. In the source portion of the leaf, the minor veins are now mature. Switching on sucrose transporters (green squares) in the minor veins signals the onset of apoplastic sucrose loading. In the source region, the class III veins function to export rather than import, which is achieved by down-regulation of the plasmodesmata (red circles) that connected them to the mesophyll during the import phase. Bidirectional phloem transport (in different sieve tubes) occurs in the class I vein, but it also may occur in class II and III veins as the 'transitional front' passes basipetally across different vein classes. From Roberts *et al.* (1997), with permission.

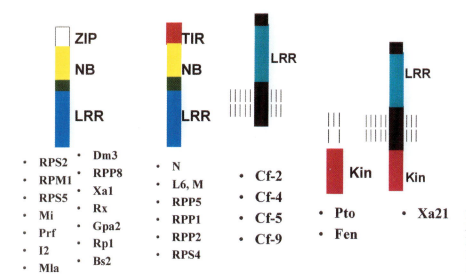

ZIP
NB
LRR

- RPS2
- RPM1
- RPS5
- Mi
- Prf
- I2
- Mla

- Dm3
- RPP8
- Xa1
- Rx
- Gpa2
- Rp1
- Bs2

TIR
NB
LRR

- N
- L6, M
- RPP5
- RPP1
- RPP2
- RPS4

LRR

- Cf-2
- Cf-4
- Cf-5
- Cf-9

Kin

- Pto
- Fen

LRR

Kin

- Xa21

Plate 10.1 The structures of 5 classes of *R* genes. LRR, leucine-rich repeat region; NB, nucleic acid binding; ZIP, leucine ZIP domain.

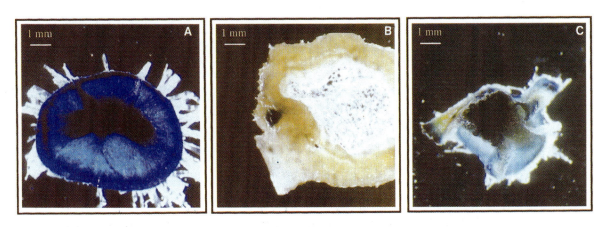

Plate 10.2 Histochemical analysis of *uidA* expression in the meristems of silenced and non-silenced tobacco plants. Histochemical GUS assays were performed on stems and shoot tips of homozygous 6b5 tobacco plants showing *uidA* PTGS and hemizygous 23b9 plants that did not show *uidA* PTGS. **(A)** Staining of stem cross-section of non-silenced 23b9 plant. **(B)** Staining of stem cross-section of silenced 6b5 plant. A non-silenced axillary meristem is visible as a dark blue spot. **(C)** Staining of the shoot apical meristem of a silenced 6b5 plant. From Béclin *et al.* (1998), with permission.

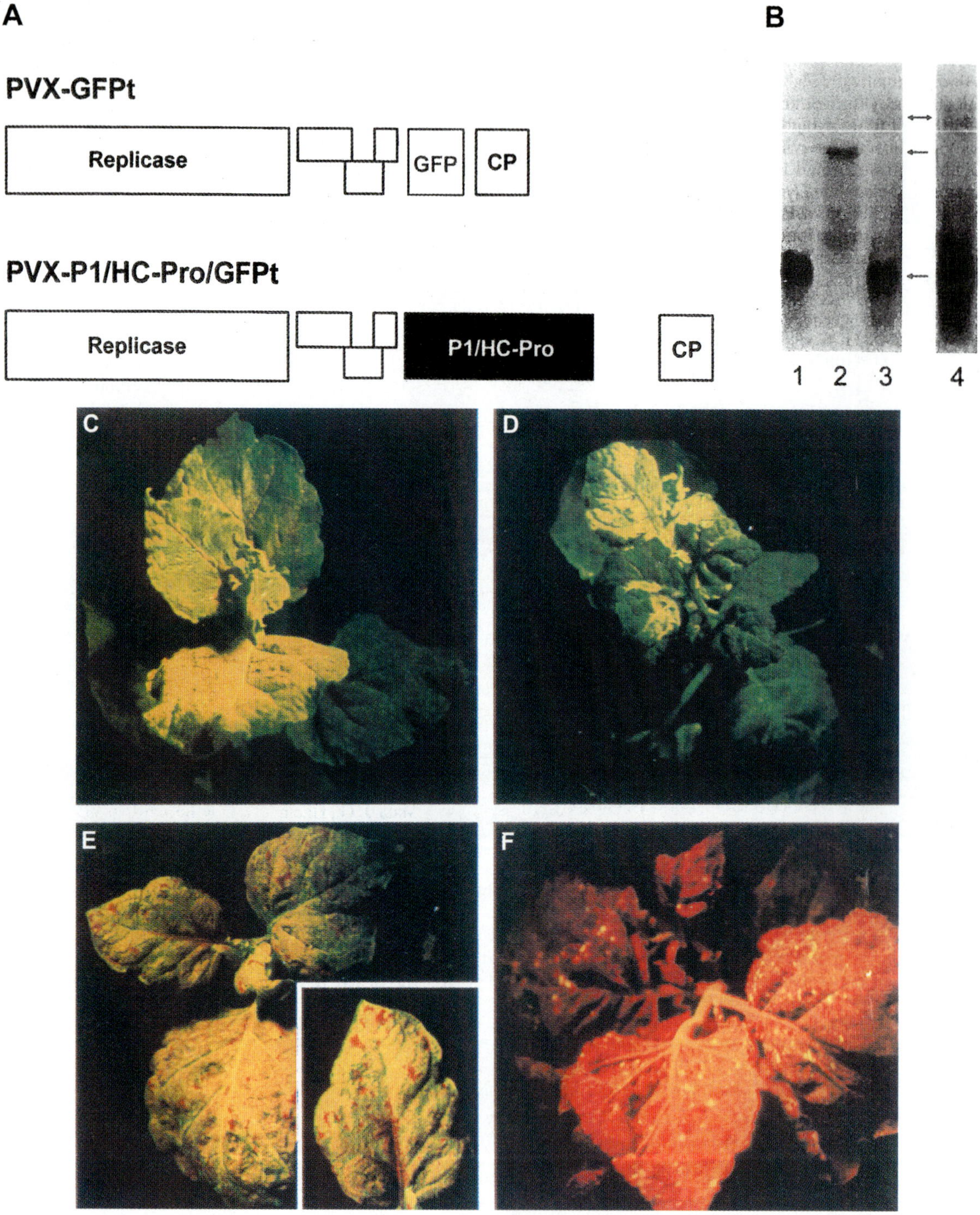

A

PVX-GFPt

| Replicase | | GFP | CP |

PVX-P1/HC-Pro/GFPt

| Replicase | | P1/HC-Pro | | CP |

B

1   2   3   4

C   D

E   F

Plate 10.3 (previous page) Virus-induced gene silencing (VIGS) of transgene-encoded GFP in mixed virus infection. **(A)** Schematic diagram of the PVX vector constructs carrying the GFPt coding sequence (PVX-GFPt) or the TEV P1/HC-Pro sequence and GFPt. GFPt is the S65T version of GFP. **(B)** Northern blot analysis of GFP transgene mRNA levels in mock-inoculated GFP transgenic plants (lane 1) or plants infected with either PVX-GFPt (lane 2) or PVX-P1/HC-Pro/GFPt (lanes 3 and 4). Lane 4 shows a 3-fold longer exposure of lane 3. Total RNA was extracted from upper leaves of the plants 20 days after inoculation; a GFP-specific probe was used. Arrows indicate the positions of PVX-P1/HC-Pro/GFPt genomic RNA, PVX-GFPt genomic RNA and GFP transgene mRNA (from top to bottom respectively). **(C–F)** Transgenic GFP *Nicotiana benthamiana* plants coinoculated with **(C)** PVX-GFP and PVX-5′TEV, showing complete suppression of virus-induced gene silencing (VIGS) of GFP; **(D)** PVX-GFP and PVX-HC, showing an almost complete suppression of VIGS of GFP. **(E)** PVX-GFP and PVX-HC, showing partial suppression of VIGS of GFP—the inset is a closer view of a partially silenced leaf showing silencing in the vicinity of the veins; **(F)** PVX-GFP and PVX-noHC, showing complete VIGS of GFP—the red fluorescence is that of chlorophyll. From Anandalakshmi *et al.* (1998) with kind permission of the copyright holders, © The National Academy of Sciences, USA.

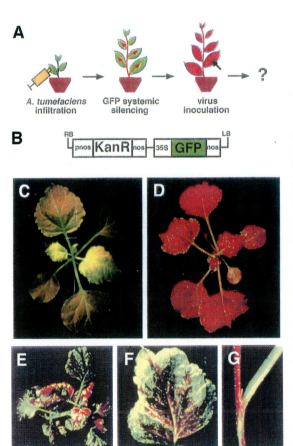

Plate 10.4 PTGS of a GFP transgene induced by infiltration with *A. tumefaciens* and subsequent suppression by PVY. **(A)** Schematic representation of the experimental system. GFP silencing was induced in transgenic *N. benthamiana* line by infiltration with a hypervirulent strain of *A. tumefaciens* carrying a binary Ti plasmid shown in **(B)**. Complete GFP silencing was achieved at 10 days post-infiltration and plants appeared uniformly red (chlorophyll fluorescence) under UV illumination. **(B)** Structure of the binary Ti plasmid cassette used to generate transgenic *N. benthamiana* plants expressing GFP and to induce GFP silencing in this line by *A. tumefaciens* (strain cor308) infiltration. The right and left borders of the T-DNA (RB and LB) flank a kanamycin resistance gene (*KanR*) in a nos promoter (pnos) and a nos terminator (nos) cassette. **(C)** Transgenic *N. benthamiana* plant showing high levels of GFP expression under UV illumination. **(D)** Transgenic *N. benthamiana* after induction of gene silencing by infiltration with *A. tumefaciens* carrying the binary Ti plasmid shown in **(B)**. **(E–G)** Suppression of PTGS by PVY. GFP-silenced transgenic *N. benthamiana*, as in **(C)**, 15 days after infection with PVY viewed under UV illumination. **(E)** Whole plant; **(B)** close-up of leaf; **(C)** close-up of stem. From Brigneti *et al.* (1998), with kind permission of the copyright holder, © The Oxford University Press.

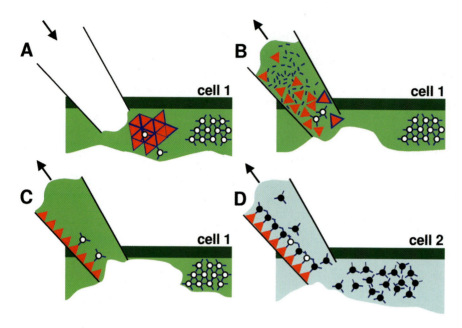

Plate 11.1 Model of the sequential acquisition of CaMV by aphids from infected cells. **(A)** In infected cells, the viral components involved in transmission are spatially separated in electron-lucent inclusion bodies (eliB) (left structure) and electron-dense inclusion bodies (ediB) (right structure). Whereas most virus particles (empty circles) complexed with P3 (blue bars) are stored in ediBs, P2 (red triangles) in loose association with P3 and some sparse viruses are located in eliBs. When the aphid stylet (at the left) pierces the cell wall, saliva is injected into the plant cell. **(B)** After salivation, the aphid sucks up some of the plant cell's contents together with viral inclusion bodies through its stylets. If an eliB is taken up, it immediately—either through the action of the aphid saliva or simply by a dilution effect—disintegrates and sets free its components P2, P3 and eventually some rare P2:P3 virion complexes. **(C)** While the liberated P3 is ingested, the released P2, perhaps together with a few virus:P3 complexes from the eliB, attaches to the aphid stylet cuticle. The aphid is now 'P2 loaded', and thus transmission competent, and ready to acquire more P3:virions (solid circles) from either the same cell or **(D)** on following feeding on other cells from the same or different host plants. From Blanc *et al.* (2001), with permission.

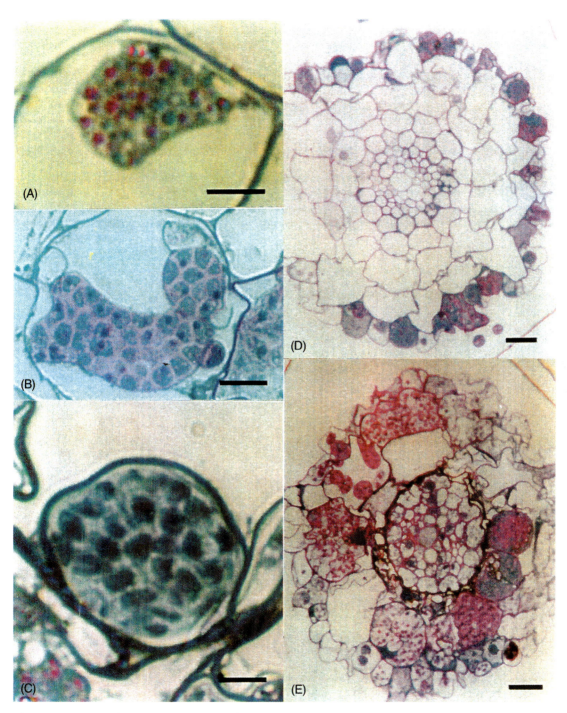

Plate 11.2 Invasion of sugar beet roots by *Polymyxa betae*. **(A)** Young plasmodium in epithelial cells with blue-stained lipid and nuclei, and pink dots of unidentified material. **(B)** Young lobed zoosporangium in cortical cell. Note pink ground colour, blue zoospores and dark blue nuclei. **(C)** Mature zoosporangium containing blue-stained zoospores. **(D)** Distribution of *P. betae* 8 days after inoculation. Plasmodia in various stages of development are restricted to epidermal cells. **(E)** Distribution of *P. betae* 18 days after inoculation. Plasmodia and zoosporangia are present throughout the root cortex but not in the endodermis or stele. Bar markers: **(A)** and **(C)**, 5 μm; **(B)**, 10 μm; **(D)** and **(E)**, 50 μm. From Barr and Asher (1996), with permission.

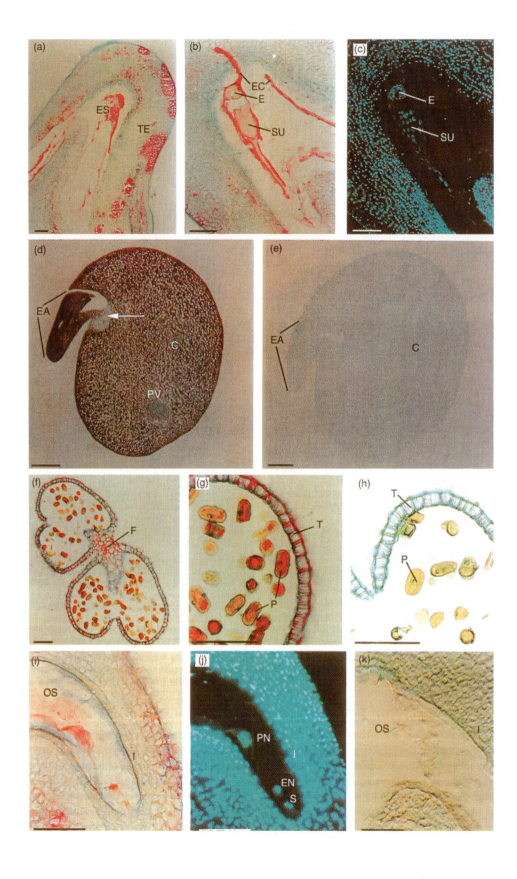

Plate 12.1 (previous page) Detection of PEBV in the embryo and gametes of pea. To show the tissue distribution of PEBV, sections of immature seeds **(a–c)**, isolated embryos **(d, e)**, anthers **(f–h)** and ovules **(i–k)** were subjected to immunohistochemistry **(a–c, f–k)** with Fast Red TR as the chromogenic substrate for detection, or *in situ* hybridization **(d, e)** with BCIP/NBT as substrate. For immature seeds, the location of PEBV **(a, b)** with respect to the testa (TE), globular embryo (E) and multinucleate suspensor (SU) was revealed after staining the section with DAPI **(c)**. Sections of Stage 5 embryos from infected **(d)** and uninfected **(e)** plants were treated with DIG-labeled RNA probes for (+)-sense PEBV RNA. Viral RNA was present uniformly throughout the embryo, except at the junction (arrow) between the axis and the cotyledon. Transverse sections through anthers were immunostained with anti-PEBV serum **(f, g)** or pre-immune serum **(h)**. PEBV was present in the vascular tissues of the anther filament (F), the tapetum (T) and the mature pollen grains (P). Longitudinal sections through unfertilized ovules were immunostained with PEBV-specific **(i, j)** or pre-immune **(k)** serum. The accumulation of PEBV in the integuments (I) and ovule sac (OS) relative to the position of the synergid (S), polar (PN and egg cell (EN) nuclei were revealed by staining the same section with DAPI **(j)**. C, cotyledons; E, globular embryo; EA, embryonic axis; EC, endospermic cytoplasm; EN, egg cell nucleus; ES, embryo sac; F, anther filament; I, integument; OS, ovule sac; P, pollen grains; PN, pollen nucleus; PV, provascular tissue; S, synergids; SU, suspensor; T, tapetum; TE, testa. Bar markers: **(d, e)** 1 mm, others 100 μm. From Wang and Maule (1997), with kind permission of the copyright holder, © Blackwell Science Ltd.

Plate 12.2 The pathway to seed transmission of PSbMV in pea. **(a, b)** Analysis of the distribution of PSbMV in longitudinal sections through immature pea seed by immunohistochemistry using a monoclonal antibody to PSbMV coat protein shows that a cultivar–virus interaction which is permissive for seed transmission (e.g. with *Pisum sativum* cv. Vedette in **(a)**) results in a widespread accumulation of the virus in the testa tissues. In contrast, in the non-permissive interaction (e.g. with cv. Progretta in **(b)**) virus enters the seed through the vascular bundle but is unable to invade the adjacent testa tissues extensively. In both cases, there is a gradual reduction in the amount of accumulated virus after invasion such that in cv. Progretta only patches (asterisks) of infected tissue remain detectable. Systematic analyzes of the immature seeds of different ages have identified the routes (red arrows) of virus invasion in the two cultivars (illustrated diagrammatically in **(c)** for cv. Vedette and in **(d)** for cv. Progretta). The most consistent observation from all these studies is that the virus must reach the micropylar area of the testa for seed transmission to occur, a location providing the closest point of contact (arrowhead in **(a)**) between the testa tissues and the embryonic suspensor. In the non-permissive interaction the virus appears to be blocked (denoted by the red squares) in its ability to spread into and/or replicate in the non-vascular testa tissues. E, embryo proper; F, funiculus; M, micropylar region; S, suspensor; T, testa; V, vascular bundle. Bar markers 500 μm in panels **(a)** and **(b)**. From Maule and Wang (1996), with kind permission of the copyright holder, © Elsevier Science.

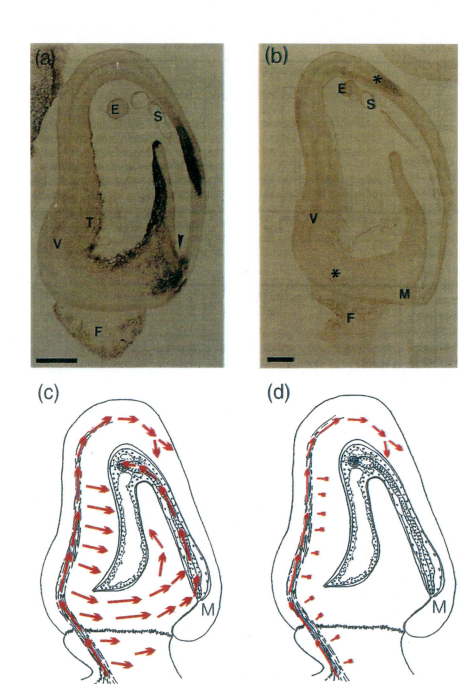

Plate 12.3 Aerial view of a sugar beet field with patches of plants infected with virus yellows (BYV and/or BMYV) and gradients of infection spreading from the patches. (Courtesy of IACR Broom's Burn Experimental Station.)

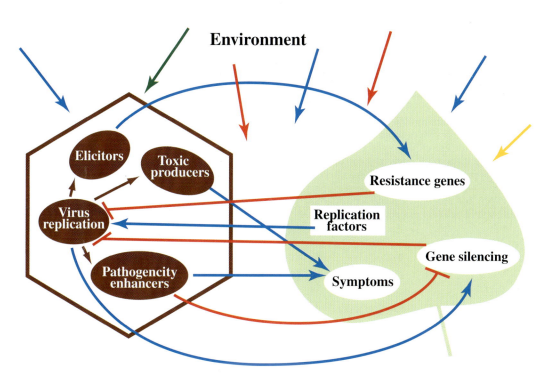

Plate 13.1 Diagram showing viral infection as the complex result of aggression, defense and counter-defense. The virus is indicated in brown on the left and the host in green on the right. (Courtesy of Dr J.A. Garcia.)

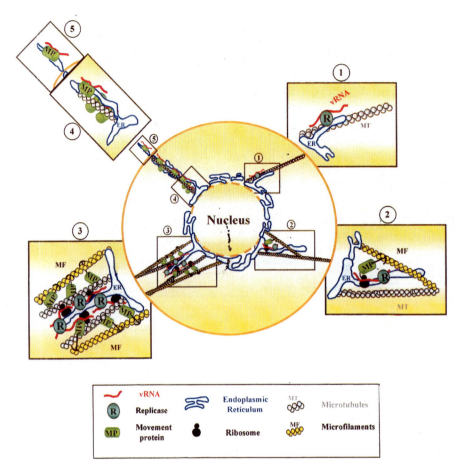

Plate 13.2 Model of TMV infection in BY2 protoplasts. (1) Early in infection TMV vRNA comes into close association with membranes of the ER, and vRNA–replicase complexes associated with ER are transported via microtubules to perinuclear positions. (2) Nascent vRNAs synthesize movement protein that remains associated with vRNA. (3) Formation and anchoring of large ER-derived structures containing movement protein, replicase and vRNA, which are stabilized by movement protein and microfilament interactions. (4) Microtubule-based transport system of vRNA–movement protein complexes towards the periphery of the cell to initiate cell-to-cell spread. (5) Protrusion of the ER containing vRNA and movement protein through the plasma membrane. The model is not drawn to scale. From Más and Beachy (1999), with kind permission of the copyright holder, © The Rockefeller University Press.

Origin of a subgroup I luteovirus by recombination

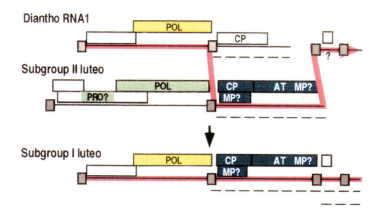

Origin of a subgroup II luteovirus by recombination

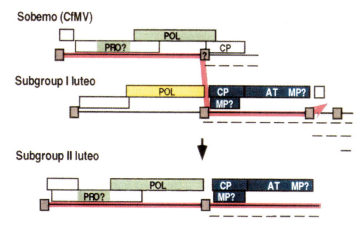

Plate 17.1 Model for origin of luteovirus subgroups. Solid black lines represent viral genomic RNA. Dashed lines indicate subgenomic RNAs. Boxes indicate genes. Blue shading, genes with sequence similarity to umbra-, diantho- and carom-viruses; green, sequence similarity to sobemoviruses. Grey boxes represent putative origins of replication and subgenomic mRNA promoters. POL, RNA-dependent RNA polymerase; PRO?, putative protease; CP, coat protein; MP?, putative movement protein; AT, read-through domain of the coat protein gene, possibly required for aphid transmission. Pink line shows the proposed path of the replicase as it switched strands during copying of viral RNAs in a mixed infection. From Miller *et al.* (1997) with kind permission of the copyright holder, © The American Phytopathological Society.

CHAPTER 4

Purification and Composition of Plant Viruses

I. INTRODUCTION

To study the basic properties of virus particles it is usually necessary to separate them from the cellular components. Since the classic studies of Stanley (1935), Bawden and Pirie (1936, 1937) and others in the 1930s, a great deal of effort has been put into devising methods for the isolation and purification of plant viruses. To ascertain the basic properties of a virus it is essential to be able to obtain purified preparations that still retain infectivity. It is not surprising that the first viruses to be isolated and studied effectively (TMV, PVX and TBSV) were among those that are fairly stable and occur in relatively high concentration in the host plant. Today, interest has extended to a range of viruses that vary widely in concentration in the host and in their stability towards various physical and chemical procedures.

In this chapter, I will describe factors to be considered in virus purification and then the basic components of virus particles.

II. ISOLATION

There are no generally applicable rules for virus purification. Procedures that are effective for one virus may not work with another apparently similar virus. Even different strains of the same virus may require different procedures for effective isolation.

A great deal has been written about purity and homogeneity as they apply to plant viruses. In a chemical sense, there is no such thing as a pure plant virus preparation. Even if

Virus acronyms are given in Appendix 1.

a preparation contained absolutely no low- or high-molecular-weight host constituents (which is most unlikely), there are other factors to be considered:

1. Most preparations almost certainly consist of a mixture of infective and non-infective virus particles. The latter will probably have one or more breaks in their nucleic acid molecules, most of which will have occurred at different places in different particles.
2. Most virus preparations almost certainly consist of a mixture of mutants even though the parent strain greatly predominates. Such mutants will differ in the base sequence of their RNA or DNA in at least one place. If the mutation is in the cistron specifying the coat protein, then this may also differ from the parent strain.
3. Purified preparations of many viruses can be shown to contain one or more classes of incomplete, non-infective particles.
4. The charged groups on viral proteins and nucleic acids will have ions associated with them. The inorganic and small organic cations found in the purified virus preparation will depend very much on the nature of the buffers and other chemicals used during isolation.
5. Some of the larger viruses appear to cover a range of particle sizes having infectivity.
6. A variable proportion of the virus particles may be altered in some way during isolation. Enzymes may attack the coat protein. For example, extracts of bean (*Phaseolus vulgaris*) contain a carboxypeptidase-like enzyme that removes the terminal threonine from TMV (Rees and Short, 1965). Proteolysis at defined sites may give rise, during isolation of the

virus, to a series of coat protein molecules of less than full size (e.g. with SNMoV: Chu and Francki, 1983; and BaYMV: Ehlers and Paul, 1986). Coat proteins may undergo chemical modification when leaf phenols are oxidized (Pierpoint *et al.*, 1977). More complex viruses, such as the reoviruses, may lose part of their structure during isolation (e.g. Hatta and Francki, 1977).

Thus, for plant viruses, purity and homogeneity are operational terms defined by the virus and the methods used. A virus preparation is pure for a particular purpose if the impurities, or variations in the particles present, do not affect the particular properties being studied or can be taken account of in the experiment. Effective isolation procedures have now been developed for many plant viruses. Rather than describe these in detail, I shall consider, in general terms, the problems involved in virus isolation. Detailed protocols for isolation procedures for a number of viruses are given in numerous publications, including Hull (1985), Walkey (1991), Stace-Smith and Martin (1993), Dijkstra and de Jager (1998), and in various chapters in Foster and Taylor (1998).

A. Choice of plant material

1. Assay host

During the development of an isolation procedure, it is useful, but not always possible, to be able to assay fractions for infectivity. Of course, this is best done with a local lesion host. Great accuracy usually is not necessary in the preliminary assays, but reliability and rapid development of lesions are a great advantage. If no local lesion host is available, then assays must be done using a systemic host. Assays by the injection of insect vectors sometimes have been used where mechanical transmission is impossible. Electron microscopy or, if antisera or probes are available, dot blot ELISA or dot blot nucleic acid hybridization, can be used to follow the progress of purification.

2. Propagation host

The choice of host plant for propagating a virus may be of critical importance for its successful isolation. Various points have to be considered in the choice of a propagating host:

1. The host plant should be easy to grow, preferably from seed. However, care should be taken that there is no seed-transmitted virus such as SoMV, which is highly transmitted in seed of *Chenopodium* spp.
2. The virus should reach a high concentration in the host.
3. The host should not contain high amounts of substances such as phenolic materials, organic acids, mucilages and gums, certain proteins, and enzymes, particularly ribonucleases that can inhibit or irreversibly precipitate the virus. For example, many stone fruit viruses were very difficult to isolate from their natural hosts, since most members of the *Rosaceae* contain high concentrations of tannins in their leaves. Discovery of alternative non-rosaceaous hosts, for example cucumber (*Cucumis sativus* L.), has allowed the isolation of several such viruses.
4. The host plant constituents should be easily separable from the virus during purification. For instance, certain legumes (e.g. *Canavalia ensiformis*) contain large amounts of Rubisco (fraction 1 protein), often in an aggregated form that can co-purify with rod-shaped viruses.

In practice, species of *Chenopodium*, *Cucumis*, *Nicotiana*, *Petunia*, *Phaseolus* and *Vigna* have been found suitable for the propagation of a large number of viruses.

The plant species used, the conditions under which it is grown, and the time at which it is harvested should be chosen to maximize the starting concentration of infectious virus. For many viruses, concentration rises to a peak after a few days or weeks and then falls quite rapidly (Fig. 4.1).

Sometimes the distribution of virus within the plant is so uneven that it is worthwhile to dissect out and use only those tissues showing prominent symptoms. Viruses frequently occur in much lower concentration in the midrib than in the lamina of the leaf. If the midrib and petiole are large, it may pay to discard them. In special situations, dissection of tissue is almost essential, for example with WTV and other

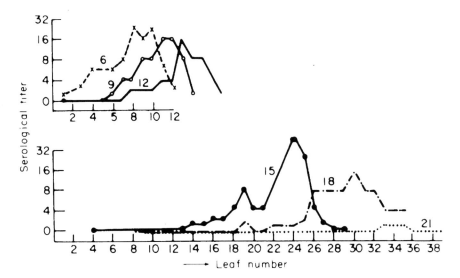

Fig. 4.1 Change in concentration of PVA with age in leaves of a single Samsun tobacco plant inoculated on leaf number 1. Virus concentration was measured serologically. From week 15 onward, older leaves had died. Numbers beside graphs show weeks after infection. From Bartels (1954), with kind permission of the copyright holders, © Blackwell Science Ltd.

viruses causing tumors where the virus is found associated with the tumor tissue. Another reason for harvesting only certain parts of the infected plant may be to avoid high concentrations of inhibitory substances or materials that adsorb to the virus and are later difficult to remove. Such materials frequently occur in lower concentration in new young growth. In certain hosts, virus can only be isolated from such tissue. Similarly, root tissue may sometimes provide more favorable starting material than leaves (e.g. Ford, 1973), although virus concentration is usually lower in roots. For some viruses, flowers may provide a suitable source of virus at high concentration

The possibility that the host used to culture a virus may already harbor another virus or become infected with one must always be borne in mind. Contamination of greenhouse-grown plants with unwanted viruses is not at all uncommon. Strains of TMV, PVX and TNV may be particularly prevalent, especially in greenhouses that have been used for virus work for some time. It is not necessarily sufficient to use a host that is only a local lesion host for such contaminating viruses (e.g. *N. glutinosa* for TMV). Very small amounts of the contaminating virus may become differentially concentrated during isolation of a second virus.

3. Extraction medium

Once infected plant cells are broken, and the contents released and mixed, the virus particles find themselves in an environment that is abnormal. Thus, it is often necessary to use an artificial extraction medium designed to preserve the virus particles in an infectious, intact and unaggregated state during the various stages of isolation. The conditions that favor stability of purified virus preparations may be different from those needed in crude extracts or partially purified preparations (e.g. Brakke, 1963). Moreover, different factors may interact strongly in the extent to which they affect virus stability. The main factors to be considered in developing a suitable medium are as follows.

a. pH and buffer system

Many viruses are stable over a rather narrow pH range, and the extract must be maintained within this range. As most viruses have an acid isoelectric point, buffer solutions with pH values in the range of 7–8 should prevent them from precipitating. However, the structural integrity of some viruses (e.g. bromoviruses) is lost on exposure to pHs in this range. The use of a buffer of pH 5 will precipitate many normal plant proteins and is advantageous if the virus itself is not precipitated. Buffers commonly used for virus extraction include borate, citrate, phosphate and Tris.

b. Metal ions and ionic strength

Some viruses require the presence of divalent metal ions (Ca^{2+} or Mg^{2+}) for the preservation of infectivity and even for the maintenance of structural integrity. Ionic strength may be important. Some viruses fall apart in media of ionic strength below about 0.2 M, while others are unstable in media above this molarity. AMV particles may be precipitated by Mg^{2+} concentrations above 1 mM and degraded by concentrations above 0.1 M (Hull and Johnson, 1968). For some viruses, NaEDTA may be included to minimize aggregation by divalent metals. On the other hand, NaEDTA disrupts certain viruses.

Decisions on the pH, presence of chelating agent and ionic strength should take account of the factors stabilizing the virus particle (see Chapter 5).

c. Reducing agents and substances protecting against phenolic compounds

Reducing agents such as ascorbic acid, cysteine hydrochloride, 2-mercaptoethanol, sodium sulfite, or sodium thioglycollate are frequently added to extraction media. Dithiothreitol (Cleland's reagent) is a useful reducing agent as it has little tendency to be oxidized by air. These materials assist in preservation of viruses that readily lose infectivity through association with products resulting from oxidation of plant extracts and also may reduce adsorption of host constituents to the virus. Phenolic materials especially may cause serious difficulties in the isolation and preservation of viruses. Several methods have been used more or less successfully to minimize the effects of phenols on plant viruses during isolation:

1. Cysteine or sodium sulfite added to the extraction medium both probably act by inhibiting the phenol oxidase and by combining with the quinone (Pierpoint, 1966).
2. Polyphenoloxidase is a copper-containing enzyme. Two chelating agents with specificity for copper, diethyldithiocarbamate, and potassium ethyl xanthate, have been used to obtain infectious preparations of several viruses (e.g. PNRSV: Barnett and Fulton, 1971); the former is also a reducing agent.
3. Materials that compete with the virus for phenols have sometimes been used. For example, Brunt and Kenten (1963) used

various soluble proteins and hide powder to obtain infective preparations of CSSV from cocoa leaves. Synthetic polymers containing the amide link required for complex formation with tannins have been used effectively to bind these materials. The most important of these is polyvinyl pyrrolidone (PVP).

d. Additives that remove plant proteins and ribosomes

Many viruses lose infectivity fairly rapidly *in vitro*. One reason for this loss may be the presence of leaf ribonucleases in extracts or partly purified preparations. Dunn and Hitchborn (1965) made a careful study of the use of magnesium bentonite as an additive in the isolation of various viruses. They found that, under appropriate conditions, contamination of the final virus product with nucleases was reduced or eliminated. In addition, ribosomes, 19S protein, and green particulate material from fragmented chloroplasts were readily adsorbed by bentonite, provided Mg^{2+} concentration was 10^{-3} M or greater. However, variation occurs in the activity of different batches of bentonite, and the material must be used with caution as some viruses are degraded in its presence. Charcoal may be used to adsorb and remove host materials, particularly pigments. Subsequent filtration to remove charcoal may lead to substantial losses of virus adsorbed in the filter cake.

NaEDTA at 0.01 M in pH 7.4 buffer will cause the disruption of most ribosomes, preventing their co-sedimentation with the virus. This substance can be used only for viruses that do not require divalent metal ions for stability.

e. Enzymes

Enzymes have been added to the initial extract for various purposes. Thus, Adomako *et al.* (1983) used pectinase to degrade mucilage in extracted sap of cocoa leaves prior to precipitation of CSSV. Improved yield of a virus limited to phloem tissue was obtained when fibrous residues were incubated with Driselase and other enzymes (Singh *et al.*, 1984). This material contains pectinase and cellulose and presumably aids in the release of virus that would otherwise remain in the fibre fraction. The enzymes also digest materials that would otherwise co-precipitate with the virus. Jones *et al.* (1991) found that digestion of rice tissue with Celluclast gave more reliable yields of the two rice tungro viruses than did digestion with Driselase. Treatment with trypsin at an optimum concentration can markedly improve the purification of TuMV (Thompson *et al.*, 1988).

f. Detergents and other additives

Non-ionic detergents such as Triton X-100 or Tween 80 are often used in the initial extraction medium to assist in release of virus particles from insoluble cell components and to dissociate cellular membranes that may contaminate or occlude virus particles. However, detergents should not be used with enveloped viruses.

The particles of some viruses, such as caulimoviruses, may be contained within inclusion bodies that have to be disrupted. Hull *et al.* (1976) showed that caulimovirus inclusion bodies were disrupted on treatment with urea and Triton X-100.

4. Extraction procedure

Freezing of the plant tissue, say by liquid N_2, before extraction enables disruption of vascular tissues releasing phloem-limited viruses and facilitates subsequent removal of host materials. For some viruses, however, freezing may have a deleterious effect.

A variety of procedures are used to crush or homogenize the virus-infected tissue. These include (1) a pestle and mortar, which are useful for small-scale preparations, (2) various batch-type food blenders and juice extractors, which are useful on an intermediate scale, and (3) roller mills, colloid mills and commercial meat mincers, which can cope with kilograms of tissue. For long, fragile, rod-shaped viruses, grinding in a pestle and mortar may be the safest procedure to minimize damage. For instance, BYV particles were broken on using a blender (Bar-Joseph and Hull, 1974). The addition of acid-washed sand greatly improves

the efficiency of extraction. If an extraction medium is used, it is often necessary to ensure immediate contact of broken cells with the medium. The crushed tissue is usually expressed through muslin or cheesecloth.

5. Preliminary isolation of the virus
a. Clarification of the extract

In the crude extract, the virus is mixed with a variety of cell constituents that lie in the same broad size range as the virus and that may have properties that are similar in some respects. These particles include ribosomes, Rubisco (fraction I) protein from chloroplasts, which has a tendency to aggregate, phytoferritin, membrane fragments, and fragments of broken chloroplasts. Also present are unbroken cells, cell wall fragments, all the smaller soluble proteins of the cell, and low-molecular-weight solutes.

The first step in virus isolation is usually designed to remove as much of the macromolecular host material as possible, leaving the virus in solution. The extraction medium may be designed to precipitate ribosomes, or to disintegrate them. The extract may be subject to some treatment such as heating to 50–60°C for a few minutes, freezing and thawing, acidification to a pH of less than 5 or the addition of K_2HPO_4, to coagulate much host material. The treatment has to be chosen on a case-by-case basis as it may damage the virus. For some viruses, organic solvents such as chloroform give very effective precipitation of host components. For others, the extract can be shaken with *n*-butanol-chloroform, which denatures much host material. The treated extract is then subjected to centrifugation at fairly low speed (e.g. 10–20 minutes at 5000–10 000 g). This treatment sediments cell debris and coagulated host material. With the butanol-chloroform system, centrifugation separates the two phases, leaving virus in the aqueous phase and much denatured protein at the interface. It should be noted that although some viruses can withstand the butanol–chloroform treatment, quite severe losses may occur with others. Chloroform alone gives a milder treatment than does a butanol–chloroform mixture. Of course, solvents and detergents cannot be used for viruses that have a membrane. It must be remembered that there are safety considerations to be taken into account in using large volumes of these organic solvents.

For many viruses, it pays to carry out the isolation procedure at 4°C and as fast as possible once the leaves have been extracted. On the other hand, some viruses occur in membrane-bound structures or other structures within the cell. It may take time for the virus to be released from these after the leaf extract is made. A low-speed centrifugation soon after extraction may then result in much virus being lost in the first pellet but, on the other hand, this can be used to concentrate the virus.

b. Concentration of the virus and removal of low-molecular-weight materials

i. High-speed sedimentation

Centrifugation at high speed for a sufficient time will sediment the virus. Provided the particular virus is not denatured by the sedimentation, it can be brought back into solution in an active form. This is a very useful step, as it serves the double purpose of concentrating the virus and leaving behind low-molecular-weight materials. However, high-speed sedimentation is a physically severe process that may damage some particles (e.g. some reoviruses; Long et al., 1976). Following high-speed sedimentation, some viruses remain as characteristic aggregates when the pellets are redissolved. The virus particles in these aggregates may be quite firmly bound together (Tremaine et al., 1976). If host membranes are involved in the binding, a non-ionic detergent may help to release virus particles. Sedimentation of viruses occurring in very low concentration may result in very poor recoveries. The major process causing losses appears to be the dissolving and redistribution of the small pellet of virus as the rotor comes to rest (McNaughton and Matthews, 1971). Redissolving particles from the surface of the pellet can lead to preferential losses of more slowly sedimenting components (Matthews, 1981).

ii. Precipitation with polyethylene glycol

Hebert (1963) showed that certain plant viruses could be preferentially precipitated in a single-phase polyethylene glycol (PEG) system, although some host DNA may also be precipitated. Since that time, precipitation with PEG has become one of the commonest procedures used in virus isolation. The exact conditions for precipitation depend on pH, ionic strength, and concentration of macromolecules. Its application to the isolation of any particular virus is empirical, although attempts have been made to develop a theory for the procedure (Juckes, 1971).

PEG precipitation is applicable to many viruses, even fragile ones. For example, the procedure gave a good yield of intact particles of CTV, a fragile rod-shaped virus (R. F. Lee et al., 1987). It has the advantage that expensive ultracentrifuges are not required, although differential centrifugation is often used as a second step in purification procedures.

iii. Density-gradient centrifugation

Many viruses, particularly rod-shaped ones, may form pellets that are very difficult to resuspend. Density gradient centrifugation offers the possibility of concentrating such viruses without pelleting. A density gradient is illustrated in Fig. 4.2. The following modification of the density gradient procedure may sometimes be used even with angle rotors for initial concentration of a virus without pelleting at the bottom of the tube. A cushion of a few millilitres of dense sucrose is placed in the bottom of the tube, this being overlaid with a column of low-density sucrose about 2 cm deep, and the rest of the tube is filled with clarified virus extract. Under appropriate conditions, virus can be collected from the region of the interface between the two sucrose layers. In a similar application, a cushion of medium-density sucrose is placed in the bottom of the tube before centrifugation; this prevents contamination of pellets with chlorophyll material. Density-gradient centrifugation is used in the isolation procedure for many viruses.

iv. Salt precipitation or crystallization

Salt precipitation was commonly employed before high-speed centrifuges became generally

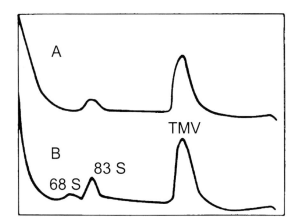

Fig. 4.2 Density-gradient centrifugation for the assay of viruses in crude extracts. An extract of 30 mg of epidermis **(A)** or the underlying tissue **(B)** from a tobacco leaf inoculated with TMV was sedimented in a 10–40% sucrose gradient at 35 000 rpm for 2 h. Absorbancy at 254 nm through the gradient was measured with an automatic scanning device. Note absence of detectable 68S ribosomes in the epidermal extracts.

available. It is still a valuable method for viruses that are not inactivated by strong salt solutions. Ammonium sulfate at concentrations up to about one-third saturation is most commonly used, although many other salts will precipitate viruses or give crystalline preparations. After standing for some hours or days, the virus is centrifuged down at low speed and redissolved in a small volume of a suitable medium.

v. Precipitation at the isoelectric point
Many proteins have low solubility at or near their isoelectric points. Isoelectric precipitation can be used for viruses that are stable under the conditions involved. The precipitate is collected by centrifugation or filtration and is redissolved in a suitable medium.

vi. Dialysis
Dialysis through cellulose membranes can be used to remove low-molecular-weight materials from an initial extract and to change the medium. It is more usually employed to remove salt following salt precipitation or crystallization, or following density gradient fractionation in salt or sucrose solutions, as discussed in a subsequent section.

6. Further purification of the virus preparation

Virus preparations taken through one step of purification and concentration will still contain some low- and high-molecular-weight host materials. Further purification steps can remove more of these. The procedure to be used will depend very much on the stability of the virus, the scale of the preparation, and the purpose for which the preparation is required.

Sometimes highly purified preparations can be obtained by repeated application of the same procedure. For example, a preparation may be subjected to repeated crystallization or precipitation from ammonium sulfate, or may be given several cycles of high- and low-speed sedimentation. The latter procedure leads to the preferential removal of host macromolecules because they remain insoluble when the pellets from a high-speed sedimentation are resuspended. Losses of virus often occur with this procedure, either because some, or all, of the virus itself also becomes insoluble or because insufficient time is allowed for the virus to resuspend. This is a particular difficulty with some rod-shaped viruses. Losses may also occur from virus resuspending before the supernatant fluid is removed. Such losses may be severe with very small pellets, where the surface-to-volume ratio is high.

Generally speaking, during an isolation it is useful to apply at least two procedures that depend on different properties of the virus. This is likely to be more effective in removing host constituents than repeated application of the same procedure.

a. Density-gradient centrifugation

One of the most useful procedures for further purification and for assay, particularly of less stable viruses, is density-gradient centrifugation which was developed by Brakke (1951, 1960). This technique has proved to be highly versatile and has been widely used in the fields of virology and molecular biology. It has largely replaced the analytical ultracentrifuge in studies on viruses. A centrifuge tube is partially filled with a solution having a decreasing density from the bottom to the top of the tube. For plant viruses, sucrose is commonly used to form the gradient, and the virus solution is layered on top of the gradient. With gradients formed with cesium salts, the virus particles may be distributed throughout the solution at the start of the sedimentation or they may be layered on top of the density gradient.

Brakke (1960) defined three ways in which density gradients may be used. *Isopycnic gradient centrifugation* occurs when centrifugation continues until all the particles in the gradient have reached a position where their density is equal to that of the medium. This type of centrifugation separates different particles based on their different densities. Sucrose alone may not provide sufficient density for isopycnic banding of many viruses. In *rate zonal sedimentation* the virus is layered over a preformed gradient before centrifugation. Each kind of particle sediments as a zone or band through the gradient, at a rate dependent on its size, shape and density. The centrifugation is stopped while the particles are still sedimenting. *Equilibrium zonal sedimentation* is like rate zonal sedimentation except that sedimentation is continued until most of the particles have reached an approximate isopycnic position. The role of the density gradient in these techniques is to prevent convectional stirring and to keep different molecular species in localized zones. The theories of density gradient centrifugation are complex. In practice, this technique is a simple and elegant method that has found widespread use in plant virology.

A high-speed preparative ultracentrifuge and appropriate swing-out or angle rotors are required. Following centrifugation, virus bands may be visualized due to their light scattering. The contents of the tube are removed in some suitable way prior to assay. The bottom of the tube can be punctured and the contents allowed to drip into a series of test tubes. Fractionating devices based on upward displacement of the contents of the tube with a dense sucrose solution are available commercially. The UV absorption of the liquid column is measured and recorded, and fractions of various sizes can be collected as required. Fig. 4.2 illustrates the sensitivity of this procedure.

Since successive fractions from a gradient can be collected, a variety of procedures can be used to identify the virus, non-infectious virus-like components and host materials. These include infectivity, UV absorption spectra, and examination in the electron microscope. Using appropriate procedures, very small differences in sedimentation rate can be detected (Matthews and Witz, 1985).

With rate zonal sedimentation, if the sedimentation coefficients of some components in a mixture are known, approximate values for other components can be estimated. If antisera are available, serological tests can be applied to the fractions, or antiserum can be mixed with the sample before application to the gradient. Components reacting with the antiserum will disappear from the sedimentation pattern.

The various forms of density gradient centrifugation can give some indication of the purity of the preparation. They also allow a correlation between particles and infectivity to be made, and frequently reveal the presence of non-infective virus-like particles or multiparticle viruses. Bands of a single component may spread more widely in the gradient than is apparent from a trace of optical density. Several cycles of density gradient centrifugation may be necessary to obtain components reasonably free of mutual contamination. Work with multicomponent viruses has shown that it may be extremely difficult to obtain one component completely free of another, even by repeated density-gradient fractionation. Density gradients prepared from the non-ionic medium Nycodenz were used effectively for the further purification of several viruses (Gugerli, 1984).

Brakke and Dayly (1965) showed that zone spreading of a major component by non-ideal sedimentation can cause zone spreading of a minor component that the major component overlaps. The way in which zones are removed from the gradient may have a marked effect on the extent of cross-contamination.

Strong solutions of salts such as CsCl are also effective gradient materials for viruses that are sufficiently stable. Successive fractionation in two different gradients may sometimes give useful results. The effective buoyant density and the stability of a virus in strong CsCl solutions may depend markedly on pH and on other ions present (Matthews, 1974). Viruses that are unstable in CsCl may be stable in Cs₂SO₄ (Hull, 1976b).

When a virus preparation is subjected to density-gradient centrifugation in CsCl or Cs₂SO₄, multiple bands may be formed. These may be due to the presence of components containing differing amounts of RNA, as is found with TYMV (Keeling *et al.*, 1979). Virus particles containing uniform amounts of nucleic acid may sometimes form multiple bands in CsCl or other salt gradients because of such factors as differential binding of ions (e.g. Hull, 1977a; Noort *et al.*, 1982).

b. Gel filtration

Filtration through agar gel or Sephadex may offer a useful step for the further purification of viruses that are unstable to the pelleting involved in high-speed centrifugation. However, such a step will dilute the virus.

c. Immunoaffinity columns

Monoclonal antiviral antibodies can be bound to a support matrix such as agarose to form a column that will specifically bind the virus from a solution passed through the column. Virus can then be eluted by lowering the pH. Clarified plant sap may destroy the reactivity of such columns (Ronald *et al.*, 1986), while low pH may damage the virus. To avoid such treatment, De Bortoli and Roggero (1985) developed an electrophoretic elution technique. The use of such columns is probably justified only in special circumstances.

d. Chromatography

Chromatographic procedures have occasionally been used to give an effective purification step for partially purified preparations. McLean and Francki (1967) used a column of calcium phosphate gel in phosphate buffer to purify LNYV. OBDV was purified using cellulose column chromatography (Banttari and Zeyen, 1969), while Smith (1987) used fast protein liquid chromatography to separate the two electrophoretic forms of CPMV.

e. Concentration of the virus and removal of low-molecular-weight materials

At various stages in the isolation of a virus, it is necessary to concentrate virus and remove salts or sucrose. For viruses that are stable to pelleting, high-speed centrifugation is commonly employed for the concentration of virus and the reduction of the amount of low-molecular-weight material. Ordinary dialysis is used for removal or exchange of salts. For unstable viruses, some other procedures are available. A dry gel powder (e.g. Sephadex) can be added to the preparation, or the preparation in a dialysis bag can be packed in the dry gel powder and placed in the refrigerator for a period of hours, while water and ions are absorbed through the tubing by the Sephadex. Ultrafiltration under pressure through a membrane can be used to concentrate larger volumes and to remove salts. Pervaporation, in which virus solution in a dialysis bag is hung in a draft of air, may lead to loss of virus due to local drying on the walls of the tube, and to the concentration of any salts that are present.

7. Storage of purified viruses

Storage of purified preparations of many plant viruses for more than a few days may present a problem. It is often best to avoid long-term storage by using the preparations as soon as possible after they are made. Under the best of conditions, most viruses except TMV lose infectivity on storage at 4°C in solution or as crystalline preparations under ammonium sulfate. Such storage allows fungi and bacteria to grow and contaminate preparations with extraneous antigens and enzymes. Addition of low concentrations of sodium azide, thymol or EDTA will prevent growth of micro-organisms, but EDTA may affect virus structure. Preparations may be stored in liquid form at low temperature by the addition of an equal volume of glycerol. By careful attention to the additives used in the medium (a suitable buffer plus some protective protein, sugar or polysaccharide), it may be possible to retain infectivity in deep-frozen solutions or as frozen dried powders for fairly long periods (e.g. Fukumoto and Tochihara, 1984). Preparations to be used for analytical studies on protein or nucleic acid are best stored as frozen solutions, after the components have been separated.

In solutions containing more than about 10 mg/mL of viruses such as TYMV, virus particles interact quite strongly and spontaneously degrade, especially if the ionic strength is low.

8. Identification of the infective particles and criteria of purity

The best methods for determining the purity of a virus preparation depend on the purpose of the experiments and the conclusions to be drawn. The three most common purposes for which virus preparations are made are: (1) to establish the general properties of a newly isolated virus—this aspect is discussed in the following sections; (2) to aid in the diagnosis of a particular disease—diagnosis is discussed in Chapter 15; and (3) to prepare specific antisera or diagnostic nucleic acid probes.

a. Identification of the characteristic virus particle or particles

With newly isolated viruses, or during attempts to isolate such viruses, the main objective is to purify the virus sufficiently to identify the infectious particles and characterize them, at least in a preliminary way. The most suitable method is sucrose density-gradient centrifugation of the purified preparation. Fractions taken from the gradient can be assayed for infectivity singly, and in various combinations if a multiparticle system is involved. Ultraviolet absorption spectra can be obtained. Samples can be examined by electron microscopy for characteristic particles. Further advantages of the sucrose gradient system are that it requires relatively small quantities of virus, and that an approximate estimate of sedimentation coefficients can be obtained.

b. Criteria of purity

There is much discussion about the purity of virus preparations, but essentially a preparation should be as pure as is necessary for its final use. Thus, if the preparation is to be used for raising an antiserum, it should not contain normal host proteins; and if it is to be used for making a virus-specific hybridization probe, it should be free from host nucleic acids. On the latter point, if the probe is to be made using a cloning procedure, it should be recognized that cloning is a purification method itself.

Many of the criteria for purity that are applied to virus preparations involve the application of the assay methods discussed in Chapter 15. Crystallization used to be taken as an important criterion of purity, but it does not distinguish components that co-crystallize with the infectious virus. A nucleoprotein-type spectrum is often put forward as evidence of purity. This is a very poor measure since contamination with large amounts of host nucleoprotein, or polysaccharides, or significant amounts of host proteins can easily go undetected. Sucrose density gradient analysis or sedimentation in the analytical ultracentrifuge will often reveal the presence of non-viral material, provided the contaminants do not sediment together with the virus and provided they are present in high enough concentrations.

Sedimentation analysis of purified preparations of some spherical viruses may reveal the presence of dimers or trimers of the virus (Markham, 1962). These might be confused with host contaminants unless other tests are applied. Likewise, preparations of rod-shaped viruses often contain a range of particles of various lengths shorter than the virus. These shorter particles and the virus may aggregate to give a range of sizes that sediment over a broad band, or as two or more discrete peaks. Sedimentation analysis is less useful for detecting the presence of non-viral materials with multiparticle viruses since more of the sedimentation profile is occupied by virus particles.

Equilibrium density-gradient sedimentation in CsCl or Cs_2SO_4 is a much more sensitive criterion of homogeneity than sedimentation velocity analysis for viruses or viral components that are stable in the strong salt solution. For example, a purified preparation of the B_0 noninfectious nucleoprotein from a TYMV preparation appeared as a single component on sedimentation velocity analysis, but after equilibrium sedimentation in CsCl it was clearly shown to be a mixture of more than one species (Matthews, 1981).

Electron microscopy is often useful in a qualitative way for revealing the presence of extraneous material, provided it is of sufficient size and differs in appearance from the virus itself. Rubisco, phytoferritin, or pieces of host membrane structures often can be detected in purified preparations. Small proteins and low-molecular-weight materials would not be detected unless they crystallized on the grids or were present in relatively large amounts.

Serological methods, such as immunodiffusion and immunoelectrophoresis, provide very sensitive methods for detecting diffusible host impurities that are antigenic, as do nucleic acid hybridization procedures for detecting nucleic acid contamination.

Where the gross composition of a particular virus such as TMV is well established, chemical analysis of major constituents, for example phosphorus nitrogen ration, would show up a nitrogen- or phosphorus-containing impurity if it were present in sufficient amount. More sophisticated chemical methods might also be used in certain circumstances, for example end-group amino acid analysis on the protein, which should show only one amino acid. Fingerprinting of the peptides from tryptic digests of well-characterized virus proteins might also be used, but again the sensitivity as a test for purity would not be very high.

Radiochemical methods can be used in various ways to test for purity. For example, a healthy plant can be labeled with ^{32}P orthophosphate or ^{35}S sulfate for several days. This material is then mixed with equivalent unlabeled virus-infected tissue before the leaf is extracted. If millicurie (10^7 bq) amounts of radioactivity are used, the sensitivity of this method can be rather depressing. Even so, it is not a complete check on purity with respect to compounds containing these two elements. For example, it is possible that virus infection leads to a kind of contaminant not found in healthy tissues. At present, the only general rule for deciding what criteria of purity to apply is that they should ensure that the virus preparation is adequate for the desired purpose.

9. Virus concentration in plants and yields of purified virus

a. Measurement of yield

There are many difficulties and ambiguities in obtaining estimates of virus concentration in a plant.

i. Non-extracted virus

The method used to extract virus from the tissue may frequently leave a variable but quite large proportion of the virus bound to cell debris or retained in vesicles or organelles. Some viruses, particularly those rod-shaped viruses that occur as large fibrous inclusions in the cell, may remain aggregated during the initial stages of isolation. The losses during virus purification can be large and variable and the extent to which various factors lead to virus loss often have not been adequately assessed.

ii. Method of measurement

The method of measuring virus yield depends upon the purpose for which the virus is being purified. If it is for serology, the criterion is the amount of virus particles, whether they are infectious or not. If it is for infectivity studies or, say, for making full-length clones of the genome, the integrity of the nucleic acid is of paramount importance. Infectivity measurements cannot give an absolute estimate of amount of virus and are influenced by many variables. For a reasonably stable virus, an estimate of such losses can be obtained by adding a small, known amount of radioactively labeled virus to the starting material. The proportion of label recovered in the final preparation gives an estimate of virus loss. This method would not account for virus bound within cell structures, however.

For viruses occurring in high enough concentrations, sucrose density-gradient analysis on clarified extracts is probably one of the most direct methods for estimating virus concentration in plant extracts. Virus concentration estimates can also be obtained from the ultraviolet spectrum of the preparation (with the caveat noted above on impurities) or by immuno-dot blots or hybridization dot blots.

iii. Tissue sampled

As discussed in Chapter 9 (Section II.J), the concentration of virus in different parts of the plant may vary widely—even in different parts of a leaf showing mosaic symptoms.

iv. Basis for expressing results

Virus concentration or virus yield is usually expressed as weight per unit of fresh weight of tissue or weight per millilitre of extract. However, different tissues and different leaves, even in the same plant, vary widely in their water content under the same conditions, reflecting, for example, the size of vacuoles or the amount of fibrous material.

b. Reported yields of virus

There is no entirely satisfactory answer to many of the preceding problems. Most workers, in reporting the isolation of a virus, if they give estimates of yield at all, express their results as weight of purified virus obtained from a given weight of starting material. This is usually whole leaves or leaf laminae without midribs. Reported yields for the same virus from different laboratories may vary quite widely because of such factors as host species, growing conditions and isolation procedure. Yields vary from several milligrams of virus per gram of fresh weight of tissue for viruses like TMV and TYMV, down to 1–2 µg/g for rice tungro viruses (Jones *et al.*, 1991) or even fractions of µg/g for luteoviruses (e.g. Matsubara *et al.*, 1985).

10. Discussion and summary

Different viruses vary over a 10 000-fold range in the amount of virus that can be extracted from infected tissue (from 0.4 or less to 4000 µg/g fresh weight). They also vary widely in their stability to various physical, chemical and enzymatic agents that may be encountered during isolation and storage. For these reasons, isolation procedures have to be optimized for each virus, or even each strain of a virus.

Important factors for the successful isolation of a virus are: (1) choice of host plant species and conditions for propagation that will maximize virus replication and minimize the formation of interfering substances; and (2) an extraction medium that will protect the virus from inactivation or irreversible aggregation.

Most viruses can be isolated by a combination of two or more of the following procedures: high-speed sedimentation, density-gradient fractionation, precipitation using polyethylene glycol, salt precipitation or crystallization, gel filtration, and dialysis. Density-gradient fractionation of the purified preparation, with combined physical and biological examination of particles in fractions from the gradient, can usually best achieve positive identification of the infectious virus particle, or particles.

When a new virus or strain is being isolated, it is essential to back-inoculate the isolated virus to the original host to demonstrate that it, and it alone, is in fact the cause of the original disease. Attention must also be paid to the precise conditions under which purified viruses are stored, as many viruses may lose infectivity and undergo other changes quite rapidly following their isolation.

As noted earlier, in a chemical sense there is no such thing as a pure virus preparation. Purity and homogeneity are operational terms. A virus preparation is pure for a particular purpose if the impurities or inhomogeneities in it do not interfere with the objectives of the experiment.

III. COMPONENTS

Most plant viruses have their genomes enclosed in either a tube-shaped or an isometric shell made up of many small protein molecules. The majority of viral genomes are RNA (Table 4.1) with most of these being messenger or (+)-sense RNA. In these small geometric viruses, there is usually only one kind of protein molecule but some have two. Some viruses have (−)-sense or dsRNA genomes while others have ssDNA or dsDNA. A few plant viruses have an outer envelope of lipoprotein. Some of the larger viruses contain several viral-coded proteins, including enzymes involved in nucleic acid synthesis. Thus, the major components of all viruses are protein and nucleic acid. A few contain lipids, while some contain small molecules such as polyamines and metal ions. This section deals with the isolation and some properties of these various components.

TABLE 4.1 Viral genomes in host from different kingdoms

Genome nucleic acid	Plants		Fungi		Animal		Bacteria	
	No.	%	No.	%	No.	%	No.	%
dsDNA	0	0	1	3.5	606	26.5	445	75.4
ssDNA	166	17.0	0	0	58	2.5	88	14.9
RT	31	3.2	0	0	112	4.9	0	0
dsRNA	45	4.6	27	93.0	383	16.7	1	0.2
(−)ssRNA	100	10.2	0	0	604	26.5	0	0
(+)ssRNA	635	65.0	1	3.5	525	22.9	57	9.6
Total	977		29		2288		591	

ds, double-stranded; ss, single-stranded; RT, viruses that replicate by reverse transcription – they may have either DNA or RNA genomes.

The genomes of, at least, one member of each virus genus has now been sequenced or will be sequenced soon. There is great interest in comparing the nucleotide and derived amino acid sequences of different viruses, from the point of view of virus evolution, virus classification, and the functional roles of viral genes and controlling elements. Reeck *et al.* (1987) have drawn attention to the terminological muddle that has arisen through the imprecise use in the literature of the term 'homology' in comparing two sequences when in fact 'similarity' is what is meant. Similarity is a fact, a measurable property of two sequences, to which measures of statistical significance can be applied. Homology describes a relationship between two things. The word has a precise meaning in biology of 'having a common evolutionary origin'. Thus, a sequence is homologous or it is not. Sequences that are homologous in an evolutionary sense may range from being highly similar to having no significant sequence similarity. In the interest of clarity, I will attempt to maintain this distinction.

A. Nucleic acids

For many viruses containing ssRNA, the RNA can act directly as an mRNA upon infection. Such molecules are called positive sense or plus- (+)-strand RNAs. The complementary sequences are called negative sense or minus- (−)-strand RNAs. The types of nucleic acid found in plant viruses are listed in Table 4.1.

In the following sections, most attention is given to ssRNAs since most plant viruses contain this type of nucleic acid. The RNAs of plant viruses contain adenylic, guanylic, cytidylic and uridylic acids—as are found in cellular RNAs. Except for the methylated guanine in 5' cap structures (Section III.A.3.c), no minor bases have been demonstrated to be present in plant viral RNAs.

The plant reoviruses, containing dsRNA, have regularities in base composition like those found in dsDNA, namely guanine = cytosine and adenine = uracil. For the viruses containing ssRNA there is no expectation that guanine should equal cytosine or that adenine should equal uracil, and many viruses show a wide deviation from any base-pairing rule. In plant viruses containing DNA, deoxyribose replaces ribose, and thymine is present instead of uracil as with other DNAs. If the DNA is ds, guanine = cytosine and adenine = thymine. Adams *et al.* (1981) in general gives a good account of the properties of nucleic acids.

1. Isolation

The usual aim in nucleic acid isolation is to obtain an undegraded and undenatured product in a state as close as possible to that existing in the virus particle. A variety of physical and chemical agents can be used to remove the protein from viruses and give infectious nucleic acid, provided that (1) the nucleic acid was infectious within the intact particle, (2) extremes of pH are avoided, and (3) the nucleic acid is protected against attack by nucleases.

Much of the pioneering work on viral RNA was carried out with TMV. However, the RNA of this virus while it is within the protein coat has exceptional stability. For most other viruses the RNA within the intact virus particle is probably subject, in varying degrees, to some degradation, with loss of infectivity. This occurs both in the intact plant and during isolation and storage of the virus. Best conditions for maintaining intact RNA within the virus particle differ from one virus to another.

Isolation of viral RNA often involves the use of phenol. Gierer and Schramm (1956) used a phenol procedure to isolate infectious TMV RNA. Phenol is an effective protein denaturant and nuclease inhibitor. There are many variations of the basic procedure. Particular variations that work well for one virus may not be effective for another. However, phenol has been reported to be unsatisfactory for some viruses (e.g. WSMV: Brakke and van Pelt, 1970; BYDV: Brakke and Rochow, 1974).

Destruction of nucleases and effective release of the viral nucleic acid may be achieved by a preliminary incubation with pronase in the presence of 1% SDS. Deoxyribonucleases require Mg^{2+} and thus are inhibited by the use of a chelating agent such as EDTA. Phenol extraction is then used to remove pronase and protein digestion products. In most currently used RNA isolation methods, SDS is included during the phenol extraction. If this is done, good nucleic acid preparations can be obtained with most viruses. If the virus preparation is at a high enough concentration the particles can be disrupted by treatment with SDS, and if necessary with heat and/or pronase, and the nucleic acid directly fractionated on an agarose or acrylamide gel.

Various procedures can be used to fractionate or to purify further the nucleic acids isolated from viruses or from infected cells. These include sucrose density gradients; equilibrium density-gradient centrifugation in solutions of CsCl or Cs_2SO_4; fractionation on columns of hydroxylapatite (Bernardi, 1971); cellulose chromatography (e.g. Jackson et al., 1971); and electrophoresis in polyacrylamide or agarose gels (e.g. Symons, 1978). This last is the method of choice for many purposes since the procedure can be fast and simple, requires very small amounts of nucleic acids, and can detect heterogeneity that would not be seen by other methods (e.g. Fowlks and Young, 1970).

In summary, for any particular virus, a protocol for nucleic acid isolation has to be devised that maximizes the required qualities in the final product. These qualities may include yield, purity, structural integrity, infectivity, or the ability to act with fidelity in *in vitro* translation systems (e.g. Brisco et al., 1985).

2. Methods for determining size

The size of a viral nucleic acid is perhaps the most important property of the virus that can be expressed as a single number. There is some uncertainty for any method of measurement of nucleic acid size except with a full sequence analysis.

There is often confusion about the terminology for describing the size of a nucleic acid (or protein) molecule. The term *molecular weight* (MW) is the sum of the atomic weights of all the atoms in the molecule. *Molecular radius* (M_r) is the radius of space occupied by a (macro)molecule and is usually expressed as relative to the M_r of a hydrogen atom; this is generally regarded as the more correct term for a nucleic acid or protein molecule. The unit *dalton* (Da) is frequently used for a molecular weight. The availability of sequence data, especially for nucleic acids, enables these terms to be avoided in most cases and units such as *bases* (b), *kilobases* (kb) and *kilobase pairs* (kbp) are much more informative. However, for protein, the unit *kilodaltons* (kDa) is in widespread use.

a. Sequence analysis

The nucleotide sequences for many viral RNAs and DNAs and viroid RNAs have been determined. These allow chemically precise determinations of molecular weight.

b. Gel electrophoresis

Electrophoresis in either a polyacrylamide (PAGE) or agarose gel is now a widely used procedure for estimating the M_r of viral nucleic acids, by reference to the mobilities of standard nucleic acids of known M_r. Standard 'RNA or DNA ladders' are available commercially. Any secondary structure in ss nucleic acids will affect the mobility and, therefore, the M_r estimate. Even formaldehyde treatment may not eliminate all secondary structure. If the analysis is performed in a Tris-EDTA buffer at pH 7.5 containing 8 M urea at 60°C, more reliable estimates of M_r may be obtained (Reijnders et al., 1974). Glyoxal has also been used as an effective denaturant (Murant et al., 1981). Estimates of M_r using gels will be approximate whatever method and markers are used.

c. Renaturation kinetics

Where it appears necessary to check whether a virus is monopartite or whether it might have, say, two RNAs or DNAs of different base sequence but of the same size, then renaturation of the nucleic acid with complementary sequences can be useful. The sequence complexity of the nucleic acid can give a clear indication as to whether one, two or more different molecules exist (e.g. Gould, *et al.*, 1978). Nucleic acid hybridization is discussed in more detail in Chapter 15 (Section V.C).

d. Electron microscopy

The size of ds nucleic acids has been estimated from length measurements made on electron micrographs of individual molecules. This method may give erroneous results with ss nucleic acids because of doubt as to the internucleotide distances (Reijnders *et al.*, 1973). Errors are considerable and reliable marker molecules must be used.

e. Physicochemical methods

Gierer (1957, 1958) estimated the size of isolated TMV RNA by measuring the sedimentation coefficient and the intrinsic viscosity of the RNA. He found the general relationship between molecular weight (m) and sedimentation coefficient (s) for this RNA to be $m = 1100s^{2.2}$. With the s equal to 31 he determined that the molecular weight of the isolated intact TMV RNA was 2.1×10^6. Empirical relationships between MW (m) and $s_{20.w}$ of the general form $m = ks^d$ have been published for RNAs by several workers (e.g. Spirin, 1961; Hull *et al.*, 1969). Such a relationship is very useful but it should be remembered that methods depending on the determination of s are much more laborious than those using PAGE and are no more accurate.

3. ssRNA genomes

a. Heterogeneity

When RNA is prepared from a purified virus preparation and subjected to some fractiona-tion procedure that separates RNA species on the basis of size, RNAs of more than one size are usually found. Apart from degradation during RNA isolation and storage, there are several possible reasons for heterogeneity.

i. Multipartite genomes

Of the 70 genera of plant viruses, 28 have their genome in two or more separate pieces of different size, as discussed in detail in Chapter 6 (see Appendix 2).

ii. Other sources of heterogeneity

1. Some degradation of the viral RNA may have occurred inside the virus before RNA isolation.
2. Less than full-length copies of the viral RNA (subgenomic pieces) synthesized as such in infected cells may be encapsulated (Palukaitis, 1984).
3. Some host RNA species may become accidentally encapsulated. For example, the empty protein shells found in preparation of EMV and several other tymoviruses contain an average of two to three small RNA molecules with tRNA activity (Bouley *et al.*, 1976). These are very probably encapsulated host tRNAs. Similarly, host RNAs may become encapsulated in TMV coat protein (Siegel, 1971).
4. A number of viruses contain small satellite RNAs unrelated to the viral genome in nucleotide sequence (see Chapter 14, Section II).
5. Small defective interfering RNAs made up of viral sequences are found in some virus infections (see Chapter 8, Section IX.C).
6. As noted in the next section (III.A.3.b), ssRNAs in solution may have a substantial degree of secondary and tertiary structure. A single RNA species may assume two or more distinct configurations under appropriate ionic conditions. These 'conformers' may be separable by electrophoresis or centrifugation. Thus, they give rise to additional heterogeneity that is not based on differences in length (Dickerson and Trim, 1978).
7. The RNA may migrate as a transient dimer under certain conditions (e.g. Asselin and Zaitlin, 1978).

b. Some physical properties of ssRNAs

i. Ultraviolet absorption

Like other nucleic acids, plant viral RNAs have an absorption spectrum in the ultraviolet region between 230 and 290 nm that is largely due to absorption by the purine and pyrimidine bases. The absorption spectra of the individual bases average out to give a strong peak of absorption near 260 nm with a trough near 235 nm (see Fig. 15.4). Ultraviolet absorption spectra are usually of little use for distinguishing between one viral RNA and another.

The absorbance of an RNA solution at 260 nm measured in a cell of 1 cm path length is a convenient measure of concentration. The absorbance per unit weight varies somewhat with base composition and also with secondary structure. In general, for ssRNA at 1 mg/mL in 0.1 M NaCl, $A_{260} = 25$; at 1 mg/mL for TMV RNA, $A_{260} = 29$; for TYMV RNA, $A_{260} = 23$ (in 0.01 M NaCl at pH 7) (Matthews, 1991).

ii. Secondary structures

In the intact virus particle, the three-dimensional arrangement of the RNA is partly or entirely determined by its association with the virus protein or proteins (see Chapter 5). Here I shall consider briefly what is known about the conformation of viral RNAs in solution.

dsDNA has a well-defined secondary structure imposed by base-pairing and base-stacking in the double helix. Single-stranded viral RNAs have no such regular structure. However, it has been shown by a variety of physical methods that an RNA such as that of TMV in solution near pH 7 at room temperature in, say, 0.1 M NaCl does not exist as an extended thread. Under appropriate conditions, ssRNAs contain numerous short helical regions of intrastrand hydrogen-bonded base-pairing interspersed with ss regions. They behave under these conditions as more or less compact molecules. The degree of secondary structure in the molecule under standard conditions will depend to some extent on the base sequence and base composition of the RNA.

The helical regions in the RNA can be abolished, making the molecule into a random, extended, disorganized coil by a variety of changes in the environment (Spirin, 1961).

These include heating, raising or lowering the pH, or lowering the concentrations or changing the nature of the counterions present (e.g. Na^+, Mg^{2+}) (Boedtker, 1960; Eecen et al., 1985). The change from helical to random coil alters a number of measurable physical properties of the viral RNA. The absorbance at 260 nm and viscosity are increased, while the sedimentation rate is decreased.

A polyribonucleotide lacking secondary structure has about 90% the absorbance of the constituent nucleotides found on hydrolysis. A fully base-paired structure (e.g. a synthetic helical polyribonucleotide composed of poly(A) plus poly(U) has about 60% of the UV absorbency of the constituent nucleotides. The changing absorption characteristics of RNA with changing pH are due in part to changes in the extent of the base-paired helical regions in the molecule, and in part to shifts in the absorption spectrum of individual bases due to changes in their ionization state. Methods for probing the structure of RNAs in solution are discussed by Ehresmann et al. (1987). Viral RNAs with amino-acid accepting activity almost certainly have a three-dimensional tRNA-like configuration near the 3′ terminus when they are in solution under appropriate conditions (see Section III.A.3.c).

iii. Pseudoknots (reviewed by Deiman and Pleij, 1997; Herman and Patel, 1999)

RNA molecules can fold into complex three-dimensional shapes and structures in order to perform their diverse biological functions. The most prevalent of these forms is the pseudoknot, which in its simplest manifestation involves the loop of a stem-loop structure base-pairing with a sequence some distance away (Fig. 4.3). The involvement of pseudoknots has been recognized in a great variety of functions of viral RNAs including control of translation by −1 frameshifting (see Chapter 7, Section V.B.10) (Giedroc et al., 2000), by read-through of stop codons, by internal ribosome entry sites (see Chapter 7, Section V.B.4), and by translational enhancers (see Chapter 7, Section V.C).

The tRNA-like structures at the 3′ termini of a variety of plant viral RNAs also form pseudoknots (Figs 4.4 and 4.5). These structures are

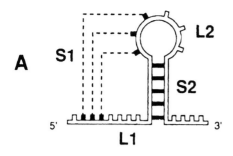

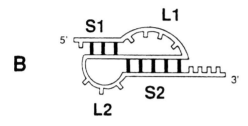

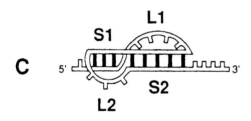

Fig. 4.3 The H(airpin) type of classical RNA pseudoknot. **(A)** Secondary structure. The dotted lines indicate the base-pair formation of the nucleotides from the hairpin loop with the complementary region at the 5′ side of the hairpin. **(B,C)** Schematic folding. **(D)** Three-dimensional folding, showing the quasi-continuous double-stranded helix. The stem regions (S1 and S2) and the loop regions (L1 and L2) are indicated. L1 crosses a deep groove and L2 a shallow groove respectively. From Deiman and Pleij (1997), with permission.

involved in the regulation of RNA replication (see Chapter 8, Section IV.H.3). The structure of the classical pseudoknot of TYMV has been solved by NMR (Kolk *et al.*, 1998). The structure displays internal mobility, which may be a general feature of RNA pseudoknots that regulates their interactions with other RNA molecules or with proteins.

iv. Effective buoyant density in solutions of cesium salts

When a dense solution of CsCl or Cs_2SO_4 is subjected to centrifugation under appropriate conditions, the dense Cs ions redistribute by sedimentation and diffusion to form a density gradient in the tube. If a nucleic acid is present, it will band at a particular position in the density gradient. This density, known as the effective buoyant density, provides a useful criterion for characterizing virus nucleic acids and for distinguishing between RNAs and DNAs. dsDNA forms a band at about 1.69–1.71 g/cm^3 in a solution of CsCl; the exact banding position depends on the G+C content of the DNA. Thus, the method can be used to discriminate between DNAs with differing base composition. Under the same conditions, RNA is pelleted from the gradient. Cs_2SO_4 solutions form gradients in which all the various kinds of nucleic acid can form bands: dsDNA banding at 1.42–1.44 g/cm^3; ssDNA at about 1.49 g/cm^3; DNA–RNA hybrids at about 1.56 g/cm^3; dsRNAs at about 1.60 g/cm^3; and ssRNAs at about 1.65 g/cm^3. If a fluorescent dye, such as ethidium bromide, is intercalated into the nucleic acid to enable it to be located more easily, the nucleic acid will band at a lower density.

c. End-group structures

Many plant viral ssRNA genomes contain specialized structures at their 5′ and 3′ termini. This section summarizes the nature of these structures, while their biological functions are discussed in Chapter 6 (Section II.D).

i. The 5′ cap

Many mammalian cellular messenger RNAs and animal virus messenger RNAs have a methylated blocked 5′ terminal group of the form:

$$m^7G^5ppp^5 X^{(m)}pY^{(m)}p$$

where $X^{(m)}$ and $Y^{(m)}$ are two methylated bases.

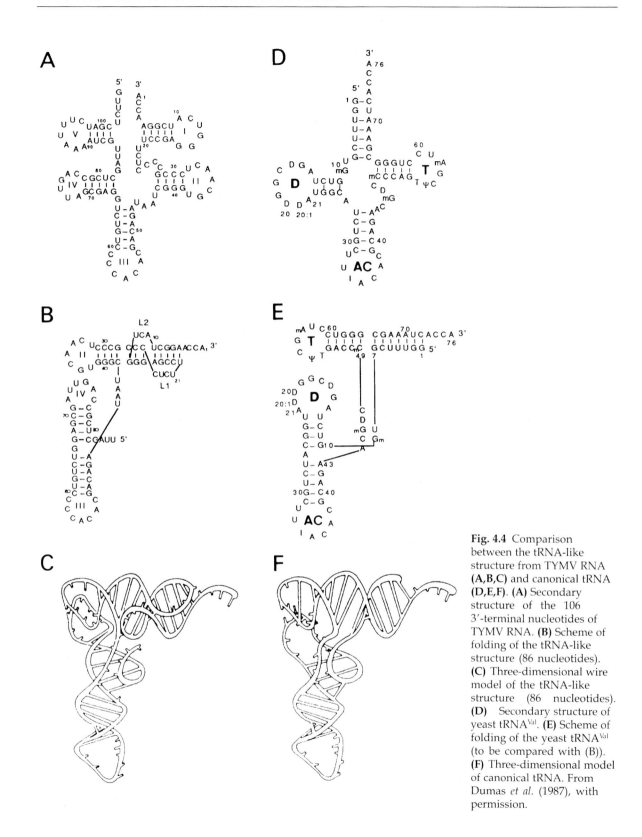

Fig. 4.4 Comparison between the tRNA-like structure from TYMV RNA **(A,B,C)** and canonical tRNA **(D,E,F)**. **(A)** Secondary structure of the 106 3'-terminal nucleotides of TYMV RNA. **(B)** Scheme of folding of the tRNA-like structure (86 nucleotides). **(C)** Three-dimensional wire model of the tRNA-like structure (86 nucleotides). **(D)** Secondary structure of yeast tRNA[Val]. **(E)** Scheme of folding of the yeast tRNA[Val] (to be compared with (B)). **(F)** Three-dimensional model of canonical tRNA. From Dumas *et al.* (1987), with permission.

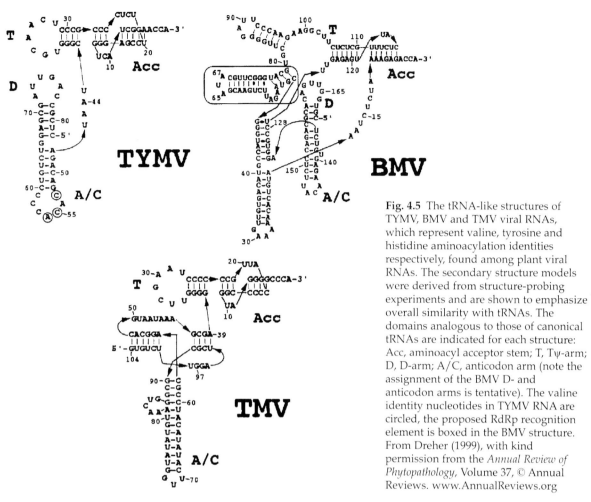

Fig. 4.5 The tRNA-like structures of TYMV, BMV and TMV viral RNAs, which represent valine, tyrosine and histidine aminoacylation identities respectively, found among plant viral RNAs. The secondary structure models were derived from structure-probing experiments and are shown to emphasize overall similarity with tRNAs. The domains analogous to those of canonical tRNAs are indicated for each structure: Acc, aminoacyl acceptor stem; T, Tψ-arm; D, D-arm; A/C, anticodon arm (note the assignment of the BMV D- and anticodon arms is tentative). The valine identity nucleotides in TYMV RNA are circled, the proposed RdRp recognition element is boxed in the BMV structure. From Dreher (1999), with kind permission from the *Annual Review of Phytopathology*, Volume 37, © Annual Reviews. www.AnnualReviews.org

Some plant viral RNAs have this type of 5'-end, known as a 'cap', but in the known plant viral RNAs the bases X and Y are not methylated. Virus groups with 5' capped RNAs are shown in the genome maps in Chapter 6.

The capping activity is virus coded (Dunigan and Zaitlin, 1990) and differs from the host capping activity (see Merits *et al.*, 1999). The viral and cellular enzymes involved in the synthesis of cap structures are reviewed in Bisaillon and Lemay (1997). Capping activity that resembles that of alphaviruses has been identified in the 126-kDa TMV protein and in protein 1a encoded by RNA1 of BMV (Merits *et al.*, 1999; Ahola and Ahlquist, 1999; Kong *et al.*, 1999). These activities methylate GTP using S-adenosylmethionine (AdoMet) as the methyl donor. This guanyl transferase and transferase activity is specific for guanine position 7 form-ing a covalent complex with m7GTP. This activity is termed methyl transferase.

ii. 5' linked protein

Members of several plant virus groups have a protein of relatively small size (*c* 3.5–24 kDa) covalently linked to the 5'-end of the genome RNA. These are known as VPgs (short for virus protein, genome-linked). All VPgs are coded for by the virus concerned and for most viruses the viral gene coding for the VPg has been identified (e.g. for CPMV: see Chapter 6, Section VIII.B.1). If a multipartite RNA genome possesses a VPg, all the genomic RNAs will have the same protein attached. The VPg is attached to the genomic RNAs by a phosphodiester bond between the β-OH group of a serine or tyrosine residue located at the NH$_2$ terminus of the VPg and the 5' terminal uridine residue of the genomic RNA(s) (Table 4.2).

TABLE 4.2 Properties of plant viral VPg

Virus genus	M_r (kDa)	Linkage		Requirement for infectivity
		VPg	RNA	
Comovirus	2	Serine	Uridine	No
Nepovirus	2–3	Serine	Uridine	Yes
Potyvirus	22–24	Tyrosine		No
Sobemovirus	8–9	–	–	Yes
Polerovirus	3	–	–	No
Enamovirus	3	–	–	No

Data from: *Comovirus*: Eggen and van Kammen (1988); *Nepovirus*: Zalloua *et al.* (1996), Wang *et al.* (1999a); *Potyvirus*: Murphy *et al.* (1991), Murphy *et al.* (1996); *Sobemovirus*: van der Wilk *et al.* (1998); *Polerovirus*: van der Wilk *et al.* (1997); *Enamovirus*: Wobus *et al.* (1998).

Figure 4.6 shows how the CPMV VPg is excised from a longer precursor by a protease. The basic nature of the protein is also indicated. Figure 4.7 shows the bond between the genomic uracil and the serine of TRSV VPg (Zalloua *et al.*, 1996).

A three-dimensional model of the potyviral VPg has been proposed based on the similarity of its hydrophobic–hydrophilic residue distribution to that of malate dehydrogenase (Plochocka *et al.*, 1996) (Fig. 4.8).

Daubert and Bruening (1984) discuss methods for the detection and characterization of VPgs. RNAs possessing a VPg can be isolated by northern-immunoblots (Margis *et al.*, 1993).

VPg are involved in virus replication (see Chapter 8, Section IV.J).

iii. 3′ poly(A) tracts (reviewed by Dreher, 1999)

Polyadenylate sequences have been identified at the 3′ terminus of the messenger RNAs of a variety of eukaryotes. Such sequences have been found at the 3′ terminus of several plant viral RNAs that can act as messengers. Virus groups with RNAs that have 3′ poly(A) tracts are shown in the genome maps in Chapter 6. The length of the poly(A) tract may vary for different RNA molecules in the same preparation, and such variation appears to be a general phenomenon. For example, in ClYMV

RNA the range was about 75–100 residues (AbouHaidar, 1983). For CPMV, numbers of residues ranged from about 25 to 170 for B RNA and 25 to 370 for M RNA (Ahlquist and Kaesberg, 1979). Internal poly(A) tracts are found in bromoviruses and hordeiviruses.

iv. 3′ tRNA-like structures (reviewed by Dreher, 1999)

Pinck *et al.* (1970) reported that TYMV RNA, when incubated with various ^{14}C-labeled amino acids in the presence of appropriate cell-free extracts from yeast or *E. coli*, bound valine, which became attached to the 3′ terminal adenosine by an ester linkage. Later work showed that several other tymoviruses accepted valine (Pinck *et al.*, 1975). The accepting activity was also present in the ds replicative form of the viral RNA, and in this state was resistant to RNase attack. These experiments demonstrated that the amino acid accepting activity was an integral part of the viral (+)-sense ssRNA.

Subsequent work has shown that the 3′ terminus of the RNAs of the type strain of TMV and some other tobamoviruses accept histidine (Litvak *et al.*, 1973; García-Arenal, 1988), those of BMV RNAs and BSMV RNAs accept tyrosine (Loesch-Fries and Hall, 1982), and CMV RNAs accept tyrosine (Kohl and Hall, 1974). BSMV RNAs are unusual in having an internal poly(A) sequence between the end of the coding region and the 3′ tRNA-like sequence that accepts tyrosine (Agranovsky *et al.*, 1981, 1982, 1983). Thus three amino acid specificities are now known: valine (tymoviruses, furoviruses, pomoviruses and the tobamovirus SHMV), histidine (most tobamoviruses), and tyrosine (bromoviruses, cucumoviruses and hordeiviruses).

Methylated purines and pseudouracil occur in cellular tRNAs and some of these bases have cytokinin activity. Claims for cytokinin activity in TMV RNA have not been confirmed (Whenham and Fraser, 1982).

The tertiary structure of these tRNA-like 3′ termini involve pseudoknots (see Section III.A.3.b). The structures and the functions of these tRNA-like structures are further discussed in Chapter 8 (Section IV.A.1).

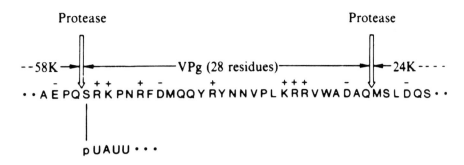

Fig. 4.6 Primary structure and properties of CPMV VPg. The VPg is released from longer precursors by the 24-kDa protease at Glu/Ser and Glu/Met cleavage sites respectively. Basic and acidic amino acids are indicated by + and − signs respectively From Jaegle *et al.* (1987), with permission.

H$_2$N-Ser-Val-Val-Ser-NH - CH - CO -Gly-Ser-Ser-Pro-Val-Ala-His-Arg-Asn...
|
CH$_2$
|
O
|
O-PO$_2$
|
H$_2$C   O   uracil
|
O   OH
|
RNA   S-CH$_2$CH$_3$
|
CH$_2$
|
H$_2$N-Ser-Val-Val-Ser-NH - C - CO -Gly-Ser-Ser-Pro-Val-Ala-His-Arg-Asn...

ethanethiol
+
NaOH

Fig. 4.7 Diagram of a VPg phosphodiester-linked to the 5′ nucleotide residue of a TRSV genomic RNA and of the expected VPg-derived product of a base-catalyzed β-elimination reaction in the presence of ethanediol. The VPg amino terminal sequence was determined by Edman degradation. An *S*-ethylcysteine residue was detected at cycle 5 for the VPg derivative obtained after treatment of VPg-5′-oligo with ethanediol. From this result, position 5 is deduced to be a serine linked by a phosphodiester to the 5′-uridylate residue of each genomic RNA of TRSV. From Zalloua *et al.* (1996), with permission.

v. Complementary 5′ and 3′ sequences

The RNA genome segments of members of the tospoviruses and tenuiviruses have complementary sequences at the 5′- and 3′-ends that enable the termini of the RNAs to anneal to form 'pan-handle structures'. These sequences are conserved across the genome segments of members of each genus. The three topsovirus genome segments have a 5′ consensus sequence of AGAGCAAU . . . and a 3′ consensus sequence of UCUCGUUA. . . . The consensus sequences of the four tenuivirus genome segments are 5′ ACACAAAGUCC . . . and 3′ UGUGUUUCAG . . . (Falk and Tsai, 1998). The RNAs of ophioviruses are circular, suggesting the presence of a pan-handle structure, but complementarity between the termini has not yet been established (Milne *et al.*, 2000).

d. Infectivity

i. Historical

Gierer and Schramm (1956) used several criteria to establish that phenol-extracted TMV RNA isolated from the virus was infectious. The infectivity in their RNA preparations was only about 0.1% that of the same amount of RNA in intact virus. Fraenkel-Conrat and Williams (1955) had shown that, while the

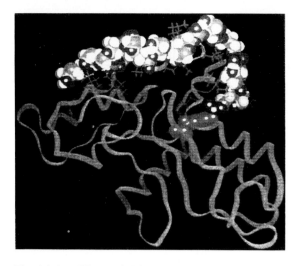

Fig. 4.8 (see Plate 4.1) Schematic ribbon drawing illustrating interactions of PVY VPg with RNA fragment. The RNA backbone and Tyr-64 of the VPg are shown in space-filling representation. The side-chain residues that are in close contact with RNA are shown in blue. From Plochocka *et al.* (1996), with kind permission of the copyright holder, © National Academy of Sciences, USA.

infectivity of isolated TMV RNA was very much lower than in intact virus, the infectivity of the RNA could be greatly enhanced by allowing viral protein subunits to reaggregate around the RNA to reform virus rods. When TMV RNA from one virus strain was reconstituted *in vitro* with protein from a distinct strain to form 'hybrid' rods, and these particles were inoculated on to plants, the progeny virus had protein of the type naturally found associated with the RNA used (Fraenkel-Conrat, 1956). This was strong evidence that the RNA alone specified the virus coat protein structure. Since these classic experiments, infectious RNA has been prepared from many plant viruses. Most viral RNA preparations are much less infectious than the intact virus on an equal RNA basis, and the infectivity of the RNA relative to intact virus varies considerably with method of preparation, with different viruses, and with inoculation medium (e.g. CCMV: Wyatt and Kuhn, 1977).

ii. Structural requirements

From experiments with various inactivating agents, it has been established that the com-

plete intact TMV RNA molecule is necessary for infectivity. A single break in the polynucleotide chain destroys infectivity. Chemical modification of only the terminal base at either the 5' or 3' terminus may inactivate a viral RNA (Kohl and Hall, 1977). On the other hand, the polyadenylic acid covalently linked to the 3' terminus of some viral RNAs may not be necessary for infectivity. For BSMV (Agranovsky *et al.*, 1978) and TEV (Hari *et al.*, 1979), RNA molecules lacking poly(A) were as infectious as the polyadenylated molecules.

The VPg found at the 5'-ends of certain viruses are necessary for the infectivity of some RNAs (e.g. those of nepoviruses: Mayo *et al.*, 1982; Hellen and Cooper, 1987) but not for others (e.g. the comoviruses: Mayo *et al.*, 1982).

Full-length transcripts of TMV RNA from cDNA lacking the 5' cap were much less infectious than capped molecules (Meshi *et al.*, 1986). Similarly, BMV RNA transcripts without a cap were much less infectious than capped molecules (Janda *et al.*, 1987).

The ds replicative RNAs of TMV (Jackson *et al.*, 1971) and CPMV (Shanks *et al.*, 1985) are not infectious. Heat denaturation renders such dsRNA preparations infectious because the infectious (+)-sense strands are released from the ds structure.

The ability to produce infectious viral RNA copies from cloned viral DNA (Ahlquist *et al.*, 1984b) has made it possible to examine the effects of adding extra non-viral nucleotides to either end of the genomic RNA. Addition of a single extra 5' G residue reduced the infectivity of BMV transcripts more than threefold, while 7 or 16 base extensions caused a marked fall in infectivity (Janda *et al.*, 1987). On the other hand, addition of non-viral bases at the 3'-ends of BMV or TMV had much less effect, and the additional bases were eliminated in progeny virus (Ahlquist *et al.*, 1984b; Meshi *et al.*, 1986).

iii. Negative-strand viruses

Members of the *Rhabdoviridae* possess an ss (–)-sense RNA genome, and virus particles contain a viral RNA polymerase (transcriptase) essential for initiating infection. Thus, the isolated RNAs of these viruses are not infectious.

Similarly, the isolated RNAs of tospoviruses and tenuiviruses, being mainly negative sense, are not infectious.

iv. Viruses with multipartite ssRNA genomes

All of these viruses require a complete set of intact genomic RNAs to establish a full infection giving rise to progeny virus. However, in several groups with bipartite genomes the RNA-dependent RNA polymerase is coded on the larger genome RNA. Thus, this RNA alone can infect cells and reproduce itself, while not being able to elicit virus particles because the coat protein gene is on the other genomic RNA (e.g. tobraviruses: Harrison and Nixon, 1959; and nepoviruses: Robinson *et al.*, 1980). While the B component of CPMV can replicate alone when inoculated to protoplasts (Goldbach *et al.*, 1980), it cannot produce symptoms or move from cell to cell when inoculated on to leaves (Rezelman *et al.*, 1982). Similarly RNAs 1 and 2 of BMV and CMV can replicate without the presence of RNA3, but cannot form virus particles or move from cell to cell (see Chapter 9, Sections II.D.2.e).

AMV and Ilarviruses have genomes made up of three RNAs (1, 2 and 3). A fourth smaller RNA containing the coat protein gene is usually encapsidated; the coat gene is also found in RNA3. The three isolated genomic RNAs alone are not infectious, because they require the presence of some coat protein molecules or RNA4 in the inoculum. Each of the three RNAs needs to bind a coat protein molecule at a specific site to initiate infection. This phenomenon is discussed further in Chapter 8 (Section IV.G).

4. dsRNA genomes

The genomes of plant reoviruses consist of 10 or 12 pieces of dsRNA (see Chapter 6, Section VI.A). The cryptoviruses that have been studied contain two segments of dsRNA (Accotto and Boccardo, 1986). The fully base-paired double-helical structure is most stable in solutions of high ionic strength (0.1 M), at pH values near neutrality, and at low temperatures. At low ionic strengths, extremes of pH, or high temperatures, the ds structure is lost (i.e. the structure melts) as described in Chapter 15

(Section V.C.1.a). These dsRNAs can be distinguished from ssRNAs by a number of physical and chemical properties. Many of these differences result from the fact that dsRNAs have a much more highly ordered structure than ssRNAs.

The following physical and chemical properties are commonly used.

a. Base composition

Base-pairing of the Watson–Crick type is present, thus base analyzes show that adenine = uracil and guanine = cytosine in these RNAs.

b. X-ray diffraction

Because of their highly ordered structure it was possible to prove the ds nature of these nucleic acids by X-ray diffraction analysis (Tomita and Rich, 1964).

c. Melting profile

Native RDV dsRNA has a sharp DNA-like melting curve, with $T_m = 80°C$. RDV RNA that had been denatured by heating followed by rapid cooling behaved like ribosomal and transfer RNA (Miura *et al.*, 1966) (Fig. 4.9).

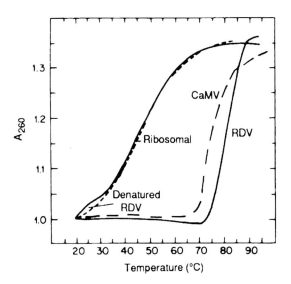

Fig. 4.9 Melting profile of RDV dsRNA, ribosomal RNA, and denatured RDV RNA, all in 0.01 × SSC and CaMV dsDNA in 0.1 × SSC (SSC = 0.15 M NaCl, 0.015 M Na citrate).

d. Electron microscopy

dsRNA has a general appearance like that of dsDNA when shadowed or stained specimens are examined in the electron microscope.

e. Resistance to RNase

The ds structure makes the phosphodiester backbone of the RNA strands resistant to attack by pancreatic RNase provided the reaction is carried out at an appropriate pH and ionic strength. If dsRNA is denatured by heating followed by rapid cooling, the individual RNA strands become susceptible to RNase. The increase in absorbance at 260 nm gives a convenient measure of the extent to which a given sample of nucleic acid is attacked by the enzyme.

f. Buoyant density in Cs_2SO_4

dsRNAs have a characteristic buoyant density in Cs_2SO_4 gradients (about 1.60 g/cm^3).

g. Reactivity with formaldehyde

Formaldehyde reacts with the free amino groups of the nucleic acid bases. The amino groups in dsRNA are involved in the hydrogen-bonded base-pairs and do not react with formaldehyde under conditions where ssRNA does. This difference can be used as an additional criterion to identify dsRNA (e.g. Ikegami and Francki, 1975).

h. Infectivity

The isolated dsRNAs of reoviruses are not infectious. This is because the virus particles contain a viral RNA polymerase that functions at an early stage in infection to produce positive sense strands from negative sense strands in the infecting virus.

5. dsDNA genomes

Members of the *Caulimoviridae* are the only plant viruses known to have a genome consisting of dsDNA. These genomes are about 7.5–8 kbp. Working with CaMV, Shepherd *et al.* (1968) provided the first good evidence that a plant virus contained DNA. The nucleic acid was identified as DNA by the following properties:

1. There was a positive diphenylamine reaction.
2. Thymine was present.
3. Infectivity of nucleic acid isolated from the virus was abolished by pancreatic DNase, but not by RNase.
4. When purified virus was subject to equilibrium density-gradient centrifugation in CsCl, the DNA as measured by the diphenylamine test (or material absorbing at 260 nm) and the infectivity all banded at the same position in the gradient.

Further work showed that the DNA was ds, because it had a melting profile typical of a dsDNA (Fig. 4.9) and because it had a base composition in which G = C and A = T (Shepherd *et al.*, 1970).

DNA isolated from CaMV, subjected to gel electrophoresis, and examined in the electron microscope reveals a number of conformations (Ménissier *et al.*, 1983). Open circles are the most abundant, but singly and multiply knotted forms also occur (Fig. 4.10).

Both strands of the DNA contain ss discontinuities or gaps. In one strand there is a unique discontinuity found in all caulimoviruses. In the second strand, different viruses or strains have one, two or three ss discontinuities. The strand with a single discontinuity is the α or (−) strand. At each of the three discontinuities, the 5' and 3' extremities of the interrupted strand overlap one another to a variable extent over a range of about 8–20 nucleotides (Richards *et al.*, 1981). These discontinuities are described in more detail in Chapter 8 (Section VII.B.3).

A small ssDNA fragment is also found in CaMV and RTBV particles with an RNA species of 75 nucleotides covalently linked to the 5'-end (Turner and Covey, 1984; Bao and Hull, 1993a). The RNA had the properties of host plant tRNA^{met}. Many of the ss strands making up the CaMV genome also possess one or more ribonucleotides covalently bound to their 5' termini (Guilley *et al.*, 1983). These RNAs are primers or remnants of primers involved in replication (Chapter 8, Section VII.B). Isolated CaMV DNA infects turnip plants (e.g. Howell *et al.*, 1980) and protoplasts (Yamaoka *et al.*, 1982a).

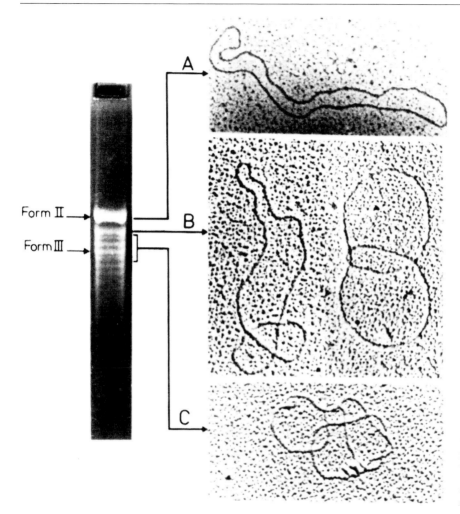

Form II

Form III

Fig. 4.10 'Knotted' CaMV DNA. DNA (1 μg) purified from virus particles was electrophoresed on a 1% agarose gel at 30 V for 16 h. The gel was then stained with ethidium bromide (0.5 μg/mL) and observed under UV illumination. DNA in different sections of the gel was electrophoresed and prepared for electron microscopy. Band A corresponds to circular DNA molecules, band B to singly knotted molecules, and band C to highly knotted circular DNA. Linear DNA molecules, sometimes knotted, were also detected (not shown). From Ménissier *et al.* (1983), with kind permission of the copyright holder, © Oxford University Press.

6. ssDNA genomes

Two families of plant viruses, the *Geminiviridae* and *Circoviridae*, have ssDNA genomes. In some geminiviruses the genome is a single covalently closed circle of about 2.5–3.0 kb. The genome of others consists of two such circles of about the same size, generally known as DNAs 1 and 2 or A and B (see Chapter 6, Section V.A.4). The *Circovirus* genus, the Nanoviruses, have six or more small circular ssDNA species, each about 1 kb (see Chapter 6, Section V.B.1).

The single circle of genomic DNA encapsulated in MSV is partially ds, having a short (70–80 bases) strand base-paired to it. The short strand is capped with one or a few ribonucleotides (Howell, 1984). ACMV has a genome consisting of two ssDNA circles. In addition to

these two molecules, DNA preparations made from the virus contain minor DNA populations of twice and approximately half the length of genomic DNA (Stanley and Townsend, 1985). The dimeric DNA is involved in the replication cycle (see Chapter 8, Section VIII.C). The half-size DNA, which is derived from only one of the genome strands, may represent DI molecules (see Chapter 8, Section IX.C.4).

Goodman (1977) showed that isolated *Geminivirus* DNA was infectious. Stanley (1983), using cloned DNA, demonstrated that both DNAs of ACMV are necessary for infectivity and movement in plants. However, one of the DNAs (DNA1) can replicate alone in protoplasts but cannot move systemically in plants (Davies *et al.*, 1987). It has been demonstrated, using cloned DNA, that the single ssDNA of

MSV is infectious and constitutes the complete genome (Lazarowitz, 1988).

7. Nucleotide sequences

The nucleotide sequences at the 5′ and 3′ termini were determined for several plant viruses in the 1970s by sequencing RNA fragments directly. The first plant viral genome to be fully sequenced was the DNA of CaMV (Franck *et al.*, 1980). The ability to generate cDNA libraries covering the full extent of viral RNA genomes allowed DNA sequencing methods to be applied to determining the complete nucleotide sequence of such RNAs. The first such sequence to be determined was that of TMV (Goelet *et al.*, 1982). Since that time many other plant viral RNA genomes have been fully sequenced (see Chapter 6). Of particular importance has been the development of a system in which cDNA clones of an RNA virus can be transcribed *in vitro* to produce RNA copies that are fully infectious *in vivo* (Ahlquist *et al.*, 1984b).

Many plant viral nucleotide sequences are available in sequence databanks. Details of sequencing procedures are given by Ausubel *et al.* (1987, 1988). Functional aspects of nucleotide sequences are discussed in Chapters 6–8.

B. Proteins

Most of the small geometric plant viruses with ssRNA as their genetic material have only one kind of protein subunit in the virus particle. In some viruses with (+)-sense ssRNA genomes the basic capsid structure is made up of two coat protein species (e.g. comoviruses). In others, some of the coat protein subunits may be variants of the major coat protein; these are frequently associated with interaction with their vector (see Chapter 11). The geminiviruses and the *Caulimoviridae* have one coat protein species but this may be end-processed to give the appearance of two or more species. The plant reoviruses, the tenuiviruses, and the much larger enveloped plant viruses contain several different proteins, some of which are enzymes; as noted in Chapter 7 (Section IV.A), these viruses require the presence of a polymerase to establish infection.

The same 20 amino acids, or a selection of them, are found in plant virus coat proteins as are present in other living material, and the average proportion of amino acids shows the same general trend. For example, cysteine, methio-nine, tryptophan, histidine and tyrosine usually occur in low amounts.

1. Isolation from virus preparations

Many of the conditions and reagents used to isolate viral RNA can also be used to give a preparation of the virus protein that is more or less free of RNA. However, the best method for obtaining intact RNA from a particular virus is usually different from that which gives satisfactory protein preparations. Protein isolation is usually performed for establishing the properties of the proteins and for studies on viral reassembly.

a. Detergents

With the advent of electrophoresis in polyacrylamide gels, disruption of virus samples in SDS became a popular procedure to prepare the monomers from non-covalently linked polypeptides in the virus.

b. Acids

Fraenkel-Conrat (1957) described an elegant and simple method for the isolation of TMV protein subunits by treating the virus with cold 67% acetic acid. The ease with which the virus is split varies with different strains of the virus. TMV protein isolated by this procedure is in a native state and can be used to reconstitute virus rods (see Chapter 5, Section IV.A). Unfortunately, the acetic acid method is not applicable to most other viruses.

c. Strong salt solutions

One molar $CaCl_2$ dissociates and solubilizes the protein coats of BBMV (Yamazaki and Kaesberg, 1963a) and BMV (Yamazaki and Kaesberg, 1963b). For BMV the product obtained appeared from sedimentation analysis to be a relatively stable dimer of the chemical subunit. A simple procedure using 1 M NaCl produced protein subunits that could be reconstituted with RNA to give infective virus particles (Hiebert *et al.*, 1968). LiCl is useful as an alternative to CaCl$_2$ for some viruses and may give a superior product (e.g. see Moghal and Francki, 1976).

2. Nature of the protein product

The chemical and physical state of the isolated protein varies widely depending on the virus and the treatment used. The types of product are discussed in the following sections.

a. Viral protein shells

These are virus particles that have not encapsidated RNA (e.g. TYMV) (empty particles).

b. Subunits

Subunits in native form and in various stages of aggregation may be present. For example, minor components (a few percent of the total protein) were found in preparations of CCMV and STNV (Rice, 1974). These components had the size and composition expected for a dimer of the coat protein. The chemical evidence suggested that the dimers were joined by covalent bonds other than disulfide bonds.

On occasions, strong bonds are formed between adjacent coat protein molecules that are not dissociable into monomer units. A good example of stable dimers of coat proteins is in the Sobemoviruses where gel electrophoresis reveals a minor band that is twice the molecular radius of the major coat protein (Sehgal and Hsu, 1997).

c. Reversibly denatured subunits

Phenol treatment of TMV gives denatured protein, but Anderer (1959) found that if the denatured protein is precipitated from the phenol by methanol it could be renatured to give fully functional subunits by heating at 60°C and pH 7.0–7.5.

d. Irreversibly denatured aggregates

Irreversibly denatured aggregates are insoluble in aqueous media. A major factor leading to the production of insoluble aggregates is the formation of cross-linking disulfide bridges between cysteine residues due to oxidation of the sulfhydryl groups. Such cross-linking may be blocked by various procedures, usually the use of reducing agents.

e. Partly degraded protein

Some of the protein subunits in a virus particle may be cleaved at specific sites during isolation from the plant or on subsequent storage. For example, Koenig *et al.* (1978) described how PVX protein can be partially degraded from the N terminus in the intact virus by reducing agent-dependent proteases in crude plant extracts, and at the C terminus by reducing agent-independent proteases that occurred in some virus preparations.

Enzymatic degradation could lead to ambiguity as to the 'true' size of the protein subunit in the virus particle. For example, most sobemoviruses have a single species of coat protein. However, proteins of more than one size are found in some members of the group. Thus, dissociated VTMoV yields proteins of approximately 37, 33 and 31.5 kDa. Chu and Francki (1983) considered that the 33-kDa protein was an unstable intermediate resulting from the loss of about 40 amino acids from the 37-kDa protein. Loss of about 15 more gave rise to the 31.5-kDa protein. The related SNMoV has a single coat protein of about 31-kDa. It is not clear whether the reduction in size from 37 kDa to 31.5 kDa is a biologically significant event or merely an artifact of the isolation process (Kibertis and Zimmern, 1984).

f. Coat protein covalently linked to a host protein

A more unusual situation has been described for the U$_1$ strain of TMV. A protein of about 26.5 kDa, occurring at about one molecule per virus particle, has most of the amino acid sequences found in TMV coat protein together with unrelated sequences (Collmer and Zaitlin, 1983). The isolated protein, called protein H, reassociated into U$_1$ TMV rods when coat protein, H protein, and RNA were mixed *in vitro* (Collmer *et al.*, 1983). The host-derived portion of H protein has been shown by amino acid sequencing and immunological cross-reactivity to be ubiquitin, probably linked to the TMV coat protein at lysine 53 (Dunigan et al., 1988). Ubiquitin is a small protein (76 amino acids) found in all eukaryotes with a high degree of amino acid conservation between groups of organisms. The significance of its binding to TMV is not known, but Dunigan et al. (1988) suggested that it may be a stress response by the plant.

3. Size determination

The comments on the terminology of size of nucleic acids (Section III.A.2) apply also to descriptions of the sizes of protein molecules. However, in practice the size of proteins is usually given in kDa.

Most plant viral coat polypeptides have M_r in the range 17–40 kDa. This size distribution is at the lower end of the range of the M_r of most polypeptides found in higher plants. Sedimentation-diffusion measurements are difficult to apply to molecules as small as the coat proteins. SDS-PAGE and calculations from amino acid sequence are now the methods most commonly used to determine MW.

a. Polyacrylamide gel electrophoresis

The most popular method in recent years has been electrophoresis of viral proteins in polyacrylamide gels containing SDS and urea. The molecular weights are estimated by reference to the mobility of marker proteins of known MW. Outstanding advantages of the method are its simplicity, the small quantities of proteins required, and the fact that mixtures of proteins are readily resolved.

However, there are several limitations that must be taken into account. First, there is an intrinsic variability of about 5–10% in estimation of molecular weight. Second, the method is based on several assumptions: (1) that the unknown proteins have the same conformation in the SDS gels as the marker proteins; (2) that all the proteins bind the same amount of SDS (about 1.4 g/g protein); and (3) that as a consequence of uniform binding of SDS all the proteins have a similar charge/mass ratio. Different proteins do not bind uniform amounts of SDS. Anomalous behavior is revealed when esti-

mates of MW made at different gel concentrations give different answers (e.g. Ghabrial and Lister, 1973; Shepherd, 1976; Hammond and Hull, 1981). Corrections can be made for anomalous migration by using a Ferguson plot (Hedrick and Smith, 1968; Hammond and Hull, 1981).

Coat proteins that are certainly single species sometimes form double bands (e.g. TYMV protein with apparent size of 20.1 and 21.7 kDa: Matthews, 1974). Substitution of certain groups on a protein may lead to unexpected changes in apparent M_r. For example, after carboxymethylation, TYMV protein had an apparent M_r of 23.2 kDa (Matthews, 1974). Such anomalies are presumably due to the existence of different conformational states of the polypeptide that bind SDS differently, but the details are not understood.

b. Chemical methods

Where the full sequence of amino acids is established the M_r is known with a high degree of precision. A combination of amino acid analysis and tryptic peptide mapping, mapping of cyanogen bromide fragments, or end-group analysis can give an estimate of size in the absence of a full sequence analysis. However, these procedures alone may give erroneous results (e.g. Gibbs and McIntyre, 1970).

4. Amino acid sequences

The primary structure of TMV coat protein was determined by Fraenkel-Conrat and colleagues (Tsugita et al., 1960) and by Anderer et al. (1960). There were many difficulties associated with techniques and interpretation of results over a period of years, and the elucidation of the full sequence of 158 amino acids represented a substantial achievement. Knowledge of the sequence of amino acids in TMV coat protein was of great value in studies that correlated base changes in the RNA with changes in amino acid sequence in the coat proteins of virus mutants. These studies confirmed the nature and universality of the genetic code.

During the 1970s, the partial or full amino acid sequences of several coat proteins were determined by direct chemical methods. However, with the development of procedures for copying RNA genomes into DNA, and the rapid sequencing of such DNA, it has proved simpler and quicker to determine the amino acid sequence of proteins from the nucleotide sequence of the genes coding for them. Use of this strategy for plant virus proteins began in the early 1980s (e.g. for STNV coat protein: Ysebaert et al., 1980; and BMV: Ahlquist et al., 1981). The full sequences of the coat protein amino acids is now known for representative members of most, if not all, genera of plant viruses.

Identification of the coat protein gene in the RNA sequence can sometimes be made from its size and its position in the genome compared with that of known related viruses. However, proof of identity can be obtained by determining chemically the sequence of 20–30 amino acids somewhere in the coat protein and matching this sequence with the appropriate nucleotide sequences (e.g. TCV coat protein: Carrington et al., 1987).

5. Secondary and tertiary structure

A variety of physical and chemical techniques that give some information about secondary and tertiary structure have been applied to plant virus coat proteins. The essential requirements for the determination of a full three-dimensional structure are (1) a knowledge of the primary sequence of amino acids, and (2) a crystalline form of the protein that allows the techniques of high-resolution X-ray crystallography to be applied. Knowledge of the chemistry of the amino acid sequence is essential for the model building that assists in the three-dimensional interpretation of the data obtained by X-ray crystallographic analysis. In addition, where sequences for several related proteins are known, residues of particular importance can be identified (see Fig. 2.5). Other physical and chemical techniques can give additional guiding or confirmatory information.

a. X-ray crystallography

X-ray analysis has been the key technique in elucidating the three-dimensional structure of the coat proteins of several viruses. These structures are discussed in Chapter 5 in relation to the virus as a whole. Where the primary sequence of the coat proteins of two or more related viruses is known, computer-based methods are available for making predictions about tertiary structures (Sawyer et al., 1987). Such predictions may give useful information but they are no substitute for detailed X-ray analysis.

b. Other physical techniques

A variety of other physical techniques has been applied to the study of viral coat protein structure. These techniques may give information on which low-resolution models can be based. For example, Kan et al. (1982) used data from proton magnetic resonance to propose a low-resolution model for AMV coat protein consisting of a rigid core and flexible N-terminal arm of approximately 36 amino acids. Further examples are described in Chapter 5.

c. Residues exposed at the surface of the protein or intact virus

It is possible to obtain evidence as to whether particular amino acids are at the surface of the protein subunit or the intact virus by using specific enzymes, chemical reagents, or antibodies. For example, carboxypeptidase A removes the C-terminal threonine residue from TMV, which must, therefore, be exposed in the intact virus (Harris and Knight, 1955). Of 15 side-chain carboxylic acid groups in TMV protein, only three (on residues 64 and 66 and the C-terminus) are readily available in the virus for reaction with a carbodiimide, and therefore are on or near the surface (King and Leberman, 1973).

Biochemical evidence together with immunological studies showed that the hydrophilic N-terminal region of the potyvirus coat protein is at the surface of the virus (Allison et al., 1985) (see Chapter 17, Section II.B.4).

The use of monoclonal antibodies in identifying residues on the surface of virus particles is discussed in Chapter 15 (Section IV.D) and is reviewed in Kekuda et al. (1995).

d. Groups added to the protein structure

Many coat proteins appear to consist entirely of unmodified polypeptides. In others, various groups are added subsequent to protein synthesis. For example, the coat protein of CaMV may be both phosphorylated (Hahn and Shepherd, 1980) and glycosylated (Du Plessis and Smith, 1981). Lee *et al.* (1972) presented indirect evidence that neuraminic acid or some related compound is present at the surface of SYVV.

Partridge *et al.* (1976) showed that carbohydrate was associated with the coat protein of BSMV. Its composition included glucose, mannose, xylose, galactosamine, and glucosamine. Pronase digestion of the coat protein of BSMV yielded a tripeptide, Gly-Asp-Ala, attached to carbohydrate through an amide bond with the asparagine (Gumpf *et al.*, 1977).

e. *The problem of polypeptide chain folding*

Following biosynthesis, the primary polypeptide chain must fold correctly to give the functionally correct three-dimensional structure. Anfinsen (1973) has pointed out that a chain of 149 amino acids would theoretically be able to assume about 4^{149}–9^{149} different conformations in solution. Thus folding to give the correct structure must proceed from a limited number of initiating events.

There is an increasing number of computer programs for the prediction of protein secondary structure. While the prediction of features such as α-helical, β-sheet and random structure regions is reasonably reliable, the prediction of the assembly of these into tertiary structures is not so advanced.

6. Enzymes and other non-coat proteins in virus particles

a. *Caulimoviruses*

CaMV has been reported to contain a DNA polymerase activity in a complex of a 76-kDa enzyme and viral DNA protected within the virus particle (Ménissier *et al.*, 1984). CaMV also contains an endogenous protein kinase activity. The main proteins phosphorylated were the coat protein and two larger proteins. Serine and threonine were the residues phosphorylated (Ménissier *et al.*, 1986; Martinez-Izquierdo and Hohn, 1987).

b. *Reoviridae*

By analogy with reoviruses infecting animals, members of the family infecting plant would be expected to contain an RNA-dependent RNA polymerase (RdRp) within the virus particle. Such activity has been detected in preparations of several plant reoviruses, including RRSV and RBSDV (Uyeda *et al.*, 1987). In addition, these viruses, for example RRSV (Hagiwara *et al.*, 1986), contain five or possibly more structural polypeptides. These are discussed in relation to the structure of the viruses in Chapter 5 (Section VII).

c. *Cryptoviruses*

CCV-1 contains an RdRp activity, which appears to be a replicase that catalyzes the synthesis of copies of the ds genomic RNAs (Marzachi *et al.*, 1988).

d. *Rhabdoviridae*

An RdRp that transcribes viral ($-$)-strand RNA into mRNA has been found in animal rhabdoviruses. Similar enzymes should also be present in plant members of the family. Such activity has been shown in preparations of LNYV (Francki and Randles, 1972) and BNYV (Toriyama and Peters, 1981). Like members of the *Rhabdoviridae* infecting animals, plant rhabdoviruses contain several structural proteins (see Chapter 5, Section VIII.A).

e. *Tomato spotted wilt* Tospovirus

TSWV particles contain three major structural proteins together with a small number of polymerase molecules (see Chapter 5, Section VIII.B).

f. *Tenuiviruses*

The filamentous nucleoprotein particles of RSV have an RdRp associated with them (Toriyama, 1986a).

C. Other components in viruses

1. Polyamines

Johnson and Markham (1962) reported the presence of a polyamine in preparations of

TYMV. On dissociating TYMV into RNA and protein, about two-thirds of the amine was found with the RNA. Johnson and Markham considered that in the intact virus it was primarily associated with the RNA. Beer and Kosuge (1970) identified the polyamine as a mixture of spermidine (I), making up about 1.0% of the TYMV by weight, and spermine (II), making up only 0.04%. This would be sufficient to neutralize about 20% of the RNA phosphate groups in the virus.

I: $H_2N-CH2-CH2-CH2-NH-CH2-CH2-CH2-CH2-NH_2$

II: $H_2N-CH2-CH2-CH2-NH-CH2-CH2-CH2-CH2-NH-CH2-CH2-CH2-NH_2$

These amines have been identified as occurring widely in plant and animal tissues. Spermidine occurs in uninfected Chinese cabbage and the amount present is increased by TYMV infection (Beer and Kosuge, 1970). Although the protein shell of TYMV can adsorb small amounts of these polyamines reversibly, the shell is impermeable to them. The virus does not exchange or leak the polyamines from within the virus particle (Cohen and Greenberg, 1981). The polyamines in the virus become associated with the RNA before or at the time it is packaged into virus particles.

Using an improved method for polyamine determination involving HPLC, Torrigiani *et al.* (1995) showed that TMV contains putrescine, spermidine and spermine which account for 0.0022% of the particle weight and possibly capable of neutralizing 0.3% of the phosphate residues of the RNA molecule.

Not all small plant viruses contain these polyamines. CPMV contained about 1% by weight of spermidine, whereas in BSMV and three bromoviruses no polyamines were detected (Nickerson and Lane, 1977).

2. Lipids

It can be safely assumed that all viruses with an outer bilayer envelope contain lipid. A few enveloped plant viruses have been analyzed for this component. Preparations of TSWV from *N. glutinosa* contained about 19% lipid (Best and Katekar, 1964). The rhabdovirus PYDV isolated from *N. rustica* contained more than 20% of lipid (Ahmed *et al.*, 1964). Lipid has also been reported in LNYV (Francki, 1973).

Selstam and Jackson (1983) found that SYNV contained about 18% of lipid composed of 62% phospholipids, 31% sterols and 7% triglycerides. Detailed analysis of these fractions showed that the lipid composition of this virus differed in several respects from that of rhabdoviruses infecting vertebrates. The detailed composition fits with the view that the virus acquires most of its lipid from a host membrane. Ahmed *et al.* (1964) considered that no more than one-fifth of the lipid in their PYDV preparations was due to contaminating host materials, but more detailed chemical characterization is required.

3. Metals

Since early work on the composition of plant viruses (Stanley, 1936) it has been recognized that the ash fraction of purified virus preparations contains various metals. We can distinguish two kinds of metal binding by viruses.

First, there are the metals bound as a consequence of the medium and method used for virus isolation. For example, Johnson (1964) found that the metal content of TYMV varied widely according to the method of preparation. Such variable binding may be trivial as far as the biological activity of the virus is concerned, but it may affect physical properties. Thus, the effective buoyant density of TYMV in CsCl gradients depends on the amount of divalent metal ions associated with the virus. In this connection, Durham *et al.* (1977) have pointed out that most virologists are not aware how many trace ions are present even in the purest distilled water.

Second, there are bound metals that may have important biological functions. Durham and colleagues have shown that several viruses (e.g. PapMV: Durham and Bancroft, 1979) bind H^+ near pH 7.0 in some way not predictable from the amino acid composition of the protein. Ca^{2+} displaces H^+, suggesting that co-ordinating oxygens are involved in the binding site. Carboxylate groups constrained in an electronegative environment are probably involved in this Ca^{2+} binding. There are probably three such binding sites per subunit for PapMV. Durham *et al.* have proposed that Ca^{2+} binding in particular may be important in the disassembly of some viruses (Durham *et al.*, 1977).

TMV in unbuffered solutions binds two Ca^{2+} ions per subunit. The two sites are non-identical and titrate independently (Gallagher and Lauffer, 1983). The stability of some viruses depends on the presence of certain ions. For example, the stability of SBMV appears to depend critically on the presence of Ca^{2+} and Mg^{2+} (Hsu *et al.*, 1976; Hull, 1977c) with the related sobemovirus, TRoV, binding one Ca^{2+} ion per protein subunit (Hull, 1978). With the structure of SBMV now established at high resolution, the binding sites for Ca^{2+} have been established (Abdel-Meguid *et al.*, 1981). The Ca^{2+} binding sites in TBSV are shown in Fig. 5.34.

Some plant viruses such as TSV have a 'zinc finger' binding domain that specifically binds an atom of zinc in a protein involved in nucleic acid binding.

4. Water

The importance of water in living cells is well recognized, but the water content of viruses and its significance have been somewhat neglected. The exact meaning to be attached to the term 'hydration' varies with the method of measurement, a problem discussed by Jaenicke and Lauffer (1969). Measurement of various hydrodynamic properties of a virus in solution allows the degree of hydration to be calculated. The use of intensity fluctuation spectroscopy of laser light has enabled the molecular weights, particle dimensions, and solvation of viruses to be measured with increased ease and accuracy (Harvey, 1973; Camerini-Otero *et al.*, 1974). However, errors concerning solvation may be large even with modern techniques.

Typical prokaryotic cells contain about 3–4 mL of water per gram of dry matter. The few enveloped viruses for which data are available have water contents in the same range (Matthews, 1975). By contrast, the small geometric viruses contain much less (e.g. 0.75 mL/g for TBSV: Camerino-Otero *et al.*, 1974; 0.78 mL/g for TYMV and about 1.1 ml/g for TSV: J.D. Harvey, personal communication,

quoted in Matthews, 1991). These amounts are similar to those reported for proteins such as bovine serum albumin (0.6 mL/g) (Kuntz and Kauzmann, 1974). 0.6 mL/g corresponds approximately to an average of three water molecules per amino acid residue.

CaMV had 1.9 mL water/g (Hull *et al.*, 1976), somewhat more than the smaller geometric viruses. Reovirus has 1.5 mL/g (Harvey *et al.*, 1974). It is probable that plant reoviruses will be found to be hydrated to a similar extent.

It seems likely that water is associated with virus particles in several different ways. The geometric viruses, and probably enveloped viruses as well, have water bound to their protein components (c 0.5–1.0 mL/g) in the same way as water binds to isolated proteins. Those viruses with osmotically functional bilayer membranes have, in addition, 2–3 mL/g water that is probably retained in the same way as cells retain water. Structures of viral coat proteins analyzed to atomic resolution show that some water molecules are bound at specific sites within the protein.

D. Discussion and summary

Knowledge of the components found in a virus and details of their properties and structure are essential to our understanding of virus architecture. For example, the detailed symmetry and the size of the protein shell of isometric viruses depend on the properties of the coat protein. The length of rod-shaped viruses depends on the length of the viral RNA, as well as the protein subunit, which determines the pitch and diameter of the helix.

The genomes of many plant virus groups consist of (+)-sense ssRNA, which can act directly as mRNA upon infecting a cell. It is interesting to note the differences in the most common genomes in the viruses that infect members of the various kingdoms (Table 4.1). In contrast to plant viruses, viruses with dsDNA genomes predominate in bacteria and are equal with those with (−)-strand ssRNA genomes in animals; viruses with dsRNA genomes are the most common in fungi. The reason for this is not known.

For the viruses with (+)-strand genomes, the isolated RNA is infectious. However, for a few viruses of this sort, a small protein covalently bound to the 5′ terminus (VPg) is necessary for infectivity. If the complete genome is split between two or three pieces of RNA, then all must be inoculated to produce a complete infection. The isolated dsDNA genomes of members of the *Caulimoviridae* and the ssDNA genomes of geminiviruses are infectious. However, the isolated (−)-sense ssRNA genomes of rhabdoviruses and the dsRNA genomes of reoviruses are not. This is because viral mRNA synthesis requires a viral-coded polymerase that is present in the virus particle, a topic discussed more fully in Chapter 7 (Section IV.A). The full nucleotide sequence has been determined for the genome of at least one member of most plant virus families and genera. This information provides the basis for our understanding of viral genome strategy and replication.

The particles of most of the small geometric viruses contain a single kind of coat protein. These proteins range in size from about 17 kDa to 60 kDa. In earlier work the amino acid sequences of several coat proteins were determined by direct chemical methods. However, today the simplest procedure is to derive the amino acid sequence from the nucleotide sequence of the coat protein gene.

The plant virus families and genera with large particles (rhabdoviruses, reoviruses, *Caulimoviridae* and tospoviruses) contain more than one kind of protein, some of which are enzymes. Others have a structural role. The two kinds of enveloped plant viruses (rhabdoviruses and tospoviruses) contain lipid as part of an outer-bounding lipoprotein envelope. This envelope is probably derived from a host bilayer membrane. All viruses contain water, with enveloped viruses containing more than geometric ones. Many viruses contain divalent metal ions, especially Ca^{2+}, and a few contain polyamines. These components may be important for stability of the intact virus particle.

CHAPTER 5

Architecture and Assembly of Virus Particles

I. INTRODUCTION

Knowledge of the detailed structure of virus particles is an essential prerequisite to our understanding of many aspects of virology; for example, how viruses survive outside the cell, how they infect and replicate within the cell, and how they are related to one another. Knowledge of virus architecture has increased greatly in recent years, owing both to more detailed chemical information and to the application of more refined electron microscopic, optical diffraction and X-ray crystallographic procedures.

The term *capsid* has been proposed for the closed shell or tube of a virus and the term *capsomere* for clusters of subunits seen in electron micrographs. The mature virus has been termed the *virion* (Caspar *et al.*, 1962). In membrane-bound viruses, the inner nucleoprotein core has been called the *nucleocapsid*. These names can, at times, cause confusion. For example, what is the capsid or the virion in a multicomponent virus such as AMV? Which is the capsomere in a virus like TBSV, where different parts of the same protein subunit are arranged with 2-fold and 3-fold symmetry? The terms *encapsulation* and *encapsidation* are now widely used to refer to the process by which a viral genome becomes encased in a shell of viral protein. These terms now have an established meaning, and I will use them. I shall also use *protein subunit* or *structural subunit* to refer to the covalently linked peptide chain. The term *morphological subunit* will refer to the groups of protein subunits revealed by electron microscopy and X-ray crystallography.

Virus acronyms are given in Appendix 1.

The implications of virus structure for virus self-assembly are discussed later in this chapter. Fig. 2.1 indicates the range of sizes and shapes found among plant viruses.

II. METHODS

A. Chemical and biochemical studies

As already pointed out in Chapter 4, knowledge of the size and nature of the viral nucleic acid and of the proteins and other components occurring in a virus is essential to an understanding of its architecture. Chemical and enzymatic studies may give various kinds of information about virus structure. For example, the fact that carboxypeptidase A removed the terminal threonine from intact TMV indicated that the C terminus of the polypeptide was exposed at the surface of the virus (Harris and Knight, 1955). Methyl picolinimidate reacts with the exposed amino group or lysine residues. Perham (1973) used this reagent with several TMV mutants to determine which lysine residues were at the surface of the virus and which were buried. For viruses with more complex structures, partial degradation by chemical or physical means—for example, removal of the outer envelope—can be used to establish where particular proteins are located within the particle (e.g. Jackson, 1978; Lu *et al.*, 1998b) (see Sections VII and VIII). Studies on the stability of a virus under various pH and ionic or other conditions may give clues as to the structure and the nature of the bonds holding the structure together. Many studies of this sort have been carried out but results are often difficult to interpret in a definitive way (Kaper, 1975). Some examples of this approach are given later in this chapter.

B. Methods for studying size of viruses

1. Hydrodynamic measurements

The classic procedure for hydrodynamic measurements is to use the Svedberg equation (see Schachman, 1959):

$$M = RTs/D(1 - v^*[rho]),$$

where R is the gas constant per mole (8.134 J mol^{-1} K^{-1}), T is the absolute temperature in kelvin, s is the sedimentation coefficient, D is the diffusion coefficient, ρ is the solvent density, and v^* is the partial specific volume for the virus being studied. D and v^* are rather troublesome to determine by classic procedures. s can be determined readily in the analytical ultracentrifuge. For this reason many workers go no further than determining s for a new virus. The term *molecular weight* is widely used with reference to viruses. I will follow this usage but particle weight would be strictly a more appropriate term.

s may be conveniently determined in a preparative ultracentrifuge by sedimenting the virus in a linear sucrose density gradient and comparing the distance it sediments to internal markers of known s. The markers and unknown should have similar sedimentation rates. If not, other methods of calculation can be used (Clark, 1976).

Laser light scattering has been used to determine the radii of several approximately spherical viruses with a high degree of precision (Harvey, 1973; Camerino-Otero et al., 1974). This method is of particular interest since it gives an estimate of the hydrated diameter whereas the Svedberg equation and electron microscopy give a diameter for the dehydrated virus. Comparison of the two approaches can give a measure of the hydration of the virus particle.

2. Electron microscopy

Measurements made on electron micrographs of isolated virus particles, or thin sections of infected cells, offer very convenient estimates of the size of viruses. For some of the large viruses and for the rod-shaped viruses such measurements may be the best available, but they are subject to significant errors. There are various magnification errors inherent in measurements on electron micrographs. These may be overcome in part, but not entirely, by using a well-characterized, stable and distinctive virus, such as TMV, as an internal standard (Bos, 1975). However, Markham et al. (1964) pointed out that TMV rods may not be absolutely uniform in length. The width of the rods may offer a better standard. Flattening of particles on the supporting film may cause significant errors. Flattening can be detected with a tilting stage (Serwer, 1977). If films not coated with carbon are used, there may be significant distortion of individual virus particles probably due to stretching of the film (Ronald et al., 1977).

When length distributions of rods are being prepared, it is necessary to combine individual measurements into size classes. If the size classes chosen are too large then the presence of more than one length of particle may be obscured. There are additional difficulties in measuring the contour lengths of flexuous rods (De Leeuw, 1975).

Hatta (1976) pointed out that measurements of interparticle distances in arrays of particles seen in thin sections of crystals of purified small isometric viruses will always be less than the true diameter of the particles because particles overlap in the arrays.

3. X-ray crystallography

For viruses that can be obtained as stable crystalline preparations, X-ray crystallography can give accurate and unambiguous estimates of the radius of the particles in the crystalline state. The technique is limited to viruses that are stable or can be made stable in the salt solutions necessary to produce crystals.

4. Neutron small-angle scattering

Neutron small-scale scattering (see Section II.E) can give accurate estimates of the radius of icosahedral viruses in solutions containing relatively low salt concentrations. The method has confirmed that particles of some small icosahedral viruses such as BMV, in solution under conditions where they are stable, may vary in size (Chauvin *et al.*, 1978).

5. Atomic force microscopy

TMV particles, when dried on glass substrates, assemble into characteristic patterns that can be studied using atomic force microscopy (Maeda, 1997). In highly orientated regions, the particle length was measured as 301 nm and the width as 14.7 nm; the latter measurement shows intercalation of packed particles. The particles are not flattened and their depth was 16.8–18.6 nm.

C. Fine structure determination: electron microscopy

Horne (1985) and Milne (1993) give accounts of the application of electron microscopy to the study of plant virus structure.

1. Metal shadowed preparations

In early electron microscope studies of virus particles, shadowing with heavy metals was used to enhance contrast. Such shadowing tended to obscure surface details but gave information on overall size and shape of the dry particle. Much more information can be obtained if specimens are freeze-dried and then shadowed in a very high vacuum (e.g. Hatta and Francki, 1977). For example, Roberts (1988), using this kind of procedure, was able to show that TRSV had a structure resembling that of models made up of 60 subunits in clusters of five arranged in a $T = 1$ structure.

2. Freeze etching

This technique can give useful information about the surfaces and substructure of larger viruses—particularly those with lipoprotein bilayer membranes. Although Steere (1969) illustrated arrays of freeze-fractured PYDV particles within an infected cell, no novel structural information has yet been obtained for plant viruses using this procedure.

3. Negative staining

The use of electron-dense stains has proved of much greater value than metal shadowing in revealing the detailed morphology of virus particles. Such stains may be positive or negative.

Positive stains react chemically with and are bound to the virus (e.g. various osmium, lead and uranyl compounds and phosphotungstic acid (PTA) under appropriate conditions). Chemical combination may lead to alteration in, or disintegration of, the virus structure. In negative staining, on the other hand, the electron-dense material does not react with the virus but penetrates available spaces on the surface or within the virus particle. This is now the preferred technique for examining virus structure by electron microscopy. Common negative stains are potassium phosphotungstate (KPT) at pH 7.0, and uranyl acetate or formate used near pH 5.0. The virus structure stands out against the surrounding electron-dense material. The carbon substrates used to mount stained specimens have a granular background. Structural detail can often be seen most clearly in images of particles in a film of stain suspended over holes in the carbon grid. The stain may or may not penetrate any hollow space within the virus particle. For recent technical improvements in negative staining, see Harris and Horne (1994).

With helical rod-shaped particles, the central hollow is frequently revealed by the stain. With the small spherical viruses, which often have associated with them empty protein shells, it was assumed by some workers that stained particles showing a dense inner region represented empty shells in the preparation. However, staining conditions may lead to loss of RNA from a proportion of the intact virus particles, allowing stain to penetrate, while stain may not enter some empty shells. Stains differ in the extent to which they destroy or alter a virus structure, and the extent of such changes depends closely on the conditions used. The particles of many viruses that are stabilized by protein:RNA bonds are disrupted by KPT at pHs above 7.0.

Even in the best electron micrographs, the finer details of virus structure tend to be obscured, firstly by noise due to minor irregularities in the actual virus structure and other factors such as irregularities in the stain, and secondly by the fact that contrast due to the stain is often developed on both sides of the particle to varying degrees. To overcome these difficulties and to extract more reliable and detailed information from images of negatively stained individual virus particles, several methods have been used.

The main procedures have been photographic superposition (Markham *et al.*, 1963, 1964; Finch and Holmes, 1967), shadowgraphs (Finch and Klug, 1966, 1967), and optical transforms (Klug and Berger, 1964; Crowther and Klug, 1975; Steven *et al.*, 1981). Vogel and Provencher (1988) developed a computational procedure for three-dimensional reconstruction from projections of a disordered collection of single particles and have applied it to TBSV. Optical diffraction has been applied to *in vitro* crystalline arrays of isometric viruses viewed under the electron microscope (see Horne and Pasquali-Ronchetti, 1974; Horne *et al.*, 1977) (Fig. 5.1). With recent developments facilitating the application of X-ray crystallographic analysis to virus structure, these procedures will probably not be widely used, except for viruses to which X-ray analysis cannot yet be applied.

4. Thin sections

Some aspects of the structure of the enveloped viruses, particularly bilayer membranes, can be studied using thin sections of infected cells or of a pellet containing the virus (see Fig. 5.41).

5. Cryoelectron microscopy

Cryoelectron microscopy, which involves the extremely rapid freezing of samples in an aqueous medium, allows the imaging of symmetrical particles in the absence of stain and under conditions that preserve their symmetry. Jeng *et al.* (1989) used TMV as a test object to develop the procedure for determining structural detail in objects with helical symmetry. They obtained data on TMV with spacing better than 10 Å from electron images of hydrated TMV frozen in vitreous ice. This technique has been widely used for isometric viruses. For these particles, the images have to be reconstructed to give a three-dimensional object from a two-dimensional image using a range of approaches (e.g. Fuller *et al.*, 1996; Lanczycki *et al.*, 1998; Rossman and Tao, 1999). This leads to

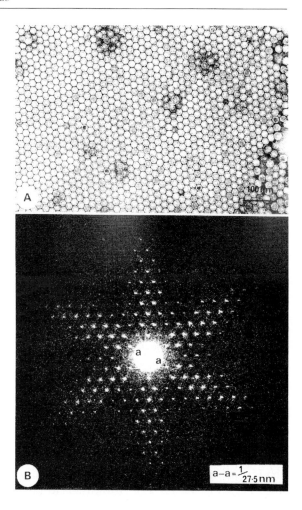

Fig. 5.1 (A) Part of a two-dimensional hexagonal array of TroV showing regular packing of the particles within the lattice. The central light region of individual capsids can be clearly seen together with a few particles where the interior has been completely penetrated by stain. The shape and size of the capsids in this orientation within the hexagonal array are more regular than those observed in a square or skewed array. **(B)** Optical diffraction pattern obtained from material in (A) showing typical spectra present in TRoV hexagonal arrays. The first-order spots a–a correspond to the center-to-center particle spacing of 27.5 nm. From Horne *et al.* (1977), with permission.

information, which complements that from X-ray crystallography and also allows the detailed analysis of viruses that are not amenable to crystal formation. For instance, the structure of CaMV has been resolved by cryo-electron microscopy to 30 Å resolution (Cheng *et al.*, 1992).

6. Cryoelectron microscopy compared with negative staining

There has been considerable discussion as to the relative advantages and disadvantages of cryoelectron microscopy and negative staining. The basic points are:

1. The contrast of unstained biological materials is low, leading to problems in focusing and in the signal to noise ratio.
2. Fully hydrated specimens are electron-beam sensitive.
3. Cryoelectron microscopy is considered to be difficult and costly.
4. The resolution of cryoelectron microscopy combined with image analyzes is often greater than with negative staining.

Adrian *et al.* (1998) have developed a technique, termed 'cryo-negative staining', that combines the advantages of the two techniques. The vitrified samples are prepared with ammonium molybdate and are blotted on to holey carbon supports.

D. X-ray crystallographic analysis

Where it can be applied, X-ray crystallography provides the most powerful means of obtaining information about structures that are regularly arrayed in three dimensions. Finch and Holmes (1967) give an introductory account of the methods involved.

Over about the past 20 years or so there have been significant advances that have allowed the application of X-ray crystallographic analysis to more viruses, and at higher resolutions. With the definition of structures at the atomic level, it has become possible to interpret structure in more detail in relation to biological function. The major advances have been as follows:

1. There has been an increase in the capacity and speed of computers, together with a reduction in cost.
2. Non-crystallographic symmetry averaging has been introduced. Icosahedral viruses such as TBSV and STNV have 60-fold symmetry in the virus particle. The degree to which this coincides with the symmetry of the crystal used for analysis depends on the crystal form that happens to be obtained. For example, in TBSV the smallest crystal repeating unit (the asymmetric unit) consists of five trimers of the coat protein. These are exactly related by crystallographic symmetry. Thus, symmetry averaging for this virus takes place around a chosen 5-fold axis. Successive recalculations from an initial electron density map remove noise and enhance detail in the map. An example of the procedure is given by Olson *et al.* (1983) for TBSV.
3. Computer graphics have now replaced the laborious manual model-building of the past, which was necessary for the refinement of a structure (Olson *et al.*, 1985). Computer graphics are also extremely useful for exploring particular aspects of a three-dimensional structure and for communicating structural ideas in a more readily comprehensible form, by such means as color coding and selective omission of detail (Namba *et al.*, 1984, 1988).
4. Site-directed mutagenesis has been introduced to protein crystallography. The possibility of exchanging one amino acid for another in any chosen site in a protein changes crystallography from a passive technique to one in which the relation between structure and function can be studied in a systematic and rational fashion.

X-ray crystallography analysis has two main limitations as far as virus structures are concerned:

1. For nearly all icosahedral viruses most of the nucleic acid is not arranged in a regular manner in relation to the symmetries in the protein shell. However, for some isometric viruses it is possible to obtain information on the distribution of nucleic acid within the particle (see Section VI.C).
2. Larger viruses such as members of the *Rhabdoviridae* cannot be crystallized. In addition, it is probable that many features of their structure are not strictly regular. For such viruses, electron microscopy provides the best information.

E. Neutron small-angle scattering

Neutron scattering by virus solutions is a method by which low-resolution information can be obtained about the structure of small isometric viruses and in particular about the radial dimensions of the RNA or DNA and the protein shell. The effects of different conditions in solution on these virus dimensions can be determined readily. The method takes advantage of the fact that $H_2O–D_2O$ mixtures can be used that match either the RNA or the protein in scattering power. Analysis of the neutron diffraction at small angles gives a set of data from which models can be built.

F. Mass spectrometry

As well as measuring the mass of viral proteins and particles, mass spectrometry can be used to identify viral protein post-translational modifications such as myristoylation, phosphorylation and disulfide bridging (reviewed by Siuzdak, 1998). When this technique is used in conjunction with X-ray crystallography, the mobility of the capsid can be studied. Similarly, nuclear magnetic-resonance spectroscopy can detect mobile elements on the surface of virus particles (Brierley et al., 1993).

G. Serological methods

The reaction of specific antibodies with intact viruses or dissociated viral coat proteins has been used to obtain information that is relevant to virus structure. For instance, the terminal location of the minor coat protein on the flexuous rod-shaped particles of closteroviruses and criniviruses was recognized using polyclonal antibodies (see Fig. 5.12).

Monoclonal antibodies (MAbs) are proving to be particularly useful for this kind of investigation, although there are significant limitations (reviewed in Kekuda et al., 1995). Some other examples of the use of MAbs are in the determination of the antigenic structure of PMTV (Pereira et al., 1994) and of PVA (Moravec et al., 1998). For a description of the type of epitopes that induce antibodies, see Chapter 15 (Section IV.A.2).

Structural interpretation of the results of cross-reactions between closely related proteins can be confused by the fact that the conformation of any antigenic determinant may be changed by an amino acid substitution occurring elsewhere in the protein. Furthermore, the ability of different MAbs to detect residue exchanges may be extremely variable as was found for a set of TMV mutants (Al-Moudallal et al., 1982). Nevertheless, serological techniques have produced some structural information, especially with viruses for which the detailed protein structure has not been determined by X-ray crystallography. For example, at least three different antigenic determinants were distinguished on the coat protein of PVX with a set of MAbs (Koenig and Torrance, 1986). The results are summarized in Fig. 5.2. The surface location of the N terminus was confirmed by Söber et al. (1988).

The N-terminal and C-terminal regions of the coat protein of TYMV were interpreted to be at the surface of the virus (Quesniaux et al., 1983a,b). Similarly immunological evidence confirmed that these N-terminal regions are also at the virus surface of potyviruses (Shukla et al., 1988b). A surface location for both N and C termini appears to be a common feature in other rod-shaped viruses.

Caution must be used in interpreting the results of ELISA tests in structural terms when the whole virus is the antigen, as the following results demonstrate. Antibodies against a synthetic peptide of TBSV making up part of the flexible amino-terminal arm (amino acids 28–40) reacted with whole virus as antigen in ELISA tests (Jaegle et al., 1988). Since the amino-terminal arm is in the interior of a very compact shell (Section VI.B.4.d), this result must mean that virus structure was opened up sufficiently on the ELISA plate to allow antibodies to react with the normally buried arm. Similar results have been obtained for SBMV (MacKenzie and Tremaine, 1986).

As was indicated by Dore et al. (1987b), a definitive delineation of antigenic sites requires knowledge of the three-dimensional structure of the polypeptides. In most cases, if the three-dimensional structure is known, what is the point, in relation to their structure, of determin-

Reactivity of determinants with MAb of group	Schematic representation of determinants (●) on undisturbed (a) and partially denatured (b) subunits	Properties of determinants

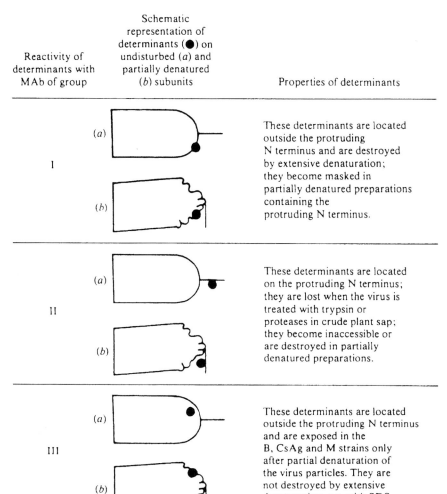

I — (a) / (b): These determinants are located outside the protruding N terminus and are destroyed by extensive denaturation; they become masked in partially denatured preparations containing the protruding N terminus.

II — (a) / (b): These determinants are located on the protruding N terminus; they are lost when the virus is treated with trypsin or proteases in crude plant sap; they become inaccessible or are destroyed in partially denatured preparations.

III — (a) / (b): These determinants are located outside the protruding N terminus and are exposed in the B, CsAg and M strains only after partial denaturation of the virus particles. They are not destroyed by extensive denaturation, e.g. with SDS.

Fig. 5.2 Schematic representation of the three kinds of antigenic determinants distinguished in the coat protein of PVX by their differential reactivity to a set of MAbs. From Koenig and Torrance (1986), with permission.

ing the antigenic sites for plant viruses? The situation is quite different for viruses infecting vertebrates, where knowledge of such sites may be very important for vaccine development. However, with recombinant DNA technology an understanding of the surface structure of a plant virus can lead to identification of sites at which foreign epitopes can be introduced and the recombinant virus used for vaccination (see Chapter 16, Section IX.C).

Dore et al. (1988) developed an elegant procedure involving ELISA reactions on electron microscope grids and gold-labeled antibody. Using this procedure, they showed that anti-TMV MAbs that reacted with both virus particles and the coat protein subunits bound to the virus rods only at one end (Fig. 5.3). Further studies have shown that the MAbs bind to the surface of the protein subunit that contains the right radial and left radial α-helices (Dore et al., 1989). This result confirms in a graphic way a fact already known from X-ray crystallographic analysis, namely that the coat protein in the TMV rod presents a chemically different aspect on its upper and lower surfaces.

Using a similar technique, Lesemann et al. (1990) distinguished three groups of MAbs reacting with BNYVV. One reacted along the entire length of the particles. The other two groups reacted with antigenic sites on opposite ends of the particles.

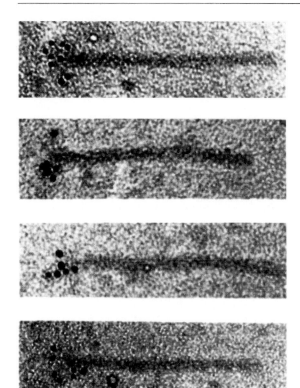

Fig. 5.3 Electron micrographs showing the binding of a gold-labeled MAb to one extremity of TMV rods. Bar = 100 nm. From Dore *et al.* (1988), with permission.

H. Methods for studying stabilizing bonds

The primary structures of viral coat proteins and nucleic acids depend on covalent bonds. In the final structure of the simple geometric viruses, these two major components are held together in a precise manner by a variety of non-covalent bonds. Three kinds of interaction are involved: protein:protein, protein:RNA, and RNA:RNA. In addition, small molecules such as divalent metal ions (Ca^{2+} in particular) may have a marked effect on the stability of some viruses. Knowledge of these interactions is important for understanding the stability of the virus in various environments, how it might be assembled during virus synthesis, and how the nucleic acid might be released following infection of a cell. The stabilizing interactions are hydrophobic bonds, hydrogen bonds, salt linkages, and various other long-

and short-range interactions. A variety of physical and chemical methods has been used in attempts to refine our understanding of the role of these bonds in virus structure.

1. X-ray crystallographic analysis

Models built to a resolution of less than 3 Å, based on X-ray crystallographic analysis and knowledge of the primary structure of the coat protein, provide a detailed understanding of the bonds in the secondary and tertiary structure of the protein subunit. In addition, the bonds between subunits that make up the quaternary structure of the virus, and the role of accessory molecules and ions, mainly water and Ca^{2+}, can be defined. A virus structure examined to atomic resolution provides a vast amount of detail that cannot be encompassed in a book of this size. The highlights of several such structures are discussed in later sections.

2. Stability to chemical and physical agents

The effects of pH, ionic strength, kind of ion, temperature, compounds such as phenol and detergents, and hydrogen bond-breaking agents such as urea on the stability of viruses have been studied in many laboratories (see Kaper, 1975). Such experiments can give us information only of a general kind about the bonds involved in virus stability.

Among the small isometric viruses there is a wide range of stabilities. Viruses like TYMV with strong protein:protein interactions are the most stable. The other end of the spectrum is illustrated by CMV and AMV, where protein:RNA interactions predominate (Kaper, 1975).

Protein:protein interactions are obviously very important for the stability of TYMV since empty protein shells are quite stable. Hydrophobic bonding between subunits is a significant factor in this stability because (1) TYMV is very stable at high ionic strengths, (2) it is readily degraded by phenol and ethanol, and (3) degradation of the virus and empty protein shell by urea, organic mercurial compounds and other chemicals can be interpreted on the basis that the predominant protein:protein interaction is via hydrophobic bonding (Kaper, 1975). Extreme conditions such as high pH (Keeling and Matthews, 1982) and freezing and thawing (Katouzian-Safadi and Haenni, 1986) have been used to study release of RNA from TYMV particles.

Alkaline conditions have been used to demonstrate the removal of protein subunits from the TMV rod, beginning at the 5'-end, and to reveal intermediates in the stripping process, these being attributed to regions of unusually strong interaction between the protein and RNA (Perham and Wilson, 1978).

3. Chemical modification of the coat protein

Particular amino acid residues in the coat protein can be modified by the chemical addition of a side-chain, and the effects of such substitution on stability of the virus can be examined (e.g. TMV: Wilson and Perham, 1985).

4. Removal of ions

For small isometric viruses whose structures are partially stabilized by Ca^{2+} ions, removal of these ions, usually by EDTA or EGTA, leads to swelling of the virus particle. A study of the swelling phenomenon can give information about the kinds of bonds important for stability. A variety of techniques has been used for monitoring swelling. For example, Krüse et al. (1982) used small-angle X-ray and neutron scattering, analytical centrifugation and fluorescence techniques to study swelling of TBSV. Swelling of STNV in the presence of EDTA was demonstrated by ultracentrifugation and X-ray crystallography (Unge et al., 1986). In swollen virus an amino-terminal peptide that was normally well buried in the structure became susceptible to trypsin. The kinetics of swelling of SBMV was studied using photon correlation spectroscopy (Brisco et al., 1986b).

Durham et al. (1984) used hydrogen-ion titration curves to follow the reversible swelling and contraction of particles of TBSV, CCMV and TCV. They drew conclusions regarding the semi-permeability of various protein shells.

The topic of virus particle swelling and of other conformational changes involved in the disassembly of viruses is discussed in Chapter 7 (Section II).

5. Circular dichroism

Circular dichroism spectra can be used to obtain estimates of the extent of α-helix and β structure in a viral protein subunit (e.g. TRoV: Denloye et al., 1978; SBMV: Odumosu et al., 1981).

6. Methods applicable to nucleic acid within the virus particle

Several methods are available to give approximate estimates of the degree of helical base-pairing or other ordered arrangement of the nucleic acid within the virus. These include relative absorbency at 260 nm (e.g. TYMV: Haselkorn, 1962), laser Raman spectroscopy (e.g. TYMV: Hartman et al., 1978), circular dichroism spectra (e.g. SBMV: Odumosu et al., 1981), and magnetic birefringence (e.g. CaMV: Torbet et al., 1986).

III. ARCHITECTURE OF ROD-SHAPED VIRUSES

A. Introduction

Crick and Watson (1956) put forward a hypothesis concerning the structure of small viruses, which has since been generally confirmed. Using the knowledge then available for TYMV and TMV, namely, that the viral RNA was enclosed in a coat of protein and (for TMV only, at that stage) that the naked RNA was infectious, they assumed that the basic structural requirement for a small virus was the provision of a shell of protein to protect its ribonucleic acid. They considered that the relatively large protein coat might be made most efficiently by the virus controlling production in the cell of a large number of identical small protein molecules, rather than in one or a few very large ones.

They pointed out that if the same bonding arrangement is to be used repeatedly in the particle the small protein molecules would then aggregate around the RNA in a regular manner. There are only a limited number of ways in which the subunits can be arranged. The structures of all the geometric viruses are based on the principles that govern either rod-shaped or spherical particles.

In rod-shaped viruses, the protein subunits are arranged in a helical manner. There is no theoretical restriction on the number of protein subunits that can pack into a helical array in rod-shaped viruses.

B. *Tobamovirus* genus

1. General features

The particle of TMV is a rigid helical rod, 300 nm long and 18 nm in diameter. The composition of the particle is approximately 95% protein and approximately 5% RNA. It is an extremely stable structure, having been reported to retain infectivity in non-sterile extracts at room temperature for at least 50 years (Silber and Burk, 1965). The stability of naked TMV RNA is no greater than that of any other ssRNA. Thus, stability of the virus with respect to infectivity is a consequence of the interactions between neighboring protein subunits and between the protein and the RNA.

X-ray diffraction analyzes have given us a detailed picture of the arrangement of the protein subunits and the RNA in the virus rod. The particle comprises approximately 2130 subunits that are closely packed in a helical array. The pitch of the helix is 2.3 nm, and the RNA chain is compactly coiled in a helix following that of the protein subunits (Fig. 5.4).

There are three nucleotides of RNA associated with each protein subunit, and there are 49 nucleotides and $16\,^1/_3$ protein subunits per turn. The phosphates of the RNA are at about 4 nm from the rod axis. By tilting negatively stained TMV particles in the electron beam and noting changes in the edge appearance of the rods, Finch (1972) established that the basic helix of TMV is right-handed. In a proportion of negatively stained particles, one end of the rod can be seen as concave, and the other end is convex. The 3'-end of the RNA is at the convex end and the 5' at the concave end (Wilson *et al.*, 1976; Butler *et al.*, 1977). A central canal with a

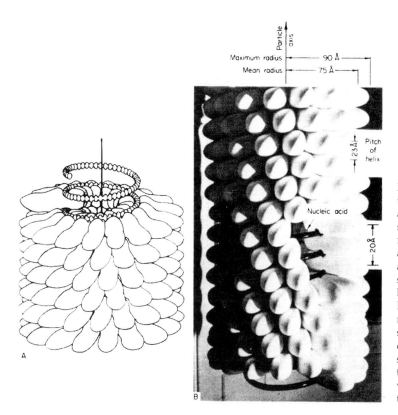

Fig. 5.4 Structure of TMV. **(A)** Drawing to show the relationship of the RNA and the protein subunits. The RNA shown free of protein could not maintain this configuration in the absence of the protein subunits. There are three nucleotides per protein subunit, or 49 per turn of the major helix spaced about 5 Å apart. **(B)** Photograph of a model of TMV with major dimensions indicated. The structure repeats after 69 Å in the axial direction, the repeat containing 49 subunits distributed over three turns of the helix. From Klug and Caspar (1960), with permission, and based largely on the work of R. E. Franklin.

radius of about 2 nm becomes filled with stain in negatively stained preparations of the virus (Fig. 5.5).

2. Short rods

Most purified preparations of TMV contain a proportion of rods that are of variable length and less than 300 nm. Study of these short rods may be complicated by the problem of aggregation. Both intact TMV rods and the shorter rods have a tendency to aggregate end to end under appropriate conditions, giving a wide range of lengths. The length distribution observed is very dependent on the precise history of the virus preparation. Some factors affecting the length distribution are pH, ionic strength, nature of ions present, temperature, and length of storage, but not all the factors involved are as yet fully understood.

Many of the shorter rods are probably of no special significance. They may be virus particles that were only partially assembled at the time of isolation, or parts of rods fractured during isolation. However, in beans infected with SHMV the RNA containing the coat protein gene ($\approx$1 kb), which is synthesized as a discrete species during virus replication, is assembled into a rod about 40 nm long (Higgins *et al.*, 1976).

3. Properties of the coat protein

The coat protein comprises 158 amino acids giving it a molecular weight of 17–18 kDa. Fibre diffraction studies have determined the structure to 2.9 Å resolution (Namba *et al.*, 1989) (Fig. 5.6).

The protein has a high proportion of secondary structure, with 50% of the residues forming four *a*-helices and 10% of the residues in β-structures, in addition to numerous reverse turns. The four closely parallel or anti-parallel *a*-helices (residues 20–32, 38–48, 74–88 and 114–134) make up the core of the subunit and the distal ends of the four helices are connected transversely by a narrow and twisted strip of β-sheet. The central part of the subunit distal to the β-sheet is a cluster of aromatic residues (Phe^{12}, Trp^{17}, Phe^{62}, Tyr^{70}, Tyr^{139}, Phe^{144}) forming

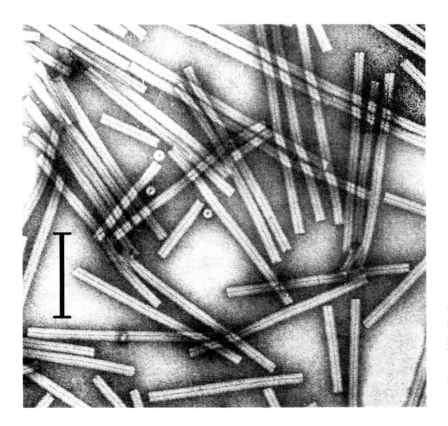

Fig. 5.5 Electron micrograph of negatively stained TMV particles. The axial hole filled with stain can be clearly seen. A few protein disks showing the central hole are also present. Bar = 100 nm. (Courtesy of R. W. Horne.)

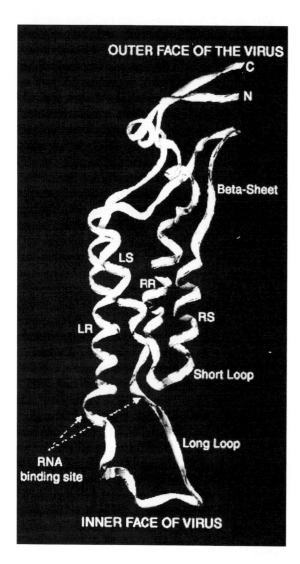

Fig. 5.6 Ribbon representation of the α-carbon tracing of TMV coat protein. At the core of the protein is a right-handed helical bundle of four α-helices, the left-slewed (LS), the left-radial (LR), the right-slewed (RS) and the right-radial (RR). The long loop connects the LR and RR helices. From Taraporewala and Culver (1997), with kind permission of the copyright holder, © The American Phytopathological Society.

a hydrophobic patch. The N and C termini of the protein are to the outside of the particle. The polypeptide chain is in a flexible or disordered state below a radius in the particle of about 4 nm so that no structure is revealed in this region.

4. Structure of the double disk

One of the reassembly products of TMV protein subunits (see Section IV.A) is a double disk containing two rings of 17 protein subunits that is of particular interest in that the details of inter-subunit contacts can be determined. Under appropriate conditions, the disks form true three-dimensional crystals. Although the repeating unit is very large, X-ray crystallographic procedures can be applied. This was the approach taken by Klug and colleagues, which after 12 years' work led to an elucidation of the structure of the protein subunit and the double disk to 2.8 Å resolution (Bloomer *et al.*, 1978).

Viewed in section, the subunits of the upper ring in the disk are flat. Those of the lower ring are tilted downward toward the center of the disk (Fig. 5.7). There are three regions of contact vertically between subunits as detailed in the figure.

The outermost contact is a polar interaction between Serines 147 and 148 and Thr[59]. At lower radius the largest contact region is a salt-bridge system involving a complex hydrogen bonded three-dimensional network in which two water molecules also participate. The innermost contact is a small hydrophobic patch where Ala[74] and Val[75] both touch the ring of Pro[54]. Thus, the two disks, in contact at the outer part, open toward the center like a pair of jaws. The flexible inner parts of the folded chains are indicated by dotted lines in Fig. 5.7. Other physical studies show that there is thermal motional disorder in this region rather than static disorder (Jardetsky *et al.*, 1978).

5. Virus structure

Intact TMV does not form three-dimensional regular crystalline arrays in solution. For this reason structural analysis has not yet proceeded to the absolute detail available for the double disk though it is approaching it. However, using fibre diffraction methods, Stubbs *et al.* (1977) solved the structure to a resolution of 4 Å. Using this, together with the other data already available, they produced a model for the virus (Fig. 5.8). Fiber diffraction data can be refined by the molecular dynamics method (Wang and Stubbs, 1993).

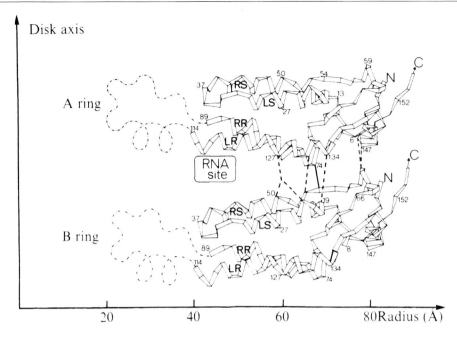

Fig. 5.7 Side view of a sector through the disk showing the relative disposition of subunits in the two rings and the axial contacts between them. There are three regions of contact indicated by a solid line for the hydrophobic contact of Pro 54 with Ala 74 and Val 75, by dashed lines for the hydrogen bonds between Thr 59 and serines 147 and 148, and further dashed lines for the extended salt-bridge system. The low-radius region of the chain, which has no ordered structure in the absence of RNA, is shown schematically with an indication of an additional 1–2 turns extending the LR helix before it turns upwards into the vertical column. The primary nucleotide-binding site is probably below the RR helix. From Bloomer *et al.* (1978), with kind permission of the copyright holder, © Macmillan Magazines Ltd.

Namba and Stubbs (1986) established a structure at 3.6 Å that refined the structure of Stubbs *et al.* (1977) in certain details that are particularly relevant to our understanding of virus assembly. Namba *et al.* (1989) produced an even more refined structure at 2.9 Å that has generated a great deal of detailed information. The following discussion is based mainly on the structure of Namba and Stubbs (1986). The following are important features.

a. The outer surface

As noted above, the N and C termini are at the virus surface. However, the very C-terminal residues 155–158 were not located on the density map and are therefore assumed to be somewhat disordered (Namba and Stubbs, 1986).

b. The inner surface

The presence of the RNA stabilizes the inner part of the protein subunit in the virus so that its position can be established. The highest peak in the radial density distribution is at about 2.3 nm. This is the region occupied by the vertical chains containing the V helices (Namba and Stubbs, 1986). These chains fill a space 0.9 nm wide by 2.3 nm high, and they are packed closely together to form a dense wall around the axial hole, protecting the RNA from the medium.

c. The RNA binding site

The binding site is in two parts, being formed by the top of one subunit and the bottom of the next. The three bases associated with each protein subunit form a claw that grips the left radial helix of the top subunit. The left radial helix has a large number of aliphatic residues between positions 117 and 128, which form three faces for the bases. The bases lie flat against the hydrophobic side-chains and each face can accommodate any base. The other part of the RNA binding site is mostly on the right-slewed helix.

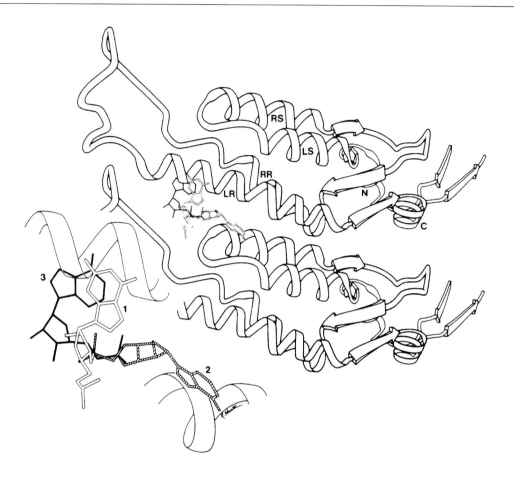

Fig. 5.8 Secondary structure in TMV coat protein. Backbone structures of two protein subunits and three RNA nucleotides (labeled 1, 2 and 3), represented as GAA, are illustrated. The RNA is enlarged at lower left with each nucleotide shaded differently. α-Helices are designated as in Fig. 5.6. N, N-terminal, substantantially obscured in this view; C, C-terminal. The viral axis is vertical, to the left of the figure. From Namba *et al.* (1989), with permission.

Two phosphate groups form ion pairs with Arg[90] and Arg[92]. The third phosphate appears to form a hydrogen bond with Thr[37] (Namba and Stubbs, 1986). Arg[41] extends toward the same phosphate as Arg[90] but does not approach as closely. Namba *et al.* (1989) suggest that the electrostatic interactions between protein and RNA are best considered as complementarity between the electrostatic surfaces of protein and RNA, rather than as simple ion pairs between arginines and phosphates.

d. Electrostatic interactions involved in assembly and disassembly

Caspar (1963) and Butler *et al.* (1972) obtained results indicating that pairs of carboxyl groups with anomalous pK values (near pH 7.0) were present in TMV. Namba and Stubbs (1986) identified two inter-subunit pairs of carboxyl groups in their model that may be those proposed by Caspar: Glu[95]–Glu[106] at 2.5 nm radius and in a side-to-side interface, and Glu[50]–Asp[77] at 5.8 nm radius in the top-to-bottom interface. These groups may play a critical role in assembly and disassembly of the virus (see Section IV.A and Chapter 7, Section II.B).

Namba *et al.* (1989) identified three sites where negative charges from different molecules are juxtaposed in subunit interfaces. These create an electrostatic potential that could be used to drive disassembly:

- a low-radius carboxyl–carboxylate pair appears to bind calcium
- a phosphate–carboxylate pair that also appears to bind calcium
- a high-radius carboxyl–carboxylate pair in the axial interface (this could not bind calcium but could bind a proton and thus titrate with an anomalous pK).

e Water structure

Water molecules are distributed throughout the surface of the protein subunit, on both the inner and outer surfaces of the virus and in the sub-unit interfaces (Namba *et al.,* 1989).

f. Specificity of TMV protein for RNA

Gallie *et al.* (1987d) showed that TMV protein does not assemble with DNA even if the origin of assembly sequence is included. Namba *et al.* (1989) concluded that this specificity must involve interactions made by the ribose hydroxyl groups, because all three base-binding sites could easily accommodate thymine.

C. *Tobravirus* genus

TRV is a rigid cylindrical rod built on the same general plan as TMV. TRV requires two parti-cles, a long rod and a shorter one, to co-operate in the production of intact virus particles. This aspect is discussed in Chapter 6 (Section VIII.H.2). Here I shall consider the structure of the long rods. The short rods are considered to be constructed on the same plan using the same protein subunit.

Most information about the structure of this virus has been obtained by electron micro-scopy of virus particles stained in uranyl formate (Offord, 1966) (Fig. 5.9). Some infor-mation was obtained by optical diffraction. For the strains studied by Offord, the length of the infectious particle was approximately 191 nm and its diameter was 25.6 nm. The pitch of the helix was 2.55 nm with 76 subunits in three turns. The radius of the central hole was about 2.7 nm. An annular feature was observed at a radius of about 8.2 nm, which might represent the RNA chain. A densely stained ring has been observed at about this radius in cross-sections of rods (Tollin and Wilson, 1971). The strain studied by Offord had about 72 turns of the helix with probably four nucleotides per protein subunit, giving 7100 nucleotides.

Harrison and Woods (1966) noted that, in many of their electron micrographs of different TRV isolates, the two ends of the particle had different shapes. One end was slightly convex, while the other end of the axial canal was slightly flared. They suggest that this appearance of the ends of the rods might be due to the protein subunits being banana-shaped with the long axes not inclined at exactly 90 degrees to the long axis of the particle. Recently, using vibrational Roman optical activity, Blanch *et al.* (2001) showed that both TMV and TRV coat proteins have a significant amount of α helix, TRV having more than TMV.

Like TMV coat protein, TRV protein can exist in various discrete states of aggregation in solu-tion. In particular, a double disk structure that sediments at 35S has been observed that may well play a part in virus assembly (Gugerli, 1976).

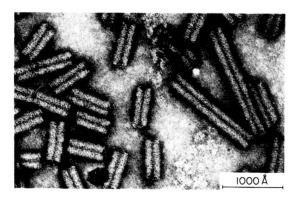

Fig. 5.9 Electron micrograph of TRV showing long and short rods negatively stained with uranyl formats. From Offord (1966), with permission.

D. Other helical viruses

Structural details of other rod-shaped viruses with helical symmetry are given in Table 5.1, and electron micrographs of some flexuous rod-shaped virus particles are shown in Fig. 5.10.

For many of these viruses there is no information on the details of the helical arrangement of the coat protein subunits as they are difficult viruses for analysis by optical and other diffraction techniques. It is likely that the rigid rod-shaped viruses have basic structures similar to those of TMV and TRV. However, notable features are that the rigid rods have a greater diameter (about 20 nm) than do the flexuous rods (about 11–13 nm) and that the helix pitch in the rigid rods (2.4–2.6 nm) is less than that of the flexuous rods (3.3–3.8 nm). The flexibility of the flexuous rods is likely to result from their looser structure.

The antigenic structures of two viruses with rigid rods, the furovirus SBWMV and the pomovirus PMTV have been determined using MAbs (Chen *et al.*, 1996; Pereira *et al.*, 1994). Both these studies showed that, as with TMV, the termini of the coat protein subunits were at or near the surface of the particle.

The most detailed structural studies of flexuous rod-shaped viruses have been on potexviruses (reviewed by Tollin and Wilson, 1988). Circular dichroism studies showed that the secondary structures of NMV and PVX coat proteins have a large element of α-helix (45%) with much less β-sheet (Wilson *et al.*, 1991). The helix pitch varied according to the humidity under which it was measured. Similarly, the length of some potyviruses can be affected by the presence or absence of divalent cations. PepMoV particles have a length of 850 nm in the presence of magnesium but when exposed to EDTA they were only about 750 nm long (Govier and Woods, 1971).

The organization of the coat protein subunits in PVX particles was studied by tritium planigraphy (Baratova *et al.*, 1992). The particles were bombarded with thermally activated tritium atoms and the intramolecular distribution of tritium label was assessed. The N-terminal region of the unmodified coat protein is the most accessible to the hot tritium, the C-terminal region being almost completely inaccessible. From these and the predicted secondary structure they built a model of the coat protein (Fig. 5.11).

TABLE 5.1 Properties of other rod-shaped viruses

Virus group	Particle length (nm)	Particle diameter (nm)	Rod type[a]	Helix pitch (nm)	Turns/ repeat	Subunits/ turn	Axial hole	Data source
Hordeivirus	110–150 (m)[b]	20	R	2.5	ND[c]	ND	Yes	1
Furovirus	140–160, 260–300 (m)	20	R	2.4–2.5	ND	ND	Yes	2
Pomovirus	65–80, 150–160, 290–310 (m)	18-20	R	2.4–2.5	ND	ND	Yes	3
Pecluvirus	190, 245 (m)	21	R	2.6	ND	ND	Yes	4
Benyvirus	85, 100, 265, 390 (m)	20	R	2.6	ND	ND	Yes	5
Potexvirus	470–580	13	F	3.3–3.6	4–11	8–9	Yes	6
Carlavirus	610–700	12–15	F	3.4	ND	ND	ND[c]	6
Potyvirus	680–900	11–13	F	3.4–3.5	ND	ND	ND	6
Capillovirus	640–700	12	F	3.4–3.8	9–10	ND	ND	6
Trichovirus	640–760	12	F	3.4–3.8	*c* 10	9.3–9.8	ND	6
Vitivirus	725–825	12	F	3.8	5	8–9	ND	6,7
Closterovirus	1250–2000	12	F	3.4–3.8	5	9–10	ND	6,7

[a]: R, rigid rod; F, flexuous rod.
[b]: (m), multicomponent virus.
[c]: ND, not determined or, for axial hole, not detected.
Data: 1, Laurence *et al.* (2000); 2, Torrance (2000); 3, Koenig and Lesemann (2000a); 4. Fritsch and Dollett (2000); 5. Koenig and Lesemann (2000b); 6, Tollin and Wilson (1988); 7, Tollin *et al.* (1992); all with permission.

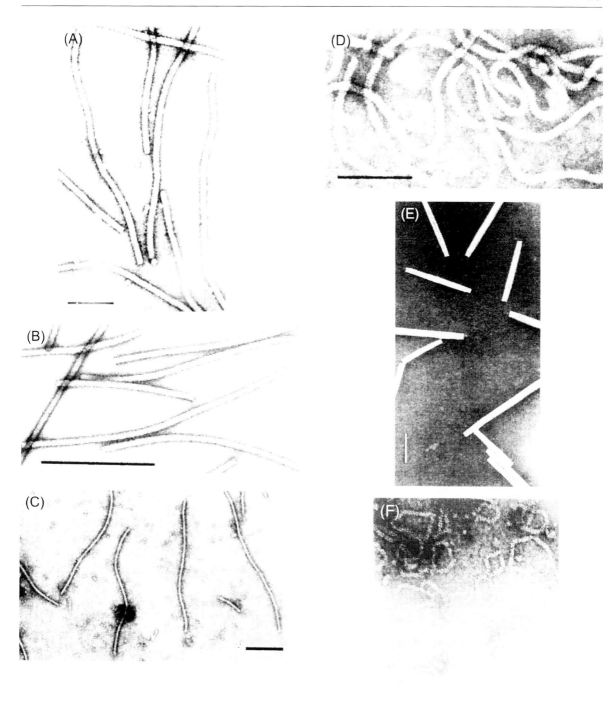

Fig.5.10 Electron micrographs of negatively stained virus particles as representatives of virus groups. **(A)** PVX potexvirus; bar = 100 nm (from Brunt *et al.*, 2000a). **(B)** CLV carlavirus; bar = 250 nm (from Brunt *et al.*, 2000b). **(C)** PPV potyvius; bar = 200 nm (from Berger *et al.*, 2000). **(D)** BSMV hordeivirus; bar = 50 nm (from Lawrence *et al.*, 2000). **(E)** CTV closterovirus; bar = 100 nm (from Martelli *et al.*, 2000). **(F)** RHBV tenuivirus; bar = 100 nm. From Mayo *et al.* (2000); all with permission.

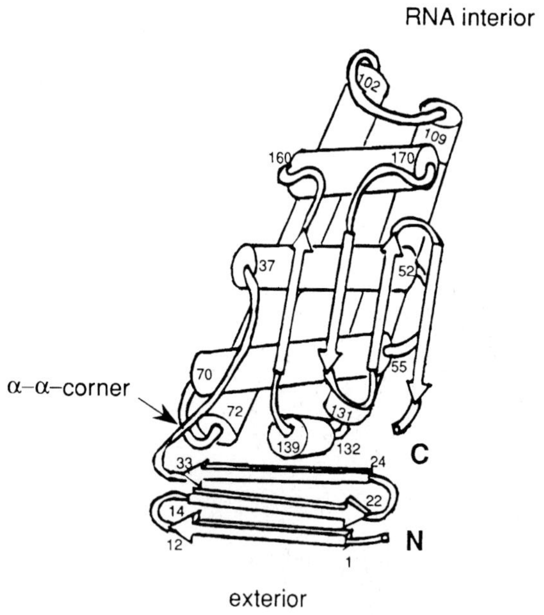

RNA interior

α–α–corner

exterior

Fig. 5.11 Schematic of the polypeptide chain fold of PVX coat protein. α-helices are shown as cylinders and β-strands as arrows. N and C are the N and C termini of the polypeptide chain, respectively. Positions of some amino acid residues are numbered. From Baratova *et al.* (1992), with permission.

The closterovirus BYV encodes a protein (p24) that is structurally related to, but somewhat larger than (with an N-terminal addition), the major coat protein (p22) (see Chapter 6, Section VIII.F.1.b). Agranovsky *et al.* (1995) probed BYV particles with coat protein (p22) antiserum and an antiserum to the N-terminal part of p24. A 75 nm segment at the 3′-end of the flexuous particle (Zinovkin *et al.*, 1999) was consistently labeled with both types of antibody and the N-terminal antiserum did not label the rest of the particle (Fig 5.12A–D), giving a 'rattlesnake' structure. A similar situation has been found with CTV (Febres *et al.*, 1996) (Fig 5.12E,F) and probably pertains to all closteroviruses; it is also found in the crinivirus LIYV (Tian *et al.*, 1999).

Mention should also be made of tenuiviruses that show folded, branched or coiled thread-like particles in electron micrographs (Toriyama, 1986b) (Fig 5.10F). Preparations of these particles are infectious.

IV. ASSEMBLY OF ROD-SHAPED VIRUSES

A. TMV

The pioneering work of Fraenkel-Conrat and Williams (1955) – who disassembled TMV into coat protein subunits and RNA by dialyzing virus preparations against alkaline buffers and separating the protein from the nucleic acid by ammonium sulfate precipitation—has led to much study of the factors involved in *in vitro* virus disassembly and reassembly. Disassembly of TMV in the plant is described in Chapter 7 (Section II.B).

1. Assembly of TMV coat protein

The protein monomer can aggregate in solution in various ways depending on pH, ionic strength and temperature. The major forms are summarized in Fig. 5.13. A kind of aggregate known as the stacked disk (not illustrated in Fig. 5.13) consists of three or more pairs of rings of subunits (Raghavendra *et al.*, 1986).

Experiments with MAbs show that both ends expose the same protein subunit surface in stacked disks. Thus, each two-layer unit in the stack must be bipolar (i.e. facing in opposite directions) (Dore *et al.*, 1989). A four-layered disk aggregate has been crystallized and its structure solved to atomic resolution (Diaz-Avalos and Caspar, 1998). This showed that it was made up of pairs of disks.

The existence of these various aggregates has been important both for our understanding of how the virus is assembled and also for the X-ray analysis that has led to a detailed understanding of the virus structure. The helical protein rods that are produced at low pH are of two kinds, one with $16\frac{1}{3}$ subunits per turn of the helix, as in the virus, and one with $17\frac{1}{3}$. In both of these forms the protein subunit structure is very similar to that in the virus. The RNA is replaced by at least one anion binding near a phosphate-binding site (Mandelkow *et al.*, 1981). Constraints on the permissible amino acid exchanges found in the coat proteins of different tobamoviruses are discussed in Chapter 2 (Section II.D).

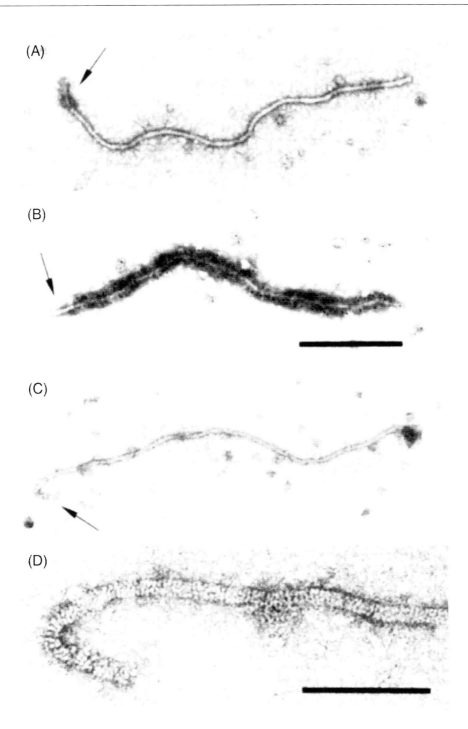

Fig. 5.12 'Rattlesnake' tails on closterovirus particles. **(A–D)** BYV particles. **(A)** Particle labeled with mouse polyclonal antiserum to the N-terminal peptide of BYV p24 protein. **(B)** Labeling with rabbit anti-BYV serum and immunogold labeling with secondary goat ant-rabbit 10-nm gold conjugate. **(C)** Unlabeled (uranyl acetate-stained) particle with distinct terminal structure. **(D)** The terminal structure shown in (C) but at a higher magnification. Arrows indicate distinct virus tail. Bars, A–C 300 nm, D 100 nm. From Agranovsky *et al.* (1995), with kind permission of the copyright holder, © National Academy of Sciences, USA.

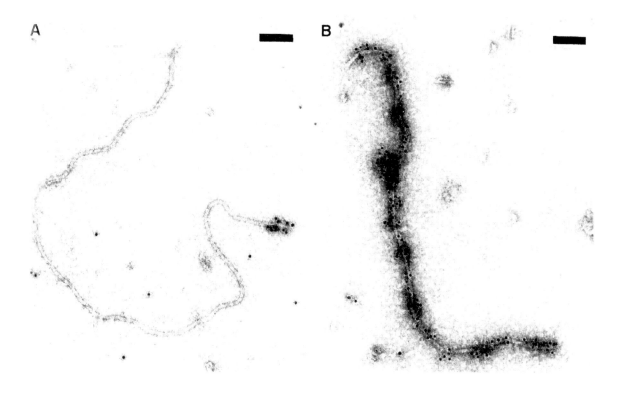

Fig. 5.12 cont. (E,F) 'Rattlesnake' tails on closterovirus particles. **(E)** Particle gold-labeled with p27 antibodies. **(F)** Particles gold-labeled with coat protein antibodies. Bars: 100 nm. From Febres *et al.* (1996), with kind permission of the copyright holder, © The American Phytopathological Society.

2. Assembly of the TMV rod

a. *Assembly* in vitro

In their classic experiments, Fraenkel-Conrat and Williams (1955) showed that it was possible to reassemble intact virus particles from TMV coat protein and TMV RNA. TMV RNA alone had an infectivity about 0.1% that of intact virus. Reconstitution of virus rods gave greatly increased specific infectivity (about 10–80% that of the native virus) and the infectivity was resistant to RNase attack. Since these early experiments, many workers have studied the mechanism of assembly of the virus rod, for there is considerable general interest in the problem. The three-dimensional structure of the coat protein is known in atomic detail (see Figs 5.6, 5.7 and 5.8) and the complete nucleotide sequence of several strains of the virus and related viruses is known. The system therefore provides a useful model for studying interactions during the formation of a macro-molecular assembly from protein and RNA.

There are four aspects of rod assembly to be considered: the site on the RNA where rod formation begins; the initial nucleating event that begins rod formation; rod extension in the 5' direction; and rod extension in the 3' direction. There is a general consensus concerning most of the details of the initiation site and the initial event, but the nature of the elongation processes remains somewhat controversial.

A central problem has been the fact that the coat protein monomer can exist in a variety of aggregation states, the existence of which is closely dependent on conditions in the medium (see Fig. 5.13). Equilibria exist between different aggregates so that, while one species may dominate under a given set of experimental conditions, others may be present in smaller amounts. The subject has been reviewed by Wilson and McNicol (1995), Butler (1999)

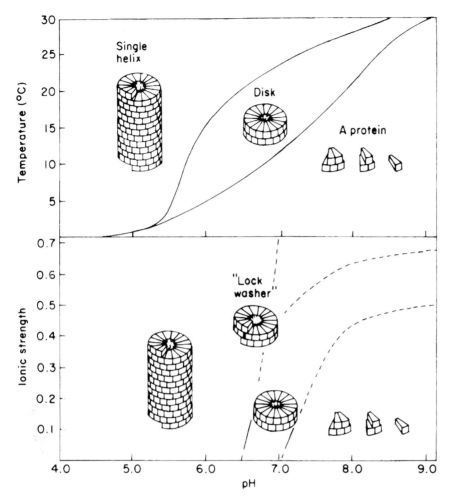

Fig. 5.13 Some aggregation states of TMV coat protein. Effects of pH, ionic strength and temperature. The helical rod of protein, the A protein, and the 20S disk have been well characterized (Champness *et al.*, 1976). The lockwasher is the proposed intermediate in the initiation of the assembly of TMV (Durham *et al.*, 1971) (see Fig. 5.15). Although the lockwasher form has never been isolated, the 'nicked' protein helices formed when the pH is lowered rapidly give direct support to the existence of such a form (Durham and Finch, 1972). Upper figure modified from Richards and Williams (1976); lower part modified from Durham *et al.* (1971), with permission.

and Klug (1999) and has the following major features.

i. The assembly origin in the RNA

Coat protein does not begin association with the RNA in a random manner. Zimmern and Wilson (1976) located the origin of assembly between 900 and 1300 nucleotides from the 3'-end. Butler *et al.* (1977) and Lebeurier *et al.* (1977) produced evidence showing that the longer RNA tail loops back down the axial hole of the rod. Lebeurier *et al.* used electron microscopy to show that, in partially assembled rods, a long and a short tail of RNA both protrude from one end of the rod. As the rod lengthens the longer tail disappears. The short tail is then incorporated, completing rod formation. These and other experiments (e.g. Otsuki *et al.*, 1977) demonstrated an internal

origin of assembly. The nucleotide sequence of the origin of assembly was established by various workers for the common strains of TMV (e.g. Zimmern, 1977; Jonard *et al.*, 1977). Nucleotide sequences near the initiating site can form quite extensive regions of internal base-pairing as is illustrated in Fig. 5.14. The origin of assembly is located elsewhere in some tobamoviruses and, when in the region that is present in the coat protein subgenomic mRNA, can give rise to short rods.

Loop 1 in Fig. 5.14, with the sequence AAGAAGUCG, combines first with a 20S aggregate of coat protein. Steckert and Schuster (1982) assayed the binding of 25 different trinucleoside diphosphates to polymerized TMV coat protein at low temperatures and low pH, and 5'-AAG-3' bound most strongly. Under conditions used for virus reconstitution, longer

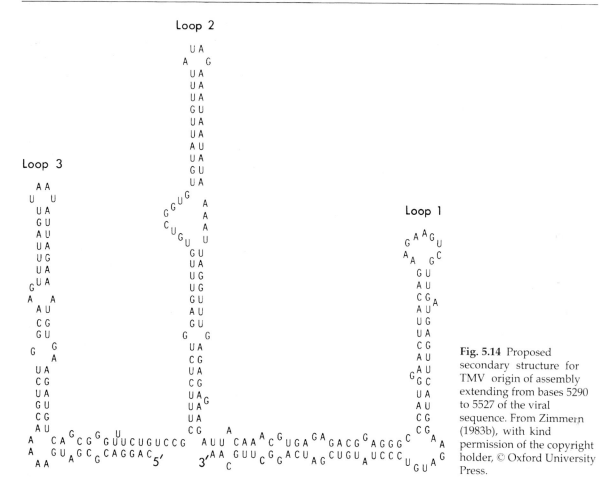

Fig. 5.14 Proposed secondary structure for TMV origin of assembly extending from bases 5290 to 5527 of the viral sequence. From Zimmern (1983b), with kind permission of the copyright holder, © Oxford University Press.

oligomers ((AAG)$_2$ or AAGAAGUUG) were required for strong binding (Turner *et al.*, 1986). The importance of various aspects of loop 1 has been established in more detail by site-directed mutagenesis (Turner and Butler, 1986; Turner *et al.*, 1988). These studies have shown that specific assembly initiation occurs in the absence of loops 2 and 3 of Fig. 5.14, but loop 1 is essential. Deletion or alteration of the unpaired sequence in loop 1 abolishes rapid packaging. The binding of loop 1 is mainly due to the regularly spaced G residues. The sequences (UUG)$_3$ and (GUG)$_3$ are as effective as the natural sequence. However, sequences such as (CCG)$_3$ and (CUG)$_3$ reduced the assembly initiation rate. Thus, there is some, but not complete, latitude in the bases in the first two positions.

Base sequence in the stem of loop 1 is not critical, except that shortening the stem reduces the rate of protein binding, as do

changes that alter the RNA folding close to the loop (Turner *et al.*, 1988). Overall stability of loop 1 is important because base changes that made the stem either more or less strongly base-paired were detrimental to protein binding. The small loop at the base of the stem is also important, but its phasing to the top loop is not critical.

The preceding discussion refers to the *vulgare* strain of TMV. Other tobamoviruses have somewhat different sequences in the stem-loop structure. From comparing these sequences, Okada (1986b) suggested a different target sequence in the loop, namely GAAGUUG. Some other tobamoviruses have the assembly origin in different sites on the genome. For instance, the assembly origin sites are in the coat protein gene of SHMV (Takamatsu *et al.*, 1983) and the sgRNA for the coat protein becomes encapsidated.

ii. The initial nucleating event

Since Butler and Klug (1971) showed that a 20S polymer of coat protein was responsible for initiating TMV rod assembly, the finding has been confirmed in various ways by many workers. The structure of the 20S double disk is known to atomic resolution (see Fig. 5.7). For many years it has been assumed that this was the configuration that initiated assembly, and that on interaction with the origin of assembly sequence in the RNA, the disk converts to a protohelical form. The structure in Fig. 5.7 was derived from 20S aggregates crystallized in solutions of high ionic strength, and it has been assumed that the double disk was also the favored structure under reconstitution conditions. Studies using sedimentation equilibrium (Correia et al., 1985), near-UV circular dichroism (Raghavendra et al., 1985), and electron microscopy (Raghavendra et al., 1986) under salt and temperature conditions used in rod reconstitution experiments suggested that the 20S species observed is a helical aggregate of 39 ± 2 subunits. However, using the rapid-freeze technique, it has been shown directly that under conditions in solution favoring most rapid TMV assembly, about 80% of the coat protein is in a structure that is compatible only with an aggregate of two rings (Butler, 1999). Thus the predominant structure in solution is likely to be very similar to that found in the crystal (Fig. 5.7).

Certainly, the model proposed by Champness et al. (1976) appeals on functional grounds. The disks can form an elongated helical rod of indefinite length at lower pH values. The transition between helix and double disk is mainly controlled by a switching mechanism involving abnormally titrating carboxyl groups. At low pH the protein can form a helix on its own because the carboxyl groups become protonated. When the protein is in the helical state either at low pH or in combination with RNA in the virus, the interlocking vertical helices are present. In this condition the abnormally titrating carboxyl groups are assumed to be forced together. Champness et al. proposed that the inner part of the two-layered disk acts as a pair of jaws (Fig. 5.7) to bind specifically to the origin-of-assembly loop in the RNA (Fig.

5.14), in the process converting each disk of the double disk into a 'lock washer' and forming a protohelix. A model for the early steps in the assembly of TMV is shown in Fig. 5.15.

iii. Rod extension in the 5' direction

Following initiation of rod assembly there is rapid growth of the rod in the 5' direction (e.g. Zimmern, 1977) 'pulling' the RNA through the central hole until about 300 nucleotides are coated. Zimmern (1983) has proposed a model for this initial rapid assembly involving two additional hairpin loops located 5' to the origin of assembly loop (Fig. 5.14). The spacing of these loops is consistent with the idea that they interact successively with three double disks. Various workers have isolated partially assembled rods of definite length classes. At least some of these classes are probably due to regions of RNA secondary structure that delay rod elongation at particular rod lengths (Godchaux and Schuster, 1987).

It is generally agreed that rod extension is faster in the 5' than in the 3' direction, but there has been considerable disagreement as to other aspects. Butler's group believed that the 20S aggregate is used in 5' extension and the A protein in the 3' direction with complete rods being formed in 5–7 minutes (see Fig. 5.13 for aggregation states of the coat protein monomer). Okada's group (e.g. Fukuda and Okada, 1985, 1987) claimed that 5' extension uses 4S protein while 3' extension uses 20S aggregates, with complete rods taking 40–60 minutes to form. Some of the earlier relevant experiments are discussed in detail by Butler (1984), Lomonossoff and Wilson (1985) and Okada (1986b). The experiments of Fukuda et al. (1978) show that for the strain of virus and assembly conditions they used, full-length rods were formed only after 40–60 minutes. Fukuda and Okada (1985) found that rod elongation in the 5' direction is complete in 5–7 minutes giving rise to a 260-nm rod, which subsequently elongates in the 3' direction to give the 300-nm rod. They used a Japanese strain of TMV and a higher ionic strength than Butler's group, which may account for some of the differences.

However, it is now generally accepted that under optimum assembly conditions, growth

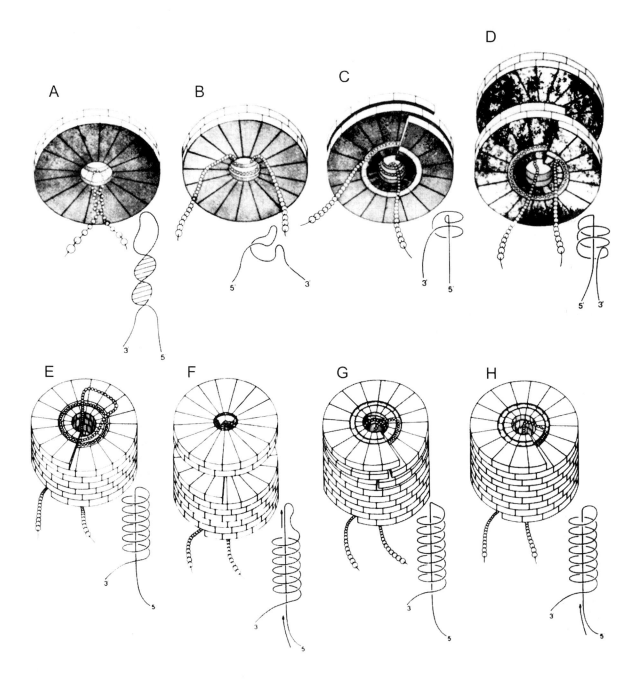

Fig. 5.15 Model for the assembly of TMV; **(A–C)** initiation; **(D–H)** elongation. **(A)** The hairpin loop inserts into the central hole of the 20S disk. This insertion is from the lower side of the disk. It is not yet apparent how the correct side for entry is chosen. **(B)** The loop opens up as it intercalates between the two layers of subunits. **(C)** This protein RNA interaction causes the disk to switch to the helical lockwasher form (a protohelix). Both RNA tails protrude from the same end. The lockwasher–RNA complex is the beginning of the helical rod. **(D)** A second double disk can add to the first on the side away from the RNA tails. As it does so it switches to helical form and two more turns of the RNA become entrapped. **(E–H)** Growth of the helical rod continues in the 5′ direction as the loop of RNA receives successive disks, and the 5′ tail of the RNA is drawn through the axial hole. In each drawing, the three-dimensional state of the RNA strand is indicated. (Courtesy of P. J. G. Butler.)

of the TMV rod in the 5′ direction is mainly by the addition of double disks, or sometimes single disks of coat protein (Turner *et al.*, 1989). They prepared *in vitro* RNA transcripts containing various heterologous non-viral RNAs 5′ to the TMV origin of assembly sequence, instead of the natural TMV sequence. There was no sequence 3′ to the origin of assembly. They then determined the lengths of RNA fragments protected from nuclease attack after allowing a short period for rod assembly. They found a series of lengths in steps of slightly over 100 nucleotides, or occasionally of just over 50 nucleotides. These are approximately the lengths expected to be protected by two (or one) turns of protein helix. They found such steps in length whatever heterologous RNA was used. This experiment rules out the possibility that the steps are due to some regularity in the RNA sequence that slows assembly at certain distances along the molecule.

There is uncertainty as to how the 5′ cap structure of the RNA is encapsidated. Disassembly by ribosomes (see Chapter 7, Section II.B.4) and *in vivo* would suggest that the structure at the extreme 5′-end might differ from that over most of the virus particle.

iv. Rod extension in the 3′ direction

Fairall *et al.* (1986) studied reassembly with RNAs that had been blocked at various sites by short lengths of hybridized and cross-linked cDNA probes. They found that even when 5′ extension was incomplete due to the blocked sequence, 3′ extension was completed. However, lengths of rods were determined after 20 minutes of incubation so the data are not relevant to the question of whether 3′ elongation is completed in 5–7 minutes.

Fukuda and Okada (1987) prepared an ss cDNA probe that extended from the origin of assembly to the 3′ terminus and that was complementary to TMV RNA. They used this to determine the length of rod extension in the 3′ direction. The results showed that significant rod extension in the 3′ direction did not occur at least in the first 4 minutes. It was first observed at 8 minutes and was still increasing between 15 and 40 minutes. At 4 minutes there was substantial encapsidation of RNA in the 5′ direction. Fukuda and Okada (1987) found a series of discrete sizes of RNA in the RNA-protected material in the 3′ direction and suggested that these differed by about 100 nucleotides in length. However, the results illustrated in their paper indicate a very uneven increment in length from about 55 to 135 nucleotides. This scatter in lengths may be due to the effects of the nucleotide sequence on assembly using A protein and on nuclease specificity rather than to the addition of double disks. Turner *et al.* (1989) carried out assembly experiments on RNA with heterologous sequences inserted 3′ to the TMV origin of assembly sequence. They found no evidence for banding in the protected RNA, giving strong support to earlier work indicating that extension of the rod in the 3′ direction is by the addition of small A protein aggregates.

The assembly of other strains of TMV and other tobamoviruses has not been studied as intensively as that of the type strain, but the available evidence indicates that the same basic assembly mechanism operates for all these viruses. However, the structures involved may differ in detail.

Studies on reassembly *in vitro* in which the TMV origin of assembly was embedded in various positions support the ideas that rod extension is much faster in the 5′ than the 3′ direction, that 5′ extension is probably completed before 3′ extension begins, and that the two extension reactions are different from each other (Gaddipatti and Siegel, 1990).

Artificial RNAs have been constructed that combine the TMV origin of assembly and the mRNA for a foreign gene. This RNA has been assembled into a rod with TMV coat protein and used to introduce the mRNA into plant cells.

b. *Assembly* in vivo

Very short rods have been seen in electron micrographs of infected leaf extracts, but the 20S aggregate has not been definitively established as occurring *in vivo*. Nevertheless, the following evidence shows that the process involved in the initiation of assembly outlined earlier is almost certainly used *in vivo*:

1. The conditions under which assembly occurs most efficiently *in vitro* (pH 7.0, 0.1 M ionic strength, and 20°C) can be regarded as reasonably physiological.
2. The correlation between the location of the origin of assembly in different tobamoviruses and the encapsulation, or not, of short rods containing the coat protein mRNA (Section III.B.2) strongly suggests that the origin of assembly found *in vitro* is used *in vivo*.
3. The mutant Ni2519, which is *ts* for viral assembly *in vivo*, has a single base change that is at position 5332 (Zimmern, 1983). This change weakens the secondary structure near the origin-of-assembly loop.
4. Tobacco plants transgenic for the CAT gene with the TMV origin of assembly inserted next to the 3' terminus give rise to RNA transcripts that can be assembled into virus-like rods with the TMV coat protein when the plants are systemically infected with TMV (Sleat *et al.*, 1988b).

It is known from *in vitro* experiments that TMV coat protein can form rods with other RNAs. *In vivo* there appears to be substantial specificity in that most rods formed contain the homologous RNA. *In vivo* this specificity may be due to (1) specific recognition of the correct RNA by the 20S disk, (2) rods assembled at an intracellular site where the homologous viral RNA predominates, and (3) coordinated RNA replication and virion assembly. Nevertheless, fidelity in *in vivo* assembly is not total (see Chapter 8, Section X).

There is no evidence that establishes the method by which the TMV rod elongates *in vivo*, but there is no reason to suppose that it differs from the mechanism that has been proposed for *in vitro* assembly.

B. Other rod-shaped viruses

There is much less understanding of the processes of virion assembly for other rod-shaped viruses. The assembly of both potexviruses (PapMV) and potyviruses (TVMV) is initiated at, or close to, the 5'-end of the RNA (Sit *et al.*, 1994; Wu and Shaw, 1998). The minimal sequence for *in vitro* initiation of assembly of PapMV is between 38 and 47 nucleotides from the 5' terminus (Sit *et al.*, 1994), a region rich in adenosine and cytidine residues, poor in uridine residues and lacking any discernable secondary structure. The initiation probably involves a 14S coat protein aggregate, thought to be a double-disk structure of nine subunits per disk, and proceeds in the 5' → 3' direction most probably by the addition of dimers and trimers of coat protein (reviewed in Abouhaidar and Erickson, 1985).

Potyvirus coat protein subunits also form stacked ring structures under certain conditions (McDonald *et al.*, 1976; Goodman *et al.*, 1976) that, on addition of viral RNA, yield virus-like particles that were shorter than native virus and were non-infectious (McDonald and Bancroft, 1977). The evidence that the origin of assembly is close to the 5'-end of the RNA came from *in vivo* studies using RNase protection (Wu and Shaw, 1998) that showed that assembly took place 30–45 minutes after inoculation of protoplasts with TVMV.

Potyvirus coat protein assembles into virus-like particles when expressed in *E. coli*, yeast and, mediated by recombinant vaccinia virus, in mammalian cells (Jagadish *et al.*, 1991; Hammond *et al.*, 1998). This gives a new tool to the studies of the factors involved in virion assembly. For instance, using expression in *E. coli*, Jagadish *et al.* (1993) showed that Arg[194] and Asp[238] of JGMV are required for the assembly process but were not necessarily involved in forming the predicted salt bridge.

V. ARCHITECTURE OF ISOMETRIC VIRUSES

A. Introduction

From crystallographic considerations, Crick and Watson concluded that the protein shell of a small 'spherical' virus could be constructed from identical protein subunits arranged with cubic symmetry, for which case the number of subunits would be a multiple of 12.

Crick and Watson (1956) pointed out that cubic symmetry was most likely to lead to an isometric virus particle. There are three types of cubic symmetry: tetrahedral (2:3), octahedral (4:3:2), and icosahedral (5:3:2). Thus, an icosahedron has 5-fold, 3-fold, and 2-fold rotational symmetry. For a virus particle, the three types of cubic symmetry would imply 12, 24 or 60 identical subunits arranged identically on the surface of a sphere. These subunits could be of any shape.

Klug and Caspar (1960) realized that many viruses probably had shells made up of subunits arranged with icosahedral symmetry. Horne and Wildy (1961) discussed possible models for the arrangement of protein subunits in icosahedral shells with 5:3:2 symmetry. They considered possible packing arrangements for clusters of five and six protein subunits. Pawley (1962) enumerated the plane groups that can be fitted on polyhedra, and he suggested that these may have application in the study of virus structure. Caspar and Klug (1962) further developed the principles of virus construction, particularly with respect to the isometric viruses based on icosahedral structures. A shell made up of many small identical protein molecules makes most efficient use of a virus's genetic material.

Fig. 5.16 shows a regular icosahedron (20 faces). With three units in identical positions on each face, this icosahedron gives 60 identical subunits. This is the largest number of subunits that can be located in identical positions in an isometric shell. Some viruses have this structure, but many have much larger numbers of subunits, so that not all subunits can be in identical environments.

B. Quasi-equivalence

Caspar and Klug (1962), with their theory of quasi-equivalence, laid the basis for our further understanding of the way in which the shells of many smaller isometric viruses are constructed of more than 60 identical subunits. In general terms, they assumed that not all the chemical subunits in the shell need be arrayed in a strictly mathematically equivalent way, but

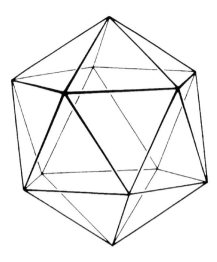

Fig. 5.16 The regular icosahedron. This solid has 12 vertices with 5-fold rotational symmetry; the center of each triangular face is on a 3-fold symmetry axis, and the midpoint of each edge is on a 2-fold symmetry axis. There are 20 identical triangular faces. Three structural units of any shape can be placed in identical positions on each face, giving 60 structural units. Some of the smallest viruses have 60 subunits arranged in this way.

only quasi-equivalently. They also assumed that the shell is held together by the same type of bonds throughout, but that the bonds may be deformed in slightly different ways in different non-symmetry-related environments. They calculated that the degree of deformation necessary would be physically acceptable. Quasi-equivalence would occur in all icosahedra except the basic structure (Fig. 5.16). Basically, pentamers give the three-dimensional curvature, 12 giving a closed icosahedron, whereas hexamers give a two-dimensional curvature (or tubular structure). Thus, the conceptual basis of quasi-equivalence is the interchangeable formation of hexamers and pentamers by the same protein subunit.

However, more recent detailed information on the arrangement of subunits in the shells of small viruses has altered the original view of quasi-equivalence (see Johnson and Speir, 1997). A series of molecular switches, often involving a segment of 10–30 residues of the subunit polypeptide, have been identified. The switches enable the subunit polypeptide to show icosahedral symmetry at some interfaces and to be disordered at similar, but not symmetrically equivalent, interfaces. The switches include specific segments of the subunit polypeptides, ss or ds RNA and divalent cations. For instance, viruses such as TBSV have multi-domain subunits that, to a large extent, adjust to the different symmetry-related positions in the shell by means other than distortion of inter-subunit bonds (Section VI.B.4.d). Even particles made up of 60 identical subunits may have some variation in the detailed structure of the subunits, possibly associated by their interactions with the genomic nucleic acid (Chapman, 1998).

Other viruses have evolved different variations on the icosahedral theme. For example, comoviruses (Section VI.B.6.a) and members of the *Reoviridae* (Section VII) have different polypeptides in different symmetry environments within the shell. Bacilliform virus particles, such as those of AMV and badnaviruses, are also based on icosahedral symmetry (Sections V.F, VI.B.2.a and VI.B.5).

C. Possible icosahedra

Caspar and Klug (1962) enumerated all the possible icosahedral surface lattices and the number of structural subunits involved. The basic icosahedron (Fig. 5.16), with 20 _ 3 = 60 structural subunits, can be sub-triangulated according to the formula:

$$T = P (f^2,$$

where *T* is called the 'triangulation number'.

a. The meaning of parameter f

The basic triangular face can be subdivided by lines joining equally spaced divisions on each side (diagram i). Thus, *f* is the number of subdivisions of each side and is the number of smaller triangles formed.

b. The meaning of parameter P

There is another way in which sub-triangulation can be made and this is represented by *P*. It is easier to consider a plane network of equilateral triangles (diagram ii). Such a sheet can

be folded down to give the basic icosahedron by cutting out one triangle from a hexagon (e.g. cross-hatching) and then joining the cut edges to give a vertex with 5-fold symmetry.

However, if each vertex is joined to another by a line not passing through the nearest vertex, other triangulations of the surface are obtained. In the simplest case the 'next but one' vertices are joined (diagram iii). This gives a new array of equilateral triangles. This plane net can be folded to give the solid in Fig. 5.17C by removing one triangle from each of the original vertices (e.g. the shaded triangle) and then folding in to give a vertex with 5-fold symmetry.

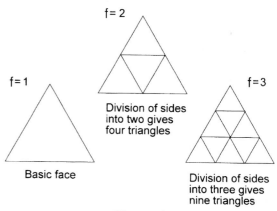

Diagram i

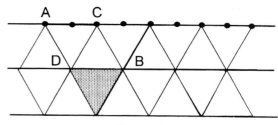

Diagram ii

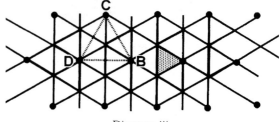

Diagram iii

It can be shown by simple trigonometry that each of the small triangles has one-third the area of the original faces. This can be seen by inspection by noting that there are six new half-triangles within one original face (dashed lines CBD on diagram iii). In this example $P = 3$. In general:

$$P = h^2 + hk + k^2,$$

where h and k are any integers having no common factor.

- For $h = 1$ and $k = 0$, $P = 1$.
- For $h = 1$ and $k = 1$, $P = 3$.
- For $h = 2$ and $k = 1$, $P = 7$.

With $p \geq 7$, the icosahedra are skew, and right-handed and left-handed versions are possible. The physical meaning of h and k in a virus structure is illustrated in the following section.

Since each of the triangles formed with the P parameter can be further subdivided into f^2 smaller triangles, T gives the total number of subdivisions of the original faces and $20T$ the total number of triangles. Fig. 5.17 gives some examples.

Thus, the number of structural subunits in an icosahedral shell is $20 \times 3 \times T = 60T$.

D. Clustering of subunits

The actual detailed structure of the virus surface will depend on how the physical subunits are packed together. For example, three clustering possibilities for the basic icosahedron are shown in diagram iv.

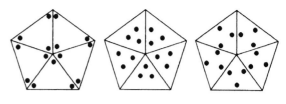

Diagram iv

In fact many smaller plant viruses are based on the $P = 3$, $f = 1$, $T = 3$ icosahedron. In this structure, the structural subunits are commonly clustered about the vertices to give pentamers and hexamers of the subunits. These are the morphological subunits seen in electron micrographs of negatively stained particles (e.g. Fig. 5.30B).

Since there are always 12 vertices with 5-fold symmetry in icosahedra, we can calculate the number of morphological subunits (M) (assuming clustering into pentamers and hexamers) as follows:

$$M = [(60T - 60)/6] \text{ hexamers}$$
$$+ (60/5) \text{ pentamers}$$
$$= 10(T - 1) \text{ hexamers} + 12 \text{ pentamers}$$

In photographs of virus particles where the pentamers and hexamers can be unambiguously recognized (e.g. one-sided images of negatively stained particles or freeze-fracture replicas of the outer faces of larger viruses), the parameters h and k might be used to establish the icosahedral class of the particle. This procedure would be particularly useful for shells containing large numbers of hexamers. h and k represent the numbers of hexamers that

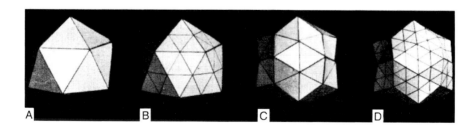

Fig. 5.17 Ways of sub-triangulation of the basic icosahedron shown in Fig, 5.16 to give a series of deltahedra with icosahedral symmetry (icosadeltahedra). **(A)** The basic icosahedron, with $T = 1$ ($P = 1$, $f = 1$). **(B)** With $T = 4$ ($P = 1$, $f = 2$). **(C)** With $T = 3$ ($P = 3$, $f = 1$). **(D)** With $T = 12$ ($P = 3$, $f = 2$). From Caspar and Klug (1962), with permission.

must be traversed to move by the shortest route from one pentamer to the next. Thus, we must identify two adjacent pentamers. For example, the structure shown in diagram v was observed on the surface of freeze-etch replicas of phage λ (Bayer and Bocharov, 1973).

$$h = 4, k = 1, T = 21$$

Diagram v

This indicates a skew icosahedron with 'right-handed' skewness.

E. 'True' and 'quasi' symmetries

In the basic icosahedron (Fig. 5.16), a feature located half way along any edge of a triangular face is positioned on an axis of rotation for the whole solid. Thus, it is on a 'true' or icosahedral symmetry axis. This is a true dyad. In any more complex icosahedron ($T > 1$) there is more than one kind of 2-fold symmetry. For example, in a $P = 3$, $T = 3$ shell (Fig. 5.17C), the center of one edge on each of the 60 triangular faces is on a true dyad axis relating to the solid as a whole. The center positions of the other two edges of a face have only local 2-fold symmetry. These are called 'quasi' dyads.

On the 3' axis of the 'quasi' symmetry, the three chemically identical but structurally independent subunits in an icosahedral asymmetric unit are designated A, B and C (Harrison *et al.*, 1978) (Fig. 5.18).

F. Bacilliform particles

Some virus particles, for instance those of AMV and badnaviruses, are bacilliform with rounded ends separated by a tubular section. Hull (1976a) suggested that the structure of these particles is based on icosahedral symmetry. The rounded ends would have constraints

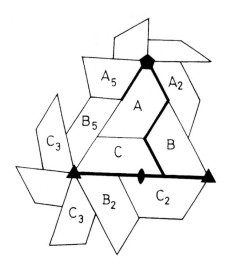

Fig. 5.18 Arrangement of protein subunits as found in several T = 3 plant and animal viruses. The nomenclature for the chemically identical subunits follows Harrison *et al.* (1978). From Krishna *et al.* (1999), with permission.

of icosahedra with the three-dimensional curvature determined by 12 pentamers, six at each end. The two-dimensional structure of the tubular section would be made up of hexamers. Hull derived various hexamer structures from icosahedra cut across the 2-fold, 3-fold, 5-fold and interlattice axes.

VI. SMALL ICOSAHEDRAL VIRUSES

A. Subunit structure

The coat protein subunits of most small icosahedral viruses are in the range of 20–40 kDa; some are larger but fold to give effective 'pseudomolecules' within this range (Section VI.B.6). In contrast to rod-shaped viruses, the subunits of most small icosahedral viruses have a relatively high proportion of β-sheet structure and a low proportion of α-helix (Denloye *et al.*, 1978; Odumosu *et al.*, 1981) and have the same basic structure. This comprises an eight-stranded antiparallel β sandwich, often termed a β barrel or 'jellyroll', which is shown schematically in Fig. 5.19.

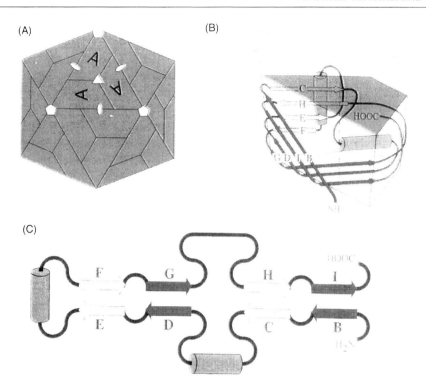

Fig. 5.19 (see Plate 5.1) (A) The icosahedral capsid contains 60 identical copies of the protein subunit – in blue, labeled A. These are related by 5-fold symmetry elements (pentagons at vertices), 3-fold symmetry elements (triangles in faces) and 2-fold symmetry elements (ellipses at edges) – shown in yellow. For a given-sized subunit, this point group symmetry generates the largest possible assembly (60 subunits) in which every protein lies in an identical environment. **(B)** Schematic of the subunit building block found in many RNA, and some DNA, viral structures. Such subunits have complementary interfacial surfaces which, when they repeatedly interact, lead to the symmetry of the icosahedron. The tertiary structure of the subunit is an eight-stranded β-barrel with the topology of the jellyroll – see part (C). In this diagram the β-strand and helix coding are identical to that in (C). Subunit sizes generally range between 20 and 40 kDa with variation between different viruses occurring at the N and C termini and in the size of insertions between strands of β-sheet. These insertions generally do not occur at the narrow end of the wedge (B–C, H–I, D–E and F–G turns). **(C)** The topology of viral β-barrel showing the connections between strands of the sheets (represented by fat arrows) and positions of the insertions between strands. The green cylinders represent helices that are usually conserved. The C–D, E–F and G–H loops often contain large insertions. From Johnson and Spier (1999), with permission.

The overall shape is a three-dimensional wedge with the B–C, H–I, D–E and F–G turns being at the narrow (interior) end. Most variation between the subunit sizes occurs at the N and C termini and between the strands of β sheet at the broad end of the subunit.

It is the detailed positioning of the elements of the β barrel and of the N and C termini that give the flexibility to overcome the quasi-equivalence problems. This is illustrated in the detailed structures described below.

The coat protein subunits form one, two or three structural domains, the S (shell) domain, the R domain (random-binding) and the P (pro-truding) domain. All viruses have the S domain. The R domain is somewhat of a misnomer but defines an N-terminal region of the polypeptide chain that associates with the viral RNA. As it is random, no structure can be determined by X-ray crystallography. The P domain gives surface protuberances on some viruses.

B. Virion structure

At present, we can distinguish seven kinds of structure among the protein shells of small icosahedral or icosahedra-based plant viruses whose architecture has been studied in sufficient detail. These are: $T = 1$ particles; bacilliform particles based on $T = 1$; geminate particles based on $T = 1$; $T = 3$ particles; bacilliform particles based on $T = 3$; pseudo $T = 3$ particles; and $T = 7$ particles.

1. $T = 1$ particles (satellite viruses)

The satellite viruses are the smallest known plant viruses having a particle diameter of about 17 nm and capsids made up of 17–21-kDa polypeptides (see Chapter 14; Section II.A, for satellite viruses).

The structure of STNV was the first to be solved and shown to be made of 60 protein subunits of 21.3 kDa arranged in a $T = 1$ icosahedral surface lattice. The structure of the protein subunit has been determined crystallographically with refinement to 2.5 Å resolution (Jones and Liljàs, 1984). The general topology of the polypeptide chain is like that of the S domains of TBSV and SBMV (Fig. 5.20) but the packing of the subunits in the $T = 1$ icosahedral structure is clearly different (Rossmann et al., 1983) and there is no P domain.

In the amino terminus of STNV, there are only 11 disordered residues followed by an ordered helical section (residues 12–22) buried in the RNA. Three different sets of metal ion binding sites (probably Ca^{2+}) have been located. These link the protein subunits together. A more detailed structure for the protein shell has been proposed by Montelius et al. (1988).

The crystal structure of STNV has been studied at 16 Å resolution using neutron diffraction in H_2O/D_2O (Bentley et al., 1987). At 40% D_2O, scattering arises largely from the RNA component. The two main RNA motifs are shown in Fig. 5.21.

These are in fact connected by regions of weaker RNA density. Each spherical motif (II) is connected to five symmetry-related extended motifs (I). They form a continuous network of RNA density on the inside surface of the protein coat. The I motifs form the edges of an icosahedron, leaving triangular holes centerd on the 3-fold axes. It is into these holes that triple-helical arms of amino-terminal regions of the protein subunits penetrate, and make close

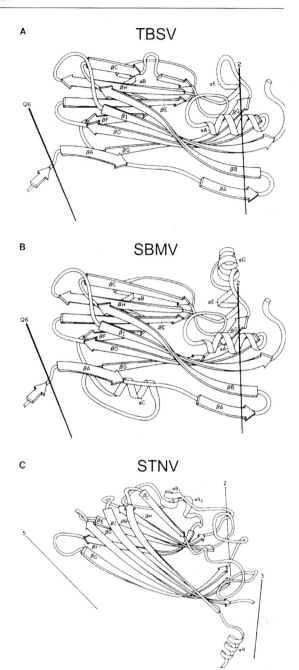

Fig. 5.20 Diagrammatic representation of the backbone folding of the coat protein of **(A)** TBSV, **(B)** SBMV, and **(C)** STNV shown in roughly comparable orientations. From Rossmann et al. (1983), with permission.

contact with the RNA. Basic amino acids are well placed to make contact with the RNA.

Except for this protein—RNA interface, the inner face of the protein shell is separated from

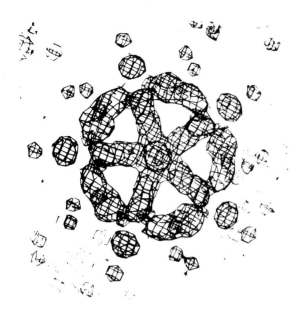

Fig. 5.21 Low-resolution structure of the RNA within STNV determined by neutron diffraction. Positive density at 40% D_2O looking down the 5-fold axis from the center of the virus. I is the RNA density motif, which lies along the edges of each triangle and forms the edges of an icosahedron. Its length is about 45 Å and its diameter is 22–25 Å. The RNA density of motif II lies along each 5-fold axis at a distance of 67 Å from the virus center. Minor regions of density (III) at higher radii correspond to positive fluctuations of protein density. See Bentley *et al.* (1987) for details of I, II and III. From Bentley *et al.* (1987), with permission.

the RNA by a thin layer of solvent. The cross-section of motif I fits well with the idea that it represents a double helix in the RNA. If this is so, then 72% of the total RNA would be in double-helical form. The fact that STNV RNA has thermal denaturation kinetics like that of a transfer RNA—indicating a high degree of secondary structure (Mossop and Francki, 1979a) – confirms this.

The structures of two other satellite viruses have been resolved, STMV and SPMV (Larson *et al.*, 1993a,b; Ban and McPherson, 1995) and have been compared to that of STNV (Ban *et al.*, 1995). In spite of all having the β-barrel structure in the subunit that the 5-fold contacts at the narrow end, the three viruses were remarkably different in the arrangements of the secondary structural elements. Also, the 5-fold protein

interactions are organized by Ca^{2+} in STNV, an anion in STMV and apparently neither of these in SPMV. Finally, nucleic acid was visible only in electron density maps of STMV and showed as double-helical RNA segments associated with each coat protein dimer.

2. Bacilliform particles based on $T = 1$ symmetry

As noted in Section V.F, it has been proposed that the structure of bacilliform particles is based on icosahedral symmetry.

a. Alfamovirus *and* Ilarvirus *genera*

Purified preparations of AMV contain four nucleoprotein components present in major amounts (bottom, B; middle, M; top b, Tb; and top a, Ta). They each contain an RNA species of definite length. The genome is split between the Tb, M and B RNAs. Three of the four major components are bacilliform particles, 19 nm in diameter. The fourth (Ta) is normally spheroidal with a diameter slightly larger than 19 nm (Fig. 5.22).

However, two forms of Ta component have been recognized (Heijtink and Jaspars, 1976). Ta^t is spheroidal and soluble in 0.3 M $MgSO_4$. Ta^b is a rodlet and insoluble in 0.3 M $MgSO_4$. These two particles appear identical in other properties. The Ta components have 120 protein subunits, and Cusack *et al.* (1983) raised the possibility that these may have a non-icosahedral structure.

From a careful study of the molecular weights of the RNAs, the protein subunit and the virus particles, Heijtink *et al.* (1977) concluded that the number of coat protein monomers in the four major components is equal to $60 + (n \times 18)$, n being 10, 7, 5 or 4.

As noted above (Section V.F), it has been proposed that such particles are based on icosahedra cut across various axes with the tubular portion made up of hexamer subunits. From optical diffraction studies on electron micrographs of negatively stained AMV particles, Hull *et al.* (1969b) suggested that the tubular structure of the bacilliform particles was based on a $T = 1$ icosahedron cut across its 3-fold axis with rings of three hexamers (18 coat protein monomers) forming the tubular portion (Fig 5.23).

Proton magnetic resonance studies provide a low-resolution model in which the coat protein consists of a rigid core and a flexible amino-terminal part of about 36 amino acid residues (Kan *et al.*, 1982). The protein behaves as a water-soluble dimer stabilized by hydrophobic interactions between the two molecules. This dimer is the morphological unit out of which the viral shells are constructed.

Under appropriate conditions of ionic strength, ionic species, pH, temperature and protein concentration, the protein dimer forms a $T = 1$ icosahedral structure built from 30 dimers (Fig. 5.24) (Driedonks *et al.*, 1977). This structure has been confirmed by X-ray crystallographic analysis (Fukuyama *et al.*, 1983) at 4.5 Å resolution that was further refined to 4.0 Å resolution together with cryoelectron microscopy and image reconstruction (Kumar

et al., 1997). This showed that the subunit structure and dimer association is structurally similar to CCMV. Large holes are observed at the pentamer axes, giving a porous particle structure.

The virus is unstable with regard to high ionic strength and SDS, and is sensitive to RNase, which might be explained by the holes in the protein coat. Conformational changes occur at mildly alkaline pH (Verhagen *et al.*, 1976). AMV nucleoproteins are readily dissociated into protein and RNA at high salt concentrations, and bacilliform particles can be reformed under appropriate conditions. Thus, the virus is mainly stabilized by protein–RNA interactions (van Vloten-Doting and Jaspars, 1977). Under a wide range of solvent conditions, AMV particles do not show the phenomenon of swelling described below for bromoviruses (Oostergetel *et al.*, 1981). Compared with the small isometric viruses,

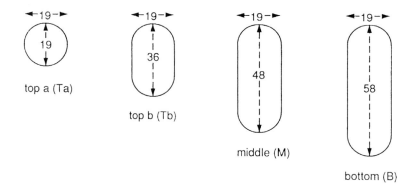

top a (Ta)

top b (Tb)

middle (M)

bottom (B)

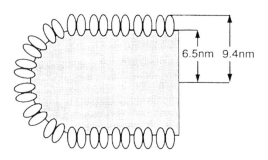

6.5nm 9.4nm

Fig. 5.22 AMV particles. **Top:** Sizes (in nm) of the four main classes of particle. **Bottom:** Schematic of the distribution of protein and RNA in AMV bottom component. RNA is indicated by the tinted area and the protein molecules are represented by ellipsoids. The model is derived from the analysis of both the 30S and the bottom component by small-angle neutron scattering. Bottom part reprinted from Cusack *et al.* (1981), with permission.

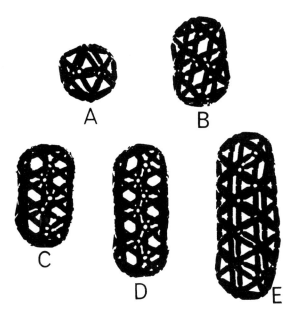

Fig. 5.23 Geodestix models showing the proposed structure of the components of AMV. **(A)** Top$_z$, (53S) component; **(B)** top$_a$ component; **(C)** top$_b$ component; **(D)** middle component; **(E)** bottom component. From Hull *et al.* (1969b), with permission.

Fig. 5.24 $T = 1$ of the 30S AMV particle composed of 30 dimers of coat protein. The 2-fold symmetry axes of the dimers coincide with the dyad positions of an icosahedron leaving large holes at 5- and 3-fold positions. For clarity, the rear-facing subunits are omitted from the model. From Driedonks *et al.* (1977), with permission.

AMV may be regarded as being in a permanently swollen state and thus resembles CMV.

Purified preparations of some AMV strains can be fractionated by polyacrylamide gel electrophoresis to reveal the presence of at least 17 nucleoprotein components (Bol and Lak-Kaashoek, 1974). These are all made up from the single viral coat protein (van Beynum *et al.*, 1977). Besides the four major RNA species, there are at least 10 minor RNA species of different lengths. The nucleoproteins occurring in minor amounts contain both major and minor species of RNA. Particles of somewhat different size may contain the same RNA, while particles of the same size may contain different RNAs. Thus, a small variation is possible in the amount of RNA encapsulated by a given amount of viral protein (Bol and Lak-Kaashoek, 1974). Although AMV is labile in gradients of CsCl, it is stable and bands isopycnically in gradients of Cs_2SO_4 and metrizamide (Hull, 1976b). In these gradients it forms two bands at very similar densities, the major band containing mainly bottom component and the minor band predominantly the other components. This indicates that the components have very similar, but not identical, protein:RNA ratios.

Some AMV strains (e.g. 15/64 and VRU) form unusually long particles (Fig 5.25) (Heijtink and Jaspars, 1974; Hull, 1970a) with particles up to 1 μm long; these particles do not contain any abnormally long RNA molecules and most likely contain several molecules of the genome segments.

Reconstitution experiments indicated that the tendency to form long particles is coat protein-directed (Hull, 1970a). In a comparison of the coat proteins of normal-length particle strains with those of long particle strains, Thole *et al.* (1998) identified two amino acid substitutions, Ser[66] and Leu[175], that were associated with the long particles. When these amino acid alterations were introduced into a normal-length particle strain, long particles were formed.

Ilarviruses such as TSV have quasi-isometric or occasionally bacilliform particles of four different size classes (van Vloten-Doting, 1975) that appear to share many properties in common with the AMV group (van Vloten-Doting,

1976). The top component of TSV has been crystallized to give a hexagonal space group, but the crystals were not amenable to X-ray diffraction as they were disordered (Senke and Johnson, 1993).

b. Ourmiaviruses

Members of the *Ourmiavirus* genus have bacilliform virions of 18 nm diameter and of three lengths, 30, 37 and 45.5 nm (Lisa *et al.*, 1988). The protein subunits cluster into dimers or trimers. The pointed ends may be formed from icosahedra cut through 3-fold axes for a dimer or 2-fold axes for a trimer. The tubular body does not form a continuous geometrical net as with AMV, but contains discontinuities marked by fissures between double disks of the protein (Fig. 5.26). Particles consisting of two, three, four or six double disks have been observed, with four- and six-disk particles being rare. Figure 5.26 also illustrates these structures diagrammatically.

3. Other particles based on T = 1 symmetry (geminiviruses)

Geminiviruses contain ssDNA and one type of coat polypeptide. Particles in purified preparations consist of twinned or geminate icosahedra (Fig. 5.27).

From a study of negatively stained particles together with models of possible structures, Francki *et al.* (1980b) suggested that CSMV consists of two *T* = 1 icosahedra joined together at a site where one morphological subunit is missing from each, giving a total of 22 morphological units in the geminate particle. This structure received support from a study of the size of the DNA and protein subunit of this virus, indicating that each geminate particle shell, consisting of 110 polypeptides of 28 kDa arranged in 22 morphological units, contains one molecule of ssDNA with molecular weight 7.1×10^5 (about 2.1 kb).

The structure of the coat protein of MSV has been deduced by modeling it on the atomic coordinates of STNV coat protein as a template

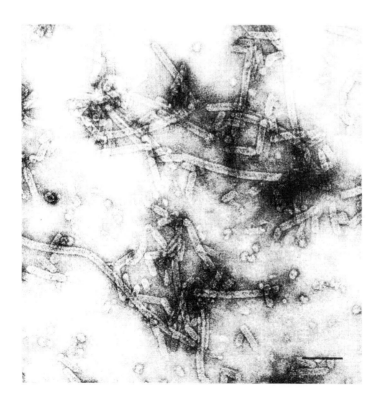

Fig. 5.25 Electron micrograph of unfractionated preparation of VRU strain of AMV negatively stained in saturated uranyl acetate. Bar = 100 nm. From Hull (1970a), with permission.

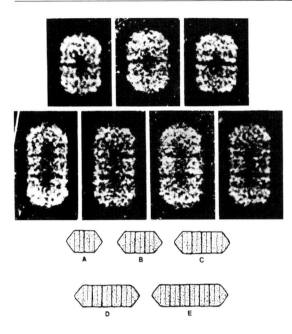

Fig. 5.26 Structure of Ourmia melon mosaic virus. **Top:** Averaged images of negatively stained particles. Each image was built up photographically by equal superimposed exposures of 10 original particle images. The top row has two double disks and the bottom row has three. **Bottom:** Sketches showing the suggested arrangement of double disks in particles of different length. Particles of type D have not yet been observed. (Courtesy of R. G. Milne.)

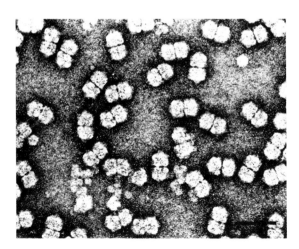

Fig. 5.27 Purified *geminivirus* virus from *Digitaria* negatively stained in 2% aqueous uranyl acetate. Bar = 50 nm. From Dollet *et al.* (1986), with permission.

motif (Zhang *et al.*, 2001) (Fig. 5.28). The fine structure of MSV particles was determined by cryoelectron microscopy and three-dimensional image reconstruction (Zhang *et al.*, 2001) (Fig. 5.28) and confirmed the model suggested by Francki *et al.* (1980b).

4. T = 3 particles

a. Tymovirus genus

The particles of tymoviruses are about 30 nm in diameter and made up of subunits of about 20 kDa.

i. Classes of particle

Purified TYMV preparations can be fractionated on CsCl density gradients into a large number of components. There are three classes of particle.

1. The empty protein shell. About one-third to one-fifth of the particles found in a TYMV preparation isolated from infected leaves are empty protein shells (the top or T component). These contain no RNA but otherwise are identical in structure to the protein shell of the infectious virus (B_1 component).

Full particles can be converted to empty ones by freeze-thaw or by alkali treatment. When TYMV B_1 is taken to pH 11.6 in 1 M KCl, the particles swell from 14.6 to 15.2 nm radius within 30 seconds (Keeling *et al.*, 1979). RNA escapes from these particles in 3–10 minutes in a partially degraded state. In addition, an amount of protein is lost within 1–3 minutes that is equivalent to the loss of one pentamer or hexamer of subunits from each particle. No such loss of protein occurs with the minor nucleoproteins (Keeling and Matthews, 1982). On return to pH 7.0, the radius of the resultant empty shells returns to normal. The nucleoproteins containing less than the full genome RNA do not swell at pH 11.6 under the same conditions, and their RNA does not escape, although it is also degraded within the particle.

2. Infectious virus nucleoprotein (B or bottom components and particles derived from them). The infectious virus fractionates to form two density classes in CsCl gradients (B_{1a} and B_{1b}) that are equally infectious (Matthews, 1974). A third B_1 fraction, B_{1c}, more dense than B_{1b}, has been

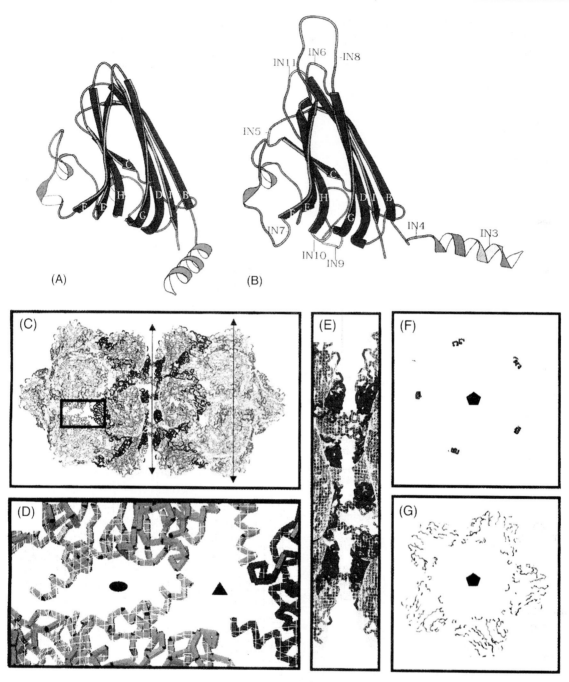

Fig. 5.28 (see Plate 5.2) Models of MSV-N coat protein and complete capsid. Ribbon drawings of polypeptide chain of **(A)** STNV coat protein and **(B)** MSV-N coat protein. In these drawings β-strands are labeled according to convention. **(C)** Fit of 110 copies of the MSV-N pseudo-atomic model, shown as Cα tracings (in yellow), the two apical capsomeres, 10 peripentonal capsomeres, and 10 equatorial capsomeres. The rectangular boxed region indicates the position of close-up views of the models shown in (D) and (E) and the double-ended arrows the cross-sectional views shown in (F) and (G). **(D)** Close-up view of the 2- and 3-fold icosahedral and **(E)** equatorial capsomere interactions. Cross-sectional slices of the MSV-N particle **(F)** through the equatorial region and **(G)** peripentonal capsomeres of the particle. The icosahedral 2-, 3- and 5-fold axes are indicated as a black ellipsoid, triangle and pentamer respectively. From Zhang *et al.* (2001), with permission.

characterized. B_{1b} and B_{1c} both contain copies of the coat protein mRNA as well as a molecule of genome RNA. These B_1 components can be converted in strong solutions of CsCl to a B_2 series with higher densities, especially if the pH is above 6.5. These are designated B_{2a}, B_{2b} and B_{2c}. Their formation is prevented by the presence of 0.1 M $MgCl_2$ in the CsCl.

3. *Nucleoprotein particles containing subgenomic RNAs and having densities in CsCl intermediate between that of the T and B components.* Mellema et al. (1979) and Keeling et al. (1979) isolated a series of eight minor components. The coat mRNA and a series of eight other subgenomic RNAs of discrete size have been isolated from these particles. These eight subgenomic RNAs have not yet been firmly allocated to particular nucleoprotein species. If the coat mRNA and the eight others are encapsidated in various combinations, and numbers of copies per particle, there may in fact be a very large number of particles of slightly differing density.

Some properties of the minor noninfective nucleoprotein fractions have been determined by Mellema et al. (1979) and Keeling et al. (1979). Their RNA content ranges from about 5% for fraction 1 to 28% for fraction 8. The proportion of total minor nucleoproteins relative to the infectious nucleoproteins is about 5% on a particle number basis. The coat protein cistron is found in most of the minor nucleoproteins along with other subgenomic RNAs (Pleij et al., 1977; Higgins et al., 1978).

ii. The protein shell

Using negative-staining procedures, Huxley and Zubay (1960) and Nixon and Gibbs (1960) showed that the protein shell of TYMV is made up of 32 protuberances, occupying two structurally distinct sites in the shell. Klug and colleagues used TYMV extensively in developing X-ray diffraction and electron microscopy as tools for the study of smaller isometric viruses. Klug et al. (1966) and Finch and Klug (1966) concluded that the protein shell has 180 scattering centers lying at a radius of about 14.5 nm. These points were identified with protuberances of the protein structure units at the surface of the particle. A higher resolution X-ray

analysis to 3.2 Å was made by Canady et al. (1995, 1996).

Each individual protein subunit is somewhat banana-shaped. Within the intact virus, each protein subunit is made up of about 9% α-helix and 43% β-sheet. About 48% of the polypeptide is in an irregular conformation (Hartman et al., 1978). This conclusion was broadly confirmed by Tamburro et al. (1978). The polypeptide chain forms an eight-stranded antiparallel β sandwich (β barrel) (Fig. 5.29) (Canady et al., 1996).

The X-ray data on the virus gave good agreement for a model of the protein shell with 32 scattering centers lying at a radius of about 121 Å and extending to a radius of 159 Å from the center of the particle. Figure 5.30A summarizes the knowledge of the external arrangement of the protein subunits, while Fig. 5.30B shows views of these particles in three different orientations obtained by the three-dimensional image reconstruction technique. Thus, the surface of the particle has the subunits in a $T = 3$ arrangement with the pentameric and hexameric protein aggregates protruding from the surface and forming deep valleys at the quasi 3-fold axes.

The N-terminal 26 residues of the A subunit are disordered, whereas those of the B and C subunits interact around the interior of the quasi 6-fold cluster where they form an annulus (Fig. 5.31).

There are extensive internal contacts between the A, B and C subunits (Fig. 5.32). The three histidine residues of each protein subunit are positioned to the interior and are accessible for interaction with the RNA genome. The appearance of the interior surface of the virus capsid suggests that a pentameric subunit is lost during virus disassembly.

The structure of the tymovirus PhyMV has been resolved to 3.0 Å (Krishna et al. 1999). The basic structure is similar to that of TYMV but there are some differences. The N-terminal 17 residues of the A subunits making up the 12 pentamers show order that is not seen in TYMV and have a conformation that is very different from that observed in the B and C subunits constituting the hexameric capsomeres. Analysis of interfacial contacts indicates that

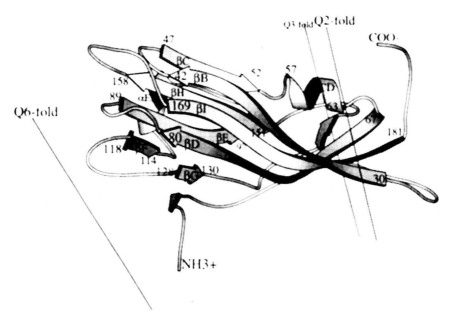

Fig. 5.29 Ribbon diagram of the TYMV coat protein with orientations of the quasi-sixfold, quasi-threefold and quasi-twofold axes indicated. The B-subunit is shown; the A- and C-subunits are virtually the same except that the N-terminal 26 residues are disordered in the A-subunit. The eight strands of the β-barrel are labeled, along with the N and C termini. Helix CD, which is a regular α-helix, is indicated. The EF helix, appearing behind, is irregular, with some qualities of a 3_{10}-helix. Two small segments of β-sheet are also formed by residues 6–8 and 140–142. From Canady *et al.* (1996), with permission.

hexamers are held together more strongly than pentamers and that hexamer–hexamer contacts are stronger than pentamer–hexamer contacts. These observations suggested an explanation for the formation of empty capsids which might be initiated by a change in the conformation of the A-subunit N-terminal arm.

iii. Location of the RNA

Finch and Klug (1966) considered that folds of the RNA in TYMV were intimately associated with each of the 32 morphological protein units. They thought that it was the presence of this RNA in and around these positions that enhanced the appearance of the 32 morphological subunits seen in electron micrographs of this virus, as compared with empty protein shells. However, neutron small-angle scattering shows that there is very little penetration of the RNA into the protein shell and that the protein subunits are densely packed. Comparison of cryoelectron microscopy images of full and empty capsids of TYMV identified strong inner features around the 3-fold axes of the full, but

not the empty, particles (Bottcher and Crowther, 1996). This suggested that substantial parts of the RNA are icosahedrally ordered. The X-ray analysis indicates density attributable to RNA to form a core or radius ~40 Å with a hollow at its center (radius ~25 Å) and pronounced projections extending to a radius of ~75 Å along the 5-fold directions (Canady *et al.*, 1996).

b. Bromovirus genus

The protein shell of these viruses is 25–28 nm in diameter, made up of 180 protein subunits of 19.4 kDa. The three genomic RNAs (1, 2 and 3) are packaged separately; the subgenomic RNA for coat protein (RNA4) is packaged with RNA3.

i. Stability of the virus

Although bromoviruses are very similar to tymoviruses in the arrangement of their viral RNA and protein shells, they have much less stable particles. As the pH is raised between 6.0 and 7.0, particles of BMV swell from a radius of

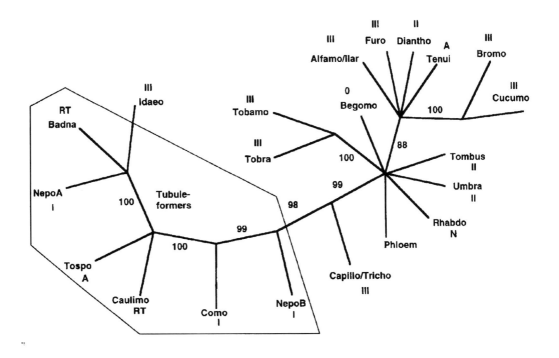

Fig. 9.12 Relationships among putative 30-kDa MP superfamily members determined by bootstrappin Branches with less than 80% support (100 bootstrap replicates) were collapsed. Labels 0, RT, N, A, I, II and the types of polymerase encoded by the viruses, which are respectively: 0; RNA-dependent DNA polymer strand virus; ambisense-strand virus; and positive-strand supergroups I, II and III RNA-dependent RNA (see Chapter 8, Section IV.B.1). The thin-lined polygon encloses those MPs known to form virion-bearing Melcher (2000), with permission.

A model for the two stages, intra- and intercellular transport, is shown in Fig. 9.13.

At the early stage, the virus replicates on the endoplasmic reticulum, leading to viral RNA accumulating in large irregular bodies. P30 co-localizes with this viral RNA, suggesting that viral replication and translation occur at the same subcellular site. The viral RNA and P30 co-localize with microtubules which are thought to direct the RNA–P30 complex to the plasmodesmata, which the P30 gates open to allow the complex to pass through (the intercellular transport stage). The mechanisms by which the MP targets and gates plasmodesmata are unknown, but it is interesting to note that P30 binds specifically to the cell wall enzyme, pectin methylesterase, which is also an RNA binding protein (Dorokhov et al., 1999). This gating is only temporary at the infection front, and behind the front plasmodesmata

appear to return to their normal S virus infection of protoplasts, high weight forms of P30 have been de accumulation of which was enhan inhibition of the 26S proteasome (I Beachy, 2000). It was suggested ubiquinated P30 is degraded by the somes. It has also been suggeste phosphorylation could act as a me sequester P30 in cell walls and in function (Citovsky et al., 1993); in thaliana P30 is N-terminally processe thought to also inactivate it (Hu; 1995).

The similarities between the RCNMV noted above, and the fac TMV, RCNMV does not require vir; tein for cell-to-cell movement (Xi 1993), suggests that these two viruse ilar local movement strategies.

13.5 nm to over 15 nm (Zulauf, 1977). The consequences of this swelling depend on many factors and particularly on the ionic conditions. The particles are readily disrupted in 1 M NaCl at pH 7.0. When swollen virus is dissociated, the protein subunits can reassemble under a variety of conditions to form a range of products, described by Bancroft and Horne (1977) (see Section IX.A). Of particular interest is the fact that, in the presence of trypsin, the coat protein of BMV loses 63 amino acids from the amino terminus, and can then self-assemble into a $T = 1$ empty shell (Cuillel *et al.*, 1981). BMV can be fully dissociated into subunits by high pressures, the formation of $T = 1$ particles being a step in the disassembly process (Silva and Weber, 1988). Only swollen particles disassemble under high pressure (Leimkühler *et al.*, 2000).

The structural polymorphism of BMV has been investigated using neutron small-angle scattering (Jacrot *et al.*, 1976; Chauvin *et al.*, 1978). At low pH (around 5.0) the virus is in a compact state under a range of ionic condi-

tions. Near pH 7.0 at moderate ionic strength, the virus particle swells and the RNA penetrates more deeply between the protein subunits; Mg^{2+} suppresses this effect.

Titration studies have shown that BMV contains several (probably two) cation-binding sites per subunit. Both sites titrate together at about pH 6.7. It is probable that both carboxylate groups of the protein and RNA phosphates contribute to this binding (Pfeiffer and Durham, 1977). When both sites bind H^+, Ca^{2+} or Mg^{2+}, the virus is compact. As these ions are released, the virus swells.

The N-terminal 25 amino acids of the coat protein are rich in basic amino acids, and structure prediction methods indicate that this sequence may interact in a helical form with the RNA (Argos, 1981) thus forming an R domain.

ii. Particle structure

Finch *et al.* (1967b) found that the morphological units of BBMV protrude at least 1.5 nm from the body of the particle. The negative stain

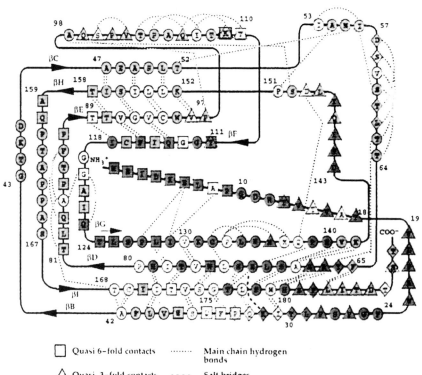

Quasi 6–fold contacts
Quasi 3–fold contacts
Quasi 2–fold contacts

Main chain hydrogen bonds
Salt bridges

Fig. 5.32 (see Plate 5.4) Hydrogen bonding, intersubunit contacts, and accessibility diagram for the B-subunit of TYMV. Atoms were assumed in contact if they were 4.11 Å apart for van der Waals interactions, 3.3 Å for hydrogen bonds, and 3.8 Å for salt bridges. Residues were considered inaccessible if they had less than 5 Å² accessible surface area. From Canady *et al.* (1996), with kind permission of the copyright holder, © John Wiley and Sons Ltd.

appeared to penetrate into the center of the virus particles, suggesting the presence of an appreciable central hole, which is not found in TYMV. The presence of this hole, about 5.5 nm in radius, was confirmed by X-ray diffraction studies (Finch *et al.*, 1967b), and for BMV by small-angle neutron scattering. Finch and Klug (1967) suggested that the absence of a central hole in TYMV may be due to the need to pack a higher proportion of RNA into the particle.

A detailed analysis has been made of the structure of native and swollen particles of CCMV using X-ray crystallography (to 3.2 Å) and cryoelectron microscopy (Speir et al., 1993, 1995). The polypeptide chains of the coat protein subunits are arranged in β barrels with the C-terminal regions of adjacent subunits being interwoven. Additional particle stability is provided by contacts between metal ion (primarily Ca^{2+})-mediated carboxyl cages on each subunit and by protein interactions with regions of ordered RNA. Swelling of the particle results in a 29 Å radial expansion due to electrostatic repulsion at the carboxylate cages. Complete disassembly of the particle is prevented by preservation of the interwoven C termini and by the protein–RNA interactions.

iii. Location of the RNA

In a cryoelectron microscopy study of CCMV, Fox *et al.* (1998) concluded that RNAs 1, 2 and 3+4 were packaged in a similar manner against the interior surface of the virion shell. The viral RNA appeared to have an ordered conformation at each of the quasi-threefold axes.

c. Cucumovirus *genus*

Purified preparations of cucumoviruses contain four RNA species housed in three particles of 30 nm diameter in the same arrangement as the bromoviruses. Some isolates of cucumoviruses have associated with them a small satellite RNA (see Chapter 14, Section II.B.2).

Using electron microscope methods similar to those employed in their study of BBMV, Finch *et al.* (1967a) showed that CMV resembled bromoviruses, both in surface structure and in the fact that there is a central hole in the particle.

There is substantial penetration of the RNA into the shell of protein (Jacrot *et al.*, 1977). The packing of the protein subunits is such that about 15% of the surface (at a radius of 11.7 nm) could be made up of holes, which would expose the RNA to inactivating agents and could explain the sensitivity of this virus to RNase.

Cryoelectron microscopy and reconstruction to 23 Å resolution shows that CMV is structurally similar to CCMV (Wikoff *et al.*, 1997). The CMV structure was confirmed by X-ray crystallography at 8 Å resolution, which also showed that the coat protein subunits have a β-barrel structure. Thus, CMV and CCMV particle structures are similar in (1) particle morphology, (2) size and orientation of their β-barrels, (3) stabilizing interactions for hexamer formation, and (4) subunit primary sequence.

However, CMV has relatively unstable particles when compared with those of bromoviruses. CMV particles do not swell at pH 7.0 under conditions where those of bromoviruses do—or in reality, they do not 'shrink' at pHs below 7. Thus, they behave essentially as swollen particles stabilized primarily by protein:RNA interactions.

d. Tombusviridae *family*

The 30-nm diameter particles of tombusviruses are composed of subunits of 38–43 kDa encapsidating a single species of genomic RNA.

The structure of TBSV has been determined crystallographically to a resolution of 2.9 Å (Harrison *et al.*, 1978; Olson *et al.*, 1983) and those of TCV and CarMV to 3.2 Å (Hogle *et al.*, 1986; Morgunova *et al.*, 1994). Each virus contains 180 protein subunits arranged to form a $T = 3$ icosahedral surface lattice, with prominent dimer clustering at the outside of the particle, the clusters extending to a radius of about 17 nm.

TBSV, CarMV and TCV have very similar structures. The essential features of the structure are summarized in Fig. 5.33. The two distinct globular P and S domains of the protein subunit are connected by a flexible hinge involving five amino acids (Fig. 5.33B).

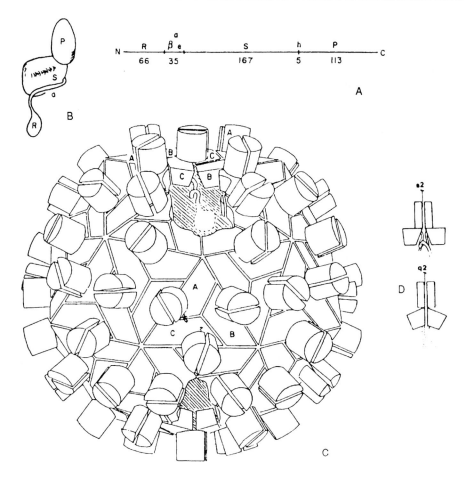

Fig. 5.33 Architecture of TBSV particle. **(A)** Order of domains in polypeptide chain from N terminus to C terminus. The number of residues in each segment is indicated below the line. The letters indicate R-domain (possible RNA binding region), arm 'a' (the connector that forms the β-annulus and extended arm structure on C-subunits and that remains disordered on A- and B-subunits), S-domain, hinge, and P-domain. **(B)** Schematic of folded polypeptide chain, showing P-, S- and R-domains. **(C)** Arrangement of subunits in particle. Here, labels A, B and C denote distinct packing environments for the subunit (see Fig. 5.18). S-domains of A-subunits pack around 5-fold axes; S-domains of B- and C-subunits alternate around 3-fold axes. The differences in local curvature can be seen at the two places where the shell has been cut away to reveal S-domain packing near strict (top) and quasi (bottom) dyads. **(D)** The two states of the TBSV subunit found in this structure, viewed as dimers about the strict (s2) and local (q2) 2-fold axes. Subunits in C positions have the interdomain hinge 'up' and a cleft between 2-fold-related S-domains into which fold parts of the N-terminal arms. Subunits in the quasi-twofold-related A and B positions have hinge 'down', S-domains abutting, and a disordered arm. Parts (A)–(C) from Olson *et al.* (1983), with permission; (D) from Harrison *et al.* (1978), with kind permission of the copyright holder, © Macmillan Magazines Ltd.

Mutations of the hinge of TCV do not affect movement of the virus through the plant but some alter the symptoms induced by the virus (Lin and Heaton, 1999). Each P domain forms one-half of the dimer-clustered protrusions on the surface of the particle occupying approximately one-third of one icosahedral triangular face. The S domain forms part of the icosahedral shell which is about 3 nm thick and from it protrude the 90 dimer clusters formed by the P domain pairs. In addition to domains P and S, each protein subunit has a flexibly linked N-terminal arm containing 102 amino acid residues comprising the R domain and the connecting arm called 'arm a' (Fig. 5.33B).

The coat protein assumes two different large-scale conformational states in the shell that differ in the angle between domains P and S by about 20 degrees (Fig. 5.33D). The conformation taken up depends on whether the subunit is near a quasi dyad or a true dyad in the $T = 3$ surface lattice. The state of the flexible N-terminal arm also depends on the symmetry position.

The N-terminal arms originating near a strict dyad follow each other in the cleft between two adjacent S domains. On reaching a 3-fold axis, such an arm winds around the axis in an anticlockwise fashion (viewed from outside the particle). Two other N-terminal arms originating at neighboring strict dyads will be at each 3-fold axis. The three polypeptides overlap with each other to form a circular structure called the β-annulus, made up of 19 amino acids from each arm, around the 3-fold axis (Fig. 5.34). Thus 60 of the 180 N-terminal arms create an interlocking network that, in principle, could form an open $T = 1$ structure without the other 120 subunits. The R domain and the 'a' connecting arm arising from the 120 subunits at quasi-dyad positions hang down into the interior of the particle in an irregular way so that their detailed position cannot be derived by X-ray analysis. The RNA of TBSV is tightly packed within the particle, and these N-terminal polypeptide arms very probably interact with the RNA.

Two divalent cation binding sites are a feature of each of the three trimer contacts between the S domains (Fig. 5.34). If the divalent cations are removed above pH 7.0 the particle swells. Mutation of the calcium-binding sites of TCV can affect cell-to-cell or long-distance movement or induce delayed mild systemic symptoms (Laasko and Heaton, 1993; Lin and Heaton, 1999).

The crystal structure of TNV has been resolved to 2.25 Å (Oda et al., 2000). The basic features are similar to other $T = 3$ icosahedral viruses, but in detail the tertiary and quaternary of this virus resembles sobemoviruses more closely than *Tombusviridae*.

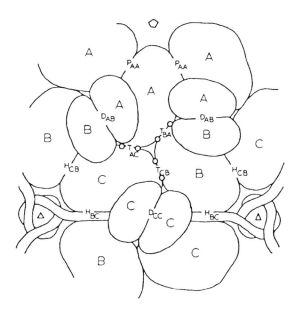

Fig. 5.34 Packing of subunits in the icosahedral asymmetric unit of TBSV and notation for interfaces. The different kinds of subunit contact are labeled D (dimer), T (trimer), P (pentamer), and H (hexamer), with a subscript showing the types of subunit interacting across the contact in question. The positions of Ca^{2+} at the T interfaces are shown by small circles. A 5-fold axis is indicated at the top of the diagram. The β-annulus with 3-fold symmetry is shown twice in the lower part of the diagram. From Olson *et al.* (1983), with permission.

e. Sobemovirus *genus*

The particles of sobemoviruses are 30 nm in diameter and are made up of subunits of about 30 kDa surrounding a single species of RNA. The virus particle shows little surface detail in negatively stained preparations.

The structure of SBMV has been determined at a resolution of 2.8 Å (Abad-Zapatero *et al.*, 1980; Silva and Rossmann, 1987). The folding of the S domain of the protein subunit is very similar to that of TBSV (Fig. 5.20B). SBMV resembles tombusviruses in having an amino-terminal R domain but differs in that it lacks the P domain. This lack of the P domain accounts for the smaller subunit (260 amino acids compared with 380), the smaller outside radius, and the smoother appearance of virus particles in electron micrographs.

The arrangement of the three quasi-equivalent subunits (A, B and C) is very similar to that in TBSV. As with TBSV, part of the R domain of the C subunits is in an ordered state forming a β-annulus around each icosahedral 3-fold axis. The R domain is shorter than that in TBSV. The first 64 residues of the A and B subunits (the R domain) and the first 38 residues of the C subunit are disordered and associated with the RNA. Cation binding sites (Ca^{2+}) are near the external surface of the shell, the virus being strongly dependent on Mg^{2+} and Ca^{2+} for its structural integrity. About 200 of these ions are bound firmly to each virus particle, and their removal with EDTA causes a reversible destabilization of the structure. Treatment with trypsin removes the N-terminal 61 amino acids from the isolated protein subunit. The resulting 22-kDa fragment can assemble into a 17.5-nm diameter $T = 1$ particle (Erickson and Rossmann, 1982). Rossmann (1985) gives a detailed account of SBMV structure.

Using electron cryomicroscopy and image reconstruction, Opalka et al. (2000) showed that SCPMV has an exterior face of deep valleys along the 2-fold axes and protrusions at the quasi-threefold axes, whereas the surface of RYMV is comparatively smooth. Particles of both viruses display two concentric shells of density beneath the capsid layer which are interpreted as ordered layers of genomic RNA.

A structural study on Sesbania mosaic virus, which is closely related to SBMV, at 3.0 Å showed that, although the basic structures were similar, there were some differences (Bhuvaneshwari et al., 1995). The polar interactions at the quasi-threefold axes are substantially less in Sesbania mosaic virus, and the positively charged residues on the RNA-facing side of the subunits and in the N-terminal arm (R domain) are not well conserved, suggesting that the protein:RNA interactions differ between the two viruses.

f. Pea enation mosaic

Pea enation mosaic disease is caused by a complex of two viruses, PEMV-1 and PEMV-2 (see Chapter 2, Section III.P.3) the genomes of which are encapsidated in the same coat protein, encoded by PEMV-1. The complex has two particle sizes, that encapsidating RNA-1 being about 28 nm in diameter and that containing RNA-2 about 25 nm in diameter. PEMV-1 is stable in CsCl but PEMV-2 is labile. Both the particles band at the same density in Cs_2SO_4 and in sucrose made up in D_2O (Hull and Lane, 1973; Hull, 1976b), indicating that they have very similar nucleic acid content. Based on estimates of the molecular weight of the particles and the coat protein, Hull and Lane (1973) suggested that PEMV-1 particles had 180 subunits consistent with a $T = 3$ icosahedral structure, but that PEMV-2 particles had 150 subunits. A quasi-icosahedral model has been proposed for PEMV-2 (Hull, 1977d).

Particles of PEMV-1 and PEMV-2 can be separated on electrophoresis in polyacrylamide gels (Hull and Lane, 1973). Some viral strains that have lost their aphid transmissibility form two homogeneous bands in gels; but in others that are aphid-transmitted, the band formed by PEMV-1 is very heterogeneous (Hull, 1977d) (see Fig. 2.7). The reason for this heterogeneity is unknown.

5. Bacilliform based on $T = 3$ symmetry

As noted in Section V.F, it has been suggested that the structure of bacilliform particles is based on icosahedral symmetry. RTBV has bacilliform particles of 130×30 nm. From the size of the coat protein and the diameter of the particle, it was proposed that these particles could be based on $T = 3$ icosahedral symmetry (Hull, 1996). Optical diffraction of electron micrographs of RTBV particles indicated that the size of the morphological subunit was 100 Å and the structure was based on a $T = 3$ icosahedron cut across its 3-fold axis.

6. Pseudo $T = 3$ symmetry

Comoviruses, and most likely nepoviruses, fabaviruses and sequiviruses, have icosahedral particles formed of one, two or three coat protein species. The structure resembles that of picornaviruses in that if all the coat protein species are considered as one protein the symmetry would appear to be $T = 1$. However, the larger polypeptides of viruses with one or two protein species form to give 'pseudomolecules' and for each virus the structure can be considered to be made up of effectively three 'species'. This then gives a pseudo $T = 3$ symmetry.

a. Comovirus *genus*

CPMV has a diameter of about 28 nm and an icosahedral structure with an unusual arrangement of subunits. The virus shell contains two proteins, a large and small one (42 and 22 kDa) with different amino acid compositions (Wu and Bruening, 1971). Purified preparations of CPMV contain three classes of particle (B component containing the larger RNA, M component with the smaller RNA, and T component being empty particles), with identical protein shells, that can be separated by centrifugation. Early studies on this virus were complicated by the essentially trivial fact that the smaller protein is susceptible *in situ* to attack by proteolytic enzymes both in the host plant and after isolation of the virus. CPMV contains about 200 spermidine molecules and a trace of spermine per particle.

From the sizes of the protein and the virus shell, Geelen *et al.* (1972) calculated that there should be 60 of each of the two structural proteins in the shell. Using this information and data they obtained from three-dimensional image reconstruction, Crowther *et al.* (1974) proposed a model in which the 60 larger proteins are clustered at the 12 positions with 5-fold symmetry; the 60 smaller proteins form 20 clusters about the positions with 3-fold symmetry (Fig. 5.35).

Although the particle of CPMV is clearly a $T = 1$ icosahedral structure, the fact that it has two proteins in different symmetry environments gives it some overall resemblance to a $T = 3$ structure (Schmidt and Johnson, 1983). The structure of CPMV has been solved by X-ray crystallography to 3.5 Å (Stauffacher *et al.*, 1985) and refined to 2.8 Å (Lin *et al.*, 1999) and shows that the two coat proteins produce three distinct β-barrel domains in the icosahedral asymmetric unit (Fig. 5.36). Although two of these β-barrels are covalently linked, their relationship is essentially like the β-barrels found in the three major separate coat proteins of the picornaviruses.

A similar structure has been deduced for BPMV (Chen *et al.*, 1989) and RCMV (Lin *et al.*,

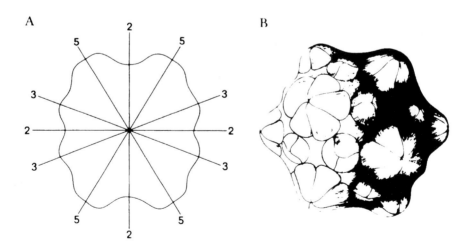

Fig. 5.35 Structure of CPMV based on image reconstruction and on known chemical composition. **(A)** Relation between symmetry axes and contour at the virus surface. Trace of the equator of the image reconstruction in a plane normal to a 2-fold axis. The positions of the 2-, 3- and 5-fold axes lying in the plane are indicated. The radii of the contour at the 5-, 3- and 2-fold positions are approximately 12.0, 10.6 and 9.3 nm respectively. The indentations at the 2-fold positions are approximately twice as long (between two 5-fold axes) as they are wide (between two 3-fold axes). **(B)** Schematic of the model. The large protein subunits are clustered about the 5-fold positions while the small subunits are clustered about the 3-fold positions. From Crowther *et al.* (1974), with permission.

(A)

(B)

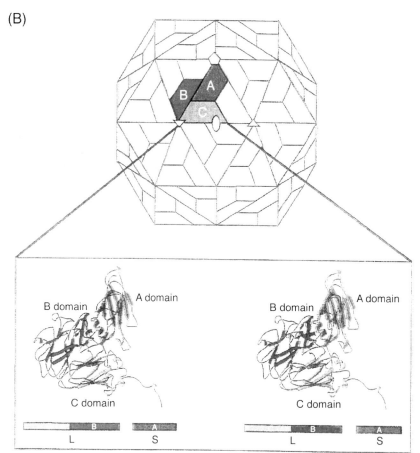

Fig. 5.36 (see Plate 5.5) The structures of viral capsid and icosahedral asymmetric unit of RCMV. Two proteins (S- and L-subunits) are in the RCMV capsid. The S-subunit forms the A domain (in blue), while the L-subunit forms the B (red) and C (green) domains. **(A)** Stereoview of a space-filling drawing of the RCMV capsid. All atoms are shown as spheres corresponding to a diameter of 1.8 Å. The pentameric S-subunits form the protrusion. **(B)** At the top, the icosahedral asymmetric unit of the capsid is color-coded in the schematic presentation of the CPMV capsid. The S-subunit occupies the A position, forming the A domain around the 5-fold axis; the two domains of the L-subunit occupy the B and C positions. Positions A, B and C are quasi-equivalent positions of identical gene products on a $T = 3$ surface lattice. At the bottom, a stereoview of a ribbon diagram of the icosahedral asymmetric unit is shown. All three domains are variants of the jellyroll β-sandwich structure. The schematic presentation of composite proteins, L- and S-subunits, is also shown. From Lin *et al.* (2000), with permission.

2000), the latter to 2.4 Å resolution. In comparing the structures of CPMV, BPMV and RCMV, Lin *et al.* (2000) identified a structural fingerprint at the N terminus of the small subunit that allowed subgrouping of comoviruses.

Nearly 20% of the RNA in BPMV particles binds to the interior of the protein shell in a manner displaying icosahedral symmetry. The RNA that binds is single-stranded, and interactions with the protein are dominated by non-bonding forces with few specific contacts (Chen *et al.*, 1989). From resolving the protein—RNA interactions in BPMV at 3.0 Å resolution, Chen *et al.* (1989) suggested that seven ribonucleotides that can be seen as ordered in the structure lie in a shallow pocket on the inner surface of the protein shell formed by the two covalently linked domains of the large coat protein.

b. Waikavirus *genus*

RTSV has isometric particles 30 nm in diameter on which no obvious structural features can be seen by electron microscopy. The three coat protein species are processed from a polyprotein and are present in the capsid in equimolar amounts (Druka *et al.*, 1996).

7. $T = 7$ particles (caulimoviruses)

CaMV has a very stable isometric particle about 50 nm in diameter containing dsDNA. The circular dsDNA is encapsidated in subunits of a protein processed from a 58-kDa precursor to several products, the major ones being approximately 37 kDa and 42 kDa. There is also evidence that minor amounts of the product of ORF III are present in the viral coat (see Chapter 11, Section III.F). Electron microscopy shows a relatively smooth protein shell with no structural features. Neutron diffraction studies revealed a central empty cavity, about 25 nm in diameter (Chauvin *et al.*, 1979; Krüse *et al.*, 1987) and suggested a structure of four concentric shells. The particle consists of an outer protein shell. Within this is a zone where both DNA and protein are present. Most of the DNA

is in a zone with very little protein, while the central region of the particle (one-eighth of the volume) is occupied only by solvent. The DNA does not appear to be associated with any significant amount of histone-like protein (Al Ani *et al.*, 1979b). Exposure of the virus to pH 11.25 leads to the release of DNA tails without total disruption of the protein shell (Al Ani *et al.*, 1979a).

Calculations from the molecular weight of the coat protein and the amount of protein in the virus suggested that it may have a $T = 7$ icosahedral structure (Krüse *et al.*, 1987). Using cryoelectron microscopy and image reconstruction procedures, the structure of CaMV was solved to about 3 nm (Cheng *et al.*, 1992). These studies showed that CaMV particles are composed of three concentric layers of solvent-excluding density surrounding a large (*c* 27-nm diameter) solvent-filled cavity (Fig. 5.37).

The outer layer (I) comprises 12 pentameric units and 60 hexameric units arranged in a $T = 7$ icosahedral symmetry, this being the first example of a $T = 7$ virus that obeys the icosahedral rules. The dsDNA genome is distributed in layers II and III together with some of the capsid protein.

C. The arrangement of nucleic acid within icosahedral viruses

For most icosahedral viruses, it is not possible to gain a detailed picture of the arrangement of the nucleic acid within the particles, as it does not form an ordered structure that can be identified by current techniques. In some cases,

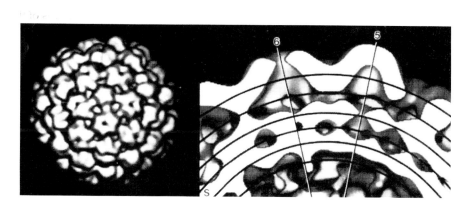

Fig. 5.37 Left: Reconstruction of CaMV surface structure showing $T = 7$ symmetry. **Right:** Cutaway surface reconstruction showing multiplayer structure. From Cheng *et al.* (1992), with permission.

there is some ordered structure that has been noted under the specific virus. However, there are some general points that can be made concerning the arrangement of nucleic acid within these particles.

1. Percentage of double-helical structure

It is probable that the RNA inside many small icosahedral viruses has some double-helical structure. For example, Haselkorn (1962) concluded from hypochromicity studies that at least two-thirds of the bases in TYMV were in an ordered structure. Laser Raman spectroscopy indicated that about 77% of the RNA was in α-helical form (Hartman et al., 1978). Circular dichroism studies confirmed that the RNA has a considerable amount of base-pairing and/or base-stacking, and that the configuration of isolated RNA is very similar to that inside the virus (Tamburro et al., 1978).

From hypochromicity experiments on intact CMV and isolated RNA, Kaper et al. (1965) concluded that the RNA inside the virus has a high degree of secondary structure, equivalent to a helical content of about 70%.

Circular dichroism studies on TRoV suggested that the RNA within the particle has considerable base-pairing (Denloye et al., 1978). In CCMV, the RNA has a high degree (c 95%) of ordered secondary structure as determined by laser Raman spectroscopy (Verduin et al., 1984).

Magnetically induced birefringence in solution for several isometric plant viruses indicated that at least part of the RNA core has a symmetry differing from that of the protein shell (Torbet, 1983). ^{31}P nuclear magnetic resonance studies showed that at 10°C the RNA within particles of AMV lacked movement, but that at 25°C there was some mobility in this component (Kan et al., 1987).

2. Contributions from base sequence

The question as to whether the base sequence in an RNA plays a role in the folding of RNA within an icosahedral virus has been approached by Yamamoto and Yoshikura (1986) using a computer program for calculating RNA secondary structure. Calculations for 20 viral RNAs indicated that genomes of icosahedral viruses had higher folding probabilities than those of helical viruses. When the folding probability of a viral sequence was compared with that of a random sequence of the same base composition, the viral sequences were more folded. These results suggest that base sequence plays some part in the way in which ssRNA genomes fold within the virus.

3. Interactions between RNA and protein in small isometric viruses

As has already been indicated in earlier sections, current knowledge suggests that there may be two types of RNA:protein interaction in the small isometric viruses, depending on whether basic amino acid side-chains or polyamines neutralize charged phosphates of the RNA.

a. Viruses in which basic amino acids bind RNA phosphates

The best studied virus in which basic amino acids bind RNA phosphates is TBSV (Olson et al., 1983). A total of 21 lysine and arginine residues are available to bind phosphates (11 on the R domain, 4 on the arm, and 6 on the inner surface of the S domain). Since there are approximately 26 nucleotides in the RNA for each protein subunit, most of the RNA charges can be neutralized by these amino acids. The flexible connection of the R arm allows it to conform to irregularities in RNA packing.

A similar situation exists for the SBMV (Silva and Rossmann, 1987). STNV has a shorter amino-terminal arm associated with the RNA (Jones and Liljàs, 1984). The positive charges on the inner surface of the protein shell and the amino-terminal arm are sufficient to neutralize 70% of the RNA phosphates. The remaining 30% are probably neutralized by bound Mg^{2+} ions (Liljàs et al., 1982).

BMV coat protein has a basic amino-terminal region that is predicted to interact in a helical form with the RNA (Argos, 1981). A coat protein mutant in which the first 25 N-terminal amino acids in the basic arm of BMV coat protein had been deleted failed to direct the packaging of RNA (Sacher and Ahlquist, 1989). A mutant giving rise to a protein lacking only the first seven amino acids packaged viral RNA *in vivo*.

Argos also predicted a similar helical domain for the amino-terminal region of the AMV coat protein. About 10–15 amino acids penetrate the RNA (Oostergetel *et al.*, 1983). These helices are considered to be stabilized by neutralization of the positive charges by RNA phosphates. The composition of these N-terminal arms bears a similarity to histones (Argos, 1981). None of these viruses contains polyamines as part of its structure.

b. TYMV

There is no evidence for significant penetration of protein into the RNA of TYMV, nor is there any accumulation of basic amino acids at the protein–RNA interface. Ehresmann *et al.* (1980) used two *in situ* cross-linking procedures to identify three regions of the coat protein that lie close to the RNA. Cytosine was the base most prominently involved in cross-linking, but there was no enrichment of basic amino acids in the cross-linked peptides. An examination of the coat protein sequences of TYMV and another tymovirus, EGMV, showed that neither protein possessed an accumulation of basic residues able to form a strong ionic interaction with the RNA (Dupin *et al.*, 1985).

The RNA within TYMV may be stabilized by two kinds of interaction: (1) by glutamyl or aspartyl side-chains hydrogen bonding to cytosine phosphate residues as suggested by Kaper (1972); and (2) by the spermine and spermidine found in this virus (see Chapter 4, Section III.C.1), which could neutralize a significant proportion of the charged RNA, together with divalent cations. It is probable that all tymoviruses are stabilized in part by polyamines, but some members of the group may lose these components quite readily during isolation. Thus, BeMV isolated in the absence of CsCl contained about 100–200 polyamine molecules and 500–900 Ca^{2+} ions per virus particle. The polyamines could readily be exchanged with other cations such as Cs, leading to a loss of particle stability (Savithri *et al.*, 1987).

c. *Comoviruses*

CPMV may be similar to TYMV with respect to neutralization of charged phosphates, since particles containing RNA also contain polyamines, and there is no evidence for a mobile protein arm (Virudachalum *et al.*, 1985).

VII. MORE COMPLEX ISOMETRIC VIRUSES

1. *Phytoreovirus*

Phytoreoviruses have distinctly angular particles about 65–70 nm in diameter that contain 12 pieces of dsRNA. There are seven different proteins in the particle of the type species, WTV (Reddy and MacLeod, 1976) and RDV (see Table 6.4). Unlike most animal reoviruses which have triple-shelled capsids, the particles of phytoreoviruses consist of an outer shell of protein and an inner core containing protein and the 12 pieces of dsRNA. However, there is no protein in close association with the RNA. The particles are readily disrupted during isolation, by various agents. Under suitable conditions, subviral particles can be produced, which lack the outer envelope and which reveal the presence of 12 projections at the 5-fold vertices of an icosahedron.

By controlled degradation and study of the products, it has been possible to locate some of the proteins within the particle. If the outer shell is removed by a gentle treatment (e.g. enzyme digestion), the resultant cores have unimpaired infectivity for insect vector monolayer cultures (Reddy and MacLeod, 1976).

The location of the RNA within RDV has been studied by small-angle neutron scattering (Inoue and Timmins, 1985). The RNA is located within a radius of 23 nm, and the protein shell lies 23–36 nm from the center of the particle. Thus, there appears to be no major interpenetration of RNA and protein. The location of the proteins within the virus has also been established for RGDV (Omura *et al.*, 1985).

Using cryoelectron microscopy and image reconstruction, Lu *et al.* (1998b) derived the structure of RDV to 25 Å resolution. This revealed two distinct icosahedral shells, a

$T = 13$ outer shell, 700 Å in diameter, composed of 260 trimeric clusters of P8 (46 kDa) and an inner $T = 1$ shell, 567 Å in diameter and 25 Å thick, made up of 60 dimers of P3 (114 kDa) (Fig. 5.38). From tilt experiments, the outer shell was shown to be left-handed ($T = 13l$). The $T = 1$ core plays a critical role in the organization of the quasi-equivalence of the $T = 13$ outer capsid (Wu *et al.*, 2000). It is suggested that the core may guide the assembly of the outer capsid.

Mizuno *et al.* (1986) presented electron microscopic evidence suggesting that the ds genomic RNA segments of RDV are packed within the viral core as supercoiled structures complexed within protein.

2. Fijiviruses

Fijiviruses have 10 pieces of RNA in spherical particles 65–70 nm diameter. Fijiviruses resemble phytoreoviruses in having double-shelled particles but differ in having spikes both on the outer shell ('A'-type spikes) and on the inner shell ('B'-type spikes).

MRDV is one of the best characterized members of the *Fijivirus* genus (Milne and Lovisolo, 1977). Milne *et al.* (1973) detected 12 'A'-type spikes projecting from the surface of intact MRDV particles, one at each 5-fold symmetry axis. These spikes were about 11 nm long. Beneath each 'A'-type spike was a 'B'-type spike about 8 nm long, revealed when the outer coat was removed. The 'B'-type spikes were associated with some differentiated structure in the inner core, which they termed a 'baseplate'. Detached 'B'-type spikes could be seen to be made up of five morphological units surrounding a central hole.

Cores without spikes (smooth cores) contain 136-kDa and 126-kDa polypeptides. The spiked cores contained, in addition, a 123-kDa polypeptide that is, therefore, probably located in the 'B'-type spikes (Boccardo and Milne, 1975). None of the other polypeptides has been unequivocally associated with particular structures seen in negatively stained images of particles (Boccardo and Milne, 1975; Luisoni *et al.*, 1975). FDV has a very similar structure to that of MRDV, as revealed by electron microscopy (Fig. 5.39A).

3. Oryzaviruses

Members of the *Oryzavirus* genus have particles of 57–65 nm diameter that contain 10 pieces of dsRNA. The particle structure differs significantly from those of the other two plant reovirus genera, in that oryzaviruses appear to lack an outer capsid and thus to consist of the inner core. There are 12 'B'-type spikes, 8–10 nm in height and 23–26 nm wide at the base and 14–17 nm at the top (Fig 5.39B).

VIII. ENVELOPED VIRUSES

A. *Rhabdoviridae*

Rhabdoviridae is a family of viruses whose members infect vertebrates, invertebrates and plants. They have a complex structure. Rhabdoviruses form widely differing organisms and are constructed on the basic plan shown in Fig. 5.40.

Some animal rhabdoviruses may be bullet-shaped, but most and perhaps all plant members are rounded at both ends to give a bacilliform shape. Electron microscopy on thin sections and negatively stained particles has been used to determine size and details of morphology (Fig. 5.41). The size measurements can be only approximate and are probably underestimates because of shrinkage taking place during dehydration for electron microscopy. Particles of these viruses readily deform and fragment *in vitro* unless pH and other conditions are closely controlled (e.g. Francki and Randles, 1978) (Fig. 5.41E).

The location of the various proteins within the particles has been established by fractionation of subviral structures, by successive removal of superficially located proteins by detergent or enzyme treatment, and by labeling of the exposed proteins with [125]I (e.g. Ziemiecki and Peters, 1976). The N protein appears to be firmly bound to the RNA to form the ribonucleoprotein core (RNP). This is usually seen in thin sections as having cross-striations 4.5–5.5 nm apart. The RNP is a single continuous strand arranged in helical fashion (e.g. Hull, 1976a). The surrounding membrane contains two proteins: the matrix or M protein (or proteins),

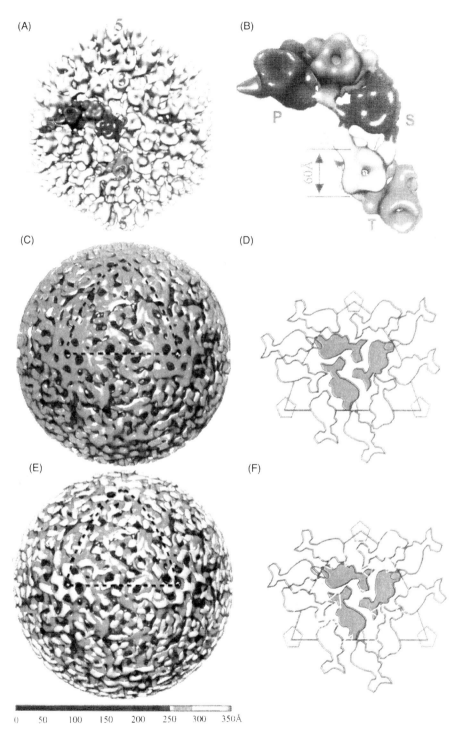

Fig. 5.38 (see Plate 5.6)
Structure of RDV. **(A)**
Shaded surface view of the
reconstruction of the full
RDV as viewed along the
icosahedral 2-fold axis. The
numbers 5, 3 and 2
designate the icosahedral 5-,
3- and 2-fold axes.
Highlighted in color are a
contiguous group of five
trimers found in each
asymmetric unit. **(B)** Blown-
up view of the group of five
trimers computationally
extracted from panel (A).
These trimers at the distinct
quasi-equivalent positions
are designated P, Q, S, R
and T, using the convention
set out for BTV (Grimes *et
al.*, 1997). **(C)** Inner shell
computationally extracted at
590 Å diameter, showing a
$T = 1$ lattice. The dashed
triangle designates one
triangular face of the
icosahedron. **(D)** Schematic
of fish-shaped density
distribution within a
triangle of a $T = 1$ lattice.
(E) Capsid computationally
extracted at 604 Å diameter,
showing the interface
between the outer shell
(yellow) and inner shell
(blue) proteins. Color bar
shows the coding as a
function of radius. **(F)**
Schematic illustrating the
interaction pattern of
trimers (yellow triangles) of
the outer shell with the fish-
shaped densities of the
inner shell. From Lu *et al.*
(1998b), with permission.

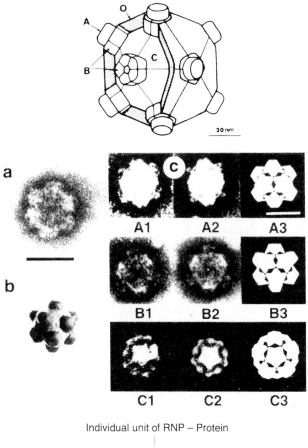

Fig. 5.39 **Top:** Scale model of the Fiji disease virus. Part of the outer shell (O) of the particle and one of the A spikes have been removed to expose the core (C) and the structure and arrangement of the B and A spikes. The arrangement of distinct morphological subunits within the A spikes, outer shell and core has not been fully established. From Hatta and Francki (1977), with permission. **Bottom:** Structure of RRSV. **(a)** Electron micrograph of RRSV, **(b)** schematic of RRSV particle, and **(c)** micrographs of the virus, showing the 2-, 3- and 5-fold symmetries (A1, B1 and C1 respectively), images of the same rotated by increments of 180 degrees (A2) or 120 degrees (B2) or 72 degrees (C2) and proposed models of the 2-, 3- and 5-fold symmetries (A3, B3 and C3 respectively). From Mertens *et al.* (2000), with permission.

which is in a hexagonal array, and the G protein (glycoprotein), which forms the surface projections (e.g. Ziemiecki and Peters, 1976).

The detailed fine structure remains to be determined. In particular, it is not clear how the hexagonally arrayed M protein is related spatially to the helical RNP. Likewise the way the rounded ends are formed and the arrangement of RNP within them is not established. Hull (1976a) suggested that both the M protein layer and the arrangement of the G protein was based on half icosahedral rounded ends as in other bacilliform particles.

B. Tospoviruses

TSWV is very unstable and difficult to purify (e.g. Joubert *et al.*, 1974). Its structure has been studied in thin sections of infected cells, and in partially purified preparations. Isolated preparations contain many deformed and damaged particles.

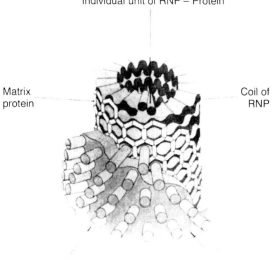

Fig. 5.40 Model for rhabdovirus structure showing the proposed three-dimensional relationship between the three proteins and the lipid layer. From Cartwright *et al.* (1972), with permission.

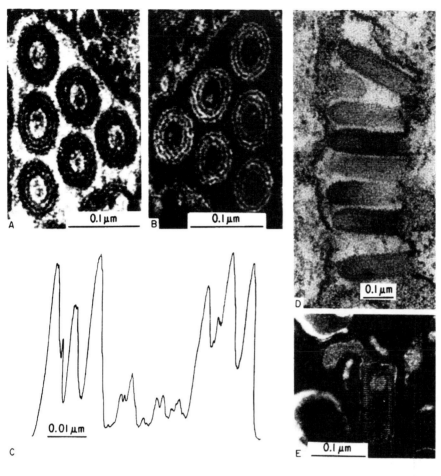

Fig. 5.41 Rhabdoviruses. **(A,B)** Transverse sections through particles of PYDV showing the lamellar nature of the viral envelope. Particles show some lateral compression. **(B)** is a photographic reversal of (A). **(C)** Microdensiometer trace through the top particle in (A). From MacLeod *et al.* (1966), with permission. **(D)** Section through a cell of *N. rustica* showing a group of PYDV particles enclosed in a membrane-bound vesicle. Courtesy of R. MacLeod. **(E)** A particle of SYVV in a negatively stained leaf dip preparation. From Richardson and Sylvester (1968), with permission.

Tospovirus particles are spherical with a diameter of 80–110 nm and comprise a lipid envelope encompassing the genomic RNAs which are associated with the N protein as a nucleoprotein complex. The viral polymerase is also contained within the particle. The lipid envelope contains two types of glycoproteins (Fig 5.42).

IX. ASSEMBLY OF ICOSAHEDRAL VIRUSES

A. Bromoviruses

The pioneering work of Bancroft and colleagues showed that the protein subunits of several bromoviruses could be reassembled *in vitro* to give a variety of structures (reviewed by Bancroft and Horne, 1977). In the presence of viral RNA, the protein subunits could reassemble to form particles indistinguishable from native virus. However, the conditions used were nonphysiological in several respects. The assembly mechanism of CCMV is thought to involve carboxyl–carboxylate pairs (Jacrot, 1975).

Of considerable interest in relation to possible *in vivo* mechanisms is the assembly of infectious CCMV under mild conditions (Adolph and Butler, 1977). A 3S protein aggregate dimer at pH 6.0, ionic strength 0.1–0.2 and 25°C, in the absence of added Mg^{2+}, combines with the RNA to form infectious virus. The reassembled particles cannot be distinguished from native virus by physicochemical or structural means. Zlotnick *et al.* (2001) reported on experiments which indicated that capsid assembly of CCMV is nucleated by

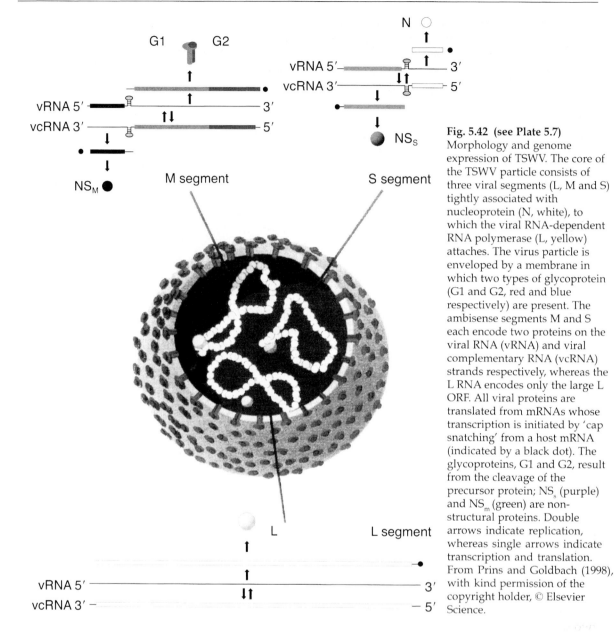

Fig. 5.42 (see Plate 5.7) Morphology and genome expression of TSWV. The core of the TSWV particle consists of three viral segments (L, M and S) tightly associated with nucleoprotein (N, white), to which the viral RNA-dependent RNA polymerase (L, yellow) attaches. The virus particle is enveloped by a membrane in which two types of glycoprotein (G1 and G2, red and blue respectively) are present. The ambisense segments M and S each encode two proteins on the viral RNA (vRNA) and viral complementary RNA (vcRNA) strands respectively, whereas the L RNA encodes only the large L ORF. All viral proteins are translated from mRNAs whose transcription is initiated by 'cap snatching' from a host mRNA (indicated by a black dot). The glycoproteins, G1 and G2, result from the cleavage of the precursor protein; NS_s (purple) and NS_m (green) are non-structural proteins. Double arrows indicate replication, whereas single arrows indicate transcription and translation. From Prins and Goldbach (1998), with kind permission of the copyright holder, © Elsevier Science.

a pentamer and proceeds by the cooperative addition of dimers.

The competition experiments of Cuillel *et al.* (1979) with various foreign RNAs showed that under appropriate conditions BMV protein can recognize its own RNA molecules to some extent. Using neutron and X-ray scattering, Cuillel *et al.* (1983) found that the assembly of BMV empty protein shells *in vitro* was a very rapid process, with forward scattering reaching half-maximum in less than a second. The

oligonucleotides $(AP)_8A$ and $(A-T)_5$ interact with empty shells of CCMV. They bind specifically to the arginine and lysine residues of the N-terminal arm of the polypeptide within the shell (Section VI.B.4.b). The binding studies support a model for virus assembly in which a random coil to α-helix transition occurs, induced by neutralization of the basic amino acid side-chains (Vriend *et al.*, 1986).

Various *in vitro* studies with BMV and CCMV indicate that the highly basic N-terminal region of the coat protein is involved in reactions with the viral RNA. Experiments with BMV variants containing known deletions in the coat protein gene confirm this view. The first seven amino acids are completely dispensable for packaging of the RNA *in vivo*, but if 25 N-terminal amino acids are missing, no virus particles are produced (Sacher and Ahlquist, 1989). Mutagenic studies on BMV showed that the *in vivo* interactions of the N-terminal arm and the three genomic RNAs are distinct (Choi *et al.*, 2000). For instance, deletion in the arginine-rich motif (ARM) specifically affected the stability of virions containing RNA1. The ARM region also contains crucial amino acids required for packaging RNA4, independent of genomic RNA3.

Although BMV and CMV are structurally and genetically very similar, there is considerably specificity for each virus's coat protein to encapsidate the cognate nucleic acid (Osman *et al.*, 1998). RNA3 chimera were constructed in which the respective coat protein genes were exchanged and the replicative competence of each chimera examined in *Nicotiana benthamiana* protoplasts and in their ability to give systemic infection in *Chenopodium quinoa* and *N. benthamiana*. Each chimera replicated to near wild-type levels and expressed the expected coat protein. However, CMV coat protein encapsidated BMV RNAs to much lower levels than the wild type, and BMV did not encapsidate the heterologous CMV RNAs. Thus, there is considerable specificity during *in vivo* packaging, with BMV coat protein being more specific than that of CMV (Osman *et al.*, 1998).

BMV coat protein can assemble *in vivo* into two distinct capsids (Krol *et al.*, 1999). Natural genomic BMV RNA was encapsidated in a $T = 3$ 180-subunit particle, but engineered mRNA containing only the BMV coat protein gene was packaged into a 120-subunit particle. The 120-subunit particle was made up of 60 coat protein dimers in distinct non-equivalent environments that differed from the quasi-equivalent environments in the 180-subunit particle. Thus, there is potential flexibility in the interactions between coat protein subunits that is potentiated by RNA features.

B. Alfalfa mosaic virus

The coat protein of AMV assembles into bacilliform particles in the presence of nucleic acids or into $T = 1$ icosahedral particles in the absence of nucleic acid (Bol and Kruseman, 1969; Hull, 1970c; Driedonks et al., 1977). Driedonks *et al.* (1980) followed polymerization in the analytical ultracentrifuge. They postulated four stages in assembly of the virus: (1) an initiation stage, (2) initial cap formation, (3) cylindrical elongation, and (4) a cap formation or closure stage. The dimer of the coat protein is a very stable configuration in solution (Driedonks *et al.*, 1977) and is likely to be involved in virus assembly. X-ray crystallography of the $T = 1$ particle indicates that dimer is formed by the C-terminal arm of one subunit hooking around the N-terminal arm of an adjacent subunit (Kumar *et al.*, 1997). The ability to form dimers is controlled by the C terminus of the coat protein (Choi and Loesch-Fries, 1999).

C. Other viruses

SBMV particles can be assembled *in vitro* at low ionic strength from isolated RNA and coat protein. The components assembled into $T = 1$ or $T = 3$ particles depending on the size of the viral RNA used and the pH (Savithri and Erickson, 1983). The coat protein of the cowpea strain of SBMV (now known as SCPMV) binds to specific sites on the viral RNA (Hacker, 1995). This region, mapped to nucleotides 1410–1436, is predicted to fold into a hairpin with a 4-base loop and a duplex stem of 24 nucleotides.

Dissociation of TCV at high pH and ionic strength gives rise to coat protein dimers and a ribonucleoprotein complex made up of the genomic RNA, six coat protein subunits, and an 80-kDa protein that is a covalent coat protein dimer. TCV particles can be reassembled *in vitro* using coat protein and either the free RNA or the ribonucleoprotein complex. For both forms of the RNA, the process is selective for viral RNA, and proceeds by continuous growth of a shell from an initiating structure (Sorger *et al.*, 1986). Analysis of the encapsidation of various mutant viral RNAs showed that a 186-nucleotide region at the 3'-end of the coat protein gene, with a bulging hairpin loop of 28 nucleotides, is indispensable for TCV RNA encapsidation (Qu and Morris, 1997).

OkMV produces large quantities of empty viral protein shells in inoculated cucumber cotyledons (Marshall and Matthews, 1981). At 1 day after inoculation, about one-half the total viral protein shells were found in the nucleus. This accumulation occurred in the presence of virus particles that were found exclusively in the cytoplasm. The most likely explanation for this active and preferential accumulation is that viral coat protein enters the nucleus in the form of monomers or pentamers and hexamers and is assembled into shells once inside.

Co-expression of the regions of CPMV RNA2 that encode the large and small coat proteins in *Spodoptera frugiperda* (sf21) insect cells using baculovirus vectors led to the formation of virus-like particles that had the sedimentation characteristics of empty particles (Shanks and Lomonossoff, 2000). However, no particles were formed when either of the proteins was expressed individually. Similarly, CPMV proteins expressed in protoplasts (Wellink *et al.*, 1996) and ArMV proteins expressed in transgenic plants and insect cells (Bertioli *et al.*, 1991) form empty virus-like particles.

Proteolysis of the small coat protein of CPMV results in the loss of the C-terminal 24 amino acids (Taylor *et al.*, 1999). Infections with mutants of cDNA of CPMV RNA2 from which the 24 C-terminal amino acids had been deleted were debilitated in virus accumulation with a much increased proportion (73%) of top component particles not containing RNA. It is suggested that the C-terminal region of the small coat protein is involved with RNA packaging.

TYMV particles have not yet been reassembled *in vitro* from RNA and protein subunits. Figure 5.43 illustrates a possible model for the assembly of TYMV particles *in vivo*.

Matthews (1981) discussed this in more detail. The model predicts that there is an accumulation of coat protein just before virus assembly begins, and that, unlike TMV

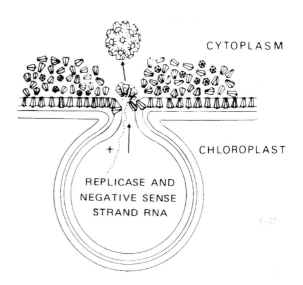

Fig. 5.43 Model for the assembly of TYMV. (i) Pentamer and hexamer clusters of coat protein subunits are synthesized by the ER and accumulate in the cytoplasm overlying clustered vesicles in the chloroplast. (ii) These become inserted into the outer chloroplast membrane in an oriented fashion—that is, with the hydrophobic sides that are normally buried in the complete protein shell lying within the lipid bilayer, with the end of the cluster that is normally inside the virus particle at the membrane surface. (iii) An RNA strand synthesized or being synthesized within a vesicle begins to emerge through the vesicle neck. (iv) At this site a specific nucleotide sequence in the RNA recognizes and binds a surface feature of a pentamer cluster lying in the outer chloroplast membrane near the vesicle neck, thus initiating virus assembly. (v) Assembly proceeds by the addition of pentamers and hexamers from the uniformly oriented supply in the membrane. (vi) The completed virus particle is released into the cytoplasm. From Matthews (1991).

replication, there would be no accumulation of complete uncoated viral genomes. As the pentamer and hexamer clusters are depleted in the membrane, others would replace them from the electron-lucent layer until the supply was exhausted. Empty protein shells presumably represent errors in virus assembly, which take place in the absence of RNA. They can form because of the strong protein–protein interactions in the shell of this virus.

The model readily explains the effect of 2-thiouracil on TYMV replication. When genome

RNA synthesis in the vesicles is blocked by the analog, empty protein shells are made in increased amounts from the accumulated coat protein and from further protein being synthesized on pre-existing coat mRNA. The model also explains the apparent requirement of illumination for virus assembly (Rohozinski and Hancock, 1996). It is suggested that the light-induced generation of low pH drives TYMV assembly.

D. RNA selection during assembly of plant reoviruses

Every WTV particle appears to contain one copy of each genome segment, because (1) RNA isolated from virus has equimolar amounts of each segment (Reddy and Black, 1973), and (2) an infection can be initiated by a single particle (Kimura and Black, 1972). Thus, there is a significant problem with this kind of virus. What are the macromolecular recognition signals that allow one, and one only, of each of 10 or 12 genome segments to appear in each particle during virus assembly? For example, the packaging of the 12 segments of WTV presumably involves 12 different and specific protein–RNA and/or RNA–RNA interactions. The first evidence that may be relevant to this problem comes from the work of Anzola *et al.* (1987) with WTV. They established the structure of a defective (DI) genomic segment 5, which was only one-fifth the length of the functional S5 RNA because of a large internal deletion. However, this DI RNA was packaged at one copy per particle, as for the normal sequence. Thus, they established that the sequence(s) involved in packaging must reside within 319 base-pairs of the 5′ end of the plus strand and 205 from the 3′ end. Reddy and Black (1974, 1977) showed that an increase in the DI RNA content in a virus population led to a corresponding decrease in the molar proportion of the normal fragment; that is, the DI fragment competes only with its parent molecule. Thus, there must be two recognition signals, one that specifies a genome segment as viral rather than host, and one that specifies each of the 12 segments.

Anzola *et al.* (1987) sequenced the 5′- and 3′-terminal domains of all 12 genome segments (see Fig. 8.21). They suggested that a fully conserved hexanucleotide sequence at the 5′ terminus and a fully conserved tetranucleotide sequence at the 3′ terminus might form the recognition signals for viral as opposed to host RNA. They also found segment-specific inverted repeats of variable length just inside the conserved segments (Fig. 8.22), which they suggested might represent the specific recognition sequence for each individual genome segment. A similar inverted repeat was found for segment 9 of RDV (Uyeda *et al.*, 1989).

Other members of the *Reoviridae* family each have conserved 5′ and 3′ sequences (Asamizu *et al.*, 1985). Influenza viruses, which have segmented ssRNA genomes, also have comparable conserved sequences at the 5′ and 3′ termini of each genome segment (Stoeckle *et al.*, 1987). These similarities strengthen the idea that the 5′- and 3′-terminal sequences have a role in the packaging of these segmented RNA genomes. However, the problem is by no means solved. If only RNA–RNA recognition is involved, how is this brought about to give a set of 12 dsRNAs for packaging? If protein:RNA recognition is important, how are 12 specific sites constructed out of the three proteins known to be in the nucleoprotein viral core?

In *Reovirus*, the (−)-sense strands are synthesized by the viral replicase on a (+)-sense template that is associated with a particulate fraction (Acs *et al.*, 1971). These and related results led to the proposal that dsRNA is formed within the nascent cores of developing virus particles, and that the dsRNA remains within these particles. If true, this mechanism almost certainly applies to the plant reoviruses. It implies that the mechanism that leads to selection of a correct set of 12 genomic RNAs involves the ss plus strand. Thus, the base-paired inverted repeats illustrated in Fig. 8.22 could be the recognition signals. It may be that other virus-coded 'scaffold' proteins transiently present in the developing core are involved in RNA recognition rather than, or as well as, the three proteins found in mature particles.

Xu *et al.* (1989) constructed a series of transcription vectors that allowed production of an exact transcript of S8 RNA and of four analogs that differed only in the immediate 3′ terminus. Their experiments provided three lines of evidence supporting the view that the 5′- and 3′-terminal domains interact in a functional way:

1. Nuclease T1 sensitivity assays showed that even a slight change in the 39-terminal sequence can affect the conformation of the 59 terminus.
2. Translation *in vitro* is slightly decreased by alterations in the 3′ terminus, which extends the potential for 3′–5′ terminal base-pairing, and is increased by changes that reduce potential base-pairing.
3. Computer modeling for minimal energy structures for six WTV transcripts predicted a conformation in which the terminal inverted repeats were base-paired.

Dall *et al.* (1990) developed a gel retardation assay with which they demonstrated selective binding of WTV transcripts by a component of extracts from infected leafhopper cell cultures. Using terminally modified and internally deleted transcripts, they established that the segment-specific inverted repeats present in the terminal domains were necessary but not sufficient for optimal binding. Some involvement of internal sequences was also necessary. There was no evidence for discrimination in binding between transcripts from different segments. The binding component or components present in extracts of infected cells, which are not present in those of healthy cells, have not yet been characterized.

X. DISCUSSION AND SUMMARY

Advances in the capacity, speed and availability of computers, the use of non-crystallographic symmetry averaging, and developments in computer graphics have allowed resolution down to atomic detail for the proteins and protein coats of several geometric plant viruses, both rod-shaped and isometric. X-ray crystallographic analysis can, of course, be applied only to viruses or virus coat proteins that can be obtained in crystalline form. For larger viruses that have not been crystallized, other techniques, especially electron microscopy, remain important for studying virus architecture.

Serological methods, even those employing monoclonal antibodies, have proved to be of limited use in delineating the structures of the small plant viruses, because of ambiguities in interpretation of the results, unless they can be related to a protein structure that is already established by other methods.

The idea of quasiequivalence in the bonding between subunits in icosahedral protein shells with $T > 1$ as put forward by Caspar and Klug in 1962 is still useful but has required modification in the light of later developments. As noted earlier, it is now even applicable to the organization of subunits in particles that have $T = 1$ symmetry (Chapman, 1998). Viruses have evolved at least two methods by which a substantial proportion of the potential nonequivalence in bonding between subunits in different symmetry related environment can be avoided: (1) by having quite different proteins in different symmetry-related positions as in the reoviruses and the comoviruses, and (2) by developing a protein subunit with two or more domains that can adjust flexibly in different symmetry positions, as in TBSV and SBMV.

The high-resolution analysis of the structures of isometric viruses is showing that, although the outer surface of the shell shows quasiequivalent icosahedral symmetry, this may not extend into the inner parts. There is frequently interweaving, especially of the C-terminal regions of the polypeptide chain, which forms an internal network.

The many negatively charged phosphate groups on the nucleic acid within a virus are mutually repelling. To produce a sufficiently stable virus particle these charges, or most of them, need to be neutralized. Structural studies to atomic resolution have revealed three solutions to this problem.

1. In TMV, the RNA is closely confined within a helical array of protein subunits. Two of the

phosphates associated with each protein subunit are close to arginine residues. However, the electrostatic interactions between protein and RNA are best considered as complementarity between two electrostatic surfaces. Similar arrangements may hold for other rod-shaped viruses, but no detailed data are available for these.

2. In TBSV and a number of other icosahedral viruses, a flexible basic amino-terminal arm with a histone-like composition projects into the interior of the virus interacting with RNA phosphates. Additional phosphates are neutralized by divalent metals, especially Ca^{2+}.

3. In TYMV, where there is little interpenetration of RNA and protein, the RNA phosphates are neutralized by polyamines and Ca^{2+} ions.

X-ray analysis has located Ca^{2+} ions in specific locations between protein subunits in several virus shells. All this detailed structural information has gone some distance in explaining the relative stability of different viruses to various agents such as chelating compounds and changes in pH or ionic strength.

In the more complex particles with structures comprising several layers of proteins, there is increasing evidence for interactions between the layers. It seems likely that these interactions drive the structural arrangements of protein subunits in adjacent layers.

The organization of the nucleic acid within icosahedral virus particles has been difficult to study because the nucleic acid does not contribute to the orientation of virus particles within a crystal. However, indirect methods show that the nucleic acids are quite highly ordered within the virus, most viral RNAs having substantial double-helical structure. The data that are becoming available indicate that, for several viruses, the organization of the RNA is related to the icosahedral symmetry. For these, and for the viruses in which the charge neutralization is primarily by polyamines or divalent cations, the strong secondary structure is essential to give the compactness required by the small isometric structure. In the simplest virus, STNV, neutron diffraction data show that much of the RNA is arranged in an icosahedral network just inside the protein shell. A high-resolution density map of BPMV shows that about 20% of the RNA is arranged inside the protein shell in an icosahedral manner.

An increasing number of the interactions involved in the structures and stabilization of both rod-shaped and isometric viruses have been explored by site-directed mutagenesis. This has revealed the sophistication of interactions that give the stability necessary to protect the viral genome on passage between hosts but also to enable the genome to be released on entry into the cell. It is likely that the application of new technologies, such as electron energy-loss spectroscopy (Leapman and Rizzo, 1999), will lead to an even more detailed understanding of these interactions.

CHAPTER 6

Genome Organization

I. INTRODUCTION

One of the main features of the last decade has been the great explosion of sequence data on plant virus genomes. The first viral genome to be sequenced was the DNA of CaMV (Franck *et al.*, 1980) followed by the RNA of TMV (Goelet *et al.*, 1982). By 1990, the genome sequences from about 40 species from about 20 groups (some of the groups have been reclassified since) had been determined. In the year 2000, the genomes of about 250 species had been fully sequenced including representatives of most plant virus genera. There are also numerous partial sequences mainly of viral coat protein genes.

The great expansion of sequences has enabled refinements to be made to the classification of plant viruses and the separation of apparently similar species into distinct genera (see Chapter 2). Good examples are the separation of the genera of the *Caulimoviridae* and of the *Luteoviridae* based on genome organization.

This plethora of data has led to much comparison between sequences often going into fine detail beyond the scope of this book. However, the comparisons, coupled with the use of infectious clones of viruses, have given valuable information for use in mutagenesis experiments to elucidate the functions of various gene products and non-coding regions.

In this chapter, I am going to describe the various gene products that viruses encode and, in general terms, the genome organizations from which they are expressed. In the next chapter, I will discuss how the genetic information is expressed from these genomes, and in Chapter

8, I will describe how these genomes are replicated.

II. GENERAL PROPERTIES OF PLANT VIRAL GENOMES

Basically, the viral genome comprises coding regions that express the proteins required for the viral infection cycle, movement through the plant, interactions with the host and movement between hosts, and non-coding regions that control the expression and replication of the genome; control sequences can also be found in the coding regions.

A. Information content

In theory, the same nucleotide sequence in a viral genome could code for up to 12 or more polypeptides. There could be an open reading frame (ORF) in each of the three reading frames of both the positive (+)- and negative (−)-sense strands, giving six polypeptides. Usually an ORF is defined as a sequence commencing with an AUG initiation codon and capable of expressing a protein of 10 kDa or more. If each of these ORFs had a leaky termination signal, they could give rise to a second read-through polypeptide; frameshift to a downstream ORF also gives a second polypeptide. However, in nature, there must be severe evolutionary constraints on such multiple use of a nucleotide sequence, because even a single base change could have consequences for several gene products. However, two overlapping genes in different reading frames do occasionally occur, as do genes on both (+)- and (−)-sense strands. Read-through

Virus acronyms are given in Appendix 1.

and frameshift proteins are quite common and are described in Chapter 7 (Sections V.B.9 and 10).

The number of genes found in plant viruses ranges from 1 for the satellite virus STNV, to 12 for some closteroviruses and reoviruses. Most of the ss (+)-sense RNA genomes code for about four to seven proteins. In addition to coding regions for proteins, genomic nucleic acids contain nucleotide sequences with recognition and control functions that are important for virus replication. These control and recognition functions are mainly found in the 5' and 3' non-coding sequences of the ssRNA viruses, but they may also occur internally, even in coding sequences.

B. Economy in the use of genomic nucleic acids

Viruses make very efficient use of the limited amount of genomic nucleic acids they possess. Eukaryote genomes may have a content of introns that is 10–30 times larger than that of the coding sequences. Like prokaryote cells, most plant viruses lack introns but there are some which are described in Chapter 7 (Section V.B.11). Plant viruses share with viruses of other host kingdoms several other features that indicate very efficient use of the genomic nucleic acids:

1. Coding sequences are usually very closely packed, with a rather small number of non-coding nucleotides between genes.
2. Coding regions for two different genes may overlap in different reading frames (e.g. in BNYVV; see Fig. 6.49) or one gene may be contained entirely within another in a different reading frame (e.g. the OCSV genome; see Fig. 6.25).
3. Read-through of a 'leaky' termination codon may give rise to a second, longer read-through polypeptide that is co-terminal at the amino end with the shorter protein. This is quite common among the virus groups with ss (+)-sense RNA genomes. Frameshift proteins in which the ribosome avoids a stop codon by switching to another reading frame have a result that is similar to a 'leaky' termination signal; these are described in Chapter 7 (Sections V.B. 9 and 10).

4. A single gene product may have more than one function. For example, the coat protein of AMV has a protective function in the virus particle, a function in insect transmission and a function in the initiation of infection by the viral RNAs. The coat protein of MSV has a protective function, is involved in insect vector specificity (Chapter 11, Section IV.D.1), cell-to-cell transport of the virus (Chapter 9, Section II.D.2.i), nuclear transport of the viral DNA (Chapter 9, Section II.C), and possibly in symptom expression and in control of replication.
5. Functional introns have been found in several geminiviruses, in RTBV and in CaMV (see Chapter 7, Sections V.H.1 and 2). Thus, mRNA splicing, a process that can increase the diversity of mRNA transcripts available and therefore the number of gene products, may be a feature common in viruses with a DNA genome.
6. A functional viral enzyme may use a host-coded protein in combination with a virus-coded polypeptide (e.g. the replicase of TMV).
7. Regulatory functions in the nucleotide sequence may overlap with coding sequences (e.g. the signals for subgenomic RNA synthesis in TMV).
8. In the 5' and 3' non-coding sequences of the ssRNA viruses, a given sequence of nucleotides may be involved in more than one function. For example, in genomic RNA, the 5'-terminal non-coding sequences may provide a ribosome recognition site, and at the same time contain the complementary sequence for a replicase recognition site in the 3' region of the (−) strand. In members of the *Potexvirus* genus the origin of assembly is also at the 5' end of the genome (Chapter 5, Section IV.B).

C. The functions of viral gene products

The known functions of plant viral gene products may be classified as follows:

1. Structural proteins

These are coat proteins of the small viruses, the matrix, core or nucleoproteins proteins of the reoviruses, tenuiviruses and those viruses with a lipoprotein membrane, and proteins found within such membranes.

2. Enzymes

a. Proteases

These are proteases coded for by those virus groups in which the whole genome or a segment of the genome is first transcribed into a single polyprotein. The properties of these enzymes have been reviewed by Wellink and van Kammen (1988). These are described in more detail in Chapter 7 (Section V.B.1).

b. Enzymes involved in nucleic acid synthesis

It is now generally accepted that all plant viruses, except some satellite sequences, code for one or more proteins that have an enzymatic function in nucleic acid synthesis, either genomic nucleic acid or mRNAs or both. The general term for these enzymes is *polymerase*. There is some inconsistency in the literature in relation to the terms used for different polymerases. I will use the various terms with the following meanings. Polymerases that catalyze transcription of RNA from an RNA template have the general name *RNA-dependent RNA polymerase* (RdRp). The enzyme complex that makes copies of an entire RNA genome and the subgenomic mRNAs is called a *replicase*. If an RdRp is found as a functional part of the virus particle as in the *Rhabdoviridae* and *Reoviridae* it is often called a *transcriptase*. The enzyme coded by members of the *Caulimoviridae*, which copies a full-length viral RNA into genomic DNA, is called an *RNA-dependent DNA polymerase* or *reverse transcriptase*. In the *Geminiviridae* the viral gene product(s) associate(s) with the host *DNA-dependent DNA polymerase*.

A further terminological difficulty is that replicase enzymes often either have various functional domains or are made up of virus-encoded subunits with different functions. For instance, RNA → RNA replication can involve methyl transferase and helicase activities as well as the actual polymerase itself. I will use the general term *replicase* to describe the activity *in toto* and the specific names of the subactivities when appropriate. Viral replication is discussed in Chapter 8.

3. Virus movement and transmission

For many plant viruses, a specific virus-coded protein has been identified as an essential requirement for cell-to-cell movement and for systemic movement within the host plant (see Chapter 9, Section II.D.2). Other gene products have been identified as essential for successful transmission by invertebrate vectors, and viral gene products may also be involved in transmission by fungi (see Chapter 11).

4. Non-enzymatic role in RNA synthesis

The 5' VPg protein found in some virus genera is thought to act as a primer in RNA synthesis (see Chapter 8, Section IV.J).

5. Coat protein of AMV

The AMV coat protein, and the corresponding protein in ilarviruses, has an essential role in the initiation of infection by the viral RNA, possibly by priming (−)-strand synthesis. This protein is discussed in more detail in Chapter 8 (Section IV.G).

6. Host cell-recognizing proteins

Bacterial cells that can be infected by viruses usually exist in a liquid medium. Cells in an animal body usually have no cytoplasmic connections between them. Thus, viruses infecting bacterial or animal cells normally release virus from infected cells to infect further cells of an appropriate type. To identify markers on the surface of appropriate host cells, most viruses infecting bacteria and animals have recognition proteins on their surface. Such recognition proteins are lacking for most plant viruses, which is probably related to two properties of the host organism. A recognition protein would be of no use to a plant virus for recognizing the surface of a suitable host plant because the virus cannot penetrate the surface layers unaided. Furthermore, a plant virus, once it infects a host cell, say in a leaf, can move from cell to cell via plasmodesmata and vascular tissue throughout almost the entire plant. It therefore has no need for a special recognition protein for the surface of cells inside the host organism. However, plant viruses that circulate in their insect vectors have to cross various barriers such as the gut and salivary gland accessory cell walls. These barriers are crossed by receptor-mediated mechanisms (see Chapter 11, Section III.H.1.a).

Within plant cells, the situation may be quite different. There is increasing evidence for the involvement of membranes in virus replication and in cell-to-cell movement. Thus, it is likely that there are receptor sites on intracellular membranes that are targeted by viral proteins.

D. Non-coding regions

1. End structures

The structures at the 5′ and 3′ ends of viral nucleic acids are described in Chapter 4 (Section III.A.3.c). As well as these specific terminal structures, there are various features of terminal and intergenic non-coding regions.

2. 5′ and 3′ non-coding regions

The 5′ and 3′ non-coding regions control both translation and replication. As will be described in Chapter 7, these two regions interact in the initiation of translation of, at least, the 5′ open reading frames (ORFs). The 3′ non-coding region is the site of initiation of (−)-strand RNA synthesis and the 5′ non-coding region (the 3′ end of (−)-strand RNA) is the site of initiation of (+)-strand synthesis (see Chapter 8).

3. Intergenic regions

Sequences in intergenic regions are also involved in both RNA synthesis and the translation of downstream ORFs. The initiation of synthesis of subgenomic RNAs is in these regions and these RNAs are the messengers for translation of non-5′ ORFs in many viruses. This is described in detail in Chapter 7.

As will become apparent in Chapter 7, there are an increasing number of interactions being recognized between terminal and internal sequence regions in the control of expression of the genomic information from (+)-strand RNAs. It is likely that similar interactions will be found that control the expression of (−)-strand and dsRNA genomes and the genomes of ss and ds DNA viruses.

III. PLANT VIRAL GENOME ORGANIZATION

In the following sections, I will describe the genome organizations of the plant virus genera and the functions of the gene products as far as they are known. Further details can be found in the reviews quoted for many of the genera and in van Regenmortel *et al.* (2000).

IV. DOUBLE-STRANDED DNA VIRUSES

A. Family *Caulimoviridae* (reviewed by Rothnie *et al.*, 1994; Hohn and Fütterer, 1997)

The basic features of the genomes of members of this family is that they are circular dsDNA molecules with one discontinuity in one strand and one or more discontinuity in the other; the discontinuities are associated with the replication of the viruses (see Chapter 8, Section VII.B.3). Transcription is asymmetric (see Chapter 7, Section V.H.1) with all the coding information on one strand.

1. Genus *Caulimovirus* (Hohn, 1999)

a. Genome structure

The DNA nucleotide sequences of the type member of the genus *Caulimovirus*, CaMV, consists of a circular dsDNA molecule of about 8 kb. The circular ds DNA of CaMV has a single gap in one strand and two in the complementary strand that are described in detail in Chapter 8 (Section VII.B.3). All caulimoviruses have a single gap in the α- (or minus) strand. Viruses in the group other than CaMV may have one, two or three discontinuities in the (+) strand. The DNA encodes six and possibly eight genes. These are closely spaced but with very little overlap except for the possible gene VIII. The arrangement of the ORFs in relation to the dsDNA is shown in Fig. 6.1.

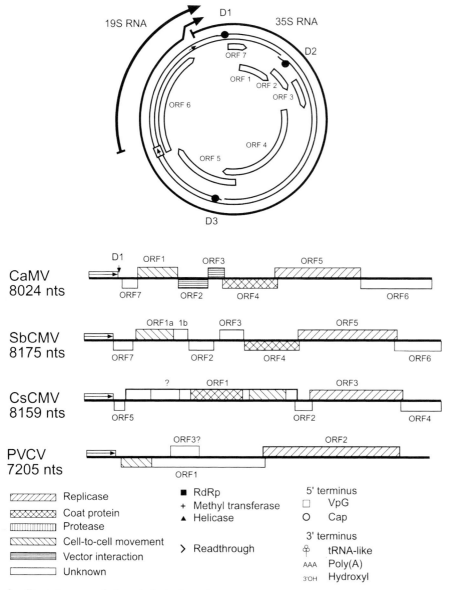

Fig. 6.1 Genome organization of some caulimoviruses. The circular map shows the genome organization of CaMV. The double complete circles represent the dsDNA genome, the positions of the discontinuities (D1, D2 and D3) being represented by ●; the boxed arrows show the positions of the promoters. Inside the double circles the arced boxes represent the ORFs, labeled ORF 1–7. Outside the double circles, the positions of the 35S and 19S RNA transcripts are shown, the arrowhead marking the 3' ends.

Below the circular map are linear representations of the genome organizations of CaMV, SbCMV, CsCMV and PVCV. The maps start at the intergenic region (the boxed arrow) and the relative positions of the ORFs are shown as boxes. The vertical arrow shows the position of D1. The infills of the boxes represent the functions of the ORF products.

b. Proteins encoded and their functions

The eight ORFs illustrated in Fig. 6.1 have been cloned, enabling *in vitro* transcription and translation experiments to be carried out. All eight ORFs could be translated *in vitro* (Gordon *et al.*, 1988). However, not all the protein products have been detected *in vivo* and not all have had functions unequivocally assigned to them. Gordon *et al.* (1988) tested translation products from all eight ORFs with an antiserum prepared against purified dissociated virus. Gene products of ORFs III, IV and V reacted with this antiserum. The others did not.

i. ORF I

The MW of the gene product expected from the DNA sequence is 37 kDa. The gene product has been produced in expression vectors and antiserum prepared. Such an antiserum detected virus-specific proteins of 46 kDa, 42 kDa and 38 kDa only in replication complexes isolated from infected tissue (Harker *et al.*, 1987b). In a similar study, Martinez-Izquierdo *et al.* (1987) found a virus-specified protein of $M_r \sim 41$ kDa. A 45 kDa product was found by Young *et al.* (1987). Given the ambiguities in MW determinations in gels, the major polypeptide found by all three groups is probably the expected 37 kDa protein, especially as the *in vitro* translation product had an $M_r = 41$ kDa (Gordon *et al.*, 1988).

The ORF I product was detected immunologically in an enriched cell wall fraction from infected leaves (Albrecht *et al.*, 1988). At early times after infection the protein was detected in replication complexes, from which it later disappeared. Immunogold cytochemistry indicated an accumulation of the ORF I protein at or in plasmodesmata between infected cells (Linstead *et al.*, 1988). It appeared to have a close association with the cell wall matrix associated with modified plasmodesmata, suggesting a role in cell-to-cell movement. Such a role is supported by some degree of sequence similarity between the ORF I protein and the TMV P30, which is involved in viral movement (Hull *et al.*, 1986). The role of this gene product in cell-to-cell movement of the virus is described in more detail in Chapter 9 (Section II.D.2.d).

ii. ORF II

A protein with $M_r \sim 19$ kDa is translated from ORF II mRNA *in vitro* (Gordon *et al.*, 1988) and an 18-kDa ORF II product can be isolated from infected leaf tissue in association with the viroplasms. Some naturally occurring strains of CaMV are transmitted by aphids, whereas others are not. Deletions made within ORF II led to loss of aphid transmissibility (Armour *et al.*, 1983; Woolston *et al.*, 1983). The Campbell isolate is a natural isolate of CaMV that is not aphid transmissible. The 18-kDa product is produced by this strain (Harker *et al.*, 1987a) but appears to be inactive in the aphid transmission function because of an amino acid change, glycine to arginine, in position 94 (Woolston *et al.*, 1987). Experiments in which various recombinant genotypes were produced between CaMV strains showed that aphid transmissibility mapped to ORF II (Daubert *et al.*, 1984). Collectively these results demonstrated a function in aphid transmission for the product of ORF II and the role of this gene product in aphid transmission is described in more detail in Chapter 11 (Section III.F.2). The ORF II protein also appears to increase the extent to which virus particles are held within the viroplasms

(Givord *et al.*, 1984) and makes up one of the types of viroplasm (see Espinoza *et al.*, 1991). This protein forms paracrystalline structures in insect cells (Fig. 6.2) (Blanc *et al.*, 1993a) that were found to include tubulin (Blanc *et al.*, 1996).

iii. ORF III

ORF III mRNA is translated *in vitro* to give a protein of the expected size (14 kDa) (Gordon *et al.*, 1988). A protein of about 15 kDa was detected *in vivo* using an antiserum raised against an NH_2-terminal 19-amino-acid peptide with a sequence corresponding to that of the ORF III (Xiong *et al.*, 1984). A fusion protein expressed in bacteria and consisting of the N terminus of β-galactosidase and the complete gene III protein (15 kDa) demonstrated DNA-binding activity with a preference for ds DNA (Giband *et al.*, 1986). Mesnard *et al.* (1990) showed that the gene III product is a non-sequence-specific DNA binding protein, the C-terminal part of the protein being responsible for this activity (Mougeot *et al.*, 1993). It was suggested that it is a structural protein within the virus particle. As described in Chapter 11 (Section III.F.2), the product of ORF III is also involved in the insect transmission of CaMV.

iv. ORF IV

The product of ORF IV is a 57-kDa precursor of the 42-kDa protein subunit of the icosahedral shell of the virus (Franck *et al.*, 1980; Daubert *et al.*, 1982). The 42-kDa protein is assumed to be derived from the 57-kDa molecule by proteolysis after formation of the virus shell (Hahn and Shepherd, 1982). This protein contains a cysteine motif (the *cys* motif) ($CX_2CX_4HX_4C$, where C = cysteine, H = histidine and X = any amino acid) that is considered to give a zinc finger structure and to be RNA binding; this motif is characteristic of the coat or *gag* proteins of most reverse transcribing elements (Covey, 1986). The coat protein is phosphorylated on serine and threonine residues by a protein kinase that is firmly bound to the virus (Ménissier de Murcia *et al.*, 1986; Martinez-lzquierdo and Hohn, 1987). The coat protein is also glycosylated to a limited extent (e.g. Du Plessis and Smith, 1981).

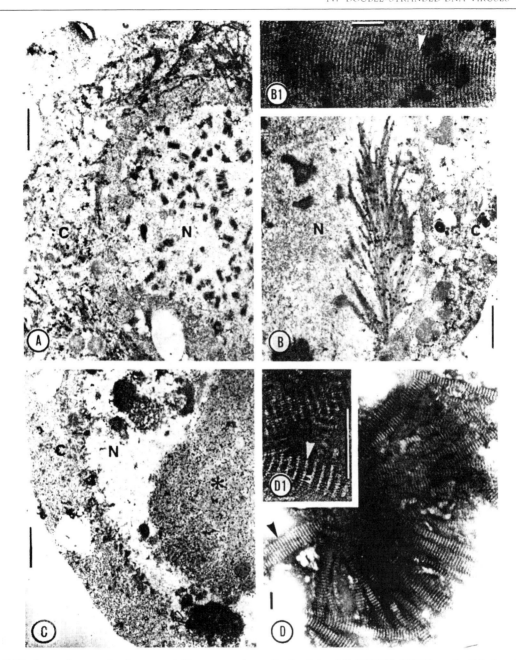

Fig. 6.2 Electron micrograph of paracrystalline array of negatively stained CaMV P18 (ORF 2 product) in baculovirus-recombinant infected insect cell. Bar corresponds to 1 μm. From Blanc *et al.* (1993), with permission.

v. ORF V

This is the largest ORF in the genome. mRNA derived from cloned DNA is translated *in vitro* to give a protein of the expected M_r of 79 kDa (Gordon *et al.*, 1988). The gene can also be expressed in *E. coli*. Antibodies raised against the protein produced in *E. coli* reacted with an 80-kDa protein in extracts from infected leaves as well as a series of smaller polypeptides (Pietrzak and Hohn, 1987). The following lines of evidence have established that ORF V is the viral reverse transcriptase gene:

1. A protein of the expected size is present both in replication complexes (Pfeiffer *et al.*, 1984) and in virus particles (Ménissier *et al.*, 1984; Gordon *et al.*, 1988). Furthermore the 80 kDa polypeptide associated with the replication complexes is recognized by antibodies against a gene V translation product (Laquel *et al.*, 1986).
2. There are regions of significant sequence similarity between gene V of CaMV and the reverse transcriptase gene of retroviruses.
3. Takatsuji *et al.* (1986) cloned gene V and expressed it in *Saccharomyces cerevisiae*. Yeast expressing the gene accumulated significant levels of reverse transcriptase activity.

It has been suggested that the CaMV reverse transcriptase gene may be translated as a large fusion protein involving the neighboring coat protein gene IV. This idea comes from analogy with the retroviruses, where there is a gene known as *gag* next to the *pol* (polymerase) gene. A *gag* polyprotein gives rise to four small structural proteins of the virus. The *pol* gene may encode a complex of four enzyme activities—reverse transcriptase, RNaseH, endonuclease and protease. The *pol* gene is translated as a large fusion polyprotein with the *gag* gene (reviewed by Mason *et al.*, 1987). The fusion protein suggestion for CaMV is supported by the fact that there is a domain in CaMV gene V with sequence similarity to the retrovirus protease. Using a dicistronic mRNA for genes IV and V transcribed from a cloned DNA construct, Gordon *et al.* (1988) could detect no such fusion protein, suggesting that the transcriptase gene is translated from its own AUG. However, Toruella *et al.* (1989) showed directly that the N-terminal domain of ORF V produces a functional aspartic protease that can process several CaMV polyproteins. Further properties of the aspartate protease and reverse transcriptase + RNaseH activities of this gene product are described in Chapters 7 and 8.

vi. ORF VI

The product of ORF VI (58 kDa) has been identified as the major protein found in CaMV viroplasms (e.g. Odell and Howell, 1980; Xiong *et al.*, 1982). RNA transcribed from cloned DNA was translated to give a polypeptide of appropriate size (Gordon *et al.*, 1988). The ORF VI product was detected immunologically in a cell fraction enriched for viroplasms (Harker *et al.*, 1987b). Various studies show that the ORF VI product plays a major role in disease induction, in symptom expression, and in controlling host range, but that other genes may also play some part (Schoelz and Shepherd, 1988; Baughman *et al.*, 1988). These topics are discussed further in Chapter 10 (Sections III.E.6 and III.O.1.c). Gene VI also functions in *trans* to activate translation of other viral genes (Chapter 7, Section V.B.7).

vii. ORFs VII and VIII

The two smallest ORFs (VII and VIII) are not present in another caulimovirus (CERV) so their significance is doubtful. However, they can be translated *in vitro* and there is indirect evidence for their existence *in vivo* (reviewed by Givord *et al.*, 1988). Gene VII has no role in aphid transmission. They have a high content of basic amino acids and maybe DNA binding proteins (Givord *et al.*, 1988). ORF VIII (not shown on Fig. 6.1) lies within ORF 4. Site-directed mutagenesis of this ORF, in which start and stop codons were removed or a new stop codon introduced, did not affect infectivity (Schultze *et al.*, 1990).

2. Genus 'Soybean chlorotic mottle virus-like' (reviewed by Reddy and Richins, 1999)

The genome of SbCMV (8175 bp) was sequenced by Hasegawa *et al.* (1989) and shown to contain nine ORFs (Fig. 6.1). The ORFs appeared to be equivalent to those of CaMV except that ORF I was divided into ORF Ia and Ib. ORF IV was identified as the coat protein based in the *cys* sequence and ORF V contained motifs suggestive of aspartate proteinase and reverse transcriptase.

3. Genus 'Cassava vein mosaic virus-like' (reviewed by de Kocho, 1999)

The 8158 bp genome of CsVMV was sequenced by Calvert *et al.* (1995) and five ORFs were identified (Fig. 6.1). ORF I (186 kDa) has a region in the middle containing the *cys* sequence characteristic of coat protein and a C-terminal region with similarity to the cell-to-cell movement proteins of caulimoviruses. The

77-kDa ORF III contained motifs indicative of aspartate proteinase, reverse transcriptase and RNaseH. The functions of the products of ORFs II, IV and V are unknown.

4. Genus 'Petunia vein clearing virus-like'

The 7205 bp genome of PVCV contains two definitive ORFs and possibly a third ORF (Fig. 6.1) (Richert-Pöggeler and Shepherd, 1997). ORF 1 of 126 kDa has an N-terminal motif similar to the CaMV cell-to-cell movement protein. In the C-terminal region of this ORF are the motifs HHCC and DD(35)E characteristic of the integrase of retroviruses and retrotransposons. The 125-kDa ORF II contains the *cys* sequence found in the RNA-binding domain of the caulimovirus coat protein, a sequence suggestive of an aspartate proteinase and sequences indicative of reverse transcriptase and RNaseH. Thus, this ORF is similar to the *gag*-pol of retroviruses. Recent sequencing indicates that there is only one large ORF comprising ORFs 1 and 2 illustrated in Fig. 6.1 (K. R. Richert-Pöggeler and T. Hohn, personal communication).

5. Genus *Badnavirus* (reviewed by Lockhart and Olszewski, 1999)

The genomes of five members of this genus have been sequenced and all have a very similar genome organization to that of the type member, ComYMV (Fig. 6.3). The genome of about 7.5 kb has two discontinuities, one in each strand. There are ORFs. The functions of the products of ORFs I and II are not known. ORF III encodes a polyprotein that is cleaved into at least four products, the capsid protein, an aspartate proteinase, reverse transcriptase + RNaseH and a putative cell-to-cell movement protein. It is thought that this processing is by the aspartate proteinase. The coat protein has the *cys* motif that is found in most reverse transcribing elements and a second cysteine motif ($CX_2CX_{11}CX_2CX_4CX_2C$) that is found only in members of the genera *Badnavirus* and rice tungro bacilliform-like viruses.

6. Genus 'Rice tungro bacilliform virus-like' (reviewed in Hull, 1996, 1999a)

The genome of the sole species in this genus, RTBV, contains four ORFs (Fig. 6.4). ORFs I to

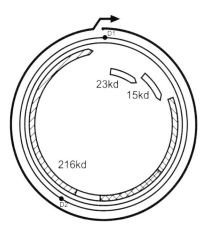

Fig. 6.3 Genome organization of CoYMV. The double complete circles represent the dsDNA genome, the positions of the discontinuities (D1 and D2) being represented by ●. Inside the double circles the arced boxes represent the ORFs, labelled ORF 1–3. Outside the double circles, the position of the 35S RNA transcript is shown, the arrowhead marking the 3' end. The infills of the boxes represent the functions of the ORF products — see Fig. 6.1 for details.

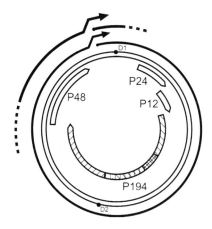

Fig. 6.4 Genome organization of RTBV. The double complete circles represent the dsDNA genome, the positions of the discontinuities (D1 and D2) being represented by ●. Inside the double circles the arced boxes represent the ORFs, labelled ORF 1–4. Outside the double circles, the positions of the 35S RNA transcript and the spliced mRNA for ORF 4 are shown, the arrowheads marking the 3' ends. The infills of the boxes represent the functions of the ORF products—see Fig. 6.1 or inside book cover for details.

III are similar to those of members of the genus *Badnavirus*. ORF IV is expressed from an mRNA spliced from the more-than-genome length 35S RNA. The product has a leucine zipper motif but its function is unknown.

V. SINGLE-STRANDED DNA VIRUSES

A. Family *Geminiviridae* (reviewed by Buck, 1999a)

1. Genome structure

As described in Chapter 2 (Section III.B), four genera have been described within the family *Geminiviridae*—the mastreviruses, the curtoviruses, the begomoviruses and the topocuviruses. The genomes of all four genera comprise circular ss DNA and are either as one species or divided between two species. The genera are distinguished on their genome organizations. However, all geminiviruses have a conserved genome sequence in common. There is a large (–200 base) non-coding inter-genic region, termed the common region, in the two-component geminiviruses, features of which are found in the similar genomic posi-tion in the single-component geminiviruses. This region has sequences capable of forming a hairpin loop. Within this loop is a conserved sequence, TAATATTAC (see Fig. 8.26) found in all geminiviruses (Lazarowitz, 1987).

2. Genus *Mastrevirus* (reviewed by Palmer and Rybicki, 1998)

The nucleotide sequence of the MSV genome (Nigerian strain) revealed a single DNA circle of 2687 nucleotides that had no detectable overall sequence similarity with genomes of geminiviruses with two DNA components (Mullineaux *et al.*, 1984). Analysis of ORFs with potential for polypeptides of 10 kDa or larger revealed four potential coding regions arranged as shown in Fig. 6.5. There are two intergenic regions, the long intergenic region (LIR) and the short intergenic region (SIR). As will be described in Chapter 7, most, if not all, of the transcription starts in the LIR and ter-minates in the SIR. A slightly different map from that found for the Nigerian isolate was

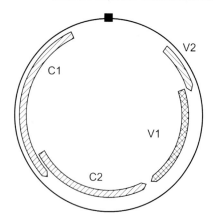

Fig. 6.5 Genome organization of MSV. The circle represents the ssDNA; the black box the common region. The arced boxes within the circle are the ORFs, the infills representing the functions of the products of each ORF—see Fig. 6.1 or inside book cover for details.

determined for a South African isolate of MSV on a clone that was infectious for maize using an *A. tumefaciens* delivery system (Lazarowitz, 1988).

As can be seen from Fig 6.5, two of the ORFs are on the viral strand (given the designation V) and two are on the complementary strand (given the designation C). ORF V1 encodes the cell-to-cell movement protein (described in more detail in Chapter 9, Section II.D.2.i), ORF V2 the virion coat protein, and ORFs C1 and C2 encode the two replication-associated proteins. The functions of these replication-associated proteins are described in Chapter 8 (Sections VIII.D.4 and 5).

3. Genus *Curtovirus*

The sequence of 2993 nucleotides in an infec-tious clone of BCTV has been established (Stanley *et al.*, 1986). The organization of the single-component genome (Fig. 6.6) reveals that curtovirus DNAs produce six or seven pro-teins. The product of ORF V1 is the virion coat protein, that of ORFs V2 is involved in regula-tion of the genomic ss and ds DNA levels, and that of ORF V3 facilitates the cell-to-cell move-ment of the virus. The four proteins encoded by the complementary-sense ORFs are involved

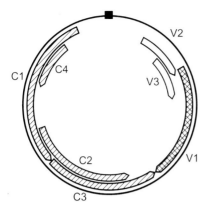

Fig. 6.6 Genome organization of BCTV. The circle represents the ssDNA, the black box the common region. The arced boxes within the circle are the ORFs, the infills representing the functions of the products of each ORF—see Fig. 6.1 or inside book cover for details.

have two component genomes, the single-component ones being found only in the 'Old World' (Eurasia, Africa and Australasia).

The two-component begomoviruses encode five or six genes on DNA A, one or two on the virion-sense strand (termed AV) and four in the complementary-sense strand (AC). DNA B encodes two proteins, one on the virion-sense strand (BV) and one on the complementary strand (BC) (Fig. 6.7).

The product of ORF AV1 is the coat protein, that of AV2 the cell-to-cell movement protein; AV2 is not present in 'New World' begomoviruses. The proteins encoded by the complementary sense of DNA A are all involved with replication and expression (see Chapter 8, Section VIII). AC1 is the Rep protein, AC2 the transcriptional activator protein, and AC3 the replication enhancer protein; the function of AC4 is unknown. BV1, encoded on the virion

in virus replication. ORF C1 encodes the Rep protein (see Chapter 8, Section VIII.D.4), C3 a protein analogous to the replication enhancer protein of begomoviruses, and C4 a protein that can initiate cell division; the function of the product of ORF C2 is unknown.

4. Genus *Begomovirus*

The full nucleotide sequence of ACMV DNA showed that the genome consists of two ss DNA circles of similar size (Stanley and Gay, 1983). Stanley (1983) demonstrated that both DNAs are needed for infectivity. Among all well-characterized DNA viruses, these and similar geminiviruses are the only type with a bipartite genome. The nucleotide sequences of the two DNAs are very different except for a 200-nucleotide non-coding region, the common region, which is almost identical on each DNA. The two DNAs have been arbitrarily labeled A and B (some workers use the terminology 1 and 2). Since these initial characterizations, several viruses that have other properties that assign them to the genus *Begomovirus* (e.g. whitefly transmission, infection of dicotyledons) have been found to have genomes comprising a single DNA component. Most of the 'Old World' and all the 'New World' begomoviruses

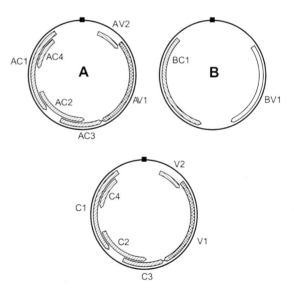

Fig. 6.7 Genome organization of begomoviruses. The upper pair of diagrams are for two-component viruses (components A and B) and the lower diagram for single-component viruses. The circle represents the ssDNA; the black box the common region. The arced boxes within the circle are the ORFs, the infills representing the functions of the products of each ORF—see Fig. 6.1 or inside book cover for details.

sense of DNA B, is the nuclear shuttle protein (Chapter 9, Section II.C) and BC1 is involved (together with AV2 when present) in cell-to-cell movement.

The genome organization of the single-component begomoviruses is essentially the same as that of DNA A of the two-component viruses (Fig. 6.7).

5. Genus *Topocuvirus*

The single-component genome of TPCTV has properties of both mastreviruses and begomoviruses (Briddon *et al.*, 1996) (Fig. 6.8). The coat protein has features more akin to the leafhopper-transmitted mastreviruses than to those transmitted by whiteflies whereas the organization of the complementary-sense genes is similar to that of the single-component begomoviruses.

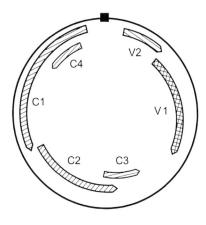

Fig. 6.8 Genome organization of TPCTV. The circle represents the ssDNA, the black box the common region. The arced boxes within the circle are the ORFs, the infills representing the functions of the products of each ORF—see Fig. 6.1 or inside book cover for details.

6. Functions of the ORF products

a. Coat protein

The ORF encoding coat protein maps to similar positions on the single-component genome or DNA A of the two-component viruses.

As well as its function in protecting the viral genome it might be expected that the properties of the coat protein might play a role in insect vector specificity. This idea is supported by the sequence homologies between the coat proteins of viruses transmitted by *Bemisia tabaci*. By contrast, the coat protein ORFs of BCTV, MSV and WDV, each of which has a different leafhopper vector, show much less homology (MacDowell *et al.*, 1985; Stanley *et al.*, 1986). Briddon *et al.* (1990) constructed chimeric clones in which the coat protein gene in DNA1 of ACMV (whitefly transmitted) was replaced with the coat protein gene of BCTV (leafhopper transmitted). The BCTV gene was expressed in plants and gave rise to particles containing the ACTV DNAs. These chimeric particles were transmitted by the BCTV leafhopper vector, when injected into the insects, whereas normal ACMV was not. These results demonstrated that vector specificity is a function of the coat protein, a topic explored in more detail in Chapter 11 (Section IV.D.1).

Genetic analysis of TGMV (whitefly transmitted) has shown that the coat protein is not essential for infectivity, for systemic spread, or for symptom development (Gardiner *et al.*, 1988). Unlike the situation with the bipartite TGMV, the coat protein gene of monopartite viruses is essential for infectivity, as shown, for example, for MSV by Boulton *et al.* (1989), Lazarowitz *et al.* (1989), and for WDV (Woolston *et al.*, 1989) and BCTV by Briddon *et al.* (1989).

b. ORFs related to DNA replication

The ORFs involved directly in DNA replication are found on the complementary sense of the single-component DNAs and of DNA A of two-component viruses. These gene products are described in more detail in Chapter 8 (Sections VIII.D.4 and 5). It should be remembered that the function of replication is closely coordinated with other functions and that many regions of the viral genome, both coding and non-coding, are likely to be involved in the synthesis of new viral DNA molecules.

c. Virus movement

Geminiviruses encode proteins that are involved in both intracellular and intercellular movement of viral genomes. Within the cell the viral genome has to enter the nucleus; this is mediated by virus-encoded products as described in Chapter 9 (Section II.C). Similarly, cell-to-cell movement is facilitated by virus-encoded proteins (see Chapter 9, Section II.D.2.i).

d. Insect transmission

Since DNA B is found only with whitefly-transmitted begomoviruses, it had been suggested that products of this DNA must be involved in insect transmission. Insertion and deletion mutagenesis experiments showed that both coding regions of ACMV DNA2 had other functions in addition to any possible role in insect transmission (Etessami *et al.*, 1988). The experiments described in Chapter 11 (Section IV.D.1) show that the coat protein, a product of a DNA A gene, controls transmission by insect vectors.

B. Family *Circoviridae*

1. Genus *Nanovirus*

The genomes of members of this genus consist of multiple circular ss DNA molecules of approximately 1 kb in size. The number of distinct DNAs varies from virus to virus; for example, the Australian isolate of BBTV has 6 DNA components (Fig. 6.9) whereas FBNYV has 11 DNA components (Table 6.1).

All the DNA components appear to be structurally similar in being (+) sense, transcribed in one direction and containing a conserved stem-loop structure and other domains in the non-coding region (Fig. 6.9).

The individual DNA components encode products involved in replication and also ones involved in non-replication functions. The replication (Rep) proteins are described in more detail in Chapter 8 (Section VIII.E) as are those that potentiate replication such as the protein that interacts with a retinoblasma-like plant protein. The functions of the nanovirus proteins as far as they are known are listed in Table 6.1. It is obvious from the table that there needs to be a better way of naming the DNA components to give consistent information on their gene products.

The number of distinct DNA components that comprise a fully functional nanovirus genome has not yet been determined. It is possible that some are 'satellite' molecules, especially when one considers that the Australian isolate of BBTV has six DNA components and that from Taiwan has eight.

VI. DOUBLE-STRANDED RNA VIRUSES

Double-stranded RNA viruses have genomes that consist of multiple linear dsRNA molecules. In most cases the dsRNA segments are monocistronic. There are some segments that apparently have two ORFs, but is has not been established that both ORFs are expressed. The genera differ in the number of RNA molecules that make up the genome.

A. Family *Reoviridae*

The genomes of plant reoviruses comprise either 10 or 12 RNA segments. These segments are numbered according to their electrophoretic mobility in gels, the slowest being segment 1. However, sequencing of the genome segments has shown that the true size is not necessarily the size deduced from its electrophoretic mobility. There are several examples below of higher numbered RNA segments being larger than smaller numbered ones.

Four types of gene products are recognized, those that make up the capsid (structural proteins), those involved with RNA replication, non-structural proteins, and those for which no function is known. In several cases, the function is attributed because of sequence similarities or common motifs with better characterized vertebrate-infecting reoviruses.

1. Genus *Fijivirus* (reviewed by Marzachi *et al.*, 1995)

The genome of Fijiviruses is divided between 10 RNA segments. The coding assignments for those that are known are summarized in Table 6.2. Also listed in the table are the coding assignments that have been determined for

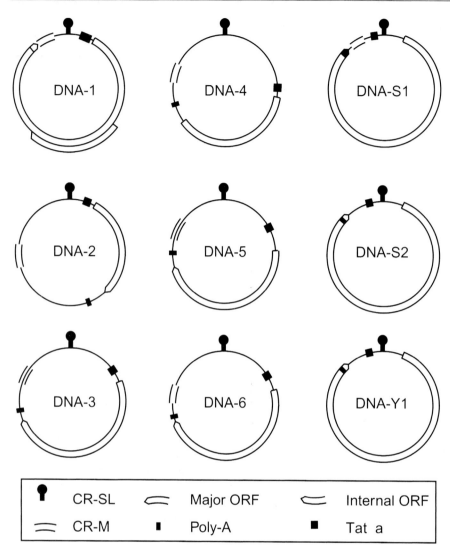

Fig. 6.9 Genome organization of BBTV showing the nine circular ssRNA components identified in this species. DNAs 1–6 occur in all species whereas components S1, S2 and Y1 are found also in Taiwan isolates. From Randles *et al.* (2000), with permission.

insect-infecting *Nilaparvata lugens reovirus* which is classified as a Fijivirus (Nakashima *et al.*, 1996). The functions of more of the RNA segments have been determined for this virus and this gives an indication of what functions may be encoded by the Fijiviruses that also infect plants. It should be remembered that these viruses replicate in their insect vectors as well as plants.

As described in Chapter 5 (Section VII.2), Fijivirus particles are double-shelled with 'A'-type spikes on the 12 vertices of the outer shell and 'B'-type spikes on the inner shell or core.

The gene product from segment 3 of NLRV is the major core protein and those from segments 1 and 7 contribute enzymatic activities also found in the core. The product from segment 2 has been attributed to the 'B' spike on the inner core. Three of the products, those from segments 8, 9 and 10, give proteins described as non-structural and no function has yet been ascribed to the products of segments 4, 5 and 6.

2. Genus *Oryzavirus*

Oryzaviruses have 10 RNA segments. The genome properties of the type member of this

TABLE 6.1 Functions of components of nanoviruses

	BBTV	FBNYV	MDV	SCSV
No. of DNA components	6 (8)[a]	11	10	7
Function				
Rep[b]	C1[f]	C2	C10	C2, C6
Rep-associated[c]		C1, C7, C9, C11	C1, C2, C3	
Rb-interacting[d]	C5	C10	C4	
Capsid protein	C3	C5	C9	C5
Putative movement protein	C4	?C4	C8	
Nuclear shuttle	C6			
Unknown	C2	C3, C6, C8	C5, C6, C7	C1, C3, C4, C7
Other functions[e]		Aphid transmission		

[a] Some isolates reported to have eight components.
[b] Protein(s) considered to be that which initiates replication.
[c] Proteins resembling Rep.
[d] Protein that interacts with a retinoblasma-like plant protein possibly controlling cell cycle.
[e] Other functions identified for which no gene product ascribed.
[f] Component number as in publication.

Data from: BBTV: Burns *et al.* (1995), Wanitchakorn *et al.* (2000); FBNYV: Katul *et al.* (1997, 1998), Franz *et al.* (1999), Timchenko *et al.* (1999); MDV: Sano *et al.* (1998); SCSV: Boevink *et al.* (1995); all with permission.

genus, RRSV, have been most studied and are summarized in Table 6.3. Unlike the other two genera, the RNA polymerase is not encoded on the largest RNA segment but on the second largest one. However, like the other two genera it is the largest gene product.

As with most other plant reoviruses, the particles of oryzaviruses are double-shelled with A-type spikes on the outer shell and B-type spikes on the inner shell. The B-type spikes are encoded on segment 1 and the A-type spikes, which are involved in vector transmission by delphacid planthoppers, are encoded by segment 9.

3. Genus *Phytoreovirus* (reviewed by Omura, 1995; Hillman and Nuss, 1999)

Phytoreoviruses have 12 genome segments. The genome of RDV has been studied in detail and the properties of the RNA segments are summarized in Table 6.4. Ten of the genome segments have one ORF, one has two ORFs and one three. However, it is not certain as to whether the additional ORFs are expressed *in vivo*.

It can be seen from Table 6.4 that segments 2, 3 and 8 encode structural proteins; the outer capsid protein encoded by segment 2 is essential for vector transmission. The products of segments 1, 5, 7 and possibly one of the products of segment 12 are involved with RNA replication, and those of segments 4, 6, 9, 10, 11 and one encoded by segment 12 are classed as non-structural proteins.

4. Discussion

As noted earlier, the RNA segments of plant reoviruses are identified by their electrophoretic mobility in gels. With information becoming available on sequence and functions of these RNAs this classification of the parts of the genome is beginning to give rise to confusion. For instance, it can be seen from Table 6.2 that similar-numbered (or -sized) RNA segments from different viruses do not necessarily encode the same function. Thus, the core protein that has NTP-binding activity is encoded by segment 8 (1927 bp) of RBSVD, possibly by segment 7 (1936 bp) of MRDV, segment 9 (1893 bp) of OSDV and

TABLE 6.2. Genome properties of members of the genus *Fijivirus*

Seg-ment no.	RBSVD			MRDV			OSDV			NLRV		
	Size (bp)	Protein size (kDa)	Protein function	Size (bp)	Protein size (kDa)	Protein function	Size (bp)	Protein size (kDa)	Protein function	Size (bp)	Protein size (kDa)	Protein function
1	ND	–	–	ND	–	–	ND	–	–	4391	165.9	Core/Pol
2	ND	–	–	ND	–	–	ND	–	–	3732	136.6	OsB
3	ND	–	–	ND	–	–	ND	–	–	3753	138.5	Mc
4	ND	–	–	ND	–	–	ND	–	–	3560	130	U
5	ND	–	–	ND	–	–	ND	–	–	3427	106.4	U
6	ND	–	–	2193	41.0 36.3	Ns Tup U	ND	–	–	2970	95.1	U
7	2193	41.2 36.4	Ns/Tup Ns	1936	68.1	?Core/NTP	1944	42.0 30.0	?Ns/TuP U	1994	73.5	Core/NTP
8	1927	68.1	Core/?NTP	1900	40 24.2	?Ns/VP U	1874	66.2	?Mos	1802	62.4	Mos
9	1900	39.9 24.2	Ns VP Ns	ND	–	–	1893	68.2	?Core/NTP	1640	33.0 23.6	Ns/VP Ns
10	1801	63.3	Mos	1802	62.9	?Mos	1761	35.7 22.7	?Ns/VP U	1430	49.4	Ns/?Tup

Data from van Regenmortel *et al.* (2000), with permission. Core, virus core protein; Mos, major outer shell protein; Mc, major core protein; ND, not determined; Ns, non-structural protein; NTP, NTP-binding protein; OsB, outer shell, B spike; Pol, RNA polymerase; TuP, protein forms tubular structure; U, unknown function; VP, protein found in viroplasm; ?, probable function indicated by equivalence to other virus species.

TABLE 6.3 Genome properties of RRSV

Segment no.	Size (bp)	Protein size (kDa)	Function
1	3849	137	OsB
2	3810	118	Core
3	3669	130	Mc
4	3823	145	Pol
		36.9	U
5	2682	90	Gt
6	2157	65.6	U
7	1938	66	Ns
8	1814	67	Pre
		25.6	Pro
		41.7	Mos
9	1132	38.6	Sp
10	1162	32.3	Ns

Data from van Regenmortel *et al.* (2000), with permission. Core, core protein; Gt, guanyltransferase activity; Mc, major core protein; Mos, major outer shell protein; Ns, non-structural protein; OsB, outer shell B spike; Pol, RNA polymerase; Pre, precursor protein; Pro, protease; Sp, spike involved in vector transmission; U, unknown.

TABLE 6.4 Genome properties of RDV, Akita isolate

Segmen. no.	Size (bp)	Protein size (kDa)	Function
1	4423	164.1	Core/Pol
2	3512	123.0	Os
3	3195	114.2	Mc
4	2468	79.8	Ns
5	2570	90.5	Core/Gt
6	1699	57.4	Ns
7	1696	55.3	Core/Nab
8	1427	46.5	Mos
9	1305	38.9	Ns
10	1321	39.2	Ns
11	1067	20.0	Ns
		20.7	U
12	1066	33.9	Ns
		10.6	Pp
		9.6	U

Data from van Regenmortel *et al.* (2000), with permission. Core, core protein; Gt, protein has guanylytransferase activity; Mc, major core protein; Mos, major outer shell protein; Nab, nucleic acid binding protein; Ns, non-structural protein; Os, outer shell protein; Pol, RNA polymerase; Pp, phosphoprotein; U, unknown.

segment 7 (1994 bp) of NLRV; however, segment 9 of OSDV is larger than segment 8. The understanding of the genome organization and functions of plant reoviruses will obviously increase, and thus there is a need to develop a less confusing and more informative system for designating the genome segments.

B. Family *Partitiviridae* (reviewed by Milne and Marzachi, 1999)

The two plant genera of this family, the genus *Alphacryptovirus* and the genus *Betacryptovirus*, each have genomes comprising two dsRNA genome segments. In alphacryptoviruses the larger segment encodes the virion-associated RNA polymerase and the smaller segment codes for the capsid protein. It is thought that betacryptoviruses have the same genome arrangement.

C. Genus *Varicosavirus*

LBVV has a bipartite genome of dsRNA segments of approximately 7 and 6.5 kbp. Currently, nothing is known about the genome organization.

VII. NEGATIVE-SENSE SINGLE-STRANDED RNA GENOMES

There are two families (*Rhabdoviridae* and *Bunyaviridae*) and two unassigned genera (*Tenuivirus* and *Ophiovirus*) of plant viruses with (−)-sense ssRNA genomes. In each of the viral genomic RNA is closely associated with a protein, termed the nucleocapsid protein. In members of the two families this is encapsidated in membrane-bound particles. The structures of these involve several virus-encoded proteins and are described in Chapter 5 (Section VIII). The genomes of tenuiviruses and ophioviruses do not appear to be contained within membranous particles though, from the relationship that tenuiviruses have with Bunyaviruses, membrane-bound particles might be expected. All the

(−)-sense RNA viruses have the virus-coded RNA polymerase associated with the virion as this is required for initial transcription as an early step in the infection cycle (see Chapter 7, Section IV.A).

A. Family *Rhabdoviridae* (reviewed by Jackson *et al.*, 1999)

As described in Chapter 2 (Section III.G), there are two genera of plant-infecting rhabdoviruses, the *Cytorhabdovirus* and the *Nucleorhabdovirus*. In many properties these resemble animal-infecting rhabdoviruses, the genome organizations of several of which are well characterized.

Thus, plant-infecting rhabdoviruses, like those infecting vertebrates, possess a genome consisting of a single piece of single-stranded negative-sense RNA, with a length in the range of 11 000–13 000 nucleotides.

The genomes of the cytorhabdovirus, LNYV (Wetzel *et al.*, 1994), and of the nucleorhabdovirus, SYNV (Heaton *et al.*, 1989a), have been sequenced and shown to have very similar organizations. The genome organization of SNYV is shown in Fig. 6.10. From the 3' end (this is a (−)-strand genome) there is a 144 nucleotide leader sequence followed by six genes, with short intergenic regions between them. Each gene is transcribed separately as described in Chapter 7 (Sections IV.A and V.F).

Four of the proteins are structural (for rhabdovirus structure see Chapter 5, Section VIII). The N gene gives the 54-kDa nucleocapsid protein, M1 and M2 are matrix proteins, M2 being phosphorylated and the G gene gives the glycosylated spikes (70 kDa) that extend through the viral membrane to the surface of the particle. The L gene encodes the RNA polymerase (241 kDa) that is incorporated in the virion. These five proteins are found in animal-infecting rhabdoviruses, but the sixth, sc4, is unique to plant rhabdoviruses. This 37-kDa protein is membrane-associated and is rich in serine and threonine residues (16%). It has four potential casein kinase II phosphorylation sites and contains motifs related to *a*-amylases and aspartate proteases (Scholthof *et al.*, 1994). It is thought that this protein either effects some control on viral replication, or facilitates cell-to-cell movement of the virus through plasmodesmata.

B. Family *Bunyaviridae*

1. Genus *Tospovirus* (reviewed by Moyer, 1999)

The particles of tospoviruses encapsidate three RNA species, the largest of which (L RNA) is (−) sense and the other two (M and S RNAs) each encode two proteins in the ambisense arrangement (see Chapter 7, Section V.B.12 for description of the ambisense arrangement) (Fig. 6.11). Thus, the viral genome codes for five proteins. The protein encoded by the largest RNA is the RNA polymerase (330 kDa). The virion-sense of the M RNA encodes a 34-kDa protein (termed NS$_M$) that is involved in cell-to-cell movement of the virus (see Chapter 9, Section II.D.2.j). The complementary sense of this RNA encodes a 127-kDa protein that is processed into two polypeptides each of which is glycosylated. The C-terminal G1 is estimated

to be 75 kDa and N-terminal G2 to be 46 kDa; because of glycosylation, these migrate as 78 kDa and 58 kDa respectively (Kormelink *et al.*, 1992b). These two glycoproteins form the spikes on the surface of the virus particles. The function of the protein (NS$_S$; 52 kDa) encoded by the virion sense of S RNA is unknown; the complementary sense ORF on this RNA encodes the 29-kDa nucleocapsid protein.

2. Genus *Tenuivirus* (reviewed by Ramires and Haenni, 1994; Falk and Tsai, 1998; and Falk, 1999)

The genome of most tenuiviruses is divided between four RNAs (Fig. 6.12). In some tenuiviruses (e.g. some isolates of RSV and RHBV) there is a fifth RNA that has a single ORF but it is not certain as to whether this is an essential part of the viral genome or a satellite-like molecule. As with tospoviruses, the largest RNA (RNA1) is (−) sense and encodes the RNA polymerase (337 kDa). The other three RNAs

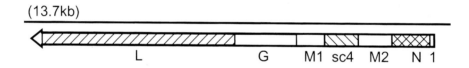

(13.7kb)

L   G   M1   sc4   M2   N   1

Fig. 6.10 Genome organization of SYNV as representative of nucleorhabdoviruses. The single line represents the (−)-strand RNA genome, and the box the coding region with the final gene products identified. The arrowhead indicates the C-terminus. The infills represent the function of the gene products—see Fig. 6.1 or inside book cover for details.

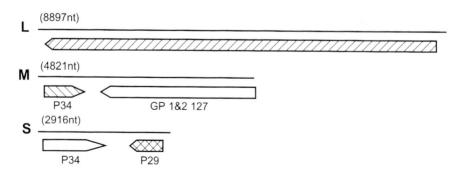

L   (8897nt)

M   (4821nt)

P34   GP 1&2 127

S   (2916nt)

P34   P29

Fig. 6.11 Genome organization of TSWV. The single lines represent the (−)-strand RNA genome segments, and the boxes the coding regions with the final gene products identified. The arrowheads indicate the C-terminus. The infills represent the function of the gene products—see Fig. 6.1 or inside book cover for details.

(RNA2-4) each have an ambisense arrangement of two ORFs (Fig. 6.12). The virion-sense ORF on RNA3 encodes the nucleocapsid protein and the virion-sense ORF on RNA 4 encodes a major non-structural protein that accumulates in infected plant cells. The functions of the products of the other four ORF on these three ambisense RNAs have not yet been determined with certainty.

RGSV differs from the other tenuiviruses in that its genome comprises six segments, all with an ambisense arrangement (Toryama *et al.*, 1998). The virus-complementary (vc) sequence on RNA1 encodes the 339-kDa RNA polymerase. The function of the product of the viral (v)-sense ORF of RNA1 (19 kDa) is unknown. The products of the v and vc ORFs of RNA2 resembled those of RNA2 of other tenuiviruses. However, the functions of the products from the ORFs of RNAs 3–6 are unknown except that the vc ORF of RNA5 encodes the nucleocapsid protein.

3. Genus *Ophiovirus*

The genome of ophioviruses is divided between three segments of 7.5–9.0, 1.6–1.8 and 1.5 kb. Nothing is known about the organization of genes within these RNAs.

VIII. POSITIVE-SENSE SINGLE-STRANDED RNA GENOMES

As noted in Table 4.1, the majority of plant viruses have genomes of positive-sense ssRNA.

This can be either as a single molecule or the genome can be divided into several molecules.

A. Family *Bromoviridae*

1. Genus *Bromovirus* (reviewed by Ahlquist, 1999)

a. Genome structure

BMV has a tripartite genome totaling 8243 nucleotides (Fig. 6.13). In addition, a subgenomic RNA containing the coat protein gene is found in infected plants and in virus particles. This coat protein gene is encoded in the sequence toward the 3′ end of RNA3. Each of the four RNAs has a 5′ cap and a highly conserved 3′-terminal sequence of about 200 nucleotides. The terminal 135 nucleotides of this sequence can be folded into a three-dimensional tRNA-like structure (Perret *et al.*, 1989), which accepts tyrosine, in a reaction similar to the aminoacylation of tRNAs, but this reaction probably needs the 3′-terminal 155 nucleotides.

The genome of BMV has been completely sequenced (Ahlquist *et al.*, 1984a). RNAs 1 and 2 each encode a single protein on what are termed ORFs 1a and 2a (Fig. 6.13). RNA3 encodes a protein of 35 kDa on ORF 3a at its 5′ end. Between this cistron and the coat protein cistron there is an intercistronic non-coding region approximately 250 nucleotides long. An internal poly(A) sequence of heterogeneous length (16–22 nucleotides) occurs in this intercistronic region ending 20 bases 5′ to the start of

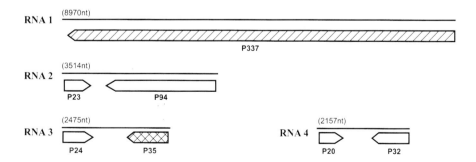

Fig. 6.12 Genome organization of RSV as representative of tenuiviruses. The single lines represent the (–)-strand RNA genome segments, and the boxes the coding regions with the final gene products identified. The arrowheads indicate the C-terminus. The infills represent the function of the gene products—see Fig. 6.1 or inside book cover for details.

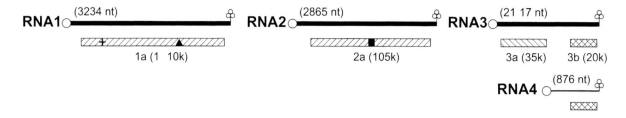

Fig. 6.13 Genome organization of BMV as representative of bromoviruses. The thick single lines represent the (+)-strand RNA genome segments, and the thin line the subgenomic RNA4. The boxes show the coding regions with the final gene products identified. The infills represent the function of the gene products—see Fig. 6.1 or inside book cover for details of these and other features.

the coat protein gene (ORF 3b). The first nine bases of the RNA4 consist of the last nine bases of the intercistronic region. The internal poly(A) sequence is not essential for RNA3 replication. However, during replication, *in vivo*, of constructs lacking this sequence, restoration of the poly(A) took place (Karpova *et al.*, 1989).

The genomes of BBMV and CCMV have also been sequenced (Allison *et al.*, 1989; Dzianott and Bujarski, 1991; Romero *et al.*, 1992). CCMV has extensive sequence similarity with the proteins of BMV and also CMV. However, the CCMV RNA3 5′ non-coding region contains a 111-base insertion not found in the other viruses.

b. Proteins encoded and their functions

BMV RNAs are efficient mRNAs in *in vitro* systems, and in particular in the wheat germ system that is derived from a host plant of the virus. In this system, RNA1 directs the synthesis of a single polypeptide of 110 kDa and RNA2 directs a single polypeptide of 105 kDa (Shih and Kaesberg, 1976; Davies, 1979). RNA3 directs the synthesis of a 35-kDa protein (Shih and Kaesberg, 1973). The coat protein cistron in RNA3 is not translated. RNA4 directs synthesis of the 20-kDa coat protein very efficiently. It is preferentially translated in the presence of the other viral RNAs in part, at least, because of more efficient binding of ribosomes (Pyne and Hall, 1979). *In vitro* the coat protein inhibits RNA synthesis by the BMV replicase, in a specific manner, possibly by partial assembly of nucleoprotein (Horikoshi *et al.*, 1987).

Four new proteins were observed in tobacco protoplasts after infection with BMV. These

had MWs of 20 kDa (coat protein), 35 kDa, 100 kDa and 107 kDa (Sakai *et al.*, 1979). Okuno and Furusawa (1979) also found four BMV-induced proteins with the same MWs (within the error of estimation in gels). They were found in infected protoplasts prepared from three plant species—a systemic host, a local lesion host and a non-host. These four proteins account for over 90% of the viral genome. They correspond well in size with the *in vitro* products noted earlier and the precise MWs in Fig. 6.13.

RNAs 1 and 2 inoculated alone to protoplasts replicate, showing that they must contain a replicase function (Kiberstis *et al.*, 1981). French *et al.* (1986) demonstrated with cloned material, where no contaminating RNA3 could be present, that RNAs 1 and 2 by themselves could replicate. Further evidence that the proteins from RNAs 1 and 2 are involved in RNA replication comes from the amino acid sequence similarities with proteins of other viruses known to function as viral RNA polymerases, or genetically implicated in replication functions (discussed in Chapter 8, Section IV.E). An active replicase can be isolated from BMV-infected tissue. Bujarski *et al.* (1982) demonstrated the presence of a 110-kDa protein in such a replicase fraction that had the same electrophoretic mobility and tryptic polypeptide pattern as the *in vitro* translation product of RNA1. Quadt *et al.* (1988) synthesized C-terminal peptides derived from the polypeptides coded by RNAs 1 and 2. Antibodies raised against these peptides could recognize the corresponding native proteins in replicase preparations. Antibodies against the polypeptide of

RNA1 but not that of RNA2 completely blocked replicase activity. Similar results were obtained by Horikoshi *et al.* (1988).

Kroner *et al.* (1989) introduced defined mutations resulting in substitutions between amino acids 451 and 484 of the BMV RNA2 protein (94 kDa), a region conserved in many (+)-strand RNA viruses. Four mutants had unconditional blocks in RNA synthesis while five others showed *ts* defects in RNA replication. Two of the *ts* mutants also showed a preferential reduction in the synthesis of genomic RNA relative to subgenomic RNA, at both permissive and non-permissive temperatures. Other studies using directed mutagenesis have shown that each of the three conserved domains in BMV gene products of RNAs 1 and 2 (109 and 94 kDa) are involved in RNA replication (Ahlquist *et al.*, 1990).

As originally suggested by Haseloff *et al.* (1984) the protein translated from RNA3 has a role in cell-to-cell movement (see Chapter 9; Section II.D.2.e). This function is supported by the fact that RNAs 1 and 2 will replicate in protoplasts in the absence of RNA3 (Kiberstis *et al.*, 1981) but this RNA is needed for systemic plant infection. Thus, functions have been assigned for all four proteins known to be translated from the BMV genome.

2. Genus *Cucumovirus* (reviewed by Roossinck, 1999a; 1999b)

The cucumoviruses have many properties in common with the bromoviruses. The 3′ termini of the three genomic RNAs and of the sgRNA4 within a species are highly similar and can form tRNA-like structures aminoacylatable with tyrosine. The genome organization is similar to that of the bromoviruses (Fig. 6.14) except that each of the three genomic RNAs and the subgenomic coat protein RNA are slightly longer and there is an additional ORF (2b) on RNA2. ORF 2b overlaps ORF 2a by 69 codons (Ding *et al.*, 1994) and is expressed from a subgenomic RNA, RNA4A. The product from ORF 2b associates with the nucleus (Mayers *et al.*, 2000). Expression of the genome gives five protein products, four of which (those from ORFs 1a, 2a, 3a and 3b) have similar functions to those of bromoviruses. ORF 2b is required

for systemic spread and disease induction (S.-W. Ding *et al.*, 1995) and an interspecific hybrid in which CMV ORF 2b was replaced by that of TAV is significantly more virulent than either parental virus (S.-W. Ding *et al.*, 1996). As discussed in Chapter 10 (Section IV.H), these properties of ORF 2b are conferred by its ability to suppress the normal host defense system.

The ability of strain Fny of CMV to induce rapid and severe disease in zucchini squash is a function of RNA1 (Roossinck and Palukaitis, 1990). Infectious RNA transcripts have been obtained from full-length cDNAs of all three CMV RNAs (Rizzo and Palukaitis, 1990).

A major difference from other members of the *Bromoviridae* is the occurrence of satellite RNAs in many CMV isolates (see Chapter 14, Section II.B.2).

3. Genus *Alfamovirus* (reviewed by Bol, 1999a)

a. Genome structure

The genome of AMV is tripartite with a fourth subgenomic coat protein mRNA (Fig. 6.15). The arrangement is very similar to that of BMV. The genome of strain 425 was the first tripartite RNA genome to be fully sequenced (Brederode *et al.*, 1980; Barker *et al.*, 1983; Cornelissen *et al.*, 1983a,b). All RNA species have an M^7G^5ppp cap structure and a 3′ terminus ending with –AUGC-OH. They cannot be aminoacylated. RNAs 1 and 2 each have a long ORF, that of RNA1 encoding a protein of 125 kDa and RNA2 encoding an 89-kDa protein. RNA3 encodes a protein of 300 amino acids and the coat protein gene. The coat protein gene is repeated in RNA4. The leader of RNA3 is longer than the others and contains a sequence of 28–30 nucleotides repeated three times. This repeat is found in all strains of the virus (Langereis *et al.*, 1986). With the exception of a few nucleotides, the last 145 at the 3′ terminus are identical in all four RNA species.

b. Proteins encoded by the genome

The four RNAs have all been translated in various cell-free systems. RNA1 directs the synthesis of a 115-kDa protein at low mRNA concentrations in the rabbit reticulocyte system,

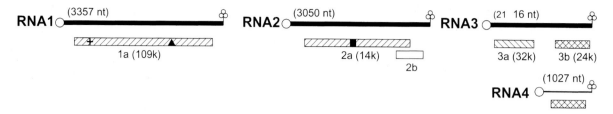

Fig. 6.14 Genome organization of CMV as representative of cucumoviruses. The thick single lines represent the (+)-strand RNA genome segments, and the thin line the subgenomic RNA4. The boxes show the coding regions with the final gene products identified. The infills represent the function of the gene products—see Fig. 6.1 or inside book cover for details of these and other features.

Fig. 6.15 Genome organization of AMV. The thick single lines represent the (+)-strand RNA genome segments, and the thin line the subgenomic RNA4. The boxes show the coding regions with the final gene products identified. The infills represent the function of the gene products—see Fig. 6.1 or inside book cover for details of these and other features.

or a mixture of 58-kDa and 62-kDa proteins at high mRNA concentrations (van Tol and van Vloten-Doting, 1979; van Tol *et al.*, 1980). They have the same N terminus as the 115-kDa protein (125 kDa in Fig. 6.15). In various cell-free systems RNA2 was translated to give an 84-kDa protein (Mohier *et al.*, 1976) corresponding to the 89-kDa protein predicted in Fig. 6.15. In a variety of *in vitro* systems, the primary translation product from RNA3 is 35 kDa (Dougherty and Hiebert, 1985) corresponding to the expected protein. RNA4 is efficiently translated *in vitro* to produce the viral coat protein (e.g. Mohier *et al.*, 1976).

All three non-structural proteins indicated in Fig. 6.15 have been detected in a crude membrane fraction from AMV-infected tobacco leaves using antisera raised against synthetic peptides corresponding to the C-terminal sequences (Berna *et al.*, 1986).

c. Functions of the non-structural proteins

i. The 125-kDa and 89-kDa proteins

The following evidence indicates that the 125-kDa and 89-kDa proteins are involved in RNA replication, presumably as part of a membrane-bound replicase complex:

1. RNAs 1 and 2 can replicate independently of RNA3 in protoplasts (Nassuth *et al.*, 1981).
2. The time course of accumulation of these two proteins following inoculation corresponded to the rise and fall in replicase activity in infected tobacco leaves (Berna *et al.*, 1986). In infected protoplasts, the 125-kDa and 89-kDa proteins were detected early and disappeared when virus production reached a plateau (van Pelt-Heerschap *et al.*, 1987b).
3. The effects of mutations in RNAs 1 and 2 suggest that they function cooperatively (Sarachu *et al.*, 1983).
4. RNA1 and RNA2 *ts* mutants were defective in the synthesis of viral minus strand RNAs (Sarachu *et al.*, 1985). A period of 6 hours immediately after inoculation at the permissive temperature was sufficient to allow normal minus strand synthesis (Huisman *et al.*, 1985).
5. There are strong amino acid sequence similarities between three regions of the 89-kDa

protein and regions in the RNA polymerase of some other viruses (Kamer and Argos, 1984).

Other aspects of the structures of these proteins and their involvement in virus replication are discussed in Chapter 8 (Section IV.G).

ii. The 35-kDa protein

Various lines of evidence suggest that the 35-kDa protein has a cell-to-cell transport function.

1. Studies with *ts* mutants of RNA2 showed that some can multiply in protoplasts but cannot move from cell to cell in tobacco leaves (Huisman *et al.*, 1986).
2. The 35-kDa protein is largely confined to the cell wall fraction of infected tobacco leaves (Godefroy-Colburn *et al.*, 1986). Using an immunocytochemical technique, Stussi-Garaud *et al.* (1987) showed that the 35-kDa protein was localized almost entirely in the middle lamella of recently infected cells. The 35-kDa protein was not detected in fully infected cells. However, no relationship of the 35-kDa protein to plasmodesmata was observed in contrast to the situation with TMV.
3. Virus particles containing only RNAs 1 and 2 can replicate the RNAs in protoplasts but cannot move from cell to cell in leaves.

The involvement of the protein in cell-to-cell movement of the virus is discussed further in Chapter 9, Section II.D.2.

4. Genus *Ilarvirus* (reviewed by Bol, 1999a)

As with other members of the *Bromoviridae*, the ilarvirus genome is divided between three RNA species (Fig. 6.16). The 3' termini of the RNAs within a species are very similar and, like AMV, form complex structures. RNA1 encodes one protein and RNA2 two proteins. The protein from RNA1 and that from the 5' ORF of RNA2 resemble those from other members of the *Bromoviridae* and are involved in RNA replication. Like cucumoviruses but unlike other members of the *Bromoviridae*, ilarvirus RNA2 has a second ORF encoding protein 2b of about 22 kDa that is expressed from sgRNA4a (Ge *et al.*, 1997; Xin *et al.*, 1998). The product of the 5' ORF of RNA3 is the cell-to-cell movement protein and that of the 3' ORF of this RNA is the viral coat protein, which is expressed from sgRNA4. As with AMV, ilarvirus coat protein is required for activation of replication and is interchangeable with that of AMV (see Chapter 8, Section IV.G).

5. Genus *Oleavirus*

OLV-2, the type and only member of this genus, also has a tripartite genome organized in a manner similar to other members of the *Bromoviridae* (Grieco *et al.*, 1995, 1996) (Fig. 6.17). However, unlike other members of this family, the sgRNA for coat protein is encapsidated to a very low level, if at all. Another virus-associated RNA, termed RNA4, of 2073 nucleotides is commonly found encapsidated (Grieco *et al.*, 1995). This additional RNA differs

Fig. 6.16 Genome organization of TSV as representative of ilarviruses. The thick single lines represent the (+)-strand RNA genome segments and the thin line, the subgenomic RNA4. The boxes show the coding regions with the final gene products identified. The infills represent the function of the gene products—see Fig. 6.1 or inside book cover for details of these and other features.

from RNA3 in only a few positions but did not produce any protein on *in vitro* translation.

B. Family *Comoviridae*

1. Genus *Comovirus* (reviewed by Goldbach and Wellink, 1996; Lomonossoff and Shanks, 1999)

a. *Genome structure*

The genome of CPMV consists of two strands of (+)-sense ssRNA called B- and M-RNAs (or RNAs 1 and 2), of 5.9 and 3.5 kb respectively. Both RNAs have a small VPg covalently linked at the 5′ end, and a poly(A) sequence at the 3′ end. The full nucleotide sequence of the M-RNA was determined by van Wezenbeek *et al.* (1983) and that of B-RNA by Lomonossoff and Shanks (1983). Each RNA has a single long open reading frame, that of B-RNA encoding a protein of 200 kDa (Fig. 6.18). In M-RNA there are three AUGs beginning at nucleotides 161, 512 and 524. Using site-directed mutagenesis, Holness *et al.* (1989) confirmed that the AUG at position 161 is used *in vitro* to direct the synthesis of a 105-kDa product. Both the 105-kDa protein initiating at nucleotide 161 and a 95-kDa product initiating at nucleotide 512 have been detected in infected protoplasts

(Rezelman *et al.* 1989). As discussed in Chapter 7 (Section V.E.8.a), these polyproteins are cleaved to give a series of functional products (see Fig. 7.28).

The first 44 nucleotides in the 5′ leader sequences of M- and B-RNA show 86% homology. The last 65 nucleotides preceding the poly(A) sequence show 83% homology. The seven nucleotides before the poly(A) sequence are the same in the two RNAs. This conservation suggests that these sequences are involved in recognition signals for viral or host proteins. Eggen *et al.* (1990) constructed a series of mutants in B-RNA involving the seven 3′-terminal nucleotides and the first four As of the poly(A) sequence and tested their effects on B-RNA replication in protoplasts and plants. Only mutants with minor modifications were able to replicate. Mutant transcripts reverted stepwise to the wild-type sequence during replication in plants.

Neither RNA has an AAUAAA polyadenylation signal preceding the 3′ poly(A) sequence. The presence of poly(U) stretches at the 5′ end of (−)-sense strands in RF molecules indicates that the poly(A) sequences are transcribed in the RNA replication process (Lomonossoff *et al.*, 1985).

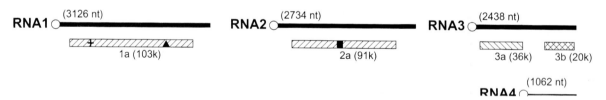

Fig. 6.17 Genome organization of OLV-2. The thick single lines represent the (+)-strand RNA genome segments, and the thin line the subgenomic RNA4. The circle at the 5′ end indicates a cap. The boxes show the coding regions with the final gene products identified. The infills represent the function of the gene products—see Fig. 6.1 or inside book cover for details of these and other features.

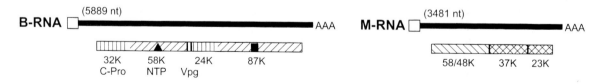

Fig. 6.18 Genome organization of CPMV as representative of comoviruses. The thick single lines represent the (+)-strand RNA genome segments. The ❑ at the 5′ end indicates a VPg and AAA shows Poly(A) at the 3′ end. . The ▲ shows the position of the helicase motif and the ■, the RdRp motif. The boxes show the coding regions with the final gene products after processing identified. The infills represent the function of the gene products – See Fig. 6.1 or inside book cover for details of these and other features.

b. The partial independence of B-RNAs

Both the B and M virus particles, or the RNA that they contain, are necessary for an infection producing progeny virus particles (van Kammen, 1968; De Jager, 1976). However, inoculation of protoplasts with B component leads to production of B-encoded polypeptides and to replication of B-component RNA. M-RNA is not replicated in the absence of B-RNA. Thus, B-RNA must code for proteins involved in replication, while M-RNA codes for the structural proteins of the virus. B-RNA inoculated to intact leaves cannot move to surrounding cells in the absence of M-RNA (Rezelman et al., 1982).

c. Proteins coded by B-RNA

B-RNA is translated both in vitro and in vivo into a 200-kDa polyprotein that is subsequently proteolytically cleaved at specific sites, in a series of steps, to give the functional proteins; the processing pathway is discussed in Chapter 7 (Section V.E.8.a). The relationships between the various polypeptides synthesized in protoplasts have been established by comparison of the proteolytic digestion patterns of the isolated proteins and by immunological methods (e.g. Rezelman et al., 1980; Goldbach et al., 1981, 1982; Zabel et al., 1982; Goldbach and Rezelman, 1983; Franssen et al., 1984a). In addition, radioactively labeled CPMV proteins synthesized in infected protoplasts were isolated and partial amino acid sequences determined (Wellink et al., 1986). This allowed precise location of the coding sequence for each protein on the B-RNA, precise determination of molecular weights, and determination of the cleavage sites at which they are released from the polyprotein precursor. The cleavage sites are described in Chapter 7 (Section V.E.8.a).

The B-RNA polyprotein yields five final products (see Fig. 7.28). Two of the products, the N-terminal 32-kDa product and the central 24-kDa protein, have proteinase activity. The 58-kDa protein has nucleotide-binding activity and the C-terminal 87-kDa contains the RNA-dependent RNA polymerase motif. As described in Chapter 8 (Section IV.J), these two proteins are found in the replication complex.

Between the 58-kDa and 24-kDa products, the polyprotein is processed to give the VPg.

As described in Chapter 13, intermediate processing products are important in the replication cycle with replication and processing being highly coordinated.

d. Proteins coded by M-RNA

A scheme for the processing of the M-RNA polyprotein is shown in Fig. 7.28. This polyprotein is processed to four final products, the identification of which is based on the in vitro translation and processing studies of Franssen et al. (1982) and later work. The two mature coat proteins are known and the 60-kDa precursor of these proteins has been detected in infected protoplasts (Wellink et al., 1987b). The in vitro studies of Franssen et al. (1982) demonstrated directly that the two coat proteins are coded on M-RNA. This virus can spread through the plant only if the RNAs are in the form of virus particles (Wellink and van Kammen, 1989).

Wellink et al. (1987b), using antisera raised against synthetic peptides, detected a 48-kDa M-RNA-encoded protein in the membrane fraction of infected cells. A 58-kDa polypeptide initiated at the AUG at nucleotide 161 and sharing a C-terminal sequence with the 48-kDa protein was detected in protoplasts inoculated with CPMV (Wellink and van Kammen, 1989). A 95-kDa protein was detected in vitro initiating from either of the AUGs at positions 512 or 524 (Holness et al., 1989) but this protein did not appear to be necessary for M-RNA replication in protoplasts.

As noted above, B-RNA is dependent on a function of M-RNA for cell-to-cell movement. Besides a requirement for the coat proteins, experiments using insertion and deletion mutants have shown that the 48-kDa protein encoded by M-RNA is needed for cell-to-cell movement (Wellink and van Kammen, 1989) (for more details see Chapter 9, Section II.D.2.c). A 43-kDa protein coded by RCMV (corresponding to the 48-kDa CPMV protein) was found by immunogold labeling to be associated with the plasmodesmata (Shanks et al., 1988, 1989).

The function of the 58-kDa protein is unknown but it may be involved in the control of replication.

2. Genus *Fabavirus* (reviewed by Lisa and Boccardo, 1996; Cooper, 1999)

The aphid-transmitted fabaviruses have two RNA components that, as with comoviruses, are each expressed as polyproteins (Fig. 6.19). Little is known about the details of the processing of the B-RNA product but it is thought to be similar to that of comoviruses. The processing of the M-RNA polyprotein is likely to be similar to that of comoviruses giving the two coat protein species and a protein thought to be involved in cell-to-cell movement.

3. Genus *Nepovirus* (reviewed by Mayo and Robinson, 1996; Mayo and Jones, 1999a)

a. Genome structure

The genome of nepoviruses is bipartite, consisting of two strands of ss (+)-sense RNA. The larger RNA (RNA1) has about 7.1–8.4 kb depending on the virus, while RNA2 is much more variable with about 3.4–7.2 nucleotides (Murant *et al.*, 1981). Definitive nepoviruses are divided into three groups on the size and encapsidation of RNA2. RNA2 of members of subgroup 'a' has 4.3–5.0 kb and is encapsidated in both B and M components; that of subgroup 'b' has 4.6–5.3 kb encapsidated in M component only; and that of subgroup 'c' has 6.3–7.3 kb encapsidated in M components that are separable only with difficulty from B component. Two tentative nepoviruses, SLRSV and SDV, do not fit with this grouping. In another method of subgrouping all those with RNA2 <5 kb are placed in subgroup I and the others in subgroup II (Mayo and Jones, 1999a).

Both RNAs have a 3' poly(A) tail, and both have a VPg linked to the 5' terminus, which is essential for infectivity but not for *in vitro* translation (Chu *et al.*, 1981; Mayo *et al.*, 1982).

b. Genome organization

As with other genera of the *Comoviridae*, each of the nepovirus RNAs encodes a polyprotein that is processed to give functional gene products (Fig. 6.20). The genome organization is similar to that of comoviruses, the major differences being in the sizes of the products and the processing of the coat proteins. For instance, the coat protein of definitive nepoviruses is a single polypeptide species of 52–60 kDa whereas that of tentative species comprises either two polypeptides (about 40–47 and 21–27 kDa) or three polypeptides of about 20–27 kDa.

C. Family *Potyviridae* (reviewed by Shukla *et al.* 1994; López-Moya and Garcia, 1999)

The flexuous particles of members of the *Potyviridae* contain (+)-sense ssRNA, about 8.5–10 kb, that of five of the genera being a single species and that of the genus *Bymovirus* being divided between two components. The genomes of all genera have a 5' VPg and are polyadenylated at the 3' end.

1. Single-component potyviruses

Most of the studies have been on members of the genus *Potyvirus*, but the more limited studies on the other genera of single-component potyviruses (*Macluravirus*, *Ipomovirus*, *Tritimovirus* and *Rymovirus*) indicate that they have a similar genome structure and organization but differ somewhat in detail.

Fig. 6.19 Genome organization of PatMMVas representative of fabaviruses. The thick single lines represent the (+)-strand RNA genome segments. The boxes show the coding regions with the final gene products after processing identified. The infills represent the function of the gene products—see Fig. 6.1 or inside book cover for details of these and other features.

Fig. 6.20 Genome organization of BRSV as representative of nepoviruses. The thick single lines represent the (+)-strand RNA genome segments. The boxes show the coding regions with the final gene products after processing identified. The infills represent the function of the gene products—see Fig. 6.1 or inside book cover for details of these and other features.

a. Genome structure

The main features of the TEV genome are: (1) a 5′ VPg (about 24 kDa); (2) a 5′ non-coding region of 144 nucleotides rich in A and U; (3) a single large ORF of 9161 nucleotides initiating at residue 145–147, which could code for a polyprotein with about 3000 amino acids (about 340 kDa); and (4) a 3′ untranslated region of 190 bases terminating in a poly(A) tract (20 to 160 adenosines).

The genome encodes a polyprotein that is cleaved to nine or ten products by virus-coded proteinases (Fig. 6.21) (see Chapter 7, Section V.B.1 for processing pathway).

b. Gene products

i. Proteins in the virus particle

The rod-shaped particle of TEV is made up of about 2000 molecules of a coat protein of 30 kDa that is processed from the C-terminus of the polyprotein (Fig. 7.16). It is involved in successful aphid transmission of potyviruses (Shaw *et al.*, 1990) (see Chapter 11, Section III.E.7). In addition, there is a single molecule of a VPg linked covalently to the 5′ terminus that is processed from the middle of the polyprotein (Fig. 6.21).

ii. Cytoplasmic pinwheel inclusion protein

A protein monomer that aggregates to form characteristic pinwheel inclusion bodies is found in the cytoplasm plants infected by all potyviruses (Edwardson, 1974) (see Fig. 3.10), including those transmitted by mites (Brakke *et al.*, 1987a). *In vitro* translation studies have shown that this protein, termed the cytoplasmic inclusion (CI) protein, is coded by the virus (Dougherty and Hiebert, 1980). The CI protein of TEV is about 70 kDa. It has been suggested that it is involved in cell-to-cell movement of virus (Langenberg, 1986a), but this seems unlikely. From nucleotide sequence similarities with similar proteins in picornaviruses and comoviruses, it is now considered to be involved in RNA replication (see Chapter 8, Section IV.I). In particular, it contains the highly conserved NTP binding motif.

iii. Cytoplasmic amorphous inclusion protein and the helper component protein

The cytoplasmic amorphous inclusion body protein has been detected in some but not all potyviruses. The helper component protein is a virus-coded protein necessary for aphid transmission of the virus (see Chapter 11, Section III.E.7). Several lines of evidence indicate they are the same proteins, in whole or in part, and

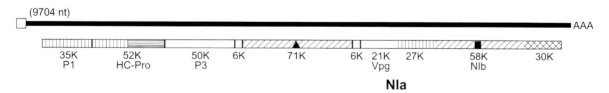

Fig. 6.21 Genome organization of TEV as representative of potyviruses. The thick single line represents the (+)-strand RNA genome segments. The boxes show the coding regions with the final gene products after processing identified. The infills represent the function of the gene products—see Fig. 6.1 or inside book cover for details of these and other features.

that the gene is located in the 5' region of the genome: (1) they comigrate in gel electrophoresis; (2) amorphous inclusion body protein antibody reacts with the main protein in purified helper protein extracts; (3) antiserum to a partially purified helper component blocked the biological activity of homologous helper component and also reacted with a cell-free 75-kDa translation product from homologous RNA; and (4) amorphous inclusion protein and 5' cell-free translation products have similar partial proteolysis patterns (de Mejia *et al.*, 1985; Thornbury *et al.*, 1985). The helper component (HC) protein of all the aphid-transmitted potyviruses that have been sequenced contains a potential 'zinc finger' metal binding site of unknown function (Robaglia *et al.*, 1989). Helper component activity is located in the 52-kDa protein coded near the 5' terminus (Fig. 6.21).

iv. Nuclear inclusion proteins

The nuclear inclusion (NI) bodies found in TEV and some other *Potyvirus* infections are composed of two proteins of approximately 48 kDa (NIa) and 58 kDa (NIb). These are coded for by all potyviruses but only some produce the inclusions. The NIa protein of TEV is a virus-coded proteinase related to the trypsin superfamily of serine proteinases that are excised from the polyprotein by autoproteolysis (Carrington and Dougherty, 1987a,b). It is also responsible for four other cleavages in the 3' two-thirds genome (see Chapter 7, Section V.B.1). The NIb protein has sequence similarity with polymerases of some other viruses (Allison *et al.*, 1986; Domier *et al.*, 1987), and is the RNA-dependent RNA polymerase (see Chapter 8, Section IV.I).

v. Proteinases

Potyviruses encode three proteinases that are involved in processing the polyprotein, the N-terminal 35-kDa P1 protein, the N-terminal part of the 52-kDa HC-Pro, and the 27-kDa C-terminal part of the NIa protein. The properties of these proteinases and the role that they play in processing the polyprotein are described in Chapter 7 (Section V.B.1.a).

2. Genus Bymovirus

The RNAs of BaYMV have been sequenced (Kashiwazaki *et al.*, 1990, 1991) and shown to encode a genome organization similar to that on single-component potyviruses but divided between the two segments (Fig. 6.22). Essentially, the polyprotein encoded by RNA2 contains the two products that are at the N terminus of the potyvirus polyprotein and that encoded by RNA1 is processed to give the other products.

D. Family *Tombusviridae*

With the exception of the genus *Dianthovirus*, the genomes of members of the *Tombusviridae* are a single molecule of (+)-sense ssRNA; the genomes of dianthoviruses are divided between two RNA species. The 3'ends of the RNAs are not polyadenylated. Cap structures have been identified at the 5' ends of CarMV, RCNMV and MCMV but, although the 5' ends of the other genera are protected, no cap structure has been identified. Furthermore, addition of a cap analog to *in vitro* RNA transcripts enhances infectivity little or not at all (Lommel *et al.*, 2000).

1. Genus Tombusvirus (reviewed by Rochon, 1999)

The genomic RNA of TBSV was sequenced by Hearne *et al.* (1990) and shown to contain four ORFs (Fig. 6.23). ORF 1 encodes a 33-kDa protein and read-through of its amber termination

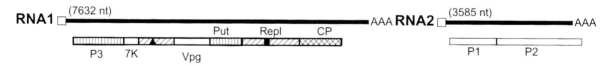

Fig. 6.22 Genome organization of BaYMV. The thick single lines represent the (+)-strand RNA genome segments. The boxes show the coding regions with the final gene products after processing identified. The infills represent the function of the gene products—see Fig. 6.1 or inside book cover for details of these and other features.

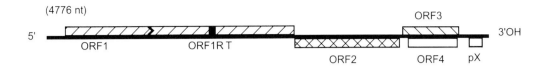

Fig. 6.23 Genome organization of TBSV as representative of tombusviruses. The thick single lines represent the (+)-strand RNA genome segments. The boxes show the coding regions with the final gene products after processing identified. The infills represent the function of the gene products—see Fig. 6.1 or inside book cover for details of these and other features.

codon give a protein of 92 kDa. These two proteins have been detected *in vivo* and are produced by *in vitro* translation of virion RNA. Together they form the viral polymerase. The product of ORF 2 is the 41-kDa coat protein, which is expressed from sgRNA1 (2.2 kb). ORFs 3 and 4 are expressed *in vivo* from a second sgRNA (sgRNA2, 0.9 kb). The product of ORF 3 (22 kDa) is membrane-associated and facilitates cell-to-cell movement of the virus (Scholthof *et al.*, 1995b; Chu *et al.*, 1999). That of ORF 4 (19 kDa) regulates necrosis formation and host-dependent induction of systemic invasion (Scholthof *et al.*, 1995b; Chu *et al.*, 2000). p19 is involved in the systemic necrotic reaction in *Nicotiana benthamiana* and *N. clevelandii*, but in spinach is required in high abundance for efficient systemic invasion. The N-terminal portion of this protein mediates the host-dependent activity and the central portion is required for all activities. As the initiation codon of ORF 3 is in a suboptimal context, ribosome scanning occurs, allowing the translation of ORF 4 from sgRNA2.

2. Genus *Aureusvirus*

The genome of PoLV has been sequenced and the genome organization determined (Fig.

6.24) (Rubino *et al.*, 1995). The viral RNA has four ORFs. ORF 1 encodes a 25-kDa protein at its N-terminal end; read-through of a weak termination codon gives an 84-kDa product that has the RNA-dependent RNA polymerase (RdRp) motif in the C-terminal portion. The two products from this ORF are involved in the viral polymerase complex. The 40-kDa product of ORF 2 is the coat protein. The ORF 3 product (27 kDa) is the cell-to-cell movement protein and that from ORF 4 (14 kDa) is responsible for symptom severity (Rubino and Russo, 1997). The coat protein regulates the expression of ORF 4 as excess production of the 14-kDa protein is lethal to plants.

ORF 2 is expressed from an sgRNA of 2.0 kb and ORFs 3 and 4 from a 0.8 sgRNA.

3. Genus *Avenavirus*

The sequence of the genomic RNA of OCSV reveals three ORFs (Boonham *et al.*, 1995) (Fig. 6.25). ORF 1 encodes a 23-kDa protein and read-through of its amber stop codon gives a protein of 84 kDa. These two proteins from ORF 1 are involved in the viral replication complex, the 84-kDa protein having the RdRp motif in its C-terminal region. ORF 2 encodes the

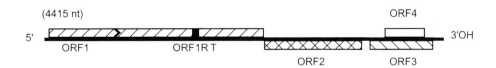

Fig. 6.24 Genome organization of PoLV. The thick single lines represent the (+)-strand RNA genome segments. The boxes show the coding regions with the final gene products after processing identified. The infills represent the function of the gene products—see Fig. 6.1 or inside book cover for details of these and other features.

48-kDa coat protein. ORF 3 is within ORF 2 in a different reading frame and its 8-kDa product is thought to be involved in cell-to-cell movement. ORFs 2 and 3 are expressed from a single subgenomic mRNA that is encapsidated.

4. Genus *Carmovirus* (reviewed by Qu and Morris, 1999)

The complete sequence of the nucleotides in CarMV has been established (Guilley *et al.*, 1985) and shown to encode four ORFs (Fig. 6.26). There is a 69-nucleotide 5' leader sequence before the first AUG. There is a UAG termination codon that would give a protein of 28 kDa, but read-through of this to a UAA codon would give a protein of 86 kDa. The 28-kDa and 86-kDa proteins are both found *in vivo* and *in vitro*. These two proteins from ORF 1 make up the viral polymerase. The product of ORF 2 (7 kDa) facilitates the cell-to-cell movement of the virus and that from ORF 3 (9 kDa) has been implicated in the systemic movement of the virus. Experiments on the localization of these two proteins of TCV that had been fused to the green fluorescent protein showed that p9 is found throughout the cytoplasm as well as in cell nuclei whereas p8 (the product of ORF 3) was confined to the cell nucleus (Cohen *et al.*, 2000). Analysis of the

sequence of p8 revealed two nuclear localization signals but these were not required for cell-to-cell movement. ORF 4 encodes the 38-kDa coat protein. This is followed by a 3' non-coding region of 290 nucleotides.

Besides genomic RNA, two sgRNA species of 1.7 and 1.45 kb are produced *in vivo* and encapsulated in virus particles (Carrington and Morris, 1984). They are 3' co-terminal with the genomic RNA and are the mRNAs for ORFs 2 and 3 and for ORF 4 respectively. Their 5' termini have been precisely mapped (Carrington and Morris, 1986).

5. Genus *Machlomovirus* (reviewed by Lommel, 1999b)

The genomic RNA of MCMV contains four ORFs (Lommel *et al.*, 1991) (Fig. 6.27). ORF 1 encodes a protein of 32 kDa of unknown function. The N-terminal part of ORF 2 codes for a 50-kDa protein, and read-through of its amber termination codon allows translation to continue to give a 111-kDa product. These two products from ORF 2 make up the viral polymerase. The N-terminal part of ORF 3 gives a protein of 9 kDa, and read-through of its opal termination codon gives a product of 33 kDa. The functions of these two products from ORF

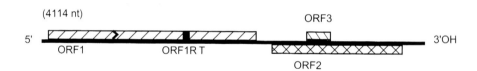

Fig. 6.25 Genome organization of OCSV. The thick single lines represent the (+)-strand RNA genome segments. The boxes show the coding regions with the final gene products after processing identified. The infills represent the function of the gene products—see Fig. 6.1 or inside book cover for details of these and other features.

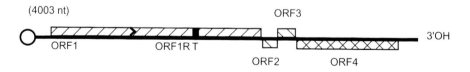

Fig. 6.26 Genome organization of CarMV as representative of carmoviruses. The thick single lines represent the (+)-strand RNA genome segments. The boxes show the coding regions with the final gene products after processing identified. The infills represent the function of the gene products—see Fig. 6.1 or inside book cover for details of these and other features.

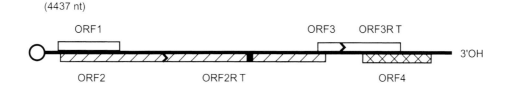

Fig. 6.27 Genome organization of MCMV. The thick single lines represent the (+)-strand RNA genome segments. The boxes show the coding regions with the final gene products after processing identified. The infills represent the function of the gene products—see Fig. 6.1 or inside book cover for details of these and other features.

3 are unknown. ORF 4 is expressed from a 1.1 kb sgRNA to give the 25-kDa coat protein.

6. Genus *Necrovirus* (reviewed by Meulewaeter, 1999)

The genomic RNAs of several strains of TNV and some other necroviruses have been sequenced (see Meuleweiter *et al.*, 1990; Coutts *et al.*, 1991; Lot *et al.*, 1996; Molnar *et al.*, 1997) and the genome organization of members of this genus exemplified by that of TNV-A (Fig. 6.28). The genome of TNV-A contains five ORFs. The N-terminal part of ORF 1 encodes a 23-kDa protein; the amber stop codon is read through to give an 82-kDa protein that has a RdRp motif in its C-terminal region. ORFs 2 and 3 are expressed from sgRNA1 (1.6 kb). The product of ORF 2 (8 kDa) is involved in cell-to-cell movement; that of ORF 3 (6 kDa) has unknown function. The 30-kDa coat protein is expressed from ORF 4 on sgRNA2 (1.3 kb), as is that of the 3′ ORF 5 (7 kDa); the 7-kDa protein has nucleic acid-binding properties (Offei *et al.*, 1995).

7. Genus *Panicovirus*

The genomic RNA of PMV has been sequenced (Turina *et al.*, 1998) and shown to contain five ORFs (Fig. 6.29). The N-terminal part of ORF 1 encodes a protein of 48 kDa and its amber termination codon is read through to give a 112-kDa protein. Both these proteins are produced by *in vitro* translation of virion RNA and function as the viral polymerase. ORF 2 encodes an 8-kDa protein that is produced by *in vitro* translation of a transcript representing sgRNA1 (1.5 kb). It is thought a −1 frameshift event from this ORF to ORF 3 (6.6 kDa) gives a 14.6-kDa protein that has also been detected upon *in vitro* translation of the sgRNA. The two products from ORFs 2 and 2+3 are thought to be involved in cell-to-cell movement. ORF 4 encodes the 26-kDa coat protein thought to be also expressed from the 1.5 kb sgRNA. ORF 5 (15 kDa) is within ORF 4 in a different reading frame and is also thought to be expressed from the sgRNA; the function of ORF 5 product is unknown. Mutagenic studies imply that the successful establishment of a PMV systemic infection in millet plants is dependent upon the concerted expression of the p8, p6.6, p15 and coat protein genes (Turina *et al.*, 1998).

8. Genus *Dianthovirus* (reviewed by Lommel, 1999a)

Dianthoviruses differ from the other members of the *Tombusviridae* in that their genome is divided between two RNA species, RNA1 and

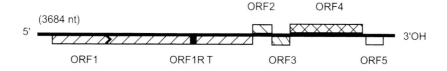

Fig. 6.28 Genome organization of TNV-A as representative of necroviruses. The thick single lines represent the (+)-strand RNA genome segments. The boxes show the coding regions with the final gene products after processing identified. The infills represent the function of the gene products—see Fig. 6.1 or inside book cover for details of these and other features.

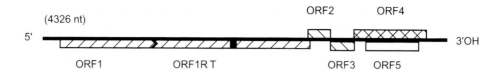

Fig. 6.29 Genome organization of PMV. The thick single lines represent the (+)-strand RNA genome segments. The boxes show the coding regions with the final gene products after processing identified. The infills represent the function of the gene products—see Fig. 6.18 or inside book cover for details of these and other features.

RNA2. The nucleotide sequence of CRSV show two large ORFs in RNA1 and one in RNA2 (Fig. 6.30). ORF 1 of RNA1 encodes a protein of 27 kDa and then, by a –1 frameshift event (see Chapter 7, Section V.B.10.a), is translated to give a protein of 88 kDa. These two proteins have been observed *in vivo* and *in vitro* and form the viral polymerase. The product of ORF 2 is the 38-kDa coat protein that is expressed from a 1.5 kb sgRNA (Xiong and Lommel, 1989). The 35-kDa product of ORF 2 is the cell-to-cell movement protein (see Chapter 9, Section II.D.2).

E. Family *Sequiviridae*

Members of the *Sequiviridae* have large RNA genomes (about 10–12.5 kb) that have a VPg at the 5′ end, the 3′ termini being polyadenylated. Their genomic information is expressed in a polyprotein that is cleaved by virus-coded proteinase(s).

1. Genus *Sequivirus* (reviewed by Mayo and Murant, 1999)

PYFV has a (+)-strand ssRNA genome of 9.9 kb (Turnbull-Ross *et al.*, 1993) that has one ORF expressing a polyprotein (Fig. 6.31). Although not much is known about the processing of this polyprotein, the sites of the three coat protein species have been identified, as have the motifs for the proteinase and RdRp.

2. Genus *Waikavirus* (reviewed by Hull, 1996; Gordon, 1999)

The genomic sequences of RTSV and MCMV have been determined (Shen *et al.*, 1993; Thole and Hull, 1996; Reddick *et al.*, 1997). The genome organization of RTSV is illustrated in Fig. 6.32. The 12.4-kb RNA encodes for a polyprotein of 393 kDa, and Shen *et al.* (1993) suggested two 3′ small ORFs. However, Thole and Hull (1996) could not detect any of the predicted products from the small ORFs. Various features of the polyprotein have been deter-

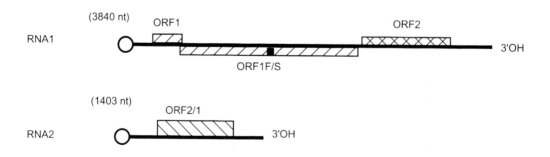

Fig. 6.30 Genome organization of CRSV as representative of dianthoviruses. The thick single lines represent the (+)-strand RNA genome segments. The boxes show the coding regions with the final gene products after processing identified. The infills represent the function of the gene products—see Fig. 6.18 or inside book cover for details of these and other features.

Fig. 6.31 Genome organization of PYFV as representative of sequiviruses. The single line represents the (+)-strand RNA genome. The boxes show the coding regions with the final gene products after processing identified for those that are known. The infills represent the function of the gene products—see Fig. 6.1 or inside book cover for details of these and other features.

mined. The N-terminal positions of the three coat protein species have been mapped to the polyprotein (Shen *et al.*, 1993; Zhang *et al.*, 1993). As these proteins are adjacent to each other, the C-termini of CP1 and CP2 were mapped but not that of CP3. RTSV CP3 has unusual behavior as the protein from the purified virus migrates in gels as a 33-kDa molecule but immunoprobing of extracts from infected plants detected several proteins in the range 40–42 kDa (Druka *et al.*, 1996). As the C-terminus could not be determined, it was not possible to say whether this discrepancy was due to processing during purification or the loss of some post-translational modification. Thole and Hull (1998) characterized the serine-like proteinase as a 35-kDa protein (see Chapter 7, Section V.B.1.a) and identified the C-terminal genomic position of the 70-kDa polymerase.

F. Family *Closteroviridae* (reviewed by German-Retana *et al.*, 1999)

Members of the *Closteroviridae* have among the largest genomes among the (+)-strand RNA plant viruses, with that of CTV being 19.3 kb.

Two genera are currently recognized, the *Closterovirus*, which has a single RNA component, and the *Crinivirus*, which has the genome divided between two RNA species.

1. Genus *Closterovirus* (reviewed by Dolja *et al.*, 1994; Bar-Joseph *et al.*, 1997; German-Retana *et al.*, 1999)

The genomes of the aphid-transmitted BYV (15.5 kb) and CTV (19.3 kb) (Agranovsky *et al.*, 1994; Karasev *et al.*, 1995), of the mealybug-transmitted LCV (16.9 kb) (Jelkmann *et al.*, 1997), and of GLRaV-1 (12.4 kb) (Fazeli and Rezaian, 2000) have been sequenced. The 5' end of the RNA is thought to be capped and the 3' end is not polyadenylated.

The sequence of the type member of this family, BYV, shows nine ORFs (Fig. 6.33) and the properties of the ORFs are summarized in Table 6.5. The sequence of CTV reveals 12 ORFs and their properties are also summarized in Table 6.5. The three additional ORFs in CTV are one encoding a 33-kDa protein immediately downstream of the pol-encoding ORF 1b, and two ORFs at the 3' end. The organization and functions of the ORFs of both viruses essentially fall

Fig. 6.32 Genome organization of RTSV as representative of waikaviruses. The single line represents the (+)-strand RNA genome. The boxes show the coding regions with the final gene products after processing identified for those that are known. The infills represent the function of the gene products—see Fig. 6.1 or inside book cover for details of these and other features.

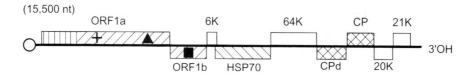

Fig. 6.33 Genome organization of BYV, the type member of the genus *Closterovirus*. The single line represents the (+)-strand RNA genome. The boxes show the coding regions with the final gene products identified. The infills represent the function of the gene products—see Fig. 6.1 or inside book cover for details of these and other features.

into four groups, those essential for replication, those essential for cell-to-cell movement, those involved in virion assembly, and those with unknown function.

a. Replicase enzymes

The product of ORF 1a has domains characteristic of a papain-like proteinase, methyl transferase and helicase; that of ORF 1b has an RdRp domain. It is thought that ORF 1b is expressed by a +1 ribosomal frameshift from ORF 1a to give a 348-kDa fusion protein.

b. Coat proteins

Two coat proteins are expressed, from ORFs 5 and 6 of BYV and 6 and 7 of CTV. The downstream protein (CP) is the major coat protein encapsidating most of the particle. The upstream protein (CPm) is a minor protein encapsidating the 5′ end of the particle (see Fig. 5.12) (Zinovkin *et al.*, 1999) and is thought to be involved in aphid transmission (see Chapter 11, Section III.F).

c. Cell-to-cell movement proteins

Five proteins are involved in the cell-to-cell movement of BYV (Alzhanova *et al.*, 2000): p6, HSP70h, p64, CPm and CP, the products of ORFs 3–7. HSP70h is a homolog of heat-shock protein 70 (Agranovsky *et al.*, 1991) and immunogold labeling locates it in association with virion aggregates and in close proximity of, and even within, plasmodesmata (Medina *et al.*, 1999). Tagging the protein with green fluorescent protein and mutagenesis show that it is a cell-to-cell movement protein (Peremyslov *et al.*, 1999). HSP70h forms physical complexes with BYV virions with non-ionic, but not covalent, interactions (Napuli *et al.*,

2000); each virion contains approximately 20 molecules of HSP70h.

d. Other functions

By analysis of subcellular fractions, the C-terminal p21 of BYV was detected in cytoplasmic and cell wall fractions (Zinovkin and Agranovsky, 1998). The C-terminal 23-kDa protein of CTV is an RNA-binding protein (Lopez *et al.*, 2000). The small protein (BYV2, CTV3, LIYV 2/1) is hydrophobic and is thought to be a membrane-binding protein.

e. Expression of ORFs

ORFs 1a and 1b are translated from the genomic RNA whereas the remaining ORFs are expressed from a nested set of sgRNAs (see Fig. 7.27) (Hilf *et al.*, 1995; He *et al.*, 1997; Karasev *et al.*, 1997).

2. Genus *Crinivirus*

The genome of criniviruses is divided between two RNA species that are separately encapsidated. The genome of LIYV has been sequenced (Klaassen *et al.*, 1995) and three ORFs identified on RNA1 and seven on RNA2 (Fig. 6.34; Table 6.5).

The 5′ ORF on RNA1 has amino acid motifs indicative of a papain-like proteinase, and methyl transferase and helicase activities. A +1 frameshift (see Chapter 7, Section V.B.10.b) leads into ORF 1b, which contains the RdRp motif. Thus, as with closteroviruses, the products of ORF 1a and 1b make up the viral polymerase. The 3′ ORF in RNA1 encodes a 31-kDa protein that is similar in size and location to CTV ORF 2, but is of unknown function.

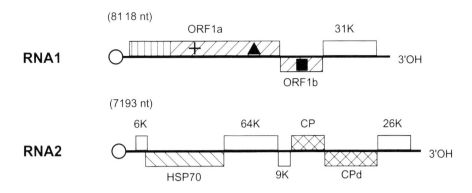

Fig. 6.34 Genome organization of LIYV as representative of criniviruses. The single lines represent the (+)-strand RNA genome. The boxes show the coding regions with the final gene products identified. The infills represent the function of the gene products—see Fig. 6.1 or inside book cover for details of these and other features.

TABLE 6.5 Properties of ORFs of closteroviruses and criniviruses

ORF	BYV		CTV		LIYV		
	Size (kDa)	Function	Size (kDa)	Function	ORF[a]	Size (kDa)	Function
1a	295	P-pro, MT, Hel	349	P-pro, MT, Hel	1/1a	217	P-pro, MT, Hel
1b	48	Pol	57	Pol	1/1b	55	Pol
2	6	Mb	33	ND	1/2	32	ND
3	65	HSP70h	6	Mb	2/1	5	Mb
4	64	(HSP90)	65	HSP70h	2/2	62	HSP70h
5	24	CPm	61	ND	2/3	59	ND
6	22	CP	27	CPd	2/4	9	ND
7	20	ND	25	CP	2/5	28	CP
8	21	ND	18	ND	2/6	52	CPd
9			13	ND	2/7	26	ND
10			20	ND			
11			23	NAb			

[a] Designation of ORF: RNA segment/ORF number.

CP, coat protein; CPm, coat protein analogue, minor coat protein; Hel, helicase; HSP90, heat-shock protein 90-like; HSP70h, heat-shock-related protein; Mb, membrane-binding protein; MT, methyltransferase; NAb, nucleic acid-binding protein; ND, not determined; P-pro, papain-like proteinase; Pol, polymerase.

Data from: BYV: Agranovsky *et al.* (1991, 1994); CTV: Pappu *et al.* (1994), Karasev *et al.* (1995); LIYV: Klaassen *et al.* (1995); all with permission.

The organization of the seven ORFs on RNA2 resembles that of the 3′ half of the closterovirus genome, except for the additional 9-kDa ORF (ORF 2/4) upstream of the coat protein gene, the order of the major and minor coat protein ORFs (ORFs 2/5 and 2/6) being reversed, and the number of ORFs downstream of the coat protein module. As with BYV, the HSP70h is found associated with virions (Tian *et al.*, 1999)

G. Family *Luteoviridae* (reviewed by Mayo and Miller, 1999; Smith and Parker, 1999; Miller, 1999)

Members of the *Luteoviridae* have genomes of (+)-sense ssRNA of 5.6–5.7 kb, the 3′ end of which is not polyadenylated and without a tRNA-like structure. The genera are distinguished on their genome organization and on the presence or absence of a 5′ VPg; the genomes of poleroviruses and enamoviruses have a 5′ VPg whereas those of luteoviruses do not. The genome organizations of representative members of the three genera are shown in Fig. 6.35. The ORFs are so numbered that ones with the same number are thought to have similar functions.

The products of ORFs 1 (~70 kDa) and 2 (~48 kDa) are involved in the viral replicase, ORF 2 being expressed by a -1 ribosomal frameshift from ORF 1 (see Chapter 7, Section V.B.10.a). The RdRp motif is in ORF 2. The ORF 1s of poleroviruses and enamoviruses contain a motif for a chymotrypsin-like (serine) proteinase and the VPg is processed from its C-terminal region, either from P1 itself or from P1 + P2 (van der Wilk *et al.*, 1997; Wobus *et al.*, 1998). As a ~27-kDa C-terminal fragment is produced on cleavage of the ORF 1 of PLRV product expressed in insect cells (Li *et al.*, 2000), it would seem likely that the VPg is produced from this protein. This arrangement of proteinase—VPg:RdRp—differs from that found in picorna-like viruses (VPg:proteinase:RdRp) (Domier *et al.*, 1987).

The product from ORF 3 is the major coat protein. As described in Chapter 7 (Section V.E.10), the termination codon of this ORF is read through to give a fusion protein with ORF 5. In most viruses the ORF 3 + ORF 5 fusion product is cleaved midway in the ORF 5 sequence. It is thought that the product of ORF 5 is involved in aphid transmission though the evidence on this is somewhat confusing. This product interacts with symbionin (see Chapter 11, Section III.H.1.a) conferring stability on the virus particles in the insect hemolymph. It appears that there is considerable structural redundancy in the read-through domain that makes dissection of functions difficult (Brault *et al.*, 2000).

ORF 4 is not found in enamoviruses. In the other two genera, it is thought to be involved in virus movement in the plant though there is some conflicting evidence for this.

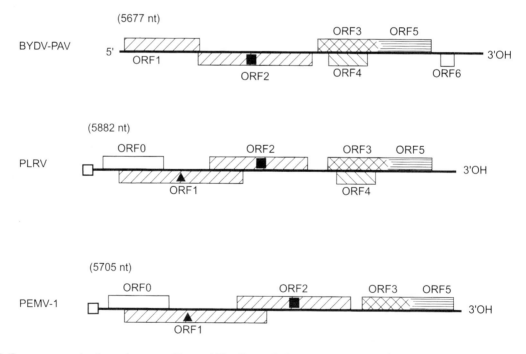

Fig. 6.35 Genome organizations of genera of *Luteoviridae*. For each the single line represents the (+)-strand RNA genome segments, and the thin line the subgenomic RNA4. The boxes show the coding regions with the final gene products identified. The infills represent the function of the gene products—see Fig. 6.1 or inside book cover for details of these and other features.

ORF 0, encoding a 28–29-kDa protein, is found only in poleroviruses and is indispensible for virus accumulation (Sadowy *et al.*, 2001a). ORF 6 (4–7 kDa) is also found only in luteoviruses and its function is unknown.

As noted above, ORFs 2 and 5 are expressed from frameshift and read-through respectively from their upstream ORF. ORFs 1 and 2 are translated from the genomic RNA, ORFs 3, 4 and 5 from sgRNA1 and, in luteoviruses, ORF 6 from sgRNA2.

H. Floating genera

1. Genus *Tobamovirus* (reviewed by Lewandowski and Dawson, 1999)

a. Genome structure

The genome of TMV was the first plant RNA virus to be sequenced (Goelet *et al.*, 1982). Since then the genomes in the (+)-sense ssRNAs of most species of the Tobamovirus genus have been sequenced (Van Regenmortel *et al.*, 2000) and the organization of TMV is shown in Fig. 6.36.

The sequence revealed several closely packed ORFs. An M⁷Gppp cap is attached to the first nucleotide (guanylic acid). This is followed by an untranslated leader sequence of 69 nucleotides (the Ω sequence; see Chapter 7, Section V.C.5). This initiates an ORF that codes for a 126-kDa protein. Experiments described next show that the termination codon for this protein (UAG) is 'leaky' and that a second larger read-through protein of 183 kDa is possible. The terminal five codons of this read-through protein overlap a third ORF coding for a 30-kDa protein. This ORF terminates two nucleotides before the initiation codon of the fourth ORF, which is closest to the 3' terminus. It codes for 17.6-kDa coat protein. As discussed in the following section, the two smaller ORFs at the 3' end of the genome are translated from sgRNAs that have been termed the I₂ and CP sgRNAs. The 3'-untranslated sequences can fold in the terminal region to give a tRNA-like structure that accepts histidine. A comparison of other sequenced or partially sequenced *Tobamovirus* genomes suggests that the genome structure outlined is common to all members of the genus.

A third subgenomic RNA, called I₁ RNA, representing approximately the 3' half of the genome, has been isolated from TMV-infected tissue. SI mapping showed that this RNA species had a distinct 5' terminus at residue 3405 in the genome (Sulzinski *et al.*, 1985). These workers proposed a model for the translation of I₁ RNA. There is an untranslated region of 90 bases followed by an AUG codon initiating a 54-kDa protein terminating at residue 4915. Thus, the amino acid sequence of the 54-kDa protein is the same as the residues at the carboxy terminus of the 183-kDa protein. The 54-kDa protein is discussed in Chapter 7 (Section V.E.1).

b. Proteins synthesized in vitro

TMV genomic RNA has been translated in several cell-free systems. Two large polypeptides were produced in reticulocyte lysates (Knowland *et al.*, 1975) and in the wheatgerm system (e.g. Bruening *et al.*, 1976), but no coat protein was found by these or later workers. The TMV genome is not large enough to code independently for the two large proteins that were produced. Using a reticulocyte lysate system, Pelham (1978) showed that the synthesis

Fig. 6.36 Genome organization of TMV as representative of tobamoviruses. The single line represents the (+)-strand RNA genome. The circle at the 5' end indicates a cap and >, readthrough of a stop codon. The + shows the position of the methyl transferase motif, the ▲, the helicase motif and the ■, the RdRp motif. The boxes show the coding regions with the final gene products identified. The infills represent the function of the gene products – see Fig. 6.1 or inside book cover for details.

of these two proteins is initiated at the same site. The larger protein is generated by partial read-through of a termination codon. The two proteins are read in the same phase, so the amino acid sequence of the smaller protein is also contained within the larger one, a conclusion fully confirmed by the nucleotide sequence discussed in the previous section. Increased production of the larger protein occurred *in vitro* at lower temperatures (Kurkinen, 1981). The extent of production of the larger protein may also depend on the kind of tRNA[tyr] present in the extract (Beier *et al.*, 1984a,b).

The I_1 subgenomic RNA isolated from infected tissue is translated *in vitro* in the rabbit reticulocyte system to produce a 54-kDa polypeptide (Sulzinski *et al.*, 1985).

The I_2 genes coding for the 30-kDa and 17.6-kDa proteins are not translated from genomic RNA but from two subgenomic RNAs. The I_2 RNA isolated from infected tissue has been studied in *in vitro* systems by many workers. It is translated to produce a protein of 30 kDa. It is uncapped (Joshi *et al.*, 1983) and appears to terminate in 5′ di- and triphosphates (Hunter *et al.*, 1983). The initiation site for transcription has been mapped at residue 1558 from the 3′ terminus (Watanabe *et al.*, 1984a). The I_2 RNA also contains the smaller 3′ gene (see Fig. 7.26) but this is not translated in *in vitro* systems.

The smallest TMV gene (the 3′ coat protein gene) is translated *in vitro* only from the mono-cistronic subgenomic RNA (see Fig. 7.26) (Knowland *et al.*, 1975; Beachy and Zaitlin, 1977). This gene can also be translated efficiently *in vitro* by prokaryotic protein-synthesizing machinery from *Escherichia coli* (Glover and Wilson, 1982).

c. *Proteins synthesized* in vivo

Proteins corresponding approximately in size to the 183-kDa and 126-kDa proteins have been found in infected tobacco leaves (e.g. Scalla *et al.*, 1976) and in infected protoplasts (e.g. Sakai and Takebe, 1974). Cyanogen bromide peptide analysis on a 110-kDa protein from infected leaves showed it to be the same as the *in vitro* translation product of similar size (Scalla *et al.*,

1978). The 30-kDa protein has been detected in both infected tobacco protoplasts (Beier *et al.*, 1980) and leaves (Joshi *et al.*, 1983).

Kiberstis *et al.* (1983) and Ooshika *et al.* (1984) raised antibodies against a synthetic peptide with the predicted sequence for the 11 or 16 C-terminal amino acids of the 30-kDa protein. The 30-kDa protein from TMV-infected protoplasts was precipitated by these antibodies, positively identifying it as the I_2 gene product.

Many workers have demonstrated the synthesis of TMV coat protein *in vivo*. Determination of the nucleotide sequence in the 3′ region of the TMV genome readily located the gene for this protein since the full amino acid sequence was already known (Anderer *et al.*, 1960; Tsugita *et al.*, 1960). *In vitro* protein synthesis primed with alkali-treated virus is discussed in Chapter 7 (Section II.A.4).

No protein has yet been detected *in vivo* that corresponds to the 54-kDa *in vitro* translation product of the I_1 subgenomic RNA. Thus, the TMV genome codes for four gene products.

d. *Functions of the virus-coded proteins*

i. *Coat protein*

The major function of the coat protein is obvious. However, several studies have shown that it is multifunctional. For instance, *Nicotiana* species containing the *N′* gene react to the common strain of TMV with systemic mosaic disease, whereas ToMV induces necrotic local lesions without mosaic disease. Saito *et al.* (1987a) showed, by constructing recombinants between the two virus species, that the viral factor causing the necrotic response lies in the coat protein gene. Using recombinants between wild-type TMV and mutants induced by nitrous acid, Knorr and Dawson (1988) identified a single point mutation at nucleotide 6157 in the coat protein gene that leads to a substitution of phenylalanine for serine at position 148 and that is responsible for the necrotic local lesion response (for further discussion see Chapter 10, Section III.E). Other studies with mutants in which the coat protein gene was partially or completely deleted showed that the coat protein can also influence symptoms in other ways (W. O. Dawson *et al.*, 1988). The coat protein plays a role in virus movement since the mutants established systemic infection less effectively than did wild-type virus (see Chapter 9, Section II.D.5.a).

ii. The 126-kDa and 183-kDa proteins

Both of these proteins are involved in viral RNA replication. They have significant amino acid sequence homology with known viral RNA-dependent RNA polymerases (replicases) (Kamer and Argos, 1984). Mutation in the 126-kDa gene caused a reduction in the synthesis of the 30-kDa protein and its mRNA, suggesting that the 126-kDa (and/or 183-kDa) proteins are involved in the synthesis of I_2 RNA (Watanabe *et al.*, 1987b). The most convincing evidence is the fact that a mutant constructed by Meshi *et al.* (1987) in which both the 30-kDa and the coat protein genes were deleted gave rise, in infected protoplasts, to a shortened viral RNA.

In vitro mutagenesis at or near the leaky termination codon of the 126-kDa gene indicates that both proteins are necessary for normal TMV replication in tobacco leaves (Ishikawa *et al.*, 1986).

The current understanding of the involvement of these proteins in the TMV replication complex is discussed in Chapter 8 (Section IV.H).

iii. The 30-kDa protein

Genetic studies with *ts* mutants first demonstrated the existence of a virus-coded function required for cell-to-cell movement of TMV (e.g. Taliansky *et al.*, 1982b; Atabekov *et al.*, 1983). Studies on the four proteins coded for by a TMV

mutant *ts* for cell-to-cell movement (LsI) showed a difference from the normal virus only in the 30-kDa protein (Leonard and Zaitlin, 1982). The mutation leading to the *ts* state in this mutant was shown to cause a change from a serine to a proline in the 30-kDa protein (Ohno *et al.*, 1983). Since then there have been detailed studies of the involvement of the 30-kDa protein in cell-to-cell movement of TMV; these are described in Chapter 9 (Section II.D.5.a).

2. Genus *Tobravirus* (reviewed by Visser *et al.*, 1999b)

The genome of tobraviruses is divided between two species of (+)-strand ssRNA (Fig. 6.37).

a. Genome structure

i. RNA1

The RNA1 of TRV (strain SYM) was the first tobravirus to be fully sequenced (Boccara *et al.*, 1986; Hamilton *et al.*, 1987). The RNA is 6791 nucleotides long and all tobraviruses have an RNA1 about this length. There are four ORFs arranged as shown in Fig. 6.37. ORFs 1 (194 kDa) and 2 (29 kDa) juxtapose in the same reading frame. ORFs 3 and 4, also termed P1a and P1b, encode proteins of 29 kDa and 12 kDa respectively. There is a 3' non-coding region of 255 nucleotides. Although TRV RNA1 cannot be aminoacylated, there is a tRNA-like feature in the 3'-terminal region (van Belkum *et al.*, 1987).

The sequence of PEBV RNA1 (MacFarlane *et al.*, 1989) showed a similar gene arrangement to that of TRV. However, the protein coded by

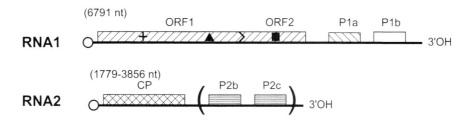

Fig. 6.37 Genome organization of TRV as representative of tobraviruses. The single lines represent the (+)-strand RNA genome. The boxes show the coding regions with the final gene products identified. The infills represent the function of the gene products—see Fig. 6.36 or inside book cover for details of these and other features. The ( ) show open reading frames on RNA2 that differ between strains of the virus (see Fig. 6.1).

ORF 4 is somewhat smaller (12 kDa). The difference in size is due to a continuous sequence of 35 amino acids absent from PEBV but present in TRV. In both proteins, there is a duplicated motif of cysteine and histidine residues. Thus, this *Tobravirus* gene product may be a 'zinc finger' protein that may be capable of zinc-dependent nucleic acid binding.

ii. RNA2

The RNA2 of tobraviruses has widely different lengths in the range of 1800–4000 nucleotides. Different strains of TRV show substantial diversity in the sequences of their RNA2. The variability in length of RNA2 is due both to the presence of additional genes in the larger molecules and to a variable sized 3'-terminal region homologous to that of the corresponding RNA1.

The only ORF consistently found in RNA2 is that for the coat protein (22–24 kDa) which is located at the 5' end. Downstream of this the sequences of RNA2 vary between viruses and strains (Fig 6.38). In some strains, for example TCM and PLB, the amount of 3' RNA homologous to RNA1 includes the 16-kDa ORF.

b. Proteins encoded

i. RNA1

The RNA1 ORF 1 of the SYM strain encodes a protein of 134 kDa, the opal termination codon (UGA) of which can be read through to give a protein of 194 kDa. The 194-kDa protein shows sequence similarities with the putative replicase genes of TMV, AMV, BMV and CMV (Hamilton *et al.*, 1987). The similarities are particularly strong with TMV. The 29-kDa protein from ORF P1a shows some sequence similarity to the TMV 30-kDa protein and thus is involved in virus movement from cell to cell. No similarities were found between the 16-kDa ORF (P1b) and other viral proteins. This protein is not required for replication, but virus from infectious transcripts of PEBV RNA1 with this ORF mutated were seed-transmitted at less than 1% of the rate of the wild-type virus.

RNA1 is translated in the rabbit reticulocyte system to give two polypeptides of 170 kDa and 120 kDa (Pelham, 1979). These two polypeptides are also produced together with many smaller products in the wheatgerm system containing added spermidine (Fritsch *et al.*, 1977). These two proteins correspond to the 194-kDa and 134-kDa ORFs (ORFs 1 and 2) shown in Fig. 6.37. Similar results have been reported for PEBV (Hughes *et al.*, 1986). A protein product for the 29-kDa ORF is translated from a subgenomic RNA (1A) (Robinson *et al.*, 1983). The 16-kDa protein is also translated from a subgenomic RNA (1B) (Guilford, 1989). A 16K protein product was found in infected protoplasts (Angenent *et al.*, 1989a), which was incorporated into a high-MW cellular component.

ii. RNA2

RNA2 of the PRN strain is translated *in vitro* to give the coat protein identified by serology, peptide mapping, and specific aggregation with authentic coat protein to form disk aggregates (Fritsch *et al.*, 1977). A second unrelated protein of 31 kDa is also translated. Different strains and viruses appear to differ in the products translated *in vitro* from RNA2 preparations, most likely due to variable contamination with sgRNAs from this genomic segment.

TRV strain PSG, which does not have additional ORFs in its RNA2, is infectious indicating that the extra genes on the RNA2s of other strains are not involved in replication. However, this strain, and also strains PLB and TCM which do contain extra ORFs on RNA2, are not transmitted by any known nematode vector. Mutations in the non-structural genes of RNA2 of strains TPA56 and PpK20 abolished their nematode transmissibility. Thus, some of the non-structural genes of RNA2 are involved in nematode transmission, a topic discussed more fully in Chapter 11 (Section XI.E).

3. Genus *Potexvirus* (reviewed by AbouHaidar and Gellatly, 1999)

a. Genome structure and strategy

Potexviruses have a (+)-sense ssRNA genome with an MW in the range 5.9–7.0 kb. The 5' end has an M7Gppp cap and the 3' end is polyadenylated. The genomes of several mem-

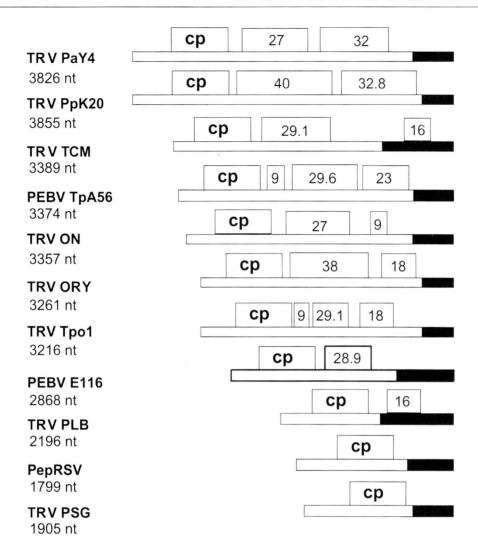

Fig. 6.38 Genome organization and length (nt) of RNA2 molecules of tobravirus isolates. Open reading frames encoding the coat protein (CP) and non-structural proteins are indicated by boxes: figures denote the P_r (kDa) of the encoded proteins. Shaded bars at the 3′ termini of RNA2 correspond to nucleotide sequences homologous with RNA1. Adapted from Visser (2000), with permission.

bers of this genus have been sequenced and most have been shown to have five ORFs, with the arrangement and coding capacity shown in Fig. 6.39. Some species (e.g. CsCMV, NMV, SMYEV and WClMV) have a sixth small ORF located completely within ORF 5, although it has no known protein product and is of unknown function.

Forty of the most 5′-terminal nucleotides show strong homology among the potexviruses and the 5′ non-coding region is rich in A and C residues. ORF 1 (147 kDa) begins 108

nucleotides from the 5′ terminus and the protein encoded by this ORF has domains of sequence similarity with the presumed polymerase genes of other RNA viruses. It corresponds to a 160-kDa protein translated *in vitro* from the genomic RNA (Forster *et al.*, 1987). ORFs 2 (25 kDa), 3 (12 kDa) and 4 (8 kDa) slightly overlap, and are commonly referred to as the 'triple gene block'. The products from these ORFs are involved in cell-to-cell movement and are described in Chapter 9 (Sections II.D.2.f and II.D.5.c). There are some apparent

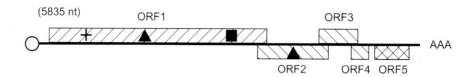

Fig. 6.39 Genome organization of PVX as representative of potexviruses. The single line represents the (+)-strand RNA genome. The boxes show the coding regions with the final gene products identified. The infills represent the function of the gene products—see Fig. 6.1 or inside book cover for details of these and other features.

exceptions to the organization of ORFs 2–4 in some potexviruses. For instance, ORF 2 of SMYEV lacks a conventional initiation codon (AUU; Jelkmann *et al.*, 1992)), as does ORF 4 of LVX. ORF 5 encodes the virion coat protein, which ranges in size from 18 kDa to 27 kDa according to virus species. The coat protein subunits may be proteolytically cleaved during purification and storage of the virus.

ORF 1 is translated from the genomic RNA, ORFs 2, 3 and 4 from sgRNA1 (1.9–2.1 kb), and the coat protein from sgRNA2 (0.9–1.0 kb).

4. Genus *Carlavirus* (reviewed by Zavriev, 1999)

The virions of most carlaviruses contain a single molecule of (+)-sense ssRNA of 7.4–7.7 kb; those of BBScV and PVM are somewhat larger at 8.5 kb. The RNA is polyadenylated at the 3′ end and is probably capped. It contains six ORFs (Fig. 6.40).

ORF 1, which follows a non-coding leader sequence of 75 nucleotides, encodes a polypeptide of 223 kDa, which is the viral replicase. ORFs 2 (25 kDa), 3 (12 kDa) and 4 (7 kDa) form a triple gene block involved in cell-to-cell movement of the virus as described above for potexviruses. ORF 5 encodes the 34-kDa coat protein and overlaps ORF 6 which codes for a cysteine-rich protein (11–16 kDa) of unknown function.

ORF 1 is translated from the genomic RNA. Two subgenomic RNAs of 2.1–3.3 kb and 1.3–1.6 kb that are possibly encapsidated are found in some species.

5. Genus *Allexivirus*

The (+)-strand ssRNA of ShVX encodes six ORFs (Fig. 6.41) (Kanyuku *et al.*, 1992). The 3′ terminus is polyadenylated, the structure of the 5′ terminus being unknown. ORF 1 follows a 5′

Fig. 6.40 Genome organization of PVM as representative of carlaviruses. The single line represents the (+)-strand RNA genome. The boxes show the coding regions with the final gene products identified. The infills represent the function of the gene products—see Fig. 6.36 or inside book cover for details of these and other features.

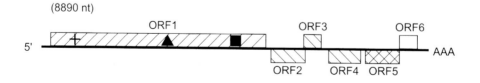

Fig. 6.41 Genome organization of ShVX as representative of allexiviruses. The single line represents the (+)-strand RNA genome. The boxes show the coding regions with the final gene products identified. The infills represent the function of the gene products—see Fig. 6.1 or inside book cover for details of these and other features.

non-coding region of 98 nucleotides and codes for a 195-kDa polypeptide that contains methyl transferase, helicase and RdRp motifs characteristic of a viral polymerase. ORFs 2 (26 kDa) and 3 (11 kDa) resemble the first two ORFs of the potexvirus and carlavirus triple gene block described above. There is a coding sequence for a small (7–8 kDa) third element of a triple gene block but this lacks the initiation AUG codon. The 42-kDa product of ORF 4 is expressed in high amounts in infected plants but its function is unknown. ORF 5 codes for the 28-kDa coat protein. Because of its high hydrophobicity, this protein migrates as a 32–36 kDa molecule on polyacrylamide gel electrophoresis. The function of the 3′ 15-kDa protein encoded by ORF 6 is unknown; like the protein from the 3′ ORF of carlaviruses, this protein has a zinc-binding finger motif and can bind nucleic acids.

6. Genus *Capillovirus*

The genomic RNA of ASGV is polyadenylated at its 3′ end, the structure of the 5′ end being unknown (Yoshikawa *et al.*, 1992). It has two ORFs (Fig. 6.42). ORF 1 encodes a putative 240–266-kDa protein with motifs at the 5′ end suggestive of a polymerase. The coat protein cistron is located at the C-terminal end of this ORF. Although it would appear that this large protein is a polyprotein, there is an sgRNA of 1.0 nucleotides that might be its mRNA. The

size of the product of ORF 2 (36–52 kDa) varies with virus species. It is likely that this protein is expressed from a 2.0 nucleotide sgRNA and that it functions as a cell-to-cell movement protein.

7. Genus *Foveavirus*

The genomes of foveaviruses are polyadenylated at their 3′ ends; that of the tentative member CGRMV is capped at the 5′ end. The type member, ASPV, encodes 5 ORFs (Fig. 6.43) (Jelkmann, 1994; Martelli and Jelkmann, 1998). ORF 1 encodes a protein of 247 kDa that has methyl transferase, helicase and RdRp motifs suggestive of it being a polymerase. ORFs 2 (25 kDa), 3 (13 kDa) and 4 (7 kDa) constitute a triple gene block and their products are thought to be involved in cell-to-cell movement (see Chapter 9, Section II.D.2.f). The 44-kDa coat protein is coded by ORF 5. CGRMV has two additional ORFs, 2a (14 kDa) and 4a (18 kDa), nested in ORFs 2 and 5 respectively, the functions of which are unknown.

ORF 1 is translated directly from the genomic RNA. There are several sgRNAs that are thought to express the other ORFs.

8. Genus *Trichovirus* (reviewed by German-Retana and Candresse, 1999).

The genomic RNA of trichoviruses is capped at the 5′ end and polyadenylated at the 3′ end. It

Fig. 6.42 Genome organization of ASGV as representative of capilloviruses. The single line represents the (+)-strand RNA genome. The boxes show the coding regions with the final gene products identified. The infills represent the function of the gene products—see Fig. 6.1 or inside book cover for details of these and other features.

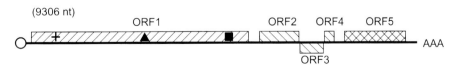

Fig. 6.43 Genome organization of ASPV as representative of foveaviruses. The single line represents the (+)-strand RNA genome. The boxes show the coding regions with the final gene products identified. The infills represent the function of the gene products—see Fig. 6.1 or inside book cover for details of these and other features.

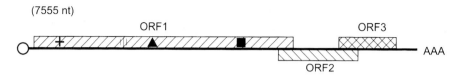

Fig. 6.44 Genome organization of ACLSV as representative of trichoviruses. The single line represents the (+)-strand RNA genome. The boxes show the coding regions with the final gene products identified. The infills represent the function of the gene products—see Fig. 6.1 or inside book cover for details of these and other features.

contains three ORFs (Fig. 6.44). ORF 1 encodes a protein of 216 kDa that has the methyl transferase, helicase and RdRp motifs characteristic of polymerases. It also contains a papain-like proteinase domain. The product of ORF 2 (50 kDa) shares distant similarities with CaMV gene 1 and TMV 30-kDa proteins and is thought to be considered with cell-to-cell movement. It has been detected in cell wall and membrane preparations from infected *Chenopodium quinoa* and is phosphorylated *in vivo*. The 22–28-kDa coat protein is coded on ORF 3.

ORF 1 is translated from the genomic RNA and ORFs 2 and 3 from sgRNAs of 2.2 and 1.1 kb respectively.

9. Genus *Vitivirus* (reviewed by German-Retana and Candresse, 1999)

The six species now grouped in the *Vitivirus* genus were originally classified as trichoviruses but they differ significantly from them in genome organization. Like trichoviruses, the genomic RNA of vitiviruses has a 5′ cap and is 3′ polyadenylated. However, unlike trichoviruses, the vitiviruses have five ORFs (Fig. 6.45). ORF 1 (194 kDa) is the viral polymerase with methyl transferase, helicase and polymerase motifs; it does not appear to have the papain-like protease motif of trichoviruses. ORF 3 (31 kDa) encodes the cell to-cell movement protein and ORF 4 the 21.5-kDa coat pro-

tein. The functions of the other two ORFs, not found in trichoviruses, ORF 2 (19 kDa) and the 3′ ORF 5 (10 kDa), are not known.

The four major bands of virus-specific dsRNAs from GVA- and GVD-infected plants are suggestive a sgRNA strategy of expression of the non-5′ ORFs, though the details are not known.

10. Genus *Furovirus* (reviewed by Shirako and Wilson, 1999)

The genome of the type member of the genus *Furovirus*, SBWMV, is divided between two RNA species (Fig. 6.46) (Shirako and Wilson, 1993). Both RNA molecules have 5′ caps and the 3′ ends can be folded into a tRNA-like structure that accepts valine. RNA1 contains two ORFs. The N-terminal section of ORF 1A encodes a 150-kDa protein that has methyl transferase and helicase motifs; read-through of the leaky UGA termination codon at the end of ORF 1A leads to a 209-kDa protein encoded by ORF 1aRT. The read-through portion contains an RdRp motif. Thus, ORF 1A codes for the viral polymerase. The product of ORF 1b is the 37-kDa putative cell-to-cell movement protein that is thought to be expressed from a sgRNA. RNA2 also contains two ORFs. ORF 2a encodes the 19-kDa coat protein from its AUG start codon and a 25-kDa protein from an upstream CUG codon (see Chapter 7, Section V.B.6).

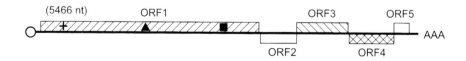

Fig. 6.45 Genome organization of GVA as representative of vitiviruses. The single line represents the (+)-strand RNA genome. The boxes show the coding regions with the final gene products identified. The infills represent the function of the gene products—see Fig. 6.1 or inside book cover for details of these and other features.

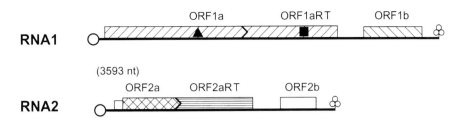

Fig. 6.46 Genome organization of SBWMV as representative of furoviruses. The single lines represent the (+)-strand RNA genome segments. The boxes show the coding regions with the final gene products identified. The infills represent the function of the gene products—see Fig. 6.1 or inside book cover for details of these and other features.

Read-through of the 19-kDa protein UGA termination codon gives an 84-kDa protein. It is thought that the read-through portion of this protein is associated with interaction of the virus with its fungal vector. ORF 2b product is a 19-kDa cysteine-rich protein that is thought to be expressed from a sgRNA; the function of this protein is unknown.

11. Genus *Pecluvirus* (reviewed by Reddy *et al.*, 1999; Shirako and Wilson, 1999)

The RNAs of the bipartite genome of PCV are capped at their 5′ ends and their 3′ ends can fold into a tRNA-like structure that accepts valine (Fig. 6.47) (Manohar *et al.*, 1993; Hertzog *et al.*, 1994). RNA1 contains two ORFs. ORF 1a encodes a 131-kDa protein that has a methyl transferase capping enzyme motif in its N-terminal region and a helicase motif in its C-terminal region. Read-through

of a weak UGA stop codon gives a 191-kDa protein, the read-through portion of which contains an RdRp motif. Thus, as with furoviruses, the product of ORF 1 is the viral moiety of the replicase. ORF 1b encodes a 15-kDa cysteine-rich protein of unknown function that is thought to be expressed from an sgRNA. PCV RNA2 contains five ORFs. The 390 nucleotide 5′ untranslated region is followed by the 23-kDa coat protein ORF 2a. ORF 2b overlaps ORF 2a by two nucleotides and is expressed as a 39-kDa protein by leaky scanning. The product is thought to be involved in the transmission of the virus by its fungal vector. The other three ORFs make up a triple gene block of 51-, 14- and 17-kDa proteins that potentiate the cell-to-cell movement of the virus. The products from ORFs 2–5 are thought to be expressed from sgRNAs though these have not been fully characterized.

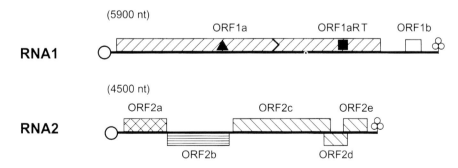

Fig. 6.47 Genome organization of PCV as representative of pecluviruses. The single lines represent the (+)-strand RNA genome segments. The boxes show the coding regions with the final gene products identified. The infills represent the function of the gene products—see Fig. 6.1 or inside book cover for details of these and other features.

12. Genus *Pomovirus* (reviewed by Shirako and Wilson, 1999; Torrance, 1999)

Members of the genus *Pomovirus* have their genomes divided between three species of (+)-strand ssRNA. The organization of the genome of BSBV is shown in Fig. 6.48 (Koenig *et al.*, 1996, 1997; Koenig and Loss, 1997). The 5' end of each RNA molecule is capped and the 3' end forms a tRNA-like structure containing an anticodon for valine. RNA1 contains a single ORF that codes for a 149-kDa protein and a read-through protein of 207 kDa. The N-terminal part of the ORF has motifs for methyl transferase and helicase activities, and the C-terminal part resulting from the read-through of a weak UAA termination codon has an RdRp motif.

The N-terminal part of RNA2 codes for the 20-kDa capsid protein, and read-through of a UAG stop codon results in a 104-kDa capsid fusion protein. RNA3 contains three ORFs that form a triple gene block of 48-, 13- and 22-kDa proteins involved in cell-to-cell spread of the virus.

13. Genus *Benyvirus* (reviewed by Tamada, 1999)

The genome of BNYVV consists of up to five molecules of ss (+)-sense RNA numbered 1–5 in decreasing order of size. The five RNAs have a 5' cap and a 3' poly(A) sequence. The nucleotide sequences of all four RNAs have been established (Bouzoubaa *et al.*, 1985, 1986, 1987; Kiguchi *et al.*, 1996) (Fig. 6.49). The 5' sequences all begin with an AAA sequence following the cap. The 3'-terminal 60 nucleotides of the five RNAs show extensive sequence homology. The infectious genome of this virus is contained in RNAs 1 and 2; RNAs 3–5 contribute to the symptom expression and vector transmission (see Chapter 14, Section II.B.3.c).

RNA1 has an ORF for one large polypeptide of 237 kDa that contains motifs for methyl transferase, helicase and RdRp, and thus can be considered to be the viral replicase protein. This protein is processed to give a 150-kDa and 66-kDa product by a papain-like protease activity located between the helicase and RdRp motifs. *In vitro* translation of RNA1 in the wheatgerm system gave two large polypeptides: 240 kDa initiating at AUG (nucleotide 154) and 220 kDa initiating at an internal AUG (nucleotide 496). In rabbit reticulocyte lysate only the 220-kDa protein was made. However, the nucleotide context around AUG (154) is much closer to the plant consensus sequence (see Chapter 7, Section V.A) than that near AUG (496) (Jupin *et al.*, 1988).

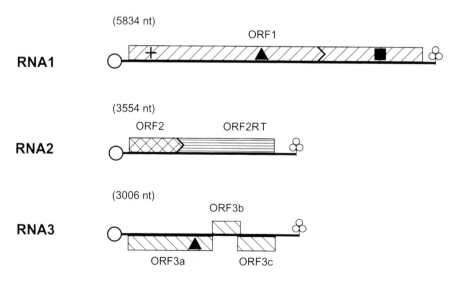

Fig. 6.48 Genome organization of BSBV as representative of pomoviruses. The single lines represent the (+)-strand RNA genome segments. The boxes show the coding regions with the final gene products identified. The infills represent the function of the gene products—see Fig. 6.1 or inside book cover for details of these and other features.

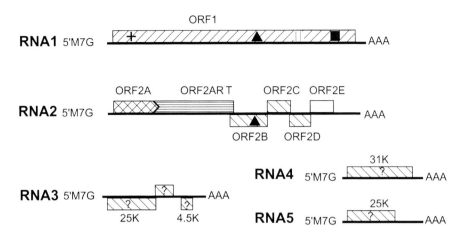

Fig. 6.49 Genome organization of BNYVV as representative of benyviruses. The single lines represent the (+)-strand RNA genome segments. The boxes show the coding regions with the final gene products identified. The infills represent the function of the gene products—see Fig. 6.1 or inside book cover for details of these and other features.

RNA2 has six ORFs. The 5′ ORF codes for the 21-kDa coat protein at the 5′ end with a weak UAG termination codon giving a read-through protein of 75 kDa. The read-through protein is involved in virus assembly and fungus transmission. ORFs 2b, 2c and 2d form a triple gene block giving proteins of 42, 13 and 15 kDa that are involved with cell-to-cell spread of the virus. This block of ORFs is expressed from two sgRNAs. The 3′ proximal ORF, 2e, encodes a zinc and nucleic acid binding protein of 14 kDa; it is expressed from a separate sgRNA.

RNA3 has three ORFs: 3a encoding a 25-kDa protein; a short 3b ORF that overlaps 3a and is involved with a necrotic response; and the 4.5-kDa ORF 3c. The latter is expressed from a subgenomic RNA. RNA4 encodes for a 31-kDa protein and RNA4 for a 25-kDa protein.

14. Genus *Hordeivirus* (reviewed by Lawrence and Jackson, 1999; Lawrence *et al.*, 2000)

a. Genome structure and the proteins encoded

The three genomic RNAs of BSMV are uniquely designated α, β and γ (Fig. 6.50). I will use these designations as they are in common practice, although the terms RNAs 1, 2 and 3 (in decreasing order of size)—which are consistent with the other viruses having tripartite genomes—are sometime used. These genomic RNAs have an M⁷Gppp cap at the 5′ end (Agranovsky *et al.*,

1979) and a tRNA-like structure at the 3′ end, which can be aminoacylated with tyrosine (Agranovsky *et al.*, 1981; Loesch-Fries and Hall, 1982). There is an internal poly(A) sequence of 8–30 residues, approximately 210 nucleotides from the 3′ terminus (Agranovsky *et al.*, 1982).

i. RNAα

A single protein of 120 kDa is translated *in vitro* from RNA1 (Dolja *et al.*, 1983). Complete sequencing data show that this is the only polypeptide likely to be coded for by this RNA with a true size of 130 kDa and termed αa (Gustafson *et al.*, 1989). A 120-kDa virus-specific protein, probably the αa product, has been found *in vivo*. The αa gene product shows amino acid sequence similarities with the large proteins of BMV RNA1, AMV RNA1, and the 126-kDa protein of TMV with methyl transferase and helicase motifs. A helicase motif is also found in the 58-kDa βb ORF of RNA2 (Gustafson *et al.*, 1987).

ii. RNAβ

RNAβ has been fully sequenced (Gustafson and Armour, 1986). There are four major ORFs as illustrated in Fig. 6.50. Three lines of evidence show that RNA2, and ORF βa in particular, codes for the 22-kDa BSMV coat protein: (1) RNA2 is translated efficiently *in vitro* to give

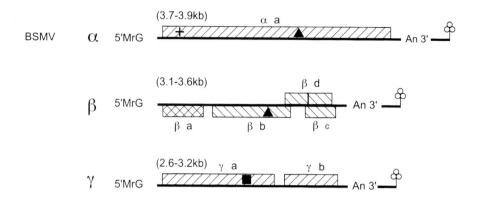

Fig. 6.50 Genome organization of BSMV as representative of hordeiviruses. The single lines represent the (+)-strand RNA genome segments. The boxes show the coding regions with the final gene products identified. The infills represent the function of the gene products—see Fig. 6.1 or inside book cover for details of these and other features.

coat protein (Dolja *et al.*, 1979); (2) the amino acid composition deduced from the nucleotide sequence is very close to that determined experimentally for the coat protein; and (3) the predicted sequence of the 30 N-terminal amino acids in the ORF 2 translation product is identical to 29 of the corresponding amino acids in BSMV coat protein determined by direct sequencing.

The next three ORFs on RNAβ comprise a triple gene block that is involved in cell-to-cell spread of the virus. ORF βb encodes a 58-kDa polypeptide, which, as noted above, contains a helicase motif. The products of ORFs βc and βd are 17 and 14 kDa respectively. In BSMV the stop codon of βd is read through to give a minor 23-kDa product, βd'. ORF βa is expressed from the genomic RNA, βb from sgRNA β1, and ORFs βc, βd and βd' from sgRNAβ2.

iii. RNAγ

Unlike RNAs α and β, which are relatively constant in length between different strains of BSMV, RNAγ varies significantly between some strains. Gustafson *et al.* (1987) sequenced two RNAγs, of different length, from the type and ND18 strains. There are two ORFs, γa and γb, separated by an intercistronic region. The larger size of the type strain is due to an in-frame 366-nucleotide direct tandem repeat located within ORF γa.

The predicted size of the translation product for type strain ORF γa is 87 kDa and for the ND18 strain it is 74 kDa. This strain-specific difference is readily seen in the polypeptides translated *in vitro* from these two strains (Gustafson *et al.*, 1981). The amino acid sequence predicted in ORF γa contains the RdRp motif. Thus, the viral contribution to the BSMV replicase complex comprises the products of ORFs αa and γa.

The 3' region of RNAγ has an ORF, γb, which encodes a polypeptide of 17 kDa; this polypeptide is translated from a subgenomic RNA (e.g. Jackson *et al.*, 1983). The initiation point for this RNA is 27 nucleotides upstream of the ORF γb initiation codon in the intercistronic region. Unlike subgenomic RNAs of some other viruses, the ORF γb RNA is not strictly co-terminal with the 3' end of the genomic RNA. Instead it contains a poly(A) terminus of variable length up to 150 nucleotides (Stanley *et al.*, 1984). Synthesis of the subgenomic RNA is presumably initiated at an internal site on (−)-strand RNAγ. The protein coded by ORF γb is rich in sulphur-containing amino acids, with no sequence similarity to any proteins coded for by other viruses with tripartite genomes. This protein is a pathogenicity determinant that is involved in regulating expression of genes encoded on RNAβ.

In summary, the genome of BSMV differs from that of the other viruses with (+)-sense tripartite RNA genomes in several ways:

1. It codes for seven proteins instead of four.
2. The coat protein gene is at the 5' end of RNAβ and is translated from the genomic RNA.
3. The RNAγ subgenomic RNA is not strictly 3' co-terminal with the genomic RNA.

15. Genus *Sobemovirus* (reviewed by Sehgal, 1999; Tamm and Truve, 2000)

The genomes of SCPMV, RYMV, CfMV, SBMV and LTSV have been sequenced and shown to consist of a single strand of (+)-sense ssRNA (Wu *et al.*, 1987; Yassi *et al.*, 1994; Mäkinen *et al.*, 1995b; Othman and Hull, 1995; Lee and Anderson, 1998). All except CfMV have a similar genome organization, and that of SCPMV is shown in Fig. 6.51. The 5' terminus is covalently linked to a small VPg (Mang *et al.*, 1982). The 48-nucleotide leader contains a sequence that is partially complementary to the 3' end of the 18S ribosomal RNA, suggesting a possible role in ribosome binding. This site is seven bases 5' to the first AUG codon. There is a 3'-terminal untranslated region 153 nucleotides long that does not have the potential to form a tRNA-like structure.

The genomic RNAs contain four ORFs. All the sobemoviruses have ORF 1 at the 5' end of the genome and ORF 4 at the 3' end. The product of ORF 1 (12–18 kDa) is involved in cell-to-cell movement of the virus (Bonneau *et al.*, 1998; Sivakumaran *et al.*, 1998) and that of ORF 4 encodes the viral coat protein (26–30 kDa). In most of the sobemoviruses ORF 2 encodes a large protein (90–110 kDa) that contains sequence motifs for a serine proteinase and for an RdRp. This ORF 2 product is processed to

give the 5' VPg (10–12 kDa) from a region between the proteinase and RdRp motifs. The ORF 2 of CfMV comprises two overlapping ORFs, 2a and 2b, the polyprotein being expressed by a –1 ribosomal frameshift mechanism (Mäkinen *et al.*, 1995a). The proteinase motif and the region from which the VPg is derived lie in ORF 2a and the RdRp motif in ORF 2b. Nested within ORF 2 of most sobemoviruses is ORF 3 capable of encoding a protein of about 18 kDa; the function of this protein is unknown. This ORF is lacking in one strain of SBMV and in CfMV

ORFs 1 and 2 are expressed from the genomic RNA, ORF 2 presumably from leaky initiation. ORF 4 is translated from a sgRNA (Ghosh *et al.*, 1981). Whether and how ORF 3 is expressed is unknown.

16. Genus *Marafivirus*

The genomic RNAs of the marafiviruses, OBDV and PnMV, have been sequenced (Edwards *et al.*, 1997; Bradel *et al.*, 2000). The genomic RNA is polyadenylated at its 3' end and capped at the 5' end. It contains two ORFs (Fig. 6.52). The 5' ORF 1 encodes a large polyprotein (221–227 kDa) the N-terminal part of which has motifs of methyl transferase, papain-like protease, helicase and RdRp activities. The C-terminal part of this large protein contains one of the two coat proteins, the 24-kDa protein. The other coat protein (21 kDa) is encoded in ORF 2 that overlaps ORF 1 at the 3' end of the genome; the gene product from ORF 2 is translated from a sgRNA.

17. Genus *Tymovirus* (reviewed by Gibbs, 1999)

The tymovirus genome comprises one species of linear (+)-sense ssRNA that is capped at the

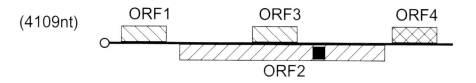

Fig. 6.51 Genome organization of SCPMV as representative of sobemoviruses. The single line represents the (+)-strand RNA genome. The boxes show the coding regions with the final gene products identified. The infills represent the function of the gene products—see Fig. 6.1 or inside book cover for details of these and other features.

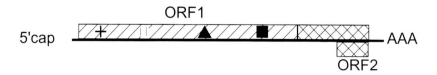

Fig. 6.52 Genome organization of OBDV as representative of marafiviruses. The single line represents the (+)-strand RNA genome. The boxes show the coding regions with the final gene products identified. The infills represent the function of the gene products—see Fig. 6.1 or inside book cover for details of these and other features.

5' end and the 3' end of which has a tRNA-like structure that accepts valine. Tymovirus RNAs are characterized by having high cytidine content (31–42%).

The genome organization is very compact (Fig. 6.53). For example, with the European TYMV only 192 (3%) of the 6318 nucleotides are non-coding. The largest ORF, ORF 1, initiates at position 95 and ends at position 5627 with a UAG codon (to give a protein of 206 kDa and 1844 amino acids). This 206-kDa protein contains sequence motifs characteristic of a methyl transferase, a papain-like protease, a helicase and RdRp. This protein is processed to give a 141-kDa and a 66-kDa product.

A second ORF, ORF 2, starts at the first AUG on the RNA (beginning at nucleotide 88) and terminates with a UGA at 1972 (69 kDa; 628 amino acids). It overlaps ORF 1 in the same reading frame as the coat protein gene. The coat protein gene (ORF 3) is at the 3' end of the vital genome There is a 105-nucleotide non-coding 3' region, which contains the tRNA-like structure.

ORFs 1 and 2 are expressed from the genomic RNA by a leaky initiation mechanism. The coat protein is translated from a sgRNA that is packaged with the genomic RNA and in a series of partially filled particles

(e.g. Pleij *et al.*, 1977). No dsRNA corresponding in length to this subgenomic RNA could be detected *in vivo*. Gargouri *et al.* (1989) detected nascent subgenomic coat protein plus strand RNAs on dsRNA of genomic length. Thus, the coat protein mRNA is synthesized *in vivo* by internal initiation on (−) strands of genomic length. Ding *et al.* (1990b) compared the available *Tymovirus* nucleotide sequences around the initiation site for the subgenomic mRNA. They found two conserved regions: CAAU/C at the initiation site, and a 16-nucleotide sequence GAGUCUGAAU-UGCUUC on the 5' side of it. This longer sequence, which they called the *tymobox*, may be an important component of the promoter for subgenomic RNA synthesis.

18. Genus *Idaeovirus* (reviewed by Mayo and Jones, 1999b)

The (+)-strand ssRNA genomes of idaeoviruses are bipartite (Fig. 6.54). RNA1 contains a large ORF that codes for a 190-kDa protein. This protein contains motifs for methyl transferase, helicase and RdRp activities and thus is the viral replicase. There is a small 12-kDa ORF overlapping the large one on RNA1; it is not known if this ORF is expressed. RNA2 contains two ORFs. The 39-kDa protein from the 5' ORF

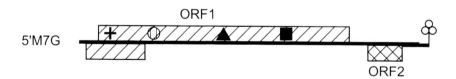

Fig. 6.53 Genome organization of TYMV as representative of tymoviruses. The single line represents the (+)-strand RNA genome. The boxes show the coding regions with the final gene products identified. The infills represent the function of the gene products—see Fig. 6.1 or inside book cover for details of these and other features.

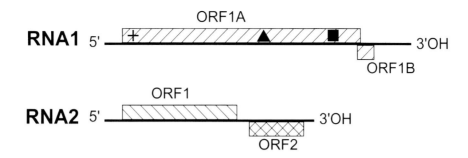

Fig. 6.54 Genome organization of the idaeovirus RBDV. The single lines represent the (+)-strand RNA genome segments. The boxes show the coding regions with the final gene products identified. The infills represent the function of the gene products—see Fig. 6.1 or inside book cover for details of these and other features.

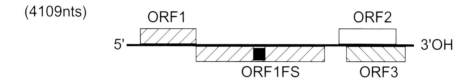

Fig. 6.55 Genome organization of GRV as representative of umbraviruses. The single line represents the (+)-strand RNA genome. The boxes show the coding regions with the final gene products identified. The infills represent the function of the gene products—see Fig. 6.1 or inside book cover for details of these and other features.

is the putative cell-to-cell movement protein, and the 3′ ORF encodes the 30-kDa coat protein. The coat protein is expressed from a 946-nucleotide sgRNA.

19. Genus *Ourmiavirus*

Ourmiaviruses have linear (+)-sense ssRNA genomes divided into three segments of approximately 3.0, 1.1 and 1.0 kb (Accotto *et al.*, 1997). Apart from the fact that the coat protein is encoded by RNA3, nothing is currently known about the genome organization.

20. Genus *Umbravirus* (reviewed by Robinson and Murant, 1999)

Umbraviruses do not form conventional virus particles unless they are associated with a helper virus (see Chapter 11, Section H.1.a). The genomes of three members have been sequenced, GRV (Taliansky *et al.*, 1996), CMoMV (Gibbs *et al.*, 1996) and PEMV-2 (Demler *et al.*, 1993). The genome of GRV contains four ORFs (Fig. 6.55). ORF 1 encodes a protein of 31 kDa and, most probably by a

frameshift event, translation continues into ORF 2 to give a 95-kDa product. The ORF 2 region of this product contains the sequence motifs for an RdRp and thus it can be considered to be the viral replicase. The function of the 27-kDa product of ORF 3 is associated with long-distance movement (Ryabov *et al.*, 1999). The 28-kDa protein encoded by ORF 4 has a cell-to-cell movement function. Umbravirus genomes lack genes for plausible coat proteins.

IX. SUMMARY AND DISCUSSION

As noted in Section II.A, the conventional definition of an ORF is that it is a coding region, starting with an AUG initiation codon, capable of expressing a protein of 10 kDa or more. Recent results indicate that there are limitations to this definition that may lead to functional ORFs being overlooked. These limitations include:

1. Several ORFs of less than 10 kDa have been found to be functional; e.g. ORF 4 of potexviruses and ORF 2 of BYV.

2. Several ORFs have been found to initiate with a non-conventional codon; e.g. AUU in RTBV ORF 1.
3. Frameshift or read-through events could produce a larger ORF from two (or more) smaller ones.

The descriptions in the above sections show the great range of variability of the genome organizations of plant viruses, which is emphasized by Table 6.6.

The variety of plant RNA virus genomes has also been discussed by Zaccomer *et al.* (1995). However, there are some basic features to genome organization. Essentially, a viral genome, be it of a higher or lower plant, a higher or lower animal or a bacterium, is composed of various cassettes. The basic cassette comprises the gene or genes and the nucleic acid sequences that replicate and express the genome. This is made up of the replicase enzymes, the sequences that initiate and/or terminate the formation of a new viral genome from the input one and sequences and/or proteins that control expression. Attached to the replication cassette are cassettes of genes and/or nucleic acid sequences that control the replication and that adapt the virus to its host. As a virus is totally dependent on its host for its multiplication, it can be considered to be advantageous for it to control its replication and to maintain its natural host in a living state for as long as possible. Viruses that successfully infect plants have to be able to move from the initially infected cell to most, if not all, parts of the plant and to control or overcome any plant defense system. They also have to be able to be moved from an infected host to a healthy one. Thus, there are individual genes or collections of genes that potentiate movement within the plant and movement between plants. The latter involves gene products that protect the viral genomic nucleic acid from degradation while it is outside the host cell, usually the viral coat protein (except umbraviruses that use the coat protein of a helper virus), and often one or more gene products that facilitate interactions with the virus vector. The genes involved in movement within and between plants will be described in more detail in Chapters 9, 11 and

12. As will be described in Chapter 10, plants have a defense system against 'foreign' nucleic acids which successful viruses have to overcome.

Thus, the 'basic' virus has a cassette of one or more proteins for replicating its genome, one for one or more coat proteins, a cassette of one or more proteins for movement within the plant and, if it has a vector, possibly a cassette of one or more proteins (which may be the coat protein) for interacting with it. Coupled with these cassettes, many viruses have proteins that overcome the inherent host defense system.

Table 6.6 summarizes the current knowledge of the proteins encoded by the various families and genera of plant viruses. Several of the virus genera have this 'basic' organization. Perhaps the most 'basic' are the cryptoviruses (*Partitiviridae*) that encode just a replicase and coat protein. They rely on their ability to infect meristematic cells to be able to infect new cells within the plant, as well as on their high rates of seed transmission to be able to move to new sites.

Cell-to-cell movement proteins have been identified for most virus groups. For those that no movement protein has yet been identified there are proteins for which no function has yet been ascribed. As is discussed in detail in Chapter 9, cell-to-cell movement is mediated either by a single protein or by a cassette of three proteins, termed the triple gene block.

As described in Chapter 11, there are specific interactions with their invertebrate or fungal vector. This is either with sequences on the surface of the major coat protein, with a minor modified coat protein species incorporated into the capsid, or by a helper protein that bridges between the viral coat protein and specific sites within the vector.

Many of the virus groups have proteins with other functions. In those that express all or part of their genetic information as a polyprotein the virus encodes one or more proteinase that gives the final products; these are described in Chapter 7. For several viruses, no function(s) has yet been ascribed to one or more of their gene functions. Possible functions could include: (1) cell-to-cell movement as suggested

TABLE 6.6 Proteins encoded by plant viruses

Family/genus	Replicase protein	Coat protein	Movement protein	Vector protein	Suppressor protein	Other protein
Caulimoviridae	+	+	+	+		+
Geminiviridae	(+)[a]	+	+	(+)[c]		+
Nanovirus	+	+	+			
Reoviridae	+	+				
Partitivirus	+	+	−	−		
Rhabdoviridae	+	(+)[b]	+	+		
Tospovirus	+	(+)[b]	+	+		+
Tenuivirus	+	(+)[b]	+			+
Bromovirus	+	+	+			
Cucumovirus	+	+	+	(+)[c]	+	
Alfamovirus	+	+	+	(+)[c]		
Ilarvirus	+	+	+		+	
Oleovirus	+	+	+			
Comovirus	+	+	+			+
Nepovirus	+	+	+	+		+
Potyviridae	+	+	+			+
Tombusvirus	+	+	+			+
Aureusvirus	+	+	+			+
Avenavirus	+	+				+
Carmovirus	+	+	+			+
Dianthovirus	+	+	+			
Machlomovirus	+	+				+
Necrovirus	+	+	+			+
Panicovirus	+	+	+			+
Closterovirus	+	+	+	+		+
Crinivirus	+	+	+	+		+
Luteovirus	+	+	+	+		+
Polerovirus	+	+	+	+		+
Enamovirus	+	+	+	+		
Tobamovirus	+	+	+			
Tobravirus	+	+	+			+
Potexvirus	+	+	+			
Carlavirus	+	+	+			+
Allexivirus	+	+	+			+
Capillovirus	+	+	+			
Foveavirus	+	+	+			
Trichovirus	+	+	+			
Vitivirus	+	+	+			+
Furovirus	+	+	+	+		+
Pecluvirus	+	+	+	+		+
Pomovirus	+	+	+			
Benyvirus	+	+	+	+		+
Hordeivirus	+	+	+			+
Sobemovirus	+	+	+			+
Marafivirus	+	+				
Tymovirus	+	+				
Idaeovirus	+	+	+			+
Umbravirus	+	−	+			+

[a] Proteins that initiate host replicase; [b] Nucleocapsid; [c] Vector specificity by coat protein.
+ = present, − = absent, blank = unknown.

above, (2) suppression of the host defense system, and (3) control of the expression of viral genomes.

This chapter has ascribed the functions to individual cassettes that might be a single gene or a group of genes. However, it is becoming increasingly apparent that few, if any, gene products operate on their own. As will be described in Chapter 13, the virus infection cycle is a highly integrated process and each gene product (and non-coding sequence) interacts with several or many others.

C H A P T E R **7**

Expression of Viral Genomes

Having described the structures of viruses and their genomes, I now move on to how these genomes function. As an introduction to the subject, I shall consider a brief outline of the probable main stages in the replication of a virus. There are many variations in detail in these stages.

1. The virus particle enters the cell. At the time of entry or shortly afterwards the genome is released from the protein coat or the structure of the particle relaxes to enable the next stages to take place.
2. The infecting genome is either translated directly if it is (+)-sense ssRNA, or mRNAs are formed and translated, to give early products such as the viral replicase, and perhaps other virus-specific proteins. This is described in this chapter.
3. The viral replicase or replication-associated protein(s) are used to synthesize subgenomic mRNAs if required by the genome strategy. This is also described in this chapter.
4. The viral replicase or replication-associated proteins are used to synthesize new viral genomes, as described in the next chapter.
5. Proteins required relatively late in the viral replication cycle, such as coat protein and cell-to-cell movement protein, are synthesized.
6. Coat protein subunits and viral genomes are assembled to give new virus particles, which accumulate within the cell, usually in the cytoplasm. This was described in Chapter 5.
7. Infectious units of the virus move from the initially infected cell to adjacent cells and possibly through the plant to initiate a systemic infection, as described in Chapter 9.

Virus acronyms are given in Appendix 1.

I. INTRODUCTION

Viral genomes are expressed from mRNAs that are either the nucleic acid of positive-sense [(+)-sense] ssRNA viruses or transcripts from negative-sense [(–)-sense] or dsRNA or from ds or ss DNA viruses. Baltimore (1971) pointed out that the expression of all viral genomes, be they RNA or DNA, ss or ds, (+)- or (–)-sense, converge on the mRNA stage (Fig. 7.1).

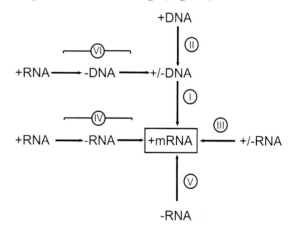

Fig. 7.1 Routing of viral genome expression through mRNA. Route I is transcription of dsDNA usually by host DNA-dependent RNA polymerase. Route II is the transcription of ssDNA to give the dsDNA template for I (e.g. geminiviruses). Route III is transcription of dsRNA, usually by virus-coded RdRp (e.g. reoviruses). Route IV is replication of (+)-strand RNA via a (–)-strand template by virus-coded RdRp — the viral (+) strand is often the template for early translation (the (+)-strand RNA viruses). Route V is transcription of (–)-strand virus genome by virus-coded RdRp (e.g. tospoviruses). Route VI is reverse transcription of RNA stage of retro- and pararetro-viruses leading to the dsDNA template for mRNA transcription. From Baltimore (1971), with permission.

When the encapsidated virus particle enters a susceptible plant cell, the genome must be released from the relatively stable capsid required for movement from host to host. Once the genome becomes available, it can be translated directly if (+)-sense ss RNA is present, or else the formation of mRNA can commence.

As will be described in Section V.A, expression of the viral mRNA faces various constraints imposed by the eukaryotic translation system.

In this chapter, I describe how viruses release their encapsidated genome on entry into the host cell, how they express their genetic information overcoming the various constraints imposed by the host translation system, and how the expression is regulated.

II. VIRUS ENTRY AND UNCOATING

A. Virus entry

As described in Chapters 11 and 12, plant viruses require damage of the cuticle and cell wall to be able to enter a plant cell. There have been various suggestions as to the mechanism of entry into the cell in which infection is initially established (reviewed by Shaw, 1999) (Fig. 7.2).

There is no evidence for a specific entry mechanism such as plasma membrane receptor sites or endocytotic uptake, and it is generally considered that 'entry is accomplished by brute force' (Shaw, 1999).

B. Uncoating of TMV (reviewed by Shaw, 1999)

1. Early events in intact leaves

The nature of the leaf surface, the requirement for wounding, the efficiency of the process, and other aspects of infection in intact leaves are discussed in Chapter 12. The uncoating process has been examined directly by applying TMV radioactively labelled in the protein or the RNA or in both components (e.g. Shaw, 1973; Hayashi, 1974). The following conclusions were drawn from such experiments:

1. Within a few minutes of inoculation, about 10% of the RNA may be released from the virus retained on the leaf.
2. Much of the RNA is in a degraded state but some full-length RNAs have been detected.
3. *In vivo* stripping of the protein from the rod begins at a minimum of two and probably many more sites along the rod (Shaw, 1973) (but see Section II.B.2).
4. The early stages of the process do not appear to depend on pre-existing or induced enzymes (Shaw, 1969).
5. The process is not host-specific, at least in the early stages. However, there is a fundamental difficulty with all such experiments. Concentrated inocula must be used to provide sufficient virus for analysis, but this means that large numbers of virus particles enter cells rapidly (Fig. 7.3). It is impossible to know which among these particles actually establish an infection.

2. Disassembly of the virus *in vitro*

To initiate infection, TMV RNA must be uncoated, at least to the extent of allowing the first ORF to be translated. Most initial *in vitro* experiments on the disassembly of TMV were carried out under non-physiological conditions. For example, alkali or detergent (1% sodium dodecyl sulfate, SDS) cause the protein subunits to be stripped from TMV RNA beginning at the 5'-end of the RNA (the concave end of the rod) (e.g. Perham and Wilson, 1976). Controlled disassembly by such reagents yields a series of subviral rods of discrete length (e.g. Hogue and Asselin, 1984). Various cations slow down or prevent the stripping process at pH 9.0 (Powell, 1975). Durham *et al.* (1977) suggested that Ca^{2+} binding sites might act as a switch controlling disassembly of TMV in the cell. Removal of Ca^{2+} would result in a change in the conformation of protein subunits, leading to their disaggregation. Durham (1978) proposed that TMV (and other small viruses) might be disassembled at or within a cell membrane. The virus might be in a medium roughly 10^{-3} M with respect to Ca^{2+} outside the cell, while inside the cell Ca^{2+} is about 10^{-7} M. The ion dilution would provide free energy to help

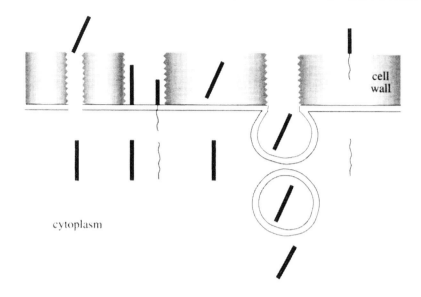

Fig. 7.2 Proposed routes for entry of TMV particles during manual inoculation of leaves. None of these routes has been demonstrated directly and all remain unproven. Left to right: direct entry of virus particles through wound; attachment of virus particle to cell membrane and passage of virus particle or viral RNA into cell; passage of virus particle through cell wall via ectodesmata or 'bleb'; attachment of virus particle to cell membrane and entry after invagination of membrane and formation of endocytotic vesicle; attachment of virus particle to outer cell wall and passage of viral RNA through wall into cell. From Shaw (1999), with kind permission of the copyright holder, © The Royal Society.

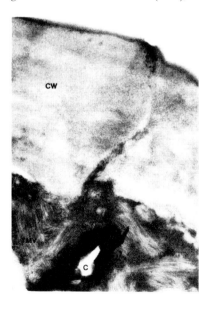

Fig. 7.3 TMV particles that have entered a tobacco leaf lower epidermal cell through a wound caused by abrasive (celite). Tissue was excised and fixed immediately after inoculation. Large numbers of virus rods (TMV) are visible in the cytoplasm. CW, cell wall; C, celite. From Plaskitt *et al.* (1987), with kind permission of the copyright holder, © The American Phytopathological Society.

break inter-subunit bonds. These ideas remain speculative.

Wilson (1984a) found that treatment of TMV rods briefly at pH 8.0 allowed some polypeptide synthesis to occur when the treated virus was incubated in an mRNA-dependent rabbit reticulocyte lysate (Fig. 7.4).

Wilson suggested that the alkali treatment destabilizes the 5'-end of the rod sufficiently to allow a ribosome to attach to the 5' leader sequence and then to move down the RNA, displacing coat protein subunits as it moves by a process termed *co-translational disassembly*. He called the ribosome-partially-stripped-rod complexes 'striposomes' (Fig. 7.5) and suggested that a similar uncoating mechanism may occur *in vivo*.

In contrast to the reticulocyte lysate system, in the wheatgerm system virus treated at pH 8.0 gave rise to three times as much polypeptide synthesis as isolated TMV RNA, presumably owing to protection of the RNA in the rod from nucleases before it was uncoated (Wilson, 1984b). In an *in vitro* protein-synthesizing

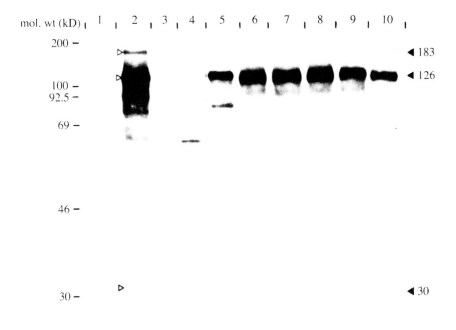

mol. wt (kD)

Fig. 7.4 *In vitro* co-translational disassembly of TMV. Electrophoretic resolution of the products of cell-free translation reactions programed with TMV RNA (lane 2) or purified TMV particles that had been pretreated at pH 8.0–8.2 (lanes 5–10). Numbers at left show positions of markers; those to the right are positions of TMV proteins. The appearance of the 126 kDa product provided evidence of a co-translational disassembly mechanism. From Wilson (1984a), with permission.

system from *E. coli*, virus treated at pH 8.0 gave rise to significant amounts of the 126-kDa protein, whereas TMV RNA gave polypeptides of 50 kDa or less, with a substantial amount of coat protein size (Wilson 1986). *Xenopus* oocytes micro-injected with TMV produced at least as much immunoreactive 126-kDa protein as did oocytes injected with TMV RNA (Ph. C. Turner *et al.*, 1987). This experiment appears to rule out a specific role for the cellulose cell wall in the uncoating of TMV in leaves. Whether it also rules out a role for the plasma membrane depends on whether intact virus particles contaminating the outside of the needle used for injection could have entered the oocytes via the cell membrane, being uncoated on the way.

Treatment of TMV *in vitro* with SDS for 15 seconds exposed a sequence of nucleotides from the 5′ terminus to beyond the first AUG codon. No more 5′ nucleotides were exposed during a further 15 minutes in SDS. Incubation of SDS-treated rods with wheatgerm extract, or rabbit reticulocyte lysate, led to the binding of one or two ribosomes in ≈20% of the particles (Mundry *et al.*, 1991). Structure predictions suggest that the exposed sequence up to the first AUG exists in an extended single-stranded configuration (see Section V.C.5), which would assist in the recruitment of ribosomes.

3. Experiments with protoplasts

To obtain infection of a reasonable proportion of protoplasts it is necessary to treat the virus and/or the protoplasts in one of several ways (see Chapter 8, Section III.A.5). Electron microscopy has been used to study the entry process, and it has been suggested that poly-L-ornithine stimulates entry of TMV either by damaging the plasmalemma (Burgess *et al.*, 1973) or by stimulating endocytotic activity (Takebe *et al.*, 1975). Estimates of the extent of uncoating of the adsorbed TMV inoculum vary from 5% (Wyatt and Shaw, 1975) to about 30% one hour after inoculation (Zhuravlev *et al.*, 1975), but the proportion of fully stripped RNA has not been determined. In view of the abnormal state of the cells—and particularly the nature of the suspension medium—the relevance of studies in protoplasts to the infection process in leaves following mechanical inoculation is open to question.

4. Co-translational disassembly

The initial experiments of Wilson and colleagues demonstrating co-translational disassembly *in vitro* suggested an attractive mechanism for a key step in the infection process. To investigate whether such a mechanism operates *in vivo*, Shaw *et al.* (1986)

extracted samples from epidermal cells of tobacco leaves inoculated with TMV and identified molecules indicative an 80S ribosome moving along the RNA from the 5'-end in the manner of co-translational disassembly suggested from *in vitro* studies (see above, Section II.B.2). Translation complexes with the expected properties of striposomes have been isolated from the epidermis of tobacco leaves shortly after inoculation (Plaskitt *et al.*, 1987). Subsequent experiments with protoplasts (Wu *et al.*, 1994; Wu and Shaw, 1996, 1997) have built up a more complete picture.

The first event in co-translational disassembly is that the structure of the virion has to relax so that the 5' terminus of the RNA is accessible to a ribosome. *In vitro* treatments, such as SDS or weak alkali, showed that the 68 nucleotide 5' leader sequence, which lacks G residues, interacts more weakly with coat protein subunits than do other regions of the genome (Mundry *et al.*, 1991). As discussed in Chapter 5 (Section III.B.5), TMV particles are stabilized by carboxylate interactions, there being two carboxyl–carboxylate bonds between adjacent subunits and one carboxylate–RNA interaction. At slightly alkaline pHs these carboxylates become protonated, leading to electrostatic repulsion. Mutagenic studies on these bonds have shown that the situation is probably more complex than initially thought but that these groups provide the key controlling mechanisms for virus disassembly (Culver *et al.*, 1995; Lu *et al.*, 1996, 1998a; Wang *et al.*, 1998a).

Having initiated translation, ribosomes proceed along TMV RNA translating the 5' ORF, the 126/183-kDa replicase protein, and displacing coat protein subunits. When the ribosomes reach the stop codon of the 126/183-kDa ORF they disengage. This raises the question of how the 3' quarter of the particle is disassembled. Wilson (1985) suggested that the replicase might perform this task in a $3' \rightarrow 5'$ direction in synthesizing the $(-)$-strand replication intermediate. This suggestion was supported by the experiments of Wu and Shaw (1997) who obtained evidence for *co-replicational disassembly* from the 3'-end.

They showed that particles containing mutations in the 126- or 183-kDa ORFs were unable to undergo $3' \rightarrow 5'$ disassembly in electroporated protoplasts, but that this disassembly could be complemented *in trans* by wild-type TMV.

Thus, TMV is uncoated in a bi-directional manner, using the co-translational mechanism for the $5' \rightarrow 3'$ direction yielding the replicase which disassembles the rest of the particle in the $3' \rightarrow 5'$ direction, showing that disassembly and replication are coupled processes. The process happens rapidly with the whole capsid uncoated within 20 minutes (Wu *et al.*, 1994; Wu and Shaw, 1996) (Fig. 7.6).

C. Uncoating of bromoviruses

The isometric particles of bromoviruses swell at pHs above 7 (see Chapter 5, Section VI.B.4.b) and it has been suggested that, under these conditions, co-translational disassembly can take place (Brisco *et al.*, 1986a). In these *in vitro* experiments, swollen BMV and CCMV particles were added to a wheatgerm extract and translation products were obtained. However, it was not possible to perform translations on unswollen virus particles as the wheatgerm extract did not translate mRNAs at pHs below 7. By analogy with similar experiments on SBMV (Section II.D), in which the swelling could be controlled by both pH and Ca^{2+}, it was concluded that swelling of the bromovirus particle was required for uncoating. Analysis by sucrose and CsCl density gradients showed that the virus–ribosome complexes contain up to four ribosomes per virus particle (Roenhorst *et al.*, 1989).

Using mutants of CCMV that did not swell under alkaline conditions, Albert *et al.* (1997) found that swelling was not necessarily required for co-translational disassembly. They suggested that there is a pH-dependent structural transition in the virion, other than swelling, which enables the RNA to be accessible to the translation system. The proposed model, which is similar to ones from some vertebrate and insect viruses, postulates that the N termini of the five subunits in the pentameric capsomere undergo a major structural

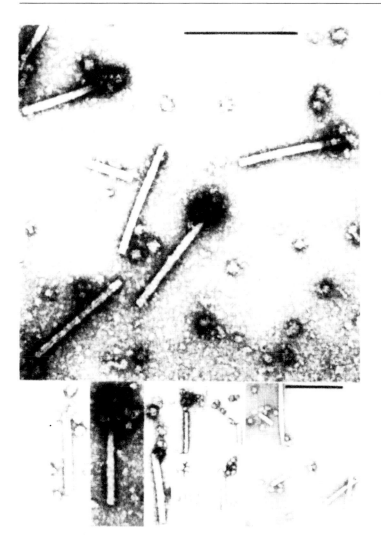

Fig. 7.5 'Striposome' complexes. Electron microscopic examination of the products of *in vitro* translation reactions programed with TMV particles. One end of some of the particles is associated with structures thought to be ribosomes. The complexes are thought to be intermediates in the co-translational disassembly process. From Wilson (1984a), with permission.

transition from the interior to the exterior of the virion. This provides a channel through which the RNA passes to be accessible for translation (Albert *et al.*, 1997). However, the 5'-end of the RNA must be released, which suggests that it is located in association with a pentameric capsomere.

D. Uncoating of SBMV

In similar experiments to those described above for bromoviruses, Brisco *et al.* (1986a) showed that co-translational disassembly takes place on swollen SBMV particles. As the stabilization of particles of SBMV is controlled by both pH-dependent and Ca^{2+}-mediated interactions (see Chapter 5, Section VI.B.4.e), they were able to

control the swelling at the alkaline pHs required for the translation system. When swollen SBMV particles were incubated with a wheatgerm extract containing [^{35}S]methionine, sucrose density gradient analysis showed that 80S ribosomes were associated with intact or almost intact virus particles (Shields *et al.*, 1989). The data suggested that translation of the viral RNA begins before it is fully released from the virus particle. It is not known whether the disassembly involved release of the RNA through 'holes' in the pentameric capsomeres as suggested above for CCMV.

E. Uncoating of TYMV

TYMV did not co-translationally disassemble in the *in vitro* translation system described

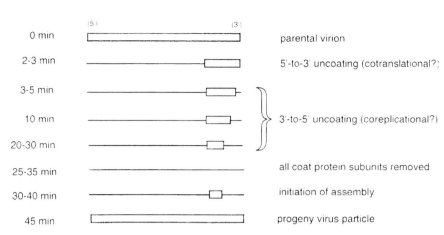

Fig. 7.6 Bidirectional disassembly of TMV particles *in vivo*. Coat protein subunits are removed in a 5′ → 3′ direction from *c.* 75% of the viral RNA in the first 2–3 minutes after inoculation of protoplasts. Uncoating the 3′-end of the RNA begins shortly thereafter and is completed by removal of subunits in the 3′ → 5′ direction. From Wu *et al.* (1994), with kind permission of the copyright holder, © The National Academy of Sciences, USA.

above for bromoviruses and SBMV (Brisco *et al.*, 1986a).

In vitro studies show that, under various non-physiological conditions, the RNA can escape from TYMV particles without disintegration of the protein shell. Thus, at pH 11.5 in 1 M KCl the RNA is rapidly released together with a cluster of 5–8 protein subunits from the shell (Keeling and Matthews, 1982). A hole corresponding to 5–7 subunits is left in the protein shell following release of RNA by freezing and thawing (Katouzian-Safadi and Berthet-Colominas, 1983). Treatment of TYMV with 3–7% butanol at pH 7.4 leads to the rapid release of RNA and five or six protein subunits in monomer form (Matthews, 1991).

In Chinese cabbage leaves, Kurtz-Fritsch and Hirth (1972) found that about 2% of the retained TYMV inoculum was uncoated after 20 minutes. They showed that empty shells and low-molecular-weight protein were formed following RNA release.

Matthews and Witz (1985) confirmed these findings and demonstrated that a significant proportion of the retained inoculum was uncoated within 45 seconds, and that the process was complete within 2 minutes. At least 80–90% of this uncoating takes place in the epidermis. Approximately 10^6 particles per epidermal cell can be uncoated (see also Chapter 12, Sections II.D and II.E). The process gives rise to empty shells that have lost about 5–6 protein subunits and to low-molecular-weight protein. At the high inoculum concentrations used, most

of the released RNA must be inactivated, presumably in the epidermal cells. Celite was used as an abrasive, so that the mechanism of entry proposed for TMV (Fig. 7.3) could account for the large numbers of particles entering each cell. The uncoating process just described was not confined to known hosts of TYMV.

F. Discussion

There is a dichotomy in the structural stabilization requirements of viruses in that the particles have to be stable enough to protect the viral genome when being transported outside the host yet they have to be able to present the genome to the cellular milieu for the first stages in replication. For at least some of the viruses with (+)-sense RNA genomes, the process of co-translational disassembly answers this problem. The coupled co-translational and co-replicational uncoating of TMV is an elegant process applicable to a rod-shaped virus. However, for other longer rod-shaped viruses, this may not be the process by which they are disassembled. The origin of assembly is at or near the 5′-end of the RNA (see Chapter 5, Section IV.B) and, as suggested for TMV, would be likely to present an obstacle to the translocation of the ribosomes.

It is likely that some form of co-translational disassembly takes place *in vivo* for the isometric viruses that swell or are permanently swollen (e.g. AMV). Whether the proposed mechanism for release of the RNA through a destabilized pentamer structure is applicable to more stable isometric viruses is still an open question as is the possible involvement of membranes in the uncoating process.

As for other viral genomes, the requirements in the first stages of infection are different to those of the (+)-sense ssRNA viruses. Viruses with dsRNA or (–)-sense ssRNA have to transcribe their genome to give mRNA. These viruses carry the viral RNA-dependent RNA polymerase in the virus particle and, presumably, transcription is an early event. It is not known whether this occurs within the virus particle, possibly in a relaxed structure, or whether the viral genome is released into the cell. However, it is most likely that this process takes place in an environment protected from cellular nucleases and that it is coupled to translation of the mRNA.

The dsDNA genomes of members of the *Caulimoviridae* have to be transported to the nucleus where they are transcribed to mRNA by the host RNA-dependent RNA polymerase (see Section IV.C.1; Chapter 8, Section VII.B). The coat protein of CaMV has a nuclear localization signal (Leclerc *et al.*, 1999) that will presumably target the particle into the nucleus. Particles of some caulimoviruses and badnaviruses are particularly stable, being able to resist phenol (Hull and Covey, 1983a; Bao and Hull, 1994) and nothing is known about how they disassemble.

The ssDNA genomes of members of the *Geminiviridae* also have to be transported to the nucleus so that they can be replicated before being transcribed to give mRNAs. As described in Chapter 9 (Section II.C), nuclear localization signals have been recognized in some geminiviral proteins. However, nothing is known about how the particles uncoat.

III. VIRAL GENOME EXPRESSION

Genome strategy is a useful but rather vague term (see Wolstenholme and O'Connor, 1971; Matthews, 1991), which could be extended to include almost every aspect of virus structure, replication and ecology. The term has been taken to include: (1) the structure of the genome (DNA or RNA; ds or ss; if ss whether it is (+)-or (–)-sense); (2) the question as to whether the nucleic acid alone is infectious; (3) general aspects of the enzymology by which the genome is replicated (e.g. the presence of nucleic acid polymerases in the virus particle, and any other enzymes concerned in nucleic acid metabolism that are coded for by the virus); and (4) the overall pattern whereby the information in the genome is transcribed and translated into viral proteins (not the detailed molecular biology of this process). However, in this chapter I will use this term to describe the kinds of strategy that have evolved among the groups of plant viruses to translate the genomic information from the mRNA stage of the infection cycle. Several selection pressures have probably been involved in this evolution. After describing the methods for studying genome strategies, I will discuss ways in which mRNAs are synthesized, the selection pressures and then the ways that viruses use to overcome these limitations.

The actual sequence of events that has led to the understanding of viral genome strategies has varied widely for different viruses. This is because of the rather haphazard manner in which a particular branch of science tends to develop. To take two examples:

- All four of the definite TMV gene products, three of which are non-structural, were identified before the nucleotide sequence of the genome of that virus had been established.
- At the other extreme, the full nucleotide sequences of several viruses are known, while only one or two of seven potential gene products had been characterized, and these are usually proteins found in the virus.

I shall attempt to present a summary of the various methods involved in a more logical sequence than that in which they have actually been applied to many viruses. First, we must understand the structure of the genome, and in particular the number of genome pieces, the arrangement of the ORFs in the genome, the deduced amino sequences for those ORFs, and the positions of any likely regulatory and recognition nucleotide sequences. As a second stage, we must define the ORFs that are actually functional by both *in vitro* and *in vivo* studies. We need to recognize the gene functions of the virus, either by direct studies on any viral proteins that can be isolated from infected cells, or by classic genetic studies that may reveal various biological activities. Finally, we need to match these viral gene activities with the functional ORFs. It is here that the techniques of reverse genetics can play a very important role. In reverse genetics, an alteration (base change, insertion or deletion of bases) is made at any preselected position in the genome. The consequences of the change are then studied with respect to its effect on the gene product and the product's biological function, a function that may not have been previously recognized by traditional methods. However, it must be recognized that the changes could induce secondary effects on other gene products.

A. Structure of the genome

There are several steps in determining the structure of a viral genome. The starting material is almost always nucleic acid isolated from purified virus preparations.

1. Kind of nucleic acid

Whether the nucleic acid is ss or ds, DNA or RNA, or linear or circular can be established by the various chemical, physical and enzymatic procedures outlined in Chapter 4.

2. Number of genome pieces

When virus particles housing separate pieces of a multi-partite genome differ sufficiently in size or density, they may be fractionated on density gradients and the nucleic acids isolated from the fractions. Alternatively, nucleic acids of differing size may be separated on density gradients or by gel electrophoresis. When the two pieces of a bipartite genome are of very similar size, as with some geminiviruses, the existence of two distinct parts of the genome may be inferred from hybridization experiments estimating sequence complexity. However, formal proof that the genome is in two pieces of nucleic acid can best be obtained by cloning the full length of each piece separately and demonstrating that

both are required for infectivity (e.g. the geminiviruses: Hamilton *et al.*, 1983).

3. Terminal structures

Chemical and enzymatic procedures can be used to establish the nature of any structures at the 5' and 3' termini of a linear nucleic acid (see Chapter 4, Section III.A.3).

4. Nucleotide sequence

Knowledge of the full nucleotide sequence of a viral genome is essential for understanding genome structure and strategy. The methods used are detailed in many publications and laboratory manuals.

5. Open reading frames (ORFs)

With the help of an appropriate computer program, the nucleotide sequence is searched for ORFs in each of the three reading frames of both (+)- and (−)-sense strands. All ORFs are tabulated, as is illustrated in Fig. 7.7 for a tymovirus.

As shown in Fig. 7.7, a large number of ORFs may be revealed. Those ORFs that could code for polypeptides of MW less than 7–10 kDa, or that would give rise to proteins of highly improbable amino acid composition, are usually not given further consideration. However, sequence similarity between small ORFs in

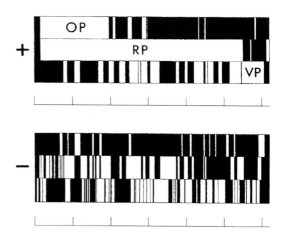

Fig. 7.7 Diagram of the three triplet codon phases of the plus and minus strand RNAs of OYMV genomic RNA. White boxes indicate all ORFs that begin with an AUG and terminate with UGA, UAG or UAA. There are three ORFs considered to be significant labeled OP (overlapping protein), RP (replicase protein) and VP (coat protein). From Ding *et al.* (1989), with permission.

several viruses may indicate that they are functional (e.g. the 7–9 kDa ORFs of potexviruses and carlaviruses).

ORFs of significant size representing possible proteins of 100 amino acids or more occur in the (−)-sense strands of several viruses that are normally regarded as being (+)-stranded (e.g. CPMV RNA2: Lomonossoff and Shanks, 1983; TMV: Goelet *et al.*, 1982; AMV RNA1 and RNA2: Cornelissen *et al.*, 1983a,b; PapMV: AbouHaidar, 1988). There is no evidence that any of these have functional significance. However, there is no reason, in principle, why functional ORFs should not occur in the (−)-sense strand. Such ORFs are found in the geminiviruses, tospoviruses and tenuiviruses (see Chapter 6, Sections V and VII).

ORFs do not necessarily start with the conventional AUG start codon. An AUU start codon has been recognized for ORF I of RTBV (Fütterer *et al.*, 1996) (see Section V.B.6) and a CUG start codon for the capsid protein of SBWMV (Shirako, 1998). This phenomenon raises the question of the definition of an ORF. Conventionally, it starts with an AUG codon and stops with one of the three stop codons. If non-AUG start codons are more widely used than at present thought, an ORF should be a largish in frame region without a stop codon.

6. Amino acid sequence

The amino acid sequence and MW of the potential polypeptide for each ORF of interest can be determined from the nucleotide sequence and the genetic code.

7. Regulatory signals

Various parts of the genome and particularly the 5′ and 3′ non-coding sequences are searched for relevant regulatory and recognition signals, as will be discussed in Section V.C. Regulatory sequences may also be found in coding regions.

8. mRNAs

The genomes of DNA viruses must be transcribed into one or more mRNAs. These must be identified in nucleic acids isolated from infected tissue and matched for sequence with the genomic DNA. Many plant viruses with ss

(+)-sense RNA genomes have some ORFs that are translated only from a subgenomic RNA (discussed in Section V.B.2). These too must be identified to establish the strategy of the genome.

In RNA preparations isolated from virus-infected tissue or from purified virus preparations, subgenomic RNAs may be present that can be translated *in vitro* to give polypeptides with a range of sizes that do not correspond to ORFs in the genomic RNA. For example, Higgins *et al.* (1978) detected RNAs of eight discrete lengths in nucleic acid isolated from preparations of TYMV. Mellema *et al.* (1979) were able to associate five of these with particular polypeptides synthesized in the reticulocyte system. The full-length translation products of these RNAs and the genomic RNA overlapped with one another and shared a common amino terminus. Mellema *et al.* concluded that these RNAs share a common translation initiation site near their 5′ termini.

In vitro translation of TMV RNAs isolated from TMV-infected tissue gave rise to a series of products with molecular weights of 45, 55, 80 and 95 kDa (Goelet and Karn, 1982). These formed a nested set of proteins sharing C-terminal sequences and having staggered N-terminal amino acids. Goelet and Karn suggested that viral RNA may be transcribed into a set of incomplete negative-sense strands that are in turn transcribed into a set of incomplete mRNAs. Since no function in viral replication has yet been ascribed to such N-terminal or C-terminal families of proteins, they will not be further discussed.

Thus, it may be a difficult task to establish whether a viral RNA of subgenomic size is a functional mRNA or merely a partly degraded or partly synthesized piece of genomic RNA. One criterion is to first isolate an active polyribosome fraction and then isolate the presumed mRNAs. The RNAs may then be fractionated by gel electrophoresis, and those with virus-specific sequences identified by the use of appropriate hybridization probes or by PCR.

Not infrequently, genuine viral subgenomic mRNAs are encapsidated along with the genomic RNAs. These can then be isolated from purified virus preparations and characterized. When the sequence of the genomic nucleic acid is known there are two techniques available to locate precisely the 5′ terminus of a presumed subgenomic RNA. In the S1 nuclease protection procedure, the mRNA is hybridized with a complementary DNA sequence that covers the 5′ region of the subgenomic RNA. The ss regions of the hybridized molecule are removed with S1 nuclease. The DNA that has been protected by the mRNA is then sequenced. In the second method, primer extension, a suitable ss primer molecule is annealed to the mRNA. Reverse transcriptase is then used to extend the primer as far as the 5′ terminus of the mRNA and the DNA produced sequenced. Carrington and Morris (1986) used both these procedures to locate the 5′ termini of the two subgenomic RNAs of CarMV. A sequence determination that reveals a single termination nucleotide rather than several is a good indication that the subgenomic RNA under study is a single distinct species and not a set of heterogeneous molecules (e.g. Sulzinski et al., 1985).

B. Defining functional ORFs

Some of the ORFs revealed by the nucleotide sequence will code for proteins *in vivo*, whereas others may not. The functional ORFs can be unequivocally identified only by *in vitro* translation studies using viral mRNAs and by finding the relevant protein in infected cells.

1. *In vitro* translation of mRNAs

The monocistronic RNA of STNV was translated with fidelity in the prokaryotic *in vitro* system derived from *Escherichia coli* (Lundquist et al., 1972). However, results with other plant viral RNAs were difficult to interpret. Three systems derived from eukaryotic sources have proven useful with plant viral RNAs. The general outline of the procedure for these three systems follows.

1. The RNA or RNAs of interest are purified to high degree, using density gradient centrifugation and/or polyacrylamide gel electrophoresis. For viruses whose particles become swollen under the conditions of the *in vitro* protein-synthesizing system, the RNA associated with the virus may act effectively as mRNA (e.g. Brisco *et al.*, 1986a). Alternatively, RNA may be transcribed from cloned viral cDNA or DNA.

2. The RNAs are then added to the protein-synthesizing system in the presence of amino acids, one or more of which is radioactively labelled.

3. After the reaction is terminated, the polypeptide products are fractionated by electrophoresis on SDS–PAGE, together with markers of known size.

4. The products are located on the gels by means of the incorporated radioactivity.

The three systems are:

- *The rabbit reticulocyte lysate system*. The cells from anaemic rabbit blood are lysed in water and centrifuged at 12 000g for 10 minutes. The supernatant fluid is then used. This is a useful system because of the virtual absence of RNase activity. Fig. 7.8 illustrates the use of this system.

- *The wheat embryo system*. In this system, the viral RNA is added in the presence of an appropriate label to a supernatant fraction from extracted wheat embryos from which the mitochondria have been removed. This system may contain plant factors not present in animal systems.

- *Toad oocytes*. These strictly do not constitute an *in vitro* system. Intact oocytes of *Xenopus* or *Bufo* are injected with the viral mRNA and incubated in a labelled medium.

Jagus (1987a,b) gives technical details of these methods and the first two systems are available in kit form from several companies. Not infrequently, when purified viral genomic RNAs are translated in cell-free systems, several viral-specific polypeptides may be produced in minor amounts in addition to those expected from the ORFs in the genomic RNA. It is unlikely that such polypeptides have any functional role *in vivo*. They are probably formed *in vitro* by one or more of the following mechanisms: (1) endonuclease cleavage of genomic RNA at specific sites; (2) proteolytic cleavage of

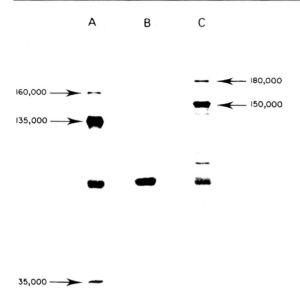

Fig. 7.8 Translation of plant viral RNAs in the rabbit reticulocyte system. The polypeptide products were fractionated by electrophoresis in a polyacrylamide gel and located by autoradiography of the gel. **(A)** Products using TMV RNA as message. **(B)** Control with no added RNA. **(C)** Products using TYMV RNA. The unmarked smaller polypeptides may be incomplete transcripts of the viral message or due to endogenous mRNA. Note that no protein the size of the viral coat protein (17.5 or 20 kDa) is produced by either viral RNA. From Briand (1978),with permission.

longer products during the incubation; (3) misreading of sense codons as termination signals; and (4) secondary structure of the genomic RNA, formed under the *in vitro* conditions, which prevents translocation of a proportion of ribosomes along the RNA. For example, multiple polypeptides of MW below 11 kDa are synthesized in the rabbit reticulocyte system with TMV RNA (Wilson and Glover, 1983). Similarly, when the I_2 subgenomic RNA of TMV is translated *in vitro*, a family of polypeptides besides the 30-kDa protein is produced, but only the 30-kDa protein could be detected *in vivo* (Ooshika *et al.*, 1984). Such polypeptides will not be discussed further in relation to virus replication.

What criteria can be used to 'optimize' conditions for *in vitro* translation? Measurement of total radioactivity incorporated is not particularly informative. Measurements of radioactivity in individual polypeptides separated on polyacrylamide gels are much more useful.

One might aim for conditions producing (1) the greatest number of polypeptides, (2) the fewest, (3) the longest, or (4) the most of a particular known gene product. It thus becomes apparent that to obtain definitive mapping of the genome from studies on the polypeptides produced *in vitro*, we must also know what polypeptides are actually synthesized *in vivo* by the virus.

2. Methods for identifying ORFs that are functional *in vivo*

Virus-coded proteins, other than those found in virus particles, may be difficult to detect *in vivo* especially if they occur in very low amounts and are only transiently during a particular phase of the virus replication cycle. However, a battery of methods is now available for detecting virus-coded proteins *in vivo* and matching these with the ORFs in a sequenced viral genome. In particular, the nucleotide sequence information gives a precise estimate of the size and amino acid composition of the expected protein. Knowledge of the expected amino acid sequence can be used to identify the *in vivo* product either from a partial amino acid sequence of that product or by reaction with antibodies raised against either a synthetic polypeptide that matches part of the expected amino acid sequence or against the ORF or part thereof expressed in, say, *E. coli*.

a. Proteins found in the virus

Coat proteins are readily allocated to a particular ORF based on several criteria: (1) amino acid composition compared with that calculated for the ORF; (2) amino acid sequence of part or all of the coat protein; (3) serological reaction of an *in vitro* translation product with an antiserum raised against the virus; and (4) for a few viruses such as TMV, assembly of an *in vitro* translation product into virus particles when mixed with authentic coat protein. Viruses such as rhabdoviruses and reoviruses may be exceptional in that several of the gene products, corresponding to various ORFs in the genome, are found in purified virus preparations (see Chapter 4, Section III.B.6.d, and Chapter 6, Section VII.A).

b. Direct isolation from infected tissue or protoplasts

Healthy and virus-infected leaves or proto-plasts are labelled with one or more radioactive amino acids. Cell extracts are fractionated by appropriate procedures and proteins separated by gel electrophoresis. Protein bands appearing in the samples from infected cells and not from healthy cells may be identified with the expected product of a particular ORF by comparing its mobility and its pattern of tryptic peptides with that of an *in vitro* translation product of the ORF (e.g. Bujarski *et al.*, 1982). With appropriate *in vivo* labelling, partial amino acid sequencing of the isolated protein may allow the precise location of its coding sequence in the genome to be established (e.g. Wellink *et al.*, 1986). In infections with some viruses such as potyviruses, large amounts of several virus-coded non-structural proteins accumulate in infected cells, facilitating the allocation of each protein to its position in the genome.

c. Serological reactions

Antisera provide a powerful set of methods for recognizing viral-coded proteins produced *in vivo* and identifying them with the appropriate ORF in the genome.

i. Antisera against synthetic peptides

A synthetic peptide can be prepared corresponding to part of the amino acid sequence predicted from an ORF. An antiserum is raised against the synthetic peptide and used to search for the expected protein in extracts of healthy or infected tissue or protoplasts. For example, Kibertis *et al.* (1983) synthesized a peptide corresponding to the C-terminal 11 amino acids of a 30-kDa ORF in the TMV genome. They were able to show that a polypeptide corresponding to this ORF was synthesized in infected protoplasts.

ii. Antisera against in vitro translation products

If an mRNA is available that is translated *in vitro* to give a polypeptide product clearly identified with a particular ORF or genome segment, antisera raised against the *in vitro* product can be used to search for the same pro-tein in extracts of infected cells or tissue. For example, such antisera have been used to identify non-structural proteins coded for by AMV RNAs in tobacco leaves. The antisera were used in conjunction with a very sensitive immunoblotting procedure.

iii. Antisera against recombinant proteins

Recombinant proteins can be derived either *in vitro* from translation of RNA transcripts of a cloned gene or by expression in various *E. coli* systems. In the latter it is convenient to attach a 'tag' to the protein to enable it to be purified. Antibodies are then raised against the recombinant protein and used to search for the corresponding protein in extracts of infected plant. An example of the *in vitro* transcript procedure was used to establish the position of the nucleocapsid protein gene in the rhabdovirus SYNV (Zuidema *et al.*, 1987). The *E. coli* procedure is exemplified by the antisera raised against the polymerase and protease gene regions in the polyprotein of RTSV (Thole and Hull, 1998). The RTSV cDNAs were placed in an *E. coli* vector that expressed them as fusion proteins with glutathione S-transferase that enabled them to be purified by absorption on to glutathione-agarose beads.

iv. Immunogold labeling

Antibodies produced against a synthetic peptide corresponding to part of a particular ORF in the genomic nucleic acid and labeled with gold can be used to probe infected cells for the presence of the putative gene product. This was done for the 30-kDa gene product of TMV (Tomenius *et al.*, 1987) and examples include antibodies produced against CaMV ORF I product expressed in *E. coli* combined with immunogold labeling demonstrated that this protein is expressed in infected leaves (Linstead *et al.*, 1988).

d. Comparison with genes known to be functional in other viruses

Size, location in the genome and nucleotide sequence similarities with known functional genes may give a strong indication that a particular ORF codes for a functional protein *in vivo*. These are frequently identified by searches of computer databases. For example, ORFs coding potential polymerases (RNA-dependent RNA polymerases and reverse transcriptases) are often recognized by the presence of characteristic motifs (see Chapter 8, Sections IV.B.1 and VII.A).

e. Presence of a well-characterized subgenomic RNA

Occasionally, a viral subgenomic RNA (sgRNA) has been well characterized but no *in vivo* protein product has been detected. Thus, the I_1 sgRNA of TMV was recognized as a functional mRNA because (1) it is located in the polyribosome fraction from infected cells, and (2) it has a precisely defined 5′ terminus (Sulzinski *et al.*, 1985). Thus, it is reasonable to suppose that the 5′ ORF of this subgenomic RNA is functional *in vivo*.

f. Presence of appropriate regulatory signals in the RNA

AUG triplets that are used to initiate protein synthesis may have a characteristic sequence of nucleotides nearby (Section V.A). Upstream of the AUG triplet there may be identifiable ribosome recognition signals. Presence of these sequences would indicate that the ORF is functional.

g. Codon usage

Frequency of codon usage has sometimes been used to indicate whether an ORF revealed in a genomic nucleotide sequence is likely to produce a functional protein (e.g. Morch *et al.*, 1988). However, an analysis of the codon usage by RTBV showed that it used many rare codons (R. Hull, unpublished), and thus this character should not be used as a firm criterion.

h. Reoviruses

The reoviruses are a special case with respect to establishing functional ORFs. Each ds genome segment is transcribed *in vitro* to give an mRNA that gives a single protein product (Nuss and Peterson, 1981). On this basis, it was considered reasonable to assume that each genome piece has a single functional ORF (but see Chapter 6, Section VI.A).

C. Recognizing activities of viral genes

Before information on the sequence of nucleotides in viral genomes became available and before the advent of *in vitro* translation systems, there were two ways of recognizing the activities of viral genes—identification of proteins in the virus particle and classic virus genetics. These approaches are still relevant.

1. Gene products in the virus

Fraenkel-Conrat and Singer (1957) reconstituted the RNA of one strain of TMV in the protein of another strain that was recognizably different. The progeny virus produced *in vivo* by this *in vitro* 'hybrid' had the coat protein corresponding to the strain that provided the RNA. Since this classic experiment, it has been universally assumed that coat proteins are encoded by viral genomes. Likewise, it has usually been assumed that other proteins found as part of the virus particle are also virus coded; for example, those found in reoviruses and rhabdoviruses.

2. Classic viral genetics

Two kinds of classic genetic study have identified many biological activities of viral genomes, and both of these procedures are still useful in appropriate circumstances.

a. Allocation of functions in multi-particle viruses

The discovery of viruses with the genome divided between two or three particles opened up the possibility of locating specific functions on particular RNA species. The requirements for, and stages in, this kind of analysis are as follows.

1. Purification of the virus.
2. Fractionation of the genome components, either as nucleoprotein particles (on density gradients of sucrose or cesium salts) or as isolated biologically active RNA species (usually by electrophoresis on polyacrylamide gels).
3. Definition of the set of RNA molecules that constitute the minimum viral genome.
4. Identification and isolation of natural strains

or artificial mutants differing in some defined biological or physical properties, which will provide suitable experimental markers. For example, Dawson (1978a) isolated a set of *ts* mutants of CMV. One group of mutations mapped on RNA3 and the rest on RNA1.

5. *In vitro* substitution of components from different strains or mutants in various combinations. These are inoculated to appropriate host plant species. The relevant biological or physical properties of the various combinations are determined. A particular property may then be allocated to a particular genome segment or segments.

6. Back-mixing experiments. In such experiments, the parental genome pieces are isolated from the artificial hybrids, mixed in the original combinations, and tested for appropriate physical or biological properties. Such tests are necessary to show that the RNAs of the hybrids retain their identity during replication.

7. Supplementation tests. These provide an alternative procedure to *in vitro* reassortment. Individual wild-type genome segments are added to a defective (mutant) inoculum. Restoration of the wild-type character in a particular mixture will indicate which segment controls the character (e.g. Dawson, 1978a). Transgenic plants expressing a single genome segment can also be used in supplementation tests.

8. Mixing of mutants. Unfractionated preparations of two different mutants may be mixed and tested. If the wild-type property is restored it can be assumed that the two mutations are on different pieces of RNA.

Supplementation tests and mixing of mutants do not require purification and fractionation of the mutant viruses. They can provide independent confirmation of results obtained by *in vitro* substitution experiments (de Jager, 1976).

Various factors may complicate the analysis of reassortment experiments:

1. A particular property may be determined by more than one gene, located on the same or on separate pieces of RNA.

2. Some genes are pleiotropic (i.e. have more than one effect). An example is the coat protein of AMV, which is involved in encapsidation of the virus, its aphid transmission, RNA replication and the spread of the virus through the plant (Tenllado and Bol, 2000).

3. If certain parts of the RNA are used to produce two proteins with different functions (e.g. by the read-through mechanism), then a single base change might induce changes in the two different functions.

4. Some amino acid replacements might be 'silent' with respect to one property of the protein but not another.

However, by using these procedures, many activities of viral genes have been attributed to one or more of the genome segments in a multipartite virus. Local or systemic symptoms in particular hosts, host range, and proteins found in the virus particle are activities that have commonly been studied. It must be recognized that these reassortment experiments have two limitations:

1. Where more than one gene is present on the RNA or DNA segment an activity cannot be allocated to a particular gene.

2. Except for structural proteins, they do not prove that the gene product is responsible for the activity. In principle, the activity could be due to some direct effect of the nucleic acid itself.

b. Natural or artificially induced virus mutants

The study of naturally occurring or artificially induced mutants of a virus has allowed various virus activities to be identified. Again, many of the activities involve biological properties of the virus.

Mutants that grow at a normal (permissive) temperature but that replicate abnormally or not at all at the non-permissive (usually higher) temperature are particularly useful. Such temperature-sensitive (*ts*) mutants are easy to score and manipulate, and most genes seem to be potentially susceptible to such mutations. They arise when a base change (or changes) in the viral nucleic acid gives rise to an amino acid substitution (or substitutions) in a protein, which results in defective function at the non-permissive temperature. Alternatively, the base change might affect the function of a non-translated part of the genome—a control element, for example. The experimental objective is to collect and study a series of *ts* mutants of a particular virus. To be useful for studies on replication, *ts* mutants must possess certain characteristics: (1) they must not be significantly 'leaky' at the non-permissive temperature; and (2) the rate of reversion to wild type must be low enough to allow extended culture of the mutant at both the permissive and non-permissive temperatures.

If the mutation studied occurred in the gene for the coat protein and the amino acid sequence of the coat protein is known, it is possible to locate the mutation within that protein. The location of mutations in other viral genes had to await information of nucleotide sequence and genome structure. The *ts* strain of TMV known as Ls1 is a good illustration. This strain replicates normally at 22°C, but at the non-permissive temperature (32°C) there is very little replication compared with the parent virus (TMV-L). Nishiguchi *et al.* (1980) studied the replication and movement of the two strains at the two temperatures, using fluorescent antibody staining to identify cells where virus had replicated. The results, illustrated in Fig. 7.9, showed clearly that Ls1 was defective in a cell-to-cell movement function. However, these results did not locate the cell-to-cell function involved. This had to await a knowledge of the genome structure of TMV and the site of the mutation within that structure (see Chapter 9, Section II.D.2.a). Viral mutants and other variants are discussed further in Chapter 17.

D. Matching gene activities with functional ORFs

A variety of methods is now available for attempting to match the *in vivo* function of particular viral gene products with a particular ORF. A few of these give unequivocal proof of function, whereas others are more or less strongly indicative of a particular function.

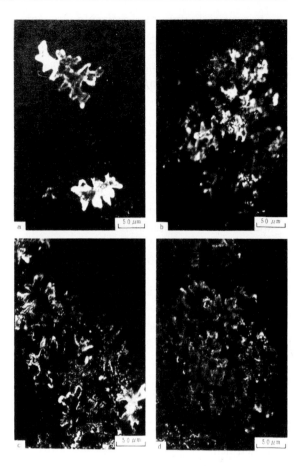

Fig. 7.9 Example of a *ts* mutant of TMV defective in cell-to-cell transport function. Fluorescent antibody staining of epidermal cells indicates the distribution of coat protein antigen 24 hours after inoculation. Tomato leaflets were infected at the permissive and non-permissive temperature with a *ts* strain (Ls1) or a wild-type strain (L) of TMV. Inoculated leaflets were cultured for 24 hours **(a)** Ls1 at 32°C; **(b)** Ls1 at 22°C; **(c)** L at 32°C; **(d)** L at 22°C. From Nishiguchi *et al.* (1980), with permission.

There are two kinds of method. In the first, which may not be generally applicable, the natural gene product produced in infected tissue is isolated, and its activity is established by direct methods. The second group of methods involves, directly or indirectly, the use of recombinant DNA technology.

1. Direct testing of protein function

Some virus-coded proteins besides coat proteins have functions that can be identified in *in vitro* tests.

a. In vitro *translation products*

Carrington and Dougherty (1987a) prepared an *in vitro* plasmid expression vector that allowed cell-free synthesis of particular segments of the TEV genome. The RNAs obtained were translated in rabbit reticulocyte lysates to give polypeptides that could be assessed for protease activity and the ability to act as protease substrates. In this way, they showed that the 40-kDa protein was a viral protease. Using an *in vitro* transcription and translation system on mutagenized cDNAs, Thole and Hull (1998) showed that a 35-kDa protein from the RTSV polyprotein has proteolytic activity, and they identified the potential cleavage sites in the polyprotein.

b. *A protein isolated from infected cells*

Thornbury *et al.* (1985) purified the protein helper component for aphid transmission from leaves infected with PVY, showing that it has a molecular weight of 58 kDa. They isolated the corresponding protein from another potyvirus and produced antisera against the two proteins. Using an *in vitro* aphid-feeding test, they showed that the antisera specifically inhibited aphid transmission of the virus that induced formation of the corresponding protein in infected plants. These tests demonstrated that the polypeptides were essential for helper component activity. Active CaMV aphid transmission factor was recovered after expression of a recombinant of the ORF II product and a baculovirus in insect cells (Blanc *et al.*, 1993b); however, this protein was not active when expressed in *E. coli*.

2. Approaches depending on recombinant DNA technology

a. *Location of spontaneous point mutations*

Knowledge of nucleotide sequences in natural virus variants allows a point mutation to be located in a particular gene, even if the protein product has not been isolated. In this way, the changed or defective function can be allocated to a particular gene. The *ts* mutant of TMV known as Ls1 can serve again as an example. At the non-permissive temperature, it replicates and forms virus particles normally in protoplasts and infected leaf cells, but is unable to move from cell to cell in leaves. A nucleotide comparison of the Ls1 mutant and the parent virus showed that the Ls1 mutant had a single base change in the 30-kDa protein gene that substituted a serine for a proline (Ohno *et al.*, 1983). This was a good indication, but not definitive proof, that the 30-kDa protein is involved in cell-to-cell movement.

b. *Introduction of point mutations, deletions or insertions*

The genomes of many DNA viruses and cDNAs to many RNA viruses have been cloned and the DNA or transcripts thereof shown to be infectious. There are numerous examples of experiments in which point mutations, deletions or insertions have been used to elucidate the function(s) of the gene produced by the modified ORF. The introduction of defined changes in particular RNA viral genes to study their biological effects, and thus define gene functions, is commonly known as reverse genetics. This approach has been of major importance in gaining understanding of the gene functions described in this book.

c. *Recombinant viruses*

Recombinant DNA technology can be used to construct viable viruses from segments of related virus strains that have differing properties, and thus to associate that property with a particular viral gene. For example, Saito *et al.* (1987a) constructed various viable recombinants containing parts of the genome of two strains of TMV, only one of which caused necrotic local lesions on plants such as *Nicotiana sylvestris*, which contain the *N′* gene. Their results indicated that the viral factor responsible for the necrotic response in *N′* plants is coded for in the coat protein gene. This response is discussed further in Chapter 10 (Section III.E.1).

Another example of the use of recombinants constructed *in vitro* is given by the work of Woolston *et al.* (1983) with CaMV. They infected plants with hybrids constructed from the genomes of an aphid-transmissible and an aphid non-transmissible strain of the virus. The results showed that aphid transmission and the synthesis of an 18-kDa protein were located in either ORF I or ORF II. Tests with a deletion indicated that ORF II was the gene involved.

It is also possible to construct viable recombinant hybrids between different viruses. Sacher *et al.* (1988) used biologically active cDNA clones to replace the natural coat protein gene of BMV RNA3 with the coat protein gene of SHMV. In SHMV the origin of assembly lies within the coat protein gene. In barley protoplasts co-inoculated with BMV RNAs 1 and 2, the hybrid RNA3 was replicated by *trans*-acting BMV factors, but was coated in TMV coat protein to give rod-shaped particles instead of the normal BMV icosahedra. However, since functions are highly integrated in a viral genome it is sure to be able to produce viable recombinants between distantly-related viruses.

One further application of recombinants is the tagging of gene products with fluorescent or other probes that report where in the plant or protoplast that gene product is being expressed or accumulates. This is described more fully in Chapter 9 (Section II.B).

d. Expression of the gene in a transgenic plant

As with mutagenesis of infectious cloned genomes of viruses, the technique of transforming plants with viral (and other) sequences has had a major impact on understanding viral genes and control functions and there are numerous examples of their expression in transgenic plants. The technique is described in detail in numerous texts, including Old and Primrose (1989) and Draper and Scott (1991). The basic features of the technique are that a construct comprising the gene of interest, a promoter, often the 35S promoter of CaMV (see Section IV.C.1) and a transcriptional terminator sequence, are introduced into suitable plant material. The plant material was originally protoplasts but now

usually embryonic cell suspensions or similar meristematic tissue from which plants can be regenerated.

There are two commonly used ways of introducing the construct into the plant material. In the biolistic approach, the construct to be introduced is coated on to small microparticles which are propelled into the plant tissues by an explosive or blast of high pressure. The other approach is to use the integrating properties of the Ti plasmid of *Agrobacterium tumefaciens*. The construct of interest is placed in a T-DNA plasmid that retains the integrating properties (see Old and Primrose, 1989) but has tumor-inducing genes deleted. This is then co-cultivated with the plant tissue, allowing the integration of the construct. In both approaches, a selection marker, usually an antibiotic resistance gene or a herbicide tolerance gene, is included so that successful transformation events can be identified and isolated.

e. Bacterial, yeast and insect cell systems

There are numerous bacterial systems that are used for the expression of proteins and many commercial kits available. Basically, the gene for the protein of interest is cloned into the appropriate site in a vector (an expression vector), which is then transformed into a bacterium, usually *E. coli*. By cloning in frame with a known sequence at the N or C terminus of the protein of interest, that protein can be 'tagged' — which facilitates its purification. The main problems with bacterial expression of plant viral proteins are:

1. They are expressed in a prokaryotic system and will not be modified (say phosphorylated or glycosylated) in the manner that they would be in a eukaryotic system.
2. They may be processed by prokaryotic enzymes (e.g. proteases) in a manner not found in eukaryotic systems.
3. They may prove toxic to the bacterium.

There are two eukaryotic systems commonly used for the expression of plant viral gene products. A frequently used vector is derived from the *Autographica californica* nuclear polyhedrosis virus (AcMNPV), which is a member of the *Baculoviridae*, a large family of occluded viruses pathogenic to arthropods. Baculoviruses occlude their virions in large protein crystals, the matrix of which is composed primarily of polyhedrin, a protein of about 29 kDa. In the baculovirus expression system, the polyhedrin gene, which is not required for viral replication, is replaced by the gene of interest (Smith *et al.*, 1983; Lucknow and Summers, 1988). The foreign gene is expressed from the polyhedrin promoter on infection of an insect cell line, such as Sf21 derived from the moth *Spodoptera frugiperda*. There are many variants on these baculovirus-based expression systems; for more details see King and Possee (1992).

The other eukaryotic system involves yeast. There is a large pool of information available on the classical and molecular genetics of *Saccharomyces cerevisiae* (see Botstein and Fink, 1988; Ausubel *et al.*, 1998) and an increasing amount on *Schizosaccharonyces pombe*. Vector systems are available for the expression of foreign genes in yeasts. As described in Chapter 8 (Section III.A.6), yeast systems are used for the analysis of interactions between proteins and also for unraveling details of viral replication (see Chapter 8, Section III.A.6).

f. Hybrid arrest and hybrid selection procedures

Hybrid arrest and hybrid selection procedures can be used to demonstrate that a particular cDNA clone contains the gene for a particular protein. In hybrid arrest, the cloned cDNA is hybridized to mRNAs, and the mRNAs are translated in an *in vitro* system. The hybrid will not be translated. Identification of the missing polypeptide defines the gene on the cDNA.

In the hybrid selection procedure, the cDNA–mRNA hybrid is isolated and dissociated. The mRNA is translated *in vitro* to define the encoded protein. In appropriate circumstances, these procedures can be used to identify gene function. For example, Hellman *et al.*

(1988) used a modified hybrid arrest procedure to obtain evidence identifying the protease gene in TVMV.

g. Sequence comparison with genes of known function

As noted in Section III.B.2.d, sequence comparisons can be used to obtain evidence that a particular ORF may be functional. The same information may also give strong indications as to actual function. For example, the gene for an RNA-dependent RNA polymerase (RdRp) was identified in poliovirus. The study by Kamer and Argos (1984) revealed amino acid sequence similarities between this poliovirus protein and proteins coded for by several plant viruses. This similarity implied quite strongly that these plant viral-coded proteins also have a polymerase function. The conserved amino acid sequences (motifs) of RdRps and many other viral gene products are described at the appropriate places in this book (e.g. for RdRps, see Chapter 8, Section IV.B.1).

h. Functional regions within a gene

Spontaneous mutations and deletions can be used to identify important functional regions within a gene. However, mutants obtained by site-directed mutagenesis, and deletions constructed *in vitro* can give similar information in a more systematic and controlled manner. For example, the construction and transcription of cDNA representing various portions of the TEV genome, and translation *in vitro* and testing of the polypeptide products, showed that the proteolytic activity of the 49-kDa viral proteinase lies in the 3' terminal region. The amino acid sequence in this region suggested that it is a thiol protease related in mechanism to papain (Carrington and Dougherty, 1987a). Proteinases are further described in Section V.B.1.a.

However, care must be taken with this approach. Many functions depend upon the three-dimensional structure of the protein, and mutations not at the active site may have a secondary effect on the protein structure.

IV. SYNTHESIS OF mRNAs

As noted earlier (Fig. 7.1), Baltimore (1971) pointed out that the expression of all viruses has to pass through an mRNA stage. The (+)-sense ssRNAs of many genera of plant viruses can act as mRNAs directly on entry into the host cell. For viruses with other types of genome, mRNAs have to be synthesized at some stage of the infection cycle.

A. Negative-sense single-stranded RNA viruses

All viruses with a (−)-sense ssRNA genome carry the viral RdRp in their virus particles. Thus, one of the early events on entry into a host cell is the transcription of the viral genome to (+)-sense RNA required for both translation of the viral genetic information and as an intermediate for replication. Replication of such viruses is described in Chapter 8 (Section V). Here I will discuss how the mRNAs are formed.

1. Plant *Rhabdoviridae*

Plant rhabdoviruses, like those infecting vertebrates, possess a genome consisting of a single piece of (–)-sense ssRNA, with a length in the range 11–13 kb and encoding six proteins, one more than animal rhabdoviruses (see Chapter 6, Section VII.A).

From patterns of hybridization with cDNA clones, Heaton *et al.* (1989a) showed that the SYNV genome is transcribed into a short 3'-terminal 'leader' RNA and six mRNAs. Thus, the plant rhabdoviruses appear to be expressed in a manner similar to animal rhabdoviruses such as vesicular stomatitis virus (VSV), which has been studied much more extensively (reviewed by Rodriguez and Nichol, 1999). For vesicular stomatitis virus (VSV), the active transcribing complex consists of the RNA genome tightly associated with the N protein, and the polymerase made up of the phosphoprotein (P) and the large (L) protein. This complex starts transcribing (+)-sense RNA at a single entry site at the 3'-end of the genome and transcribes the leader RNA that is transported to the nucleus

where it inhibits host cell transcription. The complex then transcribes the mRNA for the N protein, which is capped during synthesis by the polymerase. At the end of the N gene, and of all genes, is the sequence 5'-AGUUUUUU-3' (element I) which signals termination and polyadenylation of the mRNA. This intergenic sequence also comprises a short untranscribed sequence (element II) and the start site for transcription of the next mRNA (element III). Similar sequences are found in plant rhabdoviruses (Fig. 7.10) (see Jackson *et al.*, 1999).

Thus, the viral genes are transcribed separately from the 3'-end and they are transcribed in decreasing amounts (N>P>sc4>M>G>L) (Wagner and Jackson, 1997). This is an efficient way of regulating gene expression, as the genes that are located at the 3'-end are those that are required in greatest amounts.

	I	II	III
Consensus	AP$_y$UP$_y$UUUU	G N C	P$_y$UNNN
SYNV	AUUCUUUUU	GG	AA UUG UC
LNYV	AUUCUUUU	G(N)$_x$	AAG CU UUU
VSV	ACUUUUUUU	G U C	UUGUC
RV	ACUUUUUUU	C(N)$_x$	A UUGU G

Fig. 7.10 Alignment of the intergenic regions of selected plant and animal rhabdoviruses. The rhabdovirus consensus sequence is shown at the top followed by the sequences of SYVV, LNTV, vesicular stomatitis virus (VSV) and rabies virus (RV). The intergenic sequences ('gene–junction' sequences) are separated into three elements: element I constitutes the poly(U) tract at the 3'-end of each gene on the genomic RNA; element II is a short sequence that is not transcribed during mRNA synthesis; element III constitutes the start site for transcription of each mRNA. The bold type in the viral sequences indicates the consensus nucleotides. P$_y$ indicates pyrimidine, (N)$_x$ corresponds to a variable number of nucleotides. From Jackson *et al.* (1999), with permission.

2. Tospoviruses

As described in Chapter 6 (Section VII.B.1), the genomes of tospoviruses comprise three ssRNA segments. L RNA is (–)-sense and monocistronic encoding the viral RdRp. The mRNA is transcribed from the virion RNA by the virion-associated polymerase.

The other two RNAs have an ambisense gene arrangement (see Fig. 6.11) with one ORF in the viral strand and one in the complementary strand. The two ORFs are separated by an AU-rich intergenic region of variable length. For both RNAs the virion-sense ORF is expressed from an sgRNA transcribed from the complementary RNA and the complementary-sense ORF from an sgRNA transcribed from the virion RNA (see Fig. 5.42) (de Haan et al., 1990; Kormelink et al., 1992b, 1994).

The intergenic region between the ambisense ORFs is predicted to form stable hairpin structures that are suggested to control the termination of transcription of sgRNAs. However, as noted in the next section this should be considered with circumspection.

As described in Section V.C.4, formation of tospovirus mRNAs involves cap-snatching.

3. Tenuiviruses (reviewed by Falk and Tsai, 1998)

The genome organization of tenuiviruses is described in Chapter 6 (Section VII.B.2). Members of this genus have genomes divided between four or more ssRNA. As with the tospoviruses, the largest RNA is (–)-sense and monocistronic and is considered to be expressed from transcripts made using the virion-associated, virus-encoded RdRp.

Most of the other species in this genus have three other RNAs, each containing two ORFs in an ambisense arrangement (see previous section, and Fig. 6.12). When the MSpV and RHBV virion RNAs are translated in vitro, only a few proteins, including the NCP and N proteins, are detectable (Falk et al., 1987; Ramirez et al., 1992). RNAs corresponding to, but shorter than, RNAs 2, 3 and 4 are found in infected plants and insects. Northern hybridization analysis using strand-specific probes show that these RNA correspond in size and polarity to the ORFs on the ambisense RNAs (Falk et al., 1989; Huiet et al., 1991; Huiet et al., 1992; Estabrook et al., 1996). As at least some of these RNAs are found associated with polyribosomes (Estabrook et al., 1996), they are interpreted as being sgRNAs that arise from transcription in a manner similar to that described above for tospoviruses (Fig. 5.42).

The ambisense ORFs are separated by an AU-rich intergenic region of varying length (see de Miranda et al., 1994, 1995a). It has been suggested that the intergenic regions of RNAs with ambisense ORF arrangement fold into hairpin structures that function in transcriptional termination (Emery and Bishop, 1987; Kakutani et al., 1991); but de Miranda et al. (1994) found that the predicted folding for the RHBV RNA3 intergenic region differed according to what was being analyzed. Stable hairpin structures could be predicted if the computer-assisted folding was performed on the intergenic region alone but not if it was on the whole RNA.

As described in Section V.C.4, the formation of MSpV sgRNA involves cap-snatching. Cap-snatching has been demonstrated for the tospovirus TSWV and tenuivirus MSpV (Estabrook et al., 1998). It is likely that other tenuiviruses also cap-snatch.

B. Double-stranded RNA viruses

1. Plant Reoviridae

Plant members of the Reoviridae family are placed in three genera: Phytoreovirus with 12 dsRNA genome segments, and Fijivirus and Oryzavirus each with 10 dsRNA genome segments. The genome organizations of these genera are shown in Chapter 6 (Section VI). Most of the dsRNA segments are monocistronic but the Fijiviruses RBSDV segments 7 and 9, MRDV segments 6 and 8, OSDV segments 7 and 10, the Phytoreovirus RDV segment 11 and the Oryzavirus segment 4 are bicistronic; RDV segment 12 possibly has three ORFs. However, there is no evidence of these downstream ORFs being expressed.

The plant reoviruses, like their counterparts infecting vertebrates and insects, contain a transcriptase that uses the RNA in the particle as template to produce ssRNA copies. In animal reoviruses, this occurs in subviral particles comprising part of the capsid, the polymerase and the dsRNAs (reviewed by Joklik, 1999; Lawton et al., 2000). Early in infection only (+)-sense ssRNAs are synthesized which act as mRNAs. Later, (–)-sense strands are synthesized leading to viral replication (see Chapter 8, Section VI.A). It is likely that a similar series of events occurs in the plant reoviruses, especially when they multiply in their insect vectors.

C. DNA viruses

The synthesis of mRNAs from either the dsDNA members of the *Caulimoviridae* or the ssDNA members of the *Geminiviridae* and the nanoviruses does not involve a virus-coded enzyme but is performed by the host DNA-dependent RNA polymerase II located in the nucleus. This synthesis is initiated by viral promoter sequences, and so in this section I will consider these sequences in the plant DNA viruses. Plant viral DNA promoter sequences have been used widely in gene vectors in plants (see Chapter 16, Section IX.B.1).

1. Caulimoviridae

The genome organizations of the *Caulimoviridae* genera are described in Chapter 6 (Section IV). Most of the detailed studies have been performed on CaMV and these observations most likely pertain to all members of this family.

As described in Chapter 8 (Section VI.B), there are two phases in the nucleic acid replication cycle of CaMV, the nuclear phase of transcription and the cytoplasmic phase of gene expression and reverse transcription. In the first, the dsDNA of the infecting particle moves to the cell nucleus, where the overlapping nucleotides at the gaps are removed, and the gaps are covalently closed to form a fully dsDNA. These mini-chromosomes form the template used by the host DNA-dependent RNA polymerase to transcribe two RNAs of 19S and 35S, as indicated in Fig. 6.1. As well as promoters for these two mRNAs, the viral

DNA also has signals for the polyadenylated termination of transcription.

a. The 35S promoter (reviewed by Hull et al., 2000a)

The identification of a promoter involves mapping the 5′-end of the transcript on to the viral genome. This has been performed for the 35S RNAs of CaMV (Odell et al., 1985), CsMV (Verdaguer et al., 1996), FMV (termed the 34S promoter) (Sanger et al., 1990; Maiti et al., 1997), MMV (Day and Maiti, 1999), PClSV (Maiti and Shepherd, 1998), SoyCMV (Hasegawa et al., 1989), SVBV (Wang et al., 2000), CoYMV (Medberry et al., 1992), RTBV (Bhattacharyya-Pakrasi et al., 1993; Bao and Hull, 1993a; Chen et al., 1994),and SCBV (Tzafrir et al., 1998; Schenk et al., 1999). The approach to studying these promoters involves transgenic or transient expression of constructs comprising the promoter region coupled to a reporter gene, usually the *uidA* gene expressing β-glucuronidase (GUS). The promoter region usually consisted of several hundred nucleotides upstream of, and up to one hundred nucleotides downstream of, the transcription start site. Mutagenesis and deletion analysis was then used to dissect the regions responsible for the strength and tissue specificity of the promoter.

These studies show that the promoter sequences comprise the core promoter upstream of the transcription start site and various control elements both upstream and downstream of the start site. The core promoter is characterized by what is termed a 'TATA box' about 25 nucleotides upstream of the start site (Table 7.1).

A detailed analysis of the CaMV promoter revealed that it had a modular nature (reviewed by Benfey and Chua, 1990) with subdomains conferring patterns of tissue-specific expression. Two major domains were identified, domain A (–90 to +8) (numbering relative to transcription start site at +1) which is important for root-specific expression, and domain B (–343 to –90) mainly involved in expression in the aerial parts of the plant. The region of the A domain between –83 and –63 contains an as-1 (activation sequence-1)-like element that is

TABLE 7.1 Defined and putative promoter sequences of some caulimoviruses and badnaviruses

		as-1 sequence		TATA sequences		Transcription start		Poly(A) signal	
CaMV	7850	cacTGACGtaagggaTGACGcac	34	ctcTATATAAgca	21	ACACGCG	154	atcAATAAAttt	
CVMV	7380	tgaAGACGtaagcacTGACGaca	34	tccTATATAAgga	24	AAGAAAA			
FMV	6857	gtaTGACGaacgcacTGACGacc	13	ctcTATATAAgaa					
MMV		aaaTGACGtaagccaTGACGtct	21	tccTATATAAgga	15	GAAGAGA	186	atcAATAAAata	
BSV	7083	tagTCACGcacga—TGACCttt	181	ctcTATATAAgga	20	ACACGCA			
RTBV	7370			cagTATATAAgga	21	TCATCGA	184	atcAATAAAgct	

In the transcription start sequences, the nucleotide indicated in 'outline' is +1 of the transcript. Gaps are where reliable information is not available.
Data from: Sanger *et al.* (1990); Bao *et al.* (1993a); Verdaguer *et al.* (1996); Harper and Hull (1998); Dey and Maiti (1999); all with permission.

important for the root-specific expression. The as-1 element is present in several non-viral promoters and can be recognized in many of the caulimovirus promoters where it is important. The B domain was dissected further into five subdomains, B1 to B5, each conferring specific expression patterns in developing and mature leaves. Thus, in the full promoter these domains and subdomains act co-ordinately and synergistically to give the constitutive expression of the CaMV 35S promoter. Plant nuclear factors have been identified that bind to various regulatory regions in this promoter (Benfey and Chua, 1990; Hohn and Fütterer, 1992; Sanfaçon, 1992) and also to the RTBV promoter (Yin and Beachy, 1995). Other caulimovirus promoters also show similar modular structures. For instance, three overlapping regions have been identified in the CsVMV promoter (Verdaguer *et al.*, 1998), –222 to –173 controlling expression in green tissues and root tips, –178 to –63 giving vascular-specific expression, and –149 to –63 controlling expression in mesophyll tissues. The promoters of some viruses require sequences downstream of the transcription start site for maximum expression. An example of this is MMV promoter for which the region of +33 to +63 is essential for maximum expression (Day and Maiti, 1999). Similarly, efficient transcription from the RTBV promoter requires an enhancer located in the first 90 nucleotides of the transcript (Chen *et al.*, 1996). Two subelements were identified in this enhancer region, one being independent of position and orientation and the other being position-dependent.

The RTBV promoter has been analyzed in detail (Yin and Beachy, 1995; Yin *et al.*, 1997a,b; Klöti *et al.*, 1999) and shown to comprise several elements (Fig. 7.11). Some of these elements are upstream of the transcription initiation site but others are downstream (Chen *et al.*, 1996).

Some promoters are specific to the vascular tissue. In that of RTBV the region between –164 and –100 is essential for vascular tissue expression (Klöti *et al.*, 1999) and deletion leads to specificity in the epidermis. This tissue specificity is not surprising for RTBV as the virus itself is phloem-limited. However, the CoYMV promoter gives expression primarily in the phloem (Medberry *et al.*, 1992) but the parent virus infects most tissues.

Most of the caulimovirus promoters act in both monocot and dicot plant species even though the parent virus is restricted in host range. The CaMV promoter has been used for the expression of transgenes in many dicot and monocot plant species and is considered a good strong constitutive promoter. It has also been shown to be active in bacteria (Assad and Signer, 1990), in yeast (Pobjecky *et al.*, 1990), in animal HeLa cells (Zahm *et al.*, 1989) and in *Xenopus* oocytes (Ballas *et al.*, 1989).

b. The 19S promoter

The caulimovirus 19S promoter (reviewed by Rothnie *et al.*, 1994) has been much less studied. Only that of CaMV has been analyzed and shown to be weaker than the 35S promoter when tested in transgenic constructs. For example, expression of the α-subunit of β-conglycinin in petunia plants under control

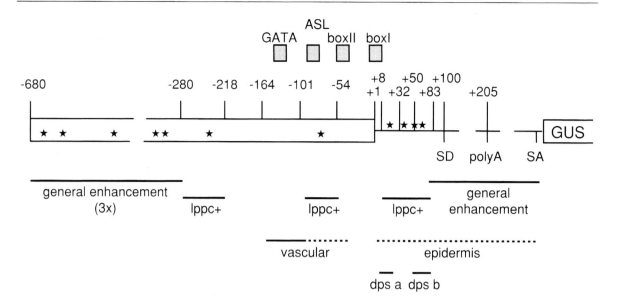

Fig. 7.11 Organization of elements influencing expression from the RTBV promoter. The location of regions containing different elements with apparent activity in transcription control that can be deduced from Klöti *et al.* (1999) are indicated below a schematic presentation of the transcription unit; stippled lines indicate supposed activity. Sequence elements defined by Yin and Beachy (1995) and Yin *et al.* (1997a,b) are indicated above. Positions are given relative to the transcription start site. GAGA-like elements are indicated by *; SD, splice donor; SA splice acceptor; lppc+, regions stimulating assembly of low-processivity polymerase complexes. From Klöti *et al.* (1999) with kind permission of the copyright holder, © Kluwer Academic Publishers.

of the 35S promoter was 10–50 times greater than from the 19S promoter (Lawton *et al.*, 1987). The 35S promoter was also found to be 10–30 times more effective than the nopaline synthase promoter from *Agrobacterium tumefaciens* (García *et al.*, 1987). This is in contrast to virus infections leading to comparable levels of the 35S and 19S RNAs and the product of the gene encoded by the 19S RNA, the ORF VI product, being the most abundant viral protein. The core 19S promoter can be strongly activated by the 35S promoter enhancer elements but no enhancer elements have been detected for the promoter itself.

c. *The polyadenylation signal* (reviewed by Rothnie *et al.*, 1994)

The caulimovirus 35S and 19S RNAs are 3' co-terminal and share a polyadenylation signal. The signal motif, AAUAAA, is found upstream of the transcription termination or cleavage site but downstream of the transcription initiation site (Table 7.1). It

can be seen from Table 7.1 and Fig. 6.1 that, to generate the more-than-genome length 35S transcript, the transcription termination and polyadenylation signal has not to be effective the first time that it is passed. Sanfaçon and Hohn (1990) suggested that it is the proximity to the transcription start signal that occludes the polyadenylation signal on the first passage. However, sequences upstream of the transcription start site are not required for the initiation of polyadenylation and all the information for efficient polyadenylation are in the repeated region. The bypass is not 100% efficient and short-stop transcripts can be detected arising from first-pass processing. Notable amounts of short-stop transcripts are found in RTBV infections (Klöti *et al.*, 1999). The RTBV polyadenylation signals are very similar to those of CaMV (Rothnie *et al.*, 2001).

2. Geminiviridae

The circular ssDNA genomes of members of the *Geminiviridae* have ORFs both in the virion-sense and complementary-sense orientations (see Figs. 6.5–6.8). All geminiviruses employ the same basic strategy to transcribe their genomes in that it is bi-directional from at or near the common region (see Chapter 6, Section V.A, for description of the common region), terminating diametrically opposite. However, there are differences between the genera in the details of transcription that are reviewed by depth by Hanley-Bowdoin *et al.* (1999). Here I will summarize the transcription, and for original references the reader is referred to the review.

a. Begomoviruses

The most detailed studies on begomoviruses have been on TGMV, the DNA A of which is transcribed into six polyadenylated RNAs and DNA B into four polyadenylated RNAs (Fig. 7.12).

The polyadenylation sites for the virion- and complementary-sense RNAs overlap so they share a few 3′ nucleotides. Each genome component gives rise to a single virion-sense RNA that is translated into either the coat protein from DNA A or BV1 from DNA B. Complementary-sense transcription is much more complex and gives rise to multiple overlapping RNAs with common 3′-ends but differing in their 5′-ends. The three complementary-sense RNAs from TGMV B DNA all translate to give protein BC1, whereas those from DNA A have different coding capacities. The largest transcript, AC61 (transcripts are designated according to their 5′-ends so AC61 is from the complementary side of DNA A and starts at nucleotide 61), encodes the entire left side of DNA A and is the only RNA giving full-length Rep protein (see Chapter 8, Section VIII.D.4, for Rep proteins). AC2540 and AC2515 may express ORF AC4, and the smallest RNAs (AC1938 and AC1629) specify AC2 from their first ORF and AC3 from their second ORF.

Upstream of the TGMV mRNAs are characteristic eukaryotic RNA polymerase II promoter sequences (reviewed by Hanley-Bowdoin *et al.*, 1999). Transcription of each of four RNAs initiates 20–30 bp downstream of the TATA box motifs, whereas the other RNAs have sequences resembling initiator elements overlapping their 5′-ends. The promoters for the complementary-sense AC61 and AC1629 mRNAs and for the virion-sense AV1 and BV1 RNAs have been studied in some detail. The AC 61 promoter maps to the TGMV A common region and supports high levels of transcription. Deletion mutagenesis showed that most of its activity resided in the 60 bp immediately preceding the AC61 transcription start site, a region that overlaps the origin of (+)-strand DNA synthesis (see Chapter 8, Section VIII.D.3, and Fig. 8.27). These mutations showed the close interactions between transcription and replication. Mutations in both the host factor binding sequences and in the G-box reduced promoter function. The AC61 promoter is autoregulated through the Rep binding site, the repression being specific for the homologous Rep protein and is thought to involve active interference with the transcription apparatus and not just steric hindrance. The AC4 protein also negatively regulates this promoter, the *cis* element involved being upstream of the G-box and being distinct from the Rep binding site. The analogous promoters of most other begomoviruses are probably regulated by similar mechanisms, but those of some (e.g. ACMV) differ in detail (reviewed by Hanley-Bowdoin *et al.*, 1999). The BC1 promoter sequences are similar to those of the AC61 promoter but differ in the transcription start site and in that the Rep protein does not regulate it.

The virion-sense promoters of TGMV are not as well characterized as those described above for the complementary-sense RNAs. The promoter regions include the common regions and the downstream sequences containing the TATA boxes. The virion-sense promoters require the AC2 protein in *trans*, the activation being independent of replication. Study of the TGMV AV1 promoter in transgenic plants indicated that its regulation is complex and is controlled differently in different tissues (Sunter and Bisaro, 1997). The promoter for the AC2 gene, which lies upstream of the AC1629 transcription start site, is not responsive to Rep, AC2 or AC3 (reviewed by Hanley-Bowdoin *et al.*, 1999). The regulation of this promoter is unclear and possibilities are discussed by Hanley-Bowdoin *et al.* (1999).

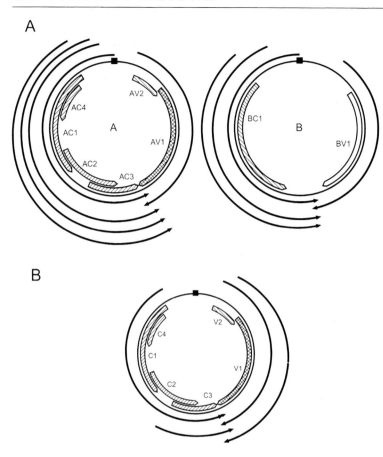

Fig. 7.12 Diagram of transcription of begomoviruses. **(a)** TGMV; **(b)** TLCV. Genome maps are described in Fig. 6.7. The outer arcs represent the transcripts with the arrowheads indicating the 3′-ends. Data from Sunter and Bisaro (1989) and Mullineaux *et al.* (1993), with permission.

Some of the begomoviruses have monopartite genomes the transcription of which has been much less studied than that of the bipartite members. Four major transcripts were identified and mapped to the TLCV genome (Mullineaux *et al.*, 1993) (Fig. 7.12). One transcript spanned the C1, C2 and C3 ORFs and a second covered C2 and C3; the C4 ORF, located within C1, was not recognized at that time. On the virion-sense side, there are two 3′ co-terminal transcripts, the 5′-ends of which map either side of the first in-frame AUG codon of the V1 ORF.

To analyze the regulation of expression of TLCV, Dry *et al.* (2000) studied the expression of fusions of the ORFs with the GUS reporter gene in both stably and transiently transformed *Nicotiana tabacum* tissues. They showed that the C2 ORF curtailed the expression of the V2 ORF, indicating that the C2 protein is involved in transactivation of virion-sense gene expression. The TLCV ORF–GUS constructs had distinctive tissue expression patterns in transgenic tobacco plants: C1, C4 and V2ΔC (deletion at C terminus of V2 ORF) were constitutive; C2 and C3 were predominantly vascular; V1ΔC reduced expression in cells associated with vascular bundles.

b. Curtoviruses

The single-component curtoviruses produce six or seven proteins in infected plants (Fig. 7.13). Frischmuth *et al.* (1993) identified an abundant virion-sense population of poly-adenylated RNA and four complementary-sense polyadenylated RNAs in BCTV-infected plants. The population of virion-sense RNAs comprises 3′ co-terminal overlapping transcripts, the 5′ termini of which are positioned so that they express ORFs V1, V2 and V3. There are two consensus TATA boxes at the appropriate position upstream of the larger two virion-sense RNAs, but there is no detailed information on these or any other promoters in curtoviruses.

c. Mastreviruses

The mastreviruses have a single component of ss DNA that codes for four proteins (Fig. 7.14), V1 and V2 being expressed from transcripts in the virion sense, while C1 and C2 are expressed from transcripts in the complementary sense.

The bi-directional transcripts have multiple initiation sites and terminate at overlapping polyadenylation signals. However, unlike the other two geminivirus genera, the production of mRNAs from both the virion and complementary senses involves splicing. The C transcripts are of low abundance and a splicing event fusing C1 to C2 has been found for MSV, DSV and TYDV (Mullineaux *et al.*, 1990; Dekker *et al.*, 1991; Morris *et al.*, 1992). In MSV about 20% of the C transcript is spliced (Wright *et al.*, 1997). For MSV there are two V-sense transcripts, the most abundant starting one nucleotide upstream of the V1 ORF, and the longer, least abundant one initiating 141 nucleotides upstream of V1 (see Wright *et al.*, 1997). About 50% of the major transcript is spliced opening up the V2 ORF and about 10% of the longer transcript is spliced to give an mRNA for V1 (Wright *et al.*, 1997). The products of C1 and C2 are early functions involved with viral replication and the C1:C2 fusion product is essential. The V1 product (movement protein) and V2 product (coat protein) are late functions, the coat protein being required in much greater amounts than the products from the other ORFs. Thus, the splicing is important in both the temporal and quantitative expression of mastreviruses.

Three TATA promoter consensus sequences may be involved in complementary-sense transcription of the MSV genome (Boulton *et al.*, 1991b). As described in Chapter 10 (Section III.O.1.d), mutation of one of these promoters affected both symptoms and host range possibly by indirectly influencing viral replication by influencing the synthesis of the Rep protein. Using a maize protoplast tran-

sient expression system, Fenoll *et al.* (1988) defined further the structure of the MSV virion-sense promoter that drives transcription of the RNA for coat protein. They identified a 122-bp sequence upstream from the start site for transcription that enhances promoter activity. The 122-bp sequence activates the MSV core promoter in a position-dependent, but orientation-independent fashion. The activating sequence specifically binds proteins in maize nuclear extracts. This 'upstream activating sequence' lies in the large intergenic region and includes the common region and was mapped to two GC-rich boxes on the distal side of the origin hairpin motif relative to the transcription start site (Fenoll *et al.*, 1990). These GC-rich boxes and the TGMV G-box element are positioned similarly with respect to the hairpins of the respective viruses, implying similar functions (Arguello-Astorga *et al.*, 1994). Thus, mastreviruses may encode a transactivator of virion-sense transcription analogous to the AL2/C2 proteins of begomoviruses.

All the MSV promoters show cell-cycle specificity (Nicovics *et al.*, 2001). The coat protein promoter had highest activity in early G2, whereas the C-sense promoter sequences produced two peaks of activity, in the S and G2 cell-cycle phases.

d. Polyadenylation

As can be seen from Figs. 7.12–7.14, the transcripts of geminiviruses terminate on the opposite side of the genome to the promoters. This is an AT-rich region that contains putative signals for polyadenylation. However, no detailed analysis has yet been made of poly(A) signals in these viruses.

3. Nanoviruses

As described in Chapter 6 (Section V.B), the genomes of nanoviruses are divided between six or more circular ssDNA molecules of about 1 kb. Each genome segment encodes at least one protein. The promoters of these genome segments have been studied in most detail in BBTV. Beetham *et al.* (1997) showed that two mRNAs are transcribed from BBTV DNA-1,

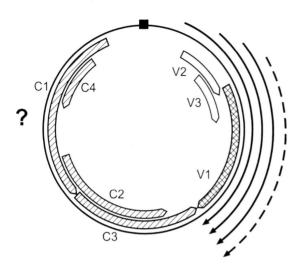

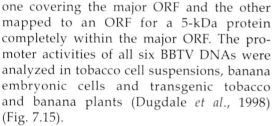

Fig. 7.13 Diagram of transcription map of BCTV. Genome map is described in Fig. 6.6. The outer arcs represent the transcripts with the arrowheads indicating the 3'-ends. The dotted arc is a minor transcript but fits with being that for V1, the coat protein. It possibly was a minor transcript at the sampling time but a major one at another time. The complementary sense transcript(s) were not mapped because deletions complicated their analysis. Data from Frischmuth *et al.* (1993), with permission.

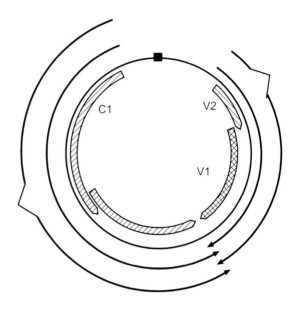

Fig. 7.14 Diagram of transcription map of MSV. Genome map is described in Fig. 6.5. The outer arcs represent the transcripts with the arrowhead indicating the 3'-end. The breaks in the outer arcs linked by angled lines show the site of introns spliced out during processing of transcripts. Data from Wright *et al.* (1997), with kind permission of the copyright holder, © Blackwell Science Ltd.

one covering the major ORF and the other mapped to an ORF for a 5-kDa protein completely within the major ORF. The promoter activities of all six BBTV DNAs were analyzed in tobacco cell suspensions, banana embryonic cells and transgenic tobacco and banana plants (Dugdale *et al.*, 1998) (Fig. 7.15).

In these experiments the intergenic region of each genome segment that contains three regions of homology were fused to the *uid*A (GUS) and the green fluorescent protein (GFP) reporter genes. Two of the three homologous regions are associated with replication (see Chapter 8, Section VIII.E) and one contains the potential TATA box of the promoter. Dugdale *et al.* (1998) showed that (1) the intergenic regions of all six BBTV DNAs have promoter activity, (2) the activities of the different intergenic regions vary considerably, and (3) the relative activities of the different intergenic

regions vary between tobacco and banana. In tobacco cell suspensions, transient expression from promoters of DNAs 2 and 6 was greater than that from the other promoters. In transgenic tobacco, the weak expression from each promoter was phloem-associated, that of the promoter of DNA 6 being the strongest. However, expression from the DNA 6 promoter became high in callus derived from transgenic tobacco. The sensitive GFP reporter could detect activity from these promoters only in banana embryonic cells, where the promoters of DNAs 4 and 5 gave the highest levels of transient activity. In transgenic banana, activity from the DNA 6 promoter was restricted to the phloem of leaves and roots, stomata and root meristems. Deletion analysis of the DNA 6 promoter suggested that the elements required for strong expression were within 239 nucleotides upstream of the translational start site.

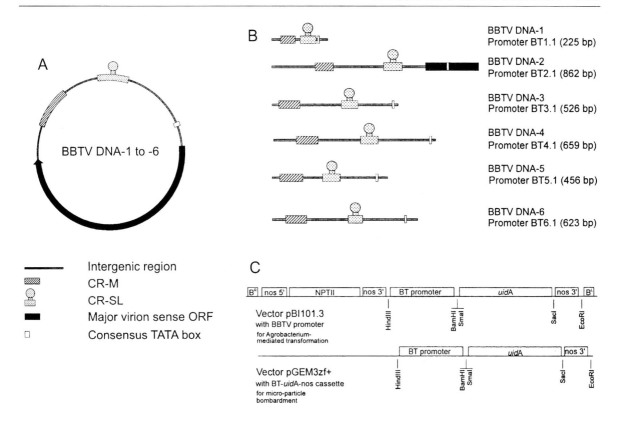

Fig. 7.15 Schematic presentation of **(A)** general BBTV circular ssDNA genome organization, **(B)** BBTV DNA 1–6 promoter fragments, and **(C)** cloning strategy. Promoter fragments incorporating the intergenic regions of BBTV DNA 1–6 were isolated by PCR or restriction digestion from cloned components and inserted upstream of the *uidA* reporter gene in pBI101.3 for *Agrobacterium*-mediated transformation of tobacco. The BT-*uidA*-nos cassettes from each construction were subsequently cloned into pGEM-zf⁺ for microparticle bombardment transient assays. From Dugdale *et al.* (1998), with permission.

V. PLANT VIRAL GENOME STRATEGIES

A. The eukaryotic protein-synthesizing system

It is generally accepted that the eukaryotic protein-synthesizing system translates the information from viral RNAs. This translation system has various features and controls (reviewed by Gallie, 1996):

1. Plant cellular mRNAs have a cap (reviewed by Furuichi and Slatkin, 2000), an inverted and methylated GTP at the 5′ terminus [m⁷G(5′)ppp(5′)N] and a poly(A) tail at the 3′ terminus.
2. In most circumstances, mRNAs contain a single open reading frame (ORF).
3. Translation is initiated at an AUG start codon, the context of which controls the efficiency of initiation (Kozak, 1989, 1992).
4. The cap, 5′ untranslated region, the coding region, the 3′ untranslated region and the poly(A) tail all have potential to influence translational efficiency and mRNA stability.

In the scanning model for translation (Kozak, 1989, 1992) the 40S ribosomal subunit binds to the 5′ cap (see Fig. 7.23), translocates to first AUG in a suitable context where it forms the 80S ribosome which translates only that ORF immediately downstream from the 5′ region of an mRNA; at the stop codon of this ORF the ribosomes dissociate. Thus, ORFs beyond this point normally remain untranslated. Viral genomes, except those of satellite viruses, encode two or more proteins, and therefore are presented with a problem of how to express their downstream proteins in the eukaryotic system. Much of the variation in the way gene products are translated from viral RNA genomes appears to have evolved to meet this constraint and at the same time to provide differential control of expression of the various ORFs.

As noted above, the nucleotide sequence surrounding the AUG start codon is important in the efficiency of initiation of translation. For plant systems, the favourable context is A<u>A</u>CAA<u>UG</u>G (Lehto *et al.*, 1990a) with a purine at the −3 position and/or a guanine at the +4 positions (position numbering is relative to the A of the AUG codon which is designated +1) playing essential roles.

B. Virus strategies to overcome eukaryotic translation constraints

On current knowledge, there are at least 12 strategies by which RNA viral genomes and transcripts from DNA viruses ensure that all their genes are accessible to the eukaryotic protein-synthesizing system; see Fütterer and Hohn (1996), Gallie (1996) and Drugeon *et al.* (1999) for reviews. The strategies fall into three groups:

1. Making the viral genomic RNA or segment thereof effectively monocistronic by bringing any downstream AUG to the 5'-end. This is done by either having a single ORF expressing a polyprotein that is subsequently cleaved to give the functional proteins (strategy 1), or by dividing up the viral genome to give monocistronic RNAs either during expression (strategies 2, 11 and 12) or permanently (strategy 3).
2. Avoiding the constraints of the 5' AUG. There are various strategies to do this (strategies 3 to 8).
3. Maximizing the information expressed from a viral RNA by bringing together two adjacent ORFs to give two proteins, one from the 5' ORF and the other from both ORFs. Thus the second protein comprises the upstream protein in its N-terminal region, the C-terminal region being from the downstream ORF (strategies 9 and 10).

Strategies 8, 9 and 10 have been termed 'recoding' (reviewed by Gesteland and Atkins, 1996).

1. Strategy 1: Polyproteins

Here the coding capacity of the RNA for more than one protein, and sometimes for the whole genome, is translated from a single ORF. The polyprotein is then cleaved at specific sites by a virus-coded proteinase, or proteinases, to give the final gene products. The use of this strategy is exemplified by the potyviruses.

a. Virus-coded proteinases (reviewed by Dougherty and Semler, 1993; Ryan and Flint, 1997; Spall *et al.*, 1997)

Four classes of virus-coded proteinases are currently recognized: serine, cysteine, aspartic and metallo-proteinases named usually after their catalytic site. Three of these four types of proteinase are found among plant viruses (Table 7.2).

Serine proteinases are also termed 3C proteases (from their expression in picornaviruses) or chymotrypsin-like proteases. Most have a catalytic triad of amino acids His, Asp, Ser, but in some the Ser is replaced by Cys; as the latter have the same overall structure as serine proteinases they are termed 'serine-like' proteinases. The serine residue is unusually active and acts as a nucleophile during catalysis by donating an electron to the carbonyl carbon of the peptide bond to be hydrolyzed. An acyl serine is formed and a proton donated to the departing amyl group by the active-site histi-

TABLE 7.2 Proteinases encoded by plant viruses

Virus group	Viral proteinase
Caulimoviridae	Aspartate
Potyviridae	
Potyvirus	Serine, cysteine, serine-like[a]
Bymovirus	Cysteine, serine-like
Comoviridae	Serine-like
Sequiviridae	Serine-like
Benyvirus	Cysteine?[b]
Marafivirus	Cysteine
Tymovirus	Cysteine
Closterovirus	Cysteine, aspartate
Polerovirus	Serine

[a] Serine-like proteinase with cysteine at its active site but with a structure similar to a serine proteinase.
[b] ? indicates doubt as to the type of proteinase.

dine residue. The acyl enzyme is then hydrolyzed, the carboxylic acid product is released, and the active site is regenerated. Serine proteases cleave primarily at Gln–Gly, Gln–Ser, Gln–Ala and Gln–Asn (reviewed by Palmenberg, 1990).

Cysteine proteinases, also known as papain-like or thiol proteinases, have a catalytic dyad comprising Cys and His residues in close proximity that interact with each other. During proteolysis, the Cys sulfhydryl group acts as a nucleophile to initiate attack on the carbonyl carbon of the peptide bond to be hydrolyzed. An acyl enzyme is formed through the carbonyl carbon of the substrate and the sulfhydryl group of the active-site His. The carbonyl carbon is then hydrolyzed from the thiol group of the proteinase and the active-site residues are regenerated.

Aspartic or acid proteinases are composed of a catalytic dyad of two Asp residues. They most likely do not form covalent enzyme–substrate intermediates and are thought to operate by an acid–base catalysis.

Viral proteinases are highly specific for their cognate substrates, a specificity that depends on the three-dimensional structures of both the proteinase and substrate. For instance, CPMV proteinase does not recognize primary translation products of M RNAs from other comoviruses (Goldbach and Krijt, 1982). This substrate specificity is further exemplified by experiments on CPMV reported by Clark *et al.* (1999) (see Section V.E.8 for processing of CPMV polyprotein). In these experiments, the regions encoding the large (L) and small (S) coat proteins on CPMV M RNA were replaced by the equivalent region of BPMV. These recombinant molecules replicated in cowpea protoplasts in the presence of CPMV B RNA. The junction between the 58/48-kDa and the L coat protein was cleaved. However, there was no processing of that between the L and S coat proteins, and thus no virus particles were formed, even when the sequence at that junction was made CPMV-like. Clark *et al.* (1999) also translated transcripts from their constructs in the *in vitro* rabbit reticulocyte system and showed that there was processing *in trans*, albeit somewhat inefficiently, when the L–S junction was made CPMV-like. This greater specificity in cleavage of the L–S junction in *cis* than in *trans* has also been shown for ToRSV (Carrier *et al.*, 1999). It is suggested that the L–S cleavage site is defined by more than just a linear amino acid sequence and probably involves interactions between the L–S loop and the β-barrels of the viral coat proteins.

A fluorometric assay for TuMV proteinase was developed by Yoon *et al.* (2000). This showed that intramolecularly quenched fluorogenic substrates can be used for the continuous assay of TuMV NIa proteinase (and presumably other proteinases).

b. Potyvirus *genus* (reviewed by Reichmann *et al.*, 1992; Shukla *et al.*, 1994)

Potyviruses have genomes of approximately 10 kb that contain a single ORF for a polyprotein of about 3000 to 3300 amino acids. Shields and Wilson (1987) found no evidence for the presence of subgenomic RNAs in their preparations of TuMV. Likewise, Dougherty (1983) could find no evidence for authentic subgenomic RNAs in total RNA preparations made from TEV-infected leaves. The polyprotein is cleaved to give 10 proteins (Fig. 7.16) using three virus-coded proteases.

The 35-kDa P1 cleaves itself from the polyprotein at Phe–Ser, the 52-kDa HC-Pro cleaves at its C-terminal Gly–Gly, and the 27-kDa protease domain of the NIa region is a serine protease responsible at most, if not all, the other cleavages at Gln–(Ser/Gly). Although there is not full information on the sequential order of the processing events, there is some evidence for some events taking place before others. For instance, the cleavage at the C terminus of HC-Pro takes place, giving P1-HC-Pro, before P1 is cleaved off (Carrington *et al.*, 1989).

i. Potyviral proteases

P1 is a serine protease that cleaves P1 from the P1-HC-Pro product (reviewed by Reichmann *et al.*, 1992; Shukla *et al.*, 1994). It contains a serine protease active site sequence ($Hx_8Dx_{30\ or\ 31}G.x.S.G$) in its C-terminal region. This protease has proved to be difficult to analyze, as it does not appear to function in rabbit reticulocyte lysate *in vitro* translation systems. However, it is active in wheatgerm *in vitro* translation systems and the cleavage site

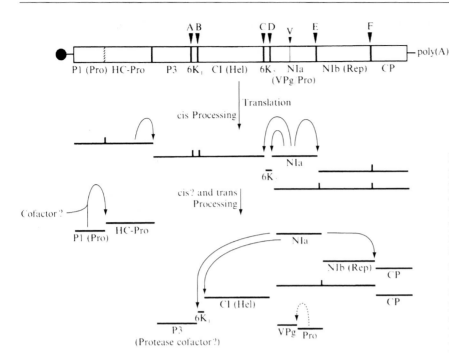

Fig. 7.16 Schematic of the processing of the potyvirus polyprotein. The primary events are probably co-translational and autocatalytic, yielding precursors and mature products. There is little information about the sequential order of these events, nor about the extent, if any, of further C-terminal processing of primary cleavage products. From Reichmann *et al.* (1992), with permission.

has been identified as between F and S in both TVMV and TEV. Comparison with homologous regions in other potyviral polyproteins indicates conservation of F or Y at the –1 position and S at the +1 position with preference for Q or H at –2 and M or I at –4. The P1 protein and the P1–HC junction appear not to be present in bymoviruses.

The enzymatic activity responsible for the HC-Pro–P3 cleavage is in a 20-kDa domain in the C-terminal half of the HC-Pro protein (Carrington *et al.*, 1989). This proteinase has cysteine and histidine residues at its active site (Oh and Carrington, 1989) and resembles papain-like cysteine proteinases. The cleavage site specificity has been studied by site-directed mutagenesis and *in vitro* expression. Positions –4, –2, –1 and +1 of the G:G cleavage site were critical while –5, –3 and +2 were not (Carrington and Herndon, 1992). These sites are conserved in the aphid-transmitted potyviruses but not in the fungus-transmitted bymovirus BaYMV which has G/Y substitution at –4 and S/G substitution at +1 (Shukla *et al.*, 1994).

The small nuclear inclusion protein, NIa, region contains the major proteinase of potyviruses. It has a two-domain structure, the

N-terminal genome-linked protein VPg (22 kDa) and the C-terminal protease (27 kDa) (Dougherty and Carrington, 1988). NIa is autocatalytically released from the polyprotein and then catalyses the cleavage of the various junctions to release P3, $6K_1$, CI $6K_2$, NIb and CP. The cleavage of the VPg domain from the NIa protein is much less efficient.

Data from experiments using site-directed mutagenesis suggest that the catalytic triad of amino acids in the 49-kDa proteinase of TEV is probably His[234], Asp[269] and Cys[339] (Dougherty *et al.*, 1989). This motif is found in all potyviruses.

The cleavage sites have been identified for several viruses (Table 7.3) and the primary sites are considered to be QS, QG and QA with the motif V-X-X-Q/(A, S, G or V) thought to be common to all potyviruses. There is some variation on this motif (Table 7.3) and the site releasing the VPg from the NIa protein obviously differs from most of the others.

A seven-residue motif, E-X-X-Y-X-Q/(S or G), has been identified for TEV (Carrington and Dougherty, 1988). The rate of processing is thought to be controlled by additional virus-specific motifs that include hydrophobic residues at –4, residues other than A, G or S at

TABLE 7.3 Amino acid sequences of the demonstrated and putative cleavage sites for the NIa and HC-Pro proteinases of four potyviruses and the NIa-like proteinase of BaYMV

HC-Pro[a]	PPV YLVG/GL	TEV YVVG/GM	TVMV YKVG/GL	PVY YRVG/GV	BaYMV
A	QVVVHQ/S	EDVLEQ/A	EIVEFQ/A	YDVRHQ/R	PKIVLQ/A
B	QAVQHQ/S	EIIYTQ/S	NNVRFQ/S	YEVRHQ/S	ASYGLQ/A
C	ECVHHQ/T	ETIYLQ/S	EAVRFQ/S	QFVHHQ/A	
D	EEVVHQ/G	EPVYFQ/G	EPVKFQ/G	ETVSHQ/G	
E	EFVYTQ/S	ELVYSQ/G	DLVRTQ/G	DVVVEQ/A	DIIHMQ/A
F	NVVVHQ/A	ENLYFQ/S	ETVRFQ/S	YEVNHQ/A	DEIWLQ/A
V	EEVDHE/S	EDLTFE/G	QEVAFE/S	QEVEHE/A	

[a] HC-Pro and letter (A–F and V) refer to sites shown in Fig. 7.16. Amino acids in bold indicate similarities between sites, and / indicates cleavage site.
Data from Reichmann *et al.* (1992), with permission.

+1, and residues further from the cleavage site. Controls on the cleavage site appear to cover at least the region of –7 to +2 (reviewed by Shukla *et al.*, 1994).

The properties of the 27-kDa protease have been studied in detail in the cloned and expressed TuMV and TEV proteins. The optimum catalytic activity of the TuMV protein is at about 15°C and pH 8.5 (Kim *et al.*, 1996). The proteins of both viruses are C-terminally processed to a 25-kDa product that has less activity than the 27-kDa protein (Kim *et al.*, 1995; Parks *et al.*, 1995); a second internal cleavage, leading to a 24-kDa product, has been identified for the TuMV protein (Kim *et al.*, 1996). The cleavage site leading to the 25-kDa protein is unusual, being between a Ser and Gly for TuMV and a Met and Ser for TEV; that leading to the 24-kDa TuMV protein is between Thr and Ser. The 25-kDa product has reduced proteolytic activity and the 24-kDa protein has no detectable activity.

ii. In vivo protein synthesis
Some of the proteins found *in vivo* have been discussed in the preceding section. Donofrio *et al.* (1986) isolated an RdRp activity from corn infected with MDMV. The activity was solubilized and attributed to a ~160-kDa protein. The subunit structure of the polymerase has not been characterized. Vance and Beachy (1984) found a genomic-length RNA associated specifically with active polyribosomes in extracts of soybean leaves infected with SMV. They concluded that this RNA is the only viral RNA translated *in vivo*. Full-length viral RNA,

the complementary strand, and a ds viral RNA have been found associated with the chloroplast fraction in tissue extracts (Gadh and Hari, 1986). However, there was no evidence that viral RNA synthesis took place in the chloroplasts.

c. Discussion

There are both advantages and disadvantages to the polyprotein processing strategy. Apart from overcoming the non-5′ start codon problem, there are advantages both in the fact that several functional proteins are produced from a minimum of genetic information and also in the potential for regulating the processing pathway. This is shown in the differences in the cleavage sites for the NIa protease and the influence that the surrounding residues can have on the rates of cleavage. Similarly, the requirement for the processing at the C terminus of the HC-Pro before P1 is cleaved from the product most probably represents a control mechanism that is not yet understood.

The major disadvantage is that it is difficult to visualize how the polyprotein strategy of the potyviruses can be efficient. The coat protein gene is at the 3′-end of the genome (see Fig. 7.16). Thus, for every molecule of the 20-kDa coat protein produced by TEV, a molecule of all the other gene products has to be made, totalling about 320 kDa. Since about 2000 molecules of coat protein are needed to encapsidate each virus particle but probably only one replicase molecule to produce it, this appears to be a very inefficient procedure. Indeed large quantities of several gene products, apparently in a non-functional state, accumulate in infected cells (see Chapter 3, Section IV.C). Nevertheless, the potyviruses are a very successful group. There are many member viruses, and they infect a wide range of host plants. Other viruses using polyprotein processing have additional devices that can avoid this problem. Comoviruses have their two coat proteins on a separate genome segment (see Section V.E.8). There does not appear to be any massive accumulation of non-coat gene products in cells infected with these viruses.

2. Strategy 2: Subgenomic RNAs

Subgenomic RNAs (sgRNA) are synthesized during viral replication from a genomic RNA that contains more than one ORF giving 5'-truncated, 3' co-terminal versions of the genome. This then places the ORFs that were originally downstream at the 5'-end of the mRNA. When several genes are present at the 3'-end of the genomic RNA, a family of 3' collinear sgRNAs may be produced; CTV has a nested set of at least nine sgRNAs (Karasev *et al.*, 1997). SgRNAs may be encapsidated (e.g. *Bromoviridae*, SHMV) and can cause uncertainty as to what comprises an infectious genome. The encapsidation of sgRNAs is dependent on the presence of the origin of assembly on that RNA, and this can differ between viruses within a genus or even between strains of a virus. For instance, the origin of assembly is present on the coat protein sgRNA of SHMV but not on the equivalent RNA of other tobamoviruses, such as TMV (see Chapter 5, Sections III.B and IV.A.2.a); the type strain of BaMV does not encapsidate any sgRNAs (Lin *et al.*, 1992) whereas the V strain encapsidates two sgRNAs (Lee *et al.*, 1998b).

a. Synthesis of subgenomic RNAs

At least four models have been proposed for the synthesis of sgRNA from the genomic RNA. These include:

1. *De novo* internal initiation on the full-length (−) strand of the genome during (+)-strand synthesis.
2. Initiation on the full-length (−) strand

primed by a short leader from the 5'-end of the genomic DNA during (+)-strand synthesis. This has been found for coronaviruses (Liao and Lai, 1994).
3. Intramolecular recombination during (−)-strand synthesis in which the replicase jumps from the subgenomic RNA start site on the full-length (+) strand and reinitiates near the 5'-end of the genome. This also has been found for coronaviruses (Sawiki and Sawiki, 1998).
4. Premature termination during (−)-strand synthesis of the genome followed by the use of the truncated nascent RNA as a template for sgRNA synthesis.

The first and fourth mechanisms have been proposed for plant viruses.

b. De novo *internal initiation*

The simplest model for *de novo* internal initiation of sgRNAs necessitates the replicase recognizing a sequence upstream of the sgRNA 5'-end. This is termed the subgenomic promoter. Most studies on subgenomic promoters have been made on BMV but subgenomic promoters have been mapped for several other viruses (e.g. CuNV: Johnston and Rochon, 1995; PVX: Kim and Hemenway, 1996).

BMV has a tripartite genome, RNAs 1 and 2 being monocistronic and RNA3 (2114 nucleotides) dicistronic; the downstream ORF on RNA3, that encoding the coat protein, is expressed from an sgRNA, RNA4 (876 nucleotides) (see Fig. 6.13 for genome organization). Since RNAs 3 and 4 are 3' co-terminal, the 5'-end of RNA4 maps to position 1238 on RNA3. Thus, a subgenomic promoter for RNA4 is likely to be upstream of position 1238, which is in an intergenic region.

The subgenomic promoter of BMV comprises a 'core' promoter, which is the smallest region capable of promoting sgRNA synthesis with low accuracy at a basal level, and 'enhancer' regions, which provide accuracy of replication initiation and control yields of sgRNAs. The fully functional subgenomic promoter encompasses about 150 nucleotides.

The BMV core promoter is the 20 nucleotides (−20 to +1) upstream of the subgenomic initiation site (the subgenomic initiation site is designated as +1, and minus numbers are 5' of that site as read on the positive-strand RNA) (Marsh et al., 1988; French and Ahlquist, 1988). Siegel et al. (1997, 1998) examined the structure of the core promoter by constructing 'proscripts' which comprised the core promoter and a template and which directed (+)-strand synthesis. By mutagenesis they showed that the nucleotides at positions –17, –14, –13 and –11 were essential for promoter activity. There was some evidence that the –17 nucleotide recognized the RdRp. The +1 and +2 nucleotides (cytidulate and adenylate) are also important for RNA synthesis (Adkins et al., 1998). The core promoter forms a stable hairpin structure (Jaspars, 1998).

In vitro studies identified three enhancer regions in the BMV promoter, the 16 nucleotides downstream of the start site (+1 to +16) which provide accurate initiation, and two upstream domains, the poly(A) stretch present in all bromoviruses (–20 to –37) and a triple repeat of UUA (–38 to –48) (Marsh et al., 1988). However, analysis in vivo indicated that the core region extended downstream and was from –20 to +16, that the poly(A) tract was essential as were three repeats of the sequence AUCUAUGUU extending the complete promoter to a site between –74 and –95 upstream of the start site (French and Ahlquist, 1988). Deletion of the poly(A) tract led to three revertants in which subgenomic RNA synthesis has been restored (Smirnyagina et al., 1994). Two of these revertants were in the subgenomic promoter and gave increased levels of genomic RNA3. In the third revertant, the mutation was upstream of the subgenomic promoter in a sequence, designated box B, which led to decreased levels of RNA3. Box B is an ICR2-like (internal control region 2) sequence similar to those found in cellular RNA polymerase III and in the TψU loop of tRNA as well as at the 5'-ends of BMV RNAs 1 and 2.

Both the BMV and CCMV RdRps recognize the BMV core subgenomic promoter requiring specific functional residues at positions –17, –14, –13 and –11 (Adkins and Kao, 1998). For CCMV sgRNA synthesis, both RdRps require the same nucleotides and four additional nucleotides at positions –20, –16, –15 and –10. The –20 nucleotides are partially responsible for the differential recognition of the two promoters.

The subgenomic promoter of AMV is more complex than that of BMV particularly in vivo, in that: (1) the sgRNA4 is the mRNA for coat protein which is required for (–)-strand replication; and (2) the intergenic region is only 49 nucleotides which is too short to accommodate the promoter. Thus, the promoter extends into the carboxy terminus of the movement protein gene (ORF 3a) which hampers the possibilities of mutational analysis for in vivo studies. However, characterization of the in vivo behaviour of the promoter was achieved making large enough insertions to contain it at the 5'-end of the sgRNA (van der Kuyl et al., 1990). This study showed that the core promoter was located between –26 and +1 and that there were two enhancer regions, one downstream (+1 to +12) and one upstream (–136 to –94). AMV core promoter has a small conserved sequence (AAU), also present in BMV, which mutations show to be important (van der Vossen et al., 1995). As with BMV, the core promoter forms a hairpin structure (Jaspars, 1998) but the structure for AMV (and Ilarviruses) is more stable than that of BMV. It is suggested that in AMV and Ilarviruses, that require coat protein for replication (see Chapter 8, Section IV.G), the coat protein is needed for the polymerase to interact with the core promoter hairpin.

Thus, AMV and BMV have a common spatial organization and some conserved sequences homologous to a consensus sequence derived from all the alphaviruses of plants and animals (Ou et al., 1982).

c. Premature termination of minus strand

Premature termination of (–)-strand synthesis is effected by either cis or trans long-distance interactions between a region just upstream of the subgenomic promoter and another region of the viral nucleic acid. This can either be on the same nucleic acid molecule as that giving rise to the sgRNA (cis interaction) or on another genomic fragment of a split genome virus (*trans* interaction).

Formation of sgRNAs by *cis* interactions have been suggested for TBSV and PVX (Miller *et al.*, 1998; Zhang *et al.*, 1999) and by *trans* interactions for RCNMV (Sit *et al.*, 1998).

The genome of TBSV has five ORFs, the 3′ three of which are expressed from two sgRNAs, sg mRNA1 and sg mRNA2 (see Fig. 6.23 for genome organization). Deletion mutagenesis identified a 12-nucleotide sequence approximately 1 kb upstream of the initiation site for sg mRNA2 that was required specifically for the accumulation of that sgRNA. The 12-nucleotide sequence can potentially base-pair with a sequences just 5′ to the sgRNA2 initiation site and mutagenesis supported this interpretation. Base-pairing sequence regions with similar stability and relative location are found in the genomes of different tombusviruses. It was proposed that the upstream sequence represents a *cis*-acting element that facilitates sgRNA promoter activity by long-distance RNA–RNA interactions.

The genome of PVX expresses five proteins, the polymerase from the genomic RNA and the triple block movement proteins and the coat protein from two or three sgRNAs (see Fig. 6.39 for genome organization). Upstream of the two promoters is a conserved octanucleotide sequence (GUUAAGUU) that is conserved between potexviruses (Kim and Hemenway, 1997). Mutagenesis indicated that this conserved sequence and its distance from the start site for sgRNA synthesis are critical for the accumulation of the sgRNAs. Other mutagenesis experiments showed that multiple elements in the 5′-end of the genomic RNA are important in both genomic and sgRNA accumulation (Kim and Hemenway, 1996). The conserved sequences upstream of the sgRNA initiation sites are also found in the 5′ region of the genomic RNA and mutation and compensatory mutation suggested that there are long-distance interactions between the 5′ and sgRNA initiation sites (Kim and Hemenway, 1999).

The genome of RCNMV is divided between two RNAs: RNA1 which is bicistronic encoding the viral polymerase and the coat protein, and the monocistronic RNA2 (for genome organization of dianthoviruses see Fig. 6.30). The coat protein gene is expressed from a subgenomic RNA from RNA1, the putative upstream promoter region predicted to form a stable stem-loop structure (Zavriev *et al.*, 1996). The coat protein sgRNA is expressed only in the presence of RNA2 (Vaewhongs and Lommel, 1995). To study the interaction between RNAs 1 and 2 in expressing the sgRNA, Sit *et al.* (1998) replaced the coat protein gene on RNA1 with the GFP (green fluorescence protein) gene showing that subgenomic GFP (sGFP) was expressed *in vivo* when RNA2 from RCNMV, and from the related CRSV and SCNMV, were present. Mutagenesis of RNA2 was difficult because identifying the *trans*-acting element(s) for subgenomic expression could interfere with the *cis*-acting element(s) for RNA2 replication. To uncouple these elements, an infectious clone of TBSV was used as a vector for fragments of RCNMV RNA2 (Table 7.4).

In this way, a 34-nucleotide segment (756–789) of RNA2 was identified as the *trans*-acting element. This 34-nucleotide element forms a stem-loop structure that is conserved between the RNA2 of RCNMV, CRSV and SCNMV, and the 8-nucleotide loop is complementary to an 8-nucleotide sequence on RNA1 two nucleotides upstream of the subgenomic RNA initiation site. Mutations of the 8-nucleotide sequence in RNA1 that did not alter the amino acid sequence of the overlapping replicase gene abolished the formation of the subgenomic RNA, but complementary mutation of the 8-nucleotide loop in RNA2 restored the ability to generate subgenomic RNAs. Thus, the 8-nucleotide loop on RNA2 transactivates the synthesis of the subgenomic RNA on RNA1. From these observations, Sit *et al.* (1998) proposed a model for the formation of RCNMV subgenomic RNAs (Fig. 7.17).

In the early phase, RNA 1 and 2 are replicated to high levels. The transactivator element on RNA2 binds to RNA1, preventing the replicase from forming full-length minus strands of that RNA. As the 5′ terminal sequences of genomic RNA1 and the subgenomic RNA are conserved (Fig. 7.17), it is suggested that the

TABLE 7.4 Experiment to show that expression of a subgenomic RNA (sGFP) from a construct in which the coat protein gene of RCNMV RNA1 is replaced by GFP (R1sGFP) depends on RNA2 sequences

R1sGFP was co-inoculated with transcripts from the TBSV replicon (pHST2 in which the coat protein region was engineered to accept and express foreign genes from an sgRNA) containing segments from RCNMV RNA2 in order to delimit the minimal *trans*-activating elements.

Construct	Position of RNA2 sequence	Length of RNA2 inserts (nucleotides)	sGFP produced[a]
RCNMV RNA2	1–448	1448	+++
pHST2	–	–	–
pHST2-RC2.3	708–1031	324	+
pHST2-RC2.4	1031–708 (-sense)	324	–
pHST2-ΔBX	708–916	209	++
pHST2-707	707–837	121	+
pHST2-828	828–918	91	–
pHST2-SL2	756–789	34	+++
pHST2-TA38	792–755 (-sense)	38	–
pHST2-20	762–782	21	+
pHST2-LT2	767–775	9	–

[a] sGFP production relative to induction by RNA2: +++ = 100%; ++ = 50%; + = 25%; – = not detected.
From Sit *et al.* (1998), with permission.

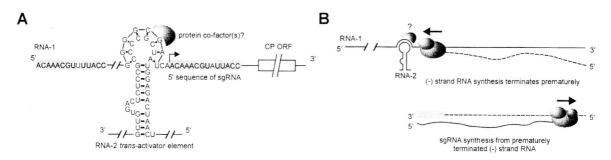

Fig. 7.17 Proposed components and model of RCNMV *trans*-activation mechanism. **(A)** Sequences involved in *trans*-activation. Grey-shaded nucleotides represent conserved sequences between genomic RNA1 and the sgRNA that are likely to be (+)-strand promoters. The loop region of the 34-nucleotide trans-activator is shown base-pairing with the complementary 8-nucleotide element, two nucleotides upstream from the sgRNA start site (right-angle arrow) on RNA1. **(B)** Model for the generation of sgRNA. Complementary strands are depicted as dashed lines. From Sit *et al.* (1998), with kind permission of the copyright holder, © The American Association for the Advancement of Science.

truncated complementary RNA serves as a template for the production of positive-sense subgenomic RNA.

d. Discussion

The use of sgRNAs is widespread in plant viruses as a strategy to obviate the limitations of eukaryotic translation (see Table 7.7). There appear to be two mechanisms by which these RNAs are synthesized, both of which involve close interlinks with viral replication and both of which have strong controlling systems that are only just beginning to be recognized. The subgenomic promoter may have elements both in intergenic and in coding regions, the latter suggesting that the position may control expression of the promoter.

Several examples have been noted above of the sgRNA promoter regions being predicted to form strong secondary structures, usually stem-loops. This is predicted for other viruses (e.g. TCV: Wang and Simon, 1997; Wang *et al.*, 1999a).

3. Strategy 3: Multi-partite genomes

Viruses with multi-partite genomes have the information required for the virus infection cycle divided between two or more nucleic acid segments. This is found for both DNA and RNA plant viruses. For the (+)-sense ssRNA viruses, this strategy places the gene at the 5'-end of each RNA segment and thus it is open for translation.

Of the 70 plant virus genera, 28 have multipartite genomes (see Table 7.7 and Appendix 2). In most of these, the genome segments are encapsidated in separate particles, such viruses being termed *multi-component*. Members of the *Reoviridae*, and possibly the *Partitiviridae*, have all their genome segments in one particle.

4. Strategy 4: Internal initiation

In the internal initiation strategy translation is initiated at a site, termed the internal ribosome entry site (IRES) or ribosome landing pad, that is not the 5' AUG start codon; see Belsham and Sonenberg (1996) for a review of IRESs. It is suggested that the IRES forms a complex secondary/tertiary structure to which ribosomes and transacting factors bind. There are various reports of the downstream ORF of bicistronic RNAs of plant viruses being expressed (Hefferson *et al.*, 1997), but few can be fully attributable to internal initiation. However, the subgenomic mRNA for the coat protein of crucifer TMV (crTMV) and the subgenomic, I_2 RNA, mRNAs for the movement proteins of both crTMV and TMV U1 strains are reported to be expressed by internal initiation (Ivanov *et al.*, 1997; Skulachev *et al.*, 1999). These RNAs are uncapped and have long 5' untranslated regions (UTRs) — 148 nt for the crTMV coat-protein and 228 nt for the movement-protein mRNAs. The problem is to show that an RNA demonstrating internal initiation *in vitro* actually does this *in vivo*. Skulachev *et al.* (1999) addressed this point and showed good evidence for it happening *in vivo*.

It is considered that the IRES strategy enables a potentially inefficient mRNA (no cap and long 5' UTR) to be translated efficiently and might also provide translational control so that gene products such as a movement protein can be expressed at the appropriate time.

5. Strategy 5: Leaky scanning

Leaky scanning is where the 40S ribosomal subunits start scanning from the 5'-end of the RNA but do not all start translation at the first AUG. Some, or all, pass the first AUG and start translation at downstream ORFs. In some cases, the 40S subunits of the 80S ribosomes fail to disengage at a stop codon and they reinitiate translation at a downstream start codon. There are three forms of leaky scanning: (1) two initiation sites on one ORF, (2) overlapping ORFs, and (3) two consecutive ORFs (Fütterer and Hohn, 1996; Maia *et al.*, 1996).

a. *Two initiation sites on one ORF (two-start)*

The genome of CPMV is divided between two RNA species, the shorter of which, M-RNA, codes for two C-terminal collinear polyproteins of 105 kDa and 95 kDa initiated from two in-frame AUG codons (see Fig. 6.18 for CPMV genome organization). From *in vitro* translation experiments, it was suggested that there might be internal initiation (Thomas *et al.*, 1991; Verver *et al.*, 1991), but *in vivo* experiments showed that there was a greater likelihood of leaky scanning being involved (Belsham and Lomonossoff, 1991). This was supported by the fact that the AUG for the 95-kDa protein is in a more favourable context (G at positions −3 and +4) than is that for the 105-kDa protein (A at −3 and U at +4).

b. *Overlapping ORFs*

The genomes of members of the *Luteovirus* and *Polerovirus* genera contain six ORFs, the 3' ones of which are expressed from sgRNAs (see Fig. 6.35 for genome organizations). ORF 4 (17 kDa) is contained in a different reading frame within ORF 3 (coat protein). The translation of ORF 4 fits very well with leaky scanning from the translation initiation of ORF 3 (Tacke *et al.*1990; Dinesh-Kumar and Miller, 1993). The context of the ORF-3 AUG (U at position −3 and A at +4) is unfavourable. There are also interactions between the translation of these two ORFs. Mutations that reduce the initiation efficiency of ORF 4 also reduce initiation at the ORF-3 AUG if the latter's flanking bases are not changed (Dinesh-Kumar and Miller, 1993).

There are two overlapping ORFs on RNA2 of PCV that are translated by leaky scanning

(Herzog *et al.*, 1995), and the second and third ORFs of the triple gene block movement protein complexes (see Chapter 9, Section II.D.2.f) of potexviruses, carlaviruses, furoviruses and hordeiviruses are also expressed by a similar leaky scanning mechanism (Zhou and Jackson, 1996; Verchot *et al.*, 1998). The upstream AUG is usually in a poor context and the upstream ORF contains few, if any, other AUGs in any of the three reading frames.

c. Two consecutive ORFs

PCV has a bipartite genome, RNA2 of which has five ORFs (see Fig. 6.47 for genome organization). The 5' ORF of RNA2 encodes the viral coat protein and overlaps the next ORF (39-kDa protein) by two nucleotides and both are translated from the same RNA. The 620-nucleotide coat-protein ORF is devoid of AUG codons apart from the initiation codon. The insertion of stem structures or of AUG codons in the upstream region inhibited the translation of the 39-kDa protein, suggesting leaky scanning (Herzog *et al.*, 1995).

The expression of ORFs I, II and III of RTBV is another example of leaky scanning due to the lack of AUG codons in upstream regions (see Section V.H.1.b).

d. Discussion

The translation of non-5' ORFs by leaky scanning usually results from the AUG start codon in the upstream ORF being in a poor context. A lack or dearth of AUG codons in the upstream region can enhance the effect. However, care must be taken in interpreting *in vitro* translation information as evidence for leaky scanning. Parameters such as translation system and conditions, especially the presence of divalent cations, can affect the expression from non-5' ORFS.

6. Strategy 6: Non-AUG start codon

Some viral ORFs appear to start with a codon that is not the conventional AUG start codon; the initiation at these ORFs is inefficient.

The genome of RTBV is expressed from a more-than-genome length RNA transcribed from the viral DNA (see Section IV.C).

Computer analyzes of the sequence of this RNA show three conventional ORFs that start with an AUG and finish with a stop codon and one region lacking stop codons but without an AUG. Using mutagenesis and expression of a coupled chlorophenical acetyl transferase (CAT) gene, Fütterer *et al.* (1996) demonstrated that this region (ORF I) was translated and that the translation was initiated at an AUU codon (Fig. 7.18). The efficiency of translation was about 10% that of a gene that had the conventional AUG start codon.

SBWMV has a bipartite genome, RNA2 of which has three ORFs, the coat protein (19 kDa), a read-through ORF from the coat protein, and a 3' 19-kDa ORF expressed from an sgRNA (see Fig. 6.46 for genome organization). RNA2 also expresses both *in vitro* and *in vivo* a 28-kDa protein that reacts with SBWMV coat protein antiserum (Hsu and Brakke, 1985; Shirako and Ehara, 1986). *In vitro* translation and site-directed mutagenesis indicated that a CUG codon initiated the translation of the 28-kDa protein in a region upstream of, and in frame with, the AUG codon for the 19-kDa coat protein giving a 40-amino-acid N-terminal extension to that protein. (Shirako and Wilson, 1993; Shirako 1998).

7. Strategy 7: Transactivation (reviewed by Fütterer and Hohn, 1996)

The dsDNA genome of CaMV has six closely spaced functional ORFs (I–IV) (see Fig. 6.1 for genome organization) and is transcribed to give a more-than-genome length RNA, the 35S RNA and an mRNA (19S) for ORF VI (see Section IV.C). Although there is some evidence for the use of spliced RNAs for expressing some of the ORFs (see Section V.B.11), it appears that some of the ORFs are expressed from the long RNA. Most of the downstream ORFs are not, or are poorly, expressed in protoplasts or transgenic plants unless the product of ORF VI is present (Bonneville *et al.*, 1989; Fütterer and Hohn, 1991). This gene product is termed a transactivator (TAV) and is thought to facilitate internal initiation. A similar function has been shown for the analogous gene product in FMV (Gowda *et al.*, 1989; Scholthof *et al.*, 1992a,b) and it is

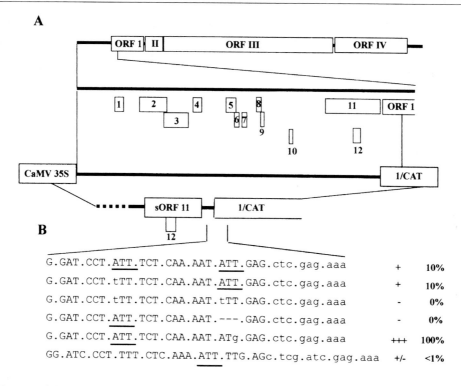

Fig. 7.18 Diagram of the experiments by Fütterer *et al.* (1996) showing the non-AUG start codon of RTBV. **(A)** The top line depicts the genome organization of RTBV displayed in a linear manner with the ORFs shown as boxes. The middle line is an enlargement of the 5′ part of the genome with the 12 short ORFs (sORF) numbered. The bottom line shows the basic construct transiently expressed in rice cells. Transcription is driven by the CaMV 35S promoter and the chloramphenicol acetyl transferase (CAT) gene without its AUG start codon is fused in frame with ORF I. **(B)** Sequences around the 5′-end of ORF I of various constructs. The RTBV sequence is shown in upper case and the CAT sequence in lower case; mutations in the RTBV sequence are shown in lower case or as –. The two ATT codons are underlined. The + and – signs to the right are an estimate of the relative CAT activity, and the percentage figures on the extreme right are the % CAT activity relative to that of constructs in which the CAT gene retained its ATG start codon.

likely that all caulimoviruses have a gene with this function in the same relative region of the genome.

Using artificial polycistronic RNAs, Fütterer and Hohn (1991, 1992) examined the parameters that control transactivation. They showed that, for reinitiation at a downstream ORF, there should not be overlapping of long ORFs and that it was particularly efficient when the first ORF was about 30 codons long. The polar effects of the insertion of stem-loop structures, that would inhibit the translocation of ribosomes, into polycistronic mRNAs and the specificity for non-overlapping ORFs indicated that transactivation causes enhanced reinitiation of ribosomes. The optimal 30-codon length for the first ORF is just long enough to emerge from a translating ribosome and suggests that the TAV acts directly or indirectly on the translating or terminating ribosome.

The transactivating function locates to the central one-third of the TAV protein (De Tapia *et al.*, 1993). Using this portion of the gene product (mini-TAV), 100-fold more DNA had to be used to produce normal levels of transactivation in transfected cells. The mini-TAV was only active in *Nicotiana plumbaginifolia* protoplasts whereas the full-length TAV was active in protoplasts from several dicot and monocot species. There is an RNA-binding activity outside the mini-TAV region that may enhance activity by increasing the TAV concentration near the RNA. TAV associates with polysomes and, as shown by an overlay binding assay, also with an 18-kDa ribosomal or ribosome-associated protein from plants and yeast

(Fütterer and Hohn, 1996). It also interacts with the 60S ribosomal subunit protein L18 (RPL18) from *Arabidopsis thaliana* (Leh *et al.*, 2000), the interacting region being defined as the mini-TAV region.

Using *N. edwardsonii* protoplasts, Edskes *et al.* (1996) analyzed the ability of the TAV from three different caulimoviruses to activate viral RNA-based reporter constructs. They found that efficient expression of polycistronic and monocistronic caulimovirus mRNAs in plant cells requires compatible interactions between ORF VI, a translational TAV and a cognate *cis*-element at the 3'-end of the mRNA.

Expression of the TAV in transgenic plants can give rise to virus-like symptoms (see Chapter 10, Section III.E.6). It is not known whether the transactivation or other activity of this gene causes this.

8. Strategy 8: Translational (ribosome) shunt

Short ORFs (sORFs) (defined arbitrarily as having less than 50 codons and no known function for the product) in a leader sequence can interfere with translation of a downstream ORF. In the translational ribosome shunt mechanism, initially scanning ribosomes are transferred directly from a donor to an acceptor site without linear scanning of the intervening region, thus avoiding sORFs.

The 35S RNAs of plant pararetroviruses, members of the *Caulimoviridae*, have long and complex leader sequences (Figs 7.18 and 7.19). The leader sequences range in length from about 350 to more than 750 nucleotides and contain between three and 19 sORFs. They fold to give complex stem-loop structures (Pooggin *et al.*, 1999). Translational shunting has been proposed as the mechanism by which these constraints to translation are bypassed (Fütterer *et al.*, 1993). For translational shunting there need to be shunt donor and acceptor sites and well-defined structure to bring these together. Most of the studies on these features have been on CaMV.

The leader sequence of CaMV 35S RNA forms an elongate hairpin conformation that contains a long-range pseudoknot and a dimer (Hemmings-Mieszczak *et al.*, 1997). Chemical modification, enzymatic probing and temperature-gradient electrophoresis showed that the hairpin is made up of three elements—termed stem sections I, II and III (Hemmings-Mieszczak *et al.*, 1997, 1998). Stem section I is the most stable and its structure, rather than its sequence, is important in ribosome shunting (Hemmings-Mieszczak and Hohn, 1999).

To study the effects of the sORFs in the leader sequence, Pooggin *et al.* (1998) mutated the start codon of each and tested the infectivity of the mutants. Mutation of the 5' sORF (A) delayed symptom development and the mutant had reverted; infectious mutants in the sORFs located in the stem-loop either reverted or there were compensatory mutations to restore the structure. Thus, sORF A as well as a polypyrimidine stretch at the 3'-end of the hairpin (Hemmings-Mieszczak and Hohn, 1999) are important in stimulating translation via the shunt. The current interpretation of the shunt structure is shown in Fig. 7.20.

The ribosome enters at the 5' cap, which is essential, translocates to sORF A, shunts across the stable hairpin, and then reaches the first true ORF. The sORF MAGDIS from the mammalian *AdoMetDC* RNA, which conditionally suppresses re-iniation at a downstream ORF, prevented shunting, indicating that in CaMV the shunt involves re-initiation (Ryabova and Hohn, 2000). There are still unknown details as to what happens at sORF A and as to the actual crossing of the shunt junction.

In CaMV the shunt 'donor' or 'take-off' site is just downstream of sORF A (Dominguez *et al.*, 1998) and the shunt 'acceptor' or 'landing' site is just upstream of ORF I (Fütterer *et al.*, 1993).

In a study of the effect of the CaMV leader sequence on the expression of a downstream *uid*A (GUS) reporter gene, Schärer-Hernández and Hohn (1998) showed that the shunt mechanism occurs *in planta* with an average efficiency of 5% compared with that of a leaderless construct. However, there are some reservations on the use of *in planta* systems to examine what are essentially artificial constructs as there may be RNA processing, transport and modifications which do not occur in the natural system. Dominguez *et al.* (1998) have developed an *in vitro* system that avoids these shortcomings and which supports the above *in vivo* data. In

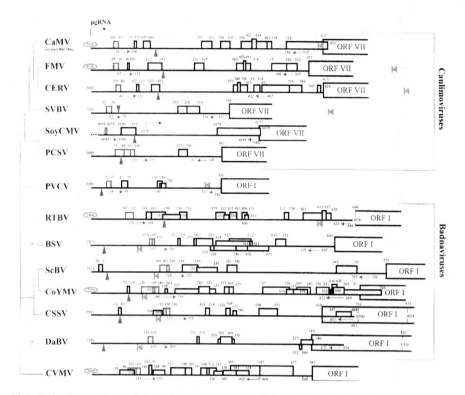

Fig. 7.19 (see Plate 7.1) Comparison of the primary structures of the pgRNA leader of plant pararetroviruses. On the left, a phylogenetic tree shows the relationship of their reverse transcriptases (according to Richert-Pöggeler and Shepherd, 1997; note that BSV and DaBV were not included in that comparison). Caulimoviruses and badnaviruses are grouped (as indicated on the right). The leader sequence preceding the first long ORF (ORF VII or ORF I) is depicted as a thick line; the sORFs are indicated by boxes, with internal start codons shown by vertical lines. The numbered genome position of the pgRNA 5′-end is enclosed within an ellipse if mapped by primer extension or not enclosed if putative. The numbering within the leader is from the 5′-end (except for SoyCMV where the latter is unclear). Red arrows under the leader define the complementary sequences that form the base of the large stem-loop structures. The conserved sORFs preceding the structures are also in red. A green triangle indicates a putative or, in the case of CaMV, FMV and RTBV (Rothnie, 1996), mapped poly(A) signal. An arrowhead adjacent to a vertical line (in blue) shows the location of the primer binding site for reverse transcription. From Pooggin *et al.* (1999), with permission.

spite of this there must be some reservations, mentioned above, in the extrapolation of *in vitro* data to the *in vivo* situation.

Three cellular proteins that bind to several sites in the CaMV leader sequence have been identified by UV cross-linking assays (Dominguez *et al.*, 1996). p35 binds to RNA non-specifically, p49 binds with low specificity, but p100 interacts specifically with viral sequences. The expression of these proteins is not induced by virus infection.

Translational shunt has also been demonstrated for RTBV with the shunt donor near sORF 1 and the shunt acceptor near the AUU start codon for ORF 1 (Fütterer *et al.*, 1996) (see Section V.H.1.b).

9. Strategy 9: Read-through proteins

The termination codon of the 5′ gene may be 'leaky' and allow a proportion of ribosomes to carry on translation to another stop codon downstream from the first, giving rise to a second longer functional polypeptide. This is termed 'read through' or 'stop-codon suppression' resulting in a protein that has the same sequence as the upstream protein in its N-terminal portion and a unique sequence in its C-terminal portion.

The read-through strategy is found in at least 17 plant virus genera (Table 7.5) and is characteristic of all three genera in the *Luteoviridae* and most genera in the *Tombusviridae*. The proteins that are produced by read through are either

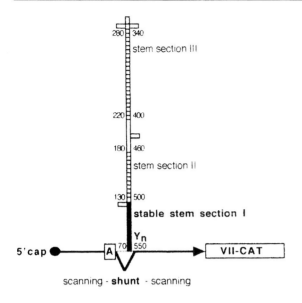

Fig. 7.20 Non-linear ribosome migration (ribosome shunt) during translation initiation on the CaMV 35S RNA leader. The leader is represented schematically as a combination of unstructured 5′ and 3′ terminal regions flanking an elongated hairpin structure (nucleotide 70–550). Three stem sections of different stability are labeled and the numbers indicate their approximate borders; stem section I (nucleotide 70–130 and 500–550), stem section II (nucleotide 130–180 and 460–500) and stem section III (nucleotide 220–280 and 340–400). The presence of a stable hairpin (stem section 1, black box) promotes translation of a downstream ORF (VII-CAT) via the ribosome shunt mechanism; all essential elements; 5′ cap, sORFa (boxed A), stable segment I, and the pyrimidine-rich sequence around nucleotide 500 (Yn) are indicated in bold. The 5′ and 3′ unidirectional ribosome migration is depicted by a thick black arrow with description below. Sequences preceding and directly following the hairpin are called shunt donor and acceptor sites respectively. The model predicts that, after the cap-dependent initial scanning, the 40S subunit of the ribosome translating sORFA is transferred (shunted) from the shunt donor to the shunt acceptor site, where the scanning is resumed and continues. Then, the 80S ribosome is assembled at the initiation codon of ORF VII, where translation starts. From Hemmings-Mieszczak *et al.* (1998), with permission.

replicases (*Tombusviridae*) or extensions to the coat protein (luteoviruses) thought to be involved in transmission vector interactions (see Chapter 11, Section III.H.1.a).

In read through, the stop codon is read by a suppressor tRNA instead of the ribosome being released by the eukaryotic release-factor complex. Essentially, there is competition between the release-factor complex and the suppressor tRNAs. Two main factors are involved in read through of stop codons, the context of the stop codon and the suppressor tRNAs involved.

a. Stop codon context

Stop codons have different efficiencies of termination (UAA>UAG>UGA) and the first, and possibly the second, nucleotide 3′ of the stop codon acts as an important efficiency determinant (Stansfield *et al.*, 1995). It can be seen from Table 7.5 that either amber (UAG) or opal (UGA) stop codons are read through; there are no examples of read through of the natural ochre (UAA) stop codon. However, when the suppressible UAG codon of TMV is replaced by a UAA codon the virus can still replicate and produce mature virions (Ishikawa *et al.*, 1986). Unlike with retroviruses, there appears to be no structural requirements for stop codon suppression in plant systems. The context of the TMV amber stop codon at the end of the 126-kDa ORF has been studied in detail by insertion into constructs containing the GUS or other genes which enable read through to be quantitated in protoplasts (Valle *et al.*, 1992). This experimental approach has defined the sequence (C/A)(A/C)A.UAG.CAR.YYA (R = purine, Y = pyrimidine) as the optimal consensus context (Skuzeski *et al.*, 1991; Hamamoto *et al.*, 1993). Efficiently recognized stop codons normally have a purine immediately downstream and avoid having a C residue (Fütterer and Hohn, 1996).

However, an analysis of Table 7.5 shows that the read-through stop codons identified for plant viruses do not necessarily conform to these contexts determined from *in vitro* systems. As the sequence context differs from that described above, the read through of UAG.G in many of the *Tombusviridae* and the *Luteoviridae* suggests that a different mechanism might be involved. It is possible that there may be a requirement for additional *cis*-acting sequences such as a conserved CCCCA motif or repeated CCXXXX motifs beginning 12 to 21 bases downstream of many of these read-through

TABLE 7.5 Read-through or stop codons

Family	Genus	Virus	Read-through stop codon	Read-through protein
Tombusviridae	Tombusvirus	TBSV	AAA **UAG** GGA GGC	Replicase
	Aureusvirus	PoLV	UAC **UAG** GGG UGC	Replicase
	Avenavirus	OCSV	AAA **UAG** GGG UGC	Replicase
	Carmovirus	CarMV	AAA **UAG** UUG GAA	Replicase
	Machlomovirus	MCMV 1	AAA **UAG** GGG UGU	Replicase
		MCMV 2	AAC **UGA** GCU GGA	Unknown
	Necrovirus	TNV	AAA **UAG** GGA GGC	Replicase
	Panicovirus	PMV	AAG **UAG** GGG UGU	Replicase
Luteoviridae	Luteovirus	BYDV	AAA **UAG** GUA GAC	Coat protein extension
	Polerovirus	PLRV	AAA **UAG** GUA GAC	Coat protein extension
	Enamovirus	PEMV	CUC **UGA** GGG GAC	Coat protein extension
No family	Benyvirus	BNYVV	CAA **UAG** CAA UUA	Coat protein extension
	Furovirus	SBWMV RNA1	AAA **UGA** CGG UUU	Replicase
		SBWMV RNA2	AGU **UGA** CGG GAC	Coat protein extension
	Pecluvirus	PCV RNA1	AAA **UGA** CGG UUU	Replicase
	Pomovirus	BSBV RNA1	AAA **UGA** CGG	Replicase
		BSBV RNA2	GAA **UGA** CAA UCA	Coat protein extension
	Tobamovirus	TMV	CAA **UAG** CAA UUA	Replicase
	Tobravirus	TRV	UUA **UGA** CGG UUU	Replicase

Data from Fütterer and Hohn (1996), amended and updated.

sites (see Fütterer and Hohn, 1996; Miller *et al.*, 1997a). It may be that other long-distance interactions are involved in the deviations from the experimentally determined optimal contexts for TMV.

b. Suppressor tRNAs

The synthesis of a read-through protein depends primarily on the presence of appropriate suppressors tRNAs. Thus, two normal tRNAs^{tyr} from tobacco plants were shown to promote UAG read-through during the translation of TMV RNA *in vitro* (Beier *et al.*, 1984a). The tRNA^{tyr} must have the appropriate anticodon, shown to be GψA, to allow effective read through. A tRNA^{tyr} from wheatgerm, with an anticodon different from that in tobacco, was ineffective (Beier *et al.*, 1984b). tRNA^{gln} will also suppress the UAG stop codon (Grimm *et al.*, 1998). Various suppressor tRNAs have been found for the UGA stop codon of TRV (Zerfass and Beier, 1992; Urban and Beier, 1995; Baum and Beier, 1998), chloroplast (chl) and cytoplasmic (cyt) tryptophan-specific tRNAs with the anticodon CmCA, and chl and cyt cysteine-specific tRNAs with the anticodon GCA and

arginine tRNA isoacceptor with a U*CG anticodon. Interestingly, the chl tRNA^{trp} suppresses the UGA codon more efficiently than does the cyt tRNA^{trp}.

The replication of some strains of TRV is associated with mitochondria (see Fig. 8.17) and it is likely that the prokaryote-like tRNAs that are found in chloroplasts will also be found in mitochondria. The tRNA^{cys} isoacceptor from tobacco chloroplasts was more efficient at reading through a UGA stop codon in the TRV (and TMV) context than was the cytoplasmic acceptor (Urban and Beier, 1995).

c. Proportion read through

The proportion of read-through protein produced may be modulated by sequence context of the termination codon (Bouzoubaa *et al.*, 1987; Miller *et al.*, 1988; Brown *et al.*, 1996a), by long-distance effects and by the availability of the suppressor tRNA. Thus, data from *in vitro* systems may not be fully relevant to the *in vivo* situation. However, it is reasonable to suggest that about 1–10% of the times that a ribosome reaches a suppressible stop codon result in read through.

d. Discussion

As well as overcoming the constraints of the eukaryotic translation system, read through of stop codons provides a mechanism for the control of the expression of gene products. Transmission helper factors have to be incorporated in the virus capsid but it is probably not necessary to have them on all coat protein subunits. Thus, it could be more efficient for the virus expression system if 1–10% of the coat protein subunits also contain the transmission factor. Similarly, the viral replicase comprises several functional domains (described in Chapter 8, Section IV.B) that are probably required in different amounts and even at different times. The production of two proteins both containing some of the domains and one, the read-through protein, also containing the other domains would give control over the availability of these functions.

10. Strategy 10: Frameshift proteins (reviewed by Brierley, 1995; Farabaugh, 1996a,b)

Another mechanism by which two proteins may commence at the same 5′ AUG is by a switch of reading frame before the termination codon of the 5′ ORF to give a second longer 'frameshift' protein. This translational frameshift event allows a ribosome to bypass the stop codon at the 3′-end in one reading frame and switch to another reading frame so that translation can continue to the next stop codon in that reading frame (Hizi et al., 1987). A frameshift is illustrated in Fig. 7.21.

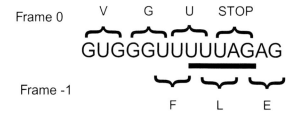

Frame 0   V   G   U   STOP

GUGGGUUUUUAGAG

Frame -1   F   L   E

Fig. 7.21 Translational frameshift. The ribosome bypasses the stop codon in frame 0 by switching back one nucleotide to frame –1 at a UUUAG sequence before continuing to read triplets in frame –1 to give the fusion or frameshift protein. From Matthews (1991).

At the frameshift site ribosomes change their reading frame either one nucleotide in the 5′ direction (–1 frameshift) or one nucleotide in the 3′ direction (+1 frameshift). This gives two proteins (the frame and frameshift proteins) identical from the N terminus to the frameshift site but different beyond that point. The frame protein is always produced in greater quantity than the frameshift protein.

Frameshift obviously occurs where ORFs overlap and may be at any place within that overlap.

The frameshift strategy is found in at least nine plant virus genera (Table 7.6) and, in all cases that are known, involves the replicase. Most instances of frameshift are in the –1 direction, only those of the *Closteroviridae* being in the +1 direction.

a. –1 Frameshifting

For a –1 frameshift three features are needed, a heptanucleotide sequence termed the 'slippery' or 'shifty' sequence at the frameshift site, a strongly structured region downstream of the frameshift site, and a spacer of four to nine nucleotides between the slippery sequence and the structured region. The involvement of the structure, stability and function of the pseudoknots making up the strongly structured region is reviewed by Giedroc et al. (2000).

The slippery sequence comprises two homopolymeric triplets of the type XXX YYY Z (X = A/G/U; Y = A/U; Z = A/C/U). Upon reaching this heptanucleotide sequence, the two ribosome-bound tRNAs that are in one reading frame (X.XXY.YYZ) shift by one nucleotide in the 5′ direction (XXX.YYY.Z) retaining two out of the three base-paired nucleotides with the viral RNA (Jacks et al., 1988). This mechanism was deduced for retroviruses but the evidence points to a similar mechanism in plant RNA viruses. The slippery sequences listed in Table 7.6 are either indicated from experimentation or predicted from similarities with other viruses.

Mäkinen at al. (1995b) noted a consensus amino acid sequence, WAD/WGD (W = Trp; A = Ala; G = Gly; D = Asp; E = Glu) followed

TABLE 7.6 Frameshift between open reading frames

Family	Genus	Virus	Frameshift sequence	Type of frameshift	Signal	Protein(s)
Tombusviridae	*Dianthovirus*	RCNMV	D F * G.GAU.UUU.UAG.GC G F L G	−1	Stem-look	Replicase
	Panicovirus	PMV	N F * C. AAU.UUC.UAG.UG Q F L V	−1	ND	Unknown
Closteroviridae	*Closterovirus*	BYV	R V * CGG.GUU.UAG.CUC G F S S	+1	Pseudoknot	Replicase
	Crinivirus	LiYV	L K D Y UUG.AAA.GAC.UAUC * K U I	+1		Replicase
Luteoviridae	*Luteovirus*	BYDV	G F * G.GGU.UUU.UAG.AG G F L E	−1	Stem-loop or pseudoknot	Replicase
	Polerovirus	PLRV	L N G Q U.UUA.AAU.GGG.CA F K W A	−1	Pseudoknot	Replicase
	Enamovirus	PEMV-1	G N G F G.GGA.AAC.GGA.UU G K R I	−1	Pseudoknot	Replicase
No family	*Sobemovirus*	SBMV	L N C L U.UUA.AAC.UGC.UU F K L L	−1	Stem-loop	
	Umbravirus	PEMV-2	F W * U.UUU.UGG.UAG.GG F L V G	−1		Replicase

by a D/E-rich domain in front of frameshift sites. The significance of this motif is unknown.

The strongly structured regions are separated by the spacer region from the frameshift point and are either hairpins or pseudoknots (see Chapter 4, Section III.A.3.b, for pseudoknots). It is considered that the structure causes the ribosome to pause, thereby initiating frameshift. The recent determination of the crystal structure of the pseudoknot in BWYV RNA (Su *et al.*, 1999) should help in gaining a detailed understanding of the mechanics of frameshifting.

The most detailed studies on –1 frameshifting have been on luteoviruses and these observations can be extended to other viruses. The overlap between the two ORFs ranges from 13 nucleotides in BYDV-PAV to several hundred nucleotides in most other luteoviruses. In BYDV-PAV and BWYV there is a stop codon close to the slippery sequence. Mutation analyzes of the slippery sequence and the structured region result in frameshift efficiencies in the range of 1–30% in *in vitro* translation systems. Slippery sequences containing A or U residues produced higher rates of frameshifting than did those containing G or C residues (reviewed by Fütterer and Hohn, 1996). A stop codon immediately downstream of the frameshift site increases frameshifting probably because it contributes to ribosome pausing.

The kinetics of ribosomal pausing during –1 framshifting have been studied for the *Saccharomyces cerevisiae* dsRNA virus that has a slippery sequence and a downstream pseudoknot (Lopinski *et al.*, 2000). About 10% of the ribosomes pause at the slippery site *in vitro* and some 60% of these continue in the –1 frame. Those that moved to the –1 frame paused for about 10 times longer than it takes to complete a peptide bond *in vitro*. Thus, there are three ways in which ribosomes pass a frameshift site: (1) without pausing leading to no frameshift, (2) pausing but no frameshift, and (3) pausing and frameshifting. As the features of the frameshift site in the *S. cerevisiae* viruses are the same as those in viruses of higher plants, presumably these observations also apply to the latter.

Translation of full-length infectious BYDV-PAV transcripts in an *in vitro* system gave higher rates of frameshift than did shorter RNAs (Di *et al.*, 1993); furthermore, the adjacent stop codon was found not to be necessary for frameshifting (Miller *et al.*, 1997a). Full-length transcripts of RCNMV behave in a similar manner (Kim and Lommel, 1994). For BYDV, sequences located in the 3′ untranslated region of the genome, 4 kb downstream of the frameshift site, were found to be essential (Wang and Miller, 1995). Thus, there is evidence for long-distance interactions but the nature of these has not yet been elucidated.

b. +1 frameshifting

Frameshifting in the +1 direction requires a run of slippery bases and a rare or 'hungry' codon or termination codon. A downstream structures region is not necessary but may be found, as has been suggested for BYV (Agranovsky *et al.*, 1994). The following mechanism has been suggested for the +1 frameshift between LIYV ORF 1a and ORF 1b (Klaassen *et al.*, 1995):

```
              K (protein 1a)                     K (protein 1b)
LIYV RNA   5′...AAAG...3′        slippage      5′...AAAG...3′
               | | |               →               | | –
tRNA^lys   3′...UUU....5′                       3′....UUU...5′
```

c. Proportion frameshifted

As with the read-through strategy, most of the studies on the proportion of translation events that result in a read-through protein involve *in vitro* systems. These can give frameshift rates as high as 30%; *in vivo*, rates of 1–5% are more likely (Brault and Miller, 1992; Prüfer *at al.*, 1992).

11. Strategy 11: Splicing

The production of mRNAs from DNA in eukaryotes involves splicing which removes internal non-coding sequences and can give various versions of an mRNA. Two of the families of plant viruses with DNA genomes, the *Caulimoviridae* and *Geminiviridae*, use splicing in the production of mRNAs, a process which, at least in the caulimoviruses, opens up downstream ORFs.

As described above, CaMV produces two major transcripts, the more-than-genome-length 35S RNA and the monocistronic 19S RNA. Several other RNA species and deleted DNA species (suggested to have arisen by reverse transcription of RNA) have been found in CaMV-infected plants and have been interpreted as arising from splicing (Hirochika *et al.*, 1985; Vaden and Melcher, 1990; Scholthof *et al.*, 1991). In an analysis of the potential for splicing in the CaMV genome, Kiss-László *et al.* (1995) identified splice donor sites in the 35S RNA leader sequence and in ORF I and a splice acceptor site in ORF II. Splicing between the leader sequence and ORF II makes ORF III the 5′ ORF, thus opening it up to conventional translation. Mutants of the splice acceptor site in ORF II are not infectious, indicating that it is essential for the virus infection cycle. It is likely that these and possibly other splicing events are involved in the controlled expression of this group of viruses.

ORF 4 of RTBV is expressed from an mRNA spliced from the 35S RNA (Fütterer *et al.*, 1994). The splice removes an intron of about 6.3 kb and brings ORF 1 into frame with ORF 4. The splice donor and acceptor sequences correspond to the plant splice consensus sequences (Fig. 7.22).

The circular ssDNA genomes of mastreviruses have four ORFs, two being expressed from transcripts, V1 and V2, in the virion sense and two, C1 and C2, from transcripts in the complementary sense (see Figs 6.5 and 7.14 for genome organization). The C transcripts are of low abundance and a splicing event fusing C1 to C2 has been found for MSV, DSV and TYDV (Mullineaux *et al.*, 1990; Dekker *et al.*, 1991; Morris *et al.*, 1992). Splicing has not yet been found in the other two geminivirus

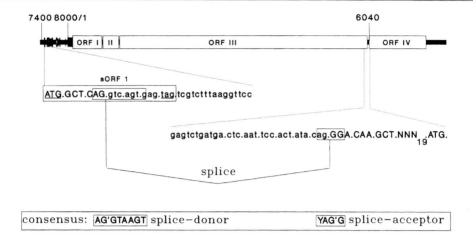

Fig. 7.22 Schematic presentation of splice junction sequences. The sequences around the supposed splice sites are shown and localized schematically on the linear RTBV map. At the 5'-end, the sequence starts with the ATG codon of the first short ORF in the leader (indicated as box; start and stop codons are underlined). At the 3'-end, the sequence ends with the ATG codon of the RTBV ORF IV. The sORF and ORF IV codons are indicated. Sequences found in the cDNA are in uppercase. On the spliced RNA, the ORF initiated at the sORF AUG is in phase with ORF IV. The consensus for splice donor and splice acceptor sites is also shown (' marks the point of splicing). From Fütterer *et al.* (1994), with permission.

genera and, apart from TYDV, appears to be restricted to geminiviruses that infect monocots.

12. Strategy 12: Translation for both viral and complementary strands (ambisense)

Some of the genome segments of the tospoviruses and tenuiviruses encode two proteins having one ORF in the virion sense and the other in the complementary sense. Thus, one of the proteins is expressed from complementary-sense RNA. This is termed the *ambisense* expression strategy and is another means by which viruses overcome the eukaryotic translation constraints (see Sections IV.A.2 and 3).

C. Control of translation

Various mechanisms for the control of expression of viral genomes have been described in the section above. These relate primarily to mechanisms that viruses have developed to overcome the problem of the limitation of translation of mRNAs in eukaryotic systems to the 5' ORF. Among other features that control or regulate the translation of eukaryotic mRNAs are the various non-coding regions

(untranslated regions or UTRs) which include the termini of the RNAs, the 5' terminus being capped (see Section V.A) and the 3' terminus having a poly(A) tail. Also involved in the control and efficiency of translation are the 5' leader sequence and the 3' non-coding region. In eukaryotic mRNAs, there is co-ordinated interaction between the 5' and 3' UTRs of these mRNAs (Fig. 7.23) (reviewed by Gallie, 1996, 1998) and even evidence for circularized mRNAs (Wells *et al.*, 1998). Only some plant viral RNAs are capped and have poly(A) tails (see Table 7.7). The majority either have a cap or a poly(A) tail or have neither, yet these may be translated very efficiently.

1. Cap but no poly(A) tail

The genome of TMV is capped but lacks a poly(A) tail. The structure of the 3' UTR is complex, being composed of five pseudoknots covering a 177 base region (Fig. 7.24B).

The 3'-terminal two pseudoknots form the tRNA-like structure that is involved in virus replication (see Chapter 8, Section IV.H). The remaining three pseudoknots make up the upstream pseudoknot domain that is conserved in the tobamoviruses and also found in the hordeiviruses. This domain appears to

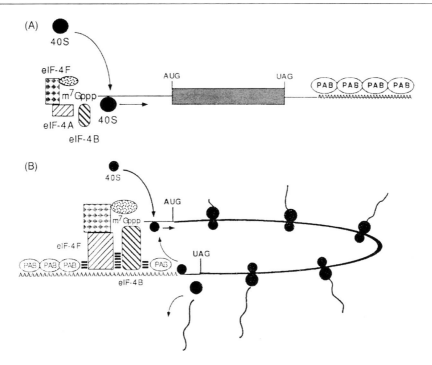

Fig. 7.23 Regulation of mRNA expression. **(A)** Schematic of a typical mRNA and the proteins that bind to the terminal regulatory elements. The initiation factors, eIF-4F, eIF-4A and eIF4B are shown associated with the 5′-terminal cap structure and the poly(A)-binding (PAB) protein is shown bound to the poly(A) tail at the 3′ terminus. After initiator factor binding, the 40S ribosomal subunit at or close to the 5′ terminus scans down the 5′ leader in search for the AUG initiation codon. **(B)** The co-dependent model of translation, the eIF-4F and eIF-4B are shown bound to both the 5′-terminal cap structure and the poly(A) tail. Protein–protein contacts between eIF-4F/eIF-4B, eIF-4A/PAB and eIF-4B/PAB that stabilize protein/mRNA binding are shown as the multiple thick lines between the proteins. This stable complex maintains the close physical proximity of the termini of the mRNA, which allows the efficient recycling of ribosomes. The 60S subunit is shown dissociating from the mRNA upon translation termination whereas the 40S subunit recycles back to the 5′ terminus to anticipate another round of initiation. From Gallie (1996), with kind permission of the copyright holder, © Kluwer Academic Publishers.

functionally substitute for a poly(A) region in promoting interactions between the 5′ and 3′ termini and enhancing translation initiation in a cap-dependent manner (Leathers *et al.*, 1993). A 102-kDa protein binds to the pseudoknot domain (Fig. 7.24) and also to the 5′ UTR (Tanguay and Gallie, 1996). It is likely that the 102-kDa protein mentioned above is involved in bringing the 5′- and 3′-ends together in the manner shown for eukaryotic mRNAs in Fig. 7.23 above.

2. Poly(A) tail but no cap

Potyviruses are polyadenylated at their 3′ termini but the 5′ terminus is attached to a VPg. The 5′ UTR of TEV confers cap-independent enhancement of translation of reporter genes by interactions between the leader and the poly(A) tail (Carrington and Freed, 1990; Gallie *et al.*, 1995). In a deletion analysis of the 143-nucleotide leader sequence of TEV, Niepel and Gallie (1999) identified two centrally located cap-independent regulatory elements that promote cap-independent translation. Placing the leader sequence into the intercistronic region of a bicistronic construct increased the expression of the second cistron, the enhancement being markedly increased by introducing a stable stem-loop structure upstream of the TEV leader sequence. It was concluded that the two elements in the TEV leader together promote internal initiation and that the function of one element is facilitated by the proximity to the 5′ terminus.

The VPg of the potyvirus TuMV interacts with the eukaryotic translational initiation

A

m⁷GpppGUAUUUUUACAACAAUUACCAACAACAACAAACAACAAACAACAUUACAAUUACUAUUUACAAUUACAUG

B

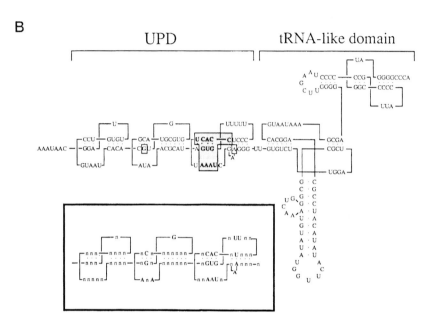

Fig. 7.24 The sequence and structure of the TMV 5′ leader and 3′ UTR showing the primary sequence required for regulating translation. **(A)** The 5′ leader sequence with the poly(CAA) region responsible for the enhancing activity of Ω underlined. Three direct repeats are indicated by arrows. **(B)** The predicted structure of the 3′ UTR. The primary sequence essential for the regulation by the upstream pseudoknot domain (UPD) is shaded and the tRNA-like domain is indicated. The sequence that is absolutely conserved in eight tobamo- and hordei-viruses is shown in the boxed UPD. From Tanguay and Gallie (1996), with permission.

factor eIF(iso)4E of *Arabidopsis thaliana* and wheat (*Triticum aestivim*) (Wittmann *et al.*, 1997; Léonard *et al.*, 2000). eIF(iso)4E binds to cap structures of mRNAs and plays an important role in regulating the initiation of translation (see McEndrick *et al.*, 1999). The interaction domain of TuMV VPGg was mapped to a stretch of 35 amino acids and the substitution of an aspartate residue in this region completely abolished the interaction (Léonard *et al.*, 2000). The cap analog m⁷GTP, but not GTP, inhibited the VPg–eIF(iso)4E complex formation, suggesting that the VPg and cellular mRNAs compete for eIF(iso)4E binding. The NIa protein of TEV that contains the VPg also binds to eIF4E but in a strain-specific manner (Schaad *et al.*, 2000) with that from the HAT but not from the Oxnard strain interacting with eIF4E from tomato and tobacco.

3. Neither cap nor poly(A) tail

Many plant viruses have neither a 5′ cap nor a 3′ poly(A) tail (see Table 7.7) but these mRNAs are expressed very effectively and thus can be considered to have translation enhancement. There is currently information on three viruses which show some variation in details.

The 4-kb genome of TCV expresses the 5′-proximal two genes from the genomic RNA (by read through) and the other three genes from sgRNAs. Two genes are translated from the sgRNA1 by leaky scanning and the 3′ coat protein gene is expressed from a 1.45-kb sgRNA. Qu and Morris (2000) examined the 5′ UTRs of the genomic and 1.45-kb sgRNAs and their common 3′-ends for roles in translation regulation. Using the firefly luciferase reporter gene, they obtained optimal translation activity when the mRNA contained both 5′ and 3′ UTRs, the synergistic effect being at least 4-fold greater than the sum of the contributions of the individual UTRs. The translational enhancement was cap-independent and was greater for the sgRNA UTRs than for the genomic RNA UTRs.

TBSV has a genome of 4.8 kb with five ORFs of known function and a 3′ sixth small ORF of unknown function (see Fig. 6.23 for genome organization). A 167-nucleotide region (segment 3.5) near the 3′-end was implicated as a determinant of translational efficiency (Oster *et al.*, 1998) and was shown to be involved in cap-independent translation (Wu and White, 1999). Segment 3.5 is part of a larger 3′ cap-independent translational enhancer (termed 3′CITE) and none of the major viral proteins was involved in 3′CITE activity. Unlike the translational enhancers of some other plant small RNA viruses (BYDV and STNV), there was no TBSV 3′CITE-dependent stimulation of translation in an *in vitro* wheatgerm extract system, which suggests that there might be differences in the enhancement mechanisms (Wu and White, 1999).

The luteovirus BYDV has a genome of 5.6 kb expressing six ORFs from the genomic and three sgRNAs (see Fig. 6.35 for genome organization). Highly efficient cap-independent translation initiation is facilitated by a 3′ translation enhancer sequence (3′TE) (Wang and Miller, 1995; Wang *et al.*, 1997c; Allen *et al.*, 1999). A 109-nucleotide 3′ sequence is sufficient for translational enhancement *in vitro*, but a larger 3′ region is required for optimal enhancement *in vivo* (Wang *et al.*, 1999c). The 5′ extremity of this larger region coincides with the 5′ terminus of sgRNA2. Competition studies showed that the 3′TE did not enhance the translation of the genomic RNA as much as that of sgRNA1, this difference being attributable to the different 5′ UTRs of these two RNAs. Thus, the 3′TE stimulates translation in *cis* but selectively inhibits translation in *trans*. As the 5′ genes on the genomic RNA, translation of which is stimulated in *cis* and inhibited in *trans*, encode an early function, the viral replicase, it is suggested that sgRNA2 is a novel regulatory RNA that switches from early to late gene expression (Fig. 7.25) (Wang *et al.*, 1999c) (see also Fig. 7.31).

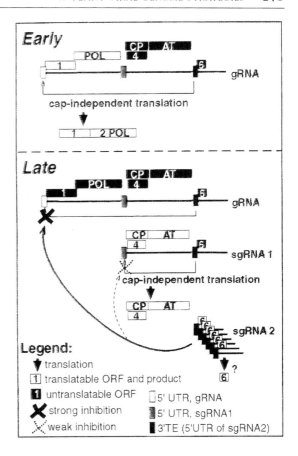

Fig. 7.25 (see Plate 7.2) Translational switch model for *trans*-regulation of BYDV gene expression by sgRNA2. Open boxes indicate translational ORFs and their translation products (below large arrows). Black boxes indicate ORFs that are not translated. **Early:** polymerase is translated from the genomic (g)RNA (the only viral RNA at this stage) via the 3′TE (red box) *in cis*. **Late:** As abundant sgRNA2 accumulates, it specifically inhibits gRNA (bold cross) in preference to sgRNA1 (dashed cross), via the 3′ TE *in trans*. This allows almost exclusive translation of late genes from sgRNA1. The different 5′ UTRs of gRNA (gold box) and sgRNA1 (green box) contribute to differential inhibition. The role of ORF 6 encoded by sgRNA2 is unknown (?) but it is not necessary for *trans*-inhibition. From Wang *et al.* (1999c, where the model is discussed in detail), with permission.

4. Cap-snatching

Negative-strand RNA viruses with segmented genomes use a mechanism, termed 'cap-snatching', to initiate transcription of their mRNAs. In this process, cap structures comprising between twelve and twenty 5′ nucleotides are cleaved from host mRNAs by a virus-encoded endonuclease and are then used to prime transcription. Cap-snatching has been demonstrated for the tospovirus TSWV and tenuivirus MSpV (Kormelink *et al.*, 1992a; van Poelwijk *et al.*, 1996; Estabrook *et al.*, 1998). Both viruses can snatch caps from positive-sense RNA viruses (Estabrook *et al.*, 1998; Duijsings *et al.*, 1999).

5. 5′ UTR

As well as being involved in the enhancement of translation initiation, the 5′ UTR of TMV also enhances the efficiency of translation. The 67-nucleotide 5′ UTR, termed the Ω sequence, dramatically enhances translation of downstream genes in both plant and animal cells (reviewed by Gallie, 1996); in constructs in transgenic plants it enhanced translation by 4- to 6-fold (Dowson Day *et al.*, 1993). The Ω sequence has reduced secondary structure (Sleat *et al.*, 1988a) and a 25 base poly(CAA) region (Fig. 7.24A) which mutagenesis indicated is the primary element for *in vivo* translational enhancement (Gallie and Walbot, 1992).

The 36-base 5′ leader sequence from AMV RNA4 also enhances translation (Jobling and Gehrke, 1987) as does the 84-nucleotide leader sequence of PVX genomic RNAS (Pooggin and Skryabin, 1992).

D. Discussion

The above descriptions show the great diversity of mechanisms that viruses use to express the information required for their function from what are often compact genomes. This diversity is summarized in Table 7.7.

The diversity of mechanisms overcomes constraints imposed by viruses using their host translational machinery. There are two other mechanisms of creating further diversity in gene products, transcriptional editing which is found in paramyxoviruses (Cattaneo, 1991; Niswender, 1998) and protein trans-splicing (reviewed by Perler, 1999), that have not yet been found in plant viruses.

These mechanisms can be viewed in two ways, specifically overcoming the constraints of the eukaryotic system, say in the requirement for a cap for translation initiation and the translation of only the 5′ ORF, and the control

of translation so that the right product is in the right place in the right quantity at the right time. The two uses of the mechanisms cannot be separated. For instance, in many cases the frameshift and read-through mechanisms provide different functions of the replication complex, the upstream one from the shorter protein containing the helicase and capping activities and the downstream one the replicase.

The regulation is not only by the recognized ORFs but can be by non-coding sequences and possibly by short ORFs that normally might not be considered. There are many examples of the coat protein gene being expressed more efficiently than that for the replicase; more coat protein is required than the replicase.

Many of the studies on virus expression systems involve the use of *in vitro* techniques, especially *in vitro* translation systems. There can be dangers in extrapolating data from these systems to the *in vivo* situation. The differences between the rabbit reticulocyte and wheatgerm translation system in the processing of the potyviral polyprotein (see Section V.B.1.b) shows that cellular factors and possibly the cellular environment is likely to be involved. Furthermore, there is no real evidence that the RNAs of viruses considered to lack a cap are not capped when in *in vivo* translation complexes.

Many of the above points occur again in the discussion below of viruses with multiple strategies.

E. Positive-sense ssRNA viruses that have more than one strategy

Most viruses use more than one of the strategies outlined above to express their genetic information. This section describes examples of these multiple strategies.

1. Two strategies: Subgenomic RNAs + read-through protein

a. Tobamovirus *genus* (reviewed by Okada, 1999)

The genomes of many species of the *Tobamovirus* genus have been sequenced (van Regenmortel

TABLE 7.7 Expression strategies of (+)-strand ssRNA plant viruses

Family	Genus	Termini		Expression strategies					
		5'	3'	Genome segments	Sub-genomic	Read-through	Frame-shift	Proteolytic cleavage	Other
Bromoviridae	Bromovirus	Cap[a]	t	3	1				
	Alfamovirus	Cap	c	3	1				
	Cucumovirus	Cap	t	3	1				
	Ilarvirus	Cap	c	3	1				
	Oleavirus	Cap	c	3	1				
Comoviridae	Comovirus	VPg	An	2				+	2-start[b]
	Fabavirus	?VPg	An	2				+	
	Nepovirus	VPg	An	2				+	
Potyviridae	Potyvirus	VPg	An	1				+	
	Ipomovirus	?VPg	An	1				+	
	Macluravirus	?VPg	An	1				+	
	Rymovirus	?VPg	An	1				?+	
	Tritimovirus	?VPg	An	1				+	
	Bymovirus	?VPg	An	2				+	
Tombusviridae	Tombusvirus		OH	1	2	+			
	Avenavirus		OH	1	1	+			
	Aureusvirus		OH	1	2	+			
	Carmovirus		OH	1	2	+			
	Dianthovirus	Cap	OH	2	1		-1		
	Machlomovirus	Cap	OH	1	+	+2			2-start
	Necrovirus		OH	1	2	+			
	Panicovirus		OH	1	1	+	-1		
Sequiviridae	Sequivirus	?VPg	An	1				+	
	Waikavirus	?VPg	An	1				+	
Closteroviridae	Closterovirus	?Cap	OH	1	7–10		+1	+	
	Crinivirus	?Cap	OH	2	+		+1		
Luteoviridae	Luteovirus		OH	1	+	+	+		
	Polerovirus	VPg	OH	1	+	+	+		
	Enamovirus	VPg	OH	1	+	+	+		
No family	Tobamovirus	Cap	t	1	2	+			
	Tobravirus	Cap	OH	2	+	+			
	Potexvirus	Cap		1	2				
	Carlavirus	?Cap	An	1	2				
	Allexivirus	?Cap	An	1	+				
	Capillovirus	?Cap	An	1	2				
	Foveavirus	Cap	An	1	+				
	Trichovirus	Cap	An	1	2				
	Vitivirus	Cap							
	Furovirus	Cap	t	2	+	+			
	Pecluvirus	?Cap		2	+	+			
	Pomovirus	Cap	t	3	+	+			
	Benyvirus	Cap	An	4	+	+		+	2-start
	Hordeivirus	Cap	t	3	+				
	Sobemovirus	VPg		1	+		+CfMV	+	
	Marafivirus	Cap		1	+			+	
	Tymovirus	Cap	t	1	1			+	2-start
	Ourmiavirus			3					
	Umbravirus		OH	1	+		+		

[a] An = poly(A); c = conserved 3' sequence between genome segments; Cap = cap sequence; OH = hydroxyl 3' terminus; t = tRNA-like sequence; VPg = genomic viral protein.
[b] Two translational starts on same ORF.

et al., 2000) and the organization of TMV is described in Chapter 6 (Section VIII.H.1). There are several closely packed ORFs, the expression of which is shown in Fig. 7.26.

An M7Gppp cap is attached to the first nucleotide (guanylic acid). The untranslated leader sequence of 69 nucleotides, the Ω sequence, potentiates efficient translation as

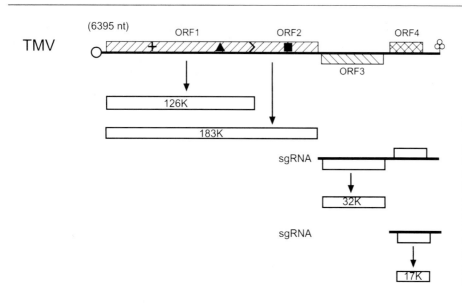

Fig. 7.26 Expression of TMV genome. The genome map is as in Fig. 6.36. The genomic RNA is the template for the 126-kDa and 183-kDa replicase proteins. ORFs III and IV are expressed from two separate 3' co-terminal sgRNAs giving the 32-kDa movement protein and the 17-kDa coat protein respectively.

described in Section V.C.5. The first ORF, coding for a 126-kDa protein, has a leaky termination codon (UAG) leading to a larger read-through protein (183-kDa). The two smaller ORFs at the 3'-end of the genome are translated from subgenomic RNAs, the 30-kDa from sgRNA I_2 and the 17.6-kDa coat protein from the smallest sgRNA. The 3' untranslated sequences can fold in the terminal region to give a tRNA-like structure that accepts histidine.

A third subgenomic RNA, called I_1 RNA, representing approximately the 3' half of the genome, has been isolated from TMV-infected tissue. SI mapping showed that this RNA species had a distinct 5' terminus at residue 3405 in the genome (Sulzinski *et al.*, 1985). These workers proposed a model for the translation of I_1 RNA. There is an untranslated region of 90 bases followed by an AUG codon initiating a 54-kDa protein terminating at residue 4915. Thus, the amino acid sequence of the 54-kDa protein is the same as the residues at the carboxy terminus of the 183-kDa protein. However, there is no evidence that this sgRNA is involved in viral expression.

The expression of TMV has been studied in great detail and the description below shows how the understanding of the processes involved was developed.

i. Proteins synthesized in vitro
TMV genomic RNA has been translated in several cell-free systems. Two large polypeptides were produced in reticulocyte lysates (Knowland *et al.*, 1975) and in the wheatgerm system (e.g. Bruening *et al.*, 1976), but no coat protein was found by these or later workers. The TMV genome is not large enough to code independently for the two large proteins that were produced. Using a reticulocyte lysate system, Pelham (1978) showed that the synthesis of these two proteins is initiated at the same site, the larger protein being generated by partial read through of a termination codon. The two proteins are read in the same phase, so the amino acid sequence of the smaller protein is also contained within the larger one. Increased production of the larger protein occurred *in vitro* at lower temperatures (Kurkinen, 1981). The extent of production of the larger protein may also depend on the kind of tRNA[tyr] present in the extract (Beier *et al.*, 1984a,b) (see Section V.B.9.b).

The 30-kDa and 17.6-kDa proteins are not translated from genomic RNA but from two subgenomic RNAs. The I_2 RNA isolated from infected tissue has been studied in *in vitro* systems by many workers. It is translated to produce a 30-kDa protein. It is uncapped (Joshi *et al.*, 1983) and appears to terminate in 5' di- and triphosphates (Hunter *et al.*, 1983). The initiation site for transcription has been mapped at residue 1558 from the 3' terminus (Watanabe *et al.*, 1984a). The I_2 RNA also contains the smaller 3' gene (Fig. 7.26) but this is not translated in *in vitro* systems.

The smallest TMV gene (the 3' coat protein gene) is translated *in vitro* only from the monocistronic subgenomic RNA (Fig. 7.26) (Knowland *et al.*, 1975; Beachy and Zaitlin, 1977). This gene can also be translated efficiently *in vitro* by prokaryotic protein-synthesizing machinery from *E. coli* (Glover and Wilson, 1982).

ii. Proteins synthesized in vivo

The data of Siegel *et al.* (1978) indicated that viral protein synthesis does not suppress total host-cell protein synthesis but occurs in addition to normal synthesis. Two days after infection, viral coat protein synthesis accounted for about 7% of total protein synthesis. Synthesis of the 126-kDa protein was about 1.4% and that of the 183-kDa protein about 0.3% as much as coat protein.

Several workers have reported the association of both full-length TMV RNA and the coat protein mRNA with cytoplasmic polyribosomes in infected tobacco leaves (e.g. Beachy and Zaitlin, 1975). Beachy and Zaitlin also found TMV dsRNAs to be associated with the membrane-bound polyribosome fraction. Confirmation that TMV proteins are synthesized on 80S ribosomes comes from the fact that cycloheximide completely inhibits TMV replication in protoplasts, whereas chloramphenicol does not (Sakai and Takebe, 1970).

Dorokhov *et al.* (1983, 1984) isolated a ribonucleoprotein fraction from infected tobacco tissue in CsCl density gradients, which had a higher buoyant density than TMV. This material can be released from polyribosomes by EDTA treatment. Genomic, I_1, I_2 and coat protein RNAs and polypeptides of various sizes were identified as components of the ribonucleoprotein complex.

From studies on viral protein synthesis in tobacco leaves using wild-type virus, a *ts* mutant held at different temperatures and a

protein synthesis inhibitor, Dawson (1983) concluded that the synthesis of 183-kDa, 126-kDa and 17.5-kDa proteins was correlated with dsRNA synthesis rather than that of ssRNAs. It is possible that nascent ssRNAs from replication complexes function as mRNAs. Dawson's results suggest that TMV mRNA is relatively transitory *in vivo*. However, Dawson and Boyd (1987b) showed that synthesis of TMV proteins in tobacco leaves was not translationally regulated under conditions of heat shock, as were most host proteins. Thus, under appropriate conditions most of the protein being synthesized was virus-coded.

As might be expected, replicative intermediates (RI) and replicative forms (RF) (see Chapter 8, Section IV.A) appear early in infections as does the 126-kDa protein. Coat protein mRNA and genomic RNA are early products. At a later stage, virus production follows closely that of coat protein synthesis. Thus, it appears that the amount of coat protein available may limit the rate at which progeny virus is produced. Watanabe *et al.* (1984b) first detected the 183-kDa, 126-kDa, 30-kDa and coat proteins 2–4 hours after infection, and before infectious virus could be found. The 183-kDa, 126-kDa and coat protein were synthesized over a period of many hours but synthesis of the 30-kDa protein and its mRNA was detected between 2 and 9 hours after inoculation.

Proteins corresponding approximately in size to the 183-kDa and 126-kDa proteins were found in infected tobacco leaves (e.g. Scalla *et al.*, 1976) and in infected protoplasts (e.g. Sakai and Takebe, 1974). Cyanogen bromide peptide analysis on the 110-kDa protein from infected leaves showed it to be the same as the *in vitro* translation product of similar size (Scalla *et al.*, 1978).

The 30-kDa protein was detected in both infected tobacco protoplasts (Beier *et al.*, 1980) and leaves (Joshi *et al.*, 1983). Kiberstis *et al.* (1983) and Ooshika *et al.* (1984) raised antibodies against a synthetic peptide with the predicted sequence for the 11 or 16 C-terminal amino acids of the 30-kDa protein. The 30-kDa protein from TMV-infected protoplasts was precipitated by these antibodies, positively identifying it as the I_2 gene product. Kiberstis *et al.* (1983) found that the 30-kDa protein was synthesized only between 8 and 16 hours after inoculation of protoplasts. However, in intact leaves, production of the 30-kDa protein continued for some days, but was maximal at around 24 hours (Lehto *et al.*, 1990b).

Many workers have demonstrated the synthesis of TMV coat protein *in vivo*. Determination of the nucleotide sequence in the 3' region of the TMV genome readily located the gene for this protein since the full amino acid sequence was already known (Anderer *et al.*, 1960; Tsugita *et al.*, 1960).

iii. Controlling elements in the viral genome
Six controlling elements have been recognized or inferred in TMV RNA:

1. The nucleotide sequence involved in initiating assembly of virus rods, as discussed in Chapter 5 (Section IV.A). This sequence may have other controlling effects. Studies using pH 8.0 treated virus in an *in vitro* protein-synthesizing system suggest that the strong coat protein–RNA interactions occurring at the origin-of-assembly nucleotide sequence may be a site where translocation of 80S ribosomes is inhibited during uncoating of the virus rod. However, the effects of such a control mechanism would differ with different strains of the virus (Wilson and Watkins, 1985). Furthermore, encapsidation most probably removes the viral genome from the translation system.
2. The replicase recognition site in the 3' noncoding region (see Chapter 8, Section IV.H.3).
3. The interactions between the 5' cap and 3' sequences described in Section V.C.
4. The Ω sequence described above which enhances translation.
5. The start codon sequence context may be one form of translational regulation. The context differs for each of the four known gene products. For example, in strain U_1 the contexts are as follows:

126K: ACA<u>AUG</u>G   54K: GAU<u>AUG</u>C
30K: UAG<u>AUG</u>G   coat protein: AAU<u>AUG</u>U

The context for the 30-kDa protein is least optimal according to Kozak's model (Kozak, 1981, 1986). This might be considered to be the reason why so little 30-kDa protein is produced compared with coat protein, but changing the start codon context for the 30-kDa AUG to the optimal strong context as defined by Kozak (1986) did not increase expression of the gene in tobacco plants (Lehto and Dawson, 1990). Furthermore, insertion of sequences containing the coat protein subgenomic RNA promoter and leader upstream from the 30-kDa ORF did not lead to increased production of the 30-kDa product (Lehto *et al.*, 1990a). In fact, the production of 30-kDa protein was delayed and virus movement impaired, suggesting that different sequences influence timing of the expression of different genes.

6. RNA promoters presumably have a role in regulating the amounts of subgenomic RNAs produced, but the exact sequences for these subgenomic promoters have not been identified. That for the coat protein lies within 100 nucleotides upstream of the ORF. The regulation of *Tobamovirus* gene expression is discussed by Dawson and Korhonen-Lehto (1990).

2. Two strategies: Subgenomic RNA + polyprotein
a. **Sobemovirus genus** (Tamm and Truve, 2000)

The monopartite sobemovirus genomes have four ORFs (see Fig. 6.51 for genome organization). The 3' ORF, that for coat protein, is expressed from an sgRNA (Rutgers *et al.*, 1980). *In vitro* translation of SBMV RNAs give four major polypeptides, P1 (100–105 kDa), P2 (60–75 kDa), P3 (28–29 kDa) and P4 (14–25 kDa) (Salerno-Rife *et al.*, 1980; Ghosh *et al.*, 1981; Brisco *et al.*, 1985). It would appear that P4 is from the 5' ORF but the attribution of the other translation products is not clear. ORF 2 of the cowpea strain of SBMV (now named SCPMV) is translated from scanning ribosomes passing ORF 1, the AUG of which is in a poor context (Sivakumaran and Hacker, 1998). The ORF 2 product is a polyprotein that is cleaved by a virus-encoded protease (Gorbalenya et al., 1988). There is also evidence for frameshift between ORFs 2a and 2b of CfMV (Mäkinen et al., 1995a; Tamm and Truve, 2000), which would suggest that this virus has three strategies.

3. Two strategies: Subgenomic RNA + multipartite genome

a. Bromovirus genus

BMV has a tripartite genome totalling 8243 nucleotides, the organization of which is described in Chapter 6 (Section VIII.A.1). RNAs 1 and 2 are monocistronic and RNA3 is bicistronic, the 3' ORF, that for the coat protein being translated from a subgenomic RNA (see Fig. 6.13). This subgenomic RNA containing the coat protein gene is found in virus particles.

BMV RNAs are efficient mRNAs in in vitro systems and, in particular, in the wheatgerm system that is derived from a host plant of the virus. In this system, RNA1 directs the synthesis of a single polypeptide of 110 kDa and RNA2 directs a single 105-kDa polypeptide (Shih and Kaesberg, 1976; Davies, 1979). RNA3 directs the synthesis of a 35-kDa protein (Shih and Kaesberg, 1973). The coat protein cistron in RNA3 is not translated. RNA4 directs synthesis of the 20-kDa coat protein very efficiently. It is preferentially translated in the presence of the other viral RNAs in part, at least, because of more efficient binding of ribosomes (Pyne and Hall, 1979). In vitro the coat protein inhibits RNA synthesis by the BMV replicase, in a specific manner, possibly by partial assembly of nucleoprotein (Horikoshi et al., 1987).

Four new proteins were observed in tobacco protoplasts after infection with BMV. These were 20 kDa (coat protein), 35 kDa, 100 kDa and 107 kDa (Sakai et al., 1979). Four BMV-induced proteins with the same molecular weights (within the error of estimation in gels) were also found by Okuno and Furusawa (1979) in infected protoplasts prepared from three plant species—a systemic host, a local lesion host and a non-host. These four proteins account for over 90% of the viral genome. They correspond well in size with the in vitro products noted earlier.

b. Hordeivirus genus

The three genomic RNAs of BSMV are designated α, β and γ and their organization is described in Chapter 6 (Section VIII.H.14). RNAα is monocistronic expressing a 130-kDa protein. RNAβ has five ORFs, the 5' of which, βa or coat protein, is expressed from the genomic RNA. The second, βb, is expressed from sgRNAβ1 and the 3' triple gene box (see Chapter 9, Section II.D.2.f) from sgRNAβ2. Genomic RNAγ is bicistronic, the 5' ORF being expressed from the genomic RNA and the 3' ORF from sgRNAγ.

4. Two strategies: Multi-partite genome + polyprotein

Species in the bymovirus genus have their genomes divided into two RNA species, each of which encodes a polyprotein (see Fig. 6.22 for genome organization). Essentially, the genome organization is the same as that of potyviruses with the gene products of RNA2 representing the 5' products from the potyviral genome. It is considered that the polyproteins are processed is a manner similar to potyviruses (see Section V.B.1.b).

5. Three strategies: Subgenomic RNA + multipartite genome + read-through protein

a. Tobravirus genus

Tobraviruses have a bipartite (+)-sense ssRNA genome (see Fig. 6.37).

RNA1 is translated in the rabbit reticulocyte system to give two polypeptides of 170 kDa and 120 kDa (Pelham, 1979). These two polypeptides are also produced together with many smaller products in the wheatgerm system containing added spermidine (Fritsch et al., 1977). These two proteins correspond to the products of ORFs 1 and 2 shown in Fig. 6.37. Similar results have been reported for PEBV (Hughes et al., 1986). A protein product for the 29-kDa ORF is translated from a subgenomic RNA (1A) (Robinson *et al.*, 1983). The 16-kDa protein is also translated from a subgenomic RNA (1B), which is not required for replication or cell-to-cell transport in leaves (Guilford, 1989). A 16-kDa protein product, which was incorporated into a high-molecular-weight cellular component, was found in infected protoplasts (Angenent *et al.*, 1989a).

RNA2 of the PRN strain is translated *in vitro* to give the coat protein identified by serology, peptide mapping, and specific aggregation with authentic coat protein to form disk aggregates (Fritsch *et al.*, 1977). A second unrelated protein of 31 kDa is also translated. Different strains and viruses appear to differ in the products translated *in vitro* from RNA2 preparations, perhaps in part due to variable contamination with sgRNAs. No messenger activity has been detected for TRV strain SYM RNA2 in *in vitro* tests (Robinson *et al.*, 1983). However, a subgenomic mRNA derived from RNA2 was shown to be the mRNA for coat protein.

In summary, RNA1 has three ORFs, the 5′ of which resembles that of TMV. It has a UGA stop codon (TMV has a UAG stop codon) at the end of the 134-kDa product which is read through to give a 194-kDa protein. The 3′ ORFs are expressed from sgRNAs. As noted in Chapter 6 (Section VIII.H.2 and Fig. 6.38), RNA2 varies in length and has variable numbers of ORFs between isolates. The 5′ ORF, that encoding the viral coat protein, is the only consistent one and is expressed from an sgRNA; it is not known how the other ORFs are expressed. However, if coat protein is the only gene product it is not clear why a subgenomic RNA should be required. Certain types of disease symptoms are specified by the short rods even when these give rise to identical coat proteins (e.g. Robinson, 1977). Thus, there may be a second protein coded for by RNA2.

6. Three strategies: Subgenomic RNA + multi-partite genome + frameshift protein

a. Dianthovirus *genus*

The genomes of dianthoviruses are bipartite (+)-sense ssRNAs (Fig. 6.30). RNA1 has three

ORFs, the 5′ two of which overlap. The second ORF is expressed by a –1 frameshift from the 5′ ORF (27 kDa) to give a protein of 88 kDa. Both the 27- and 88-kDa proteins are found *in vivo* and are made by *in vitro* translation of RNA1 (reviewed by Hamilton and Tremaine, 1996). The 3′ ORF of RNA1 is expressed from a 1.5-kb sgRNA to give the viral coat protein. RNA2 is monocistronic for the viral movement protein.

7. Three strategies: Subgenomic RNA + polyprotein + frameshift protein

a. Closterovirus *genus*

Closteroviruses have large complex genomes containing up to 12 ORFs (see Fig. 6.33). The first two ORFs overlap and the second is expressed by a +1 frameshift from the first. The first ORF contains a papain-like protease that processes at two sites. The frameshift between the first two ORFs of CTV and the proteolytic processing give rise to nine polypeptides (Fig. 7.27) (reviewed by Karasev and Hilf, 1997). The other 10 ORFs are expressed from subgenomic RNAs.

A temporal analysis of the expression of BYV genes was undertaken by tagging the genes for HSP70h, the major coat protein and the 20-kDa protein (p20, ORF 7) with the β-glucuronidase (GUS) gene (Hagiwara *et al.*, 1999). This showed that the HSP70h promoter expressed early, followed by the coat protein promoter and later the p20 promoter. The kinetics of other sg promoters was followed by northern blot analysis. These two approaches showed temporal gene expression for BYV with HSP70h, CPm, CP and p21 being expressed early and p64 and p20 later (see Table 6.5 for designation of BYV genes).

8. Three strategies: Multi-partite genome + polyprotein + two-start

a. Comovirus *genus* (reviewed by Goldbach and Wellink, 1996)

Comoviruses have bipartite genomes, RNA1 also known as B RNA and RNA2 (M RNA), each of which expresses a polyprotein. CPMV, the type member of the *Comovirus* genus, has been most studied and so this discussion will

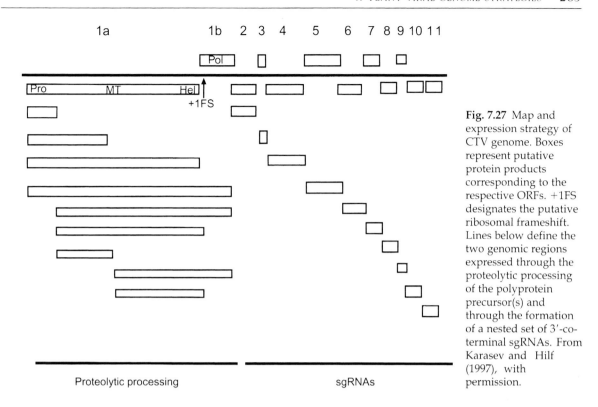

Fig. 7.27 Map and expression strategy of CTV genome. Boxes represent putative protein products corresponding to the respective ORFs. +1FS designates the putative ribosomal frameshift. Lines below define the two genomic regions expressed through the proteolytic processing of the polyprotein precursor(s) and through the formation of a nested set of 3'-co-terminal sgRNAs. From Karasev and Hilf (1997), with permission.

relate to this virus. The genome organization is described in Chapter 6 (Section VIII.B.1.a).

i. The processing products

The 200-kDa polyprotein of RNA1 is processed at four sites, which should give five products. RNA2 encodes a polyprotein of 105 kDa with two translational start sites and the two cleavage sites giving four products. However, the processing is complex and various intermediate products are also found that have been studied both *in vivo* (protoplasts) and *in vitro* (reticulocyte lysate translation system) on cloned expression vectors. In protoplasts inoculated with CPMV and labelled with 35S-methionine, virus-specific proteins of 170, 112, 110, 87, 84, 60, 58, 37, 32 and 23 kDa are readily detected (Rottier *et al.*, 1980; Rezelman *et al.*, 1980). Obviously, the combined molecular weights of these products greatly exceeds the coding capacity of the two RNAs. The 37- and 23-kDa-proteins encoded by RNA2 are the viral capsid proteins. By peptide mapping and immunological analyzes the processing pathway of the polyprotein from RNA1 was determined (Fig. 7.28) (reviewed by Goldbach and Wellink, 1996).

The recognition that there were two translation start sites in RNA2 came from *in vitro* translation and was verified on *in vivo* samples by the use of antipeptide antibodies. Furthermore, incubation of protoplasts in 2 mM $ZnCl_2$ inhibited the proteinase activity and led to the production of the 105- and 95-kDa polyproteins. From these and other data, the processing pathway of RNA2 products was determined (Fig. 7.28).

In RNA2 there are three AUGs beginning at nucleotides 161, 512 and 524. Using site-directed mutagenesis, Holness *et al.* (1989) confirmed that the AUG at position 161 is used *in vitro* to direct the synthesis of a 105-kDa product. Both the 105-kDa protein initiating at nucleotide 161 and a 95 kDa product initiating at nucleotide 512 have been detected in infected protoplasts, as have their cleavage products (Rezelman *et al.* 1989).

ii. The proteinase

The CPMV proteinase was identified as the 24-kDa protein encoded by RNA1 by two different procedures. Wellink *et al.* (1987a) synthesized a peptide corresponding to an amino

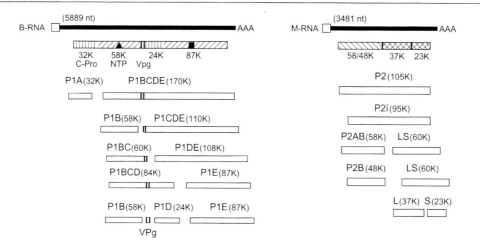

Fig. 7.28 Organization and expression of the genome of CPMV. See Chapter 6, Section VIII.B.1 for description of proteins. All intermediate and final cleavage products have been detected in infected cells. From Wellink *et al.* (2000), with permission.

acid sequence in the 200 kDa polyprotein that showed similarity to picornaviral 3C proteases. Antibodies to the peptide reacted with a 24-kDa protein found in CPMV-infected protoplasts and leaves. Verver *et al.* (1987) constructed a full-length cDNA copy of RNA1. RNA transcribed *in vitro* could be efficiently translated and proteolytically cleaved. Introduction of an 87-bp deletion into the coding region of the 24-kDa protein abolished cleavage activity, demonstrating this protein to be the viral protease. Vos *et al.* (1988a,b) extended this work to include a cDNA copy of RNA2. They constructed a series of deletion mutants in the 24-kDa reading frames of RNA1 and the results showed that the 24-kDa protein is the protease responsible for all cleavages in both B and M polyproteins. It cleaves the RNA1 polyprotein most in *cis* and that of RNA2 in *trans*. The proteinase has considerable amino acid sequence homology to the picornavirus 3C proteinase (Argos *et al.*, 1984; Franssen *et al.*, 1984c). However, unlike the cellular serine proteinases that the viral ones also show homology to, the viral proteinases have a cysteine instead of a serine at their active site (see Section V.B.1.a). The catalytic site of CPMV proteinase is formed by His[40], Glu[75] and Cys[166] (Dessens and Lomonossoff, 1991).

The 32-kDa protein is released rapidly from the 200-kDa polyprotein, occurring even as soon as ribosome finish translation of the 24-kDa coding region (Franssen *et al.*, 1984a). The processing of the remaining 170-kDa protein occurs slowly both *in vitro* and *in vivo*. However, on *in vitro* translation of mutants lacking the 32-kDa protein the processing of the 170-kDa protein is rapid (Peters *et al.*, 1992). Thus, the 32-kDa protein controls the processing of the 170-kDa product.

For efficient cleavage of the glutamine–methionine site in the RNA2 polyprotein, a second B-encoded protein (32 kDa) is essential, although this protein does not itself have proteolytic activity.

iii. The cleavage sites
N-terminal sequence analyzes have shown that cleavage of the CPMV polyproteins occurs at Gln/Met (two sites), Gln/Ser (two sites) and at Gln/Gly (two sites). Ala or Pro were present at position –2 and five of the six sites have Ala at position –4 (Wellink *et al.*, 1986). Mutational analyzes show that some of the sites have greater constraints on the surrounding residues than do others. For example, when the Gly at position +1 of the Gln/Gly cleavage site between the two coat proteins was changed to Ala, Ser or Met (amino acids present at this

position in other sites), cleavage was almost abolished (Vos *et al.*, 1988b).

9: Three strategies: Subgenomic RNA + polyprotein + two-start

a. Tymovirus *genus*

The genome organization of the tymovirus TYMV is described in Chapter 6 (Section VIII.H.17). The organization is very compact. For example, with the European TYMV only 192 (3%) of the 6318 nucleotides are noncoding. The 5' ORF runs from the first AUG codon on the RNA (beginning at nucleotide 88) and terminates with a UGA codon at position 1972 (69 kDa). The largest ORF initiates at position 95 and ends at position 5627 with a UAG codon (to give a protein of 206 kDa). Thus, it overlaps the 69-kDa gene over all its length. The 3' terminal gene encodes the coat protein. There is a 105-nucleotide non-coding 3' region, which contains a tRNA-like structure.

i. Subgenomic RNA

A small subgenomic RNA is packaged with the genomic RNA and in a series of partially filled particles (e.g. Pleij *et al.*, 1977). No dsRNA corresponding in length to this subgenomic RNA could be detected *in vivo*. Gargouri *et al.* (1989) detected nascent subgenomic coat protein (+)-strand RNAs on dsRNA of genomic length. Thus, the coat protein mRNA is synthesized *in vivo* by internal initiation on (−) strands of genomic length. Ding *et al.* (1990) compared the available *Tymovirus* nucleotide sequences around the initiation site for the subgenomic mRNA. They found two conserved regions, one at the initiation site and a 16-nucleotide sequence on the 5' side of it (Fig. 7.29). This longer sequence, which they called the *tymobox*, may be an important component of the promoter for subgenomic RNA synthesis. The elements that make up this subgenomic promoter have recently been studied in detail (Schirawski *et al.*, 2000).

The sg promoter sequence of TYMV has been located to a 494-nucleotide fragment that contains the tymobox (Schirawski *et al.*, 2000). Duplication of this fragment into the coat protein ORF led to the *in vivo* production of a second sgRNA, and mutagenesis of the tymobox

showed that it was an essential part of the promoter. The tymobox region can be folded into a hairpin formation, a feature found in other sg promoters (see Section V.B.2).

ii. In vitro *translation studies*

The small subgenomic RNA is very efficiently translated in *in vitro* systems to give coat protein (e.g. Pleij *et al.*, 1977; Higgins *et al.*, 1978). The coat protein gene in the genomic RNA is not translated *in vitro*.

Weiland and Dreher (1989) obtained infectious TYMV RNA transcripts from cloned full-length cDNA. By making mutants in the initiation codons they showed that the 69-kDa protein is expressed from the first AUG beginning at nucleotide 88, while the much larger gene product is expressed from an AUG beginning at nucleotide 95. Antibodies raised from a synthetic peptide corresponding to the C terminus of the shorter ORF are specific for the 69-kDa protein, demonstrating *in vitro* expression (C. Bozarth, J. Weiland and T. Dreher, personal communication to R. E. F. Matthews).

The large ORF with the AUG beginning at nucleotide 95 appears to be expressed in a variety of *in vitro* systems and is then cleaved *in vitro* to give a larger 5' and a smaller 3' product (Morch *et al.*, 1982, 1988, 1989; Zagorski *et al.*, 1983).

iii. *Viral proteins synthesized* in vivo

Coat protein. Biochemical studies indicate that coat protein is synthesized on 80S cytoplasmic ribosomes. Renaudin *et al.* (1975) showed that viral protein synthesis in Chinese cabbage protoplasts is inhibited by cycloheximide but not by chloramphenicol. The cytoplasmic site for viral protein synthesis has been confirmed by cytological evidence, but viral protein also accumulates in the nucleus.

The uracil analog, 2-thiouracil, blocks TYMV RNA synthesis but not protein synthesis (Francki and Matthews, 1962), implying that the mRNA for coat protein is a relatively stable molecule. The kinetics of labelling with [^{35}S]methionine of empty protein shells and viral nucleoprotein in infected protoplasts shows that these two protein shells are assembled from different pools of protein subunits (Sugimura and Matthews, 1981).

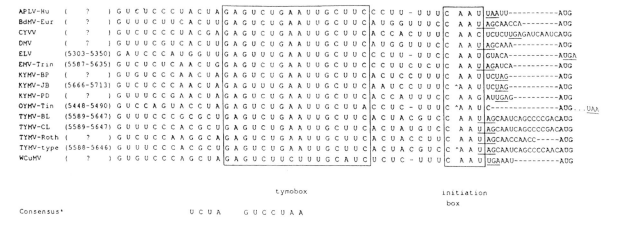

```
APLV-Hu    (    ?    ) G U C U C C C U A C U A G A G U C U G A A U U G C U U C C C U U - U U U C A A U UAAUU----------AUG
BdMV-Eur   (    ?    ) G U U U C U U C A C U U G A G U C U G A A U U G C U U C A U G G U U U C C A A U AGCAACCA--------AUG
CYVV       (    ?    ) G U C U C C C U A C G A G A G U C U G A A U U G C U U C A C C A C U U U C A A C UCUCUUGAGAUCAAUCAUG
DMV        (    ?    ) G U U U C G U C A C U U G A G U C U G A A U U G C U U C A U G G U U U C C A A U AGCAAA----------AUG
ELV        (5303-5350) G A U C C C A U G G U U G A G U U U G A A U U G C U U C C C U U - U U C C A A U GUACA----------AUGA
EMV-Trin   (5587-5635) G U C U C U C A A C U G G A G U C U G A A U U G C U U C C C U U C U C U C A A U AGAUCA----------AUG
KYMV-BP    (    ?    ) G U G U C C C A A C U A G A G U C U G A A U U G C U U C A C U C C U U U C A A U UCUAG----------AUG
KYMV-JB    (5666-5713) G U C U C C C A A C U A G A G U U U G A A U U G C U U C A A U C C U U U C ^A A U UCUAG----------AUG
KYMV-PD    (    ?    ) G U U U C C C A A C U A G A G U C U G A A U U G C U U C A C C A U U U C C A A U AUUGAG----------AUG
OYMV-Tin   (5448-5490) G U C C A G U A C C U A G A G U C U G A A U U G C U U A C C U C - U U U C ^A A U C------------AUG...UAA
TYMV-BL    (5589-5647) G U U U C C C G C G C U G A G U C U G A A U U G C U U C A C U A C G U C C A A U AGCAAUCAGCCCCGACAUG
TYMV-CL    (5589-5647) G U U U C C C A C G C U G A G U C U G A A U U G C U U C A C U A U G U C C A A U AGCAAUCAGCCCCGACAUG
TYMV-Roth  (    ?    ) G U C U C C A A G G C A G A G U C U G A A U U G C U U C A C U A C C U U C A A U AGCAACCACC-----AUG
TYMV-type  (5588-5646) G U U U C C C A C G C U G A G U C U G A A U U G C U U C A C U A C G U C C A A U AGCAAUCAGCCCCAACAUG
WCuMV      (    ?    ) G U G U C C C A G C U A G A G U C U U C U U U G C A U C U C U C - U U U C A A U UGAAAU---------AUG
```

```
                                        tymobox                              initiation
                                                                                box

Consensus*                    U C U A      G U C C U A A
```

Fig. 7.29 Aligned nucleotide sequences of tymoviral genomic RNAs in a region surrounding the initiation site (indicated by arrows in the initiation box) of subgenomic RNA transcription. The positions of these segments in their genomic RNA are shown in brackets where complete genomic sequences have been determined. The two conserved sequences are boxed. The stop codon of the replicase protein gene is underlined and the start codon of the coat protein is in bold. The consensus sequence is that of the subgenomic promoter of BMV and the possible subgenomic promoter of alphaviruses (Gargouri *et al.*, 1989). From Ding *et al.* (1990), with kind permission of the copyright holder, © Oxford University Press.

Other proteins. The 69-kDa protein has been detected *in vivo* in both Chinese cabbage and *Arabidopsis thaliana*. It is expressed at a 500× lower level than coat protein and appears to be an early nonstructural protein. The 70-kDa C-terminal fragment shown is also found *in vivo* (C. Bozarth, J. Weiland and T. Dreher, personal communication to R. E. F. Matthews).

Mouches *et al.* (1984) showed that the TYMV replicase consisted of viral-coded 115-kDa polypeptide and a host-coded protein of 45 kDa. Using immunoblotting, Candresse *et al.* (1987) showed that the virus-coded replicase subunit appears very soon after inoculation of plants or protoplasts.

iv. Polyamine synthesis

In protoplasts derived from infected leaves, or in healthy protoplasts infected *in vitro*, newly formed virus particles contained predominantly newly synthesized spermidine and spermine (Balint and Cohen, 1985a,b). When a specific inhibitor of spermidine was present, there was increased synthesis of spermine and an increase in the spermine content of virus particles. Thus, there is some flexibility in the way in which the positive charge contributed by the polyamines is conserved. The biosynthesis of polyamines and their possible roles in plants are discussed in Smith (1985).

10. Four strategies: Subgenomic RNA + read-through protein + frameshift protein + internal initiation

All three genera of the *Luteoviridae* have similar expression strategies but differ in detail (reviewed by Miller *et al.*, 1997a). The genomes of the *Luteovirus* and *Polerovirus* genera contains six ORFs and that of the *Enamovirus* genus five ORFs. Their expression strategies are shown in Fig. 7.30.

The 5' two ORFs overlap and the second ORF is expressed by a −1 frameshift from the 5' ORF (see Section V.B.10.a). The product from PLRV ORF 1 is processed by a serine proteinase that it contains (Li *et al.*, 2000). ORFs 3, 4 and 5 are expressed from sgRNA1. ORF 3, which encodes the coat protein, ends in a UAG stop codon that is read through to give read-through protein with ORF 5. ORF 4 is translated from sgRNA1 by internal initiation. ORF 6 is expressed from sgRNA2. A third sgRNA has been suggested for the putative ORF 7 of PLRV (Ashoub *et al.*, 1998). The VPg of poleroviruses is thought to be released from the ORF 1 product by a serine proteinase (see Sadowy *et al.*, 2001b)

The 5' terminus of BYDV sgRNA1 has been mapped to nucleotide 2670 on the genomic RNA (Koev *et al.*, 1999), a hotspot region of

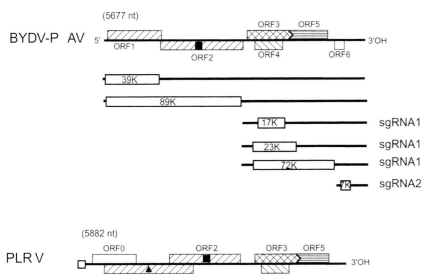

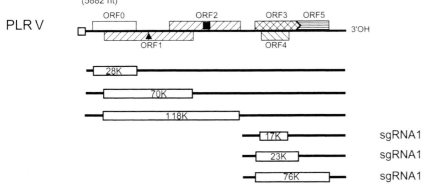

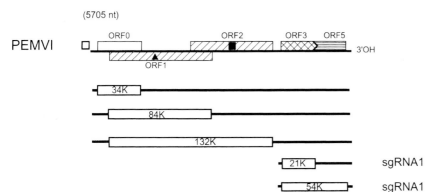

Fig. 7.30 Diagram of the genome (from Fig. 6.35) and map of the translation products typical of viruses in each genus of the family *Luteoviridae*. Solid lines represent RNA; boxes represent ORFs; thinner boxes represent translation boxes; circles represent VPgs. From D'Arcy *et al.* (2000), with permission.

recombination. The promoter for this sgRNA maps between nucleotides 2595 and 2692 and computer predictions reveal two stem-loop structures on the (−) strand of this region.

Results from *in vitro* and *in vivo* studies show that the genomic and subgenomic leader sequences of PLRV do not function as translational enhancers (Juszczuk *et al.*, 2000). In fact, deletion analyzes show that both leader sequences not only decrease translation of downstream genes but also alter the ratio of the expressed proteins.

The mechanisms involved in the expression of luteovirus and polerovirus genomes are summarized in Fig. 7.31. From this it can be seen that there are complex control mechanisms involved in the expression of these viruses which were discussed in Section V.C.3.

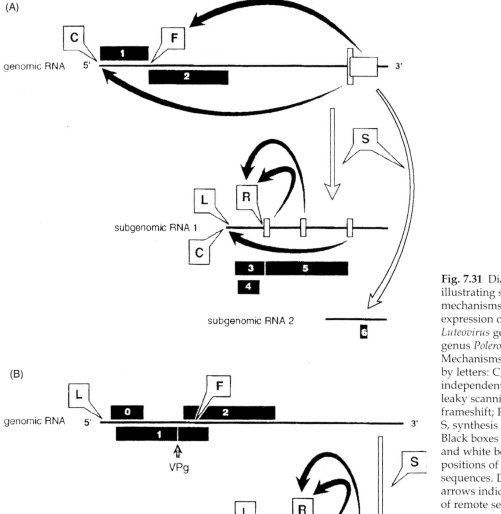

Fig. 7.31 Diagram illustrating some mechanisms used in the expression of **(A)** genus *Luteovirus* genomes and **(B)** genus *Polerovirus* genomes. Mechanisms are indicated by letters: C, cap-independent translation; L, leaky scanning; F, –1 frameshift; R, read through; S, synthesis of an sgRNA. Black boxes represent ORFs and white boxes show the positions of controlling sequences. Dark grey arrows indicate the effects of remote sequences on upstream processes, and pale grey arrows represent the synthesis of an sg mRNA. From Mayo and Miller (1999), with permission.

11. Five strategies: Multi-partite genome + subgenomic RNA + read-through protein + polyprotein + two-start

As described in Chapter 6 (Section VIII.H.13), the virions in plants naturally infected with BNYVV contain four, or sometimes five, (+)-sense ssRNA species, but only two of them, RNAs 1 and 2, are required for infection. RNA1 has a single ORF (see Fig. 6.49), the translation of which can be initiated at two sites: an AUG at position 154, which gives a 237-kDa product, and an AUG at position 496, which gives a 220-kDa product. The 220-kDa protein is processed by a papain-like protease encoded within this protein to give a 150-kDa and a 66-kDa protein (Hehn *et al.*, 1997). BNYVV RNA2 has six ORFs, the 5′ two of which are in the same frame and separated by a suppressible UAG stop codon. The 5′ ORF encodes the 19-kDa coat protein, the stop codon of which is read through to give a 75-kDa protein. This read-through protein can be detected by immunogold labeling near the extremity of the virus particles and is important in fungal transmission of the virus (see Chapter 11, Section XII) (Haberlé *et al.*, 1994; Tamada *et al.*, 1996). The other four ORFs on RNA2 are expressed from sgRNAs, two of them, sgRNA a and sgRNA c, being monocistronic for a 42-kDa and 14-kDa protein respectively, and sgRNA b being bicistronic for a 13-kDa and 15-kDa protein. The expression of the 42-kDa, 13-kDa and 15-kDa is similar to that of the triple gene block of PVX.

12. Discussion

The above examples show that most plant viruses with (+)-sense ssRNA genomes adopt more than one strategy to overcome the constraints that their host places on them when expressing their genetic information. However, as noted above this bypassing of constraints is also used to control their gene expression, and there is often an interlink between the strategies to give temporal and quantitative control.

F. Negative-sense single-stranded RNA viruses

The ways by which plant viruses with (–)-sense ssRNA genomes derive their mRNAs are described in Section IV.A. The *Rhabdovirus* genome is transcribed to give a series of monocistronic sgRNAs that differ from those of (+)-strand RNA viruses in being sequential along the genome and not 3' co-terminal. The major control of genome expression appears to be by the sequential transcription giving decreasing amounts of the mRNAs along the genome.

As described in Sections IV.A.2 and 3, the tospoviruses and tenuiviruses use two strategies; their largest RNA segment is transcribed to give a monocistronic mRNA and their other RNA genomic segments have an ambisense arrangement. Thus, mRNAs are transcribed from both the genomic- and complementary-sense RNAs. There is little information on any control mechanisms that may be involved.

G. Double-stranded RNA viruses

Transcription of the dsRNA genome segments of plant Reoviruses gives monocistronic mRNAs in most cases (see Section IV.B). Some of the genome segments potentially contain two ORFs but there is no evidence that the downstream ORF is expressed.

H. DNA viruses

1. *Caulimoviridae*

The transcription of the dsDNA genomes of members of the *Caulimoviridae*, and the promoters involved, are described in Section IV.C. For all genera in this family the transcription yields a more-than-genome-length RNA, the 35S or 34S RNA, that is both the template for expression of some of the gene products and the template for the reverse transcription phase of replication. For some of the genera other RNAs are transcribed from the viral DNA. Most of the studies on genome expression have been on CaMV and RTBV and these will be described in detail.

a. *Cauliflower mosaic virus*

The CaMV genome is transcribed from two promoters, one giving the 35S RNA and the other the 19S RNA (see Fig. 6.1). The 19S RNA is the monocistronic mRNA for ORF VI; ORFs I–V are expressed from the 35S RNA or products thereof. There have been suggestions of other transcripts such as a separate one for ORF V (Plant *et al.*, 1985; Schulze *et al.*, 1990) and for ORFs I and IV (Kobayashi *et al.*, 1998) but these have not been substantiated.

There are two major problems to be faced in translating the information from ORFs I–V in the 35S RNA. Firstly, the RNA has a long leader sequence of 600 nucleotides or more; and secondly, these ORFs are downstream of the putative ORF VII and several sORFs and therefore should be closed to the eukaryotic translation system. Two unusual mechanisms are proposed to overcome these problems together with the probability that at least some of the downstream ORFs are opened up by splicing.

The first of the unusual mechanisms is used to bypass the long leader sequence up to ORF I that contains not only ORF VII but also several sORFs. This mechanism, termed 'ribosome shunting', is described in Section V.B.8 and involves the ribosome passing from a donor to acceptor site in the highly structured leader sequence. It is considered that ribosome shunting enables much of the leader sequence to be bypassed and delivers the 40S ribosome subunit to the start codon of ORF I where it forms an 80S ribosome and starts translation.

ORFs I–V are closely appressed to each other, either overlapping by a few nucleotides or being separated by a few nucleotides. At the termination codon of ORF I it is considered that the second mechanism, that of transactivation by the product of ORF VI, takes over. This has been termed the 'relay race' model for the translation of the 35S RNA (Dixon and Hohn, 1984). In this model, a ribosome binds first to the 5'-end of the RNA and translates to the first termination codon. At this point, it does not completely leave the RNA but re-initiates protein synthesis at the nearest AUG whether just downstream or upstream from the termination. Support for the model came from site-directed mutagenesis studies in ORF VII and the region between ORF VII and ORF I, regions that are not essential for infectivity under laboratory conditions. Insertion of an AUG into either of these regions rendered the viral DNA non-infectious unless the inserted AUG was followed by an in-frame termination codon (Dixon and Hohn, 1984).

Transactivation is discussed in Section V.B.7 and is considered to prevent the ribosome completely disengaging at the stop codon and to enable it to commence translation again at the next start codon. The transactivation of CaMV is fully described in Rothnie et al. (1994) and in Hohn and Fütterer (1997).

As noted in Section V.B.11, there is increasing evidence for splicing in CaMV. It is difficult to identify spliced RNAs by northern blotting techniques of RNAs extracted from CaMV-infected tissues as there is always a smear of hybridizing RNA below the 35S RNA; this most likely arises from degradation of the 35S RNA during reverse transcription.

Using protoplasts transfected with 35S RNA-driven and promoter-less ORF I- and ORF IV-Ⅎ-glucuronidase fusion constructs, Kobayashi et al. (1998) obtained results that they interpreted as suggesting that sgRNAs were involved in the expression of ORFs I and IV.

However, no such sgRNAs have been found in CaMV infections.

b. Rice tungro bacilliform virus

The genome of RTBV contains four ORFs (see Fig. 6.4). The dsDNA genome is transcribed to give a 35S RNA which is spliced (as described in Section V.B.11) to form the mRNA for ORF IV. Thus, the expression of ORFs I–III faces similar problems to those described for CaMV above. RTBV has a long leader sequence (more than 600 nucleotides) that contains 12 sORFs. ORF I has an AUU start codon (see Fig. 7.18) and the next approximately 1000 nucleotides have only two AUG codons in any reading frame, those for ORFs II and III. ORFs I, II and III each overlap the next by one nucleotide having a 'stop/start' signal of AUGA (Hay et al., 1991).

This has led to the development of the model shown in Fig. 7.32 for the expression of RTBV ORFs I–III. As noted in Section V.B.8, RTBV resembles CaMV in that much of the long leader sequence is bypassed by a ribosome shunt mechanism. This places the 40S ribosome subunits at the AUU start codon of ORF I, a start codon that is about 10% as efficient as an AUG codon. It is suggested that only some of the 40S ribosome subunits initiate translation at the ORF I start codon, the rest translocating to the next AUG, which is the start codon for ORF II but is in a poor context. The model suggests that only some of these 40S ribosome subunits initiate here, the remainder passing to the next AUG, which is at the start of ORF III and in a good context.

2. Geminiviridae

The synthesis and regulation of mRNAs from the ssDNA genomes of geminiviruses are described in Section IV.C.2.

VI. DISCUSSION

Plant viruses have evolved strategies for effectively expressing their genetic information from a minimum of genetic material (Drugeon et al., 1999). In doing so they have to overcome

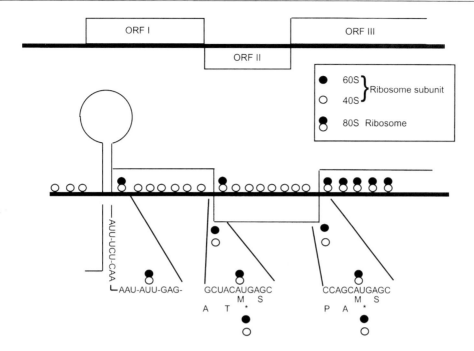

Fig. 7.32 Model for the expression of the first three ORFs of RTBV. The top line depicts the RTBV DNA genome (thick line) and ORFs I–III. The bottom line shows the genome with, above it, the stem loop formed by the 5′ region of the 35S RNA, ORFs I-III and depiction of ribosomes translating the genetic information. Note that only a proportion of ribosomes initiate at the 5′ termini of ORFs I and II, whereas initiation on ORF III is more efficient. Below the line are the sequences at the initiation sites of the three ORFs with the protein sequence below. The assembly of 80S ribosomes and disassembly into ribosomal subunits are depicted. From data reviewed in Hull (1996).

various constraints imposed by their presence in eukaryotic cells. These constraints are outlined in Sections V.A and V.B. Some emerging ideas on new mechanisms of initiating translation in eukaryotes have been reviewed by Kosak (2001). As well as overcoming these constraints, the viral mRNA(s) has to compete with host mRNAs for the translational machinery and also has to express the various gene products in the right amount, at the right time and in the right place. Thus, these strategies have a sophistication that we are only just beginning to understand.

Little is known about how viruses compete with host mRNAs. They are, in many cases, expressed to much higher levels that many plant genes. Two of the virus groups (the tospoviruses and tenuiviruses) undertake 'cap-snatching' which removes the 5′ cap structure from host mRNAs (see Section V.C.4) and is presumed to give the viral messengers an advantage over the host messengers. But, what about the plant viral mRNAs that have caps

and those that do not? Several viruses that have capped mRNAs express some of their proteins to very high levels. For instance, TMV coat protein can be one of the most highly expressed proteins in infected cells. Other uncapped viral RNAs also express proteins to a high level. To a certain extent, the level of expression is a reflection of the turnover rate of the various proteins as the comparisons of viral and host protein amounts are made on their steady-state levels.

However, in spite of these considerations, presumably there are factors that make viral mRNAs more efficient at sequestering the host translational machinery or more effective at doing so. In the latter case, it is likely that the fact they are present at much higher levels than host mRNAs that is important. It is probable that the more efficient expression of viral RNAs is due to the influence(s) of regions distant from the sites of ribosome binding and translation initiation. We are just beginning to identify these sites and to identify the interactions involved. The findings described in Chapter 9

(Section III.B.3) show that, in at least some viral infections, there is a differential effect on host gene expression with some being turned off, others enhanced and yet others not affected. It will be interesting to know if this is a widespread phenomenon and to gain an understanding of the mechanism of it.

As well as host macromolecules, such as ribosomes, being involved in the expression of viruses, there is an increasing number of host proteins and factors being recognized as being used. It is likely that the avidity of viral mRNAs for these proteins and factors is, in part, responsible for the relative efficiency of viral expression. On the other hand, the interactions between viral sequences and these proteins and factors could play a part in controlling virus expression.

It is obvious that viral gene products are required in different amounts and at different times during the infection cycle. As noted in Section V.B.1.c, just one or, at the most, a very few copies of the replicase protein(s) are required to produce a new copy of the viral genome, but 60 (for spherical viruses with $T = 1$ structure) to several thousand (for long rod-shaped viruses) copies of the coat protein to encapsidate it. Furthermore, the replicase is required early in the infection cycle whereas the coat protein is required relatively late. The other gene products, such as those required for cell-to-cell movement or interactions with vectors, are also required in different amount at different times. Viral expression strategies that involve subgenomic RNAs would give control of both the timing and amount of gene products. It is of interest to note that, for many viruses, the coat protein gene lies at the 3'-end of the genome, and possibly this is effective in producing coat protein relatively late in infection. As noted in Section V.B.1.c, the use of the polyprotein strategy does pose some questions about its efficiency. In the viruses that express their information via a polyprotein but from a divided genome (e.g. comoviruses, bymoviruses) the 'early' genes appear to be on one genome component and the 'late' genes on

another; this would allow temporal differentiation of expression.

The strategies that involve frameshift from one ORF to an overlapping one, or read through a weak stop codon, control the relative amount of the product from the downstream ORF; *in vitro* studies indicate that the frameshift or read through occurs on about 5–10% of the occasions that the upstream ORF is translated. Most, if not all, the products of these two strategies are involved either in viral replication or interaction with the virus vector. The read-through or frameshift products from the coat protein gene that are involved in virus vector interactions are incorporated into the viral capsid and it is obvious that they are required in a smaller amount than the coat protein itself. The requirement for differential expression of the replication proteins is less well understood.

It may be that, as we gather more information about the replication of viruses that our concepts of what is required early and late in infection will change. For instance, coat protein is required for the replication of AMV (see Chapter 8, Section IV.G) and thus cannot be considered to be a 'late' gene.

As discussed above, viruses compete very effectively with the host mRNA for their expression. However, there has to be control to prevent viral expression causing irreparable damage to the host cell, as the virus is totally dependent on maintaining the integrity of the cell and host plant. Thus, encapsidation plays an important role in sequestering the viral genome from the translation machinery. However, this raises a further question as to how re-initiation of virus uncoating is prevented. As discussed in Section II, for uncoating, the virus particle has to be destabilized by factors such as pH and removal of divalent cations. It would seem likely that the newly encapsidated virions are placed in a cellular compartment where this cannot occur.

CHAPTER 8

Virus Replication

I. INTRODUCTION

One of the major features of viruses is their ability to replicate their genomic nucleic acid, often to high levels, in cells in which there are normally strict limits on the production of new nucleic acid molecules. Some viruses do this by adapting the existing cellular machinery and others replicate their nucleic acid by mechanisms not widely used in host cells.

Our understanding of the ways in which plant viruses replicate has increased remarkably over the past few years. This is, in part, because the complete nucleotide sequences of many plant viral genomes have been established, allowing the number, size and amino acid sequence of putative gene products to be determined. We now have this information for representatives of most plant virus genera, genome organizations being discussed in Chapter 6. Developments in gene manipulation technology have permitted an artificial DNA step to be introduced into the infection cycle of RNA viruses; thus, infectious genomic RNA transcripts with a uniform nucleotide sequence can be produced *in vitro*. In turn, this has allowed the application of site-directed mutagenesis in experiments to determine functional regions in the non-coding regions of genomic nucleic acid, as well as the functions of gene products and putative gene products. Determination of the functions and properties of gene products has also been greatly assisted by the use of well-established *in vitro* translation systems.

This chapter will describe how plant viruses replicate. In doing this, we must bear in mind

Virus acronyms are given in Appendix 1.

that replication is not just a function on its own but is integrated and co-ordinated with many other viral functions. This integration of functions is explored in Chapter 13.

II. HOST FUNCTIONS USED BY PLANT VIRUSES

Like all other viruses, plant viruses are intimately dependent on the activities of the host cell for many aspects of replication.

A. Components for virus synthesis

Viruses use amino acids and nucleotides synthesized by host-cell metabolism to build viral proteins and nucleic acids. Certain other more specialized components found in some viruses (e.g. polyamines) are also synthesized by the host.

B. Energy

The energy required for the polymerization involved in viral protein and RNA synthesis is provided by the host cell, mainly in the form of nucleoside triphosphates.

C. Protein synthesis

Viruses use the ribosomes, tRNAs and associated enzymes and factors of the host cell's protein-synthesizing system for the synthesis of viral proteins using viral mRNAs. All plant viruses appear to use the 80S cytoplasmic ribosome system. There is no authenticated example of the chloroplast or mitochondrial ribosomes being used. Most viruses also depend on host enzymes for any post-translational modification of their proteins, such as glycosylation.

D. Nucleic acid synthesis

Almost all viruses code for an enzyme or enzymes involved in the synthesis of their nucleic acids, but they may not contribute all the polypeptides involved. For example, in the first phase of the replication of caulimoviruses, the viral DNA enters the host-cell nucleus and is transcribed into RNA form by the host's DNA-dependent RNA polymerase II. In most, if not all, RNA viruses, the replication complex comprises the viral RNA-dependent RNA polymerase (RdRp), several other virus-coded activities and various host factors. ssDNA viruses alter the cell cycle constraints on the host DNA replication system. These aspects will be developed in greater detail in subsequent sections.

E. Structural components of the cell

Structural components of the cell, particularly membranes, are involved in virus replication. For example, viral nucleic acid synthesis usually involves a membrane-bound complex. This is described in more detail in Section IV.C and in Chapter 13.

III. METHODS FOR STUDYING VIRAL REPLICATION

Because of the involvement of host systems and the close integration with other stages of the infection cycle, it is generally accepted that a full picture of viral replication can be obtained only from *in vivo* systems. However, owing to their complexity, *in vivo* systems are extremely difficult to establish and many of the questions of detailed interactions and functions can be addressed by *in vitro* systems. In this section, I describe some of the systems that have yielded information on viral replication.

A. *In vivo* systems

1. The intact plant

In Chapter 12 (Section II), some of the variables involved in sampling intact plants are discussed. It should be borne in mind that, in spite of these difficulties, there are certain aspects of virus replication that can be resolved only by study of the intact developing plant; for example, the relationship between mosaic symptoms and virus replication. The tissue that has been most commonly used in the study of virus replication is the green leaf blade. This tissue constitutes approximately 50–70% of the fresh weight of most experimental plants, and final virus concentration in the leaf blade is often 10–20 times higher than in other parts of the plant. We can distinguish four types of plant system *in vivo*: the intact plant, surviving tissue samples, cells or organs in tissue culture, and protoplasts; virus replication has also been studied in yeast cells. The advantages and difficulties of these systems are discussed next. Some plant viruses also replicate in their insect vectors; this topic is discussed in Chapter 11.

a. Inoculated leaves

Inoculated leaves have several advantages. Events can be timed more precisely from the time of inoculation than can those from systemic infections. A fairly uniform set of leaves from different plants can be selected, and half-leaves may be used as control material. There are two major disadvantages.

1. A typical leaf such as a tobacco leaf with a surface area of 200 cm^2, for example, contains about 3×10^7 cells. The upper limit for the proportion of epidermal cells that can be infected by mechanical inoculation under the best conditions is not known precisely, but is probably not more than about 10^4 cells per leaf. Thus, at the beginning of an experiment, only about 1 in 10^3 of the cells in the system has been infected. Even for those that are directly infected, the synchrony of infection may not be very sharp, especially if whole virus is used as inoculum. Thus, early changes in the small proportion of infected cells will probably be diluted out beyond

detection by the relatively enormous number of as-yet-uninfected cells. Then, as infection progresses, a mixed population of cells at different stages of infection will be produced.

2. The second major disadvantage of inoculated leaf tissue, at least for studying events over the first few hours, is that mechanical inoculation itself is a severe shock to the leaf, causing changes in respiration, water content, and probably many other things as well, including nucleic acid synthesis. Thus, the use of appropriately treated control leaves is essential.

A third difficulty applies to experiments in which radioactively labeled virus is used as inoculum. Most of the virus applied to the leaf does not infect cells, and a substantial but variable proportion cannot be washed off after inoculation. The fate of the infecting particles may well be masked by the mass of potentially infective inoculum remaining on or in the leaf.

For particular kinds of experiments, two modifications in the use of the inoculated leaf have proven useful. With some leaves grown under appropriate conditions, it is relatively easy to strip areas of epidermis from the leaf surface. Very limited amounts of tissue can be harvested in this way; but the method increases, by a factor of about eight, the proportion of cells infected at times soon after inoculation (Fry and Matthews, 1963).

Several workers have used micromanipulation methods to infect single cells on a leaf—usually leaf hair cells—and then to follow events in the living cells as they can be observed by phase or ultraviolet microscopy or in preparations stained with fluorescent antibody (see Chapter 9, Section II.B). This procedure, while it has given useful information, is limited to microscopical examination and cannot at present be used for biochemical investigations. It is mainly used for studying cell-to-cell movement of viruses but also gives information on cellular distribution of viruses and their gene products.

b. Systemically infected leaves

Moving from the inoculated leaf, a virus may invade the youngest leaves first, then successively infect the older and older leaves (see Fig. 9.2). Thus, systemically infected leaves may be in very different states with respect to virus infection. Furthermore, the time at which infectious material moves from inoculated leaves to young growth may vary significantly between individual plants in a batch. Nevertheless, it is probable that in young systemically infected leaves (perhaps about 4 cm long at the time virus enter for plants like tobacco and Chinese cabbage) most of the cells in a leaf become infected over a period of 1–2 days. Such a leaf has been used to study the replication of TMV (Nilsson-Tillgren et al., 1969) and TYMV (Hatta and Matthews, 1974; Bedbrook et al., 1974).

The synchrony of infection in the young systemically infected leaf can be greatly improved by manipulating the temperature. The lower inoculated leaves of an intact plant are maintained at normal temperatures (c. 25–30°C) while the upper leaves are kept at 5–12°C. Under these conditions, systemic infection of the young leaves occurs, but replication does not. When the upper leaves are shifted to a higher temperature, replication begins in a fairly synchronous fashion (W. O. Dawson et al., 1975). This procedure provides a very useful system that complements, in several reports, the study of virus replication in protoplasts, discussed subsequently. The technique uses intact plants, is simple, and can provide substantial amounts of material. The main requirement is for a systemic host with a habit of growth that makes it possible for upper and lower leaves to be kept at different temperatures. Its use has been extended to some other viruses, for example CMV in tobacco (Roberts and Wood, 1981).

c. Transgenic plants

The expression of viral genes in plants, often coupled with mutagenesis of the viral genome, is proving to be an increasingly useful approach. Plant transformation is discussed in Chapter 7 (Section III.D.2.d).

d. Virus mutagenesis

Infectious cDNA or DNA clones are available for many viruses and open the possibility to make specific mutations, the effects of which on virus replication can be studied in intact plants.

e. Viral reporter systems

Manipulation of cloned viral genomes enables reporter molecules, usually fluorescent proteins (see Chapter 9, Section II.B), to be attached to specific viral gene products or expressed separately from the viral genome. This enables the virus to be studied in intact plants in real time and for details to be obtained on the exact location of the gene function being studied.

2. Surviving tissue samples

a. Excised leaves

These are useful when fairly large quantities of leaf tissue are required. Petioles may be placed in water or a nutrient solution. Under these conditions, leaves vary widely in the amount of fluid they take up, and may wilt unpredictably. Tissue near the cut end of the petiole acts as a 'sink' for radioactively labeled metabolites (Pratt and Matthews, 1971). On the other hand, the method minimizes the problem of the growth of micro-organisms in the tissue during incubation. More commonly, leaves are placed in dishes covered with glass under moist conditions. Growth of bacteria, fungi and protozoa is then likely to be a problem.

b. Disks of leaf

Disks of tissue 5–20 mm in diameter cut from leaves with a cork borer and floated on distilled water or some nutrient salt solution have the advantage that pieces from many leaves can be combined in one sample to smooth out leaf-to-leaf variations. The physiological state of the leaves from which disks are taken affects uptake and metabolism of radioactively labeled materials (e.g. Kummert and Semal, 1969). There may be two serious disadvantages: (1) micro-organisms grow on the surface of the disks and in the intercellular spaces, so addition of antibiotics may not block all micro-organisms and may well alter the biochemical situation in the cells of interest; and (2) excised disks are not uniform in several ways (Pratt and Matthews, 1971). First, there is a 'geographical' gradient from the cut edge to the center of the piece of tissue. Differences involve the uptake of labeled precursors and their utilization for nucleic acid synthesis. Second,

excised tissues change with time in a complex fashion in their ability to accumulate substances from the medium. There may be a differential accumulation of labeled precursors in the cut ends of veins. Third, further variables are introduced when the excised tissue is treated with a drug such as actinomycin D, which may be distributed very unevenly in the tissue.

c. Epidermal strips

Dijkstra (1966) explored the possibility of studying TMV replication in strips of epidermis removed from leaves immediately after inoculation with TMV and floated on nutrient solutions or distilled water, but no significant progress has been made with this system.

3. Tissue culture

Plant cells can be grown in tissue culture in several ways, either as whole organs (e.g. roots or stem tips) or as solid masses of callus tissue growing in solid or liquid culture, or as cell suspensions. Amounts of virus produced in cultured tissue or cells are usually very much less than in intact green leaves, although tobacco callus cells disrupted in the presence of TMV inoculum produced high yields of virus (Murakishi et al., 1971; Pelcher et al., 1972). Various methods have been tested in the study of virus replication, but except for some microscopical studies, results have been disappointing. White et al. (1977) and Wu and Murakishi (1979) have adapted the low-temperature preincubation procedure of W. O. Dawson et al. (1975) to callus cultures infected with plant viruses. The virus growth curves obtained for TMV in tobacco callus cells were comparable to that obtained with protoplasts.

4. Cell suspensions and tissue minces

In principle, suspensions of surviving but non-dividing cells offer considerable advantages in the study of virus replication. Dissociated cells from callus tissue grown in culture and leaf cells separated enzymatically have been used. For example, Jackson et al. (1972) successfully used separated leaf cells to study the replication of TMV RNA.

5. Protoplasts

Protoplasts are isolated plant cells that lack the rigid cellulose walls found in intact tissue. Cocking (1966) first isolated protoplasts from tomato fruit by using enzymes to degrade the cell wall. Takebe *et al.* (1968), Takebe and Otsuki (1969) and Aoki and Takebe (1969) showed that metabolically active protoplasts could be isolated from tobacco leaf cells, that such protoplasts could be synchronously infected with TMV or TMV RNA, and that virus replication could be studied in them. Since then, protoplasts have been prepared from many species and infected with a range of viruses. Progress has been reviewed by Murakishi *et al.* (1984) and Sander and Mertes (1984). Methods for isolating and inoculating protoplasts are given in Dijkstra and de Jager (1998).

In outline, protoplasts from tobacco leaves are prepared as follows. The lower epidermis is stripped from the leaf tissue, which is then vacuum infiltrated with a solution of a commercial pectinase (polygalacturonidase) preparation called Macerozyme from *Rhizopus* spp. The medium contains 0.4–0.7 M mannitol plus 0.5% potassium dextran sulfate. The leaf pieces are then shaken on a waterbath. Early fractions of cells released from the tissue may be discarded. The veins and similar debris are removed by filtration and the cells collected by centrifugation. They are then treated with a cellulase preparation (from *Trichoderma viride*). On complete removal of the cellulose wall, the cells, now bounded only by the plasma membrane, assume a spherical shape (Fig. 8.1).

About 10^7 palisade cells can be obtained from 1 g of tobacco leaf in 2 hours. Many minor variations on the procedure have been developed (e.g. Kassanis and White, 1974; Beier and Bruening, 1975, 1976; Motoyoshi and Oshima, 1976; Shepard and Uyemoto, 1976; Kikkawa *et al.*, 1982). The ability to infect protoplasts with improved synchrony enables plant virologists to carry out one-step virus growth experiments (Fig. 8.2), an important kind of experiment that has long been available to those studying viruses of bacteria and mammals.

Besides improved synchrony of infection, the use of protoplasts has several other advantages. First, there can be close control of experimental conditions, and uniform sampling can be carried out by pipetting. Secondly, there is a high proportion of infected cells (often 60–90%) and a relatively high efficiency of infection. Thirdly, organelles such as chloroplasts and nuclei can be isolated in much better condition from protoplasts than from intact leaves. However, a number of actual or potential limitations and difficulties must be borne in mind.

1. Protoplasts are very fragile, both mechanically and biochemically, and their fragility may vary markedly, depending on the growing conditions of the plants, season of the year, time of day, and the particular age of leaf chosen. Defined plant growth conditions may improve the quality and reproducibility of the isolated preparations (e.g. Kubo *et al.*, 1975b).
2. Under culture conditions that favor virus

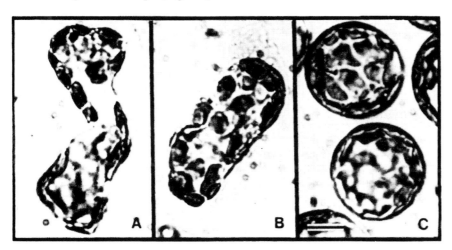

Fig. 8.1 Isolation of protoplasts from Chinese cabbage leaves. **(A)** A separated spongy mesophyll cell and **(B)** a separated palisade cell, following pectinase treatment. The cells still retain the cellulose cell wall. Following treatments with cellulase, spherical protoplasts are produced, **(C)**. Bar = 10 μm. (Courtesy of Y. Sugimura.)

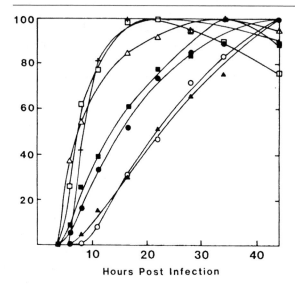

Fig. 8.2 Time course of production of TMV-related RNAs, proteins and progeny virus particles in synchronously infected protoplasts. One-half of a batch of protoplasts was incubated with [^{14}C]uridine in the presence of actinomycin D from the time of inoculation. The other half was incubated with [^{14}C]leucine under the same conditions. Samples were taken for analysis at the times indicated. Data are expressed as the percentages of the maximum values attained for each component during the time course studies. (□) RI; (△) RF; (●) TMV RNA; (■) coat protein mRNA; (+) 126-kDa (140-kDa) protein; (▲) coat protein; (○) progeny virus particles. From Ogawa and Sakai (1984), with kind permission of the copyright holder, © Blackwell Science Ltd.

replication, protoplasts survive only for 2–3 days and then decline and die.

3. To prevent growth of micro-organisms during incubation, antibiotics may be added to the medium. These may have unexpected effects on virus replication (e.g. gentamycin: Kassanis *et al.*, 1975).

4. Compared with the use of intact tissue, relatively small quantities of cells are made available.

5. Cytological effects observed in thin sections of infected leaf tissue may not be reproduced in protoplasts — probably because of the effects of changed osmotic conditions on cell membranes, for example, with TMV in tobacco (Otsuki *et al.*, 1972a), TYMV in Chinese cabbage (Sugimura and Matthews, 1981), FLSV in cowpea (van Beek *et al.*, 1985), and CaMV in turnip (Yamaoka *et al.*, 1982b).

6. The isolation procedure and the medium in which they are maintained must drastically affect the physiological state of the cells. Physical and chemical disturbances include: (i) partial dehydration; (ii) severing of plasmodesmata; (iii) loss of the cell wall compartment, which is not metabolically inert; (iv) reversal of the cell's electrical potential; (v) inhibition of leucine uptake; (vi) a large increase in RNase activity; and (vii) cellulose synthesis and wall regeneration, which begin very soon after the protoplasts are isolated. In addition, tobacco mesophyll protoplasts have been shown to synthesize six basic proteins that are undetectable in tobacco leaf. Three of these are like a 1,3-β-glucanase and two chitinases found in TMV-infected tobacco leaves (Grosset *et al.*, 1990) (see Chapter 10, Section III.K.1).

As a consequence of these changes, protoplasts vary with time in many properties during the period that they survive after isolation. Although little systematic study has yet been made of the changes, it is known that some features of virus replication differ in intact leaves and in protoplasts. Thus, Föglein *et al.* (1975) showed that, when protoplasts are prepared from leaves fully infected with TMV, vigorous viral RNA synthesis is re-initiated. Tobacco protoplasts containing the *N* gene escape necrotic cell death when infected with TMV (Otsuki *et al.*, 1972b).

In many studies using protoplasts it has been reported that yields of virus (virus particles per cell) are very similar to that found in intact plants. For example, Renaudin *et al.* (1975) found that Chinese cabbage protoplasts infected *in vitro* produced about 10^6 TYMV particles per cell. This figure is similar to the published yields of TYMV obtained with extracts of intact leaf. These estimates were based on the assumption that all the cells in the leaf were infected, and that they were of the same size. If, however, the estimates are made on the same class of cell as used for the *in vitro* studies (i.e. palisade mesophyll), and if only infected cells are considered, then yields per cell in the intact leaf are about 10 times higher (Sugimura and Matthews, 1981). Despite these limitations, protoplast systems have contributed to our knowledge and will contribute more in the future.

Many efforts have been made to improve the process of infection of protoplasts with viruses or viral RNAs. For example, Watanabe *et al.* (1982) inoculated tobacco protoplasts successfully with TMV RNA encapsulated in large unilamellar vesicle liposomes. Another technique known as 'electroporation' is used to introduce viral nucleic acids into protoplasts. It involves the application of a brief high-voltage pulse to a mixture of cells and nucleic acid. The pulse renders the cells transiently permeable to the nucleic acid. It is a widely applicable procedure but the mechanism is not well understood, and optimum conditions need to be determined for each system. For example, Nishiguchi *et al.* (1986) infected tobacco leaf protoplasts with TMV and CMV RNAs using this procedure. The method facilitated infection of protoplasts by infectious TMV RNA produced in low concentrations by transcription from TMV cDNA (Watanabe *et al.*, 1987a). A positively charged virus (BMV) could be readily induced to infect by electroporation but the negatively charged CCMV gave only a poor rate of infection (Watts *et al.*, 1987).

A polycation such as poly(L) ornithine is necessary in the medium for infection using viral RNAs or viruses with an acidic isoelectric point. Takanami *et al.* (1989) reported a greatly increased efficiency of infection using polyethyleneimine in the medium.

6. Yeast

Yeast, *Saccharomyces cerevisiae*, is a single-cell organism for which there is a considerable resource of classical and molecular genetics (see Chapter 7, Section III.D.2.e). Although yeast is the host for several viruses and virus-like agents, no ssRNA virus is known to naturally infect it. In a major advance to using this organism to advance our understanding of how RNA viruses replicate, Janda and Ahlquist (1993) showed that yeast expressing BMV RNA replication genes 1a and 2a supported the RNA-dependent replication and transcription of BMV RNA3 derivatives. In initial experiments, they were unable to detect expression of BMV coat protein on transfection of yeast spheroplasts with transcripts of RNAs 1, 2 and 3, most probably owing to the low efficiency of co-transfection with multiple RNAs. The system they used comprised yeast $2\,\mu$m-based plasmids expressing the BMV 1a and 2a genes under the control of a constitutive derivative of the yeast *ADH1* promoter and *ADH1* polyadenylation signal. Replication was studied using constructs based on RNA3 (Fig. 8.3) (Janda and Ahlquist, 1998).

On the addition of constructs of RNA3 in which the coat protein gene had been replaced by the chloramphenical acetyl transferase (CAT) or yeast *URA3* gene, they demonstrated replication of RNA3, the formation of subgenomic RNAs containing the downstream gene on RNA3 and the expression of that gene. The RNA3 derivatives were maintained through, at least several, cell division cycles of the yeast. Placing the expression of RNA3 under the galactose-inducible yeast GAL1 promoter and incorporating a self-cleaving ribozyme at or near the natural 3'-end increased expression of RNA3 by 45-fold (Ishikawa *et al.*, 1997a). The use of this system to gain an understanding of BMV replication is described in Section IV.E.

It is possible that *S. cerevisiae* will be able to support the replication and expression of other RNA viruses. However, as the organization of elements in its promoter system differs from that of most eukaryotes, it is unlikely to be of use for viruses with DNA genomes. Another yeast species, *Schizosaccharonyces pombe*, has a promoter system more similar to those of many eukaryotes and may be a more likely candidate for use with DNA viruses.

7. Radioisotopes

The use of radioactively labeled virus precursors is essential for many studies on virus replication. There are substantial difficulties and limitations in the effective use of tracer compounds for studying the replication of plant viruses. Various ways have been used to introduce the labeled material into the tissue being studied. Whole plants can be removed from their pots, the roots carefully washed free of soil, and the isotope applied to the roots. This procedure is useful for ^{32}P-labeled orthophosphate and ^{35}S-labeled sulfate. Provided there is no delay in applying the isotope after washing the roots, uptake is rapid and efficient. With

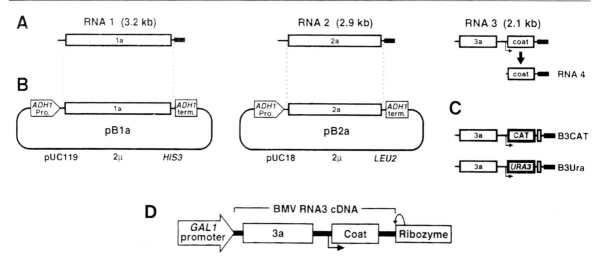

Fig. 8.3 Structure of BMV genome and derivatives used in studies on replication in yeast. **(A)** Schematic of BMV genomic RNAs 1, 2 and 3. RNA4, the 3′ co-terminal mRNA synthesized from the (−)-strand RNA3 replicative intermediate, is shown below RNA3. The open boxes represent the ORFs for the 1a, 2a, 3a and coat proteins, while the single line denotes the non-coding regions flanking these genes. The solid box at the 3′-end of each genomic RNA marks the conserved 200-base tRNA-like region, including the (−)-strand initiation signals. The bent arrow 5′ of the coat protein gene on RNA3 marks the position complementary to the RNA4 transcription start site. **(B)** Schematic of plasmids showing the BMV 1a and 2a ORFs and short flanking segments of non-coding sequence that were inserted between the *ADH1* promoter and the *ADH1* transcription terminator. The relative positions of sequences derived from pUC119 and pUC118, the yeast 2 μm origin of replication, and the *HIS3* and *Leu2* selectable markers are also shown. **(C)** Structures of BMV RNA3 derivatives carrying the CAT or yeast *URA3* genes (stippled boxes). In the B3CAT and B3URA RNAs, all but the 3′ portion of the coat protein gene (represented by a short residual open box) has been detected and replaced by the complete CAT and *URA3* genes respectively, plus small segments of 3′ non-coding sequence derived from each gene. **(D)** Schematic of BMV RNA3 cDNA in yeast centromeric plasmid B3. The *GAL1* promoter fused to the 5′-end of RNA3 allows gal-induced *in vivo* transcription of RNA3, and the hepatitis Δ ribozyme cleaves the transcript at the natural 3′-end of RNA3. Parts (A)–(C) from Janda and Ahlquist (1993), with kind permission of the copyright holder, © Elsevier Science; (D) from Janda and Ahlquist (1998), with kind permission of the copyright holder, © The National Academy of Sciences, USA.

plants such as actively growing Chinese cabbage, ^{32}P may be detected in leaves within minutes of application, and uptake into the plant may be more or less complete within a few hours. With these two isotopes, uptake into leaves through the roots is much more effective than floating intact disks of leaf tissue on solutions of the isotope, even if the disks are sliced to expose more vein ends. Placing leaves with their cut petioles in the solution can lead to a highly variable and irregular uptake of isotope. However, by careful timing and attention to growth conditions, quite high specific activities can be obtained (e.g. 1 mCi of ^{32}P (3.7×10^7 Bq) per milligram of viral RNA; see Bastin and Kaesberg, 1975). Kopp *et al.* (1981) describes a procedure in which pieces of leaf from which the lower epidermis has been stripped are floated on a solution containing the radioactive precursor. No systematic study of the best ways

to introduce such precursors as amino acids and nucleotides appears to have been made. Devices are available for injecting solutions into leaves (e.g. Hagborg, 1970; Konate and Fritig, 1983).

Most plant leaves have rather large reserves of low-molecular-weight phosphorus compounds. By various manipulations, it is possible to lower or raise the overall concentration of phosphorus compounds not more than 2- to 3-fold. Thus, in leaf tissue it has not been possible to carry out effective pulse-chase type experiments with phosphorus. With most organic compounds that can be used as labeled virus precursors, active leaves are continually providing an endogenous source of supply. Furthermore, plant tissues have the capacity to metabolize carbon compounds in many different ways, so that the labeled atom may soon appear in a wide range of low-molecular-weight compounds. For certain kinds of experiments, it is useful to be able to label purified virus chemically *in vitro* to high specific activity. A variety of procedures is available (e.g. Frost 1977; Montelaro and Rueckert, 1975).

8. Metabolic inhibitors

Inhibitors of certain specific processes in normal cellular metabolism have been widely applied to the study of virus replication. Three have been of particular importance: (1) actinomycin D, which inhibits DNA-dependent RNA synthesis but not RNA-dependent RNA synthesis; (2) cycloheximide, which is used as a specific inhibitor of protein synthesis on 80S cytoplasmic ribosomes; and (3) chloramphenicol, which inhibits protein synthesis on 70S ribosomes (e.g. in chloroplasts, mitochondria and bacteria).

Results with these inhibitors must always be treated with caution, as they may have other diverse subsidiary effects in eukaryotic cells, which may make it difficult to interpret results. For example, actinomycin D may affect the size of nucleotide pools (Semal and Kummert, 1969), can cause substantially increased uptake of metabolites by excised leaves (Pratt and Matthews, 1971), may reduce uptake by infiltrated disks (Babos and Shearer, 1969), and may not suppress synthesis of certain species of host RNA (e.g. Antignus *et al.*, 1971).

Synthesis of the large polypeptide of ribulose bisphosphate carboxylase takes place in the chloroplasts on 70S ribosomes, while the small polypeptide is synthesized on 80S ribosomes in the cytoplasm. Owens and Bruening (1975) used these two polypeptides as an elegant internal control in their examination of the effects of chloramphenicol and cycloheximide on the synthesis of CPMV proteins.

9. Metabolic compartmentation

If we count a membrane as a compartment, eukaryotic cells have at least 20 compartments. In their replication, plant viruses have adapted in a variety of ways to the opportunities provided by this intracellular metabolic diversity (see Chapter 13). In thinking about experiments on virus replication (particularly those involving the use of radioisotopes and/or metabolic inhibitors) we must take account of the fact that processes take place in cells that have a high degree of metabolic compartmentation. This exists in several forms: (1) in different cell types, which are metabolically adapted for diverse functions; (2) in membrane-bound compartments within individual cells, such as nuclei, mitochondria, chloroplasts, lysosomes, peroxisomes and vacuoles; (3) in isolatable stable complexes of enzymes; and (4) in microenvironments created without membranes, by means of weakly interacting proteins, or unstirred water layers near a surface.

a. Sites of synthesis and assembly

Two general kinds of procedure have been used in attempts to define the intracellular sites of virus synthesis and assembly: (1) fractionation of cell components from tissue extracts followed by assay for virus or virus components in the various fractions; and (2) light and electron microscopy. There are many difficulties involved in using cell fractionation procedures to locate sites of virus assembly.

1. Chloroplasts are fragile organelles, and a proportion of these are always broken. Chloroplast fragments cover a wide range of sizes and will contaminate other fractions.
2. Viruses such as TMV, occurring in high concentration, will almost certainly be distributed among all fractions, at least in small amounts.
3. Virus-specific structures may be very fragile and unable to withstand the usual cell breakage and fractionation methods.
4. If virus-specific structures are stable, they may fractionate with one or more of the normal cell organelles.
5. Virus infection may alter the way in which certain cell organelles behave on fractionation.

Considerable progress has been made with some viruses using cell fractionation procedures. However, in recent years we have learned more from ultrastructural studies and most where both kinds of technique have been applied.

Viruses belonging to many different groups induce the development in infected cells of regions of cytoplasm that differ from the surrounding normal cytoplasm in staining and ultrastructural properties. These are not bounded by a clearly defined membrane but usually include some endoplasmic reticulum and ribosomes. They vary widely in size and may be visible by light microscopy. In varying degrees for different viruses, there is evidence that these bodies are sites of synthesis of viral components and the assembly of virus particles. I use the term *viroplasm* to describe such inclusions. Some of the amorphous and 'X-body' inclusions described in the older literature are of this type. The detailed structure of the viroplasms may be highly characteristic for different virus groups (see Chapter 3, Section IV), and sometimes even for strains within a group. Proteins coded for by the viral genome probably bring about the formation of these characteristic structures. How this is accomplished in most cases is quite unknown. Immunocytochemical methods are used to locate viral coded proteins within cells and tissues (e.g. Stussi-Garaud *et al.*, 1987) (Fig. 8.4).

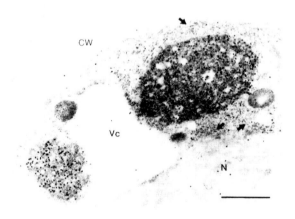

Fig. 8.4 Immunocytochemical localization of the 126 (130) kDa–180 kDa TMV proteins in a cell from a young infected tobacco leaf. Leaf sections were treated with an anti-126-kDa–anti-180-kDa antiserum, and then with a protein A–gold complex. Gold label is strongly localized in the viroplasms regions of the cytoplasm. CW, cell wall; Vc, vacuole; N, nucleus; arrows indicate ER. Bar = 1μm. From Saito *et al.* (1987b), with permission.

B. *In vitro* systems

1. *In vitro* replication systems

There have been various attempts to isolate competent replication complexes from virus-infected plant material. There are three main problems:

1. First there is the difficulty of separating complexes of proteins and nucleic acids from normal cell constituents. This is normally done by differential centrifugation and by gradient centrifugation. Solubilized TMV replication complexes were separated from cellular components in linear gradients of 10–40% glycerol (Watanabe et al., 1999) and membrane-bound complexes of ToMV in gradients of 20–60% sucrose (Osman and Buck 1996).
2. Membranes are an integral part of the replication complexes of (+)-strand RNA viruses and the technology for isolating such components is not yet well developed.
3. Uninfected plant cells contain an endogenous RdRp which is often enhanced on virus infection. Care has to be taken to separate this activity from the virus-coded activity.

2. Primer extension

Properties of replication complexes can be studied by adding nucleotide triphosphates under the appropriate conditions and assessing the resulting products from extension of primed strands on the existing template. The products can be analyzed by incorporating a labeled nucleotide triphosphate (radioactive or fluorescent label) or by probing the product with a labeled probe. This approach can be used to study the optimum condition for the replicase enzymes.

3. Enzyme activities

The enzymes involved in replication have been purified by standard protein and enzyme purification techniques including size-exclusion chromatography and ion-exchange chromatography. The properties of these enzymes have been studied by standard enzymatological techniques and other techniques such as activity gels. Description of these techniques is beyond the scope of this book but can be found in standard manuals on proteins and enzymes.

4. Protein–protein interactions

a. Yeast two-hybrid system (reviewed by Brachmann and Boeke, 1997)

In this system for investigating the interactions between two proteins, one — termed the 'bait' — is cloned so that it is fused to a DNA-binding domain (commonly derived from Gal4 or LexA). The other protein — termed the 'prey' — is fused to an activation domain (usually from Gal4 or the transactivator VP16). The two clones are introduced into a suitable yeast line and interaction between the bait and prey proteins tether the DNA-binding domain to the activation domain, allowing the activation of a downstream reporter (e.g. lacZ, HIS3, LEU2 or URA3) (Fig. 8.5).

b. Cross-linking

Proteins can be cross-linked to one another or to nucleic acids by the treatment with chemicals or with UV radiation; see Ausubel *et al.* (1998) for details. The size of the chemical cross-linker can give information on the distances separating the proteins.

c. Sandwich blots

In an adaptation of western blotting where the protein, immobilized on a membrane, is detected by an antiserum, a second protein can be allowed to bind to the immobilized protein and then detected by its specific antiserum. This allows protein–protein interactions to be studied. However, it must be recognized that the immobilization of the first protein may alter its conformation or hide binding sites.

5. Protein–nucleic acid interactions

a. Yeast three-hybrid system (reviewed by Brachmann and Boeke, 1997)

This system enables interactions between three partners to be studied, including three proteins, the functional activation of one partner through phosphorylation by a tyrosine

kinase, the extracellular domains of trans-membrane receptors, and RNA–protein interactions. The latter requires two interacting proteins and one interacting hybrid RNA. One part of the hybrid RNA acts in a known interaction and the other part is used to screen for RNA-binding proteins (Fig. 8.5). These interactions bring together the components, which allow the activation of a downstream reporter as in the two-hybrid system described above.

b. DNA and RNA footprints

The basis of this assay is that bound protein protects the DNA or RNA from DNase- or RNase-catalyzed hydrolysis. The protected DNA or RNA fragments are separated by denaturing gel electrophoresis and analyzed by techniques such as sequencing or autoradiography. Binding curves for each individual protein-binding site can give quantitative information, and sites that interact co-operatively can be identified (Ausubel *et al.*, 1998).

c. Gel mobility shift

This assay using non-denaturing polyacrylamide gel electrophoresis provides a simple, rapid and sensitive method for detecting proteins that bind to nucleic acids. The binding of the proteins retards the mobility of the nucleic acid fragment on gel electrophoresis that can be detected by comparing the mobilities of treated and untreated nucleic acid fragments. The retarded fragments correspond to individual protein–nucleic acid complexes and can identify specific purified proteins or uncharacterized proteins in crude extracts. Properties of the protein–nucleic acid interaction, such as affinity and binding specificity, can be studied by this technique (Ausubel *et al.*, 1998).

d. North-western blot

In north-western blots, either the protein or the nucleic acid under study is separated by gel electrophoresis and blotted to a membrane. The

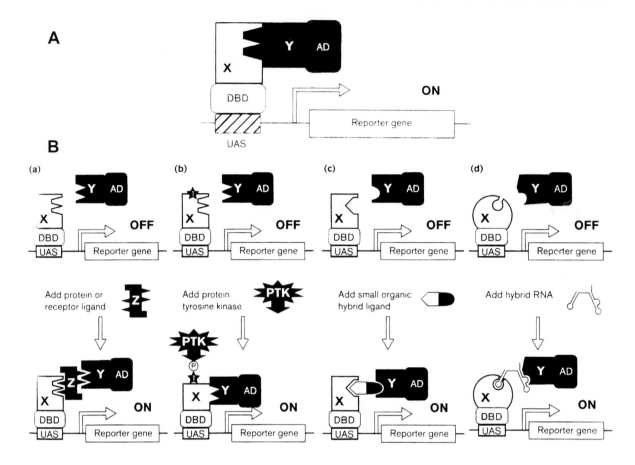

Fig. 8.5 (A) The yeast two-hybrid system. The gene for protein X, the bait, is fused to a DNA-binding domain (DBD); the gene for protein Y, the prey, is fused to an activation domain (AD). Interaction of proteins X and Y tethers the DBD to the AD, allowing activation of transcription of a downstream reporter (e.g. *lacZ*, *HIS3*, *LEU2* or *URA3*). **(B)** Some examples of yeast three-hybrid systems. All system components that can be analyzed or screened for are depicted in solid black. **(a)** The protein and peptide ligand 3HS require a third partner, Z, for proteins X and Y to form a stable complex. These partners may be bridging proteins, proteins that bind to X and Y, thus changing them to the right conformation, or peptide ligands that induce the dimerization of two extracellular receptor domains. **(b)** The kinase 3HS is related to (a): the conformational change of one of the interactive partners, X, is achieved through tyrosine (T) phosphorylation by a tyrosine kinase. **(c)** The small ligand 3HS utilizes chemical inducers of dimerization (CIDs) to study receptors and their small organic ligands. The key feature is a hybrid ligand: one part of it, FK506, is recognized by the binding protein FKBP12. The second part can be a known small ligand that allows screening for, or mutational analysis of, a receptor. The second part could also be a ligand library designed to isolate the ligand of an orphan receptor, Y. **(d)** The RNA 3HS requires two hybrid proteins and one hybrid RNA for the phenotypic readout. Conceptually similar to (c), one half of the hybrid RNA represents the bait as part of a known RNA-protein interaction, while the other half can be designed to analyze or screen for RNAs or RNA-binding proteins. From Brachmann and Boeke (1997), with kind permission of the copyright holder. © Elsevier Science.

membrane is then probed with the counterpart nucleic acid or protein under study and binding is detected by methods such as label on the nucleic acid or antiserum to the protein. As with sandwich blots, it should be remembered that immobilization might alter the conformation of the macromolecule.

IV. REPLICATION OF POSITIVE-SENSE SINGLE-STRANDED RNA VIRUSES

Nearly 50 genera of plant viruses with (+)-sense ssRNA genomes have been established, but we have a significant body of information about the replication of very few of these. However, it is likely that all viruses with this form of genome have the same basic replication mechanism though there most probably will be variations in detail.

The basic mechanism of replication of (+)-sense RNA genomes is that the virus-encoded replicase synthesizes a complementary (−) strand using the (+) strand as a template and then new (+) strands are synthesized from the (−)-strand template. Synthesis of new RNA is from the 3' to 5'-ends of the templates. Replication occurs in a replication complex that comprises the templates, newly synthesized RNA, the replicase and host factors. These will be discussed in turn and then details given of some viral systems that have attracted most study.

A. Viral templates

Two kinds of RNA structure have been isolated from viral RNA synthesizing systems. One, known as *replicative form* (RF), is a fully base-paired ds structure, whose role is not certain. For example, it may represent RNA molecules that have ceased replicating. The other, called replicative intermediate (RI), is only partly ds and contains several ss tails (nascent product strands) (Fig. 8.6).

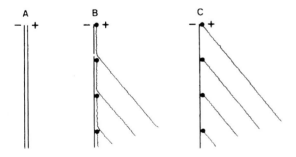

Fig. 8.6 Forms of association between positive- and negative-sense viral RNA. **(A)** Replicative form (RF)—a base-paired structure with full-length (+)- and (−)-sense strands. **(B)** Replicative intermediate (RI)—a partially base-paired structure with polymerase molecules (●) and ss tails of nascent progeny (+)-sense strands. **(C)** Probable true state of the RI *in vivo*—the progeny (+)-sense strands and the template (−)-sense strand are almost entirely ss. From Matthews (1991).

This structure is closely related to that actually replicating the viral RNA. It is thought that the RI as isolated may be derived from a structure like that in Fig. 8.6C by annealing of parts of the progeny strands to the template.

The ds nature of RFs and RIs is apparent when these molecules have been isolated from infected cells. The nature of the structures in the infected cell is unknown but is of great significance in view of the importance of dsRNA in the plant defense response (see Chapter 10, Section IV). It is likely that these molecules are essentially single-stranded *in vivo*, the strands being kept apart either by compartmentalization or by proteins bound to them. Cytological evidence suggests that the RI of TYMV is essentially ss (see Section IV.K.4).

The (+) and (−) forms of the viral genome contain the signals that control both the specificity and timing of their replication.

1. Control signals

It is likely that many of the control signals are at the 3'-ends of the template strands, though it should be recognized that there might be long-distance signals elsewhere in the template and/or the 3'-end that interact with other regions in the RNA. For (−)-strand synthesis, the 3' terminus of the (+) strand has been found to be important (reviewed in Dreher, 1999). As described in Chapter 4 (Section III.A.3.c), there are three basic structures in plant viral RNA 3' termini, tRNA-like structures, poly(A) tails and non-tRNA heteropolymeric sequences. Dreher (1999) points out that there is no real correlation between the 3' terminal structure and the supergrouping of RdRps (described below), which might indicate similar roles for dissimilar 3' termini.

The (−)-strand synthesis of viruses with a 3' tRNA-like sequence initiates with the insertion of a GTP opposite the 3'-most C residue on the (+)-strand template (reviewed in Dreher, 1999). How (−)-strand synthesis initiates on templates with other 3' termini is unknown. The general concept is that the 3' terminus provides a specific binding site for the replication enzyme complex leading to the replication complex.

The 3' untranslated regions of LMV expressed in transgenic plants served as the template for synthesis of (−)-strand RNA on infection with other potyviruses such as PepMoV, TEV or TVMV but not on infection with CMV (Teycheney et al., 2000). In plants expressing the 3'-end of CMV, (-)-strand RNA was synthesized on infection with the related TAV. Thus, the replicase complexes of potyviruses and cucumoviruses have the ability to recognize heterologous 3' UTRs of viruses in the same genus.

The 5' termini of the (+) strands (i.e. the 3' termini of the (−)-strand templates for (+)-strand synthesis) are much more variable and offer no obvious directions as to the specific priming of (+)-strand synthesis. It is generally thought that the specificity is provided by the (−)-strand initiation and synthesis and that the (−)- and (+)- strand synthesis are coupled.

B. Replicase (reviewed by Ishihama and Barbier, 1994; Buck, 1996)

Three or more virus-coded enzymatic activities can be involved in the replication of (+)-strand RNA viruses, the RNA-directed RNA polymerase (RdRp), a helicase and a methyltransferase activity. These are collectively known as the viral replicase but sometimes this term is used (incorrectly) for the RdRp. Although the ascribed function has not been formally demonstrated for most of these activities — and technically they should be termed 'RdRp-like', 'helicase-like' and 'methyltransferase-like'– for the sake of simplicity I will use the functional terms in this book.

1. RNA-dependent RNA polymerase

The RdRp catalyzes the synthesis of RNA using an RNA template. Two features have made this enzymic activity difficult to study. First, it is usually associated with membrane structures in the cell and, on isolation, the enzyme(s) often become(s) unstable. They are, therefore, difficult to purify sufficiently for positive identification of any virus-coded polypeptides. Secondly, tissues of healthy plants may contain, in the soluble fraction of the cell, low amounts of an enzyme with similar activities. The amounts of

such enzyme activity may be stimulated by virus infection.

Kamer and Argos (1984) identified various amino acid sequence motifs that were characteristic of RdRps. The most conserved of these is a Gly-Asp-Asp (GDD) motif which is flanked by segments of mainly hydrophobic amino acids and is involved with Mg^{2+} binding. This suggested a structure of two antiparallel β-strands connected by a short exposed loop containing the GDD motif. Further alignments of RdRp sequences have now resulted in the recognition of eight motifs (Poch et al., 1989; Koonin, 1991; Koonin and Dolja, 1993). This has led to the classification of RdRps into three 'supergroups'; supergroup 1 is sometime called the 'picornavirus(-like)' supergroup, supergroup 2 the 'carmovirus(-like)' supergroup, and supergroup 3 the 'alphavirus(-like)' supergroup. The supergroups extend across viruses that infect vertebrates, plants and bacteria and there are representatives of plant viruses in each supergroup (Table 8.1).

Supergroup 1 members are characterized by usually having one genome segment that has a 5' VPg and expresses its genetic information as a polyprotein. Members of supergroups 2 and 3 have one to several genome segments, the RNA is often capped, and individual genes are translated.

The three-dimensional structure of the RdRp of poliovirus (supergroup 1) has been determined by X-ray crystallography (Hansen et al., 1997). The overall structure of the enzyme is similar to those of other polymerases (DNA-dependent DNA polymerase, DNA-dependent RNA polymerase and reverse transcriptase) that have been likened to a right hand. The palm domain contains the catalytic core and is similar to that of the other three polymerases. The thumb and finger domains differ from those of the other polymerases. Using the neural net PHD (for Predict at Heidelberg) computer method (Rost et al., 1994), O'Reilly and Kao (1998) predicted the secondary structure of the RdRps of BMV, TBSV and TMV and compared the predictions with the poliovirus RdRp structure. This analysis indicated that the RdRps of these supergroup 2 and 3 viruses have a similar structure to the supergroup 1 poliovirus

TABLE 8.1 Supergroups of RdRps

Supergroup 1 (Picornavirus supergroup)	Supergroup 2 (Carmovirus supergroup)	Supergroup 3 (Alphavirus supergroup)	
Comovirus	*Aureusvirus*	*Alfamovirus*	*Ilarvirus*
Enamovirus	*Avenavirus*	*Allexivirus*	*Marafivirus*
Polerovirus	*Carmovirus*	*Bromovirus*	*Oleavirus*
Potyvirus	*Dianthovirus*	*Capillovirus*	*Pecluvirus*
Sequivirus	*Luteovirus*	*Carlavirus*	*Pomovirus*
Sobemovirus	*Machlomovirus*	*Closterovirus*	*Potexvirus*
	Necrovirus	*Crinivirus*	*Tobamovirus*
	Panicovirus	*Cucumovirus*	*Tobravirus*
	Tombusvirus	*Foveavirus*	*Trichovirus*
	Umbravirus	*Furovirus*	*Tymovirus*
		Hordeivirus	*Vitivirus*
		Ideaovirus	

enzyme, each containing a region unique to RdRps.

RdRp activity is also present in uninfected plants and, at one time, was thought to be involved in RNA virus replication. This activity has been isolated as a cDNA clone encoding a 128-kDa protein (Schiebel *et al.*, 1993, 1998).

2. Helicases (reviewed by Kadarei and Haenni, 1997; Bird *et al.*, 1998)

Helicases are polynucleotide-dependent nucleoside triphosphate (NTP) phosphatases that possess ssDNA and/or RNA-displacing activity. They play a pivotal role in genome replication and recombination by displacing complementary strands in duplex nucleic acids and possibly removing secondary structure from nucleic acid templates. Some helicases require a 3′ flanking single strand of nucleic acid and others a 5′ flanking single strand; these are known as 3′–5′ and 5′–3′ helicases respectively.

Based on conserved amino acid sequence motifs, helicases have been grouped into a number of superfamilies. Gorbalenya and Koonin (1993) recognized five superfamilies, three of which have representatives in (+)-strand RNA viruses. Superfamilies I and II have seven conserved motifs whereas superfamily III has three motifs. Two motifs common to all three superfamilies are variants of ATP-binding motifs and have the conserved sequences GXXXXGKT/S and $\phi\phi\phi\phi$D, where

X is an unspecified amino acid and ϕ is a hydrophobic residue. Most members of superfamilies I–III are 3′–5′ helicases. The superfamily designation for various plant virus genera is given in Table 8.2.

The crystal structure for the hepatitis C virus RNA helicase (superfamily II) has been determined and a mechanism for unwinding duplex RNA suggested (Cho *et al.*, 1998). The structure comprises three domains forming a Y-shaped molecule. The RNA-binding domain is separated from the NTPase and other domain by a cleft into which ssRNA could be modeled. It is suggested that a dimer form of this protein unwinds dsRNA by passing one strand through the channel formed by the clefts of the two molecules and by passing the other strand outside the dimer. Because of the conserved motifs between the various superfamilies, it is likely that many of the features determined for the hepatitis C virus helicase are applicable to this enzyme from plant viruses.

Several plant virus genera appear to lack the characteristic NTP-binding motifs of helicases (Table 8.2). Several possible reasons for this have been suggested (Buck, 1996):

1. It is possible that the NTP-binding motifs have diverged so much that they are not recognizable from the primary amino acid sequence.
2. The viral polymerase may have unwinding activity.

TABLE 8.2 Helicase superfamilies and plant virus genera

Superfamily I	Superfamily II	Superfamily III	No motif
Alfamovirus	*Bymovrus*	*Comovirus*	*Aureusvirus*
Benyvirus	*Potyvirus*	*Sequivirus*	*Avenavirus*
Bromoviris		*Waikavirus*	*Carmovirus*
Capillovirus			*Dianthovirus*
Carlavirus			*Luteovirus*
Closterovirus			*Machlomovirus*
Crinivirus			*Necrovirus*
Cucumovirus			*Panicovirus*
Furovirus			*Polerovirus*
Hordeivirus			*Sobemovirus*
Idaeovirus			*Tombusvirus*
Ilarvirus			
Potexvirus			
Tobamovirus			
Tobravirus			
Trichovirus			
Tymovirus			

3. Unwinding may be effected by a helix-destabilizing protein that uses the energy of stoichometric binding to ssRNA to melt the duplex in the absence of NTP hydrolysis.
4. The virus may co-opt a host helicase.

As can be seen from Table 8.3, some viruses have additional helicase activities located elsewhere in their genomes. It is thought that the additional helicases in benyviruses, hordeiviruses and potexviruses are involved in cell-to-cell movement (see Chapter 9, Section II.D.2).

3. Methyl transferase activity (reviewed by Schuman and Schwer, 1995)

The methyl transferase activity leading to 5′ capping of RNAs is described in Chapter 4 (Section III.A.3.c).

4. Organization of functional domains in viral ORFs

The presence or recognition of the three above function domains of viral replicases in members of the (+)-stranded plant virus genera is listed in Table 8.3. From this it can be seen that not all the genera have all three domains. There are several reasons for this:

1. Only those viruses that have a m⁷G 5′ cap would require the methyl transferase activity. There are several viruses (e.g. furoviruses and

pecluviruses) that are thought to have capped RNAs for which the methyl transferase domain has not yet been recognized. In addition, the 5′ structure of the RNA has not yet been determined for several virus genera.
2. No helicase domain has been recognized in the *Tombusviridae*, the sobemoviruses and umbraviruses (see Section IV.B.2 above).

For all the viral genomes that express as polyproteins or fused protein (frameshift or read-through), the domains appear to be in the order (N-terminal to C-terminal) methyl transferase, helicase and the RdRp; in divided genomes it is not possible to allocate the order. However, a feature of many of the virus genera is that the methyl transferase and helicase domains are separated from the RdRp domain. This can be by the methyl transferase and helicase domains being in one ORF and the RdRp being in a separately expressed ORF, by them being in two adjacent ORFs separated by either a frameshift or read-through translational event, or by them being on a polyprotein and separated by protease activity (Table 8.3). However, in the capilloviruses and marafiviruses the protease domain lies between the methyl transferase and the other two domains. In six of the genera the three domains appear not to be separated.

For many viruses, it appears that the methyl transferase and helicase domains are on a

TABLE 8.3 Organization of domains within the (+)-strand RNA replication ORFs

Genus	MT	Hel	RdRp	MT and Hel separated from RdRp			Comments
				Different ORF	RT or FS	Protease	
Bromovirus	+		+	+			
Alfamovirus	+	+	+	+			
Cucumovirus	+	+	+	+			
Ilarvirus	+	+	+	+			
Oleovirus	+	+	+	+			
Comovirus		+	+			+	
Fabavirus		+	+			+	
Nepovirus		+	+			+	
Potyvirus		+	+			+	
Bymovirus		+	+			+	
Tombusvirus	−		+				
Aureusvirus	−		+				
Avenavirus	−		+				
Carmovirus	−		+				
Dianthovirus	−		+				
Machlomovirus	−		+				
Necrovirus	−		+				
Panicovirus	−		+				
Sequivirus		+	+			+	
Waikavirus		+	+			+	
Closterovirus	+	+	+		+		
Crinivirus	+	+	+		+		
Luteovirus		+	+		+		
Polerovirus		+	+		+		
Enamovirus		+	+		+		
Tobamovirus	+	+	+		+		
Tobravirus	+	+	+		+		
Potexvirus	+	+	+				MT, Hel and RdRp not separated
Carlavirus	+	+	+				MT, Hel and RdRp not separated
Allexivirus	+	+	+				MT, Hel and RdRp not separated
Capillovirus	+	+	+			+	Protease separates MT from Hel/RdRp
Foveavirus	+	+	+				MT, Hel and RdRp not separated
Trichovirus	+	+	+				Ditto
Vitivirus	+	+	+				Ditto
Furovirus	?	+	+		+		
Pecluvirus	?	+	+		+		
Pomovirus	+	+	+		+		
Benyvirus	+	+	+			+	
Hordeivirus	+	+	+	+			Helicase also on βb
Marafivirus	+	+	+			+	Protease separates MT from Hel/RdRp
Sobemovirus		−	+				
Tymovirus	+	+	+			+	
Idaeovirus	+	+	+				MT, Hel and RdRp not separated
Umbravirus	+	−	+				

MT, methyl transferase; Hel, helicase; RdRp, RNA-dependent RNA polymerase; RT, read-through; FS, frameshift.

single protein. However, although these two activities are expressed on the same ORF of BYV, probing extracts from infected plants with monoclonal antibodies indicates that *in vivo* the 295-kDa protein is processed to a 63-kDa protein containing the methyltransferase domain and a 100-kda protein containing the helicase domain (Erokhina *et al.*, 2000).

C. Sites of replication (reviewed by Buck, 1996)

Advances in technologies, especially identifying the location of virus-specific gene products and nucleic acids in infected cells, are leading to determining the sites of viral replication. It is becoming generally assumed that the replication of (+)-strand RNA viruses involves association with cellular membranes. Although this has been proven in only a few cases (see, for example, Sections IV.E.4 and IV.H.6 below), two facts support the generality of this association.

Firstly, in only one instance, that of CMV (Hayes and Buck, 1990), has full replication been demonstrated in solubilized replicase extracts from plant tissues. Generally, only the complementary strand is synthesized in such extracts. In the case of CMV, the *in vitro* replication was very inefficient and the ratio of (+) to (-) strands was much lower than *in vivo* (Hayes and Buck, 1990).

Secondly, in many virus infections there is perturbation of membrane structures, frequently leading to the formation of vesicles, which in the proven cases have been shown to be associated with replication complexes. A variety of membranes are involved in vesicle formation (Table 8.4). It can be seen from the table that there does not appear to be any correlation between the membrane involved and the supergroups of the replicase proteins or even with the virus family.

The involvement of membranes in the virus infection cycle is discussed further in Chapter 13.

D. Mechanism of replication

The RF and RI RNAs are considered to provide evidence on the mechanism of RNA replication.

It is suggested that the RF could arise from the initial synthesis of a (−) strand on a (+)-strand template. RI RNAs usually contain more (+) strands than (−) strands (see Aoki and Takebe, 1975) which is taken to indicate that each is a single (−) strand to which is attached several (+) strands (Fig 8.7).

There are two hypotheses as to the mechanism of (+)-strand synthesis. The semi-conservative mechanism involves total displacement of the newly synthesized strand by the oncoming strand (Fig 8.7b) and, in the conservative mechanism, it is suggested that the duplex RNA is only transiently unwound at the growing end of the nascent strands (Fig. 8.7c). The majority of evidence supports the semi-conservative mechanism (reviewed in Buck, 1999b).

E. Replication of brome mosaic virus

1 BMV replication *in vivo*

a. The process of infection

In the experiments of Kurtz-Fritsch and Hirth (1972), 17% of the BMV inoculum was uncoated after 20 minutes. Uncoating led to the appearance of low-molecular-weight viral protein rather than empty protein shells. The current understanding of the mechanisms of the uncoating of BMV particles and those of related bromoviruses is described in Chapter 7 (Section II.B).

b. Time course of events

In barley protoplasts infected *in vitro*, a membrane-bound RNA polymerase activity resistant to actinomycin D was increased up to 30-fold over mock inoculated protoplasts (Okuno and Furusawa, 1979). In the same system, RNAs 1 and 2 were detected 6 hours after infection. All four RNAs were present at 10 hours. Production of ds forms of all four ssRNAs followed a time course similar to that of the ssRNAs (Loesch-Fries and Hall, 1980). Maximum RNA synthesis was from 16 to 25 hours. Virus particle formation was greatest between 10 and 25 hours after inoculation.

In young cowpea leaves in which infection with CCMV was synchronized by differential

TABLE 8.4 Example of membranes possibly associated with (+)-strand RNA virus replication

Membrane	Virus	Reference
Endoplasmic reticulum	BMV	Restrepo-Hartwig and Ahlquist (1996)
	TMV	Heinlein *et al.* (1998); Reichel and Beachy (1998)
	CPMV	Carette *et al.* (2000)
Chloroplast outer membrane	TYMV	Reviewed in Garnier *et al.* (1986)
	AMV	de Graaf *et al.* (1993)
Vacuolar membrane	CMV	Hatta and Francki (1981b)
Peroxisome	TBSV	Russo *et al.* (1983); Lupo *et al.* (1994)
Mitochondria	CIRV	Rubino *et al.* (1995)
	TRV	Harrison and Roberts (1968)
Nucleus	PEMV	Demler *et al.* (1994a)

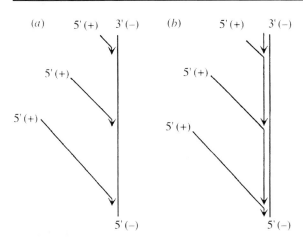

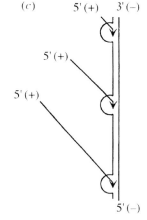

Fig. 8.7 Three possible structures of TMV replicative-intermediate (RI) RNA. From Buck (1999b), with kind permission of the copyright holder, © The Royal Society.

temperature treatment, the three largest RNAs were synthesized at relatively constant ratios throughout the infection (Dawson, 1978b), but very little RNA4 was produced early in infection. As the infection progressed the proportion of RNA4 continued to increase. The RFs of components 1, 2 and 3 were produced with kinetics similar to that for the corresponding ssRNAs.

2. Replication proteins

The genome of BMV is divided between three RNA species (see Fig. 6.13 for genome organization). RNAs 1 and 2 encode *trans*-acting functions required for RNA replication and, in barley protoplasts, can replicate independently of RNA3 (Kiberstis *et al.*, 1981; French *et al.*, 1986; Dinant *et al.*, 1993). RNAs 1 and 2 code for proteins 1a and 2a respectively and both proteins have been recognized in replicase preparations from BMV-infected plants (Horikoshi *et al.*, 1988; Quadt *et al.*, 1988). Protein 1a contains methyltransferase and helicase motifs, and protein 2a has RdRp motifs. Kao *et al.* (1992) demonstrated that proteins 1a and 2a form a complex *in vitro* by showing that: (1) they co-purified together with polymerase activity; (2) when expressed in rabbit reticulocyte lysates or in insect cells they co-precipitated with specific antisera; (3) protein 1a bound to protein 2a that had been fixed on a nylon membrane; and (4) a three-amino-acid insertion in protein 1a that abolished BMV replication *in vivo* also blocked the *in vitro* interaction with protein 2a. These observations fitted well with previous genetic observations (e.g. Allison *et al.*, 1988; Traynor and Ahlquist, 1990), suggesting that the interactions found *in vitro* occur *in vivo*. The interaction is between a 115-amino-acid region at the N terminus of protein 2a and a 50-kDa region of protein 1a encompassing the helicase domain (Kao and Ahlquist, 1992). The large interacting region of protein 1a indicated that a higher-ordered structure is involved – which was confirmed by O'Reilly *et al.* (1995) using limited proteolysis and the yeast two-hybrid system. The limited proteolysis also identified a highly structured helicase domain region in the analogous 1a proteins of CCMV and AMV. The yeast two-hybrid system has shown further features of interactions between and within the 1a and 2a proteins. The direct interaction between these two proteins is stabilized by the presence of the centrally conserved RdRp domain of 2a (O'Reilly *et al.*, 1997). There are both intramolecular interactions between the capping and helicase domains of protein 1a, and intermolecular 1a–1a interactions involving the N-terminal 515 residues of that protein (O'Reilly *et al.*, 1997, 1998). This has led to a model for the assembly of BMV replicase (Fig. 8.8) (O'Reilly *et al.*, 1998); it is suggested that this model is applicable to other members of the *Bromoviridae*.

Mutations in the helicase and capping enzyme active sites of BMV RNA 1a cause defects in template recruitment, (−)-strand RNA synthesis and viral RNA capping (Ahola *et al.*, 2000). The helicase mutants abolished all functions of the replicase system. Two groups of capping domain mutants were identified, one that increased template recruitment but allowed only low levels of (−)-strand and sgRNA synthesis, and the other that was defective in template recruitment and allowed synthesis of (−) strands to only a very low level. Deletion of the yeast chromosomal gene for the exonuclease that degrades uncapped mRNAs suppressed the second but not the first group of capping mutants. This points to the importance of the viral capping functions. There was no complementation between the helicase and capping enzyme mutants.

The replication complex also involves BMV RNA. RNA2 is recruited by the 1a protein through a 5′ signal on the RNA (Chen *et al.*, 2001). Quadt *et al.* (1995) showed that BMV RdRp activity could be extracted only from yeast expressing the 1a and 2a proteins that was also expressing certain RNA3 sequences. The minimum requirements were the 3′ untranslated region and intercistronic region, suggesting a non-template role for BMV RNA in assembly, activity or stability of the replication complex. This involvement of RNA3 with the replication complex increases the *in vivo* stability of that RNA induced by protein 1a (Janda and Ahlquist, 1998) and involves a 150- to 190-base segment of the intergenic region that has been implicated as an enhancer for RNA3 replication (Sullivan and Ahlquist, 1999).

3. *In vitro* assays

As well as the studies on *in vivo* protoplast and yeast cell systems, much has been learnt from the use of replication complexes isolated from infected cells and analyzed *in vitro*. The isolation of the replication complex was helped by the development of a detergent-based, dodecyl-β-D-maltoside, purification (Bujarski *et al.*, 1982) and the use of micrococcal nuclease to make it template-dependent (Miller and Hall, 1983). However, as noted in Section III.B, the isolation of soluble replication complexes removes them from their membrane association and usually prevents them from functioning in full.

The isolation of replication complexes that had been assembled *in vivo* in plants or yeast and the removal of the RNA template have enabled studies to be made on added RNA templates. This has demonstrated the template specificity of the BMV system (Hardy *et al.*, 1979) and (−)-strand synthesis can be initiated on the addition of (+)-strand templates (Miller and Hall, 1983; Quadt *et al.*, 1995; Sun and Kao, 1997b). Thus, it would appear that, having copied one template, the replication complex can be reused on another template—which would indicate that *cis*-acting sequences are involved in both replicase assembly and promoter recognition and the initiation of (−)-strand synthesis

In spite of the above reservations, much has been learnt from *in vitro* systems. Essentially, the signals are *cis*-acting and are located in the 5′ and 3′ untranslated regions (UTRs) and in the intercistronic region (ICR) of RNA3 (reviewed by Sullivan and Ahlquist, 1997b), sites where (−)-strand, (+)-strand and sgRNA synthesis are initiated. Most of the information on BMV applies to other bromoviruses and to cucumoviruses.

There are several stages in the synthesis of a new RNA strand, the three most obvious being initiation, extension and release. Within each of stages there are likely to be various events.

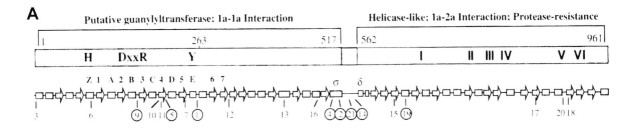

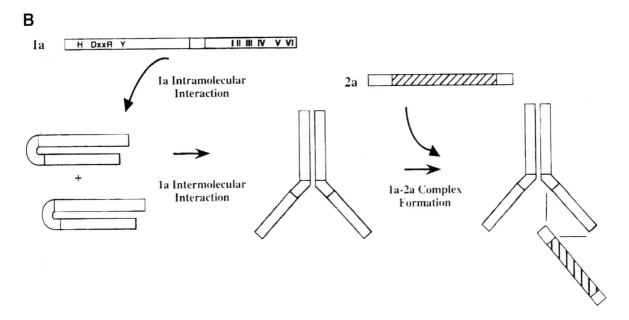

Fig. 8.8 **(A)** Diagram of BMV 1a protein and its predicted secondary structures. The sequences involved in capping and helicase function are indicated. The putative capping domain required for 1a–1a interaction is lightly shaded, while the protease-resistant helicase-like domain required for 1a–2a interaction is darkly shaded. The predicted secondary structures are shown below: boxes, α-helices; arrows, β-sheets; solid shapes, >80% accuracy; open shapes, >50% accuracy. The secondary structures that correspond to those found in DNA methyl transferase are numbered according to the system of Schluckebeier *et al.* (1995). The location of various mutants are numbered with the replication-competent ones being circled. **(B)** Working model for the assembly of the 1a–2a complex. The putative intramolecular interaction in 1a prevents the formation of the 1a–2a complex until the intermolecular 1a–1a interaction has occurred. The 2a protein interacts with the helicase-like domain of 1a through its N terminus. From O'Reilly *et al.* (1998), with permission.

a. The 3′ untranslated region

The 3′ UTR of BMV (and other bromoviruses) is highly structured, mimicking tRNAs and forming a pseudoknot (see Fig. 4.5). The 3′ 134 bases of BMV RNAs containing the tRNA-like sequence have been identified as the minimum sequence requirement for *in vitro* (−)-strand production (Miller *et al.*, 1986).

Initiation *in vitro* is independent of the primer and is at the penultimate C residue of the 3′-CCA. This indicates that the terminal A residue is not templated but is added to the (+) strands by tRNA nucleotidyl transferase after replication (Rao *et al.*, 1989). The tRNA-like sequence contains the replicase-binding site (Chapman and Kao, 1999), the specificity for the interaction with the replicase being determined by a stem-loop structure in the tRNA-like domain (Rao and Hall, 1993; Chapman and Kao, 1999; Sivakumaran *et al.*, 2000).

A transition between initiation to elongation of (−)-strand synthesis was observed by Sun and Kao (1997a). They showed that nascent RNAs of 10 nucleotides or longer remained associated with the replication complex and could be extended into full-length RNAs, whereas shorter RNAs were released from the complex. In a further analysis of this using the sensitivity of the non-templated complex to heparin, Sun and Kao (1997b) determined three steps with two transitions in the stability of the RNA synthesis complex (Fig. 8.9). They suggest that the replication proteins first bind to the 3' tRNA-like structure and, after synthesis of nascent RNA, the complex undergoes a transition to a more stable structure. The second transition occurs when the short nascent RNA chains are between 8 and 14 nucleotides in length.

b. The 5' untranslated region

The 5' UTRs of bromovirus and cucumovirus RNAs share sequence similarity, the most notable region matching the box B recognition sequence of RNA polymerase III promoters and thus also the conserved TΨC loop of tRNAs (Marsh and Hall, 1987); there is also a box B consensus in the intercistronic region of RNA3. It is suggested that this sequence might have a role in interacting with host factors (Sullivan and Ahlquist, 1999)

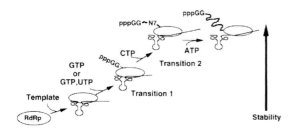

Fig. 8.9 A model for the transitions of BMV RNA synthesis from initiation to elongation by RdRp. The stability of the RNA synthesis complex increased in three distinct steps with two transitions. First, RdRp (ellipse) binds to the tRNA structure (represented by a three-leaf clover) of the viral RNA to form a binary complex of RdRp–RNA. After the synthesis of nascent RNA, the ternary complex then undergoes a transition to a structure that is more tightly associated with the template RNA, having a second stability level. A last transition occurs when further incorporation of CTP generates the short nascent RNA chains of 8 to 14 nucleotides in length. The latter ternary complex is now committed to template and will resist challenge with heparin and other template RNAs. From Sun and Kao (1997b), with permission.

As noted above, BMV replication complexes can initiate the synthesis of (−)-strand RNA from a (+)-strand template but little is known about the sequence and structural elements involved in the initiation of (+)-strand synthesis. Sivakumaran and Kao (1999) showed that initiation of (+)-strand synthesis requires the addition of a non-templated nucleotide at the 3'-end of the (−)-strand template. Mutational analysis demonstrated that this non-templated nucleotide, together with the +1 and +2 nucleotides (cytidylate and adenylate), are important for the interaction with the replicase complex. Furthermore, genomic (+)-strand RNA synthesis is affected by sequences 5' of the initiation site. This template recognition is controlled by the sequence rather than by the secondary structure at the 3'-end of the (−)-strand RNA (Sivakumaran et al., 1999).

c. The intercistronic region

Bromovirus RNA3s encode two proteins, the downstream one being expressed from a sgRNA (see Chapter 6, Section VIII.A.1). As well as the box B consensus sequence described above, bromovirus RNA3s have an oligo (A) tract in their intercistronic regions.

4. Cellular distribution

Using immunoradioautography, Lastra and Schlegel (1975) detected BBMV antigen in the nucleus and cytoplasm of broad bean leaves 1 day after inoculation. The amount of antigen associated with the nucleus remained more or less constant, while massive amounts accumulated in the cytoplasm. No antigen was associated with chloroplasts or mitochondria. Amorphous membranous inclusion bodies, which are probably viroplasms, and filamentous inclusions have been described for cells infected with CCMV (Kim, 1977).

Ultrastructural changes induced by BMV (and CCMV) involve a proliferation and modification of parts of the endoplasmic reticulum. In addition, the nuclear membrane appears to give rise to small cytoplasmic vacuoles that appear to contain nucleic acid (Burgess et al., 1974a).

As noted above, BMV replicase activity is associated with membranes. Using double-label immunofluorescence and confocal microscopy of BMV-infected barley protoplasts and in yeast cells, Restrepo-Hartwig and Ahlquist (1996, 1999) showed that 1a and 1b proteins co-localized throughout infection to defined cytoplasmic spots adjacent to, or surrounding, the nucleus, and BMV-specific RNA synthesis coincided with these spots. The BMV replication complexes were tightly associated with markers for endoplasmic reticulum and not with the medial Golgi or later compartments of the cellular secretory pathway. In yeast cells, protein 1a localizes to the endoplasmic reticulum in the absence of protein 2a (Restrepo-Hartwig and Ahlquist, 1999). As it interacts with the 2a protein, it recruits that protein to the endoplasmic reticulum, and thus it is considered to be a key organizer of the replication complex (Chen and Ahlquist, 2000).

5. Host proteins associated with replication complexes

Various host proteins have been found associated with BMV replication complexes. In characterizing the replication complex from BMV-infected barley, Quadt *et al.* (1993) detected a host protein antigenically related to the wheatgerm eukaryotic translation initiation factor 3 (eIF-3). The p41 subunit of wheatgerm eIF-3 binds strongly and specifically to BMV protein 2a, and addition of this host protein stimulated (−)-strand synthesis in an *in vitro* replication system. The development of the yeast system for studying BMV replication has led to the identification of further host proteins that appear to be involved. Study on yeast mutants in three complementation groups, *mab1-1* (*m*aintenance of *B*MV functions), *mab2-1* and *mab3-1* showed that they were due to single recessive mutations inhibiting accumulation of (+)- and (−)-strand RNA3 and sgRNA (Ishikawa *et al.*, 1997b). A similar study showed that BMV replication requires Lsm1p, a yeast protein related to the core RNA splicing factors but shown to be cytoplasmic (Diez *et al.*, 2000). The Lsm1p interaction involved BMV 1a protein and an internal viral element and the

dependence was suppressed by adding a 3' poly(A) to the normally non-polyadenylated BMV RNA. Thus, Lsm1p is involved in a specific step of BMV replication possibly through interaction with factors binding mRNA 5'-ends.

F. Replication of cucumber mosaic virus

The cucumoviruses have many properties in common with the bromoviruses although they have not been studied in such detail. The genome organization is very similar to that of the bromoviruses except that each of the three genomic RNAs and the subgenomic coat protein RNA are slightly longer. Expression of the genome gives four protein products. The replicase activity is coded for by RNAs 1 and 2 (Nitta *et al.*, 1988). Protein 1a–1a interaction through the N-terminal methyl transferase domain, described above for BMV (Section IV.E.2), has also been demonstrated for CMV (O'Reilly *et al.*, 1997, 1998). Using cloned CMV cDNA in which the termini of the 1a and 2a proteins had been tagged with histidine, Gal-On *et al.* (2000) showed that these proteins associated with a membrane fraction. They were able to elute a template-dependent RdRp when extracts from plants infected with these modified clones were passed over Ni^{2+}–NTA resin. However, only a small proportion of the viral RdRp bound to these columns.

Using an enriched replicase fraction from tobacco plants infected with the Fny isolate of CMV, Sivakumaran *et al.* (2000) identified a 3' stem-loop structure that interacted with the replicase. This structure is similar to the (−)-strand synthesis initiation site of BMV described above.

Many isolates of CMV support the replication of satellite RNA (see Chapter 14, Section II.B.2), which indicates that there must be recognition sequences for the viral replication complex on the satellite RNA.

1. *In vitro* replication

The isolation of the functional polymerase complex from CMV-infected tobacco plants by Hayes and Buck (1990) was the first for a eukaryotic virus. The membrane-bound polymerase was obtained by differential centrifugation of extracts from infected plants and the enzymatic activity was solubilized by treatment with non-ionic detergents. The polymerase was CMV RNA-dependent and synthesized both (−) and (+) strands. Although it synthesized predominantly (+) strands, the ratio of (+) to (−) did not duplicate that found *in vivo*. The replication complex comprised CMV proteins 1a and 2a and a 50-kDa host protein. Loss of the 50-kDa host protein during purification abolished the replicase activity.

A 334-nucleotide satellite RNA was fully replicated by this replication complex (Hayes *et al.*, 1992) when (+) strand was used as template but not with (−) strand as template.

2. Site of replication

Membrane-bound vesicles associated with the tonoplast may be the site of CMV RNA synthesis (Hatta and Francki, 1981b).

G. Replication of alfalfa mosaic virus
(reviewed by Bol, 1999b)

The AMV genome is divided between three RNA species. However, as was initially shown by Bol *et al.* (1971), inoculation with the three RNA species does not give infection unless either viral coat protein or the sgRNA (RNA4) for coat protein is present. The ilarviruses also have a requirement for coat protein to initiate infection, and functional equivalence has been shown between their coat protein and that of AMV by the activation of the AMV genome by TSV coat protein and *vice versa* (van Vloten-Doting, 1975).

1. Coat protein–RNA interaction

The main features of the genome activation by coat protein are as follows.

1. Some coat protein or coat mRNA, as well as the three genomic RNAs, is necessary to initiate an infection.
2. AMV infection starts with some coat protein bound to each of the three genomic RNAs (Smit *et al.*, 1981).
3. Coat protein has a high affinity for viral RNA. Free viral RNA added to a virus preparation will remove coat protein subunits from the nucleoprotein (van Vloten-Doting and Jaspars, 1972).
4. The binding and genome activation are specific to AMV coat protein and the coat proteins of the related ilarviruses (e.g. van Vloten-Doting, 1975). AMV protein does not bind specifically to RNAs of unrelated viruses (Zuidema and Jaspars, 1985).
5. Transgenic tobacco plants expressing AMV coat protein will support virus replication following inoculation with AMV RNAs 1, 2 and 3 or TSV RNAs 1, 2 and 3. Transgenic plants expressing TSV coat protein also support replication of the three AMV RNAs (van Dun *et al.*, 1988b).
6. The coat protein binds to a specific site near the 3′ terminus of RNAs 1–3 (Smit *et al.*, 1981). There is a homologous sequence at the 3′ termini of the three genomic RNAs.
7. The RNA binds either one to three coat protein dimers. Internal coat protein binding sites are also found on all the genomic RNAs, but these do not appear to have such strong affinity for the protein (Jaspars, 1985).
8. Removal of N-terminal amino acids of the coat protein destroys effective binding to RNA and the ability to make the genome RNAs infectious (Zuidema *et al.*, 1983).
9. Most of the features of RNA-coat protein interactions determined for AMV are applicable to ilarviruses.

Details of the interaction between AMV coat protein and its RNA have been elucidated using a range of techniques. Treatment of RNA fragments from the 3′-end of AMV RNAs 1–4 by RNase T1 digestion has shown that coat protein will protect a 36-nucleotide fragment but not a 28-nucleotide fragment (reviewed by Jaspars, 1985). Modeling of the secondary structures of the 3′-ends of RNAs 1–3 (as RNA4 is 3′ co-terminal with RNA3 it will have the same structure) (Fig 8.10) reveals similar structures of up to five stem-loops each flanked by the tetranucleotide AUGC box (van Rossum *et al.*, 1997a).

Band-shift assays using transcripts of wild-type and mutated cDNAs and of synthetic

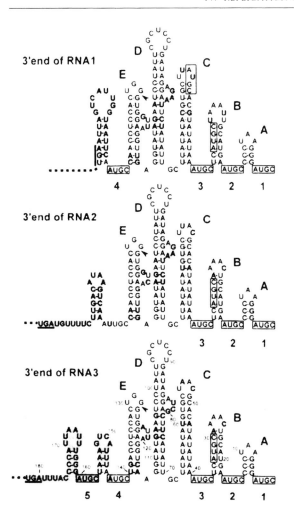

Fig. 8.10 Schematic of the proposed secondary structure of the 3' UTRs of the genomic RNA of AMV. Bases that are not identical in the three RNAs are shaded. The stop codons of the P1, P2 and CP are indicated by black bars. AUGC sequence motifs are boxed. The arrowhead indicates a unique *Dra*III site in the viral cDNAs. Nucleotides of RNA3 are numbered from the 3'-end. From van Rossum *et al.* (1997a), with permission.

deoxyribonucleotides identified four RNA–coat protein complexes involving five regions of the 3' UTR of AMV RNA (nucleotides 9–18, 29–41, 63–72, 107–121 and 133–144; numbering from the 3'-end of the RNA). Three of the complexes involved nucleotides 1–86 and one complex nucleotides 85–175 (reviewed in Bol, 1999b). The importance of the AUGC box, and especially of box 2, in coat protein binding has been revealed by mutagenesis; the UUGC

sequence between hairpins C and D also plays a role in coat protein binding (Reusken and Bol, 1996). From these mutagenesis and binding studies it is suggested that there are two independent binding sites: site I including AUGC boxes 1, 2 and 3, and site II including AUGC boxes 4 and 5 (reviewed in Bol, 1999b). It is proposed that the hairpin structures arrange the AUGC boxes and the UUGC sequence to give tandem binding sites to which the coat protein dimers can bind (Reusken and Bol, 1996). A pseudoknot conformation has been proposed for the 3' UTRs of AMV and ilarviruses (Olsthoorn *et al.*, 1999). This pseudoknot conformation resembles the tRNA-like structures at the 3' termini of bromo-, cucumo- and hordeivirus RNAs, and a weak, though specific, interaction of the AMV 3' UTR with yeast CCA-adding enzyme was detected.

As noted above, the coat protein binds as a dimer. The binding is through the N terminus, there being no binding when the N-terminal 40 amino acids are removed (Baer *et al.*, 1994). Synthetic peptides corresponding to the 38 and 25 amino acids (CP38 and CP25 respectively) bind, and a peptide corresponding to amino acids 13–26 conferred protection against RNase T2 digestion of the 39-nucleotide fragment (Ansel-McKinney *et al.*, 1996). CP25 contains seven basic residues and replacement showed that it is the arginine at position 17 (numbering from the N terminus excluding the initiating methionine) that is important (Ansel-McKinney *et al.*, 1996; Yusibov and Loesch-Fries, 1998). Ansel-McKinney and Gehrke (1998) used nucleotide substitutions, hydroxyl radical footprinting, and ethylation and chemical modification interference analysis to explore the potential phosphate and base-specific contacts in the 3' terminal 39 nucleotides of RNA3. They confirmed that the AUGC boxes and not the hairpin loops were involved and identified two contact sites (Fig. 8.11). It is suggested that these two sites are the binding sites for the N termini of the coat protein dimer, and that the G residue of AUGC box 2 may hydrogen-bond to the arginine 17 of the coat protein. However, the coat protein does not necessarily need to be in a dimer form to be able to bind (Tenllado and Bol, 2000).

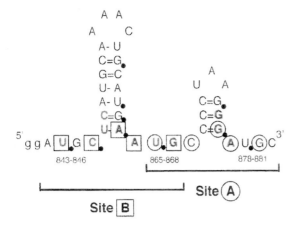

Fig. 8.11 Model for the AMV$_{843-881}$-peptide interaction. Circled and boxed letters represent putative base-specific contacts for proposed peptide binding sites A and B respectively. Filled circles represent potential phosphate contacts identified by ethylation interference. Bold letters represent base modifications that interfere with CP26 binding. From Ansel-McKinney and Gehrke (1998), with permission.

North-western and dot blot analyzes showed that the coat protein of the Ilarvirus, PNRSV, binds to the viral RNA4 even at very high salt concentrations (Pallas *et al.*, 1999).

2. AMV replication

Three ds RF RNAs corresponding to RNAs 1, 2 and 3 have been isolated from plants in which AMV was multiplying (Pinck and Hirth, 1972; Mohier *et al.*, 1974). Only mixtures of components that were infectious gave rise to any detectable RF RNA. Mohier *et al.* (1974) could detect no RF corresponding to RNA4.

Host factors may be involved in viral RNA synthesis because actinomycin D added early after inoculation of protoplasts substantially inhibited both (−)- and (+)-strand synthesis (Nassuth *et al.*, 1983b). The time course of RNA synthesis in protoplasts is shown in Fig. 8.12.

Most of the basic features of the replication of AMV as far as they have been determined appear to be similar to those described above for BMV. The major difference appeared to be the structure of the 3' terminus of the genomic RNAs and the necessity for binding of coat protein for genome activation. As noted above, recent findings indicate that the tertiary structures of AMV RNAs resemble those of bromoviruses. The pseudoknot structure is essential for AMV replication both *in vivo* and *in vitro* (Olsthoorn *et al.*, 1999). Point mutations in the five AUGC motifs had no effect on the template activity of RNA3 for (−)-strand synthesis *in vitro* (van Rossum *et al.*, 1997b) but, as noted above, could affect the coat protein binding. Binding of coat protein to the 3'-end of AMV RNAs inhibits the production of (−)-strand RNA by interfering with the formation of the pseudoknot. It is proposed that the coat protein binding has two roles, firstly in very early stages of virus replication most probably in translation, and later in infection in shutting off (−)-strand RNA synthesis (Olsthoorn *et al.*, 1999).

Using protoplasts from non-transgenic tobacco plants and from plants transformed with the P1 and P2 genes cDNA clones to RNAs 1 and 2, Neeleman and Bol (1999) showed that transgenic P1 was unable to complement mutations in RNA1 affecting the P1 gene—indicating that P1 is *cis*-acting in RNA1 replication. From the behavior of mutants of RNA3 in non-transgenic plants, it was concluded that the coat protein was required in *trans* for the replication and encapsidation of RNAs 1 and 2 but was required in *cis* for the replication and encapsidation of RNA3. Infection of protoplasts from plants co-transgenic with P1 and P2 with RNAs 1, 2 and 3 indicated that the initiation of replication of RNAs 1 and 2 required the presence of coat protein in the inoculum but that of RNA3 did not. This showed that coat protein expressed from RNA3 could not substitute for the early function of coat protein in the inoculum. Thus, coat protein in the inoculum is required to permit viral (−)-strand synthesis whereas that expressed from RNA3 after the initiation of infection is required for (+)-strand synthesis.

RNA4 is not able to influence the genetic properties of the progeny of heterologous RNAs (Bol and van Vloten-Doting, 1973). No RNA4 was found among the RNAs produced in protoplasts inoculated with virus particles containing only RNAs 1, 2 and 4 (Nassuth *et al.*, 1981). Van der Kuyl *et al.* (1990) prepared RNA transcripts from cDNA of AMV RNA3 using a virus-specific RdRp from infected plants. They

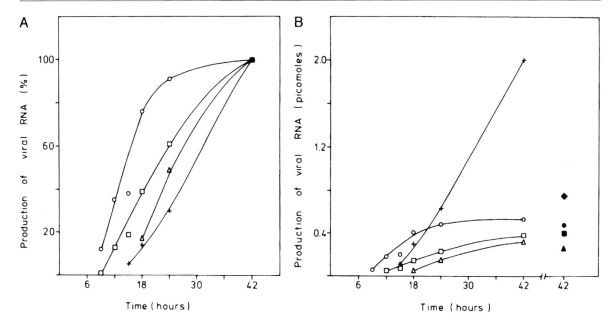

Fig. 8.12 Time course of (+)-strand AMV RNA synthesis in infected cowpea protoplasts. **(A)** The relative increase of viral RNA with the amount of each RNA present at 42 hours post-inoculation taken as 100%. **(B)** The increase of viral RNA in picomoles. RNA1 (□), RNA2 (△), RNA3(○) and RNA4 (+). Solid symbols at the right of (B) indicate the amounts of RNAs 1 (■), 2 (▲), 3 (●) and 4 (◆) present in virions 42 h after inoculation of the protoplasts. From Nassuth *et al.* (1983a), with permission.

used deletion analysis to show that a sequence located between –8 and –55 upstream of the initiation site for RNA4 synthesis was sufficient for coat-protein subgenomic RNA synthesis *in vitro*. Thus, like BMV, AMV RNA4 synthesis appears to be initiated at an internal replicase recognition site on the (−) strand of RNA3 (Nassuth *et al.*, 1983a).

3. Site of virus synthesis and assembly

The replication complex is found in the cytoplasmic fraction. Coat protein was present in the cytoplasm 6 hours after inoculation and appeared in the nucleus after 48 hours (van Pelt-Heerschap *et al.*, 1987a). Masses of AMV particles were seen in the cytoplasm of AMV-infected cells (De Zoeten and Gaard, 1969a; Hull *et al.*, 1970a) and sometimes in the nucleus (Hull *et al.*, 1970a). While the evidence is not conclusive, it is most likely that all AMV components are synthesized and assembled in the cytoplasm.

On isolating replication complexes from AMV-infected tobacco protoplasts, de Graaff *et al.* (1993) found that most of the RNA-syn-thesizing activity co-localized with intact chloroplasts in sucrose gradients. The RNA polymerase activity was sensitive to protease treatment of intact chloroplasts, indicating that enzyme activity was strongly associated with the outside of the chloroplasts.

4. *In vitro* replication

A solubilized replication complex has been isolated from AMV-infected *Nicotiana benthamiana* plants (Quadt *et al.*, 1991). This complex was AMV template-specific and the use of native dsRNA from AMV indicated that there was helicase activity. The complex comprised AMV-encoded proteins P1 and P2 from RNAs 1 and 2 together with coat protein, but (−)-strand synthesis was inhibited by AMV coat protein.

H. Replication of tobacco mosaic virus
(reviewed by Buck, 1999b)

1. RNA in infected plants

Unencapsidated forms of (−) and (+) strands of TMV RNA are found in infected plants and protoplasts both in RFs and in RIs. In protoplasts, the synthesis of (−) strands ceases 6–8 hours after inoculation, whereas (+)-strand synthesis continues for a further 10 hours (Ishikawa *et al.*, 1991a). It is considered likely that later in the infection cycle most of the (+) strands become encapsidated into virions (Aoki and Takebe, 1975; Palukaitis *et al.*, 1983).

2. Replication proteins

The 5′ ORF of TMV encodes a 126-kDa protein, the stop codon of which is read through to give a 183-kDa protein (see Chapter 6, Section VIII.H.1); thus, both proteins have the same N-terminal sequence, the 183-kDa protein having a unique C-terminal sequence. Several lines of evidence, including the following, point to both of these proteins being involved in the replication of the viral RNA:

1. Both proteins have the methyl transferase and helicase motifs and the 183-kDa protein also contains the RdRp motif.
2. Both proteins are found in *in vitro* replication complexes isolated from infected plants (Osman and Buck, 1996; Watanabe *et al.*, 1999). In these complexes, the 126- and 183-kDa proteins are in a 1:1 ratio (Watanabe *et al.*, 1999).
3. Antibodies to the 126- and 183-kDa proteins inhibited replication in the *in vitro* system (Osman and Buck, 1996).
4. A mutant engineered to produce the 126-kDa protein, but not the 183-kDa protein, could not replicate (Ishikawa *et al.*, 1986).
5. A mutant in which the UAG amber stop codon was mutated to a UAU tyrosine codon, thus producing only the 183-kDa protein, replicated in protoplasts to only 20% the level of wild-type virus. In plants, the mutation reverted to a UAA ochre stop codon, the virus then replicating more efficiently (Ishikawa *et al.*, 1986).
6. The helicase domain of the 126-kDa protein interacts with the intervening region of the 183-kDa protein (Goregaoker *et al.*, 2001).

To examine the functions of the 126- and 183-kDa proteins, Lewandowski and Dawson (2000) developed a bipartite system to express the two proteins from separate RNAs. The 183-kDa protein had all the functions expected from its motifs and it recognized promoters for (-)- and (+)-strand synthesis, transcribed sgRNAs, capped RNAs and replicated defective RNAs; it or its mRNA also moved from cell to cell within the plant. Addition of the 126-kDa protein increased the rate of replication approximately 10-fold; it functioned primarily in *cis*. Thus, efficient replication requires the 183-kDa protein to form a heterodimer with the 126-kDa protein which is probably bound to the template RNA. However, there may be still other functions for the 126-kDa protein. As noted in Chapter 7 (Section V.B.9.c), it is produced in about ten times the amount of the read-through product, the 183-kDa protein, yet the two proteins form a 1:1 heterodimer. Thus, what is the function of the other 90% of the 126-kDa protein?

3. *Cis*-acting factors on TMV RNA

The 3′ UTR of TMV RNA can be folded into three structural domains, a 3′ domain mimicking a tRNA acceptor branch, an analog of a tRNA anticodon branch, and an upstream domain comprising three pseudoknots each containing two double-helical segments (van Belkum *et al.*, 1985; Felden *et al.*, 1996). Mutational analysis indicated that one of these double-helical segments is essential for replication, the secondary structure rather than the primary structure being the important feature (Takamatsu *et al.*, 1990a).

Osman *et al.* (2000) identified four domains in the 3′ untranslated region, D1 equivalent to tRNA acceptor arm, D2 similar to a tRNA anticodon arm, an upstream domain D3 and a central core C which connects D1, D2 and D3 domains. They showed that all four domains were required for promotion of (–)-strand RNA synthesis and that the D2 and C domains bound the RNA polymerase with highest efficiency.

The tRNA-like structures of most TMV strains can be aminoacylated by the host histidyl-tRNA synthetase, but that of the cowpea strain (now known as SHMV) resembles that of TYMV in being aminoacylated with valine. Chimeras in which the 3′ UTR of TMV-L is replaced with that of TMV-OM, CGMMV or SHMV can replicate in protoplasts and plants, though the latter two replicate to a lower level than TMV-L (Ishikawa et al., 1988). When the 3′ UTR of TMV-L is replaced with that of BMV (a tyrosine acceptor) the chimera can replicate but to a much lower efficiency than TMV-L itself (Ishikawa et al., 1991b). This indicates that the TMV polymerase has some flexibility in its recognition site – which is in contrast to that of BMV which does not recognize BMV RNA3 with the 3′ UTR replaced by TMV-L 3′ UTR (Ishikawa et al., 1991b).

Sequences at the 5′-end of TMV RNA are important for replication (Takamatsu et al., 1991). Large deletions in the 5′ region and deletion of nucleotides 2–8 abolished replication but other small deletions in the 5′ UTR did not. This suggests that the 5′ replicase binding site may be complex.

Internal sequences in the TMV genome have been recognized to inhibit replication in trans (Lewandowski and Dawson, 1998a). Deletion of the region between nucleotides 3420 and 4902 (sequences encoding the RdRp domain of the 183-kDa protein) created a replication-defective RNA (dRNA) that could be replicated in trans by TMV.

The satellite (STMV) requires the helper virus for its replication, indicating that it has the necessary trans-acting sequences. STMV is described in Chapter 14 (Section II.A.4). Although STMV and various subgenomic replicons use TMV replicase in trans, the synthesis of (−) strands of genomic RNA prefers interactions in cis (Lewandowski and Dawson, 1998b).

4. In vitro assays

In vitro RNA synthesis was initially detected only in more or less crude fractions from infected cells. A virus-specific RdRp is associated with the membrane fraction (Zaitlin et al., 1973; Romaine and Zaitlin, 1978). This could catalyze the synthesis of both (+)- and (−)-sense strands.

Young and Zaitlin (1986) analyzed the RNA structures synthesized in vitro by a crude 30 000g pellet fraction from TMV-infected tobacco leaves using DNA probes of (+) or (−) sense. The major products labeled with radioactive nucleotides had the expected properties for TMV RF and RI (see Fig. 8.6). Much more label was incorporated into (+) than into (–) strands in the RF fraction. In the RI fraction

only labeled (+) strands could be detected. The system had a number of limitations: synthetic activity was short-lived; the system did not respond to added template RNA; and few, if any, free progeny viral RNA molecules were formed. These results are very similar to those reported by Watanabe and Okada (1986). They used a 2500g pellet and showed that coat protein sgRNA could also be synthesized by this fraction. They detected RF of genomic size but not of coat protein sg size. Furthermore, they detected no synthesis of (−)-strand RNA of this size, indicating that the subgenomic coat protein RNA was synthesized by internal initiation on the (−)-strand genomic RNA rather than by transcription from a subgenomic (−)-sense strand.

Osman and Buck (1996) isolated a membrane-bound replication complex from ToMV-infected tomato leaves. This replication complex was template-dependent, template-specific and primer-independent, and it was able to initiate RNA synthesis on templates containing the 3′-terminal sequences of (+)-strand RNA. The main product of the reaction with homologous template was genome-length ssRNA together with some dsRNA. Both the 126- and 183-kDa proteins were present in the replication complex, and antibodies to these proteins inhibited the template-specific replication.

Watanabe et al. (1999) immunopurified a solubilized replication complex from TMV (OM strain)-infected tobacco plants and showed that it comprised a heterodimer of the 126-and 183-kDa proteins. The complex could synthesize (−)-strand from (+)-strand templates made up from the 3′ 249 nucleotides of TMV RNA.

5. Host proteins involved with replication complex

The involvement of host protein(s) in TMV replication was first suggested by the sensitivity to actinomycin D of an early step in the virus multiplication (Dawson and Schlegel, 1976; Dawson, 1978b). However, as the inoculations were with virus particles, there was the possibility that the actinomycin D treatment could affect a stage such as virus disassembly. In two unlinked single recessive mutations of Arabidopsis thaliana, tom-1 and tom-2, TMV replication is reduced to very low levels

(Ishikawa *et al.*, 1993; Ohshima *et al.*, 1998); these mutations did not affect the replication of CMV, TCV or TYMV. The product of the *Tm-1* resistance gene, conferring resistance to TMV in tomato (see Chapter 10, Section III.A.1), has been suggested to be an altered form of a host protein involved in virus RNA replication (Meshi *et al.*, 1988). Resistance-breaking strains are mutated in the helicase region of the 126-/183-kDa proteins, which might be the region to which a putative host protein binds. However, alternate hypotheses suggest that the Tm-1 protein may interact with the viral replicase to inactivate it, or that the host response in Tm-1 plants to non-resistance-breaking TMV strains may result in non-virus-specific inhibition of virus replication (reviewed in Buck, 1999b).

A highly purified TMV replication complex contained, as well as the 126- and 183-kDa proteins, three host proteins of 56, 54 and 50 kDa (Osman and Buck, 1997). The 56-kDa protein is immunologically related to the 55-kDa (GCD10) subunit of translation initiation factor eIF-3 from yeast and wheatgerm (56 kDa). Using the yeast two-hybrid system, Taylor and Carr (2000) showed that yeast GCD10 interacts specifically with the methyl transferase domain shared by the 126- and 183-kDa TMV proteins. The association of a GCD10-like protein with the TMV replication complex suggests that, *in vivo*, replication and protein synthesis may be closely connected. Watanabe *et al.* (1999) found two putative host proteins of 34 kDa and 220 kDa in their purified replication complex; the migration position of a 55-kDa GCD10 protein was obscured by the immunoglobulin heavy chain used in purifying the replication complex.

The *Arabidopsis* gene, *tom-1*, is necessary for the efficient multiplication of TMV-Cg and TMV-L at the single-cell level, but is not implicated in the multiplication of CMV, TYMV or TCV (Ishikawa *et al.*, 1991c, 1993). Map-based cloning shows that *tom-1* protein has the structure of a multi-pass transmembrane protein (Yamanaka *et al.*, 2000). *Tom-1* interacts with the helicase domain of the 126(130)/183-kDa replicase proteins and it is suggested that it participates in the *in vivo* formation of the replication complex by serving as a membrane anchor.

6. Site of replication

The cellular site for TMV replication is associated with cytoplasmic inclusions or viroplasms that enlarge to form what are termed 'X bodies' composed of aggregates of tubules embedded in a ribosome-rich matrix (Hills *et al.*, 1987; Saito *et al.*, 1987b) (see Fig. 8.4). Initial cytological studies suggested that these were associated with the nucleus (Bald, 1964; Smith and Schlegel, 1965); but subsequently it was shown that RIs and the RdRp are associated with cellular cytoplasmic membranes (see Beachy and Zaitlin, 1975; Osman and Buck, 1996). The use of the green fluorescent protein (GFP) (see Chapter 9, Section II.B) as a reporter for TMV expression identified irregularly shaped structures in cells that contained the viral replicase (and movement protein); these structures were derived from the endoplasmic reticulum (Heinlein *et al.*, 1998; Reichel and Beachy, 1998). Using *in situ* hybridization and immunostaining, Más and Beachy (1999) showed that TMV RNA, the viral replicase and the viral movement protein co-localized at the endoplasmic reticulum, including the perinuclear endoplasmic reticulum and associated with ER-related vesicles. This is discussed further in Chapter 13.

I. Replication of potyviruses

The potyvirus genome is expressed as a polyprotein that is processed into 10 or so products (see Chapter 7, Section V.B.1). Although most of the protein products are involved in some way in viral replication, there is a set of core replication proteins, the CI helicase, the $6K_2$, the NIa VPg proteinase and the NIb RdRp. As can be seen from the genome map (see Fig. 8.13) these form a block of proteins. These proteins are analogous, in terms of gene order and sequence motifs, to the poliovirus 2C, 3A, 3B, 3C and 3D proteins respectively (Domier *et al.*, 1987). Thus, it is likely that there are similarities in functions and in the replication strategy. I will consider the properties of these proteins in the order in which they occur in the polyprotein and then bring together the information on the interactions between them.

The CI proteins of potyviruses have amino acid motifs indicative of helicases (see Section

IV.B.2). RNA binding, NTPase and helicase activities have been demonstrated in TamMV and PPV (Eagles *et al.*, 1994; Fernández *et al.*, 1995). Mutagenesis of one of the seven conserved amino acid motifs characteristic of helicases (motif V) that abolished the helicase activity of *E. coli*-expressed PPV CI protein also affected NTP hydrolysis but not RNA binding (Fernández *et al.*, 1997). These mutations affected the ability of an infectious cDNA clone to replicate in protoplasts and to infect whole plants unless they had reverted to wild-type sequence. The N-terminal 177 amino acids of PPV CI protein are involved in CI–CI binding (López *et al.*, 2001).

The TEV 6K$_2$ protein is membrane-associated (Restrepo-Hartwig and Carrington, 1994) and mutagenesis demonstrated that it is involved in virus replication. The association with membranes is mediated by a central 19-amino-acid hydrophobic domain (Fig 8.13) (Schaad *et al.*, 1997a).

Using fluorescent fusion proteins containing the 6-kDa protein, Schaad *et al.* (1997a) showed that it associated with large vesicular structures derived from endoplasmic reticulum (ER). On infection with TEV, the ER network appears to collapse into discrete aggregated structures and viral RNA in replication complexes is associated with ER-like membranes. Replication of PPV RNA has also been shown to be associated with crude membrane fractions (Martin and Garcia, 1991).

The NIa processing product comprises two regions, the N-terminal VPg and the C-terminal protease; the properties of the VPg is described in Chapter 4 (Section III.A.3.c) and the NIa protease in Chapter 7 (Section V.B.1). *E. coli*-expressed

PVA VPg and NIa both bind RNA (Merits *et al.*, 1998), as does the proteinase domain of TEV NIa (Daròs and Carrington, 1997).

The NIb protein has the amino acid sequence motifs suggestive of an RdRp, and nucleic acid binding has been demonstrated for this protein from PVA that had been expressed in *E. coli* (Merits *et al.*, 1998). Mutations to the GDD motif, the nuclear localization signal and one that debilitated the NIb N-terminal cleavage site abolished the ability of an infectious clone of TEV to replicate in protoplasts, as did the complete removal of this gene (Li and Carrington, 1995). These mutants, except for that affecting the N-terminal cleavage, could be complemented to varying degrees by the transgenic expression of NIb.

Various studies have revealed a series of interactions between several of these potyviral replication-associated proteins. Using the yeast two-hybrid system, Li *et al.* (1997) found that TEV NIb interacted with NIa, the interaction involving the protease region and not the VPg region. A similar interaction between the NIb and NIa of TVMV was found in binding studies, but in this case the VPg region was involved (Hong *et al.*, 1995; Fellers *et al.*, 1998). It is uncertain as to whether this difference reflects differences between individual potyviruses or differences in techniques. The NIa and VPg proteins (from TVMV) stimulated the NIb-associated RNA polymerase activity, the stimulation being mainly attributable to the VPg (Fellers *et al.*, 1998). Using conditional and suppressor mutants Daròs *et al.* (1999) demonstrated that the interaction between NIa and NIb is essential for TEV genome replication.

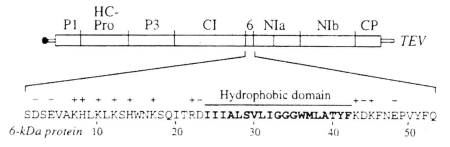

Fig. 8.13 Schematic of the TEV genome and amino acid sequence of the 6-kDa protein. The map is drawn to scale with the individual coding regions indicated above the diagram. Sequences coding for proteolytic cleavage sites are indicated by vertical lines. The VPg at the 5′ terminus is represented as a filled circle. The sequence of the 6-kDa protein is drawn with the central hydrophobic domain in bold and charged residues indicated. From Schaad *et al.* (1997), with kind permission of the copyright holder, © Oxford University Press.

The first (N-terminal) and third proteins (P1 and P3) of the PVA polyprotein interact with each other and with CI, VPg, NIa and NIb (Merits *et al.*, 1999). The significance of these interactions is not known. Using the yeast two-hybrid system, Wittmann *et al.* (1997) found that the TuMV VPg interacted with the translational eukaryotic initiation factor (iso) 4E from *Arabidopsis thaliana*.

Thus, a picture is being built up of the composition and assembly of potyviral replication complexes. This involves both interactions between the gene products and control of the processing of the polyprotein. Three polyproteins containing the 6K protein have been detected in TEV-infected cells, CI/6K, 6K/VPg and 6K/NIa (VPg plus proteinase) (Restrepo-Hartwig and Carrington, 1994). It is suggested that the replication complex assembles on ER-like membranes, initially by the binding of the 6K protein to those membranes. This also brings in the NIa protein as the membrane-binding activity of the 6K protein overrides the nuclear localization signal of NIa (Restrepo-Hartwig and Carrington, 1992). NIa brings the viral RNA into the complex by its RNA-binding ability. Then the 6K–NIa complex recruits the NIb product through the NIa–NIb interaction delivering the polymerase to the RNA. The controlled processing of the potyviral polyprotein enables the elements of the complex to be assembled and then released when they have completed their function. Thus, NIa, having recruited NIb, is released from the complex and its nuclear-localization function takes over targeting it to the nucleus.

It is thought that the VPg is involved in the initiation of RNA synthesis in a manner similar to that proposed for picornaviruses (Wimmer *et al.*, 1993; Pfister *et al.*, 1999). It is likely that host proteins and other potyviral proteins are also recruited into the complex.

J. Replication of *Comoviridae*

1. Comoviruses

As described in Chapter 6 (Section VIII.B.1), CPMV has a bipartite genome, each part of which encodes a polyprotein that is cleaved to functional proteins (Fig. 7.28). The proteins encoded by RNA1 are sufficient for replication, and the organization and sequence motifs of these proteins show a strong resemblance to those of animal picornaviruses (Frannssen *et al.*, 1984b). In this, comoviruses resemble potyviruses and it is likely that they have many features of replication in common. However, much less is known about comovirus replication than about that of potyviruses, and there will almost certainly be differences in detail.

2. General features of CPMV replication

In inoculated cowpea protoplasts there was a rapid increase in infectious virus between 9 and 24 hours after inoculation and over 10^6 progeny virus were produced per protoplast (Hibi *et al.*, 1975). Two dsRNA species have been characterized from infected *Vigna* leaves corresponding to the two ss genome RNAs (van Griensven *et al.*, 1973).

To investigate *cis*- and *trans*-acting elements in CPMV, van Bokhoven *et al.* (1993a) made heterologous sequence insertions at various places in the ORF of RNA1, leaving the 5' and 3' UTRs untouched. None of the mutant RNAs were able to replicate and the addition of wild-type RNA1 did not complement their replication. This indicates that RNA1 must replicate in *cis* and cannot be replicated in *trans*. The replication of RNA2 must be in *trans* as the replication proteins are encoded by RNA1. However, mutagenesis of RNA2 indicated that the N-terminal domain of the 58-kDa protein is required in *cis* for replication. It was suggested that the N-terminal domain of the 58-kDa protein targets RNA2 to the replication complex.

Transient expression from expression vectors of RNA1 based on the 35S promoter was able to support the replication of RNA2 in cowpea protoplasts (van Bokhoven *et al.*, 1993b). However, replication was supported only when the entire 200-kDa polyprotein was expressed and not when vectors containing the 170-kDa, 110-kDa or 87-kDa coding sequences were used.

From experiments using cycloheximide and chloramphenicol, with the polypeptides of ribulose bisphosphate carboxylase as an internal control, Owens and Bruening (1975) concluded that both the coat proteins found in CPMV are synthesized in the cytoplasm. Progeny virus particles in crystalline arrays have been seen in cytoplasm and vacuoles, but not within nuclei, chloroplasts or mitochondria (Langenberg and Schroeder, 1975). CPMV-infected cells develop quite large cytopathological structures (Fig. 8.14).

These contain groups of membranous vesicles forming a reticulum. The vesicles contain stranded material with the staining appearance of a ds nucleic acid. Virus particles are also present. From a detailed study combining electron microscopy and cell fractionation procedures, De Zoeten *et al.* (1974) concluded that the viral dsRNAs are located in these structures, which can therefore be regarded as viroplasms. These viroplasms have also been observed in infected cowpea and tobacco protoplasts (Hibi *et al.*, 1975). The vesicles are made up from a massive proliferation of a specific region of the cortical ER and this does not involve the Golgi apparatus (Carette *et al.*, 2000). CPMV replication was strongly inhibited by cerulenin, an inhibitor of *de novo* lipid synthesis, which suggests that the proliferation of membranes involved lipid biosynthesis.

This is in contrast to AMV and TMV (Carette *et al.*, 2000).

The use of green fluorescent protein tags allowed the stages in the formation of the membranous structures to be followed (Carette *et al.*, 2000). In early stages there were several fluorescent bodies at the cortical ER. These then aggregated into one large fluorescent body, usually located near the nucleus. Similar structures were observed in insect cells expressing the 200-kDa polyprotein from RNA1 using a baculovirus expression vector (van Bokhoven *et al.*, 1992).

A host-encoded RdRp activity increases some 20-fold following infection of leaves with CPMV. No such increase occurs in infected protoplasts. The activity is associated with a 130-kDa polypeptide. Although this enzyme is active in the crude membrane fraction isolated from infected leaves, it plays no role in CPMV replication (Dorssers *et al.*, 1983, 1984; van der Meer *et al.*, 1984).

3. CPMV replication complex

A CPMV RNA replication complex, containing a 110-kDa virus-coded polypeptide, has been isolated from infected cowpea leaves (Dorssers

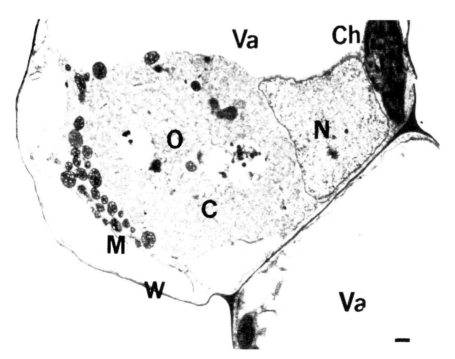

Fig. 8.14 Intracellular site of CPMV synthesis. An electron micrograph of a CPMV-infected mesophyll cell. Note the cytopathological structure (C) associated with mitochondria (M) and the nucleus (N); the large vacuoles (Va); wall (W); and chloroplasts (Ch). Bar = 0.5 μm. From De Zoeten *et al.* (1974), with permission.

et al., 1984). Other host-specific polypeptides were also present in the complex but their role, if any, in RNA replication is not established. However, actinomycin D inhibits RNA synthesis if applied immediately after inoculation, but not if supplied later (de Varennes et al., 1985). This suggests that some host-coded protein may be needed for RNA replication. This replication complex is capable of elongating nascent viral RNA strands in vitro to full-length RNA. Possible models for RNA replication are discussed by van Kammen and Eggen (1986). The initiation of viral RNA replication may be linked to processing of the polyprotein, and the VPg may play some role, but further work is needed. The position of the 110-kDa polypeptide in the polyprotein (see Fig. 7.28) requires that, for every molecule of this protein produced, the other B-RNA-encoded polypeptides must also be formed, if not fully cleaved. Other B-RNA-encoded proteins are in fact present in these cytopathic structures (Wellink et al., 1988) but their identity has not been established.

Only a small fraction of the nonstructural proteins detected in infected cowpea leaves was active in RNA replication (Eggen et al., 1988). Thus, virus replication proteins may be used only once, perhaps because of a strong coupling between polyprotein processing and replication.

4. Nepoviruses

In vivo studies show that nepovirus RNA1 can replicate in the absence of RNA2 and therefore must carry the RNA polymerase and VPg genes. The replication of nepoviruses has not been studied in detail. In cucumber cotyledons, TRSV-induced RNA polymerase activity rises rapidly to a maximum at about 3 days after inoculation and then falls (Peden et al., 1972). TRSV presumably replicates by means of an RI form of the RNA, but only heterogeneous low-molecular-weight dsRNA has been isolated from infected plants (Rezaian and Francki, 1973).

Mutagenesis of GFLV RNA2 showed that protein 2A (analogous to the CPMV 58-kDa protein described above) is necessary for RNA2 replication; other regions in RNA2 are also required (Gaire et al., 1999). Protein 2A was fused to the green fluorescent protein (2AGFP)

and expressed in tobacco-BY2 protoplasts. When expressed on its own, 2AGFP showed as punctate structures evenly distributed through the cytoplasm. When co-transfected with GFLV RNAs, 2AGFP was found located with two of the RNA1-encoded replication proteins, the protease 1D and the VPg 1C. Newly synthesized viral RNA, identified by incorporation of 5-bromouridine 5'-triphosphate, was found in the same location. This supported the suggestion for CPMV that the N-terminal part of the RNA2 polyprotein is responsible for targeting RNA2 to the replication complex.

The viral coat protein appears to be synthesized on cytoplasmic ribosomes, and both electron microscope and cell fractionation experiments indicate that TRSV replicates in the cytoplasm (Rezaian et al., 1976), probably in association with characteristic membranous vesicles.

K. Replication of turnip yellow mosaic virus

1. Virus increase in protoplasts and leaves

In infected Brassica protoplasts, virus production probably begins somewhat earlier than 12 hours after inoculation. Renaudin et al. (1975) found that $1-2 \times 10^6$ virus particles per protoplast were produced 48 hours after infection, which is similar to estimates made for cells in intact leaves (Matthews, 1970). However, these latter estimates were made on extracts of whole infected leaves, without any account being taken of variation in size and virus productivity of different cell types or of the percentage of cells actually infected in the leaf. Based on virus particles per infected mesophyll cell, protoplasts support only about one-tenth of the virus produced by similar cells in the intact leaf. In protoplasts, empty protein shell production is not reduced as much as that of virus production. The two types of particle are produced in almost equal amounts compared with about 20% of empty shells in leaves. The approximate maximum rates of particle production per infected protoplast are about the same for cells infected in vitro or in vivo. In protoplasts, virus increase ceased after about 48 hours, but continued for about 10 days in intact leaves.

Disease development in protoplasts is not synchronous as judged by rounding and clumping of chloroplasts (Sugimura and Matthews, 1981).

2. TYMV replication in relation to cytological effects

The *Tymovirus* genus produces very characteristic cytological alterations in infected cells. Some of these changes are intimately associated with virus replication and will be discussed here. Other, secondary cytological effects are described in Chapter 3.

The most characteristic cytological feature of tymovirus infection is the small peripheral vesicles on chloroplasts (Fig. 8.15). These vesicles contain viral RI molecules and viral RdRp, are present in all infected host species, and occur in all parts of the plant that have been examined (evidence summarized by Matthews, 1973). How the vesicles are formed is not established. In a formal sense, they can be considered to be derived from invaginations of the chloroplast membranes, but it is not known whether pre-existing membrane is modified or whether additional chloroplast membrane is synthesized.

Using the presence of vesicles as a marker of infection, it was possible to follow cytologically the invasion of a single systemically infected leaf as the infection spread out from the phloem elements of the small veins (Hatta and Matthews, 1974). The more important stages are illustrated in Figs. 8.16 and 3.6 and described in the legends.

The earliest stage at which TYMV particles could be detected was at stage D, when crystalline arrays were seen in the cytoplasm next to the electron-lucent zones (see Fig. 3.6A). It was concluded that the clumping of the chloroplasts is a secondary effect of infection, not essential for virus synthesis. In successive stages later than C, more and larger arrays of virus particles appeared in the cytoplasm while the electron-lucent material disappeared. The electron-lucent material is an accumulation of viral coat protein, mainly or entirely in the form of pentamers and hexamers, and these are used for the assembly of virus particles.

The role of the ER that appears transiently at stage B (Fig. 8.16) is not established. From the timing of its appearance, and its very specific position in the cell, it could be involved in any or all of the following processes: (1) the

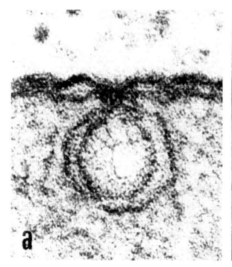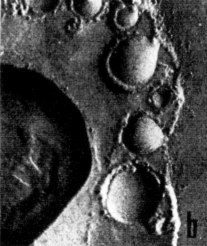

Fig. 8.15 Fine structure of TYMV-induced peripheral vesicles in the chloroplasts of infected Chinese cabbage cells. **(a)** Thin section showing continuity of inner chloroplast and outer vesicle membranes and stranded material inside the vesicle with the staining properties of ds nucleic acid (×235 000). Courtesy of S. Bullivant. **(b)** Fine structure of vesicle membranes revealed by freeze-fracturing of isolated chloroplasts (×92 000). From Hatta *et al.* (1973), with permission.

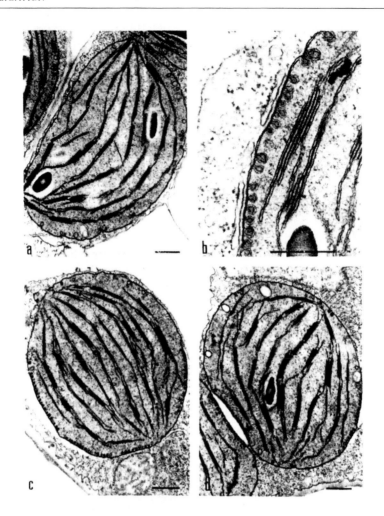

Fig. 8.16 Some steps in the sequence of cytological changes induced by TYMV infection in Chinese cabbage leaf cells. **(a)** Scattered small peripheral vesicles; chloroplasts are otherwise normal. **(b)** Chloroplasts are now swollen. Scattered vesicles are still present, but many clusters of vesicles are also present. In freeze-fracture preparations, these can be seen as hexagonal arrays. ER is present in the cytoplasm over the clustered vesicles. The zones of ER over each cluster of vesicles are connected by strands of ER, making a network over the chloroplast surface (Hatta and Matthews, 1976). **(c)** Electron-lucent areas have appeared in the cytoplasm over the clustered vesicles. This material reacts with antibodies against TYMV. Freeze-fracturing indicates that it may be in the form of pentamers and hexamers. **(d)** The chloroplasts have become clumped with electron-lucent areas in contact. At a later stage, the electron-lucent material is replaced by virus particles. From Hatta and Matthews (1974), with permission.

synthesis of a protein or proteins needed to form the peripheral vesicles, (2) synthesis of viral RNA polymerase, and (3) synthesis of viral coat protein.

Viral coat protein in the form of empty protein shells accumulates in large quantities in the nuclei of cells infected with TYMV and other tymoviruses (Hatta and Matthews, 1976). No virus particles have been identified in nuclei.

The protein can be detected in nuclei of TYMV-infected Chinese cabbage cells before the electron-lucent zones appear in the cytoplasm. The amount increases with time until the nuclei are almost filled with empty protein shells (see Fig. 3.6B). It must be assumed that the viral protein is synthesized in the cytoplasm and moves into the nucleus, either via the nuclear pores or directly through the nuclear membrane.

3. TYMV replication complexes

In extracts from Chinese cabbage plants infected with TYMV, RNA polymerase activity is associated with the chloroplast fraction (Ralph and Wojcik, 1966; Bové et al., 1967). The enzyme has been detected in the bounding membrane fraction of the chloroplasts that contains the small virus-induced vesicles (Laflèche et al., 1972; Bové et al., 1972). Mouches et al. (1984) purified the TYMV replicase from infected Chinese cabbage leaves. The enzyme was template-dependent and showed specificity for TYMV RNA. It consisted of two subunits. One of 115 kDa was virus-coded as shown by serological cross-reaction between this protein and a 115-kDa in vitro translation product of genomic RNA (Candresse et al., 1986). The other, of 45 kDa, is host-encoded, because antibodies raised against it do not react with any TYMV in vitro translation products (Bové and Bové, 1985). In vivo the viral replicase is tightly bound in a replication complex involving the chloroplast envelope membrane. By contrast, the host-encoded RdRp found in healthy and infected leaves is a soluble enzyme of different size and with different template specificity (Mouches et al., 1984).

RNA replication activity has been isolated from detergent-solubilized chloroplast membranes of TYMV-infected Chinese cabbage leaves (Deiman et al., 1997a; Singh and Dreher, 1997). This activity is template-dependent and, although it could not synthesize full-length ($-$) strands from genomic TYMV RNA, it could form shorter fragments based on the 3'-end. As described in Chapter 4 (Section III.A.3.b), the 3'end of TYMV RNA has a tRNA-like structure and forms a pseudoknot. In studies on the role of this tRNA-like structure (Figs 4.4 and 4.5), Deiman et al. (1997b) and Singh and Dreher (1997) showed its importance in the initiation of ($-$)-strand synthesis. Disruption of the pseudoknot structure reduced transcriptional efficiency by about 50%. ($-$) strands were shown to arise by de novo initiation with the insertion of GTP opposite the penultimate C residue of the 3'-terminal –CCA (Singh and Dreher, 1997). The minimum template length is the 3' nine nucleotides and transcriptional efficiency increases with template length (Deiman et al.,

2000). This observation, coupled with a mutational analysis of the 3' terminal hairpin, indicated that proper base-stacking contributes to efficient transcription initiation. Thus, it is likely that the structure at the 3'-end of the RNA targets the 3' CCA for initiation and prevents initiation at internal CCA sites.

4. In vivo replication of TYMV

Garnier et al. (1980), using a combination of electron microscopy and other techniques, demonstrated that, in vivo, the TYMV RI is mainly in ss form and that the dsRNA isolated from infected leaves (Bockstahler, 1967) is mainly an artifact generated during isolation (see Fig. 8.6). However, late in the infection cycle some dsRF may accumulate in vivo.

The idea that most of the viral RNA synthesis occurs in the small peripheral vesicles found in diseased chloroplasts (Fig. 8.15) is supported by the following additional evidence: (1) When observed in thin sections, many vesicles can be seen to contain stranded material with the staining properties expected for a ds nucleic acid (Fig. 8.15) (Ushiyama and Matthews, 1970). (2) The ds form of TYMV RNA is associated with the chloroplast fraction in extracts of infected tissue rather than with the nuclei or soluble fraction (Ralph and Wojcik, 1966; Bové et al., 1967). (3) Using autoradiography with ^{3}H-labeled uridine, Laflèche and Bové (1968) showed that radioactivity accumulated in the spaces between clumped chloroplasts in diseased cells. (4) Immunogold labeling located the viral replicase near the periphery of clumped chloroplasts in infected cells (Garnier et al., 1986). The 66 kDa protein containing the RdRp motif is located in the membrane vesicles in the chloroplast envelope (Prod'homme et al., 2001).

In spite of the close association of TYMV with the chloroplasts, the presence of chlorophyll is not necessary for TYMV replication. The virus multiplied in chlorophyll-less protoplasts from etiolated Chinese cabbage hypocotyls (Fernandez-Gonzalez et al., 1980).

TYMV infection also stimulates nucleic acid synthesis in the nucleus. Laflèche and Bové (1968) noted that ^{3}H-labeled uridine accumulated in or near the nucleoli in TYMV-infected cells treated with actinomycin D, as well as near the margins of the clumped chloroplasts. Bedbrook *et al.* (1974) showed that radioactively labeled nuclear nucleic acids prepared from TYMV-infected tissue contained two components not present in nucleic acids from equivalent healthy leaves. The significance of these events in the nuclei for TYMV replication remains to be established.

The unusually high cytidylic acid content of TYMV RNA made it possible to distinguish fairly readily between ss and ds TYMV RNA by nucleotide ratios of RNA labeled *in vivo* with ^{32}P. The results fitted with the view that the viral RNA is produced by an asymmetric semiconservative process in which the (+)-sense strands are produced more frequently than the (−)-sense strands (Ralph *et al.*, 1965). More detailed studies indicated that, at very early and very late stages of infection, (−)-sense strand synthesis predominates (Bedbrook and Matthews, 1976).

The subgenomic RNA containing the coat protein gene becomes labeled *in vivo* with ^{32}P more rapidly than genomic RNA, presumably reflecting more rapid synthesis of the coat mRNA (Matthews *et al.*, 1963). However, the site of production of this mRNA is not established.

L. Replication of other (+)-strand RNA viruses

Although the replication of the other (+)-strand RNA viruses has not been studied in the detail of those described above, there are various observations that are relevant to the understanding of this aspect of the viral infection cycle. Some of these are discussed in this section.

1. *Hordeivirus* genus

For several years, it appeared that some strains of BSMV, the type member of the genus, contained two genome pieces while others had three. This was because RNAs 2 and 3 of some strains were difficult to separate by gel electrophoresis. Full-length genomic cDNA clones of two strains of BSMV were transcribed *in vitro* in the presence of 5′ cap analogs. A combination of three transcribed RNAs was infectious for barley, demonstrating the tripartite nature

of the genome (Petty *et al.*, 1989). Hordeiviruses have been reviewed by Lawrence and Jackson (1999).

a. *Virus replication* in vivo

In barley protoplasts inoculated with BSMV, progeny virus was first detected after about 12 hours, increasing to a maximum of about 2×10^6 particles per protoplast at 48 hours (Ben-Sin and Po, 1982). In chronically infected younger leaves of barley in which virus is replicating, free viral RNA is difficult to detect. By contrast, during acute systemic infections, large amounts of uncoated RNA accumulate and this RNA is associated with a 60-kDa protein (Brakke *et al.*, 1988).

Using an immunogold cytochemical technique in which tissue was stained after sectioning, Lin and Langenberg (1985) localized dsRNA in vesicles at the periphery of proplastids in cells of infected barley roots. By following sequential cell-to-cell infection in the meristematic region of infected roots and shoots, Lin and Langenberg (1984a) established a series of cytological events. The earliest event (stage 1) was the appearance of peripheral vesicles in the proplastids of the root meristem cells. Viral protein but not virus particles was detected late in stage 1. Stage 2 was marked by the appearance of rod-shaped virus particles oriented vertically and attached to the proplastid membrane. In stage 3, rods were also seen attached to the endoplasmic reticulum. In stage 4, viral protein (but very few rods) was found in the nuclei. No virus protein was seen in heterochromatin.

b. *Viral genes and replication*

The tripartite genome of BSMV contains seven ORFs (see Fig. 6.50). The products of αa and γa ORFs comprise the essential BSMV-encoded replication components (Petty *et al.*, 1990). To analyze the *cis*-acting elements in the three RNAs required for replication, Zhou and Jackson (1996) constructed a series of deletions in infectious cDNA clones of the ND18 strain. Mixtures of transcripts of these and of wild-type α and γ components were inoculated to protoplasts with the following observations:

1. The 5′ and 3′ non-coding regions were required for replication.
2. All internal deletions in RNA α prevented replication.
3. Deletions in RNA β involving each ORF and the poly(A) region were amplified by RNAs α and γ, but for some there was reduced accumulation.
4. Deletions in the first 507 nucleotides of γa ORF abrogated expression.
5. Deletions in the central and 3′ regions of γa, in the γa–γb intergenic region and in the γb ORF could be amplified *in trans.*
6. γb protein participates in homologous interactions forming dimers, trimers and higher order complexes (Bragg *et al.*, 2001).

2. *Tombusvirus* genus

The single component of tombusvirus RNA contains five major genes and has an additional 3′ sixth small open reading frame (pX) (see Fig. 6.23 for genome organization). ORF 2 contains the GDD motif characteristic of an RdRp and is expressed by read through from ORF 1 which lacks any replicase motifs (methyltransferase or helicase). However, there are several lines of evidence pointing to the core replicase being the products of ORFs 1 and 2.

1. Transgenic plants expressing ORFs 1 and 2 of CymRSV support the replication of a DI RNA that is incapable of autonomous replication (Kollár and Burgyán, 1994).
2. Using anti-p33 (ORF 1) and anti-p92 (ORF 2) sera, Scholthof *et al.* (1995b,c) showed that these proteins are expressed early in TBSV infections and that they are associated with membranes. p33 was about 20-fold more abundant than p92. Mutants in which the UAG stop codon at the end of ORF 1 had been converted to UAC or UAU (both tyrosine) did not multiply in protoplasts.
3. The uncoupled expression of TBSV p33 and p92, as described above for TMV (Section IV.H.2), allows the replication of the viral RNA (Oster *et al.*, 1998). Both proteins are required for viral RNA amplification, but if there are interactions between them they are not particularly strict. The C-terminal ~6% of p33 is necessary for it to function.

4. Expression of p36 (ORF I) of CIRV in yeast causes membrane proliferation and altered mitochondrial morphology (Rubino et al., 2000).

Disruption of the pX ORF of TBSV by deletions or insertions gives mutants that were incapable of accumulating detectable levels of virus in cucumber or *Nicotiana benthamiana* protoplasts (Scholthof and Jackson, 1997). However, mutants in which a stop codon was introduced into the middle of this ORF were infectious, indicating that the effect was due to RNA rather than the protein product. This requirement for *cis*-acting RNA sequences is host-specific, as mutation that affected virus accumulation in cucumber or *N. benthamiana* did not do so in *Chenopodium quinoa* protoplasts or plants.

The analysis of DI RNA associated with tombusvirus infections gives some indication of viral RNA sequences involved in replication (see Section IX.C.2).

3. Tobacco rattle virus

TRV has a bipartite genome with an organization similar to that of TMV (see Fig. 6.37). An outline of the events leading to TRV replication in tobacco mesophyll protoplasts has been obtained using a variety of techniques (Harrison et al., 1976). At 22–25°C, infective RNA1 was detected 7 hours after inoculation. No other changes were seen at this stage. Nucleoprotein rods were detected by electron microscopy, fluorescent antibody, and infectivity at 9 hours. There appeared to be no accumulation of coat protein but some accumulation of viral RNA, which was incorporated into rods about 4–5 hours after synthesis. Infectious RNA synthesis was largely complete by 12 hours and infectious nucleoprotein by 24 hours.

Although some short rods appeared at early times, their synthesis lagged behind that of long rods, but by 40 hours both species had reached plateau values. A second difference between the two kinds of rod is their distribution in the cell. Short rods occur mainly scattered in the cytoplasm, whereas long rods are associated with mitochondria in a characteristic fashion (Harrison and Roberts, 1968; Kitajima and Costa, 1969) (Fig. 8.17).

A

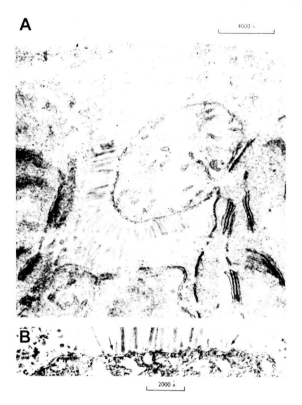

4000 Å

B

2000 Å

Fig. 8.17 Association of TRV particles with mitochondria. **(A)** Section of *Nicotiana clevelandii* leaf fixed with potassium permanganate, showing mitochondria with a fringe of virus particles. Second mitochondrion at bottom left. Parts of chloroplasts at left and right. **(B)** Enlarged view of mitochondrial membranes, showing virus particles appressed to but not penetrating the outer membrane (arrows). From Harrison and Roberts (1968), with permission.

The ends of the rods are closely appressed to the mitochondrial membrane, but do not penetrate it. No rods were associated with other cell organelles. There was no apparent association of the Californian strain of the virus with mitochondria in *N. tabacum* (de Zoeten, 1966).

4. Pea enation mosaic

Pea enation mosaic is a complex of two viruses, an enamovirus PEMV-1 and an umbravirus, PEMV-2 (see Chapter 2, Section III.P.3). However, for studying the disease both viruses have to be considered together.

Tests with fluorescent antibody showed that antigen is located in both nucleus and cytoplasm at early stages after inoculation of tobacco protoplasts, but at later stages detectable antigen was mainly confined to nuclei (Motoyoshi and Hull, 1974). In another electron microscope study with protoplasts, the first visible signs of PEMV infection at 17 hours were cytoplasmic membrane-bound bodies enclosing a series of vesicles containing fibrils. Some of these appeared to fuse with the nuclear membrane. Virus particles were seen only in the nucleus (Burgess *et al.*, 1974b).

By a variety of techniques, de Zoeten *et al.* (1976) established that PEMV dsRNA is localized in the nuclei of infected cells. The PEMV-induced RNA polymerase activity is also associated with the nuclei as well as with virus-induced vesicles in the cytoplasm (Powell *et al.*, 1977).

Powell and de Zoeten (1977) demonstrated that nuclei isolated from healthy pea plants could support the initiation of PEMV RNA replication when PEMV RNA was added to them *in vitro*. This was shown by (1) an increase in actinomycin D-resistant polymerase activity with a maximum after about 10 hours incubation, and (2) hybridization experiments, which showed that at least some of the polymerase activity led to PEMV-specific RNA synthesis. Most of the RNA made was (−)-strand in ds form but some ss (+)-sense strand was also made. This priming reaction took place with viral RNA but not intact PEMV, strongly suggesting that in the intact cell the virus RNA is uncoated before it reaches the nucleus. Several other viral RNAs were tested but only PEMV RNA stimulated viral RNA synthesis in the nuclei.

The nuclear membrane-based replication complex is found in pea protoplasts infected with PEMV-1 RNA (Demler *et al.*, 1994a). The VPg found on both RNAs is encoded by PEMV-1. Mutations of the VPg affected the ability of the RNA of PEMV-1 to replicate but not that of the RNA of PEMV-2 (Skaf *et al.*, 2000). While both PEMV-1 and -2 can replicate independently, PEMV-1 also supplies the coat protein to encapsidate both RNAs. It appears that only the RNAs that have VPg covalently attached are encapsidated, and it is suggested that for PEMV-2 the VPg is not involved with replication but provides an encapsidation signal.

There are several unanswered questions about PEMV replication, including:

Do both RNAs of the virus complex replicate in the nucleus?

On which nuclear membrane, the inner or outer, does replication take place?

Are there any interactions between the replication complexes of the two viruses?

M. Discussion

There have been rapid advances in the understanding of how (+)-strand viruses replicate. The evidence for the involvement of membranes is incontrovertible but the reasons why different viruses use different membranes (Table 8.4) is not yet understood.

The replication complexes comprise several virus-coded proteins with different functions. These are assembled on to the relevant membrane by a membrane binding protein or domain that then interacts with the other components. The co-ordination of assembly and the composition of the replication complexes are controlled not only by the interactions between the component proteins (and nucleic acids) but also by the way they are expressed. Thus, in some cases, some of the component proteins are expressed from frameshift or read through; in other cases they are expressed from a polyprotein processed in a defined manner. Some aspects are not yet fully understood, such as why frameshift and read through occurs in only about 5–10% of the times that the ribosome reaches the site of this feature, yet the products (at least of TMV) are assembled in a 1:1 ratio in the replication complex.

Host proteins involved in replication complexes are just starting to be identified. The involvement of translation initiation factors is of particular interest in indicating co-ordination between translation and replication. However, these two functions operate along the template RNA in different directions, translation being 5'>3' and replication 3'>5'; therefore there must be controls to prevent interference between the two processes.

Most of the RNA elements involved in replication operate in *cis*, showing that the template RNA is an integral part of the replication com-plex. The elements at the 3'-end of the template RNA that initiate (−)-strand synthesis appear to be well defined. This is in contrast to those at the 5'-end of the genomic RNA, which initiate (+)-strand synthesis. It seems likely that (−)-strand and (+)-strand synthesis are highly co-ordinated and that, once the (+)-strand template has been 'captured' by the *in vivo* replication complex, the full round of RNA replication will occur. The lack of (+)-strand synthesis in most *in vitro* replication systems indicates either that an important factor is lost during extraction or that there are conformational con-straints imposed by the location of the complex *in vivo*. The observations on TBSV pX (see Section IV.L.2) indicate that there might be host-specific *cis*-acting elements. It will be of interest to see whether this is a widespread fea-ture among (+)-strand viruses, and what is the nature of the host specificity.

V. REPLICATION OF NEGATIVE-SENSE SINGLE-STRANDED RNA VIRUSES

A. Plant *Rhabdoviridae* (reviewed by Jackson *et al.*, 1999)

Rhabdoviruses have large membrane-bound particles containing a single species of (−)-sense ssRNA; see Chapter 2 (Section III.G) for a general description, and Chapter 6 (Section VII.A) for genome organization. Basically, the virion RNA is associated with the nucleocap-sid protein (N) to form coiled nucleocapsid; a large protein (L), considered to be the repli-case, is also associated with the nucleocapsid. The nucleocapsid is encased in the matrix (M) protein which, in turn, is enveloped in a mem-brane to form the bacilliform particle. Virus-encoded glycoproteins (G) extend through this membrane. It has been suggested (Cartwright *et al.*, 1972) that there are structural interac-tions between the N, M and G proteins. The overall structure and replication resembles that of animal rhabdoviruses but there are some differences. For instance, all vertebrate rhab-doviruses replicate and assemble in the cytoplasm, as do some plant rhabdoviruses, but other plant rhabdoviruses replicate in the nucleus.

The (−)-strand genome of rhabdoviruses has two functions, as the template for transcription of mRNAs for individual genes (described in Chapter 7, Section IV.A.1) and as the template for replication via a full-length (+) strand. The polymerase complex undertakes both functions but the switch mechanism is not fully understood, even for the much studied animal-infecting vesicular stomatitis virus (Rodriguez and Nichol, 1999).

1. Cytological observations on replication

Because of their large size and distinctive morphology, the rhabdoviruses are particularly amenable to study in thin sections of infected cells. Morphologically they appear to fall into three groups.

First there are the nucleorhabdoviruses that accumulate in the perinuclear space with some particles scattered in the cytoplasm. With some viruses of this group, structures resembling the inner nucleoprotein cores have been seen within the nucleus. The envelopes of some particles in the perinuclear space can be seen to be continuous with the inner lamella of the nuclear membrane. Fig. 8.18 illustrates this group.

Immunogold labeling with an antiserum against the five structural proteins of PYDV showed that viral proteins accumulate mainly in the nucleus (Lin *et al.*, 1987). *In situ* hybridization demonstrated that (−)-strand genomic RNAs are found only in nuclei of infected plants, whereas the (+)-strand RNA sequences are in both nuclei and cytoplasm (Martins *et al.*, 1998). Immunofluorescence and immunogold labeling showed that N and L proteins were in viroplasms in the nucleus and the M2 protein was more generally distributed within the nucleus.

In the second group, the cytorhabdoviruses (e.g. LNYV), maturation of virus particles occurs in association with the endoplasmic reticulum (ER), and particles accumulate in vesicles in the reticulum. Biochemical evidence suggests that the nucleus might be involved in the early stages of infection by members of this group.

In the third group, there are structures that appear to be rhabdovirus nucleocapsid cores

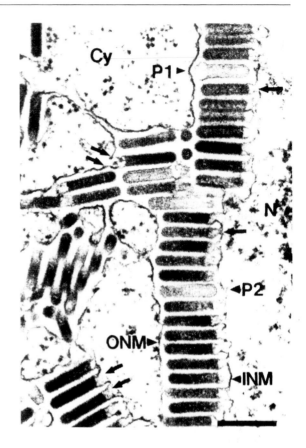

Fig. 8.18 Electron micrograph of a thin section of a maize leaf infected with a Hawaiian isolate of MMV. Virus particles apparently budding through the inner nuclear membrane (INM) and through intracytoplasmic extensions (double arrows) of the outer nuclear membrane (ONM); single arrows indicate constriction of the INM. Alignment of particles at P1 and P2 suggests budding on the ONM. Cy, cytoplasm; N, nucleus. Bar = 0.3 μm. From McDaniel *et al.* (1985), with kind permission of the copyright holder, © The American Phytopathological Society.

lacking the surrounding membrane (Francki *et al.*, 1985a).

When examining rhabdoviruses in the cell it must be remembered that the outer nuclear membrane is contiguous with the ER. Thus, nucleorhabdoviruses budding through the inner nuclear membrane into the perinuclear space may further be included in vesicles derived from the outer membrane and be found in the cytoplasm. Similarly, cytorhabdoviruses that associate with the ER may affect the outer nuclear membrane giving an appearance of a nuclear involvement.

2. Nucleorhabdoviruses

Most of the studies have been performed on SYNV for which a reasonably detailed picture of its replication has been developed.

a. In vitro studies

A salt extraction procedure proved effective in isolating an active polymerase complex from SYNV-infected *Nicotiana edwardsonii* leaf tissue (Wagner *et al.*, 1996). The products of the *in vitro* polymerase reactions included full-length, polyadenylated N and M2 mRNAs and (+)-strand leader RNA. Animal rhabdoviruses do not polyadenylate their (+)-strand leader transcript, and it is suggested that this feature of SYNV may reflect its replication in the nucleus. The polymerase complex comprises the N, M2 and L proteins (Wagner and Jackson, 1997) and addition of antibodies to L protein inhibited the *in vitro* system. The reaction condition of this system favors sgRNA transcription over (–)-sense replication, possibly owing to depletion of N protein during extraction. Some small virus-sense RNAs were formed and it is suggested that the formation of genomic (–) strand is inhibited by specific signal sequences in the (+)-strand RNA (Wagner and Jackson, 1997).

b. Replication

The following steps have been described for the replication of nucleorhabdoviruses (Fig 8.19A) (Jackson *et al.*, 1999):

1. On entry into the cell, virus particles associate with the ER and release the nucleocapsid cores into the cytoplasm.
2. It is thought that the nucleocapsid cores enter the nucleus through the nuclear pore complexes.
3. Primary transcription takes place using the L protein incorporated in the nucleocapsid core to give mRNAs that are transported to the cytoplasm and translated.
4. The core polymerase proteins, N, M2 and L, are transported back to the nucleus where

they initiate genomic RNA replication and further mRNA synthesis.
5. Granular electron-dense viroplasms, containing N, M2 and L proteins, form near the periphery of the nucleus and are the site of viral replication
6. In the late stages of replication, M protein associates with the newly synthesized nucleocapsid cores coiling them. This complex then associates with G protein that is concentrated at sites on the inner nuclear membrane.
7. Newly synthesized virus particles bud into the perinuclear space.

3. Cytorhabdoviruses

Few molecular details are available on cytorhabdoviruses, most of the information coming from cytological studies. Features of the infection cycle are shown in Fig. 8.19B and are summarized by Jackson *et al.* (1999). Early in infection, some cytorhabdoviruses (e.g. LNYV) induce blisters on the outer nuclear membrane that develop into small vesicles containing some virus particles. These are not found with other viruses (e.g. BYSMV). In both cases, masses of fibrillar viroplasms are found in the cytoplasm closely associated with dense networks of ER. Newly synthesized virus particles bud into vesicles derived from the ER.

The three major structural proteins of FLSV, the G, N and M proteins, were found at the periphery of viroplasms, but only the N protein was detected in the granular matrix of the viroplasm (Lundsgaard, 1992). In a similar immunocytological study, the FLSV G protein was detected in ER and the perinuclear membrane but not in Golgi apparatus (Lundsgaard, 1995).

B. Tospoviruses

The tospovirus genome consists of three RNA segments, L, M and S, enclosed in a membrane-bound particle (see Fig. 6.11 for genome organization). RNA L is (−)-sense and M and S have an ambisense strategy (see Chapter 7, Section IV.A.2). There are four structural polypeptides (Fig. 5.42). Two are glycosylated (G1 and G2) and are at the surface. The N protein binds to

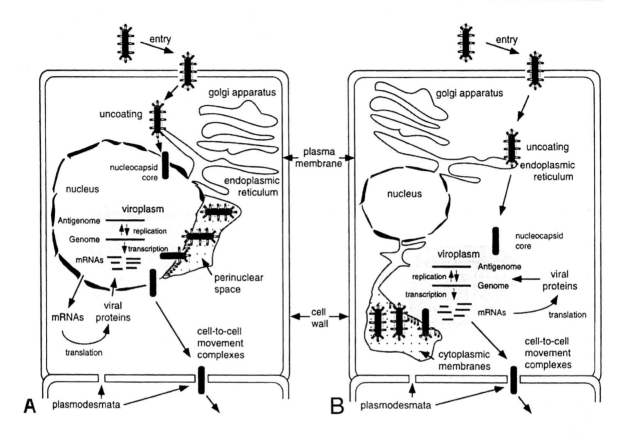

Fig. 8.19 Model for the replication cycle of **(A)** plant nucleorhabdoviruses, and **(B)** cytorhabdoviruses. From Jackson *et al.* (1999), with permission.

the RNA, and there is a large protein, the RdRp, occurring in a minor amount that is encoded by RNA L. These and other properties place tospoviruses in the *Bunyaviridae*, a large family of viruses replicating in vertebrates and invertebrates.

Little is known about the details of tospovirus replication except that it occurs in the cytoplasm in association with the Golgic stack membranes. Double enveloped particles in the ER cisternae may result from fusion of Golgi-based particles with the ER (Kitajima *et al.*, 1992; Kibbert *et al.*, 1997). It is thought that tospovirus replication is similar to that of other bunyaviruses with the concentration of N protein regulating the switch from the production of viral mRNAs to replication of the genome

(Storms, 1998). For an account of bunyavirus replication, see Elliott (1999).

VI. REPLICATION OF DOUBLE-STRANDED RNA VIRUSES

A. Plant *Reoviridae*

Plant members of the *Reoviridae* family are placed in three genera: *Phytoreovirus* with 12 dsRNA genome segments, the type member being WTV; *Fijivirus*, with 10 dsRNA genome segments, the type member being FDV; and *Oryzavirus* with 10 dsRNA genome segments, the type member being RRSV (see Chapter 6, Section VI.A, for details of genome organization). Little is known about the molecular aspects of plant reovirus replication but it is likely to be similar to that of animal reoviruses (described by Joklik, 1999).

1. Intracellular site of replication

Plant reoviruses replicate in the cytoplasm as do those infecting mammals (Wood, 1973). Following infection, densely staining viroplasms appear in the cytoplasm (Fig. 8.20).

Viroplasms were present in cells of various tissues of leafhopper vectors infected with WTV as well as infected plant cells (Shikata and Maramorosch, 1967). Immunofluorescence demonstrated the presence of viral antigen in the cytoplasm of cultured leafhopper cells (Chiu *et al.*, 1970). It is not yet possible to relate the *in vitro* studies on the replication of WTV to the structures seen cytologically.

Enzyme digestion experiments and radioautographic assay of the incorporation of ³H-labeled uridine into maize cells infected with the fijivirus, MRDV, indicate that much of the viroplasm is made up of protein—probably viral proteins. Viral RNA appears to be synthesized in the viroplasm, where the mature particles are assembled. The mature particles then

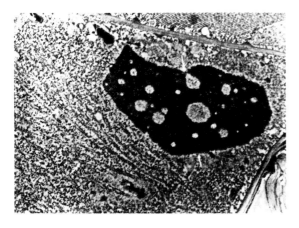

Fig. 8.20 Leaf vein tumor cells of maize experi-mentally infected with MRDV. The three different kinds of inclusions caused by MRDV are easily recognizable: viroplasm (arrows), cytoplasmic tubules along and inside which the virus particles are aligned (double arrows), and part of a virus crystal (top right). From Bassi and Favali (1972), with permission.

migrate into the cytoplasm where they may (1) remain as scattered particles, (2) form crystalline arrays, or (3) become enclosed in or associated with tube-like proteinaceous structures (Fig. 8.20) (Bassi and Favali, 1972; Favali *et al.*, 1974). The autoradiographic studies failed to implicate the nucleus, mitochondria or chloroplasts in virus replication.

A detailed electron microscopic study supported the view that the viroplasms caused by FDV in sugarcane are the sites of virus component synthesis and assembly (Hatta and Francki, 1981c). The viroplasms are composed mostly of protein and dsRNA. Some areas contained numerous isometric particles 50–60 nm in diameter. Some appeared to be empty shells while others contained densely staining centers of dsRNA. These particle types appeared to be incomplete virus particles or cores. Complete virus particles were seen only in the cytoplasm.

2. Packaging of RNAs

A problem faced by reoviruses is the selection of progeny for packaging. Each particle contains the genomic complement of one copy each of 10 or 12 dsRNA species. The evidence that every WTV particle appears to contain one copy of each genome segment is: (1) RNA isolated from virus has equimolar amounts of each segment (Reddy and Black, 1973); and (2) an infection can be initiated by a single particle (Kimura and Black, 1972). Thus, what are the macromolecular recognition signals that allow one, and one only, of each of 10 or 12 genome segments to appear in each particle during virus assembly? For example, the packaging of the 12 segments of WTV presumably involves 12 different and specific protein–RNA and/or RNA–RNA interactions.

The first evidence that may be relevant to this problem came from the work of Anzola *et al.* (1987) with WTV. They determined the structure of a defective (DI) genomic segment 5, which was only one-fifth the length of the functional S5 RNA because of a large internal deletion. However, this DI RNA was packaged at one copy per particle, as for the normal sequence. Thus, they established that the sequence(s) involved in packaging must reside

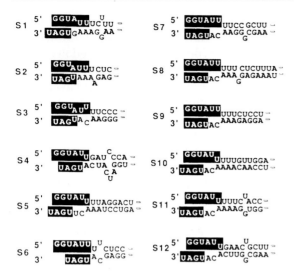

Fig. 8.21 Terminal sequence domains of the positive-sense strands of the 12 genome segments of WTV. The segment-specific inverted repeats near the 5' and 3' termini are oriented to indicate potential base-pairing interactions. The conserved 5' hexanucleotide and 3' tetranucleotide shared by all 12 genome segments are shown in white on black. From Anzola *et al.* (1987), with kind permission of the copyright holder, © The National Academy of Sciences, USA.

within 319 base-pairs of the 5'-end of the (+) strand and 205 from the 3'-end. Reddy and Black (1974, 1977) showed that an increase in the DI RNA content in a virus population led to a corresponding decrease in the molar proportion of the normal fragment; that is, the DI fragment competes only with its parent molecule. Thus, there must be two recognition signals, one that specifies a genome segment as viral rather than host, and one that specifies each of the 12 segments.

Anzola *et al.* (1987) sequenced the 5'- and 3'-terminal domains of all 12 genome segments of WTV (Fig. 8.21). They suggested that a fully conserved hexanucleotide sequence at the 5' terminus and a fully conserved tetranucleotide sequence at the 3' terminus might form the recognition signals for viral as opposed to host RNA. They also found segment-specific inverted repeats of variable length just inside the conserved segments (Fig. 8.21). They suggested that these might represent the specific recognition sequence for each individual genome segment. A similar inverted repeat was found for segment 9 of RDV (Uyeda *et al.*, 1989).

Other members of the plant reoviruses family each have conserved 5' and 3' sequences (Kudo *et al.*, 1991; Mertens *at al.*, 2000). Influenza viruses, which have segmented ssRNA genomes, also have comparable conserved sequences at the 5' and 3' termini of each genome segment (Stoeckle *et al.*, 1987). These similarities strengthen the idea that the 5'- and 3'-terminal sequences have a role in the packaging of these segmented RNA genomes. However, the problem is by no means solved. If only RNA–RNA recognition is involved, how is this brought about to give a set of 12 dsRNAs for packaging? If protein–RNA recognition is important, how are 12 specific sites constructed out of the three proteins known to be in the nucleoprotein viral core?

3. Replication

In *Reovirus* the (−)-sense strands are synthesized by the viral replicase on a (+)-sense template that is associated with a particulate fraction (Acs *et al.*, 1971). These and related results led to the proposal that dsRNA is formed within the nascent cores of developing virus particles, and the dsRNA remains within these particles. If true, this mechanism almost certainly applies to the plant reoviruses. It implies that the mechanism that leads to selection of a correct set of 12 genomic RNAs could involve the ss (+) strand. Thus, the base-paired inverted repeats illustrated in Fig. 8.21 could be the recognition signals. It may be that other virus-coded 'scaffold' proteins transiently present in the developing core are involved in RNA recognition rather than, or as well as, the three proteins found in mature particles.

Xu *et al.* (1989) constructed a series of transcription vectors that allowed production of an exact transcript of S8 RNA and of four analogs that differed only in the immediate 3' terminus. Their experiments provided three lines of evidence supporting the view that the 5'- and 3'-terminal domains interact in a functional way:

Nuclease T1 sensitivity assays showed that even a slight change in the 3'-terminal sequence can affect the conformation of the 5' terminus.

Translation *in vitro* is slightly decreased by alterations in the 3' terminus, which extends the potential for 3'–5' terminal base-pairing, and is increased by changes that reduce potential base-pairing.

Computer modeling for minimal energy structures for six WTV transcripts predicted a conformation in which the terminal inverted repeats were base-paired.

Dall *et al.* (1990) developed a gel retardation assay with which they demonstrated selective binding of WTV transcripts by a component of extracts from infected leafhopper cell cultures. Using terminally modified and internally deleted transcripts, they established that the segment-specific inverted repeats present in the terminal domains were necessary but not sufficient for optimal binding. Some involvement of internal sequences was also necessary. There was no evidence for discrimination in binding between transcripts from different segments. The binding component or components present in extracts of infected cells, which are not present in those of healthy cells, have not yet been characterized.

VII. REPLICATION OF REVERSE TRANSCRIBING VIRUSES

The *Caulimoviridae* is the only family of plant viruses with dsDNA genomes; see Chapter 2 (Section III.A) for a description of the family. In 1979 very little was known about the replication of this group, but since then progress has been very rapid. There have been two main motivating factors. First, it was hoped that these viruses, because of their dsDNA genomes, might be effective gene vectors in plants. This aspect is discussed in Chapter 16 (Section IX). Second, the realization that the DNA is replicated by a process of reverse transcription made their study a matter of wide interest.

The family comprises six genera that form two groups, the 'caulimoviruses' and the 'badnaviruses'. These two groups differ in genome organization but have essentially the same replication methods. Most experimental work has been carried out on the 'caulimovirus'

CaMV and the 'badnavirus' RTBV. Reviews include Hohn *et al.* (1985), Hull *et al.* (1987), Pfeiffer *et al.* (1987), Mason *et al.* (1987), Hull (1996) and Hohn and Fütterer (1997).

Although the replication of members of the *Caulimoviridae* is by reverse transcription and, in many respects, is similar to that of retroviruses, it does differ from that of retroviruses in several important points:

The replication does not involve integration into the host genome for transcription of the RNA. This is done from an episomal minichromosome.

The virus does not encode an integrase gene.

The virion DNA is circular dsDNA and not the linear DNA with long terminal repeats characteristic of retroviruses. (This and the previous point relate to the lack of integration.)

The DNA phase of the replication cycle is encapsidated rather than the RNA phase, which is encapsidated in retroviruses. Thus, the *Caulimoviridae* are known as pararetroviruses.

As with retroviruses, the replication cycle of pararetroviruses has two phases, a nuclear phase where the viral DNA is transcribed by host DNA-dependent RNA polymerase II; and a cytoplasmic phase where the RNA product of transcription is reverse-transcribed by virus-encoded RNA-dependent DNA polymerase or reverse transcriptase (RT) to give DNA. In retroviruses, the RT activity is part of the *pol* gene which also includes the RNase H activity that removes the RNA moiety of the RNA:DNA intermediate of replication. The *pol* gene in turn is part of the gag-pol polyprotein that is cleaved by an aspartate proteinase, the gag being analogous to coat protein. In pararetroviruses, the reverse transcriptase and RNaseH activities are closely associated. In badnaviruses the coat protein and pol are expressed from the same ORF but in caulimoviruses they are expressed from separate ORFs (see genome maps in Figs 6.1, 6.3 and 6.4). All plant pararetroviruses encode an aspartate proteinase.

A. Reverse transcriptase

Most studies have been performed on retrovirus pol. The reverse transcriptase has a characteristic motif of tyrosine–isoleucine–aspartic acid–aspartic acid (YIDD) and several amino acid motifs identify the RNase H domain. Processing of the 66-kDa pol region of human immunodeficiency virus (HIV) by the aspartate protease removes the RNase H domain giving a heterodimer of 66- and 51-kDa proteins. This enzyme complex has three activities for the conversion of ssRNA to dsDNA, RNA-dependent DNA polymerase, DNA-dependent DNA polymerase and RNase H. The crystal structure of HIV RT has been determined at 3.5 Å (Kohlstaedt *et al.*, 1992) and it was shown to have a structure resembling a right hand, as described above for RdRp (Section IV.B.1).

The first indication that the product of CaMV ORF V is analogous to the retrovirus *pol* gene came from sequence comparison (Toh *et al.*, 1983). The N-terminal domain of ORF V has an aspartate protease motif (Toruella *et al.*, 1989) that autocatalytically cleaved an N-terminal doublet of polypeptides (20 and 22 kDa) from *in vitro* translated ORF V transcript. Mutants of the protease active site were not processed. This fitted with other features of the replication cycle as will be described below. ORF V was expressed in yeast as a 60-kDa protein that had RT activity on a synthetic template (Takatsuji *et al.*, 1986). In contrast, expression of ORF V in *E. coli* gave a protein of 78 kDa, the size expected from that ORF, but it did not have any RT activity. An activity gel analysis revealed that RT activity associated with CaMV particles is also 60-kDa (Takatsuji *et al.*, 1992). Deletion analysis showed that removal of between 143 and 185 N-terminal amino acids from the *E. coli*-expressed protein gave RT activity similar to that of the yeast-expressed protein. This suggests that CaMV RT is translated as an inactive precursor form that is converted to the active form by proteolytic processing. It is presumed that this is by the N-terminal aspartate protease activity.

The pol motifs are in the C-terminal part of ORF III of 'badnaviruses' and that of RTBV has been studied in insect cells (Laco and Beachy, 1994). The predicted 87-kDa product was detected and was processed to give a 62-kDa and 55-kDa protein. Sequencing showed that these proteins were N co-terminal. Both proteins exhibited RT and DNA polymerase activities but only the 55-kDa protein had RNase H activity. The precise weights of the 62- and 55-kDa proteins were determined by mass spectrometry (Laco *et al.*, 1995) enabling the C termini to be identified. Mutagenesis of the putative active site of the aspartate protease prevented the 87-kDa protein being processed in insect cells. Using antisera to specific fragments of the RTBV ORF III product, Hay *et al.* (1994) detected a 13.5-kDa protein corresponding to the aspartate protease in extracts from infected plants. The protease antibody labeled the surface of the virus particle. Antibodies against the RT domain identified proteins of 68, 65 and 56 kDa, the latter two probably corresponding to the 62- and 55-kDa proteins found on expression of this ORF in insect cells (Laco and Beachy, 1994).

Mutation of the Y^{1339}, D^{1341} or D^{1342} residues of the RT core motif abolished RT activity, whereas that of the I^{1340} did not (S.-C. Lee and R. Hull, unpublished observation).

B. Replication of 'caulimoviruses'

There are a large number of publications concerned with CaMV nucleic acid replication and the phenomenon of reverse transcription. Many of these are referred to in the review articles noted at the beginning of this section. Here, only a few key or recent references will be given.

By 1983, various aspects of CaMV nucleic acid replication led three groups to propose that CaMV DNA is replicated by a process of reverse transcription involving an RNA intermediate (Guilley *et al.*, 1983; Hull and Covey, 1983b; Pfeiffer and Hohn, 1983). Here are some of the observations that led to the model:

A full-length RNA transcript is produced that has terminal repeats (Covey and Hull, 1981).

DNA in virus particles has discontinuities (see Fig. 6.1) while that found in the nucleus does not, but is supercoiled and is associated with histones as a minichromosome (Ménissier *et al.*, 1982; Olszewski *et al.*, 1982).

dsDNA exists in knotted forms (see Fig. 4.10).

Other forms of CaMV DNA in the cell are not encapsulated, such as an ss molecule of 625 nucleotides with the same polarity as the α-strand covalently linked to about 100 ribonucleotides (Covey *et al.*, 1983).

Since 1983, a detailed picture of the replication of CaMV has been built up.

1. Replication pathway

The replication pathway is outlined in Fig. 8.22. Essentially, the replication has two phases: transcription of an RNA template from the virion DNA and then reverse transcription of the RNA template to give dsDNA. The transcription phase occurs in the nucleus and the reverse transcription phase in the cytoplasm. In the first phase of replication, the dsDNA of the infecting particle moves to the cell nucleus, where the overlapping nucleotides at the gaps are removed, and the gaps are covalently closed to form a fully ds DNA. The covalently closed DNA associates with host histones to form minichromosomes that are the template used by the host enzyme, DNA-dependent RNA polymerase II, to transcribe two RNAs of 19S and 35S, as indicated in Fig. 6.1. This is described in more detail in Chapter 7 (Section IV.C.1).

The two polyadenylated RNA species migrate to the cytoplasm for the second phase of the replication cycle that takes place in the viroplasms (e.g. Mazzolini *et al.*, 1985). The 19S RNA is the mRNA for gene VI that is translated in large amounts to produce the viroplasm protein. Gene VI is the only *Caulimovirus* gene to be

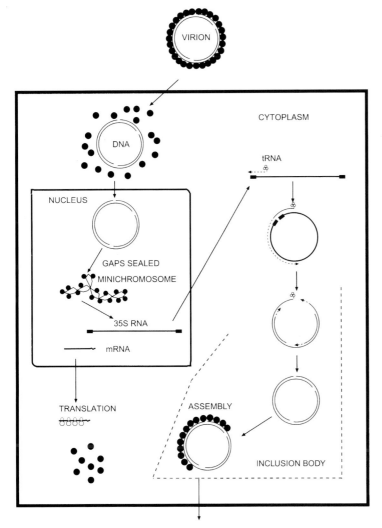

Fig. 8.22 Diagram of the replication cycle of CaMV.

transcribed as a separate transcript from its own promoter, suggesting that it may have an important role at an early stage following infection (Gowda *et al.*, 1989). Mutagenesis of the coding part of gene VI showed that it was the protein product rather than the mRNA that was responsible for transactivation which is described in more detail in Chapter 7 (Section V.B.7).

To commence viral DNA synthesis on the 35S RNA template, a plant methionyl tRNA molecule forms base-pairs over 14 nucleotides at its 3'-end with a site on the 35S RNA corresponding to a position immediately downstream from the D1 discontinuity in the α-strand DNA (see below). The viral reverse transcriptase commences synthesis of a DNA (−) strand and continues until it reaches the 5'-end of the 35S RNA with the RNase H activity removing the RNA moiety of the RNA:DNA duplex giving what is termed 'strong-stop DNA'. At this point, a switch of the enzyme to the 3'-end of the 35S RNA is needed to continue the copying. The switch is made possible by the 180-nucleotide direct repeat sequence at each end of the 35S RNA which enables the 3'-end of the strong-stop DNA to hybridize with the 3'-end of the 35S RNA. When the template switch is completed, reverse transcription of the 35S RNA continues up to the site of the tRNA primer, which is displaced and degraded to give the D1 discontinuity in the newly synthesized DNA.

The rest of the used 35S template is removed by an RNase H activity. In this process, two polypurine tracts (PPT) of the RNA are left near the position of discontinuities D2 and D3 in the second DNA strand (+ strand). Synthesis of the second (+) strand of the DNA then occurs, initiating at these two RNA primers. The growing (+) strand has to pass the D1 gap in the (−) strand, which again involves a template switch.

There are several observations that support and enhance this model for CaMV replication (reviewed in Hohn and Fütterer, 1997):

Both (−)- and (+)-strand DNA synthesis are resistant to aphidocolin, an inhibitor of DNA → DNA synthesis.

RT activity is associated with viral inclusion bodies and virus particles.

Various unencapsidated nucleic acid molecules interpreted as being replication intermediates have been isolated. These include 'strong-stop' DNAs which have ribonucleotides at the 5'-end, DNA molecules that are partially double- and partially single-stranded, compatible with being products of defective replication and hairpin structures (Turner and Covey, 1988).

Replication intermediates are associated with apparently incomplete virus particles (Thomas *et al.*, 1985; Marsh and Guifoyle, 1987; Fütterer and Hohn, 1987).

The finding of RT activity in inclusion bodies and virus particles, and of replication intermediates in virus particles, indicates that, as with retroviruses, the reverse transcription of CaMV occurs in particle-like proviral structures.

2. Inclusion bodies

As noted in an earlier section, CaMV (and other caulimoviruses) induce characteristic inclusion bodies or viroplasms in the cytoplasm of their host cells (Fig. 8.23).

There are two forms of inclusion bodies, electron-dense ones that are made up of ORF VI product, and electron-lucent ones that are made up of ORF II product (Espinoza *et al.*, 1991). Virus particles are found in both types of inclusion body. The electron-dense inclusion bodies are the site for progeny viral DNA synthesis and for the assembly of virus particles; it is not known whether virus replication takes place in the electron-lucent inclusion bodies. Viral coat protein appears to be confined to the inclusion bodies and most virus particles are retained within them.

At an early stage in their development, the ORF VI product inclusion bodies appear as very small patches of electron-dense matrix material in the cytoplasm, surrounded by numerous ribosomes. Larger inclusion bodies are probably formed by the growth and coalescence of the smaller ones, leading to mature inclusion bodies that vary quite widely in size from about 0.2 to 20 μm in diameter. They are usually spherical and are not membrane-bound. They often have ribosomes at the periphery and consist of a fine granular matrix with some electron-lucent areas not bounded by membranes. Virus particles are

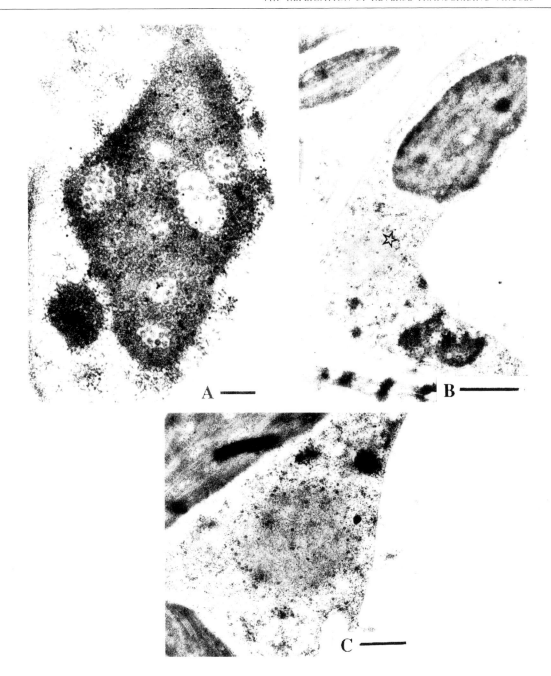

Fig. 8.23 Electron micrographs of the inclusion bodies of CaMV Cabb B-JI in infected turnip leaves immunogold labeled with anti-P62 (ORF VI product) antiserum. **(A)** Electron-dense inclusion body with gold particles preferentially labeling the inclusion body matrix (bar = 200 nm). **(B)** Cell showing an electron-dense inclusion body (filled-in star) heavily labeled and an electron-lucent inclusion body (open star) without gold particles contained within the same cell (bar = 1 μm). **(C)** An electron-lucent inclusion body showing the lack of gold particles (bar = 500 nm). From Espinoza *et al.* (1991), with permission.

present in scattered or irregular clusters in the lucent areas and the matrix.

Little is known about the way CaMV particles are assembled. No empty virus shells are found in infected tissue. These observations suggest that encapsulation may be closely linked to DNA synthesis. The role of glycosylation and phosphorylation of the coat protein remains to be determined.

3. Discontinuities

The DNA of 'caulimoviruses' has gaps or discontinuities at specific sites, one (D1) in one strand (the + or α strand) and one or more in the other strand. Those of CaMV have been studied in detail and shown to comprise an overlapping sequence with the 5'-end being in a fixed position and the 3'-end being in a variable position giving an overlap varying between 8 and 40 nucleotides (Fig. 8.24) (Richards *et al.*, 1981).

As explained above (Section VII.B.1), these discontinuities arise from replication where the advancing DNA strand reaches the priming site. Thus, D1 is at the tRNA priming site for (−)-strand synthesis and the discontinuities in the other strand are at the PPTs of RNA, generated by RNase H cleavage that give (+)-strand priming. In an analysis of the PPT-associated (+)-strand priming, Noad *et al.* (1998) showed that altering the length of the 13-base-pair PPT by ±25% significantly reduced priming efficiency but did not affect the site of the 5'-end of the new (+)-strand DNA which is 3 nucleotides from the PPT 3'-end. There is a short pyrimidine tract 5' to the PPT that plays an important role in PPT recognition *in vivo*. Noad *et al.* (1998) propose a model for pararetroviral (+)-strand priming in which the pyrimidines enhance the PPT recognition during RNase H cleavage, and suggest that the fidelity of primer maturation involves PPT length measurement and 3'-end recognition by the RNase H.

C. Replication of 'badnaviruses'

The replication of RTBV, the most studied of the 'badnaviruses', is similar to that of CaMV in most respects and is supported by the detection

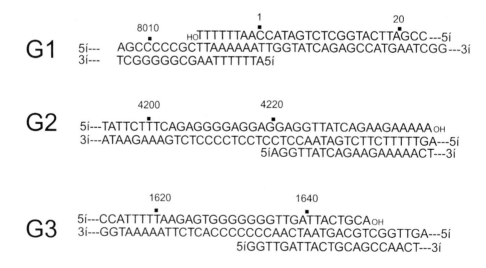

Fig. 8.24 Structure of CaMV gaps or discontinuities. G1 (also referred to as D1) is on the transcribed α strand and G2 (D2) and G3 (D3) are on the complementary strand. For each of the gaps the upper and lower sequences are those of discontinuous strand and the middle sequence the unbroken strand. These sequences are the most common found, but for each the 5' terminus is at a fixed position; the 3' terminus may vary from the shown position. The numbers above each sequence are the positions on the CaMV sequence. From Richards *et al.* (1981), with permission.

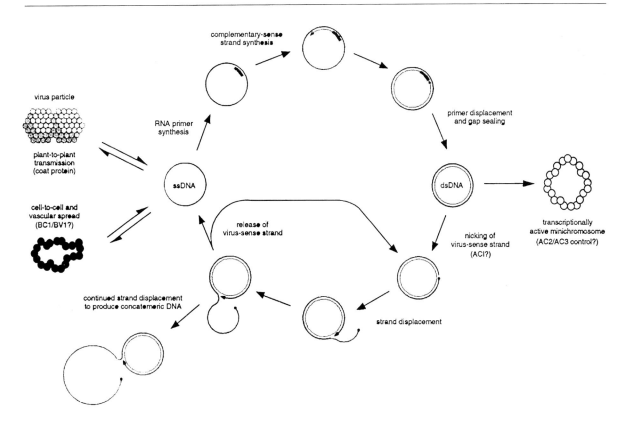

Fig. 8.25 Diagram of replication of a geminivirus. Kindly provided by J. Stanley.

a site-specific endonuclease that nicks and ligates (+)-strand viral DNA at the same position *in vitro* (Laufs *et al.*, 1995; Orozco and Hanley Bowdoin, 1996).

There are two major categories of molecular organization of the (+)-strand DNA replication origin. The replication origin for mastreviruses consists of a large *cis*-acting region where the Rep protein forms multiple complexes (Castellano *et al.*, 1999) and that for the begomoviruses contains one binding site for Rep (Fontes *et al.*, 1992, 1994b; Lazarowitz *et al.*, 1992; Orozco and Hanley-Bowdoin, 1998).

3. Plus-strand origin

The origin of TGMV (+)-strand synthesis has been studied in detail (reviewed in Hanley-Bowdoin *et al.*, 1999) and has been compared with those of other begomoviruses and of curtoviruses. The features of the TGMV (+)-strand origin is illustrated in Fig 8.26. The origin is in the left-hand side of the common region and

overlaps the promoter for AC61 (also termed AL61) (described in Chapter 7, Section IV.C.2). Six *cis* elements have been identified in this region.

1. The hairpin element is common to all geminivirus genomes (Arguello-Astorga *et al.*, 1994). This comprises a GC-rich stem and an AT-rich loop, and mutagenesis demonstrates that it is the structure of the stem rather than the sequence that is essential for its activity. The stems of mastreviruses are much longer than those of members of the other two genera. The 5'-TAATATTAC loop is conserved in all geminiviruses and is found in the (+)-strand origins of other nucleic acids that replicate by rolling circle. There is some sequence flexibility in the loop sequence but the cleavage is between the TT↓AC as noted above.

2. The binding site for Rep has several features: (i) It is virus-specific (Table 8.5) but has some conserved sequence. (ii) The GGAT repeat is an absolute requirement. (iii) The spacing

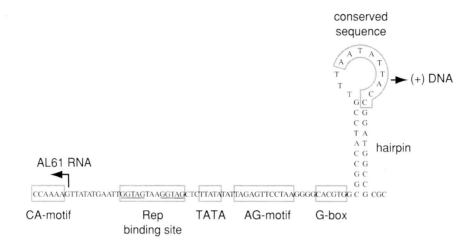

Fig. 8.26 The TGMV (+)-strand origin of replication and AC61 (AL61) promoter. The DNA sequence corresponding to the TGMV A positions 54 to 153 is shown. Only the top strand of the duplex DNA is given. The initiation sites and the directions of synthesis for (+)-strand DNA replication and AC61 transcription are indicated. Other functional elements are boxed (see the text). The hairpin structure is drawn and the conserved non-anucleotide loop sequence is marked. From Hanley-Bowdoin *et al.* (1999), with kind permission of the copyright holder, © CRC Press LLC.

TABLE 8.5 Rep binding sites

TGMV	GGTAGTAA-GGTAG
BGMV	GG-AG-ACTGG-AG
ABMV	GG-AGTATTGG-AG
Consensus	GG-AGTAYYGG-AG

between the repeats is important. (iv) The spacing between the binding site and the cleavage site is important. Less is known about how the Reps of mastreviruses recognize their origins. No sequences analogous to the binding sites described above have been found, and mastreviruses do not have the same degree of specificity, as do the other two genera. It has been suggested that mastrevirus Rep proteins bind to reiterated sequences in the stem of the hairpin (Arguello-Astorga *et al.*, 1994), but electron microscopy of WDV Rep/DNA complexes indicated that the binding site is in a similar site to those of the other two genera (Sanz-Burgos and Gutierrez, 1998).

3. Begomovirus and curtovirus (+)-strand origins share binding sites for two transcription factors, the TATA box and the G box. Neither of the host transcription factor binding sites are required for virus replication.

4. The two other elements are the AG motif and the CA motif. The AG motif is essential for virus replication (Orozco *et al.*, 1998) but has no detectable role in transcription of AC61. Deletion of the CA motif reduced TGMV replication 20-fold, and mutagenesis suggested that it acts as an efficiency element.

Although the mechanism by which these six elements operate has not been determined, it has been suggested that they may bind plant factors required for replication (Hanley-Bowdoin *et al.*, 1999).

4. Geminivirus Rep proteins

The Rep protein is the only geminiviral protein essential for replication (Elmer *et al.*, 1988a; Schalk *et al.*, 1989). In the curtoviruses and begomoviruses, Rep is encoded by ORF C1, and in mastreviruses it is expressed from ORFs C1:C2 through a spliced mRNA; unspliced RNA gives RepA from ORF C1.

Rep and RepA are multifunctional proteins:

1. They localize within the nucleus (Nagar *et al.*, 1995).
2. They have specific DNA recognition sites (Fontes *et al.*, 1994a).
3. They have site-specific endonuclease and

ligation activity for (+)-strand viral DNA (see above).

4. They have ATP/GTPase activity (Desbiez *et al.*, 1995).

5. Those of some mastreviruses and begomoviruses have been shown to activate the promoter for the coat protein gene mRNA (Hayley *et al.*, 1992; Hofer *et al.*, 1992; Sunter and Bisaro 1991; Zhan *et al.*, 1993).

6. The begomovirus Rep can repress its own promoter (Hong and Stanley, 1995; Sunter *et al.*, 1993).

7. The begomovirus Rep can stimulate the expression of the proliferating cell nuclear antigen (Nagar *et al.*, 1995).

8. They interact with retinoblastoma proteins (see below).

The domain structure of TGMV Rep protein has been determined (reviewed in Hanley-Bowdoin *et al.*, 1999) (Fig. 8.27A). It is likely that other begomovirus and curtovirus Rep proteins have a similar structure (Fige. 8.27B). Although the mastrevirus Reps differ from those of the

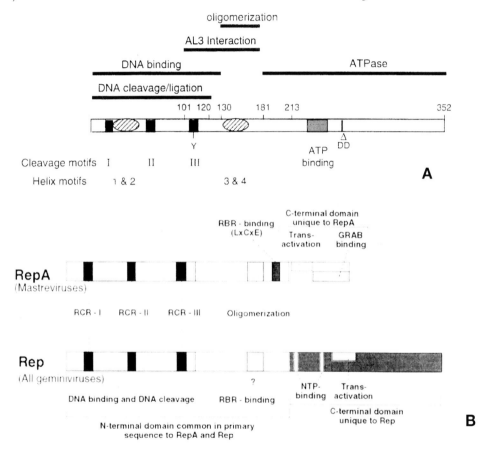

Fig. 8.27 Domain organization of Rep and RepA proteins. **(A)** TGMV Rep domains and predicted motifs. Solid boxes mark the locations of the three conserved DNA cleavage motifs in the Rep protein. The active site tyrosine residue is shown in motif III. The stippled box shows the location of the ATP binding motif which, in combination with the indicated conserved aspartic acid residues, are related to DNA helicases. The hatched circles indicate predicted sets of α-helices. Helix 2 is strongly amphipathic in character. Solid lines above the protein mark the location of the functional domains for oligomerization, AC3 (AL3) interaction, DNA binding, and DNA cleavage/ligation. The numbers correspond to amino acid positions in Rep. **(B)** Diagram comparing the organizations of RepA of mastreviruses and Rep of all geminiviruses. The different domains correspond to a composite based on data available for different geminiviruses and, therefore, the location of each domain is approximate. RCR-I, -II and -III refer to the amino acid motifs conserved in proteins involved in rolling circle replication. Part (A) from Hanley-Bowdoin *et al.* (1999) with kind permission of the copyright holder, © CRC Press LLC; part (B) from Gutierrez (2000b), with kind permission of the copyright holder, © Kluwer Academic Publishers.

other genera, they do have some motifs in common and they are likely to have domain structure similarities.

Rep proteins have several protein–protein interactions. Firstly, Rep proteins form large oligomers of about eight subunits in a virus non-specific manner (Orozco *et al.*, 1997). It is thought that the DNA-binding activity of Rep is dependent on this multi-merization (Orozco and Hanley-Bowdoin, 1998). MSV Rep dimerizes and it is suggested that, as the monomer has only one active tyrosine for cleavage, the additional tyrosine in the dimeric form would be available for a second cleavage or a ligation reaction (Horváth *et al.*, 1998). In a study on WDV Rep and RepA using DNase footprinting and chemical cross-linking, Missich *et al.* (2000) showed that these replication proteins formed large nucleoprotein complexes near the TATA boxes of the virion-sense and complementary-sense promoters. Oligomerization of both proteins was dependent on pH with octomers being formed at pH ≤ 7.0, while at pH ≥ 7.4 the predominant form was a monomer. Preformed oligomers interacted very poorly with DNA. The authors suggested that there was a step-wise assembly of the protein–DNA complex with the monomers interacting with the DNA and then with other monomers to assemble the oligomeric structure.

Secondly, Rep binds to the (A)C3 protein. The (A)C3 protein locates to the nucleus and enhances DNA accumulation of begomo- and curtoviruses. It is thought that this enhancement is through the binding to Rep (Hanley-Bowdoin *et al.*, 1999).

Thirdly, Rep binds to retinoblastoma factor from both plants and animals (see below).

Finally, RepA binds to GRAB proteins (geminivirus RepA-binding proteins) (Xie *et al.*, 1999). Using the WDV RepA protein as a bait in the yeast two-hybrid system (Fig. 8.5), a family of GRAB proteins was isolated from wheat suspension cultured cells. The 37 amino acids at the C terminus of RepA, a region conserved among other mastreviruses, is involved in the interaction with an N-terminal domain of the GRAB proteins. The expression of GRAB1 or GRAB2 proteins in wheat cells inhibits WDV replication. GRAB proteins have significant amino acid homology with the NAC domain of proteins involved in plant development and senescence.

5. Geminivirus control of the cell cycle (reviewed by Gutierrez, 2000b)

As noted above, geminiviruses replicate in differentiated plant cells in which host DNA replication has ceased. The viral replication is dependent upon host DNA replication factors, and thus the cell cycle has to be modified. There are several lines of evidence suggesting that the Rep protein is involved in this modification (reviewed in Hanley-Bowdoin *et al.*, 1999). Many Rep proteins are recalcitrant to stable constitutive expression in transgenic plants. In those plants in which expression does occur, and in plants infected with TGMV, the nucleus becomes round and migrates to the cell center, features associated with de-differentiation. Rep proteins (or RepA of mastreviruses) bind to retinoblastoma (Rb) proteins from a variety of sources including plants (Horváth *et al.*, 1998; Liu *et al.*, 1999b).

Animal Rb proteins regulate cell growth most probably through control of the transition of the G0/G1 into S phase of the cell cycle (reviewed by Gutierrez, 1998; de Jager and Murray, 1999). It is thought that the plant analogs of Rb proteins have a similar function. Various animal DNA viruses control their host cell cycle through the binding of a virus-encoded protein with the host Rb protein through a LxCxE motif (reviewed by Hanley-Bowdoin *et al.*, 1999; Gutierrez, 2000a). Curtovirus and begomovirus Rep proteins have this LxCxE motif as does the RepA/Rep protein of mastreviruses. However, CSMV does not have a LxCxE motif (Horváth *et al.*, 1998) and, although mutagenesis of this motif in BYDV reduces the Rep binding capacity, it does not abolish the replication of the virus (Liu *et al.*, 1999b). As noted above, the curtovirus and begomovirus Rep proteins and the mastrevirus RepA protein bind Rb proteins from various sources, and it is likely that this binding is in a similar manner to that of animal viral proteins to their host Rb protein. Thus, the suggestion is that the binding inhibits the Rb protein control that maintains the host cell in the G phase of the cell cycle, enabling it to return to S phase and produce the factors required for viral replication. However, for this to occur Rep must be expressed from the incoming virus. Therefore there must be enough capability in the newly infected cell to initiate (−)-strand synthesis to give the dsDNA for transcription of the mRNA for Rep. Details of this initial event have not yet been determined, but a model is shown in Fig. 8.28.

E. Nanovirus replication

As described in Chapter 6 (Section V.B.1), the genomes of nanoviruses are distributed over at least six small circular ssDNA species, each of which (with one exception) has a single ORF. Although they have not been studied in as much detail as geminiviruses, nanoviruses have several features in common with gemi-

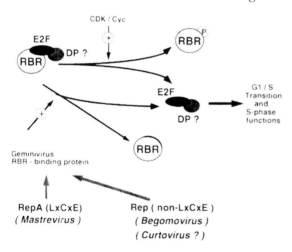

Fig. 8.28 A proposal for the two mechanisms that geminiviruses may use to interfere with the retinoblastoma-related (RBR)/E2F pathway. The G1/S transition is normally regulated by phosphorylation of RBR by CDK/cyclin complexes. RBR phosphorylation releases the RBR-bound E2F (the question mark indicates that a DP-like protein has not been identified yet in plants) transcription factors required for G1/S transition and S-phase functions. Geminivirus proteins (RepA in mastreviruses and Rep in begomoviruses) are proposed to sequester RBR and release RBR-bound factors, thus bypassing the normal cellular control. Mastrevirus RepA protein interacts with RBR through its LxCxE motif while begomovirus, and most likely curtovirus, Rep does so using a different motif. From Gutierrez (2000b), with kind permission of the copyright holder, © Kluwer Academic Publishers.

niviruses that suggest that their replication mechanisms are very similar.

Each of the nanovirus DNA species has a common region that is predicted to form a stem-loop, the loop containing the sequence 5'-TANTATTAC-3' (see Burns *et al.*, 1995) which is found in the origin of geminivirus (+)-strand synthesis (see Section VIII.D.2).

At least one of the DNA species is inferred from the amino acid sequence to encode a Rep protein (see Table 6.1) (see Harding *et al.*, 1993; Wu *et al.*, 1994b; Sano *et al.*, 1998; Timchenko *et al.*, 1999). *In vitro* tests with *E. coli*-expressed protein showed that the BBTV Rep protein has site-specific cleaving and joining activity (Hafner *et al.*, 1997a). In FBNYV, five of the DNA segments appear to encode a Rep protein (Timchenko *et al.*, 1999). Site-specific DNA cleavage and nucleotide transfer activities have been shown *in vitro* for those from DNAs 1 and 2 (Rep 1 and 2) and the essential tyrosine residue that catalyzes these reactions has been identified by mutagenesis. Rep 1 and 2 proteins hydrolyze ATP and this activity is essential for multiplication of the viral DNA. Each of the five Rep proteins initiated replication of the DNA species by which it was encoded, but only Rep2 was capable of replication of all the six DNAs that did not encode a Rep protein. Thus, only one of the Reps is a master Rep, and this is capable of triggering replication of heterologous nanovirus DNAs (Timchenko *et al.*, 2000).

Nanoviruses face the same problem as geminiviruses in that they need to start replicating their DNA in cells that are not transcriptionally active. The DNA of SCSV is able to self prime (−)-strand synthesis (Chu and Helms, 1988). Using self-primed extension with a DNA-dependent DNA polymerase, Hafner *et al.* (1997b) showed that all six DNAs of BBTV had endogenous primers bound to the genomic DNA. These primers were heterogeneous in size and appeared to be derived from DNA5, and that DNA self-primed more efficiently than the other DNAs. It is suggested that the function of the protein encoded by DNA5 is important early in the infection process.

The product of BBTV DNA5 contains the LxCxE motif characteristic of the retinoblastoma (RB)-binding protein described above

(Section VIII.D.5) (Wanitchakorn *et al.*, 2000). The yeast two-hybrid system showed that this protein has RB-binding activity, and the activity is dependent upon the LxCxE motif. None of the five Rep proteins of FBNYV contains the LxCxE motif. However, the 20-kDa protein encoded by DNA10 does contain this motif and also an F-box associated with binding to a ubiquitin-ligase (a plant SKP1 homolog) (Aronson *et al.*, 2000). The protein from DNA10, named Clink (cell cycle link), binds to RB and stimulates viral replication; the product of BBTV DNA5, described above, is a homolog of Clink. However, Clink is not an absolute requirement for infection of *Nicotiana benthamiana* and it is likely that this virus encodes one or more cell cycle-modulating activity. From its association with a constituent of the ubiquitin–protein turnover pathway, it is suggested that, as well as blocking the action of the RB protein, it targets that protein for processing.

IX. MUTATION AND RECOMBINATION

The main two ways by which faults arise in replication is by mutation and recombination. I shall discuss here the mechanisms leading to these faults, and in Chapter 17 the impact that these faults in replication have on the variation and evolution of plant viruses. Recombination is also involved in various other viral phenomena, which will be described here.

A. Mutation (reviewed by Domingo and Holland, 1997; Drake *et al.*, 1998)

Replication mutations can be either base substitutions, base additions or base deletions. In discussing mutations, one has to distinguish between *mutation frequency* and *mutation rate* (see Domingo, 1999). Mutation frequency is the proportion of mutants (averaged for an entire sequence or specific for a defined site) in a genome population. Mutation rate is the frequency of occurrence of a mutation event during genome replication. The relationship between mutation frequency and mutation rate is discussed by Drake and Holland (1999). Here

I will discuss mutation rate, but the frequency is important in the analysis of variation and evolution.

The rate of mutational errors depends on the mode of replication, the nucleotide sequence context and environmental factors. As shown in Fig 8.29, nucleic acids that replicate DNA → DNA have much lower mutation rates than those that replicate by other pathways. This is because DNA-dependent DNA polymerase has a proofreading ability that checks that the correct nucleotide has been added, whereas the other polymerases (DNA-dependent RNA polymerase, RdRp and RT) do not (Steinhauer *et al.*, 1992). The crystal structures of RNA replicases and RT do not reveal the 5′ to 3′ exonucleolytic proofreading domain present in DNA-dependent DNA polymerases (Kohlstaedt *et al.*, 1992; Joyce and Steitz, 1994; Hansen *et al.*, 1997).

Most of the studies on error rate have been undertaken *in vitro* and these have shown that parameters such as ionic composition of the medium and relative concentration of nucleoside triphosphate substrates can have significant effects (Domingo and Holland, 1997). Similarly, *in vitro* studies show that the sequence context being copied can have an effect with some regions being hypermutagenized. There is little evidence on which to judge the significance of these effects on plant viruses *in vivo*. In an attempt to minimize selection pressure, Kearney *et al.* (1993) measured the drift in two foreign sequences (dihydrofolate reductase and neomycinphosphotransferase II) cloned into a TMV vector. It was suggested that these foreign sequences would not be subject to the selection constraints of the

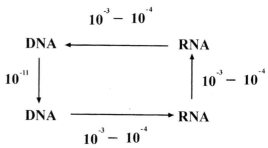

Fig. 8.29 Error rates of transcription within and between RNA and DNA.

viral sequence, and it was determined that the accumulated rate was $\leq 10^{-4}$ mutations per base per passage through *Nicotiana benthamiana*.

B. Recombination (reviewed by Bujarski, 1999; Hammond *et al.*, 1999)

Recombination is the formation of chimeric nucleic acid molecules from segments previously separated on the same molecule or present in different parental molecules. It usually, but not always, takes place during replication and can be a repair mechanism for aberrations resulting from mutation. It also is a major source of variation, as discussed in Chapter 17.

In many of the experiments on recombination, the design is to restore an important function by recombination between two nucleic acids with lost or depleted function. In this approach there is strong selection for the recombination event that may distort measurement of recombination frequency. A more realistic picture of the 'natural' situation is given if one performs the experiments under reduced or non-selective conditions. Thus, although it is recognized that the rates of recombination, especially that of RNA, are high, there are few estimates as to the actual values under 'natural' conditions.

There are differences and similarities in the mechanisms of recombination between DNA and RNA viruses.

1. DNA virus recombination

There are two basic forms of recombination in DNA viruses:

- homologous recombination occurring between two DNA sequences that are the same or very similar at the crossover point;
- non-homologous or illegitimate recombination occurring at sites where there is either microhomology or no obvious homology.

The latter usually occurs during double-strand break repair (Sargent *et al.*, 1997). In animal and bacterial viruses, non-homologous recombination is a rare event and is usually mediated by a virus- or host-encoded protein. Homologous recombination can require specific host or viral proteins but can also be due to template switching during replication.

Recombination is common among geminiviruses and is a major driving force in the evolution of this virus family (Padidam *et al.*, 1999) (see Chapter 17, Section X.D.7). Some of the early evidence came from insertion or deletion mutagenesis of the two large ORFs of ACMV DNA2, which destroyed infectivity, but infectivity was restored by co-inoculation of constructs that contained single mutations in different ORFs (Etessami *et al.*, 1988). Frequent intermolecular recombination produced dominant parental-type virus. Infection of *Nicotiana benthamiana* with uncut cloned tandem dimers of TGMV DNA components gave rise to genome-length ssDNA species of both components (Hayes *et al.*, 1988a). It has not been established whether intermolecular recombination or some replicative event led to the production of these unit-sized DNAs. Lethal mutations within the conserved stem-loop of ACMV are rapidly corrected by recombination (Roberts and Stanley, 1994).

Recombination has been found both within and between species of geminiviruses. The replicational release from tandem constructs agro-inoculated into plants (see Stenger *et al.*, 1991) is presumably due to homologous recombination. It is difficult to distinguish whether interspecies recombination is homologous or non-homologous. The apparent recombination site in the complex of begomoviruses found in cotton is often close to the origin of (+)-strand synthesis (Sanz *et al.*, 2000). As the sequence of the origin of (+)-strand synthesis is conserved between virus species, the recombination crossover could be due to homologous sequences. On the other hand, there is strong evidence for non-homologous recombination from reversion of deletion mutants to wild-type genome size, deletion of foreign sequences from geminivirus vectors, synthesis of wild-type molecules from two mutants, synthesis of subgenomic defective molecules, and release of infectious virus DNA from recombinant plasmids containing monomer genome inserts (reviewed by Bisaro, 1994).

Recombination is also common in CaMV and probably in all the *Caulimoviridae*. Both DNA and RNA recombination have been implicated in CaMV and this is discussed later in Section IX.B.3.

2. RNA virus recombination (reviewed by Lai, 1992; Nagy and Simon, 1997)

Initially, RNA recombination was categorized in the same manner as DNA recombination, into homologous and non-homologous types (King, 1988). Because of the range of variation in homologous recombination, Lai (1992) divided this category into homologous and aberrant homologous. Homologous recombinants had no sequence alterations in comparison with the parental molecules, whereas aberrant homologous recombinants contained mutations, deletions or insertions at, or close to, the insertion site. In a further analysis of RNA recombinants, Nagy and Simon (1997) proposed three classes of recombination (Fig. 8.30).

Class 1, termed *similarity-essential recombination*, has substantial sequence similarity between parental RNAs. There can be two types of products, precise and imprecise recombinants, similar to the homologous and aberrant homologous of Lai (1992). Class 2 recombination, *similarity-non-essential recombination*, occurs when there are no similar regions between the parents. It is thought that features such as transesterification, RdRp binding sites and secondary structure play a role in the recombination event (reviewed by Chetverin, 1999). Class 3 recombination, *similarity-assisted recombination*, combines features from both classes 1 and 2. In this class there are sequence similarities between the parental RNAs but additional RNA determinants on only one of the parental RNAs are required for efficient recombination.

Three mechanisms have been proposed for RNA recombination (reviewed in Nagy and Simon, 1997):

- a replicase-driven template switching model;
- a RNA breakage and ligation model;
- a breakage-induced template switching model.

The replicase-driven template switching model involves four elements, three RNAs (the primary RNA template (donor strand), the strand synthesized from the primary strand

Class 1: Similarity-Essential Recombination

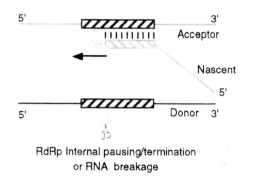

RdRp Internal pausing/termination
or RNA breakage

Class 2: Similarity-Nonessential Recombination

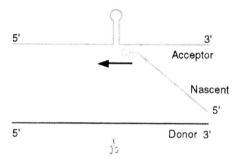

Class 3: Similarity-Assisted Recombination

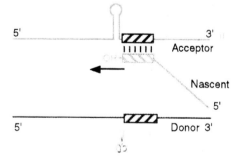

Fig. 8.30 Three classes of RNA recombination. Replicase mediated RNA synthesis after the template-switch events is shown by an arrow. The hairpin structure shown on the acceptor RNA symbolically represents various RNA determinants that are required for class-2 and class-3 recombination. From Nagy and Simon (1997), with permission.

(nascent strand), and the acceptor strand), and the replicase complex (Fig. 8.31). Synthesis of the nascent strand on the donor strand is halted or slowed temporarily, which enables either the RdRp or nascent strand to interact with the acceptor strand — leading to template switching. Thus, there are two types of signal on the donor or nascent strand: one (pausing or arrest signal) that halts the RdRp but from which it can escape; and the other (terminator signal) that releases the RdRp from the RNAs. It is thought that these signals may be similar to those involved in template switching by DNA-dependent DNA polymerases and RT, and to be the sequence and/or secondary structure of the donor or nascent RNA (see Kim and Koo, 2001). These regions that promote RdRp pausing or termination will constitute recombination hotspots. To enable template switching the RdRp must be able to bind to the acceptor RNA and use the 3'-end of the nascent RNA as a primer to re-initiate RNA synthesis.

Little is known about the actual mechanism of the switching of template by the RdRp, though there are several models. In the processive model, it is suggested that when the RdRp approaches the heteroduplex region it pauses and either switches to the acceptor sequence or slides backwards by 10–20 nucleotides, enabling the nascent strand to hybridize to the acceptor strand. In the non-processive model, the RdRp is suggested to dissociate from the donor strand and then re-associates with the nascent and acceptor RNAs.

The RNA breakage and ligation model is similar to the well-characterized DNA breakage and ligation system. It has not been formally demonstrated for RNA recombination.

The breakage-induced template switching model is similar to the replicase-driven template switching model with the switch being induced by pausing of the replicase at a break in the template RNA. Because of the lability of RNA, it can be difficult to distinguish this model from the replicase-driven template switching model.

Recombination is thought to occur in most, if not all, RNA viruses. The evidence for recombination is chimeric molecules, defective (D) and defective interfering (DI) molecules. However, it may be that all viruses recombine at similar

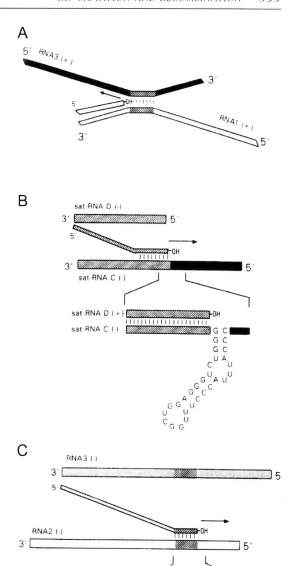

Fig. 8.31 Current replicase-mediated template-switching models of RNA recombination in BMV and TCV. **(A)** Heteroduplex-mediated recombination between (+) strands of BMV RNAs 1 and 3. **(B)** Recombination between satellite RNAs associated with TCV. The sequence of the required motif 1-hairpin is shown. **(C)** Recombination events within the identical regions of BMV RNAs 2 and 3. Recombination is favored when GC-rich and AU-rich sequences are located as shown. From Nagy and Simon (1997), with permission.

rates and these higher rates reflect a greater chance of a successful outcome of recombination. Some viruses appear to have higher rates of recombination than do others, and these have been the subjects for detailed studies. I will describe two of the viral systems: the bromoviruses and TCV that have been studied extensively for the formation of chimeric and D molecules, giving much theoretical information; and then DI molecules that can be important in the expression of disease.

a. Recombination in bromoviruses (reviewed by Bujarski and Nagy, 1996)

Recombination in plant RNA viruses was first demonstrated for BMV in an experiment that showed that a short deletion in the conserved 3′ terminal tRNA-like structure of RNA3 was repaired in plants by recombination event with the 3′-ends of either RNA1 or RNA2 (Bujarski and Kaesberg, 1986). Since then BMV has been the subject of detailed studies on both homologous and non-homologous recombination.

By testing the homologous recombination activity between various BMV-derived sequences, Nagy et al. (1999) identified five features (when all present, termed 'homology recombination activators') that can influence recombination:

- the length of sequence identity of the common region between the donor and acceptor RNA—this needs to be 15 nucleotides or more to effect efficient homologous crossovers (Nagy and Bujarski, 1995);
- the importance of the extent of sequence identity (Nagy and Bujarski, 1995);

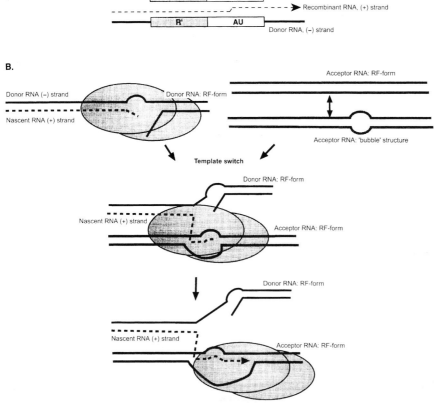

Fig. 8.32 A BMV replicase template switching model of homologous recombination. **(A)** The RNA strands and regions that are postulated to participate in the recombination events. **(B)** The proposed structure of the recombination intermediates during the homologous recombination event. The BMV replicase is thought to switch templates in the AU-rich regions that can form temporary 'bubble structures'. The GC-rich or average-content regions are postulated to stabilize the recombination intermediate formation via base-pairing between the aborted nascent strand and the acceptor strand. From Bujarski and Nagy (1996), with permission.

- the AU content of the common region, with at least 61–65% AU supporting frequent homologous recombinations (Nagy and Bujarski, 1997; Nagy *et al.*, 1999);
- the relative position of the AU-rich region (Nagy and Bujarski, 1997);
- the inhibitory effect of GC-rich sequences on upstream hot-spot regions while present on the acceptor RNA (Nagy and Bujarski, 1998).

Thus, a recombination hotspot has a GC-rich common region followed by an AU-rich region (Fig. 8.31). It is thought that the AU-rich region causes the replicase to pause and promotes RdRp slippage, leading to the incorporation of non-templated nucleotides; the latter are found in BMV recombinants (Nagy and Bujarski, 1993, 1996). The model for homologous recombination is illustrated in Fig. 8.32.

Non-homologous recombination of BMV occurs at about 10% of the frequency of homologous recombination. Nagy and Bujarski (1993) developed a BMV recombination vector system to explore non-homologous recombination. In these and subsequent experiments (Bujarski and Nagy, 1996) they demonstrated that the recombination is heteroduplex-based and the heteroduplex must be greater than 30

nucleotides. Their model (Fig. 8.33) suggests that the RdRp has difficulty in passing through the heteroduplex region and occasionally the template switches to the complementary region in the acceptor strand.

Mutagenesis of the replicase also gives some information on the mechanism. Mutations within the helicase domain of BMV 1a protein increased the frequency of recombination and shifted the recombination sites to energetically less stable parts of the heteroduplex (Nagy *et al.*, 1995). This is interpreted as suggesting that the mutations reduced the processivity of the replication complex, facilitating template switching at higher frequencies. On the other hand, mutations in the RdRp (2a protein) decreased the frequency of recombination (Bujarski and Nagy, 1996; Figlerowitz *et al.*, 1997).

Using PCR, recombinants were detected in tobacco plants that had been inoculated with CMV and the closely related TAV (Aaziz and Tepfer, 1999a) under conditions of minimum selection pressure. The recombination was detected in RNA3 in 3 of 82 plants inoculated with both viruses, and it occurred at several places at which there was high sequence similarity. In all cases, the recombination was homologous.

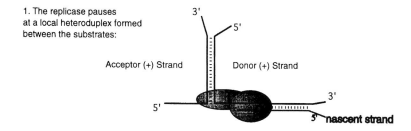

1. The replicase pauses at a local heteroduplex formed between the substrates:

Acceptor (+) Strand   Donor (+) Strand

nascent strand

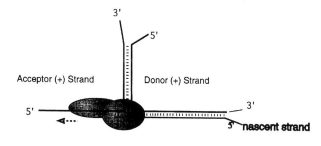

2. the replicase slides under the heteroduplexed region and restart the synthesis of the nascent strand on the acceptor template

Acceptor (+) Strand   Donor (+) Strand

nascent strand

Fig. 8.33 A template switching model of heteroduplex-mediated non-homologous recombination of BMV. From Bujarski and Nagy (1996), with permission.

b. Recombination in turnip crinkle virus
(reviewed by Simon and Nagy, 1996)

In addition to the genomic RNA, infections with TCV are often associated with satellite and DI RNAs. Sat-RNA D is a satellite RNA whose only similarity with TCV RNA is in the 3'-terminal 7 bases. Sat-RNA C is a chimera between nearly full-length Sat-RNA D and two segments from the 3' region of TCV (Fig. 8.34) (see Chapter 14, Section II.B.2).

When plants were inoculated with TCV, sat-RNA D and sat-RNA C in which there were deletions in the 5' region, a heterogeneous population of sat-RNA D/sat-RNA C recombinant molecules was formed. Recombination occurred in a region just upstream of a stable hairpin (termed motif I) (Fig. 8.35).

Similarly, recombinants were formed between TCV genomic RNA and sat-RNA D, with the crossover point again being upstream of a stable hairpin (motif IIIA/IIIB) (Fig. 8.35). About 30% of the recombinants between sat-RNA D and either sat-RNA C or TCV contained non-template nucleotides at the recombination site. There were also multimeric forms of the satellite RNAs, and these contained non-templated nucleotides. Recombination repaired truncations in the 3'-end of sat-RNA C or D when co-inoculated with TCV

These results led to the proposal of a model for recombination or TCV, shown in Fig. 8.36. In this model, the TCV RdRp copying the (−) strand of sat-RNA D either reaches the natural 5'-end or pauses at some position possibly owing to the presence of a protein binding site. The RdRp and nascent sat-RNA D switches to the acceptor (−) strand of sat-RNA C or TCV facilitated by either motif I or IIIA/IIIB hairpins. The TCV RdRp then re-initiates RNA synthesis using the 3'-end of the nascent sat-RNA D as a primer copying the acceptor RNA into the recombinant molecule.

3. Recombination in CaMV DNA
As described in Section VII.B.1, the reverse transcription mechanism used by the *Caulimoviridae* has two stages: transcription of the virion DNA in the nucleus, and reverse

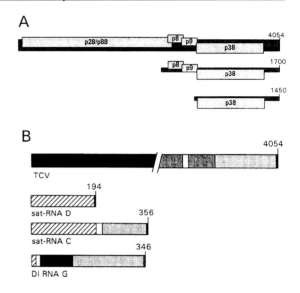

Fig. 8.34 (A) Genomic organization of TCV. The five ORFs of TCV are shown along with two known subgenomic RNAs. **(B)** Sequence similarity among subviral RNAs associated with TCV. Similar regions are shaded alike. All RNAs terminate with CCUGCCC-3'. From Simon and Nagy (1996), with permission.

transcription in the cytoplasm. Thus, there is the possibility of both DNA:DNA recombination in the nucleus and replicational recombination during the reverse transcription phase. Both forms of recombination are found.

The fact that CaMV DNA is converted to a covalently closed ds circle to allow transcription shows that there must be an early involvement of host plant DNA repair enzymes following infection. This idea is reinforced by the fact that cloned DNA, excised from the plasmid in linear form, is infectious, and that the progeny DNA is circular. Co-infection of plants with non-overlapping defective deletion mutants usually led to the production of viable virus particles (Howell *et al.*, 1981). Analysis of the progeny DNA showed that the rescue was by recombination rather than complementation. Lebeurier *et al.* (1982) showed that pairs of non-infectious recombinant full-length CaMV genomes integrated with a plasmid at different sites regained infectivity on inoculation to an appropriate host. In the progeny virus, all the plasmid DNA had been eliminated and

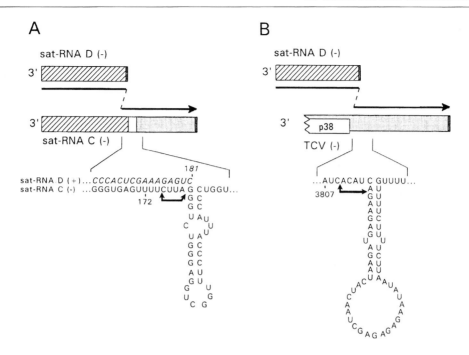

Fig. 8.35 Recombination between **(A)** TCV sat-RNA D and sat-RNA C, or **(B)** sat-RNA D and TCV. Since recombination is thought to occur during (+)-strand synthesis, the (–) strands of the parental molecules are shown. The motif I and motif IIIA/IIIB hairpins required for recombination, and their locations, are indicated. The arrowheads enclose the major hotspot crossover sites in sat-RNA C and TCV. Potential heteroduplex formation between nascent sat-RNA D strands and sequences upstream of the crossover site in sat-RNA C acceptor RNAs is shown. Similar sequences are shaded alike. From Simon and Nagy (1996), with permission.

the viral DNA had a normal structure. Walden and Howell (1982) provided further evidence for intergenomic recombination.

Based on experiments with pairs of heterologous genomes, Geldreich *et al.* (1986) proposed a model for recombination in CaMV mediated by the 35S RNA. In this model, just after inoculation two different DNAs with identical cohesive ends can be ligated together to give a dimer DNA. This dimer is then transcribed to generate a hybrid 35S RNA that is responsible for the formation of the recombinant genome by reverse transcription.

To study the mechanisms of recombination, Vaden and Melcher (1990) inoculated turnip plants with pairs of mutated CaMV DNAs and analyzed the progeny by restriction fragment patterns and sequencing. They found evidence

for both DNA:DNA and replicational recombination. Several of the chimeras had junctions between the parental sequences at, or near, the site for initiation of (−)-strand DNA synthesis or near the initiation sites for 35S or 19S RNA transcription. These were taken as being indicative of strand switching during reverse transcription. Other junctions were found that did not bear any obvious relationship with (−)-strand DNA synthesis, suggesting that they arose from DNA:DNA recombination. The deletion of inserts from the large intergenic region of CaMV DNA also suggested illegitimate recombination (Pennington and Melcher, 1993).

Similar forms of recombination have been found on the interaction of episomal CaMV with integrated viral sequences (see Section IX.B.5).

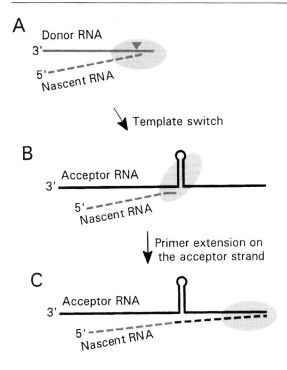

A Donor RNA

B

Acceptor RNA

Template switch

Primer extension on the acceptor strand

C

Acceptor RNA

Nascent RNA

Fig. 8.36 Model for recombination in the TCV system. **(A)** The TCV RdRp copying (−)-strand sat-RNA D either reaches the natural 5′-end or pauses (stalls) at some positions (mainly 13 nucleotides from the 5′-end, as indicated by a triangle) perhaps owing to the presence of a protein binding site. Alternatively, the nascent sat-RNA D downstream terminal sequences are removed by exonucleases, thus leading to truncation in some molecules. **(B)** The TCV RdRp, which is still associated with the nascent sat-RNA D strand, switches to the acceptor template ((−)-strand sat-RNA C or TCV) facilitated by either the motif I or motif IIIA/IIIB hairpins (see Fig. 8.35). Hybridization between the nascent strand and the acceptor strand may stabilize the recombination intermediate. **(C)** The TCV RdRp re-initiates RNA synthesis using the 3′-end of the nascent sat-RNA D as a primer. Further copying of the acceptor RNA by the RdRp results in a recombinant RNA molecule. From Simon and Nagy (1996), with permission.

4. Recombination and integrated viral sequences (reviewed in Hull *et al.*, 2000b)

An increasing number of viral sequences have been found to be naturally integrated into plant genomes. Thus far, they have all been 'DNA viruses' (*Caulimoviridae* and *Geminiviridae*) that have at least one phase of their replication cycle as DNA in the nucleus. The first reports were of multiple direct repeats of partial geminivirus sequences in *Nicotiana tabacum*, but there was no associated virus infection (Bejerano *et al.*, 1996). PVCV hybridizes to the petunia genome (Richert-Pöggler and Shepherd, 1997) and pararetrovirus-like sequences have been found in the genome of *N. tabacum* (Jakowitsch *et al.*, 1999). It is not altogether surprising that nuclear-located viral DNA sequences are inserted into the host chromosomes by illegitimate recombination. However, there are now at least two cases in which episomal infection can arise from such integrated sequences.

In certain banana (*Musa*) cultivars there has been apparently spontaneous outbreaks of BSV, especially during tissue culture and breeding programes. The best studied situation is with variety Obino l'Ewai (AAB genome) and the tetraploids following crossing with variety Calcutta 4 (AA genome) (Harper *et al.*, 1999b; Ndowora *et al.*, 1999). Obino l'Ewai itself showed occasional outbreaks of episomal BSV, whereas the male parent, Calcutta 4, never showed episomal BSV. The evidence for episomal infections arising from integrated sequences is as follows:

1. Many of the tetraploid lines had up to 100% infection after crossing symptomless parent plants. Tissue culture plantlets of Obino l'Ewai from symptomless mother plants had lower rates of infection and those of Calcutta 4 no infection.
2. PCR of total DNA from Obino l'Ewai using BSV primers and southern blotting of that DNA probed with BSV sequences gave positive results even though no virus could be detected by immuno-electron microscopy or by immune-capture PCR (Harper *et al.*, 1999a).
3. Sequencing of genomic clones from Obino l'Ewai revealed a complex insert of BSV (see below), the sequence of which was >99% homologous to that of the episomal virus. The BSV sequence interfaced with *Musa* sequence (Ndowora *et al.*, 1999).
4. The cloned products prepared by the sequence-specific amplification polymorphism (S-SAP) approach fell into three

classes, one of which comprised *Musa* sequence interfacing BSV sequence (Harper *et al.*, 1999b).

5. Fluorescent *in situ* hybridization revealed a major and a minor BSV locus (Harper *et al.*, 1999b).

6. Fibre-stretch hybridization, in which chromosomes are denatured, spread and then hybridized with fluorescent probes showed that the integrants were complex (see below).

The picture that has been derived for BSV is that there are two integration sites, as shown in Fig. 8.37. In each, there are tandem replicates of BSV sequence interspersed with *Musa* sequence. The major locus comprises six repeats over 150 kb and the minor locus three repeats over 50 kb. The interface of the BSV and *Musa* sequence is at the same site as the 5'-end of the 35S RNA transcript. From there, a segment of BSV sequence is uninterrupted for about 5 kbp. There then follows a region of short BSV-derived sequences in reverse and forward orientations followed by a segment of BSV in reverse orientation. Downstream of this is a further region of BSV in forward orientation covering the part of the genome missing from the first segment. The model is that the 7.4 kbp episomal viral genome is derived from the integrant by two recombination events (Fig. 8.38), these events being triggered by the stresses induced by crossing and/or tissue culture.

TVCV is solely seed-transmitted in the hybrid tobacco species, *Nicotiana edwardsonii* (Lockhart *et al.*, 2000). The symptoms occur only under conditions of markedly changing day length and episomal virus can only be detected in such plants. Genomic clones of *N. edwardsonii* contain sequences that have >99% homology with the episomal viral sequence. There are various interesting features of TVCV:

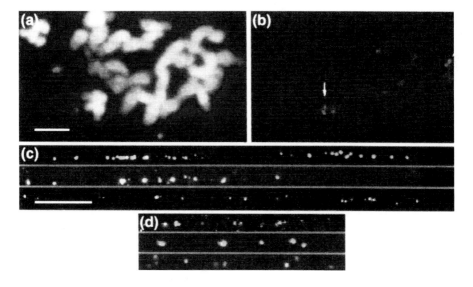

Fig. 8.37 (see Plate 8.1) *In situ* hybridization of *Musa* Obino l'Ewai chromosomes. **(a,b)** Hybridization of metaphase chromosomes. **(a)** Chromosomes stained blue with the DNA stain DAPI (4,6-diamidino-2-phenylindole). Bar = 5 μm. **(b)** Hybridization sites of BSV (red) showing one major site in each metaphase (arrowhead) and at least one minor site near the limits of hybridization sensitivity (arrow). **(c,d)** Hybridization of extended chromosomal DNA fibres. Two different-length hybridization patterns, with chains of dots representing probe hybridization sites, were detected with BSV (green) and *Musa* flanking sequence (red probes). **(c)** Three independent aligned long fibres. Both BSV and *Musa* flanking sequence are present in multiple copies in a 150-kb structure. Bar = 5 μm. **(d)** Three independent aligned short fibres in a 50-kb structure. From Harper *et al.* (1999b), with permission.

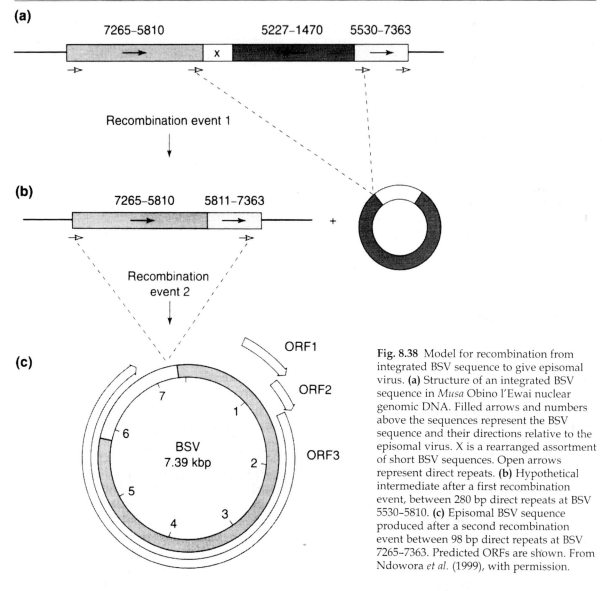

Fig. 8.38 Model for recombination from integrated BSV sequence to give episomal virus. **(a)** Structure of an integrated BSV sequence in *Musa* Obino l'Ewai nuclear genomic DNA. Filled arrows and numbers above the sequences represent the BSV sequence and their directions relative to the episomal virus. X is a rearranged assortment of short BSV sequences. Open arrows represent direct repeats. **(b)** Hypothetical intermediate after a first recombination event, between 280 bp direct repeats at BSV 5530–5810. **(c)** Episomal BSV sequence produced after a second recombination event between 98 bp direct repeats at BSV 7265–7363. Predicted ORFs are shown. From Ndowora *et al.* (1999), with permission.

1. It appears to be activated only under certain climatic conditions related to day length. In Florida where *N. edwardsonii* was developed, the presence of the virus appears not to have been detected whereas it appeared in Minnesota.

2. Southern blotting detects TVCV-related sequences in the male parent of *N. edwardsonii*, *N. glutinosa*, but not in the female parent, *N. clevelandii*. However, *N. glutinosa* does not appear to develop episomal infections of TVCV under conditions in which it becomes activated in *N. edwardsonii*.

3. TVCV sequences have been detected in

Nicotiana species of South American origin but not in those of North American or Australasian origin.

Thus, in both these viruses that appear to be activated from integrated sequences, stresses are needed for activation. Furthermore, they are more prevalent in crosses between different parent genome types. The integrated viruses are pararetroviruses, but neither the structure of the integrant nor the release of the episomal virus appears to resemble that of retroviruses.

It is likely that PVCV shares a similar phenomenon of activation of episomal virus from integrated sequences.

5. Recombination and transgenic plants

Infection of plants containing a transgene of part of the episomal virus has led to the production of recombinants. However, the interpretation of these experiments to throw light on recombination in joint infections must be taken with caution. Firstly, there is the point mentioned above about the selection pressure for recombinants. Secondly, because of the positioning of the transgene in the chromosome and transcription of it in the cytoplasm, there might be different compartmentalization to that in natural joint infections. Recombination in transgenic plants has been found for RNA, DNA and reverse transcribing viruses.

Inoculation of defective CaMV DNA into plants transgenic in the sequence of the defect has resulted in recombination producing infectious wild-type virus or recombinants between two viral strains (Gal *et al.*, 1992; Schoelz and Wintermantel, 1993). Analysis of the progeny virus indicated that both DNA:DNA and replicational recombination had taken place. Using two isolates of CaMV, Schoelz and Wintermantel (1993) identified two template switches, one at the 5'-end of the 35S RNA and the other at the 5'-end of the 19S RNA. This was taken as suggesting that, at the first strand switch in (−)-strand DNA synthesis (for CaMV replication see Section VII.B.1), the reverse transcriptase moved to the 3'-end of the 19S RNA instead of the 35S RNA; and when it reached the 5'-end of the 19S RNA it moved back to the 35S RNA template. However, this system created a strong selection pressure as only the recombinants gave infections. Under less selection pressure there was recombination but at a lower rate (3 out of 23 plants) (Wintermantel and Schoelz, 1996).

Using an integrated multimer of two different strains of CaMV, Gal *et al.* (1991) detected both types of recombination. A replicational hotspot was identified in the region of the promoter for the 19S RNA. Molecular characterization of integrants in rice of constructs driven by the 35S RNA promoter has revealed a recombinational hotspot also in that promoter (Kohli *et al.*, 1999a).

Recombination has been shown between a coat protein deletion mutant of the begomovirus ACMV and its transgenic coat protein gene (Frischmuth and Stanley, 1998). The frequency of recombination was highest when the transgene also contained the common region, suggesting that this region is a recombinational hotspot.

There can be recombination between the genome of an episomally infecting RNA virus and a transgene expressing part of that or a related virus. Transcripts in plants expressing the 3' two-thirds of CCMV coat protein gene (from RNA3) recombined with a CCMV mutant lacking the 3' one-third of the coat protein gene (Greene and Allison, 1994). Similarly, recombination has been detected between a transgene and episomally infecting virus for TBSV (Borja *et al.*, 1999) and between the two related cucumoviruses, CMV and TAV (Aaziz and Tepfer 1999a,b).

The possibility of recombination between episomal and transgenic viral sequences has obvious implications in risk assessment for the field release of transgenic plants. This is discussed in Chapter 16 (Section VII.E).

C. Defective and defective interfering nucleic acids and particles (reviewed by Graves *et al.*, 1996; White, 1996)

Recombination can lead to subviral RNA molecules derived from parent genomic RNA or DNA. These molecules comprise unmodified terminal sequences and, in some cases, some internal sequences. They depend upon the parent virus for their replication. In some cases, the defective molecules interfere with the replication and reduce symptom production by the parent virus; these are called *defective interfering* (DI) nucleic acids. If they do not interfere with the parent virus they are termed *defective* (D) nucleic acids and, if encapsidated by the parental coat protein, *defective* (D) viruses. DI and D nucleic acids are found with both RNA and DNA viruses.

The D and DI nucleic acids and viruses of RNA viruses are listed in Table 8.6. The deletion patterns in the DI and D RNAs and viruses fall into two groups. In the first group the defective RNA is derived from a single internal deletion; and in the second group it consists of a mosaic of the parental viral genome.

Single-deletion DI and D RNAs have been found in *Bromoviridae* (AMV, BBMV, CMV) family and in the potexvirus (ClYMV), tobravirus (TRV), furovirus (SBWMV), pecluvirus (PCV), benyvirus (BNYVV), phytoreovirus (WTV) and tospovirus (TSWV) genera (Table 8.6). All the defective RNAs isolated thus far from multipartite genome viruses fall into this group.

Multiple-deletion DI and D RNAs are characteristic of several members of the *Tombusviridae* family (Table 8.6) and have been studied in great detail in TBSV and TCV. Examples of the single-deletion and multiple-deletion defective molecules are described below.

As well as the natural DI and D RNAs, artificial molecules have been made that contain deletions and which can be supported by the parent virus. An example is TYMV in which molecules with small deletions in the coat protein gene were poorly supported by the parent virus but those with a large deletion in the 70-kDa ORF were replicated efficiently (Dreher and Weiland, 1994). Comparison of the replication requirement of a TMV-based D RNA and its helper virus revealed different requirements for the replication of TMV RNAs in *cis* and in *trans* (Chandrika *et al.*, 2000). Deletions of certain 3' terminal pseudoknots decreased the level of replication of full-length TNV RNA but did not affect the replication of the D RNA. However, the 3'-most pseudoknot was required for replication of both full-length and D RNAs. Homologous 3' sequences were important for the replication of the D RNA, the precise requirement appearing to involve the terminal 28 nucleotides and specifically the pseudoknot in

TABLE 8.6 Defective and defective interfering nucleic acids

Family/genus	Virus	Deleted segment	Type of defective element	Comments and reference
Group 1				
Reoviridae	WTV	RNAs 2 and 5	D virus	Loss of vector transmission (1)
Tospovirus	TSWV	L RNA	DI RNA	Encodes polymerase protein (2)
		M RNA	D virus	Loss of viral envelope and probable loss of vector transmission (2)
	PBNV	L RNA	D virus	(3)
Rhabdoviridae	SYNV			(4)
Bromoviridae	AMV	RNA3	D RNA	See Section IX.C.1
	BBMV	RNA2	DI RNA	Exacerbates symptoms in some hosts, encodes viral polymerase. See Section IX.C.1
	CMV	RNA3	D RNA	Deletion in 3a protein; encodes coat protein. See Section IX.C.1
Closteroviridae	CTV	Various	D virus	(5). Can affect aphid transmission (6)
Tobravirus	TRV	RNA2	D and DI virus	Vector transmission eliminates DI (7)
Potexvirus	ClYMV	Various	DI and D RNA	(8)
	BaMV	Central region	D virus	(8)
	CsCMV	Central region	D virus	(9)
Furovirus	SBWMV	RNA2	D virus	Loss of vector transmission (1)
		RNAs 3 and 4	D virus	Loss of vector transmission and ability to infect roots (1)
Pecluvirus	PCV	RNA2	D virus	Loss of vector transmission (1)
Benyvirus	BNYVV	RNA2	D virus	Loss of vector transmission (1)
Sobemovirus	CfMV	Central region	DI RNA	(11)
Group 2				
Tombusviridae	TBSV	Various		See Section IX.C.2
	TCV	Various		See Section IX.C.2

References: (1) Graves *et al.* (1996) (review); (2) Nagata *et al.* (2000); (3) Gowda *et al.* (1998); (4) Ismail and Milner (1988); (5) Bar-Joseph *et al.* (1997) (review); (6) Albiach-Martí *et al.* (2000a); (7) Visser *et al.* (1999b); (8) White *et al.* (1992); (9) Yeh *et al.* (1999); (10) Calvert *et al.* (1996); (11) Makinen *et al.* (2000).

the aminoacyl acceptor arm of the tRNA-like structure.

1. Defective RNAs in the *Bromoviridae* (reviewed by Graves *et al.*, 1996)

The defective RNAs associated with AMV and CMV are derived from RNA3 and have no apparent effect on either symptom production or on virus accumulation. The D RNA of CMV is derived by a single deletion that removes a segment of the 3a ORF (Fig 8.39) but maintaining the reading frame downstream of the deletion (Graves and Roossinck, 1995a).

This leaves a defective cell-to-cell movement protein and a functional coat protein. CMV D RNAs accumulate in various *Nicotiana* species; but in tomato, zucchini squash and muskmelon the D RNAs accumulated only in inoculated tissue but did not move systemically (Graves and Roossinck, 1995b). Furthermore, the D RNA accumulated and was encapsidated in both inoculated cotyledons and leaves of tomato and zucchini squash and accumulated but was not encapsidated only in the inoculated cotyledons of muskmelon. This indicates that host and tissue specificity is involved in replication, cell-to-cell movement,

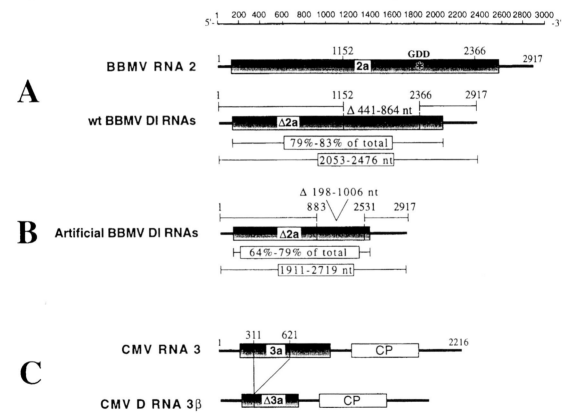

Fig. 8.39 Defective RNAs derived from members of the *Bromoviridae*. A schematic representation of BBMV and CMV defective RNAs. The 5′ and 3′ untranslated regions are shown by black lines. The 2a and 3a ORFs are represented by shaded boxes while the CP ORF is represented by a white box. The approximate location of the conserved 'GDD' motif in the 2a ORF is indicated by an asterisk. A scale bar (in nucleotides) is located at the very top of the figure. **(A)** BBMV genomic RNA2 (top) and wt DI RNAs (bottom). The numbered lines above the DI RNA indicate the minimum 5′ and 3′ sequences conserved in all the wt DI RNAs. The size range of the deletions is also given above the DI RNA. The top numbered line below the DI RNA indicates the length of the Δ2a ORF as compared to the entire length of the DI RNA. The bottom numbered line indicates the range in the total lengths of the wt DI RNAs. **(B)** BBMV artificial DI RNAs. The artificial DI RNA is labeled as in (A). The data reflect only those DI RNAs that accumulated efficiently in plants. **(C)** CMV genomic RNA3 (top) and wt D RNA 3β (bottom). The region deleted from the 3a ORF is indicated. From Graves *et al.* (1996), with permission.

systemic movement and encapsidation of CMV D RNAs.

The defective RNA of BBMV is derived from RNA2 and is a DI RNA as it decreases the concentration of RNA2. Even though it decreased the concentration of RNA2, the presence of the DI RNA exacerbated the severity of BBMV symptoms in some hosts, causing the symptoms to appear one or two days earlier than when the DI RNA was not present, and to be more severe, especially in *Pisum sativum* cv Rondo where it was lethal (Romero *et al.*, 1993). The internal deletion is in the 2a ORF and includes the GDD motif characteristic of the function of this ORF product as an RdRp (Romero *et al.*, 1993; Pogany *et al.*, 1995). As with CMV, the region of the ORF downstream of the deletion is in-frame with that upstream, suggesting that effective translation is an important feature for the production and maintenance of the defective molecules. Analysis of the sequences flanking the junction sites of BBMV DI RNAs shows the presence of either short complementary and/or similar sequences (Pogany *et al.*, 1995). An artificial DI RNA in which a 60-nucleotide sequence was duplicated in another part of the molecule produced shorter RNAs if the duplication was in the reverse orientation but not if it was in the direct orientation (Pogany and Bujarski, 1996). The further deletion was at, or close to, the base of the hairpin formed by the inverse duplication. This suggests that the DI molecules are formed by recombination during RNA replication.

The production of DI RNAs by BBMV is controlled by three factors that have been determined from both natural and artificial molecules (see Graves *et al.*, 1996):

- the presence of the terminal regions — the DI RNAs retained the 5'-terminal 1152 nucleotides and the 3'-terminal 468 nucleotides;
- the overall size — the sizes of the deleted sequences are between 15% and 30% that of the wild-type RNA2;
- the coding capacity — the DI RNA retains a deleted form of the 2a ORF that comprises at least 79% of the molecule.

BBMV DI RNAs do not accumulate and are not encapsidated in local lesion hosts or in some systemic hosts (Romero *et al.*, 1993). Thus, as with CMV D RNAs, there is host specificity involved in the replication and encapsidation of BBMV DI RNAs.

2. Defective interfering RNAs in the *Tombusviridae* (reviewed by White, 1996; White and Morris, 1999)

Some of the earliest reports of DI RNAs came from work on tombusviruses. Hillman *et al.* (1987) and Morris and Hillman (1989) described an abnormal RNA from a culture of TBSV that met all the criteria for a DI RNA. The RNA was about 396 nucleotides long and was derived from the genomic RNA by six internal deletions, the 5' and 3' sequences being conserved. Two of the deletions were large (1180 and 3000 nucleotides) while the others were much smaller. Co-inoculation of the small RNA depressed virus synthesis in whole plants and attenuated disease symptoms. Although the DI RNA could represent 60% of virus-specific RNA in leaf extracts, only about 3–4% of the encapsidated RNA was DI RNA. Experiments in protoplasts showed that the DI RNA suppresses replication of genomic TBSV RNA (Jones *et al.*, 1990). A similar DI RNA was described from a culture of CymRSV (Burgyan *et al.*, 1989). A DI RNA associated with TCV was shown to be a mosaic molecule containing sequences derived from TCV together with a block of nucleotides corresponding to a 5' sequence found in satellite RNAs associated with TCV (Li *et al.*, 1989). This DI RNA was also unusual in that its presence made the disease caused by TCV more severe.

DI RNAs have been found in several other members of the *Tombusviridae* family including CIRV and CNV. The naturally occurring DI RNAs are about 10–20% (400–800 nucleotides) of the size of the parent genomes. The DI RNAs are composed of conserved non-contiguous segments of the genome that accumulate *de novo* after serial passage of the parent virus at high multiplicity of infection. These molecules suppress the accumulation of the parent virus and usually attenuate the severe symptoms that these viruses normally induce. It is generally considered that the DI RNAs are more efficient than the parent virus at recruiting the viral replicase, and that the reduction in symptoms is a direct effect of the reduction in parent virus concentration.

The DI RNA of TBSV has been studied in detail and has features that are common to DI RNAs of other members of the *Tombusvirus* genus. The typical molecule is composed of four non-contiguous segments (regions I to IV) derived from the parent virus (Fig. 8.40).

Regions I and IV are derived from the 5' and 3' termini respectively. Region I contains the 5' non-coding region and start codon for the 5' ORF; and region IV comprises a non-coding 3' sequence. Region II is from just downstream of the stop codon for the 33-kDa protein; and region III is from the C-terminal region of the 22-kDa protein. Infected plants contain two size classes of TBSV DI RNA, the larger not being deleted between regions III and IV and which is thought to be a precursor for the smaller molecule. There are also small amounts of molecules that have extra duplications, such as of region II, or are direct repeat dimers of the entire DI RNA.

By manipulating natural DI RNAs or making artificial ones, the importance of the conserved regions for the replication of the molecules has been determined (Chang *et al.*, 1995; Havelda and Burgyan, 1995; Havelda *et al.*, 1995). Deletion of region I, II or IV abolishes the abil-ity of the DI RNA to be replicated by the parent virus. Region IV contains secondary structure that plays an important role in DI RNA accumulation (Havelda and Burgyan, 1995). Region III is critical for CymRSV DI RNA (Havelda *et al.*, 1995) but not for those of CNV or TBSV (Chang *et al.*, 1995).

The deletion events leading to the production of tombusvirus DI RNAs are described in detail by White (1996). It is noted that the polymerases of tombusviruses do not appear to have a helicase domain (see Section IV.B.2) and it may be lack of this activity that leads to replication errors.

3. Other defective RNAs associated with RNA viruses

A putative DI RNA has been described for the comovirus BPMV (Sundararaman *et al.*, 2000). This was found in a cDNA library from mRNA from apparently healthy soybean pods and appeared to be a deleted form of BPMV RNA2.

Viruses other than those with (+)-strand RNA genomes can have DI-like RNAs. For those with (−)-strand RNA genomes, Adam *et al.* (1983) described a population of DI-like particles associated with a plant rhabdovirus that arose after 30 passages. Ismail and Milner (1988) isolated DI particles from *Nicotiana edwardsonii* plants chronically infected with SYNV. Most of the DI particles were 73–86% as long as the standard virus. Alone they were

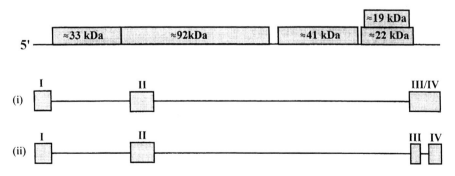

Fig. 8.40 Schematic of the RNA genome of a typical tombusvirus and of naturally occurring DI RNAs. The organization of the coding regions in the ~4.7-kb genome is shown at the top with the approximate sizes of the encoded proteins. Regions from which the TBSV DI RNAS were derived are shown below as shaded boxes with the deleted intervening regions depicted as lines. The four regions that are conserved, to some degree, in all characterized naturally occurring DI RNAs are indicated by roman numerals. **(i)** A larger size-class DI RNA containing three non-contiguous regions. Region III/IV represents a contiguous 3' terminus which includes the segment between regions III and IV. **(ii)** A proteotypical DI RNA containing four distinct regions. From White (1996), with permission.

non-infectious, but when co-inoculated with complete virus they were replicated to a greater extent than the infectious particles.

Shorter than normal dsRNA segments are associated with transmission-defective isolates of WTV. Nuss and Summers (1984) showed that these RNAs are formed by the deletion of up to 85% of a genomic RNA segment, giving rise to terminally conserved RNAs that are functional with respect to transcription, replication and packaging. These isolates interfere with standard virus production in leafhopper cell monolayers (Reddy and Black, 1977).

4. Defective DNAs associated with DNA viruses

MacDowell *et al.* (1986) characterized a less than full-sized class of DNA from preparations of the begomovirus TGMV. These DNAs, of slightly varying length, were derived only from genomic DNA B, by a deletion of about half the DNA. A DI DNA associated with ACMV interferes with the replication of both genomic components (Frischmuth and Stanley, 1991). These DNAs delayed symptom expression and to this extent were like the DI nucleic acids of animal viruses.

A population of defective DNAs is associated with AYVV (Stanley *et al.*, 1997). They are all approximately half the size of the AYVV genomic DNA and contain intergenic region and the 5′ terminus of ORF C1 (Fig. 8.41); they also contain some additional sequences unrelated to the AYVV sequence. They ameliorate the symptoms of AYVV in *Nicotiana benthamiana*, suggesting that they are DI DNAs.

X. MIXED VIRUS ASSEMBLY

Mixed virus assembly can be shown to take place *in vitro* between the RNA of one strain of a virus and the coat protein of another (e.g. Okada *et al.*, 1970; Okada, 1986b), between RNA and protein from unrelated viruses (e.g. Matthews, 1966), and between one kind of RNA and two different coat proteins (e.g. Wagner and Bancroft, 1968; Talianski *et al.*, 1977). Mixed infections are discussed by Hammond *et al.* (1999). Of more interest is the formation of mixed virus particles *in vivo*.

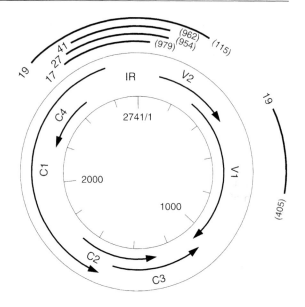

Fig. 8.41 AYVV DNA sequences retained within the defective viral DNAs. The location of the virion-sense genes (V1 and V2), complementary-sense genes (C1–C4), and the intergenic region (IR) are indicated within the genome map of AYVV. AYVV sequences retained within various clones are indicated on the outside of the map. Figures in parentheses indicate the number of non-viral nucleotides associated with each defective DNA. From Stanley *et al.* (1997), with permission.

When two viruses multiply together in the same tissue, some progeny particles may be formed that consist of the genome of one virus housed in a particle made partially or completely from the structural components of the other virus. Among enveloped viruses infecting animals, mixed infections may lead to the production of nucleoprotein cores of one virus enclosed in an envelope of the other. Such mixed particles, called *pseudotypes*, have not been observed with enveloped plant viruses. They will probably be found among the plant *Rhabdoviridae*.

Other kinds of mixed particle may be formed. Where the genome of one non-enveloped virus is encased in a protein shell made entirely of subunits of another virus (or strain) the phenomenon has been called *genomic masking, heterologous encapsidation, heteroencapsidation* or *transencapsidation*. When the protein coat consists of a mixture of proteins from the two viruses, it has been called *phenotypic mixing*.

Dodds and Hamilton (1976) gave an account of the methods used to study phenotypic mixing. Various studies on phenotypic mixing have been carried out with defective mutants of TMV whose protein will not form rods with the RNA when plants are grown at high temperature. When such strains are grown in mixed infections with type TMV (or some other strain able to form virus rods at the higher temperature), then a proportion of the progeny contain the mutant strain RNA in a rod made with the protein of the competent strain (Schaskolskaya *et al.*, 1968; Sarkar, 1969; Atabekov *et al.*, 1970b).

Such mixing may take place in leaves only under conditions where two viral RNAs are present and one functional coat protein is made (Atabekova *et al.*, 1975). On the other hand, Otsuki and Takebe (1978) showed that when protoplasts were inoculated with TMV together with ToMV, some of the individual progeny rods were coated with a mixture of the two coat proteins.

Strains of BYDV show aphid vector specificity (see Chapter 11, Section III.H.1). When a strain of the virus normally transmitted in a particular vector was grown in oats in a double infection with a serologically unrelated strain not normally transmitted by the aphid, this latter strain was transmitted. Rochow (1970) showed that this transmission was because some of the RNA of the second strain had been assembled into protein shells of the normally transmitted strain. In an analysis of mixed infections of four isolates of cereal-infecting luteoviruses, Wen and Lister (1991) demonstrated heterologous encapsidation between various combinations of CYDV-RPV, BYDV-MAV, BYDV-PAV and BYDV-RMV. In most combinations the heterologous encapsidation was in both directions, but in two of the cases (CYDV-RPV + BYDV-PAV and CYDV-RPV + BYDV-MAV) it was only in one direction with CYDV-RPV providing the capsid.

A novel immunohybridization procedure has been developed to demonstrate directly that, in mixed infections in the field, an aphid non-transmitted strain of BYDV became encapsulated in the protein of an aphid-transmitted strain (Creamer and Falk, 1990). Similarly, phenotypic mixing has been demonstrated in potyviruses.

Bourdin and Lecoq (1991) showed that an isolate of ZYMV that was not aphid-transmissible because of a defect in its coat protein was aphid-transmitted from plants co-infected with a transmissible strain of PRSV. Immunosorbent electron microscopy revealed particles that were heteroencapsidated (or transencapsidated) by the coat proteins of both viruses (Fig. 8.42).

Phenotypic mixing can occur between two unrelated helical viruses with different dimensions (TMV and BSMV in barley) as shown by Dodds and Hamilton (1974). It has even been found between a helical virus (BSMV) and an icosahedral one (BMV) (Peterson and Brakke, 1973).

The encapsidation of umbravirus genomes by luteovirus coat protein is described in Chapter 11 (Section III.H.1.a) and heteroencapsidation by transgenically expressed coat protein in Chapter 16 (Section IV.E.1).

The existence of phenotypic mixing also suggests that two unrelated viruses or two related strains can replicate together in the same cell at least under some conditions.

The formation of distinctive inclusion bodies has been used to confirm that two unrelated viruses can replicate in the same cell; for example, TMV and TEV in tobacco (Fujisawa *et al.*, 1967), TuMV and CaMV in *Brassica perviridis* (Kamei *et al.*, 1969), and SMV and BPMV in soybean (Lee and Ross, 1972).

In tobacco leaves doubly infected with TMV plus PVX or PVY plus PVX, no assembly of one viral RNA in the coat protein of another could be detected (Goodman and Ross, 1974b). A likely reason for this is that, whereas closely related strains of a virus might replicate in the same region of the cell, different viruses may be assembled from components accumulated in separate sites or viroplasms in the same cell. Such separation may not always be complete.

Efficient and specific virus assembly would be favored by the localization of the RNA and protein subunits in a compartment within the cell. There are several reasons for this:

1. If *in vivo* assembly is due to random meeting between protein subunits, then maintenance of a high local concentration of these would favor efficient assembly.

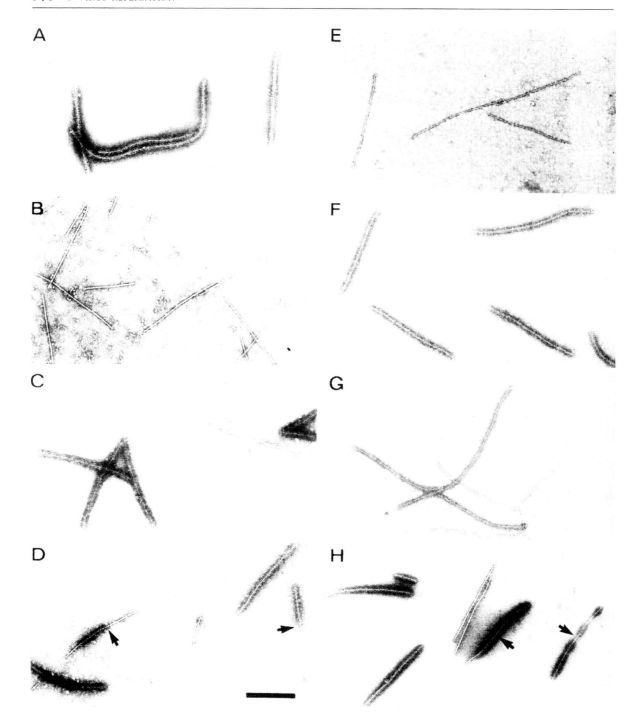

Fig. 8.42 Electron micrographs of purified preparations of: **(A,E)** PRSV (-E2); **(B,F)** ZYMV (-NAT) viruses; **(C,G)** a 1:1 mixture of these virus preparations; and **(D,H)** a purified preparation from plants co-infected with PRSV-E2 and ZYMV-NAT. Grids were coated with a mixture of PRSV and ZYMV antisera, and trapped particles were decorated with PRSV antisera (panels A–D) and ZYMV antisera (panels E–H). In artificial mixtures, particles were decorated or not decorated (C,G), whereas in preparations from doubly infected plants some particles appeared partially decorated (arrows) regardless of which antiserum was used for decoration (D,H). Bar = 400 nm. From Bourdoin and Lecoq (1991), with kind permission of the copyright holder, © The American Phytopathological Society.

2. Since subunits can pack around non-viral RNA of appropriate size, and since insignificant amounts of non-viral RNA are usually present in virus particles, free host RNA must be largely excluded from the assembly sites.

3. *In vitro* studies show that aggregation of subunits is markedly dependent on ionic environment and pH. These specific conditions differ *in vitro* for different viruses.

4. Uncoated RNA must be protected from attack by nucleases.

Nevertheless, with some viruses significant amounts of host RNA may be incorporated into virus-like particles or pseudovirions. Such particles have been reported as making up to 2.5% of preparations of various strains of TMV (Siegel, 1971; Rochon *et al.*, 1986). Most of the encapsulated host RNA is the 5′ region of 18S ribosomal RNA (Rochon *et al.*, 1986). The site for initiation of this packaging has been located within a 43-nucleotide region beginning at position 157 from the 5′ terminus of the rRNA. This sequence has limited similarity to the TMV assembly initiation sequence (see Fig. 5.14), but it can fold to give a stem-loop structure (Gaddipati *et al.*, 1988).

The mRNA for the large subunit of ribulose 1,5-bisphosphate carboxylase, which is coded by chloroplast DNA, is encapsulated in TMV coat protein in infected cells. This mRNA was found to contain at least three sites that were capable of reacting with coat protein aggregates *in vitro* to initiate rod formation (Atreya and Siegel, 1989). The most reactive site had significant sequence similarity to the initiation site in TMV.

XI. DISCUSSION

There are basically three types of viral replication mechanisms, making DNA directly from DNA, alternating between DNA and RNA, and making RNA from RNA. Each of these faces different problems as well as the overall problem of replicating to a level that would ensure propagation to a new host without causing irreparable damage to the current host.

The viruses that replicate DNA ⟶ DNA use the host machinery for this process; however, this machinery is usually active only during cell division. As described in detail in Section VIII, geminiviruses and nanoviruses have mechanisms that 'switch on' the host DNA replication enzymes, thereby overcoming this limitation.

The replication of nucleic acid by the routes DNA ⟶ RNA ⟶ DNA and RNA ⟶ RNA are not found to any great extent in uninfected plant cells. Although plant genomes contain retrotransposons, most are inactive due to mutations and the active ones appear to replicate only under certain stresses. Furthermore, their replication is thought to be controlled by host defense systems (see Chapter 10, Section IV). The *Caulimoviridae* use this route to replicate their genomes with the DNA ⟶ RNA phase being effected by a host enzymic system and the RNA ⟶ DNA phase by virus-coded enzyme activities. How these viruses overcome the host constraints directed at the unrestrained replication of retrotransposons is unknown. However, the fact that the episomal replication mechanism does not involve integration into the host genome may play a part in this. The replication of viruses such as BSV and TVCV is episomal and the integration of the viral sequences into the host genome supplies only the inoculum and is not an integral part of the replication mechanism. However, in considering the constraints on the reverse transcription route of replication it must be recognized that animal retroviruses involve an integration stage and there is little, if any, episomal replication. It may be that other host defense systems such as immune-surveillance, not found in plants, play a role.

The synthesis of RNA from RNA has not been considered to be a major mechanism in uninfected plants. As described in Chapter 10 (Section IV), this route of nucleic acid replication is used in a host defense system. Furthermore, the importance of the formation of antisense RNAs for control of gene expression is becoming increasingly recognized (see Terryn and Rouzé, 2000). However, the majority of plant viruses replicate by this route using virus-coded enzyme systems together with some host-coded factors. Why this mechanism is so relatively common in plant viruses when compared with those infecting hosts in other kingdoms (see Table 4.1) is unknown.

As noted above, there are controls on the unfettered replication of viruses that limit their detrimental effects on their hosts. As discussed in Chapter 17, there is selective pressure on viruses not to overly damage their natural hosts. In evolutionary terms, the damage that viruses cause to crops is to the disadvantage of the virus. Thus, there are controls built in to the replication of viruses that we are just beginning to understand. These include controls on gene expression so that virus-coded factors and enzymes are produced in certain amounts at certain times, as well as the sequestering of newly synthesized viral genomes by encapsidation in viral coat protein.

As well as synthesizing new genomes, viral replication also produces variants that form the basis of virus adaptation and evolution. This is discussed in Chapter 17.

Induction of Disease 1: Virus Movement through the Plant and Effects on Plant Metabolism

I. INTRODUCTION

The transmission mechanisms described in Chapters 11 and 12 lead to the virus being introduced into usually one or a few cells within the plant. Here it replicates as discussed in Chapter 8. To induce a disease, the virus has to spread through, and replicate, in much of the plant. At this stage the viral genome and the host genome confront one another, with the virus attempting to establish infection and the host attempting to resist it. In most cases, these two conflicting forces essentially reach a *quid pro quo*. It is not to the advantage of the virus to kill its host before it can be transmitted to other host plants. However, as discussed in Chapter 12, it is likely that symptom production can facilitate spread, especially by arthropods.

In this and the next chapter, I shall be discussing many of the interactions between the viral and host genomes. Here I describe first how viruses move from the site of initial infection to give a systemic infection, and the effects this has on plant metabolism.

II. MOVEMENT AND FINAL DISTRIBUTION

Some viruses are confined to the inoculated leaf while others may move systemically through the plant. If a virus replicates in the initially

Virus acronyms are given in Appendix 1.

infected cell but cannot move to neighboring cells, its replication may remain undetected. This is termed *subliminal infection* and has implications in considering the host range of viruses (see Chapter 3, Section V). It is usually identified by the ability of a virus to infect protoplasts but not apparently infect the complete host (Hull, 1989). In many host–virus combinations, cell-to-cell movement is limited, giving rise to a relatively small zone of infected tissue that may be visible as a local lesion, often consisting of chlorotic or dead cells (see Plates 3.1 and 3.2). In still others, movement within the inoculated leaf is not limited but systemic spread may not occur.

Over the last 10 years or so, there has been a rapid increase in the understanding of how viruses move around plants, giving a greater appreciation of the interactions of viruses with their hosts. This has led, amongst other things, to the characterization of the virus-encoded proteins (movement proteins: MPs) (see below) involved in movement from cell to cell and the various factors controlling that movement. There are several reviews that chart the advances in our understanding of virus movement in plants (see Hull, 1991a; Lucas and Gilbertson, 1994; Maule, 1994; McLean *et al.*, 1994; Carrington *et al.*, 1996; Lartey and Citovsky, 1997; Ding, 1998; Nelson and van Bel, 1998; Lazarowitz and Beachy, 1999; Santa Cruz, 1999); a historical perspective on cell-to-cell movement is given by Pennazio *et al.* (1999). The picture that is being built up is that virus movement is closely coupled with virus replication and that it is a dynamic regulated cascade of events (see Chapter 13).

A. Routes by which viruses move through plants

As was initially suggested by Samuel (1934), there are two basic routes by which a virus moves through the plant to give a full systemic infection, cell-to-cell movement and long-distance movement. The cell-to-cell (or short-distance) movement is from the initially infected cell(s), which are usually epidermal or mesophyll cells, to the vascular bundle. The long-distance transport is then through vascular tissue, usually the phloem sieve-tubes. Then further cell-to-cell movement establishes systemic infection of the young leaves. An exception to this is that the first stage of cell-to-cell movement may be bypassed in phloem-limited viruses, which are injected directly into the phloem by their vector (see Chapter 11, Section III.H.1). In the majority of cases there are three major barriers to movement: (1) movement from the first infected cell, (2) movement out of parenchyma cells into vascular tissues, and (3) movement out of vascular tissue into the parenchyma cells of an invaded leaf. These and various other barriers are described below.

Since the virus cannot cross the cell wall directly, it has to use plasmodesmata, which are cytoplasmic connections between adjacent cells. As was initially pointed out by Gibbs (1976), plant virus particles, or even free, folded viral nucleic acids, are too large to pass through unmodified plasmodesmata (Fig. 9.1). Thus, the plasmodesmatal size exclusion limit (SEL) has to be increased and viral MPs facilitate this.

The recognition that viral movement between cells is closely linked to viral replication raises questions about virus movement within cells. The viral genome has to be moved from the site of replication to the site of transport to the adjacent cell (the plasmodesma). For cytoplasmically replicating viruses, this involves cytoskeletal elements (see Section II.D.2.a and Chapter 13); for viruses that replicate in the nucleus, transport across nuclear membranes must also be involved (Section II.C).

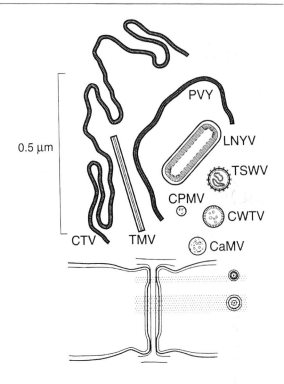

Fig. 9.1 The relative sizes of some plant viruses (top) compared with the size of a plasmodesma: CTV, 2 μm × 10 nm;   TMV,   300 nm × 18 nm;   PVY, 750 nm × 11 nm; LNYV, 220 nm × 80 nm; TSWV, 80 nm; (C)WTV, 70 nm; CaMV, 50 nm; CPMV, 28 nm. From Gibbs (1976), with kind permission of the copyright holder, © Springer-Verlag GmbH and Co. KG.

B. Methods for studying virus movement

The rapid increase in the information available on virus movement has come about by the application of a variety of methods ranging from cytological to molecular techniques; see Hull (1989) and Maule (1991) for more detailed descriptions. These include the following.

a. Classical methods

The classic experiments on movement and distribution of viruses involved dissection of plants into appropriate pieces at various times after inoculation. Extracts of these parts were then inoculated to suitable assay hosts, either immediately or after an incubation period, to allow very small amounts of virus that might be present to increase and give detectable amounts (Fig. 9.2).

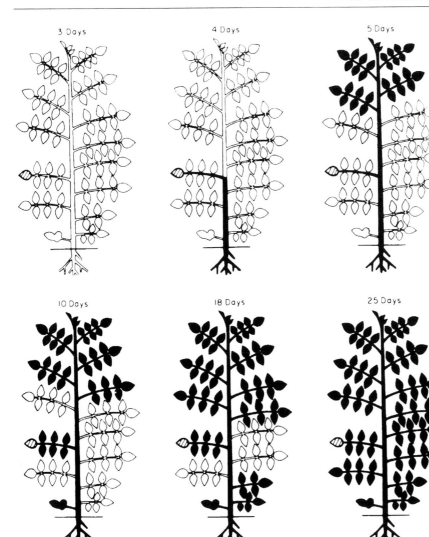

3 Days   4 Days   5 Days

10 Days   18 Days   25 Days

Fig. 9.2 Diagram showing the spread of TMV through a medium young tomato plant. The inoculated leaf on the left is marked by hatching and systemically infected leaves are shown in black. From Samuel (1934), with permission.

Fluorescent antibody methods have been applied to the detection of viruses in various parts of the plant. This method is, of course, less sensitive than infectivity but gives information on the exact location of the viral antigen. Various other procedures used for assay and diagnosis of viruses could be used to follow movement (see Chapter 15).

b. Protein expression

Molecular techniques such as western blotting are used to identify the time at which and tissues in which MPs are expressed. The properties of MPs (see below) have been studied on the gene(s) expressed in *E. coli* and in insect cells.

c. Tissue blotting

Printing of the cut surface of a leaf, petiole or stem on to a nitrocellulose membrane which is then developed by western blotting reveals which cells a protein is expressed in (Mutterer *et al.*, 1999).

d. Electron microscopy

By studying the cytology of infected plants by electron microscopy of thin sections and using electron microscopy of infected protoplasts, various features associated with cell-to-cell spread of viruses have been identified. This has been coupled with the use of gold-labeled antibodies, which reveal where the antigen is located.

e. Viral genome tagging and mutagenesis

The availability of infectious constructs of viruses has enabled the genome to be modified to include genes expressing fluorescing compounds, which allow the site of the virus, or more precisely, the viral gene product to which the fluorescing compound is attached, to be determined. The most commonly used fluorescent compound is the jellyfish (*Aequoria victorea*) green fluorescent protein (GFP) (see Baulcombe *et al.*, 1995; Oparka *et al.*, 1996; Verver *et al.*, 1998). β-glucuronidase (GUS) can also be used as a marker (Dolja *et al.*, 1992; Schmitz and Rao, 1996). However, not all tagged viral genomes behave like normal, untagged ones. For instance, the genome of PLRV, tagged with GFP and inoculated to *Nicotiana benthamiana,* only expressed the GFP in a few cells and the PLRV genome in systemically infected tissues lacked some or all of the GFP cDNA (Nurkiyanova *et al.*, 2000).

Specific sites in infectious constructs of viruses can be mutagenized to identify regions, or even amino acids, which are important to the cell-to-cell movement of the virus.

f. Microinjection procedures

Methods for micro-injecting virus are described in Chapter 12 (Section I.B). By injecting viral genomes tagged with genes expressing fluorescing compounds, the movement of virus from cell to cell can be monitored. Similarly, recombinants of a fluorescent protein gene fused either to the N or C-terminus of the MP have been micro-injected to act *in vivo* as a reporter for the location of the MP in infected tissues, in isolated cells and after transient expression in the absence of infection. Another approach is to co-inject dextrans of different molecular radii tagged with fluorochromes together with virus or movement protein (Wolf *et al.*, 1989). By using confocal laser microscopy, linked to advanced computer software (e.g. three-dimensional digital analysis packages), the movement of tagged macromolecules to adjacent cells can be visualized (Heath, 2000).

However, one has to exercise caution in the use and interpretation of microinjection techniques. Oparka and Prior (1992) showed that the application of pressure from microinjection can lead to the closure of plasmodesmata. In a comparison of two methods of microinjection, Storms *et al.* (1998) compared the plasmodesmatal size exclusion limit (SEL) (see below) in tissues of plants transgenically expressing the MP of both TSWV and TMV with nontransgenic plants. They showed that pressure-mediated injection of fluorescent probes of different sizes revealed a similar increase in the SEL of plasmodesmata in both transgenic plant lines. In contrast, iontophoretic injection of the same probes indicated that the SEL in the transgenic plants had decreased in relation to that in the non-transgenic plants. They concluded that the increase in SEL may be a consequence of the disturbance of plasmodesmatal structure owing to the application of pneumatic pressure as an additional stress to that already caused by the presence of the viral MPs.

Recombinant constructs of fusions of an MP gene with a GFP gene can be introduced into epidermal cells by biolistic bombardment where they express the fusion protein and move to adjacent cells (Itaya *et al.*, 1997). This approach can be used if there is concern that fusion proteins expressed and purified from *E. coli* might not function correctly in plants.

g. Transgenic plants expressing movement proteins

Transgenic plants expressing MPs or mutated MPs have been produced to study aspects of the function of these proteins and provided some of the early evidence for MPs. They have also been produced to confer resistance to the virus (see Chapter 16, Section VII).

h. Sequence similarities

Some MPs have been identified by amino acid sequence similarities to known MPs.

C. Transport across nuclear membranes
(reviewed by Whittaker and Helenius, 1998)

Three virus families have at least one phase of their replication in the nucleus. The replication of the single-stranded (ss) DNA genomes of

Geminiviridae and circoviruses is nuclear, as is the transcription of the double-stranded (ds) DNA of members of the *Caulimoviridae*. This necessitates import of viral nucleic acid both into the nucleus and export back to the cytoplasm.

Import and export of bipartite begomoviral DNA is mediated by the BV1 (BR1) protein (Pascal *et al.*, 1994), which has two classical nuclear location signals (Sanderfoot *et al.*, 1996). These signals would be responsible for nuclear import but those for export have not yet been identified.

The monopartite begomoviruses and the other geminiviral genera do not have a homolog of BV1. Immunofluorescent staining showed that the coat protein of the mastrevirus, MSV, accumulated in nuclei and that mutation of a potential nuclear localization signal led only to cytoplasmic accumulation (see Fig. 9.11). MSV coat protein binds both ss and ds viral DNA and, on co-microinjection, transports both these forms into the nucleus (Liu *et al.*, 1999a; Kotlisky *et al.*, 2000). Using a GUS fusion, Kunik *et al.* (1998) showed that TYLCV coat protein was imported into the nuclei of both plant and insect cells and, by mutagenesis, demonstrated a nuclear location signal in the N-terminus of the protein.

As described in Chapter 6 (Section V.B.1), nanoviruses have genomes comprising multiple small circular ssDNAs. By attaching GFP to the proteins encoded by BBTV DNAs 3, 4 and 6, Wanitchakorn *et al.* (2000) showed that proteins from DNAs 3 and 6 were found in both the cytoplasm and nucleus. Co-expression of DNA4 protein with those of DNAs 3 and 6 revealed that DNA4 protein was able to relocate the DNA6 protein, but not the DNA3 protein, to the cell periphery. This suggested that DNA6 protein is a nuclear shuttle protein, similar to those of geminiviruses.

Transient expression of constructs of CaMV coat protein identified a nuclear localization signal in the N-terminal region of this protein (Leclerc *et al.*, 1999). This signal is exposed on the surface of the virion and is regulated by the N-terminus of the precapsid, which inhibits its nuclear targeting. It is thought that this regula-

tion is important in allowing virus assembly in the cytoplasm.

Some RNA viruses or their products are found associated with the nucleus. Particles of PEMV (see Chapter 8, Section IV.L.4), of SBMV (De Zoeten and Gaard, 1969b) and of plant rhabdoviruses (see Chapter 5, Section VIII.A) are found within or associated with nuclei. Two of the proteins of TCV, the 8- and 9-kDa proteins, are involved in cell-to-cell movement (see Section D.2.g). The 8-kDa protein, tagged with GFP, locates to the nucleus (Cohen *et al.*, 2000). Mutation of either of the two nuclear localization signals found on this protein did not affect the nuclear localization but mutation of both did. However, these mutations did not affect the cell-to-cell movement of TCV.

D. Cell-to-cell movement

1. Plasmodesmata

Plasmodesmata form an important route for communication between plant cells. They regulate cell-to-cell communication, thus enabling the differentiation of plant organs and tissues. Developmental changes in their structure, frequency and SEL can lead to establishment of symplastic domains within which the metabolism and functions of all cells is probably synchronized. Because they are incorporated into the cell wall, they are one of the few plant cell organelles that it is not possible to purify by grinding up plant material and fractionating. Thus, most of the studies on them have been by electron microscopy. The interactions they have with viruses have both increased interest in studying them and provided further insights into their structure and function.

Recent ideas on the structure of plasmodesmata have been reviewed by Overall and Blackman (1996), Ghoshroy *et al.* (1997), Ding (1998) and Aaziz *et al.* (2001). The basic structure consists of two concentric membrane cylinders, the plasma membrane and the endoplasmic reticulum (appressed ER or desmotubule) that traverse the cellulose walls between adjacent plant cells (Fig. 9.3). The annulus between the two membrane cylinders gives continuity of the cytosol between cells. High-resolution electron microscopy reveals

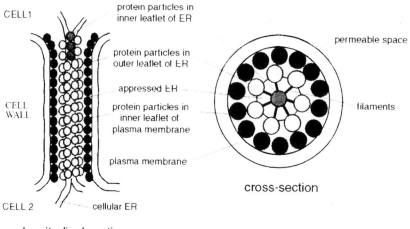

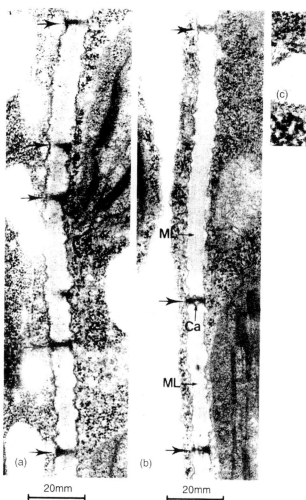

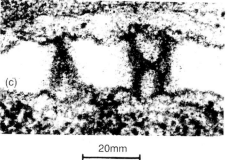

Fig. 9.3 The structure of plasmodesmata. **(A)** Diagram of the structure of a simple plasmodesma. From Ghoshroy *et al.* (1997), with kind permission of the copyright holder, © Annual Reviews. www.AnnualReviews.org **(B)** Ultrastructural details of tobacco plasmodesmata. **(a,b)** Primary plasmodesmata (unlabeled arrows) between mesophyll cells at the base and tip, respectively, of the first leaf. Note that central cavities (Ca) have developed in plasmodesmata at the tip. ML, middle lamella. **(c)** Mesophyll plasmodesmata from the tip of the first leaf, in which a more complex structure is apparent. Bars = (a) 408 nm; (b) 466 nm; (c) 140 nm. From B. Ding *et al.* (1992), with permission.

proteinaceous particles, about 3 nm in diameter, embedded in both the plasma membrane and the appressed ER and connected by spoke-like extensions. The spaces between these protein particles are thought to form tortuous micro-channels about 2.5 nm in diameter. Injection of dyes of various molecular radii indicates that these micro-channels have a basal SEL allowing passive diffusion of molecules of about 1 kDa.

There is a variety of plasmodesmata (reviewed in Nelson and van Bel, 1998). Primary plasmodesmata formed in a new cell wall are simple but then undergo modification to give complex structures (secondary plasmodesmata) with branched channels and a conspicuous central cavity (Ding *et al.*, 1992; Itaya *et al.*, 1998). Secondary plasmodesmata are formed in a basipetal pattern as leaves undergo expansion growth and, in transgenic plants, the expressed TMV MP is located at these plasmodesmata (Ding *et al.*, 1992). Of especial note to virus movement are plasmodesmata between the bundle sheath and phloem parenchyma, those between the phloem parenchyma and companion cells, and those between the companion cells (or intermediary cells: Turgeon *et al.*, 1993) and the sieve elements giving different tissue boundaries, at least for some viruses (see Fig. 9.16). Plasmodesmata between the bundle sheath and phloem parenchyma differ from mesophyll plasmodesmata in that they require additional modification before some viruses can enter the phloem of minor veins (reviewed in Ding, 1998; Derrick and Nelson, 1999). The systemic infection of several viruses, such as TAV (Thompson and Garcia-Arenal, 1998), appears to be controlled by the bundle-sheath/vascular-bundle interface as does that of BMV in monocots (Ding *et al.*, 1999) and of RCNMV in certain hosts (Wang *et al.*, 1998b). This tissue boundary also appears to be demonstrated by phloem-limited viruses (e.g. luteoviruses) that are unable to spread across the bundle sheath to mesophyll cells (see below). For several viruses it has been observed that, whereas many phloem parenchyma cells are infected, very few companion cells are (Ding, 1998). Plasmodesmata between companion cells and sieve elements have a special structure comprising a single pore on the sieve element wall and a branched arrangement in the adjoining companion cell wall (van Bel and Kempers, 1997). The SEL of the connection between companion cells and sieve elements is more than 10 kDa (Kempers and van Bel, 1997).

The initial experiments of Wolf *et al.* (1989) using microinjection of dye-coupled dextrans indicated that the SEL of unmodified plasmodesmata between mesophyll cells is about 0.75–1.0 kDa and that TMV MP increased this to 9.4–17.2 kDa (Fig. 9.4).

This was accepted as dogma for plasmodesmatal SELs for several years, but recent observations suggest that there can be major differences in the SELs of unmodified plasmodesmata. For instance, plasmodesmata in *Nicotiana clevelandii* trichomes have a basal SEL of about 7 kDa (Waigmann and Zambryski, 1995) which may be accounted for by a different structure (Waigmann *et al.*, 1997). Furthermore, it has become apparent that, during the sink–source transition of photoassimilate production and use in leaves, a major change occurs in the permeability of plasmodesmata, with molecules of up to 60 kDa being able to pass into developing sink cells (Oparka *et al.*, 1999). One of the factors thought to naturally regulate plasmodesmatal SEL is callose turnover. A β-1,3 glucanase-deficient mutant of tobacco, that had increased callose deposition in response to elevated temperatures, elicitors and wounding, also showed delayed movement of several viruses (Inglesias and Meins, 2000).

Plasmodesmata occur in groups on cell walls and it must not be assumed that those within one group are all the same. Similarly, they will vary with time, and possibly with condition, such as virus infection. Shalla *et al.* (1982) showed that the restricted movement of a *ts* mutant of TMV in tobacco leaves was associated with a substantial reduction in the numbers of plasmodesmata. During plant morphogenesis, symplastic domains are formed by plugging of plasmodesmata (Ehlers *et al.*, 1999); it is possible that there are shifting symplastic domains in leaves at various other times and under various other conditions.

Thus, the system of interconnections between plant cells is dynamic and changing

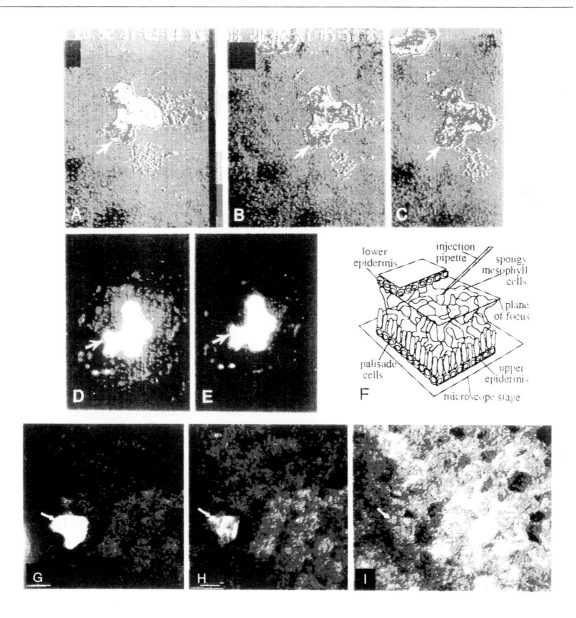

Fig. 9.4 (see Plate 9.1) Microinjection of fluorescently labeled 10-kDa Lucifer yellow dextran (**A–E**) or 20-kDa fluorescein isothiocyanate-labeled dextran (**G–I**), together with TMV MP (**A–C and G–I**) and without TMV MP (**D,E**) into spongy mesophyll cells of mature tobacco leaves. Magnifications: A–C, ×230; D and E, ×130. The time course of 10-kDa dextran movement is shown. Elapsed time after injection: (A) 1 min; (B) 4 min; (C) 5 min; (D) 1 min; (E) 30 min; (G) 3 min; (H) 14 min; (I) 34 min. Arrows point to the injected cell. Pictures are *averaged*: colors are derived by false color imaging and represent intensity of fluorescent signal ranging from blue (lowest intensity) to white (highest intensity) with black and blue representing the background; see color bar at right of fluorescent image in (A). In the top part of the picture an automatic timer gives date and time of experiment. The timer has been turned off in image (C) to allow a better view of the fluorescent cells. (**F**) Schematic view through a tobacco leaf. Spongy mesophyll cells form a three-dimensional interconnected network. To inject into spongy mesophyll cells, a small portion (5 mm²) of the lower epidermis of the leaf is peeled off. The leaf is then placed, abaxial side up, on the microscope stage. Only surface cells in the plane of focus can be injected and subsequently monitored. These cells are indicated in a lighter pattern. From Waigmann *et al.* (1994), with kind permission of the copyright holder, © The National Academy of Sciences, USA.

with the different functional demands put upon this symplastic system.

2. Movement proteins

The first evidence for a virus-coded MP came from the studies of Nishiguchi and colleagues (1978) who showed that the temperature sensitivity in the capacity of ToMV to move through its host was not due to any effect on replication. This led to the identification of the 30-kDa MP of tobamoviruses and, since then, to the identification of MPs in many viruses.

There have been various criteria suggested by which virus gene products have been identified or implicated as being MPs. These include:

- loss of cell-to-cell movement on removal of a nucleic acid component of a multicomponent virus;
- mutagenesis of a viral gene;
- complementation of movement between viruses
- sequence comparisons
- immunodetection of the gene product associated with plasmodesmata.

For genome organizations of the viruses described below, see Chapter 6; for their expression see Chapter 7.

a. The TMV '30K' protein (reviewed by Citovsky, 1999; Rhee et al., 2000)

The TMV 30-kDa protein (P30) was identified as an MP by the demonstration that, at restrictive temperatures, the temperature-sensitive mutant, Ls1, could replicate at the single-cell level but could not move out of initially infected cells (Nishiguchi et al., 1978, 1980), by the mapping of the Ls1 defect to P30 (Ohno et al., 1983), and by the complementation of movement under non-permissive conditions in P30 transgenic plants (Deom et al., 1987; Meshi et al., 1987). Since then this protein has been studied in detail and shown to bind single-stranded nucleic acids (Citovsky et al., 1990), to localize to plasmodesmata in infected and transgenic plants (Tomenius et al., 1987; Atkins et al., 1991b) (Fig 9.5), to increase the plasmodesmatal SEL (Wolf et al., 1989; Waigman et al., 1994) (Fig. 9.4), and to be phosphorylated by a

developmentally regulated plant cell wall-associated, and possibly another, protein kinase (Atkins et al., 1991a,b; Watanabe et al., 1992; Citovsky et al., 1993; Haley et al., 1995).

Atomic force microscopy shows that P30 binds all the way along the RNA chain (Kiselyev et al., 2001). The plasmodesmatal association of P30 is controlled by amino acids at the very N-terminus (Boyko et al., 2000c).

As phosphorylated P30 occurs in amounts too small for purification and analysis, Citovsky et al. (1993) used an in vitro system to determine that the phosphorylation was at the C-terminal end of the protein at Ser-258, Thr-261 and Ser-265. Matsushita et al. (2000) showed that a cellular kinase that phosphorylated some or all of these C-terminal residues bound tightly to immobilized ToMV P30. However, mutagenesis indicates that Ser-37 and Ser-238 of ToMV P30 are phosphorylated with the position 37 being important for infectivity and localization in protoplasts (Kawakami et al., 1999). Whether the difference in these reports is due to different, but closely related, viruses being studied, different systems being used or other factors are not known. Phosphorylation and/or the presence of Ser-37 is essential for the intracellular localization of ToMV P30 (Kawakami et al., 1999). The regulation of plasmodesmatal transport by TMV P30 is dependent on phosphorylation, suggested to be of the C-terminal Ser-258, Thr-261 and Ser-265, in a host-dependent manner (Waigmann et al., 2000). Mutants, in which sites for phosphorylation were mimicked by negatively charged amino acid substitutions, were unable to spread from cell to cell in Nicotiana tabacum but were able to spread in N. benthamiana.

The nucleic acid binding property of P30 melts the nucleic acid secondary structure, giving ribonucleoprotein complexes that are very long and thin (1.5–2.0 nm diameter) (Citovsky et al., 1992). The domains on P30 conserved between tobamoviruses and those associated with these activities have been mapped (Saito et al., 1988; Waigman et al., 1994) (Fig. 9.6A).

P30 also associates with cytoskeletal structures. In experiments involving antibody to P30 and using infectious TMV clones with GFP fused to P30, co-localization of P30 with

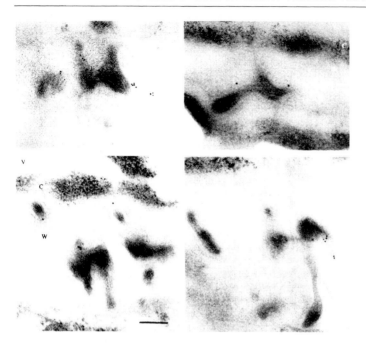

Fig. 9.5 Representative micrographs of plasmodesmata in the cell walls of leaf cells from transgenic tobacco plants (277) that express the MP of TMV. The material has not been osmicated so as to preserve antigenicity, and so the contrast, particularly in membranes, is low. Immunogold labeling of the MP shows that gold grains are localized over plasmodesmata. The signal-to-noise ratio is high, with negligible gold found over the cell wall, vacuole and other cell organelles. V, vacuole; C, cytoplasm; W, wall. Bar = 200 nm. From Atkins *et al.* (1991b), with permission.

microtubules was demonstrated in protoplasts and whole plants (Heinlein *et al.*, 1995). The microtubule association is effected by the C-terminal 66 amino acids (Boyko *et al.*, 2000c). P30 also co-localized with actin fibres and, in *in vitro* tests, bound to both actin and tubulin (McLean *et al.*, 1995). However, microtubule association of P30 was observed in cells undergoing late infections rather than those at the leading edge of expanding infection sites where virus cell-to-cell movement occurs (Boyko *et al.*, 2000a). The intercellular transport of TMV RNA is temperature-dependent with higher efficiency at elevated temperatures. This corresponds to an increased association of P30 with microtubules early in infection (Boyko *et al.*, 2000a). Antibody to P30, and the use of infectious TMV clones with GFP fused to P30, have also revealed that P30 associates with cortical endoplasmic reticulum (Heinlein *et al.*, 1998; Reichel and Beachy, 1998).

The dynamics of the association of P30 with microtubules was explored by Más and Beachy (2000) using the fact that microtubules are disrupted at low temperatures. At early stages of infection of protoplasts, cold treatment caused the accumulation of P30 fused to GFP in large virus replication bodies localized in the perinuclear region. Cold treatment (4°C) at mid stages of infection disrupted the association of P30 with microtubules. Rewarming the protoplasts to 29°C re-established the association of P30 and replication bodies with microtubules and led to the subsequent spread of the replication bodies throughout the cytoplasm and to the periphery of the cell. A mutant of P30, TAD5, in which amino acids 49–51 were deleted (Kahn *et al.*, 1998), associated with the endoplasmic reticulum, but not with microtubules. It was concluded that the role of microtubules was the intracellular distribution of P30 to peripheral sites within the cell. This is further discussed in Chapter 13.

P30 also interacts with pectin methylesterase (PME) isolated from tobacco leaf cell walls (Chen *et al.*, 2000). Deletion mutagenesis identified the interaction domain in P30 and showed the PME-binding region was necessary for TMV cell-to-cell movement.

Circular dichroism spectroscopy showed that *E. coli*-expressed P30 had ~70% α-helical conformation in the presence of urea and SDS (Brill *et al.*, 2000). P30 has two trypsin-resistant hydrophobic regions forming a tightly filled core domain. Based on these observations, Brill *et al.* (2000) proposed a topological model for TMV P30 (Fig. 9.7).

b. RCNMV

RCNMV has a two-component genome, the RNA1 of which gives subliminal infection; the RNA2, which encodes a single 35-kDa protein, confers the ability for the virus to infect plants systemically. Thus, the 35-kDa protein was identified as the MP (Lommel *et al.*, 1988). This protein binds single-stranded nucleic acids co-operatively *in vitro*, locates to the cell wall fraction, can increase plasmodesmatal SELs and transport RNA from cell to cell (Osman and Buck, 1992; Osman *et al.*, 1992; Fujiwara *et al.*, 1993). Three functional domains have been identified in the 35-kDa protein (Giesman-Cookmeyer and Lommel, 1993), an RNA binding domain, a co-operative RNA binding domain, and a domain necessary for cell-to-cell movement *in vivo* (Fig. 9.6B). RCNMV MP resembles that of TMV in these properties but there is little amino acid sequence similarity between these two proteins. However, they can complement each other's functions, indicating that they are functionally homologous (Giesman-Cookmeyer *et al.*, 1995). They have similar predicted secondary structures (Fig. 9.8).

c. CPMV

The MP of CPMV, other comoviruses and nepoviruses is encoded by the RNA2 and was identified by mutagenesis to be the 58–48-kDa product of the polyprotein expressed from that RNA (Wellink and van Kammen, 1989) (see Figs. 6.18 and 7.28 for CPMV genome organization). Electron microscopic observations of CPMV-infected plant material revealed structural modifications of plasmodesmata (reviewed in Kasteel, 1999), and especially the presence of virion-containing tubules (35 nm in diameter) replacing the desmotubules normally present in the plasmodesmata. Similar tubules, reacting with antiserum to the 58–48-kDa protein, were also found on the surface of CPMV-infected cowpea protoplasts (van Lent *et al.*, 1991) (Fig. 9.9) and of insect cells in which the protein is expressed using a baculovirus vector (Kasteel *et al.*, 1996). These tubules extend up to 50 µm, are 35 nm in diameter and are enveloped by the plasma membrane. By mutagenesis it has been shown

that it is the 48-kDa moiety of the 58–48-kDa protein that makes up the tubule, there being no evidence for host proteins being involved (Kasteel, 1999). The domain structure of the 48-kDa protein is shown in Figure 9.6D.

d. CaMV

CaMV MP was identified by mutagenesis as the product of ORF I (P1), a 38-kDa protein (Thomas *et al.*, 1993). As with como- and nepoviruses, CaMV infection gives rise to plasmodesmata modified by tubules passing through them (Fig. 9.10). Tubules are also found on the surface of CaMV-infected protoplasts and insect cells (Fig. 9.9C) (Kasteel *et al.*, 1996). Cytoskeletal assembly inhibitors did not affect tubule structure; but brenfeldin A, which disrupts the plant cell endomembrane system, interfered with tubule formation (Huang *et al.*, 2000). CaMV P1 has nucleic acid binding activity *in vitro* (Citovsky *et al.*, 1991; Thomas and Maule, 1995a). By using alanine scanning mutagenesis, epitope tagging, sequence comparison with other caulimoviruses, infection of plants and expression of P1 in insect cells, Thomas and Maule (1995b) and Huang *et al.* (2000) have built up a picture of functional domains of the CaMV MP (Fig. 9.6C). The epitope tagging indicated that the N-terminus of the protein was presented on the outer face of the tubule and the C-terminus on the inner face; they suggested that the C-terminus might be held out into the lumen of the tubule by a hypervariable upstream spacer sequence.

The MP of SbCMV is expressed from ORF 1a (see Fig. 6.1 for genome map) (Takemoto and Hibi, 2001).

e. Bromoviridae

The MPs of AMV, bromoviruses, cucumoviruses and ilarviruses are encoded by the 3a gene on RNA3 giving products ranging from 32 to 36 kDa (see Figs 6.13–6.16 for genome organizations). These proteins accumulate early in virus infection, bind RNA co-operatively, associate with the cell wall fraction, locate to the plasmodesmatal region, and increase plasmodesmatal SEL (Zheng *at al.*, 1997; Canto and Palukaitis 1999a). Cell-to-cell

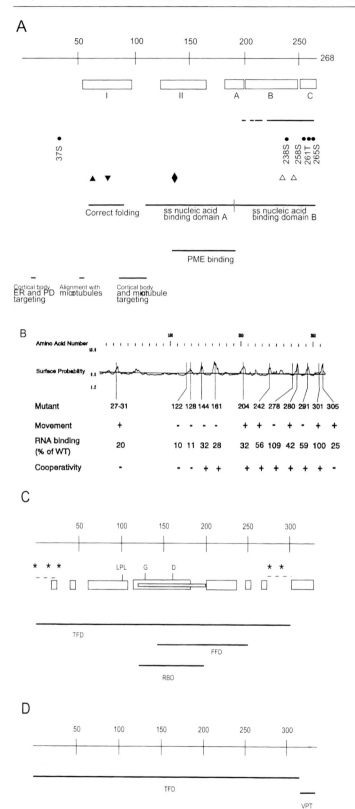

Fig. 9.6 Diagrams of functional domains on movement proteins. **(A)** TMV P30. The top line represents the gene product with amino acid positions indicated. The boxes in the second line are the regions of homology (I and II) and acidic (A and C) and basic (B) amino acids identified by Saito *et al.* (1988). The third line shows the effects of deletions in the C-terminal region on cell-to-cell movement, with the complete line showing no effect, the long dashed line slight effects, the short dashed line more effects, and no line abolition of movement (Gafney *et al.*, 1992). The black spots on the next line show the positions of the phosphorylated amino acids as discussed in the text. The triangles on the next line show the positions of the variant residues in the Tm2 resistance breaking strains: ▼ strain Ltb1, ▲ C52, and the Tm2² resistance breaking strain, △. The next line shows the domains identified by Citovsky *et al.* (1992). The next line shows the PME-binding domain (Chen *et al.*, 2000). The bottom line shows the domains identified by Kahn *et al.* (1998) and Boyko *et al.* (2000c). **(B)** Domains in the movement protein of RCNMV. The top line indicates the gene product with the amino acid positions. The second line shows the prediction for surface probability with the positions of alanine-scanning amino acid mutants. The bottom three lines show respectively the effects of these mutations of cell-to-cell movement, RNA binding and co-operativity. **(C)** Domains in the movement protein of CaMV. The top line indicates the gene product with the amino acid positions. The second line shows the results from the deletion analysis of Thomas and Maule (1995b) with the major regions of sequence conservation with other caulimoviruses shown as open boxes and the RNA-binding domain as a hatched box; the N-terminal (dots) and hypervariable spacer regions (dashes), the locations of the viable scanning deletion mutations (stars) and the positions of the conserved LPL and G- and D-boxes are indicated. Below this is shown the results of the analysis by Huang *et al.* (2000) with the tubule-forming domain (TFD), RNA-binding domain (RBD) and fluorescent focus-forming domain (FFD) indicated. **(D)** Functional domains of the 448-kDa movement protein of CPMV. The top line indicates the gene product with the amino acid positions; below are shown the regions identified by Lekkerkerker *et al.* (1996); TDF, tubule-forming domain; VPT, domain thought to be involved in incorporation of virus particles into tubules. From Bertens *et al.* (2000), with permission.

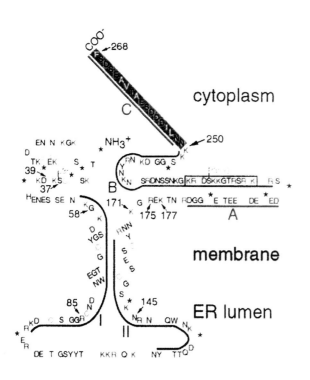

Fig. 9.7 (see Plate 9.2) Topological model of the TMV MP. The amino acid sequence was deduced from the ORF encoding the MP. Hydrophobic amino acids are in yellow, basic residues are blue, acidic residues are red and Cys residues are green. The trypsin-resistant core domain contains the first 249 or 250 residues. The C-terminus (grey) was rapidly removed by trypsin. Domains I (residues 56–96) and II (residues 125–164) are regions that are conserved among tobamovirus MPs and are outlined in black; domains A (residues 183–200) and C (residues 252–268) are acidic and are outlined in red; and domain B (residues 206–250) is basic and is outlined in blue. Cytoplasmic, transmembrane, ER luminal and core domains were inferred from western blots of proteinase K-treated microsomes, hydropathy analysis, fluorescent microscopy, trypsin susceptibility, and mass spectrometry. Transmembrane domains are presumed to be α-helical. Hydrophobic peptides (38–85 vs 58–85) differed in length between the monomer and core, suggesting that the C-terminus of the monomer may interact with its N-terminus. Serine-37 is phosphorylated (p³-) and S S 218 may be phosphorylated (P³?) based on Prosite prediction and sequence conservation among tobamoviruses. Met residues (*) have been reported to generally be involved in protein–protein interactions. From Brill *et al.* (2000), with kind permission of the copyright holder, © The National Academy of Sciences, USA.

movement of members of the *Bromoviridae* requires the viral coat protein in most situations (Rao and Grantham, 1996; Kaplan *et al.*, 1998) and is associated with the formation of tubules (Zheng *et al.*, 1997; Canto and Palukaitis, 1999a) (Fig 9.9E). However, although cell-to-cell movement of CCMV in most cell types requires coat protein, it does not in epidermal cells of several hosts (Rao, 1997) or for movement in cowpea (Schneider *et al.*, 1997). Although cell-to-cell movement of CMV needs coat protein, the virus does not need to be encapsidated as virions (Kaplan *et al.* 1998) whereas BMV requires an encapsidation-competent coat protein (Schmitz and Rao, 1996). The 3a gene of BMV is not required for cell-to-cell movement in yeast (Ishikawa *et al.*, 2000).

f. Triple-gene block movement proteins

The proteins which mediate cell-to-cell movement of several groups of plant viruses, the potex-, carla-, alexi-, fovea-, peclu-, pomo-, beny- and hordei-viruses, are translated from a set of three overlapping ORFs, termed the *triple gene block*. The genes are named TGB1, TGB2 and TGB3 from the 5′-end and all three act in concert to mediate cell-to-cell spread of the virus (Lawrence and Jackson, 2001). In the hordeiviruses, TGB 1, 2 and 3 are termed βb, βc and βd. The triple gene block of potexviruses has been most studied (see Fig. 6.39 for potexvirus genome organization). There is no requirement for synthesis in *cis* of the three TCB proteins (Lough *et al.*, 1998). The product of TGB1 has ATPase and RNA binding properties (Rouleau *et al.*, 1994) as well as domains present in RNA helicases (Habili and Symons, 1989). This protein is also associated with an increase in plasmodesmatal SEL (Angell *et al.*, 1996) and, for WClMV, functions as the cell-to-cell MP (Lough *et al.*, 1998), the viral genome being transported as a ribonucleoprotein complex of single-stranded RNA, TGB1 and coat protein. The product of TGB2 has two potential membrane-spanning hydrophobic domains

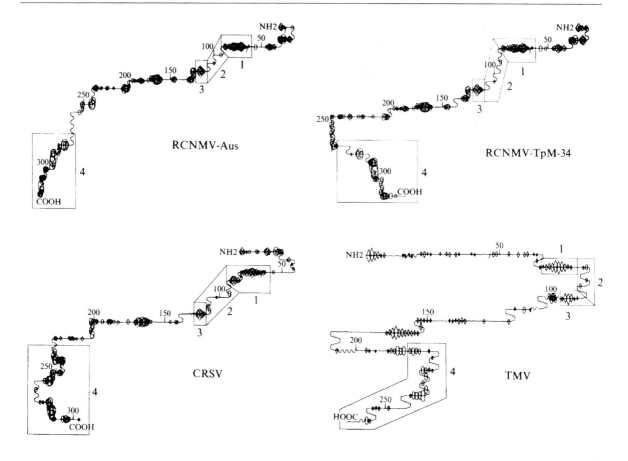

Fig. 9.8 Secondary structure predictions presented as 'Plotstructures' for CRSV, RCNMV-Aus and -TpM-34 dianthoviruses and TMV cell-to-cell movement proteins. The two-dimensional plot helices are shown as sine waves, β-sheets with a sharp sawtooth wave, 180-degree turns and coils with a dull sawtooth wave. Superimposed over the waveforms are diamonds representing hydrophobic and hexagons representing hydrophilic residues. The size of the symbol represents the extent of hydrophobicity or hydrophilicity. Boxed regions with identical numbers identify structural domains conserved among the four compared movement proteins. From Kendall and Lommel (1992), with permission.

flanking a conserved sequence of unknown function; that of TGB3 is variable in sequence but is always rather hydrophobic (Koonin and Dolja, 1993). Even though TGB1 is classed as the MP, there is an absolute requirement for TGB2 and TGB3 (Lough *et al.*, 1998).

Two classes of TGB have now been recognized: class 1 found in the hordeiviruses having a TGB1 encoding a product of 42–63 kDa; and class 2 (potex- and carlaviruses) with a TGB1 product of ~25 kDa (Solovyev *et al.*, 1996). Viral coat protein is required for the cell-to-cell movement of class-2 potexviruses (Forster *et al.*, 1992; Baulcombe *et al.*, 1995) whereas it

is not for the class-1 hordeiviruses (see Donald and Jackson, 1996).

Defective mutants of the TGB of BNYVV can be complemented in *trans* by TMV P30 but not by the tubule-forming MPs of AMV or GFLV. Although BNYVV movement could be *trans*-complemented by the three TGB proteins of PCV supplied together, it could not when the TGB proteins were substituted one by one (Lauber *et al.*, 1998b).

g. 'Double-gene' block movement proteins

Carmoviruses have two small internal overlapping ORFs that are required for cell-to-cell

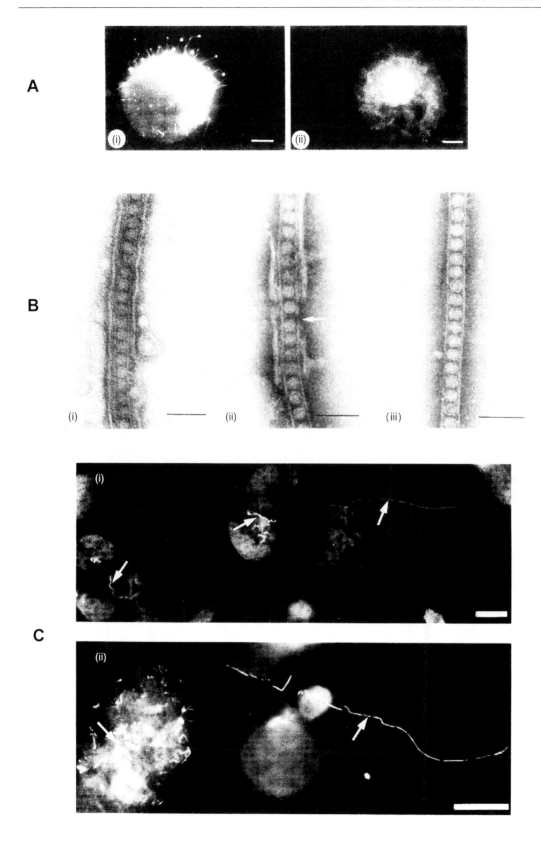

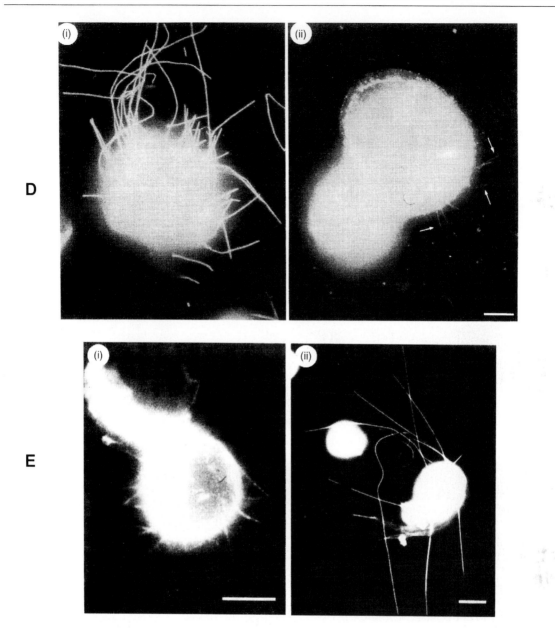

Fig. 9.9 Tubules extending from virus-infected protoplasts. **(A)** Immunofluorescent images of CPMV-infected cells treated with a peptide anti-48/58-kDa serum: **(i)** tubular structures extending from surface of the cell; **(ii)** fluorescence in nucleus. **(B)** Electron micrographs of CPMV tubules in partially purified fractions. Tubules are encased by the plasmamembrane. **(i)** After 3 weeks of refrigerated storage the tubular structure is still intact, but **(ii)** the plasma membrane has partially disintegrated (arrow). **(iii)** The plasma membrane was then removed by treatment with Nonidet P40. Bar = 100 nm. **(C)** Transient expression of CaMV MP in protoplasts. **(i)** Immunofluorescent assay of MP-expressing *A. thaliana* protoplasts showed MP-specific tubules (white arrows) extending from surface of some protoplasts, similar to those formed after culture of protoplasts isolated from CaMV-infected tissue (ii). Bar = 20 nm. **(D)** Immunofluorescent image of TSWV-infected *N. rustica* protoplasts 22 hours post-infection probed **(i)** with anti-NSm serum and **(ii)** with anti-N serum. Arrows indicate thin tubule threads visualized with anti-N serum. Bar = 5 nm. **(E)** Immunofluorescent images of cowpea protoplasts 42 hours after inoculation with **(i)** AMV or **(ii)** BMV, labeled with homologous anti-MP serum. Bar = 10 nm. Parts (A), (B) and (E) from Kasteel (1999); part (C) from Kasteel *et al.* (1996) and (D) from Storms (1998), all with permission.

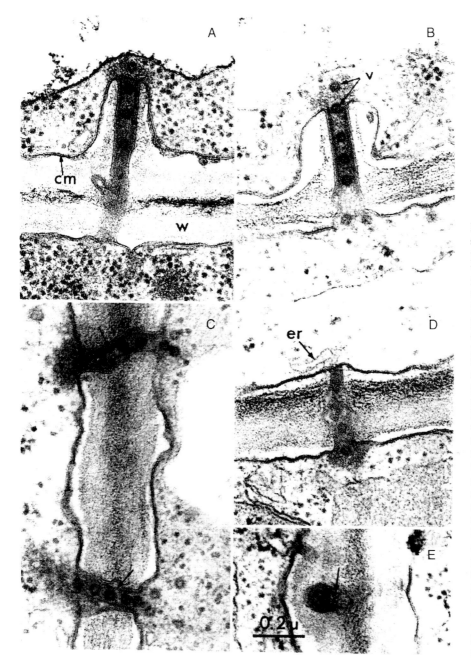

Fig. 9.10 Longitudinal sections of several plasmodesmata containing virus particles, in zinnia foliar cells infected with the caulimovirus DMV. In **(A)** and **(B)**, cell wall projects into the cytoplasm around one end of the plasmodesma. In **(C)**, a thin electron-dense layer is noticeable underneath the plasma lemma, within the plasmodesmata (arrows). In addition, the material within each plasmodesma appears to be flowing in the same direction. In **(D)**, the plasmodesma is half normal (top) and half modified (bottom). **(E)** A quasi-transverse section of a modified plasmodesma. Besides the virion, the thin dense layer below the plasma lemma (arrow) is clearly depicted in the plasmodesma. cm, cell membrane; er, endoplasmic reticulum; v, virion; w, cell wall. From Kitajima and Lauritis (1969), with permission.

movement. The 5′ ORF of CarMV, encoding p7, has RNA-binding properties (Marcos *et al.*, 1999). Cell-to-cell movement of TCV is mediated by the two small proteins, p8 and p9, acting in *trans* at the same time but not requiring the viral coat protein (Li *et al.*, 1998). A single amino acid change in TCV p8 affects the ability of the virus to move in *Arabidopsis* (Wobbe *et al.*, 1998), this ability being inversely correlated with the p8 RNA binding affinity. Two regions of amino acids are required for RNA binding (Abgoz *et al.*, 2001). As described in Section II.C, p8 localizes to the nucleus; p9 is found both in the cytoplasm and nucleus (Cohen *et al.*,

2000). Unlike the MP of TMV and other viruses, the GFP fusions of p8 and p9 did not produce punctate patterns in the cell walls of infected cells or appear to be associated with the cytoskeleton (Cohen *et al.*, 2000).

TCV P8 interacts with a protein from *Arabidopsis* that has two possible transmembrane holices, several potential phosphorylation sites and two 'RGD' sequences (Lin and Heaton, 2001).

Tombusviruses have two internal overlapping ORFs encoding, for TBSV, proteins of 22 and 19 kDa; p19 encodes a soluble protein whereas the product of p22 is membrane-associated (Scholthof *et al.*, 1995b). At least p22 is required for cell-to-cell movement (Scholthof *et al.*, 1995b; Chu *et al.*, 1999); p19 influences lesion diameter in some hosts, which suggests that it has an auxiliary host-dependent role in movement.

h. *Luteovirus movement proteins* (reviewed by Taliansky and Barker, 1999)

A rapidly growing body of evidence suggests that the 17-kDa product of the luteovirus ORF 4 is an MP. This ORF is absent in PEMV-1, which requires the umbravirus PEMV-2 for cell-to-cell movement (see Chapter 2, Section III.P.3). The 17-kDa protein has nucleic acid binding properties, accumulates at plasmodesmata and becomes phosphorylated.

The luteovirus coat protein gene has a weak termination codon leading to the production of molecules that have a read-through portion (see Chapter 7, Section V.B.9). Deletion of this read-through portion restricts accumulation of virus in agro-infected plants, suggesting that it might be involved in short-distance movement of the virus (Mutterer *et al.*, 1999).

i. *Geminivirus movement proteins*

The cell-to-cell movement of bipartite geminiviruses is facilitated by the BC1 (BL1) gene (reviewed in Carrington *et al.*, 1996) that interacts with BV1 (BR1), preventing it from re-entering the nucleus (Fig 9.11). Thus, in this group of viruses, two proteins are involved in the movement of newly synthe-

sized DNA from the nucleus to adjacent cells: BV1 for nuclear transport (see section II.C) and BC1 for movement through the plasmodesmata. A similar situation is thought to pertain in monopartite geminiviruses with the coat protein providing the nuclear shuttle function and the V1 protein cell-to-cell movement and preventing the coat protein from re-entering the nucleus (Boulton *et al.*, 1993; Liu *et al.*, 1999a) (Fig. 9.11).

j. *Tospoviral movement proteins*

The NS_M protein has been identified as the MP of TSWV (reviewed in Storms, 1998) (see Fig. 6.11 for genome organization). This protein is expressed early and temporally in infected plants and assembles into tubular structures that extend from the plasma membranes of protoplasts and insect cells (Storms *et al.*, 1995) (Fig. 9.9D). In infected plants, these tubules extend through plasmodesmata and are suggested to conduct presumed non-enveloped nucleocapsids to the adjacent cell.

TSWV NS_M, expressed in *E. coli*, specifically interacts with TSWV N protein and binds ssRNA in a manner that is not sequence-specific. Using the yeast two-hybrid system, two NS_M-binding proteins, belonging to the DnaJ family, were isolated from *Nicotiana tabacum* and *Arabidopsis thaliana* (Soellinck *et al.*, 2000).

TSWV also infects its thrips vector, *Frankinella occidentalis* (see Chapter 11, Section VI). NS_M protein was detected in only in the midgut epithelium in the L2 development stage and in salivary glands and midgut muscle cells in the adult development stage (Storms, 1998). The protein was associated with electron-dense inclusions thought to be part of an autophagic pathway for degradation of recycling proteins, and was considered not to be involved in cell-to-cell movement in that host.

k. *Other movement proteins*

The potyviral MP is the cylindrical inclusion (CI) protein (see Chapter 6, Section VIII.C.1.b). Alanine scanning mutagenesis showed that this protein is multi-functional with roles in

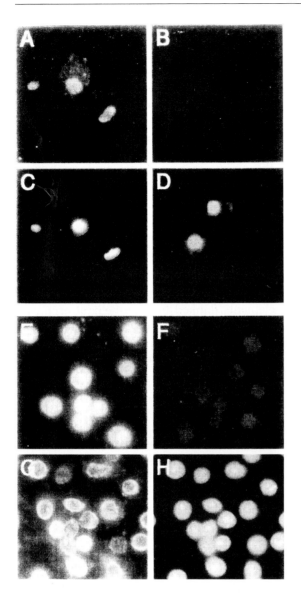

Fig. 9.11 (see Plate 9.3) Subcellular localization of MSV coat protein in **(A–D)** tobacco protoplasts and **(E–H)** insect cells determined by immunofluorescent staining. BY2 tobacco cells were transfected either (A and C) with pJITMSVCP (the MSV coat protein fragment cloned into a pUC-based transient expression vector, pJIT163), or (B and D) with pJIT163, and sampled after 48 hours' incubation. **(A,B)** Fluorescein isothiocyanate (FITC) fluorescence is identified by excitation at 488 nm, and indicates location of coat protein. **(C,D)** Images of the same cells, photographed after excitation at 366 nm, to show 4,6-diamidine-2-phenylindole dihydrochloride (DAPI) fluorescence indicating the location of nuclei. **(E–H)** Sf21 insect cells were infected with recombinant virus containing (E) the wild-type (wt) coat protein sequence, or (G and H) mutant CP201, which lacks the 20 N-terminal amino acids (Liu *et al.*, 1997a), or (F) with wt baculovirus. Cells were sampled 30 hours post-inoculation. Presence of wt or mutant coat protein is identified by fluorescence at 488 nm (E and G). For wt coat protein, most of the FITC signal was present in the nuclei, whereas for mutant CP201, comparison of the images produced at 488 and 366 nm (G and H, which depict the same cells) showed that the mutant protein did not accumulate in the nuclei. Nuclei were identified by DAPI staining and excitation at 366 nm (H), the nuclei staining intensely, but less intensive staining was always seen in surrounding insect cell cytoplasm. From Liu *et al.* (1999a), with kind permission of the copyright holder, © The American Phytopathological Society.

virus replication, short- and long-distance movement (Carrington *et al.*, 1998). CIs are associated with plasmodesmata (see below). Potyviral coat protein is also required for cell-to-cell movement (Dolja *et al.*, 1995).

Mutation of the 5′ ORF of sobemoviruses (encoding proteins of 18–21 kDa) abolishes the ability of the virus to move from cell to cell while not affecting virus replication (Bonneau *et al.*, 1998; Sivakumaran *et al.*, 1998). This identified the P1 as an MP.

The putative MP of TYMV, a 69-kDa protein, is much larger than other MPs. It does not have any sequence similarity to other MPs and is very rich in proline (19.4%). It is involved with systemic spread of the virus and possibly cell-to-cell movement (Bozarth *et al.*, 1992).

Closteroviruses have large complex genomes with that of BYV being the most studied. BYV involves five proteins in its cell-to-cell movement (Alzhanova *et al.*, 2000): a hydrophobic protein p6; p64; a homolog of heat shock protein 70 (HSP70h); and the minor and major capsid proteins (see Fig. 6.33

and Table 6.5 for closterovirus genome organization). HSP70h binds tightly to BYV particles with a non-ionic interaction and approximately 10 molecules of it are found in each BYV virion (Napuli *et al.*, 2000). Inactivation of HSP70h completely arrests translocation of BYV from cell to cell (Peremyslov *et al.*, 1999). This protein has two domains, an N-terminal ATPase domain and a C-terminal domain harboring a peptide-binding pocket; it also binds microtubules (Karasev *et al.*, 1992) and locates to plasmodesmata (Medina *et al.*, 1999).

The 28 kDa protein encoded by ORF 4 of the umbravirus GRV can replace the PVX triple-gene block proteins for some of their functions, indicating that it might be involved in cell-to-cell movement (Ryabov *et al.*, 1998).

LIYV RNA2-encoded p26 is found associated with plasmodesmata (Medina *et al.*, 2001).

3. Cell-to-cell movement of viroids

In a study on PSTVd, Ding *et al.* (1998) microinjected infectious transcripts labeled with a fluorescent dye into mesophyll cells where the nucleic acid moved rapidly in contrast to a 1.4-kb RNA comprising vector sequences. When PSTVd was fused to the vector RNA, the fusion RNA moved from cell to cell. This indicated that the movement of PSTVd through plasmodesmata was mediated by a sequence-specific or structural motif (see Chapter 14, Section I.B.5).

4. Movement protein superfamilies

Four 'superfamilies' of MP have now been recognized: the '30K' superfamily; the products of the triple-gene block of potexviruses and related viruses; the tymoviral MPs; and a series of small polypeptides, less than 10 kDa, encoded by carmo-like viruses and some geminiviruses. The triple-gene block, the carmovirus double-gene block and the tymoviral MPs are described above.

i The '30K' superfamily

The '30K' superfamily, based on the well-characterized TMV 30-kDa MP, is a collection of viral gene products for which a variety of activities has been demonstrated (reviewed by Melcher, 2000). These include:

- the ability to bind nucleic acids;
- the ability to increase plasmodesmatal SELs, localization and accumulation in plasmodesmata;
- the ability to facilitate movement of RNA to neighboring cells on microinjection;
- the ability in some cases to form tubular structures and to interact with cytoskeletal elements.

The members of the '30K' superfamily have very few amino-acid conserved motifs, and they are grouped on their predicted secondary structure (Melcher, 2000). These predictions suggest a common core structure of a series of β-elements flanked by an α-helix at each end; the core structure is flanked by variable N- and C-terminal domains. Parsimony analysis grouped the tubule-forming MP separate from the others (Fig. 9.12).

5. Movement strategies

Five movement strategies have been recognized thus far, in each of which there is a strong co-ordination between viral replication and movement. Cell-to-cell movement has been divided into two separate, but closely linked, phases: intracellular transport that is linked closely to viral replication; and intercellular transport through plasmodesmata (Carrington *et al.*, 1996). These are discussed further in Chapter 13.

a. TMV strategy (reviewed by Más and Beachy, 1999; Reichel *et al.*, 1999)

The linking of GFP to P30 in infectious TMV constructs has led to a detailed picture of many of the aspects of cell-to-cell movement of this virus. Two approaches have been used. As the gating of plasmodesmata is restricted to the leading edge of expanding infection sites (Oparka *et al.*, 1997), a radial section across such a site should give a time scan of events (Padgett *et al.*, 1996). The other approach was to make real time observations on single living protoplasts immobilized in agarose (Más and Beachy, 1998).

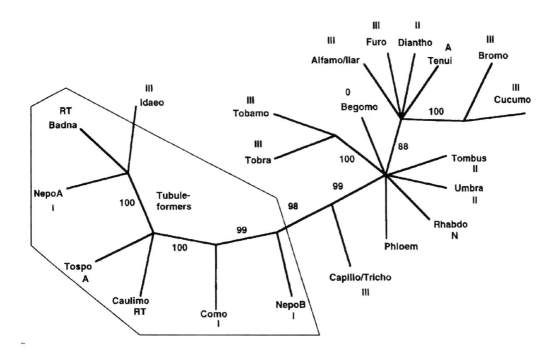

Fig. 9.12 Relationships among putative 30-kDa MP superfamily members determined by bootstrapping parsimony. Branches with less than 80% support (100 bootstrap replicates) were collapsed. Labels 0, RT, N, A, I, II and III represent the types of polymerase encoded by the viruses, which are respectively: 0; RNA-dependent DNA polymerase; negative-strand virus; ambisense-strand virus; and positive-strand supergroups I, II and III RNA-dependent RNA polymerases (see Chapter 8, Section IV.B.1). The thin-lined polygon encloses those MPs known to form virion-bearing tubules. From Melcher (2000), with permission.

A model for the two stages, intra- and intercellular transport, is shown in Fig. 9.13.

At the early stage, the virus replicates on the endoplasmic reticulum, leading to viral RNA accumulating in large irregular bodies. P30 co-localizes with this viral RNA, suggesting that viral replication and translation occur at the same subcellular site. The viral RNA and P30 co-localize with microtubules which are thought to direct the RNA–P30 complex to the plasmodesmata, which the P30 gates open to allow the complex to pass through (the intercellular transport stage). The mechanisms by which the MP targets and gates plasmodesmata are unknown, but it is interesting to note that P30 binds specifically to the cell wall enzyme, pectin methylesterase, which is also an RNA binding protein (Dorokhov et al., 1999). This gating is only temporary at the infection front, and behind the front plasmodesmata appear to return to their normal SEL. During virus infection of protoplasts, high-molecular-weight forms of P30 have been detected, the accumulation of which was enhanced by the inhibition of the 26S proteasome (Reichel and Beachy, 2000). It was suggested that poly-ubiquinated P30 is degraded by the 26S proteasomes. It has also been suggested that the phosphorylation could act as a mechanism to sequester P30 in cell walls and inactivate its function (Citovsky et al., 1993); in Arabidopsis thaliana P30 is N-terminally processed, which is thought to also inactivate it (Hughes et al., 1995).

The similarities between the TMV and RCNMV noted above, and the fact that, like TMV, RCNMV does not require viral coat protein for cell-to-cell movement (Xiong et al., 1993), suggests that these two viruses have similar local movement strategies.

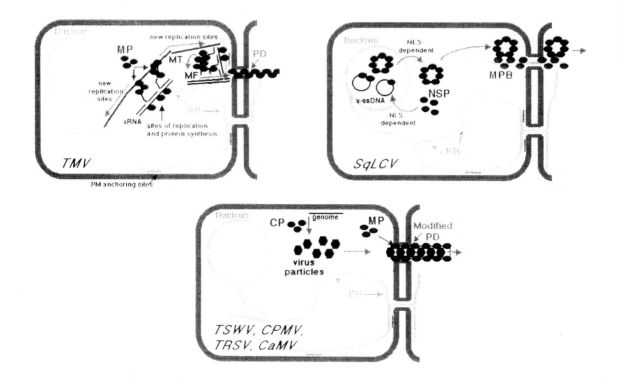

Fig. 9.13 (see Plate 9.4) Models for plant virus intracellular movement. **(Top left)** TMV. MP complexed with viral genomic RNA is proposed to move along microtubules from ER-associated sites of viral replication and protein synthesis ('viral factories') to establish additional viral factories at other ER sites. From viral factories associated with cortical ER, actin microfilaments may deliver the MP-genome complexes to putative cell-wall adhesion sites and plasmodesmata. **(Top right)** The geminivirus SqLCV. NSP is a nuclear shuttle protein that moves newly replicated viral ssDNA genomes from the nucleus to the cytoplasm. MPB, associated with ER-derived tubules, traps the NSP-ssDNA complexes in the cytoplasm and is proposed to guide these along the tubules and through the cell wall into adjacent cells. In these adjacent cells, NSP-ssDNA complexes would be released and target the viral DNA to the nucleus to initiate a new round of replication and infection. **(Bottom)** Viruses that form MP tubules: TSWV, CPMV, TRSV and CaMV. These viruses require coat protein or equivalent as well as MP for cell-to-cell movement. The viral genome, encapsidated in virus particles, moves through MP-containing tubules that appear to emerge from highly modified plasmodesmata. CP, coat protein; ER, endoplasmic reticulum; MF, microfilament; MP, movement protein; MT microtubule; NES, nuclear export signal; NLS, nuclear localization signal; PD, plasmodesma; vRNA, viral RNA genome; v-ssDNA, viral ssDNA genome. From Lazarowitz and Beachy (1999), with permission.

b. The tubule strategy

At least four virus families, *Bromoviridae*, *Comoviridae*, *Caulimoviridae* and *Nepoviridae*, together with the *Tospovirus*, *Badnavirus* (Cheng *et al.*, 1998) and *Trichovirus* (Satoh *et al.*, 2000) genera, produce tubules that extend through plasmodesmata or cell walls. Electron microscopy shows that those of bromo-, como-, nepo- and caulimoviruses contain virus particles, and those of tospoviruses are presumed non-enveloped nucleocapsids. The model for this form of virus movement

(Carrington *et al.*, 1996; see Fig. 9.13) is that MPs localize to the plasmodesmata where they induce the removal of the desmotubule and assemble into tubules extending unidirectionally into the adjacent plant cell. Virions assembled in the cytoplasm are escorted to the tubular structures through interactions with their MP and are then transported to the adjacent cell.

Three host factors in *A. thaliana* that suppress mutations in CaMV MP have been recognized

(Callaway *et al.*, 2000). One, termed *asc1* (acceleration of symptoms by CaMV N7) was mapped to chromosome 1 and others were mapped to chromosomes 3 and 4.

The situation with AMV appears to differ somewhat from that of other viruses that form tubules. AMV MP associates with the endoplasmic reticulum in tobacco cells onion bulb scales (Huang and Zhang, 1999). Infection of protoplasts produced tubules extending from the plasma membrane (similar to those in Fig. 9.9E) (Kasteel *et al.*, 1997), but tubules are not seen in plasmodesmata of infected *N. benthamiana* plants (van der Wel, 2000). However, the MP is found associated with enlarged plasmodesmata (van der Wel *et al.*, 1998) at the infection front where the number of plasmodesmata per unit area was three to four times higher than in uninfected tissue (van der Wel, 2000); the modified plasmodesmata appeared to be at least twice the diameter of unmodified plasmodesmata and lacked the desmotubule. Virus particles were seen within these modified plasmodesmata. Furthermore, the size of the plasmodesmata returns to apparently normal behind the infection front. These observations have led to the model shown in Fig. 9.14.

c. Triple-gene block strategy

Although the TGB1 product of potexviruses can increase the plasmodesmatal SEL for virus movement, the products of all three TGBs together with the viral coat protein are required for cell-to-cell movement (Lough *et al.*, 1998). The movement of the TGB1 (25-kDa) PVX protein is thought to be regulated by interactions with the TGB2, TGB3 and coat proteins (Yang *et al.*, 2000). This interaction is most likely with the C-terminus of the coat protein (Fedorkin *et al.*, 2001). The current hypothesis is that TGB1 and coat protein form a complex with the viral RNA, which is transported intracellularly. For intercellular transport and plasmodesmatal location, the presence of products from TGB 2 and 3 are required (Lough *et al.*, 1998; Erhardt *et al.*, 2000). There is some debate as to whether the viral coat protein forms a conventional capsid (Lough *et al.*, 1998; Santa Cruz *et al.*, 1998).

d. Potyviral strategy

By observing cells at the advancing infection front of PSbMV by electron microscopy, Roberts *et al.* (1998) showed that the CI pinwheel inclusions (see Chapter 6, Section VIII.C.1.b) were positioned directly over plasmodesmatal apertures. This event was linked with an apparent transient reduction of callose in the vicinity of the plasmodesma. Viral coat protein was seen in a continuous channel that passed along the axis of the pinwheel and through the plasmodesma. Behind the infection front, the CI was no longer associated with cell

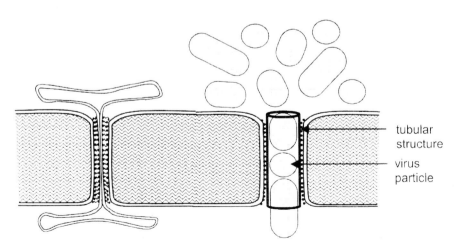

Fig. 9.14 Model for cell-to-cell movement of AMV. A short MP tubular structure is inserted into a plasmodesmata pore. Immunological studies revealed that the modified plasmodesmata still contain compounds typically present in plasmodesmata. From van der Wel (2000), with permission.

tubular structure

virus particle

walls. These observations suggested that the CIs lined up viral particles so that they could pass through the plasmodesmata. Whether there was any increase of SEL and the relevance of the decrease in callose are not known.

e. Geminiviral strategy

As noted above (Section II.C), geminiviruses replicate in the nucleus and their genomes are transported across the nuclear membrane by virus-coded proteins and cell-to-cell movement is mediated by other proteins. Thus, the model for geminiviral movement (Fig. 9.13) involves transport out of the nucleus by the BV1 or coat protein, interaction of this gene product with BC1 or V2 protein, and then passage through the plasmodesma.

6. Complementation

The presence of a virus with a compatible MP can complement the cell-to-cell spread of another virus in an apparent non-permissive host. The cowpea strain of SBMV (now known as SCPMV) did not accumulate in leaves of bean (*Phaseolus vulgaris*) when inoculated by itself, but did so when co-inoculated with SHMV. In the joint inoculations SCPMV invaded the non-vascular leaf tissue of inoculated leaves but did not spread into the vascular tissue or go systemic (Fuentes and Hamilton, 1991). As SCPMV could replicate in bean protoplasts it appeared to give a subliminal infection of bean plants and, in the joint infection, the MP of SHMV complemented its lack of movement activity. However, it could complement movement only up to the vascular tissue. SBMV complements the systemic movement of SCPMV in bean (Hacker and Fowler, 2000), it being suggested that the limitation on the systemic spread of SCPMV might be the inability of the virus to assemble in this host. SBMV infects *P. vulgaris* but does not accumulate in cowpea. When it is co-inoculated with SCPMV in cowpea it spreads locally but not systemically (Hacker and Fowler, 2000). The inverse results were obtained by Othman (1994) who observed complementation of long-distance movement of SBMV by SCPMV in cowpea but not of SCPMV by SBMV in bean.

The reasons for these differences are not known, but they could be due to either different cultivars of the hosts or different strains (isolates) of the viruses.

Complementation can also lead to the invasion of cells outside the bundle sheath by phloem-limited viruses. For instance, joint infection with a mesophyll cell-infecting virus can lead to luteoviruses also infecting mesophyll cells (Barker, 1989; Taliansky and Barker, 1999).

Recently, it has been shown that TMV and RCNMV MPs will complement the movement of an insect virus, flock house virus, in *N. benthamiana* (Dasgupta et al., 2001).

E. Time of movement from first infected cells

Uppal (1934) stripped the epidermis from *N. sylvestris* at various times after inoculation with TMV and assayed the underlying mesophyll by inoculating the local lesion hosts, *P. vulgaris* and *N. glutinosa*. He found that virus moved into the mesophyll in 4 hours at 24–30°C. Estimates for the time taken for virus to move into the underlying mesophyll using similar methods have been made for several host–virus combinations: 2 hours at 28°C for CMV in cowpea (Welkie and Pound, 1958); 10 hours at 20°C for TMV in *N. glutinosa* (Dijkstra, 1962); and 4 hours at 27°C for TMV in tobacco (Fry and Matthews, 1963).

Using microinjection of *N. clevelandii* trichome cells with TRV, Derrick et al. (1992) also found that 4 hours was required for the virus to move out of primarily inoculated cells.

F. Rate of cell-to-cell movement

The rate of movement from cell to cell has been estimated in several ways. The radius of necrotic local lesions produced by three tobamoviruses in *N. glutinosa* increases in a linear fashion with time (Rappaport and Wildman, 1957). Each of the viruses examined had a different rate of spread ranging from about 6 to 13 µm/h. Uppal (1934) estimated that TMV moved from the upper epidermis to the lower at the rate of about

8 µm/h, a value in agreement with the estimate of lateral spread. Rates of cell-to-cell spread may vary with leaf age, with different cell types, and in different directions within the leaf. TYMV infection spreads outward from the phloem of the small veins of young systemically infected Chinese cabbage leaves at about 4 µm/h; that is, roughly one cell every 3 hours (Hatta and Matthews, 1974). The rate and extent of cell-to-cell spread in inoculated Chinese cabbage leaves differed widely with different strains of TYMV (Matthews, 1981). Experiments with *ts* mutants of TMV show that rate of cell-to-cell movement of this virus is also influenced by the viral genome (Nishiguchi *et al.*, 1978).

There are generally fewer plasmodesmata per unit area on the vertical walls of mesophyll cells than on the walls that are more or less parallel to the leaf surface. Furthermore, there tend to be lines of mesophyll cells linked efficiently together and ending in contact with a minor vein. Viruses may spread more rapidly along such routes than in other directions within the mesophyll. In local lesions caused by TYMV in Chinese cabbage, not all the cells become infected even after many days. Also spread of the virus can be affected by host defense mechanisms (see Chapter 10, Section IV). For these various reasons, estimates of rates of cell-to-cell movement based on numbers of infected cells and those based on estimates of virus production can give only average values.

G. Long-distance movement

Evidence that viruses can move over long distances in the phloem came from several kinds of experiments:

Virus spread is influenced by the flow of metabolites in the plant (Bennett, 1940a). When a section of stem is killed or ringed, movement of most viruses is prevented or substantially delayed (e.g. Helms and Wardlaw, 1976). The presence of virus particles in sieve elements has been demonstrated for such phloem-restricted viruses as BYV and BWYV in *Beta vulgaris* (Esau *et al.*, 1967; Esau and Hoefert, 1972), BYSV in *Sonchus* (Esau and Hoefert,

1981), and for TRSV in soybean, which is not phloem-restricted (Halk and McGuire, 1973). When TYMV is translocated into a young Chinese cabbage leaf, the first mesophyll cells to show cytological signs of infection are always next to phloem elements and not xylem (Hatta and Matthews, 1974).

The involvement of phloem in long-distance movement has been confirmed by more recent direct observations (Leisner *et al.*, 1992a; Roberts *et al.*, 1997) and has led to detailed analysis of the phloem vascular system. Long-distance transport has recently been reviewed by Nelson and van Bel (1998), Santa Cruz (1999) and Derrick and Nelson (1999). Derrick and Nelson (1999) distinguish between long-distance movement, which is the movement of the virus in the phloem, and vascular-dependent accumulation of virus, which is the accumulation of the virus in the sink tissue after vascular movement.

Up to the early 1990s, it was considered that the phloem was involved primarily in the transport of sugars and other photoassimilates. However, it is becoming increasingly apparent that proteins and other macromolecules are actively transported in the phloem (reviewed by Thompson and Schulz, 1999). This has led to the 'Fisher model' for protein translocation in the phloem, which suggests that proteins are synthesized and processed in the companion cells, loaded into sieve elements of minor veins of source tissues and then unloaded into companion cells in sink tissues (Fisher *et al.*, 1992).

Some viruses, however, use the xylem for long-distance movement (see Section II.G.3).

1. Vein and phloem structure (reviewed by Derrick and Nelson, 1999)

Veins are defined as minor or major based on their structure, location, branching pattern and function. For dicotyledons, major veins are enclosed in parenchyma tissue forming a rib rising above the leaf surface and branch usually no more than twice. They function in long-distance transport of water, inorganic nutrients, photoassimilates and other organic matter, and are thought to be involved with photoassimilate unloading. Minor veins are embedded in mesophyll cells, are the result of three or more branchings from the first-order veins, and function in loading of photoassimilates in mature leaves. Veins have also been divided into five classes, classes I–III being major veins and IV–V being minor veins (Roberts *et al.*, 1997; see Fig. 9.15). The difference between major and minor veins is less apparent in monocotyledons. The structure of veins and vein-associated cells changes as the leaf develops from a sink to a source. This can have a significant effect on virus movement.

The vascular tissue is surrounded by the bundle sheath and comprises parenchyma (in major veins), sieve tubes with companion cells, and xylem elements (Fig. 9.16). The plasmodesmatal connections between these cells are discussed in Section II.D.1.

Many plant species have external and internal phloem in their stems. There is evidence for photoassimilate transport out of source leaves in abaxial phloem (connected to the external phloem) but not in the adaxial phloem (connected to the internal phloem).

2. Virus entry into phloem

It is suggested that most systemically invading viruses enter the vascular system through the minor veins (Roberts *et al.*, 1997; Ding *et al.*, 1998) (Fig. 9.17). Whether viruses enter veins through the termini of minor veins, branch points or over the length of the vein remains to be determined. However, detailed studies indicated that TMV enters both minor and major veins of *N. benthamiana* directly from non-vascular cells (Cheng *et al.*, 2000).

Plant species vary in the route of transport of photoassimilates across the vascular tissue to loading into the sieve element. The movement in some species is apoplastic and in others is symplastic. All the evidence points to virus movement being symplastic even in plant species that load photoassimilates apoplastically (Santa Cruz, 1999). Thus, the entry of the virus into the companion cell from the bundle sheath or parenchyma and the exit from the companion cell to the sieve element were thought to represent significant barriers. Recent studies (reviewed in Santa Cruz, 1999; Derrick and Nelson, 1999) show that the SEL of plasmodesmata in the sieve-element/companion-cell complex in transport phloem is significantly larger than that of other plasmodesmata and therefore phloem loading of virus may not be hindered at this site.

3. Virus transport through sieve elements

Using tissue blots, Andrianifahanana *et al.* (1997) showed that the potyvirus, PeMoV, moved first through the external phloem of *Capsicum annuum* towards the roots and, at or near the cotyledonary node, entered the internal phloem through which it moved to the shoot apex. Similarly, TMV first accumulated in the external phloem in the major veins and petioles of inoculated leaves and the stem beneath that leaf (Cheng *et al.*, 2000). Above the inoculated leaf the virus was mainly in the internal phloem of the stem and petioles of sink leaves. Thus, the connections between the external and internal phloem may have a role in regulating movement.

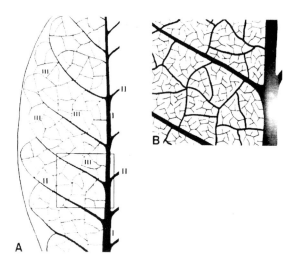

Fig. 9.15 Diagramatic representation of vein classes in the *N. benthamiana* leaf. **(A)** The midrib class (I) gives rise at regular intervals to class II veins. The class III veinal network, a branched veinal system that forms discrete islands on the lamina, lies between the class II veins. **(B)** Detail of boxed region in (A) showing the position of the minor veins (classes IV and V) within the islands of the class III veinal network. From Roberts *et al.* (1997), with permission.

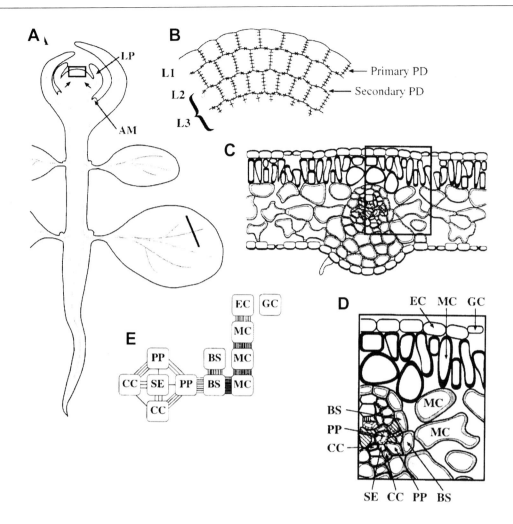

Fig. 9.16 Schematic of the supracellular nature of higher plants. **(A)** General arrangement of plant axis involving the shoot apical meristem (boxed region represents the central zone of the meristem), leaf primordia (LP), axillary meristem (AM), expanding leaves, the stem and mature leaves, and the root with its associated root apical meristem. Vascular (phloem and xylem) tissues (in grey) interconnect all regions of the plant. Nutrients essential for the growth of the meristem are delivered by the phloem into the region behind the meristem proper. Transport of these nutrients, and communication signals delivered via the phloem, are thought to move from this region via the symplasm; this route is indicated by arrows. **(B)** Central zone of the shoot apical meristem (boxed region in part A) made up of three layers, L1, L2 and L3, illustrating the arrangement of the cells and the presence of primary and secondary plasmodesmata that interconnect cells within and between layers, respectively. **(C)** Transverse section through the mature region of a leaf (solid line on leaf in part A). **(D)** Higher magnification of boxed area in (C) indicating the cell types associated with the vascular tissue and surrounding mesophyll: BS, bundle sheath; CC, companion cell; EC, epidermal cell; GC, guard cell; MC, mesophyll cell; PP, phloem parenchyma cell; SE, sieve element. **(E)** Diagrammatic representation of the spatial distribution of plasmodesmata between cell types illustrated in (D). Relative density of plasmodesmata indicated by the number of lines connecting specific cell types (cont.).

Mature sieve elements are devoid of nuclei and cytoplasm, which suggests that viruses do not replicate in them. Direct evidence has been obtained on this by using an experimental sys-tem designed to study the effects of low tem-perature on the vascular transport of TMV (Susi, 1999) and by grafting experiments with plants transgenically expressing TMV MP

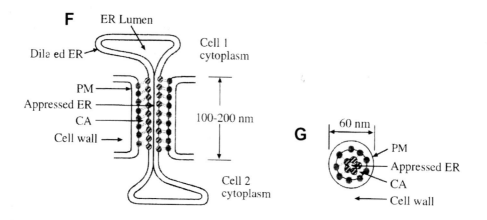

Fig. 9.16 cont. Schematic of the supracellular nature of higher plants. **(F)** Diagrammatic representation of a longitudinal section through a primary plasmodesma. Globular integral membrane proteins (3 nm in diameter) are located on the inner and outer leaflets of the plasmamembrane and appressed ER, respectively, and these proteins are thought to be interconnected by specific linking proteins. CA, cytoplasmic annulus; PM, plasma membrane. **(G)** Transverse section of primary plasmodesma. Note that the CA is divided into a number of microchannels with diameters of about 2–3 nm. From Mezitt and Lucas (1996), with kind permission of the copyright holder, © Kluwer Academic Press.

(Gera *et al.*, 1995). Both experimental systems showed that vascular transport did not depend on replication in the phloem.

4. Virus exit from sieve elements

Similar considerations pertain to the exiting of viruses from the sieve elements to those involved with entry into the elements. As noted in Section II.D.1, the SEL of plasmodesmata in sink tissues appears to be much greater than that in source tissues and may be sufficient for a virus to exit the sieve element and establish infection in new tissue. This non-specific trafficking is through simple plasmodesmata (Oparka *et al.*, 1999).

In a comparison of the phloem unloading of a fluorescent solute with fluorescent PVX, Roberts *et al.* (1997) showed that both predominantly exited in class-III veins of *Nicotiana benthamiana* (Fig. 9.17).

5. The form in which virus is transported

Most viruses require coat protein for long-distance transport (reviewed in Santa Cruz, 1999; Derrick and Nelson, 1999). However, some (e.g. GRV and BSMV) are capable of long-distance movement without a coat protein. Both these viruses form nucleoprotein complexed with a non-structural protein, the product of ORF 3 of GRV (Ryabov *et al.*, 1999) and βb of BSMV (Donald and Jackson, 1996). Whether all the viruses that require coat protein for long-distance transport move in the sieve elements as virus particles is uncertain. There is good evidence for virus particle formation for some such as TMV and TRV (Derrick and Nelson, 1999). Blackman *et al.* (1998) showed that CMV enters the sieve element as a ribonucleoprotein complex and assembles to form virions on the sieve element plasma membrane. For other viruses, it is difficult to distinguish between a requirement of coat protein for cell-to-cell movement leading to phloem loading and after phloem unloading, from that for actual sieve element transport.

Other viral proteins are important for vascular-dependent accumulation (Derrick and Nelson, 1999; see Table 9.1). Furthermore, the vascular movement of BNYVV in *Beta macrocarpa* is probably dependent on an RNA3 sequence domain rather than a gene product (Lauber *et al.*, 1998a). However, care must be

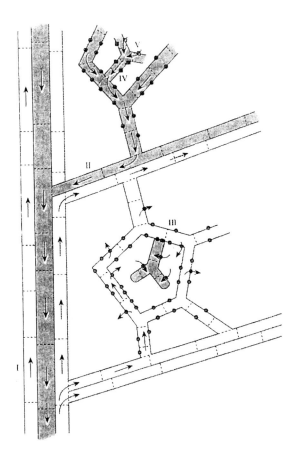

Fig. 9.17 (see Plate 9.5) Schematic model of the sink/source transition in a developing *Nicotiana benthamiana* leaf. At the top, the diagram depicts the exporting (source; blue) region of the leaf, and the lower part depicts the importing (sink; yellow) region of the leaf. Symplastic phloem unloading of solute and virus occurs from the class III vein network (yellow circles). In the sink portion of the leaf, the minor veins are immature (or non-functional; green shading) and receive assimilate (and virus) directly from the mesophyll by cell-to-cell movement. In the source portion of the leaf, the minor veins are now mature. Switching on sucrose transporters (green squares) in the minor veins signals the onset of apoplastic sucrose loading. In the source region, the class III veins function to export rather than import, which is achieved by down-regulation of the plasmodesmata (red circles) that connected them to the mesophyll during the import phase. Bidirectional phloem transport (in different sieve tubes) occurs in the class I vein, but it also may occur in class II and III veins as the 'transitional front' passes basipetally across different vein classes. From Roberts *et al.* (1997), with permission.

taken in attributing protein function directly to long-distance spread as they may be involved in, for example, suppression of host defense mechanisms. Thus, CMV gene 2b, identified as facilitating long-distance virus movement (Ding *et al.*, 1995a), is a suppressor of gene silencing (Brigneti *et al.*, 1998) (see Chapter 10, Section IV.H).

H. Rate of systemic movement

The time at which infectious material moves out of the inoculated leaf into the rest of the plant varies widely depending on such factors as host species and virus, age of host, method of inoculation, and temperature. After transmission by aphids, BYDV may move out of the inoculated leaf within 12 hours (Jensen, 1973). TMV moved out of tobacco leaves 32–48 hours after mechanical inoculation (Oxelfelt, 1970). SYNV moved systemically into leaves and roots within 24 hours of inoculation (Ismail *et al.*, 1987). However, in many early studies the rate of movement was assessed by systemic symptom production. A better indication of the rate of virus movement out of inoculated leaves is by leaf detachment experiments. Using such an approach, Gal-On *et al.* (1994) showed that CMV could spread systemically 24–30 hours after inoculation. In the classic experiments of Samuel (1934), one terminal leaflet of a tomato plant was inoculated with TMV and the spread of virus with time then followed by cutting up sets of plants into many pieces at various times. He incubated the pieces to allow any small amount of virus present to increase and then tested for the presence of virus by infectivity. Fig. 9.2 shows the course of virus spread in a medium-sized tomato plant. Virus moved first to the roots, then to the young leaves. It was some time before the middle-aged and older leaves became infected. In very young plants, older leaves did not become infected, even after several months.

Once virus enters the phloem, movement can be very rapid. Values of about 1.5 cm/h have been recorded for TMV (Bennett, 1940a). Capoor (1949) obtained a value of about 8 cm/h for the movement of TMV and PVX in tobacco stems. Helms and Wardlaw (1976)

TABLE 9.1 Viral factors that affect vascular-dependent accumulation of the virus in the host. From Derrick and Nelson (1999), with kind permission of the copyright holders, © Springer-Verlag GmbH and Co. KG

Viral factors	Virus	Known or possible site of effect	Reference
Coat protein	TMV TRV RCNMV	For TMV, possibly at entrance to SE/CC complex or within SE	Siegel et al. (1962) Sänger (1969) Takamatsu et al. (1987) Dawson et al. (1988) Hamilton and Baulcombe (1989) Vaewhongs and Lommel (1995) Ding et al. (1996b)
HC-Pro	TEV	Possibly for entry into or exit from SES	Cronin et al. (1995)
VPg domain of NIa	TEV	Possibly for entry into or exit from SES	Schaad et al. (1997)
126/183-kDa proteins	TMV	Entrance of virus into VPCs and CC or possibly for entrance of virus into SES, passage through sieve tubes, or exit from SES	Watanabe et al. (1987b) Lewandowski and Dawson (1993) Ding et al. (1995b) X.S. Ding and Nelson (unpubl.)
129/186-kDa protein	SHMV		Deom et al. (1994, 1997)
1a	CMV		Lakshman and Gonsalves (1983) Roossinck and Palukaitis (1990) Gal-On et al. (1994)
1a	BMV		De Jong and Ahlquist (1995)
1	CCMV		Wyatt and Kuhn (1980)
αa	BSMV		Weiland and Edwards (1994, 1996)
2	BMV		Traynor et al. (1991)
2	CMV	Possibly host resistance to cell-to-cell spread	Edwards et al. (1983)
2b	CMV		Ding et al. (1995a)
P19	TBSV		Scholthof et al. (1995a)
Gene VI	CaMV		Wintermantel et al. (1993)
RNA3	BNYVV		Lauber et al. (1998a)
5′ leader of RNA γ	BSMV		Petty et al. (1990)
30-kDa MP	TMV	(a) Entry to or exit from vasculature (b) Transport protein	(a) Hilf and Dawson (1993) Fenczik et al. (1995) (b) Arce-Johnson et al. (1997)

reported rates of up to 3.5 cm/h for TMV in *N. glutinosa* compared with 60 cm/h for ^{14}C assimilates. SBMV appears to move in the phloem of pinto beans, and in these cells it moves 10–100 times as fast as the cell-to-cell movement through parenchyma (Worley, 1965). However, when examined in detail, systemic invasion by a virus is affected by a complex of factors (Roberts *et al.*, 1997) and especially the source–sink status of individual leaves and parts of leaves.

Long-distance transport does not appear to depend on a concentration gradient of virus along the route of translocation, but rather on the rapid and random transport of infective mat-

erial. In the early stages of systemic infection, virus may pass through infectible tissues without causing infection there (e.g. Capoor, 1949).

The time at which unrelated viruses move from the inoculated leaf of the same individual host plant may be different. For example, Smith (1931) found that in tobacco plants inoculated with a mixture of PVX and PVY, PVY moved ahead of PVX, and could be isolated alone from the tip of the plant.

Systemic invasion of a leaf may result in a much more even movement of virus into mesophyll cells than that following mechanical inoculation. The differential temperature system that leads to infection of the cells in a young tobacco leaf with TMV under conditions where TMV replication does not occur (see Chapter 8, Section III.A.1.b) indicates that, in this leaf, cell-to-cell spread from the vascular elements is not dependent on the complete cycle of virus replication.

I. Movement in the xylem

Few viruses move long distances through the xylem. From girdling experiments, SBMV and other sobemoviruses were shown to move through dead tissue, which implicated xylem vessels (Schneider and Worley, 1959a,b; Gergerich and Scott, 1988a,b; Urban et al., 1989). The lack of normal virion formation was linked to the failure of long-distance movement of SBMV (Fuentes and Hamilton, 1993), indicating that virions are required for that type of movement. Most of the virions of RYMV accumulated in large crystalline patches in xylem parenchyma cells and sieve elements (Oparka et al., 1998). Anti-RYMV antibodies co-localized with a cell wall marker for cellulosic β-(1-4)-D-glucans over vessel pit membranes, suggesting that the vessel pits might be a pathway for virus migration between vessels.

For the sobemoviruses, movement in the xylem has been correlated with transmission by beetles (Gergerich and Scott, 1988b) (see Chapter 11, Section VIII).

The characteristic particles of LNYV were observed in young xylem cells of leaf veins (Chambers and Francki, 1966). Infective virus could also be recovered from xylem sap in the stem. However, there was no evidence that LNYV particles found in xylem played any role in systemic movement. Similarly, particles of RTBV are found in the xylem (Sta Cruz et al., 1993). PMTV is defective, in that the protein helices of its particles are mostly uncoiled at one end (Harrison and Jones, 1970). It moves rather slowly through tobacco plants, a pattern of movement consistent with transport in the xylem (Jones, 1975). The defective TMV mutants PM_1 and PM_2 may move by the same route. Some viruses normally found in other tissues may, in certain circumstances, be able to move in the xylem.

J. Final distribution in the plant

Whether or not a virus will move systemically at all depends on events in or near the local infection. It is often assumed that viruses that do move systemically become fairly evenly distributed throughout the plant, but in fact this hardly ever happens. There are several factors that can result in a very unequal distribution, including host genes, viral genes, the host defense system and environmental factors, all of which affect the rate and extent of virus movement through the plant. The effects of viral genes and host genes are discussed in Chapter 10 (Section III), those of the host defense system in Chapter 10 (Section IV), and those of environmental factors in Chapter 12 (Section II).

1. Limitation of local lesion size
a. Necrotic local lesions

With many host–virus combinations, virus moves into and multiplies in only a small group of cells around the point of infection. It has been suggested that, where infection results in rapid necrosis and death of cells, further spread of the virus is limited by the fact that no movement of virus from the dead cells can occur. However, examples are known where limited local infection occurs without death of cells (e.g. PVY in potato: Gebre Selassie et al., 1985). Limitation of spread may be a complex phenomenon associated with development of resistance to virus invasion in surrounding tissue (see Chapter 10).

A virus that normally produces only necrotic local lesions with no systemic spread when inoculated by mechanical means may spread through the plant systemically if infection is achieved by grafting. For example, SBMV, which on mechanical inoculation to P. vulgaris var. Pinto gives only necrotic local lesions, moves both upward and downward in the stem if introduced by grafting, and causes scattered limited areas of infection (Schneider and Worley, 1959a). Environmental factors as they affect extent and rate of virus movement are discussed in Chapter 12 (Section II).

b. Chlorotic local lesions

For many host–virus combinations giving rise to chlorotic local lesions, increase in lesion diameter does not continue indefinitely. This limitation may be due in part to increasing age of the inoculated leaf, but it bears no necessary relation to the question of systemic movement.

TYMV moves out of inoculated Chinese cabbage leaves about the time local lesions first become visible, usually at 4–5 days. Local lesions do not increase significantly in size after about 8 days. With TYMV, it is possible to assess the proportion of infected cells using light microscopy of protoplasts to determine whether the chloroplasts are rounded and clumped (Fig. 9.18).

Using this procedure, it has been determined that even in the central zone of a well-established yellow lesion there are three kinds of cell. First there are those showing full infection at the time protoplasts are prepared. Secondly there are those showing no sign of infection by light microscopy when first prepared, but which within about 6 hours develop rounding and clumping of chloroplasts. Electron microscopic examination showed that this group was infected in the intact lesion, as judged by the presence of characteristic peripheral vesicles (see Fig. 8.15), but further development of cytological effects had been suppressed in the intact leaf. Thirdly, in the central zone there are those cells showing no sign of infection by light or electron microscopy at any stage. These constituted 25–50% of cells in different experiments (Matthews, 1991). Tissue samples further from the center of lesions showed progressively higher proportions of cells that appeared to be uninfected. Whether the cells showing no evidence of infection in a local lesion are equivalent to dark green tissue in the systemic mosaic remains to be determined.

2. Escaping infection

The efficiency with which localized infection resulting from mechanical inoculation leads to systemic invasion varies quite widely for different viruses and hosts. Thus, in a rapidly growing tobacco or tomato plant a single local infection with TMV is sufficient to give systemic invasion in a high proportion of plants. In contrast, Bennett (1960) found that only about 10% of *Chenopodium capitatum* plants with 25 lesions or less became infected systemically with BYV. Two hundred or more local lesions were needed to ensure systemic infection in every plant.

Certain viruses may move with difficulty into particular leaves. SBMV and TRSV both multiply in an inoculated primary leaf of Black Valentine bean, and both readily move from this leaf to younger leaves on the plant. However, while SBMV readily moves to the uninoculated primary leaf in as short a time as 4 days, TRSV rarely moves into this leaf (Schneider, 1964). Schneider suggested that TRSV, for some unknown reason, becomes dependent on slow cell-to-cell invasion in the petiole of the uninoculated primary leaf.

Asymmetric infection may be even more marked with viruses that move systemically rather inefficiently. Inoculation of one primary leaf of soybean seedlings with TNV results in an asymmetric infection of the entire plant. In most trifoliate leaves, half the center leaflet and the leaflet on the same side as the inoculated primary leaf became infected. This asymmetry extended to the root system (Resconich, 1963). Some viruses that move rapidly into the root system when inoculated into leaves may not move out of the root system when the roots are inoculated, or may move only after a long delay (Roberts, 1950).

With viruses infecting woody perennials, distribution may be very uneven through the plant. For example, Gilmer and Brase (1963) found a marked non-uniform distribution of PDV in mature sweet and sour cherry trees that had been infected for at least 5 years.

3. Systemic distribution

Most viruses that give a general systemic infection do not infect and give symptoms in all the leaves. A common situation is that described by Leisner et al. (1992b) who showed that the systemic distribution of CaMV in turnip plants (Fig. 9.19) was influenced by both plant phyllotaxis and leaf development

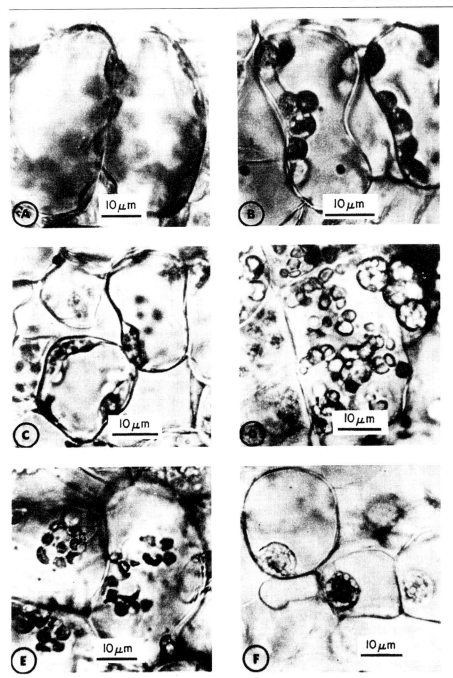

Fig. 9.18 Chloroplast abnormalities in different tissue types from Chinese cabbage leaves showing mosaic caused by TYMV. (A) Healthy leaf. (B) Pale green. (C) Yellow–green. (D) Yellow–green tissue with fragmenting chloroplasts. (E) Yellow–green tissue with highly vesiculate chloroplasts. (F) White tissue with a single spherical mass of chloroplastic material in each cell. From Chalcroft and Matthews (1967b), with permission.

stage. This is a reflection of the source–sink long-distance movement of the virus and the phloem connections in relation to the inoculated (source) leaf. As leaves undergo sink–source transition in photoassimilate import, there is a progressive basipetal decline in the amount of photoassimilate and virus entering the lamina so that, in more mature sink leaves, only the base of the leaf becomes infected.

Many of the viruses giving a general systemic infection apparently recover from infection in newly produced young leaves or even go through cycles of recovery and re-infection.

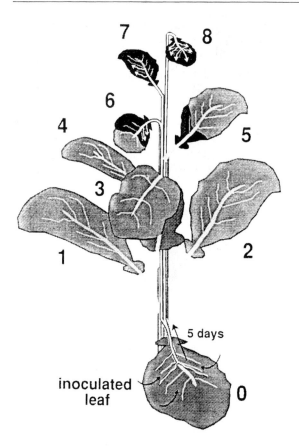

Fig. 9.19 Exploded diagram of turnip plant to show parameters that influence the bilateral and basipetal accumulation of virus in leaves of systemically-infected plant. Inoculated leaf is indicated as leaf number 0. White lines indicate the inferred vascular bundles through which the virus moves from the inoculated leaf to younger systemically infected leaves. Shading of leaves indicates the source-to-sink transition of photoassimilates, with the light shading representing the mature (source) part of the leaf and the dark shading representing the immature (sink) part of the leaf. From Leisner *et al.* (1992b), with kind permission of the copyright holder, © The American Phytopathological Society.

It is now considered that this is due to the normal plant defense system (see Chapter 10, Section IV).

4. Rise and fall in virus concentration with age

The amount of some viruses rises rapidly in infected leaves and then falls off. Considering the plant as a whole, this can lead to a very uneven distribution of virus in different leaves at any given time during growth.

Fig. 9.20 illustrates the distribution of PVA in leaves of field-grown tobacco plants over a 21-week period. This change of virus concentration is likely to be due to a combination of the normal plant defense system (Chapter 10, Section IV) and normal plant protein turnover systems.

5. Uneven concentration in different organs and tissues

In a plant infected for some time with a systemic virus, the concentration of virus may not be uniform in different organs. With most mosaic-type viruses that have been investigated, virus reaches a much higher concentration in the leaf lamina than in other parts of the plant. For example, TYMV concentration in the stem and root system and the midrib and petiole of expanded leaves is only one-tenth to one-twentieth of that found in the leaf lamina.

6. Viruses confined to certain tissues

CTV is not mechanically transmitted, but is transmitted by a phloem-feeding aphid. Price (1966) found phloem cells in infected lime plants to be packed with virus-like rods. These rods were absent from parenchyma tissues, and it seems likely that the virus is confined to the phloem. Luteoviruses, and other phloem-limited viruses, are found usually only in phloem parenchyma, companion cells and sieve elements (Taliansky and Barker, 1999). BYV appears to be confined to phloem at early stages of infection but, at later stages, it is found throughout the mesophyll and in the epidermis (Esau *et al.*, 1967). As judged by staining with specific fluorescent antibody, the viral antigen (protein) of WTV is confined to the pseudo-phloem tissue of root and stem tumors in sweet clover and to a few thick-walled cells in the xylem region (Nagaraj and Black, 1961). Similarly, curly top virus antigen was detected only in the phloem of several hosts (Mumford and Thornley, 1977).

7. Uneven distribution in leaves showing mosaic patterns

Dark green areas in the mosaic patterns of diseased leaves usually contain very little virus compared with yellow or yellow–green areas. This has been found for viruses that differ widely in structure, and infecting both mono- and dicotyledonous hosts. The phenomenon may therefore be a fairly general one for diseases of the mosaic type. Fig. 9.21 illustrates TMV distribution at the junction between a dark green and a yellow–green area in a leaf showing a mosaic pattern. Possible reasons for the low concentration of virus in dark green areas are discussed in Chapter 10 (Section III.O.4). This uneven distribution of virus in mosaic diseases may also extend to the petals to broken flowers. Thus, TMV produces white sectors in pink tobacco flowers to which virus appears to be confined.

8. Distribution of virus near apical meristems

There is evidence that some viruses may invade the primary meristematic tissues. For example, Walkey and Webb (1968) examined squashes of excised apical meristem tissue by electron microscopy and were able to show the presence of several viruses in the apices of various hosts. Virus particles have been observed in the apical initials of tobacco shoots (Roberts *et al.*, 1970). PepRSV was identified in meristematic cells of both root and shoot apices (Kitajima and Costa, 1969). Rods of PVX have been detected within or close to the potato shoot apex (Appiano and Pennazio, 1972). Rods of ORSV were observed in mitotically active cells in the apical meristem

of *Cymbidium* (Toussaint *et al.*, 1984). Clusters of BSMV-infected cells were distributed quite unevenly in both infected root and shoot apices of wheat (Lin and Langenberg, 1984b).

However, with many virus–shoot combinations, there appears to be a zone of variable length (usually about 100 µm but up to 1000 µm) near the shoot or root tip that is free of virus or that contains very little virus (e.g. Mori *et al.*, 1982; Faccioli *et al.*, 1988). This situation has been exploited to obtain virus-free clones by growing excised shoot tips in tissue culture (see Chapter 16, Sections II.C.2.c and d). However, with some virus-infected plants, one can start with meristems containing virus, and after a period in tissue culture these may give rise to virus-free plantlets (Walkey *et al.*, 1969).

Root tips of plants infected with one of several viruses have been found to be free of detectable virus (e.g. Appiano and D'Agostino, 1983). Smith and Schlegel (1964) studied the distribution of ClYVMV (described as a tymovirus in Matthews, 1991) in root tip of *Vicia faba*. They cut serial sections and assayed for infectivity (Fig. 9.22). Within the limits of the assay method, the first 400 µm of the root tip, which included the root cap and the meristem, were virus-free.

Viruses, such as members of the *Geminiviridae* and possibly of the *Caulimoviridae*, which might be expected to require dividing cells for their own replication, are not found in

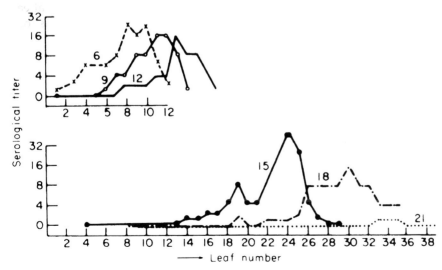

Fig. 9.20 Change in concentration of PVA with age in the leaves of a single Samsun tobacco plant inoculated on leaf number 1. Virus concentration was measured serologically. From week 15 onward, some of the older leaves had died. Numbers beside graphs show weeks after infection. From Bartels (1954), with kind permission of the copyright holder, © Blackwell Science Ltd.

Fig. 9.21 Distribution of TMV in the region of a junction between a dark green area (right) and a yellow–green area (left) of a tobacco leaf showing mosaic, determined by electron microscopic examination of a thin section across the macroscopically visible junction. To the left of the heavy line, all cells showed at least one crystalline inclusion of TMV. To the right, no crystalline inclusions were seen, and cells appeared cytologically normal. Numbers indicate the numbers of virus-like rods seen in the section of each cell. X, crystalline TMV inclusion seen; •, plasmodesmata observed between two cells. From Atkinson and Matthews (1970), with permission.

meristematic tissues (Lucy *et al.*, 1996). In fact, the replication of CaMV appears to be shut off in dividing cells (Covey and Turner, 1993). The *Geminiviridae* encode protein(s) which adjust the mature cell to support DNA replication (see Chapter 8, Section VIII.D.5).

The reasons why the meristematic zone of cells frequently fails to support virus growth are becoming clearer. It seems likely that the plant defense system of gene silencing is enhanced in meristematic cells and this inhibits the presence and replication of the virus.

K. Host factors

It is obvious from the above descriptions that host proteins are involved in viral movement, and some interactions between MPs and host proteins are mentioned above. The direct evidence comes from the interactions that viral MPs have with the cytoskele-

ton, presumably with its proteins (see Section II.D.2.a). It is likely that they will also interact with plasmodesmatal and, in the cases of *Geminiviridae* and *Caulimoviridae*, with nuclear pore proteins.

There are various other lines of indirect evidence suggesting the involvement of host proteins. Hull (1989) identified some of the lines of evidence including the fact that viruses can cause subliminal infection of some plants yet systemically infect other species. Furthermore, plant resistance such as that conferred against TMV by the dominant *Tm2* gene in tomato operates by preventing cell-to-cell spread. This resistance is overcome by simple changes in the TMV movement protein (see Chapter 10, Section III.E.3). Whether this resistance controls the interaction of the MP with a host protein is unknown, as are any potential proteins that may be involved.

However, plasmodesmatal and nuclear pore proteins are not the only host proteins

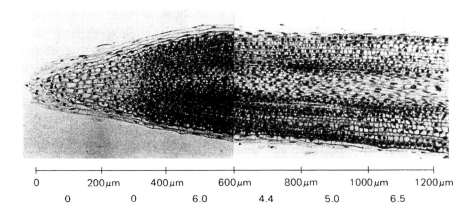

Fig. 9.22 Distribution of ClYVMV in the root tips of *Vicia faba*. Longitudinal section, showing the number of local lesions obtained on inoculation with extracts of 200-nm sections. From Smith and Schlegel (1964), with kind permission of the copyright holder, © The American Phytopathological Society.

involved in determining the spread of viruses in plants. Virus movement can be constrained by proteins in host defense systems, either those involved in local lesion formation, those involved in systemic acquired resistance or those in the gene silencing.

The formation of tubules by CPMV, CaMV and TSWV MPs in both protoplasts and insect cells (see Section II.D.4.b) indicates the lack of requirement of a specific plant protein in generating these structures, but the recent finding that AMV MP interacts with a host protein (van der Wel, 2000) indicates this need to be considered further.

a. Nicotiana benthamiana

There are several examples of viruses spreading systemically in *N. benthamiana* and which are limited to the inoculated leaf of other hosts. For example, the ability of TMV mutated in P30 to spread in *N. benthamiana* but not in *N. tabacum* was noted earlier (in Section II.D.2.a). ORSV spreads systemically in *N. benthamiana* but is limited in *N. tabacum* (Hilf and Dawson, 1993). Similarly, four mutants of RCNMV, mutated in the MP, were able to infect *N. benthamiana* systemically but were restricted in *N. edwardsonii* and *Vigna unguiculata* (Wang *et al.*, 1998b). In both these cases, the restriction to long-distance movement appears to operate at or near the bundle sheath and to be associated with the MP. These observations would suggest that the MP interacted with a host factor at, or close to, the

bundle sheath and that this factor differed between species.

b. Rate of movement

Mutants of RCNMV differed not only in the ability to systemically infect *N. benthamiana* but also in the rates of systemic movement in cowpea. It was suggested that, for these 'slow movers', the impaired long-distance transport makes it easier for the plant to mount a defense response and thus decrease the likelihood of establishing a systemic infection (Wang *et al.*, 1998b).

c. Long-distance movement

The entry of TEV was restricted at the companion-cell/sieve-tube boundary of the non-susceptible *N. tabacum* line V20, blocking either entry into, or exit from, the sieve elements (Schaad and Carrington, 1996). By analysing crosses between line V20 and the susceptible line, Havana 425, the restriction to movement could be attributed to the interaction of recessive genes at two unlinked loci.

A mutant of *A. thaliana* (*vsm1*) does not support the systemic spread of the tobamovirus, TVCV, and the tomato strain of TMV, the local movement probably being stopped at the step of viral entry into the vasculature (Lartey *et al.*, 1998); however, the carmovirus, TCV, moved systemically in this mutant host.

Four ecotypes (e.g. C24) of *A. thaliana* can support full systemic infection of TEV whereas spread of the virus in other ecotypes (e.g. Col-0 and Col-3) was limited to inoculated leaves (Mahajan *et al.*, 1998). By analysing crosses and backcrosses between C24 and Col-3, a dominant locus, RTM1, conditioning the restricted TEV infection phenotype was identified on chromosome 1. A mutational analysis of Col-0 showed that the restricted infection phenotype is a multi-gene trait that requires at least three loci, RTM1, RTM2 and RTM3 (Whitham *et al.*, 1999). The RTM1 gene has been isolated by map-based cloning and the product shown to be similar to the α-chain, the *Artocarpus integrifolia* lectin, jacalin, and to several proteins that contain multiple repeats of a jacalin-like sequence (Chisholm *et al.*, 2000).

A recessive gene, *ra*, in diploid crosses involving *Solanum tuberosum* subsp. *andigena* fully blocks vascular transport of PVA (Hämäläinen *et al.*, 2000).

L. Discussion

The recent increases in our understanding of how viruses move through plants have revealed information not only about the viruses themselves, but also about how plants function. The mechanisms involved are complex and integrated but essentially result in a plant infected by one of the mechanisms described in Chapters 11 and 12 to be able to become a source for infection of other plants.

The studies on plant virus movement have given several new insights into plant function but there are also some major gaps in the detailed understanding. Viruses which are introduced into epidermal or mesophyll cells have developed various mechanisms for moving from cell to cell so that they can be transported to more distant parts of the plant. They have acquired genes, the products of which will gate plasmodesmata to enable the infectious unit to pass to an uninfected cell. Only a few years ago it was considered that, apart from viruses, large macromolecules did not move from cell to cell. However, an increasing number of proteins are now known to move, albeit selectively, from cell to cell. These include the product of the homeobox gene, KNOTTED1 (KN1), and its mRNA, which

move selectively from cell to cell in developing meristems (Lucas *et al.*, 1995) and the pathogenesis-related PRms (Murillo *et al.*, 1997). Furthermore, a plant paralogue to TMV P30 potentiates transport of mRNA into the phloem (Xoconostle-Cazares *et al.*, 1999). Even nuclei appear to move through plasmodesmata in developing endosperm (Zhang *et al.*, 1990). It is becoming recognized that, in the processes of development, there is selective macromolecular communication between cells. It would seem likely that, as suggested by Hull (1991a), host genes which potentiate plasmodesmatal gating during development have been acquired by viruses, and been modified to facilitate movement in mature leaf cells. However, presumably communication between these leaf cells is controlled by the SEL of their plasmodesmata, and any long-term increase of the SEL could be seriously deleterious. Therefore, there appear to be mechanisms which, at least for some viruses, make the gating short term at the infection front. The phosphorylation and location at the plasmodesmatal neck of TMV P30 in tobacco and the N-terminal cleavage of this protein in *Arabidopsis* would appear to be mechanisms to inactivate this protein. Although we are beginning to gain a picture as to how viruses are transported across the cell to plasmodesmata, there is little information on how MPs increase the plasmodesmatal SEL.

The study of movement of viruses has identified various tissue barriers that have to be passed. Notable are those at the inner surface of the bundle sheath and for entry into companion cells. It was thought that there would be barriers for entry into and especially exit from sieve elements; since there is little or no active cytoplasm in the sieve elements, viral genomes would not be able to express any proteins to facilitate exit. However, recent studies, noted above, have shown that the SELs of some plasmodesmata are large and probably do not form a significant barrier.

Experiments on the long-distance movement of viruses have revealed much about the details of photoassimilate sources and sinks and phloem loading and unloading. They can also explain the mosaic or mottle (or in monocotyledons, streak) symptoms that so many viruses produce. However, there are still various questions to be answered, such as how viruses move from the external phloem to the internal phloem.

The final distribution of a virus, both on the macro-scale of the plant itself and on the micro-scale at the tissue and cellular level, is conditioned by a wide variety of factors that will differ between plant species. These will determine if a plant species is a host and, if it is, whether the virus will go systemic and to which parts of the plant. The factors include the compatibility of the viral MP(s) with the host, the developmental stages of the host, the nutritional and water status of the host, and the interactions that the virus will have with host defense systems. The final picture of the infected plant is a summation of all these, and other, dynamic variables.

It is becoming increasingly apparent that, as well as integrating and co-ordinating with plant functions, the separate functions of the viruses themselves are closely integrated (see Chapter 13). For instance, the replication and cell-to-cell movement of TMV are closely linked, as are these functions of geminiviruses (Más and Beachy, 1999; Lazarowitz and Beachy, 1999). This is not altogether surprising and it is likely that it occurs in all viruses.

III. EFFECTS ON PLANT METABOLISM

All the various macroscopic and microscopic symptoms of disease discussed in Chapter 3 (Sections II, III and IV) must originate in biochemical aberrations induced directly or indirectly by the virus. Many early workers described differences in composition or in rate of some process between healthy and virus-infected tissues. Aspects commonly investigated were total carbohydrates and sugars, total nitrogen and various nitrogen fractions, the carbon–nitrogen ratio, total ash or various ash components, and rates of photosynthesis, respiration and transpiration. Estimations were often made on fractions containing many different compounds, such as 'soluble nitrogen'.

Most analytical work has been carried out on fully infected plants, perhaps many weeks after inoculation. In such plants, we may well expect to observe changes in the amounts of many substances and in the rates of major biochemical processes. For obtaining any detailed understanding of the effects of virus infection on host metabolism and the initiation of disease processes, this earlier work is virtually useless.

A. Experimental variables

The experimental systems available for studying effects on host plant metabolism are very much the same as those described in Chapter 8 for virus replication. An effective cell culture system would be most useful for investigating some aspects of the metabolism of infected cells. However, there is increasing evidence to show that, following their isolation, leaf protoplasts are in a disturbed and changing metabolic state and results of metabolic studies on the effects of virus infection on such cells need to be interpreted with caution.

There are many variables to be taken into account when using intact plants or organs. The following discussion is concerned mainly with changes taking place in leaves, because these constitute most of the herbaceous host plant. More virus is usually produced in them, and they are most often used for experimental work. There are three kinds of variable factors to consider when designing experiments to study the effects of virus infection on host metabolism:

- the basis on which results are to be expressed;
- factors in the material that vary with time;
- non-uniformity in the material being sampled at any given time.

1. Basis for expressing results

Measurements or units that can be used as a basis for expressing results include fresh weight of tissue, dry weight, protein nitrogen, DNA, leaf area, or on a per-plant, per-leaf or per-cell basis. Fresh weight is the most convenient measurement and has been most widely used. It may be satisfactory where big differences are being sought at fairly early stages after infection. However, virus infection may alter water content of tissues soon after infection. At later times, when stunting of growth may have occurred, serious difficulties arise in the use of fresh weight. Virus-infected leaves, because they are smaller than healthy leaves of the same age, may have a higher concentration of many components per unit fresh weight. On the other hand, when compared with healthy leaves of the same size, they may have a lower concentration of these same components per unit fresh weight. Dry-weight measurement may likewise introduce serious ambiguities. The virus itself may come to represent about 10% of the dry weight of a leaf. Stunting may increase the proportion of inert cell wall materials, and starch accumulation due to virus infection may increase dry weight. Similarly, protein nitrogen content may be significantly affected by the presence of the virus and by general stunting. Where DNA content per cell remains constant, it forms a satisfactory basis for making comparisons.

Leaf area is a satisfactory basis for comparison where virus infection has not altered the area of the leaves—for example, at early stages after inoculation. Used on stunted leaves, it could be quite misleading. Measurements on a per-plant basis may be quite satisfactory where chronically infected material is examined, such as potatoes grown from infected tubers. Where the information sought is relevant to the economic aspects of disease, the whole plant, or the parts used commercially, will be the most relevant unit. However, when whole inoculated plants are used, the sample will include tissues that have been affected for different periods of time.

For many purposes, the leaf is the most satisfactory basis for calculating results, particularly where measurements can be begun before infection and continue as the disease progresses, and where leaf expansion is taking place. The cell may be the unit of choice in experiments involving tissue minces or cells grown in tissue culture. There may be no ideal method for expressing results, so that, generally speaking, it is highly desirable to use two or more different bases on the same experimental material.

2. Factors that vary with time

a. Leaf age

For most of the herbaceous species used in experiments with viruses, individual leaves are never in a steady state. They normally pass through four stages, each of which merges into the next. The first stage, up to about 2 cm long for leaves like those of tobacco, is one of intensive cell division. This is followed by a stage in which cell expansion and protein synthesis become the predominant activities. As the leaf approaches full size, photosynthesis and the export of metabolites become the major activity (although it has been going on from an early stage), and finally the processes of senescence take over. Virus infecting leaves at different stages in their development will thus meet different conditions and may have different effects on cell processes.

b. Diurnal variations

Many basic processes in leaves of plants grown in daylight follow a diurnal rhythm (reviewed by Millar, 1999). For example, the proportion of ribosomes in the polyribosome form, which presumably reflects the rate of protein synthesis, follows a diurnal cycle with a minimum at the end of the night period and a maximum near the middle of the day (Clark et al., 1964). Infection by virus at different times of day might produce different immediate effects. Diurnal rhythms can vary between species (Oberschmidt et al., 1995).

c. Seasonal variations

Plants grown at different times of the year under uncontrolled or partly controlled greenhouse conditions vary in many properties and may be affected differently by virus infection. Growing plants from seed under fully controlled environmental conditions can reduce this source of variation.

d. Wound responses

Metabolic changes, which begin as soon as healthy tissue is excised or wounded, alter the rate of uptake of isotopically labeled metabolites (Pratt and Matthews, 1971) and the pathways concerned with respiration (Macnicol, 1976).

3. Non-uniformity at a given sampling time

a. Between leaves

Variation is minimal between the two primary leaves, and these form useful experimental material in such plants as *P. vulgaris*. Apart from the primary leaves, no two leaves on many of the plants used in virus studies will be in exactly the same stage of development. This variation can be overcome to some extent by using groups of plants and selecting one leaf of a particular size for study. Even here, very few leaves in a group will be just at the same stage, since leaf initiation does not proceed in step in different plants. In fast-growing plants such as Chinese cabbage, where young leaves may double in size in 30–40 hours, it may require selection of a few plants from a large group to give a reasonably uniform set of younger leaves.

b. Within leaves

Positions in the leaf blade symmetrically placed with respect to the midrib are usually very similar, so that half-leaves cut longitudinally down each side of the midrib give useful samples for comparative studies. Within half-leaves, however, tissue is not uniform. The tip of the leaf is older than the base. Lamina thickness and venation pattern may vary. Rates of uptake of radioactively labeled compounds may not be constant between different parts of the blade. Furthermore, as discussed in Section II.J.3, infection may not be uniform over the whole leaf.

c. Cell type

Mesophyll cells usually make up a high proportion of leaf tissue, but epidermal cells and epidermal appendages may constitute 10–15% of the fresh weight. If the midrib and major veins are large, they are usually discarded in biochemical studies on virus infection, unless they are of special interest.

d. Mosaic symptoms

As discussed in Chapter 10 (Section III.P) and in Section IV.D of this chapter, blocks of tissue in leaves showing mosaic symptoms may be very heterogeneous with respect to virus infection and such heterogeneity may be detectable only by microscopic examination. With mosaic patterns such as those induced by TYMV, the dark green areas can be used with advantage as control tissue in studying some of the effects of chronic virus infection on cell components and processes. A dark green area is dissected from one-half of a leaf and a corresponding virus-bearing area is taken from the opposite half-leaf.

e. Cell organelles

In higher plants, isolated and apparently intact chloroplasts frequently fractionate into two or more bands in density gradients. In some species of higher plants, two morphological types of chloroplast have been observed that fix carbon by different routes (Slack, 1969). Thus, normal variation between chloroplasts could complicate observations on the metabolic effects of virus infection.

f. Changes in excised leaf pieces

After short periods of incubation, the marginal areas of excised leaf disks, even very small ones, vary markedly from the central region in various metabolic activities (Pratt and Matthews, 1971; Macnicol, 1976).

B. Nucleic acids and proteins

The synthesis and expression of viral nucleic acids and proteins is discussed in Chapters 7 and 8. Here I will consider the general effects of virus infection on host-cell nucleic acids and protein synthesis.

1. DNA

It is widely assumed that the small RNA viruses have little effect on host-cell DNA synthesis, but there are very few, if any, definitive experiments bearing on the question. Virus infection may well have some effect on host-cell DNA synthesis, but such effects are likely to be fairly small and difficult to establish, because (1) DNA content per cell may increase for some time in a normal expanding leaf; (2) minor DNA fractions, which might be affected by virus infection, may be difficult to isolate and identify; and (3) any effect might be very transitory and, therefore, difficult to detect in asynchronous infections. Using a radioautographic technique to assay for DNA synthesis in individual cells, Atchison (1973) found that there was a drop in DNA synthesis in the terminal 1 mm of French bean roots about the time they were invaded by TRSV. This was soon followed by a transient drop in the mitotic index.

2. Ribosomes and ribosomal RNA

Effects of virus infection on ribosomal RNA synthesis and the concentration of ribosomes may differ with the virus, strain of virus, time after infection, and the host and tissue concerned. In addition, 70S and 80S ribosomes may be affected differently.

In TMV-infected leaves, viral RNA may come to represent about 75% of the total nucleic acids without having any marked effects on the main host RNA fractions except to cause a reduction in 16S and 23S chloroplast ribosomal RNAs (Fraser, 1987b). However, under some conditions cytoplasmic ribosomal RNA synthesis is inhibited too. A reduction in chloroplast ribosomes without a marked effect on cytoplasmic ribosomes is a fairly common feature for mosaic diseases (e.g. BSMV in barley: Brakke et al., 1987b; TYMV: see the following discussion).

The variables that affect the outcome of TMV infection on cytoplasmic and chloroplast ribosomes have been discussed by Fraser (1987b). Experiments with TYMV will illustrate the difficulties in making generalizations.

In Chinese cabbage leaves chronically infected with TYMV, the concentration of 70S ribosomes in the yellow–green islands in the mosaic is greatly reduced compared with that in dark green islands in the same leaf (Reid and Matthews, 1966). There is little effect on the concentration of cytoplasmic ribosomes in such yellow–green islands of tissue. The extent of this reduction depends very much on the strain of TYMV, and it also becomes more severe with time after infection. Loss of 70S ribosomes more or less parallels the loss of chlorophyll, 'white' strains causing the most severe loss.

A somewhat different result is obtained if the effect of TYMV infection with time in a young systemically infected leaf is followed. Chloroplast ribosome concentration falls markedly about the time virus concentration reaches a maximum. About the same time, there is a significant increase in cytoplasmic ribosome concentration, which is mainly due to the stunting effect of infection. On the other hand, if the effects of virus infection on these components for the plant as a whole are considered, a different picture emerges. Infection reduces both cytoplasmic and chloroplast ribosomes.

These results emphasize the fact that infection of a growing plant with a virus introduces an additional time-dependent variable into a system in which many normal interacting components are changing with time. Analyzes made on only one or two components of the system, or at some particular time, are unlikely to give much insight into virus replication and the nature of the disease process. Very little is known about any effects of virus infection on host tRNAs, nuclear RNAs or mitochondrial ribosomal RNAs. Effects on host mRNAs are discussed in the next section in relation to protein synthesis.

3. Proteins

The coat protein of a virus such as TMV can come to represent about half the total protein in the diseased leaf. This can occur without marked effects on the overall content of host proteins. Most other viruses multiply to a much more limited extent. Effects on host protein synthesis are not necessarily correlated with amounts of virus produced. A reduction in the amount of the most abundant host protein—ribulose bisphosphate carboxylase-oxygenase (rbcs or rubisco)—is one of the commonest effects of viruses that cause mosaic and yellowing diseases (e.g. TYMV:

Reid and Matthews, 1966; WSMV: White and Brakke, 1983).

Fraser (1987b) estimated that TMV infection reduced host protein synthesis by up to 75% during the period of virus replication. Infection did not alter the concentration of host polyadenylated RNA, or its size distribution. This suggested that infection may alter host protein synthesis at the translation stage rather than interfering with transcription. Stratford and Covey (1988) found that there were changes in the levels of specific translatable mRNAs in response to infection of turnip leaves with CaMV. More such changes were found with a severe strain. In particular, the mRNA encoding the precursor to the small subunit of rbcs was markedly decreased following infection with the severe strain. Saunders *et al.* (1989) used another approach to the same problem. They generated a library of cDNA clones corresponding to the host RNAs isolated from turnip leaves infected with CaMV during the early vein-clearing stage. Hybridization was used to select clones that represented RNAs whose levels had been raised or lowered by infection. For example, one RNA that was greatly reduced in amount was identified as the mRNA for the rbcs polypeptide. Overall, the findings of Stratford and Covey (1988) and Saunders *et al.* (1989) were taken to suggest that there are few major changes in host gene expression during infection with CaMV.

In detailed studies with PSbMV, Wang and Maule (1995), Aranda *et al.* (1996) and Maule et al. (2000) identified the infection front in pea cotyledons by *in situ* hybridization using probes for both (+)- and (−)-sense of the viral genome. The detection of (−)-sense viral RNA was taken as being indicative of viral replication. They then probed for the presence of mRNAs for various host genes (Fig. 9.23) and identified three situations:

1. Expression of at least 11 host genes was suppressed; e.g. lipoxigenase 1 and HSC70 (Fig. 9.23F).
2. Expression of the heat shock protein HSP70 and of polyubiquitin was induced in association with viral replication (Figs 9.23D and E);

however, the induction of HSP70 by viral replication differed from that by heat (Aranda *et al.*, 1999).

3. There was no effect on host gene expression with, for example, actin and β-tubulin.

Probing the infection front with antibodies showed that host protein accumulation was also affected in a similar manner. In this plant viruses resemble, in some respects, various animal viruses (Aranda and Maule, 1998) but differ in that all three phenomena are associated with the same virus. These observations raise many questions (Aranda *et al.*, 1996), including:

1. Does the virus shut off expression of genes which could be considered as being 'competitive' and enhance others which may be 'helpful'?
2. Is this phenomenon restricted to cotyledons or is it a general mechanism at the infection front throughout the plant?
3. If it is general, could it be associated with symptom expression?

C. Lipids

The sites of virus synthesis within the cell almost always contain membrane structures (see Chapter 8). TYMV infection alters the ultrastructure of chloroplast membranes, and rhabdovirus and tospovirus particles obtain their outer membrane by budding through some host-cell membrane. There have been a few studies of the effects of virus infection on lipid metabolism (e.g. Trevathan *et al.*, 1982) but none of these has illuminated the mechanisms by which viruses change and use plant membrane systems.

The expression of lipid transfer proteins, which are implicated in defense responses to bacterial and fungal infection, are stimulated on CaMV infection of *Arabidopsis* (Sohal *et al.*, 1999).

D. Carbohydrates

Some viruses appear to have little effect on carbohydrates in the leaves, while others may

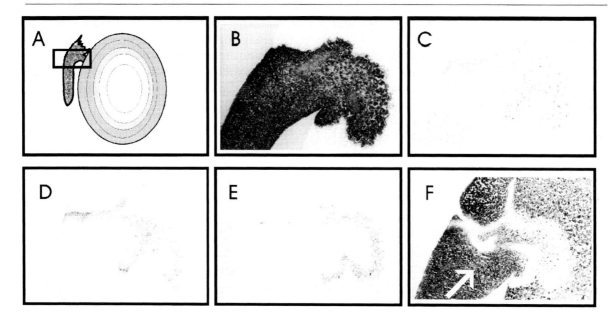

Fig. 9.23 Shutoff of cellular genes and induction of HSP70 and polyubiquitin associated with PSbMV replication. The figure shows the results of *in situ* hybridization on near-consecutive sections of PSbMV-infected pea tissues. Positive detection of specific RNAs is shown by dark staining on the tissue section. **(A)** Diagrammatic representation of a section of a pea seed; the portion of the tissue which appears in the other panels is identified (box). **(B)** Detection of genomic viral positive-strand RNA. **(C)** Detection of viral negative-strand RNA equating to cells where the viral RNA is, or has recently been, replicating. **(D–F)** Detection of HSP70, polyubiquitin, and lipoxigenase RNA respectively. The lipoxigenase probe records the shutoff effect. Induction of HSP70 and polyubiquitin genes coincides within the shutoff area and the area of viral RNA replication (Aranda *et al.*, 1996). In this example, both phenomena, shutoff and induction, occur in a tissue of complex structure where cell types corresponding to different vegetative components of the mature plant are equally affected. Note the recovery from shutoff (restoration of lipoxigenase expression; arrow) within the affected area. From Aranda and Maule (1998), with permission.

alter both their rate of synthesis and rate of translocation. These changes may be illustrated in a simple manner.

Leaves that have been inoculated several days previously with a virus that does not cause necrotic local lesions are harvested in the morning or after some hours in darkness, decolorized and treated with iodine. The local lesions may show up as dark-staining areas against a pale background, indicating a block in carbohydrate translocation. On the other hand, if the inoculated leaves are harvested in the afternoon after a period of photosynthesis, decolorized and stained with iodine, the local lesions may show up as pale spots against the dark-staining background of uninfected tissue (Holmes, 1931). Thus, virus infection can decrease the rate of accumulation of starch when leaves are exposed to light.

Infection of cotyledons of *Cucurbita pepo* with CMV results in the formation of these starch lesions. Técsi *et al.* (1994a) examined the distribution of virus across such lesions by immunolocalization and correlated this with starch accumulation, $^{14}CO_2$ assimilation and chlorophyll *a* quenching. This analysis revealed an arrangement of cell types with diverse physiology that was complex both spatially and temporally (Fig. 9.24).

The maximum virus accumulation was at the edge of the lesion which physiologically comprised four concentric zones (see legend to Fig. 9.24).

As the concentration distribution of virus did not correlate with the four zones it seemed unlikely that the physiological changes were a direct consequence of replication.

Infection with some viruses, such as BYV,

can induce damage to the phloem that restricts translocation of photoassimilates. These accumulate in the leaf lamina and possibly give rise to some of the symptoms such as thickened leaves and leaf rolling.

From the few diseases that have been examined in any detail, it is not possible to make very firmly based generalizations about other carbohydrate changes, but the following may be fairly common effects: (1) a rise in glucose, fructose and sucrose in virus-infected leaves; (2) a greater rise in these sugars caused by mild strains of a given virus compared with severe strains. Further, effects of infection on mesophyll cells, not yet understood, may reduce translocation of carbohydrates out of the leaves.

E. Cell wall compounds

Although cytological studies have demonstrated ultrastructural changes in the cell walls in many virus infections, the biochemical basis of such changes would be difficult to study. Changes to plasmodesmata associated

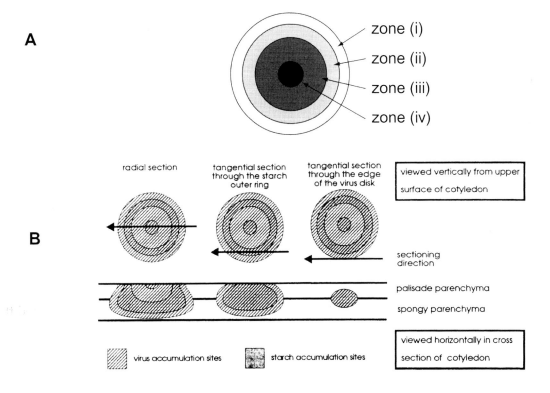

Fig. 9.24 Schematic diagrams of the distribution of CMV, starch and plant compounds and metabolites across a lesion in *Cucurbita pepo*. **(A)** Distribution of CMV, starch, chlorophyll *a* fluorescence, ^{14}C incorporation and chlorosis in a lesion on day 4 post-infection. From the outside in, the infected region consists of (i) a zone of cells which may have an elevated level of photosynthesis but which does not yet have high levels of starch; (ii) a zone which probably consists of cells with high photosynthetic activity (fast chlorophyll fluorescence quenching, high CO_2 assimilation) and low starch accumulation at its outer edge and cells of relatively low photosynthetic activity but high starch content at its inner edge; (iii) a zone of largely starchless cells which are starting to become chlorotic and which have low photosynthetic activity (slow chlorophyll fluorescence quenching, low CO_2 assimilation), and (iv) a relatively stable group of infected cells which have high photosynthetic activity and high starch content. **(B)** Diagram of the distribution of starch and virus in a lesion. Virus accumulation precedes starch accumulation in the lesions. In radial section both the central dot and ring can be observed. In tangential section through the starch outer ring only the starch ring is observed, while in tangential section through the edge of the virus disc no starch accumulation is observed. From Técsi *et al.* (1994a), with kind permission of the copyright holder, © Blackwell Science Ltd.

with virus cell-to-cell movement are described in Section II.D. Future work may show that virus infection has effects on various activities in the cell wall compartment, which is not metabolically inert. Eighty-five percent of detectable peroxidase activity and 22% of the acid phosphatase are located in the cell wall of healthy tobacco leaves (Yung and Northcote, 1975). Elevated peroxidase activity has been reported as a response of tobacco and many other hosts to virus infection (see Matthews, 1981).

F. Respiration

Many studies have been made of the effects of virus infection on rates and pathways of respiration, but it is not possible to relate the results to the processes involved in virus replication. In summary, for many host–virus combinations where necrosis does not occur, there is a rise in respiration rate, which may begin before symptoms appear and continue for a time as disease develops. In chronically infected plants, respiration is often lower than normal. In the one systemic disease so far examined in detail, there is no detectable change in the pathway of respiration. In host–virus combinations where necrotic local lesions develop, there is an increase in respiration as necrosis develops. This increase is accounted for, at least in part, by activation of the hexose monophosphate shunt pathway (see Matthews, 1981; Fraser, 1987b).

Usually, respiration and other areas of metabolism have been examined separately. Técsi *et al.* (1994b) made a detailed examination of the infection of *Cucurbita pepo* cotyledons by CMV to provide an assessment of the influence of virus infection on the rates of respiration and photosynthesis together with the activities of enzymes of primary metabolism and changes in the synthesis of soluble and insoluble carbohydrates, amino and organic acids and proteins (Figs 9.25 and 9.26).

A description of photosynthesis and starch accumulation in starch lesions is given in Section III.D above. On studying whole cotyledons, Técsi *et al.* (1994b) showed that the characteristic starch accumulation in early

stages of infection is reversed in later stages, a decline that is correlated with a reduced capacity for starch synthesis (ADP-glucose pyrophorylase) and a rise in the capacity for starch degradation (total starch hydrolase, starch phosphorylase). Newly assimilated carbon was lost at a lower rate from infected cotyledons and less was incorporated into structural carbohydrates, phosphorylated intermediates including organic acids, and more into soluble sugars, amino acids and proteins. Later in infection, there was a marked increase in respiratory capacity and a substantial alteration in carbohydrate metabolism, with the capacity being stimulated for the oxidative pentose–phosphate pathway, glycolysis, the tri-carboxylic acid cycle, anaplerotic reactions and oxidative electron transport. While there were no overall changes in photosynthetic rate, infection either reduced (rubisco and glycerate kinase) or did not affect (chloroplastic fructose *bis*-phosphate and hydroxypyruvate kinase) the capacities of the photosynthetic carbon reduction pathway or the photosynthetic carbon oxidative pathway.

G. Photosynthesis

In a tobacco mutant in which some islands of leaf tissue had no chlorophyll, TMV replication occurred in white leaf areas in the intact plant. However, replication did not occur if the white tissue was detached and floated on water immediately after inoculation (Matthews, 1991). Detached white tissue supplied with glucose supported TMV replication, indicating that the process of photosynthesis itself is not necessary for replication of this virus. Nevertheless, virus infection usually affects the process of photosynthesis. Reduction in carbon fixation is the most commonly reported effect in leaves showing mosaic or yellows diseases. This reduction usually becomes detectable some days after infection.

Photosynthetic activity can be reduced by changes in chloroplast structure, by reduced content of photosynthetic pigments or rubisco, or by reduction in specific proteins associated with the particles of photosystem II (Naidu

A

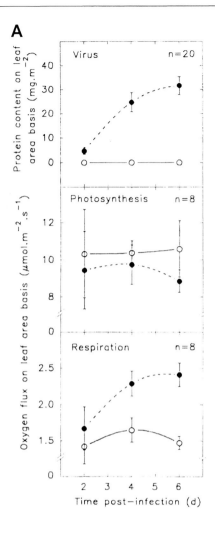

B

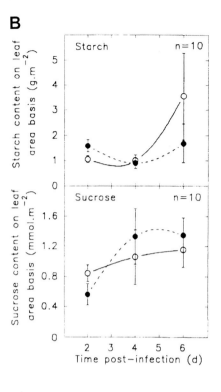

Fig. 9.25 Effects of CMV infection on the metabolism of marrow (*C. pepo*) cotyledons. **(A)** Virus content, photosynthetic oxygen evolution and respiratory oxygen uptake. **(B)** Starch and sucrose contents (cont.).

et al., 1986). However, such changes appear to be secondary, occurring some time after infection when much virus synthesis has already taken place. In tobacco plants infected with various strains of TMV, electron transport rates were reduced when loss of chlorophyll occurred. In inoculated leaves, photosystem II appeared to be irreversibly damaged in inoculated leaves even when no macroscopic symptoms were apparent (van Kooten *et al.*, 1990). A variety of effects of localized and systemic TMV infection in tobacco were observed in experiments with isolated chloroplasts. However, some enzyme activities were little affected (Montalbini and Lupattelli, 1989).

Some effects on photosynthesis are known that appear to be closely linked in time to the early period of maximum virus production. In chloroplasts isolated from Chinese cabbage leaves infected with TYMV, the Hill reaction and cyclic and non-cyclic photophosphorylation were all increased compared with healthy leaves during the phase of active virus multiplication (on an equal chlorophyll basis) (Goffeau and Bové, 1965). At a late stage of infection, photosynthetic activity was lower than in controls measured on chloroplasts isolated from whole plants. In young Chinese cabbage leaves infected with TYMV, there was a substantial diversion of the products of photosynthetic carbon fixation away from sugars and into organic

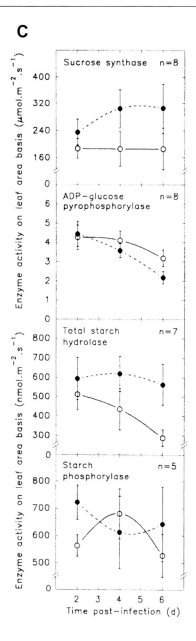

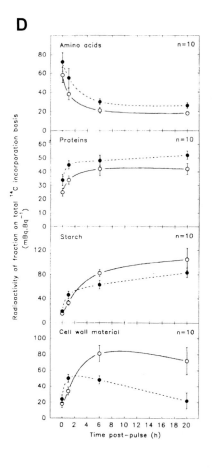

Fig. 9.25 cont. Effects of CMV infection on the metabolism of marrow cotyledons. **(C)** Maximal catalytic activities of enzymes involved in sucrose metabolism (sucrose synthase) and starch metabolism (ADP-glucose pyrophosphorylase, total starch hydrolase, starch phosphorylase). **(D)** $^{14}CO_2$ incorporation into amino acids, proteins, starch and cell-wall material in a pulse-chase experiment. ○ uninfected cotyledon; ● infected cotyledon; bars on graphs indicate confidence intervals at 90% level, n denotes the number of replicates of separate experiments. From Técsi *et al.* (1994b), with permission.

acids and amino acids. This change was most marked during the period of rapid virus increase and returned to the normal pattern when virus replication was near completion (Bedbrook and Matthews, 1973). An increase in

the activity of the enzymes phosphoenolpyruvate carboxylase and aspartate aminotransferase followed a similar time course.

Magyarosy *et al.* (1973) found a similar shift from the production of sugars to amino acids

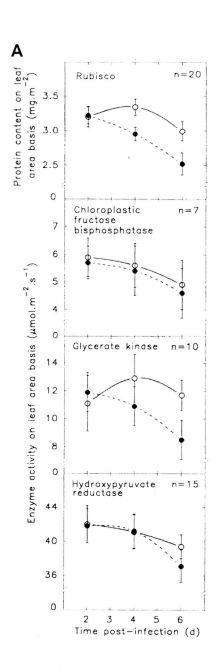

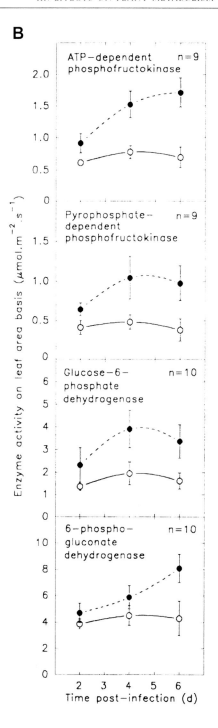

Fig. 9.26 Further effects of CMV infection on the metabolism of marrow cotyledons. **(A)** Capacity for photosynthetic carbon reduction pathway (rubisco protein content, maximal catalytic activity of chloroplastic fructose bisphosphatase) and carbon oxidation pathway (activities of glycerate kinase and hydroxypyruvate reductase). **(B)** Maximal catalytic activities of enzymes involved in glycolysis (ATP-dependent and pyrophosphate-dependent phosphofructokinases) and oxidative pentose-phosphate pathway (glucose-6-phosphate dehydro-genase, 6-phosphogluconate dehydrogenase) (cont.).

C

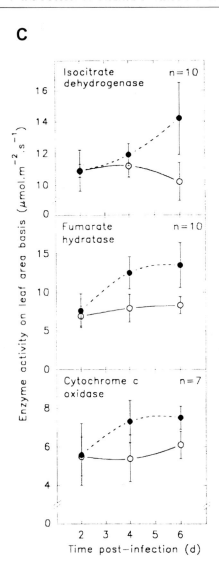

D

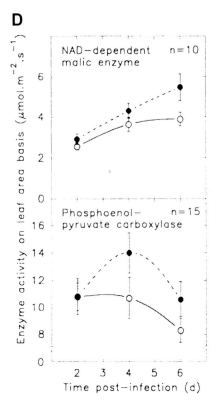

Fig. 9.26 cont. Further effects of CMV infection on the metabolism of marrow cotyledons. (C) Maximal catalytic activities of enzymes involved in tricarboxylic acid cycle (isocitrate dehydrogenase, fumarase hydratase) and in oxidative electron transport (cytochrome c oxidase). (D) Maximal catalytic activities of enzymes of anaplerotic reactions for tricarboxylic acid cycle (NAD-dependent malic enzyme, phosphoenolpyruvate carboxylase) (cont.).

and organic acids in squash plants systemically infected with SqMV. They isolated chloroplasts from healthy and diseased leaves and showed that both produced a similar pattern of carbon fixation products and that the total carbon fixed was about the same. They concluded that the virus-induced production of amino acids was taking place in the cytoplasm.

Changes in photosynthesis in starch lesions in cotyledons of *C. pepo* infected with CMV are described in Section III.D and those in whole cotyledons in Section III.F above.

The effects of infection by TMV (orchid strain) and CymMV on the crassulacean acid metabolism (CAM) in two orchid species are described by Izaguirre-Mayoral *et al.* (1993). In both species, TMV significantly reduced CAM activity in leaves, and in one species CymMV almost completely inhibited the diurnal change in titratable acidity. These changes were correlated with ultrastructural changes in chloroplasts of virus-infected leaves.

In summary, during the period of rapid replication, virus infection may cause a diversion of the early products of carbon fixation away from sugars and into pathways that lead more directly to the production of building blocks for the synthesis of nucleic acids

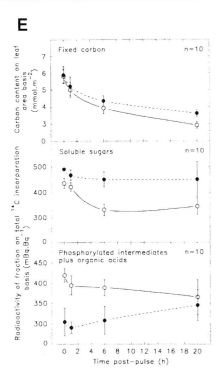

E

Fig. 9.26 cont. Further effects of CMV infection on the metabolism of marrow cotyledons. **(E)** Changes of total fixed carbon and $^{14}CO_2$ into soluble sugars and phosphorylated intermediates plus organic acids in a pulse-chase experiment. ○ uninfected cotyledon; ● infected cotyledon; bars on graphs indicate confidence intervals at 90% level, *n* denotes the number of replicates of separate experiments. From Técsi *et al.* (1994b), with permission.

and proteins. The most general result of virus infection is a reduction in photosynthetic activity but, as shown in CMV infection of *C. pepo* cotyledons, there may be no overall effect in relatively early stages of infection. Any reduction in photosynthesis is likely to arise from a variety of biochemical and physical changes. The relative importance of different factors varies with the disease.

H. Transpiration

In chronically virus-infected leaves, transpiration rate and water content have been found to be generally lower than in corresponding healthy tissues. The reported

effects over the first 1–2 weeks after inoculation vary. Results are difficult to compare and interpret because different viruses and host species have been used together with different conditions of growth and different tissue sampling procedures.

Bedbrook (1972) used the cobalt chloride paper method (Stahl, 1894) to estimate relative transpiration rates and to give a measure of stomatal opening. He compared, in intact Chinese cabbage plants, dark green islands in leaves showing mosaic patterns due to TYMV infection and various islands of tissue fully invaded by the virus. In darkness or low light intensity, stomata in darker green and pale green islands were closed, while those in islands of more severely affected lamina were open. In plants that had been held in full daylight, the dark and pale green islands were transpiring rapidly. Transpiration from severely affected islands was much less. These and other experiments showed that TYMV infection lowers the responsiveness of the stomata to changes in light intensity, the lowered response being most marked with strains causing the greatest reduction in chlorophyll. Because of diminished transpiration, the temperature of sugarbeet leaves in susceptible plants infected with BNYVV was 2–3°C higher than that of a tolerant variety (Keller *et al.*, 1989).

I. Activities of specific enzymes

Much of the work dealing with the effect of virus infection on specific enzymes is difficult to interpret for the following reasons:

1. Where differences have been found, it has usually been assumed that virus infection alters the amount of enzyme present and little consideration has been given to the possibility that infection may affect enzyme activities through changes in the amount of enzyme inhibitors or activators released when cells are disrupted.
2. The difficulty of deciding on an appropriate basis for expressing enzyme activity has often been ignored.
3. Much of the work was done before the

widespread existence of isoenzymes was recognized.

4. In most cases enzymatic activities were not studied in relation to related or associated activities.

There have been various studies involving the use of polyacrylamide gel electrophoresis to fractionate and assay isoenzymes and to study the consequences of virus infection on these patterns. It is relatively easy to generate data by this means. It is much less easy to provide meaningful interpretations of any observed changes. There are several reasons for these difficulties:

1. In the healthy plant there may be a continually changing developmental sequence of isoenzymes (Denna and Alexander, 1975).
2. Electrophoretically distinct isoenzymes may be determined genetically or may be different conformational forms derived by post-translational modification from the same primary structure.
3. The pattern of isoenzymes in the normal host may differ in closely related genotypes, and the effects of virus infection may differ with these (e.g. Eşanu and Dumitrescu, 1971).
4. Isoenzymes may be distributed in several subcellular sites. For example, a different set of peroxidase isoenzymes was associated with the cell wall and with the soluble fraction in extracts of normal maize root tips (Parish, 1975). Virus infection may affect various sites in different ways.
5. Virus-induced cell death may lead to changes in isoenzyme patterns that do not differ significantly from those induced by entirely unrelated causes of necrosis.
6. The observed effect of virus infection may depend on the substrate used for isoenzyme assay.
7. Aggregation states (e.g. monomer–dimer) may affect the kinetic properties of an enzyme, e.g. aspartate aminotransferase (Melander, 1975).

Integrated studies of the effect of CMV infection on enzymatic and other activities in *C. pepo* cotyledons are described in Sections III.D and III.F. Virus infection can also affect the expression of host genes encoding enzymes (Section III.B.3).

Enzymes involved in the phenomenon of the hypersensitive response and acquired resistance are discussed in Chapter 10 (Sections III.C, III.F and III.G).

J. Hormones

There is little doubt that virus infection influences hormone activities in infected plants and that hormones play some part in the induction of disease. This aspect is discussed in Section IV.B.1. Quantitative effects of infection on concentration have been shown for all the major groups of plant hormones (reviewed by Fraser, 1987b). Systemic virus infections tend to impair the steady-state auxin levels with consequent morphological alterations (Pennazio and Roggero, 1996). Usually the auxin level is reduced but a substantial increase in auxin activity has sometimes been observed associated with severe symptoms. Treatment of virus-infected plants with auxins may inhibit virus replication, leading to a reduction in symptom severity. Gibberellin concentrations are often reduced on virus infection whereas those of abscisic acid are increased. Cytokinin activity may be either reduced or increased (Pennazio and Roggero, 1998). Stimulation of ethylene production is associated with necrotic or chlorotic local responses.

Hormones are also involved in the hypersensitive response which is discussed in Chapter 10 (Section III.G).

K. Low-molecular-weight compounds

There are numerous reports on the effects of virus infection on concentration of low-molecular-weight compounds in various parts of virus-infected plants. The analyzes give rise to large amounts of data, which vary with different hosts and viruses, and which are impossible to interpret in relation to virus replication. In this section, some of these effects will be briefly noted.

1. Amino acids and related compounds

The most consistent change observed has been an increase in one or both of the amides, glutamine and asparagine. The amino acid, pipecolic acid, has been reported to occur in relatively high concentrations in several virus-infected tissues (e.g. Welkie *et al.*, 1967). A general deficiency in soluble nitrogen compounds compared with healthy leaves may occur during periods of rapid virus synthesis.

2. Compounds containing phosphorus

Phosphorus is a vital component of all viruses and as such may come to represent about one-fifth of the total phosphorus in the leaf. In spite of this, we still have no clear picture for any virus of the source of virus phosphorus, or the effects of infection on host phosphorus metabolism.

In Chinese cabbage leaves infected with TYMV, sampled 12–20 days after inoculation, a rise in virus phosphorus was accompanied by a corresponding fall in non-virus-insoluble phosphorus, suggesting that this virus uses phosphorus at the expense of (but not necessarily directly from) some insoluble source of phosphate in the leaf (Matthews *et al.*, 1963).

3. Leaf pigments

Virus infection frequently involves yellow mosaic mottling or a generalized yellowing of the leaves. Such changes are obviously due to a reduction in leaf pigments. Many workers have measured the effects of virus infection on the amounts of pigments in leaves. Frequently it appears to involve a loss of the chlorophylls, giving the yellowish coloration due to carotene and xanthophyll, but the latter pigments are also decreased in some diseases. Changes in chloroplast pigments are probably often secondary changes, because many viruses appear to multiply and accumulate in other parts of the cell, and closely related strains of the same virus may have markedly different effects on chloroplast pigments even though they multiply to the same extent.

The reduction in amount of leaf pigments can be due either to an inhibition of chloroplast development or to the destruction of pigments in mature chloroplasts. The first effect probably predominates in young leaves that are developing as virus infection proceeds. The rapidly developing chlorosis frequently observed in local lesions when mature green leaves are inoculated with a virus must be due to destruction of pigments already present. In systemically infected leaves, TYMV reduced the concentration of all six photosynthetic pigments to a similar extent. This was due to a cessation of net synthesis, and subsequent dilution by leaf expansion (Crosbie and Matthews, 1974a).

Dark green islands of tissue in Chinese cabbage leaves showing mosaic symptoms had essentially normal concentrations of pigments (Crosbie and Matthews, 1974a). Small leaves near the apex of large Chinese cabbage plants are shielded from light and contain little chlorophyll. When such cream-colored leaves, about 2 cm long, were excised and exposed to light, those from healthy plants became uniformly dark green, while those from TYMV-infected plants developed a prominent mosaic pattern of dark green islands and yellow areas within 24 hours. Thus, in young expanding leaves chlorophyll synthesis is inhibited in those islands of tissue in which TYMV is replicating.

Not all viruses cause a loss of chlorophyll. Infection of turnip with the Bari 1 strain of CaMV is characterized by dark green symptoms in contrast to the severe mosaic and chlorosis caused by strains such as Cabb B-JI (Stratford and Covey, 1988). Analysis of the chlorophyll content showed that leaves infected with Bari 1 had 2.55 ± 0.22 mg/g fresh weight; those infected with Cabb B-JI had 0.88 ± 0.04 mg/g; and the uninoculated controls had 2.12 ± 0.32 mg/g. Thus, plants infected with Bari 1 had 1.2-fold more chlorophyll than the uninoculated control plants.

4. Flower pigments

In view of the work that has been done on the genetics and biochemistry of normal flower coloration, surprisingly little is known about the biochemistry of the flower color-breaking process, which is such a conspicuous feature of many virus diseases. In tobacco plants infected with TMV, the normal pink color of the petals may be broken by white stripes or sectors. We have found the virus present only in the white areas. However, presence or absence of virus may not be the only cause for color breaks. In sweet peas (*Lathrus odoratus*) infected with what was presumably a single virus—BYMV or PEMV—a pale pink flower sometimes became flecked with both darker pink and white areas.

Virus infection usually appears to affect only the vacuolar anthocyanin pigments. The pigments residing in chromoplasts may not be affected. For example, the brown wallflower (*Cheiranthus cheiri*), which contains an anthocyanin, cyanin and a yellow plastid pigment (Gairdner, 1936), breaks to a yellow color when infected by TuMV (Plate 3.12). A preliminary chromatographic examination of broken and normal parts of petals infected with several viruses showed that the absence of color was due to the absence of particular pigments rather than to other factors, such as change in pH within the vacuole (Matthews, 1991).

Kruckelmann and Seyffert (1970) examined the effect of TuMV infection on several genotypes of *Matthiola incana*. Infection brought about both white stripes and pigment intensification. Observations on a set of known host genotypes have shown that virus infection affected only the activities of genes controlling the quantities of pigments produced. It appeared to have no effect on the activities of genes modifying anthocyanin structure.

L. Summary

The physiological and biochemical changes most commonly found in virus-infected plants are: (1) a decrease in rate of photosynthesis, often associated with a decrease in photosynthetic pigments, chloroplast ribosomes, and rbcs; (2) an increase in respiratory rate; (3) an increase in the activity of certain enzymes, particularly polyphenoloxidases; and (4) decreased or increased activity of plant growth regulators.

There is no reason to suppose that major disturbances of host plant metabolism are necessarily determined by the major processes directly concerned with virus replication. Some

minor initial effect of virus invasion and replication may lead to profound secondary changes in the host cell. Such changes may obscure important primary effects even at an early stage after infection.

Many of the changes in host plant metabolism noted earlier are probably secondary consequences of virus infection and not essential for virus replication. A single gene change in the host, or a single mutation in the virus, may change an almost symptomless infection into a severe disease. Furthermore, metabolic changes induced by virus infection are often non-specific. Similar changes may occur in disease caused by cellular pathogens or following mechanical or chemical injury. In many virus diseases, the general pattern of metabolic change appears to resemble an accelerated aging process. Because of these similarities, rapid diagnostic procedures based on altered chemical composition of the virus-infected plant must be used with considerable caution.

IV. PROCESSES INVOLVED IN SYMPTOM INDUCTION

In this section, I discuss some of the processes involved in the induction of disease. At the present time we are unable to implicate specific viral- or host-coded proteins with the initiation of any of these processes.

A. Sequestration of raw materials

The diversion of supplies of raw materials into virus production, thus making host cells deficient in some respect, is an obvious mechanism by which a virus could induce disease symptoms. This mechanism is very probably a factor when the host plant is under nutritional stress. For example, in mildly nitrogen-deficient Chinese cabbage plants the local lesions produced by TYMV have a purple halo, the purple coloration being characteristic of nitrogen starvation (Diener and Jenifer, 1964). Specific nutritional stress can also be induced by environmental conditions. For example, Ketellapper (1963) showed that adverse effects of unfavorable temperature on plant growth could be partly or completely prevented by providing the plants with some essential metabolites which, applied under normal temperature conditions, had no effect. The effects were to some extent species-specific. Thus, at least part of the effect of adverse temperature on plant growth may be due to deficiencies in specific metabolites. Virus infection may well aggravate such deficiencies.

Sometimes an increase in severity of symptoms is associated with increased virus production. Thus, with BPMV in soybean, Gillaspie and Bancroft (1965) observed flushes of severe symptoms, followed by a recovery period at 1.5 and 4.5 weeks after inoculation. Assays on leaf extracts showed that highest infectivity per unit of tissue occurred at the first symptom flush. Infectivity then fell, but rose again during the second symptom flush. However, in well-nourished plants there is no general correlation between amount of virus produced and severity of disease (Palomar and Brakke, 1976). Similarly there is not necessarily a correlation between the severity of macroscopic disease and various physiological and biochemical changes brought about by infection (Ziemiecki and Wood, 1976), although some correlations of this sort have been found.

Using 26 different strains of TMV in tobacco and tomato, Fraser *et al.* (1986) showed by multiple regression analysis that variation in virus multiplication and symptom severity could account for most of the variation of growth in tobacco. The same situation did not apply in tomato. Furthermore, it is in the nature of such experiments that they cannot demonstrate a relationship of cause and effect.

Except under conditions of specific pre-existing nutritional stress, as indicated, it is unlikely that the actual sequestration of amino acids and other materials into virus particles has any direct connection with the induction of symptoms. The following considerations support this view:

1. Viruses are made up of the same building blocks in roughly the same proportions as are found in the cell's proteins and nucleic acids.

Even with viruses, such as TMV and TYMV, which reach a relatively high concentration in the diseased tissue (>1.0 mg/g fresh weight of lamina), the amount of virus formed may be quite small relative to the reduction in other macromolecules caused by infection. In Chinese cabbage plants infected with TYMV, the reduction in normal proteins and ribosomes was more than 20 times as great as the amount of virus produced (Crosbie and Matthews, 1974b).

2. Closely related strains of the same virus may multiply in a particular host to give the same final concentration of virus, and yet have markedly different effects on host cell constituents.

3. The type strain of TMV multiplying in White Burley tobacco produced chlorotic lesions at 35°C but none at 20°C. About one-tenth as much virus is made at 35°C as at 20°C (Kassanis and Bastow, 1971).

4. A single gene change in the tobacco plant may result in a change from the typical mosaic disease produced by TMV to the hypersensitive reaction. F_1 hybrids between the two genetic plant types may respond to TMV with a lethal systemic necrotic disease with greatly reduced virus production.

5. The changes in metabolism of *Cucurbita pepo* cotyledons on infection with CMV could not be related directly to the diversion of carbon from normal metabolism into the synthesis of viral components (Técsi *et al.*, 1994b).

In the discussion so far I have been considering the amount of virus produced as measured in tissue extracts. However, the rate of virus replication in individual cells could be an important factor in influencing the course of events. Very high demand for key amino acids or other materials over a very short period, perhaps of a few hours, could lead to irreversible changes with major long-term effects on the cell, and subsequently on tissues and organs. Another consideration is the possibility that the 'switching off' of host genes at the infection front of PSbMV in pea cotyledons (see Section II.B.3) is a general phenomenon and that this could lead to depletion of key proteins at an important time. However, there is no unequivocal experimental evidence for such effects, measured on individual cells.

B. Effects on growth

1. Stunting

There appear to be three biochemical mechanisms by which virus infection could cause stunting of growth: (1) a change in the activity of growth hormones, (2) a reduction in the availability of the products of carbon fixation, and (3) a reduction in uptake of nutrients.

a. Changes in growth hormones

There is little doubt that a virus-induced change in hormone concentration is one of the ways in which virus infection causes stunting.

It should be remembered that plant hormones have been defined chemically and biologically in controlled growth tests on excised tissues or organs. In the intact plant, however, each of the groups of hormones induced many growth and physiological effects. Their functions overlap to some extent and their interactions are complex. For a given process, their effects may be similar, synergistic or antagonistic. In the intact plant, members of all or most of the groups are involved in any particular developmental process. There are many possible ways in which virus infection could influence plant growth by increasing or decreasing the synthesis, translocation or effectiveness of these various hormones in different organs and at different stages of growth. Thus, it is not surprising that our understanding of the interactions between viruses and hormones is extremely sketchy.

In most situations, virus infection decreases the concentration of auxin and gibberellin and increases that of abscisic acid. Ethylene production is stimulated by virus-induced necrosis and by development of chlorotic local lesions, but not where virus moves systemically without necrosis. Reported effects on cytokinins are variable (Fraser and Whenham, 1982; Whenham, 1989). Three examples of studies on virus-hormone interactions will be considered.

Ethylene production was enhanced in cucumber cotyledons infected with CMV, and the increase began just before hypocotyl elongation rate slowed down (Marco et al., 1976). Artificially reducing the ethylene content enhanced elongation of infected seedlings, but not as much as it did healthy seedlings, suggesting that some other factors were involved. This was borne out by further work showing that suppression of hypocotyl elongation was also accompanied by a reduction in gibberellin-like substances and an increase in abscisic acid (Aharoni et al., 1977). This work illustrates well the complex effects of infection on hormone balance and the difficulty in establishing relationships of cause and effect.

Whenham et al. (1985) found that TMV infection of tobacco increased the concentration of abscisic acid outside the chloroplasts, while having little effect on the amount of the hormone inside the chloroplasts. The increase in abscisic acid concentration was strongly and positively correlated with increasing severity of disease for six strains of TMV. However, infection of tomatoes with TMV did not increase abscisic acid (Fraser and Whenham, 1989).

Fraser and Matthews (1981) found that inoculation of the cotyledons of small Chinese cabbage seedlings with TYMV was followed by a rapid transient inhibition of leaf initiation in the shoot apex, long before infectious virus moved out of the inoculated leaves. The idea that abscisic acid might be responsible for this effect was supported by the fact that it could be reproduced by applying the acid in a single dose to the cotyledon leaves. The magnitude of the effect was independent of acid concentration over the range 10^{-5} to 3×10^{-9} M (Fraser and Matthews, 1983). Further support for the involvement of abscisic acid came from the following experiment. When the petioles of freshly inoculated Chinese cabbage leaves are placed in water, a substance diffuses from the cut ends that can mimic the rapid effects of virus inoculation on leaf initiation. This effect of the eluate can be abolished by the addition of gibberellic acid (GA_3 at 10^{-4} M) (Fraser et al., 1984). Thus, it seems probable that virus inoculation brings about a rapid change in the balance of abscisic acid and GA_3 in the apical dome.

b. Reduction in the availability of fixed carbon

Apart from any effects on hormone balance, plants will become stunted (on a dry weight basis, at least) if the availability of carbon fixed in photosynthesis is limiting. A reduction in available fixed carbon could be brought about in several ways.

i. Direct effects on the photosynthetic apparatus

This is the most obvious and perhaps the most common way by which infection reduces plant size.

Infection of Chinese cabbage leaves with TYMV reduced the three major chloroplast components, chlorophyll *a*, rubisco and 68S ribosomes, to about the same extent on a per-plant basis (Crosbie and Matthews, 1974b). These changes took place after most of the virus replication was completed, and reduction in the first two began before the ribosomes were affected.

The initial events that lead to reduced carbon fixation in the chloroplasts are not known for any host–virus combination. Hormones may play a role in the initiation of chlorophyll degradation. In leaves of *Tetragona expansa* inoculated with BYMV, chlorotic local lesions developed, and shortly after their appearance there was a substantial increase in ethylene release from the leaves (Gáborjányi *et al.*, 1971). In cucumber cotyledons infected with CMV, Marco and Levy (1979) showed that ethylene is produced before local lesion appearance. They were able to delay the appearance of chlorotic local lesions by removing the ethylene by hypobaric ventilation.

ii. Starch accumulation in chloroplasts

The accumulation of starch in chloroplasts commonly seen in virus infections must deprive the growing parts of the plant of some newly fixed carbon. This accumulation may be due to reduced permeability of the chloroplast membrane or to changes in enzyme activities within the chloroplast.

iii. Stomatal opening

Lowered photosynthesis in yellows-infected sugarbeet could be accounted for in part by a virus-induced reduction in stomatal opening (Hall and Loomis, 1972). The reduced responsive-ness of stomata to changes in light intensity might be a factor limiting carbon fixation in TYMV-infected leaves during the earlier part of the day.

iv. Translocation of fixed carbon

Any effect of virus infection such as necrosis that reduces the efficiency of phloem tissues must limit translocation of fixed carbon from mature leaves to growing tissues. However, reduced permeability of leaf cells to the migration of sugars into the phloem may be more commonly the limiting factor.

v. Leaf posture

Leaf posture may affect the overall efficiency of photosynthesis in the field. For example, a wheat variety with an erect habit fixed more CO_2 than one with a lax posture (Austin *et al.*, 1976). Virus infection can affect growth habit, but no studies have been made of this factor in relation to carbon fixation per plant.

c. Reduced leaf initiation

Reduction in leaf numbers is a very small factor in the stunting of herbaceous plants following virus infection. For example, 100 days after infection, tobacco plants infected with TMV possessed about one leaf less than healthy plants (Takahashi, 1972b). Careful analysis of leaf initiation following inoculation of very young Chinese cabbage plants with TYMV has shown that there is a rapid transient reduction in the rate of leaf initiation. The rate of leaf initiation is reduced to about one-third to one-quarter that of control plants for a period of 1–2 days. The rate of initiation in inoculated plants then recovers. Plants then have approximately one leaf less than healthy plants for the rest of their life (Fraser and Matthews, 1981). The same effect is produced by TuMV and by CaMV in Chinese cabbage. Thus, the effect may be a general one.

d. Reduction in uptake of nutrients

Very little experimental work has been done on the effects of virus infection on the capacity of roots to take up mineral nutrients from the soil and to transport them to other parts of the plant. However, nitrogen fixation may be adversely affected. Reduction in nodulation is described in Chapter 3 (Section II.B.10). TRSV may impair growth of soybean, at least in part by suppressing leghemoglobin synthesis and nitrogen fixation during the early stages of growth (Orellana et al., 1978). Nodulation of soybean plants by *Rhizobium* was reduced by infection with SMV, BPMV, or both viruses. Greatest reduction (about 80%) occurred when plants were inoculated with virus at an early stage of growth (Tu et al., 1970). The first phase in the infection of soybean root cells by *Rhizobium* (i.e. development of an infection thread and release of rhizobia into the cytoplasm) appears not to be affected by infection of the plant with SMV (Tu, 1973). In the second stage of infection, a membrane envelope forms around the bacterial cell to form a bacteroid. Structural differences such as decreased vesiculation of this membrane envelope were observed in virus-infected roots (Tu, 1977). Similarly, there was a decrease in nodulation of *P. vulgaris* on infection with BCMV and BYMV, which corresponded to symptom severity (Tu, 1997). On the other hand, infection of beans with BRMV, of *Medicago truncatula* and *Lupinus angustifolius* with CMV and of *M. polymorpha* with AMV did not affect nodule number but reduced nodule biomass (Wahyuni and Randles, 1993; Wroth et al., 1993; Izaguirre-Mayoral et al., 1994). In the case of seed-transmitted AMV in *M. polymorpha*, a reduction in nodule mass of 23% led to a decrease of 21% in shoot dry weight and of 24% in fixed nitrogen at 53 days post-germination (Wroth et al., 1993). Nodulated *M. truncatula* and *L. angustifolius* were less susceptible to CMV than were un-nodulated plants; however, un-nodulated plants which had been given nitrogen were more susceptible than those which had no additional nitrogen (Wahani and Randles, 1993).

e. Growth analysis of the whole plant

Analysis of the stunting process in a herbaceous plant during the development of a mosaic disease is an extremely complex problem, which has not been adequately investigated. A beginning was made with Chinese cabbage plants infected with a 'white' strain of TYMV. Such strains cause a very marked

reduction in chlorophyll content in diseased tissue (Crosbie and Matthews, 1974b).

Healthy plant growth was approximately logarithmic over the period studied (as was the amount of chlorophyll per plant). Virus infection caused a virtual cessation of chlorophyll *a* production for several weeks. Growth of the diseased plant was almost linear over this period. It was assumed that this was because chlorophyll content increased very little for several weeks.

The diseased plant appeared to respond to the reduction in chlorophyll in two ways. First, a higher proportion of the material available for growth was used to make leaf lamina rather than midrib, petiole, stem or roots. Secondly, the proportion of dark green island tissue increased in leaves formed later, so that chlorophyll in these islands came to form a significant proportion of the total.

The 'white' strain of TYMV used in these experiments causes a much more severe disease than mild 'pale green' strains of the virus when judged by eye inspection. Closer analysis showed there was little difference in total fresh weight of aerial parts or of roots between plants infected with the two kinds of strain. The gross differences were mainly due to: (1) different distribution of total chlorophyll within the plant, and (2) a more extreme alteration in growth form with the severe strain (i.e. reduction in stem, petiole and midrib). Takahashi (1972a,b) studied the effect of TMV infection on tobacco leaf size and shape and internode length, but he did not relate the effects to chlorophyll loss.

2. Epinasty and leaf abscission

The experiments of Ross and Williamson (1951) indicated that the epinastic response and leaf abscission in *Physalis floridiana* infected with PVY were associated with the evolution of a physiologically active gas, which was probably ethylene.

Beginning about 48 hours after inoculation of cucumber cotyledons with CMV, there was a decrease in ethylene emanation compared with controls (Levy and Marco, 1976). Over the same period, there was an increase in the internal concentration of ethylene. Infection gave rise to a marked epinastic response in the cotyledons. The following evidence strongly suggests that the epinasty was due to the increased ethylene concentration:

1. Ethylene reached a peak concentration before the epinastic response began.
2. Treatment of healthy cucumbers with ethylene caused epinasty.
3. Epinasty could be prevented in infected cotyledons by hypobaric ventilation, which facilitates removal of the internal ethylene.

This is the best evidence showing that a hormone is directly involved in the production of symptoms induced by virus infection. How virus infection stimulates ethylene production and how the hormone produces its effects are not known.

3. Abnormal growth

a. Virus-induced tumors

All the plant viruses belonging to the *Reoviridae*, except RDV, induce galls or tumors in their plant hosts (see Chapter 3, Section III.C), but not in the insect vectors in which they also multiply.

There is a clear organ or tissue specificity for the different viruses. For example, tumors caused by WTV predominate on roots and, to a lesser extent, stems. FDV causes neoplastic growths on veins of stems and leaves. Thus, we can be reasonably certain that some function of the viral genome induces tumor formation, but we are quite ignorant as to how this is brought about. Wounding plays an important role as an inducer or promoter of tumors caused by WTV. Hormones released on wounding may play some part in this process. Microscopic tumor initials, which normally do not develop into macroscopic tumors, occur in the stem apices of sweet clover infected with WTV. Application of IAA stimulates the growth of these tumors to macroscopic size (Black, 1972).

Some viruses, such as PEMV and curtoviruses, cause enations on leaves (see Chapter 3, Section II.B.7). Little is known about the processes involved in their induction.

b. Distortion of tissues

In leaves showing mosaic disease, the dark green islands of tissue frequently show blistering or distortion. This is due to the reduced cell size in the surrounding tissues, and the reduced size of the leaf as a whole. The cells in the dark green island are much less affected and may not have room to expand in the plane of the lamina. The lamina then becomes convex or concave to accommodate this expansion. Other effects on leaf shape are discussed in Chapter 3 (Section II.B).

C. Effects on chloroplasts

1. Chlorotic local lesions

Marco and Levy (1979) have suggested that the chlorotic local lesions formed in cucumber cotyledons following inoculation with CMV may be caused by the virus-induced release of endogenous ethylene. They found a substantial rise in ethylene production in the few hours before local lesions appeared. Furthermore, application of ethylene caused yellowing of the leaves, but suppression of ethylene production delayed lesion appearance.

2. Structural changes

TYMV infection in Chinese cabbage provides particularly favorable material for the study of virus-induced structural changes of chloroplasts. This is because virus-induced clumping of chloroplasts and some other changes such as the formation of large vesicles (sickling) and fragmentation can be monitored in large numbers of cells by light microscopy. Protoplasts isolated from leaves inoculated with TYMV can be used to study the effects of various factors on these virus-induced structural changes.

For example, in the sickling process a large clear vesicle appears in chloroplasts of an infected protoplast, the chlorophyll-bearing structures being confined to a crescent-shaped fraction of the chloroplast volume. The vesicle is bounded by a membrane that appears to arise from stroma lamellae (Fraser and Matthews, 1979b). Red and blue light are equally effective inducers of sickling, which does not occur in the dark (Matthews and Sarkar, 1976). At 6–8 days after inoculation of leaves, very few protoplasts show sickling when freshly isolated, although the light intensity received by the leaves is more than adequate to induce sickling in isolated protoplasts. Thus, some factor partially represses the sickling process in intact leaves.

D. Mosaic symptoms

Mosaic patterns are described in Chapter 3 (Section II.B.2). Mosaics develop only in sink leaves to which the virus has spread by long-distance movement. Here I shall describe various factors involved in mosaic symptoms. In Chapter 10 (Section III), I discuss aspects of the molecular biology of viruses involved in inducing mosaic symptoms and the current ideas on how the symptomatology develops.

1. Role of virus strains and dark green tissue in mosaic disease

In some infections, such as TMV in tobacco, the disease in individual plants appears to be produced largely by a single strain of the virus. However, it has been known for many years that occasional bright yellow islands of tissue in the mosaic contain different strains of the virus. Such strains probably arise by mutation and, during leaf development, come to exclude the original mild strain from a block of tissue. In Chinese cabbage plants infected with TYMV there may be many islands of tissue of slightly different color from which different strains of the virus can be isolated (Chalcroft and Matthews, 1967a,b).

Dark green islands of tissue that superficially appear normal occur in these two diseases and in many other mosaics. They are a prominent component of the mosaics and may be important in the growth of the diseased plant.

2. The structure of the mosaic symptom

a. Leaf age at time of infection

By inoculating plants at various stages of growth, it has been demonstrated for TYMV (Chalcroft and Matthews, 1966) and TMV

(Atkinson and Matthews, 1970; Takahashi, 1971) that the type of mosaic pattern developing in a leaf at a particular position on the plant depends not on its position but on its stage of development when infected by the virus. There is a critical leaf size at the time of infection above which mosaic disease does not develop. This critical size is about 1.5 cm (length) for tobacco leaves infected with TMV (Atkinson and Matthews, 1970; Gianinazzi et al., 1977). Symptoms of African cassava mosaic were determined early in leaf development and subsequently change little (Fargette et al., 1987).

b. Gradients in size of islands

Although the size of macroscopic islands in a mosaic are very variable, there tends to be a definite gradient up the plant. Leaves that were younger when systemically infected tend to have a mosaic made up of larger islands of tissue. Even within one leaf there may be a relationship between both the number and size of islands and the age of different parts of the lamina as determined by frequency of cell division.

c. Presence of virus, and the mosaic pattern in very young leaves

Some viruses can invade the apical dome (see Section II.J.8). However, the mosaic pattern is already laid in very small TYMV-infected Chinese cabbage leaves that have not yet developed significant amounts of chlorophyll (Matthews, 1973).

d. Patterns in the macroscopic mosaic

Patterns in the macroscopic mosaic are often so jumbled that an ontogenetic origin for them cannot be deduced. Occasionally patterns that are clearly derived from the apical initials have been observed. For example, Chinese cabbage leaves (or several successive leaves) have been observed to be divided about the midrib into two islands of tissue, one dark green and the other containing a uniform virus infection (Matthews, 1973). These observations have been taken to show that individual apical initials might have been infected either with virus or with the agent that induces dark green tissue.

e. The microscopic mosaic

The overall mosaic pattern may be made up of microscopic mosaics. In Chinese cabbage leaves infected with TYMV, the microscopic mosaic develops only on leaves that were less than about 1–22 mm long at the time of infection. In such leaves, areas in the mosaic pattern that macroscopically appear to be a uniform color may be found on microscopic examination to consist of mixed tissue in which different horizontal layers of the mesophyll have different chloroplast types. A wide variety of mixed tissues can be found. For example, in some areas, both palisade and the lower layers of the spongy tissue may consist of dark green tissue while the central zone of cells in the lamina is white or yellow–green (Fig. 9.27A). This situation may be reversed, with the central layer being dark green and the upper and lower layers consisting of chlorotic cells (Fig. 9.27B).

As seen in fresh leaf sections, these areas of horizontal layering may extend for several millimeters, or they may be quite small, grading down to islands of a few cells or even one cell of a different type. The junction between islands of dark green cells and abnormal cells in the microscopic mosaic is often very sharp.

Such arrangements are not confined to dark green versus virus-infected layers. The microscopic mosaic also includes cell layers infected by different strains of the virus with distinctive effects on chloroplasts.

3. Genetic control of mosaics

In spite of the fact that the most striking effects of TYMV infection in *Brassica* spp. are on the chloroplasts, the response of these organelles to infection is under some degree of nuclear control. Certain varieties of *Brassica rapa* respond to all the strains isolated from *B. pekinensis* with a mild diffuse mottle. In reciprocal crosses between *B. rapa* and *B. chinensis* all the progeny gave a *B. chinensis* type of response to the strains (Matthews, 1973).

4. The nature of dark green tissue

Dark green islands in the mosaic pattern are cytologically and biochemically normal as far as has been tested. They contain low or zero

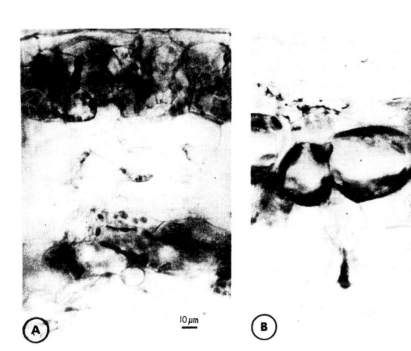

Fig. 9.27 Examples of layers in the microscopic mosaic in Chinese cabbage leaves infected with TYMV. **(A)** Upper and lower cell layers (LII) dark green; central zone (LIII), white; **(B)** Reverse of (A). From Chalcroft and Matthews (1967b), with permission.

amounts of infectious virus and no detectable viral protein or viral dsRNA. In pumpkin leaves showing mosaic disease caused by WMV, the cylindrical inclusion and amorphous inclusion proteins as well as viral coat protein were absent from dark green areas of the mosaic, or present in trace amounts (Suzuki et al., 1989).

Various lines of evidence show that dark green islands are resistant to superinfection with the same virus or closely related viruses (e.g. Chalcroft and Matthews, 1967a,b; Atkinson and Matthews, 1970; Loebenstein et al., 1977).

Various factors can influence the proportion of leaf tissue that develops as green islands in a mosaic. These include leaf age, strain of virus, season of the year, and removal of the lower leaves on the plant (Crosbie and Matthews, 1974b). The dark green islands of tissue may not persist in an essentially virus-free state for the life of the leaf. 'Breakdown' leading to virus replication usually takes place after a period of weeks, or after a sudden elevation in temperature (Atkinson and Matthews, 1970; Matthews, 1973; Loebenstein et al., 1977).

Various theories have been put forward to explain the nature of dark green islands. They certainly do not consist merely of tissue that has escaped infection. There is no evidence for the presence of very mild strains of virus in dark green islands. It is possible that the cells in dark green islands are occupied by defective strains, but such strains would have to produce little or no intact virus or viral antigen, and would have to replicate without detectable cytological effects. These cannot be ruled out at present, but on the other hand, there is no positive evidence for their presence.

Gera and Loebenstein (1988) isolated an inhibitor of virus multiplication from dark green island tissue of tobacco leaves infected with CMV. However, they do not appear to have assayed for the presence of inhibitors in the yellow virus-bearing tissues. The role of such inhibitors in dark green tissue remains to be established. When transgenic tobacco plants expressing the 30-kDa movement protein are infected with TMV, the mosaic disease which develops appears normal in all respects. Thus, we can assume that the 30-kDa protein is not involved in the formation or maintenance of dark green islands.

As discussed in Chapter 10, Section III, it is likely that at least some dark green islands are due to local host defense effects.

E. The role of membranes

Typical leaf cells are highly compartmentalized and there is an increasing awareness of the vital roles that membranes play in the functioning of normal cells. Virus replication involves the induction of new or modified membrane systems within infected cells (see Chapters 8 and 13). It is known that both animal and bacterial viruses alter the structure and permeability properties of host cell membranes. Infection of insect vector cells by plant viruses may alter the physical properties of the plasma membrane (Hsu, 1978).

Various virus-induced responses, some of which were discussed in earlier sections, indicate ways in which changes in membrane function may have far-reaching consequences for the plant. For example:

1. Virus infection may affect stomatal opening. Net photosynthesis may be limited by stomatal opening, which in turn depends on guard cell membrane function.
2. A different consequence of virus-induced stomatal closure is indicated by the work of Levy and Marco (1976). A virus-induced increase in stomatal resistance led to increased internal ethylene concentration followed by an epinastic response. The change in stomatal opening was probably induced by changes in properties of the guard cell plasma membranes.
3. Sucrose accumulation in infected leaves appears to depend on membrane permeability changes rather than altered phloem function.
4. Virus-induced starch accumulation in chloroplasts may be due in part at least to alterations in permeability of chloroplast membranes.

5. There is little doubt that ethylene is involved in the induction of some symptoms of virus disease. Ethylene has been found to cause a very rapid decrease in the incorporation of [1B^{14}C]glycerol into phospholipids of pea stems (Irvine and Osborne, 1973).

6. Living cells maintain an electrochemical potential difference across their plasma membrane, which is internally negative. There is an interdependence between this potential difference and ion transport across the membrane. Stack and Tattar (1978) showed that infection of *Vigna sinensis* cells by TRSV altered their transmembrane electropotentials.

Thus, a more detailed knowledge of the ways in which viruses alter the properties of membranes will be a prerequisite to our fuller understanding of disease induction.

V. DISCUSSION

There have been major advances in our understanding of how viruses move about plants, both from cell to cell and from the inoculated leaf to other parts of the plant. These studies have also stimulated more general investigations on the movement of macromolecules throughout the plant.

As described in Section II.D.5, five strategies have been proposed for cell-to-cell movement through plasmodesmata. However, recent results have blurred the distinctions between at least two of them. The MP of AMV has features in common with those of TMV and the '30K superfamily', yet it forms tubules on a temporary basis through plasmodesmata. There are still many questions to be answered about cell-to-cell movement of viruses, including:

1. How do MPs gate plasmodesmata open at the infection front?
2. Do MPs themselves interact with plasmodesmata proteins or other macromolecules, or are they involved in initiating a cascade of events leading to an increase in SEL?
3. What are the forms of infection unit that pass through plasmodesmata, and what are the

forces that move them from the infected to the uninfected cell?

4. How do plasmodesmata close or return to normal signaling control after the passage of the infection unit?
5. Do viruses pass through existing plasmodesmata, or are new ones formed? If they pass through existing plasmodesmata, are there certain ones amenable to gating?
6. What are the features of plasmodesmata that determine tissue specificity and symplastic domains?

There are also many questions to be answered about the details of long-distance transport, including:

1. What are the exact routes of movement from the phloem parenchyma to the sieve elements? Are there different routes specific for individual viruses or groups of viruses? If so, what are the factors that determine individual routes?
2. What is the sieve element route from the source leaf to the sink leaf? As noted in Section II.G.3, there is some evidence for switching from the external phloem from the source leaf to the internal phloem leading to the sink leaf. Does this occur for all systemically moving viruses in all plant species?
3. What are the forms in which the infection units are transported? Many viruses require coat protein for long-distance transport (see Section II.G.5), but do they form particles? If particles are seen, are these the unloaded form?
4. What are the factors that lead to the infection units being unloaded from the sieve elements?

The last two questions are difficult to answer, as it may be only a minority form of the infection unit that is unloaded and establishes systemic infection. Furthermore, for individual viruses there may be differences in the preferred unloading (and loading) form under different conditions.

In spite of many years work, relatively little is known about the detailed biochemistry of symptom production. Most of the information is essentially descriptive. However, with new techniques and approaches that are becoming available, it is likely that this will be a profitable area of research. As with the studies on virus movement, the understanding of the virus situation will help our understanding of the uninfected situation.

Study of the perturbation of the biochemical pathways leading to symptom production should increase our understanding of the controls in these pathways in the uninfected condition.

CHAPTER 10

Induction of Disease 2: Virus–Plant Interactions

I. INTRODUCTION

In various earlier chapters, I have summarized the present knowledge about viral replication and the symptoms, both macroscopic and microscopic, physiological and biochemical that are associated with this replication. Over the past 10 years, there have been major advances in the understanding of the molecular interactions between viruses and plants that I will examine in this chapter. Ultimately we would like to explain, in the terms of molecular biology and biochemistry, how a single virus particle containing genetic material sufficient to specify a few polypeptides can infect and cause disease in its host plant. This is a major task when we remember that the host plant is growing continually and is organized into a variety of tissues and organs with specific structures and functions (see Goldberg, 1988).

We are still far from attaining this ultimate objective for any host–virus combination. Nevertheless, the application of methods based on recombinant DNA technology for studying the role of viral gene products in disease induction is providing increasing amounts of information on the interactions involved in the full virus infection of a plant. Important procedures include the construction of infectious clones of viruses, site-directed mutagenesis of the viral genome, switching genes between viruses and virus strains, the construction of transgenic plants that express only one or a few viral genes, and the isolation and sequencing of genes from

Virus acronyms are given in Appendix 1.

host plants, especially *Arabidopsis thaliana*. These procedures are providing new information on the role of viral genes in disease induction and plant responses to them, but they have limitations, as discussed in the following sections.

Even more important are the recent conceptual advances in the interplay between the viral and host genomes. Essentially, the host is attempting to restrict the viral infection and the virus is attempting to overcome these restrictions. In the light of these new concepts, we are having to readjust some previous ideas. In this chapter, I discuss the interplay between the two genomes and various factors that can affect the outcome of this interplay.

II. DEFINITIONS AND TERMINOLOGY OF HOST RESPONSES TO INOCULATION

The term host is defined in Chapter 3 (Section V), but the terms for describing the various kinds of response made by plants to inoculation with a virus have been used in ambiguous and sometimes inconsistent ways. Thus, over many years, there has been confusion about the meaning of certain terms. Cooper and Jones (1983) discussed this problem and suggested a standardized usage. In particular, they introduced the term *infectible* to mean the opposite of *immune*.

In the third edition of *Plant Virology*, Matthews (1991) defined the relevant terms in the light of current knowledge (Table 10.1), taking account of the suggestion made by Cooper and Jones (1983). The use of some of

TABLE 10.1 Types of response by plants to inoculation with a virus

---

IMMUNE (non-host). Virus does not replicate in protoplasts, nor in cells of the intact plant, even in initially inoculated cells. Inoculum virus may be uncoated, but no progeny viral genomes are produced.

INFECTIBLE (host). Virus can infect and replicate in protoplasts.

　　Resistant (extreme hypersensitivity). Virus multiplication is limited to initially infected cells because of an ineffectual virus-coded movement protein, giving rise to *subliminal infection.*Plants are *field resistant.*

　　Resistant (hypersensitivity). Infection limited by a host response to a zone of cells around the initially infected cell, usually with the formation of visible necrotic local lesions. Plants are *field resistant.*

　　Susceptible (systemic movement and replication)
　　　　Sensitive. Plants react with more or less severe disease.
　　　　Tolerant. There is little or no apparent effect on the plant, giving rise to *latent* infection.

---

these terms differs from that in other branches of virology. For example, *latent* used in reference to bacterial or vertebrate viruses usually indicates that the viral genome is integrated into the host genome. There are very few plant viruses that integrate into the host genome, the integrant being activated to give an episomal infection (Chapter 8, Section IX.B.4), and these have been recognized only recently. Thus, we may have to reconsider the term 'latent' in the light of these findings.

It had been assumed for many years that virus–host cell interactions must involve specific recognition or lack of recognition between host and viral macromolecules. Interactions might involve activities of viral nucleic acids, or specific virus-coded proteins, or host proteins that are induced or repressed by viral infection. The effects of virus infection on host proteins are discussed in Chapter 9 (Section III.B.3). An increasing understanding of interactions involved in the induction of disease is accruing from the study of various forms of virus resistance and the realization of how similar some of these are to responses to attack by other microbes and by pests. These responses can involve specific host resistance (*R*) genes and/or generalized host defense systems.

A. R genes

In his studies on the inheritance of the resistance of flax (*Linum usitatissimum*) to the flax rust pathogen (*Melampsora lini*), Flor (1971) revealed the classical 'gene-for-gene' model that proposes that, for resistance to occur, complementary pairs of dominant genes, one in the host and the other in the pathogen, are required. A loss or alteration in the host resistance gene or in the pathogen avirulence (*Avr*) gene leads to disease or compatibility (Fig. 10.1 and Table 10.3). Over the past few years, there has been an increasing understanding of the interactions between the *R* and *Avr* genes. Basically, this interaction leads to both a local and a systemic signal cascade. The local signaling cascade triggers a host response that contains the pathogen infection to the primary site; the systemic cascade primes defense systems in other parts of the plant.

Over the past decade, numerous plant *R* genes have been identified, cloned and sequenced (for reviews see Hammond-Kosack and Jones, 1997; Richter and Ronald, 2000). Many of them share striking structural similarities and they are now placed into five groups based on their structures (Fig. 10.2). Many have a leucine-rich region (LRR) and a nucleotide-binding site (NBS). Other common features include a serine–threonine kinase domain, a leucine zipper (LZ) or structures found in insects (the developmental gene *Toll* of *Drosophila*) and the immune response gene *TIR* from mammals (Baker *et al.*, 1997). The detailed structures of these genes evolve, which allows the host to generate novel resistance specificities (reviewed by Richter and Ronald, 2000).

Plant *R* genes seem to encode receptors that interact directly or indirectly with elicitors produced by the pathogen *Avr* genes. It is likely that the LRR is the pathogen recognition domain (see Parker *et al.*, 2000). As noted above, this recognition prompts a signal transduction cascade, the precise mechanisms of which are poorly understood but which possibly involve salicylic acid, jasmonic acid and ethylene (for reviews see Yang *et al.*, 1997b; Innes, 1998; Parker *et al.* 2000).

Host phenotype	Pathogen phenotype		
	Avirulent AA	Avirulent Aa	Virulent Aa
Susceptible Rr	+	+	+
Resistant Rr	-	-	+
Resistant RR	-	-	+

Fig. 10.1 The gene-for-gene hypothesis of Flor (1971). The reaction of the host to the pathogen is indicated as: +, susceptible; −, resistant.

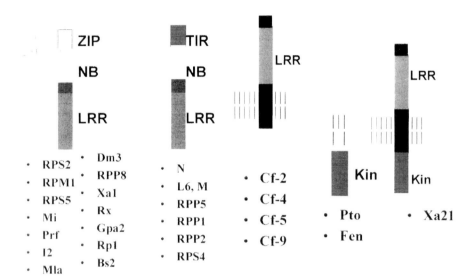

- RPS2
- RPM1
- RPS5
- Mi
- Prf
- I2
- Mla

- Dm3
- RPP8
- Xa1
- Rx
- Gpa2
- Rp1
- Bs2

- N
- L6, M
- RPP5
- RPP1
- RPP2
- RPS4

- Cf-2
- Cf-4
- Cf-5
- Cf-9

- Pto
- Fen

- Xa21

Fig. 10.2 (see Plate 10.1) The structures of 5 classes of *R* genes. LRR, leucine-rich repeat region; NB, nucleic acid binding; ZIP, leucine ZIP domain. Below are listed the genes that fall into each class.

III. STEPS IN THE INDUCTION OF DISEASE

Various lines of evidence have shown that viral genes must be involved in quite specific ways in the induction of disease. For example, closely related strains of TYMV produce two distinct and mutually exclusive pathways of change in the chloroplasts of TYMV-infected cells. In one, the chloroplasts first become rounded and clumped, and then develop a large vacuole. In the other, they become angular before clumping, and then fragment to yield many small pieces. These different pathways must be activated by a viral gene or genes (Fraser and Matthews, 1979b).

The existence of viruses with genomes consisting of two or more pieces of nucleic acid has allowed the production of pseudorecombinants in the laboratory. Reassortment of genomic segments enables determinants of specific disease symptoms to be located on particular genome segments. Many such experiments have been reported both for viruses with DNA genomes and for those with RNA genomes. For example, it was established that at least some determinants of symptom production are located on DNA1 of ACMV (Stanley *et al.*, 1985). Experiments with only a few virus strains tested on only a few hosts may not reveal the full range of genome segments involved. Thus, Rao and Francki (1982) constructed 18 pseudorecombinants *in vitro* by exchanging the three genomic RNA segments between pairs of three strains of CMV. These were tested on 10 host plant species. Some host reactions were determined by RNA2 or RNA3 alone, but others resulted from an interaction between both RNAs. Others probably involved all three genome segments. While such reassortment experiments have given useful information, they do not always pinpoint the gene or genes involved as genome segments often code for more than one protein product.

The ability to be able to produce cloned infectious genomes of viruses and to manipulate these genomes by mutagenesis has led to major advances in identifying the gene(s) or non-coding regions involved symptom determination. Examples of this approach will be discussed in this chapter.

In considering the induction of disease, it is useful to make a distinction between the functions of viral genes in the virus infection cycle and the effects of the genes on the host. In normal circumstances, all the gene products of a viral genome will have one or more functions in the complete virus infection cycle. There is no evidence for the existence of viral genes that have a role only in inducing disease.

If we consider a range of host species and environmental conditions (Chapter 12, Section II), a given virus can cause far more different kinds of symptoms than it has different gene products or combinations of two or more gene products acting together. We must conclude that a particular viral gene may have a variety of effects on the kind of disease produced depending on the host plant involved, the environmental conditions, and possible interactions with other viral genes. At present, we can distinguish three kinds of effect of viral genes. (1) Those based on a specific requirement for an essential virus function. The small vesicles induced by TYMV in chloroplast membranes (see Fig. 8.15) may be an example of this sort. (2) Those based on a defect in a viral gene, for example a cell-to-cell transport function, that results in limitation of infection to the point of entry. (3) Those based on effects of virus infection that are quite inconsequential as far as viral replication or movement is concerned, unless they so damage host cells that further replication is inhibited.

Many viral infections show no observable macroscopic disease symptoms. However, apart from the cryptic viruses, most produce observable and often characteristic cytological effects in at least one of their hosts (see Chapter 3, Section IV). Thus, most macroscopic diseases induced by viruses may be inconsequential in the sense noted in (3) above, whereas some cytological effects may represent essential requirements for viral replication and movement. Essentially macroscopic disease symptoms should be regarded as effects of the virus on its cellular environment, rather than as part of the viral phenotype.

It can be seen from Table 10.1 that there are various hurdles that a virus has to overcome in establishing a full systemic disease. These are discussed below from the initial infection to the full systemic manifestation of the viral disease to demonstrate the interactions between the aggression of the virus and the resistance of the host.

A. Ability of virus to replicate in initial cell

There have been few studies of the factors involved in controlling or inhibiting the replication of viruses in non-hosts. This is because these experiments would have to be conducted at the single-cell level and it is natural to establish protoplast systems for hosts of a virus rather than for non-hosts. Thus, it is not known what proportion of apparent non-hosts at the whole-plant level cannot support viral replication at the single-cell level.

As described in Chapter 7 (Section II), one of the earliest events in viral replication in an initially infected cell is the uncoating of the virus particle. Various lines of evidence suggest that there is little or no host specificity in this process. Thus, for TMV (Kiho et al., 1972) and TYMV (Matthews and Witz, 1985), virus was uncoated as readily in non-host as in host species. However, it is uncertain as to whether the non-host plant species used in these experiments could support the replication of the viruses and that what was being observed was subliminal infection with the virus unable to move to adjacent cells.

For RNA viruses there is some evidence for the involvement of host proteins or factors in their replication (see Chapter 8, Section IV.E.5). Similarly, geminiviruses have to involve the host DNA polymerase enzyme systems in their replication (see Chapter 8, Section VIII.D.4). Thus, it is likely that incompatibility between the plant and the virus at this step of virus multiplication could prevent the virus from replicating.

Certain resistance genes operate at the single-cell level as is shown by various forms of resistance to infection of protoplasts. This is termed *extreme resistance*.

1. TMV and the *Tm-1* gene in tomato

The multiplication of TMV in tomato plants is inhibited by the presence of the *Tm-1* gene, the inhibition being more effective in the homozygote (*Tm-1/Tm-1*) than in the heterozygote (*Tm-1/+*) (Fraser *et al.*, 1980; Watanabe *et al.*, 1987c). *Tm-1* resistance is expressed in protoplasts even in the presence of actinomycin D (Motoyoshi and Oshima, 1977; Watanabe *et al.*, 1987c) and thus can be classed as being extreme.

Comparison of the sequences of a virulent with an avirulent strain of TMV revealed two base substitutions resulting in amino acid changes in the replicase 130- and 180-kDa proteins (Gln to Glu at position 979 and His to Tyr at 984) (Meshi *et al.*, 1988). Mutagenesis of infectious transcripts suggested that the two concomitant base substitutions, and possibly also the resulting amino acid changes, were involved in the recognition of this *Avr* gene by the tomato *Tm-1* gene.

2. PVX and the *Rx* genes in potato

Extreme resistance to PVX in potato is provided by the *Rx1* and *Rx2* genes, which are located on chromosomes XII and V, respectively (Ritter *et al.*, 1991; Bendahmane *et al.*, 1997a). The *Rx1* gene is tightly linked to the nematode resistance gene *Gpa2* (Rouppe van der Voort *et al.*, 1999). *Rx* resistance operates in protoplasts by suppressing the accumulation of PVX (Adams *et al.*, 1986b; Bendahmane *et al.*, 1995). In joint infections of protoplasts, the levels of TMV and CMV production were also suppressed (Köhm *et al.*, 1993; Bendahmane *et al.*, 1995). The resistance is elicited by the PVX coat protein in a strain-specific manner (see Kavanagh *et al.*, 1992; Bendahmane *et al.*, 1995) and, by comparison of PVX sequences and mutagenesis, amino acid residue 121 in the coat protein gene was recognized as the major determinant of resistance-breaking activity (Querci *et al.*, 1995). Resistance is expressed prior to virus inoculation (Gilbert *et al.*, 1998). This extreme resistance is not associated with a hypersensitive response (HR).

The *Rx1* gene has been isolated and operates when transformed into *Nicotiana benthamiana* and *N. tabacum* (Bendahmane *et al.*, 1999). The sequence of the gene shows that it comprises three exons and two introns, and contains a single open reading frame (ORF) encoding a protein of 107.5 kDa (Bendahmane *et al.*, 1999). The primary structure is similar to the LZ-NBS-LRR class of *R* genes (Fig. 10.2). Thus, *Rx* is similar to *R* genes that confer an HR, but phenotypic analysis showed that *Rx*-mediated resistance is independent of HR. However, when the coat protein from a virulent strain of PVX was expressed constitutively in *Rx*-gene transgenic *N. benthamiana*, there was a necrotic response causing complete death of the treated region within 72 hours. This experiment showed that there is the potential for an *Rx*-mediated HR but that this potential is not realized when the coat protein is expressed from the PVX genome during viral infection. To examine the relationship between extreme resistance and the HR, experiments were conducted with *N. tabacum* plants carrying the *N* gene (see Section III.D.3) either alone or in combination with transgenic *Rx* gene (Bendehmane *et al.*, 1999). These plants were challenged with recombinant TMV expressing the coat protein from either a virulent PVX strain (TMV-KR) or an avirulent strain (TMV-TK); thus, TMV-TK contains the elicitors for both the *Rx*- and *N*-mediated resistances. Both TMV-KR and TMV-TK induced an HR on tobacco plants expressing the *N* gene, and TMV-KR gave an HR on tobacco of the *N + Rx* genotype. However, there was no HR when the *N + Rx* plants were inoculated with TMV-TK. These results show that the *Rx*-mediated extreme resistance was activated before the *N*-mediated resistance and, therefore, that the extreme resistance is epstatic to an HR.

The *Rx2* gene has also been cloned and sequenced (Bendahmane *et al.*, 2000) and shown to encode a protein very similar to the products of *Rx1* and *Gpa2*.

3. PSbMV and the *sbm* genes in pea

Plants containing genes *sbm-1*, *sbm-3* and *sbm-4*, closely linked on chromosome 6 of peas (*Pisum sativum*), are unable to support the replication of PSbMV pathotypes 1, 3 and 4 respectively; *sbm-2*, has the same properties as *sbm-3* but is on a different linkage group (Hagedorn and Gritton, 1973; Provvidenti and Alconero, 1988a,b); PSbMV resistance gene *sbm-2*, conferring resistance to pathotype 2, is located on chromosome 2 (Provvidenti and Alconero, 1988a). Pathotype 1 could not replicate in *sbm-1* protoplasts or plants. In joint infections of *sbm-1* plants, there was no inhibition of the replication of pathotype 4, nor was there complementation of the replication of pathotype 1 (Keller *et al.*, 1998).

Using recombinant viruses between pathotypes 1 and 4, Keller *et al.* (1998) showed that the virulence of PSbMV was associated with the 21-kDa genome-linked protein (VPg). Because of the likely involvement of the VPg in viral replication, it is considered possible that the *sbm-1* gene interferes with the VPg's role in replication.

B. Ability of virus to move out of first cell

The main limitation to a virus moving from the cell that is initially infected to adjacent cells is its ability to gate the plasmodesmata. This is considered in detail in Chapter 9 (Section II.D).

However, there may be other factors that could influence the ability for viruses to move. As will be discussed below, viruses can be contained to a small number of cells by a hypersensitive response (HR) giving a visible lesion. The induction of an HR at the single-cell level would not be easily visible and thus may not have been recognized.

C. Hypersensitive local response

The HR may be induced if the virus can move from the initially infected cells. In some cases this may only happen if the inoculum is heavy and continuous, say after grafting (see *RSV-1*-mediated HR in soybeans induced by SMV (Hajimorad and Hill, 2001).

The structure of local lesions is described in Chapter 3 (Section II.A). Some viruses that cause necrotic local lesions induce a nonspecific host response that includes the *de novo* synthesis of host-coded proteins. The development of both localized and systemic resistance to superinfection follows the development of the necrotic local lesions. The local lesion reaction of various virus-host combinations is used for virus diagnosis and for assay of the concentration of infectious virus (see Chapter 15, Section II.A.1). The localization of viral replication in tissue near the site of infection, the HR (Table 10.1), is important in agriculture and horticulture as the basis for field resistance to viral infection. These phenomena have been the subject of many studies and are reviewed by Ponz and Bruening (1986), Bol (1988), Bol and van Kan (1988), van Loon (1989) and Culver (1997).

A number of virus–host combinations produce necrotic local lesions and, in an increasing number of cases, the host and viral genetic determinants have been identified (Table 10.2).

Genes that induce an HR in intact plants or excised leaf pieces fail to do so when isolated protoplasts are infected. This has been found for the N gene in tobacco (Otsuki *et al.*, 1972b) and for Tm-2 and Tm-2^2 in tomato (Motoyoshi

TABLE 10.2 Virus and host genes involved in hypersensitive responses

Virus	Virus gene	Host	Host gene
Tobacco mosaic virus	Coat protein	Tobacco	*N'*
		Pepper	*L^1, L^2, L^3*
		Eggplant	?
	30-kDa movement protein	Tomato	*Tm-2, Tm-2^2*
	Replicase (helicase domain)	Tobacco	*N*
Cauliflower mosaic virus	Gene VI	Tobacco/*Datura*	?
Tomato bushy stunt virus	p19 and p22	Tobacco	?
Cucumber mosaic virus	2a polymerase	Cowpea	?
Potato virus X	Coat protein	Potato	*Nx*
	25-kDa movement protein	Potato	*Nb*
Barley stripe mosaic virus	RNAγ	*Chenopodium amaranticolor*	?

and Oshima, 1975, 1977). Protoplasts from plants carrying these genes in the homozygous condition allowed replication of TMV without death of the cells. It has been suggested that this effect might be due to the epidermis being involved in the necrotic response (Motoyoshi and Oshima, 1975, 1976), or that cell-to-cell connections are necessary for the hypersensitive phenotype to be expressed (Motoyoshi and Oshima, 1977). Another, and possibly more likely, explanation could be that the presence of the cell wall is necessary for expression of the N gene. Peroxidase enzymes appear to be involved in the necrotic response. These enzymes may be located mainly in the cell wall compartment. Thus, removal of the wall may cause a gap in the chain of biochemical events that lead to cell death.

D. HR induced by TMV in N-gene tobacco

1. Properties of lesions formed by TMV infection of N-gene tobacco

Some varieties of tobacco respond to infection with all known tobamoviruses except the ob strain (see Tobiás et al., 1982; Padgett and Beachy, 1993) (reviewed in Culver et al., 1991) at normal temperatures by producing necrotic local lesions and no systemic spread, instead of the usual chlorotic local lesions followed by mosaic disease. This HR is overcome at temperatures above 28°C (Samuel, 1931) and occasionally at lower temperatures (White and Sugars, 1996). In several Nicotiana varieties the reaction is under the control of a single dominant gene, the N gene, found naturally in N. glutinosa (Holmes, 1938). This has been incorporated into N. tabacum cultivar Samsun NN (Holmes, 1938), N. tabacum cultivar Burley NN (Valleau, 1952) and N. tabacum cultivar Xanthi nc (Takahashi, 1956). In very young seedlings, plants containing the N gene (either NN or Nn) develop systemic necrosis. In older plants, only two responses to infection are usually observed: localized necrotic lesions in NN and Nn, and systemic mosaic disease in nn (Holmes, 1938). However, systemic necrosis may occur in some older plants with the N gene (Dijkstra et al., 1977).

TMV infection of N gene-containing tobacco results in cell death at the site of infection, and virus particles are found in, but restricted to, the region immediately surrounding the necrotic lesion (Da Graça and Martin, 1976). Micro-injection of a fluorescent dye into mesophyll cells surrounding local lesions induced by TMV in N. tabacum cv. Xanthi nc indicated that defects in plasmodesmatal function accounted for at least some of late local defense reactions restricting viral spread (Susi, 2000).

2. TMV helicase domain induces necrosis in N-gene plants

The current hypothesis for the HR reaction to TMV in tobacco containing the N gene is that a TMV protein interacts either directly or indirectly with the N protein, thereby activating the signal transduction pathway leading to the HR. Two approaches have been used to identify the elicitor of the HR. In the first, analysis of HR-inducing mutants of the normally virulent ob strain of TMV suggested that the replicase region (the 126/183-kDa genes) of the genome was involved (Padgett and Beachy, 1993). By use of chimeric viral genomes and further mutagenesis, this was narrowed down to the helicase domain of the replicase region (Padgett et al., 1997). In the second approach, TMV ORFs were expressed transiently in Samsun NN plants (Abbink et al., 1998; Erickson et al., 1999). Frameshift mutation showed that it was the protein, and not the mRNA, that was responsible. Expression of the helicase domain of the 126-kDa gene also induced the expression of the PR-1 genes (Abbink et al., 1998). Necrosis was not induced by the 183-kDa protein, the movement protein (P30) or the coat protein. As the 183-kDa gene is expressed from read through of the 126-kDa gene, it might be expected to induce necrosis; however, it was probably not expressed at a sufficiently high level to give the response.

The helicase domain also encodes ATPase activity but, although mutation of this domain abolished ATPase activity, it did not affect HR induction (Erickson et al., 1999).

The HR response of the N gene to TMV is temperature sensitive, being inactivated above 28°C. However, mutants in the helicase domain of TMV ob strain overcame the HR at lower temperatures (Fig. 10.3) (Padgett *et al.*, 1997). It was suggested that, at higher temperatures, the interaction between the viral elicitor and the host surveillance mechanism that leads to HR is weakened.

3. The *N* gene

The *N* gene has been isolated by transposon tagging using the maize activator transposon (Whitham *et al.*, 1994; Dinesh-Kumar *et al.*, 1995), and characterized. The sequence of the *N* gene shows that it encodes a protein of 131.4 kDa with an N-terminal domain similar that of the cytoplasmic domain of the *Drosophila*

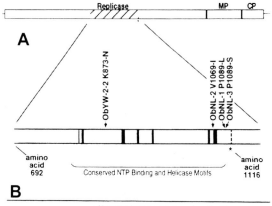

B

Virus[a]	HR elicitation and virus localization					
	18°C	20°C	22°C	24°C	26°C	28°C
TMV	+[b]	+	+	+	+	–[b]
ObNL-1.1	+	+	+	+	–	–
ObNL-3.1	+	+	+	+	–	–
ObNL-2.1	+	+	–	–	–	–
ObYW-2-2	+	+	–	–	–	–
Ob	–	–	–	–	–	–

[a] Viruses were tested for induction of the hypersensitive response (HR) and localization to inoculated leaves of tobacco cv. Xanthi NN.
[b] (+), HR accompanied by virus localizations; (–), systemic virus spread.

Fig. 10.3 Identification of HR-inducing region in TMV. **(A)** Mutations of the 126/183-kDa gene of TMV strain Ob that lead to induction of the HR. Hatched region is the genomic map representing the portion of the TMV replicase (amino acids 692–1116) that is required for induction of the *N* gene-mediated HR. At the bottom is an expanded view of this region of the Ob replicase, indicating the positions in Ob that result in HR. Shaded bars represent the nucleotide triphosphate-binding domain and helicase consensus motifs. Asterisk denotes the amber termination codon in the 126/183-kDa protein. **(B)** Temperature sensitivity of N against TMV Ob and ObNL mutants. From Padgett *et al.* (1997), with kind permission of the copyright holder, © The American Phytopathological Society.

Toll protein and the interleukin-1 receptor in mammals, a nucleotide binding site and four imperfect leucine-rich regions (Fig. 10.4). Thus, it belongs to the TIR-NB-LRR class of *R* genes (see Fig. 10.2).

A deletion analysis suggested that the TIR, NB and LRR domains all play an indispensable role in the induction of HR (Dinesh-Kumar *et al.*, 2000).

The *N* gene is expressed from two transcripts, N_S and N_L, via alternative splicing pathways (Dinesh-Kumar and Baker, 2000). The N_S transcript codes for the full-length N protein and is more prevalent before, and for 3 hours after, TMV infection. The N_L transcript codes for a truncated N protein (N^{tr}), lacking 13 of the 14 leucine-rich repeats, and is more prevalent 4–8 hours after infection. A TMV-sensitive tobacco variety transformed to express the N protein but not the N^{tr} protein is susceptible to TMV, whereas transgenic plants expressing both N_S and N_L transcripts are completely resistant.

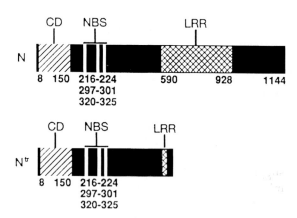

Fig. 10.4 Schematic diagram of the N and N^{tr} proteins. An analysis of N protein amino acid sequence identified three domains of possible functional significance. These domains are indicated and shown to scale within the full-length N protein. CD, putative cytoplasmic domain of N with sequence similarity to the cytoplasmic domains of Toll, interleukin-1R and MyD88; NBS, putative nucleotide-binding site comprising three motifs from amino acid 216 to 325; LRR, leucine-rich repeat region consisting of 14 imperfect tandem leucine-rich repeats. Alternative splicing yields a truncated protein, N^{tr}. N^{tr} is identical to the N-terminus of N except for the C-terminal 36 amino acids, indicated by the black box next to the LRR on the right. From Dinesh-Kumar and Baker (2000), with kind permission of the copyright holder, © The National Academy of Sciences, USA.

However, the ratio of N_S to N_L mRNAs before and after TMV infection is critical as the expression of either one mRNA alone or the two at a 1:1 ratio gives incomplete resistance. It is suggested that the relative ratio of the two N messages is regulated by TMV signals (Dinesh-Kumar and Baker, 2000).

The N gene has been transferred to tomato, where it confers resistance to TMV (Whitham et al., 1996).

E. Other viral–host hypersensitive responses

1. TMV and the N' gene (reviewed in Culver et al., 1991)

The N' gene, originating from *Nicotiana sylvestris*, controls the HR directed against most tobamoviruses, except U1 (vulgare) and OM strains that move systemically and produce mosaic symptoms in N'-containing plants. However, mutants that induce necrosis can easily be isolated as spontaneous mutants of virus strains causing systemic symptoms. The TMV coat protein gene is involved in the induction of the N' gene HR (Saito et al., 1987a). Five independent amino acid substitutions have been identified as being involved in the HR elicitation (Culver and Dawson, 1989; Culver et al., 1991), but the HR response varied for each mutation, from strong elicitors producing visible necrosis in 2–3 days to weak elicitors requiring at least 6 days for necrosis to appear. This suggested that the structure of the coat protein might influence the response. To investigate this, Culver et al. (1994) made a set of amino acid substitutions that would have predicted structural effects on the coat protein, and examined their HR. Substitutions eliciting the HR were within, or would predictably interfere with, interface regions between adjacent subunits in ordered aggregates of coat protein (see Chapter 5, Section III.B.5, for TMV structure). Substitutions that did not elicit the HR were either conservative or located outside the interface regions. The HR-inducing substitutions formed rod-shaped particles with reduced quaternary stability, and the strength with which a coat protein elicited the HR correlated with the degree of destabilization of quaternary

structure. However, mutations that affected the tertiary structure of the coat protein did not elicit the HR. It was suggested that, to elicit the HR, the weakened quaternary structure exposed a receptor binding site (Culver et al., 1994). It is likely that the size distribution and/or lifetime of small coat protein aggregates in elicitors allow the N' gene to recognize the invading virus (Toedt et al., 1999).

To identify the elicitor site further, Taraporewala and Culver (1996) studied various amino acid substitutions in relation to the known tertiary structure of the coat protein (see Chapter 5, Section III.B.3, for details of TMV coat protein tertiary structure). They showed that substitutions that disrupted the right face of the coat protein α-helical bundle interfered with N' gene recognition. The elicitor active site covers approximately 600 Å^2 and comprises 30% polar, 50% non-polar and 20% charged residues. Comparison of the coat proteins of various tobamoviruses and the effects of substitutions in these coat proteins on the N' gene HR revealed the presence of a conserved central hydrophobic cavity surrounded by surface features that were less conserved (Taraporewala and Culver, 1997). These findings suggested that the N'-gene specificity is dependent upon the three-dimensional fold of the coat protein as well as upon specific surface features within the elicitor active site (Fig. 10.5).

2. TMV and the L genes of Capsicum

Capsicum spp. carry genes, L^1 (*C. annuum*), L^2 (*C. frutescens*) and L^3 (*C. chinense*) that confer HR resistance to TMV; there is also a genetically uncharacterized HR in eggplant, *Solanum melongena* (Berzal-Herranz et al., 1995; Dardick and Culver, 1997; de la Cruz et al., 1997; Dardick et al., 1999). The HR of each of these genes is induced by TMV coat protein.

Chimeric constructs in which the coat protein gene of TMV U1 was substituted by those of other TMV strains showed that U1, U2, ORSV and CGMMV coat proteins elicited phenotypically similar HRs in pepper (L^1 gene). U1 and CGMMV coat proteins did not elicit the HR in tobacco and eggplant respectively (Dardick and Culver 1997; Taraporewala and Culver, 1997). In a comparison of the elicitation of the

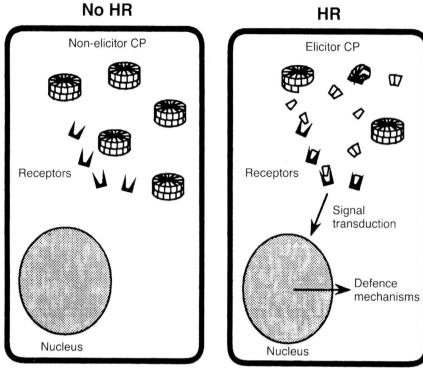

No HR

HR

Prime gene host cells

Fig. 10.5 Model explaining the effects of elicitor and non-elicitor TMV coat proteins (CPs) on induction of the N'-gene hypersensitive response. From Culver (1997), with permission.

HR in tobacco, pepper and eggplant, Dardick *et al.* (1999) showed that the α-helical bundle was essential in all cases. Differences in recognition were considered to result from how these hosts perceived the coat protein surface features and/or quaternary configurations. This suggests that these resistance genes are functionally related and may be structurally homologous.

3. TMV and the *Tm-2* genes of tomato (reviewed in Culver *et al.*, 1991)

There are two allelic genes in tomato, *Tm-2* and *Tm-2²*, that give an HR to certain strains of ToMV (see Table 10.3). As well as being determined by virus strain, the HR is also dependent on genotype of tomato and on environmental conditions, especially temperature. The response can vary from a very mild necrotic lesion giving apparent subliminal infection, through the normal necrotic local lesion to systemic necrosis.

By comparison of sequences of HR-inducing and non-inducing isolates of TMV, and use of

mutagenesis and chimeric viruses, the 30-kDa movement protein (MP) has been identified as being the inducer of the HR in both *Tm-2* and *Tm-2²* plants (Culver *et al.*, 1991; Calder and Palukaitis, 1992; Weber *et al.*, 1993). Sequence comparison of *Tm-2* resistance-breaking isolates (Ltb1 and C32) with wild-type isolates identified two amino acid differences for Ltb1 (Cys to Phe at position 68 and Glu to Lys at position 133) and also two in C32 (Glu to Lys at position 52 and Glu to Lys at position 133) (see Fig. 9.6). In *Tm-2²* resistance-breaking isolates, amino acid substitutions are at the C-terminus of the MP (Ser to Arg at position 238 and Lys to Glu at position 244) (Weber *et al.*, 1993). The C-terminal region of the MP, including the sites of the two substitutions, is not required for cell-to-cell movement (Weber and Pfitzner, 1998), although the N-terminus is (see Fig. 9.6). Even so, it now appears that both these resistance genes operate via an HR rather than inhibition of movement out of the initially infected cell due to incompatibility of the MP (Weber and Pfitzner, 1998).

TABLE 10.3 Genetic interactions between ToMV-resistant tomato plants and strains of the virus

Host genotype[a]	Virus genotype					
	0	1	2	2^2	1.2	1.2^2
Wild type	M	M	M	M	M	M
Tm-1	R	M	R	R	M	M
Tm-2*	R	R	M	R	M	R
Tm-2²*	R	R	R	M	R	M
Tm-1/Tm-2	R	R	R	R	M	R
Tm-1/Tm-2²	R	R	R	R	R	M
Tm-2/Tm-2²	R	R	R	R	R	R
Tm-1/Tm-2/Tm-2²	R	R	R	R	R	R

Plants with genotype marked with an (*) may show local and variable systemic necrosis rather than mosaic when inoculated with virulent strains.
M, systemic mosaic; R, resistance.
[a] Modified from Fraser (1985), with permission.

A sequence-characterized amplified region (SCAR) marker linked to the Tm-2^2 resistance gene in tomato has been produced (Dax *et al.*, 1998).

4. PVX and the N genes of potato

HR in potato to PVX is controlled by two genes, Nb and Nx. Nb has been mapped to a resistance gene cluster in the upper arm of chromosome V (De Jong *et al.*, 1997; Rouppe van der Voort *et al.*, 1998) and Nx to a region of chromosome IX similar to that containing the gene Sw-5 for resistance to TSWV (Tommiska *et al.*, 1998).

To identify the viral elicitor of the Nb HR, hybrid viruses were constructed between avirulent PVX strains and virulent strains (Malcuit *et al.*, 1999). The Nb avirulence determinant was mapped to the PVX 25-kDa gene encoding the MP. The isoleucine at position 6 in this protein was shown to be involved in the elicitor function. However, this amino acid is present in the resistance-breaking strain HB and may act as a determinant of the three-dimensional structure of the avirulence domain that is specifically recognized by the Nb gene product (Malcuit *et al.*, 1999).

The Nx-mediated resistance is elicited by the PVX coat protein gene (Kavanagh *et al.*, 1992), a single amino acid at position 78 being an important determinant (Santa Cruz and Baulcombe, 1993).

It is suggested that the multiple virulence–avirulence determinants were acquired by PVX through convergent evolution rather than through recombination (Malcuit *et al.*, 2000).

5. TCV and the HRT and rrt genes of Arabidopsis

Inoculation of the *Arabidopsis* ecotypes Di-0 or Di-17 with TCV results in an HR on the inoculated leaves (Simon *et al.*, 1992; Dempsey *et al.*, 1993, 1997). The HR development is conferred by a single dominant gene termed HRT (Dempsey *et al.*, 1997), located on chromosome 5 and encoding a classical leucine zipper-NBS-LRR protein (Cooley *et al.*, 2000). HRT shares extensive homology with the $RPP8$ gene family that confers resistance to the oomycete, *Peronospora parasitica*. The TCV coat protein is the Avr factor recognized by HRT and mutations in the very N-terminus of the coat protein produced hypervirulent strains of TCV that failed to induce an HR (Oh *et al.*, 1995; Wang and Simon, 1999; Cooley *et al.*, 2000). HRT may not be sufficient for complete resistance as many of the HR^+ progeny become fully infected (Kachroo *et al.*, 2000). A recessive allele, rrt, which regulates resistance to TCV, has now been identified (Kachroo *et al.*, 2000).

6. CaMV and solanaceous hosts

The D4 and W260 strains of CaMV induce chlorotic local lesions and a systemic mosaic in *Datura stramonium*, *Nicotiana bigelovii* and *N. edwardsonii*, whereas strain CM1841 induces necrotic local lesions limiting it to the inoculated leaf (Schoelz *et al.*, 1986; Schoelz and

Shepherd, 1988). Using chimeric viruses constructed between these two strains it was shown that the HR was elicited by gene VI (see Chapter 6, Section IV.A.1, for details of the CaMV genome). This deduction was confirmed by infiltration of *Agrobacterium tumefaciens* containing a binary vector expressing gene VI of WD260 strain into leaves (Palanichelvam *et al.*, 2000). Mutational analysis and chimeras identified the N-terminal one third of gene VI as being involved (Wintermantel *et al.*, 1993; Broglio, 1995) and gene II, the aphid transmission factor, also has a light-dependent influence (Qiu *et al.*, 1997). A point mutant at position 1628 (amino acid 94 in gene II) of strain W260 could systemically infect *N. bigelovii* at low but not at high light intensity, causing necrotic local lesions at under both conditions; wild-type W260 gave systemic infection at both light intensities. Thus, the HR and containment of the virus within the HR region is conditioned by host, virus strain and growing conditions. The translational transactivation of ORF VI can be uncoupled from its ability to elicit HR (Palanichelvam and Schoelz, 2001).

7. Other viruses

By expression of individual genes of TBSV, Scholthof *et al.* (1995a) showed that the MP, p22, was responsible for the HR on *N. edwardsonii*. The p22 residues responsible for cell-to-cell movement were separable from those eliciting the HR (Chu *et al.*, 1999). Three mutants of p22 impaired in cell-to-cell movement elicited necrotic local lesions on *N. edwardsonii*, whereas two mutants, capable of cell-to-cell movement, gave chlorotic lesions and systemic infection.

The resistance to infection of cowpea by strains (e.g. Fny) of CMV involves an HR and a localization of infection (Kim and Palukaitis, 1997), responses that can be separated by mutation at two sites (nucleotides 1978 and 2007, codons 631 and 641) in the viral 2a polymerase gene. Changes to both sites of Fny strain allowed systemic infection without an HR and increase of viral synthesis in protoplasts, whereas changing position 1978 alone resulted in systemic infection, systemic HR and an increase of viral RNA accumulation in protoplasts. It is suggested that the inhibition of RNA accumulation in protoplasts, where an HR does not occur, leads to localization of infection in whole plants and that different plant genes are involved in eliciting the HR and the localization response (Kim and Palukaitis, 1997). An *R* gene, *Cry*, has been recognized in cowpea cultivar Kurodane-Sanjaku which confers an HR resistance to CMV strain Y. This resistance is overcome by a legume strain of CMV (CMV-L) and its elicitation was shown by point mutation to be associated with amino acid 631 in the 2b replicase gene ((Karasawa *et al.*, 1999).

The HR caused by CMV in *Chenopodium amaranticolor* is affected by mutation of both the 3a movement protein and the 3b coat protein. It was considered that movement of the virus from the initially infected cell was required to elicit the HR (Canto and Palukaitis, 1999b).

The begomoviruses, BDMV and BGYMV, differentially infect certain bean (*Phaseolus vulgaris*) varieties, with some cultivars giving an HR to BDMV. A series of hybrid DNA-B components, containing BDMV and BGYMV sequences, were co-inoculated with BDMV DNA-A or BDMVA-green fluorescent protein into seedlings of cv. Topcrop (susceptible to BDMV and BGYMV) and cvs Othello and Black Turtle Soup T-39 (BDMV resistant) (Garrido-Ramirez *et al.*, 2000). The BDMV avirulence determinant was mapped to BV1 ORF and most likely BV1 protein. The product of BV1 is the nuclear shuttle protein (Chapter 9, Section II.C), and thus represents a new class of viral avirulence determinant.

F. Host protein changes in the hypersensitive response

The HR involves a series of complex biochemical changes at and near the infection site that include the accumulation of cytotoxic phytoalexins, the deposition of callose and lignin in the cell walls and the rapid death of plant cells forming the necrotic lesion (Dixon and Harrison, 1990). The regulation of HR is equally complex, involving the interplay of many potential signal-transducing molecules including reactive oxygen species, ion fluxes, G proteins, jasmonic and salicylic acids, protein phosphorylation cascades, activation of transcription factors and protein recycling by the polyubiquitin system (Dangl et al., 1996; Hammond-Kosack and Jones, 1996). To identify some of the tobacco genes eliciting the HR, Karrer et al. (1998) made a cDNA library from tobacco leaves undergoing HR and cloned the cDNAs into a TMV-based expression vector. Infectious transcripts from this TMV vector were inoculated to Xanthi nn tobacco (lacking the N gene) and those giving an HR were characterized. One of the 12 unique clones that were sequenced encoded ubiquitin, which was considered to elicit the necrotic response by a co-suppression mechanism.

Increases in the activity of a variety of enzymes other than those designated as PR proteins have been observed during the HR to viruses. Peroxidase, polyphenoloxidase and ribonuclease activities are increased (e.g. Wagih and Coutts, 1982). The metabolism of phenyl-propanoid compounds is strongly activated by infection with various pathogens, including viruses that induce an HR. This activation leads to the accumulation of compounds derived from phenylalanine, such as flavonoids and lignin. Activation of the pathway involves de novo enzyme synthesis, for example O-methyl-transferase (Collendavelloo et al., 1982). Van Kan et al. (1988) identified two genes in Samsun NN tobacco coding for a glycine-rich protein that is strongly induced during the HR to TMV infection.

Among other host proteins that are affected by the HR in tobacco, and which may be important in the reaction, are the myb onco-gene homolog which is increased (Yang and Klessig, 1996), catalase which is decreased (Yi et al., 1999) and a glycine-rich RNA-binding protein which decreases in the early stages of HR but increases in the later stages (Naqvi et al., 1998).

One of the earliest detectable events in the interaction between a plant host and a pathogen that induces necrosis is a rapid increase in the production of ethylene, which is a gaseous plant stress hormone. In the HR to viruses, there is an increased release of ethylene from leaves (e.g. Gáborjányi et al., 1971). The fact that ethepon (a substance releasing ethylene), introduced into leaves with a needle,

can mimic the changes associated with the response of Samsun NN to TMV is good evidence that ethylene is involved in the initiation of this HR (van Loon, 1977). An early burst of ethylene production is associated with the virus-localizing reaction, although the increase in ethylene production is not determined by the onset of necrosis but by a much earlier event (De Laat and van Loon, 1983).

The outcome of the interaction between a virus and a host resistance gene may be controlled by a quantitative interaction between the viral avirulence determinant(s) and the host resistance determinant(s) (Collmer et al., 2000). The resistance to BCMV conferred by the I allele in cultivars of P. vulgaris varied according to I allele dosage and temperature, giving a range or responses from immunity, to hypersensitive resistance, to systemic phloem necrosis leading to plant death.

In assessing the role(s) of all these changes in host proteins during the HR, it should be remembered that some of the changes are actually involved in the response and some are secondary to the response. At present, it is difficult to distinguish between these roles.

G. Other biochemical changes during the hypersensitive response

Many other biochemical changes have been observed during the HR. For example, Uegaki et al. (1988) detected 19 sesquiterpenoids that were considered to be stress-induced compounds in Nicotiana undulata inoculated with TMV. An enhanced NADPH-dependent oxygen-generating system was found in a membrane-rich cellular fraction from tobacco leaves reacting hypersensitively to TMV infection. Early electrolyte leakage occurred from cells of cowpea leaves during an HR to virus infection and other stress stimuli (Pennazio and Sapetti, 1981). Free abscisic acid concentration was raised up to 18-fold of normal in tissue near or within necrotic local lesions caused by TMV in tobacco (Whenham and Fraser, 1981). The cytokinin content of Xanthi NN tobacco leaves with systemic acquired resistance was increased (Balázs et al., 1977).

H. Systemic necrosis

1. Specific necrotic response

On occasions the necrosis induced by virus infection is not limited to local lesions but spreads. This is usually from expanding local lesions that reach veins and result in systemic cell death. The systemic necrosis can range from necrosis in a few areas of upper leaves or sporadic necrotic spots mixed with mosaic symptoms to widespread necrosis leading to death of the plant. The systemic necrotic symptoms are dependent on host genotype, virus strain and environmental conditions.

Some *Nicotiana* spp. that contain the *N* gene may develop systemic cell death in response to TMV infection, even when the plants are maintained at 17–20°C (Dijkstra *et al.*, 1977). A single amino acid substitution may alter the coat protein of TMV from being a strong elicitor of the HR in *N'* gene-containing tobacco to being a weak elicitor resulting in systemic spread of necrosis (Culver and Dawson, 1989).

ToMV-2^2 infection of tomato variety GCR 267, which is homozygous for the *Tm-2^2* gene, results in systemic necrosis 2–3 weeks after infection. Infection with ToMV-30.2^2, in which the wild-type ToMV 30-kDa gene had been replaced by the *Tm-2^2* 30-kDa gene, led to much milder symptoms and later development of systemic necrosis (Weber *et al.*, 1993).

Infection of *N. clevelandii* with CaMV strains D4 and W260 leads to systemic cell death elicited by the CaMV gene VI (Király *et al.*, 1999). The F1 generation of crosses between *N. clevelandii* with *N. bigelovii*, which does not give systemic necrosis with these CaMV strains, developed systemic mosaic symptoms on inoculation with W260, whereas the F2 plants segregated 3:1 for systemic mosaic versus systemic death. The plant gene responsible for cell death has been named *ccd1*. Thus, the systemic death phenotype is induced by the interaction between *ccd1* and CaMV gene VI (Király *et al.*, 1999).

2. Non-specific necrosis

Some other types of necrotic response are less specific and can occur in hosts that are different genera and apparently not in a gene-for-gene manner. Several deletion mutants of TMV coat protein cause an apparent non-specific necrosis (Dawson *et al.*, 1988; Dawson and Bubrick, 1989).

The 5′ non-coding region of GCMV RNA2, cloned into a PVX vector, induces a systemic necrotic response in *N. benthamiana*, *N. clevelandii* and *N. tabacum* (Fernandez *et al.*, 1999). GCMV itself does not infect *N. clevelandii* and is symptomless or gives very mild symptoms in the other two *Nicotiana* spp.

Joint infection of tomato with TMV and PVX can cause systemic necrosis (Plate 3.20).

Ringspot symptoms, in which necrotic rings spread in an apparently diurnal manner, are described in Chapter 3 (Section II.B.5).

I. Programed cell death and plant viruses

Multicellular organisms have mechanisms for eliminating developmentally misplaced or unwanted cells, or for sacrificing cells to prevent the spread of pathogens. This is termed programed cell death (PCD) or apoptosis; apoptosis is a specific case of PCD with a distinct set of physiological and morphological features (reviewed in Mittler and Lam, 1996; Gilchrist, 1998). Although much of the work on PCD has been done in animal systems, there is increasing interest in this process in plants. The HR to plant pathogens has various features in common with PCD (see Mittler and Lam, 1996; Mitsuhara *et al.*, 1999; Pontier *et al.*, 1999).

Certain animal viruses can inhibit PCD (Osborne and Schwartz, 1994). It will be of interest to determine whether plant viruses have similar properties.

J. Local acquired resistance

Most experimental work on local acquired resistance has been carried out with tobacco varieties containing the *N* gene. Ross (1961a) showed that a high degree of resistance to TMV developed in a 1–2-mm zone surrounding TMV local lesions in Samsun NN tobacco (Fig. 10.6). The zone increased in size and resistance for about 6 days after inoculation. Greatest resist-

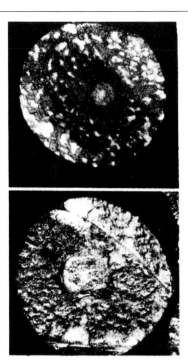

Fig. 10.6 Acquired resistance to infection. **Upper:** Disc cut from a Samsun NN tobacco leaf inoculated first with TMV (large lesion in center) and 7 days later given a challenge inoculation with a concentrated TMV inoculum. Note absence of lesions from the second inoculation in a zone around the original lesion. **Lower:** Similar experiment in which PVX was the challenge virus. No zone free of lesions is present. From Ross (1961a), with permission.

ance developed in plants grown at 20–24°C. Resistance was not found in plants grown at 30°C.

There appears to be no local acquired resistance without the HR (Costet *et al.*, 1999).

K. Systemic acquired resistance (see Ryals *et al.*, 1996 for review)

Our initial knowledge about systemic acquired resistance came mainly from the work of Ross and colleagues (e.g. Ross, 1961b, 1966) on TMV in the tobacco variety Samsun NN and on TNV in pinto bean. In tests with tobacco, lower leaves are inoculated with TMV and then some days later the same leaves or upper leaves may be challenged by a second inoculation with TMV (Fig. 10.7). Acquired resistance is measured by the reduction in diameter of the lesions (and with some viruses, reduction in number). With bean, one primary leaf is inoculated and

the opposite primary leaf challenged by inoculation some days later.

Lesions were about one-fifth to one-third the size found in control leaves, but lesion number was not reduced with TMV in Samsun NN tobacco. Resistance was detectable in 2–3 days, rose to a maximum in about 7 days, and persisted for about 20 days. Leaves that developed resistance were free of virus before the challenge inoculation. No conditions have been found that would give complete resistance. In plants held at 30°C no resistance developed. Mechanical or chemical injury that killed cells did not lead to resistance, nor did infection with viruses that do not cause necrotic local lesions. On the other hand, many other non-specific agents applied to leaves will induce the phenomenon (e.g. Gupta *et al.*, 1974). In such experiments it is not possible to be sure that the same phenomenon is being studied because many treatments affect lesions size.

The resistance induced by TMV was not specific for TMV, but was effective for TNV and several other viruses. A similar lack of specificity in the resistance acquired following the development of necrotic local lesions was found with various other host–virus combinations giving the HR. However, virus-specific factors may regulate the extent of the resistance (van Loon and Dijkstra, 1976).

Ross (1966) pointed out that effects on lesion number tend to be more variable and develop later. He considered that a fall in lesion number merely means that in highly resistant leaves lesions do not become large enough to be countable.

1. Pathogenesis-related (PR) proteins (reviewed by van Loon and van Strien, 1999)

Gianinazzi *et al.* (1970) and van Loon and van Kammen (1970) showed that changes in the pattern of soluble leaf proteins occurred in tobacco leaves responding hypersensitively to infection with TMV. They were termed pathogenesis-related proteins, or 'PR proteins', by Antoniw *et al.* (1980). These proteins have been studied extensively and their activities characterized (see Linthorst, 1991; Shewry and Lucas, 1997; Dempsey *et al.*, 1999). There are 14 families of PR protein (PR1–14), five of which are listed in

Fig. 10.7 Resistance acquired at a distance from the site of inoculation. **Right:** Samsun NN leaf inoculated first on the apical half with TMV and 7 days later given a challenge inoculation over its whole surface with TMV. **Left:** Control leaf given only the second inoculation. From Ross (1961b), with permission.

TABLE 10.4 Some types of PR proteins

Type	Biological activity	Inhibit
1	Not known	Oomycetes
2	β-1,3-Gluconases	Fungi
3	Chitinases	Fungi
4	Chitin binding	Fungi
5	Membrane permeabilization	Fungi

Table 10.4, and for many there are two classes: acidic PR proteins and their basic homologs. They have been identified in a range of plant species and shown to be induced by a variety of microbial infections (viruses, viroids, bacteria and fungi) and by treatment with certain chemical elicitors, notably salicylic acid (SA) and acetyl salicylic acid (aspirin) (White, 1979). These various PR proteins are not induced following infection with a virus that does not cause necrotic local lesions in the host used (Fig. 10.8, lane 4).

The kinetics of the HR-induced expression of a tobacco peroxidase gene differ from those of the induction of PR proteins especially in being insensitive to inducers of PR genes such as SA, methyl jasmonate and ethephron (Hiraga *et al.*, 2000).

It is not yet established whether any of the PR proteins play a role in the limitation of virus

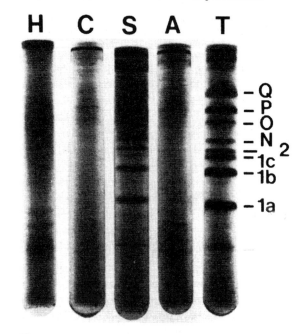

Fig. 10.8 Accumulation of PR proteins in the intercellular fluid of tobacco after various treatments. Plants were sprayed with water (H), *p*-coumaric acid (C) or salicylic acid (S), or inoculated with AMV (A) or TMV (T). Samples of the intercellular fluid were electrophoresed in non-denaturing polyacrylamide gels. The position of the major PR proteins is indicated in the margin. From Bol and van Kan (1988), with kind permission of the copyright holder, © Blackwell Science Ltd.

spread and acquired resistance. If any of the PR proteins are, in fact, active in the antiviral response, they may be found among those proteins for which a function has not yet been established. However, when expressed to high levels in transgenic tobacco plants with the NN constitution, individual PR proteins did not affect the necrotic response to inoculation with TMV, for example the 1b gene (Cutt *et al.*, 1989) and the 1a and s genes (Linthorst *et al.*, 1989).

2. Effect of salicylic acid on viruses

Application of SA suppresses the replication of TMV and PVX in inoculated tissue (Chivasa *et al.*, 1997; Naylor *et al.*, 1998) by not only decreasing the accumulation of virus, but also changing the ratio of viral genomic RNA to mRNA accumulation. It also affects the replication of AMV in cowpea protoplasts but not in tobacco leaves (Hooft van Huijsduijnen *et al.*, 1986; Murphy *et al.*, 1999a).

SA treatment appears to have no effect on the replication of CMV but causes a significant delay in the development of systemic symptoms, suggesting that this treatment affects the entry of the virus into the vasculature (Naylor *et al.*, 1998). It also affects the long-distance movement of AMV (Murphy *et al.*, 1999a). At least in the case of AMV, this treatment appears to have different effects in different hosts.

For other viruses, for example PVY, SA appears to have no effect on virus infection (Pennazio *et al.*, 1985).

3. Pathway for systemic acquired resistance to viruses

The characterization of PR proteins presented a conundrum in that, although they were induced by viral infection and associated with systemic acquired resistance, their activities, as far as is known (Table 10.4), do not appear to be related to virus inhibition but function in fungal and bacterial resistance. However, the HR induced by TMV infection of *N*-gene tobacco has many features in common with HRs caused by fungi and bacteria (reviewed in Murphy *et al.*, 1999a). The reaction is mediated by a sustained burst of reactive oxygen (ROS) followed first by local and then by systemic accumulation of SA. Since the production of SA induces metabolic heating, its local accumulation can be detected very early on by thermography (Chaerle *et al.*, 1999). SA is essential for the localization of the virus to the vicinity of the necrotic lesion and for the establishment of systemic acquired resistance.

The SA-induced resistance to TMV and PVX replication and to CMV movement in tobacco can be inhibited by salicylhydroxamic acid (SHAM) (Chivasa *et al.*, 1997; Naylor *et al.*, 1998). However, SHAM does not inhibit the SA-induced synthesis of PR proteins (Chivasa *et al.*, 1997; Chivasa and Carr, 1998), which suggests that there are two branches in the pathway to SA-induced resistance (reviewed by Murphy *et al.*, 1999a) (Fig. 10.9). One branch leads to the production of PR proteins that confer resistance to fungi and bacteria, and the other induces resistance to viral replication and movement.

SHAM is a relatively selective competitive inhibitor of the alternative oxidase (AOX) in the mitochondrial electron flow pathway in plants (reviewed in Murphy *et al.*, 1999a). The SHAM-sensitive pathway, induced by SA and potentially by AOX, is critical in the early stages of *N* gene-mediated resistance to TMV in tobacco. However, it does not explain all the observations, and other antiviral mechanism(s) must also play a role in the HR.

Various other chemicals are involved in the signal transduction pathway for defense response against pathogens including nitric oxide, mitogen-activated protein (MAP) kinases, jasmonic acid and ethylene (see Dempsey *et al.*, 1999; Klessig *et al.*, 2000; Zhang and Klessig, 2000). However, it is becoming recognized that, although there are many common features in the defense response, there is not just one single response pathway. For instance, the HR formation and TMV/*N* gene and TCV/*HRT* gene resistance are dependent upon SA but not on ethylene or jasmonic acid, and are unaffected by mutations in *NRP1* (Knoester *et al.*, 1998; Murphy *et al.*, 1999a; Kachroo *et al.*, 2000), whereas they are all required for resistance to other pathogens (see Dong, 1998; Shah *et al.*, 1999).

As noted in Section II.A, there is not yet a detailed understanding of the signal transduction

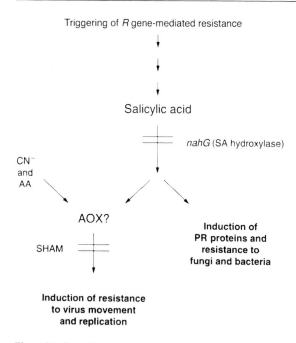

Triggering of *R* gene-mediated resistance

Salicylic acid

nahG (SA hydroxylase)

CN⁻
and
AA

AOX?

SHAM

Induction of
PR proteins and
resistance to
fungi and bacteria

Induction of resistance
to virus movement
and replication

Fig. 10.9 Possible model to explain the induction of resistance to viruses and other pathogens in tobacco. Recognition of a pathogen by the product of a resistance (*R*) gene results in localized cell death and the activation of a defense signal transduction pathway which includes salicylic acid (SA) as one of its components. Steps subsequent to SA in the defense signal transduction pathway are prevented in transgenic plants that express SA hydroxylase, the product of the bacterial *nahG* gene. Downstream of SA, the defense signal transduction pathway appears to divide. One branch (right) leads to the induction of extracellular pathogenesis-related (PR) proteins and systemic acquired resistance (SAR) to bacteria and fungi. The other (left) branch leads to the induction of resistance to the replication or long-distance movement of viruses. The virus-specific branch can be activated by cyanide (CN⁻) and antinomycin A (AA), or be inhibited with salicylhydroxamic acid (SHAM), independently of the other branch. These observations have led to the suggestion that alternative oxidase activity (AOX) might play a role in the induction of SAR to viruses. This suggestion is consistent with the finding that AOX protein and transcript levels are raised in tobacco tissue expressing SAR. From Murphy *et al.* (1999a), with kind permission of the copyright holder, © Elsevier Science.

pathway following the interaction of an *Avr* gene with an *R* gene. The action at a distance involved in SAR presumably requires the translocation of some substance or substances. Ross (1966) has presented good evidence that transport of a resistance-inducing material is involved. For example, when the midrib of an

upper tobacco leaf was cut, resistance did not develop in the portion of the lamina distal to the cut. Similarly, killing sections of petiole of inoculated leaves with boiling water, while allowing the leaf to remain turgid, prevented development of resistance in other leaves. Other experiments showed that, in large tobacco plants, the material moved equally well both up and down the stems.

The nature of the material that migrates is unknown, as is the actual mechanism of resistance in the resistant uninfected leaves. The migrating material might involve SA, ethylene, jasmonic acid, nitrous oxide or even small peptides such as systemin (see Howe and Ryan, 1999). This mechanism may or may not be the same as that in the zone of tissue around necrotic lesions. Systemic acquired resistance can be induced by non-necrotic localized viral infection (Roberts, 1982). Systemic acquired resistance is not effective when the challenge virus is one that moves systemically (Pennazio and Roggero, 1988). Thus, one also has to consider an inherent host response (see Section IV).

L. Wound healing responses

Wounds involving necrosis caused by mechanical injury, insects and various pathogens, including viruses, frequently result in a series of wound healing responses by the plant. These responses must involve the non-specific induction of many host-coded proteins. The most complex response is the development of a wound periderm and cell wall changes, including lignification, suberization and the deposition of callose. Virus-induced necrosis may lead to such wound responses. A periderm was formed in young bean leaves inoculated with the VM strain of TMV, which gives very small lesions, but not in old leaves or in leaves inoculated with the U1 strain giving large lesions (Wu, 1973).

Various workers have noted a deposition of callose in cells around necrotic local lesions, leading to thickening of cell walls and probably blocking of plasmodesmata (e.g. Hiruki and Tu, 1972; Wu, 1973). Stobbs *et al.* (1977) found that callose deposition in live cells extended beyond the margin of detectable virus, while remaining within the zone of fluorescent metabolite accumulation in pinto bean leaves infected with TMV. Cell wall glycoproteins were determined chemically following extraction from leaves by Kimmins and Brown (1973). They found an accumulation of glycoproteins following inoculation of hypersensitive hosts with TMV or TNV. An identical response occurred in leaves that had been mock inoculated.

Other observations suggest that cell wall modifications may not be a factor limiting spread. Appiano *et al.* (1977) considered that the conspicuous cell wall lignification seen in lesions caused by TBSV in *Gomphrena* leaves was not a barrier to spread of virus because lignification did not follow the whole cell perimeter, and because virus could be detected beyond the cells with modified walls.

There may be several systemic signaling pathways for wound response (see Rhodes *et al.*, 1999) but recent research suggests that the SAR and wound response pathways may not be completely independent (Maleck and Dietrich, 1999).

M. Antiviral factors

The lesion response in tobacco varieties containing the *N* gene has been associated with the presence of a protein with antiviral properties named 'inhibitor of virus replication' (IVR) (see Loebenstein and Gera, 1981; Loebenstein *et al.*, 1990). IVR, released into the medium from protoplasts of tobacco Samsun NN, inhibits the replication of TMV in protoplasts from both local lesion-responding and systemically infectible tobacco varieties. A 23-kDa protein with antiviral properties has been purified from these extracts (Gera *et al.*, 1990) and a cDNA to a protein recognized by anti-AVR antiserum has been isolated and cloned (Akad *et al.*, 1999). It was suggested that the N-gene products serve as a signal for the induction of IVR-like proteins or that IVR is similar to PR proteins (Loebenstein *et al.*, 1990; Akad *et al.*, 1999).

Sela and colleagues have partially purified and characterized 'antiviral factors' (AVF) from *Nicotiana* cultures with the *N* gene (reviewed by Fraser, 1987a). Parallels have been drawn between the AVFs and human interferons (e.g.

Sela *et al.*, 1987; Edelbaum *et al.*, 1990). However, the significance of this work in relation to host plant resistance remains to be established, as does the relationship of the AVFs and the IVR described above.

N. Ability of virus to spread through various barriers

The factors that control the ability of a virus to pass various tissue boundaries in reaching the sieve elements of an inoculated leaf, to spread through the vascular system and to exit in photosynthetic product sink leaves are discussed in Chapter 9 (Sections II.E and II.G).

One form of resistance that is not similar to the gene-for-gene-type resistance (see Section II.A) for the gene-for-gene hypothesis) is shown by the *Arabidopsis RTM1* gene (Chisholm *et al.*, 2000). This gene is necessary for restriction of long-distance movement of TEV without causing an HR or SAR. The gene product is similar to the *a*-chain of the lectin, jacalin, from *Artocarpus integrifolia*; jacalin belongs to a family of proteins with members that are implicated in defense against insects and fungi.

O. Systemic host response

As described in Chapter 3 (Section II.B), there is a wide range of systemic symptoms induced by viruses, the most common and characteristic of which is the mosaic symptom. The mosaic symptom comprises areas of the leaf showing various degrees of chlorosis together with areas that remain green and are termed 'dark green islands'. The dark green, light green and even chlorotic patches that make up mosaics range from relatively large (e.g. TMV, TYMV and AbMV) to relatively small, giving a fine mosaic (e.g. CPMV in cowpea, AMV in tobacco). These areas are often delimited by the vein structure of the leaf, giving streak or stripe symptoms in monocotyledons. The development of mosaic symptoms is described in more detail in Chapter 9 (Section IV.D). Two approaches are used to study the mosaic symptomatology: mutagenesis of the viral genome and microscopic examination of the infected leaf. Most of the work has been performed on TMV and TYMV, although there is an increasing amount of information accruing from studies on other viral genomes.

1. Chlorosis

In most mosaic symptoms there are regions (domains) with different levels of chlorosis which can vary from almost completely white to very pale green-yellow. The chlorotic response is obviously the effect of the virus either directly or indirectly on the chloroplasts in that region of the leaf causing loss of chlorophyll by perturbation of chloroplast structure and function.

a. TYMV and TMV

The effects of virus infection on the cytology of chloroplasts, especially in relation to TYMV, are described in Chapter 9 (Sections III.K.3 and IV.C). In understanding the effects that the virus has on causing the chlorotic element of a mosaic symptom it is important to consider both the developmental stage of the leaf at the time of infection and the possibility of variants of the virus in different chlorotic domains. If the leaf is almost fully grown when it becomes infected, chlorosis must result from the breakdown of mature chloroplasts. In systemic infections where leaves are very young when they become infected, it is likely that the virus affects chloroplast development.

In some infections, such as TMV in tobacco, the disease in individual plants appears to be produced largely by a single strain of the virus. However, it has been known for many years that occasional bright yellow islands of tissue in the mosaic contain different strains of the virus. Such strains probably arise by mutation, and during leaf development come to exclude the original mild strain from a block of tissue. In Chinese cabbage plants infected with TYMV there may be many islands of tissue of slightly different color from which different strains of the virus can be isolated (Chalcroft and Matthews, 1967a,b).

Specific amino acid substitutions and deletions in the coat protein of TMV affect the production of chlorotic symptoms (Dawson *et al.*,

1988). Mutants that retained the C-terminus of the TMV coat protein induced the strongest chlorotic symptoms in tobacco (Dawson *et al.*, 1988) in both expanded and developing leaves (Lindbeck *et al.*, 1991, 1992). The chlorotic symptom formation is related to the concentration of TMV capsid proteins forming aggregates in infected cells but not accumulating in chloroplasts. In contrast to this, in infections with YSI/1, a naturally occurring chlorotic mutant of the U1 strain of TMV, coat protein is found in the chloroplasts associated with the thylacoid membrane fraction (Banerjee *et al.*, 1995). Reinero and Beachy (1986) detected TMV coat protein in both the stroma and membrane fractions of the chloroplasts of infected cells. The coat protein of YSI/1 differs from that of U1 in two nucleotides, one of which, giving an Asp to Val change at amino acid 19, is responsible for the chlorotic phenotype (Banerjee *et al.*, 1995). The coat protein of another natural chlorotic mutant, flavum, also has a substitution at amino acid 19, this time Asp to Ala (Wittmann *et al.*, 1965). However, these severe chlorotic symptoms are unusual in TMV infections and are associated only with mutants. In natural infections, the chlorotic element of the mosaic is usually light green and is not accompanied by the accumulation of coat protein bodies or with coat protein in chloroplasts.

b. CMV

Chlorosis is a strain-specific symptom of CMV with strains CMV-M and CMV-Y inducing severe systemic chlorosis in tobacco. Pseudorecombinants between CMV-M and a green mosaic-inducing strain, CMV-Fny, located the gene responsible on CMV-M RNA3. Further experiments with recombinant RNA3 transcribed from engineered cDNAs showed that the symptom in tobacco was controlled by the coat protein gene (Shintaku and Palukaitis, 1990). The determinant was further narrowed down to the amino acid at position 129 in the coat protein (Shintaku *et al.*, 1992), which suggested that the local secondary structure influenced the symptom type. The induction of severe chlorotic spots on tobacco leaves inoculated with CMV-Y is also associated with amino acid 129 and with two nuclear-coded recessive host genes (Takahashi and Ehara, 1993). In mutagenesis experiments with CMV-Y and a green mosaic strain, CMV-O, Suzuki *et al.* (1995) showed that the coat protein amino acid 129 was involved not only in the induction of chlorosis but also in severe veinal necrosis and necrotic local lesion production (Table 10.5). They concluded that the symptomatology was associated with the tertiary structure of the coat protein molecule as well as the local secondary structure surrounding amino acid 129.

c. CaMV

Two approaches have been used to dissect the symptom determinants of CaMV: infecting plants with chimeric viral genomes and transforming plants with viral genes. Both of these approaches identify gene VI, encoding the protein P6 as the major factor in inducing chloro-

sis. Stratford and Covey (1989) constructed a series of hybrid CaMV genomes between two strains that cause severe (strain Cabb B-JI) or mild (strain Bari-1) disease in turnips. The determinants for the degree of leaf chlorosis were located in a domain consisting of part of gene VI together with the large intergenic region and nucleotides 6103–6190 of gene VII (see Fig. 10.10).

The transformation of tobacco to express segments of CaMV DNA containing gene VI resulted in transgenic plants showing virus-like symptoms (Baughman *et al.*, 1988; Balázs, 1990; Goldberg *et al.*, 1991). Gene VI from two different virus isolates produced different symptoms—either mosaic-like or a bleaching of the leaves. Symptom production was blocked by deletions or frameshift mutations in gene VI. Production of symptoms was closely correlated with the appearance in the leaves of the 66-kDa gene product, as shown by immunoblotting. However, it was estimated that the amount of

TABLE 10.5 Coat protein involvement in symptom production by CMV

Inoculum[a]	Base		Amino Acid		Symptoms on tobacco plants
	Position[b]	Change(s)	Position	Substitution	
A. Summary of coat protein mutants					
Y3		None		None	Chlorosis
O3		None		None	Green mosaic
Y(SP)	1644	U to C		Ser to Pro	Green mosaic
O(PS)	1645	C to U		Pro to Ser	Chlorosis
Y(SL)	1644, 1645	UC to CU		Ser to Leu	NLL, veinal necrosis
O(PL)	1646	C to U		Pro to Leu	NLL, veinal necrosis
Y(SF)	1645	C to U		Ser to Phe	NLL
O(PF)	1645, 1646	CC to UU		Pro to Phe	NLL
Y(SG)	1645 to 1646	UCU to GGC		Ser to Gly	Green mosaic
B. Summary of Y(SF) revertants					
Y(SF)		None		None	NLL
Y(SF)R1	1645	U to C	129	Phe to Ser	Chlorosis
Y(SF)R2	1698	G to U	147	Ala to Ser	Chlorosis
Y(SF)R3	1698	G to U	147	Ala to Ser	Chlorosis
Y(SF)R4	1690	C to A	144	Ala to Glu	Green mosaic
Y(SF)R5	1672	C to A	138	Ala to Asp	Slight vein necrosis

[a] Designation refers to transcripts from wild-type and mutant RNA3 cDNA. Each transcript was inoculated with transcripts from cDNAs of CMV-Y RNAs 1 and 2.
[b] Nucleotide number (CMV-Y refer to Nitta *et al.*, 1988; CMV-O, Hayakawa *et al.*, 1989) of altered base(s).
NLL, necrotic local lesions on inoculated leaves.
From Suzuki *et al.* (1995), with permission.

the 66-kDa protein produced in transgenic tobacco plants was only about one-twentieth of that found in infected turnips. Furthermore, the fact that tobacco is not a host of CaMV and that transformation with gene VI of FMV, which infects tobacco, gave no symptoms (Goldberg *et al.*, 1991) raised the possibility that this was a non-host effect. To investigate this, Cecchini *et al.* (1997a) constructed a collection of transgenic *Arabidopsis* lines expressing gene VI sequences from CaMV isolates Cabb B-JI, Bari-1 and a recombinant, Baji-31, which causes very severe infections of *Arabidopsis* (Cecchini *et al.*, 1997b). They showed that the symptom character elicited in the gene VI-expressing host plants was dependent on the level of P6 expression and also upon the P6 sequence itself. Using differential display PCR, the changes in abundance of mRNA species in P6 transgenic and CaMV-infected *Arabidopsis* plants were similar when compared with those in uninfected plants (Geri *et al.*, 1999). mRNA species that were down-regulated in transformed and infected plants included those for a phenol-like sulfotransferase and a glycine-rich RNA-binding protein; up-regulated species included a *myb* protein, glycine-rich and stress-inducible proteins.

In further studies on the interactions between CaMV and *Arabidopsis*, Cecchini *et al.* (1998) inoculated *Arabidopsis* ecotype Col-O *gl1* with 30 CaMV isolates. Thirteen isolates failed to cause symptoms, the remainder inducing symptoms that varied between mild and very severe. Some of the symptoms differed markedly from those produced by that isolate in turnip (*Brassica rapa*). A greater variety of symptoms was observed in a single *Arabidopsis* ecotype infected by a range of CaMV isolates than was produced by a single isolate in a range of *Arabidopsis* ecotypes. An EMS-mutagenized *Arabidopsis* (Col-O *dv1*) gave altered symptoms to CaMV, including an uncharacteristic necrotic response. With the detailed molecular understanding of the *Arabidopsis* genome, this approach should reveal plant genes involved in interactions leading to symptom production (see Schenk *et al.*, 2000). The ability of CaMV isolate, W260, to overcome resistance in *A. thaliana* ecotype Tsu-O is associated with ORF VI (Leisner *et al.*, 2001).

d. MSV

Immuno-histochemical and *in situ* hybridization techniques have been used to localize MSV in infected maize plants (Lucy *et al.*, 1996). The virus did not invade the apical meristem and was present only in areas of leaves displaying the characteristic chlorotic streak symptoms. Strains of MSV differ in the width of the streaks and amount of chlorosis that they cause. Strain Ns produces earlier symptoms with broader and more chlorotic streaks than does strain Nm (Boulton *et al.*, 1991a); it also has a broader host range. Site-directed mutagenesis and construction of chimeric viruses showed that genome nucleotide 40 affected the streak width and that nucleotide 2473 determined the severity of chlorosis, the length of the streaks, the latency and host range (Table 10.6). Nucleotide 40 is in gene *V1*, which controls cell-to-cell movement; nucleotide 2473 is in the large intergenic region and the nucleotide change alters a potential promoter sequence 101 nucleotides upstream of the initiation codon of the *C1* gene (see Chapter 7, Section IV.E.3 for mastrevirus promoters). Mutagenesis of promoters downstream of that at −101 showed that the −101 promoter alone conferred the chlorosis, streak length, latency and host range phenotype (Boulton *et al.*, 1991b).

2. Vein clearing

An early systemic symptom of mosaic-causing viruses is vein clearing, which occurs temporarily when the first flush of virus reaches the young leaves. Dawson and colleagues have been trying to explain this phenomenon (reviewed in Dawson, 1999). In TMV infections vein clearing occurs at temperatures above 25°C and its intensity increases with increasing temperature up to 40°C. It is light dependent and can be manifest within 5 minutes of permissive conditions being reached. It precedes most of the virus replication in the leaf and is not associated with any chloroplast abnormalities. The conclusion is that it is an optical illusion.

3. Vein banding

As well as patches of chlorosis and of light and dark green, some mosaic-inducing viruses cause characteristic chlorotic vein banding.

TABLE 10.6 Pathogenicity characteristics of wild-type and mutant MSV genomes in infected maize plants

Virus or hybrid	Symptoms (days pi)[b]	Streak morphology[d]	Chlorosis[e]	Viral DNA concentration[f]	Host range[g]
Experiments 1 and 2					
MSV-Nm[a]	9	N, S	Mild	33	R
MSV-Ns	6	W, L	Severe	100	B
MSV-Nm (s40)	9	W, S	Mild	76	R
MSV-Nm (s2473)	6	N, L	Severe	95	B
MSV-Nm (s40, 2473)	6	W, L	Severe	108	B
Experiments 3 and 4					
MSV-Nm	8	N, S	Mild	ND	R
MSV-Ns	6	W, L	Severe	ND	B
MSV-Nm (TATA, 101)	6[c]	N, L	Severe	ND	ND
MSV-Ns (TATA, 101)	6	W, L	Severe	ND	B

[a] MSV-Ns, Nigerian strain of MSV inducing wide, long, severely chlorotic streaks; MSV-Nm, variant of MSV-Ns giving narrow, discontinuous, mildly chlorotic streaks; (s40) and (s2473), substitutions at nucleotides 40 and 2473 respectively; (TATA, 101), TATA sequences at −57 and −62 modified to GATA and TACA.
[b] Earliest day of symptom appearance following agroinoculation of at least 50 plants with each construct.
[c] In experiment 4, symptoms were first seen on MSV-Nm (TATA, 101)-infected plants 8 days following inoculation.
[d] N, narrow; W, wide; L, long; S, short.
[e] Assessed visually as degree of streak chlorosis.
[f] Relative concentration, percentage values expressed relative to MSV-Ns-infected tissue = 100; mean value from two experiments.
[g] B, broad host range; R, restricted host range.
ND, not determined; pi, post-inoculation.
From Boulton *et al.* (1991b), with permission.

This is especially marked in many strains of CaMV. In a study on the distribution of CaMV, Al-Kaff and Covey (1996) found that there was more virus in the vein bands in systemically invaded expanded leaves than in the interveinal areas, but in younger systemically infected leaves there was less virus in chlorotic vein borders than in the interveinal green islands.

4. Dark green islands

Dark green islands in the mosaic pattern are cytologically and biochemically normal as far as has been tested. In most cases, for example tobamoviruses (Atkinson and Matthews, 1970), cucumoviruses (Loebenstein *et al.*, 1977) and potyviruses (Suzuki *et al.*, 1989), they contain low or zero amounts of infectious virus and no detectable viral protein or viral dsRNA. About 50% of the plants regenerated from dark green island material were virus free (Murakishi and Carlson, 1976). However, in CaMV infections, the dark green islands contain unusual accumulations of virus (Al-Kaff and Covey, 1996). The interveinal areas of CaMV-infected turnip,

which developed as dark green islands, had some of the highest virus concentrations in the leaf.

In a detailed study of the dark green islands in TMV-infected tobacco, Atkinson and Matthews (1970) distinguished two types of island: the 'true' island which remained dark green for the life of the leaf, and the 'pseudo' island in which virus production began at a late stage of leaf development. The pseudo islands had fuzzy boundaries whereas those around true islands were sharp. These authors made a close analysis of the virus content of cells on either side of the chlorotic–dark green boundary (see Fig. 9.21). The chlorotic cells contained large numbers of crystalline TMV inclusions whereas none of the dark green cells had any inclusions. TMV particles were observed in dark green cells close to the boundary but decreased in number away from the boundary until no virus particles were observed in the center of the dark green island.

Several lines of evidence show that dark green islands are resistant to superinfection with the same virus or closely related viruses (e.g. Chalcroft and Matthews, 1967a,b; Atkinson and Matthews, 1970; Loebenstein *et al.*, 1977).

Various factors can influence the proportion of leaf tissue that develops as green islands in a mosaic. These include leaf age, strain of virus, season of the year, and removal of the lower leaves on the plant (Crosbie and Matthews, 1974b). The dark green islands of tissue may not persist in an essentially virus-free state for the life of the leaf. 'Breakdown' leading to virus replication usually takes place after a period of weeks, as in the case of pseudo islands, or after a sudden increase in temperature (Atkinson and Matthews, 1970; Matthews, 1973; Loebenstein *et al.*, 1977).

Dark green islands may be part of a more general phenomenon in plant virus infections. In local lesions caused by TYMV in Chinese cabbage, a proportion of cells show no evidence of virus infection and remain in that state for considerable periods even though neighboring cells are fully infected. It may be that these cells are in a resistant state, like that found in dark green islands of the mosaic. Perhaps a proportion of cells in all infected leaves develop the resistant state, but only cells still retaining the potential to divide can give rise to microscopic or macroscopic islands of dark green tissue. Dark green islands may also represent domains within the leaf into which virus had not entered before host defense mechanisms became activated (see below).

P. Development of mosaic disease

The structure of mosaic symptoms is described in Chapter 9 (Section IV.D). Various hypotheses have been proposed as to how these patterns develop. As noted previously, they develop only in sink leaves to which the virus is transported systemically and are most obvious in leaves that develop at or after the time of systemic invasion.

When the third edition of this book was written (Matthews, 1991), it was thought that mosaic patterns were laid down in the shoot apex. It was assumed that, in a plant infected with a mixture of virus strains, the first virus particle to establish itself in a dividing cell pre-empted that cell and all or almost all its progeny, giving rise in the mature leaf to a macroscopic or microscopic island of tissue occupied by the initial strain. The main observations to support this hypothesis were:

- Mosaics caused by TMV and TYMV are most apparent in leaves that were in the developmental stage at the time of systemic invasion; in the case of TMV, this is tobacco leaves less than 1.5 cm long.
- There is often a gradient in sizes of the patches making up the mosaic, with younger leaves having larger islands.
- In some cases, one half of the leaf would be dark green and the other half light green or chlorotic.
- Detailed examination showed microscopic mosaics of cells with different levels of virus infection. In these, different horizontal layers of mesophyll had different type of virus-induced chloroplast damage (see Fig. 9.27).

However, there are several points that do not support this hypothesis:

1. Only some viruses reach the meristematic region (see Chapter 9, Section II.J.8) where they would have to be to infect cells during the division stage.
2. The recent understanding of long-distance movement of viruses in relation to source–sink translocation of photoassimilates (see Chapter 9, Section II.G) shows that systemically moving viruses unload from the vascular system some distance from the meristematic region. They would then have to move from cell to cell to reach dividing leaf primordial.
3. The recent studies on cell-to-cell communication indicate that there are symplastic domains within which there is good communication and between which the communication is poor.
4. There is an increasing realization that the gene-silencing host defense system (see Section IV.A) plays a very important role in the overall picture of virus infection. Ratcliff *et al.* (1997) suggested that dark green islands might be the result of gene silencing.

Thus, the picture that is emerging is that there is a complex of factors involved in the development of mosaic symptoms including systemic and local movement of the virus, the ability of the virus to invade meristematic tissues, the strain of the virus and its propensity to mutate and, probably above all, the conflict between the invasiveness of the virus and the response of the host defense system(s). It is quite likely that there is a different balance between these, and other, factors in the development of mosaic symptoms in different virus–host combinations.

Q. Symptom severity

A combination of factors leads to differences in the severity of symptoms caused by different isolates or strains of a virus. Some of these have been noted above in the discussion of chlorosis (Section III.O). The factors controlling symptom severity are both viral and host. For instance, Martin *et al.* (1997a) examined the responses of 116 ecotypes of *Arabidopsis* to inoculation with YoMV. They defined five symptom groups of ecotypes based on stunting, abnormal flower or seed formation, and plant death. Lee *et al.* (1996) have identified a single locus, *TTR1*, in *Arabidopsis* that confers tolerance to TRSV.

In some cases, differences in symptom severity have been attributable to individual viral genes. For instance, comparison of the nucleotide sequences of two almost symptomless mutants of TMV (LII and LIIA) with the parent-type strain showed that a change from Cys to Tyr at amino acid position 348 of the 126-kDa and 183-kDa protein(s) was involved in loss of symptom production. Two other amino acid changes in these proteins may also have been involved (Nishiguchi *et al.*, 1985). The differential symptom determinants of the Holmes' masked (M) and U1 strains of TMV have been mapped to the 126/183-kDa protein, this time to the N-terminal region (Shintaku *et al.*, 1996). Single or multiple substitutions between eight nucleotides in non-conserved domains in the two strains altered the symptomatology but did not always induce complementary visual symptoms. In some cases, there were spontaneous second-site mutations that also influenced the symptom phenotype. There was no correlation between the severity of systemic symptoms and virus accumulation.

Pseudorecombinants and recombinants between CMV and TAV identified the 3′ region of RNA3 (encoding the coat protein) as the major determinant of symptom severity (Salánki *et al.*, 1997). One combination of RNA1 and RNA2 from TAV and RNA3 from CMV gave more severe symptoms in *N. benthamiana* than either of the parental viruses or any other pseudorecombinant. On the other hand, RNA2 of a subgroup I CMV strain was identified as being involved in symptom severity in tomato (Hellwald *et al.*, 2000).

In an increasing number of cases the transgenic expression of viral genes has been used to explore symptom expression. For instance, Ghorbel *et al.* (2001) suggest that CTV p23 is responsible for some symptom expression in Mexican lime. However, transgene expression does not necessarily identify true symptom determinants.

Not all severity determinants are proteins.

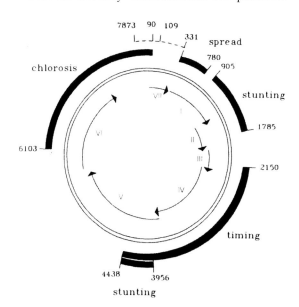

Fig. 10.10 Location of CaMV genome domains containing strain-specific symptom determinants. Arrows with roman numerals represent viral genes. Domains containing sequences involved in determining different symptom characters between CaMV strains Cabb B-JI and Bari 1 infections are labeled: spread, rate of spread of systemic vein clearing symptoms on younger leaves; stunting, whole plant stunting; timing, period between inoculation and appearance of first systemic vein clearing symptoms; chlorosis, change in chlorophyll content of leaves. The dashed line represents a domain that could contain loci influencing stunting and timing (nucleotides 7873–109) and symptom spread (nucleotides 109–331), but its specific role has not been determined. From Stratford and Covey (1989), with permission.

The 3'-terminal non-coding region of TVMV determines the symptom severity in *N. tabacum* (Rodríguez-Cerezo *et al.*, 1991a).

However, for many viruses, it is the combination of different viral genes (and possibly non-coding regions) that determines the severity of the symptoms. In their analysis of the effect of various CaMV genes on the symptomatology of the virus in turnip, Stratford and Covey (1988) detected a variety of loci affecting disease development (Fig. 10.10). As well as the effect of gene VI on chlorosis (see Section III.O.1.c), plant stunting was affected by at least two separate loci, one containing parts of genes I and II and the second within the reverse transcriptase gene (V). Thus, different aspects of the disease process can be assigned to specific parts of the genome, and much of the viral genome appears to be involved.

An example of more than one RNA viral gene being responsible is shown in experiments with hybrids between the tombusviruses Cym-RSV and ClRV. Burgyán *et al.* (2000) demonstrated that the necrotic response of *N. benthamiana* is associated with the produce of ORF 1 (p35) as well as p19.

As members of the plant reovirus group cannot be cloned by single local lesion selection, isolation of mutants has been difficult. However, Kimura *et al.* (1987) injected dilute inoculum of the type strain (0) of RDV into insect vectors, which were then fed on rice plants. They repeatedly selected for rice plants that showed unusually severe symptoms. By these means, they obtained a severe strain (S) (see Fig. 10.11). The fourth largest genome segment of strain S had an apparent MW about 20 kDa larger than that of strain 0. The corresponding gene product in strain 0 had an M_r of about 43 kDa, and in strain S the M_r was about 44 kDa. This protein is located in the outer envelope of the virus. The idea that this gene product is involved in producing the more severe symptoms receives support from the fact that neurovirulence in a reovirus infecting mice has been shown to be controlled by the outer envelope protein (Weiner *et al.*, 1977).

We have to distinguish between symptom severity and symptom phenotype. For instance, the common strain of the begomovirus, TGMV, induces extensive chlorosis in *N. benthamiana*,

Fig. 10.11 Stunting effect of rice dwarf *Phytoreovirus* infection in rice. Healthy plant on right. Plants infected with a standard strain (O) and a severe mutant (S). From Kimura *et al.* (1987), with permission.

whereas the yellow strain produces vein chlorosis on systemically-infected leaves. The symptoms of these two strains also differ on *Datura stramonium*. The difference in phenotype expression of these two strains is determined by a single nucleotide in the 3' region of the gene encoding the movement protein (Saunders *et al.*, 2001a).

R. Recovery

For several viruses, infected plants apparently recover after the initial outbreak of systemic symptoms. This is especially marked in nepoviruses (see Section IV.G) (Wingard, 1928; Lister and Murant, 1967), tobraviruses (Cadman and Harrison, 1959) and caulimoviruses (Al-Kaff and Covey, 1995). Virus can be isolated from the symptomless young leaves of plants that have recovered from nepovirus infection

(Wingard, 1928) and also from symptomless seedlings in which nepoviruses are seed transmitted (Lister and Murant, 1967).

AMV can also show recovery in *N. tabacum* but not in *Chenopodium amaranticolor* (Ross, 1941; Gibbs and Tinsley, 1961); virus can still be isolated from recovered tobacco leaves (Gibbs and Tinsley, 1961). The virus concentrations in tobacco plants rises and falls in a cyclical manner (reviewed in Hull, 1969) (Fig. 10.12).

Ratcliff *et al.* (1999) note that many of the viruses that appear to recover (except CaMV) have the ability to invade the meristem; this is likely to be associated with pollen transmissibility of the virus.

IV. INHERENT HOST RESPONSE

A. Gene silencing

The application of genetic engineering to confer resistance to viruses in plants described in

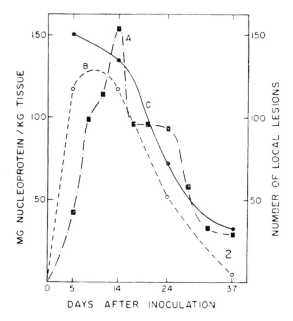

Fig. 10.12 Concentration and specific infectivity of AMV harvested from whole tobacco plants at different times following inoculation. Zero lesions are assumed for time 0. Curve A, amount of purified virus nucleoprotein (mg/kg total leaf wet weight); curve B, numbers of local lesions induced by sap inoculation; curve C, number of local lesions when purified virus samples were equalized spectrophotometrically. From Kuhn and Bancroft (1961), with permission.

Chapter 16 (Section VII) was directed primarily at the expression of viral sequences that would interfere with the normal functioning of the target virus. The initial approaches were to express wild-type or mutated viral genes in the expectation that the gene products would block crucial steps in viral replication and/or propagation. However, various observations on several systems suggested that the mechanism(s) by which the transgenic plants were being protected against viral infection were, in some cases, not as predicted. These observations included:

- In many cases, it was the transcript and not the protein that was involved in effecting protection. For instance, some lines of potato plants, transgenic for the PLRV coat protein and showing high levels of resistance to the virus, did not contain detectable levels of coat protein (Kawchuk *et al.*, 1991). Resistance to TSWV and TEV was induced by transforming tobacco plants with translationally deficient versions of the *N* gene and of the coat protein respectively (de Haan *et al.*, 1992; Lindbo and Dougherty, 1992b).
- Often plants expressing a viral transgene at low level were more resistant than those expressing the transgene at high levels. Thus, in plants transgenic for the TSWV *N* gene, the highest levels of resistance were found in plants accumulating the lowest levels of transcript (Pang *et al.*, 1993).
- Some tobacco lines transformed with either the complete or truncated coat protein gene of TEV were initially susceptible to TEV infection but the plants 'recovered' from the infection after about 3–5 weeks. The recovered tissue could not be reinfected with TEV, nor could protoplasts from it. There was a significant decline in the steady-state level of the transgene RNA in the recovered tissue, although nuclear run-off studies showed no differences in transgene transcription rates between recovered and uninoculated tissue (Lindbo *et al.*, 1993).

These and other observations suggested that the resistance might be due to homology-dependent gene silencing. To test this possibility, crosses were made between lines of tobacco

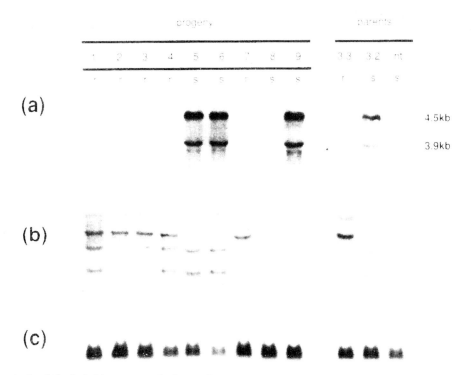

Fig. 10.13 Analysis of the hybrid progeny of tobacco lines, transgenic in PVX replicase, 3.2 (PVX susceptible) and 3.3 (PVX resistant). Leaves were harvested for RNA analysis when the plants were at the five-leaf stage. The probe for gel blot analysis of total RNA (8 μg) samples was a 1.6-kb fragment of the PVX RdRp DNA **(a)** or a rubisco small subunit probe **(c)** to confirm equal loading of the RNA samples. The genotype of each line was determined by DNA gel blot analysis of *Eco*RV-digested DNA (8 μg per track) **(b)**. After removal of the leaf sample for RNA analysis the plants were inoculated with PVX and assigned as either resistant (r) or susceptible (s) depending on the absence or presence of systemic mosaic symptoms at 14 days post-inoculation. Samples of the lines 3.2.2 and 3.3.4 and non-transformed tobacco were included at each stage of the analysis for comparison and labeled as 3.2, 3.3 and nt respectively. From Mueller *et al.* (1995), with kind permission of the copyright holder, © Blackwell Science Ltd.

transformed with the replicase enzyme gene of PVX (Mueller *et al.*, 1995) (Fig. 10.13).

Some of the lines (e.g. line 3.3) were virus resistant and expressed the transgene at low level; other lines (e.g. 3.2) were susceptible and expressed the transgene at relatively high levels. All the progeny from crosses between these two lines that were virus resistant expressed the transgene at low levels. The progeny of crosses with non-transgenic tobaccos showed the transgenic parent phenotype. These observations were in accord with an increasing number of cases in which transformation with homologs of endogenous plant genes led to both the transgene and endogenous gene expression being co-suppressed (e.g. Napoli *et al.*, 1990; van der Krol *et al.*, 1990). This co-suppression is due to either transcriptional gene silencing (TGS) or post-transcriptional gene silencing (PTGS) or, possibly, a combination of the two (for reviews

see Depicker and van Montagu, 1997; Vaucheret *et al.*, 1998; Matzke *et al.*, 2000).

B. Transcriptional and post-transcriptional gene silencing

The main features of TGS and PTGS are given in Table 10.7.

As most plant viruses have RNA genomes that replicate in the cytoplasm, it is likely that the transgenic resistance operates by a PTGS mechanism involving RNA. In an important experiment, English *et al.* (1996) showed that PTGS operates against the whole RNA in which a target sequence is located. PVX vectors in which the β-glucuronidase (GUS) or green fluorescent protein (GFP) gene had been inserted between the triple gene block and the coat protein gene (see Fig. 6.39 for PVX genome organization) were inoculated to tobacco plants

TABLE 10.7 General features of gene silencing mechanisms

Feature	Position-dependent gene silencing	Homology-dependent TGS	Homology-dependent PTGS
Homology of silencing sequence and silenced genes	No	Yes	Yes
Properties of the silenced target locus	*Cis*-located sequence	Homology with silencing locus in the promoter	Homology with silencing locus in the transcribed region
Properties of the silencing locus	Not applicable	Frequently direct or inverted repeated sequences	Frequently direct or inverted repeated sequences
Silencing step	Transcriptional	Transcriptional	Post-transcriptional
Localization	Nuclear	Nuclear	Cytoplasmic
Trans-inactivation	No	Yes; paramutagenic	Yes
Phenotype	Recessive	Dominant, sometimes epistatic	Dominant and epistatic
Protein levels	Strongly reduced	Strongly reduced	Reduced
RNA levels	Strongly reduced	Strongly reduced	Reduced
Silencing capacity of a locus			
Influence of position effects	Not applicable	Strong	Variable
Influence of sequence organization	Not applicable	Strong	Strong
Silencing susceptibility of a locus			
Influence of position effects	Strong	Variable	No
Influence of locus organization	Variable	Variable	No
Silencing inducer signals	Developmental regulation by thresholds of non-coding RNAs and silencing proteins	*In cis* and *in trans* inactivation by DNA–DNA pairing	RNA thresholds
Silencing effectors	Silencing proteins and/or non-coding RNAs mediating heterochromatinization	Methylation and packing of the chromatin	Degradosomes and anti-sense RNA
Methylation of gene	Present in promoter or complete locus	Present in promoter or complete locus	When present, found in transcribed and downstream regions
Stability/persistence	Meiotic resetting Developmentally controlled	Persistent in different degrees	Meiotic resetting Infrequent somatic resetting Developmentally controlled
Environmental susceptibility	Yes (gene regulation) No (genetic programs)	Variable	Yes (weak silencing loci) No (strong silencing loci)
Examples	*Cis*-regulated elements and enhancers Position–effect variegation X-chromosome dosage compensation	Natural paramutation Homology-dependent silencing of transgenes, endogens and transposons	Natural *in trans* silencing Variation of transgene expression Co-suppression Virus-induced gene silencing

From Depicker and van Montagu (1997), with kind permission of the copyright holder, © Elsevier Science.

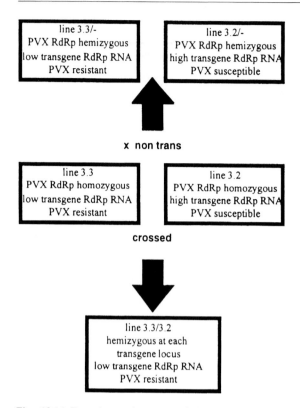

Fig. 10.14 Experimental testing of the relationship of virus resistance and PTGS. Lines 3.3 and 3.2 are tobacco lines carrying a PVX RdRp transgene. The diagram illustrates the phenotypes of plants carrying these genes either individually or in combination, and shows how it was concluded that the PVX RdRp transgenes conferring PVX resistance were also able to confer PTGS. The primary data were presented by Mueller *et al.* (1995). The crosses with non-transformed plants (upper panel) revealed that the transgene phenotype with these lines was not affected by transgene dosage. The crosses between the two lines showed that the transgene in line 3.3 could suppress expression of the transgene in line 3.2. The results also show that PVX resistance and low transgene expression were epistatic to high-level expression and PVX susceptibility. From Baulcombe (1999a), with kind permission of the copyright holder, © Springer-Verlag GmbH and Co. KG.

that were transformed with, and expressed, the GUS gene at low and high levels; the low-level expressers showed PTGS whereas the high-level expressers did not. The low-level expressers were resistant to both the GUS and to PVX in the PVX:GUS construct but not to the PVX:GFP construct. Plants expressing high levels of GUS transgene were susceptible to both constructs (Fig. 10.14).

Various experiments have shown that PTGS is highly nucleotide sequence specific. For

instance, the above example demonstrated that it targeted GUS but not GFP inserts in the PVX vector inoculated to GUS transgenic plants. A PTGS transgene based on the TEV coat protein confers resistance against TEV but not against other potyviruses (Lindbo *et al.*, 1993).

C. Genes involved in post-transcriptional gene silencing

PTGS has been found in other eukaryotic organisms and has been studied in detail in fungi where it is termed 'quelling', and in nematodes and *Drosophila* where it is termed RNA interference (RNAi) (Gura, 2000; Marx, 2000). By studying mutants of these organisms and of *Arabidopsis* that were defective in PTGS, several genes thought to be involved in the defense pathway have been identified (see Morel and Vaucheret, 2000).

An RNA-dependent RNA polymerase (RdRp) (QDE-1) was shown to be required for PTGS in the fungus *Neurospora crassa* (Cogoni and Macino, 1999a) and in *Arabidopsis* (Dalmay *et al.*, 2000). Mutagenesis of *Arabidopsis* identified four silencing defective (*sde*) loci, one of which, *sde1*, is a homolog of the RdRP, QDE-1, from *N. crassa*. Although *sde* mutations affect transgene silencing, they do not have any effect on virus-induced silencing (Dalmay *et al.*, 2000). In a similar study, Mourrain *et al.* (2000) isolated two *Arabidopsis* mutants, *sgs2* and *sgs3*, impaired in PTGS. SGS2 protein is similar to *N. crassa* QDE-1 and to the nematode, *Caenorhabditis elegans* EGO-1, both RdRps; EGO-1 functions in RNAi (Smardon *et al.*, 2000). The functions of SGS3 and of the other sde proteins are unknown but these proteins are likely to be involved in the PTGS pathway in plants. Analysis of a further *Arabidopsis* mutant impaired in PTGS, AGO-1, revealed one amino acid essential for PTGS that is also present in QDE-2 and RDE-1 in a highly conserved motif (Fagard *et al.*, 2000).

A mutant of the alga, *Chlamydomonas reinhardtii*, *Mut6*, required for the silencing of a transgene and two transposon families, is homologous to the DEAH-box family of RNA helicases (Wu-Scharf *et al.*, 2000). A different helicase, encoded by the gene *smg-2*, is involved in PTGS in *C. elegans* (Domeier *et al.*, 2000) and a further one (QDE-3), related to RecQ DNA helicase, in *N. crassa* (Cogoni and Macino, 1999b). Mutants have also identified an RNaseD-like protein (Ketting *et al.*, 1999) and a protein encoded by the *piwi/sting/argonaute/zwille/eIF2C* gene family (Tabara *et al.*, 1999); the domain organization of the latter, which includes the QDE-2 and RDE-1 proteins, is described by Cerutti *et al.* (2000).

D. Mechanism of post-transcriptional gene silencing

PTGS operates against transgenes, retro-elements and RNA viruses. The target for PTGS in most cases is sense RNA (see Baulcombe, 1996b), suggesting that the mechanism oper-ates via an antisense RNA (Dougherty and Parks, 1995). Gene silencing can be induced by the simultaneous expression of sense and anti-sense RNA (Waterhouse *et al.*, 1998) and efficient silencing is induced by hairpin RNAs into which introns have been spliced (Fig. 10.15) (Smith *et al.*, 2000b).

Much of the early search was for antisense RNA molecules in the size range 200–500 bp, which were not found. However, when Hamilton and Baulcombe (1999) examined four classes of PTGS in plants — (1) tomato lines co-suppressing endogenous 1-aminocyclopropane-1-carboxylate oxidase (ACO), (2) a tobacco line expressing GUS transgenes from the CaMV 35S promoter crossed with a line containing a trans-gene suppressor of the 35S promoter, (3)

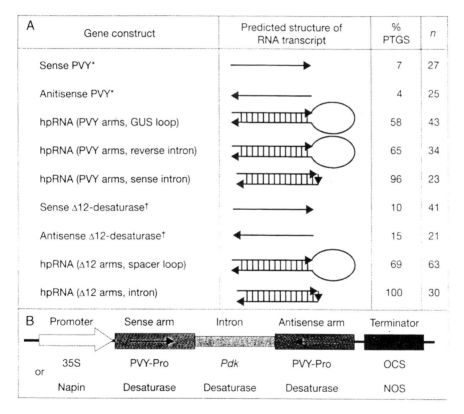

Fig. 10.15 Efficiency of induction of PTGS by different gene constructs and predicted structure of RNA transcribed from the transgenes. **(A)** PTGS efficiency measured for PVY and Δ12-desaturase genes as the percentage of independent transgenic plants immune to PVY and the percentage of plants with enzyme activity reduced by more than 20% compared with wild type respectively. In the predicted structures of RNA transcripts, right- and left-pointing arrows represent sense and anti-sense orientation of sequences respectively; small vertical arrows represent splice-junction sequences remaining after the intron has been spliced out. Vertical lines in the predicted structures indicate duplex formation. *Data from Waterhouse *et al.* (1998); †data from Cartea *et al.* (1998); hpRNA, hairpin RNA; n, number of independent transformants; GUS, β-glucuronidase. **(B)** Design of intron-containing hairpin constructs. OCS, octopine synthase; NOS, nopaline synthase. From Smith *et al.* (2000b) with kind permission of the copyright holder, © Macmillan Magazines Ltd.

Nicotiana benthamiana plants transformed to express GFP infiltrated with *Agrobacterium tumefaciens* containing GFP sequences in a binary vector, (4) PVX-infected plants—they detected in all four cases sense and antisense RNA molecules to the transgene or virus of about 25 nt. These RNA species accumulated in plants showing PTGS but were never found in the absence of PTGS. Similarly, in *Drosophila,* RNAi results in the formation of mRNA fragments of 21–23 nucleotides (Sharp and Zamore, 2000; Zamore *et al.*, 2000). The involvement of RNA degradation in PTGS is reviewed by Meins (2000) and Waterhouse (2001).

Thus, the target for PTGS is dsRNA (see Bass, 2000). Various models have been proposed to explain how these antisense RNAs can be produced, two of which have been considered in depth (Baulcombe, 1999a). In the first, it is suggested that the insertion site of the transgene places it adjacent to an endogenous promoter that would transcribe antisense RNA. However, this seems unlikely on several counts. There is no correlation between the direct transcription of an antisense RNA and PTGS (van Blokland *et al.*, 1994); the suppression of the transgene promoter leads to loss of PTGS (English *et al.*, 1997). Furthermore, it would not explain the production of the 25-nt antisense RNAs in PVX infection (class (4) above). The

second model suggests that the antisense RNA is produced indirectly from a sense transcript from the transgene. This model seems the more likely as it accords with the available data. Various templates for the formation of dsRNA have been suggested. For instance, transcription of integrated DNA has been suggested to give aberrant RNAs; the replication of RNA viruses involves the synthesis of complementary strands. The transcription of antisense RNA from an RNA template (say the aberrant RNAs) requires an RNA-dependent RNA polymerase (RdRp) and such an enzyme was shown to be required for PTGS in *Neurospora crassa* (Cogoni and Macino, 1999a) and in *Arabidopsis* (Dalmay *et al.*, 2000). As described in the previous section, various other gene products have been implicated as being involved in PTGS. Several of these proteins are also involved in the normal molecular network that regulates RNA processing and stability in eukaryotes (see Baulcombe, 2000; Wu-Scharf *et al.*, 2000).

Not all transgenes induce PTGS, as novel traits can be introduced by transformation. Various models have been proposed to explain this (reviewed by Waterhouse *et al.*, 1999), several of which suggest that PTGS is induced when the transcripts of the transgene exceed a certain level. This 'threshold' model, initially suggested by Lindbo and Dougherty (1992a)

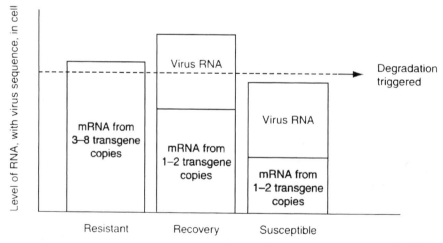

Fig. 10.16 Initiation of RNA-mediated virus resistance (RMVR) according to the threshold model. A threshold of RNA containing virus sequences is required to initiate RMVR. Plants with between three and eight transgenes meet this level and show resistance. Some plants with one or two transgene copies exceed this level only in conjunction with RNA from the virus, and resistance develops some time after infection. Other plants with one or two transgene copies express insufficient mRNA to exceed the threshold even in conjunction with viral RNA, and show no resistance. From Waterhouse *et al.* (1999), with kind permission of the copyright holder, © Elsevier Science.

and by Dougherty and Parks (1995), for the induction of PTGS is illustrated in Fig. 10.16.

Multiple copies of transgenes can induce their methylation, which, it is suggested, might lead to the production of short aberrant RNA species that induce PTGS (see Waterhouse *et al.*, 1999; Mette *et al.*, 2000), although not all cases of silencing involve DNA methylation (Wang and Waterhouse, 2000).

E. PTGS systemic signaling (reviewed by Fagard and Vaucheret, 2000)

In grafting experiments, plants showing PTGS transmit this character 100% from silenced stocks to non-silenced scions expressing the corresponding transgene (Palauqui *et al.*, 1997), even when the stock and scion are separated by 30 cm of stem of non-target wild-type plant. This suggested that a transgene-specific diffusible messenger mediates the propagation of *de novo* PTGS through the plant. Systemic silencing has also been demonstrated by Voinnet *et al.* (1998), who infiltrated lower leaves of *N. benthamiana*, transgenic with a GFP construct, with *A. tumefaciens* carrying a 35S:GFP construct. After 7–14 days the GFP expression was lost from the upper leaves, especially around the veins. The identity of the signal molecule is unknown but all the evidence (sequence specificity, systemic translocation) suggests a small antisense RNA.

Staining tobacco plants showing PTGS of the β-glucuronidase (*uidA*) transgene for Gus expression revealed no GUS staining of leaves or cross-sections of stems, but dark blue GUS staining of the shoot apical and axillary meristems (Fig. 10.17) (Béclin *et al.*, 1998). This indicates that silencing does not affect meristems and thus takes place during the development of each leaf.

F. Induction and maintenance

In a virus-induced gene silencing (VIGS) approach, Ruiz *et al.* (1998) inoculated *N. benthamiana* plants with PVX vectors carrying the exons of the endogenous phytoene desaturase (PDS) gene. The PDS mRNA was affected in all the green tissue. When PVX:GFP was inoculated to GFP transgenic plants, the PVX:GFP was silenced. From their analysis of the silencing in different parts of these plants, Ruiz *et al.* (1998) suggested that there are two stages in PTGS: initiation of silencing and maintenance of silencing. The initiation of silencing is an RNA-mediated defense reaction as a transgene is not necessary, and the maintenance requires the presence of the transgene and probably involves methylation of the transgene dependent on an RNA–DNA interaction (Jones *et al.*, 1999). Thus, for PTGS in transgenic plants there are three stages: initiation, systemic signaling and maintenance.

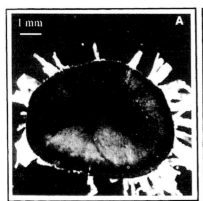 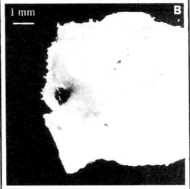 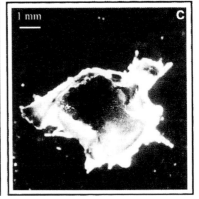

Fig. 10.17 (see Plate 10.2) Histochemical analysis of *uidA* expression in the meristems of silenced and non-silenced tobacco plants. Histochemical GUS assays were performed on stems and shoot tips of homozygous 6b5 tobacco plants showing *uidA* PTGS and hemizygous 23b9 plants that did not show *uidA* PTGS. **(A)** Staining of stem cross-section of non-silenced 23b9 plant. **(B)** Staining of stem cross-section of silenced 6b5 plant. A non-silenced axillary meristem is visible as a dark blue spot. **(C)** Staining of the shoot apical meristem of a silenced 6b5 plant. From Béclin *et al.* (1998), with permission.

G. PTGS in virus-infected plants

(reviewed by Carrington and Whitham, 1998; Marathe *et al.*, 2000a; Carrington *et al.*, 2001)

The finding that viruses can both initiate and be the targets of gene silencing in transgenic plants raised the speculation that PTGS is part of a defense system (also termed RNA-mediated defense, RMD; Voinnet *et al.*, 1999) in plants against viruses and other 'foreign' nucleic acids (Baulcombe, 1996a; Pruss *et al.*, 1997). The similarities between plant defense against viruses and gene silencing were enhanced by several further observations including: (1) various experiments similar to that shown in Fig. 10.18 indicated that homologous genes carried by virus infection of a transgenic plant silenced the expression of the transgene. (2) In nepovirus infection of non-transgenic plants there are severe viral symptoms on the inoculated and first systemically infected leaves; however, upper leaves developing after systemic infection are symptom-free and contain lesser amounts of virus than do the systemic leaves (Wingard, 1928) (see Section III.R). These recovered leaves are resistant to reinoculation of the virus. When the recovered leaves of *N. clevelandii* plants inoculated with the TBRV strain W22 were challenged with a PVX:W22 construct, the PVX:W22 RNA could not be detected, in contrast to the high levels of this RNA is control plants (Ratcliff *et al.*, 1997). Thus, the PVX:W22 was suppressed by the infection of the plants by TBRV W22 in a manner similar to PTGS in transgenic plants. (3) Kohlrabi (*Brassica oleracea gongylodes*) plants inoculated with CaMV initially develop systemic symptoms from which they recover completely. This recovery coincides with marked changes in intermediates of viral replication (see Chapter 8, Section VII.B, for CaMV replication cycle). The viral minichromosome accumulates but the levels of both of the main viral polyadenylated RNAs decline rapidly, although non-polyadenylated fragments persist (Covey *et al.*, 1997). These changes in the levels of viral replication intermediates are consistent with gene silencing halting the replication of the virus.

However, the difference between these two examples of host defense against viral infection

Systemic accumulation ofTMV::GFP and PVX::GUS

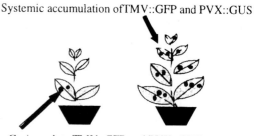

Co-inoculate TMV::GFP and PVX::GUS

Systemic accumulation PVX::GUSGF **only**

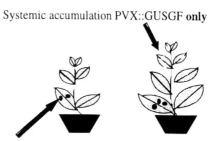

Co-inoculate TMV::GFP and PVX::GUSGF

Fig. 10.18 RNA sequence specificity in cross-protection (PTGS). *Nicotiana benthamiana* plants were inoculated with mixed TMV and PVX vector constructs. The TMV vector was TMV:GFP. The PVX vectors were either PVX:GUS or PVX:GUSGF in which the GF component represents the 5′ part of the GFP reporter gene. Cross-protection between the two vector constructs was assessed in systemically infected tissue (Ratcliff *et al.*, 1999). In the plants inoculated with PVX:GUSGF and TMV:GFP there was an interaction involving the GFP sequence in the TMV construct and the GEP-derived GF in PVX:GUSGF. The consequence of this interaction was cross-protection and subsequent suppression of TMV:GFP in the systemic parts of the plant, which is suggested to involve a PTGS-like mechanism. From Baulcombe (1999a), with kind permission of the copyright holder, © Springer-Verlag GmbH and Co. KG.

is that the replication of the RNA of TBRV is cytoplasmic whereas that of the DNA of the pararetrovirus CaMV occurs in both the nucleus and cytoplasm. The silencing of TBRV is consistent with the PTGS mechanism, whereas there is evidence that silencing of CaMV involves both PTGS and TGS (Al-Kaff *et al.*, 1998; Covey and Al-Kaff, 2000).

Thus, PTGS can be considered to be a generalized response to infection of plants by viruses. This raises the questions of how this defense is initiated and maintained. As RNA viruses replicate via a complementary strand, the formation of dsRNA is an integral part of the replication cycle. Furthermore, host RdRp activity is induced in virus-infected plants (Xie et al., 2001). The replication intermediate, or dsRNA formed by transcription by the host RdRp, would be a target for the pathway giving the 22–25 nucleotide characteristic degradation products (see Ruiz et al., 1998). Presumably the systemic movement of the signal would be similar to that described above for the transgenic situation. If PTGS operated more rapidly than the virus spread, the virus would not be able to move from the site of infection. If the virus moved more rapidly than the PTGS, it would establish systemic infection, the effectiveness of which would depend upon the 'aggressiveness' of the virus and the 'responsiveness' of the PTGS. Factors involved in virus aggressiveness are discussed in the next section.

H. Suppression of gene silencing

(reviewed in Carrington et al., 2001)

Obviously, if PTGS is a normal defense system in plants against 'foreign' nucleic acids, for a virus to establish infection successfully this defense has to be overcome. In studying the

mediator of synergistic interactions (Section V.F.4) between PVX and PVY, Pruss et al. (1997) identified that the TEV HC-Pro gene potentiated the synergism and suggested that this might be due to it interfering with a host defense system. Experiments reported by Anandalakshmi et al. (1998) and by Kasschau and Carrington (1998) confirmed that the potyviral HC-Pro did indeed suppress the action of PTGS. Both groups essentially used the same approach of inoculating transgenic plants with virus vectors containing various inserts. The experiments are summarized in Fig. 10.19 and Table 10.8.

Since this initial demonstration of viral suppression of gene silencing, the phenomenon has been shown for several other viruses. Voinnet et al. (1999) examined 16 viruses using the PVX vector system and showed suppression of PTGS in 12 of them (Table 10.9).

The experiments of Brigneti et al. (1998) (Fig. 10.20), which confirmed the suppression of gene silencing by TEV and CMV (Béclin et al., 1998), also showed that it was the 2b gene of CMV that was involved and that the suppression was

TABLE 10.8 Experiment demonstrating suppression of gene silencing by TEV P1/HC-Pro

Plants	Transgene genotype		Infection	
	Gus	P1/HC-Pro	TEV-GFP	TEV-GUS
Non-transgenic	–	–	+++	+++
407	2n	–	++	–
F3 generation				
U-6B × 407#34	–	–	+++	+++
U-6B-407#7	2n	–	+++	–
U-6B × 407#17				
P1/HC-Pro (+)	2n	1n.2n	+++	+++
P1/HC-Pro (−)	2n	–	+++	–
407 × U-6B#13	–	–	+++	+++
407 × U-6B#9	2n	–	+++	=
407 × U-6B#25				
P1/HC-Pro (+)	2n	1n/2n	++	+++
P1/HC-Pro (−)	2n	=	+++	—

Experimental system:
Transgenic *Nicotiana benthamiana* expressing a non-translatable GUS gene (line 407).
Two infectious cDNAs to TEV, TEV-GUS containing the GUS gene and TEV-GFP containing the GFP gene (inserted between P1 and HC-Pro). 407 is immune to TEV-GUS and susceptible to TEV-GFP.
N. benthamiana line U-6B expressing TEV P1/HC-Pro genes.
407 crossed with U-6B, F1 selfed to give F2; F3 generation as in Table.
There were three phenotypes: homozygous GUS:hemizygous P1/HC-Pro (#17, #25); homozygous GUS:nul P1/HC-Pro (#7, #9); nul GUS:nul P1/HC-Pro (#34, #13).
From Kasschau and Carrington (1998), with kind permission of the copyright holder, © Elsevier Science.

A

B

PVX-GFPt

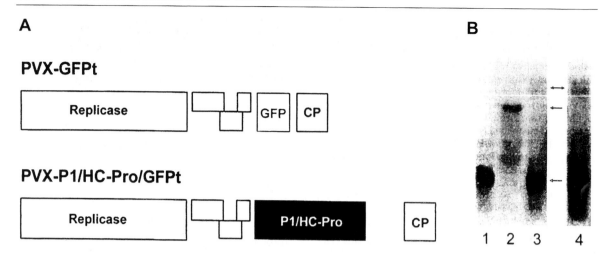

PVX-P1/HC-Pro/GFPt

1   2   3   4

Fig. 10.19 See caption on the following page.

mediated by the protein rather than a nucleic acid. This protein localizes to the nucleus via an arginine-rich nuclear localization signal ([22]KRRRRR[27]) (Lucy et al., 2000). The nuclear targeting is required for the suppression of PTGS.

Voinnet et al. (1999) identified three further viral gene products that are involved in gene silencing suppression (Table 10.9) and observed that there was no common feature between them except that they were frequently identified as 'pathogenicity determinants'.

TABLE 10.9 Suppression of PTGS of GFP mRNA caused by various plant viruses

Virus genus	Virus	Suppression of PTGS	Old or new leaves	Whole leaf or vein centric	Protein	Other known functions
Alfamovirus	AMV	0/9	–	–	–	–
Comovirus	CPMV	5/6	OL and NL	Vein centric	?	–
Cucumovirus	CMV	20/20	NL only	Whole leaf	2b	Host-specific long-distance movement
Geminivirus	ACMV	6/6	OL and NL	Whole leaf	AC2	Virion-sense gene expression transactivator
Nepovirus	TBRV	0/6	–	–	–	–
Potexvirus	PVX	0/9	–	–	–	–
	FoMV	0/9	–	–	–	–
	NMV	8/9	OL and NL	Whole leaf	?	–
	NVX	7/9	OL and NL	Whole leaf	?	–
	VMV	7/9	OL and NL	Whole leaf	?	–
Potyvirus	PVY/TEV TEV	10/10	OL and NL	Whole leaf	HC-Pro	Genome amplification Viral synergy Long-distance movement Polyprotein processing Aphid transmission
Sobemovirus	RYMV	–	–	–	P1	Virus accumulation Long-distance movement
Tobamovirus	TMV	4/6	OL and NL	Vein centric	?	–
Tobravirus	TRV	7/9	OL and NL	Whole leaf	?	–
Tombusvirus	TBSV	7/9	NL only	Vein centric	19 kDa	Host-specific spread and symptom determinant

PTGS of the GFP was induced in transgenic *N. benthamiana* by *Agrobacterium* infiltration as described in Voinnet et al. (1999), with kind permission of the copyright holder, © The National Academy of Sciences, USA. OL, old leaves; NL, new leaves.

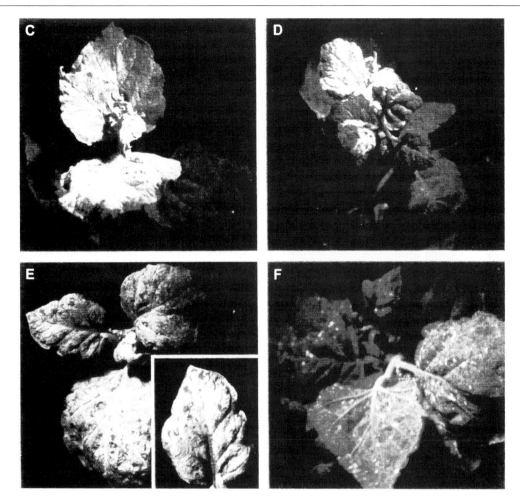

Fig. 10.19 (see Plate 10.3) Virus-induced gene silencing (VIGS) of transgene-encoded GFP in mixed virus infection. **(A)** Schematic diagram of the PVX vector constructs carrying the GFPt coding sequence (PVX-GFPt) or the TEV P1/HC-Pro sequence and GFPt. GFPt is the S65T version of GFP. **(B)** Northern blot analysis of GFP transgene mRNA levels in mock-inoculated GFP transgenic plants (lane 1) or plants infected with either PVX-GFPt (lane 2) or PVX-P1/HC-Pro/GFPt (lanes 3 and 4). Lane 4 shows a 3-fold longer exposure of lane 3. Total RNA was extracted from upper leaves of the plants 20 days after inoculation; a GFP-specific probe was used. Arrows indicate the positions of PVX-P1/HC-Pro/GFPt genomic RNA, PVX-GFPt genomic RNA and GFP transgene mRNA (from top to bottom respectively). **(C–F)** Transgenic GFP *Nicotiana benthamiana* plants coinoculated with **(C)** PVX-GFP and PVX-5'TEV, showing complete suppression of virus-induced gene silencing (VIGS) of GFP; **(D)** PVX-GFP and PVX-HC, showing an almost complete suppression of VIGS of GFP. **(E)** PVX-GFP and PVX-HC, showing partial suppression of VIGS of GFP—the inset is a closer view of a partially silenced leaf showing silencing in the vicinity of the veins; **(F)** PVX-GFP and PVX-noHC, showing complete VIGS of GFP—the red fluorescence is that of chlorophyll. From Anandalakshmi *et al.* (1998), with kind permission of the copyright holder, © The National Academy of Sciences, USA.

There were also differences in the degrees and spatial details of the suppression, ranging from suppression in all tissues of all infected leaves to suppression only in the veins of newly emerged leaves. This suggests that suppressors might be targeted to different parts of the gene silencing mechanism. For instance, if the suppressor blocks the initiation stage of silencing it would show in newly emerging leaves where virus was just starting to be synthesized, whereas if it suppressed the maintenance stage it would show in both old and young leaves (Brigneti *et al.*, 1998; Voinnet *et al.*, 1999). The evidence points to potyvirus HC-Pro suppressing at the maintenance stage and the cucumovirus 2b protein blocking at

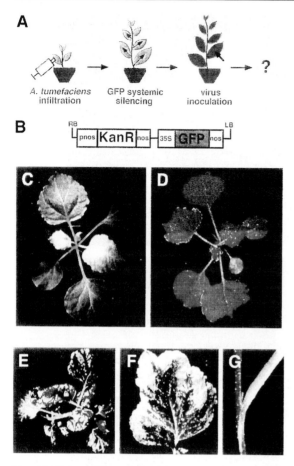

Fig. 10.20 (see Plate 10.4) PTGS of a GFP transgene induced by infiltration with *A. tumefaciens* and subsequent suppression by PVY. **(A)** Schematic representation of the experimental system. GFP silencing was induced in transgenic *N. benthamiana* line by infiltration with a hypervirulent strain of *A. tumefaciens* carrying a binary Ti plasmid shown in **(B)**. Complete GFP silencing was achieved at 10 days post-infiltration and plants appeared uniformly red (chlorophyll fluorescence) under UV illumination. **(B)** Structure of the binary Ti plasmid cassette used to generate transgenic *N. benthamiana* plants expressing GFP and to induce GFP silencing in this line by *A. tumefaciens* (strain cor308) infiltration. The right and left borders of the T-DNA (RB and LB) flank a kanamycin resistance gene (*KanR*) in a nos promoter (pnos) and a nos terminator (nos) cassette. **(C)** Transgenic *N. benthamiana* plant showing high levels of GFP expression under UV illumination. **(D)** Transgenic *N. benthamiana* after induction of gene silencing by infiltration with *A. tumefaciens* carrying the binary Ti plasmid shown in **(B)**. **(E–G)** Suppression of PTGS by PVY. GFP-silenced transgenic *N. benthamiana*, as in **(C)**, 15 days after infection with PVY viewed under UV illumination. **(E)** Whole plant; **(B)** close-up of leaf; **(C)** close-up of stem. From Brigneti *et al.* (1998), with kind permission of the copyright holder, © The Oxford University Press.

the initiation stage (Brigneti *et al.*, 1998). Furthermore, there were differences within suppression at the maintenance stage, with TMV, CPMV and TBSV suppressing around the veins and the other viruses suppressing over all the old and young leaves (Table 10.9). This is taken as suggesting that, in the vein-oriented phenotype, the suppression is targeted against the systemic signal of silencing (Voinnet *et al.*, 1999). Virus-specific spatial differences with the interference of silencing have also been reported by Teycheney and Tepfor (2001).

To analyze the point of potyviral HC-Pro-mediated suppression of the PTGS pathway, Llave *et al.* (2000) transiently delivered HC-Pro by *Agrobacterium* injection into tissue of a plant with a silenced GUS transgene. They confirmed that HC-Pro suppresses one or more maintenance steps by targeting a factor that is required on a continual basis or is relatively labile. They showed that HC-Pro inhibits a step required for the accumulation of the small RNAs and that it reduces the level of cytosine methylation of the transgene sequence. It is possible that the methylation of the PTGS transgene locus is guided by small RNAs diffusing from the cytoplasm and interacting with chromosomal DNA. HC-Pro appears to prevent plants from responding to the mobile silencing signal but does not eliminate the ability to produce or send the signal (Mallory *et al.*, 2001).

A suppressor of gene silencing in one host may evoke a different reaction in another. For instance, the 2b gene of TAV, placed in a TMV vector, functions as a virulence determinant suppressing PTGS in *N. benthamiana* (Li *et al.*, 1999). However, the same gene evokes an HR in *N. tabacum*.

The plant viral suppressors of PTGS, P1/HC-Pro and the 2b gene do not suppress transcriptional gene silencing (Marathe *et al.*, 2000b).

I. Other mechanisms of avoiding PTGS

Table 10.9 shows that not all viruses appear to be able to suppress gene silencing, at least in the vector system and host used. Thus, there may be other mechanisms by which some viruses can avoid the host defense systems. Brigneti *et al.* (1998) suggest that PVX, which does not appear to have a suppression mechanism, might out-compete the defense system by very rapid replication and spread, or it might avoid the defense by compartmentalization.

J. Discussion

The realization that plants have a general defense system against viruses has given potential answers to many aspects of the interactions of viruses with plants, but has also raised many questions. The common features of the defenses in plants (PTGS), fungi (quelling) (Cogoni and Macino, 1999c) and animals (RNAi) (Sharp, 2000) are indicating that this is likely to be an ancient system to control 'foreign' nucleic acids and is probably involved in the regulation of gene expression during development (see Fagard et al., 2000; Matzke et al., 2000).

The picture that is beginning to appear in relation to PTGS and plant virus infection is one of balance and counter-balance between the virus establishing itself and moving through the plant and the plant system responding to the invasion of a foreign nucleic acid. The results of these interactions can explain why some cell regions become infected and others do not, as in the case of the common mosaic symptom. External factors, such as environmental conditions, and internal factors, such as the plant genome, also play a significant part in the outcome of the initial infection of the plant by the virus.

The mechanism(s) of PTGS are beginning to be unraveled. It is becoming more apparent that PTGS and TGS (through methylation) are interlinked (Wang et al., 2001). The overall outline and the similarities to the centre of expression of transgenes are shown in Fig. 10.21.

It is likely that there are several pathways by which the final event of silencing occurs. Observations such as the difference in suppression of PTGS by potyviruses and cucumoviruses suggest that there are several routes to the RNA degradation stage. Mutagenesis of model plants such as Arabidopsis will reveal the pathway(s) and gene products involved in PTGS. However, it should be recognized that there may be differences in plants from other families. Similarly, it appears that there are two mechanisms by which viruses can suppress PTGS and it is likely that others will be found.

The ability to use viruses to induce sequence-specific suppression of gene expression (virus-induced gene silencing, or VIGS) is leading to a high throughput procedure for functional genomics in plants (Baulcombe, 1999b).

V. INFLUENCE OF OTHER AGENTS

Most viruses mutate frequently to give new strains that may have a marked effect on the disease produced (see Chapter 17). The disease produced by a particular virus in a given host and the extent to which it replicates are sometimes influenced markedly by the presence of a second independent and unrelated virus or by infection with cellular parasites. These latter effects are discussed below.

A. Viroids and satellite RNAs

The molecular basis for disease induction by viroids and satellite RNAs is discussed in Sections I.E and II.B.5 of Chapter 14.

B. Defective interfering nucleic acids

Virus-dependent defective nucleic acids are biologically similar to satellite nucleic acids in that they require a helper virus for replication and encapsidation, but they differ from satellites in being deletion and/or recombination mutants of the helper virus (reviewed in Bruening, 2000). They are found associated with both RNA and DNA viruses including the following genera: RNA viruses—Alfamovirus, Bromovirus, Closterovirus, Carmovirus, Cucumovirus, Potexvirus, Rhabdovirus, Tobamovirus, Tombusvirus and Tospovirus; DNA viruses—Begomovirus and Curtovirus. As they frequently interfere with the replication of the helper virus, they are termed defective interfering (DI) nucleic acids. The properties of DI particles and nucleic acids are discussed as a manifestation of faults in virus replication (Chapter 8, Section IX.C).

The presence of DI nucleic acids in the plant tends to make disease symptoms milder. Thus Hillman et al. (1987) described a DI RNA derived from a culture of TBSV (see Fig. 8.40). In N. clevelandii, TBSV alone causes fatal necrosis, but addition of increasing amounts of the DI RNA to

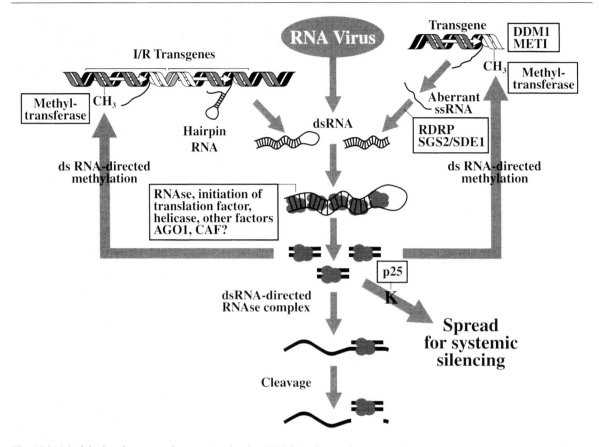

Fig. 10.21 Model of pathways and genes involved in PTGS in plants. dsRNA can be produced in three ways; as a hairpin RNA produced by read-through transcription of two transgenes integrated into the genome as an inverted repeat; as the replicative form of an RNA virus; or as a product of a host-encoded RdRp acting off a ssRNA that is in some way aberrant (and hence recognized as a template). A complex of proteins binds to and then cleaves the dsRNA into short 22–25-nucleotide segments. These dsRNA–protein complexes cleave ssRNA molecules containing the same sequences as the oligomers. The same or modified versions of complexes travel to the nucleus, where they direct methylation of DNA sequences homologous to the oligomers, and also travel to other cells where they propagate the RNA degradation mechanism and direct sequence-specific DNA methylation in their nuclei. From Finnegan *et al.* (2001), with kind permission of the copyright holder, © The American Association for the Advancement of Science.

the inoculum resulted in increasingly milder symptoms, which was accompanied by a reduced production of TBSV. Ismail and Milner (1988) isolated a DI particle of SYNV whose RNA was 77% as long as that of standard virus. Particles containing this RNA were only about 80% as long as the standard virus. They were non-infectious on their own but, when mixed with standard virus, the resulting symptoms in *N. edwardsonii* were chlorotic mottling instead of the normal vein-clearing.

The DI DNA of ACMV interferes with the replication of the parent virus in *N. benthamiana* causing a reduction in infected plants, a delay in symptom development and symptom attenuation (Stanley and Townsend, 1985).

C. Other associated nucleic acids

Examination of the causal agents for the whitefly-transmitted diseases inducing yellow veins in *Ageratum conyzoides* and leafcurl in cotton has revealed three DNA species. One of these resembled DNA A of begomoviruses, one had the characteristics of a satellite (DNA β) and one had features suggestive of a recombinant between DNA A and a nanovirus (Stanley *et al.*, 1997; Mansoor *et al.*, 1999; Saunders and Stanley, 1999). Although the DNA A of AYVV systemically infects *A. conyzoides* and *N. benthamiana*, it causes no symptoms in either host. DNA A + β gives the wild-type symptoms, which are not affected by the presence of the nanovirus-like DNA (J. Stanley, unpublished data). It is thought possible that other apparently single-component begomoviruses might have a similar complex etiology.

D. Cross-protection

Cross-protection is the protection conferred on a host by infection with one strain of a virus that prevents infection by a closely related strain of that virus. It was demonstrated by McKinney (1929), who showed that tobacco plants infected with a green mosaic virus (TMV) developed no further symptoms when inoculated with a yellow mosaic virus. Salaman (1933) found that tobaccos inoculated with a mild strain of PVX were immune from subsequent inoculation with severe strains of the virus, even if inoculated after only 5 days. They were not immune to infection with the unrelated viruses, TMV and PVY. This phenomenon, which has also been called antagonism, or interference, was soon shown to occur very commonly among related virus strains. It is most readily demonstrated when the first strain inoculated causes a fairly mild systemic disease and the second strain causes necrotic local lesions or a severe disease. Development of such lesions can be readily observed and a quantitative assessment made. Interference between related strains can also be demonstrated by mixing the two viruses in the same inoculum and inoculating to a host that gives distinctive lesions for one or both of the two viruses or strains. Interference by type TMV with the formation of yellow local lesions by another strain is shown in Fig. 10.22.

Strains of AMV differ in the aggregation bodies that their particles form in infected cells (see Chapter 17, Section II.C.1.b). Hull and Plaskitt (1970) used electron microscopy to investigate the cross-protection interactions between two strains of AMV. When the challenging strain was inoculated at the same time as the protecting strain or a short time after (about 4 hours), the two aggregation types were found side by side, merging into each other in the same cell. When there was a longer interval between inoculation of the protecting and challenging strains (about 7 hours) the two strains were found in separate parts of the cytoplasm of the same cell and, after an interval of about 10 hours, in separate cells. Only when

Fig. 10.22 Tobacco leaf inoculated with a mutant strain of TMV that produces large slowly developing yellow local lesions. The isolate is not completely freed of the type strain (causing no obvious local lesions). The sharp boundaries to the yellow (light) areas in the lower left quarter of the leaf delineate areas in which type TMV has multiplied and prevented invasion by the yellow strain. (Courtesy of B. Kassanis.)

cross-protection was complete as assessed by back inoculation to an indicator host could the aggregation bodies of the challenging strain not be found.

There have been several mechanisms proposed for cross-protection, which were reviewed in the previous edition of this book (Matthews, 1991). However, the recent recognition of gene silencing mechanisms being induced by plant virus infection (see Section IV.G) gives a much more rational explanation of this phenomenon. Ratcliff et al. (1999) have produced convincing evidence that PTGS is responsible for cross-protection in a tobravirus and a potexvirus system.

Cross-protection with mild virus strains is

used as a control measure (see Chapter 16, Section IV.A).

E. Concurrent protection

Concurrent protection is the reduction in challenging virus infection rate and/or titer due to the co-inoculation with a protecting virus that does not accumulate or induce symptoms in that host plant (Ponz and Bruening, 1986). The cowpea line Arlington shows no symptoms after inoculation with CPMV and no infectivity or accumulation of capsid antigen can be detected (Beier et al., 1977). Co-inoculation of Arlington with CPMV and CPSMV reduced the numbers of CPSMV-induced lesions. Inoculation of an isogenic line derived from Arlington with CPMV also protected it against infection with CLRV and SCPMV (Bruening et al., 2000). This protection was elicited by CPMV RNA1.

F. Interactions between unrelated viruses

1. Complete dependence for disease

For some virus combinations, there is complete dependence of one virus on the other for its replication. The dependent virus is termed a satellite virus and is described in fully Chapter 14 (Section II.A).

2. Incomplete dependence for disease

This situation exists between two viruses where both are normally associated with a recognized disease in the field. For example, the important tungro disease of rice is normally caused by a mixture of RTBV, a reverse transcribing DNA virus, and RTSV, an RNA virus (reviewed by Anjaneyulu et al., 1994). RTSV on its own is transmitted by the rice green leafhopper, but causes no symptoms. RTBV causes severe symptoms in rice but no vectors are known for this virus on its own; it requires the presence of RTSV for transmission (see Chapter 2, Section III.A.6). Thus, in the disease complex RTSV gives the transmission and RTBV most of the symptoms.

Most, if not all, umbraviruses are associated with luteoviruses, which provide their insect transmission (see Chapter 11, Section III.H.1.a).

An even more complex system is that of groundnut rosette disease which involves three agents: an umbravirus (GRV), a luteovirus (GRAV) and a satellite RNA (GRV sat-RNA). The sat-RNA is the major cause of symptoms (Murant et al., 1988a; Murant and Kumar, 1990) and both it and GRAV are required for aphid transmission (Murant, 1990). The GRAV coat protein encapsidates its own genome and also that of GRV and sat-RNA, the presence of sat-RNA being essential for encapsidation (Robinson et al., 1999).

3. Effects on numbers of local lesions

Thomson (1961) found that, when PVX was inoculated on to tobacco leaves together with TMV or PVY, the PVX lesions were more numerous. The increase was highly variable, ranging from 2- to 80-fold depending on conditions. From studies on the effect of dilution and of changing ratios of the two viruses, Close (1962) concluded that the maximum stimulation by TMV of PVX lesion formation occurs at a definite concentration of TMV in the inoculum and does not depend on the ratio of amounts of the two viruses. This can possibly be explained by virus suppression of PTGS (see Section IV.H).

4. Synergistic effects on virus replication

Joint infection of tobacco plants with PVX and PVY is characterized by severe veinal necrosis in the first systemically infected leaves. Leaves showing this synergistic reaction contain up to 10 times as much PVX as with single infections but only the same amount of PVY (e.g. Stouffer and Ross, 1961). Ultrastructural studies and fluorescent antibody staining showed that both viruses were replicating in the same cells and that the increased production of PVX was due to an increase in virus production per cell rather than an increase in the number of cells supporting PVX replication (Goodman and Ross, 1974a). The level of PVX (−)-strand RNA increased disproportionately to that of (+)-strand RNA in doubly infected tissues, suggesting that the synergism involves an alteration in the normal regulation of the relative levels of the two RNA strand polarities during viral replication (Vance, 1991). TVMV, TEV and PepMoV also give a synergistic reaction with

PVX (Vance *et al.*, 1995) and various other combinations of potyviruses and unrelated viruses have synergistic interactions. The joint infection of VlCMV and CMV causes cowpea stunt (Anderson *et al.*, 1996) and corn lethal necrosis is due to joint infections of the machlovirus, MCMV, with the potyvirus, MDMV or SCMV (see Scheets, 1998). Joint infections of potyvirus SPFMV and the crinivirus SPCSV lead to severe symptoms in sweet potato (*Ipomoea batatas*), whereas the individual viruses cause mild symptoms (Karyeija *et al.*, 2000). In all these cases, the concentration of the potyvirus is similar to that in a single infection, whereas that of the other virus is increased markedly; in the sweet potato disease complex, SPCSV remained limited to the phloem. However, corn lethal necrosis is also caused by joint infection of MCMV with the rymovirus WSMV (Scheets, 1998). In this case, the concentrations of both viruses were enhanced by the dual infection.

To determine which region of the potyviral genome enhanced PVX multiplication, Vance *et al.* (1995) inoculated tobacco plants transgenically expressing various regions of the TEV genome (Fig. 10.23). The synergistic reaction was only in plants expressing P1-(HC-Pro)-P3. This region also enhanced the replication of CMV and TMV and, for all three enhanced viruses, it was the expression of the HC-Pro protein that was required for synergism (Pruss *et al.*, 1997). The recognition that potyviral HC-Pro suppresses the host gene silencing defense mechanism (see Section IV.H) suggests that the synergy might be due to reduction of the host inhibition of the replication of the other virus.

A striking synergistic effect was found with BRNV in herbaceous hosts. In mixed infections with the sobemovirus, SNMoV, the normally very low concentration of the raspberry virus was increased 1000-fold (Jones and Mitchell, 1986).

In rice tungro disease, the presence of RTSV enhances the symptoms caused by RTBV. Unlike the potyviral synergisms, there was no significant increase of either RTBV or RTSV in joint infection when compared with single infections (F. Sta Cruz, O. Azzam and R. Hull, unpublished data).

The presence of a satellite nucleic acid can either reduce or enhance the helper virus repli-

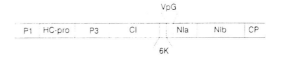

Line	TVMV gene	Km segregation	Protein level	Synergism
1	coat protein (CP)	98:1	0.30%	-
2	coat protein (CP)	65:22	0.25%	-
3	CI	66:26	0.15%	-
4	CI	64:22	0.10%	-
5	NIa	73:22	#	-
6	NIa-glnH	59:20	0.20%	-
8	CI-6K-NIa-NIb-CP	*	0.02%	-
9	CI-6K-NIa-NIb-CP	*	0.02%	-
10	P1-(HC-Pro)-P3	40:1	1%*	+
11	P1-(HC-Pro)-P3	2:1	1%*	+

Fig. 10.23 Diagrammatic representation of the TVMV polyprotein showing the localization of the individual genes and a listing of the TVMV transgenic lines indicating the genes included in the construct, the segregation characteristics of the transgene as measured by selection on kanamycin, and the protein level of the encoded gene product(s) in plant leaves. Km, kanamycin; CI, cytoplasmic inclusion body protein; NIa, nuclear inclusion body protein a; glnH, *Escherichia coli* glutamine-binding protein; NIb, nuclear inclusion body protein b. *Indicates that these seeds are homozygous and do not segregate as measured on kanamycin plates. #Indicates that no appropriate antibodies were available for this measurement. Expression levels for lines 1–9 refer to percentage of transgene-derived protein/total soluble protein. ᵃIndicates that expression levels for lines 10 and 11 refer to percentage of transgene expression compared with a wild-type virus infection. In all cases, plants used to determine the level of gene expression were grown from the same generation of seed used to produce plants for the studies on synergism with PVX. From Vance *et al.* (1995), with permission.

cation and symptom production (see Chapter 14, Section II.B.5).

5. Effects on virus movement

As discussed in Chapter 9 (Section II.D.6), infection and systemic movement by one virus in a particular host may allow the cell-to-cell and systemic movement of an unrelated virus that normally would not move from the initially infected cells in that host. Similarly, a fully systemically infecting virus can complement the movement of a tissue-restricted virus (e.g. phloem-limited virus) out of that tissue.

6. Effects on macroscopic disease symptoms

Not uncommonly, a mixed infection with two viruses produces a more severe disease than either alone. The classical example of this synergistic effect is the mixture of PVX and PVY in tobacco described above, which produces a severe veinal necrosis instead of the milder mottling or vein-banding diseases seen with either virus separately (Smith, 1931). In many potato varieties, these two viruses together produce a 'rugose mosaic' that is more severe than the disease produced by either virus alone. Many strains of ToMV produce only a mosaic disease in tomatoes but, in combination with PVX, a severe 'streak' disease ensues and usually kills the plant (Plate 3.20). Double infections with PVY and the viroid PSTVd caused severe necrotic disease in some potato cultivars in the field (Singh and Somerville, 1987).

Isolated protoplasts have been used to study mixed infections. For example, Otsuki and Takebe (1976a) inoculated tobacco mesophyll protoplasts with TMV and CMV. Some 70–80% of protoplasts supported both viruses without any synergistic or antagonistic effects. On the other hand, in mixed infections of BMV and CCMV in tobacco protoplasts, both types of virus particle are produced, but only those of BMV are infectious. Only RNA3 of CCMV was synthesized in the mixed infections (Sakai *et al.*, 1983).

7. Cytological effects in mixed infections

Mixed infections leading to necrosis must clearly have marked effects on individual cells. In mixed infections where cell death does not occur, two different viruses may replicate in the same cell producing their characteristic inclusion bodies or arrays of virus particles without any significant indications of mutual interference (e.g. Poolpol and Inouye, 1986; Langenberg, 1987).

G. Interactions between viruses and fungi

1. Effect of virus infection on fungal diseases

a. Increased resistance

Observations in the field and glasshouse tests have shown that infection with *Phytophthora* *infestans* developed less rapidly in potato plants infected with one of a number of viruses (Müller and Munro, 1951). Numerous other examples are known where virus infection reduces susceptibility to, or development of, fungal and bacterial parasites. For example, infection of a hypersensitive tobacco cultivar with TMV induced systemic and long-lived resistance against *Phytophthora parasitica*, *Peronospora parasitica* and *Pseudomonas tabaci* (McIntyre *et al.*, 1981). Similarly, systemic resistance to anthracnose in cucumber was induced by inoculation with TMV (Jenns and Kuc, 1980), as was resistance to *Peronospora tabacina* in tobacco (Ye *et al.*, 1989) and *Erysiphe chicoracearum* (Marte *et al.*, 1993). The protection against *E. chicoracearum* was related to an accumulation of hydroxyproline-rich glycoproteins induced by the TMV infection (Raggi, 1998).

The development of resistance of this sort probably involves the PR proteins discussed in Section III.K.1. Indeed, fungicidal compounds have been isolated from plants reacting with necrosis to virus infection (e.g. Burden *et al.*, 1985). Generalized defense reactions may not be involved in some other virus-fungus interactions. Thus, infection of faba bean with BYMV decreased pustule density on leaves subsequently inoculated with *Uromyces viciafabae*. Changes were most marked on leaves showing yellowing symptoms (Omar *et al.*, 1986).

b. Increased susceptibility

Russell (1966) showed that sugar beet plants in the field that were infected with BMYV had greatly increased susceptibility to *Alternaria* infection. BYV had no such effect. BMYV increased and BYV decreased susceptibility to another fungus, *Erisphye polygoni*. Plants infected with both viruses had about the same susceptibility as healthy plants. The precise extent of the interaction depended on the genetic constitution of the host plant and on the environmental conditions.

Many other instances of increased susceptibility to fungi have been reported following virus infection. For example, prior infection of wheat or barley with BYDV predisposed the ears to infection by *Cladosporium* spp. and

Verticillum spp. (Ajayi and Dewar, 1983). Sporulation of *Helminthosporium maydis* on corn leaves began sooner and was more abundant in lesions formed on leaves infected with MDMV (Stevens and Gudauskas, 1983). Asparagus infected with AV-2 had increased susceptibility to *Fusarium* crown and root rot (Evans and Stephens, 1989).

2. Effects of fungal infection on susceptibility to viruses

Infection by a rust fungus may greatly increase the susceptibility of leaves to several viruses (Yarwood, 1951). For example, pinto bean leaves were heavily inoculated with the uredinial stage of *Uromyces phaseoli* on one half-leaf and then later with TMV over the whole leaf. Subsequent estimations of the amounts of TMV showed the presence of up to 1000 times as much virus infectivity in the rusted as in the non-rusted half-leaves. Rust infection partially suppressed the development of visible necrotic lesions. This suppression made it impossible to determine whether the increase in virus content was due to an increase in the number of successful entry points or to increased virus multiplication or to a combination of both of these factors.

Other fungi may induce resistance or apparent resistance to viral infection. Xanthi tobacco plants that had been injected in the stem with a suspension of spores of *Peronospora tabacina* produced fewer and smaller necrotic local lesions when inoculated with TMV 3 weeks later (Mandryk, 1963). The fungus *Thielaviopsis basicola* causes necrotic local lesions in tobacco. Hecht and Bateman (1964) found that, if tobacco plants were infected on the lower leaves with this fungus, upper leaves became resistant to infection with TNV or TMV, as judged by size of the necrotic viral lesions. Inoculation of cucumber with *Colletotrichum lagenarium* or *Pseudomonas lachrymans* induced systemic resistance to CMV in an upper leaf (Bergström *et al.*, 1982). Some of these inhibitory effects at a distance may be due to SAR or possibly PR proteins discussed in Section III.K. On the other hand, a glucan preparation from *Phytophthora megasperma* protected a number of *Nicotiana* spp. from infection

by several viruses without the induction of pathogenesis-related proteins (Kopp *et al.*, 1989).

3. Effects of virus infection on non-vector insects

Infection by TMV improved the suitability of tomato for survival of Colorado potato beetle larvae (Hare and Dodds, 1987). It was suggested that virus infection may facilitate the adaptation of phytophagous insects to 'marginal' host plant species.

4. Interactions between virus infection and air pollutants

There is increasing concern about the effects of air pollutants such as ozone on plant growth. There have been several investigations of the effect of virus infection on the severity of ozone leaf damage. For most host–virus combinations, viral infection reduced damage. For example, infection with TMV reduced leaf damage in tobacco leaves due to ozone from 11% to 5% (Bisessar and Temple, 1977). The effect was seen in both glasshouse and field trials. Subacute doses of sulfur dioxide caused small but consistent increases in the content of SBMV in beans and MDMV in maize (Laurence *et al.*, 1981).

VI. DISCUSSION AND SUMMARY

The effects of a virus entering a cell in an uninfected plant depend on the interplay of two genomes, that of the virus and that of the plant. The activities of these genomes can be affected by influences such as the environment, the growth stage of the plant, the site of virus entry and the effects of other pathogens.

The interplay between viral and host genes in disease induction is currently one of the most biologically interesting and practically important areas of plant virus research. Site-directed mutagenesis, transgenic plants, non-destructive fluorescent probes and other techniques together with *in vitro* or protoplast systems, and sometimes whole plants, can identify the main function or functions of a gene product, for example that it is a replicase or a protease. However, to study disease induction intact plants or parts of plants must be used, and a full virus replication cycle studied. For these reasons there may be significant difficulties ahead.

The *in planta* roles of some of the protein products of viral genes involved in disease induction may be very difficult to study for several reasons including:

1. The proteins may be present in very low concentration, as a very few molecules per cell of a virus-specific protein could block or derepress some host-cell functions.
2. It is quite possible that such proteins would be present in the infected cells for a short period relative to that required for the completion of virus synthesis.
3. The virus-specified polypeptide may form only part of the active molecule in the cell.
4. The virus-specified polypeptide may be biologically active only *in situ*, for example in the membrane of some particular organelle.

As noted in Section III, many macroscopic disease symptoms may be due to quite unexpected side effects of virus replication. The following hypothetical example illustrates this kind of possibility. Consider the NIa proteinase coded for by TEV (Carrington and Dougherty, 1988). The amino acid sequences that function as substrate recognition signals have been identified. In the usual host under normal conditions, this proteinase can accumulate to high levels within infected cells without causing cell death. This must mean that the proteinase does not significantly deplete the amount of any vital host protein. Suppose that we change to another host species, or to different environmental conditions. In the new situation, some host-coded protein essential for cell function might be sensitive to cleavage, with a new pattern of disease developing as a consequence. In such circumstances, it might be difficult to identify the host protein involved, and its functions. In attempting to understand disease induction in molecular terms it may be most profitable to concentrate initially on effects that are known to be a direct consequence of and essential for virus replication, or for virus movement from cell to cell.

Although molecular approaches will no doubt continue to increase our understanding of the role of viral genes in the induction of disease, difficulties are emerging. For example, site-directed mutagenesis, which is in principle a powerful technique, has several limitations. (1) The number of possible permutations and combinations of base sequence alteration is enormous. (2) It will be difficult to find changes that are not fatal for the virus when a full infectious cycle is required. (3) Because of the high mutation rate in RNA viruses and the existence of recombination (Chapter 17), it is often necessary to check the complete base sequence of the engineered mutant culture to ensure that spontaneous changes have not caused reversion to the original sequence or altered the sequence in some other way. (4) A more general difficulty may be that, as increasingly detailed experiments are carried out with a particular virus, most or all viral genes may be found to have interacting roles in various aspects of disease induction.

The new techniques being developed in genomic research should provide powerful tools for unraveling details of the host responses to viral infections. Not only will they increase the understanding of virus disease resistance (see Michelmore, 2000), but also they should give information on the permissive situation. For instance, microarray analysis of infection of *Arabidopsis* with the fungal pathogen *Alternaria brassicicola* (albeit an incompatible situation) revealed substantial changes in the steady-state abundance of 705 host mRNAs (Schenk *et al.*, 2000).

Depending on the response to inoculation with a virus, plants have been described as either *immune* or *infectible*. If a plant appeared immune it is considered a non-host for the virus, and the virus does not replicate in any cells of the intact plant or in isolated protoplasts. This extreme resistance is discussed in detail in Section III.A.

In infectible species or cultivars, the virus can infect and replicate in isolated protoplasts. The plant may be either resistant or susceptible to infection. Until recently, two kinds of resistance were recognized. In resistance involving subliminal infection, virus multiplication is limited to the initially infected cells because the viral-coded protein necessary for cell-to-cell movement of virus is non-functional in the particular host. In the past, many examples of this type of resistance were described as immune. In the second kind of resistance, infection was considered to be limited by a host response to a zone of cells around the initially infected cell, usually resulting in necrotic local lesions. Uninfected tissue surrounding these lesions becomes resistant to infection. This was called *acquired resistance.*

However, at least two forms of host resistance response have now been recognized. These can be termed a *specific resistance* response and a *generalized resistance* response. The specific resistance response is directed by one or more host genes, the products of which interact with certain viral determinants and limit the spread of the virus from the site of initial infection. The limitation is usually by an HR. It is likely that the containment by the specific response is not always by an HR and may sometimes be by a mechanism that does give a visible symptom. The generalized resistance is that mounted by the plant against 'foreign' nucleic acids, in the case of viruses usually by PTGS (see Section IV) (reviewed by Covey, 2000; Ding, 2000). Very recently, it has been demonstrated the begomovirus AC2 and curtovirus C2 genes supress a general stress defence response in plants (Sunter *et al.*, 2001). This points to at least two forms of generalized resistance, both of which viruses can overcome.

A virus that does not cause systemic disease in a particular plant has been termed as *non-pathogenic* for that plant. If a virus or virus strain causes systemic disease in a particular species or cultivar, it is *pathogenic*. A gene for resistance introduced into such a susceptible species or cultivar may make the virus *avirulent*. However, the virus may mutate and overcome the host resistance to become a *virulent* strain. Thus, both host and viral genes interact to determine the outcome of inoculation. The change from an avirulent to a virulent virus strain may involve no more than a single amino acid change in a virus-coded protein (for examples see Sections III.A–E).

In species or cultivars that are susceptible, the virus replicates and moves systemically. In a *sensitive* reaction, disease ensues. If the plant is *tolerant*, there is no obvious effect on the plant, giving rise to a *latent* infection. The consequences of infection for a susceptible plant are determined by both host and viral genes. For example, a single base change in the TMV coat protein gene may be sufficient to alter the nature of the resulting disease.

In the light of these new findings on host responses to virus infection, it is necessary to modify the ideas of the events leading to the establishment of fully systemic virus infection. If the virus is not contained by extreme resistance or a local hypersensitive response, or if it overcomes the HR, the outcome of the infection is dependent on the 'aggressiveness' of the virus overcoming the generalized resistance response of the host. Several viruses have been found to encode genes that suppress the generalized PTGS defense (see Section IV.H), and two basic systems have been recognized that operate at different times in the infection process. It is likely that other viral genes with be found that suppress PTGS and that there will be other variants of the suppression process. However, for other viruses such as PVX, no PTGS suppression system has been detected and it is suggested that the rapid multiplication and spread of the virus outcompetes the host defense response.

As noted earlier for viruses that overcome the defense mechanisms, many environmental factors influence the course of infection and disease. These include light, temperature, water supply, nutrition and the interactions between these factors during the growing season. Complex interactions may occur when plants are infected with two unrelated viruses or with a virus and a cellular pathogen. Factors such as these would affect the expression of either the plant or viral genome or both and thus, the outcome of an infection depends upon the individual circumstances.

The actual processes involved in the induction of disease symptoms are not well understood. For strains of a given virus we should distinguish between the severity of a particular symptom and different symptom phenotypes.

Many of the biochemical changes involved may not be directly connected with virus replication. Stunting probably involves changes in the balance of growth hormones. The formation of mosaic patterns in virus-infected leaves involves events that occur in the early stages of leaf ontogeny and the local balance between the aggressiveness of the virus and the host defense systems.

CHAPTER 11

Transmission 1:
By Invertebrates, Nematodes and Fungi

I. INTRODUCTION

One could envisage a virus surviving for hundreds of years in an individual tree of a long-lived species. However, being obligate parasites, viruses usually depend for survival on being able to spread from one susceptible individual to another fairly frequently. Knowledge of the ways in which viruses are transmitted from plant to plant is important for several reasons:

- From the experimental point of view, we can recognize a particular disease as being caused by a virus only if we can transmit the virus to healthy individuals by some means and reproduce the disease.
- Viruses are important economically only if they can spread from plant to plant fairly rapidly in relation to the normal commercial lifetime of the crop.
- Knowledge of the ways in which a virus maintains itself and spreads in the field is usually essential for the development of satisfactory control measures.
- The interactions between viruses and their invertebrate and fungal vectors are of considerable interest, both from the scientific point of view and in developing new approaches to the control of viruses.
- Certain methods, particularly mechanical transmission, are very important for the effective laboratory study of viruses.

Viruses cannot penetrate the intact plant cuti-

cle and the cellulose cell wall. This problem is overcome either by avoiding the need to penetrate the intact outer surface (e.g. in seed transmission or by vegetative propagation) or by some method involving penetration through a wound in the surface layers, such as in mechanical inoculation and transmission by insects. There is considerable specificity in the mechanism by which any one virus is naturally transmitted (see Appendix 2). In this chapter, I will discuss how their biological vectors transmit viruses. Mechanical and seed transmission of viruses are described in Chapter 12 together with how viruses spread through crops and natural populations.

II. TRANSMISSION BY INVERTEBRATES

Many plant viruses are transmitted from plant to plant in nature by invertebrate vectors (see Appendix 3). Of some 22 phyla in the invertebrates, only three have many members that feed on living green land plants. These are the *Urinamia* and *Crustacea* phyla of the *Arthropoda* and the *Nematoda*. Both of the *Arthropoda* phyla, and the *Nematoda*, contain vectors of plant viruses. Two additional phyla, the *Annelida* and *Mollusca*, have a few plant feeders, and it these that may contain potential vectors of a strictly mechanical sort.

A. *Arthropoda*

I have followed the taxonomy used by Webb *et al.* (1978) and by Richards and Davies (1994). Of the three subphyla of the *Uniramia*, one, the

Hexapoda (*Insecta*), contains members feeding on living green land plants; one class of the *Crustacea*, the *Arachnida*, has members feeding on green land plants. Both of these groups contain virus vectors.

1. *Insecta*

Among 29 orders in the living *Insecta* there are nine with members feeding on living green land plants and that might, therefore, be possible vectors. These are listed below with the approximate number of vector species at present known:

1. *Collembola*—chewing insects; some feed on green plants (0)
2. *Orthoptera*—chewing insects; some feed on green plants (27)
3. *Dermaptera*—chewing insects; a few feed on green plants (1)
4. *Coleoptera*—chewing insects; many feed on green plants; see Table 11.1 for vectors
5. *Lepidoptera*—chewing insects; larvae of many feed on green plants (4)
6. *Diptera*—larvae of a few feed on green plants (2)
7. *Hymenoptera*—larvae of a few feed on green plants (0)
8. *Thysanoptera* (thrips)—some are rasping and sucking plant feeders (10)
9. *Hemiptera*—feed by sucking on green plants
      Suborder *Heteroptera*, families *Myridae* and *Piesmatidae* (~4)
      Suborder *Homoptera*, see Table 11.1.

The first seven orders listed are all chewing insects, and representatives of these orders feed on living green plants as larvae or adults, or both. Vectors of a strictly mechanical sort have been found among the *Orthoptera*, *Dermaptera* and larvae of *Lepidoptera* and *Diptera*. Except for a few viruses, vectors in these orders are of minor importance. Important vectors occur in the *Coleoptera*. The *Collembola* and *Hymenoptera* contain relatively few species that are common pests of agricultural plants. There may be potential vectors among them. The *Thysanoptera* contains vector species for one important group of plant viruses. The *Homoptera* is numerically the most important suborder containing plant virus vectors. There

are still uncertainties about the classification within the *Homoptera* (Campbell *et al.*, 1995).

2. *Arachnida*

Only one of 12 orders in this class, the *Acari* (mites and ticks), contains members feeding on living green land plants. The *Acari* have four families containing mites that are green plant feeders: the *Tetranychidae*, *Tarsonemidae*, *Eriophyidae* and *Acaridae*. Virus vectors are known in the third and possibly the first of these families.

B. Nematoda

There are 10 orders in the *Nematoda* (Goodey, 1963). Most of the nematodes parasitic in living green plants belong to the *Tylenchida*, but none from this group has yet been found to be a virus vector. Vectors known so far are confined to the *Dorylaimida* group containing only a few plant parasites.

C. Relationships between plant viruses and invertebrates

The transmission of viruses from plant to plant by invertebrate animals is of considerable interest from two points of view. First, such vectors provide the main method of spread in the field for many viruses that cause severe economic loss. Second, there is considerable biological and molecular interest in the relationships between vectors and viruses, especially as some viruses have been shown to multiply in the vector. Such viruses can be regarded as both plant and animal viruses. Even for those that do not multiply in the animal vector, the relationship is usually more than just a simple one involving passive transport of virus on some external surface of the animal. Transmission by invertebrate vectors is usually a complex phenomenon involving specific interactions between the virus, the vector and the host plant coupled with the effects of environmental conditions. In this chapter I consider the groups involved as virus vectors and outline the kinds of relationship that have been found to exist between virus and vector. Vectors are considered in relation to the ecology of viruses in

TABLE 11.1 Distribution of plant virus vectors among selected *Homoptera* and *Coleoptera* families

Order, suborder, family	Common name of insect group	Approx. no. species described	No. vector species	No. viruses transmitted
Homoptera				
Auchenorrhyncha				
Cicadidae	Cicada	3 200	0	0
Membracidae	Treehopper	4 500	1	1
Cercopidae	Spittlebug	3 600	0	0
Cicadellidae	Leafhopper	15 000	49	31
Fulgoroidea	Planthopper	19 000	28	24
Sternorrhyncha				
Psyllidae	Psyllid	2 000	0	0
Aleyroididae	Whitefly	1 200	3	43
Aphididae	Aphid	4 000	192	275
Pseudococcidae	Mealybug	6 000	19	10
Coleoptera				
Chrysomelidae	Leaf beetle	20 000	48	30
Coccinellidae	Ladybird beetle	3 500	2	7
Cucurlionidae	Weevil	36 000	10	4
Meloidae	Blister beetle	2 100	1	1

From Nault (1997), with permission.

Chapter 12 and in relation to disease control in Chapter 16.

As a general rule, viruses that are transmitted by one type of vector are not transmitted by any of the others. This specificity is not only at the level of vector type, family, genus or species but can be even at the level of biotype. There are two basic interactions between viruses and their biological vector. They may be taken up internally within the vector, termed *internally borne* or *circulative,* or they may not pass to the vector's interior, in which case they are termed *externally borne* or *non-circulative* (Hull, 1994a; Gray, 1996). Virus–vector relationships have been studied in most detail in insects, and especially aphids. The basic features of these relationships are described in Section III.D.

III. APHIDS (*APHIDIDAE*)

A. Aphid life cycle and feeding habits

Among insects, the aphids have evolved to be the most successful exploiters of higher plants as a food source, particularly flora of temperate regions. It is therefore not surprising that they have also developed into the most important group of virus vectors.

1. Life cycle

In temperate climates, aphids frequently alternate between a primary and a secondary host. They are remarkable for the number of forms produced. A complete cycle is shown in Fig. 11.1.

There are many variations in this cycle, depending on aphid species and on climate. For example, some may overwinter as parthenogenetic viviparous forms. Some may pass through their life cycle on one host species or several species within one genus. *Myzus persicae,* an important vector aphid, has *Prunus* spp. as its primary host, alternating with secondary hosts in over 50 plant families.

There are three kinds of variability in aphids that may affect their ability to transmit a virus:

- An aphid species may contain different clones or races (beotypes), with or without obvious morphological differences.
- Aphids can exist in different forms, as noted earlier.
- Successive moults by the developing insect define the number of stages or instars.

2. Mouthparts

The mouthparts of aphids consist of two pairs of flexible stylets, held within a groove of the

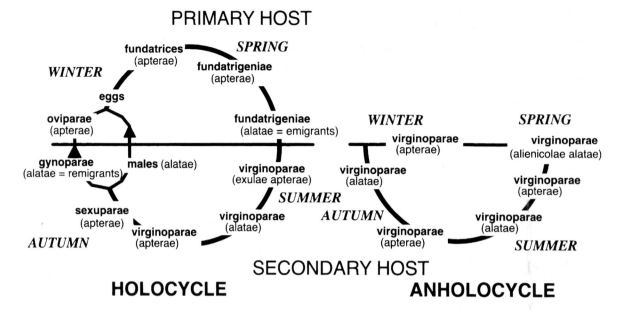

Fig. 11.1 Two different types of aphid life cycle. In the holocycle, the aphid alternates between two hosts, overwintering usually as eggs on the primary host. In the anholocycle, the aphid overwinters as apterae, which usually find shelter. From Robert and Lemaire (1999b), with permission.

labium. They are extended from the labium during feeding. The maxillary stylets have a series of toothlike projections near their tips (Fig. 11.2). Details of the feeding mechanism such as the sucking pump and the esophageal valve have been described by Forbes (1977) and McLean and Kinsey (1984) (Fig. 11.3).

3. Feeding habits

At the beginning of feeding a drop of gelling saliva is secreted. The stylets then rapidly penetrate the epidermis and, in exploratory probes, the aphid may feed there temporarily. Penetration usually continues into the deeper layers with a sheath of gelled saliva forming during penetration. The stylets usually move between cells until they reach a phloem sieve tube (a process that may take minutes or hours) (Fig. 11.5A).

Only the maxillary stylets enter the sieve tube. Compression by the cell wall causes the tips to open, exposing the end of the food and salivary canals. Electronic monitoring of insects while they are feeding can give information on their feeding behavior and sites of feeding (for further information see Ellsbury *et al.*, 1994). It

should be recognized that there are two systems of electronic monitoring, AC and DC, which give different waveform patterns; also there have been differences in the terminology used by European and American scientists. Evidence from electronic monitoring of aphids (Fig. 11.5A) while they feed in the phloem suggests that they salivate, but it is the watery, enzyme-bearing, non-gelling saliva. Indeed, if aphids did not salivate in the phloem, the circulative and propagative viruses that are restricted to the phloem would have no access to this tissue from an aphid carrying the virus. No gelling sheath saliva is secreted during feeding in a sieve tube, but, on withdrawal, such saliva is used to seal the lumen in the salivary sheath that had been occupied by the stylets. This feeding process causes minimal damage to the sieve tube and to surrounding cells in the stylet path. Penetration by aphids has been reviewed by Pollard (1977).

4. Role of the host plant

Both physical and chemical features of the plant may markedly affect aphid feeding behavior. For example, resistance of certain

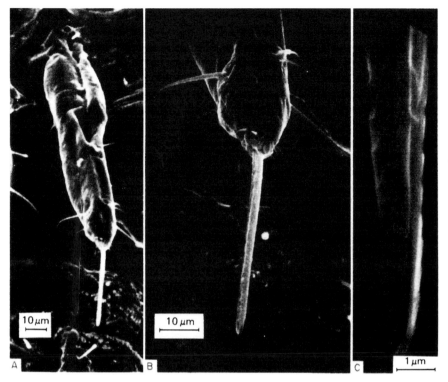

Fig. 11.2 Mouthparts of *Myzus persicae* revealed by scanning electron microscopy. **(A)** Labium with joint area and bristles. Mandibular stylets protrude from the labium. (The aphid was frozen in liquid nitrogen immediately after it had withdrawn its stylets from a leaf.) **(B)** Tip of labium and mandibular stylets at higher magnification. **(C)** Tip of mandibular stylets showing ridges in both stylets, and the overlap of the tip of one stylet. From de Zoeten (1968), with permission.

brassicas to *Brevicoryne brassicae* (L.) (but not to *M. persicae*) has been shown to depend on the physical state of the wax on the leaf surface (Jadot and Roland, 1971). The density of trichomes on soybean leaves influenced probing behavior by several aphid species. Spread of SMV in the field correlated negatively with density of pubescence (Gunasinghe *et al.*, 1988). Specific chemicals may either attract or inhibit feeding by particular aphid species. For example, sinigrin, a mustard oil glucoside found in the *Brassicaceae*, stimulates feeding by aphids that normally feed on brassicas, but inhibits uptake by species that do not feed on members of this family (Nault and Styer, 1972).

Several examples are known where infection of a plant with a virus makes the plant more suitable for the insect vector to grow and reproduce. *Aphis fabae* produced more young per mother on beet plants infected with BtMV than on healthy plants (Kennedy, 1951). Because overcrowding began sooner on virus-infected plants, emigration of aphids began sooner. The cumulative increase of the aphid numbers in a set of plants would be substantial. In other experiments with individual leaves, this differ-

ence appeared on leaves of all ages on the plant. Baker (1960) found somewhat similar effects with four species of aphids on beet. *Myzus persicae* preferentially selected plants infected with BYV for feeding, and subsequently bred more rapidly and lived longer than on normal green plants.

Individuals of two aphid species excreted fewer droplets of honeydew when feeding on plants infected with BYDV compared with healthy plants (Ajayi and Dewar, 1982). Maturation of aphids as alatae on oats was favored by BYDV infection (85%) compared with that on healthy plants (35%) (Gildow, 1983). On the other hand, aphids reared on BYDV-infected wheat had a shorter lifespan (Araya and Foster, 1987).

In transmission experiments, plants are involved in three ways: (1) for breeding the aphids, (2) for providing virus-infected material, and (3) for providing healthy plants to test the ability of aphids to infect. It is common practice to rear virus-free aphids on a plant species 'immune' to the virus under study. However, when aphids are placed on the virus-infected plant, the change of species

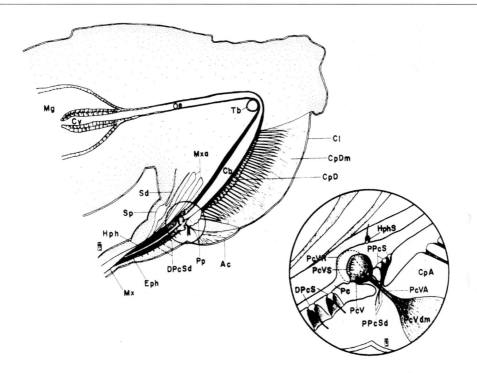

Fig. 11.3 Stylized sagittal view of head of an adult pea aphid (*Acyrthosiphon pisum*), showing the primary structure of the anterior alimentary canal. The epipharyngeal pharynx protuberance (Pp) (actually offset 3 μm from the midline), included here, demonstrates precisely its relative location with respect to the precibarial valve. An enlarged sagittal view, at the juncture of the precibarial canal and cibarium (Cb), is shown within the circle. The dashed line that scribes an arc between the precibarial valve piston (PcV) and the precibarial valve receptacle (PcVR) denotes a closed valve. Ac, antecylpeus; Cl, clypeus; CpA, cybarial pump apodemes; CpD, cibarial pump diaphragm; CpDm, cibarial valve dilator muscle; Cv, cardiac valve, DPcS, distal precibarial sensilla; DPcSd, distal precibarial dendrites; Eph, epipharynx; Hph, hypopharynx; HphS, hypopharyngeal sensillum; Mg, midgut; Mx, maxillary stylets; Mxa, maxillary stylets apodemes; Oe, esophagus; Pc, precibarium; PcVA, precibarial valve apodemes; PcVdm, precibarial valve dilator muscle; PcVS, precibarial valve suture; PPcS, proximal precibarial sensilla; PPcSd, proximal precibarial dendrites; Sd, salivary duct; SP, salivary pump; Tb; tentorial bar. From McLean and Kinsey (1984), with permission.

may influence their feeding behavior. The species and even the variety of plant used as a source of virus or as a test plant may affect the efficiency of transmission.

Source plants may change in their efficiency with time after inoculation if virus concentration changes. As discussed in Chapter 9 (Section II.J), the concentration of virus in a systemically infected plant may vary widely even in adjacent areas of tissue. This can affect the efficiency with which an aphid acquires virus. Species and varieties may differ in their susceptibility to a given virus more when one species of aphid is used as a vector than with another.

5. Environmental conditions

Environmental conditions, particularly temperature, humidity and wind, may have marked effects on aphid movement and feeding. For example, both transmission and acquisition efficiency for SDV by an aphid vector were greater at 20–22°C than at 10–11°C or 29°C (Damsteegt and Hewings, 1987). High humidity favored the transmission of PVY and PLRV by various aphid species (Singh *et al.*, 1988a). Light quality influenced the translocation of two luteoviruses in *Lycopersicon* and affected the recovery of virus by the aphid *Myzus persicae* from the plants (Thomas *et al.*, 1988). These various effects are discussed in relation to virus ecology and control of virus diseases in Chapters 12 and 16. Environmental factors may also affect transmission through effects on plant susceptibility and on the concentration of virus in source plants.

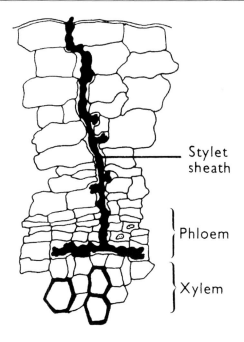

Fig. 11.4 Stylet sheath laid down in the tissues of a plant by an aphid feeding on the phloem sieve tubes. With kind permission from Dixon (1973).

B. The vector groups of aphids

There is no agreed classification for the 3700 aphid species that have been described. Eastop (1977) uses the superfamily *Aphidoidea*, which contains three families. The *Adelgidae* consists of about 45 species living only on *Coniferae*. They are of no known significance for viruses. The *Phyloxeridae*, with about 60 species, have been little studied with respect to viruses except for *Phylloxera* on vines. The *Aphididae* contain 10 subfamilies, of which the *Aphidinae* contain more than one-half of the species of aphids and most of the important virus vectors. About 50% of the approximately 600 or so viruses with invertebrate vectors are transmitted by aphids.

Data concerning vector groups can be regarded only as indicative, because the numbers are biased in various ways. For example, in virus–vector studies there has been a marked preference for trials with certain aphid species such as *M. persicae* and *Aphis gossipii* that are widespread and easily reared on a range of plant species. Negative results in transmission trials are often not reported. Only a very small proportion of the possible virus–vector combinations has actually been tested. Twenty-two viruses affecting the *Solanaceae* are transmitted by aphid species, most of which had not encountered potatoes until about 400 years ago (Eastop, 1977). Thus, it seems certain that many more actual and potential vectors exist. For example, in a 3-year study in Holland eight new vector species for PVY-N were discovered (Piron, 1986).

C. Aphid transmission by cell injury

Unstable viruses occurring in low concentration may be readily transmitted by aphid feeding, but some stable viruses such as TMV, TYMV and SBMV are not. This curious fact has not yet been fully explained. Most experimental work on this problem has been done with TMV, with the following results: (1) aphids cannot transmit TMV via their stylets (Harris and Bradley, 1973); (2) under laboratory conditions they can transmit by making small wounds when they claw the surface of the leaf (Harris and Bradley, 1973); (3) they can ingest TMV from infected plants and through membranes, and release virus again in an infectious state (Pirone, 1967); (4) aphid saliva does not inhibit TMV infection; and (5) when purified TMV is mixed with poly(L)-ornithine and potassium chloride, aphids can acquire TMV through a membrane and transmit it to plants via their stylets (e.g. Pirone, 1977). Perhaps the poly(L)-ornithine in some way makes the cell penetrated by the stylets susceptible to infection or facilitates the retention of the virus in the stylets.

D. Types of aphid–virus relationship

The basic concepts of virus–vector interactions were introduced by Watson and Roberts (1939), who coined the terms 'persistent' and 'non-persistent' to describe the length of time for which the aphid vector was able to transmit the virus after acquisition. Since then, the terminology has been modified and refined

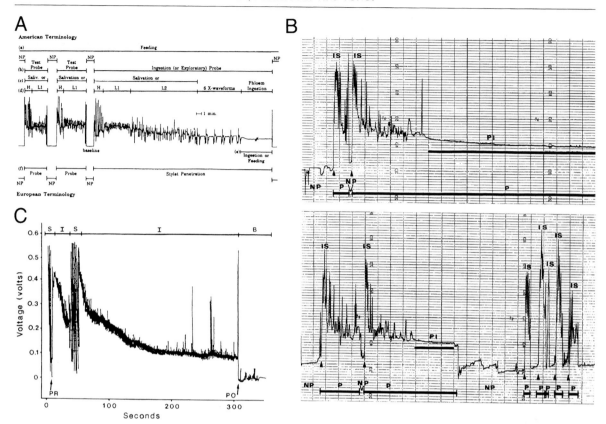

Fig. 11.5 Electronic monitoring of insect feeding. **(A)** Three composite probes made by a pea aphid (*Acyrthosiphon pisum*) on broad bean (*Vicia faba*) leaf. Lines above the waveform explain the hierarchical interpretation (from the top of the waveforms) of the terminology used by North American researchers. Line (a) defines feeding behavior encompassing all others, line (b) defines probing as the durations of all stylet insertions into a plant, whether as short test probes or as long ingestion (or exploratory) probes. Line (c) defines the categories or waveforms within the probes according to the classical AC interpretations for aphids (salivation or S, X-waveforms, and Ingestion or I). Salivation can also be split, as on line (d), into high-voltage salivation (H) and two types of low-voltage salivation (L1 and L2), although the biological meanings of these patterns are not known. Lines below the waveform tracing explain the terminology used by the Europeans, where line (e) defines ingestion as synonymous with feeding, and line (f) defines the differences between the probes and stylet penetrations. European waveform interpretations are not listed because they are the same when an AC monitor is used; the DC monitor produced waveforms, which are not yet convertible into AC waveform terms. NP, non-probing or non-penetrating. From Backus (1994). **(B)** Representative waveforms made during the electronic monitoring of the leafhopper, *Cicadulina mbila*, on a host, *Digitaria sanguinalis* (upper) and a non-host, rice (lower). IS, initial salivation; PI, phloem sieve element ingestion; P, probe (period from start of feed to end); NP, non-probing activities; ?, beginning of probe. Strip chart to be read from left to right. Each section between two vertical lines represents 20 seconds. From Mesfin *et al.* (1995). **(C)** Strip-chart recording feeding of the thrips, *Frankliniella occidentalis*, on broad bean. PR, probe initiation; S, salivation and stylet movement; I, ingestion; PO, probe termination and pull out spike; B, baseline. From Hunter *et al.* (1994), with permission.

many times to take account of new findings, but often causing controversy (reviewed in Hull, 1994a; Nault, 1997; Gray and Banerjee, 1999; Blanc *et al.*, 2001). The current, most widely accepted, terminology is given in Table 11.2, which differentiates both between and within externally and internally borne interactions. This differentiation is based on the region(s) of the vector in which the interaction(s) occurs and also takes into account the virus gene product(s) involved in the interaction.

Table 11.2 summarizes the main properties of the different kinds of relationships. Essentially, there are three stages in the transmission cycle:

TABLE 11.2 Relationships between plant viruses and their vectors

Virus transmission group			Transmission characteristics						
Site in vector	Type of transmission	Virus product inter-acting with vector	Aquisition time (max dose)	Retention time (half-life)	Transtadial passage	Virus in vector hemo-lymph	Latent period	Virus multiples in vector	Transovarial transmission
Externally borne	Non-persistently transmitted, stylet-borne	Capsid Helper factor	Seconds to minutes	Minutes	No	No	No	No	No
	Non-persistently transmitted, foregut-borne (semi-persistent)	Capsid Helper factor	Minutes to hours	Hours	No	No	No	No	No
Internally borne	Persistent, circulative		Hours to days	Days to weeks	Yes	Yes	Hours to days	No	No
	Persistent, propagative		Hours to days	Weeks to months	Yes	Yes	Weeks	Yes	Often

1. The *acquisition phase* in which the vector feeds on the infected plant and acquires sufficient virus for it to be able to transmit it.
2. The *latent period* in which the vector has acquired sufficient virus but is not able to transmit it. For externally borne viruses there is little or no latent period.
3. The *retention* (transmission) *period* is the length of time during which the vector can transmit the virus to a healthy host.

Some further definitions are needed. *Inoculativity* is the ability of an aphid or other insect to deliver infectious virus into a healthy plant. The *acquisition feed* is the feeding process by which the insect acquires virus from an infected plant. The *inoculative* (transmission) *feed* is the feed during which virus is delivered into a healthy plant.

These relationships have been studied in most detail with aphids but are applicable to most other viruses that are transmitted by arthropods with piercing–sucking mouthparts. The basic features of the interactions between viruses and their arthropod vectors are shown diagrammatically in Fig. 11.6.

E. Non-persistent transmission

Because of its role in the field transmission of many economically important viruses, non-persistent transmission by aphids has been studied in many laboratories over seven decades, and molecular details of the interactions are becoming more clearly understood.

1. The non-persistent viruses

Of the approximately 290 or so known aphid-borne viruses, most are non-persistent (see Appendix 2). The following virus genera have definite members transmitted in a non-persistent manner: *Alfamovirus*, *Caulimovirus* (by *M. persicae*), *Cucumovirus*, *Fabavirus*, *Macluravirus* and *Potyvirus*. These genera include viruses with helical and isometric particles, and with DNA and RNA mono-, bi- and tri-partite genomes.

2. Acquisition time

Non-persistently transmitted viruses are acquired rapidly from plants, usually in a matter of seconds. During this time, aphids stylets do not usually penetrate beyond the epidermal cells and when they penetrate beyond the epidermis

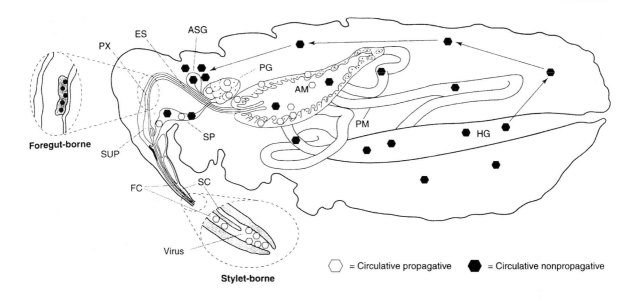

Fig. 11.6 Mechanism of transmission of plant viruses by arthropods with piercing–sucking mouthparts. The general anatomy of the alimentary system and the salivary system is shown; the areas relevant to virus transmission are labeled. One inset shows a detailed view of the distal end of the mouthparts where the food canal (FC) and the salivary canal (SC) empty into a common space (see Fig. 11.3). One current model of transmission of stylet-borne (non-persistent, non-circulative) viruses suggests that the transmissible virus is retained at the distal tip of the stylets and then released by salivary secretions as the insect salivates during feeding. A second inset shows a detailed view of the foregut-borne (semi-persistent, non-circulative) viruses attached to the cuticle lining of the foregut, a region that would include the sucking pump (SUP), pharynx (PX) and esophagus (ES). Note the virus is embedded in a matrix material attached to the cuticle. The origin or composition of the matrix material is unknown. The circulative non-propagative viruses will pass through the foregut into the anterior midgut (AM), the posterior midgut (PM) and then into the hindgut (HG). They do not infect the gut cells but are transported through the posterior midgut and hindgut cells and released into the hemocoel (body cavity). Current information indicates that these viruses specifically associate with the accessory salivary glands (ASG) and are transported across the ASG cells and then released into the salivary canal (SC) (see Fig. 11.11). The circulative propagative viruses will infect midgut cells and subsequently infect other tissues. These viruses ultimately associate with the principal salivary glands (PG) and possibly the ASG before their release into the SC. SP, salivary pump. From Gray and Rochon (1999), with permission.

into the mesophyll and vascular tissue the transmission rate declines rapidly (see Nault, 1997). The initial host-finding behavior of aphids—short probing, which is thought to involve sampling of the epidermal cells sap—fits very well with this mechanism. As the sampling is especially brief on non-hosts, the vectors of non-persistent viruses are often non-colonizers of that species. Acquisition can also be increased by starving the aphids, which can affect their initial feed behavior.

3. Retention time

With a non-persistent virus, aphids begin to lose the ability to infect immediately after the acquisition feed. The rate at which infectivity is lost depends on many factors, including tem-

perature and whether they are held on plants or under some artificial condition. With many non-persistent viruses, aphids become non-infective very rapidly when they are allowed to feed on test plants—often in a matter of minutes. The rate at which they become non-infective is about the same on plants that are, and plants that are not, susceptible to the virus (Bradley, 1959).

The main reason for attempting to determine how long aphids retain the ability to infect is in relation to the spread of virus in the field. In most experiments, conditions have not been particularly close to those that might exist under field conditions. However, Cockbain *et al.* (1963) simulated field conditions by allowing tethered aphids to fly for

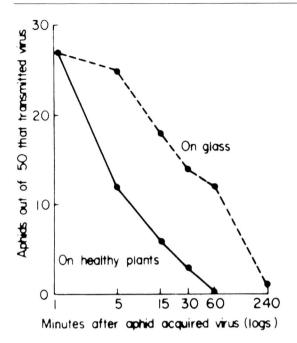

Fig. 11.7 Loss of ability of infective *Myzus persicae* to transmit PVY after various periods on healthy tobacco leaves or in a glass dish without food. From Bradley (1959), with permission.

various times in an air current. The infectivity of alatae of *M. persicae* and *A. fabae* for PMV and BtMV fell off at about the same rate whether they were flying or kept fasting in a glass container. Figure 11.7 illustrates the rate at which *M. persicae* lost the ability to transmit PVY when held on healthy plants or in a glass container.

4. Virus and virus strain specificity

The V strain of TAV is transmitted by *M. persicae*, whereas the M strain of CMV is not. The RNAs of these two viruses and also that of TMV were reassembled *in vitro* using either V or M strain coat protein. All of the three viral RNAs coated in V strain protein were transmitted by *M. persicae*, following acquisition through membranes. Reassembly with M strain protein did not give aphid transmission (Chen and Francki, 1990). These results show that aphid transmission of cucumoviruses depends solely on the coat protein (see Section III.E.7.a for molecular interactions).

Different strains of the same virus may differ in the efficiency with which they are transmitted by a particular aphid species. Some strains may not be transmitted by aphids at all. Examples are also known where a given virus strain either gained or lost the ability to be transmitted by a particular aphid. Again, properties of the viral coat protein appear to be important in determining specificity of transmission (e.g. Gera *et al.*, 1979).

Different strains of the same non-persistent virus do not usually interfere with each other's transmission, as is sometimes found with propagative viruses. Thus, Castillo and Orlob (1966), in experiments with two strains of CMV and two of AMV, found that each strain was acquired independently of the other. The proportion of aphids transmitting one or both strains was in accord with the probability expected for independent transmission. By contrast, such interference has been reported between strains of PVY (Katis *et al.*, 1986).

5. Aphid specificity

a. Aphid species

The conditions of the aphid culture may affect the efficiency with which individuals transmit and thus the results of experiments on aphid vector specificity. This factor may well account for some of the discrepancies in the literature. Young colonies will provide active individuals, whereas older colonies will give less efficient and more variable transmission.

Aphid species vary widely in the number of different viruses they can transmit. At one extreme, *M. persicae* is known to be able to transmit a large number of non-persistent viruses. Other aphids have been found to transmit only one virus. These differences in part reflect the extent to which different aphid species have been tested, but there is no doubt that real differences in versatility occur.

Among species that transmit a given virus, one species may be very much more efficient than another. In some experiments on this problem, adequate care was not taken to ensure that factors such as age and vigour of colony, as well as host plant, were controlled properly. However, even when considerable care is taken in the design of the experiment, big differences in vector efficiency may be observed. Thus, Bradley and Rideout (1953) found marked differences in the efficiency with which PVY was transmitted by different species even when acquisition feed and test feed times were standardized (Fig. 11.8). This can reflect the initial feeding behavior of different aphid species on the test plant species.

b. Non-transmitters in a single population

When individual aphids of a vector species are tested for their ability to transmit a given virus, not all individuals will transmit in a first trial. However, using two vector species for PVY, Gibson *et al.* (1988) showed that individual aphids that did not transmit in a first trial were as likely to transmit virus in a second test as those that did on the first occasion. Thus, there was no evidence for non-uniformity in a given vector population. For any given virus, aphid vector and host plant combination there appears to be a statistical probability that both the acquisition and infection feeds will be successful.

By feeding aphids on solutions containing a virus and a radioisotope marker, Pirone and Thornbury (1988) calculated that individual *M. persicae* aphids given a 10-minute access feeding acquired between 10 and 4000 particles of a *Potyvirus*. Although these estimates were based

on an artificial feeding experiment, they indicate the kind of variability involved in virus uptake by individuals in the same populations.

c. Aphid form

Alate or apterous viviparous females have been used in most experimental work. The different forms of an aphid species adapted to different seasons and different host plants may vary in the efficiency with which they transmit a virus, although not many species have been studied carefully from this point of view (e.g. Gill, 1970).

d. Aphid clones

Aphid species may exist as clones, perhaps showing minor anatomical differences. Such clones may differ in the efficiency with which they transmit. For example, Jurík *et al.* (1980) found that 36 clones of the pea aphid *Acyrthosiphon pisum* varied widely in their ability to transmit BYMV. Successful transmissions, which ranged from 3% to 37%, were not correlated with the color of the clone or its geographical origin within the Czech Republic.

e. Geographically diverse populations

Populations of *Acyrthosiphon solani* originating in the USA and New Zealand transmitted strains of SDV less efficiently than did a population from Japan (Damsteegt and Hewings, 1986).

6. Site of virus retention

As noted above, when they alight on a leaf, aphids may make brief probes into the leaf (usually less than 30 seconds) (Fig. 11.5A). Thus, the initial behavior of such aphids on reaching a leaf is ideally suited to rapid acquisition of a non-persistent virus.

Sap sampling on a virus-infected plant will contaminate the stylet tips, the food canal and the foregut. These sites have been favored for the retention of virus that will be injected subsequently into a healthy plant following another exploratory probing feed. The early workers focused on the stylet tips as being the site of retention. However, experiments in

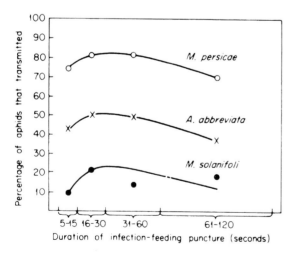

Fig. 11.8 Relative efficiency of three aphid species in transmitting PVY after defined acquisition feeding times. From Bradley and Rideout (1953), with permission.

which treatment of the stylets with formalin or UV irradiation eliminated or reduced virus transmission (Bradley, 1966) were criticized for the effects that these treatments had on the aphid's probing behavior. Electron microscopy has shown TVMV and TEV particles in a matrix attached primarily to the epicuticle of the maxillary food canal (Ammar *et al.*, 1994), with fewer particles attached to the precibarium and cibarium of the foregut. Radioactively labeled virus was detected on the surfaces of the distal third of the maxillary food canal (Wang and Pirone, 1996; Wang *et al.*, 1996). Thus, the weight of evidence now favors the food canal in the maxillae as the site where infective virus is retained during nonpersistent transmission, although it must be remembered that this evidence identifies sites of accumulation but does not give any indication as to whether that virus is transmitted.

7. Virus–vector molecular interactions

There are two phases to the interaction involved in non-persistent virus transmission: retention of the virus at a specific site and release of the virus (reviewed by Gray and Banerjee, 1999). All non-persistently transmitted viruses have a simple structure of nucleic acid encapsidated in simple icosahedral or rod-shaped particles by one or more coat protein species. Thus, it is the capsid protein that is available for any interactions. Two forms of interaction have been identified in the retention phase, one in which there appears to be direct interaction between the virus capsid and the site of retention in the aphid and the other in which a non-structural virus-encoded protein is involved. This non-structural protein is termed a *helper component, helper factor* or *aphid transmission factor* (reviewed by Pirone and Blanc, 1996). As well as these virus gene products that directly control virus transmission, other viral genes can affect transmission indirectly. The efficiency of transmission can be influenced by virus concentration, which is controlled by virus replication and turnover (Atreya *et al.*, 1992). Similarly, acquisition of a virus can be affected by the localization of the virus in the host in relation to the site(s) of interaction of the vector with the host. Viral genes, such as those involved with cell-to-cell movement, can influence virus distribution in the host.

a. Direct capsid interaction

It is thought that AMV and CMV transmission involves direct links between the capsid and the binding site within the aphid vector. The prime evidence is that purified virus can be transmitted from artificial feeding systems without the addition of other proteins or factors. Most of the detailed evidence is for CMV. Gera *et al.* (1979) made heterologously assembled particles between the genomes and capsid proteins of a highly aphid-transmitted (HAT) and a poorly aphid-transmitted (PAT) strain of CMV and showed that the efficiency of aphid transmission segregated with the source of coat protein. There were no differences in sizes of coat protein and, thus, no indication of read through to a helper component in the transmissible strain. Chen and Francki (1990) obtained similar results with other cucumo-viruses and showed that the heteroencapsidant of TMV RNA in TAV coat protein was aphid transmissible. The amino acid changes in M-CMV (a PAT strain) coat protein required to restore transmission by *A. gossypii* have been identified (Perry *et al.*, 1994). However, in contrast to the amino acid positions 129 and 162 identified as important in *A. gossypii* transmission, further amino acids (positions 25, 168 and 214) proved to be important in establishing transmission of the PAT strain by *M. persicae* (Perry *et al.*, 1998). Surface residues are important for interacting with the vector (Perry *et al.*, 2001) but the amino acids identified above are not necessarily those directly involved in the interaction(s) with the vectors. As well as the possibility of pleiotropic effects on the availability of the virus for acquisition from the host plant, these amino acids may be important for the folding of the coat protein and/or the physical stability of the virus in the vector.

The stability of the aphid transmission phenotype after mechanical transmission of seven field isolates of CMV was studied using the aphids *A. gossypii* and *M. persicae* (Ng and Perry, 1999). After more than 24 mechanical passages, one of the isolates lost its transmissibility by *M. persicae* but not by *A. gossypii*, which illustrates differences in the interaction between the virus and the two aphid species.

b. Indirect interaction involving helper component

It has been known for many years that certain viruses could be transmitted by aphids in a non-persistent fashion only when another virus was also present in the source plant. For example, transmission of PAMV required the presence of PVY. Kassanis and Govier (1971) made the important observation that PAMV need not be present in a mixed infection with PVY in order to be aphid transmitted. Aphids first fed on plants infected with PVY then transferred to plants infected with PAMV, transmitted PAMV; this led to the concept that a helper component was required for aphid transmission. Subsequently it has been shown that helper factors are specific gene products of the helper virus and are required for the transmission of aphid-transmitted potyviruses. Their properties are reviewed by Raccah *et al.* (2001) and are summarized here:

1. The helper factor of one potyvirus may or may not permit the aphid transmission of another potyvirus when tested in an *in vitro* acquisition system. Thus, there is some specificity in the phenomenon (e.g. Sako and Ogata, 1980; Lecoq and Pitrat, 1985).
2. The helper component has to be acquired by the aphid either during or before virus acquisition. If it is provided after the virus, there is no transmission.
3. Potyvirus helper components have MWs in the range of 53 kDa (TVMV) and 58 kDa (PVY). They are cleaved from the polyprotein encoded by the virus (Chapter 7, Section V.B.1) as a product that, as well as helper component activity, has various other activities including being a protease. Thus, it is termed HC-Pro.
4. Some potyvirus helper components (e.g. PRSV, WMV2, ZYMV and TuMV) can be purified on Ni-charged resins, whereas others (e.g. PVY and TVMV) cannot.

5. Purified helper component can be used to facilitate the transmission of potyviruses by feeding through artificial membranes.
6. The biologically active form of helper component appears to be a dimer with molecular weight ranging between 100 and 160 kDa.
7. Potyviral HC-Pro binds non-specifically to single-stranded nucleic acids, with a preference for RNA (Urcuqui-Inchima *et al.*, 2000). The protein contains two nucleic acid-binding regions.

By studying the effects of mutations in the coat protein and helper component on aphid transmission of various potyviruses, a picture has been built up of some of the molecular interactions involved. The current hypothesis, reviewed by Raccah *et al.* (2001), is that the helper component forms a bridge between the virus capsid and the aphid stylet (Fig. 11.9). Close to the N-terminus of the coat protein is the amino acid triplet DAG (Asp-Ala-Gly) which is important for transmissibility. Mutagenesis shows that both the DAG sequence itself and the context of the surrounding amino acids affects transmissibility (see Lopez-Moya *et al.*, 1999). Biochemical and immunological analyzes (Allison *et al.*, 1985; Dougherty *et al.*, 1985) indicate that this N-terminal region is located on the external surface of the virus particle. Two important regions have been identified in HC-Pro. One, characterized by the amino acid sequence PKT (Pro-Lys-Thr), appears to be involved in the interaction with the capsid protein. The other, termed the KITC (Lys-Ile-Thr-Cys) region, appears to be involved with the HC-Pro retention on the aphid's stylets.

c. Release

Little is known about the mechanisms of release of non-persistent viruses from the site of binding in the aphid's stylet but three theories have been proposed (reviewed by Gray and Banerjee, 1999). (1) The mechanical transmission theory suggests that the virus is simply inoculated by the stylet (Kennedy *et al.*, 1962). (2) In the ingestion–egestion theory, release is effected by regurgitation and salivation (Harris, 1977). (3) As the food and salivary canals of

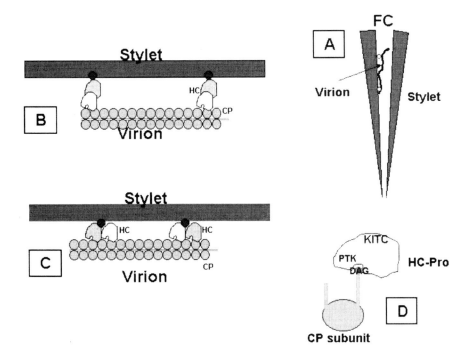

Fig. 11.9 Models depicting possible interactions between the HC-Pro, the aphid's stylets and the potyviral coat protein. **(A)** Position of the virion particles close to the apical section of the food canal. **(B)** Model assuming an association between two molecules of HC-Pro. Note that one molecule of the HC-Pro is bound to a 'receptor' on the stylet, whereas the second HC-Pro molecule is bound to the coat protein subunit. **(C)** Model assuming that a dimer is needed to bind to the 'receptor' on the stylet. Both HC-Pro molecules are linked to coat protein subunits. **(D)** A proposed structural binding between the PTK motif of HC-Pro and the DAG motif found on the N-terminus of the coat protein subunit. FC, food canal; KITC, lysine, isoleucine, threonine, cysteine; PTK, proline, threonine, lysine; DAG, aspartic acid, alanine, glycine; HC, helper component; CP, coat protein. From Raccah *et al.* (2001), with permission.

the stylets fuse near the tip of the maxillary stylet, non-persistently stylet-borne viruses could be released by saliva alone (Martin *et al.*, 1997b).

F. Semi-persistent transmission

1. Semi-persistently transmitted viruses

The aphid transmission of members of five virus genera has properties that are intermediate between non-persistent and circulative systems, although as a group these viruses do not show very uniform transmission properties.

The best studied of the semi-persistent viruses are the caulimoviruses (reviewed by Pirone and Blanc, 1996; Blanc *et al.*, 2001) and the closteroviruses BYV and CTV (Raccah *et al.*, 2001). The closteroviruses are found particularly in the phloem, whereas the caulimo-viruses are found in most cell types. The caulimoviruses have been studied extensively and their transmission shown to involve a helper component that has been characterized in great detail.

Helper components have been implicated in sequiviruses (reviewed in Harrison and Murant, 1984) but have not been formally identified. The aphid *Cavariella aegopodii* transmits both AYV and PYFV in a semi-persistent manner. PYFV is a small isometric RNA virus with a diameter of about 30 nm, as is AYV (Hemida *et al.*, 1989). The aphid transmits PYFV only when it is carrying AYV, a virus that is not sap transmitted. Unlike the 'helper' for some non-persistent viruses, the AYV itself is the helper agent for PYFV (Elnagar and Murant, 1976). However, PYFV can be acquired from plants infected singly with this virus only if the aphids are already carrying the helper AYV. This suggests that a helper protein may be involved as with some non-persistent viruses. AYV appears to have a specific retention site in the vector foregut (Harrison and Murant, 1984).

There have been suggestions that carla-viruses and closteroviruses might also involve helper components in their transmission (Murant *et al.*, 1988b). In another helper relationship between two semi-persistent viruses, a novel mechanism operates: one closterovirus (HLV, with 730-nm rods) depends on another (HV-6, with 1600-nm rods) for transmission by aphids. In this system the two virus rods become attached to each other at one end (Murant, 1984; Murant *et al.*, 1988b). The anomalous coat protein encapsidating the 5′ end of closterovirus particles (see Fig. 5.12) has been suggested to be involved in aphid transmission (Agranovsky *et al.*, 1995; Febres *et al.*, 1996).

2. Virus–vector molecular interactions

CaMV, and presumably other caulimoviruses, requires a helper component (or aphid transmission factor) when being transmitted in a semi-persistent manner by *M. persicae*. The CaMV helper component system has the following properties (reviewed by Blanc *et al.*, 2001):

- As with potyviruses, the helper component has to be acquired by the aphid either during or before virus acquisition.
- Helper components of other caulimoviruses can complement defects in CaMV helper component (Markham and Hull, 1985).
- The CaMV helper component system involves two non-capsid proteins, the 18-kDa product of ORF II (P2 or P18) and the 15-kDa product of ORF III (P3) (for CaMV gene map see Chapter 6, Section IV.A.1).
- In infected cells, P2 is found in crystalline inclusion bodies (Fig. 6.1) and P3 in association with virus particles.
- P2 interacts very strongly with microtubules with binding domains in two regions, one near the N-terminus and the other near the C-terminus.
- Active P2 is not post-translationally modified and is 23% α-helical mainly in the C-terminal domain. The active form is a large

soluble oligomer of 200–300 subunits (Hebrard *et al.*, 2001).

Thus, the CaMV helper component system is more complex that that of potyviruses. The virus can be transmitted from an *in vitro* acquisition system containing baculovirus-expressed P2 and sap from a plant infected with a non-transmissible isolate but not from P2 plus purified virus (see Blanc *et al.*, 2001). However, the virus can be transmitted when P3 is added to the purified virus (Leh *et al.*, 1999). Secondary structure predictions of P2 suggest two domains, the N-terminus being predominantly β sheet and the C-terminus predominantly α helix; the two domains are separated by predicted random structure. The 61-amino-acid C-terminal domain interacts with partially purified virus and with the 30 N-terminal amino acids of P3 (Leh *et al.*, 1999). Mutations of the N-terminal domain abolish its ability to facilitate transmission but do not affect its ability to bind to semi-purified virus. This leads to a model for how the CaMV helper system operates (Fig. 11.10). P3, which forms a tetramer (Leclerc *et al.*, 1998), binds to the virus capsid composed mainly of P4, with the C-terminus of P2 binding to P3. The P2 interaction is thought to be via coiled-coil structures but self interaction of those structures could possibly interfere with the P3 interaction (Hebrard *et al.*, 2001). The bridge is completed by the N-terminus of P2 binding to a site in the aphid foregut. The role of the microtubule binding activity of P2 is unknown but it is noted that the microtubule binding domains overlap the P3 and aphid binding domains.

As with non-persistent viruses, nothing is known about the molecular details of virus release from the vector.

G. Bimodal transmission

Some aphid transmission studies (e.g. Chalfant and Chapman, 1962) have shown that a virus could apparently be transmitted both non-persistently and semi-persistently by the same aphid species. This was termed bimodal transmission. However, there have

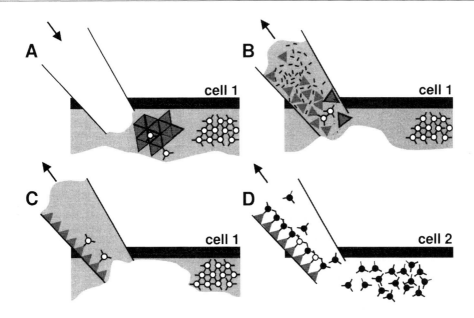

Fig. 11.10 (see Plate 11.1) Model of the sequential acquisition of CaMV by aphids from infected cells. **(A)** In infected cells, the viral components involved in transmission are spatially separated in electron-lucent inclusion bodies (eliB) (left structure) and electron-dense inclusion bodies (ediB) (right structure). Whereas most virus particles (empty circles) complexed with P3 (blue bars) are stored in ediBs, P2 (red triangles) in loose association with P3 and some sparse viruses are located in eliBs. When the aphid stylet (at the left) pierces the cell wall, saliva is injected into the plant cell. **(B)** After salivation, the aphid sucks up some of the plant cell's contents together with viral inclusion bodies through its stylets. If an eliB is taken up, it immediately—either through the action of the aphid saliva or simply by a dilution effect—disintegrates and sets free its components P2, P3 and eventually some rare P2:P3 virion complexes. **(C)** While the liberated P3 is ingested, the released P2, perhaps together with a few virus:P3 complexes from the eliB, attaches to the aphid stylet cuticle. The aphid is now 'P2 loaded', and thus transmission competent, and ready to acquire more P3:virions (solid circles) from either the same cell or **(D)** on following feeding on other cells from the same or different host plants. From Blanc *et al.* (2001), with permission.

been various inconsistencies in these reports, most of which might be due to other factors such as aphid–plant interactions. Blanc *et al.* (2001) argue that the molecular strategy of transmission should be used in classifying the virus–vector interaction and the transmission system thus making the term bimodal transmission unnecessary.

H. Persistent transmission

The main features of persistent transmission are summarized in Table 11.2. Viruses transmitted in this manner are usually transmitted by one or a few species of aphid. Yellowing and leaf-rolling symptoms are commonly produced by infection with persistently transmitted viruses. Viruses that are internally borne in their aphid vectors may replicate in the vector (propagative) or may not (circulative). For an aphid to become a transmitter by either type of relationship the virus has to be ingested from the infected plant and reach the salivary glands, usually via the hemolymph to be egested into the healthy plant. Thus, it has to pass at least two barriers: the gut wall and the wall of the salivary glands.

1. Circulative viruses

Circulative (persistent) viruses for which there is no demonstration of replication in the vector are the luteoviruses (among which BYDV and PLRV have been the most studied), PEMV which is a complex between a luteovirus and an umbravirus (see Chapter 2; Section III.P.3) and nanoviruses. Some other less well-characterized viruses are also transmitted in this manner.

a. Luteoviruses

i General features of transmission

The minimum acquisition time of luteoviruses can be a little as 5 minutes (for BYDV; Duffus, 1972) but is usually several hours. This is followed by a latent period of at least 12 hours, after which the virus can be transmitted with an inoculation access time of 10–30 minutes. The aphids then remain capable of transmitting for at least several days. Wide differences were found in the ability of the five stages of *Schizaphis graminum* to transmit BYDV. Percentage transmission progressively declined from 36% for the first instar to 2% for adults (Zhou and Rochow, 1984).

ii Uptake route in vector

As noted above, persistently transmitted viruses have to cross at least the gut and salivary gland barriers. The most detailed studies on circulative viruses have been made on luteoviruses (reviewed by Gildow, 1987, 1999) (Fig. 11.11). Gildow (1985) showed that particles of CYDV-RPV (formerly BYDV-RPV) associate only with the cell membranes of the hindgut of the aphid vector *Rhopalosiphon padi*. He suggested that the particles entered the hindgut cells by endocytosis into coated pits and coated vesicles, and accumulated in tubular vesicles and lysosomes (Fig. 11.11A). Interaction with the gut membrane or passage across gut cells regulated the transmission effi-

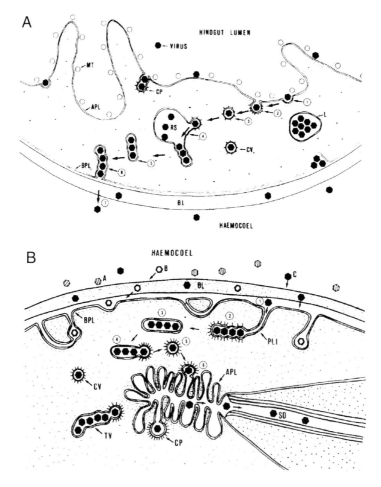

Fig. 11.11 Models for interactions and transcellular transport of luteoviruses in aphid vectors. **(A)** Transcellular transport through aphid gut epithelium. Visualization of endocytotic- and exocytotic-associated ultrastructure supports receptor-mediated endocytosis as a mechanism regulating vector-specific luteovirus acquisition. Based on this model,

luteoviruses recognized at the gut-cell apical plasmalemma (APL) bind to the membrane (1) initiating virus invagination (2) into coated pits (CP). Coated pits bud off the APL as virus-containing coated vesicles (CV), which transport the virus (3) to larger uncoated vesicles, called receptosomes (RS), which act to concentrate the virus (4). Tubular vesicles containing linear aggregates of virus form on the receptosomes (5) and transport the virus to the basal plasmalemma (BPL). Tubular vesicles containing virus are observed to fuse with the BPL, allowing release of the virus from the cell (6). Luteoviruses can then diffuse through the gut basal lamina (BL) and into the hemocoel (7). Eventually, receptosomes (endosomes) mature into lysosomes (L) and any virus particles remaining in the lysosome are probably degraded. MT, microtubules. **(B)** Luteovirus interactions with accessory salivary glands (ASG) determining vector-specific transmission. Luteoviruses suspended in hemolymph first encounter the extracellular basal lamina (BL) surrounding the ASG. The BL acts as a selective barrier to luteovirus transmission. Depending on the aphid biotype and the specific luteovirus, the virus particles may be prevented from penetrating the BL (A) or may diffuse through (B, C) to the basal plasmalemma (BPL). A second selective barrier occurs at the BPL. Luteovirus particles not recognized at the BPL remain outside the cell in the pericellular space (B). Luteoviruses recognized by putative virus receptors (C) on the BPL (1) are endocytosed by coated pits (2) and accumulated into tubular vesicles (TV) in the cytoplasm (3). The TV adjacent to the microvilli-lined canals formed by the apical plasmalemma (APL) bud off coated vesicles (CVs) (4) containing individual virions. These CVs transport the virus (5) to the canals, fuse with the APL (6), forming coated pits (CP), and release the virus into the canal lumen allowing transport of luteovirus out of the aphid along with salivary secretions. PLI, plasmalemma invagination; SD, salivary duct. From Gildow (1999), with permission.

ciency of a strain of PLRV (Rouzé-Jouan *et al.,* 2001).

Particles were then released into the hemocoel by fusion of the tubular vesicles with the basal plasmalemma. Aphid salivary glands comprise two principal glands and two accessory glands (Ponsen, 1977). In a study (Gildow, 1982) on the interactions between BWYV and PLRV and the salivary glands of their vector, *M. persicae*, virus particles were seen in the basal lamina and plasmalemma invaginations of accessory salivary cells (Fig. 11.11B). Particles were also found in tubular vesicles in the cytoplasm near salivary canals and in coated pits connected to the canal membrane. The basal lamina and the basal plasmalemma function as independent barriers to transmission of different luteovirus–aphid combinations (reviewed by Gray and Banerjee, 1999). From these studies the route that luteoviruses take across the two barrier tissues in their aphid vector would appear to be by incorporation into coated vesicles and transport across the cell(s). Thus, the main sites of interaction for the virus particles are with the plasma membrane on the gut side of the gut wall cells and with two plasma membranes on the hemocoel side of the salivary gland accessory cells, which suggests a receptor-mediated interaction.

iii Luteovirus–vector molecular interactions (reviewed by Gildow, 1999)

As purified luteoviruses can be aphid transmit-

ted from *in vitro* acquisition it is likely that no non-capsid proteins are involved. The capsid comprises the major capsid protein and a minor amount of a larger protein translated via a read through of the coat protein stop codon (see Chapter 7, Section V.B.9). This read-through portion is C-terminally processed to a certain extent in most luteoviruses except PEMV, and the C-terminal portion has been shown not to be required for aphid transmission (reviewed in Gray and Banerjee, 1999). However, particles containing just the major coat protein without any read-through protein are not transmissible, which led to the widespread assumption that the read-through portion was required for aphid transmissibility. Particles without read-through protein can reach the hemocoel, which shows that they can cross the gut wall barrier. Furthermore, virus-like particles from coat protein lacking read-through protein expressed in insect cells using a baculovirus system and either injected into or fed to aphids have been observed in the accessory salivary gland cells and in the salivary ducts (reviewed by Gray and Banerjee, 1999).

Point mutations introduced into or near conserved domains in the read-through domain of BWYV minor capsid protein affected aphid transmissibility (Brault *et al.,* 2000). Aphid transmissibility was restored in some of the mutants by pseudo-reversion or second-site mutations. One of the second-site mutations restored the ability of the virus to move from the hemocoel through the accessory salivary gland following micro-injection of the mutant virus into the hemocoel but did not permit virus movement across epithelium separating the gut from the hemocoel. It was suggested that distinct features of the read-through domain operate at different stages of the transmission process.

Also to be taken into account are the interactions of luteoviruses with symbionin (see below). Thus, there is no clear picture of the luteovirus component of the receptor-mediated recognition.

Several aphid proteins of M_r ranging from 31 to 85 kDa have been shown to interact with purified luteoviruses *in vitro* (reviewed by van den Heuvel, 1999; van den Heuvel *et al.*, 1999). Antisera raised against two of these proteins, P31 and P44, react specifically with extracts of accessory salivary glands from vector aphids, suggesting that these proteins might be involved in luteovirus-specific recognition at this site.

iv Symbionin

Another binding aphid protein, which interacts with many luteoviruses and other viruses, is the 60-kDa symbionin or GroEL from the endosymbiotic bacterium *Buchneria* spp. This protein, found readily in aphid hemolymph, is a member of the molecular chaperone family responsible for stabilizing the structure of proteins. The interaction of luteoviruses with symbionin is determined by the read-through domain of the minor capsid protein. Injection of BWYV mutants devoid of the read-through domain persist for much shorter times in the aphid hemolymph than do unmutated virions (van den Heuvel *et al.*, 1997). Furthermore, treatment of *M. persicae* with antibiotics that affect *Buchneria* and reduce symbionin levels in the hemolymph inhibits transmission and leads to loss of capsid integrity in the hemolymph (van den Heuvel *et al.*, 1994). Taken together, these observations indicate that the luteovirus–symbionin interaction is essential for retention of the virus in the hemolymph (van den Heuvel *et al.*, 1999).

v Dependent transmission (for a reviews see Falk and Tian, 1999)

Within luteoviruses. As with certain non-persistent viruses, some persistent viruses require a helper virus to be present in the plant before aphid transmission can occur. For persistent viruses that are dependent on another virus, it is the presence of the virus itself in a mixed infection that provides the assistance. BYDV is the best studied of the persistent viruses with dependent transmission.

There is a high degree of vector specificity among the luteoviruses, including BYDV. For example, CYDV-RPV is transmitted efficiently by *Rhopalosiphum padi* but not by *Sitobion avenae*. For BYDV-MAV, the reverse is true. Virus can enter the hemolymph of inefficient vectors. Thus, the block in transmission appears to lie in the inability of a virus to move from the hemolymph, via the salivary glands, to the plant. *R. padi* very rarely transmits BYDV-MAV from oats when plants are infected only with that virus. However, it regularly transmits both BYDV-MAV and CYDV-RPV from doubly infected plants. When aphids are fed through membranes on concentrated mixtures of the two viruses purified from separate infections, they transmit only the virus that they transmit specifically from singly infected leaves. Transmission from purified virus is blocked only by homologous antiserum.

When aphids were fed through membranes on virus purified from mixed infections, transmission of BYDV-MAV by *M. avenae* was blocked by BYDV-MAV antiserum. From the same mixture, however, *R. padi* infected plants that often developed BYDV-MAV as well as CYDV-RPV (Rochow, 1970). Rochow and Muller (1975) injected aphids with strain-specific antisera before they fed on leaves infected with both viruses. Injection of BYDV-MAV antiserum prevented transmission of BYDV-MAV by control aphids. However, *R. padi* injected with this antiserum transmitted BYDV-MAV along with CYDV-RPV from mixed infections.

These and other experiments are best interpreted by assuming that this type of dependent transmission is due to phenotypic mixing during replication of the two viruses together in the plant (Chapter 8, Section X). Some copies of the BYDV-MAV genome are presumed to become encapsulated in a coat of CYDV-RPV protein and are thus protected from inactivation by BYDV-MAV serum.

Mixed infections with BYDV strains and with CYDV are common in the field. The fact that dependent transmission by *R. padi* occurred with combinations of several BYDV-MAV-like and CYDV-RPV-like isolates collected in various parts of the United States and Canada indicates that the phenomenon can influence virus spread in the field.

Phenotypic mixing may occur in other important members of the luteovirus group, such as BWYV.

Umbraviruses. In Chapter 6 (Section VIII.H.20), I described how umbraviruses do not encode a coat protein. For their aphid transmission, they associate with a helper luteovirus which is presumed to supply the coat protein and thus aphid transmission properties. There are seven definitive umbravirus species, each associating with a luteovirus (Table 11.3) (Robinson and Murant, 1999). These systems have the following characteristics: (1) both viruses are transmitted in a circulative non-propagative manner; (2) the dependent virus is sap transmissible, but the helper is not; (3) the dependent virus is transmitted only by aphids from source plants that contain both viruses; in other words, aphids already carrying helper virus cannot transmit the dependent virus from plants infected only with this virus; (4) evidence from a variety of experiments indicates that the dependent virus is transmitted by the aphid vector only when its RNA is packaged in a protein shell comprising the helper virus protein (Harrison and Murant, 1984); and (5) the dependent virus may be helped by several different luteoviruses. This phenotypic mixing can take place in doubly infected plants. GRV depends on its satellite RNA as well as on GRAV for transmission by *Aphis craccivora* (Murant, 1990).

Viroids. PSTVd is transmitted from plants that are also infected with PLRV (Salazar *et al.*, 1995; Singh and Kurz, 1997), although there is some doubt as to whether the viroid is encapsidated in PLRV coat protein.

b. Nanoviruses

Nanoviruses are transmitted by aphids in the persistent circulative manner. The minimum acquisition feed for FBNYV is 15–30 minutes (Franz *et al.*, 1998) and for BBTV is within 4 hours (Hu *et al.*, 1996). The inoculation access period for both viruses is 5–15 minutes, and individuals transmitted virus throughout their lifespan. The persistent transmission of nanoviruses can be erratic, which has been attributed to their multicomponent nature (Franz *et al.*, 1998). Purified FBNYV is not aphid transmissible, which suggests that it needs a helper factor (Franz *et al.*, 1999).

2. Propagative viruses

Whereas aphids are vectors of many of the persistent circulative viruses, most of the persistent propagative viruses are transmitted by leafhoppers or planthoppers. However, several members of the *Rhabdoviridae* replicate in their aphid vector, including SYVV and LNYV, both in the vector *Hyperomyzus lactucae* (reviewed by Jackson *et al.*, 1999).

The latent period of SYVV in the vector is long and depends strongly on temperature. Characteristic bacilliform particles have been observed in the nucleus and cytoplasm of cells in the brain, subesophageal ganglion, salivary glands, ovaries, fat body, mycetome and muscle (Sylvester and Richardson, 1970). Virus

TABLE 11.3 Umbravirus–helper luteovirus complexes that are naturally transmitted[a]

Umbravirus species	Helper luteovirus	Main aphid vector
Bean yellow veinbanding virus	*Bean leaf roll virus*	*Acyrthosiphon pisum*
Carrot mottle virus	*Carrot red leaf virus*	*Cavariella aegopodii*
Carrot mottle mimic virus	*Carrot red leaf virus*	*C. aegopodii*
Groundnut rosette virus	*Groundnut rosette assistor virus*	*Aphis craccivora*
Lettuce speckles mottle virus	*Beet western yellows virus*	*Myzus persicae*
Pea enation mosaic virus 2	*Pea enation mosaic virus 1*	*A. pisum*
Tobacco mottle virus	*Tobacco vein distorting virus*	*M. persicae*

[a] Modified from Robinson and Murant (1999), with permission.

particles appear to be assembled in the nucleus. The virus can be transmitted serially from aphid to aphid by injection of hemolymph, and infection is associated with increased mortality of the aphids. Decreased lifespan varied with different virus isolates. However, as infected aphids lived through the period of maximum larviposition, the intrinsic rate of population growth was little affected (Sylvester, 1973). The virus is transmitted through the egg of *H. lactucae*, about 1% of larvae produced being able to infect plants (Sylvester, 1969). The virus had been shown to multiply in primary cultures of aphid cells (Peters and Black, 1970).

Continuous passage of SYVV in the aphid by mechanical inoculation gives rise to isolates that have lost the ability to infect the plant host (Sylvester and Richardson, 1971).

Similar kinds of evidence have shown that LNYV replicates in its aphid vector (O'Loughlin and Chambers, 1967), as does BNYV (Garrett and O'Loughlin, 1977). It is assumed that virus particles produced in aphid cells are released into the hemolymph, find their way to the salivary glands, and are injected into the plant along with saliva. There is no evidence for interference or cross-protection when aphids are allowed to acquire two different rhabdoviruses (Sylvester and Richardson, 1981).

SCV has been maintained through six consecutive serial passages by needle inoculation of the aphid vector *Chaetosiphon jacobi* (Sylvester *et al.*, 1974). As each transmission involved a 10^{-3} dilution factor of the original inoculum, this is strong evidence for replication. In spite of good survival of and a long lifespan for injected aphids, their ability to transmit was greatest at 10 days and then declined rapidly (Fig. 11.12). This rapid fall-off might be due either to a drop in concentration of virus arriving at the salivary glands or to reduced feeding by aging aphids.

IV. LEAFHOPPERS AND PLANTHOPPERS (AUCHENORRHYNCHA)

A. Structure and life cycle

Unlike aphids, leafhoppers have a simple life cycle in which the egg hatches to a nymph,

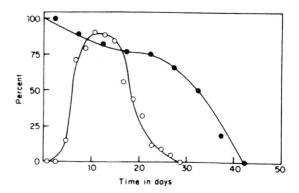

Fig. 11.12 Survival and transmission curves of *Chaetosiphon jacobi* aphids injected with SCV during serial passage experiments. After injection, the aphids were caged on healthy strawberry seedlings in a growth chamber at 25°C and constant light, and thereafter moved at 24–48-hour intervals to fresh seedlings until death. The survival curve represents the percentage of 212 insects that was alive at the beginning of each successive 4-day period. The transmission curve represents the proportion of the test seedlings fed upon by the insects during each successive 48-hour interval that developed symptoms. ○—○, Transmission; ●—●, survival. From Sylvester *et al.* (1974), with permission.

which feeds by sucking and passes through a number of moults before becoming an adult. There may be one or several generations per year. Different species overwinter as the egg, as the adult or as immature forms.

Only a few comments can be made here on structures important for vector leafhopper relationships with viruses. Figure 11.13 shows in diagrammatic sagittal section the arrangement of major organs in the head and thorax. The salivary glands, which are important in virus transmission, consist of a principal four-lobed gland and an accessory gland. It contains five different types of acini.

The mycetome occurs as an isolated mass on each side of the abdomen. The fat body, which probably has a storage function, surrounds all the organs of the body and is present in head, thorax and abdomen. The alimentary tract is in three main regions: foregut, midgut and hindgut (Fig. 11.14). Methods for studying the feeding behavior of hoppers have been described by Markham *et al.* (1988) and Mesfin *et al.* (1995). Studying the feeding behavior of the leafhopper, *Cicadulina mbila*, Mesfin *et al.*

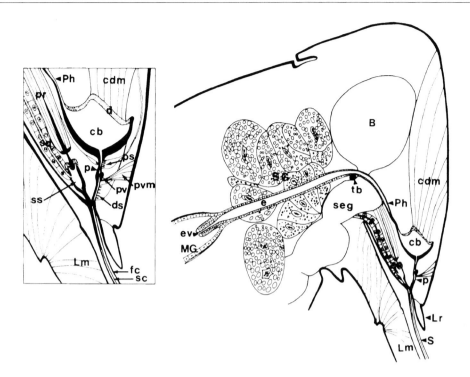

Fig. 11.13 Sagittal view of the head of the leafhopper *Graminella nigrifrons* showing the anterior alimentary canal, salivary system and surrounding structures. (Inset) Details of the precibarium and salivary syringe areas. B, brain; cb, cibarium; cdm, cibarial dilator muscle; d, cibarial diaphragm; ds, distal sensilia; e, esophagus, ev, esophageal valve; fc, food canal; Lm, labium; Lr, labrum; MG, midgut; p, precibarium; Ph, pharynx; pr, piston retractor muscle; ps, proximal sensilla; pv, precibarial valve; pvm, precibarial valve muscle; S, stylets; sc, salivary canal; sd, salivary duct; seg, subesophageal ganglion; SG, salivary gland; ss, salivary syringe; tb, tentorial bar. From Nault and Ammar (1989), with kind permission of the copyright holder, © Annual Reviews. www.AnnualReviews.org

(1995) showed that it was similar to that of aphids, with the mouthparts, surrounded by the salivary sheath, penetrating to the phloem in preferred hosts (see Fig. 11.5B).

B. Kinds of virus–vector relationship

No viruses have been found to be transmitted in a non-persistent manner by hoppers, or by purely mechanical means. Some have the characteristics of semi-persistent transmission with virus-like particles found attached to the cuticular linings of the anterior alimentary canal or foregut (Nault and Ammar, 1989). Hopper-borne viruses are often transmitted by only one or a few closely related species of hopper.

There are about 60 subfamilies in the leafhopper family (*Cicadellidae*) and two of these, the *Agalliinae* and the *Deltocephalinae*, contain species that are virus vectors. The *Agalliinae* have herbaceous dicotyledonous

hosts, while most *Deltocephalinae* feed on mono-cotyledons. There are about 15 000 described species of leafhopper in about 2000 genera. Of these, only 49 species from 21 genera have been reported as being virus vectors (Table 11.1) (Nault and Ammar, 1989).

There are about 20 families of planthoppers (*Fulgoroidea*) but only the *Delphacidae* have definite virus vector species. Members of this family feed on monocotyledons, primarily members of the *Poaceae*. Thus, all the viruses known to be transmitted by members of this family have hosts in the *Poaceae*. These cause important diseases of cereal crops, including rice, wheat and maize. Table 11.4 lists examples of representative hopper-borne viruses and their vectors. Hopper transmission of plant viruses has been reviewed by Nault and Ammar (1989) and by Nault (1997). Evolution of viruses and vectors is discussed in Chapter 17 (Section XIV).

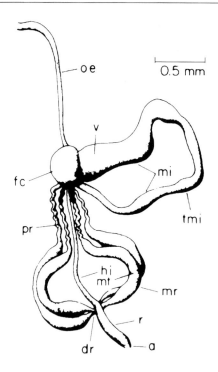

Fig. 11.14 Digestive system of *Agallia constricta*. a, Anus; r, rectum; dr, distal region; mr, middle region; mt, malpighian tubules; hi, hind intestine; pr, proximal region; tmi, tubular mid-intestine; mi, mid-intestine; v, ventriculus; fc, filter chamber; oe, esophagus. From Gil-Fernandez and Black (1965); modified by Black to conform with the findings of Bharadwaj *et al.* (1966); all with permission. (Diagram courtesy of L.M. Black.)

C. Semi-persistent transmission

Two viruses, MCDV and RTSV, are transmitted in the semi-persistent manner: the RTSV transmission is associated with that of RTBV, the two viruses causing rice tungro disease (see Chapter 2, Section III.A.6). The minimum acquisition and inoculation times for MCDV by *Graminella nigrifrons* are both 15 minutes (Choudhury and Rosenkranz, 1983), and for RTSV by *Nephotettix virescens* are 5–30 minutes and 5–10 minutes respectively (reviewed in Anjaneyulu *et al.*, 1994). Longer acquisition times increased the frequency of transmission of both viruses. The minimum time is thought to be the time the hopper takes to penetrate to phloem cells where most virus is found. Infectious MCDV is lost from the insect in less than 24 hours at 25°C, but is retained for several

days at lower temperatures (Nault and Ammar, 1989). Nymphs lose transmissible MCDV following a moult. Male and female adults are vectors of both MCDV and RTSV, with females being more efficient (Choudhury and Rosenkranz, 1983; Anjaneyulu *et al.*, 1994). MCDV particles attached to the lining of the vector's foregut, and particles have been observed by electron microscopy adhering to the cuticular lining in the precibarium, cibarium, pharynx and fore esophagus (reviewed by Nault, 1997) (Fig. 11.15). None was observed in the alimentary canal beyond the cardiac valve, or in other tissues or the stylets and stylet tips (Childress and Harris, 1989). Hoppers egest material (extravasation) from the foregut from time to time during feeding and it is considered that this is the process by which plants are inoculated (Wayadande and Nault, 1992). Thus, semi-persistent leafhopper transmission of viruses may involve an ingestion–egestion mechanism (Harris *et al.*, 1981).

A virus-coded helper protein like that for potyviruses may be needed for the hopper transmission of MCDV and RTSV, but the evidence is indirect (Hibino and Cabauatan, 1987; Hunt *et al.*, 1988; Creamer *et al.*, 1993). It is thought that the RTSV helper component facilitates the co-transmission of RTBV.

D. Persistent transmission

Persistent transmission involves movement of ingested virus to the salivary glands. As with the aphid vectors, some persistently transmitted viruses replicate in the hopper vector (propagative) and some do not (circulative).

1. Circulative viruses

Two genera of geminiviruses (mastreviruses and curtoviruses) are leafhopper transmitted in the persistent circulative manner; the curtovirus, TPCTV, is transmitted in this manner by treehoppers. Each leafhopper-transmitted geminivirus has one principal vector; for example, the principal vectors for BCTV, MSV and WDV are *Circulifer tenellus*, *Cicadulina mbila* and *Psammotettix alienus* respectively.

Acquisition times range from a few seconds to an hour. Longer feeding times give higher

TABLE 11.4 Auchenorrhyncha-borne plant viruses and vector genera grouped according to mode of transmission, virus taxonomy and vector family

Transmission mode and virus group	Vector family	Virus	Vector genera
Foregut-borne			
Machlomovirus	*Cicadellidae*	MCDV	*Graminella* and six others
Sequivirus	*Cicadellidae*	RTSV	*Nephotettix, Recilia*
Circulative			
Mastrevirus	*Cicadellidae*	MSV	*Cicadulina*
		CSMV	*Nesoclutha*
		PSMV	*Orosius*
		TYDV	*Nesoclutha*
		WDV	*Psammotettix*
Curtovirus	*Cicadellidae*	BCTV	*Circulifera*
Topocuvirus	*Membracidae*	TPCTV	*Micrutalis*
Propagative			
Marafivirus	*Cicadellidae*	MRFV	*Dalbulus* and three others
		OBDV	*Macrosteles*
Nucleorhabdovirus	*Cicadellidae*	CCMoV	*Nesoclutha, Cicadulina*
		OSMV	*Graminella*
		PYDV	*Aceratagallia[b], Agallia[b], Agalliopsis[b]*
		RTYV	*Nephotettix*
	Delphacidae	CBDV	*Tarophagus*
		FMMV	*Sogatella, Peregrinus*
		MMV	*Peregrinus*
		MSSV	*Sogatella, Peregrinus*
		WCSV[a]	*Laodelphax*
		WRSV	*Laodelphax*
Cyttorhabdovirus	*Cicadellidae*	WASMV	*Endria, Elymana*
	Delphacidae	BYSMV	*Laodelphax*
		NCMV	*Laodelphax, Muellerianella, Ribautodelphax, Unkanodes*
Unclassified rhabdovirus	*Cicadellidae*	SSMV	*Graminella*
		WWMV[a]	*Psammotettix*
	Delphacidae	CynCSV	*Toya*
		DiSV	*Sogatella*
		IMMV	*Ribautodelphus*
Tenuivirus	*Delphacidae*	RSV[a]	*Laodelphax, Terthron, Unkanodes*
		MSpV[a]	*Peregrinus*
		RGSV	*Niloparvata*
		RHBV[a]	*Sogatodes*
		EWSMV[a]	*Javesella*
Phytoreovirus	*Cicadellidae*	WTV[a]	*Agallia[b], Agalliopsis[b], Aceratagallia[b]*
		RDV[a]	*Nephotettix, Recilia*
		RGDV[a]	*Nephotettix, Recilia*
Fijivirus	*Delphacidae*	FDV[a]	*Laodelphax, Delphacodes, Javesella*
		MRDV[a]	*Sogatella*
		OSDV[a]	*Javesella, Dicranotropis, Ribautodelphax*
		PaSV	*Sogatella*
		RBSDV	*Laodelphax, Unkanodes*
Oryzavirus	*Delphacidae*	RRSV	*Niloparvata*
		ERSV	*Sogatella*

[a] Transovarial transmission has been reported in some vectors.
[b] Genera from subfamily *Agalliinae;* all other cicadellids from subfamily *Deltocephalinae.*
From Nault and Ammar, (1989), with kind permission of the copyright holder, © Annual Reviews.
www.AnnualReviews.org

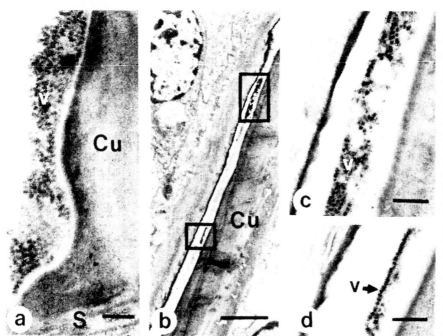

Fig. 11.15 Electron micrographs of MCMV-like particles (v), embedded in a dense matrix, in inoculative *Graminella nigrifrons.* **(a)** Particles attached to the cuticular lining (Cu) of the precibarium. **(b–d)** Particles attached to parts of the pharynx. Boxed areas in b are enlarged in **(c)** and **(d)**. S, preciberial sensillum. See Fig. 11.13 for the identification of these structures in the leafhopper's head. Scale bars in **(a)**, **(c)** and **(d)** represent 200 nm, and in **(b)** 2 μm. From Nault and Ammar (1989), with kind permission of the copyright holder, © Annual Reviews. www.AnnualReviews.org

transmission rates and longer persistence in the vector (Goodman, 1981; Harrison, 1985). The latent period for 10 circulative viruses was 23±4.1 hours (Nault, 1991), which is presumably the time taken for virus to reach the salivary glands.

MSV occurs in the mesophyll of infected plants and cannot be acquired from the phloem. Its vector, *C. mbila*, is usually a phloem feeder but prefers to feed in the mesophyll of MSV-infected plants, feeding behavior that is critical for transmission of this virus.

The movement of MSV through its vector *C. mbila* has been studied by Markham (1992a,b), who showed that the virus passes to the hemocoel via the filter chamber and anterior cells of the ventriculus of the midgut by receptor-mediated endocytosis. The receptor-mediated uptake into the salivary glands differs from that of the midgut. Some *Cicadulina* species lack the gut receptors but retain those in the salivary glands. They cannot acquire the virus orally but can transmit after injection of the virus into the hemocoel.

The main virus determinant for leafhopper transmission is the viral coat protein. Briddon *et al.* (1990) replaced the coat protein gene of the whitefly-transmitted ACMV with that of the leafhopper-transmitted BCTV. Plants were infected by agroinoculation and the resultant virus was injected into leafhoppers, which transmitted it to produce ACMV symptoms.

2. Propagative transmission (reviewed by Ammar, 1994)

Four families and genera of plant viruses have members that replicate in their hopper vectors (Table 11.4). It is notable that the *Rhabdoviridae* and *Reoviridae* contain animal-infecting members and the plant members have been considered to be plant-infecting arboviruses or phytoarboviruses (Whitcombe and Davies, 1970). The latent period for 13 propagative viruses from four families and groups of viruses was 368±41 hours (Nault, 1991). The reasons for this very much longer latent period compared with that of viruses that are merely circulative is not fully understood but probably relates to a requirement for the virus to replicate before it becomes transmissible. Both enter the plant via saliva.

a. Reoviridae

The classical experiments demonstrating that some plant viruses can also replicate in their animal vectors were carried out with plant reoviruses and their leafhopper vectors. Early work in this area generated considerable controversy, a subject that has been reviewed by Black (1984). The main kinds of evidence that replication in the hopper vector took place were: (1) transovarial transmission and dilution experiments; (2) serial injection experiments showing transmission after effective dilution of about 10^{-18}; (3) growth curves of virus and viral antigen; and (4) localization of virus within the insect.

Phytoreoviruses are transmitted by leafhoppers, and fijiviruses and oryzaviruses are transmitted by planthoppers. Plant reoviruses have a latent, or incubation, period of usually 1–3 weeks. During this period the virus replicates and invades most tissues. Ability to transmit appears to coincide with arrival of the virus in the salivary glands. Hoppers then usually retain infectivity for the rest of their lifespan. Minimum efficient times for the three stages of transmission are given by Conti (1984). Rates of transmission may be quite low. For example, only 15% of *Perkinsiella saccharicida* reared on infected sugarcane contained FDV antigen and only 6% transmitted the virus (Francki *et al.*, 1986c). Some, but not all, plant reoviruses have been shown to be transmitted through the eggs of vectors, often with a low frequency. Nevertheless, transovarial transmission is important in the ecology of these viruses (see Chapter 12).

Omura and Yan (1999) reviewed the virus–vector interactions involved in phytoreovirus transmission. RDV has been studied both in vector cell monolayers and in its leafhopper vector, *Nephotettix cincticeps*. Treatment of virus particles with carbon tetrachloride removed the minor outer capsid protein P2 (see Chapter 6, Section VI.A, for reovirus proteins) and abolished the ability to infect vector cell monolayers and also the ability of the vector to acquire the virus by *in vitro* feeding. P2 plays a role in the attachment of virus particles to insect cells, and lack of this protein did not affect the initial round of replication. When RDV particles lacking P2 were injected into vector leafhoppers, they became infected and were able to transmit the virus.

These observations suggest that P2 is responsible for the viral interaction with cells of the insects' intestinal tract. If this is bypassed by injection the virus can establish itself in susceptible cells within the insect and produce fully competent particles containing P2. Prolonged vegetative propagation of RDV and WTV leads to loss of vector transmissibility, the particles of transmission-defective RDV losing P2 and those of WTV losing both P2 and P5. There is also some evidence to implicate RDV P8, a major capsid protein, in virus–vector interactions. Transmission of this virus is blocked by treatment with P8 antiserum.

The viral spike protein, a 39-kDa protein, is implicated in the virus–vector interactions of the oryzavirus, RRSV (Zhou *et al.*, 1999). This protein was expressed in bacteria and fed to the insect planthopper vector, *Niloparvata lugens*, before they were fed on infected rice plants. This pre-feeding with the 39-kDa protein completely inhibited the ability of the planthopper to transmit the virus. The 39-kDa protein bound to a 32-kDa insect cell membrane protein, suggesting that this might be involved in the receptor site.

b. Rhabdoviridae

Work on the plant members of the *Rhabdoviridae* family has been reviewed by Jackson *et al.* (1999). Fifteen members have hopper vectors. Each virus has cicadellid or delphacid vectors, but not both.

The same kinds of experiment described for the *Reoviridae* have been used to establish that plant members of the *Rhabdoviridae* replicate in their hopper vectors. Thus Chiu *et al.* (1970) used fluorescent antibody staining to show that soon after infection PYDV antigen accumulates in the nucleus of *Agallia constricta* cells in culture. Association of maturing virus particles with the nuclear envelope was established by electron microscopy.

The growth curve of PYDV in cultured vector cell monolayers showed an eclipse period of 9 hours at 28°C. Between 9 and 29 hours, virus concentration doubled every 80 minutes. A plateau of virus concentration was then maintained (Hsu and Black, 1974).

Minimum acquisition times range from less than 1 minute to about 15 minutes for different viruses. The longer times are associated with rhabdoviruses confined to the phloem and nearby cells. The latent period may be days or months. For instance, SSMV has a minimum acquisition period of 6 hours, a minimum latent period of 9 days at 30°C and a minimum inoculation time of 1 hour (Creamer *et al.*, 1993).

There may be a high degree of vector specificity, even between strains of the same virus. It is probable that the G (glycosylated) protein, which is exposed at the virus surface, is involved in this specificity (Jackson *et al.*, 1987). In rhabdoviruses infecting vertebrates this protein functions in recognition and attachment to a host-cell recognition site. Such a role is supported for SYDV by experiments with its G protein. G-specific antibodies blocked infection of insect vector monolayers (Gaedigk *et al.*, 1986).

Long-term propagation of rhabdoviruses in plants can lead to loss of hopper transmissibility.

c. Tenuiviruses

Tenuiviruses are transmitted by Delphacid planthoppers in the persistent propagative manner (reviewed by Falk and Tsai, 1998). The virus can be acquired from infected plants by feeding times of 15 minutes to 4 hours. There then follows a latent period of 4–31 days during which the vector is not able to transmit the virus. After that, the vector can usually transmit the virus for the rest of its life, with inoculation feeds of as short as 30 seconds but more usually a few minutes to hours. Nymphs are usually more efficient vectors of tenuiviruses than adults, and females are more efficient than males. Transmission efficiency can vary with population or biotype and by selective interbreeding colonies with transmission efficiencies of up to 100% can be obtained. All tenuiviruses, except RGSV, are transmitted transovarially with rates ranging from about 20% to 100% and for up to 40 generations. RHBV has been reported to be paternally transmitted.

Tenuiviruses can infect various planthopper organs including the brain, digestive, respira-

tory and reproductive tracts, salivary glands, Malpigian tubules, leg muscles and fat bodies, causing mild to lethal effects on the vector. These effects include reduction in the fecundity and longevity of viruliferous females and lethality to eggs and early instar nymphs.

Little is known about molecular aspects of tenuivirus–vector interactions. Genome similarities to tospoviruses and bunyaviruses (see Chapter 2, Section III.H) suggest that tenuivirus particles should be associated with membranes and that receptor-mediated interactions should involve virus-coded glycoproteins. However, no enveloped particles have been found and nothing has been reported about the predicted glycoproteins. There are some differences in the expression of MSpV genes in plants and insects. Whereas the N protein and its mRNA are expressed equally in plants and insects, the NCP and its mRNA are expressed only in plants. It is not known whether this reflects a plant-specific function for the NCP.

d. Marafiviruses

The main vector of MRFV is the deltocephaline leafhopper *Dalbulus maidis*, in which the virus has been shown to replicate. Although about 80% of insects may be infected as revealed by ELISA tests, only about 10–34% of an ordinary population can transmit the virus. The ability to transmit appears to be under genetic control, because the proportion of transmitters can be rapidly increased by selective breeding (Gamez and Léon, 1988). After a few generations of outcrossing by random mating, transmission reverts to normal rates. Thus active transmitters may be recessive homozygotes for rare alleles that are rapidly diluted in outcrosses.

The virus has been shown to be widespread in the organs of infected insects. There is no evidence for cytopathological effects caused by the virus (Gamez and Léon, 1988).

e. Factors affecting multiplication and transmission

Virus can multiply in hoppers feeding on an immune host. Eggs may overwinter and provide a source of virus for spring crops in the absence of diseased plants. Thus, persistence of virus in the insect and transovarial transmission, and the factors that affect its efficiency, may be of considerable economic importance.

i Age of vector when infected

For several leafhopper vectors under experimental conditions, nymphs are more efficient vectors than adults, and adults decrease in efficiency as they age. For example, Sinha (1967) followed the appearance of WTV antigen in nymphs and adults of *A. constricta* after a 1-day acquisition feed. By 32 days after infection, 50% of the individuals infected as nymphs had antigen in their hemolymph and salivary glands. For individuals infected as adults, only 5% showed antigen at these sites. With the remaining insects, antigen was confined to the primary site of infection in the filter chamber of the intestine. Antigen had spread over a more limited region of intestine than with the nymphs. These experiments suggest that, as the insect ages, the intestine becomes more refractory to infection, and that virus present in the filter chamber of many adults may not be able to pass into the body cavity.

ii Time after infection

Slykhuis (1963) showed that the leafhopper *Endria inimica* lost its ability to transmit WASMV after a variable number of days. Some transmitted intermittently and none was infective after 72 days.

iii Temperature

Sinha (1967) examined the effect of increased temperature on the spread of WTV in *A. constricta*. Groups of nymphs were given an acquisition feed of 1 day at 27°C, and then held at 27°C for three further days to allow virus infection to become established in the filter chamber. Virus antigen was found in the filter chamber at this stage. One group continued at 27°C and the other was held at 36°C. High temperature prevented the spread of virus from the intestine to the hemolymph, salivary glands and other parts of the intestine. By day 6, antigen had disappeared from the filter chamber of 60% of the insects.

Temperature had a marked effect on the rate of transovarial passage of RSV in its planthopper vector *Laodelphax striatellus*. At 17.5°C, 83% of viruliferous females passed virus to their progeny at a rate greater than 90%. At 32.5°C, only 12.5% of females reached this frequency of transovarial transmission (Raga *et al.*, 1988).

iv Genetic variation in the leafhopper

Different lines or races within a vector species may vary widely in their efficiency as vectors. Thus, to demonstrate with a reasonable degree of confidence that a species is not a vector, it may be necessary to test populations from various regions where the virus and insect occur.

v Change in properties of the virus

Long-term culture of a virus in plants without recourse to leafhopper transmission may lead to loss of ability to be transmitted, as noted earlier for WTV and rhabdoviruses.

V. WHITEFLIES (*ALEYRODIDAE*)

Three genera of viruses are transmitted by whiteflies, the begomoviruses of the *Geminiviridae* family and the criniviruses and some closteroviruses of the *Closteroviridae*. Most of these viruses are found in the tropics and subtropics where they can be of substantial importance.

A. Whiteflies

Begomoviruses are transmitted exclusively by *Bemisia tabaci* whereas the whitefly-transmitted closteroviruses and the criniviruses are transmitted by the glasshouse whitefly, *Trialeuroides vaporariorum* (BPYV, TICV and CCSV), *T. abutilonea* (DVCV and AYV), *B. tabaci* (CYSDV and SPSVV) and *B. argentifolia* (LCV).

The most studied vector is *B. tabaci*, which is important in the epidemiology of several major diseases. Only the first instar of the larva is mobile, and it does not move far. Adults are winged, and many generations may be produced in a year. The nymphs of *B. tabaci* are phloem feeders. The B type of *B. tabaci*, sometimes named *B. argentifolia*, is of increasing importance as a vector.

The general morphology of whiteflies is reviewed by Gill (1990) and details of the ultrastructure of the mouthparts of *B. tabaci* have been studied by Rosell *et al.* (1995). Adult whitefly mouthparts are similar to those of other homopterans, especially aphids, and comprise the labrum, the lamium and the stylets. The stylet bundle, made up of two mandibular and two maxillary stylets, is the feeding organ. The feeding behavior of whiteflies resembles that of aphids in being piercing and sucking and involving a salivary sheath (Janssen *et al.*, 1989).

B. Begomoviruses

Most, if not all, bipartite begomoviruses are transmitted in the persistent circulative manner. Virus particles have been observed in the gut epithelial cells and associated with salivary glands of whitefly vectors and are thought to follow a similar circulative route to those of luteoviruses (Cohen and Antignus, 1994; Hunter *et al.*, 1998). As with circulative aphid-transmitted viruses (Section III.H.1.a), the GroEL homolog, symbionin, is implicated in the circulative transmission in whiteflies of TYLCV (Morin *et al.*, 1999). However, the non-transmissibility of AbMV is not the result of lack of binding of the coat protein to symbionin (Morin *et al.*, 2000).

However, certain complexities have been described for some begomoviruses (reviewed in Gray and Banerjee, 1999). TYLCV appears to persist in its whitefly vector for longer than expected from its infectivity, and is reported to be transovarially transmitted. SCLV particles have been associated with cytopathological abnormalities in some vector tissues and this virus can have detrimental effects on the vector biology and reproduction. These observations have been taken to suggest that the viruses are propagative but no replication intermediates have yet been detected. Sexual transmission of TYLCV-Is from male to female and from female to male insects has been suggested (Ghanim and Czosnek, 2000).

As with the persistent circulative transmission of luteoviruses (Section III.H.1.a), begomoviruses have to cross at least two barriers, the gut wall and membranes, to enter and exit the salivary glands; it is likely that these involve receptor-mediated strategies. 'Squash-blot' experiments show that ACMV can be acquired by two non-vectors, the whitefly, *T. vaporariorum*, and the aphid, *M. persicae* (Liu *et al.*, 1997b), which suggests that the virus can cross the gut wall but not enter the salivary glands. The coat protein is shown to be essential for the acquisition process because unencapsidated cloned nucleic acid is not acquired (Azzam *et al.*, 1994). By making chimeras between the genomes of a whitefly-transmissible isolate of ACMV-NOg and a non-transmissible isolate ACMV-K, Liu *et al.* (1997b, 1998) demonstrated that the defects responsible for lack of transmissibility were in the coat protein and DNA-BC1 gene of ACMV-K. The BC1 gene is thought to mediate cell-to-cell spread of the virus (see Chapter 9, Section II.D.2.i) and thus may influence the virus distribution in plant tissues. Comparison of coat protein sequences of transmissible and non-transmissible isolates of TYLCV and then mutagenesis showed that the region of the coat protein between amino acids 129 and 134 was essential for the correct assembly of virion and for whitefly transmission (Norris *et al.*, 1998). In contrast with these observations with ACMV, SLCV does not cross the gut wall barrier of non-vector *B. tabaci* (Rosell *et al.*, 1999).

C. Closteroviruses and criniviruses

Both the monopartite whitefly-transmitted closteroviruses and the bipartite criniviruses are transmitted in a foregut-borne, semi-persistent manner. LIYV is retained in the vector for a maximum of 3 days whereas LCV and CYSDV persist for 4 and 9 days respectively. As with the aphid-transmitted closteroviruses, BYV and CTV, (see Section III.F.1), LIYV encodes two capsid proteins: the major protein (CP) and a minor protein (CPm) (see Chapter 6, Section VIII.F.2). Also as with the aphid-transmitted closteroviruses, CPm is found at one end of the particles (see Fig. 5.12) (Tian *et al.*, 1999). Purified LIYV virions could be transmitted by *B. tabaci* after *in vitro* acquisition, and transmission was neutralized by antiserum to CPm but not by antiserum to CP. Thus, CPm is involved in the transmission of LIYV (Tian *et al.*, 1999).

VI. THRIPS (*THYSANOPTERA*)

(reviewed by Ullman *et al.*, 1997)

Of the 5000 or so species of thrips, only 10 species, all in the family *Thripidae*, are vectors of plant viruses (Table 11.5). Most of these vector species are extremely polyphagous and able to reproduce on a broad range of host plants. *Thrips tabaci* is cosmopolitan, feeding on at least 140 species from over 40 families of plants. It reproduces mainly parthenogenetically. The larvae are rather inactive but the adults are winged and very active. *Thrips tabaci* feeds by sucking the contents of the subepidermal cells of the host plant. Adults live up to about 20 days. Several generations can develop in a year.

Viruses from four plant virus families or groups are transmitted by thrips (Table 11.5). The ilarviruses, sobemoviruses and carmoviruses are pollen transmitted, the thrips carrying the pollen and inoculating it by mechanical damage during feeding. Tospoviruses are transmitted in a persistent propagative manner.

A. Thrip anatomy (reviewed by Nagata, 1999)

Although detailed studies have been made on the anatomy of *Hercinothrips femoralis*, most of the information on the internal anatomy is applicable to other members of the *Thripidae*. Thrip feeding apparatus consists of one mandible that punches a hole in the leaf and two maxillae that are inserted into the plant cell and through which the cell contents are sucked (Fig. 11.16). The mouthparts lead into the foregut and then the midgut of which there are three regions, anterior (Mg1), middle (Mg2) and posterior (Mg3). The foregut and hindgut are of ectodermal origin lined by a thick impermeable cuticle and the midgut is endodermal with a soft inner epithelial cell layer. The

TABLE 11.5 Transmission of viruses by thrips

Thrip species	Virus	Virus family (group)	Virus–vector relationship
Frankiniella occidentalis	GRSV[a]	*Tospovirus*	PP
	INSV	*Tospovirus*	PP
	PDV	*Ilarvirus*	Pollen
	PFBV	*Carmovirus*	Pollen
	PNRSV	*Ilarvirus*	Pollen
	TCSV	*Tospovirus*	PP
	TSWV	*Tospovirus*	PP
F. fusca	TSWV	*Tospovirus*	PP
F. intensa	GRSV	*Tospovirus*	PP
	TCSV	*Tospovirus*	PP
	TSWV	*Tospovirus*	PP
F. schulzei	GRSV	*Tospovirus*	PP
	TCSV	*Tospovirus*	PP
	TSWV	*Tospovirus*	PP
Microcephalothrips abdominalis	PNRSV	*Ilarvirus*	Pollen
	TSV	*Ilarvirus*	Pollen
Thrips australis	PNRSV	*Ilarvirus*	Pollen
T. imaginis	PNRSV	*Ilarvirus*	Pollen
T. palmi	GBNV	*Tospovirus*	PP
	WSMV	*Tospovirus*	PP
T. setosus	TSWV	*Tospovirus*	PP
T. tabaci	PNRSV	*Ilarvirus*	Pollen
	SoMV	*Sobemovirus*	Pollen
	TSV	*Ilarvirus*	Pollen
	TSWV	*Tospovirus*	PP

[a] See Appendix 1 for virus acronym.
PP, persistent propagative.
Adapted from Ullman *et al.* (1997), with permission.

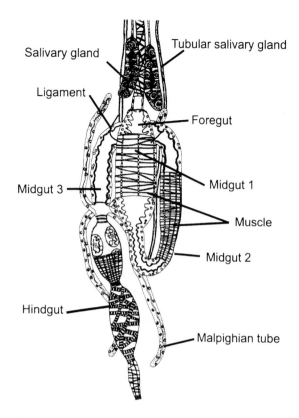

Salivary gland

Tubular salivary gland

Ligament

Foregut

Midgut 3

Midgut 1

Muscle

Midgut 2

Hindgut

Malpighian tube

Fig. 11.16 Composite drawing of the alimentary tract and associated organs of the thrips, *Hercinothrips femoralis*. From Nagata (1999), with permission.

salivary gland complex comprises two lobular and two tubular glands, which may correspond to the accessory salivary glands of other insects. It has been suggested that the tubular glands connect to Mg1 (Ullman *et al.*, 1989) but no direct connection could be found (Del Bene *et al.*, 1991). Mg1 is also connected to the salivary glands, this time the lobular glands by thin thread-like structures, termed ligaments (Ullman *et al.*, 1989).

B. Tospovirus transmission (reviewed by Nagata, 1999)

Transmission of tospoviruses by thrips has several unusual features (Fig. 11.17). Only the larvae, and not adults, can acquire the virus, and the competence to acquire decreases with age of the larvae. Effectively, the virus has to be acquired by a first instar nymph for realistic

transmission (van de Wetering *et al.*, 1996). The virus could be acquired or transmitted by first instar nymphs of *Frankliniella occidentalis* in feeding periods of as short as 5 minutes but the median acquisition access period on infected *Impatiens* plants was 106 minutes and the median inoculation access period to petunia leaf discs was 58 minutes. The median latent period varied with temperature, being 84 hours at 27°C and 171 hours at 20°C. Individuals may retain infectivity for life, but their ability to transmit may be erratic. The virus is not passed through the egg.

Although the titer of virus was higher in females, males were more efficient transmitters, probably because of their feeding behavior (Sakurai *et al.*, 1998; van der Wetering *et al.*, 1999). There are distinct levels of species and biotype specificity (Wijkamp *et al.*, 1995). *Frankliniella occidentalis* was the most efficient vector for four tospoviruses, TSWV, INSV, TCSV and GRSV. The dark form of *F. schultzei* transmitted TSWV, TCSV and GRSV, whereas the light form of this species transmitted TSWV and TCSV poorly. Only one of four populations of *T. tabaci* from different geographical regions transmitted only TSWV of the four viruses tested, and that with low efficiency.

If TSWV is cultured by successive transfers only in plants, the isolate loses the ability to be transmitted by thrips.

C. Virus–vector relationship

The accumulation of the nucleocapsid and a non-structural protein of TSWV was studied in developing nymphs and adults of *F. occidentalis* (Wijkamp *et al.*, 1993) and both proteins were shown to increase in the vector. The proteins and virus particles accumulated in the salivary glands and other tissues. These observations taken with the times for virus acquisition and the latent period indicated that TSWV multiplied in its vector.

As with other internally-borne persistently transmitted viruses, tospoviruses have to pass several barriers in the vector, which suggests that there is a receptor-mediated mechanism(s). TSWV is enveloped with spikes, made up from two virus-coded glycoproteins, extending from the envelope (see Chapter 5, Section VIII.B). Feeding *F. occidentalis* on plants infected with wild type and an envelope-deficient isolate showed that the thrips became infected only when they acquired intact virus particles (Nagata, 1999). Similarly the envelope-deficient isolate did not infect primary *F. occidentalis* cell cultures, nor did the nucleocapsids of wild type which had had the envelope removed. These observations suggested that the viral glycoprotein(s) contain the binding site for a receptor in the vector's midgut. Two proteins from *F. occidentalis* have been shown to bind to TSWV glycoproteins. Gel overlay assays and immuno-labeling identified a 50-kDa thrips protein and anti-idiotype antibodies against each of the two TSWV glycoproteins labeled a 50-kDa thrip protein which was localized to the larval thrip midgut (Bandla *et al.*, 1998). A 94-kDa protein that bound to TSWV was identified in *F. occidentalis* and *T. tabaci* but not in the non-vector aphid *M. persicae*. This protein bound to the TSWV G2 glycoprotein and was present throughout the thrips body (Kikkert *et al.*, 1998). The binding properties of these two thrip proteins would suggest that they may be associated with receptor sites but the detailed sites have not yet been identified.

D. Route through the thrips

A detailed study of the route that TSWV takes through *F. occidentalis* has been made by Nagata (1999), who used immunofluorescent staining of nymphs at various times after virus acquisition. The first infections were found in the Mg1 region (see Fig. 11.16) about 24 hours post-acquisition (hpa). These infections increased in intensity but remained restricted to the Mg1 epithelium for some time. In late larval stage, it spread to the circular and longitudinal midgut muscle tissues. By the adult stage the visceral muscle tissues of the midgut and foregut were infected. Infection of the salivary glands was first observed 72 hpa and at the same time the ligaments connecting the midgut with the salivary glands became infected. There was no evidence for TSWV in either the hemocoel or the midgut basal lamina. It appeared that the virus reached the salivary glands through the ligaments connecting Mg1 to the salivary glands. This is a different route to

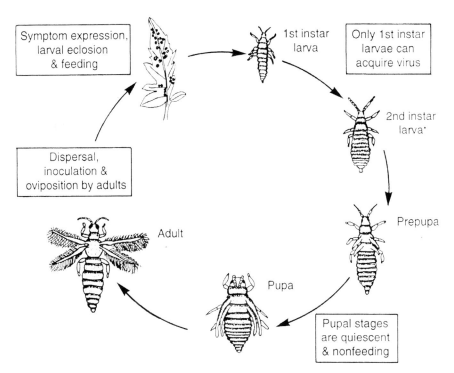

Fig. 11.17 Line drawing depicting the relationship between thrips and tospoviruses. Transmission by adults occurs only if larvae acquire the virus. *Second instar larvae occasionally transmit tospoviruses, but are wingless and seldom disperse widely. The adult is the primary dispersal stage of the insect and thought to be of greatest epidemiological significance. From Ullman *et al.* (1997), with permission.

Symptom expression, larval eclosion & feeding

1st instar larva

Only 1st instar larvae can acquire virus

2nd instar larva*

Dispersal, inoculation & oviposition by adults

Prepupa

Adult

Pupa

Pupal stages are quiescent & nonfeeding

that conventionally proposed for persistent viruses, which is movement through the hemocoel from the gut cells to the salivary glands.

Thrips that ingested TSWV at the second instar stage had less infection of the midgut and rarely infection of the salivary glands.

VII. OTHER SUCKING AND PIERCING VECTOR GROUPS

A. Mealybugs (Coccoidea and Pseudococcoidea)

Mealybugs are much less mobile on the plant than other groups of vectors such as aphids and leafhoppers, a feature that makes them relatively inefficient as virus vectors. They spread from one plant to another in contact with it, and the crawling nymphs move more readily than adults. Ants that tend the mealybugs may move them from one plant to another (Sether et al., 1998) and occasional long-distance dispersal by wind may occur. Mealybugs feed on the phloem.

They have been established as the vectors of many of the badnaviruses and several closteroviruses (GLRaV-3, LCV and PMWaV) and trichoviruses (GVA and GVB). Six vector species of differing abundance for CSSV were recorded by Bigger (1981) but there are transmitting and non-transmitting races of certain species (Posnette, 1950). CSSV could be acquired by Pseudococcus njalensis within 20 minutes but the mealybug took at least 16 minutes to penetrate the leaf tissue. The virus persisted in the mealybug for less that 3 hours (Posnette and Robertson, 1950). The relationship between the CSSV and mealybugs has some similarities to the non-persistent aphid-transmitted viruses. Presumably, the virus is carried on or near the stylets of the mealybug.

The closterovirus PMWaV is transmitted by the pink pineapple mealybug Dysmicoccus brevipes and the grey pineapple mealybug D. neobrevipes, the second and third instars being more effective at acquiring the virus than first instars and gravid females (Sether et al., 1998).

GLRaV-3 is retained in its vector, Planococcus citri, for approximately 24 hours (Cabaleiro and Segura, 1997), which suggests a semi-persistent relationship.

B. Bugs (Miridae and Piesmatidae)

The mirid bugs feed by means of stylets but their biology and taxonomy are not well understood. Cyrtopeltis nicotianae has been shown to be a vector of VTMoV, SBMV and several other, but not all, sobemoviruses (Gibb and Randles, 1988). Minimum acquisition time was 1 minute, a characteristic of non-persistent viruses. However, rate of transmission increased with increasing acquisition feeding time, a property characteristic of semi-persistent or circulative transmission. Other characteristics were like those of semi-persistent or circulative transmission. There was no evidence for virus replication in the vector. The viral antigen of VTMoV was detected in the gut, hemolymph and feces of its vector, Cyrtopeltis nicotianae, but not in the salivary glands, and infective virus was found in feces 6 days after acquisition (Gibb and Randles, 1990). Non-infective myrids were able to inoculate plants from infectious sap deposits on the upper epidermis and an ingestion–defecation model, not involving salivary glands, is suggested for this virus–vector association. Thus, the transmission was like that of beetles in several respects. Mirids are important crop pests, so it is possible that other virus vectors will be found.

BLCV is transmitted in a persistent, propagative manner by the piesmatid bug Piesma quadratum. There is no evidence for transmission through the egg (Proesler, 1980).

VIII. INSECTS WITH BITING MOUTHPARTS

A. Vector groups and feeding habits

A few vectors have been reported from the orders Orthoptera and Dermaptera. Important vectors are found in the Coleoptera (beetles). Almost all the vectors belong in a few families. Interest centers on the family Chrysomelidae, which consists of 55 000 species of plant-eating beetles. About 30 of these are known to transmit plant viruses, and each species feeds on a limited range of host plants. Twenty vector species are found in the subfamilies *Galerucinae* and *Halticinae* (flea-beetles), two are in the *Crysomelinae* and two in *Criocerinae*. As pointed out by Selman (1973), this distribution almost certainly reflects the interests of investigators rather than the actual situation in nature. Many beetle vectors probably remain to be discovered.

In the *Curculionidae* and *Apionidae* a few species of weevils have been shown to be virus vectors.

Leaf-feeding beetles do not have salivary glands. The chrysomelid beetles tend to eat the parenchyma tissues between vascular bundles, thus leaving holes in the leaf but, with heavy infestation, damage may be more severe. They regurgitate during feeding, which bathes the mouthparts with plant sap, as well as with viruses, if the plant fed upon is infected. It was once thought that transmission by beetles involved simply a mechanical process of wounding in the presence of virus. This is not so because: (1) some very stable sap-transmissible viruses such as TMV are not easily transmitted by beetles (but see Orlob, 1963); (2) some transmitted viruses may be retained by beetle vectors for long periods; and (3) there is a substantial degree of specificity between viruses and vector beetles. Beetle transmission has been reviewed by Fulton *et al.* (1987).

B. Viruses transmitted by beetles

The viruses transmitted by beetles belong to the *Tymovirus*, *Comovirus*, *Bromovirus* and *Sobemovirus* groups. Most viruses in these groups are not transmitted by members of other arthropod groups (but see transmission of sobemoviruses by myrids; Section VII.B above) and are usually quite stable, reaching high concentrations in infected tissues. They have small isometric particles (25–30 nm diameter) and are readily transmitted by mechanical inoculation. The viruses tend to have relatively narrow host ranges, as do their beetle vectors.

C. Beetle–virus relationships

Beetles can acquire virus very quickly—even after a single bite—but efficiency of transmission increases with longer feeding, as does retention time (Fulton *et al.*, 1987). Some viruses appear quickly in the hemolymph after certain beetle species have fed on an infected plant; others do not (Wang *et al.*, 1992, 1994). Insects become viruliferous after injection of virus into the hemocoel. Retention time varies between about 1 and 10 days with different beetles. However, under dormant, overwintering conditions, beetles may stay viruliferous for periods of months. Beetles can transmit the virus with their first bite on a susceptible plant. There is no good evidence for the existence of a latent period following virus acquisition, and no evidence for virus replication in beetle vectors. These observations suggest that beetle-transmitted viruses are externally borne.

It has been established that the regurgitant fluid is a key factor in determining whether a virus will or will not have beetle vectors (Gergerich *et al.*, 1983). This discovery was made possible by a gross wounding technique, which involved cutting discs from a leaf with a glass cylinder contaminated with virus-regurgitant mixture, thus mimicking the kind of wounds made by feeding beetles. When virus was mixed with regurgitant, only viruses normally transmitted by beetles were transmitted by the gross wounding technique. In ordinary mechanical inoculation using abrasives, all mixtures were non-infectious (Gergerich *et al.*, 1983; Monis *et al.*, 1986). Regurgitant does not irreversibly inactivate viruses not transmitted by beetles, because infectious virus could be recovered by dilution of the regurgitant–virus mixture or by isolation of the virus.

Regurgitant from several leaf-feeding beetle species was found to contain an RNase activity equivalent to 0.1–1.0 mg/ml of pancreatic RNase. This enzyme, used in this concentration range, inactivated beetle-non-transmitted viruses such as TMV when inoculated by the gross wounding technique (Gergerich *et al.*, 1986), but the transmission of viruses that are normally transmitted by beetles was not affected. Thus, pancreatic RNase could mimic the effect of beetle regurgitant. In further work, three kinds of RNase differing in the way they cleave RNA were found to act with the same discrimination as pancreatic RNase. Other basic proteins did not inhibit transmission of viruses not transmissible by beetles. Thus, it appears to be the enzymatic activity of the proteins that affects transmissibility (Gergerich and Scott, 1988a). Why this should be so is not clear. Perhaps the RNase activity affects establishment of the beetle-non-transmitted viruses in the initially inoculated cells as suggested by Gergerich and Scott. Alternatively, RNases may bind more firmly to viruses such as TMV, preventing uncoating in the cell.

Gergerich and Scott (1988b) showed that several beetle-transmitted viruses could move through cut stems. Furthermore, such viruses, inoculated below a steam-killed section of stem in an intact plant, could move and infect the upper parts of the plant, whereas beetle-non-transmitted viruses could not. These results suggest that the ability to be translocated in the xylem and to infect non-wounded tissue is a feature of beetle-transmitted viruses. However, TYMV, which is beetle transmitted, can move into a leaf from a cut petiole but cannot infect the leaf (Matthews, 1970).

When sodium azide was included in the inoculum mixture, cells in a zone around the gross wounding site were rapidly killed but infection by SBMV still occurred. Transmission of a beetle-non-transmitted virus was severely affected (Fulton et al., 1987). It was suggested that the ability of beetle-transmitted viruses to move in the xylem might take them to cells unaffected by the sodium azide treatment.

The apparently simple transmission of viruses by beetles is now seen to be quite a complex process. For example, the experiments summarized above shed little light on the problem of specificity among beetle species—why some species are highly efficient vectors of a particular virus and others are not.

IX. MITES (ARACHNIDA)

Members of the mite families *Eriophyidae* and *Tetranychidae* feed by piercing plant cells and sucking the contents, but they differ in many other respects. Several members of the first of these families are vectors for rymoviruses. One unassigned virus (*Peach mosaic virus*) is transmitted by the *Tetranychidae*.

A. *Eriophyidae*

The eriophyid mites are not closely related to other groups of mites. Members are known to transmit at least six plant viruses, and they feed by puncturing plant cells with stylets and sucking in the cell contents. The stylets are held inside the groove of the rostrum, which has two pads that act as ducts for the saliva (Fig. 11.18).

Eriophyid mites are very small arthropods (about 0.2 mm in length) and have limited powers of independent movement. They frequently infest buds and young leaves, where they often cause little damage and may quite easily be overlooked. They are readily killed by desiccation. In spite of this, their main method of spread from plant to plant is by wind (Slykhuis, 1955).

Most species are quite specific for the host plant on which they feed, usually being confined to one plant genus or, at most, the members of a single family. These mites cannot survive for long periods away from a host plant and, thus, most of the plant species on which they feed are perennials. They have a relatively simple developmental history that may be completed in 6–14 days. There are two nymphal instars followed by a resting 'pseudopupa'. Males are not often seen. Some species have two kinds of female, one being specialized for hibernation.

One of the best-studied mite vectors is *Aceria tulipae* (Keifer), which can transmit two viruses simultaneously: WSMV and wheat spot mosaic virus (Slykhuis, 1962). Figure 11.19 illustrates the main internal structures of an adult of this species.

The relationship of this vector with WSMV has been studied in considerable detail. Like non-persistent aphid-transmitted viruses, it has a rod-shaped particle and is readily sap transmitted. *Aceria tulipae* can acquire the virus in a 15-minute period on an infected leaf. A similar minimum period is required to transmit the

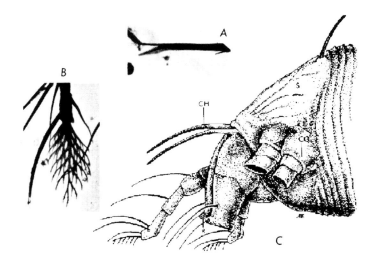

Fig. 11.18 Head of an adult mite *Aceria tulipae*. S, shield; CO, coxa; R, rostrum; CH, stylets. **(A)** Electron micrograph of stylets. A hair lies along the upper side of stylets. Stylets are about 5 μm long, and only about one-third of this length penetrates during feeding. **(B)** Electron micrograph of feather claw. From Orlob (1966), with kind permission of the copyright holder, © Blackwell Science Ltd..

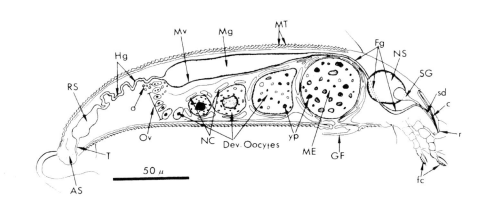

Fig. 11.19 Diagram of female adult *Aceria tulipae*. fc, Feather claws; r, rostrum; sd, salivary duct; c, chelicera; SG, salivary gland; NS, neurosynganglion; Fg, foregut; GF, genital flap; ME, mature egg; yp, yolk platelets; MT, microtubercles; Mg, midgut; Mv, microvilli; Dev. Oocytes, developing oocytes; NC, nurse cells; o, ovariole; Ov, oviduct region; Hg, hindgut; RS, rectal sac; T, rectal tube; AS, anal sucker. From Whitmoyer *et al.* (1972), with permission.

virus. The mite retains infectivity through moults and may remain infective, at least in the greenhouse, for 6–9 days after removal from an infected plant. On a 'virus-immune' host held at 3°C, they remained infective for over 2 months. The mites become infective as nymphs but not as adults when fed on an infected plant. About 30% of mites from a diseased wheat plant tested singly will transmit the virus (Slykhuis, 1965; Orlob, 1966). The uncharacterized wheat spot mosaic virus has a similar relationship with its *A. tulipae* vector (Nault and Styer, 1970).

Virus–vector relationships are difficult to study because of the small size of the mite. There is no good evidence for replication of viruses in mite vectors. Paliwal (1980) found particles of WSMV in the midgut, body cavity and salivary glands of the mite *Eriophyes tulipae*, suggesting that the virus is circulative in this vector.

B. *Tetranychidae*

The *Tetranychidae* consists of medium-sized mites (approximately 0.8 mm) that are all plant feeders, usually with a wide host range. The spider mite *Tetranychus urticae* (Koch) was claimed to be a vector of PVY, but this was not confirmed by Orlob (1968). Orlob showed that *T. urticae* was unable to transmit nine viruses, but could acquire several viruses as revealed by presence of virus within the insect. Several non-transmitted viruses reached high concentrations in the mite (Orlob and Takahashi, 1971).

BaYSMV is transmitted by the brown wheat mite (*Petrobia latens*) (Robertson and Carroll, 1988). Pre-adult mites readily acquired BaYSMV and both they and adult mites efficiently transmitted the virus to barley plants (Smidansky and Carroll, 1996). There is indirect evidence for transovarial passage of BaYSMV (Robertson and Carroll, 1988; Smidansky and Carroll, 1996). However, the virus was lost on keeping an originally viruliferous mite colony on a cucurbit (a non-host for BaYSMV), which is not in accord with a propagative relationship between the virus and vector.

X. POLLINATING INSECTS

RBDV (Murant *et al.*, 1974) and other viruses transmitted through infected pollen, and having insect-pollinated host plants, probably have the infecting pollen distributed by pollinating insects. Field experiments showed that BLMV is transmitted via pollen carried by foraging honeybees. About half the honeybees trapped in a field containing infected blueberry bushes had the virus in pollen from their pollen baskets, as determined by ELISA (Childress and Ramsdell, 1987). Caging experiments demonstrated that bees and infected pollen were both essential for new infections to occur.

As noted in Section VI, plants can be infected with some carmoviruses, ilarviruses and sobemoviruses through pollen carried by thrips.

XI. NEMATODES (NEMATODA)
(reviewed by Brown *et al.*, 1995, 1996b)

Since Hewitt *et al.* (1958) demonstrated that the fanleaf virus of grapes is transmitted by a dagger nematode, several widespread and important viruses have been shown to be transmitted through the soil by nematodes. Vector nematodes belong to the order *Dorylaimida*, family *Longidoridae*, in which three closely-related genera, *Xiphinema*, *Longidorus* and *Paralongidorus*, transmit viruses, and to the order *Triplonchida*, family *Trichodoridae*, in which the genera *Paratrichodorus* and *Trichodorus* transmit viruses. The three genera are all ectoparasitic and feed on epidermal cells of the root (Fig. 11.20) with feeding punctures occurring frequently near the root cap.

Of the approximately 375 species in the *Longidoridae* only eight *Longidorus*, one *Paralongidorus* and seven *Xiphenema* species are virus vectors. Similarly only seven *Paratrichodorus* and four *Trichodorus* species are vectors out of the 80 or so species in the *Trichodoridae*.

Two genera of plant viruses are transmitted by nematodes. Nepoviruses are transmitted by species in the genera *Xiphinema* and *Longidorus*, and tobraviruses are transmitted by species of *Trichodorus* and *Paratrichodorus*. All three tobraviruses are nematode transmitted but only about one-third of the nepoviruses are transmitted by these vectors. With the exception of TRSV, none of the viruses in these two genera is known to have invertebrate vectors other than nematodes; some nepoviruses are pollen transmitted (see Chapter 12, Section III.A.4).

Fig. 11.20 Nematodes of the species *Trichodorus christei* feeding on a root of blueberry (*Vaccinium corymbosum*). (Courtesy of B.M. Zuckermann.)

A. Criteria for demonstrating nematode transmission

Nematodes are difficult vectors to deal with experimentally because of their small size and their somewhat critical requirements with respect to soil moisture content, type of soil and, to a lesser extent, temperature. To overcome these problems five criteria have been proposed for establishing the nematode vectoring of viruses (Trudgill *et al.*, 1983; Brown *et al.*, 1989a).

- Infection of a bait plant must be demonstrated.
- Experiments should be done with handpicked nematodes.
- Appropriate controls should be included to show unequivocally that the nematode is the vector.
- The nematode should be fully identified.
- The virus should be fully characterized.

A common method for detecting nematode transmission has been to set out suitable 'bait' plants (such as cucumber) in a sample of the test soil. These plants are grown for a time to allow any viruliferous nematodes to feed on the roots and transmit the virus, and for any transmitted virus to replicate. Extracts from the roots and leaves of the bait plants are then inoculated mechanically to a suitable range of indicator species. Various modifications of the procedure can be used. For example, Valdez (1972) described a small-scale procedure for testing individual handpicked nematodes, while van Hoof (1976) showed that detached tobacco leaves buried in soil could become infected with TRV by nematodes within a few hours.

The proportion of nematodes ingesting virus can be determined by crushing whole nematodes and examining the extract by immunosorbent electron microscopy (Roberts and Brown, 1980) or by inoculating the suspension to suitable test plants Yassin, 1968). TRV can be detected in trichodorids by an RT-PCR method (van der Wilk *et al.*, 1994). However, it must be remembered that, although these methods show that the nematode ingested the virus, they do not show that it transmits it.

Estimation of the extent of transmission of a particular virus in the field by examination of the nematodes present may be complicated by various factors. For example, the distribution of individuals carrying the virus may be different from that of the population as a whole. Subpopulations within a given area, for example those in the surface layer and those in subsoil, may be of differing importance (Gugerli, 1977). Populations from different geographical areas may also differ in efficiency of transmission (Brown, 1986).

B. Nematode feeding

Longidorids are large nematodes (2–12 mm long) with long hollow feeding stylets (60–250 µm) which enable them to penetrate deep into root tips. The stylet comprises the anterior odontostyle, which penetrates the root cells, and the posterior odontophore in which there is nerve tissue adjacent to the food canal. The esophagus connects the stylet to a muscular pump which withdraws the plant cell contents and forces them through a one-way valve into the gut. Most longidorids induce the formation of galls when feeding at root tips, this being the first of two feeding phases. The second phase is exploitive involving repeated bouts of salivation causing liquefaction of the cytoplasm followed by ingestion. *Longidorus* spp. have long periods of ingestion during which a volume approximately equivalent to 40 normal root-tip cells is removed each hour.

Trichodorids are much smaller, with adults about 1 mm long. They typically feed on root hairs and epidermal and subepidermal cells of the root elongation zone. Their feeding has five phases: exploration, perforation of the cell wall and penetration to a depth of 2–3 µm, salivation, ingestion and withdrawal from the cell.

C. Virus–nematode relationships

Brown and Weischer (1998) divided the nematode transmission of a virus into seven discrete but inter-related processes: ingestion, acquisition, adsorption, retention, release, transfer and establishment (reviewed by Visser, 2000). Ingestion is the intake of virus particles from the infected plant and, although it does not require a specific interaction between nematode and virus, it needs a specific interaction between the nematode and plant. In the acquisition phase the ingested viral particles are retained in an intact state and specific features on the surface of the particle are recognized by receptor sites in the nematode feeding apparatus leading to adsorption (Fig. 11.21).

Once adsorbed, infectious particles can be retained in the nematode for months or even years, but not after moulting. Release of the viral particles is thought to occur by a change in pH caused by saliva flow when the nematode commences feeding on a new plant. In the transfer and establishment phases, the viral particles are placed in the plant cell, and start replicating and causing infection.

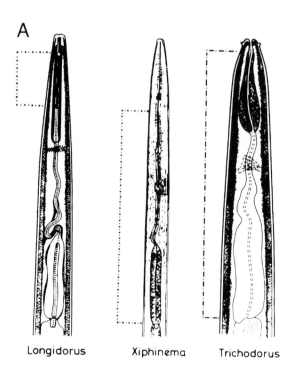

Longidorus   Xiphinema   Trichodorus

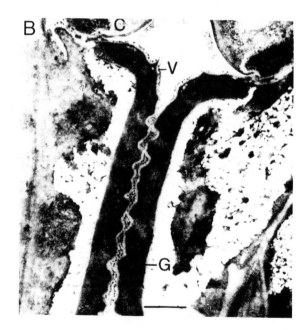

Fig. 11.21 Localization of viruses within nematode vectors. **(A)** Diagram of anterior portion of vectors. Broken lines indicate portions of the alimentary tract where virus particles are retained. **(B)** Longitudinal section of the buccal region of *Longidorus elongatus* carrying the nepovirus, RRSV. Note numerous virus-like particles (V) lining the guide sheath (G) and that none is associated with the stoma cuticle (C). Bar marker 1 µm. From Harrison *et al.* (1974), with permission.

There is specificity in the relationships between nematodes and the viruses they transmit with often an apparent unique association between the virus isolate and the vector species. There are some cases of different virus isolates sharing the same vector species or, conversely, one particular virus isolate being transmitted by several nematode species (Brown *et al.*, 1995; Vassilakos *et al.*, 1997; Brown and Weischer, 1998). There are 13 trichodorid species known to be tobravirus vectors, but only one or two of these transmits each tobravirus. By undertaking virus transmission bait tests with single trichodorid nematodes from England, the Netherlands, Scotland and Sweden, Ploeg *et al.* (1992) showed that there was a substantial degree of specificity between the nematode vector and the tobravirus serotype. This specificity was more marked with *Paratrichodorus* species than with *Trichodorus* species (Table 11.6).

Several nepoviruses are transmitted by more than one vector species but there can be differences in the observations under laboratory and field conditions. The Scottish and English isolates of RRSV are each transmitted in the laboratory by both *Longidorus elongatus* and *L. macrosoma*. However, under field conditions the Scottish isolate is only associated with *L. elongatus* and the English isolate with *L. macrosoma* (Brown *et al.*, 1995).

Once acquired, viruses may persist in transmissible form in starved *Longidorus* for up to 12 weeks, in *Xiphinema* for about a year, and much more than a year in *Trichodorus* (van Hoof, 1970). Transmission does not appear to involve replication of the virus in the vector. Plant virus particles have never been observed within nematode cells. Consistent with this is the fact that no evidence has been obtained for virus transmission through eggs of nematode vectors.

Specificity of transmission does not appear to involve the ability to ingest active virus since both transmitted and non-transmitted viruses have been detected within individuals of the same nematode species (Harrison *et al.*, 1974). Sites of retention of virus particles within nematodes have been identified by electron microscopy of thin sections (reviewed by Brown *et al.*, 1995). Nepovirus particles are associated with the inner surface of the odontostyle of various *Longidorus* species and with the cuticular lining of the odontophore and esophagus of *Xiphinema* species. Tobravirus particles have been observed absorbed to the cuticular lining of the esophageal lumen.

D. Virus–vector molecular interactions

i Nepoviruses

The genetic determinants for the transmissibility of RRSV and TBRV are encoded by RNA2 which expresses, among other proteins, the viral coat protein (reviewed by Brown *et al.*, 1996b). ArMV particles were found associated with carbohydrate-like material on the food canal walls of *Xiphinema diversicaudatum* (Robertson and Henry, 1986).

ii Tobraviruses

By making reciprocal pseudo-recombinants between a nematode-transmissible and a non-transmissible isolate of TRV, Ploeg *et al.* (1993b) showed that transmissibility segregated with RNA2. As noted in Chapter 6 (Section VIII.H.2.a), tobravirus RNA2 is variable in size and, as well as encoding the viral coat protein, encodes one to three non-structural proteins. A recombinant virus, in which the coat protein gene of a nematode non-transmissible isolate of

TABLE 11.6 Complementarity of tobravirus transmission by trichodorid nematodes

Virus	Serotype	Sequenced isolates	Vectors[a]
PEBV	English	TpA56	*P. anemones* *T. cylindricus* *T. primitivus* *T. viruliferus*
PEBV	Dutch	–	*P. pachydermus* *P. teres*
TRV	PaY4	PaY4	*P. anemones* *P. pachydermus*
TRV	PRN	PpK20	*P. nanus* *P. pachydermus*
TRV	RQ	TpO1	*T. cylindricus* *T. primitivus* *T. viruliferus*

[a] Generic names: *P.*, *Paratrichodorus*; *T.*, *Trichodorus*.
From Visser (2000), with permission.

PEBV was replaced with that of a highly nematode transmissible isolate of TRV, was not transmitted by nematodes indicating that more than one of the RNA2 genes was involved (MacFarlane *et al.*, 1995). Mutations in both the 29-kDa and 23-kDa non-structural genes of PEBV abolished nematode transmission without affecting particle formation, as did removal of the C-terminal mobile region of the coat protein (MacFarlane *et al.*, 1996). However, only mutation in the 40-kDa (formerly designated 29.5 kDa) non-structural gene of TRV isolate PpK20 abolished transmission by *Paratrichodorus pachydermis*, whereas that of the 32.8-kDa gene did not (Hernández *et al.*, 1997); it was suggested that the 32.8-kDa protein might be involved in transmission by other vector nematode species. Using the yeast two-hybrid system an interaction was detected between the TRV-PpK20 coat protein (CP) and both the 32.8- and 40-kDa proteins (Visser and Bol, 1999). Deletion of the C-terminal 19 amino acids interfered with the CP-40K interaction but not with the CP-32.8K interaction, whereas deletion of the C-terminal 79 amino acids affected both interactions. It is suggested that the non-structural proteins may be transmission helper components analogous to those in some aphid and leafhopper virus transmission systems (reviewed by Visser, 2000).

XII. FUNGI (reviewed by Campbell, 1996)

Several viruses have been shown to be transmitted by soil-inhabiting fungi. The known vectors are members of the class *Plasmodiophoromycetes* in the division *Myxomycota*, or in the class *Chytridiomycetes* in the division *Eumycota*. Both classes include endoparasites of higher plants. Species in the chytrid genus *Olpidium* transmit viruses with isometric particles, while species in two plasmodiophorus genera (*Polymyxa* and *Spongospora*) transmit rod-shaped or filamentous viruses (Table 11.7).

The two chytrid vectors, *Olpidium brassicae* and *O. bornavirus*, are characterized by having posteriorly uniflagellate zoospores, whereas those of the three plasmodiophoral vectors,

Polymyxa graminis, P. betae and *Spongospora subterranean*, are biflagellate. All five species are obligate parasites of plant roots and have similar development stages (Fig. 11.22).

They survive between crops by resting spores that produce zoospores, which infect the host. The zoospores form thalli in the host cytoplasm. In the early stages of infection the cytoplasm of thalli is separated from the host cytoplasm by a membrane, but later the thalli form a cell wall. The entire thallus is converted into vegetative sporangia or resting spores. A detailed study of the infection of sugar beet roots by *P. betae* is described by Barr and Asher (1996) (Fig. 11.23). The invasion of a root cell by *Olpidium* is illustrated in Fig. 11.24.

Various degrees of host specificity exist in both the chytrid and plasmodiophoral vectors, with some isolates having a wide host range and others a narrow host range. The wide host range isolates tend to be better vectors than do the narrow ones.

Two types of virus–fungal vector relationships have been recognized, termed *in vitro* and *in vivo* (Campbell, 1996).

A. *In vitro* fungal transmission

The *in vitro* virus–vector relationship is found between the isometric viruses of the *Tombusviridae* and two *Olpidium* species (Table 11.7). Virions from the soil water adsorb on to the surface of the zoospore membrane and are thought to enter the zoospore cytoplasm when the flagellum is 'reeled in'. It is unknown how the virus passes from the zoospore cytoplasm to the host cytoplasm, but it is thought that this occurs early in fungal infection of the root. Reciprocal exchange of the coat proteins of TBSV (not transmitted by *O. bornavarus*) and CNV (transmitted by *O. bornavarus*) showed that the coat protein is involved in the uptake of the virus by the zoospore (McLean *et al.*, 1994). One amino acid in the coat protein of CNV was identified as being important for transmissibility, and binding studies showed that this was associated with recognition of the virus by *O. bornavarus* zoospores (Robbins *et al.*, 1997).

PLASMODIOPHORA

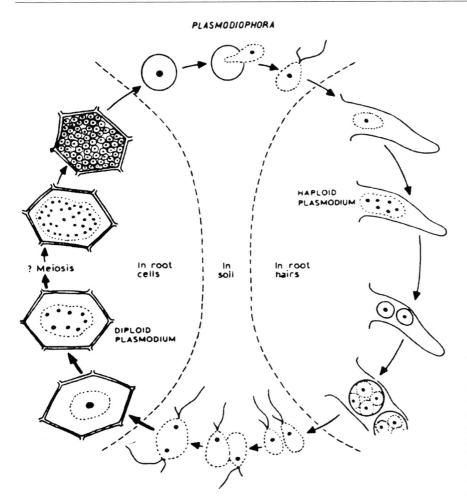

Fig. 11.22 Life cycle of a plasmodiophoral fungus. On the left-hand side is the diploid stage in root cells; on the right-hand side is the haploid stage in root hairs. Between are the phases in the soil where plant-to-plant transmission of viruses can occur.

B. *In vivo* fungal transmission

The *in vivo* virus–vector relationship occurs between the rod-shaped viruses of the *Bymovirus*, *Furovirus* and *Varicosavirus* genera and *O. brassicae* and the three plasmodiophoral species. The model for this relationship is based on observations on *O. brassica* and LBVV, *P. graminis* and SBWMV, and *P. betae* and BNYVV (Campbell, 1996). The virus is within the zoospores when they are released from the vegetative sporangia or resting spores and infects the new host when these zoospores establish their own infection of the root. The processes of virus acquisition and release by the zoospores are unknown. Studies on various bymoviruses and benyviruses suggest that a read-through domain from the coat protein (for genome structure of bymoviruses and benyviruses see Chapter 6, Sections VIII.C.2 and H.13 respectively) is implicated in the fungal transmission of these viruses (reviewed by Campbell, 1996). BNYVV RNAs 3 and 4 also have an indirect effect on the transmission, most likely through controlling factors such as spread and accumulation of the virus in the root system.

XIII. DISCUSSION AND SUMMARY

Study of the invertebrate vectors that transmit plant viruses is important for two reasons. First, these vectors play a major role in disseminating virus diseases of economic importance in all countries. Second, virus–vector relationships are of considerable biological interest, especially those where the virus replicates in the animal vector as well as in its plant host.

TABLE 11.7 Viruses and virus-like agents for which fungal vectors have been proven or suggested

Virus genus or group	Virus	Fungal vector[a]					
		Obr	Obo	Pgr	Pbe	Sss	Ssn
Polyhedral virions, *in vitro* acquisition							
Tombusvirus	CNV		+u				
Carmovirus	MNSV		+u				
	CLSV		+u				
	CSBV		+u				
	SqNV		+u				
Necrovirus	TNV	+u					
	ChNV	+					
	LNV	+					
Dianthovirus	RCNMV		+				
Satellite virus	STNV	+u					
Polyhedral virions, acquisition unknown							
	WYSV						+
Virion not characterized, acquisition unknown	WCLA						+
Virion rod-shaped, *in vivo* acquisition							
Furovirus	SBWMV			+			
	OGSV			+			
	RSNV			+			
Pecluvirus	PCV			+			
	IPCV			+			
Benyvirus	BNYVV				+u		
	BSBV				+		
Pomovirus	PMTV					+	
Bymovirus	BaMMV			+u			
	BaYMV			+			
	OMV			+			
	RNMV			+			
	WSSMV			+u			
Varicosavirus	LBVV	+u					
	TSV	+					
	FLNV	+					
Other rod-shaped, not characterized, *in vivo* acquisition	LRNA	+u					
	PYVA	+u					

[a] Vectors: Obr, *Olpidium brassicae*; Obo, *O. bornavanus*; Pgr, *Polymyxa graminis*; Pbe, *P. betae*; Sss, *Spongospora subterranea* f.sp. *subterranea*; Ssn, *S. subterranea* f.sp. *nasturtii*.

+, Specific fungus associated with virus transmission; +u, unifungal or equivalent culture of fungus demonstrated to transmit virus, not necessarily the association of vector with virus; WCLA, watercress chlorotic leafspot agent; LRNA, lettuce ring necrosis agent; PYVA, pepper yellow vein agent.

From Campbell (1996), with kind permission of the copyright holder, © Annual Reivews. www.AnnualReviews.org

Only two invertebrate phyla have members that feed on living green land plants and both of these—the *Nematoda* and *Arthropoda*—contain vectors of plant viruses.

Three genera of nematodes are vectors. Two, *Xiphinema* and *Longidorus*, contain vec-

tors for the polyhedral viruses of the *Nepovirus* group. Rod-shaped tobraviruses are transmitted by species of *Trichodorus*. There is a substantial degree of specificity in the virus–nematode relationship, almost certainly involving attachment of viral particles

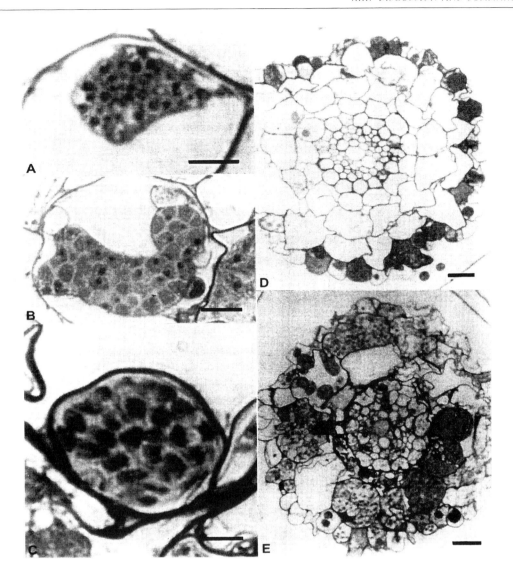

Fig. 11.23 (see Plate 11.2) Invasion of sugar beet roots by *Polymyxa betae*. **(A)** Young plasmodium in epithelial cells with blue-stained lipid and nuclei, and pink dots of unidentified material. **(B)** Young lobed zoosporangium in cortical cell. Note pink ground color, blue zoospores and dark blue nuclei. **(C)** Mature zoosporangium containing blue-stained zoospores. **(D)** Distribution of *P. betae* 8 days after inoculation. Plasmodia in various stages of development are restricted to epidermal cells. **(E)** Distribution of *P. betae* 18 days after inoculation. Plasmodia and zoosporangia are present throughout the root cortex but not in the endodermis or stele. Bar markers: **(A)** and **(C)**, 5 μm; **(B)**, 10 μm; **(D)** and **(E)**, 50 μm. From Barr and Asher (1996), with permission.

to specific zones in the lining of the nematode gut.

Among the classes of the *Arthropoda* phylum, two have members that feed on living green land plants—the *Arachnida* and *Insecta*—and both of these contain viral vectors. The most important group of vectors numerically is to be found in the insect order *Homoptera*, which includes the aphids, leafhoppers, planthoppers, whiteflies and mealybugs. As these groups have sucking mouthparts that penetrate leaf cells and tissues, they are ideally suited to transmit viruses from diseased to healthy plants.

About 66% of the arthropod-borne viruses are transmitted by aphids. There are various

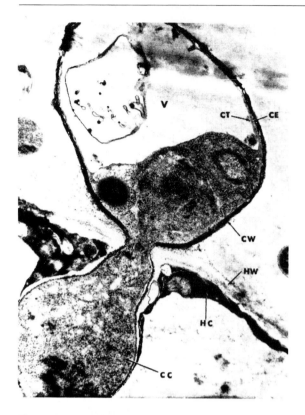

Fig. 11.24 Fungal transmission. Infection of a root cell by *Olpidium*. Electron micrograph showing contents of an encysted zoospore entering the host cell. CC, cyst cytoplasm; HC, host cytoplasm; HW, host wall; CW, cyst wall; CT, cyst tonoplast; CE, cyst ectoplast; V, vacuole. From Temminck and Campbell (1969), with permission.

types of virus–aphid relationship, as summarized in Table 11.2. None of these involves a 'flying pin' kind of transmission. Several members of the *Rhabdoviridae* replicate in their aphid vectors, and vectors may remain infective for their lifetime. Even when the aphid remains infective for a relatively brief period, there is specificity in the virus–vector relationship. This specificity involves virus-coded proteins. (1) Potyviruses and caulimoviruses each produce a virus-coded protein, a *helper factor*, that is essential for aphid transmission. They probably act by facilitating the binding of virus to some internal surface of the aphid. (2) Some semi-persistent viruses cannot be transmitted by an aphid vector when the RNA is coated with its own virus-coded coat protein. They can, however, be transmitted when the RNA is encapsulated in the coat protein of some unrelated

virus, a *helper virus*, in a plant infected with both viruses. (3) Viral coat proteins are also involved in the specific retention of particular viruses by particular aphids. (4) Specificity of aphid vector transmission may involve different strains of the same virus, as occurs with luteoviruses. Virus strains will be transmitted by a particular vector only if the RNA is in the protein coat appropriate for that vector. Details of the viral side of the interactions of some of the the helper component and coat protein systems are now understood at the molecular level but little is known about the vector side of the interaction.

The leafhoppers and planthoppers constitute a second important group of vectors. No viruses are transmitted in a non-persistent manner by hoppers. Two important viruses, MCDV and RTSV, are transmitted in a semi-persistent manner but most are circulative, either being non-propagative or propagative. Viruses transmitted by hoppers are not as numerous as those transmitted by aphid vectors, but they include a number of economically important viruses, especially those infecting food crops belonging to the *Poaceae*. There is a substantial degree of vector specificity between virus and hopper, and frequently both have somewhat narrow host ranges.

Four families and genera—*Reoviridae*, *Rhabdoviridae*, and the *Tenuivirus* and *Marafivirus* genera—contain members that replicate in their leafhopper vectors. Such replication usually has little effect on the hoppers. However, from the point of view of the virus, replication in the vector has two important consequences: (1) once they acquire virus the vectors normally remain infective for the rest of their lives, and (2) replication in the vector is often associated with transovarial passage of the virus, thus giving it a means of survival over winter that is quite independent of the host plant. In the plant reoviruses, particular genome segments code for gene products required for replication in the insect, but not in the plant. With viruses that replicate in their vectors, there may be a high degree of specificity between vector and virus, or even strains of a virus.

Leaf-feeding beetles have chewing mouthparts and do not possess salivary glands. They regurgitate during feeding, which bathes the mouthparts in sap. This regurgitant will contain virus if the beetle has fed on an infected plant. Beetles can acquire virus after a single bite and can infect a healthy plant with one bite. However, beetle transmission is not a purely mechanical process. There is a high degree of specificity between beetle vector and virus, and some very stable viruses such as TMV are not transmitted by beetles. The viruses that are transmitted belong to the *Tymovirus*, *Comovirus*, *Bromovirus* and *Sobemovirus* genera.

Sometimes one beetle species will transmit a particular virus with high efficiency while a related species does so inefficiently. The reasons for this sort of specificity are not understood. However, we are beginning to understand why some stable viruses are not beetle transmitted. The regurgitant fluid of the beetles contains an inhibitor that prevents the transmission of non-beetle-transmitted viruses but does not affect those that are transmitted. There is good evidence that the inhibitor is an RNase.

Other insect vectors are found among the mealybugs, whiteflies, mirid bugs and thrips. The viruses transmitted by these groups are not numerous, but the first two are vectors for some viruses causing important diseases in tropical crops. In the *Arachnida*, eriophyid mites are vectors for several viruses.

For viruses transmitted through the pollen, pollinating insects can transfer infected pollen to healthy plants, thus transmitting the virus in an indirect manner.

Other biological vectors include plasmodiophoral fungi with which viruses have specific interactions. Little is known about the detailed molecular interactions involved in this form of transmission.

Thus, for biological vectors, analyzes of the viral determinants of the specificity of virus–vector interactions have given much detailed information. This is relatively easy because of the relatively small size of the viral genome and the ability to manipulate it. However, the genomes and molecular biology of the vectors are much more complex and there has been little research on them. The application of modern technologies should enable these aspects to be studied in the future. Findings from such research could lead to new approaches in generic control of important viruses of crops by interfering with key aspects of the virus–vector interaction.

Transmission 2:
Mechanical, Seed, Pollen and Epidemiology

In the previous chapter, I described how viruses are transmitted by 'biological' vectors: arthropods, nematodes and fungi. This chapter covers three further aspects of virus movement between plants. As viruses need mechanical damage to enter a host, mechanical inoculation is a widely used technique for infecting plants experimentally; some viruses are also transmitted naturally by mechanical means. In this chapter, I discuss various considerations in experimental mechanical inoculation of viruses and also some other means (seed and pollen) that do not involve a biological vector by which viruses can move from the initially infected plant. Then I will describe how viruses move in the field and the various factors involved in the occurrence of epidemics.

I. MECHANICAL TRANSMISSION

Mechanical inoculation involves the introduction of infective virus or viral RNA into a wound made through the plant surface. When virus establishes itself successfully in the cell, infection occurs. This method of transmission is of great importance for many aspects of experimental plant virology, particularly the assay of viruses often by local lesion production (see Chapter 15) in the propagation of viruses for purification (see Chapter 4) and in the study of the early events in the interaction between a virus and susceptible cells (see Chapter 10). When intact virus is used as inoculum, the viral nucleic acid must be partly or entirely uncoated at an early stage. This process is discussed in Chapter 7 (Section II).

Virus acronyms are given in Appendix 1.

Many factors have to be taken into account in performing efficient experimental mechanical inoculation of viruses; these are described in detail in Walkey (1991) and Dijkstra and de Jager (1998), and are outlined below.

A. Source and preparation of inoculum

The most common source of virus for mechanical inoculation is infected leaf tissue. Since the virus has to be released from cells in infected plants, many of the considerations for obtaining the best quality inoculum are similar to those in the extraction of viruses from plants for purification (Chapter 4). It is usual that plant material likely to contain the highest amount of virus is used as a source but one also has to consider the possible presence of inhibitors of virus infection. In many cases, infected young leaves showing strong symptoms are used but for some viruses, other tissues may be better. For instance, roots of plants infected with TNV contain greater amounts of virus than do leaves (Smith, 1937). CMV is transmitted more efficiently from cucumber flower petals than from leaves which contain more inhibitors (Sill and Walker, 1952).

The infected plant material is sometimes ground in tap water (Dijkstra and de Jager, 1998) but, more frequently, an extraction buffer is used. It has been recognized for some years that phosphate buffers enhance infectivity of many viruses (Yarwood, 1952; Fulton, 1964). Breaking the plant cells exposes the virus to secondary metabolites that can affect infectivity and that are more prevalent in some hosts than in others. As well as plant products such as nucleases and the products from the oxidation of polyphenols which may affect virus structure and stability, there are many other substances in crude sap that could affect the number of successful infections produced by a virus. The former can be countered by additives similar to those used in virus purification (see Chapter 4), the latter often by dilution of the sap. Buffer mixtures with additives may be specially designed for successful inoculation from particular hosts (e.g. Martin and Converse, 1982).

An alternative method for preparing inoculum which overcomes inhibitor problems is to grind infected tissue frozen in liquid nitrogen to a fine powder and inoculate that directly, for instance with a fine brush (Lawson and Taconis, 1965: Ragetli et al., 1973).

B. Applying the inoculum

The efficiency of mechanical inoculation is greatly increased when some abrasive material is added to the inoculum or sprinkled over the leaves before inoculation. The most commonly used abrasives are carborundum (400–500 mesh) or diatomaceous earths such as Celite. The increase in number of local lesions obtained by the use of abrasives varies with different hosts and viruses, but may be 100-fold or more. The time of addition of these materials may be important. Celite added after grinding and dilution was much more effective than when added before grinding (Yarwood, 1968).

The early method of placing drops of the inoculum on the leaf and scratching or pricking the leaf surface with a needle to cause wounding was very inefficient and has been largely superseded by gently rubbing the leaf surface with some suitable object wetted with the inoculum to give more efficient transmission. A wide variety of objects has been used, depending on the preference of the operator and the volume of inoculum available (see Walkey, 1991; Dijkstra and de Jager, 1998). The objective in mechanical inoculation is to make numerous small wounds in the leaf surface without causing death of the cells. The pressure required to do this depends on many factors, such as plant species, age and condition of leaf, and additives present in the inoculum.

Macroscopic areas of dead tissue appearing on the inoculated leaf within a day or so indicate that the wounding was excessive. With a few viruses and hosts severe abrasion is more effective (Louie and Lorbeer, 1966). For some pathogens, for example CTV, which is probably confined to the phloem, TBRV and CEVd, cutting or slashing the plant stem with a contaminated blade is the most effective method of mechanical transmission (Garnsey and Whidden, 1973; Garnsey et al., 1977; Bitterlin et al., 1987). The vascular puncture technique in which the virus is inoculated to kernels of maize and barley is successful for MWLMV and MRFV (Louie, 1995; Madriz-Ordeñana et al., 2000). Particle bombardment has been used for infection with viruses (Franz et al., 1999) and with viral RNA (Klein et al., 1987), cDNA clones of RNA viruses (see Gal-On et al., 1995, 1997; Fakhfakh et al., 1996) and cloned DNA viruses (see Gilbertson et al; 1991; Garzon-Tiznado et al., 1993; Hagen et al., 1994). The cDNA constructs of RNA viruses included a promoter, usually the CaMV 35S promoter (see Chapter 7, Section IV.C.1), to express an infectious transcript; those of DNA viruses use the viral promoter but have to be such that the introduced genome can effectively replicate (see below). Particle bombardment inoculation has resulted in infection with several viruses that had proved to be difficult, if not impossible, to transmit mechanically but some, for instance RTBV, are not transmissible by this procedure (Dasgupta et al., 1991).

Holmes (1929) considered that washing inoculated leaves with water immediately after inoculation increases the number of local lesions formed, and this has become a fairly widespread practice. However, washing leaves after inoculation, spraying with water or dipping leaves in water may substantially reduce the number of local lesions produced by several viruses or have variable effects depending on other conditions (Yarwood, 1973). The effect of washing or dipping leaves in water on the number of lesions probably depends on many factors, and particularly on whether inhibitors of infection are present in the inoculum. If such inhibitors are present, washing may minimize their effect.

If the leaves are dried rapidly after inoculation, either by blotting or with an air jet, there may be a marked increase in the number of local lesions, but again the effect is variable (Yarwood, 1973).

Polson and von Wechmar (1980) used electro-endosmosis to introduce MSV into leaves through cut petioles in a way that gave rise to infection even though this virus is normally transmitted only by leafhoppers. Konate and Fritig (1984) described an efficient microinoculation procedure that allowed early events following mechanical inoculation to be studied at predetermined individual infection sites on the leaves. Injection of virus into petioles or stems with a hypodermic syringe has been used occasionally but, apart from the use of a high-pressure medical serum injector to transmit BCTV (Mumford, 1972), is generally a very inefficient method. although micro-injection of trichome cells has proved useful in studying cell-to-cell spread of viruses (Derrick *et al.*, 1992). However, the injection of infective constructs of viruses cloned into *Agrobacterium tumefaciens* has proved to be very effective, especially with viruses that are difficult to transmit mechanically. In this process, termed *agro-inoculation*, the viral construct, cloned into the T-DNA of *A. tumefaciens* Ti plasmid is transformed into *A. tumefaciens*, which is then injected into the host. The system was initially developed for CaMV, which infects dicotyledonous species (Grimsley *at al.*, 1986). *A. tumefaciens* is thought not to infect monocotyledons, but Grimsley *et al.* (1987) showed that, when cultures of the bacterium containing a plasmid with tandemly repeated copies of MSV DNA were inoculated to whole maize plants, the plants developed symptoms caused by MSV. Agro-inoculation has been used for several other geminiviruses (e.g. Donson *et al.*, 1988; Kheyr-Pour *et al.*, 1994) and for RTBV (Dasgupta *et al.*, 1991), which are not mechanically transmissible. As with particle bombardment, agro-inoculation constructs of these DNA viruses has to be such that transcription and replication can be initiated on their introduction into the plant. For geminiviruses, dimers of the viral genome are usually used; for reverse transcribing viruses (e.g. CaMV, RTBV) a 'one and a

bitmer' construct that can transcribe the 35S RNA (see Chapter 7, Section IV.C.1) is produced. There appears to be some specificity on the strains of *A. tumefaciens* that give efficient transmission of viruses to monocotyledons (Boulton *et al.*, 1989; Dasgupta *et al.*, 1991). cDNAs of RNA viruses have also been agro-inoculated effectively (see Leisner *et al.*, 1992a; Turpen *et al.*, 1993); as with particle bombardment inoculation, these constructs have to include a suitable promoter.

Inoculating large numbers of plants can be a time-consuming process, and various procedures have been adopted to reduce the time involved. For example, dipping and moving the leaves of seedling plants in the inoculum may provide a rapid method for inoculating large numbers of seedlings at the time of transplantation. Where large numbers of leaves are to be inoculated with the same inoculum, airbrushes of the type used by artists may prove useful (Whitham *et al.*, 1999). Alternatively, the inoculum may be applied in a solid stream (Louie *et al.*, 1983). A sensitive and rapid method using a specially designed airgun has been described by Laidlaw (1987). To avoid washing pestles and mortars or other glassware where large numbers of individual tests have to be made, it may be possible to rub a piece of diseased leaf directly on a leaf of the test plant (Murakishi, 1963).

II. FACTORS INFLUENCING THE COURSE OF INFECTION AND DISEASE

For a given species of plant and a given virus, there are many factors that can influence the course of infection and the disease that develops. The way in which a virus is inoculated into the plant may be important. Sometimes many different strains of a virus occur and these may cause quite different kinds of disease in the same host plant under the same conditions (see Chapter 17). In this section, I shall discuss inherent variables in the host plant itself, and environmental factors. Interactions between unrelated viruses and between viruses and some other agents of disease are described in Chapter 10, Section V.

A. The plant being inoculated

Sometimes a virus can be transmitted mechanically by inoculating the cotyledon leaves, but not the first true leaves, for example with a virus of sweet potato (Alconero, 1973). Mechanical inoculation of roots is inconvenient and is often less successful than with leaves. However, transmission by this means has been achieved for several viruses (e.g. Moline and Ford, 1974).

1. Susceptibility to infection

Most commonly, small young leaves and old leaves are less susceptible than well-expanded younger leaves. There may be a marked gradient of susceptibility with age (Fig. 12.1). The curves in Fig. 12.1 can be taken to indicate the changes in susceptibility that individual leaves undergo with time. One has to consider the recent

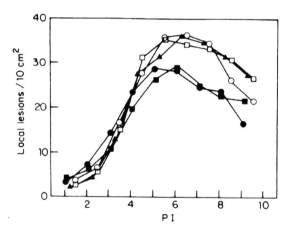

Fig. 12.1 Effect of leaf age on number of local lesions produced by TMV in leaves of Samsun NN tobacco. Since younger leaves are expanding, the number is expressed per unit leaf area at the time of inoculation. A *plastochron* is the time interval between corresponding developmental stages of successive leaves. The *leaf plastochron index* (PI) is an arbitrary measure that allows plants at slightly different developmental stages to be compared. Each curve represents successive leaves from a single plant. PI ●, 15.10; ■, 15.12; ▲, 15.30; ○, 15.52; □ 15.53. From Takahashi (1972b), with kind permission of the copyright holder, © Blackwell Science Ltd.

understanding of host response to virus infection (see Chapter 10).

The gradient may not always be of the form shown in Fig. 12.1. Thus, on *Nicotiana glutinosa* plants with 8 to 10 well-developed leaves, TMV will produce more local lesions on the middle and lower leaves than on the younger ones. In contrast, TBSV may produce no lesions on the oldest leaves and most on the youngest (Bawden, 1964).

Abscisic acid accelerates senescence processes in leaves. Exogenously applied abscisic acid increased susceptibility of Samsun tobacco leaves to TMV (Balázs *et al.*, 1973), indicating a possible role for this hormone in age-dependent changes in leaf susceptibility.

2. Environmental factors

The environmental conditions under which plants are grown before inoculation, at the time of inoculation, and during the development of disease can have profound effects on the course of infection. A plant that is highly susceptible to a given virus under one set of conditions may be completely resistant under another. If infection occurs, the plant may support a high or low concentration of virus and develop severe disease or remain almost symptomless, depending on the conditions.

3. Factors affecting susceptibility to infection

Any factor that alters the ease with which the surface of the leaf is wounded will alter the probability of successful entry of virus introduced by mechanical inoculation, while physiological changes in the leaf may make the cell more or less suitable for virus establishment. As a broad generalization, greenhouse-grown plants will have greatest susceptibility when they are grown and used under the following conditions: (1) mineral nutrition and water supply that do not limit growth; (2) moderate to low light intensities; (3) a temperature in the range of 18–30°C, depending on virus and host; and (4) inoculation carried out in the afternoon. These plants are termed horticulturally 'soft' and may be different to plants grown for normal horticultural purposes.

a. Light

There are two general situations in which light affects susceptibility: short-term changes in light intensity or deviation over a period of no more than a day or two, and long-term effects that may markedly influence the growth of the plant.

Bawden and Roberts (1947) found that in summer reducing light intensity or giving plants complete darkness for periods of days increased the susceptibility of several hosts to certain viruses. This effect of shading or darkening plants for about a day before inoculation is often used as a practical measure to increase the susceptibility of test plants with viruses that are difficult to transmit.

Various experiments suggest that light may have two opposing kinds of effects. Excessive illumination may reduce susceptibility. On the other hand, after a period of darkness, even a short burst of light may increase the number of local lesions. For example, when bean plants that had been in the dark for 18 hours were given a 1-minute exposure to 800 foot-candles before inoculation, the number of local lesions produced by a TNV inoculum was double that of plants inoculated under minimal illumination required to see the leaves (Matthews, 1953b). In almost all studies on the effects of light, the experimental plants have been raised under ordinary greenhouse conditions. Thus, any experimentally imposed change will have been superimposed on the natural daily cycle of variation in susceptibility, discussed in a subsequent section. The time of day at which an experiment is performed can have a marked effect on results (Matthews, 1953b).

As far as the long-term effects of light on susceptibility are concerned, it is generally found that high light intensities over the period of growth of the test plant give rise to 'hard' plants that may have very low susceptibility compared with 'soft' plants raised under lower light intensities.

A further consideration is the quality of light. Plant grown in the spring and autumn in temperate climates are often more susceptible and produce more virus than those grown in the summer or winter. Even with supplementary lighting, it appears that the quality of natural light and possibly the changing of natural day length might have an influence on virus yield.

b. Temperature

In general, pre-incubation of plants at slightly higher temperatures than normal before inoculation increases susceptibility. The effect of treating plants at a higher temperature after inoculation may vary with the virus or the strain of virus tested (e.g. Kassanis, 1952). A brief shock treatment, for example at 50°C, after inoculation may affect the kind of lesion that subsequently develops (e.g. Foster and Ross, 1975). Where necrotic ring spot local lesions are produced, light and temperature conditions interact in the production of rings (Harrison and Jones, 1971). Temperature may also affect the speed and efficiency of graft transmission (Fridlund, 1967a).

c. Water supply and humidity

The immediate effect of washing inoculated leaves with water was discussed in Section I.B. Generally speaking, if plants are grown with a minimal supply of water they become more or less stunted, leaf texture is 'hard', and they will give greatly reduced numbers of local lesions compared with plants raised under conditions where water supply is not limiting growth. For example, Tinsley (1953) found that as many as 10 times more local lesions were produced by several viruses on well-watered N. glutinosa and tobacco plants as on poorly watered plants.

Most reports suggest that moderate wilting near the time of inoculation increases susceptibility. For example, when detached bean leaves were wilted before inoculation with TMV, water deficits of 0–15% of the green weight of the leaf increased infection. More severe deficits in the range of 15–29% decreased infection. When leaves were wilted after inoculation, numbers of local lesions increased with increasing water deficit over the range of 0–35% (Yarwood, 1955). Maximum increase was 4–8-fold over that of unwilted leaves.

d. Nutrition

The nutrition of the host plant may have a marked effect on the numbers of local lesions produced by viruses. As would be expected, interactions between different nutrients are quite complex. For example, Bawden and Kassanis (1950) investigated the effects of various levels of nitrogen, phosphorus and potassium on susceptibility of tobacco and *N. glutinosa* plants to strains of TMV. The range of fertilizer treatments used produced large effects on plant growth and many significant differences in susceptibility. In general, nutritional conditions that were most favorable for plant growth were also those giving greatest susceptibility. There was no evidence that one particular element increased susceptibility on its own.

Trace elements may also affect transmissibility. For example, increased manganese supply caused a marked increase in the number of local lesions produced by PSTVd in *Scopolia* leaves (Singh *et al.*, 1974).

e. Time of day

Since many basic processes in leaves are influenced by the diurnal cycle of night and day, it is not surprising that the susceptibility of leaves to mechanical inoculation varies systematically with time of day. The number of local lesions produced rises to a maximum in the afternoon and falls to a minimum during the night, usually just before dawn (Fig. 12.2).

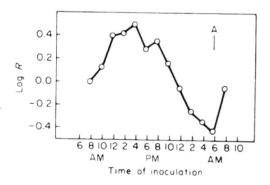

Fig. 12.2 Effect of time of day of inoculation on the numbers of local lesions produced by TNV in beans. Log *R* relates the number produced to the number produced at a base time (8 a.m.). Bar A represents the difference between any two points required for significance at *P* = 1%. From Matthews (1953a), with permission.

A diurnal variation has been found for various host–virus combinations. This diurnal change is not dependent on immediate changes in environmental conditions, but appears to be 'built in', as are some other physiological processes in leaves. Thus, bean plants placed in darkness at constant temperature at 4 p.m. and maintained in darkness through the following day showed a very similar cycle of susceptibility to plants that were exposed to the normal daylight cycle (Matthews, 1953b).

f. Time of year

Where there are large climatic differences between the seasons, plants may vary widely in their susceptibility to a virus at different times of the year. For example, certain varieties of bean (*Phaseolus vulgaris*) grown at Rothamsted in England during the summer appeared to be almost immune to CMV. In the winter, they produced substantial numbers of necrotic local lesions (Bhargava, 1951). Similar, but less extreme, differences have been noted with other viruses producing local lesions. The major environmental factor concerned in greenhouse-grown plants is probably light, as Bawden and Roberts (1947) showed that typical winter reactions of plants can be reproduced in summer by appropriate shading.

g. Chemical and mechanical injury

Nematicides and fungicides may increase susceptibility to virus infection. For example, atrazine increased susceptibility of maize hybrids to MDMV (MacKenzie *et al.*, 1970).

The mechanical injury involved in transplanting wheat plants led to increased transmission of WSSMV from infected soil (Slykhuis, 1976).

B. Development of disease

1. Virus concentration in the inoculum

The amount of infecting virus may influence the extent to which growth is subsequently depressed (compare (A) and (B) in Fig. 12.3). This may in fact be an age-of-plant effect because it is probable that virus moves systemically through the plant sooner after a heavy

Fig. 12.3 Interaction between time of infection and number of infecting aphids on subsequent depression in size of barley plants by BYDV. **(A)** About eight aphids per plant. **(B)** About 150 aphids per plant. The inoculative feed was 2 days, so that feeding damage was insignificant. Stages of growth when infected: (left to right) control; flowering; boot; tillering; three-leaf; one-leaf. From Smith (1967), with permission.

inoculation, either mechanically or by insects. However, other workers have not found a decrease in plant yield with increasing numbers of aphids (e.g. Skaria *et al.*, 1984). The apparent discrepancy may be in the number of aphids used. To obtain effects such as that shown in Fig. 12.3, 100 or more aphids were used per plant for the heavy inoculation.

2. Environmental factors

a. Light

Light intensity and duration affect virus production and disease expression in different ways with different viruses, but, generally speaking, high light intensities and long days favor replication. For instance, local lesions formed in 17 hours under continuous illumination at 25°C on detached *P. vulgaris* leaves inoculated with AMV maintained in moist Petri dishes, whereas they took nearly 30 h under low lighting on the laboratory bench (R. Hull, unpublished observation).

b. Temperature

Over the range of temperatures at which plants are normally grown, increasing temperature usually increases the rate at which viruses replicate and move through the plant. Like other biological phenomena, however, increase in temperature above a certain point leads to a reduction in the rate of replication. The species of host plant, strain of virus, and age of the leaf in which the virus is replicating may have a major effect on the way the virus behaves with changes in temperature. Figure 12.4 illustrates some effects of temperature on TMV replication.

The temperature at which plants are grown frequently affects the kind of disease that develops. For example, the type of pigmentation developing in subterranean clover infected with red-leaf virus was dependent on the temperature at which the plants were grown (Helms *et al.*, 1985). A severe stem tip necrosis developed in certain soybean cultivars held at 24°C following infection with SMV. At 28°C, most plants developed typical mosaic disease (Tu and Buzzell, 1987). Plantains and bananas infected with BSV exhibited marked symptoms when grown at 22°C, but the symptoms were reduced or even disappeared when the plants were transferred to 28–35°C (Dahal *et al.*, 1998); symptoms reappeared when the plants were returned to the lower temperature. The viral titer was significantly higher in the symptomatic plants. However, this may not necessarily be a temperature effect: it could be due to changes in light quantity or quality.

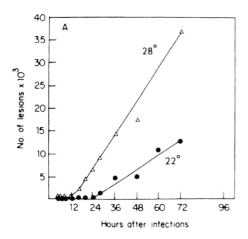
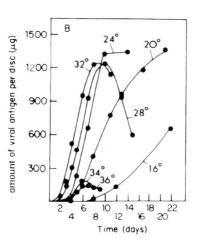

Fig. 12.4 Effect of temperature on TMV replication. **(A)** A linear plot of the time course of TMV replication in tobacco mesophyll protoplasts. From Dawson *et al.* (1975), with permission. **(B)** Effect of temperature on the replication of TMV (common strain) in tobacco leaf discs. Concentration of virus measured serologically. From Lebeurier and Hirth (1966), with permission.

High growth temperatures may annul the effect of some hypersensitivity genes. For example, in *Nicotiana* cultivars with the *N* gene, TMV causes systemic disease at 36°C, whereas at 28°C or lower it gave a hypersensitive response (Kassanis, 1952). On the other hand, the activity of genes *Nx* and *Nb* in potato, causing hypersensitivity to PVX, were not affected by higher temperatures (Adams *et al.*, 1986b).

Increasing temperature, up to a certain point, increases the rate of systemic movement and decreases the time before the first appearance of systemic symptoms (e.g. Jensen, 1973). The shortest incubation period for systemic invasion does not necessarily occur at the temperature that leads to maximum production of virus.

In plants kept at different temperatures there may or may not be an approximate correlation between severity of disease symptoms and virus concentration reached. For example, such a correlation is observed with TMV and PVX in tobacco in the range 16–28°C (Bancroft and Pound, 1954).

Changes in the temperature at which plants are grown may lead to a selective multiplication of certain strains adapted to the particular conditions (see Chapter 17, Sections III and IV.A.3) . Practical applications of heat therapy are discussed in Chapter 16 (Section II.C.2.b).

c. Water supply

The effects of water supply on virus replication have not been studied systematically. A chronic deficiency of water giving stunted 'hard' plants will usually give rise to less obvious symptoms. A liberal supply of soil water increases the incidence of internal browning disease in tomatoes due to ToMV infection (Boyle and Bergman, 1967).

d. Nutrition

A number of investigations have been made into the effects of host plant nutrition on virus replication, particularly of TMV in tobacco plants. Many of the results are conflicting, probably due to variations in methods of growing the plants, the effective concentration of nutrients supplied, strain of virus used, method of estimating virus, and the basis for expressing the results. Supply of the major elements that give best plant growth usually allows for the greatest replication of virus. For example, Bawden and Kassanis (1950) concluded that the effects of nitrogen and phosphorus supply on TMV replication were fairly closely correlated with effects on plant growth. Treatments giving better plant growth led to a greater increase in virus production, both per unit fresh weight and per plant. Pea plants provided with low calcium nutrition developed severe stunting when infected with pea leafroll virus (Thompson and Ferguson, 1976).

Minor element nutrition (zinc, molybdenum, manganese, iron, boron and copper) has a variable effect on the capacity of plants to support virus replication. Effects on virus accumulation often parallel effects on plant growth, but exceptions have been observed. For example, manganese deficiency led to an increase in TMV concentration in the leaves. By contrast, toxic concentrations of this element in cowpea led to greatly enhanced replication of CCMV (Dawson, 1972).

e. Time of year

The complex factors including day length, light intensity and quality, air and soil temperatures, and water supply that change during the seasonal cycle will affect plant growth, and thus the disease produced by a given virus, and the extent to which a virus replicates. For example, there was a seasonal variation in the concentration of AMV in alfalfa, the highest concentrations being recorded in spring and the lowest in autumn (Matisová, 1971). Some viruses, such as PLRV in potato and BYV in sugar beet, cause much more distinct disease symptoms in summer than in winter. Others, such as PVX in potato, may cause more severe disease under winter conditions. Chloroplasts in the leaves of *Abutilon* infected with AbMV showed a marked seasonal variation in the severity of ultrastructural changes, being most severe in the summer (Schuchalter-Eicke and Jeske, 1983).

3. Systemic spread

In a fairly mature tobacco or tomato plant inoculated with TMV in the younger leaves, the lower leaves may never become infected with virus unless directly inoculated. With increasing age there is a greater tendency for infection to remain localized. Physiological age rather than actual age is the significant factor as discussed under virus movement (see Chapter 9). When potato plants are infected with PVX, either naturally in the field or by inoculation, some tubers may be free of the virus at the end of the first season. Beemster (1966) found that the proportion of tubers and eyes infected with PVY is closely related to the time at which the plants are inoculated. For example, the percentage of infected tubers was 100% for plants 8 weeks old on inoculation and only 25% for plants that were 13 weeks old when inoculated. Time of infection is often an important factor in determining loss of yield in economically important crops. For example, loss of yield is generally more severe when cereals are infected

at an early stage of growth with BYDV (Smith, 1967) (Fig. 12.3).

In field experiments with *Capsicum annuum*, plant growth and fruit yield improved almost in direct proportion to the lateness of inoculating the plants with CMV (Agrios *et al.*, 1985). The term 'mature plant resistance' has been used for these age-of-plant effects, but the mechanism for them is unknown. Many metabolic changes occur as leaves age. For example, ribosome content decreases (Venekamp and Beemster, 1980), but no causal relationship has been established between such changes and increased resistance to the effects of inoculation with a virus. Barker and Woodford (1987) describe a longer-term effect of late infection. Potato plants grown from tubers of plants infected with PLRV late in the previous season developed unusually mild symptoms, although they contained as much virus as plants with severe symptoms. Perhaps a strain selection process was at work or there is a change in the natural plant infection response (see Chapter 17).

4. The kinds of host response

Genetic and molecular aspects of resistance to viruses are discussed in Chapters 16 and 10 respectively.

C. Viral nucleic acid as inoculum

For any virus that has (+)-sense ssRNA, and many with ssDNA or dsDNA, as its genetic material it should be possible, in principle, to prepare an extract of total nucleic acids from infected tissue and use this to inoculate healthy plants. This procedure may allow mechanical transmission when whole-leaf extracts are ineffective. Success may be due to removal of virus inhibitors into the phenol phase or interface (or by another method), or it may be due to the existence of unstable or incomplete virus that is inactivated by nucleases unless these are removed by the phenol. With some tissues and viruses, grinding or blending in the presence of phenol may release virus from sites where it remains bound in normal sap extracts. Any infectivity in phenol extracts or RNA preparations will be fully susceptible to nucleases on contaminated glassware once the phenol is removed, as it must be before inoculation using the extract.

D. Nature and number of infectible sites

1. Nature of the leaf surface

On the upper surface of leaves commonly used for mechanical inoculation, there will be about $1\text{-}5 \times 10^6$ cells of various types. The structure of the leaf surface is shown in Fig. 12.5 and details can be found in Esau (1953). The outermost layer is the epicuticular wax, which in different species may be very different in amount, fine structure and chemical constitution. The epicuticular wax layer will have a strong influence on the wettability of the leaf surface during mechanical inoculation.

2. Nature of the infectible sites

Virus placed on an intact leaf surface cannot infect directly and wounds must be made that break through the inert leaf surface. It is possible that some types of cell on the leaf surface are more susceptible to wounding than others. It has been shown directly, by microinfection methods that virus can be introduced into leaf hairs, but the infection rate was low (Zech, 1952). It seems probable that all the cell types making up the epidermis are potentially capable of being infected by mechanical inoculation, the cuticle being probably the major barrier to infection. Wounds that penetrate right through the cuticle and the cell wall are probably effective in allowing virus to enter, and broken hairs are probably a common route of entry (Fig. 12.6). There have been some suggestions of direct contacts between plant cells and the environment through ectodesmata and hydathodes. Ectodesmata, retermed ectoteicoides, are not analogous to plasmodesmata (Franke, 1971) and it is considered that, in most cases, they are not sites of virus entry. Experiments with several viruses indicate that, following mechanical inoculation, some underlying mesophyll cells may be infected directly with the inoculum (Salinas Calvete and Wieringa-Brants, 1984; Matthews and Witz, 1985).

Hydathodes are structures containing water pores located at leaf margins (Cook *et al.*, 1952) that connect to the intracellular spaces and to the xylem vascular system. Under conditions of water uptake and limited transpiration, such as warm soils and high humidity in the dark, liquid is expelled through the hydathodes in a process termed guttation. Particles of TMV have been found in the guttant of tomato (Johnson, 1937), of ToMV in tomato and *Gomphrena globosa*, of PPMV from *Capsicum annuum* (French *et al.*, 1993), of 10 genera of viruses in the guttant of cucumber (French and Elder, 1999) and of BMV in barley (Ding *et al.*, 2000). BMV was found in the intercellular spaces (Ding *et al.*, 2000) and ZYMV in the xylem (French and Elder, 1999) but how they got there has not yet been determined.

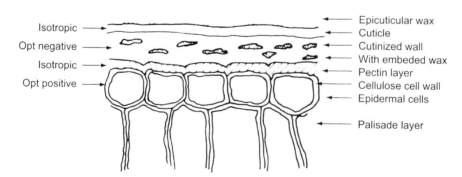

Fig. 12.5 The barriers to virus infection. Diagrammatic representation of the epicuticle of the plant seen in cross-section. The lines dividing the layers above the epidermal cells indicate regions of major change in the construction of components rather than sharp boundaries. Individual plant species may depart greatly from this general arrangement. From Eglinton and Hamilton (1967), with kind permission of the copyright holder, © The American Association for the Advancement of Science.

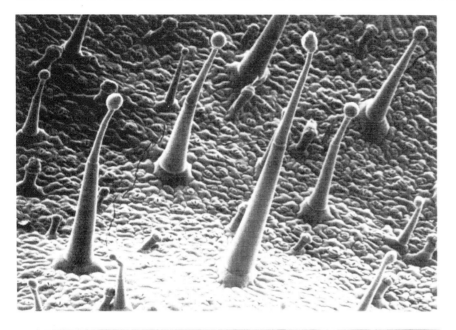

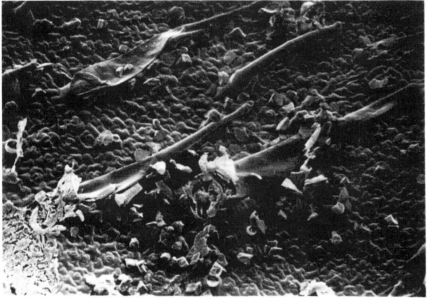

Fig. 12.6 Scanning electron micrographs of the surface of *Nicotiana glutinosa* leaves before and after mechanical inoculation. (*Top*) An untreated leaf showing intact leaf hairs and epidermis. (*Bottom*) the broken hairs following mechanical inoculation. The particles of the carborundum abrasive may be clearly seen. Bar marker 0.1 nm. (Courtesy of M.J.W. Webb.)

3. Evidence from the infectivity dilution curve

The relation between numbers of local lesions produced and the dilution of the inoculum is discussed in relation to the assay of viruses in Chapter 15. Various theoretical models have been developed in attempts to explain the nature of the dilution curve (Fulton, 1962; Furumoto and Mickey, 1967; Gokhale and Bald, 1987). Dilution curves for AMV (requiring three particles) and TNV (assumed to require only one particle) are shown in Fig. 15.2 for data obtained in the same host species. Although the difference between the two curves in Fig. 15.2 is quite clear, a dilution curve on its own cannot be used to decide whether a viral genome is housed in one, two or three particles. There are many other factors that cause the slope of the curve to vary from experiment to experiment (see Section II.A).

4. Lifetime of infectible sites following wounding

The time for which wounds remain infectible has usually been studied by dipping leaves into inoculum at various times after abrasion and counting the number of lesions that subsequently develop. Generally speaking, the number of infectible sites falls off very rapidly after abrasion. For example, Furumoto and Wildman (1963a) found that about 70% of the sites on *N. glutinosa* leaves that were susceptible to infection with TMV by dipping 2 seconds after wounding had lost their susceptibility by 90 seconds. The remaining sites lost their susceptibility much more slowly over a period of about 1 hour. However, not all viruses and hosts follow this pattern. Jedlinski (1964) found that during the first 10 minutes after abrasion the number of infectible sites could either increase, decrease or remain constant, depending on the host–virus system tested. There are many treatments that, applied to the leaf before or after wounding, can alter the course of events (see Section II.A).

E. Number of particles required to give an infection

There are three aspects to consider concerning the number of particles required for infection.

1. Number of particles to give one successful infection

Mechanical inoculation of leaves has been widely regarded as a very inefficient process. Various estimates suggested that between about 10^4 and 10^7 virus particles have to be applied to the leaf for each local lesion that subsequently develops (e.g. Walker and Pirone, 1972a). With multi-particle viruses the requirement for more than one particle makes the inoculation process even less efficient than for single-particle viruses. For CPMV (two component) 10^6 to 10^8 particles were applied mechanically for each lesion produced (van Kammen, 1968) and for AMV (three component) this number was 10^8 to 10^{10} (van Vloten-Doting, 1968).

In most estimates of the efficiency of mechanical inoculation, relatively large volumes of inoculum were applied to the test leaves and the calculation involved determining the number of particles applied per lesion produced.

However, if a very small volume of inoculum is applied using a series of dilutions of the virus, the limiting dilution at which an infection is obtained gives a much lower estimate. Thus, Walker and Pirone (1972b) found that about 450 TMV particles in 2.5 µl of inoculum were sufficient to infect a tobacco plant. Using even smaller volumes of inoculum (0.1–1.0 µl) as few as 10 to 30 particles of TYMV were required to produce a single local lesion in Chinese cabbage (Fraser and Matthews, 1979a). These very substantial increases in efficiency of the mechanical inoculation process may be due to three effects: (1) for a given number of virus particles applied, the smaller the volume, the higher the virus concentration; (2) a virus particle in a very small volume will have a greater probability of finding an infectible site in a short time than one in a large volume; and (3) the fact that rapid drying increases the number of local lesions (see Section I.B) may also be a factor.

Two other methods have been used to determine efficiencies of infection: inoculation of protoplast suspensions and micro-injection of cells. The number of particles adsorbed to each viable cell for each successful infection can be measured for inoculated protoplasts, and is usually about 0.1–1.0% of the applied inoculum. Efficiency of infection of protoplasts can be expressed as the average number of virus particles adsorbed per protoplast to give infection in one-half of the protoplasts (ID_{50}). Values of the ID_{50} calculated from published data for several viruses fell in the range 50–500 (Fraser and Matthews, 1979a). The numbers of particles actually supplied in the inoculum were of the order of 100 to 1000 times these numbers. Thus, the efficiency of inoculation of protoplasts may not be intrinsically any higher than mechanical inoculation of leaves. Furthermore, the amount of virus binding to protoplasts is influenced by factors such as temperature, pH, ionic strength and the presence of added compounds such as polyethylene glycol (Roenhorst et al., 1988).

The number of particles actually entering a cell for each successful infection can be measured only with micro-injection. For example, Halliwell and Gazaway (1975) obtained an ID_{50} of 310 TMV particles injected in 1 pL into single tobacco cells.

2. Proportion of particles containing an infectious genome

In the absence of an efficient method for inoculation, it is not possible to obtain unequivocal estimates of the proportion of infectious particles in a virus preparation, but Furumoto and Wildman (1963b) concluded that at least one in 10 of the TMV particles in a purified preparation was infectious. The use of infectious constructs or infectious transcripts should provide a higher proportion of infectious molecules, but these unencapsidated nucleic acids are prone to degradation by nucleases before they can reach the site of initial replication.

3. Can one infectious genome give an infection?

It is generally agreed that for many viruses only a single infectious particle or infection unit is needed actually to infect a cell and give rise to the visible lesion (e.g. Boxall and MacNeil, 1974). Theoretical consideration of the dilution curve is consistent with these data. Experiments with protoplasts indicate that about one TMV particle is sufficient to infect one protoplast (Takebe, 1977). Reddy and Black (1973) found that two to three WTV particles were sufficient to infect a cell in insect vector cell monolayers. However, the results from protoplasts may not be directly applicable to inoculation of whole leaves.

In practice, a large number of viral particles is needed for each successful infection, and a major factor is almost certainly the inefficiency of the process as it is usually performed. There are probably several reasons for this: (1) it is very likely that only a very small proportion of epidermal cells have potentially infectible wounds made in the leaf surface above them; (2) the lifetime of the infectible wounds is short, and many of the viral particles in the fluid above a wound never make contact with the site; (3) the distribution of the virus applied over the leaf surface is probably very uneven in terms of surface areas of the size of single cells; (4) much of the virus may be adsorbed to inactive sites on the leaf surface and remain trapped there; and (5) some viral particles may enter potentially infectible cells, but not become successfully established, unless 'rescued' in some way. The existence of such centers has been shown for some host–virus combinations (Rappaport and Wu, 1963).

F. Mechanical transmission in the field

Compared with transmission by invertebrate vectors or vegetative propagation, field spread by mechanical means is usually of minor importance. However, with some viruses it is of considerable practical significance. TMV can readily contaminate hands, clothing and implements, and be spread by workers and, for instance, birds in tobacco and tomato crops (Broadbent, 1963, 1965a; Broadbent and Fletcher, 1963). This is particularly important during the early growth of the crop, for example during the setting out of plants. Plants infected early act as sources of infection for further spread either during cultural operations as disbudding or by rubbing together of healthy and infected leaves by wind. TMV may be spread mechanically by tobacco smokers, because the virus is commonly present in processed tobacco leaf. For example, all 37 brands of cigarette sold in West Germany contained TMV (Wetter, 1975) and the virus was detected in 60 of 64 smoking tobaccos (Broadbent, 1962).

PVX can also be readily transmitted by contaminated implements or machines and by workers or animals that have been in contact with diseased plants (Todd, 1958). On some materials, the virus persists for several weeks.

PVX can spread either by direct leaf contact between neighboring plants (Clinch et al., 1938) or between neighbors when leaves are not in contact. This has been assumed to be due to mechanical inoculation by contact between roots, but a soil-inhabiting vector has not been excluded. Some viruses of fruit trees have been shown to be spread in orchards by cutting tools (e.g. Wutscher and Shull, 1975). Field trials have suggested that RCMV can be readily transmitted by mowing machines (Rydén and Gerhardson, 1978). On the other hand, Heard and Chapman (1986) concluded that in mown ryegrass swards most of the local spread of RGMV was due to transmission by mites rather than the mowing operation.

G. Abiotic transmission in soil

1. Above ground

Allen (1981) showed that, in glasshouses, TMV could be transmitted from soil containing the virus coming into contact with leaves. Some other stable viruses may well be transmitted in this way.

2. Below ground

Several viruses may infect roots from virus-contaminated soil apparently without fungi or arthropods being involved, for example TMV (Broadbent, 1965b), TBSV (Kleinhempel and Kegler, 1982), CRSV (Brown and Trudgill, 1984) and SBMV (Teakle, 1986).

Levels of infection of glasshouse tomatoes from ToMV-containing plant debris was affected by inoculum concentration (Pares et al., 1996). As the inoculum concentration decreased, infection levels decreased and a greater proportion of infections were symptomless or were restricted to the roots.

H. Summary and discussion

Mechanical inoculation is used for many purposes in plant virology such as host range studies, assessing the infectivity of a virus preparation and diagnosis of a virus. The outcome of the inoculation is, to a great extent, dependent upon the condition of the plants being inoculated. As a generality, the inoculated plants should be horticulturally 'soft'. This does not necessarily apply to natural conditions, but it is only a few viruses that are transmitted mechanically in the field. These viruses, such as TMV and PVX, occur at high concentrations in infected plants and can be spread to adjacent plants through broken hairs. We should also remember that humans are very effective at transmitting viruses (and viroids) through cutting tools and through various agronomic practices.

As described in Section II.A.3 there is diurnal variation in the susceptibility of plants to viral infection. Plants have circadian expression of some genes (see Beator and Kloppstech, 1996) and it would be interesting to know whether there was any link between this and their diurnal susceptibility.

The excretion of viruses in the guttant from a range of plants points to another possible route of virus infection. McKinney (1953) suggested that human activity in pastures could transmit BMV between plants. The recent finding (Ding et al., 2000) that BMV is present in guttation fluid from infected barley plants suggests a source of virus for mechanical transmission by the activities of humans and other animals. Plant pathogenic bacteria are transmitted through guttation fluid that is withdrawn into the plant through hydathodes (see Carlton et al., 1998; Hugouvieux et al., 1998). However, it is not known whether virus infection could result from entry by this route as the virus particles would have to find a means of movement from the apoplast or xylem vessels to the cytoplasm of susceptible cells.

III. DIRECT PASSAGE IN LIVING HIGHER PLANT MATERIAL

A. Through the seed

About one-seventh of the known plant viruses are transmitted through the seed of at least one of their infected host plants. Seed transmission provides a very effective means of introducing virus into a crop at an early stage, giving randomized foci of primary infection throughout the planting. Thus, when some other method of transmission can operate to spread the virus within the growing crop, seed transmission may be of considerable economic importance. Viruses may persist in seed for long periods so that commercial distribution of a seed-borne virus over long distances may occur.

Table 12.1 lists the approximate frequency with which seed transmission has been found among some viruses of various groups and among viroids.

Two general types of seed transmission can be distinguished. With TMV in tomato, seed transmission is largely due to contamination of the seedling by mechanical means. This type of transmission may occur with other tobamoviruses. The external virus can be readily inactivated by certain treatments eliminating all, or almost all, seed-borne infection. A small but

TABLE 12.1 Relative importance of seed transmission for viruses of various virus groups

| Virus group | No. of members | | Type of potential injury[a] | | | | | | Seed transmission (%)[b] |
	In group	Seed-borne	A	B	C	D	E	F	
Alfamovirus	1	1	+	+	+				1–23
Bromovirus	6	1	+	+	+				+
Capillovirus	4	1							1–60
Carlavirus	60	2							2–90
Carmovirus	18	2							10–40
Caulimoviridae	34	1[c]							Up to 100
Closterovirus	28	1							+
Comovirus	15	6	+	+	+				1–90
Cryptovirus	31	31							100
Cucumovirus	3	3	+	+	+				<1–1
Dianthovirus	5	0							
Enamovirus	1	1							1–2
Fabavirus	4	0							
Geminivirus	102	1							+
Hordeivirus	4	1	+	+	+		+		+
Ilarvirus	17	8				+			1–90
Luteovirus	7	0							
Marafivirus	3	0							
Nepovirus	40	17	+	+	+				3–100
Plant reovirus	14	0							
Potexvirus	36	4							1–6
Potyvirus	179	16	+	+	+	+	+	+	<1–80
Rhabdovirus	15	1							+
Sobemovirus	14	4							1–80
Tenuivirus	11	0							
Tobamovirus	17	7	+	+	+				1–20[d]
Tobravirus	3	3	+	+	+				1–35
Tospovirus	13	1							Up to 95
Tombusvirus	13	1							+
Tymovirus	23	3							+
Viroids	15	5				+	+		+

Data from Stace-Smith and Hamilton (1988), with permission, and from AAB Descriptions of Plant Viruses (see Appendix 3); note that not all members of groups tested for seed transmissibility.
[a] A, survival of inoculum; B, dispersal of inoculum; C, primary inoculum source; D, contamination of germplasm lines; E, contamination of virus-free planting material; F, direct crop losses due to plants arising from infected seed.
[b] A plus sign (+) indicates that no percentage value was given.
[c] BSV, PVCV and TVCV are apparently seed-transmitted in their respective hosts, but probably by activation of integrated viral sequences.
[d] Seed transmission of TMV probably due to contamination.

variable proportion of the seed may be infected in the endosperm where virus may persist for many years. No TMV has been detected in the embryo of tomato (Broadbent, 1965c; Lartey et al., 1997) or of Arabidopsis (Filho and Sherwood, 2000). The seed transmission of MNSV is assisted by the fungal vector Olpidium bornovanus (Campbell et al., 1996).

In the second and more common type of seed transmission the virus is found within the tissues of the embryo. The developing embryo can become infected either before fertilization by infection of the gametes (indirect embryo invasion or gametic transmission) or by direct invasion after fertilization (Johansen at al., 1994: Maule and Wang, 1996). Many viruses use both processes in the production of infected seed.

Some infections cause disease symptoms in the seed (see Chapter 3) but there is not necessarily a correlation between seeds showing symptoms and those transmitting the virus. Small seeds have been found to transmit PSbMV at a much higher rate than larger seeds (Khetarpal et al., 1988).

1. Factors affecting the proportion of infected seed

a. Virus

The proportion of infected seed from infected plants varies quite widely with different viruses (Table 12.1). It may be as much as 100%, for example TRSV in soybean (Athow and Bancroft, 1959) or SLRSV in celery (Walkey and Whittingham-Jones, 1970). By contrast, 1% transmission was found for APLV in potato (Jones and Fribourg, 1977), and lettuce plants infected with LMV may produce about 3–15% of seed giving rise to infected plants (Couch, 1955).

b. Viral determinants

Strain specificity of seed transmissibility is shown by the observation that different isolates of a seed-borne virus may have different seed transmission rates in the same host (e.g. Bennett, 1969; Shepherd, 1972; Frosheiser, 1974; Adams and Kuhn, 1977; Hanada and Harrison, 1977; Hampton and Francki, 1992). This enables experiments on the viral determinants to be conducted by reassortment or pseudo-recombination of divided genome segments or by molecular techniques such as sequence comparison and mutagenesis or the formation of chimeric viruses. These experiments indicate that the determinants of seed transmissibility are complex.

Pseudo-recombination experiments with the two RNAs of some nepoviruses showed that seed transmissibility in *Stellaria media* was markedly dependent on some virus function carried by RNA1. RNA2 had an additional but smaller influence (Hanada and Harrison, 1977). In similar experiments with PEBV, RNA1 was also the major determinant, with RNA2 or its products playing a minor role in determining seed transmissibility (Wang *et al.*, 1997a). The removal of the 12-kDa gene from RNA1 almost completely abolished seed transmission. This 12-kDa deletion mutant accumulated poorly in anthers and carpels, and could not be detected in pollen grains or ovules suggesting that this gene is involved in the infection of the gametic cells (Wang *et al.*, 1997a).

Pseudo-recombinants between the ND18 (seed transmitted) and CV17 (not seed trans-mitted) strains of BSMV showed that RNAγ had a predominant effect on seed transmission of this virus; there were also more subtle effects from RNAβ (Edwards, 1995). The major determinants of seed transmission in RNAγ included the 5'- untranslated leader, a 369 nt repeat in the γa and γb genes. There is a complex interaction between the RNAγ leader, the γb gene and RNAs α and β. These results indicated that viral replication and movement play a pivotal role in the seed transmission of BSMV.

Construction of hybrids between PSbMV isolate DPD-1 (highly seed transmitted) and NY (poorly seed transmitted) showed that the 5' NTR, the HC-Pro and the coat protein regions of the genome affected seed transmission, with the HC-Pro having the major influence (Johansen *et al.*, 1996). As with BSMV, these regions are important for virus replication and movement.

c. Host plant

Some viruses are seed transmitted by a wide range of host species. For example, TBRV was seed transmitted by all nine species tested in six families (Lister, 1960). Some viruses are transmitted at different rates in different hosts, for example AMV in 10% of the seed of *Melilotus indica*, 2% in *Stachys arvensis* and 0.1% in *Ornithopus compressus* (McKirdy and Jones, 1994). Other viruses may be seed transmitted in one host but not in another. Thus, *Dodder latent mosaic virus* was transmitted through 5% of seed from infected *Cuscuta campestris*, but not through seed from cantaloupe, buckwheat or pokeweed.

Different varieties of the same host species often vary widely in the rate at which seed transmission of a particular virus occurs. For example, LMV does not appear to be transmitted through the seed in the variety Cheshunt Early Giant (Couch, 1955). Grogan and Bardin (1950) found rates of transmission ranging from 1% to 8% in other varieties. Seed transmission rates reported for BSMV in different barley cultivars have varied from 0% to 75% (Carroll and Chapman, 1970). There is also variation between plants of a given variety.

d. Time at which plant is infected

Generally speaking, for infection of the embryo from the mother plant, the earlier that the plant is infected, the higher the percentage of seed that will transmit the virus (e.g. Owusu *et al.*, 1968; Ren *et al.*, 1997). One exception to this trend appears to be BSMV in barley, where the percentage of infected seed rose steadily as time of infection was delayed, reaching a maximum at 10 days before heading. After this time the percentage declined (Eslick and Afanasiev, 1955).

Crowley (1959) examined the effect of time of flowering in relation to time of inoculation on the proportion of bean seeds (*P. vulgaris*) infected with SBMV. In the bean plant, self-fertilization takes place as the flower opens, and timing of this event in relation to inoculation time was recorded. Embryos were removed from the seed before they were fully mature, and tested for infectivity. The virus was able to infect the embryo both by infecting the gametes and by infecting the embryo, but only during the early stages of its development (up to about 4 days after fertilization). Crowley found a similar timing for TRSV in soybean and for BSMV in barley.

Obviously, for indirect embryo invasion by pollen the infection takes place at pollination.

e. Location of seed on plant

There appears to be no consistent pattern in the way infected and non-infected seeds are distributed on the plant. For example, Athow and Laviolette (1962) found that position of seed in the pod and location of pod on the plant did not affect the proportion of soybean seeds that were infected with TRSV. In these experiments, the plants had been infected for a long time. For plants that set seed in succession over a period, infection occurring near or during the flowering period might give a distribution in which older seeds would have less infection than younger seeds.

There was no difference in the rates of seed transmission of SMV in mottled and non-mottled seeds (Pacumbaba, 1995).

f. Age of seed

Some viruses appear to be lost quite rapidly from seed on storage, while others persist for years. Fulton (1964) described a loss of virus from the seed of *Prunus pennsylvanica* carrying Cherry necrotic ring spot virus. For the first 4 years of storage at 2°C the percentage remained fairly constant at 60–70%. By the sixth year, less than 5% of the seed was infected, while loss of viability in the seed was minor. There was little loss of AMV in infected alfalfa seeds after 5 years at –18°C or room temperature (Frosheiser, 1974).

g. High temperature

Well-dried seed is much more resistant to high temperatures than most other plant parts. Some seed-borne viruses appear to tolerate about as high a temperature as the seed they infect. TRSV survived for 5 years in soybean seed at 16–32°C as effectively as at 1–2°C, although seed germination was greatly reduced at the higher temperature (Laviolette and Athow, 1971). The reason for the resistance of these viruses in the seed is quite obscure. It may be due to the general stabilizing effect on intact virus particles of low water and high protein content.

h. Host resistance

Although there are several examples of varietal differences in the ability to seed transmit viruses, there have been few studies on the inheritance of the characters involved. A single recessive gene conditions the resistance to seed transmission of BSMV in barley (Carroll *et al.*, 1979). However, as was pointed out by Maule and Wang (1996), this study did not assess whether the resistance was due to affecting gametic transmission or direct embryo invasion, both of which mechanisms operate with this virus. Seed transmission of PSbMV is by direct embryo invasion and, as the potential resistance or susceptibility to transmission introduced into progeny by cross-pollination had no effect on the final seed transmission efficiency, it is controlled in maternal tissues rather than the progeny (Wang and Maule, 1994). The characters involved in seed transmissibility of this virus in peas are incompletely dominant and segregate in the F2 generation as a quantitative trait, probably involving only a few genes.

Although no host genes controlling seed transmissibility have been identified, various predictions have been made on the properties of their products (Maule and Wang, 1996). These include: (1) genes that control the ability to invade gametic tissues being related to the ability to infect meristematic tissues (e.g. cryptic viruses, which are ubiquitously distributed throughout the host plant are very efficiently seed transmitted; Kassanis *et al.*, 1978); (2) genes that control the survival of gametes and embryos in the presence of the virus; (3) host factors controlling virus multiplication and movement; and (4) host factors involved in embryo maturation which affect virus longevity.

Obviously, the genetics of the factors controlling seed transmissibility are complex. They must relate to the mechanisms by which the embryo acquires the virus and probably vary between different virus–host combinations.

2. Distribution of virus within the seed

Viral transmission due to surface contamination of the seed or virus in the seed coat is rare and is essentially found only with viruses, such as TMV, which are very stable (reviewed in Johansen *et al.*, 1994). Thus, most seed transmission occurs when the embryo becomes infected (Johansen *et al.*, 1994).

3. Mechanism of seed transmission

The embryo can be infected by two routes: directly from the mother plant or by pollen (reviewed by Johansen *et al.*, 1994; Maule and

Wang, 1996). The direct route from the mother plant poses problems in that symplastic connection is severed at meiosis. By either route, the pathways leading to embryo infection are complex and involve a range of genetic and environmental interactions (Maule, 2000). To infect the embryo, the virus has to reach either the floral meristems, which are beyond the limits of normal long-distance movement in the phloem (see Chapter 9), or the embryo itself. Using *in situ* hybridization and immunohistochemistry, Wang and Maule (1997) compared the temporal and spatial accumulation of PEBV and PSbMV in pea embryos and distinguished two routes of infection.

Gametic infection. Some viruses, for instance PEBV, nepoviruses and cryptic viruses, can infect floral meristems and, hence, the gametes (Fig. 12.7). Details of the mechanism are unknown.

Direct embryo infection. This route has been examined for PSbMV in detail (Wang and Maule, 1994). The virus moves through the testa of the immature seed after fertilization and must reach the micropylar region of the seed for embryo infection to occur. The micropyle is in close contact with the base of the embryonic suspensor, which functions as a conduit for nutrient flow to support growth of the embryo. The suspensor is the route by which the virus invades the embryo itself (Fig 12.8) but it degrades as part of the seed development program. This leaves a 'window of opportunity' for embryo infection. However, there is no symplastic connection between

Fig. 12.7 (page opposite) (see Plate 12.1) Detection of PEBV in the embryo and gametes of pea. To show the tissue distribution of PEBV, sections of immature seeds **(a–c)**, isolated embryos **(d, e)**, anthers **(f–h)** and ovules **(i–k)** were subjected to immunohistochemistry **(a–c, f–k)** with Fast Red TR as the chromogenic substrate for detection, or *in situ* hybridization **(d, e)** with BCIP/NBT as substrate. For immature seeds, the location of PEBV **(a, b)** with respect to the testa (TE), globular embryo (E) and multinucleate suspensor (SU) was revealed after staining the section with DAPI **(c)**. Sections of Stage 5 embryos from infected **(d)** and uninfected **(e)** plants were treated with DIG-labeled RNA probes for (+)-sense PEBV RNA. Viral RNA was present uniformly throughout the embryo, except at the junction (arrow) between the axis and the cotyledon. Transverse sections through anthers were immunostained with anti-PEBV serum **(f, g)** or pre-immune serum **(h)**. PEBV was present in the vascular tissues of the anther filament (F), the tapetum (T) and the mature pollen grains (P). Longitudinal sections through unfertilized ovules were immunostained with PEBV-specific **(i, j)** or pre-immune **(k)** serum. The accumulation of PEBV in the integuments (I) and ovule sac (OS) relative to the position of the synergid (S), polar (PN and egg cell (EN) nuclei were revealed by staining the same section with DAPI **(j)**. C, cotyledons; E, globular embryo; EA, embryonic axis; EC, endospermic cytoplasm; EN, egg cell nucleus; ES, embryo sac; F, anther filament; I, integument; OS, ovule sac; P, pollen grains; PN, pollen nucleus; PV, provascular tissue; S, synergids; SU, suspensor; T, tapetum; TE, testa. Bar markers: **(d, e)** 1 mm, others 100 μm. From Wang and Maule (1997), with kind permission of the copyright holder, © Blackwell Science Ltd.

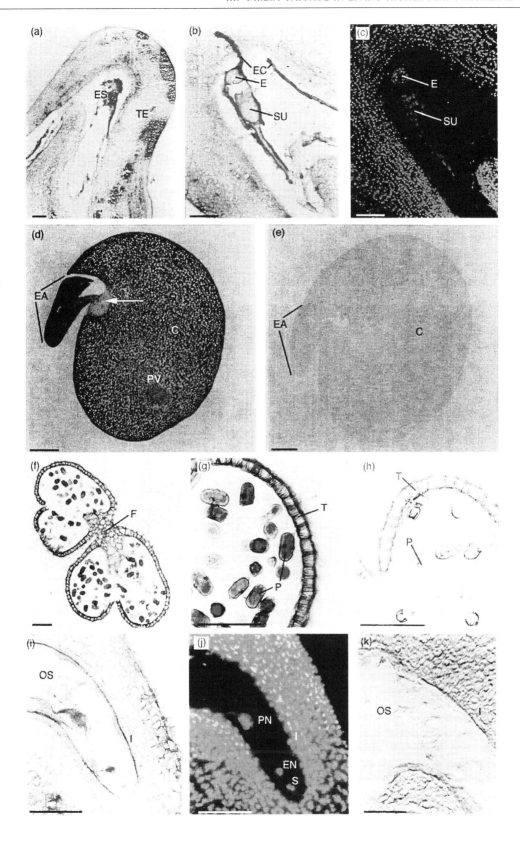

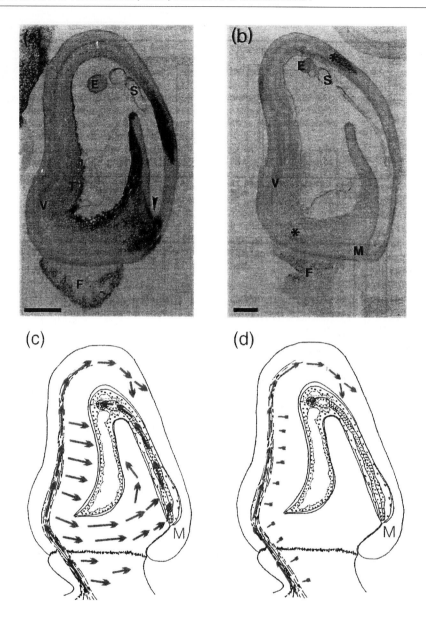

Fig. 12.8 (see Plate 12.2) The pathway to seed transmission of PSbMV in pea. **(a, b)** Analysis of the distribution of PSbMV in longitudinal sections through immature pea seed by immunohistochemistry using a monoclonal antibody to PSbMV coat protein shows that a cultivar–virus interaction which is permissive for seed transmission (e.g. with *Pisum sativum* cv. Vedette in **(a)**) results in a widespread accumulation of the virus in the testa tissues. In contrast, in the non-permissive interaction (e.g. with cv. Progretta in **(b)**) virus enters the seed through the vascular bundle but is unable to invade the adjacent testa tissues extensively. In both cases, there is a gradual reduction in the amount of accumulated virus after invasion such that in cv. Progretta only patches (asterisks) of infected tissue remain detectable. Systematic analyzes of the immature seeds of different ages have identified the routes (red arrows) of virus invasion in the two cultivars (illustrated diagrammatically in **(c)** for cv. Vedette and in **(d)** for cv. Progretta). The most consistent observation from all these studies is that the virus must reach the micropylar area of the testa for seed transmission to occur, a location providing the closest point of contact (arrowhead in **(a)**) between the testa tissues and the embryonic suspensor. In the non-permissive interaction the virus appears to be blocked (denoted by the red squares) in its ability to spread into and/or replicate in the non-vascular testa tissues. E, embryo proper; F, funiculus; M, micropylar region; S, suspensor; T, testa; V, vascular bundle. Bar markers 500 μm in panels **(a)** and **(b)**. From Maule and Wang (1996), with kind permission of the copyright holder, © Elsevier Science.

maternal and embryonic tissue, and it is still unknown how the virus crosses from the maternal testa cells to the embryonic suspensor.

Comparison of the spatial and temporal distribution of PsbMV in pea variety Vedette (60–80% seed transmission) with that in variety Progretta (0% seed transmission) indicated that, in Progretta the distribution of the virus did not take advantage of this window of opportunity (Wang and Maule, 1994) (Fig. 12.8)

4. Transmission through infected pollen

Some viruses are transmitted from plant to plant via pollen (Table 12.2). AMV is more efficiently transmitted through pollen than the ovules (Frosheiser, 1974). In contrast, while there was 5% transmission of LMV through the ovule in lettuce, there was less than 0.5% infected seed produced by pollen transmission (Ryder, 1964). Self-pollination of infected plants presumably can result in a higher percentage of infected seed than when only one of the gametes comes from an infected individual. Crosses between healthy and infected *Arabidopsis* plants indicated that TYMV can invade the seed from either the male or the female parent but that invasion from maternal tissue was the only route for seed invasion by TMV (de Assis and Sherwood, 2000). As with seed transmission, two mechanisms appear to operate in pollen transmission: gametic infection of the embryo and direct infection of the mother plant.

BSMV (Carroll, 1974) and TRSV (Yang and Hamilton, 1974) particles have been found by electron microscopy within infected pollen grains, and those of PNRSV by RT-PCR, dot blot and *in situ* hybridization (Aparicio *et al.*, 1999). Particles of BSMV were seen within the sperm cells in both the cytoplasm and nuclei, and were found to be widely distributed in cells and tissues associated with embryogenesis (Brlansky *et al.*, 1986). In another study, immuno-gold labeling showed AMV particles to be widely distributed in ovules, pollen and anthers of alfalfa (Pesic *et al.*, 1988). For such viruses, the way in which the egg cell is infected is presumably via the infected sperm, or perhaps the sperm nucleus alone. On the other hand, PNRSV particles were found in the

TABLE 12.2 Viruses and viroids that appear to be spread to other plants by pollen

	Viruses or viroid	Plant
Virus group		
Ilarvirus	BlShV	Blueberry
	PDV	Stone fruit
	PNRSV	Stone fruit
	TSV	Various
Nepovirus		
	AYRSV	Artichoke
	BLMoV	Blueberry
	CLRV	Walnut, betula
Sobemovirus		
	SoMV	Chenopodium
Idaeovirus		
	RBDV	Raspberry
Viroid	ASSVd	Apple
	ASBVd	Avocado
	CSVd	Chrysanthemum, tomato
	CEVd	Citrus, tomato
	CCCVd	Coconut
	CbVd-1	Coleus
	GYSVd-1	Grapevine
	HSVd	Cucumber, grapevine, hop, tomato[a]
	PSTVd	Potato, tomato

[a] Depends on strain of viroid.
From Mink (1993), with permission.

cytoplasm of the vegetative cell but not in the generative cell, which indicates that infection of the embryo is not by the sperm cell. The exine of mature pollen from infected plants has been shown to carry several viruses, such as TMV, occurring in high concentration (Hamilton *et al.*, 1977). These observations point to a second mechanism for pollen transmission. The germ tubes growing from the infected pollen grain may pick up virus particles, or actually become infected by mechanical means, and thus carry active virus to the ovule. Cryptoviruses are unusual in that they are transmitted with high efficiency through pollen and seed, but not by mechanical transmission, grafting or invertebrate vectors (Lisa *et al.*, 1986). It is suggested that the ability of viruses to infect pollen is related to their ability to invade meristematic regions (see Maule and Wang, 1996).

In some cases, the mother plant itself may become infected. For example, Gilmer (1965) recorded an experimental tree-to-tree transmission of Sour cherry yellows virus by pollen. The pollen may be carried by humans, wind or honey bees (e.g. Converse and Lister, 1969). Natural transmission in the field may be via infected pollen and no other means, as was found for RBDV in raspberry by Murant *et al.* (1974). Francki and Miles (1985) showed that if leaves on to which pollen from plants infected with SoMV had fallen were subjected to mechanical abrasion, then virus transmission occurred. However, *Thrips tabaci* and other thrips species have been implicated in the pollen transmission of several viruses (Hardy and Teakle, 1992; Sdoodee and Teakle, 1993; Klose *et al.*, 1996). Sdoodee and Teakle (1993) showed that pollen infected with TSV is carried both externally and internally in the thrips and is thought to infect through feeding wounds.

The extent to which infected pollen is a significant factor in the spread of viruses in the field has not been thoroughly assessed. It may well be more important economically with cross-pollinated woody perennials than with annual crops. With certain viruses, infected pollen may cause only the resulting seed to become infected when a healthy plant is pollinated.

B. By vegetative propagation

Vegetative propagation is an important horticultural practice, but is unfortunately a very effective method for perpetuating and spreading viruses. Economically important viruses spread systemically through most vegetative parts of the plant. A plant once systemically infected with a virus usually remains infected for its lifetime. Thus, any vegetative parts taken for propagation, such as tubers, bulbs, corms, runners and cuttings, will normally be infected. There are many instances where every individual of a particular cultivar tested has been found infected with a particular virus, for example some potato cultivars infected with PVX. However, when a healthy plant of a vegetatively reproducing species is infected, even at a fairly early stage of growth, the virus may not move throughout the plant in the first growing season.

C. By grafting

Grafting is essentially a form of vegetative propagation in which part of one plant grows on the roots of another individual. Once organic union has been established, the stock and scion become effectively a single plant. Where either the rootstock or the individual from which the scion is taken is infected systemically with a virus, the grafted plant as a whole will become infected if both partners in the graft are susceptible. Early descriptions of graft transmission are noted in Chapter 1. Since the early days of work with plant viruses, the demonstration that a disease was transmissible by grafting, together with the absence of a pathogen visible by light microscopy, has been taken as an indication that the disease is due to a virus. Many viruses once thought to be transmissible only by grafting are now known to be transmitted by other means as well.

Procedures for grafting can be found in Dijkstra and de Jager (1998).

Grafting transmission may lead to a different disease from that appearing after, say, mechanical inoculation. For example, *N. glutinosa* normally gives necrotic local lesions with no systemic movement of virus following mechanical inoculation with TMV. However, healthy plants grafted with tobacco plants infected systemically with TMV die from a systemic necrotic disease which is probably due to the introduction of virus into the vascular elements of the hypersensitive host (Zaitlin, 1962).

Grafting may succeed in transmitting a virus where other methods fail. Nevertheless, it is not always an efficient process. This may sometimes be due to lack of complete systemic invasion in the plant supplying the supposedly diseased scions. Where tissue of the healthy plant material being grafted gives a necrotic local reaction to the virus, transmission of the disease may not be accomplished. For example, Chamberlain *et al.* (1951) found that when healthy buds from Burbank and Sultan plums were grafted on to plum rootstocks infected with the vein-banding type of plum mosaic, the buds reacted with necrosis and died. This is probably the reason why these varieties were very rarely found infected with the virus in the field. Under standard conditions, different viruses are transmitted in different minimal times after grafting. Minimum times of bud contact to give 100% transmission varied from 74 to 152 hours for 12 *Prunus* viruses studied by Fridlund (1967b). Unaided graft formation is a somewhat uncommon occurrence, and this method of virus transmission is not of wide importance in nature. However, the formation of viable unions between roots may be significant in the spread of viruses, especially in some perennial hosts.

D. By dodder

Dodder (*Cuscuta* spp.) (*Convolvulaceae*) is a parasitic vine on higher plants. There are many different species with different host ranges, some of which are extensive. Bennett (1940b) showed that dodder would transmit viruses from plant to plant. The parasite forms haustoria, which connect with the vascular tissues of the host. Viruses are probably transmitted via the plasmodesmata which transiently connect the parasite's hyphal tips with host-cell cytoplasm.

Transmission by dodder is in some respects similar to grafting. However, graft compatibility is limited to quite closely related plants — usually within a genus. Dodder, on the other hand, can be used to transmit a virus between distantly related plants (e.g. Desjardins *et al.*, 1969). The virus being transmitted experimentally may not multiply in the dodder, which then appears to act as a passive pipeline connecting two plants. Transmission of TMV was substantially increased by conditions (such as pruning the dodder and shading the healthy plant) that might be expected to lead to a flow of food materials through the dodder from the diseased to the healthy plant (Cochran, 1946). Bennett (1940b) was able to separate CMV from TMV because it persisted in the dodder when the parasite was grown on hosts immune to both viruses, whereas the TMV was lost.

Dodder used in transmission studies may sometimes harbor an unsuspected virus. Thus, Bennett (1944) found that symptomless *Cuscuta californica* was frequently infected with a virus he called *Dodder latent mosaic virus*, which caused serious disease in several unrelated plant species.

One of the main experimental uses of dodder transmission has been to transfer viruses from hosts where they are difficult to study to useful experimental plants. Dodder is probably an insignificant factor in the transmission of economically important viruses in the field, and has rarely been used in experimental work in recent times.

E. Summary and discussion

Seed transmission is important in the epidemiology of some viruses, notably some with relatively immobile vectors such as nematodes. As will be described in Section IV.A.2, several nematode-transmitted viruses are seed transmitted in annual weeds, giving a means of longer distance dispersal. Seed transmission is also significant in providing sites of primary infections for some aphid-transmitted viruses such as LMV and SMV.

Only certain viruses are seed transmitted and then often only in certain host species. The work of Maule and colleagues described in Section III.A.3 has provided an understanding of some of the factors that limit the entry of a virus into the embryo. It is likely that other factors include the rate and route of systemic spread of the virus and the interaction of the virus with the host's general defense system.

IV. ECOLOGY AND EPIDEMIOLOGY

In order to survive, a plant virus must have (1) one or more host plant species in which it can multiply, (2) an effective means of spreading to and infecting fresh individual host plants, and (3) a supply of suitable healthy host plants to which it can spread. The actual situation that exists for any given virus in a particular locality, or on the global scale, will be the result of complex interactions between many physical and biological factors, the major ones of which are shown in Fig. 12.9.

In this section, I shall consider briefly the more important of these, with illustrations of the ways they can interact to affect the survival and distribution of plant viruses. An understanding

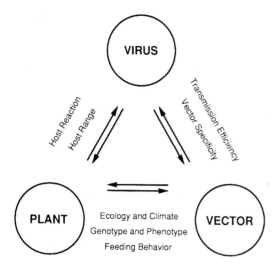

Fig. 12.9 Some of the interactions involved in the natural infection of a plant by a virus. From Hull (1991b), with permission.

of the ecology and epidemiology of a virus in a particular crop and locality is essential for the development of appropriate methods for the control of the disease it causes. As with most other obligate parasites, the dominant ecological factors to be considered are usually the way viruses spread from plant to plant and the ways in which other factors influence such spread. Aspects of ecology and epidemiology are discussed by Maramarosch and Harris (1981) and Harrison (1981) and for specific virus groups by Irwin and Thresh (1990), Wisler *et al.* (1998), Robert and Lemaire (1999a), Robert (1999), Dewar and Smith (1999), Lecoq (1999) and Burgess *et al.* (1999); points raised for specific viruses are applicable to many other viral situations.

We should distinguish between the terms epidemiology and ecology, which in many papers have been used interchangeably. *Ecology* describes the factors influencing the behavior of a virus in a given physical situation. These factors include host range, tissue tropism, pathogenesis and host responses. It is a fundamental concept based on the relational properties of the virus and is not a property of the environment. *Epidemiology* is the study of the determinants, dynamics and distribution of viral diseases in host populations. It includes a dimensional aspect of the factors determining the spread of a virus into a given situation. The

two terms often interlink but each has a specific meaning. In this section, I will consider them both together.

A. Biological factors

1. Properties of viruses and their host plants

a. Physical stability of viruses and concentrations reached

For viruses that depend on mechanical transmission, a virus that is stable, both inside and outside the plant, and that reaches a high concentration in the tissues is more likely to survive and spread than one that is highly unstable. The survival and spread of certain viruses appear to depend largely on a high degree of stability and the large amounts produced in infected tissues. For example, TMV may survive for long periods in dead plant material in the soil, which is then a source of infection for subsequent crops (Johnson and Ogden, 1929). TMV was recovered from 42 brands of cigarettes in Germany by Wetter and Bernard (1977); the virus yield was 0.1–0.3 mg/g tobacco. TMV from cigarette tobacco had about one-half the specific infectivity of that isolated from fresh leaf.

High concentrations of virus at a particular season may be important. *Stellaria media* is an overwintering weed host for CMV. Walkey and Cooper (1976) found that CMV reached its highest concentration in this host at low temperatures. Thus, the plant would provide a very good source for acquisition of virus by aphids in the spring.

b. Rate of movement and distribution within host plants

Viruses or virus strains that move slowly through the host plant from the point of infection are less likely to survive and spread efficiently than those that move rapidly. Speed of movement will be important as measured relative to the lifetime of individual host plants. Viruses that infect long-lived shrubs or trees can afford to move much more slowly through their hosts than those affecting annual plants. Viruses that can move into the seed and survive there have an important advantage in spread and survival. TNV is confined to the roots in most hosts and, in nature, it depends on transmission through the soil and infection of new hosts with the aid of fungal vectors.

c. Severity of the disease

A virus that kills its host plant with a rapidly developing systemic disease is much less likely to survive than one that causes only a mild or moderate disease which allows the host plant to survive and reproduce effectively. There is probably a natural selection in the field against strains that cause rapid death of the host plant. As discussed in Chapter 17, it is likely that viruses have co-evolved with their natural hosts and that many crop diseases are severe because there has not been sufficient time for a stable relationship to have evolved. Leaf-hoppers living on desert plants of the western United States are infective primarily with strains of BCTV that cause mild symptoms in beet. Virulent strains of BCTV kill certain desert plant species before they can allow a generation of leafhoppers to mature. Thus, virulent strains in the desert tend to be self-eliminating (Bennett, 1963).

However, disease severity has a very different effect in beet plantings. Sugar beet is a good host for the leafhopper if the plants are small and exposed to full sunlight. They are poor hosts when large, providing a lot of shade. For this reason, severe strains of BCTV facilitate their own spread in beet by producing small, stunted plants, which favor vector multiplication (Bennett, 1963). The spread of severe strains is further assisted by the fact that mild strains do not protect against infection. Leafhoppers overwintering near beet fields carry strains of higher virulence than hoppers found in the desert.

d. Mutability and strain selection

The extent to which a virus is able to mutate to produce strains that can cope effectively with changes in the environment may well affect the survival and dispersal of the virus. It is difficult to make valid comparisons of rates of mutation between different viruses, but they probably vary quite considerably. For example, PLRV seems relatively stable (although this may merely reflect lack of experimentation or appropriate techniques for this virus), while many others such as CMV and TSWV exist in nature as numerous strains. There is fairly good evidence that different strains of the same virus may vary in the rate at which they produce a certain type of mutant. Mild strains of PVX isolated from potato frequently give rise to ringspot strains in tobacco, but the TBR strain originally isolated from tomato was never observed to do so (Matthews, 1949a). The common TMV can give rise to a wide range of mutant types (Bawden, 1964).

As discussed in Chapter 17 (Section III), various examples are known where particular host plants allow the selective multiplication of certain strains of a virus when presented with a mixture. Likewise, invertebrate vectors have sometimes been shown to transmit some strains of a virus more efficiently than others (Chapter 11). These experiments illustrate ways in which strains might become selected under natural conditions.

Different tobacco species and varieties in the field may be infected naturally with different strains of TMV. It is probable that a major factor leading to the dominance of a particular strain in a particular host is the rate at which it can invade the plant systemically and thus exclude other strains (Chapter 17).

In districts where an annual crop such as tobacco has been farmed for many years, characteristic strains may come to dominate. Thus, Johnson and Valleau (1946) described how different dominant strains of TMV tended to be characteristic of different tobacco farms in old tobacco-growing areas. On a wider scale, geographical isolation may tend to lead to divergence of strains, particularly where climatic conditions are different. Such geographical variation in the dominant strain type is not uncommon (however, see Chapter 17, Section XIII.B).

One strain of a virus may remain dominant for many seasons in a particular crop and region if there is a stable natural reservoir host for the strain in that region. In regions with high year-round temperatures, strains of viruses may be adapted to grow at such temperatures. However, where summer temperatures are very high, viruses may actually be inactivated *in vivo*. For example, SCV was eliminated from strawberry plants grown in the Imperial Valley of California (Frazier *et al.*, 1965). Season appeared to be a factor in the dominance of two strains of CMV in tomato and pepper crops in south-eastern France. A thermosensitive strain dominated in spring, whereas a thermoresistant strain was prevalent in summer (Quiot *et al.*, 1979a). BYDV-related viruses (then considered to be strains) can be distinguished by severity of disease and by the species of aphid that transmits them. The dominant 'strains' in New York State changed gradually over 20 years from isolates of the vector-specific MAV type to those similar to PAV 'strains' (Rochow, 1979). By contrast, in a given crop and season there may be no detectable variation in the pathogenicity of the virus over wide areas, for example BPMV in soybean (Ross and Butler, 1985).

Agricultural practices may influence in many different ways the strains of virus that become dominant in a crop. For example, on the estate in Scotland where the Arran varieties of potato were bred, cultural and selection practices unintentionally led to virtually all promising seedlings being infected with PVX at an early stage through contact with old commercial varieties heavily infected with the virus. As seedlings were multiplied by vegetative propagation, individuals showing mosaic symptoms were discarded, ensuring that the variety came to contain a predominantly mild strain of PVX (Matthews, 1949b). An outbreak of rapidly lethal virus disease may well indicate a new and unstable relationship between host and virus. When potato plants free of PVX were inoculated with virus obtained from another variety showing no obvious disease, they showed no symptoms in the year of infection. In the following year, the disease produced varied widely, even in different shoots from the same tuber. Some shoots became invaded by strains causing severe necrotic disease and death, while others showed mild mottling or no symptoms (Matthews, 1949a). A period of natural and artificial selection would no doubt have given a line of potatoes again showing only a mild type of infection.

The use of an avirulent strain of TMV (derived from strain 1) for providing cross-protection against ToMV in commercial tomato plantings (Chapter 16, Section III.A) led to a marked increase in the appearance of strain 1 infections in susceptible cultivars (Fletcher and Butler, 1975).

Where there is a source of virus for an annual agricultural crop in perennial weeds and wild hosts, successive crops may become infected with strains that never have the opportunity to become well adapted to the crop plant.

When agricultural operations or other factors do not bring about sudden changes, we can envisage, for any particular host species and environment, that selection of strains of a virus will occur until those best adapted for survival will dominate. Factors important for survival of a strain will include (1) efficient transmission by insects or other means, (2) more rapid multiplication and movement in the plant than for any competing strains, and (3) the production of mild or only moderately severe disease. These factors have been studied experimentally for BSMV in barley by Timian (1974).

e. *Plant host range of viruses*

The general distribution of viruses in the plant kingdom is considered in Chapters 2 and 17. Here I shall consider the ways in which the host range of a virus may affect its survival and distribution in the field.

Viruses vary greatly in the range of species they are able to infect. Some viruses affecting strawberries appear to be confined to the genus *Fragaria*. Other viruses may be able to infect a wide range of plants. For example, the host range of CMV includes over 1000 host species in more than 85 families. Other viruses with very narrow host ranges presumably survive either because their host is perennial, or because it is vegetatively propagated, or because the virus is transmitted efficiently through the seed.

A diversity of hosts gives a virus much greater opportunities to maintain itself and spread widely. Viruses that have perennial ornamental plants as hosts as well as other agricultural and horticultural species have become widespread around the world.

Important examples among ornamental flowering species are (1) BYMV and CMV carried in gladioli, (2) TSWV carried in dahlia and other species, which is frequently infected and is a major reservoir of this virus in many countries (dahlia may also carry CMV, often without symptoms) and (3) CMV carried in lilies, often without symptoms.

Weeds, wild plants, hedgerows, and ornamental trees and shrubs may also act as virus reservoirs. The actual importance of these various sorts of host for neighboring crops will depend on circumstances, particularly on the presence of active invertebrate vectors. For example, the presence of OkMV in three malvaceous weeds in Nigeria appeared to be an important source of the virus for crop plants, because infective beetle vectors were shown to be active (Atiri, 1984). On the other hand, although ACLSV was found in a significant proportion of hedgerow hawthorn plants in England, they appeared to be of little significance for spread of the virus to fruit trees (Sweet, 1980). Among the weed hosts, *Plantago* species may be an important potential virus reservoir. They are efficient and adaptable perennial weeds with a worldwide distribution. They have been found infected naturally with at least 26 viruses from 19 groups and families (Hammond, 1982). The natural hosts of EHBV and RHBV occupy different ecological niches and thus, although EHBV can infect rice, it does not under field conditions (Madriz *et al.*, 1998). Different strains of a virus may preferentially infect certain weed species. For example, in the south-east of France, thermo-sensitive strains of CMV were found preferentially in *Rubia peregrina* while thermoresistant strains predominated in *Portulaca oleracea* (Quiot *et al.*, 1979b). The important and complex role of weeds in the incidence of virus diseases is discussed by Duffus (1971).

Many nematodes and the viruses transmitted by them have wide host ranges, including woody perennial plants. Such viruses and their vectors can often survive in woody plants in hedges and forests in the absence of suitable agricultural crop plants. GFLV and its nematode vector *Xiphinema index* are unusual in that they are both confined largely to the grape plant in the field. Because the grape vine is long-lived, alternative hosts may not be necessary for survival. Furthermore, vector and virus can survive for several years in viable roots that may remain in the soil after the grapevine tops have been removed.

2. Dispersal

Dispersal of viruses by air-borne or soil-inhabiting vectors, by seed and pollen, and over long distances by human activities plays a key role in the ecology and epidemiology of viruses.

In many infections of annual crops and in some cases with perennial crops, there are two phases in the epidemiology or spread of a virus. Primary infections are brought in by winged vectors, or are due to a few plants infected by, say, seed transmission. This is followed by secondary spread, which may be by winged vectors flying locally between plants or by wingless vectors walking from plant to plant when leaves are in contact. Frequently, secondary spread of viruses with non- or semi-persistent relationships with their vectors is by winged insects on their host-seeking flights, whereas spread of persistent viruses is by insects that colonize that plant species. Primary infections followed by secondary spread often leads to patches of infected plants within a crop with gradients of infection from the primary foci (Fig. 12.10).

a. Air-borne vectors

Taking plant viruses as a whole, the flying, sap-sucking groups of insect vectors, particularly the aphids, are by far the most important agents of spread and survival. The pattern of spread in the crop and the rate and extent of spread will depend on many factors, including (1) the source of the inoculum—whether it comes from outside the crop, from diseased individuals within the crop arising from seed transmission or through vegetative propagation, from weeds or other plants within the crop, or from crop debris; (2) the amount of potential inoculum available; (3) the nature and habits of the vector—for example, with aphids, whether they are winged transients or colonizers; (4) whether virus is non-persistent, semi-persistent or persistent in the vector; (5) the

Fig. 12.10 (see Plate 12.3) Aerial view of a sugar beet field with patches of plants infected with virus yellows (BYV and/or BMYV) and gradients of infection spreading from the patches. (Courtesy of IACR Broom's Barn Experimental Station.)

time at which vectors become active in relation to the lifetime of the crop; and (6) weather conditions.

In early work attempting to relate aphid numbers to virus spread, counts were made of aphids actually on plants at different times during the season. Relationships between such numbers and the spread of virus were often not apparent. Doncaster and Gregory (1948) made a significant contribution when they pointed out the importance of migrant winged aphids, particularly those that move through the crop early in the season. The size of the largely static populations that build up on individual plants depends to a great extent on the local weather and other conditions within the crop. Populations can increase at a very rapid rate (about 10-fold in 7 days) and population density may vary widely even in a small region within a crop. Subsequent work has confirmed the general importance of aphid migrations early in the season. For example, Heathcote and Broadbent (1961) placed potatoes growing in pots and infected with PLRV or PVY in plots or in the field for successive periods through the season and noted aphid numbers and subsequent virus incidence. The viruses spread from the infected plants early in the season when there were few aphids colonizing plants, but not later in midseason when wingless aphids were numerous. See Fig. 12.10 for another example of the importance of early infestations.

With different viruses, even those transmitted by members of the same group of vectors, rates and patterns of spread in a crop can vary widely. Two different patterns of spread are illustrated in Fig. 12.11. A virus brought into a crop from outside by vectors does not necessarily spread from the first infected plant in a crop to other plants. Little spread within the crop may occur if the invasion occurs late in the growing season, or if the presence of the vectors in the crop is very transient. This sort of situation is illustrated by WMV-2 spread in Fig. 12.11. This represents part of a melon field that was initially free of WMV-2, and was then infected from a source several hundred meters away.

The rapid accumulation of CABYV-infected plants (Fig. 12.11) shows that plant-to-plant spread within the crop had taken place with this virus.

Van der Plank (1946) developed a method for testing whether virus was spreading from diseased plants within a crop. It is based on the assumption that virus coming from outside the crop will infect plants at random. On this basis, there will be a certain expectation of the frequency with which infected plants will occur side by side as pairs:

$$p = X[(X - 1)/n]$$

where p is the expected number of infected pairs, n is the number of consecutive plants

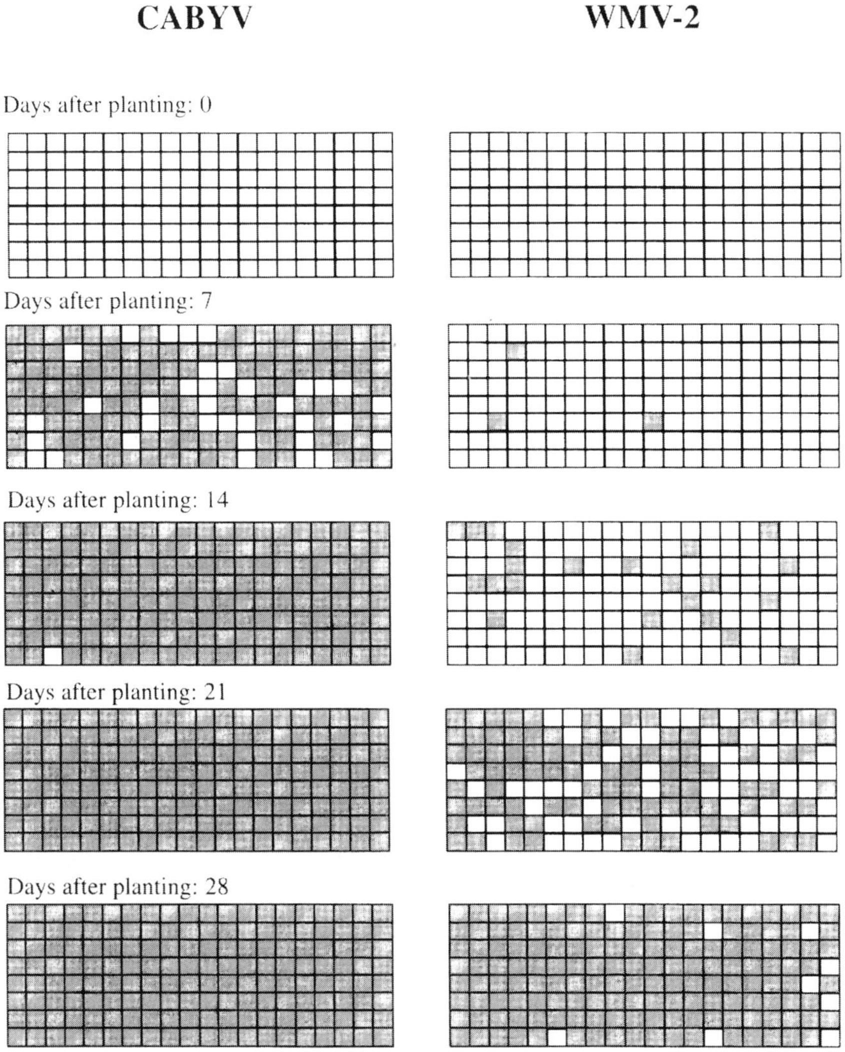

Fig. 12.11 Comparative spatial spread of two viruses transmitted by aphids either persistently (CABYV) or non-persistently (WMV-2) in a melon field (Avignon, 1992). Every plant was tested at weekly intervals for infection by CABYV or WMV-2 by ELISA tests. Individual melon plants are represented by white squares, which are shaded when found infected by one of these two viruses. After 4 weeks, most plants were doubly infected. From Lecoq (1999), with permission.

examined, and X is the number of infected plants observed. For high values of n, the standard error of p is $\sqrt{p}$.

If the observed number of pairs (counting three plants together as two pairs) is significantly greater than expectation, then spread from infected plants within the field can be assumed. Spread within a crop need not necessarily favor neighboring plants, so that an apparently random distribution cannot exclude the possibility of such spread. The pattern of infection brought about in a field by one vector species may be confused if two forms of the vector are active at the same time, one causing spread to neighbors and the other 'jump' trans-

mission to plants at a distance. More sophisticated procedures are now available for analysing the extent of randomness or clumping (patchiness) in the distribution of infected plants during an epidemic (e.g. Madden and Campbell, 1986; Madden et al., 1987a,b).

Various kinds of traps have been developed for assessing the number of flying aphids, including yellow pan traps, vertical sticky traps, conical nets and suction traps. The horizontal mosaic green pan trap designed by Irwin is proving useful (Irwin and Ruesink, 1986). Different types of traps may give somewhat different answers with different aphid species, and the height at which the traps are set will affect the data obtained. In general, the frequency of flying aphids decreases with increasing height, and relative numbers of vector species caught in traps may not correspond to relative numbers found on the crop (e.g. Tatchell *et al.*, 1988). Furthermore, trapping alone merely gives an estimate of numbers of aphids of various species and the morphs involved and not the numbers actually infective for particular viruses. Transient winged forms of aphids that do not colonize the crop at all, but move from plant to plant, may be especially important with stylet-borne viruses. Although they may form a small proportion of the total population, they may introduce virus from outside or acquire it from infected plants within the crop and spread it rapidly as they move about seeking suitable food plants. Colonizers will also be important if they move about within the crop.

Ideally, from the point of view of virus transmission, we want to know the numbers of infective aphids that are flying and that will land on the crop of interest. Sensitive techniques, such as PCR, can detect virus in individual aphids (see Stevens *et al.*, 1997) but even these do not tell us whether the aphid is able to transmit the virus. To determine this may be a very tedious and labor-intensive task, especially when non-persistent viruses are involved. Trapped aphids then have to be placed on healthy test plants as soon as possible after capture. Nevertheless, as they land in a crop, winged aphids can be collected, identified and tested on appropriate plants to determine whether they are carrying particular viruses. For example, Ashby *et al.* (1979) used wind sock traps to assess numbers of viruliferous vectors of subterranean Clover red-leaf virus. On average over the season 1972–1977, 40–50% of alate *Aulacorthum solani* (Kltb.) were found to be carrying this virus.

Using similar traps in a potato field, Piron (1986) collected 101 aphid species. Twenty-three species transmitted PVY^N, and 22 of these were recorded as new vectors of the virus. This experiment illustrates the fact that a lot more information than is presently available may be needed about vectors active in the field in relation to the ecology of particular viruses in particular locations.

The 'infection pressure' to which a crop is subjected can also be assessed by placing sets of bait plants in pots within the field for successive periods of a few days. The sets of plants are then maintained in the glasshouse and observed for infection. Figure 12.12 illustrates such a trial in relation to the prevalence of potential vector species. The frequency with which aphids were trapped clearly implicates the early flights of *Cavariella aegopodii* as being of prime importance in the spread of PVY in this particular field and season. Measurement of vector activity has been discussed by Irwin and Ruesink (1986). If particular virus strains are of interest it is important to know whether the bait plant being used is preferentially infected by one out of two or more strains that are present (Marrou *et al.*, 1979).

Most of the preceding discussion has been concerned with variation in the number of plants infected with a virus. In addition to this aspect, the severity of the disease in individual plants may be substantially increased by an increased dose of virus provided by numerous aphids (Smith, 1967) (see Fig. 12.3). Disease is also generally more severe when younger plants are infected.

The number of aphid species known to transmit different viruses varies widely. For example, SCV has two vector species, while more than 60 vector species have been recorded for CMV.

The fact that the majority of epidemiological studies have been carried out on aphid-borne viruses reflects the importance of this group of vectors. Other groups of air-borne vectors such as hoppers and beetles are, of course, also of crucial importance in both the epidemiology and ecology of the viruses they transmit. Beetles may differ from aphids and hoppers in the pattern of virus incidence they bring about. For example, flea beetles differ significantly from the aphids in their patterns of movement within a crop. They are generally much more active than aphids. They tend to move from plant to plant over short distances much more frequently than wingless aphids, but they do not usually travel as far as winged forms. Figure 12.13 shows the pattern of spread of TYMV by flea beetles in a field of turnips

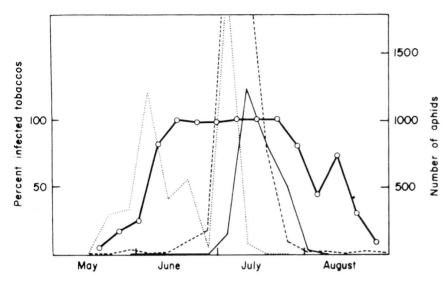

Fig. 12.12 Infection pressure of PVY in a field of potatoes. Batches of 100 tobacco plants in pots were set out in the field for 7 days. The left-hand axis and the solid line show the percentage of these plants that subsequently developed infection with PVY in the glasshouse. The number of aphids of three species trapped each week is given by the axis on the right. ○ — ○, PVY^N in tobacco as a percentage; —, *Myzus persicae*; -----, *Rhopalosiphum padi*; ......, *Cavariella aegopodii*. From van Hoof (1977), with kind permission of the copyright holder, © Kluwer Academic Publishers.

growing adjacent to a block of plants infected mechanically with the virus. The initially healthy field was divided into strips parallel to the block of infected plants. A strong gradient of infection developed in the field (Markham and Smith, 1949). Similar gradients of infection have been observed for spread by aphids under some conditions.

b. Soil-borne viruses

i. Viruses with no known vector

TMV is one of the few viruses transmitted throughout the soil to any significant extent without the aid of any known vector. The stability of this virus allows it to survive from season to season in plant remains, provided conditions are suitable. A susceptible crop planted in contaminated soil will become infected, presumably through small wounds made in the roots during transplanting cultivation, or root growth. Viruses with no known vectors, such as members of the *Potexvirus* and *Tobamovirus* genera, are widespread in soils of forest ecosystems in Germany (Büttner and Nienhaus, 1989).

ii. Viruses with fungal vectors

The ecological implications of transmission by fungal vectors depend on the way in which the fungus carries the virus. Viruses such as SBWMV and PMTV are carried in the resting spores of their plasmodiophoromycete vectors.

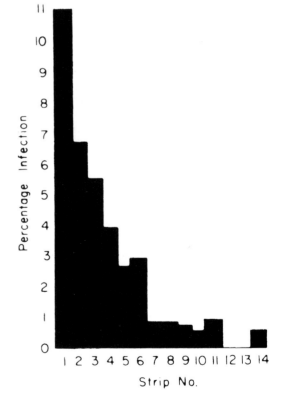

Fig. 12.13 Gradient of infection due to spread by flea-beetles. Percentage infection in each strip of plants is plotted against distance from the source of infection, which was next to strip no. 1. Natural spread of TYMV in turnips. Each strip was 2 m wide. From Markham and Smith (1949), with kind permission of the copyright holder, © Elsevier Science.

In these spores, the viruses may survive in air-dried soil or in soil that has been stored for many years. Build-up of infection may take several seasons but, once established in a field, viruses with this type of vector may persist for many years even in the absence of suitable plant hosts (e.g. Jones and Harrison, 1972). Localized spread of these viruses is by zoospores and resting spores, soil water being a major factor influencing local movement. Viruses with this type of transmission tend to have rather narrow host ranges.

Viruses transmitted by *Olpidium brassicae*, such as TNV, STNV and CuNV, are carried on the surface of the spores. Zoospores carrying virus probably survive for only a few hours. However, virus that is free in the soil may be picked up and transmitted by newly released zoospores. If the soil is moist, *Olpidium* zoospores can recover TNV from soil for at least 11 weeks.

Viruses transmitted by *Olpidium* do not survive in air-dried soil. In general, viruses of this type have wide host ranges and probably survive in the soil by frequent transmission to successive hosts. Drainage water and movement of soil and root fragments are probably important in the spread of these viruses from one site to another. In model experiments in which viruses were added to a soil-less recirculating water supply, several viruses, including some such as TMV with no known vectors were transmitted (Paludan, 1985).

Polymyxa betae, the fungal vector of BNYVV, forms very persistent cytosores. These were transmitted to infected soil, and both cytosores and BNYVV could pass through the digestive system of the sheep (Heijbroek, 1988).

iii. Viruses with nematode vectors

The ecology and epidemiology of viruses with nematode vectors (tobraviruses and nepoviruses) differs considerably from that of viruses with air-borne vectors. Nematodes are long lived, may have wide host ranges, and are capable of surviving in adverse conditions and in the absence of host plants for considerable periods.

Vector nematodes do not have a resistant resting stage, but can survive adverse soil conditions by movement through the soil profile. As soils become dry in summer or cold in winter, they move to the subsoil and return when conditions are favorable. Some of the viruses (e.g. ArMV) may persist through the winter in the nematode vector.

Nematode-transmitted viruses usually have two other characteristics in common: a wide host range, especially among annual weed species, and seed transmission in many of their hosts. Thus, nematode vectors that lose infectivity during the winter may regain it in the spring from germinating infected weeds (Murant and Taylor, 1965).

The spread of nematodes in undisturbed soil may be rather slow. Harrison and Winslow (1961) calculated that a population of *Xiphinema diversicaudatum* invaded uncultivated woodland at the rate of about 30 cm per year. Agricultural practices will increase distribution of the vectors during cultivation and probably in drainage or floodwater. Farm workers' footwear and machinery contaminated with infested soil may transmit nematodes and their associated viruses over both short and long distances (Boag, 1985). However, the pattern of infection observed in a crop may often depend largely on the vector and virus situation in the soil before the crop was planted. In a field already infected with nematodes and planted with a biennial or perennial crop, it may take 1 or 2 years before the initial pattern of infection becomes apparent, since leaf symptoms may not show until about a year after infection. Subsequent spread may give rise to slowly expanding patches of infected plants within the field (Fig. 12.14).

Individual woody trees or hedges of long-lived woody species have been implicated as a reservoir of vector nematodes (e.g. Fig. 12.15). Sometimes such species may harbor a virus as well.

c. Seed and pollen transmission

The occurrence of and mechanisms for seed and pollen transmission were discussed in Section III.A. These two methods of transmission may be of crucial importance in the ecology and epidemiology of certain viruses. Survival in the seed can be particularly important for

Fig. 12.14 Picture of an outbreak of RRSV in Talisman strawberry. The patchy distribution of stunted plants reflects the patchy distribution of the vector *Longidorus elongatus*. (Courtesy of Scottish Horticultural Research Institute.)

viruses that have only annual plants as hosts, and for those that have invertebrate vectors, such as nematodes that normally move only slowly.

CMV persists in seeds of *Stellaria media* buried in soil for at least 12 months (Tomlinson and Walker, 1973). If 10^7 seeds were present per hectare, with 1% infected with CMV, then a 10% emergence of seedlings would give rise to about one infected seedling per square meter. Under such conditions, a rapid build-up of CMV infection in a crop might result. LMV in lettuce is an outstanding example of the importance of seed transmission in a commercial crop.

Human activities in the dispersal of virus-infected seed are noted in Section IV.A.2.d. Natural dispersal of seeds by wind or water may also be a factor. Seed output per plant may be very large for wind-borne seeds, but wastage is also high.

Transmission by pollen may be the major, and perhaps only, method of natural spread in the field for certain viruses, for example PNRSV, On the other hand, pollen dispersal is probably of little or no ecological significance when the host plant is mainly or entirely self-pollinated, for example, barley. However, Shepherd (1972) pointed out that some barley varieties produce large amounts of pollen, which if infected with BSMV could lead to mechanical transmission by abrasion between leaves.

d. Dispersal over long distances

i. Human dispersal

Even in countries that have had a highly developed agricultural technology for some time, it may be very difficult or impossible to document the arrival and spread of particular viruses. However, it is widely accepted that over the past few centuries humans have been mainly responsible for the wide distribution around the world of many viruses that were previously localized in one or a few geographical areas.

Viruses have been transported in plants or plant parts and perhaps occasionally in invertebrate vectors. There is little doubt that many of the virus diseases of potato and some of their vectors were brought to Europe with the potato from America and have since been spread to many other countries in tubers (e.g. Jones and Harrison, 1972). LMV, because it is seed transmitted, has probably been distributed wherever this crop is grown. The fact that TMV can survive in infectious form in prepared smoking tobacco is probably sufficient to account for its presence wherever tobacco is grown commercially.

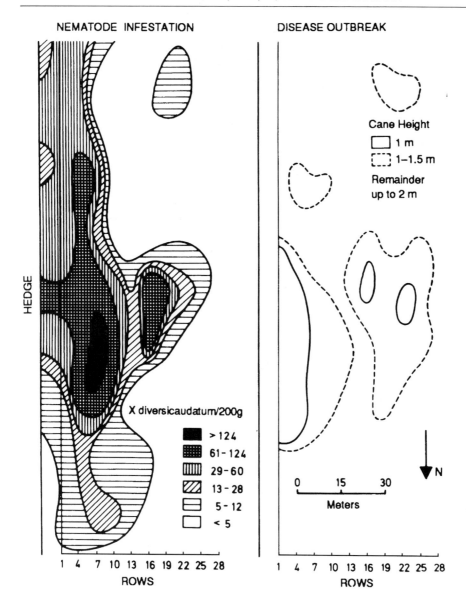

Fig. 12.15 Relationship between density of nematode infestation and an outbreak of virus disease. *Left*: Population contours of *Xiphinema diversicaudatum*. *Right*: Outbreak of SLRSV in a 10-year-old raspberry plantation. The virus causes a reduction in cane height. Plants were spaced 1.8 m between rows and 9.6 m within rows. From Taylor and Thomas (1968), with permission.

The spread of some viruses may be a more complex problem, involving the necessity for spread of a suitable invertebrate or fungal vector as well as the virus and appropriate host plants. For example, Raski and Hewitt (1963) consider that humans have been largely responsible for the dispersal of GFLV (and its nematode vector) around the world as grape vines were taken from place to place.

In relation to worldwide dispersal of viruses, New Zealand is an interesting example, becuase it is geographically perhaps the most isolated of the countries that have a diverse and modern agriculture and horticulture. Polynesian immigrants introduced a few food plants, for example taro (*Colocasia esculenta*) and kumara (*Ipomea batatas*). Over the past 170 years or so, European immigrants have introduced a wide range of agricultural and horticultural crop plants, together with a large selection of weed species. By 1990, 139 viruses had been recorded in the country, all present in introduced species (see Chapter 16, Section II.E.6). These have almost all been identified as viruses occurring elsewhere, mainly Europe or North America, with a few from Australia.

It is easy to see how most of these could have arrived in tubers, corms, runners, rootstocks, seed, and so on. For example, 38 of the 139 viruses infect vegetatively propagated fruit trees and vines.

Most of the vectors for these viruses are introduced species of aphids. New Zealand is very deficient in endemic aphids, their place in the fauna being taken by *Psyllidae*. It is a curious fact that few or no viruses with leafhopper vectors are yet known in New Zealand, although representatives of some potential vector genera occur.

There are other examples where the effects of human activities can be more precisely dated. PPV infects *Prunus* spp. and is transmitted in a non-persistent manner by aphids. The disease was reported first in Bulgaria in 1915. It has since spread through Europe, as indicated in Fig. 12.16. Aphids spread the virus to nearby orchards. Long-distance spread is most probably by means of vegetative plant material distributed by humans (Walkey, 1985).

The more recent movement of BNYVV, which causes the important rhizomania disease in sugar beet, is summarized in Table 16.1. A virus arriving in a new environment may sometimes find conditions suitable for very rapid spread. Stubbs (1956) pointed out that *Carrot motley dwarf virus* disease spread very rapidly in Australia, where the aphid vector *Cavariella aegopodii* was already plentiful as an introduced species. It also lacked natural enemies. By

contrast, spread was slow in California, where the aphid vector was heavily parasitized.

Human activities may spread both virus and vector within a country. For example, soils of nurseries in southern England frequently contain nepoviruses and their vectors, which can be transported with the host plants, especially if the infection is latent (Sweet, 1975).

ii. Airborne vectors

Aphid vectors are important in the long-distance as well as local spread of viruses. This may be true for non-persistent as well as persistent viruses. For example, detailed studies of aphid movements in the UK show that many important aphid species re-colonize the whole island each year, as long as appropriate host plants are available (Taylor, 1986). This movement involves distances of up to 1000 km, although not usually in one flight. Several successive colonizations may be involved.

On the other hand, under appropriate climatic conditions, a continuous long-distance journey may not be uncommon. Geostrophic airstreams are air movements at altitudes of about 1000 m or more moving along the isobars of a relatively large-scale weather system. Such airstreams have almost certainly led to the mass transport of winged aphids from Australia to New Zealand, a distance over sea of about 2000 km (Close and Tomlinson, 1975). LNYV and several vector species of aphids have probably been introduced in this way. An example of the potential effects of weather on aphid movement is illustrated in Fig. 12.17. Not only the viruses but also potential aphid vectors are still moving into new continents. Thus *Metopalophium dirhodum*, a vector for BYDV, was first recorded in Australia in 1985 and was almost certainly a fairly recent arrival (Waterhouse and Helms, 1985).

Leafhopper vectors may also travel long distances. For example, large numbers of *Macrosteles fascifrons* may be blown each spring from an overwintering region about 300 km north of the Gulf of Mexico through the midwestern United States and into the prairie provinces of Canada (Chiykowski and Chapman, 1965). BYDV and cereal aphids follow similar routes.

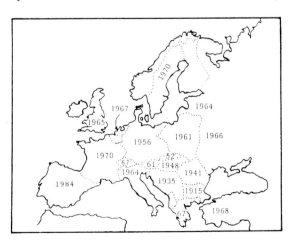

Fig. 12.16 First reports of plum pox *Potyvirus* in Europe and western Asia. Dates signify the year the virus was first recorded. (Courtesy of J.M. Thresh.)

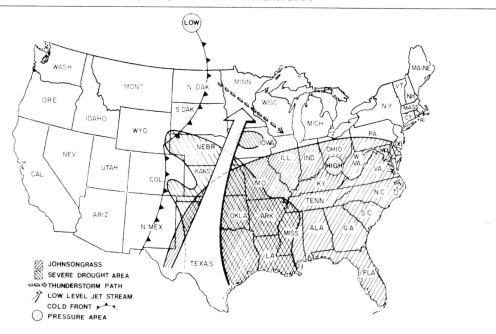

Fig. 12.17 Hypothesis to account for long-distance aphid vector movement and a massive outbreak of MDMV in corn in Minnesota in 1977. The continental United States with the mid-continental and southern distributions of Johnsongrass, the major wild host for MDMV (western distributions not shown), relative to the severe drought during the months of May and June 1977. The large arrow indicates the path of low-level jet winds (up to 80 km/hour) on 2 July 1977. The smaller blocked arrow indicates the path of a cold front that caused the low-level jet winds to become diffuse and triggered thunderstorm activity, which moved in a south-southeasterly direction through Minnesota and into Wisconsin in the late evening of 2 July and early morning of 3 July 1977. From Zeyen *et al.* (1987), with permission.

Pollen and seed dispersal. Pollen and wind-borne seed transmission may sometimes be important over shorter distances. Seed-transmitted viruses might be transported over considerable distances by birds, but there is no documented example. Proctor (1968) has shown that viable seeds may persist in the alimentary tract of migratory seabirds for as long as 340 hours, long enough to be transported several thousand kilometers.

Water dispersal. Over many years, methods have been developed for screening water for the presence of human and animal viruses. These have also been applied to a search for water-borne plant viruses, especially in Europe, with some surprising results. The work has been reviewed by Koenig (1986). Most sampling has been done in rivers and lakes. Infectious viruses isolated from such waters include TMV, CGMV and other tobamoviruses; a potexvirus; TNV and STNV; TBSV, CIRV and other tombusviruses; CarMV; CMV; and CRSV. In

addition, CarMV was isolated from Baltic Sea water (Kontzog *et al.*, 1988). Many of these viruses share several properties. Most of them are very stable and lack air-borne vectors that would allow spread over long distances. They occur in high concentrations in infected plants and are released from infected roots and can infect roots without vectors. Many have a wide host range. Most of the infectious virus probably moves in water while adsorbed on to organic and inorganic colloidal particles, especially clays (Koenig, 1986). In this state, they would be substantially more resistant to inactivation than as free virus.

The viruses found in water originate by release of virus from living roots, or from decaying plant material, or from sewage. Tomlinson and Faithfull (1984) isolated TBSV from waters of the River Thames in England. This virus and others were isolated from a large number of tomato plants growing in semi-solid or dried sludge in the primary settlement beds of various sewage plants. They concluded that these tomato plants and at least some of the viruses were derived from infected tomatoes eaten raw by humans. It is known that TBSV can pass in an infectious state through the human alimentary tract (Tomlinson *et al.*, 1982). CRSV was isolated from water of a canal near a sewage plant in northern Germany, again implicating sewage in possible long-distance dissemination (Koenig *et al.*, 1988b). These water-borne viruses may be a factor in the forest decline occurring in Europe, but further research is needed on this question (Koenig, 1986; Büttner *et al.*, 1987).

3. Cultural practices

The way a particular crop is grown and cultivated in a particular locality and the ways land is used through the year in the area may have a marked effect on the incidence of a virus disease in the crop. This may, of course, offer the opportunity to prevent infection by appropriate cultural practices. Many diverse situations arise. Some examples will illustrate the kinds of factors involved.

a. Planting date

A clear-cut example of the way time of sowing seed affects virus incidence occurs with the winter wheat crop in southern Alberta. In a normal season, the percentage infection with streak mosaic is markedly dependent on the time at which the seed is sown (Slykhuis *et al.*, 1957). For sowing dates earlier than September, the spring-sown crop, carrying disease, overlaps with the autumn-sown crop. High air temperatures may cause a dramatic reduction in some aphid vector populations. If the planting date for the autumn crop is delayed until such conditions prevail, much less virus spread may take place in a crop. Another example of the importance of planting date is given in Fig. 12.18. The importance of the planting date in relation to vector migrations and control measures is illustrated in Fig. 16.6.

Changes in planting practice that provide an overwintering crop may cause an increased virus incidence. Thus, the increased incidence of BaYMV in the UK has been attributed, in part, to the growing of winter malting barley as

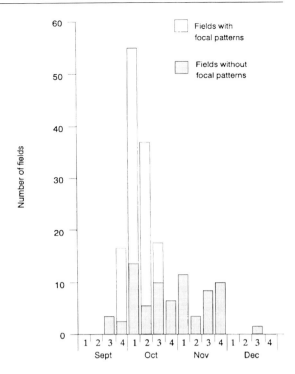

Fig. 12.18 Effect of time of sowing on BYDV incidence in wheat fields in the southern UK in 1976–1977, as estimated from aerial observation of foci of infection. Unshaded columns represent fields with obvious foci of infection; shaded columns are fields apparently free from infection. From Hill (1987), with kind permission of the copyright holder, © Blackwell Science Ltd.

a profitable spring crop, leading to increased perpetuation of the virus through its fungal vector (Coutts, 1986). The increase in BaYMV has also been associated with a switch from spring-sown to autumn-sown barley in various parts of the UK. In Washington State, the highest incidence of BYDV in winter wheat crops occurred in irrigated areas that supported aphid vectors during the summer, and with early planting dates (Wyatt *et al.*, 1988).

Planting date is also important in tropical situations. For example, from studies on planting dates of rice in Maros and Lanrang (Indonesia) for 7 years, Manwan *et al.*, (1987) identified the best time to plant to avoid Tungro virus disease. These were based on the times that the leafhopper vector populations were low and climatic conditions were suitable for planting rice.

b. Crop rotation

The kind of crop rotation practiced may have a marked effect on the incidence of viruses that can survive the winter in weeds or volunteer plants. With certain crops, volunteer plants that can carry viruses may survive in high numbers for considerable periods. Doncaster and Gregory (1948) showed that it may take 5 or 6 years to eliminate volunteer potatoes from a field in which potatoes had been grown.

For a crop such as carrots, fields for seed production and for root production together with volunteer plants may form a continuous yearly cycle enabling viruses to be perpetuated (Howell and Mink, 1977). The importance of volunteer plants in the spread of beet viruses is illustrated in Fig. 12.19. With perennial crops, virus incidence will tend to increase as the crop ages.

c. Soil cultivation

Soil cultivation practices may affect the spread and survival of viruses in the soil or in plant remains. Nematode and fungal vectors may be spread by movement of soil during cultivation. The extent to which soil is aerated and kept moist may affect the survival of TMV in plant remains. Cultivation during frosty weather of

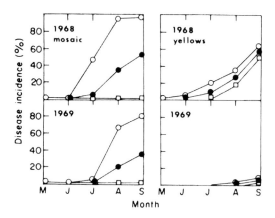

Fig. 12.19 Relationship between volunteer sugar beets and the occurrence of BtMV and BWYV in beet fields in Washington State during 1968 and 1969. ○, fields containing volunteer plants; ●, fields adjacent to volunteer plants; □ fields at least 1 mile from known volunteer plants. From Howell and Mink (1971), with kind permission of the copyright holder, © The American Phytopathological Society.

fields previously in potatoes may greatly reduce the survival of volunteer potato plants (Doncaster and Gregory, 1948).

It has been thought that BNYVV was spread at a significant level by irrigation. However, a comparison of the spread of the virus by irrigation with spread by the physical movement of soil tillage and harvesting operations showed that the latter was the major means of spread of the virus from a point source (Harveson et al., 1996).

d. Field size

The influence of the size of field on the spread of a virus will depend to a great extent on the source of initial infection. If virus is coming from outside the crop, then, as pointed out by Van der Plank (1948), concentration of the crop into large fields of compact shape will reduce infection from outside the crop to a minimum. For example, this situation has been shown to occur with AMV in lucerne (Gibbs, 1962).

e. Population density and plant size

Air-borne vectors bringing a virus into a crop from outside will infect a greater proportion of the plants in a given area when they are widely spaced than when they are close together. For example, the incidence of beet yellows was reduced where the distances between plants or between rows was reduced (Blencowe and Tinsley, 1951). Crop spacing may affect the landing response of flying aphids. Closely spaced groundnuts were not visited by alate *Aphis cracivora* as frequently as widely spaced plants (Hull, 1964). The frequency of alates being found in the crop decreased significantly after the plants had met within and between rows. Some species were trapped more frequently over widely spaced crops of cocksfoot and kale (A'Brook, 1973).

Large plants in a crop might be expected to become infected more readily with insect-borne viruses than small ones, because they are more likely to be visited by a vector. Broadbent (1957) found that this held in cauliflower seedbeds. In 1 year, 30% of large seedlings 15% of medium-sized seedlings and 5% of small seedlings were infected with CaMV.

f. Effects of glasshouses

It might be expected that the use of glasshouses and polythene tunnels would favor the survival of a stable virus such as TMV, because the structures remain at one site and are in use for intensive cultivation. On the other hand, they might provide some protection against aphid-borne viruses. Thus, Conti and Masenga (1977) reported that in the Piedmont, in pepper crops under plastic tunnels, 84% of virus-infected plants contained TMV. The remaining 16% were infected with aphid-borne viruses. By contrast, 88% of virus-infected pepper plants grown outdoors were infected by viruses transmitted by aphids in a non-persistent manner. Glasshouses are normally used in regions with cool winters and will therefore favor the introduction of viruses more adapted to tropical and subtropical climates, such as with TSWV.

g. Pollination practices

The horticultural practice of planting mixtures of varieties or pollinators in orchards may favor the spread of pollen-borne viruses. Howell and Mink (1988) present circumstantial evidence suggesting that the annual appearance of new sites of infection with pollen-borne PNRSV in a sweet cherry orchard was caused by the practice of moving commercial beehives directly from earlier blooming orchards.

h. Nurseries and production fields as sources of infection

Nurseries, especially where they have been used for some years, may act for themselves as important sources of virus infection. For example, Hampton (1988) found in several US states no evidence of infection in wild *Humulus lupulus* plants tested for the following viruses: two ilarviruses and three carlaviruses common in cultivated hops, and in particular AHLV. Several hundred hop plants, primarily from breeding nurseries in Oregon, had substantial infection rates with these viruses. Fifty-three non-*Humulus* species growing around infected hop yards and nurseries were found to be free of AHLV. Thus, the commercial yards and nurseries appeared to be the only source of infection for this virus.

Nursery beds for transplanted rice can be important in the epidemiology of rice tungro disease. The young plants are very susceptible to virus infection and, furthermore, the process of uprooting the plants and transplanting them disturbs the leafhopper vectors, which will move to other plants (see Mukhopadhyay and Chowdhury, 1973).

i. Movement of crop plants into new areas

As well as distributing viruses around the world (Section IV.A.2.d), humans have moved crop species to new countries, often with disastrous consequences as far as viral infection is concerned. Plant species that were relatively virus-free in their native land may become infected with viruses that have long been present in the countries to which they were moved for commercial purposes. It is often difficult to prove a sequence of events of this sort, especially if the movement took place before virologists could investigate and record events. Thresh (1980) has detailed several instances. CSSV is a significant example because of the importance of the cacao crop in several West African economies. Cacao was transferred from the Amazonian jungle to West Africa late last century, and since then major commercial production has developed there. The swollen shoot disease, first reported in cacao in 1936, was probably transmitted from natural West African tree hosts of the virus by the mealybug vectors that are indigenous to the region.

Movement of plant species between countries and continents has been carried out with increasing frequency during the last two centuries. Agriculture in India, North America and Australia is almost totally dependent on introduced crop plants (Thresh, 1980). Thus, there has been ample opportunity for events such as that outlined for cacao to occur.

i. Monocropping

Cultivation of a single crop, or at least a very dominant crop, over a wide area continuously for many years may lead to major epidemics of virus disease, especially if an air-borne vector is involved. The development of such an epidemic is shown in Fig. 12.20. Soil-borne vectors

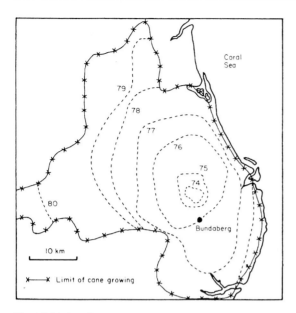

Fig. 12.20 Development of an epidemic of planthopper-transmitted virus disease in sugarcane over 7 years. The incidence of FDV in the Bundaberg district of Queensland, Australia, 1974–1980. Disease contours indicate the boundary of the area with an average of at least 1% affected cane. The solid line is the outer limit of the main cane-growing area. Modified from Egan and Hall (1983), with kind permission of the copyright holder, © Blackwell Science Ltd.

may also be important from this point of view, for example with GFLV in vineyards, where the vines are cropped for many years. Mono-cropping may also lead to a build-up of crop debris and the proliferation of weeds that become associated with the particular crop.

As pointed out by Diener (1987a), most viroid diseases in crops have appeared only in recent decades. He suggested that viroids themselves have existed for a very long period in wild host species, causing little disease there and escaping from time to time into agricultural crops that were grown on a small scale with varying genetic composition of the crop plants. Under these conditions, they would cause no widespread disease. However, with the advent of modern large-scale monocultures of geneti-cally uniform plants, the opportunities for serious outbreaks of viroid diseases have greatly increased. This is particularly so with crops such as chrysanthemums, where propa-gation of vegetative stocks may be highly centralized.

An outstanding example of the effects of monocropping is the rise in importance of cer-tain virus diseases of rice and of their hopper vectors as a consequence of the 'green revolu-tion' that began about 35 years ago. New rice varieties introduced as part of the green revolu-tion in tropical and subtropical areas are a major factor in the greatly improved yields that have been achieved. However, the gains have been seriously impaired by the increased prevalence of several serious virus diseases of rice, and also of their hopper vectors (Thresh, 1988a). An important example is RGSV and RRSV and their rice brown planthopper vector (*Nilaparvata lugens*) (Thresh, 1989a). As well as transmitting these viruses, the hopper can itself cause serious crop damage. The first new rice cultivars released during 1966–1971 were all susceptible to both vector and RGSV. Effective sources of resistance to both were found and incorporated into new varieties that were released in 1974–1975. These were at first grown widely and successfully in the Philippines, Indonesia and elsewhere. However, severe infestations with the plant-hopper were reported within 2–3 years, and in 1982–1983 a resistance breaking strain of RGSV was reported. No new source of virus resist-ance has been found. A new source of hopper resistance was identified and incorporated into new varieties, and it was successful for a few years until it, too, broke down with the emer-gence of a new hopper biotype (Thresh, 1989a).

Even within one country, the viruses infect-ing a particular host plant may well be influ-enced by a new development in agricultural practice. For example, *Solanum laciniatum*, an indigenous plant in New Zealand, had no virus infections recorded from natural habitats. Shortly after commercial plantings were begun, field infections with PVX, PVY, TMV and CMV were recorded (Thomson, 1976).

B. Physical factors

1. Rainfall

Rainfall may influence both air- and soil-borne virus vectors. The timing and extent of rainfall may alter the influence it has on vector popu-lations. For example, some rainfall or high humidity is necessary for the build-up of whitefly populations, while continuous heavy rainfall may be a factor in reducing the size of such populations (Vetten and Allen, 1983). Similarly, heavy rainfall just after air-borne aphids have arrived in a crop may kill many potential vectors, thus reducing subsequent virus incidence (Wallin and Loonan, 1971).

PMTV has its highest incidence in Scotland in areas with the highest rainfall (Cooper and Harrison, 1973), which was correlated with increased incidence of the fungal vector (*Spongospora subterranea*) in wetter soils. Similarly, in Peru this virus is widely distributed in *Solanum* spp. in the highlands of the Andes, where rainfall and temperature favor the fungal vector (Salazar and Jones, 1975). Symptoms of the virus were not seen in plantings in the dry coastal region.

2. Wind

Wind may be an important factor not only in assisting or inhibiting spread of viruses by air-borne vectors, but also in determining the predominant direction of spread. Windbreaks may affect the local incidence of vectors and viruses in complex ways (Lewis and Dibley, 1970). Winged aphids tend not to fly when wind speed is too great, although their direction of flight can be influenced by the prevailing wind. At low wind speeds, some species may fly with the wind and others against it (Fig. 12.21).

Freak wind conditions may cause annual epidemics. In 1977, there was a massive and unexpected epidemic of MDMV in the corn crop of the northern US state of Minnesota. This virus had been usually confined to southern states. From studies of the continental weather patterns in 1977, and from the fact that aphid vectors could retain the virus for more than 19 hours, Zeyen *et al.* (1987) proposed that low-level jet winds rapidly transported infective aphid vectors from drought-stricken southern areas north to Minnesota (Fig. 12.16).

The direction of movement of leafhoppers may also be influenced markedly by wind speed and direction. The beet hopper *C. tenellus* cannot make progress against a headwind stronger than about 3 km per hour. Movement

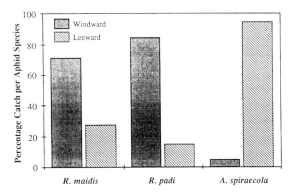

Fig. 12.21 Variation in flying direction by different aphid species under low wind velocities. The percentage catch of selected aphid species on the windward and leeward sides of a sticky cylinder set vertically just above a soybean canopy, Urbana, Illinois, USA, 1978. Modified from Irwin and Ruesink (1986), with permission.

of whiteflies is also substantially influenced by wind (Fig. 12.22). Mealybugs, which transmit swollen shoot disease of cacao, are relatively inactive and move only short distances on the plant, although they are carried about by ants. The young stages may, however, be carried some distance by wind.

The spread of blackcurrant reversion virus by its gall mite vector (*Phytoptus ribis*) is influenced to a substantial extent by the prevailing

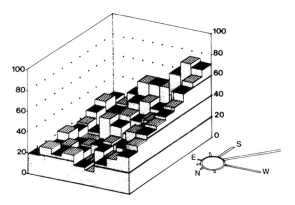

Fig. 12.22 Effect of wind direction on the incidence of ACMV infection in a 0.8-ha field of cassava surrounded by a sugarcane windbreak at a coastal site in the Côte d'Ivoire. 0–100 indicates the percentage infection in each block of plants. Wind frequency and direction are indicated by the length of the lines to the right of the block graph. The pattern of catch of whitefly vectors (*Bemisia*) was very similar to that shown for virus incidence. From Fargette *et al.* (1985), with permission.

wind over distances of about 15 m from infected plants. However, other factors, such as air turbulence caused by nearby buildings and trees, may influence the passive movement of this vector and spread of the virus (Thresh, 1966). With certain climates and soil types, nematodes of many genera may also be passively carried in wind-borne dust (Orr and Newton, 1971). However, virus vector species have not yet been identified in such dust.

3. Air temperature

Air temperature may have marked effects on the rate of multiplication and movement of air-borne virus vectors. For example, winged aphids tend to fly only when conditions are reasonably warm. However, very high temperatures may be particularly effective in reducing certain aphid populations.

Exposure of the planthopper vector of MRDV to 36°C prevented virus replication in the vector and suppressed transmission (Klein and Harpaz, 1970). Such air temperatures occur during the hot season in Israel.

4. Soil

Conditions in the soil can influence the incidence of virus disease in various ways. Highly fertile soils tend to increase the incidence of virus disease. For example, animal manure and several inorganic fertilizers increased the incidence of leaf roll and rugose mosaic diseases in potato crops (Broadbent et al., 1952). Some species of aphid vectors also multiplied faster on such plants. Plant nutrition may also influence the apparent amount of virus infection by accentuating or suppressing symptom expression.

Soil conditions can have a marked effect on the survival of TMV in plant debris. Moist well-aerated soils favor inactivation of the virus compared with dry, compacted or waterlogged soils (Johnson and Ogden, 1929).

Soil temperature may have a marked effect on the transmission of viruses by nematodes. The optimum temperature and temperature range may vary with different viruses, hosts and nematode species (e.g. Cooper and Harrison, 1973; Douthit and McGuire, 1975). However, in spite of seasonal fluctuations in

temperature, populations of nematodes in the field may remain quite stable for years. WSSMV in Canada is transmitted by its fungal vector with an optimal soil temperature of 10°C (Slykhuis, 1974). This may be a geographical adaptation since SBWMV from Illinois develops optimally at about 16°C.

The physical soil type may also affect vector distribution, and thus the incidence of viral disease. Nepoviruses predominated in cruciferous species on light soils in the former German Democratic Republic (Shukla and Schmelzer, 1975). In Scotland, TRV in potatoes was found in freely draining podzols but not on heavy, badly drained soils (Cooper, 1971). Soil type may influence virus incidence in yet another way: by the extent to which moisture is retained. For example, in Tasmania, BYDV infection was consistently higher in pastures on heavier soils, probably because plant growth was better on such soils under drought conditions (Guy, 1988).

5. Seasonal variation in weather and the development of epidemics

The wide annual variation that can occur in the incidence of virus disease in an annual crop is shown in Fig. 12.22 for beet yellowing viruses in sugar beet. Watson and Heathcote (1965) emphasized the importance of early migration of M. persicae for subsequent disease spread. However, other factors must be involved. For example, in 1945, there was a low incidence of M. persicae in June followed by a severe outbreak of virus infection. The year 1946 had the highest count of M. persicae in June for the 8 years studied, but this was followed by a low incidence of disease. During the 8-year period, the extent of infection in nearby seed crops was almost certainly a major factor affecting incidence of virus. Given similar sources of virus inoculum, seasonal variations in virus incidence in a crop such as sugar beet are probably due to continuing effects of weather conditions on the aphid vector population through most of the season, affecting timing and size of early migrations into the crop, build-up of population within the crop, and the mobility of this population (Fig. 12.23).

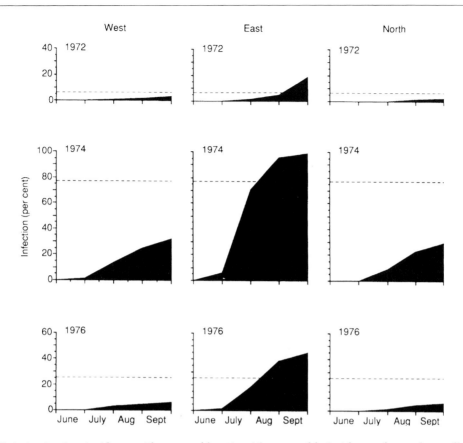

Fig. 12.23 Variation in virus incidence with year and location. Mean monthly incidence of sugar beet yellows disease in different regions of England in years with greatly differing amounts of spread. Data supplied by G.D. Heathcote (Brooms Barn Experimental Station) for representative crops sown annually in March or April in eastern England (Suffolk, Essex and Cambridgeshire), in wetter areas to the west (Shropshire and Worcestershire) and in cooler areas to the north (Yorkshire). The dashed lines indicate the mean incidence of yellows at the end of September as calculated for the entire English crop. From Thresh (1983), with permission.

A sequence of unusual weather conditions can lead to a severe outbreak of a disease that is normally present at fairly low levels in a crop in a particular region each year. The history of a severe epidemic of WSMV that developed in southern Alberta in the autumn of 1963 will serve to illustrate the kinds of interactions involved. Winter wheat is normally sown in the first 2 weeks of September. This sowing date normally allows the crop to escape infection from maturing spring-sown wheat (see Fig. 16.5). In 1963, an unusual sequence of weather conditions prevailed (Atkinson and Slykhuis, 1963). Spring rainfall was much below normal. Crops of spring-sown grain were sparse and not much seed germinated. This situation was suddenly changed in late June by very heavy rains. In the area that had previously suffered from drought, seed of spring wheat and barley that had lain dormant germinated, now about a month late, and previously stunted plants tillered profusely. Above normal or near normal rainfall in June and July assured further vigorous development of the crop and, by reducing the effectiveness of summer fallow operations, allowed extensive and profuse development of volunteer wheat. In this zone where spring rainfall had been low, extensive acreages of wheat and barley did not mature until late September. The late-developing wheat and barley crops became massive reservoirs of inoculum for the virus and for its mite vector.

Winter wheat sown at the normally recommended time in the first 2 weeks of September then became heavily infected. The autumn weather was also abnormal and facilitated massive spread of virus by the mite. The mean September temperature was 62°F (10°F above the 30-year average). These unusually high temperatures continued well into October. The 1964 season was not conducive to spread of the virus but, because of the very extensive spread that had occurred in the winter wheat before the freeze in 1963, very large acreages of grain were lost. The distribution of this epidemic outbreak in 1964 showed a clear relationship with the lack of spring rainfall in 1963 (Atkinson and Grant, 1967).

Three major epidemics of PLRV in potatoes lasting 5–10 years have occurred in the northern seed-growing areas of eastern North America and western Europe during the twentieth century. The climax years for these epidemics were 1912, 1944 and 1976, and were near the minima of every third sunspot cycle. The dry warm summers led to a northward movement of the aphid vector *Myzus persicae* (Bagnall, 1988).

An epidemic in which unusual wind patterns were thought to be a factor is illustrated in Fig. 12.16. The problem of forecasting disease outbreaks is discussed in Section IV.D below.

C. Survival through the seasonal cycle

Viruses can survive a cold winter period or a dry summer season in various ways. Some have more than one means of survival:

1. Many viruses readily survive from season to season in the same host plant or propagating material from it. These include viruses in perennial hosts, in tubers, runners, and so on, and viruses that are transmitted through the seed. Similarly, crops stored over winter in the field, such as mangolds, may act as a source of both viruses and vectors.
2. Viruses with a wide host range are well adapted to survive, provided their hosts include perennial species, a group of annual species with overlapping growing seasons, or species in which seed transmission occurs.
3. Biennial or perennial wild plants, ornamental trees and shrubs, or weed hosts such as *Plantago* spp. may be important sources for

overwintering or oversummering of many viruses (Hammond, 1982).

4. Leafhopper-transmitted viruses that pass through the egg of the vector may overwinter in the egg, or in young nymphs (Fig. 12.24).
5. TMV and a few other viruses can survive through the winter under appropriate conditions in plant refuse, and perhaps free in the soil. TMV can also survive in plant litter left in curing sheds, and so on. In the past, attention has been focused on virus surviving as free virus particles. However, it is known that many viruses infecting vertebrates may adsorb to inanimate surfaces and survive there (Gerpa, 1984). More attention should be paid to the possible survival of plant viruses adsorbed to minerals. Piazzolla *et al.* (1989b) have shown that CMV adsorbs readily to

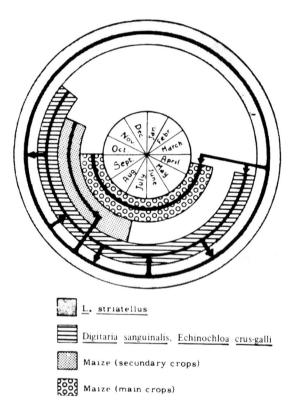

L. striatellus

Digitaria sanguinalis, Echinochloa crus-galli

Maize (secondary crops)

Maize (main crops)

Fig. 12.24 Survival of MRDV through the seasonal cycle in northern Italy. *Laodelphax stratellus* overwinters as a young nymph in diapause. This vector carries the virus through the winter when no known plant hosts are available. The black line and arrows indicate movement of the virus to and from its various hosts. From Conti (1972), with permission.

montmorillonite (an expandable layer silicate). In this state, it was resistant to inactivation by a strong salt solution.

6. Viruses carried within the resting spores of their fungal vectors may survive for long periods in soil.

7. Agricultural practices may allow the virus to survive in successive crops of the same plant grown in the same locality throughout the year. This may happen where the climate is suitable for production cropping throughout the year, where seed crops of the species are grown through the winter and overlap with production crops, or with wheat, where spring and winter crops are grown in the same area. This is shown in the cultivation cycle of crops in some regions of Europe and North America that allows a continuous food source for polyphagous aphids (Fig. 12.25). Another example is the substantial increases in the planting of winter oilseed rape in the UK, which pose a threat to vegetable crops because a range of viruses has been identified in these winter crops (Walsh and Tomlinson, 1985).

D. Disease forecasting

An understanding of the epidemiology and ecology of some major crop virus diseases has led to procedures for forecasting potential epidemics. This is very useful in implementing control measures (see Chapter 16). There are two main approaches to forecasting, monitoring the progress of a disease and developing mathematical models.

1. Monitoring virus disease progress

The monitoring of virus vectors to enable efficient insecticide application is described in Chapter 16 (Section II.A.1). Many large-scale farmers routinely monitor their crops and apply control measures at an appropriate time. However, as viral diseases take several days or even weeks to show symptoms after infection, the application of control measures based on symptom appearance can be too late. It also depends on the correct diagnosis of the disease and knowledge of how it is spread.

2. Mathematical modeling

There are an increasing number of mathematical models directed at forecasting the outcome of the spread of a disease into an agronomic situation. Basically, there are two types of model: prediction models to predict a possible epidemic and simulation models to understand the factors that give rise to and control a given situation. A model is developed to answer specific questions and there is no general model to predict the potential and outcome of all potential viral epidemics. In developing a model, as many factors as can be predicted are taken into account. These include knowledge of the virus, its vector, the virus–vector interaction, type of crop, the cropping system and various environmental factors that can impact on these biological factors. A good model enables strategic management decisions to be taken on whether the problem is going to be significant and, if so, when and how to deal with it (Jeger and Chan, 1995; Irwin, 1999). Chan and Jeger (1994) and Jeger *et al.* (1998) describe a model based on

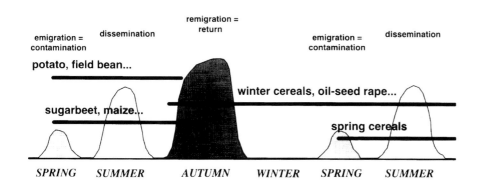

Fig. 12.25 Temporal relations between aphid flights and periods of cultivation of various crops. From Robert (1999), with permission.

transmission characteristics of arthropod virus vectors to analyze the effects of roguing of diseased plants and/or reduction of the vector population size by insecticide treatment. From this model, they suggest that roguing would be effective for non-persistently transmitted viruses only at low population densities and for propagative viruses only at high population densities. There would be a clear advantage for reducing vector population densities for propagative viruses but not for non-persistent viruses. Similarly, a model has been developed assessing the potential effects of the virus–vector interaction on the development of plant virus epidemics (Madden *et al.*, 2000).

There are numerous models for specific disease situations. Some examples to show the range are:

- Predicting the effects of using healthy and infected cassava cuttings on the spread of ACMV (Holt *et al.*, 1997). If virus-free cuttings were used then either high rates of virus transmission by the whitefly vector or large populations of vectors could lead to persistent cycles of disease incidence. If some of the cuttings were infected, the model predicts three possible outcomes: disease elimination, persistence of both healthy and infected plants, or total infection, depending on various parameters.
- A model for predicting virus yellows disease of sugar beet (BMYV and BYV) in the UK is based on the preceding winter weather (Watson *et al.*, 1975) (especially the number of frost-days) (Fig. 12.26A), the dates when the aphid vectors begin their spring migration (Harrington *at al.*, 1989) (Fig. 12.26B) and region in which the beet is being grown (eastern, western or northern region). This model has been refined to allow for the numbers of migrating *Myzus persicae*, the major vector (Werker *et al.*, 1998).
- A model based on vector preference for healthy plants or those infected by BYDV showed that the effect of vector preference for infected plants depended on the frequency of the infected plants in the population (McElhany *et al.*, 1995). The effect of vector preference depended on the persistence of

the virus in the vector. The results of the analysis contrasted with the assumption that preference for diseased plants automatically leads to increase in disease spread.
- To investigate the effects of competition between viruses, Power (1996) analyzed the interactions between BYDV-MAV and BYDV-PAV. Transmission rate plays an important role in determining the outcome of competition between the viruses, as does the interaction between transmission rate and vector behavior.
- Many annual crops in tropical regions are grown in continuous, contiguous cultivation without a seasonal break. Holt and Chancellor (1997) developed a model for this situation. The model predicted that, in a given region, disease incidence depended on infection efficiency, the dispersal gradient and the variance in planting date. This model is discussed in relation to the spread of Rice tungro virus disease.

The above examples show some of the complexities that have to be taken into account in developing meaningful models. Thus, the temporal dynamics (Nutter, 1997) and the spatial patterns both of the plant host and of the vector(s) and its hosts have to be analyzed (see Real and McElhany, 1996; Perry, 1998; Raybould *et al.*, 1999). The spatial patterns have to be large scale and take account of the landscape ecology (Barnes *et al.*, 1999).

E. Conclusions

We can see from this brief account that the factors and interactions involved in the ecology and epidemiology of viruses in crops are extremely complex. This can be illustrated by the processes involved in the spread of luteoviruses within and between crops (Fig. 12.27). The luteoviruses are among the most studied of virus groups as far as epidemiology is concerned but the complexities shown in Fig. 12.27 are just as applicable to other crop situations. There are virtually no data on the epidemiology of viruses in non-crop natural systems.

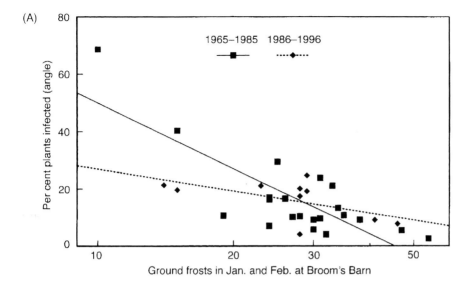

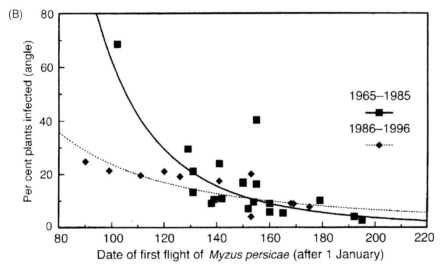

Fig. 12.26 Factors for predicting occurrence of virus yellows of sugar beet in England. **(A)** Effect of winter weather on virus yellows incidence in sugar beet in the eastern region of the UK: 1965–1996. Ground frosts were measured at Broom's Barn Experimental Station in eastern England. **(B)** Relationship between aphid migrations and virus yellows incidence in sugar beet in the eastern region of the UK: 1965–1996. From Dewar and Smith (1999), with permission.

We can see from this brief account that all viruses do not possess all characteristics that are generally favorable to survival and spread, which is perhaps just as well. Each virus has some combination of properties that allows it to persist more or less effectively. TMV has no significant invertebrate vector, but survives because of its high resistance to inactivation, the high concentration it generally reaches in its host, the ease with which it can be mechanically transmitted, and its wide host range. By contrast, viruses that multiply in the insect vector and are transmitted through the egg have a complete alternative to the plant host for survival.

TSWV is an example of a very unstable virus that is extremely successful in the field. It can retain infectivity only for periods of hours under normal conditions outside the plant. It has a specialized type of vector (species of thrips that can acquire virus only as nymphs), and yet it is widespread around the world and common within individual countries. Its success probably depends on several factors (Best, 1968): (1) its very wide host range, including many perennial symptomless carriers that provide year-round reservoirs of the virus; (2) the widespread distribution of species of thrips capable of transmitting the virus; and (3) the existence of a wide range of strains in strain

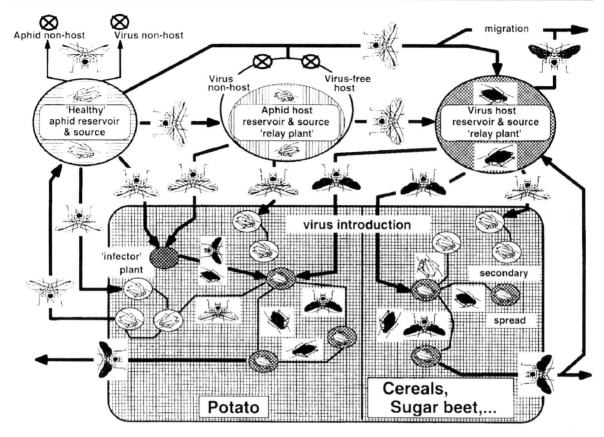

Fig. 12.27 Epidemiological process of luteovirus spread showing the various routes of infection and the relevant differences between potato and other crops. From Robert (1999), with permission.

mixtures, and the probable occurrence of recombination or gene re-assortment allowing rapid adaptation of the virus to new hosts to give relatively mild disease.

Many nematode-transmitted viruses are also transmitted through the seed in a range of weed and crop plant hosts. Infected seedlings often appear to be healthy. The combination of nematode and seed transmission may be important for the survival and dispersal of many of these viruses (Murant and Lister, 1967). Germination of infected weed seeds may allow re-infection of the nematode population in a field, and infected seeds may be transferred over longer distances than nematodes to set up new centers of infection. Murant and Lister noted that those viruses such as TBRV, which are commonly found in soils in infected weed seeds, do not persist more than about 9 weeks

in their vector *Longidorus elongatus*. In contrast, two nepoviruses (ArMV and GFLV), which can persist for 8 months or more in the vectors *Xiphinema* spp., were rarely or never found in seed in infected soil.

Viruses existing in completely natural or almost natural habitats and their ecology have been very little studied. Most of our knowledge is a by-product of the investigation of commercially important diseases. It seems probable that, under conditions where natural selection had operated on the virus and the host plant genotype over long periods undisturbed by the activities of humans, viruses would be closely adapted to their plant hosts and invertebrate or other vectors. The effects of virus infection on plant growth would probably be minimal. Severe disease in individual plants might occur occasionally, but epidemics would be rare.

Several investigations suggest that in wild plants, symptomless infection is the most common situation. Thus, weed hosts of CMV in both England (Tomlinson *et al.*, 1970) and the former German Democratic Republic (Shukla and Schmelzer, 1973) were infected without symptoms.

There may be regional adaptation between virus and host genotype to give mild disease. Tomlinson and Walker (1973) found that *Stellaria media* plants grown from seed from Britain, North America and Australia were least severely affected by the CMV strain obtained from their country of origin. Thus, natural selection of tolerant races of the species and mild strains of the virus had probably taken place.

A passionfruit virus causes disease in the introduced *Passiflora edulis* in the Ivory Coast. However, infections of indigenous *Adenia* spp. in remote areas were often symptomless (De Wijs, 1975). On the other hand, intensive agriculture may bring about complex ecological changes in nearby wild plant areas. Such changes may lead to more severe strains of a virus appearing in the wild flora, for example with BCTV in California (Magyarosy and Duffus, 1977).

Agricultural practices, even in fairly early primitive stages, must have changed the environment in which viruses existed in several important ways:

1. Plant selection and later plant breeding have given rise to new host genotypes with differing reactions to the existing viruses.
2. Cultivation brought new communities of plants together, both the useful species and the weeds associated with cultivation.
3. These agricultural communities of plants have been transported about the world and have become neighbors of the indigenous flora. Viruses present in these have moved into the agricultural crops.
4. Along with the plants, insect vectors and potential vectors have been transported into the new areas, and insects already present in the new areas, may have been able to spread into the agricultural crops.
5. Culture of a single species as the major plant over wide areas may allow the development of enormous populations of insect vectors and the consequent large-scale outbreaks of disease. This possibility may be made more likely by the maintenance of a particular species throughout the year in the same locality (e.g. the growing of seed crops of biennial plants such as sugar beet) or by the introduction of weeds that harbor the virus.
6. Agricultural practices such as cultivation and drainage may change soil conditions so as to affect markedly the population of soil-inhabiting virus vectors.
7. Grafting procedures, apart from their role in spreading viruses, may have allowed the expression of new diseases by selection of virus strains and by the introduction of viruses into previously uninfected species and varieties of plants.

A few agricultural practices may have played a part in unconscious selection of mild strains of a virus but, generally speaking, the foregoing factors combine in various ways continually to produce new and unstable ecological situations in which outbreaks of disease can occur in agricultural and horticultural crops. On the other hand, in regions where a crop has been grown continually for long periods under conditions favoring infection with a virus, resistant cultivars may have developed without a deliberate selection procedure. For example, many cultivars of barley from the Ethiopian plateau have been found to have a high degree of resistance to BYDV and BSMV (Harlan, 1976). It is probable that barley has been grown in this region for millennia.

CHAPTER 13

New Understanding of the Functions of Plant Viruses

I. INTRODUCTION

Two recurring observations in many of the preceding chapters have been the integration of viral features and functions and the interactions between viruses, their hosts and vectors. The latter is illustrated in the diagram in Fig. 13.1.

These are areas in which the knowledge and understanding of the mechanisms involved are likely to develop over the next few years. In the

first part of this chapter, I propose to bring together some of the threads from the previous chapters to emphasize the form and function of plant viruses. To do this, I will consider the various stages of the viral infection cycle, although it must be remembered that these are not necessarily distinct and that there are points of interaction between them. As stated by Revers *et al.* (1999): 'almost certainly, we will find the processes of RNA replication and translation

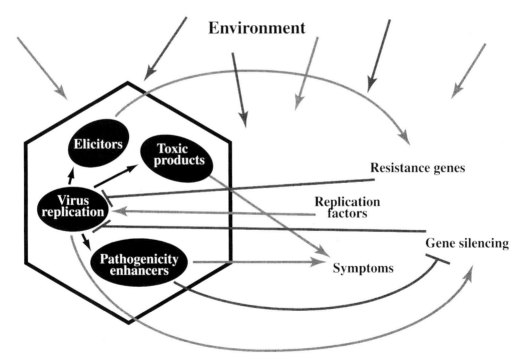

Fig. 13.1 (see Plate 13.1) Diagram showing viral infection as the complex result of aggression, defense and counter-defense. The virus is indicated in brown on the left and the host in green on the right. (Courtesy of Dr J.A. Garcia.)

Virus acronyms are given in Appendix 1.

are spatially and temporally integrated'. This statement related to a discussion on potyviruses but it is equally applicable to most, if not all, other groups of viruses.

In the discussion here, the infection cycle is divided into early events (entry, uncoating and initial translation), mid events (translation and replication) and late events (cell-to-cell movement and virus assembly). The division of functions into early, mid and late is not strictly correct as the timing of many of the functions overlaps and some of the 'early' functions may also occur late in infection. However, there is a natural progression from the initial entry of the virus into the cell to its exit from that cell, its entry into an adjacent cell and the formation of new virus particles. Thus, for the sake of simplifying the discussion, I will use these arbitrary divisions.

Most of the technologies involved in unravelling these interactions have been described in preceding chapters. Basically, they are *in vitro* techniques, such as the yeast two-hybrid system, that tell you what potentially can happen and *in vivo* techniques, such as the use of fluorescent reporters and confocal microscopy, that tell you what is happening in the cell.

II. EARLY EVENTS

The entry of RNA viruses into the cell, the uncoating of the particles and the initial translation of genomic RNA are discussed in Chapter 7. The integration of these events is illustrated by TMV for which co-translational disassembly of the virus particles yields the 5' gene products, which are those required to replicate the viral genome. Co-translational disassembly has been demonstrated for some other groups of viruses, such as bromoviruses and sobemoviruses. However, the mechanisms of disassembly of isometric viruses whose capsids are stabilized by protein–protein interactions (e.g. comoviruses) and rod-shaped viruses whose origin of assembly is thought to be near the 5' end of the viral genome (e.g. potexviruses) are unknown. Although these may differ from the TMV type of co-translational disassembly mechanism, it is likely that uncoating and early translation are co-ordinated.

A differing view on the early events of virus entry and uncoating is presented by de Zoeten (1995). He suggested that there was an initial destabilization of the virus particle at the cell wall on entry into the cell. His argument was based on the fact that the response to wounding a cell is the production of phenolic compounds that would inhibit a virus and that the plugging of the plasmodesmata leading from the wounded cell would prevent the virus moving to adjacent cells. Thus, he suggested that a virus would not function in such a wounded cell and that the successful infection events occurred in cells in which the plasma membrane was not damaged. In his hypothesis, the cytosol does not provide the environment for destabilizing a virus particle and infection of other cells in the plant is by progeny virus that has not 'restabilized'. A difficulty of reconciling this view with the more generally accepted one is that initial infection requires just one viable infection unit and that many virus particles enter the plant. With the current technology it is not possible to distinguish the particle(s) that initiate infection from all those that accompany it (or them).

As described in Chapter 7 (Section IV.A), viruses with (–)-sense single-stranded RNA genomes have to transcribe their encapsidated RNA to give (+)-sense mRNA; this is effected by the RdRp that is contained within the virus particles. Steps in the pathway for uncoating, transcription and translation of nucleorhabdoviruses have been described by Jackson *et al.* (1999) (Chapter 8, Section V.A.2; Fig. 8.19). However, details of the interactions involved have still to be worked out. It is likely that other (–)-strand viruses have similar pathways. Since these viruses also infect insects, it would be interesting to determine whether the interactions involved in their uncoating, initial transcription and translation are the same in insect and plant hosts. There is more information on these stages of the infection cycle of animal rhabdoviruses and bunyaviruses in animal cell systems that may be applicable to the plant-infecting counterparts in insects. However, it would be unwise to extend the analogy to the behavior of these viruses in plant cells until further information is forthcoming.

The information on the early stages of infection of plant cells by dsRNA viruses is similar to that of the (–)-sense RNA viruses described above.

The current understanding of the replication of reverse transcribing plant viruses and those with ssDNA genomes is described Chapter 8 (Sections VII.B and VIII). As with the (–)-sense and dsRNA viruses, there are many gaps in the details of interactions involved at various stages. For instance, it is not known how the particles of CaMV are uncoated. Since the coat protein has a nuclear localization signal, it is likely that this takes place in the nucleus but, as noted in Chapter 7 (Section II), CaMV particles are particularly stable.

III. MID-STAGE EVENTS

The main mid-stage events are full translation and the replication of the viral genome. For most, if not all, viruses there is considerable host involvement with these functions which are closely integrated both temporally and spatially.

A. Host and virus translation

The expression of plant viral genomes is described in detail in Chapter 7. Viral mRNAs use the host translation system. In certain cases, there is evidence for the virus controlling precedence for the translation of its own RNAs over at least some of the host mRNAs. The best understood situation is that of the potyvirus PSbMV described in Chapter 9 (Section III.B.3), in which the expression of the virus is associated with decreases in some host mRNA, increases in others, and no change in the steady state of yet others. This could reflect viral control of expression of host proteins to make the cellular environment more favorable. Another way in which some viruses could affect host protein synthesis is the 'cap snatching' shown by tospoviruses and tenuiviruses (Chapter 7, Section V.C.4). Cap snatching occurs in vertebrate-infecting viruses in the families *Orthomyxoviridae*, *Arenaviridae* and *Bunyaviridae*, and is thought to be a mechanism by which translation of viral mRNAs gains precedence over that of the host mRNAs.

For the correct functioning of an organism—whether based on a plant or animal cell, or a bacterium or virus—it is important that the correct proteins are expressed in the right place at the right time and in the right quantity and form. With cellular organisms, much of this control is effected by the transcription of mRNA from the chromosomal DNA. There is some control, as described in Chapter 7 (Section V.C), by the structure and interaction between the 5′ and 3′ ends of the mRNA. With cytoplasmically replicating and expressing RNA viruses, the control of expression of final gene products is by a range of mechanisms described in Chapter 7 (Section V.B). This wide variety of mechanisms has evolved, not only to overcome the constraints of eukaryotic ribosomes translating only 5′ ORFs but also to give effective temporal and quantitative production of final gene products.

B. Host and virus replication

The replication of plant virus genomes is described in detail in Chapter 8. The normal replication of genetic material in host cells is DNA → DNA, and the DNA viruses (geminiviruses and nanoviruses) use this pathway. However, these viruses have to overcome the problem of host DNA replication taking place only at a specific phase (the G phase) of the cell cycle, and they encode a gene product, similar to a normal host protein, that converts the cell from S phase to G phase (Chapter 8, Section VIII.D.5). The genome of MSV first moves to the nucleus with the coat protein providing the nuclear localization signal (Chapter 9, Section II.C). However, the coat protein also interacts with the viral movement protein which inhibits nuclear transport of the coat protein–DNA complex (Liu *et al.*, 2001). This would direct newly-synthesized DNA for the nucleus toward cell-to-cell transport.

The reverse transcription pathway of virus replication (DNA → RNA → DNA) (caulimoviruses and badnaviruses) is found in retroelements that occur in plant chromosomes. The host is directly involved in the transcription of the viral DNA genome by DNA-dependent RNA polymerase II, but the RNA → DNA phase is catalysed by the virus-encoded reverse transcriptase. It is not known whether there is any direct host involvement in this phase.

The production of RNA directly from RNA is an unusual situation in normal plant cells, although the recent work on the involvement of host RNA-dependent RNA polymerase in gene silencing (Chapter 10, Section IV) indicates that it may not be as unusual as originally thought. As described in Chapter 8 (Sections IV.E.5, IV.F.1 and IV.H.5), host proteins are involved in RNA virus replication complexes. It is of especial note that some of the host proteins are translation initiation factors, which might indicate co-ordination of translation and replication. As the two processes operate in different directions along the template, it may be that they (and other host proteins) prevent interference between the two processes (Chapter 8, Section IV.M).

C. Spatial factors in virus expression and replication

There is increasing evidence that the expression and replication of plant viruses occurs at specific sites within plant cells and that these sites are related to membrane systems and cytoskeletal elements. This is also the case with animal viruses (reviewed in Ploubidou and Way, 2001).

1. Plant endomembrane and cytoskeletal systems

It is not known whether the early events of entry and co-translational disassembly involve the plant endomembrane and cytoskeletal systems, but it is likely that co-translational disassembly takes place on rough endoplasmic reticulum which contains protein-synthesizing ribosomes. There is increasing evidence that most of the mid-stage events of the virus infection cycle are associated with the endomembrane system and cytoskeleton, and so a general understanding of these structures is important.

a. Endoplasmic reticulum

The endoplasmic reticulum (ER) system is a pleomorphic and multifunctional organelle found in all eukaryotic cells, and offers a large membrane surface with different functional domains (for a review see Staehelin, 1997). At least 16 types of ER domain have been recognized (Fig. 13.2).

Of these, seven are currently recognized as having, or potentially having, an involvement with plant viruses. It is quite possible that interactions with other ER domains will be recognized in the future.

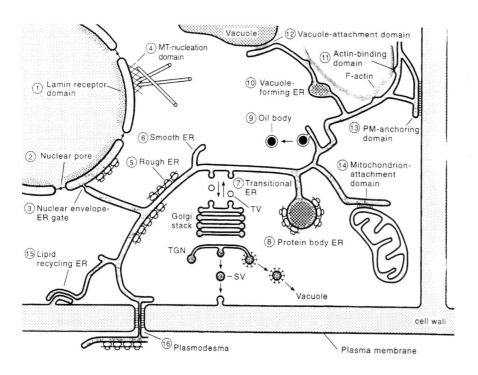

Fig. 13.2 Schematic diagram of a plant cell depicting the 16 types of ER domain. PM, plasma membrane; TV, transport vesicle; SV secretory vesicle; TGN, trans-Golgi network. From Staehelin (1997), with kind permission of the copyright holder, © Blackwell Science Ltd.

The outer membrane of the nuclear envelope is continuous with the ER and has membrane-bound ribosomes. The outer nuclear membrane joins the inner membrane at nuclear pore complexes (no. 2 in Fig. 13.2) that mediate the directed transport of proteins and nucleic acids between the nucleus and cytoplasm. Although little is known about the biochemical and molecular aspects of plant nuclear pores, it is thought that they have some unique features when compared with the better-studied structures in other eukaryotes (Panté and Aebi, 1994; Heese-Peck and Raikhel, 1998). As described in Chapter 8 (Sections V, VII and VIII), transport of the genomes of DNA viruses (caulimoviruses, geminiviruses and nanoviruses) and the nucleorhabdoviruses into the nucleus is an essential part of their replication cycle; nuclear localization signals have been identified on proteins of some of these viruses (Chapter 9, Section II.C). Particles of some viruses with (+)-strand RNA genomes are also found in the nucleus.

The outer membrane of the nuclear envelope is also the major microtubule organizing center (MTOC) in plant cells (no. 4 in Fig. 13.2). Microtubules are discussed in a subsequent section (Section III.C.1.b).

The two classical types of ER are the rough (sheet) and smooth ER, which are distinguished by the presence (rough) or absence (smooth) of attached ribosomes (nos 5 and 6 in Fig. 13.2). The rough ER is the site of translation of mRNAs. It is suggested that there is reversible conversion between rough ER and smooth tubular ER, which could be due to the binding of new, and the loss of old, polysomes (Knebel *et al.*, 1990); however, the generation and maintenance of tubular ER also requires a functional cytoskeleton (Quader, 1990). By analogy with vertebrate cells, it is suggested that one of the principal functions of the smooth ER is the synthesis of membrane lipids.

As described in Chapter 8 (Section V.B), the membrane of tospoviruses is derived by budding into the Golgi cisternae. although the replication and expression of the viral genome takes place in the cytoplasm. The transport of proteins from the ER to the Golgi bodies is by transport vesicles from the transitional ER (no. 7 in Fig. 13.2).

The sensitivity of the ER network to cytochalasin D has been taken as indicating that its movement and deployment is dependent on an actinomyosin-based mobility system (Liebe and Quader, 1994). In animal cells there is reasonable evidence that microtubules and motor proteins are involved in this mobility and the dependence on actin may be indirect. There is also electron microscopic evidence for a structural association between the ER and actin filaments in plant cells, suggesting actin-binding domains (no. 11 in Fig. 13.2). Actin microfilaments are discussed further in Section III.C.1.b.

The seventh domain with viral associations is the part of the ER that passes through plasmodesmata (no. 16 in Fig. 13.2). The ER central tubular cylinder in plasmodesmata is termed the desmotubule or appressed ER and is shown in Fig. 9.3.

b. *The cytoskeleton*

The two main elements of the plant cytoskeleton are the microtubules and the microfilaments. There are dynamic interactions between each of these entities in response to extracellular and intracellular signals (see Baluška *et al.*, 2001).

Microtubules are made up of tubulin, seven distinct families of which have been recognized recently (Oakley, 2000; Dutcher, 2001). The most abundant are the α and β tubulins that form the α/β dimers, which are the monomers for constructing microtubules. Flowering plants contain multiple copies of functional α and β tubulin genes, the products of which can be post-translationally modified to give a large array of isoforms (reviewed in Ludueña, 1998). In plant cells, γ tubulin is found in association with all microtubule arrays as well as being involved with MTOCs, as in animal and fungal cells. The other tubulins, δ, ε, ζ and η have recently been recognized in animal and fungal cells but not yet in plant cells. It is thought that these various tubulins have distinct functions associated with the placement and dynamic behavior of microtubules.

Associated with microtubules are specific proteins, microtubule-associated proteins (MAPs), some of which are mechanochemical motors that move various components along microtubules. Other MAPs modulate the assembly of microtubules at the MTOCs (no. 4 in Fig. 13.2) and the disassembly at their distal end; this forms a 'treadmill' that moves any protein associated with the microtubule from the MTOC end to the distal end (see Schroeder (2001) for a review of microtubule end dynamics). These functions are probably involved in cytoplasmic streaming and intracellular transport of molecules and macromolecules. In interphase cells the microtubule arrays comprise microtubules cross-linked by MAPs to the plasma membrane. Plant MAPs have been reviewed by Lloyd and Hussey (2001).

Microfilaments are composed primarily of actin, a helical assembly of globular units (G-actin) that form a rope-like filament (F-actin). The endomembrane system is surrounded by a sheath of cytoskeleton, a primary constituent of which is F-actin (for a review see Reuzeau et al., 1997). This endomembrane sheath confers both basic structure and the structural dynamism to the ER, and is suggested to be the framework on, and within which, many of the metabolic reactions within the cell are organized and operate. The sheath links the ER to various organelles such as the plasma membrane via various molecules. Actin is associated with plasmodesmata (White et al., 1994) and is suggested to be involved in regulation of the size exclusion limit.

Various other proteins have been found to be involved in the plant cytoskeleton. Myosin-like proteins co-localize with the ER and are suggested to function as membrane-bound motor proteins, which translocate organelles and the ER along actin filaments (Liebe and Quader, 1994). A myosin-like protein is also found in plasmodesmata, and the application of an inhibitor of actin–myosin activity (2,3-butane-dione monoxime) resulted in strong constriction of the neck region of plasmodesmata (Radford and White, 1998). A protein similar to centrin, a calcium-binding contractile protein, also localizes to plasmodesmata (Blackman et al., 1999). It is suggested that this centrin-like protein is a component of the calcium-sensitive contractile microfilaments in the neck region of plasmodesmata. The protein translation factor, EF-1, and the initiation factor-(iso)4F bind to tubulin and are associated with microtubules (Hugdahl et al., 1995; Moore and Cyr, 2000). The heat shock protein (HSP) 90, which has homologs in some viruses (e.g. closteroviruses) is also found associated with microtubules (Freudenrich and Nick, 1998).

2. The cytoskeleton and movement of macromolecules

There is increasing evidence that both intracellular and intercellular movement of many macromolecules involves the cytoskeleton and the associated endomembrane system. In some animal cell types, such as oocytes and neurons, microtubules are involved in the long-distance travel of mRNAs (see Bassell et al., 1994a). In other animal cell types, for instance fibroblasts, long-distance transport of mRNA involves actin filaments (see Bassell et al., 1994b). The general conclusions are that cytoskeletal elements play a major role in the regulation, movement and localization of mRNAs (Bassell and Singer, 1997). It is likely that this also pertains in plant cells, but the evidence for this is much less firm.

As well as the involvement of cytoskeletal elements in the movement of proteins described above, other mechanisms for protein movement involving the cytoskeleton have been described. For instance, protein complexes made up of the chaperone heat-shock proteins HSP90 and HSP70, together with some other proteins, are involved in the targeted movement of signaling and other proteins along microtubule tracts (Pratt et al., 2001).

D. Plant viruses and cytoskeletal elements

The replication of an increasing number of RNA plant viruses is being shown to be associated with intracellular membranes (see Table 8.4). It is not clear whether the association is with the membrane itself or with the related cytoskeletal element. The most detailed studies have been made on TMV.

The work of Más and Beachy (1999) develops a picture of the structural associations of TMV with the ER and cytoskeletal elements during the replication of the virus (Fig. 13.3).

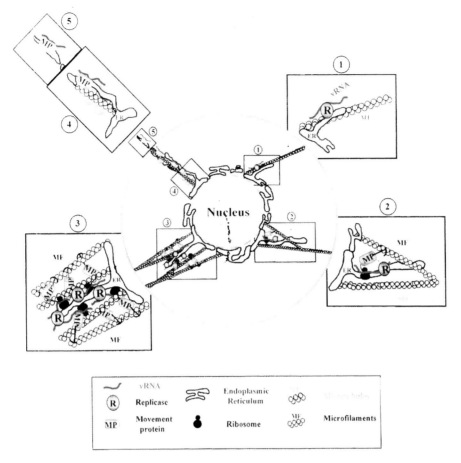

Fig. 13.3 (see Plate 13.2) Model of TMV infection in BY2 protoplasts. (1) Early in infection TMV vRNA comes into close association with membranes of the ER, and vRNA–replicase complexes associated with ER are transported via microtubules to perinuclear positions. (2) Nascent vRNAs synthesize movement protein that remains associated with vRNA. (3) Formation and anchoring of large ER-derived structures containing movement protein, replicase and vRNA, which are stabilized by movement protein and microfilament interactions. (4) Microtubule-based transport system of vRNA–movement protein complexes towards the periphery of the cell to initiate cell-to-cell spread. (5) Protrusion of the ER containing vRNA and movement protein through the plasma membrane. The model is not drawn to scale. From Más and Beachy (1999), with kind permission of the copyright holder, © The Rockefeller University Press.

Using *in situ* hybridization to localize TMV viral RNA (vRNA), immunostaining to locate the ER luminal binding protein (BiP), tubulin, actin and the TMV replicase (126/183-kDa protein), cytoskeleton depolymerizing agents (oryzalin for microtubules and cytochalasin D for actin microfilaments) and a TMV construct with GFP tagged to the 30-kDa movement protein, coupled with confocal microscopy, Más and Beachy (1999) studied the spatial and temporal changes on infection of tobacco protoplasts. One of the earliest events was the association of the vRNA with the cytoskeleton and ER; it is suggested that the vRNA is transported by microtubules to perinuclear positions (no. 1 in Fig. 13.3). The (+)- and (−)-strands of TMV RNA are both present in the perinuclear region, indicating that viral replication takes place here. However, they appear to be compartmentalized in this region, which is taken as suggesting that their synthesis may require different types of factors or molecular interactions. Movement protein was synthesized in the perinuclear region (no. 2 in Fig. 13.3).

IV. LATE EVENTS

The late events of the virus replication cycle require the cell-to-cell movement to infect adjacent cells and encapsidation of newly synthesized viral genomes. Cell-to-cell movement of viruses is described in Chapter 9, and viral encapsidation in Chapter 5. As with the mid-stage events, most of the detailed studies have been on TMV.

The observations of Más and Beachy (1999) described in the previous section extend to the intracellular movement of the viral replication complex from the perinuclear site to the plasmodesmata. The perinuclear replication sites develop into large ER-derived structures containing the vRNA, the replicase and movement protein, and which it is suggested are stabilized by movement protein and microfilament interactions (no. 3 in Fig. 13.3); these structures may correspond to previously described viroplasms or amorphous inclusions (Martelli and Russo, 1977). The complexes of vRNA and movement protein are transported towards the cell periphery by the microtubule-based transport system (no. 4 in Fig. 13.3). The TMV movement protein has a conserved sequence found in a region of tubulins that are thought to mediate lateral contacts between microtubule protofilaments (Boyko *et al.*, 2000b). Point mutations of this motif affect the association of movement protein with microtubules and it is suggested that the movement mimics the tubulin assembly surfaces, enabling the complex with vRNA to move along the microfilament 'treadmill'. The final stage in protoplasts that lack plasmodesmata is the protrusion of ER containing vRNA and movement protein through the plasma membrane (no. 5 in Fig. 13.3). It is not known whether these protrusions represent desmotubules or some other structure. The details of the movement of the complexes of vRNA, movement protein, ER and microtubules exactly to the plasmodesmata and the movement of the vRNA through the plasmodesmata are not known. It may be of significance that actin microfilaments extend through plasmodesmata (White *et al.*, 1994), whereas microtubules appear not to do so.

The final products of replication go into one of two pathways. First, there is the transfer of the infection unit to initiate infection of adjacent cells. Second, there is encapsidation of progeny genomes to give virus particles. This encapsidation provides protection of the genome for systemic spread of the virus (for those viruses that require particles for long-distance spread) and for spread to other host plants. It also sequesters viral genomes away from the replication and translation machinery in the plant cell, thereby limiting damaging overproduction of the virus. Little is known about the factors that control the directing of the progeny into these two pathways.

V. SYSTEMIC INTERACTIONS WITH PLANTS

The systemic movement of viruses within the plant is described in Chapter 9 and many of the factors that lead to symptom production in Chapter 10. In considering aspects of interactions and co-ordination on the outcome of virus replication and movement from the initially infected cell, there are three major points to bear in mind.

First, plants have a generalized defense system that recognizes 'foreign' nucleic acids (Chapter 10, Section IV). This system is activated by double-stranded RNA and it is likely that the replication of RNA viruses causes this activation. As described in Chapter 10 (Section IV.H), several viruses have been shown to have mechanisms that suppress or overrun this defense system. Thus, the establishment of systemic infection and the infection of individual cells depends upon the balance between the defense system and its suppression. Hence, in systemic infections one should view the interactions of a virus with its host on a cell basis rather than a whole-plant basis.

Second, as well as the generalized defense system, many plant have genes, the products of which interact with viruses to limit infection. The most obvious of these, which control the ability of the virus to move out of the initially infected cell (subliminal infection) or result in a response that limits virus spread (hypersensitive reaction), are described in Chapter 10 (Section III).

Third, subliminal infection and hypersensitive reaction are obvious responses but there are also likely to be less obvious and varying interactions of the viral and plant genomes. For instance, environmental factors impacting on the host genome can affect the detailed outcome of a virus infection (see Chapter 12). Factors that affect the detailed metabolism of the plant could also have an impact on spatial and quantitative aspects of infection of the plant. For instance, transcription of several *Arabidopsis* key pathway genes has been shown to be orchestrated by the circadian clock (Harmer *et al.*, 2000). It is quite possible that these effects on the host could influence the outcome of the infections of individual cells during the systemic spread of a virus. This, in turn, could affect the phenotypic expression of the virus-infected plant.

VI. DISCUSSION

The above descriptions show some of the interactions both within the viral genome and between the viral and plant host genomes that occur during an infection. There is an increasing number of other individual observations, not only for RNA viruses but also for viruses with DNA genomes. There are several reasons

for these interactions and the coordination of functions including:

1. Protection of the viral genome. For many years, the functioning of the viral genome was considered on the basis of its nucleic acid. It is now thought likely that nucleic acids do not exist alone at any stage in the cell, as they would be rapidly degraded in that state. Thus, they are associated with proteins and membranes that protect them at the various stages of replication.

2. As noted above, for each stage in the replication cycle and for efficient sequential expression it is necessary to have the required materials in the right place, at the right time, in the right amount and in the right conformation. Thus, it is necessary to co-ordinate the stages of the replication cycle.

3. The use of plant functions reduces the number of genes that the virus has to carry in its infectious genome. However, the virus has to be in the right place in the cell and at the right time to capitalize on the use of plant functions. Also its gene products have to be compatible with, and interact with, the plant gene products.

4. The strict co-ordination of functions gives control over the level of replication. The virus has to control the amount of progeny produced to maximize the chance of being transmitted to another host, but reduce the chance of causing irreversible damage to the current host.

CHAPTER 14

Viroids, Satellite Viruses and Satellite RNAs

There are various small RNAs associated with plant pathogenic situations. Essentially, most fall into two groups: those that can replicate autonomously and those that require a functional virus for their replication. The former group is termed viroids and the latter include satellites and defective nucleic acids. Defective nucleic acids are described in Chapter 8 (Section IX.C). In this chapter, I will describe viroids and satellites.

I. VIROIDS

Various important virus-like diseases in plants have been shown to be caused by pathogenic RNAs known as viroids, a term first introduced by Diener (1971) to describe the infectious agent causing the spindle tuber disease of potato. They are small circular molecules, a few hundred nucleotides long, with a high degree of secondary structure. They do not code for any polypeptides and replicate independently of any associated plant virus. Viroids are of practical importance as the cause of several economically significant diseases and are of general biological interest as being among the smallest known agents of infectious disease. Work on viroids has been reviewed many times, for example by Diener (1987b), Diener (1993), Flores *et al.* (1997), Symons (1997) and Diener (1999). The most studied viroid is potato spindle tuber viroid (PSTVd). Viroid names are abbreviated to initials with a 'd' added to distinguish them

Virus acronyms are given in Appendix 1.

from abbreviations for virus names. Acronyms for viroids are listed in Appendix 1.

A. Classification of viroids

Based on the sequence and predicted structures of their RNAs, viroids are classified into two families, the *Pospiviroidae* and the *Avsunviroidae* (reviewed by Flores *et al.*, 2000); each family has several genera (Table 14.1).

Carnation small viroid-like (CarSV) RNA does not fit with the above classification. It resembles viroids in that both its (+) and (–) strands self-cleave via a hammerhead structure (Hernández *et al.*, 1992) (see below). However, purified RNA is not infectious to carnations, and CarSV sequences have been isolated from 'infected' carnations (Daròs and Flores, 1995). Furthermore, CarSV DNA has been found fused directly to DNA sequences of the caulimovirus CERV (Daròs and Flores, 1995). CarSV DNA is organized as a series of head-to-tail multimers fusing to distinct regions of the CERV genome (Vera *et al.*, 2000). Each junction had short nucleotide stretches common to both CarSV and CERV DNAs, suggesting a polymerase-driven mechanism for their origin. As well as the most abundant CarSV RNA, there was a series of deletions, some of which had corresponding DNA forms.

B. Pathology of viroids

1. Macroscopic disease symptoms

Viroids infect both dicotyledonous and monocotyledonous plants. As a group, there is

TABLE 14.1 Classification of viroids

Family	Genus	Viroid	Abbreviation[a]	Size (nt)	Reference[b]
Pospiviroidae	Pospiviroid	Potato spindle tuber	PSTVd	356–360	1
		Chrysanthemum stunt	CSVd	354–356	1
		Citrus exocortis	CEVd	370–375	1
		Columna latent	CLVd	370–373	1
		Iresine viroid 1	IrVd-1	370	1
		Mexican papita	MPVd	361	2
		Tomato apical stunt	TASVd	360–363	1
		Tomato planta macho	TPMVd	360	1
	Hostuviroid	Hop stunt	HSVd	297–303	1
	Cocadviroid	Coconut cadang-cadang	CCCVd	246–247	1
		Citrus viroid IV	CVd-IV	284	1
		Coconut tinangaja	CtiVd	254	1
		Hop latent	HLVd	256	1
	Apscaviroid	Apple scar skin	ASSVd	329–330	1
		Apple dimple fruit	ADFVd	306	1
		Australian grapevine	AGVd	369	1
		Citrus viroid III	CVd-III	294–297	1
		Citrus bent leaf	CBLVd	318	1
		Grapevine yellow speckle viroid 1	GYSVd-1	366–368	1
		Grapevine yellow speckle viroid 2	GYSVd-2	363	1
		Pear blister canker	PBCVd	315–316	1
	Coleviroid	Coleus blumei viroid 1	CbVd-1	248–251	1
		Coleus blumei viroid 2	CbVd-2	301	1
		Coleus blumei viroid 3	CbVd-3	361–364	1
Avsunviroidae	Avsunviroid	Avocado sunblotch	ASBVd	246–250	1
	Pelamoviroid	Peach latent mosaic	PLMVd	336–339	1
		Chrysanthemum chlorotic mottle	CChMVd	399	3

[a] See Appendix 1.
[b] 1, Flores *et al.* (1997); 2, Martinez-Soriano *et al.* (1996); 3, Navarro and Flores (1997); all with permission.

nothing that distinguishes the disease symptoms produced by them from those caused by viruses, which are described in Chapter 3. They include stunting, mottling, leaf distortion and necrosis. From an agricultural crop point of view, symptoms cover a wide range from the slowly developing lethal disease in coconut palms caused by CCCVd (e.g. Haseloff *et al.*, 1982) to the worldwide symptomless infection of HLVd (Puchta *et al.*, 1988). It is probable that many more symptomless viroid infections remain to be discovered.

2. Cytopathic effects

Various effects of viroid infection on cellular structures have been reported. For example, in some infections changes have been observed in membranous structures called 'plasmalemmasomes'. Several workers have described pronounced corrugations and irregular thickness in cell walls of viroid-infected tissue (e.g. Momma and Takahashi, 1983). Various degenerative abnormalities have been found in the chloroplasts of viroid-infected cells (e.g. da Graça and Martin, 1981).

3. Subcellular location of viroids

Using confocal laser scanning microscopy and transmission electron microscopy in conjunction with *in situ* hybridization, both CEVd and CCCVd were found in vascular tissues as well as mesophyll cells (Bonfiglioli *et al.*, 1996).

From experiments involving fractionating components of viroid-infected cells it has become generally accepted that most viroids are located in the nucleus (reviewed by Riesner, 1987). The main exception is ASBVd, which is found in chloroplasts (Bonfiglioli *et al.*, 1994; Lima *et al.*, 1994). Within nuclei, PSTVd and CCCVd are located in nucleoli whereas CEVd accumulates to higher concentrations in the nucleoplasm (Harders *et al.*, 1989; Bonfiglioli *et al.*, 1996).

4. Biochemical changes

Viroid infection appears to cause no gross changes in host nucleic acid metabolism. By contrast, marked changes in the amounts of various host proteins have been described in infected tissue, for example a 14-kDa protein (Diener, 1987a). Perhaps the most dramatic effect is the increase in a 140-kDa host protein in tomatoes infected with several different viroids (Camacho-Henriquez and Sänger, 1984). Induction of this protein was not specific to viroid infection. Increases in other pathogenesis-related proteins have also been noted following infections with viroids, and some of these proteins appear to bind to viroid molecules *in vivo* (Grannel *et al.*, 1987; Hadidi, 1988). Hiddinga *et al.* (1988) identified a host-encoded 68-kDa protein that is differentially phosphorylated in extracts from viroid-infected tissue and mock-inoculated tissue. The protein appeared to be a dsRNA-dependent protein kinase immunologically related to a similar protein in mammalian cells that has been implicated in the regulation of virus synthesis. Significant changes in the composition of cell walls have been found in viroid-infected tissues (Wang *et al.*, 1986).

5. Movement in the plant

Viruses with defective coat proteins and naked RNAs move slowly through the plant by cell-to-cell movement. By contrast, viroids move rapidly from cell to cell of a host plant in the manner of competent viruses. When infectious transcripts produced from PSTVd cDNA clones and labeled with the nucleotide-specific fluorescent dye, TOTO-1 iodide, were injected into symplastically isolated guard cells, the viroid remained in the injected cell (Ding *et al.*, 1997). However, when the labeled RNA was injected into symplastically connected mesophyll cells, it moved rapidly from cell to cell. A 1400-nucleotide RNA transcript of the vector was unable to move from the injected cell but when the transcript included PSTVd it moved from cell to cell. From these experiments, it was concluded that viroids move from cell to cell via plasmodesmata and that the movement is mediated by a specific sequence or structural motif.

Long distance movement of viroids is almost certainly through the phloem (Palukaitis, 1987; Zhu *et al.*, 2001). The relative resistance of viroid RNA to nuclease attack probably facilitates their long-distance movement. It is also possible that viroid 'particles' undergo translocation while bound to some host protein.

6. Transmission

Viroids are readily transmitted by mechanical means in most of their hosts. Transmission in the field is probably mainly by contaminated tools and similar means. This ease of transmission in the presence of nucleases is probably due to viroid secondary structure and to the complexing of viroids to host components during the transmission process.

PSTVd is transmitted through the pollen and true seed of potato plants (Grasmick and Slack, 1986) and can survive in infected seed for long periods. However, this route of transmission is not likely to be of great commercial importance in potato. Several viroids have been shown to be transmitted by pollen and seed in tomato (Kryczynski *et al.*, 1988) and in grapes (Wah and Symons, 1999).

PSTVd was reported to be transmitted at low frequency in a non-persistent manner by the aphid *Macrosiphum euphorbiae* (De Bokx and Piron, 1981) and to be aphid transmitted to up to 100% of test plants from source plants in which it was in joint infection with PLRV (Salazar *et al.*, 1995; Syller *et al.*, 1997). However, in a detailed analysis of this transmission route, Singh and Kurz (1997) found only low rates of transmission (7%) in association with PLRV and as, discussed in Chapter 11 (Section III.H.1.a), there are some open questions concerning aphid transmission of viroids. Thus, it is doubtful whether aphid transmission of these agents is of any significance in the field.

7. Epidemiology

The main methods by which viroids are spread through crops are by vegetative propagation, mechanical contamination, and through pollen and seed. The relative importance of these methods varies with different viroids and hosts. For example, vegetative propagation is dominant for PSTVd in potatoes and for CSVd in chrysanthemums. Mechanical transmission is a significant factor for others such as CEVd in citrus and HSVd in hops. Seed and pollen transmission are factors in the spread of ASBVd in avocados.

For most viroid diseases, the reservoir of inoculum appears to be within the crop itself, which raises the question as to where the viroid diseases came from. The evidence suggests that many viroid diseases are of relatively recent origin (Diener, 1987a). As Diener pointed out, none of the recognized viroid diseases was known to exist before 1900. In fact, many of them were first described after 1940. The sudden appearance and rapid spread of a new viroid disease can probably be accounted for by the following factors. Viroids are readily trans-mitted by mechanical means. Many modern crops are grown as large-scale monocultures. Thus, from time to time a viroid present in a natural host and probably causing no disease might escape into a nearby susceptible com-mercial crop and spread rapidly within it. If the viroid and crop plant had not evolved together, disease would be a likely outcome. There is direct evidence for such a sequence of events with the tomato planta macho disease in Mexico (Diener, 1987a). A study of HLVd in the UK suggests that the current prevalence of this viroid in hops is a consequence of infection becoming established in the hop propagation system during the late 1970s (Barbara *et al.*, 1990).

C. Structure of viroids

1. Circular nature of viroid RNA

The circular nature of many viroids was first shown directly by electron microscopy. Under non-denaturing conditions, the molecules appear as small rods with an axial ratio of about 20:1, and for PSTVd an average length of about 37 nm. When spread under denaturing conditions, the molecules can be seen to be covalently closed circles of about 100 nm con-tour length (Sänger *et al.*, 1976; McClements and Kaesberg, 1977; Riesner *et al.*, 1979). All viroid preparations contain a variable propor-tion of linear molecules.

2. Nucleotide sequences

The nucleotide sequences of about 27 members of the viroid group and those of numerous viroid variants are now known. They range in size from 246 to 375 nucleotides (Table 14.1). These sequences have confirmed the circular nature of the molecules (Fig. 14.1). All viroids have some degree of sequence similarity. Based on the degree of overall sequence similarity, 27 distinct viroids have been recognized, and these have been placed in seven genera (Table 14.1). Other distinct viroids continue to be discovered.

Most viroids have a relatively high G+C con-tent (53–60%), but ASBVd is rich in A+U (62%).

A database of sequences of viroids and viroid-like RNAs is published annually and can be found at http://www.callisto.si.usherb.ca/~jpperra (see Pelchat *et al.*, 2000).

The use of cDNA clones to sequence field iso-lates of viroids has revealed that a single isolate may contain a range of closely related sequence variants (e.g. Visvader and Symons, 1985).

3. Secondary structure

From the primary sequence it is possible to predict a secondary structure that maximizes the number of base pairs. Based on this viroids fall into two groups, the families mentioned above. The RNAs of members of the *Pospiviroidae* are predicted to form rod-like molecules with base-paired regions inter-spersed with unpaired loops, whereas those of the *Avsunviroidae* are less structured (Fig. 14.2).

Melting curves for double-stranded nucleic acids are discussed in Chapter 15 (Section V.C.1). Detailed studies of the melting curves of PSTVd, together with the sequence data, allowed Riesner *et al.* (1979) to develop a model for the stages in the denaturation of PSTVd RNA as the temperature is raised. A more refined model has been proposed by Steger *et al.* (1984) (Fig. 14.3).

It should be remembered that these struc-tures have been derived either from computer predictions or from *in vitro* experiments and that, *in vivo*, viroids may be associated with host proteins and have other structures. However, there is some evidence pointing to at least a partial rod-shaped structure *in vivo* in that viable duplications or deletions preserve this type of structure (Haseloff *et al.*, 1982; Semancik *et al.*, 1994; Wassenegger *et al.*, 1994).

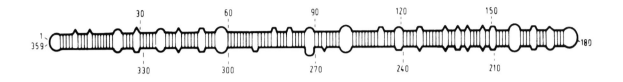

Fig. 14.1 Structure of PSTVd. **Lower:** Nucleotide sequence as determined by Gross *et al.* (1978); **centre:** proposed secondary structure in outline; **upper:** three-dimensional representation of the viroid molecule. From H. Sänger; used in Matthews (1991).

Viroids have tertiary structure that is thought to be important in interactions with host proteins (Branch *et al.*, 1985; Gast *et al.*, 1996).

a. Structure of Pospiviroidae

The predicted rod-like structures of the *Pospiviroidae* have five structural–functional domains that are common to all members (Fig. 14.2) (Keese and Symons, 1985). The five domains are: C (central), P (pathogenic), V (variable), T_R and T_L (terminal right and left respectively). These structures were originally thought to have specific functions but the situation is now considered to be more complex. For instance, symptom expression is thought to be controlled by determinants located within the T_L, P, V and T_R domains (Sano *et al.*, 1992).

The C domain contains the central conserved region (CCR) of about 95 nucleotides that comprises two sets of conserved nucleotides

located in the upper and lower strand flanked by an inverted repeat in the upper strand (Fig. 14.4). Based on the core nucleotides, three type of CCR have been recognized, exemplified by PSTVd, ASSVd and CbVd-1, and several subclasses of these can be distinguished.

The upper strands of the CCR of the five subgroups of the non-self-cleaving viroids are predicted to form similar hairpin structures (hairpin 1) that have conserved nucleotides in similar positions (Fig. 14.5).

Two further sequences are conserved between members of the *Pospiviroidae*. The terminal conserved sequence (TCR) is conserved between *Pospiviroid* and *Apscaviroid* genera and the two largest members (CbVd-2 and CbVd-3) of the *Coleoviroid* genus (Fig. 14.4). The conservation of this sequence, CNNGNGGUUCCUGUGG, and its similar location in viroids from three genera suggest that it has an important function. The TCR is

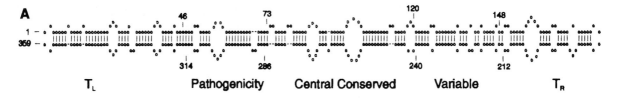

A

T_L   Pathogenicity   Central Conserved   Variable   T_R

Potato spindle tuber viroid

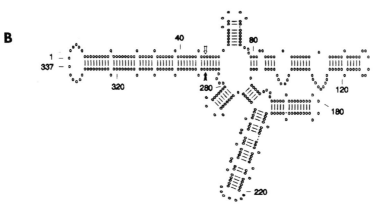

B

Peach latent mosaic viroid

Fig. 14.2 Secondary structures of PSTVd and PLMVd. **(A)** A rod-like secondary structure for PSTVd is supported by a variety of physical studies as well as chemical and enzymatic mapping data. Boundaries of the terminal-left (T_L), pathogenicity, central conserved, variable and terminal-right (T_R) domains are indicated by vertical lines. **(B)** Proposed lowest-free-energy structure of PLMVd. Predicted self-cleavage sites in the (+) and (–) strands are indicated by filled and open arrows respectively. From Owens (1999), redrawn from Hernández and Flores (1991), with kind permission of the copyright holders, © The National Academy of Sciences, USA.

not found in viroids of about 300 nucleotides or less but another sequence (CCCCUCUGGGGAA) is conserved in most of these smaller viroids. This conserved sequence forms the left terminal conserved hairpin (TCH) (Fig. 14.4).

The P domain contains an adenine-dominated purine-rich sequence in one strand and an oligo(U) sequence in the opposite strand. It has been implicated in pathogenesis (Section I.E).

The V domain is the most variable region in the molecule and may show less than 50% homology between otherwise closely related viroids.

The main sequence homologies between PSTVd-like viroids are in the two terminal domains shown in Figs 14.2 and 14.4. They have been implicated in viroid replication (Section I.D).

b. Structure of Avsunviroidae

Members of this family differ markedly from other viroids in that they lack a CCR and their RNAs of both polarities contain hammer-head ribozymes (see Section I.D.4.b). The RNAs of CChMVd and PLMVd adopt a complex branched secondary structure (Bussière *et al.*, 2000).

D. Replication of viroids

Studies on viroid replication have been reviewed in detail by Sänger (1987), Robertson and Branch (1987), Flores *et al.* (1997) and Symons (1997).

Even if it was assumed that the three out-of-phase potential reading frames were fully utilized, viroids do not contain sufficient

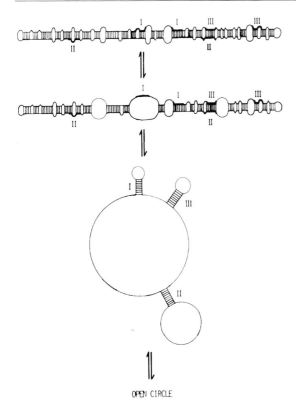

OPEN CIRCLE

Fig. 14.3 Mechanism of denaturation and renaturation of PSTVd. Schematic structures are represented at temperatures of approximately 25, 70, 75 and greater than 95°C. The formation of two or three stable hairpins (I, II, III) is possible with PSTVd during the main transition. The depicted hairpins are formed by base pairing from the following regions: (I) 79–87/110–102, (II) 227–236/328–319 and (III) 127–135/168–160. Analogous hairpins to hairpins I and II of PSTVd are found in CEVd and CSVd. In CCCVd, only hairpin 1 is found. From Steger *et al.* (1984), with permission.

information to code for an RNA replicase. In principle, however, they could code for a relatively small polypeptide that could combine with host proteins to form a viroid-specific replicase, in the way that the small bacteriophage Qβ operates in its host. Nevertheless, it is now generally accepted that viroids are not translated to give any polypeptides. The main lines of evidence are: (1) a few viroids contain one or more AUG codons, but many, such as PSTVd and its variants, contain none; however, one must not ignore non-AUG start codons (see Chapter 7, Section V.B.6); (2) attempts to demonstrate mRNA function for viroid RNAs *in vitro* using various systems under a range of

conditions have all failed; and (3) analysis of the leaf proteins present in healthy and viroid-infected tissues has failed to reveal any new viroid-specified polypeptides.

1. Site of replication

As noted above, the RNAs of members of the *Pospiviridae* are found in the nucleus and the available evidence is that this family of viroids replicates there. On the other hand, members of the *Avsunviroidae* are associated with chloroplasts. Navarro *et al.* (1999) found RNA forms of ASBVd indicative of being replication intermediates in chloroplasts and, as described below, there is evidence for a chloroplast polymerase being involved in the replication of this viroid. Similarly, both (+) and (–) strands of PLMVd were found in chloroplasts (Bussière *et al.*, 1999).

Entry of viruses into nuclei and plastids is usually controlled by a signal often on a protein associated with the viral genome (see Chapter 9, Section II.C). The import of PSTVd into nuclei or permeabilized tobacco BY2 cells was studied using fluorescein-labeled or GFP-labeled transcripts (Woo *et al.*, 1999; Zhao *et al.*, 2001). Nuclear import was observed for PSTVd but not for mRNA fragments of the same size or for two viroids, ASBVd or PLMVd, thought to replicate in chloroplasts. Import was inhibited by the addition of a 10-fold excess of non-fluorescent PSTVd but not by similar amounts of non-imported RNAs. These and other observations were taken to indicate that (1) PSTVd has a sequence and/or structure motif(s) for nuclear import and (2) the import is independent of cytoskeleton and is mediated by a specific and saturatable receptor.

It is likely that members of the *Avsunviroidae* have signals for chloroplast import, but these have not yet been determined.

2. The template for replication

RNA strands complementary to viroid RNA (defined for viroids as (–) strands) were first described by Grill and Semancik (1978) in tissue infected with CEVd. Owens and Cress (1980) showed that the PSTVd (–) strands present in RNase-treated dsPSTVd RNA had the same electrophoretic mobility as linear 359-nucleotide (+) strands. Zelcer *et al.* (1982)

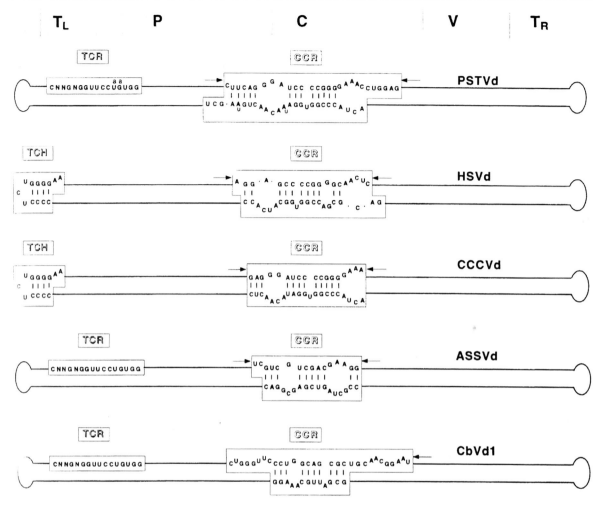

Fig. 14.4 Rod-like structure models for the different subgroups of non-self-cleaving viroids. The type member of each genus is indicated to the right and the approximate locations of the five domains C (central), P (pathogenic), V (variable) and T$_L$ (terminal left) and T$_R$ (terminal right) are on the top of the figure. The core nucleotides of the central conserved region (CCR), terminal conserved region (TCR) and terminal conserved hairpin (TCH) are shown. Arrows indicate flanking sequences which form, together with the core nucleotides of the CCR upper strand, imperfect inverted repeats. The core nucleotides of HSVd CCR have been defined by comparison with CLVd, which, based on other criteria (the presence of the TCR, the absence of the TCH and overall sequence similarity), is included in the PSTVd genus. Substitutions found in the CCR and TCR of IrVd, a member of the PSTVd genus, are indicated by lower-case letters. In the CbVd genus the TCR exists only in the two largest members CbVd2 and 3. From Flores *et al.* (1997), with permission.

found that PSTVd-infected tissues contained RNA complementary to the entire viroid RNA. Later studies showed that the PSTVd (–)-strand RNA existed as a tandem multimer of several unit-length monomers (e.g. Branch *et al.*, 1981). These were present as complexes with extensive double-stranded regions. Further studies showed that monomeric PSTVd (+) strands, both circular and linear, are complexed with

long multimeric (–) strands. Synthesis of the ds complex increases simultaneously with that of ss (+)-sense RNA (Owens and Diener, 1982). Time-course studies with PSTVd-infected protoplasts have shown that synthesis of oligomeric linear (–)-strand RNA preceded that of the (+)-strand oligomers, suggesting that the (–)-strand oligomers represent transient intermediates (Faustmann *et al.*, 1986).

```
PSTVd       HSVd       CCCVd      ASBVd      CbVd1

  C  G        C  G        C  G        C  G        C  G
 C    G      C    G      C    G      U    A      G    C
 C · G       C · G       C · G       G · C       A · U
 G · G       C · G       C · G       G · G       G · G
 U · A       G · C       U · A       U · A       G · C
A      A    A      A    A      A    G      A    G      A
G      A    G      A    G      A     C · G      U      A
 G · C       G · C       G · C       U · G       C      C
 G · C       A · U       G · C       G · C       C · G
 A · U       A · U  C    A · U      U      C     C · G
 C · G       A · U       G · C       C · G       U · A
 U · G       G · C       U · A       C · G       U · A
 U · A       A · U       U · A       A · U       G · U
 C · G       G · C       C · G       C · G       G · C
 G · C       U · A       G · C       U · A       G · C
 G · G       G · G       G · G       G · G       U · A
                                                 G · G
```

Fig. 14.5 Hairpin 1 structures that can be formed by the CCR upper strand of the five subgroups of non-self-cleaving viroids. Outlined fonts indicate conserved nucleotides in similar positions in all the structures. From Flores *et al.* (1997), with permission.

3. Mechanism for viroid RNA replication

The experiments noted briefly in the previous section and many others (e.g. Steger *et al.*, 1986) demonstrated that viroids replicate via an RNA template. The existence of closed circular monomers complexed to long linear multimers makes it highly probable that a rolling circle type of replication gives rise to progeny viroid RNA and it is generally accepted that this is the mechanism. Several models of the rolling circle mechanism have been put forward (e.g. Branch *et al.*, 1981; Owens and Diener, 1982; Branch and Robertson, 1984; Ishikawa *et al.* 1984; Hutchins *et al.*, 1985). Figure 14.6 (left-hand side) illustrates two rolling circle models. In the asymmetrical pathway, the infecting circular (+)-strand monomer is transcribed into linear multimeric (–) strands, which then are the template for the synthesis of linear multimeric (+) strands. In the symmetrical pathway, the linear multimeric (–) strands are processed and ligated to give (–) monomer circles which are the template for linear multimeric (+)-strand synthesis. In both cases, the multimeric (+) strand is processed to give monomeric circles. As the symmetrical pathway involves both (+)- and (–)-strand circular forms and the asymmetrical pathway only (+)-strand circular forms, the two mechanisms can be distinguished by the presence or absence of the (–)-strand circular form. This RNA species has not been found in

plants infected with PSTVd (see Branch *et al.*, 1988a) and thus the replication of this viroid is considered to follow the asymmetrical pathway. In contrast, (–)-strand circular monomer RNA forms have been found in ASBVd infections (Hutchins *et al.*, 1985; Daròs *et al.*, 1994), which suggests that replication of this viroid follows the symmetrical pathway.

4. Enzymes involved viroid replication

Given that viroids code for no polypeptides of their own, we must assume that they use pre-existing host nucleic acid-synthesizing enzymes. Healthy plants contain two kinds of RNA-synthesizing enzymes: DNA-dependent RNA polymerases I, II and III, and an RNA-dependent RNA polymerase. Early experiments on viroid replication, particularly the finding that actinomycin D inhibited viroid replication in leaf tissue, suggested that viroids might replicate via a DNA template. However, no viroid-specific DNA has been found in infected tissue (Branch and Dickson, 1980; Zaitlin *et al.*, 1980; Hadidi *et al.*, 1981), and it is now accepted that viroids replicate via a (–)-strand RNA template (see above).

Three enzymatic activities are required for viroid replication by the rolling circle mechanism: an RNA polymerase, an RNase for processing the multimeric products of rolling circle replication and an RNA ligase to produce circular monomers.

a. RNA polymerases

The host polymerases involved in viroid RNA replication have not been identified beyond doubt and there has been considerable controversy on this matter.

Mühlbach and Sänger (1979) studied the effect of α-amanitin on replication of cucumber pale fruit viroid (now known as HSVd) in tomato protoplasts. They found that viroid synthesis was inhibited by concentrations around 10^{-8} M, which is characteristic for the inhibition of DNA-dependent RNA polymerase II by this drug. Rackwitz *et al.* (1981) purified the enzyme from healthy plant tissues and showed that it is capable of synthesizing full-length linear (–)-strand viroid copies from

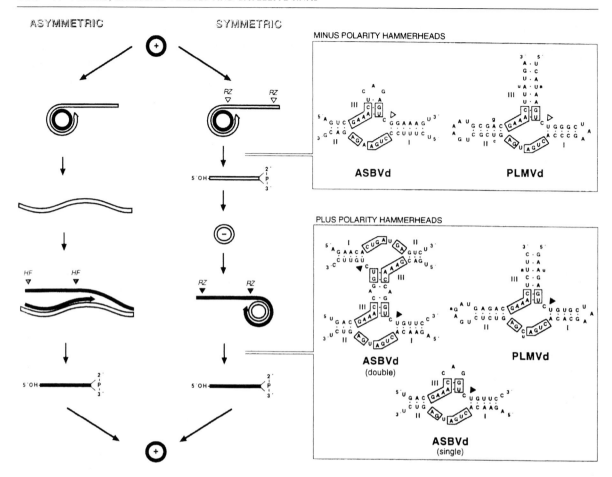

Fig. 14.6 Models for viroid replication. The (+) polarity (solid lines) is assigned by convention to the most abundant infectious RNA, and the (–) polarity (open lines) to its complementary strand. The alternative asymmetrical and symmetrical pathways involve one or two rolling circles respectively. In the symmetrical variant, cleavage of (+) and (–) multimeric strands is mediated by hammerhead ribozymes (RZ), which lead to linear monomeric RNAs with 5'-hydroxyl and 2',3'-cyclic phosphate termini. Arrowheads denote cleavage sites. The hammerhead structures that can be formed by ASBVd and PLMVd RNAs are shown to the right; conserved nucleotides are boxed and substitutions found in two PLMVd cDNA clones are indicated by lower-case letters. In the asymmetrical variant, cleavage of (+) multimeric strands is mediated by a host factor (HF), which generates linear monomeric RNAs containing probably 5'-hydroxyl and 2',3'-cyclic phosphate termini. From Flores *et al.* (1997), with permission.

(+)-strand templates *in vitro*. As other enzymes can copy viroid RNAs *in vitro*, this work does not demonstrate a role for the polymerase II enzymes (Symons *et al.*, 1985). Polymerase II initiates synthesis of different viroid (–) strands at specific sites (Sänger, 1987). Various workers have shown that viroids replicate and accumulate in the nucleus, a location that would fit with a role for DNA polymerase II (e.g. Takahashi and Diener, 1975). In addition, Rivera-Bustamente and Semancik (1989) prepared a subnuclear fraction from nuclei of

CEVd-infected tissue that was able to synthesize CEV RNA.

Other work suggested a possible role for DNA-dependent RNA polymerase I, which normally carries out ribosomal RNA synthesis in the nucleolus. Riesner *et al.* (1983) showed that PSTVd RNA accumulates in the nucleolus, while Palukaitis and Zaitlin (1987) described regions of sequence similarity between regions of PSTVd and an rDNA promoter sequence. However, DNA-dependent polymerase III transcribes viroid RNA *in vitro* into complete copies whereas polymerase I produces low yields of smaller products (Sänger, 1987).

As a third possibility, it was found that an RNA-dependent RNA polymerase from healthy tomato plants can use PSTVd RNA as a template to produce full-length copies in low yield (Boege et al., 1982). Synthesis was not inhibited by relatively high concentrations of α-amanitin. The viroid copy was the first well-defined homogeneous product reported to be synthesized *in vitro* by an RNA-dependent RNA polymerase from healthy plants (Sänger, 1987).

Studies with isolated nuclei strongly suggest that the transcription of (–) strands to (+) strands of *Pospiviroidae* is carried out by the DNA-dependent RNA polymerase II found in the nucleoplasm and it is now generally accepted that this is the enzyme involved in this phase of replication (see Schindler and Mühlbach, 1992). However, it is uncertain as to whether this enzyme is involved in the (+) strand to (–) strand phase of replication.

The replication of ASBVd, and particularly of its (+) strand, is insensitive to α-amanitin (Marcos and Flores, 1992). This suggests that ASBVd either uses a nuclear α-amanitin-insensitive polymerase or that it replicates in the chloroplasts where the RNA polymerase is α-amanitin-insensitive. Two main chloroplastic RNA polymerases have been described, one plastid encoded and the other nuclear encoded. The replication of ASBVd is resistant to the inhibitor tagetitoxin, whereas the transcription of several chloroplast genes is blocked by this inhibitor (Navarro et al., 2000). Taken with the evidence that ASBVd replication occurs in the cytoplasm, it seems likely that it is catalyzed by the nuclear-encoded polymerase. However, the possibility of an unrecognized tagetitoxin-resistant chloroplast-encoded polymerase cannot be ruled out.

b. RNA cleavage

Hutchins et al. (1986) prepared tandem dimeric cDNA clones of ASBVd, and RNA dimers were transcribed from these clones. They found that self-cleavage of both (+)- and (–)-RNA transcripts occurred at specific sites in each tran-

script to give (+)- and (–)-viroid monomers. They proposed a hammerhead-like secondary structure around the cleavage site (Fig. 14.6) similar to those proposed for satellite RNAs (see Fig. 14.15). However, the ASBVd hammerhead structure proposed is unlikely to be sufficiently stable. Forster et al. (1988) therefore proposed a double hammerhead structure involving the association of two (+)-strand or two (–) hammerheads of ASBVd (Fig. 14.7). This model has two advantages: (1) the proposed cleavage site has sufficient stability and (2) monomer-length RNAs could not cleave further. Only dimers or multimers could form the double hammerhead structure and thus be self-cleaved. As with the satellite RNAs discussed in Section II.B.5, self-cleavage requires a divalent ion and gives rise to 5'-hydroxyl and 2',3'-cyclic phosphate 3' termini. The sequences required to form the (+)- and (–)-strand self-cleavage structures in ASBVd are situated side by side in the middle of the molecule. About

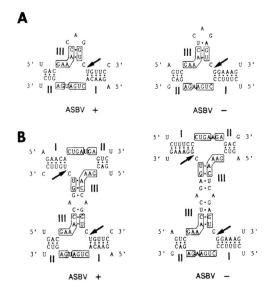

Fig. 14.7 Hammerhead structures proposed for the self-cleaving ASBVd RNA. **(A)** Single hammerhead structures proposed by Hutchins *et al.* (1986), which are similar to those proposed for self-cleaving satellite RNAs (see Fig. 14.15). **(B)** Double hammerhead structure proposed for ASBVd RNA by Forster *et al.* (1988). Sites of cleavage are indicated by arrows. Nucleotides conserved between ASBVd, and newt self-cleaving RNA and plant satellite RNAs are boxed. From Forster *et al.* (1988), with kind permission of the copyright holders, © Macmillan Magazines Ltd.

one-third of the total viroid sequence appears to be devoted to the self-cleavage function (Symons *et al.*, 1987).

Not all strands of self-cleaving viroid and satellite RNAs are cleaved by hammerhead structures. These *in vitro* self-cleaving RNAs fall into three groups (Table 14.2) (Symons, 1997): those in which both the (+) and (–) strands are cleaved via a hammerhead structure, those in which just the (+) strand is cleaved by a hammerhead structure, and those

in which the (+) strand is cleaved by a hammerhead structure and the (–) strand by a hairpin structure. These two self-cleavage structures are illustrated in Fig. 14.8.

Secondary structures such as those proposed for the processing of ASBVd and satellite RNAs have not been found in other viroids such as PSTVd and CEVd. Furthermore, Tsagris *et al.* (1987a) tested oligomeric linear PSTVd RNAs under a wide range of conditions *in vitro*, but could find no evidence for autolytic processing,

TABLE 14.2 Plant pathogenic RNAs that cleave *in vitro*

| | Size (nt) | RNA self-cleavage structure | |
		(+) RNA	(−) RNA
Viroids			
ASBVd	246–251	Hammerhead	Hammerhead
PLMVd	337–338	Hammerhead	Hammerhead
Viroid-like RNA			
CarSV RNA	275	Hammerhead	Hammerhead
Satellite RNAs			
sLSTV	322–324	Hammerhead	Hammerhead
sSNMV	377	Hammerhead	–
sSCMoV	322 and 328	Hammerhead	–
sVTMoV	365–366	Hammerhead	–
SARMV	300	Hammerhead	Hairpin
sCYMV	457	Hammerhead	Hairpin
sTRSV	359–360	Hammerhead	Hairpin
sBYDV	322	Hammerhead	Hammerhead

From Symons (1997), with permission.

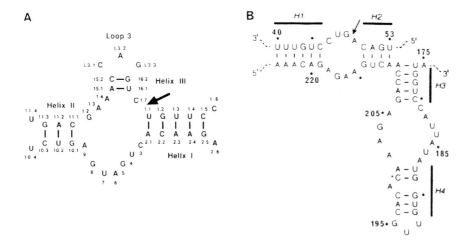

Fig. 14.8 Hammerhead and hairpin self-cleavage structures. **(A)** Hammerhead self-cleavage structure of (+) ASBVd. The residue numbering system was introduced in order to simplify the comparisons of data from different laboratories. **(B)** Hairpin self-cleavage structure of (–) sTRSV. Residue numbers are those of the (+) RNA and hence run in the 3' to 5' direction. H1–H4 are the four helices. The arrow indicates the self-cleavage site in both structures. From Symons (1997), with kind permission of the copyright holders, © Oxford University Press.

as has been found for ASBVd. On the other hand, such oligomers were accurately processed to monomers when incubated with a nuclear extract from healthy potato cells (Tsagris *et al.*, 1987b). Various treatments of the extracts showed that excision of the linear monomers required the presence of catalytically active protein in the extract. The central conserved domain plays an important role in the generation of monomers from multimers for viroids such as PSTVd, but other less efficient sites may be used (Hammond *et al.*, 1989). There are other possible mechanisms of cleavage, as discussed by Flores *et al.* (1997) and Liu and Symons (1998).

c. RNA *ligases*

The last step in viroid replication must be the ligation of linear monomers to give circular molecules. Linear PSTVd monomers were efficiently converted to the circular form by an RNA ligase isolated from wheatgerm (Branch *et al.*, 1982). The nuclear extract from potato cells shown by Tsagris *et al.* (1987b) to cleave PSTVd multimer RNAs into linear monomers also ligated these monomers to give (+)-strand circles. The ligation process is presumably assisted by the 5'-hydroxyl and 2',3'-cyclic phosphate termini formed during cleavage of the multimers. Kikuchi *et al.* (1982) described a novel type of RNA ligase from wheatgerm that circularized linear viroid RNA to form 2' phosphomonoester and 3',5' phosphodiester bonds. It has been suggested that the self-cleaving ribozymes can self-ligate (Lafontaine *et al.*, 1995) but, as discussed by Flores *et al.* (1997), it is thought likely that a host enzyme is involved.

5. Initiation of replication

There is accumulating evidence that replication of viroids is initiated at specific sites on the circular template. Linear (+) and (–) strands of ASBVd were studied using primer extension and terminal capping, and were shown to begin with the sequence UAAAA that maps to A+U-rich terminal loops in the predicted secondary structure (Navarro and Flores, 2000). The sequences around the initiation sites of this viroid are similar to promoters of a nuclear-

encoded chloroplastic RNA polymerase. Using *in vitro* transcription with nuclear extracts from potato cells, the initiation of PSTVd strands was proposed to occur at two different positions, neither in a terminal loop (Riesner *et al.*, 1999).

6. Evidence for recombination between viroids

The available nucleotide sequence data make it highly probable that recombination events have taken place in the past between different viroids, presumably during replication in mixed infections (Keese and Symons, 1985). For example, TASVd appears to be a recombinant viroid made up of most of the sequence of a CEVd-like viroid but with the T2 domain replaced by a T2 domain from a PSTVd-like viroid. Other examples are known (Keese and Symons, 1985; Hammond *et al.*, 1990). AGVd appears to have originated from extensive RNA recombination between four other viroids. Its sequence can be divided into regions, each with high sequence similarity with sequences of CEVd, PSTVd, ASSVd or GYSVd, the central conserved domain being that of ASSVd (Rezaian, 1990).

7. Interference between viroids

Fernow (1967) showed that inoculation of tomato plants with a very mild strain of PSTVd gave substantial protection against a second inoculation with a severe strain applied 2 weeks after the first. However, the mild strain has to be established first. In experiments in which a mild and a severe strain of PSTVd were inoculated simultaneously, the severe strain dominated and most plants developed severe disease, even when the mild strain was in 100-fold excess in the inoculum (Branch *et al.*, 1988b). Cross-protection occurs not only between strains of a particular viroid but also between different viroids (Niblett *et al.*, 1978). When PSTVd RNA transcripts from a cloned PSTVd DNA were inoculated together with HSVd RNA, the PSTVd reduced the level of HSVd RNA in infected plants. Plants inoculated with dual transcripts (two copies of a severe PSTVd strain linked to two of HSVd) developed PSTVd symptoms and only PSTVd progeny RNA could be detected. The molecular basis for this interference is not understood but it would be interesting to know whether the post-translational gene silencing defense system was involved (see Chapter 10, Section IV).

With CEVd there was variation in the protection given against a severe strain by a mild strain (Duran-Vila and Semancik, 1990). The level of protection was dependent on the length of interval between inoculation of the mild and severe strains.

E. Molecular basis for biological activity

Because of their very small size, their autonomous replication, the known structure of many variants, and the lack of any viroid-specific polypeptides, it has been a hope of many workers that viroids might provide a simple model system that would provide insights as to how variations in the structure of a pathogen modulate disease expression. This hope has not yet been realized. As discussed in the following paragraphs, very small changes in nucleotide sequence may give rise to dramatic changes in the kind of disease induced by a viroid. Therefore, disease induction must involve specific recognition of the viroid sequence by some host macromolecule. Until the nature of this host macromolecule (or molecules) is known, the interpretation of correlations between nucleotide sequence and biological properties of viroids will remain speculative. At present, the only practicable biological properties that can be observed are infectivity and severity of disease, although viroid-binding proteins are starting to be recognized. Five approaches are being used and these will be briefly summarized.

1. Sequence comparisons of naturally occurring viroid variants

The sequence comparison approach takes advantage of the fact that many sequence variants of a particular viroid may exist in nature, which provide a range of viable mutants. A major difficulty is that many natural infections contain a mixture of viroid variants. This problem has been overcome by the preparation of cDNA clones of isolates each with a single defined nucleotide sequence. The work of Visvader and Symons (1985) is an example of this approach. By cDNA cloning, they isolated 11 new variants of CEVd. These variants, together with six already known, fell into two classes, severe and mild, with respect to symptom severity in tomatoes. Sequence differences lay in the pathogenicity and variable domains indicated in Fig. 14.2. Later work using cDNA chimeric clones located pathogenicity in the P_L domain (Visvader and Symons, 1986).

For PSTVd, only the P domain has been shown to be involved in pathogenicity (Schnölzer et al., 1985). They placed PSTVd variants in four classes with respect to disease severity in tomato: mild, intermediate, severe and lethal. Increasing severity of disease was correlated with decreasing stability in vitro of the secondary structure in the virulence-modulating domain. Such a correlation was not observed with the CEVd variants described above (Visvader and Symons, 1985).

An interesting potential application of viroids is as a dwarfing agent for citrus trees (see Semancik et al., 1997). Sequence changes in the putative pathogenicity domain of CVd-III can lead to a marked reduction in symptom expression in Etrog citrus (Owens et al., 1999). It is suggested that, with a suitable screening assay, it should be possible to select stable variants with different dwarfing properties from populations of mutagenized CVd-III synthesized in vitro.

2. Site-directed mutagenesis

In principle, the ability to prepare infectious cDNA clones of viroids should allow the application of site-directed mutagenesis to study of the effects of structural changes on disease induction. In practice, it has been found that many base changes in a viroid sequence render it non-infectious, even those changes that might be expected to have a minimal effect on secondary structure. When a PSTVd cDNA containing a single base change was introduced into the host plant in an Agrobacterium vector, wild-type viroid was recovered from the inoculated plant (Owens et al., 1986; Hammond and Owens, 1987). This reversion may well be a consequence of the high error frequency characteristic of RNA copying (Chapter 8, Section

IX.A; see Fig. 8.29). Because of this reversion, the progeny arising from inoculations with mutant viroid cDNAs must be examined carefully to determine whether or not they have maintained the alteration(s) present in the cDNA inoculum.

3. Chimeric viroids

Construction of chimeric viroid cDNAs provides another possible approach to the study of disease induction, but this has so far met with little success. For example, Owens *et al.* (1986) constructed mixed tandem dimers of full-length PSTVd and TASVd cDNAs. These were infectious, but analysis of the progeny showed that individual plants contained only normal PSTVd or TASVd; none of the expected chimeras was found. In another approach, Owens *et al.* (1986) constructed chimeric cDNA monomers between parts of the PSTVd and TASVd cDNA molecules, but these were not infectious. Hammond *et al.* (1989) made a chimera between PSTVd and TPMVd that produced infectious progeny. Owens *et al.* (1990) constructed infectious chimeras between TASVd and CEVd. Symptoms in tomato were milder than those produced by TASVd, and resembled those produced by CEVd, which was the source of the pathogenicity domain in the construct.

4. Sequence comparisons with host RNAs

Another possible way to attempt to identify viroid nucleotide sequences that function in pathogenesis is to search for possible host RNA sequences to which viroid sequences might bind. For example, during ribosomal RNA processing there is an intron-like sequence between 5.8S RNA and 25S RNA that is excised. U3 small nuclear RNA is thought to bind to the intron-like sequence in the reaction leading to excision. Jakab *et al.* (1986) found sequence similarities between two domains of PSTVd and broad bean U3 snRNA. On this basis they suggested that viroids (except ASBVd) exert their pathogenic effects by interfering with normal pre-rRNA processing in nucleoli. Haas *et al.* (1988) showed that the 7S RNA from tomato leaf tissue resembles a sequence recognition particle RNA. It also has a notable sequence complementarity to the part of PSTVd RNA known to modulate virulence. Thus, this 7S RNA is another possible host target for viroid RNA.

5. Viroid-binding proteins

A 43-kDa host protein was found to bind to PSTVd when the viroid was mixed *in vitro* with nuclear extracts (Klaff *et al.*, 1989). A similar complex was found *in vivo* in the nuclei of infected cells.

Using a screening assay for detecting and isolating RNA-binding protein, Sägesser *et al.* (1997) detected a protein from a tomato cDNA expression library that bound PSTVd.

6. Summary

All of these approaches are a long way from explaining how pathogens as small as viroids cause such a variety of diseases, ranging from symptomless to lethal. Further significant progress in this important area will have several prerequisites, including the following: (1) specific *in vitro* and *in vivo* assays for different viroid functions, (2) a knowledge of viroid structure (or structures) *in vivo,* and (3) identification of the host macromolecules with which viroids interact during the processes of infection, replication and movement in the plant.

F. Diagnostic procedures for viroids

As viroids produce no specific proteins, the immunological methods applied so successfully to viruses cannot be used for the diagnosis of diseases caused by viroids. Similarly, because no characteristic particles can be detected, electron microscopic techniques are inappropriate. For these reasons diagnostic procedures have been confined to biological tests, gel electrophoresis, PCR and nucleic acid hybridization tests.

1. Biological tests

Biological tests for viroid detection and diagnosis have been important where suitable diagnostic test plants have been identified, and they remain important for some viroids, for example where strains of different severity exist. Some indicator hosts have been found that give severe symptoms with strains causing both mild and severe disease in the crop of interest (e.g. Singh, 1984). However, no suitable indicator hosts have been found for some viroid diseases such as cadang-cadang disease. Mild isolates of viroids may produce barely detectable symptoms. Environmental conditions may affect markedly the disease produced by other isolates. For such reasons, *in vitro* tests based on the properties of the viroid RNA have assumed considerable importance.

2. Gel electrophoresis

Viroids generally occur in very low concentration in the infected host. Thus, some partial purification and concentration procedure must be used before the nucleic acids are run in an appropriate polyacrylamide gel electrophoresis system. Various modifications have dramatically increased the sensitivity of viroid detection by electrophoresis. For example, Schumacher *et al.* (1986) devised a 'return' gel technique in which the samples were first subjected to electrophoresis under non-denaturing conditions. Denaturing conditions to produce open viroid circles are then applied, and the polarity of the electric field is reversed. Viroid molecules then lag behind, well clear of all host RNA species. Such a procedure has been used to detect PSTVd in single true seeds of potato (Singh *et al.*, 1988b).

3. Nucleic acid hybridization

Methods for detection and diagnosis based on nucleic acid hybridization are important. For application in routine testing it was necessary to develop methods for the large-scale production of cloned, highly labeled, viroid cDNA and to develop procedures where clarified tissue extracts could be used in the tests (e.g. Owens and Diener, 1981). Methods for preparing labeled nucleic acid probes are discussed in Chapter 15 (Section V.C.5). Extraction of nucleic acids from fruit trees can pose problems because of the high levels of tannins. An extraction procedure based on non-organic solvents has been suggested by Astruc *et al.* (1996). RNA probes are now used for routine detection of PSTVd in tuber flesh and sprouts (Salazar *et al.*, 1988a). Sano *et al.* (1988) used synthetic

oligonucleotide probes to diagnose HSVd strains and CEVd. The dot blot hybridization procedure is now being widely used (e.g. for CSVd, see Candresse *et al.*, 1988). This procedure has been coupled with the use of non-radioactive DNA probes to provide a sensitive procedure for the routine diagnosis of viroids in plant extracts (McInnes *et al.*, 1989; Roy *et al.*, 1989). Several viroids including PSTVd and citrus viroids have been detected by tissue print hybridization (see Romero-Durbán et al., 1995; StarkLorenzen *et al.*, 1997; Palacio-Bielsa *et al.*, 1999).

4. Reverse transcription–polymerase chain reaction

Viroid RNAs can be amplified by first converting them to DNA by reverse transcription and then using the DNA for the polymerase chain reaction; this is known as RT-PCR. RT-PCR has been applied in the diagnosis of several viroids, for example ASBVd, ASSVd and CEVd (Hadidi and Yang, 1990; Yang *et al.*, 1992; Schnell *et al.*, 1997).

5. General

With present procedures, it is relatively simple to verify a positive test result for viroid infection. It is much more difficult to ensure that a negative result means a viroid-free plant. This may be an important practical issue where vegetative crops are concerned. Methods for the control of viroid diseases are discussed in Chapter 16 (Section II).

II. SATELLITE VIRUSES AND SATELLITE RNAs

Purified virus preparations isolated from infected plants may contain a variety of nucleic acids other than the genomic nucleic acids. Some of these, such as subgenomic RNAs and defective RNAs or DNAs, were discussed in Chapters 7 and 8. In addition, some isolates of certain plant viruses contain satellite agents, which are subviral agents that lack genes that would encode the enzymes needed for their replication. Two major classes of satellite agents can be distinguished according to the source of the coat protein used to encapsulate the nucleic acid. In *satellite viruses*, the satellite nucleic acid codes for its own coat protein. In *satellite RNAs* or *DNAs*, the nucleic acid becomes packaged in protein shells made from coat protein of the helper virus. Satellite viruses and satellite nucleic acids have the following properties in common:

- Their genetic material is a nucleic acid molecule of small size. The nucleic acid is not part of the helper virus genome and usually has little sequence similarity to it apart from the terminal regions.
- Replication of the nucleic acid is dependent on a specific helper virus.
- The agent affects disease symptoms, at least in some hosts.
- Replication of the satellite interferes to some degree with replication of the helper.
- Satellites are replicated on their own nucleic acid template.

Both RNA and DNA viruses have satellites associated with them, those of RNA viruses being RNA and those of DNA viruses being DNA. Satellite viruses and nucleic acids can be categorized into six groups based on their properties (Table 14.3) (Bruening, 2000). Not included in the table are defective RNAs which have sequences derived from those of the helper virus; these are described in detail in Chapter 8 (Section IX.C). The coat-dependent RNA replicons are included in this section as their sequences differ from those of the helper virus.

Several satellite RNAs associated with a particular group of viruses have been shown to have viroid-like structural properties. These agents have been termed 'virusoids', but this term is no longer favored. Satellite viruses and satellite nucleic acids have been reviewed by Scholthof *et al.* (1999b) and Bruening (2000).

There appears to be no taxonomic correlation between the viruses that are associated with satellites, and satellism seems to have arisen a number of times during virus evolution. Some viruses are associated with more than one satellite, and satellites can even require a second satellite as well as the helper virus for replication.

This section summarizes the properties of satellite viruses and satellite RNAs and DNAs. Their methods of replication, and the ways in which they induce or modulate expression of disease, are also discussed.

A. Satellite plant viruses

Four definite satellite plant viruses have been described and there is one other possible satellite virus. They are the smallest known viruses. As shown in Table 14.3 these are known as type A satellites.

1. Satellite tobacco necrosis virus (STNV) (reviewed by Scholthof et al., 1999b)

STNV was the first satellite virus to be recognized, the term 'satellite virus' being coined to denote the 17-nm isometric particles associated with the 30-nm TNV isometric particles (Kassanis, 1962) (Table 14.4). The structure of STNV particles is described in Chapter 5 (Section VI.B.1). The helper *Necrovirus*, TNV (see Chapter 6, Section VIII.D.6), replicates

TABLE 14.3 Satellite viruses, nucleic acids and coat-dependent RNA replicons. From Bruening (2000), with permission.

Nucleic acid description[a]	Example	Independent replication[b]	Directs protein synthesis[c]	Coat protein[d]
Non-messenger satellite RNA (C type)	sCMV RNA	No	No	Helper virus
Non-messenger satellite RNA (D type)	sTRSV RNA	No	No	Helper virus
Messenger satellite RNA (B type)	BsatTomBRV RNA	No	Yes	Helper virus
Satellite virus genomic RNA (A type)	STNV genomic RNA	No	Yes	Own
Satellite DNA		No	No	Helper virus
Coat-dependent RNA replicon	ST9aRNA	Yes	Yes	Helper virus

[a] Satellite types defined by Mayo *et al.* (1995).
[b] Replication independent of helper virus.
[c] Does the satellite nucleic acid express a protein?
[d] The coat protein encapsidating the satellite nucleic acid.

TABLE 14.4 Properties of satellite and helper viruses

	Satellite				Helper		
Virus	Particle[a]	Genome (nt)	CP (kDa)	Virus	Particle[a]	Genome (nt)	CP (kDa)
STNV-1	Sphere (18)	1239	21	TNV	Sphere (28)	3684	30
STNV-2	Sphere (18)	1245	21	TNV	Sphere (28)	3684	30
SPMV	Sphere (16)	826	17	PMV	Sphere (30)	4326	26
SMWLMV	Sphere	1168	24	MWLMV	Sphere (35)	–	33
STMV	Sphere (18)	1059	17.5	TMV	Rod (300 × 18)	6395	17

[a] Values in parentheses denote the size in nanometers, where known.

independently of other viruses and normally infects plant roots in the field. It had been known for some time that certain cultures of TNV contained substantial amounts of a smaller virus-like particle. Kassanis and colleagues (e.g. Kassanis, 1962) showed that the smaller particle with a diameter of about 17 nm, now know as STNV, depended for its replication on the larger virus. There is significant specificity in the relationship between satellite and helper. Strains of both viruses have been isolated, and only certain strains of the helper virus will activate particular strains of the satellite (Uyemoto et al., 1968). The ability of particular TNV isolates to activate the satellites was correlated with the ability of TNV to infect bean or tobacco plants but not with their serological relatedness (Kassanis and Phillips, 1970).

Three major strains of STNV have been defined by serological analysis, amino acid composition of the coat proteins, host range, fungal vector specificity and the TNV isolates that support their replication. STNV-1 and STNV-2 are supported by strains of TNV that readily infect tobacco and bean following mechanical inoculation. STNV-C does not multiply in the inoculated leaves but must be maintained in roots of tobacco plants (Rees et al., 1970).

Both STNV and TNV are transmitted by the zoospores of the fungus *Olpidium brassicae* (Chapter 11, Section XII). Transmission depends on an appropriate combination of four factors: satellite and helper virus strains, race of fungus, and species of host plant (e.g. Kassanis and Macfarlane, 1968).

The complete nucleotide sequence of STNV RNA was one of the first viral sequences to be determined (Fig. 14.9A). STNV RNA has no significant sequence similarity with the TNV genome. The arrangement of the RNA within STNV is shown in Fig. 5.21. The amino acid sequence of the coat protein was deduced from the nucleotide sequence (Fig. 14.9B) and later confirmed by direct sequencing of the STNV coat protein. These results demonstrated that STNV codes for only one gene product: its coat protein (Fig. 14.10). Although there is considerable sequence variation between isolates of STNV-1 and STNV-2 (Danthinne et al., 1991), each encodes only the coat protein. However, the leader sequences of STNV-1 and STNV-2 are identical for 21 of the first 27 nucleotides from the 5′ end. Similarly, the predicted secondary structures of the 3′ UTR of the two strains are nearly identical (Danthinne et al., 1991).

STNV RNA has a high degree of secondary structure and it has been suggested that this region may be involved in the stability of the molecule (Ysebaert et al., 1980). The RNA is remarkably stable *in vivo*, having been shown to survive in inoculated leaves for at least 10 days in the absence of helper virus (Mossop and Francki, 1979a). This stability may have evolved to allow the satellite to survive a period within a cell after uncoating, but before the cell becomes infected with helper virus.

Relatively little is known about the replication of STNV *in vivo*, but it is widely assumed that STNV RNA replication must be carried out by an RNA-dependent RNA polymerase coded for, at least in part, by the helper virus. There is no evidence for rolling circle replication used by type D satellites (see Section II.B.5) and it is assumed that replication is through a (–) strand

A

```
AGUAAAGACAGGAAACUUUACUGACUAACAUGGCAAAACAACAGAACAACAGGCGAAAAUCCGCAACAAUGCGUGCAGUGAAGCGCAUGAUAAAAUACACA  100
CUUGGAGCAUAAAAGGUUUGCACUGAUCAACUCAGGGAACACCAAUGCAACUGCUGGUACAGUACAAAAUCUGUCCAACGGUAUAAUCCAAGGAGAUGAU  200
AUCAACCAGAGAAGUGGUGAUCAAGUGCGUAUAGUUUCACAUAAACUUCACGUACGAGGCACUGCCAUCACCGUCAGCCAGACCUUUAGAUUUAUCUGGU  300
UUCGUGAUAAACAUGAACCGUGGGACCACUCCCACAGUUCUUGAGGUGUUGAACACUGCGAAUUUCAUGUCGCAGUAUAACCCAAUCACGUUGCAGCAAAA  400
GAGAUUUACAAUACUCAAGGAUGUAACUCUCAAUUGUUCGCUGACAGGGGAGAGCAUUAAAGAUCGGAUAAUUAACCUUCCAGGACAACUGGUGAACUAU  500
AAUGGAGCGACGGCUGUAGCAGCCUCCAAUGGUCCCGGCGCAAUAUUUAUGUUGCAGAUUGGCGACUCCUUUGGUUGGGUCUGUGGGGACUCCUCUUAUGAGG  600
CUGUGUACACAGAUGCAUAAUCCCAGAGGUUCACAAUGUUAUGUGAUGGGGCGCUGAAAGAUGCGUAGCUACCCUUCUGGAGCCACUUCCUGGUGGUAAGC  700
AGAAAUCCAAGGGUACGGUGGUACGGUGGAAAGCAGUCCCAGCUCUGCAUUGGGAACCGGCUUACACCCAGCUUAGGGCUAAAGUGUACUACUUGCUCAU  800
UUGUAGUCUAAAUGAGACGUUGGCCUCGACGUGUCGAGGUGGCCUAAAGGGAUUGGAACCCCUGAUGGUCGUAGUCGAAUUUCCGUGUUUCAUUCCGAGU  900
CUCUUGGUCAUAAUGCCAUUAGUAGGUCUAGCACUCAACGUAACUUCAAAGAUAUCCUCCUUGCAACAAGAAAUAUGUGCGCCGUCUGUGUGUUUAAAGCGGU  1000
AUAUUAAGUGCGCCGGCAUAUCGUUGUUUGGGACCAGGGCCCCACGCCGGUUGGUACCCGGGUGGCUUCCCCUCGUUCACAGGGCUUUAGGAGAUGAUAAG  1100
GUAUAGUUAUUAGACAAAUGCGGACAAACCUGAAAAGCUCGCUAGUGGUGGGGCUGGCCAAGCGAAGAACCUCAUCCAGGUAUAGUUCUACAUGGGAAAUU  1200
UGGUACCAUCCAAACUUCUAUGAAGUCCUCGACUACCCC
```

B

```
ALA-LYS-GLN-GLN-ASN-ASN-ARG-ARG-LYS-SER-ALA-THR-MET-ARG-ALA-VAL-LYS-ARG-MET-ILE-  20
ASN-THR-HIS-LEU-GLU-HIS-LYS-ARG-PHE-ALA-LEU- ILE-ASN-SER-GLY-ASN-THR-ASN-ALA-THR-  40
ALA-GLY-THR-VAL-GLN-ASN-LEU-SER-ASN-GLY- ILE- ILE-GLN-GLY-ASP-ASP- ILE-ASN-GLN-ARG-  60
SER-GLY-ASP-GLN-VAL-ARG- ILE-VAL-SER-HIS-LYS-LEU-HIS-VAL-ARG-GLY-THR-ALA- ILE-THR-  80
VAL-SER-GLN-THR-PHE-ARG-PHE- ILE-TRP-PHE-ARG-ASP-ASN-MET-ASN-ARG-GLY-THR-THR-PRO-  100
THR-VAL-LEU-GLU-VAL-LEU-ASN-THR-ALA-ASN-PHE-MET-SER-GLN-TYR-ASN-PRO- ILE-THR-LEU-  120
GLN-GLN-LYS-ARG-PHE-THR- ILE-LEU-LYS-ASP-VAL-THR-LEU-ASN-CYS-SER-LEU-THR-GLY-GLU-  140
SER- ILE-LYS-ASP-ARG- ILE- ILE-ASN-LEU-PRO-GLY-GLN-LEU-VAL-ASN-TYR-ASN-GLY-ALA-THR-  160
ALA-VAL-ALA-ALA-SER-ASN-GLY-PRO-GLY-ALA- ILE-PHE-MET-LEU-GLN- ILE-GLY-ASP-SER-LEU-  180
VAL-GLY-LEU-TRP-ASP-SER-SER-TYR-GLU-ALA-VAL-TYR-THR-ASP-ALA  195
```

Fig. 14.9 (A) Nucleotide sequence of STNV RNA. The sequence, represented here as viral RNA, combines information obtained from cloned cDNA and the 5'-terminal sequence reported by Leung *et al.* (1976). The initiation and termination codons of the STNV coat protein gene are boxed. **(B)** Amino acid sequence of STNV coat protein as deduced from the nucleotide sequence. From Ysebaert *et al.* (1980), with permission.

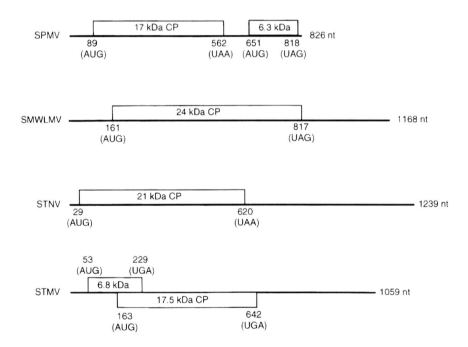

Fig. 14.10 Schematic organization of the plant virus satellite virus genomes. SPMV RNA encodes the 17-kDa capsid protein (CP) in the 5'-proximal region of the genome and has a 3'-proximal 6.4-kDa ORF of unknown function. Two small ORFs are also located on the (–) strand (not shown). SMWLV encodes a 24-kDa CP, and the TNV satellite viruses, STNV-1 and STNV-2, encode CPs of 21 kDa. The 3' non-coding regions of the satellite RNAs are highly structured and this extensive base pairing probably reflects *cis* requirements for replication, translation of the encoded genes and packaging by the capsid protein. STMV is also included for comparison. The genome lengths of the satellite virus RNAs are indicated in nucleotides (nt). From Scholthof *et al.* (1999b), with kind permission of the copyright holders, © Springer-Verlag GmbH and Co. KG.

in a manner similar to that of (+)-strand RNA viruses (see Chapter 8, Section IV.D). Expression of TNV-A replicase proteins (p23 and p82) from the CaMV 35S promoter either transiently in protoplasts or in transgenic plants supported STNV-2 replication (Andriessen *et al.*, 1995). However, the satellite accumulated to only about 1% of the level found when wild-type TNV-A RNA and STNV-2 were electroporated into protoplasts. This could be due to the low levels of polymerase proteins being expressed or to a requirement for other TNV-A encoded proteins.

Replication of STNV substantially suppresses TNV replication (Jones and Reichmann, 1973), and it is thought likely that this involves competition for the replicase. The presence of STNV in the inoculum reduces the size of the local lesions produced by the helper virus, possibly by reduction in TNV replication.

STNV is readily translated *in vitro* in both prokaryotic and eukaryotic systems, the only product being the coat protein. In the wheatgerm system, where both TNV and STNV RNAs were translated in a mixture, STNV RNA was preferentially translated even in the presence of an excess of TNV RNA (Salvato and Fraenkel-Conrat, 1977). Using mRNAs transcribed from recombinant plasmids containing DNA copies of various lengths of the 5' region of STNV RNA, sequences that affect the requirement for various initiation factors have been delineated in the wheatgerm protein synthesis system (Browning *et al.*, 1988).

The 5' terminus of STNV is unlike that of most ss (+)-sense RNA plant viruses. There is neither a 5' cap nor a VPg, the 5' termination being 5'-ppApGpUp- (Ysebaert *et al.*, 1980). The 3'-terminal region can be folded to give a tRNA-like structure with an AUG anticodon, but there is no evidence that this structure can accept methionine. The fact that STNV RNA, compared with other plant viral RNAs, is efficiently translated *in vitro* in prokaryotic systems may be explained by the 5' non-coding region containing a sequence, -AGGA-, which is part of the Shine–Dalgarno prototype sequence complementary to a region at the 3' end of the 16S ribosomal RNA of *Escherichia coli* (Ysebaert *et al.*, 1980). A cDNA copy of STNV

RNA inserted into a plasmid gave rise to efficient translation of STNV coat protein in *E. coli*, in line with its efficient translation in a bacterial system *in vitro* (van Emmelo *et al.*, 1984).

2. Satellite panicum mosaic virus (reviewed by Scholthof *et al.*, 1999b)

A relationship very similar to that between TNV and STNV exists between the isometric viruses, PMV and a smaller satellite virus (SPMV) (Fig. 14.11) (Niblett and Paulsen, 1975; Buzen *et al.*, 1984) (Table 14.4). PMV induces mild symptoms in pearl millet but co-infection with SPMV usually results in severe mosaic, stunting and reduced seed-set (Masuta *et al.*, 1987).This is associated with an increased rate of systemic infection of PMV together with increases of PMV RNA and its p8 and coat protein (Scholthof, 1999).

SPMV RNA has two ORFs on the (+) sense, the 5' ORF encoding the 17-kDa coat protein and the downstream ORF potentially coding for a 6.3-kDa protein (Masuta *et al.*, 1987) (Fig. 14.10). Whether the latter protein is expressed and, if so, what its function is, are unknown. The (–)-sense SPMV strand has two ORFs potentially encoding polypeptides of 7.1 and 11 kDa. Neither PMV nor SPMV has sequence similarity to TNV or STNV.

The 5' and 3' termini of PMV and SPMV have limited sequence similarities.

3. Maize white-line mosaic satellite virus (reviewed by Scholthof *et al.*, 1999b)

MWLMV is an ungrouped virus similar in size to TNV but with no evidence of relationship to it. Small satellite virus particles have been found in the roots of infected plants (Gingery and Louie, 1985) (Table 14.4).

The 1168 nucleotide RNA genome of SMWLMV has a single ORF encoding a 24-kDa protein (Zhang *et al.*, 1991). It resembles other satellite virus RNAs in lacking a 5' cap structure and with the 3' terminus being 'CCC'. The limited data available on the genomic sequence of MWLMV do not allow sequence comparisons with the satellite virus.

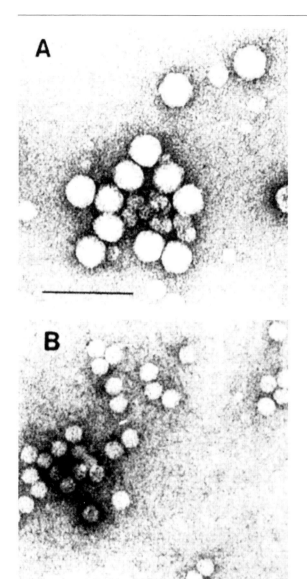

Fig. 14.11 Electron micrographs of PMV and its satellite sPMV. **(A)** A mixture of 30-nm PMV and 17-nm SPMV particles that were present in preparations obtained from a mixed infection of pearl millet. **(B)** The 42S SPMV particles following separation from the 109S PMV particles on sucrose density gradients. Bar marker 100 nm. From Scholthof *et al.* (1999), with kind permission of the copyright holders, © Springer-Verlag GmbH and Co. KG.

4. Satellite tobacco mosaic virus (reviewed by Dodds, 1998, 1999)

Valverde and Dodds (1986) described a small RNA, unrelated in sequence to TMV, that they found in some natural isolates of TMV strain U5 (now known as TMGMV). At first, they considered this to be a satellite RNA. However, when the TMGMV isolation procedure was altered, a 17-nm-diameter icosahedral virus particle was isolated that contained the satellite RNA (Fig. 14.12) (Valverde and Dodds, 1987) (Table 14.4). The satellite virus had no serological relationship with the helper virus or with other satellite viruses. These experiments are a good example of the need for care when attempting to delineate a newly isolated satellite RNA. STMV can be adapted to replicate experimentally using other tobamoviruses as helpers, but has only been found naturally in association with TMGMV. It does not appear to have any effects on the symptoms produced by the helper virus.

STMV RNA has been sequenced (Mirkov *et al.*, 1989) (Fig. 14.10). Comparison of this sequence with that of TMGMV showed that similarity (approximately 60%) was limited to the 3' non-coding region of TMGMV (Solis and García-Arenal, 1990). The 3' end of the satellite RNA can fold to give a tRNA-like structure involving two pseudo-knots, as has been proposed for TMV (see Fig. 4.5) and TMGMV. Like TMV, this can be aminoacylated with histidine (Felden *et al.*, 1994; Gultyaev *et al.*, 1994) and STMV can be adapted to be helped by the tobamoviruses that aminoacylate with histidine. These similarities are assumed to reflect dependence of the satellite on a replicase coded for by the helper virus.

The satellite virus RNA contains two ORFs: the 5' one coding for a 6.8-kDa protein of unknown function and the 3' ORF encoding the 17.5-kDa coat protein. The 6.8-kDa protein ORF is dispensable for infection.

The adaptation of STMV to other helper tobamoviruses involves the deletion of a single G residue in a series of five G residues (positions 61–65) in the 5' leader sequence together with a change in the 5' nucleotide (Yassi and Dodds, 1998). This is taken as suggesting that the adaptation does not involve the ability of the TMV replicase to make (–)-sense RNA from the (+)-sense genome, but does involve the ability to make (+) strand from (–) strand (Dodds, 1998).

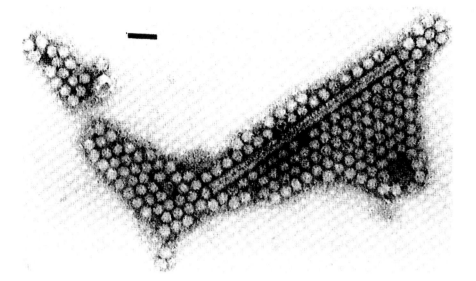

Fig. 14.12 Electron micrograph of a negatively stained purified preparation of TMGMV and STMV from doubly infected tobacco showing a full-length rod-shaped virion of TMGMV (300 nm) surrounded by spherical virions of STMV (16–17 nm). Bar marker 50 nm. From Valverde and Dodds (1987), with permission.

B. Satellite RNAs (satRNAs)

As can be seen from Table 14.3, there are three classes of satellite RNAs, the A type being a satellite virus. These are grouped according to size, the large ones with mRNA activity being termed B type (Mayo *et al.*, 1995) or subgroup 1 (Mayo *et al.*, 2000b), and the smaller ones without mRNA activity being divided with respect to form, those with linear molecules (C type or subgroup 2) and those with circular molecules (D type or subgroup 3).

White and Kaper (1989) describe a simple method for the detection satellite RNAs in small samples of plant tissue. Such RNAs have been found in association with members of four plant virus groups.

1. Large satellite RNAs (B type or subgroup 1) (reviewed by Mayo *et al.*, 1999)

As with the other satellite RNAs, B-type satellites are dependent upon the helper virus for replication and encapsidation but, in contrast to the other types, they direct protein synthesis both *in vitro* and *in vivo* from a single ORF. The RNAs of the B-type satellites thus far studied are 0.7 kb or larger. Most of the B-type satellites (given the prefix 'Bsat') are found associated with nepoviruses, and one has been found associated with the potexvirus BaMV.

a. Satellite RNA of Tomato black ring virus (BsatTBRV)

Some isolates of TBRV have associated with them a small RNA of about 1.4 kb that has the properties of a satellite RNA (Murant *et al.*, 1973). The satellite RNA is packaged in varying numbers in particles made of the helper virus coat protein, giving rise to a series of components of differing buoyant density in solutions of cesium chloride. Like the helper virus, the satellite RNA is transmitted through the seed and also by the nematode vector of the virus (Murant and Mayo, 1982). Various strains of the satellite have been isolated in the field, and there is some specificity between strains of helper virus and strains of the satellite RNA. BsatTBRV has a VPg at the 5' terminus and is polyadenylated at the 3' end. The satellite does not appear to affect replication of the helper virus or to modify symptoms, except that the number of local lesions on *Chenopodium amaranticolor* is reduced (Murant *et al.*, 1973).

The first complete nucleotide sequence of an isolate of BsatTBRV revealed a molecule 1375 nucleotides long. The sequence contained one ORF, with 5' and 3' non-coding sequences. The 3' region contained a sequence resembling a polyadenylation signal (Meyer *et al.*, 1984). On the basis of sequence homology, five BsatTBRV isolates fell into two groups with about 60%

homology between the groups. Within groups, homology was much stronger (Hemmer *et al.*, 1987). All isolates contained a single ORF with a coding capacity for a protein of 419–424 amino acids. Several regions of potential amino acid sequence were identical in all isolates, including a cationic N-terminal region and several potential phosphorylation sites. Mutagenesis studies suggest that this protein region and the 5′ and 3′ UTRs are all necessary for the replication of the Bsat.

Satellite RNAs with general properties quite similar to those of BsatTBRV have been described for other nepoviruses. A small satellite-like RNA has been found in some isolates of ArMV from hops. The evidence suggests that, when present together with the helper virus, it may be the cause of the important nettlehead disease of hops (Davies and Clark, 1983).

An interesting situation with respect to satellite RNAs has been described for ChYMV, a nepovirus from southern Italy. Strain ChYMV-T contains two major RNA components with the properties of satellite RNAs. Hybridization experiments showed no sequence homology between the two RNAs, or between these RNAs and the helper virus genomic RNAs (Piazzolla *et al.*, 1989a). Both satellite RNAs have been sequenced (Rubino *et al.*, 1990). One has only linear molecules and is 1145 nucleotides long with a 3′ poly(A) sequence. There is one ORF capable of coding for a polypeptide of c40 kDa. There is a 5′ leader sequence of 16 nucleotides and a 3′ non-coding region of 40 nucleotides. The ORF is translated *in vitro* to give the expected product. The other satellite is 457 nucleotides long. There is no ORF of significant length and no mRNA activity *in vitro*. Both linear and circular forms are found in infected tissue. There are large regions of sequence similarity with other small *Nepovirus* satellites (see Section II.B.4.a).

b. *Satellite RNA of* Bamboo mosaic virus (*BsatBaMV*)

BsatBaMV is 836 nucleotides long and has a 5′ hexanucleotide sequence and a 3′ polyadenylate in common with its helper virus, the potexvirus BaMV (Lin *et al.*, 1994; Yang *et al.*, 1997a). The single ORF of BsatBaMV encodes a 20-kDa protein that binds tightly to BsatBaMV RNA and with less affinity to other ssRNA molecules (Tsai *et al.*, 1999). However, this 20-kDa protein is not required for replication of the satellite RNA as it can be replaced by a reporter gene (Lin *et al.*, 1996).

Different isolates of BsatBaMV have different effects on the accumulation of the genomic and subgenomic RNAs, and on the symptom production of the helper virus (Hsu *et al.*, 1998). These effects also differ depending on the host plant.

2. Small linear satellite RNAs (C type or subgroup 2)

C-type satellites have linear RNA molecules that are generally smaller than 0.7 kb and do not have mRNA activity.

a. *Satellite RNAs of cucumoviruses* (reviewed by Garcia-Arenal and Palukaitis, 1999)

In 1972, a devastating outbreak of a lethal necrotic disease of field-grown tomatoes occurred in the Alsace region of France. It was realized that CMV was involved in this outbreak but it was not clear why necrosis occurred instead of the usual fern-leaf symptoms (e.g. Putz *et al.*, 1974). Kaper and colleagues (e.g. Lot and Kaper, 1976) described the presence of a fifth small RNA component present in some isolates of CMV in addition to the three genomic RNAs and the subgenomic coat protein mRNA, and that was not part of the viral genome. Kaper and Waterworth (1977) showed that the additional small RNA present in cultures of CMV strain S caused lethal necrotic disease in tomatoes when added to the CMV genomic RNAs. They called this satellite RNA CARNA5 and proposed that it had been responsible for the lethal necrotic disease in Alsace. Similar outbreaks in tomatoes in southern Italy have been shown to be due to a necrogenic isolate of satCMV (CARNA5) (Kaper *et al.*, 1990). However, most CMV satellites attenuate the symptom (see Palukaitis, 1988) but the overall effect depends on the combination of satellite RNA and helper virus strain.

Several different satellite RNAs have been described but CARNA5 is the best characterized (Fig. 14.13). It is found in all CMV density fractions following equilibrium density gradient centrifugation. Therefore, it is probably packaged in particles containing RNAs 1 and 2, and also, in varying numbers of copies per particle in particles not containing any CMV RNA (Kaper *et al.*, 1976). This RNA is dependent on, but not significantly related in overall nucleotide sequence to, the CMV genome. Packaging in CMV protein enables the satellite RNA to be transmitted by aphids that transmit CMV (Mossop and Francki, 1977).

Many natural variants related in nucleotide sequence to satCMV have been described (e.g. García-Luque *et al.*, 1984; Palukaitis and Zaitlin, 1984; Hidaka *et al.*, 1988). They vary in length from about 330 to 390 nucleotides. It is probable that most satellite RNA preparations consist of populations of molecules with closely related sequences.

Like the helper genomic CMV RNAs, they are all capped with M7Gppp at their 5' termini and have a hydroxyl group at the 3' end. Ten residues appear to be conserved at the 5' terminus and six at the 3' terminus, with other conserved regions in between. The satellite RNAs have a potential for substantial secondary structure, and models have been proposed (e.g. Gordon and Symons, 1983). A 3'-terminal tRNA-like structure is possible but attempts to amino-acylate the RNA have been unsuccessful. The high degree of secondary structure may account for the stability of the RNA both *in vitro* and *in vivo*, and for its relatively high specific infectivity (Mossop and Francki, 1978).

With many CMV isolates, the amount of satellite RNA is small, and some workers could not eliminate the satellite from CMV cultures (e.g. Kaper *et al.*, 1976). This may result from the

```
                     met   glu   asn   cys   ala   glu   gly   leu   tyr
m7GpppGUUUUGUUUG     AUG   GAG   AAU   UGC   GCA   GAG   GGG   UUA   UAU


   leu   arg   glu   asp   leu   ser   leu   gly   gly   val   gly   tyr
   CUG   CGU   GAG   GAU   CUG   UCA   CUC   GGC   GGU   GUG   GGA   UAC


   leu   pro   ala   lys   ala   gly
   CUC   CCU   GCU   AAG   GCG   GGU   UGAGUGAUGUUCCCUCGGACUGG


                           met   ser   ala   thr   leu   ser   thr
   GGACCGCUGGCUUGCGAGCU    AUG   UCC   GCU   ACU   CUC   AGU   ACU


   thr   leu   ser   phe   glu   pro   pro   leu   ser   leu   leu   ala
   ACA   CUC   UCA   UUU   GAG   CCC   CCG   CUC   AGU   UUG   CUA   GCA


   glu   pro   gly   thr   trp   phe   ala   asp   thr   met   asp   phe
   GAA   CCC   GGC   ACA   UGG   UUC   GCC   GAU   ACU   AUG   GAU   UUU


   leu   lys   lys   his   ser   val   arg   trp   tyr   glu   ser
   CUA   AAG   AAA   CAC   UCU   GUU   AGG   UGG   UAU   GAG   UCA   UGA


CGCACGCAGGGAGAGGCUAAGGCUUAUGCUAUGCUGAUCUCCGUGAA


UGUCUAUCAUUCCUCUGCAGGACCCOH
```

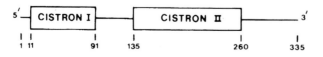

Fig. 14.13 The sequence of 335 nucleotides in CARNA5 (strain 7). The possible coding scheme shown below is that proposed by Richards *et al.* (1978), with permission.

high stability and specific infectivity of the small RNA. The effects of the presence of satellite RNA on disease symptoms are discussed in Section II.B.5.

Using RNA transcripts from a cDNA clone of a CMV satellite inducing yellowing symptoms (strain Y), Masuta *et al.* (1988b) showed that extra non-satellite nucleotides at the 5' terminus greatly reduced infectivity. This parallels the effect seen with transcripts from cloned viral RNAs.

The amounts of satellite RNA produced are highly variable depending on both the host species used and the strain of helper virus. RNAs 1 and 2 of CMV have highly conserved 5' leader sequences that form part of a hairpin structure. This conserved 5' sequence appears in the complementary sequence of CMV satellite RNA. Base-pairing of this sequence to the 5' leaders of genomic RNAs 1 and 2 may be one mechanism whereby the satellite RNA influences both viral and satellite RNA production (Rezaian *et al.*, 1985). On the other hand, it has been shown that a CMV satellite RNA binds specifically to a 33-nucleotide region in the coat protein gene of CMV, probably forming an unusual knot-like structure (Rezaian and Symons, 1986). This finding provides another possible mechanism whereby the satellite RNA could regulate CMV replication.

PSV isolates may contain a satellite RNA known as PARNA5. It is more than 15% longer than any CMV satellite RNA (393 nucleotides), has the same 5'- and 3'-terminal structures as CMV satellites, and shows short stretches of sequence similarity near these termini (Collmer *et al.*, 1985). However, PARNA5 has several regions of high sequence similarity with some viroids. There are also striking similarities with host intron sequences essential for correct RNA processing.

b. Satellite RNAs of tombusviruses

Particles of most members of the *Tombusvirus* genus contain a small satellite RNA. These RNAs have extensive sequence similarity with one another but not with the helper virus genomic RNA (Gallitelli and Hull, 1985). In most other systems, replication of satellite RNAs is strain specific with respect to the helper virus. By contrast, replication of these *Tombusvirus* satellite RNAs can be supported by heterologous viruses that are not normally associated with the satellite. The presence of the satellite RNA may alter disease expression (e.g. Hillman *et al.*, 1985).

c. Satellite RNAs of carmoviruses (reviewed by Simon, 1999)

TCV supports a family of satellite RNAs (Altenbach and Howell, 1981) (see Fig. 8.34). Two of the satellites (D and F) do not affect symptoms, while C is virulent, intensifying TCV symptoms in turnip. Simon and Howell (1987) synthesized RNA copies of the virulent satellite (RNA C) *in vitro* using full-length cDNA copies in an expression vector. The RNA was infectious only when inoculated together with helper TCV. cDNA or (−)-sense RNA copies of the satellite were not infectious. Using a cloned cDNA probe of the satellite RNA, Altenbach and Howell (1984) detected a series of at least six multimeric forms of the RNA. Most of this RNA was of the same polarity as the encapsulated monomer, suggesting a role in RNA replication for the multimers. They also found some sequence relatedness between the satellite and host DNA, but the significance of this remains to be determined.

The virulent satellite C of TCV is of particular interest because, unlike other known satellites, it has a substantial region of homology with the helper TCV genome (Simon and Howell, 1986; Simon *et al.*, 1988). The 189 bases at the 5' end of RNA C are homologous to the entire sequence of a smaller non-virulent TCV satellite D RNA. The rest of the RNA C molecule (166 bases in the 3' region) is nearly identical in sequence to two regions at the 5' end of the TCV helper genome. This interesting composite molecule is illustrated in Fig. 14.14 along with the functional domains in the molecule established by Simon *et al.* (1988) by means of insertion and deletion analysis. A 30 nucleotide stem-loop region and several motifs in RNS C enhance replication, showing that the replication enhancer has a modular nature (Simon and Zhang, 2001; Zhang and Simon, 2001). Thus, in a structural sense at least, TCV satellite RNA C is a hybrid between a satellite RNA and a DI

RNA (see Chapter 8, Section IX.B.2.b). When plants were inoculated with a mixture of TCV, satellite RNA D and satellite RNA C transcripts containing non-viable mutations in the 5' domain, recombinant satellite RNAs were recovered (Cascone *et al.*, 1990). Sequence analysis around 20 recombinant junctions supported a copy-choice model for this recombination. In this model, while in the process of replicating (–) strands, the replicase can leave the template together with the nascent (+) strand and can reinitiate synthesis at one of two recognition sequences on the same or a different template.

3. Satellite-like RNAs

Most satellites are not essential for the biological success of the helper virus. However, it is a relatively small evolutionary step from this situation to one where the satellite is involved in the disease spread and expression. Since these molecules cannot be described as inessential for the helper virus, it cannot technically be considered to be a satellite; hence, they are grouped together as 'satellite-like RNAs' (Mayo *et al.*, 1999).

a. A *Satellite RNA of* Groundnut rosette virus

The satellite associated with GRV (satRNA or sGRV) is a linear ssRNA of 895–903 nucleotides which relies on GRV for its replication (Murant *et al.*, 1988a; Murant and Kumar, 1990). Although different variants of sGRV contain up to five potential ORFs in either the (+)- or (–)-sense strands (Blok *et al.*, 1994), none of these ORFs is essential for replication of the satellite RNA (Taliansky and Robinson, 1997).

As discussed in Section II.D, sGRV is involved with the two viruses causing groundnut rosette

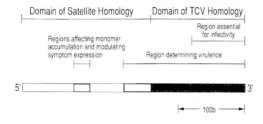

Fig. 14.14 Domains in the virulent composite satellite RNA C of TCV. Modified from Simon *et al.* (1988), with kind permission of the copyright holders, © Oxford University Press.

disease and is essential for encapsidation and symptom production within the complex.

b. A *Satellite RNA of* Pea enation mosaic virus (reviewed by Demler *et al.*, 1996a)

In addition to the two genomic RNAs, a third small RNA (717 nucleotides) is found in some PEMV preparations and has been characterized as a satellite RNA (Demler and de Zoeten, 1989; Demler *et al.*, 1994b). sPEMV is not infectious on its own and requires PEMV RNA2 for its replication. The 5'- and 3'-terminal sequences of sPEMV have limited homology to those of PEMV RNA2. sPEMV replication can be supported by GRV (de Zoeten *et al.*, 1994).

The presence of the RNA does not affect symptom expression, aphid transmission or particle morphology. Thus, sPEMV is fully dispensable, like a true satellite, but is placed here because of its close relationship with sGRV (Demler *et al.*, 1996b).

c. *Ancillary RNAs of* Beet necrotic yellow vein virus

Field isolates of BNYVV contain four or five RNA components but laboratory isolates maintained by mechanical inoculation of *Chenopodium quinoa* or *Tetragonia expansa* may lack any or all of the three smaller RNAs. Thus, the BNYVV genome comprises RNAs 1 and 2 (see Chapter 6, Section VIII.H.13), and RNAs 3, 4 and 5 resemble satellite RNAs. They each encode one ORF and contribute to the field pathology of the parent virus. RNA3 contributes to a large degree to the pathology of the virus (Lauber *et al.*, 1998a) whereas RNA4 greatly increases the efficiency of virus transmission by its fungal vector. RNA5 seems to play a role in both the symptom production and fungus transmission (Mayo *et al.*, 1999).

4. Small circular satellite RNAs (D type or subgroup 3)

D-type satellite RNAs are small, about 350–450 nucleotides, and occur as circular as well as linear molecules.

a. *D-type satellite RNAs of nepoviruses*

Several nepoviruses support the replication of satellite RNAs, which become encapsulated in particles made of the helper virus coat protein.

Some of these satellite RNAs are relatively large and are the B-type described in Section II.B.1. They are most likely replicated by the helper virus replicase in the manner described for (+)-strand RNA viruses (see Chapter 8, Section IV.D). Others appear to replicate by a quite different rolling circle mechanism.

i. Satellite RNA of Tobacco ringspot virus (sTRSV)

sTRSV consists of a small RNA species about 359 nucleotides long (Buzayan et al., 1986d) in a protein shell identical to that of the helper virus. Twelve to 25 satellite RNA molecules become packaged in a single particle (Schneider et al., 1972a). Different field isolates of the satellite produce different lesion types (Schneider et al., 1972b). The satellite cannot replicate on its own, and it interferes with the replication of TRSV. There is an absence of ORFs, and the RNA is not translated in vitro (Owens and Schneider, 1977). Schneider and Thompson (1977) isolated a population of dsRNA molecules from leaves infected with sTRSV that were absent from leaves infected with TRSV alone, or from healthy tissue. The MWs found in this multi-component dsRNA preparation were much greater than that expected from double-stranded sTRSV RNA. Electrophoretic analysis showed that the monomer RNA was the smallest and most abundant of at least 10 size classes with the sTRSV sequence. A dsRNA fraction from infected tissue, when denatured, gave a similar series of up to 12 zones containing both (+)- and (–)-sense strands. (The polarity of the encapsulated single-stranded satellite RNA is defined as (+) sense). The MW increment of each zone corresponded approximately to the size of the monomer satellite RNA (Kiefer et al., 1982). No polyadenylated was detected at the 3′ terminus and no VPg at the 5′ terminus; these are present in the helper virus. The 3′ terminus was a cytosine 2′,3′-cyclic phosphate, and adenosine was at the 5′ terminus. Circular monomeric and dimeric forms of sTRSV RNA have also been isolated (Linthorst and Kaper, 1984). sTRSV RNA may have some base-paired viroid-like secondary structure (Buzayan et al., 1987).

Similar satellite RNAs are found with some other nepoviruses.

b. Satellite RNA associated with Cereal yellow dwarf virus

sRPV is found associated with the luteovirus, CYDV-RPV, where it reduces the accumulation and symptom production of the helper virus in oats (Rasochova and Miller, 1996). The satellite has similarities to the D-type satellites in that there are multimeric linear and circular forms together with a hammerhead ribozyme sequence (see Section II.B.5.b). Association with CYDV-RPV leads to encapsidation, mainly of linear molecules. Replication of sRPV is also supported by the polerovirus, BWYV, and encapsidated satellite molecules are mainly circular (Rasochova et al., 1997).

c. Satellite RNAs associated with sobemoviruses (reviewed by Symons and Randles, 1999)

Four viruses described from Australia, which belong in the Sobemovirus genus, have been shown to contain small RNAs that occur in both circular and linear forms. Two of the viruses, VTMoV and SNMoV, infect Solanaceae species, while the other two, LTSV and SCMoV, infect legumes. The small RNAs have been shown to have the biological properties of satellite RNAs. The helper virus can replicate independently, whereas the small RNA cannot (Jones et al., 1983; Jones and Mayo, 1984; Francki et al., 1986a). The satellite RNAs, which range in size from about 325 to 390 nucleotides, are encapsidated in both linear and closed circular forms. They show no sequence similarity with the helper RNAs. Replication of these satellite RNAs can be supported by other sobemoviruses, including viruses for which no satellite has been discovered.

A smaller satellite RNA (220 nucleotides) has been found associated with another sobemovirus, RYMV (Collins et al., 1998).

The properties of these satellite RNAs are somewhat like those of viroids (reviewed by Francki, 1985) and they are sometimes termed 'virusoids'. However, their thermodynamic properties and their ability to self-cleave are quite different from those of viroids, except possibly ASBVd. They do not show the co-operativity during melting that is found in viroids, but behave like random base sequences (Steger et al., 1984). All four satellite RNAs share a common sequence, GAUUUU, which is also in the same position and loop structure for all four RNAs (Keese *et al.*, 1983). Canadian and Australian isolates of the LTSV satellite have 80% sequence homology (AbouHaidar and Paliwal, 1988).

While in biological properties these RNAs are undoubtedly satellites, in some properties, particularly their small size, circularity, high degree of base-pairing, and lack of mRNA activity, they are like viroids (Francki, 1985). A sequence, GAAAC, is found in the central region of viroids and also in a similar position in the satellites associated with viruses infecting solanaceous plants (Keese and Symons, 1987).

5. Replication of satellite RNAs

In the previous sections, the various satellite RNAs were classified according to the size and shape of the RNA, its coding capacity and the kind of helper virus involved. This may not be the most functional way to consider these small RNAs. It is becoming increasingly apparent that satellite RNAs differ in the ways in which they are replicated and that most of them may fall into two groups on this basis. Piazzolla *et al.* (1989a) suggested that one group may consist of satellite RNAs that have 5'- and 3'-terminal structures like those of the helper virus, and whose replication depends on the helper virus RNA replicase. The second group uses a rolling circle means of replication and depends on some other function of the helper virus. This grouping cuts across the classification based on the taxonomic position of the helper virus. Indeed, Piazzolla *et al.* (1989a) and Rubino *et al.* (1990) have described one satellite of each type, both of which are dependent on ChYMV (see Section II.B.1.a).

The presence of a replication enhancer in TCV RNA C is noted in Section II.B.2.b.

a. Satellite RNAs with terminal structures like those of the helper virus

This group is typified by the satellite RNAs of CMV and TBRV. Their properties with respect to replication may be summarized as described below.

i. Terminal structures

CMV satellite RNAs have a 5' cap structure and a 3' hydroxyl group, as does CMV. sTBRV has a 5' VPg and 3' polyadenylation, as does TBRV. These similarities to the corresponding helper virus suggest a common method for replication of the satellite and helper RNAs.

A short RNA element located 41 nucleotides from the 5' end of (–) strands of TCV satRNA C has been identified as important for (+)-strand synthesis (Guan *et al.*, 1997, 2000). There is a strict requirement for 10 of the 14 nucleotides in this element, indicating that the primary sequence is essential for RNA accumulation.

TCV and its associated satRNA have similar, but not identical, hairpins near their 3' ends and terminate with CCUGCCC-OH, which forms a single-stranded tail (Carpenter *et al.*, 1995; Carpenter and Simon,1996a,b; Nagy *et al.*, 1997). The hairpin structure is necessary for the replication of the satRNA but it does not need to be as stable as that of the wild type (Stupina and Symon, 1997).

ii. The presence of ORFs

Satellite RNAs in this group that have been sequenced contain one or two ORFs. Some satellites of CMV contain two ORFs (Fig. 14.13), while others contain one (Avila-Rincon *et al.*, 1986). The isolate of CARNA5 used by Owens and Kaper (1977) was translated in the *in vitro* wheatgerm system to give two polypeptides of about 5.2 and 3.8 kDa. These polypeptides have not yet been found *in vivo*, so their significance is uncertain. This *in vitro* translation is inhibited by 7-methylguanosine-5'-monophosphate, which would be expected because the 5' end of CARNA5 is capped.

All variants of sTBRV contained a single large ORF coding for a putative protein containing 419–424 amino acids with several regions of identical amino acid sequence (Hemmer *et al.*, 1987). These features strongly suggest a functional role for the ORF, but it is doubtful whether they play any role in symptom induction. These satellite RNAs were translated in the wheatgerm system to give a single polypeptide of approximately 48 kDa requiring most of the satellite's coding potential (Fristch *et al.*, 1980). The protein was produced in both the absence and the presence of helper viral RNAs 1 and 2 (Fristch *et al.*, 1978). A protein of the same size was detected in extracts of protoplasts infected with TBRV preparations containing the satellite, but not with isolates that lacked it.

iii. RNA replication via a unit-length (−)-sense template

The following lines of evidence strongly suggest that the CMV satRNAs replicate via a unit-length (−)-sense template:

- Kaper and Diaz-Ruiz (1977) isolated from infected tobacco four dsRNA species corresponding to the four CMV ssRNA species and in addition a ds species of MW approximately 220 000 Da corresponding to an RF of CARNA5. Similar results were obtained in protoplasts by Takanami *et al.* (1977), where large amounts of unit-length dsRNA accumulated.
- The kinetics of labeling the two strands in the dsRNA with ^{32}P fit with the production of an excess of (+)-sense strands early in infection, with more ds satRNA accumulating later (Piazzolla *et al.*, 1982).
- The ds forms of both CARNA5 and genomic CMV RNA3 contain an unpaired guanosine at the 3′ end of the (-) strand. This is a feature of the RFs of some other viruses and suggests that the viral and satellite RNAs share a common replicative mechanism (Collmer and Kaper, 1985).
- Baulcombe *et al.* (1986) and Masuta *et al.* (1989) constructed plasmids containing a single cDNA copy of the RNA of a CMV satellite and introduced this into tobacco plants using *Agrobacterium tumefaciens*. Transformed plants could produce biologically active satRNA when inoculated with CMV. Following CMV inoculation there was no delay in symptom appearance in plants transformed with a monomer compared with those containing a dimer DNA copy. Thus, the monomer form appeared to be as effective in replication as the dimer.

iv. The nature of dependence on the helper virus

Linthorst and Kaper (1985) showed that CARNA5 did not replicate in protoplasts unless the helper virus was present. This result suggests that the satellite depends on the helper for some replication function, rather than just for encapsulation or movement through the plant. Inoculation of tobacco mesophyll protoplasts with RNAs 1 and 2 of CMV induces the synthesis of viral RNA replicase activity. Satellite RNA was replicated when added to such protoplasts (Nitta *et al.*, 1988).

The preceding evidence strongly suggests that replication of satRNAs of this type depends on a replicase function of the helper virus. This raises the question of the function of the polypeptide potentially coded for by these satellites. Fritsch *et al.* (1978) suggested that this polypeptide might modulate the activity of the host replicase so that it preferentially replicated the satellite RNA.

b. Satellite replication by rolling circle

This group comprises the type D satellites. sTRSV has no detectable mRNA activity (Owens and Schneider, 1977), and the nucleotide sequences do not indicate ORFs of significant length. There is no clear evidence as to why satellite RNAs of this group are dependent for their replication on a helper virus. However, there is quite strong evidence that they are replicated by a rolling circle mechanism involving intermediates that are multimeric tandem repeats of the sTRSV RNA (Kiefer *et al.*, 1982; Branch and Robertson, 1984).

The circular and multimer forms of sTRSV RNA found in infected tissue have been described in Section II.B.4.a. Large amounts of the circular monomer form may be present in nucleic acids extracted from infected tissue (Linthorst and Kaper, 1984) but only linear molecules are packaged in virus particles.

In vitro, autolytic processing of dimeric and trimeric forms of sTRSV takes place to give rise to biologically active monomers of 359 nucleotides (Prody *et al.*, 1986). The reaction is promoted by Mg^{2+} and some other divalent ions. The autolytic processing appears to be a phospho-transfer reaction between an adenylate residue and the 2′ hydroxyl of a neighboring cytidylate, to give rise to a 5′-hydroxyl terminal adenosine and a 3′ cytosine 2′,3′-

cyclic phosphodiester group. As noted earlier, these termini are characteristic of the monomeric RNA. Under appropriate conditions *in vitro*, the process is reversible in a non-enzymatic reaction. Ligation restores the original bonds. RNA self-cleavage by satellite RNAs has been reviewed by Sheldon *et al.* (1990).

The viroid-like satellite RNAs are predicted to form hammerhead-shaped self-cleavage domains (Fig. 14.15) which resemble those of ASBVd (see Fig. 14.6).

Multimeric polyribonucleotides with sTRSV (−) RNA sequences are also able to process autolytically at an A–G bond 48 residues distant from the autolytic site in the (+) strand (Buzayan *et al.*, 1986b). Non-enzymatic ligation *in vitro* of sTRSV (−) RNA also occurs, and at a far higher rate than ligation of the (+) strand. The ligation reaction is the precise reversal of autolytic processing (Buzayan *et al.*, 1986c).

The cleavage reaction of the (+) sTRSV RNA was shown to be a property of a core sequence no longer than 64 satellite nucleotides that contains the bond that is cleaved (Buzayan *et al.*, 1986a). Comparisons have been made of the nucleotide sequences of related satellite RNAs with viroid-like replication, for example several isolates of sTRSV RNA (Buzayan *et al.*, 1987) and the satellite RNA of ArMV with an sTRSV sequence (Kaper *et al.*, 1988b). These show that the sequences near the 5′ and 3′ termini of the linear forms that are involved in ligation, and in cleavage of circular forms, are highly conserved. In the more central region of the sequence, various differences are apparent, including single-base deletions and insertions, base substitutions and inverted blocks of

sequences, indicating past recombinational events.

The kinetics of labeling of the circular and linear forms of the satellite RNA of VTMoV indicates that the linear form is a precursor of the circular form (Hanada and Francki, 1989). A large pool of unencapsidated satellite RNA appears to be present in infected cells.

6. Molecular basis for symptom modulation (reviewed by Collmer and Howell, 1992; Roossinck et al., 1992)

As already noted in Section II.B.2.a, satellite RNAs may be responsible for outbreaks of severe disease in the field. In addition, because of their small size and the availability of many sequence variants, either isolated from nature or produced in the laboratory, they are amenable to detailed molecular study. For these reasons it was hoped that studies correlating molecular structure of the satellite RNAs with their effects on disease in the plant would give us some insight into the molecular basis of disease induction. Results so far have been difficult to interpret. This is not surprising, because, as discussed below, multiple factors are involved.

The coat protein of TCV is involved in the symptom changes induced by satRNA C dependent on the TCV derivative tested. Normally, satRNA C enhances the symptoms of TCV infections. However, in plants infected with a mutant of TCV in which the coat protein ORF is replaced by that of the related virus CCFV, the presence of satRNA C attenuates the symptoms (Kong *et al.*, 1995). In the inoculated leaves at 7 or 10 days post-inoculation there is

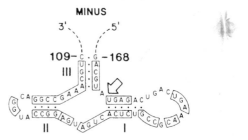

Fig. 14.15 Location of self-cleavage sites, indicated by arrows, in the hammerhead-shaped self-cleavage domain proposed for the viroid-like satellite RNA of LTSV. Boxed nucleotides are conserved between the (+) and (−) hammerhead structures. From Forster and Symons (1987), with kind permission of the coyright holders, © Elsevier Science.

an 80% reduction in the level of TCV-CP$_{ccfv}$ in the presence of satRNA C; no viral RNA can be detected in uninoculated leaves (Kong *et al.*, 1995, 1997b). The symptom of mutants of TCV in which the the coat protein ORF initiation codon is mutated from AUG to ACG are also ameliorated by satRNA C (Kong *et al.*, 1997b). Thus, it appears that the coat protein, and not the RNA encoding it, is the viral determinant that affects the symptoms induced by satRNA C. The region of satRNA C involved in symptom modulation has been localized to the 3'-terminal 53 bases (Kong *et al.*, 1997a), which forms a stem–loop stucture that binds coat protein (Wang and Simon, 2000).

a. Variation in the disease symptoms induced by satellite RNAs

The biological effects of CARNA5 depend on the host. In tobacco, large amounts of satellite RNA are produced, the yield of CMV is markedly depressed and symptoms become milder (Kaper and Tousignant, 1977). Disease symptoms may also be suppressed in other hosts (Waterworth *et al.*, 1979). By contrast, the addition of CARNA5 to CMV in tomato leads to a lethal necrotic disease, as noted in Section II.B.2.a. Figure 14.16 shows the contrasting effects of CARNA5 in tomato and pepper.

Another satellite RNA associated with CMV in the field causes the tomato white-leaf disease (Gonsalves *et al.*, 1982). A different satellite RNA of CMV caused disease attenuation when inoculated together with some strains of CMV (Mossop and Francki, 1979b). With other strains of CMV, this satellite had no effect on symptoms. Yet another CMV satellite RNA (Y strain) caused a bright yellowing of the leaves when inoculated with the helper CMV to some, but not all, of the *Nicotiana* species tested (Takanami, 1981). On the other hand, in tomato it induced lethal necrosis in combination with one strain of CMV but not with another (Masuta *et al.*, 1988a).

White and Kaper (1987) made a systematic study of the effects of CMV strain D and CARNA5 on 52 *Lycopersicon* spp. accessions. This combination, which causes lethal necrosis in *L. esculentum* cv. Rutgers, caused only mild mosaic symptoms in many species. There was wide variation in different accessions of the same species. Thus, while most accessions of *L. hirsutum* gave no necrotic plants, one gave rise to 100% necrotic plants.

Symptom modulation is not confined to satellite RNAs associated with CMV. For example, a satellite RNA of GRV is largely responsible for the symptoms of groundnut rosette disease (Murant *et al.*, 1988a). This is also true for

Fig. 14.16 Modulation of CMV symptoms by its satellite CARNA5. Upper row of plants was infected by virus alone; lower row of plants by virus plus CARNA5. In Tabasco pepper plants on the left, CARNA5 attenuates disease symptoms; in tomato plants on the right, CARNA5 induces lethal necrosis. From Kaper and Collmer (1988), with kind permission of the copyright holder, © CRC Press LLC.

VTMoV and its satellite. The groundnut rosette disease has two main forms, 'chlorotic mottle' and 'green rosette'. Sometimes mild chlorosis or mottle symptoms are found. These variations depend on the satellite RNAs or mixture of satellite RNAs that are present in the plant (Murant and Kumar, 1990).

In summary, these and other results demonstrate that, where satellite RNAs occur, the disease outcome depends on an interaction between the helper virus strain, the strain of satellite RNA, and the species and cultivar of host plant. Environmental conditions no doubt provide a fourth factor.

b. Interference between satellite RNAs

Jacquemond (1982) showed that infection of young tomato plants with CMV together with a satellite RNA giving mild symptoms protected plants against a second inoculation with a satellite RNA that normally caused lethal necrosis. Similar results have been reported by Yoshida *et al.* (1985). The use of satellite RNAs in disease control is discussed in Chapter 16 (Section IV.B).

c. Nucleotide sequences in relation to disease modulation

Kaper *et al.* (1988a) compared the nucleotide sequences of nine CARNA5 variants that were necrogenic in tomato with five that were not. The sequence of ORF 1 (see Fig. 14.13) was completely conserved in all nine necrogenic strains but not in any of the non-necrogenic strains, suggesting a possible role for the putative product of ORF 1 in causing necrosis. However, in a direct approach to this question, Collmer and Kaper (1988) used site-directed mutagenesis to eliminate the AUG initiation codon for ORF 1. Such mutants retained the ability to cause necrosis, demonstrating that a protein product of ORF 1 is not involved. Comparison of the nucleotide sequences of six CMV satellite RNAs suggested that only a few nucleotide changes may be necessary to change the host response and that different kinds of disease response (e.g. yellowing or necrosis) may be associated with different domains of the satellite sequence (Palukaitis, 1988).

Several workers have used cDNA clones to construct recombinant RNA genomes *in vitro* to investigate functional domains in satellite RNAs. Thus, Kurath and Palukaitis (1989a) constructed six recombinant satellite RNA genomes from infectious cDNA clones of three CMV satellite RNAs. The results showed clearly that the domain for chlorosis in tomato is in the 5' 185 nucleotides of these RNAs, while the domain for necrosis is in the 3' 150 nucleotides. Devic *et al.* (1989), using a similar approach, showed that the determinant for symptom production by the CMV satellite RNA Y lay between nucleotides 1 and 219, while the domain for necrosis in tomato lay in the 3' region, beyond nucleotide 219. The experiments of Masuta and Takanami (1989) suggest that the formation of a secondary structure between nucleotides 100 and 200 in the Y satellite may be involved in the induction of chlorosis.

The sequence of the WL1-sat, which attenuates CMV symptoms on tomatoes, differs from all necrogenic satellite RNAs at three nucleotides positions in the conserved 3' region of the RNA. Sleat and Palukaitis (1990a) used site-directed mutagenesis to show that when all three nucleotide positions in the attenuating strain were mutated to the nucleotides found in necrogenic strains, lethal necrosis was induced in tomato. If only two of the three nucleotides were mutated to those present in a necrogenic satellite, no necrosis ensued.

As noted above, the Y-satRNA of CMV elicits bright yellow mosaics on tobacco. When compared with other satellite strains, Y-satRNA has a unique domain (approximately residues 100–200; the Y region) that is responsible for the yellow symptom expression (Masuta and Takanami, 1989; Kuwata *et al.*, 1991). The Y region has a sequence motif exhibiting extensive sequence complementarity to a specific tRNA^{Glu} that functions as a co-factor in the first step of chlorophyll biosynthesis (Masuta *et al.*, 1992). It is suggested that the chlorosis might result from this sequence acting as an antisense RNA.

A satellite RNA of particular interest in relation to disease modulation is the virulent satellite RNA of TCV, illustrated in Fig. 14.14. Other avirulent strains of the satellite (e.g.

strain F) lack the 3' sequence derived from the helper TCV genome. To demonstrate that this domain determines virulence, Simon *et al.* (1988) constructed a chimeric satellite composed mostly of the 5' sequence of strains F and the 3' domain of the virulent satellite C. Other constructs contained small insertions or deletions at various sites. Tests with these RNAs led to the conclusion that the 3' sequence derived from TCV contains a region essential for infectivity and a larger overlapping region determining virulence (see Fig. 14.14). The domain of satellite homology contains regions affecting monomer accumulation and modulating symptom expression.

The only general conclusion to be drawn from studies of this kind is that changes in a disease induced by the presence of a satellite RNA depend on changes in nucleotide sequence in the RNA. So far, there is no convincing evidence that such changes are mediated by a polypeptide translated from the satellite RNA. Indeed, differences in disease modulation occur between RNAs that have no significant ORFs. Disease modulation is most probably brought about by specific macromolecular interactions between the satellite RNA and (1) helper virus RNAs, (2) host RNAs, (3) virus-coded proteins, (4) host proteins, or (5) any combination of (1) to (4). Symptom modulation of CMV in tomato by CARNA5 involves the function in the helper virus that supports satellite replication, as well as some function of the satellite itself (Matthews, 1991). A similar conclusion can be drawn concerning CMV satellite RNAs causing systemic chlorosis in tobacco (Sleat and Palukaitis, 1990b). Nucleic acid hybridization studies have enabled CMV strains to be placed in two groups. Induction of chlorosis by satellite RNAs occurred only with subgroup II strains of the virus, and also appeared to be associated with CMV RNA2.

Until the kinds of interaction that are important in disease induction have been established on a molecular basis, differences in nucleotide sequence between related satellites will remain largely uninterpretable.

7. Chimeras between satellite and defective RNAs

The classical distinction between satellite and defective RNAs is that the latter are derived from the helper virus sequence whereas the former have little or no sequence similarity to that of the helper virus. However, some viral infections are associated with helper virus-dependent RNAs that appear in part to be derived from the helper virus and in part to have unique sequence. As noted above, the 356-nucleotide RNA C of TCV has a 5' portion corresponding to a 194-nucleotide true satellite RNA of TCV and a 3' portion corresponding in sequence to the 3' end of TCV genomic RNA (Simon 1999). Between the 5' and 3' sequences is another distinct segment of TCV genomic RNA.

Replicating artificial satellite RNA–D-RNA hybrids have been constructed from tombusvirus sequences (Burgyan *et al.*, 1992). Both these artificial sequences and RNA C alter the titer and symptoms of the helper virus.

C. Satellite DNAs

The first satellite to be recognized to be associated with a DNA virus was a 262-nucleotide circular ssDNA associated with the begomovirus ToLCV from northern Australia (Dry *et al.*, 1997). Replication of SToLCV is dependent upon the helper virus replication-associated protein and is encapsidated by the helper virus coat protein. It has no significant ORFs and the only significant sequence similarity to that of the helper virus is in two short motifs in separate putative stem–loop structures: TAATATTAC, which is conserved in all geminiviruses, and AATCGGTGTC, which is identical to a putative replication-associated protein-binding motif in TLCV (see Chapter 8, Section VIII). Replication of SToLCV can be supported by other geminiviruses such as the begomoviruses, ACMV and TYLCV, and the curtovirus, BCTV.

Other single-component begomoviruses have several non-viral ssDNAs associated with them. Both CLCuV and AYVV have an associated nanovirus-like component (Mansoor *et al.*, 1999; Saunders and Stanley, 1999). This component resembles the nanovirus components that encode Reps (see Chapter 8, Section VIII.E) and requires the helper virus for its replication; it contains the conserved geminivirus sequence described above. Plants infected with AYVV also contain another circular ssDNA, DNAβ, which is approximately half the size (1347 nucleotides) of the parent DNA A (Saunders *et al.*, 2000). Apart from the TAATATTAC sequence common to all geminiviruses, this has negligible sequence similarity to the AYVV sequence. DNAβ is dependent upon AYVV DNA A for its replication, is encapsidated in the coat protein encoded by AYVV DNA A and, thus, is characteristic of a satellite. The yellow vein symptoms in *Ageratum conyzoides* are dependent on joint infection with AYVV DNA A and DNAβ; the nanovirus-like satellite has no effect on symptom production (Saunders *et al.*, 2000).

D. Complex-dependent viruses

As noted above, satellite viruses and RNAs depend upon a helper virus for their replication and satellite RNAs also require the helper for their encapsidation. Umbraviruses can replicate autonomously but require another virus to effect encapsidation (see Chapter 2, Section III.Q.20). In the case of the complex causing groundnut rosette disease, a satellite RNA also plays an important role. Groundnut rosette disease is caused by the complex of two viruses, the luteovirus GRAV and the umbravirus GRV, together with a satellite. Both GRV and sGRV are encapsidated in the coat protein of GRAV, which contributes the aphid transmission to the disease. sGRV is essential for the encapsidation of GRV (Robinson *et al.*, 1999) and thus for its aphid transmission. Furthermore, the expression of chlorotic or green symptoms shown by different 'strains' of rosette is controlled by the strain of sGRV (Murant *et al.*, 1988a; Murant and Kumar, 1990).

Pea enation disease is also caused by a complex of a luteovirus, PEMV-1, and an umbravirus, PEMV-2 (see Chapter 2, Section III.P.3). PEMV-2 has a satellite RNA, sPEMV (see Section II.B.3.b), associated with it but this does not appear to contribute to the encapsidation of PEMV-2, even though it has sequence similarity to sGRV and helper viruses can support the two satellites (Demler *et al.*, 1996b).

E. Discussion

Satellite viruses and RNAs form specific associations with their helper viruses. This raises two questions: how did they evolve and what is their natural function? Ideas on evolution of satellites are discussed in Chapter 17 (Section XII.C). As to function, there are two possibilities: satellites are either 'molecular parasites' or they have a beneficial role in the biology of the helper viruses. As molecular parasites, satellites compete for the replication machinery with the helper virus. However, this may not be to the disadvantage of the helper virus. It is possible that the attenuation of symptoms by many satellites could enhance the survival of the host to the benefit of the helper virus. The variants that induce more severe symptoms, such as some CARNA5 isolates, would be selected against under natural conditions as they kill their host thus limiting the spread of the helper virus. Kassanis (1962) suggested that satellite viruses (and satellite nucleic acids) may alter the metabolism of host cells so that they become more tolerant to the replication of the helper virus.

If the complex associations of viruses (and satellites) described in Section II.D above reflects an early stage in the evolution of satellites, one can see other advantages accruing. For instance, PEMV-2 facilitates the full systemic spread of phloem-limited PEMV-1, which could help in the insect transmission of the disease complex. One could envisage PEMV-2 losing its ability to replicate autonomously and that property being provided by PEMV-1. Similarly, the groundnut rosette disease complex requires the presence of sGRV for encapsidation of its component members, this being essential for insect transmission.

Methods for Assay, Detection and Diagnosis

I. INTRODUCTION

The ability to assay viruses is basic to almost all kinds of virological investigation. Simultaneous application of two or more assay methods that depend on different properties of the virus to separate samples of the same material is useful, and is often essential for many kinds of experiments. The problems of detecting viruses and of diagnosing viral disease involve the use of assay techniques, so these topics are also considered in this chapter. Some of the methods are used for assay, some for diagnosis, and some for both. The advances in molecular biology over the past 20 years or so have had a major impact on the range of tools available for virus detection and diagnosis. Diagnosis of a viral disease on a routine basis is a requirement for the development of satisfactory measures to control that disease (see Chapter 16).

We can distinguish at least five situations where the procedures outlined in this chapter are used. First, we may wish to assay a virus, for example during the step in a procedure designed to purify the virus, or in experiments to investigate the stability of the virus *in vitro*. Second, diagnostic and assay techniques are needed when investigating the effects of mutagenesis on cloned infectious genomes of a virus. The third situation is when we can anticipate from previous experience that one or more known viruses may be present in a crop, and we wish to detect and identify these and assess their prevalence through a growing season. The fourth occurs when we isolate a new virus in a particular crop or region, and we wish to deter-

mine whether it is a new virus or a variant or strain of a known virus, or identical to a known virus. In the fifth situation, assay and diagnostic techniques are necessary in the study of the epidemiology of a virus. In the past, descriptions of viruses have often been published without sufficient data to distinguish between these possibilities. Thus, the literature has become cluttered with partial descriptions of viruses that are essentially useless.

The problem of delineating strains of viruses is discussed in Chapter 17, while the criteria that can be used to describe and classify an unknown virus are summarized in Chapter 2. Other chapters include material relevant to the problem of disease diagnosis: Chapter 4, viral components; Chapter 5, viral architecture; Chapters 7 and 8, expression and replication; and Chapter 14, agents causing virus-like symptoms.

The choice of plant material to be sampled is of great importance for successful assay, detection and diagnosis. The distribution of virus in an infected plant may be very uneven (see Chapter 9). Many factors influence the concentration of virus in a plant and therefore its ease of detection (see Chapter 10). These problems may be particularly important for crops, where tissue samples are taken from bulbs, corms or tubers, and for woody perennial crops.

When appraising the relative merits of different methods, the following important factors need to be considered:

1. What question is being addressed? Is one just determining whether the plant is virus-infected or what virus is infecting the plant, or what strain of the virus is infecting the plant?

Virus acronyms are given in Appendix 1.

2. Sensitivity—how small an amount of virus can be measured or detected.
3. Accuracy and reproducibility.
4. Numbers of samples that can be processed in a given time by one operator.
5. Cost and sophistication of the apparatus and materials needed.
6. The degree of operator training required.
7. Adaptability to field conditions.

The last three considerations may be particularly important in the developing countries of the tropics. For this reason, adequate biological tests may be especially important in these countries (Lana, 1981).

Finally, it must always be remembered that diseased plants in the field may be infected by more than one virus. Thus, an early step in diagnosis for an unknown disease must be to determine whether more than one virus is involved.

The methods involved in assay, detection and diagnosis can be placed in four groups according to the properties of the virus upon which they depend: biological activities; physical properties of the virus particle; properties of viral proteins; and properties of the viral nucleic acid.

II. METHODS INVOLVING BIOLOGICAL ACTIVITIES OF THE VIRUS

Biological methods for the assay, detection and diagnosis of viruses are much more time consuming than most other methods now available. Nevertheless, they remain very important. Only measurements of infectivity give us relative estimates of the concentration of viable virus particles. As far as diagnosis is concerned, in most circumstances only inoculation to an appropriate host species can determine whether a particular virus isolate causes severe or mild disease. However, this group of methods does have a major drawback, especially when used on a new virus or in quarantine situations. It raises the possibility that the production of infected plants might be a source of infection for local crops, even in spite of the

strictest containment conditions. Some practical aspects of biological testing are described by Hill (1984), Walkey (1991), Matthews (1993) and Dijkstra and de Jager (1998).

A. Infectivity assays

1. Quantitative assay based on local lesions

Holmes (1929) showed that the necrotic local lesions produced in leaves of *Nicotiana glutinosa* following mechanical inoculation with TMV could be used for assay of relative infectivity. The method was more precise and used fewer plants than the older procedure of estimating the number of systemically infected plants in an inoculated group. Since that time, much effort has been devoted to seeking local lesion hosts for particular viruses. Various aspects of mechanical transmission are discussed in Chapter 12.

Whenever possible, hosts that give a clear-cut necrotic or ring spot type of local lesion are used for local lesion assays (Plates 3.1A,B, 3.2 and 3.3). Some viruses give reproducible chlorotic lesions, but for others chlorotic lesions may grade from clear-cut spots to faint yellow areas that require arbitrary and subjective assessment. With such plants, it is sometimes possible to take advantage of the fact that the starch content of the cells in the lesion may differ from that in the uninfected cells. At the end of a photosynthetic period, virus-infected cells may contain less starch. At the end of a dark period, they may contain more starch. Leaves are decolored in ethanol and stained with iodine. For satisfactory and reproducible results with starch–iodine lesions, environmental conditions and sampling times need to be carefully controlled. Necrotic local lesions induced by heat treatment of the leaves have been used for certain host–virus combinations (Foster and Ross, 1975).

The nutritional state of the plant may affect the distinctness of the local lesions formed. For example, in nitrogen-deficient Chinese cabbage plants, TYMV may produce well-defined purple local lesions (Diener and Jenifer, 1964).

With due care and an appropriate experimental design, local lesion assays can distinguish a difference in infectivity of as little as 10–20% between two preparations. However, there are many examples in the literature where unwarranted conclusions are drawn from local lesion assays. The four major aspects to be borne in mind are: (1) the wide variation in number of local lesions produced in different leaves by a standard inoculum; (2) the general nature of the curve relating dilution of inoculum and lesion number; (3) the hit kinetics of multicomponent viruses; and (4) the statistical requirements for making valid comparisons.

a. Variation between leaves

The environmental and physiological factors that influence susceptibility to infection are discussed in Chapter 12. They may include the age of the plant, genetic variation in the host, position of the leaf on the plant, nutrition of the plant, water supply, temperature, light intensity, season of the year and time of day. Samuel and Bald (1933) recognized that there was much less variation between opposite halves of the same leaf than between different leaves. Since that time, most experimental designs for local lesion assays have used half-leaf comparisons. For plants like *Phaseolus vulgaris*, which have two usable primary leaves in the same position on the plant, four fairly equivalent half-leaves are available. Close attention to the inoculation procedure, and to the conditions under which inoculated leaves are held, may improve both the sensitivity and reproducibility of local lesion assays (e.g. McKlusky and Stobbs, 1985).

The number of half-leaf comparisons that are necessary depends on the accuracy required, the uniformity of the test plants, and the number of samples to be compared. With an assay plant grown under fairly standard conditions, it is usually possible after some experience to pick individual plants that will have a susceptibility differing markedly from the group, and to discard them. For a single comparison between two samples, a minimum number of six to eight leaves should be used. When more than two samples are compared, a variety of experimental layouts is possible. One-half of every leaf can be inoculated with the same standard preparation, and the various test solutions to the other half-leaves. This is a simple design but rather wasteful of plants.

Where appropriate, a latin square design is effective. For example, with a plant such as *N. glutinosa*, where about four to eight leaves may be available on each plant, it is possible, for a limited number of samples, to arrange that each sample is compared on a leaf at each position on the plant.

Where the number of treatments exceeds the number of leaves, an appropriate design is one in which each test inoculum appears on each leaf position the same number of times. Kleczkowski (1950), Fry and Taylor (1954) and Preece (1967) give some examples of more complex experimental designs. However, there may be a useful limit to the size and complexity of local lesion assays. In large experiments the risk of error in the inoculations or labeling is increased. Because of the long time required to carry out inoculations, changes in the susceptibility of plants with time of day might influence results unless the experimental design is further complicated to take account of this effect, which may be quite large.

b. Relation between dilution of inoculum and lesion number

Best (1937b) examined the nature of the curve relating dilution of TMV inoculum and numbers of local lesions produced in *N. glutinosa*. He found that the curve could be divided into three parts: (1) a section at high concentrations where a change in concentration is accompanied by very little change in local lesion number; (2) a section in the middle of the curve where a change in concentration is accompanied by a more or less equivalent change in lesion number; and (3) a section at low virus concentration where change in concentration has little effect on lesion number. This general situation holds for many viruses, but not for all. Two dilutions curves are shown in Fig. 15.1.

Virus dilution curves are interesting from both the theoretical and the practical points of view. Infectivity is considered in relation to the processes of infection in Chapter 12. From the practical point of view, the form of the dilution curve has several important implications. First, comparisons of two samples where local lesion numbers are very high or very low are quite useless. They may be in error by several orders

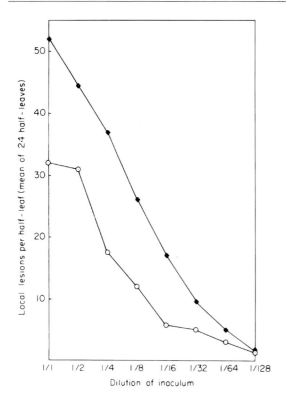

Fig. 15.1 Infectivity dilution curves. Effect of dilution of inoculum on number of local lesions produced by two plant viruses. ◆ TBSV in *Nicotiana glutinosa*; ○ TMV in *N. glutinosa*. Data from Kleczkowski (1950), with permission.

of magnitude. Second, valid comparisons can be made only in the region where lesion number is responding more or less proportionally to dilution. Third, the exact slope of the dilution curve is variable and unpredictable from one experiment to another. This means that two samples must be compared at several dilutions (usually 2-, 5- or 10-fold). Generally speaking, for leaves about the size of *N. glutinosa* leaves, mean values in the range of 10–100 local lesions per half-leaf give useful estimates. For much larger or much smaller leaves, the range would be different.

Factors that may affect the slope of the dilution curve include: (1) the presence of inhibitors in the inoculum—dilution of the inhibitor may give rise to a curve that is flatter than expected; (2) virus that is in an aggregated state but becomes disaggregated on dilution—this also would give a flattened curve; (3) the need for

more than one virus particle to give a local lesion—this would give a curve steeper than expected (Fulton, 1962) and has been found for several viruses that require two or more particles for infectivity (Fig. 15.2); and (4) changes in susceptibility of test plants during the time taken to carry out inoculations—this could affect the slope of the curve either way depending on order of inoculation and time of day.

c. Relationship between lesion number and concentration of infective virus

Even when account has been taken of leaf-to-leaf variation (by replication and experimental design) and when the actual lesion numbers fall in the middle range of the dilution curve, lesion number cannot be translated directly into relative infective virus content. The simplest practical way to overcome this problem is to arrange the dilutions (on the basis of a preliminary test) so that the samples compared give nearly equal numbers of local lesions (within the useful range of about 10–100 per leaf or half leaf) in one of the comparisons.

Proper use of local lesion data requires some statistical analysis. Kleczkowski (1949, 1953) drew attention to the fact that neither lesion

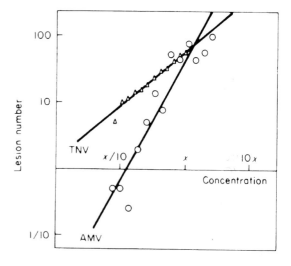

Fig. 15.2 Comparison of dilution curves for a single-particle virus (TNV) and a multi-particle virus (AMV). Local lesions in *Phaseolus vulgaris* for both viruses. For TNV, $X = 0.15 \, \mu g/mL$; for AMV, $X = 9 \, \mu g/mL$. From van Vloten-Doting *et al.* (1968), with permission.

numbers nor logarithms of these numbers (a transformation used by earlier workers) are satisfactory for statistical analysis. Numbers of local lesions are not normally distributed, and the variance of the mean increases with increasing mean. Kleczkowski derived a transformation that is satisfactory when the mean number of local lesions is greater than about 10. This transformation is:

$$y = \log_{10}(X + C)$$

where X is the number of local lesions and C is a constant. C can be assessed for each experiment, but any value of C between about 5 and 20 could be used satisfactorily.

d. Some general considerations

The many factors that can influence the infectivity of viruses and susceptibility of plants should be borne in mind when devising and using an assay system. When studying the effect of some treatment on the infectivity of a preparation it is important to consider whether the treatment may be altering the state of the medium in some way (e.g. pH, chelating agents) rather than having an effect directly on the virus.

The size and complexity of an assay should be appropriate to the needs of the experiment. It is a waste of labor to set up an elaborate randomized design when a very approximate estimate of infectivity will give the required answer. The more common failing is to draw conclusions from inadequately designed and analyzed experiments. In this connection, one factor that has been widely neglected is the influence of time of day on susceptibility of test plants (Chapter 9, Section III.A.3). The magnitude of this effect varies a great deal and, with assays that take only an hour or so to carry out, probably can be neglected. However, with complex experimental designs and the larger numbers of plants required for maximum accuracy, or with a large number of samples to be assayed, this factor can influence results in a systematic way.

Practical advice on local lesion assays is given in Dijkstra and de Jager (1998).

e. Assay in insect vector cell monolayers
(see Uyeda et al., 1995)

L.M. Black and colleagues have developed an assay technique for PYDV and WTV in which the virus is applied to insect vector cells growing as a monolayer on coverslips. The method provides an assay for these two viruses that is, in principle, the same as the plaque methods available for bacterial and vertebrate viruses. In a comparative study with PYDV, Hsu and Black (1973) found that on the basis of the number of cells per unit area of monolayer, or of leaf epidermis, an assay using insect cells was $10^{3.7}$ times more sensitive than that using local lesions on Nicotiana rustica. Assay on vector cells was also much less variable than when leaves were used. Kimura (1986) developed conditions for infectivity assays of RDV using a focus counting technique on vector cell monolayers. This method was about 100 times more sensitive than vector injection. RGDV has also been assayed in cell monolayers (Omura et al., 1988).

2. Quantal assay based on number of individuals infected

Before Holmes (1929) introduced the local lesion assay, the only method available for measuring infectivity was to inoculate the sample at various dilutions to groups of plants and to record the number of plants that became systemically infected. This type of test takes longer and requires many more plants to obtain a reliable answer. Nevertheless, there are still occasions when this type of test has to be used, for example with viruses having no suitable local lesion host or with viruses that have to be assayed by the use of insect vectors. Statistical aspects of quantal assays are discussed by Brakke (1970).

a. Mechanical inoculation of whole plants

Groups of plants are inoculated with a series of dilutions of the inocula to be compared. The dilutions must span the range over which only a proportion of the test plants become infected. Presence or absence of systemic symptoms are subsequently recorded. This procedure is known as a quantal assay, distinct from a quantitative assay based on local lesions. Various means have been used to improve the precision of quantal assays. When large numbers of plants can be grown and inoculated easily, these can give useful data (Fig. 15.3). Probit analysis can be used to estimate the LD_{50} and to give a statistical estimate of precision with such data.

When plants are inoculated with small amounts of virus, systemic symptom development takes longer than with heavy inocula. A record of the time taken for systemic symptoms to appear, combined by some arithmetic manipulation with the proportion of plant infected, may give increased precision to a quantal assay (Diener and Hadidi, 1977).

b. Incubated tissue samples

It is sometimes possible to estimate the rate at which a virus moves from one part of the plant to another by sampling many small pieces of tissue that may contain no virus or very low amounts of virus. The tissue samples are incubated in isolation to allow any virus present to multiply to give a detectable amount, and then are assayed for the presence or absence of virus either by infectivity or some other method. Such a procedure was used by Fry and Matthews (1963) to determine the time after

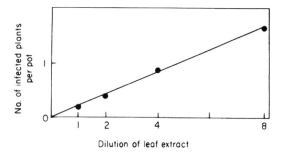

Fig. 15.3 Quantal assays. Use of large numbers of plants to obtain precise assays of infectious virus. Four dilutions of crude sorghum leaf extract from plants infected with SCMV were prepared and each dilution was inoculated to 525 sorghum seedlings (25 to a pot). This experiment was repeated seven times. Thus, each point is based on 147 pots. 1, 1/160 000; 2, 1/80 000; 4, 1/40 000; 8, 1/20 000. As the straight line passes through the origin, it was concluded that SCMV probably has a monopartite genome. From Dean (1979), with kind permission of the copyright holders, © The American Phytopathological Society.

inoculation at which cells beneath the epidermis of tobacco leaves became infected with TMV.

c. Insect vectors

For viruses that are not mechanically transmissible but have an insect vector, it may be possible to use the percentage of successful insect transmissions to estimate relative amounts of virus. Insects might feed on the test plants or through membranes on solutions containing the virus, or they might be injected with such solutions.

Results obtained by feeding insects on infected tissue may reflect differences in the availability of virus to the insect rather than concentration in the tissue or organ. These methods are all laborious and involve biological variability in the insect as well as in the host plant. For example, the length of time insects are allowed to feed on the test plants may affect the results. Insects may die on the test plants at various times. If more than one insect is used on each test plant, more elaborate statistical analysis is needed. Because of these difficulties, the methods have not been widely used, but they have provided valuable information for certain viruses such as WTV (Whitcomb and Black, 1961).

B. Indicator hosts for diagnosis

Disease symptoms on plants in the field are frequently inadequate on their own to give a positive identification. This is particularly so when several viruses cause similar symptoms, as do yellowing diseases in beet (Duffus, 1973), when a single virus, such as CMV, is very variable in the symptoms it causes (Francki and Hatta, 1980), or when both of these factors are relevant in a single host (Francki et al., 1980a). Thus, since the early days of plant virology, searches have been made for suitable species or varieties of host plant that will give clear, characteristic and consistent symptoms for the virus or viruses being studied, usually under glasshouse conditions. Such indicator hosts provide one of the most basic tools for routine diagnosis. Many good indicator species have been found in the genera *Nicotiana, Solanum, Chenopodium, Cucumus, Phaseolus, Vicia* and *Brassica*. Certain plant species, such as *Chenopodium amaranticolor, C. quinoa* and *Nicotiana benthamiana,* react to a wide range of viruses. Indicator hosts for plant virus diagnosis are reviewed by Horváth (1993). Sometimes the usefulness of a particular plant can be enhanced by changing the growing conditions (e.g. Lee and Singh, 1972).

Our knowledge of virus structure and replication has increased greatly in recent years, and the taxonomy of plant viruses has developed to the stage where most of the known viruses have been allocated to families or genera. Nevertheless, the search for improved indicator hosts continues. For example, Schmidt and Zobywalski (1984) surveyed a set of 33 French bean cultivars and lines for their ability to discriminate between a range of isolates of BYMV. Van Dijk *et al.* (1987) screened some 200 species and accessions within the genus *Nicotiana* to find several new and useful indicator hosts. Lists of useful indicator species for specific viruses are given in the Association of Applied Biologists Descriptions of Plant Viruses (see Appendix 3).

When comparing the results of viral diagnosis from different laboratories using indicator species, it must always be borne in mind that different lines of the same named variety may vary quite markedly in their symptom response to a given virus (e.g. van der Want *et al.*, 1975).

Testing large numbers of samples using indicator hosts requires glasshouse facilities that may be occupied for weeks or longer. However, the actual manipulations involved in mechanical inoculation may take less time per sample than many other methods of testing, and many attempts have been made to streamline the procedures for sap extraction and inoculation (e.g. Laidlaw, 1986). As noted above, care must be taken to keep inoculated test plants under suitable containment to minimize escape of the virus(es) to the outside environment.

For some groups of diseases, international co-operation has led to the definition of standard sets of indicator plants for particular diseases. Thus, the working group on fruit tree virus diseases of the International Society for Horticultural Science has developed a list of indicator hosts for virus and virus-like diseases of eight major woody fruit species. Conditions for field or greenhouse indexing are defined. Of the 137 diseases listed, diagnosis depended solely on the indicator hosts for 87. Antisera were available for testing the remaining 50 by ELISA to complement the indicator results (Dunez, 1983).

C. Host range in diagnosis

In earlier work on plant viruses, host range was used as an important criterion in diagnosis. Such information may still be important, or even crucial in certain circumstances. However, apart from the fact that only a relatively small proportion of possible host–virus combinations has been tested, our knowledge of host ranges is limited, or must be qualified, in several ways:

1. In many of the reported host range studies, only positive results have been recorded.
2. Absence of symptoms following inoculation of a test plant has not always been followed by back-inoculation to an indicator species to test for masked infection. Such infections may not be uncommon. In testing the susceptibility of 456 plant species to 24 viruses, Horváth (1983) discovered 1312 new susceptible host–virus combinations. Of these, 13% were latent infections. Where inoculation tests are carried out, an arbitrary decision may have to be made about what constitutes multiplication of virus over and above inoculum remaining on the leaf. Thus, Holmes (1946) took 10 lesions per leaf on back-inoculation to *N. glutinosa* to mean that multiplication of TMV or TEV had occurred in the test species. Multiplication of the virus in one or a few cells of a host near the sites of infection may well go undetected by current procedures. Inoculation with infectious RNA would reduce or eliminate background infectivity due to residual inoculum.
3. The manner of inoculation may well affect the results. Mechanical inoculation has almost always been used in extensive host range studies because of its convenience, but

many plants contain inhibitors of infection that prevent mechanical inoculation to the species, or from it, or both.

4. In studying large numbers of species it is usually practicable to make tests only under one set of conditions, but it is known that a given species may vary widely in susceptibility to a virus depending on the conditions of growth (Chapter 12, Section II.B.2).

5. Quite closely related strains of a virus may differ in the range of plants they will infect. Host range data may apply only to the virus strain studied.

6. Mesophyll protoplasts may be readily infected with a virus that gives little or no infection when applied to intact leaves (e.g. Huber *et al.*, 1977). (See also Table 10.1.)

In spite of these qualifications, it may still be worthwhile to seek sets of host and non-host plants that will distinguish between certain viruses in particular crop species. For example, an international study of persistent aphid-transmitted viruses causing diseases in legumes identified a set of 12 leguminous and non-leguminous test plants that could discriminate between the eight viruses in the study (Johnstone *et al.*, 1984).

Sometimes host range is more discriminating than other tests. For example, raspberry plantings in Scotland have been protected against RBDV by the use of cultivars resistant to the common strain of the virus. A resistance-breaking strain could not be distinguished from the common strain either serologically or by indicator hosts (Murant *et al.*, 1986). Different viruses vary widely in the taxonomic breadth of their host range. Host range studies for diagnosis will usually be most useful for those infecting a relatively narrow range of plants. Because of the gene-for-gene resistance (see Chapter 16, Section V.B.1) specific genotypes of plants can be useful in identifying strains of a virus.

D. Symptom-related methods

Virus infection often affects the chlorophyll levels of plants. Daley (1995) suggested the use of chlorophyll fluorescence analysis for long-term, non-destructive field studies of plant virus infections. Presumably this could be used in remote sensing systems for detection of viral infections but would probably not have much potential in the identification of the virus.

As described in Chapter 10, Section III.K, salicylic acid (SA) is produced in certain virus–host interactions. SA induces metabolic heating and the response to TMV infection of tobacco leaves was detected by thermography before any symptoms became visible (Chaerle *et al.*, 1999).

E. Methods of transmission in diagnosis

The different methods of virus transmission, discussed in Chapters 11 and 12, may be useful diagnostic criteria. Their usefulness may depend on the particular stances. For example, a virus with an icosahedral particle that is transmitted through the seed and by nematodes is very probably a nepovirus. On the other hand, such a virus transmitted mechanically and by the aphid *Myzus persicae* might belong to any one of several groups.

F. Cytological effects for diagnosis

The cytological effects of infection, described in Chapter 3, have been used to assist in diagnosis. Cytological effects detectable by light microscopy can sometimes be used effectively to supplement macroscopic symptoms in diagnosis. Christie and Edwardson (1986) provide an illustrated catalog of virus-induced inclusions and discuss the problems involved, and Edwardson *et al.* (1993) review this approach to diagnosis. The light microscope has a number of advantages when the inclusions are large enough to be easily observed: (1) it is a readily available instrument; (2) specimen preparation techniques can be simple, for example inclusions seen in epidermal strips can assist in the rapid diagnosis of some viral diseases in red clover (Khan and Maxwell, 1977); (3) there is a wide field of view, thus allowing many cells to be readily examined; and (4) a variety of cytochemical procedures is available.

Nevertheless, electron microscopy of thin sections is necessary for some types of inclusion to provide information for use in diagnosis. Hamilton *et al.* (1981) list nine virus groups that induce inclusions that are diagnostic for the group. Presence of characteristic inclusions may be diagnostic for a particular virus when a specific host plant is involved, for example the CTV in citrus trees (Brlansky, 1987). Individual strains of a virus may produce distinctive cytological effects, as occurs with CaMV (Shalla *et al.*, 1980). Strains of AMV produce different aggregation forms of their virus particles (Hull *et al.*, 1970).

G. Mixed infections

Simultaneous infection with two or even more viruses is not uncommon. Such mixed infections may complicate a diagnosis based on biological properties alone, especially if the host response is variable, as with the internal rib necrosis disease of lettuce in California (Zink and Duffus, 1972). However, several possible differences in biological properties may be used to separate the viruses: (1) if one virus is confined to the inoculated leaf in a particular host, while the other moves systemically; (2) if a host can be found that only one of the viruses infects; (3) if the two viruses cause distinctive local lesions in a single host; and (4) if the two viruses have different methods of transmission, for example by different species of invertebrate vectors.

On the other hand, certain diseases in the field may be the result of a more or less stable association between two or more viruses. For instance, two viruses are involved in pea enation mosaic disease (Demler *et al.*, 1996a) and two viruses and a satellite in groundnut rosette disease (Murant, 1990; Murant and Kumar, 1990). At least three viruses may be implicated in the carrot 'motley dwarf' disease complex (see Vercruysse *et al.*, 2000) and in the lettuce speckles disease (Falk *et al.*, 1979).

H. Preservation of virus inoculum

To aid in diagnosis of new diseases and for other virus studies, it is often useful to store virus inocula rather than maintain stock cultures in plants in the glasshouse. Storage saves space and minimizes the risk of cross-contamination or changes in the virus isolate. Preservation of purified viruses is discussed in Chapter 4 (Section II.A.7).

Many strains of TMV can be stored for long periods in air-dried leaf or in non-sterile aqueous media. Inoculum for most other viruses loses infectivity more or less rapidly unless special conditions are met. Most procedures involve removal of water from the tissue or liquid, the addition of protectant materials, storage at low temperature, or a combination of these procedures. Hollings and Stone (1970) found that a high proportion of the 74 viruses they tested survived for at least a year after lyophilization of infective leaf sap. D-Glucose and peptone were added before lyophilization, and the ampoules were stored at room temperature. Some remained infective for over 10 years.

Skimmed milk has been used as a protectant for LNYV and for some other unstable viruses stored in a dehydrated state at 4°C (Grivell *et al.*, 1971).

Deep-frozen liquid inocula are satisfactory for some viruses, for example WMV (De Wijs and Suda-Bachmann, 1979), but may be unsatisfactory for others, for example RCMV (Marcinka and Musil, 1977). Many of the more unstable viruses have been stored for periods of years in pieces of chemically dehydrated tissue held at about 10°C (McKinney *et al.*, 1965). Leaf tissue taken from young, actively growing, infected plants and held in sealed vials gave the longest storage of infective virus. However, care must be taken that the leaf material being preserved contains only the virus of interest. For example, a leaf sample indicated to contain SCPMV dried in 1962 was found to also contain CPMV and CCMV; this was 2 years before CCMV was first reported! (R. Hull, personal observation).

For air transport between countries, fresh leaf samples can be placed between pieces of dry paper, sealed in polyethylene bags, and transported at 4°C in a 'cool bag'; if wet paper is used, fungal rots may set in, especially on leaves collected in tropical climates. For particularly perishable material or where 4°C storage is not available, leaves may be soaked in 50% glycerol for a few hours before being sealed in bags (Alhubaishi *et al.*, 1987).

III. METHODS DEPENDING ON PHYSICAL PROPERTIES OF THE VIRUS PARTICLE

A. Stability and physicochemical properties

1. Measurement of stability

Historically, stability of the virus as measured by infectivity (often in crude extracts) was an important criterion in attempting to establish groups of viruses. Such properties as thermal inactivation, aging at room temperature and the effect of dilution were used. The main justification for their use was the absence of alternative physical data. It is unfortunate that they are still sometimes invoked, because they have been shown to be far too variable to provide a sound basis for the identification of viruses or the placing of viruses into groups (Francki, 1980). Publication of a virus description based solely on biological properties together with stability in a crude extract will usually do no more than clutter the literature. However, useful comparative information can sometimes be obtained from the kinds of reagent that degrade a virus. For example, those whose subunits appear to be held together mainly by ionic bonds are degraded by strong salt, which tends to stabilize those held together mainly by hydrophobic forces. The latter type may be readily degraded by reagents such as urea. This is useful information in designing purification procedures.

2. Physicochemical properties

A virus has certain independently measurable properties of the particle that depend on its detailed composition and architecture. These properties can be useful for identification and as criteria for establishing relationships. The most commonly measured properties are described below.

Density. Density is measured in water, sucrose or cesium chloride solutions (see Chapter 4, Section II.A.5). For non-enveloped viruses, this property reflects mainly the proportion of nucleic acid in the virus. The buffer used may affect the results obtained in cesium chloride (see Scotti, 1985).

Sedimentation coefficient and diffusion coefficient. These properties reflect the mass, density and shape of the virus particle and are discussed further in Chapter 5 (Section II.B.1).

Ultraviolet absorption spectrum. The ultraviolet absorption spectrum of a virus is a combination of the absorption spectra of the nucleic acid and the coat protein (Fig. 15.4A). Nucleic acids have maximum absorption at about 260 nm and a trough at about 230 nm, whereas the absorption spectrum for proteins peaks at about 280 nm and troughs at about 250 nm. The specific absorption for nucleic acids (20–25 OD units per mg per mL at 260 nm) is much greater than that for proteins (about 1 OD unit per mg per mL at 280 nm). The combination gives an absorption spectrum for a virus that peaks at about 260 nm and troughs at about 230–240 nm (Fig. 15.4A,B).

Thus, this property depends mainly on the ratio of nucleic acid to protein in the virus; the amino acid composition of the protein can also influence the spectrum. Many unrelated viruses with similar compositions have similar absorption spectra. Ultraviolet absorption provides a useful assay for purified virus preparations, provided other criteria have eliminated the possibility of contamination with non-viral nucleic acids or proteins, especially ribosomes. Measurements of A_{260} may be unreliable for rod-shaped viruses that vary substantially in their degree of aggregation from one preparation to another, thus leading to changes in the amount of light scattering for a given virus concentration.

Electrophoresis. Electrophoretic mobility is now usually determined by migration in polyacrylamide gels. Thus, this property depends both on the size of the particle and on the net charge at its surface. While strains of the same virus usually have very similar mobilities, different viruses within a group may be distinguished,

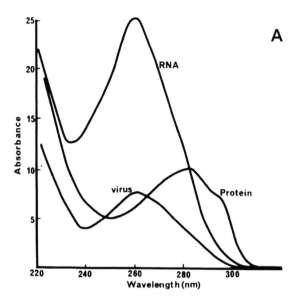

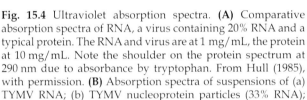

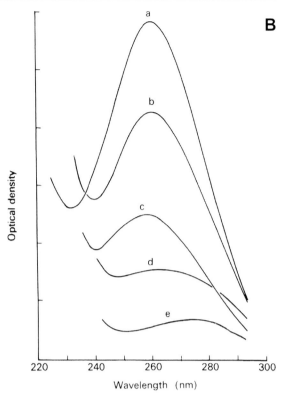

Fig. 15.4 Ultraviolet absorption spectra. **(A)** Comparative absorption spectra of RNA, a virus containing 20% RNA and a typical protein. The RNA and virus are at 1 mg/mL, the protein at 10 mg/mL. Note the shoulder on the protein spectrum at 290 nm due to absorbance by tryptophan. From Hull (1985), with permission. **(B)** Absorption spectra of suspensions of (a) TYMV RNA; (b) TYMV nucleoprotein particles (33% RNA); (c) CMV particles (19% RNA); (d) TMV particles (5% RNA); (e) TYMV protein particles (top component). All suspensions contained the same weight of material except (a), which contained half as much as the others. From Gibbs and Harrison (1976), with permission.

as was shown for the *Cucumovirus* group by Hanada (1984) and for PEMV (Hull, 1977b).

B. Ultracentrifugation

1. Analytical ultracentrifugation

Plant viruses fall in a size range that makes them very suitable for assay using the analytical ultracentrifuge. The technique has been particularly useful for monitoring the progress of a virus purification procedure, for studying the effects of various treatments on the physical state of a virus and for assay of amounts of virus in crude tissue samples. With Schlieren optics, virus at 100 µg/mL could usually be measured, and it can be detected down to about 50 µg/mL. The best measurements have been made at concentrations of single components in the range 0.5–2.0 mg/mL. One of the big advantages of the analytical centrifuge for virus assay has been that, as well as giving a measure of the amount of material present, it

can provide a physical criterion of identity, the sedimentation coefficient.

Analytical centrifugation has proved a valuable method for studying the purity of virus preparations with respect to (1) the presence of virus-like particles with different sedimentation properties, (2) the presence of a contaminating virus, (3) the presence of contaminating host macromolecules, and (4) the state of aggregation of the virus. Although this technique was used widely in the 1970s and 1980s, it is now used less frequently. Even so, it still remains a useful tool.

2. Density gradient centrifugation

Density gradient centrifugation, developed by Brakke (1951, 1960), is a method that can be used for both isolation and assay of plant viruses. It has proved to be a highly versatile technique and has been widely used in the fields of virology and molecular biology. A centrifuge tube is partially filled with a solution having a decreasing density from the bottom to the top of the tube. For plant viruses, sucrose is commonly used to form the gradient, and the virus solution is layered on top of the gradient. With gradients formed with cesium salts, the viral particles may be distributed throughout the solution at the start of the sedimentation or they may be layered on top of the density gradient.

Brakke (1960) defined three ways in which density gradients might be used. *Isopycnic gradient centrifugation* occurs when centrifugation continues until all the particles in the gradient have reached a position where their density is equal to that of the medium. This type of centrifugation separates different particles based on their different densities. Sucrose alone may not provide sufficient density for isopycnic banding of many viruses. In *rate zonal sedimentation* the virus is layered over a pre-formed gradient before centrifugation. Each kind of particle sediments as a zone or band through the gradient, at a rate dependent on its size, shape and density. The centrifugation is stopped while the particles are still sedimenting. *Equilibrium zonal sedimentation* is like rate zonal sedimentation except that sedimentation is continued until most of the particles have reached an approximate isopycnic position. The role of the density gradient in these techniques is to prevent convectional stirring and to keep different molecular species in localized zones. The theories of density gradient centrifugation are complex. In practice, this technique is a simple and elegant method that has found widespread use in plant virology.

A high-speed preparative ultracentrifuge and appropriate swing-out or angle rotors are required. Following centrifugation, virus bands may be visualized as a result of their light scattering. The contents of the tube are removed in some suitable way before assay. The bottom of the tube can be punctured and the contents allowed to drip into a series of test tubes. Fractionating devices based on upward displacement of the contents of the tube with a dense sucrose solution are available commercially. The UV absorption of the liquid column is measured and recorded, and fractions of various sizes can be collected as required. Figure 4.2 illustrates the sensitivity of this procedure.

As successive fractions from a gradient can be collected, a variety of procedures can be used to identify the virus, non-infectious virus-like components and host materials. These include infectivity, UV absorption spectra and examination in the electron microscope. Using appropriate procedures, very small differences in sedimentation rate can be detected (Matthews and Witz, 1985).

With rate zonal sedimentation, if the sedimentation coefficients of some components in a mixture are known, approximate values for other components can be estimated. If antisera are available, serological tests can be applied to the fractions, or antiserum can be mixed with the sample before application to the gradient. Components reacting with the antiserum will disappear from the sedimentation pattern.

C. Electron microscopy

1. Number of virus particles per unit volume

A very crude but rapid indication of relative numbers of virus particles can be obtained for reasonably concentrated preparations by mixing a drop of the solutions with an appropriate amount of a negative stain such as phosphotungstic acid or uranyl acetate, placing a small amount of the mixture on an electron microscope grid and examining directly in the microscope for characteristic virus particles. It should be remembered that the particles of some viruses, especially those stabilized by ionic bonds, may be unstable in phosphotungstic acid. To obtain an accurate estimate of the number of virus particles by electron microscopy it is necessary to know the volume of solution being examined and to be able to count all the particles that were in that volume. Backus and Williams (1950) described a method in which the virus samples were diluted in a solution containing volatile salts (ammonium acetate or ammonium carbonate). The sample was then mixed with a solution containing a known weight of polystyrene latex particles of known and uniform size. The mixture was sprayed on to electron microscope grids using an atomizer.

The number of polystyrene latex particles present can be counted in photographs of drops, and the number present gives an estimate of drop size. The number of characteristic virus particles in the drop also is counted. The ratio of number of virus particles to number of latex particles will vary in different drops. A number of drops must be counted to give a reliable estimate of particle number, and appropriate statistical procedures must be applied (e.g. Williamson and Taylor, 1958). The method is laborious, but gives valuable information in some kinds of experiments.

2. Virus identification

Knowledge of the size, shape and any surface features of the virus particle is a basic requirement for virus identification. Electron microscopy can provide this information quickly and, in general, reliably. For the examination of virus particles in crude extracts or purified preparations, a negative-staining procedure is now used almost universally. Commonly used negative stains are sodium phosphotungstate, ammonium molybdate or uranyl acetate, depending on the stability of the virus to these stains.

Approximate particle dimensions can be determined. This is particularly useful for rod-shaped viruses for which particle length distributions can be obtained. Measured particle size may depend very much on how the specimen is prepared and stained (Fig. 15.5). Measurements of particle diameters made on crystalline arrays seen in thin sections of infected cells may also be in error (Hatta, 1976). Diameters of icosahedral viruses are sometimes best determined by measuring linear arrays of particles on the grid. Surface features may be seen best in isolated particles on the grid. Depending on size and morphology, a virus may be tentatively assigned to a particular taxonomic group. However, some small icosahedral viruses cannot be distinguished from members of unrelated groups on morphology alone (Hatta and Francki, 1984). Good results depend very much on optimizing the extraction method and the staining procedure for the particular virus and host plant. Details of current methods are given by Milne (1984), Roberts (1986), Christie et al. (1987) and Milne (1993).

For many viruses, examination of thin sections by electron microscopy is a valuable procedure for detecting virus within cells and tissues, but this has its limitations. The large enveloped viruses, the plant reoviruses, and the rod-shaped viruses can usually be readily distinguished because their appearance in thin sections generally differs from that of any normal structures. However, the concentration of virus in the cell and the distribution of particles

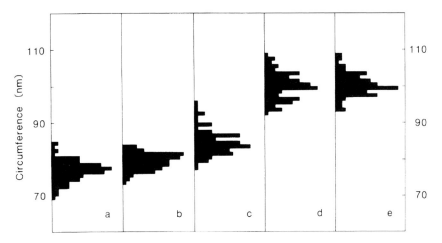

Fig. 15.5 Effects of various treatments on the apparent diameter of an icosahedral virus. Histograms of the circumference measurements (nm) of TRSV particles treated in different ways: particles freeze-dried and shadowed after washing grids with ammonium formate (a) and ammonium acetate (b) buffers; grids negatively stained with 2% ammonium molybdate, pH 7.0 (c), 2% uranyl acetate, pH 3.5 (d), and 2% methylamine tungstate, pH 7.0 (e). Only particles negatively stained with ammonium molybdate (c) have dimensions similar to those from freeze-dried preparations. From Roberts (1986), with permission.

must be such that there is a reasonable probability of observing virus particles if they are present, in a random section through the cell. It should be remembered that it would require 1000 serial sections 50 nm thick to examine completely the contents of a single cell 50 μm in diameter.

Most of the small isometric viruses have staining properties and apparent diameters that make it very difficult to distinguish scattered individual virus particles from cytoplasmic ribosomes. Some of these viruses form crystalline arrays within the cell and can thus be recognized as virus particles rather than ribosomes.

Some isometric viruses can be induced to form readily recognizable intracellular crystalline arrays if the water content of the tissue is reduced either by wilting (Milne, 1967; Ushiyama and Matthews, 1970) or by plasmolysis with sucrose solutions (Hatta and Matthews, 1974). The pattern of particles seen in the arrays will depend on the relationship between the plane of the section and the crystal lattice (Hatta, 1976).

Ribosomes are susceptible to digestion with RNase, while a small isometric virus may be resistant. This characteristic has been developed as a method for detecting such virus particles in cells (Hatta and Francki, 1981a). If isometric virus particles occur in cells or organelles where 80S ribosomes are absent, scattered virus particles may be recognized, for example in nuclei (Esau and Hoefert, 1973), plasmodesmata (de Zoeten and Gaard, 1969b), sieve elements (Esau and Hoefert, 1972) and vacuoles. However, nuclei of healthy cells sometimes contain crystalline structures that might be mistaken for viral inclusions (e.g. Lawson et al., 1971).

The identification technique that combines serological diagnosis with electron microscopy is discussed later in this chapter. Electron microscopy in relation to the architecture of virus particles is discussed further in Chapter 5.

D. Chemical assays for purified viruses

Procedures used in chemical analysis may be applied to the assay of viruses that can be obtained in a sufficiently purified condition. The simplest procedure, and one that is often overlooked, is to measure the dry weight content of a given volume of the solution. However, dry weights do not distinguish between virus and non-infectious particles containing less than the full amount of nucleic acid. Dry weight measurements form the basis for determining the nitrogen and phosphorus content of the virus and of other components, such as ribose or particular amino acids. Measurement of one of these components then could provide a method of virus assay in purified solutions. A solution containing a known weight of purified virus can be used to determine the absorbency at 260 nm for a given weight per milliliter of the virus, and also factors for making refractive index measurements with a differential refractometer, for converting the areas under Schlieren peaks, or absorbency peaks from sucrose gradients to weight of virus. These measures can then be used for the assay of other preparations of the virus.

E. Assay using radioisotopes

1. In vivo experiments

When the radioactive isotope ^{32}P or ^{33}P as orthophosphate is introduced into tissue where a virus is multiplying, either through the roots of intact plants or in tissue floated on a solution of the isotope, the viral RNA becomes labeled. Similarly, ^{35}S-labeled sulfate can be used to label virus protein. These two isotopes are fairly cheap, and are readily available with high specific activities. In certain circumstances, they can be used to detect very small amounts of virus. ^{35}S-labeled methionine is a very convenient material for labeling viral proteins in experiments using protoplasts to study viral replication.

2. In vitro experiments

For viruses such as TMV and TYMV that can be fairly readily freed of most contaminating material, use of radioactive virus provides a sensitive and accurate assay for certain kinds of in vitro experiments. To obtain the best yield of labeled virus, plants are fed the isotope for a period of days during the time of maximum virus increase. For intact plants, the highest practicable amount of ^{32}P- or ^{33}P-labeled orthophosphate or ^{35}S-labeled sulfate is about 3.7×10^8 Bq (10 mCi) per plant. Using these amounts, a purified virus is obtained containing roughly 1000 cpm/µg counting with an efficiency of about 5% for both labels.

Most methods of viral isolation from infected tissue lead to losses of some virus. The extent of such losses can be estimated by adding a very small known amount of radioactive virus to the starting material and estimating the loss of radioactivity as the isolation proceeds (e.g. Fraser and Gerwitz, 1985).

IV. METHODS DEPENDING ON PROPERTIES OF VIRAL PROTEINS

Some of the most important and widely used methods for assay, detection and diagnosis depend on the surface properties of viral proteins. For most plant viruses, this means the protein or proteins that make up the viral coat. Different procedures may use the protein in the intact virus, the protein subunits from disrupted virus or proteins expressed from cloned cDNA or DNA in a system such as *Escherichia coli* or insect cells. More recently non-structural proteins coded for by a virus have been used in diagnosis.

A. Serological procedures

Serological procedures are based on the interaction between a protein or proteins (termed the *antigen*) in the pathogen with antibodies raised against them in a vertebrate. The theories of the immune response and the practical applications of serology have been reviewed by Harlow and Lane (1988), Hampton *et al.* (1990) and van Regenmortel and Dubs (1993).

1. Antibodies

The term 'immunoglobulin' is often used interchangeably with 'antibody'. However, strictly an antibody is a molecule that binds to a known antigen, whereas immunoglobulin refers to this group of proteins irrespective or not of whether their binding target is known. Antibodies are secreted by B lymphocytes. They are a large family of glycoproteins that share key structural and functional features. Structurally they are composed of one or more copies of a characteristic unit that can be visualized as forming a Y shape (Fig. 15.6).

Each Y contains four polypeptides: two identical copies of the heavy chain and two identical copies of the light chain joined by disulfide bonds. Antibodies are divided into five classes, IgG, IgM, IgA, IgE and IgD, based on the number of Y-like units and the type of heavy-chain polypeptide they contain (Table 15.1)

IgG molecules have three protein domains. Two of the domains, forming the arms of the Y, are identical and are termed the Fab domain. They each contain an antigen-binding site at the end, making the IgG molecule bivalent. The third domain, the Fc domain, forms the stem of the Y. The three domains may be separated from one another by cleavage with the protease papain. The Fc region binds protein A, a protein obtained from the cell wall of *Staphylococcus aureus*, with very high affinity. This property is used in several serological procedures.

TABLE 15.1 The five classes of immunoglobulins

Characteristics	IgG	IgM	IgA	IgE	IgD
Heavy chain	γ	μ	α	ε	δ
Light chain	κ or λ	κ or λ	κ or λ	κ or λ	κ or λ
Molecular formula	$\gamma_2 \kappa_2$ or $\gamma_2 \lambda_2$	$(\mu_2 \kappa_2)_5$ or $(\mu_2 \lambda_2)_5$	$(\alpha_2 \kappa_2)_n$ or $(\alpha_2 \lambda_2)_n$[a]	$\varepsilon_2 \kappa_2$ or $\varepsilon_2 \lambda_2$	$\delta_2 \kappa_2$ or $\delta_2 \lambda_2$
Valency	2	10	2, 4 or 6	2	2
Serum concentration (mg/mL)	8–16	0.5–2	1–4	0.01–0.4	0–0.4
Function	Secondary response	Primary response	Protects mucous membranes	Protects against parasites?	?

[a] n = 1, 2 or 3.

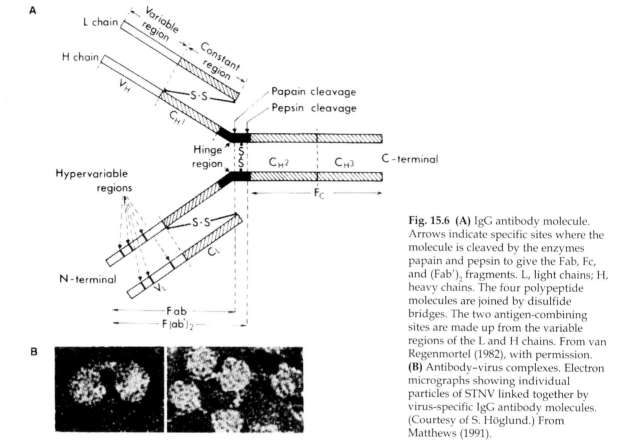

Fig. 15.6 **(A)** IgG antibody molecule. Arrows indicate specific sites where the molecule is cleaved by the enzymes papain and pepsin to give the Fab, Fc, and $(Fab')_2$ fragments. L, light chains; H, heavy chains. The four polypeptide molecules are joined by disulfide bridges. The two antigen-combining sites are made up from the variable regions of the L and H chains. From van Regenmortel (1982), with permission. **(B)** Antibody–virus complexes. Electron micrographs showing individual particles of STNV linked together by virus-specific IgG antibody molecules. (Courtesy of S. Höglund.) From Matthews (1991).

The N-terminal regions of both the light and heavy chains in the arms of the Y-shaped IgG molecule comprise very heterogeneous sequences. This is known as the variable (V) region. The C-terminal region of the light chains and the rest of the heavy chains form the constant (C) region. The V regions of one heavy and one light chain combine to form one antigen-binding site. The heterogeneity of the V regions provides the structural basis for the large repertoire of binding sites used in an effective immune response. For more details of the variation and how it arises see Harlow and Lane (1988).

2. Antigens

Antigens are usually fairly large molecules or particles consisting of, or containing, protein or polysaccharides that are foreign to the vertebrate species into which they are introduced. Most have a molecular weight greater than 10 kDa, although smaller peptides can elicit antibody production. There are two aspects to the activity of an antigen. First, the antigen can stimulate the animal to produce antibody proteins that will react specifically with the antigen. This aspect is known as the immunogenicity of the antigen. Second, the antigen must be able to combine with the specific antibody produced. This is generally referred to as the antigenicity of the molecule. Some small molecules with a specific structure such as amino acids may not be immunogenic by themselves but may be able to combine with antibodies produced in response to a larger antigen containing the small molecule as part of its structure. Such small molecules are known as *haptens*.

Large molecules are usually more effective immunogens than small ones. Thus, plant viruses containing protein macromolecules are often very effective in stimulating specific antibody production. The subunits of a viral protein coat are much less efficient.

It is specific regions on antigens, termed *epi-topes*, that induce and interact with specific antibodies. Epitopes can be composed of amino acids, carbohydrates, lipids, nucleic acids and a wide range of synthetic organic chemicals. About 7–15 amino acids at the surface of a protein may be involved in an antigenic site. However, there are difficulties in defining such sites precisely. These are discussed by van Regenmortel (1989b).

a. Type of epitope

There are several different grouping of epitopes. In one, the *continuous epitope* is a linear stretch of amino acids that is distinguished from the *discontinuous epitope*, which is formed from a group of spatially adjacent surface residues brought together by the folding of the polypeptide chain or from the juxtaposition of residues from two or more separate peptide chains. In another grouping system there are *cryptotopes*, which are hidden epitopes revealed only on dissociation or denaturation of the antigen, *neotopes* formed by the juxtaposition of adjacent polypeptides (e.g. adjacent viral coat protein subunits), *metatopes*, which are epitopes present on both the dissociated and polymerized forms of the antigen, and *neutralization epitopes*, which are specifically recognized by antibody molecules able to neutralize the infectivity of a virus. The conformational specificity of viral epitopes is discussed by van Regenmortel (1992). The epitope type giving rise to a monoclonal antibody (MAb) and the relative proportions of different epitopes recognized by a polyclonal antiserum can affect the outcome of certain serological tests. For instance, an antibody to a cryptotope is unlikely to recognize an antigen in DAS-ELISA but is likely to in a Western blot.

It has been suggested that if the surface area of a protein that is recognized by an antibody molecule is about 2.0×2.5 nm, then all protein antigenic determinants are likely to be discontinuous (Barlow *et al.*, 1986). Antigenic determinants specific for quaternary structure in the virus particle have been demonstrated for all the groups of viruses (van Regenmortel, 1982). The existence of such determinants in the TMV

rod was confirmed by the finding that, of 18 MAbs raised against the virus, eight did not react with viral subunits in ELISA tests (Altschuh *et al.*, 1985).

The molecular dissection of antigens by monoclonal antibodies has been discussed by van Regenmortel (1984b). Three methods have been used to localize antigenic determinants: (1) peptides obtained from the protein by chemical or enzymatic cleavage are screened for their reactivity with antibodies; (2) short synthetic peptides representing known amino acid sequences in the protein can be similarly screened; and (3) immunological cross-reactivity is assayed between closely related proteins having one or a few amino acid substitutions at known sites.

There are substantial limitations for all these procedures. Peptides derived by cleavage or synthesis may not maintain in solution the conformation that the sequence possessed in the intact molecule. Furthermore, such peptides will only rarely represent all the amino acids in the original antigenic determinant. For these reasons reactivity of peptides with antibodies to the intact protein is usually very low.

The limitations of the peptide approach can be further illustrated from work with TMV. Tests with a set of synthetic and cleavage peptides indicated that the dissociated TMV protein possesses seven antigenic determinants (Altschuh *et al.*, 1983). Work with a set of synthetic peptides representing almost the full length of the protein showed that almost the entire sequence possesses antigenic activity (Al-Moudallal *et al.*, 1985).

3. Interaction between antibodies and antigens

The interaction between an epitope and antibody is affected by both affinity and avidity.

a. Affinity

Affinity is a measure of the strength of the binding of an epitope to an antibody. As this binding is a reversible bimolecular interaction, affinity describes the amount of antibody–antigen complex that will be found at equilibrium. High-affinity antibodies perform better in all immunochemical techniques, not only because of their higher capacity but also because of the stability of the complex.

b. Avidity

Avidity is a measure of the overall stability of the complex between antibodies and antigens, and is governed by three factors: the intrinsic affinity of the antibody for the epitope, the valency of the antibody and antigen, and the geometric arrangement of the interacting components.

c. Titer

Titer is a relative measure of the concentration of a specific antibody in an antiserum. It is often used to define the dilution endpoint of the antiserum for detection of an antigen. Thus, as the sensitivities of various serological tests differ, the apparent titer is applicable only to the test under discussion.

d. Serological differentiation index

The serological differentiation index (SDI) is a measure of the serological cross-reactivity of two antigens. It is the number of 2-fold dilution steps separating the homologous and heterologous titers. The homologous titer is that of the antiserum with respect to the antigen used for immunizing the animal, while the heterologous titer is that with respect to another related antigen. Because of the variation in the response of animals to antigens, the SDI is reliable only if it is the average of several measurements. It is measured by a variety of techniques such as ELISA and by single radial immunodiffusion (Krajacic *et al.*, 1992).

4. Types of antisera

There are two basic types of antisera: polyclonal, which contain antibodies to all the available epitopes on the antigen, and monoclonal, which contain antibodies to one epitope. There is much discussion as to which is the best for diagnosis but this will depend on what question the diagnostician is addressing. Monoclonal antisera are much more specific than polyclonal antisera and can be used to differentiate strains of many pathogens. There are some examples of a monoclonal antiserum that is virus group specific (e.g. potyviruses). On the other hand specificity can be a disadvantage and a variant of the pathogen may not be detected.

5. Production of antisera

Antisera have been produced against plant viruses in a variety of animals. Rabbits have been used most often because they respond well to plant virus antigens, are easy to handle and produce useful volumes of serum. Individual animals may vary quite widely in their response to a particular antigen. The amount and specificity of antibody produced in response to a given plant virus antigen may be determined genetically. Unstable plant viruses may have their immunogenicity increased by chemical stabilization (e.g. Francki and Habili, 1972). Protocols for the production of antisera and for the removal of any antibodies reacting with host antigens are given by van Regenmortel (1982), Hampton et al. (1990) and van Regenmortel and Dubs (1993). A recent new approach is to inject a cDNA to the antigen protein sequence cloned in a mammalian expressing vector. Using this approach, an antiserum was made to BYDV (Pal *et al.*, 2000).

Although mice provide relatively small volumes of antiserum, their use may be an advantage when very small amounts of viral antigen are available. Highly inbred strains of mice also minimize variation in response between animals. Mice are, of course, used in the production of monoclonal antisera, as are rats. Chickens are convenient animals to use for the production of polyclonal antisera. When laying hens are immunized, large quantities of purified IgG can be obtained from the egg yolks in a relatively short time (van Regenmortel, 1982; Polson *et al.*, 1985; Bollen *et al.*, 1996; Schade *et al.*, 1997).

Purified IgG from rabbit antisera can be obtained by isolating a specific precipitate of virus and IgG, dissociating virus and IgG at low pH, and removing the virus by high-speed centrifugation. Alternatively, to avoid lengthy acid treatment, antibodies can be bound to virus made insoluble by polymerization with glutaraldehyde. This procedure requires only a short low-speed centrifugation after acid treatment to remove the virus (Maeda and Inouye, 1985).

It was assumed for a long time that plants could not produce antibodies. However, as described in Chapter 16 (Section VII.C.3), antibodies can be produced in transgenic plants.

6. Monoclonal antibodies

B lymphocytes cannot be cultured *in vitro*. To overcome this problem, Köhler and Milstein (1975) took B lymphocytes from an immunized mouse and fused these *in vitro* with an 'immortal' mouse myeloma cell line. Selection of appropriate single fused cells gave 'hybridomas' producing only a single kind of antibody — a monoclonal antibody.

Since monoclonal antibodies against TMV were described (Briand *et al.*, 1982; Dietzgen and Sander, 1982), there has been an explosive growth of interest in this type of antibody for many aspects of plant virus research, and particularly for detection and diagnosis. The term monoclonal antibody has been variously abbreviated as MAb, McAb, MA and McA. I will use the first of these. By 1986, MAbs had been prepared against more than 30 different plant viruses (van Regenmortel, 1986), and many additions to the list have since been reported.

a. The nature of antibody specificity in relation to MAbs

The binding site of a MAb, or any individual antibody in a polyclonal serum, is able to bind to different antigenic determinants with varying degrees of affinity, that is, an individual binding site on an antibody is polyfunctional. Furthermore, antibodies against a single antigenic determinant (or epitope) may vary in affinities from one that is scarcely measurable with standard techniques to one that is 100 000-fold higher. Thus, it is quite possible for an antibody to bind more strongly to an antigenic determinant differing from the one that stimulated its production. Such a phenomenon may well be obscured by the many different antibodies in a polyclonal antiserum, but with MAbs it will show up clearly. For example, when MAbs were produced following immunization of a mouse with a particular strain of TMV it was found that some of the MAbs reacted more strongly with other strains of the virus (Al-Moudallal *et al.*, 1982; van Regenmortel, 1982). This has important implications for the use of MAbs in the delineation of virus strains (Chapter 17, Section II.B.3).

As individual MAbs react with a specific epitope, a panel of MAbs will have ones that react against the different types of epitope. For instance, Dore *et al.* (1988) distinguished MAbs that reacted against dissociated TMV rods (probably cryptotopes) and those that reacted against assembled virions (probably neotopes). Electron microscopy showed that the neotopes reacted with the entire surface of the virion, whereas the cryptotopes reacted only with the extremities of the viral rod and with disc aggregates of the protein.

Another aspect of the specificity of MAbs must be borne in mind. A particular protein antigen may have only one site for binding a particular MAb. If this single site is shared with another protein, a significant cross-reaction could occur even in the absence of general structural similarity between the two proteins. This may be an uncommon phenomenon. Nevertheless, cross-reactions detected with MAbs cannot be taken to indicate significant structural or functional similarity between two proteins without other supporting evidence (Carter and ter Meulen, 1984).

b. Production of MAbs

Technical details for the production of MAbs directed against plant viruses are given by Sander and Dietzgen (1984), Harlow and Lane (1988) and Hampton *et al.* (1990). The use of rats is described by Torrance *et al.* (1986a,b). An outline of the procedure using mice is shown in Fig. 15.7.

The kind of screening test used to test for MAb production is of critical importance (van Regenmortel, 1986). Most workers use some form of ELISA. Multi-layered sandwich procedures, especially using chicken antibody in one of the layers, appeared to be the most sensitive (Al-Moudallal *et al.*, 1984). The results obtained depend very much on the quality of the reagents used and the exact conditions of pH and other factors in the medium.

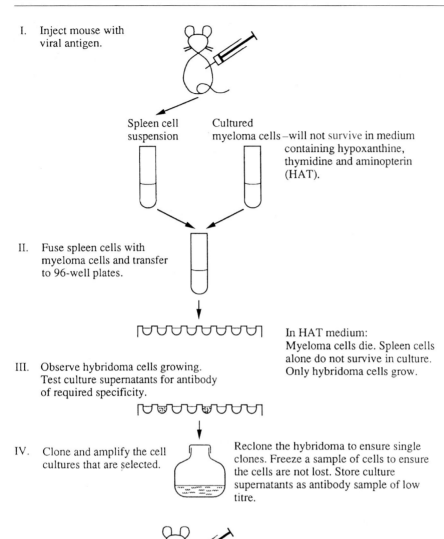

I.   Inject mouse with viral antigen.

Spleen cell suspension

Cultured myeloma cells –will not survive in medium containing hypoxanthine, thymidine and aminopterin (HAT).

II. Fuse spleen cells with myeloma cells and transfer to 96-well plates.

In HAT medium: Myeloma cells die. Spleen cells alone do not survive in culture. Only hybridoma cells grow.

III. Observe hybridoma cells growing. Test culture supernatants for antibody of required specificity.

IV. Clone and amplify the cell cultures that are selected.

Reclone the hybridoma to ensure single clones. Freeze a sample of cells to ensure the cells are not lost. Store culture supernatants as antibody sample of low titre.

V.   Grow cells as ascitic tumour.

The ascitic fluid contains monoclonal antibody of very high titer.

Fig. 15.7 Steps in the production of monoclonal antibodies. From Matthews (1991).

7. Single-chain antibodies and phage displays

The bringing together of two technologies has resulted in the ability to produce a large range of molecules with the properties of single-type antibodies similar to MAbs without involving injection of animals. These technologies are the production of single-chain antibodies and the display of recombinant proteins on the surface of bacteriophage (phage display).

The antibody repertoire of a vertebrate is estimated to consist of more than 10^8 different antibodies, the range being given by variation in the sequence of the variable regions of the heavy (V_H) and light (V_L) immunoglobulin chain (Fig. 15.6). Forced cloning of cDNA produced by RT-PCR of the sequences encoding the VH and VJ chain domains allowed amplification of the antibody repertoire encoding sequences (Orlandi *et al.*, 1989; Sastry *et al.*, 1989). The two variable chains could be expressed in *E. coli* as a single-chain variable fragment (scFv) fusion protein by joining them with a flexible linker protein (Huston *et al.*, 1988). Cloning of randomly combined VH and VL sequences gave a pool of scFv-encoding genes, allowing the generation of large combinatorial antibody libraries with a diversity comparable to the natural immune repertoire (Huse *et al.*, 1989; Marks *et al.*, 1991).

Phage display depends upon the insertion of coding sequences into a phage structural gene which, if it does not disrupt an essential function of the gene product, will be displayed on the surface of the viral particle. If the insert is a set of random sequences, the resulting particles will present a library of peptides, each displayed on a viral scaffold (reviewed in Rodi and Makowski, 1999). The most commonly used phage for display are the filamentous M13 and fd. These phage have minor structural proteins, usually at the ends of the particles as well as the major coat proteins (see Gao *et al.*, 1999). Both these type of protein have been used to display recombinant proteins.

In bringing these two technologies together, the repertoire of scFv molecules is displayed on the surface of the phage (see McCafferty *et al.*, 1990). Those phage displaying antibodies to the antigen of interest, such as a viral coat protein, are selected by binding to the antigen immobilized in, say, a microtiter plate (termed panning). The bound phage are then released by elution and used to infect *E. coli* (Fig. 15.8). This cycle of panning, elution and infection is repeated several times until antibodies with the desired affinity are obtained. The Fv cDNA library does not necessarily have to be from animals that have an immune response to the antigen of interest; if they are from unimmunized animals they are termed naive.

B. Methods for detecting antibody–virus combination

A wide variety of methods has been developed for demonstrating and estimating combination

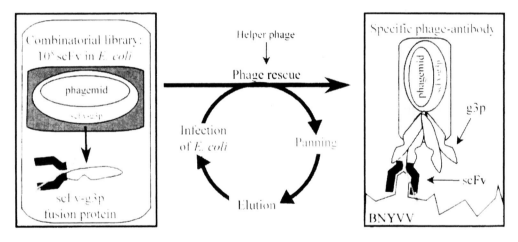

Fig. 15.8 Outline of antibody engineering and selection of antigen-specific antibodies by phage display technology. The minimum requirements of an antibody for antigen recognition are located in the antigen-binding site. The V_H and V_L domains, joined by a linker peptide-encoding DNA fragment, are inserted into a phagemid vector as a single-chain Fv segment. In the scFv-encoding phagemid an ampicillin resistance gene, an origin of replication for *E. coli* and a packaging signal which is required for phage assembly, are present. Depending on the *E. coli* strain used, the scFvs can be expressed as soluble proteins in HB2151 bacteria, wherein translation halts at an amber stop codon (located between scFv and g3p). Fusion proteins of scFv with the minor coat protein of filamentous phage Fd are obtained from TG1 bacteria in which translation proceeds at the amber codon. Selection of BNYVV-specific scFV-antibodies from a combinatorial antibody library can be achieved using phage display. Helper phage, which contain the entire phage genome but lack an efficient packaging signal, are used to 'rescue' phagemids from a combinatorial antibody library. When both helper phage and phagemid are present within the same bacterium, phage antibodies are assembled which carry scFv antibodies on their surface and contain the scFv-encoding phagemid vector. Consequently, within PhAbs, the genotype and phenotype are linked. To select for antigen specificity, PhAbs, rescued from a combinatorial antibody library, are allowed to bind to immobilized BNYVV (panning). Washing removes the PhAbs that lack affinity for BNYVV. Bound PhAbs are eluted and the selected PhAbs are applied to sequential rounds of panning until the desired affinity is obtained. From Griep *et al.* (1999), with kind permission of the copyright holders, © Kluwer Academic Publishers.

between antibodies and antigens. Some of these, such as complement fixation and anaphylaxis, have rarely been used with plant viruses and will not be further discussed. Most traditional methods for using antisera with plant viruses involved direct observation of specific precipitates of virus and antibody, either in liquid media or in agar gels. Over about the past 20 years or so, these methods have been progressively superseded by the use of the enzyme-linked immunosorbent assay (ELISA), immunosorbent electron microscopy (ISEM) and 'dot blots' employing either polyclonal or monoclonal antibodies. These are described in more detail in subsequent sections.

1. Advantages and limitations

Serological tests have provided rapid and convenient methods for the identification and estimation of plant viruses, the main advantages being: (1) the specificity of the reaction allows virus to be measured in the presence of host material or other impurities; (2) results are obtained in a few hours or overnight compared with days, or even weeks, for infectivity assays; (3) the methods give an answer that is directly proportional to viral concentration over a wide range of concentrations; (4) some serological detection and assay procedures are more sensitive than infectivity measurements; (5) serological tests are particularly useful with viruses that have no good local lesion host or that are not sap transmissible; and (6) antisera can be stored and comparable tests made over periods of years and in different laboratories.

It must be borne in mind that serological tests detect and measure the virus protein antigen, not the amount of infective virus. This fact may, of course, be used to advantage in some situations.

The possibility of chance cross-reactions must be considered. Just a few amino acids in a protein sequence are able to elicit an antibody combining with that sequence, provided it is suitably displayed at the surface of the protein. Thus, it is theoretically possible that two quite unrelated protein antigens might elicit cross-reacting antibodies purely because of a chance short amino acid sequence similarity. Such similarity has been established between the coat protein of the U1 strain of TMV and the large subunit of the host protein ribulose bisphosphate carboxylase (Dietzgen and Zaitlin, 1986). There was no cross-reaction with distantly related TMV strains or other viruses.

Aside from forming the basis for a range of assay and diagnostic methods, serological reactions can be used in a variety of other ways, such as investigating virus structure (see Chapter 5), virus relationships (see Chapters 2 and 17) and virus activities in cells (see Chapter 10). Technical details of methods are given by van Regenmortel (1982), and van Regenmortel and Dubs (1993).

2. Precipitation methods
a. Precipitation in liquid media

The formation of a visible specific precipitate between the antigen and antibody is one of the most direct ways of observing the combination between antibody and virus, but relatively high concentrations of reagents are needed.

Plant viruses are polyvalent, that is, each particle can combine with many antibody protein molecules. The actual number that can combine depends on the size of the virus antigen. Polyvalent antigen combines with divalent antibody to form a lattice-structured precipitate. The precipitation reaction of a virus with a particular antiserum is clearly delineated if a series of tests are made with 2-fold dilutions of both reagents in all combinations, under standard conditions of temperature and mixing (Matthews, 1957).

In a modification of the precipitation reaction in tubes, known as a ring test, a small volume of undiluted or slightly diluted serum is placed in a small glass tube and overlaid carefully with the virus antigen solution. With time, antibody diffuses into the virus solution and virus diffuses into the antiserum. Somewhere near the boundary, a zone of specific precipitate will form, provided both reagents are sufficiently concentrated. It is a useful quick test, but somewhat insensitive.

When a drop of freshly expressed leaf sap from plants infected with viruses occurring in high concentration is mixed on a microscope slide with a drop of antiserum, clumping of small particles of host material occurs. This may be seen with the naked eye but is viewed more readily with a hand lens or low-powered microscope. Chloroplasts and chloroplast fragments are prominent in the clumped aggregates. However, some viruses such as WSMV may undergo rapid antigenic modification in leaf extracts unless glutaraldehyde fixation is used (Langenberg, 1989).

b. Precipitation methods using virus bound to larger objects

The sensitivity of the precipitation reaction depends on the smallest amount of antigen that will form a visible precipitate. The smaller the antigen, the greater the weight of antigen required. Various methods have been developed in which the virus is adsorbed to particles substantially larger than viruses, such as latex or red blood cells, before reaction with antiserum. Compared with simple precipitation tests, these procedures allow the detection of 100- to 1000-fold smaller quantities of virus (van Regenmortel, 1982). Latex agglutination allowed the detection of between 0.1 and 0.5 μg/mL of six rod-shaped viruses infecting legumes (Demski et al., 1986). A simple field kit using latex agglutination for three viruses infecting potatoes allowed results to be obtained within 3–10 minutes (Talley et al., 1980). Coating of latex particles with protein A has been used to increase the sensitivity of the procedure (Torrance, 1980).

Reverse passive hemagglutination (RPH) is a procedure in which antibody is linked non-specifically to red blood cells by some appropriate treatment. A solution containing virus is added to a suspension of the antibody-linked red blood cells in round-bottom microtiter plates. In a negative test, the cells settle as a compact button. Agglutination occurs in a positive test, giving a shield of cells over the bottom of the tube. Sander et al. (1989), using TMV as a model system, have shown that RPH is equal in sensitivity and specificity to double-antibody sandwich ELISA. The method involves only one step; readings can be made by eye within 90 minutes without the need of equipment; and

appropriately stabilized red blood cells can be stored for long periods.

The protocol for a chloroplast agglutination test is described by Dijkstra and de Jager (1998).

c. Immunodiffusion reactions in gels

The great advantages of immunodiffusion reactions carried out in gels are: (1) mixtures of antigenic molecules and their corresponding antibodies may be physically separated, either because of differing rates of diffusion in the gel or because of differing rates of migration in an electric field (in immunoelectrophoresis), or by a combination of these factors; and (2) direct comparisons can be made of two antigens by placing them in neighboring wells on the same plate. The method may not be as sensitive as the tube precipitation method in terms of a detectable concentration of virus, but, of course, much smaller volumes of fluid are required. In early work, immunodiffusion tests were carried out in small tubes (i.e. in one dimension). However, this type of test has been superseded by double diffusion tests in two dimensions on glass slides. Ouchterlony (1962) gives a general account of these methods.

Wells are punched in the agar or agarose gel in a defined geometrical arrangement. It is usual in double diffusion tests to have the antiserum in a central well and the antigen solutions being tested in a series of wells surrounding the central well. However, if a single antigen sample is being tested against several antisera, say in virus identification, the sample can be placed in the central well. Antigen and antibody diffuse toward each other in the gel, and after a time a zone will form where the two reagents are in suitable proportions to form a precipitating complex. Both reactants leave the solution at this point and more diffuses in to build up a visible line of precipitation that traps related antigen and antibody. Unrelated antigens or antibodies can pass through the band of precipitation. Bands can be recorded by direct visual observation with appropriate lighting, by the use of protein stains or by photography. When radioactive virus is used, radioautography can be used to detect the bands.

When comparing bands formed by two antigens in neighboring wells, several types of patterns have been distinguished. Movement of antigen in the agar gel is strongly dependent on the size and shape of the virus. For small isometric plant viruses, the methods are satisfactory. Rods may diffuse slowly or not at all. For routine tests, a suitable detergent in the immunodiffusion system may allow rapid migration in the gel of virus degradation products. Elongated virus particles can also be made to diffuse by sonication. This has the advantage that the original antigenic surface of the virus survives in the sonicated fragments (e.g. Moghal and Francki, 1976).

Serological relationships between viruses can be determined by the interactions of bands from adjacent wells. The bands from serologically identical or very closely related viruses fuse, whereas those from more distantly related viruses can form spurs (Fig 15.9).

A modification of gel diffusion uses an agar-like polysaccharide from *Pseudomonas elodea* (Gelrite) instead of agar. Antigens were detected more rapidly in Gelrite, and with about a 100-fold increase in sensitivity. This would make the test almost comparable with ELISA in this respect (Ohki and Inouye, 1987).

d. Immunoelectrophoresis in gels

In immunoelectrophoresis, a mixture of antigens is first separated by migration in an electric field in an agar gel containing an appropriate buffer. Antiserum is then placed in a trough parallel to the path of electrophoretic migration and an immunodiffusion test is carried out. This is a powerful method for resolving mixtures of antigens as two independent criteria are involved: electrophoretic mobility and antigenic specificity.

In immuno-osmophoresis two sets of wells are cut in agar buffered near pH 7. Antiserum is placed in one well of each pair and a virus dilution in the other, and a voltage gradient is applied. Antibody protein moves because of endosmotic flow; the virus moves because of a net negative charge. Thus, the reagents are brought together in the gel much more rapidly than by simple gel diffusion (e.g. John, 1965).

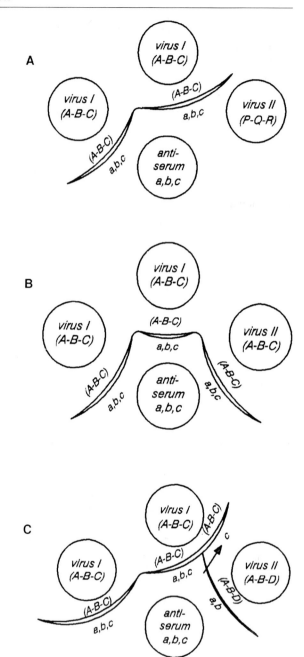

Fig. 15.9 Possible precipitin line patterns in gel double-diffusion tests indicating the degree of serological relationship between two viruses, I and II. Arrangement of reactants is indicated in the wells. Epitopes of viral coat proteins are A, B, C, D, P, Q or R. Their linkage in coat protein molecules and virus particles is indicated by hyphen and parentheses. The antiserum is against virus I (epitopes A-B-C) and contains antibodies a, b and c, which occur separately. From Dijkstra and de Jager (1998), with kind permission of the copyright holders, © Springer-Verlag GmbH and Co. KG.

In rocket immunoelectrophoresis, antigen migrates in an electric field in a layer of agarose containing the appropriate antibody. The migration of the antigen toward the anode gives rise to rocket-shaped patterns of precipitation. The area under the rocket is proportional to antigen concentration. This procedure has been adapted to the assay of 25-ng to 10-µg amounts of CaMV in leaf extracts from about 25 mg of tissue (Hagen *et al.*, 1982).

3. Radioimmune assay

Radioimmune assay, using radioactively labeled viral antigens, provides a sensitive and specific method for measuring plant viral antigens (Ball, 1973). In this method, viral antibody is irreversibly attached to the walls of disposable tubes. The amounts of labeled virus that bind to the surface of such tubes can then be measured. In view of the sensitivity now available with ELISA procedures, this method of assay is unlikely to be widely used.

4. ELISA procedures

Clark and Adams (1977) showed that the microplate method of ELISA could be very effectively applied to the detection and assay of plant viruses. Since that time the method has come to be used more and more widely. Many variations of the basic procedure have been described, with the objective of optimizing the tests for particular purposes.

The method is very economical in the use of reactants and readily adapted to quantitative measurement. It can be applied to viruses of various morphological types in both purified preparations and crude extracts. It is particularly convenient when large numbers of tests are needed. It is very sensitive, detecting concentrations as low as 1–10 ng/mL. Detailed protocols are given in many manuals and journals, including Clark and Bar-Joseph (1984), Hamilton *et al.* (1990), van Regenmortel and Dubs (1993) and Dijkstra and de Jager (1998). Two general procedures have been developed.

a. Direct double-antibody sandwich method

The principle of the direct double-antibody sandwich procedure is summarized in Fig. 15.10. The kinds of data obtained in ELISA tests are illustrated in Fig. 15.11. This is the method described by Clark and Adams (1977). It has been widely used, but suffers two limitations. First, it may be very strain specific. For discrimination between virus strains, this can be a useful feature; however, for routine diagnostic tests it means that different viral serotypes may escape detection. This high specificity is almost certainly due to the fact that the coupling of the enzyme to the antibody interferes with weaker combining reactions with strains that are not closely related. Second, this procedure requires a different antivirus enzyme–antibody complex to be prepared for each virus to be tested.

b. Indirect double-antibody sandwich methods

In the indirect procedure, the enzyme used in the final detection and assay step is conjugated to an anti-globulin antibody. For example, if the

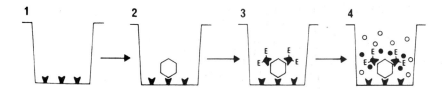

Fig. 15.10 Principle of the ELISA technique for plant viruses (direct double-antibody sandwich method). (1) The γ globulin fraction from an antiserum is allowed to coat the surface of wells in a polystyrene microtiter plate. The plates are then washed. (2) The test sample containing virus is added and combination with the fixed antibody is allowed to occur. (3) After washing again, enzyme-labeled specific antibody is allowed to combine with any virus attached to the fixed antibody. (Alkaline phosphatase is linked to the antibody with glutaraldehyde.) (4) The plate is again washed and enzyme substrate is added. The colorless substrate *p*-nitrophenyl phosphate (○) gives rise to a yellow product (●), which can be observed visually in field applications or measured at 405 nm using an automated spectrophotometer. An example of the kind of data obtained is given in Fig. 15.11. Modified from Clark and Adams (1977), with permission.

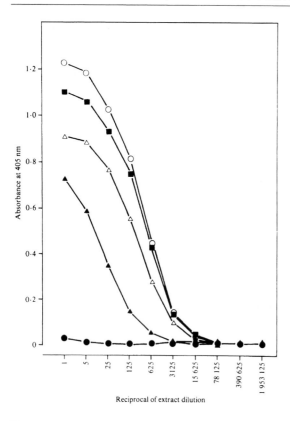

Fig. 15.11 Example of data obtained in an ELISA test. Detection of lettuce LNYV in *Nicotiana glutinosa* plants systemically infected for various periods. ●, Uninfected plants; plants infected for 7 days (no systemic symptoms, ▲), 15 days (prominent systemic symptoms, ■), 19 days (prominent systemic symptoms, ○) and 30 days (severe chlorosis and stunting, △). From Chu and Francki (1982), with permission.

virus antibodies were raised in a rabbit, a chicken anti-rabbit globulin might be used. Thus, one conjugated globulin preparation can be used to assay bound rabbit antibody for a range of viruses. Furthermore, indirect methods detect a broader range of related viruses with a single antiserum (Koenig, 1981).

Many variations of these procedures are possible. Koenig and Paul (1982) studied nine of them with the aim of optimizing the tests for different objectives (Fig. 15.12).

Their results emphasized the versatility of the assay for different applications. They concluded that the direct double-antibody sandwich method is the most convenient for the routine detection of plant viruses in situations where strain specificity and very low virus concentrations cause no problems. The broadest range of serologically related viruses is detected by indirect ELISA using unpre-coated plates, but this procedure is open to interference by crude plant sap. Pre-coating of plates with antibodies or their F(ab')$_2$ fragments (Barbara and Clark, 1982) eliminated the interference problem but narrowed the specificity. Other forms of interference may occur. For example, roots of herbaceous plants may contain a factor that makes the use of protein A–horseradish peroxidase unsatisfactory as an enzyme conjugate (Jones and Mitchell, 1987). Other virus-coated proteins such as the cylindrical inclusion body protein produced by potyviruses can be used for diagnosis with ELISA methods (Yeh and Gonsalves, 1984).

c. Dot ELISA

Several procedures have been developed that use nitrocellulose or nylon membranes as the solid substrate for ELISA tests. In one procedure, the virus in a plant extract is electroblotted on to the membrane as the first step (Rybicki and von Wechmar, 1982). As an alternative first step, the membrane is coated with antiviral IgG by soaking in an appropriate solution (Banttari and Goodwin, 1985). For the final color development, a substrate is added that the enzyme linked to the IgG converts to an insoluble colored material. The intensity of the colored spot can be assessed by eye or by using a reflectance densitometer. This kind of assay was twice as sensitive as the test carried out in microliter plates for PLRV (Smith and Banttari, 1987). An example of a dot blot assay is given in Fig. 15.13.

The dot immunobinding procedure for TMV used by Hibi and Saito (1985) was only about one-tenth as sensitive as a similar ELISA test in microliter plates. Nevertheless, 1.0 ng of virus could be detected, and its simplicity and speed (a few hours) make the test useful as a practical diagnostic technique.

Graddon and Randles (1986) described a single antibody dot immunoassay for the rapid detection of SCMoV. The method was found to be 12 times as sensitive as ELISA in terms of total antigen detectable. The main advantages

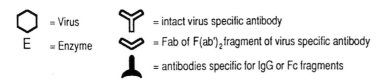

○ = Virus

E = Enzyme

Y = intact virus specific antibody

⋁ = Fab of F(ab')₂ fragment of virus specific antibody

♀ = antibodies specific for IgG or Fc fragments

(i) Black and white symbols indicate antibodies derived from two different animal species.
(ii) Shaded areas indicate a reaction between the Fc portion of a virus-specific antibody and an Fc-specific antibody.
(iii) Reagents added in order from bottom of diagram except where vertical bars indicate two reagents are preincubated together before addition to the plate.

Direct procedures

Indirect procedures

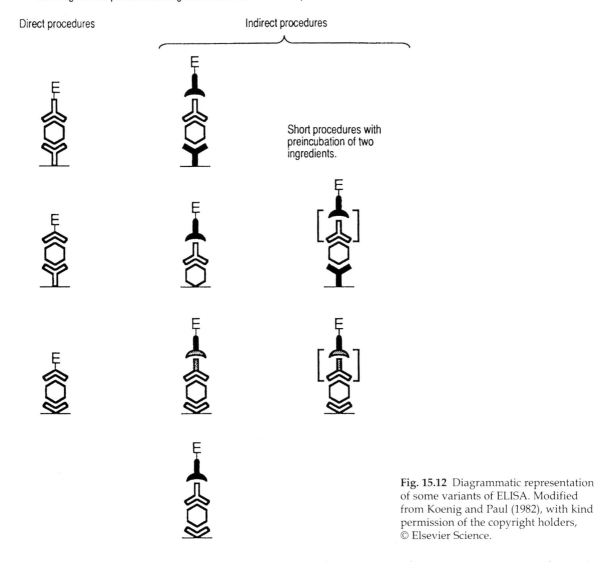

Short procedures with preincubation of two ingredients.

Fig. 15.12 Diagrammatic representation of some variants of ELISA. Modified from Koenig and Paul (1982), with kind permission of the copyright holders, © Elsevier Science.

of the method were speed (3 hours to complete a test), low cost and the small amount of reagents required. Dot blot immunoassays may be particularly useful for routine detection of virus in seeds or seed samples, especially for laboratories where an inexpensive and simple test is needed (Lange and Heide, 1986). However, endogenous insect enzymes may interfere with tests using insect extracts (Berger *et al.*, 1985). Dot blot immunobinding using

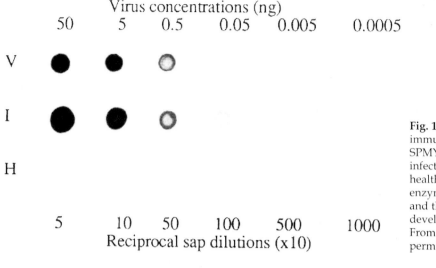

Fig. 15.13 The sensitivity of a dot immunobinding assay. V, purified SPMYEV; I , crude sap from infected leaves; H, crude sap from healthy leaves. The conjugated enzyme was alkaline phosphatase and the substrate for color development was Fast Red TR salt. From Yoshikawa *et al.* (1986), with permission.

plain paper has been developed as a diagnostic method for five potato viruses using direct extraction from green leaves (Heide and Lange, 1988).

The technique of dotting the antigen on to nitrocellulose or other papers can be used with methods other than ELISA for detection of the antigen. For example, Hsu (1984) used gold-labeled goat antirabbit IgG to detect rabbit antibodies bound to TMV on nitrocellulose paper. The gold-labeled antibody is directly visible because of its pink color. The method could detect 1–5 pg of TMV protein and is also simple and economical. The use of electrophoretic separation of proteins followed by immunoblotting (western blotting) is discussed in Section IV.I.2. Koenig and Burgermeister (1986) discuss the application of various immunoblotting techniques and consider the problem of unexplained or false-positive reactions that sometimes occur. These reactions may be due to non-specific binding of immunoglobulins to coat proteins of certain plant viruses (Dietzgen and Francki, 1987).

d. Immuno-tissue printing

As an extension to dot ELISA, immuno-tissue printing involves applying the cut surface of a leaf or stem (or any other plant organ) to a nitrocellulose membrane and revealing the presence of an antigen (say viral coat protein) by immunoprobing. Standard protocols for tissue printing have been published (Holt, 1992) and its use has been described in several papers (see Knapp *et al.*, 1995).

Tissue printing has several advantages. (1) It gives detailed information on the tissue distribution of a virus. (2) Extraction of sap from leaves in which a virus has limited tissue distribution leads to the dilution of the virus with sap from uninfected cells. Since this technique samples the contents of each cell on the cut surface individually, there is increased sensitivity. (3) The technique is easily applicable to field sampling: tissue blots can be made in the field and there is no need to collect leaf samples for sap extraction in the laboratory.

e. Other modifications

The following examples illustrate the versatility of the basic ELISA procedure.

i. Biotin–avidin

This detection system is based on the very high affinity of the protein avidin for biotin. The biotin is chemically coupled to the globulin fraction of the antiserum, while the enzyme to be used for detection is coupled to the avidin. Biotin coupling does not interfere with the binding capacity of the antiviral antibody so that a broad range of viral serotypes can be detected using the direct double-antibody sandwich procedure, and the assays have increased sensitivity (Zrein *et al.*, 1986). Further advantages are speed of development and versatility (Hewish *et al.*, 1986).

ii. Use of a fluorogenic substrate

Torrance and Jones (1982) showed that a fluorogenic substrate of alkaline phosphatase (4-methyl-umbelliferyl phosphate) gave a more sensitive ELISA test than the standard chromogenic substrate, but the technique requires an expensive plate-reading machine. Reichenbächer *et al.* (1984) developed an ultra-micro ELISA test using the same fluorogenic substrate and plates with a 10-µL test volume. Time-resolved fluoroimmunoassay has been developed to increase sensitivity by reducing background (Siitari and Kurppa, 1987). This is done by using a europium chelate that produces an intense fluorescence with a very long decay time, during which background fluorescence has decayed. Using this procedure with MAbs, Sinijärv *et al.* (1988) were able to detect PVX at 5 pg/mL in potato tuber extracts diluted 7×10^4-fold, and in leaf extracts diluted 2×10^7-fold.

iii. Enzyme amplification

Torrance (1987) described a procedure for increasing the sensitivity of ELISA assays in which the enzyme bound to the antibody catalyses the conversion of NADP to NAD, which then takes part in a second enzyme-mediated cyclic reaction to produce a red-colored end product. The method could be used for rapid diagnosis of a luteovirus occurring in very low concentrations in plants and also for detecting the virus in individual aphid vectors.

iv. Use of F(ab')₂ and a protein A–enzyme conjugate

Barbara and Clark (1982) described a simple indirect ELISA test in which the virus was trapped by F(ab')₂ fragments of homologous antibody bound to the plate. Trapped virus was then allowed to react with intact antivirus immunoglobulin. Thus, only one viral antiserum was needed. Bound immunoglobulin was then assayed by a conjugate of protein A and enzyme. Protein A will bind only to the Fc portion of the intact immunoglobulin. The procedure has been used effectively with several viruses in small fruits (Converse and Martin, 1982).

v. Radioimmune ELISA

Ghabrial and Shepherd (1980) developed a simple and highly sensitive radioimmunosorbent assay using the principle of the direct double-antibody sandwich technique in microliter plates, except that ¹²⁵I-labeled IgG is substituted for the enzyme-linked IgG in an ELISA assay. The bound and labeled IgG is dissociated from the sandwich before being assayed.

vi. Measurement of specific activity of viruses in crude extracts

Konate and Fritig (1983) combined two procedures to enable the specific activity of virus radioactively labeled *in vivo* to be determined. Indirect double-antibody sandwich ELISA was used to estimate the amount of virus in a well. Then, as a second step, the radioactivity of the virus trapped in the well was assayed.

vii. ELISA on polystyrene beads

Polystyrene beads 6.5 mm in diameter have been used as the solid phase (one bead per test) instead of microtiter plates (Chen *et al.*, 1982). This system was more discriminating for detecting differences among isolates of SMV than the standard plate procedure.

C. Collection, preparation and storage of samples

As with other diagnostic tests, it is necessary to define and optimize time of sampling and tissue to be sampled to achieve a reliable routine detection procedure (e.g. Torrance and Dolby, 1984; Rowhani *et al.*, 1985; Pérez de San Román *et al.*, 1988). The main factor limiting the number of tests that can be done with ELISA procedures is the preparation of tissue extracts. Various procedures have been tested to minimize this problem (e.g. Mathon *et al.*, 1987). On the other hand, with some viruses at least, it may be possible to avoid the extraction step by placing several small discs of leaf tissue, cut from the leaves to be tested, directly in the well of a microtiter plate (Romaine *et al.*, 1981).

When many hundreds of field samples have to be processed, it is often necessary to store them for a time before ELISA tests are carried out. Storage conditions may be critical for reliable results. For this reason conditions need to be optimized for each virus and host (e.g. Ward *et al.*, 1987).

D. Monoclonal antibodies

1. Tests using MAbs

Many MAbs do not provide a precipitation reaction, especially with protein antigen monomers, for which there may be only one antibody combining site per molecule. However, the use of a MAb and an anti-MAb is a convenient way of extracting or precipitating a peptide or protein antigen with a single determinant. For detection and diagnosis, MAbs have been used most commonly in conjunction with ELISA tests or sometimes with radioimmune assays. Again, exact conditions with respect to pH and other factors may be vitally important. For many applications it may be best to use the same ELISA protocol that was used to screen for MAbs during the isolation procedure because quite different MAbs may be selected depending on the ELISA procedure used (van Regenmortel, 1986). If highly concentrated preparations of MAbs are used in ELISA tests, spurious cross-reactions between viruses in different groups may be detected. Such effects can be avoided by the use of milk proteins instead of bovine albumin as a blocking agent and as a diluent (Zimmermann and van Regenmortel, 1989). The reactivity of a MAb with a given antigen may differ greatly depending on the kind of ELISA procedure used (Dekker *et al.*, 1989).

Diaco *et al.* (1985) used MAbs in a biotin–avidin ELISA for the detection of SMV in soybean seed, while Omura *et al.* (1986) used MAbs with latex flocculation to detect RSV in plants and insects. Sherwood *et al.* (1989) obtained a MAb that reacted in ELISA and dot blot tests with the inner core of TSWV, a virus with a lipoprotein envelope. For diagnostic purposes it may be useful to select for MAbs that react with a common antigenic determinant on several strains, for example PVY (Gugerli and Fries, 1983), PVX (Torrance *et al.*, 1986a) and CTV (Vela *et al.*, 1986). Alternatively, several strain-specific or virus-specific MAbs may be pooled to give the required reagent (e.g. Dore *et al.*, 1987a; Pérez de San Román *et al.*,

1988). Alternatively, MAbs can be used to distinguish between strains or even isolates with single amino acids differences in their coat proteins (e.g. Chen *et al.*, 1997). Epitope profiling using a panel of MAbs to distinguish virus strains is described in Chapter 17 (Section II.B.4). MAbs can also be used to establish relationships between viruses (e.g. Seddas *et al.*, 2000).

2. Advantages of MAbs

i. Requirements for immunization

Mice and rats can be immunized with small amounts of antigen (approximately 100 μg or less). If the virus preparation used is contaminated with host material or other viruses it is still possible to select for MAbs that react only with the virus of interest.

ii. Standardization

MAbs provide a uniform reagent that can be distributed to different laboratories, eliminating much of the confusion that has arisen in the past from the use of variable polyclonal antisera. Furthermore, MAbs can be obtained in almost unlimited quantities in suitable circumstances.

iii. Specificity

MAbs combine with only one antigenic site on the antigen. Thus, they may have very high specificity and can provide a refined tool for distinguishing between virus strains (see Chapter 17). They can also be used to investigate aspects of virus architecture (see Chapter 5) and transmission by vectors (see Chapter 11).

iv. High affinity

The screening procedure for detecting MAbs allows for the possibility of obtaining antibodies with very high affinity for the antigen. High-affinity antibodies may be used at very high dilutions, minimizing problems of background in the assays. They can also be used for virus purification by affinity chromatography.

v. Storage of cells

Hybridomas can be stored in liquid nitrogen to provide a source of MAb-producing cells over a long period.

3. Disadvantages of MAbs

a. Preparation

Polyclonal antisera are relatively easy to prepare. The isolation of new MAbs is labor intensive, time consuming and relatively expensive. For any particular project, these realities must be weighed against the substantial advantages discussed above.

b. Specificity

MAbs may be too specific for some applications, especially in diagnosis. This is an important consideration.

c. Sensitivity to conformational changes

Because of their high specificity, MAbs may be very sensitive to conformational changes in the antigen brought about by binding to the solid phase or by other conditions in the assay.

4. Summary

Although MAbs provide an excellent tool for many aspects of plant virus research, it is unlikely that they will replace the use of polyclonal antisera in all applications. As far as detection and diagnosis are concerned, appropriate MAbs are available commercially.

E. Phage-displayed single-chain antibodies

The theory and use of phage display for the production of antibodies comprising scFv regions of immunoglobulins is described in Section IV.A.7. This technology has been used to provide epitope-specific antibodies for a range of viruses, e.g. BNYVV (Griep et al., 1999; Uhde et al., 2000), PLRV (Harper et al., 1999c; Toth et al., 1999) and TSWV (Griep et al., 2000). A phage library displaying random nonapeptides bound to CMV coat protein (Gough et al., 1999). This shows that proteins other than antibodies that bind to viruses can be discovered.

This technique has several advantages over the production of MAbs by hybridomas including: (1) it is more rapid, taking weeks rather than months to produce the panel of antibodies; (2) it does not require the use of animals; (3)

it is much less expensive; and (4) it overcomes the problem of immunological epitopes being lost by proteolysis during the production of antibodies. For instance, the important C-terminal epitope of the coat protein of BNYVV is readily lost by proteolysis during the preparation of MAbs in hybridomas. Phage display obviates this problem (Uhde et al., 2000).

F. Serologically specific electron microscopy

A combination of electron microscopy and serology was first used by Larson et al. (1950) and also illustrated by Matthews (1957), but the technique of negative staining was needed before the procedure could be developed as an effective and widely applicable diagnostic tool. Derrick (1973) described a procedure in which the support film on an electron microscope grid was first coated with specific antibody for the virus being studied. Grids were then floated on appropriate dilutions of the virus solution for 1 hour. They were then washed, dried, shadowed and examined in the electron microscope. Under appropriate conditions, many more viral particles are trapped on a grid coated with antiserum than on one coated with normal serum. Thus, the method offers a diagnostic procedure based on two properties of the virus: serological reactivity with the antiserum used and particle morphology. The method was not popular for some years but is now widely used as a diagnostic tool in appropriate circumstances. Various terms have been used to describe the process: serologically specific electron microscopy (SSEM), immunosorbent electron microscopy (ISEM), solid-phase immune electron microscopy and electron microscope serology. The subject has been reviewed by Milne (1991, 1993).

1. Decoration technique

Milne and Luisoni (1977) introduced to plant virology a modification of the general procedure in which, after being adsorbed on to the EM grid, viral particles are coated with virus-specific antibody. This produces a halo of IgG molecules around the virus particles that can be readily visualized in negatively stained preparations. The phenomenon was termed 'decoration'. This procedure probably offers the most convincing demonstration by electron microscopy of specific combination between virus and IgG. For a critical test, a second serologically unrelated virus of similar morphology should be added to the preparation (Fig. 15.14).

Technical details of protocols are given and difficulties with the procedure are discussed by Milne (1984), Milne and Lesemann (1984) and Katz and Kohn (1984). Particles of some viruses may disintegrate following decoration (Langenberg, 1986b). Nevertheless, the unstable rod-shaped virus-like particles associated with lettuce big vein disease have been recognized using the technique and shown to be related to TStV (Vetten et al., 1987).

The main advantages of the method are: (1) the result is usually clear, in the form of virus particles of a particular morphology, and thus false-positive results are rare; (2) sensitivity may be of the same order as with ELISA procedures and may be 1000 times more sensitive for the detection of some viruses than conventional EM (Roberts and Harrison, 1979); (3) when the support film is coated with antibody, much less host background material is bound to the grid; (4) antisera can be used without fractionation or conjugation, low-titer sera can be satisfactory, and only small volumes are required; (5) very small volumes (about 1 µL) of virus extract may be sufficient; (6) antibodies against host components are not a problem inasmuch as they do not bind to virus; (7) one antiserum may detect a range of serological variants (on the other hand, the use of monoclonal antibodies may greatly increase the specificity of the test) (e.g. Diaco et al., 1986); (8) results may be obtained within 30 minutes; (9) when decoration is used, unrelated undecorated virus particles on the grid are readily detected; (10) different proteins on a particle can be decorated

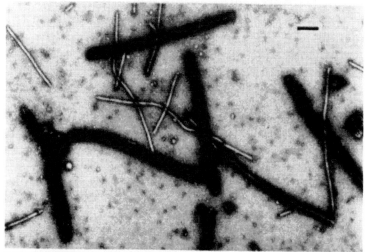
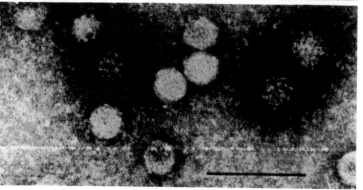

Fig. 15.14 Immune specific electron microscopy (ISEM). **Bottom**: Mixture of purified TNV and TBSV adsorbed to the grid and then treated with saturating levels of antibodies to TBSV. TBSV particles are 'decorated'. TNV particles remain clean, with sharp outlines. Antibody molecules appear in the background. Bar marker 100 nm. **Top**: A natural mixture of two potyviruses from a perennial cucurbit, Bryonia cretica. The antiserum used has decorated particles of only one of the viruses in the mixture. One particle near the center is longer than normal and decorated for only part of its length. This particle probably arose by end-to-end aggregation between a particle of each of the two viruses. Bar marker 100 nm. From Milne (1990), with kind permission of the copyright holders, © Springer-Verlag GmbH and Co. KG.

(e.g. the two coat proteins on closteroviruses and criniviruses; see Fig. 5.12); and (11) prepared grids may be sent to a distant laboratory for application of virus extracts and returned to a base laboratory for further steps and EM examination.

Some disadvantages of the procedure are: (1) it will not detect virus structures too small to be resolved in the EM (e.g. coat protein monomers); (2) sometimes the method works inconsistently or not at all for reasons that are not well understood (Milne and Lesemann, 1984); (3) it involves the use of expensive EM equipment, which requires skilled technical work and is labor intensive—for these reasons it cannot compete with, say, ELISA for large-scale routine testing; and (4) when quantitative results are required, particle counting is laborious, variability of particle numbers per grid square may be high and control grids are required.

In summary, SSEM and ISEM cannot replace ELISA tests when large numbers of samples have to be tested. Their main uses are (1) in the identification of an unknown virus, (2) in situations where only a few diagnostic tests are needed; and (3) when ELISA tests are equivocal, or not sufficiently sensitive and need direct confirmation (e.g. detection of BSV; Lockhart, 1986; Thottappilly *et al.*, 1998).

2. Immuno-gold labeling

Gold particles are highly electron dense and thus show the position on an EM grid of any particle to which they are attached. Protein A forms a reasonably stable complex with gold particles. This complex can then be used to locate any IgG molecules bound to virus coat protein on the EM grid.

Van Lent and Verduin (1985, 1986) prepared gold particles of two average diameters (7 and 16 nm). The size of these particles in relation to virus particles and the specificity of the reaction are illustrated in Fig. 15.15A.

Immuno-gold labeling is particularly valuable in locating viral antigens in thin sections of infected cells (see Fig. 8.24). Van Lent and Verduin found that the larger gold particles were more readily seen in stained thin sections. The technique has been adopted for localiza-

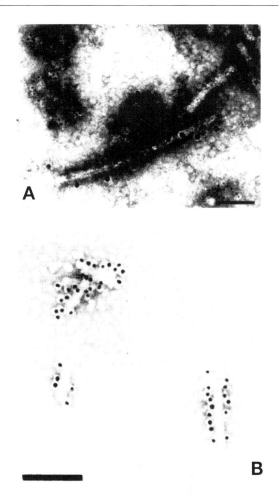

Fig. 15.15 Gold particle labeling of virus particles. **(A)** Protein A labeled with gold particles combining with virus-specific IgG near the virus surface. Virus particles were labeled in suspension with protein A–gold complexes. A mixture of purified CCMV labeled with 7-nm protein A–gold and TMV labeled with 16-nm protein A–gold. Bar marker 100 nm. From van Lent and Verduin (1986). **(B)** RTBV particle treated first with anti-RTBV protease rabbit antiserum and then with gold-labeled goat anti-rabbit serum. Bar marker 200 nm. From Hay *et al.* (1994), with permission.

tion of viral antigens by light microscopy (van Lent and Verduin, 1987) and has been reviewed by Patterson and Verduin (1987).

G. Fluorescent antibody

The fluorescent antibody method has been applied to the study of the intracellular location and distribution of plant viruses within tissues of the host plant and in insect vectors. Nagaraj (1965) gives the detailed procedures that must be carried out in the preparation and storage of TMV antibody conjugated with fluorescein to avoid non-specific staining. Appropriate controls are essential. Fluorescent antibody staining is particularly suited to detecting viral antigens in isolated infected protoplasts (Otsuki and Takebe, 1969). Chiu and Black (1969) and others have used standard fluorescent antibody methods very successfully for detecting virus antigens in insect vectors.

H. Neutralization of infectivity

When viruses are mixed with a specific antiserum, infectivity is reduced to a greater extent than when mixed with a non-immune serum. At high dilutions of serum, inactivation may occur only with the specific antiserum (Rappaport and Siegel, 1955). This phenomenon has been little used in plant virology mainly because of the lack of precision in assays for infectivity. Neutralization tests have been used to demonstrate serological relationships between viruses that are not transmissible mechanically (e.g. Rochow and Duffus, 1978). With a sufficiently stable virus such as TMV, infectivity of the virus can be restored by removal of the antibody at low pH. The mechanism of neutralization has not been established. The binding of IgG to the virus protein may prevent effective release of the nucleic acid from the protein coat. Alternatively, it may block virus from attaching to some site within the cell at an early stage of infection. Dietzgen (1986) selected a monoclonal antibody specific for a C-terminal antigenic determinant of the TMV coat protein using a chemically synthesized tetrapeptide. This MAb neutralized the infectivity of TMV. A systematic study using these procedures may illuminate the mechanism of neutralization.

I. Electrophoretic procedures

Electrophoresis in a suitable substrate separates proteins according to size and net electric charge at the pH used. Polyacrylamide gel electrophoresis in a medium containing sodium dodecyl sulfate (SDS-PAGE) is commonly used.

The position and amount of the proteins can then be visualized by a non-specific procedure such as staining or by a specific procedure such as immunoassay (termed western blotting).

1. Gel electrophoresis followed by staining

Provided a virus occurs in sufficient concentration, and provided it can be freed sufficiently from any interfering host proteins by some simple preliminary procedure, gel electrophoresis is a rapid method for detecting viral coat proteins. Provided they occur in high enough concentration, other virus-coded proteins may be detected.

2. Electrophoresis followed by electroblot immunoassay

O'Donnell et al. (1982) and Rybicki and von Wechmar (1982) were the first to use the protein-fractionating power of electrophoresis together with the sensitivity and specificity of solid-phase immunoassay to identify and assay viral proteins. The main stages in the technique are: (1) fractionation of the proteins in infected plant sap by SDS-PAGE; (2) electrophoretic transfer of the protein bands from the gel to a membrane (usually nitro-cellulose); (3) blocking of remaining free binding sites on the membrane with protein, usually serum albumin or fat-free dried milk powder; (4) probing for the viral coat protein band with specific antiserum; and (5) detection of the antibody–antigen complex with ^{125}I-labeled protein A or by some ELISA procedure. A great advantage of this technique is that it identifies the virus by two independent properties of its coat protein: molecular weight and serological specificity. Shukla et al. (1983) found that amino-phenylthioether (APF) paper was more efficient than nitrocellulose in binding coat proteins. As little as 0.5 ng of viral protein could be detected, compared with 500 ng for Coomassie blue staining. Furthermore, APT paper binds the proteins very stably so that a sheet can be cleaned of antibody and re-probed sequentially with a number of antisera. Electroblot immunoassay can be used following electrophoresis of intact virus particles in agarose gels and transfer of the virus by diffusion blotting to the paper (Koenig and Burgermeister, 1986).

As noted in Section III.A.2, virus particles can be separated by gel electrophoresis. In a technique termed native electrophoresis and western blot analysis (NEWeB), particles of PPV were electrophoresed directly from plant extracts in agarose or mixed acrylamide-agarose gels and blotted on to nitrocellulose membranes (Manoussopoulos *et al.*, 2000). The position of the particles was then identified by immunoprobing. Two different strains of PPV could be distinguished in double infections.

V. METHODS INVOLVING PROPERTIES OF THE VIRAL NUCLEIC ACID

General properties of a viral nucleic acid, such as whether it is DNA or RNA, double stranded (ds) or single stranded (ss), or consists of one or more pieces, are fundamental for allocating an unknown virus to a particular family or group. However, with the exception of dsRNA, these properties are usually of little use for routine diagnosis, detection or assay. The ability to make DNA copies (cDNA) of parts or all of a plant viral RNA genome has opened up many new possibilities. The nucleotide sequence of the DNA copy can be determined, but this is far too time consuming to be considered as a diagnostic procedure, except in special circumstances.

Basically, there are four approaches to the use of nucleic acids for detection and diagnosis of viruses:

1. the type and molecular sizes of the virion-associated nucleic acids
2. the cleavage pattern of viral DNA or cDNA
3. hybridization between nucleic acids
4. the polymerase chain reaction.

In this section, I discuss the principles of these approaches, and then their applications.

A. Type and size of nucleic acid

As noted above, the properties of dsRNAs associated with RNA viral infections have been used for diagnosis (reviewed by Dodds, 1993).

Double-stranded RNAs are associated with plant RNA viruses in two ways: (1) the plant reoviruses (Chapter 6, Section VI) and cryptoviruses have genomes consisting of dsRNA pieces; and (2) in tissues infected with ssRNA viruses, a double-stranded form of the genome RNA accumulates that is twice the size of the genomic RNA. This is known as the replicative form (see Fig. 8.6). These dsRNAs have been used for diagnosis either following characterization of the double-stranded form by PAGE or by the use of antibodies reacting with dsRNA.

a. Electrophoresis of isolated dsRNAs

Dodds *et al.* (1984) and Dodds (1993) summarize procedures for the isolation of dsRNAs from infected tissue, and for their separation and characterization by PAGE or an agarose gel, using appropriate molecular weight markers. Bands of dsRNA are normally revealed by staining with ethidium bromide or by hybridization (Bar-Joseph *et al.*, 1983). The dsRNA bands can also be eluted from the gel and used to make radioactive probes for use in dot blot analyzes.

In principle, each RNA virus should give rise to a distinctive band of dsRNA. This has been found for some virus groups, for example a series of rod-shaped viruses with monopartite genomes (Valverde *et al.*, 1986). However, a series of dsRNA molecules smaller than the full-length dsRNA are almost always present. The pattern of these smaller virus-specific RNAs is characteristic for the virus. Sometimes strains of the same virus can be distinguished by the pattern of smaller bands.

The use of dsRNAs in diagnosis is complicated by the fact that some uninoculated and apparently healthy plants contain a series of dsRNA species. For example, Nameth and Dodds (1985) found that 40 of 50 glasshouse-grown uninoculated cucurbit cultivars contained readily detectable dsRNA species in the molecular weight range $0.5–11.0 \times 10^6$ Da. Each cultivar had a characteristic pattern, indicating seed transmission. The nature of these dsRNAs has not been established, but in view of the high molecular weight of the largest species, they may be associated with undescribed cryptic viruses. On the other hand, Wakarchuk and Hamilton (1985) described high-molecular-weight dsRNAs from one variety of *Phaseolus vulgaris*. These dsRNAs had sequence similarity to the genome DNA of the bean variety in which it was found, as well as that of other bean varieties that contained no detectable dsRNA. Other large dsRNA species from several plants are noted in Chapter 17. Another source of dsRNA could be fungi adventitiously associated with the plant being tested. Many of the viruses of fungi have dsRNA genomes and there are dsRNA molecules associated with plant pathogenic fungi (see Nuss and Koltin, 1990).

In spite of these difficulties, the dsRNA method has a role to play in virus diagnosis for some host species. Thus TRSV, RBDV and Raspberry leaf spot virus could be readily detected in infected *Rubus* spp. Field samples of diseased plants often contained two or more viruses that could be readily identified from the dsRNA patterns (Martin, 1986). The procedure provided an alternative to diagnosis by grafting or aphid transmission to indicator hosts. Similarly, the method was valuable for diagnosis of CTV and strains of the virus, in various citrus species, provided that the tissue sampled was in optimal condition (Dodds *et al.*, 1987). A low-molecular-weight dsRNA associated with groundnut rosette disease has been used as a diagnostic tool (Breyel *et al.*, 1988).

b. Antibodies against dsRNA

Antibodies reacting non-specifically against dsRNAs were found in antisera prepared against plant reoviruses (Ikegami and Francki, 1973; van der Lubbe *et al.*, 1979). Such non-specific antisera can be prepared using synthetic poly(I):poly(C) antigen (Stollar, 1975), but the procedure has not received wide application for plant viruses. MAbs have been produced that recognize dsRNA and have been used in immuno-blots to detect this form of RNA in extracts from plants infected with 18 viruses (Lukács, 1994). Immuno-blot analyzes in combination with temperature gradient electrophoresis could distinguish between the dsRNA genomes of BCV-1 and BCV-2 from one another. As MAbs to dsRNAs do not react with double-stranded sequences of less than 11 bp, they can be used as probes of RNA structure in crude nucleic acid extracts (Schonborn *et al.*, 1991).

c. Polymorphisms in heteroduplex RNAs

The electrophoretic migration of single- and double-stranded nucleic acids in gels differs and can be used as a basis for diagnostic techniques. The technique of single-stranded conformation polymorphism (SSCP) was originally based on the migration rates of ssDNA molecules conferred by their particular primary and tertiary structures (Orita *et al.*, 1989a). As described in Chapter 17 (Section II.A.1.b), SSCP was applied to the ssRNA of BNYVV in an attempt to differentiate strains (Koenig *et al.*, 1995) but proved to be variable in detecting mutations. By adapting the method using heterologous duplexes of RNA transcripts polymorphisms were detected between strains of PNRSV (Rosner *et al.*, 1998; 1999).

B. Cleavage patterns of DNA

Cleaving cDNAs of RNA genomes and the genomes of DNA viruses at specific sites with restriction enzymes and determining the sizes of the fragments by PAGE is a possible procedure for distinguishing viruses in a particular group. For instance, Hull (1980) showed that isolates of CaMV could be distinguished on the restriction endonuclease patterns (subsequently termed restriction length polymorphism mapping; RFLP) of the virion DNA. However, there is considerable variation in viral populations (see Chapter 17, Section I.A) as shown by RFLPs between individual clones of RTBV DNA from a single virus preparation (Villegas *et al.*, 1997). For the *Geminivirus* group, Haber *et al.* (1987) used labeled DNA from one virus of the group to identify a range of group members by probing restriction enzyme fragments that gave a pattern of bands after PAGE characteristic for each member. An example of the application of RFLP analysis to cDNA of an RNA virus is given by Kruse *et al.* (1994), who used the technique to distinguish two strains of BNYVV.

This technique may be of use in specific situations but it has not yet achieved any wide acceptance in plant virology.

C. Hybridization procedures

These procedures depend on the fact that single-stranded nucleic acid molecules of opposite polarity and with sufficient similarity in their nucleotide sequence will hybridize to form a double-stranded molecule. The motivation to develop hybridization methods came first with the viroids, which have no associated proteins that can be used for diagnosis (Palukatis *et al.*, 1981; Allen and Dale, 1981).

1. Basis for hybridization procedures

The theory concerning nucleic acid hybridization is complex. It has been discussed by Britten *et al.* (1974) and Hull (1993, 2000). The Watson and Crick model for the structure of dsDNA showed that the two strands were held together by hydrogen bonds between specific (complementary) bases, namely adenine and thymidine, cytosine and guanine. This interaction of base pairing is the basis of all molecular hybridization. The early studies revealed several important features of this process. The ability to use various physical and chemical procedures to disrupt base pairing and hence separate the strands (termed *melting* or *denaturing* the nucleic acid) and then to reinstate base pairing and thus renature the double-stranded nucleic acid (termed *hybridization*) enabled the various factors controlling the stability of the duplex to be examined. Most of these basic studies were performed using both the target and probe DNAs in solution (liquid–liquid hybridization); there are also some data available for RNA:RNA and DNA:RNA interactions in solution. Although most of plant pathogen diagnosis now involves mixed-phase hybridization with the target immobilized on a solid matrix, the theory developed for liquid–liquid hybridization is still very relevant.

Below is a general account of some of the major factors that have to be considered in hybridization experiments. More details on the theory of hybridization can be found in Ausubel *et al.* (1998), Britten *et al.* (1974), Britten and Davidson (1985) and Young and Anderson (1985).

a. Denaturation

There are various factors that affect denaturation.

i. Temperature

Double-stranded nucleic acid will denature at high temperatures. Strand separation takes place over a relatively narrow range of temperatures, the temperature at which 50% of the sequences are denatured being called the melting temperature (T_m). The major factors affecting the T_m are the composition of the nucleic acid, the concentration of salt in, and the pH of, the solution, and the presence of materials that can disrupt hydrogen bonding such as formamide.

ii. Nucleic acid composition

Guanosine + cytosine (G+C) base pairs, which have three hydrogen bonds, are more stable than adenosine + thymidine (uridine) (A+T) base pairs, which have two. For perfectly base-paired DNA in 1× SSC (SSC = 0.15 M sodium chloride, 0.015 M sodium citrate), T_m is related to the G+C content by:

$$T_m = 0.41 \, (\%G+C) + 69.3$$

iii. Salt

The salt concentration has a marked effect on the T_m of a duplex. The T_m increases by almost 16°C for each 10-fold increase in the concentration of monovalent cation over the range of 0.01–0.1 M; the effect is less at higher concentrations. Over the lower concentration range the T_m is given by:

$$T_m = (\%GC/2.44) + 81.5 + 16.6 \, \mathrm{logM}$$

where M is the molarity of the monovalent cation.

Divalent cations have an even greater effect and should be removed by the use of chelators such as EDTA.

iv. pH

The T_m is insensitive to pH in the range 5–9. Below pH 5, depuration will start and will become more rapid as the pH is lowered. This can be of use as a method for introducing nicks into DNA as apurinic acid is alkali labile. Above pH 9 denaturation of double-stranded nucleic acid sets in, first in A+T-rich regions. Most duplex DNAs are fully denatured at pH 12. The phosphodiester bonds of RNA are degraded above pH 8, the higher the pH and temperature, the more rapid the degradation.

v. Organic solvents

Some organic solvents such as formamide, dimethyl formamide and dimethylsulfoxide, and also urea, lower the T_m. The T_m is reduced by 0.7°C for each percent of formamide.

vi. Base-pair mismatch

Mismatched sequences are less stable than perfectly base-paired duplexes. In nucleic acids of more than 100 bp, a mismatch of 1% reduces the T_m by about 1°C. Thus, mismatching can be assessed by varying the hybridization conditions (see stringency below).

vii. RNA:RNA and RNA:DNA duplexes

The T_m of dsRNA is significantly higher than that of dsDNA. A general value for the T_m of DNA in 1× SSC is about 85°C and of RNA is close to 100°C.

The T_m of RNA:DNA hybrids is about 4–5°C higher than that for DNA:DNA duplexes under the same conditions.

b. Re-association

The two main considerations relevant to re-association are the rate at which it occurs and the stability of the products. The stability of the products is affected by the same factors that control denaturation. The main factors affecting rate of re-association are described below.

i. Temperature

For a typical DNA:DNA re-association reaction, the graph relating the rate of formation of duplexes to temperature is a broad curve with the maximum rate at about 25°C below the T_m. This is taken to be the optimum temperature for re-association. At temperatures well below the optimum, re-association may effectively cease. Thus, it is possible to maintain the two strands of a duplex separated for considerable periods of time by melting at high temperature and then lowering the temperature rapidly (by plunging the tube into ice) to well below the T_m (termed quenching).

If the T_m is lowered, for example by the use of formamide, the optimum rate of re-association is at the new T_m-25°C. However, because the temperature is lower, the overall rate of hybridization will be slower than in the absence of formamide.

ii. Salt concentration

The concentration of monovalent cations affects the rate of re-association markedly. Below 0.1 M sodium chloride, a 2-fold increase in salt concentration increases the rate by 5–10-fold or more. The rate continues to rise with increasing salt concentration, but becomes constant above 1.2 M sodium chloride. Divalent cations have very pronounced effects on the rate and should be removed by the use of chelators such as EDTA.

iii. Base mismatch

The precision with which base-pairing occurs also affects the rate of re-association. For each 10% mismatch the rate is halved when the reaction is under optimal conditions (e.g. T_m-25°C).

iv. Fragment length

If the cDNA strands are the same length, the rate of re-association rises with the square root of the length. When the strands are of different length, the rate depends upon which fragment is in excess and interpretation can be very complicated.

v. Concentration of nucleic acid

The time required for the formation of duplexes is directly proportional to the initial concentration of the interacting single-stranded molecules. Reactions are normally measured as the product of the initial concentration (C^0) of the nucleic acid in moles of nucleotide per liter and time (t) of reaction in seconds (C^0t expressed as moles per second per liter). C^0t values are of great use in liquid–liquid hybridization in determining features such as amounts of unique sequence (complexity) and repeated sequences in nucleic acids.

vi. Polymers

The anionic polymer, dextran sulfate, accelerates the rate of re-association both in solution and on filters. A 10% solution of dextran sulfate (MW 500 000 Da) increases the rate of hybridization in solution by about 10-fold and of hybridization to immobilized nucleic acids by up to 100-fold. This increase is thought to be due to the exclusion of the nucleic acid from the volume occupied by the polymer, thus effectively increasing its concentration.

vii. RNA:DNA hybridization

The factors affecting hybridization of RNA to cDNA are somewhat different from those affecting DNA:DNA interactions, probably because of the greater amount of strong secondary structure in the RNA. If RNA is in excess under moderate salt conditions (0.18 M sodium chloride), the rate is almost the same as with DNA:DNA hybridization. However, the rate does not rise as rapidly as does DNA:DNA hybridization with increasing salt concentration.

If DNA is in excess, the rate is four to five times slower than that expected for DNA:DNA re-association.

c. Mixed-phase hybridization

The kinetics of mixed-phase hybridization have been less well studied than those of liquid–liquid hybridization. It is generally assumed that the effects of the reaction conditions are qualitatively, and in most cases quantitatively, the same for filter hybridization as they are for liquid–liquid hybridization (for details see Anderson and Young, 1985). However, with the common practice in filter hybridization of using double-stranded probes, the strandedness of the probe can have effects that may be important if kinetics are being studied.

i. Probe strandedness

When double-stranded probes are used in filter hybridization, two sets of hybridization take place, that between the probe and the immobilized target sequence and the self-annealing of the probe. The latter can have two effects:

removal of the effective probe, which reduces sensitivity, and the formation of concatenates, which may then hybridize to the target nucleic acid and thus increase sensitivity but decrease specificity. To minimize these effects, the sequences in the probe complementary to the target nucleic acid should be relatively short, the probe should be at low concentration in a small reaction volume, and the reaction should be at as high a temperature as possible.

Single-stranded probes overcome many of these problems. However if ^{32}P-labeled single-stranded probes are used at concentrations above 100 ng/mL, non-specific binding to the membrane may occur.

ii. Length of probe

For short oligonucleotide probes (less than 30 base pairs) the T_m can be estimated from:

$$T_m = 2(A+T) + 4(G+C)$$

where A, T, G and C are the numbers of adenosine, thymidine, guanosine and cytosine bases respectively.

For filter hybridization the dissociation temperature (T_d) is calculated:

$$T_d = T_m - k$$

where the constant k has been determined experimentally to be 7.6°C.

The T_m of DNA longer than 50 bp can be calculated from:

$$T_m = 81.5 + 16.6 \times \log[\text{Na}] + 0.41 \times (\%G+C) - 675/\text{length} - 0.65 \times (\%\text{formamide})$$

iii. Stringency (see Anderson and Young, 1985; Meinkoth and Wahl, 1984)

The term 'stringency' is often used imprecisely. It relates to the effect of hybridization and/or wash conditions on the interaction between complementary nucleic acids, which may be incompletely matched. The use of different stringencies is one of the more powerful tools of the hybridization technology.

Filter hybridization can be used to determine the degree of relatedness between sequences. To achieve this, one has to be able to estimate approximately the proportion of mismatches in the hybrid. This is done by adjusting the reaction conditions so that the desired interaction can be examined. If close relationships are to be distinguished from distant ones, more stringent conditions are used; if distant relationships are to be detected, the conditions should be less stringent. The stringency can be varied at two stages in the procedure: at hybridization and at the post-hybridization wash. As a general rule, for distantly related sequences the hybridization conditions should be relaxed and the washes carried out under increasing stringency; for closely related sequences hybridization and washing should be under stringent conditions. Stringency can be varied by changing the temperature and/or salt concentration and/or formamide concentration.

Also to be considered in experiments to determine the relationship between nucleic acids are the relative concentrations of probe and target nucleic acid and the size of the probe. The probe nucleic acid should never be in excess, to ensure that it does not saturate the target nucleic acid. For probes larger than 100 bases, the T_m of a DNA duplex is decreased by approximately 1°C for every 1% mismatch; for hybrids shorter than 20 bp the T_m decreases by about 5°C for each mismatched base pair.

As a general guideline for nucleic acids of more than 200 bp and 40–50% G+C content, the following are conditions for various stringencies:

Low stringency	50°C	5× SSC	allowing approx. 25% mismatch
Stringent	65°C	2× SSC	allowing approx. 10% mismatch
High stringency	65°C	0.1× SSC	allowing <1% mismatch

d. *Types of hybridization format*

Liquid–liquid hybridization has been used to examine the relationships between plant viruses. Essentially, the technique is to hybridize the target nucleic acid with the probe that has been radioactively labeled (see below), then to digest away any unhybridized ss probe, precipitate the ds hybrid on to a filter membrane and count the radioactivity.

Most of the approaches to the use of nucleic acid hybridization for plant pathogen diagnosis now involve the use of mixed-phase hybridization. There are several basic forms of membranes, the most common being nitrocellulose and nylon. Nitrocellulose membranes are widely used but have two main disadvantages. As nucleic acids are attached to them by hydrophobic rather than covalent bonds, they can be released slowly during hybridization and washing. Nitrocellulose membranes also become brittle when dry and are frequently damaged if one wishes to strip a probe off and reprobe. Nylon membranes bind nucleic acids irreversibly and, as they are more flexible, can be used for stripping and reprobing experiments with fewer breakage problems. Nucleic acids can be immobilized on to nylon membranes in buffers of low ionic strength. The main disadvantage of nylon membranes is the tendency for high background hybridization, especially with RNA probes and with some non-radioactive reporter systems (see below). There are two basic types of nylon membrane, unmodified and charge-modified. The latter, which has a positively charged surface, has a greater binding capacity for nucleic acids. There are numerous brands of nylon membranes available that differ in detail in their properties (see Khandjian, 1987; Twomey and Krawetz, 1990).

As noted above, nucleic acid has to be denatured to bind to membranes. This can be effected by alkali (but remember alkali treatment will degrade RNA), heating and quenching the sample or by treating the sample with formamide, formaldehyde or glyoxal. The latter two compounds form complexes with nucleic acid which can inhibit hybridization, and so the compound has to be removed by mild alkali treatment. Details of the various denaturation procedures are given in Sambrook *et al.* (1989).

After transfer to either nitrocellulose or nylon membranes the nucleic acid must be immobilized. This can be effected either by baking under vacuum for 0.5–2 hours at 80°C or by exposing the side of the membrane carrying the nucleic acid to UV radiation (254 nm).

The simplest format is the dot blot in which the target sample is spotted on to the membrane. This can be either directly or by using a template in which the samples are placed into holes or slots (for slot blots) in a piece of plastic placed over the membrane. In a more sophisticated version, the template is incorporated in an apparatus that draws the sample through the membrane by suction; this enables dilute samples to be concentrated on the membrane.

The dot blot does not give any information on the size or number of species of the target nucleic acid. Such information is gained by electrophoresing the nucleic acid in a gel and then transferring it to the membrane. Details of protocols for transferring DNA (southern blots) and RNA (northern blots) are given in Sambrook *et al.* (1989).

Various other hybridization formats have been described. The one that shows most promise is sandwich hybridization (Ranki *et al.*, 1983). In this procedure two sequences complementary to adjacent parts of the target nucleic acid are used. One, the capture sequence, is immobilized on to a solid matrix and is hybridized to the target nucleic acid to which the other complementary strand, the probe, is also being hybridized (liquid–liquid hybridization). In a modification of the technique (Syvanen *et al.*, 1986), the capture strand is modified, for example by being sulfonated or by the attachment of a protein. Hybridization of the capture and probe strands to the target is performed in liquid and the hybrid is captured in the wells of a microtiter plate by an antiserum to the capture strand modifying molecule. Sandwich hybridization is relatively fast as at least part of it involves liquid–liquid hybridization. It enables contaminating materials in crude sap extracts to be washed away and, in the modified form, ELISA technology and equipment can be used.

2. Dot blot hybridization

Dot blot hybridization is now the most commonly used procedure for testing of large numbers of samples. The main steps in dot blot hybridization are: (1) a small amount of sap is extracted from the plant under test; (2) the viral nucleic acid is denatured by heating or, if it is DNA, by alkali treatment; (3) a spot of the extract is applied to a membrane; (4) the membrane is baked or exposed to ultraviolet light to bind the nucleic acid firmly to it; (5) nonspecific binding sites on the membrane are blocked by incubation in a prehybridization solution containing a protein, usually bovine serum albumin or non-fat dried milk (Johnson *et al.*, 1984), and small single-stranded fragments of an unrelated DNA, together with salt, etc.; (6) hybridization of a labeled probe nucleic acid to the test nucleic acid bound to the substrate; and (7) washing off excess (unhybridized) probe and estimation of the amount of probe bound by a method appropriate to the kind of label used for the probe. The prehybridization step (about 2 hours) and the hybridization step (overnight) are often carried out in heat-sealable plastic bags in a water bath at about 65°C or in specialized equipment. Detailed procedures have been given by Hull (1993) and Dijkstra and de Jager (1998).

Technical details for blotting methods using labeled nucleic acid probes are given by Owens and Diener (1984) and Hull (1985, 1986). The technique is now widely used in plant virology. For instance, a dot blot technique has been successfully applied to screen large numbers of potato plants in a programe of breeding potatoes for resistance to several viruses (Boulton *et al.*, 1986). A non-radioactive dot blot system using minimal equipment was developed for the routine diagnosis of a range of insect-transmitted viruses (Harper and Creamer, 1995). The method has been used to assess relationships between tombusviruses, but some unexpected cross-hybridizations were observed (Koenig *et al.*, 1988a). A sensitive non-radioactive procedure has been developed for detecting BaMV and its associated satellite RNA in meristen-tip cultured plants (Hsu *et al.*, 2000).

3. Tissue print hybridization

Immunoprobing of tissue prints is described in Section IV.B.4.d above. Tissue prints can be hybridized with labeled probes. Guilfoyle *et al.*

(1993) described the methods for tissue print hybridization using ^{32}P-labeled probes. Más and Pallás (1995) showed that, although non-radioactive probes with colorigenic reporter systems were less sensitive those with chemilu-minescent reporters, the colorigenic systems gave anatomical information (see Section V.C.5 for reporter groups on probes).

Tissue print hybridization is of especial use for viroids that do not have proteins that can be detected immunologically. Viroids have been detected by molecular hybridization of imprinted membranes (see Podleckis *et al.*, 1993; Romero-Durbán *et al.*, 1995; Duran-Vila *et al.*, 1996; Palacio-Bielsa *et al.*, 1999).

Boulton and Markham (1986) used an adap-tation of the procedure they termed 'squash-blot' (or 'swat blot') to assay MSV in single leafhopper vectors squashed directly on to the nitrocellulose filter. Their probe was ^{32}P-labeled and the extent of hybridization was determined by autoradiography. The method is illustrated in Fig. 15.16.

4. *In situ* hybridization

In situ hybridization can give information of the distribution of the target nucleic acid within a cell. The loci at which BSV is integrated into the banana chromosomes were identified by *in situ* hybridization (Harper *et al.*, 1999c). As an extension of this procedure, fibre stretch hybridization (Fransz *et al.*, 1996; Brandes *et al.*, 1997), in which denatured chromosomal DNA is stretched out on slides before hybridization, gives information on the detailed structure of the integrated locus; this is described for BSV in Chapter 8 (Section IX.B.4).

Increased sensitivity can be achieved by use of the *in situ* PCR (Muro-Cacho, 1999).

5. Preparation and labeling of probes

The probe comprises sequences complemen-tary to the target sequences to which a reporter system it attached to reveal when hybridization has taken place.

Labeled viral RNA has occasionally been used as a probe and has been shown to have several advantages (e.g. Lakshman *et al.*, 1986), but in most experiments, with either DNA or RNA viruses, DNA probes have been used. However, RNA probes have the advantage of forming stronger duplexes with their target, especially if it is RNA and thus more sensitive. Sensitivity of detection may be related to probe

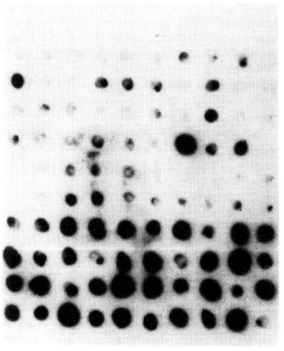
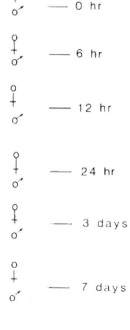

Fig. 15.16 Use of the squash-blot nucleic acid hybridization technique to estimate the time course of uptake of MSV by a leafhopper vector. Autoradiograph of squash-blots of male (♂) and female (♀) *Cicadulina mbila* showing the time course of MSV acquisition. Each blot is from one insect. From Boulton and Markham (1986), with permission.

size, with larger probes giving more sensitive assays (e.g. Barbara *et al.*, 1987).

Where appropriate information is available, it is possible to make diagnostic cDNA probes to regions of the viral genome where the extent of sequence similarity has been shown to distinguish between different viruses in a group and to identify strains within a group (see Chapter 17, Section II.A).

The probe may be labeled with a radioactive marker (usually ^{32}P for total nucleic acids bound to paper or ^{3}H for cytological experiments). Alternatively, a non-radioactive reporter group may be used. Various procedures can be used to radioactively label nucleic acid probes, or to make non-radioactive probes.

a. Reverse transcription

The viral RNA is first 'random' primed with short DNA oligomers. Then, DNA copies of the RNA are made using a retrovirus reverse transcriptase. The strand formed is complementary to the viral RNA (copy or cDNA). It can be used directly as a probe if radioactive or non-radioactive nucleotides are incorporated during reverse transcription.

b. Cloned probes

In a more commonly used procedure, a ds cDNA is made, inserted into an appropriate bacterial plasmid, and grown in the bacterial host. (See Sambrook *et al.* (1989) for practical aspects, and Old and Primrose (1989) for theory.) This procedure can provide a long-term source of a standard probe. Batches of the probe can then be labeled by nick translation, an enzymatic process in which some nucleotides are excised from the dsDNA and the gaps filled in with nucleotides, some of which are labeled. Alternatively, the ds cDNA is made single stranded by heating and then random primed with short synthetic oligomers. DNA polymerase may then be used to complete new double-stranded molecules, again using some labeled nucleotides.

c. RNA probes

Although theoretically dsRNA could be used as a probe, this is impractical because of preparation difficulties. The more usual approach is to synthesize *in vitro* complementary RNA from a cloned cDNA using nucleotides, some of which are labeled. This method has provided probes that are significantly more sensitive than DNA probes (see Varveri *et al.*, 1988; Robinson and Romero, 1991).

d. Strand-specific probes

For some experiments, it is necessary to have DNA probes that are either homologous to or complementary to the viral RNA. By cloning with the bacteriophage M13, which packages ssDNA into virus particles, it is possible to make probes of either positive or negative polarity.

e. Synthetic probes

When the nucleotide sequence of part or all of the viral genome is known, it is possible chemically to synthesize oligonucleotides of the desired sequence with a chain length of 15–20 bases. In principle, at least, this procedure has a number of advantages: (1) large amounts of single-stranded probe can be made and readily end-labeled with a ^{32}P-labeled nucleotide by means of a polynucleotide kinase; (2) strand-specific probes can be made; (3) several oligomers 15–20 bases long can be joined in tandem and cloned as outlined above; and (4) it is possible to construct a library of probes specifically designed to detect particular parts of the viral genome.

Bar-Joseph *et al.* (1986) discuss details of the application of synthetic probes. One limitation is sensitivity. Using such probes they could detect 4 ng of TMV RNA, whereas Sela *et al.* (1984), using randomly ^{32}P-labeled TMV cDNA, could detect as little as 25 pg.

f. PCR probes

The polymerase chain reaction (PCR) is described in Section V.D.

g. Non-radioactive probes (reviewed by Kricka, 1993)

There are two reasons why various workers have sought alternatives to radioactive labels for nucleic acid probes. ^{32}P, the isotope most commonly used, has a short half-life, and many diagnostic laboratories are not well equipped to handle this isotope. Thus, there is increasing interest in using non-radioactive reporter systems. Some examples of the use of non-radioactive probes, not only for virus detection but also for analysing transgenic plants, are given by Accotto *et al.* (1998).

There is an increasing number of non-radioactive reporter systems, which fall into three basic types: those that directly modify bases in the probe DNA, those that attach precursors to the probe DNA or RNA, and those that incorporate labeled precursors into the probe.

i. Direct modification of DNA

Various modified bases are specifically immunogenic and thus can be detected by antibodies to which an enzyme has been complexed.

ii. Attachment of precursors to DNA

One of the major approaches to non-radioactive reporters has been to cross-link compounds to DNA. These compounds can either be an enzyme that directly gives a color or luminescent reaction, or can be a molecule that reacts to an antibody or other molecules carrying an enzyme. An example of the former is the cross-linking of horseradish peroxidase (HRP) or alkaline phosphatase (AP) to DNA, both of which form the basis of commercially available products, for example the enhanced chemiluminescence (ECL) system. The disadvantage of HRP is that it is sensitive to higher temperatures and therefore hybridization has to be at 42°C in the presence of urea. Photobiotin is a photoreactive derivative of biotin that cross-links to single- or double-stranded DNA or RNA on irradiation with visible light.

iii. Incorporation of labeled precursors into DNA

This approach is essentially the same as that used for the radioactive labeling of nucleic acids described above. There are several forms of biotinylated nucleotides, for instance pyrimidine labeled at the 5 position, adenine at the N6 position, or cytidine at the N4 position. The method used for incorporation of biotinylated dNTPs can influence the sensitivity of the probe.

A commercially available kit is based upon the incorporation of digoxigenin–dUTP into the probe. The digoxigenin is then detected by an antibody–enzyme complex.

iv. Labeling of oligonucleotides

The above methods can also be used to label oligonucleotide probes. For example, biotin and AP have been incorporated during the synthesis of oligonucleotide, and the same molecules have been cross-linked to oligonucleotides.

v. Detection of reporter molecules

The actual detection is usually by an enzyme that gives a colored product or a luminescent compound on reaction with a substrate. The colored product has to be insoluble (unlike the colored product of the ELISA reaction). The most commonly used enzymes are AP and HRP for which the detection systems are the same as in western blots.

Luminescent systems have several advantages over the color systems. They give very rapid results of which a permanent record can easily be made as light released by the luminescence reaction can be detected by exposure of the membrane to photographic film. As it is easy to destroy most luminescent compounds, it is possible to reprobe with several different probes. With the right equipment, one can quantitate the amount of hybridized probe and thus, as with radioactive probes, determine relationships. Luminescence systems based on both HRP and AP have been developed. In the ECL system, mentioned above, positive-charge modified, polymerized HRP, cross-linked to ss DNA or RNA, catalyses the production of oxygen from hydrogen peroxide, which then oxidizes luminol. This gives a chemiluminescent reaction, the light output being enhanced by the presence of various phenolic compounds. This, and various other HRP-based systems, are reviewed by Durrant (1990). Among the AP-based systems is one in which the enzyme acts on D-luciferin-O-phosphate to liberate D-luciferin, which then emits light on oxidation by luciferase. In another system, AP deprotects phosphorylated phenyl dioxetane. Light output is enhanced by intermolecular energy transfer to a micelle-solubilized fluorophore. In the digoxigenin system the substrate is 3-(2'-spiroadamantane)-4-methoxy-4-(3''-phosphoryloxy)-phenyl-1,2-dioxetane (AMPPD) which, upon de-phosphorylation at alkaline pH, releases light.

Biotin can be detected by its close affinity for avidin (or streptavidin), which is one of the strongest associations between biological macromolecules; the avidin usually has an enzyme conjugated to it, though gold-labeled avidin has been used. Biotin on probes can also be detected by an anti-biotin antibody conjugate. One conjugate of interest involves coating the antibody with colloidal gold, the presence of which can be further enhanced by silver treatment.

Although biotin is widely used as a reporter molecule, it does have some disadvantages. Many sap samples contain significant amounts of endogenous biotin, which can give false positives. Avidin frequently binds non-specifically to nylon membranes giving high backgrounds, although this can be reduced by pre-treating the membrane with a protease. Both these problems are overcome with sandwich hybridization.

6. Southern blots

Nucleic acids that have been separated by polyacrylamide gel electrophoresis (see Chapter 4, Section III.A.2.b) can be blotted on to membranes and detected by the probing systems described above. This has the advantage over dot blot hybridization in that it gives information on the numbers and sizes of nucleic molecules.

D. Polymerase chain reaction

DNA fragments of interest can be amplified enzymatically *in vitro* by the polymerase chain reaction (PCR) (Mullis *et al.*, 1986; Saiki *et al.*, 1988). The technique involves the hybridization of synthetic complementary oligonucleotide primers to the target sequence and synthesis of multiple copies of cDNA of the sequence between the primers using heat-stable DNA polymerase. The process goes through a series of amplification cycles, each consisting of melting the double-stranded template DNA molecules in the presence of the oligonucleotide primers and the four deoxyribonucleotide triphosphates at high temperature (melting), hybridization of the primers with the complementary sequences in the template DNAs at lower temperature (annealing) and extension of the primers with DNA polymerase (DNA synthesis). This is illustrated in Fig. 15.17. During each cycle, the sequence between the primers is doubled so that after n cycles a 2^n amplification should be obtained. Usually the reaction is of 30 to 50 cycles. Details of this technique can be found in various manuals, including Innis *et al.* (1990).

The selection of primers depends on the target sequence and it is absolutely essential that the 3' nucleotide is complementary to the desired nucleotide on the target sequence. Usually primers are between 18 and 25 bases long and have a T_m of between 55 and 65°C (see Section V.C.1.a. for T_m of short sequences).

As PCR is based on DNA, it is not directly applicable to most plant viruses that have RNA genomes. However, a cDNA can be made to the desired region of the RNA genome using a primer and reverse transcriptase, and this used as the initial template. This procedure, now widely used, is termed RT-PCR. A further refinement is to couple PCR with the capture of the virus particles by immobilized antibodies, termed immune-capture PCR (IC-PCR or IC-RT-PCR).

PCR (and RT-PCR) has proved to be a very powerful tool for virus detection and diagnosis. It can be used directly to produce a DNA product of predicted size that can be confirmed by gel electrophoresis. The choice of primers can be used to distinguish between strains of a virus or, with primers containing a variety of nucleotides at specific positions, for more generic determinations. Strains can also be distinguished by amplifying a region that has differences in restriction endonuclease sites. As noted above, PCR is widely used to produce probes for hybridization by incorporating reporter nucleotides in the reaction. To illustrate the widespread applicability to the detection and diagnosis of plant viruses, some

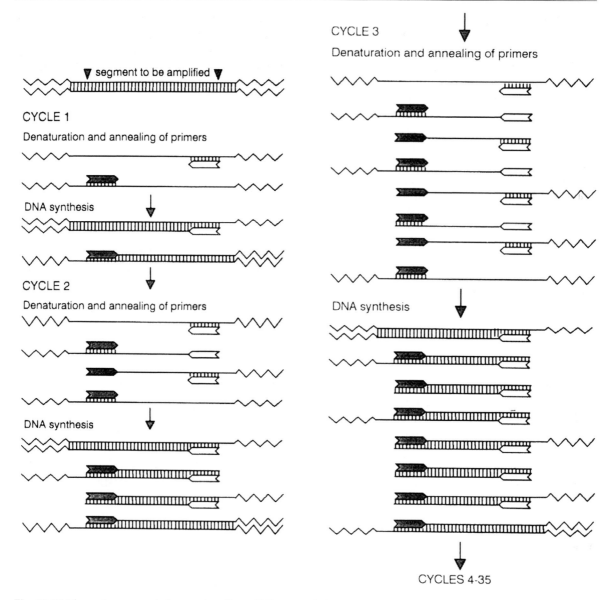

Fig. 15.17 The polymerase chain reaction. From Dijkstra and de Jager (1998), with kind permission of the copyright holders, © Springer-Verlag GmbH and Co. KG.

examples are provided; these are by no means a comprehensive collection of the uses to which this technique can be put.

As described in Chapter 8 (Section IX.B.4), BSV sequences are frequently found integrated in the host genome. This precludes the use of straightforward PCR for diagnosis of episomal infections of this virus. Harper *et al.* (1999a) described an IC-PCR technique that is specific for episomal BSV. Using an IC-RT-PCR method

in a single closed tube, TSWV can be detected in leaf and root tissue (Jain *et al.*, 1998), and CTV and PPV can be detected in plant tissue and single aphid samples (Olmos *et al.*, 1999).

In an RFLP analysis of 10 PVY isolates representative of four symptomatic groups, the whole genomes were each amplified in two fragments by RT-PCR (Glais *et al.*, 1998). Using seven restriction enzymes, three RFLP groups were determined in the 5′ fragment and two in the 3' fragment that correlated with the biological characters. One group of isolates appeared to have resulted from a recombination event.

The coat proteins of many potyviruses have two regions of conserved amino acids (MVW-CIENG and QMKAAA). These conserved motifs have been used to produce group-specific primers, albeit with some redundant nucleotides (Langeveld *et al.*, 1991; Pappu *et al.*, 1993). Using these primers, Marie-Jeanne *et al.* (2000) amplified by RT-PCR a 327-nucleotide fragment from several potyviruses that infect *Poaceae* and then differentiated between the viruses by RFLP. Other conserved regions in potyvirus coat proteins have also been used (see Colinet *et al.*, 1994, 1998).

Closely related viruses can be distinguished by multiplex RT-PCR (Clover and Henry, 1999) using primers specific to either WSSMV or WYMV. No PCR product was obtained with the closely related Ba MMV, BaYMV or OMV.

Degenerate primers to highly conserved regions have been used to detect whitefly-transmitted geminiviruses (Rojas *et al.*, 1993). Strains of WDV can be differentiated by using universal and strain-specific promoters (Commandeur and Huth, 1999).

A highly sensitive procedure for the early detection of BNYVV in plant, soil and vector samples involves PCR and digoxigenin labeling (Fenby *et al.*, 1995).

PCR is being used widely in the detection and identification of phytoplasmas in both plant and insect samples (see Lee *et al.*, 1993; Marzachi *et al.*, 1998; Seemüller *et al.*, 1998).

VI. DISCUSSION AND SUMMARY

The application of biological, physical and molecular techniques has given a large 'tool-bag' for the detection and diagnosis of plant viruses. This emphasizes the point made in the introduction that it is important to identify the question to be addressed. If one wishes to determine whether a plant is virus infected, say for quarantine purposes, one does not necessarily need a sophisticated technique that identifies a virus strain. On the other hand, if one is studying the durability of a potential resistance gene (or transgene) it is very useful to have an understanding of the range of variation of the virus. Thus, one has to select the best technique for what is wanted. In making this decision, various points have to be taken into account, as outlined below.

1. Sensitivity required

There is much discussion about the relative sensitivity of detection procedures. However, the sensitivity of many of the serological and nucleic acid-based tests is adequate for most purposes, and so the system that is most convenient should be used. For ease and speed of operation, and low cost, dot blots based on either an immunological test or nucleic acid hybridization have a lot to offer. There are field kits available for testing for, say, potato viruses, that are based on the same technology as home pregnancy kits.

However, for each virus situation it is advisable to compare tests to see which is the most appropriate. For instance, in comparative tests on serological and nucleic acid methods, using MAbs for the routine diagnosis of TSWV in the field, ELISA was found to be the most suitable (Huguenot *et al.*, 1990). On the other hand, the gain in sensitivity of 1000-fold by IC-RT-PCR over ELISA enabled PVY to be detected in individual aphids (Varveri, 2000). Similarly, BYV could be detected in individual aphids by PCR but not by TAS-ELISA (Stevens *et al.*, 1997). Higher sensitivity also enables samples, for example of seeds, to be pooled, and infection of only a small proportion to be detected.

2. Number of samples

Where large numbers of samples have to be handled the following factors are important in choosing a test procedure: (1) specificity; (2) sensitivity; (3) ease and speed of operation; and (4) cost of equipment and consumable supplies. Serological tests offer a range of specificities from exquisitely specific MAbs to broadly cross-reacting polyclonal antisera. However, nucleic acid hybridization tests may sometimes provide a better test across a range of related viruses (Harrison *et al.*, 1983).

3. The material being sampled

In many cases, and especially in trees, the distribution of virus is not uniform. Thus, samples need to be taken carefully and it is advisable to take several samples from each tree. Pretreatment of the plant material can influence the distribution of the virus. For instance, storage of tubers of some varieties of potatoes at raised temperatures for 4 weeks increased the reliability of detection of PMTV by ELISA (Sokmen et al., 1998).

4. Reliability of the technique

A technique can be unreliable in two ways. False negatives can result from sampling part of the plant not containing virus, inhibition of the reaction by a plant constituent or limitation in the materials being used (e.g. mismatch in a primer for PCR). False positives can be due to plant constituents, especially when testing new plant species (see Murphy et al., 1999b).

5. Equipment and expertise available

The routine reliable detection of most viruses usually does not require expensive equipment. There are some exceptions, such as the detection of BSV described in Section V.D, which requires either ISEM (Lockhart, 1986) or IC-PCR. However, it is important to have a good reliable supply of consumables and means for storing them without deterioration. Similarly, most of the basic techniques are relatively easy to learn, although it is important that they are learnt properly so that potential sources of error can be recognized.

For many kinds of experiments, for example studies on virus inactivation, interest centers on the results of measurements made by different methods that depend on different properties of the virus. When two or more methods are used on the same samples, the difficulties of interpretation that may arise because of the different sensitivities of the methods used must always be borne in mind.

Apart from any technical problems associated with any particular testing methods, certain general difficulties may be encountered in routine diagnosis. First, various factors that influence virus concentration in the plant can affect reliability. This may relate to variation throughout the plant, or to very localized variations, as described above. Second is the possibility of mixed infections with more than one virus. Some recognized disease entities are caused by multiple infections (e.g. the lettuce speckles disease; Falk et al., 1979). Third, other kinds of agent sometimes combine with a virus infection to produce a disease state. Thus, the internal browning disorder in tomato appears to depend on infection by TMV at a particular stage of fruit maturity (Taylor et al., 1969).

Perhaps the only generalizations that can be made are that the biological methods of detection will always be important for validating any application of the newer technologies, especially when weak reactions are observed, and that symptoms and host range, tested on an appropriate set of species, still give essential information for a reliable diagnosis.

Finally, with respect to diagnosis, there are international guidelines (FAO/IPGRI Technical Guidelines for the Safe Movement of Germplasm) that have been drawn up by expert panels to assist with international germ plasm movement. These detail the current ideas on the safest and simplest tests for assuring that plant propagules do not contain the viruses that are known to infect that species. It should always be remembered that one cannot, of course, test for unknown viruses.

Control and Uses of Plant Viruses

I. INTRODUCTION

As outlined in Chapter 3, viruses cause considerable losses and there is a range of control measures aimed at trying to mitigate these losses. However, as well as being detrimental, viruses also have their uses. In this chapter, I describe the main approaches to controlling viruses and also discuss some of their uses.

The use of fungicidal chemicals that, when applied to crop plants, protect them from infection or minimize invasion is an important method for the control of many fungal diseases. No such direct method for the control of virus diseases is yet available. Most of the procedures that can be used effectively involve measures designed to reduce sources of infection inside and outside the crop, to limit spread by vectors and to minimize the effect of infection on yield. Generally speaking, such measures offer no permanent solution to a virus disease problem in a particular area. Control of virus disease is usually a running battle in which organization of control procedures, care by individual growers and co-operation among them is necessary year by year. The few exceptions are where a source of resistance to a particular virus has been found in, or successfully incorporated into, an agriculturally useful cultivar. This is becoming of increasing importance with the development of transgenic protection of plants against viruses. Even with conventional and transgenic resistance, protection may not be permanent when new strains of the virus arise that can cause disease in a previously resistant cultivar.

Virus acronyms are given in Appendix 1.

I shall consider here the kinds of measures that have been tried for the control of virus diseases. It should be borne in mind that virus infection can sometimes increase the incidence of some other kind of disease. In such situations, different sorts of control may be needed. For example, yellowing viruses in sugar beet increase susceptibility to *Alternaria* infection. Spraying with appropriate fungicides reduced this secondary effect of virus infection in some seasons (Russell, 1966).

Correct identification of the virus or viruses infecting a particular crop is essential for effective control measures to be applied. Disease symptoms alone may be very misleading. For example, virus disease in lettuce can be caused by some 14 viruses with an aphid, leafhopper, thrip, nematode or fungus vector (Cock, 1968). Many of these viruses produce brown necrotic spots or bronzing on leaves, and later chlorotic stunting. Another example is that of the yellowing diseases of beets in the UK (Russell, 1958) and the western United States.

Of major importance in designing a strategy for control of a virus in a specific crop is an understanding of the epidemiology of that virus. This enables disease outbreaks to be forecasted; these aspects are discussed in Chapter 12 (Section IV).

It has become apparent that most of the serious virus disease problems around the world are the direct or indirect result of human activities (Thresh, 1982). Important activities leading to epidemics include:

- introduction of viruses into new areas through transport of infected seed or vegetative material

- introduction of virus vectors into new areas
- introduction of a new variety of a crop into an area when that variety is especially susceptible to a virus already present there
- use of monocultures, that is, the planting of genetically uniform crops over large areas replacing traditional polycultures
- use of irrigation to prolong the cropping season with overlapping crops
- repeated use of the same fields for the same crop
- increased use of fertilizer and herbicides or other forms of weed control.

In recent years the most active areas of research into the control of virus diseases have been: (1) the breeding of resistant or immune cultivars by classical genetic procedures; (2) the control of vectors by various strategies; (3) the production of virus-free stocks of seed and vegetative propagules; and (4) the production of transgenic plants containing viral genes that confer resistance to the virus. General approaches to control are discussed in Harris and Maramorosch (1982) and Walkey (1991), and control of specific viruses is reviewed in several places, for example control of luteoviruses (Robert and Lemaire, 1999b), of ACMV (Thresh and Otim-Nape, 1994) and of PRSV (Gonsalves, 1998), and in several chapters in Hadidi et al. (1998).

Data for many of the control measures discussed in this chapter are derived from laboratory and field trials. Because of the many variables, and the large number of countries involved, it is often difficult to assess the extent to which any particular procedure or set of procedures has actually been adopted on a regular basis into commercial practice.

More and more attention is being given to the possibilities of integrated control involving several strategies. Several of these are noted in this chapter. Again, it is often difficult to assess whether these are actually being effectively used in the field, or whether they remain optimistic dreams. When adequate facilities and expertise are available, a multi-disciplinary approach may be useful. The attempts to control diseases caused by TSWV in Hawaii are an example (Cho et al., 1989). Control of plant pathogens integrated with control of pests in integrated pest management (IPM) systems is discussed in Jacobsen (1997).

II. REMOVAL OR AVOIDANCE OF SOURCES OF INFECTION

A. Removal of sources of infection in or near the crop

It is obvious that there will be no virus problem if the crop is free of virus when planted and when there is no source of infection in the field, or none near enough to allow it to spread into the crop.

The extent to which it will be worthwhile to attempt to eliminate sources of infection in the field can be decided only on the basis of a detailed knowledge of such sources and of the ways in which the virus is spreading from them into a crop. Eradication as a control measure has been reviewed by Thresh (1988b) and eradication schemes directed against CSSV and CTV are described by Ollennu et al. (1989) and Garnsey et al. (1998).

1. Living hosts for the virus

Living hosts as sources of infection may include: (1) perennial weed hosts, annual weed hosts in which the virus is seed transmitted, or annual weed hosts that have several overlapping generations throughout the year; (2) perennial ornamental plants that often harbor infection in a mild form. For instance, gladioli are often infected with BYMV that can spread to adjacent annual legume crops (Hull, 1965); (3) unrelated crops; (4) plants of the same species remaining from a previous crop (these may be groundkeepers, as with potatoes, or seedling volunteers); and (5) seed crops of biennial plants that may be approaching maturity about the time the annual crop is emerging.

In theory, it should be possible to eliminate most of such sources of infection. In practice, it is usually difficult and often impossible, particularly in cropping areas that also contain private gardens. Private household gardens and adjacent farms in temperate and subtropical regions often contain a diverse collection of plants, many of which can carry economically important viruses. It is usually difficult or impossible to control such gardens effectively. In tropical countries, the diversity of cropping makes control of living sources of infection very difficult.

The extent to which attempts to remove other hosts of a virus from an area may succeed will depend largely on the host range of the virus. It may be practicable to control alternative hosts where the virus has a narrow host range, but with others, such as CMV and TSWV, the task is usually impossible.

2. Plant remains

Plant remains in the soil, or attached to structures such as greenhouses, may harbor a mechanically transmitted virus and act as a source of infection for the next crop. With a very stable virus such as TMV, general hygiene is very important for control, particularly where susceptible crops are grown in the same area every year. ToMV may be very difficult to eliminate completely from greenhouse soil using commercially practicable methods of partial soil sterilization (Broadbent et al., 1965; Broadbent, 1976; Lanter et al., 1982). A major development has been the replacement of soil by sand/peat substrates that are renewed every year.

3. Roguing and eradication schemes

Sometimes it may be worthwhile to remove infected plants from a crop. If the spread is occurring rapidly from sources outside the crop, roguing the crop will have no beneficial effect. If virus spread is relatively slow and mainly from within the crop, then roguing may be worthwhile, especially early in the season. Even with a perennial crop, if a disease spreads slowly, roguing and replanting with healthy plants may maintain a relatively productive stand. A study of the distribution of infected plants within the field using the formula developed by van der Plank (Chapter 12, Section V.A.2) may give an indication as to whether spread within the crop is taking place.

In certain situations, roguing may increase disease incidence by disturbing vectors on infected plants (Rose, 1974). In many crops, newly infected plants may be acting as sources of virus for further vector infection before they show visible signs of disease (e.g. Beemster, 1979).

Most of the successful eradication schemes have been on tree crops. Among the factors that dictate success are: (1) relatively small numbers of infected trees and infection foci; (2) low rate of natural spread; (3) good data on extent and distribution of infection; (4) rapid, reliable and inexpensive diagnostic procedure for the virus; and (5) resources for rapid and extensive surveys and tree removal.

Regular roguing of infected plants has been effective in the control of ACMV in cassava in trials carried out in tropical Africa (Robertson, 1987), but the method is not widely used. One of the most successful examples of disease control by roguing of infected crop plants has been the reduction in incidence of the BBTV in bananas in eastern Australia. Legislation to enforce destruction of diseased plants and abandoned plantations was enacted in the late 1920s. Within about 10 years, the campaign was effective to the point where bunchy top disease was no longer a limiting factor in production. Dale (1987) and Dale and Harding (1998) attributed the success of the scheme to the following main factors: (1) absence of virus reservoirs other than bananas, together with a small number of wild bananas; (2) knowledge that the primary source of virus was planting material, and that spread was by aphids; (3) cultivation of the crop in small, discrete plantations, rather than as a scattered subsistence crop; (4) strict enforcement of strong government legislation; and (5) co-operation of most farmers.

A model has shown that extensive eradication will control CTV (Fig. 16.1). An analytical model of the virus disease dynamics with roguing and replanting has been put forward by Chan and Jeger (1994). They showed that roguing only in the post-infectious category confers no advantage. At low contact rates, roguing only when plants became infectious is sufficient to eradicate the disease but at high contact rates, roguing latently infected as well as infectious plants is advisable. The model indicates that eradication is achievable with realistic roguing intensities. Smith et al. (1998) used a model to explore a variable roguing rate across blocks of bananas in the Philippines infected to varying degrees with BBTV. This model indicated that small gains would accrue from constant roguing across all blocks but changing the emphasis to the most infected blocks where the disease was spreading in from outside could result in a dramatic increase in disease.

4. Hygiene

For some mechanically transmitted viruses, and particularly for TMV or ToMV, human activities during cultivation and tending of a crop are a major means by which the virus is spread. Once TMV or ToMV enters a crop such as tobacco or tomato, it is very difficult to prevent its spread during cultivation and particularly during such processes as lateraling and tying of plants. Control measures consist of treatment of implements and washing of the hands. For this, Broadbent (1963) recommends a 3% solution of trisodium orthophosphate.

Workers' clothing may become heavily contaminated with TMV and thus spread the virus by contact. TMV persisted for over 3 years on clothing stored in a dark enclosed space (Broadbent and Fletcher, 1963), but was inactivated in a few weeks in daylight. Clothing was largely decontaminated by dry cleaning or washing in detergents with hot water.

While TMV is the most stable of the mechanically transmitted viruses, others can be transmitted more or less readily on cutting knives and other tools. These include TBV, CymMV, PVX and PSTVd. These mechanically transmitted agents may be a particular problem in glasshouse crops, where lush growth, close contact between plants, high temperatures and frequent handling are important factors in facilitating virus transmission.

Since mechanical transmission is an important means whereby viroids are spread in the field, decontamination of tools and hands is an

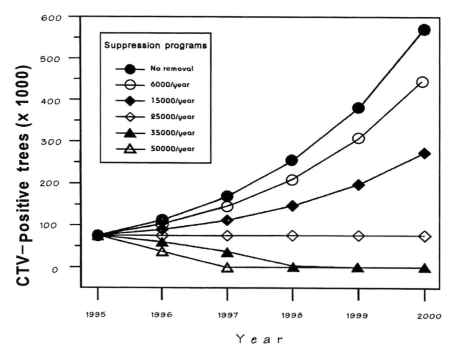

Fig. 16.1 Effects of various tree removal rates on suppression of CTV infection. Figure is from a model developed for a tree removal project in California. Surveys indicated approximately 75 000 CTV-infected trees in an area of approximately 80 000 ha. An infection increase of 50% per year of existing trees is assumed. If 25 000 of the existing trees are removed, a 50% increase in the remaining 50 000 creates essentially a static situation. Increasing the initial tree removal to 50 000 rapidly reduces the population of infected trees and lowers the overall number of tree removals compared with less aggressive initial approaches. From Garnsey *et al.* (1998), with kind permission of the copyright holder, © The American Phytopathological Society.

important control measure. However, this presents difficulties because of the stability of viroids to heat and many decontaminating agents as normally used. Brief exposure to 0.25% sodium or calcium hypochlorite is probably the best procedure (Garnsey and Randles, 1987). One of the earliest and most effective methods for the control of viroids has been the avoidance of sources of infection. This method has been particularly successful for vegetatively propagated crops susceptible to viroids, such as potatoes and chrysanthemums. This success is probably due to the absence of vectors in the field apart from humans.

B. Virus-free seed (reviewed in Maury et al., 1998)

Where a virus is transmitted through the seed, such transmission may be an important source of infection because it introduces the virus into the crop at a very early stage, allowing infection to be spread to other plants while they are still young. In addition, seed transmission introduces scattered foci of infection throughout the crop. Where seed infection is the main or only source of virus, and where the crop can be grown in reasonable isolation from outside sources of infection, virus-free seed may provide a very effective means for control of a disease. Factors that affect seed transmissibility of viruses are discussed in Chapter 12 (Section III.A.1).

LMV is perhaps a good example of controlling a virus problem through clean seed (see Dinant and Lot, 1992). Grogan et al. (1958) found that crops grown from virus-free seed in California had a much lower percentage of mosaic at harvest than adjacent plots grown from standard commercial seed. For a time, control was unsatisfactory, until it was realized that even a small percentage of seed infection could give much infection within the crop if the aphid vector was active (Zink et al., 1956).

Tomlinson (1962) obtained similar results under English conditions. To obtain effective control by the use of virus-free or low-virus seed, a certification scheme may be necessary, with seed plants being grown in appropriate isolation. More recent work indicated that even

0.1% seed transmission did not give effective control of lettuce mosaic (Kimble et al., 1975). The use of seedstocks that tested <0.003% infection in the Salinas Valley area of California has given consistent control in practice.

In the 5 years prior to the introduction of the virus-free seed program in the Salinas Valley area of California, lettuce yields were about 140 cartons per hectare. In the 5 years immediately after introduction of the scheme, the average yield was 190 cartons per hectare. Most of this increase can be attributed to the reduction in losses due to LMV (Kimble et al., 1975).

ELISA is used to test for the presence of a virus in batches of seed. If there is a significant amount of virus in the seed coats that is not involved in seed transmission it may be necessary to remove the seed coats before testing the embryos (Maury et al., 1988). Seed certification schemes against BSMV have been credited with avoiding millions of dollars in losses due to this disease in barley crops in some US states (Carroll, 1983).

To improve the quality control of analyzing large quantities of seed being tested for PSbMV, statistical and serological approaches were taken to identify thresholds for identification (Masmoudi et al., 1994). Large groups (200 to 500 seeds per group) were tested by ELISA and account was taken of detecting one infected embryo in such a group size.

Tomato seed from ToMV-infected tomatoes carries the virus on the surface of the seed coat. As the seed germinates, virus contaminates the cotyledons and is inoculated into the plant by handling during pricking out. Seed can be cleaned by extraction in hydrochloric acid, heating dried seed in 0.1 N hydrochloric acid, or treatment with trisodium orthophosphate or sodium hypochlorite (e.g. Gooding, 1975).

It may be particularly important to attempt to eliminate seed-transmitted viruses from national or international germplasm collections. Among 207 Phaseolus vulgaris accessions in the USDA germ plasm collection, about 60% contained some seed infected with BCMV (Klein et al., 1988). In at least one instance, that of PSbMV in peas, elimination of infected individuals led to a loss of genetic diversity, as judged by seed coat color and isoenzyme genotypes (Alconero *et al.*, 1985). In this circumstance, methods that do not involve selective loss of particular plant types might be used to eliminate the virus (see below).

C. Virus-free vegetative stocks

For many vegetatively propagated plants, the main source of virus is chronic infection in the plant itself. With such crops, one of the most successful forms of control has involved the development of virus-free clones, that is, clones free of the particular virus under consideration. Two problems are involved. First, a virus-free line of the desired variety with good horticultural characteristics must be found. When the variety is 100% infected, attempts must be made to free a plant or part of a plant from the virus. Second, having obtained a virus-free clone, a foundation stock or 'mother' line must be maintained virus free, while other material is grown up on a sufficiently large scale under conditions where reinfection with the virus is minimal or does not take place. These stocks are then used for commercial planting.

1. Methods for identification of virus-free material

Visual inspection for symptoms of virus disease is usually quite inadequate when selecting virus-free plants. Appropriate indexing methods are essential. A variety of methods is available, and the most suitable will depend on the host plant and virus. For many viruses, especially those of woody plants, the rather laborious process of graft indexing to one or more indicator hosts is essential. Distribution of a virus within the tree may be uneven, especially in the early stages after infection, so that tests repeated in successive seasons may be necessary to ensure freedom from virus. For example, Hampton (1966) found, using four buds for indexing per tree, that first-year infection with PDV was not detected in a high proportion of cherry trees (29–63%, depending on the variety). The probability of detection improved substantially in trees that had been infected for 3 years. Distribution may also be uneven in herbaceous plants (see Chapter 9, Section II.J). Thus, Beemster (1967) found that in potatoes

inoculated with PVY, not all tubers were infected and not all parts of a single tuber might be infected. The heel end of tubers was less frequently infected than the rose end. The rose end was, therefore, a more reliable source material for testing.

Mechanical inoculation to indicator hosts can be used with some viruses, but other methods of diagnosis now rival infectivity tests in their sensitivity for the detection of viruses (see Chapter 15).

2. Methods for obtaining virus-free plants

a. Naturally occurring virus-free material

Occasionally, individual plants of a variety, or plants in a particular location, may be found to be free of the virus. If all plants are infected, advantage may sometimes be taken of uneven distribution of the virus in the plant. This is not uncommon with some viruses in fruit trees. Budwood can be taken from uninfected parts of the tree. The shoot tips of rapidly growing stems may sometimes be free of a virus that is systemic through the rest of the plant. Thus, Holmes (1955) was able to obtain dahlia cuttings free of TSWV. This sort of procedure has been used successfully for several viruses in certain hosts. However, many vegetatively reproduced varieties appear to be virtually 100% infected with a virus, and, with these, one or more of the special treatments and methods described next have to be used to obtain a nucleus of virus-free material. Some examples have been reported where natural elimination of a virus occurred. When TRV-infected tulip bulbs were grown for several seasons in soil free of the nematode vector, a proportion of the bulbs were found to be free of the virus (van Hoof and Silver, 1976).

b. Heat therapy (reviewed in Mink et al., 1998)

Heat treatment has been a most useful method for freeing plant material from viruses. Many viruses have been eliminated from at least one host plant by heat treatment (Walkey, 1991). Two kinds of plant material have been used. Dormant plant parts such as tubers or budwood can generally stand higher temperatures than growing tissues, and the method probably depends on direct heat inactivation of the virus. Temperatures and times of treatment vary widely (35–54°C for minutes or hours). Hot water treatments are often used, as hot air tends to give uneven heating during short treatments. Unless tissues are thoroughly hydrated, dry heat is much less effective than wet heat.

Growing plants are much more generally treated, and hot air rather than hot water is applied. Temperatures in the range of 35–40°C for periods of weeks are commonly employed. This form of treatment gives a better survival rate for growing plant material. Details of the treatment vary widely and have to be worked out empirically for each host–virus combination. Very frequently, small cuttings are taken from the shoot tips immediately after the heat treatment, as these may be free of virus when the rest of the plant is not. For example, culture of shoot tips from heat-treated sprouting potatoes gave a useful proportion of plantlets free of PVX (Faccioli and Rubies-Autonell, 1982). The percentage of garlic plants free of three viruses that were regenerated from meristem tip culture increased from 25–50% to 85% when infected plants were treated at 38°C before tip culture (Walkey *et al.*, 1987).

Alternatively, apical explants established in culture may be given the heat treatment. Thus, strawberry mild yellow edge disease was eliminated from 4-mm-long strawberry stolon explants held at 38°C (Converse and Tanne, 1984). These authors suggest that heat therapy and stolon apex culture contribute independently to the process.

A regime of alternating high (approximately 40°C) and normal (about 20°C) temperatures may help to reduce the damaging effects of high temperatures on the plant tissues while still allowing some shoot tips to be freed of virus. Detailed regimes have to be developed for each host and virus. This procedure has been used for CCMV in cowpea (Lozoya-Saldaña and Dawson, 1982) and for GFLV and ArMV in grapes (Monette, 1986). Reduction and elimination of virus may depend on the total hours held at the high temperature, as illustrated for CMV in Fig. 16.2. The actual temperature within plant tissues may be

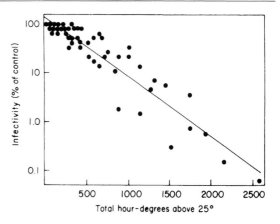

Fig. 16.2 Relationship between inactivation of CMV in infected tissue cultures of *Nicotiana rustica* and the total hour-degrees above 25°C resulting from various alternating diurnal temperature regimes. From Walkey and Freeman (1977), with permission.

several degrees below the measured ambient temperature.

At present, there is no basis for predicting that tissue from a certain plant species can, or cannot, be freed of a particular virus. The mechanisms underlying the preferential elimination of virus are not yet understood, but they presumably involve inactivation both of intact virus already synthesized and of the means for making more.

A protocol suitable for demonstrating heat treatment curing of a plant virus is given by Dijkstra and de Jager (1998).

c. Meristem tip culture (reviewed in Walkey, 1991; Faccioli and Marani, 1998)

The distribution of virus in apical meristems was considered in Chapter 9 (Section II.J.8). Culture of meristem tips has proved an effective way of obtaining vegetatively propagated plants free from certain viruses. Hollings (1965) defined meristem tip culture as aseptic culture of the apical meristem dome plus the first pair of leaf primordia. This piece of tissue is about 0.1–0.5 mm long in different plants. The minimum size of tip that will survive varies with different species. For example, it was necessary to use tips at least 0.3 mm long to obtain survival of rhubarb (Walkey, 1968). The kind of meristem tip usually taken is illustrated in Fig. 16.3. (The smaller the excised tips are at the

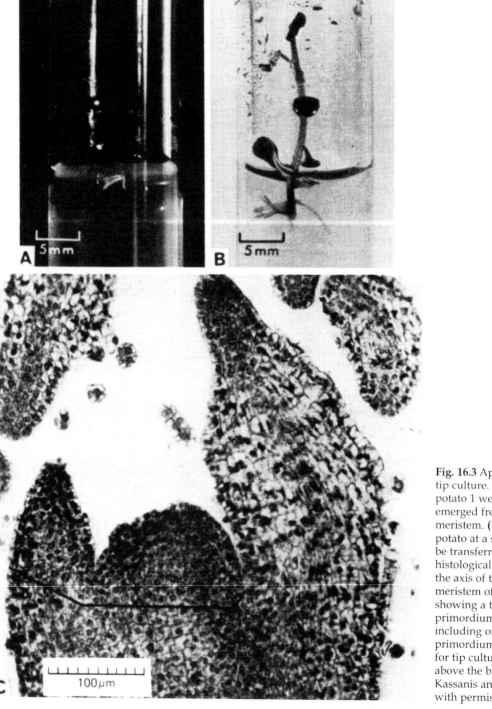

Fig. 16.3 Apical meristem tip culture. **(A)** Stem of potato 1 week after it emerged from a dormant meristem. **(B)** Plantlet of potato at a stage ready to be transferred to soil. **(C)** A histological section along the axis of the apical meristem of a potato sprout showing a two-leaf primordium. The piece including one leaf primordium that is excised for tip culture is shown above the black line. From Kassanis and Varma (1967), with permission.

time of removal, the better the chance that they will give rise to virus-free plants, although the more difficult they are to regenerate.) For many plants, at least with the culture methods currently used, one leaf primordium needs to be included to get regeneration of a complete plant.

A wide range of nutrient media has been used by different workers. The basic ingredients are an appropriate selection of mineral salts (macro- and micro-nutrients), sucrose, and one or more growth-stimulating factors such as indole acetic acid or gibberellic acid, sometimes in agar.

Only a proportion of meristem tip cultures yield virus-free plants. It is not always clear to what extent the success of the method depends on: (1) the regular absence of virus from meristem tissue, some tips being accidentally contaminated; (2) some meristematic regions in the plant containing virus and others containing none; or (3) virus present in the meristem being inactivated during culture on the synthetic medium.

Some viruses, for example, those present in members of the *Araceae* (Hartman, 1974), appear to be readily eliminated by culture in a suitable medium at 20–25°C. For others, such as TRSV and PVX, most or all such cultures remain infected. In this situation, it is now common practice to combine meristem tip culture with heat therapy, as discussed in the preceding section. Figure 16.4 illustrates a protocol for the combined use of thermotherapy and meristem tip culture.

As an alternative to direct culture on a synthetic medium, shoot tips up to 1.0 mm in length may be grafted aseptically on to virus-free seedling plants growing *in vitro*. Several viruses have been eliminated from peaches by this procedure (Navarro *et al.*, 1983).

The additional treatment of 'electrotherapy' before meristem culture has been used to free potatoes from PVX (Lozoya-Saldaña *et al.*, 1996). In this treatment potato stems were exposed to either 5, 10 or 15 mA for 5–10 minutes before the axillary bud was taken. Organogenesis and virus elimination were both stimulated by the electrical treatment.

Where the virus occurs in the apical meristem it may be difficult to obtain virus-free plants by tip culture (Toussaint *et al.*, 1984). Virus-free plants of several important tropical food crops have been obtained using meristem tip culture. These include cassava (Adejare and Coutts, 1981), sweet potato (Frison and Ng, 1981) and various aroids (Zettler and Hartman, 1987).

A protocol suitable for demonstrating curing of a plant virus by meristem culturing is given by Dijkstra and de Jager (1998).

d. Tissue culture

Culture of single cells or small clumps of cells from virus-infected plants may sometimes give rise to virus-free plants. For example, two viruses were eliminated from *Euphorbia pulcherrima* by cell suspension culture followed by regeneration of plants *in vitro* (Preil *et al.*, 1982). Plants regenerated from shake subculture of small pieces of tobacco tissue originally from TMV-infected plants were free of TMV (Toyoda *et al.*, 1985). A significant proportion of calli obtained from yellow-green areas of TMV-infected *Nicotiana tomentosa* gave rise to regenerated plants that were free of virus (White, 1982). This may mean that some cells in TMV-infected zones of the leaf are in fact free of virus, as appears to be the situation for some cells in TYMV local lesions.

In coupling tissue culture with heat treatment, BBTV-infected plantlets of Cavendish banana were exposed to 40°C for 16 hours per day for 12 weeks, but this failed to eliminate the virus (Wu and Su, 1991). Culturing for 3 months at 35°C gave some healthy buds (Wu and Su, 1991), as did culturing at 28°C for 9 months (Thomas *et al.*, 1995).

e. Low temperatures

The effect of holding plants at lower than normal temperatures on virus survival has not been widely investigated. Low temperatures might be expected to have little effect on viruses that are stable *in vitro*. In a few instances, growth at low temperatures has given virus-free plant material. Selsky and Black (1962) grew cuttings from sweet clover

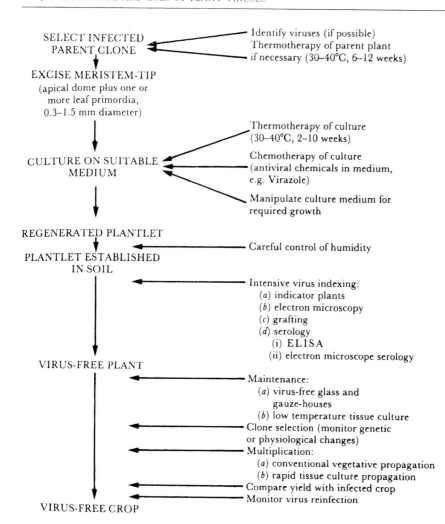

Fig. 16.4 Scheme for virus-free plant production by meristem tip culture. From Walkey (1985), with permission.

plants infected with WTV at 14°C, and no tumors developed even after several vegetative generations. After three generations, cuttings were taken from the plants and grown under normal greenhouse conditions. Some 95% of these gave rise to a second generation of greenhouse-grown cuttings that were 90% free of virus as indicated by absence of tumors. However, maintenance of three peach cultivars at 4°C for long periods did not free them of PNRSV (Heuss *et al.*, 1999).

Potato and chrysanthemum plants freed from four different viruses were regenerated from meristem tip cultures that had been held for 6 months in the dark at 6–7°C (Paduch-Cichal and Krycyznski, 1987).

f. Chemotherapy

Attempts to free infected plant material of a virus solely by the application of antiviral chemicals have been disappointing. There have been several reports of such cures, but they have often been based on very small numbers of plants, or the results have been open to other interpretations. Chemical treatment by itself has not yet found practical use. However, chemical treatment in combination with heat treatment or meristem tip culture may have been an advantage in a few instances.

Several compounds have been used. For example, Hansen and Lane (1985) used ribavirin (also called virazole), an analog of guanosine, in combination with *in vitro* culture to eliminate ACLSV from apple shoot cultures, and Toussaint *et al.* (1993) used it to eradicate ORSV from Cymbidium cultures. 2,4-Dioxohexahydro-1,3,5-triazine was used to eliminate potato viruses in potato meristem cultures (Borissenko *et al.*, 1985) and potato stem cuttings (Bittner *et al.*, 1987); for a review, see Tomlinson (1981).

3. The importance of using selected clones

The selection of horticulturally desirable clonal material free of virus infection is an important aspect of any program. Selection may need to be carried out both before and after the material has been rendered virus free by any of the preceding procedures. For example, in New Zealand the planting of apple trees freed from virus infection by heat therapy has resulted in higher yields and better trees, but problems have occurred with poor skin color in colored varieties. This may have been due to poor clone selection prior to treatment (Wood, 1983). The problems caused by variation are even greater if clones are regenerated from single cells.

Similarly, there may be horticultural variability in the material after it has been freed of virus. For example, van Oosten *et al.* (1983) noted variation in fruit skin properties in Golden Delicious apples that had been rendered virus free.

4. The importance of adequate virus testing

Plants found to be apparently free of the virus at an early stage of growth may develop infection after quite a long incubation, so that, in practice, it is very important that apparently virus-free plants obtained by meristem tip culture be tested over a period before release. For example, Mullin *et al.* (1974) monitored the progeny of meristem-cultured strawberry plants and found them to be still free of graft-transmissible disease after 7 years.

The recent finding that sequences of certain reverse transcribing plant viruses can be activated by stresses such as tissue culture and crossing (see Harper *et al.*, 1999b; Ndowora *et al.*, 1999; Hull *et al.*, 2000) (see Chapter 8, Section IX.B.4) throw an even greater emphasis on the need for a robust diagnostic procedure.

D. Propagation and maintenance of virus-free stocks (see Hadidi *et al.*, 1998)

Once suitable virus-free material of a variety is obtained, it has to be multiplied under conditions that preclude reinfection of the nucleus stock and that allow the horticultural value of the material to be checked with respect to trueness to type. Nuclear stock is then further multiplied for commercial use. This multiplication and distribution phase requires a continuing organization for checking on all aspects of the growth and sale of the material. A classical example is the potato certification scheme in Great Britain, which, over many decades, led to a 2- to 3-fold increase in yield, much of which was due to decreased incidence of virus infection. Tested Foundation Stocks, which are virtually free of virus, are grown in isolation in parts of Scotland that are unfavorable to early aphid migration and colonization (Todd, 1961). High-grade stocks are grown from this seed elsewhere in Britain, in areas selected because of the low incidence of aphid vectors. Health of the stocks is regularly checked. The use of systemic insecticides has extended the areas in which seed potato crops can be grown (for reviews, see Ebbels, 1979; Slack and Singh, 1998).

Many such schemes are now in operation around the world for a variety of agricultural and horticultural plants, including stone and pome fruits, grapevines and berry fruits, as well as potatoes (see Hadidi *at al.*, 1998). For example, the Australian Fruit Variety Foundation supplies virus-free trees to Australian growers (Smith, 1983). On the other hand, for some groups of plants, particularly those grown for cut flowers and bulbs, lack of co-operation by individual growers may limit the effective application of virus eradication programs.

It may be possible to obtain rapid initial multiplication of virus-free material. For example, Logan and Zettler (1985) describe a procedure involving repeated rapid shoot proliferation on an appropriate medium that has the potential to produce more than 50 000 virus-free gladioli within 30 weeks from a single shoot tip. As virus-free material is introduced into commercial planting, grower cooperation is essential for the implementation of measures to minimize reinfection. Avoidance of soil containing nematode vectors may be important in the propagation of virus-free crops such as hops (Cotten, 1979).

One problem that often arises as a certification scheme develops is that many plants need to be checked for infection before release. Markham *et al.* (1948) suggested a group testing procedure to save labor. Using appropriate sampling conditions, the number of plants infected in a field can be estimated from the proportion of groups found to be infected. The reliability of the test of course increases as the number of plants tested increases.

Any sampling and testing scheme should be considered from two points of view: first, the probability required that a crop of a certain 'high' level of infection will be rejected and, second, the probability required that a crop of a certain 'low' level of infection will be accepted. It is possible to construct schemes having various probability levels of acceptance of rejection. This sort of procedure has been developed by Marrou *et al.* (1967) for testing lettuce seeds for freedom from LMV. In their test, several hundred seeds are extracted together and inocu-

lated to a sensitive indicator host; results are interpreted in relation to graphs or tables based on the binomial distribution.

When large numbers of virus-free plants or seeds are being produced, it is necessary to have an 'audit trail' so that any problems that may arise can be traced back to source. One way of doing this is to use bar coding coupled to computer records that would enable individual or batches to be followed through all the stages of propagation and release.

E. Modified planting and harvesting procedures

1. Breaks in infection cycle

Where one major susceptible annual crop or group of related crops is grown in an area, and where these are the main hosts for a virus in that area, it may be possible to reduce infection very greatly by ensuring that there is a period when none of the crop is grown. A good example of this is the control of planting date of the winter wheat crops in Alberta to avoid overlap with the previous spring- or winter-sown crop (Fig. 16.5). This procedure, together with elimination of volunteer wheat and barley plants and grass hosts of WSMV before the new win-

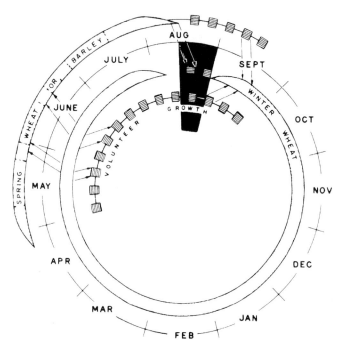

Fig. 16.5 Wheat streak mosaic disease cycle. Preventing the infection of winter wheat in the autumn is the key to controlling this disease in southern Alberta. Dark area, period during which effective control can normally be achieved; broken hatched bands, problems presented by volunteer seedlings, early-seeded winter wheat and/or late-maturing spring wheat or barley; arrows, transfer of virus by wind-blown mites. (Diagram courtesy of T.G. Atkinson.) From Matthews (1991).

ter crop emerges, can give good control in most seasons.

A break during the year where no susceptible plants are grown has proved effective in the control of certain other viruses with limited host ranges; even though they have efficient air-borne vectors. Control measures of this sort may be difficult to implement in developing countries where a major food plant is traditionally grown in an overlapping succession. Rice is the major example. The increased use of irrigation in the tropics and protected cropping in cold temperate regions limits the options available to growers for breaking the infection cycle.

2. Changed planting dates

The effect of infection on yield is usually much greater when young plants are infected. Furthermore, older plants may be more resistant to infection and virus moves more slowly through them. Thus, with viruses that have an air-borne vector, the choice of sowing or planting date may influence the time and amount of infection. The best time to sow will depend on the time of migration of the vector. If it migrates early, late sowing may be advisable. If it is a late migration, early sowing may allow the plants to become quite large and probably less susceptible before they are infected. An example of the need for an early sowing date is given in Fig. 16.6.

To reduce the incidence of rice tungro disease, which is transmitted by the rice green leafhopper, in irrigated rice grown under highly intensive conditions, farmers are encouraged to plant synchronously, to allow a break between crops and to avoid very late planting. In Indonesia they are advised to adopt recommended planting dates so that the plants are not exposed to infection when they are most vulnerable to incoming vectors (Manwan *et al.*, 1985; Sama *et al.*, 1991).

For any particular crop, the effectiveness of changed planting or harvesting dates in minimizing virus infection has to be considered in relation to other economic factors. Thus, Broadbent *et al.* (1957) found that potatoes planted early and lifted early had reduced virus infection, but a quite uneconomic reduction in yield also resulted.

3. Plant spacing

The fact that a higher percentage of more closely spaced plants tend to escape infection than widely spaced ones was discussed in Chapter 12 (Section IV.A.3.e). The practical effects of planting density on incidence of a virus and its aphid vectors and on plant yield are well illustrated from the work of Hull (1964) and A'Brook (1964, 1968) on groundnuts and rosette virus disease. A'Brook tested a wide range of planting densities over several seasons. Aphid densities were higher over well-spaced plants. Figure 16.7 shows the marked reduction in rosette infection with higher plant densities (number of infections were transformed to allow for multiple infections).

Although larger populations decreased rosette incidence, plant competition tended to decrease yield with the very high densities, and seed costs were greatly increased. The objective is to use a planting rate that will achieve complete ground cover as soon as possible without reducing yield due to competition.

Grain yields were significantly higher for some rice varieties planted in a close spacing, which reduced the incidence of rice tungro disease (Shukla and Anjaneyulu, 1981). Incidence of yellows in sugar beet was also reduced by increasing plant density (Johnstone *et al.*, 1982). Thus, the phenomenon may be a fairly general one. For possible explanations see Chapter 12 (Section IV.A.3.e).

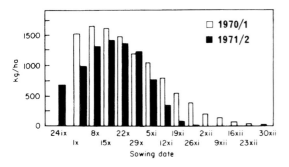

Fig. 16.6 Effect of sowing date on yield of field beans (*Vicia faba*) over two seasons in the Sudan. Losses were due to Sudanese broad bean mosaic virus transmitted by aphids. Plots sown after the end of October suffered increasingly severe losses. From Abu Salih *et al.* (1973), with permission.

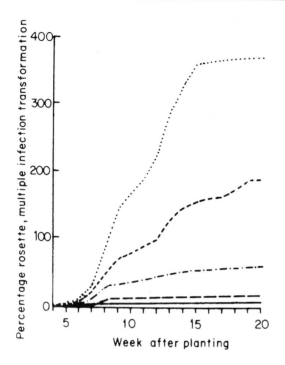

Fig. 16.7 Effect of planting density on the percentage of groundnut plants infected with rosette virus. Plants per acre: ⋯⋯, 9600; – –, 19 050; –·–·, 38 550; —, 78 750; —, 160 200. From A'Brook (1964). For derivation of multiple-infection transformation, see Gregory (1948), with permission.

4. Destruction of aerial parts of the plant

To limit virus spread late in the season, some certification schemes for virus-free vegetative stocks require the crop to be harvested before a certain date. This applies to seed potatoes in Holland, where lifting of the crop or killing of the haulms is required before a date determined each season from aphid trapping data (de Bokx, 1972).

Fodder-beet and mangolds are stored over-winter in clamps to supply animal food. The retention of these into the spring when the roots start resprouting can cause severe infections of BMYV and BYV (Smith, 1986).

5. Isolation of plantings

Where land availability and other factors permit, isolation of plantings from a large source of aphid-borne infection might give a useful reduction in disease incidence. Thus, isolation of beet fields from a large source of beets

infected with two viruses markedly reduced infection (Fig. 16.8). Production of virus-free seed potatoes is frequently carried out in areas that are well separated from crops being grown for food. Distances may be controlled by legislation, and the planting of home garden potatoes forbidden within a prescribed area.

6. Prevention of long-distance spread

Most agriculturally advanced countries have regulations controlling the entry of plant material to prevent the entry of diseases and pests not already present. Many countries now have regulations aimed at excluding specific viruses and their vectors, sometimes from specific countries or areas. The setting up of quarantine regulations and providing effective means for administering them is a complex problem. Economic and political factors frequently have to be considered. Quarantine measures may be well worthwhile with certain viruses, such as those transmitted through seed, or in dormant vegetative parts such as fruit trees and bud wood.

There is good evidence indicating that infected rootstocks were important in the worldwide distribution of grapevine viruses (Luhn and Goheen, 1970). Similarly, the spread

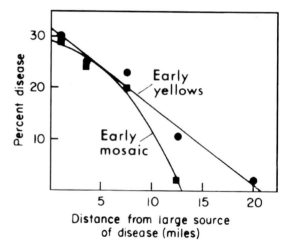

Fig. 16.8 Effect of distance from a large source of infection with two aphid-borne viruses on the percentage of sugar beet plants infected early in the season. The large source was about 30 000 acres of infected beets in the Sacramento Valley, California. From Shepherd and Hills (1970), with kind permission of the copyright holder, © The American Phytopathological Society.

of CTV worldwide is associated with infected budwood and exacerbated by efficient transmission by its aphid vector (Fig. 16.9).

Kahn (1976) suggested the use of aseptic plantlet culture for the transfer of vegetatively propagated material between countries. Co-operation between importing and exporting countries or areas may greatly improve the effectiveness of quarantine regulations. Quarantine in relation to plant health is reviewed in Hewitt and Chairappa (1977), Khan (1989), Foster and Hadidi (1998) and Waterworth (1998), and there are guidelines produced by FAO/IPGRI for the safe movement of germ plasm (see Diekmann and Putter, 1996).

The value of quarantine regulations will depend to a significant degree on the previous history of plant movements in a region. For example, active exchanges of ornamental plants between the countries of Europe has been going on for a long period, leading to an already fairly uniform geographical distribution of viruses infecting this type of plant (Lovisolo, 1985). On the other hand, the European Plant Protection Organization found it worthwhile to set up quarantine regulations against fruit tree viruses not already recorded in Europe (Rønde Kristensen, 1983; Krczal, 1998).

It is difficult to obtain any objective assessment as to how effective quarantine regulations have been in limiting long-distance spread of viruses. Because of its long-standing geographical isolation and the fact that almost all crop species have been imported, it is almost certain that all the plant viruses recorded in New Zealand have arrived from other

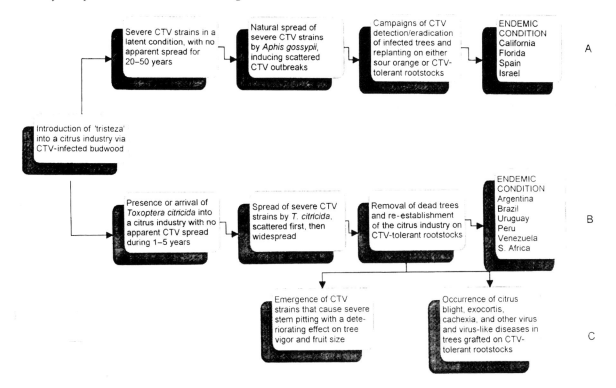

Fig. 16.9 Introduction of CTV and epidemics in several parts of the world. **(A)** CTV epidemics in California, Florida, Spain and Israel with the melon aphid *Aphis gossypii* as primary vector. **(B)** CTV epidemics in Argentina, Brazil, Uruguay, etc. after the arrival and co-localization of the brown citrus aphid, *Toxoptera citricida*. In both cases, CTV was present in a latent condition in CTV-infected propagative material brought from abroad. The simultaneous presence of CTV strains and *T. citricida* notably shortens the time of occurrence of CTV epidemics. **(C)** After 10–20 years of CTV epidemics and establishment of *T. citricida*, the citrus industries contend with the occurrence of severe CTV stem-pitting strains, citrus blight, and other virus and virus-like diseases. From Rocha-Peña *et al.* (1995), with kind permission of the copyright holder, © The American Phytopathological Society.

countries within the past 200 years. Viruses were first listed in New Zealand quarantine regulations in 1952, and by 1990 about 130 viruses had been found in that country (Matthews, 1991). One example is the introduction of RgMV, which is thought to have been introduced to New Zealand, together with its mite vector, on perennial ryegrass vegetative material (Guy et al., 1998). Studies on the coat protein of the virus suggest that there are two strains present, one in the south and one in the north, indicating two separate introductions.

There is also good evidence in particular instances that quarantine measures can limit further long-distance dispersal of viruses. For example, 37 of 61 importations of cuttings of *Datura* species from South America grown under quarantine in the United States were found to be infected with a range of viruses (Kahn and Monroe, 1970). South American potato viruses have been found in potatoes in quarantine in Europe. An effective quarantine system requires an effective technological infrastructure that is capable of detecting viruses under a variety of circumstances. Furthermore, quarantine must be concerned not only with important crop species but also with species, unimportant in themselves, that may harbor viruses infecting major crops. However, natural spread of some viruses over very long distances by invertebrate vectors may negate the effects of quarantine measures.

In spite of many countries having regulations designed to prevent the entry of damaging viruses, they can spread internationally very rapidly. A good example is the rhizomania disease of sugar beet, which was first recognized in the Po Valley in northern Italy in 1952. Table 16.1 plots the spread of this disease to many of the major sugar beet-growing countries (Asher, 1999). Attempts to prevent viruses spreading into a country may involve more than regulations controlling the importation of particular plant species. For example, since rhizomania reached the Netherlands in 1983, close surveys of British sugar beet crops have been made, especially those in areas closest to mainland Europe. BNYVV was first found in 1987 in a beet crop in Suffolk (Hill and Torrance, 1989). Steps were immediately taken in an attempt to contain the disease. The crop was destroyed, the land was sown in pasture, and strict hygiene measures were imposed. However, in 1989 when some 3000 sugar beet crops were surveyed, two further outbreaks of rhizomania, one near Norwich and one near King's Lynn, were found. These outbreaks appear to be separate from that of 1987 in Suffolk and since then more than 1600 hectares have been found infected and further sugar beet cropping on them has been prohibited (M. Asher, personal communication). In addition, rhizomania was found for the first time in the western hemisphere in the early 1980s (Table 16.1). Thus, the prospects for preventing worldwide spread of this disease do not appear to be good.

III. CONTROL OR AVOIDANCE OF VECTORS

Before control of virus spread by vectors can be attempted, it is necessary to identify the vector. This information has sometimes been difficult

TABLE 16.1 First records of rhizomania in different countries

1950s and 1960s		1970s		1980s and 1990s	
Italy	1952	Yugoslavia	1971	Hungary	1982
Japan	1965	Greece	1972	USA	1983
		France	1973	Switzerland	1983
		Germany	1974	Bulgaria	1983
		Czech. Republic	1978	Netherlands	1983
		China	1978	Belgium	1984
		Austria	1979	UK	1987
		Romania	1979	Sweden	1997
		USSR	1979		

From Asher (1999), with permission.

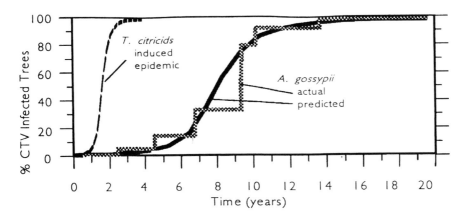

Fig. 16.10 Comparative increase of CTV infection in field situations when vectored by the brown citrus aphid (*Toxoptera citricida*), an efficient vector, and by the melon aphid (*Aphis gossypii*), a less efficient vector. Data for the brown citrus aphid were taken from test plots in Costa Rica and the Dominican Republic. Data for the melon aphid were taken from surveys and experimental plots in Spain, Florida and California. Initial infection levels were less than 1%. Note the 'stair-step' progression in infection with the melon aphid, which is believed to correspond to periodic heavy aphid migrations. From Garnsey *et al.* (1998), with kind permission of the copyright holder, © The American Phytopathological Society.

to obtain. Not uncommonly, it is an occasional visitor rather than a regular colonizer that is the main or even the only vector of a virus. Furthermore, some aphid species are more efficient vectors than others. For instance the brown citrus aphid (*Toxoptera citricida*) is a much more efficient vector of CTV than is the melon aphid (*Aphis gossypii*) (Fig. 16.10).

Control or avoidance of invertebrate or fungal vectors is of prime importance for the limitation of crop damage by viruses that have such vectors. For this reason more than one method for control may be used simultaneously for a particular crop, location, virus and vector.

A. Air-borne vectors

1. Insecticides (reviewed by Satapathy, 1998; Perring *et al.*, 1999)

A wide range of insecticides is available for the control of insect pests on plants. To prevent an insect from causing direct damage to a crop, it is necessary only to reduce the population below a damaging level. Control of insect vectors to prevent infection by viruses is a much more difficult problem, as relatively few winged individuals may cause substantial spread of virus. Contact insecticides would be expected to be of little use unless they were applied very frequently. Persistent insecticides,

especially those that move systemically through the plant, offer more hope for virus control. Viruses are often brought into crops by winged aphids, and these may infect a plant during their first feeding, before any insecticide can kill them. When the virus is non-persistent, the incoming aphid when feeding rapidly loses infectivity anyway, so that killing it with insecticide will not make much difference to infection of the crop from outside. On the other hand, an aphid bringing in a persistent virus is normally able to infect many plants, so that killing it on the first plant will reduce spread.

As far as subsequent spread within the crop is concerned, similar factors should operate. Spread of a virus that is non-persistent should not be reduced as much by insecticide treatment as a persistent virus where the insect requires a fairly long feed on an infected plant. Thus, spread of the persistent PLRV in potato crops was substantially reduced by appropriate application of insecticides but spread of the non-persistent PVY was not (Burt *et al.*, 1964; Webley and Stone, 1972).

Burt *et al.* (1964) used systemic insecticides, applied as sprays at different times during the season, to estimate the stage at which most spread of PLRV occurred in potatoes in England. Their results emphasized the importance of spread early in the season by winged aphids and the need for plants to be made lethal for aphids as early as possible. Like PVY, other non-persistent viruses, such as LMV, have not been controlled in any useful way by insecticide treatments.

A screen or 'trap crop' of plants sprayed with a systemic insecticide may reduce virus infection in plants growing within the screened field (Gay et al., 1973). Use of inappropriate insecticides may cause an increase in virus infection, either by disturbing the aphids present in the crop or by destruction of predators (e.g. Broadbent et al., 1963).

Spraying as a farming operation has certain disadvantages. It is an extra operation to be performed, tractor damage to the crop occurs, and spraying may not be practicable at the time required. Drifting spray can also lead to damage in other crops. Persistent systemic insecticides applied in granular form at time of planting overcome many of these difficulties. With a crop such as potatoes, granules can be metered into the furrow through applicators attached to the planting machines. For example, various systemic granular pesticides such as Aldicarb and Thiofanox have given control of aphids and PLRV in potato crops, applied in furrow at planting, or as side bands some weeks later (Woodford et al., 1988). The side-band treatments were generally less effective than those in furrow.

The photostable synthetic pyrethroid PP321 was effective in field trials in the control of BYDV in winter barley, of rice tungro disease in rice, and of the non-persistent PVY in potatoes (Perrin and Gibson, 1985), in contrast to the preceding experiments. This same material was effective in controlling BWYV in winter oilseed rape (Walsh, 1986).

The effectiveness of systemic insecticide treatment in controlling the non-persistent PVY in tobacco varied with season, probably depending on climatic factors that affect the numbers and movement of the aphid vector (Pirone et al., 1988).

With leafhopper vectors, the speed at which they are killed after feeding on a treated plant may be an important factor for virus control. For example, several synthetic pyrethroids that killed the vector Nephotettix virescens within 7 minutes prevented spread of the rice tungro disease (Anjaneyulu and Bhaktavatsalam, 1987). Slower-acting chemicals were ineffective in reducing virus incidence.

From the point of view of the economics of insecticide use, it may be important to forecast whether applications are necessary and, if so, to define optimal times for treatment. For example, in China two crops of rice followed by winter barley and wheat give the planthopper Laodelphax striatella a year-round succession of hosts. The best timing for the application of insecticides has been found to be at the winter seedling stage of barley and wheat, and at the early paddy stage of the second crop of rice (Yi-Li et al., 1981). In India, where an overlapping cropping pattern is also common, application of systemic insecticides under rice nursery conditions appears to be the most effective for the control of tungro disease (Shukla and Anjaneyulu, 1982).

As discussed in Section III.A.1, disease forecasting data can be an important factor in the economic use of insecticides. Sometimes a long-term program of insecticide use aimed primarily at one group of viruses will help in the control of another virus. Thus, the well-timed use of insecticides in beet crops in England, aimed mainly at reducing or delaying the incidence of yellows diseases (BYV and BMYV), has also been a major factor in the decline in the importance of BtMV in this crop (Heathcote, 1978). A warning scheme to spray against the vectors of beet yellows viruses was initiated in the UK in 1959 and is based on monitoring populations of aphids in crops from May until early July (reviewed in Dewar and Smith, 1999).

As well as the problems described above, there may be other adverse biological and economic consequences related to the use of insecticides. These include: (1) development of resistance by the target insect to the insecticide (see Georghiou and Saito, 1983); (2) resurgence of the pest once the insecticide activity has worn off; and (3) possible effects on humans and other animals in the food chain, leading to the banning of effective chemicals such as some of those listed above.

In spite of these problems, the use of insecticides for virus control is widespread. This is because, among other reasons: it is the way that growers control pests and, thus, they think it will control viruses; there are few alternative strategies; it is difficult to predict epidemics and, thus, prophylactic applications are used; and the cost of materials is low relative to the total production costs.

2. Oil sprays (reviewed by Perring et al., 1999)

Bradley (1956) and Bradley et al. (1962) showed that when aphids carrying PVY probed a membrane containing oil their subsequent ability to transmit the virus was greatly reduced. These observations led to considerable interest in the possibility of using oil sprays in the field to control viruses spread by aphids. Compared with synthetic insecticides, oil sprays have considerable appeal because of their lack of toxicity for humans and other animals. However, various limitations have emerged and their commercial use is not widespread.

In some reports, oil sprays have given useful results in field trials against a range of non-persistent viruses (e.g. Hein, 1971) and also BYV, which is semi-persistent. However, oil does not prevent the spread of persistent viruses. Other limitations involve possible plant toxicity, volatility or viscosity of the oil (de Wijs, 1980), adequate coverage of the leaves, and removal of the oil cover by rain or irrigation water, and effects on the marketable quality of the food produced.

Mineral oil sprays have proved useful in controlling viruses in bulb crops such as lilies and hyacinths in the Netherlands (Asjes, 1978, 1980). Similar sprays, used routinely on potato seed crops in Brittany, have greatly reduced the spread of PVY (Kerlan et al., 1987). Promising results have been obtained using oil sprays on tomatoes in the Sudan to limit the spread of a whitefly-transmitted geminivirus (Yassin, 1983). On the other hand, no control of several non-persistent viruses was obtained with oil sprays in England (Walkey and Dance, 1979), or of MDMV in sweet corn in Ohio (Szatmari-Goodmann and Nault, 1983). The combined use of mineral oils and pyrethroids gave better control in trials with PVY and BtMV than either component alone (Gibson and Rice, 1986).

Oil sprays do not appear to affect significantly the susceptibility of the plant, aphid behavior or virus infectivity. When aphids probe leaves sprayed with oil, the oil spreads readily over the whole length of the stylets. This observation and various experiments summarized by Vanderveken (1977) and Wang and Pirone (1996) support the idea that oil alters the surface structure or charge on the stylets, thus limiting the ability to adsorb (or release) virus particles. In the light of the wide range of non-persistently and semi-persistently transmitted viruses that oil interferes with, it is likely that the mechanism is non-specific and is not targeted to any specific virus–vector interaction.

The normal feeding behavior of the leafhopper *Nephotettix virescens* was disrupted on rice leaves sprayed with oil. There was reduced phloem feeding and increased xylem feeding, with restless behavior, repeated probing and profuse salivation. These effects may explain the reduced survival time of the insect and the decrease in virus transmission (Saxena and Khan, 1985).

The mechanism by which the application of oils controls virus spread is unknown. There are two hypotheses: (1) that it affects the acquisition or inoculation of the virus by the vector; and (2) that it affects the infection process of the virus.

3. Pheromone derivatives (reviewed by Perring et al., 1999)

The process of virus transmission is influenced by a variety of insect behaviors, such as leaving and alighting on plants, probing and feeding, interactions with adjacent insects and response to alarms. These behaviors are influenced by a wide range of chemicals and it is suggested that manipulating them may be a way of effecting virus control (Irwin and Nault, 1996).

Derivatives prepared from the pheromone (E)-β-farnesene and related compounds interfered with the transmission of PVY by *Myzus persicae* in glasshouse experiments (Gibson et al., 1984). It is possible that these substances act in a manner similar to that of mineral oils, because they possess aliphatic carbon chains. While new compounds continue to be developed (Dawson et al., 1988), this type of vector control is still very much at the experimental stage.

4. Non-chemical barriers against infestation

Several kinds of possible barriers to, or repellents against, vector movement into a crop have been investigated.

A tall cover crop will sometimes protect an undersown crop from insect-borne viruses. For example, cucurbits are sometimes grown intermixed with maize. Broadbent (1957) found that surrounding cauliflower seedbeds with quite narrow strips of barley (about three rows 0.3 m apart) could reduce virus incidence in seedlings to about one-fifth. Barley is not attacked by crucifer viruses. Many incoming aphids were assumed to land on the barrier crops, feed briefly, and either stay there or fly off. If they then land on the *Brassica* crop they may have lost any non-persistent virus they were carrying during probes on the barrier crop.

The reported action against aphids of aluminum strips laid on the ground has been tested for several crops. As the aphids come in to land, the reflected UV light is thought to act as a repellent. Reflective aluminum polythene mulches reduced the incidence of WMV and increased yields of cucurbits in Western Australia under conditions where both insecticide and oil sprays proved ineffective (McLean *et al.*, 1982). However, these reflective mulches are expensive, and where they have come into regular use difficulties in disposal at the end of the season may occur, at least when disposal by burning is forbidden (Nameth *et al.*, 1986).

A strip of sticky yellow polythene, 0.5 m wide and 0.7 m above the soil, surrounding the trial plots reduced the incidence of aphid-transmitted viruses in peppers (Cohen and Marco, 1973). A yellow polythene mulch significantly delayed the appearance of a yellow vein mosaic virus in okra in India (Khan and Mukhopadhyay, 1985). Nets spread above the crop may also reduce the winged aphid population and virus infection, while allowing normal plant growth. Coarse white nets may be the most effective under some conditions (Cohen, 1981). However, most of the findings noted in this section have not moved beyond the experimental stage into commercial practice.

5. Plant resistance to vectors

There has been substantially increased interest in breeding crops for resistance to insect pests as an alternative to the use of pesticide chemicals. This has been due to various factors, including emergence of resistance to insecticides in insects, the costs of developing new pesticides, and increasing concern regarding environmental hazards and the effects on natural enemies. Along with these developments, there has been increased activity in breeding for resistance to invertebrates that are virus vectors. Some virus vectors are not pests in their own right but others, especially leafhoppers and planthoppers, may be severe pests. In this situation there is a double benefit in achieving a resistant cultivar, sometimes with striking improvements in performance. The subject has been reviewed by Jones (1987; 1998), and for luteoviruses by Barker and Waterhouse (1999).

Sources of resistance have been found among most of the air-borne vector groups. Some examples are given in Table 16.2.

The basis for resistance to the vectors is not always clearly understood, but some factors have been defined. In general terms, there are two kinds of resistance relevant to the control of vectors. First, *non-preference* involves an adverse effect on vector behavior, resulting in decreased colonization, and, second, antibiosis involves an adverse effect on vector growth, reproduction and survival after colonization has occurred. These two kinds of factor may not always be readily distinguished. Some specific mechanisms for resistance are: (1) sticky material exuded by glandular trichomes such as those in tomato (Berlinger and Dahan, 1987); (2) heavy leaf pubescence in soybean (Gunasinghe *et al.*, 1988); (3) A-type hairs on *Solanum betrhaultii* which, when ruptured, entrap aphids with their contents, and B-type hairs on the same host, which entangle aphids making them struggle more and so rupture more A-type hairs (Tingey and Laubengayer, 1981); (4) inability of the vector to find the phloem in *Agropyron* species (Shukle *et al.*, 1987) — although this effect was not operative with an aphid vector of BYDV in barley (Ullman *et al.*, 1988); and (5) interference with the ability of the vector to locate the host plant.

TABLE 16.2 Examples of plants with resistance to air-borne vectors that have been associated with a decreased incidence of virus infection

Vector	Crop	Virus	Reference
Aphids			
Aphis gossypii	Musk melon	CMV	Lecoq *et al.* (1981)
Myzus persicae	Potato	PLRV	Rizvi and Raman (1983)
Rhopalosiphon maidis	Soybean	SMV	Gunasinghe *et al.* (1988)
Leafhopper			
Nephotettix virescens	Rice	Rice tungro viruses	Hibino *et al.* (1987)
Planthopper			
Niloparvata lugens	Rice	RRSV	Parejarlarn *et al.* (1984)
Whitefly			
Bemisia tabaci	Tomato	TYLCV	Berlinger and Dahan (1987)
Thrips			
Frankiniella schiltzei	Groundnut	TSWV	Amin (1985)
Mites			
Aceria tulipae	Wheat	WSMV	Martin *et al.* (1984)

For example, in cucurbits with silvery leaves there was a delay of several weeks in the development of 100% infection in the field with CMV and ClYVV (Davis and Shifriss, 1983). This effect may be due to aphids visiting plants with silvery leaves less frequently because of their different light-reflecting properties.

Combining resistance to a vector with some other control measure may sometimes be useful. For example, in field trials, rice tungro disease was effectively controlled by a combination of insecticide application and moderate resistance of the rice cultivar to the leafhopper vectors (Heinrichs *et al.*, 1986). Sprays would be unnecessary with fully resistant cultivars.

There may be various limitations on the use of vector-resistant cultivars. Sometimes such resistance provides no protection against viruses. For example, resistance to aphid infestation in cowpea did not provide any protection against CABMV (Atiri *et al.*, 1984). In addition, if a particular virus has several vector species, or if the crop is subject to infection with several viruses, breeding effective resistance against all the possibilities may not be practicable, unless a non-specific mechanism is used (e.g. tomentose leaves). Perhaps the most serious problem is the potential for new vector biotypes to emerge following widespread cultivation of a resistant cultivar, as may happen following the use of insecticides.

This difficulty is well illustrated by the recent history of the rice brown planthopper (*Nilaparvata lugens*) (BPH). With the advent of high-yielding rice varieties in south-east Asia in the 1960s and 1970s, the rice BPH and RGSV, which it transmits, became serious problems. Cultivars containing a dominant gene (*Bph1*) for resistance to the hopper were released about 1974. Within about 3 years, resistance-breaking populations of the hopper emerged. A new, recessive, resistance gene (*bph2*) was exploited in cultivars released between 1975 and 1983. They were grown successfully for a few years until a new hopper biotype emerged that overcame the resistance (Thresh, 1988a, 1989a,b). In an experiment to study the adaptation of three colonies of *N. lugens* to rice cultivars containing different resistance genes, Alam and Cohen (1998) showed that the *bhp1* and *bhp3* resistance genes were overcome more readily by colonies that had been exposed for about 10 years to those genes. However, cultivar IR64 which contained *bph1* and some minor resistance genes showed a greater durability of resistance than other cultivars. DNA markers to BPH resistance genes are being mapped to the rice genome and are being used in breeding programs (for a review see Yencho *et al.*, 2000).

In spite of these difficulties, and the problems associated with the identification of plants with resistance to vectors, it seems certain that substantial efforts will continue to be made to improve and extend the range of crop cultivars with resistance to virus vectors. A combination of resistance to the vector and to the virus will frequently be the goal.

6. Control by predators or parasites

Parasites and predators undoubtedly play a major role in limiting the population growth of aphids and other insects. As well at using integrated pest management strategies such as selection of variety, planting date, and the planned and controlled use of insecticides, a potential strategy for increasing the effectiveness of natural enemies is to incorporate natural enemy-enhancing traits into crop plants by breeding (reviewed by Bottrell and Barbosa, 1998).

This approach has not been used much for virus control. However, it does have some promise under certain circumstances, as exemplified by Figure 16.11, which shows the results of an experiment by Evans (1954) in Tanzania. He established small colonies of *Aphis craccivora* (one adult and five nymphs) on single plants in various parts of a field of groundnuts. He then followed the growth of the aphid colonies and the appearance of predators.

Under some circumstances, predators might play a part in limiting spread of a virus, but generally they will have little effect if they arrive after the early migratory aphids, which are so important for virus spread. In west Africa, introduction of fungal and hymenopter-

ous parasites of the mealybug vectors of cacao viruses to control CSSV were unsuccessful, even though the vector is relatively immobile (Thresh, 1958).

B. Soil-borne vectors

Most work on the control of viruses transmitted by nematodes and fungi has centered on the use of soil sterilization with chemicals. However, several factors make general and long-term success unlikely: (1) huge volumes of soil may have to be treated; (2) a mortality rate of 99.99% still leaves many viable vectors; and (3) use of some of the chemicals involved has been banned in certain countries, and such bans are likely to be extended. In any event, chemical control will be justified economically only for high-return crops, or crops that can remain in the ground for many years. However, some recent advances in nematode control procedures may be applicable to the control of viruses that they transmit and may be adaptable to the control of fungus-transmitted viruses.

1. Nematodes (reviewed by Barker and Koenning, 1998; Satapathy, 1998)

There are four basic strategies for nematode control:

1. Exclusion or avoidance
2. Reduction of the initial population density
3. Suppression of nematode reproduction
4. Restriction of the current or future crop damage.

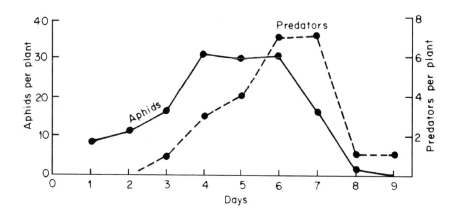

Fig. 16.11 Growth of initially small aphid colonies on groundnuts and their subsequent destruction by predators. From Evans (1954), with permission.

Within each strategy, there are various tactics that reduce the nematode population or minimize the interaction that the nematode has with the plant. As with other vectors, it is important to have a method for identification of the potential virus vector species and there have been considerable advances in the development of molecular diagnostics (see Hyman, 1996). In this section, I briefly describe these strategies and summarize the approaches already used to control nematode-borne viruses.

Exclusion or avoidance of nematodes is usually by quarantine. However, this can be difficult to implement as large quantities of root vegetables and rooted plants are traded worldwide.

There area several tactics to reduce the initial population density. These include cultural approaches such as use of clean planting stock, crop rotation with a break crop of a species that is not a host for the target nematode, weed host control, which is important as several nematode-transmitted viruses are also seed transmitted in weed species, and the use of plant species that are antagonistic to the target nematode species. Population densities can also be reduced by chemical nematicides, by biological tactics such as introducing biological agents antagonistic to nematodes and by organic amendments. The use of nematode-resistant crop varieties will reduce nematode populations. Recently, transgenic resistance has been developed against nematodes (see Lilley et al., 1999).

Nematode reproduction can be reduced by chemicals, organic amendments and by certain natural and transgenic resistance traits.

Tolerant cultivars will reduce crop damage due to nematode feeding but will not reduce the chances of virus infection.

Overall, the best way to minimize the impact of nematodes is a combination of the above strategies and tactics in an integrated manner to discourage the nematode and to encourage organisms and soil conditions antagonistic to them (see Stirling, 1999). The reduction of nematode populations should lead to a reduction of the chances of plant infection by the viruses that they may carry.

The main approach to controlling nematode-transmitted viruses has been by the use of nematicides, which fall into two groups: fumigant (e.g. methyl bromide) and non-fumigant (e.g. Aldicarb). In principle, this control method should be effective. Movement and dispersal of nematodes are generally slow, so that one treatment might be expected to remain effective for longer periods than with air-borne vectors. On the other hand, as pointed out by Sol (1963), infective nematodes may occur at considerable depths. Nematodes in soil samples taken up to 80–100 cm deep were able to infect plants with TRV. Thus, it is possible that fumigated soil could be re-infested by nematodes moving up from deep sites where they had escaped the effects of fumigation.

Harrison et al. (1963), in field trials at several sites in southern Britain, showed that the fumigants dichloropropane-dichloropropene (D-D) or methyl bromide at $1 \, kg/10 \, m^2$, applied in summer, killed over 99% of *Xiphinema diversicaudatum* in the soil. The treatments very largely prevented infection of strawberry crops with ArMV. The crops were planted after the fumigation and examined after 1–2 years. Both chemicals killed the nematode down to 70 cm, the greatest depth sampled. Virus incidence following fumigation was closely related to the numbers of surviving nematodes.

Murant and Taylor (1965) and Taylor and Murant (1968) showed that a single soil fumigation with D-D or pentachloronitrobenzene (quintozine) prevented infection of strawberries by TBRV and RpRSV transmitted by *Longidorus elongatus*. Quintozine was preferred because it had no marked effect on plant growth. Because *L. elongatus* has a wide host range among cultivated plants and weeds, and because it can survive for long periods in the soil without food, its elimination by crop rotation is most unlikely. Chemical control of the vector in the soil probably offers the best solution for viruses transmitted by this nematode.

At present, soil fumigation appears to be the only effective measure for the control of *Xiphinema index*, the vector of GFLV. If proper care is taken with the fumigation process, vineyards may be grown successfully for 15–20 years (Raski et al., 1983). With vines, it is very important to kill all nematodes before replanting. The process may be assisted by treating old vines with a herbicide so that deep roots, and the nematodes they harbor, are killed before fumigation treatment of the upper soil.

Little work has been done on seeking resistance to nematode vectors. However, Bouquet (1981) reported resistance to *X. index* and protection from grapevine fanleaf infection in *Vitis rotundifolia*. Furthermore, there are transgenic approaches to protecting plants against nematodes described above.

2. Fungi (reviewed by Walsh, 1998)

There are two agronomic situations in which fungally transmitted viruses are important: nutrient or aquatic systems and fields. The major control measures are chemical; they divide into three types: (1) surfactants, heavy metals and sometimes fungicides for use in nutrient or aquatic systems; (2) soil amendments and fungicides to control the fungal vector in the soil; and (3) soil partial sterilants or disinfectants to reduce the active and resting spore stages of fungal vectors in the field. Some examples of the three approaches are given in Table 16.3.

In general, attempts to control infection with viruses having fungal vectors by application of chemicals to the soil have usually not been successful. For example, various fungicides failed to control transmission of BaYMV to winter

barley by *Polymyxa graminis* (Proeseler and Kastirr, 1988). Several fungicides prevented infection of sugar beet by BNYVV in glasshouse trials, but were ineffective in the field, even at high rates of application (Hein, 1987). The zoospores of *Olpidium brassicae* were susceptible to several fungicides and surfactants in laboratory tests (Tomlinson and Faithfull, 1979). The chemical control of fungal vectors has been reviewed by Walsh (1998).

Some cultivars of wheat show field resistance to SBWMV. The mechanism for such resistance is not established, but it may involve, in part, resistance to the fungal vector *P. graminis* (Lapierre *et al.*, 1985). Culturing this vector is difficult, which probably accounts for the fact that little work has been done on the development of vector-resistant cereals, in spite of the important viruses this fungus transmits.

IV. PROTECTING THE PLANT FROM SYSTEMIC DISEASE

Even if sources of infection are available, and the vectors are active, there is a third kind of control measure available: protecting inoculated plants from developing systemic disease. There are essentially three approaches that have been used to protect plants: use of a mild strain of the virus (termed cross-protection or mild strain protection), the use of chemicals

TABLE 16.3 Examples of chemical treatments to control fungal vectors of plant viruses

Virus	Vector	Agent of control	Effective concentration	Reference
Nutrient or aquatic culture				
WYSV	*S. subterranea*	Zinc	> 0.05 ppm	Walsh (1992)
LBVV	*O. brassicae*	Bavastin	0.06 g/L	Tomlinson and Faithfull (1979)
LBVV and lettuce ring necrosis	*O. brassicae*	Ultraviolet light	Dependent on flow rate	Gharbi and Verhoyen (1993)
Soil amendments and fungicides				
PMTV	*S. subterranea*	Zinc sulfate	1320 kg/ha	Cooper *et al.* (1976)
TNV	*O. brassicae*	Captan	1 g/kg soil	Thomas (1973)
BNYVV	*P. betae*	Fluazinam	10–50 ppm	Uchino *et al.* (1993)
Soil treatments				
LBVV	*O. brassicae*	Mercuric chloride	0.05%	Grogan *et al.*, (1958)
LBVV	*O. brassicae*	Jet 5	2%	Walsh (1998)
WSSMV	*P. graminis*	Methyl bromide	1.5 ml/L	Slykhuis (1970)

and genetic protection (conventional resistance and transgenic resistance). The first two of these approaches are discussed in this section and genetic protection in the next two sections.

A. Mild strain protection (cross-protection) (reviewed by Urban *et al.*, 1990; Lecoq, 1998)

Infection of a plant with a strain of virus causing only mild disease symptoms (the protecting strain; also known as the mild, attenuated, hypovirulent or avirulent strain) may protect it from infection with severe strains (the challenging strain) (Chapter 17, Section II.C.4). Thus, plants might be purposely infected with a mild strain as a protective measure against severe disease. This was first reported by McKinney (1929), who observed that tobacco plants systemically infected by a 'green' strain of TMV were protected from infection by another strain that induced yellow mosaic symptoms.

While such a procedure could be worthwhile as an expedient in very difficult situations, it is not to be recommended as a general practice, for the following reasons:

1. So-called mild strains often reduce yield by about 5–10%.
2. The infected crop may act as a reservoir of virus from which other more sensitive species or varieties can become infected.
3. The dominant strain of virus may change to a more severe type in some plants.
4. Serious disease may result from mixed infection when an unrelated virus is introduced into the crop.
5. For annual crops, introduction of a mild strain is a labor-intensive procedure.
6. The genome of the mild strain may recombine with that of another virus, leading to the production of a new virus.

In spite of these difficulties, the procedure has been used successfully, at least for a time, with some crops. A suitable mild isolate should have the following properties:

- It should induce milder symptoms in all the cultivated hosts than isolates commonly encountered and should not alter the marketable properties of the crop products.
- It should give fully systemic infections and invade most, if not all, tissues.
- It should be genetically stable and not give rise to severe forms.
- It should not be easily disseminated by vectors to limit unintentional spread to other crops.
- It should provide protection against the widest possible range of strains of the challenging virus.
- The protective inoculum should be easy and inexpensive to produce, check for purity, provide to farmers and apply to the target crops.

Mild protecting strains are produced from naturally occurring variants, from random mutagenesis or from directed mutagenesis of severe strains.

Broadbent (1964) suggested that, because late infection of greenhouse tomatoes with ToMV often causes a severe reduction in quality of the fruit compared with early infection, growers who regularly suffer those losses should inoculate their plants at an early stage with a mild strain of the virus. Rast obtained a nitrous acid mutant of ToMV that was symptomless in tomato (Rast, 1975). Seedling inoculation with this strain was widely practiced for a time in some western European countries and in Canada, New Zealand and Japan. However, cultivars with resistance genes or with multiple genes for tolerance to ToMV have largely replaced seedling inoculation with the attenuated strain in commercial practice.

CTV provides the most successful example for the use of cross-protection. Worldwide, this is the most important virus in citrus orchards. In the 1920s, after its introduction to South America from South Africa, the virus virtually destroyed the citrus industry in many parts of Argentina, Brazil and Uruguay. The successful application of cross-protection by inoculation with mild CTV isolates in Brazil has been detailed by Costa and Muller (1980). The method has been particularly successful with Pera oranges, with more than eight million trees being planted in Brazil by 1980. Protection continues in most individual plants through successive clonal generations. However, in an 8-year assessment of the ability of four mild isolates to suppress severe CTV isolates in Valencia sweet orange on sour orange rootstock in Florida, about 75% of the mild-strain protected trees had severe symptoms compared with about 85% of the unprotected trees (Powell *et al.*, 1992). The use of the same isolates gave better protection of Ruby Red grapefruit on sour orange rootstock (Powell *et al.*, 1999). Thus, there are differences in the responses of the scion–rootstock combination, but it is also important to have a compatible mild strain. The search for improved attenuated strains of the virus continues (e.g. Muller and Costa, 1987; Roistacher *et al.*, 1987), and the technique is being adopted in other countries.

Other viruses and crops for which attempts are being made to develop effective cross-protection include PRSV in papaya (Yeh *et al.*, 1988; Gonsalves, 1998), TMV in pepper (Tanzi *et al.*, 1986) and ZYMV in courgette (Walkey *et al.*, 1992). In principle, the cacao swollen shoot disease of cacao in Ghana should be controllable to some degree by cross-protection with mild strains of the virus (Posnette and Todd, 1955; Hughes and Ollennu, 1994), but various difficulties have prevented its effective application (Fulton, 1986). In particular, the use of the technique was incompatible with the objective of treating all known outbreaks by removal of infected trees. This is no longer feasible, so that there is now scope for using mild strain protection in the worst affected areas of Ghana, where other control measures have been abandoned (Matthews, 1991).

In Chapter 10 (Section IV), I describe the recent developments in understanding host response to virus infection by post-transcriptional gene silencing. It is most likely that mild strain protection operates by the mild strain 'priming' this defense system so that it operates against the superinfecting severe strain. With this in mind, it is important that the sequence of the protecting mild strain is similar to that of the superinfecting severe strain. An analysis of sequence of CTV isolates from different sites and collected at different times showed that the mild strains used in Florida and Spain were very similar to a

diverse range of isolates (Albiach-Martí *et al.*, 2000a). This was unexpected as the first two CTV isolates to be sequenced had up to 60% sequence divergence.

However, it is possible that, in some instance of mild strain protection, other mechanisms, such as competition for replication sites, operate.

B. Satellite-mediated protection
(reviewed by Tien and Wu, 1991; Jacquemond and Tepfer, 1998)

Satellite viruses and RNAs are described in Chapter 14 (Section II) and, as far as potential biocontrol agents, fall into three categories: those that enhance the helper virus symptoms, those that have no effect and those that reduce the helper virus symptoms.. It is the latter that have potential as control agents. Most of the work has focused on the satellites of CMV.

Tien *et al.* (1987) obtained mild strains of CMV by adding selected satellite RNAs to a CMV isolate. This procedure has been tested, with increased yields, in many areas of China for the control of CMV in peppers. The extent of the protective effect depended on such factors as inoculation time, percentage of plants inoculated, the nature of virulent CMV strains already in the field, and variety of pepper.

Wu *et al.* (1989) compared the degree of protection obtained by pre-inoculating tobacco and pepper plants with a mild strain of CMV with or without satellite RNA. The presence of satellite RNA increased the protection obtained. Similar results were obtained in greenhouse experiments with CMV in tomato. In field trials, protection was maintained when the virus was introduced by aphids (Montasser *et al.*, 1991). This work has been taken a stage further by Gallitelli *et al.* (1991), who inoculated several hundred young tomato seedlings using several varieties, with CMV strain S carrying a non-necrogenic satellite called S-CARNA5. These were planted out at the seedling stage in spring on a farm in southern Italy, where a severe epidemic of the tomato necrosis disease was expected. The epidemic occurred, with 100% of plants being destroyed in some fields. In the field containing the inoculated plants, protection against necrosis was almost 100%, while 40% of uninoculated plants developed lethal necrosis. Fruit yields were about doubled in the protected plants. Satellite protection against CMV infection of tomato has also been used in Japan (Sayama *et al.*, 1993) and China (Tien and Wu, 1991), and against CMV in pepper and melon plants in the USA (Montasser *et al.*, 1998).

A model suggested that inoculation with CMV containing a mild satellite RNA prior to challenge with a severe satellite isolate with or without CMV interfered with the replication and symptom production of the severe strain (Smith *et al.*, 1992).

However, there has been concern over the durability of using satellites as biocontrol agents. There is a wide range of necrogenic and other virulent strains of satCMV (Jacquemond and Tepfer, 1998). Passage of a benign satellite of CMV through *Nicotiana tabacum* led to the satellite rapidly mutating to a pathogenic form (Palukaitis and Roossinck, 1996) and mutations of a single or a few bases can change a non-necrogenic variant to a necrogenic one (Fig. 16.12). Necrogenic variants of the CMV satellite have a greater virulence than non-necrogenic variants (Escriu *et al.*, 2000), but, as they depress the accumulation of the helper virus more than do non-necrogenic variants, the necrogenic variants are not so efficiently aphid transmitted.

C. Antiviral chemicals

1. Chemical control of viruses (reviewed by Hansen, 1989)

Considerable effort has gone into a search for inhibitors of virus infection and multiplication that could be used to give direct protection to a crop against virus infection in the way that fungicides protect against fungi.

There has been no successful control on a commercial scale by the application of antiviral chemicals. The major difficulties are:

1. An effective compound must inhibit virus infection and multiplication without damaging the plant. This is the first, and major, problem. Virus replication is so intimately bound up with cell processes that any compound blocking virus replication is likely to have damaging effects on the host.
2. An effective antiviral compound would need to move systemically through the plant if it were to prevent virus infection by invertebrate vectors.
3. A compound acting systemically would need to retain its activity for a reasonable period. Frequent protective treatments would be impracticable. Many compounds that have some antiviral activity are inactivated in the plant after a time.

```
Y Sat-RNA(a)      CUAAGGCUUAUGCUAUGCUGAUCUCCGUGAAUGUCUAUACAUUCCUCUACAGGACCC      necrogenic
                             ⇓                                                     ⇓
                             U                                               ameliorative


Y Sat-RNA(b)      CUAAGGCUUAUGCUAUGCUGAUCUCCGUGAAUGUCUAUACAUUCCUCUACAGGACCC      necrogenic
                          ⇓    ⇓ ⇓                                                ⇓
                          A    G U                                           ameliorative


W11 Sat-RNA(c)    CUUAGACUUAGGUUAUGCUGAUCUCCGUGAAUGUCUACACAUUCCUCUACAGGACCC      ameliorative
                          ⇓    ⇓ ⇓                                                ⇓
                          G    U C                                            necrogenic


R Sat-RNA(d)      CUAAGGCUUAUGCUACGCUGAUCUCCGUGAAUGUCUA.UCAUUCCUC.ACAGGACCC      ameliorative
                             ⇓                                                     ⇓
                             U                                               necrogenic
```

Fig. 16.12 Alignment of the 55 3′-terminal residues of the satellite RNA variants mutated from a necrogenic form towards a non-necrogenic one, or *vice versa*. Arrows indicate the positions found to be determinant for necrogenicity. From Jaquemond and Tepfer (1998), with kind permission of the copyright holder, © The American Phytopathological Society.

4. For most crops and viruses, the compound would need to be able to be produced on a large scale at an economic price. This might not apply to certain relatively small-scale, high-value crops, such as greenhouse orchids.

5. For use with many crops, the compound would have to pass food and drug regulations. Many of the compounds that have been used experimentally would not be approved under such regulations.

Many substances isolated from plants and other organisms, as well as synthetic organic chemicals, have been tested for activity against plant viruses. Almost all the substances showing some inhibition of virus infectivity do so only if applied to the leaves before inoculation or very shortly afterwards. An example is a glucan preparation obtained from *Phytophthora megasperma f.s.p. glycinea*, which appears to inhibit infection by several viruses by a novel, but unknown mechanism (Kopp *et al.*, 1989). Work in this field has been reviewed by Matthews (1981), Tomlinson (1981), White and Antoniw (1983) and Verma *et al.* (1998).

Synthetic analogs of the purine and pyrimidine bases found in nucleic acids have been widely studied, and the search for inhibitory compounds of this type continues (Dawson and Boyd, 1987a). The substituted triazole, 4(5)-amino-1H-1,2,3-triazole-5(4)-carboxyamide, can be regarded as an aza analog of the substituted imidazole compound which is known to be a precursor of the purine ring in some systems. This compound showed some plant virus inhibitory activity but was less effective than 8-azaguanine. The riboside of this compound was suggested as a possible antiviral agent by Matthews (1953c). It was later found to have broad-spectrum activity in experimental animal virus systems and given the name virazole (Sidwell *et al.*, 1972). Virazole, also known as ribavirin, has been studied in a variety of plant–virus systems. For example, pretreatment of tobacco plants with the compound delayed or prevented systemic infection with TSWV (De Fazio *et al.*, 1980). Virazole reduced the concentration of CMV and AMV in cultured plant tissues. However, virus-free plants were obtained from meristem tip cultures whether or not virazole had been in the medium (Simpkins *et al.*, 1981). Other virus-inhibitory compounds have been included in culture media in attempts to improve the efficiency of meristem tip culture for obtaining virus-free plants. None has found widespread use.

2. Suppression of disease symptoms by chemicals

For a time, there was considerable interest in the use of certain systemic fungicides to suppress the symptoms of virus infection without necessarily having any effect on the amount of virus produced in the leaves. The fungicides concerned (e.g. benlate and bavistin) break down in aqueous solution to give methylbenzimidazole-2-yl-carbamate (MBC or carbendazim). It has been reported that these systemic fungicides possess cytokinin-like activity, although at a low level compared with kinetin. Application of MBC as a soil drench caused a substantial reduction in leaf symptoms caused by TMV in tobacco (Tomlinson *et al.*, 1976). However, this compound or others with similar effects have not found commercial application.

The severity of infection of tomato plants by ToMoV was reduced by treatment with plant growth-promoting rhizobacteria (PGPR) (Murphy *et al.*, 2000). It was suggested that the use of PGPR could become a component of an integrated program for management of this virus in tomato.

It should be noted that, in some cases, high applications of nitrogen fertilizers can suppress virus disease symptoms. However, virus concentrations are not diminished.

V. CONVENTIONAL RESISTANCE TO PLANT VIRUSES

A. Kinds of host response

The genetic makeup of the host plant has a profound influence on the outcome following inoculation with a particular virus. The kinds of host response are defined in Table 10.1 and molecular aspects are discussed in Chapter 10. Figure 16.13 illustrates the dramatic effect that a single mendelian gene can have on disease

Fig. 16.13 Reactions of two varieties of tobacco to a strain of TMV from tomato. **Left:** Var. White Burley showing typical systemic mosaic. **Center:** Var. Warnes, necrotic local lesions with no systemic spread of virus. **Right:** An F₁ hybrid (Warnes ♀ × White Burley ♂) showing necrotic local lesions and a severe systemic necrosis and stunting. From Matthews (1991).

response. Defining the response of a particular host species or cultivar to a particular virus must always be regarded as provisional. A new mutant of the virus may develop in the stock culture, or a new strain of the virus may be found in nature that causes a different response in the plant. This applies particularly to non-host immunity. For example, for many decades the potato seedling USDA 41956 was considered immune to PVX. However, in due course a strain of this virus was discovered that can infect this genotype (Moreira *et al.*, 1978).

Nevertheless, immunity with long-term durability must occur frequently in nature. It has been suggested with respect to cellular parasites that long-lived plants such as trees, and species naturally dominant over large areas, such as prairie grasses and aggressive weeds, must have developed greater resistance to lethal parasites than other species. Similar considerations may apply as far as virus infection is concerned, but little relevant experimental evidence is available.

It is probable that many of the plants described in the past as immune to a particular virus were in fact infectible, but resistant and showing extreme hypersensitivity as defined in Table 10.1. The viruses concerned may have had cell-to-cell movement proteins that were defec-

tive in the particular plant, resulting in multiplication only in the initially infected cells. This phenomenon may be detectable only by special procedures (for example, see Fig. 9.4).

The following points concerning the effects of host genes on the plant's response to infection emerge from many different studies:

1. Both dominant and recessive mendelian genes may have effects. However, while most genes known to affect host responses are inherited in a mendelian manner, cytoplasmically transferred factors may sometimes be involved (Nagaich *et al.*, 1968).
2. There may or may not be a gene dose effect.
3. Genes at different loci may have similar effects.
4. The genetic background of the host may affect the activity of a resistance gene.
5. Genes may have their effect with all strains of a virus, or with only some.
6. Some genes influence the response to more than one virus.
7. Plant age and environmental conditions may interact strongly with host genotype to produce the final response.
8. Route of infection may affect the host response. Systemic necrosis may develop following introduction of a virus by grafting

into a high-resistant host that does not allow systemic spread of the same virus following mechanical inoculation (e.g. ToMV in resistant tomato lines; Stobbs and McNeill, 1980).

9. Resistance originally thought to be to the virus may be really to the vector (Dahal et al., 1990a,b).

As pointed out by Fraser (1988, 1992), there are three main types of resistance and immunity to a particular virus, considered from the point of view of the complexity of the host population involved: (1) *immunity* involves every individual of the species; little is known about the basis for immunity, but it is related to the question of the host range of viruses discussed in Chapter 3 (Section V); (2) *cultivar resistance* describes the situation where one or more cultivars or breeding lines within a species show resistance while others do not; and (3) *acquired or induced resistance* is present where resistance is conferred on otherwise susceptible individual plants following inoculation with a virus. This last phenomenon was discussed in Chapter 10 (Section III.K). Some authors have considered that immunity and cultivar resistance are based on quite different underlying mechanisms. However, studies with a bacterial pathogen in which only one pathogen gene was used show that, for this class of pathogen at least, the two phenomena have the same basis (Whalen et al., 1988).

B. Genetics of resistance to viruses

This section briefly outlines the kinds of genetics involved in cultivar resistance to virus infection that can be used for virus control. The possible mechanisms for such resistance are discussed in Chapter 10. Plant resistance to viruses has been reviewed by Ponz and Bruening (1986),

Fraser (1987a, 1988, 1992), Evered and Hamett (1987), Khetarpal et al. (1998) and Hammond (1998).

The summary in Table 16.4 shows that resistance to viruses in most crop virus combinations is controlled by a single dominant gene. However, this may merely reflect the fact that most resistant cultivars were developed in breeding programs aimed at the introduction of a single resistance gene. Furthermore, incomplete dominance may be a reflection of gene dosage or be due to environmental factors (Fischer and Rufty, 1993). There have not been many studies of the inheritance of resistance in wild species. Some specific examples of dominant, incompletely dominant, and apparently recessive genes for resistance are given in Table 16.5.

Sometimes response to virus infection is associated with resistance to some other kind of disease agent. Thus, the necrotic response of tobacco to infection by a strain of PVY may be a pleiotropic effect of the gene controlling resistance to a root-knot nematode (Rufty et al., 1983).

1. The gene-for-gene hypothesis

Gene-for-gene relationships are well known between host plant and fungal or bacterial pathogens. They have been established primarily based on genetic analyzes of both plants and pathogens. With these parasites, each allele in the host that confers resistance may be reflected in a complementary virulence locus in the parasite that can overcome the resistance. Virulence and avirulence genes and plant resistance are discussed in Chapter 10.

Viruses and virus strains may be described in relation to the various host responses defined in Table 10.1. Thus, if a particular plant species or cultivar shows immunity or resistance to a

TABLE 16.4 Summary of number of virus resistance genes reported

Resistance gene	Monogenic	Oligogenic or polygenic
Dominant	81	10
Recessive	43	20
Incompletely dominant	15	6
Nature unknown	–	4
Total number of resistance genes	139	40

From Khertarpal et al. (1998), with kind permission of the copyright holder, © The American Phytopathological Society.

TABLE 16.5. Examples of host genes for resistance to plant viruses

Gene	Host species	Virus	Virulent virus isolates known	Selected references[a]
Controlled by dominant genes				
N	*Nicotiana glutinosa*	TMV	No	Holmes (1929)
N'	*N. sylvestris*	TMV	Yes	Melchers *et al.* (1966)
Zym[a]	*Cucurbita moschata*	ZYMV	No[b]	Paris *et al.* (1988)
Tm-2	*Lycopersicon esculentum*	TMV	Yes	Pilowski *et al.* (1981)
Tm-2²	*L. esculentum*	TMV	Yes	Cirulli and Alexander (1969)
Nx, Nb	*Solanum tuberosum*	PVX	Yes	Jones (1982)
By, By-2	*Phaseolus vulgaris*	BYMV	Yes	Schroeder and Provvidenti (1968)
Rsv₁, Rsv₂	*Glycine max*	SbMV	Yes	Buzzell and Tu (1984)
Controlled by incompletely dominant gene				
Tm-1	*L. esculentum*	TMV	Yes	Fraser *et al.* (1980)
L¹, L, L3	*Capsicum* spp.	TMV	Yes	Berzal-Herranz *et al.* (1995), Dardick and Culver (1997), Dardick *et al.* (1999)
Two genes	*Hordeum vulgare*	BSMV	Yes	Sisler and Timian (1956)
Multiple genes	*Vigna sinensis*	SCPMV	?	Hobbs *et al.* (1987)
Apparently recessive genes				
By-3	*P. vulgaris*	BYMV	No	Provvidenti and Schroeder (1973)
Sw₂, Sw₃, Sw₄	*L. esculentum*	TSWV	Yes	Finlay (1953)

[a] Resistance considered to be conferred by three genes, *zym-1*, *zym-2* and *zym-3*, of which *zym-1* is essential and *zym-2* and *zym-3* lower the degree of susceptibility (Paris and Cohen, 2000), with permission.

virus, that virus is said to be non-pathogenic for that species or cultivar. A virus is pathogenic if it usually causes systemic disease in a species or cultivar. A gene for immunity or resistance introduced into such a species or cultivar may make the virus avirulent. However, a mutant strain may overcome the host gene resistance and it would then be virulent.

Gene-for-gene relationships have been proposed by some authors for virus–host interactions (e.g. Fraser, 1987a). A well-studied example of the gene-for-gene hypothesis applied to a plant virus and its host is the resistance of tomato to ToMV (Table 10.3). There are three resistance genes: *Tm-1*, *Tm-2* and *Tm-2²*. The virus has evolved variants that can overcome all three host resistance genes. The virulent virus strains are numbered according to the host genes they can overcome. For example, strain 0 is avirulent. Strain 2 overcomes gene *Tm-2*; strain *1.2²* overcomes genes 1 and 2². No virus strains are yet known that overcome host genes 2 plus 2² or 1, 2 plus 2.2². The host genotypes differ in their 'durability' in the field.

Thus, strain 1 isolates appeared within a year in commercial crops containing only the *Tm-1* gene. By contrast, most commercial ToMV-resistant tomato cultivars now contain the genes *Tm-1/Tm-1:Tm-2/Tm-2²* or *Tm-1/+:Tm-2/Tm-2²*, and these appear to be highly durable in their resistance to TMV (Fraser, 1985).

The relationships summarized in Table 10.3 bear a superficial similarity to the gene-for-gene relationships seen between bacterial and fungal pathogens and their hosts. Such pathogens contain large numbers of genes and could easily maintain or develop a suite of genes that either allow or overcome host resistance. By contrast, ToMV contains four genes, all with functions involved in virus replication or movement. The change to virulence by the virus must involve mutational events in one or more of the four genes or in the controlling elements of the viral genome. An example is provided by the work of Meshi *et al.* (1988). They examined the nucleotide sequence of a resistance breaking strain of ToMV, Ltal, which is able to replicate in tomatoes with the *Tm-1* gene (i.e. a strain of genotype 1 in Table 10.3). They found two base substitutions resulting in amino acid changes Gln979 → Glu and His984 → Tyr in the 130- and 180-kDa viral proteins. They demonstrated that these were indeed the changes responsible for resistance breaking by introducing these mutations singly or together into the parent strain. All three constructed mutants replicated in tomato protoplasts with the *Tm-1* gene, but the change His984 → Tyr did so to a greatly reduced extent.

Another resistance breaking strain of ToMV, Ltbl, is able to multiply in tomatoes with the *Tm-2* gene. Nucleotide sequence analysis revealed two changes in the 30-kDa protein compared with the parent ToMV (Cys66 → Phe and Glu133 → Lys) (Meshi *et al.*, 1989). Thus, the *Tm-2* resistance in tomato may involve some aspect of cell-to-cell movement.

Mutations in virus genotypes 2 and 2² of Table 10.3 involve nucleotide changes in the TMV gene coding for the movement protein, as has been found for the *ts* mutant LsI (see Chapter 9, Section II.D) (Nishiguchi and Motoyoshi, 1987). Thus, for viruses, virulence or avirulence appears to be controlled by single point mutations in genes essential for virus replication or movement, rather than some change in genes dedicated to maintaining or overcoming host resistance.

2. Mechanisms of host immunity and resistance (reviewed in Fraser, 1998)

Details of what is known of some molecular mechanisms of host immunity and resistance are described in Chapter 10. Here I discuss some aspects related to the control of plant viruses.

a. Resistance of cowpea to CPMV

Seedlings of the cowpea cultivar Arlington are resistant to CPMV, and isolated protoplasts are also resistant. This cultivar is therefore immune, as defined in Table 10.1. This immunity is governed by a single dominant gene as determined by crosses with the susceptible Black Eye variety. Ponz *et al.* (1988a) found three inhibitory activities in extracts of Arlington cowpea protoplasts that were at higher levels than those in Black Eye extracts. These were: (1) inhibitor(s) of the translation of CPMV RNAs; (2) proteinase(s) that degrades CPMV proteins; and (3) an inhibitor of proteolytic processing of a CPMV polyprotein. The proteinases were not specific for CPMV proteins and were not coinherited with the immunity to CPMV.

The inhibitor of polyprotein processing was specific for CPMV and had the co-inheritance expected for an agent mediating the immunity to CPMV. It can be reasonably concluded that the proteinase inhibitor is the host-coded gene product responsible for immunity to CPMV. This is the first such gene product identified for plant viruses and the first example of immunity for which a clear molecular mechanism has been established. The inheritance of the translation inhibitor activities was complex but one or more of these may play an accessory role in immunity. In another study, Ponz *et al.* (1988b) found that the Arlington cowpea line also had a single dominant gene for resistance to TRSV, but that this was quite distinct from the gene for immunity to CPMV.

b. Impaired systemic movement of the virus

In some cultivars that have a genotype conferring resistance to a particular virus, movement through the vascular tissue is affected in a manner that appears not to involve a virus-coded cell-to-cell movement protein in the way discussed in a subsequent section. For example, the spread of PLRV infection within the phloem of resistant potato cultivars appears to be impaired, leading to much less efficient acquisition of the virus by the aphid vector *Myzus persicae* (Barker and Harrison, 1986). Similarly, in corn (*Zea mays*) resistance to MDMV, the pattern of virus spread in inoculated leaves suggested that the plant inhibited the virus from moving through the vascular system (Lei and Agrios, 1986). The molecular mechanism for such host effects is unknown.

Expression of the TMV 30-kDa movement protein is strongly and selectively enhanced in tobacco protoplasts by actinomycin (Blum *et al.*, 1989). It was suggested that the drug may act by selectively inhibiting a host factor that normally suppresses the expression of the 30-kDa viral protein.

c. Non-specific virus inhibitors

Extracts of many plants contain substances that inhibit infection by viruses when mixed with the inoculum. Some may inhibit virus replication in experimental systems, and many are known to be proteins or glycoproteins. The relevance of any of these substances to plant resistance to viruses has not been established. One of the most studied is a 29-kDa protein isolated from *Phytolacca americana* that is known to inhibit a wide range of eukaryotic ribosomes. Ready *et al.* (1986) showed that the *Phytolacca* protein is located mainly in the cell wall. They suggested that when injury to a cell occurs during the process of virus infection, the inhibitor enters the cytoplasm and shuts off virus synthesis.

d. Ineffective viral genes

As discussed in Chapter 9 (Section II.D), many viruses code for a gene product that is necessary for cell-to-cell movement. If a movement protein is ineffective in a particular host plant then virus replication is confined to the initially infected cells. Consequently, the plant shows extreme hypersensitivity (see Table 10.1) (although not necessarily a necrotic response) and is effectively resistant to the virus. A virus that does not normally move systemically in a particular plant species may do so in the presence of an unrelated virus that does infect systemically and complements the movement of the defective virus (see Chapter 9, Section II.D.6). This indicates that a mismatch between viral movement protein and plant species may be a quite common mechanism for plant resistance.

3. Clustering of resistance genes

In several plant species, the resistance virus resistance genes are clustered to specific loci on the chromosomes; in this, virus resistance resembles that for fungi (Boller and Meins, 1992; Pryor and Ellis, 1993). For instance, in *Pisum sativum*, resistances to the lentil strain of PSbMV, BYMV, WMV-2, CYVV and BCMV

NL-8 strain are controlled by tightly linked recessive genes on chromosome 2 (Provvidenti and Alconero; 1988a; Provvidenti, 1990). Other examples include the close linkages between resistance to WMV and ZYMV in melon (*Cucumis melo*) (Anagnostou *et al.*, 2000) and between resistance to PRSV-W, ZYMV, WMV and MWMV in cucumber (*C. sativa*) (Grumet *et al.*, 2000).

C. Tolerance

The classical example of genetically controlled tolerance is the Ambalema tobacco variety. TMV infects and multiplies through the plant, but in the field, infected plants remain almost normal in appearance. This tolerance is due to a pair of independently segregating recessive genes r_{m1} and r_{m2}, and perhaps to others as well, with minor effects.

Other examples are known where either a single gene or many genes control tolerance. For example, tolerance of a set of barley genotypes to BYDV was controlled by a single major gene probably with different alleles giving differing degrees of tolerance (Catherall *et al.*, 1970). From a study of the relative abundance of DNA forms and viral RNAs in plants infected with CaMV, Covey *et al.* (1990) concluded that the expression of the CaMV minichromosome is a key phase in the replication cycle that is regulated by the host. Kohlrabi is a host that is tolerant of infection compared with turnip. In this host, high levels of supercoiled DNA accumulated, with very little generation of RNA transcripts, viral products or virus.

D. Use of conventional resistance for control

A review of the consideration in a breeding program for resistance to an important virus, that causing rhizomania of sugar beet, is given in Scholten and Lange (2000). Many of the aspects that they discuss are applicable to breeding programs for resistance to other viruses. In this section, I describe some examples of the application of conventional resistance to the control of viruses.

1. Immunity

The apparent immunity of seven raspberry cultivars to RVCV was confirmed by the fact that they could not be infected by graft inoculation (Jennings and Jones, 1986). Certain cultivars of swede (*Brassica napus*) appear to be immune to TuMV (Tomlinson and Ward, 1982), and certain cultivars of barley appear to be immune to BaYMV. Plants remained uninfected with the virus following root inoculation with virus-bearing cultures of the fungal vector *Polymyxa graminis*. Transmission by zoospores from the fungus growing on roots of an immune cultivar was rare or absent.

However, although many searches have been made, true immunity against viruses and viroids, which can be incorporated into useful crop cultivars, is a rather uncommon phenomenon. For example, Singh and Slack (1984) found no immunity to PSTVd among 555 introductions belonging to 81 tuber-bearing *Solanum* species.

2. Resistance

Where suitable genes can be introduced into agriculturally satisfactory cultivars, breeding for resistance to a virus provides one of the best solutions to the problem of virus disease. Attempts to achieve this objective have been made with many of the virus diseases of major importance. Genes for resistance or hypersensitivity have often been found, but it has frequently proved difficult to incorporate such factors into useful cultivars.

A hypersensitive reaction to a virus involving necrotic local lesions with no systemic spread, as discussed in Chapter 10 (Section III), can give effective resistance in the field. For example, the necrotic reaction of *N. glutinosa* to TMV was bred into some commercial lines of *N. tabacum* (Valleau, 1946). It has since been used widely with Virginia-type tobacco cultivars, but not with flue-cured types.

Occasionally, very high resistance to a virus has been discovered even in a plant where most known cultivars were highly susceptible. However, the designation of a variety or genotype as very highly resistant to a particular virus must always be provisional, because it is always possible that a mutant of the virus will arise that can overcome the plant resistance.

The major problem with resistance of any sort as a control measure is its durability. How long can it be deployed successfully before a resistance breaking (virulent) strain of the virus emerges? Fraser (1992) listed 87 host–virus combinations for which resistance genes have been found. Of those tested, virulent virus isolates able to overcome these resistances are known for more than 75% of these (Table 16.6). Fewer than 10% of the resistance genes listed have remained effective when tested against a wide range of virus isolates over a long period. However, some of the virulent isolates were found only in laboratory tests rather than field outbreaks.

The costs of a breeding program must be weighed against the possible gains in crop

TABLE 16.6. Summary of occurrence of resistance-breaking isolates

	Immunity or subliminal infection	Local lesion	Partial localization	Systemically effective	Not known
Resistance phenotype					
Dominant alleles	5	22	1	3	8
Incompletely dominant	0	0	4	11	0
Recessive dominant	6	0	3	9	4
Virulent isolates reported	Yes	No	Not tested	–	–
Dominant alleles	20	4	16	–	–
Incompletely dominant	9	3	3	–	–
Recessive	9	4	8	–	–
Total	38	11	27	–	–

From Fraser (1992), with kind permission of the copyright holder, © Kluwer Academic Publishers.

yield (see Buddenhagen, 1981). Many factors are involved, such as: (1) the severity of the viral disease in relation to other yield-limiting factors; (2) the 'quality' of the available resistance genes (for example, resistance genes against CMV are usually 'weak' and short lived, which may be due, at least in part, to the many strains of CMV that exist in the field); (3) the importance of the crop (compare, for instance, a minor ornamental species with a staple food crop such as rice); and (4) crop quality. Good virus resistance that gives increased yields may be accompanied by poorer quality in the product, as happened with some TMV-resistant tobacco cultivars (Johnson and Main, 1983).

a. Sources of resistant genotypes

Quite frequently, no resistance could be found for particular crops and viruses. For example, no resistance to BWYV was found in 70 cultivars and 500 breeding lines of lettuce (Watts, 1975) but some was found later (Maisonneuve et al., 1991). Russell (1960) found no resistance to BYV in 100 000 beet seedlings. In such circumstances, it may be necessary to search for resistance among wild species. In general, there is certainly a need to broaden the search for resistance genes. For example, in the UK, 22 horticultural crops have no known source of resistance to a total of 25 viruses known to affect them (Fraser, 1988). Nevertheless, many effective sources of resistance have been found and are currently in use (for list see Khetarpal et al., 1998). Among the main sources of genetic resistance to pest and pathogens are the centers of origin and regions of diversification of the crop species. In these places plants have been exposed to selective pressure from the pest or pathogen over long periods of time and thus have developed resistance ((Leppik, 1970). This is well illustrated by the Solanum species that contain sources of resistance to PLRV (Table 16.7).

Sometimes certain existing lines or varieties have been found to have a useful degree of resistance, as with the resistance of Corbett refugee bean to BCMV (Zaumeyer and Meiners, 1975). Resistance may not always be uniform, however. For example, even within inbred lines of corn there was variation in resistance to MDMV (Louie et al., 1976). In cabbage, variation in degree of resistance to TuMV was found between different lines of the same variety (Polak, 1983).

Occasionally, useful sources of resistance can be identified by making initial selections from plants showing good growth in otherwise severely infected fields (e.g. Cope et al., 1978). With important crops, searches for sources of resistance have been made on a worldwide

TABLE 16.7 *Solanum* species containing sources of resistance to PLRV

Species	Series	Ploidy	Country of origin	Cultivated or wild	Tuber-bearing
S. etuberosum	Etuberosa	2×	Chile	Wild	No
S. brevidens	Etuberosa	2×	Argentina, Chile	Wild	No
S. raphanifolium	Megistacroloba	2×	Peru	Wild	Yes
S. chacoense	Yungasensa	2×	Argentina, Bolivia, Paraguay, Uruguay	Wild	Yes
S. acaule	Acaulia	4×	Argentina, Bolivia, Peru	Wild	Yes
S. demissum	Demissa	6×	Mexico	Wild	Yes
S. phureja	Tuberosa	2×	Bolivia, Colombia, Ecuador, Peru, Venezuela	Cultivated	Yes
S. tuberosum ssp. andigena	Tuberosa	4×	Argentina, Bolivia, Colombia, Ecuador, Peru, Venezuela	Cultivated	Yes
S. tuberosum ssp. tuberosum	Tuberosa	4×	Chile	Cultivated	Yes

From Barker and Waterhouse (1999), with permission.

basis. Timian (1975) tested 4889 entries in the
Barley World collection and found 44 that
showed no symptoms of infection with BSMV
in the field. As another example, sources of
resistance to CSSV in Ghana have been found
in the upper Amazon region (Legg and
Lockwood, 1977), and attempts have been
made to combine resistance from various
sources (Kenten and Lockwood, 1977).

Useful resistance has sometimes been found
among a collection of mutants induced by
physical or chemical means. For example, Ukai
and Yamashita (1984) identified such a barley
mutant that was highly resistant to BaYMV.

In principle, culture of plant cells as protoplasts offers several possibilities for obtaining
new sources of resistance to viruses. However,
no commercially successful virus-resistant cultivars have yet been derived by this means. The
topic has been reviewed by Shepard (1981).
When plant cells are grown in culture for a time
and then plants regenerated from single cells or
small clusters of cells, considerable genetic
variation or *somaclonal* variation may be
observed in various properties, including
resistance to disease. Attempts have been made
to increase the frequency and range of variations by treatment of the tissue in culture with
mutagens. Somaclonal variants of tomatoes
produced by adventitious shoot formation
from leaf discs showed some resistance to
ToMV (Smith and Murakishi, 1993).

Protoplast fusion can take place *in vitro* even
between protoplasts belonging to different genera. In principle, this offers the possibility of
introducing virus resistance genes into a crop
cultivar from a quite distantly related species.
The donor cells are often irradiated before
fusion to fragment the chromosomes.
Following fusion, chromosomes and parts of
chromosomes are eliminated in a fairly random
manner, until a more or less stable set of chromosomes is achieved. The hope is that a cell
line will arise that contains a functional set of
chromosomes from the crop cultivar, together
with a minimal amount of alien genetic material containing the desired resistance gene or
genes.

In model experiments, transfer of methotrexate and 5-methyltryptophan resistances from
carrot to tobacco was achieved by fusion
between leaf mesophyll protoplasts of tobacco
and irradiated cell culture protoplasts of carrot
(Dudits *et al.*, 1987). Some of the regenerated
plants had tobacco morphology and independently segregating genes for the two resistance
markers from carrot.

Intergeneric hybridization in a search for
resistance to BYDV in wheat has been reviewed
by Comeau and Plourde (1987). Substantial
resistance has been found in about 20 species in
the tribe *Triticeae*, which could be, or have been,
hybridized to cultivated wheat or close
relatives.

b. Low seed transmission

The discussion in the previous section was concerned with resistance of the growing plant to
virus infection. For those viruses affecting
annual crops that are transmitted through the
seed, resistance to seed transmission may be an
important method for limiting infection in the
field. For example, a single recessive gene conditions resistance to seed transmission of BSMV
in barley (Carroll *et al.*, 1979). The resistance
gene, found in an Ethiopian barley called
Modjo, was introduced into a new variety,
Mobet, with good agronomic qualities and high
resistance to seed transmission of Montana isolates of BSMV (Carroll *et al.*, 1983). Resistance to
seed transmission has also been found for LMV
in lettuce and for some legume viruses.

c. Adequate testing of resistant material

Varieties or lines showing resistance in preliminary trials must then be tested under a range of
conditions. Important factors to be considered
are strains of the virus, climatic conditions (e.g.
Thompson and Hebert, 1970) and inoculum
pressure (e.g. Kenten and Lockwood, 1977).

d. The need for resistance to multiple pathogens

The difficulties in finding suitable breeding
material are compounded when there are
strains of not one but several viruses to consider. Cowpeas in tropical Africa are infected to

a significant extent by at least seven different viruses. In such circumstances, a breeding program may utilize any form of genetic protection that can be found. Sources of resistance, hypersensitivity or tolerance have been found for five of the viruses (Matthews, 1991). However, several of these viruses have different strains or isolates that may break resistance to other isolates (van Boxtel *et al.*, 2000). There is of course the further problem of combining these factors with multiple resistance to fungal and bacterial diseases. For example, genetic resistance to TMV, cyst nematodes, root-knot nematodes, and wildfire from *Nicotiana repanda* has been incorporated into *N. tabacum* (Gwynn *et al.*, 1986).

e. Durability of resistance and the emergence of resistance-breaking strains

For some crop plants and viruses, resistance has proved to be remarkably durable. Thus, the resistance to BCMV found in Corbett Refugee bean has been bred into most varieties of dry and snap beans in the United States, and the resistance had not broken down after 45 years (Zaumeyer and Meiners, 1975). One of the most noteworthy examples of durability has been the resistance in sugar beet to the BCTV (Duffus, 1987), the original selections for which were made in the 1920s. The resistance is multigenic and appears to involve a lower concentration of virus in resistant plants that do become infected and a much longer incubation period in resistant varieties. However, over a period of years a series of more aggressive strains of the virus has emerged. Resistance to TuMV in lettuce described more than 40 years ago has been tested with strains of the virus from many countries, but the resistance has not been broken (Duffus, 1987; Robbins *et al.*, 1994).

A hypersensitive type of resistance to CCMV in cowpeas was overcome by certain strains of the virus (Paguio *et al.*, 1988). Resistance breaking strains of PVX have been described in the UK for resistant potato varieties, but these have not yet become a serious practical problem (Jones, 1985). RBDV has been controlled in

Scotland by the use of cultivars that are immune to the prevalent strain of the virus. This situation has been threatened by the discovery of a resistance-breaking strain (Murant *et al.*, 1986).

Once a substantial population of resistant plants is exposed in the field there is a good probability that a new strain of the virus will evolve, or be introduced, that can overcome the resistance.

The problem of virus strains in the development of a breeding program is well illustrated by the tests carried out by Rast (1967a,b) with 64 different isolates of ToMV on 30 clones of *Lycopersicum peruvianum*. Different isolates (even from the same strain of ToMV as judged by symptoms on tomato or tobacco) differed in the range of *L. peruvianum* clones they would infect. Every clone could be infected by at least one isolate.

Even when an apparently successful resistant variety has been developed, it may be important to maintain other measures to minimize contact of the resistant variety with the virus concerned. For example, in tomatoes resistant to ToMV, the virus did not move systemically for several weeks and reached only a low concentration (Dawson, 1967). Extracts from infected resistant plants were more infective for healthy resistant plants than virus from susceptible lines. Resistant plants infected with such virus showed more obvious symptoms.

The $Tm2^2$ gene in tomato was very useful for protection against ToMV for more than 10 years. However, ToMV strains have now been found that can, in the laboratory, overcome the resistance due to the $Tm2^2$ gene. It is probably only a matter of time before these become prevalent in commercial glasshouses.

We must conclude that, on present knowledge for most crops and most viruses, the search for new sources of resistance and their incorporation into useful cultivars will be a continuing and very long-term process, as it is with many fungal and other parasitic agents. The interrelations between host genetic factors and other epidemiological aspects of virus diseases are discussed by Buddenhagen (1983).

3. Tolerance

Where no source of genetic resistance can be found in the host plant, a search for tolerant varieties or races is sometimes made. However, tolerance is not nearly as satisfactory a solution as genetic resistance for several reasons.

1. The infected tolerant plants may act as a reservoir of infection for other hosts. Thus, it is bad practice to grow tolerant and sensitive varieties together under conditions where spread of virus may be rapid.
2. Large numbers of virus-infected plants may come into cultivation. The genetic constitution of host or virus may change to give a breakdown in the tolerant reaction.
3. The deployment of tolerant varieties removes the incentive to find immunity to the virus until the tolerance breaks down in an 'out of sight, out of mind' attitude.
4. Virus infection may increase susceptibility to a fungal disease (see Chapter 10, Section V.G).

However, tolerant varieties may yield very much better than standard varieties where virus infection causes severe crop losses and where large reservoirs of virus exist under conditions where they cannot be eradicated. Thus, tolerance has, in fact, been widely used (see Posnette, 1969). Cultivars of wheat and oats commonly grown in the US Midwest have probably been selected for tolerance to BYDV in an incidental manner, because of the prevalence of the virus (Clement *et al.*, 1986). Tolerance to MSV has been found in maize and rapidly incorporated into high-yielding maize populations for use in tropical Africa (Soto *et al.*, 1982). Walkey and Antill (1989) obtained an unusual result with a variety of garlic called Fructidor. The yield of selected stocks was significantly less than that of unselected infected stocks.

Salomon (1999) argues that breeding for tolerance has an advantage over breeding for resistance in that the selection pressures for the development of resistance-breaking strains are less. This may be so for fungal and bacterial pathogens but it is open to question as to whether this consideration pertains to viruses as well.

VI. TRANSGENIC PROTECTION AGAINST PLANT VIRUSES

A. Introduction

It is now possible to introduce almost any foreign sequence into a plant and obtain expression of that sequence. In principle, this should make it possible to transfer genes for resistance or immunity to a particular virus, across species, genus and family boundaries. Furthermore, genes can be designed to interfere with directly, or induce the host to interfere with, the virus replication cycle. Several approaches to producing transgenic plants resistant to virus infection are being actively explored.

There are essentially three sources of transgenes for protecting plants against viruses: (1) natural resistance genes; (2) genes derived from viral sequences, giving what is termed *pathogen-derived resistance* (PDR); and (3) genes from various other sources that interfere with the target virus. These are discussed in the following sections. There have been numerous reviews on the subject, including Beachy (1993, 1997), Fitchen and Beachy (1993), Wilson (1993), Scholthof *et al.* (1993), Baulcombe (1994), Hull (1994b), Wilson and Davies (1994), Lomonossoff (1995), Prins and Goldbach (1996), Palukaitis and Zaitlin (1997), Kaniewski and Lawson (1998), Bendahmane and Beachy (1999), Hammond *et al.* (1999), Waterhouse and Upadhyaya (1999). This selection of reviews, over 7 years, illustrates the rapid development of understanding of this subject.

B. Natural resistance genes

Molecular aspects of genes found in plant species that confer resistance to various viruses are discussed in Chapter 10. When a resistance gene has been identified, it can be isolated and transferred to another plant species.

The *Rx1* gene, which gives extreme resistance to PVX, has been isolated from potato and transformed into *Nicotiana benthamiana* and *N. tabacum* (Bendahmane *et al.*, 1999) where it gives resistance to the virus. Similarly, the *N* gene, found naturally in *N. glutinosa*, and which confers hypersensitive resistance to

TMV, gives resistance to TMV when transferred to tomato (Whitham *et al.*, 1996).

VII. PATHOGEN-DERIVED RESISTANCE

The ideas leading up to the concept of pathogen-derived resistance for plant viruses were first postulated by Hamilton (1980) and are encapsulated as a general concept in a paper by Sanford and Johnson (1985). They suggested that the transgenic expression of pathogen sequences might interfere with the pathogen itself terming this concept 'parasite-derived resistance'. Since then, several names have been used for this approach including 'non-conventional protection', 'transgenic resistance' and 'engineered virus resistance', but the generally accepted term is now pathogen-derived resistance (PDR). Since the mid-1980s, this approach has attracted major interest and is the main one by which transgenic protection is being produced against viruses in plants. The first demonstration of PDR against plant viruses was by Powell-Abel *et al.* (1986), who showed that the expression of TMV coat protein in tobacco plants protected those plants against TMV. This opened the floodgates for extensive research both on protecting crop species against viruses and on the mechanisms involved.

The basic idea arising out of Sanford and Johnson's (1985) concept is that, if one understands the molecular interactions involved in the functioning of a pathogen, mechanisms can be devised for interfering with them. Although this concept applies to all pathogens and invertebrate pests, it has mainly been used against viruses because of their relatively simple genomes. In developing the concept it was recognized that the interactions of interest occur at all stages of the virus infection cycle and that they can potentially be interfered with in various ways, for example by blocking the interaction or by decoying one or more of the molecules involved in the interaction. This then led to the idea that the overall strategy as being one of attacking specific viral 'targets' with specific molecular 'bullets'. Some examples of targets and attacking mechanisms, or bullets, are given in Table 16.8. However, in practice, much of the development of this approach was done without detailed knowledge of the precise molecular mechanisms involved, and analysis of these results has thrown light on several new mechanisms. Perhaps the most important is the gene-silencing phenomenon described in Chapter 10 (Section IV) and which is further discussed in Section VII.B.3 below.

In this rapidly expanding subject, there are various terminological problems. The main one, whether to term this phenomenon resistance or protection, is discussed in Section

TABLE 16.8 Examples of 'targets' and 'bullets' for pathogen-derived resistance

Targets	Bullets
Viral gene products	Molecular blockers
Coat protein	Viral gene products
Replicase	Mutated viral gene products
Cell-to-cell spread function	Antisense nucleic acid $\pm$ ribozymes
Protease	Sense nucleic acid
	Antibodies
Control sequences	
Replication control sequences	Decoys
Origins of replication	Nucleic acid control sequences
Primer binding sites	Satellites
Expression control sequences	Protease sites
Subgenomic RNA promoters	
Translational leader sequences	Other
Splice sequences	Non-host resistance

From Hull (1994b), with kind permission of the copyright holder, © Kluwer Academic Publishers.

VII.B.7. I will use the term protection wherever possible, but in situations where it has been used widely (such as pathogen-derived resistance) I will retain the term resistance.

Currently, there are two basic molecular mechanisms by with PDR is thought to operate. In some systems the expression of an unmodified or a modified viral gene product interferes with the viral infection cycle – this I will term *protein-based protection*. The second mechanism does not involve the expression of a protein product, and I will call this *nucleic acid-based protection*.

A. Protein-based protection

As noted above, the first demonstration of PDR involved the expression of TMV coat protein (Powell-Abel *et al.*, 1986). Since then, there have been many examples of the use of this coat protein-mediated protection. The expression of other viral gene products also gives protection to a greater or lesser extent against the target virus.

1. Transgenic plants expressing a viral coat protein

The sequences encoding viral coat proteins are the most widely used for conferring protection in plants (Fitchen and Beachy, 1993); coat protein genes from at least 35 viruses, representing 15 viral taxonomic groups, have been transformed into many different plant species (Palukaitis and Zaitlin, 1997). This is because this gene was used in the first example of this approach and because coat protein genes are relatively easy to identify and clone. The phenomenon is often referred to as 'coat protein-mediated resistance' (CP-MR).

a. Tobacco mosaic virus *coat protein* (reviewed by Beachy, 1999)

Beachy *et al.* (1986) and Bevan *et al.* (1985) first reported the expression of the coat protein of TMV in tobacco plants into which a cDNA containing the coat protein gene had been incorporated. This was quickly followed by reports on the expression of the coat protein gene of AMV by several groups (e.g. Loesch-Fries *et al.*, 1987),

that of TRV by van Dun *et al.* (1987) and that of PVX by Hemenway *et al.* (1988). Some laboratories used the CaMV 35S promoter and others the 19S promoter to obtain expression. The 35S promoter was considerably more effective.

Powell-Abel *et al.* (1986) showed that transgenic plants expressing TMV coat protein either escaped infection following inoculation or developed systemic disease symptoms significantly later than plants not expressing the gene. Plants that showed no systemic disease did not accumulate TMV in uninoculated leaves (Nelson *et al.*, 1987). Transgenic plants produced only 10–20% as many local lesions as controls when inoculated with a strain of TMV causing local lesions. The idea that transgenic plants resist initial infection rather than subsequent replication was suggested by results obtained using transgenic Xanthi nc tobacco plants, in which fewer local lesions were produced than on control plants. However, the lesions that did develop were just as big as those on control leaves, indicating that once infection was initiated there was no further block in the infection cycle.

In transgenic plants it would be possible, in principle, for the resistance to infection to be due to the coat protein mRNAs transcribed from the cDNA, to the coat protein itself, or to both of these molecules. To test these possibilities, Register *et al.* (1988) constructed a series of cDNAs generating mRNA sequences that would produce no coat protein, or mRNA that lacked the replicase recognition site but that would produce coat protein. These experiments conclusively implicated the coat protein rather than the mRNA in causing resistance to superinfection. Further experiments with TMV confirmed the earlier results and showed that the 3' tRNA-like sequence was not necessary to generate resistance (Powell *et al.*, 1990).

Register *et al.* (1988) and Register and Beachy (1988) showed that protoplasts made from transgenic plants expressing coat protein were specifically protected against infection with TMV. When tobacco protoplasts took up coat protein, they were transiently protected from infection with TMV (Register and Beachy, 1989). Thus, coat protein outside the cell is probably not involved in coat protein-mediated protection. Pathogenesis-related proteins (see Chapter 10, Section III.K.1) also do not appear to be involved in this resistance (Carr *et al.*, 1989).

b. Dose and sequence dependency of protection by TMV coat protein

The greater the amount of virus inoculum, the lower is the protection afforded by TMV coat protein (Nejidat and Beachy, 1990; Bendahmane *et al.*, 1997b).

There is a positive correlation between the level of protections and the sequence similarity between the transgene coat protein and that of the challenge virus (Nejidat and Beachy, 1990). For instance, the coat proteins of ToMV and TMGMV have 82% and 72% sequence identity, respectively, with that of TMV, whereas that of RMV is only 45% identical; TMV coat protein gave much better protection against ToMV and TMGMV than against RMV. It appears likely that it is the structure of the coat protein and possibly differences in the sites of carboxyl–carboxylate interactions (see Chapter 5, Section III.B.5) that influence the protection given against different tobamoviruses (Bendahmane and Beachy, 1999). There was little or no protection against viruses from other families or genera (e.g. AMV, CMV, PVX or PVY) (Anderson *et al.*, 1989).

Transgenic expression of TMV coat protein does not protect against inoculation with viral RNA (Nelson *et al.*, 1987).

c. Mechanism of TMV coat protein protection
(reviewed by Reimann-Philipp and Beachy, 1993; Bendahmane and Beachy, 1999)

A careful analysis of virus spread in single lesions in transgenic and control tobacco plants showed that, when local infection did take place in tissues expressing the coat protein, there was no inhibition in subsequent cell-to-cell movement (Register *et al.*, 1988). On the other hand, when a leaf-bearing stem segment from a transgenic plant was grafted between lower and upper sections of a non-transgenic plant, systemic movement of TMV into the leaves above the graft was inhibited. Transgenic tobacco plants expressing the 30-kDa movement protein of TMV were not pro-

tected against infection or disease development (Deom *et al.*, 1987).

When TMV is treated at pH 8.0, translatability *in vitro* is greatly enhanced (see Chapter 7, Section II.B.4). Register *et al.* (1988) and Register and Beachy (1988) found that TMV treated in this manner was able to overcome the resistance of transgenic plants in the same way as RNA. This result certainly supports an early event as being important in the resistance of transgenic plants. Osbourn *et al.* (1989b) showed that TMV-like pseudo-particles containing the GUS reporter gene expressed this gene 100 times less efficiently in protoplasts from coat protein-transgenic tobacco plants than in control protoplasts. The data suggested that about 97% of the GUS pseudo-particles remained uncoated. However, other experiments indicated that inhibition of virus disassembly is insufficient to account entirely for coat protein-mediated resistance, and that some later event or events in virus replication must be involved.

The TMV mutant DT-IG produces no coat protein and does not normally move systemically. When this mutant is inoculated into transgenic tobaccos expressing the coat protein gene, rod-shaped particles were found in the systemically invaded leaves (Matthews, 1991). However, the viral rods isolated from the systemic infection were unable to infect fresh transgenic plants, supporting the view that some early uncoating event is involved in resistance.

Osbourn *et al.* (1989a) tested the possibility that coat protein expressed in transgenic plants inhibits virus replication by recoating uncoated viral RNA from a challenge inoculum. They produced double-transformed tobacco plants that were expressing TMV coat protein and a reporter gene (CAT) whose transcripts contained a copy of the TMV origin of assembly sequence. No rods could be detected in cell extracts of these plants by electron microscopy, and there was no significant reduction in CAT activity. However, transformed plants retained their ability to resist infection by TMV. Thus, it seems unlikely that re-encapsidation of uncoated RNA of challenge virus by endogenous coat protein is involved in the resistance of transgenic plants expressing coat proteins.

In a comparison of the co-translational disassembly (see Chapter 7, Section II.B.4, for a description of co-translational disassembly) of TMV in CP(+) and CP(–) protoplasts, Wu *et al.* (1990) showed that TMV recruited polyribosomes within 5–10 minutes and that the input virus was largely undetectable within 60 minutes. In contrast, in the CP(+) protoplasts input virus was not recruited to polyribosomes and was largely intact at 60 minutes. Thus, it appeared most likely that coat protein-mediated resistance is an early event after virus entry into the cell and it was suggested that the transgenically expressed coat protein blocks disassembly (Wu *et al.*, 1990).

There are several hypotheses as to how the coat protein can block disassembly. The first suggests an inhibition of viral uncoating (Register and Nelson, 1992; Bendahmane *et al.*, 1997b) with the transgenic coat protein driving the disassembly reaction towards assembly. The major point in this hypothesis is that the balance between uncoating of the virus particle and particle assembly is controlled by the amount of free coat protein in the relevant cellular compartment. Thus, in the non-transgenic situation, there is little or no free coat protein in the cell during disassembly, and assembly occurs when the amount of 'local' free coat protein becomes significant. In the transgenic plant, the expression of the transgene produces significant amounts of free coat protein, thus switching the balance to particle assembly.

The second hypothesis proposes that a cellular site for TMV disassembly is blocked by the transgenic coat protein. To examine the second hypothesis, Clark *et al.* (1995) postulated that, if the recognition binding transgenic coat protein to a putative receptor site was on the radial surface of the coat protein, the level of resistance against a challenging virus containing SHMV sequences on its surface would be similar to the resistance against SHMV itself. However, plants expressing TMV coat protein were just as resistant to the chimeric virus as they were to TMV. Plants challenged with a TMV mutant in which the coat protein was replaced by that of SHMV showed the same low level of resistance as that to SHMV. This experiment gave a

strong indication that the binding site hypothesis was not tenable.

The structure of TMV coat protein and factors involved with the assembly of TMV particles are discussed in Chapter 5 (Section IV). A series of experiments on the effects of mutations of TMV coat protein designed to affect virus assembly are reviewed in Bendahmane and Beachy (1999). Although the experiments showed differences in protection, they did not fully resolve the mechanism of the protection.

d. Other viral coat proteins

Coat protein-mediated resistance (protection) has been demonstrated for many other virus families and genera (Waterhouse and Upadhyaya, 1999).

Plants that were expressing AMV coat protein were also resistant to infection with AMV (e.g. Loesch-Fries *et al.*, 1987; Turner *et al.*, 1987a; van Dun *et al.*, 1987), even though this protein is required for virus replication (see Chapter 8, Section IV.G). Similar results were obtained with a mutated AMV coat protein (van Dun *et al.*, 1988b).

Transgenic tobacco plants expressing the PVX coat protein gene were significantly protected against PVX infection, as shown by a reduced number of local lesions on inoculated leaves, delayed or no systemic symptom development, and a reduction in virus accumulation in both inoculated and systemically infected leaves. The higher the level of coat protein expression, the higher was the level of protection (Hemenway *et al.*, 1988). Plants expressing an antisense coat protein transcript were resistant to infection with PVX, but only with low concentrations of virus in the inoculum.

The extent of protection provided by such transgenic plants is correlated with the degree of expression of the coat protein gene. Figure 16.14 shows the striking protection obtained with the transgenic expression of CMV coat protein. The extent of protection is reduced as inoculum concentration is increased.

Like TMV, the coat proteins of AMV and TRV do not protect against RNA inoculum (Loesch-Fries *et al.*, 1987; van Dun *et al.*, 1987; Angenent *et al.*, 1990) but, unlike TMV, plants expressing

Fig. 16.14 Protection of transgenic tobacco plants expressing the CMV coat protein gene against mechanical inoculation with CMV at 25 µg/mL. CP$^+$ plants (back row) and CP$^-$ plants (front row) were photographed 1 month after inoculation. From Cuozzo *et al.* (1988), with permission.

high levels of PVX coat protein were resistant to infection with PVX RNA (Hemenway *et al.*, 1988).

The expressed coat protein does not necessarily have to be that of the target virus, but it has to be sufficiently closely related. For instance, field resistance in tobacco to PVY was conferred by by expression of LMV coat protein (Dinant *et al.*, 1998).

Two coat proteins can be expressed from one construct. Marcos and Beachy (1997) designed a construct comprising the coat proteins of the tobamovirus TMV and the potyvirus SMV together with the highly specific TEV NIa proteinase. In plants transformed with this construct, the proteinase processed the multifunctional polypeptide to give accumulation of the two viral coat proteins. These plants were protected against both TMV and PVY. Similarly, *Nicotiana benthamiana* plants expressing sequences of the *N* gene of TSWV and the coat protein of TuMV were protected against both viruses (Jan *et al.*, 2000a). Squash lines expressing the coat protein genes from CMV, WMV 2 and ZYMV were resistant to all three viruses (Fuchs *et al.*, 1998), indicating that these genes can be pyramided.

Of interest is 'coat' protein-mediated protection against viruses either that have several coat protein species or that do not have conventional capsids. Rice plants transformed with one of three structural proteins (S5, S8 or S9) of the oryzavirus RRSV showed resistance to the virus (Waterhouse and Upadhyaya, 1999) but it was not noted whether any of these genes were expressed as proteins. The particles of tenuiviruses are ribonucleoproteins, the main protein being the nucleocapsid (N) protein (see Chapter 2, Section III.I). Rice plants transformed with, and expressing, the N protein of RSV were protected against the virus (Haykawa *et al.*, 1992).

2. Viral movement proteins

The proteins encoded by viruses that facilitate their cell-to-cell movement are described in Chapter 9 (Section II.D.2). The relationship between both structure and function of movement proteins indicated that, if one could block the function by, for instance, using a defective mutant protein, a broader resistance might result (Cooper *et al.*, 1996). Expression of the BMV movement protein gave partial resistance to TMV (Malyshenko *et al.*, 1993) and that of the potexvirus triple gene block will confer protection against other viruses with a similar genome organization (Beck *et al.*, 1994; Seppänen *et al.*, 1997). However, a functional movement protein can complement defective proteins; when the 30-kDa movement protein gene was expressed in transgenic tobacco plants, movement of the transport-defective LSI mutant of TMV was possible (Deom *et al.*, 1987). As noted in Chapter 9, Section II.D, TMV and RCNMV movement proteins will complement the movement of the insect virus, flock house virus, in transgenic plants (Dasgupta *et al.*, 2001).

The transgenic expression of the movement proteins of BDMV had a deleterious effect on the plant development (Hou *et al.*, 2000).

3. Viral replicase proteins (reviewed by Palukaitis and Zaitlin, 1997)

Transformed tobacco plants containing cDNA copies of AMV RNAs 1 or 2 showed no resistance to AMV (van Dun *et al.*, 1988a). It is not certain whether this result was due to low levels of expression or whether both genes in the same plant might confer some resistance. No resistance was detectable in tobacco plants transformed with the non-structural 13- and 16-kDa genes of strain PLB of TRV, or the 29-kDa gene unique to strain TCM (Angenent *et al.*, 1990).

Replicase sequences expressing proteins were shown to provide protection against three plant viruses, AMV (Brederode *et al.*, 1995), CMV (Palukaitis and Zaitlin, 1997) and TMV (Golemboski *et al.*, 1990). Transformation with the P1 or P2 replicase genes of AMV did not give protection and neither did mutants of the P2 gene, N-truncated to resemble TMV 54-kDa gene or with the GDD motif (the RNA-dependent RNA polymerase catalytic site; see Chapter 8, Section IV.B.1) changed to VDD (Brederode *et al.*, 1995). However, mutation of the GDD motif to GGD, GVD or DDD gave high levels of protection against AMV.

As described in Section V.B below, in most cases use of replicase sequences give RNA-mediated protection. However, the AMV, CMV and TMV systems show relatively high steady-state levels of accumulation of transgene mRNA. In each case, the construct comprised only part of the replicase protein, the 54-kDa moiety of the TMV replicase (see Chapter 7, Section V.E.1), truncated CMV 2a protein and the AMV 2a protein. The truncated 2a protein encoded by the transgene was

detected for CMV (Carr *et al.*, 1994) but not for the other two viruses, and transgene translatability increased the effectiveness of the protection (Wintermantel and Zaitlin, 2000). Mutations in the sequence encoding the TMV and AMV proteins interfered with protection, suggesting that the protein itself was involved (Carr and Zaitlin, 1992; Brederode *et al.*, 1995). Expression of PVY NIb gene is possibly protein mediated (Audy *et al.*, 1994).

The transgenic plants expressing the TMV 54-kDa protein were resistant to infection with TMVU1 or its RNA at high concentrations. They were also resistant to a U1 mutant, but not to two other tobamoviruses or to CMV. Experiments in protoplasts derived from plants transgenic for the 54-kDa protein have shown that a very early event is involved. Carr *et al.* (1993) demonstrated that expression of the 54-kDa protein was required for protection, but Marano and Baulcombe (1998) suggested that the situation is more complicated, with at least part of the protection being nucleic acid based.

Various observations discussed by Patukaitis and Zaitlin (1997) indicate that this form of protection is not a dominant negative mutant effect but point to complex and subtle mechanisms. It is suggested that the interactions between the transgenic replicase proteins and other virus-encoded proteins may affect the processes of replication and cell-to-cell movement at the wrong time in the infection cycle, leading to the arrest of some stage of the replication process.

When a construct of ACMV replication-associated gene (*AC1*) mutated to alter the putative NTP-binding site was expressed in *Nicotiana benthamiana*, the plants showed a delay in symptom appearance and/or mild symptoms (Sangare *et al.*, 1999). A high level of the mutated gene was necessary for protection.

B. Nucleic acid-based protection

Four potential forms of protection based on the expression of viral RNA sequences have been recognized: (1) that induced by the viral RNA sequence expressed in (+) sense; (2) that induced by the viral RNA expressed as anti-sense molecules; (3) that induced by the expression of satellite RNAs; (4) that in which ribozymes are targeted to viral genomes.

Early in the development of coat protein-mediated protection, there were some unexpected observations. For instance, there was no correlation between resistance and the expression of potyvirus, PVY or luteovirus, PLRV coat proteins in potato (e.g. Kawchuk *et al.*, 1990; Lawson *et al.*, 1990). Lindbo and Dougherty (1992a,b) found that the untranslatable coat protein gene of TEV gave higher levels of protection than either full-length or truncated translatable constructs. Similar observations were made for protection given by untranslatable sequences of TSWV (de Haan *et al.*, 1992) and PVY (van der Vlugt *et al.*, 1992). These and other observations suggested that the protection, at least in these cases, was mediated by nucleic acid rather than by protein.

1. RNA-mediated protection (reviewed by Prins and Goldbach, 1996)

As no promoterless transgenes have been shown to confer protection (Lomonossoff, 1995), it must be assumed that either the RNA transcript or a protein that is encoded give the protection. In a plant transformed with a construct that does not give a protein, any protection is obviously due to the RNA. However, not all the plant lines expressing a non-coding RNA show protection. When a plant is transformed with a construct designed to produce a viral protein, it can often be difficult to distinguish between protection due to expression of the protein itself or that due to the RNA transcript. However, there are various features of the protection that tend to be characteristic for RNA-mediated protection (Smith *et al.*, 1994):

- There is no direct correlation between RNA expression levels and the level of protection (see Pang *et al.*, 1993).
- Usually, no transgene-encoded protein can be detected and the steady state of the transcript in inoculated plants is often in low amount.
- The protection is usually narrow and against strains of the virus that have very similar sequences to that of the transgene.

- Unlike coat protein-mediated protection, the protection is not overcome by inoculating RNA.
- Also, unlike coat protein-mediated protection, RNA-mediated protection is not dose dependent and operates at high levels of inoculum.
- The insert in the host genome comprises multiple copies of the transgene, particularly with direct repeats of coding regions (Sijen *et al.*, 1996).
- Copies of the transgene may be truncated and/or in an antisense orientation (Waterhouse *et al.*, 1998; Kohli *et al.*, 1999b).
- Transgene sequences and sometimes their promoter(s) may be methylated (Jones *et al.*, 1999; Kohli *et al.*, 1999b; Sonoda *et al.*, 1999).

When transcript levels have been examined, three general classes of resistance phenotype have been recognized:

1. Plants that are fully susceptible. These plants have low to moderate levels of transgene transcription and steady-state RNA.
2. Plants that become infected and then recover. These have moderate to high levels of transgene transcription and steady-state RNA in uninfected plants, but low-level steady-state RNA in recovered tissues.
3. Plants that are highly resistant. These plants have high levels of transgene expression but low steady-state levels.

Since the recognition of the differences between protein- and RNA-mediated protection, there have been many examples of plants transformed with viral sequences that show the properties of RNA-mediated protection.

2. Sequences for RNA-mediated protection

RNA-mediated protection has been given by a range of sequences from viral genomes. In many cases, it has resulted from attempts to transform plants with the viral genes described above.

Cowpea plants transformed with the CPMV movement protein gene are protected against CPMV by an RNA-based mechanism (Sijen *et al.*, 1995). The resistance also operates against PVX RNA containing the CPMV movement protein gene. This protection against PVX is initiated using sequences as small as 60 nucleotides, particularly situated in the 3' part of the transcribed region of the movement protein transgene (Sijen *et al.*, 1996). Transformation with a direct repeat of the movement protein sequences increased the frequency of resistant lines by 20–60%.

Transgenic expression in rice of four separate constructs derived from the BMV genome, an artificial DI RNA from RNA2, a sense tRNA-like structure corresponding to the 3' end of RNA2, an antisense sequence to the intergenic region of RNA3, and RNA encoding the virus coat protein, all gave protection against BMV (Huntley and Hall, 1996).

Protection is given by the expression of the NIa, NIb and coat protein regions of potyviruses. To determine whether combinations of these genes would give additively greater protection, Maiti *et al.* (1999) compared transgenic lines expressing TVMV coat protein with those expressing NIa, NIb and coat protein. The plants with the combination of three genes were invariably less resistant to TVMV than those expressing the coat protein gene alone. Furthermore, plants with the three-gene combination had virtually no resistance to TEV in contrast to the coat protein lines.

Immunity to BYDV-PAV was conferred on barley plants transformed to express hairpin (hp)RNA (see Chapter 10, Section IV.D) containing the polymerase gene of BYDV (Wang *et al.*, 2001a). Previous attempts to produce transgenic protection against this luteovirus have had only limited success (McGrath *et al.*, 1997; Koev *et al.*, 1998; Wang *et al.*, 2001b). This indicates that the hpRNA construct is important in obtaining high levels of protection.

3. Molecular basis of RNA-mediated protection

The recovery phenomenon associated with low steady states of transgene RNA in recovered tissues and the low steady-state RNA levels in highly resistant plants, coupled with the narrow range of protection against viruses with homologous sequences to the transgene, are all suggestive of homology-dependent or post-transcriptional gene silencing. This is described in detail in Chapter 10 (Section IV).

4. Protein- and RNA-mediated protection

A single type of construct may confer protection by both protein- and RNA-mediated mechanisms. For instance, transformation with the TSWV N gene sequence conferred resistance to heterologous tospoviruses in plant lines with the highest levels of expressed protein but the most effective protection to the homologous virus in plant lines that had the lowest steady-state levels of RNA and little protein (Pang *et al.*, 1993). Similarly, when barley plants were transformed with constructs to express the coat protein of BYDV-PAV, some resistant lines had detectable levels of coat protein whereas others did not (McGrath *et al.* 1997).

From an analysis of protection by TMV replicase sequences, Goregaoker *et al.* (2000) concluded that both RNA and protein sequences were involved in conjunction with the speed of the infecting challenging virus.

Nicotiana benthamiana transformed with PMTV coat protein gene showed strong resistance to the virus irrespective of the amount of transcript RNA or coat protein detected (Barker *et al.*, 1998). Lines transformed with a non-translatable form of PMTV coat protein gene were either not resistant or showed low levels of resistance. Barker *et al.* (1998) suggested that the form of protection given by PMTV coat protein is unique because, although it depends on coat protein translation to be effective, it mediates very strong resistance.

Transformation of potato with constructs expressing a mutant form of PLRV ORF 4, the gene for the movement protein gave transgenic lines that showed broad-range protection against PLRV and also against PVY and PVX (Tacke *et al.*, 1996). One of the transgenic lines had strong protection against PLRV but was susceptible to PVY and PVX. From an analysis of the RNA levels associated with these two phenotypes, it was concluded that the protection against PLRV was RNA mediated and that against PVY and PVX was mediated by the mutant movement protein binding to important sites in the plasmodesmata (Tacke *et al.*, 1996).

Palukaitis and Zaitlin (1997) draw attention to the need to test large numbers of independent transformants, not only to obtain lines with the best protection characteristics, but also to rule out the possibility that protection is not given by a particular construct.

5. Transgenic plants expressing antisense RNAs (reviewed by Tabler et al., 1998)

One method of gene regulation in organisms is by complementary RNA molecules that are able to bind to the RNA transcripts of specific genes and thus prevent their translation. Such RNA has been called antisense or micRNA (*m*essenger-RNA-*i*nterfering *c*omplementary RNA). Appropriate antisense sequences incorporated into a plant genome have been shown to block the activity of specific genes (e.g. Delauney *et al.*, 1988; van der Krol *et al.*, 1988). The possibility of using this strategy for the control of plant viruses has been explored.

Various laboratories have carried out *in vitro* studies with oligonucleotides complementary to some plant virus RNA sequences. Oligodeoxynucleotides complementary to genomic PVX RNA caused translation arrest in a Krebs-2 cell-free system. This was thought to be due to endogenous RNase H activity in the cell-free system (Miroshnichenko *et al.*, 1988). Antisense sequences complementary to sequences near the 5' end of TMV RNA inhibited *in vitro* translation of this RNA in a rabbit reticulocyte lysate. The inhibition was probably due to direct interference with ribosome attachment (Crum *et al.*, 1988). Morch *et al.* (1987) found that the 'sense' nucleotide sequences corresponding to the replicase recognition site near the 3' end of genomic TYMV RNA specifically inhibited *in vitro* the activity of the TYMV replicase isolated from virus-infected plants. The relevance of these various experiments to possible virus inhibition *in vivo* remains to be determined.

Among transgenic tobacco plants containing genes for the production of antisense RNAs for three regions of the CMV genome, only one showed some resistance to the virus (Rezaian *et al.*, 1988). Cuozzo *et al.* (1988) compared the extent of protection provided in transgenic tobacco plants by the coat protein gene of CMV or its antisense transcript. Symptom development and virus accumulation were reduced or absent in plants transgenic for the sense gene,

and this was unaffected by inoculum concentration over the range used. By contrast, antisense plants were protected only at low inoculum concentrations. Transgenic tobacco plants expressing RNA sequences complementary to the coat protein gene of TMV were not protected as strongly from TMV infection as were plants expressing the coat protein gene itself (Powell *et al.*, 1989). In a few cases expression of antisense RNA has given significant protection, for example to BYMV (Hammond and Kamo, 1995). Similarly, both sense and antisense constructs of TRSV coat protein gave protection against the virus, apparently by an RNA-mediated mechanism (Yepes *et al.*, 1996).

Antisense RNAs also give protection against geminiviruses (Frischmuth and Stanley, 1993). The targets have ranged from the rare mRNA of the Rep protein of TGMV and TYLCV (Day *et al.*, 1991; Bendahmane and Gronenborn, 1997) to the coat protein of ToMoV (Sinisterra *et al.*, 1999).

Antisense RNA has given some protection against viroids (reviewed by Tabler *et al.*, 1998).

It is likely that antisense protection operates by mechanisms similar to those of RNA-mediated protection described above.

6. Ribozymes (reviewed by Tabler et al., 1998)

As described in Chapter 14 (Sections I.C.3.b and II.B.4.b), ribozymes are catalytic RNAs that can cleave at specific sites in complementary target RNAs. Since the ribozyme has to be complementary to the target viral sequence, it can be considered to be an antisense RNA. Incorporation of a ribozyme into an antisense RNA to TMV gave no significant advantage over the antisense RNA itself (de Feyter *et al.*, 1996), but constructs directing PPV that contained a hammerhead ribozyme gave stronger protection than did the ordinary antisense RNA construct (Liu *et al.*, 2000).

7. Levels of protection

The reactions of various forms of transgenic protection give a great range of responses. These vary from delay in symptom production for just a few days to complete immunity. This raises the question of definitions of resistance and protection. Many transgenic plant responses do not fit the definitions outlined in Table 10.1. For instance, can the delay of symptom expression by a few days really be called resistance? Hull and Davies (1992) discuss this point, and suggest the following definitions:

> Resistance to a virus is a property of the plant that reduces virus multiplication and reduces or prevents virus spread within the plant and/or symptom expression.

> Protection is a property conferred to a plant that interferes with the virus infection cycle (the virus infection cycle includes transmission from an infected to a healthy plant and the full systemic infection of the healthy plant).

Thus, most of the phenotypes described above should be classed as protection.

Hull and Davies (1992) also go on to suggest that there should be categorization of field protection, with the seven levels describing the behavior of the transgenic plant (Table 16.9). This would enable the reader or listener to understand the level of protection afforded by the 'gene' being discussed. However, there are some problems in adopting such a system. First, the level of protection may vary between siblings in a transgenic line. In many reports on protection, results are quoted as percent or number of plants showing (or not showing) symptoms. Second, the level of protection may vary according to conditions such as temperature (Nejidat and Beachy, 1989). Third, the level of protection may vary with the generation of progeny from the transformant.

Thus, there are certain difficulties in categorizing levels of protection. For instance, the level of protection may vary according to the plant developmental stage (Table 16.10) (Jan et al., 2000b) or possibly due to environmental factors.

8. Relationship between natural cross-protection and protection in transgenic plants

The mechanism for transgenic protection against a viral infection, especially coat protein-mediated protection, has been compared with natural cross-protection or mild strain protection (see Section IV.A). There are several similarities that have been used to support the idea. (1) In both situations the degree of resistance depends on the inoculum concentration, with high concentrations reducing the observed resistance. (2) Both are effective against closely related strains of a virus, less against distantly related strains, and not against unrelated viruses. (3) In some circumstances cross-protection can be substantially overcome when RNA is used as inoculum rather than whole virus (Sherwood and Fulton, 1982; Dodds et al., 1985). Similarly, the resistance of transgenic plants expressing the coat protein of several viruses is substantially but not completely overcome when RNA is used as inoculum (see Section VII.A.1). (4) In classic cross-protection experiments, no cross-protection was observed between two rather similar viruses, AMV and TSV. In experiments with transgenic plants expressing their viral coat proteins, high resistance to infection was observed against the homologous virus and none against the heterologous virus (van Dun et al., 1988b).

On the other hand, there appear to be some differences between natural cross-protection and coat protein-induced resistance. When cross-protection between related strains of a virus is incomplete, the local lesions produced may be much smaller than in control leaves (illustrated for PVX in Fig. 17.6). This indicates reduced movement and/or replication of the superinfecting strain. Local lesions that formed in transgenic tobacco plants expressing the PVX coat were smaller than those of the controls (Hemenway et al., 1988), in line with the result for PVX shown in Fig. 17.6. However, the reduced number of local lesions that do form on transgenic plants infected with TMV became as large as those of controls, indicating no block in replication or local movement once infection was successful.

It is quite possible that there are several mechanisms that give cross-protection. One of them is likely to involve the post-transcriptional gene silencing host defense system and, thus, to resemble RNA-mediated protection.

TABLE 16.9 Proposed levels of protection

Level	Finding
1	Full immunity to a range of viruses
2	Full immunity to a range of strains of a virus
3	Full immunity to a few closely related strains of a virus
4	Subliminal infection with the virus unable to spread from initially infected cells
5	Delay in systemic infection with a description of duration
6	Reduction in severity of systemic symptoms associated with a reduction in virus titer
7	Reduction of systemic symptoms with no reduction in virus titer

From Hull and Davies (1992), with kind permission of the copyright holder, © CRC Press LLC.

TABLE 16.10 Reactions of transgenic R1 squash plants to inoculation with SqMV at different developmental stages under field conditions

Genotype	Percentage of plants showing resistance after inoculation at:[a]		
	17 DAG	31 DAG	45 DAG
Control	0	0	0
SqMV-22	0	7	7
SqMV-3	0 (21)	21	17 (8)
SqMV-127	79 (7)	93	92

[a] Three upper leaves of each transgenic plant were inoculated at 1, 3 or 5 weeks after transplanting (17, 31 or 45 DAG) (DAG, days after germination). With 1:15 diluted extract of infected squash leaves; 12 to 14 plants were inoculated in each treatment.
Values in parentheses indicate plants that displayed the recovery phenotype.
From Jan *et al.* (2000b), with permission.

9. Satellite-mediated protection

The general nature of satellite RNAs is described in Chapter 14 (Section II.B), including the ability of some satellite RNAs to attenuate the symptoms of the helper virus. It has been shown for two satellite RNAs that transgenic plants expressing the satellite RNA are less severely diseased when inoculated with the helper virus. Harrison *et al.* (1987) showed that, when transgenic tobacco plants containing DNA copies of a CMV satellite RNA were inoculated with a satellite-free CMV isolate, satellite replication occurred. At the same time, CMV replication was reduced and disease symptoms were greatly attenuated. In untransformed plants the CMV isolate caused mosaic disease and stunting. In the transformed plants, no mosaic appeared and plants grew almost as well as healthy ones. These differences persisted for 14 weeks, the longest period tested. Furthermore, the same result was obtained in plants raised from seed of the transformed plants. When transformed plants were inoculated with TAV, there was a similar attenuation of disease symptoms but without a marked decrease in TAV genome synthesis. Jacquemond *et al.* (1988) showed that tobacco plants transgenic for a CMV satRNA were tolerant to infection by aphids, the main method of field transmission for CMV.

Results similar to those of Harrison *et al.* (1987) were reported at the same time for another satellite–virus combination, satTRSV and TRSV (Gerlach *et al.*, 1987). Tobacco plants that expressed full-length satTRSV or its complementary sequence as RNA transcripts increased their synthesis of satTRSV RNA following inoculation with TRSV, but virus replication was reduced and disease symptoms were greatly ameliorated. This protection was maintained for the life of the plants.

Two distinct mechanisms of resistance were found in *N. benthamiana* with full-length of sequences of a mild variant of satGRV inoculated with GRV plus severe satRNA (Taliansky *et al.*, 1998). In one set of transformed lines, there were high levels of transcript RNA and the replication of both severe satRNA and GRV genomic RNA was inhibited. In the second set of plants there were low levels of transcript RNA, and replication of severe satRNA, but not that of GRV genomic RNA, was inhibited. It was concluded that in the first set of plants both GRV genomic and severe satRNA replication was down-regulated by the mild satRNA and in the second there was homology-dependent gene silencing of the severe satRNA. Resistant plants were also produced using only the 5' terminal one-third of the mild satRNA.

The use of satellite RNAs in transgenic plants to protect against the effect of virus infection has both advantages and disadvantages. The protection afforded is not affected by the inoculum concentration, as it is with viral coat protein transformants. The losses that do occur in transgenic plants because of slight stunting will affect only the plants that become naturally infected in the field, whereas if all plants are deliberately infected with a mild CMV–satellite combination they will all suffer some loss (Section IV.B). Furthermore, the resistance may be stronger in transgenic plants than in plants inoculated with the satellite. Inoculation is not needed each season, and the mutation frequency is lower.

Nevertheless, there are distinct risks and limitations with the satellite control strategy. The satellite RNA could cause virulent disease in another crop species or could mutate to a form that enhances disease rather than causing attenuation (see Chapter 14, Section II.B.5). Another risk is the reservoir of virus available to vectors in the protected plants. Lastly, the satellite approach will be limited to those viruses for which satellite RNAs are known.

10. DI nucleic acid-mediated protection

Defective interfering (DI) nucleic acids are described in Chapter 8 (Section IX.C). They are mutants of viral genomes that are incapable of autonomous replication but contain sequences that enable them to be replicated in the presence of the parent helper virus. In many cases, they are amplified at the expense of the parent virus, ameliorating the symptoms induced by that virus. When such nucleic acids are transgenically expressed, infection with the parent virus mobilizes and amplifies them.

Transgenic expression of DI RNAs was shown to protect *N. benthamiana* plants against the apical necrosis and death usually caused by CymRSV (Kollár *et al.*, 1993). *N. benthamiana* plants transformed to express the TBSV DI RNA were protected against that virus and closely related tombusviruses (CNV and CIRV) but were susceptible to a distantly related tombus-like virus (CymRSV) and unrelated viruses (BDMV, PVX, TMV) (Rubio *et al.*, 1999).

Transformation of *N. benthamiana* to express the DI DNA of ACMV interferes with the replication of both genomic components of that geminivirus (Frischmuth and Stanley, 1991, 1993). Serial transmission from the transgenic plants led to increasing numbers of asymptomatic plants with undetectable levels of viral DNA. The protection afforded by the DI DNA is confined to closely related strains of the virus. Similarly, the accumulation of the Logan strain of BCTV is reduced in *N. benthamiana* plants transgenically expressing the DI DNA of that virus strain (Stenger, 1994); however, there was no effect on other BCTV strains.

C. Other forms of transgenic protection

Various forms of protection against viruses have been shown for a variety of transgenes that are not derived from viruses themselves. Some of these are described in this section.

1. Transgenic plants expressing PR proteins

The PR host proteins induced following infection with viruses causing necrotic local lesions were discussed in Chapter 10 (Section III.K.1). These proteins, which are part of a non-specific host defense reaction, are involved in the phenomenon of local acquired resistance. Treatment of leaves with salicylic acid induces certain PR proteins and inhibits AMV replication in such leaves. Hooft van Huijsduijnen *et al.* (1986) have isolated and cloned the mRNAs for some of these proteins. In principle, it might be possible to provide protection against certain viruses by using 'transgenic' plants in which PR protein genes are expressed constitutively under the control of a suitable promoter, but this has not yet been proven. The evidence is against this, with no protection arising from transgenic expression of PR proteins (see Lusso and Kuc, 1996).

2. Antisense to β-1,3-glucanase

The β-1,3-glucanases are proteins believed to be part of the constitutive and induced defense system of plants against fungal infection. Unexpectedly plants deficient in these enzymes as a result of the expression of an antisense RNA show markedly reduced lesion size and number in the local lesion response of *N. tabacum* Havana 425 to TMV and in *N. sylvestris* to TNV (Beffa *et al.*, 1996). The mutant plants also showed reduced severity and delay of symptoms of TMV in *N. sylvestris*.

3. Transgenic plants expressing virus-specific antibodies

Plants do not have an immune system like that of animals in which specific antibody proteins are formed in response to an infection, and it has long been assumed that plants could not produce such proteins. However, the work of Hiatt *et al.* (1989) demonstrates that this is possible. They obtained cDNAs derived from mouse hybridoma mRNA, transformed tobacco leaf segments, and regenerated plants. Plants expressing single γ (heavy) or κ (light) chains were crossed to produce plants in which both chains were expressed simultaneously. A functional antibody made up over 1% of leaf proteins. Production of antibodies in plants (plantibodies) has been reviewed by Smith (1996) and by Zhang and Wu (1998).

The expression in plants of a single-chain Fv antibody, derived from a panel of monoclonal antibodies against AMCV coat protein, reduced the incidence of infection and delayed symptom development (Tavladoraki *et al.*, 1993).

A further suggestion is to express anti-idiotypic antibodies to important binding sites (Martin, 1998). Anti-idiotypic antibodies are antibodies made against the antigen-interacting portion of an antibody and thus have the same surface conformation as the antigen. Such an antibody should bind to a virus-specific domain in, say, a coat protein, movement protein or replicase.

The transgenic expression of plantibodies has potential for the control of plant viruses and also for determining functions of plant proteins (Fig. 16.15) (de Jaeger *et al.*, 2000), including those involved in disease determination.

Antibodies directed against other antigens, e.g. lymphoma-associated protein, have been produced in plants (McCormick *et al.*, 1999).

4. Transgenic plants expressing 2′,5′-oligoadenylate synthetase

In mammalian systems, interferons are effective antiviral molecules. When one of the components of the virus-inhibiting pathway, 2′,5′-oligoadenylate synthetase, was expressed in potato plants, it gave protection against PVX (Truve *et al.*, 1993). The virus concentration in transgenic plants was lower than that in plants expressing PVX coat protein.

5. Transgenic plants expressing ribosome-inactivating proteins (reviewed by Wang and Tumer, 2000)

Ribosome-inactivating proteins (RIPs) deglycosylate a specific base in the 28S rRNA and prevent binding of elongation factor 2. RIPs have been isolated from several plant species. Transgenic plants expressing the pokeweed (*Phytolacca americana*) RIP (termed pokeweed antiviral protein; PAP) showed protection against PVX, PVY and PLRV (Lodge *et al.*, 1993). Plants expressing a low level of a C-terminal deletion mutant of PAP were resistant to PVX (Tumer *et al.*, 1997). As the intact C-terminus is required for toxicity and depurination of tobacco ribosomes *in vivo*, it was concluded that the antiviral activity of PAP can be dissociated from its toxicity. Another RIP from pokeweed, PAP-II, is less toxic than PAP and plants expressing it are protected against TMV and PVX (Wang *et al.*, 1998c); a similar protein, PIP-2 from *P. insularis*, exhibits antiviral activity against TMV (Song *et al.*, 2000). The pokeweed antiviral protein and its applications are reviewed in Tumer *et al.* (1999).

Other RIPs that show antiviral activity include BAP from *Bougainvillea spectabilis* (Balasaraswathi *et al.*, 1998) which protected against TSWV, trichosanthin (Lam *et al.*, 1996) which gave protection against TuMV, and dianthin (Hong *et al.*, 1996) giving protection against ACMV.

6. Transgenic plants expressing ribonuclease gene pac-1

RNA viruses replicate via a complementary strand and, thus, are thought to have a dsRNA

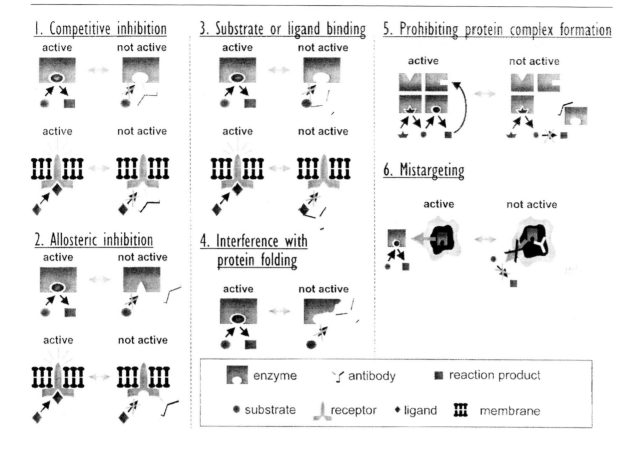

Fig. 16.15 Potential mechanisms of antibody-mediated *in vivo* modulation of protein or signal molecule activity. From de Jaeger *et al.* (2000), with kind permission of the copyright holder, © Kluwer Academic Publishers.

stage in their infection cycle. To attack these replication intermediates the yeast-derived dsRNA-specific RNase gene, *pac-1*, was transformed into tobacco plants (Watanabe *et al.*, 1995). Transformed plants showed a decrease in lesion numbers when inoculated with TMV and a delay in symptom appearance when inoculated with CMV or PVY.

7. Human cystatin C

As described in Chapter 7 (Section V.B.1), potyviruses express their genetic information as a polyprotein that is cleaved by virus-specific proteases to functional proteins. One of the viral enzymes, HCPro, is a papain-like cysteine protease. Human cystatin C, an inhibitor of cysteine proteases, interfered with the autoprocessing of PPV polyprotein by HCPro and, unexpectedly had an inhibitory effect on the NIa protease, a serine protease

(Garcia *et al.*, 1993). It was suggested that the transgenic expression of such a protease inhibitor might protect plants against viruses that had polyprotein processing in their infection cycle.

8. Transgenic plants expressing an insect toxin

The bacterium *Bacillus thuringiensis* produces a polypeptide that is toxic to insects. Different strains of this bacterium produce toxic polypeptides with specificity for different insect groups. Vaeck *et al.* (1987) produced transgenic plants expressing the toxin gene that were protected against insect attack. Among the virus vector groups of insects, only the *Coleoptera* appear to be affected by the toxin from some strains of the bacterium. In principle, it might be possible to produce transgenic plants protected against beetle vectors of some plant viruses.

D. Field releases of transgenic plants

(reviewed by Kaniewski and Thomas, 1993)

Concerns have been expressed about the release and use of plants modified by genetic manipulation. This has led to plants being produced by this means being treated in a different manner to those produced by conventional breeding techniques and being subject to specific regulatory structures.

In most countries that have regulatory structures, there are two stages in the field releases of transgenic plants. In the first stage, the plants are released under contained conditions directed by the provisions of the country's biosafety regulatory structure. This stage of release essentially has two purposes: (1) to address any potential problems that a risk assessment might identify; and (2) to assess the field performance of the transgenic lines being tested. Plant lines that satisfy both the regulators and the releasers then go to the second stage of more general release, termed commercial release or farmer release. In this section, I will discuss field performance of transgenic plants and in the next section the possible risks of transgenes that protect plants against viruses.

Testing the field performance of transgenic plants is essentially no different to testing plant lines that have been obtained by traditional breeding (Delannay *et al.*, 1989). The testing objectives include evaluating the plant appearance, typeness, growth vigor, yield and quality. Of especial importance is to assess the stability and durability of the protecting transgene under these conditions. Two main factors can affect stability and durability: (1) possible climatic effects on the expression of the transgene; and (2) the presence of protection-breaking strains or isolates of the virus that are present in the viral ecosystem but were not recognized in the initial glasshouse tests.

There have been numerous contained field trials of virus-protected transgenic plants, and lines of some crops such as potatoes and papaya are in more general release (see Perlak *et al.*, 1995; Thomas *et al.*, 1997; Gonsalves, 1998). Here I will give some examples of the results obtained.

Field experiments with tomatoes suggest that transgenic expression of TMV coat protein may be commercially useful in this host (Fig. 16.16). The transgenic plants were partially resistant to TMV and to strains L, 2, and 2^2 of ToMV. In the field, no more than 5% of transgenic plants showed systemic disease symptoms, compared with 99% for the control plants. Lack of visual symptoms was correlated

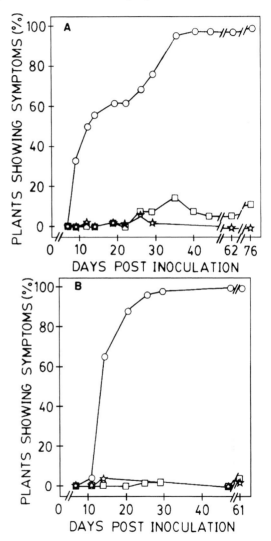

Fig. 16.16 Development of systemic symptoms of TMV infection under field conditions in tomatoes non-transgenic for coat protein. ○, Non-transformed plants; □ and ★, plants of two lines of transgenic tomatoes expressing TMV coat protein. Plants were inoculated on terminal leaflets of three successive leaves with TMV strain U1 at 10 µg/mL, 8 days after planting out in the field. Observations were made on 48 plants from each line. From Nelson *et al.* (1988), with permission.

with an absence of ToMV. In inoculated control plants, fruit yields were depressed by 26–35%. There was no evidence that expression of the coat protein gene reduced plant growth or fruit yield compared with uninoculated non-transgenic plants (Nelson *et al.*, 1988).

In Australia, most lines of potato cvs Kennebec and Atlantic containing the PLRV coat protein gene showed no measurable differences in agronomic performance when compared with non-transgenic lines in the absence of PLRV (Graham *et al.*, 1995). However, under conditions where PLRV was prevalent, one transgenic Kennebec line gave approximately 30% increased tuber yield over its non-transgenic counterpart (Barker and Waterhouse, 1999). In the USA, some Russet Burbank potato lines containing the PLRV coat protein gene showed low levels of protection against primary spread of the virus but marked reductions in the secondary spread (Thomas *et al.*, 1997). Tomato lines expressing CMV coat protein showed field resistance to the target virus but at a lower level to that found in growth chamber experiments (Tomassolit *et al.*, 1999).

The expression of CMV mild satRNA in tomato confers field tolerance to CMV (Stommel *et al.*, 1998). Yie *et al.* (1995) described the rapid production of homozygous tobacco lines protected against CMV by mild satRNA. Three of these lines were highly protected against CMV under field conditions.

Transgenic lines that are protected against mechanically inoculated target virus may not be protected when the virus is inoculated by natural vectors. For example, one line of potatoes expressing the PVY coat protein gave promising results on mechanical inoculation but was not protected against aphid inoculation (Lawson *et al.*, 1990). Similarly, transgenic plants expressing TRV coat protein were not protected against nematode transmission (Ploeg *et al.*, 1993a). Of particular concern is the stability of the protection conferred by the transgene. There are several ways in which the protection can be overcome or broken down, including:

1. There are many factors that may affect the stability of a transgene including rearrange-

ment during meiosis, and methylation. These usually show up in early generations of a transgenic line and would be unlikely to occur in later generations that were released to the field.

2. As noted in Section VII.B.7, environmental conditions may affect the expression of a transgene.

3. Protection may not always be effective with different strains of a virus. For example, tobacco plants transformed to express the coat protein of strain PLB or TCM of TRV were resistant only to the strain whose coat protein was being expressed (Angenent *et al.*, 1990). On the other hand, Nejidat and Beachy (1990) found that transgenic tobacco plants expressing the coat protein gene of TMV were resistant to ToMV and TMGMV but not to RMV. The protective effect against TMV in transgenic plants expressing coat protein was greatly reduced when tobacco plants were held at 35°C instead of 25°C. However, plants held in a 25°C–35°C night–day cycle retained resistance (Nejidat and Beachy, 1989).

4. As described in Chapter 10 (Section IV.H), some viruses encode gene products that suppress PTGS. There is concern that the suppression of PTGS by one virus may overcome the transgenic protection against another. However, Wang *et al.* (2001a) reported that coinfection by CYDV did not compromise the transgenic protection against BYDV-PAV.

5. The CaMV 35S promoter is widely used in constructs to express virus-protecting transgenes. However, in oilseed rape plants in which this promoter is being used to express transgenes, it can be silenced on infection with CaMV (Covey *et al.*, 1997; Al-Kaff *et al.*, 1998, 2000).

E. Potential risks associated with field release of virus transgenic plants (reviewed by Hammond *et al.* 1999)

A major consideration for the use of plant lines transgenically protected against viruses involves the possible risks that might arise from their release to the general environment. This has been discussed in many places (e.g. Hull, 1990a,b; de Zoeten, 1991; Hull and Davies, 1992; Tepfer, 1993; Miller *et al.*,

1997b). Essentially, there are three areas of potential risk: risk to humans and other animals, risks to the environment, and commercial risks. Risks to humans are basically any potential deleterious effects that the transgenic product may have to food and whether it has any allergenic properties. For instance, the potential risks of long-term consumption of food expressing relatively high levels of RIPs would need special consideration. However, it must be remembered that humans have been eating virus-infected plants for millenia without any obvious deleterious effects. The potential commercial risks involve the agronomic properties of the transgenic line, the durability of the protection and any possible effects of spread of the transgene to other crops. These considerations are not limited to viral transgenes but apply to most, if not all transgenes.

The area of virus-protecting transgenes that has attracted specific interest is the use of virus sequences. The basic question that is asked is: what is the risk of any interactions that might arise between a virus or virus-related sequence integrated in the host genome and another virus superinfecting that plant? Three scenarios are considered: heteroencapsidation, recombination and synergism.

1. Heteroencapsidation

Heteroencapsidation involves the superinfection of a plant expressing the coat protein of virus A by the unrelated virus B, the expression of the virus A coat protein not protecting the plant against virus B. The risk is that the coat protein of virus A might encapsidate the genome of virus B, thereby conferring on it other properties such as different transmission characteristics. Heteroencapsidation by transgenically expressed coat protein has been reported for closely related viruses, such as CMV and AMV (Candelier-Harvey and Hull, 1993), and between potyviruses (Lecoq et al., 1993). In a broader study using unrelated viruses, the particles of viruses which have different forms of stabilization, P. Candelier-Harvey and R. Hull (unpublished results) showed that heteroencapsidation occurs only between closely related viruses, the particles of which have similar forms of stabilization.

2. Recombination

The concern here is that recombination between the transgene and superinfecting virus might lead to a new virus. Recombination involving integrated sequences is discussed in Chapter 8 (Section IX.B.5). A hybrid virus, made in vitro by replacing the 2b gene of CMV by its homolog from TAV, is significantly more virulent than either of its parents (Ding et al., 1996). This potential risk is particularly pertinent to luteoviruses (Miller et al., 1997b).

3. Synergism

As described in Chapter 10 (Section V.F), synergistic interactions between two unrelated viruses are potentiated by distinct virus sequences. Thus, there is a possibility that the effect of a superinfecting virus could be exacerbated by a transgene expressing a synergism-inducing sequence.

4. Avoiding risk

Understanding the molecular interactions involved in the potential risk situations can lead to methods for the 'sanitizing' of the transgene to avoid that risk. For example, as described in Chapter 11 (Section III.E.7), aphid transmission of potyviruses involves an amino acid triplet (Asp, Ala, Gly; DAG) in the coat protein. Mutation of this in a PPV coat protein transgene abolished the aphid transmissibility but did not affect the protection offered by the transgene (Jacquet et al., 1998). Similarly, mutation in the PPV coat protein gene suppressed particle assembly, heterologous encapsidation and complementation in transgenic N. benthamiana but retained protection against ChiVMV and PVY (Varrelmann and Maiss, 2000). The understanding of the factors involved in recombination (Chapter 8, Section IX.B) will lead to transgene constructs that lessen the possibility of new molecules being formed between the transgene and a superinfecting virus. Similarly, sequences that potentiate synergism, such as the potyvirus HCPro or the cucumovirus 2b genes (see Chapter 10, Section IV.H), can be avoided.

In all these risk assessments, it is important to compare the transgenic situation with the non-transgenic situation. Thus, there is the possibility of the above potential risks occurring in mixed infections between viruses. This is discussed in Hammond *et al.* (1999).

VIII. DISCUSSION AND CONCLUSIONS

As discussed in Chapter 3, it is impossible to give precise measures of crop losses due to viruses. For a given virus, losses may vary widely with season, crop, country and locality. Nevertheless, there are sufficient data to show that continuing effort is needed to prevent losses from becoming more and more extensive. Essentially, once a plant becomes infected with a virus, in most cases it is not feasible to try to cure it. The approaches of heat treatment and meristem culturing (Sections II.C.2.b and II.C.2.c) are in practicality applicable only to high-value crops. Three kinds of situation are of particular importance: (1) annual crops of staple foods such as grains and sugar beet that are either grown on a large scale or are subsistence crops and that, under certain seasonal conditions, may be subject to epidemics of viral disease; (2) perennial crops, mainly fruit trees with a big investment in time and land, where spread of a virus disease, such as citrus tristeza or plum pox, may be particularly damaging; and (3) high-value cash crops such as tobacco, tomato, cucurbits, peppers and a number of ornamental plants that are subject to widespread virus infections.

Possible control measures have been classified under three headings: (1) removal or avoidance of sources of infection; (2) control or avoidance of vectors; and (3) protecting the plant from systemic disease. In principle, by far the best method for control would be the development of cultivars that resist a particular virus on a permanent basis. Experience has shown that viruses continually mutate in the field with respect to both virulence and the range of crops or cultivars they can infect. Thus, it has been generally considered that breeding for resistance or the development of transgenic plants is unlikely to give a permanent solution for any particular virus and crop. It has been suggested for conventional resistance to pests and diseases that one should not put too strong a selection pressure on a viral population or else a resistance-breaking strain will arise. However, Hull (1994b) has argued that, as variation arises from the replication of a virus, the more that one can inhibit the replication the less chance there is that variants will arise that can overcome the protection. Furthermore, if transgenic protection is targeted at highly conserved sequences or motifs that are essential for replication, it is highly unlikely that any variant would be able to compensate for the defect.

With almost all crops affected by viruses, an integrated and continuing program of control measures is necessary to reduce crop losses to acceptable levels. Such programs will usually need to include elements of all three kinds of control measure identified at the beginning of this discussion. As noted above, the deployment of resistant or protected cultivars or lines is one of the most promising approaches. Genetic technology is, and will increasingly, contribute to this approach. Breeding programs are being speeded up by the use of molecular marker techniques, and the co-linearity of the organization of many plant genomes is enabling the prediction of loci of potential resistance genes. Transgenic technology is providing a range of protecting genes, some of which should be relatively durable. Thus, future breeding programs for virus resistance should bring together the conventional and transgenic approaches with stacking, pyramiding and/or deployment of protecting genes to tackle the problem being addressed. In developing strategies for the integrated approach it is essential to have a full understanding of the disease, its epidemiology and ecology, and of the pathogen, its genetic make-up and functioning, and its potential for variation.

However, as well as the science involved in improving agriculture, the use of transgene technology has raised many public issues such as the impact of agricultural practices on the environment and on the food that this industry provides, and the ethics of overcoming the species barrier. Thus, the application of this technology to improving the world's food security will involve many more issues than just the science and its application.

IX. POSSIBLE USES OF VIRUSES FOR GENE TECHNOLOGY

A. Viruses as gene vectors (reviewed in Porta and Lomonossoff, 1996; Scholthof et al., 1996)

In the early 1980s there was considerable interest in the possibility of developing plant viruses as vectors for introducing foreign genes into plants. At first, interest centered on the caulimoviruses, the only plant viruses with dsDNA genomes, because cloned DNA of the viruses was shown to be infectious (Howell et al., 1980). Interest later extended to the ssDNA geminiviruses, and then to RNA viruses when it became possible to reverse-transcribe these into dsDNA, which could produce infectious RNA transcripts.

The main potential advantages of a plant virus as a gene vector were seen to be: (1) the virus or infectious nucleic acid could be applied directly to leaves, thus avoiding the need to use protoplasts and the consequent difficulties in plant regeneration; (2) it could replicate to high copy number; (3) there would be no 'position effects' of a site in the plant chromosomal DNA; and (4) the virus could move throughout the plant, thus offering the potential to introduce a gene into an existing perennial crop such as orchard trees.

Such a virus vector would have to be able to carry a non-viral gene (or genes) in a way that did not interfere with replication or movement of the genomic viral nucleic acid. Ideally, it would also have the following properties: (1) inability to spread from plant to plant in the field, providing a natural containment system; (2) induction of very mild or no disease; (3) a broad host range, which would allow one vector to be used for many species, but would be a potential disadvantage in terms of safety; and (4) maintenance of continuous infection for the lifetime of the host plant.

The major general limitations in the use of plant viruses as gene vectors are: (1) they are not inherited in the DNA of the host plant, and therefore genes introduced by viruses cannot be used in conventional breeding programs; (2) plants of annual crops would have to be inoculated every season, unless there was a very high rate of seed transmission; (3) by recombination or other means the foreign gene introduced with the viral genome may be lost quite rapidly with the virus reverting to wild type; and (4) it would be necessary to use a virus that caused minimal disease in the crop cultivar. The virus used as vector might mutate to produce significant disease, or be transmitted to other crops that were susceptible. Infection in the field with an unrelated second virus might cause very severe disease.

1. Caulimoviruses

Howell et al. (1981) inserted an eight-base-pair EcoRI linker molecule into the large intergenic region of cloned CaMV DNA (see Chapter 6, Section IV.A.1, for CaMV genome organization) and showed there was no impairment of infectivity. Gronenborn et al. (1981) reported the successful propagation of foreign DNA in turnip plants using CaMV as a vector. However, they found that the size of the DNA insert that could be successfully propagated was limited to about 250 base pairs or less.

Brisson et al. (1984) reported the first successful expression of a foreign gene in plants using CaMV as a vector. They used the 234-bp dihydrofolate reductase gene from Escherichia coli, which confers methotrexate resistance in the bacterium. They inserted the gene into a derived strain of CaMV from which most of gene II had been deleted. The chimeric DNA was stably propagated in turnip plants and the bacterial gene was expressed, as shown by assays for methotrexate resistance. De Zoeten et al. (1989) replaced the ORF II of CaMV DNA with the human interferon (IFNaD) gene. They obtained a stable strain of CaMV that replicated in Brassica rapa and led to the production of IFNaD. The interferon was located in the CaMV viroplasms and it had antiviral activity in an animal cell assay.

Paszkowski et al. (1986) constructed a hybrid CaMV genome containing the selectable marker gene neomycin phosphotransferase type II, which replaced the gene VI coding region. This construct was not viable in plants and could not be complemented in trans by wild-type CaMV. However, inoculation of

Brassica campestris protoplasts under DNA-uptake conditions gave rise to stable cell lines genetically transformed for the marker gene. This occurred only when the hybrid was coinoculated with wild-type CaMV. The mechanism for this effect is not understood. CaMV as a probe for studying gene expression in plants is discussed by Pfeiffer and Hohn (1989).

Thus, there appear to be several constraints to the use of CaMV as a gene vector. These include the packing capacity of the CaMV particle, the amount of viral DNA that can be removed without affecting the functioning of the genome, and the interactions between different parts of the genome in expression and replication. Removal of non-essential regions of the genome should enable about 1000 bp of sequence to be inserted (Fütterer *et al.*, 1990), but it is not certain whether all this sequence is really non-essential. The use of pairs of vectors with overlapping deletions has been suggested (Hirochika and Hayashi, 1991).

2. Geminiviruses

Much attention has been focused on the geminiviruses as potential gene vectors because of their DNA genomes and the fact that the small size of the genomes makes them convenient for *in vitro* manipulations (reviewed by Mullineaux *et al.*, 1992; Stanley, 1993; Timmermans *et al.*, 1994; Palmer and Rybicki, 1997). Nevertheless, this small size may restrict the amount of viral DNA that can be deleted (see Davies *et al.*, 1987). However, this is counterbalanced by the fact that for some geminiviruses a viable coat protein and encapsidation are not necessary for successful inoculation by mechanical means, or for systemic movement through the plant (Gardiner *et al.*, 1988).

There are other potential difficulties (Davies *et al.*, 1987). Recombination can occur to give parent-type molecules. Most geminiviruses are restricted mainly to the phloem and associated cells. However, the wide host range of the geminiviruses (compared with the caulimoviruses) makes them of considerable interest. The ability of some members to infect cereal crops would be particularly useful, except for the fact that they are not seed transmitted and are mechanically transmitted only with difficulty. In any event, inoculations on the scale needed for cereal crops would be impractical.

Nevertheless, model experiments have shown that a geminivirus can be used to introduce and express foreign genes in plants. Various workers have shown that an intact coat protein gene is not essential for TGMV replication and movement. Hayes *et al.* (1988d) constructed a chimeric TGMV DNA in which most of the coat protein gene had been deleted and replaced by the bacterial neomycin phosphotransferase gene (*neo* gene). They used *Agrobacterium* inoculation (agro-infection) to introduce this construct into tobacco plants. The TGMV DNA was replicated in the transgenic plants and the *neo* gene expressed. Ward *et al.* (1988) isolated a coat protein mutant of ACMV DNA1 in which 727 nucleotides in the coat protein gene were deleted. This mutant was non-infectious. However, when the coat protein ORF was replaced by the coding region of the bacterial chloramphenicol acetyltransferase (*CAT*) gene under control of the coat protein promoter, infectivity was restored, virus spread through the plant, and the *CAT* gene was expressed. Nevertheless, systemic movement of infectious constructs with larger than normal DNAs in the coat protein position exerted a strong selection pressure favoring derivatives of normal size (Elmer and Rogers, 1990).

Transient expression systems allow for the rapid screening of DNA constructs designed to study the activities of promoter sequences, RNA-processing signals, and so on, in cells, and as a preliminary screen in the construction of transgenic plants. In principle, viral DNA might be altered to provide plasmid-type vectors for high copy number and rapid expression of modified or foreign genes. The work of Elmer *et al.* (1988a) with a geminivirus progressed some distance toward the construction of such a plant plasmid. Deletion derivatives of TGMV DNA A defined the minimal DNA fragment capable of self-replication as about 1640 bp or 60% of the A sequence.

Hanley-Bowdoin *et al.* (1988) showed that following agro-inoculation of petunia leaf discs with TGMV DNA A, the viral coat protein gene was transiently expressed, beginning about 2 days after agroinfection. Constructs were then made in which the coat protein ORF was replaced by the bacterial *CAT* or β-glucuronidase (*GUS*) genes, under the control of the coat protein promoter. Following agro-inoculation, these foreign genes were also transiently expressed in petunia leaf discs.

Gröning *et al.* (1987) found that the DNA of a geminivirus infecting *Abutilon sellovianum* was located in the chloroplasts, raising the possibility of constructing a chloroplast-specific transformation vector.

As noted above, some geminiviruses such as mastreviruses have the potential for expressing genes in monocots. However, although constructs of MSV with the V2 gene substituted by a *CAT* gene or with the *GUS* gene inserted into the short intergenic region expressed the inserted gene, they did not move systemically (Lazarowitz *et al.*, 1989; Shen and Hohn, 1994). A 3.7-kbp plant cell–*E. coli* shuttle vector based on WDV has been developed (Ugaki *et al.*, 1991). Although this shuttle vector has been useful in studying viral replication maize cells (Timmermans *et al.*, 1992), its use in whole plants has yet to be demonstrated.

3. RNA viruses

The ability to manipulate RNA virus genomes by means of a cloned cDNA intermediate has opened up the possibility of using RNA as well as DNA viruses as gene vectors. In principle the known high error rate in RNA replication (see Chapter 17, Section IV.A) might place a limitation on the use of RNA viruses (Siegel, 1985; van Vloten-Doting *et al.*, 1985). The experimental evidence to date suggests that mutation may not be a major limiting factor, at least in the short term. Viruses with isometric and rod-shaped particles have been studied as potential vectors but those with rod-shaped particles have better potential as there are fewer constraints on the amount of nucleic acid that can be inserted.

a. TMV

Takamatsu *et al.* (1987) prepared a TMV cDNA construct in which most of the coat protein gene was removed and the *CAT* gene was in its place.

When *in vitro* transcripts from this construct were inoculated on to tobacco leaves, the local lesions that formed were smaller than normal, but biologically active CAT was produced, and this activity increased in the inoculated leaves for 2 weeks. RNA was extracted from these leaves and re-inoculated on to tobacco after being encapsulated in TMV protein *in vitro*. CAT activity was again detected, indicating that this replicating RNA had some degree of stability *in vivo*. Yamaya *et al.* (1988), using a disarmed Ti plasmid vector, introduced a cDNA copy of the TMV genome under the control of the 35S CaMV promoter into the genomic DNA of tobacco plants. This experiment demonstrated that a non-seed-transmitted RNA virus can be made seed transmissible.

Dawson *et al.* (1989) constructed a hybrid TMV in which the *CAT* gene was inserted between the coat protein and the 30-kDa genes (Fig. 16.17A). This construct replicated efficiently, produced an additional subgenomic RNA and CAT activity, and assembled into 350-nm virus rods. However, during systemic infection the insert was precisely deleted, giving rise to wild-type virus; it was thought that this deletion resulted from homologous recombination between the two copies of the coat protein gene subgenomic promoter. Takamatsu *et al.* (1990b) constructed, via cDNA, a TMV RNA in which an additional sequence encoding Leu-enkephalin, a pentapeptide with opiate-like activity, was incorporated just 5' to the termination codon for coat protein. The pentapeptide was expressed in protoplasts as a fusion product with coat protein.

To overcome the problem of loss of the insert through recombination between homologous sequences, Donson *et al.* (1991) developed a hybrid vector containing sequences from two tobamoviruses (TMV-U1 and ORSV (Fig. 16.17B)). In this vector, the gene of interest is expressed from the TMV coat protein subgenomic promoter and the coat protein from the ORSV promoter. There was a problem with partial deletion of one of the inserts, NPTII, but this was considered to be insert specific. This vector system has been used to express α-trichosanthin (Kumagai *et al.*, 1993), and a derivative of this vector expressing the coat

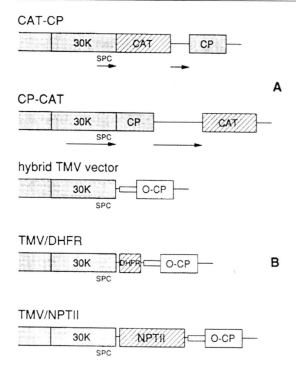

Fig. 16.17 Schematic diagrams of TMV expression vectors. **(A)** TMV mutants in which the gene coding for CAT is added into the complete TMV-U1 genome. The *CAT* gene is fused behind the subgenomic promoter for the coat protein in order to produce an additional sgRNA. The resulting sequence duplications are indicated by arrows beneath each construct. **(B)** Hybrid TMV expression vector in which all ORFs are TMV-U1 sequences, except the coat protein sequence, which originates from ORSV and is transcribed into mRNA from its own ORSV-derived, subgenomic promoter (shown as a small open box). Foreign genes coding for DHFR and NPTII are expressed via subgenomic mRNAs transcribed from the heterologous TMV-U1 promoter located in the 30-kDa ORF. SPC, subgenomic promoter for the TMV-U1 coat protein. From Porta and Lomonossoff (1998), with permission.

protein of ToMV instead of ORSV has been used to manipulate a biosynthetic pathway *in planta* (Kumagai *et al.*, 1995).

b. PVX

A similar construct to that of TMV shown in Fig. 16.17A was made with PVX with a duplication of the coat protein promoter (Chapman *et al.*, 1992). The PVX construct with the *GUS* gene being expressed from one of the promoters gave systemic infection of various *Nicotiana*

species and retained the inserted gene especially in *N. clevelandii* plants.

c. BMV

French *et al.* (1986) constructed variants of BMV RNA3 in which the coat protein gene was replaced with the bacterial *CAT* gene, or in which the *CAT* gene was inserted near the 5' end of the coat gene. When inoculated on to barley protoplasts together with normal BMV RNAs 1 and 2, these RNA3 constructs replicated and produced subgenomic RNAs equivalent to the normal coat protein subgenomic RNA (see Chapter 7, Section V.E.3.a). When the *CAT* gene was inserted in-frame with the upstream coat protein initiation codon, CAT expression exceeded that in plant cells transformed by Ti plasmid-based vectors.

Pacha *et al.* (1990) showed that all the *cis*-acting elements required for the replication of the 2.1-kb CCMV RNA3 can be contained in a 454-base replicon made of 5'- and 3'-terminal sequences. Thus, it may be possible to express a foreign gene of significant size using this replicon.

d. TRV

Angenent *et al.* (1989c) found that sequences of 340 nucleotides at the 5' end and of 405 at the 3' end of the RNA2 of TRV (strain PLB) were sufficient for replication. Constructs in which the deleted viral nucleotides were replaced with a 1401-nucleotide sequence from plasmid DNA were replicated in protoplasts co-inoculated with the complete genome of strain PLB, indicating that this virus may have potential as a gene vector.

Tobravirus vectors express foreign proteins efficiently in roots (MacFarlane and Popovich, 2000).

e. TBSV

The coat protein of TBSV is dispensable for systemic infection of certain *Nicotiana* species (Scholthof *et al.*, 1993) and hence can be replaced by foreign sequences such as the *GUS* gene (see Scholthof *et al.*, 1993; Scholthof, 1999). By placing a cDNA copy of the TBSV genome between the CaMV 35S promoter and the hepatitis Δ ribozyme, followed by the nopaline synthase gene polyadenylation signal, Sholthof (1999) was able to gain infections by directly rub-inoculating the construct. The 35S promoter and the NOS terminator controlled transcription of the construct and the cleavage of the transcript by the ribozyme delimited the 3′ end of the infectious genome. This vector enables expression of inserts into the coat protein gene region.

B. Viruses as sources of control elements for transgenic plants

As explained in Chapter 7 (Section III.D.2.d), constructs for transforming plants contain various control elements. Certain plant viral nucleic acid sequences have been found to have useful activity in these chimeric gene constructs as promoters of DNA and RNA transcription and as enhancers of mRNA translation.

1. Promoters
a. Caulimovirus promoters

Transcription of CaMV DNA gives rise to a 19S and a 35S mRNA (see Chapter 7, Section IV.C.1). Shewmaker et al. (1985) introduced a full-length copy of CaMV DNA into the T DNA of the Agrobacterium Ti plasmid, and this was integrated into various plant genomes. They showed that the 19S and 35S promoters were functional in this form in several plant species. They are both strong constitutive promoters and have found wide application for the expression of a range of heterologous genes (e.g. Balázs et al., 1985; Bevan et al., 1985; Odell et al., 1985, Lloyd et al., 1986; Nagy et al., 1986).

The 35S promoter has been found to be much more effective than the 19S promoter in several systems. For example, expression of the α-subunit of β-conglycinin in petunia plants under control of the 35S promoter was 10 to 50 times greater than that from the 19S promoter (Lawton et al., 1987). The 35S promoter was also found to be 10 to 30 times more effective than the nopaline synthase promoter from Agrobacterium tumefaciens (García et al., 1987).

Kay et al. (1987) constructed a variant 35S promoter that contained a tandem duplication of 250 bp of upstream sequences. This modification gave about a 10-fold increase in transcriptional activity. The '34S' promoter from FMV had about the same activity as the 35S CaMV promoter (Sanger et al., 1990).

The 35S promoter is nominally a constitutive one. However, Benfey and Chua (1989) showed that there was marked histological localization of expression of GUS activity in petunia under control of this promoter. By contrast, the 19S promoter directed expression of the CAT gene in a range of tissues in tobacco plants (Morris et al., 1988b).

b. Other DNA virus promoters

Properties of promoters from other DNA viruses, such as badnaviruses and geminiviruses, are discussed in Chapter 7 (Sections IV.C.2). Several of these promoters have been shown to have activity in transformation constructs but are not used as widely as the CaMV 35S promoter.

2. Untranslated leader sequences as enhancers of translation

Untranslated leader sequences of several viruses have been shown to act as very efficient enhancers of mRNA translational efficiency both in vitro and in vivo, and in prokaryotic and eukaryotic systems (see Chapter 7, Section V.C.5). AMV RNA4 is known to be a well-translated message for AMV coat protein. Jobling and Gehrke (1987) replaced the natural leader sequences of a barley and a human gene with AMV RNA4 leader sequence. These constructs showed up to a 35-fold increase in mRNA translational efficiency in the rabbit reticulocyte and wheatgerm systems. Sleat et al. (1987) made similar constructs involving the uncapped mRNAs for two vertebrate genes and the bacterial GUS gene with or without a 5′-terminal 67-nucleotide sequence derived from the untranslated region of TMV RNA (the Ω sequence). These were tested in vitro in the rabbit reticulocyte, wheatgerm and E. coli systems. The TMV leader sequence enhanced translation of almost every mRNA in every system. Gallie et al. (1987a,b) extended these results to show that the 67-nucleotide sequence was also a potentially useful enhancer *in vivo* in mesophyll protoplasts and *Xenopus* oocytes. A deletion derivative of Ω appeared to be functionally equivalent to a Shine-Dalgarno sequence in several bacterial systems (Gallie and Kado, 1989). The translational enhancement brought about by the TMV Ω sequence is mediated by the ribosomal fraction of the *in vitro* system used (Gallie *et al.*, 1988). The experiments of Sleat *et al.* (1988c) support the view that the untranslated viral leader sequences reduce RNA secondary structure, making the 5' terminus more accessible to scanning by ribosomal subunits or by interaction with initiation factors.

3. *In vitro* studies on transcription

In vitro translation systems which faithfully reproduce *in vivo* gene expression have proved very useful in animal and other systems for developing an understanding of nucleotide sequences and protein factors involved in the control of transcription. Cooke and Penon (1990) have taken a step in this direction for plant systems. They obtained a partially purified extract from tobacco cell suspensions that contained all the factors necessary for transcription from the 19S promoter.

4. Use of the TMV origin of assembly (OAS) for the introduction of foreign RNAs

Sleat *et al.* (1986) showed that a correctly oriented OAS sequence located 3' to a foreign RNA sequence could initiate the efficient encapsulation of the foreign RNA *in vitro*. In an extension of these experiments, Gallie *et al.* (1987c) prepared transcripts that encoded the OAS together with the RNA sequence for the CAT enzyme. When these were encapsulated in TMV coat protein *in vitro* and inoculated to a wide range of cell types, the CAT mRNA was transiently expressed. Immuno-gold labeling located the site of disassembly and transient gene expression in epidermal cells of inoculated tobacco leaves (Plaskitt *et al.*, 1987).

Sleat *et al.* (1988c) constructed a plasmid derivative containing the 5' leader sequence of TMV followed by the CAT sequence and the OAS. This was introduced into the DNA of tobacco plants using an *Agrobacterium* vector. Transcripts from this nuclear DNA were encap-

sulated into TMV-like rods when the transgenic plants were infected with TMV. These experiments demonstrated an efficient complementation between functions encoded in the host genome and those of the infecting virus.

5. Other possible RNA virus sequences

The subgenomic promoters for several RNA viruses have now been defined (see Chapter 7, Section V.B.2). These are used in the RNA virus vectors described above and may prove useful for gene amplification, as might 5'- or 3'-terminal sequences that have not yet been tested. The potential of simple and specific RNA enzymes (ribozymes) from satellite RNA sequences is discussed in Section VII.B.6.

C. Viruses for presenting heterologous peptides (reviewed by Lomonossoff and Johnson, 1996; Porta and Lomonossoff, 1996, 1998; Johnson *et al.*, 1997)

The production of small peptides is required for several reasons, including acting as epitopes for vaccines and biologically active peptides. As described in Chapter 15 (Section IV.A.2), epitopes are patches of amino acids that adopt specific conformations. Free peptides can act as epitopes but the immunogenicity is enhanced by presentation on the surface of a macromolecular assembly. One approach to presenting the peptide sequence in the correct conformation is to incorporate them into a viral coat protein sequence in such a way that they are exposed on the surface of the virus particle. The virus particle can then be used as a vaccine. There are several advantages to doing this with plant viruses, including: (1) the virus can be produced in large amounts and in less developed countries where the technology for animal virus vaccine production may be limited; (2) such vaccines may be given orally as part of the normal food supply; (3) the virus will not infect the human or other animal and thus is completely inactive; and (4) the system is not subject to contamination by other virulent animal pathogens. A potential disadvantage is that the high rate of mutation of RNA viruses could result in the deletion or loss of inserted sequences, especially as they would not be

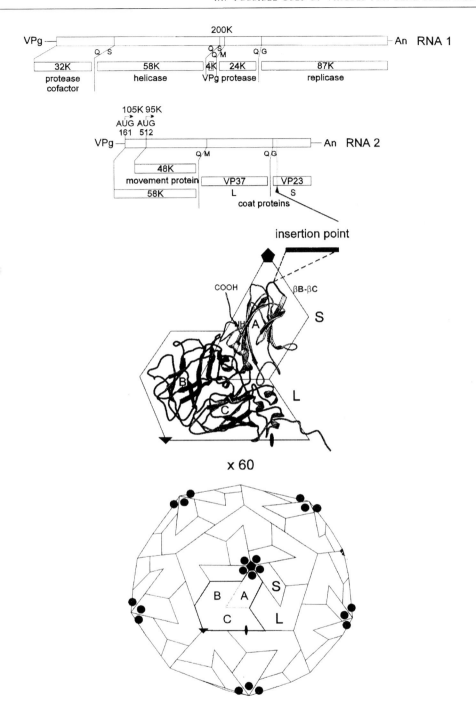

Fig. 16.18 Generation of chimeric CPMV particles. Foreign sequences are inserted into the gene for the S coat protein borne by RNA2. Both RNA1 and RNA2 are translated into polyproteins and undergo a cascade of cleavages whose sites and final products are shown. RNA2 (bearing the heterologous sequence) needs to be co-inoculated with RNA1 (unmodified) to initiate an infection in cowpea plants. S protein harboring a foreign epitope in its βB–βC loop and native L protein assemble at 60 copies each into icosahedral virus particles on which the foreign insert is expressed around the 5-fold axes of symmetry. From Porta and Lomonossoff (1998), with kind permission of the copyright holder, © John Wiley and Sons Ltd.

under selection pressure (van Vloten Doting *et al.*, 1985); however, experience is indicating that this problem may not be as significant as at first feared (Johnson *et al.*, 1997). Several plant viruses have been used for the presentation of foreign peptides.

1. CPMV (reviewed by Lomonossoff and Hamilton, 1999)

The structure of CPMV has been solved to atomic resolution (see Chapter 5, Section VI.B.6.a) (Lomonossoff and Johnson, 1991). The capsid comprises two types of protein: the L protein, which has two β-barrel domain domains, and the S protein, which has one β-barrel domain. Analysis of the three-dimensional structure suggested that loops between the β strands would be suitable for the insertion of sequences to be expressed as epitopes, as these loops are not involved in contacts between protein subunits. The βB–βC loop of the S protein is highly exposed (Lomonossoff and Johnson, 1995) and was used for most of the insertions (Fig. 16.18); some insertions have been made in other loops (Lomonossoff and Hamilton, 1999).

Early studies on inserting sequences at the βB–βC loop site gave guidelines for construction of viable, genetically stable, chimeras (Porta *et al.*, 1994). These included: (1) foreign sequences should be inserted as additions and not replacements of the CPMV sequence; (2) sequence duplication should be avoided as this led to loss of insert by recombination; and (3) the precise site of insertion was important for maximizing growth of chimeras. Understanding of these guidelines gave a standard procedure for inserting foreign DNA into the βB–βC loop of the S protein (Spall *et al.*, 1997).

Chimeras with inserts for the sequence for up to 38 amino acids have been successfully made in which the presence of foreign DNA did not significantly affect the ability of the modified virus to replicate. Various epitopes have been inserted, including ones from *Human rhinovirus 14*, *Human immunodeficiency virus type 1* and *Canine parvovirus* (Lomonossoff and Hamilton, 1999).

2. TMV

The initial attempts to express foreign peptides on TMV coat protein used an *E. coli* expression vector to produce a self-assembling chimera of the viral coat protein and a poliovirus epitope (Haynes *et al.*, 1986). With the development of infectious cDNA clones to TMV, it was possible to use a self-replicating system in plants. Fusion of a foreign sequence to the C-terminus of the coat protein prevented particle assembly (Takamatsu *et al.*, 1990b). To overcome this problem, Hamamoto *et al.* (1993) placed the insert after an amber stop codon at the C-terminus of the coat protein gene (Fig. 16.19A) so that it could be expressed as a read-through protein. Particles were assembled with about 5% of the coat protein subunits expressing the inserted sequence (Sugiyama *et al.*, 1995). Replacement of two amino acids on a surface loop near the C-terminus of the coat protein gave particles with 100% of the subunits containing the insert (Fig. 16.19B) (Turpen *et al.*, 1995), as did replacement into another part of the C-terminal region not involved in particle assembly (Fig. 16.19C) (Fitchen *et al.*, 1995).

3. Other viruses

Chimeric fusions to the coat protein of the potyvirus JGMV have been expressed in *E. coli* (Jagedish *et al.*, 1993, 1996). The extracted protein could be polymerized into potyvirus-like particles. These particles could be formed with inserts at the N-terminus or replacing some of the C-terminal sequence.

A construct that gave both fused and unfused versions of the green fluorescence protein at the N-terminus of PVX coat protein was produced by Santa Cruz *et al.* (1996). In this construct the GFP gene was separated from the coat protein by the *foot-and-mouth disease virus* 2A sequence which mediates processing at its C-terminal end. The presence of some unfused coat protein molecules was essential for the assembly of virus particles. In subsequent experiments, proteins ranging in size from 8.5 to 31 kDa were expressed as 'overcoats' on the surface of PVX particles (Santa Cruz *et al.*, 1996).

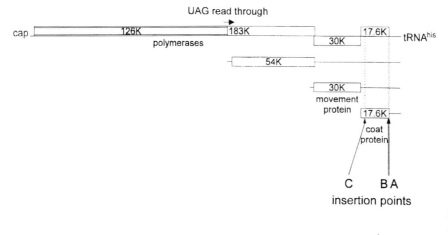

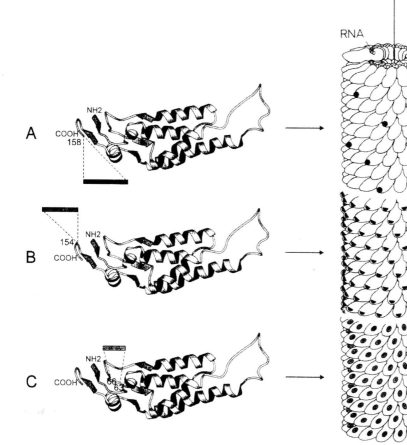

Fig. 16.19 Production of chimeric TMV particles *in planta*. Foreign oligonucleotide sequences are introduced at one of three positions, labeled A, B and C, in the gene of the TMV coat protein, which is expressed from the most 3' of the three viral subgenomic mRNAs. *In vitro* transcripts of the altered full-length genomic cDNA are inoculated on to tobacco. The resulting recombinant TMV coat proteins are represented as ribbon drawings with the numbers indicating each insertion site. Upon assembly of these coat proteins, chimeric rod-shaped virions are formed on which the foreign peptides are differentially displayed and distributed; a maximum of 5% of the coat proteins present an insert in position A. All of the coat proteins are modified in positions B and C. From Porta and Lomonossoff (1998), with kind permission of the copyright holder, © John Wiley and Sons Ltd.

TBSV particles were assembled from the construct shown in Fig. 16.20, in which an insert was made at the C-terminus of the coat protein gene (Joelson *et al.*, 1997). Similarly, the HIV p24 ORF was expressed as an in-frame fusion with a 5'-terminal portion of the TBSV coat protein ORF (Zhang *et al.*, 2000). Thus, it would appear that TBSV can tolerate insertions in both the N- and C-terminal regions of the coat protein gene.

D. Viruses in functional genomics of plants (reviewed in Lindbo *et al.*, 2001)

The gene silencing induced by virus infection of plants is described in Chapter 10 (Section

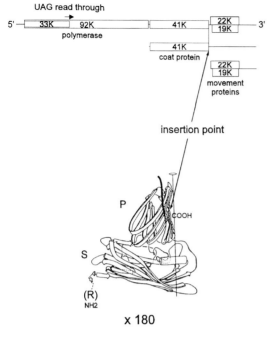

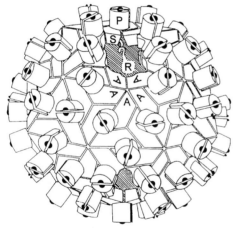

Fig. 16.20 Generation of chimeric TBSV particles. Fusions of heterologous sequences are made at the 3′ end of the coat protein gene in a full-length cDNA clone of the viral genome. The coat protein is translated from the larger of the two viral subgenomic mRNAs and comprises three domains, designated R, S and P, with P bearing the foreign amino acids (shown in black). When 180 copies of the viral coat protein, labeled A, assemble to form icosahedral particles, the fused peptides, each represented by a black half-circle, are well exposed at the surface of the virions. From Porta and Lomonossoff (1998), with kind permission of the copyright holder, © John Wiley and Sons Ltd.

IV.G). In virus-induced gene silencing (VIGS) (see Fig. 10.18) a gene incorporated into a virus vector, for instance a TMV-based vector (Kumagai *et al.*, 1995), a PVX-based vector (Ruiz *et al.*, 1998) or a TGMV-based vector (Kjemtrup *et al.*, 1998), will silence a homologous gene in a plant. As described in Chapter 10 (Section IV.H), some viruses can suppress PTGS; these would not be effective for VIGS. The VIGS phenomenon was initially shown for transgenes and has now been demonstrated for various plant genes such as phytoene desaturase (Ruiz *et al.*, 1998) and a magnesium chelatase gene (Kjemtrup *et al.*, 1998). Baulcombe (1999b) proposed that a VIGS approach could be an approach for determining the function(s) of genes identified from genome sequencing. The current approach of insertional mutagenesis has various drawbacks, such as the inability to identify genes, whose disruption is lethal before the plant has developed. With the VIGS approach, the lethality would be apparent from the death of the mature plant that had been inoculated. Thus, virus vectors could be a powerful tool in functional genomics.

E. Summary and discussion

The discussion in this section demonstrates that viruses do not always cause problems and some have their uses. As well as the uses in gene technology, some viruses have a positive horticultural importance, with cultivars being identified and named according to their virus symptoms. Thus, *Abutilon striatum* cv. Tompsonii is infected with AbMV and derives its attractiveness from the variegation induced by that virus (similar to Plate 3.7). Similarly, *Lonicera japonica* cv. Aureo-reticulata shows symptoms of HYVMV.

A wide range of gene technology has been suggested and proven for plant viruses, and several of these basic approaches have been described above. The ones that are making the most impact are the use of viral control sequences in transgenic constructs and the use of viral vectors for expressing foreign sequences. The use of VIGS in functional genomics is not yet widespread, but has great potential.

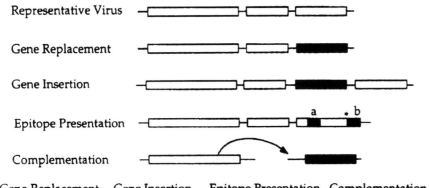

Gene Replacement	Gene Insertion	Epitope Presentation	Complementation
Bromovirus	*Caulimovirus*	*Comovirus*	*Alfamovirus*
Caulimovirus	*Geminivirus*	*Tobamovirus*	*Caulimovirus*
Furovirus	*Potexvirus*	*Tombusvirus*	*Dianthovirus*
Geminivirus	*Potyvirus*		*Geminivirus*
Hordeivirus	*Tobamovirus*		*Potexvirus*
Potexvirus			*Tobamovirus*
Tobamovirus			*Tombusvirus*
Tobravirus			
Tombusvirus			

Fig. 16.21 Comparison of strategies used to express foreign genes (black box) from different viruses. White boxes indicate viral genes. The epitope presentation method (a) involves translational fusion of a small sequence inside the coat protein gene or (b) translational read through of an amber stop codon (*) at the 3′ end. From Scholthof *et al.* (1996), with kind permission of the copyright holder, © Annual Reviews. www.AnnualReviews.org

Approaches to using viruses as expression vectors have been grouped under four headings by Scholthof *et al.* (1996) (Fig. 16.21). Most of these uses have been discussed above and can be found in more detail in the review of Scholthof *et al.* (1996).

The DNA plant viruses did not realize their initial expectations as gene vectors. There are several reasons for this including their high rates of recombination, the close integration of viral gene functions and packaging constraints. However, these viruses provide important control sequences for transformation constructs and will probably continue to do so, especially with those that give tissue-specific expression.

Although it was initially suggested that RNA viruses would have problems with accumulation of errors due to replication under non-selecting conditions, this has not proved to be so. One major advantage that rod-shaped viruses have is that they do not have the con-straints on packaging large increases in the size of the viral genome that isometric viruses have. Theoretically, there are no limits on the size of nucleic acid that could be packaged in a rod-shaped particle, but it is likely that limits will be found in the future.

Plant viruses have great potential in the production of biologically active peptides and vaccines. However, there are various constraints that will have to be borne in mind if they are to be used on a commercial scale. These include the need to reduce the symptoms caused by the virus so that the host is not overly damaged, prevention of spread of the modified virus to other crops where it could cause severe infections, control of spread of any food products from the crop bearing the modified virus into the food supplies of humans and other animals, and, of course, quality control of the modified virus to detect any non-functional mutants.

Variation, Evolution and Origins of Plant Viruses

Like other living entities, viruses substantially resemble the parent during their replication, but can change to give rise to new types or 'strains'. This inherent variation enables viruses to adapt to new and changing situations. Over longer periods of time new viruses arise and there must have been a time at which the archetypical virus arose. For many groups of organisms an understanding of the evolutionary pathways can be obtained from the fossil record over geological time, but viruses do not leave the conventional form of fossils. However, the increasing molecular information that is accruing on plant (and other) viruses is revealing what can be termed 'molecular fossil' information. Even so, our knowledge of the pathways of virus evolution is quite fragmentary, although there is no doubt that viruses have undergone, and continue to undergo, evolutionary change that is sometimes rapid. New strains provide the raw material for such change.

From the plant pathologist's point of view, the existence of viral strains in the field that cause different kinds of disease is often a matter of considerable practical importance. Reliable criteria are needed for distinguishing and identifying strains. For these reasons the study of virus strains is an interesting and important aspect of plant virology.

In this chapter, I discuss the variation of plant viruses and how this information is giving an understanding of viral origins and evolution.

Virus acronyms are given in Appendix 1.

I. STRAINS OF VIRUSES

A. Quasi-species

As pointed out in Chapter 2 (Section I.C.2), a virus species is not a uniform population but is a quasi-species. The concept of quasi-species is fundamental to the understanding of virus variation and evolution; it is discussed in detail by Eigen (1993), Domingo et al. (1995), Smith et al. (1997), Domingo (1999) and Domingo et al. (1999a). The term quasi-species describes a type of population structure in which collections of closely related genomes are subjected to a continuous process of genetic variation, competition and selection. Usually, the distribution of mutants or variants is centerd on one or several master sequences. The selection equilibrium is meta-stable and may collapse or change when an advantageous mutant appears in the distribution. In this case, the previous quasi-species will be substituted by a new one characterized by a new master sequence and a new mutant spectrum. The stability of a quasi-species depends upon the complexity of the genetic information in the viral genome, the copy fidelity on replication of the genome and the superiority of the master sequence.

A quasi-species has a physical, chemical and biological definition. In the physical definition, a quasi-species can be regarded as a cloud in sequence space which is the theoretical representation of all the possible variants of a genomic sequence. For a ssRNA virus of 10 kb the sequence space is $4^{10\,000}$. Thus, the quasi-species cloud represents only a very small proportion of the sequence space and is constrained by the requirements of gene and nucleic acid functions. Chemically, the quasi-species is a rated distribution of related non-identical genomes. Biologically, a quasi-species is the phenotypic expression of the population, most likely dominated by that of the master sequence.

B. Virus strains

The definition of a virus species is given in Chapter 2 (Section I.C) and, as discussed in the previous section, is essentially a cloud in sequence space. This makes a strict definition of a strain difficult, if not impossible. However, one has to describe variants within a species and, in reality, take a pragmatic approach. Characters have to be weighed up as to how they would contribute to making subdivisions and to communication, not only between virologists, but also to plant pathologists, extension workers, farmers and many other groups. This can be illustrated by two examples.

BWYV is a luteovirus with a wide host range including sugar beet in the USA and for many years BMYV, which infected sugar beet in Europe, was regarded as a strain of BWYV (see Waterhouse et al., 1988). Confusion arose when it was realized that the European luteovirus that was most closely related to BWYV did not infect sugar beet but was common in the oilseed rape crop. This caused many problems in explaining to farmers that the BWYV (*Beet western yellows virus*) in their overwintering oilseed rape crop would not infect their beet crop the next year. In an analysis of nucleotide sequences of 38 isolates of BYMV and BMYV from Europe, Iran and the USA, de Miranda et al. (1995b) identified three distinct sequence groups. The first contained isolates that could infect both oilseed rape and beet, the second infected only oilseed rape and the third only sugar beet. It was suggested that groups 1 and 2 be named BWYV (possibly BWYV-1 and BWYV-2) and the third group, BMYV.

The second example is RTBV that has two major strains or isolate groups, one in southeast Asia and the other in the Indian subcontinent (Druka and Hull, 1998). These two strains differ in nucleotide sequence by about 25% and at the amino acid level by about 12–40% depending on ORF (Hull, 1996). Thus, using

the criteria used for potyviruses (see below), these should be considered as two distinct viruses. However, the strains cause the same symptoms and cross-interact with their transmission helper virus, RTSV, in a similar manner. Thus, it would be confusing to farmers and extension workers to give them different names, although useful to virologists to indicate that they are different strains.

II. CRITERIA FOR THE RECOGNITION OF STRAINS

A virus species might be defined simply as a collection of strains with similar properties. Sometimes we wish to ask whether two similar virus isolates are identical or not; on other occasions we have to decide whether two isolates are different virus species or strains of the same species. Two kinds of properties are available for the recognition and delineation of virus strains: (1) structural criteria based on the properties of the virus particle itself and its components; and (2) biological criteria based on various interactions between the virus and its host plant and its insect or other vectors. These criteria were discussed in Chapter 2 in relation to the general problem of virus classification. Serological properties are based on the structure of the viral protein or proteins, but, because of their practical importance, serological criteria are considered here in a separate section.

A. Structural criteria

1. Nucleic acids

a. Heterogeneity in viral RNA genomes

As described above, all the RNA genomes that have been examined have been found to exist not as a single nucleotide sequence but as a distribution of sequence variants around a consensus sequence. Most of the variants in a culture of a particular virus strain will normally consist of base substitutions at various sites, perhaps with some deletions or additions of nucleotides. However, more substantial variation can occur when one or more segments of a multi-partite genome are not under selection pressure. Variants may arise quite rapidly, and these could lead to confusion in strain identification. For example, genomic RNAs 3 and 4 of BNYVV code for functions required for fungal transmission in the soil (Chapter 11, Section XII). When virus was isolated from leaves of infected sugar beets, where there is presumably no selection pressure to maintain the integrity of these genes, a wide range of deletion mutants was recovered (Burgermeister *et al.*, 1986). This potential for rapid change must be borne in mind when considering nucleic acid differences as criteria for recognizing virus strains.

b. Methods for assessing nucleic acid relationships

i. Nucleotide sequences

Nucleotide sequences give valuable information about the extent of relationships between viruses. Determination of even relatively short sequences can provide useful information. However, to study the relationships between strains of a virus effectively it is necessary to determine the full nucleotide sequence of at least one, or preferably several, isolates. In earlier studies on nucleotide sequences, where sequence was determined directly on the labeled viral RNA, the microheterogeneity noted above was not detected. The sequence determined would normally be the consensus sequence for that particular culture (e.g. Swinkels and Bol, 1980). When cDNA clones of RNA viruses are used for sequencing, the resulting sequence may, by chance, contain sequences that are not typical of the consensus sequence. This possibility is checked by the sequencing of several clones covering the same section of the genome. Care should also be taken on possible sequence errors contributed by the reverse transcriptase in making cDNA from virion RNA (Bracho *et al.*, 1998). To ensure that a functional genome has been sequenced, it is necessary to use a cloned DNA that has been shown to be infectious (e.g. Lazarowitz, 1988) or a cloned DNA that can be transcribed into infectious RNA (e.g. Dawson *et al.*, 1986).

In considering nucleotide sequences as a measure of relatedness between virus strains, it must be remembered that the extent and distribution of such differences may vary quite widely depending on the virus and the part of the genome examined. To take some examples: (1) the coat protein gene and 3' untranslated region of three strains of PVY and two strains of WMV2 had 83% and 92% identity respectively (Frenkel *et al.*, 1989). There were a few clusters of non-identical nucleotides but most were distributed more or less randomly throughout the sequence studied. (2) By contrast, the coat protein and 3' untranslated regions of four strains of PPV and two strains of TEV had 97–99% identity (Frenkel *et al.*, 1989). Similarly, the 39-nucleotide 5' leader sequences of the RNA4 of seven strains of AMV were identical except at position 26, where A, G or U residues were found (Swinkels and Bol, 1980). Sequences with one or more recognition function(s) may often be highly conserved. For example, based on sequence homology in the 3' non-translated region, it has been proposed that PepMoV should be regarded as a strain of PVY (van der Vlugt *et al.*, 1989). However, in considering these examples it must be remembered that certain sequences, especially the 5'- and 3'-terminal regions are important for the control of replication and expression of the genome, and hence may be less variable than other regions.

ii. Hybridization

Nucleic acid hybridization experiments can give valuable information concerning the degree of base sequence homology between the nucleic acids of different virus isolates, but interpretation of the data may not be straightforward.

The basis for nucleic acid hybridization was discussed in Chapter 15 (Section V.C). A significant advantage of hybridization procedures is that a comparison can be made between total genome RNAs, or RNA segments. This contrasts with serological tests, for example, which usually compare antigenic sites on the coat protein, which, for many viruses, represents only about one-tenth or less of the total genome. However, the degree of nucleic acid homology estimated between different strains will depend on various experimental factors such as the stringency of the hybridization conditions and the conditions for enzymatic removal of unpaired nucleotides (see Fig. 17.1). Figure 17.1 also illustrates the use of the dot blot method in a semi-quantitative manner.

Koenig *et al.* (1988a) found that quantitative dot blot hybridization provided a very sensitive method for distinguishing closely related viruses that could barely be distinguished in serological tests. On the other hand, they observed some unexpected cross-hybridization between viruses belonging to different taxonomic groups. These cross-reactions may have resulted from real base sequence similarities, as has been found between some viruses in separate groups, and may also have resulted from the use of random-primed cDNAs.

iii. Heterogeneity mapping
Heterogeneity mapping is a method based on RNA hybridization that allows the detection of single point mutations provided they occur in a significant proportion of the molecules. The method takes advantage of the ability of RNase A to recognize and excise single-base mismatches in RNA heteroduplexes (Winter *et al.*, 1985). Labeled (–)-sense RNA probes are transcribed

from cDNA clones of the RNAs. Following hybridization with the test RNA, and digestion by RNase A and T, fragments are separated and sized by PAGE. The method has been used to assess the extent and location of heterogeneity among strains of CMV (Owen and Palukaitis, 1988) and satellite RNAs of this virus (Kurath and Palukaitis, 1989b), and makes possible the assessment of entire populations of RNA molecules for major sites of heterogeneity. By contrast, the sequencing of individual clones gives precise and detailed information on a few molecules that may make up a tiny fraction of the virus population. Polymorphisms between heterologous duplexes from cRNA transcripts from different virus isolates can be recognized by their electrophoretic mobilities (Rosner *et al.*, 1999). This method distinguished isolates of PNRSV.

iv. Restriction fragment length polymorphism
The cutting of dsDNA with restriction endonucleases can reveal polymorphisms (RFLPs). dsDNAs can be obtained directly from DNA viruses or as cDNA produced from RNA viruses by reverse transcription. The usual application

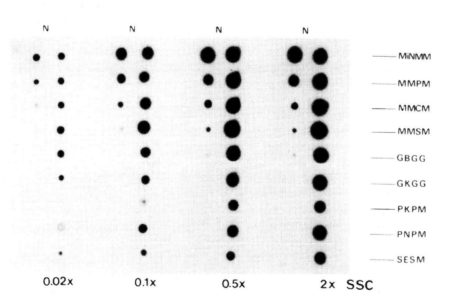

Fig. 17.1 Use of the dot blot technique to estimate degree of relationship between strains of a virus. Autoradiograph showing the extent of sequence similarity between MSV-MNM and other MSV strains. Four identical filters were each spotted, on the right with 2 ng of DNA from different MSV strains under test and on the left with doubling dilutions (2 ng → 7.8 pg) of MSV-MnM (N) as controls. Filters were hybridized with a full-length nick-translated clone of MSV-MnM (N) prior to washing under conditions of different stringency (0.02, 0.1, 0.5 and 2 × SSC at 65°C). From Boulton and Markham (1986), with permission.

of this technique is to detect differences between viral genomes (e.g. Hull, 1980; Kruse *et al.*, 1994; Villegas *et al.*, 1997), but it can also be used to show similarities between genomes.

v. Single-strand conformation polymorphisms

Analysis of single-strand conformation polymorphisms (SSCPs) is a powerful method for detecting differences (or similarities) between genomes. When the (+) and (–) strands of a dsDNA are separated by heat treatment, they attain metastable sequence-specific folded structures that can be detected by their electrophoretic mobilities in non-denaturing polyacrylamide gels (Orita *et al.*, 1989b); even single nucleotide changes are detectable under the right conditions. A non-isotopic variant of SSCP analysis was used for comparing genomes of BNYVV (Koenig *et al.*, 1995).

Both SSCP and RFLP can be determined on PCR products after reverse transcription of viral RNAs.

vi. Other properties of the genomic nucleic acid

Size of the viral nucleic acid may differ with different virus strains, for example with various TRV isolates (e.g. Cooper and Mayo, 1972).

c. Additional genes

The RNA2 of the TCM strain of TRV is considerably larger than that of other strains that have been sequenced, for example PSG. Angenent *et al.* (1986; 1989b) showed that the greater length was due partly to a repetition of 1099 3′ nucleotides from RNA1, which includes a 16-kDa ORF, and partly to a 29-kDa ORF that was unique to this RNA2 (see Fig. 6.38). The 29-kDa ORF has no significant homology with the 28.8-kDa ORF of PSG RNA1, and its origin and function, if any, are not established.

d. Subgroups of strains

When a sufficient number of strains has been examined, nucleotide sequence relationships may be used to delineate subgroups of strains of a virus. For example, based on competition hybridization tests, 30 strains of PSV could be divided into two groups with little homology between them, but extensive homology within

groups (Diaz-Ruiz and Kaper, 1983). Similarly, CMV strains have been subgrouped (see Section VII.C.4).

e. Some limitations concerning base sequence data

A given base substitution, deletion or addition may have very different effects in the protein resulting, depending on a number of circumstances. The following factors may be important:

1. Because the genetic code is degenerate, many base substitutions cause no change in the amino acid being coded for. For example, in the TMV strain $L_{II}A$, derived from the L strain, there were 10 base substitutions, seven of which occurred in the third position of in-phase codons and did not influence amino acid sequence (Nishiguchi *et al.*, 1985).

2. A given base substitution may result in change to an amino acid of very similar properties, which causes very little change in the protein. Alternatively, the change may be to a very different kind of amino acid (e.g. from an aliphatic side-chain to an aromatic one), giving rise to a viable protein with changed physical properties or to a non-functional mutant that does not survive.

3. A deletion or addition of one or two bases will cause a frameshift mutation with greater or lesser effect depending on whether it is near the beginning or the end of the gene, whether a second change (addition or deletion) brings the reading frame back to the original, and how many proteins are coded for by the section of nucleic acid in question.

4. A more general problem in using base sequence data for classification is that some parts of the genome, and some products, may have multiple functions. Some parts of the genome may code for a single polypeptide, but others may code for more than one; in addition, some polypeptides may have more than one function. Some parts of the genome may have both coding and control or recognition functions; other parts may have just control functions. Furthermore, mutations in one gene may affect the production of another. For example, mutations in the presumed polymerase gene of TMV

mutant $L_{II}A$ cause reduced synthesis of the cell-to-cell movement protein (30 kDa), thus reducing efficiency of movement of this strain (Watanabe *et al.*, 1987b).

Thus, even if we knew the full base sequences for the nucleic acids of a set of viral strains, it would be unwise to use these sequences to establish degrees of relationship without other information. It was once thought that a virus classification scheme based only on nucleotide sequence would be the ultimate aim (Gibbs, 1969). It is now apparent that the significance to be placed on nucleic acid base sequence data can be judged from a biological point of view only in conjunction with knowledge of the organization of the genome and the functions and interactions of its parts and products. The use of infectious clones and molecular biological techniques is helping in the determination of the association between sequence and biological variation.

2. Structural proteins

The coat protein or proteins and other structural proteins found in viruses are very important, both for the viruses and for virologists wishing to delineate viruses and virus strains. The coat proteins of the small RNA viruses must have evolved to give a satisfactory balance between three important functions.

1. The ability to self-assemble around the RNA; mutants are known in which this function is defective even at normal temperatures. For example, strain PM2 of TMV cannot form virus rods with RNA. The protein aggregates *in vitro* at pH 5.2 to form long, open, flexuous, helical structures rather than compact rods (Zaitlin and Ferris, 1964);
2. Stability of the intact particle inside the cell, and during transmission to a fresh host plant.
3. The ability to disassemble to the extent necessary to free the RNA for transcription and translation. A variant that could not carry out this function would not survive in nature. For example, Bancroft *et al.* (1971) described a mutant of CCMV induced by nitrous acid that was unable to be uncoated in the cell, and was therefore non-infectious

in spite of the fact that RNA isolated by the phenol procedure from the virus was highly infectious.

For the small RNA viruses, the coat protein is of particular importance for the delineation of viruses and virus strains. Besides the intrinsic properties of this protein (size, amino acid sequence, and secondary and tertiary structure), many other measurable structural properties of the virus depend largely or entirely on the coat protein. These include serological specificity, architecture of the virus, electrophoretic mobility, cation binding and stability to various agents. Thus, ideas on relationships within groups of virus strains, based on properties dependent on the coat protein, may be rather heavily biased. On the other hand, if mutations in the non-coat protein genes have occurred more or less at the same rate as in the coat protein during the evolution of strains in nature, then such views on relationships may be reasonably well based.

The *Potyvirus* genus will serve to illustrate the use of the properties of coat proteins in the delineation of virus strains. Potyviruses have been one of the most difficult virus groups to study taxonomically. The group contains about one-tenth of all the known plant viruses. The viruses infect a wide range of host plants and exist in nature as many strains or pathotypes differing in biological properties such as host range and disease severity. It has been considered by some workers that strains of potyviruses may form a continuous spectrum between two or more otherwise distinct viruses, making delineation of viruses and groups of strains difficult or impossible. However, comparisons between the amino acid sequences of the coat proteins of several viruses and many strains indicate that this approach may provide a useful basis for taxonomy within the group (Shukla and Ward, 1989a,b). Analysis of the 136 possible pairings between a set of viruses and strains revealed a clear-cut bimodal distribution, with distinct viruses having an average sequence homology of 54%, while strains averaged 95% (see Fig. 2.2). These data give no support for the 'continuous spectrum' idea among the potyviruses. Distinct viruses showed major differences in length of their coat proteins (Fig. 17.2). Major differences in amino acid sequence were near the N-termini, with high homology in the C-terminal half of the proteins. On the other hand, strains have very similar N-termini.

Two exceptions to this pattern appear to reflect the misplacing of certain potyvirus isolates on the basis of previous data. Serological tests suggested that PeMV and PVY were only distantly related, yet the sequence data shown in Fig. 17.2 clearly indicate that PeMV should be considered a strain of PVY. SMV-N and SMV-V, formerly considered to be strains of SMV but, when shown to have a sequence homology of 58%, were considered as two distinct viruses (Shukla and Ward, 1988). In both cases and in others a pragmatic approach has to be taken in assigning species status.

High-performance liquid chromatography of tryptic peptides may be useful in differentiating potyviruses and their strains (Shukla *et al.*, 1988a). This technique does not provide the detailed information obtained by amino acid sequencing, but its greatest value may lie in the ease with which the method can be applied.

The projecting (P) domain of *Tombusvirus* coat proteins has a more variable amino acid sequence than the structural (S) domain (Hearne *et al.*, 1990). Thus, greater variability at the exposed surface may be a feature of both rod-shaped and icosahedral plant viruses.

3. Non-structural proteins

The non-structural virus-coded proteins are not much used in delineating strains. Mayo *et al.* (1982) could detect no difference in the tryptic peptides obtained from the VPg of different strains of RpRSV or TBRV. Some strains of TMV may differ widely in the amino acid sequence of their 30-kDa proteins, as discussed by Atabekov and Dorokhov (1984). Four structural proteins of the two serotypes of PYDV (SYDV and CYDV) fell into two groups based on peptide mapping. Proteins M and N differed little between the strains, whereas M_2 and G were significantly different (Adam and Hsu, 1984).

4. Proportion of particle classes

The proportion of particles with differing sedimentation rates found in purified virus preparations or in crude extracts may vary quite widely with different strains of a virus or members of a virus genus. Variation of three kinds can be distinguished:

1. In relative amounts of top component (empty protein shells). For example, among the tymoviruses, the proportion of empty shells to viral nucleoprotein is usually in the range of 1:2 to 1:5 for TYMV and 10:1 to 15:1 for OkMV. Even quite closely related strains may vary in this property, for example strains of RCMV (Oxelfelt, 1976). For some multipartite viruses the proportion of top component

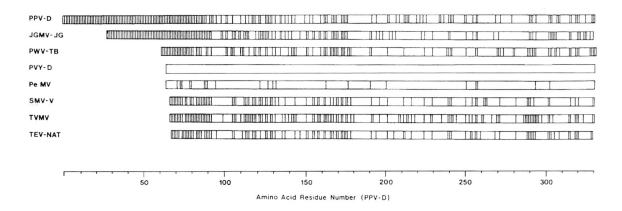

Fig. 17.2 Schematic diagram showing the location of amino acid sequence differences between seven distinct members of the *Potyvirus* genus and PeMV. The sequences were compared with strain D of PVY, the type member. PeMV is very similar to PVY in its coat protein sequence. From Shukla and Ward (1988), with permission.

has been shown to depend on a function of one RNA species.

2. The proportion of nucleic acid components encapsidated may vary in different strains of viruses with multipartite genomes, for example AMV (van Vloten-Doting *et al.*, 1968). Again, nucleoprotein proportions may be under the control of a particular RNA species.

3. Abnormal particle classes may be produced by particular strains. Thus, Hull (1970a) described an isolate of AMV producing considerable amounts of particles longer than the B component.

It should be remembered that the proportion of particle classes can be affected by factors other than the strain of virus. These include: (1) time after infection; (2) host species; (3) environmental conditions; (4) system of culture, for example the proportion of TYMV empty protein shells is higher in infected protoplasts than in whole leaf tissue; and (5) isolation procedure.

5. Other structural features

a. Architecture of the virus particle

Related viruses will be expected to have very similar size, shape and geometrical arrangement of subunits. However, significant differences in particle morphology have been found within groups of related strains. Differences in rod length are frequent between strains of helical viruses such as TRV (e.g. Cooper and Mayo, 1972) and BSMV (Chiko, 1975). Sometimes the variation in architecture appears to be 'abnormal' even though the strain of virus is a viable one. Thus, the packing of the coat protein of the Dahlemense strain of TMV involves a periodic perturbation of the helix (Caspar and Holmes, 1969). Some AMV strains contain abnormally long particles that have the normal diameter but contain more than one RNA molecule (Hull, 1970a: Heijtink and Jaspars, 1974).

b. Electrophoretic mobility

The electrophoretic mobility of a virus depends in the first place on the amino acid composition of the protein and second on the three-dimensional structure, which affects the availability of ionizing groups. Mobility is also dependent on the ions present in the buffer used. Isolates of PEMV that differed in aphid transmissibility also differed in electrophoretic behavior (Fig. 2.7) (Hull, 1977b).

c. Stability and density

Among the small RNA viruses, differences in stability and density have been used to differentiate virus strains. The RNA content of the virus may vary with strain and thus affect buoyant density in strong salt solutions (e.g. Lot and Kaper, 1976). However, differences in the coat protein most commonly lead to a difference in stability or density.

B. Serological criteria

The nature of antigens and antibodies, the basis for serological tests, and their advantages and limitations are discussed in Chapter 15 (Section IV). This section considers the use of serological criteria to delineate viruses and virus strains.

1. Some general considerations

a. Presence or absence of serological relationship

Serological tests provide a useful criterion for establishing whether two virus isolates are related or not. Any of the tests described in Chapter 15 can be applied, but most commonly some modification of the precipitation reaction or ELISA tests is used. Provided adequate precautions are taken, serological tests can be valuable for placing viruses into groups.

If two virus isolates show some degree of serological relationship, it is highly probable that they will have many other properties in common and belong in the same virus group. There are a few unexplained exceptions. Various examples are known of viruses that undoubtedly belong in the same group but that show no serological cross-reactivity, for example TYMV and EMV in the *Tymovirus* genus. In making tests for serological relationships, there are several potential sources of error:

- Presence in viral antisera of antibodies reacting with host constituents such as the abundant protein ribulose 1′,5′-bisphosphate carboxylase.

- Non-specific precipitation of host materials in crude extracts.
- Non-specific precipitation of viral antigens, especially at high concentrations.
- Contamination of antigen preparations with other viruses.
- Virus altered during isolation. It should always be borne in mind that virus may be altered during isolation in a way that can affect its serological specificity.
- Non-reciprocal positive reactions. To demonstrate that two viruses are serologically unrelated, reactive antisera must be prepared against each of the viruses under test. It must be shown that each reacts with its own antiserum, but gives no reaction with the heterologous antiserum. This reciprocal test is necessary because the viruses might in fact be related, but one may occur in too low a concentration in the extracts to give any positive reaction. Negative one-way tests are of little value. As discussed in the next section, it is preferable to use high-titer antisera to demonstrate a lack of serological relationship.
- Isolates taken from the field may be mixtures of several different serotypes (e.g. Dekker et al., 1988).

These considerations apply particularly to the use of polyclonal antisera. The use of MAbs avoids several of these problems, but they have

limitations of their own as discussed in Section II.B.3.b.

b. Degrees of serological relationship

i. Among a group of virus strains

A considerable amount of experimental work has been directed toward determining degrees of relatedness within groups of strains and in attempts to correlate serological properties with other biological and chemical characteristics. Delineation of virus strains is a particularly important aspect of any programe designed to produce resistant varieties of a host species.

If two isolates of a virus are identical, they will respond identically when cross-reacted with each other's antisera, whatever form of serological test is applied. If, however, they are related but distinct, some degree of cross-reaction will be observed, at least with polyclonal antisera, although the reactions will not be identical. Various types of serological test can be used to identify and distinguish virus strains. Examples are given in Table 17.1.

When a group of only two or three virus isolates is to be considered, it is a relatively simple matter, provided technical precautions are observed, to determine whether the isolates are unrelated serologically, whether they are identical, or whether they show differing degrees of relationship. Using the same set of isolates and the same antisera, quite reproducible results can be obtained, to indicate, for example, that

Table 17.1 Some serological procedures used for the delineation of viruses and virus strains

Procedure	Virus or virus group	Reference
Indirect ELISA	Tymo-, tombus-, como-, tobamo-, potex-, carla-, poty-viruses	Koenig (1981)
F(ab′)$_2$ ELISA	Carlaviruses	Adams and Barbara (1982)
Radial double diffusion in agar, ELISA and SSEM	Cucumoviruses	Rao et al. (1982)
Quantitative rocket immuno-electrophoresis	PMV	Berger and Toler (1983)
Electroblot immuno-assay	Tymo-, tombus-, como-, nepo-, tobamo-, potex-, carla-, poty-viruses	Burgermeister and Koenig (1984)
Direct or indirect ELISA	Ilarviruses and AMV	Halk et al. (1984)
Various ELISA procedures	GFLV strains	Huss et al. (1987)
Indirect protein A–sandwich ELISA	Tobamoviruses and virus strains	Hughes and Thomas (1988)
Indirect ELISA	MSV isolates	Dekker et al. (1988)
Immunocapture–PCR	PVY^{NTN} and CSSV	Weidemann and Maiss (1996) Hoffmann et al. (1997)

strains A and B are closely related and that both are more distantly related to strain C. However, when large numbers of related strains are tested, the situation may become quite complex and less and less meaningful as more strains are considered in relation to one another.

ii. Experimental variables

There are a number of important experimental variables that can affect the estimated degree of serological relationship between viruses and strains. (1) A major source of experimental variation is the variability in antisera, both in successive bleedings from the same animal and in sera from different individuals. The proportion of cross-reacting antibody present in a series of bleeding taken over a period of months from a single animal may vary widely (Koenig and Bercks, 1968). (2) The extent to which antisera to two virus strains cross-react is usually correlated with the antibody content of the serum. Sera of low titer show lower cross-reactivity, and those with high titers show greater cross-reactivity. Thus, to detect serological differences between closely related strains using polyclonal antisera it is preferable to use antisera of fairly low titer. To demonstrate distant serological relationships, it may be necessary to use high-titer antisera. (3) Many virus preparations used for immunization and for antibody assay may contain varying amounts of free coat protein or coat protein in various intermediate states of aggregation or in a denatured state. Coat protein in the intact virus may lose amino acids through proteolysis. Antibodies reactive with coat protein in these various forms may or may not indicate the same sort of relationships as antibodies against intact virus. An example of this sort is discussed in relation to the *Potyvirus* genus in Section II.B.4. The method used to detect and assay cross-reacting and strain-specific antibodies may affect the apparent degree of relationship. Examples are given in the references listed in Table 17.1.

c. The serological differentiation index

In spite of all the variables, useful assessment of degrees of serological relationship can be obtained by testing successive bleedings from many animals and pooling the results. Most quantitative measurements of degrees of serological relationship have been carried out using precipitation titers. In such tests, the extent of serological cross-reactivity can be expressed by a serological differentiation index (SDI) (van Regenmortel and von Wechmar, 1970) (see Chapter 15, Section IV.A.3.d). The SDI values are equal to the difference in those titers expressed as negative log2. For example, such replicated comparisons have been made for sets of tobamoviruses (van Regenmortel, 1975) and tymoviruses (Koenig, 1976).

ELISA can also be used to calculate SDIs as a measure of the extent of serological relatedness between viruses or virus strains (Jaegle and van Regenmortel, 1985; Clark and Barbara, 1987). Table 17.2 shows a comparison of the SDIs obtained from ELISA and precipitin tests. There were differences in the reciprocal SDIs found by ELISA for pairs of viruses, and the average of these values did not correspond closely to those found by precipitin SDIs. Nevertheless, both kinds of test show clearly that CGMMV is substantially different from the other tobamoviruses.

Clark and Barbara (1987) describe a more refined statistical procedure for calculating SDIs from ELISA tests that is capable of discriminating reliably among virus strains that differ by as little as 0.2 SDI.

2. Role of virus components in serological reactions

There is no good evidence that plant viral ssRNA can elicit RNA-specific antibodies. Antibodies formed in response to injection with a plant virus react only with the virus protein, either in the intact virus or as various partial degradation products of the intact protein shell. A formal demonstration of the role of the protein was made by Fraenkel-Conrat and Singer (1957). They carried out mixed reconstitution experiments between serologically distinct strains of TMV. The artificial hybrid virus had the serological type of the protein used to coat the RNA, but the progeny following infection had protein of the type from which the RNA was obtained.

Nevertheless, the viral RNA may play some secondary role in stimulating production of

Table 17.2   Serological differentiation indices (SDIs) for pairs of tobamoviruses calculated from ELISA and precipitin tests[a]

Tobamoviruses		SDI from ELISA			Average SDI from precipitin test	Sequence similarity in viral coat proteins(%)
x	y	y–anti-x	x–anti-y	Average value		
TMV	ToMV	1.4	0.5	0.9	1.2	82
U2	ToMV	0.6	1.9	1.3	1.9	70
TMV	RMV	2.0	1.2	1.6	2.1	44
TMV	U2	2.4	1.9	2.1	2.7	74
ToMV	RMV	1.9	0.7	1.3	4	47
U2	RMV	3.5	2.5	3.0	4.5	46
RMV	CGMMV	7.0	6.4	6.7	5	30
U2	CGMMV	2.6	8	5.3	6	33
TMV	CGMMV	5.4	6.9	6.2	6.8	36
ToMV	CGMMV	5.7	6.9	6.3	7	33

From Jaegle and van Regenmortel (1985), with kind permission of the copyright holder, © Elsevier Science.

antibodies against the viral protein shell. Intact TYMV is substantially more immunogenic than the apparently identical empty protein shell, which contains no RNA (Marbrook and Matthews, 1966). This difference was found in rabbits and mice using several routes of injection. The difference persisted throughout the time course of the primary response and was also found in the response to a second injection. Isolated TYMV RNA injected at the same time did not augment the immunogenicity of the empty protein shells. Artificial empty protein shells produced from the infectious virus *in vitro* were no more immunogenic than the natural empty shells.

Non-infectious TYMV nucleoprotein was just as immunogenic as infectious virus. Thus, it was concluded that the enhanced immunogenicity of the nucleoprotein must be due to the physical presence of the RNA inside the particle. Whether this enhanced immunogenicity of the viral nucleoprotein is a general feature of plant viruses remains to be determined. TMV appears to be more immunogenic than either protein rods or subunits (Marbrook and Matthews, 1966; Loor, 1967). The mechanism by which the ssRNA within the virus stimulates immunogenicity is not yet understood.

dsRNAs can be immunogenic. dsRNA antisera react with dsRNA but not dsDNA or single-stranded nucleic acids. The antisera lack specificity for particular double-stranded nucleic acids (see Chapter 15, Section IV.A.2).

There are several reasons why we would expect intact viruses (such as TMV and TYMV) and protein subunits or subviral aggregates prepared from them to differ in the antigenic sites they possess:

- Some antibody-combining sites on the intact virus may be made up of parts of the exposed surface of two or more subunits. Such a site would not exist in isolated subunits.
- Subunits probably have characteristic combining sites, which are masked when the subunits are packed into the intact shell.
- Conformational changes occur when the subunits aggregate, so that the configuration of the exposed surface of the packed subunit may not be the same as when it exists as a monomer.

Examples of all these phenomena are known among plant viruses and their protein subunits.

3. Procedures used for delineating viruses and strains

a. Assay methods

The various serological methods that are used in the detection and assay of viruses are discussed in Chapter 15 (Section IV.B). Most of these procedures have been used for delineating viruses and virus strains. Some examples are listed in Table 17.1. This list reflects the fact that ELISA procedures have become the most popular for the delineation of viruses and strains.

b. Monoclonal antibodies

The advantages and disadvantages of using MAbs for assay, detection and diagnosis of viruses were summarized in Chapter 15 (Section IV.D). The outstanding value of MAbs in the delineation of virus strains is that their molecular homogeneity ensures that only one antigenic determinant is involved in a particular reaction.

The high specificity of this single interaction is not swamped in a large number of other interactions as with a polyclonal antiserum. Provided a MAb can be found that recognizes a small antigenic change between two virus strains, then very fine distinctions can be made in a reproducible manner. Although the ability of different MAbs to distinguish single amino acid exchanges may vary widely, some may be able to do so (Al-Moudallal *et al.*, 1982).

However, there are several limitations in the use of Mabs. (1) There is usually no immunoprecipitation between MAbs and viral protein monomers. (2) MAbs are often sensitive to minor conformation changes in the antigen such as may be caused by detergent or by binding of antigen to an ELISA plate (e.g. Dekker *et al.*, 1987). (3) Among a set of virus strains the relative reactivity of different MAbs may vary considerably. For example, TMV strain 06 differs from TMV by residue exchanges at positions 9, 65 and 129 of the coat protein. In the tests shown in Fig. 17.3, this strain reacted like TMV with MAb a, more strongly than TMV with MAb c, and not at all with MAb b. (4) MAbs may be heterospecific, that is, they may frequently react more strongly with other antigens than with the virus used for immunization (see Chapter 15, Section IV.D). The reaction of strain 06 with MAb c, which is stronger than that with TMV, the strain used as the immunogen, illustrates this phenomenon (Fig. 17.3).

If, during the selection of hybridoma clones for the isolation of MAbs, the clones are tested only with the strain of virus used as immunogen, MAbs with low affinity for this strain may go undetected and be discarded. Among such MAbs may be clones that would be very useful for the detection of other strains. Thus, when searching for strain-specific MAbs it is important to screen clones against a panel of structural relatives of the immunogen.

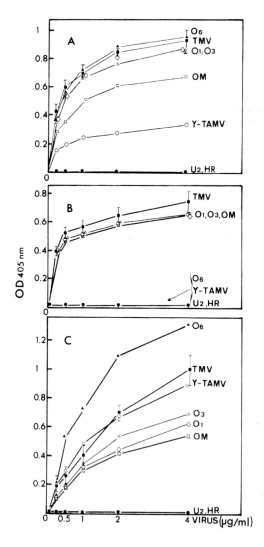

Fig. 17.3 Detection of tobamoviruses by indirect ELISA using three MAbs (a, b and c) obtained using TMV as immunogen. Strains 01, 03 and 06 are orchid strains and 0M is the common Japanese strain of TMV; Y-TAMV is a strain of tomato mosaic virus; U2 and HR are two distinct tobamoviruses. From van Regenmortel (1984c), with kind permission of the copyright holder, © Blackwell Science Ltd.

Another potential limitation of MAbs in the delineation of strains can occur if two strains have an identical antigenic determinant in common. If, by chance, the MAb specificity is directed against this determinant, the strains will appear identical even though they may have substantially different determinants elsewhere in the molecule. For example, strain Y-TAMV is a member of the ToMV group of strains that has an 18% difference in coat protein amino acid sequence compared with TMV. The two viruses are readily distinguished by polyclonal antisera but not by some MAbs (e.g. antibody c in Fig. 17.3).

These limitations highlight the importance of generating diverse panels of MAbs for the delineation of viruses and strains. Experiments with ToMV will further illustrate the problem. Strains of this virus are considered to be serologically quite uniform. Ten MAbs raised against the virus reacted in an identical manner with 15 ToMV strains and isolates. However, two of them cross-reacted with TMV and RMV (Dekker *et al.*, 1987).

4. Antigenic sites involved in the serological delineation of viruses and strains

Because of the crucial role they play in inter-subunit bonding, the sides of protein subunits that make up the shells or rods of a particular virus might be expected to be fairly constrained in the extent to which amino acid replacements would allow the subunit to remain functional. This would be particularly so with rod-shaped viruses (and also certain isometric viruses) in which RNA–protein interactions are also important. One might expect much less constraint on that part of the protein subunit that makes up the surface of the virus, and that would therefore also provide the antigenic sites of the intact virus. This expectation has been confirmed for members of the *Potyvirus* genus.

Biochemical and immunological evidence suggested that the N-terminal 29 amino acids of TEV were hydrophilic and located at or near the virus surface (Allison *et al.*, 1985). Mild proteolysis by trypsin of the particles of six distinct potyviruses showed that the N- and C-terminal regions of the coat proteins are exposed at the particle surface. Trypsinization removed 30–67

residues from the N-terminus and 18–20 from the C-terminus, the length removed depending on the virus (Shukla *et al.*, 1988b). This proteolysis left a fully assembled, infectious virus particle containing protein cores consisting of 216 or 218 amino acids. Electroblot immunoassays with polyclonal antisera showed that potyvirus-specific antigenic sites are located in the trypsin-resistant core protein region. Thus, antibodies to the dissociated core protein should react with most potyviruses. On the other hand, the surface-located N-terminus is the only large region in the coat protein that is unique to a particular potyvirus, and most virus-specific antibodies should react with this region. This fits with the amino acid sequence data, which showed that the N-terminal region is the most variable in potyvirus coat proteins (see Fig. 17.2). It has been known for some time that potyviruses become partly degraded on storage. The use of partially degraded virus as an immunogen or in antigenic analyzes may account for many of the contradictory reports in the literature concerning serological relationships among the potyviruses (Shukla *et al.*, 1988b) (see also Fig. 2.3).

Shukla *et al.* (1989a) developed the following simple procedure to remove cross-reacting group-specific antibodies. The virus-specific N-terminal region of the coat protein of one potyvirus was removed using lysylendopeptidase. The truncated coat protein was then coupled to cyanogen bromide-activated Sepharose. By passing antisera to different potyviruses through such a column, the cross-reacting antibodies were bound. Antibodies that did not bind reacted only with the homologous virus and its strains, as judged by electroblot immunoassays.

The practical utility of this procedure has been demonstrated for a group of 17 potyvirus isolates infecting maize, sorghum and sugarcane in Australia and the United States whose taxonomy was in a confused state (Shukla *et al.*, 1989b). The results demonstrated that the 17 strains belong to four distinct potyviruses, for which the names JGMV, MDMV, SCMV and SrMV were proposed.

Electroblot immunoassays using native and truncated coat proteins (minus the N-terminus) can be used to screen MAbs to determine whether they are group specific or virus specific. This procedure was used to distinguish MAbs that were virus specific from those that reacted with two or more, and sometimes all 15, of the potyviruses tested (Hewish and Shukla, reported in Shukla and Ward, 1989b).

Using panels of MAbs to ACMV, ICMV and OLCV, epitope profiles have been obtained for various whitefly-transmitted geminiviruses (see Swanson et al., 1992; Konaté et al., 1995; Harrison et al., 1997). These profiles reveal MAbs that react with several viruses and those that are virus or even strain specific. The epitope profiles of 12 begomoviruses is illustrated in Fig. 17.4.

5. Production of antibodies against defined antigenic determinants

Geysen et al. (1984) described a procedure for the rapid concurrent synthesis of hundreds of peptides on solid supports. These had sufficient purity to be used for ELISA. Using sets of such peptides and antisera against a virus, immunologically important amino acid sequences on the virus could be closely defined. This procedure has been applied successfully to the analysis of both polyclonal sera and MAbs raised against potyviruses (Shukla et al., 1989c). This work opened up the possibility of using synthetic peptides corresponding to defined antigenic sites as immunogens to generate group-specific, virus-specific and perhaps some strain-specific serological probes. Intrinsic and extrinsic factors affecting antigenic sites in relation to the prediction of important amino acid sequences are discussed in Berzofsky (1985). The ability to be able to express single-chain variable fragments from cloned cDNAs on the surface of phage particles (see Chapter 15, Section IV.E) gives another approach to obtaining highly specific antibodies.

6. Antibodies against non-structural proteins

Variation in non-structural proteins can sometimes be used to define strains. For instance, Chang et al. (1988) used the serological reactions of nuclear inclusion proteins to study relationships between a set of potyviruses and potyvirus strains.

7. Other uses of strain-specific antisera

Besides the use of serological methods for establishing relationships between plant viruses, strain-specific antisera provide very useful reagents for various kinds of experiments. For example, antisera specific for ToMV strains have been used to monitor the effectiveness of the protection given by infection of tomatoes with mild strains of ToMV to superinfection with wild strains (Cassells and Herrick, 1977), and to study the mechanism of cross-protection (Barker and Harrison, 1978).

Strain-specific antisera were used to show that, when tobacco leaf protoplasts were

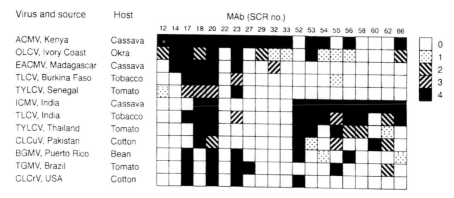

Fig. 17.4 Epitope profiles of 12 begomoviruses from six hosts, illustrating the differences between viruses from the same host in different continents, and similarities among viruses from different hosts in the same continent. Strengths of reaction range from imperceptible (0) to very strong (4). From Harrison and Robinson (1999), with kind permission of the copyright holder, © Annual Reviews. www.AnnualReviews.org

doubly infected with two TMV strains, some progeny rods contained a mixture of both coat proteins (Otsuki and Takebe, 1976b). Antibodies specific for TMV strains were used to study the conditions under which phenotypically mixed rods of TMV could be formed *in vivo* and *in vitro* (Atabekova *et al.*, 1975; Taliansky *et al.*, 1977). Purcifull *et al.* (1973) used strain-specific antisera for several potyviruses to show that the protein found in the inclusion bodies induced by each strain was distinct, unrelated to the viral coat protein, and independent of host species in which the virus was grown. The site of initiation and direction of TMV assembly were elegantly confirmed by Otsuki *et al.* (1977) using strain-specific antibody.

C. Biological criteria

1. Symptoms

a. Macroscopic symptoms

As noted above, symptom differences are of prime importance in the recognition of mutant strains. However, the extent of differences in disease symptoms may be a quite unreliable measure of the degree of relatedness between different members of a group of strains. Symptoms produced by different virus strains in the same species and variety of host plant may range from the symptomless 'carrier' state, through mosaic diseases of varying degrees of severity, to lethal necrotic disease. Figure 17.5 illustrates the range of systemic symptom types produced by four strains of TSV in tobacco. The strains are sufficiently closely related that experimental reassortment experiments are possible between them.

The diseases produced by a given set of strains in one host plant may not be correlated at all with the kinds of disease produced in another host species. Most viruses, including many of widespread occurrences such as TMV, PVX, PVY, AMV and CMV, occur as numerous strains in nature. Many 'new' viruses have been described primarily based on symptoms and other biological properties, which have turned out later to be a strain of one of these com-

monly occurring viruses. Some viruses appear to have given rise to relatively few strains as judged by symptoms, for example PLRV in potato varieties.

A set of defined cultivars that give differential local lesion responses may provide a particularly useful and rapid method for delineating strains among field isolates of a virus. However, the important influence of environmental conditions on local lesion responses must be controlled.

A virus causing severe disease is often said to be more 'virulent' than one causing mild disease. From what has been said in other sections, it should be apparent that the description can be applied only to a given strain of the virus inoculated into a particular variety of host plant in a specific manner and growing under particular environmental conditions.

A named variety of host plant, especially a long-established one, may come to vary considerably in its reaction to a given strain of virus, due, for example, to the fact that seed merchants in different localities may make different selections for propagation. This may add a further complication to the identification of strains by means of symptoms produced on named cultivars. Nevertheless, a systematic study of symptoms produced on several host species or varieties under standard conditions may help considerably to delineate strains among large numbers of field isolates of a virus.

b. Cytological effects

The cytological changes induced by different strains of a virus are often readily distinguished. Differences are of three kinds: (1) in the effects on cell organelles; (2) in the virus-induced structures within the cell; or (3) in the distribution or aggregation state of virus particles within the cell. Such differences may be of increasing importance in the delineation of viruses and virus strains. However, other factors may cause variation in the extent of differences between strains. For example, various strains in the stock culture of TYMV have markedly different effects on chloroplasts in cells of systemically infected leaves (see Fig. 9.18), but these differences may be much less

Fig. 17.5 Control of disease expression by the viral genome. Variation in chronic disease symptom type caused by four TSV isolates in tobacco. **(a)** The 'Standard' North American strain. Tobaccos became more or less symptomless. **(b)** A strain causing toothed margins on the leaves. **(c)** A strain in which tobaccos continue to show mosaic and necrotic symptoms. **(d)** A strain causing severe chronic stunting. These symptom types can be artificially re-assorted by making crosses between top, middle and bottom components of the various strains (see Fulton, 1975). (Courtesy of R.W. Fulton.) From Matthews (1991).

marked or non-existent in the infected cells of local lesions.

Different strains of TuMV show differences in the morphology of their cylindrical inclusions (McDonald and Hiebert, 1975).

Ultrastructural changes in both nucleus and cytoplasm of oat cells infected with BYDV strains differed between strains that were specific for a particular aphid vector and those that were not (Gill and Chong, 1979).

Different strains of AMV may differ markedly in the way in which virus particles form aggregates within infected cells (e.g. Hull *et al.*, 1970; Wilcoxson *et al.*, 1975). The characteristic viroplasms found in cells infected with caulimoviruses (see Fig. 8.23) may vary with different strains (Givord *et al.*, 1984; Stratford *et al.*, 1988). The variation may be associated with differences in gene II and the proportions of the products of genes II and IV. Mixed infections with two variants of BYDV in oats gave rise to altered patterns of effects in vascular tissue, including a predisposition for the xylem to become infected (Gill and Chong, 1981).

2. Host range and host plant genotype

Host ranges of viruses generally are discussed in Chapter 3 (Section V). Many strains of a virus may have very similar host ranges, but others may differ considerably.

Similar responses of a set of host plant genotypes to two viruses may provide good evidence that they are related strains (e.g. Schroeder and Provvidenti, 1971). On the other hand, a loss in ability to infect a particular host may be brought about by a single mutation. Dahl and Knight (1963) studied 12 mutants isolated from ToMV that had been treated with nitrous acid. One of these strains had lost the capacity to infect tomato.

Strains of a virus that have different host ranges often produce different disease symptoms on some common host. This is not always so. For example, four strains of TMV that were not clearly distinguishable by symptoms on *N. tabacum* or on common varieties of *Lycopersicon esculentum* could be differentiated by their host ranges on a set of *Lycopersicon* hosts, including two varieties of *L. esculentum* and three selections of *L. peruvianum* (McRitchie and Alexander, 1963). Strains of PVX have been grouped according to their reactions to a range of host plant genotypes (Cockerham, 1970).

3. Methods of transmission

Different arthropod vector species or different races of a single species may differ in their transmission of various strains of the same virus. Differences may be of the following kinds.

- In the percentage of successful transmissions, for example MDMV strains by aphid species (Louie and Knoke, 1975).
- In minimum acquisition time by the vector, for example MDMV in aphid vectors (Thongmeearkom *et al.*, 1976).
- In the length of the latent period, for example strains of PEMV in its aphid vector (Bath and Tsai, 1969).
- In the time that the vectors remain infective (e.g. Thongmeearkom *et al.*, 1976).
- Some strains may not be transmitted at all by particular vectors, for example strains of PYDV and leafhopper species (Black, 1941).

Patterns of transmissibility by three aphid species have allowed large numbers of field variants of BYDV found in North America to be placed into five main groups (Rochow, 1979), which are now recognized as different species. The quite stable groupings have facilitated studies on the distribution of virus variants both geographically and in successive seasons.

If one strain of a virus is transmissible by mechanical means all others usually are too. However, there are reports of marked variation in mechanical transmissibility depending on both host clone and virus strain, for example AMV in alfalfa (Frosheiser, 1969). Defective strains may occur in which the RNA is not coated or is incompletely coated with protein. Such strains will not be mechanically transmissible except under conditions where they are protected from attack by nucleases.

4. Cross-protection

The mechanism of cross-protection is discussed in Chapter 10 (Section V.D).

Early observations on the interactions between virus strains led to the development of the concept of cross-protection. It was shown by McKinney (1929) that tobacco plants infected with a green mosaic virus (TMV) developed no further symptoms when inoculated with a yellow mosaic virus.

Salaman (1933) found that tobaccos inoculated with a mild strain of PVX were immune from subsequent inoculation with severe strains of the virus, even if inoculated after only 5 days. They were not immune to infection with the unrelated viruses, TMV and PVY. This phenomenon, which has been variously called cross-protection, antagonism or interference, was soon found to occur very commonly among related virus strains. It is most readily demonstrated when the first strain inoculated causes a fairly mild systemic disease and the second strain causes severe symptoms or necrotic local lesions. Development of such lesions can be readily observed and a quantitative assessment can be made. Interference between related strains can also be demonstrated by mixing the two viruses in the same inoculum and inoculating to a host that gives distinctive lesions for one or both of the two viruses or strains. Interference by type TMV with the formation of yellow local lesions by another strain is shown in Fig. 10.22.

For a time, cross-protection tests were given considerable weight in establishing whether two virus isolates were related strains or not, but subsequent developments have indicated the need for caution. Among a group of strains that on other grounds are undoubtedly related, some may give complete cross-protection, while with other combinations protection may be only partial. This is illustrated in Fig. 17.6 for strains of PVX in *Datura*.

Some virus strains do not appear to cross-protect at all. Thus, none of the strains of BCTV protects against the others in water pimpernel *Samolus parviflorus* (Raf) (Bennett, 1955).

Within a set of isolates that are undoubtedly related strains, all possibilities may exist: reciprocal cross-protection of varying degrees of completeness, unilateral cross-protection and no cross-protection, as was found for strains of TSV in tobacco (Fulton, 1978). The other factor that may make cross-protection tests ambiguous is that there can be quite strong interference between some unrelated viruses (Bos, 1970).

Most experiments on cross-protection have been carried out using mechanical transmission. Cross-protection may also occur in the plant with viruses transmitted in a persistent manner by insect vectors. Thus, Harrison (1958) found that infection with a mild strain of PLRV protected plants against infection with a severe strain introduced by the aphid vector *Myzus persicae* (Sulz). Cross-protection also occurs in viroids (Chapter 14, Section I.D.7).

5. Productivity

Different strains of a virus may vary widely in the amount of virus produced in a given host under standard conditions. For example, the common strain of TMV was the most productive, and other naturally occurring strains varied over a range down to about one-tenth that of common TMV when productivity was measured as the number of local lesions produced from inocula made from extracts of single local lesions produced in *N. tabacum* cultivar Xanthi nc (Veldee and Fraenkel-Conrat, 1962).

Chemically induced mutants also varied widely in productivity, and all were less productive than common TMV. Some of these strains caused severe symptoms in certain hosts, but there was no correlation between severity of disease and productivity. Productivity appeared to be a genetically stable character as it remained fairly constant for a given mutant when tested after successive transfers. Chemical mutation quite frequently increased the severity of disease produced, but

Fig. 17.6 Cross-protection by strains of PVX in *Datura tatula*. **(A)** Healthy leaf inoculated with a strain giving necrotic local lesions. **(B)** Leaf previously systemically infected with a very mild strain of the virus, and showing complete protection against inoculation with the necrotic strain. **(C)** Leaf previously systemically infected with a mottling strain, and showing only partial protection. From Matthews (1949b), with permission.

rarely if ever increased the productivity. From a type culture of TMV, B. Kassanis (quoted in Matthews, 1991) isolated strains causing slowly spreading bright yellow local lesions, usually without systemic spread, in White Burley tobacco (see Fig. 10.22). Virus content of these yellow lesions was extremely low. Such strains are difficult to maintain in the laboratory and would never survive in nature.

6. Specific infectivity

Bawden and Pirie (1956) showed that infectivity per unit weight of purified type TMV was greater than that of a *Datura* strain when they were tested in *N. glutinosa*. There is some evidence suggesting that the protein coat of a virus may be involved in differences in specific infectivity at least between different viruses. Thus, Fraenkel-Conrat and Singer (1957) found that RMV had only about 5% of the specific infectivity of common TMV. However, when RMV RNA was reconstituted with common TMV protein, the specific infectivity was about four times higher than the RMV preparation that provided the RNA. Reconstituted TMV usually has a lower specific infectivity than the intact virus. The reason for the increase when the RMV RNA was coated with type TMV protein is not known, but might be due to the relative ease with which intact RMV and the RMV RNA reconstituted with type protein are uncoated *in vivo*.

7. Genome compatibility

The possibility of carrying out viability tests with mixtures of components from different isolates of viruses with multi-partite genomes provides a functional biological test of relationship. Such tests were carried out with TRV strains by Sänger (1969). Only two of the 20 combinations he tested gave a functional interaction. Members of the *Nepovirus* genus show various degrees of compatibility in genetic reassortment experiments (Randles *et al.*, 1977).

Rao and Francki (1981) found that the RNAs 1, 2 and 3 of three strains of CMV were interchangeable in all combinations. However, with TAV, a distinct virus in the *Cucumovirus* genus, only RNA3 could be exchanged with those of the CMV strains. Similarly, only RNA3 could be successfully exchanged between two members of the *Bromovirus* genus, BMV and CCMV (Allison *et al.*, 1988). The incompatibility of RNAs 1 and 2 of these viruses is presumably due to the way in which their gene products interact (see Chapter 8, Section IV).

Genome compatibility can be tested in a more direct fashion when the gene products can be isolated and their function is known. For example, Goldbach and Krijt (1982) showed that the protease coded for by CPMV did not process the primary translation products of other comoviruses. The transcriptase activities found in the particles of two rhabdoviruses (LNYV and BNYV) did not carry out transcription with the heterologous virus (Toriyama and Peters, 1981).

8. Activation of satellites

Particular isolates of TNV will support the replication of some STNVs but not others (Uyemoto and Gilmer, 1972). Similarly, among the cucumoviruses and the small satellite RNAs found in association with them, some viruses support the replication of particular satellites while others do not (Chapter 14, Section II).

D. Discussion

In considering use of the various possible criteria for the delineation of virus strains, we must bear in mind that, from a strictly genetic point of view, complete nucleotide sequence data would be sufficient to establish relationships between strains. Nevertheless, small changes in nucleotide sequence could have very different phenotypic effects. At one extreme a single base change in the coat protein gene could give rise to changes in several of the phenotypic properties noted earlier. On the other hand, several base changes might give rise to no phenotypic effects at all. For practical purposes, phenotypic characters such as host range, disease symptoms and insect vectors must usually be given some weight in delineating and grouping virus strains. One further consideration in the delimitation of virus strains is how to differentiate them from virus isolates. A common mistake is to name different virus isolates as strains when there are no real differences between them. This is discussed further in Section III.

III. ISOLATION OF STRAINS

The property of a new strain that first allows it to be distinguished from other known strains of a virus has been almost always biological — usually a difference in disease symptoms in some particular host. There are several ways in which new strains may be obtained. Where the virus is mechanically transmissible, it is usual to pass new isolates through several successive single local lesion cultures, if a suitable host is available. This is done to ensure as far as possible that a single strain is being dealt with, and that no unrelated contaminating virus is present in the culture. There is good evidence that a single virus particle or infection unit can give rise to a local lesion (see Chapter 12, Section II.E). On the other hand, there is ample evidence that new mutants soon appear.

As discussed in Section I.A, a virus culture actually consists of a large range of variants with usually one being dominant. Many mutants arise even during the development of a single local lesion. For example, such mutants have been found in U1 TMV passaged through single lesions (García-Arenal et al., 1984). The various chemical and physical methods described in this chapter give useful information only because they are not sufficiently sensitive to detect the small proportion of any particular mutant or variant present in a culture.

A. Strains occurring naturally in particular hosts

Different strains of a virus frequently occur in nature, either in particular host species or varieties or in particular locations. These can be cultured in appropriate host plants in the greenhouse.

B. Isolation from systemically infected plants

Plants systemically infected with a virus frequently show atypical areas of tissue that contain strains differing from the major strain in the culture. These areas of tissue may be different parts of a mosaic disease pattern (see Chapter 9, Section IV.D) or they may be merely small necrotic or yellow spots in systemically infected leaves, for example in tobacco plants infected with mild mottling strains of PVX. When such areas or spots are dissected out and inoculated to fresh plants, they may be shown to contain distinctive strains.

On occasions, routine passage of a virus through a host either by mechanical inoculation or by vector transmission can result in symptom variants. An example of this is the separation of variants of RTBV during routine transmission in rice with the leafhopper vector, *Nephotettix virescens* (Cabauatan et al., 1995). Four symptom variants (named strains) were isolated and their genomes have subsequently been sequenced, revealing various nucleotide substitutions, insertions and deletions (Cabauatan et al., 1999).

The preparation of protoplasts from systemically infected leaves, even when these are showing apparently uniform symptoms, offers the possibility of a very fine 'dissection' of the leaves. Natural mutants may be revealed among such protoplasts either by regenerating plants from them (Shepard, 1975) or by inoculating test plants with virus from a single cell (Fraser and Matthews, 1979a).

C. Selection by particular hosts or conditions of growth

A particular strain may multiply and move ahead of others in a certain plant. Such a host can be used to isolate the strain. Similarly, strains may differ in the rate at which they multiply and move at different temperatures in a given host. Holmes (1934) found that tomato and tobacco stems inoculated with a severe strain of TMV and incubated at 35°C subsequently contained mild strains, which were able to multiply readily at 35°C, although the original severe strain was not. This result has since been amply confirmed by others. For example, Lebeurier and Hirth (1966) used serial passage in tobacco at successively higher temperatures to isolate a strain of TMV that grew effectively at 36°C.

Low temperatures have also been used to isolate strains (McGovern and Kuhn, 1984). It seems reasonable to suppose that selection of strains by particular hosts or conditions of growth may sometimes involve selection of a strain with a cell-to-cell movement protein that is better adapted to the host or conditions than the previously dominant strain. A small sub-population of virus exists in a U1 TMV culture that can cause the hypersensitive reaction in *Nicotiana sylvestris*. This subpopulation moves upward rapidly and is selected for in the upper parts of the plant during rapid growth (bolting) of the shoot axis (Khan and Jones, 1989).

The new concepts of host defense systems based on gene silencing and the ability of viruses to suppress this defense system (see Chapter 10, Section IV) raise issues of strains differing in suppression ability. These issues have yet to be considered.

D. Isolation by means of vectors

Vectors may be used in three ways to isolate strains. First, by using short feeding periods on the plants to be infected, only one strain out of a mixture may be transmitted. This occurred with BCTV transmitted by a leafhopper vector following a 15-minute infection feeding (Thomas, 1970). Second, particular vectors may preferentially transmit certain strains of a virus (see Chapter 11, Section III.E.4). Third, inoculation of insect vectors with diluted inoculum followed by repeated selection of infected plants for a particular type of symptom can result in the isolation of a variant virus (Kimura *et al.*, 1987).

E. Isolation of artificially induced mutants

Experimentally induced mutants have been used for two important kinds of investigation in plant virology. Nitrous acid-induced mutations in the coat protein gene of TMV were of considerable importance in determining the nature of the genetic code and in confirming the nature of mutation. Nitrous acid has also been used to induce temperature-sensitive (*ts*) mutants, mainly in TMV, to aid in the delin-

eation of the *in vivo* functions carried by the viral genome. Nitrous acid mutants have been used as a source of mild virus strains to give disease control by cross-protection.

There is a spontaneous background mutation rate giving the quasi-species described in Section I.A, but many of the natural mutants will be suppressed in inoculated leaves by a dominant strain or master sequence. Mutagens such as HNO_2 inactivate much of the treated virus. Thus, to show that a mutagen is increasing the mutation rate rather than selecting for a minor sequence, it is necessary for treated and untreated virus to be assayed at fairly high dilutions that give about the same number of total infections for treated and control samples (e.g. Melchers, 1968).

1. Coat protein and other mutants of TMV

To isolate mutants, the reactions of different hosts containing the *N* or *NN* genes (see Chapter 10, Sections III.D and III.E) have been exploited. One convenient method for isolating mutants is to inoculate a necrotic local lesion host under conditions where most lesions will have arisen from infection with single virus particles. Sometimes mutants give a recognizably different necrotic local lesion, very frequently smaller than normal (Siegel, 1960). To detect other symptom differences, single lesions are dissected out and inoculated to hosts giving systemic symptoms. Mutants may then be recognized by the different symptoms that they produce.

Another method for isolating mutants and estimating their frequency is to use a parental strain of virus in a host that gives systemic symptoms without necrotic or other conspicuous local lesions. One then looks for mutants producing necrotic, yellow or other characteristic local lesions in the host. This method selects one class of mutants, but it is sometimes difficult to eliminate the parental virus from the culture, even by repeated single local lesion culture (see Fig. 10.22).

A third method for isolating mutants depends on diluting the inoculum to a point where only about one-half (or less) of the plants inoculated become infected. A host giving systemic infection is used. Under these conditions, it can be assumed that most of infected plants were infected by a single virus particle or, for multi-component viruses, a single infection unit.

2. ts Mutants

In isolating ts mutants following treatment of virus with nitrous acid, the objective is to obtain a range of independent mutants that are defective in one virus function at the non-permissive temperature. The procedure used to select for such mutants should not select against mutants in any particular cistron.

Early methods selected for ts mutants with amino acid substitutions giving coat proteins that were defective at the non-permissive temperature (Jockusch, 1964; Wittmann and Wittmann-Liebold, 1966).

Dawson and Jones (1976) described a selection procedure for ts mutants of TMV based on the idea that all ts mutants should infect and begin replication more slowly at the non-permissive temperature (32°C). U1 TMV replicates in Xanthi nc leaves at 32°C but no necrotic lesions appear. On return to 25°C, necrotic lesions develop rapidly. If an infection was due to a ts mutant, development of necrosis on return to 25°C would be significantly delayed. Using this procedure, Dawson and Jones screened approximately 50 000 lesions formed by nitrous acid-treated TMV. They eventually obtained 25 ts mutants. The reversion to wild type by many of these mutants is a sufficiently rare event that they can be used for biochemical experiments at 25°C (Jones and Dawson, 1978).

In an attempt to obtain ts mutants of TRV with the mutation located in the long-rod RNA rather than the coat protein, Robinson (1973) added a large excess of untreated short rods to nitrous acid-treated virus. The mixture was inoculated to a local lesion host at 20°C. Single lesions were isolated and inoculated to a local lesion host at 20°C and 30°C as a preliminary screen for mutants.

F. Isolation of strains by molecular cloning

For many viruses, infectious genomes of DNA viruses and cDNAs of RNA viruses have been cloned. The process of cloning means that individual genome molecules of the quasi-species

population are separated and thus infection arises from just one component of that population. This is the ultimate way of separating strains. An example of this is the separation of two distinct symptom variants of MSV from plants infected with the Nigerian strain (MSV-N) (Boulton et al., 1991a). One variant (MSV-Nm) gave narrow, mildly chlorotic, discontinuous streaks after agro-inoculation to maize whereas the other variant (MSV-Ns) gave wide, severely chlorotic streaks. Some molecular aspects of these variants are described in Chapter 10 (Section III.O.1.d).

IV. THE MOLECULAR BASIS OF VARIATION

A. Mutation (nucleotide changes)

1. Chemical mutagens

Mutations involving single nucleotides consist of the replacement of one base by another at a particular site, or the deletion or the addition of a nucleotide. Single base changes occurring in a coding region may lead to: (1) replacement of one amino acid by another in the protein product; (2) the introduction of a new stop codon that results in early termination of translation and a shorter polypeptide; or (3) replacement of a codon that has either greater or lesser usage in the particular host. Deletion or addition of a single nucleotide in a coding region will give rise to a frameshift, with consequent amino acid changes downstream of the deletion or addition. Such deletions or additions will usually be lethal unless compensated for by a second change (addition or deletion) that restores the original reading frame. Nucleotide changes in non-coding regions will vary in their effects depending on the regulatory or recognition functions of the sequence involved.

Treatments that cause mutation also inactivate viruses. Quantitative studies on inactivation and on the appearance of necrotic local lesions in a culture normally giving chlorotic lesions fitted quite well with a theoretical curve based on the assumptions that a single chemical event can result in inactivation and that a single event can result in mutation.

Gierer and Mundry (1958) first demonstrated the high efficiency of nitrous acid as an *in vitro* mutagen for TMV. The mutagenic action was considered to be through deamination of cytosine to give uracil, and deamination of adenine to give hypoxanthine. The hypoxanthine acts like guanine during replication of the treated RNA and base-pairs with a cytosine. In the next copying event this will pair with a guanine and, thus, an adenine is replaced by a guanine at the original site. The amino acid exchanges found in the coat protein of mutants of TMV induced by nitrous acid confirmed these deaminations as the basis for the induced mutations. Such studies on TMV mutants made an important contribution to our understanding of the genetic code. They confirmed that the code is non-overlapping and degenerate, and gave strong support to the idea that the code is universal (Wittmann and Wittmann-Liebold, 1966; Sarkar, 1986). The single base changes brought about by nitrous acid have also been amply confirmed by later amino acid and nucleotide sequence data (e.g. Rees and Short, 1982; Knorr and Dawson, 1988).

5-Fluorouracil is an analog of uracil that is incorporated into the RNA of viruses that are replicating in plants supplied with the analog. It replaces uracil residues in the RNA and can lead to the changes uracil → cytosine and adenine → guanine (Wittmann and Wittmann-Liebold, 1966). Some other chemicals have less clearly defined mutagenic effects on RNA viruses. For the experimental induction of mutations, nitrous acid will usually be the most useful mutagen.

2. X-irradiation and ultraviolet irradiation

Ionizing radiation and UV irradiation inactivate viruses containing dsDNA, and these agents also cause mutations in such viruses as well as in ssRNA viruses. For instance, temperature-sensitive mutants of AMV have been isolated following irradiation of purified Tb component of AMV with UV light (van Vloten-Doting *et al.*, 1980).

3. Raised temperature

Several workers have noted that an increased number of variant strains could be isolated when plants were grown at higher temperatures (e.g. Mundry, 1957). However, there is good evidence that the multiplication of and invasion by particular strains may be favored by certain temperatures. It seems probable that a major effect of heat treatment is to favor certain types of spontaneous mutant or a minor member of the quasi-species population.

4. Natural mutations

There is no doubt that a single base change giving rise to a single amino acid substitution in the protein concerned is a frequent source of virus variability under natural conditions *in vivo*. Thus, of 16 spontaneous TMV mutants listed by Wittmann and Wittmann-Liebold (1966), six had one amino acid exchange in the coat protein, one had two, and one had three such exchanges. Many of the coat protein exchanges represented base changes different from those induced by nitrous acid. Eight of the mutants had normal coat proteins and, therefore, must have had one or more base changes in the RNA outside the coat protein cistron. Most of the 15 amino acid substitutions found between the coat proteins of two naturally occurring strains of AMV could be explained by single point mutations (Castel *et al.*, 1979). Presumably, most of these mutants would have arisen from copying errors made by the RNA-dependent RNA polymerase during viral RNA replication. The differences between naturally occurring sequence variants of viroids usually consist mainly of a series of base substitutions (e.g. Visvader and Symons, 1985), although additions and deletions of nucleotides also occur.

Natural mutation is discussed in more detail in Chapter 8 (Section IX.A).

B. Recombination

For many years, it was considered that recombination was a genetic mechanism confined almost entirely to organisms with DNA as their genetic material. It is now known that recombination occurs in plant viruses with genomes consisting of either DNA or RNA. Recombination in both DNA and RNA viruses is discussed in detail in Chapter 8 (Section IX.B).

C. Deletions and additions

Recombinational events can lead to deletions in the genomes of both DNA and RNA viruses. These often give rise to defective or defective interfering nucleic acids, which are discussed in Chapter 8 (Section IX.C).

In the literature, many examples are described of mutant viruses with additional base sequences that have been generated by *in vitro* modification of recombinant DNA plasmids. Naturally occurring examples of such additions are much less common. In a group of naturally occurring variants, it may sometimes be difficult to establish whether a difference in length is due to an addition or a deletion of nucleotides. Kimura *et al.* (1987) repeatedly selected rice plants for severe symptoms after inoculation with RDV by leafhoppers injected with a dilute inoculum of the stock culture (strain 0). This procedure allowed the isolation of a severe strain (S). The fourth largest RNA of strain S had an M_r about 20 kDa greater than that of strain 0. The M_r of the protein corresponding to the 43-kDa protein of strain 0 was 44 kDa in strain S. It has not been demonstrated that S was derived from 0 by an addition of nucleotides. Strain 0 could have been derived from S by a deletion event, with the parent strain maintained at a low level in the culture.

Among the bromoviruses, the CCMV RNA3 5′ non-coding sequence contains a clearly demarcated 111-base insertion not present in BMV, which must represent a sequence re-arrangement in one of the two viruses (Allison *et al.*, 1989).

Repetition of blocks of sequences is known in some RNA viruses. For example, sequences of 56 and 75 nucleotides are duplicated in the leaders of the RNAs 3 of strain S and strain L of AMV (Langereis *et al.*, 1986). The duplications are next to one another in the leader sequences and do not appear to be essential for replication. Langereis *et al.* suggested that these duplications may have been generated by polymerase molecules releasing prematurely from a (−)-strand RNA3 template and re-initiating again on the same template with the nascent strand still attached.

D. Nucleotide sequence re-arrangement

An example of nucleotide re-arrangement has been described among satellite RNAs of TRSV (Buzayan *et al.*, 1987). STRSV RNAs from strains 62L and NC-87 of TRSV have the same 360-residue sequence. The budblight satellite RNA with 359 residues differs from these mainly in the nucleotides from 100 to 140. The differences in sequences in this region are consistent with re-arrangements of blocks of nucleotide residues. as indicated in Fig. 17.7.

E. Re-assortment of multi-particle genomes

Since the classical experiments with the long and short rods of strains of TRV (Lister, 1966, 1968; Frost *et al.*, 1967), genetic assortment experiments have been carried out with most of the known multi-partite viruses. These experiments have demonstrated beyond doubt that new variants can arise by re-assortment of the pre-existing segments of the viral genome, both in the laboratory and in nature (Fulton, 1980).

Successful re-assortment may not always be mutually effective. For example, Rao and Hiruki (1987) found with two strains of RCNMV that a mixture of RNA1 of strain TpM34 plus RNA2 of strain TpM48 was infectious, while the reciprocal mixture was not.

Although BMV and CCMV have been regarded as distinct bromoviruses, their individual RNA components will complement one another in certain combinations (Allison *et al.*, 1988). Capped *in vitro* transcripts were made from complete cDNA copies of genomic RNAs

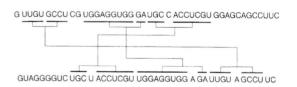

Fig. 17.7 Possible nucleotide sequence re-arrangements between two satellite RNAs of TobRSV. **Upper line**: strain 62L, nucleotides 100–145. **Lower line**: budblight strain, nucleotides 100–143. Heavy underlines indicate blocks of sequences that may have been re-arranged. Connecting lines indicate identical sequences in the two strains. (Courtesy of G. Bruening.) From Matthews (1991).

1, 2 and 3 of each virus. No viral replication was detected with any heterologous combination of RNAs 1 and 2, which code for proteins involved in RNA replication (discussed in Section II.C.7). By contrast, heterologous RNA3 was viable in both combinations, replicating in protoplasts and giving local lesions in *Chenopodium*. However, neither hybrid virus systemically infected the natural parental host plants.

F. The origin of strains in nature

All the kinds of molecular change noted above will contribute to the evolution of strains in nature. A single base change resulting in a single amino acid change in a protein is probably one of the most common events giving rise to natural variation. The primary structure of the coat proteins of naturally occurring strains of TMV supports this view (Wittmann and Wittmann-Liebold, 1966).

Viruses appear to vary quite widely in the rate at which they give rise to new strains. Large numbers of TMV strains are known, while only a few have been isolated for TBSV. Different strains of a virus may also vary quite markedly in the rates at which they give rise to mutants of a particular symptom type. Thus, some strains of PVX producing chlorotic local lesions in tobacco frequently gave rise to ring spot local lesion strains, and other strains gave none at all (Matthews, 1949a). Strains of TNV that produced white lesions on cowpea frequently gave rise to strains giving red lesions (Fulton, 1952). These red lesions always first appeared as spots resembling sectors in association with a white lesion. Various white strains produced red mutants at different rates over a 5-fold range. The reverse process (red strains giving rise to white) was not observed.

These various differences in the rate of appearance of strains may not necessarily be due to differences in actual mutation rate. Some viruses or strains may produce a much higher proportion of defective or completely non-viable mutants than others. Some of the apparent differences probably reflect our ability to detect mutants.

We can envisage strains diverging further and further from the parent type as changes brought about by various mechanisms accumulate in the various proteins specified by the virus. The survival and spread of such strains will often depend on their competitive advantage within the host in which they happen to arise, and in others that they subsequently infect.

The survival of new strains may not always depend on immediate selective advantage, although continued survival would of course require an adequate combination of properties. For example, Reddy and Black (1977) pointed out how deletion variants of WTV could by chance come to dominate in particular shoots of a growing clover plant. If the deletion event occurred in a cell near the apical meristem where the virus concentration is low, the mutant may come to replace the parent virus entirely in a particular shoot. They point out that the process can be regarded as an example of evolution by geographical isolation in miniature, with the branches of the plant providing the geographical isolation. The evolution of virus strains is further discussed in Section IX.

V. CONSTRAINTS ON VARIATION

As discussed in Chapter 8 (Section IX.A), in most viral genomes there is no proofreading on replication; this leads to much of the variation described in previous sections in this chapter. In light of this high level of variation, there must be some constraints that control the preservation of the identity of a virus species or strain. It is likely that most mutations in a viral genome will be either neutral or deleterious. Mutations that lead to loss of a critical function would not be propagated in a population unless they were complemented by other members of that population. However, mutations that caused only a decrease in 'fitness' would be more likely to be retained.

A. Muller's ratchet

Theoretical considerations indicate that high mutation rates can have significant impact on populations, especially if they are small in size.

The concept of 'Muller's ratchet' is that, if the average mutation is deleterious, there will be a drift to decrease of population fitness leading to 'mutational meltdown' (see Muller, 1964; Lynch *et al.*, 1993). Muller's ratchet is particularly effective in small populations, and for many viruses transmission and infection forms a bottleneck in which the population is small (Duarte *et al.*, 1993; Bergstrom *et al.*, 1999).

Back mutations at the specific site of a deleterious mutation or compensatory mutations are likely to occur at a lower rate than forward mutations (Haigh, 1978; Maynard Smith, 1978). In populations of higher organisms, this drift is limited by sex which re-creates, through genetic exchange, genomes with fewer or no mutations. Obviously, this process does not occur in the conventional sense in viruses but it is likely that recombination or genetic re-assortment within the quasi-species could play a part in controlling the effects of Muller's ratchet (see Chao, 1997).

B. Does Muller's ratchet operate with plant viruses?

In a study on the tobamoviruses that infected *Nicotiana glauca* in Australia over a 100-year period, Fraile *et al.* (1997a) found an example that could be interpreted as being due to Muller's ratchet. The tobamoviruses in herbarium samples and living samples of New South Wales *N. glauca* covering a period from 1899 to 1993 were analyzed. Before 1950, many plants were infected with both TMV and TMGMV but after that date only TMGMV was found. In experimental joint infections of *N. glauca* TMV accumulated to about 10% of the level of that in single inoculations; the level of TMGMV was not affected. Fraile *et al.* (1997a) concluded that TMV colonized *N. glauca* in New South Wales earlier or faster than TMGMV, but in joint infections the latter virus caused a decrease of the TMV population below the threshold at which deleterious mutations were eliminated.

However, nematode transmission of TRV serves as a bottleneck to clear the virus population of DI RNAs (Visser *et al.*, 1999a). These DI RNAs, derived from RNA2 of TRV isolate PpK20 (see Fig. 6.38) have a modified coat protein gene and interfere with viral replication. It was suggested that TRV RNA2 and the DI RNA are encapsidated in *cis* by their encoded coat proteins which are, respectively, functional and non-functional in transmission, eliminating the DI RNA at the transmission bottleneck. This contrasts with heterologous encapsidation of potyvirus and luteovirus genomes (see Chapter 8, Section X) leading to isolates defective in vector transmission becoming transmissible.

Similarly, aphid transmission often sorts the populations of genomic RNA variants and D-RNAs present in CTV isolates (Albiach-Martí *et al.*, 2000b).

VI. VIRUS STRAINS IN THE PLANT

In the previous sections, we have considered ways of isolating virus strains, the molecular mechanisms by which they originate, and the criteria that can be used for distinguishing them. Here certain activities of strains in the infected plant are discussed.

A. Cross-protection

It is generally accepted that protection of a plant by one strain of a virus against infection with a second depends on the presence of the protecting virus in the protected tissue. This is discussed in detail in Section II.C.4. If, for any reason, a plant or part of a plant becomes freed of the protecting virus, it is often susceptible to re-infection with the first strain or other strains. Exceptions may occur with systemic spread of the response to the host general defense system (see Chapter 10, Section IV.E).

B. Selective survival in specific hosts

When a virus culture that has been maintained in an apparently stable state in one host species is transferred to another species and then inoculated back to the original host, it is sometimes found that the dominant strain in the culture has been changed. Carsner (1925) showed that a culture of BCTV could be altered by transmission to *Chenopodium murale*. When the virus was returned to beet from this plant, it produced only mild symptoms. Lackey (1932) found that the change was reversible and that the virus culture could be returned to its original condition, with respect to symptoms in beet, by passage through *Stellaria media*. According to Salaman (1938), a strain of PVX, which caused ring spot symptoms, when inoculated into seedling beet, produced small necrotic rings only. When virus from the local lesions was returned to tobacco, only a faint mottle developed. A reverse situation was described by Matthews (1949c). When PVX cultures giving a mild mottle in tobacco were passed through *Cyphomandra betacea* and re-inoculation was made to tobacco, only local and systemic ring spot-type disease was produced. *Cyphomandra* was the only one of 19 solanaceous species tested to cause this change. The mild cultures that were used contained a small proportion of ring spot strains that had presumably arisen by mutation. It was suggested that these ring spot strains multiplied more effectively in *Cyphomandra* than did the mild strains and that they thus came to dominate in this host. Ring spot strains alone reached several times the virus concentration in *Cyphomandra* compared with tobacco. Mild strains were four to eight times as concentrated in tobacco as the ring spot strains in this host. This is an example of the host selecting a different member of a quasi-species population to become the dominant member.

Johnson (1947) found that passage of the ordinary severe-type TMV through sea holly (*Eryngium aquaticum*) resulted in mild symptoms on tobacco. He showed that severe strains moved more slowly in sea holly, thus accounting for the 'filtering' action of this host.

Inoculation of a culture of CMV that did not cause systemic infection in Blackeye cowpea to the variety Catjang led to the appearance of some abnormally large local lesions. Inoculation from these large local lesions to the variety Blackeye was followed by systemic necrotic disease (Lakshman *et al.*, 1985).

These various selection phenomena may well involve differences in the cell-to-cell movement protein coded for by strains of the virus or possibly in the interaction that the virus has with the host defense system. For example, the mild cultures of PVX in tobacco just described may have been giving rise continually to mutants that could not compete in tobacco with the parent strain based on their cell-to-cell movement proteins. However, the movement protein of some of these mutants may have been better adapted to systemic movement in *Cyphomandra* than the parental strain. The back mutation rate must be very low since the strain or strains selected in *Cyphomandra* appeared quite stable when cultured in tobacco. Now that the nucleotide sequences of the genomic RNAs are known it should be possible to establish whether strain selection by particular hosts involves mutations in the movement protein. Satellite RNAs such as those associated with CMV may undergo differential replication in particular hosts. This may provide another basis for variation in symptoms following culture of a virus in a given species (Waterworth *et al.*, 1978). For example, CARNA5 exists as a series of closely related sequences (Richards *et al.*, 1978), which could provide for a rapid response to changed conditions for replication.

C. Loss of infectivity for one host following passage through another

Loss of infectivity for one host may develop following repeated passage through another species. For example, several strains of PVX lost infectivity for potato during continued propagation in tobacco (Matthews, 1949a). No change in symptoms produced in tobacco, *N. glutinosa* or *Datura tatula* could be observed over the period that the AP strain lost its infectivity for potato. The nature of the change is not understood, but possibly reflects gradual selection of a strain or strains better adapted to growth in tobacco. Alternatively, a minor member of the quasi-species in potato might have been able to multiply only in that host through complementation by the master sequence or other more major sequence.

An immediate loss of infectivity for one host following passage through another can be due to a virus inhibitor that is effective only in certain hosts, rather than to any change in the virus itself.

D. Double infections *in vivo*

The existence of phenotypic mixing (see Chapter 8, Section X) suggests that two virus strains can replicate together in the same cell. The following evidence confirms this view:

- Various observations have demonstrated the presence of two strains of a virus in the same cell in intact plants. Thus, Hull and Plaskitt (1970) could recognize characteristic aggregates of particles of two AMV isolates in the same cell.
- When the *ts* mutant of TMV, Ni118, was inoculated in a mixture to tobacco with common TMV and grown at 35°C, some mutant RNA was found to be coated with common strain protein (Takebe, 1977).
- Protoplasts inoculated with TMV show a multitarget response to inactivation by UV light (Takebe, 1977), indicating that more than one particle can initiate infection in the same cell.

Thus, while proof is lacking, the weight of evidence suggests that two strains can infect and replicate simultaneously in the same cell in the intact plant. The quasi-species concept (Section I.A) with a master sequence and a multitude of minor ones suggests that this is an unusual situation. However, on some occasions it is likely that there are two or more master sequences, which would reflect the situation with co-infection by two strains.

E. Selective multiplication under different environmental conditions

Temperature is probably the environmental factor that most commonly influences the survival or predominance of strains occurring in nature. Experimental use of temperature to isolate strains is noted in Sections III.C and III.E.2. In parts of the world with hot climates, strains of viruses surviving at these temperatures are selected naturally. Some understanding of the basis for the effects of temperature on naturally occurring strains can be gained from the results of the studies on artificially induced *ts* mutants (see Section III.E.2).

VII. CORRELATIONS BETWEEN CRITERIA FOR CHARACTERIZING VIRUSES AND VIRUS STRAINS

In the preceding sections, we have surveyed the various criteria that can be used to delineate variation among virus isolates. How can we use these criteria to decide whether a particular isolate is identical to another isolate or a related variant or strain, or whether it is a distinct virus? This is a question of considerable practical importance, because the recognition and identification of virus strains may be most important for effective virus control. In addition, the virological literature is cluttered with inadequate descriptions of virus isolates. These are frequently described as new viruses or new strains, particularly if they are found in a new host or a different country, when adequate study might show they were very probably identical to some virus already described. The definition of a virus species is given in Chapter 2 (Section I.C).

A. Criteria for identity

There is only one criterion that will establish that two virus isolates are identical: the identity of the complete base sequence of their genome nucleic acids. In spite of recent rapid advances in nucleic acid sequencing techniques, it is usually impractical for this to become a routine procedure. For most practical purposes, the following criteria would be sufficient to establish provisional identity of two virus isolates: (1) identity of size, shape and any substructure of the virus particle as revealed by appropriate electron microscopy; (2) serological identity in adequate tests; (3) identical disease symptoms and host ranges for a set of indicator hosts and genotypes; and (4) identical transmission, especially with respect to any arthropod, nematode or fungal vectors. The presumption of identity would be greatly reinforced by information on some aspect of nucleic acid sequence, for example identical sequences in a particular region of the genome or identity as judged by heterogeneity mapping.

B. Strains and viruses

The broad questions of virus classification are dealt with in Chapter 2. Here we will consider the problems involved in using the various properties outlined earlier in this chapter to define virus strains, to group them and to decide whether an isolate is a strain or a different virus.

One method is to take a quantitatively determined set of characteristics such as the amino acid composition of the coat protein. Statistical procedures and computer analysis are then used to derive a classification with degrees of relationship indicated. Computer analysis is particularly useful for handling large amounts of numerical data as was used, for example, to derive Fig. 2.3 from amino acid sequences. However, a classification based on computer analysis is no more objective than other ways of making a classification. It will depend on the personal judgments and selections made by the taxonomist providing the data.

In the adansonian approach, all the known characters are given weight in determining groupings. This approach has become popular with the widespread availability of computers, but there are significant limitations. For example, as noted earlier, many of the fairly easily measured characters of a small virus depend on properties of the coat protein. Thus, differences in the coat protein may be given undue weight. Similarly, symptom differences between two strains could be emphasized, merely by recording differences on an extended host range.

On the other hand, the hierarchical system involves making arbitrary decisions about which characters are the most important. There are serious objections to applying such a system, without some modification, especially when we are considering classification within a group of related viruses. The most useful characters will be different within different virus groups. Thus, the coat protein of STNV, being the only gene product of this virus, should be given more substantial weight than would the coat protein of a virus with, say, 10 genes. Similarly, particle morphology may be most useful for those groups such as the *Rhabdoviridae* that possess a complex structure.

The best course is probably the pragmatic one of considering all known properties within a group of variants and weighting them in a commonsense manner in relation to the overall properties of the group in question.

When strains arise in the stock culture of a virus in the laboratory, as they do with such viruses as TMV and TYMV, we can be reasonably sure that they will be closely related to the parental strain—usually arising from a single mutation. Phenotypic differences in most properties will usually be small, but may sometimes be large as with the TMV strains, such as PM1, that produce defective coat proteins and no intact virus.

Virus isolates collected in the field, perhaps from different host species in different countries, may appear to be related on the basis of some properties and unrelated on others. The only generalization that can be made at present is that closely related strains will differ in only a few properties, while distantly related strains will differ in many. The extent to which different properties show correlations varies widely in the different groups of viruses.

C. Correlations for various criteria

From a purely genetic point of view, the relationships between a set of virus strains can be assessed precisely if we know the differences in nucleotide sequences between their genomes. However, from the virological point of view, other factors must be taken into consideration. For example: (1) nucleotide changes that are silent, that is, lead to no change in the structure or function of the virus, are usually of little interest. However, it must be remembered that, with accumulating knowledge, what are considered as non-functional nucleotides at one time may in the future be highly significant. (2) Particular gene functions may be of particular ecological and therefore practical significance, for example mutations in a viral gene that affects insect vector specificity. (3) When large numbers of field isolates have to be typed over a short time interval, only rapid diagnostic methods are practicable. The confidence with which particular criteria can be used depends in part on the extent to which they correlate with other criteria. This section gives a brief overview of these problems.

1. Host responses

Where a group of strains are fairly closely related, host responses may provide the best, or even only practicable, criteria for establishing strain types. For example, Mosch et al. (1973) found that 18 isolates of TMV from glasshouse tomato crops could be placed in three groups depending on their pathogenicity for a set of *Lycopersicon esculentum* clones. There were no differences in certain physical properties (buoyant densities and $S_{20,W}$) and only small individual differences in coat protein composition. These did not correlate with the pathogenicity groups. When a virus of economic importance such as AMV is highly variable, the classification of large numbers of field isolates must usually depend primarily on symptoms and host range on a standard set of indicators (e.g. Crill et al., 1971; Hajimorad and Francki, 1988).

2. Vector transmission

Among three isolates of BYDV there was a correlation between closeness of serological relationship and transmission by aphid vectors (Aapola and Rochow, 1971). Pead and Torrance (1989) found that MAbs could be used to type the three major vector-specific strain groups (now species) of BYDV. On the other hand, an isolate of PLRV that was poorly transmitted by aphids was indistinguishable serologically from readily transmitted isolates (Tamada et al., 1984). There was no correlation between serological relatedness and the ability of English populations of *Longidorus attenuatus* to transmit different isolates of TBRV (Brown et al., 1989b).

3. Multi-partite genomes

The ability of multi-particle viruses to complement one another successfully provides a powerful functional criterion indicating relationship. However, this property may not correlate closely with the physical properties of the virus particle or other properties of the virus. For example, certain viruses that have been considered as strains of CPMV (Swaans and

van Kammen, 1973) did not successfully complement one another in mixed infection experiments (van Kammen, 1968). Successful complementation has been shown to occur not only between already well-recognized strains but also between viruses thought to be distinct members of the same group. Such results further complicate the use of complementation tests as a criterion of relationship. For example, Bancroft (1972) demonstrated successful complementation between BMV and CCMV. These are both in the *Bromovirus* genus, although they have almost totally different host ranges and appear unrelated serologically.

At present two genera of viruses are known with a tripartite genome and a separately encapsulated coat protein cistron: the *Ilarvirus* and *Alfamovirus* genera. With members of both these groups, if the three-genome RNAs are used for infection, the coat protein RNA, or some coat protein itself, is required for infectivity (discussed in Chapter 8, Section IV.G). The coat protein or the coat protein RNA of some ilarviruses, for example TSV, will activate the RNAs of AMV. The reverse combination is also active. However, mixtures of the three-genome RNAs from the two viruses do not complement one another (van Vloten-Doting, 1975; Gonsalves and Fulton, 1977), and there is no sequence similarity between the corresponding RNA segments.

Transgenic tobacco plants expressing the TSV coat protein gene were resistant to infection with TSV but susceptible to AMV. They could be infected with AMV RNAs 1, 2 and 3, demonstrating that the endogenously produced TSV coat protein can activate the AMV genome, even though it does not protect against this virus (van Dun et al., 1988b). The coat protein of AMV nucleoproteins is specifically removed by the addition of AMV RNA. Similarly, the nucleoproteins of ilarviruses may lose their protein when free viral RNA is added. There is reciprocity in this reaction between certain ilarviruses and AMV (van Vloten-Doting, 1975; Gonsalves and Fulton, 1977). These results have led some workers to suggest that AMV should be placed in the *Ilarvirus* genus (see Chapter 2, Section III.J.4).

4. General nucleotide sequence similarities

Using hybridization techniques, there may be complete lack of detectable base sequence homology between viruses that on other grounds, such as morphology of the particle and serology, are certainly related (Zaitlin *et al.*, 1977). At the other extreme, Bol *et al.* (1975) described four strains of AMV with well-characterized differences in biological tests that were virtually indistinguishable in nucleic acid hybridization tests.

Strains of CMV are divided into two subgroups, based on serology and nucleic acid hybridization, as discussed by Rizzo and Palukaitis (1988). Of 39 strains examined by nucleic acid hybridization, 30 belong to subgroup I and nine to subgroup II. RNAs belonging to the two subgroups can be re-assorted to yield viable recombinants. The RNAs 1 and 2 of representatives of the two groups have been sequenced and compared (Rizzo and Palukaitis, 1988, 1989). Different regions of the RNAs varied in the extent of sequence homology (from 62% to 81%). Strains within the two subgroups cannot be distinguished by the usual nucleic acid hybridization techniques. However, Owen and Palukaitis (1988) used molecular heterogeneity mapping to place 13 of the CMV strains into two groups based on their ability to hybridize to two representative strains. Molecular heterogeneity mapping could distinguish strains within the two groups.

5. 3' Non-coding nucleotide sequences

Another approach for discriminating between distinct potyviruses and strains has been explored by Frenkel *et al.* (1989). They compared the 3' non-coding nucleotide sequences of 13 potyviruses and found that viruses that were distinct on other grounds had 3' non-coding sequences of different lengths (189–475 nucleotides). The degree of sequence similarity ranged from 39% to 53%. Such values are comparable to that obtained when the 3'-untranslated regions from unrelated potyviruses are compared, and they are probably in the range expected for chance matching. By contrast, the 3'-untranslated regions of sets of viruses recognized on other criteria as related strains were very similar in length and in nucleotide sequence homology (83–99%). WMV-2 and SGMV-N were found to have 78% homology and on this basis were considered to be strains of the same virus.

6. Serological relationships

Relationships determined by serological methods might, by chance, correlate quite well with any other properties. However, it is reasonable to expect that they may show some correlation with those criteria that also depend on some property of the coat protein.

Correlations have been reported between degree of relatedness, measured by cross-protection tests, and serological relatedness (e.g. PVX strains—Matthews, 1949b; BYDV isolates—Aapola and Rochow, 1971). On the other hand, there was no correlation between serological relatedness within a group of TNV isolates and their ability to support the replication of three differing isolates of STNV (Kassanis and Phillips, 1970), nor was there any correlation between serological relatedness and symptoms in tobacco for TRSV (Gooding, 1970).

MAbs raised against strains of PVX reacted in a complex manner with the strains in different groups based on the reaction of host varieties (Torrance *et al.*, 1986a). Nevertheless, a resistance-breaking strain could be identified. A panel of 10 MAbs raised against PLRV failed to differentiate between strains that caused different symptoms in indicator hosts (Massalski and Harrison, 1987).

Traditionally, viruses and strains within the *Potyvirus* genus have been very difficult to delineate. This has been not only because of the large number of viruses involved but because different tests for relationship gave different answers. The work of Shukla and colleagues has gone some way toward establishing a sound basis for classification of virus isolates belonging to this group. For example, earlier work suggested that cross-protection did not distinguish some isolates that, on other criteria such as serology, were considered to be separate viruses. Shukla *et al.* (1989b), using antibodies directed against the N-terminal part of the coat protein, showed that potyviruses infecting maize, sorghum and sugarcane in Australia and the United States comprised four distinct viruses. The earlier cross-protection tests fell neatly into place on this basis (Shukla and Ward, 1989a).

Similarly, the kind of cytoplasmic inclusions found with these isolates supported the idea of four distinct viruses. Difficulties remain, however, because some unexpected serological cross-reactions occur between viruses that on other soundly based criteria are regarded as distant members of the *Potyvirus* genus. The antigenic site for these cross-reactions may reside with a few common amino acids close to the N-terminus of the coat protein (Shukla and Ward, 1989a,b).

7. Non-structural proteins

Yeh and Gonsalves (1984) used antisera raised against the inclusion body proteins of two potyviruses to confirm that they were related strains of one virus rather than two distinct viruses. Thornbury and Pirone (1983) showed that the helper component protein of two different potyviruses were serologically distinct. There was no serological relationship between the 35-kDa protein coded for by AMV and the corresponding proteins of three other viruses with a tripartite genome (van Tol and van Vloten-Doting, 1981).

VIII. DISCUSSION AND SUMMARY

The study of variability is one of the most important aspects of plant virology. It is important from the practical point of view because strains vary in the severity of disease they cause in the field, and because strains can mutate to break crop plant resistance to a virus. It is important also for developing our understanding of how viruses have evolved in the past, and how they are evolving at present.

A range of procedures is available for isolating virus variants either from nature or following some form of mutagenesis or other manipulation outside the plant. Mutants of the *ts* type have been particularly useful in studying various aspects of virus structure and replication.

Because of the very high mutation rate, it is probable that all cultures of plant viruses consist of a mixture of numerous strains even after single lesion passage. However, a master genome sequence will usually dominate in the culture and many variants will not be detected. The selection of the master sequence will depend on many factors, including the host genotype and the environment in which it is growing.

The molecular mechanisms by which variation within a virus population is produced are like many of those found in cellular organisms, except that for many plant virus groups the material upon which variation operates is RNA rather than DNA. Mechanisms include mutations involving single nucleotide changes or the addition or deletion of one or a few nucleotides, recombination, deletions or additions of blocks of nucleotide sequences, rearrangement of nucleotide sequences, and re-assortment among multi-partite genomes.

A range of structural, serological and biological criteria is available for delineating viruses and virus strains within a group or family of viruses. The kind of criteria to be used will depend on the purpose of the study. If we are studying evolutionary relationships within a virus group or family, or among the variants of a single virus, then the full nucleotide sequences of the viruses concerned will be of prime importance, but knowledge of the functional products of the viral genome will often be needed as well. If the full nucleotide sequences are known for representative viruses, then other methods, such as various forms of nucleic acid hybridization or PCR, can be usefully interpreted for additional viruses and strains. If we are interested in developing methods for reliably and rapidly diagnosing viruses and virus strains from the field, then other methods will be appropriate. Dot blot serological assays using some form of ELISA are an important type of test. Polyclonal antisera of wide specificity or MAbs of very narrow specificity can be used in such tests as appropriate. Biological criteria such as disease symptoms, host range, methods of transmission and cross-protection may be important in defining viruses and virus strains.

The extent to which virus species have been clearly delineated varies widely among the different genera and families of viruses. There are dangers in formalizing virus species or virus groups before a sufficient number and diversity of strains have been investigated. For example, at a stage when only about seven tymoviruses were known, two subgroups were suggested on the basis of serological relationships and RNA base composition (Gibbs, 1969; Harrison *et al.*, 1971). Since then, further tymoviruses have been discovered with intermediate characteristics (Koenig and Givord, 1974). For some groups, such as the potyviruses, 'a common set or pattern of correlating stable properties' has emerged that can allow the grouping of virus strains into species with some degree of confidence.

The relative importance, or weight, to be placed on different properties of a virus for purposes of classification remains a difficult problem. An adequate understanding of the significance to be placed on the various properties may come only when we have a detailed knowledge of the structure of the viral genome, the polypeptides it codes for and their functions, and the regulatory or other roles of any translated or untranslated regions in the genome. Even with such knowledge difficulties will remain. For example:

- Disease induction, which is a complex process, has been shown for some viruses to depend on the functions of two or more viral genes.
- Various possible mechanisms are now known whereby a single mutation could have effects on two or more functions.
- A single gene product may have two or more functions, differing in importance, for the virus infection cycle.

Thus from a practical point of view it may be an oversimplification to establish relationships between viruses and strains within a family or group solely on the basis of nucleotide sequences.

IX. SPECULATIONS ON ORIGINS AND EVOLUTION

The most fundamental single property of an organism is the size of its genome. In this respect, viruses infecting all kinds of organisms span a range of almost three orders of magnitude (see Fig. 1.4). The genome of the smallest virus, STNV, consists of a monocistronic mRNA coding for the coat protein of the virus shell. The largest viruses (infecting animals) have genomes about as large as those of the smallest cells. Where and when did this great diversity of agents arise? When and how did they evolve?

Although much relevant information has become available from studies on the structure and replication of viruses and on the molecular biology of normal cells, the origin and evolution of viruses is only now beginning to emerge from the realm of speculation. Nevertheless, the topic is one of general interest, and one that will be relevant to the problem of classification. For a meaningful discussion of these topics, we must also consider examples from other groups of viruses besides those infecting plants.

In discussing evolution of viruses, we must recognize that it is distinct from the evolution of virus diseases. A new disease may be the consequence of the 'evolution' of the causal virus, but can also result from no change in the virus (Nathanson *et al.*, 1995). For instance, a new disease can result from the movement of an 'old' virus into a new situation. It is likely that the epidemics of swollen shoot in cacao in west Africa and of tungro in rice in south-east Asia were due to the spread of the viruses from asymptomatic natural hosts into either a new species in that area or a changed agronomic situation.

When considering virus origins, it must be remembered that there are three basic types of plant viral genome: those that replicate RNA → RNA, those that replicate DNA → RNA → DNA (reverse transcribing) and those that replicate DNA → DNA. It is likely that these three genome types have different evolutionary pathways but it is not known whether they originate from the same type of macromolecule. Similarities in polymerase structure (see Chapter 8, Section IV.B) suggest a possible common basic origin, although they could be the result of convergent evolution. Thus, there is no compelling reason to suppose that all viruses arose in the same way. Furthermore, it is possible that viruses that originated in one major group of organisms may now exist primarily or solely as agents infecting another group.

The origin and evolution of plant viruses is reviewed in Gibbs (1999b) and that of viruses in general by Domingo *et al.* (1999b).

X. TYPES OF EVOLUTION

A. Microevolution and macroevolution

In earlier sections of this chapter, I discussed the variation of plant viruses and its molecular basis. This variation gives the material on which selection pressures can act which results in virus evolution. The different forms of variation have different importance in the level or type of evolution (Fig. 17.8).

Thus, strains are differentiated mainly by mutations and small insertions or deletions that are selected, changing the master sequence in the quasi-species cloud. This can be termed *microevolution*. Larger and more radical changes caused by recombination and/or acquisition of new genes, termed *macroevolution*, lead to the generation of new genera or families. This essentially starts a new quasi-species cloud that

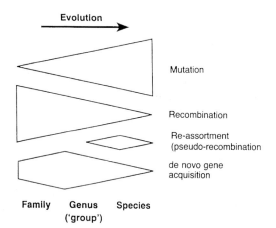

Fig. 17.8 The apparent relative importance of some sources of genetic novelty that influence virus evolution. From Gibbs and Keese (1995), with kind permission of the copyright holder, © The Cambridge University Press.

is selected upon in further microevolutionary diversification.

Because of the population structure of a virus isolate, microevolution can be a continuous process. However, as discussed by Holland and Domingo (1998), extremely high mutation rates do not necessitate rapid evolution. As described below (Section X.D), there is evidence that some plant virus genera appear to be diversifying rapidly at present whereas others appear to be more stable. It is likely that, as with the evolution of higher organisms, viruses go through a stage of relatively rapid diversification when presented with a changing environment and then enter a relatively stable phase where they have adapted to the new environment(s). The selection pressures are discussed in Section XIII.

Macroevolution is a much rarer and a stepwise process. The great majority of recombination events between viral sequences or those leading to the acquisition of new genes will be lethal or deleterious to the virus. However, it is the very rare event that leads to the formation of a viral genome with new properties that enable it to be more successful than its progenitors. Examples of this will be discussed later in this section.

Microevolution and macroevolution are not necessarily independent evolutionary systems. It is quite possible that microevolutionary changes of one virus could lead to the formation of 'hotspots' for recombination with sequences in other viruses. Similarly, not all recombination events will lead to major changes. Recombination between near homologous sequences will create new sequence combinations that have only minor differences to the parental sequences. Thus, microevolution and macroevolution are all a matter of degree.

B. Sequence divergence or convergence

Sequence similarity between two genes does not necessarily indicate evolutionary relationship (homology). Without other evidence, it may be impossible to establish whether sequence similarity between two genes is due to a common evolutionary origin or to convergence. The lysozyme enzymes of foregut fermenters constitute a good example of sequence convergence (Stewart and Wilson, 1987). Foregut fermentation has evolved independently twice in placental mammals, first in ruminants and later in colobine monkeys. In both instances, lysozyme C was recruited to function in the true stomach. About half the amino acid replacements along the langur lysozyme lineage were in parallel or were convergent with those that had evolved earlier along the cow stomach lysozyme lineage. This convergence was driven by selection for adaptation to the acidic pepsin-containing environment of the stomach. Similar convergence has probably occurred from time to time during the evolution of viral genes. However, with such genes it is difficult to obtain independent evidence demonstrating convergence, as has been possible with the stomach lysozymes. Amino acid sequence similarities that are sufficient to lead to a serological cross-reaction may sometimes arise by chance, as appears to have happened between the TMV coat protein and the large subunit of ribulose bisphosphate carboxylase (Dietzgen and Zaitlin, 1986).

Zanotto *et al.* (1996) argue for the convergent evolution of RNA-dependent RNA polymerases. However, this is debated by others (see Koonin and Dolja, 1993; Gorbalenya, 1995), who suggest that the evolution of this gene is divergent.

C. Modular evolution

A process of modular evolution was proposed by Botstein (1980) for DNA bacteriophage but is now considered to also apply to RNA viruses. It is suggested that viruses have evolved by recombinational re-arrangements or re-assortments of interchangeable elements or modules. Modules are defined as interchangeable genetic elements, each of which carries out a particular biological function; examples are replicase proteins, capsid proteins and regulatory systems. This interchange enables independent evolution of the modules under a wide variety of selective conditions. Such modular mobility can overcome the evolutionary constraints that would occur if all

the modules had to co-evolve within a single genomic unit.

The essentials of modular evolution include:

1. The product of evolution is a favorable combination of modules selected to work optimally individually and together to fill a particular niche.
2. Joint infection of the host by two or more viruses is essential for the assembly of new combinations of modules. The viruses do not necessarily have to give full systemic infection of the host; they just have to replicate in the same cell. This can lead to changes in virus host ranges.
3. Viruses in the same 'interbreeding' population can differ widely in any characteristic as these are aspects of the function of individual modules.
4. Evolution acts primarily at the level of the individual module and not at the level of the intact virus. Selection upon modules is for a good execution of function, retention of the appropriate regulatory sequences and functional compatibility with most, if not all, other modules in that genome.

There is increasing evidence for a modular mechanism of macroevolution of plant viruses, which will be illustrated below. The question of where the modules came from is discussed in Section XI.

D. Evidence for virus evolution

No fossil viruses have yet been discovered. Some may await discovery, for example in insects preserved in amber or PCR of long-preserved material. In the meantime, evidence for virus evolution must come from the study of present-day viruses in present-day hosts. Most of the current ideas and concepts about viral evolution are based on molecular data ('molecular fossils') that are accumulating at a considerable rate. It is the comparisons of genome organizations and the details of sequences that are giving evidence for the evolutionary pathways of many of the plant viruses. Below I describe how the data are analyzed and the evidence for six virus groups, which illustrates some current ideas on plant virus evolution.

1. Phylogenetic analyzes

There is a variety of methods used for inferring phylogenetic or other relationships of sequences (reviewed by Weiller *et al.*, 1995). The main aim of these methods is to analyze the nucleic acid or amino acid sequence data in such a way as to reveal relationships. Thus, important steps are in the recording and presentation of the data for analysis and in the interpretation of the analysis. The theory, methods and practice of analysing molecular sequence information is the subject of several monographs (e.g. Waterman, 1988; Doolittle, 1990; Gribskov and Devereux, 1991).

a. Comparison of sequences

The simplest method for comparing two sequences to determine whether there are regions of sequence similarity is the dot plot (dot diagram) (Gibbs and McIntyre, 1971). In this, the two sequences are placed at right angles to form the adjacent areas of a rectangular matrix and a dot is placed in the matrix wherever a row and column with the same sequence element intersect (Fig. 17.9). In practice, overlapping windows of a pre-determined size of sequence are compared and the value above which the similarity is recorded as a dot is selected beforehand to reduce the 'background' dots. Sequence similarities appear as a diagonal run of dots that can easily be seen by the eye. Deviations in this dot matrix reveal features of the two sequences being compared. For instance, mutations give gaps in the diagonal run, insertions or deletions cause the run to change from one diagonal to another, and sequence repetitions give parallel diagonal runs of dots.

Sequences can also be compared by using various alignment computer programs. These attempt to give the optimum sequence alignment by putting blanks in one or other of the sequences. This is best used for pairs of sequences that a dot plot has shown to be related. There are two types of method for aligning sequences. *Global* methods attempt to find an 'optimal' alignment throughout the length of the sequence, and *local* methods only attempt to identify short regions of similarity and do not

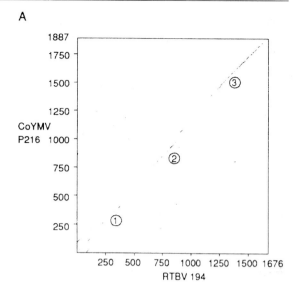

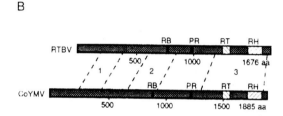

Fig. 17.9 (A) Comparison of RTBV P194 and CoYMV, P216 by DIAGON. Three major regions of homology were identified as indicated. **(B)** Comparison of the spatial distribution of various functional domains in RTBV P194 and CoYMV P216. RB, RNA-binding domain; PR, proteinase domain; RT, reverse transcriptase domain; RH, RNase H domain; aa, amino acids. Also shown are the regions of homology identified in (A). From Hay *et al.* (1991), with kind permission of the copyright holder, © The Oxford University Press.

attempt to align the sequences between these regions. There is a wide variety of sequence alignment programs (see McClure *et al.*, 1993) and a useful one color-codes amino acids according to their properties. This enables the manual alignment of amino acids on their properties (e.g. basic or acidic or polar or non-polar) without the need to align specific residues.

From multiple alignments of amino acid sequences, sequence motifs that are characteristic of specific proteins such as enzymes can be identified. Various motifs found in viral proteins are described in previous chapters (e.g. those involved in RNA replication; Chapter 8, Section IV.B). Identification of sequence motifs and comparison of the sequence(s) with databases can enable the function of the protein to be identified and the closeness of relationship to known proteins to be determined. Although such searches up to now have not revealed functions for viral proteins in many cases, the rapid increase in accessions to databases, especially of plant genome sequences, coupled with advanced computer programs that enable reiterative searches, will reveal more relationships.

b. Reconstructing phylogeny

Multiple alignments of sequences and/or lists of numbers giving details of sequence similarities do not enable the easy interpretation of evolutionary relationships or pathways. There are two groups of methods for reconstructing phylogenies. The phenetic methods use phenotypic data in attempts to reconstruct phylogeny without completely understanding the evolutionary pathway. The cladistic methods concentrate on the evolutionary pathway by attempting to predict the ancestors. Initially, it would appear that the cladistic approach is superior but the criteria used to predict ancestors may be invalid and lead to wrong conclusions. In practice, one has to use a pragmatic approach, and methods that involve a combination of phenetic and cladistic approaches can be equally useful.

Phylogenetic relationships are often presented as a structure resembling a tree or dendrogram made up of various parts termed root, stem, branch, node and leaf. There are various styles of dendrograms (see Fig. 17.10) and various methods for their construction. These are discussed in detail in Weiller *et al.* (1995) and their advantages and disadvantages are beyond the scope of this book.

However, in determining evolutionary relationships, it is important to recognize the limitations of the various types of dendrogram; it is also important to test the statistical significance of the branching within the tree. These statistical methods are discussed by Weiller *et al.* (1995), who comment that no one method of tree construction is superior to others and advise the use of more than one method for each data set and then comparison of the results.

2. Bromoviridae

Members of the *Bromoviridae* have genomes that are divided between three (+)-strand ssRNA segments (see Chapter 6, Section VIII.A). There is evidence for both genome re-assortment and recombination in the evolution of members of this family. A phylogenetic analysis of cucumoviruses using aligned amino acids revealed different relationships among species when the three genome segments were compared (White *et al.*, 1995). This suggested that re-assortment events have given rise to the current isolates. Furthermore, an interspecies pseudo-recombinant between CMV and PSV was found. However, a study in which 217 field isolates of CMV from 11 natural populations were typed by RNase protection assay (Fraile *et al.*, 1997b) provided evidence of selection pressure against re-assortment.

CMV is divided into two subgroups based on serological data, and sequence data suggest a further subdivision of subgroup I (Palukaitis *et al.*, 1992; Chaumpluk *et al.*, 1996). Recombination has been found within RNA3 of two cucumoviruses, CMV and TAV, co-inoculated to tobacco plants and grown under minimal selection pressure (Aaziz and Tepfer, 1999a). Alignment of the 5′ non-translated regions of RNA3 of 26 isolates identified possible re-arrangements, deletions and insertions in this region that may have been the precursors of the subsequent radiation of each subgroup (Roossinck *et al.*, 1999). Phylogenetic analyzes indicated that the three subgroups evolved radially, each from a single origin. Roossinck (2001) describes CMV as a model for RNA virus evolution.

3. Closteroviruses (reviewed by Karasev, 2000)

The genome organizations of members of the *Closteroviridae* are described in Chapter 6 (Section VIII.F) and are illustrated in Figs 6.33 and 6.34. These viruses have the largest (+)-strand ssRNA genomes among plant viruses. Dolja *et al.* (1994) describe the modular organization of closterovirus genomes, identifying

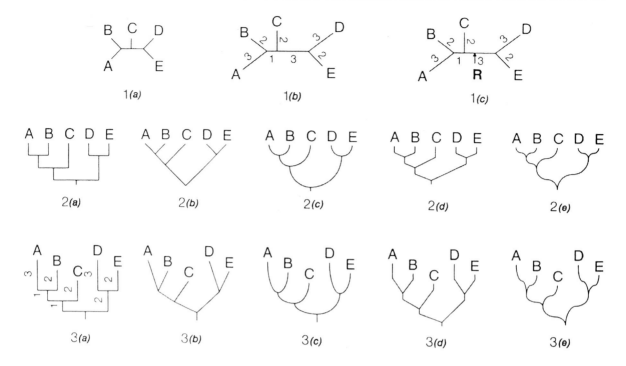

Fig. 17.10 Styles of dendrograms. Series 1 (a–c) represents unrooted trees (networks) as undirected graphs: 1(a) is unweighted and merely indicates relationships; 1(b) is weighted and gives the relative edge lengths; 1(c) also shows the possible root 'R', which is at the midpoint of the path between the most dissimilar OTUs (operational taxonomic units, a generic term that can represent many types of comparable taxa), A and D. Series 2 (unweighted) and 3 (weighted) are rooted dendrograms representing 1(c); (a)s are phenograms, (b)s are cladograms, (c)s are curvograms, (d)s are eurograms and (e)s are swoopograms. From Weiller *et al.* (1995), with kind permission of the copyright holder, © The Cambridge University Press.

four modules. (At that time of writing, some viruses were classified as closteroviruses but have now been moved to other genera; therefore, these comments apply to closteroviruses *sensu stricto*.) For BYV the core module consists of the key domains for RNA replication (the methyl transferase (MTR), the helicase (HEL) and the RNA-dependent RNA polymerase (POL) (in ORFs 1a and 1b; Fig. 6.33) and belongs to the alphavirus supergroup (see Chapter 8, Section IV.B.1). The upstream accessory module is the papain protease (P-PRO) at the N-terminus of ORF 1a. The chaperone module, which is separated from the core module by a short intergenic spacer, combines a small hydrophobic protein (ORF 2), the HSP70 homolog (ORF 3) and the 64-kDa protein with similarities to HSP90 (ORF 4). The fourth module contains the 3' ORFs 5, 6, 7 and 8, which include the major and minor coat proteins.

Obviously, there would be a slightly different composition of each module for CTV (see Table 6.5).

Dolja *et al* (1994) suggest that the closterovirus genome arose from a common ancestor, with re-arrangement of that genome and acquisition of other modules by recombination. The progenitor of the alphavirus supergroup has been proposed to comprise a complex of genes encoding MTR, HEL, P-PRO, POL and capsid protein. They propose that the evolutionary pathway was as shown in Fig. 17.11.

This evolutionary pathway comprised the following steps:

1. Deletion of P-PRO from the core of the replication genes.
2. Substitution of the postulated alphavirus-type capsid protein by a capsid protein capable of forming elongated virions.

3. Invention of the frameshift mechanism of POL expression (see Chapter 7, Section V.B.10.b).
4. Acquisition of the HSP70 from the cellular genome.
5. Duplication of the capsid protein and the functional switch for one of the tandem copies to facilitate aphid transmission.

6. Insertion of long coding sequences between the MET and HEL cistrons.
7. Secondary acquisition of the leader P-PRO, perhaps from a potyvirus or a related virus.
8. Additional diversification and acquisition of the 3'-terminal genes.
9. Split of the genome into two components giving the crinivirus genome organization.

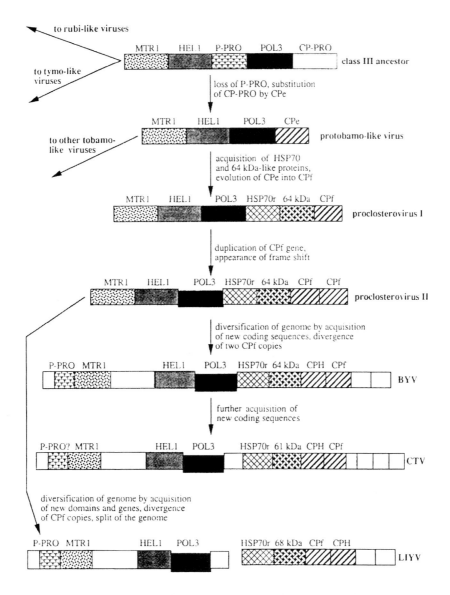

Fig. 17.11 A tentative scenario for the evolution of closteroviruses. Cpe designates an elongated particle capsid protein (a possible ancestor of capsid protein of both rod-shaped and filamentous viruses). CPH, capsid protein homolog; HEL 1, RNA helicase of superfamily 1; HSP70r, HSP70-related protein; MTR 1, methyltransferase of type 1; POL 3, polymerase of supergroup 3; P-PRO, papain-like proteinase. From Dolja *et al.* (1994), with kind permission of the copyright holder, © Annual Reviews. www.AnnualReviews.org

It is suggested that steps 1 and 2 occurred early in evolution and gave a common ancestor of the whole tobamovirus cluster. The order of the other events is rather arbitrary.

Recombination is an essential part of this proposed evolutionary pathway. Analyzes of multiple species of CTV defective RNAs show that they appear to have arisen by recombination of a subgenomic RNA (sgRNA) with distant parts from the 5′ end of the CTV genome (Bar-Joseph et al., 1997). It is suggested that closteroviruses are able to use the sgRNA and/or their promoter signals for modular exchange and rearrangement of their genomes.

In spite of the potential for variation, a study showed that mild CTV isolates maintained in different citrus hosts, from several geographical locations (Spain, Taiwan, Colombia, Florida and California) and isolated at different times were remarkably similar (Albiach-Martí et al., 2000a). This indicated a high degree of evolutionary stasis in some CTV populations.

4. Luteoviruses

The organization of the (+)-sense ssRNA genomes of the three genera of the *Luteoviridae* is described in Chapter 6 (Section VIII.G) and shown in Fig. 6.35.

Gibbs (1995) recognized a 'supergroup' of small icosahedral viruses that have (+)-sense ssRNA genomes, which now comprise the *Luteoviridae*, *Tombusviridae* and *Sobemovirus* taxa. The genome organization of these viruses comprises two basic modules: the 5′ replicase proteins and the 3′ proteins, which include the virion coat protein. Phylogenetic analyzes of the RdRp and the coat protein (Fig. 17.12) suggest that there have been gene transfer events between the modules of members of the supergroup.

The phylogenetic analyzes indicated that this supergroup fell into two main clusters termed the 'enamo' and 'carmo' clusters. Likely gene transfers were recognized both between and within these main clusters.

The current classification of the *Luteoviridae* recognizes three genera: the *Luteovirus*, which falls in the 'carmo' cluster, and the *Polerovirus* and the *Enamovirus*, which are placed in the 'enamo' cluster (Fig. 17.12).

Various models have been proposed for the evolution of the luteovirus and polerovirus genome organizations. There is a clear relatedness among the P3 sequences of both genera, and the organizations of ORFs 3, 4 and 5 are obviously similar (see Fig. 6.35). However, the arrangement and composition of the 5′ ORFs of the two genera are clearly very different. The most likely model for the origin of the genomes of the two genera is shown in Fig. 17.13 (Miller et al., 1997). In this model, it is suggested that recombination arose by strand switching at subgenomic RNA start sites during RNA replication in cells jointly infected by the two parental viruses. For the derivation of the luteovirus genome, the sgRNA start site on diantho-like viruses has homology to that of poleroviruses. Recombination at this site would create a hybrid virus with dianthovirus polymerase and polerovirus coat protein and neighboring genes. A recombination event at the sgRNA start site downstream of ORF 5 would give the complete luteovirus genome organization. A single recombination between the 5′ region of a sobemovirus and the 3′ part of a luteovirus followed by premature termination would give the polerovirus genome organization (Fig. 17.13).

There is further considerable evidence for recombination within the *Luteoviridae*. One can distinguish three types of event: recombination within a gene, recombination of large parts of the genome within a genus, and recombination between large parts of the genome between genera. An example of recombination within a gene can be found in the analysis the luteovirus readthrough proteins (ORFs 3 and 5) (Gibbs and Cooper, 1995). There appeared to be a recombinational event between the ancestors of CABYV and PEMV that led to the transfer of the RNA encoding the 5′ part of this region to CABYV. As a result of the second type of event, BMYV appears to be a chimera between two members of the genus *Polerovirus*, with the 5′ sequence encompassing ORFs 0, 1 and 2 resembling CABV and the 3′ region covering ORFs 3, 4 and 5 being similar to BWYV (Guilley et al., 1995). For at least two viruses, there is evidence for intergeneric recombina-

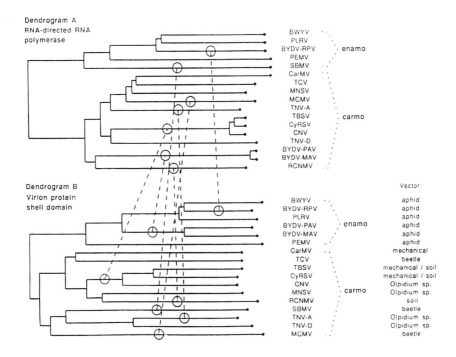

Fig. 17.12 Classification of RNA-directed RNA polymerase and virion protein shell domain amino acid sequences from 17 sequenced members of the luteovirus 'supergroup' (dendrograms A and B respectively). Dotted lines linking the two dendrograms represent likely gene transfer events between the two main clusters of the 'supergroup'. The dendrograms were deduced by the neighbor-joining method from estimates of the evolutionary distance between each pair of sequences calculated after the progressive alignment of each complete set of sequences. The pattern of transfer events was assessed from the incongruity of lineages given by different genes. The links are drawn arbitrarily at the midpoints of lines joining the nodes where incongruities are evident, as it is not possible to assess at what point the transfer event occurred within an incongruent lineage. The lineages leading to PLRV and BWYV are assumed to be congruent as this minimizes a rate anomaly between the polymerase and virion protein trees within this subcluster. The lineages leading to CarMV, TCV, MNSV and TNV-D are assumed to be congruent within the carom cluster. The means of transmission of each virus is indicated next to the virion protein tree. From Gibbs (1995), with kind permission of the copyright holder, © The Cambridge University Press.

tion between luteoviruses and poleroviruses. The genome arrangement of SbDV (Rathjen *et al.*, 1994) suggests that it should be classified in the genus *Luteovirus*, as (1) there is no ORF 0; (2) ORF 1 is relatively small and overlaps ORF 2 by only a few nucleotides; (3) the RdRp is carmovirus-like; and (4) the 5' and 3' UTRs contain sequences typical of that genus. However, the products of SbDV ORFs 3, 4 and 5 are more typical of the genus *Polerovirus*. Thus, the SbDV genome resembles a hybrid between those of the two genera. Phylogenetic analysis of the genomic nucleotide sequence of ScYLV shows that ORFs 1 and 2 most closely resemble polerovirus counterparts, ORFs 3 and 4 are most closely related to counterparts in the

luteovirus genomes, and ORF 5 is related to the read-through protein gene of the only known enamovirus (Moonan *et al.*, 2000; Smith *et al.*, 2000a). The two recombination sites on the SCYLV genome map to the known sites for the transcription of subgenomic RNAs.

As described in Chapter 11 (Section III.H.1.a), luteoviruses are involved in close biological associations with other viruses, especially umbraviruses. In these complexes, the luteovirus provides the coat protein that enables aphid transmission of the umbravirus together with the luteovirus. Gibbs (1995) suggested that close associations of viruses such as these complexes could give a greater chance of interviral recombination.

Thus, recombination appears to be rampant both within the *Luteoviridae* and between members of this family and those of some other groups of small ssRNA viruses. There is also evidence that PLRV can recombine with host sequences (Mayo and Jolly, 1991); the 5'-terminal 119 nucleotides of some RNAs of a Scottish isolate of PLRV are very similar to an exon of tobacco chloroplast DNA.

5. Potyviruses

Potyviruses have a (+)-strand ssRNA genome, which encodes a polyprotein that is processed to the final gene products (see Chapter 7,

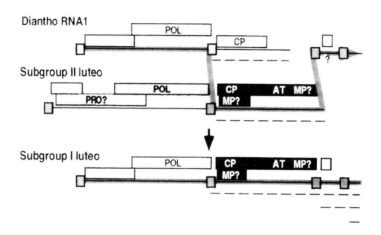

Origin of a subgroup I luteovirus by recombination

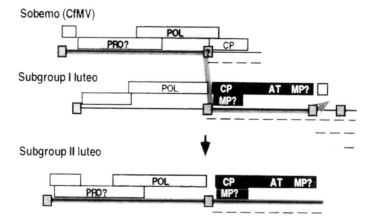

Origin of a subgroup II luteovirus by recombination

Fig. 17.13 (see Plate 17.1) Model for origin of luteovirus subgroups. Solid black lines represent viral genomic RNA. Dashed lines indicate subgenomic RNAs. Boxes indicate genes. Blue shading, genes with sequence similarity to umbra-, diantho- and carom-viruses; green, sequence similarity to sobemoviruses. Grey boxes represent putative origins of replication and subgenomic mRNA promoters. POL, RNA-dependent RNA polymerase; PRO?, putative protease; CP, coat protein; MP?, putative movement protein; AT, read-through domain of the coat protein gene, possibly required for aphid transmission. Pink line shows the proposed path of the replicase as it switched strands during copying of viral RNAs in a mixed infection. From Miller *et al.* (1997b), with kind permission of the copyright holder, © The American Phytopathological Society.

Section V.B.1.b). Phylogenetic relationships have been assessed using the coat protein sequence (reviewed by Ward *et al.*, 1995). The coat protein comprises three domains: a N-terminal region, a conserved core, and a much smaller C-terminal region. Various phylogenetic analyzes have been performed using the coat protein core sequences (Fig. 17.14).

In both dendrograms shown in Fig. 17.14, the basal node is on the branch connecting the bymovirus, BaYMV, coat protein to the others. The second branch is the tritimovirus WSMV and the third branch connects all the aphid-transmitted potyviruses. It is suggested that these taxonomic features correlate with the taxonomy of the hosts and with the vector specificities. The bymoviruses and tritimoviruses are found naturally only in monocotyledonous species, with bymoviruses being transmitted by plasmodiophoraceous fungi and tritimoviruses by eriophyid mites. These are separated by the dicotyledonous plant-infecting viruses that are aphid-transmitted. However, a more recent analysis (Fig. 17.15) (Berger *et al.*, 2000) does not fully support this, with dicot-infecting ipomoviruses and both dicot- and monocot-infecting macluraviruses being relatively close to the tritimovirus and bymovirus branches respectively. The phylogenetic tree shown in Fig. 17.15 does place the dicot-infecting aphid-transmitted viruses together (except for macluraviruses) and separates these from the virus genera with other vectors.

Analysis of the potyviral coat protein N-terminal domain sequences reveals evidence for partial gene duplication. For instance, two strains of SCMV have core domains that are 92% identical but N-terminal domains that differ in length and are only 22% identical (Frenkel *et al.*, 1991). Comparison of the N-terminal domain sequences show that they appear to come from different sources and also that there is duplication in the larger sequence. These N-terminal sequences are on the exposed surface of the coat protein and changes in them would not affect the basic coat protein structure important for assembly of rod-shaped particles. The DAG recognition site for the aphid-transmission helper protein

(see Chapter 11, Section III.E.7.b) is in this domain.

Recombination also appears to have played a role in the evolution of YMV (Bousalem *et al.*, 2000). Analysis of a region covering the C-terminal part of the NIb, the coat protein and the 3′ untranslated region of 27 YMV isolates showed that it had the most variable coat protein compared with eight other potyviruses. There was no correlation between the coat protein and the 3′-UTR diversities and phylogenies. There was a geographical distinction between isolates from the Caribbean, South America and Africa, with most variation occurring in the African isolates. There was also evidence for host selection of particular isolates. Bousalem *et al.* (2000) hypothesize an African origin of YMV in the *Dioscorea cayenensis–D. rotunda* complex and independent transfers to *D. alata* and *D. trifida*.

Geographical distinction has also been recognized between isolates of the peanut stripe strain of BCMV collected from China and Indonesia (Higgins *et al.*, 1999), with variation in the coat protein nucleotide sequence reaching up to 3.5% within and 7.3% between geographical groups. It is suggested that this strain appears to have arisen independently in different locations.

6. Tobamoviruses

Tobamoviruses are among the most studied of plant viruses. Relationships have been assessed from a wide variety of data including nucleic acid sequences, amino acid sequences, peptide mapping, serological relationships and biological properties. The evolution of this virus genus has been reviewed by Fraile *et al.* (1995) and Gibbs (1999b).

Estimates of the average number of amino acid substitutions per site between pairs of tobamoviruses for four coding regions (Table 17.3) showed that the 54-kDa region (ORF 2; see Fig. 6.36) was the most conserved and that the 30-kDa movement protein was the most divergent. Most of the nucleotide substitutions were synonymous (not causing a change in amino acid) and did not differ between the four ORFs. Thus, the whole genome appears to have diverged as an entity, with little evidence of recombination playing a significant part in diversification. One possible exception is SHMV, which appears to have acquired its 3' non-coding region from a tymo-like virus.

Various studies on tobamoviruses have

revealed very stable populations. For example, RNA fingerprinting of 26 isolates of PMMoV from epidemic outbreaks in Spain and Sicily in the 1980s revealed high stability (Rodríguez-

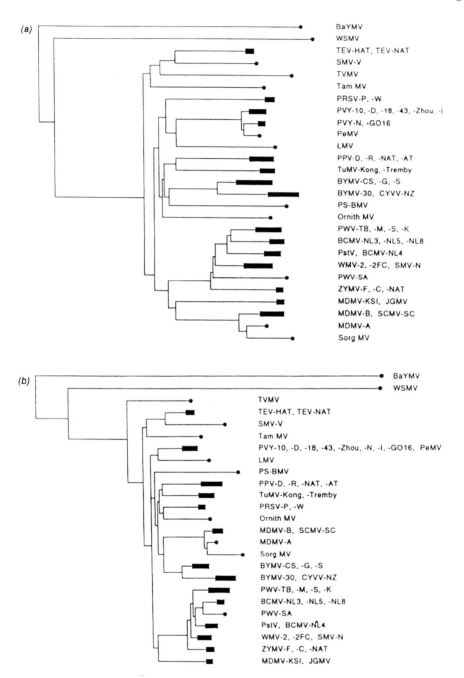

Fig. 17.14 Neighbor-joining trees showing the relationships of potyviruses calculated from the '% non-identity' (panel (a)) or 'FJD distance' (Feng *et al.*, 1985) (panel (b)) of the aligned core domains of their coat proteins. From Ward *et al.* (1995), with kind permission of the copyright holder, © The Cambridge University Press.

Cerezo *et al.*, 1989). Although there was variation within the population, presumably reflecting its quasi-species status, the master sequence was relatively constant in different geographical sites and over a period of time. The genetic divergence between five populations of TMGMV naturally infecting *Nicotiana glauca* in southern Spain was very small (Moya *et al.*, 1993), as was that from isolates of this virus from *N. glauca* plants from Australia, California, Spain and the east Mediterranean Basin (Fraile *et al.*, 1996).

In vitro studies on TMV show that homologous recombination can occur frequently. For instance, repeated sequences inserted into cloned cDNA are eliminated rapidly on infection of plants (see Dawson *et al.*, 1989; Beck and Dawson, 1990). However, under natural conditions there is little evidence for recombination. Similarly, quantification of mutational error rates of TMV replication, using bacterial genes inserted into the viral genome, give values of 10^{-3} to 10^{-5}, which are very similar to those of other RNA viruses (Donson *et al.*, 1991; Kearney *et al.*, 1999). However, the passage of cloned TMV DNA through seven plant host species over a period of 413–515 days revealed a small mutation rate of 3.1×10^{-4} nucleotide substitutions per base-year (Kearney *et al.*, 1999). Thus, the inherent variability of tobamoviruses is similar to that of other RNA viruses and the sequence conservation in the natural situation must reflect considerable constraints on the accumulation of variants.

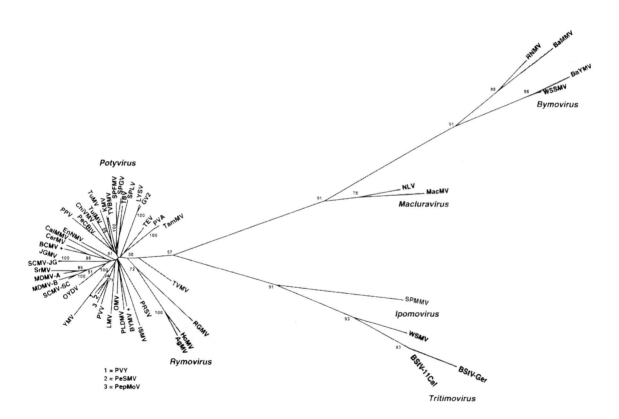

Fig. 17.15 Phylogenetic tree of the family *Potyviridae* using amino acid relationships of the coat protein. Inference based on the Fitch and Margoliash (1967) least-squares method. The sequences were aligned using PILEUP (Devereux *et al.*, 1984). Branch lengths are proportional to sequence distances. The dendrogram was bootstrapped 100 times (percentage scores shown at nodes) and are not rooted. Many branches are condensed and may contain multiple viruses. BCMV+, for example, includes BCNMV, CABMV, DMV, PWV, SAPV, SMV, WMV and ZYMV, as well as all the viruses contained in the BCMV subgroup (AzMV, BlCMV, DeMV and PStV). Similarly, the branch labeled BYMV+ includes CYVV, PMV, PSbMV and PeMotV. From Berger *et al.* (2000), with permission.

Table 17.3   Estimates of the average number of amino acid substitutions per site, d for different encoded proteins[a], and estimates of the nucleotide substitution per site for the 3' ncr[b]

Gene	ToMV	TMGMV	PMMV	ORSV	SHMV	CGMMV	RMV
126K							
TMV	0.0929	0.4246	0.3064	–	–	0.7967	–
ToMV		0.4108	0.2871	–	–	0.7824	–
TMGMV			0.4480	–	–	0.7899	–
PMMV				–	–	0.8007	–
54K							
TMV	0.0622	0.1999	0.2211	–	–	0.5236	–
ToMV		0.2992	0.1961	–	–	0.5477	–
TMGMV			0.2985	–	–	0.5667	–
PMMV				–	–	0.5327	–
MP							
TMV	0.2384	0.5307	0.3729	0.4397	1.1865	0.9848	–
ToMV		0.4979	0.4054	0.4761	1.0303	0.9822	–
TMGMV			0.4214	0.5017	1.0026	0.9822	–
PMMV				0.3539	1.0671	0.9636	–
ORSV					0.9899	0.9899	–
SHMV						0.9004	–
CP							
TMV	0.1785	0.3415	0.3558	0.3136	0.6424	0.7820	0.7459
ToMV		0.3264	0.3289	0.2792	0.8077	0.9618	0.7657
TMGMV			0.3835	0.3249	0.7722	0.9618	0.6868
PMMV				0.3585	0.8092	0.8625	0.7668
ORSV					0.8518	0.8625	0.7744
SHMV						0.7910	0.8692
CGMMV							0.9556
3'ncr							
TMV	0.3199	1.0578	0.6785	0.8080	1.3837	0.8084	–
ToMV		1.0692	0.6833	0.8317	1.3817	0.8996	–
TMGMV			1.0333	1.1382	1.7285	0.9948	–
PMMV				0.8863	1.7461	0.9192	–
ORSV					1.6126	0.7687	–
SHMV						1.7319	–

[a] $d = -\log_e n_i/n$, where n = average number of amino acids in the protein, n_i = number of identical amino acids in the two proteins (Nei, 1987).

[b] Nucleotide substitutions per site in the 3' non-coding region (ncr) calculated by Kimura's two-parameter method (Kimura, 1980).

126k and 54k, 126- and 54-kDa parts of the replicase; MP, cell-to-cell movement protein; CP, coat protein.

From Fraile *et al.* (1995), with kind permission of the copyright holder, © The Cambridge University Press.

7. Geminiviruses

The DNA → DNA replication of geminiviruses should result in less mutagenic variation than RNA → RNA replication (see Fig. 8.29). However, there is evidence for significant genomic variation in at least two of the geminivirus genera, the begomoviruses and the mastreviruses. There is widespread genomic and serological variation in the begomoviruses (reviewed by Harrison and Robinson, 1999), much of which represents geographically related lineages that have little relation to host range (Fig. 17.16).

Using short acquisition and inoculation feeding periods of single *Cicadulina mbila* leafhopper vector, transmission from an initial MSV infection of a wild perennial host and serial passage on almost completely resistant cultivars yielded three MSV isolates (Isnard *et al.*, 1998). Sequence analysis of these isolates indicated that the original infection had a quasi-species structure with mutations distributed throughout the genome. Mutation frequencies were estimated to be between 3.8×1^{-4} and 10.5×10^{-4}, levels similar to those found with RNA → RNA replication. This variation was highest in synonymous positions in AC1 and AV1 ORFs of CLCuV (Sanz *et al.*, 1999). The ratio of non-synonymous to synonymous substitutions varied for the different ORFs, being higher for AV1 than for AC1 and lower still in the AC4 and AV2 ORFs. It is suggested that the evolution of the AC4 and AV2

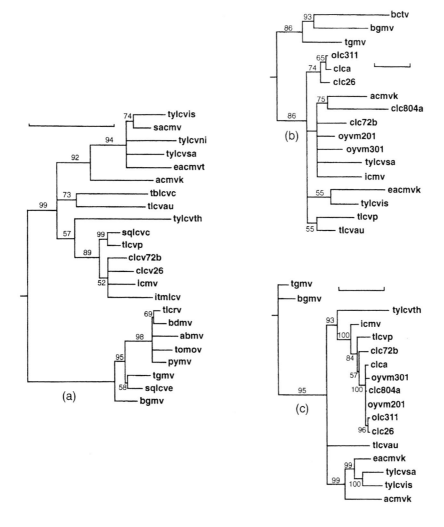

Fig. 17.16 Phylogenetic trees obtained from alignments of deduced amino acid sequences of begomoviruses using the PUZZLE program. Horizontal scale bars represent 0.2 replacements per position. **(a)** CP, showing two main branches containing Old World (upper branch) and New World (lower branch) viruses. Three subclusters of Old World viruses are evident. **(b)** and **(c)** Rep and CP, respectively, of seven begomovirus isolates from cotton (clc) or okra (olc, oyvm) in Pakistan together with selected other viruses from five continents. Note differences in the relative positions of individual cotton and okra isolates in **(b)** and **(c)**. abmv, *Abutilon mosaic virus*; acmvk, eacmvt, icmv, sacmv, African, East African, Indian and South American *Cassava mosaic virus* respectively; bctv, *Beet curly top virus*; pymv, *Potato yellow mosaic virus*; sqlcvc, sqlcve, *Squash leaf curlvirus* from China and USA respectively; thlcvc, *Tobacco leaf curl virus* from China; tgmv, *Tomato golden mosaic virus*; tlcrv, *Tomato leaf crumple virus*; tlcvau, tlcvp, itmlcv, *Tomato leaf curl virus* from Australia, New Delhi and Bangalore respectively; tomov, *Tomato mottle virus*; tylcvis, tylcvni, tylcvsa, tylcvth, *Tomato yellow leaf curl virus* from Israel, Nigeria, Sardinia and Thailand respectively. From Harrison and Robinson (1999), with kind permission of the copyright holder, © Annual Reviews. www.AnnualReviews.org

ORFs is constrained by AC1 and AV1 respectively, and that AC4 and AV2 are relatively new genes originating by overprinting of preexistent AC1 and AV1 genes.

The mutagenic variation of geminiviruses is unexpected in view of the proofreading normally associated with DNA → DNA replication. This point is discussed by Roossinck (1997), who suggests that it may reflect a lack of post-replicative repair. Geminivirus DNA does not appear to be methylated, as its replication is inhibited by methylation. Thus, although these viruses use the host system for their replication, the normal host mechanisms for mismatch repair probably do not function.

There is increasing evidence for natural recombination within geminiviruses, on occasions leading to new diseases. For example, three geminivirus species have been recognized as causing mosaic in cassava: ACMV, EACMV and ICMV (Hong et al., 1993). Comparison of the sequence of a geminivirus causing an epidemic of severe cassava mosaic disease in Uganda (the Uganda variant, UgV) (see Plate 3.21) with the sequences of the three cassava viruses showed that it was similar to a Tanzanian isolate of EACMV except in the coat protein (Zhou et al., 1997). The 5' part of the UgV coat protein gene was 99% identical to EACMV (only 79% to ACMV), the middle region of the gene was 99% identical to ACMV (75% to EACMV) and the 3' part was 98% identical to EACMV (76% to ACMV). This was taken as strong evidence that the UgV arose by recombination between ACMV and EACMV. Pita et al. (2001) paint a more complex picture of severe epidemics in Uganda involving recombination, pseudorecombination and synergism. Sequence analysis of cassava mosaic disease agents from Cameroon provides further evidence of recombination, this time in the AC2–AC3 region of DNA-A and the BC-1 region of DNA-B (Fondong et al., 2000) (see Fig. 6.7 for begomovirus genome organization). This is also evidence for recombination in South African cassava mosaic virus (Berrie et al., 2001).

The severe outbreaks of cotton leaf curl disease in Pakistan are associated with multiple infections of begomoviruses (see Zhou et al., 1998; Sanz et al., 1999, 2000). The 5' and 3' halves of the intergenic region of some isolates had different affinities and occurred in several combinations, suggesting that recombination had occurred and that the origin of replication was a favored recombination site (Zhou et al., 1998). Evidence for recombination was also found in the AC1 and AV1 ORFs, indicating that the origin of replication is not the only recombination site but that it is the major one (Sanz et al., 1999). It is suggested that recombination, following multiple infection, explains the network of relationships among many of the begomoviruses found in the Indian subcontinent, and their evolutionary divergence, as a group, from begomoviruses causing similar diseases in other geographical regions (Sanz et al., 2000).

Evidence has also been found for recombination of geminiviruses infecting pepper in Mexico (Torres-Pacheco et al., 1993) and tomato in central America (Umaharan et al., 1998) and Ageratum conyzoides in many countries (Saunders et al., 2001b). An analysis of 64 distinct geminivirus species revealed 420 statistically significant recombinant fragments distributed across the genome (Padidam et al., 1999). It is suggested that this high rate of recombination contributes to the recent emergence of new geminivirus diseases.

New variants of bipartite geminiviruses can arise from pseudo-recombination between the components. Unseld et al. (2000) demonstrated the gradual increase in infectivity of pseudo-recombinants between two isolates of SIGMV associated with virus-specific adaptation of the components. Serial passage of a pseudo-recombinant between ToMoV DNA-A and BDMV DNA-B led to a recombination event in which the common region of the DNA-B was replaced by that of DNA-A (Hou and Gilbertson, 1996). The recombinant had increased pathogenicity.

8. Discussion and summary

There is extensive molecular evidence that has been used to suggest evolutionary pathways for several groups of viruses. The examples given above are not the only ones and there is similar evidence for many of the other plant virus groups.

As described in Section IV, there are three ways in which variation is generated: mutation, re-assortment and recombination. Re-assortment of the segments of multi-partite viral genomes (pseudo-recombination) has been used in laboratory experiments to study gene functions, but there are few examples of new pathogens arising naturally by this means. As mentioned in Section X.D.2, there appears to be selection against pseudo-recombinants, at least in the *Bromoviridae*, but there is evidence for re-assortment in various natural isolates of the tenuivirus, RGSV, from the Philippines (Miranda *et al.*, 2000). Similarly, genome re-assortment between two isolates of TSWV overcame transgenic protection conferred by the *N* gene (Qiu and Moyer, 1999). In both these latter cases, it is likely that the selection pressure for the re-assorted isolate overcame any constraints of incompatibilities between genome segments.

The examples described above show that there is evidence for extensive mutation and recombination among plant viruses. The high levels of mutation can be expected among viruses that do not have proofreading in their replication system, that is those that replicate RNA → RNA and by reverse transcription. The relatively high levels of variation among geminiviruses is unexpected but, as discussed in Section X.D.7, there would appear to be a molecular explanation for this. Mutation is considered to be the major source of variation upon which microevolution operates to generate new species (see Fig. 17.8). The analyzes of the sequences of viral genomes is giving increasing evidence for recombination being the major factor in macroevolution, the generation of new genera and families.

Laboratory experiments have shown that viruses have a great potential for both mutation and recombination. This potential does not appear to be fully realized in the natural situation and thus there would seem to be some constraints on the adoption of both mutants and recombinants. These constraints are discussed in Section XIII.

The examples above indicate that the theory, proposed initially for DNA bacteriophage (Botstein, 1980) (Section X.C), that macroevolution proceeds by the exchange of modules by

recombination would seem to apply equally to RNA viruses. This has also been suggested for reverse-transcribing viruses (Hull, 1992; Hull and Covey, 1996). However, when one considers the modular theory in detail in light of the above examples, there is a problem in the original definition of 'module', which was that they are 'interchangeable genetic elements, each of which carries out a particular biological function'. In the examples given above, modules varied from large, with several functions to small, with perhaps one function. Luteoviruses are suggested to have evolved by recombination of two half genomes (see Fig. 17.13), one half containing the replicase and the other all the other biological functions (e.g. coat protein, insect transmission factor, cell-to-cell movement protein, and transcriptional and translational control factors). In contrast, it is suggested that closteroviruses have evolved by acquisition of modules that effect one, or only a few, biological functions and the genesis of CABYV involved the exchange of the 5' portion of the ORF 3 + 5 read-through protein. Thus, if the original definition of module is to be retained, for which there is a strong case as it is very descriptive, there is the need for terms to describe collections of modules and parts of modules.

Two problems arise from the recognition of the important part that exchange of modules plays in virus evolution. On the practical side it would indicate that there may be risks associated with the release of transgenic plants to the field; this is discussed in Chapter 16 (Section VII.E). Modular exchange also raises problems in virus classification, for instance with SbDV and SCYLV, which incorporate characters of two, or even three, genera of the *Luteoviridae*.

XI. SOURCES OF VIRAL GENES

The discussion in the previous section showed that, on current understanding, a significant part of viral evolution involves the acquisition and exchange of functional modules that make up the virus genome. In this section, I discuss some ideas on the sources of some of these modules. As to the overall origins of genes, there are two theories: a common origin by a molecular 'big bang' and continuous creation (Keese and Gibbs, 1992).

One mechanism by which new coding sequences can be generated in a continuous creation scenario is by 'overprinting' in which an existing nucleotide sequence is translated *de novo* in a different reading frame or from non-coding ORFs. There is some evidence for over-printing, as discussed by Gibbs and Keese (1995). A suggested example is in the genome of TYMV (see Fig. 6.53). The largest ORF encodes the replicase with the characteristic GDD motif. The 5' terminus of this ORF also encompasses an overlapping out-of-frame gene that begins at the first AUG of the genome, which is four nucleotides upstream of the start of the replicase ORF. The suggestion is that the replicase was the original ORF and the overlapping one was formed by overprinting.

A. Replicases

1. RNA replicases

The structure of core RNA replicases is discussed in Chapter 8 (Section IV.B), where it is pointed out that they comprise several functional units, the RNA-dependent RNA polymerase (RdRp), a helicase and a methyl-transferase; the RdRps and helicases fall into several supergroups or superfamilies (see Tables 8.1 and 8.2). There is disagreement as to whether there is a common origin for each of these basic activities giving a monophyletic evolutionary pathway or whether their evolution is polyphyletic from several origins (Gorbalenya, 1995; Zanotto *et al.*, 1996). However, it has been suggested that the earliest form of nucleic acid was RNA (see Cech and Bass, 1986), which would suggest RNA replication is very ancient.

Notwithstanding the divergence of opinion about the monophyletic or polyphyletic origins of the component parts of core RNA replicases, there is reasonable evidence that their arrangement in modern viral genomes has arisen by modular shuffling of these components (Fig. 17.17).

2. Reverse transcriptase

The replication of members of the *Caulimoviridae* is by reverse transcriptase (RT), which converts

an RNA template to DNA (see Chapter 8, Section VII.A). There are several arguments that point to the basic elements of RT being very ancient. The suggestion that RNA preceded DNA (see above),

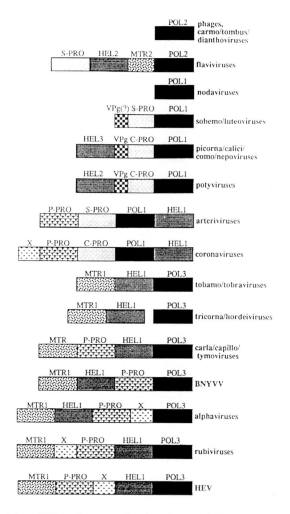

Fig. 17.17 Conservation and variability in the arrangement of the 'core' genes of (+)-strand RNA virus principal genes (domains) encoding proteins involved in genome replication and expression. Omission of less well-conserved domains is not specified. POL 1, 2, 3, RNA-dependent RNA polymerase of supergroups 1, 2 and 3 respectively; HEL 1, 2, 3, (putative) RNA helicase of superfamily I, II and III respectively (for the supergroups and superfamilies see Chapter 8, Section IV.B); S-PRO, serine chymotrypsin-like proteinase; C-PRO, cysteine chymotrypsin-like proteinase; P-PRO, papain-like proteinase; MTR 1, 2, methyltransferases of types 1 and 2 respectively. The 'shell' genes coding for virion 'accessory' proteins are not shown. Boxes are not to scale. From Koonin and Dolja (1993), with kind permission of the copyright holder, © CRC Press LLC.

coupled with sequence similarities between RdRps and RTs (see Argos, 1988; Doolittle *et al.*, 1989), has been taken to indicate that RT is amongst the earliest of enzymes (Cech and Bass, 1986). The amino acid homologies between the RTs of all retro-elements suggest that the enzyme has evolved only once (Doolittle *et al.*, 1989; Xiong and Eichbush, 1990).

3. DNA replicases

The DNA → DNA replication of geminiviruses and nanoviruses is described in Chapter 8 (Section VIII). These viruses use host DNA replicase functions and, as they replicate in differentiated cells, the cell cycle has to be modified. Thus, one of the viral gene functions is to bind to retinoblastoma (Rb) proteins, which directs the cell into the S phase. It is likely that the Rb-binding protein(s) has been acquired by the viruses from host sources.

B. Proteinases

Three classes of virus-coded proteinases are found among plant viruses (see Table 7.2). The functions of these were identified by comparison with amino acid sequences from plants and animals — hence the terms 'chymotrypsin-like' and 'papain-like'. This would suggest a host origin.

In most cases, the protease is a unique gene product, but it is sometimes associated with a protein with another function, for example the HC-Pro of potyviruses in which it is associated with the insect transmission helper factor and also the suppressor of host gene silencing. Whether this represents a multiple function for the original proteinase or the acquisition of various functional modules into the proteinase molecule is unknown.

C. Coat proteins

The structure of the virus particle is determined by the structure of the coat protein subunit (see Chapter 5). Because of packing considerations, proteins making up rod-shaped particles have to be wedge shaped, whereas those making up spherical particles have to be conical or have a three-dimensional wedge shape. The shape of these proteins is determined by the secondary and tertiary structures of their polypeptide chains. The coat proteins of most, if not all, viruses with isometric particles have a basic structure comprising eight anti-parallel β sheets, termed the β barrel or jellyroll motif (see Chapter 5, Section VI, and Fig. 5.19). From a comparison of the coat protein structures of 25 isometric viruses, Chelvanayagam *et al.* (1992) noted that the structure was best viewed as a two-sheet wedge rather than a barrel. They suggested an evolutionary pathway from subunits capable of forming T = 1 particles to those capable of forming T = 3 and higher-order particles (see Chapter 5, Section V, for a description of the architecture of isometric viruses). This evolution involved changes in the protein loop structures surrounding the basic β barrel to accommodate quasi-equivalent interactions between adjacent protein subunits. These considerations might indicate a single origin for the isometric virus coat protein. However, the coat protein of Sindbis virus has close relationships to chymotrypsin-like proteinases (Choi *et al.*, 1991), which would suggest a separate origin.

From structural considerations, McLachlan *et al.* (1980) suggested that the ancestral coat protein of TMV was a dimer of two smaller units. A tandem gene duplication event was proposed, followed by drift in amino acid sequence, arising from point mutations, to give rise to the present coat protein subunit. Gorbalenya *et al.* (1986) have invoked extensive gene duplications to explain internal repeating homologies in the poliovirus polyprotein. They also suggested that the VPg of this genome was built up by multi-step amplification of a gene coding for an 11-membered peptide followed by divergence in individual copies. The present 22-amino-acid VPg would be derived from a dimer of the original gene.

D. Cell-to-cell movement proteins

The cell-to-cell movement of many viruses is potentiated by one or more movement proteins (MP) (see Chapter 9, Section II.D.2). At least one type of these proteins, the '30K superfamily' that includes the TMV MP, appears to function by gating the plasmodesma open to enable the viral infection unit to pass to an uninfected cell. As discussed in Chapter 9 (Section II.L), plant proteins such as the KNOTTED1 homeodomain protein (Lucas *et al.*, 1995) selectively traffick through plasmodesmata, and a plant paralog of TMV MP has been recognized (Xoconostle-Cázares *et al.*, 1999). Thus, it is likely that this functional module has been acquired by viruses from the plant genome. As there are several ways in which cell-to-cell movement is effected, it is likely that different genes have been acquired by various viruses. For instance, the triple gene block would appear to be a combination of three different genes.

E. Suppressors of gene silencing

The plant (and other organisms) post-transcriptional gene silencing (PTGS) defense system against 'foreign' nucleic acids is described in Chapter 10 (Section IV). At least some viruses have genes (e.g. the potyvirus HC-Pro and the cucumovirus 2b genes) that suppress this defense system. The origin of these suppression genes is not known but there is a possibility that they have been acquired from the plant genome. One time that that plants have to tolerate 'foreign' nucleic acids is during the fertilization process. Proteins involved in the fertilization process might be the source of the viral PTGS suppression gene(s).

XII. ORIGINS OF VIRUSES, VIROIDS AND SATELLITES

Much of the evidence for the origins and evolutionary pathways of higher organisms comes from the fossil record of macroscopic (and microscopic) remains. As noted above, viruses do not form conventional fossils, although insects preserved in amber and seeds preserved in archeological sites may have some evidence of viruses. Another source of evolutionary data is what has been termed 'molecular fossils'. Analysis of nucleic acids and proteins of higher organisms and even bacterial and fungi can give much information on relatedness and even on evolutionary pathways. The high turnover rate of viral genomes coupled with their potential for great variation would seem to preclude this source of information on viral evolution. However, most basic functions are controlled by highly conserved sequences that are essential, say, for enzyme catalytic sites; significant variation in these motifs would be lethal for the organism. Such motifs are conserved in viruses and form an important source of information for gaining an insight into viral evolution and even origins.

A. Origins of viruses

There are numerous suggestions for the origins of viruses. At the time of the previous edition of this book three were generally considered:

1. Viruses are descended from primitive precellular life forms. This hypothesis was based on the ideas that the earliest prebiotic polymers were RNA, which had enzymatic properties, such as that shown by ribozymes (see Chapter 14, Section I.C.3.b). These pre-biotic RNAs later parasitized the earliest cells.
2. Viruses developed from the normal constituents of cells. The suggestion is that viruses arose from some cell constituent that escaped the normal control mechanism and became self-replicating entities. Examples of the normal cell constituent include transposable elements and mRNAs.
3. Viruses are derived from degenerate cells that eventually parasitized normal cells.

From the discussion in the sections above, it is evident that, on current thinking, much of virus evolution has taken place through the acquisition and exchange of functional modules. The basic module is the replicon, the minimal unit capable of replicating the viral genome, and attached to this are modules that enable the virus to fill its 'niche'. The source of the basic replicon probably differs for different genome types. It would seem likely that reverse-transcribing viruses arose from retroelement-like sequences that acquired modules enabling horizontal transmission. However, it is possible that retroelements are degenerate viruses that have lost their horizontal transmission genes.

The source of the basic replicon for RNA viruses is more difficult to assess and this module could be from either (1) or (2) above. As discussed by Schuster and Stadler (1999), the development of early replicons involved the transition from simple chemical reaction networks to autocatalytic processes capable of forming self-organized systems that undertook replication.

It is possible that DNA viruses such as geminiviruses and nanoviruses arose from plasmid-like DNA molecules. Possible origins and sources of 'niche-adapting' modules have been discussed in the previous section.

It is unlikely that any plant viruses originated from degenerate cells, though this has been suggested as a possible origin for large animal viruses, such as poxviruses (Fenner, 1979).

1. Viruses of prokaryotes and eukaryotes

Is there a fundamental distinction between viruses infecting prokaryotes and those infecting eukaryotes? There is no authenticated example of a virus of one sort completing its infection cycle successfully in cells of the other type. There are a few reports of a short sequence similarity between prokaryotic viruses and various eukaryotic viruses. Bacteriophage MS2 had the Asp-Asp sequence in its RNA-dependent RNA polymerase (RdRp) (Kamer and Argos, 1984), which is part of the conserved motif of viruses that infect eukaryotes (Chapter 8, Section IV.B.1). Rogers et al. (1986) drew attention to similarities between a nucleotide sequence in the stem and loop structure in the common region of various Geminivirus DNAs and that in the recognition and cleavage site for the ΦXI74 coliphage gene A protein and suggested similar mechanisms for the replication of their DNA (see Chapter 8, Section VIII.C). There is also another observation that may be relevant. It is now well established that chloroplasts evolved from prokaryotic cells. ssDNA from the AbMV is present in the chloroplasts of infected Abutilon (Gröning et al., 1987). However, on present evidence it seems likely that viruses infecting prokaryotes, and most of those infecting eukaryotes, had separate evolutionary origins, or that, if there was a common ancestral virus, then it existed in the very distant past.

2. Viruses infecting photosynthetic eukaryotes below the angiosperms

Virus-like particles (VLPs) have been observed in thin sections of many eukaryotic algal species belonging to the Chlorophyceae, Rhodophyceae and Phaeophyceae. These are described in Chapter 2 (Section V.A). Basically, there are two types of viruses, isometric ones that have dsDNA genomes and most closely resemble phage, and a rod-shaped virus with an RNA genome. The rod-shaped virus from the eukaryotic alga Chara australis (CAV) has some properties like those of tobamoviruses. However, its genome is much larger (11 kb, rather than 6.4 kb for TMV), and about 7 kb of the genome has been sequenced, revealing other relationships (A.J. Gibbs, quoted in Matthews, 1991): (1) the coat protein of CAV has a composition closer to BNYVV, and is closer to TRV than to TMV; (2) the GDD-polymerase motif of CAV is closest to that of BNYVV; and (3) the two GKT nucleotide-binding motifs found in CAV are arranged in a manner similar to that found in potexviruses. Thus, CAV appears to share features of genome organization and sequences found in several groups of rod-shaped viruses infecting angiosperms. It appears to have strongest affinity with the Furovirus group and has no known angiosperm host. For these reasons it is most unlikely that CAV originated in a recent transfer of some rod-shaped virus from an angiosperm host to Chara. In its morphology, Chara is one of the most complex types of Charophyceae, which is a well-defined group with a very long geographical history (Round, 1984). Based on cytological and chemical similarities, land plants (embryophytes) are considered to have evolved from a charophycean green alga. Coleochaete, another of the more complex types among the Charophyceae, has been shown to contain lignin, a substance thought to be absent from green algae (Delwiche et al., 1989). Molecular genetic evidence confirms a charophycean origin for land plants. Group II introns have been found in the tRNA^ala and tRNA^ile genes of all land plant chloroplast DNAs examined. All the algae and eubacteria examined have uninterrupted genes but Manhart and Palmer (1990) have shown that introns are present in three members of the Charophyceae in the same arrangement as in Marchantia, giving strong support to the view that they are related to the lineage that gave rise to land plants. Tree construction suggests that the Charophyceae may have acquired the introns

400 to 500 million years ago. Thus, the virus described in *Chara* may be the oldest recorded virus infecting a plant on or near the lineage that ultimately gave rise to the angiosperms.

No viruses have been reported from bryophytes. A virus with particles like those of a *Tobravirus* were found in hart's tongue fern (*Phyllitis scolopendrium*) (see Chapter 2, Section V.C). Some viruses have been found in gymnosperms (see Chapter 2, Section V.D), one from *Cycas revoluta* resembling a nepovirus.

In summary, the existence of a (+)-sense ssRNA virus infecting the genus *Chara* suggests an ancient origin for this type of virus. Other than this example, the meager information about viruses infecting photosynthetic eukaryotes below the angiosperms can tell us very little about the age and course of evolution among the plant viruses. The cycads are regarded as living fossils, being in the record since early Mesozoic times. However, the *Nepovirus* found in *Cycas revoluta* is quite likely to have originated in a modern angiosperm, as it readily infects *Chenopodium* spp. The *Phycodnavirus* PBCV-1 infecting a *Chlorella*-like alga is much more likely to be of ancient origin. However, based on structure they do not appear to be primitive viruses. They are much larger and more complex than any known viruses infecting angiosperms, with a genome of about 300 kbp and at least 50 structural proteins (Meints *et al.*, 1986).

B. Origin of viroids (reviewed by Diener, 1996; Semancik and Duran-Vila, 1999)

As far as the evolution of viroids is concerned, it is important to remember that their viability is readily destroyed by relatively minor sequence alterations (Chapter 14, Section I.E.2). Nevertheless, sequence comparisons of sets of closely related viroids show clearly that single base changes, insertions and deletions have played a part in viroid evolution. Of more general interest is the evidence that intermolecular re-arrangements may also have played a role in viroid evolution as discussed in Chapter 14 (Section I.D.6).

1. Possible origins from other nucleic acids

a. From small nuclear RNAs

Small nuclear RNAs (snRNAs) are believed to play a role in the processing of the primary transcription products of split genes, thus allowing for precise alignment and correct excision of introns. Some, such as U1 snRNA, have been shown to have base complementarity with the ends of introns. Kiss *et al.* (1983) found significant sequence similarity between PSTVd RNA and the U3B snRNA of Novikoff hepatoma cells, suggesting that the viroid might be related to the snRNA.

b. From introns

Group 1 introns are widely distributed among the genes for mitochondrial mRNA and tRNA genes, chloroplast tRNA genes and nuclear tRNA genes. They are characterized by a highly conserved 16-nucleotide sequence and three sets of complementary sequences. The phenomenon of self-splicing has been described for several group 1 introns.

Diener (1981) noted a 23-nucleotide region of high base complementarity between the (−)-strand RNA of PSTVd and the 5′-terminal nucleotides of mammalian U1 snRNA (snRNAs reviewed by Filipowicz and Kiss, 1993). On this basis, Diener suggested that viroids might have arisen as escaped introns. Dinter-Gottlieb (1986) pointed out that viroids contain a consensus sequence and three sets of complementary bases. In addition, there were stretches of sequence similarity between viroids and a self-splicing intron from *Tetrahymena*. Dinter-Gottlieb showed that pairing of these complementary sequences within a viroid could generate a structure resembling the self-splicing group 1 intron from *Tetrahymena*. This would be a more open structure than a maximally base-paired viroid, perhaps stabilized by protein binding. Hadidi (1986) compared sequence similarities between viroids, other RNA pathogens, introns and exons. As the statistical stringency of the comparisons was increased, only similarities between introns and viroids remained significant.

The escaped intron concept includes the idea of self-processing of the RNAs. However, current evidence for viroids does not fit this model. ASBV is able to self-cleave (Chapter 14, Section I.D) but this viroid does not have the introns-like sequence motifs noted earlier for other viroids. With PSTVd and related viroids, Tsagris *et al.* (1987b) have shown that autocatalytic processing does not operate *in vitro* as with introns. In all the discussions on relationships between snRNA or introns and viroids, it has been assumed that plant snRNAs have sequences similar to those known from animal systems. Furthermore, it is assumed that viroids were derived from the host element, ignoring the possibility that the host element is a 'captured' viroid.

c. From hypothetical signal RNA

Zimmern (1982) proposed that some control elements between cells might consist of small mobile RNAs and that viroids might have arisen from these.

d. From transposable genetic elements

Sequence analysis of a group of viroids showed some striking similarities with transposable genetic elements, including the proviruses of retroviruses (Kiefer *et al.*, 1983). The similarities included the presence of inverted repeats often ending in the dinucleotides U-G and C-A and flanking imperfect direct repeats. Kiefer *et al.* suggested that viroids might have arisen by deletion of interior nucleotide sequences from transposable elements or retroviruses. ASBVd sequences do not fit this model.

e. From mitochondrial plasmids

Analogies have been drawn between viroids and certain mitochondrial plasmids from the fungus, *Neurospora* spp. (reviewed in Diener, 1996). Although the Mauriceville and Varkud plasmids consist of small circular DNA molecules, their RNA transcripts contain conserved sequences characteristic of group 1 introns. The Varkud Small Plasmid comprises linear and circular ssRNA, the linear form being generated 'by a viroid-like, RNA catalyzed cleavage reaction, which leaves 5'-OH and 2',3'-cyclic phosphate termini' (Lambowitz and Belford, 1993).

f. From satellite RNAs

Although viroids and the viroid-like satellite RNAs show little sequence similarity, they have several properties in common: small size, circular ssRNA and lack of mRNA activity. In addition, some satellite RNAs have a rolling circle model for replication involving greater than unit-length RNAs. On the other hand, viroids replicate in the nucleus, whereas viroid-like satellites do so in the cytoplasm. Nevertheless, Francki *et al.* (1986b) showed that PSTVd could be encapsidated *in vivo* in particles of VTMoV. Tomato plants, which are immune to this virus, became infected with PSTVd when inoculated with virus containing the viroid. In principle, a viroid might have arisen from a viroid-like satellite by becoming independent of the helper virus for replication.

As noted in Chapter 14 (Section II.B.2.a), the satellite of peanut stunt virus (PARNA5) has several regions of high sequence similarity with various viroids, including sequences of the conserved central region of most viroids. However, the replication of PARNA5 appears to be like that of the helper virus RNA rather than viroid-like. There are also conserved intron sequences in PARNA5 (Collmer *et al.*, 1985).

2. Viroids as living fossils of pre-biotic evolution

It has been proposed by a number of authors (see Cech and Bass, 1986) that RNA preceded DNA in pre-biotic evolution. The discovery of RNA molecules with splicing and polymerase activity has led to models for the RNA-catalysed replication of RNA (Cech, 1986). Thus, it is conceivable that modern viroids, because of their small size, circular conformation, inherent stability, RNA → RNA replication cycle, and lack of protein coding capacity, may be descended from a type of RNA molecule that existed before the evolution of DNA. This view is discussed by Diener (1989, 1996).

3. Origin and evolution of viroids and viroid-like satellites

A model for the current ideas on the origin and evolution of viroids has been proposed by Semancik and Duran-Vila (1999) (Fig. 17.18).

C. Origin of satellite viruses and nucleic acids

1. Satellite viruses

Apart from their dependence on a helper virus and their small size, satellite viruses appear to belong in the same category of agents as other viruses. This has been taken as suggesting that they most probably arose from an independent virus by degenerative loss of functions in a mixed infection that provided the helper virus (Francki, 1985).

2. Satellite RNAs

As discussed in Chapter 14 (Section II.B), most satellite RNAs fall into one of two groups based on their replication mechanism. The first, typified by satellite RNAs of CMV and TBRV, has terminal structures like that of the helper virus RNAs, replicate via a unit-length (−)-sense template, depend on the helper virus replicase, and appear to code for one to three proteins. The second group, typified by sTRSV, appears to have no mRNA function and is replicated by a viroid-like rolling circle mechanism. The predicted secondary structure of the viroid-like satellite RNAs mimics the rod-shaped configuration of the viroids, but there is little base sequence homology between them except for a conserved GAAAC occurring in some members of each group of agents (Keese and Symons, 1987).

Thus, it is possible that satellite RNAs have arisen from at least three different lineages (Fig. 17.19). The viral and viroid origins are mentioned above. The 3′ termini of satellites that are replicated in a virus-like manner show similarities to the helper virus; this is necessary for recognition by helper virus replication machinery. Thus, such satellites could have arisen by a possible third route involving recombination between the 3′-terminal region of the helper virus with host sequences. The lack of selection

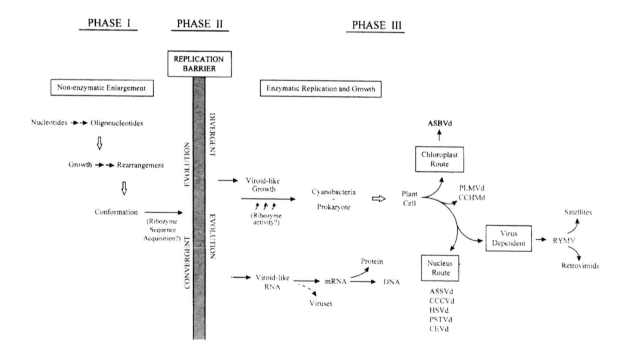

Fig. 17.18 Considerations for the evolution of viroids and viroid-like molecules as related to viruses and other nucleic acid species. From Semancik and Duran-Vila (1999), with permission.

pressure on the host sequence would enable it to diverge so that its host origin was no longer apparent. The satellite RNA C of TCV (see Fig. 14.14) is of particular interest from the evolutionary point of view. Its structure appears to demonstrate that recombination can take place between an RNA of unknown origin and the helper virus genome to generate a new satellite with different biological properties.

As with viruses, satellites can vary both by mutation and by recombination. In studying the nucleotide sequence of 17 variants of CMV-satRNA isolated from field-infected tomato plants in 1989, 1990 and 1991, Aranda *et al.* (1997) concluded that recombination may make an important contribution to the generation of new variants.

XIII. SELECTION PRESSURES FOR EVOLUTION

Virus evolution, like that of other organisms, is a tradeoff between costs and benefits and is driven by the impact of selection pressures on the inherent variation. Much of the molecular information on viruses comes from laboratory experiments, which although they can provide an insight on the mechanisms involved cannot give a picture of what happens in the 'natural situation'. In this section, I discuss some of the factors that drive and limit the evolution of plant viruses.

A. Maximizing the variation

In the sections above, I have described the sources of variation, mutation, pseudo-recombination and recombination that provide the material upon which selection pressures can act. The rapid replication of viruses and the lack of proofreading or repair mechanisms give the potential for much mutagenic variation in both RNA and DNA viruses. Thus, selection pressures operate on quasi-species, which comprise a dominant sequence and a multitude of minor variant sequences.

As pointed out by Padidam *et al.* (1999), three factors contribute to the success of recombination giving rise to new viruses, namely mixed infections, high levels of viral replication and increased host range of the vector.

B. Controlling the variation

1. Factors controlling variation

For the survival of a virus, this rampant variation has to be controlled for various reasons, including:

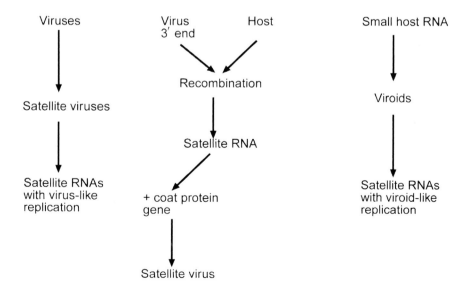

Fig. 17.19 Hypothetical schemes for the derivation of satellite RNAs.

- Variants that are lethal to replication have to be removed. It is possible in a quasi-species population that such variants could be complemented by other variants but this does not appear to be common.

- Variants that reduce the fitness of the virus population have to be removed. Muller's ratchet results in loss of fitness on passage of a virus through a bottleneck such as transmission that limits virus numbers (see Section V.A). From their success, it is obvious that plant viruses can overcome this constraint, although the mechanism is not known for most viruses.

- Viral genomes are packaged in nucleoprotein particles for transmission between hosts and there may be packing constraints. For many viruses, packaging is effected by the direct interaction of the nucleic acid with the coat protein subunits, which could be affected by base changes. The secondary and tertiary structures of the nucleic acid are important in packaging, as rod-shaped viruses require little or no secondary structure in contrast to isometric viruses, which require highly structured nucleic acid. Furthermore, there are physical limitations on the amount of nucleic acid that can be encapsidated to provide a stable isometric particle. An example of this is bromoviruses, which have three nucleoprotein components, one encapsidating RNA1 (about 3.3 kb), one encapsidating RNA2 (about 3.0 kb) and the third encapsidating RNA3 + RNA4 (2.1 + 1.0 kb); thus, each particle contains about 3 kb of RNA.

- Viruses are absolutely dependent upon their hosts for their propagation and survival. Thus, any viral variant that damages the host significantly would be selected against. However, a variant, say by recombination, that changes the host range may be advantageous.

- Viruses are dependent on various parts of the host machinery for their expression. Thus, a mutation might change the codon usage of an ORF from one that was advantageous in that host to one that was disadvantageous. As described in Chapter 7 (Section III.B.2.g), codon usage by a virus is not necessarily the most common one for mRNAs in that species.

- Viruses are transmitted from an infected to a healthy host in a specific manner, often by an insect vector with which it has a specific interaction. Any variant that reduces the transmission efficiency will be selected against. However, as with host range, a variant that changed the vector might be advantageous.

2. Proteins approaching the limits of change

As two proteins with a common evolutionary origin diverge, they may approach a limit of change that retains common amino acid positions in excess of that expected for random sequences. This limit on change would be due to functional requirements for the protein. When this limit is reached, convergence or back mutations and parallel mutations become as common as divergent mutations. As two diverging proteins approach and remain in this steady-state condition, sequence differences no longer reflect evolutionary distance, and therefore such sequences should not be placed in evolutionary trees. Some bacterial amino acid sequence data are considered to be of this sort (Meyer et al., 1986). Some coat proteins in certain groups of plant viruses may also have approached this state (e.g. the tobamoviruses). The limit for change may vary even within one protein. Thus, the surface domains of some plant viral coat proteins are more variable (or less constrained) than domains with structural roles. The three-dimensional structure of proteins tends to be conserved for longer periods of evolutionary time than primary sequences. Therefore, they might be useful for evolutionary comparisons long after amino acid sequences have markedly diverged. Rossmann and Rueckert (1987) pointed out that all the proteins of icosahedral viruses examined up to that time had an eight-stranded β barrel as their core. By contrast, TMV and probably other rod-shaped viruses have as their core a bundle of four α helices. This led to the idea that perhaps all icosahedral virus coat proteins had a common origin, and the coat proteins of all rod-shaped viruses had another common origin. An alternative explanation, where no significant amino acid sequence similarity exists, could be that convergent evolution has occurred.

Perhaps a functional coat protein for a virus with icosahedral symmetry requires a β-barrel core, and similarly for rod-shaped viruses and α helices.

3. Role of selection pressure

Functional viral genes will be retained only if they are needed for survival of the virus. For example, maintenance of WTV exclusively in vegetative plant hosts leads to loss of the ability of the virus to replicate in, and be transmitted by, leafhoppers. This functional loss is due to deletions in certain genome segments (see Fig. 17.20). Similarly, RNAs 3 and 4 of BNYVV appear to suffer deletions when not under the requirement to be transmitted by the fungal vector (see Chapter 14, Section II.B.3.c). The reversion of viroid point mutants to wild type provides another example of the effect of selection pressure (Owens *et al.*, 1986).

4. Selection pressure by host plants

Conditions within a given host species or variety will exert pressure on an infecting virus against rapid and drastic change. Viral genomes and gene products must interact in highly specific ways with host macromolecules during virus replication and movement. These host molecules, changing at a rate that is slow compared with the potential for change in a virus, will act as a brake on virus evolution. This general idea led Koonin and Gorbalenya (1989) to propose that RNA viral proteins evolve with an amino acid substitution rate that is similar to that of their host. However, Sala and Wain-Hobson (2000) conclude that RNA virus evolution appears conservative in comparison to that of their hosts. They suggest that this is due to the size limit imposed on RNA viral genomes by the lack of proofreading.

Selection pressure in an experimental system is illustrated by the work of Aldaoud *et al.* (1989). They studied the evolution of variants in populations of TMV originating from an *in vitro* transcript of a cDNA clone. One of the markers measured was the ability to cause necrotic lesions in *Nicotiana sylvestris* (nl). The proportion of nl variants in tobacco, tomato, *Solanum nigrum* and *Petunia hybrida* was similar.

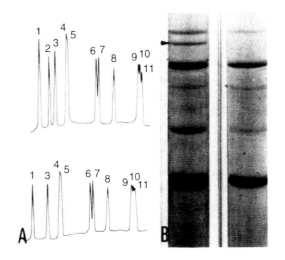

Fig. 17.20 The genome of wild WTV and a mutant that had lost the ability to be transmitted by leafhoppers. **(A)** Scan of a gel electrophoresis separation of the dsRNA genome segments (RNA no. 12 is not shown). Upper scan is of wild-type virus. In the mutant (lower scan), RNA2 is missing. **(B)** The polypeptides of a wild-type virus are shown on the left. On the right the mutant has one polypeptide missing. This polypeptide is therefore coded for by RNA2. It is involved in leafhopper transmission (see text). From Reddy and Black (1977), with permission.

However, in *Physalis floridiana* there was strong selection pressure against nl variants, which were reduced to almost undetectable levels. Nevertheless, in all hosts tested there were large and apparently random changes in the proportion of nl variants in individual plants, showing that viral populations can evolve rapidly on a timescale of days. Perhaps these variations in a strain population occur during the early stages of infection in new host species. Over a period of many transfers in the same species, the populations may stabilize. This idea is supported by the fact that TMV obtained from W.M. Stanley and independently subcultured many times over decades in tobacco varieties in the United States and Germany was shown in both countries to have the same coat protein amino acid sequence (Zaitlin and Israel, 1975).

Further support for stability of the dominant strain in a single host comes from the experiments of Rodriguez-Cerezo and García-Arenal (1989) with TMGMV (formerly known as

U5-TMV) in Samsun tobacco, a systemic host. They passaged the virus culture twice through single lesions in Xanthi nc tobacco and then followed heterogeneity in isolates after four series of 20 passages in Samsun tobacco. T1 fingerprint characterization of the population of genomic RNAs showed no detectable variation.

The few field studies that are available support the idea that a given host species tends to stabilize a virus population. Rodriguez-Cerezo *et al.* (1989) studied genome variations in isolates of PMMoV isolated from peppers in the field over the period 1983–1987. They concluded that there was a highly stable virus population from which variants continually arose without replacing the parent population.

A quantitative study of variation in a virus in a natural host in the wild was carried out on variants of TMGMV in *Nicotiana glauca* in southern Spain and Australia (Rodriguez-Cerezo *et al.*, 1991b). Their two main conclusions were: (1) that RNA sequence divergence between isolates was low, no matter what the distance was between the sites where they were isolated; and (2) there was no correlation between closeness of sequence relationship and geographical proximity of the plants from which they had been isolated. This is difficult to explain because the only known natural method of transmission for TMGMV is by contact between plants.

C. Adaptation to niches

Using the analogy that higher organisms evolve to adapt to specific niches, we should consider what is the ecosystem and the niches therein for viruses. Hull and Covey (1996) identified two levels of niches for viruses that involve different forms of adaptation. At the 'major' level is the host type that the virus infects. Thus, viruses that infect animals require different genes (modules) for spreading from cell to cell to those that infect plants. Viruses require living cells for their existence and so the living cells can be regarded as the ecosystem. Thus, at the 'fine-tuning' level there are sites within the cell that can be regarded as 'niches'. The organelles such as nucleus and plastids are

obvious niches. However, as discussed in Chapter 13, many of the functions of viruses in cells, such as replication and cell-to-cell movement, are associated with specific membranes. Thus, one could consider the individual membranes or cellular components as being niches within a cell to which a virus can adapt.

D. Rates of evolution

1. Non-constant rates of evolution

The molecular clock hypothesis, which assumes a constant rate of change in sequence over evolutionary time, has often been used in the interpretation of 'trees'. However, comparisons of organelle DNA in both vertebrates (Vawter and Brown, 1986) and plants (Wolfe *et al.*, 1987) reveal extreme rate variation in the molecular clock between nuclear, mitochondrial and chloroplast DNAs. It is now known that rates of change in different genes and lineages may vary widely (Li *et al.*, 1987). Ubiquitin is an example of a very highly conserved protein. It is composed of 76 amino acids with only three substitutions between animals, higher plants and yeast (Dunigan *et al.*, 1988). By contrast, there has been very rapid evolution of the reactive center regions of serine protease inhibitors of rodents (Hill and Hastie, 1987). The molecular phylogeny of globin genes is particularly instructive in this respect because amino acid sequences are known for several hundred globin chains (Goodman *et al.*, 1987). Furthermore, the molecular clock runs at different rates among closely related members of a gene family (Gibbs *et al.*, 1998). There is little doubt that different proteins, or parts of a protein coded by a viral genome, may evolve at different rates. In addition, some non-coding sequences in the genome may be highly conserved, particularly those recognition sequences essential for genome replication. Thus, there are considerable doubts as to the use and applicability of molecular clocks (see Ayala, 1999). It has been concluded that 'molecular evolution is dependent on the fickle process of natural selection', but it is a time-dependent process with the accumulation of empirical data often giving an approximate clock (Ayala, 1999).

2. Estimated rates of evolution

The rate of point mutation for RNA viruses has been estimated to be approximately 10^{-3}–10^{-4} per nucleotide per round of replication with some variation between different viruses (reviewed by Holland *et al.*, 1982). This contrasts with estimates of about 10^{-11} for DNA polymerases (see Fig. 8.29). However, Smith and Inglis (1987) have questioned the idea that the error rate in eukaryotic viral DNA synthesis is significantly lower than that for RNA viruses. Also, as noted in Section X.D.7, the observed rate of variation of geminiviruses is greater than predicted. In theory, the measured mutation rates would allow for very rapid change in viral genomes either RNA or DNA.

Virus cultures contain large numbers of sequence variants (Domingo and Holland, 1988). Thus, there is no doubt concerning the reality of the quasi-species concept (see Section I.A).

It is very difficult to relate mutation rates to the actual rates of change in viruses that might be occurring in the field at present, or over past evolutionary time. The reasons for this include the following.

i. Variation in rates of change in different parts of a viral genome

As noted in Chapters 6–8, non-coding regions of viral genomes, particularly at the 5' and 3' termini, which function as recognition sites in viral RNA translation and replication, may be highly conserved in the members of a virus family or group. On the other hand, in viruses with multi-partite genomes, one genome segment may be conserved and the other highly variable, as with the RNAs 1 and 2 of tobraviruses (see Fig. 6.38). Coat protein genes may be much more strongly conserved in some regions than in others by functional requirements in the proteins, for example with TMV coat protein (see Fig. 2.5) and potyvirus coat proteins (see Fig. 17.2).

ii. Uneven rates of change over a time period

The environment that dictates the selection pressure on the replication and movement of a virus within a plant consists almost entirely of the internal milieu of the host. Other selection pressures on survival involve transmission from plant to plant, for example by invertebrate or other vectors. It is possible that a switch to a new host plant species may induce rapid evolution of a virus over quite a short period. Recombination between viral genomes with mutations in different parts of the genome may speed this process. This stage may be followed by a further period of stability for the new virus in the new host. The situation may be somewhat analogous, on a very much shorter time scale, to the episodic evolution or 'punctuated equilibria' deduced from the fossil record for the evolution of some groups of invertebrates (Eldredge, 1985).

One way in which a virus may gain a foothold in a new host species is through co-infection with another virus that normally infects that species. For example, TMV does not normally systemically infect wheat, but can do so if the plants are already infected with BSMV; it is assumed that the BSMV movement protein complements the movement of TMV (see Chapter 9, Section II.D.6).

iii. Lack of precise historical information

Until recently the type member of the *Tymovirus* genus, TYMV, had been found occurring naturally only in western Europe. However, a very closely related virus has been found in an isolated area of Australia infecting the endemic brassicaceous species *Cardamine lilacina* (Keese *et al.*, 1989). From the genomic sequences of the European and Australian TYMV populations, Blok *et al.* (1987) calculated that the two strains have diverged by about 1%, at most, over the past 10 000 years. This is the only estimate we have of the possible rate of evolution of a plant virus over a geological timespan. However, the estimate is based mainly on biogeographical data, concerning the *Cardamine* host, which are themselves subject to considerable error (Guy and Gibbs, 1985). Furthermore, the 1% difference could be due to a very rapid process of adaptation to different host genera over a relatively short period at some time in the past rather than to slow change over thousands of years.

The sequence of ToMV recovered from glacial ice cores approximately 140 000 years old in Greenland was nearly identical to that of contemporary isolates (Castello *et al.*, 1999). However, there is some uncertainty as to whether the virus was contemporary with the ice.

XIV. CO-EVOLUTION OF VIRUSES WITH THEIR HOSTS AND VECTORS

A. Co-evolution of viruses, host plants and invertebrate vectors

Fahrenholtz's rule postulates that parasites and their hosts speciate in synchrony (Eichler, 1948). Thus, there is a prediction that phylogenetic trees of parasites and their hosts should be topologically identical. Using protein electrophoretic data, Hafner and Nadler (1988) obtained phylogenetic trees for rodents and their ectoparasites that confirm a history of co-speciation. In view of the known wide host ranges of many present-day plant viruses, it is not to be expected that Fahrenholtz's rule will be followed closely for viruses and their hosts. Nevertheless, it is now widely accepted that viruses have had a long evolutionary history and have co-evolved with their host organisms. In this section, I consider aspects of this idea as they relate to plant viruses. In discussing this idea I do not wish to imply that viruses have evolved to a state of higher complexity following their plant hosts in this respect. On the contrary, the evidence available at present shows that the largest and most complex virus infecting photosynthetic organisms is found in the simplest host, a *Chlorella*-like green alga.

B. Evolution of angiosperms and insects

The most recent major evolutionary explosion, as evidenced in the fossil record, took place about the beginning of the Cretaceous (approximately 135 million years ago). Monocotyledons and dicotyledons emerged and various orders of mammals and birds appeared. Some orders of insects had evolved by the Devonian (400 million years ago) but several important orders produced large num-

bers of new types as the higher plants emerged. Many of these co-evolved with their angiosperm food plants. Thus, some viruses that were already present in insects may have adapted to replicate in the evolving mammals and angiosperms during Cretaceous times.

The angiosperms appear in the fossil record no earlier than the early Cretaceous, that is, about 130 million years ago, with the major early diversification of the group occurring during the mid-Cretaceous (see Crane and Lidgard, 1989; Crane *et al.*, 1995; Qiu *et al.*, 1999). However, chloroplast DNA sequence data suggest that the monocotyledons and dicotyledons diverged from a common stock about 200–40 million years ago (Wolfe *et al.*, 1989). Thus, over this period of divergence, the present-day viruses infecting angiosperms presumably also evolved, at least with respect to their main host specificities. This is assuming that viruses spread to angiosperms during the early stages of angiosperm evolution.

C. Horizontal transmission through plants of viruses infecting only insects

A virus of the leafhopper, *Niloparvata lugens* (NLRV), has particles similar to those of fijiviruses infecting plants. NLRV infects and multiplies in the leafhopper and is transmitted through its eggs. It does not multiply in the maize host plants of the insect. However, when infected hoppers feed on a plant, virus is injected into and circulates transiently in the plant. This virus can infect healthy hoppers that feed simultaneously on the same plant. Thus, the rice plant can be regarded as a circulative but non-persistent vector of an insect virus (Nakashima and Noda, 1995). A similar situation probably exists with leafhopper A virus (Boccardo *et al.*, 1980; Ofori and Francki, 1985) and *Perigrinus maidis* virus (Falk *et al.*, 1988).

A virus that infects the aphid *Rhapalosiphum padi* (RhPV) has a 27-nm icosahedral particle containing a single molecule of ssRNA. It is transmitted through the aphid eggs and does not replicate in the host plants of the aphid. However, when virus-infected and healthy aphids were fed simultaneously on the same barley leaf, healthy aphids became infected

(Gildow and D'Arcy, 1988). It is not unreasonable to suppose that over a long period of time viruses such as LAV and RhPV might occasionally acquire a gene or genes that would allow them to establish and replicate in the plant as well as the insect.

D. Affinities of viruses that replicate in both insects and plants

The *Reoviridae* family of viruses has members that infect both vertebrates and invertebrates, and others that infect both invertebrates and plants. This taxonomic distribution of hosts suggests that the more ancient invertebrates may have been the original source of this virus family. A number of features support the view that plant reoviruses originated in the leaf-hopper vectors (Conti, 1984; Nault, 1987; Nault and Ammar, 1989):

- Fijiviruses are morphologically similar to viruses such as LAV that replicate only in the insect and to the *Peregrinus maidis* virus (Falk *et al.*, 1988).
- All known plant reoviruses replicate in their hopper vectors.
- Plant reoviruses are not seed-borne, nor are they transmitted by mechanical means, except in special circumstances. Thus, they are entirely dependent on the hopper vectors for survival.
- Plant species infected by reoviruses are usually the prime food and breeding hosts of their hopper vectors.
- Plant reoviruses appear more closely adapted to their hopper hosts because: (1) they replicate to higher titer in the insects; (2) several plant reoviruses are transmitted through insect eggs but none is transmitted through plant seed; (3) the percentage of virus-carrying insects in a given vector population is higher than the percentage of plants that can be infected by feeding single hoppers on them; and (4) some cause cytopathic effects in the hopper vectors, but in general these viruses have less severe pathogenic effects in the insects than in their plant hosts. In fact, most can be considered as causing latent infections.

Like the *Reoviridae*, members of the *Rhabdoviridae* family all have a very similar particle morphology and genome strategy, and infect either vertebrates and invertebrates or invertebrates and plants. Thus, a common origin for this family among the insect vectors is indicated. The viruses are not seed transmitted but are transmitted through the eggs of hopper vectors. Most do not appear to cause disease in their insect vectors. The situation is somewhat more complex than with the plant reoviruses because different plant rhabdoviruses have hopper, aphid, piesmid or mite vectors. Perhaps, as a rare event in the past, one rhabdovirus could transfer from a vector in one family to a vector in another where the vectors had a common host plant (Nault, 1991). It may be relevant to this idea that vesicular stomatitis virus, a rhabdovirus infecting vertebrates, replicates when injected into the leafhopper vector of a plant virus (Lastra and Esparza, 1976).

Based on evidence discussed in Chapter 2 (Section III.H), tospoviruses now belong in the *Bunyaviridae*. This family contains over 200 viruses that infect warm- and cold-blooded vertebrates and arthropods. Different viruses are transmitted by mosquitoes, sandflies or ticks. Tospoviruses are transmitted by, and replicate in, thrips. It seems very probable that, like the *Reoviridae* and *Rhabdoviridae* families, the bunyaviruses originated in their insect hosts. The tenuiviruses almost certainly originated in insects. The same may be true for the marafiviruses but the evidence is less compelling.

Flock house virus (FHV) is an insect virus belonging to a family of small RNA viruses, the *Nodaviridae*. It has no known relationship with plants in nature. However, Selling *et al.* (1990) have shown that the virus replicated to low levels in the leaves of several plant species following mechanical inoculation with FHV RNA. No replication could be detected following inoculation with whole virus. However, inoculation of barley protoplasts with intact FHV resulted in the synthesis of small amounts of progeny virus particles, indicating that the virus could be uncoated in plant cells. FHV particles moved systemically in *Nicotiana benthamiana* but no replication in systemic leaves could be detected. The virus showed no symptoms in plants it infected. As noted in Chapter 9, Section II.D.6, TMV and RCNMV movement proteins will complement the movement of FHV in *N. benthamiana* (Dasgupta *et al.*, 2001). These results suggest that the internal milieu of diverse organisms may be sufficiently similar to be able to support the replication of simple RNA viruses once inside a cell. This may have been a factor in past evolutionary processes.

The vertebrate-infecting circoviruses and the plant-infecting nanoviruses each have small circular ssDNA genomes with several features in common. In an analysis of circovirus and nanovirus Rep protein sequences, Gibbs and Weiller (1999) noted similarities between the N-terminal regions whereas the C-terminal region of calicivirus Rep is related to protein 2C of picorna-like viruses. They concluded that circoviruses evolved from nanoviruses by transference of nanovirus DNA to an animal followed by recombination.

E. Adaptation of plant viruses to their present invertebrate vectors

Fossil records indicate that the main taxonomic groups of the *Homoptera* had diverged by the Upper Triassic about 180 million years ago (Hennig, 1981). As discussed in Chapter 11, there are some aphid species among the vectors of plant viruses that transmit only one virus, and others such as *Myzus persicae* that transmit large numbers. Some viruses are transmitted by only one aphid species, whereas others are transmitted by many. There is some evidence that among a phylogenetically diverse array of aphid vectors, some groups are better vectors of a particular virus than others. For example, Zettler (1967) showed that aphids from the subfamily *Aphidinae* were better vectors of BCMV than were aphids from the subfamilies *Callaphidinae* or *Chaitophorinae*. The differences were attributed to differences in probing behavior. Nault and Madden (1988) compared the efficiency of transmission of MCDV by 25 leafhopper species from 13 genera. They concluded that leafhopper species from the tribes *Deltocephalini* or recent (advanced) *Euscelini*

that use maize as a feeding and breeding host have a higher probability of being MCDV vectors, whereas species from other taxa even if they feed well on maize, have a lower probability of being vectors.

Among the geminiviruses, the whitefly-transmitted viruses all have the same vector species, *Bemisia tabaci*. In immunoabsorbent electron microscopy tests, strong relationships were detected between five viruses transmitted by *Bemisia* (Roberts *et al.*, 1984). Similar relationships were found using monoclonal antibodies (see Section II.B.4) and nucleic acid hybridization, using DNA1, which contains the coat protein gene. No relationship was detected between any whitefly-transmitted and any leafhopper-transmitted virus. The whitefly-transmitted viruses had diverse host ranges, came from different countries, and either were or were not sap transmissible. Roberts *et al.* (1984) suggested that the relationships found may be due to a key role played by the coat protein in vector transmission by these viruses.

Assuming this association by descent, and assuming an insect origin for plant reoviruses and rhabdoviruses, Nault (1987) determined the latest stages during the evolution of vector groups at which the associations could have taken place (Fig. 17.21). However, an insect origin for geminiviruses may be doubtful because they appear not to replicate in their vector.

Plant viruses and their beetle vectors may have evolved together even though virus replication in the vector does not occur. The *Tymovirus* genus provides an interesting example. Most members of this group have a relatively restricted host range, usually within a single plant family when only natural hosts are considered. The host families do not occur at random among the six subclasses of the dicotyledons. None is found in the *Magnoliidae* or the *Hammaelidae*, subclasses that contain the older families among the dicotyledons.

The chrysomelid and curculionid beetle vectors of tymoviruses show specificity with respect to the viruses they transmit. They feed on a fairly narrow range of plant species, usually within the family constituting the host range of the virus. The earliest direct fossil evidence for the *Curculionaidae* goes back to the

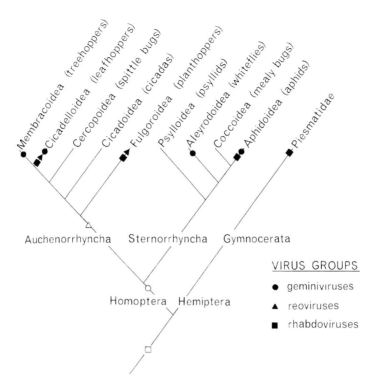

Fig. 17.21 Possible co-evolution of plant virus families and vector groups. Phylogeny of superfamilies of *Homoptera* with common names of groups in parentheses. Shown are homopteran groups (and one hemipteran) that are vectors for the geminiviruses, reoviruses and rhabdoviruses. Assuming these plant viruses are associated by descent with the *Homoptera*, open symbols show the latest time at which associations could have taken place. From Nault (1987). *Notes*: (i) The genus *Phytoreovirus* is transmitted by leafhoppers, and the genus *Fijivirus* by planthoppers. (ii) TPCTV, a *Geminivirus*, is transmitted by a treehopper, while another possible *Geminivirus*, Oat chlorotic stripe virus, is transmitted by an aphid. From Matthews (1991).

Upper Jurassic and for the *Crysomelidae* to the Middle Eocene (Crowson *et al.*, 1967). Thus, although the fossil record is very scanty, it is not inconsistent with the idea that host plants, beetle vectors and the tymoviruses underwent substantial evolutionary diversification together during the Eocene period.

Adaptation of plant viruses to their invertebrate vectors has probably involved several different processes on different timescales, and these may be very difficult to unravel. These processes include: (1) evolutionary origins by descent on a geological time scale as suggested in Fig. 17.21; (2) adaptations to particular vector species that may be of quite recent origin, for example through mutational changes in the viral coat protein or in a helper protein, in a group such as the potyviruses; and (3) evolutionary origins that involve both co-evolution and descent, and direct colonization of new

vector groups. For example, such colonizations may explain the fact that among the plant rhabdoviruses there are both insect and mite vectors (e.g. for CoRSV) (Nault, 1987).

XV. DISCUSSION AND SUMMARY

The accumulating evidence is supporting the modular structure of the viral genome, consisting of the basic core replicon and modules added to adapt the replicon to the niche that the virus occupies. The cryptic viruses (see Chapter 2, Section III.E) would appear to be basic replicons plus coat protein which could be part of the replication complex. Their dsRNA genomes encode a replicase and the viral coat protein but no factors that facilitate movement from cell to cell or from plant to plant. These viruses could represent either viruses that had lost various functions or precursors of modern plant viruses yet to acquire other functional modules (Roossinck, 1997). Large dsRNA species have been found in several plants including cultivated rice (*Oryza sativa*) (Moriyama *et al.*, 1995), *O. rufipogon* (Moriyama *et al.*, 1999), broad bean (*Vicia faba*) (Pfeiffer, 1998), bean (*Phaseolus vulgaris*) (Wakarchuk and Hamilton, 1990) and barley (Zabalgogeazcoa and Gildow, 1992; Zabalgogeazcoa *et al.*, 1993). These so-called 'endogenous dsRNAs' do not appear to be horizontally transmitted and have been considered to be a kind of RNA plasmid. The genomes are much larger than those of cryptic viruses (10–17 versus 2–3 kbp) but, unlike cryptic viruses, these endogenous dsRNAs do not appear to be associated with particles. In their size and lack of particles they resemble the fungus-infecting *Hypoviridae* (see Chapter 2, Section V.B) but they have a different genome organization. Analysis of the RNA-dependent RNA polymerase of plant-associated endogenous dsRNAs has led to the suggestion that they have evolved from a 'defective' ssRNA virus (Gibbs *et al.*, 2000). Some dsRNAs from distantly related plants are closely related, indicating that they have not strictly co-speciated with their hosts.

Much of the discussion on virus evolution is directed at (+)-sense RNA viruses. This is not surprising because the majority of plant viruses have this genome type and most of the conclusions can be applied also to (–)-strand ssRNA and dsRNA viruses. Koonin and Dolja (1993) formulated five principles to be taken into account in discussing RNA virus evolution:

1. RNA viruses evolve rapidly. Hence, only important functional motifs are conserved over the wide range of virus groups.
2. (+)-Sense RNA virus genomes (and those of other viruses) are made from a limited number of building blocks (modules). The universal blocks are genes for the RdRp and coat proteins. Only the RdRp contains the universal sequence motifs that are conserved in all (+)-strand RNA viruses. Therefore, phylogenetic analysis of the RdRp sequences inevitably forms the basis for producing a coherent picture of the evolution of this group of viruses in general. This is applicable to the groups of viruses with other polymerases.
3. Evolution of (+)-strand RNA viruses is shaped by two opposite trends: (1) conservation of distinct arrays of genes, primarily those encoding proteins involved in RNA replication; and (2) the recombinational shuffling of genes and gene blocks.
4. The frequency of recombination, even among distantly related viruses makes it impossible in principle to show the evolutionary history of (+)-strand RNA viruses as a single phylogenetic tree. The description has to be a complex combination of both vertical ('tree-like') and horizontal flows of genetic information.
5. Correlation between phylogeny and strategy of genome replication and expression is only limited, suggesting fundamental expression and replication mechanisms could have evolved more than once.

The question of monophyletic or polyphyletic origins of viruses is discussed extensively. In view of the modular arrangement and presumed modular evolution of most, if not all, viruses we really should consider whether the modules have single or multiple origins. In most cases the discussion is based on the replicon, for which, for RNA viruses, there are diverging opinions on the origins (see Section XI.A.1). There are stronger suggestions for single origins for reverse transcriptase and the module(s) involved in the replication of ssDNA plant viruses. As to the other modules in plant viruses, the suggestion is that most have been acquired from the host and adapted to the virus's specific use. Because of the variation due to lack of proofreading in nucleic acid replication, and because of the adaptation to virus use, it will probably not be possible to find sequence similarities to host genes. However, it may be possible to determine functional similarities; these are becoming apparent for cell-to-cell movement proteins (see Section XI.D).

Koonin and Dolja (1993) suggested that the basis of phylogenetic taxonomy of (+)-strand RNA viruses should be the 'core' gene complex of the replicase and the coat protein. Based on this they proposed three classes into which these viruses could be grouped. While there are considerable reservations about the use of these three classes in a formal taxonomy, analysis of the inter-relationships of the core genome organizations provides material for future discussions on virus evolutionary pathways. They suggested that the ancestors of the three classes arose from a common ancestor (Fig. 17.22).

From these three class ancestors they created tentative scenarios for the evolution of the virus groups (Fig. 17.23). These are based on the classification extant in 1993 and do not include all the current groups. However, these scenarios show potential evolutionary pathways for (+)-strand RNA viruses.

The acquisition of the non-core modules is to a large extent directed by the hosts and vectors.

There is strong evidence that viruses, hosts and vectors have co-evolved. A good example of co-evolution with vectors is shown in the closteroviruses (Karasev, 2000). Phylogenetic analysis of the replication genes and the HSP70 shows that the evolution has resulted in three major lineages relating to the aphid, mealybug and whitefly vectors.

However, it is important to remember that

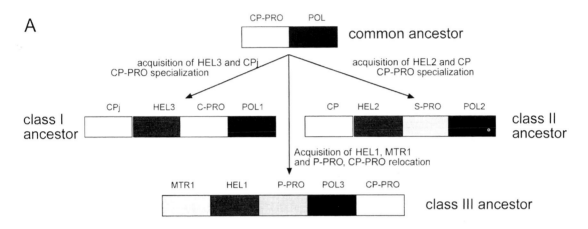

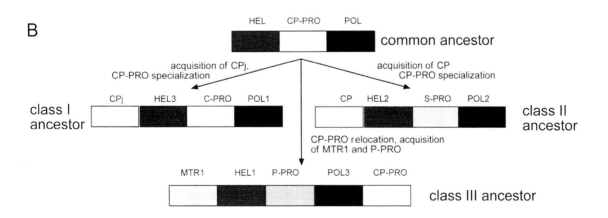

Fig. 17.22 Two alternative scenarios for the evolution of the ancestors of the three virus classes from the hypothetical common ancestor virus. **(A)** 'Gene capture' scenario. It is postulated that the ultimate ancestor contained two genes and the capsid autoprotease performed only one cleavage at its own C-terminus, like the contemporary alphavirus capsid protease. **(B)** 'Primordial' scenario. The ultimate ancestor is postulated to contain three genes, with the primitive capsid autoprotease performing two cleavages at its N- and C-termini. Symbols as in Fig. 17.17. From Koonin and Dolja (1993), with kind permission of the copyright holder, © CRC Press LLC.

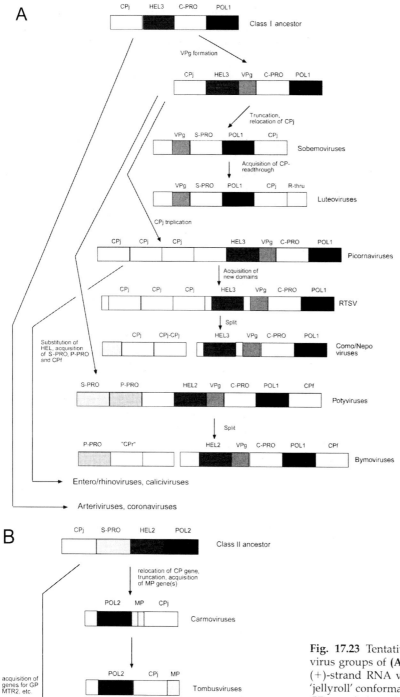

Fig. 17.23 Tentative evolutionary scenarios for some virus groups of **(A)** class 1, **(B)** class II and **(C)** class III (+)-strand RNA viruses. CPj, icosahedral capsid with 'jellyroll' conformation; CPf, filamentous capsid protein; CPr, rod-shaped capsid protein; CP-PRO, capsid autoprotease; MP, movement protein; MP-TGB, triple gene block movement protein; RBP, (putative) RNA-binding protein; R-thru, read-through protein; VPg, genome-linked viral protein; other abbreviations are given in Fig. 17.17. Adapted from Koonin and Dolja (1993), with kind permission of the copyright holder, © CRC Press LLC.

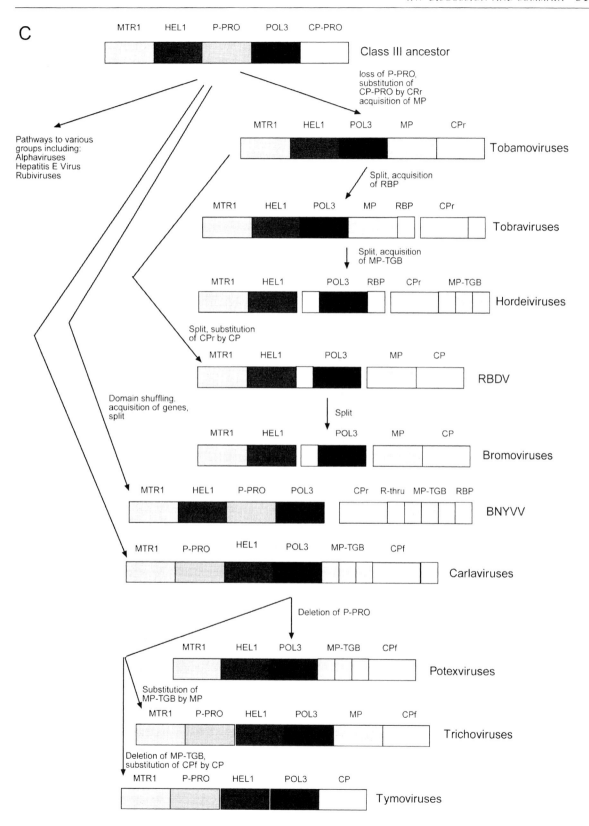

the most studied viruses are those that cause important diseases of crops. Modern crop varieties have relatively short, rapid and complex evolutionary histories (see Gaut *et al.*, 2000). Deriving concepts of the co-evolutiion of viruses with such plant species should be regarded with caution.

Another feature favoring an ancient origin for present-day viruses is the geographical distribution of some groups. Members of the *Tymovirus* genus of plant viruses have no known means of intercontinental spread except by humans. Different tymoviruses occur in Europe, North and South America, Africa, Asia and Australia. These continents were still in one landmass at the end of the Jurassic period. The present-day tymoviruses may have diverged from a common ancestral stock: the continents drifted apart between 138 and 80 million years ago. The strain of TYMV found in *Cardamine* in Australia may be anomalous with respect to this general idea. However, the fact that both the European and Australian TYMVs are found in brassicaceous hosts may have restrained their evolutionary divergence.

Phylogenetic trees based on the amino acid sequences of *Geminivirus* coat proteins or replication-associated proteins suggest that the geographical isolation between Old World and New World geminiviruses may have played a role in the evolution of this group (Howarth and Vandemark, 1989).

There is another aspect of present-day virus and host distribution that is relevant to the problem of virus evolution. An area containing many species and varieties of a plant genus is taken to represent a site where evolution within that genus has occurred. For example, the *Solanum* species related to the potato probably evolved in the Andean region, with a center in the Lake Titicaca area of southern Peru and northern Bolivia, and were first domesticated there (Hawkes, 1967). Most of the viruses infecting potatoes are restricted in nature to the genus *Solanum*, for example the carlavirus PVS, the potexvirus PVX, the comovirus APMoV, the

tymovirus APLV and the pomovirus PMTV. These viruses have almost certainly been spread from the Andes to other parts of the world in potato tubers. To the extent that they have been examined in detail, these viruses show marked variation within the Andean region. For example, 26 isolates of APLV fell into three serological groups. One group was found only in the south of the region, one only in the north, and the third was distributed through the whole region (Koenig *et al.*, 1979). APLV and other viruses such as APMoV that are widely distributed in the Andean region multiply well only under the cool conditions prevailing in the potato areas at 2000–4000 m above sea level (Fribourg *et al.*, 1977). This diversity of virus strain, climatic adaptation and geographical restriction of some strains to parts of the region support the idea that the section of the genus *Solanum* comprising the potatoes and the major viruses of potatoes have co-evolved and are still co-evolving in the Andean region.

Similar examples of the effects of geography on evolution were described above for YMV (see Section X.D.5) and CLCuV (see Section X.D.7).

There is ample experimental evidence from the study of existing viruses to demonstrate that viruses continue to evolve, but rates of evolution for most viruses and virus genes are very difficult to establish with any degree of certainty. Several new factors may affect the rates at which plant viruses evolve. Changes in agricultural systems are giving rise to large areas of monoculture, the overlapping cropping of the same species and the introduction of new species into a given region. Increased international travel and trade is introducing new viruses in many regions and also possibly new vectors. Thus, the selection pressures on virus variants are changing and the possibilities of joint infections of new combinations of viruses are providing further opportunities for recombination. It will be interesting to view the impact of these and other societal changes on the emergence of new viruses.

Appendix 1A

TABLE I Alphabetical list of plant virus names and their abbreviations, used in the Seventh Report of the International Committee on the Taxonomy of Viruses. Virus names in italics are of virus Species, and those not in italics are of Tentative species. Synonyms do not have abbreviations allocated and are therefore not listed here. Data from Fauquet and Mayo (1999).

Plant virus name	Abbreviation	Family	Genus[a]
Abutilon mosaic virus	AbMV	*Geminiviridae*	*Begomovirus*
Abutilon yellows virus	AbYV	*Closteroviridae*	*Crinivirus*
Acalypha yellow mosaic virus	AYMV	*Geminiviridae*	*Begomovirus*
African cassava mosaic virus	ACMV	*Geminiviridae*	*Begomovirus*
Ageratum yellow vein virus	AYVV	*Geminiviridae*	*Begomovirus*
Aglaonema bacilliform virus	ABV	*Caulimoviridae*	*Badnavirus*
Agropyron mosaic virus	AgMV	*Potyviridae*	*Rymovirus*
Ahlum waterborne virus	AWBV	*Tombusviridae*	*Carmovirus*
Alfalfa cryptic virus 1	ACV-1	*Partitiviridae*	*Alphacryptovirus*
Alfalfa cryptic virus 2	ACV-2	*Partitiviridae*	*Betacryptovirus*
Alfalfa mosaic virus	AMV	*Bromoviridae*	*Alfamovirus*
Alligatorweed stunting virus	AWSV	*Closteroviridae*	*Closterovirus*
Alpinia mosaic virus	AlpMV	*Potyviridae*	*Potyvirus*
Alstroemeria mosaic virus	AlMV	*Potyviridae*	*Potyvirus*
Alstroemeria streak virus	AlStV	*Potyviridae*	*Potyvirus*
Alternanthera mosaic virus	AltMV		*Potexvirus*
Althea rosea enation virus	AREV	*Geminiviridae*	*Begomovirus*
Amaranthus leaf mottle virus	AmLMV	*Potyviridae*	*Potyvirus*
Amazon lily mosaic virus	ALiMV	*Potyviridae*	*Potyvirus*
American hop latent virus	AHLV		*Carlavirus*
American plum line pattern virus	APLPV	*Bromoviridae*	*Ilarvirus*
Andean potato latent virus	APLV		*Tymovirus*
Andean potato mottle virus	APMoV	*Comoviridae*	*Comovirus*
Aneilema mosaic virus	AneMV	*Potyviridae*	*Potyvirus*
Anthoxanthum latent blanching virus	ALBV		*Hordeivirus*
Anthoxanthum mosaic virus	AntMV	*Potyviridae*	*Potyvirus*
Anthriscus latent virus	AntLV		*Carlavirus*
Anthriscus yellows virus	AYV	*Sequiviridae*	*Waikavirus*
Apple chlorotic leafspot virus	ACLSV		*Trichovirus*
Apple dimple fruit viroid	ADFVd	*Pospiviroidae*	*Apscaviroid*
Apple fruit crinkle viroid	AFCVd	Viroid	Unassigned

(continued)

TABLE I Alphabetical list of plant virus names and their abbreviations (*Continued*)

Plant virus name	Abbreviation	Family	Genus[a]
Apple mosaic virus	ApMV	*Bromoviridae*	*Ilarvirus*
Apple scar skin viroid	ASSVd	*Pospiviroidae*	*Apscaviroid*
Apple stem grooving virus	ASGV		*Capillovirus*
Apple stem pitting virus	ASPV		*Foveavirus*
Aquilegia necrotic mosaic virus	ANMV	*Caulimoviridae*	*Caulimovirus*
Aquilegia necrotic ringspot virus	AqNRSV	*Potyviridae*	*Potyvirus*
Arabis mosaic virus	ArMV	*Comoviridae*	*Nepovirus*
Araujia mosaic virus	ArjMV	*Potyviridae*	*Potyvirus*
Arracacha latent virus	ALV		*Carlavirus*
Arracacha virus A	AVA	*Comoviridae*	*Nepovirus*
Arracacha virus B	AVB	*Comoviridae*	*Nepovirus*
Arracacha virus Y	AVY	*Potyviridae*	*Potyvirus*
Artichoke Aegean ringspot virus	AARSV	*Comoviridae*	*Nepovirus*
Artichoke curly dwarf virus	ACDV		*Potexvirus*
Artichoke Italian latent virus	AILV	*Comoviridae*	*Nepovirus*
Artichoke latent virus	ArLV	*Potyviridae*	*Potyvirus*
Artichoke latent virus M	ArLVM		*Carlavirus*
Artichoke latent virus S	ArLVS		*Carlavirus*
Artichoke mottled crinkle virus	AMCV	*Tombusviridae*	*Tombusvirus*
Artichoke vein banding virus	AVBV	*Comoviridae*	*Nepovirus*
Artichoke yellow ringspot virus	AYRSV	*Comoviridae*	*Nepovirus*
Asparagus virus 1	AV-1	*Potyviridae*	*Potyvirus*
Asparagus virus 2	AV-2	*Bromoviridae*	*Ilarvirus*
Asparagus virus 3	AV-3		*Potexvirus*
Asystasia gangetica mottle virus	AGMoV	*Potyviridae*	*Potyvirus*
Asystasia golden mosaic virus	AGMV	*Geminiviridae*	*Begomovirus*
Atropa belladonna virus	AtBV	*Rhabdoviridae*	*Nucleorhabdovirus*
Aucuba bacilliform virus	AuBV	*Caulimoviridae*	*Badnavirus*
Australian grapevine viroid	AGVd	*Pospiviroidae*	*Apscaviroid*
Avocado sunblotch viroid	ASBVd	*Avsunviroidae*	*Avsunviroid*
Bajra streak virus	BaSV	*Geminiviridae*	*Mastrevirus*
Bamboo mosaic virus	BaMV		*Potexvirus*
Banana bract mosaic virus	BBrMV	*Potyviridae*	*Potyvirus*
Banana bunchy top virus	BBTV		*Nanovirus*
Banana streak virus	BSV	*Caulimoviridae*	*Badnavirus*
Barley mild mosaic virus	BaMMV	*Potyviridae*	*Bymovirus*
Barley stripe mosaic virus	BSMV		*Hordeivirus*
Barley virus B1	BarV-B1		*Potexvirus*
Barley yellow dwarf virus – GPV	BYDV-GPV	*Luteoviridae*	Unassigned
Barley yellow dwarf virus – MAV	BYDV-MAV	*Luteoviridae*	*Luteovirus*
Barley yellow dwarf virus – PAV	BYDV-PAV	*Luteoviridae*	*Luteovirus*
Barley yellow dwarf virus – RMV	BYDV-RMV	*Luteoviridae*	Unassigned
Barley yellow dwarf virus – SGV	BYDV-SGV	*Luteoviridae*	Unassigned
Barley yellow mosaic virus	BaYMV	*Potyviridae*	*Bymovirus*
Barley yellow striate mosaic virus	BYSMV	*Rhabdoviridae*	*Cytorhabdovirus*
Bean calico mosaic virus	BCaMV	*Geminiviridae*	*Begomovirus*
Bean common mosaic necrosis virus	BCMNV	*Potyviridae*	*Potyvirus*

Plant virus name	Abbreviation	Family	Genus[a]
Bean common mosaic virus	BCMV	*Potyviridae*	*Potyvirus*
Bean dwarf mosaic virus	BDMV	*Geminiviridae*	*Begomovirus*
Bean golden mosaic virus – Brazil	BGMV-Br	*Geminiviridae*	*Begomovirus*
Bean golden mosaic virus – Puerto Rico	BGMV-PR	*Geminiviridae*	*Begomovirus*
Bean leafroll virus	BLRV	*Luteoviridae*	Unassigned
Bean mild mosaic virus	BMMV	*Tombusviridae*	*Carmovirus*
Bean pod mottle virus	BPMV	*Comoviridae*	*Comovirus*
Bean rugose mosaic virus	BRMV	*Comoviridae*	*Comovirus*
Bean yellow dwarf virus	BeYDV	*Geminiviridae*	*Mastrevirus*
Bean yellow mosaic virus	BYMV	*Potyviridae*	*Potyvirus*
Bean yellow vein-banding virus	BYVBV		*Umbravirus*
Beet cryptic virus 1	BCV-1	*Partitiviridae*	*Alphacryptovirus*
Beet cryptic virus 2	BCV-2	*Partitiviridae*	*Alphacryptovirus*
Beet cryptic virus 3	BCV-3	*Partitiviridae*	*Alphacryptovirus*
Beet curly top virus – California/Logan	BCTV-Cal	*Geminiviridae*	*Curtovirus*
Beet curly top virus	BCTV	*Geminiviridae*	*Curtovirus*
Beet curly top virus – Iran/CFH	BCTV-CFH	*Geminiviridae*	*Curtovirus*
Beet curly top virus – Worland	BCTV-Wor	*Geminiviridae*	*Curtovirus*
Beet leaf curl virus	BLCV	*Rhabdoviridae*	*Nucleorhabdovirus*
Beet mild yellowing virus	BMYV	*Luteoviridae*	*Luteovirus*
Beet mosaic virus	BtMV	*Potyviridae*	*Potyvirus*
Beet necrotic yellow vein virus	BNYVV		*Benyvirus*
Beet pseudoyellows virus	BPYV	*Closteroviridae*	*Closterovirus*
Beet ringspot virus	BRSV	*Comoviridae*	*Nepovirus*
Beet soil-borne mosaic virus	BSBMV		*Benyvirus*
Beet soil-borne virus	BSBV		*Pomovirus*
Beet virus Q	BVQ		*Pomovirus*
Beet western yellows virus	BWYV	*Luteoviridae*	*Luteovirus*
Beet yellow stunt virus	BYSV	*Closteroviridae*	*Closterovirus*
Beet yellows virus	BYV	*Closteroviridae*	*Closterovirus*
Belladonna mottle virus	BeMV		*Tymovirus*
Bermuda grass etched-line virus	BELV		*Marafivirus*
Bhendi yellow vein mosaic virus	BYVMV	*Geminiviridae*	*Begomovirus*
Bidens mosaic virus	BiMV	*Potyviridae*	*Potyvirus*
Bidens mottle virus	BiMoV	*Potyviridae*	*Potyvirus*
Black raspberry necrosis virus	BRNV		Unassigned
Blackcurrant reversion associated virus	BRAV	*Comoviridae*	*Nepovirus*
Blackgram mottle virus	BMoV	*Tombusviridae*	*Carmovirus*
Blueberry leaf mottle virus	BLMoV	*Comoviridae*	*Nepovirus*
Blueberry mosaic viroid-like RNA	BlMVd-RNA	Viroid	Tentative
Blueberry red ringspot virus	BRRV	*Caulimoviridae*	*Caulimovirus*
Blueberry scorch virus	BlScV		*Carlavirus*
Blueberry shock virus	BlShV	*Bromoviridae*	*Ilarvirus*
Blueberry shoestring virus	BSSV		*Sobemovirus*
Boletus virus X	BolVX		*Potexvirus*
Brachypodium yellow streak virus	BraYSV		Unassigned
Bramble yellow mosaic virus	BrmYMV	*Potyviridae*	*Potyvirus*
Brazilian wheat spike virus	BWSpV		*Tenuivirus*

(continued)

TABLE I Alphabetical list of plant virus names and their abbreviations (*Continued*)

Plant virus name	Abbreviation	Family	Genus[a]
Broad bean mottle virus	BBMV	*Bromoviridae*	*Bromovirus*
Broad bean necrosis virus	BBNV		*Pomovirus*
Broad bean stain virus	BBSV	*Comoviridae*	*Comovirus*
Broad bean true mosaic virus	BBTMV	*Comoviridae*	*Comovirus*
Broad bean wilt virus 1	BBWV-1	*Comoviridae*	*Fabavirus*
Broad bean wilt virus 2	BBWV-2	*Comoviridae*	*Fabavirus*
Broccoli necrotic yellows virus	BNYV	*Rhabdoviridae*	*Cytorhabdovirus*
Brome mosaic virus	BMV	*Bromoviridae*	*Bromovirus*
Brome streak virus	BStV	*Potyviridae*	*Tritimovirus*
Bromus striate mosaic virus	BrSMV	*Geminiviridae*	*Mastrevirus*
Bryonia mottle virus	BryMoV	*Potyviridae*	*Potyvirus*
Burdock stunt viroid	BuSVd	Viroid	Unassigned
Burdock yellows virus	BuYV	*Closteroviridae*	*Closterovirus*
Butterbur mosaic virus	ButMV		*Carlavirus*
Cacao swollen shoot virus	CSSV	*Caulimoviridae*	*Badnavirus*
Cacao yellow mosaic virus	CYMV		*Tymovirus*
Cactus virus 2	CV-2		*Carlavirus*
Cactus virus X	CVX		*Potexvirus*
Calanthe mild mosaic virus	CalMMV	*Potyviridae*	*Potyvirus*
Callistephus chinensis chlorosis virus	CCCV	*Rhabdoviridae*	*Nucleorhabdovirus*
Calopogonium yellow vein virus	CalYVV		*Tymovirus*
Camellia yellow mottle virus	CYMoV		*Varicosavirus*
Canary reed mosaic virus	CRMV	*Potyviridae*	*Potyvirus*
Canavalia maritima mosaic virus	CnMMV	*Potyviridae*	*Potyvirus*
Canna yellow mottle virus	CaYMV	*Caulimoviridae*	*Badnavirus*
Caper latent virus	CapLV		*Carlavirus*
Caraway latent virus	CawLV		*Carlavirus*
Cardamine chlorotic fleck virus	CCFV	*Tombusviridae*	*Carmovirus*
Cardamine latent virus	CaLV		*Carlavirus*
Cardamom mosaic virus	CdMV	*Potyviridae*	*Potyvirus*
Carnation bacilliform virus	CBV	*Rhabdoviridae*	*Nucleorhabdovirus*
Carnation cryptic virus 1	CCV-1	*Partitiviridae*	*Alphacryptovirus*
Carnation cryptic virus 2	CCV-2	*Partitiviridae*	*Alphacryptovirus*
Carnation etched ring virus	CERV	*Caulimoviridae*	*Caulimovirus*
Carnation Italian ringspot virus	CIRV	*Tombusviridae*	*Tombusvirus*
Carnation latent virus	CLV		*Carlavirus*
Carnation mottle virus	CarMV	*Tombusviridae*	*Carmovirus*
Carnation necrotic fleck virus	CNFV	*Closteroviridae*	*Closterovirus*
Carnation ringspot virus	CRSV	*Tombusviridae*	*Dianthovirus*
Carnation vein mottle virus	CVMoV	*Potyviridae*	*Potyvirus*
Carnation yellow stripe virus	CYSV	*Tombusviridae*	*Necrovirus*
Carrot latent virus	CtLV	*Rhabdoviridae*	*Nucleorhabdovirus*
Carrot mosaic virus	CtMV	*Potyviridae*	*Potyvirus*
Carrot mottle mimic virus	CMoMV		*Umbravirus*
Carrot mottle virus	CMoV		*Umbravirus*
Carrot red leaf virus	CtRLV	*Luteoviridae*	Unassigned
Carrot temperate virus 1	CTeV-1	*Partitiviridae*	*Alphacryptovirus*

Plant virus name	Abbreviation	Family	Genus[a]
Carrot temperate virus 2	CTeV-2	*Partitiviridae*	*Betacryptovirus*
Carrot temperate virus 3	CTeV-3	*Partitiviridae*	*Alphacryptovirus*
Carrot temperate virus 4	CTeV-4	*Partitiviridae*	*Alphacryptovirus*
Carrot thin leaf virus	CTLV	*Potyviridae*	*Potyvirus*
Carrot yellow leaf virus	CYLV	*Closteroviridae*	*Closterovirus*
Cassava American latent virus	CsALV	*Comoviridae*	*Nepovirus*
Cassava Ivorian bacilliform virus	CsIBV		Unassigned
Cassava common mosaic virus	CsCMV		*Potexvirus*
Cassava green mottle virus	CsGMV	*Comoviridae*	*Nepovirus*
Cassava symptomless virus	CsSLV	*Rhabdoviridae*	*Nucleorhabdovirus*
Cassava vein mosaic virus	CsVMV	*Caulimoviridae*	'Cassava vein mosaic-like viruses'
Cassava virus C	CsVC		*Ourmiavirus*
Cassava virus X	CsVX		*Potexvirus*
Cassia mild mosaic virus	CasMMV		*Carlavirus*
Cassia yellow blotch virus	CYBV	*Bromoviridae*	*Bromovirus*
Cassia yellow spot virus	CasYSV	*Potyviridae*	*Potyvirus*
Cauliflower mosaic virus	CaMV	*Caulimoviridae*	*Caulimovirus*
Celery mosaic virus	CeMV	*Potyviridae*	*Potyvirus*
Celery yellow mosaic virus	CeYMV	*Potyviridae*	*Potyvirus*
Centrosema mosaic virus	CenMV		*Potexvirus*
Ceratobium mosaic virus	CerMV	*Potyviridae*	*Potyvirus*
Cereal chlorotic mottle virus	CCMoV	*Rhabdoviridae*	*Nucleorhabdovirus*
Cereal yellow dwarf virus – RMV	CYDV-RMV	*Luteoviridae*	Unassigned
Cereal yellow dwarf virus – RPV	CYDV-RPV	*Luteoviridae*	*Luteovirus*
Cestrum virus	CV	*Caulimoviridae*	*Caulimovirus*
Chara corallina virus	ChaCV		*Tobamovirus*
Chayote mosaic virus	ChMV		*Tymovirus*
Chenopodium necrosis virus	ChNV	*Tombusviridae*	*Necrovirus*
Cherry green ring mottle virus	CGRMV		*Foveavirus*
Cherry leafroll virus	CLRV	*Comoviridae*	*Nepovirus*
Cherry rasp leaf virus	CRLV	*Comoviridae*	*Nepovirus*
Cherry rosette virus	CRV	*Comoviridae*	*Nepovirus*
Cherry virus A	CVA		*Capillovirus*
Chickpea bushy dwarf virus	CpBDV	*Potyviridae*	*Potyvirus*
Chickpea chlorotic dwarf virus	CpCDV	*Geminiviridae*	*Mastrevirus*
Chickpea filiform virus	CpFV	*Potyviridae*	*Potyvirus*
Chickpea stunt disease-associated virus	CpSDaV	*Luteoviridae*	Unassigned
Chicory yellow blotch virus	ChYBV		*Carlavirus*
Chicory yellow mottle virus	ChYMV	*Comoviridae*	*Nepovirus*
Chilli veinal mottle virus	ChiVMV	*Potyviridae*	*Potyvirus*
Chinese yam necrotic mosaic virus	ChYNMV		*Carlavirus*
Chino del tomate virus	CdTV	*Geminiviridae*	*Begomovirus*
Chloris striate mosaic virus	CSMV	*Geminiviridae*	*Mastrevirus*
Chrysanthemum chlorotic mottle viroid	CChMVd	*Avsunviroidae*	*Pelamoviroid*
Chrysanthemum frutescens virus	CFV	*Rhabdoviridae*	*Nucleorhabdovirus*
Chrysanthemum stem necrosis virus	CSNV	*Bunyaviridae*	*Tospovirus*

(continued)

TABLE I Alphabetical list of plant virus names and their abbreviations (*Continued*)

Plant virus name	Abbreviation	Family	Genus[a]
Chrysanthemum stunt viroid	CSVd	*Pospiviroidae*	*Pospiviroid*
Chrysanthemum vein chlorosis virus	CVCV	*Rhabdoviridae*	*Nucleorhabdovirus*
Chrysanthemum virus B	CVB		*Carlavirus*
Citrus bent leaf viroid	CBLVd	*Pospiviroidae*	*Apscaviroid*
Citrus exocortis viroid	CEVd	*Pospiviroidae*	*Pospiviroid*
Citrus III viroid	CVd-III	*Pospiviroidae*	*Apscaviroid*
Citrus IV viroid	CVd-IV	*Pospiviroidae*	*Cocadviroid*
Citrus leaf rugose virus	CiLRV	*Bromoviridae*	*Ilarvirus*
Citrus leprosis virus	CiLV	*Rhabdoviridae*	*Nucleorhabdovirus*
Citrus mosaic virus	CMBV	*Caulimoviridae*	*Badnavirus*
Citrus psorosis virus	CPsV		*Ophiovirus*
Citrus tristeza virus	CTV	*Closteroviridae*	*Closterovirus*
Citrus variegation virus	CVV	*Bromoviridae*	*Ilarvirus*
Clitoria yellow mosaic virus	CtYMV	*Potyviridae*	*Potyvirus*
Clitoria yellow vein virus	CYVV		*Tymovirus*
Clover enation virus	ClEV	*Rhabdoviridae*	*Nucleorhabdovirus*
Clover yellow mosaic virus	ClYMV		*Potexvirus*
Clover yellow vein virus	ClYVV	*Potyviridae*	*Potyvirus*
Clover yellows virus	CYV	*Closteroviridae*	*Closterovirus*
Cocksfoot mild mosaic virus	CMMV		*Sobemovirus*
Cocksfoot mottle virus	CoMV		*Sobemovirus*
Cocksfoot streak virus	CSV	*Potyviridae*	*Potyvirus*
Cocoa necrosis virus	CoNV	*Comoviridae*	*Nepovirus*
Coconut tinangaja viroid	CTiVd	*Pospiviroidae*	*Cocadviroid*
Coconut cadang-cadang viroid	CCCVd	*Pospiviroidae*	*Cocadviroid*
Coconut foliar decay virus	CFDV		*Nanovirus*
Coffee ringspot virus	CoRSV	*Rhabdoviridae*	*Nucleorhabdovirus*
Cole latent virus	CoLV		*Carlavirus*
Coleus blumei viroid 1	CbVd-1	*Pospiviroidae*	*Coleviroid*
Coleus blumei viroid 2	CbVd-2	*Pospiviroidae*	*Coleviroid*
Coleus blumei viroid 3	CbVd-3	*Pospiviroidae*	*Coleviroid*
Colocasia bobone disease virus	CBDV	*Rhabdoviridae*	*Nucleorhabdovirus*
Colombian datura virus	CDV	*Potyviridae*	*Potyvirus*
Columnea latent viroid	CLVd	*Pospiviroidae*	*Pospiviroid*
Commelina mosaic virus	ComMV	*Potyviridae*	*Potyvirus*
Commelina virus X	ComVX		*Potexvirus*
Commelina yellow mottle virus	ComYMV	*Caulimoviridae*	*Badnavirus*
Coriander feathery red vein virus	CFRVV	*Rhabdoviridae*	*Nucleorhabdovirus*
Cotton leaf crumple virus	CLCrV	*Geminiviridae*	*Begomovirus*
Cotton leaf curl virus – Pakistan 1	CLCuV-Pk1	*Geminiviridae*	*Begomovirus*
Cotton leaf curl virus – Pakistan 2	CLCuV-Pk2	*Geminiviridae*	*Begomovirus*
Cow parsnip mosaic virus	CPaMV	*Rhabdoviridae*	*Nucleorhabdovirus*
Cowpea aphid-borne mosaic virus	CABMV	*Potyviridae*	*Potyvirus*
Cowpea chlorotic mottle virus	CCMV	*Bromoviridae*	*Bromovirus*
Cowpea golden mosaic virus	CPGMV	*Geminiviridae*	*Begomovirus*
Cowpea green vein banding virus	CGVBV	*Potyviridae*	*Potyvirus*
Cowpea mild mottle virus	CPMMV		*Carlavirus*

Plant virus name	Abbreviation	Family	Genus[a]
Cowpea mosaic virus	CPMV	*Comoviridae*	*Comovirus*
Cowpea mottle virus	CPMoV	*Tombusviridae*	*Carmovirus*
Cowpea rugose mosaic virus	CPRMV	*Potyviridae*	*Potyvirus*
Cowpea severe mosaic virus	CPSMV	*Comoviridae*	*Comovirus*
Crimson clover latent virus	CCLV	*Comoviridae*	*Nepovirus*
Crinum mosaic virus	CriMV	*Potyviridae*	*Potyvirus*
Croatian clover virus	CroCV	*Potyviridae*	*Potyvirus*
Croton yellow vein mosaic virus	CYVMV	*Geminiviridae*	*Begomovirus*
Crucifer strain of TMV	TMV-Cg		*Tobamovirus*
Crucifer-infecting TMV	crTMV		*Tobamovirus*
Cucumber leafspot virus	CLSV	*Tombusviridae*	Unassigned
Cucumber chlorotic spot virus	CCSV	*Closteroviridae*	*Closterovirus*
Cucumber cryptic virus	CuCV	*Partitiviridae*	*Alphacryptovirus*
Cucumber green mottle mosaic virus	CGMMV		*Tobamovirus*
Cucumber mosaic virus	CMV	*Bromoviridae*	*Cucumovirus*
Cucumber necrosis virus	CuNV	*Tombusviridae*	*Tombusvirus*
Cucumber soil-borne virus	CuSBV	*Tombusviridae*	*Carmovirus*
Cucurbit aphid-borne yellows virus	CABYV	*Luteoviridae*	*Luteovirus*
Cucurbit yellow stunting disorder virus	CYSDV	*Closteroviridae*	*Crinivirus*
Cycas necrotic stunt virus	CNSV	*Comoviridae*	*Nepovirus*
Cymbidium mosaic virus	CymMV		*Potexvirus*
Cymbidium ringspot virus	CymRSV	*Tombusviridae*	*Tombusvirus*
Cynara virus	CraV	*Rhabdoviridae*	*Nucleorhabdovirus*
Cynodon mosaic virus	CynMV		*Carlavirus*
Cynosurus mottle virus	CnMoV		*Sobemovirus*
Cypripedium chlorotic streak virus	CypCSV	*Potyviridae*	*Potyvirus*
Dahlia mosaic virus	DMV	*Caulimoviridae*	*Caulimovirus*
Dandelion latent virus	DaLV		*Carlavirus*
Dandelion yellow mosaic virus	DaYMV	*Sequiviridae*	*Sequivirus*
Daphne virus S	DVS		*Carlavirus*
Daphne virus X	DVX		*Potexvirus*
Daphne virus Y	DVY	*Potyviridae*	*Potyvirus*
Dasheen mosaic virus	DsMV	*Potyviridae*	*Potyvirus*
Datura distortion mosaic virus	DDMV	*Potyviridae*	*Potyvirus*
Datura mosaic virus	DTMV	*Potyviridae*	*Potyvirus*
Datura necrosis virus	DNV	*Potyviridae*	*Potyvirus*
Datura shoestring virus	DSSV	*Potyviridae*	*Potyvirus*
Datura virus 437	DV-437	*Potyviridae*	*Potyvirus*
Datura yellow vein virus	DYVV	*Rhabdoviridae*	*Nucleorhabdovirus*
Dendrobium leaf streak virus	DLSV	*Rhabdoviridae*	*Nucleorhabdovirus*
Dendrobium vein necrosis virus	DVNV	*Closteroviridae*	*Closterovirus*
Desmodium mosaic virus	DesMV	*Potyviridae*	*Potyvirus*
Desmodium yellow mottle virus	DYMoV		*Tymovirus*
Digitaria streak virus	DSV	*Geminiviridae*	*Mastrevirus*
Digitaria striate mosaic virus	DiSMV	*Geminiviridae*	*Mastrevirus*
Digitaria striate virus	DiSV	*Rhabdoviridae*	*Nucleorhabdovirus*
Diodea vein chlorosis virus	DVCV	*Closteroviridae*	*Closterovirus*
Dioscorea bacilliform virus	DBV	*Caulimoviridae*	*Badnavirus*

(continued)

TABLE I Alphabetical list of plant virus names and their abbreviations (*Continued*)

Plant virus name	Abbreviation	Family	Genus[a]
Dioscorea latent virus	DLV		*Potexvirus*
Dioscorea trifida virus	DTV	*Potyviridae*	*Potyvirus*
Dipladenia mosaic virus	DipMV	*Potyviridae*	*Potyvirus*
Dock mottling mosaic virus	DMMV	*Potyviridae*	*Potyvirus*
Dolichos yellow mosaic virus	DoYMV	*Geminiviridae*	*Begomovirus*
Dulcamara mottle virus	DuMV		*Tymovirus*
Dulcamara virus A	DuVA		*Carlavirus*
Dulcamara virus B	DuVB		*Carlavirus*
East African cassava mosaic virus	EACMV	*Geminiviridae*	*Begomovirus*
Echinochloa hoja blanca virus	EHBV		*Tenuivirus*
Echinochloa ragged stunt virus	ERSV	*Reoviridae*	*Oryzavirus*
Eclipta yellow vein virus	EYVV	*Geminiviridae*	*Begomovirus*
Eggplant green mosaic virus	EGMV	*Potyviridae*	*Potyvirus*
Eggplant latent viroid	ELVd	Viroid	Unassigned
Eggplant mild mottle virus	EMMV		*Carlavirus*
Eggplant mosaic virus	EMV		*Tymovirus*
Eggplant mottled crinkle virus	EMCV	*Tombusviridae*	*Tombusvirus*
Eggplant mottled dwarf virus	EMDV	*Rhabdoviridae*	*Nucleorhabdovirus*
Eggplant severe mottle virus	ESMoV	*Potyviridae*	*Potyvirus*
Eggplant yellow mosaic virus	EYMV	*Geminiviridae*	*Begomovirus*
Elderberry latent virus	ElLV	*Tombusviridae*	*Carmovirus*
Elderberry symptomless virus	ElSLV		*Carlavirus*
Elm mottle virus	EMoV	*Bromoviridae*	*Ilarvirus*
Endive necrotic mosaic virus	ENMV	*Potyviridae*	*Potyvirus*
Epirus cherry virus	EpCV		*Ourmiavirus*
Erysimum latent virus	ErLV		*Tymovirus*
Euonymus fasciation virus	EFV	*Rhabdoviridae*	*Nucleorhabdovirus*
Euonymus mosaic virus	EuoMV		*Carlavirus*
Eupatorium yellow vein virus	EpYVV	*Geminiviridae*	*Begomovirus*
Euphorbia mosaic virus	EuMV	*Geminiviridae*	*Begomovirus*
Euphorbia ringspot virus	EuRSV	*Potyviridae*	*Potyvirus*
European wheat striate mosaic virus	EWSMV		*Tenuivirus*
Faba bean necrotic yellows virus	FBNYV		*Nanovirus*
Fescue cryptic virus	FCV	*Partitiviridae*	*Alphacryptovirus*
Festuca leaf streak virus	FLSV	*Rhabdoviridae*	*Cytorhabdovirus*
Festuca necrosis virus	FNV	*Closteroviridae*	*Closterovirus*
Fig leaf chlorosis virus	FigLCV	*Potyviridae*	*Potyvirus*
Fig virus S	FVS		*Carlavirus*
Figwort mosaic virus	FMV	*Caulimoviridae*	*Caulimovirus*
Fiji disease virus	FDV	*Reoviridae*	*Fijivirus*
Finger millet mosaic virus	FMMV	*Rhabdoviridae*	*Nucleorhabdovirus*
Flame chlorosis virus	FlCV		Unassigned
Foxtail mosaic virus	FoMV		*Potexvirus*
Fragaria chiloensis latent virus	FClLV	*Bromoviridae*	*Ilarvirus*
Frangipani mosaic virus	FrMV		*Tobamovirus*
Freesia leaf necrosis virus	FLNV		*Varicosavirus*
Freesia mosaic virus	FreMV	*Potyviridae*	*Potyvirus*

Plant virus name	Abbreviation	Family	Genus[a]
Fuchsia latent virus	FLV		*Carlavirus*
Furcraea necrotic streak virus	FNSV	*Tombusviridae*	*Dianthovirus*
Galinsoga mosaic virus	GaMV	*Tombusviridae*	*Carmovirus*
Garland chrysanthemum temperate virus	GCTV	*Partitiviridae*	*Alphacryptovirus*
Garlic common latent virus	GarCLV		*Carlavirus*
Garlic latent virus	GarLV		*Carlavirus*
Garlic mite-borne filamentous virus	GarMbFV		*Allexivirus*
Garlic mite-borne latent virus	GarMbLV		*Allexivirus*
Garlic mosaic virus	GarMV		*Carlavirus*
Garlic virus A	GarV-A		*Allexivirus*
Garlic virus B	GarV-B		*Allexivirus*
Garlic virus C	GarV-C		*Allexivirus*
Garlic virus D	GarV-D		*Allexivirus*
Garlic virus X	GarV-X		*Allexivirus*
Gentiana latent virus	GenLV		*Carlavirus*
Gerbera symptomless virus	GeSLV	*Rhabdoviridae*	*Nucleorhabdovirus*
Ginger chlorotic fleck virus	GCFV		*Sobemovirus*
Gloriosa stripe mosaic virus	GSMV	*Potyviridae*	*Potyvirus*
Glycine mosaic virus	GMV	*Comoviridae*	*Comovirus*
Glycine mottle virus	GMoV	*Tombusviridae*	*Carmovirus*
Gomphrena virus	GoV	*Rhabdoviridae*	*Nucleorhabdovirus*
Grapevine Algerian latent virus	GALV	*Tombusviridae*	*Tombusvirus*
Grapevine berry inner necrosis virus	GINV		*Trichovirus*
Grapevine Bulgarian latent virus	GBLV	*Comoviridae*	*Nepovirus*
Grapevine chrome mosaic virus	GCMV	*Comoviridae*	*Nepovirus*
Grapevine fanleaf virus	GFLV	*Comoviridae*	*Nepovirus*
Grapevine fleck virus	GFkV		Unassigned
Grapevine leafroll-associated virus 1	GLRaV-1	*Closteroviridae*	*Closterovirus*
Grapevine leafroll-associated virus 2	GLRaV-2	*Closteroviridae*	*Closterovirus*
Grapevine leafroll-associated virus 3	GLRaV-3	*Closteroviridae*	*Closterovirus*
Grapevine leafroll-associated virus 4	GLRaV-4	*Closteroviridae*	*Closterovirus*
Grapevine leafroll-associated virus 5	GLRaV-5	*Closteroviridae*	*Closterovirus*
Grapevine leafroll-associated virus 6	GLRaV-6	*Closteroviridae*	*Closterovirus*
Grapevine leafroll-associated virus 7	GLRaV-7	*Closteroviridae*	*Closterovirus*
Grapevine Tunisian ringspot virus	GTRSV	*Comoviridae*	*Nepovirus*
Grapevine virus A	GVA		*Vitivirus*
Grapevine virus B	GVB		*Vitivirus*
Grapevine virus C	GVC		*Vitivirus*
Grapevine virus D	GVD		*Vitivirus*
Grapevine yellow speckle viroid 1	GYSVd-1	*Pospiviroidae*	*Apscaviroid*
Grapevine yellow speckle viroid 2	GYSVd-2	*Pospiviroidae*	*Apscaviroid*
Groundnut bud necrosis virus	GBNV	*Bunyaviridae*	*Tospovirus*
Groundnut chlorotic fan-spot virus	GCFSV	*Bunyaviridae*	*Tospovirus*
Groundnut eyespot virus	GEV	*Potyviridae*	*Potyvirus*
Groundnut ringspot virus	GRSV	*Bunyaviridae*	*Tospovirus*
Groundnut rosette assistor virus	GRAV	*Luteoviridae*	*Luteovirus*
Groundnut rosette virus	GRV		*Umbravirus*

(*continued*)

TABLE I Alphabetical list of plant virus names and their abbreviations (*Continued*)

Plant virus name	Abbreviation	Family	Genus[a]
Groundnut yellow spot virus	GYSV	*Bunyaviridae*	*Tospovirus*
Guar symptomless virus	GSLV	*Potyviridae*	*Potyvirus*
Guinea grass mosaic virus	GGMV	*Potyviridae*	*Potyvirus*
Gynura latent virus	GyLV		*Carlavirus*
Habenaria mosaic virus	HaMV	*Potyviridae*	*Potyvirus*
Hart's tongue fern mottle virus	HTFMoV		Unassigned
Helenium virus S	HVS		*Carlavirus*
Helenium virus Y	HVY	*Potyviridae*	*Potyvirus*
Helleborus mosaic virus	HeMV		*Carlavirus*
Henbane mosaic virus	HMV	*Potyviridae*	*Potyvirus*
Heracleum latent virus	HLV		*Vitivirus*
Heracleum virus 6	HV-6	*Closteroviridae*	*Closterovirus*
Hibiscus chlorotic ringspot virus	HCRSV	*Tombusviridae*	*Carmovirus*
Hibiscus latent ringspot virus	HLRSV	*Comoviridae*	*Nepovirus*
Hippeastrum mosaic virus	HiMV	*Potyviridae*	*Potyvirus*
Holcus lanatus yellowing virus	HLYV	*Rhabdoviridae*	*Nucleorhabdovirus*
Holcus streak virus	HSV	*Potyviridae*	*Potyvirus*
Hollyhock leaf curl virus	HLCV	*Geminiviridae*	*Begomovirus*
Honeysuckle latent virus	HnLV		*Carlavirus*
Honeysuckle yellow vein mosaic virus	HYVMV	*Geminiviridae*	*Begomovirus*
Hop latent viroid	HpLVd	*Pospiviroidae*	*Cocadviroid*
Hop latent virus	HpLV		*Carlavirus*
Hop mosaic virus	HpMV		*Carlavirus*
Hop stunt viroid	HSVd	*Pospiviroidae*	*Hostuviroid*
Hop trefoil cryptic virus 1	HTCV-1	*Partitiviridae*	*Alphacryptovirus*
Hop trefoil cryptic virus 2	HTCV-2	*Partitiviridae*	*Betacryptovirus*
Hop trefoil cryptic virus 3	HTCV-3	*Partitiviridae*	*Alphacryptovirus*
Hordeum mosaic virus	HoMV	*Potyviridae*	*Rymovirus*
Horsegram yellow mosaic virus	HgYMV	*Geminiviridae*	*Begomovirus*
Horseradish curly top virus	HrCTV	*Geminiviridae*	*Curtovirus*
Horseradish latent virus	HrLV	*Caulimoviridae*	*Caulimovirus*
Hosta virus X	HVX		*Potexvirus*
Humulus japonicus latent virus	HJLV	*Bromoviridae*	*Ilarvirus*
Hungarian datura innoxia virus	HDIV	*Potyviridae*	*Potyvirus*
Hyacinth mosaic virus	HyaMV	*Potyviridae*	*Potyvirus*
Hydrangea latent virus	HdLV		*Carlavirus*
Hydrangea mosaic virus	HdMV	*Bromoviridae*	*Ilarvirus*
Hydrangea ringspot virus	HdRSV		*Potexvirus*
Hypochoeris mosaic virus	HyMV		*Furovirus*
Impatiens latent virus	ILV		*Carlavirus*
Impatiens necrotic spot virus	INSV	*Bunyaviridae*	*Tospovirus*
Indian cassava mosaic virus	ICMV	*Geminiviridae*	*Begomovirus*
Indian peanut clump virus	IPCV		*Pecluvirus*
Indian pepper mottle virus	IPMoV	*Potyviridae*	*Potyvirus*
Indian tomato leaf curl virus	IToLCV	*Geminiviridae*	*Begomovirus*
Indonesian soybean dwarf virus	ISDV	*Luteoviridae*	Unassigned
Iranian wheat stripe virus	IWSV		*Tenuivirus*

Plant virus name	Abbreviation	Family	Genus[a]
Iresine viroid 1	IrVd-1	*Pospiviroidae*	*Pospiviroid*
Iris fulva mosaic virus	IFMV	*Potyviridae*	*Potyvirus*
Iris germanica leaf stripe virus	IGLSV	*Rhabdoviridae*	*Nucleorhabdovirus*
Iris mild mosaic virus	IMMV	*Potyviridae*	*Potyvirus*
Iris severe mosaic virus	ISMV	*Potyviridae*	*Potyvirus*
Iris yellow spot virus	IYSV	*Bunyaviridae*	*Tospovirus*
Isachne mosaic virus	IsaMV	*Potyviridae*	*Potyvirus*
Ivy vein clearing virus	IVCV	*Rhabdoviridae*	*Nucleorhabdovirus*
Jatropha mosaic virus	JMV	*Geminiviridae*	*Begomovirus*
Johnsongrass mosaic virus	JGMV	*Potyviridae*	*Potyvirus*
Kalanchoë latent virus	KLV		*Carlavirus*
Kalanchoë mosaic virus	KMV	*Potyviridae*	*Potyvirus*
Kalanchoë top-spotting virus	KTSV	*Caulimoviridae*	*Badnavirus*
Kennedya virus Y	KVY	*Potyviridae*	*Potyvirus*
Kennedya yellow mosaic virus	KYMV		*Tymovirus*
Konjac mosaic virus	KoMV	*Potyviridae*	*Potyvirus*
Kyuri green mottle mosaic virus	KGMMV		*Tobamovirus*
Laelia red leafspot virus	LRLV	*Rhabdoviridae*	*Nucleorhabdovirus*
Lamium mild mosaic virus	LMMV	*Comoviridae*	*Fabavirus*
Lato river virus	LRV	*Tombusviridae*	*Tombusvirus*
Launea arborescens stunt virus	LArSV	*Rhabdoviridae*	*Nucleorhabdovirus*
Leek white stripe virus	LWSV	*Tombusviridae*	*Necrovirus*
Leek yellow stripe virus	LYSV	*Potyviridae*	*Potyvirus*
Lemon scented thyme leaf chlorosis virus	LSTCV	*Rhabdoviridae*	*Nucleorhabdovirus*
Leonurus mosaic virus	LeMV	*Geminiviridae*	*Begomovirus*
Lettuce big-vein virus	LBVV		*Varicosavirus*
Lettuce chlorosis virus	LCV	*Closteroviridae*	*Crinivirus*
Lettuce infectious yellows virus	LIYV	*Closteroviridae*	*Crinivirus*
Lettuce mosaic virus	LMV	*Potyviridae*	*Potyvirus*
Lettuce necrotic yellows virus	LNYV	*Rhabdoviridae*	*Cytorhabdovirus*
Lettuce speckles mottle virus	LSMV		*Umbravirus*
Lilac chlorotic leafspot virus	LiCLV		*Capillovirus*
Lilac mottle virus	LiMoV		*Carlavirus*
Lilac ring mottle virus	LiRMoV	*Bromoviridae*	*Ilarvirus*
Lilac ringspot virus	LiRSV		*Carlavirus*
Lily mild mottle virus	LMMoV	*Potyviridae*	*Potyvirus*
Lily mottle virus	LMoV	*Potyviridae*	*Potyvirus*
Lily symptomless virus	LSLV		*Carlavirus*
Lily virus X	LVX		*Potexvirus*
Limabean golden mosaic virus	LGMV	*Geminiviridae*	*Begomovirus*
Lisianthus necrosis virus	LNV	*Tombusviridae*	*Necrovirus*
Little cherry virus	LChV	*Closteroviridae*	*Closterovirus*
Lolium ryegrass virus	LoRV	*Rhabdoviridae*	*Nucleorhabdovirus*
Lotus stem necrosis	LoSNV	*Rhabdoviridae*	*Nucleorhabdovirus*
Lucerne Australian latent virus	LALV	*Comoviridae*	*Nepovirus*
Lucerne Australian symptomless virus	LASV	*Comoviridae*	*Nepovirus*

(continued)

TABLE I Alphabetical list of plant virus names and their abbreviations (*Continued*)

Plant virus name	Abbreviation	Family	Genus[a]
Lucerne enation virus	LEV	*Rhabdoviridae*	*Nucleorhabdovirus*
Lucerne transient streak virus	LTSV		*Sobemovirus*
Lupin leaf curl virus	LLCV	*Geminiviridae*	*Begomovirus*
Lupin yellow vein virus	LYVV	*Rhabdoviridae*	*Nucleorhabdovirus*
Lychnis ringspot virus	LRSV		*Hordeivirus*
Lychnis symptomless virus	LycSLV		*Potexvirus*
Maclura mosaic virus	MacMV	*Potyviridae*	*Macluravirus*
Macroptilium golden mosaic virus	MGMV	*Geminiviridae*	*Begomovirus*
Macrotyloma mosaic virus	MaMV	*Geminiviridae*	*Begomovirus*
Maize chlorotic dwarf virus	MCDV	*Sequiviridae*	*Waikavirus*
Maize chlorotic mottle virus	MCMV	*Tombusviridae*	*Machlomovirus*
Maize dwarf mosaic virus	MDMV	*Potyviridae*	*Potyvirus*
Maize mosaic virus	MMV	*Rhabdoviridae*	*Nucleorhabdovirus*
Maize rayado fino virus	MRFV		*Marafivirus*
Maize rough dwarf virus	MRDV	*Reoviridae*	*Fijivirus*
Maize sterile stunt virus	MSSV	*Rhabdoviridae*	*Nucleorhabdovirus*
Maize streak virus	MSV	*Geminiviridae*	*Mastrevirus*
Maize stripe virus	MSpV		*Tenuivirus*
Maize white line mosaic virus	MWLMV		Unassigned
Malva silvestris virus	MaSV	*Rhabdoviridae*	*Nucleorhabdovirus*
Malva vein clearing virus	MVCV	*Potyviridae*	*Potyvirus*
Malva veinal necrosis virus	MVNV		*Potexvirus*
Malvaceous chlorosis virus	MCV	*Geminiviridae*	*Begomovirus*
Maracuja mosaic virus	MarMV		*Tobamovirus*
Marigold mottle virus	MaMoV	*Potyviridae*	*Potyvirus*
Megakepasma mosaic virus	MegMV	*Closteroviridae*	*Closterovirus*
Melandrium yellow fleck virus	MYFV	*Bromoviridae*	*Bromovirus*
Melilotus latent virus	MeLV	*Rhabdoviridae*	*Nucleorhabdovirus*
Melilotus mosaic virus	MeMV	*Potyviridae*	*Potyvirus*
Melon leaf curl virus	MLCV	*Geminiviridae*	*Begomovirus*
Melon necrotic spot virus	MNSV	*Tombusviridae*	*Carmovirus*
Melon rugose mosaic virus	MRMV		*Tymovirus*
Melon variegation virus	MVV	*Rhabdoviridae*	*Nucleorhabdovirus*
Melon vein-banding mosaic virus	MVBMV	*Potyviridae*	*Potyvirus*
Mexican papita viroid	MPVd	*Pospiviroidae*	*Pospiviroid*
Mibuna temperate virus	MTV	*Partitiviridae*	*Alphacryptovirus*
Milk vetch dwarf virus	MDV		*Nanovirus*
Mimosa bacilliform virus	MBV	*Caulimoviridae*	*Badnavirus*
Mirabilis mosaic virus	MiMV	*Caulimoviridae*	*Caulimovirus*
Miscanthus streak virus	MiSV	*Geminiviridae*	*Mastrevirus*
Molinia streak virus	MoSV	*Tombusviridae*	*Panicovirus*
Moroccan pepper virus	MPV	*Tombusviridae*	*Tombusvirus*
Moroccan watermelon mosaic virus	MWMV	*Potyviridae*	*Potyvirus*
Mulberry latent virus	MLV		*Carlavirus*
Mulberry ringspot virus	MRSV	*Comoviridae*	*Nepovirus*
Mungbean mosaic virus	MbMV	*Potyviridae*	*Potyvirus*
Mungbean mottle virus	MMoV	*Potyviridae*	*Potyvirus*

Plant virus name	Abbreviation	Family	Genus[a]
Mungbean yellow mosaic virus	MYMV	*Geminiviridae*	*Begomovirus*
Muskmelon vein necrosis virus	MuVNV		*Carlavirus*
Nandina mosaic virus	NaMV		*Potexvirus*
Nandina stem pitting virus	NSPV		*Capillovirus*
Narcissus degeneration virus	NDV	*Potyviridae*	*Potyvirus*
Narcissus late season yellows virus	NLSYV	*Potyviridae*	*Potyvirus*
Narcissus latent virus	NLV	*Potyviridae*	*Macluravirus*
Narcissus mosaic virus	NMV		*Potexvirus*
Narcissus tip necrosis virus	NTNV	*Tombusviridae*	*Carmovirus*
Narcissus yellow stripe virus	NYSV	*Potyviridae*	*Potyvirus*
Neckar river virus	NRV	*Tombusviridae*	*Tombusvirus*
Negro coffee mosaic virus	NeCMV		*Potexvirus*
Nerine latent virus	NeLV		*Carlavirus*
Nerine virus	NV	*Potyviridae*	*Potyvirus*
Nerine virus X	NVX		*Potexvirus*
Nerine virus Y	NVY	*Potyviridae*	*Potyvirus*
Nerine yellow stripe virus	NeYSV	*Potyviridae*	*Potyvirus*
Nicotiana glutinosa stunt viroid	NGSVd	*Viroid*	Unassigned
Nicotiana velutina mosaic virus	NVMV		Unassigned
Northern cereal mosaic virus	NCMV	*Rhabdoviridae*	*Cytorhabdovirus*
Nothoscordum mosaic virus	NoMV	*Potyviridae*	*Potyvirus*
Oat blue dwarf virus	OBDV		*Marafivirus*
Oat chlorotic stunt virus	OCSV	*Tombusviridae*	*Avenavirus*
Oat mosaic virus	OMV	*Potyviridae*	*Bymovirus*
Oat necrotic mottle virus	ONMV	*Potyviridae*	*Rymovirus*
Oat sterile dwarf virus	OSDV	*Reoviridae*	*Fijivirus*
Oat striate mosaic virus	OSMV	*Rhabdoviridae*	*Nucleorhabdovirus*
Obuda pepper virus	ObPV		*Tobamovirus*
Odontoglossum ringspot virus	ORSV		*Tobamovirus*
Okra leaf curl virus	OLCV	*Geminiviridae*	*Begomovirus*
Okra mosaic virus	OkMV		*Tymovirus*
Olive latent ringspot virus	OLRSV	*Comoviridae*	*Nepovirus*
Olive latent virus 1	OLV-1	*Tombusviridae*	*Necrovirus*
Olive latent virus 2	OLV-2	*Bromoviridae*	*Oleavirus*
Onion mite-borne latent virus	OMbLV		*Allexivirus*
Onion yellow dwarf virus	OYDV	*Potyviridae*	*Potyvirus*
Ononis yellow mosaic virus	OYMV		*Tymovirus*
Orchid fleck virus	OFV	*Rhabdoviridae*	*Nucleorhabdovirus*
Orchid fleck virus	OFV		Unassigned
Ornithogalum mosaic virus	OrMV	*Potyviridae*	*Potyvirus*
Ourmia melon virus	OuMV		*Ourmiavirus*
Palm mosaic virus	PalMV	*Potyviridae*	*Potyvirus*
Pangola stunt virus	PaSV	*Reoviridae*	*Fijivirus*
Panicum mosaic virus	PMV	*Tombusviridae*	*Panicovirus*
Panicum streak virus	PanSV	*Geminiviridae*	*Mastrevirus*
Papaya leaf curl virus	PaLCV	*Geminiviridae*	*Begomovirus*
Papaya leaf distortion mosaic virus	PLDMV	*Potyviridae*	*Potyvirus*
Papaya mosaic virus	PapMV		*Potexvirus*

(continued)

TABLE I Alphabetical list of plant virus names and their abbreviations (*Continued*)

Plant virus name	Abbreviation	Family	Genus[a]
Papaya ringspot virus	PRSV	*Potyviridae*	*Potyvirus*
Paprika mild mottle virus	PaMMV		*Tobamovirus*
Parietaria mottle virus	PMoV	*Bromoviridae*	*Ilarvirus*
Parsley latent virus	PaLV		Unassigned
Parsley virus	PaV	*Rhabdoviridae*	*Nucleorhabdovirus*
Parsley virus 5	PaV-5		*Potexvirus*
Parsnip mosaic virus	ParMV	*Potyviridae*	*Potyvirus*
Parsnip virus 3	ParV-3		*Potexvirus*
Parsnip virus 5	ParV-5		*Potexvirus*
Parsnip yellow fleck virus	PYFV	*Sequiviridae*	*Sequivirus*
Paspalum striate mosaic virus	PSMV	*Geminiviridae*	*Mastrevirus*
Passiflora latent virus	PLV		*Carlavirus*
Passion fruit mottle virus	PFMoV	*Potyviridae*	*Potyvirus*
Passion fruit ringspot virus	PFRSV	*Potyviridae*	*Potyvirus*
Passion fruit woodiness virus	PWV	*Potyviridae*	*Potyvirus*
Passion fruit yellow mosaic virus	PFYMV		*Tymovirus*
Patchouli mild mosaic virus	PatMMV	*Comoviridae*	*Fabavirus*
Patchouli mottle virus	PatMoV	*Potyviridae*	*Potyvirus*
Patchouli virus X	PatVX		*Potexvirus*
Pea early-browning virus	PEBV		*Tobravirus*
Pea enation mosaic virus 2	PEMV-2		*Umbravirus*
Pea enation mosaic virus 1	PEMV-1	*Luteoviridae*	*Enamovirus*
Pea green mottle virus	PGMV	*Comoviridae*	*Comovirus*
Pea mild mosaic virus	PMiMV	*Comoviridae*	*Comovirus*
Pea seed-borne mosaic virus	PSbMV	*Potyviridae*	*Potyvirus*
Pea streak virus	PeSV		*Carlavirus*
Peach latent mosaic viroid	PLMVd	*Avsunviroidae*	*Pelamoviroid*
Peach rosette mosaic virus	PRMV	*Comoviridae*	*Nepovirus*
Peanut chlorotic streak virus	PCSV	*Caulimoviridae*	'Soybean chlorotic mottle-like viruses'
Peanut clump virus	PCV		*Pecluvirus*
Peanut green mottle virus	PeGMoV	*Potyviridae*	*Potyvirus*
Peanut mottle virus	PeMoV	*Potyviridae*	*Potyvirus*
Peanut stunt virus	PSV	*Bromoviridae*	*Cucumovirus*
Peanut yellow mosaic virus	PeYMV		*Tymovirus*
Pear blister canker viroid	PBCVd	*Pospiviroidae*	*Apscaviroid*
Pecteilis mosaic virus	PcMV	*Potyviridae*	*Potyvirus*
Pelargonium flower break virus	PFBV	*Tombusviridae*	*Carmovirus*
Pelargonium leaf curl virus	PLCV	*Tombusviridae*	*Tombusvirus*
Pelargonium zonate spot virus	PZSV		Unassigned
Pepino mosaic virus	PepMV		*Potexvirus*
Pepper huasteco virus	PHV	*Geminiviridae*	*Begomovirus*
Pepper leaf curl virus	PepLCV	*Geminiviridae*	*Begomovirus*
Pepper mild mosaic virus	PMMV	*Potyviridae*	*Potyvirus*
Pepper mild mottle virus	PMMoV		*Tobamovirus*

Plant virus name	Abbreviation	Family	Genus[a]
Pepper mild tigré virus	PepMTV	*Geminiviridae*	*Begomovirus*
Pepper mottle virus	PepMoV	*Potyviridae*	*Potyvirus*
Pepper ringspot virus	PepRSV		*Tobravirus*
Pepper severe mosaic virus	PepSMV	*Potyviridae*	*Potyvirus*
Pepper vein banding virus	PVBV	*Potyviridae*	*Potyvirus*
Pepper veinal mottle virus	PVMV	*Potyviridae*	*Potyvirus*
Perilla mottle virus	PerMoV	*Potyviridae*	*Potyvirus*
Peru tomato mosaic virus	PTV	*Potyviridae*	*Potyvirus*
Petunia asteroid mosaic virus	PetAMV	*Tombusviridae*	*Tombusvirus*
Petunia flower mottle virus	PetFMV	*Potyviridae*	*Potyvirus*
Petunia vein clearing virus	PVCV	*Caulimoviridae*	'Petunia vein clearing-like viruses'
Phalaenopsis chlorotic spot virus	PhCSV	*Rhabdoviridae*	*Nucleorhabdovirus*
Physalis mottle virus	PhyMV		*Tymovirus*
Physalis severe mottle virus	PhySMV	*Bunyaviridae*	*Tospovirus*
Pigeon pea mosaic mottle viroid	PPMMoVd	*Viroid*	Unassigned
Pigeon pea proliferation virus	PPPV	*Rhabdoviridae*	*Nucleorhabdovirus*
Pineapple bacilliform virus	PBV	*Caulimoviridae*	*Badnavirus*
Pineapple chlorotic leaf streak virus	PCLSV	*Rhabdoviridae*	*Nucleorhabdovirus*
Pineapple mealybug wilt-associated virus 1	PMWaV-1	*Closteroviridae*	*Closterovirus*
Pineapple mealybug wilt-associated virus 2	PMWaV-2	*Closteroviridae*	*Closterovirus*
Piper yellow mottle virus	PYMoV	*Caulimoviridae*	*Badnavirus*
Pisum virus	PisV	*Rhabdoviridae*	*Nucleorhabdovirus*
Plantago asiatica mosaic virus	PlAMV		*Potexvirus*
Plantago mottle virus	PlMoV		*Tymovirus*
Plantago severe mottle virus	PlSMoV		*Potexvirus*
Plantago virus 4	PlV-4	*Caulimoviridae*	*Caulimovirus*
Plantain mottle virus	PlMV	*Rhabdoviridae*	*Nucleorhabdovirus*
Plantain virus 6	PlV-6	*Tombusviridae*	*Carmovirus*
Plantain virus 7	PlV-7	*Potyviridae*	*Potyvirus*
Plantain virus 8	PlV-8		*Carlavirus*
Plantain virus X	PlVX		*Potexvirus*
Pleioblastus mosaic virus	PleMV	*Potyviridae*	*Potyvirus*
Plum pox virus	PPV	*Potyviridae*	*Potyvirus*
Poa semilatent virus	PSLV		*Hordeivirus*
Poinsettia cryptic virus	PnCV	*Partitiviridae*	*Alphacryptovirus*
Poinsettia mosaic virus	PnMV		*Tymovirus*
Pokeweed mosaic virus	PkMV	*Potyviridae*	*Potyvirus*
Poplar mosaic virus	PopMV		*Carlavirus*
Poplar decline virus	PDV	*Potyviridae*	*Potyvirus*
Potato aucuba mosaic virus	PAMV		*Potexvirus*
Potato black ringspot virus	PBRSV	*Comoviridae*	*Nepovirus*
Potato leafroll virus	PLRV	*Luteoviridae*	*Luteovirus*
Potato mop-top virus	PMTV		*Pomovirus*
Potato spindle tuber viroid	PSTVd	*Pospiviroidae*	*Pospiviroid*

(continued)

TABLE I Alphabetical list of plant virus names and their abbreviations (*Continued*)

Plant virus name	Abbreviation	Family	Genus[a]
Potato virus A	PVA	*Potyviridae*	*Potyvirus*
Potato virus M	PVM		*Carlavirus*
Potato virus S	PVS		*Carlavirus*
Potato virus T	PVT		*Trichovirus*
Potato virus U	PVU	*Comoviridae*	*Nepovirus*
Potato virus V	PVV	*Potyviridae*	*Potyvirus*
Potato virus X	PVX		*Potexvirus*
Potato virus Y	PVY	*Potyviridae*	*Potyvirus*
Potato yellow dwarf virus	PYDV	*Rhabdoviridae*	*Nucleorhabdovirus*
Potato yellow mosaic virus	PYMV	*Geminiviridae*	*Begomovirus*
Pothos latent virus	PoLV	*Tombusviridae*	*Aureusvirus*
Primula mosaic virus	PrMV	*Potyviridae*	*Potyvirus*
Primula mottle virus	PrMoV	*Potyviridae*	*Potyvirus*
Prune dwarf virus	PDV	*Bromoviridae*	*Ilarvirus*
Prunus necrotic ringspot virus	PNRSV	*Bromoviridae*	*Ilarvirus*
Prunus virus S	PruVS		*Carlavirus*
Pseuderanthemum yellow vein virus	PYVV	*Geminiviridae*	*Begomovirus*
Quail pea mosaic virus	QPMV	*Comoviridae*	*Comovirus*
Radish mosaic virus	RaMV	*Comoviridae*	*Comovirus*
Radish vein clearing virus	RaVCV	*Potyviridae*	*Potyvirus*
Radish yellow edge virus	RYEV	*Partitiviridae*	*Alphacryptovirus*
Ranunculus mottle virus	RanMoV	*Potyviridae*	*Potyvirus*
Ranunculus repens symptomless virus	RaRSV	*Rhabdoviridae*	*Nucleorhabdovirus*
Ranunculus white mottle virus	RWMV		*Ophiovirus*
Raphanus virus	RaV	*Rhabdoviridae*	*Nucleorhabdovirus*
Raspberry bushy dwarf virus	RBDV		*Idaeovirus*
Raspberry ringspot virus	RpRSV	*Comoviridae*	*Nepovirus*
Raspberry vein chlorosis virus	RVCV	*Rhabdoviridae*	*Nucleorhabdovirus*
Red clover cryptic virus 2	RCCV-2	*Partitiviridae*	*Betacryptovirus*
Red clover mosaic virus	RClMV	*Rhabdoviridae*	*Nucleorhabdovirus*
Red clover mottle virus	RCMV	*Comoviridae*	*Comovirus*
Red clover necrotic mosaic virus	RCNMV	*Tombusviridae*	*Dianthovirus*
Red clover vein mosaic virus	RCVMV		*Carlavirus*
Red pepper cryptic virus 1	RPCV-1	*Partitiviridae*	*Alphacryptovirus*
Red pepper cryptic virus 2	RPCV-2	*Partitiviridae*	*Alphacryptovirus*
Rembrandt tulip breaking virus	ReTBV	*Potyviridae*	*Potyvirus*
Rhododendron necrotic ringspot virus	RoNRSV		*Potexvirus*
Rhubarb temperate virus	RTV	*Partitiviridae*	*Alphacryptovirus*
Rhubarb virus 1	RV-1		*Potexvirus*
Rhynchosia mosaic virus	RhMV	*Geminiviridae*	*Begomovirus*
Ribgrass mosaic virus	RMV		*Tobamovirus*
Rice black streaked dwarf virus	RBSDV	*Reoviridae*	*Fijivirus*
Rice dwarf virus	RDV	*Reoviridae*	*Phytoreovirus*
Rice gall dwarf virus	RGDV	*Reoviridae*	*Phytoreovirus*
Rice grassy stunt virus	RGSV		*Tenuivirus*

Plant virus name	Abbreviation	Family	Genus[a]
Rice hoja blanca virus	RHBV		*Tenuivirus*
Rice necrosis mosaic virus	RNMV	*Potyviridae*	*Bymovirus*
Rice ragged stunt virus	RRSV	*Reoviridae*	*Oryzavirus*
Rice stripe necrosis virus	RSNV		*Furovirus*
Rice stripe virus	RSV		*Tenuivirus*
Rice tungro bacilliform virus	RTBV	*Caulimoviridae*	'Rice tungro bacilliform-like viruses'
Rice tungro spherical virus	RTSV	*Sequiviridae*	*Waikavirus*
Rice wilted stunt virus	RWSV		*Tenuivirus*
Rice yellow mottle virus	RYMV		*Sobemovirus*
Rice yellow stunt virus	RYSV	*Rhabdoviridae*	*Nucleorhabdovirus*
Rubus Chinese seed-borne virus	RCSV	*Comoviridae*	*Nepovirus*
Rudbeckia mosaic virus	RuMV	*Potyviridae*	*Potyvirus*
Rupestris stem pitting-associated virus	RSPaV		*Foveavirus*
Ryegrass cryptic virus	RGCV	*Partitiviridae*	*Alphacryptovirus*
Ryegrass mosaic virus	RGMV	*Potyviridae*	*Rymovirus*
Saguaro cactus virus	SgCV	*Tombusviridae*	*Carmovirus*
Sainpaulia leaf necrosis virus	SLNV	*Rhabdoviridae*	*Nucleorhabdovirus*
Sambucus vein clearing virus	SVCV	*Rhabdoviridae*	*Nucleorhabdovirus*
Sammons's Opuntia virus	SOV		*Tobamovirus*
Santosai temperate virus	STV	*Partitiviridae*	*Alphacryptovirus*
Sarracenia purpurea virus	SPV	*Rhabdoviridae*	*Nucleorhabdovirus*
Satsuma dwarf virus	SDV	*Comoviridae*	*Nepovirus*
Schefflera ringspot virus	SRV	*Caulimoviridae*	*Badnavirus*
Scrophularia mottle virus	ScrMV		*Tymovirus*
Serrano golden mosaic virus	SGMV	*Geminiviridae*	*Begomovirus*
Shallot latent virus	SLV		*Carlavirus*
Shallot mite-borne latent virus	ShMbLV		*Allexivirus*
Shallot virus X	ShVX		*Allexivirus*
Shallot yellow stripe virus	SYSV	*Potyviridae*	*Potyvirus*
Sida golden mosaic virus	SiGMV	*Geminiviridae*	*Begomovirus*
Sida yellow vein virus	SiYVV	*Geminiviridae*	*Begomovirus*
Sikte waterborne virus	SWBV	*Tombusviridae*	*Tombusvirus*
Sinaloa tomato leaf curl virus	STLCV	*Geminiviridae*	*Begomovirus*
Sint-Jem's onion latent virus	SJOLV		*Carlavirus*
Smithiantha latent virus	SmiLV		*Potexvirus*
Soil-borne rye mosaic virus	SBRMV		*Furovirus*
Soil-borne wheat mosaic virus	SBWMV		*Furovirus*
Solanum apical leaf curl virus	SALCV	*Geminiviridae*	*Begomovirus*
Solanum nodiflorum mottle virus	SNMoV		*Sobemovirus*
Solanum tomato leaf curl virus	SToLCV	*Geminiviridae*	*Begomovirus*
Solanum yellow leaf curl virus	SYLCV	*Geminiviridae*	*Begomovirus*
Sonchus mottle virus	SMoV	*Caulimoviridae*	*Caulimovirus*
Sonchus virus	SonV	*Rhabdoviridae*	*Cytorhabdovirus*
Sonchus yellow net virus	SYNV	*Rhabdoviridae*	*Nucleorhabdovirus*
Sorghum chlorotic spot virus	SrCSV		*Furovirus*
Sorghum mosaic virus	SrMV	*Potyviridae*	*Potyvirus*

(continued)

TABLE I Alphabetical list of plant virus names and their abbreviations (*Continued*)

Plant virus name	Abbreviation	Family	Genus[a]
Sorghum virus	SrV	*Rhabdoviridae*	*Nucleorhabdovirus*
Soursop yellow blotch virus	SYBV	*Rhabdoviridae*	*Nucleorhabdovirus*
South African cassava mosaic virus	SACMV	*Geminiviridae*	*Begomovirus*
Southern bean mosaic virus	SBMV		*Sobemovirus*
Southern cowpea mosaic virus	SCPMV		*Sobemovirus*
Southern potato latent virus	SoPLV		*Carlavirus*
Sowbane mosaic virus	SoMV		*Sobemovirus*
Sowthistle yellow vein virus	SYVV	*Rhabdoviridae*	*Nucleorhabdovirus*
Soybean chlorotic mottle virus	SbCMV	*Caulimoviridae*	'Soybean chlorotic mottle-like viruses'
Soybean crinkle leaf virus	SCLV	*Geminiviridae*	*Begomovirus*
Soybean dwarf virus	SbDV	*Luteoviridae*	Unassigned
Soybean mosaic virus	SMV	*Potyviridae*	*Potyvirus*
Spartina mottle virus	SpMV	*Potyviridae*	*Rymovirus*
Spinach latent virus	SpLV	*Bromoviridae*	*Ilarvirus*
Spinach temperate virus	SpTV	*Partitiviridae*	*Alphacryptovirus*
Spring beauty latent virus	SBLV	*Bromoviridae*	*Bromovirus*
Squash leaf curl virus	SLCV	*Geminiviridae*	*Begomovirus*
Squash leaf curl virus – China	SLCV-Ch	*Geminiviridae*	*Begomovirus*
Squash mosaic virus	SqMV	*Comoviridae*	*Comovirus*
Sri Lankan passion fruit mottle virus	SLPMoV	*Potyviridae*	*Potyvirus*
Strawberry crinkle virus	SCV	*Rhabdoviridae*	*Cytorhabdovirus*
Strawberry latent ringspot virus	SLRSV	*Comoviridae*	*Nepovirus*
Strawberry mild yellow edge virus	SMYEV		*Potexvirus*
Strawberry pseudo mild yellow edge virus	SPMYEV		*Carlavirus*
Strawberry vein banding virus	SVBV	*Caulimoviridae*	*Caulimovirus*
Subterranean clover mottle virus	SCMoV		*Sobemovirus*
Subterranean clover stunt virus	SCSV		*Nanovirus*
Sugarcane bacilliform virus	SCBV	*Caulimoviridae*	*Badnavirus*
Sugarcane mild mosaic virus	SMMV	*Closteroviridae*	*Closterovirus*
Sugarcane mosaic virus	SCMV	*Potyviridae*	*Potyvirus*
Sugarcane streak virus	SSV	*Geminiviridae*	*Mastrevirus*
Sunflower crinkle virus	SuCV		*Umbravirus*
Sunflower mosaic virus	SuMV	*Potyviridae*	*Potyvirus*
Sunflower yellow blotch virus	SuYBV		*Umbravirus*
Sunn-hemp mosaic virus	SHMV		*Tobamovirus*
Sweet clover necrotic mosaic virus	SCNMV	*Tombusviridae*	*Dianthovirus*
Sweet potato chlorotic stunt virus	SPCSV	*Closteroviridae*	*Crinivirus*
Sweet potato feathery mottle virus	SPFMV	*Potyviridae*	*Potyvirus*
Sweet potato latent virus	SPLV	*Potyviridae*	*Potyvirus*
Sweet potato leaf speckling virus	SPLSV	*Luteoviridae*	Unassigned
Sweet potato mild mottle virus	SPMMV	*Potyviridae*	*Ipomovirus*
Sweet potato mild speckling virus	SPMSV	*Potyviridae*	*Potyvirus*
Sweet potato vein mosaic virus	SPVMV	*Potyviridae*	*Potyvirus*

Plant virus name	Abbreviation	Family	Genus[a]
Sweet potato yellow dwarf virus	SPYDV	*Potyviridae*	*Ipomovirus*
Sword bean distortion mosaic virus	SBDMV	*Potyviridae*	*Potyvirus*
Taino tomato mottle virus	TToMoV	*Geminiviridae*	*Begomovirus*
Tamarillo mosaic virus	TamMV	*Potyviridae*	*Potyvirus*
Tamus red mosaic virus	TRMV		*Potexvirus*
Taro bacilliform virus	TaBV	*Caulimoviridae*	*Badnavirus*
Taro feathery mottle virus	TFMoV	*Potyviridae*	*Potyvirus*
Teasel mosaic virus	TeaMV	*Potyviridae*	*Potyvirus*
Telfairia mosaic virus	TeMV	*Potyviridae*	*Potyvirus*
Tephrosia symptomless virus	TeSV	*Tombusviridae*	*Carmovirus*
Thistle mottle virus	ThMoV	*Caulimoviridae*	*Caulimovirus*
Tobacco bushy top virus	TBTV		*Umbravirus*
Tobacco etch virus	TEV	*Potyviridae*	*Potyvirus*
Tobacco leaf curl virus	TLCV	*Geminiviridae*	*Begomovirus*
Tobacco mild green mosaic virus	TMGMV		*Tobamovirus*
Tobacco mosaic virus	TMV		*Tobamovirus*
Tobacco mottle virus	TMoV		*Umbravirus*
Tobacco necrosis virus A	TNV-A	*Tombusviridae*	*Necrovirus*
Tobacco necrosis virus D	TNV-D	*Tombusviridae*	*Necrovirus*
Tobacco necrotic dwarf virus	TNDV	*Luteoviridae*	Unassigned
Tobacco rattle virus	TRV		*Tobravirus*
Tobacco ringspot virus	TRSV	*Comoviridae*	*Nepovirus*
Tobacco streak virus	TSV	*Bromoviridae*	*Ilarvirus*
Tobacco stunt virus	TStV		*Varicosavirus*
Tobacco vein banding mosaic virus	TVBMV	*Potyviridae*	*Potyvirus*
Tobacco vein mottling virus	TVMV	*Potyviridae*	*Potyvirus*
Tobacco wilt virus	TWV	*Potyviridae*	*Potyvirus*
Tobacco yellow dwarf virus	TYDV	*Geminiviridae*	*Mastrevirus*
Tobacco yellow vein virus	TYVV		*Umbravirus*
Tomato apical stunt viroid	TASVd	*Pospiviroidae*	*Pospiviroid*
Tomato aspermy virus	TAV	*Bromoviridae*	*Cucumovirus*
Tomato black ring virus	TBRV	*Comoviridae*	*Nepovirus*
Tomato bunchy top viroid	ToBTVd	Viroid	Unassigned
Tomato bushy stunt virus	TBSV	*Tombusviridae*	*Tombusvirus*
Tomato chlorosis virus	ToCV	*Closteroviridae*	*Crinivirus*
Tomato chlorotic spot virus	TCSV	*Bunyaviridae*	*Tospovirus*
Tomato golden mosaic virus	TGMV	*Geminiviridae*	*Begomovirus*
Tomato infectious chlorosis virus	TICV	*Closteroviridae*	*Crinivirus*
Tomato leaf curl virus – Australia	ToLCV-Au	*Geminiviridae*	*Begomovirus*
Tomato leaf curl virus – Bangalore I	ToLCV-BanI	*Geminiviridae*	*Begomovirus*
Tomato leaf curl virus – Bangalore II	ToLCV-BanII	*Geminiviridae*	*Begomovirus*
Tomato leaf curl virus – New Delhi	ToLCV-NDe	*Geminiviridae*	*Begomovirus*
Tomato leaf curl virus – Senegal	ToLCV-Sn	*Geminiviridae*	*Begomovirus*
Tomato leaf curl virus – Taiwan	ToLCV-Tw	*Geminiviridae*	*Begomovirus*
Tomato leaf curl virus – Tanzania	ToLCV-Tz	*Geminiviridae*	*Begomovirus*
Tomato leafroll virus	TLRV	*Geminiviridae*	*Curtovirus*
Tomato mosaic virus	ToMV		*Tobamovirus*
Tomato mottle virus	ToMoV	*Geminiviridae*	*Begomovirus*

(continued)

TABLE I Alphabetical list of plant virus names and their abbreviations (*Continued*)

Plant virus name	Abbreviation	Family	Genus[a]
Tomato planta macho viroid	TPMVd	*Pospiviroidae*	*Pospiviroid*
Tomato pseudo-curly top virus	TPCTV	*Geminiviridae*	*Curtovirus*
Tomato ringspot virus	ToRSV	*Comoviridae*	*Nepovirus*
Tomato severe leaf curl virus	ToSLCV	*Geminiviridae*	*Begomovirus*
Tomato spotted wilt virus	TSWV	*Bunyaviridae*	*Tospovirus*
Tomato top necrosis virus	ToTNV	*Comoviridae*	*Nepovirus*
Tomato yellow dwarf virus	ToYDV	*Geminiviridae*	*Begomovirus*
Tomato yellow leaf curl virus – China	TYLCV-Ch	*Geminiviridae*	*Begomovirus*
Tomato yellow leaf curl virus – Israel	TYLCV-Is	*Geminiviridae*	*Begomovirus*
Tomato yellow leaf curl virus – Nigeria	TYLCV-Ng	*Geminiviridae*	*Begomovirus*
Tomato yellow leaf curl virus – Sardinia	TYLCV-Sar	*Geminiviridae*	*Begomovirus*
Tomato yellow leaf curl virus – Southern Saudi Arabia	TYLCV-SSA	*Geminiviridae*	*Begomovirus*
Tomato yellow leaf curl virus – Tanzania	TYLCV-Tz	*Geminiviridae*	*Begomovirus*
Tomato yellow leaf curl virus – Thailand	TYLCV-Th	*Geminiviridae*	*Begomovirus*
Tomato yellow leaf curl virus – Yemen	TYLCV-Ye	*Geminiviridae*	*Begomovirus*
Tomato yellow mosaic virus	ToYMV	*Geminiviridae*	*Begomovirus*
Tomato yellow mottle virus	ToYMoV	*Geminiviridae*	*Begomovirus*
Tomato yellow vein streak virus	ToYVSV	*Geminiviridae*	*Begomovirus*
Tongan vanilla virus	TVV	*Potyviridae*	*Potyvirus*
Tradescantia mosaic virus	TraMV	*Potyviridae*	*Potyvirus*
Trichosanthes mottle virus	TrMoV	*Potyviridae*	*Potyvirus*
Triticum aestivum chlorotic spot virus	TACSV	*Rhabdoviridae*	*Nucleorhabdovirus*
Tropaeolum mosaic virus	TrMV	*Potyviridae*	*Potyvirus*
Tropaeolum virus 1	TV-1	*Potyviridae*	*Potyvirus*
Tropaeolum virus 2	TV-2	*Potyviridae*	*Potyvirus*
Tuberose mild mosaic virus	TuMMV	*Potyviridae*	*Potyvirus*
Tulare apple mosaic virus	TAMV	*Bromoviridae*	*Ilarvirus*
Tulip band breaking virus	TBBV	*Potyviridae*	*Potyvirus*
Tulip breaking virus	TBV	*Potyviridae*	*Potyvirus*
Tulip mild mottle mosaic virus	TMMMV		*Ophiovirus*
Tulip virus X	TVX		*Potexvirus*
Turnip crinkle virus	TCV	*Tombusviridae*	*Carmovirus*
Turnip mosaic virus	TuMV	*Potyviridae*	*Potyvirus*
Turnip rosette virus	TRoV		*Sobemovirus*
Turnip vein clearing virus	TVCV		*Tobamovirus*
Turnip yellow mosaic virus	TYMV		*Tymovirus*
Ullucus mild mottle virus	UMMV		*Tobamovirus*
Ullucus mosaic virus	UMV	*Potyviridae*	*Potyvirus*
Ullucus virus C	UVC	*Comoviridae*	*Comovirus*
Urochloa hoja blanca virus	UHBV		*Tenuivirus*
Vallota mosaic virus	ValMV	*Potyviridae*	*Potyvirus*
Vanilla mosaic virus	VanMV	*Potyviridae*	*Potyvirus*
Velvet tobacco mottle virus	VTMoV		*Sobemovirus*
Vicia cryptic virus	VCV	*Partitiviridae*	*Alphacryptovirus*
Vigna sinensis mosaic virus	VSMV	*Rhabdoviridae*	*Nucleorhabdovirus*
Viola mottle virus	VMoV		*Potexvirus*

Plant virus name	Abbreviation	Family	Genus[a]
Voandzeia necrotic mosaic virus	VNMV		*Tymovirus*
Watercress yellow spot virus	WYSV		Unassigned
Watermelon bud necrosis virus	WBNV	*Bunyaviridae*	*Tospovirus*
Watermelon chlorotic stunt virus	WmCSV	*Geminiviridae*	*Begomovirus*
Watermelon curly mottle virus	WmCMV	*Geminiviridae*	*Begomovirus*
Watermelon mosaic virus	WMV	*Potyviridae*	*Potyvirus*
Watermelon silver mottle virus	WSMoV	*Bunyaviridae*	*Tospovirus*
Weddel waterborne virus	WWBV	*Tombusviridae*	*Carmovirus*
Wheat American striate mosaic virus	WASMV	*Rhabdoviridae*	*Cytorhabdovirus*
Wheat chlorotic streak virus	WCSV	*Rhabdoviridae*	*Nucleorhabdovirus*
Wheat dwarf virus	WDV	*Geminiviridae*	*Mastrevirus*
Wheat rosette stunt virus	WRSV	*Rhabdoviridae*	*Nucleorhabdovirus*
Wheat spindle streak mosaic virus	WSSMV	*Potyviridae*	*Bymovirus*
Wheat streak mosaic virus	WSMV	*Potyviridae*	*Tritimovirus*
Wheat yellow leaf virus	WYLV	*Closteroviridae*	*Closterovirus*
Wheat yellow mosaic virus	WYMV	*Potyviridae*	*Bymovirus*
White bryony mosaic virus	WBMV		*Carlavirus*
White bryony virus	WBV	*Potyviridae*	*Potyvirus*
White clover cryptic virus 1	WCCV-1	*Partitiviridae*	*Alphacryptovirus*
White clover cryptic virus 2	WCCV-2	*Partitiviridae*	*Betacryptovirus*
White clover cryptic virus 3	WCCV-3	*Partitiviridae*	*Alphacryptovirus*
White clover mosaic virus	WClMV		*Potexvirus*
White clover virus L	WClVL		Unassigned
Wild cucumber mosaic virus	WCMV		*Tymovirus*
Wild potato mosaic virus	WPMV	*Potyviridae*	*Potyvirus*
Winter wheat mosaic virus	WWMV		*Tenuivirus*
Winter wheat Russian mosaic virus	WWRMV	*Rhabdoviridae*	*Nucleorhabdovirus*
Wissadula golden mosaic virus	WGMV	*Geminiviridae*	*Begomovirus*
Wisteria vein mosaic virus	WVMV	*Potyviridae*	*Potyvirus*
Wound tumor virus	WTV	*Reoviridae*	*Phytoreovirus*
Yam mosaic virus	YMV	*Potyviridae*	*Potyvirus*
Youcai mosaic virus	YoMV		*Tobamovirus*
Yucca bacilliform virus	YBV	*Caulimoviridae*	*Badnavirus*
Zea mays virus	ZMV	*Rhabdoviridae*	*Nucleorhabdovirus*
Zinnia leaf curl virus	ZiLCV	*Geminiviridae*	*Begomovirus*
Zoysia mosaic virus	ZoMV	*Potyviridae*	*Potyvirus*
Zucchini lethal chlorosis virus	ZLCV	*Bunyaviridae*	*Tospovirus*
Zucchini yellow fleck virus	ZYFV	*Potyviridae*	*Potyvirus*
Zucchini yellow mosaic virus	ZYMV	*Potyviridae*	*Potyvirus*
Zygocactus symptomless virus	ZSLV		*Potexvirus*

[a]Species or agents not assigned to a particular genus are shown as 'unassigned' with an indication under 'family' as to their broader classification.

TABLE II Discrepancies in plant name abbreviations used in plant virus names

A and Ar for artichoke

A and Ant for *Anthoxanthum*

A and Ant for *Anthriscus*

A and Ap (once) for apple

A and Aq for *Aquilegia*

B, Ba and Bar (once) for barley

B and Be (once) for bean

B and Bt (once) for beet

B and Bl for blueberry

C and Car (once) for carnation. Car is used only once, for the type species of the genus *Carmovirus*

C and Ct for carrot

C and Cas for *Casssia*

C and Ch for chicory

C and Ci for citrus

C and Cl for *Clitoria*

C and Co for cocksfoot (as Co is also used for coffee and cole, Ck could be the new abbreviation)

C and CP for cowpea

C and Cu for cucumber

D and Des (once) for *Desmodium*

D (once) and Di for *Digitaria*

E (once) and Euo for *Euonimus*

F and Fre for *Freesia*

H and Hn (once) for honeysuckle

H and Hp for hop

L and Lyc for *Lychnis*

M and Ma for *Malva*

M and Mb for mungbean

N and Na (once) for *Nandina*

N and Ne (once) for *Nerine*

P and Pan (once) for *Panicum*

P, Pa (once) and Pap (once) for papaya

P (once) and Par for parsnip

P (once) and PF for passion fruit

P and Pe (once) for pea

P and Pe for peanut

P and Pel for *Pelargonium*

P and Pep for pepper

P and Pet for *Petunia*

P and PP for pigeon pea

Pl used for *Plantago* and plantain (i.e. the scientific and vernacular name of the same host)

P and Pop for poplar

P and Pru (once) for *Prunus*

R (once) and Ra for radish

R (once) and Ran (once) for *Ranunculus*

R and Rp (once) for raspberry

S and Sh for shallot

S and Son (once) for *Sonchus*

S and So (once) for southern

S and Sb for soybean

S and Sq (once) for squash

T and To for tomato

T and Tra (once) for *Tradescantia*

T and Tu (once) for turnip

Wm and W for watermelon

TABLE III List of international country abbreviation names

Afghanistan	AF	Comoros	KM
Albania	AL	Congo	CG
Algeria	DZ	Cook Islands	CK
American Samoa	AS	Costa Rica	CR
Andorra	AD	Cote d'Ivoire (Ivory Coast)	CI
Angola	AO	Croatia (Hrvatska)	HR
Anguilla	Al	Cuba	Cu
Antarctica	AQ	Cyprus	CY
Antigua and Barbuda	AG	Czech Republic	CZ
Argentina	AR	Denmark	DK
Armenia	AM	Djibouti	DJ
Aruba	AW	Dominica	OM
Australia	AU	Dominican Republic	DO
Austria	AT	East Timor	TP
Azerbaijan	AZ	Ecuador	EC
Bahamas	BS	Egypt	EG
Bahrain	BH	El Salvador	SV
Bangladesh	BD	Equatorial Guinea	GO
Barbados	BB	Eritrea	ER
Belarus	BY	Estonia	EE
Belgium	BE	Ethiopia	ET
Belize	BZ	Falkland Islands (Malvinas)	FK
Benin	BJ	Faroe Islands	FO
Bermuda	BM	Fiji	FJ
Bhutan	BT	Finland	Fl
Bolivia	BO	France	FR
Bosnia and Herzegovina	BA	France, Metropolitan	FX
Botswana	EW	French Guiana	GF
Bouvet Island	BV	French Polynesia	PF
Brazil	BR	French Southern Territories	TF
British Indian Ocean Territory	IO	Gabon	GA
Brunei Darussalam	BN	Gambia	GM
Bulgaria	BG	Georgia	GE
Burkina Faso	BF	Germany	DE
Burundi	BI	Ghana	GH
Cambodia	KH	Gibraltar	GI
Cameroon	CM	Greece	GR
Canada	CA	Greenland	GL
Cape Verde	CV	Grenada	GD
Cayman Islands	KY	Guadeloupe	GP
Central African Republic	CF	Guam	GU
Chad	TD	Guatemala	GT
Chile	CL	Guinea	GN
China	CN	Guinea-Bissau	GW
Christmas Island	CX	Guyana	GY
Cocos (Keeling Islands)	CC	Haiti	HT
Colombia	CO	Heard and McDonald Islands	HM

(*continued*)

TABLE III List of international country abbreviation names (*Continued*)

Honduras	HN	Montserrat	MS
Hong Kong	HK	Morocco	MA
Hungary	HU	Mozambique	MZ
Iceland	IS	Myanmar	MM
India	IN	Namibia	NA
Indonesia	ID	Nauru	NR
Iran	IR	Nepal	NP
Iraq	10	Netherlands	NL
Ireland	IE	Netherlands Antilles	AN
Israel	IL	New Caledonia	NC
Italy	IT	New Zealand	NZ
Jamaica	JM	Nicaragua	NI
Japan	JR	Niger	NE
Jordan	JO	Nigeria	NG
Kazakhstan	KZ	Niue	NU
Kenya	KE	Norfolk Island	NF
Kiribati	KI	Northern Mariana Islands	MR
Korea (North)	KP	Norway	NO
Korea (South)	KR	Oman	OM
Kuwait	KW	Pakistan	PK
Kyrgyzstan	KG	Palau	RW
Laos	LA	Panama	RA
Latvia	LV	Papua New Guinea	PG
Lebanon	LB	Paraguay	PY
Lesotho	LS	Peru	RE
Liberia	LR	Philippines	PH
Libya	LY	Pitcairn	PN
Liechtenstein	LI	Poland	PL
Lithuania	LT	Portugal	PT
Luxembourg	LU	Puerto Rico	PR
Macau	MO	Qatar	QA
Macedonia	MK	Reunion	RE
Madagascar	MG	Romania	RO
Malawi	MW	Russian Federation	RU
Malaysia	MY	Rwanda	RW
Maldives	MV	Saint Kitts and Nevis	KN
Mali	ML	Saint Lucia	LC
Malta	MT	Saint Vincent and The Grenadines	VC
Marshall Islands	MH	Samoa	WS
Martinique	MO	San Marino	SM
Mauritania	MR	Sao Tome and Principe	ST
Mauritius	MU	Saudi Arabia	SA
Mayotte	YT	Senegal	SN
Mexico	MX	Seychelles	SC
Micronesia	FM	Sierra Leone	SL
Moldova	MD	Singapore	SG
Monaco	MC	Slovak Republic	SK
Mongolia	MN	Slovenia	SI

Solomon Islands	SB	Turkmenistan	TM
Somalia	SO	Turks and Caicos Islands	TC
South Africa	ZA	Tuvalu	TV
S. Georgia and S. Sandwich Islands	GS	Uganda	UG
Spain	ES	Ukraine	UA
Sri Lanka	LK	United Arab Emirates	AE
St. Helena	SH	United Kingdom	UK
St. Pierre and Miquelon	PM	United States	US
Sudan	SD	US Minor Outlying Islands	UM
Suriname	SR	Uruguay	UY
Svalbard and Jan Mayen Islands	SJ	Uzbekistan	UZ
Swaziland	SZ	Vanuatu	VU
Sweden	SE	Vatican City State (Holy City)	VA
Switzerland	CH	Venezuela	VE
Syria	SY	Vietnam	VN
Taiwan	TW	Virgin Islands (British)	VG
Tajikistan	TJ	Virgin Islands (US)	VI
Tanzania	TZ	Wallis and Futuna Islands	WF
Thailand	TH	Western Sahara	EH
Togo	TG	Yemen	YE
Tokelau	TK	Yugoslavia	YU
Tonga	TO	Zaire	ZR
Trinidad and Tobago	TT	Zambia	ZM
Tunisia	TN	Zimbabwe	ZW
Turkey	TR		

Appendix 1B

Alphabetical list of plant virus abbreviations and viruses

Abbreviation	Virus	Abbreviation	Virus
AARSV	*Artichoke Aegean ringspot virus*	APLPV	*American plum line pattern virus*
AbMV	*Abutilon mosaic virus*	APLV	*Andean potato latent virus*
AbYV	*Abutilon yellows virus*	ApMV	*Apple mosaic virus*
ABV	*Aglaonema bacilliform virus*	APMoV	*Andean potato mottle virus*
ACDV	Artichoke curly dwarf virus	AqNRSV	Aquilegia necrotic ringspot virus
ACLSV	*Apple chlorotic leafspot virus*		
ACMV	*African cassava mosaic virus*	AREV	*Althea rosea enation virus*
ACV-1	*Alfalfa cryptic virus 1*	ArjMV	*Araujia mosaic virus*
ACV-2	Alfalfa cryptic virus 2	ArLV	*Artichoke latent virus*
ADFVd	*Apple dimple fruit viroid*	ArLVM	Artichoke latent virus M
AFCVd	*Apple fruit crinkle viroid*	ArLVS	Artichoke latent virus S
AGMV	*Asystasia golden mosaic virus*	ArMV	*Arabis mosaic virus*
AGMoV	Asystasia gangetica mottle virus	ASBVd	*Avocado sunblotch viroid*
		ASGV	*Apple stem grooving virus*
AgMV	*Agropyron mosaic virus*	ASPV	*Apple stem pitting virus*
AGVd	*Australian grapevine viroid*	ASSVd	*Apple scar skin viroid*
AHLV	*American hop latent virus*	AtBV	Atropa belladonna virus
AILV	*Artichoke Italian latent virus*	AuBV	Aucuba bacilliform virus
ALBV	*Anthoxanthum latent blanching virus*	AV-1	*Asparagus virus 1*
		AV-2	*Asparagus virus 2*
AliMV	Amazon lily mosaic virus	AV-3	*Asparagus virus 3*
AlMV	*Alstroemeria mosaic virus*	AVA	*Arracacha virus A*
AlpMV	Alpinia mosaic virus	AVB	Arracacha virus B
AlStV	Alstroemeria streak virus	AVBV	Artichoke vein banding virus
AltMV	Alternanthera mosaic virus	AVY	Arracacha virus Y
ALV	Arracacha latent virus	AYRSV	*Artichoke yellow ringspot virus*
AMCV	*Artichoke mottled crinkle virus*	AWBV	*Ahlum waterborne virus*
AmLMV	*Amaranthus leaf mottle virus*	AWSV	Alligatorweed stunting virus
AMV	*Alfalfa mosaic virusx*	AYMV	*Acalypha yellow mosaic virus*
AneMV	Aneilema mosaic virus	AYV	*Anthriscus yellows virus*
ANMV	Aquilegia necrotic mosaic virus	AYVV	*Ageratum yellow vein virus*
		BaMMV	*Barley mild mosaic virus*
AntLV	Anthriscus latent virus	BaMV	*Bamboo mosaic virus*
AntMV	Anthoxanthum mosaic virus	BarV-B1	*Barley virus B1*

Abbreviation	Virus	Abbreviation	Virus
BaSV	Bajra streak virus	BRAV	*Blackcurrant reversion associated virus*
BaYMV	*Barley yellow mosaic virus*		
BBMV	*Broad bean mottle virus*	BraYSV	Brachypodium yellow streak virus
BBNV	*Broad bean necrosis virus*		
BBrMV	*Banana bract mosaic virus*	BRMV	*Bean rugose mosaic virus*
BBSV	*Broad bean stain virus*	BrmYMV	Bramble yellow mosaic virus
BBTMV	*Broad bean true mosaic virus*	BRNV	Black raspberry necrosis virus
BBTV	*Banana bunchy top virus*	BRRV	*Blueberry red ringspot virus*
BBWV-1	*Broad bean wilt virus 1*	BrSMV	*Bromus striate mosaic virus*
BBWV-2	*Broad bean wilt virus 2*	BRSV	*Beet ringspot virus*
BcaMV	*Bean calico mosaic virus*	BryMoV	Bryonia mottle virus
BCMNV	*Bean common mosaic necrosis virus*	BSBMV	*Beet soil-borne mosaic virus*
		BSBV	*Beet soil-borne virus*
BCMV	*Bean common mosaic virus*	BSMV	*Barley stripe mosaic virus*
BCTV-Cal	*Beet curly top virus – California/Logan*	BSSV	*Blueberry shoestring virus*
		BStV	*Brome streak virus*
BCTV	*Beet curly top virus*	BSV	*Banana streak virus*
BCTV-CFH	*Beet curly top virus – Iran/CFH*	BtMV	*Beet mosaic virus*
BCTV-Wor	*Beet curly top virus – Worland*	BuSVd	*Burdock stunt viroid*
BCV-1	*Beet cryptic virus 1*	ButMV	Butterbur mosaic virus
BCV-2	*Beet cryptic virus 2*	BuYV	*Burdock yellows virus*
BCV-3	*Beet cryptic virus 3*	BVQ	*Beet virus Q*
BDMV	*Bean dwarf mosaic virus*	BWSpV	Brazilian wheat spike virus
BELV	*Bermuda grass etched-line virus*	BWYV	*Beet western yellows virus*
BeMV	*Belladonna mottle virus*	BYDV-GPV	*Barley yellow dwarf virus – GPV*
BeYDV	*Bean yellow dwarf virus*	BYDV-MAV	*Barley yellow dwarf virus – MAV*
BGMV-Br	*Bean golden mosaic virus – Brazil*	BYDV-PAV	*Barley yellow dwarf virus – PAV*
BGMV-PR	*Bean golden mosaic virus – Puerto Rico*	BYDV-RMV	*Barley yellow dwarf virus – RMV*
BiMoV	*Bidens mottle virus*	BYDV-SGV	*Barley yellow dwarf virus – SGV*
BiMV	Bidens mosaic virus	BYMV	*Bean yellow mosaic virus*
BLCV	Beet leaf curl virus	BYSMV	*Barley yellow striate mosaic virus*
BLMoV	*Blueberry leaf mottle virus*		
BlMVd-RNA	Blueberry mosaic viroid-like RNA	BYSV	*Beet yellow stunt virus*
		BYV	*Beet yellows virus*
BlScV	*Blueberry scorch virus*	BYVBV	*Bean yellow vein-banding virus*
BlShV	*Blueberry shock virus*	BYVMV	*Bhendi yellow vein mosaic virus*
BLRV	*Bean leafroll virus*	CABMV	*Cowpea aphid-borne mosaic virus*
BMV	*Brome mosaic virus*	CABYV	*Cucurbit aphid-borne yellows virus*
BMMV	*Bean mild mosaic virus*		
BmoV	Blackgram mottle virus	CalMMV	*Calanthe mild mosaic virus*
BMYV	*Beet mild yellowing virus*	CaLV	Cardamine latent virus
BNYV	*Broccoli necrotic yellows virus*	CalYVV	*Calopogonium yellow vein virus*
BNYVV	*Beet necrotic yellow vein virus*	CaMV	*Cauliflower mosaic virus*
BolVX	Boletus virus X	CapLV	*Caper latent virus*
BPMV	*Bean pod mottle virus*	CarMV	*Carnation mottle virus*
BPYV	*Beet pseudoyellows virus*	CasMMV	Cassia mild mosaic virus

(continued)

Alphabetical list of plant virus abbreviations and viruses (*Continued*)

Abbreviation	Virus	Abbreviation	Virus
CasYSV	Cassia yellow spot virus	ChYNMV	Chinese yam necrotic mosaic virus
CawLV	Caraway latent virus		
CaYMV	*Canna yellow mottle virus*	ClEV	Clover enation virus
CBDV	Colocasia bobone disease virus	CiLV	Citrus leprosis virus
		CIRV	*Carnation Italian ringspot virus*
CBLVd	*Citrus bent leaf viroid*	CLCrV	*Cotton leaf crumple virus*
CBV	Carnation bacilliform virus	CLCuV-Pk1	*Cotton leaf curl virus – Pakistan 1*
CbVd-1	*Coleus blumei viroid 1*		
CbVd-2	*Coleus blumei viroid 2*	CLCuV-Pk2	*Cotton leaf curl virus – Pakistan 2*
CbVd-3	*Coleus blumei viroid 3*		
CCCV	Callistephus chinensis chlorosis virus	CiLRV	*Citrus leaf rugose virus*
		CLRV	*Cherry leafroll virus*
CCCVd	*Coconut cadang-cadang viroid*	CLSV	Cucumber leafspot virus
CCFV	*Cardamine chlorotic fleck virus*	CLV	*Carnation latent virus*
CChMVd	*Chrysanthemum chlorotic mottle viroid*	CLVd	*Columnea latent viroid*
		ClYMV	*Clover yellow mosaic virus*
CCLV	*Crimson clover latent virus*	ClYVV	*Clover yellow vein virus*
CCMoV	Cereal chlorotic mottle virus	CMBV	*Citrus mosaic virus*
CCMV	*Cowpea chlorotic mottle virus*	CMMV	Cocksfoot mild mosaic virus
CCSV	Cucumber chlorotic spot virus	CMoMV	*Carrot mottle mimic virus*
CCV-1	*Carnation cryptic virus 1*	CMoV	*Carrot mottle virus*
CCV-2	Carnation cryptic virus 2	CMV	*Cucumber mosaic virus*
CdMV	*Cardamom mosaic virus*	CNFV	*Carnation necrotic fleck virus*
CdTV	*Chino del tomate virus*	CnMMV	Canavalia maritima mosaic virus
CDV	*Colombian datura virus*		
CeMV	*Celery mosaic virus*	CnMoV	Cynosurus mottle virus
CenMV	Centrosema mosaic virus	CNSV	*Cycas necrotic stunt virus*
CerMV	*Ceratobium mosaic virus*	CoLV	Cole latent virus
CERV	*Carnation etched ring virus*	ComMV	*Commelina mosaic virus*
CEVd	*Citrus exocortis viroid*	CoMV	*Cocksfoot mottle virus*
CeYMV	Celery yellow mosaic virus	ComVX	*Commelina virus X*
CFDV	Coconut foliar decay virus	ComYMV	*Commelina yellow mottle virus*
CFRVV	Coriander feathery red vein virus	CoNV	*Cocoa necrosis virus*
		CoRSV	Coffee ringspot virus
CFV	Chrysanthemum frutescens virus	CPaMV	Cow parsnip mosaic virus
		CpBDV	Chickpea bushy dwarf virus
CGMMV	*Cucumber green mottle mosaic virus*	CpCDV	Chickpea chlorotic dwarf virus
CGRMV	Cherry green ring mottle virus	CpFV	Chickpea filiform virus
CGVBV	*Cowpea green vein banding virus*	CPGMV	*Cowpea golden mosaic virus*
ChaCV	Chara corallina virus	CPMMV	*Cowpea mild mottle virus*
ChiVMV	*Chilli veinal mottle virus*	CPMoV	*Cowpea mottle virus*
ChMV	Chayote mosaic virus	CPMV	*Cowpea mosaic virus*
ChNV	*Chenopodium necrosis virus*	CPRMV	Cowpea rugose mosaic virus
ChYBV	Chicory yellow blotch virus	CpSDaV	*Chickpea stunt disease-associated virus*
ChYMV	*Chicory yellow mottle virus*		

Abbreviation	Virus	Abbreviation	Virus
CPSMV	*Cowpea severe mosaic virus*	CYBV	*Cassia yellow blotch virus*
CPsV	*Citrus psorosis virus*	CYDV-RMV	*Cereal yellow dwarf virus – RMV*
CraV	Cynara virus		
CriMV	Crinum mosaic virus	CYDV-RPV	*Cereal yellow dwarf virus – RPV*
CRLV	Cherry rasp leaf virus	CYLV	*Carrot yellow leaf virus*
CRMV	Canary reed mosaic virus	CymMV	*Cymbidium mosaic virus*
CroCV	Croatian clover virus	CYMoV	Camellia yellow mottle virus
CRSV	*Carnation ringspot virus*	CymRSV	*Cymbidium ringspot virus*
CRV	Cherry rosette virus	CYMV	*Cacao yellow mosaic virus*
CsALV	*Cassava American latent virus*	CynMV	Cynodon mosaic virus
CsCMV	*Cassava common mosaic virus*	CypCSV	Cypripedium chlorotic streak virus
CsGMV	*Cassava green mottle virus*		
CsIBV	*Cassava Ivorian bacilliform virus*	CYSDV	*Cucurbit yellow stunting disorder virus*
CSMV	*Chloris striate mosaic virus*		
CSNV	*Chrysanthemum stem necrosis virus*	CYSV	Carnation yellow stripe virus
		CYV	Clover yellows virus
CsSLV	Cassava symptomless virus	CYVMV	*Croton yellow vein mosaic virus*
CSSV	*Cacao swollen shoot virus*	CYVV	*Clitoria yellow vein virus*
CSV	*Cocksfoot streak virus*	DaLV	*Dandelion latent virus*
CsVC	*Cassava virus C*	DaYMV	*Dandelion yellow mosaic virus*
CsVMV	*Cassava vein mosaic virus*	DBV	*Dioscorea bacilliform virus*
CSVd	*Chrysanthemum stunt viroid*	DDMV	Datura distortion mosaic virus
CsVX	*Cassava virus X*	DesMV	Desmodium mosaic virus
CTeV-1	*Carrot temperate virus 1*	DipMV	Dipladenia mosaic virus
CTeV-2	*Carrot temperate virus 2*	DiSMV	*Digitaria striate mosaic virus*
CTeV-3	*Carrot temperate virus 3*	DiSV	Digitaria striate virus
CTeV-4	*Carrot temperate virus 4*	DLSV	Dendrobium leaf streak virus
CTiVd	*Coconut tinangaja viroid*	DLV	Dioscorea latent virus
CTLV	*Carrot thin leaf virus*	DMMV	Dock mottling mosaic virus
CtLV	Carrot latent virus	DMV	*Dahlia mosaic virus*
CtMV	Carrot mosaic virus	DNV	Datura necrosis virus
CtRLV	*Carrot red leaf virus*	DoYMV	*Dolichos yellow mosaic virus*
CTV	*Citrus tristeza virus*	DsMV	*Dasheen mosaic virus*
CtYMV	Clitoria yellow mosaic virus	DSSV	*Datura shoestring virus*
CuCV	Cucumber cryptic virus	DSV	*Digitaria streak virus*
CuNV	*Cucumber necrosis virus*	DTMV	Datura mosaic virus
CuSBV	*Cucumber soil-borne virus*	DTV	Dioscorea trifida virus
CV	Cestrum virus	DuMV	*Dulcamara mottle virus*
CV-2	*Cactus virus 2*	DuVA	Dulcamara virus A
CVA	*Cherry virus A*	DuVB	Dulcamara virus B
CVB	*Chrysanthemum virus B*	DV-437	Datura virus 437
CVCV	Chrysanthemum vein chlorosis virus	DVCV	Diodea vein chlorosis virus
		DYMoV	*Desmodium yellow mottle virus*
CVd-III	*Citrus III viroid*	DVNV	Dendrobium vein necrosis virus
CVd-IV	*Citrus IV viroid*		
CVMoV	*Carnation vein mottle virus*	DVS	Daphne virus S
CVV	*Citrus variegation virus*	DVX	*Daphne virus X*
CVX	*Cactus virus X*	DVY	Daphne virus Y

(*continued*)

Alphabetical list of plant virus abbreviations and viruses (*Continued*)

Abbreviation	Virus	Abbreviation	Virus
DYVV	*Datura yellow vein virus*	GarCLV	*Garlic common latent virus*
EACMV	*East African cassava mosaic virus*	GarLV	*Garlic latent virus*
		GarMbFV	*Garlic mite-borne filamentous virus*
EFV	Euonymus fasciation virus		
EGMV	Eggplant green mosaic virus	GarMbLV	Garlic mite-borne latent virus
EHBV	*Echinochloa hoja blanca virus*	GarMV	Garlic mosaic virus
ElLV	Elderberry latent virus	GarV-A	*Garlic virus A*
ElSLV	*Elderberry symptomless virus*	GarV-B	*Garlic virus B*
ELVd	*Eggplant latent viroid*	GarV-C	*Garlic virus C*
EMCV	*Eggplant mottled crinkle virus*	GarV-D	*Garlic virus D*
EMDV	*Eggplant mottled dwarf virus*	GarV-X	*Garlic virus X*
EMMV	Eggplant mild mottle virus	GBLV	*Grapevine Bulgarian latent virus*
EMoV	*Elm mottle virus*	GBNV	*Groundnut bud necrosis virus*
EMV	*Eggplant mosaic virus*	GCFSV	*Groundnut chlorotic fan-spot virus*
ENMV	*Endive necrotic mosaic virus*		
EpCV	*Epirus cherry virus*	GCFV	Ginger chlorotic fleck virus
EpYVV	Eupatorium yellow vein virus	GCMV	*Grapevine chrome mosaic virus*
ErLV	*Erysimum latent virus*	GCTV	Garland chrysanthemum temperate virus
ERSV	*Echinochloa ragged stunt virus*		
ESMoV	Eggplant severe mottle virus	GenLV	Gentiana latent virus
EuMV	*Euphorbia mosaic virus*	GeSLV	Gerbera symptomless virus
EuoMV	Euonymus mosaic virus	GEV	*Groundnut eyespot virus*
EuRSV	Euphorbia ringspot virus	GFkV	Grapevine fleck virus
EWSMV	European wheat striate mosaic virus	GFLV	*Grapevine fanleaf virus*
		GGMV	*Guinea grass mosaic virus*
EYMV	Eggplant yellow mosaic virus	GINV	*Grapevine berry inner necrosis virus*
EYVV	*Eclipta yellow vein virus*		
FBNYV	*Faba bean necrotic yellows virus*	GLRaV-1	Grapevine leafroll-associated virus 1
FClLV	*Fragaria chiloensis latent virus*		
FCV	Fescue cryptic virus	GLRaV-2	*Grapevine leafroll-associated virus 2*
FDV	*Fiji disease virus*		
FigLCV	Fig leaf chlorosis virus	GLRaV-3	*Grapevine leafroll-associated virus 3*
FlCV	Flame chlorosis virus		
FLNV	Freesia leaf necrosis virus	GLRaV-4	Grapevine leafroll-associated virus 4
FLSV	*Festuca leaf streak virus*		
FLV	Fuchsia latent virus	GLRaV-5	Grapevine leafroll-associated virus 5
FMMV	Finger millet mosaic virus		
FMV	*Figwort mosaic virus*	GLRaV-6	Grapevine leafroll-associated virus 6
FNSV	Furcraea necrotic streak virus		
FNV	Festuca necrosis virus	GLRaV-7	Grapevine leafroll-associated virus 7
FoMV	*Foxtail mosaic virus*		
FreMV	*Freesia mosaic virus*	GMoV	Glycine mottle virus
FrMV	*Frangipani mosaic virus*	GMV	*Glycine mosaic virus*
FVS	Fig virus S	GoV	Gomphrena virus
GALV	*Grapevine Algerian latent virus*	GRAV	*Groundnut rosette assistor virus*
GaMV	*Galinsoga mosaic virus*	GRSV	*Groundnut ringspot virus*

Abbreviation	Virus	Abbreviation	Virus
GRV	*Groundnut rosette virus*	HyaMV	*Hyacinth mosaic virus*
GSLV	Guar symptomless virus	HyMV	Hypochoeris mosaic virus
GSMV	*Gloriosa stripe mosaic virus*	HYVMV	*Honeysuckle yellow vein mosaic virus*
GTRSV	*Grapevine Tunisian ringspot virus*	ICMV	*Indian cassava mosaic virus*
GVA	*Grapevine virus A*	IFMV	*Iris fulva mosaic virus*
GVB	*Grapevine virus B*	IGLSV	Iris germanica leaf stripe virus
GVC	Grapevine virus C	ILV	Impatiens latent virus
GVD	*Grapevine virus D*	IMMV	*Iris mild mosaic virus*
GyLV	Gynura latent virus	INSV	*Impatiens necrotic spot virus*
GYSV	*Groundnut yellow spot virus*	IPCV	*Indian peanut clump virus*
GYSVd-1	*Grapevine yellow speckle viroid 1*	IPMoV	Indian pepper mottle virus
GYSVd-2	*Grapevine yellow speckle viroid 2*	IrVd-1	*Iresine viroid 1*
HaMV	Habenaria mosaic virus	IsaMV	*Isachne mosaic virus*
HCRSV	*Hibiscus chlorotic ringspot virus*	ISDV	*Indonesian soybean dwarf virus*
HDIV	Hungarian datura innoxia virus	ISMV	*Iris severe mosaic virus*
HdLV	*Hydrangea latent virus*	IToLCV	*Indian tomato leaf curl virus*
HdMV	*Hydrangea mosaic virus*	IVCV	Ivy vein clearing virus
HdRSV	*Hydrangea ringspot virus*	IWSV	Iranian wheat stripe virus
HeMV	Helleborus mosaic virus	IYSV	*Iris yellow spot virus*
HgYMV	*Horsegram yellow mosaic virus*	JGMV	*Johnsongrass mosaic virus*
HiMV	Hippeastrum mosaic virus	JMV	*Jatropha mosaic virus*
HJLV	*Humulus japonicus latent virus*	KGMMV	*Kyuri green mottle mosaic virus*
HLCV	*Hollyhock leaf curl virus*	KLV	*Kalanchoë latent virus*
HLRSV	*Hibiscus latent ringspot virus*	KMV	*Kalanchoë mosaic virus*
HLV	*Heracleum latent virus*	KoMV	*Konjac mosaic virus*
HLYV	Holcus lanatus yellowing virus	KTSV	*Kalanchoë top-spotting virus*
HMV	*Henbane mosaic virus*	KVY	Kennedya virus Y
HnLV	*Honeysuckle latent virus*	KYMV	*Kennedya yellow mosaic virus*
HoMV	*Hordeum mosaic virus*	LALV	*Lucerne Australian latent virus*
HpLV	*Hop latent virus*	LArSV	Launea arborescens stunt virus
HpLVd	*Hop latent viroid*	LASV	Lucerne Australian symptomless virus
HpMV	*Hop mosaic virus*	LBVV	*Lettuce big-vein virus*
HrCTV	*Horseradish curly top virus*	LChV	*Little cherry virus*
HrLV	*Horseradish latent virus*	LCV	*Lettuce chlorosis virus*
HSV	Holcus streak virus	LeMV	*Leonurus mosaic virus*
HSVd	*Hop stunt viroid*	LEV	Lucerne enation virus
HTCV-1	*Hop trefoil cryptic virus 1*	LGMV	*Limabean golden mosaic virus*
HTCV-2	*Hop trefoil cryptic virus 2*	LiCLV	*Lilac chlorotic leafspot virus*
HTCV-3	*Hop trefoil cryptic virus 3*	LiMoV	*Lilac mottle virus*
HTFMoV	Hart's tongue fern mottle virus	LiRMoV	*Lilac ring mottle virus*
HV-6	Heracleum virus 6	LiRSV	Lilac ringspot virus
HVS	*Helenium virus S*	LIYV	*Lettuce infectious yellows virus*
HVX	*Hosta virus X*	LLCV	Lupin leaf curl virus
HVY	*Helenium virus Y*	LMMV	*Lamium mild mosaic virus*
		LMMoV	Lily mild mottle virus

(continued)

Alphabetical list of plant virus abbreviations and viruses (*Continued*)

Abbreviation	Virus	Abbreviation	Virus
LMoV	*Lily mottle virus*	MPVd	*Mexican papita viroid*
LMV	*Lettuce mosaic virus*	MRDV	*Maize rough dwarf virus*
LNV	Lisianthus necrosis virus	MRFV	*Maize rayado fino virus*
LNYV	*Lettuce necrotic yellows virus*	MRMV	*Melon rugose mosaic virus*
LoRV	Lolium ryegrass virus	MRSV	*Mulberry ringspot virus*
LoSNV	Lotus stem necrosis	MSpV	*Maize stripe virus*
LRLV	Laelia red leafspot virus	MSSV	Maize sterile stunt virus
LRSV	*Lychnis ringspot virus*	MSV	*Maize streak virus*
LRV	*Lato river virus*	MTV	Mibuna temperate virus
LSLV	*Lily symptomless virus*	MVBMV	Melon vein-banding mosaic virus
LSMV	*Lettuce speckles mottle virus*		
LSTCV	Lemon scented thyme leaf chlorosis virus	MVCV	*Malva vein clearing virus*
		MuVNV	*Muskmelon vein necrosis virus*
LTSV	*Lucerne transient streak virus*	MVNV	Malva veinal necrosis virus
LVX	*Lily virus X*	MVV	Melon variegation virus
LWSV	*Leek white stripe virus*	MWLMV	Maize white line mosaic virus
LycSLV	Lychnis symptomless virus	MWMV	*Moroccan watermelon mosaic virus*
LYSV	*Leek yellow stripe virus*		
LYVV	Lupin yellow vein virus	MYFV	*Melandrium yellow fleck virus*
MacMV	*Maclura mosaic virus*	MYMV	*Mungbean yellow mosaic virus*
MaMoV	Marigold mottle virus	NaMV	Nandina mosaic virus
MaMV	*Macrotyloma mosaic virus*	NCMV	*Northern cereal mosaic virus*
MarMV	Maracuja mosaic virus	NDV	*Narcissus degeneration virus*
MaSV	Malva silvestris virus	NeCMV	Negro coffee mosaic virus
MbMV	Mungbean mosaic virus	NeLV	*Nerine latent virus*
MBV	Mimosa bacilliform virus	NeYSV	*Nerine yellow stripe virus*
MCDV	*Maize chlorotic dwarf virus*	NGSVd	*Nicotiana glutinosa stunt viroid*
MCMV	*Maize chlorotic mottle virus*	NLSYV	*Narcissus late season yellows virus*
MCV	*Malvaceous chlorosis virus*		
MDMV	*Maize dwarf mosaic virus*	NLV	*Narcissus latent virus*
MDV	*Milk vetch dwarf virus*	NMV	*Narcissus mosaic virus*
MegMV	Megakepasma mosaic virus	NoMV	*Nothoscordum mosaic virus*
MeLV	Melilotus latent virus	NRV	*Neckar river virus*
MeMV	Melilotus mosaic virus	NSPV	Nandina stem pitting virus
MGMV	*Macroptilium golden mosaic virus*	NTNV	Narcissus tip necrosis virus
		NVMV	Nicotiana velutina mosaic virus
MiMV	*Mirabilis mosaic virus*		
MiSV	*Miscanthus streak virus*	NVX	*Nerine virus X*
MLCV	*Melon leaf curl virus*	NVY	Nerine virus Y
MLRSV	*Myrobalan latent ringspot virus*	NV	Nerine virus
MLV	*Mulberry latent virus*	NYSV	*Narcissus yellow stripe virus*
MMoV	Mungbean mottle virus	OBDV	*Oat blue dwarf virus*
MMV	*Maize mosaic virus*	ObPV	Obuda pepper virus
MNSV	*Melon necrotic spot virus*	OCSV	*Oat chlorotic stunt virus*
MoSV	Molinia streak virus	OFV	Orchid fleck virus
MPV	*Moroccan pepper virus*	OkMV	*Okra mosaic virus*

Abbreviation	Virus	Abbreviation	Virus
OLCV	Okra leaf curl virus	PepRSV	Pepper ringspot virus
OLRSV	Olive latent ringspot virus	PepSMV	Pepper severe mosaic virus
OLV-1	Olive latent virus 1	PerMoV	Perilla mottle virus
OLV-2	Olive latent virus 2	PeSV	Pea streak virus
OMbLV	Onion mite-borne latent virus	PetAMV	Petunia asteroid mosaic virus
OMV	Oat mosaic virus	PetFMV	Petunia flower mottle virus
ONMV	Oat necrotic mottle virus	PeYMV	Peanut yellow mosaic virus
OrMV	Ornithogalum mosaic virus	PFBV	Pelargonium flower break virus
ORSV	Odontoglossum ringspot virus	PFMoV	Passion fruit mottle virus
OSDV	Oat sterile dwarf virus	PFRSV	Passion fruit ringspot virus
OSMV	Oat striate mosaic virus	PFYMV	Passion fruit yellow mosaic virus
OuMV	Ourmia melon virus	PGMV	Pea green mottle virus
OYDV	Onion yellow dwarf virus	PhCSV	Phalaenopsis chlorotic spot virus
OYMV	Ononis yellow mosaic virus		
PaLCV	Papaya leaf curl virus	PHV	Pepper huasteco virus
PalMV	Palm mosaic virus	PhyMV	Physalis mottle virus
PaLV	Parsley latent virus	PhySMV	Physalis severe mottle virus
PaMMV	Paprika mild mottle virus	PkMV	Pokeweed mosaic virus
PAMV	Potato aucuba mosaic virus	PlAMV	Plantago asiatica mosaic virus
PanSV	Panicum streak virus	PLCV	Pelargonium leaf curl virus
PapMV	Papaya mosaic virus	PLDMV	Papaya leaf distortion mosaic virus
ParMV	Parsnip mosaic virus		
ParV-3	Parsnip virus 3	PleMV	Pleioblastus mosaic virus
ParV-5	Parsnip virus 5	PlMoV	Plantago mottle virus
PaSV	Pangola stunt virus	PlMV	Plantain mottle virus
PatMMV	Patchouli mild mosaic virus	PLMVd	Peach latent mosaic viroid
PatMoV	Patchouli mottle virus	PLRV	Potato leafroll virus
PatVX	Patchouli virus X	PlSMoV	Plantago severe mottle virus
PaV	Parsley virus	PLV	Passiflora latent virus
PaV-5	Parsley virus 5	PlV-4	Plantago virus 4
PBCVd	Pear blister canker viroid	PlV-6	Plantain virus 6
PBRSV	Potato black ringspot virus	PlV-7	Plantain virus 7
PBV	Pineapple bacilliform virus	PlV-8	Plantain virus 8
PCLSV	Pineapple chlorotic leaf streak virus	PlVX	Plantain virus X
		PMiMV	Pea mild mosaic virus
PcMV	Pecteilis mosaic virus	PMMoV	Pepper mild mottle virus
PCSV	Peanut chlorotic streak virus	PMMV	Pepper mild mosaic virus
PCV	Peanut clump virus	PMoV	Parietaria mottle virus
PDV	Prune dwarf virus	PMTV	Potato mop-top virus
PEBV	Pea early-browning virus	PMV	Panicum mosaic virus
PeGMoV	Peanut green mottle virus	PMWaV-1	Pineapple mealybug wilt-associated virus 1
PeMoV	Peanut mottle virus		
PEMV-1	Pea enation mosaic virus 1	PMWaV-2	Pineapple mealybug wilt-associated virus 2
PEMV-2	Pea enation mosaic virus 2		
PepLCV	Pepper leaf curl virus	PnCV	Poinsettia cryptic virus
PepMoV	Pepper mottle virus	PnMV	Poinsettia mosaic virus
PepMTV	Pepper mild tigré virus	PNRSV	Prunus necrotic ringspot virus
PepMV	Pepino mosaic virus	PoLV	Pothos latent virus

(continued)

Alphabetical list of plant virus abbreviations and viruses (*Continued*)

Abbreviation	Virus	Abbreviation	Virus
PopDV	Poplar decline virus	RCMV	*Red clover mottle virus*
PopMV	*Poplar mosaic virus*	RCNMV	*Red clover necrotic mosaic virus*
PPMMoVd	*Pigeon pea mosaic mottle viroid*	RCSV	Rubus Chinese seed-borne virus
PPPV	Pigeon pea proliferation virus		
PPV	*Plum pox virus*	RCVMV	*Red clover vein mosaic virus*
PRMV	*Peach rosette mosaic virus*	RDV	*Rice dwarf virus*
PrMoV	Primula mottle virus	ReTBV	*Rembrandt tulip breaking virus*
PrMV	Primula mosaic virus	RGCV	*Ryegrass cryptic virus*
PRSV	*Papaya ringspot virus*	RGDV	*Rice gall dwarf virus*
PruVS	Prunus virus S	RGMV	*Ryegrass mosaic virus*
PSbMV	*Pea seed-borne mosaic virus*	RGSV	*Rice grassy stunt virus*
PSLV	*Poa semilatent virus*	RHBV	*Rice hoja blanca virus*
PSMV	*Paspalum striate mosaic virus*	RhMV	*Rhynchosia mosaic virus*
PSTVd	*Potato spindle tuber viroid*	RMV	*Ribgrass mosaic virus*
PSV	*Peanut stunt virus*	RNMV	*Rice necrosis mosaic virus*
PTV	*Peru tomato mosaic virus*	RoNRSV	Rhododendron necrotic ringspot virus
PVA	*Potato virus A*		
PVBV	Pepper vein banding virus	RPCV-1	Red pepper cryptic virus 1
PVCV	*Petunia vein clearing virus*	RPCV-2	Red pepper cryptic virus 2
PVM	*Potato virus M*	RpRSV	*Raspberry ringspot virus*
PVMV	*Pepper veinal mottle virus*	RRSV	*Rice ragged stunt virus*
PVS	*Potato virus S*	RSNV	Rice stripe necrosis virus
PVT	*Potato virus T*	RSV	*Rice stripe virus*
PVU	*Potato virus U*	RTBV	*Rice tungro bacilliform virus*
PVV	*Potato virus V*	RSPaV	*Rupestris stem pitting-associated virus*
PVX	*Potato virus X*		
PVY	*Potato virus Y*	RTSV	*Rice tungro spherical virus*
PWV	*Passion fruit woodiness virus*	RTV	Rhubarb temperate virus
PYDV	*Potato yellow dwarf virus*	RuMV	Rudbeckia mosaic virus
PYFV	*Parsnip yellow fleck virus*	RV-1	Rhubarb virus 1
PYMoV	*Piper yellow mottle virus*	RVCV	Raspberry vein chlorosis virus
PYMV	*Potato yellow mosaic virus*	RWMV	*Ranunculus white mottle virus*
PYVV	*Pseuderanthemum yellow vein virus*	RWSV	Rice wilted stunt virus
		RYEV	*Radish yellow edge virus*
PZSV	Pelargonium zonate spot virus	RYMV	*Rice yellow mottle virus*
QPMV	*Quail pea mosaic virus*	RYSV	*Rice yellow stunt virus*
RaMV	*Radish mosaic virus*	SACMV	*South African cassava mosaic virus*
RanMoV	Ranunculus mottle virus		
RaRSV	Ranunculus repens symptomless virus	SALCV	Solanum apical leaf curl virus
		SbCMV	*Soybean chlorotic mottle virus*
RaV	Raphanus virus	SBDMV	Sword bean distortion mosaic virus
RaVCV	Radish vein clearing virus		
RBDV	*Raspberry bushy dwarf virus*	SbDV	*Soybean dwarf virus*
RBSDV	*Rice black streaked dwarf virus*	SBLV	*Spring beauty latent virus*
RCCV-2	*Red clover cryptic virus 2*	SBMV	*Southern bean mosaic virus*
RCIMV	Red clover mosaic virus	SBRMV	Soil-borne rye mosaic virus

Abbreviation	Virus	Abbreviation	Virus
SBWMV	*Soil-borne wheat mosaic virus*	SpMV	Spartina mottle virus
SCBV	*Sugarcane bacilliform virus*	SPMYEV	*Strawberry pseudo mild yellow edge virus*
SCLV	Soybean crinkle leaf virus		
SCMoV	*Subterranean clover mottle virus*	SpTV	*Spinach temperate virus*
SCMV	*Sugarcane mosaic virus*	SPVMV	Sweet potato vein mosaic virus
SCNMV	*Sweet clover necrotic mosaic virus*		
		SPV	Sarracenia purpurea virus
SCPMV	*Southern cowpea mosaic virus*	SPYDV	*Sweet potato yellow dwarf virus*
ScrMV	*Scrophularia mottle virus*	SqMV	*Squash mosaic virus*
SCSV	*Subterranean clover stunt virus*	SrCSV	*Sorghum chlorotic spot virus*
SCV	*Strawberry crinkle virus*	SrMV	*Sorghum mosaic virus*
SDV	Satsuma dwarf virus	SRV	*Schefflera ringspot virus*
SgCV	*Saguaro cactus virus*	SrV	Sorghum virus
SGMV	*Serrano golden mosaic virus*	SSV	*Sugarcane streak virus*
ShMbLV	Shallot mite-borne latent virus	STLCV	*Sinaloa tomato leaf curl virus*
SHMV	*Sunn-hemp mosaic virus*	SToLCV	*Solanum tomato leaf curl virus*
ShVX	*Shallot virus X*	STV	Santosai temperate virus
SiGMV	*Sida golden mosaic virus*	SuCV	Sunflower crinkle virus
SiYVV	Sida yellow vein virus	SuMV	Sunflower mosaic virus
SJOLV	*Sint-Jem's onion latent virus*	SuYBV	Sunflower yellow blotch virus
SLCV	*Squash leaf curl virus*	SVBV	*Strawberry vein banding virus*
SLCV-Ch	*Squash leaf curl virus – China*	SVCV	Sambucus vein clearing virus
SLNV	Sainpaulia leaf necrosis virus	SWBV	*Sikte waterborne virus*
SLPMoV	Sri Lankan passion fruit mottle virus	SYBV	Soursop yellow blotch virus
		SYLCV	*Solanum yellow leaf curl virus*
SLRSV	Strawberry latent ringspot virus	SYNV	*Sonchus yellow net virus*
		SYSV	*Shallot yellow stripe virus*
SmiLV	Smithiantha latent virus	SYVV	*Sowthistle yellow vein virus*
SMMV	Sugarcane mild mosaic virus	TaBV	*Taro bacilliform virus*
SMoV	Sonchus mottle virus	TACSV	Triticum aestivum chlorotic spot virus
SMV	*Soybean mosaic virus*		
SMYEV	*Strawberry mild yellow edge virus*	TamMV	*Tamarillo mosaic virus*
		TAMV	*Tulare apple mosaic virus*
SNMoV	*Solanum nodiflorum mottle virus*	TASVd	*Tomato apical stunt viroid*
SoMV	*Sowbane mosaic virus*	TAV	*Tomato aspermy virus*
SonV	*Sonchus virus*	TBBV	Tulip band breaking virus
SoPLV	Southern potato latent virus	TBRV	*Tomato black ring virus*
SOV	*Sammons's Opuntia virus*	TBSV	*Tomato bushy stunt virus*
SPCSV	*Sweet potato chlorotic stunt virus*	TBTV	Tobacco bushy top virus
		TBV	*Tulip breaking virus*
SPFMV	*Sweet potato feathery mottle virus*	TCSV	*Tomato chlorotic spot virus*
		TCV	*Turnip crinkle virus*
SPLSV	*Sweet potato leaf speckling virus*	TeaMV	Teasel mosaic virus
SPLV	*Sweet potato latent virus*	TeMV	*Telfairia mosaic virus*
SpLV	*Spinach latent virus*	TeSV	Tephrosia symptomless virus
SPMMV	*Sweet potato mild mottle virus*	TEV	*Tobacco etch virus*
SPMSV	Sweet potato mild speckling virus	TFMoV	Taro feathery mottle virus
		TGMV	*Tomato golden mosaic virus*

(continued)

Alphabetical list of plant virus abbreviations and viruses (*Continued*)

Abbreviation	Virus	Abbreviation	Virus
ThMoV	*Thistle mottle virus*	TSWV	*Tomato spotted wilt virus*
TICV	*Tomato infectious chlorosis virus*	TToMoV	*Taino tomato mottle virus*
TLCV	*Tobacco leaf curl virus*	TuMMV	*Tuberose mild mosaic virus*
TLRV	Tomato leafroll virus	TuMV	*Turnip mosaic virus*
TMGMV	*Tobacco mild green mosaic virus*	TV-1	Tropaeolum virus 1
TMMMV	*Tulip mild mottle mosaic virus*	TV-2	Tropaeolum virus 2
TMoV	*Tobacco mottle virus*	TVBMV	*Tobacco vein banding mosaic virus*
TMV	*Tobacco mosaic virus*		
TMV-Cg	Crucifer strain of TMV	TVCV	Turnip vein-clearing virus
crTMV	Crucifer-infecting TMV	TVMV	*Tobacco vein mottling virus*
TNV-A	*Tobacco necrosis virus A*	TVV	Tongan vanilla virus
TNV-D	*Tobacco necrosis virus D*	TVX	*Tulip virus X*
TNDV	*Tobacco necrotic dwarf virus*	TWV	Tobacco wilt virus
ToBTVd	*Tomato bunchy top viroid*	TYDV	*Tobacco yellow dwarf virus*
ToCV	*Tomato chlorosis virus*	TYLCV-Ch	*Tomato yellow leaf curl virus – China*
ToLCV-Au	*Tomato leaf curl virus – Australia*	TYLCV-Is	*Tomato yellow leaf curl virus – Israel*
ToLCV-BanI	*Tomato leaf curl virus – Bangalore I*	TYLCV-Ng	*Tomato yellow leaf curl virus – Nigeria*
ToLCV-BanII	*Tomato leaf curl virus – Bangalore II*	TYLCV-Sar	*Tomato yellow leaf curl virus – Sardinia*
ToLCV-NDe	*Tomato leaf curl virus – New Delhi*	TYLCV-SSA	*Tomato yellow leaf curl virus – Southern Saudi Arabia*
ToLCV-Sn	*Tomato leaf curl virus – Senegal*	TYLCV-Th	*Tomato yellow leaf curl virus – Thailand*
ToLCV-Tw	*Tomato leaf curl virus – Taiwan*		
ToLCV-Tz	*Tomato leaf curl virus – Tanzania*	TYLCV-Tz	*Tomato yellow leaf curl virus – Tanzania*
ToMoV	*Tomato mottle virus*	TYLCV-Ye	*Tomato yellow leaf curl virus – Yemen*
ToMV	*Tomato mosaic virus*		
ToRSV	*Tomato ringspot virus*	TYMV	*Turnip yellow mosaic virus*
ToSLCV	*Tomato severe leaf curl virus*	TYVV	Tobacco yellow vein virus
ToTNV	*Tomato top necrosis virus*	UHBV	*Urochloa hoja blanca virus*
ToYDV	*Tomato yellow dwarf virus*	UMMV	*Ullucus mild mottle virus*
ToYMoV	*Tomato yellow mottle virus*	UMV	Ullucus mosaic virus
ToYMV	*Tomato yellow mosaic virus*	UVC	*Ullucus virus C*
ToYVSV	*Tomato yellow vein streak virus*	ValMV	Vallota mosaic virus
TPCTV	*Tomato pseudo-curly top virus*	VanMV	Vanilla mosaic virus
TPMVd	*Tomato planta macho viroid*	VCV	*Vicia cryptic virus*
TraMV	Tradescantia mosaic virus	VMoV	Viola mottle virus
TrMoV	*Trichosanthes mottle virus*	VNMV	*Voandzeia necrotic mosaic virus*
TRMV	*Tamus red mosaic virus*	VSMV	Vigna sinensis mosaic virus
TrMV	*Tropaeolum mosaic virus*	VTMoV	*Velvet tobacco mottle virus*
TRoV	*Turnip rosette virus*	WASMV	*Wheat American striate mosaic virus*
TRSV	*Tobacco ringspot virus*		
TRV	*Tobacco rattle virus*		
TStV	Tobacco stunt virus	WBMV	White bryony mosaic virus
TSV	*Tobacco streak virus*		

Abbreviation	Virus	Abbreviation	Virus
WBNV	Watermelon bud necrosis virus	WTV	Wound tumor virus
WBV	White bryony virus	WVMV	Wisteria vein mosaic virus
WCCV-1	White clover cryptic virus 1	WWBV	Weddel waterborne virus
WCCV-2	White clover cryptic virus 2	WWMV	Winter wheat mosaic virus
WCCV-3	White clover cryptic virus 3	WWRMV	Winter wheat Russian mosaic virus
WClMV	White clover mosaic virus		
WClVL	White clover virus L	WYLV	Wheat yellow leaf virus
WCMV	Wild cucumber mosaic virus	WYMV	Wheat yellow mosaic virus
WCSV	Wheat chlorotic streak virus	WYSV	Watercress yellow spot virus
WDV	Wheat dwarf virus	YBV	Yucca bacilliform virus
WGMV	Wissadula golden mosaic virus	YMV	Yam mosaic virus
WmCMV	Watermelon curly mottle virus	YoMV	Youcai mosaic virus
WmCSV	Watermelon chlorotic stunt virus	ZiLCV	Zinnia leaf curl virus
WMV	Watermelon mosaic virus	ZLCV	Zucchini lethal chlorosis virus
WPMV	Wild potato mosaic virus	ZMV	Zea mays virus
WRSV	Wheat rosette stunt virus	ZoMV	Zoysia mosaic virus
WSMoV	Watermelon silver mottle virus	ZSLV	Zygocactus symptomless virus
WSMV	Wheat streak mosaic virus		
WSSMV	Wheat spindle streak mosaic virus	ZYFV	Zucchini yellow fleck virus
		ZYMV	Zucchini yellow mosaic virus

Appendix 2A

Classification of Plant Viruses

Genome[a]	Family	Genus	Type member	No. of spp.	No. of tentative spp.
dsDNA (RT)	Caulimoviridae	Badnavirus	Commelina yellow mottle virus	12	4
		Caulimovirus	Cauliflower mosaic virus	9	4
		'Soybean chlorotic mottle virus-like'	Soybean chlorotic mottle virus	2	0
		'Cassava vein mosaic virus-like'	Cassava vein mosaic virus	1	0
		'Petunia vein clearing virus-like'	Petunia vein clearing virus	1	0
		'Rice tungro bacilliform virus-like'	Rice tungro bacilliform virus	1	0
ssDNA	Geminiviridae	Mastrevirus	Maize streak virus	12	2
		Curtovirus	Beet curly top virus	3	1
		Begomovirus	Bean golden mosaic virus	76	8
	Circoviridae	Nanovirus	Subterranean clover stunt virus	4	1
dsRNA	Reoviridae	Fijivirus	Fiji disease virus	8	0
		Oryzavirus	Rice ragged stunt virus	2	0
		Phytoreovirus	Wound tumor virus	3	1
	Partitiviridae	Alphacryptovirus	White clover cryptic virus 1	16	10
		Betacryptovirus	White clover cryptic virus 2	4	1
	No family	Varicosavirus	Lettuce big-vein virus	1	3
ssRNA (−)	Rhabdoviridae	Cytorhabdovirus	Lettuce necrotic yellows virus	8	0
		Nucleorhabdovirus	Potato yellow dwarf virus	7	0
		Unassigned		58	0
	Bunyaviridae	Tospovirus	Tomato spotted wilt virus	8	5
	No family	Tenuivirus	Rice stripe virus	6	5
		Ophiovirus	Citrus psorosis virus	3	0
ssRNA (+)	Bromoviridae	Bromovirus	Brome mosaic virus	6	0
		Alfamovirus	Alfalfa mosaic virus	1	0
		Cucumovirus	Cucumber mosaic virus	3	0
		Ilarvirus	Tobacco streak virus	17	0
		Oleavirus	Olive latent virus 2	1	0
	Comoviridae	Comovirus	Cowpea mosaic virus	15	0
		Fabavirus	Broad bean wilt virus	4	0
		Nepovirus	Tobacco ringspot virus	31	9
	Potyviridae	Potyvirus	Potato virus Y	91	88
		Ipomovirus	Sweet potato mild mottle virus	1	1
		Macluravirus	Maclura mosaic virus	2	0
		Rymovirus	Ryegrass mosaic virus	4	1
		Tritimovirus	Wheat streak mosaic virus	2	0
		Bymovirus	Barley yellow mosaic virus	6	0
	Tombusviridae	Tombusvirus	Tomato bushy stunt virus	13	0
		Avenavirus	Oat chlorotic stunt virus	1	0
		Aureusvirus	Pothos latent virus	1	0

(continued)

Classification of Plant Viruses (*Continued*)

Genome[a]	Family	Genus	Type member	No. of spp.	No. of tentative spp.
		Carmovirus	*Carnation mottle virus*	12	6
		Dianthovirus	*Carnation ringspot virus*	4	1
		Machlomovirus	*Maize chlorotic mottle virus*	1	0
		Necrovirus	*Tobacco necrosis virus A*	5	2
		Panicovirus	*Panicum mosaic virus*	1	1
	Sequiviridae	*Sequivirus*	*Parsnip yellow fleck virus*	2	0
		Waikavirus	*Rice tungro spherical virus*	3	0
	Closteroviridae	*Closterovirus*	*Beet yellows virus*	11	16
		Crinivirus	*Lettuce infectious yellows virus*	1	0
	Luteoviridae	*Luteovirus*	*Barley yellow dwarf virus* (MAV strain)	2	0
		Polerovirus	*Potato leafroll virus*	5	0
		Enamovirus	*Pea enation mosaic virus 1*	1	0
		Unassigned		11	0
	No family	*Tobamovirus*	*Tobacco mosaic virus*	16	1
		Tobravirus	*Tobacco rattle virus*	3	0
		Potexvirus	*Potato virus X*	26	19
		Carlavirus	*Carnation latent virus*	31	29
		Allexivirus	*Shallot virus X*	7	3
		Capillovirus	*Apple stem grooving virus*	3	1
	No family	*Foveavirus*	*Apple stem pitting virus*	2	1
		Trichovirus	*Apple chlorotic leafspot virus*	3	0
		Vitivirus	*Grapevine virus A*	4	1
		Furovirus	*Soil-borne wheat mosaic virus*	1	4
		Pecluvirus	*Peanut clump virus*	2	0
		Pomovirus	*Potato mop-top virus*	4	0
		Benyvirus	*Beet necrotic yellow vein virus*	2	0
		Hordeivirus	*Barley stripe mosaic virus*	4	0
		Marafivirus	*Maize rayado fino virus*	3	0
		Sobemovirus	*Southern bean mosaic virus*	11	3
		Tymovirus	*Turnip yellow mosaic virus*	21	2
		Idaeovirus	*Raspberry bushy dwarf virus*	1	0
		Ourmiavirus	*Ourmia melon virus*	3	0
		Umbravirus	*Carrot mottle virus*	7	4

[a]Genome type: dsDNA (RT), double-stranded DNA (replicating by reverse transcription); ssDNA, single-stranded DNA; dsRNA, double-stranded RNA; ssRNA (−), (−)-sense RNA; ssRNA (+), (+)-sense RNA (mRNA).
Data from van Regenmortel *et al.* (2000) with permission.

Appendix 2B

Properties of plant viruses

Family	Genus	Host range[a]	Vector[b]	Particle shape[c]	No. of nucleoprotein particles[d]
Caulimoviridae	*Badnavirus*	n	m	b	1
	Caulimovirus	n	as	i	1
	'Soybean chlorotic mottle virus-like'	n	u	i	1
	'Cassava vein mosaic virus-like'	n	u	i	1
	'Petunia vein clearing virus-like'	n	u	i	1
	'Rice tungro bacilliform virus-like'	n	lh	b	1
Geminiviridae	*Mastrevirus*	n	lc	g	1
	Curtovirus	w–n	lc	g	1
	Begomovirus	n	w	g	1 or 2
Circoviridae	*Nanovirus*	n	ac	i	6 to 7
Reoviridae	*Fijivirus*	n	pp	ic	1
	Oryzavirus	n	pp	ic	1
	Phytoreovirus	w–n	lp	ic	1
Partitiviridae	*Alphacryptovirus*	n	s	i	2
	Betacryptovirus	n	s	i	2
No family	*Varicavirus*	n–m	f	r	2
Rhabdoviridae	*Cytorhabdovirus*	n	lp, ap	bc	1
	Nucleorhabdovirus	n	lp, ap	bc	1
	Unassigned	n	lp, ap, mtp	bc	1
Bunyaviridae	*Tospovirus*	w	tp	ic	3
No family	*Tenuivirus*	n	pp	fl*	4
	Ophiovirus	n	u	fl	3
Bromoviridae	*Bromovirus*	n	u	i	3
	Alfalfamovirus	w	an	b	3
	Cucumovirus	n–w	an	i	3
	Ilarvirus	w	s	(i)	3
	Oleavirus	n	u	b	3
Comoviridae	*Comovirus*	n	b	i	2
	Fabavirus	w	an	i	2
	Nepovirus	n–w	n	i	2
Potyviridae	*Potyvirus*	n–m	an, s	f	1
	Ipomovirus	w	w	f	1
	Macluravirus	n	an	f	1
	Rymovirus	n	mt	f	1
	Tritimovirus	n	mt	f	1
	Bymovirus	n	f	f	1
Tombusviridae	*Tombusvirus*	n	u, s, f	i	1
	Avenavirus	n	u	i	1
	Aureusvirus	n	u	i	1

(continued)

Properties of plant viruses (*Continued*)

Family	Genus	Host range[a]	Vector[b]	Particle shape[c]	No. of nucleoprotein particles[d]
	Carmovirus	n	u, s, f	i	1
	Dianthovirus	m	u	i	1
	Machlomovirus	n	s	i	1
	Necrovirus	w	f	i	1
	Panicovirus	n	u	i	1
Sequiviridae	*Sequivirus*	n	as	i	1
	Waikavirus	n	ls	1	1
Closteroviridae	*Closterovirus*	n	as, w, m	f	1
	Crinivirus	w	w	f	2
Luteoviridae	*Luteovirus*	n	ac	i	1
	Polerovirus	n	ac	i	1
	Enamovirus	n	ac	i	1
	Unassigned	n	ac	i	1
No family	*Tobamovirus*	m	c	r	1
	Tobravirus	w	n	r	2
	Potexvirus	n	c	f	1
	Carlavirus	n-m	as, w, s, c	f	1
	Allexivirus	n	mt	f	1
	Capillovirus	n	s	f	1
No family	*Foveavirus*	n	u	f	1
	Trichovirus	n	s	f	1
	Vitivirus	n–m	as, h, m, sc	f	1
	Furovirus	n	f	r	2
	Pecluvirus	n	f, s	r	2
	Pomovirus	n	f	r	2
	Benyvirus	n	f	r	4
	Hordeivirus	n	c, s	r	3
	Marafivirus	n	lp	i	1
	Sobemovirus	n	b, s	i	1
	Tymovirus	n	b	i	1
	Idaeovirus	n–w	s	i	2
	Ourmiavirus	m	u	b	3
	Ourmiavirus	m	u	b	3
	Umbravirus	n	h	ih	1

[a]n, Narrow (1–5 natural hosts); m, medium (6–20 natural hosts); w, wide (more than 20 natural hosts).

[b]ac, Aphid, circulative; an, aphid, non-persistent; ap, aphid, propagative; as, aphid, semi-persistent; b, beetle; c, contact (mechanical); f, fungus; h, requires a helper virus, usually a luteovirus; lc, leafhopper circulative; lh, leafhopper but requiring helper virus; lp, leafhopper propagative; ls, leafhopper, semi-persistent; m, mealybug; mt, mite; mtp, mite, propagative; n, nematode; pp, planthopper propagative; s, seed; sc, scale insect; tp, thrip, propagative; u, unknown; w, whitefly.

[c]b, Bacilliform; bc, complex bacilliform; f, filamentous rod; fl*, flexuous particle, but may be a complex particle; g, geminate; i, isometric; (i), possibly isometric; ic, complex isometric; ih, encapsidated by helper virus giving isometric particle; r, rigid rod.

[d]The number of nucleoprotein particle types containing different genome segments required for full infection.

Data from van Regenmortel *et al.* (2000) with permission.

Appendix 3

Sources of information on plant viruses

JOURNALS

The following list includes the major publications that regularly contain original papers or reviews relevant to plant virology:

- Core journals

  Intervirology
  Journal of General Virology
  Journal of Virology
  Journal of Virological Methods
  Virology

- Journals that publish papers mainly on applied aspects

  Annals of Applied Biology
  Annals of the Phytopathological Society of Japan
  European Journal of Plant Pathology (formerly *Netherlands Journal of Plant Pathology*)
  Phytopathology
  Plant Disease
  Plant Pathology

- Journals that publish papers from time to time mainly on basic plant virology

  Archives of Virology
  Cell
  EMBO Journal
  FEBS Letters
  Journal of Molecular Biology
  Molecular Plant–Microbe Interactions
  Molecular Plant Pathology
  Nature
  Physiological and Molecular Plant Pathology

  Plant Cell
  Plant Journal
  Plant Molecular Biology
  Proceedings of the National Academy of Sciences of the USA
  Science
  Virus Genes
  Virus Research

- Review journals

  Advances in Disease Vector Research
  Advances in Virus Research
  Annual Review of Biochemistry
  Annual Review of Entomology
  Annual Review of Microbiology
  Annual Review of Phytopathology
  Annual Review of Plant Physiology and Molecular Biology

COMPUTER WEBSITES

Many computer websites are ephemeral. However, 'All the Virology on the WWW' (http://www.tulare.edu/~dmsander/garry-favweb.html) has being running for several years and links to many useful sites

COMPENDIA OF INFORMATION

1. The reports of the International Committee on Taxonomy of Viruses collate information on all viruses including plant viruses. These are published at approximately 4-year intervals, the most recent being van Regenmortel *et al.* (2000) (ISBN 0–12–370200–3)

2. The Association of Applied Biologists published Descriptions of Plant Viruses from 1970 to 1989. These are now being updated and are available on CD ROM. Information on this source can be obtained from the Association of Applied Biologists, c/o Horticulture Research International, Wellesbourne, Warwick CV35 9EF, UK (e-mail: dpv.aab@hri.ac.uk).

3. A collection of digital images of crop diseases is available from The American Phytopathological Society (www.shopapspress.org).

References

A'Brook, J. (1964). The effect of planting date and spacing on the incidence of groundnut rosette disease and of the vector *Aphis craccivora*, Koch, at Mokawa, Northern Nigeria. *Ann. Appl. Biol.* **54**, 199–208.

A'Brook, J. (1968). The effect of plant spacing on the numbers of aphids trapped over the groundnut crop. *Ann. Appl. Biol.* **61**, 289–294.

A'Brook, J. (1973). The effect of plant spacing on the number of aphids trapped over cocksfoot and kale crops. *Ann. Appl. Biol.* **74**, 279–285.

Aapola, A.I.E. and Rochow, W.F. (1971). Relationships among three isolates of barley yellow dwarf virus. *Virology* **46**, 127–141.

Aaziz, R. and Tepfer, M. (1999a). Recombination between genomic RNAs of two cucumoviruses under conditions of minimal selection pressure. *Virology* **263**, 282–289.

Aaziz, R. and Tepfer, M. (1999b). Recombination in RNA viruses and in virus-resistant transgenic plants. *J. Gen. Virol.* **80**, 1339–1346.

Aaziz, R., Dinant, S. and Epel, B.L. (2001). Plasmodesmata and plant cytoskeleton. *Trends Plant Sci.* **6**, 326–330.

Abad-Zapatero, C., Abdef-Meguid, S.S., Johnson, J.E. *et al.* (1980). The structure of Southern bean mosaic virus at 2.8 Å resolution. *Nature (London)* **286**, 33–39.

Abbink, T.E.M., Tjernberg, P.A., Bol, J.F. and Linthorst, H.J.M. (1998). Tobacco mosaic virus helicase domain induces necrosis in N gene-carrying tobacco in the absence of virus replication. *Mol. Plant Microbe Interact.* **11**, 1242–1246.

Abdel-Meguid, S.S., Yamane, T., Fukuyama, K. and Rossmann, M.G. (1981). The location of calcium ions in southern bean mosaic virus. *Virology* **114**, 81–85.

AbouHaidar, M.G. (1983). The structure of the 5' and 3' ends of clover yellow mosaic virus RNA. *Can. J. Microbiol.* **29**, 151–156.

AbouHaidar, M.G. (1988). Nucleotide sequence of the capsid protein gene and 3' non-coding region of papaya mosaic virus RNA. *J. Gen. Virol.* **69**, 219–226.

AbouHaidar, M.G. and Erickson, J.W. (1985). Structure and *in vitro* assembly of papaya mosaic virus. In: J.W. Davies (ed.) *Molecular Plant Virology*, Vol. 1, pp. 85–121. CRC Press, Boca Raton.

AbouHaidar, M.G. and Gellatly, D. (1999). Potexviruses. In: A. Granoff and R.G. Webster (eds) *Encyclopedia of Virology*, 2nd edn, pp. 1364–1368. Academic Press, San Diego.

AbouHaidar, M.G. and Paliwal, Y.C. (1988). Comparison of the nucleotide sequences of viroid-like satellite RNA of the Canadian and Australasian strains of lucerne transient streak virus. *J. Gen. Virol.* **69**, 2369–2373.

Abu Salih, H.S., Ishag, H.M. and Siddig, S.A. (1973). Effect of sowing date on incidence of Sudanese broad bean mosaic virus in and yield of, *Vicia faba*. *Ann. Appl. Biol.* **74**, 371–378.

Accotto, G.P. and Boccardo, G. (1986). The coat proteins and nucleic acids of two beet cryptic viruses. *J. Gen. Virol.* **67**, 363–366.

Accotto, G.P., Riccioni, L., Barba, M. and Boccardo, G. (1997). Comparison of some molecular properties of Ourmia melon and Epirus cherry viruses, two representatives of a proposed new virus group. *J. Plant Pathol.* **78**, 87–91.

Accotto, G.P., Vaira, A.M., Noria, E. and Vecchiati, M. (1998). Using non-radioactive probes on plants: a few examples. *J. Biolum.Chemilum.* **13**, 295–301.

Acs, G., Klett, H., Schonberg, M., Christman, J., Levin, D.H. and Silverstein, S.C. (1971). Mechanism of reovirus double-stranded ribonucleic acid synthesis *in vivo* and *in vitro*. *J. Virol.* **8**, 684–689.

Adam, G. and Hsu, H.T. (1984). Comparison of structural proteins from two potato yellow dwarf viruses. *J. Gen. Virol.* **65**, 991–994.

Adam, G., Gaedigk, K. and Mundry, K.W. (1983). Alterations of a plant rhabdovirus during successive mechanical transfers. *Z. Pflanzenkr. Pflanzenschutz* **90**, 28–35.

Adams, A.N. and Barbara, D.J. (1982). The use of F(ab')$_2$-based ELISA to detect serological relationships among carlaviruses. *Ann. Appl. Biol.* **101**, 495–500.

Adams, D.B. and Kuhn, C.W. (1977). Seed transmission of peanut mottle virus in peanuts. *Phytopathology* **67**, 1126–1129.

Adams, R.L.P., Burdon, R.H., Campbell, A.M., Leader, D.P. and Smellie, R.M.S. (1981). *The Biochemistry of the Nucleic Acids*, 9th edn. Chapman & Hall, London.

Adams, S.E., Jones, R.A.C. and Coutts, R.H.A. (1986a). Expression of potato virus X resistance gene *Rx* in potato leaf protoplasts. *J. Gen. Virol.* **67**, 2341–2345.

Adams, S.E., Jones, R.A.C. and Coutts, R.H.A. (1986b). Effect of temperature on potato virus X infection in potato cultivars carrying different combinations of hypersensitivity genes. *Plant Pathol.* **35**, 517–526.

Adanson, M. (1763). *Familles des Plantes*, Vol. 1. Vincent, Paris.

Adejare, G.O. and Coutts, R.H.A. (1981). Eradication of cassava mosaic disease from Nigerian cassava clones by meristem-tip culture. *Plant Cell Tissue Organ Cult.* **1**, 25–32.

Adkins, S. and Kao, C.C. (1998). Subgenomic RNA promoters dictate the mode of recognition by bromoviral RNA-dependent RNA polymerases. *Virology* **252**, 1–8.

Adkins, S., Stawicki, S.S., Faurote, G., Siegel, R.W. and Kao, C.C. (1998). Mechanistic analysis of RNA synthesis by RNA-dependent RNA polymerase from two promoters reveals similarities to DNA-dependent RNA polymerase. *RNA* **4**, 455–470.

Adolph, K.W. and Butler, P.J.G. (1977). Studies on the assembly of a spherical plant virus: III. Reassembly of infectious virus under mild conditions. *J. Mol. Biol.* **109**, 345–357.

Adomako, D., Lesemann, D.E., Paul, H.L. and Owusu, G.K. (1983). Improved methods for the purification and detection of cocoa swollen shoot virus. *Ann. Appl. Virol.* **103**, 109–116.

Adrian, M., Dubochet, J., Fuller, S.D. and Harris, J.R. (1998). Cryo-negative staining. *Micron* **29**, 145–160.

Agranovsky, A.A., Dolja, V.V., Kavsan, V.M. and Atabekov, J.G. (1978). Detection of polyadenylate sequences in RNA components of barley stripe mosaic virus. *Virology* **91**, 95–105.

Agranovsky, A.A., Dolja, V.V., Kagramanova, V.K. and Atabekov, J.G. (1979). The presence of a cap structure at the 5' end of barley stripe mosaic virus RNA. *Virology* **95**, 208–210.

Agranovsky, A.A., Dolja, V.V., Gorbulev, V.G., Kozlov, Y.V. and Atabekov, J.G. (1981). Aminoacylation of barley stripe mosaic virus RNA: Polyadenylate-containing RNA has a 3' terminal tyrosine-accepting structure. *Virology* **113**, 174–187.

Agranovsky, A.A., Dolja, V.V. and Atabekov, J.G. (1982). Structure of the 3' extremity of barley stripe mosaic virus RNA: evidence for internal poly(A) and a 3'-terminal tRNA-like structure. *Virology* **119**, 51–58.

Agranovsky, A.A., Dolja, V.V. and Atabekov, J.G. (1983). Differences in polyadenylate length between individual barley stripe mosaic virus RNA species. *Virology* **129**, 344–349.

Agranovsky, A.A., Boyko, V.P., Karasev, A.V., Lunina, N.A., Koonin, E.V. and Dolja, V.V. (1991). Nucleotide sequence of the 3'-terminal half of beet yellows closterovirus RNA genome: unique arrangement of eight virus genes. *J. Gen. Virol.* **72**, 15–23.

Agranovsky, A.A., Koonin, E.V., Boyko *et al.* (1994). Beet yellows closterovirus: complete genome structure and identification of a leader papain-like third protease. *Virology* **195**, 311–324.

Agranovsky, A.A., Lesemann, D., Maiss, E., Hull, R. and Atabekov, J.J. (1995). 'Rattlesnake' structure of a filamentous plant RNA virus built of two capsids. *Proc. Natl Acad. Sci. USA* **92**, 2471–2473.

Agrios, G.N., Walker, M.E. and Feffo, D.N. (1985). Effect of cucumber mosaic virus inoculation at successive weekly intervals on growth and yield of pepper (*Capsicum annuum*) plants. *Plant Dis.* **69**, 52–55.

Aharoni, N., Marco, S. and Levy, D. (1977). Involvement of gibberellins and abscisic acid in the suppression of hypocotyl elongation in CMV-infected cucumbers. *Physiol. Plant Pathol.* **11**, 189–194.

Ahlquist, P. (1999). Bromoviruses (*Bromoviridae*). In: A. Granoff and R.G. Webster (eds) *Encyclopedia of Virology*, 2nd edn, pp. 198–204. Academic Press, San Diego.

Ahlquist, P. and Kaesberg, P. (1979). Determination of the length distribution of poly(A) at the 3' terminus of the virion RNAs of EMC virus, poliovirus, rhinovirus, RAV-61 and CPMV and of mouse globin MRNA. *Nucl. Acids Res.* **7**, 1195–1204.

Ahlquist, P., Luckow, V. and Kaesberg, P. (1981). Complete nucleotide sequence of brome mosaic virus RNA3. *J. Mol. Biol.* **153**, 23–38.

Ahlquist, P., Dasgupta, R. and Kaesberg, P. (1984a). Nucleotide sequence of brome mosaic virus genome and its implications for virus replication. *J. Mol. Biol.* **172**, 369–383.

Ahlquist, P., French, R., Janda, M. and Loesch-Fries, L.S. (1984b). Multicomponent RNA plant virus infection derived from cloned viral cDNA. *Proc. Natl Acad. Sci. USA* **81**, 7066–7070.

Ahlquist, P., Allison, R., Dejong, W. *et al.* (1990). Molecular biology of bromovirus replication and host specificity. In: T.P. Pirone and J.G. Shaw (eds) *Viral Genes and Plant Pathogenesis*, pp. 144–155. Springer-Verlag, New York.

Ahmed, M.E., Black, L.M., Perkins, E.G., Walker, B.L. and Kummerow, F.A. (1964). Lipid in potato yellow dwarf virus. *Biochem. Biophys. Res. Commun.* **17**, 103–107.

Ahola, T. and Ahlquist, P. (1999). Putative RNA capping activities encoded by brome mosaic virus: methylation and covalent binding of guanylate by replicase protein 1a. *J. Virol.* **73**, 10061–10069.

Ahola, T., den Boon, J.A. and Ahlquist, P. (2000). Helicase and capping enzyme active site mutations in brome mosaic virus protein 1a causes defects in template recruitment, negative-strand RNA synthesis and viral RNA capping. *J. Virol.* **74**, 8803–8811.

Ajayi, O. and Dewar, A.M. (1982). The effect of barley yellow dwarf virus on honeydew production by the cereal aphids, *Sitobium avenae* and *Metolophium dirhodium*. *Ann. Appl. Biol.* **100**, 203–212.

Ajayi, O. and Dewar, A.M. (1983). The effects of barley yellow dwarf virus, aphids and honeydew on *Cladosporium* infection of winter wheat and barley. *Ann. Appl. Biol.* **102**, 57–65.

Akad, F., Teverovsky, E., David, A. *et al.* (1999). A cDNA from tobacco codes for an inhibitor of virus replication (IVR)-like protein. *Plant Mol. Biol.* **40**, 969–976.

Akgoz, M., Nguyen, Q.N., Talmadge, A.E., Drainville, K.E. and Wobbe, K.K. (2001). Mutational analysis of turnip crinkle virus movement protein p8. *Molec. Plant Pathol.* **2**, 37–48.

Al Ani, R., Pfeiffer, P., Lebeurier, G. and Hirth, L. (1979a). The structure of cauliflower mosaic virus: I. pH-induced structural changes. *Virology* **93**, 175–187.

Al Ani, R., Pfeiffer, P. and Lebeurier, G. (1979b). The structure of cauliflower mosaic virus: II. Identity and location of the viral polypeptides. *Virology* **93**, 188–197.

Alam, S.N. and Cohen, M.B. (1998). Durability of brown planthopper, *Niloparvata lugens*, resistance in rice variety IR64 in greenhouse selection studies. *Entomol. Exp. Appl.* **89**, 71–78.

Albert, F.G., Fox, J.M. and Young, M.J. (1997). Virion swelling is not required for cotranslational disassembly of cowpea chlorotic mottle virus *in vitro*. *J. Virol.* **71**, 4296–4299.

Albiach-Martí, M., Mawassi, M., Gowda, S. *et al.* (2000a). Sequences of *Citrus tristeza virus* separated in time and space are essentially identical. *J. Virol.* **74**, 6856–6865.

Albiach-Martí, M.R., Guerri, J., de Mendoza, H., Laigret, F., Ballester-Olmos, J.F. and Moreno, P. (2000b). Aphid transmission alters the genomic and defective RNA populations of citrus tristeza virus isolates. *Phytopathology* **90**, 134–138.

Albrecht, H., Geldreich, A., Menissier de Murcia, J., Kirchherr, D., Mesnard, J.-M. and Lebeurier, G. (1988). Cauliflower mosaic virus gene 1 product detected in a cell-wall-enriched fraction. *Virology* **163**, 503–508.

Alconero, R. (1973). Mechanical transmission of viruses from sweet potato. *Phytopathology* **63**, 377–380.

Alconero, R., Weeden, N.F., Gonsalves, D. and Fox, D.T. (1985). Loss of genetic diversity in pea germplasm by the elimination of individuals infected by pea seedborne mosaic virus. *Ann. Appl. Biol.* **106**, 357–364.

Aldaoud, R., Dawson, W.O. and Jones, G.E. (1989). Rapid random evolution of the genetic structure of replicating tobacco mosaic virus populations. *Intervirology* **30**, 227–233.

Alhubaishi, A.A., Walkey, D.G.A., Webb, M.J.W., Bolland, C.J. and Cook, A.A. (1987). A survey of horticultural plant virus diseases in the Yemen Arab Republic. *FAO Plant Prot. Bull.* **35**, 135–143.

Al-Kaff, N.S. and Covey, S.N (1995). Biological diversity of cauliflower mosaic virus isolates expressed in two *Brassica* species. *Plant Pathol.* **44**, 516–526.

Al-Kaff, N.S. and Covey, S.N. (1996). Unusual accumulations of cauliflower mosaic virus in local lesions, dark green leaf tissue and roots of infected plants. *Mol. Plant Microbe Interact.* **5**, 357–363.

Al-Kaff, N.S., Turner, D.S. and Covey, S.N. (1997). Multiple pathogenic factors influence aphid transmission of cauliflower mosaic virus from infected plants. *Eur. J. Plant Pathol.* **103**, 65–71.

Al-Kaff, N.S., Covey, S.N., Kreike, M.M., Page, A.M., Pinder, R. and Dale, P.J. (1998). Transcriptional and posttranscriptional plant gene silencing in response to a pathogen. *Science* **279**, 2113–2115.

Al-Kaff, N.S., Kreike, M.M., Covey, S.N., Pitcher, R., Page, A.M. and Dale, P.J. (2000). Plants rendered herbicide-susceptible by cauliflower mosaic virus-elicited suppression of a 35S promoter-regulated transgene. *Nature Biotechnol.* **18**, 995–999.

Allen, E., Wang, S. and Miller, W.A. (1999). Barley yellow dwarf virus TNA requires a cap-independent translation sequence because it lacks a 5′ cap. *Virology* **253**, 139–144.

Allen, R.N. and Dale, J.L. (1981). Application of rapid biochemical methods for detecting avocado sunblotch disease. *Ann. Appl. Biol.* **98**, 451–461.

Allen, W.R. (1981). Dissemination of tobacco mosaic virus from soil to plant leaves under glasshouse conditions. *Can. J. Plant Pathol.* **3**, 163–168.

Allison, R.F., Dougherty, W.G., Parks, T.D. *et al.* (1985). Biochemical analysis of the capsid protein gene and capsid protein of tobacco etch virus: N-terminal amino acids are located on the virion's surface. *Virology* **147**, 309–316.

Allison, R.F., Johnston, R.E. and Dougherty, W.G. (1986). The nucleotide sequence of the coding region of tobacco etch virus genomic RNA: evidence for the synthesis of a single polyprotein. *Virology* **154**, 9–20.

Allison, R.F., Janda, M. and Ahlquist, P. (1988). Infectious *in vitro* transcripts from cowpea chlorotic mottle virus cDNA clones and exchange of individual RNA components with brome mosaic virus. *J. Virol.* **62**, 3581–3588.

Allison, R.F., Janda, M. and Ahlquist, P. (1989). Sequence of cowpea chlorotic mottle virus RNAs 2 and 3 and evidence of a recombination event during bromovirus evolution. *Virology* **172**, 321–330.

Al-Moudallal, Z., Briand, J.P. and van Regenmortel, M.H.V. (1982). Monoclonal antibodies as probes of the antigenic structure of tobacco mosaic virus. *EMBO J.* **1**, 1005–1010.

Al-Moudallal, Z., Altschuh, D., Briand, J.P. and van Regenmortel, M.H.V. (1984). Comparative sensitivity of different ELISA procedures for detecting monoclonal antibodies. *J. Immunol. Meth.* **68**, 35–43.

Al-Moudallal, Z., Briand, J.P. and van Regenmortel, M.H.V. (1985). A major part of the polypeptide chain of tobacco mosaic virus protein is antigenic. *EMBO J.* **4**, 1231–1235.

Altenbach, S.B. and Howell, S.H. (1981). Identification of a satellite RNA associated with turnip crinkle virus. *Virology* **112**, 25–33.

Altenbach, S.B. and Howell, S.H. (1984). Nucleic acid species related to the satellite RNA of turnip crinkle virus in turnip plants and virus particles. *Virology* **134**, 72–77.

Altschuh, D., Hartman, D., Reinbolt, J. and van Regenmortel, M.H.V. (1983). Immunological studies of tobacco mosaic virus: IV. Localization of four epitopes in the protein subunit by inhibition tests with synthetic peptides and cleavage peptides from three strains. *Mol. Immunol.* **20**, 271–278.

Altschuh, D., Al Moudallal, Z., Briand, J.P. and van Regenmortel, M.H.V. (1985). Immunochemical studies of tobacco mosaic virus: VI. Attempts to localize viral epitopes with monoclonal antibodies. *Mol. Immunol.* **22**, 329–337.

Altschuh, D., Lesk, A.M., Bloomer, A.C. and King, A. (1987). Correlation of coordinated amino acid substitutions with function in viruses related to tobacco mosaic virus. *J. Mol. Biol.* **193**, 693–707.

Alzhanova, D.V., Hagiwara, Y., Peremyslov, V.V. and Dolja, V.V. (2000). Genetic analysis of the cell-to-cell movement of beet yellows closterovirus. *Virology* **268**, 192-200.

Amin, P.W. (1985). Apparent resistance of groundnut cultivar Robut 33-1 to bud necrosis disease. *Plant Dis.* **69**, 718–719.

Ammar, E.D. (1994). Propagative transmission of plant and animal viruses by insects: Factors affecting vector specificity and competence. *Adv. Dis. Vector Res.* **10**, 289–359.

Ammar, E.D., Jarlfors, U. and Pirone, T.P. (1994). Association of potyvirus helper component protein with virions and the cuticle lining of the maxillary food canal and foregut of an aphid vector. *Phytopathology* **84**, 1054–1060.

Anagnostou, K., Jahn, M. and Perl-Treves, R. (2000). Inheritance and linkage analysis of resistance to zucchini yellow mosaic virus, papaya ringspot virus and powdery mildew in melon. *Euphytica* **116**, 265–270.

Anandalakshmi, R., Pruss, G.J., Ge, X. *et al.* (1998). A viral suppressor of gene silencing in plants. *Proc. Natl. Acad. Sci. USA* **95**, 13079–13084.

Anderer, F.A. (1959). Reversible Denaturierung des Proteins aus Tabakmosaikvirus. *Z. Naturforsch. B: Anorg. Chem., Org. Chem., Biochem., Biophys., Biol.* **14**, 642–647.

Anderer, F.A., Uhlig, H., Weber, E. and Schramm, G. (1960). Primary structure of the protein of tobacco mosaic virus. *Nature (London)* **186**, 922–925.

Anderson, E.J., Stark, D.M., Nelson, R.S., Powell, P.A., Tumer, N.E. and Beachy, R.N. (1989). Transgenic plants that express the coat protein genes of tobacco mosaic virus or alfalfa mosaic virus interfere with disease development of some nonrelated viruses. *Phytopathology* **79**, 1284–1290.

Anderson, E.J., Kline, A.S., Morelock, T.E. and McNew, R.W. (1996). Tolerance to blackeye cowpea mosaic potyvirus not correlated with decreased virus accumulation or protection from cowpea stunt disease. *Plant Dis.* **80**, 847–852.

Anderson, M.L.M. and Young, B.D. (1985). Quantitative filter hybridization. In: B.D. Hames and S.J. Higgins (eds) *Nucleic Acid Hybridization: A Practical Approach*, pp. 73–111. IRL Press, Oxford.

Andrews, J.H. and Shalla, T.A. (1974). The origin, development and conformation of amorphous inclusion body components in tobacco etch virus-infected cells. *Phytopathology* **64**, 1234–1243.

Andrianifahanana, M., Lovins, K., Dute, R., Sikora, E. and Murphy, J.F. (1997). Pathway for phloem-dependent movement of pepper mottle potyvirus in the stem of *Capsicum annuum*. *Phytopathology* **87**, 892–898.

Andriessen, M., Meulewaeter, F. and Cornelissen, M. (1995). Expression of tobacco necrosis virus open reading frames 1 and 2 is sufficient for the replication of satellite tobacco necrosis virus. *Virology* **212**, 222–224.

Anfinsen, C.B. (1973). Principles that govern the folding of protein chains. *Science* **181**, 223–230.

Angell, S.M., Davies, C. and Baulcombe, D.C. (1996). Cell-to-cell movement of potato virus X is associated with a change in the size-exclusion limit of plasmodesmata in trichome cells of *Nicotiana clevelandii*. *Virology* **216**, 197–201.

Angenent, G.C., Linthorst, H.J.M., van Belkum, A.F., Cornelissen, B.J.C. and Bol, J.F. (1986). RNA2 of tobacco rattle virus strain TCM encodes an unexpected gene. *Nucl. Acids Res.* **14**, 4673–4682.

Angenent, G.C., Verbeek, H.B.M. and Bol, J.F. (1989a). Expression of the 16K cistron of tobacco rattle virus in protoplasts. *Virology* **169**, 305–311.

Angenent, G.C., Posthumus, E., Brederode, F.T. and Bol, J.F. (1989b). Genome structure of tobacco rattle virus strain PLB: further evidence on the occurrence of RNA recombination among tobraviruses. *Virology* **171**, 271–274.

Angenent, G.C., Posthumus, E. and Bol, J.F. (1989c). Biological activity of transcripts synthesised *in vitro* from full-length and mutated DNA copies of tobacco rattle virus RNA2. *Virology* **173**, 68–76.

Angenent, G.C., van den Ouweland, J.M.W. and Bol, J.F. (1990). Susceptibility to virus infection of transgenic tobacco plants expression structural and non-structural genes of tobacco rattle virus. *Virology* **175**, 191–198.

Anjaneyulu, A. and Bhaktavatsalam, G. (1987). Evaluation of synthetic pyrethroid insecticides for tungro management. *Trop. Pest Manag.* **33**, 323–326.

Anjaneyulu, A., Satapathy, M.K and Shukla, V.D. (1994). *Rice Tungro*. Oxford and IHB Publishing, New Delhi.

Ansel-McKinney, P. and Gehrke, L. (1998). RNA determinants of a specific RNA-coat protein peptide interaction in alfalfa mosaic virus: conservation of homologous features in ilarvirus RNAs. *J. Mol. Biol.* **278**, 767–785.

Ansel-McKinney, P., Scott, S.W., Swanson, M., Ge, X. and Gehrke, L. (1996). A plant viral coat protein RNA binding consensus sequence contains a crucial arginine. *EMBO J.* **15**, 5077–5084.

Antignus, Y., Sela, I. and Harpaz, I. (1971). Species of RNA extracted from tobacco and *Datura* plants and their differential sensitivity to actinomycin D. *Biochem. Biophys. Res. Commun.* **44**, 78–88.

Antoniw, J.F., Ritter, C.E., Pierpoint, W.S. and van Loon, L.C. (1980). Comparison of three pathogenesis-related proteins from plants of two cultivars of tobacco infected with TMV. *J. Gen. Virol.* **47**, 79–87.

Anton-Lamprecht, I. (1966). Beitrage zum Problem der Plastidenabiinderung: III. Über das Vorkomnien von 'Rückmutationen' in einer spontanentstandenen Plastidenschecke von Epilobium hirsutum. *Z. Pflanzenphysiol.* **54**, 417–445.

Anzola, J.V., Xu, Z., Asamizu, T. and Nuss, D.L. (1987). Segment-specific inverted repeats found adjacent to conserved terminal sequences in wound tumor virus genome and defective interfering RNAs. *Proc. Natl Acad. Sci. USA* **84**, 8301–8305.

Aoki, S. and Takebe, I. (1969). Infection of tobacco mesophyll protoplasts by tobacco mosaic virus ribonucleic acid. *Virology* **39**, 439–448.

Aoki, S. and Takebe, I. (1975). Replication of tobacco mosaic virus RNA in tobacco mesophyll protoplasts *in vitro*. *Virology* **65**, 343–354.

Aparicio, F., Sánchez-Pina, M.A., Sánchez-Navarro, J.A. and Pallás, V. (1999). Location of prunus necrotic ringspot ilarvirus within pollen grains of infected nectarine trees: evidence from RT-PCR, dot-blot and *in situ* hybridisation. *Eur. J. Plant Pathol.* **105**, 623–627.

Appiano, A. and D'Agostino, G. (1983). Distribution of tomato bushy stunt virus in root tips of systemically infected *Gomphrena globosa*. *J. Ultrastruct. Res.* **85**, 239–248.

Appiano, A. and Pennazio, S. (1972). Electron microscopy of potato meristem tips infected with potato virus X. *J. Gen. Virol.* **14**, 273–276.

Appiano, A., Pennazio, S., D'Agostino, G. and Redolfi, P. (1977). Fine structure of necrotic local lesions induced by tomato bushy stunt virus *Gomphrena globosa* leaves. *Physiol. Plant Pathol.* **11**, 327–332.

Aranda, M. and Maule, A. (1998). Virus-induced host gene shutoff in animals and plants. *Virology* **243**, 261–267.

Aranda, M., Escaler, M., Wang, D. and Maule, A.J. (1996). Induction of HSP70 and polyubiquitin expression associated with plant virus replication. *Proc. Natl Acad. Sci. USA* **93**, 15289–15293.

Aranda, M.A., Fraile, A., Dopazo, J., Malpica, J.M. and García-Arenal, F. (1997). Contribution of mutation and RNA recombination to the evolution of a plant pathogenic RNA. *J. Mol. Evol.* **44**, 81–88.

Aranda, M.A., Escaler, M., Thomas, C.L. and Maule, A.J. (1999). A heat shock transcription factor in pea is differentially controlled by heat and virus replication. *Plant J.* **20**, 153–161.

Araya, J.E. and Foster, J.E. (1987). Laboratory study on the effects of barley yellow dwarf virus on the life cycle of *Rhopalosiphum padi* (L). *J. Plant Dis. Prot.* **94**, 578–583.

Arce-Johnson, P., Reimann-Philipp, U., Padgett, H.S., Rivera-Bustamante, R. and Beachy, R.N. (1997). Requirement of the movement protein for long distance spread of tobacco mosaic virus in grafted plants. *Mol. Plant-Microbe Interact.* **10**, 691–699.

Argos, P. (1981). Secondary structure prediction of plant virus coat proteins. *Virology* **110**, 55–62.

Argos, P. (1988). A sequence motif in many polymerases. *Nucl. Acids Res.* **16**, 9909–9916.

Argos, P., Kamer, G., Nicklin, M.J.H. and Wimmer, E. (1984). Similarity in gene organization and homology between proteins of animal picornaviruses and a plant comovirus suggest common ancestry of these virus families. *Nucl. Acids Res.* **12**, 7251–7267.

Arguello-Astorga, G.R., Guevara-Gonzalez, R.G., Herrera-Estrella, L.R. and Rivera-Bustamante, R.F. (1994). Geminivirus replication origins have a group-specific organization of iterative elements: a model for replication. *Virology* **203**, 90–100.

Armour, S.L., Melcher, U., Pirone, T.P., Lyttle, D.J. and Essenberg, R.C. (1983). Helper component for aphid transmission encoded by region II of cauliflower mosaic virus DNA. *Virology* **129**, 25–30.

Aronson, M.N., Meyer, A.D., Györgyey, J. *et al.* (2000). Clink, a nanovirus-encoded protein, binds both pRB and SKP1. *J. Virol.* **74**, 2967–2972.

Asamizu, T., Summers, D., Motika, M.B., Anzola, J.V. and Nuss, D.L. (1985). Molecular cloning and characterization of the genome of wound turnout virus: a tumor-inducing plant reovirus. *Virology* **144**, 398–409.

Ashby, J.W., Teh, P.B. and Close, R.C. (1979). Symptomatology of subterranean clover red leaf virus and its incidence in some legume crops, weed hosts and certain alate aphids in Canterbury, New Zealand. *N. Z. J. Agric. Res.* **22**, 361–365.

Asher, M. (1999). Sugar-beet rhizomania: the spread of a soilborne disease. *Microbiol. Today* **26**, 120–122.

Ashoub, A., Rohde, W. and Prufer, D. (1998). In planta transcription of a second subgenomic RNA increases the complexity of the subgroup 2 luteovirus genome. *Nucl. Acids Res.* **26**, 420–426.

Asjes, C.J. (1978). Minerale oliën op lelies om virusverspreiding tegen te gaan. *Bloembollencultuur* **88**, 1046–1047.

Asjes, C.J. (1980). Toepassing van minerale olie om verspreiding van hyacintemozaiekvirus in hyacinten tegen te gaan. *Bloemboliencultuur* **90**, 1396–1397.

Asjes, C.J., de Vos, P. and von Slogteren, D.H.M. (1973). Brown ring formation and streak mottle, two distinct syndromes in lilies associated with complex infections of lily symptomless virus and tulip breaking virus. *Neth. J. Plant Pathol.* **79**, 23–35.

Assad, F.F. and Signer, E.R. (1990). Cauliflower mosaic virus P35S promoter activity in *Escherichia coli*. *Mol. Gen. Genet.* **223**, 517–520.

Asselin, A. and Zaitlin, M. (1978). An anomalous form of tobacco mosaic virus RNA observed upon polyacrylamide gel electrophoresis. *Virology* **88**, 191–193.

Astruc, N., Marcos, J.F., Macquaire, G., Candresse, T. and Pallas, V. (1996). Studies on the diagnosis of hop stunt viroid in fruit trees: identification of new hosts and application of a nucleic acid extraction procedure based on non-organic solvents. *Eur. J. Plant Pathol.* **102**, 837–846.

Atabekov, J.G. (1975). Host specificity of plant viruses. *Annu. Rev. Phytopathol.* **13**, 127–145.

Atabekov, J.G. and Dorokhov, Yu.L. (1984). Plant virus-specific transport function and resistance of plants to viruses. *Adv. Virus Res.* **29**, 313–364.

Atabekov, J.G., Novikov, V.K., Vishnichenko, V.K. and Javakhia, V.G. (1970a). A study of the mechanisms controlling the host range of plant viruses: II. The host range of hybrid viruses reconstituted *in vitro*, and of free viral RNA. *Virology* **41**, 108–115.

Atabekov, J.G., Schaskolskaya, N.D., Atabekova, T.I. and Sacharovskaya, G.A. (1970b). Reproduction of temperature-sensitive strains of TMV under restrictive conditions in the presence of temperature-resistant helper strain. *Virology* **41**, 397–407.

Atabekov, J.G., Dorokhov, Yu.L. and Taliansky, M.E. (1983). A virus-coded function responsible for the transport of virus genome from infected to healthy cells. *Tag. ungs ber.—Akad. Landwirtschafts wiss. DDR* **216**, 53–58.

Atabekov, J.G., Malyshenko, S.I., Morozov, S.Yu. *et al.* (1999). Identification and study of tobacco mosaic virus movement function by complementation tests. *Phil. Trans. R. Soc. Lond. B* **354**, 629–635.

Atabekova, T.I., Taliansky, M.E. and Atabekov, J.G. (1975). Specificity of protein–RNA and protein–protein interaction upon assembly of TMV *in vivo* and *in vitro*. *Virology* **67**, 1–13.

Atchison, B.A. (1973). Division, expansion and DNA synthesis in meristematic cells of French bean (*Phaseolus vulgaris* L.) root-tips invaded by tobacco ringspot virus. *Physiol. Plant Pathol.* **3**, 1–8.

Athow, K.L. and Bancroft, J.B. (1959). Development and transmission of tobacco ringspot virus in soybean. *Phytopathology* **49**, 697–701.

Athow, K.L. and Laviolette, F.A. (1962). Relation of seed position and pod location to tobacco ringspot virus seed transmission in soybean. *Phytopathology* **52**, 714–715.

Atiri, G.I. (1984). The occurrence and importance of okra mosaic virus in Nigerian weeds. *Ann. Appl. Biol.* **104**, 261–265.

Atiri, G.I., Ekpo, E.J.A. and Thottappilly, G. (1984). The effect of aphid-resistance in cowpea on infestation and development of *Aphis craccivora* and the transmission of cowpea aphid-borne mosaic virus. *Ann. Appl. Biol.* **104**, 339–346.

Atkins, D., Roberts, K., Hull, R. and Bishop, D.H.L. (1991a). Expression of tobacco mosaic virus movement protein using a baculovirus expression vector. *J. Gen. Virol.* **72**, 2831–2835.

Atkins, D., Hull, R., Wells, B., Roberts, K., Moore, P. and Beachy, R.N. (1991b). The tobacco mosaic virus movement protein in transgenic plants is localised to plasmodesmata. *J. Gen. Virol.* **72**, 209–211.

Atkinson, P.H. and Matthews, R.E.F. (1970). On the origin of dark green tissue in tobacco leaves infected with tobacco mosaic virus. *Virology* **40**, 344–356.

Atkinson, T.G. and Grant, M.N. (1967). An evaluation of streak mosaic losses in winter wheat. *Phytopathology* **57**, 188–192.

Atkinson, T.G. and Slykhuis, J.T. (1963). Relation of spring drought, summer rains and high fall temperatures to the wheat streak mosaic epiphytotic in southern Alberta, 1963. *Can. Plant Dis. Surv.* **43**, 154–159.

Atreya, C.D. and Siegel, A. (1989). Localization of multiple TMV encapsulation initiation sites in rbcL gene transcripts. *Virology* **168**, 388–392.

Atreya, C.D., Atreya, P.L., Thornbury, D.W. and Pirone, T.P. (1992). Site-directed mutations in the potyvirus HC-Pro gene affect helper component activity, virus accumulation and symptom expression in infected tobacco plants. *Virology* **191**, 106–111.

Audy, P., Palukaitis, P., Slack, S.A. and Zaitlin, M. (1994). Replicase-mediated resistance to potato virus Y in transgenic tobacco plants. *Mol. Plant Microb. Interact.* **7**, 15–22.

Austin, R.B., Ford, M.A., Edrich, J.A. and Hooper, B.E. (1976). Some effects of leaf posture on photosynthesis and yield in wheat. *Ann. Appl. Biol.* **83**, 425–446.

Ausubel, F.M., Brent, R., Kingston, R.E. *et al.* (eds) (1987). *Current Protocols in Molecular Biology*, Vol. 1. Wiley, New York.

Ausubel, F.M., Brent, R., Kingston, R.E. *et al.* (eds) (1988). *Current Protocols in Molecular Biology*, Vol. 2. Wiley, New York.

Ausubel, F.M., Brent, R., Kingston, R.E. *et al.* (eds) (1998). *Current Protocols in Molecular Biology*, Vols 1 and 2. John Wiley, New York.

Avila-Rincon, M.J., Collmer, C.W. and Kaper, J.M. (1986). In vitro translation of cucumoviral satellites. I. Purification and nucleotide sequence of cucumber mosaic virus-associated RNA5 from cucumber mosaic virus strain S. Virology 152, 446–454.

Ayala, F.J. (1999). Molecular clock mirages. *Bioessays* **21**, 71–75.

Azzam, O., Frazer, J., Delarosa, D., Beaver, J.S., Ahlquist, P. and Maxwell, D.P. (1994). Whitefly transmission and efficient ssDNA accumulation of bean golden mosaic geminivirus require functional coat protein. *Virology* **204**, 289–296.

Babos, P. and Shearer, G.B. (1969). RNA synthesis in tobacco leaves infected with tobacco mosaic virus. *Virology* **39**, 286–295.

Backus, E.A. (1994). History, development and applications of the AC electronic monitoring system for insect feeding. In: M.M. Ellsbury, E.A. Backus and D.E. Ullman (eds) *History, Development and Application of AC Electronic Feeding Monitors*, pp. 1–51. Entomological Society of America, Lanham, Maryland.

Backus, R.C. and Williams, R.C. (1950). Use of spraying methods and of volatile suspending media in the preparation of specimens for electron microscopy. *J. Appl. Physiol.* **21**, 11–15.

Baer, M., Houser, F., Loesch-Fries, L.S. and Gehrke, L. (1994). Specific RNA binding by amino-terminal peptides of alfalfa mosaic virus coat protein. *EMBO J.* **13**, 727–735.

Bagnall, R.H. (1988). Epidemics of potato leaf roll in North America and Europe linked to drought and sunspot cycles. *Can. J. Plant. Pathol.* **10**, 193–202.

Baker, B., Zambryski, P., Staskawicz, B. and Dinesh-Kumar, S.P. (1997). Signaling in plant–microbe interactions. *Science* **276**, 726-733.

Baker, P.F. (1960). Aphid behaviour on healthy and on yellow virus-infected sugar beet. *Ann. Appl. Biol.* **48**, 384–391.

Balachandran, S., Hurry, V.M., Kelley, S.E. *et al.* (1997). Concepts of biotic stress: some insights into the stress physiology of virus-infected plants, from the perspective of photosynthesis. *Physiol. Plant* **100**, 203–213.

Balasaraswathi, R., Sadasivam, S., Ward, M. and Walker, J.M. (1998). An antiviral protein from *Bougainvillea spectabilis* roots: purification and characterization. *Phytochemistry* **47**, 1561–1565.

Balázs, E. (1990). Disease symptoms in transgenic tobacco induced by integrated gene VI of cauliflower mosaic virus. *Virus Genes* **3**, 205–211.

Balázs, E., Gáborjányi, R. and Kirdly, Z. (1973). Leaf senescence and increased virus susceptibility in tobacco. The effect of abscisic acid. *Physiol. Plant Pathol.* **3**, 341–346.

Balázs, E., Sziráki, I. and Király, Z. (1977). The role of cytokinins in the systemic acquired resistance of tobacco hypersensitive to tobacco mosaic virus. *Physiol. Plant Pathol.* **11**, 29–37.

Balázs, E., Bouzoubaa, S., Guilley, H., Jenard, G., Paszkowski, J. and Richards, K. (1985). Chimeric vector construction for higher plant transformation. *Gene* **40**, 343–348.

Bald, J.G. (1964). Cytological evidence for the replication of plant virus nucleic acid in the nucleus. *Virology* **22**, 377–387.

Bald, J.G. and Tinsley, T.W. (1967). A quasi-genetic model for plant virus host ranges: III. Congruence and relatedness. *Virology* **32**, 328–336.

Bald, J.G. and Tinsley, T.W. (1970). A quasi-genetic model for plant virus host ranges: IV. Cacao swollen-shoot and mottle-leaf viruses. *Virology* **40**, 369–378.

Balint, R. and Cohen, S.S. (1985a). The incorporation of radiolabelled polyamines and methionine into turnip yellow mosaic virus in protoplasts from infected plants. *Virology* **144**, 181–193.

Balint, R. and Cohen, S.S. (1985b). The effects of dicyclohexylamine on polyamine biosynthesis and incorporation into turnip yellow mosaic virus in chinese cabbage protoplasts infected *in vitro*. *Virology* **144**, 194–203.

Ball, E.M. (1973). Solid phase radioimmunoassay for plant viruses. *Virology* **55**, 516–520.

Ballas, N., Shimson, B., Hermona, S. and Abraham, L. (1989). Efficient functioning of plant promoters and polyadenylation sites in *Xenopus* oocytes. *Nucl. Acids Res.* **17**, 7891–7904.

Baltimore, D. (1971). Expression of animal virus genomes. *Bacteriol. Rev.* **35**, 235–241.

Baluška, F., Volkmann, D. and Barlow, P.W. (2001). Mobile plant cell body: a 'bug' within a 'cage'. *Trends Plant Sci.* **6**, 104–111.

Ban, N. and McPherson, A. (1995). The structure of satellite panicum mosaic virus at 1.9 Å resolution. *Nat. Struct. Biol.* **2**, 882–890.

Ban, N., Larson, S.B. and McPherson, A. (1995). Structural comparisons of the plant satellite viruses. *Virology* **214**, 571–583.

Bancroft, J.B. (1972). A virus made from parts of the genomes of brome mosaic and cowpea chlorotic mottle viruses. *J. Gen. Virol.* **14**, 223–228.

Bancroft, J.B. and Horne, R.W. (1977). Bromoviruses (brome mosaic virus) group. In: K. Maramorosch (ed.) *The Atlas of Insect and Plant Viruses*, pp. 287–302. Academic Press, New York.

Bancroft, J.B. and Pound, G.S. (1954). Effect of air temperature on the multiplication of tobacco mosaic virus in tobacco. *Phytopathology* **44**, 481–482.

Bancroft, J.B., McLean, J.D., Rees, M.W. and Short. M.N. (1971). The effect of an arginyl to a cysteinyl replacement on the uncoating behaviour of a spherical plant virus. *Virology* **45**, 707–715.

Bandla, M.D., Campbell, L.R., Ullman, D.E. and Sherwood, J.L. (1998). Interaction of tomato spotted wilt tospovirus (TSWV) glycoproteins with a thrips midgut protein, a potential cellular receptor for TSWV. *Phytopathology* **88**, 98–104.

Banerjee, N., Wang, J.-Y. and Zaitlin, M. (1995). A single nucleotide change in the coat protein gene of tobacco mosaic virus is involved in the induction of severe chlorosis. *Virology* **207**, 234–239.

Banttari, E.E. and Goodwin, P.H. (1985). Detection of potato viruses S, X and Y by enzyme-linked immunosorbent assay on nitro-cellulose membranes (Dot-ELISA). *Plant Dis.* **69**, 202–205.

Banttari, E.E. and Zeyen, R.J. (1969). Chromatographic purification of the oat blue dwarf virus. *Phytopathology* **59**, 183–186.

Banttari, E.E. and Zeyen, R.J. (1972). Ultrastructure of flax with a simultaneous virus and mycoplasma-like infection. *Virology* **49**, 305–308.

Bao, Y. (1994). Replication intermediates of rice tungro bacilliform virus DNA support a replication mechanism involving reverse transcription. *Virology* **204**, 626–633.

Bao, Y. and Hull, R. (1992). Characterization of the discontinuities in rice tungro bacilliform virus DNA. *J. Gen. Virol.* **73**, 1297–1301.

Bao, Y. and Hull, R. (1993a). Mapping the 5′ terminus of rice tungro bacilliform viral genomic RNA. *Virology* **197**, 445–448.

Bao, Y. and Hull, R. (1993b). A strong-stop DNA in rice plants infected with rice tungro bacilliform virus. *J. Gen. Virol.* **74**, 1611–1616.

Bao, Y. and Hull, R. (1994). Replication intermediates of rice tungro bacilliform virus DNA support a replication mechanism involving reverse transcription. *Virology* **204**, 626–633.

Baratova, L.A., Grebenshcikov, N.I., Dobrov, E.N. *et al.* (1992). The organization of potato virus X coat proteins in virus particles studied by tritium planigraphy and model building. *Virology* **188**, 175–180.

Barbara, D.J. and Clark, M.F. (1982). A simple indirect ELISA using F(ab')$_2$ fragments of immunoglobulin. *J. Gen. Virol.* **58**, 315–322.

Barbara, D.J., Kawata, E.E., Ueng, P.P., Lister, R.M. and Larkins, B.A. (1987). Production of cDNA clones from the MAV isolate of barley yellow dwarf virus. *J. Gen. Virol.* **68**, 2419–2427.

Barbara, D.J., Morton, A. and Adams, A.N. (1990). Assessment of UK hops for the occurrence of hop latent and hop stunt viroids. *Ann. Appl. Biol.* **116**, 265–272.

Bar-Joseph, M and Hull, R. (1974). Purification and partial characterization of sugar beet yellows virus. *Virology* **62**, 552–562.

Bar-Joseph, M., Rosner, A., Moscovitz, M. and Hull, R. (1983). A simple procedure for the extraction of double-stranded RNA from infected plants. *J. Virol. Methods* **6**, 1–8.

Bar-Joseph, M., Segev, D., Blickle, V., Yesodi, V., Franck, A. and Rosner, A. (1986). Application of synthetic DNA probes for the detection of viroids and viruses. In: R.A.C. Jones and L. Torrance (eds) *Developments and Applications in Virus Testing*, Dev. Appl. Biol. I, pp. 13–23. Association of Applied Biologists, Wellesbourne.

Bar-Joseph, M., Yanmg, G., Gafney, R. and Mawassi, M. (1997). Subgenomic RNAs: the possible building blocks for modular recombination of *Closteroviridae* genomes. *Sem. Virol.* **8**, 113–119.

Barker, H. (1989). Specificity of the effect of sap-transmissible viruses in increasing the accumulation of luteoviruses in co-infected plants. *Ann. Appl. Biol.* **115**, 71–78.

Barker, H. and Harrison, B.D. (1978). Double infection, interference and superinfection in protoplasts exposed to two strains of raspberry ringspot virus. *J. Gen. Virol.* **40**, 647–658.

Barker, H. and Harrison, B.D. (1986). Restricted distribution of potato leafroll virus antigen in resistant potato genotypes and its effect on transmission of the virus by aphids. *Ann. Appl. Biol.* **109**, 595–604.

Barker, H. and Waterhouse, P.M. (1999). The development of resistance to luteoviruses mediated by host genes and pathogen-derived transgenes. In: H.G. Smith and H. Barker (eds) *The Luteoviridae*, pp. 169–210. CAB International, Wallingford, UK.

Barker, H. and Woodford, J.A.T. (1987). Unusually mild symptoms of potato leafroll virus in the progeny of late-infected mother plants. *Potato Res.* **30**, 345–348.

Barker, H., Reavy, B., McGeachy, K.D. and Dawson, S. (1998). Transformation of *Nicotiana benthamiana* with potato mop-top virus coat protein gene produces a novel phenotype mediated by the coat protein. *Mol. Plant Microb. Interact.* **11**, 626–633.

Barker, K.R. and Koenning, S.R. (1998). Developing sustainable systems for nematode management. *Annu. Rev. Phytopathol.* **36**, 165–205.

Barker, R.F., Jarvis, N.P., Thompson, D.V., Loesch-Fries, L.S. and Hall, T.C. (1983). Complete nucleotide sequence of alfalfa mosaic virus RNA3. *Nucl. Acids Res.* **11**, 2881–2891.

Barlow, D.J., Edwards, M.S. and Thornton, J.M. (1986). Continuous and discontinuous protein antigenic determinants. *Nature (Lond.)* **322**, 747–752.

Barnes, J.M., Trinidad-Correa, R., Orum, T.V., Felix-Gastelum, R. and Nelson, M.R. (1999). Landscape ecology as a new infrastructure for improving management of plant viruses and their insect vectors in agroecosystems. *Ecosystem Health* **5**, 26–35.

Barnett, O.W. and Fulton, R.W. (1971). Differential response of *Prunus* necrotic ringspot and tulare apple mosaic viruses to stabilizing agents. *Virology* **46**, 613–619.

Barr, K.J. and Asher, M.J.C. (1996). Studies on the life-cycle of *Polymyxa betae* in sugar beet roots. *Mycol. Res.* **100**, 203–208.

Bartels, R. (1954). Serologische Untersuchungen über das Verhalten des Kartoffel–A–virus in Tabakpflanzen. *Phytopathol. Z.* **21**, 395–406.

Bass, B.L. (2000). Double-stranded RNA as a template for gene silencing. *Cell* **101**, 235–238.

Bassell, G. and Singer, R.H. (1997). MRNA and cytoskeleton filaments. *Curr. Opin.Cell Biol.* **9**, 109–115.

Bassell, G.J., Powers, C.M., Taneja, K.L. and Singer, R.H. (1994a). Single mRNAs visualized on ultrastructural *in situ* hybridization are principally localized at actin filament intersections in fibroblasts. *J. Cell Biol.* **126**, 863–876.

Bassell, G.J., Singer, R.H. and Kosik, K.S. (1994b). Association of poly(A) mRNA with microtubules in cultured neurons. *Neuron* **12**, 571–582.

Bassi, M. and Favali, M.A. (1972). Electron microscopy of maize rough dwarf virus assembly sites in maize: cytochemical and autoradiographic observations. *J. Gen. Virol.* **16**, 153–160.

Bassi, M., Favali, M.A. and Conti, G.G. (1974). Cell wall protrusions induced by cauliflower mosaic virus in Chinese cabbage leaves: a cytotechnical and autoradiographic study. *Virology* **60**, 353–358.

Bassi, M., Appiano, A., Barbieri, N. and D'Agostino, G. (1985). Chloroplast alterations induced by tomato bushy stunt virus in *Datura* leaves. *Protoplasma* **126**, 233–235.

Bastin, M. and Kaesberg, P. (1975). Radioactive labelling of brome mosaic virus. *J. Gen. Virol.* **26**, 321–325.

Bath, J.E. and Tsai, J.H. (1969). The use of aphids to separate two strains of pea enation mosaic virus. *Phytopathology* **59**, 1377–1380.

Baughman, G.A., Jacobs, J.D. and Howell, S.H. (1988). Cauliflower mosaic virus gene VI produces a symptomatic phenotype in transgenic tobacco plants. *Proc. Natl Acad. Sci. USA* **85**, 733–737.

Baulcombe, D.C. (1994). Novel strategies for engineering virus resistance in plants *Curr. Opin. Biotechnol.* **5**, 117–124.

Baulcombe, D.C. (1996a). Mechanisms of pathogen-derived resistance to viruses in transgenic plants. *Plant Cell* **8**, 1833–1844.

Baulcombe, D.C. (1996b). RNA as a target and initiator of post-transcriptional gene silencing in transgenic plants. *Plant Mol. Biol.* **32**, 79–88.

Baulcombe, D.C. (1999a). Viruses and gene silencing in plants. *Arch. Virol.* **15** (Suppl.), 189–201.

Baulcombe, D.C. (1999b). Fast forward genetics based on virus-induced gene silencing. *Curr. Opin. Plant Biol.* **2**, 109–113.

Baulcombe, D.C. (2000). Unwinding RNA silencing. *Science* **290**, 1108–1109.

Baulcombe, D.C., Saunders, G.R., Bevan, M.W., Mayo, M.A. and Harrison, B.D. (1986). Expression of biologically active viral satellite RNA from the nuclear genome of transformed plants. *Nature (Lond.)* **321**, 446–449.

Baulcombe, D.C., Chapman, S. and Santa Cruz, S. (1995). Jellyfish green fluorescent protein as a reporter for virus infections. *Plant J.* **7**, 1045–1053.

Baum, M. and Beier, H. (1998). Wheat cytoplasmic arginine tRNA isoacceptor with a U*CG anticodon is an efficient UGA suppressor *in vitro. Nucl. Acids Res.* **26**, 1390–1395.

Baur, E. (1904). Zur Aetiologie der infectiösen Panachierung. *Ber. Dtsch. Bot. Ges.* **22**, 453–460.

Bawden, F.C. (1964). *Plant Viruses and Virus Diseases*, 4th edn. Ronald Press, New York.

Bawden, F.C. and Kassanis, B. (1950). Some effects of host nutrition on the susceptibility of plants to infection by certain viruses. *Ann. Appl. Biol.* **37**, 46–57.

Bawden, F.C. and Pirie, N.W. (1936). Experiments on the chemical behaviour of potato virus X. *Br. J. Exp. Pathol.* **17**, 64–74.

Bawden, F.C. and Pirie, N.W. (1937). The isolation and some properties of liquid crystalline substances from solanaceous plants infected with three strains of tobacco mosaic virus. *Proc. R. Soc. London B* **123**, 274–320.

Bawden, F.C. and Pirie, N.W. (1956). Observations on the anomalous proteins occurring in extracts of plants infected with strains of tobacco mosaic virus. *J. Gen. Microbiol.* **14**, 460–477.

Bawden, F.C. and Roberts, F.M. (1947). The influence of light intensity on the susceptibility of plants to certain viruses. *Ann. Appl. Biol.* **34**, 286–296.

Bawden, F.C., Pirie, N.W., Bernal, J.D., and Fankuchen, I. (1936). Liquid crystalline substances from virus-infected plants. *Nature (London)* **138**, 1051–1052.

Bayer, M.E. and Bocharov, A.F. (1973). The capsid structure of bacteriophage lambda. *Virology* **54**, 465–475.

Beachy, R.N. (1993). Transgenic resistance to plant viruses. *Semin. Virol.* **4**, 327–416.

Beachy, R.N. (1997). Mechanisms and applications of pathogen-derived resistance in transgenic plants. *Curr. Topics Biotechnol.* **8**, 215–220.

Beachy, R.N. (1999). Coat protein-mediated resistance to tobacco mosaic virus: discovery, mechanisms and exploitation. *Phil Trans. R. Soc. Lond. B* **354**, 659–664.

Beachy, R.N. and Zaitlin, M. (1975). Replication of tobacco mosaic virus: VI. Replicative intermediate and TMV–RNA-related RNAs associated with polyribosomes. *Virology* **63**, 84–97.

Beachy, R.N. and Zaitlin, M. (1977). Characterization and *in vitro* translation of the RNAs from less-than-full-length, virus-related, nucleoprotein rods present in tobacco mosaic vines preparations. *Virology* **81**, 160–169.

Beachy, R.N., Abel, P., Oliver, M.J. *et al.* (1986). Potential for applying genetic information to studies of virus pathogenesis and cross-protection. In: M. Zaitlin, P. Day and A. Hollaender (eds) *Biotechnology in Plant Science: Relevance to Agriculture in the Eighties*, pp. 265–275. Academic Press, Orlando, FL.

Beachy, R.N., Abel, P.P., Nelson, R.S. *et al.* (1996). Expression of genes encoding TMV coat protein and 30-kDa protein in transformed plants. *J. Cell. Biochem. Suppl.* **10C**, 38.

Beale, H.P. (1928). Immunologic reactions with tobacco mosaic virus. *Proc. Soc. Exp. Biol. Med.* **25**, 602.

Beator, J. and Kloppstech, K. (1996). Significance of circadian gene expression in higher plants. *Chronobiol. Int.* **13**, 319–339.

Beck, D.L. and Dawson, W.O. (1990). Deletion of repeated sequences from tobacco mosaic virus mutants with two coat protein genes. *Virology* **177**, 462–469.

Beck, D.L., van Dolleweerd, C.J., Lough, T.J. *et al.* (1994). Disruption of virus movement confers broad-spectrum resistance against systemic infection by plant viruses with a triple gene block. *Proc. Natl Acad. Sci. USA* **91**, 10310–10314.

Béclin, C., Berthomé, R., Palauqui, J.-C., Tepfer, M. and Vaucheret, H. (1998). Infection of tobacco and *Arabidopsis* plants by CMV counteracts systemic post-transcriptional silencing of nonviral (trans)genes. *Virology* **252**, 313–317.

Beczner, L., Horváth, J., Romhanyi, J. and Förster, H. (1984). Studies on the etiology of tuber necrotic ringspot disease in potato. *Potato Res.* **27**, 339–352.

Bedbrook, J.R. (1972). The effects of TYMV infection on photosynthetic carbon metabolism. MSc thesis, University of Auckland, New Zealand.

Bedbrook, J.R. and Matthews, R.E.F. (1973). Changes in the flow of early products of photosynthetic carbon fixation associated with the replication of TYMV. *Virology* **53**, 84–91.

Bedbrook, J.R. and Matthews, R.E.F. (1976). Location, rate and asymmetry of ds-RNA synthesis during replication of TYMV in Chinese cabbage. *Ann. Microbiol. (Paris)* **127A**, 55–60.

Bedbrook, J.R., Douglas, J., and Matthews, R.E.F. (1974). Evidence for TYMV-induced RNA and DNA synthesis in the nuclear fraction from infected Chinese cabbage leaves. *Virology* **58**, 334–344.

Beemster, A.B.R. (1966). The rate of infection of potato tubers with potato virus YN in relation to position of the inoculated leaf. In: A.B.R. Beemster and J. Dijkstra (eds) *Viruses of Plants*, pp. 44–47. North-Holland, Amsterdam.

Beemster, A.B.R. (1967). Partial infection with potato virus YN of tubers from primarily infected potato plants. *Neth. J. Plant Pathol.* **73**, 161–164.

Beemster, A.B.R. (1979). Acquisition of potato virus YN by *Myzus persicae* from primarily infected 'Bintje' potato plants. *Neth. J. Plant Pathol.* **85**, 75–81.

Beer, S.V. and Kosuge, T. (1970). Spermidine and spermine: polyamine components of turnip yellow mosaic virus. *Virology* **40**, 930–938.

Beetham, P.R., Hafner, G.J., Harding, R.M and Dale, J.L. (1997). Two mRNAs are transcribed from banana bunchy top virus DNA-1. *J. Gen. Virol.* **78**, 229–236.

Beffa, R.S., Hofer, R.-M., Thomas, M. and Meins, F., Jr. (1996). Decrease susceptibility to viral disease of β-1,3-gluconase-deficient plants generated by antisense transformation. *Plant Cell* **8**, 1001–1011.

Beier, H. and Bruening, G. (1975). The use of an abrasive in the isolation of cowpea leaf protoplasts which support the multiplication of cowpea mosaic virus. *Virology* **64**, 272–276.

Beier, H. and Bruening, G. (1976). Factors influencing the infection of cowpea protoplasts by cowpea mosaic virus RNA. *Virology* **72**, 363–369.

Beier, H., Siler, D.J., Russell, M.L. and Bruening, G. (1977). Survey of susceptibility to cowpea mosaic virus among protoplasts and intact plants from *Vigna sinensis* lines. *Phytopathology* **67**, 917–921.

Beier, H., Mundry, K.W. and Issinger, O.-G. (1980). *In vivo* and *in vitro* translation of the RNAs of four tobamoviruses. *Intervirology* **14**, 292–299.

Beier, H., Barciszewska, M., Krupp, G., Mitnacht, R. and Gross, H.J. (1984a). UAG readthrough during TMV RNA translation: Isolation and sequence of two tRNAs^{tyr} with suppressor activity from tobacco plants. *EMBO J.* **3**, 351–356.

Beier, H., Barciszewska, M. and Sickinger, H.-D. (1984b). The molecular basis for the differential translation of TMV RNA in tobacco protoplasts and wheat germ extracts. *EMBO J.* **3**, 1091–1096.

Beijerinck, M.W. (1898). Over een contagium vivum fluidum als oorzaak van de vlekziekte der tabaksbladen. *Versl. Gewone Vergad. Wis. Natuurkd. Afd., K. Akad. Wet. Amsterdam* **7**, 229–235.

Bejarano, E.R., Khashoggi, A., Witty, M. and Lichtenstein, C. (1996). Integration of multiple repeats of geminiviral DNA into the nuclear genome of tobacco during evolution. *Proc. Natl Acad. Sci. USA* **93**, 759–764.

Belin, C., Schmitt, C., Gaire, F., Walter, B., Demangeat, G. and Pinck, L. (1999). The nine-C-terminal residues of the grapevine fanleaf nepovirus movement protein are critical for systemic virus spread. *J. Gen. Virol.* **80**, 1347–1356.

Belsham, G.J. and Lomonossoff, G.P. (1991). The mechanism of translation of cowpea mosaic virus middle component RNA: no evidence for internal initiation from experiments in an animal cell transient expression system. *J. Gen Virol.* **72**, 3109–3113.

Belsham, G.J. and Sonenberg, N. (1996). RNA-protein interactions in regulation of picornaviral RNA translation. *Microbiol. Rev.* **60**, 499–513.

Benda, G.T.A and Bennett, C.W. (1964). Effect of curly top virus on tobacco seedlings: infection without obvious symptoms. *Virology* **24**, 97–101.

Bendahmane, A., Köhm, B.A., Dedi, C. and Baulcombe, D.C. (1995). The coat protein of potato virus X is a strain-specific elicitor of *Rx1*-mediated virus resistance in potato. *Plant J.* **8**, 933–941.

Bendahmane, A., Kanyuka, K.V and Baulcombe, D.C. (1997a). High resolution and physical mapping of the *Rx* gene for extreme resistance to potato virus X in tetraploid potato. *Theor. Appl. Genet.* **95**, 153–162.

Bendahmane, A., Kanyuka, K. and Baulcombe, D.C. (1999). The *Rx* gene from potato controls separate virus resistance and cell death responses. *Plant Cell* **11**, 781–791.

Bendahmane, A., Querci, M., Kanyuka, K. and Baulcombe, D.C. (2000). *Agrobacterium* transient expression system as a tool for the isolation of disease resistance genes: application to the *Rx2* locus in potato. *Plant J.* **21**, 73–81.

Bendahmane, M. and Beachy, R.N. (1999). Control of tobamovirus infections via pathogen-derived resistance. *Adv. Virus Res.* **53**, 369–386.

Bendahmane, M. and Gronenborn, B. (1997). Engineering resistance against tomato yellow leaf curl virus (TYLCV) using antisense RNA. *Plant Mol. Biol.* **33**, 351–357.

Bendahmane, M., Fitchen, J.H., Zang, G.H. and Beachy, R.N. (1997b). Studies of coat protein-mediated resistance to tobacco mosaic tobamovirus: correlation between assembly of mutant coat proteins and resistance. *J. Virol.* **71**, 7942–7950.

Benfey, P.N. and Chua, N.-H. (1989). Regulated genes in transgenic plants. *Science* **244**, 174–181.

Benfey, P.N. and Chua, N.H. (1990). The cauliflower mosaic virus 35S promoter: combinatorial regulation of transcription in plants. *Science* **250**, 959–966.

Bennett, C.W. (1940a). Relation of food translocation to movement of virus of tobacco mosaic. *J. Agric. Res.* **60**, 361–390.

Bennett, C.W. (1940b). Acquisition and transmission of viruses by dodder (*Cuscuta subinclusa*). *Phytopathology* **30**, 2 (abstract).

Bennett, C.W. (1944). Latent virus of dodder and its effect on sugar beet plants. *Phytopathology* **34**, 77–91.

Bennett, C.W. (1955). Recovery of water pimpernel from curly top and the reaction of recovered plants to reinoculation with different virus strains. *Phytopathology* **45**, 531–536.

Bennett, C.W. (1960). Sugar beet yellows disease in the United States. *US Dep. Agric., Tech. Bull.* **1218**, 1–63.

Bennett, C.W. (1963). Highly virulent strains of curly top virus in sugar beet in western United States. *J. Am. Soc. Sugar Beet Technol.* **12**, 515–520.

Bennett, C.W. (1969). Seed transmission of plant viruses. *Adv. Virus Res.* **14**, 221–261.

Ben-Sin, C. and Po, T. (1982). Infection of barley protoplasts with barley stripe mosaic virus detected and assayed by immunoperoxidase. *J. Gen. Virol.* **58**, 323–327.

Bentley, G.A., Lewit-Bentley, A., Liljas, L., Skoglund, U., Roth, M. and Unge, T. (1987). Structure of RNA in satellite tobacco necrosis virus: a low-resolution neutron diffraction study using $H_2O/^2H_2O$ solvent contrast variation. *J. Mol. Biol.* **194**, 129–141.

Berger, P.H. and Toler, R.W. (1983). Quantitative immunoelectrophoresis of panicum mosaic virus and strains of St. Augustine decline. *Phytopathology* **73**, 185–189.

Berger, P.H., Thornbury, D.W. and Pirone, T.P. (1985). Detection of picogram quantities of potyviruses using a dot blot immunobinding assay. *J. Virol. Methods* **12**, 31–39.

Berger, P.H., Barnett, O.W., Brunt, A.A. *et al.* (2000). Family *Potyviridae*. In: M.H.V. van Regenmortel, C.M. Fauquet, D.H.L. Bishop *et al.* (eds) *Virus Taxonomy: Classification and Nomenclature of Viruses. Seventh Report of the International Committee on Taxonomy of Viruses*, pp. 703–724. Academic Press, San Diego.

Bergstrom, C.T., McElhany, P. and Real, L.A. (1999). Transmission bottlenecks as determinants of virulence in rapidly evolving pathogens. *Proc. Natl. Acad. Sci. USA* **96**, 5095–5100.

Bergström, G.C., Johnson, M.C. and Ku , J. (1982). Effects of local infection of cucumber by *Colletotrichum lagenarium, Pseudomonas lachrymans* or tobacco necrosis virus on systemic resistance to cucumber mosaic virus. *Phytopathology* **72**, 922–926.

Berlinger, M.J. and Daban, R. (1987). Breeding for resistance to virus transmission by whiteflies in tomatoes. *Insect Sci. Appl.* **8**, 783–784.

Berna, A., Briand, J.-P., Stussi-Garaud, C. and Godefroy-Colburn, T. (1986). Kinetics of accumulation of the three non-structural proteins of alfalfa mosaic virus in tobacco plants. *J. Gen. Virol.* **67**, 1135–1147.

Bernal, J.D. and Fankuchen, I. (1937). Structure types of protein 'crystals' from virus-infected plants. *Nature (London)* **139**, 923–924.

Bernardi, G. (1971). Chromatography of nucleic acids on hydroxyapatite columns. In: L. Grossman and K. Moldave (eds) *Methods in Enzymology* Vol. 21, Part D, pp. 95–139. Academic Press, New York.

Berrie, L.C., Rybicki, E.P. and Rey, M.E.C. (2001). Complete nucleotide sequence and host range of South African cassava mosaic virus: further evidence for recombination among begomoviruses. *J. Gen. Virol.* **82**, 53–58.

Bertens, P., Wellink, J., Goldbach, R. and van Kammen, A. (2000). Mutational analysis of the cowpea mosaic virus movement protein. *Virology* **267**, 199–208.

Bertioli, D.J., Harris, R.D., Edwards, M.I., Cooper, J.I. and Hawes, W.S. (1991). Transgenic plants and insect cells expressing the coat proteins of arabis mosaic virus produce empty virus-like particles. *J. Gen. Virol.* **72**, 1801–1809.

Berzal-Herranz, A., de la Cruz, A., Tenllado, F. *et al.* (1995). The *Capsicum* L^3 gene-mediated resistance against tobamoviruses is elicited by the coat protein. *Virology* **209**, 498–505.

Berzofsky, J.A. (1985). Intrinsic and extrinsic factors in protein antigenic structure. *Science* **229**, 932–940.

Best, R.J. (1936). Precipitation of the tobacco mosaic virus complex at its isoelectric point. *Aust. J. Exp. Biol. Med. Sci.* **14**, 1–13.

Best, R.J. (1937a). Artificially prepared visible paracrystalline fibres of tobacco mosaic virus nucleoprotein. *Nature (London)* **140**, 547–548.

Best, R.J. (1937b). The quantitative estimation of relative concentrations of the viruses of ordinary and yellow tobacco mosaics and of tomato spotted wilt by the primary lesion method. *Aust. J. Exp. Biol. Med. Sci.* **14**, 1–13.

Best, R.J. (1968). Tomato spotted wilt virus. *Adv. Virus Res.* **13**, 65–146.

Best, R.J. and Katekar, G.F. (1964). Lipid in a purified preparation of tomato spotted wilt virus. *Nature (London)* **203**, 671–672.

Bevan, M.W., Mason, S.E. and Goelet, P. (1985). Expression of tobacco mosaic virus coat protein by a cauliflower mosaic virus promoter in plants transformed by *Agrobacterium*. *EMBO J.* **4**, 1921–1926.

Bharadwaj, R.K., Reddy, D.V.R. and Sinha, R.C. (1966). A reinvestigation of the alimentary canal in the leaf-hopper *Agallia constricta* (Homoptera: Cicadellidae). *Ann. Entomol. Soc. Am.* **59**, 616–617.

Bhargava, K.S. (1951). Some properties of four strains of cucumber mosaic virus. *Ann. Appl. Biol.* **38**, 377–388.

Bhattacharyya-Pakrasi, M., Pen, J., Elmer, J.S. *et al.* (1993). Specificity of a promoter from the rice tngro bacilliform virus for expression in phloem tissues. *Plant* 4, 71–79.

Bhuvaneshwara, M., Subramanya, H.S., Gopinath, K., Savithri, H.S., Nayudu, M.V. and Murthy, M.R.N. (1995). Structure of sesbania mosaic virus at 3 angstrom resolution. *Structure* **3**, 1021–1030.

Bigger, M. (1981). The relative abundance of the mealy bug vectors (Hemiptera, Coccidae and Pseudococcidae) of cocoa swollen shoot disease in Ghana. *Bull. Entomol. Res.* **71**, 435–448.

Bird, L.E., Subramanya, H.S. and Wigley, D.B. (1998). Helicases: a unifying structural theme. *Curr. Opin. Struct. Biol.* **8**, 14–18.

Bisaillon, M. and Lemay, G. (1997). Viral and cellular enzymes involved in the synthesis of mRNA cap structure. *Virology* **238**, 1–7.

Bisaro, D.M. (1994). Recombination in geminiviruses: mechanisms for maintaining genome size and generating genomic diversity. In: J. Paszkowski (ed.) *Homologous Recombination in Plants*, pp. 39–60. Kluwer, Amsterdam.

Bisessar, S. and Temple, P.J. (1977). Reduced ozone injury on virus-infected tobacco in the field. *Plant Dis. Rep.* **61**, 961–963.

Bitterlin, M.W., Gonsalves, D. and Scorza, R. (1987). Improved mechanical transmission of tomato ringspot virus to Prunus seedlings. *Phytopathology* **77**, 560–563.

Bittner, H., Schenk, G. and Schuster, G. (1987). Chemotherapeutical elimination of potato virus X from potato stem cuttings. *J. Phytopathol.* **120**, 90–92.

Black, L.M. (1941). Specific transmission of varieties of potato yellow-dwarf virus by related insects. *Am. Potato J.* **18**, 231–233.

Black, L.M. (1972). Plant tumors of viral origin. *Prog. Exp. Tumor Res.* **15**, 110–137.

Black, L.M. (1984). The controversy regarding multiplication of some plant viruses in their insect vectors. *Curr. Top. Vector Res.* **2**, 1–30.

Blackman, I.M., Boevink, P., Santa Cruz, S., Palukaitis, P. and Oparka, K.J. (1998). The movement protein of cucumber mosaic virus traffics into sieve elements in minor veins of *Nicotiana clevelandii*. *Plant Cell* **10**, 525–537.

Blackman, L.M., Harper, J.D.I. and Overall, R.L. (1999). Localization of a centrin-like protein to higher plant plasmodesmata. *Eur. J. Cell Biol.* **78**, 297–304.

Blair, P. (1719). *Botanik Essays*. W.&J. Innys, London.

Blanc, S., Schmidt, I., Kuhl, G. *et al.* (1993a). Paracrystalline structure of cauliflower mosaic virus aphid transmission factor produced both in plants and in a heterologous system and relationship with a solubilized active form. *Virology* **197**, 283–292.

Blanc, S., Cerutti, M., Usmany, M., Vlak, J.M. and Hull, R. (1993b). Biological activity of cauliflower mosaic virus aphid transmission factor expressed in a heterologous system. *Virology* **192**, 643–650.

Blanc, S., Schmidt, I., Vantard, M. *et al.* (1996). A plant virus protein decorates cellular microtubules, yielding a highly stable complex. *Proc. Natl Acad. Sci. USA* **93**, 15158–15163.

Blanc, S. Hébrard, E., Drucker, M. and Froissart, R. (2001). Molecular basis of vector transmission: caulimovirus. In: K. Harris, J.E. Duffus and O.P. Smith (eds) *Virus–Insect–Plant Interactions*. Academic Press, San Diego (in press).

Blanch, E.W., Robinson, D.J., Hecht, L. and Barron, L.D. (2001). A comparison of the solution structures of tobacco rattle and tobacco mosaic viruses from Raman optical activity. *J. Gen. Virol.* **82**, 1499–1502.

Blencowe, J.W. and Tinsley, T.W. (1951). The influence of density of plant population on the incidence of yellows in sugar beet crops. *Ann. Appl. Biol.* **38**, 395–401.

Blok, J., Mackenzie, A., Guy, P. and Gibbs, A. (1987). Nucleotide sequence comparisons of turnip yellow mosaic virus isolates from Australia and Europe. *Arch. Virol.* **97**, 283–295.

Blok, V.C., Ziegler, A., Robinson, D.J. and Murant, A.F. (1994). Sequences of 10 variants of the satleeite-like RNA3 of groundnut rosette virus. *Virology* **202**, 25–32.

Bloomer, A.C., Champness, J.N., Bricogne, G., Staden, R. and Klug, A. (1978). Protein disk of tobacco mosaic virus at 2.8 Å resolution showing the interactions within and between subunits. *Nature (London)* **276**, 362–368.

Blum, H., Gross, H.J. and Beier, H. (1989). The expression of the TMV-specific 30-kDa protein in tobacco protoplasts is strongly and selectively enhanced by actinomycin. *Virology* **169**, 51–61.

Blunt, W. (1950). *Tulipomania*. Penguin Books, Harmondsworth, England.

Boag, B. (1985). The localised spread of virus-vector nematodes adhering to farm machinery. *Nematologica* **31**, 234–235.

Boccara, M., Hamilton, W.D.O. and Baulcombe, D.C. (1986). The organisation and interviral homologies of genes at the 3' end of tobacco rattle virus RNA1. *EMBO J.* **5**, 223–229.

Boccardo, G. and Milne, R.G. (1975). The maize rough dwarf virion: I. Protein composition and distribution of RNA in different viral fractions. *Virology* **68**, 79–85.

Boccardo, G. Hatta, T., Francki, R.I.B. and Grivell, C.J. (1980). Purification and some properties of reovirus-like particles from leafhoppers and their possible involvement in wallaby ear disease of maize. *Virology* **100**, 300–313.

Boccardo, G., Lisa, V., Lusoni, E. and Milne, R.G. (1987). Cryptic plant viruses. *Adv. Virus Res.* **32**, 171–214.

Bockstahler, L.E. (1967). Biophysical studies on double-stranded RNA from turnip yellow mosaic virus-infected plants. *Mol. Gen. Genet.* **100**, 337–348.

Boedtker, H. (1960). Configurational properties of tobacco mosaic virus ribonucleic acid. *J. Mol. Biol.* **2**, 171–188.

Boege, F., Rohde, W. and Sänger, H.L. (1982). *In vitro* transcription of viroid RNA into full-length copies by RNA-dependent RNA polymerase from healthy tomato leaf tissue. *Biosci. Rep.* **2**, 185–194.

Boevink, P., Chu, P.W.G. and Keese, P. (1995). Sequence of subterranean clover stunt virus DNA: affinities with the geminiviruses. *Virology* **207**, 354–361.

Bol, J.F. (1988). Structure and expression of plant genes encoding pathogenesis related proteins. In: D.P.S. Verina and R.B. Goldberg (eds) *Plant Gene Research*, pp. 201–221. Springer, New York.

Bol, J.F. (1999a). Alfamovirus and Ilarviruses (Bromoviridae). In: A. Granoff and R.G. Webster (eds) *Encyclopedia of Virology*, 2nd edn, pp. 38–43 Academic Press, San Diego.

Bol, J.F. (1999b). Alfalfa mosaic virus and ilarviruses: involvement of coat protein in multiple steps of the replication cycle. *J. Gen. Virol.* **80**, 1089–1102.

Bol, J.F. and Kruseman, J. (1969). The reversible dissociation of alfalfa mosaic virus. *Virology* **37**, 485–489.

Bol, J.F. and Lak-Kaashoek, M. (1974). Composition of alfalfa mosaic virus nucleoproteins. *Virology* **60**, 476–484.

Bol, J.F. and van Kan, J.A.L. (1988). The synthesis and possible functions of virus-induced proteins in plants. *Microbiol. Sci.* **5**, 47–52.

Bol, J.F. and van Vloten-Doting, L. (1973). Function of top component a RNA in the initiation of infection by alfalfa mosaic virus. *Virology* **51**, 102–108.

Bol, J.F., van Vloten-Doting, L. and Jaspars, E.M.J (1971). A functional equivalence of top component a RNA in the initiation of infection by alfalfa mosaic virus. *Virology* **46**, 73–85.

Bol, J.F., Brederode, F.T., Janze, G.C. and Rauh, D.K. (1975). Studies on sequence homology between the RNAs of alfalfa mosaic virus. *Virology* **65**, 1–15.

Bollen, L.S., Crowley, A., Stodulski, G. and Hau, J. (1996). Antibody production in rabbits and chickens immunized with human IgG —a comparison of titre and avidity development in rabbit serum, chicken serum and egg yolk using three different adjuvants. *J. Immunol. Methods* **191**, 113–120.

Boller, T. and Meins, F. (eds) (1992). *Genes Involved in Plant Defense*. Springer, Vienna.

Bonfiglioli, R., McFadden, G.I. and Symons, R.H. (1994). *In situ* hybridization localizes avocado sunblotch viroid on chloroplast thylacoid membranes and coconut cadang cadang viroid in the nucleus. *Plant J.* **6**, 99–103.

Bonfiglioli, R., Webb, D.R. and Symons, R.H. (1996). Tissue and intra-cellular distribution of coconut cadang cadang viroid and citrus exocortis viroid determined by *in situ* hybridization and confocal laser scanning and transmission electron microscopy. *Plant J.* **9**, 457–465.

Bonneau, C., Brugidou, C., Chen, L., Beachy, R.N. and Fauquet, C. (1998). Expression of the rice yellow mottle virus P1 protein *in vitro* and *in vivo* and its involvement in virus spread. *Virology* **244**, 79–86.

Bonneville, J.-M., Sanfaçon, H., Fütterer, J. and Hohn, T. (1989). Posttranslational transactivation in cauliflower mosaic virus. *Cell* **59**, 1135–1143.

Boonham, N., Henry, C.R. and Wood, K.R. (1995). The nucleotide sequence and proposed genome organization of oat chlorotic stunt virus, a new soil-borne virus of cereals. *J. Gen. Virol.* **76**, 2025–2034.

Borissensko, S., Schuster, G. and Schmygla, W. (1985). Obtaining a high percentage of explants with negative serological reactions against viruses by combining potato meristem culture with phytoviral chemotherapy. *Phytopathol. Z.* **114**, 185–188.

Borja, M., Rubio, T., Scholthof, H.B. and Jackson, A.O. (1999). Restoration of wild-type virus by double recombination of tombusvirus mutants with a host transgene. *Mol. Plant–Microbe Interact.* **12**, 153–162.

Bos, L. (1970). The identification of three new viruses isolated from *Wisteria* and *Pisum* in The Netherlands and the problem of variation within the potato virus Y group. *Neth. J. Plant Pathol.* **76**, 8–46.

Bos, L. (1975). The application of TMV particles as an internal magnification standard for determining virus particle sizes with the electron microscope. *Neth. J. Plant Pathol.* **81**, 168–175.

Bos, L. (1978). *Symptoms of Virus Diseases in Plants*, 3rd edn. Pudoc, Wageningen.

Bos, L.(1982). Crop losses caused by viruses. *Crop Protect.* **1**, 263–282.

Bos, L. (1999a). Beijerinck's work on tobacco mosaic virus: historical context and legacy. *Phil. Trans. R. Soc.Lond. B* **354**, 675–685.

Bos, L. (1999b). The naming of viruses: an urgent call to order. *Arch. Virol.* **144**, 631–636.

Bos, L. (2000a). 100 years of virology: from vitalism via molecular biology to genetic engineering. *Trends Microbiol.* **8**, 82–87.

Bos, L. (2000b). Structure and typography of virus names. *Arch. Virol.* **145**, 429–432.

Botstein, D. (1980). A theory of modular evolution for bacteriophages. *Ann. N. Y. Acad. Sci.* **354**, 484–491.

Botstein, D. and Fink, G.R. (1988). Yeast: an experimental organism for modern biology. *Science* **240**, 1439–1443.

Bottcher, B. and Crowther, R.A. (1996). Difference imaging reveals ordered regions of RNA in turnip yellow mosaic virus. *Structure* **4**, 387–394.

Bottrell, D.G. and Barbosa, P. (1998). Manipulating natural enemies by plant varietal selection and modification: a realistic strategy? *Annu. Rev. Entomol.* **43**, 347–367.

Bouley, J.-P., Briand, J.-P., Genevaux, M., Pinck, M. and Witz, J. (1976). The structure of eggplant mosaic virus: evidence for the presence of low molecular weight RNA in top component. *Virology* **69**, 775–781.

Boulton, M.I. and Markham, P.G. (1986). The use of squash-blotting to detect plant pathogens in insect vectors. In: R.A.C. Jones and L. Torrance (eds) *Developments and Applications in Virus Testing*, Dev. Appl. Biol. I, pp. 55–69. Association of Applied Biologists, Wellesbourne.

Boulton, M.I., Buchholz, W.G., Marks, M.S., Markham, P.G. and Davies, J.W. (1989) Specificity of *Agrobacterium*-mediated delivery of maize streak virus DNA to members of the Graminae. *Plant Mol. Biol.* **12**, 31–40.

Boulton, M.I., King, D.I., Markham, P.G., Pinner, M.S. and Davies, J.W. (1991a). Host range and symptoms are determined by specific domains of the maize streak virus genome. *Virology* **181**, 312–318.

Boulton, M.I., King, D.I., Donson, J. and Davies, J.W. (1991b). Point substitutions in a promoter-like region and the V1 gene affect the host range and symptoms of maize streak virus. *Virology* **183**, 114–121.

Boulton, M.I., Pallaghy, C.K., Chatani, M., MacFarlane, S. and Davies, J.W. (1993). Replication of maize streak virus mutants in maize protoplasts: evidence for a movement protein. *Virology* **192**, 85–93.

Boulton, R.E. (1996). Pea early-browning tobravirus. *Plant Pathol.* **45**, 13–28.

Boulton, R.E., Jellis, G.J., Baulcombe, D.C. and Squire, A.M. (1986). The application of complementary DNA probes to routine virus detection, with particular reference to potato viruses. In: R.A.C. Jones and L. Torrance (eds) *Developments and Applications in Virus Testing*, Dev. Appl. Biol. I, pp. 41–53. Association of Applied Biologists, Wellesbourne.

Bouquet, A. (1981). Resistance to grapevine fanleaf virus in muscadine grape inoculated with *Xiphinema index*. *Plant Dis.* **65**, 791–793.

Bourdin, D. and Lecoq, H. (1991). Evidence that heteroencapsidation between two potyviruses is involved in aphid transmission of a non-aphid-transmissible isolate from mixed infections. *Phytopathology* **81**, 1459–1464.

Bousalem, M., Douzery, E.J.P. and Fargette, D. (2000). High genetic diversity, distant phylogenetic relationships and intraspecies recombination events among natural populations of yam mosaic virus: a contribution to understanding potyvirus evolution. *J. Gen. Virol.* **81**, 243–255.

Bouzoubaa, S., Guilley, H., Jenard, G., Richards, K. and Putz, C. (1985). Nucleotide sequence analysis of RNA3 and RNA4 of beet necrotic yellow vein virus, isolates F2 and G1. *J. Gen. Virol.* **66**, 1553–1564.

Bouzoubaa, S., Ziegler, V., Beck, D., Guilley, H., Richards, K. and Jenard, G. (1986). Nucleotide sequence of beet necrotic yellow vein virus RNA2. *J. Gen. Virol.* **67**, 1689–1700.

Bouzoubaa, S., Quillet, L., Guilley, H., Jonard, G. and Richards, K. (1987). Nucleotide sequence of beet necrotic yellow vein virus RNA1. *J. Gen. Virol.* **68**, 615–626.

Bové, C., Mocquot, B. and Bové, J.M. (1972). Turnip yellow mosaic virus RNA synthesis in the plastids: partial purification of a virus-specific, DNA-independent enzyme–template complex. *Symp. Biol. Hung.* **13**, 43–59.

Bové, J.M. and Bové, C. (1985). Turnip yellow mosaic virus RNA replication on the chloroplast envelope. *Physiol. Vég.* **23**, 741–748.

Bové, J.M., Bové, C., Rondot, M.-J. and Morel, G. (1967). Chloroplasts and virus-RNA synthesis. In: T.W. Goodwin (ed.) *The Biochemistry of Chloroplasts*, Vol. 2, pp. 329–339. Academic Press, New York.

Boxall, M. and MacNeill, B.H. (1974). Local lesions as sources of biologically pure strains of tobacco mosaic virus. *Can. J. Bot.* **52**, 23–25.

Boyko, V., Ferralli, J. and Heinlein, M. (2000a). Cell-to-cell movement of TMV RNA is temperature-dependent and corresponds to the association of movement protein with microtubules. *Plant J.* **22**, 315–325.

Boyko, V., Ferralli, J., Ashby, J., Schellenbaum, P. and Heinlein, M. (2000b). Function of microtubules in intercellular transport of plant virus RNA. *Nature Cell Biol.* **2**, 826–832.

Boyko, V., van der Laak, J., Ferralli, J., Suslova, E., Kwon, M.-O. and Heinlein, M. (2000c). Cellular targets of functional and dysfunctional mutants of tobacco mosaic virus movement protein fused to green fluorescent protein. *J. Virol.* **74**, 11339–11346.

Boyle, J.S. and Bergman, E.L. (1967). Factors affecting incidence and severity of internal browning of tomato induced by tobacco mosaic virus. *Phytopathology* **57**, 354–362.

Bozarth, C.S., Weiland, J.J. and Dreher, T.E. (1992). Expression of ORF-69 of turnip yellow mosaic virus is necessary for viral spread in plants. *Virology* **187**, 124–130.

Brachmann, R.K. and Boeke, J.D. (1997). Tag games in yeast: the two hybrid system and beyond. *Curr. Opin. Biotech.* **8**, 561–568.

Bracho, M.A., Moya, A. and Barrio, E. (1998). Contribution of *Taq* polymerase-induced errors to the estimation of RNA diversity. *J. Gen. Virol.* **79**, 2921–2928.

Bradel, B.G., Preil, W. and Jeske, H. (2000). Sequence analysis and genome organization of Poinsettia mosaic virus (PnMV) reveal closer relationship to marafiviruses than to tymoviruses. *Virology* **271**, 289–297.

Bradley, R.H.E. (1956). Effects of depth of stylet penetration on aphid transmission of potato virus Y. *Can. J. Microbiol.* **2**, 539–547.

Bradley, R.H.E. (1959). Loss of virus from the stylets of aphids. *Virology* **8**, 308–318.

Bradley, R.H.E. (1966). Which of an aphid's stylets carry transmissible virus? *Virology* **29**, 396–401.

Bradley, R.H.E. and Rideout, D.W. (1953). Comparative transmission of potato virus Y by four aphid species that infest potato. *Can. J. Zool.* **31**, 333–341.

Bradley, R.H.E., Wade, C.V. and Wood, F.A. (1962). Aphid transmission of potato virus Y inhibited by oils. *Virology* **18**, 327–328.

Bragg, J.N., Lawrence, D.M. and Jackson, A.O. (2001). Interactions of barley stripe mosaic virus B protein. Abstract W42-3, 20th Annual Meeting of the American Society for Virology.

Brakke, M.K. (1951). Density gradient centrifugation: a new separation technique. *J. Am. Chem. Soc.* **73**, 1847–1848.

Brakke, M.K. (1953). Zonal separations by density gradient centrifugation. *Arch. Biochem. Biophys.* **45**, 275–290.

Brakke, M.K. (1960). Density gradient centrifugation and its application to plant viruses. *Adv. Virus Res.* **7**, 193–224.

Brakke, M.K. (1963). Stabilization of brome grass mosaic virus by magnesium and calcium. *Virology* **19**, 367–374.

Brakke, M.K. (1970). Systemic infections for the assay of plant viruses. *Annu. Rev. Phytopathol.* **8**, 61–84.

Brakke, M.K. (1984). Mutations, the aberrant ratio phenomenon, and virus infection of maize. *Annu. Rev. Phytopathol.* **22**, 77–94.

Brakke, M.K. and Daly, J.M. (1965). Density-gradient centrifugation: non-ideal sedimentation and the interaction of major and minor components. *Science* **148**, 387–389.

Brakke, M.K. and Rochow, W.F. (1974). Ribonucleic acid of barley yellow dwarf virus. *Virology* **61**, 240–248.

Brakke, M.K. and van Pelt, N. (1970). Properties of infectious ribonucleic acid from wheat streak mosaic virus. *Virology* **42**, 699–706.

Brakke, M.K., Ball, E.M., Hsu, Y.H. and Langenberg, W.G. (1987a). Wheat streak mosaic virus cylindrical inclusion body protein. *J. Gen. Virol.* **68**, 281–287.

Brakke, M.K., White, J.L., Samson, R.G. and Joshi, J. (1987b). Chlorophyll, chloroplast ribosomal RNA, and DNA are reduced by barley stripe mosaic virus systemic infection. *Phytopathology* **78**, 570–574.

Brakke, M.K., Ball, E.M. and Langenberg, W.G. (1988). A non-capsid protein associated with unencapsidated virus RNA in barley infected with barley stripe mosaic virus. *J. Gen. Virol.* **69**, 481–491.

Branch, A.D. and Dickson, E. (1980). Tomato DNA contains no detectable regions complementary to potato spindle tuber viroid as assayed by Southern hybridization. *Virology* **104**, 10–26.

Branch, A.D. and Robertson, H.D. (1984). A replication cycle for viroids and other small infectious RNAs. *Science* **223**, 450–455.

Branch, A.D., Robertson, H.D. and Dickson, E. (1981). Longer-than-unit-length viroid minus strands are present in RNA from infected plants. *Proc. Natl Acad. Sci. USA* **78**, 6381–6385.

Branch, A.D., Robertson, H.D., Greer, C., Gegenheimer, P., Peebles, C. and Abelson, J. (1982). Cell-free circularization of viroid progeny RNA by an RNA ligase from wheat germ. *Science* **217**, 1147–1149.

Branch, A.D., Benenfeld, B.J. and Robertson, H.D. (1985). Ultraviolet light-induced crosslinking reveals a unique region of local tertiary structure in potato spindle tuber viroid and HeLa 5S RNA. *Proc. Natl Acad. Sci. USA* **82**, 6590–6594.

Branch, A.D., Benenfeld, B.J. and Robertson, H.D. (1988a). Evidence for a single rolling circle in the replication of potato spindle tuber viroid. *Proc. Natl Acad. Sci. USA* **85**, 9128–9132.

Branch, A.D., Benenfeld, B.J., Franck, E.R. *et al.* (1988b). Interference between coinoculated viroids. *Virology* **163**, 538–546.

Brandes, A., Thompson, H., Dean, C. and Heslop-Harrison, J.S. (1997). Multiple repetitive DNA sequences in the paracentromeric regions of *Arabidopsis thaliana* L. *Chromosome Res.* **5**, 238–246.

Brault, V. and Miller, W.A. (1992). Translational frameshifting mediated by a viral sequence in plant cells. *Proc. Natl Acad. Sci. USA* **89**, 2262–2266.

Brault, V., Mutterer, J., Scheidecker, D. *et al.* (2000). Effects of point mutations in the readthrough domain of the beet western yellows virus minor capsid protein on virus accumulation in planta and on transmission by aphids. *J. Virol.* **74**, 1140–1148.

Brederode, F.T., Koper-Zwarthoff, E.C. and Bol, J.F. (1980). Complete nucleotide sequence of alfalfa mosaic virus RNA4. *Nucl. Acids Res.* **8**, 2213–2223.

Brederode, F.T., Taschner, P.E.M., Posthumus, E. and Bol, J.F. (1995). Replicase-mediated resistance to alfalfa mosaic virus. *Virology* **207**, 467–474.

Breyel, E., Casper, R., Ansa, O.A., Kuhn, C.W., Misari, S.M. and Demski, J.W. (1988). A simple procedure to detect a dsRNA associated with groundnut rosette. *J. Phytopathol.* **121**, 118–124.

Briand, J.-P. (1978). Contribution à l'étude de la structure des tymovirus et à l'organisation de leut génome. Thèse DSc, l'Université Louis Pasteur de Strasbourg.

Briand, J.-P., Al-Moudallal, Z. and van Regenmortel, M.H.V. (1982). Serological differentiation of tobamoviruses by means of monoclonal antibodies. *J. Virol. Meth.* **5**, 293–300.

Briddon, R.W., Watts, J., Markham, P.G. and Stanley, J. (1989). The coat protein of beet curly top virus is essential for infectivity. *Virology* **172**, 628–633.

Briddon, R.W., Pinner, M.S., Stanley, J. and Markham, P.G. (1990). Geminivirus coat protein gene replacement alters insect specificity. *Virology* **177**, 85–94.

Briddon, R.W., Bedford, I.D., Tsai, J.H. and Markham, P.G. (1996). Analysis of the nucleotide sequence of the treehopper-transmitted geminivirus, tomato pseudo-curly top virus, suggests a recombinant origin. *Virology* **219**, 387–394.

Brierley, I. (1995). Ribosomal frameshifting on viral RNAs. *J. Gen. Virol.* **76**, 1885–1892.

Brierley, K.M., Goodman, B.A. and Mayo, M.A. (1993). A mobile element on a virus particle identified by nuclear-magnetic-resonance spectroscopy. *Biochem. J.* **293**, 657–659.

Brigneti, G., Voinnet, O., Li, W.-X., Ji, L.-H., Ding, S.-W. and Baulcombe, D.C. (1998). Viral pathogenicity determinants are suppressors of transgene silencing in *Nicotiana benthamiana*. *EMBO J.* **17**, 6739–6746.

Brill, L.M., Nunn, R.S., Kahn, T.W., Yeager, M. and Beachy, R.N. (2000). Recombinant tobacco mosaic virus movement protein is an RNA-binding, α-helical membrane protein. *Proc. Natl Acad. Sci. USA* **97**, 7112–7117.

Brisco, M.J., Hull, R. and Wilson, T.M.A. (1985). The effect of extraction protocol on the yield, purity, and translation products of RNA from an isometric plant virus. *J. Virol. Meth.* **10**, 195–202.

Brisco, M.J., Hull, R. and Wilson, T.M.A. (1986a). Swelling of isometric and of bacilliform plant virus nucleocapsids is required for virus-specific protein synthesis *in vitro*. *Virology* **148**, 210–217.

Brisco, M.J., Haniff, C., Hull, R., Wilson, T.M.A. and Sattelle, D.B. (1986b). The kinetics of swelling of Southern bean mosaic virus: a study using photon correlation spectroscopy. *Virology* **148**, 218–220.

Brisson, N., Paszkowski, J., Penswick, J.R., Gronenborn, B., Potrykus, I. and Hohn, T. (1984). Expression of a bacterial gene in plants by using a viral vector. *Nature (Lond.)* **310**, 511–514.

Britten, R.J. and Davidson, E.H. (1985). Hybridization strategy. In: B.D. Hames and S.J. Higgins (eds) *Nucleic Acid Hybridization: A Practical Approach*, pp. 3–15. IRL Press, Oxford.

Britten, R.J., Graham, D.E. and Neufeld, B.R. (1974). Analysis of repeating DNA sequences by reassociation. *Methods Enzymol.* **29**, 363–418.

Brlansky, R.H. (1987). Inclusion bodies produced in *Citrus* spp. by citrus tristeza virus. *Phytophylactica* **19**, 211–213.

Brlansky, R.H., Carroll, T.W. and Zaske, S.K. (1986). Some ultrastructural aspects of the pollen transmission of barley stripe mosaic virus in barley. *Can. J. Bot.* **64**, 853–858.

Broadbent, L. (1957). *Investigation of Virus Diseases of Brassica Crops*, A.R.C. Rep. Ser. No. 14. Cambridge University Press, London.

Broadbent, L. (1976). Epidemiology and control of tomato mosaic virus. *Annu. Rev. Phytopathol.* **14**, 75–96.

Broadbent, L. (1962). The epidemiology of tomato mosaic. II. Smoking tobacco as a source of virus. *Ann. Appl. Biol.* **50**, 461–466.

Broadbent, L. (1963). The epidemiology of tomato mosaic. III. Cleaning virus from hands and tools. *Ann. Appl. Biol.* **52**, 225–232.

Broadbent, L. (1964). The epidemiology of tomato mosaic. VII. The effect of TMV on tomato fruit yield and quality under glass. *Ann. Appl. Biol.* **54**, 209–224.

Broadbent, L. (1965a). The epidemiology of tomato mosaic. IX. Transmission of TMV by birds. *Ann. Appl. Biol.* **55**, 67–69.

Broadbent, L. (1965b). The epidemiology of tomato mosaic. VII. Virus infection through tomato roots. *Ann. Appl. Biol.* **55**, 57–66.

Broadbent, L. (1965c). The epidemiology of tomato mosaic. XI. Seed transmission of TMV. *Ann. Appl. Biol.* **56**, 177–205.

Broadbent, L. and Fletcher, J.T. (1963). The epidemiology of tomato mosaic. IV. Persistence of the virus on clothing and glasshouse structures. *Ann. Appl. Biol.* **52**, 233–241.

Broadbent, L., Gregory, P.H. and Tinsley, T.W. (1952). The influence of planting date and manuring on the incidence of virus diseases in potato crops. *Ann. Appl. Biol.* **39**, 509–524.

Broadbent, L., Heathcote, G.D., McDermott, N. and Taylor, C.E. (1957). The effect of date of planting and of harvesting potatoes on virus infection and on yield. *Ann. Appl. Biol.* **45**, 603–622.

Broadbent, L., Green, D.E. and Walker, P. (1963). Narcissus virus diseases. *Daffodil Tulip Yearbook* **28**, 154–160.

Broadbent, L., Read, W.H. and Last, F.T. (1965). The epidemiology of tomato mosaic X. Persistence of TMV infected debris in soil and the effects of soil partial sterilisation. *Ann. Appl. Biol.* **55**, 471–483.

Broglio, E.P. (1995). Mutational analysis of cauliflower mosaic virus gene VI: changes in host range, symptoms, and discovery of transactivation-positive, non-infectious mutants. *Mol. Plant Microbe Interact.* **8**, 755–760.

Brown, C.M., Dinesh-Kumar, S.P. and Miller, W.A. (1996a). Local and distant sequences are required for efficient readthrough of the barley yellow dwarf PAV coat protein gene stop codon. *J. Virol.* **70**, 5884–5892.

Brown, D.J.F. (1986). The transmission of two strains of arabis mosaic virus from England by populations of *Xiphinema diversicaudatum* (Nematoda: Dorylaimoidea) from ten countries. *Rev. Nematol.* **9**, 83–87.

Brown, D.J.F. and Trudgill, D.L. (1984). The spread of carnation ringspot virus in soil with or without nematodes. *Nematologica* **30**, 102–104.

Brown, D.J.F. and Weischer, B. (1998). Specificity, exclusivity and complementarity in the transmission of plant viruses by plant parasitic nematodes: an annotated terminology. *Fund. Appl. Nematol.* **21**, 1–11.

Brown, D.J.F., Ploeg, A.T. and Robinson, D.J. (1989a). A review of reported association between *Trichodorus* and *Paratrichodorus* species (Nematoda: Trichodoridae) and tobraviruses with a description of laboratory methods for examining virus transmission by trichodorids. *Rev. Nematol.* **12**, 235–241.

Brown, D.J.F., Murant, A.F. and Trudgill, D.L. (1989b). Differences between isolates of the English serotype of tomato black ring virus in their transmissibility by an English population of *Longidorus attennatus* (Nematoda: Dorylaimoidea). *Rev. Nematol.* **12**, 51–56.

Brown, D.J.F., Robertson, W.M. and Trudgill, D.L. (1995). Transmission of viruses by plant nematodes. *Annu. Rev. Phytopathol.* **33**, 223–249.

Brown, D.J.F., Trudgill, D.L. and Robertson, W.M. (1996b). Nepoviruses: transmission by nematodes. In: B.D. Harrison and A.F. Murant (eds) *The Plant Viruses*, Vol. 5: *Polyhedral Virions and Bipartite RNA Genomes*, pp. 187–209. Plenum Press, New York.

Browning, K.S., Fletcher, L. and Ravel, J.M. (1988). Evidence that requirements for ATP and wheat germ initiation factors 4A and 4F are affected by a region of satellite tobacco necrosis virus RNA that is 3' to the ribosomal binding site. *J. Biol. Chem.* **263**, 8380–8383.

Bruening, G. (2000). Virus-dependent RNA agents. In: O.C. Malloy and T.D. Murray (eds) *The Encyclopedia of Plant Pathology*. John Wiley, New York (in press).

Bruening, G., Beachy, R.N., Scalla, R. and Zaitlin, M. (1976). *In vitro* and *in vivo* translation of the ribonucleic acids of a cowpea strain of tobacco mosaic virus. *Virology* **71**, 498–517.

Bruening, G., Buzayan, J.M., Ferreiro, C. and Lim, W. (2000). Evidence for participation of RNA 1-encoded elicitor in cowpea mosaic virus-mediated concurrent protection. *Virology* **266**, 299–309.

Brunt, A.A. and Kenten, R.H. (1963). The use of protein in the extraction of cocoa swollen-shoot virus from cocoa leaves. *Virology* **19**, 388–392.

Brunt, A.A., Foster, G.D., Martelli, G.P. and Zavriev, S.K. (2000a). Genus Potexvirus. In: M.H.V. van Regenmortel, C.M Fauquet, D.H.L. Bishop, E.B. Carstens, M.K. Estes, S.M. Lemon, J. Maniloff, M.A. Mayo, D.J. McGeoch, C.R. Pringle and R.B. Wickner (eds), *Virus Taxonomy. Seventh Report of the International Committee on Taxonomy of Viruses*, pp. 975–981. Academic Press, San Diego.

Brunt, A.A., Foster, G.D., Morozov, S. Yu. and Zavriev, S.K. (2000b). Genus Carlavirus. In: M.H.V. van Regenmortel, C.M Fauquet, D.H.L. Bishop, E.B. Carstens, M.K. Estes, S.M. Lemon, J. Maniloff, M.A. Mayo, D.J. McGeoch, C.R. Pringle and R.B. Wickner (eds) *Virus Taxonomy. Seventh Report of the International Committee on Taxonomy of Viruses*, pp. 969–975. Academic Press, San Diego.

Buchter, H., Hartmann, W. and Stoesser, R. (1987). Anatomical–histological changes in sharka-infected shoots and roots. *Z. Pflanzenkr. Pflanzenschutz* **94**, 46–57.

Buck, K.W. (1996). Comparison of the replication of positive-stranded RNA viruses of plants and animals. *Adv. Virus Res.* **47**, 159–251.

Buck, K.W. (1999a). Geminiviruses (*Geminiviridae*). In: A. Granoff and R.G. Webster (eds) *Encyclopedia of Virology*, 2nd edn, pp. 597–606. Academic Press, San Diego.

Buck, K.W. (1999b). Replication of tobacco mosaic virus RNA. *Phil. Trans. R. Soc. Lond. B* **54**, 613–627.

Buddenhagen, I.W. (1981). Conceptual and practical considerations when breeding for tolerance or resistance. In: R.C. Staples and G.H. Toenniessen (eds) *Plant Disease Control: Resistance and Susceptibility*, pp. 221–234. Wiley, New York.

Buddenhagen, I.W. (1983). Crop improvement in relation to virus diseases and their epidemiology. In: R.T. Plumb and J.M. Thresh (eds) *Plant Virus Epidemiology*, pp. 25–37. Blackwell, Oxford.

Bujarski, J.J. (1999). Recombination of viruses. In: A. Granoff and R.G. Webster (eds) *Encyclopedia of Virology*, pp. 1446–1454. Academic Press, San Diego.

Bujarski, J.J. and Kaesberg, P. (1986). Genetic recombination between RNA components of a multipartite plant virus. *Nature (London)* **321**, 528–531.

Bujarski, J.J. and Nagy, P.D. (1996). Different mechanisms of homologous and nonhomologous recombination in brome mosaic virus: role of RNA sequences and replicase proteins. *Sem. Virol.* **7**, 363–372.

Bujarski, J.J., Hardy, S.F., Miller, W.A. and Hall, T.C. (1982). Use of dodecyl-β-D-maltoside in the purification and stabilization of RNA polymerase from brome mosaic virus-infected barley. *Virology* **119**, 465–473.

Burden, R.S., Rowell, P.M., Bailey, J.A., Loeffler, R.S.T., Kemp, M.S. and Brown, C.A. (1985). Debneyol, a fungicidal sesquiterpene from TMV-infected *Nicotiana debneyi*. *Phytochemistry* **24**, 2191–2194.

Burgermeister, W. and Koenig, R. (1984). Electro-blot immunoassay — a means for studying serological relationships among plant viruses? *Phytopathol. Z.* **111**, 15–25.

Burgermeister, W., Koenig, R., Weich, H., Sebald, W. and Lesemann, D.-E. (1986). Diversity of the RNAs in thirteen isolates of beet necrotic yellow vein virus in *Chenopodium quinoa* detected by means of cloned cDNAs. *J. Phytopathol.* **115**, 229–242.

Burgess, A.J., Harrington, R. and Plumb, R.T. (1999). Barley and cereal yellow dwarf virus epidemiology and control strategies. In: H.G Smith and H. Barker (eds) *The Luteoviridae*, pp. 248–279. CAB International, Wallingford, UK.

Burgess, J., Motoyoshi, F. and Fleming, E.N. (1973). The mechanism of infection of plant protoplasts by viruses. *Planta* **112**, 323–332.

Burgess, J., Motoyoshi, F. and Fleming, E.N. (1974a). Structural changes accompanying infection of tobacco protoplasts with two spherical viruses. *Planta* **117**, 133–144.

Burgess, J., Motoyoshi, F. and Fleming, E.N. (1974b). Structural and autoradiographic observations of the infection of tobacco protoplasts with pea enation mosaic virus. *Planta* **119**, 247–256.

Burgyán, J., Grieco, F. and Russo, M. (1989). A defective interfering RNA molecule in *Cymbidium* ringspot virus infections. *J. Gen. Virol.* **70**, 235–239.

Burgyán, J., Dalmay, T., Rubino, L. and Russo, M. (1992). The replication of cymbidium ringspot tombusvirus defective interfering-satellite RNA hybrid molecules. *Virology* **190**, 579–586.

Burgyán, J., Hornyik, C., Szittya, G., Silhavy, D. and Bisztray, G. (2000). The ORF1 products of tombusviruses play a crucial role in lethal necrosis of virus-infected plants. *J. Virol.* **74**, 10873–10881.

Burns, T.M., Harding, R.M and Dale, J.L. (1995). The genome organization of banana bunchy top virus: analysis of six ssDNA components. *J. Gen. Virol.* **76**, 1471–1482.

Burt, P.E., Heathcote, G.D. and Broadbent, L. (1964). The use of insecticides to find when leaf roll and virus Y spread within potato crops. *Ann. Appl. Biol.* **54**, 13–22.

Bussière, F., Lehoux, J., Thompson, D.A., Skrzeczowski, L.J. and Perreault, J.-P. (1999). Subcellular localization and rolling circle replication of peach latent mosaic viroid: hallmarks of group A viroids. *J. Virol.* **73**, 6353–6360.

Bussière, F., Ouellet, J., Côté, F., Lévesque, D. and Perreault, J.P. (2000). Mapping in solution shows the peach latent mosaic viroid to possess a new pseudoknot in a complex, branched secondary structure. *J. Virol.* **74**, 2647–2654.

Butler, P.J.G. (1984). The current picture of the structure and assembly of tobacco mosaic virus. *J. Gen. Virol.* **65**, 253–279.

Butler, P.J.G. (1999). Self-assembly of tobacco mosaic virus: the role of an intermediate aggregate in generating both specificity and speed. *Phil. Trans. R. Soc. Lond. B* **354**, 537–550.

Butler, P.J.G. and Klug, A. (1971). Assembly of the particle of tobacco mosaic virus from RNA and disks of protein. *Nature (London)* **229**, 47–50.

Butler, P.J.G., Durham, A.C.H. and Klug, A. (1972). Structures and roles of the polymorphic forms of tobacco mosaic virus protein: IV. Control of mode of aggregation of tobacco mosaic virus protein by proton binding. *J. Mol. Biol.* **72**, 1–18.

Butler, P.J.G., Finch, J.T. and Zimmern, D. (1977). Configuration of tobacco mosaic virus RNA during virus assembly. *Nature (London)* **265**, 217–219.

Büttner, C. and Nienhaus, F. (1989). Virus contamination of soils in forest ecosystems of the Federal Republic of Germany. *Eur. J. For. Pathol.* **19**, 47–53.

Büttner, C., Jacobi, V. and Koenig, R. (1987). Isolation of carnation Italian ringspot virus from a creek in a forested area southwest of Bonn. *J. Phytopathol.* **118**, 131–134.

Buzayan, J.M., Gerlach, W.L. and Bruening, G. (1986a). Satellite tobacco ringspot virus RNA: a subset of the RNA sequence is sufficient for autolytic processing. *Proc. Natl Acad. Sci. USA* **83**, 8859–8862.

Buzayan, J.M., Gerlach, W.L. and Bruening, G. (1986b). Nonenzymatic cleavage and ligation of RNAs complementary to a plant virus satellite RNA. *Nature (Lond.)* **323**, 349–353.

Buzayan, J.M., Hampel, A. and Bruening, G. (1986c). Nucleotide sequence and newly formed phosphodiester bond of spontaneously ligated satellite tobacco ringspot virus RNA. *Nucleic Acids Res.* **14**, 9729–9743.

Buzayan, J.M., Gerlach, W.L., Bruening, G., Keese, P. and Gould, A.R. (1986d). Nucleotide sequence of satellite tobacco ringspot RNA and its relationship to multimeric forms. *Virology* **151**, 186–199.

Buzayan, J.M., McNinch, J.S., Schneider, I.R. and Bruening, G. (1987). A nucleotide sequence rearrangement distinguishes two isolates of satellite tobacco ringspot virus RNA. *Virology* **160**, 95–99.

Buzen, F.G., Jr., Niblett, C.L., Hooper, G.R., Hubbard, J. and Newman, M.A. (1984). Further characterisation of panicum mosaic virus and its associated satellite virus. *Phytopathology* **74**, 313–318.

Buzzell, R.I. and Tu, J.C. (1984). Inheritance of soybean resistance to soybean mosaic virus. *J. Hered.* **75**, 82.

Cabaleiro, C. and Segura, A. (1997). Field transmission of grapevine leafroll associated virus 3 (GLRaV-3) by the mealybug *Planococcus citri*. *Plant Dis.* **81**, 283–287.

Cabauatan, P.Q., Cabunagan, R.C and Koganezawa, H. (1995). Biological variants of rice tungro viruses in the Philippines. *Phytopathology* **85**, 77–81.

Cabauatan, P.Q., Melcher, U., Ishikawa, K. *et al.* (1999). Sequence changes in six variants of rice tungro bacilliform virus and their phylogenetic relationships. *J. Gen. Virol.* **80**, 2229–2237.

Cadman, C.H. and Harrison, B.D. (1959). Studies on the properties of soil-borne viruses of the tobacco-rattle type occurring in Scotland. *Ann. Appl. Biol.* **47**, 542–556.

Calder, V.L and Palukaitis, P. (1992). Nucleotide sequence analysis of the movement genes of resistance breaking strains of tomato mosaic virus. *J. Gen. Virol.* **73**, 165–168.

Callaway, A.S., Huang, Z. and Howell, S.H. (2000). Host suppressors in *Arabidopsis thaliana* of mutations in the movement protein gene of cauliflower mosaic virus. *Mol. Plant–Microb. Interact.* **13**, 512–519.

Calvert, L.A., Ospina, M.D. and Shepherd, R.J. (1995). Characterization of cassava vein mosaic virus: a distinct plant pararetrovirus. *J. Gen. Virol.* **76**, 1271–1278.

Calvert, L.A., Cuervo, M.I., Ospina, M.D., Fauquet, C.M. and Ramirez, B.C. (1996). Characterization of cassava common mosaic virus and a defectiveRNA species. *J. Gen. Virol.* **77**, 525–530.

Camacho-Henriquez, A. and Sänger, H.L. (1984). Purification and partial characterisation of the major pathogenesis-related tomato leaf protein P14 from potato spindle tuber viroid (PSTV)-infected tomato leaves. *Arch. Virol.* **81**, 263–284.

Camerini-Otero, R.D., Posey, P.N., Koppel, D.E., Schaefer, D.W. and Franklin, R.M. (1974). Intensity fluctuation spectroscopy of laser light scattered by solutions of spherical viruses: R17, Q, BSV, PM2 and T7: II. Diffusion coefficients, molecular weights, salvation, and particle dimensions. *Biochemistry* **13**, 960–970.

Campbell, B.C., Steffen-Campbell, J.D., Sorensen, J.T. and Gill, R.J. (1995). Paraphyly of Homoptera and Auchenorrhyncha inferred from 18S nucleotide sequences. *Syst. Entomol.* **20**, 175–194.

Campbell, R.N. (1996). Fungal transmission of plant viruses. *Annu. Rev. Phytopathol.* **34**, 87–108.

Campbell, R.N., Wipf-Scheibel, C. and Lecoq, H. (1996). Vector-assisted seed transmission of melon necrotic spot virus in melon. *Phytopathology* **86**, 1294–1298.

Canady, M.A., Day, J. and McPherson, A. (1995). Preliminary X-ray diffraction analysis of crystals of turnip yellow mosaic virus (TYMV). *Proteins* **21**, 78–81.

Canady, M.A., Larson, S.B., Day, J. and McPherson, A. (1996). Crystal structure of turnip yellow mosaic virus. *Nat. Struct. Biol.* **3**, 771–781.

Candelier-Harvey, P. and Hull, R. (1993). Cucumber mosaic virus genome is encapsidated in alfalfa mosaic virus coat protein expressed in transgenic plants. *Transgenic Res.* **2**, 277–285.

Candresse, T, Mouches, C. and Bové, J.M. (1986). Characterisation of the virus encoded subunit of turnip yellow mosaic virus RNA replicase. *Virology* **152**, 322–330.

Candresse, T., Batisti, M., Renaudin, J., Mouches, C. and Bové, J.M. (1987). Immunodetection of turnip yellow mosaic virus non-structural proteins in infected chinese cabbage leaves and protoplasts. *Ann. Inst. Pasteur./Virol.* **138**, 217–227.

Candresse, T., Macquaire, G., Monsion, M. and Dunez, J. (1988). Detection of chrysanthemum stunt viroid (CSVd) using nick translated probes in a dot blot hybridization assay. *J. Virol. Methods* **20**, 185–193.

Canto, T. and Palukaitis, P. (1999a). Are tubules generated by the 3a protein necessary for cucumber mosaic virus movement? *Mol. Plant–Microb. Interact.* **11**, 985–993.

Canto, T. and Palukaitis, P. (1999b). The hypersensitive response to cucumber mosaic virus in *Chenopodium amaranticolor* requires virus movement outside the initially infected cell. *Virology* **265**, 74–82.

Capoor, S.P. (1949). The movement of tobacco mosaic viruses and potato virus X through tomato plants. *Ann. Appl. Biol.* **36**, 307–319.

Carette, J.E., Stuiver, M., van Lent, J., Wellink, J. and van Kammen, A. (2000). Cowpea mosaic virus infection induces a massive proliferation of endoplasmic reticulum but not golgi membranes and is dependent on de novo membrane synthesis. *J. Virol.* **74**, 6556–6563.

Carlton, W.M., Braun, E.J. and Gleason, M.L. (1998). Ingress of *Clavibacter michiganensis* subsp. *michigensis* into tomato leaves through hydothodes. *Phytopathology* **99**, 525–529.

Carpenter, C.D. and Simon, A.E. (1996a). *In vivo* repair of 3′ end deletions in a TCV satellite RNA may involve two abortive synthesis and priming events. *Virology* **226**, 153–160.

Carpenter, C.D. and Simon, A.E. (1996b). *In vivo* restoration of biologically active 3′ ends of virus-associated RNAs by non-homologous RNA recombination and replacement of a terminal motif. *J. Virol.* **70**, 478–486.

Carpenter, C.D., Oh, J.-W., Zhang, C. and Simon, A.E. (1995). Involvement of a stem-loop structure in the location of junction sites in viral RNA recombination. *J. Mol. Biol.* **245**, 608–622.

Carr, J.P. and Zaitlin, M. (1993). Replicase-mediated resistance. *Semin. Virol.* **4**, 339–347.

Carr, J.P., Beachy, R.N. and Klessig, D.F. (1989). Are the PR1 proteins involved in genetically engineered resistance to TMV? *Virology* **169**, 470–473.

Carr, J.P., Marsh, L.E., Lomonossoff, G.P., Sekiya, M.E. and Zaitlin, M. (1992). Resistance to tobacco mosaic virus induced by the 54-kDa gene sequence requires expression of the 54-kDa protein. *Mol. Plant Microb Interact.* **5**, 397–404.

Carr, J.P., Gal-On, A., Palukaitis, P. and Zaitlin, M. (1994). Replicase-mediated resistance to cucumber mosaic virus in transgenic plants involves suppression of both virus replication in inoculated leaves and long-distance movement. *Virology* **199**, 439–447.

Carr, R.J. and Kim, K.S. (1983). Evidence that bean golden mosaic virus invades non-phloem tissue in double infections with tobacco mosaic virus. *J. Gen. Virol.* **64**, 2489–2492.

Carrier, K., Hans, F. and Sanfaçon, H. (1999). Mutagenesis of amino acids at two tomato ringspot nepovirus cleavage sites: effect on proteolytic processing in *cis* and in *trans* by the 3C-like protease. *Virology* **256**, 161–175.

Carrington, J.A. and Freed, D.D. (1990). Cap-dependent enhancement of translation by a plant potyvirus 5′ nontranslated region. *J. Virol.* **64**, 1590–1597.

Carrington, J.C. and Dougherty, W.G. (1987a). Small nuclear inclusion protein encoded by a plant potyvirus genome is a protease. *J. Virol.* **61**, 2540–2548.

Carrington, J.C. and Dougherty, W.G. (1987b). Processing of the tobacco etch virus 49K protease requires autoproteolysis. *Virology* **160**, 355–362.

Carrington, J.C. and Dougherty, W.G. (1988). A viral cleavage site cassette: identification of amino acid sequences required for tobacco etch virus polyprotein processing. *Proc. Natl Acad. Sci. USA* **85**, 3391–3395.

Carrington, J.C. and Herndon, K.L. (1992). Characterization of the potyviral HC-Pro autoproteolytic cleavage site. *Virology* **187**, 308–315.

Carrington, J.C. and Morris, T.J. (1984). Complementary DNA cloning and analysis of carnation mottle virus RNA. *Virology* **139**, 22–31.

Carrington, J.C. and Morris, T.J. (1986). High resolution mapping of carnation mottle virus-associated RNAs. *Virology* **150**, 196–206.

Carrington, J.C. and Whitham, S.A. (1998). Viral invasion and host defense: strategies and counter-strategies. *Curr. Opin. Plant Biol.* **1**, 336–341.

Carrington, J.C., Morris, T.J., Stockley, P.G. and Harrison, S.C. (1987). Structure and assembly of turnip crinkle virus: IV. Analysis of the coat protein gene and implications of the subunit primary structure. *J. Mol. Biol.* **194**, 265–276.

Carrington, J.C., Cary, S.M., Parks, T.D. and Dougherty, W.G. (1989). A second proteinase encoded by a plant potyvirus genome. *EMBO J.* **8**, 365–370.

Carrington, J.C., Kasschau, K.D., Mahajan, S.K. and Schaad, M.C. (1996). Cell-to-cell and long-distance transport of viruses in plants. *Plant Cell* **8**, 1669–1681.

Carrington, J.C., Jensen, P.E. and Schaad, M.C. (1998). Genetic evidence for an essential role for potyvirus CI protein in cell-to-cell movement. *Plant J.* **14**, 393–400.

Carrington, J.C., Kasschau, K.D. and Johansen, L.K. (2001). Activation and suppression of RNA silencing by plant viruses. *Virology* **281**, 1–5.

Carroll, T.W. (1974). Barley stripe mosaic virus in sperm and vegetative cells of barley pollen. *Virology* **60**, 21–28.

Carroll, T.W. (1983). Certification schemes against barley stripe mosaic. *Seed Sci. Technol.* **11**, 1033–1042.

Carroll, T.W. and Chapman, S.R. (1970). Variation in embryo infection and seed transmission of barley ctripe mosaic virus within and between two cultivars of barley. *Phytopathology* **60**, 1079–1081.

Carroll, T.W., Gossel, P.L. and Hockett, E.A. (1979). Inheritance of resistance to seed transmission of barley stripe mosaic virus in barley. *Phytopathology* **69**, 431–433.

Carroll, T.W., Hockett, E.A. and Zaske, S.K. (1983). Registration of mobet barley germplasm. *Crop Sci.* **23**, 599–600.

Carsner, E. (1925). Attenuation of the virus of sugar beet curly-top. *Phytopathology* **15**, 745–757.

Cartea, M.E., Migdal, M., Galle, A.M., Pelletier, G. and Guerche, P. (1998). Comparison of sense and antisense methodologies for modifying the fatty acid composition of *Arabidopsis thaliana* oilseed. *Plant Sci.* **136**, 181–194.

Carter, M.J. and ter Meulen, V. (1984). The application of monoclonal antibodies in the study of viruses. *Adv. Virus Res.* **29**, 95–130.

Cartwright, B., Smale, C.J., Brown, F. and Hull, R. (1972) Model for vesicular stomatitis virus. *J. Virol.* **10**, 256–260.

Cascone, P.J., Carpenter, C.D., Li, X.H. and Simon, A.E. (1990). Recombination between satellite RNAs of turnip crinkle virus. *EMBO J.* **9**, 1709–1715.

Caspar, D.L.D. (1963). Assembly and stability of the tobacco mosaic virus particle. *Adv. Protein Chem.* **18**, 37–121.

Caspar, D.L.D. and Holmes, K.C. (1969). Structure of dahlemense strain of tobacco mosaic virus: a periodically derformed helix. *J. Mol. Biol.* **46**, 99–133.

Caspar, D.L.D. and Klug, A. (1962). Physical principles in the construction of regular viruses. *Cold Spring Harbor Symp. Quant. Biol.* **27**, 1–24.

Caspar, D.L.D., Dulbecco, R., Klug, A., Lwoff, A., Stoker, M.G.P., Tournier, P. and Wildy, P. (1962). Proposals. *Cold Spring Harbor Symp. Quant. Biol.* **27**, 49.

Cassells, A.C. and Herrick, C.C. (1977). The identification of mild and severe strains of tobacco mosaic virus in double inoculated tomato plants. *Ann. Appl. Biol.* **86**, 37–46.

Castel, A., Kraal, B., De Graaf, J.M. and Bosch, L. (1979). The primary structure of the coat protein of alfalfa mosaic virus strain VRU. A hypothesis on the occurrence of two conformations in the assembly of the protein shell. *Eur. J. Biochem.* **102**, 125–138.

Castellano, M.M., Sanz-Burgos, A.P. and Gutierrez, C. (1999). Initiation of DNA replication in an eukaryotic rolling-circle replicon: identification of multiple DNA-protein complexes at the geminivirus origin. *J. Mol. Biol.* **290**, 639–652.

Castello, J.D., Rogers, S.O., Starmer, W.T. *et al.* (1999). Detection of tomato mosaic tobamovirus RNA in ancient glacial ice. *Polar Biol.* **22**, 207–212.

Castillo, M.B. and Orlob, G.B. (1966). Transmission of two strains of cucumber mosaic and alfalfa mosaic viruses by single aphids of *Myzus persicae*. *Phytopathology* **56**, 1028–1030.

Catherall, P.L., Jones, A.T. and Hayes, J.D. (1970). Inheritance and effectiveness of genes in barley that condition tolerance to barley yellow dwarf virus. *Ann. Appl. Biol.* **65**, 153–161.

Cattaneo (1991). Different types of messenger RNA editing. *Annu. Rev. Genet.* **25**, 71–88.

Cecchini, E., Gong, Z., Geri, C., Covey, S.N. and Milner, J.J. (1997a). Transgenic *Arabidopsis* lines expressing gene VI from cauliflower mosaic virus variants exhibit a range of symptom-like phenotypes and accumulate inclusion bodies. *Mol. Plant Microbe Interact.* **10**, 1094–1101.

Cecchini, E., Al-Kaff, N., Covey, S.N. and Milner, J.J. (1997b). Variation in symptom expression in *Arabidopsis* ecotypes and mutants infected with cauliflower mosaic virus. *J. Exp. Bot.* **48**, S18.

Cecchini, E., Al-Kaff, N.S., Bannister, A. *et al.* (1998). Pathogenic interactions between variants of cauliflower mosaic virus and *Arabidopsis thaliana*. *J. Exp. Bot.* **49**, 731–737.

Cech, T.R. (1986). A model for the RNA-catalysed replication of RNA. *Proc. Natl. Acad. Sci. USA* **83**, 4360–4363.

Cech, T.R. and Bass, B.L. (1986). Biological catalysis by RNA. *Annu. Rev. Biochem.* **55**, 599–629.

Cerutti, L., Mian, N. and Bateman, A. (2000). Domains in gene silencing and cell differentuiation proteins: novel PAZ domain and redefinition of the Piwi domain. *Trends Biochem. Sci.* **25**, 481–482.

Chaerle, L., van Caeneghem, W., Messens, E., Lambers, H., van Montagu, M. and van der Straeten, D. (1999). Presymptomatic visualization of plant–virus interactions by thermography. *Nature Biotechnol.* **17**, 813–816.

Chalcroft, J.P. and Matthews, R.E.F. (1966). Cytological changes induced by turnip yellow mosaic virus infections. *Virology* **28**, 555–562.

Chalcroft, J.P. and Matthews, R.E.F. (1967a). Virus strains and leaf ontogeny as factors in the production of leaf mosaic patterns by turnip yellow mosaic virus. *Virology* **33**, 167–171.

Chalcroft, J.P. and Matthews, R.E.F. (1967b). Role of virus strains and leaf ontogeny in the production of mosaic patterns by turnip yellow mosaic virus. *Virology* **33**, 659–673.

Chalfant, R.B. and Chapman, R.K. (1962). Transmission of cabbage viruses A and B by the cabbage aphid and the green peach aphid. *J. Econ. Entomol.* **55**, 584–590.

Chamberlain, E.E., Atkinson, J.D. and Hunter, J.A. (1951). Plum mosaic, a virus disease of plums, peaches and apricots in New Zealand. *N. Z. J. Sci. Technol.* **33**, 1–16.

Chambers, T.C. and Francki, R.I.B. (1966). Localization and recovery of lettuce necrotic yellows virus from xylem tissues of *Nicotiana glutinosa*. *Virology* **29**, 673–676.

Champness, J.N., Bloomer, A.C., Bricogne, G., Butler, P.J.G. and Klug, A. (1976). The structure of the protein disk of tobacco mosaic virus to 5 Å resolution. *Nature (London)* **259**, 20–24.

Chan, M.S. and Jeger, M.J. (1994). An analytical model of plant virus disease dynamics with roguing and replanting. *J. Appl. Ecol.* **31**, 413–427.

Chandrika, R., Rabindran, S., Lewandowski, D.J., Manjunath, K.L. and Dawson, W.O. (2000). Full-length tobacco mosaic virus RNAs and defective RNAs have different 3' replication signals. *Virology* **273**, 198–209.

Chang, C.-A., Hiebert, E. and Purcifull, D.E. (1988). Purification, characterisation and immunological analysis of nuclear inclusions induced by bean yellow mosaic virus and clover yellow vein, potyviruses. *Phytopathology* **78**, 1266–1275.

Chang, Y.C., Borja, M., Scholthof, H.B., Jackson, A.O. and Morris, T.J. (1995). Host effects and sequences essential for accumulation of defective interfering RNAs of cucumber necrosis and tomato bushy stunt tombusviruses. *Virology* **210**, 41–53.

Chao, L. (1997). Evolution of sex and the molecular clock in RNA viruses. *Gene* **205**, 301–308.

Chapman, M. and Kao, C. (1999). A minimal RNA promoter for minus-strand RNA synthesis by brome mosaic virus polymerase complex. *J. Mol. Biol.* **286**, 709–720.

Chapman, M.S. (1998). Watching ones P's and Q's: promiscuity, plasticity and quasiequivalence in a T = 1 virus. *Biophys. J.* **74**, 639–644.

Chapman, S., Kavanagh, T. and Baulcombe, D. (1992). Potato virus X as a vector for gene expression in plants. *Plant J.* **2**, 549–557.

Chaumpluk, P., Sasaki, Y., Nakajima, N. *et al.* (1996). Six new subgroup I members of Japanese cucumber mosaic virus as determined by nucleotide sequence analysis of RNA3's cDNAs. *Ann. Phytopathol. Soc. Jpn.* **62**, 40–44.

Chauvin, C., Pfeiffer, P., Witz, J. and Jacrot, B. (1978). Structural polymorphism of bromegrass mosaic virus: a neutron small angle scattering investigation. *Virology* **88**, 138–148.

Chauvin, C., Jacrot, B., Lebeurier, G. and Hirth, L. (1979). The structure of cauliflower mosaic virus: a neutron diffraction study. *Virology* **96**, 640–641.

Chelvanayagam, G., Heringa, J. and Argos, P. (1992). Anatomy and evolution of proteins displaying the viral capsid jellyroll topology. *J. Mol. Biol.* **228**, 220–242.

Chen, B. and Francki, R.B. (1990). Cucumovirus transmission by the aphid *Myzus persicae* is determined solely by the viral coat protein. *J. Gen. Virol.* **71**, 939–944.

Chen, G., Müller, M., Potrykus, I., Hohn, T. and Fütterer, J. (1994). Rice tungro bacilliform virus: transcription and translation in protoplasts. *Virology* **204**, 91–100.

Chen, G., Rothnie, H.M., He, X., Hohn, T. and Fütterer, J. (1996). Efficient transcription from the rice tungro bacilliform virus promoter requires elements downstream of the transcription start site. *J. Virol.* **70**, 8411–8421.

Chen, H. and Ahlquist, P. (2000). Brome mosaic virus polymerase-like protein 2a is directed to the endoplasmic reticulum by helicase-like viral protein 1a. *J. Virol.* **74**, 4310–4318.

Chen, J., Torrance, L., Cowan, G.H., MacFarlane, S.A., Stubbs, G. and Wilson, T.M.A. (1997). Monoclonal antibodies detect a single amino acid difference between the coat proteins of soilborne wheat mosaic virus isolates: implications for virus structure. *Phytopathology* **87**, 295–301.

Chen, J., Noueiry, A. and Ahlquist, P. (2001). Brome mosaic virus protein 1a recruits viral RNA2 to RNA replication through a 5' proximal RNA2 signal. *J. Virol.* **75**, 3207–3219.

Chen, L.-C., Durand, D.P. and Hill, J.H. (1982). Detection of pathogenic strains of soybean mosaic virus by enzyme-linked immunosorbent assay with polystyrene plates and beads as the solid phase. *Phytopathology* **72**, 1177–1181.

Chen, M.-H., Sheng, J., Hind, G., Handa, A.K. and Citovsky, V. (2000). Interaction between the tobacco mosaic virus movement protein and host cell pectin methylesterases is required for viral cell-to-cell movement. *EMBO J.* **19**, 913–920.

Chen, Z., Stauffacher, C., Li, Y. *et al.* (1989). Protein-RNA interactions in an icosahedral virus at 3.0 Å resolution. *Science* **245**, 154–159.

Cheng, C.-P., Tzafrir, I., Lockhart, B.E.L. and Olszewski, N.E. (1998). Tubules containing virions are present in plant tissues infected with *Commelina* yellow mottle badnavirus. *J. Gen. Virol.* **79**, 925–929.

Cheng, N.-H., Su, C.-L., Carter, S.A. and Nelson, R.S. (2000). Vascular invasion routes and systemic accumulation patterns of tobacco mosaic virus in *Nicotiana benthamiana*. *Plant J.* **23**, 349–362.

Cheng, R.H., Olson, N.H. and Baker, T.S. (1992). Cauliflower mosaic virus: a 420 subunit (T = 7), multiplayer structure. *Virology* **186**, 655–668.

Chester, K.S. (1935). A serological estimate of the absolute concentration of tobacco mosaic virus. *Science* **82**, 17.

Chester, K.S. (1936). Separation and analysis of virus strains by means of precipitin tests. *Phytopathology* **26**, 778–785.

Chetverin, A.B. (1999). The puzzle of RNA recombination. *FEBS Lett.* **460**, 1–5.

Chiarappa, L. (ed.) (1971). Crop loss assessment methods. In: *Manual on the Evaluation and Prevention of Pests, Disease and Weeds*. Alden Press, Oxford.

Chiko, A.W. (1975). Evidence of multiple virion components in leaf-dip preparations of barley stripe mosaic virus. *Virology* **63**, 115–122.

Childress, A.M. and Ramsdell, D.C. (1987). Bee mediated transmission of blueberry leaf mottle virus via infected pollen in highbush blueberry. *Phytopathology* **77**, 167–172.

Childress, S.A. and Harris, K.F. (1989). Localization of virus-like particles in the foreguts of viruliferous *Graminella nigrifrons* leafhoppers carrying the semi-persistent maize chlorotic dwarf virus. *J. Gen. Virol.* **70**, 247–251.

Chisholm, S.T., Mahajan, S., Whitham, S.A., Yamamoto, M.L and Carrington, J.C. (2000). Cloning of the *Arabidopsis RTM1* gene, which controls restriction of long-distance movement of tobacco etch virus. *Proc. Natl Acad. Sci. USA* **97**, 489–494.

Chiu, R.-J. and Black, L.M. (1969). Assay of wound tumor virus by the fluorescent cell counting technique. *Virology* **37**, 667–677.

Chiu, R.-J., Liu, H.-Y., MacLeod, R. and Black, L.M. (1970). Potato yellow dwarf virus in leafhopper cell culture. *Virology* **40**, 387–396.

Chivasa, S. and Carr, J.P. (1998). Cyanide restores N gene-mediated resistance to tobacco mosaic virus in transgenic plants expressing salicylic acid hydrolase. *Plant Cell* **10**, 1489–1498.

Chivasa, S., Murphy, A.M., Naylor, M. and Carr, J.P. (1997). Salicylic acid interferes with tobacco mosaic virus replication via a novel salicylhydroxamic acid-sensitive mechanism. *Plant Cell* **9**, 547–557.

Chiykowski, L.N. and Chapman, R.K. (1965). Migration of the six-spotted leaf hopper in central North America. *Res. Bull. Wis. Agric. Exp. Stn.* **261**, 21–45.

Cho, H.-S., Ha, N.-C., Kang, L.-W. *et al.* (1998). Crystal structure of RNA helicase from genotype 1b hepatitis C virus. *J. Biol. Chem.* **273**, 15045–15052.

Cho, J.J., Mau, R.F.L., German, T.L. *et al.* (1989). A multidisciplinary approach to management of tomato spotted wilt virus in Hawaii. *Plant Dis.* **73**, 375–383.

Choi, H.-K., Tong, L., Minor, W. *et al.* (1991). Structure of *Sindbis virus* core protein reveals a chymotrypsin-like serine proteinase and the organization of the virion. *Nature* **354**, 37–43.

Choi, J. and Loesch-Fries, S. (1999). Effect of C-terminal mutations of alfalfa mosaic virus coat protein on dimer formation and assembly *in vitro*. *Virology* **260**, 182–189.

Choi, Y.G., Grantham, G.L. and Rao, A.L.N. (2000). Molecular studies on bromovirus capsid protein: VI. Contributions of the N-terminal arginine-rich motif of BMV capsid protein in virion stability and RNA packaging. *Virology* **270**, 377–385.

Choudhury, M.M. and Rosenkranz, E. (1983). Vector relationship of *Graminella nigrifrons* to maize chlorotic dwarf virus. *Phytopathology* **73**, 685–690.

Christie, R.G. and Edwardson, J.R. (1977). Light and electron microscopy of plant virus inclusions. *Fla. Agric. Exp. Sta. Monogr. Ser. 9.* 150 pp.

Christie, R.G. and Edwardson, J.R. (1986). Light microscopic techniques for detection of plant virus inclusions. *Plant Dis.* **70**, 273–279.

Christie, S.R., Purcifull, D.E., Crawford, W.E. and Ahmed, N.A. (1987). Electron microscopy of negatively stained clarified viral concentrates obtained from small tissue samples with appendices on negative staining techniques. *Bull. Fla. Agric. Exp. Stn.* **872**, 1–45.

Chu, M., Park, J.-W. and Scholthof, H.B. (1999). Separate regions on the tomato bushy stunt virus p22 protein mediate cell-to-cell movement versus elicitation of effective resistance responses. *Mol. Plant–Microb. Interact.* **12**, 285–292.

Chu, M., Desvoyes, B., Turina, M., Noad, R. and Scholthof, K.B.G. (2000). Genetic dissection of tomato bushy stunt virus p19-protein-mediated host-dependent symptom induction and systemic invasion. *Virology* **266**, 79–87.

Chu, P.W.G. and Francki, R.I.B. (1982). Detection of lettuce necrotic yellows virus by an enzyme-linked immunosorbent assay in plant hosts and the insect vector. *Ann. Appl. Biol.* **100**, 149–156.

Chu, P.W.G. and Francki, R.I.B. (1983). Chemical and serological comparison of the coat proteins of velvet tobacco mottle and *Solanum nodiflorum* mottle viruses. *Virology* **129**, 350–356.

Chu, P.W.G. and Helms, K. (1988). Novel virus-like particles containing single-stranded DNAs associated with subterranean clover stunt disease. *Virology* **167**, 38–49.

Chu, P.W.G., Boccardo, G. and Francki, R.I.B. (1981). Requirement of a genome associated protein of tobacco ringspot virus for infectivity but not for *in vitro* translation. *Virology* **109**, 428–430.

Cireli, B. (1976). Observations on the effects of some air pollutants on *Zea mays* leaf tissue. *Phytopathol. Z.* **86**, 233–239.

Cirulli, M. and Alexander, L. (1969). Influence of temperature and strain of tobacco mosaic virus on resistance of a tomato breeding line derived from *Lycopersicon peruvianum*. *Phytopathology* **59**, 1287–1297.

Citovsky, V. (1999). Tobacco mosaic virus: a pioneer of cell-to-cell movement. *Phil. Trans. R. Soc. Lond. B* **354**, 637–643.

Citovsky, V., Knorr, D., Schuster, G. and Zambryski, P. (1990). The P30 movement protein of tobacco mosaic virus is a single-strand nucleic acid binding protein. *Cell* **60**, 637–647.

Citovsky, V., Knorr, D. and Zambryski, P. (1991). Gene 1: a potential cell-to-cell movement locus of cauliflower mosaic virus, encodes an RNA-binding protein. *Proc. Natl Acad. Sci. USA* **88**, 2476–2480.

Citovsky, V., Wong, M.L., Shaw, A.L., Prasad, B.V.V. and Zambryski, P. (1992). Visualization and characterization of tobacco mosaic virus movement protein binding to single-stranded nucleic acids. *Plant Cell* **4**, 397–411.

Citovsky, V., McLean, B.G., Zupan, J.R. and Zambryski, P. (1993). Phosphorylation of tobacco mosaic virus cell-to-cell movement protein by a developmentally regulated cell wall-associated protein kinase. *Genes Dev.* **7**, 904–910.

Clark, A.J., Bertens, P., Wellink, J., Shanks, M. and Lomonossoff, G.P. (1999). Studies on hybrid comoviruses reveal the importance of three-dimensional structure for processing of the viral coat proteins and show that the specificity of cleavage is greater in *trans* than in *cis*. *Virology* **263**, 184–194.

Clark, M.F. and Adams, A.N. (1977). Characteristics of the microplate method of enzyme-linked immunosorbent assay for the detection of plant viruses. *J. Gen. Virol.* **34**, 475–483.

Clark, M.F. and Barbara, D.J. (1987). A method for the quantitative analysis of ELISA data. *J. Virol. Methods* **15**, 213–222.

Clark, M.F. and Bar-Joseph, M. (1984). Enzyme immunosorbent assays in plant virology. *Methods Virol.* **7**, 51–85.

Clark, M.F., Matthews, R.E.F. and Ralph, R.K. (1964). Ribosomes and polyribosomes in Brassica pekinensis. *Biochim. Biophys. Acta* **91**, 289–304.

Clark, R.W. (1976). Calculation of S_{20w} values using ultracentrifuge sedimentation data from linear sucrose gradients: an improved, simplified method. *Biochim. Biophys. Acta* **428**, 269–274.

Clark, W.G., Fitchen, J.H., Nejidat, A., Deom, C.M. and Beachy, R.N. (1995). Studies of coat protein-mediated resistance to tobacco mosaic virus (TMV). 2. Challenge by a mutant with altered virion surface does not overcome resistance conferred by TMV coat protein. *J. Gen. Virol.* **76**, 2613–2617.

Clement, D.L., Lister, R.M. and Foster, J.E. (1986). ELISA-based studies on the ecology and epidemiology of barley yellow dwarf virus in Indiana. *Phytopathology* **76**, 86–92.

Clinch, P., Loughnane, J.B. and Murphy, P.A. (1938). A study of the infiltration of viruses into seed potato stocks in the field. *Sci. Proc. R. Dublin Soc.* **22**, 18–31.

Close, R.C. (1962) Some interactions between viruses when multiplying together in plants. PhD thesis, University of London.

Close, R.C. and Tomlinson, A.I. (1975). Dispersal of the grain aphid *Macrosiphum miscanthi* from Australia to New Zealand. *N. Z. Entomol.* **6**, 62–65.

Clover, G. and Henry, C. (1999). Detection and discrimination of wheat spindle streak mosaic virus and wheat yellow mosaic virus using multiplex RT-PCR. *Eur. J. Plant Pathol.* **105**, 891–896.

Coakley, S.M., Campbell, R.N. and Kimble, K.A. (1973). Internal rib necrosis and rusty brown discoloration of climax lettuce induced by lettuce mosaic virus. *Phytopathology* **63**, 1191–1197.

Cochran, G.W. (1946). Effect of shading techniques on transmission of tobacco mosaic virus through dodder. *Phytopathology* **36**, 396.

Cock, L.J. (1968). Virus diseases of lettuce. *NAAS Q. Rev.* **79**, 126–138.

Cockbain, A.J., Gibbs, A.J. and Heathcote, G.D. (1963). Some factors affecting the transmission of sugar-beet mosaic and pea mosaic viruses by *Aphis fabae* and *Myzus persicae*. *Ann. Appl. Biol.* **52**, 133–143.

Cockerham, G. (1970). Genetical studies on resistance to potato viruses X and Y. *Heredity* **25**, 309–348.

Cocking, C. (1966). An electron microscope study of the initial stages of infection of isolated tomato fruit protoplasts by tobacco mosaic virus. *Planta* **68**, 206–214.

Cogoni, C. and Macino, G. (1999a). Gene silencing in *Neurospora crassa* requires a protein homologous to RNA-dependent RNA polymerase. *Nature* **399**, 166–169.

Cogoni, C. and Macino, G. (1999b). Posttrancriptional gene silencing in *Neurospora* by a RecQ DNA helicase. *Science* **286**, 2342–2344.

Cogoni, C. and Macino, G. (1999c). Homology-dependent gene silencing in plants and fungi: a number of variations on the same theme. *Curr. Opin. Microbiol.* **2**, 657–662.

Cohen, J. and Loebenstein, G. (1975). An electron microscope study of starch lesions in cucumber cotyledons infected with tobacco mosaic virus. *Phytopathology* **65**, 32–39.

Cohen, J., Loebenstein, G. and Spiegel, S. (1988). Infection of sweet potato by cucumber mosaic virus depends on the presence of sweet potato feathery mottle virus. *Plant Dis.* **72**, 583–585.

Cohen, S. and Antignus, Y. (1994). Tomato yellow leafcurl virus, a whitefly-borne geminivirus of tomatoes. *Adv. Dis. Vector Res.* **10**, 259–288.

Cohen, S.S. (1981). Reducing the spread of aphid-transmitted viruses in peppers by coarse-net cover. *Phytoparasitica* **9**, 69–76.

Cohen, S.S. and Greenberg, M.L. (1981). Spermidine, an intrinsic component of turnip yellow mosaic virus. *Proc. Natl Acad. Sci. USA* **78**, 5470–5474.

Cohen, S.S. and Marco, S. (1973). Reducing the spread of aphid-transmitted viruses in peppers by trapping the aphids on sticky yellow polyethylene sheets. *Phytopathology* **63**, 1207–1209.

Cohen, Y., Qu, F., Gisel, A., Morris, T.J. and Zambryski, P.C. (2000). Nuclear localization of turnip crinkle virus movement protein p8. *Virology* **273**, 276–285.

Colinet, D., Kummert, J., Lepoivre, P. and Semal, J. (1994). Identification of distinct potyviruses in mixedly-infected sweet-potato by the polymerase chain reaction with degenerate primers. *Phytopathology* **84**, 65–69.

Colinet, D., Nguyen, M., Kummert, J. and Lepoivre, P. (1998). Differentiation among potyviruses infecting sweet potato based on genus-specific and virus-specific reverse transcription poly-merase chain reaction. *Plant Dis.* **82**, 223–229.

Collendavelloo, J., Legrand, M. and Fritig, B. (1982). Plant disease and regulation of enzyriies involved in lignification. *De novo* synthesis controls *o*-methyltransferase activity in hypersensitive tobacco leaves infected by tobacco mosaic virus. *Physiol. Plant Pathol.* **21**, 271–281.

Collins, R.F., Gellatly, D.L., Sehgal, O.P. and Abouhaidar, M.G. (1998). Self-cleaving circular RNA associated with rice yellow mottle virus is the smallest viroid-like RNA. *Virology* **241**, 269–275.

Collmer, C.W. and Howell, S.H. (1992). Role of satellite RNA in the expression of symptoms caused by plant viruses. *Annu. Rev. Phytopathol.* **30**, 419–442.

Collmer, C.W. and Kaper, J.M. (1985). Double-stranded RNAs of cucumber mosaic virus and its satellite contain an unpaired ter-minal guanosine: implications for replication. *Virology* **145**, 249–259.

Collmer, C.W. and Kaper, J.M. (1988). Site-directed mutagenesis of potential protein-coding regions in expressible cloned cDNAs of cucumber mosaic viral satellites. *Virology* **163**, 293–298.

Collmer, C.W. and Zaitlin, M. (1983). The H protein isolated from tobacco mosaic virus reassociates with virions reconstituted *in vitro*. *Virology* **126**, 449–458.

Collmer, C.W., Vogt, V.M. and Zaitlin, M. (1983). H protein, a minor protein of TMV virions, contains sequences of the viral coat pro-tein. *Virology* **126**, 429–448.

Collmer, C.W., Hadidi, A. and Kaper, J.M. (1985). Nucleotide sequence of the satellite of peanut stunt virus reveals structural homologies with viroids and certain nuclear and mitochondrial introns. *Proc. Natl Acad. Sci. USA* **82**, 3110–3114.

Collmer, C.W., Marston, M.F., Taylor, J.C. and Jahn, M. (2000). The *I* gene of bean: a dosage-dependent allele conferring extreme resistance, hypersensitive resistance, or spreading vascular necrosis in response to the potyvirus bean common mosaic virus. *Mol. Plant Microb. Interact.* **13**, 1266–1270.

Comeau, A. and Plourde, A. (1987). Cell, tissue culture and inter-generic hybridization for barley yellow dwarf virus resistance in wheat. *Can. J. Plant Pathol.* **9**, 188–192.

Commandeur, U. and Huth, W. (1999). Differentiation of strains of wheat dwarf virus in infected wheat and barley plants by means of polymerase chain reaction. *J. Plant Dis. Protect.* **106**, 550–552.

Conti, M. (1972). Investigations on the epidemiology of maize rough dwarf virus. I. Overwintering of virus in its planthopper vector. *Actas Congr. Uniao Fitopatol. Mediterr.* **3**, 11–17.

Conti, M. (1984). Epidemiology and vectors of plant reo-like viruses. *Curr. Top. Vector Res.* **2**, 112–139.

Conti, M. and Masenga, V. (1977). Identification and prevalence of pepper viruses in northwest Italy. *Phytopathol. Z.* **90**, 212–222.

Converse, R.H. and Lister, R.M. (1969). The occurrence and some properties of black raspberry latent virus. *Phytopathology* **59**, 325–333.

Converse, R.H. and Martin, R.R. (1982). Use of the Barbara–Clark (F(ab')₂) indirect ELISA test for detection of viruses in small fruits. *Acta Hortic.* **129**, 21.

Converse, R.H. and Tanne, E. (1984). Heat therapy and stolon apex culture to eliminate mild yello-wedge virus from Hood straw-berry. *Phytopathology* **74**, 1315–1316.

Cook, A.A., Walker, J.C. and Larson, R.H. (1952). Studies on the disease cycle of black rot of crucifers. *Phytopathology* **42**, 162–167.

Cooke, R. and Penon, P. (1990). *In vitro* transcription from cauli-flower mosaic virus promoters by a cell free extract from tobacco cells. *Plant Mol. Biol.* **14**, 391–405.

Cooley, M.B., Pathirana, S., Wu, H.-J., Kachroo, P. and Klessig, D.F. (2000). Members of the *Arabidopsis HRT/RPP8* family of resist-ance genes confer resistance to both viral and oomycete pathogens. *Plant Cell* **12**, 663–676.

Cooper, B., Schmitz, I., Rao, A.L.N., Beachy, R.N. and Dodds, J.A. (1996). Cell-to-cell transport of movement-defective cucumber mosaic and tobacco mosaic viruses in transgenic plants express-ing heterologous movement protein genes. *Virology* **216**, 208–213.

Cooper, J.I. (1971). The distribution in Scotland of tobacco rattle virus and its nematode vectors in relation to soil type. *Plant Pathol.* **20**, 51–58.

Cooper, J.I. (1999). Fabaviruses (*Comoviridae*). In: A. Granoff and R.G. Webster (eds) *Encyclopedia of Virology*, 2nd edn, pp. 531–534. Academic Press, San Diego.

Cooper, J.I. and Harrison, B.D. (1973). Distribution of potato mop-top virus in Scotland in relation to soil and climate. *Plant Pathol.* **22**, 73–78.

Cooper, J.I. and Jones, A.T. (1983). Responses of plants to viruses: proposals for the use of terms. *Phytopathology* **73**, 127–128.

Cooper, J.I. and Mayo, M.A. (1972). Some properties of the particles of three tobravirus isolates. *J. Gen. Virol.* **16**, 285–297.

Cooper, J.L., Jones, R.A.C. and Harrison, B.D. (1976). Field and glasshouse experiments on the control of potato mop top virus. *Ann. Appl. Biol.* **83**, 215–230.

Cope, W.A., Walker, S.K. and Lucas, L.T. (1978). Evaluation of selected white clover clones for resistance to viruses in the field. *Plant Dis. Rep.* **62**, 267–270.

Cornelissen, B.J.C. and Bol, J.F. (1984). Homology between the proteins encoded by tobacco mosaic virus and two tricornaviruses. *Plant Mol. Biol.* **3**, 379–384.

Cornelissen, B.J.C., Brederode, F.T., Moorinann, R.J.M. and Bol, J.F. (1983a). Complete nucleotide sequence of alfalfa mosaic virus RNA1. *Nucl. Acids Res.* **11**, 1253–1265.

Cornelissen, B.J.C., Brederode, F.T., Veeneman, G.H., van Boom, J.H. and Bol, J.F. (1983b). Complete nucleotide sequence of alfalfa mosaic virus RNA2. *Nucl. Acids Res.* **11**, 3019–3025.

Correia, J.-J., Shire, S., Yphantis, D.A. and Schuster, T.M. (1985). Sedimentation equilibrium measurements of the intermediate-size tobacco mosaic virus protein polymers. *Biochemistry* **24**, 3292–3297.

Costa, A.S. and Muller, G.W. (1980). *Tristeza* control by cross-protection: a U.S.-Brazil cooperative success. *Plant Dis.* **64**, 538–541.

Costet, L., Cordelier, S., Dorey, S., Baillieul, F., Fritig, B. and Kauffmann, S. (1999). Relationship between localized acquired resistance (LAR) and the hypersensitive response: HR is necessary for LAR to occur and salicylic acid is not sufficient to trigger LAR. *Mol. Plant Microbe Interact.* **12**, 655–662.

Cotten, J. (1979). The effectiveness of soil sampling for virus-vector nematodes in MAFF certification schemes for fruit and hops. *Plant Pathol.* **28**, 40–44.

Couch, H.B. (1955). Studies on the seed transmission of lettuce mosaic virus. *Phytopathology* **45**, 63–70.

Coutts, R.H.A. (1986). New threats to crops from changing farm practices. *Nature (Lond.)* **322**, 594.

Coutts, R.H.A. and Buck, K.W. (1985). DNA and RNA polymerase activities of nuclei and hypotonic extracts of nuclei isolated from tomato golden mosaic virus infected tobacco leaves. *Nucl. Acids Res.* **13**, 7881–7897.

Coutts, R.H.A., Rigden, J.E., Slabas, A.R., Lomonossoff, G.P. and Wise, P.J. (1991). The complete nucleotide sequence of tobacco necrosis virus strain D. *J. Gen. Virol.* **72**, 1521–1529.

Covey, S.N. (1986). Amino acid sequence homology in gag region of reverse transcribing elements and the coat protein gene of cauliflower mosaic virus. *Nucl. Acids Res.* **14**, 623–633.

Covey, S.N. (2000). Silencing gene silencing genes. *Trends Plant Sci.* **5**, 405–406.

Covey, S.N. and Al-Kaff, N.S. (2000). Plant DNA viruses and gene silencing. *Plant Mol. Biol.* **43**, 307–322.

Covey, S.N. and Hull, R. (1981). Transcription of cauliflower mosaic virus DNA: detection of transcripts, properties, and location of the gene encoding the virus inclusion body protein. *Virology* **111**, 463–474.

Covey, S.N. and Turner, D.S. (1993). Changes in populations of cauliflower mosaic virus DNA and RNA forms during turnip callus proliferation *J. Gen. Virol.* **74**, 1887–1893.

Covey, S.N., Turner, D. and Mulder, G. (1983). A small DNA molecule containing covalently linked ribonucleotides originates from the large intergenic region of the cauliflower mosaic virus genome. *Nucl. Acids Res.* **11**, 251–264.

Covey, S.N., Turner, D.S., Lucy, A.P. and Saunders, K. (1990). Host regulation of the cauliflower mosaic virus multiplication cycle. *Proc. Natl Acad. Sci. USA* **87**, 1633–1637.

Covey, S.N., Al-Kaff, N.S., Lángara, A. and Turner, D.S. (1997). Plants combat infection by gene silencing. *Nature* **385**, 781–782.

Crane, P.R. and Lidgard, S. (1989). Angiosperm diversification and paleolatitudinal gradients in cretaceous floristic diversity. *Science* **246**, 675–678.

Crane, P.R., Friis, E.M. and Pedersen, K.R. (1995). The origin and early diversification of angiosperms. *Nature* **374**, 27–33.

Creamer, R. and Falk, B.W. (1990). Direct detection of transcapsidated barley yellow dwarf luteovirus in doubly infected plants. *J. Gen. Virol.* **71**, 211–217.

Creamer, R., He, X. and Styer, W.E. (1993). Transmission of sorghum stunt mosaic rhabdovirus by the leafhopper vector, *Graminella sonora* (Homoptera: Cicadellidae). *Plant Dis.* **81**, 63–65.

Creaser, E.H., Gibbs, A.J. and Pares, R.D. (1987). The amino acid composition of the coat protein of a tobamovirus from an Australian *Capsicum* crop. *Aust. Plant Pathol.* **16**, 85–87.

Crick, F.H.C. and Watson, J.D. (1956). Structure of small viruses. *Nature (London)* **177**, 473–475.

Crill, P., Hagedorn, D.J. and Hanson, E.W. (1971). An artificial system for differentiating strains of alfalfa mosaic virus. *Plant Dis. Rep.* **55**, 127–130.

Cronin, S., Verchot, J., Haldeman-Cahill, R., Schaad, M.C. and Carrington, J.C. (1995). Long-distance movement factor: a transport function of the potyvirus helper component proteinase. *Plant Cell* **7**, 549–559.

Cronshaw, J., Hoefert, L. and Esau, K. (1966). Ultrastructural features of *Beta* leaves infected with beet yellows virus. *J. Cell Biol.* **31**, 429–443.

Crosbie, E.S. and Matthews, R.E.F. (1974a). Effects of TYMV infection on leaf pigments in *Brassica pekinensis* Rupr. *Physiol. Plant Pathol.* **4**, 379–387.

Crosbie, E.S. and Matthews, R.E.F. (1974b). Effects of TYMV infection on growth of *Brassica pekinensis* Rupr. *Physiol. Plant Pathol.* **4**, 389–400.

Crowley, N.C. (1959). Studies on the time of embryo infection by seed-transmitted viruses. *Virology* **8**, 116-123.

Crowson, R.A., Ralfe, W.D.I., Smart, J., Waterson, C.D., Willey, E.C. and Wootton, R.J. (1967). *Arthropoda: Chelicerata, Pyconogonida, Palaeoisopus, Myriapoda and Insecta. The Fossil Record*, pp. 499–534. Geological Society, London.

Crowther, R.A. and Klug, A. (1975). Structural analysis of macromolecular assemblies by image reconstruction from electron micrographs. *Annu. Rev. Biochem.* **44**, 161–182.

Crowther, R.A., Geelen, J.L.M.C. and Mellema, J.E. (1974). A three-dimensional image reconstruction of cowpea mosaic virus. *Virology* **57**, 20–27.

Crum, C., Johnson, J.D., Nelson, A. and Roth, D. (1988). Complementary oligodeoxynucleotide mediated inhibition of tobacco mosaic virus RNA translation *in vitro*. *Nucl. Acids Res.* **16**, 4569–4581.

Cuillel, M., Herzog, M., and Hirth, L. (1979). Specificity of *in vitro* reconstitution of bromegrass mosaic virus. *Virology* **95**, 146–153.

Cuillel, M., Jacrot, B. and Zulauf, M. (1981). A T = 1 capsid formed by protein of brome mosaic virus in the presence of trysin. *Virology* **110**, 63–72.

Cuillel, M., Berthet-Colominas, C., Krop, B., Tardieu, A., Vachette, P. and Jacrot, B. (1983). Self-assembly of brome mosaic virus capsids: kinetic studies using neutron and X-ray solution scattering. *J. Mol. Biol.* **164**, 645–650.

Culver, J.N. (1997). Viral avirulence genes. In: G. Stacey and N.T. Keen (eds) *Plant–Microbe Interactions*, Vol. 2, pp. 196–219. Chapman and Hall, New York.

Culver, J.N. and Dawson, W.O. (1989). Point mutations in the coat protein gene of tobacco mosaic virus induce hypersensitivity in *Nicotiana sylvestris*. *Mol. Plant Microbe Interact.* **2**, 209–213.

Culver, J.N., Lindebeck, A.G.C. and Dawson, W.O. (1991). Virus–host interactions: induction of chlorotic and necrotic responses in plants by tobamoviruses. *Annu. Rev. Phytopathol.* **29**, 193–217.

Culver, J.N., Stubbs, G. and Dawson, W.O. (1994). Structure–function relationship between tobacco mosaic virus coat protein and hypersensitivity in *Nicotiana sylvestris*. *J. Mol. Biol.* **242**, 130–138.

Culver, J.N., Dawson, W.O., Plonk, K. and Stubbs, G. (1995). Site-directed mutagenesis confirms the involvement of carboxylate groups in the disassembly of tobacco mosaic virus. *Virology* **206**, 724–730.

Cuozzo, M., O'Connell, K.M., Kaniewski, W., Fang, R.-X., Chua, N.-H. and Turner, N.E. (1988). Viral protection in transgenic tobacco plants expressing the cucumber mosaic virus coat protein or its antisense RNA. *Biotechnology* **6**, 549–557.

Cusack, S., Miller, A., Krijgsmann, P.C.J. and Mellema, J.E. (1981). An investigation of the structure of alfalfa mosaic virus by small-angle neutron scattering. *J. Mol. Biol.* **145**, 525–543.

Cusack, S., Oostergetel, G.T., Krijgsman, P.C.J. and Mellema, J.E. (1983). Structure of the Top a–t component of alfalfa mosaic virus: a non-icosahedral virion. *J. Mol. Biol.* **171**, 139–155.

Cutt, J.R., Harpster, M.H., Dixon, D.C., Carr, J.P., Dunsmuir, P. and Klessig, D.F. (1989). Disease response to tobacco mosaic virus in transgenic tobacco plants that constitutively express the pathogenesis-related PR1b gene. *Virology* **173**, 89–97.

da Graça, J.V. and Martin, M.M. (1976). An electron microscope study of hypersensitive tobacco infected with tobacco mosaic virus at 32 ∝ C. *Physiol. Plant Pathol.* **8**, 215–219.

da Graça, J.V. and Martin M.M. (1981). Ultrstructural changes in avocado leaf tissue infected with advocado sunblotch. *Phytopathol. Z.* **102**, 185–194.

Dahal, G., Hibino, H. and Saxena, R.C. (1990a). Association of leafhopper feeding behaviour with transmission of rice tungro to susceptible and resistant rice cultivars. *Phytopathology* **80**, 371–377.

Dahal, G., Hibino, H., Cabunagan, R.C., Tionco, E.M., Flores, Z.M. and Aguiero, V.M. (1990b). Changes in cultivar reaction due to changes in 'virulence' of the leafhopper vector. *Phytopathology* **80**, 659–665.

Dahal, G., Hughes, J.d'A., Thottappilly, G. and Lockhart, B.E.L. (1998). Effect of temperature on symptom expression and reliability of banana streak badnavirus detection in naturally infected plantain and banana (*Musa* spp.). *Plant Dis.* **82**, 16–21.

Dahl, D. and Knight, C.A. (1963). Some nitrous acid-induced mutants of tomato atypical mosaic virus. *Virology* **21**, 580–586.

Dale, J.L. (1987). Banana bunchy top: an economically important tropical plant virus disease. *Adv. Virus Res.* **33**, 301–325.

Dale, J.L. and Harding, R.M (1998). Banana bunchy top disease: current and future strategies for control. In: A. Hadidi, R.K. Khetarpal and H. Koganezawa (eds) *Plant Virus Disease Control*, pp. 659–669. APS Press, St. Paul, MN.

Daley, P.F. (1995). Chlorophyll fluorescence analysis and imaging plant stress and disease. *Can. J. Plant Pathol.* **17**, 167–173.

Dall, D.J., Randles, J.W. and Francki, R.I.B. (1989). The effect of alfalfa mosaic virus on productivity of annual barrel medic, *Medicago truncatula*. *Aust. J. Agric. Res.* **40**, 807–815.

Dall, D.J., Anzola, J.V. Xu, Z. and Nuss, D.L. (1990). Structure-specific binding of wound tumor virus transcript by a host factor. *Virology* 179, 599–608.

Dalmay, T., Hamilton, A., Rudd, S., Angell, S. and Baulcombe, D.C. (2000). An RNA-dependent RNA polymerase gene in *Arabidopsis* is required for posttranscriptional gene silencing mediated by a transgene but not by a virus. *Cell* **101**, 543–553.

Damsteegt, V.D. and Hewings, A.D. (1986). Comparative transmission of soybean dwarf virus by three geographically diverse populations of *Aulacorthum* (= *Acyrthosiphon*) *solani*. *Ann. Appl. Biol.* **109**, 453–463.

Damsteegt, V.D. and Hewings, A.D. (1987). Relationships between *Aulacorthum solani* and soybean dwarf virus: effect of temperature on transmission. *Phytophatholology* **77**, 515–518.

Dangl, J.L., Dietrich, R.A. and Richberg, M.H. (1996). Death don't have no mercy: cell death programs in plant–microbe interactions. *Plant Cell* **8**, 1793–1807.

Danthinne, X., Seurinck, J., van Montagu, M., Pleij, C.W.A. and van Emmelo, J. (1991). Structural similarities between the RNAs of two satellites of tobacco necrosis virus. *Virology* **185**, 605–614.

D'Arcy, C.J. and Mayo, M.A. (1997). Proposals for changes in luteovirus taxonomy and nomenclature. *Arch. Virol.* **142**, 1285–1287.

D'Arcy, C.J., Domier, L.L, and Mayo, M.A. (2000). Family *Luteoviridae*. In: M.H.V. van Regenmortel *et al.* (eds) *Virus Taxonomy: Seventh Report of the International Committee on the Taxonomy of Viruses*, pp. 775–784. Academic Press, San Diego.

Dardick, C.D. and Culver, J.N. (1997). Tobamovirus coat proteins: elicitors of the hypersensitive response in *Solanum melongena* (eggplant). *Mol. Plant Microbe Interact.* **10**, 776–778.

Dardick, C.D., Taraporewala, Z., Lu, B. and Culver, J.N. (1999). Comparison of tobamovirus coat protein structural features that affect elicitor activity in pepper, eggplant and tobacco. *Mol. Plant Microbe Interact.* **12**, 247–251.

Daròs, J.A. and Carrington, J.C. (1997). RNA-binding activity of NIa proteinase of tobacco etch potyvirus. *Virology* 237, 327–336

Daròs, J.A. and Flores, R. (1995). Identification of a retroviroid-like element from plants. *Proc. Natl Acad. Sci. USA* **92**, 6856–6860.

Daròs, J.A., Marcos, J.F., Hernández, C. and Flores, R. (1994). Replication of avocado sunblotch viroid: evidence for a symmetric pathway with two rolling circles and hammerhead ribozyme processing. *Proc. Natl Acad. Sci. USA* **91**, 12813–12817.

Daròs, J.-A., Schaad, M.C. and Carrington, J.C. (1999). Functional analysis of the interaction between VPg-proteinase (NIa) and RNA polymerase (NIb) of tobacco etch potyvirus, using conditional and suppressor mutants. *J. Virol.* **73**, 8732–8740.

Dasgupta, I., Hull, R., Eastop, S., Poggi-Pollini, C., Blakebrough, M., Boulton, M.I. and Davies, J.W. (1991). Rice tungro bacilliform virus DNA independently infects rice after *Agrobacterium*-mediated transfer. *J. Gen. Virol.* **72**, 1215–1221.

Dasgupta, R., Garcia II, B.H. and Goodman, R.M. (2001). Systemic spread of an RNA insect virus in plants expressing plant viral movement protein genes. *Proc. Natl Acad. Sci. USA* **98**, 4910–4915.

Datema, K.P., Spruijt, R.B., Verduin, J.M. and Hemminga, M.A. (1987). Interaction of plant viruses and viral coat proteins with model membranes. *Biochemistry* **26**, 6217–6223.

Daubert, S.D. and Bruening, G. (1984). Detection of genome-linked proteins of plant and animal viruses. In: K. Maramorosch and H. Koprowski (eds) *Methods in Virology*, pp. 347–379. Academic Press, Orlando, FL.

Daubert, S.D., Richins, R., Shepherd, R.J. and Gardner, R.C. (1982). Mapping of the coat protein gene of cauliflower mosaic virus by its expression in a prokaryotic system. *Virology* **122**, 444–449.

Daubert, S.D., Schoeiz, J., Debao, L. and Shepherd, R.J. (1984). Expression of disease symptoms in cauliflower mosaic virus genomic hybrids. *J. Mol. Appl. Genet.* **2**, 537–547.

Davies, D.L. and Clark, M.F. (1983). A satellite-like nucleic acid of arabis mosaic virus associated with hop nettlehead disease. *Ann. Appl. Biol.* **103**, 439–448.

Davies, J.W. (1979). Translation of plant virus ribonucleic acids in extracts from eukaryotic cells. In T.C. Hall and J.W. Davies (eds) *Nucleic Acids in Plants*, Vol. 2, pp. 111–149. CRC Press, West Palm Beach, FL.

Davies, J.W., Townsend, R. and Stanley, J. (1987). The structure, expression, functions and possible exploitation of geminivirus genomes. *Adv. Plant Sci.* **4**, 31–52.

Davis, P.H. and Heywood, V.H. (1963). *Principles of Angiosperm Taxonomy.* Oliver & Boyd, Edinburgh.

Davis, R.E. and Sinclair, W.A. (1998). Phytoplasma identity and disease etiology. *Phytopathology* **88**, 1372-1376.

Davis, R.F. and Shifriss, O. (1983). Natural virus infection in silvery and non-silvery lines of *Cucurbita pepo. Plant Dis.* **67**, 379–380.

Dawe, V.H. and Kuhn, C.W. (1983). Isolation and characterization of a double-stranded DNA mycovirus infecting the aquatic fungus *Rhizidiomyces. Virology* **130**, 21–28.

Dawson, G.W., Griffiths, D.C., Pickett, J.A., Plumb, R.T., Woodcock, C.M. and Zhong-Ning, Z. (1988). Structure/activity studies on aphid alarm pheromone derivatives and their field use against transmission of barley yellow dwarf virus. *Pestic. Sci.* **22**, 17–30.

Dawson, J.R.O. (1967). The adaptation of tomato mosaic virus to resistant tomato plants. *Ann. Appl. Biol.* **60**, 209–214.

Dawson, W.O. (1972). Enhancement of the infectivity, nucleoprotein concentration and multiplication rate of cowpea chlorotic mottle virus in manganese-treated cowpea. *Phytopathology* **62**, 1206–1209.

Dawson. W.O. (1976). The sequence of inhibition of tobacco mosaic virus synthesis by actinomycin D, 2-thiouracil and cyclohexamide in a synchronous infection. *Phytopathology* **66**, 177–181.

Dawson, W.O. (1978a). Isolation and mapping of replication-deficient, temperature-sensitive mutants of cowpea chlorotic mottle virus. *Virology* **90**, 112–118.

Dawson, W.O. (1978b). Time-course of actinomycin D inhibition of tobacco mosaic virus multiplication relative to the rate of spread of the infection. *Intervirology* **9**, 304–309.

Dawson, W.O. (1983). Tobacco mosaic virus protein synthesis is correlated with double-stranded RNA synthesis and not single-stranded RNA synthesis. *Virology* **125**, 314–323.

Dawson, W.O. (1999). Tobacco mosaic virus virulence and avirulence. *Proc. R. Soc. Lond. B* **354**, 645–651.

Dawson, W.O. and Boyd, C. (1987a). Modifications of nucleic acid precursors that inhibit plant virus multiplication. *Phytopathology* **77**, 477–480.

Dawson, W.O. and Boyd, C. (1987b). TMV protein synthesis is not translationally regulated by heat shock. *Plant Mol. Biol.* **8**, 145–149.

Dawson, W.O. and Bubrick, P. (1989). The interactions of modified viruses with their hosts. In: *Molecular Biology of Plant–Pathogen Interactions*, pp. 117–127. Alan R. Liss, New York.

Dawson, W.O. and Jones, G.E. (1976). A procedure for specifically selecting temperature-sensitive mutants of tobacco mosaic virus. *Mol. Gen. Genet.* **145**, 307–309.

Dawson, W.O. and Korhonen-Lehto, K.M. (1990). Regulation of tobamovirus gene expression. *Adv. Virus Res.* **38**, 307–342.

Dawson, W.O. and Schlegel, D.E. (1976). The sequence of inhibition of TMV synthesis by actinomycin D, 2-thiouracil and cycloheximide in a synchronous infection. *Phytopathology* **66**, 177–181.

Dawson, W.O., Schlegel, D.E., and Lung, M.C.Y. (1975). Synthesis of tobacco mosaic virus in intact tobacco leaves systemically inoculated by differential temperature treatment. *Virology* **65**, 565–573.

Dawson, W.O., Beck, D.L., Knorr, D.A. and Grantham, G.L. (1986). cDNA cloning of the complete genome of tobacco mosaic virus and production of infectious transcripts. *Proc. Natl Acad. Sci. USA* **83**, 1832–1836.

Dawson, W.O., Bubrick, P. and Grantham, G.L. (1988). Modification of the tobacco mosaic virus coat protein gene affecting replication movement and symptomatology. *Phytopathology* **78**, 783–789.

Dawson, W.O., Lewandowski, D.J., Hilf, M.E. *et al.* (1989). A tobacco mosaic virus-hybrid expresses and loses an added gene. *Virology* **172**, 285–292.

Dax, E., Livneh, O., Aliskevicius, E. *et al.* (1998). A SCAR marker linked to the ToMV resistance gene, *Tm2²*, in tomato. *Euphytica* **101**, 73–77.

Day, A.G., Bejarano, E.R., Buck, K.W., Burrell, M. and Lichtenstein, C.P. (1991). Expression of an antisense viral gene in transgenic tobacco confers resistance to the DNA virus tomato golden mosaic virus. *Proc. Natl Acad. Sci. USA* **88**, 6721–6725.

Day, N. and Maiti, I.B. (1999). Structure and promoter/leader deletion analysis of mirabilis mosaic virus (MMV) full-length promoter in transgenic plants. *Plant Mol. Biol.* **40**, 771–782.

de Assis, F.M. and Sherwood, J.L. (2000). Evaluation of seed transmission of turnip yellow mosaic virus and tobacco mosaic virus in *Arabidopsis thaliana. Phytopathology* **90**, 1233–1238.

de Bokx, J.A. (ed.) (1972). *Viruses of Potatoes and Seed-Potato Production.* Pudoc, Wageningen.

de Bokx, J.A. and Piron, P.G.M. (1981). Transmission of potato spindle tuber viroid by aphids. *Neth. J. Plant Pathol.* **87**, 31–34.

de Bortoli, M. and Roggero, P. (1985). Electrophoretic desorption of intact virus from immunoadsorbent. *Microbiologica* **8**, 113–121.

de Fazio, G., Kudamatsu, M. and Vicente, M. (1980). Virazole pretreatments for the prevention of tomato spotted wilt virus (TSWV) systemic infection in tobacco plants *Nicotiana tabacum*, L. 'White Burley'. *Fitopatol. Bras.* **5**, 343–394.

de Feyter, R., Young, M., Schroeder, K., Dennis, E.S. and Gerlach, W. (1996). A ribozyme gene and an antisense gene are equally effective in conferring resistance to tobacco mosaic virus on transgenic tobacco. *Mol. Gen. Genet.* **250**, 329–338.

de Graaf, M., Coscoy, L. and Jaspars, E.M.J. (1993). Localization and biochemical characterization of alfalfa mosaic virus replication complexes. *Virology* **194**, 878–881.

de Haan, P., Wagemakers, L., Peters, D. and Goldbach, R. (1990). The S RNA segment of tomato spotted wilt virus has an ambisense character. *J. Gen. Virol.* **71**, 1001–1007.

de Haan, P., Gielen, J.J.L., Prins, M. *et al.* (1992). Characterization of RNA-mediated resistance to tomato spotted wilt virus in transgenic tobacco plants. *Biotechnology* **10**, 1133–1137.

de Jaeger, G., De Wilde, C., Eeckhout, D., Fiers, E. and Depicker, A. (2000). The plantibody approach: expression of antibody genes in plants to modulate plant metabolism or to obtain pathogen resistance. *Plant Mol. Biol.* **43**, 419–428.

de Jager, C.P. (1976). Genetic analysis of cowpea mosaic virus mutants by supplementation and reassortment tests. *Virology* **70**, 151–163.

de Jager, S.M. and Murray, J.A.H. (1999). Retinoblastoma proteins in plants. *Plant Mol. Biol.* **41**, 295–299.

de Jong, W. and Ahlquist, P. (1995). Host-specific alterations in viral RNA accumulation and infection spread in a brome mosaic virus isolate with an expanded host range. *J. Virol.* **69**, 1485–1492.

de Jong, W., Forsyth, A., Leister, D., Gebhardt, C. and Baulcombe, D.C. (1997). A potato hypersensitive resistance gene against potato virus X maps to a resistance gene cluster on chromosome V. *Theor. Appl. Genet.* **95**, 153–162.

de Kochko, A. (1999). Plant pararetroviruses—cassava vein mosaic virus. In: A. Granoff and R.G. Webster (eds) *Encyclopedia of Virology*, 2nd edn, pp. 1285–1289. Academic Press, San Diego.

de la Cruz, A., Lopez, L., Tenllado, F. *et al.* (1997). The coat protein is required for elicitation of *Capsicum L²* gene-mediated resistance against the tobamoviruses. *Mol. Plant Microbe Interact.* **10**, 107–113.

de Laat, A.M.M. and van Loon, L.C. (1983). The relationship between stimulated ethylene production and symptom expression in virus-infected tobacco leaves. *Physiol. Plant Pathol.* **22**, 261–273.

de Leeuw, G.T.N. (1975). An easy and precise method to measure the length of flexuous virus particles from electron micrographs. *Phytopathol. Z.* **82**, 347–351.

de Mejia, M.V.G., Hiebert, E., Purcifull, D.E., Thornbury, D.W. and Pirone, T.P. (1985). Identification of potyviral amorphous inclusion protein as a non-structurlal virus-specific protein related to helper component. *Virology* **142**, 34–43.

de Miranda, J., Hernandez, M., Hull, R. and Espinoza, A.M. (1994). Sequence analysis of rice hoja blanca virus RNA 3. *J. Gen. Virol.* **75**, 2127–2132.

de Miranda, J., Hull, R. and Espinoza, A.M. (1995a). Sequence of the PV2 gene of rice hoja blanca tenuivirus RNA-2. *Virus Genes* **10**, 205–209.

de Miranda, J., Stevens, M., de Bruyne, E., Smith, H.G., Bird, C. and Hull, R. (1995b). Sequence comparison and classification of beet luteovirus isolates. *Arch. Virol.* **140**, 2183–2200.

de Tapia, M., Himmelbach, A. and Hohn, T. (1993). Molecular dissection of the cauliflower mosaic virus translation transactivator. *EMBO J.* **12**, 3305–3314.

de Varennes, A., Davies, J.W., Shaw, J.G. and Maule, A.J. (1985). A reappraisal of the effect of actinomycin D and cordycepin on the multiplication of cowpea mosaic virus in cowpea protoplasts. *J. Gen. Virol.* **66**, 817–825.

de Wijs, J.J. (1975). The distribution of passionfruit ringspot virus in its main host plants in Ivory Coast. *Neth. J. Plant Pathol.* **81**, 144–148.

de Wijs, J.J. (1980). The characteristics of mineral oils in relation to their inhibitory activity on the aphid transmission of potato virus Y. *Neth. J. Plant Pathol.* **86**, 291–300.

de Wijs, J.J. and Suda-Bachmann, F. (1979). The long-term preservation of potato virus Y and watermelon mosaic virus in liquid nitrogen in comparison to other preservation methods. *Neth. J. Plant Pathol.* **85**, 23–29.

de Zoeten, G.A. (1966). California tobacco rattle virus, its intracellular appearance, and the cytology of the infected cell. *Phytopathology* **56**, 744–754.

de Zoeten, G.A. (1968). Application of scanning microscopy in the study of virus transmission of aphids. *J. Virol.* **2**, 745–751.

de Zoeten, G.A. (1991). Risk assessment: do we let history repeat itself? *Phytopathology* **81**, 585–586.

de Zoeten, G.A. (1995). Plant virus infection: another point of view. *Adv. Bot. Res.* **21**, 105–124.

de Zoeten, G.A. and Gaard, G. (1969a). Distribution and appearance of alfalfa mosaic virus in infected plant cells. *Virology* **39**, 768–774.

de Zoeten, G.A. and Gaard, G. (1969b). Possibilities for inter- and intracellular translocation of some icosahedral plant viruses. *J. Cell Biol.* **40**, 814–823.

de Zoeten, G.A., Gaard, G. and Diez, F.B. (1972) Nuclear vesiculation associated with pea enation mosaic virus-infected plant tissue. *Virology* **48**, 638–647.

de Zoeten, G.A., Assink, A.M. and van Kanimen, A. (1974). Association of cowpea mosaic virus-induced double-stranded RNA with a cytopathological structure in infected cells. *Virology* **59**, 341–355.

de Zoeten, G.A., Powell, C.A., Gaard, G. and German, T.L. (1976). *In situ* localization of pea enation mosaic virus double-stranded ribonucleic acid. *Virology* **70**, 459–469.

de Zoeten, G.A., Penswick, J.R., Horisberger, M.A., Ahl, P., Schultze, M. and Hohn, T. (1989). The expression localization and effect of a human interferon in plants. *Virology* **172**, 213–222.

de Zoeten, G.A., Demler, S.A., Rucker, D.G., Ziegler, A., Robinson, D.J. and Murant, A.F. (1994). Replicative cross-support of the satellite RNAs of the groundnut rosette disease complex and pea enation mosaic virus. *Abstracts of the Annual Meeting of the American Society of Virology*, 1994, Madison, Wisconsin, USA, p. 206.

Dean, J.L. (1979). Sugarcane mosaic virus: shape of the inoculum–infection curve near the origin. *Phytopathology* **69**, 179–181.

Deiman, B.A.L.M. and Pleij, C.W.A. (1997). Pseudoknots: a vital feature in viral RNA. *Semin. Virol.* **8**, 166–175.

Deiman, B.A.L.M., Séron, K., Jaspars, E.M.J. and Pleij, C.W.A. (1997a). Efficient transcription of the tRNA-like structure of turnip yellow mosaic virus by a template-dependent and specific RNA polymerase obtained by a new procedure. *J. Virol. Meth.* **64**, 181–195.

Deiman, B.A.L.M., Kortlever, R.M. and Pleij, C.W.A. (1997b). The role of the pseudoknot at the 3′ end of turnip yellow mosaic virus RNA in minus-strand synthesis by the viral RNA-dependent RNA polymerase. *J. Virol.* **71**, 5990–5996.

Deiman, B.A.L.M., Verlaan, W.G. and Pleij, C.W.A. (2000). *In vitro* transcription of the turnip yellow mosaic virus RNA polymerase: a comparison with the alfalfa mosaic virus and brome mosaic virus replicases. *J. Virol.* **74**, 264–271.

Dekker, E.L., Doré, I., Porta, C. and van Regenmortel, M.H.V. (1987). Conformational specificity of monoclonal antibodies used in the diagnosis of tomato mosaic virus. *Arch. Virol.* **94**, 191–203.

Dekker, E.L., Pinner, M.S., Markham, P.G. and van Regenmortel, M.H.V. (1988). Characterisation of maize streak virus isolates from different plant species by polyclonal and monoclonal antibodies. *J. Gen. Virol.* **69**, 983–990.

Dekker, E.L., Porta, C. and van Regenmortel, M.H.V. (1989). Limitations of different ELISA procedures for localizing epitopes in viral coat protein subunits. *Arch. Virol.* **105**, 269–286.

Dekker, E.L., Woolston, C.J., Xue, Y., Cox, B. and Mullineaux, P.M. (1991). Transcript mapping reveals different expression strategies for bicistronic RNAs of the geminivirus wheat dwarf virus. *Nucl. Acids Res.* **19**, 4075–4081.

Del Bene, G., Dallai, R. and Marchini, D. (1991). Ultrastructure of the midgut and the adhering tubular salivary glands of *Frankliniella occidentalis* (Pergande) (Thysanoptera: *Thripidae*). *Int. J. Insect Morphol. Embryol.* **20**, 15–24.

Delannay, X., Fraley, R.T., Rogers, S.G. *et al.* (1989). Development and field testing of crops improved through genetic engineering. In: J.J. Cohen (ed.) *Strengthening Collaboration in Biotechnology. International Agricultural Research and the Private Sector*, pp. 185–195. Agency for International Development, Washington, DC.

Delauney, A.J., Tabaeizadeh, Z. and Verma, D.P.S. (1988). A stable bifunctional antisense transcript inhibiting gene expression in transgenic plants. *Proc. Natl Acad. Sci. USA* **85**, 4300–4304.

Delwiche, C.F., Graham, L.E. and Thomson, N. (1989). Lignin-like compounds and sporopollenin in Coleochaete, an algal model for land plant ancestry. *Science* **245**, 399–401.

Demler, S.A. and de Zoeten, G.A. (1989). Characterization of a satellite RNA associated with pea enation mosaic virus. *J. Gen. Virol.* **70**, 1075–1084.

Demler, S.A., Rucker, D.G. and de Zoeten, G.A. (1993). The chimeric nature of the genome of pea enation mosaic virus: the independent replication of RNA 2. *J. Gen. Virol.* **74**, 1–14.

Demler, S.A., Borkhsenious, O.N., Rucker, D.G. and de Zoeten, G.A. (1994a). Assessment of the autonomy of replicative and structural functions encoded by the luteo-phase of pea enation mosaic virus. *J. Gen. Virol.* **75**, 997–1007.

Demler, S.A., Rucker, D.G., Nooruddin, L. and de Zoeten, G.A. (1994b). Replication of the satellite RNA of pea enation mosaic virus is controlled by RNA 2-encoded functions. *J. Gen. Virol.* **75**, 1339–1406.

Demler, S.A., de Zoeten, G.A., Adam, G. and Harris, K.F. (1996a). Pea enation mosaic virus: properties and aphid transmission. In: B.D. Harrison and A.F Murant (eds) *The Plant Viruses*, Vol. 5, *Polyhedral Virions and Bipartite RNA Genomes*, pp. 303–344. Plenum Press, New York.

Demler, S.A., Rucker, D.G., de Zoeten, G.A., Ziegler, A., Robinson, D.J. and Murant, A.F. (1996b). The satellite RNAs associated with the groundnut rosette disease complex and pea enation mosaic virus: sequence similarities and ability of each other's helper to support their replication. *J. Gen. Virol.* **77**, 2847–2855.

Dempsey, D.A., Wobbe, K.K. and Klessig, D.F. (1993). Resistance and susceptible responses of *Arabidopsis thaliana* to turnip crinkle virus. *Phytopathology* **83**, 1021–1029.

Dempsey, D.A., Pathirana, M.S., Wobbe, K.K. and Lessig, D.F. (1997). Identification of an *Arabidopsis* locus required for resistance to turnip crinkle virus. *Plant J.* **11**, 301–311.

Dempsey, D'M.A., Shah, J. and Klessig, D.F. (1999). Salicylic acid and disease resistance in plants. *Crit. Rev. Plant Sci.* **18**, 547–575.

Demski, J.W., Bays, D.C. and Kahn, M.A. (1986). Simple latex agglutination test for detecting flexuous rod-shaped viruses in forage legumes. *Plant Dis.* **70**, 777–779.

Denloye, A.O., Homer, R.B. and Hull, R. (1978). Circular dichroism studies on turnip rosette virus. *J. Gen. Virol.* **41**, 77–85.

Denna, D.W. and Alexander, M.B. (1975). The isoperoxidases of *Cucurbita pepo* L. In: C.L. Markert (ed.) *Isozymes*, Vol. 2, pp. 851–864. Academic Press, New York.

Deom, C.M., Oliver, M.J. and Beachy, R.M. (1987). The 30-kilodalton gene product of tobacco mosaic virus potentiates virus movement. *Science* **237**, 389–394.

Deom, C.M., He, X.Z., Beachy, R.N. and Weissinger, A.K. (1994). Influence of heterologous tobamovirus movement protein and chimeric-movement protein genes on cell-to-cell and long-distance movement. *Virology* **205**, 198–209.

Deom, C.M., Quan, S. and He, X.Z. (1997). Replicase proteins as determinants of phloem-dependent long-distance movement of tobamoviruses in tobacco. *Protoplasma* **199**, 1–8.

Depicker, A. and van Montagu, M. (1997). Post-transcriptional gene silencing in plants. *Curr. Opin. Cell Biol.* **9**, 373–382.

Derrick, K.S. (1973). Quantitative assay for plant viruses using serologically specific electron microscopy. *Virology* **56**, 652–653.

Derrick, P.M. and Nelson, R.S. (1999). Plasmodesmata and long-distance virus movement. In: A.J.E. van Bel and W.J.P. van Kesteren (eds) *Plasmodesmata: Structure, Function, Role in Cell Communications*, pp. 315–339. Springer-Verlag, Berlin.

Derrick, P.M., Barker, H. and Oparka, K.J. (1992). Increase in plasmodesmatal permeability during cell-to-cell spread of tobacco rattle virus from individually inoculated cells. *Plant Cell* **4**, 1405–1412.

Desbiez, C., David, C., Mettouchi, A., Laufs, J. and Gronenborn, B. (1995). Rep protein of tomato yellow leafcurl geminivirus has an ATPase activity required for viral DNA replication. *Proc. Natl Acad. Sci. USA* **92**, 5640–5644.

Desjardins, P.R., Drake, R.J. and French, J.V. (1969). Transmission of citrus ringspot virus to citrus and non-citrus hosts by dodder (*Campestris subinclusa*). *Plant Dis. Rep.* **53**, 947–948.

Dessens, J.T. and Lomonossoff, G.P. (1991). Mutational analysis of the putative catalytic triad of the cowpea mosaic virus 24K protease. *Virology* **184**, 738–746.

Devereux, J., Haeberli, P. and Smithies, O. (1984). A comprehensive set of sequence analysis programs for the VAX. *Nucl. Acids Res.* **12**, 387–395.

Devic, M., Jaegle, M. and Baulcombe, D. (1989). Symptom production on tobacco and tomato is determined by two distinct domains of the satellite RNA of cucumber mosaic virus (strain Y). *J. Gen. Virol.* **70**, 2765–2774.

Dewar, A.M. and Smith, H.G. (1999). Forty years of forecasting virus yellow incidence in sugar beet. In: H.G. Smith and H. Barker (eds) *The Luteoviridae*, pp. 231–243. CAB International, Wallingford, UK.

Dey, N. and Maiti, I.B. (1999). Structure and promoter/leader deletion analysis of mirabilis mosaic virus (MMV) full-length transcript promoter in transgenic plants. *Plant Mol. Biol.* **40**, 771–782.

Di, R., Dinesh-Kumar, S.P. and Miller, W.A. (1993). Translational frameshifting by barley yellow dwarf RNA (PAV serotype) in *Escherichia coli* and in eukaryotic cell-free extracts. *Molec. Plant-Microbe Interact.* **6**, 444–452.

Di Franco, A. and Martelli, G.P. (1987). Some observations on the ultrastructure of galinsoga mosaic virus infections. *Phytopathol. Mediterr.* **26**, 54–56.

Di Franco, A., Russo, M. and Martelli, G.P. (1984). Ultrastructure and origin of cytoplasmic multivesicular bodies induced by carnation Italian ringspot virus. *J. Gen. Virol.* **65**, 1233–1237.

Diaco, R., Hill, J.H., Hill, G.K., Tachibana, H. and Durand, D.P. (1985). Monoclonal antibody-based biotin–avidin ELISA for the detection of soybean mosaic virus in soybean seeds. *J. Gen. Virol.* **66**, 2089–2094.

Diaco, R., Lister, R.M., Hill, J.H. and Durand, D.P. (1986). Detection of homologous and heterologous barley yellow dwarf virus isolates with monoclonal antibodies in serologically specific electron microscopy. *Phytopathology* **76**, 225–230.

Diaz-Avalos, R. and Caspar, D.L.D. (1998). Structure of the stacked disk aggregate of tobacco mosaic virus protein. *Biophys J.* **74**, 595–603.

Diaz-Ruiz, J.R. and Kaper, J.M. (1983). Nucleotide sequence relationships among thirty peanut stunt virus isolates determined by competition hybridization. *Arch. Virol.* **75**, 277–281.

Dickerson, P.E. and Trim, A.R. (1978). Conformational states of cowpea chlorotic mottle virus ribonucleic acid components. *Nucl. Acids Res.* **5**, 987–998.

Diekmann, M. and Putter, C.A. (1996). *FAO/IPGRI Technical Guidelines for the Safe Movement of Germplasm*, No. 15, Musa, 2nd ed. Food and Agriculture Organization of the United Nations/International Plant Genetic Resource Institute, Rome.

Diener, T.O. (1971). Potato spindle tuber 'virus'. IV. A replicating, low molecular weight RNA. *Virology* **45**, 411–428.

Diener, T.O. (1981). Are viroids escaped introns? *Proc. Natl Acad. Sci. USA* **78**, 5014–5015.

Diener, T.O. (1987a). Biological properties. In: T.O. Diener (ed.) *The Viroids*, pp. 9–35. Plenum, New York.

Diener, T.O. (ed.) (1987b). *The Viroids*. Plenum, New York.

Diener, T.O. (1989). Circular RNAs: relics of precellular evolution? *Proc. Natl Acad. Sci. USA* **86**, 9370–9374.

Diener, T.O. (1993). The viroid: big punch in a small package. *Trends Microbiol.* **1**, 289–294.

Diener, T.O. (1996). Origin and evolution of viroids and viroid-like satellite RNAs. *Virus Genes* **11**, 119–131.

Diener, T.O. (1999). Viroids and the nature of viroid diseases. *Arch Virol.* **15**, 203–220.

Diener, T.O. and Hadidi, A. (1977). Viroids. *Compr. Virol.* **11**, 285–337.

Diener, T.O. and Jenifer, F.G. (1964). A dependable local lesion assay for turnip yellow mosaic virus. *Phytopathology* **54**, 1258–1260.

Dietzgen, R.G. (1986). Immunological properties and biological function of monoclonal antibodies to tobacco mosaic virus. *Arch. Virol.* **87**, 73–86.

Dietzgen, R.G. and Francki, R.I.B. (1987). Nonspecific binding of immunoglobulins to coat proteins of certain plant viruses in immunoblots and indirect ELISA. *J. Virol. Methods* **15**, 159–164.

Dietzgen, R.G. and Sander, E.-M. (1982). Monoclonal antibodies against a plant virus. *Arch. Virol.* **74**, 197–204.

Dietzgen, R.G. and Zaitlin, M. (1986). Tobacco mosaic virus coat protein and the large subunit of the host protein ribulose-1,5-bisphosphate carboxylase share a common antigenic determinant. *Virology* **155**, 262–266.

Diez, J., Ishikawa, M., Kaido, M. and Ahlquist, P. (2000). Identification and characterization of a host protein required for efficient template selection in viral RNA replication. *Proc. Natl Acad. Sci. USA* **97**, 3913–3918.

Dijkstra, J. (1962). On the early stages of infection by tobacco mosaic virus in *Nicotiana glutinosa* L. *Virology* **18**, 142–143.

Dijkstra, J. (1966). Multiplication of TMV in isolated epidermal tissue of tobacco and *Nicotiana glutinosa* leaves. In: A.B.R. Beemster and J. Dijkstra (eds) *Viruses of Plants*, pp. 19–21. North-Holland, Amsterdam.

Dijkstra, J. and de Jager, C.P. (1998). *Practical Plant Pathology: Protocols and Exercises*. Springer-Verlag, Berlin

Dijkstra, J., Bruin, G.C.A., Burgers, A.C. *et al.* (1977). Systemic infection of some N-gene-carrying *Nicotiana* species and cultivars after inoculation with tobacco mosaic virus. *Neth. J. Plant Pathol.* **83**, 41–59.

Dinant, S. and Lot, H. (1992). Lettuce mosaic virus. *Plant Pathol.* **41**, 528–542.

Dinant, S., Janda, M., Kroner, P. and Ahlquist, P. (1993). Bromovirus RNA replication and transcription require compatability between the polymerase- and helicase-like RNA synthesis proteins. *J. Virol.* **67**, 7181–7189.

Dinant, S., Kusiak, C., Cailleteau, B. *et al.* (1998). Field resistance against potato virus Y infection using natural and genetically engineered resistance genes. *Eur. J. Plant Pathol.* **104**, 377–382.

Dinesh-Kumar, S.P. and Baker, B.J. (2000). Alternatively spliced N resistance gene transcripts: their possible role in tobacco mosaic virus resistance. *Proc. Natl Acad. Sci. USA* **97**, 1908–1913.

Dinesh-Kumar, S.P. and Miller, W.A. (1993). Control of start codon choice on a plant viral RNA encoding overlapping genes. *Plant Cell* **5**, 679–692.

Dinesh-Kumar, S.P., Whitham, S., Choi, D., Hehl, R., Corr, C. and Baker, B. (1995). Transposon tagging of tobacco mosaic virus resistance gene N: its possible role in TMV-N-mediated signal transduction pathway. *Proc. Natl Acad. Sci. USA* **92**, 4175–4180.

Dinesh-Kumar, S.P., Tham, W.-H. and Baker, B.J. (2000). Structure–function analysis of the tobacco mosaic virus resistance gene N. *Proc. Natl Acad. Sci. USA* **97**, 14789–14794.

Ding, B. (1998). Intercellular protein trafficking through plasmodesmata. *Plant Mol. Biol.* **38**, 279–310.

Ding, B., Haudenshield, J.S., Hull, R.J., Wolf, S., Beachy, R.N. and Lucas, W.J. (1992). Secondary plasmodesmata are specific sites of localization of the tobacco mosaic virus movement protein in transgenic tobacco plants. *Plant Cell* **4**, 915–928.

Ding, B., Kwon, M.-O., Hammond, R. and Owens, R. (1997). Cell-to-cell movement of potato spindle tuber viroid. *Plant J.* **12**, 931–936.

Ding, S.-W. (2000). RNA silencing. *Curr. Opin. Biotechnol.* **11**, 152–156.

Ding, S.-W., Keese, P. and Gibbs, A. (1989). Nucleotide sequence of the ononis yellow mosaic tymovirus genome. *Virology* **172**, 555–563.

Ding, S.-W., Howe, J., Keese, P. *et al.* (1990). The tymobox, a sequence shared by most tymoviruses: its use in molecular studies of tymoviruses. *Nucl. Acids Res.* **18**, 1181–1187.

Ding, S.-W., Anderson, B.J., Haase, H.R. and Symons, R.H. (1994). New overlapping gene encoded by the cucumber mosaic virus genome. *Virology* **198**, 593–601.

Ding, S.-W., Li, W.X. and Symons, R.H. (1995a). A novel naturally occurring hybrid gene encoded by a plant virus facilitates long-distance virus movement. *EMBO J.* **14**, 5762–5772.

Ding, S.-W., Shi, B.J., Li, W.X. and Symons, R.H. (1996a). An interspecific species hybrid RNA virus is significantly more virulent than either parental virus. *Proc. Natl Acad. Sci. USA* **93**, 7470–7474.

Ding, X.S., Shintaku, M.H., Arnold, S.A. and Nelson, R.S. (1995b). Accumulation of mild and severe strains of tobacco mosaic virus in minor veins of tobacco. *Mol. Plant-Microbe Interact.* **8**, 32–40.

Ding, X.S., Shintaku, M.H., Carter, S.A. and Nelson, R.S. (1996b). Invasion of minor veins of tobacco leaves inoculated with tobacco mosaic virus mutants defective in phloem-dependent movement. *Proc. Natl Acad. Sci. USA* **93**, 11155–11160.

Ding, X.S., Carter, S.A., Deom, C.M. and Nelson, R.S. (1998). Tobamovirus and potyvirus accumulation in minor veins of inoculated leaves from representatives of the *Solanaceae* and *Fabaceae*. *Plant Physiol.* **116**, 125–136.

Ding, X.S., Flasinski, S. and Nelson, R.S. (1999). Infection of barley by brome mosaic virus is restricted predominantly to cells in and associated with veins through a temperature-dependent mechanism. *Mol. Plant–Microb. Interact.* **12**, 615–623.

Ding, X.S., Boydiston, C.M. and Nelson, R.S. (2000). Potential roles of plant guttation in BMV transmission in nature. *Phytopathology* **90**, S19.

Dinter-Gottlieb, G. (1986). Viroids and virusoids are related to group I introns. *Proc. Natl Acad. Sci. USA* **83**, 6250–6254.

Dixon, A.F.G. (1973). *The Biology of Aphids.* Edward Arnold (Publishers) Ltd., London.

Dixon, E.A. and Harrison, M.J. (1990). Activation, structure and organization of genes involved in microbial defense in plants. *Adv. Genet.* **28**, 166–233.

Dixon, L.K. and Hohn, T. (1984). Initiation of translation of the cauliflower mosaic virus genome from a polycistronic mRNA: evidence from deletion mutagenesis. *EMBO J.* **3**, 2731–2736.

Dodds, J.A. (1993). dsRNA in diagnosis. In: R.E.F. Matthews (ed.) *Diagnosis of Plant Virus Diseases*, pp. 273–294. CRC Press, Boca Raton, FL.

Dodds, J.A. (1998). Satellite tobacco mosaic virus. *Annu. Rev. Phytopathol.* **36**, 295–310.

Dodds, J.A. (1999). Satellite tobacco mosaic virus. In: P.K. Vogt and A.O. Jackson (eds) *Satellites and Defective RNAs*, pp. 145–157. Springer, Berlin.

Dodds, J.A. and Hamilton, R.I. (1974). Masking of the RNA genome of tobacco mosaic virus by the protein of barley stripe mosaic virus in doubly infected barley. *Virology* **59**, 418–427.

Dodds, J.A. and Hamilton, R.I. (1976). Structural interactions between viruses as a consequence of mixed infections. *Adv. Virus Res.* **20**, 33–86.

Dodds, J.A., Morris, T.J., and Jordan, R.L. (1984). Plant viral double-stranded RNA. *Annu. Rev. Phytopathol.* **22**, 151–168.

Dodds, J.A., Lee, S.Q. and Tiffany, M. (1985). Cross protection between strains of cucumber mosaic virus: effect of host and type of inoculum on accumulation of virions and double-stranded RNA of the challenge strain. *Virology* **144**, 301–309.

Dodds, J.A., Jarupat, T., Lee, J.G. and Roistacher, C.N. (1987). Effects of strain, host, time of harvest and virus concentration on double-stranded RNA analysis of citrus tristeza virus. *Phytopathology* **77**, 442–447.

Doi, Y., Teranaka, M., Yora, K. and Asuyama, H. (1967). Mycoplasma or PLT group-like microorganisms found in the phloem elements of plants infected with mulberry dwarf, potato witches broom, aster yellows, or Paulownia witches broom. *Ann. Phytopathol. Soc. Jpn* **33**, 259–266.

Dolenc, J., Vihar, B. and Dermastia, M. (2000). Systemic infection with potato virus Y-NTN alters the structure and activity of the shoot meristem in a susceptible potato cultivar. *Physiol. Molec. Plant Pathol.* **56**, 33–38.

Dolja, V.V., Sokolova, N.A., Tjulkina, L.G. and Atabekov, J.G. (1979). A study of barley stripe mosaic virus (BSMV) genome: II. Translation of individual RNA species of two BSMV strains in a homologous cell-free system. *Mol. Gen. Genet.* **175**, 93–97.

Dolja, V.V., Lunina, N.A., Leiser, R.-M., Stanarius, T., Belzhelarskaya, S.N., Koziov, Yu.V., and Atabekov, J.G. (1983). A comparative study on the *in vitro* translation products of individual RNAs from two-, three-, and four-component strains of barley stripe mosaic virus. *Virology* **127**, 1–14.

Dolja, V.V., McBride, H.J. and Carrington, J.C. (1992). Tagging of plant potyvirus replication and movement by insertion of β-glucuronidase into the viral polyprotein. *Proc. Natl Acad. Sci. USA* **89**, 10208–10212.

Dolja, V.V., Karasev, A.V. and Koonin, E.V. (1994). Molecular biology and evolution of closteroviruses: sophisticated build-up of large RNA genomes. *Annu. Rev. Phytopathol.* **32**, 261–285.

Dolja, V.V., Haldeman-Cahill, R., Montgomery, A.E., Vandenbosch, K.A. and Carrington, J.C. (1995). Capsid protein determinants involved in cell-to-cell and long distance movement of tobacco etch potyvirus. *Virology* **206**, 1007–1016.

Dollet, M., Accotto, G.P., Lisa, V., Menissier, J. and Boccardo, G. (1986). A geminivirus, serologically related to maize streak virus, from *Digitaria sanguinalis* from Vanuatu. *J. Gen. Virol.* **67**, 933–937.

Domeier, M.E., Morse, D.P., Knight, S.W., Portereiko, M., Bass, B.L. and Mango, S.E. (2000). A link between RNA interference and nonsense-mediated decay in *Caenorhabditis elegans*. *Science* **289**, 1928–1930.

Domier, L.L., Shaw, J.G. and Rhoads, R.E. (1987). Potyviral proteins share amino acid sequence homology with picorna-, como-, and caulimoviral proteins. *Virology* **158**, 20–27.

Domingo, E. (1999). Quasispecies. In: A. Granoff and R.G. Webster (eds) *Encyclopedia of Virology*, pp. 1431–1436. Academic Press, San Diego.

Domingo, E. and Holland, J.-J. (1988). High error rates, population equilibrium and evolution of RNA replication systems. In: E. Domingo, J.-J. Holland and P. Ahlquist (eds) *RNA Genetics*, Vol. 3, pp. 3–36. CRC Press, Boca Raton, FL.

Domingo, E. and Holland, J.-J. (1997). RNA virus mutations and fitness for survival. *Annu. Rev. Microbiol.* **51**, 151–178.

Domingo, E., Holland, J., Biebricher, C. and Eigen, M. (1995). Quasi-species: the concept and the word. In: A.J. Gibbs, C.H. Calisher and F. Garcia-Arenal (eds) *Molecular Basis for Virus Evolution*, pp. 181–191. Cambridge University Press, Cambridge.

Domingo, E., Escarmis, C., Menéndez-Arias, L. and Holland, J. – J. (1999a). Viral quasispecies and fitness variations. In: E. Domingo, R. Webster and J. Holland (eds) *Origin and Evolution of Viruses*, pp. 141–161. Academic Press, London.

Domingo, E., Webster, R. and Holland, J. (1999b). Origin and evolution of viruses. Academic Press, London.

Dominguez, D.I., Hohn, T. and Schmidt-Puchta, W. (1996). Cellular proteins bind to multiple sites of the leader region of cauliflower mosaic virus 35S RNA. *Virology* **226**, 374–383.

Dominguez, D.I., Ryabova, L.A., Pooggin, M.M., Schmidt-Puchta, W., Fütterer, J. and Hohn, T. (1998). Ribosome shunting in cauliflower mosaic virus: identification of an essential and sufficient structural element. *J. Biol. Chem.* **273**, 3669–3678.

Donald, R.G.K. and Jackson, A.O. (1996). RNA-binding activities of barley stripe mosaic virus gamma b fusion proteins. *J. Gen. Virol.* **77**, 879–888.

Donald, R.G.K., Petty, I.D.T., Zhou, H. and Jackson, A.O. (1985). Properties of genes influencing barley stripe mosaic virus movement phenotypes. In: D.D. Bills *et al.* (eds) *Biotechnology and Plant Protection: Viral Pathogenesis and Disease Resistance*, pp. 115–150. World Scientific Publishing, Singapore.

Doncaster, J.P. and Gregory, P.H. (1948). The spread of virus diseases in the potato crop. *G. B. Agric. Res. Counc. Rep. Ser.* **7**, 1–189.

Dong, X. (1998). SA, JA, ethylene and disease resistance in plants. *Curr. Opin. Plant Biol.* **1**, 316–323.

Donofrio, J.C., Kuchta, J., Moore, R. and Kacymarczyk, W. (1986). Properties of a solubilized replicase isolated from corn infected with maize dwarf mosaic virus. *Can. J. Microbiol.* **32**, 637–644.

Donson, J., Morris-Krsinich, B.A.M., Mullineaux, P.M., Boulton, M.I. and Davies, J.W. (1984). A putative primer for second strand DNA synthesis of maize streak virus is virion associated. *EMBO J.* **3**, 3069–3073.

Donson, J., Gunn, H.V., Woolston, C.J. *et al.* (1988). *Agrobacterium*-mediated infectivity of cloned digitaria streak virus DNA. *Virology* **162**, 248–250.

Donson, J., Kearney, C.M., Hilf, M.E. and Dawson, W.O. (1991). Systemic infection of a bacterial gene by tobacco mosaic virus-based vector. *Proc. Natl Acad. Sci. USA* **88**, 7204–7208.

Doolittle, R.F. (ed.) (1990). *Molecular Evolution: Computer Analysis of Protein and Nucleic Acid Sequences*. Academic Press, San Diego.

Doolittle, R.F., Feng, D.-F., Johnson, M.S. and McClure, M.A. (1989) Origins and evolutionary relationships of retroviruses. *Q. Rev. Biol.* **64**, 1–30.

Dore, I., Dekker, E.L., Porta, C. and van Regenmortel, M.H.V. (1987a). Detection by ELISA of two tobamoviruses in orchids using monoclonal antibodies. *J. Phytopathol.* **20**, 317–326.

Dore, I., Altschuh, D., Al-Moudallal, Z. and van Regenmortel, M.H.V. (1987b). Immunochemical studies of tobacco mosaic virus: VII. Use of comparative surface accessibility of residues in antigenically related viruses for delineating epitopes recognised by monoclonal antibodies. *Mol. Immunol.* **24**, 1351–1358.

Dore, I., Weiss, E., Altschuh, D. and van Regenmortel, M.H.V. (1988). Visualization by electron microscopy of the location of tobacco mosaic virus epitopes reacting with monoclonal antibodies in enzyme immunoassay. *Virology* **162**, 279–289.

Dore, I., Ruhlmann, C., Oudet, P., Cahoon, M., Caspar, D.L.D. and van Regenmortel, M.H.V. (1989). Polarity of binding of mono-clonal antibodies to tobacco mosaic virus rods and stacked disks. *Virology* **176**, 25–29.

Dorokhov, Y.L., Alexandrova, N.M., Miroshnichenko, N.A. and Atabekov, J.G. (1983). Isolation and analysis of virus-specific ribonucleoprotein of tobacco mosaic virus-infected tobacco. *Virology* **127**, 237–252.

Dorokhov, Y.L., Alexandrova, N.M., Miroshnichenko, N.A. and Atabekov, J.G. (1984). Stimulation by aurintricarboxylic acid of tobacco mosaic virus-specific RNA synthesis and production of informosome-like infection-specific ribonucleoprotein. *Virology* **135**, 395–405.

Dorokhov, Y.L., Mäkinen, K., Frolova, O.Y. *et al.* (1999). A novel function for a ubiquitous plant enzyme pectin methylesterase: the host receptor for the tobacco mosaic virus movement protein. *FEBS Lett.* **461**, 223–228.

Dorssers, L., van der Meer, J., van Kammen, A. and Zabel, P. (1983). The cowpea mosaic virus RNA replication complex and the host-encoded RNA-dependent RNA polymerase–template complex are functionally different. *Virology* **125**, 155–174.

Dorssers, L., van der Krol, S., van der Meer, J., van Kammen, A. and Zabel, P. (1984). Purification of cowpea mosaic virus RNA replication complex: identification of a viral-encoded 110,000-dalton polypeptide responsible for RNA chain elongation. *Proc. Natl Acad. Sci. USA* **81**, 1951–1955.

Dougherty, W.G. (1983). Analysis of viral RNA isolated from leaf tissue infected with tobacco etch virus. *Virology* **131**, 473–481.

Dougherty, W.G. and Carrington, J.C. (1988). Expression and function of potyviral gene products. *Annu. Rev. Phytopathol.* **26**, 123–143.

Dougherty, W.G. and Hiebert, E. (1980). Translation of potyvirus RNA in a rabbit reticulocyte lysate: identification of nuclear inclusion proteins as products of tobacco etch virus RNA translation and cylindrical inclusion protein as a product of the potyvirus genome. *Virology* **104**, 174–182.

Dougherty, W.G. and Hiebert, E. (1985). Genome structure and gene expression of plant RNA viruses. In: J.W. Davies (ed.) *Plant Molecular Biology*, Vol. 2, pp. 23–81. CRC Press, Boca Raton, FL.

Dougherty, W.G. and Parks, T.D. (1995). Transgenes and gene silencing: telling us something new? *Curr. Opin. Cell Biol.* **7**, 399–405.

Dougherty, W.G. and Semler, B.L. (1993). Expression of virus-encoded proteinases: functional and structural similarities with cellular enzymes. *Microbiol. Rev.* **57**, 781–822

Dougherty, W.G., Wilis, L. and Johnston, R.F. (1985). Topographic analysis of tobacco etch virus capsid protein epitopes. *Virology* **144**, 66–72.

Dougherty, W.G., Parks, T.D., Cary, S.M., Bazan, J.F. and Fletterick, R.J. (1989). Characterization of the catalytic residues of the tobacco etch virus 49-kDa proteinase. *Virology* **172**, 302–310.

Douthit, L.B. and McGuire, J.M. (1975). Some effects of temperature on *Xiphinema americanum* and infection of cucumber by tobacco ringspot virus. *Phytopathology* **65**, 134–138.

Dowson Day, M.J., Ashurst, J.L., Mathias, S.F., Watts, J.W., Wilson, T.M.A. and Dixon, R.A. (1993). Plant viral leaders influence expression of a reporter gene in tobacco. *Plant. Mol. Biol.* **23**, 97–109.

Drake, J.W. and Holland, J.J. (1999). Mutation rates among RNA viruses. *Proc. Natl Acad. Sci. USA* **96**, 13910–13913.

Drake, J.R., Charlesworth, B., Charlesworth, D. and Crow, J.F. (1998). Rates of spontaneous mutation. *Genetics* **148**, 1667–1686.

Draper, J. and Scott, R. (1991). Gene transfer to plants. In: D. Grierson (ed.) *Plant Genetic Engineering*, pp. 38–81. Blackie, Glasgow.

Dreher, T.W. (1999). Functions of the 3'-untranslated regions of positive strand RNA viral genomes. *Annu. Rev. Phytopathol.* **37**, 151–174.

Dreher, T.W. and Weiland, J.J. (1994). Preferential replication of defective turnip yellow mosaic virus RNAs that express the 150-kDa protein in *cis. Archiv. Virol.* **S9**, 195–204.

Driedonks, R.A., Krijgsman, P.C.J. and Mellema, J.E. (1977). Alfalfa mosaic virus protein polymerization. *J. Mol. Biol.* **113**, 123–140.

Driedonks, R.A., Tjok Joe, M.K. K. and Mellema, J.E. (1980). Application of band centrifugation to the study of the assembly of alfalfa mosaic virus. *Biopolymers* **19**, 575–595.

Drugeon, G., Urcuqui-Inchima, S., Milner, M. *et al.* (1999). The strategies of plant virus gene expression: models of economy. *Plant Sci.* **148**, 77–88.

Druka, A. and Hull, R. (1998). Variation in rice tungro viruses: further evidence for two rice tungro bacilliform virus strains and possibly several rice tungro spherical virus variants. *J. Phytopathol.* **146**, 175–178.

Druka, A., Burns, T., Zhang, S.L. and Hull, R. (1996). Immunological characterization of rice tungro spherical virus coat proteins and differentiation of isolates from the Philippines and India. *J. Gen. Virol.* **77**, 1975–1983.

Dry, I., Krake, L.R., Rigden, J.E. and Rezaian, M.A. (1997). A novel subviral agent associated with a geminivirus: the first report of a DNA satellite. *Proc. Natl Acad. Sc. USA* **94**, 7088–7093.

Dry, I., Krake, L., Mullineaux, P. and Rezaian, A. (2000). Regulation of tomato leaf curl viral gene expression in host tissues. *Mol. Plant Microb. Interact.* **13**, 529–537.

Du Plessis, D.H. and Smith, P. (1981). Glycosylation of the cauliflower mosaic virus capsid polypeptide. *Virology* **109**, 403–408.

Duarte, E.A., Clarke, D.K., Moya, A., Elena, S.F., Domingo, E. and Holland, J. (1993). Million-trillionfold amplification of single RNA virus particles fails to overcome the Muller's ratchet effect. *J. Virol.* **67**, 3620–3623.

Dudits, D., Marcy, E., Praznovszky, T., Olah, Z., Gyorgyey, J. and Cella, R. (1987). Transfer of resistance traits from carrot into tobacco by asymmetric somatic hybridization: regeneration of fertile plants. *Proc. Natl Acad. Sci. USA* **84**, 8434–8438.

Duffus, J.E. (1971). Role of weeds in the incidence of virus diseases. *Annu. Rev. Phytopathol.* **9**, 319–340.

Duffus, J.E. (1973). The yellowing virus diseases of beet. *Adv. Virus Res.* 18, 347–386.

Duffus, J.E. (1987). Durability of resistance. *Ciba Found. Symp.* **133**, 196–199.

Dugdale, B., Beethan, P.R., Becker, D.K., Harding, R.M. and Dale, J.L. (1998). Promoter activity associated with the intergenic regions of banana bunchy top virus DNA-1 to -6 in transgenic tobacco and banana cells. *J. Gen. Virol.* **79**, 2301–2311.

Duijsings, D., Kormelink, R. and Goldbach, R. (1999). Alfalfa mosaic virus RNAs serve as cap donors for tomato spotted wilt virus transcription during coinfection of *Nicotiana benthamiana. J. Virol.* **73**, 5172–5175.

Dumas, P., Moras, D., Florentz, C., Giegé, R., Verlaan, P., van Belkum, A. and Pleij, C.W.A. (1987). 3D graphics modelling of the tRNA-like 3'-end of turnip yellow mosaic virus RNA: Structural and functional implications. *J. Biomol. Struct. Dyn.* **4**, 707–728.

Dunez, J. (1983). *Indexing of Virus and Virus-like Diseases of Fruit Trees.* International Society of Horticultural Science, Castet IMP, Bordeaux.

Dunigan, D.D. and Zaitlin, M. (1990). Capping activity of tobacco mosaic virus RNA: analysis of viral-coded guanyltransferase-like activity. *J. Biol. Chem.* **265**, 7770–1186.

Dunigan, D.D., Dietzen, R.G., Schoelz, J.E. and Zaitlin, M. (1988). Tobacco mosaic virus particles contain ubiquitinated coat protein subunits. *Virology* **165**, 310–312.

Dunn, D.B. and Hitchborn, J.H. (1965). The use of bentonite in the purification of plant viruses. *Virology* **25**, 171–192.

Dupin, A., Collot, D., Peter, R. and Witz, J. (1985). Comparison between the primary structure of the coat proteins of turnip yellow mosaic virus and eggplant mosaic virus. *J. Gen. Virol.* **66**, 2571–2579.

Duran-Vila, N. and Semancik, J.S. (1990). Variations in the 'cross protection' effect between two strains of citrus exocortis viroid. *Ann. Appl. Biol.* **117**, 367–377.

Duran-Vila, N., Romero-Durbán, J. and Hernández, M. (1996). Detection and eradication of chrysanthemum stunt in Spain. *OEPP Bull.* **26**, 399–405.

Durham, A.C.H. (1978). The roles of small ions especially calcium in virus disassembly takeover and transformation. *Biomedicine* **28**, 307–313.

Durham, A.C.H. and Bancroft, J.B. (1979). Cation binding by papaya mosaic virus and its protein. *Virology* **93**, 246–252.

Durham, A.C.H. and Finch, J.T. (1972). Structures and roles of the polymorphic forms of tobacco mosaic virus protein. II. Electron microscope observations of the larger polymers. *J. Mol. Biol.* **67**, 307–314.

Durham, A.C.H., Finch, J.T. and Klug, A. (1971). States of aggregation of tobacco mosaic virus protein. *Nature (London), New Biol.* **229**, 37–42.

Durham, A.C.H., Hendry, D.A. and von Wechmar, M.B. (1977). Does calcium ion binding control plant virus disassembly. *Virology* **77**, 524–533.

Durham, A.C.H., Witz, J. and Bancroft, J.B. (1984). The semipermeability of simple spherical virus capsids. *Virology* **133**, 1–8.

Durrant, I. (1990). Light-based detection of biomolecules. *Nature* **346**, 297–300.

Dutcher, S.K. (2001). The tubulin fraternity: alpha to eta. *Curr. Opin. Cell Biol.* **13**, 49–54.

Dzianott, A.M. and Bujarski, J.J. (1991). The nucleotide sequence and genome organization of RNA-1 segment in two bromoviruses—broad bean mottle virus and cowpea chlorotic mottle virus. *Virology* **185**, 553–562.

Eagles, R.M., Balmori-Melián, E., Beck, D.L., Gardner, R.C. and Forster, R.L.S. (1994). Characterization of NTPase, RNA-binding and RNA-helicase activities of the cytoplasmic inclusion protein of tamarillo mosaic potyvirus. *Eur. J. Biochem.* **224**, 677–684.

Easton, L.M., Lewis, G.D. and Pearson, M.N. (1997). Virus-like particles associated with dyeback symptoms in the brown alga *Ecklonia radiata. Dis. Aquatic Org.* **30**, 217–222.

Eastop, V.F. (1977). Worldwide importance of aphids as virus vectors. In: K.F. Harris and K. Maramorosch (eds) *Aphids as Virus Vectors*, pp. 3–62. Academic Press, New York.

Ebbels, D.L. (1979). A historical review of certification schemes for vegetatively propagated crops in England and Wales. *ADAS Q. Rev.* **32**, 21–58.

Edelbaum, O., Ilan, N., Grafi, G. *et al.* (1990). Two antiviral proteins from tobacco: purification and characterization by monoclonal antibodies to human β-interferon. *Proc. Natl Acad. Sci. USA* **87**, 588–592.

Edskes, H.K., Kiernan, J.M. and Shepherd, R.J. (1996). Efficient translation of distal cistrons of a polycistronic mRNA of a plant pararetrovirus requires a compatible interaction between the mRNA 3′ end and the proteinaceous *trans*-activator. *Virology* **224**, 564–567.

Edwards, M.C. (1995). Mapping of the seed transmission determinants of barley stripe mosaic virus. *Mol. Plant Microbe Interact.* **8**, 906–915.

Edwards, M.C., Zhang, Z. and Weiland, J.J. (1997). Oat blue dwarf marafivirus resembles the tymoviruses in sequence, genome organization and expression strategy. *Virology* **232**, 217–229.

Edwards, M.C., Gonsalves, D. and Provvidenti, R. (1983). Genetic analysis of cucumber mosaic virus in relation to host resistance location of determinants for pathogenicity to certain legumes and *Lactuca saligno*. *Phytopathology* **73**, 269–273.

Edwardson, J.R. (1974a). Some properties of the potato virus Y-group. *Fla. Agric. Exp. Sta. Monogr.* **4**, 398 pp.

Edwardson, J.R. (1974b). Host ranges of viruses in the PVY group. *Fla. Agric. Exp. Stn. Monogr.* **5**, 225 pp.

Edwardson, J.R. and Christie, R.G. (1983). Cytoplasmic cylindrical and nucleolar inclusions induced by potato virus-A. *Phytopathology* **73**, 290–293.

Edwardson, J.R. and Corbett, M.K. (1962). A virus-like syndrome in tomato caused by a mutation. *Am. J. Bot.* **49**, 571–574.

Edwardson, J.R., Christie, R.G. and Ko, N.J. (1984). Potyvirus cylindrical inclusions: subdivision-IV. *Phytopathology* **74**, 1111–1114.

Edwardson, J.R., Christie, R.G., Purcifull, D.E. and Petersen, M.A. (1993). Inclusions in diagnosing plant virus diseases. In: R.E.F. Matthews (ed.) *Diagnosis of Plant Virus Diseases*, pp. 101–128. CRC Press, Boca Raton, FL.

Eecen, H.G., van Dierendonck, J.H., Pleij, C.W.A., Mandel, M. and Bosch, L. (1985). Hydrodynamic properties of RNA: effect of multivalent cations on the sedimentation behaviour of turnip yellow mosaic virus RNA. *Biochemistry* **24**, 3610–3617.

Egan, B.T. and Hall, P. (1983). Monitoring the Fiji disease epidemic in sugar cane at Bundaberg, Australia. In: R.T. Plumb and J.M. Thresh (eds) *Plant Virus Epidemiology*, pp. 287–296. Blackwell, Oxford.

Egan, B.T., Ryan, C.C. and Franki, R.I.B. (1989). Fiji disease. In: C. Ricaud, B.T. Egan, A.G. Gillespie and C.G. Hughes (eds) *Diseases of Sugar Cane: Major Diseases*, pp. 263–287. Elsevier, Amsterdam.

Eggen, R. and van Kammen, A. (1988). RNA replication in comoviruses. In: P. Ahlquist, J. Holland and E. Dimingo (eds) *RNA Genetics*, Vol. 1, pp. 49–69. CRC Press, Boca Raton.

Eggen, R., Kaan, A., Goldbach, R. and van Kammen, A. (1988). Cowpea mosaic virus RNA replication in crude membrane fractions from infected cowpea and *Chenopodium amaranticolor*. *J. Gen. Virol.* **69**, 2711–2720.

Eggen, R., Verver, J., Wellink, J., Pleij, K., van Kammen, A. and Goidbach, R. (1990). Analysis of sequences involved in cowpea mosaic virus RNA replication using site specific mutants. *Virology* **173**, 456–464.

Eglinton, G. and Hamilton, R.J. (1967). Leaf epicuticular waxes. *Science* **156**, 1322–1335.

Ehlers, K., Binding, H. and Kollmann, R. (1999). The formation of symplastic domains by plugging of plasmodesmata: a general event in plant morphogenesis? *Protoplasma* **209**, 181–192.

Ehlers, U. and Paul, H.-L. (1986). Characterisation of the coat proteins of different types of barley yellow mosaic virus by polyacrylamide gel electrophoresis and electro-blot immunoassay. *J. Phytopathol.* **115**, 294–304.

Ehresmann, B., Briand, J.-P., Reinbolt, J. and Witz, J. (1980). Identification of binding sites of turnip yellow mosaic virus protein and RNA by crosslinks induced *in situ*. *Eur. J. Biochem.* **108**, 123–129.

Ehresmann, C., Baudin, F., Mougel, M., Romby, P., Ebel, J.-P. and Ehresmann, B. (1987). Probing the structure of RNAs in solution. *Nucl. Acids Res.* **15**, 9109–9128.

Eichler, W. (1948). Some rules on ectoparasitism. *Ann. Mag. Nat. Hist. Ser.* **12**, 588–598.

Eigen, M. (1993). Viral quasispecies. *Sci. Am.* **July,** 32–39.

Eldredge, N. (1985). *Time Frames. The Rethinking of Darwinian Evolution and the Theory of Punctuated Equilibria*. Simon and Schuster, New York.

Elliot, M.S., Cromray, H.L., Zettler, F.W. and Carpenter, W.R. (1987). A mosaic disease of wax myrtle associated with a new species of eryophyid mite. *HortScience* **22**, 258–260.

Elliott, R.M. (1999). *Bunyaviridae*: replication. In: A. Granoff and R.G. Webster (eds) *Encyclopedia of Virology*, 2nd edn, pp. 212–216. Academic Press, San Diego.

Ellsbury, M.M., Backus, E.A. and Ullman, D.E. (eds) (1994). *History, Development and Application of AC Electronic Feeding Monitors*. Entomological Society of America, Lanham, Maryland.

Elmer, J.S., Brand, L., Sunter, G., Gardiner, W.E., Bisaro, D.M. and Rogers, S.G. (1988a). Genetic analysis of tomato golden mosaic virus: II. The product of the AL1 coding sequence is required for replication. *Nucl. Acids Res.* **16**, 7043–7060.

Elmer, J.S., Sunter, G., Gardiner, W.E., Brand, L., Browning, C.K., Bisaro, D.M. and Rogers, S.G. (1988b). *Agrobacterium*-mediated inoculation of plants with tomato golden mosaic virus DNAs. *Plant Mol. Biol.* **10**, 225–234.

Elmer, S. and Rogers, S.G. (1990). Selection for wild type size derivatives of tomato golden mosaic virus during systemic infection. *Nucl. Acids Res.* **18**, 2001–2006.

Elnagar, S. and Murant, A.F. (1976). The role of the helper virus, anthriscus yellows, in the transmission of parsnip yellow fleck virus by the aphid *Cavariella aegopodii*. *Ann. Appl. Biol.* **84**, 169–181.

Emery, V.C. and Bishop, D.H.L. (1987). Characterization of Punta Toro S mRNA species and identification of an inverted complementary sequence in the intergenic region of Punta Toro phlebovirus ambisense S RNA that is involved in mRNA transcription termination. *Virology* **156**, 1–11.

English, J.J., Mueller, E. and Baulcombe, D.C. (1996). Suppression of virus accumulation in transgenic plants exhibiting silencing of nuclear genes. *Plant Cell* **8**, 179–188.

English, J.J., Davenport, G.F., Elmayan, T., Vaucheret, H. and Baulcombe, D.C. (1997). Requirement of sense transcription for homology-dependent virus resistance and trans-inactivation. *Plant J.* **12**, 597–603.

Erhardt, M., Morant, M., Ritzenhaler, C. *et al.* (2000). P42 movement protein of beet necrotic yellow vein virus is targeted by the movement proteins p13 and p15 to punctate bodies associated with plasmodesmata. *Mol. Plant–Microb. Interact.* **13**, 520–528.

Erickson, F.L., Holzberg, S., Calderon-Urrea, A. *et al.* (1999). The helicase domain of the TMV replicase proteins induces the *N*-mediated defense response in tobacco. *Plant J.* **18**, 67–75.

Erickson, J.W. and Rossmann, M.G. (1982). Assembly and crystallization of a *T* = 1 icosahedral particle from trypsinized Southern bean mosaic virus coat protein. *Virology* **116**, 128–136.

Erokhina, T.N., Zinovkin, R.A., Vitushkina, M.V., Jelkmann, W. and Agranovsky, A.A. (2000). Detection of beet yellows closterovirus methyltransferase-like and helicase-like proteins *in vivo* using monoclonal antibodies. *J. Gen. Virol.* **81**, 597–603.

Eşanu, V., and Dumitrescu, M. (1971). A comparative study of isoperoxidases of tobacco as influenced by TMV infection and genetic constitution. *Acta Phytopathol. Acad. Sci. Hung.* **6**, 31–35.

Esau, K. (1953). *Plant Anatomy.* John Wiley, New York.

Esau, K. (1956). An anatomist's view of virus diseases. *Am. J. Bot.* **43**, 739–748.

Esau, K. (1979). Beet yellow stunt virus in cells of *Sonchus oleraceus* L. and its relation to host mitochondria. *Virology* **98**, 1–8.

Esau, K. and Cronshaw, J. (1967). Relation of tobacco mosaic virus with host cells. *J. Cell Biol.* **33**, 665–678.

Esau, K. and Hoefert, L.L. (1972). Ultrastructure of sugarbeet leaves infected with beet western yellows virus. *J. Ultrastruct. Res.* **40**, 556–571.

Esau, K. and Hoefert, L.L. (1973). Particles and associated inclusions in sugarbeet infected with the curly top virus. *Virology* **56**, 454–464.

Esau, K. and Hoefert, L.L. (1978). Hyperplastic phloem in sugarbeet leaves infected with the beet curly top virus. *Am. J. Bot.* **65**, 772–783.

Esau, K. and Hoefert, L.L. (1981). Beet yellow stunt virus in the phloem of *Sonchus oleraceus* L. *J. Ultrastruct. Res.* **75**, 326–338.

Esau, K., Cronshaw, J. and Hoefert, L.L. (1966). Organisation of beet-yellows inclusions in leaf cells of *Beta. Proc. Natl Acad. Sci. USA* **55**, 486–493.

Esau, K., Cronshaw, J. and Hoefert, L.L. (1967). Relation of beet yellows virus to the phloem and to movement in the sieve tube. *J. Cell Biol.* **32**, 71–87.

Escriu, F., Fraile, A. and Garcia-Arenal, F. (2000). Evolution of virulence in natural populations of the satellite RNA of cucumber mosaic virus. *Phytopathology* **90**, 480–485.

Eslick, R.F. and Afanasiev, M.M. (1955). Influence of time of infection with barley stripe mosaic on symptoms, plant yield and seed infection of barley. *Plant Dis. Rep.* **39**, 722–724.

Espinoza, A.M., Medina, V., Hull, R. and Markham, P.G. (1991). Cauliflower mosaic virus gene II product forms distinct inclusion bodies in infected plant cells. *Virology* **185**, 337–344.

Estabrook, E.M., Suyenaga, K., Tsai, J.H. and Falk, B.W. (1996). Maize stripe tenuivirus RNA2 transcripts in plant and insect hosts and analysis of pvc2, a protein similar to the *Phlebovirus* virion membrane glycoproteins. *Virus Genes* **12**, 239–247.

Estabrook, E., Tsai, J. and Falk, B.W. (1998). *In vivo* transfer of barley stripe mosaic hordeivirus ribonucleotides to the 5′ terminus of maize stripe tenuivirus RNAs. *Proc. Natl Acad. Sci. USA* **95**, 8304–8309.

Etessami, P., Callis, R., Ellwood, S. and Stanley, J. (1988). Delimitation of essential genes of cassava latent virus DNA2. *Nucl. Acids Res.* **16**, 4811–4829.

Evans, A.C. (1954). Groundnut rosette disease in Tanganyika. I. Field studies. *Ann. Appl. Biol.* **41**, 189–206.

Evans, D. (1985). Isolation of a mutant of cowpea mosaic virus which is unable to grow in cowpeas. *J. Gen. Virol.* **66**, 339–343.

Evans, T.A. and Stephens, C.T. (1989). Increased susceptibility to fusarium crown and root rot in virus-infected asparagus. *Phytopathology* **79**, 253–258.

Evered, D. and Hamett, S. (eds) (1987). *Plant Resistance to Viruses*, Ciba Foundation Symposium, No. 133. Wiley, Chichester, UK.

Faccioli, G. and Marani, F. (1998). Virus elimination by meristem tip culture and tip micrografting. Breeding for resistance to plant viruses. In: A. Hadidi, R.K. Khetarpal and H. Koganezawa (eds) *Plant Virus Disease Control*, pp. 346–380. APS Press, St. Paul, MN.

Faccioli, G. and Rubies-Autonell, C. (1982). PVX and PVY distribution in potato meristem tips and their eradication by the use of thermotherapy and meristem-tip culture. *Phytopathol. Z.* **103**, 66–75.

Faccioli, G., Rubies-Autonell, C. and Resca, R. (1988). Potato leafroll virus distribution in potato meristem tips and production of virus-free plants. *Potato Res.* **31**, 511–520.

Fagard, M. and Vaucheret, H. (2000). Systemic silencing signal(s). *Plant Mol. Biol.* **43**, 285–293.

Fagard, M., Boutet, S., Morel, J.-B., Bellini, C. and Vaucheret, H. (2000). AGO1, QDE-2, and RDE-1 are related proteins required for post-transcriptional gene silencing in plants, quelling in fungi and RNA interference in animals. *Proc. Natl Acad. Sci. USA* **97**, 11650–11654.

Fairall, L., Finch, J.I., Hui, C.-F. and Cantor, C.R. and Butler, P.J.G. (1986). Studies of tobacco mosaic virus reassembly with an RNA tail blocked by a hydridised and cross-linked probe. *Eur. J. Biochem.* **156**, 459–465.

Fakhfakh, H., Vilaine, F., Makni, M. and Robaglia, C. (1996). Cell-free cloning and biolistic inoculation of an infectious cDNA of potato virus Y. *J. Gen. Virol.* **77**, 519–523.

Falk, B.W. (1999). Tenuiviruses. In: A. Granoff and R.G. Webster (eds) *Encyclopedia of Virology*, 2nd edn, pp. 1756–1764. Academic Press, San Diego.

Falk, B.W. and Tian, T. (1999). Transcapsidation interactions and dependent aphid transmission among luteoviruses, and luteovirus-associated RNAs. In: H.G. Smith and H. Barker (eds) *The Luteoviridae*, pp. 125–164. CAB International, Wallingford, UK.

Falk, B.W. and Tsai, J.H. (1998). Biology and molecular biology of viruses in the genus *Tenuivirus*. *Annu. Rev. Phytopathol.* **36**, 139–163.

Falk, B.W., Duffus, J.E. and Morris, T.J. (1979). Transmission, host range, and serological properties of the viruses that cause lettuce speckles disease. *Phytopathology* **69**, 612–617.

Falk, B.W., Tsai, J.H. and Lommel, S.A. (1987). Differences in levels of detection for the maize stripe virus capsid and major non-capsid proteins in plant and insect hosts. *J. Gen. Virol.* **68**, 1801–1811.

Falk, B.W., Kim, K.S. and Tsai, J.H. (1988). Electron microscopic and physicochemical analysis of a reo-like virus of the planthopper *Peregrinus maidis*. *Intervirology* **29**, 195–206.

Falk, B.W., Klaassen, V.A. and Tsai, J.H. (1989). Complementary DNA cloning and hybridization analysis of maize stripe virus RNAs. *Virology* **173**, 338–342.

Farabaugh, P.J. (1996a). Programmed translational frameshifting. *Microbiol. Rev.* **60**, 103–134.

Farabaugh, P.J. (1996b). Programmed translational frameshifting. *Annu. Rev. Genet.* **30**, 507–528.

Fargette, D., Fauquet, C. and Thouvenel, J.-C. (1985). Field studies on the spread of African cassava mosaic. *Ann. Appl. Biol.* **106**, 285–294.

Fargette, D., Thouvenel, J.-C. and Fauquet, C. (1987). Virus content of leaves of cassava infected by African cassava mosaic virus. *Ann. Appl. Biol.* **110**, 65–73.

Faulkner, G. and Kimmins, W.C. (1975). Staining reactions of the tissue bordering lesions induced by wounding, tobacco mosaic virus, and tobacco necrosis virus in bean. *Phytopathology* **65**, 1396–1400.

Fauquet, C.M. (1999). Taxonomy, classification and nomenclature of viruses. In: A. Granoff and R.G. Webster (eds) *Encyclopedia of Virology*, 2nd edn, pp. 1730–1756. Academic Press, New York.

Fauquet, C.M., and Martelli, G.P. (1995). Up-dated ICTV list of names and abbreviations of viruses, viroids and satellites infecting plants. *Arch. Virol.* **140**, 393–413.

Fauquet, C.M. and Mayo, M.A. (1999). Abbreviations for plant virus names—1999. *Arch. Virol.* **144**, 1249–1272.

Fauquet, C.M., Dejardin, J. and Thouvenel, J.-C. (1986). Evidence that the amino acid composition of the particle proteins of plant viruses is characteristic of the virus group: I. Multidimensional classification of plant viruses. *Intervirology* **25**, 1–13.

Faustmann, O., Kern, R., Sänger, H.L. and Muehlbach, H.-P. (1986). Potato spindle tuber viroid (PSTV) RNA oligomers of (+) and (–) strand polarity are synthesised in potato protoplasts after liposome-mediated infection with PSTV. *Virus Res.* **4**, 213–227.

Favali, M.A., Bassi, M. and Appiano, A. (1974). Synthesis and migration of maize rough dwarf virus in the host cell: an autoradiographic study. *J. Gen. Virol.* **24**, 563–565.

Fazeli, C.F. and Rezaian, M.A. (2000). Nucleotide sequence and organization of ten ORFs in the genome of Grapevine leafroll-associated virus 1 and identification of 3 subgenomic RNAs. *J. Gen. Virol.* **81**, 605–615.

Febres, V.J., Mawassi, A.M., Frank, A. *et al.* (1996). The p27 protein is present at one end of citrus tristeza virus particles. *Phytopathology* **86**, 1331–1335.

Fedorkin, O.N., Solovyev, A.G., Yelina, N.E. *et al.* (2001). Cell-to-cell movement of potato virus X involves distinct functions of the coat protein. *J. Gen. Virol.* **82**, 449–458.

Felden, B., Florentz, C., McPherson, A. and Giege, R. (1994). A histidine accepting tRNA-like fold at the 3′ end of satellite tobacco mosaic virus RNA. *Nucl. Acids Res.* **22**, 2882–2886.

Felden, B., Florenz, C., Giegé, R. and Westhof, E. (1996). A central pseudoknotted three-way junction imposes tRNA-like mimicry and the orientation of three 5′ upstream pseudoknots in the 3′ terminus of tobacco mosaic virus RNA. *RNA* **2**, 201–212.

Fellers, J., Wan, J., Hong, Y., Collins, G.B. and Hunt, A.G. (1998). *In vitro* interactions between a potyvirus-encoded, genome linked protein and RNA-dependent RNA polymerase. *J. Gen. Virol.* **79**, 2043–2049.

Fenby, N.S., Scott, N.W., Slater, A. and Elliott, M.C. (1995). PCR and nonisotopic labelling: techniques for plant virus detection. *Cell. Mol. Biol.* **41**, 639–652.

Fenczik, C.A., Padgett, H.S., Holt, C.A., Casper, S.J. and Beachy, R.N. (1995). Mutational analysis of the movement protein of odontoglossum ringspot virus to identify a host-range determinant. *Mol. Plant-Microbe Interact.* **8**, 666–673.

Feng, D.F., Johnson, M.S. and Doolittle, R.F. (1985). Aligning amino acid sequences—comparison of commonly used methods. *J. Mol. Evol.* **21**, 112–125.

Fenner, F. (1976). Classification and Nomenclature of Viruses. Second Report on the International Committee on Taxonomy of Viruses. *Intervirology* **7**, 1–115.

Fenner, F. (1979). Portraits of viruses: the poxviruses. *Intervirology* **11**, 137–157.

Fenoll, C., Black, D.M. and Howell, S.H. (1988). The intergenic region of maize streak virus (MSV) contains promoter elements involved in rightward transcription of the viral genome. *EMBO J.* **7**, 1589–1596.

Fenoll, C., Schwarz, J.J., Black, D.M., Schneider, M. and Howell, S.H. (1990). The intergenic region of maize streak virus contains a GC-rich element that activates rightward transcription and binds maize nuclear factors. *Plant Mol. Biol.* **15**, 865–877.

Fernández, A., Laín, S. and García, J.A. (1995). RNA helicase activity of the plum pox potyvirus CI protein expressed in *Escherichia coli*: mapping of an RNA binding domain. *Nucl. Acids Res.* **23**, 1327–1332.

Fernández, A., Guo, H.S., Sáenz, P., Simón-Buela, L., Gómez de Cedrón, M. and García, J.A. (1997). The motif V of plum pox potyvirus CI RNA helicase is involved in NTP hydrolysis and is essential for virus replication. *Nucl. Acids Res.* **25**, 4474–4480.

Fernandez, I., Candresse, T., Le Gall, O. and Dunez, J. (1999). The 5′ noncoding region of grapevine chrome mosaic nepovirus RNA-2 triggers a necrotic response on three *Nicotiana* spp. *Mol. Plant–Microbe Interact.* **12**, 337–344.

Fernandez-Gonzalez, O., Renaudin, J. and Bové, J.M. (1980). Infection of chlorophyll-less protoplasts from etiolated chinese cabbage hypocotyls by turnip yellow mosaic virus. *Virology* **104**, 262–265.

Fernow, K.H. (1967). Tomato as a test plant for detecting mild strains of potato spindle tuber virus. *Phytopathology* **57**, 1347–1352.

Figlerowitz, M., Nagy, P.D. and Bujarski, J.J. (1997). A mutation in the putative RNA polymerase gene inhibits nonhomologous, but not homologous, genetic recombination in an RNA virus. *Proc. Natl Acad. Sci. USA* **94**, 2073–2078.

Filho, F.M. de A. and Sherwood, J.L. (2000). Evaluation of seed transmission of turnip yellow mosaic virus and tobacco mosaic virus in *Arabidopsis thaliana*. *Phytopathology* **90**, 1233–1238.

Filipowicz, W and Kiss, T. (1993). Structure and function of nucleolar snRNPs. *Mol. Biol. Rep.* 18, 149–156.

Finch, J.T. (1972). The hand of the helix of tobacco mosaic virus. *J. Mol. Biol.* **66**, 291–294.

Finch, J.T. and Holmes, K.C. (1967). Structural studies of viruses. In: K. Maramorosch and H. Koprowski (eds) *Methods in Virology*, Vol. 3, pp. 351–474. Academic Press, New York.

Finch, J.T. and Klug, A. (1966). Arrangement of protein subunits and the distribution of nucleic acid in turnip yellow mosaic virus: II. Electron microscopic studies. *J. Mol. Biol.* **15**, 344–364.

Finch, J.T. and Klug, A. (1967). Structure of broad bean mottle virus: 1. Analysis of electron micrographs and comparison with turnip yellow mosaic virus and its top component. *J. Mol. Biol.* **24**, 289–302.

Finch, J.T., Klug, A. and van Regenmortel, M.H.V. (1967a). The structure of cucumber mosaic virus. *J. Mol. Biol.* **24**, 303–305.

Finch, J.T., Leberman, R. and Berger, J.E. (1967b). Structure of broad bean mottle virus: II. X-ray diffraction studies. *J. Mol. Biol.* **27**, 17–24.

Finlay, K.W. (1953). Inheritance of spotted wilt resistance in the tomato. II. Five genes controlling spotted wilt resistance in four tomato types. *Aust. J. Biol. Sci.* **6**, 153–163.

Finnegan, E.J., Wang, M.-B. and Waterhouse, P. (2001). Gene silencing: fleshing out the bones. *Curr. Biol.* (in press).

Fischer, D.B. and Rufty, R.C. (1993). Inheritance of partial resistance to tobacco etch virus and tobacco vein mottling virus in burley tobacco cultivar sota 6505. *Plant Dis.* **77**, 662–666.

Fisher, D.B., Wu, Y. and Ku, M.S.B. (1992). Turnover of soluble proteins in the wheat sieve tube. *Plant Physiol.* **100**, 1433–1441.

Fitch, W.M. and Margoliash, E. (1967). Construction of phylogenetic trees. *Science* **155**, 279–284.

Fitchen, J.H. and Beachy, R.N. (1993). Genetically engineered protection against viruses in transgenic plants. *Annu. Rev. Microbiol.* **47**, 739–763.

Fitchen, J., Beachy, R.N. and Hein, M.B. (1995). Plant virus expressing hybrid coat protein with added murine epitope elicits autoantibody response. *Vaccine* **13**, 1051–1057.

Fletcher, J.T. and Butler, D. (1975). Strain changes in populations of tobacco mosaic virus from tomato crops. *Ann. Appl. Biol.* **81**, 409–412.

Flor, H.H. (1971). Current status of the gene-for-gene concept. *Annu. Rev. Phytopathol.* **9**, 275–296.

Flores, R., Di Serio, F. and Hernández, C. (1997). Viroids: the non-coding genomes. *Sem. Virol.* **8**, 65–73.

Flores, R., Daròs, J.-A. and Hernández, C. (2000). Asunviroidae family: viroids containing hammerhead ribozymes. *Adv. Virus Res.* **55**, 271–323.

Föglein, F.J., Kalpagam, C., Bates, D.C., Premecz, G., Nyitrai, A. and Farkas, G.L. (1975). Viral RNA synthesis is renewed in protoplasts isolated from TMV-infected Xanthi tobacco leaves in an advanced stage of virus infection. *Virology* **67**, 74–79.

Föglein, F.J., Nyitrai, A., Gulyás, A., Premecz, G., Oláh, T. and Farkas, G.L. (1976). Crystalline inclusion bodies are degraded in protoplasts isolated from TMV-infected tobacco leaves. *Phytopathol. Z.* **86**, 266–269.

Fondong, V.N., Pita, J.S., Rey, M.E.C., de Kochko, A., Beachy, R.N. and Fauquet, C.M. (2000). Evidence for synergism between African cassava mosaic virus and a new double-recombinant geminivirus infecting cassava in Cameroon. *J. Gen. Virol.* **81**, 287–297.

Fontes, E.P.B., Luckow, V.A. and Hanley-Bowdoin, L. (1992). A geminivirus replication protein is a sequence-specific DNA binding protein. *Plant Cell* **4**, 597–608.

Fontes, E.P.B., Eagle, P.A., Sipe, P.S., Luckow, V.A. and Hanley-Bowdoin, L. (1994a). Interaction between a geminivirus replication protein and origin DNA is essential for viral replication. *J. Biol. Chem.* **11**, 8459–8465.

Fontes, E.P.B., Gladfelter, H.J., Schaffer, R.L., Petty, I.T.D. and Hanley-Bowdoin, L. (1994b). Geminivirus replication origins have a molecular organization. *Plant Cell* **6**, 405–416.

Forbes, A.R. (1977). The mouthparts and feeding mechanism of aphids. In: K.F. Harris and K. Maramorosch (eds) *Aphids as Virus Vectors*, pp. 83–103. Academic Press, New York.

Ford, R.E. (1973). Concentration and purification of clover yellow mosaic virus from pea roots and leaves. *Phytopathology* **63**, 926–930.

Forster, A.C. and Symons, R.H. (1987). Self-cleavage of plus and minus RNAs of a virusoid and a structural model for the active sites. *Cell (Camb.)* **49**, 211–220.

Forster, A.C., Davies, C., Sheldon, C.C., Jeffries, A.C. and Symons, R.H. (1988). Self-cleaving viroid and newt RNAs may only be active as dimers. *Nature (Lond.)* **334**, 265–267.

Forster, R.L.S., Guilford, P.J. and Faulds, D.V. (1987). Characterisation of the coat protein subgenomic RNA of white clover mosaic virus. *J. Gen. Virol.* **68**, 181–190.

Forster, R.L.S., Beck, D.L., Guilford, P.J., Voot, T.M., Vandolleweerd, C.J. and Andersen, M.T. (1992). The coat protein of white clover mosaic potexvirus has a role in facilitating cell-to-cell transport in the plant. *Virology* **191**, 480–484.

Foster, G.D. and Taylor, S.C. (eds) (1998). *Plant Virology Protocols: From Virus Isolation to Transgenic Resistance.* Humana Press, Totowa, NJ.

Foster, J.A. and Hadidi, A. (1998). Exclusion of viruses. In: A. Hadidi, R.K. Khetarpal and H. Koganezawa (eds) *Plant Virus Disease Control*, pp. 208-229. APS Press, St. Paul, MN.

Foster, J.A. and Ross, A.F. (1975). Properties of the initial tobacco mosaic virus infection sites revealed by heating symptomless inoculated tobacco leaves. *Phytopathology* **65**, 610–616.

Fowlks, E. and Young, R.J. (1970). Detection of heterogeneity in plant viral RNA by polyacrylamide gel electrophoresis. *Virology* **42**, 548–550.

Fox, J.M., Wang, G.J., Speir, J.A. *et al.* (1998). Comparison of the native CCMV virion with *in vitro* assembled CCMV virions by cryoelectron microscopy and image reconstruction. *Virology* **244**, 212–218.

Fraenkel-Conrat, H. (1956). The role of nucleic acid in the reconstitution of active tobacco mosaic virus. *J. Am. Chem. Soc.* **78**, 882–883.

Fraenkel-Conrat, H. (1957). Degradation of tobacco mosaic virus with acetic acid. *Virology* **4**, 1–4.

Fraenkel-Conrat, H. and Singer, B. (1957). Virus reconstitution: II. Combination of protein and nucleic acid from different strains. *Biochim. Biophys. Acta* **24**, 540–548.

Fraenkel-Conrat, H. and Williams, R.C. (1955). Reconstitution of active tobacco mosaic virus from its inactive protein and nucleic acid components. *Proc. Natl Acad. Sci. USA* **41**, 690–698.

Fraile, A. and García-Arenal, F. (1990). A classification of the tobamoviruses based on the comparison of their '126K' proteins. *J. Gen. Virol.* **71**, 2223–2228.

Fraile, A., Aranda, M.A. and García-Arenal, F. (1995). Evolution of the tobamoviruses. In: A.J. Gibbs, C.H. Calisher and F. Garcia-Arenal (eds) *Molecular Basis for Virus Evolution*, pp. 338–349. Cambridge University Press, Cambridge.

Fraile, A., Malpica, J.M., Aranda, M.A., Rodriguez-Cerezo, E. and García-Arenal, F. (1996). Genetic diversity in tobacco mild green mosaic tobamovirus infecting the wild plant *Nicotiana glauca*. *Virology* **223**, 148–155.

Fraile, A., Escriu, F., Aranda, M.A., Malpica, J.M., Gibbs, A.J. and García-Arenal, F. (1997a). A century of tobamovirus evolution in an Australian population of *Nicotiana glauca*. *J. Virol.* **71**, 8316–8320.

Fraile, A., Alonso-Prados, J.L., Aranda, M.A., Bernal, J.J., Malpica, J.M. and García-Arenal, F. (1997b). Genetic exchange by recombination or reassortment is infrequent in natural populations of a tripartite RNA plant virus. *J. Virol.* **71**, 934–940.

Franck, A., Guilley, H., Jonard, G., Richards, K. and Hirth, L. (1980). Nucleotide sequence of cauliflower mosaic virus DNA. *Cell* **21**, 285–294.

Francki, R.I.B. (1973). Plant rhabdoviruses. *Adv. Virus Res.* **18**, 257–345.

Francki, R.I.B. (1980). Limited value of the termal inactivation point, longevity *in vitro* and dilution endpoint as criteria for the characterization, identification and classification of plant viruses. *Inter-virology* **13**, 91–98.

Francki, R.I.B. (1985). Plant virus satellites. *Annu. Rev. Microbiol.* **39**, 151–174.

Francki, R.I.B. (1987). Responses of plant cells to virus infection with special references to the sites of RNA replication. In: M.A. Brinton and R.R. Rueckert (eds) *Positive Strand RNA Viruses*, pp. 423–436. Alan R. Liss, New York.

Francki, R.I.B. and Habili, N. (1972). Stabilization of capsid structure and enhancement of immunogenicity of cucumber mosaic virus (Q strain) by formaldehyde. *Virology* **48**, 309–315.

Francki, R.I.B. and Hatta, T. (1980). Cucumber mosaic virus—variation and problems of identification. *Acta Hortic.* **110**, 167–174.

Francki, R.I.B. and Matthews, R.E.F. (1962). Some effects of 2-thiouracil on the multiplication of turnip yellow mosaic virus. *Virology* **17**, 367–380.

Francki, R.I.B. and Miles, R. (1985). Mechanical transmission of sowbane mosaic virus carried on pollen from infected plants. *Plant Pathol.* **34**, 11–19.

Francki, R.I.B. and Randles, J.W. (1972). RNA-dependent RNA polymerase associated with particles of lettuce necrotic yellows virus. *Virology* **47**, 270–275.

Francki, R.I.B. and Randles, J.W. (1978). Composition of the plant rhabdovirus lettuce necrotic yellows virus in relation to its biological properties. In: B.W.J. Mahy and R.D. Barry (eds) *Negative Strand Viruses and the Host Cell*, pp. 223–242.

Francki, R.I.B., Gould, A.R. and Hatta, T. (1980a). Variation in the pathogenicity of three viruses of tomato. *Ann. Appl. Biol.* **96**, 219–226.

Francki, R.I.B., Hatta, T., Boccardo, G. and Randles, J.W. (1980b). The composition of chloris striate mosaic virus, a geminivirus. *Virology* **101**, 233–241.

Francki, R.I.B., Milne, R.G. and Hatta, T. (1985a). *Atlas of Plant Viruses*, Vol. 1. CRC Press, Boca Raton, Florida.

Francki, R.I.B., Milne, R.G. and Hatta, T. (eds) (1985b). *Atlas of Plant Viruses*, Vol. 2. CRC Press, Boca Raton, FL.

Francki, R.I.B., Milne, R.G. and Hatta, T. (1985c). Plant *Rhabdoviridae*. In: R.I.B. Francki, R.G. Milne and T. Hatta (eds) *Atlas of Plant Viruses*, Vol. 1, pp. 73–100. CRC Press, Boca Raton, FL.

Francki, R.I.B., Grivell, C.J. and Gibb, K.S. (1986a). Isolation of velvet tobacco mottle virus capable of replication with and without a viroid-like RNA. *Virology* **148**, 381–384.

Francki, R.I.B., Zaitlin, M. and Palukaitis, P. (1986b). *In vivo* encapsulation of potato spindle tuber viroid by velvet tobacco mottle virus particles. *Virology* **155**, 469–473.

Francki, R.I.B., Ryan, C.C., Hatta, T., Rohozinski, J. and Grivell, C.J. (1986c). Serological detection of Fiji disease virus antigens in the planthopper *Perkinsiella saccharricida* and its inefficient ability to transmit the virus. *Plant Pathol.* **35**, 324–328.

Francki, R.I.B., Fauquet, C.M., Knudson, D.I. and Brown, F. (1991). Classification and Nomenclature of Viruses. Fifth Report on thee International Committee on Taxonomy of Viruses. Springer-Verlag, Wien, New York.

Franke, W. (1971). Über die Natur der Ektodesmen und einen Vorschlag zur terminologie. *Ber. Dtsch. Ges.* **84**, 533–537.

Franssen, H., Goldbach, R., Broekhuijsen, M., Moerman, M. and van Kammen, A. (1982). Expression of middle-component RNA of cowpea mosaic virus: *in vitro* generation of a precursor to both capsid proteins by a bottom-component RNA-encoded protease from infected cells. *J. Virol.* **41**, 8–17.

Franssen, H., Moerman, M., Rezelman, G. and Goldbach, R. (1984a). Evidence that the 32,000-dalton protein encoded by bottom-component RNA of cowpea mosaic virus is a proteolytic processing enzyme. *J. Virol.* **50**, 183–190.

Franssen, H., Leunissen, J., Golbach, R., Lomonossoff, G. and Zimmern, D. (1984b). Homologous sequences in non-structural proteins from cowpea mosaic virus and picornaviruses. *EMBO J.* **3**, 855–861.

Franssen, H., Goldbach, R. and van Kammen, A. (1984c). Translation of bottom component RNA of cowpea mosaic virus in reticulocyte lysate: faithful proteolytic processing of primary translation product. *Virus Res.* **1**, 39–49.

Fransz, P.F., Alonso-Blanco, C., Liharska, T.B., Peeters, A.J.M., Zabel, P. and de Jong, J.H. (1996). High resolution physical mapping in *Arabidopsis thaliana* and tomato by fluorescence *in situ* hybridisation to extended DNA fibers. *Plant J.* **9**, 421–430.

Franz, A., Makkouk, K.M and Vetten, H.J. (1998). Acquisition, retention and transmission of faba bean necrotic yellows virus by two of its aphid vectors, *Aphis craccivora* (Koch) and *Acyrthosiphon pisum* (Harris). *J. Phytopathol.* **146**, 347–355

Franz, A.W.E., van der Wilk, F., Verbeek, M., Dullemans, A.M. and van den Heuvel, J.F.J.M. (1999). Faba bean necrotic yellows virus (genus *Nanovirus*) requires a helper factor for its aphid transmission. *Virology* **262**, 210–219.

Fraser, L.G. and Matthews, R.E.F. (1979a). Efficient mechanical inoculation of turnip yellow mosaic virus using small volumes of inoculum. *J. Gen. Virol.* **44**, 565–568.

Fraser, L.G. and Matthews, R.E.F. (1979b). Strain-specific pathways of cytological change in individual Chinese cabbage protoplasts infected with turnip yellow mosaic virus. *J. Gen. Virol.* **45**, 623–630.

Fraser, L.G. and Matthews, R.E.F. (1981). A rapid transient inhibition of leaf initiation induced by turnip yellow mosaic virus infection. *Physiol. Plant Pathol.* **19**, 325–336.

Fraser, L.G. and Matthews, R.E.F. (1983). A rapid transient inhibition of leaf initiation by abscisic acid. *Plant Sci. Lett.* **29**, 67–72.

Fraser, L.G., Keeling, J. and Matthews, R.E.F. (1984). A reduction in starch accumulation in the apical dome of chinese cabbage seedlings following inoculation with turnip yellow mosaic virus. *Physiol. Plant Pathol.* **24**, 157–162.

Fraser, R.S.S. (1985). Genetics of host resistance to viruses and of virulence. In: R.S.S. Fraser (ed.) *Mechanisms of Resistance to Plant Disease*, pp. 62–79. Martinus Nijhoff/W. Junk, Dordrecht.

Fraser, R.S.S. (1987a). Resistance to plant viruses. *Oxford Surv. Plant Mol. Cell Biol.* **4**, 1–45.

Fraser, R.S.S. (1987b). *Biochemistry of Virus-Infected Plants*. Research Studies Press, Letchworth, UK.

Fraser, R.S.S. (1988). Virus recognition and pathogenicity: implications for resistance mechanisms and breeding. *Pestic. Sci.* **23**, 267–275.

Fraser, R.S.S. (1992). The genetics of plant–virus interactions: implications for plant breeding. *Euphytica* **63**, 175–185.

Fraser, R.S.S. (1998) Biochemistry of resistance to plant viruses. Breeding for resistance to plant viruses. In: A. Hadidi, R.K. Khetarpal and H. Koganezawa (eds) *Plant Virus Disease Control*, pp. 56–64. APS Press, St. Paul, MN.

Fraser, R.S.S. and Gerwitz, A. (1985). A new physical assay method for tobacco mosaic virus using a radioactive virus recovery standard and the first derivative of the ultraviolet absorption spectrum. *J. Virol. Methods* **11**, 289–298.

Fraser, R.S.S. and Whenham, R.J. (1982). Plant growth regulators and virus infection: a critical review. *Plant Growth Regul.* **1**, 37–59.

Fraser, R.S.S. and Whenham, R.J. (1989). Abscisic acid metabolism in tomato plants infected with tobacco mosaic virus: relationships with growth, symptoms, and the Tm-1 gene for TMV resistance. *Physiol. Mol. Plant Pathol.* **34**, 215–226.

Fraser, R.S.S., Loughlin, S.A.R. and Connor, J.C. (1980). Resistance to tobacco mosaic virus in tomato: effects of the *Tm-1* gene on symptom formation and multiplication of virus strain 1. *J. Gen. Virol.* **50**, 221–224.

Fraser, R.S.S., Gerwitz, A. and Morris, G.E.L. (1986). Multiple regression analysis of the relationships between tobacco mosaic virus multiplication, the severity of mosaic symptoms, and the growth of tobacco and tomato. *Physiol. Mol. Plant Pathol.* **29**, 239–249.

Frazier, N.W., Voth, V. and Bringhurst, R.S. (1965). Inactivation of two strawberry viruses in plants grown in a natural high temperature environment. *Phytopathology* **53**, 1203–1205.

French, C.J. and Elder, M. (1999). Virus particles in guttate and xylem of infected cucumber (*Cucumis sativus* L.). *Ann. Appl. Biol.* **134**, 81–87.

French, C.J., Elder, M. and Skelton, F. (1993). Recovering and identifying infectious plant viruses in guttation fluid. *HortSci.* **28**, 746–747.

French, R. and Ahlquist, P. (1988). Characterisation and engineering of sequences controlling *in vivo* synthesis of brome mosaic virus subgenomic RNA. *J. Virol.* **62**, 2411–2420.

French, R., Janda, M. and Ahlquist, P. (1986). Bacterial gene inserted in an engineered RNA virus: efficient expression in monocotyledonous plant cells. *Science* **231**, 1294–1297.

Frenkel, M.J., Ward, C.W. and Shukla, D.D. (1989). The use of 3′-noncoding nucleotide sequences in the taxonomy of potyviruses: application to watermelon mosaic virus 2 and soybean mosaic virus N. *J. Gen. Virol.* **70**, 2775–2783.

Frenkel, M.J., Jilka, J., McKern, N.M. *et al.* (1991). Unexpected sequence diversity in the amino-terminal ends of the coat proteins of strains of sugarcane mosaic virus. *J. Gen. Virol.* **72**, 237–242.

Freudenrich, A and Nick, P. (1998). Microtubule organization in tobacco cells: heat-shock protein 90 can bind to tubulin *in vitro*. *Bot Acta* **111**, 273–279.

Fribourg, C.E., Jones, R.A.C. and Koenig, R. (1977). Andean potato mottle, a new member of the cowpea mosaic virus group. *Phytopathology* **67**, 969–974.

Fridlund, P.R. (1967a). Effect of time and temperature on the rate of graft transmission of the *Prunus* ringspot virus. *Phytopathology* **57**, 230–231.

Fridlund, P.R. (1967b). The relationship of inoculum–receptor contact period to the rate of graft transmission of twelve *Prunus* viruses. *Phytopathology* **57**, 1296–1299.

Frischmuth, S., Frischmuth, T., Latham, J.R. and Stanley, J. (1993). Transcriptional analysis of the virion-sense genes of the geminivirus beet curly top virus. *Virology* **197**, 312–319.

Frischmuth, T. and Stanley, J. (1991). African cassava mosaic virus DI DNA interferes with the replication of both genomic components. *Virology* **183**, 539–544.

Frischmuth, T. and Stanley, J. (1993). Strategies for the control of geminivirus diseases. *Semin. Virol.* **4**, 329–337.

Frischmuth, T. and Stanley, J. (1998). Recombination between viral DNA and the transgenic coat protein gene of African cassava mosaic geminivirus. *J. Gen. Virol.* **79**, 1265–1271.

Frison, E.A. and Ng, S.Y. (1981). Elimination of sweet potato virus disease agents by meristem tip culture. *Trop. Pest Manag.* **27**, 452–454.

Fritsch, C. and Dollett, M. (2000). Genus *Pecluvirus*. In: M.H.V. van Regenmortel *et al.* (eds) *Seventh Report of the International Committee on the Taxonomy of Viruses*, pp. 913–917. Academic Press, San Diego.

Fritsch, C., Mayo, M.A. and Hirth, L. (1977). Rirther studies on the translation products of tobacco rattle virus RNA *in vitro*. *Virology* **77**, 722–732.

Fritsch, C., Mayo, M.A. and Murant, A.F. (1978). Translation of the satellite RNA of tomato black ring virus *in vitro* and in tobacco protoplasts. *J. Gen. Virol.* **40**, 587–593.

Fritsch, C., Mayo, M.A. and Murant, A.F. (1980). Translation products of genome and satellite RNAs of tomato black ring virus. *J. Gen. Virol.* **46**, 381–389.

Frosheiser, F.I. (1969). Variable influence of alfalfa mosaic virus strains on growth and survival of alfalfa and on mechanical and aphid transmission. *Phytopathology* **59**, 857–862.

Frosheiser, F.I. (1974). Alfalfa mosaic virus transmission to seed through alfalfa gametes and longevity in alfalfa seed. *Phytopathology* **64**, 102–105.

Frost, E.H.E. (1977). Radioactive labelling of viruses: An iodination technique preserving biological properties. *J. Gen. Virol.* **35**, 181–185.

Frost, R.R., Harrison, B.D. and Woods, R.D. (1967). Apparent symbiotic interaction between particles of tobacco rattle virus. *J. Gen. Virol.* **1**, 57–70.

Fry, P.R. and Matthews, R.E.F. (1963). Timing of some early events following inoculation with tobacco mosaic virus. *Virology* **19**, 461–469.

Fry, P.R. and Taylor, W.B. (1954). Analysis of virus local lesion experiments. *Ann. Appl. Biol.* **41**, 664–674.

Fuchs, M., Tricoli, D.M., Carney, K.J., Schesser, M., McPherson, J.R. and Gonsalves, D. (1998). Comparative virus resistance and fruit yield of transgenic squash with single and multiple coat protein genes. *Plant Dis.* **82**, 1350–1356.

Fudl-Allah, A.E.A., Calavan, E.C. and Desjardins, P.R. (1971). Comparative anatomy of healthy and exocortis virus-infected citron plants. *Phytopathology* **61**, 990–993.

Fuentes, A.L. and Hamilton, R.I. (1991). Sunn-hemp mosaic virus facilitates cell-to-cell spread of southern bean mosaic virus in a nonpermissive host. *Phytopathology* **81**, 1302–1305.

Fuentes, A.L. and Hamilton, R.I. (1993). Failure of long-distance movement of southern bean mosaic virus in a resistant host is correlated with lack of normal virion formation. *J. Gen. Virol.* **74**, 1903–1910.

Fujisawa, I., Hayashi, T. and Matsui, C. (1967). Electron microscopy of mixed infections with two plant viruses: I. Intracellular interactions between tobacco mosaic virus and tobacco etch virus. *Virology* **33**, 70–76.

Fujiwara, T., Giesman_Cookmeyer, D., Ding, B., Lommel, S.A. and Lucas, W.J. (1993). Cell-to-cell trafficking of macromolecules through plasmodesmata potentiated by red clover necrotic mosaic virus movement protein. *Plant Cell* **5**, 1783–1794.

Fukuda, M. and Okada, Y. (1985). Elongation in the major direction of tobacco mosaic virus assembly. *Proc. Natl Acad. Sci. USA* **82**, 3631–3634.

Fukuda, M. and Okada, Y. (1987). Bidirectional assembly of tobacco mosaic virus *in vitro*. *Proc. Natl Acad. Sci. USA* **84**, 4035–4038.

Fukuda, M., Ohno, T., Okada, Y., Otsuki, Y. and Takebe, I. (1978). Kinetics of biphasic reconstitution of tobacco mosaic virus *in vitro*. *Proc. Natl Acad. Sci. USA* **75**, 1727–1730.

Fukumoto, F. and Tochihara, H. (1984). Effect of various additives on the long-term preservation of tobacco ringspot and radish mosaic viruses. *Ann. Phytopathol. Soc. Jpn* **50**, 158–165.

Fukushi, T. (1940). Further studies on the dwarf disease of rice plant. *J. Fac. Agric., Hokkaido Univ.* **45**, 83–154.

Fukushi, T. (1969). Relationships between propagative rice viruses and their vectors. In: K. Maramorosch (ed.) *Viruses, Vectors, and Vegetation*, pp. 279–301. Wiley, New York and London.

Fukuyama, K., Abdel-Meguid, S.S., Johnson, J.E. and Rossmann, M.G. (1983). Structure of a *T* = 1 aggregate of alfalfa mosaic virus coat protein seen at 4.5 Å resolution. *J. Mol. Biol.* **167**, 873–894.

Fuller, S.D., Butcher, S.J., Cheng, R.H. and Baker, T.S. (1996). Three-dimensional reconstruction of icosahedral particles: the uncommon line. *J. Struct. Biol.* **116**, 48–55.

Fulton, J.P. (1969). Transmission of tobacco ringspot virus to the roots of a conifer by a nematode. *Phytopathology* **59**, 236.

Fulton, J.P., Gergerich, R.C. and Scott, H.A. (1987). Beetle transmission of plant viruses. *Annu. Rev. Phytopathol.* **25**, 111–123.

Fulton, R.W. (1952). Mutation in a tobacco necrosis virus strain. *Phytopathology* **42**, 156–158.

Fulton, R.W. (1962). The effect of dilution on necrotic ringspot virus infectivity and the enhancement of infectivity by noninfective virus. *Virology* **18**, 477–485.

Fulton, R.W. (1964). Transmission of plant viruses by grafting, dodder, seed and mechanical inoculation. In: M.K. Corbett and H.D. Sisler (eds) *Plant Virology*, pp. 39–67. University of Florida Press, Gainesville, FL.

Fulton, R.W. (1975). The role of top particles in recombination of some characters of tobacco streak virus. *Virology* **67**, 188–196.

Fulton, R.W. (1978). Superinfection by strains of tobacco streak virus. *Virology* **85**, 1–8.

Fulton, R.W. (1980). Biological significance of multicomponent viruses. *Annu. Rev. Phytopathol.* **18**, 131–146.

Fulton, R.W. (1986). Practices and precautions in the use of cross protection for plant virus disease control. *Annu. Rev. Phytopathol.* **24**, 67–81.

Furuichi, Y. and Shatkin, A.J. (2000). Viral and cellular mRNA capping: past and prospects. *Adv. Virus Res.* **55**, 135–184.

Furumoto, W.A and Mickey, R. (1967). A mathematical model for the infectivity–dilution curve of tobacco mosaic virus: experimental tests. *Virology* **32**, 224–233.

Furumoto, W.A. and Wildman, S.G. (1963a). Studies on the mode of attachment of tobacco mosaic virus. *Virology* **20**, 45–52.

Furumoto, W.A. and Wildman, S.G. (1963b). The specific infectivity of tobacco mosaic virus. *Virology* **20**, 53–61.

Fütterer, J. and Hohn, T. (1987). Involvement of nucleocapsids in reverse transcription: A general phenomenon? *Trends Biochem. Res.* **12**, 92–95.

Fütterer, J. and Hohn, T. (1991). Translation of a polycistronic mRNA in the presence of the cauliflower mosaic virus transactivator protein. *EMBO J.* **10**, 3887–3896.

Fütterer, J. and Hohn, T. (1992). Role of an upstream open reading frame in the translation of polycistronic mRNAs in plant cells. *Nucl. Acids Res.* **20**, 3851–3857.

Fütterer, J. and Hohn, T. (1996). Translation in plants: rules and exceptions. *Plant Mol. Biol.* **32**, 159–189.

Fütterer, J., Bonneville, J.M. and Hohn, T. (1990). Cauliflower mosaic virus as a gene expression vector for plants. *Physiol. Plantar.* **79**. 154–157.

Fütterer, J., Kiss-László, Z. and Hohn, T. (1993) Nonlinear ribosome migration on cauliflower mosaic virus 35S RNA. *Cell* **73**, 789–802.

Fütterer, J., Potrykus, I., Brau, M.P.V., Dasgupta, I., Hull, R. and Hohn, T. (1994). Splicing in a pararetrovirus. *Virology* **198**, 663–670.

Fütterer, J. Potrykus, I., Bao, Y. *et al.* (1996). Position-dependent ATT initiation during plant pararetrovirus rice tungro bacilliform virus translation. *J. Virol.* **70**, 2999–3010.

Gáborjányi, R., Balázs, E. and Király, Z. (1971). Ethylene production, tissue senescence and local virus infections. *Acta Phytopathol. Acad. Sci. Hung.* **6**, 51–55.

Gaddipati, J.P. and Siegel, A. (1990). Study of TMV assembly with heterologous RNA containing the origin-of-assembly sequence. *Virology* **174**, 337–344.

Gaddipati, J.P., Atreya, C.D., Rochon, D'A. and Seigel, A. (1988). Characterisation of the TMV encapsidation initiation site on 18S rRNA. *Nucl. Acids Res.* **16**, 7303–7313.

Gadh, I.P.S. and Hari, V. (1986). Association of tobacco etch virus related RNA with chloroplasts in extracts of infected plants. *Virology* **150**, 304–307.

Gaedigk, K., Adam, G. and Mundry, K.-W. (1986). The spike protein of potato yellow dwarf virus and its functional role in the infection of insect vector cells. *J. Gen. Virol.* **67**, 2763–2773.

Gafney, R., Lapidot, M., Berna, A., Holt, C.A., Deom, C.M. and Beachy, R.N. (1992). Effects of terminal deletion mutants on function of the movement protein of tobacco mosaic virus. *Virology* **187**, 499–507.

Gairdner, A.E. (1936). The inheritance of factors in *Cheiranthus cheiri*. *J. Genet.* **32**, 479–486.

Gaire, F., Schmitt, C., Stussi-Garaud, C., Pinck, L. and Ritzenthaler, C. (1999). Protein 2A of grapevine fanleaf nepovirus is implicated in RNA2 replication and colocalizes to the replication site. *Virology* **264**, 25–36.

Gal, S., Pisan, B., Hohn, T., Grimsley, N. and Hohn, B. (1991). Genomic homologous recombination *in planta*. *EMBO J.* **10**, 1571–1578.

Gal, S., Pisan, B., Hohn, T., Grimsley, N. and Hohn, B. (1992). Agroinfection of transgenic plants leads to viable cauliflower mosaic virus by intermolecular recombination. *Virology* **197**, 525–533.

Gallagher, W.H. and Lauffer, M.A. (1983). Calcium ion binding by tobacco mosaic virus. *J. Mol. Biol.* **170**, 905–919.

Gallie, D.R. (1996). Translational control of cellular and viral mRNAs. *Plant Mol. Biol.* **32**, 145–158.

Gallie, D.R. (1998). A tale of two termini: a functional interaction between the termini of an mRNA is a prerequisite for efficient translation initiation. *Gene* **216**, 1–11.

Gallie, D.R. and Kado, C.I. (1989). A translational enhancer derived from tobacco mosaic virus is functionally equivalent to a Shine–Dalgarno sequence. *Proc. Natl Acad. Sci. USA* **86**, 129–132.

Gallie, D.R. and Walbot, V. (1992). Identification of the motifs within the tobacco mosaic virus 5′-leader responsible for enhancing translation. *Nucl. Acids Res.* **20**, 4631–4638.

Gallie, D.R., Sleat, D.E., Watts, J.W., Turner, P.C. and Wilson, T.M.A. (1987a). The 5′ leader sequence of tobacco mosaic virus RNA enhances the expression of foreign gene transcripts *in vitro* and *in vivo*. *Nucleic Acids Res.* **15**, 3257–3273.

Gallie, D.R., Sleat, D.E., Watts, J.W., Turner, P.C. and Wilson, T.M.A. (1987b). A comparison of eukaryotic viral 5′-leader sequences as enhancers of mRNA expression *in vivo*. *Nucl. Acids Res.* **15**, 8693–8711.

Gallie, D.R., Sleat, D.E., Watts, J.W., Turner, P.C. and Wilson, T.M.A. (1987c). *In vivo* uncoating and efficient expression of foreign mRNAs packaged in TMV-like particles. *Science* **236**, 1122–1124.

Gallie, D.R., Plaskitt, K.A. and Wilson, T.M.A. (1987d). The effect of multiple dispersed copies of the origin-of-assembly sequence from TMV RNA on the morphology of pseudovirus particles assembled *in vitro*. *Virology* **158**, 473–476.

Gallie, D.R., Walbot, V. and Hershey, J.W.B. (1988). The ribosomal fraction mediates the translational enhancement associated with the 5′-leader of tobacco mosaic virus. *Nucl. Acids Res.* **16**, 8675–8694.

Gallie, D.R., Tanguay, R. and Leathers, V. (1995). The tobacco etch viral 5′ leader and poly(A) tail are functionally synergistic regulators of translation. *Gene* **165**, 233–238.

Gallitelli, D. and Hull, R. (1985). Characterisation of satellite RNAs associated with tomato bushy stunt virus and five other definitive tombusviruses. *J. Gen. Virol.* **66**, 1533–1543.

Gallitelli, D., Vovlas, C., Martelli, G., Montasser, M.S., Tousignant, M.E. and Kaper, J.M. (1991). Satellite-mediated protection of tomato against cucumber mosaic virus: II. Field test under natural epidemic conditions in southern Italy. *Plant Dis.* **75**, 93–95.

Gal-On, A., Kaplan, I., Roossinck, M.J. and Palukaitis, P. (1994). The kinetics of infection of zucchini squash by cucumber mosaic virus indicate a function for RNA-1 in virus movement. *Virology* **205**, 280–289.

Gal-On, A., Meiri, E., Huet, H., Hua, W.J., Raccah, B. and Gaba, V. (1995). Particle bombardment drastically increases the infectivity of cloned DNA of zucchini yellow mosaic potyvirus. *J. Gen Virol.* **76**, 3223–3227.

Gal-On, A., Meiri, E., Elman, C., Gray, D.J and Gaba, V. (1997). Simple hand-held devices for the efficient infection of plants with viral-encoding constructs by particle bombardment. *J. Virol. Methods* **64**, 103–110.

Gal-On, A., Canto, T. and Palukaitis, P. (2000). Characterisation of genetically modified cucumber mosaic virus expressing histidine-tagged 1a and 2a proteins. *Archiv. Virol.* **145**, 37–50.

Gamez, R. and Léon, P. (1988). Maize raydo fino and related viruses. In: R. Koenig (ed.) *The Plant Viruses*, Vol. 3, pp. 213–233. Plenum, New York.

Gao, C., Mao, S., Lo, C.-H., Wirsching, P., Lerner, R.A. and Janda, K.D. (1999). Making artificial antibodies: a format for phage display of combinatorial heterodimeric arrays. *Proc. Natl Acad. Sci. USA* **96**, 6025–6030.

García, J.A., Schrijvers, L., Tan, A., Vos, P., Wellink, J. and Goldbach, R. (1987). Proteolytic activity of the cowpea mosaic virus encoded 24K protein synthesised in *Escherichia coli*. *Virology* **159**, 67–75.

García, J.A., Cervera, M.T., Reichmann, J.L. and López-Otin, C. (1993). Inhibitory effects of human cyctetin C. on plum pox potyvirus proteases. *Plant Mol. Biol.* **22**, 697–701.

García-Arenal, F. (1988). Sequence and structure at the genome 3′-end of the U2 strain of tobacco mosaic virus, a histidine accepting tobamovirus. *Virology* **167**, 201–206.

García-Arenal, F. and Palukaitis, P. (1999). Structure and functional relationships of satellite RNAs of cucumber mosaic virus. In: P.K. Vogt and A.O. Jackson (eds) *Satellites and Defective RNAs*, pp. 37–63. Springer, Berlin.

García-Arenal, F., Palukaitis, P. and Zaitlin, M. (1984). Strains and mutants of tobacco mosaic virus are both found in virus derived from single-lesion-passaged inoculum. *Virology* **132**, 131–137.

García-Luque, I., Kaper, J.M., Diaz-Ruiz, J.R. and Rubio-Huertos, M. (1984). Emergence and characterization of satellite RNAs associated with Spanish cucumber mosaic virus isolates. *J. Gen. Virol.* **65**, 539–547.

Gardiner, W.E., Sunter, G., Brand, L., Elmer, J.S., Rogers, S.G. and Bisaro, D.M. (1988). Genetic analysis of tomato golden mosaic virus: the coat protein is not required for systemic spread or symptom development. *EMBO J.* **7**, 899–904.

Gargouri, R., Joshi, R.L., Bol, J.F., Astier-Manifacier, S. and Haenni, A.-L. (1989). Mechanism of synthesis of turnip yellow mosaic virus coat protein subgenomic RNA *in vivo*. *Virology* **171**, 386–393.

Garnier, M., Mamoun, R. and Bové, J.M. (1980). TYMV RNA replication *in vivo*: Replicative intermediate is mainly single stranded. *Virology* **104**, 357–374.

Garnier, M., Candresse, T. and Bové, J.M. (1986). Immunocytochemical localization of TYMV coded structural and non-structural proteins by the protein A–gold technique. *Virology* **151**, 100–109.

Garnsey, S.M. and Randles, J.W. (1987). Biological interactions and agricultural implications of viroids. In: J.S. Semancik (ed.) *Viroids and Viroid-Like Pathogens*, pp. 127–160. CRC Press, Boca Raton, FL.

Garnsey, S.M. and Whidden, R. (1973). Efficiency of mechanical inoculation procedures for citrus exocortis virus. *Plant Dis. Rep.* **57**, 886–889.

Garnsey, S.M., Gonsalves, D. and Purcifull, D.E. (1977). Mechanical transmission of citrus tristeza virus. *Phytopathology* **67**, 965–968.

Garnsey, S.M., Gottwald, T.R. and Yokomi, R.K. (1998). Control strategies for citrus tristeza virus. In: A. Hadidi, R.K. Khetarpal and H. Koganezawa (eds) *Plant Virus Disease Control*, pp. 639–658. APS Press, St. Paul, MN.

Garrett, R.G. and O'Loughlin, G.T. (1977). Broccoli necrotic yellows virus in cauliflower and in the aphid *Brevicoryne brassicae* L. *Virology* **76**, 653–663.

Garrido-Ramirez, E.R., Sudarshana, M.R., Lucas, W.J. and Gilbertson, R.L. (2000). Bean dwarf mosaic virus BV1 protein is a determinant of the hypersensitive response and avirulence in *Phaseolus vulgaris*. *Mol. Plant Microb. Interact.* **13**, 1184–1194.

Garzon-Tiznado, J.A., Torres-Pacheco, I., Ascencio-Ibanez, J.T., Herrera-Estrella, L. and Rivera-Bustamante, R.F. (1993). Inoculation of peppers with infectious clones of a new gemini-virus by a biolistic procedure. *Phytopathology* **83**, 514–521.

Gast, F.U., Kempe, D., Spieker, R.L. and Sänger, H.L. (1996). Secondary structure probing of potato spindle tuber viroid (PSTVd) and sequence comparison with other small pathogenic RNA replicons provides evidence central non-canonical base-pairs, large A-rich loops and a terminal branch. *J. Mol. Biol.* **263**, 652–670.

Gaut, B.S., d'Ennequin, M. Le T., Peek, A.S. and Sawkins, M.C. (2000). Maize as a model for the evolution of plant nuclear genomes. *Proc. Natl Acad. Sci. USA* **97**, 7008–7015.

Gay, J.D., Johnson, A.W. and Chalfant, R.B. (1973). Effects of a trapcrop on the introduction and distribution of cowpea virus by soil and insect vectors. *Plant Dis. Rep.* **57**, 684–688.

Ge, X., Scott, S.W. and Zimmerman, M.T. (1997). The complete sequence of the genomic RNAs of spinach latent virus. *Archiv. Virol.* **142**, 1213–1226.

Gebre Selassie, K., Marchoux, G., Delecolle, B. and Pochard, E. (1985). Varabilité naturelle des souches du virus Y de la pomme de terre dans le cultures du piment du sud-est de la France: caractérisation et classification en pathotypes. *Agronomie* **5**, 621–630.

Geelen, J.L.M.C., van Kammen, A. and Verduin, B.J.M. (1972). Structure of the capsid of cowpea mosaic virus. The chemical subunit: molecular weight and number of subunits per particle. *Virology* **49**, 205–213.

Geldreich, A., Lebeurier, G. and Hirth, L. (1986). *In vivo* dimerisation of cauliflower mosaic virus DNA can explain recombination. *Gene* **48**, 277–286.

Georghiou, G.P. and Saito, T. (eds) (1983). *Pest Resistance to Pesticides*. Plenum, New York.

Gera, A. and Loebenstein, G. (1988). An inhibitor of virus replication associated with green island tissue of tobacco infected with cucumber mosaic virus. *Physiol. Mol. Plant Pathol.* **32**, 373–385.

Gera, A. Loebenstein, G. and Raccah, B. (1979). Protein coats of two strains of cucumber mosaic virus affect transmission by *Aphis gossypii*. *Phytopathology* **69**, 396–399.

Gera, A., Loebenstein, G., Salomon, R. and Frank, A. (1990). An inhibitor of virus replication (IVR) from protoplasts of a hypersensitive tobacco cultivar infected with tobacco mosaic virus is associated with a 23 K protein species. *Phytopathology* **80**, 78–81.

Gera, A., Deom, C.M., Shaw, J.J., Lewandowski, D.J. and Dawson, W.O. (1995). Tobacco mosaic tobamovirus does not require concomitant synthesis of movement protein during vascular transport. *Mol. Plant–Microb. Interact.* **8**, 784–787.

Gergerich, R.C. and Scott, H.A. (1988a). The enzymatic function of ribonuclease determines plant virus transmission by leaf-feeding beetles. *Phytopathology* **78**, 270–272.

Gergerich, R.C. and Scott, H.A. (1988b). Evidence that virus translation and virus infection of nonwounded cells are associated by transmissibility by leaf-feeding beetles. *J. Gen. Virol.* **69**, 2935–2938.

Gergerich, R.C., Scott, H.A. and Fulton, J.P. (1983). Regurgitant as a determinant of specificity in the transmission of plant viruses by beetles. *Phytopathology* **73**, 936–938.

Gergerich, R.C., Scott, H.A. and Fulton, J.P. (1986). Evidence that ribonuclease in beetle regurgitant determines the transmission of plant viruses. *J. Gen. Virol.* **67**, 367–370.

Geri, C., Cecchini, E., Giannakou, M.E., Covey, S.N. and Milner, J.J. (1999). Altered patterns of gene expression in *Arabidopsis* elicited by cauliflower mosaic virus (CaMV) infection and by a CaMV gene VI transgene. *Mol. Plant–Microb. Interact.* **12**, 377–384.

Gerlach, W.L., Llewellyn, D. and Haseloff, J. (1987). Construction of a plant resistance gene from the satellite RNA of tobacco ringspot virus. *Nature (Lond.)* **328**, 802–805.

German-Retana, S. and Candresse, T. (1999). Trichoviruses. . In: A. Granoff and R.G. Webster (eds) *Encyclopedia of Virology*, 2nd edn, pp. 1837–1842. Academic Press, San Diego.

German-Retana, S., Candresse, T. and Martelli, G. (1999). Closteroviruses (*Closteroviridae*). In: A. Granoff and R.G. Webster (eds) *Encyclopedia of Virology*, 2nd edn, pp. 266–273. Academic Press, San Diego.

Gerpa, C.P. (1984). Applied and theoretical aspects of virus adsorption to surfaces. *Adv. Appl. Microbiol.* **30**, 133–168.

Gesteland, R.F. and Atkins, J.F. (1996). RECODING: dynamic reprogramming of translation. *Annu. Rev. Biochem.* **65**, 741–768.

Geysen, H.M., Meloen, R.H. and Barteling, S.J. (1984). Use of peptide synthesis to probe viral antigens for epitopes to a resolution of a single amino acid. *Proc. Natl. Acad. Sci. USA* **81**, 3998–4002.

Ghabrial, S.A. and Hillman, B.I. (1999). Partitiviruses—fungal (*Partitiviridae*). In: A. Granoff and R.G. Webster (eds) *Encyclopedia of Virology*, 2nd edn, pp. 1147–1151. Academic Press, San Diego.

Ghabrial, S.A. and Lister, R.M. (1973). Anomalies in molecular weight determinations of tobacco rattle virus protein by SDS–polyacrylamide gel electrophoresis. *Virology* **51**, 485–488.

Ghabrial, S.A. and Patterson, J.L. (1999). Totiviruses (*Totiviridae*). In: A. Granoff and R.G. Webster (eds) *Encyclopedia of Virology*, 2nd edn, pp. 1808–1812.. Academic Press, San Diego.

Ghabrial, S.A. and Shepherd, R.J. (1980). A sensitive radioimmunosorbent assay for the detection of plant viruses. *J. Gen. Virol.* **48**, 311–317.

Ghanim, M. and Czosnek, H. (2000). Tomato yellow leafcurl geminivirus (TYLCV-Is) is transmitted among whiteflies (*Bemisia tabaci*) in a sex-related manner. *J. Virol.* **74**, 4738–4745.

Gharbi, S. and Verhoyen, M. (1993). Sterilisation par irradiation UV des solutions nutritives en vue d'eviter les infections virales propagees par *O. brassicae* en culture hydroponique de laitue. *Med. Fac. Landbouww. Rijksuniv. Gent* **53/3a**, 1113–1124.

Ghorbel, R., López, C., Fagoaga, C., Moreno, P., Navarro, L., Flores, R. and Peña, L. (2001). Transgenic citrus plants expressing the citrus tristeza p23 protein exhibit viral-like symptoms. *Molec. Plant Pathol.* **2**, 27–36.

Ghosh, A., Rutgers, T., Ke-Qiang, M. and Kaesberg, P. (1981). Characterisation of the coat protein mRNA of Southern bean mosaic virus and its relationship to the genomic RNA. *J. Virol.* **39**, 87–92.

Ghoshroy, S., Lartey, R., Sheng, J. and Citovsky, V. (1997). Transport of proteins and nucleic acids through plasmodesmata. *Annu. Rev. Plant Physiol. Plant Mol. Biol.* **48**, 27–50.

Gianinazzi, S., Martin, C. and Vallée, J.-C. (1970). Hypersensibilité aux virus, température et protéines solubles chez le *Nicotiana* Xanthi n.c. Apparition de nouvelles macromolécules lors de la répression de la synthèse virale. *C. R. Hebd. Seances Acad. Sci. Ser. D* **270**, 2383–2386.

Gianinazzi, S., Deshayes, A., Martin, C. and Vernoy, R. (1977). Differential reactions to tobacco mosaic virus infection in Samsun 'nn' tobacco plants: I. Necrosis, mosaic symptoms and symptomless leaves following the ontogenic gradient. *Phytopathol. Z.* **88**, 347–354.

Giband, M., Mensard, J.M. and Lebeurier, G. (1986). The gene III product (P15) of cauliflower mosaic virus is a DNA-binding protein while an immunologically related P11 polypeptide is associated with virions. *EMBO J.* **5**, 2433–2438.

Gibb, K.S. and Randles, J.W. (1988). Studies on the transmission of velvet tobacco mottle virus by the mirid *Cyrtopeltis nicotianae*. *Ann. Appl. Biol.* **112**, 427–437.

Gibb, K.S. and Randles, J.W. (1990). Distribution of velvet tobacco mottle virus in its mired vector and its relationship to transmissibility. *Ann. Appl. Biol.* **116**, 513–521.

Gibbs, A. (1976) Viruses and plasmodesmata. In: B.E.S. Gunning and R.W. Robards (eds) *Intercellular Communication in Plants: Studies on Plasmodesmata*, pp. 149–164. Springer-Verlag, Berlin.

Gibbs, A. (1986). Tobamovirus classification. In: M.H.V. van Regenmortel and H. Fraenkel-Conrat (eds) *The Plant Viruses*, Vol. 2, pp. 167–180. Plenum, New York.

Gibbs, A. (1999a). Tymoviruses. In: A. Granoff and R.G. Webster (eds) *Encyclopedia of Virology*, 2nd edn, pp. 1859–1853. Academic Press, San Diego.

Gibbs, A.J. (1962). Lucerne mosaic virus in British lucerne crops. *Plant Pathol.* **11**, 167–171.

Gibbs, A.J. (1969). Plant virus classification. *Adv. Virus Res.* **14**, 263–328.

Gibbs, A.J. (1999a). Evolution and origins of tobamoviruses. *Phil Trans. R. Soc. Lond. B* **354**, 593–602.

Gibbs, A.J. and Harrison, B. (1976). *Plant Virology: The Principles.* Edward Arnold, London.

Gibbs, A.J. and Keese, P.K. (1995). In search of the origins of viral genes. In: A.J. Gibbs, C.H. Calisher and F. Garcia-Arenal (eds) *Molecular Basis for Virus Evolution*, pp. 76–90. Cambridge University Press, Cambridge.

Gibbs, A.J. and McIntyre, G.A. (1970). A method for assessing the size of a protein from its composition: its use in evaluating data on the size of the protein subunits of plant virus particles. *J. Gen. Virol.* **9**, 51–67.

Gibbs, A.J. and McIntyre, G.A. (1971). The diagram, a method for comparing sequences. *Eur. J. Biochem.* **16**, 1–11.

Gibbs, A.J. and Tinsley, T.W. (1961). Lucerne mosaic virus in Great Britain. *Plant Pathol.* **10**, 61–62.

Gibbs, A.J., Harrison, B.D., Watson, D.H. and Wildy, P. (1966). What's in a virus name? *Nature (London)* **209**, 450–454.

Gibbs, A.J., Keese, P.L., Gibbs, M.J. and Garcia-Arenal, F. (1999). Plant virus evolution: past, present and future. In: E. Domingo, R. Webster and J. Holland (eds) *Origin and evolution of viruses*, pp. 263–285. Academic Press, London.

Gibbs, M. (1995). The luteovirus supergroup: rampant recombination and persistent partnerships. In: A.J. Gibbs, C.H. Calisher and F. Garcia-Arenal (eds) *Molecular Basis for Virus Evolution*, pp. 351–368. Cambridge University Press, Cambridge.

Gibbs, M.J. and Cooper, J.I. (1995). A recombinational event in the history of luteoviruses probably induced by base-pairing between the genomes of two distinct viruses. *Virology* **206**, 1129–1132.

Gibbs, M.J. and Weiller, G.F. (1999). Evidence that a plant virus switched hosts to infect a vertebrate and then recombined with a vertebrate-infecting virus. *Proc. Natl. Acad. Sci. USA* **96**, 8022–8027.

Gibbs, M.J., Cooper, J.I. and Waterhouse, P.M. (1996). The genome organization and affinities of an Australian isolate of carrot mottle umbravirus. *Virology* **224**, 310–313.

Gibbs, M.J., Koga, R., Moriyama, H., Pfeiffer, P. and Fukuhara, T. (2000). Phylogenetic analysis of some large double-stranded RNA replicons from plants suggests they evolved from a defective single-stranded virus. *J. Gen. Virol.* **81**, 227–233.

Gibbs, P.E.M., Witke, W.F. and Dugaiczyk, A. (1998). The molecular clock runs at different rates among closely related members of a gene family. *J. Mol. Evol.* **46**, 552–561.

Gibson, R.W. and Rice, A.D. (1986). The combined use of mineral oils and pyrethroids to control plant viruses transmitted non- and semi-persistently by *Myzus persicae*. *Ann. Appl. Biol.* **109**, 465–472.

Gibson, R.W., Pickett, J.A., Dawson, G.W., Rice, A.D. and Stribley, M.F. (1984). Effects of aphid alarm pheromone derivatives and related compounds on non- and semi-persistent plant virus transmission by *Myzus persicae*. *Ann. Appl. Biol.* **104**, 203–209.

Gibson, R.W., Payne, R.W. and Katis, N. (1988). The transmission of potato virus Y by aphids of differing vectoring abilities. *Ann. Appl. Biol.* **113**, 35–43.

Giedroc, D.P., Theimer, C.A. and Nixon, P.L. (2000). Structure, stability and function of RNA pseudoknots involved in stimulating ribosomal frameshifting. *J. Mol. Biol.* **298**, 167–185.

Gierer, A. (1957). Structure and biological function of ribonucleic acid from tobacco mosaic virus. *Nature (London)* **179**, 1297–1299.

Gierer, A. (1958). Die Gross der biologisch aktiven Einheit der Ribosenucleinsdure des Tabakmosaikvirus. *Z. Naturforsch. B: Anorg. Chem., Org. Chem., Biochem., Biophys., Biol.* **13**, 485–488.

Gierer, A. and Mundry, K.W. (1958). Production of mutants of tobacco mosaic virus by chemical alteration of its ribonucleic acid *in vitro*. *Nature (Lond.)* **182**, 1457–1458.

Gierer, A. and Schramm, G. (1956). Infectivity of ribonucleic acid from tobacco mosaic virus. *Nature (London)* **177**, 702–703.

Giesman-Cookmeyer, D. and Lommel, S.A. (1993). Alanine scanning mutagenesis of a plant virus movement protein identifies three functional domains. *Plant Cell* **5**, 973–982.

Giesman-Cookmeyer, D., Silver, S., Vaewhongs, A., Lommel, S.A. and Deom, C.M. (1995). Tobamovirus and dianthovirus movement proteins are functionally homologous. *Virology* **213**, 38–46.

Gilbert, J., Spillane, C., Kavanagh, T.A. and Baulcombe, D.C. (1998). Elicitation of *Rx*-mediated resistance to PVX in potato does not require new RNA synthesis and may involve a latent hypersensitive response. *Mol. Plant Microbe Interact.* **11**, 833–835.

Gilbertson, R.L., Faria, J.C., Hanson, S.F. et al. (1991). Cloning of the complete DNA genomes of four bean-infecting geminiviruses and determining their infectivity by electric discharge particle acceleration. *Phytopathology* **81**, 980–985.

Gilchrist, D.G. (1998). Programmed cell death in plant disease: the purpose and promise of cellular suicide. *Annu. Rev. Phytopathol.* **36**, 393–414.

Gildow, F.E. (1982). Coated-vesicle transport of luteoviruses through salivary glands of *Myzus persicae*. *Phytopathology* **72**, 1289–1296.

Gildow, F.E. (1983). Influence of barley yellow dwarf virus-infected oats and barley on morphology of aphid vectors. *Phytopathology* **73**, 1196–1199.

Gildow, F.E. (1985). Transcellular transport of barley yellow dwarf virus into the hemocoel of the aphid vector, *Rhopalosiphum padi*. *Phytopathology* **75**, 292–297.

Gildow, F.E. (1987). Virus–membrane interactions involved in circulative transmission of luteoviruses by aphids. *Curr. Top. Vector Res.* **4**, 93–120.

Gildow, F.E. (1999). Luteovirus transmission and mechanisms regulating vector specificity. In: H.G. Smith and H. Barker (eds) *The Luteoviridae*, pp.88–112. CAB International, Wallingford, UK.

Gildow, F.E. and D'Arcy, C.J. (1988). Barley and oats as reservoirs for an aphid virus and the influence on barley yellow dwarf transmission. *Phytopathology* **78**, 811–816.

Gil-Fernandez, C. and Black, L.M. (1965). Some aspects of the internal anatomy of the leafhopper *Agallia constricta* (Homoptera: Cicadellidae). *Ann. Ent. Soc. Am.* **58**, 275–284.

Gill, C.C. (1970). Aphid nymphs transmit an isolate of barley yellow dwarf virus more efficiently than do adults. *Phytopathology* **60**, 1747–1752.

Gill, C.C. (1974). Inclusions and wall deposits in cells of plants infected with oat necrotic mottle virus. *Can. J. Bot.* **52**, 621–626.

Gill, C.C. and Chong, J. (1979). Cytological alterations in cells infected with corn leaf aphid-specific isolates of barley yellow dwarf virus. *Phytopathology* **69**, 363–368.

Gill, C.C. and Chong, J. (1981). Vascular cell alterations and predisposed xylem infection in oats by inoculations with paired barkey yellow dwarf viruses. *Virology* **114**, 405–414.

Gill, R.J. (1990). The morphology of whiteflies. In: D. Gerling (ed.) *Whiteflies: Their Bionomics, Pest Status and Management*, pp. 13–46. Intercept, Andover, UK.

Gillaspie, A.G. and Bancroft, J.B. (1965). The rate of accumulation, specific infectivity and electrophoretic characteristics of bean pod mottle virus in bean and soybean. *Phytopathology* **55**, 906–908.

Gilmer, R.M. (1965). Additional evidence of tree-to-tree transmission of sour cherry yellows virus by pollen. *Phytopathology* **55**, 482–483.

Gilmer, R.M. and Brase, K.D. (1963). Nonuniform distribution of prune dwarf virus in sweet and sour cherry trees. *Phytopathology* **53**, 819–821.

Gilmer, R.M., Uyemoto, J.K. and Kelts, L.J. (1970). A new grapevine disease induced by tobacco ringspot virus. *Phytopathology* **60**, 619–627.

Gingery, R.E. and Louie, R. (1985). A satellite-like virus particle associated with maize white line mosaic virus. *Phytopathology* **75**, 870–874.

Givord, L., Xiong, C., Giband, M. et al. (1984). A second cauliflower mosaic virus gene product influences the structure of the viral inclusion body. *EMBO J.* **3**, 1423–1427.

Givord, L., Dixon, L., Rauseo-Koenig, I. and Hohn, T. (1988). Cauliflower mosaic virus ORF VII is not required for aphid transmissibility. *Ann. Inst. Pasteur Virol.* **139**, 227–231.

Glais, L., Tribodet, M., Gauthier, J.P., Astier-Manifacier, S., Robaglia, C. and Kerlan, C. (1998). RFLP mapping of the whole genome of ten viral isolates representative of different biological groups of potato virus Y. *Arch. Virol.* **143**, 2077–2091.

Glover, J.F. and Wilson, T.M.A. (1982). Efficient translation of the coat protein cistron of tobacco mosaic virus in a cell free system from *Escherichia coli*. *Eur. J. Biochem.* **122**, 485–492.

Godchaux, W. and Schuster, T.M. (1987). Isolation and characterisation of nucleoprotein assembly intermediates of tobacco mosaic virus. *Biochemistry* **26**, 454–461.

Godefroy-Colburn, T., Gagey, M.-J., Berna, A. and Stussi-Garaud, C. (1986). A non-structural protein of alfalfa mosaic virus in the walls of infected tobacco cells. *J. Gen. Virol.* **67**, 2233–2239.

Goelet, P. and Karn, J. (1982). Tobacco mosaic virus induces the synthesis of a family of 3' coterminal messenger RNAs and their complements. *J. Mol. Biol.* **154**, 541–550.

Goelet, P., Lomonossoff, G.P., Butler, P.J.G., Akam, M.E., Gait, M.J. and Karn, J. (1982). Nucleotide sequence of tobacco mosaic virus RNA. *Proc. Natl Acad. Sci. USA* **79**, 5818–5822.

Goffeau, A. and Bové, J.M. (1965). Virus infection and photosynthesis: I. Increased photophosphorylation by chloroplasts from Chinese cabbage infected with turnip yellow mosaic virus. *Virology* **27**, 243–252.

Gokhale, D.V. and Bald, J.G. (1987). Relationship between plant virus concentration and infectivity: a 'growth curve' model. *J. Virol. Methods* **18**, 225–232.

Goldbach, R. and Krijt, J. (1982). Cowpea mosaic virus-encoded protease does not recognise primary translation products of mRNAs from other comoviruses. *J. Virol.* **43**, 1151–1154.

Goldbach, R. and Rezelman, G. (1983). Orientation of the cleavage map of the 200-kilodalton polypeptide encoded by the bottom-component RNA of cowpea mosaic virus. *J. Virol.* **46**, 614–619.

Goldbach, R. and Wellink, J. (1996). Comoviruses: molecular biology and replication. In: B.D. Harrison and A.F. Murant (eds) *The Plant Viruses. Vol. 5: Polyhedral Virions and Bipartite Genomes*, pp. 35–76. Plenum Press, New York.

Goldbach, R., Rezelman, G. and van Kammen, A. (1980). Independent replication and expression of B-component RNA of cowpea mosaic virus. *Nature (London)* **285**, 297–300.

Goldbach, R., Schilthuis, J.G. and Rezelman, G. (1981). Comparison of *in vivo* and *in vitro* translation of cowpea mosaic virus RNAs. *Biochem. Biophys. Res. Commun.* **99**, 89–94.

Goldberg, K.-B., Kiernan, J. and Shepherd, R.J. (1991). A disease syndrome associated with expression of gene VI of caulimoviruses may be a nonhost reaction. *Mol. Plant Microbe Interact.* **4**, 182–189.

Goldberg, R.B. (1988). Plants. Novel developmental processes. *Science* **240**, 1460–1467.

Golemboski, D.B., Lomonossoff, G.P. and Zaitlin, M. (1990). Plants transformed with a tobacco mosaic virus non-structural gene sequence are resistant to the virus. *Proc. Natl Acad. Sci. USA* **87**, 6311–6315.

Gonsalves, D. (1998). Control of papaya ringspot virus in papaya: a case study. *Annu. Rev. Phytopathol.* **36**, 415–437.

Gonsalves, D. and Fulton, R.W. (1977). Activation of *Prunus* necrotic ringspot virus and rose mosaic virus by RNA4 components of some ilarviruses. *Virology* **81**, 398–407.

Gonsalves, D., Provvidenti, R. and Edwards, M.C. (1982). Tomato white leaf: the relation of an apparent satellite RNA and cucumber mosaic virus. *Phytopathology* **72**, 1533–1538.

Goodey, T. (1963). *Soil and Freshwater Nematodes*, 2nd rev. ed. Methuen, London.

Gooding, G.V. (1970). Natural serological strains of tobacco ringspot virus. *Phytopathology* **60**, 708–713.

Gooding, G.V. (1975). Inactivation of tobacco mosaic virus on tomato seed with trisodium orthophosphate and sodium hypochlorite. *Plant Dis. Rep.* **59**, 770–772.

Goodman, M., Czelusniak, J., Koop, B.F., Tagle, D.A. and Slightom, J.L. (1987). Globins: a case study in molecular phylogeny. *Cold Spring Harbor Symp. Quant. Biol.* **52**, 875–890.

Goodman, R.M. (1977). Single-stranded DNA genome in a whitefly-transmitted plant virus. *Virology* **83**, 171–179.

Goodman, R.M. (1981). Geminiviruses. In: E. Kurstak (ed.) *Handbook of Plant Virus Infections and Comparative Diagnosis*, pp. 883–910. Elsevier, New York.

Goodman, R.M. and Ross, A.F. (1974a). Enhancement of potato virus X synthesis in doubly infected tobacco occurs in doubly infected cells. *Virology* **58**, 16–24.

Goodman, R.M. and Ross, A.F. (1974b). Independent assembly of virions in tobacco doubly infected by potato virus X and Virus Y or tobacco mosaic virus. *Virology* **59**, 314–318.

Goodman, R.M., McDonald, J.G., Horne, R.W. and Bancroft, J.B. (1976). Assembly of flexuous plant viruses and their proteins. *Phil. Trans. R. Soc. Lond. B* **276**, 173–179.

Gorbalenya, A.E. (1995). Origin of RNA viral genomes: approaching the problem by comparative sequence analysis. In: A.J. Gibbs, C.H. Calisher and F. García-Arenal (eds) *Molecular Basis for Virus Evolution*, pp. 49–66. Cambridge University Press, Cambridge.

Gorbalenya, A.E. and Koonin, E.V. (1993). Helicases: amino acid sequence comparisons and structure-function relationships. *Curr. Opin. Struct. Biol.* **3**, 419–429.

Gorbalenya, A.E., Donchenko, A.P. and Blinov, V.M. (1986). A possible common origin of poliovirus proteins with different functions. *Mol. Genet. Mikrobiol. Virusol.* **1**, 36–41.

Gorbalenya, A.E., Koonin, E.V., Blinov, V.M. and Donchenko, A.P. (1988). Sobemovirus genome appears to encode a serine protease related to cysteine proteases of picornaviruses. *FEBS Lett.* **236**, 287–290.

Gordon, D.T. (1999). Waikaviruses (*Sequiviridae*). In: A. Granoff and R.G. Webster (eds) *Encyclopedia of Virology*, 2nd edn, pp. 1965–1970. Academic Press, San Diego.

Gordon, K.H.J. and Symons, R.H. (1983). Satellite RNA of cucumber mosaic virus forms a secondary structure with partial 3'-terminal homology to genomal RNAs. *Nucl. Acids Res.* **11**, 947–960.

Gordon, K.H.J., Pfeiffer, P., Fütterer, J. and Hohn, T. (1988). *In vitro* expression of cauliflower mosaic virus genes. *EMBO J.* **7**, 309–317.

Goregaoker, S.P., Eckhardt, L.G. and Culver, J.N. (2000). Tobacco mosaic virus replicase-mediated cross-protection: contributions of RNA and protein-mediated mechanisms. *Virology* **273**, 267–275.

Goregaoker, S.P., Lewandowski, D.J. and Culver, J.N. (2001). Interactions between the 126/183 kDa replicase associated proteins of tobacco mosaic virus. Abstract W42–2, 20th Annual Meeting of the American Society for Virology.

Gough, K.C., Cockburn, W. and Whitelam, G.C. (1999). Selection of phage display peptides that bind to cucumber mosaic virus coat protein. *J. Virol Methods* **79**, 169–180.

Gould, A.R., Palukaitis, P., Symons, R.H. and Mossop, D.W. (1978). Characterization of a satellite RNA associated with cucumber mosaic virus. *Virology* **84**, 443–455.

Govier, D.A. and Woods, R.D. (1971). Changes induced by magnesium ions in the morphology of some plant viruses with filamentous particles. *J. Gen. Virol.* **13**, 127–132

Gowda, S., Wu, F.C., Scholthof, H.B. and Shepherd, R.J. (1989). Gene VI of figwort mosaic virus (caulimovirus group) functions in post transcriptional expression of genes on the full-length RNA transcript. *Proc. Natl Acad. Sci. USA* **86**, 9203–9207.

Gowda, S., Satyanarayana, T., Naidu, R.A., Mushegian, A., Dawson, W.O. and Reddy, D.V.R. (1998). Characterization of the large (L) RNA of peanut bud necrosis tospovirus. *Arch. Virol.* **143**, 2381–2390.

Gracía, O. and Shepherd, R.J. (1985). Cauliflower mosaic virus in the nucleus of *Nicotiana*. *Virology* **146**, 141–145.

Graddon, D.J. and Randles, J.W. (1986). Single antibody dot immunoassay—a simple technique for rapid detection of a plant virus. *J. Virol. Methods* **13**, 63–69.

Graham, M.W., Keese, P. and Waterhouse, P.M. (1995). The search for the perfect potato. *Today's Life Sciences* **7**, 34–41.

Granell, A., Belles, J.M. and Conejero, V. (1987). Induction of pathogenesis-related proteins in tomato by citrus exocortis viroid, silver ion, and ethephon. *Physiol. Mol. Plant Pathol.* **31**, 83–90.

Grasmick, M.E. and Slack, S.A. (1986). Effect of potato spindle tuber viroid on sexual reproduction and viroid transmission in true potato seed. *Can. J. Bot.* **64**, 336–340.

Gratia, A. (1933). Pluralité antigénique et identification sérologique des virus de plantes. *CR Seances Soc. Biol. Ses Fil.* **114**, 923.

Graves, M.V. and Roossinck, M.J. (1995a). Characterization of defective RNAs derived from RNA 3 of the Fny strain of cucumber mosaic cucumovirus. *J. Virol.* **69**, 4746–4751.

Graves, M.V. and Roossinck, M.J. (1995b). Host specific maintenance of a cucumovirus defective RNA. Presented at the 14th Annual Meeting of the American Society for Virology, University of Texas, Austin, TX.

Graves, M.V., Pogany, J. and Romero, J. (1996). Defective interfering RNAs and defective viruses associated with multipartite RNA viruses of plants. *Sem. Virol.* **7**, 399–408.

Gray, S.M. (1996). Plant virus proteins involved in natural vector transmission. *Trends Microbiol.* **4**, 253–294.

Gray, S.M. and Banerjee, N. (1999). Mechanisms of arthropod transmission of plant and animal viruses. *Microbiol. Mol. Biol. Rev.* **63**, 128–148.

Gray, S.M. and Rochon, D.'A. (1999). Vector transmission of plant viruses. In: A. Granoff and R.G. Webster (eds) *Encyclopedia of Virology*, 2nd edn, pp. 1899–1910. Academic Press, San Diego.

Greene, A.E. and Allison, R.F. (1994). Recombination between viral RNA and transgenic plant transcripts. *Science* **263**, 1423–1425.

Gregory, P.H. (1948). The multiple-infection transformation. *Ann. Appl. Biol.* **35**, 412–417.

Gribskov, M. and Devereux, J. (eds) (1991). *Sequence Analysis Primer*. M Stockton Press, New York.

Grieco, F., Martelli, G.P. and Savino, V. (1995). The nucleotide sequence of RNA3 and RNA4 of olive latent virus—2. *J. Gen. Virol.* **76**, 929–937.

Grieco, F., DellOrco, M. and Martelli, G.P. (1996). The nucleotide sequence of RNA1 and RNA2 of olive latent virus –2 and its relationship to the family *Bromoviridae*. *J. Gen. Virol.* **77**, 2637–2644.

Griep, R.A., van Twisk, C. and Schots, A. (1999). Selection of beet necrotic yellow vein virus specific single-chain Fv antibodies from a semi-synthetic combinatorial antibody library. *Eur. J. Plant Pathol.* **105**, 147–156.

Griep, R.A., Prins, M., van Twisk, C. *et al.* (2000). Application of phage display in selecting tomato spotted wilt virus-specific single-chain antibodies (scFv) for sensitive diagnosis in ELISA. *Phytopathology* **90**, 183–190.

Grill, L.K. and Semancik, J.S. (1978). RNA sequences complementary to citrus exocortis viroid in nucleic acid preparations from infected *Gynura aurantiaca*. *Proc. Natl Acad. Sci. USA* **75**, 896–900.

Grimes, J.M., Kakana, J., Ghosh, M. *et al.* (1997). An atomic model of the outer layer of the bluetongue virus core derived from X-ray crystallography and electron cryo-microscopy. *Structure* **5**, 885–893.

Grimm, M., Nass, A., Schull, C. and Beier, H. (1998). Nucleotide sequences and functional characterization of two tobacco UAG suppressor tRNA(Gln) isoacceptors and their genes. *Plant Mol. Biol.* **38**, 689–697.

Grimsley, N., Hohn, B., Hohn, T. and Walden, R. (1986). 'Agroinfection', an alternative route for viral infection of plants by using the Ti plasmid. *Proc. Natl Acad. Sci. USA* **83**, 3282–3286.

Grimsley, N., Hohn, T., Davies, J.W. and Hohn, B. (1987). *Agrobacterium*-mediated delivery of infectious maize streak virus into maize plants. *Nature (London)* **325**, 177–179.

Grivell, A.R., Grivell, C.J., Jackson, J.F. and Nicholas, D.J.D. (1971). Preservation of lettuce necrotic yellows and some other plant viruses by dehydration with silica gel. *J. Gen. Virol.* **12**, 55–58.

Grogan, R.G. and Bardin, R. (1950). Some aspects concerning the seed transmission of lettuce mosaic virus. *Phytopathology* **40**, 965.

Grogan, R.G., Zink, F.W., Hewitt, W.B. and Kimble, K.A. (1958). The association of *Olpidium* with the big-vein disease of lettuce. *Phytopathology* **48**, 292–297.

Gronenborn, B., Gardner, R.C., Schaefer, S. and Shepherd, R.J. (1981). Propagation of foreign DNA in plants using cauliflower mosaic virus as vector. *Nature (Lond.)* **294**, 773–776.

Gröning, B.R., Abouzid, A. and Jeske, H. (1987). Single-stranded DNA from abutilon mosaic virus is present in the plastids of infected *Abutilon sellovianum*. *Proc. Natl Acad. Sci. USA* **84**, 8996–9000.

Gross, H.J., Domdey, H., Lossow, C., Jank, P., Raba, M., Alberty, H. and Sänger, H.L. (1978). Nucleotide sequence and secondary structure of potato spindle tuber viroid. *Nature (London)* **273**, 203–208.

Grosset, J., Meyer, I., Chartier, Y., Kauffmann, S., Legrand, M. and Fritig, B. (1990). Tobacco mesophyll protoplasts synthesise 1,3-β-glucanase chitinases and 'Osmotins' during *in vitro* culture. *Plant Physiol.* **92**, 520–527.

Grumet, R., Kabelka, E., McQueen, S., Wai, T. and Humphrey, R. (2000). Characterization of sources of resistance to the watermelon strain of papaya ringspot virus in cucumber: allelism and co-segregation with other potyvirus resistances. *Theor. Appl. Genet.* **101**, 463–472.

Guan, H., Song, C. and Symon, A.E. (1997). RNA promoters located on minus –strands of a subviral RNA associated with turnip crinkle virus. *RNA* **3**, 1401–1412.

Guan, H., Carpenter, C.D. and Simon, A.E. (2000). Requirement of a 5'-proximal linear sequence on minus strands for plus-strand synthesis of a satellite RNA associated with turnip crinkle virus. *Virology* **268**, 355–363.

Gugerli, P. (1976). Different states of aggregation of tobacco rattle vines coat protein. *J. Gen. Virol.* **33**, 297–307.

Gugerli, P. (1977). Untersuchungen über die großräumige und lokale Verbreitung des Tabakrattlevirus (TRV) und seiner Vektoren in der Schweiz. *Phytopathol. Z.* **89**, 1–24.

Gugerli, P. (1984). Isopycnic centrifugation of plant viruses in Nycodenz® density gradients. *J. Virol. Meth.* **9**, 249–258.

Gugerli, P. and Fries, P. (1983). Characterization of monoclonal antibodies to potato virus Y and their use for virus detection. *J. Gen. Virol.* **64**, 2471–2477.

Guilford, P.J. (1989). *A Molecular Analysis of Tobacco Rattle Mosaic Virus RNA1*. PhD thesis, University of Cambridge, UK.

Guilfoyle, T.J., McClure, B.A., Gee, M.A. and Hagen, G. (1993). Tissue-print hybridisation for detecting DNA directly. *Methods Enzymol.* **218**, 688–695.

Guilley, H., Richards, K.E. and Jonard, G. (1983). Observations concerning the discontinuous DNAs of cauliflower mosaic virus. *EMBO J.* **2**, 277–282.

Guilley, H., Carrington, J.C., Balàzs, E., Jonard, G., Richards, K. and Morris, T.J. (1985). Nucleotide sequence and genome organisation of carnation mottle virus RNA. *Nucl. Acids Res.* **13**, 6663–6677.

Guilley, H., Richards, K.E. and Jonard, G. (1995). Nucleotide sequence of beet mild yellowing virus RNA. *Arch. Virol.* **140**, 1109–1118.

Gultyaev, A., van Barenburg, E. and Plaij, C. (1994). Similarities between the secondary structure of satellite tobacco mosaic virus and tobamovirus RNAs. *J. Gen. Virol.* **75**, 2851–2856.

Gumpf, D.J., Cunningham, D.S., Heick, J.A. and Shannon, L.M. (1977). Amino acid sequence in the proteolytic glycopeptide of barley stripe mosaic virus. *Virology* **78**, 328–330.

Gunasinghe, U.B., Irwin, M.E. and Karnpmeier, G.E. (1988). Soybean leaf pubescence affects aphid vector transmission and field spread of soybean mosaic virus. *Ann. Appl. Biol.* **112**, 259–272.

Gupta, B.M., Chandra, K., Verma, H.N. and Verma, G.S. (1974). Induction of antiviral resistance in *Nicotiana glutinosa* plants by treatment with *Trichothecium* polysaccharide and its reversal by actinomycin D. *J. Gen. Virol.* **24**, 211–213.

Gura, T. (2000). A silence that speaks volumes. *Nature* **404**, 804–808.

Gustafson, F.G. and Armour, S.L. (1986). The complete nucleotide sequence of RNA β from the type strain of barley stripe mosaic virus. *Nucl. Acids Res.* **14**, 3895–3909.

Gustafson, G.D., Larkins, B.A. and Jackson, A.O. (1981). Comparative analysis of polypeptides synthesised *in vivo* and *in vitro* by two strains of barley stripe mosaic virus. *Virology* **111**, 579–587.

Gustafson, G.D., Hunter, B., Hanau, R., Armour, S.L. and Jackson, A.O. (1987). Nucleotide sequence and genetic organisation of barley stripe mosaic virus RNAλ. *Virology* **158**, 394–406.

Gustafson, G.D., Armour, S.L., Gamboa, G.C., Burgett, S.G. and Shepherd, J.W. (1989). Nucleotide sequence of barley stripe mosaic virus RNAα: RNAα encodes a single polypeptide with homology to corresponding proteins from other viruses. *Virology* **170**, 370–377.

Gutierrez, C. (1998). The retinoblastoma pathway in plant cell cycle and development. *Curr. Opin. Plant Biol.* **1**, 492–497.

Gutierrez, C. (1999). Geminivirus DNA replication. *Cell. Mol. Life Sci.* **56**, 313–329.

Gutierrez, C. (2000a). DNA replication and cell cycle in plants: learning from geminiviruses. *EMBO J.* **19**, 792–799.

Gutierrez, C. (2000b). Geminiviruses and the plant cell cycle. *Plant Mol. Biol.* **43**, 763–772.

Guy, P.L. (1988). Pasture ecology of barley yellow dwarf viruses at Sanford, Tasmania. *Plant Pathol.* **37**, 546–550.

Guy, P.L. and Gibbs, A.J. (1985). Further studies on turnip yellow mosaic tymovirus isolates from an endemic Australian Cardamine. *Plant Pathol.* **34**, 532–544.

Guy, P.L., Webster, D.E. and Davis, L. (1998). Pests of non-indigenous organisms: hidden costs of introduction. *Trends Ecol. Evol.* **13**, 111.

Gwynn, G.R., Reilly, K.R., Komn, J.J., Burk, L.G. and Reed, S.M. (1986). Genetic resistance to tobacco mosaic virus, cyst nematodes root-knot nematodes and wildfire from *Nicotiana repanda* incorporated into *N. tabacum*. *Plant Dis.* **70**, 958–962.

Haas, B., Klanner, A., Ramm, K. and Sänger, H.L. (1988). The 7S RNA from tomato leaf tissue resembles a signal recognition particle RNA and exhibits a remarkable sequence complementarity to viroids. *EMBO J.* **7**, 4063–4074.

Haber, S., Polston, J.E. and Bird, J. (1987). Use of DNA to diagnose plant diseases caused by single-stranded DNA plant viruses. *Can. J. Plant Pathol.* **9**, 156–161.

Haberlé, A.M., Stussi-Garaud, C., Schmitt, C. *et al.* (1994). Detection by immunogold labeling of P75 readthrough protein near an extremity of beet necrotic yellow vein virus particles. *Arch. Virol.* **134**, 195–203.

Habili, N. and Symons, R.H. (1989). Evolutionary relationship between luteoviruses and other RNA plant viruses based on the sequence motifs in their putative RNA polymerases and nucleic acid helicases. *Nucl. Acids Res.* **17**, 9543–9555.

Hacker, D.L. (1995). Identification of a coat protein-binding site on southern bean mosaic virus RNA *Virology* **207**, 562–565.

Hacker, D.L. and Fowler, B.C. (2000). Complementation of the host range restriction of southern cowpea mosaic virus in bean by southern bean mosaic virus. *Virology* **266**, 140–149.

Hadidi, A. (1986). Relationship of viroids and certain other plant pathogenic nucleic acids to group I and II introns. *Plant Mol. Biol.* **7**, 129–142.

Hadidi, A. (1988). Synthesis of disease associated proteins in viroid-infected tomato leaves and binding of viroid to host proteins. *Phytopathology* **78**, 575–578.

Hadidi, A. and Yang, X. (1990). Detection of pome fruit viroids by enzymatic cDNA amplification. *J. Virol. Methods* **30**, 261–270.

Hadidi, A., Cress, D.E. and Diener, T.O. (1981). Nuclear DNA from uninfected or potato spindle tuber viroid-infected tomato plants contains no detectable sequences complementary to cloned double-stranded viroid cDNA. *Proc. Natl Acad. Sci. USA* **78**, 6932–6935.

Hadidi, A., Khetarpal, R.K. and Koganezawa, H. (eds) (1998). *Plant Virus Disease Control.* APS Press, St. Paul, MN.

Hafner, G.J., Stafford, M.R., Wolter, L.C., Harding, R.M. and Dale, J.L. (1997a). Nicking and joining activity of the banana bunchy top virus replication protein *in vitro*. *J. Gen. Virol.* **78**, 1795–1799.

Hafner, G.J., Harding, R.M. and Dale, J.L. (1997b). A DNA primer associated with banana bunchy top virus. *J. Gen. Virol.* **78**, 479–486.

Hafner, M.S. and Nadler, S.A. (1988). Phylogenetic trees support the coevolution of parasites and their hosts. *Nature (Lond.)* **332**, 258–259.

Hagborg, W.A.F. (1970). A device for injecting solutions and suspensions into thin leaves of plants. *Can. J. Bot.* **48**, 1135–1136.

Hagedorn, D.J. and Gritton, E.T. (1973). Inheritance of resistance to pea seed-borne mosaic virus. *Phytopathology* **63**, 1130–1133.

Hagen, L.S., Lot, H., Godon, C., Tepfer, M. and Jacquemond, M. (1994). Infection of *Theobroma cacao* using cloned DNA of cacao swollen shoot virus and particle bombardment. *Phytopathology* **84**, 1239–1243.

Hagen, T.J., Taylor, D.B. and Meagher, R.B. (1982). Rocket immuno-electrophoresis assay for cauliflower mosaic virus. *Phytopathology* **72**, 239–242.

Hagiwara, K., Minobe, Y., Nozu, Y., Hibino, H., Kimura, I. and Omura, T. (1986). Component proteins and structure of rice ragged stunt virus. *J. Gen. Virol.* **67**, 1711–1715.

Hagiwara, Y., Peremyslov, V.V. and Dolja, V.V. (1999). Regulation of closterovirus gene expression examined by insertion of a self-processing reporter and by northern hybridization. *J. Virol.* **73**, 7988–7993.

Hahn, P. and Shepherd, R.J. (1980). Phosphorylated proteins in cauliflower mosaic virus. *Virology* **107**, 295–297.

Hahn, P. and Shepherd, R.J. (1982). Evidence for a 58-kilodalton polypeptide as precursor of the coat protein of cauliflower mosaic virus. *Virology* **116**, 480–488.

Haigh, J. (1978). The accumulation of deleterious genes in a population—Muller's ratchet. *Theor. Popul. Biol.* **14**, 251–267.

Haight, E. and Gibbs, A. (1983). Effect of viruses on pollen morphology. *Plant Pathol.* **32**, 369–372.

Hajimorad, M.R. and Francki, R.I.B. (1988). Alfalfa mosaic virus isolates from lucerne in South Australia: biological variability and antigenic similarity. *Ann. Appl. Biol.* **113**, 45–54.

Hajimorad, M.R. and Hill, J.H. (2001). Rsv1–mediated resistance against soybean mosaic virus-N is hypersensitive response-independent at inoculation site, but has the potential to initiate a hypersensitive response-like mechanism. *Molec. Plant-Microbe Interact.* **14**, 587–598.

Haley, A., Hunter, T., Kiberstis, P. and Zimmern, D. (1995). Multiple serine phosphorylation sites on the 30 kDa TMV cell-to-cell movement protein synthesised in tobacco protoplasts. *Plant J.* **8**, 715–724.

Halk, E.L. and McGuire, J.M. (1973). Translocation of tobacco ringspot virus in soybean. *Phytopathology* **63**, 1291–1300.

Halk, E.L., Hus, H.T., Aebig, J. and Franke, J. (1984). Production of monoclonal antibodies against three ilarviruses and alfalfa mosaic virus and their use in serotyping. *Phytopathology* **74**, 367–372.

Hall, A.E. and Loomis, R.S. (1972). An explanation for the difference in photosynthetic capabilities of healthy and beet yellows virus-infected sugar beets (*Beta vulgaris* L.). *Plant Physiol.* **50**, 576–580.

Hall, J.D., Barr, R., Al-Abbas, A.H. and Crane, F.L. (1972). The ultrastructure of chloroplasts in mineral-deficient maize leaves. *Plant Physiol.* **50**, 404–409.

Halliwell, R.S. and Gazaway, W.S. (1975). Quantity of microinjected tobacco mosaic virus required for infection of single cultured tobacco cells. *Virology* **65**, 583–587.

Hämäläinen, J.H., Kekarainen, T., Gebhardt, C., Watanabe, K.N. and Valkonen, J.P.T. (2000). Recessive and dominant genes interfere with the vascular transport of potato virus A in diploid potatoes. *Mol. Plant–Microb. Interact.* **13**, 402–412.

Hamamoto, H., Sugiyama, Y., Nakagawa, N. *et al.* (1993). A new tobacco mosaic virus vector and its use for the systemic production of angiotensin-1-converting enzyme inhibitor in transgenic tobacco and tomato. *Bio/Technol.* **11**, 930–932.

Hamilton, A.J. and Baulcombe, D.C. (1999). A species of small antisense RNA in posttranscriptional gene silencing in plants. *Science* **286**, 950–952.

Hamilton, R.I. (1980). Defenses triggered by previous invaders: viruses. In: J.G. Horsfall and E.B. Cowling (eds) *Plant Disease: An Advanced Treatise*, Vol. 5, pp. 279–303. Academic Press, New York.

Hamilton, R.I. and Tremaine, J.H. (1996). Dianthoviruses: properties, molecular biology, ecology and control. In: B.D. Harrison and A.F. Murant (eds) *The Plant Viruses. Vol. 5: Polyhedral Virions and Bipartite RNA genomes*, pp. 251–282. Plenum Press, New York.

Hamilton, R.I., Leung, E. and Nichols, C. (1977). Surface contamination of pollen by plant viruses. *Phytopathology* **67**, 395–399.

Hamilton, R.I., Edwardson, J.R., Francki, R.I.B. *et al.* (1981). Guidelines for the identification and characterization of plant viruses. *J. Gen. Virol.* **54**, 223–241.

Hamilton, W.D.O. and Baulcombe, D.L. (1989). Infectious RNA produced by *in vitro* transcription of a full-length tobacco rattle virus RNA1 cDNA. *J. Gen. Virol.* **70**, 963–968.

Hamilton, W.D.O., Bisaro, D.M., Coutts, R.H.A. and Buck, K.W. (1983). Demonstration of the bipartite nature of the genome of a single-stranded DNA plant virus by infection with the cloned DNA components. *Nucl. Acids Res.* **11**, 7387–7396.

Hamilton, W.D.O., Boccara, M., Robinson, D.J. and Baulcombe, D.C. (1987). The complete nucleotide sequence of tobacco rattle virus RNA1. *J. Gen. Virol.* **68**, 2563–2575.

Hammond, J. (1982). Plantago as a host of economically important viruses. *Adv. Virus Res.* **27**, 103–140.

Hammond, J. (1998). Resistance to plant viruses: an overview. Breeding for resistance to plant viruses. In: A. Hadidi, R.K. Khetarpal and H. Koganezawa (eds) *Plant Virus Disease Control*, pp. 163–171. APS Press, St. Paul, MN.

Hammond, J. and Hull, R. (1981). Plantain virus X: a new potexvirus from *Plantago lanceolata*. *J. Gen. Virol.* **54**, 75–90.

Hammond, J. and Kamo, K.K. (1995). Effective resistance to potyvirus infection in transgenic plants expressing antisense RNA. *Mol. Plant-Microb. Interact.* **8**, 674–682.

Hammond, J., Lecoq, H. and Raccah, B. (1999). Epidemiological risks from mixed infections and transgenic crops expressing viral genes. *Adv. Virus Res.* **54**, 180–314.

Hammond, J.M., Sproat, K.W., Wise, T.G., Hyatt, A.D., Jagadish, M.N. and Coupar, B.E.H. (1998). Expression of the potyvirus coat protein mediated by recombinant vaccinia virus and assembly of potyvirus-like particles in mammalian cells. *Arch. Virol.* **143**, 1433–1439

Hammond, R.W. and Owens, R.A. (1987). Mutational analysis of potato spindle tuber viroid reveals complex relationships between structure and infectivity. *Proc. Natl Acad. Sci. USA* **84**, 3967–3971.

Hammond, R.W., Diener, T.O. and Owens, R.A. (1989). Infectivity of chimeric viroid transcripts reveals the presence of alternative processing sites in potato spindle tuber viroid. *Virology* **170**, 486–495.

Hammond, R.W., Smith, D.R. and Diener, T.O. (1990). Nucleotide sequence and proposed secondary structure of *Columnea* latent viroid: a natural mosaic of viroid sequences. *Nucl. Acids Res.* **17**, 10083–10094.

Hammond-Kosack, K.E. and Jones, J.D.G. (1996). Resistance gene-dependent plant defense responses. *Plant Cell* **8**, 1773–1791.

Hammond-Kosack, K.E. and Jones, J.D.G. (1997). Plant disease resistance genes. *Annu. Rev. Plant Physiol. Plant Mol. Biol.* **48**, 575–607.

Hampton, R., Ball, E. and de Boer, S. (eds) (1990). *Serological Methods for Detection and Identification of Viral and Bacterial Plant Pathogens: A Laboratory Manual*. APS Press, St. Paul, MN.

Hampton, R.O. (1966). Probabilities of failing to detect prune dwarf virus in cherry trees by bud indexing. *Phytopathology* **56**, 650–652.

Hampton, R.O. (1975). The nature of bean yield reduction by bean yellow and bean common mosaic viruses. *Phytopathology* **65**, 1342–1346.

Hampton, R.O. (1988). Health status (virus) of native North American *Humulus lupulus* in the natural habitat. *J. Phytopathol.* **123**, 353–362.

Hampton, R.O. and Francki, R.I.B. (1992). RNA-1 dependent seed transmissibility of cucumber mosaic virus in *Phaseolus vulgaris*. *Phytopathology* **82**, 127–130.

Hanada, K. (1984). Electrophoretic analysis of virus particles of fourteen cucmovirus isolates. *Ann. Phytopathol. Soc. Jpn.* **50**, 361–367.

Hanada, K. and Francki, R.I.B. (1989). Kinetics of velvet tobacco mottle virus satellite RNA synthesis and encapsulation. *Virology* **170**, 48–54.

Hanada, K. and Harrison, B.D. (1977). Effects of virus genotype and temperature on seed transmission of nepoviruses. *Ann. Appl. Biol.* **85**, 79–92.

Hanada, K., Kusunoki, M. and Iwaki, M. (1986). Properties of virus particles, nucleic acid and coat protein of cycas necrotic stunt virus. *Ann. Phytopathol. Soc. Jpn* **52**, 422–427.

Hanley-Bowdoin, L., Elmer, J.S. and Rogers, S.G. (1988). Transient expression of heterologous RNAs using tomato golden mosaic virus. *Nucl. Acids Res.* **16**, 10511–10528.

Hanley-Bowdoin, L., Settlage, S.B., Orozco, B.M., Nagar, S. and Robertson, D. (1999). Geminiviruses: models for plant DNA replication, transcription, and cell cycle regulation. *Crit. Rev. Plant Sci.* **18**, 71–106.

Hansen, A.J. (1989). Antiviral chemicals for plant disease control. *Crit. Rev. Plant Sci.* **8**, 45–88.

Hansen, A.J. and Lane, W.D. (1985). Elimination of apple chlorotic leafspot virus from apple shoot cultures by ribavirin. *Plant Dis.* **69**, 134–135.

Hansen, J.L., Long, A.M. and Schultz, S.C. (1997). Structure of the RNA-dependent RNA polymerase of poliovirus. *Structure* **5**, 1109–1122.

Harders, J., Lukacs, N., Robert-Nicoud, M., Jovin, J.M. and Riesner, D. (1989). Imaging of viroids in nuclei from tomato leaf tissue by *in situ* hybridization and confocal laser scanning microscopy. *EMBO J.* **8**, 3941–3949.

Harding, R.M., Burns, T.M., Hafner, G.J., Dietzgen, R. and Dale, J.L. (1993). Nucleotide sequence of the banana bunchy top genome contains a putative replicase gene. *J. Gen. Virol.* **74**, 323–328.

Hardy, S.F., German, T.L., Loesch-Fries, L.S. and Hall, T.C. (1979). Highly active template-specific RNA-dependent RNA polymerase from barley leaves infected with brome mosaic virus. *Proc. Natl Acad. Sci. USA* **76**, 4956–4960.

Hardy, V.G. and Teakle, D.S. (1992). Transmission of sowbane mosaic virus by *Thrips tabaci* in the presence and absence of virus-carrying pollen. *Ann. Appl. Biol.* **121**, 315–320.

Hare, J.D. and Dodds, J.A. (1987). Survival of the Colorado potato beetle on virus-infected tomato in relation to plant nitrogen and alkaloid content. *Entomol. Exp. Appl.* **44**, 31–35.

Hari, V., Siegel, A., Rozek, C. and Timberlake, W.E. (1979). The RNA of tobacco etch virus contains poly(A). *Virology* **92**, 568–571.

Harker, C.L., Woolston, C.J., Markham, P.G. and Maule, A.J. (1987a). Cauliflower mosaic virus aphid transmission factor protein is expressed in cells infected with some aphid non-transmissible isolates. *Virology* **160**, 252–254.

Harker, C.L., Mullineaux, P.M., Bryant, J.A. and Maule, A.J. (1987b). Detection of CaMV gene I and gene VI protein products *in vivo* using antisera raised to COOH-terminal β-galactosidase fusion proteins. *Plant Mol. Biol.* **8**, 275–287.

Harlan, J.R. (1976). Diseases as a factor in plant evolution. *Annu. Rev. Phytopathol.* 14, 31–51.

Harlow, E. and Lane, D. (1988). *Antibodies: A Laboratory Manual.* Cold Spring Harbor Laboratory, Cold Spring Harbor, NY.

Harmer, S.L., Hogenesch, J.B., Straume, M. *et al.* (2000). Orchestrated transcription of key pathways in *Arabidopsis* by the circadian clock. *Science* **290**, 2110–2113.

Harper, G. and Hull, R. (1998). Cloning and sequence analysis of banana streak virus DNA. *Virus Genes* **17**, 271–278.

Harper, G., Dahal, G., Thottappilly, G. and Hull, R. (1999a). Detection of episomal banana streak badnavirus by IC-PCR. *J. Virol. Meths.* **79**, 1–8.

Harper, G., Osuji, J.O., Heslop-Harrison, J.S. and Hull, R. (1999b). Integration of banana streak badnavirus into the *Musa* genome: molecular and cytological evidence. *Virology* **255**, 207–213.

Harper, K. and Creamer, R. (1995). Hybridization detection of insect-transmitted plant viruse with digoxigenin-labelled probes. *Plant Dis.* **79**, 563–567.

Harper, K., Toth, R.L., Mayo, M.A. and Torrance, L. (1999c). Properties of a panel of single chain variable fragments against potato leafroll virus obtained from two phage display libraries. *J. Virol. Methods* **81**, 159–168.

Harrington, R., Dewar, A.M. and George, B. (1989). Forecasting the incidence of virus yellows in sugar beet in England. *Ann. Appl. Biol.* **114**, 459–469.

Harris, J.I. and Knight, C.A. (1952). Action of carboxypeptidase on tobacco mosaic virus. *Nature (London)* **170**, 613–614.

Harris, J.I. and Knight, C.A. (1955). Studies on the action of carboxypeptidase on tobacco mosaic virus. *J. Biol. Chem.* **214**, 215–230.

Harris, J.R. and Horne, R.W. (1994). Negative staining: a brief assessment of current technical benefits, limitations and future possibilities. *Micron* **25**, 5–13.

Harris, K.F. (1977). An ingestion–egestion hypothesis of non-circulative virus transmission. In: K.F. Harris, and K. Maramorosch (eds) *Aphids as Virus Vectors*, pp. 166–208. Acadenic Press, New York.

Harris, K.F. and Bradley, R.H.E. (1973). Tobacco mosaic virus: can aphids inoculate it into plants with their mouthparts? *Phytopathology* **63**, 1343–1345.

Harris, K.F. and Maramorosch, K. (eds) (1982). *Pathogens, Vectors and Plant Diseases: Approaches to Control.* Academic Press, New York.

Harris, K.F., Treur, B., Tsai, J. and Toler, R. (1981). Observations on leafhopper ingestion–egestion behaviour: its likely role in the transmission of non-circulative viruses and other plant pathogens. *J. Econ. Entomol.* **74**, 446–453.

Harrison, B.D. (1958). Ability of single aphids to transmit both avirulent and virulent strains of potato leaf roll virus. *Virology* **6**, 278–286.

Harrison. B.D. (1981). Plant virus ecology: ingredients, interactions and environmental influences. *Ann. Appl. Biol.* **99**, 195–209.

Harrison, B.D. (1985). Advances in geminivirus researrh. *Annu. Rev. Phytopathol.* **23**, 55–82.

Harrison, B.D. and Jones, R.A.C. (1970). Host range and some properties of potato mop-top virus. *Ann. Appl. Biol.* **65**, 393–402.

Harrison, B.D. and Jones, R.A.C. (1971). Effects of light and temperature on symptom development and virus content of tobacco leaves inoculated with potato mop-top virus. *Ann. Appl. Biol.* **67**, 377–387.

Harrison, B.D. and Murant, A.F. (1984). Involvement of plant virus-coded proteins in transmission of plant viruses by vectors. In: M.A. Mayo and K.R. Harrap (eds) *Vectors in Virus Biology*, pp. 1–36. Academic Press, London.

Harrison, B.D. and Nixon, H.L. (1959). Separation and properties of particles of tobacco rattle virus with different lengths. *J. Gen. Microbiol.* **21**, 569–581.

Harrison, B.D. and Roberts, I.M. (1968). Association of tobacco rattle virus with mitochondria. *J. Gen. Virol.* **3**, 121–124.

Harrison, B.D. and Robinson, D.J. (1999). Natural genomic and antigenic variation in whitefly-transmitted geminiviruses (Begomoviruses). *Annu. Rev. Phytopathol.* **37**, 369–398.

Harrison, B.D. and Wilson, T.M.A. (1999). Milestones in the research on tobacco mosaic virus. *Phil. Trans.R. Soc. Lond. B* **354**, 521–529.

Harrison, B.D. and Winslow, R.D. (1961). Laboratory and field studies on the relation of arabis mosaic virus to its nematode vector *Xiphinema diversicaudatum* Micoletzky. *Ann. Appl. Biol.* **49**, 621–633.

Harrison, B.D. and Woods, R.D. (1966). Serotypes and particle dimensions of tobacco rattle viruses from Europe and America. *Virology* **28**, 610–620.

Harrison, B.D., Peachey, J.E. and Winslow, R.D. (1963). The use of nematicides to control the spread of arabis mosaic virus by *Xiphinema diversicaudatum* (Micol). *Ann. Appl. Biol.* **52**, 243–255.

Harrison, B.D., Finch, J.T., Gibbs, A.J. *et al.* (1971). Sixteen groups of plant viruses. *Virology* **45**, 356–363.

Harrison, B.D., Robertson, W.M. and Taylor, C.E. (1974). Specificity of retention and transmission of viruses by nematodes. *J. Nematol.* **6**, 155–164.

Harrison, B.D., Kubo, S., Robinson, D.J. and Hutcheson, A.M. (1976). The multiplication cycle of tobacco rattle virus in tobacco mesophyll protoplasts. *J. Gen. Virol.* **33**, 237–248.

Harrison, B.D., Robinson, D.J., Mowat, W.P. and Duncan, G.H. (1983). Comparison of nucleic acid hybridisation and other tests for detecting tobacco rattle virus in narcissus plants and potato tubers. *Ann. Appl. Biol.* **102**, 331–338.

Harrison, B.D., Mayo, M.A. and Baulcombe, D.C. (1987). Virus resistance in transgenic plants that express cucumber mosaic virus satellite RNA. *Nature (Lond.)* **328**, 799–802.

Harrison, B.D., Liu, Y.L., Khalid, S., Hameed, S., Otim-Nape, G.W. and Robinson, D.J. (1997). Detection and relationships of cotton leaf curl virus and allied whitefly-transmitted geminiviruses occurring in Pakistan. *Ann. Appl. Biol.* **130**, 61–75.

Harrison, S.C., Olson, A.J., Schutt, C.E., Winkler, F.K. and Bricogne, G. (1978). Tomato bushy stunt virus at 2.9 Å resolution. *Nature (London)* **276**, 368–373.

Hartman, K.A., McDonald-Ordzie, P.E., Kaper, J.M., Prescott, B. and Thomas, G.J. (1978). Studies of virus structure by laser Raman spectroscopy: turnip yellow mosaic virus and capsids. *Biochemistry* **17**, 2118–2123.

Hartman, R.D. (1974). Dasheen mosaic virus and other phytopathogens eliminated from caladium, taro and cocoyam by culture of shoot tips. *Phytopathology* **64**, 237–240.

Harveson, R.M., Rush, C.M. and Wheeler, T.A. (1996). The spread of beet necrotic yellow vein virus from point source inoculations as influenced by irrigation and tillage. *Phytopathology* **86**, 1242–1247.

Harvey, J.D. (1973). Diffusion coefficients and hydrodynamic radii of three spherical RNA viruses by laser light scattering. *Virology* **56**, 365–368.

Harvey, J.D., Farrell, J.A. and Bellamy, A.R. (1974). Biophysical studies of reovirus type 3: II. Properties of the hydrated particle. *Virology* **62**, 154–160.

Hasegawa, A., Verver, J., Shimada, A. *et al.* (1989). The complete nucleotide sequence of soybean chlorotic mottle virus DNA and the identification of a novel promoter. *Nucl. Acids Res.* **17**, 9993–10013.

Haselkorn, R. (1962). Studies on infectious RNA from turnip yellow mosaic virus. *J. Mol. Biol.* **4**, 357–367.

Haseloff, J., Mohamed, N.A. and Symons, R.H. (1982). Viroid RNAs of cadang-cadang disease of coconuts. *Nature* **229**, 316–321.

Haseloff, J., Goelet, P., Zimmern, D., Ahlquist, P., Dasgupta, R. and Kaesberg, P. (1984). Striking similarities in amino acid sequence among non-structural proteins encoded by RNA viruses that have dissimilar genomic organisation. *Proc. Natl Acad. Sci. USA* **81**, 4358–4362.

Hatta, T. (1976). Recognition and measurement of small isometric virus particles in thin sections. *Virology* **69**, 237–245.

Hatta, T. and Francki, R.I.B. (1976). Anatomy of virus-induced galls on leaves of sugarcane infected with Fiji disease virus and the cellular distribution of virus particles. *Physiol. Plant Pathol.* **9**, 321–330.

Hatta, T. and Francki, R.I.B. (1977). Morphology of Fiji disease virus. *Virology* **76**, 797–807.

Hatta, T. and Francki, R.I.B. (1981a). Identification of small polyhedral virus particles in thin sections of plant cells by an enzyme cytochemical technique. *J. Ultrastruct. Res.* **74**, 116–129.

Hatta, T. and Francki, R.I.B. (1981b). Cytopathic structures associated with tonoplasts of plant cells infected with cucumber mosaic and tomato aspermy viruses. *J. Gen. Virol.* **53**, 343–346.

Hatta, T. and Francki, R.I.B. (1981c). Development and cytopathology of virus-induced galls on leaves of sugarcane infected with Fiji disease virus. *Physiol. Plant Pathol.* **19**, 337–346.

Hatta, T. and Francki, R.I.B. (1984). Differences in the morphology of isometric particles of some plant viruses stained with uranyl acetate as an aid to their identification. *J. Virol. Methods* **9**, 237–247.

Hatta, T. and Matthews, R.E.F. (1974). The sequence of early cytological changes in Chinese cabbage leaf cells following systemic infection with turnip yellow mosaic virus. *Virology* **59**, 383–396.

Hatta, T. and Matthews, R.E.F. (1976). Sites of coat protein accumulation in turnip yellow mosaic virus-infected cells. *Virology* **73**, 1–16.

Hatta, T., Nakamoto, T., Takagi, Y. and Ushiyama, R. (1971). Cytological abnormalities of mitochondria induced by infection with cucumber green mottle mosaic virus. *Virology* **45**, 292–297.

Hatta, T., Bullivant, S. and Matthews, R.E.F. (1973). Fine structure of vesicles induced in chloroplasts of Chinese cabbage leaves by infection with turnip yellow mosaic virus. *J. Gen. Virol.* **20**, 37–50.

Havelda, Z. and Burgyán, J. (1995). 3′ terminal putative stem-loop structure required for the accumulation of cymbidium ringspot virus RNA. *Virology* **214**, 269–272.

Havelda, Z., Dalmay, T. and Burgyán, J. (1995). Localization of the *cis*-acting sequences essential for cymbidium ringspot tombusvirus defective interfering RNA replication. *J. Gen. Virol.* **76**, 2311–2316.

Hawkes, J.G. (1967). The history of the potato. Part III. *J. R. Hortic. Soc.* **92**, 288–302.

Hay, J.M., Jones, M.C., Blakebrough, M., Dasgupta, I., Davies, J.W. and Hull, R. (1991). An analysis of the sequence of an infectious clone of rice tungro bacilliform virus, a plant pararetrovirus. *Nucl. Acids Res.* **19**, 2615–2621.

Hay, J.M., Grieco, F., Druka, A., Pinner, M., Lee, S. and Hull, R. (1994). Detection of rice tungro bacilliform virus gene products *in vivo*. *Virology* **205**, 430–437.

Hayakawa, T., Mizukami, M., Nakamura, I. and Suzuki, M. (1989). Cloning and sequencing RNA1 cDNA from cucumber mosaic virus strain O. *Gene* **85**, 533–540.

Hayakawa, T., Zhu, Y, Itoh, K. *et al.* (1992). Genetically engineered rice resistant to rice stripe virus, an insect-transmitted virus. *Proc. Natl Acad. Sci. USA* **89**, 9865–9869.

Hayashi, T. (1974). Fate of tobacco mosaic virus after entering the host cell. *Jpn J. Microbiol.* **18**(4), 279–286.

Hayes, R.J. and Buck, K.W. (1990). Complete replication of a eukaryotic virus RNA in vitro by a purified RNA-dependent RNA polymerase. *Cell* **63**, 363–368.

Hayes, R.J., Brough, C.L., Prince, V.E., Coutts, R.H.A and Buck, K.W. (1988a). Infection of *Nicotiana benthamiana* with uncut cloned tandem dimers of tomato golden mosaic virus DNA. *J. Gen. Virol.* **69**, 209–218.

Hayes, R.J., MacDonald, H., Coutts, R.H.A. and Buck, K.W. (1988b). Agroinfection of *Tritium aestivum* with cloned DNA of wheat dwarf virus. *J. Gen. Virol.* **69**, 891–896.

Hayes, R.J., MacDonald, H., Coutts, R.H.A. and Buck, K.W. (1988c). Priming of complimentary DNA synthesis *in vitro* by small DNA molecules tightly bound to virion DNA of wheat dwalf virus. *J. Gen. Virol.* **69**, 1345–1350.

Hayes, R.J., Petty, I.T.D., Coutts, R.H.A. and Buck, K.W. (1988d). Gene amplification and expression in plants by a replicating geminivirus vector. *Nature (Lond.)* **334**, 179–182.

Hayes, R.J., Coutts, R.H.A. and Buck, K.W. (1988e). Agroinfection of *Nicotiana* spp. with cloned DNA of tomato golden mosaic virus. *J. Gen. Virol.* **69**, 1487–1496.

Hayes, R.J., Tousch, D., Jacquemond, M., Pereira, V.C., Buck, K.W. and Tepfer, M. (1992). Complete replication of a satellite RNA in vitro by a purified RNA-dependent RNA polymerase. *J. Gen. Virol.* **73**, 1597–1600.

Hayley, A., Zhan, X., Richardson, K.A., Head, K. and Morris, B.A.M. (1992). Regulation of the activation of African cassava mosaic virus promoters by the AC1, AC2 and AC3 gene products. *Virology* **188**, 905–909.

Haynes, J.R., Cunningham, J., von Seefried, A., Lennick, M., Garvin, R.T. and Shen S.-H. (1986). Development of a genetically engineered, candidate polio vaccine employing the self-assembling properties of tobacco mosaic virus coat protein. *Bio/Technology* **4**, 637–641.

He, X.-H., Rao, A.L.N. and Creamer, R. (1997). Characterization of beet yellows closterovirus-specific RNAs in infected plants and protoplasts. *Phytopathology* **87**, 347–352.

Heard, A.J. and Chapman, P.F. (1986). A field study of the pattern of local spread of ryegrass mosaic virus in mown grassland. *Ann. Appl. Biol.* **108**, 341–345.

Hearne, P.Q., Knorr, D.A., Hillman, B. and Morris, T.J. (1990). The complete genome structure and synthesis of infectious RNA from clones of tomato bushy stunt virus. *Virology* **177**, 141–151.

Heath, M.C. (2000). Advances in imaging the cell biology of plant–microbe interactions. *Annu. Rev. Phytopathol.* **38**, 443–459.

Heathcote, G.D. (1978). Review of losses caused by virus yellows in English sugar beet crops and the cost of partial control with insecticides. *Plant Pathol.* **27**, 12–17.

Heathcote, G.D. and Broadbent, L. (1961). Local spread of potato leaf roll and Y viruses. *Eur. Potato J.* **4**, 138–143.

Heaton, L.A., Hillman, B.I., Hunter, B.G., Zuidema, D. and Jackson, A.O. (1989a). A physical map of the genome of Sonchus yellow net virus, a plant rhabdovirus with six genes and conserved genejunction sequences. *Proc. Natl Acad. Sci. USA* **86**, 8665–8668.

Heaton, L.A., Hillman, B.I., Hunter, B.G., Zuidema, D. and Jackson, A.O. (1989b). Physical map of the genome of Sonchus yellow net virus, a plant rhabdovirus with six genes and conserved junction sequences. *Proc. Natl Acad. Sci. USA* **86**, 8665–8668.

Hebers, K., Takahata, Y., Melzer, M., Mock, H.-P., Hajirezaei, M. and Sonnewald, U. (2000). Regulation of carbohydrate partitioning during the interaction of potato virus Y with tobacco. *Mol. Plant Pathol.* **1**, 51–59.

Hebert, T.T. (1963). Precipitation of plant viruses by polyethylene glycol. *Phytopathology* **53**, 362.

Hebrard, E., Drucker, M., Leclerc, D., Hohn, T. *et al.* (2001). Biochemical characterization of the helper component of cauliflower mosaic virus. *J. Virol.* In press.

Hecht, E.I. and Bateman, D.F. (1964). Non-specific acquired resistance to pathogens resulting from localised infections by *Thielaviopsis basicola* or viruses in tobacco leaves. *Phytopathology* **54**, 523–530.

Hedrick, J.L. and Smith, A.J. (1968). Size and charge isomer separation and estimation of molecular weights of proteins by disc gel electrophoresis. *Archiv. Biochem. Biophys.* **126**, 155–164.

Heese-Peck, A. and Raikhel, N.V. (1998). The nuclear pore complex. *Plant Mol. Biol.* **38**, 145–162.

Hefferson, K.L., Khalilian, H., Xu, H. and AbouHaidar, M.G. (1997). Expression of the coat protein of potato virus X from a dicistronic mRNA in transgenic potato plants. *J. Gen. Virol.* **78**, 3051–3059.

Hehn, A., Fritsch, C., Richards, K.E., Guilley, H. and Jonard, G. (1997). Evidence for *in vitro* and *in vivo* autolytic processing of the primary translation product of beet necrotic yellow vein virus RNA 1 by a papain-like proteinase. *Arch. Virol.* **142**, 1051–1058.

Heide, M. and Lange, L. (1988). Detection of potato leaf roll virus and potato viruses M, S, X and Y by dot immunobinding on plain paper. *Potato Res.* **31**, 367–373.

Heijbroek, W. (1988). Dissemination of rhizomania by soil, beet seeds and stable manure. *Neth. J. Plant Pathol.* **94**, 9–15.

Heijtink, R.A. and Jaspars, E.M.J. (1974). RNA contents of abnormally long particles of certain strains of alfalfa mosaic virus. *Virology* **59**, 371–382.

Heijtink, R.A. and Jaspars, E.M.J. (1976). Characterization of two morphologically distinct top component a particles from alfalfa mosaic virus. *Virology* **69**, 75-80.

Heijtink, R.A., Houwing, C.J. and Jaspars, E.M.J. (1977). Molecular weights of particles and RNAs of alfalfa mosaic virus: number of subunits in protein capsids. *Biochemistry* **16**, 4684–4693.

Hein, A. (1971). Zur Wirkung von Öl auf die Virusübertragung durch Blattäiuse. *Phytopathol. Z.* **71**, 42–48.

Hein, A. (1987). A contribution to the effect of fungicides and additives for formulation on the rhizomania infection of sugar beets (beet necrotic yellow vein virus). *J. Plant Dis. Prot.* **94**, 250–259.

Heinlein, M., Epel, B.L., Padgett, H.S. and Beachy, R.N. (1995). Interaction of tobamovirus movement protein with the plant cytoskeleton. *Science* **270**, 1983–1985.

Heinlein, M., Padgett, H.S., Gens, J.S. *et al.* (1998). Changing patterns of localization of TMV movement protein and replicase to endoplasmic reticulum and microtubules during infection. *Plant Cell* **10**, 1107–1120.

Heinrichs, E.A., Rapusas, H.R., Aquino, G.B. and Palis, F. (1986). Integration of host plant resistance and insecticides in the control of *Nephotettix virescens* (Homoptera: Cicadellidae), a vector of rice tungro virus. *J. Econ. Entomol.* **79**, 437–443.

Hellen, C.U.T. and Cooper, J.I. (1987). The genome-linked protein of cherry leaf roll virus. *J. Gen. Virol.* **68**, 2913–2917.

Hellmann, G.M., Shaw, J.G. and Rhoads, R.E. (1988). *In vitro* analysis of tobacco vein mottling virus NI$_a$ cistron: evidence for a virus-encoded protease. *Virology* **163**, 554–562.

Hellwald, K.H., Zimmermann, C. and Buchenauer, H. (2000). RNA 2 of cucumber mosaic virus subgroup 1 strain NT-CMV is involved in the induction of severe symptoms in tomato. *Eur. J. Plant Pathol.* **106**, 95–99.

Helms, K. and Wardlaw, I.F. (1976). Movement of viruses in plants: long distance movement of tobacco mosaic virus in *Nicotiana glutinosa*. In: I.F. Wardlaw and J.B. Passioura (eds) *Transport and Transfer Processes in Plants*, pp. 283–293. Academic Press, New York.

Helms, K., Waterhouse, P.M. and Muller, W.J. (1985). Subterranean clover red leaf virus disease: effects of temperature on plant symptoms, growth and virus content. *Phytopathology* **75**, 337–341.

Hemenway, C., Fang, R.-X., Kaniewski, W.K., Chua, N.-H. and Turner, N.E. (1988). Analysis of the mechanism of protection in transgenic plants expressing the potato virus X coat protein or its antisense RNA. *EMBO J.* **7**, 1273–1280.

Hemida, S.K., Murant, A.F. and Duncan, G.H. (1989). Purification and some particle properties of anthricus yellows virus, a phloem-limited semipersistent aphid-borne virus. *Ann. Appl. Biol.* **114**, 71–86.

Hemmati, K. and McLean, D.L. (1977). Gamete–seed transmission of alfalfa mosaic virus and its effect on seed germination and yield in alfalfa plants. *Phytopathology* **67**, 576–579.

Hemmer, O., Meyer, M., Greif, C. and Fritsch, C. (1987). Comparison of the nucleotide sequences of five tomato black ring virus satellite RNAs. *J. Gen. Virol.* **68**, 1823–1833.

Hemmings-Mieszczak, M. and Hohn, T. (1999). A stable hairpin preceded by a short open reading frame promotes nonlinear ribosome migration on a synthetic mRNA leader. *RNA* **5**, 1149–1157.

Hemmings-Mieszczak, M.W., Steger, G. and Hohn, T. (1997). Alternative structures of the cauliflower mosaic virus 35S RNA leader: implications for viral expression and replication. *J. Mol. Biol.* **267**, 1075–1088.

Hemmings-Mieszczak, M.W., Steger, G. and Hohn, T. (1998). Regulation of CaMV translation is mediated by a stable hairpin in the leader. *RNA* **4**, 101–111.

Hennig, W. (1981). *Insect Phylogeny*. Wiley, New York.

Herman, T. and Patel, D.J. (1999). Stitching together RNA tertiary architectures. *J. Mol. Biol.* **294**, 829–849.

Hernández, C. and Flores, R. (1991). Plus and minus RNAs of peach latent mosaic viroid self-cleave *in vitro* via hammerhead structures. *Proc. Natl Acad. Sci. USA* **89**, 3711–3715.

Hernández, C. Daros, J.A., Elena, S.F., Moya, A. and Flores, R. (1992). The strands of both polarities of a small circular RNA from carnation self-cleave *in vitro* through alternative double- and single-hammerhead structures. *Nucl. Acids Res.* **20**, 6323–6329.

Hernández, C, Visser, P.B., Brown, D.J.F. and Bol, J.F. (1997). Transmission of tobacco rattle virus isolate PpK20 by its nematode vector requires one of the two non-structural genes in the viral RNA2. *J. Gen. Virol.* **78**, 465–467.

Hershey, A.D. and Chase, M. (1952). Independent functions of viral protein and nucleic acid in growth of bacteriophage. *J. Gen. Physiol.* **36**, 39–56.

Hertzog, E., Guilley, H., Manohar, S.K. *et al.* (1994). Complete nucleotide sequence of peanut clump virus RNA 1 and relationships with other fungus-transmitted rod-shaped viruses. *J. Gen. Virol.* **75**, 3147–3155.

Herzog, E., Guilley, H. and Fritsch, C. (1995). Translation of the second gene of peanut clump virus RNA 2 occurs by leaky scanning *in vitro*. *Virology* **208**, 215–225.

Heuss, K., Liu, Q., Hammerschlag, F.A. and Hammond, R.W. (1999). A cDNA probe detects *Prunus* necrotic ringspot virus in three peach cultivars after micrografting and in peach shots following long-term culture at 4 degrees C. *HortScience* **34**, 346–347.

Hewish, D.R., Shukla, D.D. and Gough, K.H. (1986). The use of biotin-conjugated antisera in immunoassays for plant viruses. *J. Virol. Methods* **13**, 79–85.

Hewitt, W.B. and Chiarappa, L. (eds) (1977). *Plant Health and Quarantine in International Transfer of Genetic Resources*. CRC Press, Boca Raton, FL.

Hewitt, W.B., Raski, D.J. and Goheen, A.C. (1958). Nematode vector of soilborne fan-leaf virus of grapevines. *Phytopathology* **48**, 586–595.

Heyraud, F., Matzeit, V., Kammann, M., Schaefer, S., Schell, J. and Gronenborn, B. (1993a). Identification of the initiation sequence for viral-strand DNA synthesis of wheat dwarf virus. *EMBO J.* **12**, 4445–4452.

Heyraud, F., Matzeit, V., Schaefer, S., Schell, J. and Gronenborn, B. (1993b). The conserved nonaucleotide motif of the geminivirus stem-loop sequence promotes replicational release of virus molecules from redundant copies. *Biochimie* **75**, 605–615.

Heywood, V.H. (1978). *Flowering Plants of the World*. Oxford University Press, London.

Hiatt, A. Cafferkey, R. and Bowdish, K. (1989). Production of antibodies in transgenic plants. *Nature (Lond.)* **342**, 76–78.

Hibi, T. and Saito, Y. (1985). A dot immunobinding assay for the detection of tobacco mosaic virus in infected tissues. *J. Gen. Virol.* **66**, 1191–1194.

Hibi, T., Rezelman, G. and van Kammen, A. (1975). Infection of cowpea mesophyll protoplasts with cowpea mosaic virus. *Virology* **64**, 308–318.

Hibino, H. and Cabauatan, P.Q. (1987). Infectivity neutralization of rice tungro-associated viruses acquired by vector leafhoppers. *Phytopathology* **77**, 473–476.

Hibino, H., Tiongco, E.R., Cabunagan, R.C. and Flores, Z.M. (1987). Resistance to rice tungro-associated virus in rice under experimental and natural conditions. *Phytopathology* **77**, 871–875.

Hidaka, S., Hanada, K., Ishikawa, K. and Miura, K.-I. (1988). Complete nucleotide sequence of two new satellite RNAs associated with cucumber mosaic virus. *Virology* **164**, 326–333.

Hiddinga, H.J., Crum, C.J., Hu, J. and Roth, D.A. (1988). Viroid-induced phosphorylation of a host protein related to a dsRNA-dependent protein kinase. *Science* **241**, 451–453.

Hiebert, E. and McDonald, J.G. (1973). Characterisation of some proteins associated with viruses in the potato Y group. *Virology* **56**, 349–361.

Hiebert, E., Bancroft, J.B. and Bracker, C.E. (1968). The assembly *in vitro* of some small spherical viruses, hybrid viruses and other nucleoproteins. *Virology* **34**, 492–508.

Higgins, C.M., Dietzgen, R.G., Akin, H.M., Sudarsono, Chen, K. and Xu, Z. (1999). Biological and molecular variability of peanut stripe potyvirus. *Curr. Topics Virol.* **1**, 1–26.

Higgins, T.J.V., Goodwin, P.B. and Whitfeld, P.R. (1976). Occurrence of short particles in beans infected with the cowpea strain of TMV: II. Evidence that short particles contain the cistron for coat-protein. *Virology* **71**, 486–497.

Higgins, T.J.V., Whitfeld, P.R. and Matthews, R.E.F. (1978). Size distribution and *in vitro* translation of the RNAs isolated from turnip yellow mosaic virus nucleoproteins. *Virology* **84**, 153–161.

Hilborn, M.T., Hyland, F. and McCrum, R.C. (1965). Pathological anatomy of apple trees affected by the stem-pitting virus. *Phytopathology* **55**, 34–39.

Hilf, M.E. and Dawson, W.O. (1993). The tobamovirus capsid protein functions as a host-specific determinant of long-distance movement. *Virology* **193**, 106–114.

Hilf, M.E., Karasev, A.V., Pappu, H.R., Gumpf, S.M., Niblett, C.L. and Garnsey, S.M. (1995). Characterization of citrus tristeza virus subgenomic RNAs in infected tissue. *Virology* **208**, 576–582.

Hill, R.E. and Hastie, N.D. (1987). Accelerated evolution in the reactive centre regions of serine protease inhibitors. *Nature (Lond.)* **326**, 96–99.

Hill, S.A. (1984). *Methods in Plant Virology*. Blackwell, Oxford.

Hill, S.A. (1987). Cereal virus diseases: contrasting experience. In: M.S. Wolfe and C.E. Caten (eds) *Populations of Plant Pathogens: Their Dynamics and Genetics*, pp. 149–159. Blackwell, Oxford.

Hill, S.A. and Torrance, L. (1989). Rhizomania disease of sugar beet in England. *Plant Pathol.* **38**, 114–122.

Hillman, B.I. and Nuss, D.L. (1999). Phytoreoviruses (*Reoviridae*). . In: A. Granoff and R.G. Webster (eds) *Encyclopedia of Virology*, 2nd edn, pp. 1262–1267. Academic Press, San Diego.

Hillman, B.I., Morris, T.J. and Schlegel, D.E. (1985). Effects of low-molecular-weight RNA and temperature on tomato bushy stunt virus symptom expression. *Phytopathology* **75**, 361–365.

Hillman, B.I., Carrington, J.C. and Morris, T.J. (1987). A defective interfering RNA that contains a mosaic of a plant virus genome. *Cell* **51**, 427–433.

Hills, G.J., Plaskitt, K.A., Young, N.D. *et al.* (1987). Immunogold localization of the intracellular sites of structural and nonstructural tobacco mosaic virus proteins. *Virology* **161**, 488–496.

Hinegardner, R. (1976). Evolution of genome size. In: F.J. Ayala (ed.) *Molecular Evolution*, pp. 179–199. Sinauer, Sunderland, MA.

Hiraga, S., Iti, H., Yamakawa, H. *et al.* (2000). An HR-induced tobacco peroxidase gene is responsive to spermine, but not to salicylate, methyl jasmonate and ethephon. *Mol. Plant Microbe Interact.* **13**, 210–216.

Hirikoshi, M., Nakayama, M., Yamaoka, N., Furusawa, I. and Shishiyama, J. (1987). Brome mosaic virus coat protein inhibits viral RNA synthesis *in vitro*. *Virology* **158**, 15–19.

Hirochika, H. and Hayashi, K. (1991). A new strategy to improve a cauliflower mosaic virus vector. *Gene* **105**, 239–241.

Hirochika, H., Takatsuji, H., Ubasawa, A. and Ikeda, J. (1985). Site-specific deletion in cauliflower mosaic virus DNA: possible involvement of RNA splicing and reverse transcription. *EMBO J.* **4**, 1673–1680.

Hiruki, C. and Tu, J.C. (1972). Light and electron microscopy of potato virus M lesions and marginal tissue in red kidney bean. *Phytopathology* **62**, 77–85.

Hizi, A., Henderson, L.E., Copeland, T.D., Sowder, R.C., Hixon, C.V. and Oroszlan, S. (1987). Characterisation of mouse mammary tumor virus gag-pro gene products and the ribosomal frameshift site by protein sequencing. *Proc. Natl Acad. Sci. USA* **84**, 7041–7045.

Hobbs, H.A., Kuhn, C.W., Papa, K.E. and Brantley, B.B. (1987). Inheritance of non necrotic resistance to southern bean mosaic virus in cowpea. *Phytopathology* **77**, 1624–1629.

Hofer, J.M.I., Dekker, E.L., Reynolds, H.V., Woolston, C.J., Cox, B.S and Mullineaux, P.M. (1992). Coordinate regulation of replication and virion sense gene expression in wheat dwarf virus. *Plant Cell* **4**, 213–223.

Hoffmann, K., Sackey, S.T., Maiss, E., Adomako, D. and Vetten, H.J. (1997). Immunocapture polymerase chain reaction for the detection and characterization of cacao swollen shoot virus 1A isolates. *J. Phytopathol.* **145**, 205–212.

Hogle, J.M., Maeda, A. and Harrison, S.C. (1986). Structure and assembly of turnip crinkle virus 1: X-ray crystallographic structure analysis at 3.2 Å resolution. *J. Mol. Biol.* **191**, 625–638.

Hogue, R. and Asselin, A. (1984). Polyacrylamide–agarose gel electrophoretic analysis of tobacco mosaic virus disassembly intermediates. *Can. J. Bot.* **62**, 2336–2339.

Hohn, T. (1999). Plant Pararetroviruses—Caulimoviruses: molecular biology. . In: A. Granoff and R.G. Webster (eds) *Encyclopedia of Virology*, 2nd edn, pp. 1281–1285.. Academic Press, San Diego.

Hohn, T. and Fütterer, J. (1992). Transcriptional and translational control of gene expression in cauliflower mosaic virus. *Curr. Opin. Genetic. Devel.* **2**, 90–96.

Hohn, T. and Fütterer, J. (1997). The proteins and functions of plant pararetroviruses: known and unknowns. *Crit. Rev. Plant Sci.* **16**, 133–161.

Hohn, T., Hohn, B. and Pfeiffer, P. (1985). Reverse transcription in CaMV. *Trends Biochem. Sci.* **10**, 205–209.

Holland, J. and Domingo, E. (1998). Origin and evolution of viruses. *Virus Genes* **16**, 13–21.

Holland, J., Spindler, K., Horodyski, F., Grabau, E., Nichol, S. and Vande Pol, S. (1982). Rapid evolution of RNA genomes. *Science* **215**, 1577–1585.

Hollings, M. (1965). Disease control through virus-free stock. *Annu. Rev. Phytopathol.* **3**, 367–396.

Hollings, M. and Stone, O.M. (1970). The long-term survival of some plant viruses preserved by lyophilization. *Ann. Appl. Biol.* **65**, 411–418.

Holmes, F.O. (1929). Local lesions in tobacco mosaic. *Bot. Gaz. (Chicago)* **87**, 39–55.

Holmes, F.O. (1931). Local lesions of mosaic in *Nicotiana tabacum* L. *Contrib. Boyce Thompson Inst.* **3**, 163–172.

Holmes, F.O. (1934). A masked strain of tobacco mosaic virus. *Phytopathology* **24**, 845–873.

Holmes, F.O. (1938). Inheritance of resistance to tobacco mosaic disease in tobacco. *Phytopathology* **28**, 553–561.

Holmes, F.O. (1939). *Handbook of Phytopathogenic Viruses*. Burgess, Minneapolis, MI.

Holmes, F.O. (1946). A comparison of experimental host ranges of tobacco-etch and tobacco mosaic viruses. *Phytopathology* **36**, 643–659.

Holmes, F.O. (1955). Elimination of spotted wilt virus from dahlias by propagation of tip cuttings. *Phytopathology* **45**, 224–226.

Holmes, F.O. (1964). Symptomology of viral diseases in plants. In: M.K. Corbett and H.D. Sisler (eds) *Plant Virology*, pp. 17–38. University of Florida Press, Gainesville.

Holness, C.L., Lomonossoff, G.P., Evans, D. and Maule, A.J. (1989). Identification of the initiation codons for translation of cowpea mosaic virus middle component RNA using site-directed mutagenesis of an infectious cDNA clone. *Virology* **172**, 311–320.

Holt, C.A. (1992). Localization of plant pathogens. In: P.D. Reid and R.F. Pont-Lezica (eds) *Tissue Printing. Tools for the Study of Anatomy, Histochemistry, and Gene Expression*, pp. 125–137. Academic Press, San Diego.

Holt, J. and Chancellor, T.C.B. (1997). A model of plant virus disease epidemics in asynchronously-planted cropping systems. *Plant Pathol.* **46**, 490–501.

Holt, J. Jeger, M.J., Thresh, J.M. and Otim-Nape, G.W. (1997). An epidemiological model incorporating vector population dynamics applied to African cassava mosaic virus disease. *J. Appl. Ecol.* **34**, 793–806.

Hong, Y and Stanley, J. (1995). Regulation of African cassava mosaic virus complementary-sense gene expression by N-terminal sequences of the replication-associated protein AC1. *J. Gen. Virol.* **76**, 2415–2422.

Hong, Y., Robinson, D.J. and Harrison, B.D. (1993). Nucleotide sequence evidence for the occurrence of three distinct whitefly-transmitted geminiviruses in cassava. *J. Gen. Virol.* **74**, 2437–2443.

Hong, Y., Levay, K., Murphy, J.F., Klein, P.G., Shaw, J.G. and Hunt, A.G. (1995). A potyvirus-encoded polymerase interacts with the viral coat protein and VPg in yeast cells. *Virology* **214**, 159–166.

Hong, Y., Saunders, K., Hartley, M.R., and Stanley, J. (1996). Resistance to geminivirus infection by virus-induced expression of dianthin in transgenic plants. *Virology* **220**, 119–127.

Hooft van Huijsduijnen, R.A.M., van Loon, L.C. and Bol, J.F. (1986). cDNA cloning of six mRNAs induced by TMV infection of tobacco and a characterization of their translation products. *EMBO J.* **5**, 2057–2061.

Horikoshi, M., Nakayama, M., Yamaoka, N., Furusawa, I. and Shishiyama, J. (1987). Brome mosaic virus coat protein inhibits viral RNA synthesis *in vitro*. *Virology* **158**, 15–19.

Horikoshi, M., Mise, I., Furusawa, I. And Shishiyama, J. (1988). Immunological analysis of brome mosaic virus replicase. *J. Gen. Virol.* **69**, 3081–3087.

Horne, R.W. (1985). The development and application of electron microscopy to the structure of isolated plant viruses. In: J.W. Davies (ed.) *Molecular Plant Virology*, Vol. 1, pp. 1–41. CRC Press, Boca Raton.

Horne, R.W. and Pasquali-Ronchetti, I. (1974). A negative staining-carbon film technique for studying viruses in the electron microscope: 1. Preparative procedures for examining icosahedral and filamentous viruses. *J. Ultrastruct. Res.* **47**, 381–383.

Horne, R.W. and Wildy, P. (1961). Symmetry in virus architecture. *Virology* **15**, 348–373.

Horne, R.W., Harnden, J.M. and Hull, R. (1977). The *in vitro* crystalline formations of turnip rosette virus: 1. Electron microscopy of two- and three-dimensional arrays. *Virology* **82**, 150–162.

Horst, R.K., Langhans, R.W. and Smith, S.H. (1977). Effects of chrysanthemum stunt, chlorotic mottle, aspermy, and mosaic on flowering and rooting of chrysanthemums. *Phytopathology* **67**, 9–14.

Horváth, G.V., Pettkó-Szandtner, A., Nikovics, K. *et al.* (1998). Prediction of functional regions of the maize streak virus replication-associated proteins by protein–protein interaction analysis. *Plant Mol. Biol.* **38**, 699–712.

Horváth, J. (1983). New artificial hosts and non-hosts of plant viruses and their role in the identification and separation of viruses: XVIII. Concluding remarks. *Acta Phylopathol. Acad. Sci. Hung.* **18**, 121–161.

Horváth, J. (1993). Host plants in diagnosis. In: R.E.F. Matthews (ed.) *Diagnosis of Plant Virus Diseases*, pp. 18–48. CRC Press, Boca Raton, FL.

Hou, Y.-M. and Gilbertson, R.L. (1996). Increased pathogenicity in a pseudorecombinants bipartite geminivirus correlates with intermolecular recombination. *J. Virol.* **70**, 5430–5436.

Hou, Y.-M., Sanders, R., Ursin, V.M. and Gilbertson, R.L. (2000). Transgenic plants expressing geminivirus movement protein: abnormal phenotypes and delayed infection by tomato mottle virus in transgenic tomatoes expressing bean dwarf virus BV1 or BC1 proteins. *Mol. Plant Microb. Interact.* **13**, 297–308.

Howarth, A.J. and Vandemark, G.J. (1989). Phylogeny of geminiviruses. *J. Gen. Virol.* **70**, 2717–2727.

Howe, G.A. and Ryan, C.A. (1999). Suppressors of systemin signaling identify gene in the tomato wound response pathway. *Genetics* **153**, 1411–1421.

Howell, S.H. (1984). Physical structure and genetic organisation of the genome of maize streak virus (Kenya isolate). *Nucl. Acids Res.* **12**, 7359–7375.

Howell, S.H., Walker, L.L. and Dudley, R.K. (1980). Cloned cauliflower mosaic virus DNA infects turnips (*Brassica rapa*). *Science* **208**, 1265–1267.

Howell, S.H., Walker, L.L. and Walden, R.M. (1981). Rescue of *in vitro* generated mutants of cloned cauliflower mosaic virus genome in infected plants. *Nature (London)* **293**, 483–486.

Howell, W.E. and Mink, G.I. (1971). The relationship between volunteer sugarbeets and occurrence of beet mosaic and beet western yellow viruses in Washington beet fields. *Plant Dis. Rep.* **55**, 676–678.

Howell, W.E. and Mink, G.I. (1988). Natural spread of cherry rugose mosaic disease and two prunus necrotic ringspot virus biotypes in a central Washington sweet cherry orchard. *Plant Dis.* **72**, 636–640.

Hsu, C.H., Sehgal, O.P. and Pickett, E.E. (1976). Stabilizing effect of divalent metal ions on virions of southern bean mosaic virus. *Virology* **69**, 587–595.

Hsu, H.T. (1978). Cell fusion induced by a plant virus. *Virology* **84**, 9–18.

Hsu, H.T. and Black, L.M. (1973). Comparative efficiencies of assays of a plant virus by lesions on leaves and on vector cell monolayers. *Virology* **52**, 284–286.

Hsu, H.T. and Black, L.M. (1974). Multiplication of potato yellow dwarf virus on vector cell monolayers. *Virology* **59**, 331–334.

Hsu, Y.-H. (1984). Immunogold for detection of antigen on nitrocellulose paper. *Anal. Biochem.* **142**, 221–225.

Hsu, Y.-H. and Brakke, M.K. (1985). Cell-free translation of soilborne wheat mosaic virus RNAs. *Virology* **143**, 272–279.

Hsu, Y.-H., Lee, Y.-S., Liu, J.-S. and Lin, N.-S. (1998). Differential interactions of bamboo mosaic potexvirus satellite RNAs, helper virus and host plants. *Mol. Plant Microb. Interact.* **12**, 1207–1213.

Hsu, Y.-H., Annamalai, P., Lin, C.S., Chen, Y.Y., Chang, W.C. and Lin, N.S. (2000). A sensitive method for detecting bamboo mosaic virus (BaMV) and establishment of BsaMV-free meristem-tip cultures. *Plant Pathol.* **49**, 101–107.

Hu, J.S., Wang, M., Sether, D., Xie, W. and Leonhardt, K.W. (1996). Use of polymerase chain reaction (PCR) to study transmission of banana bunchy top virus by the banana aphid (*Pentolonia nigronervosa*). *Ann. Appl. Biol.* **128**, 55–64.

Huang, M. and Zhang, L. (1999). Association of the movement protein of alfalfa mosaic virus with the endoplasmic reticulum and its trafficking in epidermal cells of onion bulb scales. *Mol. Plant–Microb. Interact.* **12**, 680–690.

Huang, Z., Han, Y. and Howell, S.H. (2000). Formation of surface tubules and fluorescent foci in *Arabidopsis thaliana* protoplasts expressing a fusion between the green fluorescent protein and cauliflower mosaic virus movement protein. *Virology* **271**, 58–64.

Huber, R. Rezelman, G., Hibit, T. and van Kammen, A. (1977). Cowpea mosaic virus infection of protoplasts from Samsun tobacco leaves. *J. Gen. Virol.* **34**, 315–323.

Hugdahl, J.D., Bokros, C.L. and Morejohn, L.C. (1995). End-to-end annealing of plant microtubules by the p86 subunit of eukaryotic initiation factor-(iso)4F. *Plant Cell* **7**, 2129–2138.

Hughes, G., Davies, J.W. and Wood, K.R. (1986). *In vitro* translation of the bipartite genomic RNA of pea early browning virus. *J. Gen. Virol.* **67**, 2125–2133.

Hughes, J.d'A. and Ollennu, L.A.A. (1994). Mild strain protection of cocoa in Ghana against cocoa swollen shoot virus—a review. *Plant Pathol.* **43**, 442–457.

Hughes, J.d'A. and Thomas, B.J. (1988). The use of protein A–sandwich ELISA as a means for quantifying serological relationships between members of the tobamovirus group. *Ann. Appl. Biol.* **112**, 117–126.

Hughes, R.K., Perbal, M.-C., Maule, A.J. and Hull, R. (1995). Evidence for proteolytic processing of tobacco mosaic virus movement protein in *Arabidopsis thaliana*. *Mol. Plant–Microb. Interact.* **5**, 658–665.

Hugouvieux, V., Barber, C.E. and Daniels, M.J. (1998). Entry of *Xanthomonas campestris* pv. *campestris* into hydathodes of *Arabidopsis thaliana* leaves: a system for studying early events in bacterial pathogenesis. *Mol. Plant Microb. Interact.* **11**, 537–543.

Huguenot, C., van den Dobbelsteen, G., de Haan, P. *et al.* (1990). Detection of tomato spotted wilt virus using monoclonal antibodies and riboprobes. *Arch. Virol.* **110**, 47–62.

Huiet, L., Klaassen, V.A., Tsai, J.H. and Falk, B.W. (1991). Nucleotide sequence and RNA hybridization analyses reveal an ambisense coding strategy for maize stripe virus RNA-3. *Virology* **182**, 47–53.

Huiet, L., Tsai, J.H. and Falk, B.W. (1992). Complete sequence of maize stripe virus RNA4 and mapping of its subgenomic RNAs. *J. Gen. Virol.* **73**, 1603–1607.

Huisman, M.J., Sarachu, A.N., Alblas, F. and Bol, J.F. (1985). Alfalfa mosaic virus temperature-sensitive mutants. II. Early functions encoded by RNAs 1 and 2. *Virology* **141**, 23–29.

Huisman, M.J., Sarachu, A.N., Alblas, F. Broxterman, H.J.G., van Vloten-Doting, L. and Bol, J.F. (1986). Alfalfa mosaic temperature sensitive mutants: III. Mutants with a putative defect in cell-to-cell transport. *Virology* **154**, 401–404.

Hull, R. (1964). Spread of groundnut rosette virus by *Aphis craccivora* (Koch). *Nature* **202**, 213–214.

Hull, R. (1965). Virus diseases of sweet peas in England. *Plant Pathol.* **14**, 150–153.

Hull, R. (1968). A virus disease of Hart's tongue fern. *Virology* **35**, 333–335.

Hull, R. (1969). Alfalfa mosaic virus. *Adv. Virus Res.* **15**, 365–433.

Hull, R. (1970a). Studies on alfalfa mosaic virus: IV. An unusual strain. *Virology* **42**, 283–292.

Hull, R. (1970b). Large RNA plant-infecting viruses. In: R.D. Barry and B.W.J. Mahy (eds) *The Biology of Large RNA Viruses*, pp. 153–164. Academic Press, New York.

Hull, R. (1970c). Studies on alfalfa mosaic virus: III. Reversible dissociation and reconstruction studies. *Virology* **40**, 34–47.

Hull, R. (1976a). The structure of tubular viruses. *Adv. Virus Res.* **20**, 1–32.

Hull, R. (1976b). The behavior of salt-labile plant viruses in gradients of cesium sulphate. *Virology* **75**, 18–25.

Hull, R. (1977a). The banding behaviour of the viruses of southern bean mosaic virus group in gradients of caesium sulphate. *Virology* **79**, 50–57.

Hull, R. (1977b). Particle differences related to aphid transmissibility of a plant virus. *J. Gen. Virol.* **34**, 183–187.

Hull, R. (1977c). The stabilization of the particles of turnip rosette virus and of other members of the southern bean mosaic virus group. *Virology* **79**, 58–66.

Hull, R. (1977d). Properties of an aphid-borne virus: pea enation mosaic virus. In: K.F. Harris and K. Maramorosch (eds) *Aphids as Virus Vectors*, pp. 137–162. Academic Press, New York.

Hull, R. (1978). The stabilization of the particles of turnip rosette virus: III. Divalent cations. *Virology* **89**, 418–422.

Hull, R. (1980). Structure of the cauliflower mosaic virus genome. III. Restriction endonuclease mapping of thirty three isolates. *Virology* **100**, 76–90.

Hull, R. (1985). Purification, biophysical and biochemical characterisation of viruses with especial reference to plant viruses. In B.W.J. Mahy (ed.) *Virology: A Practical Approach*, pp. 1–14. IRL Press, Oxford.

Hull, R. (1986). The potential for using dot-blot hydridisation in the detection of plant viruses. In: R.A.C. Jones and L. Torrance (eds) *Developments and Applications in Virus Testing*, Appl. Biol. I, pp. 3–12. Association of Applied Biologists, Wellesbourne.

Hull, R. (1989). The movement of viruses in plants. *Annu. Rev. Phytopathol.* **27**, 213–240.

Hull, R. (1990a). Non-conventional resistance to viruses in plants—concepts and risks. In: G.P. Gustafson (ed.) *Proceedings of the 19th Stadler Conference*, pp. 289–303. Plenum Press, New York.

Hull, R. (1990b). Virus resistant plants: potential and risks. *Chemistry and Industry* **17**, 543–546.

Hull, R. (1991a). The movement of viruses within plants. *Semin. Virol.* **2**, 89–95.

Hull, R. (1991b). Introduction: plant viral pathogenesis. *Semin. Virol.* **2**, 79–80.

Hull, R. (1992). Genome organization of retroviruses and retroelements: evolutionary considerations and implications. *Semin. Virol.* **3**, 373–382.

Hull, R. (1993). Nucleic acid hybridization procedures. In: R.E.F. Matthews (ed.) *Diagnosis of Plant Virus Diseases*, pp. 253–272. CRC Press, Boca Raton, FL.

Hull, R. (1994a). Molecular biology of plant virus–vector interactions. *Adv. Dis. Vector Res.* **10**, 361–386.

Hull, R. (1994b). Resistance to plant viruses: obtaining genes by non-conventional approaches. *Euphytica* **75**, 195–205.

Hull, R. (1996). Molecular biology of rice tungro viruses. *Annu. Rev. Phytopathol.* **34**, 275–297.

Hull, R. (1999a). Plant pararetroviruses.– rice tungro bacilliform virus. . In: A. Granoff and R.G. Webster (eds) *Encyclopedia of Virology*, 2nd edn, pp. 1292–1296. Academic Press, San Diego.

Hull, R. (1999b). Classification of reverse transcribing elements: a discussion document. *Arch. Virol.* **144**, 209–214.

Hull, R. (2000). Virus-detection—nucleic acid hybridization. In: O.C. Maloy and T.D. Murray (eds) *Encyclopedia of Plant Pathology*, pp. 1092–1095. John Wiley, New York.

Hull, R. and Covey, S.N. (1983a). Characterization of cauliflower mosaic virus DNA forms isolated from infected turnip leaves. *Nucl. Acids Res.* **11**, 1881–1895.

Hull, R. and Covey, S.N. (1983b). Does cauliflower mosaic virus replicate by reverse transcription? *Trends Biochem. Sci.* **8**, 119–121.

Hull, R. and Covey, S.N. (1996). Retroelements: propagation and adaptation. *Virus Genes* **11**, 105–118.

Hull, R. and Davies, J.W. (1992). Approaches to nonconventional control of plant virus diseases. *Crit. Rev. Plant Sci.* **11**, 17–33.

Hull, R. and Donson, J. (1982). Physical mapping of the DNAs of carnation etched ring and figwort mosaic viruses. *J. Gen. Virol.* **60**, 125–134.

Hull, R. and Johnson, M.W. (1968). The precipitation of alfalfa mosaic virus by magnesium. *Virology* **34**, 388–390.

Hull, R. and Lane, L.C. (1973). The unusual nature of the components of a strain of pea enation mosaic virus. *Virology* **55**, 1–13.

Hull, R. and Plaskitt, A. (1970). Electron microscopy on the behaviour of two strains of alfalfa mosaic virus in mixed infections. *Virology* **42**, 773–776.

Hull, R. and Plaskitt, A. (1974). The *in vivo* behaviour of broad bean wilt virus and three of its strains. *Intervirology* **2**, 352–359.

Hull, R., Rees, M. and Short, M.N. (1969a). Studies on alfalfa mosaic virus: I. The protein and nucleic acid. *Virology* **37**, 404–415.

Hull, R., Hills, G.J. and Markham, R. (1969b). Studies on alfalfa mosaic virus: II. The structure of the virus components. *Virology* **37**, 416–428.

Hull, R., Hills, G.J. and Plaskitt, A. (1970). The *in vivo* behaviour of twenty-four strains of alfalfa mosaic virus. *Virology* **42**, 753–772.

Hull, R., Shepherd, R.J. and Harvey, J.D. (1976). Cauliflower mosaic virus: an improved purification procedure and some properties of the virus particles. *J. Gen. Virol.* **31**, 93–100.

Hull, R., Sadler, J. and Longstaff, M. (1986). The sequence of carnation etched ring virus DNA: comparison with cauliflower mosaic virus and retroviruses. *EMBO J.* **5**, 3083–3090.

Hull, R., Covey, S.N. and Maule, A.J. (1987). Structure and replication of caulimovirus genomes. *J. Cell Sci., Suppl.* **7**, 213–229.

Hull, R., Brown, F. and Payne, C. (1989). *Virology: Directory and Dictionary of Animal Bacterial and Plant Viruses*. Macmillan, London.

Hull, R., Milne, R.G. and van Regenmortel, M.H.V. (1991). A list of proposed standard acronyms for plant viruses and viroids. *Arch. Virol.* **120**, 151–164.

Hull, R., Covey, S.N. and Dale, P. (2000a). Genetically modified plants and the 35S promoter: assessing the risks and enhancing the debate. *Microb. Ecol. Health and Dis.* **12**, 1–5

Hull, R., Lockhart, B. and Harper, G. (2000b). Viral sequences integrated into plant genomes. *Trends Plant Sci.* **5**, 362–365.

Hunt, R.E., Nault, L.R. and Gingery, R.E. (1988). Evidence for infectivity of maize chlorotic dwarf virus and for a helper component in its leafhopper transmission. *Phytopathology* **78**, 499–504.

Hunter, T., Jackson, R. and Zimmern, D. (1983). Multiple proteins and subgenomic mRNAs may be derived from a single open reading frame on tobacco mosaic virus RNA. *Nucl. Acids Res.* **11**, 801–821.

Hunter, W.B., Ullman, D.E. and Moore, A. (1994). Electronic monitoring: Characterizing the feeding behavior of western flower thrips (Thysanoptera: Thripae). In: M.M. Ellsbury, E.A. Backus and D.E. Ullman (eds) *History, Development and Application of AC Electronic Feeding Monitors*, pp. 73–85. Entomological Society of America, Lanham, Maryland.

Hunter, W.B., Hiebert, E., Webb, E., Tsai, J.H. and Polston, J.E. (1998). Location of geminiviruses in the whitefly *Bemisia tabaci* (Homoptera: Aleyrodidae). *Plant Dis.* **82**, 1147–1151.

Huntley, C.C. and Hall, T.C. (1996). Interference with brome mosaic virus replication in transgenic rice. *Mol. Plant Microb. Interact.* **9**, 164–170.

Huse, W.D., Sastry, L., Iverson, S.A. *et al.* (1989). Generation of a large combinatorial library of the immunoglobulin repertoire in phage lambda. *Science* **246**, 1275–1281.

Huss, B., Muller, S., Sommermeyer, G., Walter, B. and van Regenmortel, M.H.V. (1987). Grapevine fanleaf virus monoclonal antibodies: their use to distinguish different isolates. *J. Phytopathol.* **119**, 358–370.

Huston, J.S., Levinson, D., Mudgett-Hunter, M. *et al.* (1988). Protein engineering of antibody binding sites: recovery of specific activity in an anti-digoxin single-chain Fv analogue produced in *Escherichia coli*. *Proc. Natl Acad. Sci. USA* **85**, 5879–5883.

Hutchins, C.J., Keese, P., Visvader, J.E., Rathjen, P.D., McInnes, J.L. and Symons, R.H. (1985). Comparison of multimeric plus and minus forms of viroids and virusoids. *Plant Mol. Biol.* **4**, 293–304.

Hutchins, C.J., Rathgen, P.D., Forster, A.C. and Symons, R.H. (1986). Self-cleavage of plus and minus transcripts of avocado sunblotch viroid. *Nucl. Acids Res.* **14**, 3627–3640.

Huxley, H.E. and Zubay, G. (1960). The structure of the protein shell of turnip yellow mosaic virus. *J. Mol. Biol.* **2**, 189–196.

Hyman, B.C. (1996). Molecular systematic and population biology of phytonematodes: some unifying principles. *Fund. Appl. Nematol.* **19**, 309–313.

Iizuka, N., Chen, C., Yang, Q., Johannes, G. and Sarnow, P. (1995). Cap-independent translation and internal initiation of translation in eukaryotic cellular mRNA molecules. *Curr. Topics Microbiol. Immunol.* **203**, 155–177.

Ikegami, M. and Francki, R.I.B. (1973). Presence of antibodies to double-stranded RNA in sera of rabbits immunized with rice dwarf and maize rough dwarf viruses. *Virology* **56**, 404–406.

Ikegami, M. and Francki, R.I.B. (1975). Some properties of RNA from Fiji disease subviral particles. *Virology* **64**, 464–470.

Inglesias, V.A. and Meins, F. (2000). Movement of plant viruses is delayed in a β-1,3-glucanose-deficient mutant showing a reduced plasmodesmatal size exclusion limit and enhanced callose deposition. *Plant J.* **21**, 157–166.

Innes, R.W. (1998). Genetic dissection of *R* gene signal transduction pathways. *Curr. Opin. Plant Biol.* **1**, 229–304.

Innis, M.A., Gelfand, D.H., Sninsky, J.J. and White, T.J. (eds) (1990). *PCR Protocols: A Guide to Methods and Applications*. Academic Press, San Diego.

Inoue, H. and Timmins, P.A. (1985). The structure of rice dwarf virus determined by small-angle neutron scattering measurements. *Virology* **147**, 214–216.

Irvine, R.F. and Osborne, D.J. (1973). The effect of ethylene on [1B¹⁴C] glycerol incorporation into phospholipids of etiolated pea stems. *Biochem. J.* **136**, 1133–1135.

Irwin, M.E. (1999). Implications of movement in developing and deploying integrated pest management strategies. *Agric. Forest Meteorol.* **97**, 235–248.

Irwin, M.E. and Nault, L.R. (1996). Virus/vector control. In: G.J. Persley (ed.) *Biotechnology and Integrated Pest Management*, pp. 304–322. CAB International, London.

Irwin, M.E. and Ruesink, W.G. (1986). Vector intensity: a product of propensity and activity. In: G.D. McClean, R.G. Garrett and W.G. Ruesink (eds) *Plant Virus Epidemics*, pp. 13–33. Academic Press, Orlando, FL.

Irwin, M.E. and Thresh, J.M. (1990). Epidemiology of barley yellow dwarf: a study in ecological complexity. *Annu. Rev Phytopathol.* **28**, 393–424.

Ishihama, A. and Barbier, P. (1994). Molecular anatomy of viral RNA-directed RNA polymerases. *Arch. Virol.* **134**, 235–258.

Ishiie, T., Doi, Y., Yora, K. and Asuyama, H. (1967). Suppressive effects of antibiotics of tetracycline group on symptom development of mulberry dwarf disease. *Ann. Phytopathol. Soc. Jpn* **33**, 267–275.

Ishikawa, M., Meshi, T., Ohno, T. *et al.* (1984). A revised replication cycle for viroids: the role of longer than unit length RNA in viroid replication. *Mol. Gen. Genet.* **196**, 421–428.

Ishikawa, M., Meshi, T., Motoyoshi, F., Takamatsu, N. and Okada, Y. (1986). *In vitro* mutagenesis of the putative replicase genes of tobacco mosaic virus. *Nucl. Acids Res.* **14**, 8291–8305.

Ishikawa, M., Meshi, T., Watanabe, Y. and Okada, Y. (1988). Replication of chimeric tobacco mosaic viruses which carry heterologous combinations of replicase genes and 3' noncoding regions. *Virology* **164**, 290–293.

Ishikawa, M., Meshi, T., Ohno, T. and Okada, Y. (1991a). Specific cessation of minus-strand RNA accumulation at an early stage of tobacco mosaic virus infection. *J. Virol.* **65**, 861–868.

Ishikawa, M., Kroner, P., Ahlquist, P. and Meshi, T. (1991b). Biological activities of hybrid RNAs generated by 3'-end exchanges between tobacco mosaic and brome mosaic viruses. *J. Virol.* **65**, 3451–3459.

Ishikawa, M., Obata, F., Kumagai, T. and Ohno, T. (1991c). Isolation of mutants of *Arabidopsis thaliana* in which accumulation of tobacco mosaic virus coat protein is reduced to low levels. *Mol. Gen. Genet.* **230**, 33–38.

Ishikawa, M., Naito, S and Ohno, T. (1993). Effects of the *tom1* mutation of *Arabidopsis thaliana* on the multiplication of tobacco mosaic virus RNA in protoplasts. *J. Virol.* **67**, 5328–5338.

Ishikawa, M., Janda, M., Krol, M.A. and Ahlquist, P. (1997a). In vivo DNA expression of functional brome mosaic virus RNA replicons in *Saccharomyces cerevisiae. J. Virol.* **71**, 7781–7790.

Ishikawa, M., Diez, J., Restrepo-Hartwig, M. and Ahlquist, P. (1997b). Yeast mutations in multiple complementation groups inhibit brome mosaic virus RNA replication and transcription and perturb regulated expression of the viral polymerase-like gene. *Proc. Natl Acad. Sci. USA* **94**, 13810–13815.

Ishikawa, M., Janda, M. and Ahlquist, P. (2000). The 3a cell-to-cell movement gene is dispensable for cell-to-cell transmission of brome mosaic virus RNA replicons in yeast but retained over $10^{4\text{-}5}$-fold amplification. *J. Gen. Virol.* **81**, 2307–2311.

Ismail, I.D. and Milner, J.J. (1988). Isolation of defective interfering particles of Sonchus yellow net virus from chronically infected plants. *J. Gen. Virol.* **69**, 999–1006.

Ismail, I.D., Hamilton, I.D., Robertson, E. and Milner, J.J. (1987). Movement and intracellular location of Sonchus yellow net virus within infected *Nicotiana edwardsonii. J. Gen. Virol.* **68**, 2429–2438.

Isnard, M., Granier, M., Frutos, R., Reynaud, B. and Peterschmidtt, M. (1998). Quasispecies nature of three maize streak virus isolates obtained through different modes of selection from a population used to assess response to infection of maize cultivars. *J. Gen. Virol.* **79**, 3091–3099.

Itaya, A., Hickman, H., Bao, Y., Nelson, R. and Ding, B. (1997). Cell-to-cell trafficking of cucumber mosaic virus movement protein:green fluorescent protein fusion produced by biolistic gene bombardment in tobacco. *Plant J.* **12**, 1223–1230.

Itaya, A., Woo, Y.-M., Matsuta, C., Bao, Y., Nelson, R. and Ding, B. (1998). Development regulation of intercellular protein trafficking through plasmodesmata in tobacco leaf epidermis. *Plant Physiol.* **118**, 373–385.

Ivanov, P.A., Karpova, O.V., Skulachev, M.V., Tomashevskaya, O.L., Rodionova, N.P., Dorokhov, Yu.L. and Atabekov, J.G. (1997). A tobamovirus genome that contains an internal ribosome entry site functional *in vitro. Virology* **232**, 32–43.

Iwanowski, D. (1892). Ueber die Mosaikkrankheit der Tabakspflanze. *Bull. Acad. Imp. Sci. St.-Petersbourg [N.W.]* **3**, 65–70.

Izaguirre-Mayoral, M.L., de Uzcategui, R.C. and Carballo, O. (1993). Crassulacean acid metabolism in two species of orchids infected by tobacco mosaic virus-orchid strain and/or cymbidium mosaic virus. *J. Phytopathol.* **137**, 272–282.

Izaguirre-Mayoral, M.L., Carballo, O., Demallorca, M.S., Marys, E. and Gil, F. (1994). Symbiotic nitrogen-fixation and physiological performance of bean (*Phaseolus vulgaris* L.) plants as affected by *Rhizobium*-inoculum position and bean rugose mosaic virus infection. *J. Exp. Bot.* **45**, 373–383.

Jacks, T., Madhani, H.D., Masiarz, F.R. and Varmus, H.E. (1988). Signals for ribosomal frameshifting in Rous sarcoma virus gag-pol region. *Cell* **55**, 447–458.

Jackson, A.O. (1978). Partial characterization of the structural proteins of sonchus yellow net virus. *Virology* **87**, 172–181.

Jackson, A.O., Mitchell, D.M. and Siegel, A. (1971). Replication of tobacco mosaic virus: I. Isolation and characterization of double-stranded forms of ribonucleic acid. *Virology* **45**, 182–191.

Jackson, A.O., Zaitlin, M., Siegel, A. and Francki, R.I.B. (1972). Replication of tobacco mosaic virus: III. Viral RNA metabolism in separated leaf cells. *Virology* **48**, 655–665.

Jackson, A.O., Dawson, J.R.O., Covey, S.N. *et al.* (1983). Sequence relations and coding properties of a subgenomic RNA isolated from barley stripe mosaic virus. *Virology* **127**, 37–44.

Jackson, A.O., Francki, R.I.B. and Zuidema, D. (1987). Biology structure and replication of plant rhabdoviruses. In: R.R. Wagner (ed.) *The Rhabdoviruses*, pp. 427–508. Plenum, New York.

Jackson, A.O., Goodin, M., Moreno, I., Johnson, J. and Lawrence, D.M. (1999). Rhabdoviruses (*Rhabdoviridae*): plant rhabdoviruses. In: A. Granoff and R.G. Webster (eds) *Encyclopedia of Virology*, pp. 1531–1541. Academic Press, San Diego.

Jacobi, V., Castello, J.D. and Flachmann, M. (1992). Isolation of tomato mosaic virus from red spruce. *Plant Dis.* **76**, 518–522.

Jacobsen, B.J. (1997). Role of plant pathology in integrated pest management. *Annu. Rev. Phytopathol.* **35**, 373–391.

Jacquemond, M. (1982). Phenomena of interferences between the two types of satellite RNA of cucumber mosaic virus. Protection of tomato plants against lethal necrosis. *C. R. Seances Acad. Sci. Ser. 3* **294**, 991–994.

Jacquemond, M. and Tepfer, M. (1998). Breeding for resistance to plant viruses. In: A. Hadidi, R.K. Khetarpal and H. Koganezawa (eds) *Plant Virus Disease Control*, pp. 94–120. APS Press, St. Paul, MN.

Jacquemond, M., Amselem, J. and Tepfer, M. (1988). A gene coding for a monomeric form of cucumber mosaic virus satellite RNA confers tolerance to CMV. *Mol. Plant Microbe Interact.* **1**, 311–316.

Jacquet, C., Delecolle, B., Raccah, B., Lecoq, H., Dunez, J. and Ravelonandro, M. (1998). Use of modified plum pox virus coat protein genes developed to limit heteroencapsidation-associated risks in transgenic plants. *J. Gen. Virol.* **79**, 1509–1517.

Jacrot, B. (1975). Studies on the assembly of a spherical plant virus: II. The mechanism of protein aggregation and virus swelling. *J. Mol. Biol.* **95**, 433–446.

Jacrot, B., Pfeiffer, P. and Witz, J. (1976). The structure of a spherical plant virus (bromegrass mosaic virus) established by neutron diffraction. *Phil. Trans. R. Soc. Lond. B* **276**, 109–112.

Jacrot, B., Chauvin, C. and Witz, J. (1977). Comparative neutron small-angle scattering study of small spherical RNA viruses. *Nature (London)* **266**, 417–421.

Jadot, R. and Roland, G. (1971). Observations sur les deplacements des aphides à partir des plantes adventices marquées dans un champ de betteraves. *Meded. Fac. Landbouwwet., Rijksuniv. Gent* **36**, 940–944.

Jaegle, M. and van Regenmortel, M.H.V. (1985). Use of ELISA for measuring the extent of serological cross-reactivity between plant viruses. *J. Virol. Methods* **11**, 189–198.

Jaegle, M., Wellink, J. and Goldbach, R. (1987). The genome-linked protein of cowpea mosaic virus is bound to the 5– terminus of virus RNA by a phosphodiester linkage to serine. *J. Gen. Virol.* **68**, 627–632.

Jaegle, M., Briand, J.P., Burckard, J. and van Regenmortel, M.H.V. (1988). Accessibility of three continuous epitopes in tomato bushy stunt virus. *Ann. Inst. Pasteur Virol.* **139**, 39–50.

Jaenicke, R. and Lauffer, M.A. (1969). Determination of hydration and partial specific volume of proteins with the spring balance. *Biochemistry* **8**, 3077–3082.

Jagadish, M.N., Ward, C.W., Gough, K.H., Tulloch, P.A., Whittaker, L.A. and Shukla, D.D. (1991). Expression of potyvirus coat protein in *Escherichia coli* and yeast and its assembly into virus-like particles. *J. Gen. Virol.* **72**, 1543–1550.

Jagadish, M.N., Huang, D.X. and Ward, C.W. (1993). Site-directed mutagenesis of a potyvirus coat protein and its assembly in *Escherichia coli*. *J. Gen. Virol.* **74**, 893–896.

Jagadish, M.N., Edwards, S.J., Hayden, M.B., Grusovin, J., Vandenberg, K., Schoofs, P., Hamilton, R.V., Shukla, D.D., Kalins, H., McNamara, M., Haynes, J., Nisbet, I.T., Ward, C.W. and Pye, D. (1996). Chimeric potyvirus-like particles as vaccine carriers. *Intervirology* **39**, 85–92.

Jagus, R. (1987a). Translation in cell-free systems. In: S.L. Berger and A.R. Kimmel (eds) *Methods in Enzymology*, Vol. 152, pp. 267–296. Academic Press, San Diego.

Jagus, R. (1987b). Characterisation of *in vitro* translation products. In: S.L. Berger and A.R. Kimmel (eds) *Methods in Enzymology*, Vol. 152, pp. 296–304. Academic Press, San Diego.

Jain, R.K., Pappu, S.S., Pappu, H.R., Culbreath, A.K. and Todd, J.W. (1998). Molecular diagnosis of tomato spotted wilt tospovirus infection of peanut and other field and greenhouse crops. *Plant Dis.* **82**, 900–904.

Jakab, G., Kiss, T. and Solymosy, F. (1986). Viroid pathogenicity and pre-rRNA processing: a model amenable to experimental testing. *Biochim. Biophys. Acta* **868**, 190–197.

Jakowitsch, J., Mette, M.F., van der Winden, J., Matzke, M,A. and Matzke, A.J.M. (1999). Integrated pararetroviral sequences define a unique class of dispersed repetitive DNA in plants. *Proc. Natl Acad. Sci. USA* **96**, 13241–13246.

James, W.C. (1974). Assessment of plant diseases and losses. *Annu. Rev. Phytopathol.* **12**, 27–48.

Jan, F.-J., Fagoaga, C., Pang, S.-Z. and Gonsalves, D. (2000a). A single chimeric transgene derived from two distinct viruses confers multi-virus resistance in transgenic plants through homology-dependent gene silencing. *J. Gen. Virol.* **81**, 2103–2109.

Jan, F.-J., Pang, S.-Z., Tricoli, D.M. and Gonsalves, D. (2000b). Evidence for resistance in squash mosaic comovirus coat protein-transgenic plants is affected by developmental stage and enhanced by combination of transgenes from different lines. *J. Gen. Virol.* **81**, 2299–2306.

Janda, M. and Ahlquist, P. (1993). RNA-dependent replication, transcription and persistence of brome mosaic virus RNA replicons in *S. cerevisiae*. *Cell* **72**, 961–970.

Janda, M. and Ahlquist, P. (1998). Brome mosaic virus RNA replication protein 1a dramatically increases *in vivo* stability but not translation of viral genomic RNA3. *Proc. Natl Acad. Sci. USA* **95**, 2227–2232.

Janda, M., French, R. and Ahlquist, P. (1987). High efficiency T7 polymerase synthesis of infectious RNA from cloned brome mosaic virus cDNA and effects of 5' extensions of transcript infectivity. *Virology* **158**, 259–262.

Janssen, J.A.M., Tjallingii, W.F. and van Lenteren, J.C. (1989). Electrical recording and ultrastructure of stylet penetration by the greenhouse whitefly. *Entomol. Exp. Appl.* **52**, 69–81.

Jardetzky, O., Akasaka, K., Vogel, D., Morris, S. and Holmes, K.C. (1978). Unusual segmental flexibility in a region of tobacco mosaic virus coat protein. *Nature (London)* **273**, 564–566.

Jaspars, E.M.J. (1985). Interaction of alfalfa mosaic virus nucleic acid and protein. In: J.W. Davies (ed.) *Plant Molecular Biology*, Vol. 1, pp. 155–221. CRC Press, Boca Raton, FL.

Jaspars, E.M.J. (1998). A core promoter hairpin is essential for subgenomic RNA synthesis in alfalfa mosaic alfamovirus and is conserved in other *Bromoviridae*. *Virus Genes* **17**, 233–242.

Jedlinski, H. (1964). Initial infection processes by certain mechanically transmitted plant viruses. *Virology* **22**, 331–341.

Jeger, M.J. and Chan, M.S. (1995). Theoretical aspects of epidemics—uses of analytical models to make strategic management decisions. *Can. J. Plant Pathol.* **17**, 109–114.

Jeger, M.J., van den Bosch, F., Madden, L.V. and Holt, J. (1998). A model for analyzing plant–virus transmission characteristics and the epidemic development. *J. Math. Appl. Med. Biol.* **15**, 1–18.

Jelkmann, W. (1994). Nucleotide sequence of apple stem pitting virus and of the coat protein gene of a similar virus from pear associated with vein yellows disease and their relationship with potex- and carlaviruses. *J. Gen. Virol.* **75**, 1535–1542.

Jelkmann, W., Maiss, E. and Martin, R.R. (1992). The nucleotide sequence and genome organization of strawberry mild yellow edge-associated potexvirus. *J. Gen. Virol.* **73**, 475–479.

Jelkmann, W., Fetchner, B. and Agronovsky, A.A. (1997). Complete genome structure and phylogenetic analysis of little cherry virus, a mealybug-transmissible closterovirus. *J. Gen. Virol.* **78**, 2067–2071.

Jeng, T.-W., Crowther, R.A., Stubbs, G. and Chiu, W. (1989). Visualization of alpha-helices in tobacco mosaic virus by cryoprotection microscopy. *J. Mol. Biol.* **205**, 251–257.

Jennings, D.L. and Jones, A.T. (1986). Immunity from raspberry vein chlorosis virus in raspberry and its potential for control of the virus through plant breeding. *Ann. Appl. Biol.* **108**, 417–422.

Jenns, A.E. and Kuc, J. (1980). Characteristics of anthracnose resistance induced by localized infection of cucumber with tobacco necrosis virus. *Physiol. Plant Pathol.* **17**, 81–91.

Jensen, S.G. (1973). Systemic movement of barley yellow dwarf virus in small grains. *Phytopathology* **63**, 854–856.

Jobling, S.A. and Gehrke, L. (1987). Enhanced translation of chimeric messenger RNAs containing a plant viral untranslated leader sequence. *Nature (London)* **325**, 622–625.

Jockusch, H. (1964). *In vivo-* and *in vitro-*Verhalten temperature sensitiver Mutanten des tabakmosaikvirus. *Z. Vererbungsl.* **95**, 379–382.

Joelson, T., Åkerblom, L., Oxelfelt, P., Strandberg, B., Tomenius, K. and Morris, T.J. (1997). Presentation of a foreign peptide on the surface of tomato bushy stunt virus. *J. Gen. Virol.* **78**, 1213–1217.

Johansen, E., Edwards, M.C. and Hampton, R.O. (1994). Seed transmission of viruses: current perspectives. *Annu. Rev. Phytopathol.* **32**, 363–386.

Johansen, E., Dougherty, W.G., Keller, K.E., Wang, D. and Hampton, R.O. (1996). Multiple viral determinants affect seed transmission of pea seedborne mosaic virus in *Pisum sativum*. *J. Gen. Virol.* **77**, 3149–3154.

John, V.T. (1965). A micro-immuno-osmophoretic technique for assay of tobacco mosaic virus. *Virology* **27**, 121–123.

John, V.T. and Weintraub, M. (1966). Symptoms resembling virus infection induced by high temperature in *Nicotiana glutinosa*. *Phytopathology* **56**, 502–506.

Johnson, C.S. and Main, C.E. (1983). Yield/quality tradeoffs of tobacco mosaic virus-resistant tobacco cultivars in relation to disease management. *Plant Dis.* **67**, 886–890.

Johnson, D.A., Gautsch, J.W., Sportsman, J.R. and Elder, J.H. (1984). Improved technique utilizing nonfat dry milk for analysis of proteins and nucleic acids transferred to nitrocellulose. *Gene Anal. Tech.* **1**, 3–8.

Johnson, E.M. and Valleau, W.D. (1946). Field strains of tobacco mosaic virus. *Phytopathology* **36**, 112–116.

Johnson, J. (1937). Factors relating to the control of ordinary tobacco mosaic virus. *J. Agr. Res.* **54**, 239–273.

Johnson, J. (1942). Translations of: 1. Concerning the mosaic disease of tobacco: Adolf Mayer; 2. Concerning the mosaic disease of the tobacco plant: Dmitrii Ivanowski; 3. Concerning a contagium vivum fluidum as cause of the spot disease of tobacco leaves: Martinus W. Beijerinck; 4. On the etiology of infectious variagation: Erwin Baur. *Phytopathology Classics Number 7.*

Johnson, J. (1947). Virus attenuation and mutation. *Phytopathology* **37**, 12.

Johnson, J. and Hoggan, I.A. (1935). A descriptive key for plant viruses. *Phytopathology* **25**, 328–343.

Johnson, J. and Ogden, W.B. (1929). The overwintering of tobacco mosaic virus. *Wis. Agric. Exp. Stn. Bull.* **95**, 1–25.

Johnson, J., Lin, T. and Lomonossoff, G. (1997). Presentation of heterologous peptides on plant viruses: Genetics, structure and function. *Annu. Rev. Phytopathol.* **35**, 67–86.

Johnson, J.E. and Speir, J.A. (1997). Quasi-equivalent viruses: a paradigm for protein assemblies. *J. Mol. Biol.* **269**, 665–675.

Johnson, J.E. and Spier, J.A. (1999). Principles of virus structure. In: A. Granoff and R.G. Webster (eds) *Encyclopedia of Virology*, pp. 1946–1956. Academic Press, San Diego.

Johnson, M.W. (1964). The binding of metal ions by turnip yellow mosaic virus. *Virology* **24**, 26–35.

Johnson, M.W. and Markham, R. (1962). Nature of the polyamine in plant viruses. *Virology* **17**, 276–281.

Johnston, J.C. and Rochon, D.M. (1995). Deletion analysis of the promoter for cucumber necrosis virus 0.9-kb subgenomic RNA. *Virology* **214**, 100–109.

Johnstone, G.R., Koen, T.B. and Conley, H.L. (1982). Incidence of yellows in sugar beet as affected by variation in plant density and arrangement. *Bull. Entomol. Res.* **72**, 289–294.

Johnstone, G.R., Ashby, J.W., Gibbs, A.J., Duffus, J.E., Thottappilly, G. and Fletcher, J.D. (1984). The host ranges, classification and identification of eight persistent aphid-transmitted viruses causing disease in legumes. *Neth. J. Plant Pathol.* **90**, 225–245.

Joklik, W.K. (1999). Reoviruses (*Reoviridae*): molecular biology. In: A. Granoff and R.G. Webster (eds) *Encyclopedia of Virology*, 2nd edn, pp. 1464–1471. Academic Press, San Diego.

Jonard, G., Richards, K.E., Guilley, H. and Hirth, L. (1977). Sequence from the assembly nucleation region of TMV RNA. *Cell* **11**, 483–493.

Jones, A.T. (1987). Control of virus infection in crop plants through vector resistance: a review of achievements, prospects and problems. *Ann. Appl. Biol.* **111**, 745–772.

Jones, A.T. (1998). Control of virus infection in crops through breeding plants for vector resistance. In: A. Hadidi, R.K. Khetarpal and H. Koganezawa (eds) *Plant Virus Disease Control*, pp. 41–55. APS Press, St. Paul, MN.

Jones, A.T. and Mayo, M.A. (1984). Satellite nature of the viroid-like RNA2 of *Solanum nodiflorum* mottle virus and the ability of other plant viruses to support the replication of viroid-like RNA molecules. *J. Gen. Virol.* **65**, 1713–1721.

Jones, A.T. and Mitchell, M.J. (1986). Propagation of black raspberry necrosis virus (BRNV) in mixed culture with *Solanum nodiflorum* mottle virus and the production and use of BRNV antiserum. *Ann. Appl. Biol.* **109**, 323–336.

Jones, A.T. and Mitchell, M.J. (1987). Oxidising activity in root extracts from plants inoculated with virus or buffer that interferes with ELISA when using the substrate 3,3′, 5,5′ tetramethyl-benzidine. *Ann. Appl. Biol.* **111**, 359–364.

Jones, A.T., Mayo, M.A. and Duncan, G.H. (1983). Satellite-like properties of small circular RNA molecules in particles of lucerne transient streak virus. *J. Gen. Virol.* **64**, 1167–1173.

Jones, G.E. and Dawson, W.O. (1978). Stability of mutations conferring temperature sensitivity on tobacco mosaic virus. *Intervirology* **9**, 149–155.

Jones, I.M. and Reichmann, M.E. (1973). The proteins synthesized in tobacco leaves infected with tobacco necrosis virus and satellite tobacco necrosis virus. *Virology* **52**, 49–56.

Jones, L., Hamilton, A.J., Voinnet, O., Thomas, C.L., Maule, A.J. and Baulcombe, D.C. (1999). RNA–DNA interactions and DNA methylation in post-transcriptional gene silencing. *Plant Cell* **11**, 2291–2301.

Jones, M.C., Gough, K., Dasgupta, I. *et al.* (1991). Rice tungro disease is caused by an RNA and a DNA virus. *J. Gen. Virol.* **72**, 757–761.

Jones, R.A.C. (1975). Systemic movement of potato mop-top virus in tobacco may occur through the xylem. *Phytopathol. Z.* **82**, 352–355.

Jones, R.A.C. (1982). Breakdown of potato virus X resistance gene Nx: selection of a group four strain from strain group three. *Plant Pathol.* **31**, 325–331.

Jones, R.A.C. (1985). Further studies on resistance-breaking strains of potato virus X. *Plant Pathol.* **34**, 182–189.

Jones, R.A.C. and Fribourg, C.E. (1977). Beetle, contact and potato true seed transmission of Andean potato latent virus. *Ann. Appl. Biol.* **86**, 123–128.

Jones, R.A.C. and Harrison, B.D. (1972). Ecological studies on potato mop-top virus in Scotland. *Ann. Appl. Biol.* **71**, 47–57.

Jones, R.W., Jackson, A.O. and Morris, T.J. (1990). Defective-interfering RNAs and elevated temperatures inhibit replication of tomato bushy stunt virus in inoculated protoplasts. *Virology* **176**, 539–545.

Jones, T.A. and Liljàs, L. (1984). Structure of satellite tobacco necrosis virus after crystallographic refinement at 2.5 Å resolution. *J. Mol. Biol.* **177**, 735–767.

Joshi, S., Pleij, C.W.A., Haenni, A.-L., Chapeville, F. and Bosch, L. (1983). Properties of the tobacco mosaic virus intermediate length RNA2 and its translocation. *Virology* **127**, 100–111.

Joubert, J.J., Hahn, J.S., von Wechmar, M.B. and van Regenmortel, M.H.V. (1974). Purification and properties of tomato spotted wilt virus. *Virology* **57**, 11–19.

Joyce, C.M. and Steitz, T.A. (1994). Function and structure relationships in DNA polymerases. *Annu. Rev. Biochem.* **63**, 777–822.

Juckes, I.R.M. (1971). Fractionation of proteins and viruses with polyethylene glycol. *Biochim. Biophys. Acta* **229**, 535–546.

Jupin, I., Quillet, L., Ziegler-Graff, V., Guilley, H., Richards, K. and Jonard, G. (1988). *In vitro* translation of natural and synthetic beet necrotic yellow vein virus RNA1. *J. Gen. Virol.* **69**, 2359–2367.

Jurík, M., Mucha, V. and Valenta, V. (1980). Intraspecies variability in transmission efficiency of stylet-borne viruses by the pea aphid (*Acyrthosiphon pisum*). *Acta Virol.* **24**, 351–357.

Juszczuk, M., Paczkowski, E., Sadowy, E., Zagórski, W. and Hulanicka, D.M. (2000). Effect of genomic and subgenomic leader sequences of potato leafroll virus on gene expression. *FEBS Lett.* **484**, 33–36.

Kachroo, P., Yoshioka, K., Shah, J., Dooner, H. and Klessig, D.F. (2000). Resistance to turnip crinkle virus in *Arabidopsis* is affected by two host genes, is salicylic acid dependent but *NRP1*, ethylene and jasmonate independent. *Plant Cell* **12**, 677–690.

Kadarei, G. and Haenni, A.-L. (1997). Virus-encoded RNA helicases. *J. Virol.* **71**, 2583–2590.

Kahn, R.P. (1976). Aseptic plantlet culture to improve the phytosanitary aspects of plant introduction for asparagus. *Plant Dis. Rep.* **60**, 459–461.

Kahn, R.P. (1989). *Plant Protection and Quarantine*, Vols 1–3. CRC Press, Boca Raton, FL.

Kahn, R.P. and Monroe, R.L. (1970). Viruses isolated from arborescent *Datura* species from Bolivia, Ecuador and Columbia. *Plant Dis. Rep.* **54**, 675–677.

Kahn, T.W., Lapidot, M., Heinlein, M. *et al.* (1998). Domains of the TMV movement protein involved in subcellular localization. *Plant J.* **15**, 15–25.

Kaiser, W.J. and Danesh, D. (1971). Etiology of virus-induced wilt of *Cicer arietinum*. *Phytopathology* **61**, 453–457.

Kakutani, T., Hayano, Y., Hayashi, T. and Minobe, Y. (1991). Ambisense segment 3 of rice stripe virus: the first instance of a virus containing two ambisense segments. *J. Gen. Virol.* **72**, 465–468.

Kamei, T., Goto, T. and Matsui, C. (1969). Turnip virus multiplication in leaves infected with cauliflower mosaic virus. *Phytopathology* **59**, 1795–1797.

Kamer, G. and Argos, P. (1984). Primary structural comparison of RNA-dependent polymerases from plant, animal and bacterial viruses. *Nucl. Acids Res.* **12**, 7269–7282.

Kammann, M., Schalk, H.-J., Matzeit, V., Schaefer, S., Schell, J. and Gronenborn, B. (1991). DNA replication of wheat dwarf virus, a geminivirus, requires two cis-acting signals. *Virology* **184**, 786–790.

Kan, J.H., Andree, P.-J., Kouijzer, L.C. and Mellema, J.E. (1982). Proton-magnetic-resonance studies on the coat protein of alfalfa mosaic virus. *Eur. J. Biochem.* **126**, 29–33.

Kan, J.H., Cremers, A.F.M., Haasnoot, C.A.G. and Hilbers, C.W. (1987). The dynamical structure of the RNA of alfalfa mosaic virus studied by ^{31}P nuclear magnetic resonance. *Eur. J. Biochem.* **168**, 635–639.

Kaniewski, W. and Lawson, C. (1998). Coat protein and replicase-mediated resistance to plant viruses. In: A. Hadidi, R.K. Khetarpal and H. Koganezawa (eds) *Plant Virus Disease Control*, pp. 65–78. APS Press, St. Paul, MN.

Kaniewski, W.K. and Thomas, P.E. (1993). Field testing of virus resistant transgenic plants. *Semin Virol.* **4**, 389–396.

Kanyuka, K.V., Vishnichenko, V.K., Levay, K.E. *et al.* (1992). Nucleotide sequence of shallot virus X RNA reveals a 5′-proximal cistron closely related to those of potexviruses and a unique arrangement of the 3′-proximal cistrons. *J. Gen. Virol.* **73**, 2553–2560.

Kao, C.C. and Ahlquist, P. (1992). Identification of the domains required for direct interaction of the helicase-like and polymerase-like RNA replication proteins of brome mosaic virus. *J. Virol.* **66**, 7293–7302.

Kao, C.C., Quadt, R., Hershberger, R.P. and Ahlquist, P. (1992). Brome mosaic virus RNA replication proteins 1a and 2a form a complex *in vitro*. *J. Virol.* **66**, 6322–6329.

Kaper, J.M. (1972). Experimental analysis of the stabilising interactions of simple RNA viruses. *Proc. FEBS Meet.* **27**, 19–41.

Kaper, J.M. (1975). The chemical basis of virus structure, dissociation and reassembly. *Front. Biol.* **39**, 1–485.

Kaper, J.M. and Collmer, C.W. (1988). Modulation of viral plant diseases by secondary RNA agents. In: E. Domingo, J.J. Holland and P. Ahlquist (eds) *RNA Genetics*, Vol. 3, pp. 171–193. CRC Press, Boca Raton, FL.

Kaper, J.M. and Diaz-Ruiz, J.R. (1977). Molecular weights of the double-stranded RNAs of cucumber mosaic virus strain S and its associated RNA5. *Virology* **80**, 214–217.

Kaper, J.M. and Tousignant, M.E. (1977). Cucumber mosaic virus-associated RNA5. I. Role of host plant and helper strain in determining amount of associated RNA5 with virions. *Virology* **80**, 186–195.

Kaper, J.M. and Waterworth, H.E. (1977). Cucumber mosaic virus associated RNA5. Causal agent for tomato necrosis. *Science* **196**, 429–431.

Kaper, J.M., Diener, T.O. and Scott, H.A. (1965). Some physical and chemical properties of cucumber mosaic virus (strain Y) and of its isolated ribonucleic acid. *Virology* **27**, 54–72.

Kaper, J.M., Tousignant, M.E. and Lot, H. (1976). A low-molecular-weight replicating RNA associated with a divided genome plant virus: defective or satellite RNA? *Biochem. Biophys. Res. Commun.* **72**, 1237–1243.

Kaper, J.M., Tousignant, M.E. and Steen, M.T. (1988a). Cucumber mosaic virus-associated RNA5. XI. Comparison of 14 CARNA5 variants relates ability to induce tomato necrosis to a conserved nucleotide sequence. *Virology* **163**, 284–292.

Kaper, J.M., Tousignant, M.E. and Steger, G. (1988b). Nucleotide sequence predicts circularity and self cleavage of 300-ribonucleotide satellite of arabis mosaic virus. *Biochem. Biophys. Res. Commun.* **154**, 318–325.

Kaper, J.M., Gallitelli, D. and Tousignant, M.E. (1990). Identification of a 334-ribonucleotide viral satellite as principal aetiological agent in a tomato necrosis epidemic. *Res. Virol.* **141**, 81–95.

Kaplan, I.B., Zhang, L. and Palukaitis, P. (1998). Characterization of cucumber mosaic virus: V. Cell-to-cell movement requires capsid protein but not virions. *Virology* **246**, 221–231.

Karasawa, A., Okada, I., Akashi, K. *et al.* (1999). One amino acid change in cucumber mosaic virus RNA polymerase determines virulent/avirulent phenotypes on cowpea. *Phytopathology* **89**, 1186–1192.

Karasev, A.V. (2000). Genetic diversity and evolution of closteroviruses. *Annu. Rev. Phytopathol.* **38**, 293–324.

Karasev, A.V. and Hilf, M.E. (1997). Molecular biology of the citrus tristeza virus. In: P.L. Monette (ed) *Filamentous Viruses of Woody Plants*, pp. 121–131. Research Signpost, Trivandrum.

Karasev, A.V., Kashina, A.S., Gelfand, V.I. and Dolja, V.V. (1992). HSP70-related 65 kDa protein of beet yellows closterovirus is a microtubule-binding protein. *FEBS Lett.* **304**, 12–14.

Karasev, A.V., Boyko, V.P., Nikolaeva, O.V. *et al.* (1995). Complete sequence of the citrus tristeza virus RNA genome. *Virology* **208**, 511–520.

Karasev, A.V., Hilf, M.E., Garnsey, S.M. and Dawson, W.O. (1997). Transcriptional strategy of closteroviruses: mapping the 5´ termini of the citrus tristeza virus subgenomic RNAs. *J. Virol.* **71**, 6233–6236.

Karpova, O.V., Tyulkina, L.G., Atabekov, K.J., Rodionova, N.P. and Atabekov, J.G. (1989). Deletion of intercistronic poly(A) tract from brome mosaic virus RNA3 by ribonuclease H and its restoration in progeny of the relegated RNA3. *J. Gen. Virol.* **70**, 2287–2297.

Karrer, E.K., Beachy, R.N. and Holt, C.A. (1998). Cloning of tobacco genes that elicit the hypersensitive response. *Plant Mol. Biol.* **36**, 681–690.

Karyeija, R.F., Kreuze, J.F., Gibson, R.W. and Valkonen, J.P.T. (2000). Synergistic interactions of a potyvirus anmd a phloem-limited crinivirus in sweet potato plants. *Virology* **269**, 26–36.

Kashiwazaki, S., Minobe, Y., Omura, T. and Hibino, H. (1990). Nucleotide sequence of barley yellow mosaic virus RNA 1: a close evolutionary relationship with potyviruses. *J. Gen. Virol.* **71**, 2781–2790.

Kashiwazaki, S., Minobe, Y. and Hibino, H. (1991). Nucleotide sequence of barley yellow mosaic virus RNA 2. *J. Gen. Virol.* **72**, 995–999.

Kassanis, B. (1939). Intranuclear inclusions in virus-infected plants. *Ann. Appl. Biol.* **26**, 705–709.

Kassanis, B. (1952). Some effects of high temperature on the susceptibility of plants to infection with viruses. *Ann. Appl. Biol.* **39**, 358–369.

Kassanis, B. (1962). Properties and behaviour of a virus depending for its multiplication on another. *J. Gen. Microbiol.* **27**, 477–488.

Kassanis, B. and Bastow, C. (1971). The relative concentration of infective intact virus and RNA of four strains of tobacco mosaic virus as influenced by temperature. *J. Gen. Virol.* **11**, 157–170.

Kassanis, B. and Govier, D.A. (1971). New evidence on the mechanism of aphid transmission of potato C and potato aucuba mosaic viruses. *J. Gen. Virol.* **10**, 99–101.

Kassanis, B. and Macfarlane, I. (1968). The transmission of satellite viruses of tobacco necrosis virus by *Olpidium brassicae*. *J. Gen. Virol.* **3**, 227–232.

Kassanis, B. and Phillips, M.P. (1970). Serological relationship of strains of tobacco necrosis virus and their ability to activate strains of satellite virus. *J. Gen. Virol.* **9**, 119–126.

Kassanis, B. and Varma, A.(1967). The production of virus-free clones of some British potato varieties. *Ann. Appl. Biol.* **59**, 447–450.

Kassanis, B. and White, R.F. (1974). A simplified method of obtaining tobacco protoplasts for infection with tobacco mosaic virus. *J. Gen. Virol.* **24**, 447–452.

Kassanis, B., Vince, D.A. and Woods, R.D. (1970). Light and electron microscopy of cells infected with tobacco necrosis and satellite viruses. *J. Gen. Virol.* **7**, 143–151.

Kassanis, B., White, R.F. and Woods, R.D. (1975). Inhibition of multiplication of tobacco mosaic virus in protoplasts by antibiotics and its prevention by divalent metals. *J. Gen. Virol.* **28**, 185–191.

Kassanis, B., Russell, G.E and White, R.F. (1978). Seed and pollen transmission of beet cryptic virus in sugar beet plants. *Phytopathol. Z.* **91**, 76–79.

Kasschau, K.D and Carrington, J.C. (1998). A counterdefense strategy of plant viruses: suppression of posttranscriptional gene silencing. *Cell* **95**, 461–470.

Kasteel, D. (1999). Structure, morphogenesis and function of tubular structures induced by cowpea mosaic virus. PhD thesis, University of Wageningen, Netherlands.

Kasteel, D.T.J., Perbal, M.-C., Boyer, J.-C. *et al.* (1996). The movement proteins of cowpea mosaic virus and cauliflower mosaic virus induce tubular structures in plant and insect cells. *J. Gen. Virol.* **77**, 2857–2864.

Kasteel, D.T.J., van der Wel, N.N., Jansen, K.A.L., Goldbach, R.W. and van Lent, J.W.M. (1997). Tubule forming capacity of the movement proteins of AlMV and BMV. *J. Gen. Virol.* **78**, 2089–2093.

Katis, N., Carpenter, J.M. and Gibson, R.W. (1986). Interference between potyviruses during aphid transmission. *Plant Pathol.* **35**, 152–157.

Katouzian-Safadi, M. and Berthet-Colominas, C. (1983). Evidence for the presence of a hole in the capsid of turnip yellow mosaic virus after RNA release by freezing and thawing: decapsidation of turnip yellow mosaic virus *in vitro*. *Eur. J. Biochem.* **137**, 47–55.

Katouzian-Safadi, M. and Haenni, A.-L. (1986). Studies on the phenomenon of turnip yellow mosaic virus RNA release by freezing and thawing. *J. Gen. Virol.* **67**, 557–565.

Katul, L., Maiss, E., Morozov, S.Y. and Vetten, H.J. (1997). Analysis of six DNA components of the faba bean necrotic yellows virus genome and their structural affinity to related plant virus genomes. *Virology* **233**, 247–259.

Katul, L., Timchenko, T., Gronenborn, B. and Vetten, H.J. (1998). Ten distinct circular ssDNA components, four of which encode putative replication associated proteins, are associated with the faba bean necrotic yellows virus genome. *J. Gen. Virol.* **79**, 3101–3109.

Katz, D. and Kohn, A. (1984). Immunoabsorbent electron microscopy for detection of vinises. *Adv. Virus Res.* **29**, 169–194.

Kausche, G.A., Pfankuch, E. and Ruska, A. (1939). Die Sichtbormachung von pflanzlichem Virus im Übermikroskop. *Naturwissenschaften* **27**, 292–299.

Kavanagh, T., Goulden, M., Santa Cruz, S., Chapman, S., Barker, I. and Baulcombe, D. (1992). Molecular analysis of a resistance-breaking strain of potato virus X. *Virology* **189**, 609–617.

Kawakami, S., Padgett, H.S., Hosokawa, D., Okada, Y., Beachy, R.N. and Watanabe, Y. (1999). Phosphorylation and/or presence of serine 37 in the movement protein of tomato mosaic tobamovirus is essential for intracellular localization and stability *in vivo*. *J. Virol.* **73**, 6831–6840.

Kawchuk, L.M., Martin, R.R. and McPherson, J. (1990). Resistance in transgenic potato expressing the potato leafroll virus coat protein gene. *Mol. Plant Microb. Interact.* **3**, 301–307.

Kawchuk, L.M., Martin, R.R. and McPherson, J. (1991). Sense and antisense RNA-mediated resistance to potato leafroll virus in Russet Burbank potato plants. *Mol. Plant Microbe Interact.* **4**, 247–253.

Kay, R., Chan, A., Daly, M. and McPherson, J. (1987). Duplication of CaMV 35S promoter sequences creates a strong enhancer for plant genes. *Science* **236**, 1299–1302.

Kearney, C.M., Donson, J., Jones, G.E. and Dawson, W.O. (1993). Low level of genetic drift in foreign sequences replicating in an RNA virus in plants. *Virology* **192**, 11–17.

Kearney, C.M., Thomson, M.J. and Roland, K.E. (1999). Genome evolution of tobacco mosaic virus populations during long-term passaging in a diverse range of hosts. *Arch. Virol.* **144**, 1513–1526.

Keeling, J. and Matthews, R.E.F. (1982). Mechanism for release of RNA from turnip yellow mosaic virus at high pH. *Virology* **119**, 214–218.

Keeling, J., Collins, E.R. and Matthews, R.E.F. (1979). Behaviour of turnip yellow mosaic virus nucleoproteins under alkaline conditions. *Virology* **97**, 100–111.

Keese, P. and Gibbs, A. (1992). Origins of genes: 'big band' or continuous creation? *Proc. Natl Acad. Sci. USA* **89**, 9489–9493.

Keese, P. and Symons, R.H. (1985). Domains in viroids: evidence of intermolecular RNA rearrangements and their contribution to viroid evolution. *Proc. Natl Acad. Sci. USA* **82**, 4582–4586.

Keese, P. and Symons, R.H. (1987). The structure of viroids and virusoids. In: J.S. Semancik (ed.) *Viroids and Viroid-like Pathogens*, pp. 1–47. CRC Press, Boca Raton, FL.

Keese, P., Bruening, G. and Symons, R.H. (1983). Comparative sequence and structure of circular RNAs from two isolates of lucerne transient streak virus. *FEBS Lett.* **159**, 185–190.

Keese, P., Mackenzie, A. and Gibbs, A. (1989). Nucleotide sequence of the genome of an Australian isolate of turnip yellow mosaic tymovirus. *Virology* **172**, 536–546.

Kekuda, R., Sundareshan, S., Karande, A. and Savithri, H.S. (1995). Monoclonal antibodies in the study of architechture of plant viruses. *Curr. Sci.* **68**, 611–617.

Keller, K. E., Johnasen, E., Martin, R.R. and Hampton, R.O. (1998). Potyvirus genome-linked protein (VPg) determines pea seed-borne mosaic virus pathotype-specific virulence in *Pisum sativum*. *Mol. Plant Microbe Interact.* **11**, 124–130.

Keller, P., Lüttge, U., Wang, X.-C. and Büttner, G. (1989). Influence of rhizomania disease on gas exchange and water relations of a susceptible and a tolerant sugar beet variety. *Physiol. Mol. Plant Pathol.* **34**, 379–392.

Kempers R. and van Bel, A.J.E. (1997). Symplastic connections between sieve elements and companion cell in the stem phloem of *Vicia faba* L. have a molecular exclusion limit of at least 10 kDa. *Planta* **201**, 195–201.

Kendall, T.L. and Lommel, S.A. (1992). Nucleotide sequence of carnation ringspot dianthovirus RNA-2. *J. Gen. Virol.* **73**, 2479–2488.

Kennedy, J.S. (1951). Benefits to aphids from feeding on galled and virus-infected leaves. *Nature (Lond.)* **168**, 825–826.

Kennedy, J.S., Day, M.F. and Eastop, V.F. (1962). *A Conspectus of Aphids as Vectors of Plant Viruses*. Commonwealth Institute of Entomology, London.

Kenten, R.H. and Lockwood, G. (1977). Studies on the possibility of increasing resistance to cocoa swollen-shoot virus by breeding. *Ann. Appl. Biol.* **85**, 71–78.

Kerlan, C., Robert, Y., Perennec, P. and Guillery, E. (1987). Survey of the level of infection by PVY-O and control methods developed in France for potato seed production. *Potato Res.* **30**, 651–667.

Kessler, C. (1993). Nonradioactive labeling methods for nucleic acids. In: L.J. Kricka (ed.) *Nonisotopic DNA Probe Techniques*, pp. 29–92. Academic Press, San Diego.

Ketellapper, H.J. (1963). Temperature-induced chemical defects in higher plants. *Plants Physiol.* **38**, 175–179.

Ketting, R.F., Haverkamp, T.H.A., van Luenen, H.G.A.M. and Plasterk, R.H.A. (1999). *mut-7* of *C. elegans*, required for transposon silencing and RNA interference, is a homolog of Werner Syndrome helicase and RNaseD. *Cell* **99**, 133–141.

Khan, I.A. and Jones, G.E. (1989). Selection for a specific tobacco mosaic virus variant during bolting of *Nicotiana sylvestris*. *Can. J. Bot.* **67**, 88–94.

Khan, M.A. and Maxwell, D.P. (1977). Use of inclusions in the rapid diagnosis of virus diseases of red clover. *Plant Dis. Rep.* **61**, 679–683.

Khan, M.A. and Mukhopadhyay, S. (1985). Studies on the effect of some alternative cultural methods on the incidence of yellow vein mosaic virus (YVMV) disease of okra (*Abelmoschus esculentus* (L.) Moench.). *Indian J. Virol.* **1**, 69–72.

Khan, M.A., Maxwell, D.P. and Maxwell, M.D. (1977). Light microscopic cytochemistry and ultrastructure of red clover vein mosaic virus-induced inclusions. *Virology* **78**, 173–182.

Khandjian, E.W. (1987) Optimized hybridization of DNA blotted and fixed to nitrocellulose and nylon membranes *Biotechnology* **5**, 165–167.

Khetarpal, R.K., Bossennec, J.-M., Burghofer, A., Cousin, A. and Maury, Y. (1988). Effect of pea seed-borne mosaic virus on yield of field pea. *Agronomie* **8**, 811–815.

Khetarpal, R.K., Maisonneuve, B., Maury, Y. *et al.* (1998). Breeding for resistance to plant viruses. In: A. Hadidi, R.K. Khetarpal and H. Koganezawa (eds) *Plant Virus Disease Control*, pp. 14–32. APS Press, St. Paul, MN.

Kheyr-Pour, A., Gronenborn, B. and Czosnek, H. (1994). Agroinoculation of tomato yellow leaf curl virus (TYLCV) overcomes the virus resistance of wild *Lycopersicon* species. *Plant Breeding* **112**, 228–233.

Kiberstis, P.A. and Zimmern, D. (1984). Translational strategy of *Solanum nodiflorum* mottle virus RNA: synthesis of a coat protein precursor *in vitro* and *in vivo*. *Nucl. Acids Res.* **12**, 933–943.

Kiberstis, P.A., Loesch-Fries, L.S. and Hall, T.C. (1981). Viral protein synthesis in barley protoplasts inoculated with native and fractionated brome mosaic virus RNA. *Virology* **112**, 804–808.

Kiberstis, P.A., Pessi, A., Atherton, E., Jackson, R., Hunter, T. and Zimmern, D. (1983). Analysis of *in vitro* and *in vivo* products of the TMV 30 kDa open reading frame using antisera raised against a synthetic peptide. *FEBS Lett.* **164**, 355–360.

Kiefer, M.C., Daubert, S.D., Schneider, I.R. and Bruening, G. (1982). Multimeric forms of satellite of tobacco ringspot virus RNA. *Virology* **121**, 262–273.

Kiefer, M.C., Owens, R.A. and Diener, T.O. (1983). Structural similarities between viroids and transposable genetic elements. *Proc. Natl Acad. Sci. USA* **80**, 6234–6238.

Kiguchi, T., Saito, M. and Tamada, T. (1996). Nucleotide sequence analysis of RNA-5 of five isolates of beet necrotic yellow vein virus and the identity of a deletion mutant. *J. Gen. Virol.* **77**, 575–580.

Kiho, Y., Machida, H. and Oshima, N. (1972). Mechanism determining the host specificity of tobacco mosaic virus: I. Formation of polysomes containing infecting viral genome in various plants. *Jpn J. Microbiol.* **16**, 451–459.

Kikkawa, H., Nagata, T., Matsui, C. and Takebe, I. (1982). Infection of protoplasts from tobacco suspension cultures by tobacco mosaic virus. *J. Gen. Virol.* **63**, 451–456.

Kikkert, M., Van Poelwijk, F., Storms, M. *et al.* (1997). A protoplast system for studying tomato spotted wilt infection. *J. Gen. Virol.* **78**, 1775–1763.

Kikkert, M., Meurs, C., van de Wetering, F. *et al.* (1998). Binding of tomato spotted wilt virus to a 94-kDa thrips protein. *Phytopathology* **88**, 63–69.

Kikuchi, Y., Tye, K., Filipowicz, W., Sänger, H.L. and Gross, H.J. (1982). Circularization of linear viroid RNA via 2'-phosphomonoester, 3',5'-phosphodiester bonds by a novel type of RNA ligase from wheat germ and *Chlamydomonas*. *Nucl. Acids Res.* **10**, 7521–7529.

Kim, C.-H. and Palukaitis, P. (1997). The plant defense response to cucumber mosaic virus in *Coppea* is elicited by the viral polymerase gene and affects virus accumulation in single cells. *EMBO J.* **16**, 4060–4068.

Kim, D.-H., Park, Y.S., Kim, S.S., Lew, J., Nam, H.G. and Choi, K.Y. (1995). Expression, purification and identification of a novel self-cleavage site of the NIa C-terminal 27-kDa protease of turnip mosaic potyvirus C5. *Virology* **213**, 517–525.

Kim, D.-H., Hwang, D.C., Kang, B.H. *et al.* (1996). Effects of internal cleavages and mutations in the C-terminal region of NIa protease of turnip mosaic potyvirus on the catalytic activity. *Virology* **226**, 183–190.

Kim, H.-H. and Lommel, S.A. (1994). Identification and analysis of the site of −1 ribosomal frameshifting in red clover necrotic mosaic virus. *Virology* **200**, 574–582.

Kim, K.-H. and Hemenway, C. (1996). The 5' nontranslated region of potato virus X RNA affects both genomic and subgenomic RNA synthesis. *J. Virol.* **70**, 5533–5540.

Kim, K.-H. and Hemenway, C. (1997). Mutations that alter a conserved element upstream of the potato virus X triple block and coat protein genes affect subgenomic RNA accumulation *Virology* **232**, 187–197.

Kim, K.-H. and Hemenway, C. (1999).Long-distance RNA–RNA interactions and conserved sequence elements affect potato virus X plus-strand RNA accumulation. *RNA* **5**, 636–645.

Kim, K.S. (1977). An ultrastructural study of inclusions and disease development in plant cells infected by cowpea chlorotic mottle virus. *J. Gen. Virol.* **35**, 535–543.

Kim, K.S., Fulton, J.P. and Scott, H.A. (1974). Osmiophilic globules and myelinic bodies in cells infected with two comoviruses. *J. Gen. Virol.* **25**, 445–452.

Kim, M.-J. and Kao, C. (2001). Factors regulating template switch *in vitro* by viral RNA-dependent RNA polymerases: implications for RNA-RNA recombination. *Proc. Natl Acad. Sci. USA* **98**, 4972–4977.

Kimble, K.A., Grogan, R.G., Greathead, A.S., Paulus, A.O. and House, J.K. (1975). Development, application and comparison of methods for indexing lettuce seed for mosaic virus in California. *Plant Dis. Rep.* **59**, 461–464.

Kimmins, W.C. and Brown, R.G. (1973). Hypersensitive resistance. The role of cell wall glycoproteins in virus localization. *Can. J. Bot.* **51**, 1923–1926.

Kimura, I. (1986). A study of rice dwarf virus in vector cell monolayers by fluorescent antibody focus counting. *J. Gen. Virol.* **67**, 2119–2124.

Kimura, I. and Black, L.M. (1972). The cell-infecting unit of wound tumor virus. *Virology* **49**, 549–561.

Kimura, I., Minobe, Y. and Omura, T. (1987). Changes in a nucleic acid and a protein component of rice dwarf virus particles associated with an increase in symptom severity. *J. Gen. Virol.* **68**, 3211–3215.

Kimura, M. (1990). A simple method for estimating evolutionary rates of base substitutions through comparative studies of nucleotide sequences. *J. Mol. Evol.* **16**, 111–120.

King, A.M.Q. (1998). Genetic recombination in positive strand RNA viruses. In: E. Domingo, J.J. Holland and P. Ahlquist (eds) *RNA Genetics*, Vol. II, pp. 149–165. CRC Press, Boca Raton, FL.

King, L. and Leberman, R. (1973). Derivatisation of carboxyl groups of tobacco mosaic virus with cystamine. *Biochim. Biophys. Acta* **322**, 279–293.

King, L.A. and Possee, R.D. (1992). *The Baculovirus Expression System: A Laboratory Guide*. Chapman & Hall, London.

Király, L., Cole, A.B., Bourque, J.E. and Schoelz, J.E. (1999). Systemic cell death is elicited by the interaction of a single gene in *Nicotiana clevelandii* and gene VI of cauliflower mosaic virus. *Mol. Plant Microbe Interact.* **12**, 919–925.

Kiselyova, O.I., Yaminsky, I.V., Karger, E.M., Frolova, O.Yu., Dorokhov, Y.L. and Atabekov, J.G. (2001). Visualization by atomic force microscopy of tobacco mosaic virus movement protein-RNA complexes formed *in vitro*. *J. Gen. Virol.* **82**, 1503–1508.

Kiss, T., Pósfai, J. and Solymosy, F. (1983). Sequence homology between potato spindle tuber viroid and U3B snRNA. *FEBS Lett.* **163**, 217–220.

Kiss-László, Z, Blanc, S. and Hohn, T. (1995). Splicing of cauliflower mosaic virus 35S RNA is essential for viral infectivity. *EMBO J.* **14**, 3552–3562.

Kitajima, E.W. and Costa, A.S. (1969). Association of pepper ringspot virus (Brazilian tobacco rattle virus) and host cell mitochondria. *J. Gen. Virol.* **4**, 177–181.

Kitajima, E.W. and Costa, A.S. (1973). Aggregates of chloroplasts in local lesions induced in *Chenopodium quinoa* wild. by turnip mosaic virus. *J. Gen. Virol.* **20**, 413–416.

Kitajima, E.W. and Lauritis, J.A. (1969). Plant virions in plasmodesmata. *Virology* **37**, 681–685.

Kitajima, E.W. and Lovisolo, O. (1972). Mitochondrial aggregates in *Datura* leaf cells infected with henbane mosaic virus. *J. Gen. Virol.* **16**, 265–271.

Kitajima, E.W., De Avila, A.C., Resende, R. de O., *et al.* (1992). Comparative cytological and immunogold labeling studies on different isolates of tomato spotted wilt virus. *J. Submicrosc. Cytol. Pathol.* **24**, 1014.

Kjemtrup, S., Sampson, K.S., Peele, C.G., Nguygen, L.V., Conkling, M.A., Thompson, W.F. and Robertson, D. (1998). Gene silencing from plant DNA carried by a geminivirus. *Plant J.* **14**, 91–100.

Klaassen, V.A., Boeshore, M.L., Koonin, E.V., Tian, T. and Falk, B.W. (1995). Genome structure and phylogenetic analysis of lettuce infectious yellows virus, a whitefly-transmitted, bipartite crinivirus. *Virology* **208**, 99–110.

Klaff, P., Gruner, R., Hecker, R., Sättler, A., Theissen, G. and Riesner, D. (1989). Reconstituted and cellular viroid–protein complexes. *J. Gen. Virol.* **70**, 2257–2270.

Kleczkowski, A. (1949). The transformation of local lesion counts for statistical analysis. *Ann. Appl. Biol.* **36**, 139–152.

Kleczkowski, A. (1950). Interpreting relationships between concentrations of plant viruses and numbers of local lesions. *J. Gen. Microbiol.* **4**, 53–69.

Kleczkowski, A. (1953). A method for testing results of infectivity tests for plant viruses for compatibility with hypotheses. *J. Gen. Microbiol.* **8**, 295–301.

Klein, M. and Harpaz, I. (1970). Heat suppression of plant-virus propagation in the insect vector's body, *Virology* **41**, 72–76.

Klein, R.E., Wyatt, S.D. and Kaiser, W.J. (1988). Incidence of bean common mosaic virus in USDA *Phaseolus* germplasm collection. *Plant Dis.* **72**, 301–302.

Klein, T.M., Wolf, E.D., Wu, R. and Sanford, J.C. (1987). High-velocity microprojectiles for delivering nucleic acids into living cells. *Nature* **327**, 70–73.

Klein, T.M., Arentzen, R., Lewis, P.A. and Fitzpatrick-McElligott, S. (1992). Transformation of microbes, plants and animals by particle bombardment. *Biotechnology* **10**, 286–291.

Kleinhempel, H. and Kegler, G. (1982). Transmission of tomato bushy stunt virus without vectors. *Acta Phytopathol. Acad. Sci. Hung.* **17**, 17–21.

Klessig, D.F., Durner, J., Navarre, D.A. *et al.* (2000). Nitric oxide and salicylic acid signaling in plant defense. *Proc. Natl Acad. Sci. USA* **97**, 8849–8855.

Klinkenberg, F.A., Ellwood, S. and Stanley, J. (1989). Fate of African cassava mosaic virus coat protein deletion mutants after agroinoculation. *J. Gen. Virol.* **70**, 1837–1844.

Klose, M.J., Sdoodee, R., Teakle, D.S., Milne, J.R., Greber, R.S. and Walter, G.H. (1996). Transmission of three strains of tobacco streak ilarvirus by different thrips species using virus-infected pollen. *J. Phytopathol.* **144**, 281–284.

Klöti, A., Heinrich, C., Bieri, S. *et al.* (1999). Upstream and downstream sequence elements determine the specificity of rice tungro bacilliform virus promoter and influence RNA production after transcription initiation. *Plant Mol. Biol.* **40**, 249–266.

Klug, A. (1999). The tobacco mosaic virus particle: structure and assembly. *Phil. Trans. R. Soc. Lond. B* **354**, 531–535.

Klug, A. and Berger, J.E. (1964). An optical method for the analysis of periodicities in electron micrographs, and some observations on the mechanism of negative staining. *J. Mol. Biol.* **10**, 565–569.

Klug, A. and Caspar, D.L.D. (1960). The structure of small viruses. *Adv. Virus Res.* **7**, 225–325.

Klug, A., Longley, W. and Leberman, R. (1966). Arrangement of protein subunits and the distribution of nucleic acid in turnip yellow mosaic virus: I. X-ray diffraction studies. *J. Mol. Biol.* **15**, 315–343.

Knapp, E. and Lewandowski, D.J. (2001). Tobacco mosaic virus, not just a single component virus anymore. *Molec. Plant Pathol.* **2**, 117–123.

Knapp, E., da Câmara Machado, A., Pühringer, H. *et al.* (1995). Localization of fruit tree viruses by immuno-tissue printing in infected shoots of *Malus* sp. and *Prunus* sp. *J. Virol. Methods* **55**, 157–173.

Knebel, W., Quader, H. and Schnepf, E. (1990). Mobile and immobile endoplasmic reticulum in onion bulb epidermis cells: short and long-term observations with a confocal laser scanning microscope. *Eur. J. Cell Biol.* **52**, 328–340.

Knoester, M., van Loon, L.C., Heuvel, J.V.D., Hennig, J., Bol, J.F. and Linthorst, H.J.M. (1998). Ethylene-insensitive tobacco lacks non-host resistance against soil-borne fungi. *Proc. Natl Acad. Sci. USA* **95**, 1933–1937.

Knorr, D.A. and Dawson, W.O. (1988). A point mutation in the tobacco mosaic virus capsid protein gene induces hypersensitivity in *Nicotiana sylvestris*. *Proc. Natl Acad. Sci. USA* **85**, 170–174.

Knowland, J., Hunter, T., Hunt, T. and Zimmern, D. (1975). Translation of tobacco mosaic virus RNA and isolation of the messenger for TMV coat protein. *Colloq. Inst. Natl Sante Rech. Med.* **47**, 211–216.

Kobayashi, K., Tsuge, S., Nakayashiki, H., Mise, K. and Furusawa, I. (1998). Evidence for a dual strategy in the expression of cauliflower mosaic virus open reading frames I and IV. *Microbiol. Immunol.* **42**, 329–334.

Koenig, R. (1976). A loop-structure in the serological classification system of tymoviruses. *Virology* **72**, 1–5.

Koenig, R. (1981). Indirect ELISA methods for the broad specificity detection of plant viruses. *J. Gen. Virol.* **55**, 53–62.

Koenig, R. (1986). Plant viruses in rivers and takes. *Adv. Virus Res.* **31**, 321–333.

Koenig, R. and Bercks, R. (1968). Änderungen in heterologen Reaktionsvermögen von Antiseren gegen Vertroter der potato virus X-Gruppe im Laufe des Immunisierungsprozesses. *Phytopathol. Z.* **61**, 382–398.

Koenig, R. and Burgermeister, W. (1986). Applications of immunoblotting in plant virus diagnosis. In: R.A.C. Jones and L. Torrance (eds) *Developments and Applications in Virus Testing*, Dev. Appl. Biol. 1, pp. 121–137. Association of Applied Biologists, Wellsbourne.

Koenig, R. and Gibbs, A. (1986). Serological relationships among tombusviruses. *J. Gen. Virol.* **67**, 75–82.

Koenig, R. and Givord, L. (1974). Serological interrelationships in the turnip yellow mosaic virus group. *Virology* **58**, 119–125.

Koenig, R. and Lesemann, D.E. (2000a). Genus *Pomovirus*. In: van Regenmortel *et al.* (eds) *Seventh Report of the International Committee on the Taxonomy of Viruses*, pp. 908–913. Academic Press, San Diego.

Koenig, R. and Lesemann, D.E. (2000b). Genus *Benyvirus*. In: van Regenmortel *et al.* (eds) *Seventh Report of the International Committee on the Taxonomy of Viruses*, pp. 917–922. Academic Press, San Diego.

Koenig, R. and Loss, S. (1997). Beet soil-borne virus RNA 1: genetic analysis enabled by a starting sequence generated with primers to highly conserved helicase-encoding domains. *J. Gen. Virol.* **78**, 3161–3165.

Koenig, R. and Paul, H.L. (1982). Variants of ELISA in plant virus diagnosis. *J. Virol. Methods* **5**, 113–125.

Koenig, R. and Torrance, L. (1986). Antigenic analysis of potato virus X by means of monoclonal antibodies. *J. Gen. Virol.* **67**, 2145–2151.

Koenig, R., Tremaine, J.H. and Shepard, J.F. (1978). *In situ* degradation of the protein chain of potato virus X at the N- and C-termini. *J. Gen. Virol.* **38**, 329–337.

Koenig, R., Fribourg, C.E. and Jones, R.A.C. (1979). Symptomatological, serological and electrophoretic diversity of isolates of Andean potato latent virus from different regions of the Andes. *Phytopathology* **69**, 748–752.

Koenig, R., An, D. and Burgermeister, W. (1988a). The use of filter hybridisation techniques for the identification, differentiation and classification of plant viruses. *J. Virol. Methods* **19**, 57–68.

Koenig, R., An, D., Lesemann, D.E. and Burgermeister, W. (1988b). Isolation of carnation ringspot virus from a canal near a sewage plant: cDNA hybridization analysis, serology, and cytopathology. *J. Phytopathol.* **121**, 346-356.

Koenig, R., Luddecke, P. and Haberle, A.M. (1995). Detection of beet necrotic yellow vein virus strains, variants and mixed infections by examining single-stranded conformation polymorphisms of immunocapture RT-PCR products. *J. Gen. Virol.* **76**, 2051–2055.

Koenig, R., Beier, C., Commandeur, U., Lüth, U., Kaufmann, A. and Lüddecke, P. (1996). Beet soil-borne virus RNA 3: a further example of the heterogeneity of the gene content of furovirus genomes and of triple gene block-carrying RNAs. *Virology* **216**, 202–207.

Koenig, R., Commandeur, U., Loss, S., Beier, C., Kaufmann, A. and Lesemann, D.-E. (1997). Beet soil-borne virus RNA-2: similarities and dissimilarities to the coat protein gene-carrying RNAs of other furoviruses. *J. Gen. Virol.* **78**, 469–477.

Koev, G., Mohan, B.R., Dinesh-Kumar, S.P., Torbert, K.A., Somers, D.A. and Miller, W.A. (1998). Extreme reduction of disease in oats transformed with the 5′ half of the barley yellow dwarf virus-PAV genome. *Phytopathology* **88**, 1013–1019.

Koev, G., Mohan, B.R. and Miller, W.A. (1999). Primary and secondary structural elements required for synthesis of barley yellow dwarf virus subgenomic RNA-1. *J. Virol.* **73**, 2867–2885.

Kohl, R.J. and Hall, T.C. (1974). Aminoacylation of RNA from several viruses: amino acid specificity and differential activity of plant, yeast and bacterial synthetases. *J. Gen. Virol.* **25**, 257–261.

Kohl, R.J. and Hall, T.C. (1977). Loss of infectivity of brome mosaic virus RNA after chemical modification of the 3′ or 5′ terminus. *Proc. Natl Acad. Sci. USA* **74**, 2682–2686.

Köhler, G. and Milstein, C. (1975). Continuous cultures of fused cells secreting antibody of predefined specificity. *Nature (Lond.)* **256**, 495–497.

Kohli, A., Griffiths, S., Palacios, N. *et al.* (1999a). Molecular characterization of transformed plasmid rearrangements in transgenic rice reveals a recombinational hotspot in the CaMV 35S promoter and confirms the predominance of microhomology mediated recombination. *Plant J.* **17**, 591–601.

Kohli, A., Gahakawa, D., Vain, P., Laurie, D.A. and Christou, P. (1999b). Transgene expression in rice engineered through particle bombardment: molecular factors controlling stable expression and transgene silencing. *Planta* **208**, 88–97.

Kohlstaedt, L.A., Wang, J., Friedman, J.M., Rice, P.A. and Steitz, T.A. (1992). Crystal structure at 3.5 Å resolution of HIV-1 reverse transcriptase complexed with an inhibitor. *Science* **256**, 1783–1790.

Köhm, B.A., Goulden, M.G., Gilbert, J.E., Kavanagh, T.A. and Baulcombe, D.C. (1993). A potato X resistance gene mediates an induced nonspecific resistance in protoplasts. *Plant Cell* **5**, 913–920.

Kolk, M.H., van der Graaf, M., Wijmenga, S.S., Pleij, C.W.A., Heus. H.A. and Hilbers, C.W. (1998). NMR structure of a classical pseudoknot: interplay of single- and double-stranded RNA. *Science* **280**, 434–438.

Kollár, A. and Burgyán, J. (1994). Evidence that ORF 1 and 2 are the only virus-encoded replicase genes of cymbidium ringspot tombusvirus. *Virology* **201**, 169–172.

Kollár, A., Dalmay, T. and Burgyán, J. (1993). Defective interfering RNA-mediated resistance against cymbidium ringspoy tombusvirus in transgenic plants. *Virology* **193**, 313–318.

Konaté, G. and Fritig, B. (1983). Extension of the ELISA method to the measurement of the specific radioactivity of viruses in crude cellular extracts. *J. Virol. Meth.* **6**, 347–356.

Konaté, G. and Fritig, B. (1984). An efficient microinoculation procedure to study plant virus multiplication at predetermined individual infection sites on the leaves. *Phytopathol. Z.* **109**, 131–138.

Konaté, G., Barro, N., Fargette, D., Swanson, M.M. and Harrison, B.D. (1995). Occurrence of whitefly-transmitted geminiviruses in crops in Burkino Faso, and their serological detection and differentiation. *Ann. Appl. Biol.* **126**, 121–129.

Kong, F., Sivakumaran, K. and Kao, C. (1999). The N-terminal half of brome mosaic virus 1a protein has RNA capping-associated activities: specificity for GTP and S-adenosylmethionine. *Virology* **259**, 200–210.

Kong, Q., Oh, J.-W. and Simon, A.E. (1995). Symptom attenuation by a normally virulent satellite RNA of turnip crinkle virus is associated with the coat protein open reading frame. *Plant Cell* **7**, 1625–1634.

Kong, Q., Oh, J.-W., Carpenter, C.D. and Simon, A.E. (1997a). The coat protein of turnip crinkle virus is involved in subviral RNA-mediated symptom modulation and accumulation. *Virology* **238**, 478–485.

Kong, Q., Wang, J. and Simon, A.E. (1997b). Satellite RNA-mediated resistance to turnip crinkle virus in *Arabidopsis* involves a reduction in virus movement. *Plant Cell* **9**, 2051–2063.

Kontzog, H.G., Kleinhempel, H. and Kegler, H. (1988). Detection of plant pathogenic viruses in waters. *Arch. Phytopathol. Pflanzenschutz.* **24**, 171–172.

Koonin, E.V. (1991). The phylogeny of RNA-dependent RNA polymerases of positive-strand RNA viruses. *J. Gen. Virol.* **72**, 2197–2206.

Koonin, E.V. and Dolja, V.V. (1993). Evolution and taxonomy of positive strand RNA viruses: implications of comparative analysis of amino acid sequences. *Biochem. Mol. Biol.* **28**, 357–430.

Koonin, E.V. and Gorbalenya, A.E. (1989). Evolution of RNA genomes: does the high mutation rate necessitate high rate of evolution of viral proteins? *J. Mol. Evol.* **28**, 524–527.

Kopp, M., Geoffrey, P. and Fritig, B. (1981). Studies on tobacco mosaic virus replication by means of radiolabelling the virus under isotonic conditions. *Ann. Phytopathol.* **12**, 314.

Kopp, M., Rouster, I., Fritig, B., Darvill, A. and Albersheim, P. (1989). Host–pathogen interactions. XXXII. A fungal glucan preparation protects *Nicotiana* against infection by viruses. *Plant Physiol.* **90**, 208–216.

Kormelink, R., van Poelwijk, F., Peters, D. and Goldbach, R. (1992a). Non-viral heterogeneous sequences at the 5' ends of tomato spotted wilt virus mRNAs. *J. Gen. Virol.* **73**, 2125–2128.

Kormelink, R., de Haan, P., Meurs, C., Peters, D. and Goldbach, R. (1992b). The nucleotide sequence of the mRNA segment of tomato spotted wilt virus, a bunyavirus with two ambisense RNA segments. *J. Gen. Virol.* **73**, 2795–2804.

Kormelink, R., Storms, M., van Lent, J., Peters, D. and Goldbach, R. (1994). Expression and subcellular location of the NS_M protein of tomato spotted wilt virus (TSWV), a putative viral movement protein. *Virology* **200**, 56–65.

Kotlisky, G., Boulton, M.I., Pitaksutheepong, C., Davies, J.W. and Epel, B.L. (2000). Intracellular and intercellular movement of maize streak geminivirus V1 and V2 proteins transiently expressed as green fluorescent protein fusions. *Virology* **274**, 32–38.

Kozak, M. (1981). Possible role of flanking nucleotides in recognition of the AUG initiator codon by eukaryotic ribosomes. *Nucl. Acids Res.* **9**, 5233–5252.

Kozak, M. (1986). Point mutations define a sequence flanking the AUG initiator codon that modulates translation by eukaryotic ribosomes. *Cell* **44**, 283–292.

Kozak, M. (1989). The scanning model for translation: an update. *J. Cell Biol.* **108**, 229–241.

Kozak, M. (1992). Regulation of translation in eukaryotic systems. *Annu. Rev. Cell Biol.* **8**, 197–225.

Kozak, M. (2001). New ways of initiating translation in eukaryotes? *Molec. Cell. Biol.* **21**, 1899–1907.

Krajacic, M., Mamula, D and Juretic, N. (1992). Determination of serological differentiation index values of turnip yellow mosaic virus (TYMV) strains by single radial immunodiffusion. *Periodicum Biologorum* **94**, 221–225.

Krczal, G. (1998). Virus certification of ornamental plants—the European strategy. Breeding for resistance to plant viruses. In: A. Hadidi, R.K. Khetarpal and H. Koganezawa (eds) *Plant Virus Disease Control*, pp. 277–287.. APS Press, St. Paul, MN.

Kricka, L.J. (ed.) (1993). *Nonisotopic DNA Probe Techniques*. Academic Press, San Diego.

Krishna, S., Hiremath, C.N., Munshi, S.K. *et al.* (1999). Three-dimensional structure of physalis mottle virus: implications for the viral assembly. *J. Mol. Biol.* **289**, 919–934.

Krol, M.A., Olson, N.H., Tate, J., Johnson, J.E., Baker, T.S. and Ahlquist, P. (1999). RNA-controlled polymorphism in the *in vivo* assembly of 180-subunit and 120-subunit virions from a single capsid protein. *Proc. Natl Acad. Sci. USA* **96**, 13650–13655.

Kroner, P., Richards, D., Traynor, P. and Ahlquist, P. (1989). Defined mutations in a small region of the brome mosaic virus 2a gene cause diverse temperature-sensitive RNA replication phenotypes. *J. Virol.* **63**, 5302–5309.

Kruckelmann, H.-W. and Seyffert, W. (1970). Wechselwirkungen zwischen einem turnip-mosaik-Virus und dem Genom des Wirtes. *Theor. Appl. Genet.* **40**, 121–123.

Krüse, J., Krüse, K.M., Witz, J., Chauvin, C., Jacrot, B. and Tardieu, A. (1982). Divalent ion-dependent reversible swelling of tomato bushy stunt virus and organisation of the expanded virion. *J. Mol. Biol.* **162**, 393–417.

Krüse, J., Timmins, P. and Witz, J. (1987). The spherically averaged structure of a DNA isometric plant virus: cauliflower mosaic virus. *Virology* **159**, 166–168.

Krüse, M., Koenig, R., Hoffmann, A. *et al.* (1994). Restriction fragment length polymorphism analysis of revese transcription-PCR products reveals the existence of two major strain groups of beet necrotic yellow vein virus. *J. Gen. Virol.* **75**, 1835–1842.

Kryczynski, S., Paduch-Cichal, E. and Skrzeczkowski, L.J. (1988). Transmission of three viroids through seed and pollen of tomato plants. *J. Phytopathol.* **121**, 51–57.

Kubo, S., Harrison, B.D., Robinson, D.J. and Mayo, M.A. (1975a). Tobacco rattle virus in tobacco mesophyll protoplasts: infection and virus multiplication. *J. Gen. Virol.* **27**, 293–304.

Kubo, S., Harrison, B.D. and Barker, H. (1975b). Defined conditions for growth of tobacco plants as sources of protoplasts for virus infection. *J. Gen. Virol.* **28**, 255–257.

Kudo, H., Uyeda, I. And Shikata, E. (1991). Viruses in the *Phytoreovirus* genus of the *Reoviridae* family have the same conserved terminal sequences. *J. Gen. Virol.* **72**, 2857–2866.

Kuhn, C.W. and Bancroft, J.B. (1961). Concentration and specific infectivity changes of alfalfa mosaic virus during systemic infection. *Virology* **18**, 281–288.

Kumar, A., Reddy, V.S., Yusibov, V. *et al.* (1997). The structure of alfalfa mosaic virus capsid protein assembled as a $T = 1$ icosahedral particles at 4.0 Å resolution. *J. Virol.* **71**, 7911–7916.

Kumagai, M.H., Turpen, T.H., Weinzettl, N., Della-Cioppa, G., Turpen, A.M., Donson, J., Hilf, M.E., Grantham, G.L., Dawson, W.O., Chow, T.P., Piatak, M. Jr. and Grill, L.K. (1993). Rapid, high-level expression of biologically active —trichosanthin in transfected plants by an RNA viral vector. *Proc. Natl Acad. Sci. USA* **90**, 427–430.

Kumagai, M.H., Donson, J., Dell-Cioppa, G., Harvey, D., Hanley, K. and Grill, L.K. (1995). Cytoplasmic inhibition of carotinoid biosynthesis with virus-derived RNA. *Proc. Natl Acad. Sci. USA* **92**, 1679–1683.

Kummert, J. and Semal, J. (1969). Study of the incorporation of radioactive uridine into virus-infected leaf fragments. *Phytopathol. Z.* **65**, 101–123.

Kunik, T., Palanichelvam, K., Czosnek, H., Citovsky, V. and Gafni, Y. (1998). Nuclear import of the capsid protein of tomato yellow leaf curl virus (TYLCV) in plants and insect cells. *Plant J.* **13**, 393–399.

Kunkel, L.O. (1922). Insect transmission of yellow stripe disease. *Hawaii Plant. Rec.* **26**, 58–64.

Kuntz, I.D. and Kauzmann, W. (1974). Hydration of proteins and polypeptides. *Adv. Protein Chem.* **28**, 239–345.

Kurath, G. and Palukaitis, P. (1989a). Satellite RNAs of cucumber mosaic virus: recombinants constructed in vitro reveal independent functional domains for chlorosis and necrosis in tomato. *Mol. Plant Microbe Interact.* **2**, 91–96.

Kurath, G. and Palukaitis, P. (1989b). RNA sequence heterogeneity in natural populations of three satellite RNAs of cucumber mosaic virus. *Virology* **173**, 231–240.

Kurkinen, M. (1981). Fidelity of protein synthesis affects the read through translation of tobacco mosaic virus RNA. *FEBS Lett.* **124**, 79–83.

Kurtz-Fritsch, C. and Hirth, L. (1972). Uncoating of two spherical plant viruses. *Virology* **47**, 385–396.

Kuwata, S., Masuta, C. and Takanami, Y. (1991). Reciprocal phenotype alterations between two satellite RNAs of cucumber mosaic virus. *J. Gen. Virol.* **72**, 2385–2389.

Laasko, M.M. and Heaton, L.A. (1993). Asp–Asn substitutions in the putative calcium binding site of the turnip crinkle coat protein affect virus movement in plants. *Virology* **197**, 774–777.

Lackey, C.F. (1932). Restoration of virulence of attenuated curleytop virus by passage through *Stellaria media*. *J. Agric. Res.* **44**, 755–765.

Laco, G.S. and Beachy, R.N. (1994). Rice tungro bacilliform virus encodes reverse transcriptase, DNA polymerase and ribonucleases H activities. *Proc. Natl Acad. Sci. USA* **91**, 2654–2658.

Laco, G.S., Kent, S.B.H. and Beachy, R.N. (1995). Analysis of proteolytic processing and activation of the rice tungro bacilliform virus reverse transcriptase. *Virology* **208**, 207–214.

Laflèche, D. and Bové, J.M. (1968). Sites d'incorporation de l'uridine tritiée dans les cellules du parenchyma foliate de *Brassica chinensis*, saines on infectées par le virus de la mosaique jaune du navet. *C.R. Hebd. Seances Acad. Sci.* **266**, 1839–1841.

Laflèche, D., Bové, C., Dupont, G. *et al.* (1972). Site of viral RNA replication in the cells of higher plants: TYMV–RNA synthesis on the chloroplast outer member system. *Proc. FEBS Meet.* **27**, 43–71.

Lafontaine, D., Beaudry, D., Marquis, P. and Perreault, J.P. (1995). Intra- and intermolecular nonenzymatic ligations occur within transcripts derived from peach latent mosaic viroid. *Virology* **212**, 705–709.

Lai, M.M.C. (1992). RNA recombination in animal and plant viruses. *Microbiol. Rev.* **56**, 61–79.

Laidlaw, W.M.R. (1986). Mechanical aids to improve the speed and sensitivity of plant virus diagnosis by the biological test method. *Ann. Appl. Biol.* **108**, 309–318.

Laidlaw, W.M.R. (1987). A new method for mechanical virus transmission and factors affecting its sensitivity. *OEPP/EPPO Bull.* **17**, 81–89.

Lakshman, D.K. and Gonsalves, D. (1985). Genetic analysis of two large-lesion isolates of cucumber mosaic virus. *Phytopathology* **75**, 758–762.

Lakshman, D.K., Gonsalves, D. and Fulton, R.W. (1985). Role of *Vigna* species in the appearance of pathogenic variants of cucumber mosaic virus. *Phytopathology* **75**, 751–757.

Lakshman, D.K., Hiruki, C., Wu, X.N. and Leung, W.C. (1986). Use of [^{32}P]RNA probes for the dot-hybridization detection of potato spindle tuber viroid. *J. Virol. Methods* **14**, 309–319.

Lam, Y.H., Wong, Y.S., Wang, B., Wong, R.N.S., Yeung, H.W. and Shaw, P.C. (1996). Use of trichosanthin to reduce infection by turnip mosaic virus. *Plant Sci.* **114**, 111–117.

Lambowitz, A.M. and Belford, M. (1993). Introns as mobile genetic elements. *Annu. Rev. Biochem.* **62**, 587–622.

Lana, A.F. (1981). Prospects of infectivity tests as a tool in plant virus disease diagnosis in the third world. *Trop. Pest Manage.* **27**, 24–28.

Lanczycki, C.J., Johnson, C.A., Trus, B.L., Conway, J.F., Steven, A.C. and Martino, R.L. (1998). Parallel computing strategies for determining viral capsid structure by cryoelectron microscopy. *IEEE Comput. Sci. Eng.* **5**, 76–91.

Lange, L. (1975). Infection of *Daucus carota* by tobacco necrosis virus. *Phytopathol. Z.* **83**, 136–143.

Lange, L. and Heide, M. (1986). Dot immunobinding (DIB) for detection of virus in seed. *Can. J. Plant Pathol.* **8**, 373–379.

Langenberg, W.G. (1979). Chilling of tissue before glutaraldehyde fixation preserves fragile inclusions of several plant viruses. *J. Ultrastruct. Res.* **66**, 120–131.

Langenberg, W.G. (1982). Fixation of plant inclusions under conditions designed for freeze-fracture. *J. Ultrastruct. Res.* **81**, 184–188.

Langenberg, W.G. (1986a). Virus protein associated with the cylindrical inclusions of two viruses that infect wheat. *J. Gen. Virol.* **67**, 1161–1168.

Langenberg, W.G. (1986b). Deterioration of several rod-shaped wheat viruses following antibody decoration. *Phytopathology* **76**, 339–341.

Langenberg, W.G. (1987). Barley stripe mosaic virus but not brome mosaic virus binds to wheat streak mosaic virus cylindrical inclusions *in vivo*. *Phytopathology* **78**, 589–594.

Langenberg, W.G. (1989). Rapid antigenic modification of wheat streak mosaic virus *in vitro* is prevented in glutaraldehyde fixed tissue. *J. Gen. Virol.* **70**, 969–973.

Langenberg, W.G. and Schroeder, H.F. (1975). The ultrastructural appearance of cowpea mosaic virus in cowpea. *J. Ultrastruct. Res.* **51**, 166–175.

Langereis, K., Mugnier, M.-A., Cornelissen, B.J.C., Pinck, L. and Bol, J.F. (1986). Variable repeats and poly (A)-stretches in the leader sequence of alfalfa mosaic virus RNA3. *Virology* **154**, 409–414.

Langeveld, S.A., Dore, J.M., Memelink, J. *et al.* (1991). Identification of potyviruses using polymerase chain reaction with degenerate primers. *J. Gen. Virol.* **72**, 1531–1541.

Lanter, J.M., McQuire, J.M. and Goode, M.J. (1982). Persistence of tomato mosaic virus in tomato debris and soil under field conditions. *Plant Dis. Rep.* **66**, 552–555.

Lapierre, H., Cortillot, M., Kusiak, C. and Hariri, D. (1985). Field resistance of autumn-sown wheat to wheat soil–borne mosaic virus (WSBMV) *Agronomie* **5**, 565–572.

Laquel, P., Ziegler, V. and Hirth, L. (1986). The 80K polypeptide associated with the replication complexes of cauliflower mosaic virus is recognised by antibodies to gene V translation product. *J. Gen. Virol.* **67**, 197–201.

Larson, R.H., Matthews, R.E.F. and Walker, J.C. (1950). Relationships between certain viruses affecting the genus *Brassica*. *Phytopathology* **40**, 955–962.

Larson, S.B., Koszelak, S., Day, J., Greenwood, A., Dodds, J.A. and McPherson, A. (1993a). Double helical RNA in satellite tobacco mosaic virus. *Nature (London)* **361**, 179–182.

Larson, S.B., Koszelak, S., Day, J., Greenwood, A., Dodds, J.A. and McPherson, A. (1993b). Three-dimensional structure of satellite tobacco mosaic virus at 2.9 Å resolution. *J. Mol. Biol.* **231**, 375–391.

Lartey, R. and Citovsky, V. (1997). Nucleic acid transport in plant–pathogen interactions. *Genet. Eng. (NY)* **19**, 201–214.

Lartey, R., Ghoshroy, S., Ho J. and Citovsky, V. (1997). Movement and subcellular localization of a tobamovirus in *Arabidopsis*. *Plant J.* **12**, 537–545.

Lartey, R.T., Ghoshroy, S. and Citovsky, V. (1998). Identification of an *Arabidopsis thaliana* mutation (*vsm1*) that restricts systemic movement of tobamoviruses. *Mol. Plant–Microb. Interact.* **11**, 706–709.

Lastra, J.R. and Esparza, J. (1976). Multiplication of vesicular stomatitis virus in the leafhopper *Peregrinus maidis* (Ashm), a vector of a plant rhabdovirus. *J. Gen. Virol.* **32**, 139–142.

Lastra, J.R. and Schlegel, D.E. (1975). Viral protein synthesis in plants infected with broadbean mottle virus. *Virology* **65**, 16–26.

Latham, J.R., Saunders, K., Pinner, M.S. and Stanley, J. (1997). Induction of plant cell division by beet curly top virus gene C4. *Plant J.* **11**, 1273–1283.

Lauber, E., Guilley, H., Tamada, T., Richards, K.E. and Jonard, G. (1998a). Vascular movement of beet necrotic yellow vein virus in *Beta macrocarpa* is probably dependent on an RNA3 sequence domain rather than a gene product. *J. Gen. Virol.* **79**, 385–393.

Lauber, E., Bleykasten-Grosshans, C., Erhardt, M. *et al.* (1998b). Cell-to-cell movement of beet necrotic yellow vein virus: 1. Heterologous complementation experiments provide evidence for specific interactions among triple gene block proteins. *Mol. Plant–Microb. Interact.* **11**, 618–625.

Laufs, J., Traut, W., Heyraud, F. *et al.* (1995). *In vitro* cleavage and joining at the viral origin of replication by the replication initiator protein of tomato yellow leaf curl virus. *Proc. Natl Acad. Sci. USA* **92**, 3879–3883.

Laurence, J.A., Aluiso, A.L., Weinstein, L.H. and McCune, D.C. (1981). Effect of sulphur dioxide on southern bean mosaic and maize dwarf mosaic. *Environ. Poll. Ser. A* **24**, 185–191.

Laviolette, F.A. and Athow, K.L. (1971). Longevity of tobacco ringspot virus in soybean seed. *Phytopathology* **61**, 755.

Lawrence, D.M. and Jackson, A.O. (1999). Hordeiviruses. In: A. Granoff and R.G. Webster (eds) *Encyclopedia of Virology*, 2nd edn, pp.749–753. Academic Press, San Diego.

Lawrence, D.M. and Jackson, A.O. (2001). Requirements for cell-to-cell movement of barley stripe mosaic virus in monocot and dicot hosts. *Molec. Plant Pathol.* **2**, 65–75.

Lawrence, D.M., Solovyer, A.G., Morozov, S. *et al.* (2000). Genus *Hordeivirus*. In: van Regenmortel *et al.* (eds) *Seventh Report of the International Committee on the Taxonomy of Viruses*, pp. 899–904. Academic Press, San Diego.

Lawrence, J. (1714). *The Clergyman's Recreation*, 2nd edn. B. Lintott, London.

Lawson, C., Kaniewski, W., Haley, L. *et al.* (1990). Engineering resistance to mixed virus infection in a commercial potato cultivar: resistance to potato virus X and potato virus Y in transgenic Russet Burbank. *Biotechnology* **8**, 127–134.

Lawson, R.H. and Taconis, P.J. (1965). Transfer of dahlia mosaic virus with liquid nitrogen and relation of transfer to symptoms and inclusions. *Phytopathology* **55**, 715–718.

Lawson, R.H., Hearon, S.S. and Smith, F.F. (1971). Development of pinwheel inclusions associated with sweet potato russet crack virus. *Virology* **46**, 453–463.

Lawton, J.A., Estes, M.K. and Prasad, B.V.V. (2000). Mechanism of genome transcription in segmented dsRNA viruses. *Adv. Virus Res.* **55**, 185–229.

Lawton, M.A., Tierney, M.A., Nakamura, I. *et al.* (1987). Expression of a soybean β-conglycinin gene under the control of the cauliflower mosaic virus 35S and 19S promoters in transformed petunia tissue. *Plant Mol. Biol.* **9**, 315–324.

Lazarowitz, S.G. (1987). The molecular characterisation of geminiviruses. *Plant Mol. Biol. Rep.* **4**, 177–192.

Lazarowitz, S.G. (1988). Infectivity and complete nucleotide sequence of the genome of a South African isolate of maize streak virus. *Nucl. Acids Res.* **16**, 229–249.

Lazarowitz, S.G. (1991). Molecular characterisation of two bipartite geminiviruses causing squash leaf curl disease: role of transactivation and defective genomic components in determining host range. *Virology* **180**, 70–80.

Lazarowitz, S.G. and Beachy, R.N. (1999). Viral movement proteins as probes for investigating intracellular and intercellular trafficking in plants. *Plant Cell* **11**, 535–548.

Lazarowitz, S.G., Pinder, A.J., Damsteegt, V.D. and Rogers, S.G. (1989). Maize streak virus genes essential for systemic spread and symptom development. *EMBO J.* **8**, 1023–1032.

Lazarowitz, S.G., Wu, L.C., Rogers, S.G and Elmer, J.S. (1992). Sequence-specific interaction with the viral AL1 protein identifies a geminivirus DNA replication origin. *Plant Cell* **4**, 799–809.

Leapman, R.D. and Rizzo, N.W. (1999). Towards single atom analysis of biological structures. *Ultramicroscopy* **78**, 251–268.

Leathers, V., Tanguay, R., Kobayashi, M and Gallie, D.R. (1993). A phylogenetically conserved sequence within the viral 3′ untranslated RNA pseudoknots regulates translation. *Mol. Cell Biol.* **13**, 5331–5347.

Lebeurier, G. and Hirth, L. (1966). Effect of elevated temperatures on the development of two strains of tobacco mosaic virus. *Virology* **29**, 385–395.

Lebeurier, G., Nicolaieff, A. and Richards, K.E. (1977). Inside-out model for self-assembly of tobacco mosaic virus. *Proc. Natl Acad. Sci. USA* **74**, 149–153.

Lebeurier, G., Hirth, L., Hohn, B. and Hohn, T. (1982). *In vivo* recombination of cauliflower mosaic virus DNA. *Proc. Natl Acad. Sci. USA* **79**, 2932–2936.

Leclerc, D., Burri, L., Kajava, A.V., Mougeot, J.L., Hess, D., Lustig, A., Kleemann, G. and Hohn T. (1998). The open reading frame III product of cauliflower mosaic virus forms a tetramer through a N-terminal coiled-coil. *J. Biol. Chem.* **273**, 29015–29021.

Leclerc, D., Chapdelaine, Y. and Hohn, T. (1999). Nuclear targeting of the cauliflower mosaic virus coat protein. *J. Virol.* **73**, 553–560.

Lecoq, H. (1998). Control of plant virus diseases by cross protection. In: A. Hadidi, R.K. Khetarpal and H. Koganezawa (eds) *Plant Virus Disease Control*, pp. 33–40. APS Press, St. Paul, MN.

Lecoq, H. (1999). Epidemiology of cucurbit aphid-borne yellows virus. In: H.G. Smith and H. Barker (eds) *The Luteoviridae*, pp. 243–248. CAB International, Wallingford, UK.

Lecoq, H. and Pitrat, M. (1985). Specificity of the helper-component-mediated aphid transmission of three potyviruses infecting muskmelon. *Phytopathology* **75**, 890–893.

Lecoq, H., Pitrat, M. and Labonne, G. (1981). Resistance to virus transmission by aphids in a *Cucumis melo* lines presenting non-acceptance to *Aphis gossypii*. *Bull. SROP* **4**, 147–151.

Lecoq, H., Ravelonandro, M., Wopf-Scheibel, C., Monsion, M., Raccah, B. and Dunez, J. (1993). Aphid transmission of a non-aphid-transmissible strain of zucchini yellow mosaic potyvirus from transgenic plants expressing the capsid protein of plum pox potyvirus. *Mol. Plant Microbe Interact.* **6**, 403–406.

Lee, C.L. and Black, L.M. (1955). Anatomical studies of *Trifolium incarnartum* infected by wound tumor virus. *Am. J. Bot.* **42**, 160–168.

Lee, C.R. and Singh, R.P. (1972). Enhancement of diagnostic symptoms of potato spindle tuber virus by manganese. *Phtopathology* **62**, 516–520.

Lee, I.M., Hammond, R.W. Davis, R.E. and Gundersen, D.E. (1993). Universal amplification and analysis of pathogen 16S rRNA for classification and identification of mycoplasma-like organisms. *Phytopathology* **83**, 834–842.

Lee, I.M., Gundersen-Rindal, D.E. and Bertaccini, A. (1998a). Phytoplasma: ecology and genomic diversity. *Phytopathology* **88**, 1359–1366.

Lee, J.-M., Hartman, G.L., Domier, L.L. and Bent, A.F. (1996). Identification and map location of *TTR1*, a single locus in *Arabidopsis thaliana* that confers tolerance to tobacco ringspot nepovirus. *Mol. Plant Microbe Interact.* **9**, 729–735.

Lee, L. and Anderson, E.J. (1998). Nucleotide sequence of a resistance breaking mutant of southern bean mosaic virus. *Arch Virol.* **143**, 2189–2201.

Lee, P.E., Boerjan, M. and Peters, D. (1972). Electron microscopic evidence for a neuraminic acid in sowthistle yellow vein virus. *Virology* **50**, 309–311.

Lee, R.F., Garnsey, S.M., Briansky, R.H. and Goheen, A.C. (1987). A purification procedure for enhancement of citrus tristeza virus yields and its application to other phloem limited viruses. *Phytopathology* **77**, 543–549.

Lee, Y.-S. and Ross, J.P. (1972). Top necrosis and cellular changes in soybean doubly infected by soybean mosaic and bean pod mottle viruses. *Phytopathology* **62**, 839–845.

Lee, Y.-S., Lin, B.Y., Hsu, Y.H., Chang, B.Y. and Lin, N.S. (1998b). Subgenomic RNAs if bamboo mosaic potexvirus-V isolate are packaged into virions. *J. Gen. Virol.* **79**, 1825–1832.

Legg, J.T. and Lockwood, G. (1977). Evaluation and use of a screening method to aid selection of cocoa (*Theobroma cacao*) with field resistance to cocoa swollen-shoot virus in Ghana. *Ann. Appl. Biol.* **86**, 241–248.

Leh, V., Jacquot, E., Geldreich, A. *et al.* (1999). Aphid transmission of cauliflower mosaic virus requires the viral PIII protein. *EMBO J.* **18**, 7077–7085.

Leh, V., Yot, P. and Keller, M. (2000). The cauliflower mosaic virus translation transactivator interacts with the 60S ribosomal subunit protein L 18 of Arabidopsis thaliana. *Virology* **266**, 1–7.

Lehto, K. and Dawson, W.O. (1990). Changing the start codon context of the 30K gene of tobacco mosaic virus from 'weak' to 'strong' does not increase expression. *Virology* **174**, 169–176.

Lehto, K., Grantham, G.L. and Dawson, W.O. (1990a). Insertion of sequences containing the coat protein subgenomic RNA promoter and leader in front of the tobacco mosaic virus 30K ORF delays its expression and causes defective cell-to-cell movement. *Virology* **174**, 145–157.

Lehto, K., Bubrick, P. and Dawson, W.O. (1990b). Time course of TMV 30K protein accumulation in intact leaves. *Virology* **174**, 290–293.

Lei, J.D. and Agrios, G.N. (1986). Mechanisms of resistance in corn to maize dwarf mosaic virus. *Phytopathology* **76**, 1034–1040.

Leimkühler, M., Goldbeck, A., Lechner, M.D. and Witz, J. (2000). Conformational changes preceding decapsidation of bromegrass mosaic virus under hydrostatic pressure: a small-angle neutron scattering study. *J. Mol. Biol.* **296**, 1295–1305.

Leisner, R.-M., Ziegler-Graff, V., Reutenauer, A. *et al.* (1992a). Agroinfection as an alternative to insects for infecting plants with beet western yellows luteovirus. *Proc. Natl Acad. Sci. USA* **89**, 9136–9140.

Leisner, S.M., Turgeon, R. and Howell, S.H. (1992b). Long-distance movement of cauliflower mosaic virus in infected turnip plants. *Mol. Plant–Microb. Interact.* **5**, 41–47.

Leisner, S.M., Agama, K., Hapiak, M. and Li, Y. (2001). Identification of cauliflower mosaic virus gene VI protein regions involved in resistance breakage and self association. Abstract W21–6, 20th Annual Meeting of the American Society for Virology.

Lekkerkerker, A.-M., Wellink, J., Yuan, P., van Lent, J., Goldbach, R. and van Kammen, A. (1996). Distinct functional domains in the cowpea mosaic virus movement protein. *J. Virol.* **70**, 5658–5661.

Leonard, D.A. and Zaitlin, M. (1982). A temperature-sensitive strain of tobacco mosaic defective in cell-to-cell movement generates an altered viral-coded protein. *Virology* **117**, 416–424.

Léonard, S., Plante, D., Wittmann,S., Daigneault, N. Fortin, M.G. and Laliberté J.-F. (2000). Complex formation between potyvirus VPg and translation eukaryotic initiation factor 4E correlates with virus infectivity. *J. Virol.* **74**, 7730–7737.

Leppik, E.E. (1970). Gene centers of plants as sources of disease resistance. *Annu. Rev. Phytopathol.* **8**, 323–344.

Lesemann, D.-E. (1988). Cytopathology. In: R.G. Milne (ed.) *The Plant Viruses. Vol. 4: The Filamentous Plant Viruses*, pp. 179–235. Plenum Press, New York.

Lesemann, D.E., Koenig, R., Torrance, L. *et al.* (1990). Electron-microscopical demonstration of different binding sites for monoclonal antibodies on particles of beet necrotic yellow vein virus. *J. Gen. Virol.* **71**, 731–733.

Leung, D.W., Gilbert, C.W., Smith, R.E., Sasavage, N.L. and Clark, J.M., Jr. (1976). Translation of satellite tobacco necrosis virus ribonucleic acid by an *in vitro* system from wheat germ. *Biochemistry* **15**, 4943–4950.

Levy, D. and Marco, S. (1976). Involvement of ethylene in epinasty of CMV-infected cucumber cotyledons which exhibit increased resistance to gaseous diffusion. *Physiol. Plant Pathol.* **9**, 121–126.

Lewandowski, D.J. and Dawson, W.O. (1993). A single amino acid change in tobacco mosaic virus replicase prevents symptom production. *Mol. Plant-Microbe Interact* **6**, 157–160.

Lewandowski, D.J. and Dawson, W.O. (1998a). Deletion of internal sequences results in tobacco mosaic virus defective RNAs that accumulate to high levels without interfering with replication of the helper virus. *Virology* **251**, 427–437.

Lewandowski, D.J. and Dawson, W.O. (1998b). The *cis*-preferential nature of tobacco mosaic virus min-strand synthesis can be uncoupled from translation. In *Abstracts of the American Society for Virology*, Vancouver, P2-06.

Lewandowski, D.J. and Dawson, W.O. (1999). Tobamoviruses. In: A. Granoff and R.G. Webster (eds) *Encyclopedia of Virology*, 2nd edn, pp. 1780–1783. Academic Press, San Diego.

Lewandowski, D.J. and Dawson, W.O. (2000). Functions of the 126- and 183-kDa proteins of tobacco mosaic virus. *Virology* **271**, 90–98.

Lewis, T. and Dibley, G.C. (1970). Air movement near windbreaks and a hypothesis of the mechanism of the accumulation of airborne insects. *Ann. Appl. Biol.* **66**, 477–484.

Li, H.-W., Lucy, A.P., Guo, H.-S. *et al.* (1999). Strong host resistance targeted against a viral suppressor of the plant gene silencing defence mechanism. *EMBO J.* **18**, 2683–2691.

Li, W.-H., Wolfe, K.H., Sourdis, J. and Sharp, P.M. (1987). Reconstruction of phylogenetic trees and estimation of divergence times under non-constant rates of evolution. *Cold Spring Harbor Symp. Quant. Biol.* **52**, 847–856.

Li, W.-Z., Qu, F. and Morris, T.J. (1998). Cell-to-cell movement of turnip crinkle virus is controlled by two small open reading frames that function *in trans*. *Virology* **244**, 405–418.

Li, X. and Carrington, J.C. (1995). Complementation of tobacco etch potyvirus mutants by active RNA polymerase expressed in transgenic cells. *Proc. Natl Acad. Sci. USA* **92**, 457–461.

Li, X., Heaton, L.A., Morris, J. and Simon, A.E. (1989). Turnip crinkle defective interfering RNAs intensify viral symptoms and are generated *de novo*. *Proc. Natl Acad. Sci. USA* **86**, 9173–9177.

Li, X., Valdez, P., Olvera, R.E. and Carrington, J.C. (1997). Functions of the tobacco etch virus RNA polymerase (NIb): subcellular transport and protein-protein interaction with VPg/proteinase (NIa). *J. Virol.* **71**, 1598–1607.

Li, X., Ryan, M.D. and Lamb, J.W. (2000). Potato leafroll virus protein P1 contains a serine proteinase domain. *J. Gen. Virol.* **81**, 1857–1864.

Liao, C.L. and Lai, M.M. (1994). Requirement of the 5'-end genomic sequence as an upstream *cis*-acting element for coronavirus subgenomic mRNA transcription. *J. Virol.* **68**, 4727–4737.

Liebe, S. and Quader, H. (1994). Myosin in onion (*Allium cepa*) bulb scale epidermal cells: involvement in the dynamics of organelles and endoplasmic reticulum. *Physiol. Plant* **90**, 114–124.

Liljàs, L., Unge, T., Alwyn Jones, T. *et al.* (1982). Structure of satellite tobacco necrosis virus at 3.0 Å resolution. *J. Mol. Biol.* **159**, 93–108.

Lilley, C.J., Devlin, F., Urwin, P.E. and Atkinson, H.J. (1999). Parasitic nematodes, proteinases and transgenic plants. *Parasitol. Today* **15**, 414–417.

Lim, T.M., Chng, C.G. and Wong, S.M. (1996). Study of the three-dimensional images of potyvirus-induced cytoplasmic inclusions using confocal laser scanning microscopy. *J. Virol. Meth.* **60**, 139–145.

Lima, M.I., Fonseca, M.E.N., Flores, R. and Kitajima, E.W. (1994). Detection of avocado sunblotch viroid in chloroplasts of avocado leaves by *in situ* hybridization. *Arch Virol.* **138**, 385–390.

Lin, B. and Heaton, L.A. (1999). Mutational analyses of the putative calcium binding site and hinge of the turnip crinkle virus coat protein. *Virology* **259**, 34–42.

Lin, B. and Heaton, L.A. (2001). An *Arabidopsis thaliana* protein interacts with a movement protein of turnip crinkle virus in yeast cells and *in vitro*. *J. Gen. Virol.* **82**, 1245–1251.

Lin, N.-S. and Langenberg, W.G. (1984a). Chronology of appearance of barley stripe mosaic virus protein in infected wheat cells. *J. Ultrastruct. Res.* **89**, 309–323.

Lin, N.-S. and Langenberg, W.G. (1984b). Distribution of barley stripe mosaic virus protein in infected wheat root and shoot tips. *J. Gen. Virol.* **65**, 2217–2224.

Lin, N.-S. and Langenberg, W.G. (1985). Peripheral vesicles in proplastids of barley stripe mosaic virus-infected wheat cells contain double-stranded RNA. *Virology* **142**, 291–298.

Lin, N.-S., Hsu, Y.-H. and Chiu, R.-J. (1987). Identification of viral structural proteins in the nucleoplasm of potato yellow dwarf virus-infected cells. *J. Gen. Virol.* **68**, 2723–2728.

Lin, N.-S., Lin, F.Z., Huang, T.Y. and Hsu, Y.H. (1992). Genome properties of bamboo mosaic virus. *Phytopathology* **82**, 731–734.

Lin, N.-S., Lin, B.-Y., Lo, N.-W., Hu, C.-C., Chow, T.-W. and Hsu, Y.-H. (1994). Nucleotide sequence of the genomic RNA of bamboo mosaic potexvirus. *J. Gen. Virol.* **75**, 2513–2518.

Lin, N.-S., Lee, Y.-S., Lin, B.-Y., Lee, C.W. and Hsu, Y.-H. (1996). The open reading frame of bamboo mosaic potexvirus satellite RNA is not essential for its replication and can be replaced by a bacterial gene. *Proc. Natl Acad. Sci. USA* **93**, 3138–3142.

Lin, T.W., Chen, Z.G., Usha, R. *et al.* (1999). The refined crystal structure of cowpea mosaic virus at 2.8 angstrom resolution. *Virology* **265**, 20–34.

Lin, T.W., Clark, A.J., Chen, Z.G. *et al.* (2000). Structural fingerprinting: subgrouping of comoviruses by structural studies of red clover mottle virus to 2.4-Å resolution and comparisons with other comoviruses. *J. Virol.* **74**, 493–504.

Lindbeck, A.G.C., Dawson, W.O. and Thomson, W.W. (1991). Coat protein-related polypeptides from *in vitro* tobacco mosaic virus coat protein mutants do not accumulate in the chloroplasts of directly inoculated leaves. *Mol. Plant Microbe Interact.* **4**, 89–94.

Lindbeck, A.G.C., Lewandowski, D.J., Culver, J.N., Thomson, W.W. and Dawson, W.O. (1992). Mutant coat protein of tobacco mosaic virus induces acute chlorosis in expanded and developing tobacco leaves. *Mol. Plant Microbe Interact.* **5**, 235–241.

Lindbo, J.A. and Dougherty, W.G. (1992a). Pathogen-derived resistance to a potyvirus: immune and resistant phenotypes in transgenic tobacco expressing altered forms of a potyvirus coat protein nucleotide sequence. *Mol. Plant Microb. Interact.* **5**, 144–153.

Lindbo, J.A. and Dougherty, W.G. (1992b). Untranslatable transcripts of the tobacco etch virus coat protein gene sequence can interfere with tobacco etch virus replication in transgenic plants and protoplasts. *Virology* **189**, 725–733.

Lindbo, J.A., Silva-Rosales, L., Proebsting, W.M. and Dougherty, W.G. (1993). Induction of a highly specific antiviral state in transgenic plants: implications for regulation of gene expression and virus resistance. *Plant Cell* **5**, 1743–1759.

Lindbo, J.A., Fitzmaurice, W.P. and della-Cioppa, G. (2001). Virus-mediated reprogramming of gene expression in plants. *Curr. Opin. Plant Biol.* **4**, 181–185.

Linstead, P.J., Hills, G.J., Plaskitt, K.A., Wilson, I.G., Harker, C.L. and Maule, A.J. (1988). The subcellular location of the gene I product of cauliflower mosaic virus is consistent with a function associated with virus spread. *J. Gen. Virol.* **69**, 1809–1818.

Linthorst, H.J.M. (1991). Pathogenesis-related proteins of plants. *Crit. Rev. Plant Sci.* **10**, 123–150.

Linthorst, H.J.M. and Kaper, J.M. (1984). Circular satellite-RNA molecules in satellite of tobacco ringspot virus-infected tissue. *Virology* **137**, 206–210.

Linthorst, H.J.M. and Kaper, J.M. (1985). Cucumovirus satellite RNAs cannot replicate autonomously in cowpea protoplasts. *J. Gen. Virol.* **66**, 1839–1842.

Linthorst, H.J.M. Meuwissen, R.L.J., Kayffmann, S. and Bol., J.F. (1989). Constitutive expression of pathogenesis-related PR-1, GRP, and PR-S in tobacco has no effect on virus infection. *Plant Cell* **1**, 285–291.

Lisa, V. and Boccardo, G. (1996). Fabaviruses: Broad bean wilt and allied viruses. In: B.D. Harrison and A.F. Murant (eds) *The Plant Viruses. Vol. 5: Polyhedral Virions and Bipartite Genomes*, pp. 229–250. Plenum Press, New York.

Lisa, V., Luisoni, E. and Milne, R.G. (1986). Carnation cryptic virus. AAB Descriptions Plant Viruses No. 315.

Lisa, V., Milne, R.G., Accotto, G.P., Boccardo, G., Caciagli, P. and Parvizy, R. (1988). Ourmia melon virus, a virus from Iran with novel properties. *Ann. Appl. Biol.* **112**, 291–302.

Lister, R.M. (1960). Transmission of soilborne virus through seed. *Virology* **10**, 547-549.

Lister, R.M. (1966). Possible relationship of virus specific products of tobacco rattle virus infections. *Virology* **28**, 350–353.

Lister, R.M. (1968). Functional relationships between virus-specific products of infection by viruses of the tobacco rattle type. *J. Gen. Virol.* **2**, 43–58.

Lister, R.M. and Murant, A.F. (1967). Seed transmission of nematode-borne viruses. *Ann. Appl. Biol.* **59**, 49–62.

Litvak, S., Tarragó, A., Tarragó-Litvak, L. and Allende, J.E. (1973). Elongation factor viral genome interaction dependent on, the aminoacylation of TYMV and TMV RNAs. *Nature (London)* **241**, 88–90.

Liu, B.L., Tabler, M. and Tsagris, M. (2000). Episomal expression of a hammerhead ribozyme directed against plum pox virus. *Virus Res.* **68**, 15–23.

Liu, H., Boulton, M.I. and Davies, J.W. (1997a). Maize streak virus coat protein binds single- and double-stranded DNA *in vitro*. *J. Gen Virol.* **78**, 1265–1270.

Liu, H., Boulton, M.I., Thomas, C.L., Ptior, D.A.M, Oparka, K.J. and Davies, J.W. (1999a). Maize streak virus coat protein is karyophyllic and facilitates nuclear transport of viral DNA. *Mol. Plant–Microb. Interact.* **12**, 894–900.

Liu, H., Boulton, M.I., Oparka, K.J. and Davies, J.W. (2001). Interaction of the movement and coat proteins of maize streak virus: implications for the transport of viral DNA. *J. Gen. Virol.* **82**, 35–44.

Liu, L., Sanders, K., Thomas, C.L., Davies, J.W. and Stanley, J. (1999b). Bean yellow dwarf virus RepA, but not Rep, binds to maize retinoblastoma protein, and the virus tolerates mutations in the consensus binding motif. *Virology*. **256**, 270–279.

Liu, S., Bedford, I.D., Briddon, R.W. and Markham, P.G. (1997b). Efficient whitefly transmission of African cassava mosaic geminivirus requires sequences from both genomic components. *J. Gen. Virol.* **78**, 1791–1794.

Liu, S., Briddon, R.W., Bedford, I.D., Pinner, M.S. and Markham, P.G. (1998). Identification of genes directly and indirectly involved in the insect transmission of African cassava mosaic geminivirus by *Bemisia tabaci*. *Virus Genes* **18**, 5–11.

Liu, Y.H. and Symons, R.H. (1998). Specific RNA self-cleavage in coconut cadang cadang viroid: potential for a role in rolling circle replication. *RNA* **4**, 418–429.

Llave, C., Kasschau, K.D. and Carrington, J.C. (2000). Virus-encoded suppressor of posttranscriptional gene silencing targets a maintenance step in the silencing pathway. *Proc. Natl Acad. Sci. USA* **97**, 13401–13406.

Lloyd, A.M., Bamason, A.R., Rogers, S.G., Byme, M.C. Fraley, R.T. and Horsch, R.B. (1986). Transformation of *Arabidopsis thaliana* with *Agrobacterium tumefaciens*. *Science* **234**, 464–466.

Lloyd, C. and Hussey, P. (2001). Microtubule-associated proteins in plants—why we need a map. *Nature Rev. Mol. Cell Biol.* **2**, 40–47.

Lockhart, B.E. and Olszewski, N.E. (1999). Plant pararetroviruses—badnaviruses. In: A. Granoff and R.G. Webster (eds) *Encyclopedia of Virology*, 2nd edn, pp. 1296–1300. Academic Press, San Diego.

Lockhart, B.E., Menke, J., Dahal, G. and Olszewski, N.E. (2000). Characterization and genomic analysis of tobacco vein-clearing virus, a plant pararetrovirus that is transmitted vertically and related to sequences integrated in the host genome. *J. Gen. Virol.* **81**, 1579–1585.

Lockhart, B.E.L. (1986). Purification and serology of a bacilliform virus associated with streak disease of banana. *Phytopathology* **80**, 127–131.

Lodge, J.K., Kaniewski, W.K. and Tumer, N.E. (1993). Broad-spectrum virus resistance in transgenic plants expressing pokeweed antiviral protein. *Proc. Natl Acad. Sci. USA* **90**, 7089–7093.

Loebenstein, G. and Gera, A. (1981). Inhibitor of virus replication released from tobacco mosaic virus infected protoplasts of local lesion-responding tobacco cultivar. *Virology* **114**, 132–139.

Loebenstein, G., Cohen, J., Shabtai, S., Coutts, R.H.A. and Wood, K.R. (1977). Distribution of cucumber mosaic virus in systemically infected tobacco leaves. *Virology* **81**, 117–125.

Loebenstein, G., Gera, A. and Gianinazzi, S. (1990). Constitutive production of an inhibitor of virus replication in the interspecific hybrid *Nicotiana glutinosa* × *N. debneyi*. *Physiol. Mol. Plant Pathol.* **37**, 145–151.

Loesch-Fries, L.S and Hall, T.C. (1980). Synthesis, accumulation and encapsulation of individual brome mosaic virus RNA components in barley protoplasts. *J. Gen. Virol.* **47**, 323–332.

Loesch-Fries, L.S. and Hall, T.C. (1982). *In vivo* aminoacylation of brome mosaic and barley stripe mosaic virus RNAs. *Nature (London)* **298**, 771–773.

Loesch-Fries, L.S., Merlo, D., Zinnen, T. *et al.* (1987). Expression of alfalfa mosaic virus RNA4 in transgenic plants confers resistance. *EMBO J.* **6**, 1845–1852.

Logan, A.E. and Zettler, F.W. (1985). Rapid in vitro propagation of virus-indexed gladioli. *Acta Hortic.* **164**, 169–180.

Lommel, S.A. (1999a). Dianthoviruses (*Tombusviridae*). In: A. Granoff and R.G. Webster (eds) *Encyclopedia of Virology*, 2nd edn, pp. 403–409. Academic Press, San Diego.

Lommel, S.A. (1999b). Machlomoviruses (*Tombusviridae*). In: A. Granoff and R.G. Webster (eds) *Encyclopedia of Virology*, 2nd edn, pp. 935–939. Academic Press, San Diego.

Lommel, S.A., Weston-Fina, M., Xiong, Z. and Lomonossoff, G.P. (1988). The nucleotide sequence and gene organisation of red clover necrotic mosaic virus RNA2. *Nucl. Acids Res.* **16**, 8587–8602.

Lommel, S.A., Kendall, R., Sui, N.F. and Nutter, R.C. (1991). Characterization of maize chlorotic mottle virus. *Phytopathology* **81**, 819–823.

Lommel, S.A., Martelli, G.P. and Russo, M. (2000). Family *Tombusviridae*. In: M.H.V. van Regenmortel *et al.* (eds) *Virus Taxonomy. Seventh Report of the International Committee on Taxonomy of Viruses*. Academic Press, New York.

Lomonossoff, G.P. (1995). Pathogen-derived resistance to plant viruses. *Annu. Rev. Phytopathol.* **33**, 323–343.

Lomonossoff, G.P. and Hamilton, W.D.O. (1999). Cowpea mosaic virus-based vaccines. *Curr. Topic. Microbiol. Immunol.* **240**, 177–189.

Lomonossoff, G.P. and Johnson, J.E. (1991). The synthesis and structure of comovirus capsids. *Prog. Biophys. Mol. Biol.* **55**, 107–137.

Lomonossoff, G.P. and Johnson, J.E. (1995). Eukaryotic viral expression systems for polypeptides. *Semin. Virol.* **6**, 257–267.

Lomonossoff, G.P. and Johnson, J.E. (1996). Use of macromolecular assemblies as expression systems for peptides and synthetic vaccines. *Curr. Opin. Struct. Biol.* **6**, 176–182.

Lomonossoff, G.P. and Shanks, M. (1983). The nucleotide sequence of cowpea mosaic virus B RNA. *EMBO J.* **2**, 2253–2258.

Lomonossoff, G.P. and Shanks, M. (1999). Comoviruses (*Comoviridae*). In: A. Granoff and R.G. Webster (eds) *Encyclopedia of Virology*, 2nd edn, pp. 285–291. Academic Press, San Diego.

Lomonossoff, G.P. and Wilson, T.M.A. (1985). Structure and *in vitro* assembly of tobacco mosaic virus. In: J.W. Davies (ed.) *Molecular Plant Virology*, Vol. 1, pp. 43–83. CRC Press, Boca Raton.

Lomonossoff, G.P., Shanks, M. and Evans, D. (1985). The structure of cowpea mosaic virus replicative form RNA. *Virology* **144**, 351–362.

Long, D.G., Borsa, J. and Sargent, M.D. (1976). A potential artifact generated by pelleting viral particles during preparative ultra-centrifugation. *Biochim. Biophys. Acta* **451**, 639–642.

Loor, F. (1967). Comparative immunogenicities of tobacco mosaic virus protein subunits and reaggregated protein subunits. *Virology* **33**, 215–220.

López , G., Navas-Castello, J., Gowda, S., Moreno, P. and Flores, R. (2000). The 25-kDa protein encoded by the 3′ terminal gene of citrus tristeza virus is an RNA-binding protein. *Virology* **269**, 462–470.

López, L., Urzainqui, A., Domínguez, E. and García, J.A. (2001). Identification of an N-terminal domain of the plum pox potyvirus CI RNA helicase involved in self-interaction in a yeast two-hybrid system. *J. Gen. Virol.* **82**, 677–686.

López-Moya, J.J. and Garcia, J.A. (1999). Potyviruses (*Potyviridae*). In: A. Granoff and R.G. Webster (eds) *Encyclopedia of Virology*, 2nd edn, pp. 1369–1375. Academic Press, San Diego.

López-Moya, J.J., Wang, R.Y. and Pirone, T.P. (1999). Context of the coat protein DAG motif affects potyvirus transmissibility by aphids. *J. Gen. Virol.* **80**, 3281–3288.

Lopinski, J.D., Dinman, J.D. and Bruenn, J.A. (2000). Kinetics of ribosomal pausing during programmed −1 translational frameshifting. *Molec. Cell. Biol.* **20**, 1095–1103.

Lot, H. and Kaper, J.M. (1976). Physical and chemical differentiation of three strains of cucumber mosaic virus and peanut stunt virus. *Virology* **74**, 209–222.

Lot, H., Rubino, L., Delecolle, B., Jaquemond, M., Turturo, C. and Russo, M. (1996). Characterization, nucleotide sequence and genome organization of leek stripe virus, a putative new species of the genus Necrovirus. *Arch. Virol.* **141**, 2375–2386.

Lough, T.J., Shash, K., Xoconostle-Càzares, B. *et al.* (1998). Molecular dissection of the mechanism by which potexvirus triple gene block proteins mediate cell-to-cell transport of infectious RNA. *Mol. Plant–Microb. Interact.* **8**, 801–814.

Louie, R. (1995). Vascular puncture of maize kernels for the mechanical transmission of maize white line mosaic virus and other viruses of maize. *Phytopathology* **85**, 139–143.

Louie, R. and Knoke, J.K. (1975). Strains of maize dwarf mosaic virus. *Plant Dis. Rep.* **59**, 518–522.

Louie, R. and Lorbeer, J.W. (1966). Mechanical transmission of onion yellow dwarf virus. *Phytopathology* **56**, 1020–1023.

Louie, R., Findley, W.R. and Knoke, J.K. (1976). Variation in resistance within corn inbred lines to infection by maize dwarf mosaic virus. *Plant Dis. Rep.* **60**, 838–842.

Louie, R., Knoke, J.K. and Reichard, D.L. (1983). Transmission of maize dwarf mosaic virus with solid-stream inoculum. *Plant Dis.* **67**, 1328–1331.

Lovisolo, O. (1985). International transport of flowers, foliage, nursery stock and ornamental plants in Europe and the Mediterranean basin. *Acta Hortic.* **164**, 139–151.

Lozoya-Saldaña, H. and Dawson, W.O. (1982). Effect of alternating temperature regimes on reduction or elimination of viruses in plant tissues. *Phytopathology* **72**, 1059–1064.

Lozoya-Saldaña, H., Abello, F. and de la Garcia, G. (1996). Electrotherapy ands shoot tip culture eliminate potato virus X in potatoes. *Am. Potato J.* **73**, 149–154.

Lu, B., Stubbs, G. and Culver, J.N. (1996). Carboxylate interactions involved in the disassembly of tobacco mosaic tobamovirus. *Virology* **225**, 11–20.

Lu, B., Taraporewala, Z.F., Stubbs, G. and Culver, J.N. (1998a). Intersubunit interactions allowing a carboxylate mutant coat protein to inhibit tobamovirus disassembly. *Virology* **244**, 13–19.

Lu, G., Zhou, Z.H., Baker, M.L. *et al.* (1998b). Structure of double-shelled rice dwarf virus. *J. Virol.* **72**, 8541–8549.

Lucas, W.J. and Gilbertson, R.L. (1994). Plasmodesmata in relation to viral movement within leaves. *Annu. Rev. Phytopathol.* **32**, 387–411.

Lucas, W.J., Bouché-Pillon, S., Jackson, D.P. *et al.* (1995). Selective trafficking of KNOTTED1 homeodomain protein and its mRNA through plasmodesmata. *Science* **270**, 1980–1983.

Lucknow, V.A. and Summers, M.D. (1988). Signals important for high-level expression of foreign genes in *Autographa californica* nuclear polyhedrosis virus expression vectors. *Virology* **167**, 56–71.

Lucy, A.P., Boulton, M.I., Davies, J.W. and Maule, A.J. (1996). Tissue specificity of *Zea mays* infection by maize streak virus. *Mol. Plant-Microb. Interact.* **9**, 22–31.

Lucy, A.P., Guo, H.-S., Li, W.-X. and Ding, S.-W. (2000). Suppression of post-transcriptional gene silencing by a plant viral protein localized in the nucleus. *EMBO J.* **19**, 1672–1680.

Ludueña, R.F. (1998). Multiple forms of tubulin: different gene products and covalent modifications. *Int. Rev. Cytol.* **178**, 207–275.

Luhn, C.F. and Goheen, A.C. (1970). Viruses in early California grapevines. *Plant Dis. Rep.* **54**, 1055–1056.

Luisoni, E., Milne, R.G. and Boccardo, G. (1975). The maize rough dwalf virion: II. serological analysis. *Virology* **68**, 86–96.

Lukács, N. (1994). Detection of virus infection in plants and differentiation between coexisting viruses by monoclonal antibodies to double-stranded RNA. *J. Virol. Methods* **47**, 255–272.

Lundquist, R.E., Lazar, J.M., Klein, W.H. and Clark, J.M. (1972). Translation of satellite tobacco necrosis virus ribonucleic acid: II. Initiation of *in vitro* translation in prokaryotic and eukaryotic systems. *Biochemistry* **11**, 2014–2019.

Lundsgaard, T. (1992). N protein of festuca leaf streak virus (*Rhabdoviridae*) detected in cytoplasmic viroplasms by immunogold labeling. *J. Phytopathol.* **134**, 27–32.

Lundsgaard, T. (1995). Routing of the G protein during maturation of festuca leaf streak rhabdovirus in its plant host. *J. Phytopathol.* **143**, 479–483.

Lupo, R., Robino, L. and Russo, M. (1994). Immunodetection of the 33K/98K polymerase proteins in cymbidium ringspot virus-infected and in transgenic plant- tissue extracts. *Arch. Virol.* **138**, 135–142.

Lusso, M. and Kuc, J. (1996). The effect of sense and antisense expression of the PR-N gene for beta-1,3–glucanase on disease resistance of tobacco to fungi and viruses. *Phys. Mol. Plant Pathol.* **49**, 267–283.

Lynch, M., Bürger, R., Butcher, D. and Gabriel, W. (1993). The mutational meltdown in asexual populations. *J. Hered.* **84**, 339–344.

MacDonald, R.J.H., Coutts, R.H.A. and Buck, K.W. (1988a). Characterization of a subgenomic DNA isolated from *Triticum aestivium* plants infected with wheat dwalf virus. *J. Gen. Virol.* **69**, 1339–1344.

MacDonald, R.J.H., Coutts, R.H.A. and Buck, K.W. (1988b). Priming of complementary DNA synthesis *in vitro* by small DNA molecules tightly bound to virion DNA of wheat dwarf virus. *J. Gen. Virol.* **69**, 1345–1350.

MacDowell, S.W., MacDonald, R.J.H., Hamilton, W.D.O., Coutts, R.H.A. and Buck, K.W. (1985). The nucleotide sequence of cloned wheat dwarf virus DNA. *EMBO J.* **4**, 2173–2180.

MacDowell, S.W., Coutts, R.H.A. and Buck, K.W. (1986). Molecular characterisation of subgenomic single-stranded and double-stranded DNA forms isolated from plants infected with tomato golden mosaic virus. *Nucl. Acids Res.* **14**, 7967–7984.

MacFarlane, S.A. and Popovich, A.H. (2000). Efficient expression of foreign proteins in roots from tobravirus vectors. *Virology* **267**, 29–35.

MacFarlane, S.A., Taylor, S.C., King, D.I., Hughes, G. and Davies, J.W. (1989). Pea early browning virus RNA1 encodes four polypeptides including a putative zinc-finger protein. *Nucl. Acids Res.* **17**, 2245–2260.

MacFarlane, S.A., Brown, D.J.F. and Bol, J.F. (1995). The transmission by nematodes of tobraviruses is not determined exclusively by the viral coat protein. *Eur. J. Plant Pathol.* **101**, 535–539.

MacFarlane, S.A., Wallis, C.V. and Brown, D.J.F. (1996). Multiple virus genes involved in the nematode transmission of pea early browning virus. *Virology* **219**, 417–422.

MacKenzie, D.J. and Tremaine, J.H. (1986). The use of a monoclonal antibody specific for the N-terminal region of southern bean mosaic virus as a probe of virus structure. *J. Gen. Virol.* **67**, 727–735.

MacKenzie, D.R., Cole, H., Smith, C.B. and Ercegovich, C. (1970). Effects of atrazine and maize dwarf mosaic virus infection on weight and macro and micro element constituents of maize seedlings in the greenhouse. *Phytopathology* **60**, 272–279.

MacLeod, R., Black, L.M. and Moyer, F.H. (1966). The fine structure and intracellular localisation of potato yellow dwarf virus. *Virology* **29**, 540–552.

Macnicol, P.K. (1976). Rapid metabolic changes in the wounding response of leaf discs following excision. *Plant Physiol.* **57**, 80–84.

Madden, L.V. and Campbell, C.L. (1986). Descriptions of virus disease epidemics in time and space. In: G.D. McLean, R.G. Garrett and W.G. Ruesink (eds) *Plant Virus Epidemics*, pp. 273–293. Academic Press, Orlando, FL.

Madden, L.V., Louie, R. and Knoke, J.K. (1987a). Temporal and spatial analysis of maize dwarf mosaic epidemics. *Phytopathology* **77**, 148–156.

Madden, L.V., Pirone, T.P. and Raccah, B. (1987b). Analysis of spatial patterns of virus-diseased tobacco plants. *Phytopathology* **77**, 1409–1417.

Madden, L.V., Jeger, M.J. and van den Bosch, F. (2000). A theoretical assessment of the effects of vector–virus transmission mechanism on plant virus disease epidemics. *Phytopathology* **90**, 576–594.

Madriz, J.A. de Miranda, J.R., Cabezas, E., Oliva, M., Hernandez, M. and Espinoza, A.M. (1998). Echinocloa hoja blanca and rice hoja blanca viruses occupy distinct ecological niches. *J. Phytopathol.* **146**, 305–308.

Madriz-Ordeñana, K., Rojas-Montero, R., Lundsgaard, T., Ramirez, P., Thordal-Christensen, H. and Collinge, D.B. (2000). Mechanical transmission of maize rayado fino marafivirus (MRFV) to maize and barley by means of the vascular puncture technique. *Plant Pathol.* **49**, 302–307.

Maeda, H. (1997). An atomic force microscopy study for the assembly structures of tobacco mosaic virus and their size evaluation. *Langmuir* **13**, 4150–4161.

Maeda, T. and Inouye, N. (1985). Insolubilization of cucumber mosaic virus with glutaraldehyde and its use for isolation of specific antibody. *Ann. Phytopathol. Soc. Jpn.* **51**, 312–314.

Magyarosy, A.C. and Duffus, J.E. (1977). The occurrence of highly virulent strains of the beet curly top virus in California. *Plant Dis. Rep.* **61**, 248–251.

Magyarosy, A.C., Buchanan, B.B. and Schürmann, P. (1973). Effect of a systemic virus infection on chloroplast function and structure. *Virology* **55**, 426–438.

Mahajan, S.K., Chisholm, S.T., Whitham, S.A. and Carrington, J.C. (1998). Identification and characterization of a locus (*RTM1*) that restricts long-distance movement of tobacco etch virus in *Arabidopsis thaliana*. *Plant J.* **14**, 177–186.

Maia, I.G., Séron, K., Haenni, A.-L. and Bernardi, F. (1996). Gene expression from viral RNA genomes. *Plant Mol. Biol.* **32**, 367–391.

Main, C.E. (1977). Crop destruction: the raison d'être of plant pathology. In: J.G. Horsfall and E.B. Cowling (eds) *Plant Disease: An Advanced Treatise*, Vol. 1, pp. 55–78. Academic Press, New York.

Main, C.E. (1983). Nature of crop losses: an overview. In: T. Kommedahl and P.H. Williams (eds) *Challenging Problems in Plant Health*, pp. 61–68. American Phytopathology Society Press, St Paul, MN.

Maisonneuve, B., Chovelon, V. and Lot, H. (1991). Inheritance of resistance to beet western yellows virus in *Lactuca virosa* L. *HortScience* **26**, 1543–1545.

Maiti, I.B. and Shepherd, R.J. (1998). Isolation and expression analysis of peanut chlorotic streak caulimovirus (PCISV) full-length transcript (FLt) promoter in transgenic plants. *Biochem. Biophys. Res. Comm.* **244**, 440–444.

Maiti, I.B., Gowda, S., Kiernan, J., Ghosh, S.K. and Shepherd, R.J. (1997). Promoter/leader deletion analysis and plant expression vectors with the figwort mosaic virus (FMV) full-length transcript (FLt) promoter containing single and double enhancer domains. *Transgen. Res.* **6**, 143–156.

Maiti, I.B., von Lanken, C., Hong, Y.L., Dey, N. and Hunt, A.G. (1999). Expression of multiple virus-derived resistance determinants in transgenic plants does not lead to additive resistance properties. *J. Plant Biochem. Biotechnol.* **8**, 67–73.

Mäkinen K., Næss, V., Tamm, T., Truve, E., Aaspõllu, A. and Saarma, M. (1995a). The putative replicase of the cocksfoot mottle sobemovirus is translated as part of a polyprotein by −1 ribosomal frameshift. *Virology* **207**, 566–571.

Mäkinen K., Tamm, T., Næss, V. *et al.* (1995b). Characterization of cocksfoot mottle sobemovirus genomic RNA and sequence comparison with related viruses. *J. Gen. Virol.* **76**, 2817–2825.

Mäkinen, K., Generozov, E., Arshava, N., Kaloshin, A., Morozov, S. and Zavriev, S. (2000). Detection and characterization of defective interfering RNAs associated with cocksfoot mottle sobemovirus. *Mol. Biol.* **34**, 291–296.

Malcuit, I., Marano, M.R., Kavanagh, T.A., de Jong, W., Forsyth, A. and Baulcombe, D.C. (1999). The 25-kDa movement protein of PVX elicits *Nb*-mediated hypersensitive cell death in potato. *Mol. Plant Microbe Interact.* **12**, 536–543.

Malcuit, I., de Jong, W., Baulcombe, D.C., Shields, D.C. and Kavanagh, T.A. (2000). Acquisition of multiple virulence/avirulence determinants by potato virus X (PVX) has occurred through convergent evolution rather than recombination. *Virus Genes* **20**, 165–172.

Maleck, K. and Dietrich, R.A. (1999). Defense on multiple fronts: How do plants cope with diverse enemies? *Trends Plant Sci.* **4**, 215–219.

Mallory, A.C., Ely, L., Smith, T.H. et al. (2001). HC-Pro suppression of transgene silencing eliminates the small RNAs but not transgene or the mobile signal. Abstract W10–1, 20th Annual Meeting of the American Society for Virology.

Malyshenko, S.I., Kondakova, O.A., Nazarova, J.V., Kaplan, I.B., Taliansky, M.E. and Atabekov, J.G. (1993). Reduction in tobacco mosaic virus accumulation in transgenic plants producing non-functional viral transport proteins. *J. Gen. Virol.* **74**, 1149–1156.

Mandelkow, E., Stubbs, G. and Warren, S. (1981). Structures of the helical aggregates of tobacco mosaic virus protein. *J. Mol. Biol.* **152**, 375–386.

Mandryk, M. (1963). Acquired systemic resistance to tobacco mosaic virus in *Nicotiana tabacum* evoked by stem injection with *Perenospora tabacina*. *Adam. Aust. J. Agric. Res.* **14**, 315–318.

Mang, K., Gosh, A. and Kaesberg, P. (1982). A comparative study of the cowpea and bean strains of southern bean mosaic virus. *Virology* **116**, 264–274.

Manhart, J.R. and Palmer, J.D. (1990). The gain of two chloroplast tRNA introns marks the green algal ancestors of land plants. *Nature (London)* **345**, 268–270.

Manohar, S.K., Guilley, H., Dollet, M., Richards, K. and Jonard, G. (1993). Nucleotide sequence and genetic organization of peanut clump virus RNA-2 and partial characterization of deleted forms. *Virology* **195**, 33–41.

Manoussopoulos, I.N., Maiss, E. and Tsagris, M. (2000). Native electrophoresis and Western blot analysis (NEWeB): a method for characterization of different forms of potyvirus particles and similar nucleoprotein complexes in extracts of infected plant tissues. *J. Gen. Virol.* **81**, 2295–2298.

Mansoor, S., Khan, S.H., Bashir, A. *et al.* (1999). Identification of a novel circular single-stranded DNA associated with cotton leaf curl disease in Pakistan. *Virology* **259**, 190–199.

Manwan, I., Sama, S. and Rizvi, S.A. (1985). Use of varietal rotation in the management of tungro disease in Indonesia. *Indones. Agric. Res. Dev. J.* **7**, 43–48.

Manwan, I., Sama, S. and Rizvi, S.A. (1987). Management strategy to control rice tungro in Indonesia. In: *Proceeding of the Workshop on Rice Tungro Virus*, pp. 91–97. Ministry of Agriculture, AARD-Maros Research Institution for Food Crops, Maros, Indonesia.

Maramorosch, K. and Harris, K.F. (eds) (1981). *Plant Diseases and Vectors: Ecology and Epidemiology*. Academic Press, New York.

Maramorosch, K. and Raychaudhuri, S.P. (eds) (1988). *Mycoplasma Diseases of Crops: Basic and Applied Aspects*. Springer-Verlag, New York.

Marano, M.R. and Baulcombe, D.C. (1998). Pathogen-derived resistance targeted against the negative-strand RNA of tobacco mosaic virus: RNA strand-specific gene silencing? *Plant J.* **13**, 537–546.

Marathe, R., Anandalakshmi, R., Smith, T.H., Pruss, G.J. and Vance, V.B. (2000a). RNA viruses as inducers, suppressors and targets of post-transcriptional gene silencing. *Plant Mol. Biol.* **43**, 295–306.

Marathe, R., Smith, T.H., Anandalakshmi, R. *et al.* (2000b). Plant viral suppressors of post-transcriptional silencing do not suppress transcriptional silencing. *Plant J.* **22**, 51–59.

Marbrook, J. and Matthews, R.E.F. (1966). The differential immunogenicity of plant virus proteins and nucleoproteins. *Virology* **28**, 219–228.

Marcinka, K. and Musil, M. (1977). Disintegration of red clover mottle virus virions under different conditions of storage *in vitro*. *Acta Virol.* **21**, 71–78.

Marco, S. and Levy, D. (1979). Involvement of ethylene in the development of cucumber mosaic virus-induced chlorotic lesions in cucumber cotyledons. *Physiol. Plant Pathol.* **14**, 235–244.

Marco, S., Levy, D. and Aharoni, N. (1976). Involvement of ethylene in the suppression of hypocotyl elongation in CMV-infected cucumbers. *Physiol. Plant Pathol.* **8**, 1–7.

Marcone, C., Neimark, H., Ragozzino, A. *et al.* (1999). Chromosome sizes of phytoplasmas composing major phylogenetic groups and subgroups. *Phytopathology* **89**, 805–810.

Marcos, J.F. and Beachy, R.N. (1997). Transgenic accumulation of two plant virus coat proteins on a single self-processing polypeptide. *J. Gen. Virol.* **78**, 1771–1778.

Marcos, J.F. and Flores, R. (1992). Characterization of RNA specific to avocado sunblotch viroid synthesized in vitro by a cell-free system from infected avocado leaves. *Virology* **186**, 481–488.

Marcos, J.F., Vilar, M., Pérez-Payá, E. and Pallás, V. (1999). *In vivo* detection, RNA-binding properties and characterization of the RNA-binding domain of the p7 putative movement protein from carnation mottle carmovirus (CarMV). *Virology* **255**, 354–365.

Margis, R., Hans, F. and Pinck, L. (1993). VPg northernimmunoblots as a means for detection of viral RNAs in protoplasts or plants infected with grapevine fanleaf nepovirus. *Archiv. Virol.* **131**, 225–232.

Marie-Jeanne, V., Ioos, R., Peyre, J., Alliot, B. and Signoret, P. (2000). Differentiation of *Poaceae* potyviruses by reverse transcription-polymerase chain reaction and restriction analysis. *J. Phytopathol.* **148**, 141–151.

Marinos, N.G. (1967). Multifuntional plastids in the meristematic region of potato tuber bulbs. *J. Ultrastruct. Res.* **17**, 91–113.

Markham, P.G. (1992a). Transmission of maize streak virus by *Cicadulina* species. *XIX International Congress of Entomology*, 28 June to 4 July 1992, Beijing, China, p. 345.

Markham, P.G. (1992b). AC-electronic monitoring and its role in virus–vector relations of cicallids. *International Congress of Entomology*, 28 June to 4 July 1992, Beijing, China, p. 213.

Markham, P.G. and Hull, R. (1985). Cauliflower mosaic virus aphid transmission facilitated by transmission factors from other caulimoviruses. *J. Gen. Virol.* **66**, 921–923.

Markham, R. (1951). Physicochemical studies on the turnip yellow mosaic virus. *Discuss. Faraday Soc.* **11**, 221–227.

Markham, R. (1962). The analytical centrifuge as a tool for the investigation of plant viruses. *Adv. Virus Res.* **9**, 241–270.

Markham, R. and Smith, K.M. (1949). Studies on the virus of turnip yellow mosaic. *Parasitology* **39**, 330–342.

Markham, R. and Smith, J.D. (1951). Chromatographic studies on nucleic acids. IV. The nucleic acid of the turnip yellow mosaic virus, including a note on the nucleic acid of tomato bushy stunt virus. *Biochem. J.* **49**, 401–406.

Markham, R., Matthews, R.E.F. and Smith, K.M. (1948). Testing potato stocks for virus X. *Farming* February, pp. 40–46.

Markham, R., Frey, S. and Hills, G.J. (1963). Methods for enhancement of image detail and accentuation of structure in electron microscopy. *Virology* **20**, 88–102.

Markham, R., Hitchborn, J.H., Hills, G.J. and Frey, S. (1964). The anatomy of tobacco mosaic virus. *Virology* **22**, 342–359.

Marks, J.D., Hoogenboom, H.R., Bonnert, T.P., McCafferty, J., Griffiths, A.D. and Winter, G. (1991). By-passing immunization: human antibodies from V-gene libraries displayed on phage. *J. Mol. Biol.* **222**, 581–597.

Marrou, J., Messiaen, C.-M. and Migliori, A. (1967). Méthode de contrôle de l'état sanitaire des graines de laitice. *Ann. Epiphyt.* **18**, 227–248.

Marrou, J., Quiot, J.B., Duteil, M., Labonne, G., Leclant, F. and Renoust, M. (1979). Ecology and epidemiology of cucumber mosaic virus. III. Interest of the exposure of bait plants in the study of cucumber mosaic virus dissemination. *Ann. Phytopathol.* **11**, 291-306.

Marsh, L.E. and Guilfoyle, T.J. (1987). Cauliflower mosaic virus replication intermediates are encapsidated into virion-like particles. *Virology* **161**, 129–137.

Marsh, L.E. and Hall, T.C. (1987). Evidence implicating a tRNA heritage for promoters of positive-strand RNA synthesis in brome mosaic and related viruses. *Cold Spring Harbor Symp. Quant. Biol.* **52**, 331–341.

Marsh, L.E., Dreher, T.W. and Hall, T.C. (1988). Mutational analysis of the core and modulator sequences of the BMV RNA3 subgenomic promoter. *Nucl. Acids Res.* **16**, 981–995.

Marshall, B. and Matthews, R.E.F. (1981). Okra mosaic virus protein shells in nuclei. *Virology* **110**, 253–256.

Marte, M., Buonaurio, R. and Della Torre, G. (1993). Induction of systemic resistance to tobacco powdery mildew by tobacco mosaic virus, tobacco necrosis virus and Ethephon. *J. Phytopathol.* **138**, 137–144.

Martelli, G.P. and Jelkmann, W. (1998). Foveavirus, a new plant genus. *Arch. Virol.* **143**, 1245–1249.

Martelli, G.P. and Russo, M. (1977). Plant virus inclusion bodies. *Adv. Virus Res.* **21**, 175–266.

Martelli, G.P., Di Franco, A. and Russo, M. (1984). The origin of multivesicular bodies in tomato bushy stunt virus-infected *Gomphrena globosa* plants. *J. Ultrastruct. Res.* **88**, 275–281.

Martelli, G.P., Agranovsky, A.A., Bar-Joseph, M. et al. (2000). Family Closteroviridae. In: M.H.V. van Regenmortel, C.M. Fauquet, D.H.L. Bishop, E.B. Carstens, M.K. Estes, S.M. Lemon, J. Maniloff, M.A. Mayo, D.J. McGeoch, C.R. Pringle and R.B. Wickner (eds) *Virus Taxonomy. Seventh Report of the International Committee on Taxonomy of Viruses*, pp. 943–952. Academic Press, San Diego.

Martin, A.M., Martinez-Herrera, D., Poch, H.L.C.Y. and Ponz, F. (1997a). Variability in the interactions between *Arabidopsis thaliana* ecotypes and oilseed rape mosaic tobamovirus. *Aust. J. Plant Physiol.* **24**, 275–281.

Martin, B., Collar, L.J., Tjallingii, W.F. and Fereres, A. (1997b). Intracellular ingestion and salivation by aphids may cause the acquisition and inoculation of non-persistently transmitted plant viruses. *J. Gen. Virol.* **78**, 2701–2705.

Martin, M.T. and Garcia, J.A. (1991). Plum pox potyvirus RNA replication in a crude membrane fraction from infected *Nicotiana clevelandii* leaves. *J. Gen. Virol.* **72**, 785–790.

Martin, R. (1986). Use of double-stranded RNA for detection and identification of virus diseases of *Rubus* species. *Acta Hortic.* **186**, 51–62.

Martin, R.R. (1998). Alternative strategies for engineering virus resistance in plants. Breeding for resistance to plant viruses. In: A. Hadidi, R.K. Khetarpal and H. Koganezawa (eds) *Plant Virus Disease Control*, pp. 121–128. APS Press, St. Paul, MN.

Martin, R.R. and Converse, R.H. (1982). An improved buffer for mechanical transmission of viruses from *Fragaria* and *Rubus*. *Acta Hortic.* **129**, 69–72.

Martin, T.J., Harvey, T.L., Bender, C.G. and Seifers, D.L. (1984). Control of wheat streak mosaic virus with vector resistance in wheat. *Phytopathology* **74**, 963–964.

Martinez-Izquierdo, J. and Hohn, T. (1987). Cauliflower mosaic virus coat protein is phosphorylated in vitro by a virion-associated protein kinase. *Proc. Natl Acad. Sci. USA* **84**, 1824–1828.

Martinez-Izquierdo, J., Fütterer, J. and Hohn, T. (1987). Protein encoded by ORF1 of cauliflower mosaic virus is part of the viral inclusion body. *Virology* **160**, 527–530.

Martinez-Soriano, J.P., Galindo-Alonso, J., Maroon, C.J.M., Yucel, I., Smith, D.R. and Diener, T.O. (1996). Mexican papita viroid: putative ancestor of crop viroids. *Proc. Natl Acad. Sci. USA* **93**, 9397–9401.

Martins, C.R.F., Johnson, J.A., Lawrence, D.M. et al. (1998). Sonchus yellow net rhabdovirus nuclear viroplasms contain polymerase-associated proteins. *J. Virol.* **72**, 5669–5679.

Marx, J. (2000). Interfering with gene expression. *Science* **288**, 1370–1372.

Marzachi, C., Milne, R.G. and Boccardo, G. (1988). *In vitro* synthesis of double-stranded RNA by carnation cryptic virus-associated RNA-dependent RNA polymerase. *Virology* **165**, 115–121.

Marzachi, C., Boccardo, G., Milne, R., Isogai, M. and Uyeda, I. (1995). Genome structure and variability of Fijiviruses. *Semin. Virol.* **6**, 103–108.

Marzachi, C., Veratti, F. and Bosco, D. (1998). Direct PCR detection of phytoplasmas in experimentally infected insects. *Ann. Appl. Biol.* **133**, 45–54.

Más, P. and Beachy, R.N. (1998). Distribution of TMV movement protein in single living protoplasts immobilized in agarose. *Plant J.* **15**, 835–842.

Más, P. and Beachy, R.N. (1999). Replication of tobacco mosaic virus on endoplasmic reticulum and role of the cytoskeleton and virus movement protein in intracellular distribution of the viral RNA. *J. Cell Biol.* **147**, 945–958.

Más, P., and Beachy, R.N. (2000). Role of microtubules in the intracellular distribution of tobacco mosaic virus movement protein. *Proc. Natl Acad. Sci. USA* **97**, 12345–12349.

Más, P. and Pallás, V. (1995). Non-isotopic tissue-printing hybridization: a new technique to study long-distance plant virus movement. *J. Virol. Methods* **52**, 317–326.

Masmoudi, K., Duby, C., Suhas, M. et al. (1994). Quality-control of pea seed for pea seed-borne mosaic virus. *Seed Sci. Technol.* **22**, 407–414.

Mason, W.S., Taylor, J.M. and Hull, R. (1987). Retroid virus genome replication. *Adv. Virus Res.* **32**, 35–96.

Massalski, P.R. and Harrison, B.D. (1987). Properties of monoclonal antibodies to potato leafroll luteovirus and their use to distinguish virus isolates differing in aphid transmissibility. *J. Gen. Virol.* **68**, 1813–1821.

Masuta, C. and Takanami, Y. (1989). Determination of sequence and structural requirements for pathogenicity of a cucumber mosaic virus satellite RNA (Y-satRNA). *Plant Cell* **1**, 1165–1173.

Masuta, C., Zuidema, D., Hunter, B.G., Heaton, L.A., Sopher, D.S. and Jackson, A.O. (1987). Analysis of the genome of satellite panicum mosaic virus. *Virology* **159**, 329–338.

Masuta, C., Kuwata, S. and Takanami, Y. (1988a). Disease modulation on several plants by cucumber mosaic virus satellite RNA (Y strain). *Ann. Phytopathol. Soc. Jpn.* **54**, 332–336.

Masuta, C., Kuwata, S. and Takanami, Y. (1988b). Effects of extra 5′ non-viral bases on the infectivity of transcripts from a cDNA clone of satellite RNA (strain Y) of cucumber mosaic virus. *J. Biochem. (Tokyo)* **104**, 841–846.

Masuta, C., Komari, T. and Takanami, Y. (1989). Expression of cucumber mosaic virus satellite RNA from cDNA copies in transgenic tobacco plants. *Ann. Phytopathol. Soc. Jpn.* **55**, 49–55.

Masuta, C., Kuwata, S., Matzuzaki, T., Takanami, Y. and Koiwai, A. (1992). A plant virus satellite RNA exhibits a significant sequence complementarity to a chloroplast tRNA. *Nucl. Acids Res.* **20**, 2885.

Mathon, M.P., Tavert, G. and Malato, G. (1987). Comparison of three methods for homogenising samples of plant material prior to ELISA testing. *OEPP/EPPO Bull.* **17**, 97–103.

Matisová, J. (1971). Alfalfa mosaic virus in lucerne plants and its transmission by aphids in the course of the vegetation period. *Acta Virol.* **15**, 411–420.

Matsubara, A., Kojima, M., Kawano, S. *et al.* (1985). Purification and serology of a Japanese isolate of barley yellow dwarf virus. *Ann. Phytopathol. Soc. Jpn* **51**, 152–158.

Matsushita, Y., Hanazawa, K., Yoshioka, K. *et al.* (2000). *In vitro* phosphorylation of the movement protein of tomato mosaic tobravirus by a cellular kinase. *J. Gen. Virol.* **81**, 2095–2102.

Matthews, R.E.F. (1949a). Studies on potato virus X. I. Types of change in potato virus X infections. *Ann. Appl. Biol.* **36**, 448–459.

Matthews, R.E.F. (1949b). Studies on potato virus X. II. Criteria of relationships between strains. *Ann. Appl. Biol.* **36**, 460–474.

Matthews, R.E.F. (1949c). Reactions of *Cyphomandra betacea* to strains of potato virus X. *Parasitology* **39**, 241–244.

Matthews, R.E.F. (1949d). *Studies on Two Plant Viruses*. PhD thesis, University of Cambridge.

Matthews, R.E.F. (1953a). Factors affecting the production of local lesions by plant viruses. I. Effect of time of day of inoculation. *Ann. Appl. Biol.* **40**, 377–383.

Matthews, R.E.F. (1953b). Factors affecting the production of local lesions by plant viruses. II. Some effects of light, darkness and temperature. *Ann. Appl. Biol.* **40**, 556–565.

Matthews, R.E.F. (1953c). Incorporation of 8-azaguanine into nucleic acid of tobacco mosaic virus. *Nature (London)* **171**, 1065–1066.

Matthews, R.E.F. (1957). *Plant Virus Serology*. Cambridge University Press, London.

Matthews, R.E.F. (1966). Reconstitution of turnip yellow mosaic virus RNA with TMV protein subunits. *Virology* **30**, 82–96.

Matthews, R.E.F. (1970). *Plant Virology*. Academic Press, New York.

Matthews, R.E.F. (1973). Induction of disease by viruses, with special reference to turnip yellow mosaic virus. *Annu. Rev. Phytopathol.* **11**, 147–170.

Matthews, R.E.F. (1974). Some properties of TYMV nucleoproteins isolated in cesium chloride density gradients. *Virology* **60**, 54–64.

Matthews, R.E.F. (1975). A classification of virus groups based on the size of the particle in relation to genome size. *J. Gen. Virol.* **27**, 135–149.

Matthews, R.E.F. (ed.) (1979). Classification and Nomenclature of Viruses. Third Report of the International Committee on Taxonomy of Viruses. *Intervirology* **12**, 132–296.

Matthews, R.E.F. (1981). *Plant Virology*, 2nd edn. Academic Press, New York.

Matthews, R.E.F. (ed.) (1982). Classification and Nomenclature of Viruses. Fourth Report of the International Committee on Taxonomy of Viruses. *Intervirology* **17**, 1–199.

Matthews, R.E.F. (1983a). The history of viral taxonomy. In: R.E.F. Mathews (ed.) *A Critical Appraisal of Viral Taxonomy*, pp. 1–35. CRC Press, Boca Raton, FL.

Matthews, R.E.F. (1983b). Future prospects for viral taxonomy. In: R.E.F. Matthews (ed.) *A Critical Appraisal of Viral Taxonomy*, pp. 219–245. CRC Press, Boca Raton, FL.

Matthews, R.E.F. (1985a). Viral taxonomy. *Microbiol. Sci.* **2**, 74–75.

Matthews, R.E.F. (1985b). Viral taxonomy for the non-virologist. *Annu. Rev. Microbiol.* **39**, 451–474.

Matthews, R.E.F. (1991). *Plant Virology*, 3rd edn. Academic Press, London.

Matthews, R.E.F. (ed.) (1993). *Diagnosis of Plant Virus Diseases*. CRC Press, Boca Raton, FL.

Matthews, R.E.F. and Sarkar, S. (1976). A light-induced structural change in chloroplasts of Chinese cabbage cells infected with turnip yellow mosaic virus. *J. Gen. Virol.* **33**, 435–446.

Matthews, R.E.F. and Witz, J. (1985). Uncoating of turnip yellow mosaic virus *in vivo*. *Virology* **144**, 318–327.

Matthews, R.E.F., Bolton, E.T. and Thompson, H.R. (1963). Kinetics of labelling of turnip yellow mosaic virus with ^{32}P and ^{35}S. *Virology* **19**, 179–189.

Matzke, M.A., Mette, M.F. and Matzke, A.J.M. (2000). Transgene silencing by the host genome defense: implications for the evolution of epigenetic control mechanisms in plants and vertebrates. *Plant Mol. Biol.* **43**, 401–415.

Maule, A.J. (1991). Virus movement in infected plants. *Crit. Rev. Plant Sci.* **9**, 457–473.

Maule, A.J. (1994). Plant-virus movement: *de novo* processing or redeployed machinery? *Trends Microbiol.* **2**, 305–306.

Maule, A.J. (2000). Virus transmission—seeds. In O.C. Maloy and T.D. Murray. (eds) *Encyclopedia of Plant Pathology*, pp. 1168–1170. John Wiley and Sons, Inc.

Maule, A.J. and Wang, D. (1996). Seed transmission of plant viruses: a lesson in biological complexity. *Trends Microbiol.* **4**, 153–158.

Maule, A.J., Escaler, M. and Aranda, M.A. (2000). Programmed responses to virus replication in plants. *Mol. Plant Pathol.* **1**, 9–15.

Maurin, J., Ackermann, H.W., Lebeurier, G. and Lwoff, A. (1984). Un système des virus—1983. *Ann. Inst. Pasteur Virol.* **135E**, 105–110.

Maury, Y., Bossennec, J.-M., Boudazin, G., Hampton, R., Pietersen, G. and Macguire, J. (1987). Factors influencing ELISA evaluation of transmission of pea seed-borne mosaic virus in infected pea seed: seed-group size and seed decortication. *Agronomie* **7**, 225–230.

Maury, Y., Duby, C. and Khetarpal, R.K. (1998). Seed certification for viruses. Breeding for resistance to plant viruses. In: A. Hadidi, R.K. Khetarpal and H. Koganezawa (eds) *Plant Virus Disease Control*, pp. 237–248. APS Press, St. Paul, MN.

Mayer, A. (1886). Ueber die Mosaikkrankheit des Tabaks. *Landwirtsch. Vers.-Stn.* **32**, 451–467.

Mayers, C.N., Palukaitis, P. and Carr, J.P. (2000). Subcellular distribution analysis of the cucumber mosaic virus 2b protein. *J. Gen. Virol.* **81**, 219–226.

Maynard Smith, J. (1978). *The Evolution of Sex*. Cambridge University Press, Cambridge.

Mayo, M.A. and Brunt, A.A. (2001). The current state of plant virus taxonomy. *Molec. Plant Pathol.* **2**, 97–100.

Mayo, M.A. and D'Arcy, C.J. (1999). Family *Luteoviridae*: a reclassification of luteoviruses. In: H.G. Smith and H. Barker (eds) *The Luteoviridae*, pp. 15–22. CAB International, Wallingford, UK.

Mayo, M.A. and Jolly, C.A. (1991). The 5′-terminal sequence of potato leafroll virus RNA: evidence of recombination between virus and host RNA. *J. Gen. Virol.* **72**, 2591–2595.

Mayo, M.A. and Jones, A.T. (1999a). Nepoviruses (*Comoviridae*). In: A. Granoff and R.G. Webster (eds) *Encyclopedia of Virology*, 2nd edn, pp. 1007–1013. Academic Press, San Diego.

Mayo, M.A. and Jones, A.T. (1999b). Idaeovirus. In: A. Granoff and R.G. Webster (eds) *Encyclopedia of Virology*, 2nd edn, pp. 809–811. Academic Press, San Diego.

Mayo, M.A. and Miller, W.A. (1999). The structure and expression of luteovirus genomes. In: H.G. Smith and H. Barker (eds) *The Luteoviridae*, pp. 23–67. CAB International, Wallingford, UK.

Mayo, M.A. and Murant, A.F. (1999). Sequiviruses (*Sequiviridae*). In: A. Granoff and R.G. Webster (eds) *Encyclopedia of Virology*, 2nd edn, pp. 1622–1625. Academic Press, San Diego.

Mayo, M.A. and Robinson, D.J. (1996). Nepoviruses: molecular biology and replication. In: B.D. Harrison and A.F. Murant (eds) *The Plant Viruses. Vol. 5: Polyhedral Virions and Bipartite Genomes*, pp. 139–185. Plenum Press, New York.

Mayo, M.A., Barker, H. and Harrison, B.D. (1982). Specificity and properties of the genome-linked proteins of nepoviruses. *J. Gen. Virol.* **59**, 149–162.

Mayo, M.A., Berns, K.L., Fritsch, C. *et al.* (1995). Subviral agents: satellites. *Arch. Virol. Suppl.* **10**, 487–492.

Mayo, M.A., Taliansky, M.E. and Fritsch, C. (1999). Large satellite RNA: molecular parasitism or molecular symbiosis. In: P.K. Vogt and A.O. Jackson (eds) *Satellites and Defective RNAs*, pp. 65–79. Springer, Berlin.

Mayo, M.A., de Miranda, J.R., Falk, B.W., Goldbach, R., Haenni, A.-L. and Toriyama, S.(2000a). Genus Tenuivirus. In: M.H.V. van Regenmortel, C.M. Fauquet, D.H.L. Bishop, E.B. Carstens, M.K. Estes, S.M. Lemon, J. Maniloff, M.A. Mayo, D.J. McGeoch, C.R. Pringle and R.B. Wickner (eds) *Virus Taxonomy. Seventh Report of the International Committee on Taxonomy of Viruses*, pp. 622–627. Academic Press, San Diego.

Mayo, M.A., Fritsch, C., Leibowitz, M.J. *et al.* (2000b). Satellite nucleic acids. In: M.H.V. van Regenmortel, C.M. Fauquet, D.H.L. Bishop *et al.* (eds) *Virus Taxonomy: Seventh Report of the International Committee on Taxonomy of Viruses*, pp. 1028–1032. Academic Press, New York.

Mazzolini, L., Bonneville, J.M., Volovitch, M., Magazin, M. and Yot, P. (1985). Strand-specific viral DNA synthesis in purified viroplasms isolated from turnip leaves infected with cauliflower mosaic virus. *Virology* **145**, 293–303.

McCabe, P.M. and van Alfen, N.K. (1999). The influence of dsRNA viruses on the biology of plant pathogenic fungi. *Trends Microbiol.* **7**, 377–381.

McCafferty, J., Griffiths, A.D., Winter, G. and Chiswell, D.J. (1990). Phage antibodies—filamentous phage displaying antibody variable domains. *Nature* **348**, 552–554.

McClements, W.L. and Kaesberg, P. (1977). Size and secondary structure of potato spindle tuber viroid. *Virology* **76**, 477–484.

McClure, M.A., Vasi, T.K. and Fitch, W.M. (1993). Comparative analysis of multiple protein sequence alignment methods. *Mol. Biol. Evol.* **11**, 571–592.

McCormick, A.A., Kumagai, M.H., Hanley, K. *et al.* (1999). Rapid production of specific vaccines for lymphoma by expression of the tumor-derived single-chain Fv epitopes in tobacco plants. *Proc. Natl Acad. Sci. USA* **96**, 703–708.

McDaniel, L.L., Ammar, E.-D. and Gordon, D.T. (1985) Assembly, morphology and accumulation of a Hawaiian isolate of maize mosaic virus. *Phytopathology* **75**, 1167–1172.

McDonald, J.G. and Bancroft, J.B. (1977). Assembly studies on potato virus Y and its coat protein. *J. Gen. Virol.* **35**, 251–263.

McDonald, J.G. and Hiebert, E. (1974). Ultrastructure of cylindrical inclusions induced by viruses of the potato Y group as visualised by freeze-etching. *Virology* **58**, 200–208.

McDonald, J.G. and Hiebert, E. (1975). Characterization of the capsid and cylindrical inclusion proteins of three strains of turnip mosaic virus. *Virology* **63**, 295–303.

McDonald, J.G., Beveridge, T.J. and Bancroft, J.B. (1976). Self assembly of protein from a flexuous virus. *Virology* **69**, 327–331.

McElhany, P., Real, L.A. and Power, A.G. (1995). Vector preference and disease dynamics—a study of barley yellow dwarf virus. *Ecology* **76**, 444–457.

McGovern, M.H. and Kuhn, C.W. (1984). A new strain of southern bean mosaic virus derived at low temperatures. *Phytopathology* **74**, 95–99.

McGrath, P.F., Vincent, J.R., Lei, C.H. *et al.* (1997). Coat protein-mediated resistance to isolates of barley yellow dwarf in oats and barley. *Eur. J. Plant Pathol.* **103**, 695–710.

McInnes, J.L., Habili, N. and Symons, R.H. (1989). Nonradioactive, photobiotin-labelled DNA probes for routine diagnosis of viroids in plant extracts. *J. Virol. Methods* **23**, 299–312.

McIntyre, J.L., Dodds, J.A. and Hare, J.D. (1981). Effects of localized infections of *Nicotiana tabacum* by tobacco mosaic virus on systemic resistance against diverse pathogens and an insect. *Phytopathology* **71**, 297–301.

McKendrick, L., Pain, V.P. and Morley, S.J. (1999). Translation initiation factor 4E. *Int. J. Biochem. Cell Biol.* **31**, 31–35.

McKenzie, D.R. (1983). Toward the management of crop losses. In: T. Kommedahl and P.H. Williams (eds) *Challenging Problems in Plant Health*, pp. 82–92. American Phytopathology Society Press, St Paul, MN.

McKinney, H.H. (1929). Mosaic diseases in the Canary Islands, West Africa and Gibraltar. *J. Agric. Res.* **39**, 557–578.

McKinney, H.H. (1953). New evidence on virus disease in barley. *Plant Dis. Reptr.* **37**, 292–295.

McKinney, H.H., Silber, G. and Greeley, L.W. (1965). Longevity of some plant viruses stored in chemically dehydrated tissues. *Phytopathology* **65**, 1043–1044.

McKirdy, S.J. and Jones, R.A.C. (1994). Infection of alternate hosts associated with annual medics (*Medicago* spp.) by alfalfa mosaic virus and its persistence between growing seasons. *Aust. J. Agric. Res.* **45**, 1413–1426.

McKlusky, D.J. and Stobbs, L.W. (1985). A modified local lesion assay procedure with improved sensitivity and reproducibility. *Cvan. J. Plant Pathol.* **7**, 347–350.

McLachlan, A.D., Bloomer, A.C. and Butler, P.J.G. (1980). Structural repeats and evolution of tobacco mosaic virus coat protein and RNA. *J. Mol. Biol.* **136**, 203–224.

McLean, B.G., Zupan, J. and Zambryski, P. (1995). Tobacco mosaic virus movement protein associates with the cytoskeleton in tobacco cells. *Plant Cell* **7**, 2101–2114.

McLean, D.L. and Kinsey, M.G. (1984). The precibarial valve and its role in the feeding behaviour of the pea aphid, *Acyrthosiphon pisum. Bull. Entomol. Soc. Am.* **30**, 26–31.

McLean, G.B., Hempel, F.D. and Zambryski, P.C. (1997). Plant intercellular communication via plasmodesmata. *Plant Cell* **9**, 1043–1054.

McLean, G.D. and Francki, R.I.B. (1967). Purification of lettuce necrotic yellows virus by column chromatography on calcium phosphate gel. *Virology* **31**, 585–591.

McLean, G.D., Burl, J.R., Thomas, D.W. and Sproul, A.N. (1982). The use of reflective mulch to reduce the incidence of watermelon mosaic virus in Western Australia. *Crop Prot.* **1**, 491–496.

McLean, M.A., Campbell, R.N., Hamilton, R.I. and Rochon, D.M. (1994). Involvement of cucumber necrosis virus coat protein in the specificity of fungal transmission by *Olpidium bornavirus. Virology* **204**, 840-842.

McMullen, C.R., Gardner, W.S. and Myers, G.A. (1977). Ultrastructure of cell-wall thickenings and paramural bodies induced by barley stripe mosaic virus. *Phytopathology* **67**, 462–467.

McNaughton, P. and Matthews, R.E.F. (1971). Sedimentation of small viruses at very low concentrations. *Virology* **45**, 1–9.

McRitchie, J.J. and Alexander, L.J. (1963). Host-specific *Lycopersicon* strains of tobacco mosaic virus. *Phytopathology* **53**, 394–398.

Medberry, S.L., Lockhart, B.E.L. and Olszewski, N.E. (1992). The *Commelina* yellow mottle virus promoter is a strong promoter in vascular and reproductive tissues. *Plant Cell* **4**, 185–192.

Medina, V., Peremyslov, V.V., Hagiwara, Y. and Dolja, V.V. (1999). Subcellular localization of the HSP70-homolog encoded by beet yellows closterovirus. *Virology* **260**, 173–181.

Medina, V., Tian, T., Yeh, H.-H., Livieratos, C. and Falk, B.W. (2001). Lettuce infectious yellows virus-encoded P20 protein accumulates in plasmalemma deposits. Abstract W42–10, 20th Annual Meeting of the American Society for Virology.

Meinkoth, J., and Wahl, G. (1984). Hybridization of nucleic acids immobilized on solid supports. *Anal. Biochem.* **138**, 267–284.

Meins, F. (2000). RNA degradation and models for post-transcriptional gene silencing. *Plant Mol. Biol.* **43**, 261–273.

Meints, R.H., Lee, K. and van Etten, J.L. (1986). Assembly site of the virus PBCV-1 in a *Chlorella*-like green alga: ultrastructural studies. *Virology* **154**, 240–245.

Melander, W.R. (1975). Effect of aggregation on the kinetic properties of aspartate aminotransferase. *Biochim. Biophys. Acta* **410**, 74–86.

Melcher, U. (2000). The '30K' superfamily of viral movement proteins. *J. Gen. Virol.* **81**, 257–266.

Melchers, G. (1968). Techniques for the quantitative study of mutation in plant viruses. *Theor. Appl. Genet.* **38**, 275–279.

Melchers, G., Jockusch, H. and Von Sengbusch, P. (1966). A tobacco mutant with a dominant allele for hypersensitivity against some TMV strains. *Phytopathol. Z.* **55**, 86–88.

Mellema, J.E. and Amos, L.A. (1972). Three-dimensional image reconstruction of turnip yellow mosaic virus. *J. Mol. Biol.* **72**, 819-822.

Mellema, J.-R., Benicourt, C., Haenni, A.-L., Noort, A., Pleij, C.W.A. and Bosch, L. (1979). Translational studies with turnip yellow mosaic virus RNAs isolated from major and minor virus particles. *Virology* **96**, 38–46.

Ménissier, J., Lebeurier, G. and Hirth, L. (1982). Free cauliflower mosaic virus supercoiled DNA in infected plants. *Virology* **117**, 322–328.

Ménissier, J., de Murcia, G., Lebeurier, G. and Hirth, L. (1983). Electron microscopic studies of the different topological forms of the cauliflower mosaic virus DNA: knotted encapsulated DNA and nuclear minichromosome. *EMBO J.* **2**, 1067–1071.

Ménissier, J., Laquel, P., Lebeurier, G. and Hirth, L. (1984). A DNA polymerase activity is associated with cauliflower mosaic virus. *Nucl. Acids Res.* **12**, 8769–8778.

Ménissier, J., de Murcia, G., Geldreich, A. and Lebeurier, G. (1986). Evidence for a protein kinase activity associated with purified particles of cauliflower mosaic virus. *J. Gen. Virol.* **67**, 1885–1891.

Merits, A., Guo, D. and Saarma, M. (1998). VPg, coat protein and five non-structural proteins of potato A potyvirus bind RNA in a sequence-unspecific manner. *J. Gen. Virol.* **79**, 3123–3127.

Merits, A., Guo, D., Järvekülg, L. and Saarma, M. (1999). Biochemical and genetic evidence for interactions between potato A potyvirus-encoded proteins P1 and P3 and proteins of the putative replication complex. *Virology* **263**, 15–22.

Mernaugh, R.L., Gardner, W.S. and Yocom, K.L. (1980). Three dimensional structure of pinwheel inclusions as determined by analytical geometry. *Virology* **106**, 273–281.

Mertens, P.P.C., Arella, M., Attoui, H. *et al.* (2000). Genus *Fijivirus*, genus *Phytoreovirus*, genus *Oryzavirus*. In: M. van Regenmortel *et al.* (eds) *Virus Taxonomy: Seventh Report of the International Committee on Taxonomy of Viruses*, pp. 455–480. Academic Press, New York.

Mesfin, T., Den Hollander, J. and Markham, P.G. (1995). Feeding activities of *Cicadulina mbila* (Hemiptera: Cicadellidae) on different host-plants. *Bull. Entomol. Res.* **85**, 387–396.

Meshi, T., Ishikawa, M., Motoyoshi, F., Semba, K. and Okada, Y. (1986). *In vitro* transcription of infectious RNAs from full-length cDNAs of tobacco mosaic virus. *Proc. Natl Acad. Sci. USA* **83**, 5043–5047.

Meshi, T., Watanabe, Y., Saito, T., Sugimoto, A., Maeda, T. and Okada, Y. (1987). Function of the 30 kD protein of tobacco mosaic virus: involvement in cell-to-cell movement and dispensability for replication. *EMBO J.* **6**, 2557–2563.

Meshi, T., Motoyoshi, F., Adachi, A., Watanabe, Y., Takamatsu, N. and Okada, Y. (1988). Two concomitant base substitutions in the putative replicase genes of tobacco mosaic virus confer the ability to overcome the effects of a tomato resistance gene, *Tm-1. EMBO J.* **7**, 1575–1581.

Meshi, T., Motoyoshi, F., Maeda, T., Yoshiwoka, S., Watanabe, H. and Okada, Y. (1989). Mutations in the tobacco mosaic virus 30–kD protein gene overcome Tm-2 resistance in tomato. *Plant Cell* **1**, 515–522.

Mesnard, J.-M., Kirchherr, D., Wurrh, T. and Lebeurier, G. (1990). The cauliflower mosaic virus gene III product is a non-sequence-specific DNA binding protein. *Virology* **174**, 622–624.

Mette, M.F., Aufsatz, W., van der Winden, J., Matske, M.A. and Matske, A.J.M. (2000). Transcriptional silencing and promoter methy-lation triggered by double-stranded RNA. *EMBO J.* **19**, 5194–5201.

Meulewaeter, F. (1999). Necroviruses (*Tombusviridae*). In: A. Granoff and R.G. Webster (eds) *Encyclopedia of Virology*, 2nd edn, pp. 1003–1007. Academic Press, San Diego.

Meuleweiter, F., Seurinck, J. and van Emmelo, J.(1990). Genome structure of tobacco necrosis virus A. *Virology* **117**, 699–709.

Meyer, M., Hemmer, O. and Fritsch, C. (1984). Complete nucleotide sequence of a satellite RNA of tomato black ring virus. *J. Gen. Virol.* **65**, 1575–1583.

Meyer, T.E., Cusanovich, M.A. and Kamen, M.D. (1986). Evidence against use of bacterial amino acid sequence data for construction of all-inclusive phylogenetic trees. *Proc. Natl Acad. Sci. USA* **83**, 217–220.

Mezitt, L.A. and Lucas, W.J. (1996). Plasmodesmatal cell-to-cell transport of proteins and nucleic acids. *Plant Molec. Biol.* **32**, 251–273.

Michelmore, R. (2000). Genomic approaches to plant disease resistance. *Curr. Opin. Plant Biol.* **3**, 125–131.

Millar, A.J. (1999). Biological clocks in *Arabidopsis thaliana*. *New Phytol.* **141**, 175–197.

Miller, E.D., Plante, C.A., Kim, K.-H., Brown, J.W. and Hemenway, C. (1998). Stem-loop structure in the 5′ region of potato virus X genome required for plus-strand RNA accumulation. *J. Mol. Biol.* **284**, 591–608.

Miller, W.A. (1999). Luteovirus (Luteoviridae). In: A. Granoff and R.G. Webster (eds) *Encyclopedia of Virology*, 2nd edn, pp. 901–908. Academic Press, San Diego.

Miller, W.A. and Hall, T.C. (1983). Use of micrococcal nuclease in the purification of highly template dependent RNA-dependent RNA polymerase from brome mosaic virus-infected barley. *Virology* **125**, 236–241.

Miller, W.A., Bujarski, J.J., Dreher, T.W. and Hall, T.C. (1986). Minus-strand initiation of brome mosaic virus replicase within the 3′ tRNA-like structure of native and modified RNA templates. *J. Mol. Biol.* **187**, 537–546.

Miller, W.A., Waterhouse, P.M. and Gerlach, W.L. (1988). Sequence and organisation of barley yellow dwarf virus genomic RNA. *Nucl. Acids Res.* **16**, 6097–6111.

Miller, W.A., Brown, C.M. and Wang, S. (1997a). New punctuation for the genetic code: luteovirus gene expression. *Semin. Virol.* **8**, 3–13.

Miller, W.A., Koev, G. and Mohan, B.R. (1997b). Are there risks associated with transgenic resistance to luteoviruses? *Plant Dis.* **81**, 700–710.

Milne, R.G. (1967). Plant viruses inside cells. *Sci. Prog. (Oxford)* **55**, 203–222.

Milne, R.G. (1984). Electron microscopy for the identification of plant viruses in *in vitro* preparations. *Methods Virol.* **7**, 87–120.

Milne, R.G. (1988). Species concept should not be universally applied to virus taxonomy—But what to do instead? *Intervirology* **29**, 254–259.

Milne, R.G. (1991). Immunoelectron microscopy for virus identification. In: K. Mendgen and D.E. Lesemann, (eds) *Electron Microscopy of Plant Pathogens,* pp. 87–102. Springer-Verlag, New York.

Milne, R.G. (1993). Electron microscopy of *in vitro* preparations. In: R.E.F. Matthews (ed.) *Diagnosis of Plant Virus Diseases*, pp. 215–251. CRC Press, Boca Raton.

Milne, R.G. and Lesemann, D.-E. (1984). Immunoabsorbent electron microscopy in plant virus studies. *Methods Virol.* **8**, 85–101.

Milne, R.G. and Lovisolo, O. (1977). Maize rough dwarf and related viruses. *Adv. Virus Res.* **21**, 267–341.

Milne, R.G. and Luisoni, E. (1977). Rapid immune electron microscopy of virus preparations. *Methods Virol.* **6**, 265–281.

Milne, R.G. and Marzachi, C. (1999). Cryptoviruses (*Partitiviridae*). In: A. Granoff and R.G. Webster (eds) *Encyclopedia of Virology*, 2nd edn, pp. 312–315. Academic Press, San Diego.

Milne, R.G., Conti, M. and Lisa, V. (1973). Partial purification, structure and infectivity of complete maize rough dwarf virus particles. *Virology* **53**, 130–141.

Milne, R.G., Garcia, M.L. and Grau, O. (2000). Genus Ophiovirus. In: M.H.V. van Regenmortel, C.M. Fauquet, D.H.L. Bishop *et al.* (eds) *Virus Taxonomy: Seventh Report of the International Committee on Taxonomy of Viruses*, pp. 627–631. Academic Press, New York.

Mink, G.I. (1993). Pollen- and seed - transmitted viruses and viroids. *Ann. Rev. Phytopathol.* **31**, 375–402.

Mink, G.I., Wample, R. and Howell, W.E. (1998). Heat treatment of perennial plants to eliminate phytoplasmas, viruses and viroids while maintaining plant survival. Breeding for resistance to plant viruses. In: A. Hadidi, R.K. Khetarpal and H. Koganezawa (eds) *Plant Virus Disease Control*, pp. 332–345. APS Press, St. Paul, MN.

Miranda, G.J., Azzam, O. and Shirako, Y. (2000). Comparison of nucleotide sequences between northern and southern Philippine isolates of rice grassy stunt virus indicates occurrence of natural genetic reassortment. *Virology* **266**, 26–32.

Mirkov, T.E., Mathews, D.M., Duplessis, D.H. and Dodds, J.A. (1989). Nucleotide sequence and translation of satellite tobacco mosaic virus RNA. *Virology* **170**, 139–146.

Miroshnichenko, N.A., Karpova, O.V., Morozov, W.Y., Rodionova, N.P. and Atabekov, J.G. (1988). Translation arrest of potato virus X RNA in Krebs-2 cell-free system: RNase H cleavage promoted by complementary oligodeoxynucleotides. *FEBS Lett.* **234**, 65–68.

Missich, R., Ramirez-Parra, E. and Gutierrez, C. (2000). Relationship of oligomerization to DNA binding of wheat dwarf virus RepA and Rep proteins. *Virology* **273**, 178–188.

Mitsuhara, I., Malik, K.A., Miura, M. and Ohashi, Y. (1999). Animal cell-death suppressors Bcl-x$_l$ and Ced-9 inhibit cell death in tobacco plants. *Curr. Biol.* **9**, 775–778.

Mittler, R. and Lam, E. (1996). Sacrifice in the face of foes: pathogen-induced programmed cell death in plants. *Trends Microbiol.* **4**, 10–15.

Miura, K.-I., Kimura, I. and Suzuki, N. (1966). Double-stranded ribonucleic acid from rice dwarf virus. *Virology* **28**, 571–579.

Mizuno, A., Sano, T., Fujii, H., Miura, K. and Yazaki, K. (1986). Supercoiling of the genomic doublestranded RNA of rice dwarf virus. *J. Gen. Virol.* **67**, 2749–2755.

Moghal, S.M. and Francki, R.I.B. (1976). Towards a system for the identification and classification of potyviruses: I. Serology and amino acid composition of six distinct viruses. *Virology* **73**, 350–362.

Mohier, E., Pinck, L. and Hirth, L. (1974). Replication of alfalfa mosaic virus RNAs. *Virology* **58**, 915.

Mohier, E., Hirth, L., LeMeur, M.-A. and Gerlinger, P. (1976). Analysis of alfalfa mosaic virus 17S RNA translational products. *Virology* **71**, 615–618.

Moline, H.E. and Ford, R.E. (1974). Clover yellow mosaic virus infection of seedling roots of *Pisum sativum*. *Physiol. Plant Pathol.* **4**, 219–228.

Molnar, A., Havelda, Z., Dalmay, T., Szutorisz, H. and Burgyan, J. (1997). Complete nucleotide sequence of tobacco necrosis virus strain D^H and genes required for RNA replication and virus movement. *J. Gen. Virol.* **78**, 1235–1239.

Momma, T. and Takahashi, T. (1983). Cytopathology of shoot apical meristem of hop plants infected with hop stunt viroid. *Phytopathol. Z.* **106**, 272–280.

Monette, P.L. (1986). Elimination in vitro of two grapevine nepoviruses by an alternating temperature regime. *J. Phytopathol.* **116**, 88–91.

Monis, J., Scott, H.A. and Gergerich, R.C. (1986). Effect of beetle regurgitant on plant virus transmission using the gross wounding technique. *Phytopathology* **76**, 808–811.

Montalbini, P. and Lupattelli, M. (1989). Effect of localised and systemic tobacco mosaic virus infection on some photochemical and enzymatic activities of isolated chloroplasts. *Physiol. Mol. Plant Pathol.* **34**, 147–162.

Montasser, M.S., Tousignant, M. and Kaper, J.M. (1991). Satellite-mediated protection against cucumber mosaic virus: I. Greenhouse experiments and simulated epidemic conditions in the field. *Plant Dis.* **75**, 86–92.

Montasser, M.S., Tousignant, M.E. and Kaper, J.M. (1998). Viral satellite RNAs for the prevention of cucumber mosaic virus (CMV) disease in field-grown pepper and melon plants. *Plant Dis.* **82**, 1298–1303.

Montelaro, R.C. and Rueckert, R.R. (1975). Radiolabeling of proteins and viruses *in vitro* by acetylation with radioactive acetic anhydride. *J. Biol. Chem.* **250**, 1413–1421.

Montelius, I., Liljàs, L. and Unge, T. (1988). Structure of EDTA-treated satellite tobacco necrosis virus at pH 6.5. *J. Mol. Biol.* **201**, 353–363.

Moonan, F., Molina, J. and Mirkov, T.E. (2000). Sugarcane yellow leaf virus: an emerging virus that has evolved by recombination between luteoviral and poleroviral ancestors. *Virology* **269**, 156–171.

Moore, R.C. and Cyr, R.J. (2000). Association between elongation factor-1a and microtubules *in vivo* is domain dependent and conditional. *Cell Motil. Cytoskeleton* **45**, 279–292.

Moravec, T., Cerovska, N. and Pavlicek, A. (1998). Electron microscopic observation of potato virus A using murine monoclonal antibodies. *Acta. Virol.* **42**, 341–345.

Morch, M.-D., Zagorski, W. and Haenni, A.-L. (1982). Proteolytic maturation of the turnip-yellow-mosaic-virus polyprotein coded *in vitro* occurs by internal catalysis. *Eur. J. Biochem.* **127**, 259–265.

Morch, M.-D., Joshi, R.L., Denial, T.M. and Haenni, A.-L. (1987). A new 'sense' RNA approach to block viral RNA replication *in vitro*. *Nucl. Acids Res.* **15**, 4123–4130.

Morch, M.-D., Boyer, J.-C. and Haenni, A.-L. (1988). Overlapping open reading frames revealed by complete nucleotide sequencing of turnip yellow mosaic virus genomic RNA. *Nucl. Acids Res.* **16**, 6157–6173.

Morch, M.-D., Drugeon, G., Szafranski, P. and Haenni, A.-L. (1989). Proteolytic origin of the 15-kilodalton protein encoded by

turnip yellow mosaic virus genomic RNA. *J. Virol.* **63**, 5153–5158.

Moreira, A., Jones, R.A.C. and Fribourg, C.E. (1978). A resistance breaking strain of potato virus X that does not cause local lesions in *Gomphrena globose*. *Proc. Int. Congr. Plant Pathol.*, 3rd, 1978 Abstract, p. 56.

Morel, J.-B. and Vaucheret, H. (2000). Post-transcriptional gene silencing mutants. *Plant Mol. Biol.* **43**, 275–284.

Morgunova, E.Y., Dauter, Z., Fry, E. *et al.* (1994). The atomic structure of carnation mottle virus capsid protein. *FEBS Lett.* **338**, 267–271.

Mori, K., Hosokawa, D. and Watanabe, M. (1982). Studies on multiplication and distribution of viruses in plants by the use of fluorescent antibody technique: I. Multiplication and distribution of viruses in shoot apices. *Ann. Phytopathol. Soc. Jpn* **48**, 433–443.

Morin, S., Ghanim, M., Zeidan, M., Czosnek, H., Verbeek, M. and van den Heuvel, F.J.M. (1999). A GroEL homologue from endosymbiotic bacteria of the whitefly *Bemisia tabaci* is implicated in the circulative transmission of tomato yellow leaf curl virus. *Virology* **256**, 75–84.

Morin, S., Ghanim, M., Sobol, I. And Czosnek, H. (2000). The GroEL protein of the whitefly *Bemisia tabaci* interacts with the coat protein of transmissible and nontransmissible begomoviruses in the yeast two-hybrid system. *Virology* **276**, 404–416.

Moriyama, H., Nitta, T. and Fukuhara, T. (1995). Double-stranded RNA in rice: a novel RNA replicon in plants. *Mol. Gen. Genet.* **248**, 364–369.

Moriyama, H., Horiuchi, H., Koga, R. and Fukuhara, T. (1999). Molecular characterization of two endogenous double-stranded RNAs in rice and their inheritance by interspecific hybrids. *J. Biol. Chem.* **274**, 6882–6888.

Morris, B.A.M., Richardson, K.A., Anderson, M.T. and Gardner, R.C. (1988a). Cassava latent virus infections mediated by the Ti plasmids of *Agrobacterium tumefaciens* containing either monomeric or dimeric viral DNA. *Plant Mol. Biol.* **11**, 795–803.

Morris, B.A.M., Richardson, K.A., Haley, A., Zhan, X. and Thomas, J.E. (1992). The nucleotide sequence of the infectious cloned DNA component of tobacco yellow dwarf virus reveals features of geminiviruses infecting monocotyledonous plants. *Virology* **187**, 633–642.

Morris, C., Gallois, P., Copley, J. and Kreis, M. (1988b). The 5' flanking region of a barley B hordein gene controls tissue and developmental specific CAT expression in tobacco plants. *Plant Mol. Biol.* **10**, 359–366.

Morris, T.J. and Hillman, B.I. (1989). Defective interfering RNAs of a plant virus. In: *Molecular Biology of Plant–Pathogen Interactions*, pp. 185–197. Alan R. Liss, New York.

Mosch, W.H.M., Huttings, H. and Rast, A.T.B. (1973). Some chemical and physical properties of 18 tobacco mosaic virus isolates from tomato. *Neth. J. Plant Pathol.* **79**, 104–111.

Mossop, D.W. and Francki, R.I.B. (1977). Association of RNA3 with aphid transmission of cucumber mosaic virus. *Virology* **81**, 177–181.

Mossop, D.W. and Francki, R.I.B. (1978). Survival of a satellite RNA *in vivo* and its dependence on cucumber mosaic virus for replication. *Virology* **86**, 562–566.

Mossop, D.W. and Francki, R.I.B. (1979a). The stability of satellite viral RNAs *in vivo* and *in vitro*. *Virology* **94**, 243–253.

Mossop, D.W. and Francki, R.I.B. (1979b). Comparative studies on two satellite RNAs of cucumber mosaic virus. *Virology* **95**, 395–404.

Motoyoshi, F. and Hull, R. (1974). The infection of tobacco protoplasts with pea enation mosaic virus. *J. Gen. Virol.* **24**, 89–99.

Motoyoshi, F. and Oshima, N. (1975). Infection with tobacco mosaic virus of leaf mesophyll protoplasts from susceptible and resistant lines of tomato. *J. Gen. Virol.* **29**, 81–91.

Motoyoshi, F. and Oshima, N. (1976). The use of tris-HCl buffer for inoculation of tomato protoplasts with tobacco mosaic virus. *J. Gen. Virol.* **32**, 311–314.

Motoyoshi, F. and Oshima, N. (1977). Expression of genetically controlled resistance to tobacco mosaic virus infection in isolated tomato leaf mesophyll protoplasts. *J. Gen. Virol.* **34**, 499–506.

Mouches, C., Candresse, T. and Bové, J.M. (1984). Turnip yellow mosaic virus RNA-replicase contains host and virus-encoded subunits. *Virology* **134**, 78–90.

Mougeot, J., Guidasci, T., Wurch, T., Lebeurier, G. and Mesnard, J. (1993). Identification of C-terminal amino acid residues of cauliflower mosaic virus open reading frame III protein responsible for its DNA binding activity. *Proc. Natl Acad. Sci. USA* **90**, 1470–1473.

Mourrain, P., Béclin, C., Elmayan, T. *et al.* (2000). *Arabidopsis SGS2* and *SGS3* genes are required for posttranscriptional gene silencing and natural virus resistance. *Cell* **101**, 533–542.

Moya, A., Rodríguez-Cerezo, E. and García-Arenal, F. (1993). Genetic structure of natural populations of the plant RNA virus tobacco mild green mosaic virus. *Mol. Biol. Evol.* **10**, 449–456.

Moyer, J.W. (1999). Tospoviruses (*Bunyaviridae*). In: A. Granoff and R.G. Webster (eds) *Encyclopedia of Virology*, 2nd edn, pp. 1803–1807. Academic Press, San Diego.

Mueller, E., Gilbert, J.E., Davenport, G., Brigneti, G. and Baulcombe, D.C. (1995). Homology-dependent resistance: transgenic virus resistance in plants related to homology-dependent gene silencing. *Plant J.* **7**, 1011–1013.

Mühlbach, H.-P. and Sänger, H.L. (1979). Viroid replication is inhibited by α-amantin. *Nature (Lond.)* **278**, 185–187.

Mukhopadhyay, S. and Chowdhury, A.K., (1973). Some epidemiological aspects of tungro virus disease of rice in West Bengal. *Int. Rice Commun. Newsl.* **22**, 44–57.

Muller, G.W. and Costa, A.S. (1987). Search for outstanding plants in tristeza infected citrus orchards: the best approach to control the disease by preimmunization. *Phytophylactica* **19**, 197–198.

Muller, H.J. (1964). The relation of recombination to mutational advance. *Mutat. Res.* **1**, 2–9.

Müller, H.O. (1942). Die Ausmessung der Tiefe übermikroskopischer Objekte. *Kolloid-Z.* **99**(1), 6–28; *Chem. Abstr.* **37**, 3991 (1943).

Müller, K.O. and Munro, J. (1951). The reaction of virus-infected potato plants to *Phytophthora infestans*. *Ann. Appl. Biol.* **38**, 765–773.

Mullin, R.H., Smith, S.H., Frazier, N.W., Schlegel, D.E. and McCall, S.R. (1974). Meristem culture frees strawberries of mild yellow edge, pallidosis and mottle diseases. *Phytopathology* **64**, 1425–1429.

Mullineaux, P.M., Donson, J., Morris-Krsinich, B.A.M., Boulton, M.I. and Davies, J.W. (1984). The nucleotide sequence of maize streak virus DNA. *EMBO J.* **3**, 3063–3068.

Mullineaux, P.M., Donson, J., Stanley, J. *et al.* (1985). Computer analysis identifies sequence homologies between potential gene products of maize streak virus and those of cassava latent virus and tomato golden mosaic virus. *Plant Mol. Biol.* **5**, 125–131.

Mullineaux, P.M., Guerineau, F. and accotto, G.P. (1990). Processing of complementary sense RNAs of *Digitaria* streak virus in its host and in transgenic tobacco. *Nucl. Acids Res.* **18**, 7259–7265.

Mullineaux, P.M., Davies, J.W. and Woolston, C.J. (1992). Geminiviruses as gene vectors. In: T.M.A. Wilson and J.W. Davies (eds) *Genetic Engineering with Plant Viruses*, pp. 187–215. CRC Press, Boca Raton, FL.

Mullineaux, P.M., Rigden, J.E., Dry, I.B., Krake, L.R. and Rezaian, M.A. (1993). Mapping of the polycistronic RNAs of tomato leaf curl geminivirus. *Virology* **193**, 414–423.

Mullis, K.B., Faloona, F., Scharf, S.J., Saiki, R.K., Horn, G.T. and Erlich, H.A. (1986). Specific enzymatic amplification of DNA *in vitro*; the polymerase chain reaction. *Cold Spring Harbor Symp. Quant. Biol.* **51**, 263–273.

Mumford, D.L. (1972). A new method of mechanically transmitting curly top virus. *Phytopathology* **62**, 1217–1218.

Mumford, D.L. and Thornley, W.R. (1977). Location of curly top virus antigen in bean, sugarbeet, tobacco, and tomato by fluorescent antibody staining. *Phytopathology* **67**, 1313–1316.

Mundry, K.W. (1957). Die abhängigkeit des auftretens neuer virusstämme von der kulturetemperatur der wirtspflanzen. *Z. Indukt. Abstamm. Vererbungsi.* **88**, 407–426.

Mundry, K.W., Watkins, P.A.C., Ashfield, T., Plaskitt, K.A., Eiselawalter, S. and Wilson, T.M.A. (1991). Complete uncoating of the 5′ leader sequence of tobacco mosaic virus RNA occurs rapidly and is required to initiate cotranslational disassembly *in vitro*. *J. Gen. Virol.* **72**, 769–777.

Murakishi, H.H. (1963). Transfer of virus by a direct leaf-to-leaf method. *Nature (Lond.)* **198**, 312–313.

Murakishi, H.H. and Carlson, P.S. (1976). Regeneration of virus-free plants from dark-green islands of tobacco mosaic virus-infected tobacco leaves. *Phytopathology* **66**, 931–932.

Murakishi, H.H., Hartmann, J.X., Beachy, R.N. and Pelcher, L.E. (1971). Growth curve and yield of tobacco mosaic virus in tobacco callus cells. *Virology* **43**, 62–68.

Murakishi, H.H., Lesney, M. and Carlson, P. (1984). Protoplasts and plant viruses. *Adv. Cell Cult.* **3**, 1–55.

Murant, A.F. (1981). Nepoviruses. In: E. Kurstak (ed.) *Handbook of Plant Virus Infections: Comparative Diagnosis*, pp. 197–238. Elsevier/North-Holland, Amsterdam.

Murant, A.F. (1984). Helper dependence among persistent and semipersistent aphid-borne viruses. *Phytoparasitica* **12**, 207.

Murant, A.F. (1990). Dependence of groundnut rosette virus on its satellite RNA as well as on groundnut rosette assistor luteovirus for transmission by *Aphis craccivora*. *J. Gen. Virol.* **71**, 2163–2166.

Murant, A.F. and Kumar, I.K. (1990). Different variants of the satellite RNA from groundnut rosette virus are responsible for the chlorotic and green forms of groundnut rosette disease. *Ann. Appl. Biol.* **117**, 85–92.

Murant, A.F. and Lister, R.M. (1967). Seed-transmission in the ecology of nematode-borne viruses. *Ann. Appl. Biol.* **59**, 63–76.

Murant, A.F. and Mayo, M.A. (1982). Satellites of plant viruses. *Annu. Rev. Phytopathol.* **20**, 49–70.

Murant, A.F. and Taylor, C.E. (1965). Treatment of soil with chemicals to prevent transmission of tomato black ring and raspberry ringspot viruses by *Longidorus elongatus* (de Man). *Ann. Appl. Biol.* **55**, 227–237.

Murant, A.F., Mayo, M.A., Harrison, B.D. and Goold, R.A. (1973). Evidence for two functional RNA species and a 'satellite' RNA in tomato blackring virus. *J. Gen. Virol.* **19**, 275–278.

Murant, A.F., Chambers, J. and Jones, A.T. (1974). Spread of raspberry bushy dwarf virus by pollination, its association with crumbly fruit and problems of control. *Ann. Appl. Biol.* **77**, 271–281.

Murant, A.F., Taylor, M., Duncan, G.H. and Raschké, J.H. (1981). Improved estimates of molecular weight of plant virus RNA by agarose gel electrophoresis and electron microscopy after denaturation with glyoxal. *J. Gen. Virol.* **53**, 321–332.

Murant, A.F., Mayo, M.A. and Raschké, J.H. (1986). Some biochemical properties of raspberry bushy dwarf virus. *Acta Hortic.* **186**, 23–30.

Murant, A.F., Rajeshwari, R., Robinson, D.J. and Raschké, J.H. (1988a). A satellite RNA of groundnut rosette virus that is largely responsible for symptoms of groundnut rosette disease. *J. Gen. Virol.* **69**, 1479–1486.

Murant, A.F., Raccah, B. and Pirone, T.P. (1988b). Transmission by vectors. In: R.G. Milne (ed.) *The Plant Viruses*, Vol. 4, *The Filamentous Plant Viruses*, pp. 237–273. Plenum, New York.

Murant, A.F., Jones, A.T., Martelli, G.P. and Stace-Smith, R. (1996). Nepoviruses: general properties, diseases, and virus identification. In: B.D. Harrison and A.F. Murant (eds) *The Plant Viruses. Vol. 5: Polyhedral Viruses and Bipartite RNA Genomes*, pp. 99–137. Plenum Press, New York.

Murdock, D.J., Nelson, P.E. and Smith, S.H. (1976). Histopathological examination of *Pelargonium* infected with tomato ringspot virus. *Phytopathology* **66**, 844–850.

Murillo, I., Cavallarin, L. and San Segundo, B. (1997). The maize pathogenesis-related PRms protein localizes to plasmodesmata in maize radicles. *Plant Cell* **9**, 145–156.

Muro-Cacho, C.A. (1999). *In situ* polymerase chain reaction: overview of procedures and applications. *Pediatr. Pathol. Mol. Med.* **18**, 231–253.

Murphy, A.M., Chivasa, S., Singh, D.P. and Carr, J.P. (1999a). Salicylic acid-induced resistance to viruses and other pathogens: a parting of the ways? *Trends Plant Sci.* **4**, 155–160.

Murphy, F.A., Fauquet, C.M., Bishop *et al.* (eds) (1995). Classification and Nomenclature of Viruses. Sixth Report of the International Committee on Taxonomy of Viruses. Springer-Verlag, Wien, New York.

Murphy, J.F., Rychlik, W., Rhoads, R.E., Hunt, A.G. and Shaw, J.G. (1991). A tyrosine residue in the small nuclear inclusion protein of tobacco vein mottling virus links to the viral RNA. *J. Virol.* **65**, 511–513.

Murphy, J.F., Klein, P.G., Hunt, A.G. and Shaw, J.G. (1996). Replacement of the tyrosine residue that links a potyviral VPg to the viral RNA is lethal. *Virology* **220**, 535–538.

Murphy, J.F., Andrianifahanana, M. and Sikora, E.J. (1999b). Detection of cucumber mosaic cucumovirus in weed species: a cautionary report on nonspecific reactions in ELISA. *Can J. Plant Pathol.* **21**, 338–344.

Murphy, J.F., Zehnder, G.W., Schuster, D.J., Sikora, E.J., Polston, J.E. and Kloepper, J.W. (2000). Plant growth-promoting rhizobacterial mediated protection of tomato against tomato mottle virus. *Plant Dis.* **84**, 779–784.

Mutterer, J.D., Stussi-Garaud, C., Milcher, P., Richards, K.E., Jonard, G. and Ziegler-Graff, V. (1999). Role of the beet western yellows virus readthrough protein in virus movement in *Nicotaina clevelandii*. *J. Gen. Virol.* **80**, 2771–2778.

Nagaich, B.B., Upadhya, M.D., Prakash, O. and Singh, S.J. (1968). Cytoplasmically determined expression of symptoms of potato virus X crosses between species of *Capsicum*. *Nature (Lond.)* **220**, 1341–1342.

Nagar, S., Pedersen, T.J., Carrick, K., Hanley-Bowdoin, L. and Robertson, D. (1995). A geminivirus induces expression of a host DNA replication protein in terminally differentiated plant cells. *Plant Cell* **7**, 705–719.

Nagaraj, A.N. (1965). Immunofluorescence studies on synthesis and distribution of tobacco mosaic virus antigen in tobacco. *Virology* **25**, 133–142.

Nagaraj, A.N. and Black, L.M. (1961). Localisation of wound-tumor virus antigen in plant tumors by the use of fluorescent antibodies. *Virology* **15**, 289–294.

Nagata, T. (1999). *Competence and Specificity of Thrips in the Transmission of Tomato Spotted Wilt Virus*. PhD thesis, University of Wageningen.

Nagata, T., InoueNagata, A.K., Prins, M., Goldbach, R. and Peters, D. (2000). Impeded thrips transmission of defective tomato spotted wilt virus isolates. *Phytopathology*. **90**, 454–459.

Nagy, F., Odell, J., Morelli, G. and Chua, N. (1986). Properties of expression of the 35S promoter from CaMV in transgenic tobacco plants. In: M. Zaitlin, P. Day and A. Hollaender (eds) *Biotechnology in Plant Science*, pp. 227–236. Academic Press, Orlando, FL.

Nagy, P.D. and Bujarski, J.J. (1993). Targetting the site of RNA–RNA recombination in brome mosaic virus with antisense sequences. *Proc. Natl Acad. Sci. USA* **90**, 6390–6394.

Nagy, P.D. and Bujarski, J.J. (1995). Efficient system of homologous RNA recombination in brome mosaic virus: sequence and structure requirements and accuracy of crossovers. *J. Virol.* **69**, 131–140.

Nagy, P.D. and Bujarski, J.J. (1996). Homologous RNA recombination of brome mosaic virus: AU-rich regions decrease the accuracy of crossovers. *J. Virol.* **70**, 415–426.

Nagy, P.D. and Bujarski, J.J. (1997). Engineering of homologous recombination hotspots with AU-rich sequences in brome mosaic virus: AU-rich sequences decrease the accuracy of crossovers. *J. Virol.* **71**, 3799–3810.

Nagy, P.D. and Bujarski, J.J. (1998). Silencing homologous recombination hotspots with GC-rich sequences in brome mosaic virus. *J. Virol.* **72**, 1122–1130.

Nagy, P.D. and Simon, A.E. (1997). New insights into the mechanisms of RNA recombination. *Virology* **235**, 1–9.

Nagy, P.D., Dzianott, A., Ahlquist, P.G. and Bujarski, J.J. (1995). Mutations in the helicase-like domain of protein 1a alter the sites of RNA-RNA recombination in brome mosaic virus. *J. Virol.* **69**, 2547–2556.

Nagy, P.D., Carpenter, C.D. and Symon, A.E. (1997). A novel 3′ end repair mechanism in an RNA virus. *Proc. Natl. Acad. Sci. USA* **94**, 1113–1118.

Nagy, P.D., Ogiela, C. and Bujarski, J.J. (1999). Mapping sequences active in homologous RNA recombination in brome mosaic virus: prediction of recombination hot spots. *Virology* **254**, 92–104.

Naidu, R.A., Krishnan, M., Nayudu, M.V. and Gnanam, A. (1986) Studies on peanut green mosaic virus infected peanut (*Arachis hypogaea* L.) leaves: III. Changes in the polypeptides of photosystem II particles. *Physiol. Mol. Plant Pathol.* **29**, 53–58.

Nakashima, N. and Noda, H. (1995). Nonpathogenic *Niloparvata lugens* reovirus is transmitted to the brown planthopper through rice plant. *Virology* **207**, 303–307.

Nakashima, N., Koizumi, M., Watanabe, H. and Noda, H. (1996). Complete nucleotide sequence of the Niloparvata-lugens reovirus—a putative member of the genus Fijivirus. *J. Gen. Virol.* **77**, 139–146.

Namba, K. and Stubbs, G. (1986). Structure of tobacco mosaic virus at 3.6 Å resolution: implications for assembly. *Science* **231**, 1401–1406.

Namba, K., Caspar, D.L.D. and Stubbs, G. (1984). Computer graphics representation of levels of organization in tobacco mosaic virus structure. *Science* **227**, 773–776.

Namba, K., Caspar, D.L.D. and Stubbs, G. (1988). Enhancement and simplification of macromolecular images. *Biophys. J.* **53**, 469–475.

Namba, K., Pattanayek, R. and Stubbs, G. (1989). Visualization of protein-nucleic acid interactions in a virus: refined structure of intact tobacco mosaic virus at 2.9 Å resolution by X-ray fibre diffraction. *J. Mol. Biol.* **208**, 307–325.

Nameth, S.T. and Dodds, J.A. (1985). Double-stranded RNAs detected in cucurbit varieties not inoculated with viruses. *Phytopathology* **75**, 1293.

Nameth, S.T., Dodds, J.A., Paulus, A.O. and Laemmien, F.F. (1986). Cucurbit viruses of California: an ever-changing problem. *Plant Dis.* **70**, 8–11.

Napoli, C., Lemieux, C. and Jorgensen, R.A. (1990). Introduction of a chimeric chalcone synthase gene into *Petunia* results in reversible co-suppression of homologous genes in trans. *Plant Cell* **2**, 279–289.

Napuli, A.J., Falk, B.W. and Dolja, V.V. (2000). Interaction between the HSP70 homolog and filamentous virion of beet yellows virus. *Virology* **274**, 232–239.

Naqvi, S.M.S., Park, K.S., Yi, S.Y., Lee, H.W., Bok, S.H. and Choi, D. (1998). A glycine-rich RNA-binding protein gene is differentially expressed during acute hypersensitive response following tobacco mosaic virus infection in tobacco. *Plant Mol. Biol.* **37**, 571–576.

Nassuth, A., Alblas, F. and Bol, J.F. (1981). Localisation of genetic information involved in the replication of alfalfa mosaic virus. *J. Gen. Virol.* **53**, 207–214.

Nassuth, A., Ten Bruggencate, G. and Bol, J.F. (1983a). Time course of alfalfa mosaic virus RNA and coat protein synthesis in cowpea protoplasts. *Virology* **125**, 75–84.

Nassuth, A., Alblas, F., van der Geest, A.J.M. and Bol, J.F. (1983b). Inhibition of alfalfa mosaic virus RNA and protein synthesis by actinomycin D and cycloheximide. *Virology* **126**, 517–524.

Nathanson, N., McGann, K.A. and Wilesmith, J. (1995). The evolution of virus diseases: their emergence, epidemicity and control. In: A.J. Gibbs, C.H. Calisher and F. García-Arenal (eds) *Molecular Basis for Virus Evolution*, pp.31–46. Cambridge University Press, Cambridge.

Nault, L.R. (1987). Origin and evolution of *Auchenorrhyncha*-transmitted, plant infecting viruses. In: M.R. Wilson and L.R. Nault (eds) *Leafhoppers and Plant Hoppers of Economic Importance*, Proceedings of the Second Workshop, pp. 131–149. CIE, London.

Nault, L.R. (1991). Transmission biology, vector specificity and evolution of planthopper transmitted plant viruses. In: R.F. Denno and T.J. Perfect (eds) *Planthoppers. Their Ecology, Genetics and Management,*. Chapman and Hall, New York.

Nault, L.R. (1997). Arthropod transmission of plant viruses: a new synthesis. *Ann. Entomol. Soc. Am.* **90**, 521–541.

Nault, L.R. and Ammar, E.D. (1989). Leafhopper and planthopper transmission of plant viruses. *Annu. Rev. Entomol.* **34**, 503–529.

Nault, L.R. and Madden, L.V. (1988). Phylogenetic relatedness of maize chlorotic dwarf virus leafhopper vectors. *Phytopathology* **78**, 1683–1687.

Nault, L.R. and Styer, W.E. (1970). An *Aceria tulipae*-borne disease agent producing a virus-like disease of Graminae. *Phytopathology* **59**, 1042.

Nault, L.R. and Styer, W.E. (1972). Effects of sinigrin on host selection by aphids. *Entomol. Exp. Appl.* **15**, 423–437.

Navarro, B. and Flores, R. (1997). Chrysanthemum chlorotic mottle viroid: unusual structural properties of a subgroup of self-cleaving viroids with hammerhead ribozymes. *Proc. Natl Acad. Sci. USA* **94**, 11262–11267.

Navarro, J.A. and Flores, R. (2000). Characterization of the initiation sites of both polarity strands of a viroid RNA reveals a motif conserved in sequence and structure. *EMBO J.* **19**, 2662–2670.

Navarro, J.A., Daros, J.A. and Flores, R. (1999). Complexes containing both polarity strands of avocado sunblotch viroid: identification in chloroplasts and characterization. *Virology* **253**, 77–85.

Navarro, J.A., Vera, A. and Flores, R. (2000). A chloroplastic RNA polymerase resistant to tagetitoxin is involved in replication of *Avocado sunblotch viroid*. *Virology* **268**, 218–225.

Navarro, L., Llacer, G., Cambra, M., Arregui, J.M. and Juarez, J. (1983). Shoot-tip grafting *in vitro* for elimination of viruses in peach plants (*Prunus persica* Batsch). *Acta Hortic.* **130**, 185–192.

Naylor, M., Murphy, A.M., Berry, J.O. and Carr, J.P. (1998). Salicylic acid can induce resistance to plant virus movement. *Mol. Plant Microbe Interact.* **11**, 860–868.

Ndowora, T., Dahal, G., LaFleur, D. *et al.* (1999). Evidence that badnavirus infection in *Musa* can originate from integrated pararetroviral sequences. *Virology* **255**, 214–220.

Neeleman, L. and Bol, J.F. (1999). *Cis*-acting functions of alfalfa mosaic virus proteins involved in replication and encapsidation of viral RNA. *Virology* **254**, 324–333.

Nei, M. (1987). *Molecular Evolutionary Genetics*. Columbia University Press, New York.

Nejidat, A. and Beachy, R.N. (1989). Decreased levels of TMV coat protein in transgenic tobacco plants at elevated temperatures reduce resistance to TMV infection. *Virology* **173**, 531–538.

Nejidat, A. and Beachy, R.N. (1990). Transgenic tobacco plants expressing a coat protein gene of tobacco mosaic virus are resistant to some other tobamoviruses. *Mol. Plant Microbe Interact.* **3**, 247–251.

Nelson, R.S. and van Bel, A.J.E. (1998). The mystery of virus trafficking into, through and out of vascular tissue. *Progr. Bot.* **59**, 476–533.

Nelson, R.S., Powell Abel, P. and Beachy, R.N. (1987). Lesions and virus accumulation in inoculated transgenic tobacco plants expressing the coat protein gene of tobacco mosaic virus. *Virology* **158**, 126–132.

Nelson, R.S., McCormick, S.M., Delannay, X. et al. (1988). Virus tolerance, plant growth and field performance of transgenic tomato plants expressing coat protein from tobacco mosaic virus. *Biotechnology* **6**, 403–409.

Newcomb, E.H. (1967). Fine structure of protein storing plastids in bean root tips. *J. Cell Biol.* **33**, 143–163.

Ng, J. and Perry, K.L. (1999). Stability of the aphid transmission phenotype in cucumber mosaic virus. *Plant Pathol.* **48**, 388–394.

Niblett, C.L. and Paulsen, A.Q. (1975). Purification and further characterization of panicum mosaic virus. *Phytopathology* **65**, 1157–1160.

Niblett, C.L., Dickson, E., Fernow, K.H., Horst, R.K. and Zaitlin, M. (1978). Cross protection among four viroids. *Virology* **91**, 198–203.

Nickerson, K.W. and Lane, L.C. (1977). Polyamine content of several RNA plant viruses. *Virology* **81**, 455–459.

Niepel, M. and Gallie, D.R. (1999). Identification and characterization of the functional elements within the tobacco etch virus 5′ leader required for cap-independent translation. *J. Virol.* **73**, 9080–9088.

Nikovics, K., Simidjieva, J., Peres, A., et al. (2001). Cell-cycle, phase-specific activation of maize streak virus promoters. *Molec. Plant Microb. Interact.* **14**, 609–617.

Nilsson-Tiligren, T., Kolehmainen-Sevéus, L., and von Wettstein, D. (1969). Studies on the biosynthesis of TMV: I. A system approaching a synchronized virus synthesis in a tobacco leaf. *Mol. Gen. Genet.* **104**, 124–141.

Nishiguchi, M. and Motoyoshi, F. (1987). Resistance mechanisms of tobacco mosaic virus strains in tomato and tobacco. In: D. Evered and S. Harnett (eds) *Plant Resistance to Viruses*, pp. 38–56. Wiley, Chichester.

Nishiguchi, M., Motoyoshi, F. and Oshima, N. (1978). Behavior of a temperature-sensitive strain of tobacco mosaic virus in tomato leaves and protoplasts. *J. Gen. Virol.* **39**, 53–61.

Nishiguchi, M., Motoyoshi, F. and Oshima, N. (1980). Further investigation of a temperature-sensitive strain of tobacco mosaic virus: its behaviour in tomato leaf epidermis. *J. Gen. Virol.* **46**, 497–5000.

Nishiguchi, M., Kikuchi, S., Kiho, Y., Ohno, T., Meshi, T. and Okada, Y. (1985). Molecular basis of plant viral virulence: the complete nucleotide sequence of an attenuated strain of tobacco mosaic virus. *Nucl. Acids Res.* **13**, 5585–5590.

Nishiguchi, M., Langridge, W.H.R., Szalay, A.A. and Zaitlin, M. (1986). Electroporation-mediated infection of tobacco leaf protoplasts with tobacco mosaic virus RNA and cucumber mosaic virus RNA. *Plant Cell Rep.* **5**, 57–60.

Niswender, C.M. (1998). Recent advances in mammalian RNA editing. *Cell. Molec. Life Sci.* **54**, 946–964.

Nitta, N., Takanami, Y., Kuwata, S. and Kubo, S. (1988). Inoculation with RNAs 1 and 2 of cucumber mosaic virus induces viral RNA replicase activity in tobacco mesophyll protoplasts. *J. Gen. Virol.* **69**, 2695–2700.

Nixon, H.L. and Gibbs, A.J. (1960). Electron microscope observations on the structure of turnip yellow mosaic virus. *J. Mol. Biol.* **2**, 197–200.

Noad, R.J., Al-Kaff, N.S., Turner, D.S. and Covey, S.N. (1998). Analysis of polypurine tract-associated DNA plus-strand priming *in vivo* utilizing a plant pararetroviral vector carrying redundant ectopic priming elements. *J. Biol. Chem.* **273**, 32568–32575.

Noort, A., van den Drics, C.L.A.M., Pleij, C.W.A., Jaspars, E.M.J. and Bosch, L. (1982). Properties of turnip yellow mosaic virus in cesium chloride solutions: the formation of high-density components. *Virology* **120**, 412–421.

Norris, E., Vaira, A.M., Caciagli, P., Masenga, V., Gronenborn, B. and Accotto, G.P. (1998). Amino acids in the capsid protein of tomato yellow leaf curl virus that are crucial for systemic infection, particle formation, and insect transmission. *J. Virol.* **72**, 10050–10057.

Novik, R.P. (1998). Contrasting lifestyles of rolling-circle phages and plasmids. *Trends Biochem. Sci.* **23**, 434–438.

Nuorteva, P. (1962). Studies on the causes of the phytopathogenicity of *Calligypona pellucida* (F). Hom. Araeopidae). *Ann. Zool. Soc. Zool. Bot. Fenn. Vanamo* **23**(4), 1–58.

Nurkiyanova, K.M., Ryabov, E.V., Commandeur, U., Duncan, G.H., Canto, T., Gray, S.M., Mayo, M.A. and Taliansky, M.E. (2000). Tagging potato leafroll virus with the jellyfish green fluorescent protein gene. *J. Gen. Virol.* **81**, 617–626.

Nuss, D.L. (1999). Hypoviruses (*Hypoviridae*). In: A. Granoff and R.G. Webster (eds) *Encyclopedia of Virology*, 2nd edn, pp. 804–807. Academic Press, San Diego.

Nuss, D.L. and Koltin, V. (1990). Significance of dsRNA genetic elements in plant pathogenic fungi. *Annu. Rev. Phytopathol.* **28**, 37–58

Nuss, D.L. and Peterson, A.J. (1981). Resolution and genome assignment of mRNA transcripts synthetized *in vitro* by wound tumor virus. *Virology* **114**, 399–404.

Nuss, D.L. and Summers, D. (1984). Variant dsRNAs associated with transmission-defective isolates of wound tumor virus represent terminally conserved remnants of genome segments. *Virology* **133**, 276–288.

Nutter, F.W. (1997). Quantifying the temporal dynamics of plant virus epidemics: a review. *Crop Protect.* **16**, 603–618.

Nutter, F.W., Teng, P.S. and Royer, M.H. (1993). Terms and concepts for yield, crop loss, and disease thresholds. *Plant Dis.* **77**, 211–215.

O'Donnell, I.J., Shukla, D.D. and Gough, K.H. (1982). Electro-blot radioimmunoassay of virus-infected plant sap—a powerful new technique for detecting plant viruses. *J. Virol. Methods* **4**, 19–26.

O'Loughlin, G.T. and Chambers, T.C. (1967). The systemic infection of an aphid by a plant virus. *Virology* **33**, 262–271.

O'Reilly, E.K. and Kao, C.C. (1998). Analysis of RNA-dependent RNA polymerase structure and function as guided by known polymerase structures and computer prediction of secondary structure. *Virology* **252**, 287–303.

O'Reilly, E.K., Tang, N., Ahlquist, P. and Kao, C.C. (1995). Biochemical and genetic analysis of the interaction between the helicase-like and polymerase-like proteins of brome mosaic virus. *Virology* **214**, 59–71.

O'Reilly, E.K., Paul, J.D. and Kao, C.C. (1997). Analysis of the interaction of viral RNA replication proteins by using the yeast two-hybrid assay. *J. Virol.* **71**, 7526–7532.

O'Reilly, E.K., Wang, Z., French, R. and Kao, C.C. (1998). Interactions between the structural domains of the RNA replication proteins of plant-infecting RNA viruses. *J. Virol.* **72**, 7160–7169.

Oakley, B.R. (2000). An abundance of tubulins. *Trends Cell Biol.* **10**, 537–542.

Oberschmidt, O., Hücking, C. and Piechulla, B. (1995). Diurnal *Lhc* gene expression is present in many but not all species of the plant kingdom. *Plant Mol. Biol.* **27**, 147–153.

Oda, Y., Saeki, K., Takahashi, Y. *et al.* (2000). Crystal structure of tobacco necrosis virus at 2.25 Å resolution. *J. Mol. Biol.* **300**, 153–169.

Odell, J.T. and Howell, S.H. (1980). The identification, mapping, and characterization of mRNA for P66, a cauliflower mosaic virus-coded protein. *Virology* **102**, 349–359.

Odell, J.T., Nagy, F. and Chua, N.-H. (1985). Identification of DNA sequences required for activity of the cauliflower mosaic virus 35S promoter. *Nature (London)* **313**, 810–812.

Odumosu, A.O., Homer, R.B. and Hull, R. (1981). Circular dichroism studies on Southern bean mosaic virus. *J. Gen. Virol.* **53**, 193–196.

Offei, S.K., Coffin, R.S. and Coutts, R.H.A. (1995). The tobacco necrosis virus p7a protein is a nucleic acid-binding protein. *J. Gen. Virol.* **76**, 1493–1496.

Offord, R.E. (1966). Electron microscopic observations on the substructure of tobacco rattle virus. *J. Mol. Biol.* **17**, 370–375.

Ofori, F.A. and Francki, R.I.B. (1983). Evidence that maize wallaby ear disease is caused by an insect toxin. *Ann. Appl. Biol.* **103**, 185–189.

Ofori,, F.A. and Francki, R.I.B. (1985). Transmission of leafhopper A virus, vertically through eggs and horizontally through maize in which it does not multiply. *Virology* **144**, 152–157.

Ogawa, M. and Sakai, F. (1984). A messenger RNA for tobacco mosaic virus coat protein in infected tobacco mesophyll protoplasts. *Phytopathol. Z.* **109**, 193–203.

Oh, C-S. and Carrington, J.C. (1989). Identification of essential residues in potyvirus proteinase HC-Pro by site-directed mutagenesis. *Virology* **173**, 692–699.

Oh, J.-W., Kong, Q., Song, C., Carpenter, C.D. and Simon, A.E. (1995). Open reading frames of turnip crinkle virus involved in satellite symptom expression and incompatibility with *Arabidopsis thaliana* ecotype Dijon. *Mol. Plant Microbe Interact.* **8**, 979–987.

Ohki, S.T. and Inouye, T. (1987). Use of Gelrite as a gelling agent in immunodiffusion test for identification of plant viral antigens. *Ann. Phytopathol. Soc. Jpn.* **53**, 557–561.

Ohki, S.T., Leps, W.T. and Hiruki, C. (1986). Effects of alfalfa mosaic virus infection on factors associated with symbiotic N_2 fixation in alfalfa. *Can. J. Plant Pathol.* **8**, 277–281.

Ohno, T., Takamatsu, N., Meshi, T., Okada, Y., Nishiguchi, M. and Kiho, Y. (1983). Single amino acid substitution in 30K protein of TMV defective in virus transport function. *Virology* **131**, 255–258.

Okada, Y. (1986a). Cucumber green mottle mosaic virus. In: M.H.V. van Regenmortel and H. Fraenkel-Conrat (eds) *The Plant Viruses*, Vol. 1, pp. 267–281. Plenum, New York.

Okada, Y. (1986b). Molecular assembly of tobacco mosaic virus *in vitro*. *Adv. Biophys.* **22**, 95–149.

Okada, Y. (1999). Historical overview of research on tobacco mosaic virus genome: genome organization, infectivity and gene manipulation. *Phil. Trans. R. Soc. Lond.* **354**, 569–582.

Okada, Y., Ohashi, Y., Ohno, T. and Nozu, Y. (1970). Sequential reconstitution of tobacco mosaic virus. *Virology* **42**, 243–245.

Okuno, T. and Furusawa, I. (1979). RNA polymerase activity and protein synthesis in brome mosaic virus-infected protoplasts. *Virology* **99**, 218–225.

Old, R.W. and Primrose, S.B. (1989). Principles of genetic manipulation. 4th Edn. Blackwell Scientific Publications.

Ollennu, N.E., Owusa, G.K. and Thresh, J.M. (1989). Spread of cacao swollen shoot virus to recent plantings in Ghana. *Crop Protect.* **8**, 251–264.

Olmos, A., Cambra, M., Esteban, O., Gorris, M.T. and Terrada, E. (1999). New device and method for capture, reverse transcription and nested PCR in a single closed-tube. *Nucl. Acids Res.* **27**, 1564–1565.

Olson, A.J., Bricogne, G. and Harrison, S.C. (1983). Structure of tomato bushy stunt virus: IV. The virus particle at 2.9 Å resolution. *J. Mol. Biol.* **171**, 61–93.

Olson, A.J., Tainer, J.A. and Getsoff, E.D. (1985). Computer graphics in the study of macromolecular interactions. In: D. Moras, J. Drenth, B. Strandberg, D. Suck and K. Wilson (eds) *Crystallography in Molecular Biology*, pp. 131–139. Plenum, New York.

Olsthoorn, R.C.L., Mertens, S., Brederode, F.T. and Bol, J.F. (1999). A conformational switch at the 3′-end of a plant virus RNA regulates viral replication. *EMBO J.* **18**, 4856–4864.

Olszewski, N., Hagen, G. and Guilfoyle, T.J. (1982). A transcriptionally active, covalently closed minichromosome of cauliflower mosaic virus DNA isolated from infected turnip leaves. *Cell* **29**, 395–402.

Omar, S.A.M., Bailiss, K.W., Chapman, G.P. and Mansfield, J.W. (1986). Effects of virus infection of faba bean on subsequent infection by *Uromyces viciae-fabae*. *Plant Pathol.* **35**, 535–543.

Omura, T. (1995). Genomes and primary protein structures of phytoreoviruses. *Semin. Virol.* **6**, 97–102.

Omura, T. and Yan, J. (1999). Role of outer capsid proteins in transmission of phytoreovirus in insect vectors. *Adv. Virus Res.* **54**, 15–43.

Omura, T., Minobe, Y., Matsuoka, M., Nozu, Y., Tsuchizaki, T. and Saito, Y. (1985). Location of structural proteins in particles of rice gall dwarf virus. *J. Gen. Virol.* **66**, 811–815.

Omura, T., Takahashi, Y., Shohara, K., Minobe, Y., Tsuchizaki, T. and Nozu, Y. (1986). Production of monoclonal antibodies against rice stripe virus for the detection of virus antigen in infected plants and viruliferous insects. *Ann. Phytopathol. Soc. Jpn.* **52**, 270–277.

Omura, T., Kimura, I., Tsuchizaki, T. and Saito, Y. (1988). Infection by rice gall dwarf virus of cultured monolayers of leafhopper cells. *J. Gen. Virol.* **69**, 429–432.

Ooshika, I., Watanabe, Y., Meshi, T. *et al.* (1984). Identification of the 30K protein of TMV by immunoprecipitation with antibodies directed against a synthetic peptide. *Virology* **132**, 71–78.

Oostergetel, G.T., Krijgsman, P.C.J., Mellema, J.E., Cusack, S. and Miller, A. (1981). Evidence for the absence of swelling of alfalfa mosaic virions. *Virology* **109**, 206–210.

Oostergetel, G.T., Mellema, J.E. and Cusack, S. (1983). Solution scattering study on the structure of alfalfa mosaic virus strain VRU. *J. Mol. Biol.* **171**, 157–173.

Opalka, N., Brugidou, C., Bonneau, C. *et al.* (1998). Movement of rice yellow mottle virus between xylem cells through pit membranes. *Proc. Natl Acad. Sci. USA* **95**, 3323–3328.

Opalka, N., Tihova, M., Brugidou, C. *et al.* (2000). Structure of native and expanded sobemoviruses by electron cryo-microscopy and image reconstruction. *J. Molec. Biol.* **303**, 197–211.

Oparka, K.J. and Prior, D.A.M. (1992). Direct evidence for pressure-generated closure of plasmodesmata. *Plant J.* **2**, 741–750.

Oparka, K.J., Boevink, P. and Santa Cruz, S. (1992) Studying the movement of plant viruses using green fluorescent protein. *Trends Plant. Sci.* **1**, 412–418.

Oparka, K.J., Boevink, P. and Santa Cruz, S. (1996). Studying the movement of plant viruses using green fluorescent protein. *Trends Plant Sci.* **1**, 412–418.

Oparka, K.J., Prior, D.A.M., Santa Cruz, S., Padgett, H.S. and Beachy, R.N. (1997). Gating of epidermal plasmodesmata is restricted to the leading edge of expanding infection sites of tobacco mosaic virus (TMV). *Plant J.* **12**, 781–789.

Oparka, K.J., Roberts, A.G., Boevink, P. *et al.* (1999). Simple, but not branched, plasmodesmata allow the non-specific trafficking of proteins in developing tobacco leaves. *Cell* **97**, 743–754.

Orellana, R.G., Fan, F. and Sloger, C. (1978). Tobacco ringspot virus and *Rhizobium* interactions in soybean: Impairment of leghemoglobin accumulation and nitrogen fixation. *Phytopathology* **68**, 577–582.

Orita, M., Iwahana, H., Kanazawa, H., Hayashi, K. and Sekiya, T. (1989a). Detection of polymorphisms of human DNA by gel electrophoresis as single-stranded conformation polymorphisms. *Proc. Natl Acad. Sci. USA* **86**, 2766–2770.

Orita, M., Suzuki, Y., Sekiya, T. and Hayashi, K. (1989b). Rapid and sensitive detection of point mutations and DNA polymorphisms using the polymerase chain reaction. *Genomics* **5**, 874–879.

Orlandi, O., Gussow, D.H., Jones, P.T. and Winter, G. (1989). Cloning immunoglobulin variable domains for the expression by the polymerase chain reaction. *Proc. Natl Acad. Sci. USA* **86**, 3833–3837.

Orlob, G.B. (1963). Reappraisal of transmission of tobacco mosaic virus by insects. *Phytopathology* **53**, 822–830.

Orlob, G.B. (1966). Feeding and transmission characteristics of *Aceria tulipae* Keifer as a vector of wheat streak mosaic virus. *Phytopathol. Z.* **55**, 218–238.

Orlob, G.B. (1968). Relationships between *Tetranychus urticae* Koch and some plant viruses. *Virology* **35**, 121–133.

Orlob, G.B. and Takahashi, Y. (1971). Location of plant viruses in two-spotted spider mite, *Tetranychus urticae* Koch. *Phytopathol. Z.* **72**, 21–28.

Orozco, B.M. and Hanley-Bowdoin, L. (1996). A DNA structure is required for geminivirus origin function. *J. Virol.* **270**, 148–158.

Orozco, B.M. and Hanley-Bowdoin, L. (1998). Conserved sequence and structural motifs contribute to the DNA binding and cleavage activities of a geminivirus replication protein. *J. Biol. Chem.* **273**, 24448–24456.

Orozco, B.M., Miller, A.R., Settlage, S.B. and Hanley-Bowdoin, L. (1997). Functional domains of a geminivirus replication protein. *J. Biol. Chem.* **272**, 9840–9846.

Orozco, B.M., Gladfelter, H.J., Settlage, S.B., Eagle, P.A., Gentry, R. and Hanley-Bowdoin, L. (1998). Multiple *cis* elements contribute to geminivirus origin functions. *Virology* **242**, 346–356.

Orr, C.C. and Newton, O.H. (1971). Distribution of nematodes by wind. *Plant Dis. Rep.* **55**, 61–63.

Osaki, T., Yamada, M. and Inouye, T. (1985). Whitefly-transmitted viruses from three plant species (abstract in Japanese). *Ann. Phytopathol. Soc. Jpn* **51**, 82–83.

Osborne, A.B. and Schwartz, M.I. (1994). Essential genes that regulate apoptosis. *Trends Cell Biol.* **4**, 394–403.

Osbourn, J.K., Plaskitt, K.A., Watts, J.W. and Wilson, T.M.A. (1989a). Tobacco mosaic virus coat protein and reporter gene transcripts containing the TMV origin-of-assembly sequence do not interact in double-transgenic tobacco plants: implications for coat protein-mediated protection. *Mol. Plant Microbe Interact.* **2**, 340–345.

Osbourn, J.K., Watts, J.W., Beachy, R.N. and Wilson, T.M.A. (1989b). Evidence that nucleocapsid disassembly and a later step in virus replication are inhibited in transgenic tobacco protoplasts expressing the TMV coat protein. *Virology* **172**, 370–373.

Oshima, K., Taniyama, T., Yamanaka, T., Ishikawa, M and Naito, S. (1998). Isolation of a mutant of *Arabidopsis thaliana* carrying two simultaneous mutations affecting tobacco mosaic virus multiplication within a single cell. *Virology* **243**, 472–481.

Osman, F., Choi, Y.G., Grantham, G.L. and Rao, A.L.N. (1998). Molecular studies on bromovirus capsid protein: V. Evidence for the specificity of brome mosaic virus encapsidation using RNA3 chimera of brome mosaic and cucumber mosaic viruses expressing heterologous coat proteins. *Virology* **251**, 438–448.

Osman, T.A.M. and Buck, K.W. (1992). Detection of the movement protein of red clover necrotic mosaic virus in a cell wall fraction from infected *Nicotaina clevelandii* plants. *J. Gen. Virol.* **72**, 2853–2856.

Osman, T.A.M. and Buck, K.W. (1996). Complete replication *in vitro* of tobacco mosaic virus RNA by a template-dependent membrane-bound RNA polymerase. *J. Virol.* **70**, 6227–6234.

Osman, T.A.M. and Buck, K.W. (1997). The tobacco mosaic virus RNA polymerase complex contains a plant protein related to the RNA-binding subunit of yeast eIF-3. *J. Virol.* **71**, 6075–6082.

Osman, T.A.M., Hayes, R.J. and Buck, K.W. (1992). Cooperative binding of the red clover necrotic mosaic virus movement protein to single-stranded nucleic acids. *J. Gen. Virol.* **73**, 223–227.

Osman, T.A.M., Hemenway, C.L. and Buck, K.W. (2000). Role of the 3' tRNA-like structure in tobacco mosaic virus minus-strand RNA synthesis by the viral RNA-dependent RNA polymerase *in vitro*. *J. Virol.* **74**, 11671–11680.

Oster, S.K., Wu, B. and White, K.A. (1998). Uncoupled expression of p33 and p92 permits amplification of tomato bushy stunt virus RNAs. *J. Virol.* **72**, 5845–5851.

Othman, R.Y. (1994). Molecular studies on southern bean mosaic virus. PhD thesis, University of East Anglia, UK.

Othman, Y. and Hull, R. (1995). Nucleotide sequence of the bean strain of southern bean mosaic virus. *Virology* **206**, 287–297.

Otsuki, Y. and Takebe, I. (1969). Fluorescent antibody staining of tobacco mosaic virus antigen in tobacco mesophyll protoplasts. *Virology* **38**, 497–499.

Otsuki, Y. and Takebe, I. (1976a). Double infection of isolated tobacco mesophyll protoplasts by unrelated plant viruses. *J. Gen. Virol.* **30**, 309–316.

Otsuki, Y. and Takebe, I. (1976b). Interaction of tobacco mosaic virus strains in doubly infected tobacco protoplasts. *Ann. Microbiol. (Paris)* **127**, 21 (abstract).

Otsuki, Y. and Takebe, I. (1978). Production of mixedly coated particles in tobacco mesophyll protoplasts doubly infected by strains of tobacco mosaic virus. *Virology* **84**, 162–171.

Otsuki, Y., Takebe, I., Honda, Y. and Matsui, C. (1972a). Ultrastructure of infection of tobacco mesophyll protoplasts by tobacco mosaic virus. *Virology* **49**, 188–194.

Otsuki, Y., Shimomura, T. and Takebe, I. (1972b). Tobacco mosaic virus multiplication and expression of the N gene in necrotic responding tobacco varieties. *Virology* **50**, 45–50.

Otsuki, Y., Takebe, I., Ohno, T., Fukuda, M. and Okada, Y. (1977). Reconstitution of tobacco mosaic virus rods occurs bidirectionally from an internal initiation region: demonstration by electron microscopic serology. *Proc. Natl Acad. Sci. USA* **74**, 1913–1917.

Ou, J.H., Rice, C.M., Dalgarno, L., Strauss, E.G. and Strauss, J.H. (1982). Sequence studies of several alphavirus genomic RNAs in the region containing the start of the subgenomic RNA. *Proc. Natl Acad. Sci. USA* **79**, 5235–5239.

Ouchterlony, O. (1962). Diffusion-in-gel methods for immunological analysis. II. *Prog. Allergy* **6**, 30–154.

Overall, R.L. and Blackman, L.M. (1996). A model if the macromolecular structure of plasmodesmata. *Trends Plant Sci.* **1**, 307–311.

Owen, J. and Palukaitis, P. (1988). Characterisation of cucumber mosaic virus. I. Molecular heterogeneity mapping of RNA3 in eight CMV strains. *Virology* **166**, 495–502.

Owens R.A. (1999). Viroids. In: A. Granoff and R.G. Webster (eds) *Encyclopedia of Virology*, pp. 1928–1937. Academic Press, San Diego.

Owens, R.A. and Bruening, G. (1975). The pattern of amino acid incorporation into two cowpea mosaic virus proteins in the presence of ribosome-specific protein synthesis inhibitors. *Virology* **64**, 520–530.

Owens, R.A. and Cress, D.E. (1980). Molecular cloning and characterisation of potato spindle tuber viroid cDNA sequences. *Proc. Natl Acad. Sci. USA* **77**, 5302–5306.

Owens, R.A. and Diener, T.O. (1981). Sensitive and rapid diagnosis of potato spindle tuber viroid disease by nucleic acid hybridization. *Science* **213**, 670–672.

Owens, R.A. and Diener, T.O. (1982). RNA intermediates in potato spindle tuber viroid replication. *Proc. Natl Acad. Sci. USA* **79**, 113–117.

Owens, R.A. and Diener, T.O. (1984). Spot hybridization for detection of viroids and viruses. *Methods Virol.* **7**, 173–187.

Owens, R.A. and Kaper, J.M. (1977). Cucumber mosaic virus-associated RNA5. II. *In vitro* translation in a wheat germ protein-synthesis system. *Virology* **80**, 196–203.

Owens, R.A. and Schneider, I.R. (1977). Satellite of tobacco ringspot virus RNA lacks detectable mRNA activity. *Virology* **80**, 222–224.

Owens, R.A., Hammond, R.W., Gardner, R.C., Kiefer, M.C., Thompson S.M. and Cress, D.E (1986). Site-specific mutagenesis of potato spindle tuber viroid cDNA: alterations within pre-melting region 2 that abolish infectivity. *Plant Mol. Biol.* **6**, 179–192.

Owens, R.A., Candresse, T. and Diener, T.O. (1990). Construction of novel viroid chimeras containing portions of tomato apical stunt and citrus exocortis viroids. *Virology* **175**, 238–246.

Owens, R.A., Thompson, S.M., Feldstein, P.A. and Garnsey, S.M. (1999). Effects of natural sequence variation on symptom induction by citrus viroid III. *Ann. Appl. Biol.* **134**, 73–80.

Owusu, G.K., Crowley, N.C. and Francki, R.I.B. (1968). Studies of the seed-transmission of tobacco ringspot virus. *Ann. Appl. Biol.* **61**, 195–202.

Oxelfelt, P. (1970). Development of systemic tobacco mosaic virus infection: I. Initiation of infection and time course of virus multiplication. *Phytopathol. Z.* **69**, 202–211.

Oxelfelt, P. (1976) Biological and physicochemical characteristics of three strains of red clover mottle virus. *Virology* **74**, 73–80.

Pacha, R.F., Allison, R.F. and Ahlquist, P. (1990). Cis-acting sequences required for *in vivo* amplification of genomic RNA3 are organised differently in related bromoviruses. *Virology* **174**, 436–443.

Pacumbaba, R.P. (1995). Seed transmission of soybean mosaic virus in mottled and nonmottled soybean seeds. *Plant Dis.* **79**, 193–195.

Padgett, H.S. and Beachy, R.N. (1993). Analysis of a tobacco mosaic virus strain capable of overcoming N gene-mediated resistance. *Plant Cell* **5**, 577–586.

Padgett, H.S., Epel, B.L., Kahn, T.W., Heinlein, M., Watanabe, Y. and Beachy, R.N. (1996). Distribution of tobamovirus movement protein in infected cells and implications for cell-to-cell spread of infection. *Plant J.* **10**, 1079–1088.

Padgett, H.S., Watanabe, Y. and Beachy, R.N. (1997). Identification of the TMV replicase sequence that activates the *N*-gene mediated hypersensitive response. *Mol. Plant Microbe Interact.* **10**, 709–715.

Padidam, M., Sawyer, S. and Fauquet, C.M. (1999). Possible emergence of new geminiviruses by frequent recombination. *Virology* **265**, 218–225.

Paduch-Cichal, E. and Kryczynski, S. (1987). A low-temperature therapy and meristem-tip culture for eliminating four viroids from infected plants. *J. Phytopathol.* **118**, 341–346.

Paguio, O.R., Kuhn, C.W. and Boerma, H.R. (1988). Resistance-breaking variants of cowpea chlorotic mottle virus in soybean. *Plant Dis.* **72**, 768–770.

Pal, N., Moon, J.S., Sandhu, J., Domier, L.L. and D'Arcy, C.J. (2000). Production of barley yellow dwarf virus antisera by DNA immunization. *Can. J. Plant Pathol.* **22**, 410–415.

Palacio-Bielsa, A., Foissac, X. and Duran-Vila, N. (1999). Indexing of citrus viroids by imprint hybridization. *Eur. J. Plant Pathol.* **105**, 897–903.

Palanichelvam, K. and Schoelz, J.E. (2001). Uncoupling sequences within CaMV gene VI required for translational transactivation from those that elicit a hypersensitive defense response. Abstract W21–5, 20th Annual Meeting of the American Society for Virology.

Palanichelvam, K., Cole, A.B., Shababi, M. and Schoelz, J.E. (2000). Agroinfiltration of cauliflower mosaic virus gene VI elicits hypersensitive response in *Nicotiana* species. *Mol. Plant Microb. Interact.* **13**, 1275–1279.

Palauqui, J.-C., Elmayan, T., Pollien, J.-M. and Vaucheret, H. (1997). Systemic acquired silencing: transgene-specific post-transcriptional silencing is transmitted by grafting from silenced stocks to non-silenced scions. *EMBO J.* **16**, 4738–4745.

Paliwal, Y.C. (1980). Relationship of wheat streak mosaic and barley stripe mosaic viruses to vector and nonvector eriophyid mites. *Arch. Virol.* **63**, 123–132.

Pallas, V., Sanchez-Navarro, J.A. and Diez, J. (1999). *In vitro* evidence for RNA binding properties of the coat protein of prunus necrotic ringspot Ilarvirus and their comparison to related and unrelated viruses. *Archiv. Virol.* **144**, 797–803.

Palmenberg, A. (1990). Proteolytic processing of picornaviral polyproteins. *Annu. Rev. Microbiol.* **44**, 603–623.

Palmer, K.E. and Rybicki, E.P. (1997). The use of geminiviruses in biotechnology and plant molecular biology, with particular focus on Mastreviruses. *Plant Sci.* **129**, 115–130.

Palmer, K.E. and Rybicki, E.P. (1998). The molecular biology of mastreviruses. *Adv. Virus Res.* **50**, 183–234.

Palomar, M.K. and Brakke, M.K. (1976). Concentration and infectivity of barley stripe mosaic virus in barley. *Phytopathology* **66**, 1422–1426.

Paludan, N. (1985). Spread of viruses by a recirculating water supply in soilless culture. *Phytoparasitica* **13**, 276.

Palukaitis, P. (1984). Detection and characterisation of subgenomic RNA in plant viruses. *Methods Virol.* **7**, 259–317.

Palukaitis, P. (1987). Potato spindle tuber viroid: investigation of the long-distance, intra-plant transport route. *Virology* **158**, 239–241.

Palukaitis, P. (1988). Pathogenicity regulation by satellite RNAs of cucumber mosaic virus: minor nucleotide sequence changes alter host responses. *Mol. Plant Microbe Interact.* **1**, 175–181.

Palukaitis, P. and Roossinck, M.J. (1996). Spontaneous change of a benign satellite RNA of cucumber mosaic virus to a pathogenic variant. *Nature Biotech.* **14**, 1264–1268.

Palukaitis, P. and Zaitlin, M. (1984). Satellite RNAs of cucumber mosaic virus: characterization of two new satellites. *Virology* **132**, 426–435.

Palukaitis, P. and Zaitlin, M. (1987). The nature and biological significance of linear potato spindle tuber viroid molecules. *Virology* **157**, 199–210.

Palukaitis, P. and Zaitlin, M. (1997). Replicase-mediated resistance to plant virus disease. *Adv. Virus Res.* **48**, 349–377.

Palukaitis, P., Rakowski, A.G., Alexander, D.McE. and Symons, R.H. (1981). Rapid indexing of the sunblotch disease of avocados using a complementary DNA probe to avocado sunblotch viroid. *Ann. Appl. Biol.* **98**, 439–449.

Palukaitis, P., García-Arenal, F., Sulzinski, M.A. and Zaitlin, M. (1983). Replication of tobacco mosaic virus: VII. Further characterization of single- and double-stranded virus-related RNAs from TMV-infected plants. *Virology* **131**, 533–545.

Palukaitis, P., Roossinck, M.J., Dietzgen, R.G. and Francki, R.I.B. (1992). Cucumber mosaic virus. *Adv. Virus Res.* **41**, 281–348.

Pang, S.-Z., Slightom, J.L. and Gonsalves, D. (1993). Different mechanisms protect transgenic tobacco against tomato spotted wilt and impatiens necrotic spot tospoviruses. *Biotechnology* **11**, 819–824.

Panté, N. and Aebi, U. (1994). Towards understanding the three dimensional structure of the nuclear pore complex at the molecular level. *Curr. Opin. Cell Biol.* **4**, 187–196.

Pappu, H.R., Karasev, A.V., Anderson, E.J. *et al.* (1994). Nucleotide sequence and organization of eight 3' open reading frames of the citrus tristeza closterovirus genome. *Virology* **199**, 35–46.

Pappu, S.S., Brand, R., Pappu, H.R. *et al.* (1993). A polymerase chain reaction method adapted for selective amplification and cloning of 3' sequences of potyviral genomes: application to dasheen mosaic virus. *J. Virol. Methods* **41**, 9–20.

Parejarlarn, A., Lapis, D.B. and Hibino, H. (1984). Reaction of rice varieties to rice ragged stunt virus (RSV) infection by three known plant hopper (BPH) biotypes. *Int. Rice Res. Newsl.* **9**, 7–8.

Pares, R.D. (1988). Serological comparison of an Australian isolate of capsicum mosaic virus with capsicum tobamovirus isolates from Europe and America. *Ann. Appl. Biol.* **112**, 609–612.

Pares, R.D., Gunn, L.V. and Keskula, E.N. (1996). The role of infective plant debris, and its concentration in soil, in the ecology of tomato mosaic tobamovirus—a non-vectored plant virus. *J. Phytopathol.* **144**, 147–150.

Paris, H.S. and Cohen, S. (2000). Oligogenic inheritance for resistance to *Zucchini yellow mosaic virus* in *Cucurbita pepo*. *Ann. Appl. Biol.* **136**, 209–214.

Paris, H.S., Cohen, S., Burger, Y. and Yoseph, R. (1988). Single-gene resistance to zucchini yellow mosaic virus in *Cucurbita moschata*. *Euphytica* **37**, 27–29.

Parish, R.W. (1975). The lysosome concept in plants. I. Peroxidases associated subcellular and wall functions of maize root tips: implications for vacuole development. *Planta* **123**, 1–13.

Parker, J.E., Feys, B.J., van der Biezen, E.A. *et al.* (2000). Unravelling R gene-mediated disease resistance pathways in *Arabidopsis*. *Mol. Plant Pathol.* **1**, 17–24.

Parkinson, J. (1656). 'Paradisi in sole paradisus teffestris, or a garden of all sorts of pleasant flowers . . . with a kitchen garden of all manner of herbes etc.' London.

Parks, T.D., Howard, E.D., Wolpert, T.J., Arp, D.J. and Dougherty, W.G. (1995). Expression and purification of a recombinant tobacco etch virus NIa proteinase: biochemical analyses of the full-length and a naturally occurring truncated proteinase form. *Virology* **210**, 194–201.

Partridge, J.E., Shannon, L. and Gumpf, D. (1976). A barley lectin that binds free amino sugars: I. Purification and characterization. *Biochim. Biophys. Acta* **451**, 470–483.

Pascal, E., Sanderfoot, A.A., Ward, B.W., Medville, R., Turgeon, R. and Lazarowitz, S.G. (1994). The geminivirus BR1 movement protein binds single-stranded DNA and localizes to the cell nucleus. *Plant Cell* **6**, 995–1006.

Paszkowski, J., Pisan, B., Shillito, R.D., Hohn, T., Hohn, B. and Potrykus, I. (1986). Genetic transformation of *Brassica compestris* var. *rapa* protoplasts with an engineered cauliflower mosaic virus genome. *Plant Mol. Biol.* **6**, 303–312.

Patterson, S. and Verduin, B.J.M. (1987). Applications of immunogold labelling in animal and plant virology. *Arch. Virol.* **97**, 1–26.

Pavord, A. (1999). *The Tulip*. Bloomsbury, London.

Pawley, G.S. (1962). Plane groups on polyhedra. *Acta Crystallogr.* **15**, 49–53.

Pead, M.T. and Torrance, L. (1989). Some characteristics of monoclonal antibodies to a British MAV-like isolate of barley yellow dwarf virus. *Ann. Appl. Biol.* **113**, 639–644.

Peden, K.W.C., May, J.T. and Symons, R.H. (1972). A comparison of two plant virus-induced RNA polymerases. *Virology* **47**, 498–501.

Pelchat, M., Deschênes, P. and Perreault, J.-P. (2000). The database of the smallest known auto-replicable RNA species: viroids and viroid-like RNAs. *Nucl. Acids Res.* **28**, 179–180.

Pelcher, L.E., Murakishi, H.H. and Hartmann, J.X. (1972). Kinetics of TMV-RNA synthesis and its correlation with virus accumulation and crystalline viral inclusion formation in tobacco tissue culture. *Virology* **47**, 787–796.

Pelham, H.R.B. (1978). Leaky UAG termination codon in tobacco mosaic virus RNA. *Nature (London)* **272**, 469–471.

Pelham, H.R.B. (1979). Translation of tobacco rattle virus RNAs *in vitro*: four proteins from three RNAs. *Virology* **97**, 256–265.

Pennazio, S. and Roggero, P. (1988). Systemic acquired resistance induced in tobacco plants by localized virus infection does not operate against challenging viruses that infect systemically. *J. Phyopathol.* **121**, 255–266.

Pennazio, S. and Roggero, P. (1996). Plant hormones and plant virus diseases: the auxins. *Microbiologica* **19**, 369–378.

Pennazio, S. and Roggero, P. (1998). Endogenous changes in cytokinin activity in systemically virus-infected plants. *Microbiologica* **21**, 419–426.

Pennazio, S. and Sapetti, C. (1981). Electrolyte leakage in relation to viral and abiotic stresses inducing necrosis in cowpea leaves. *Biol. Plant.* **24**, 218–255.

Pennazio, S., Roggero, P. and Gentile, I.A. (1985). Effects of salicylate on virus-infected tobacco plants. *Phytopathol. Z.* **114**, 203–213.

Pennazio, S., Roggero, P. and Conti, M. (1999). Plasmodesmata and plant viruses: a centenary story. *Microbiologica* **22**, 389–404.

Pennington, R.E. and Melcher, U. (1993). *In planta* deletion of DNA inserts from the large intergenic region of cauliflower mosaic virus DNA. *Virology* **192**, 188–196.

Pereira, L.G., Torrance, L., Roberts, I.M. and Harrison, B.D. (1994). Antigenic structure of the coat protein of potato mop-top furovirus. *Virology* **203**, 277–285.

Peremyslov, V.V., Hagiwara, Y. and Dolja, V.V. (1999). HSP70 homolog functions in cell-to-cell movement of a plant virus. *Proc. Natl Acad. Sci. USA* **96**, 14771–14776.

Pérez de san Román, C., Legorburu, F.J., Pascualena, J. and Gil, A. (1988). Simultaneous detection of potato viruses Y, leaf roll, X and S by DAS-ELISA technique with artificial polyvalent antibodies (APAs). *Potato Res.* **31**, 151–158.

Perham, R.N. (1973). The reactivity of functional groups as a probe for investigating the topography of tobacco mosaic virus: the use of mutants with additional lysine residues in the coat protein. *Biochem. J.* **131**, 119–126.

Perham, R.N. and Wilson, T.M.A. (1976). The polarity of stripping of coat protein subunits from the RNA in tobacco mosaic virus under alkaline conditions. *FEBS Lett.* **62**, 11–15.

Perham, R.N. and Wilson, T.M.A. (1978). The characterization of intermediates formed during the disassembly of tobacco mosaic virus at alkaline pH. *Virology* **84**, 293–302.

Perlak, F.J., Kaniewski, W.K., Lawson, E.C., Vincent, M.N. and Feldman, J. (1995). Genetically improved potatoes: their potential role in integrated pest management. In: M. Manka (ed.) *Environmental Biotic Factors in Integrated Plant Disease Control*, pp. 451–454. Polish Phytopathological Society, Poznan, Poland.

Perler, F.B. (1999). A natural example of protein trans-splicing. *Trends Biochem. Sci.* **24**, 209–211.

Perret, V., Florentz, C., Dreher, T. and Giege, R. (1989). Structural analogies between the 3– tRNA-like structure of brome mosaic virus RNA and yeast tRNA^{tyr}, revealed by protection studies with yeast tyrosyl-tRNA synthetase. *Eur. J. Biochem.* **185**, 331–339.

Perrin, R.M. and Gibson, R.W. (1985). Control of some insect-borne plant viruses with the pyrethroid PP321 (Karate). *Int. Pest Control* **Nov/Dec**, 142–143.

Perring, T.M., Gruenhagen, N.M. and Farrar, C.A. (1999). Management of plant viral diseases through chemical control of insect vectors. *Annu. Rev. Entomol.* **44**, 457–481.

Perry, J.N. (1998). Measures of spatial pattern for counts. *Ecology* **79**, 1008–1017.

Perry, K.L., Zhang, L., Shintaku, M.H. and Palukaitis, P. (1994). Mapping determinants in cucumber mosaic virus for transmission by *Aphis gossypii*. *Virology* **205**, 591–595.

Perry, K.L., Zhang, L. and Palukaitis, P. (1998). Amino acid changes in the coat protein of cucumber mosaic virus differentially affect transmission by the aphids *Myzus persicae* and *Aphis gossypii*. *Virology* **242**, 204–210.

Perry, K.L., Ng, J., Liu, S. and Smith, T.J. (2001). Structural determinants in the vector transmission of cucumber mosaic virus. Abstract W33–5, 20th Annual Meeting of the American Society for Virology.

Pesic, Z., Hiruki, C. and Chen, M.H. (1988). Detection of viral antigen by immunogold cytochemistry in ovules, pollen and anthers of alfalfa infected with alfalfa mosaic virus. *Phytopathology* **78**, 1027–1032.

Peters, D. and Black, L.M. (1970). Infection of primary cultures of aphid cells with a plant virus. *Virology* **40**, 847–853.

Peters, S.A., Voorhorst, W.G.B., Wery, J., Wellink, J. and van Kammen, A. (1992). A regulatory role for the 32K protein in proteolytic processing of cowpea mosaic virus polyproteins. *Virology* **191**, 81–89.

Peterson, J.F. and Brakke, M.K. (1973). Genomic masking in mixed infections with brome mosaic and barley stripe mosaic viruses. *Virology* **51**, 174–182.

Petty, I.T.D., Hunter, B.G., Wei, N. and Jackson, A.O. (1989). Infectious barley stripe mosaic virus RNA transcribed *in vitro* from full-length genomic cDNA clones. *Virology* **171**, 342–349.

Petty, I.T.D., French, R., Jones, R.W. and Jackson, A.O. (1990). Identification of barley stripe mosaic virus genes involved in viral replication and systemic movement. *EMBO J.* **9**, 3453–3457.

Pfeiffer, P. (1998). Nucleotide sequence, genetic organization and expression strategy of the double-stranded RNA associated with the '447' cytoplasmic male sterility trait in *Vicia faba*. *J. Gen. Virol.* **79**, 2349–2358.

Pfeiffer, P. and Durham, A.C.H. (1977). The cation binding associated with structural transitions in bromegrass mosaic virus. *Virology* **81**, 419–432.

Pfeiffer, P. and Hohn, T. (1983). Involvement of reverse transcription in the replication of cauliflower mosaic virus: a detailed model and test of some aspects. *Cell* **33**, 781–789.

Pfeiffer, P. and Hohn, T. (1989). Cauliflower mosaic virus as a probe for studying gene expression in plants. *Physiol. Plant.* **77**, 625–632.

Pfeiffer, P., Laquel, P. and Hohn, T. (1984). Cauliflower mosaic virus replication complexes: characterisation of the associated enzymes and the polarity of DNA synthesised *in vitro*. *Plant Mol. Biol.* **3**, 261–270.

Pfeiffer, P., Gordon, K., Filtterer, J. and Hohn, T. (1987). The life cycle of cauliflower mosaic virus. In: D. von Wettstein and N.-H. Chua (eds) *Plant Molecular Biology*, pp. 443–458. Plenum, New York.

Pfister, T., Mirzayan, C. and Wimmer, E. (1999). Polioviruses (*Picornaviridae*): molecular biology. In: A. Granoff and R.G. Webster (eds) *Encyclopedia of Virology*, 2nd edn, pp. 1330–1348. Academic Press, San Diego.

Piazzolla, P., Tousignant, M.E. and Kaper, J.M. (1982). Cucumber mosaic virus-associated RNA5. IX. The overtaking of viral RNA synthesis by CARNA5 and dsCARNA5 in tobacco. *Virology* **122**, 147–157.

Piazzolla, P., Rubino, L., Tousignant, M.E. and Kaper, J.M. (1989a). Two different types of satellite RNA associated with chicory yellow mottle virus. *J. Gen. Virol.* **70**, 949–954.

Piazzolla, P., Palmieri, F. and Nuzzaci, M. (1989b). Infectivity studies on cucumber mosaic virus treated with a clay material. *J. Phytopathol.* **127**, 291–295.

Pierpoint, W.S. (1966). The enzymic oxidation of chlorogenic acid and some reactions of the quinone produced. *Biochem. J.* **98**, 567–580.

Pierpoint, W.S., Ireland, R.J. and Carpenter, J.M. (1977). Modification of proteins during the oxidation of leaf phenols: reaction of potato virus X with chlorogenoquinone. *Phytochemistry* **16**, 29–34.

Pietrzak, M. and Hohn, T. (1987). Translation products of cauliflower mosaic virus ORF V, the coding region corresponding to the retrovirus pol gene. *Virus Genes* **1**, 83–96.

Pilowsky, M., Frankel, R. and Cohen, S. (1981). Studies of the variable reaction at high temperature of F1 hybrid tomato plants resistant to tobacco mosaic virus. *Phytopathology* **71**, 319–323.

Pinck, L. and Hirth, L. (1972). The replicative RNA and the viral RNA synthesis rate in tobacco infected with alfalfa mosaic virus. *Virology* **49**, 413–425.

Pinck, L., Genevaux, M., Bouley, J.P. and Pinck, M. (1975). Amino acid accepter activity of replicative form from some tymovirus RNAs. *Virology* **63**, 589–590.

Pinck, M., Yot, P., Chapeville, F. and Duranton, H.M. (1970). Enzymatic binding of valine to the 3′ end of TYMV-RNA. *Nature (London)* **226**, 954–956.

Piron, P.G.M. (1986). New aphid vectors of potato virus YN. *Neth. J. Plant Pathol.* **92**, 223–229.

Pirone, T.P. (1967). Acquisition and release of infectious tobacco mosaic virus by aphids. *Virology* **31**, 569–571.

Pirone, T.P. (1977). Accessory factors in nonpersistent virus transmission. In: K.F. Harris and K. Maramorosch (eds) *Aphids as Virus Vectors*, pp. 221–235. Academic Press, New York.

Pirone, T.P. and Blanc, S. (1996). Helper-dependent vector transmission of plant viruses. *Annu Rev. Phytopathol.* **34**, 227–247.

Pirone, T.P. and Thornbury, D.W. (1988). Quantity of virus required for aphid transmission of a potyvirus. *Phytopathology* **7**, 104–107.

Pirone, T.P., Raccah, B. and Madden, L.V. (1988). Suppression of aphid colonization by insecticides: effect on the incidence of potyviruses in tobacco. *Plant Dis.* **72**, 350–353.

Pita, J.S., Fondong, V.N., Sangaré, A., Otim-Nape, G.W., Ogwal, S. and Fauquet, C.M. (2001). Recombination, pseudorecombination and synergism of geminiviruses are determinant keys to the epidemic of severe cassava mosaic disease in Uganda. *J. Gen. Virol.* **82**, 655–665.

Plant, A.L., Covey, S.N. and Grierson, D. (1985). Detection of a subgenomic mRNA for gene V, the putative reverse transcriptase gene of cauliflower mosaic virus. *Nucl. Acids Res.* **13**, 8305–8321.

Plaskitt, K.A., Watkins, P.A.C., Sleat, D.E., Gallie, D.R., Shaw, J.G. and Wilson, T.M.A. (1987). Immunogold labelling locates the site of disassembly and transient gene expression of tobacco mosaic virus-like pseudovirus particles *in vivo*. *Mol. Plant–Microbe Interact.* **1**, 10–16.

Pleij, C.W.A., Mellema, J.R., Noort, A. and Bosch, L. (1977). The occurrence of the coat protein messenger RNA in the minor components of turnip yellow mosaic virus. *FEBS Lett.* **80**, 19–22.

Pleše, N., Hoxha, E. and Miličić, D. (1975). Pathological anatomy of trees affected with apple stem grooving virus. *Phytopathol. Z.* **82**, 315–325.

Plochocka, D., Welnicki, M., Zielenkiewicz, P. and Ostoja-Zagórski, W. (1996). Three-dimensional model of the potyviral genome-linked protein. *Proc. Natl Acad. Sci. USA* **93**, 12150–12154.

Ploeg, A.T., Brown, D.J.F. and Robinson, D.J. (1992). The association between species of *Trichodorus* and *Paratrichodorus* vector nematodes and serotypes of tobacco rattle tobravirus. *Ann. Appl. Biol.* **121**, 619–630.

Ploeg, A.T., Mathis, A., Bol, J.F., Brown, D.J.F and Robinson, D.J. (1993a). Susceptibility of transgenic tobacco plants expressing tobacco rattle virus coat protein to nematode-transmitted and mechanically inoculated tobacco rattle virus. *J. Gen. Virol.* **74**, 2709–2715.

Ploeg, A.T., Robinson, D.J. and Brown, D.J.F. (1993b). RNA-2 of tobacco rattle virus encodes the determinants of transmissibility by trichodorid vector nematodes. *J. Gen. Virol.* **74**, 1463–1466.

Ploubidou, A. and Way, M. (2001). Viral transport and the cytoskeleton. *Curr. Opin. Cell Biol.* **13**, 97–105

Pobjecky, N., Rosenberg, G.H., Dinter-Gottlieb, G. and Kaufer, N.F. (1990). Expression of the β-glucoronidase gene under the control of the CaMV 35S promoter in *Schizosaccharomyces pombe*. *Mol. Gen. Genet.* **220**, 314–316.

Poch, O., Sauvaget, I., Delarue, M. and Tordo, N. (1989). Identification of four conserved motifs among RNA-dependent polymerase encoding elements. *EMBO J.* **8**, 3867–3874.

Podleckis, E.V., Hammond, R.W., Hurtt, S.S. and Hadidi, A. (1993). Chemilumiscent detection of potato and pome fruit viroids by digoxigenin-labeled dot-blot and tissue-blot hybridization. *J. Virol. Methods* **43**, 147–158.

Pogany, J. and Bujarski, J.J. (1996). Complementary sequences facilitate deletions in defective (D) RNAs associated with broad bean mottle bromovirus (BBMV). Presented at the 15th Annual Meeting of the American Society for Virology, University of Western Ontario, Canada.

Pogany, J., Romero, J., Huang, Q., Sgro, J.-Y., Shang, H. and Bujarski, J.J. (1995). *De novo* generation of defective interfering-like RNAs in broad bean mottle bromovirus. *Virology* **212**, 574–586.

Polak, J. (1983). Variability of resistance of white cabbage to turnip mosaic virus. *Tagungsber. Akad. Landwirtschaftswiss. D.D.R.* **216**, 331–335.

Pollard, D.G. (1977). Aphid penetration of plant tissues. In: K.F. Harris and K. Maramorosch (eds) *Aphids as Virus Vectors*, pp. 105–118. Academic Press, New York.

Polson, A. and von Wechmar, M.B. (1980). A novel way to transmit plant viruses. *J. Gen. Virol.* **51**, 179–181.

Polson, A., Coetzer, T., Kruger, J., von Maltzahn, E. and van der Merwe, K.J. (1985). Improvements in the isolation of IgY from yolks of eggs laid by immunized hens. *Immunol. Invest.* **14**: 323–327.

Ponsen, M.B. (1977). Anatomy of an aphid vector: *Myzus persicae*. In: K.F. Harris and K. Maramorosch (eds) *Aphids as Virus Vectors*, pp. 63–82. Academic Press, New York.

Pontier, D., Gan, S., Amasino, R.M., Roby, D. and Lam, E. (1999). Markers for hypersensitive response and senescence show distinct patterns of expression. *Plant Mol. Biol.* **39**, 1243–1255.

Ponz, F. and Bruening, G. (1986). Mechanisms of resistance to plant viruses. *Annu. Rev. Phytopathol.* **24**, 355–381.

Ponz, F., Glascock, C.B. and Bruening, G. (1988a). An inhibitor of polyprotein processing with the characteristics of a natural virus resistance factor. *Mol. Plant Microbe Interact.* **1**, 25–31.

Ponz, F., Russell, M.L., Rowhani, A. and Bruening, G. (1988b). A cowpea line has distinct genes for resistance to tobacco ringspot virus and cowpea mosaic virus. *Phytopathology* **78**, 1124–1128.

Pooggin, M.M. and Skryabin, K.G. (1992). The 5′-untranslated leader sequence of potato virus X RNA enhances the expression of a heterologous gene *in vivo*. *Mol. Gen. Genet.* **234**, 329–331.

Pooggin, M.M., Hohn, T. and Fütterer, J. (1998). Forced evolution reveals the importance of short open reading frame A and secondary structure in the cauliflower mosaic virus 35S RNA leader. *J. Virol.* **72**, 4157–4169.

Pooggin, M.M., Fütterer, J., Skryabin, K.G. and Hohn, T. (1999). A short open reading frame terminating in front of a stable hairpin is the conserved feature in pregenomic RNA leaders of plant pararetroviruses. *J. Gen. Virol.* **80**, 2217–2228.

Poolpol, P. and Inouye, T. (1986). Ultrastructure of plant cells doubly infected with potyviruses and other unrelated viruses. *Bull. Univ. Osaka Prefect. Ser. B* **38**, 13–23.

Porta, C. and Lomonossoff, G.P. (1996). Use of viral replicons for the expression of genes in plants. *Molec. Biotech.* **5**, 209–221.

Porta, C. and Lomonossoff, G.P. (1998). Scope for using plant viruses to present epitopes from animal pathogens. *Rev. Mol. Virol.* **8**, 25–41.

Porta, C., Spall, V.E., Loveland, J., Johnson, J.E., Barker, P.J. and Lomonossoff, G.P. (1994). Development of cowpea mosaic virus as a high-yielding system for the presentation of foreign peptides. *Virology* **202**, 949–955.

Posnette, A.F. (1947). Virus diseases of cacao in West Africa: I. Cacao viruses 1A, 1B, 1C, and 1D. *Ann. Appl. Biol.* **34**, 388–402.

Posnette, A.F. (1950). Virus diseases of cacao in West Africa. VII. Virus transmission by different vector species. *Ann. Appl. Biol.* **37**, 378–384.

Posnette, A.F. (1969). Tolerance of virus infection in crop plants. *Rev. Appl. Mycol.* **48**, 113–118.

Posnette, A.F. and Robertson, N.F. (1950). Virus diseases of cacao in West Africa. VI. Vector investigations. *Ann. Appl. Biol.* **37**, 363–377.

Posnette, A.F. and Todd, J.M.C.A. (1955). Virus diseases of cacao in West Africa. IX. Strain variation and interference in virus 1A. *Ann. Appl. Biol.* **43**, 433–453.

Powell-Abel, P., Nelson, R.S., De, B. *et al.* (1986). Delay of disease development in transgenic plants that express the tobacco virus coat protein gene. *Science* **232**, 738–743.

Powell, C.A. (1975). The effect of cations on the alkaline dissociation of tobacco mosaic virus. *Virology* **64**, 75–85.

Powell, C.A. and de Zoeten, G.A. (1977). Replication of pea enation mosaic virus RNA in isolated pea nuclei. *Proc. Natl Acad. Sci. USA* **74**, 2919–2922.

Powell, C.A., de Zoeten, G.A. and Gaard, G. (1977). The localization of pea enation mosaic virus-induced RNA-dependent RNA polymerase in infected peas. *Virology* **78**, 135–143.

Powell, C.A., Pelosi, R.R. and Cohen, M. (1992). Superinfection of orange trees containing mild isolates of citrus tristeza virus with severe Florida isolates of citrus tristeza virus. *Plant Dis.* **76**, 141–144.

Powell, C.A., Pelosi, R.R., Rundell, P.A., Stover, E. and Cohen, M. (1999). Cross-protection of grapefruit from decline-inducing isolates of citrus tristeza virus. *Plant Dis.* **83**, 989–991.

Powell, P.A., Stark, D.M., Sanders, P.R. and Beachy, R.N. (1989). Protection against tobacco mosaic virus in transgenic plants that express tobacco mosaic virus antisense RNA. *Proc. Natl Acad. Sci. USA* **86**, 6949–6952.

Powell, P.A., Saunders, P.R., Turner, N., Frayley, R.T. and Beachy R.N. (1990). Protection against tobacco mosaic virus infection in transgenic plants requires accumulation of coat protein rather than coat protein RNA sequences. *Virology* **175**, 124–130.

Power, A.G. (1996). Competition between viruses in a complex plant–pathogen. *Ecology* **77**, 1004–1010.

Pratt, M.J. and Matthews, R.E.F. (1971). Non-uniformities in the metabolism of excised leaves and leaf discs. *Planta* **99**, 21–36.

Pratt, W.B., Krishna, P and Olsen, L.J. (2001). Hsp90-binding immunophilins in plants: the protein movers. *Trends Plant Sci.* **6**, 54–58.

Preece, D.A. (1967). Nested balanced incomplete block designs. *Biometrika* **54**, 479–486.

Preil, W., Koenig, R., Engelhardt, M. and Meier-Dinkel, A. (1982). Elimination of poinsettia mosaic virus (PoiMV) I and poinsettia cryptic virus (PoiCV) from *Euphorbia pulcherrima* Willd by cell suspension culture. *Phytopathol. Z.* **105**, 193–197.

Price, W.C. (1966). Flexuous rods in phloem cells of lime plants infected with citrus tristeza virus. *Virology* **29**, 285–294.

Pringle, C.R. (1998). The universal system of virus taxonomy of the International Committee on Taxonomy of Viruses (ICTV), including new proposals ratified since publication of the sixth report in 1995. *Arch. Virol.* **143**, 203–210.

Pringle, C.R. (1999a). Virus taxonomy at the XIth International Congress of Virology, Sydney, Australia, 1999. *Archiv. Virol.* **144**, 2065–2070.

Pringle, C.R. (1999b). Editorial: Virus nomenclature. *Arch. Virol.* **144**, 1464–1466.

Prins, M. and Goldbach, R. (1996). RNA-mediated virus resistance in transgenic plants. *Arch. Virol.* **141**, 2259–2276.

Prins, M. and Goldbach, R. (1998). The emerging problem of tospovirus infection and nonconventional methods of control. *Trends Microbiol.* **6**, 31–35.

Proctor, V.W. (1968). Long-distance dispersal of seeds by retention in digestive tracts of birds. *Science* **160**, 321–322.

Prod'homme, D., Le Panse, S., Drugeon, G. and Jupin, I. (2001). Detection and subcellular localization of the turnip yellow mosaic virus 66K replication protein in infected cells. *Virology* **281**, 88–101.

Prody, G.A., Bakes, J.T., Buzayan, J.M., Schneider, I.R. and Bruening, G. (1986). Autolytic processing of dimeric plant virus satellite RNA. *Science* **231**, 1577–1580.

Proeseler, G. (1980). Piesmids. In: K.F. Harris and K. Maramorosch (eds) *Vectors of Plant Pathogens*, pp. 97–113. Academic Press, New York.

Proeseler, G. and Kastirr, U. (1988). Research into the effect of fungicides against *Polymyxa graminis* Led. as vector of barley yellow mosaic virus. *Nachrichten bl. Pflanzenschutzdienst DDR* **42**, 116–117.

Provvidenti, R. (1990). Inheritance of resistance to pea mosaic virus in *Pisum sativum. J. Hered.* **81**, 143–145

Provvidenti, R. and Alconero, R. (1988a). Inheritance of resistance to a lentil strain of pea seed-borne mosaic virus in *Pisum sativum. J. Hered.* **79**, 45–47.

Provvidenti, R. and Alconero, R. (1988b). Inheritance of resistance to a third pathotype of pea seed-borne mosaic virus in *Pisum sativum. J. Hered.* **79**, 76–77.

Provvidenti, R. and Schroeder, W.T. (1973). Resistance in *Phaseolus vulgaris* to the severe strain of bean yellow mosaic virus. *Phytopathology* **63**, 196–197.

Prüfer, D., Tacke, E., Schmitz, J., Kull, B., Kaufmann, A. and Rohde, W. (1992). Ribosomal frameshifting in plants: a novel signal directs the −1 frameshift in the synthesis of the putative replicase of potato leafroll luteovirus. *EMBO J.* **11**, 1111–1117.

Pruss, G., Ge, X., Shi, X.M., Carrington, J.C. and Vance, V.B. (1997). Plant viral synergism: the potyviral genome encodes a broad-range pathogenicity enhancer that transactivates replication of heterologous viruses. *Plant Cell* **9**, 859–868.

Pryor, T. and Ellis, J. (1993). The genetic complexity of fungal resistance genes in plants. *Adv. Plant Pathol.* **10**, 281–305.

Puchta, H., Ramm, K. and Sänger, H.L. (1988). The molecular structure of hop latent viroid (HLV) a new viroid occurring worldwide in hops. *Nucl. Acids Res.* **16**, 4197–4216.

Purcifull, D.E., Hiebert, E. and McDonald, J.G. (1973). Immunochemical specificity of cytoplasmic inclusions induced by viruses in the potato Y group. *Virology* **55**, 275–279.

Putz, C., Kuszala, J., Kuszala, M. and Spindler, C. (1974). Variation de pouvoir pathogène des isolats du virus de la mosaïque du concombre associée à la necrose de la tomate. *Ann. Phytopathol.* **6**, 139–154.

Pyne, J.W. and Hall, T C. (1979). Efficient ribosome binding of brome mosaic virus (BMV) RNA4 contributes to its ability to outcompete the other BMV RNAs for translation. *Intervirology* **11**, 23–29.

Qiu, S.G., Wintermantel, W.M., Sha, Y. and Schoelz, J.E. (1997). Light-dependent systemic infection of solanaceous species by cauliflower mosaic virus can be conditioned by a viral gene encoding an aphid transmission factor. *Virology* **227**, 180–188.

Qiu, W. and Moyer, J.W. (1999). Tomato spotted wilt tospovirus adapts to the TSWV *N* gene-derived resistance by genomic reassortment. *Phytopathology* **89**, 575–582.

Qiu, Y.-L., Lee, J., Bernasconi-Quadroni, F. *et al.* (1999). The earliest angiosperms: evidence from mitochondrial, plastid and nuclear genomes. *Nature* **402**, 404–407.

Qu, F. and Morris, T.J. (1997). Encapsidation of turnip crinkle virus is defined by a specific packaging signal and RNA size. *J. Virol.* **71**, 1428–1435.

Qu, F. and Morris, T.J. (1999). Carmoviruses (*Tombusviridae*). In: A. Granoff and R.G. Webster (eds) *Encyclopedia of Virology*, 2nd edn, pp. 243–247. Academic Press, San Diego.

Qu, F. and Morris, T.J. (2000). Cap-independent translation enhancement of turnip crinkle virus genomic and subgenomic RNAs. *J. Virol.* **74**, 1085–1093.

Quader, H. (1990). Formation and disintegration of cisternae of the endoplasmic reticulum visualized in live cells by conventional fluorescence and confocal laser scanning microscopy: role of calcium and the cytoskeleton. *Protoplasma* **151**, 167–170.

Quadt, R., Verbeek, H.J.M. and Jaspars, E.M.J. (1988). Involvement of a non-structural protein in the RNA-synthesis of brome mosaic virus. *Virology* **165**, 256–261.

Quadt, R., Rosdorff, H.J.M., Hunt, T.W. and Jaspars, E.M.J. (1991). Analysis of the protein composition of alfalfa mosaic virus RNA-dependent RNA polymerase. *Virology* **182**, 309–315.

Quadt, R., Kao, C.C., Browning, K.S., Hershberger, R.P. and Ahlquist, P. (1993). Characterization of a host protein associated with brome mosaic virus RNA-dependent RNA polymerase. *Proc. Natl Acad. Sci. USA* **90**, 1498–1502.

Quadt, R., Ishikawa, M., Janda, M. and Ahlquist, P. (1995). Formation of brome mosaic virus RNA-dependent RNA polymerase in yeast requires coexpression of viral proteins and viral RNA. *Proc. Natl Acad. Sci. USA* **92**, 4892–4896.

Querci, M., Baulcombe, D.C., Goldbach, R.W. and Salazar, L.F. (1995). Analysis of the resistance-breaking determinants of potato virus X (PVX) strain HB on different potato genotypes expressing extreme resistance to PVX. *Phytopathology* **85**, 1003–1010.

Quesniaux, V., Briand, J.-P. and van Regenmortel, M.H.V. (1983a). Immunochemical studies of turnip yellow mosaic virus: II. Localisation of a viral epitope in the N-terminal residues of the coat protein. *Mol. Immunol.* **20**, 179–185.

Quesniaux, V., Jaegle, M. and van Regenmortel, M.H.V. (1983b). Immunological studies of turnip yellow mosaic virus: III. Localisation of two viral epitopes in residues 57–64 and 183–189 of the coat protein. *Biochim. Biophys. Acta* **743**, 226–231.

Quiot, J.B., Devergne, J.C., Marchoux, G., Cardin, L. and Douine, L. (1979a). Ecology and epidemiology of cucumber mosaic virus (CMV) in South East France. VI. Distribution of two CMV groups in weeds. *Ann. Phytopathol.* **11**, 349–357.

Quiot, J.B., Devergne, J.C., Cardin, L., Verbrugge, M., Marchoux, G. and Labonne, G. (1979b). Ecology and epidemiology of cucumber mosaic virus in the south-east of France. VII. Occurrence of two virus populations in various crops. *Ann. Phytopathol.* **11**, 359–373.

Raccah, B., Blanc, S. and Huet, H. (2001). Molecular basis of vector transmission: Potyvirus. In: K. Harris, J.E. Duffus and O.P. Smith (eds) *Virus–Insect–Plant Interactions*. Academic Press, San Diego (in press).

Rackwitz, H.-R., Rohde, W. and Sänger, H.L. (1981). DNA-dependent RNA polymerase II of plant origin transcribes viroid RNA into full-length copies. *Nature (Lond.)* **291**, 297–301.

Radford, J.E. and White, R.G. (1998). Localization of a myosin-like protein to plasmodesmata. *Plant J.* **14**, 743–750.

Raga, I.N., Ito, K., Matsui, M. and Okada, M. (1988). Effects of temperature on adult longevity, fertility and rate of transovarial passage of rice stripe virus in the small brown planthopper, *Laodelphax striatellus* Fallen (Homoptera: Delphacidae). *Appl. Entomol. Zool.* **23**, 67–75.

Ragetli, H.W.J., Weintraub, M. and Elder, M. (1973). Effective mechanical inoculation of plant viruses in the absence of water. *Can. J. Bot.* **51**, 1977–1981.

Raggi, V. (1998). Hydroxyproline-rich glycoprotein accumulation in TMV-infected tobacco showing systemic acquired resistance to powdery mildew. *J. Phytopathol.* **146**, 321–325.

Raghavendra, K., Adams, M.L. and Schuster, T.M. (1985). Tobacco mosaic virus protein aggregates in solution: structural comparison of 20S aggregates with those near conditions for disk crystallisation. *Biochemistry* **24**, 3298–3304.

Raghavendra, K., Salunke, D.M., Caspar, D.L.D. and Schuster, T.M. (1986). Disk aggregates of tobacco mosaic virus protein in solution: electron microscopy observations. *Biochemistry* **25**, 6276–6279.

Ralph, R.K. and Wojcik, S.J. (1966). Synthesis of double-stranded viral RNA by cell-free extracts from turnip yellow mosaic virus-infected leaves. *Biochim. Biophys. Acta* **119**, 347–361.

Ralph, R.K., Matthews, R.E.F., Matus, A.I. and Mandel, H.G. (1965). Isolation and properties of double-stranded RNA from virus infected plants. *J. Mol. Biol.* **11**, 202–212.

Ramírez, B.-C. and Haenni, A.-L. (1994). Molecular biology of tenuiviruses, a remarkable group of plant viruses. *J. Gen. Virol.* **75**, 467–475.

Ramírez, B.-C., Macaya, G., Calvert, L.A. and Haenni, A.-L. (1992). Rice hoja blanca virus genome characterization and expression *in vitro*. *J. Gen. Virol.* **74**, 2463–2468.

Randles, I.W. and Francki, R.I.B. (1972). Infectious nucleocapsid particles of lettuce necrotic yellows virus with RNA-dependent RNA polymerase activity. *Virology* **50**, 297–300.

Randles, J.W., Harrison, B.D., Murant, A.F. and Mayo, M.A. (1977). Packaging and biological activity of the two essential RNA species of tomato black ring virus. *J. Gen. Virol.* **36**, 187–193.

Randles, J.W., Chu, P.W.G., Dale, J.L. *et al.* (2000). Genus *Nanovirus*. In: M.H.V. van Regenmortel, C.M. Fauquet, D.H.L. Bishop, E.B. Carstens, M.K. Estes, S.M. Lemon, J. Maniloff, M.A. Mayo, D.J. McGeoch, C.R. Pringle and R.B. Wickner (eds) *Virus Taxonomy. Seventh Report of the International Committee on Taxonomy of Viruses*, pp. 303–309. Academic Press, San Diego.

Ranki, M., Palva, A., Virtanen, M., Laaksonen, M. and Soderlund, H. (1983). Sandwich hybridization as a convenient method for detection of nucleic acids in crude samples. *Gene* **21**, 77–85.

Rao, A.L.N. (1997). Molecular studies on bromovirus capsid protein: III. Analysis of cell-to-cell movement competence of coat protein defective variants of cowpea chlorotic mottle virus. *Virology* **232**, 385–395.

Rao, A.L.N. and Francki, R.I.B. (1981). Comparative studies on tomato aspermy and cucumber mosaic viruses. VI. Partial compatibility of genome segments from the two viruses. *Virology* **114**, 573–575.

Rao, A.L.N. and Francki, R.I.B. (1982). Distribution of determinants for symptom production and host range on the three RNA components of cucumber mosaic virus. *J. Gen. Virol.* **61**, 197–205.

Rao, A.L.N. and Grantham, G.L. (1996). Molecular studies on bromovirus capsid protein: II. Functional analysis of the amino-terminal arginine-rich motif and its role in encapsidation, movement and pathology. *Virology* **226**, 294–305.

Rao, A.L.N. and Hall, T.C. (1993). Recombination and polymerase error facilitates restoration of infectivity in brome mosaic virus. *J. Virol.* **67**, 969–979.

Rao, A.L.N. and Hiruki, C. (1987). Unilateral compatibility of genome segments from two distinct strains of red clover necrotic mosaic virus. *J. Gen. Virol.* **68**, 191–194.

Rao, A.L.N., Hatta, T. and Francki, R.I.B. (1982). Comparative studies on tomato aspermy and cucumber mosaic viruses. VII. Serological relationships reinvestigated. *Virology* **116**, 318–326.

Rao, A.L.N., Dreher, T.W., Marsh, L.E. and Hall, T.C. (1989). Telomeric function of the tRNA-like structure of brome mosaic virus RNA. *Proc. Natl Acad. Sci. USA* **86**, 5335–5339.

Rao, G.P., Shukla, K. and Gupta, S.N. (1987). Effect of cucumber mosaic virus infection on modulation, nodular physiology and nitrogen fixation of pea plants. *J. Plant Dis. Prot.* **94**, 606–613.

Rao, R.D.V.J.P. and Yaraguntiah, R.C. (1976). Natural occurrence of potato virus Y on *Datura metel*. *Curr. Sci.* **45**, 467.

Rappaport, I. and Siegel, A. (1955). Inactivation of tobacco mosaic virus by rabbit antiserum. *J. Immunol.* **74**, 106–116.

Rappaport, I. and Wildman, S.G. (1957). A kinetic study of local lesion growth on *Nicotiana glutinosa* resulting from tobacco mosaic virus infection. *Virology* **4**, 265–274.

Rappaport, I. and Wu, J.-H. (1963). Activation of latent virus infection by heat. *Virology* **20**, 472–476.

Raski, D.J. and Hewitt, W.B. (1963). Plant-parasitic nematodes as vectors of plant viruses. *Phytopathology* **53**, 39–47.

Raski, D.J., Goheen, A.C., Lider, L.A. and Meredith, C.P. (1983). Strategies against grapevine fanleaf virus and its nematode vector. *Plant Dis.* **67**, 335–339.

Rasochová, L. and Miller, W.A. (1996). Satellite RNA of barley yellow dwarf-RPV virus reduces accumulation of RPV helper virus RNA and attenuates RPV symptoms in oats. *Mol. Plant Microb. Interact.* **9**, 646–650.

Rasochová, L., Passmore, B.K., Falk, B.W. and Miller, W.A. (1997). The satellite RNA of barley yellow dwarf virus-RPV is supported by beet western yellows virus in dicotyledonous protoplasts and plants. *Virology* **231**, 182–191.

Rast, A.T.B. (1967a). Yield of glasshouse tomatoes as affected by strains of tobacco mosaic virus. *Neth. J. Plant Pathol.* **73**, 147–156.

Rast, A.T.B. (1967b). Differences in aggressiveness between TMV-isolates from tomato on clones of *Lycopersicum peruvianum*. *Neth. J. Plant Pathol.* **73**, 186–189.

Rast, A.T.B. (1975). Variability of tobacco mosaic vines in relation to control of tomato mosaic in glasshouse tomato crops by resistance breeding and cross protection. *Agric. Res. Rep. (Wageningen)* **834**, 1–76.

Ratcliff, F., Harrison, B.D. and Baulcombe, D.C. (1997). A similarity between viral defense and gene silencing in plants. *Science* **276**, 1558–1560.

Ratcliff, F.G., MacFarlane, S.A. and Baulcombe, D.C. (1999). Gene silencing without DNA: RNA-mediated cross-protection between viruses. *Plant Cell* **11**, 1207–1215.

Rathjen, J.P., Karageirgos, L.E., Habili, N., Waterhouse, P.M. and Symons, R.H. (1994). Soybean dwarf virus contains the third variant genome type in the luteovirus group. *Virology* **198**, 671–679.

Raybould, A.F., Maskell, A.F., Edwards, M.L., Cooper, J.I. and Gray, A.J. (1999). The prevalence and spatial distribution of viruses in natural populations of *Brassica oleracea*. *New Phytol.* **141**, 265–275.

Ready, M.P., Brown, D.T. and Robertus, J.D. (1986). Extracellular localization of pokeweed antiviral protein. *Proc. Natl Acad. Sci. USA* **83**, 5053–5056.

Real, L.A. and McElhany, P. (1996). Spatial pattern and process in plant–pathogen interactions. *Ecology* **77**, 1011–1025.

Reddick, B.B., Habera, L.F. and Law, M.D. (1997). Nucleotide sequence and taxonomy of maize chlorotic dwarf virus within the Sequiviridae. *J. Gen. Virol.* **78**, 1165–1174.

Reddy, D.V.R. and Black, L.M. (1973). Electrophoretic separation of all components of the double-stranded RNA of wound tumor virus. *Virology* **54**, 557–562.

Reddy, D.V.R. and Black, L.M. (1974). Deletion mutations of the genome segments of wound tumor virus. *Virology* **61**, 458–473.

Reddy, D.V.R. and Black, L.M. (1977). Isolation and replication of mutant populations of wound tumor virions lacking certain genome segments. *Virology* **80**, 336–346.

Reddy, D.V.R. and MacLeod, R. (1976). Polypeptide components of wound tumor virus. *Virology* **70**, 274–282.

Reddy, D.V.R. and Richins, D.R. (1999). Plant pararetroviruses— legume caulimoviruses. In: A. Granoff and R.G. Webster (eds) *Encyclopedia of Virology*, 2nd edn, pp. 1289–1292. Academic Press, San Diego.

Reddy, D.V.R., Delfosse, P. and Mayo, M.A. (1999). Pecluviruses. In: A. Granoff and R.G. Webster (eds) *Encyclopedia of Virology*, 2nd edn, pp. 1196–1200. Academic Press, San Diego.

Reeck, G.R., de Hahn, C., Teller, D.C. *et al.* (1987). 'Homology' in proteins and nucleic acids: a terminology muddle and a way out of it. *Cell* **50**, 667.

Rees, M.W. and Short, M.N. (1965). Variations in the composition of two strains of tobacco mosaic virus in relation to their host. *Virology* **26**, 596–602.

Rees, M.W. and Short, M.N. (1982). The primary structure of cowpea chlorotic mottle virus coat protein. *Virology* **119**, 500–503.

Rees, M.W., Short, M.N. and Kassanis, B. (1970). The amino acid composition, antigenicity, and other characteristics of the satellite viruses of tobacco necrosis virus. *Virology* **40**, 448–461.

Register, J.C. and Beachy, R.N. (1988). Resistance to TMV in transgenic plants results from interference with an early event in infection. *Virology* **166**, 524–532.

Register, J.C., III, and Beachy, R.N. (1989). Effects of protein aggregation state on coat proteinmediated protection against tobacco mosaic virus using a transient protoplast assay. *Virology* **173**, 656–663.

Register, J.C. and Nelson, R.S. (1992). Early events in plant virus infections: relationships with genetically engineered protection and host gene resistance. *Semin. Virol.* **3**, 441–451.

Register, J.C., Powell, P.A., Nelson, R.S. and Beachy, R.N. (1988). Genetically engineered cross protection against TMV interferes with initial infection and long distance spread of the virus. In: B. Staskawicz, P. Ahlquist and O. Yoder (eds) *Molecular Biology of Plant–Pathogen Interactions*, pp. 269–282. Alan R. Liss, New York.

Reichel, C. and Beachy, R.N. (1998). Tobacco mosaic virus infection induces severe morphological changes in the endoplasmic reticulum. *Proc. Natl Acad. Sci. USA* **95**, 11169–11174.

Reichel, C. and Beachy, R.N. (2000). Degradation of tobacco mosaic virus movement protein by the 26S proteasome. *J. Virol.* **74**, 3330–3337.

Reichel, C., Más, P. and Beachy, R.N. (1999). The role of ER and cytoskeleton in plant viral trafficking. *Trends Plant Sci.* **4**, 458–462.

Reichenbächer, D., Kalinina, I., Schulze, M., Hom, A. and Kleinhempel, H. (1984). Ultramicro ELISA with a fluorogenic substrate for detection of potato viruses. *Potato Res.* **27**, 353–364.

Reichmann, J.L., Lain, S. and García, J.A. (1992). Highlights and prospects of potyvirus molecular biology. *J. Gen. Virol.* **73**, 1–16.

Reid, M.S. and Matthews, R.E.F. (1966). On the origin of the mosaic induced by turnip yellow mosaic virus. *Virology* **28**, 563–570.

Reijnders, L., Sloof, P. and Borst, P. (1973). The molecular weights of the mitochondrial–ribosomal RNAs of *Saccharomyces carlsbergensis. Eur. J. Biochem.* **35**, 266–269.

Reijnders, L., Aalbers, A.M.J., van Kammen, A. and Thuring, R.W.J. (1974). Molecular weights of plant viral RNAs determined by gel electrophoresis under denaturing conditions. *Virology* **60**, 515–521.

Reimann-Philipp, U. and Beachy, R,N, (1993). The mechanism(s) of coat protein-mediated resistance against tobacco mosaic virus. *Semin. Virol.* **4**, 349–356.

Reinero, A. and Beachy, R.N. (1986). Association of TMV coat protein with chloroplast membranes in virus-infected leaves. *Plant Mol. Biol.* **6**, 291–301.

Ren, Q., Pfeiffer, T.W. and Ghabrial, S.A. (1997). Soybean mosaic virus incidence level and infection times: interaction effects on soybean. *Crop Sci.* **37**, 1706–1711.

Renaudin, J., Bové, J.M., Otsuki, Y. and Takebe, I. (1975). Infection of *Brassica* leaf protoplasts by turnip yellow mosaic virus. *Mol. Gen. Genet.* **141**, 59–68.

Resconich, E.C. (1963). Movement of tobacco necrosis virus in systemically infected soybeans. *Phytopathology* **53**, 913–916.

Restrepo-Hartwig, M.A. and Ahlquist, P. (1996). Brome mosaic virus helicase- and polymerase-like proteins colocalize on the endoplasmic reticulum at sites of viral RNA synthesis. *J. Virol.* **70**, 8908–8916.

Restrepo-Hartwig, M.A. and Ahlquist, P. (1999). Brome mosaic virus RNA replication proteins 1a and 2a colocalize and 1a independently localizes on the yeast endoplasmic reticulum. *J. Virol.* **73**, 10303–10309.

Restrepo-Hartwig, M.A. and Carrington, J.C. (1992). Regulation of nuclear transport of a plant potyvirus protein by autoproteolysis. *J. Virol.* **66**, 5662–5666.

Restrepo-Hartwig, M.A. and Carrington, J.C. (1994). The tobacco etch potyvirus 6-kilodalton protein is membrane associated and involved in viral replication. *J. Virol.* **68**, 2388–2397.

Reusken, C.B.E.M. and Bol, J.F. (1996). Structural elements of the 3'-terminal coat protein binding site in alfalfa mosaic virus RNAs. *Nucl. Acids Res.* **24**, 2660–2665.

Reuzeau, C., McNally, J.G. and Pickard, B.G. (1997). The endomembrane sheath: a key structure for understanding the plant cell? *Protoplasma* **200**, 1–9.

Revers, F., Le Gall, O., Candresse, T. and Maule, A.J. (1999). New advances in understanding the molecular biology of plant/potyvirus interactions. *Mol. Plant Microb. Interact.* **12**, 367–376.

Revill, P.A., Davidson, A.D. and Wright, P.J. (1994). The nucleotide sequence and genome organization of mushroom bacilliform virus: a single-stranded RNA virus of *Agaricus bisporus* (Lange) Inbach. *Virology* **202**, 904–911.

Revill, P.A., Davidson, A.D. and Wright, P.J. (1998). Mushroom bacilliform virus: the initiation of translation at the 5' end of the genome and identification of the VPg. *Virology* **249**, 231–237.

Revill, P.A., Davidson, A.D. and Wright, P.J. (1999). Identification of a subgenomic mRNA encoding the capsid protein of mushroom bacilliform virus, a single-stranded RNA mycovirus. *Virology* **260**, 273–276.

Rezaian, M. (1990). Australian grapevine viroid—evidence for extensive recombination between viroids. *Nucleic Acids Res.* **18**, 1813–1818.

Rezaian, M.A. and Francki, R.I.B. (1973). Replication of tobacco ringspot virus: I. Detection of a low-molecular-weight double-stranded RNA from infected plants. *Virology* **56**, 238–249.

Rezaian, M.A. and Symons, R.H. (1986). Anti-sense regions in satellite RNA of cucumber mosaic virus form stable complexes with the viral coat protein gene. *Nucl. Acids Res.* **14**, 3229–3239.

Rezaian, M.A., Francki, R.I.B., Chu, P.W.G. and Hatta, T. (1976). Replication of tobacco ringspot virus: III. Site of virus synthesis in cucumber cotyledon cells. *Virology* **74**, 481–488.

Rezaian, M.A., Williams, R.H.V. and Symons, R.H. (1985). Nucleotide sequence of cucumber mosaic virus RNA1: presence of a sequence complementary to part of the viral satellite RNA and homologies with other viral RNAs. *Eur. J. Biochem.* **150**, 331–339.

Rezaian, M.A., Skene, K.G.M. and Ellis, J.G. (1988). Anti-sense RNAs of cucumber mosaic virus in transgenic plants assessed for control of the virus. *Plant. Mol. Biol.* **11**, 463–471.

Rezelman, G., Goldbach, R. and van Kammen, A. (1980). Expression of bottom component RNA of cowpea mosaic virus in cowpea protoplasts. *J. Virol.* **36**, 366–373.

Rezelman, G., Franssen, H.J., Goldbach, R.W., Ie, T.S. and van Kammen, A. (1982). Limits to the independence of bottom component RNA of cowpea mosaic virus. *J. Gen. Virol.* **60**, 335–342.

Rezelman, G., van Kammen, A. and Wellink, J. (1989). Expression of cowpea mosaic virus mRNA in cowpea protoplasts. *J. Gen. Virol.* **70**, 3043–3050.

Rhee, Y., Tzfira, T., Chen, M.-H., Waigmann, E. and Citovsky, V. (2000). Cell-to-cell movement of tobacco mosaic virus: enigmas and explanations. *Mol. Plant Pathol.* **1**, 33–39.

Rhodes, J.D., Thain, J.F. and Wilden, D.C. (1999). Evidence for physically distinct signalling pathways in the wounded plant. *Ann. Bot.* **84**, 109–116.

Rice, R.H. (1974). Minor protein components in cowpea chlorotic mottle virus and satellite of tobacco necrosis virus. *Virology* **61**, 249–255.

Richards, K.E., and Williams, R.C. (1976). Assembly of tobacco mosaic virus *in vitro. Compr. Virol.* **6**, 1–37.

Richards, K.E., Jonard, G., Jacquemond, M. and Lot, H. (1978). Nucleotide sequence of cucumber mosaic virus-associated RNA5. *Virology* **89**, 395–408.

Richards, K.E., Guilley, H. and Jonard, G. (1981). Further characterization of the discontinuities in cauliflower mosaic virus DNA. *FEBS Lett.* **134**, 67–70.

Richards, O.W. and Davies, R.G. (1994). *Imms General Textbook of Entomology,* 10th edn. Chapman and Hall, London.

Richardson, J. and Sylvester, E.S. (1968). Further evidence of multiplication of sowthistle yellow vein virus in its aphid vector *Hyperomyzus lactucae. Virology* **35**, 347–355.

Richert-Pöggeler, K.R. and Shepherd, R.J. (1997). Petunia vein-clearing virus: a plant pararetrovirus with the core sequences of an integrase function. *Virology* **236**, 137–146.

Richert-Pöggeler, K.R. and Shepherd, R.J. (1998). Petunia vein-clearing virus: a plant pararetrovirus with the core sequences for an integrase function. *Virology* **236**, 137–146.

Richter, T.E. and Ronald, P.C. (2000). The evolution of disease resistance genes. *Plant Mol. Biol.* **42**, 195–204.

Riesner, D. (1987). Viroid function: subcellular location and *in situ* association with cellular components. In: T.O. Diener (ed.) *The Viroids,* pp. 99–116. Plenum Press, New York.

Riesner, D., Henco, K., Rokohl, U. *et al.* (1979). Structure and structure formation of viroids. *J. Mol. Biol.* **133**, 85–115.

Riesner, D., Colpan, M., Goodman, T.C., Nagel, L., Schumacher, J. and Steger, G. (1983). Dynamics and interactions of viroids. *J. Biomol. Struct. Dyn.* **1**, 669–688.

Riesner, D., Fels, A., Repsilber, D. *et al.* (1999) Structural motifs involved in replication and pathogenicity of potato spindle tuber viroid. Abstract in XI International Congress of Virology, Sydney, Australia. p. 29.

Ritter, E., Debener, T., Barone, A., Salamini, F. and Gebhardt, C. (1991). RFLP mapping of potato chromosomes of two gene controlling extreme resistance to potato virus X (PVX). *Mol. Gen. Genet.* **227**, 81–85.

Rivera-Bustamante, R.F. and Semancik, J.S. (1989). Properties of a viroid-replicating complex solubilized from nuclei. *J. Gen. Virol.* **70**, 2707–2716.

Rizvi, S.A.H. and Raman, K.V. (1983). Effect of glandular trichomes on the spread of potato virus Y (PVY) and potato leaf roll virus (PLRV) in the field. In: W.J. Hooker (ed.) *Proceeding of the International Congress on Research for the Potato in the Year 2000,* pp. 162–163. International Potato Centre, Lima, Peru.

Rizzo, T.M. and Palukaitis, P. (1988). Nucleotide sequence and evolutionary relationships of cucumber mosaic virus (CMV) strains: CMV RNA2. *J. Gen. Virol.* **69**, 1777–1787.

Rizzo, T.M. and Palukaitis, P. (1989). Nucleotide sequence and evolutionary relationships of cucumber mosaic virus (CMV) strains: CMV RNA1. *J. Gen. Virol.* **70**, 1–11.

Rizzo, T.M. and Palukaitis, P. (1990). Construction of full-length cDNA clones of cucumber mosaic virus RNAs 1, 2, and 3: generation of infectious transcripts. *Mol. Gen. Genet.* **222**, 249–256.

Robaglia, C., Durand-Tardif, M., Tronchet, M., Boudazin, G., Astier-Manifacier, S. and CasseDelbart, F. (1989). Nucleotide sequence of potato virus Y (N strain) genomic RNA. *J. Gen. Virol.* **70**, 935–947.

Robbins, M.A., Witsenboer, H., Michelmore, R.W., Laliberte, J.-F. and Fortin, M.G. (1994). Genetic mapping of turnip mosaic virus resistance in *Lactuca sativa. Theor. Appl. Genet.* **89**, 583–589.

Robbins, M.A., Reade, R.D. and Rochon, D.M. (1997). A cucumber necrosis virus variant deficient in fungal transmissibility contains an altered coat protein shell domain. *Virology* **234**, 138–146.

Robert, Y. (1999). Epidemiology of potato leafroll disease. In: H.G. Smith and H. Barker (eds) *The Luteoviridae,* pp. 221-231. CAB International, Wallingford, UK.

Robert, Y. and Lemaire, O. (1999a). Introduction to luteovirus disease epidemiology. In: H.G. Smith and H. Barker (eds) *The Luteoviridae,* pp. 211–220. CAB International, Wallingford, UK.

Robert, Y. and Lemaire, O. (1999b). Epidemiology and control strategies. In: H.G. Smith and H. Barker (eds) *The Luteoviridae,* pp. 211–279. CAB International, Wallingford, UK

Roberts, A., Santa Cruz, S., Roberts, I.M., Prior, D.A.M., Turgeon, R. and Oparka, K. (1997). Phloem-unloading in sink leaves of *Nicotaina benthamiana*: comparison of fluorescent solute with fluorescent virus. *Plant Cell* **9**, 1381–1396.

Roberts, D.A. (1982). Systemic acquired resistance induced in hypersensitive plants by non-necrotic localized viral infections. *Virology* **122**, 207–209.

Roberts, D.A., Christie, R.G. and Archer, M.C. (1970). Infection of apical initials in tobacco shoot meristems by tobacco ringspot virus. *Virology* **42**, 217–220.

Roberts, F.M. (1950). The infection of plants by viruses through roots. *Ann. Appl. Biol.* **37**, 385–396.

Roberts, I.M. (1986). Practical aspects of handling, preparing and staining samples containing planta virus particles for electron microscopy. In: R.A.C. Jones and L. Torrance (eds) *Developments and Applications in Virus Testing. Dev. Appl. Biol.,* **I**, pp. 213–243. Association of Applied Biologists, Wellesbourne.

Roberts, I.M. (1988). The structure of particles of tobacco ringspot nepovirus: evidence from electron microscopy. *J. Gen. Virol.* **69**, 1831–1840.

Roberts, I.M. and Brown, D.J.F. (1980). Detection of six nepoviruses in their nematode vectors by immunosorbent electron microscopy. *Ann. Appl. Biol.* **96**, 187–192.

Roberts, I.M. and Harrison, B.D. (1979). Detection of potato leafroll and potato mop top viruses by immunosorbent electron microscopy. *Ann. Appl. Biol.* **93**, 289–297.

Roberts, I.M., Robinson, D.J. and Harrison, B.D. (1984). Serological relationships and genome homologies among geminiviruses. *J. Gen. Virol.* **65**, 1723–1730.

Roberts, I.M., Wang, D., Findlay, K. and Maule, A.J. (1998). Ultrastructural and temporal observations of the potyvirus cylindrical inclusions (CIs) show that the CI protein acts transiently in aiding virus movement. *Virology* **245**, 173–181.

Roberts, P.L. and Wood, K.R. (1981). Methods for enhancing the synchrony of cucumber mosaic virus replication in tobacco plants. *Phytopathol. Z.* **102**, 114–121.

Roberts, S. and Stanley, J. (1994). Lethal mutations within the conserved stem-loop of African cassava mosaic virus DNA are rapidly corrected by genomic recombination. *J. Gen. Virol.* **75**, 3203–3209.

Robertson, A.D. (1987). The whitefly *Bemisia tabaci* (Gennadius) as a vector of African cassava mosaic virus at the Kenya coast and ways in which the yield losses in cassava, *Manihot esculenta* Crantz, caused by the virus can be reduced. *Insect Sci. Appl.* **8**, 797–801.

Robertson, H.D. and Branch, A.D. (1987). The viroid replication process. In: J.S. Semancik (ed.) *Viroids and Viroid-like Pathogens*, pp. 49–69. CRC Press, Boca Raton, FL.

Robertson, N.L. and Carroll, T.W. (1988). Virus-like particles and spider mite intimately associated with a new disease of barley. *Science* **240**, 1188–1190.

Robertson, W.M. and Henry, C.E. (1986). An association of carbohydrates with particles of arabis mosaic virus retained within *Xiphinema diversicaudatum*. *Ann. Appl. Biol.* **109**, 299–305.

Robinson, D.J. (1973). Inactivation and mutagenesis of tobacco rattle virus by nitrous acid. *J. Gen. Virol.* **18**, 215–222.

Robinson, D.J. (1977). A variant of tobacco rattle virus: evidence for a second gene in RNA2. *J. Gen. Virol.* **35**, 37–43.

Robinson, D.J. and Murant, A.F. (1999). Umbraviruses. In: A. Granoff and R.G. Webster (eds) *Encyclopedia of Virology*, 2nd edn, pp. 1855–1859. Academic Press, San Diego.

Robinson, D.J. and Romero, J. (1991). Sensitivity and specificity of nucleic acid probes for potato leafroll detection. *J. Virol. Methods* **34**, 209–219.

Robinson, D.J., Barker, H., Harrison, B.D. and Mayo, M.A. (1980). Replication of RNA1 of tomato black ring virus independently of RNA2. *J. Gen. Virol.* **51**, 317–326.

Robinson, D.J., Mayo, M.A., Fritsch, C., Jones, A.T. and Raschke, J.H. (1983). Origin and messenger activity of two small RNA species found in particles of tobacco rattle virus strain SYM. *J. Gen. Virol.* **64**, 1591–1599.

Robinson, D.J., Ryabov, E.V., Raj, S.K., Roberts, I.M. and Taliansky, M.E. (1999). Satellite RNA is essential for encapsidation of groundnut rosette umbravirus RNA by groundnut rosette assistor luteovirus coat protein. *Virology* **254**, 106–114.

Rocha-Peña, M.A., Lee, R.F., Lastra, R. *et al.* (1995). Citrus tristeza virus and its aphid vector *Toxoptera citricida*: threats to citrus production in the Caribbean and Central and North America. *Plant Dis.* **79**, 437–444.

Rochon, D'A.M. (1999). Tombusviruses. In: A. Granoff and R.G. Webster (eds) *Encyclopedia of Virology*, 2nd edn, pp. 1789–1798. Academic Press, San Diego.

Rochon, D'A.M., Kelly, R. and Siegel, A. (1986). Encapsidation of 18S RNA by tobacco mosaic virus coat protein. *Virology* **150**, 140–148.

Rochow, W.F. (1970). Barley yellow dwarf virus: phenotypic mixing and vector specificity. *Science* **167**, 875–878.

Rochow, W.F. (1979). Field variants of barley yellow dwarf virus: detection and fluctuation during twenty years. *Phytopathology* **69**, 655–660.

Rochow, W.F. and Duffus, J.E. (1978). Relationship between barley yellow dwarf and beet western yellows viruses. *Phytopathology* **68**, 51–58.

Rochow, W.F. and Muller, I. (1975). Use of aphids injected with virus-specific antiserum for study of plant viruses that circulate in vectors. *Virology* **63**, 282–286.

Rodi, D.J. and Makowski, L. (1999). Phage-display technology—finding a needle in a vast molecular haystack. *Curr. Opin. Biotechnol.* **10**, 87–93.

Rodríguez, L.L. and Nichol, S.T. (1999). Vesicular stomatitis viruses (*Rhabdoviridae*). In: A. Granoff and R.G. Webster (eds) *Encyclopedia of Virology*, pp. 1910–1919. Academic Press, San Diego.

Rodríguez-Cerezo, E. and García-Arenal, F. (1989). Genetic heterogeneity of the RNA genome population of the plant virus U5-TMV. *Virology* **170**, 418–423.

Rodríguez-Cerezo, E., Moya, A. and García-Arenal, F. (1989). Variability and evolution of the plant RNA virus pepper mild mottle virus. *J. Virol.* **63**, 2189–2203.

Rodríguez-Cerezo, E., Klein, P.G. and Shawe, J.G. (1991a). A determinant of disease symptom severity is located in the 3'-terminal noncoding region of the RNA of a plant virus. *Proc. Natl Acad. Sci. USA* **88**, 9863–9867.

Rodríguez-Cerezo, E., Fernandez-Elena, S., Moya, A. and García-Arenal, F. (1991b). High genetic stability in natural populations of the plant virus tobacco mild green mosaic virus. *J. Mol. Evol.* **32**, 328–332.

Roenhorst, J.W., van Lent, J.W.M. and Verdium, B.J.M. (1988). Binding of cowpea chlorotic mottle virus to cowpea protoplasts and relation of binding to virus entry and expression. *Virology* **164**, 91–98.

Roenhorst, J.W., Verdiun, B.J.M. and Goldbach, R.W. (1989). Virus–ribosome complexes from cell-free translation systems supplemented with cowpea chlorotic mottle virus particles. *Virology* **168**, 138–146.

Rogers, S.G., Bisaro, D.M., Horsch, R.B. *et al.* (1986). Tomato golden mosaic virus: a component DNA replicates autonomously in transgenic plants. *Cell (Camb.)* **45**, 593–600.

Rohozinski, J. and Hancock, J.M. (1996). Do light-induced pH changes within the chloroplast drive turnip yellow mosaic virus assembly? *J. Gen. Virol.* **77**, 163–165.

Roistacher, C.N., Dodds, J.A. and Bash, J.A. (1987). Means of obtaining and testing protective strains of seedling yellows and stem pitting tristeza virus: a preliminary report. *Phytophylactica* **19**, 199–203.

Rojas, M.R., Gilbertson, R.L., Russell, D.R. and Maxwell, D.P. (1993). Use of degenerate primers in the polymerase chain reaction to detect whitefly-transmitted geminiviruses. *Plant Dis.* **77**, 340–347.

Romaine, C.P. (1999). Barnaviruses (*Barnaviridae*). In: A. Granoff and R.G. Webster (eds) *Encyclopedia of Virology*, 2nd edn, pp. 152–154. Academic Press, San Diego.

Romaine, C.P. and Zaitlin, M. (1978). RNA-dependent RNA polymerases in uninfected and tobacco mosaic virus-infected tobacco leaves: viral-induced stimulation of a host polymerase activity. *Virology* **86**, 241–253.

Romaine, C.P., Newhart, S.R. and Anzola, D. (1981). Enzyme-linked immunosorbent assay for plant viruses in intact leaf tissue disks. *Phytopathology* **71**, 308–312.

Romero, J., Dzianott, A.M. and Bujarski, J.J. (1992). The nucleotide sequence and genome organization of RNA2 and RNA3 segments of broad bean mottle virus. *Virology* **187**, 671–681.

Romero, J., Huang, Q., Pogany, J. and Bujarski, J.J. (1993). Characterization of defective interfering components that increase symptom severity of broad bean mottle virus infections. *Virology* **194**, 576–584.

Romero-Durbán, J., Cambra, M. and Duran-Vila, N. (1995). A simple imprint-hybridization method for detection of viroids. *J. Virol. Methods* **55**, 37–47.

Ronald, W.P., Schroeder, B., Tremaine, J.H. and Paliwal, Y.C. (1977). Distorted virus particles in electron microscopy: an artifact of grid films. *Virology* **76**, 416–419.

Ronald, W.P., Tremaine, J.H. and MacKenzie, D.J. (1986). Assessment of southern bean mosaic virus monoclonal antibodies for affinity chromatography. *Phytopathology* **76**, 491–494.

Rønde Kristensen, H. (1983). European fights against fruit tree viruses as organised by EPPO and EEC. *Acta Hortic.* **130**, 19–29.

Roossinck, M. (1997). Mechanisms of plant virus evolution. *Annu. Rev. Phytopathol.* **35**, 191–209.

Roossinck, M.J. (1999a). Cucumoviruses (*Bromoviridae*)—general features. In: A. Granoff and R.G. Webster (eds) *Encyclopedia of Virology*, 2nd edn, pp. 315–320. Academic Press, San Diego.

Roossinck, M.J. (1999b). Cucumoviruses (*Bromoviridae*)—molecular biology. In: A. Granoff and R.G. Webster (eds) *Encyclopedia of Virology*, 2nd edn, pp. 320–324. Academic Press, San Diego.

Roossinck, M.J. (2001). Cucumber mosaic virus, a model for RNA virus evolution. *Molec. Plant Pathol.* **2**, 59–63.

Roossinck, M.J. and Palukaitis, P. (1990). Rapid induction and severity of symptoms in zucchini squash (*Cucurbita pepo*) map to RNA1 of cucumber mosaic virus. *Mol. Plant Microb. Interact.* **3**, 188–192.

Roossinck, M.J., Sleat, D. and Palukaitis, P. (1992). Satellite RNAs of plant viruses: Structures and biological effects. *Microbiol. Rev.* **56**, 265–279.

Roossinck, M.J., Zhang, L. and Hellwald, K.-H. (1999). Rearrangements in the 5′ nontranslated region and phylogenetic analysis of cucumber mosaic virus RNA 3 indicate radial evolution of three subgroups. *J. Virol.* **73**, 6752–6758.

Rose, D.J.W. (1974). The epidemiology of maize streak disease in relation to population densities of *Cicadulina* spp. *Ann. Appl. Biol.* **76**, 199–207.

Rosell, R.C., Lichty, J.E. and Brown, J.K. (1995). Ultrastructure of the mouthparts of adult sweetpotato whitefly, *Bemisia tabaci* Gennadius (Homoptera: Aleyrodidae). *Int. J. Insect Morphol. Embryol.* **24**, 297–306.

Rosell, R.C., Torres-Jerez, I. and Brown, J.K. (1999). Tracing the geminivirus–whitefly transmission pathway by polymerase chain reaction in whitefly extracts, saliva, hemolymph and honeydew. *Phytopathology* **89**, 239–246.

Rosner, A., Maslenin, L. and Spiegel, S. (1998). Differentiation among isolates of prunus necrotic ringspot virus by transcript conformation polymorphism. *J. Virol. Methods* **74**, 109–115.

Rosner, A., Maslenin, L. and Spiegel, S. (1999). Double-stranded conformation polymorphism of heterologous RNA transcripts and its use for virus strain differentiation. *Plant Pathol.* **48**, 235–239.

Ross, A.F. (1941). The concentration of alfalfa mosaic virus in tobacco plants at different periods of time after inoculation. *Phytopathology* **31**, 410–421.

Ross, A.F. (1961a). Localized acquired resistance to plant virus infection in hypersensitive hosts. *Virology* **14**, 329–339.

Ross, A.F. (1961b). Systemic acquired resistance induced by localized virus infections in plants. *Virology* **14**, 340–358.

Ross, A.F. (1966). Systemic effects of local lesion formation. In: A.B.R. Beemster and J. Dijkstra (eds) *Viruses of Plants*, pp. 127–150. North-Holland, Amsterdam.

Ross, A.F. and Williamson, C.E. (1951). Physiologically active emanations from virus-infected plants. *Phytopathology* **41**, 431–438.

Ross, J.P. and Butler, A.K. (1985). Distribution of bean pod mottle virus in soybeans in North Carolina. *Plant Dis.* **69**, 101–103.

Rossmann, M.G. (1985). The structure and *in vitro* assembly of southern bean mosaic virus in relation to that of other small spherical plant viruses. In: J.W. Davies (ed.) *Molecular Plant Virology*, Vol. 1, pp. 123–153. CRC Press, Boca Raton.

Rossmann, M.G. and Rueckert, R.R. (1987). What does the molecular structure of viruses tell us about viral functions? *Microbiol. Sci.* **4**, 206–214.

Rossmann, M.G. and Tao, Y.Z. (1999). Cryoelectron-microscopy reconstruction of partially symmetric objects. *J. Struct. Biol.* **125**, 196–208.

Rossmann, M.G., Abad-Zapatero, C., Murthy, M.R.N., Liljàs, L., Alwyn-Jones, T. and Strandberg, B. (1983). Structural comparisons of some small spherical plant viruses. *J. Mol. Biol.* **165**, 711–736.

Rost, B., Sander, C. and Schneider, R. (1994). PHD: a mail server for the protein secondary structure prediction. *CA BIOS* **10**, 53–60.

Rothnie, H.M. (1996). Plant mRNA 3′-end formation. *Plant Molec. Biol.* **32**, 43–61.

Rothnie, H.M., Chapdelaine, Y. and Hohn, T. (1994). Pararetroviruses and retroviruses: a comparative review of viral structure and gene expression strategies. *Adv. Virus Res.* **44**, 1–67.

Rothnie, H.M., Chen, G., Fütterer, J. and Hohn, T. (2001). Polyadenylation in rice tungro bacilliform virus: *cis*-acting signals and regulation. *J. Virol.* **75**, 4148–4194.

Rottier, P.J.M., Rezelman, G. and van Kammen, A. (1980). Protein synthesis in cowpea mosaic virus-infected cowpea protoplasts: detection of virus-related proteins. *J. Gen. Virol.* **51**, 359–379.

Rouleau, M., Smith, R.J., Bancroft, J.B. and Mackie, G.A. (1994). Purification, properties, and subcellular localization of foxtail mosaic potexvirus 26-kDa protein. *Virology* **204**, 254–265.

Round, F.E. (1984). The systematics of the Chlorophyta: an historical review leading to some modern concepts (Taxonomy of the Chlorophyta)—III. In: D.E.G. Irvine and D.M. John (eds) *Systematics of the Green Algae*, pp. 1–27. Academic Press, Orlando, FL.

Rouppe van der Voort, J., Lindeman, W., Folkertsma, R. *et al.* (1998). A QTL for broad-spectrum resistance to cyst nematode species (*Globodera* spp.) maps to a resistance gene cluster in potato. *Theor. Appl. Genet.* **96**, 654–661.

Rouppe van der Voort, J., Kanyuka, K., van der Vossen, E. *et al.* (1999). Tight physical linkage of the nematode resistance gene *Gpa2* and the virus resistance gene *Rx* on a single segment introgressed from the wild species *Solanum tuberosum* subsp. *andigena* CPC 1673 into cultivated potato. *Mol. Plant Microbe Interact.* **12**, 197–206.

Rouzé-Jouan, J., Terradot, L., Pasquer, F., Tanguy, S. and Ducray-Bourdin, D.G. (2001). The passage of potato leafroll virus through *Myzus persicae* gut membrane regulates transmission efficiency. *J. Gen. Virol.* **82**, 17–23.

Rowhani, A., Mircetich, S.M., Shepherd, R.J. and Cucuzza, J.D. (1985). Serological detection of cherry leafroll virus in English walnut trees. *Phytopathology* **75**, 48–52.

Roy, B.P., AbouHaidar, M.G. and Alexander, A. (1989). Biotinylated RNA probes for the detection of potato spindle tuber viroid (PSTV) in plants. *J. Virol. Methods* **23**, 149–156.

Rubino, L. and Russo, M. (1997). Molecular analysis of Pothos latent virus genome. *J. Gen. Virol.* **78**, 1219–1226.

Rubino, L., Tousignant, M.E. and Steger, G. and Kaper, J.M. (1990). Nucleotide sequence and structural analysis of two satellite RNAs associated with chicory yellow mottle virus (CYMV). *J. Gen. Virol.* **71**, 1897–1903.

Rubino, L., Russo, M. and Martelli, G.P. (1995). Sequence analysis of the Pothos latent virus genomic RNA. *J. Gen. Virol.* **76**, 2835–2839.

Rubino, L., DiFranco, A. and Russo, M. (2000). Expression of a plant virus non-structural protein in *Saccharomyces cerevisiae* causes membrane proliferation and altered mitochondrial morphology. *J. Gen. Virol.* **81**, 279–286.

Rubino, L., Weber-Lotfi, F., Dietrich, A., Stussi-Garroud, C. and Russo, M. (2001). The open reading frame 1– encoded (36K) protein of carnation etched ring virus localizes to mitochondria. *J. Gen. Virol.* **82**, 29–34.

Rubio, T., Borja, M., Scholthof, H.B., Feldstein, P.A., Morris, T.J. and Jackson, A.O. (1999). Broad-spectrum protection against tombusviruses elicited by defective interfering RNAs in transgenic plants. *J. Virol.* **73**, 5070–5078.

Rufty, R.C., Powell, N.T. and Gooding, G.V., Jr. (1983). Relationship between resistance to *Meloidogyne incognita* and a necrotic response to infection by a strain of potato virus Y in tobacco. *Phytopathology* **73**, 1418–1423.

Ruiz, M.T., Voinnet, O. and Baulcombe, D.C. (1998). Initiation and maintenance of virus-induced gene silencing. *Plant Cell* **10**, 937–946.

Rushing, R.E., Sunter, G., Gardiner, W.E., Dute, R.R. and Bisaro, D.M. (1987). Ultrastructural aspects of tomato golden mosaic virus infection in tobacco. *Phytopathology* **77**, 1231–1236.

Russell, G.E. (1958). Sugar beet yellows: a preliminary study of the distribution and interrelationships of viruses and virus strains found in East Anglia 1955–57. *Ann. Appl. Biol.* **46**, 393–398.

Russell, G.E. (1960). Breeding for resistance to sugar beet yellows. *Br. Sugar Beet Rev.* **28**, 163–170.

Russell, G.E. (1966). The control of *Alternaria* species on leaves of sugar beet infected with yellowing viruses. II. Experiments with two yellowing viruses and virus-tolerant sugar beet. *Ann. Appl. Biol.* **57**, 425–434.

Russo, M. and Martelli, G.P. (1972). Ultrastructural observations on tomato bushy stunt virus in plant cells. *Virology* **49**, 122–129.

Russo, M., Kishtah, A.A. and Martelli, G.P. (1979). Unusual intracellular aggregates of broad bean wilt virus particles. *J. Gen. Virol.* **43**, 453–456.

Russo, M., Di Franco, S. and Martelli, G.P. (1983). The fine structure of cymbidium ringspot virus infection in host tissues: 3. Role of peroxisomes in the genesis of multivesicular bodies. J. Ultrastuct. Res. **82**, 52–63.

Russo, M., Di Franco, A. and Martelli, G.P. (1987). Cytopathology in the identification and classification of tombusviruses. *Intervirology* **28**, 134–143.

Rutgers, T., Salerno-Rife, T. and Kaesberg, P. (1980). Messenger RNA for the coat protein of southern bean mosaic virus. *Virology* **104**, 506–509.

Ryabov, E.V., Oparka, K.J., Santa Cruz, A., Robinson, D.J. and Taliansky, M.E. (1998). Intracellular location of two groundnut rosette umbravirus proteins delivered by PVX and TMV vectors. *Virology* **242**, 303–313.

Ryabov, E.V., Robinson, D.J. and Taliansky, M.E. (1999). A plant virus-encoded protein facilitates long-distance movement of heterologous viral RNA. *Proc. Natl Acad. Sci. USA* **96**, 1212–1217.

Ryabova, L.A. and Hohn, T. (2000). Ribosome shunting in the cauliflower mosaic virus 35S RNA leader is a special case of reinitiation of translation functioning in plant and animal systems. *Genes & Dev.* **14**, 817–829.

Ryals, J.A., Neuenschwander, U.H., Willits, M.G., Molina, A., Steiner, H.-Y. and Hunt, M.D. (1996). Systemic acquired resistance. *Plant Cell* **8**, 1809–1819.

Ryan, M.D. and Flint, M. (1997). Virus-encoded proteinases of the picornavirus super-group. *J. Gen. Virol.* **78**, 699–723.

Rybicki, E.P. (1994). A phylogenetic and evolutionary justification for three genera of *Geminiviridae*. *Arch. Virol.* **139**, 49–77.

Rybicki, E.P. and von Wechmar, M.B. (1982). Enzyme-assisted immune detection of plant virus proteins electroblotted onto nitrocellulose paper. *J. Virol. Methods* **5**, 267–278.

Rydén, K. and Gerhardson, B. (1978). Rödklövermosaikvirus sprids med slättermaskiner. *Vaextskyddsnotiser* **42**, 112–115.

Ryder, E.J. (1964). Transmission of common lettuce mosaic virus through the gametes of the lettuce plant. *Plant Dis. Rep.* **48**, 522–523.

Sacher, R. and Ahlquist, P. (1989). Effect of deletions in the N-terminal basic arm of brome mosaic virus coat protein on RNA packaging and systemic infection. *J. Virol.* **63**, 4545–4552.

Sacher, R., French, R. and Ahlquist, P. (1988). Hybrid brome mosaic virus RNAs express and are packaged in tobacco mosaic virus coat protein *in vivo*. *Virology* **167**, 15–24.

Sadowy, E., Maasen, A., Juszczuk, M. *et al.* (2001a). The ORF0 product of potato leafroll virus is indispensable for virus accumulation. *J. Gen. Virol.* **82**, 1529–1532.

Sadowy, E., Juszczuk, M., David, C., Gronenborn, B. and Hulanicka, M.D. (2001b). Mutational analysis of the proteinase function of potato leafroll virus. *J. Gen. Virol.* **82**, 1517–1527.

Sägesser, R., Martinez, E., Tsagris, M. and Tabler, M. (1997). Detection and isolation of RNA-binding proteins by RNA-ligand screening of a cDNA expression library. *Nucl. Acids Res.* **25**, 3816–3822.

Saiki, R.K., Gelfand, D.H., Stoffel, S. *et al.* (1988). Primer-directed enzymatic amplification of DNA with a thermostable DNA polymerase. *Science* **239**, 487–491.

Saito, T., Meshi, T., Takamatsu, N. and Okada, Y. (1987a). Coat protein gene sequence of tobacco mosaic virus encodes a host response determinant. *Proc. Natl Acad. Sci. USA* **84**, 6074–6077.

Saito, T., Hosokawa, D., Meshi, T. and Okada, Y. (1987b). Immunocytochemical localization of the 130K and 180K proteins (putative replicase components) of tobacco mosaic virus. *Virology* **160**, 477–481.

Saito, T., Imai, Y., Meshi, T. and Okada, Y. (1988). Interviral homologies of the 30K proteins of tobamoviruses. *Virology* **167**, 653–656.

Saitou, N. and Nei, M. (1986). Polymorphism and evolution of influenza virus genes. *Mol. Biol. Evol.* **3**, 57–74.

Saitou, N. and Nei, M. (1987). The neighbor-joining method—a new method for reconstructing phylogenetic trees. *Molec. Biol. Evol.* **4**, 406–425.

Sakai, F. and Takebe, I. (1970). RNA and protein synthesis in protoplasts isolated from tobacco leaves. *Biochim. Biophys. Acta* **224**, 531–540.

Sakai, F. and Takebe, I. (1974). Protein synthesis in tobacco mesophyll protoplasts induced by tobacco mosaic virus infection. *Virology* **62**, 426–433.

Sakai, F., Dawson, J.R.O. and Watts, J.W. (1979). Synthesis of proteins in tobacco protoplasts infected with brome mosaic virus. *J. Gen. Virol.* **42**, 323–328.

Sakai, F., Dawson, J.R.O. and Watts, J.W. (1983). Interference in infections of tobacco protoplasts with two bromoviruses. *J. Gen. Virol.* **64**, 1347–1354.

Sako, N. and Ogata, K. (1980). Different helper factors associated with aphid transmission of some potyviruses. *Virology* **112**, 762–765.

Sakurai, T., Murai, T., Maeda, T. and Tsumuki, H. (1998). Sexual differences in transmission and accumulation of tomato spotted wilt virus in its insect vector *Frankliniella occidentalis* (Thysanoptera: Thripidae). *Appl. Entomol. Zool.* **33**, 583–588.

Sala, M. and Wain-Hobson, S. (2000). Are RNA viruses adapting or merely changing? *J. Mol. Evol.* **51**, 12–20.

Salaman, R.N. (1933). Protective inoculation against a plant virus. *Nature (Lond.)* **131**, 468.

Salaman, R.N. (1938). The potato virus 'X', its strains and reactions. *Philos. Trans. R. Soc. Lond. Ser. B* **229**, 137–217.

Salánki, K., Carrère, I., Jacquemond, M., Balázs, E. and Tepfer, M. (1997). Biological properties of pseudorecombinant and recombinant strains created with cucumber mosaic virus and tomato aspermy virus. *J. Virol.* **71**, 3597–3602.

Salazar, L.F. (1999). Capilloviruses. In: A. Granoff and R.G. Webster (eds) *Encyclopedia of Virology*, 2nd edn, p. 222. Academic Press, San Diego.

Salazar, L.F. and Jones, R.A.C. (1975). Some studies on the distribution and incidence of potato mop-top virus in Peru. *Am. Potato J.* **52**, 143–150.

Salazar, L.F., Balbo, I. and Owens, R.A. (1988a). Comparison of four radioactive probes for the diagnosis of potato spindle tuber viroid by nucleic acid spot hybridization. *Potato Res.* **31**, 431–442.

Salazar, L.F., Hammond, R.W., Diener, T.O. and Owens, R.A. (1988b). Analysis of viroid replication following *Agrobacterium*-mediated inoculation of non-host species with potato spindle tuber viroid cDNA. *J. Gen. Virol.* **69**, 879–889.

Salazar, L.F., Querci, M., Bartolini, I. And Lazarte, V. (1995). Aphid transmission of potato spindle tuber viroid assisted by potato leafroll virus. *Fitopatologia* **30**, 56–58.

Salerno-Rife, T., Rutgers, T. and Kaesberg, P. (1980). Translation of southern bean mosaic virus RNA in wheat germ embryo and rabbit reticulocyte extracts. *J. Virol.* **34**, 51–58.

Salinas Calvete, J. and Wieringa-Brants, D.H. (1984). Infection and necrosis of cowpea mesophyll cells by tobacco necrosis virus and two strains of tobacco mosaic virus. *Neth. J. Plant Pathol.* **90**, 71–78.

Salomon, R. (1999). The evolutionary advantage of breeding for tolerance over resistance against viral plant disease. *Israel J. Plant Sci.* **47**, 135–139.

Salvato, M.S. and Fraenkel-Conrat, H. (1977). Translation of tobacco necrosis virus and its satellite in a cell-free wheat germ system. *Proc. Natl Acad. Sci. USA* **74**, 2288–2292.

Sama, S., Hasanuddin, A., Manwan, I., Cabunagan, R.C. and Hibino, H. (1991). Integrated management of rice tungro disease in South Sulawesi. *Crop Protect.* **10**, 34–40.

Sambrook, J., Fritsch, E.F. and Maniatis, T. (1989) *Molecular Cloning: A Laboratory Manual,* Vols 1–3, 2nd edn. Cold Spring Harbor Laboratory Press, Cold Spring Harbor, NY.

Samuel, G. (1931). Some experiments on inoculating methods with plant viruses and on local lesions. *Ann. Appl. Biol.* **18**, 494–507.

Samuel, G. (1934). The movement of tobacco mosaic virus within the plant. *Ann. Appl. Biol.* **21**, 90–111.

Samuel, G. and Bald, J.G. (1933). On the use of primary lesions in queantitative work with two plant viruses. *Ann. Appl. Biol.* **20**, 70–99.

Sander, E.-M. and Dietzgen, R.G. (1984). Monoclonal antibodies against plant viruses. *Adv. Virus Res.* **29**, 131–168.

Sander, E.-M. and Mertes, G. (1984). Use of protoplasts and separate cells in plant virus research. *Adv. Virus Res.* **29**, 215–262.

Sander, E.-M., Dietzgen, R.G., Cranage, M.P. and Coombs, R.R.A. (1989). Rapid and simple detection of plant virus by reverse passive haemagglutination. I. Comparison of ELISA (enzymelinked immunosorbent assay) and RPH (reverse passive haemagglutination) for plant virus diagnosis. *J. Plant Dis. Prot.* **96**, 113–123.

Sanderfoot, A.A., Ingham, D.J and Lazarowitz, S.G. (1996). A viral movement protein as a nuclear shuttle: the geminiviral BR1 movement protein contains domains essential for interaction with BL1 and nuclear localization. *Plant Physiol.* **110**, 23–33.

Sanders, P.R., Winter, J.A., Barnason, A.R., Rogers, S.G. and Fraley, R.T. (1987). Comparison of cauliflower mosaic virus S35 and nopaline synthase promoters in transgenic plants. *Nucl. Acids Res.* **15**, 1543–1558.

Sanfaçon, H. (1992). Regulation of messenger RNA: lessons from the cauliflower mosaic virus transcription signals. *Can. J. Bot.* **70**, 885–899.

Sanfaçon, H. and Hohn, T. (1990). Proximity of the promoter inhibits recognition of cauliflower mosaic virus polyadenylation signal. *Nature* **346**, 81–84.

Sanford, J.C. and Johnson, S.A. (1985). The concept of parasite-derived resistance: deriving resistance genes from the parasites own genome. *J. Theor. Biol.* **115**, 395–405.

Sangare, A., Deng, D., Fauquet, C.M. and Beachy, R.N. (1999). Resistance to African cassava mosaic virus conferred by a mutant of the putative NTP-binding domain of the *Rep* gene (AC1) in *Nicotiana benthamiana. Mol. Breeding* **5**, 95–102.

Sänger, H.L. (1969). Functions of the two particles of tobacco rattle virus. *J. Virol.* **3**, 304–312.

Sänger, H.L. (1987). Viroid function. Viroid replication. In: T.O. Diener (ed.) *The Viroids*, pp. 117–166. Plenum, New York.

Sänger, H.L., Klotz, G., Riesner, D., Gross, H.J. and Kleinschmidt, A.K. (1976). Viroids are single stranded covalently closed circular RNA molecules existing as highly base-paired rod–like structures. *Proc. Natl Acad. Sci. USA* **73**, 3852–3856.

Sanger, M., Daubert, S. and Goodman, R.M. (1990). Characteristics of a strong promoter from figwort mosaic virus: comparison with the analogous 35S promoter from cauliflower mosaic virus and the regulated mannopine synthase promoter. *Plant. Mol. Biol.* **14**, 433–443.

Sano, T., Kudo, H., Sugimoto, T. and Shikata, E. (1988). Synthetic oligonucleotide hybridization probes to diagnose hop stunt viroid strains and citrus exocortis viroid. *J. Virol. Methods* **19**, 109–120.

Sano, T., Candresse, T., Hammond, R.W., Diener, T.O. and Owens, R.A. (1992). Identification of multiple structural domains regulating viroid pathogenicity. *Proc. Natl Acad. Sci. USA* **89**, 10104–10108.

Sano, Y., Wada, M., Hashimoto, Y., Matsumoto, T. and Kojima, M. (1998). Sequences of ten circular ssDNA components associated with the milk vetch dwarf virus genome. *J. Gen. Virol.* **79**, 3111–3118.

Santa Cruz, S. (1999). Perspective: phloem transport of viruses and macromolecules—what goes in must come out. *Trends Microbiol.* **7**, 237–241.

Santa Cruz, S. and Baulcombe, D.C. (1993). Molecular analysis of potato virus X isolates in relation to the potato hypersensitivity gene *Nx*. *Mol. Plant Microbe Interact.* **6**, 707–714.

Santa Cruz, S., Chapman, S., Roberts, A.G., Roberts, I.M., Prior, D.A.M. and Oparka, K.K. (1996). Assembly and movement of a plant virus carrying a green fluorescent protein overcoat. *Proc. Natl Acad. Sci. USA* **93**, 6286–6290.

Santa Cruz, S., Roberts, A.G., Prior, D.A.M., Chapman, S. and Oparka, K.J. (1998). Cell-to-cell and phloem-mediated transport of potato virus X. *Plant Cell* **10**, 495–510.

Sanz, A.I., Fraile, A., Gallego, J.M., Malpica, J.M. and García-Arenal, F. (1999). Genetic variability of natural populations of cotton leaf curl geminivirus, a single-stranded DNA virus. *J. Mol. Evol.* **49**, 672–681.

Sanz, A.I., Fraile, A., García-Arenal, F. *et al.* (2000). Multiple infection, recombination and genome relationships among begomovirus isolates found in cotton and other plants in Pakistan. *J. Gen. Virol.* **81**, 1839–1849.

Sanz-Burgos, A.P. and Gutierrez, C. (1998). Organization of the *cis*-acting element required for wheat dwarf geminivirus DNA replication and visualization of a Rep protein-DNA complex. *Virology* **243**, 119–129.

Sarachu, A.N., Nassuth, A., Roosien, J., van Vloten-Doting, L. and Bol, J.F. (1983). Replication of temperature-sensitive mutants of alfalfa mosaic virus in protoplasts. *Virology* **125**, 64–74.

Sarachu, A.N., Huisman, M.J., van Vloten-Doting, L. and Bol, J.F. (1985). Alfalfa mosaic virus temperature-sensitive mutants: I. Mutants defective in viral RNA and protein synthesis. *Virology* **141**, 14–22.

Sargent, R.G., Brenneman, M.A and Wilson, J.H. (1997). Repair of site-specific double-strand breaks in a mammalian chromosome by homologous and illegitimate recombination. *Mol. Cell. Biol.* **17**, 267–277.

Sarkar, S. (1969). Evidence of phenotypic mixing between two strains of tobacco mosaic virus. *Mol. Gen. Genet.* **105**, 87–90.

Sarkar, S. (1986). Tobacco mosaic virus: mutants and strains. In: M.H.V. van Regenmortel and H. Fraenkel-Conrat (eds) *The Plant Viruses*, Vol. 2, pp. 59–77. Plenum, New York.

Sastry, L., Alting-Mees, H., Huse, W.D. *et al.* (1989). Cloning of the immunological repertoire in *Escherichia coli* for generation of monoclonal catalytic antibodies: construction of a heavy chain variable region-specific cDNA library. *Proc. Natl Acad. Sci. USA* **86**, 5728–5732.

Satapathy, M.K. (1998). Chemical control of insect and nematode vectors of plant viruses. In: A. Hadidi, R.K. Khetarpal and H. Koganezawa (eds) *Plant Virus Disease Control*, pp. 188–195. APS Press, St. Paul, MN.

Satoh, H., Matsuda, H., Kawamura, T., Isogai, M., Yoshikawa, N. and Takahashi, T. (2000). Intracellular distribution, cell-to-cell trafficking and tubule-inducing activity of the 50 kDa movement protein of apple chlorotic leafspot virus fused to green fluorescent protein. *J. Gen. Virol.* **81**, 2085–2093.

Saunders, K. and Stanley, J. (1999). A nanovirus-like DNA component associated with the yellow vein disease of *Ageratum conyzoides*: evidence for interfamilial recombination between plant DNA viruses. *Virology* **264**, 142–152.

Saunders, K., Lucy, A.P. and Covey, S.N. (1989). Characterization of cDNA clones of host RNAs isolated from cauliflower mosaic virus-infected turnip leaves. *Physiol. Mol. Plant Pathol.* **35**, 339–346.

Saunders, K., Lucy, A. and Stanley, J. (1991). DNA forms of the geminivirus African cassava mosaic virus consistent with a rolling circle mechanism of replication. *Nucl. Acids Res.* **19**, 2325–2330.

Saunders, K., Bedford, I.D., Braden, R.W., Markham, P.G., Wong, S.M. and Stanley, J. (2000). A unique virus complex causes *Ageratum* yellow vein disease. *Proc. Natl Acad. Sci. USA* **97**, 6890–6895.

Saunders, K., Wege, C., Veluthambi, K., Jeske, H. and Stanley, J. (2001a). The distinct disease phenotypes of the common and yellow vein strains of tomato golden mosaic virus are determined by nucleotide differences in the 3′-terminal region of the gene encoding the movement protein. *J. Gen. Virol.* **82**, 45–51.

Saunders, K. Bedford, I.D. and Stanley, J. (2001b). Pathogenicity of a natural recombinant associated with *Ageratum* yellow vein disease: implication for geminivirus evolution and disease aetiology. *Virology* **282**, 38–47.

Savithri, H.S. and Erickson, J.W. (1983). The self-assembly of the cowpea strain of southern bean mosaic virus: formation of T = 1 and T = 3 nucleoprotein particles. *Virology* **126**, 328–335.

Savithri, H.S., Munshi, S.K., Suryanarayana, S., Divakar, S. and Murthy, M.R.N. (1987). Stability of belladonna mottle virus particles: the role of polyamines and calcium. *J. Gen. Virol.* **68**, 1533–1542.

Sawiki, S.G. and Sawiki, D.L. (1998). A new model for coronavirus transcription. *Adv. Exp. Med. Biol.* **440**, 215–219.

Sawyer, L., Tollin, P. and Wilson, W.R. (1987). A comparison between the predicted secondary structures of potato virus X and papaya mosaic virus coat proteins. *J. Gen. Virol.* **68**, 1229–1232.

Saxena, R.C. and Khan, Z.R. (1985). Electronically recorded disturbances in feeding behaviour of *Nephotettix virescens* (Homoptera: Cicadellidae) on neem oil-treated rice plants. *J. Econ. Entomol.* **78**, 222–226.

Sayama, H., Sato, T., Kominato, M., Natsuaki, T. and Kaper, J.M. (1993). Field testing of a satellite-containing attenuated strain of cucumber mosaic virus for tomato protection in Japan. *Phytopathology* **83**, 405–410.

Scalla, R., Boudon, E. and Rigaud, J. (1976) Sodium dodecyl sulphate–polyacrylamide gel electrophoretic detection of two high-molecular-weight proteins associated with tobacco mosaic virus infection in tobacco. *Virology* **69**, 339–345.

Scalla, R., Romaine, P., Asselin, A., Rigaud, J. and Zaitlin, M. (1978). An *in vivo* study of a nonstructural polypeptide synthesized upon TMV infection and its identification with a polypeptide synthesized *in vitro* from TMV RNA. *Virology* **91**, 182–193.

Schaad, M.C. and Carrington, J.C. (1996). Suppression of long-distance movement of tobacco etch virus in a nonsusceptible host. *J. Virol.* **70**, 2556–2561.

Schaad, M.C., Jensen, P.E. and Carrington, J.C. (1997a). Formation of plant RNA replication complexes on membranes: role of an endoplasmic reticulum-targeted viral protein. *EMBO J.* **16**, 4049–4059.

Schaad, M.C., Lellis, A.D. and Carrington, J.C. (1997b). VPg of tobacco etch potyvirus is a host genotype-specific determinant of long-distance movement. *J. Virol.* **71**, 8624–8631.

Schaad, M.C., Anderberg, R.J. and Carrington, J.C. (2000). Strain-specific interaction of tobacco etch virus NIa protein with the translation initiation factor eIF4E in the yeast two-hybrid system. *Virology* **273**, 300–306.

Schachman, H.K. (1959). *Ultracentrifugation in Biochemistry*. Academic Press, New York.

Schade, R., Hlinak, A., Marburger, A. *et al.* (1997). Advantages of using egg yolk antibodies in life sciences: the results of five studies. *ALTA - Alternatives to Laboratory Animals* **25**, 555–586.

Schalk, H.-J., Matzeit, V., Schiller, B., Schell, J. and Gronenborn, B. (1989). Wheat dwarf virus, a geminivirus of graminaceous plants needs splicing for replication. *EMBO J.* **8**, 359–364.

Schärer-Hernández, N and Hohn, T. (1998). Nonlinear ribosome migration on cauliflower mosaic virus 35S RNA in transgenic tobacco plants. *Virology* 242, 403–413.

Schaskolskaya, N.D., Atabekov, J.G., Sacharovskaya, G.N. and Javachia, V.G. (1968). Replication of temperature-sensitive strain of tobacco mosaic virus under nonpermissive conditions in the presence of helper strain. *Biol. Sci. USSR* **8**, 101–105.

Scheets, K. (1998). Maize chlorotic mottle machlovirus and wheat streak mosaic rymovirus concentrations increase in the synergistic disease corn lethal necrosis. *Virology* 242, 28–38.

Schenk, P.M., Sagi, L., Remans, T. *et al.* (1999). A promoter from sugarcane bacilliform badnavirus drives transgene expression in banana and other monocot and dicot plants. *Plant Mol. Biol.* **39**, 1221–1230.

Schenk, P.M., Kazan, K., Wilson, I. *et al.* (2000). Coordinated plant defense responses in *Arabidopsis* revealed by microarray analysis. *Proc. Natl Acad. Sci. USA* **97**, 11655–11660.

Schiebel, W., Haas, B., Marinkovic, S., Klanner, A. and Sänger, H. (1993). RNA-directed RNA polymerase from tomato leaves: 1. Purification and physical properties. *J. Biol. Chem.* **263**, 11851–11857.

Schiebel, W., Pélissier, T., Riedel, L. *et al.* (1998). Isolation of an RNA-directed RNA polymerase-specific cDNA clone from tomato. *Plant Cell* **10**, 2087–2101.

Schindler, I.M. and Mühlbach, H.P (1992). Involvement of nuclear DNA-dependent RNA polymerase in potato spindle tuber viroid replication: a reevaluation. *Plant Sci.* **84**, 221–229.

Schirawski, J., Voyatzakis, A., Zaccomer, B., Bernardi, F. and Haenni, A.-L. (2000). Identification and functional analysis of the subgenomic promoter of turnip yellow mosaic tymovirus. *J. Virol.* **74**, 11073–11080.

Schluckebeier, G., O'Gara, M., Saenger, W. and Cheng, X. (1995). Universal catalytic domain structure of AdoMet-dependent methyltransferase. *J. Mol. Biol.* **247**, 16–20.

Schmelzer, K., Schmidt, H.E. and Schmidt, H.B. (1966). Viruskrankheiten und Virusverdächtige Erscheinungen an Forstgehölzen. *Arch. Forstwes.* **15**, 107–120.

Schmidt, H.E. and Zobywalski, S. (1984). Determination of pathotypes of bean yellow mosaic virus using *Phaseolus vulgaris* L. as a differential host. *Arch. Phytopathol. Pflanzenschutz* **20**, 95–96.

Schmidt, T. and Johnson, J.E. (1983). The spherically averaged structures of cowpea mosaic virus components by X-ray solution scattering. *Virology* **127**, 65–73.

Schmitz, I. and Rao, A.L.N. (1996). Molecular studies on bromovirus capsid protein: 1. Characterization of cell-to-cell movement-defective RNA3 variants of brome mosaic virus. *Virology* **226**, 281–293.

Schneider, I.R. (1964). Difference in the translocatability of tobacco ringspot and southern bean mosaic viruses in bean. *Phytopathology* **54**, 701–705.

Schneider, I.R. and Thompson, S.M. (1977). Double-stranded nucleic acids found in tissue infected with the satellite of tobacco ringspot virus. *Virology* **78**, 453–462.

Schneider, I.R. and Worley, J.F. (1959a). Upward and downward transport of infectious particles of southern bean mosaic virus through steamed portions of bean stems. *Virology* **8**, 230–242.

Schneider, I.R. and Worley, J.F. (1959b). Rapid entry of infectious particles of southern bean mosaic virus into living cells following transport of the particles in the water stream. *Virology* **8**, 243–249.

Schneider, I.R., Hull, R. and Markham, R. (1972a). Multidense satellite of tobacco ringspot virus: a regular series of components of different densities. *Virology* **47**, 320–330.

Schneider, I.R., White, R.M. and Gooding, G.V. (1972b). Two new isolates of the satellite of tobacco ringspot virus. *Virology* **50**, 902–905.

Schneider, W.L., Greene, A.E. and Allison, R.F. (1997). The carboxy-terminal two-thirds of cowpea chlorotic mottle bromovirus capsid protein is incapable of virion formation yet supports systemic movement. *J. Virol.* **71**, 4862–4865.

Schnell, R.J., Kuhn, D.N., Ronning, C.M. and Harkins, D. (1997). Application of RT-PCR for indexing avocado sunblotch viroid. *Plant Dis.* **81**, 1023–1026.

Schnölzer, M., Haas, B., Ramm, K., Hofmann, H. and Sänger, H.L. (1985). Correlation between structure and pathogenicity of potato spindle tuber viroid (PSTV). *EMBO J.* **4**, 2181–2190.

Schoelz, J.E. and Bourque, J.E. (1999). Plant pararetroviruses—caulimoviruses: general features. In: A. Granoff and R.G. Webster (eds) *Encyclopedia of Virology*, 2nd edn, pp. 1275–1286. Academic Press, San Diego.

Schoelz, J.E. and Shepherd, R.J. (1988). Host range control of cauliflower mosaic virus. *Virology* **162**, 30–37.

Schoelz, J.E. and Wintermantel, W.M. (1993). Expansion of viral host range through complementation and recombination in transgenic plants. *Plant Cell* **5**, 1669–1679.

Schoelz, J.E., Shepherd, R.J. and Daubert, S.D. (1986). Region VI of cauliflower mosaic virus encodes a host range determinant. *Mol. Cell. Biol.* **6**, 2632–3627.

Scholten, O.E. and Lange, W. (2000). Breeding for resistance to rhizomania in sugar beet: a review. *Euphytica* **112**, 219–231.

Scholthof, H.B. and Jackson, A.O. (1997). The enigma of pX: a host dependent *cis*-acting element with variable effects on tombusvirus RNA accumulation. *Virology* **237**, 56–65.

Scholthof, H.B., Wu, F.C., Richins, R.D. and Shepherd, R.J. (1991). A naturally occurring deletion mutant of figwort mosaic virus (caulimovirus) is generated by RNA splicing. *Virology* **184**, 290–298.

Scholthof, H.B., Gowda, S., Wu, F.C. and Shepherd, R.J. (1992a). The full-length transcript of caulimovirus is a polycistronic mRNA whose genes are transactivated by the product of gene VI. *J. Virol.* **66**, 3131–3139.

Scholthof, H.B., Wu, F.C., Gowda, S. and Shepherd, R.J. (1992b). Regulation of caulimovirus gene expression and involvement of cis-acting elements on both viral transcripts. *Virology* **190**, 403–412.

Scholthof, H.B., Scholthof, K.-B.G. and Jackson, A.O. (1995a). Identification of tomato bushy stunt virus host-specific symptom determinants by expression of individual genes from a potato virus X vector. *Plant Cell* **7**, 1157–1172.

Scholthof, H.B., Scholthof, K.-B.G., Kikkert, M. and Jackson, A.O. (1995b). Tomato bushy stunt virus spread is regulated by two nested genes that function in cell-to-cell movement and host-dependent systemic invasion. *Virology* **213**, 425–438.

Scholthof, H.B., Scholthof, K.-B.G. and Jackson, A.O. (1996). Plant virus gene vectors for transient expression of foreign proteins in plants. *Annu. Rev. Phytopathol.* **34**, 299–323.

Scholthof, K.-B.G. (1999). A synergism induced by satellite panicum mosaic virus. *Mol. Plant. Microb. Interact.* **12**, 163–166.

Scholthof, K.-B.G., Scholthof, H.B. and Jackson, A.O. (1993). Control of plant diseases by pathogen-derived resistance in transgenic plants. *Plant Physiol.* **102**, 7–12.

Scholthof, K.-B.G., Hillman, B.I., Modrell, B., Heaton, L.A. and Jackson, A.O. (1994). Characterization and detection of sc4—a 6th gene encoded by Sonchus yellow net virus. *Virology* **204**, 279–288.

Scholthof, K.-B.G., Scholthof, H.B and Jackson, A.O. (1995c). The tomato bushy stunt virus replicase proteins are coordinately expressed and membrane associated, *Virology* **208**, 265–269.

Scholthof, K.-B., G., Shaw, J.G. and Zaitlin, M. (eds) (1999a). *Tobacco Mosaic Virus: One Hundred Years of Contributions to Virology.* American Phytopathology Society Press, St Paul, MN.

Scholthof, K.-B.G., Jones, R.W. and Jackson, A.O. (1999b). Biology and structure of plant satellite viruses activated by icosahedral helper viruses. *Curr. Topics Microbiol. Immunol.* **239**, 123–143.

Schonborn, J., Oberstrass, J., Breyel, E., Tittgen, J., Schumacher, J. and Lukacs, N. (1991). Monoclonal antibodies to double-stranded RNA as probes of RNA structure in crude nucleic acid extracts. *Nucl. Acids Res.* **19**, 2993–3000.

Schroeder, T.A. (2001). Microtubules don and doff their caps: dynamic attachments at the plus and minus ends. *Curr. Opin Cell Biol.* **13**, 92–96.

Schroeder, W.T. and Provvidenti, R. (1968). Resistance of bean (*Phaseolus vulgaris*) to the PV2 strain of bean yellow mosaic virus conditioned by the single dominant gene *By*. *Phytopathology* **58**, 1710.

Schroeder, W.T. and Provvidenti, R. (1971). A common gene for resistance to bean yellow mosaic virus and watermelon mosaic virus 2 in *Pisum sativum*. *Phytopathology* **61**, 846–848.

Schuchalter-Eicke, G. and Jeske, H. (1983). Seasonal changes in the chloroplast ultrastructure in *Abutilon* mosaic virus (ABMV) infected *Abutilon* spec. (Malvaceae). *Phytopathol. Z.* **108**, 172–184.

Schultze, M., Jiricny, J. and Hohn, T. (1990). Open reading frame VIII is not required for viability of cauliflower mosaic virus. *Virology* **176**, 662–664.

Schumacher, J., Meyer, N., Riesner, D. and Weidemann, H.L. (1986). Diagnostic procedure for detection of viroids and viruses with circular RNAs by 'return'-gel electrophoresis. *J. Phytopathol.* **115**, 332–343.

Schuman, S. and Schwer, B. (1995). RNA capping enzymes and DNA ligases, a superfamily of covalent nucleotidyl transferases. *Mol. Microbiol.* **17**, 405–410.

Schuster, P. and Stadler, P.F. (1999). Nature and evolution of early replicons. In: E. Domingo, R. Webster and J. Holland (eds) *Origin and evolution of viruses*, pp. 1–24. Academic Press, London.

Scott, S.W., Zimmerman, M.T. and Ge, X. (1998). The sequence of RNA 1 and RNA 2 of tobacco streak virus: additional evidence for the inclusion of alfalfa mosaic virus in the genus Ilarvirus. *Archiv. Virol.* **143**, 1187–1198.

Scotti, P.D. (1985). The estimation of virus density in isopycnic cesium chloride gradients. *J. Virol. Methods* **12**, 149–160.

Sdoodee, R. and Teakle, D.S. (1993). Studies on the mechanism of transmission of pollen-associated tobacco streak ilarvirus by *Thrips tabaci*. *Plant Pathol.* **42**, 88–92.

Seddas, A., Haidar, M.M., Greif, C., Jacquet, C., Cloquemin, G. and Walter, B. (2000). Establishment of a relationship between grapevine leafroll closteroviruses 1 and 3 by use of monoclonal antibodies. *Plant Pathol.* **49**, 80–85.

Seemüller, E., Marcone, C., Lauer, U., Ragozzino, A. and Göschl, M. (1998), Current status of molecular classification of the phytoplasmas. *J. Plant Pathol.* **80**, 3–27.

Sehgal, O.P. (1999). Sobemoviruses. In: A. Granoff and R.G. Webster (eds) *Encyclopedia of Virology*, 2nd edn, pp. 1674–1680. Academic Press, San Diego.

Sehgal, O.P. and Hsu, C.H. (1977). Identity and location of a minor protein component in virions of southern bean mosaic virus. *Virology* **77**, 1–11.

Sela, I., Reichman, M. and Weissbach, A. (1984). Comparison of dot molecular hybridization and enzyme-linked immunosorbent assay for detecting tobacco mosaic virus in plant tissues and protoplasts. *Phytopathology* **74**, 385–389.

Sela, I., Grafi, G., Sher, N., Edelbaum, O., Yagev, H., and Gerassi, E. (1987). Resistance systems related to the N gene and their comparison with interferon. In: D. Evered and S. Hamell, (eds) *Plant Resistance to Viruses*, pp. 109–119. Wiley, Chichester.

Selling, B.H., Allison, R.F. and Kaesberg, P. (1990). Genomic RNA of an insect virus directs synthesis of infectious virions in plants. *Proc. Natl Acad. Sci. USA* **87**, 434–438.

Selman, B.J. (1973). Beetles—phytophagous Coleoptera. In: A.J. Gibbs (ed.) *Viruses and Invertebrates*, pp. 157–177. North-Holland, Amsterdam.

Selsky, M.I. and Black, L.M. (1962). Effect of high and low temperatures on the survival of wound-tumor virus in sweet clover. *Virology* **16**, 190–198.

Selstam, E. and Jackson, A.O. (1983). Lipid composition of sonchus yellow net virus. *J. Gen. Virol.* **64**, 1607–1613.

Semal, J. and Kummert, J. (1969). Effects of actinomycin D on the incorporation of uridine into virus-infected leaf fragments. *Phytopathol. Z.* **65**, 364–372.

Semancik, J.S. and Duran-Vila, N. (1999). Viroids in plants: shadows and footprints of a primitive RNA. In: E. Domingo, R. Webster and J. Holland (eds) *Origin and evolution of viruses*, pp. 37–64. Academic Press, London.

Semancik, J.S. and Vanderwoude, W.J. (1976). Exocortis viroid: cytopathic effects at the plasma membrane in association with pathogenic RNA. *Virology* **69**, 719–726.

Semancik, J.S., Szychowski, J.A., Rakowski, A.C. and Symons, R.H. (1994). A stable 463 nucleotide variant of citrus exocortis viroid produced by terminal repeats. *J. Gen. Virol.* **75**, 727–732.

Semancik, J.S., Rakowski, A.G., Bash, J.A. and Gumpf, D.J. (1997). Application of selected viroids for dwarfing and enhancement of production of 'Valencia' orange. *J. Hort. Sci.* **72**, 563–570.

Senke, P.C. and Johnson, J.E. (1993). Crystallization and preliminary X-ray characterization of tobacco streak virus and a proteolytically modified form of the capsid protein. *Virology* **196**, 328–331.

Seppänen, P., Puska, R., Honkanen, J. *et al.* (1997). Movement protein-derived resistance to triple gene block-containing plant viruses. *J. Gen. Virol.* **78**, 1241–1246.

Séron, K. and Haenni, A.-L. (1996). Vascular movement of plant viruses. *Mol. Plant-Microb. Interact.* **6**, 435–442.

Serwer, P. (1977). Flattening and shrinkage of bacteriophage T7 after preparation for electron microscopy by negative staining. *J. Ultrastruct. Res.* **58**, 235–243.

Sether, D.M., Ullman, D.E. and Hu, J.S. (1998) Transmission of pineapple mealybug wilt-associated virus by two species of mealybug (*Dysmicoccus* spp.). *Phytopathology* **88**, 1224–1230.

Shah, J., Kachroo, P. and Klessig, D.F. (1999). The *Arabidopsis ssi1* mutation restores pathogenesis-related gene expression in *npr1* plants and renders defensin gene expression salicylic acid dependent. *Plant Cell* **11**, 191–206.

Shalla, T.A., Shepherd, R.J. and Petersen, L.J. (1980). Comparative cytology of nine isolates of cauliflower mosaic virus. *Virology* **102**, 381–388.

Shalla, T.A., Petersen, L.-J. and Zaitlin, M. (1982). Restricted movement of a temperature-sensitive virus in tobacco leaves is associated with a reduction in numbers of plasmodesmata. *J. Gen. Virol.* **60**, 355–358.

Shanks, M. and Lomonossoff, G.P. (2000). Co-expression of the capsid proteins of cowpea mosaic virus in insect cells leads to the formation of virus-like particles. *J. Gen. Virol.* **81**, 3093–3097.

Shanks, M., Lomonossoff, G.P. and Evans, D. (1985). Double-stranded, replicative form RNA molecules of cowpea mosaic virus are not infectious. *J. Gen. Virol.* **66**, 925–930.

Shanks, M., Maule, A.J., Wilson, I.G., Lomonossoff, G.P., Huskison, N. and Tomenius, K. (1988). RCMV gene expression. *Annu. Rep. John Innes Inst.* **73**.

Shanks, M., Tomenius, K., Chapham, D. *et al.* (1989). Identification and subcellular localization of a putative cell-to-cell transport protein from red clover mottle virus. *Virology* **173**, 400–407.

Sharp, P. (2000). RNAi and double-stranded RNA. *Genes Devel.* **13**, 139–141.

Sharp, P.A. and Zamore, P.D. (2000). RNA interference. *Science* **287**, 2431–2433.

Shaw, J.G. (1969). *In vivo* removal of protein from tobacco mosaic virus after inoculation of tobacco leaves: II. Some characteristics of the reaction. *Virology* **37**, 109–116.

Shaw, J.G. (1973). *In vivo* removal of protein from tobacco mosaic virus after inoculation of tobacco leaves: III. Studies on the location on virus particles for the initial removal of protein. *Virology* **53**, 337–342.

Shaw, J.G. (1999). Tobacco mosaic virus and the study of early events in virus infection. *Phil. Trans. R. Soc. Lond. B* **354**, 603–611.

Shaw, J.G., Plaskitt, K.A. and Wilson, T.M.A. (1986). Evidence that tobacco mosaic virus particles disassemble cotranslationally *in vivo*. *Virology* **148**, 326–336.

Shaw, J.G., Hunt, A.G., Pirone, T.P. and Rhodes, R.E. (1990). Organisation and expression of potyviral genes. In: T.P. Pirone and J.G. Shaw (eds) *Viral Genes and Plant Pathogens*, pp. 107–123. Springer-Verlag, New York.

Sheldon, C.C., Jeffries, A.C., Davies, C. and Symons, R.H. (1990). RNA self-cleavage by the hammerhead structure. *Nucleic Acids Mol. Biol.* **4**, 227–242.

Shen W.-H. and Hohn, B. (1994). Amplification and expression of the —glucuronidase gene in maize by vectors based on maize streak virus. *Plant J.* **5**, 227–236.

Shen, P., Kaniewski, M.B., Smith, C. and Beachy, R.N. (1993). Nucleotide sequence and genomic organization of rice tungro spherical virus. *Virology* **193**, 621–630.

Shenk, P.M., Sagi, L., Remans, T. *et al.* (1999). A promoter from sugarcane bacilliform badnavirus drives transgene expression in banana and other monocot and dicot plants. *Plant Mol. Biol.* **39**, 1221–1230.

Shepard, J.F. (1975). Regeneration of plants from protoplasts of potato virus X-infected tobacco leaves. *Virology* **66**, 492–501.

Shepard, J.F. (1981). Protoplasts as sources of disease resistance in plants. *Annu. Rev. Phytopathol.* **19**, 145–166.

Shepard, J.F. and Uyemoto, J.K. (1976). Influence of elevated temperatures on the isolation and proliferation of mesophyll protoplasts from PVX- and PVY-infected tobacco tissue. *Virology* **70**, 558–560.

Shepardson, S., Esau, K. and McCrum, R. (1980). Ultrastructure of potato leaf phloem infected with potato leafroll virus. *Virology* **105**, 379–392.

Shepherd, R.J. (1972). Transmission of viruses through seed and pollen. In: C.I. Kado and H.O. Agrawal (eds) *Principles and Techniques in Plant Virology*, pp. 267–292. Van Nostrand-Reinhold, Princeton, NJ.

Shepherd, R.J. (1976). DNA viruses of higher plants. *Adv. Virus Res.* **20**, 305–339.

Shepherd, R.J. and Hills, F.J. (1970). Dispersal of beet yellows and beet mosaic viruses in the inland valleys of California. *Phytopathology* **60**, 798–804.

Shepherd, R.J., Wakeman, R.J. and Romanko, R.R. (1968). DNA in cauliflower mosaic virus. *Virology* **36**, 150–152.

Shepherd, R.J., Bruening, G.E. and Wakeman, R.J. (1970). Double-stranded DNA from cauliflower mosaic virus. *Virology* **41**, 339–347.

Sherwood, J.L. and Fulton, R.W. (1982). The specific involvement of coat protein in tobacco mosaic virus cross protection. *Virology* **119**, 150–158.

Sherwood, J.L., Sanborn, M.R., Keyser, G.C. and Meyers, L.D. (1989). Use of monoclonal antibodies in detection of tomato spotted wilt virus. *Phytopathology* **79**, 61–64.

Shewmaker, C.K., Caton, J.R., Houck, C.M. and Gardner, R.C. (1985). Transcription of cauliflower mosaic virus integrated into plant genomes. *Virology* **140**, 281–288.

Shewry, P.R. and Lucas, J.A. (1997). Plant proteins that confer resistance to pests and pathogens. *Adv. Bot. Res.* **26**, 135–192.

Shields, S.A. and Wilson, T.M.A. (1987). Cell-free translation of turnip mosaic virus RNA. *J. Gen. Virol.* **68**, 169–180.

Shields, S.A., Brisco, M.J., Wilson, T.M.A. and Hull, R. (1989). Southern bean mosaic virus RNA remains associated with swollen virions during translation in wheat germ cell-free extracts. *Virology* **171**, 602–606.

Shih, D.-S. and Kaesberg, P. (1973). Translation of brome mosaic viral ribonucleic acid in a cell-free system derived from wheat embryo. *Proc. Natl Acad. Sci. USA* **70**, 1799–1803.

Shih, D.-S. and Kaesberg, P. (1976). Translation of the RNAs of brome mosaic virus: the monocistronic nature of RNA1 and RNA2. *J. Mol. Biol.* **103**, 77–88.

Shikata, E. and Maramorosch, K. (1966). Electron microscopy of pea enation mosaic virus in plant cell nuclei. *Virology* **30**, 439–454.

Shikata, E. and Maramorosch, K. (1967). Electron microscopy of wound tumor virus assembly sites in insect vectors and plants. *Virology* **32**, 363–377.

Shintaku, M. and Palukaitis, P. (1990). Genetic mapping of cucumber mosaic virus. In: T.P. Pirone and J.G. Shaw (eds) *Viral Genes and Plant Pathogenesis*, pp. 156–164. Springer, New York.

Shintaku, M.H., Zhang, L. and Palukaitis, P. (1992). A single amino acid substitution in the coat protein of cucumber mosaic virus induces chlorosis in tobacco. *Plant Cell* **4**, 751–757.

Shintaku, M.H., Carter, S.A., Bao, Y.M. and Nelson, R.S. (1996). Mapping nucleotides in the 126-kDa protein gene that control the differential symptoms induced by two strains of tobacco mosaic virus. *Virology* **221**, 218–225.

Shirako, Y. (1998). Non-AUG translation initiation in a plant RNA virus: a forty-amino-acid extension is added to the N terminus of the soil-borne wheat mosaic virus capsid protein. *J. Virol.* **72**, 1677–1682.

Shirako, Y. and Ehara, Y. (1986). Comparison of the *in vitro* translation products of wild-type and deletion mutant of soil-borne wheat mosaic virus. *J. Gen. Virol.* **67**, 1237–1245.

Shirako, Y. and Ehara, Y. (1990). Comparison of the *in vitro* translation products of wild-type and deletion mutant of soil-borne wheat mosaic virus. *J. Gen. Virol.* **67**, 1237–1245.

Shirako, Y. and Wilson, T.M.A. (1993). Complete nucleotide sequence and organization of the bipartite RNA genome of soil-borne wheat mosaic virus. *Virology* **195**, 16–32.

Shirako, Y. and Wilson, T.M.A. (1999). Furoviruses. In: A. Granoff and R.G. Webster (eds) *Encyclopedia of Virology*, 2nd edn, pp. 587–596. Academic Press, San Diego.

Shukla, D.D. and Schmelzer, K. (1973). Studies on viruses and virus disease of cruciferous plants. XIV. Cucumber mosaic virus in ornamental and wild species. *Acta Phytopathol. Acad. Sci. Hung.* **8**, 149–155.

Shukla, D.D. and Schmelzer, K. (1975). Studies on viruses and virus diseases of cruciferous plants. XIX. Analysis of the results obtained with ornamental and wild species. *Acta Phytopathol. Acad. Sci. Hung.* **10**, 217–229.

Shukla, D.D. and Ward, C.W. (1988). Amino acid sequence homology of coat proteins as a basis for identification and classification of the potyvirus group. *J. Gen. Virol.* **69**, 2703–2710.

Shukla, D.D. and Ward, C.W. (1989a). Structure of potyvirus and coat proteins and its application in the taxonomy of the potyvirus group. *Adv. Virus Res.* **36**, 273–314.

Shukla, D.D. and Ward, C.W. (1989b). Identification and classification of potyviruses on the basis of coat protein sequence data and serology. *Arch. Virol.* **106**, 171–200.

Shukla, D.D., O'Donnell, I.J. and Gough, K.H. (1983). Characteristics of the electro-blot radioimmunoassay (EBRIA) in relation to the identification of plant viruses. *Acta Phytopathol. Acad. Sci.* **18**, 79–84.

Shukla, D.D., McKern, N.M., Gough, K.H., Tracey, S.L. and Letho, S.G. (1988a). Differentiation of potyviruses and their strains by high performance liquid chromatographic peptide profiling of coat proteins. *J. Gen. Virol.* **69**, 493–502.

Shukla, D.D., Strike, P.M., Tracy, S.L., Gough, K.H. and Ward, C.W. (1988b). The N and C termini of the coat proteins of potyviruses are surface located and the N terminus contains the major virus-specific epitopes. *J. Gen. Virol.* **69**, 1497–1508.

Shukla, D.D., Jilka, J., Tosic, M. and Ford, R.E. (1989a). A novel approach to the serology of potyviruses involving affinity-purified polygonal antibodies directed towards virus specific N-termini of coat proteins. *J. Gen. Virol.* **70**, 13–23.

Shukla, D.D., Tosic, M., Jilka, J., Ford, R.E., Toler, R.W. and Langham, M.A.C. (1989b). Taxonomy of potyviruses infecting maize, sorghum and sugarcane in Australia and the United States as determined by reactivities of polyclonal antibodies directed towards virus-specific N-termini of coat proteins. *Phytopathology* **79**, 223–229.

Shukla, D.D., Tribbick, G., Mason, T.-J., Hewish, D.R., Geysen, H.M. and Ward, C.W. (1989c). Localization of virus-specific and group-specific epitopes of plant potyviruses by systematic immunochemical analysis of overlapping peptide fragments. *Proc. Natl Acad. Sci. USA* **86**, 8192–8196.

Shukla, D.D., Ford, R.E., Tosic, M., Jilka, J. and Ward, C.W. (1989d). Possible members of the potyvirus group transmitted by mites or whiteflies share epitopes with aphid-transmitted definitive members of the group. *Arch. Virol.* **105**, 143–151.

Shukla, D.D., Ward, C.W. and Brunt, A.A. (1994). *The Potyviridae*. CAB International, Wallingford, UK.

Shukla, V.D. and Anjaneyulu, A. (1981). Plant spacing to reduce rice tungro incidence. *Plant Dis.* **65**, 584–586.

Shukla, V.D. and Anjaneyulu, A. (1982). Evaluation of systemic insecticides to reduce tungro disease incidence in rice nursery. *Indian Phytopathol.* **35**, 502–504.

Shukle, R.H., Lampe, D.J., Lister, R.M. and Foster, J.E. (1987). Aphid feeding behaviour: relationship to barley yellow drawf virus resistance in *Agropyron* species. *Phytopathology* **77**, 725–729.

Sidwell, R.W., Huffman, J.H., Khare, G.P., Allen, L.B., Witkowski, J.T. and Robins, R.K. (1972). Broad-spectrum antiviral activity of virazole: 1-β-D-ribofuranosyl-1,2,4-triazole-3-carboxamide. *Science* **177**, 705–706.

Siegel, A. (1960). Studies on the induction of tobacco mosaic virus mutants with nitrous acid. *Virology* **11**, 156–167.

Siegel, A. (1971). Pseudovirions of tobacco mosaic virus. *Virology* **46**, 50–59.

Siegel, A. (1985). Plant-virus-based vectors for gene transfer may be of considerable use despite a presumed high error frequency during RNA synthesis. *Plant Mol. Biol.* **4**, 327–329.

Siegel, A., Zaitlin, M. and Sehgal, O.P. (1962). The isolation of defective tobacco mosaic virus strains. *Proc. Natl Acad. Sci.* **48**, 1845–1851.

Siegel, A., Hari, V. and Kolacz, K. (1978). The effect of tobacco mosaic virus infection on host and virus-specific protein synthesis in protoplasts. *Virology* **85**, 494–503.

Siegel, R.W., Adkins, S. and Kao, C.C. (1997). Sequence-specific recognition of a subgenomic RNA promoter by a viral RNA polymerase. *Proc. Natl Acad. Sci. USA* **94**, 11238–11243.

Siegel, R.W., Bellon, L., Beigelman, L. and Kao, C.C. (1998). Moieties in an RNA promoter specifically recognized by a viral RNA-dependent RNA polymerase. *Proc. Natl Acad. Sci. USA* **95**, 11613–11618.

Siitari, H. and Kurppa, A. (1987). Time resolved fluoroimmunoassay in the detection of plant viruses. *J. Gen. Virol.* **68**, 1423–1428.

Sijen, T., Wellink, J., Hendriks, J., Verver, J. and van Kammen, A. (1995). Replication of cowpea mosaic virus RNA1 and RNA2 is specifically blocked in transgenic *Nicotiana benthamiana* plants expressing the full-length replicase or movement protein genes. *Mol. Plant Microb. Interact.* **8**, 340–347.

Sijen, T., Wellink, J., Hiriart, J.-B. and van Kammen, A. (1996). RNA-mediated virus resistance: role of repeated transgenes and delineation of targeted regions. *Plant Cell* **8**, 2277–2294.

Silber, G. and Burk, L.G. (1965). Infectivity of tobacco mosaic virus stored for fifty years in extracted 'unpreserved' plant juice. *Nature (London)* **206**, 740–741.

Sill, W.H. and Walker, J.C. (1952). A virus inhibitor in cucumber in relation to mosaic resistance. *Phytopathology* **42**, 349–352.

Silva, A.M. and Rossmann, M.G. (1987). Refined structure of southern bean mosaic virus at 2.9 Å resolution. *J. Mol. Biol.* **197**, 69–87.

Silva, J.L. and Weber, G. (1988). Pressure-induced dissociation of brome mosaic virus. *J. Mol. Biol.* **199**, 149–159.

Simon, A.E. (1999). Replication, recombination and symptom-modulation properties of the satellite RNAs of turnip crinkle virus. In: P.K. Vogt and A.O. Jackson (eds) *Satellites and Defective Viral RNAs*, pp. 19–36. Springer, Berlin.

Simon, A.E. and Howell, S. (1986). The virulent satellite RNA of turnip crinkle virus has a major domain homologous to the 3′ end of the helper virus genome. *EMBO J.* **5**, 3423–3428.

Simon, A.E. and Howell, S. (1987). Synthesis *in vitro* of infectious RNA copies of the virulent satellite of turnip crinkle virus. *Virology* **156**, 146–152.

Simon, A.E. and Nagy, P.D. (1996). RNA recombination in turnip crinkle virus: its role in formation of chimeric RNAs, multimers, and in 3′-end repair. *Sem. Virol.* **7**, 373–379.

Simon, A.E. and Zhang, G. (2001). The modular nature of viral replicase enhancers. Abstract W16–1, 20th Annual Meeting of the American Society for Virology.

Simon, A.E., Engel, H., Johnson, R.P. and Howell, S.H. (1988). Identification of regions affecting virulence, RNA processing and infectivity in the virulent satellite of turnip crinkle virus. *EMBO J.* **7**, 2645–2651.

Simon, A.E., Li, X.H., Lew, J.E., Strange, R., Zhang, C., Polacco, M. and Carpenter, C.D. (1992). Susceptibility and resistance of *Arabidopsis thaliana* to turnip crinkle virus. *Mol. Plant Microbe Interact.* **5**, 496–503.

Simpkins, I., Walkey, D.G.A. and Neely, H.A. (1981). Chemical suppression of virus in cultured plant tissues. *Ann. Appl. Biol.* **99**, 161–169.

Singh, M.N., Khurana, S.M.P., Nagaich, B.B. and Agrawal, H.O. (1988a). Environmental factors influencing aphid transmission of potato virus Y and potato leafroll virus. *Potato Res.* **31**, 501–509.

Singh, R.N. and Dreher, T.W. (1997). Turnip yellow mosaic virus RNA-dependent RNA polymerase initiation of minus strand synthesis *in vitro*. *Virology* **233**, 430–439.

Singh, R.P. (1984). *Solanum X berthaultii*, a sensitive host for indexing potato spindle tuber viroid from dormant tubers. *Potato Res.* **27**, 163–172.

Singh, R.P. and Kurz, J. (1997). RT-PCR analysis of PSTVd aphid transmission in association with PLRV. *Can. J. Plant Pathol.* **19**, 418–424.

Singh, R.P. and Slack, S.A. (1984). Reactions of tuber-bearing *Solanum* species to infection with potato spindle tuber viroid. *Plant Dis.* **68**, 784–787.

Singh, R.P. and Somerville, T.H. (1987). New disease symptoms observed on field-grown potato plants with potato spindle tuber viroid and potato virus Y infections. *Potato Res.* **30**, 127–132.

Singh, R.P., Lee, C.R. and Clark, M.C. (1974). Manganese effect on the local lesion symptom of potato spindle tuber 'virus' in *Scopolia sinensis*. *Phytopathology* **64**, 1015–1018.

Singh, R.P., Boucher, A. and Seabrook, J.E.A. (1988b). Detection of the mild strains of potato spindle tuber viroid from single true potato seed by return electrophoresis. *Phytopathology* **78**, 663–667.

Singh, S.K., Anjaneyula, A. and Lapierre, H. (1984). Use of pectinocellulolytic enzymes for improving extraction of phloem-limited plant viruses as exemplified by the rice tungro virus complex. *Agronomie* **4**, 479–484.

Sinha, R.C. (1967). Response of wound tumor virus infection in insects to vector age and temperature. *Virology* **31**, 746–748.

Sinijärv, R., Järvekülg, L. andreeva, E. and Saarma, M. (1988). Detection of potato virus X by one incubation europium time-resolved fluoroimmunoassay and ELISA. *J. Gen. Virol.* **69**, 991–998.

Sinisterra, X.H., Polston, J.E., Abouzid, A.M. and Hiebert, E. (1999). Tobacco plants transformed with a modified coat protein of tomato mottle begomovirus show resistance to virus infection. *Phytopathology* **89**, 701–706.

Sisler, W.W. and Timian, R.G. (1956) Inheritance of barley stripe mosaic virus resistance of Modjo (C.I. 3212) and C.I. 3212-1. *Plant Dis. Rep.* **40**, 1106–1108.

Sit, T.L., Leclerc, D. and AbouHaidar, M.G. (1994). The minimal 5′ sequence for *in vitro* initiation of papaya mosaic potexvirus assembly. *Virology* **199**, 238–242.

Sit, T.L., Vaewhongs, A.A. and Lommel, S.A. (1998). RNA-mediated trans-activation of transcription from a viral RNA. *Science* **281**, 829–832.

Siuzdak, G. (1998). Probing viruses with mass spectrometry. *J. Mass Spectrom.* **33**, 203–211.

Sivakumaran, K. and Hacker, D.L. (1998). The 105-kDa polyprotein of southern bean mosaic virus is translated by scanning ribosomes. *Virology* **246**, 34–44.

Sivakumaran, K. and Kao, C.C. (1999). Initiation of genomic plus-strand RNA synthesis from DNA and RNA templates by a viral RNA-dependent RNA polymerase. *J. Virol.* **73**, 6415–6423.

Sivakumaran, K., Fowler, B.C. and Hacker, D.L. (1998). Identification of viral genes required for cell-to-cell movement of southern bean mosaic virus. *Virology* **252**, 376–386.

Sivakumaran, K., Kim, C.-H., Tayon, R. and Kao, C.C. (1999). RNA sequence and secondary structural determinants in a minimal viral promoter that directs replicase recognition and initiation of genomic plus-strand synthesis. *J. Mol. Biol.* **294**, 667–682.

Sivakumaran, K., Bao, Y., Roossinck, M.J. and Kao, C.C. (2000). Recognition of the core RNA promoter for minus-strand RNA synthesis by the replicase of brome mosaic virus and cucumber mosaic virus. *J. Virol.* **74**, 10323–10331.

Skaf, J.S., Schultz, M.H., Hirata, H. and de Zoeten, G.A. (2000). Mutational evidence that the VPg is involved in the replication but not the movement of pea enation mosaic virus-1. *J. Gen. Virol.* **81**, 1103–1109.

Skaria, M., Lister, R.M. and Foster, J.E. (1984). Lack of barley yellow dwarf virus dosage effects on virus content in cereals. *Plant. Dis.* **68**, 759–761.

Skotnicki, A., Gibbs, A. and Wrigley, N.G. (1976). Further studies on *Chara corallina* virus. *Virology* **75**, 457–468.

Skulachev, M.V., Ivanov, P.A., Karpova, O.V. *et al.* (1999). Internal initiation of translation directed by the 5′-untranslated region of the tobamovirus subgenomic RNA I₂. *Virology* **263**, 139–154.

Skuzeski, J.M., Nichols, L.M., Gesteland, R.F. and Atkins, J.F. (1991). The signal for a leaky UAG stop codon in several plant viruses includes the two downstream codons. *J. Mol. Biol.* **218**, 365–373.

Slack, C.R. (1969). Localization of certain photosynthetic enzymes in mesophyll and parenchyma sheath chloroplasts of maize and *Amaranthus palmeri*. *Photochemistry* **8**, 1387–1391.

Slack, S.A. and Singh, R.P. (1998). Control of viruses affecting potatoes through seed potato certification programs. Breeding for resistance to plant viruses. In: A. Hadidi, R.K. Khetarpal and H. Koganezawa (eds) *Plant Virus Disease Control*, pp. 249–260. APS Press, St. Paul, MN.

Sleat, D.E. and Palukaitis, P. (1990a). Site-directed mutagenesis of a plant viral satellite RNA changes its phenotype from ameliorative to necrogenic. *Proc. Natl Acad. Sci. USA* **87**, 2946–2950.

Sleat, D.E. and Palukaitis, P. (1990b). Induction of tobacco chlorosis by cucumber mosaic virus satellite RNAs is specific to subgroup II helper strains. *Virology* **176**, 292–295.

Sleat, D.E., Turner, P.C., Finch, J.T., Butler, P.J.G. and Wilson, T.M.A. (1986). Packaging of recombinant RNA molecules into pseudovirus particles directed by the origin-of-assembly sequences from tobacco mosaic virus RNA. *Virology* **155**, 299–308.

Sleat, D.E., Gallie, D.R., Jefferson, R.A., Bevan, M.W., Turner, P.C. and Wilson, T.M.A. (1987). Characterisation of the 5′ leader sequence of tobacco mosaic virus RNA as a general enhancer of translation *in vitro*. *Gene* **60**, 217–225.

Sleat, D.E., Hull, R., Turner, P.C. and Wilson, T.M.A. (1988a). Studies on the mechanism of translational enhancement by the 5′-leader sequence of tobacco mosaic virus RNA. *Eur. J. Biochem.* **175**, 75–86.

Sleat, D.E., Plaskitt, K.A. and Wilson, T.M.A. (1988b). Selective encapsulation of CAT gene transcripts in TMV-infected transgenic tobacco inhibits CAT synthesis. *Virology* **165**, 609–612.

Sleat, D.E., Gallie, D.R., Watts, J.W. *et al.* (1988c). Selective recovery of foreign gene transcripts as virus-like particles in TMV-infected transgenic tobaccos. *Nucleic Acids Res.* **16**, 3127–3140.

Slykhuis, J.T. (1955). *Aceria tulipae* Keifer (Acarina: Eriophyidae) in relation to the spread of wheat streak mosaic. *Phytopathology* **45**, 116–128.

Slykhuis, J.T. (1962). Mite transmission of plant viruses. In: K. Maramorosch (ed.) *Biological Transmission of Disease Agents*, pp. 41–61. Academic Press, New York.

Slykhuis, J.T. (1963). Vector and host relations of North American wheat striate mosaic virus. *Can. J. Bot.* **41**, 1171–1185.

Slykhuis, J.T. (1965). Mite transmission of plant viruses. *Adv. Virus Res.* **11**, 97–137.

Slykhuis, J.T. (1970). Factors determining the development of wheat spindle streak mosaic caused by a soil-borne virus in Ontario. *Phytopathology* **60**, 319–331.

Slykhuis, J.T. (1974). Differentiation of transmission and incubation temperatures for wheat spindle streak mosaic virus. *Phytopathology* **64**, 554–557.

Slykhuis, J.T. (1976). Stimulating effects of transplanting on the development of wheat spindle streak mosaic virus in wheat plants infected from soil. *Phytopathology* **66**, 130–131.

Slykhuis, J.T., Andrews, J.E. and Pittmann, U.J. (1957). Relation of date of seeding winter wheat in southern Alberta to losses from wheat streak mosaic, root rot and rust. *Can. J. Plant Sci.* **37**, 113–127.

Smardon, A., Spoerke, J.M., Stacey, S.C., Klein, M.E., Mackin, N. and Maine, E.M. (2000). EGO-1 is related to RNA-directed RNA polymerase and functions in germline development and RNA interference in *C. elegans*. *Curr. Biol.* **10**, 169–178.

Smidansky, E.D. and Carroll, T.W. (1996). Factors influencing the outcome of barley yellow streak mosaic virus-brown wheat mite-barley interactions. *Plant Dis.* **80**, 186–193.

Smirnyagina, E., Hsu, Y.-H., Chua, N. and Ahlquist, P. (1994). Second-site mutations in the brome mosaic virus RNA3 intercistronic region partially suppress a defect in coat protein mRNA transcription. *Virology* **198**, 427–436.

Smit, C.H., Roosien, J., van Vloten-Doting, L. and Jaspars, E.M.J. (1981). Evidence that alfalfa mosaic virus infection starts with three RNA–protein complexes. *Virology* **112**, 169–173.

Smith, C.R., Tousignant, M.E., Geletka, L.M. and Kaper, J.M. (1992). Competition between cucumber mosaic virus satellite RNAs in tomato seedlings and protoplasts: a model for satellite-mediated control of tomato necrosis. *Plant Dis.* **76**, 1270–1274.

Smith, D.B. and Inglis, S.C. (1987). The mutation rate and variability of eukaryotic viruses: an analytical review. *J. Gen. Virol.* **68**, 2729–2740.

Smith, D.B., McAllister, J., Casino, C. and Simmonds, P. (1997). Virus 'quasispecies': making a mountain out of a molehill? *J. Gen. Virol.* **78**, 1511–1519.

Smith, F.D. and Banttari, E.E. (1987). Dot-ELISA on nitrocellulose membranes for detection of potato leafroll virus. *Plant. Dis.* **71**, 795–799.

Smith, G.E., Vlak, J.M and Summers, M.D. (1983). Physical analysis of *Autographa californica* nuclear polyhedrosis virus transcripts for polyhedrin and 10,000-molecular-weight protein. *J. Virol.* **45**, 215–225.

Smith, G.R., Borg, Z., Lockhart, B.E., Braithwaite, K.S. and Gibbs, M.J. (2000a). Sugarcane yellow leaf virus: a novel member of the *Luteoviridae* that probably arose by inter-species recombination. *J. Gen. Virol.* **81**, 1865–1869.

Smith, H.A., Swaney, S.L., Parks, T.D., Wernsman, E.A. and Dougherty, W.G. (1994). Transgenic plant resistance mediated by untranslatable RNAs: expression, regulation and fate of nonessential RNAs. *Plant Cell* **6**, 1441–1453.

Smith, H.C. (1967). The effect of aphid numbers and stage of plant growth in determining tolerance to barley yellow dwarf virus in cereals. *N. Z. J. Agric. Res.* **10**, 445–466.

Smith, H.G. (1986). Fodder beet clamps as a source of virus yellows. *British Sugar Beet Rev.* **54**, 38–39.

Smith, H.G. and Barker, H. (eds) (1999). The *Luteoviridae*. CAB International, Wallingford, UK.

Smith, K.M. (1931). On the composite nature of certain potato virus diseases of the mosaic group as revealed by the use of plant indicators and selective methods of transmission. *Proc. R. Soc. London B* **109**, 251–266.

Smith, K.M. (1937). *A Text Book of Plant Virus Diseases*. Churchill, London.

Smith, K.M. (1972). *A Textbook of Plant Virus Diseases*, 3rd edn. Longmans Green, New York.

Smith, M.C., Holt, J., Kenyon, L. and Foot, C. (1998). Quantitative epidemiology of banana bunchy top virus disease and its control. *Plant Pathol.* **47**, 177–187.

Smith, M.D. (1996). Antibody production in plants. *Biotechnol. Adv.* **14**, 267–281.

Smith, N.A., Singh, S.P., Wang, M.-B., Stoutjesdijk, P.A., Green, A.G. and Waterhouse, P.M. (2000b). Total silencing by intron-spliced hairpin RNAs. *Nature* **407**, 319–320.

Smith, P.R. (1983). The Australian Fruit Variety Foundation and its role in supplying virus-tested planting material to the fruit industry. *Acta Hortic.* **130**, 263–266.

Smith, S.H. and Schlegel, D.E. (1964). The distribution of clover yellow mosaic virus in *Vicia faba* root tips. *Phytopathology* **54**, 1273–1274.

Smith, S.H. and Schlegel, D.E. (1965). The incorporation of nucleic acid precursors in healthy and virus-infected plant. *Virology* **26**, 180–189.

Smith, S.S. and Murakishi, H.M. (1993). Restricted virus multiplication and movement of tomato mosaic virus in resistant tomato somaclones. *Plant Sci.* **89**, 113–122.

Smith, T.A. (1985). Polyamines. *Annu. Rev. Plant Physiol.* **36**, 117–143.

Smith, T.A. (1987). The isolation of the two electrophoretic forms of cowpea mosaic virus using fast protein liquid chromatography. *J. Virol. Meth.* **16**, 263–269.

Sneath, P.H.A. (1962). The construction of taxonomic groups. *Symp. Soc. Gen. Microbiol.* **12**, 289–332.

Söber, J., Järvekülg, L., Toots, I., Radavsky, J., Villems, R. and Saarma, M. (1988). Antigenic characterisation of potato virus X with monoclonal antibodies. *J. Gen. Virol.* **69**, 1799–1807.

Soellick, T.-R., Uhrig, J.F., Butcher, G.L., Kellmann, J.-W. and Schreier, P.H. (2000). The movement protein NSm of tomato spotted wilt tospovirus (TSWV): RNA binding, interaction with the TSWV N protein, and identification of interacting plant proteins. *Proc. Natl Acad. Sci. USA* **97**, 2373–2378.

Sohal, A.K., Love, A.J., Cecchini, E., Covey, S.N., Jenkins, G.I. and Milner, J.J. (1999). Cauliflower mosaic virus infection stimulates lipid transfer protein gene expression in *Arabidopsis*. *J. Exp. Bot.* **50**, 1727–1733.

Sokmen, M.A., Barker, H. and Torrance, L. (1998). Factors affecting the detection of potato mop-top virus in potato tubers and improvement of test procedures for more reliable assays. *Ann. Appl. Biol.* **133**, 55–63.

Sol, H.H. (1963). Some data on the occurrence of rattle virus at various depths in the soil and on its transmission. *Tijdschr. Plantenziekten* **69**, 208–214.

Solis, I. and García-Arenal, F. (1990). The complete nucleotide sequence of the genomic RNA of the tobamovirus tobacco mild green mosaic virus. *Virology* **177**, 553–558.

Solovyev, A.G., Savenkov, E.I., Agranovsky, A.A. and Morozov, S.Y. (1996). Comparisons of the genomic *cis*-elements and coding regions in RNA components of the hordeiviruses barley stripe mosaic virus, lychnis ringspot virus, and poa semilatent virus. *Virology* **219**, 9–18.

Song, S.-K., Choi, Y., Moon, Y.H., Kim, S.-G., Choi, Y.D. and Lee, J.S. (2000). Systemic induction of a *Phytolacca insularis* antiviral protein gene by mechanical wounding, jasmonic acid, and abscisic acid. *Plant Mol. Biol.* **43**, 439–450.

Sonoda, S., Mori, M. and Nishiguchi, M. (1999). Homology-dependent virus resistance in transgenic plants with the coat protein gene of sweet potato feathery mottle potyvirus: target specificity and transgene methylation. *Phytopathology* **89**, 385–391.

Sorger, P.K., Stockley, P.G. and Harrison, S.C. (1986). Structure and assembly of turnip crinkle virus: II. Mechanism of reassembly *in vitro*. *J. Mol. Biol.* **191**, 639–658.

Soto, P.E., Buddenhagen, I.W. and Asnani, V.L. (1982). Development of streak virus-resistant maize populations through improved challenge and selection methods. *Ann. Appl. Biol.* **100**, 539–546.

Spall, V.E., Shanks, M. and Lomonossoff, G.P. (1997). Polyprotein processing as a strategy for gene expression in RNA viruses. *Sem. Virol.* **8**, 15–23.

Speir, J.A., Munshi, S., Baker, T.S. and Johnson, J.E. (1993). Preliminary X-ray analysis of crystalline cowpea chlorotic mottle virus. *Virology* **193**, 234–241.

Speir, J.A., Munshi, S., Wang, G., Baker, T.S. and Johnson, J.E. (1995). Structures of the native and swollen forms of cowpea chlorotic mottle virus determined by X-ray crystallography and cryo-electron microscopy. *Structure* **3**, 63–78.

Spirin, A.S. (1961). The 'temperature effect' and macromolecular species of high polymer ribonucleic acids of various origins. *Biokhimiya* **26**, 454–463.

Sprague, G.F. and McKinney, H.H. (1966). Aberrant ratio: an anomaly in maize associated with virus infection. *Genetics* **54**, 1287–1296.

Sprague, G.F. and McKinney, H.H. (1971). Further evidence on the genetic behavior of AR in maize. *Genetics* **67**, 533–542.

Sprague, G.F., McKinney, H.H. and Greeley, L.W. (1963). Virus as a mutagenic agent in maize. *Science* **141**, 1052–1053.

Sta Cruz, F.C., Koganezawa, H. and Hibino, H. (1993). Comparative cytology of rice tungro viruses in selected rice cultivars. *J. Phytopathol.* **138**, 274–282.

Stace-Smith, R. and Hamilton, R.I. (1988). Inoculum thresholds of seed borne pathogens: viruses. *Phytopathology* **78**, 875–880.

Stace-Smith, R. and Martin, R.R. (1993). Virus purification in relation to virus diagnosis. In: R.E.F. Matthews (ed.) *Diagnosis of Plant Viruses*, pp. 129–158. CRC Press, Boca Raton.

Stack, J.P. and Tattar, T.A. (1978). Measurement of transmembrane electropotentials of *Vigna sinensis* leaf cells infected with tobacco ringspot virus. *Physiol. Plant Pathol.* **12**, 173–178.

Staehelin, A. (1997). The plant ER: a dynamic organelle composed of a large number of discrete functional domains. *Plant J.* **11**, 1151–1165

Stahl, E. (1894). Einige Versuche über Transpiration und Assimilation. *Bot. Z.* **6/7**, 117–145.

Stanley, J. (1983). Infectivity of the cloned geminivirus genome requires sequences from both DNAs. *Nature (London)* **305**, 643–645.

Stanley, J. (1995). Analysis of African cassava mosaic virus recombinants suggests strand nicking occurs within the conserved nonanucleotide motif during initiation of rolling circle DNA replication. *Virology* **206**, 707–712.

Stanley, J. and Gay, M.R. (1983). Nucleotide sequence of cassava latent virus DNA. *Nature (London)* **301**, 260–262.

Stanley, J. and Townsend, R. (1985). Characterisation of DNA forms associated with cassava latent virus infection. *Nucl. Acids Res.* **13**, 2189–2206.

Stanley, J., Hanau, R. and Jackson, A.O. (1984). Sequence comparison of the 3' ends of a subgenomic RNA and the genomic RNAs of barley stripe mosaic virus. *Virology* **139**, 375–383.

Stanley, J., Townsend, R. and Curson, S.J. (1985). Pseudorecombinants between cloned DNAs of two isolates of cassava latent virus. *J. Gen. Virol.* **66**, 1055–1061.

Stanley, J., Markham, P.G., Callis, R.J. and Pinner, M.S. (1986). The nucleotide sequence of an infectious clone of the geminivirus beet curly top virus. *EMBO J.* **5**, 1761–1767.

Stanley, J., Saunders, K., Pinner, M.S. and Wong, S.M. (1997). Novel defective interfering DNAs associated with Ageratum yellow vein geminivirus infection of *Ageratum conyzoides*. *Virology* **239**, 87–96.

Stanley, W.M. (1935). Isolation of a crystalline protein possessing the properties of tobacco-mosaic virus. *Science* **81**, 644–645.

Stanley, W.M. (1936). Chemical studies on the virus of tobacco mosaic virus. VI. The isolation from diseased turkish tobacco plants of a crystalline protein possessing the properties of tobacco mosaic virus. *Phytopathology* **26**, 305–320.

Stansfield, I., Jones, K.M. and Tuite, M.F. (1995). The end is in sight: terminating translation in eukaryotes. *Trends Biol. Sci.* **20**, 489–491.

StarkLorenzen, P., Guitton, M.C., Werner, R. and Muhlbach, H.P. (1997). Detection and tissue distribution of potato spindle tuber viroid in infected tomato plants by tissue print hybridization. *Arch. Virol.* **142**, 1289–1296.

Stauffacher, C.V., Usha, R., Harrington, M., Schmidt, T., Hosur, M.V. and Johnson, J.E. (1985). The structure of cowpea mosaic virus at 3.5 Å resolution. *NATO Adv. Sci. Inst. Ser., Ser. A* **126**, 293–308.

Steckert, J.J. and Schuster, T.M. (1982). Sequence specificity of trinucleoside diphosphate binding to polymerised tobacco mosaic virus protein. *Nature (London)* **299**, 32–36.

Steere, R.L. (1969). Freeze-etching simplified.*Cryobiology* **5**, 306–323.

Steger, G., Hofmann, H. Förtsch, J. *et al.* (1984). Conformational transitions in viroids and virusoids: comparison of results from energy minimization algorithms and from experimental data. *J. Biomol. Struct. Dyn.* **2**, 543–571.

Steger, G., Tabler, M., Brüggermann, W. *et al.* (1986). Structure of viroid replicative intermediates: physico-chemical studies on SP6 transcripts of cloned oligomeric potato spindle tuber viroid. *Nucl. Acids Res.* **14**, 9613–9630.

Steinhauer, D., Domingo, E. and Holland, J.J. (1992). Lack of evidence for proofreading mechanisms associated with an RNA virus polymerase. *Gene* **122**, 281–288.

Stenger, D.C. (1994). Strain-specific mobilization and amplification of a transgenic defective-interfering DNA of the geminivirus beet curly top virus. *Virology* **203**, 397–402.

Stenger, D.C., Revington, G.N., Stevenson, M.C. and Bisaro, D.M. (1991). Replicational release of geminivirus genomes from tandemly repeated copies: evidence for rolling-circle replication of a plant viral DNA. *Proc. Natl Acad. Sci. USA* **88**, 8029–8033.

Steven, A.C., Trus, B.L., Putz, C. and Wurtz, M. (1981). The molecular organisation of beet necrotic yellow vein virus. *Virology* **113**, 428–438.

Stevens, C. and Gudauskas, R.T. (1983). Effects of maize dwarf mosaic virus infection of corn on inoculum potential of *Helminthosporium maydis* race 0. *Phytopathology* **73**, 439–441.

Stevens, M., Hull, R. and Smith, H.G. (1997). Comparison of ELISA and RT-PCR for the detection of beet yellows closterovirus in plants and aphids. *J. Virol. Methods* **68**, 9–16.

Stewart, C.-B. and Wilson, A.C. (1987). Sequence convergence and functional adaptation of stomach lysozymes from foregut fermenters. *Cold Spring Harbor Symp. Quant. Biol.* **52**, 891–899.

Stirling, G.R. (1999). Increasing the adoption of sustainable, integrated management strategies for soilborne diseases of high-value annual crops. *Aust. J. Plant Pathol.* **28**, 72–79.

Stobbs, L.W. and McNeill, B.H. (1980). Increase of tobacco mosaic virus in graft-inoculated TMV resistant tomatoes. *Can. J. Plant Pathol.* **2**, 217–221.

Stobbs, L.W., Manocha, M.S. and Dias, H.F. (1977). Histological changes associated with virus localization in TMV-infected pinto bean leaves. *Physiol. Plant Pathol.* **11**, 87–94.

Stoeckle, M.Y., Shaw, M.W. and Choppin, P.W. (1987). Segment-specific and common nucleotide sequences in the non-coding regions of influenza B virus genome RNAs. *Proc. Natl Acad. Sci. USA* **84**, 2703–2707.

Stollar, B.D. (1975). The specificity and applications of antibodies to helical nucleic acids. *CRC Crit. Rev. Biochem.* **3**, 45–69.

Stommel, J.R., Tousignant, M.E., Wai, T., Pasini, R. and Kaper, J.M. (1998). Viral satellite RNA expression in transgenic tomato confers field tolerance to cucumber mosaic virus. *Plant Dis.* **82**, 391–396.

Storms, M.M.H. (1998). The role of NS_M during tomato spotted wilt virus infection. PhD thesis, University of Wageningen, Netherlands.

Storms, M.M.H., Kormelink, R., Peters, D., van Lent, J.W.M. and Goldbach, R.W. (1995). The non-structural NSm protein of tomato spotted wilt virus induces tubular structures in plant and animal cells. *Virology* **214**, 485–493.

Storms, M.M.H., van der Schoot, C., Prins, M., Kormelink, R., van Lent, J.W.M. and Goldbach, R.W. (1998). A comparison of two methods of microinjection for assessing altered plasmodesmatal gating in tissues expressing viral proteins. *Plant J.* **13**, 131–140.

Stouffer, R.F. and Ross, A.F. (1961). Effect of infection by potato virus Y on the concentration of potato virus X in tobacco plants. *Phytopathology* **51**, 740–744.

Stratford, R. and Covey, S.N. (1988). Changes in turnip leaf messenger RNA populations during systemic infections by severe and mild strains of cauliflower mosaic virus. *Mol. Plant–Microbe Interact.* **1**, 243–249.

Stratford, R. and Covey, S. (1989). Segregation of cauliflower mosaic virus symptom genetic determinants. *Virology* **172**, 451–459.

Stratford, R., Plastkitt, K.A., Turner, D.S., Markham, P.G. and Covey, S.N. (1988). Molecular properties of Bari 1, a mild strain of cauliflower mosaic virus. *J. Gen. Virol.* **69**, 2375–2386.

Stubbs, G. (1999). Tobacco mosaic virus particle structure and the initiation of disassembly. *Phil. Trans. R. Soc. Lond.* **354**, 551–557.

Stubbs, G., Warren, S. and Holmes, K. (1977). Structure of RNA and RNA binding site in tobacco mosaic virus from a 4 Å map calculated from X-ray fibre diagrams. *Nature (London)* **267**, 216–221.

Stubbs, L.L. (1956). Motley dwarf virus disease of carrot in California. *Plant Dis. Rep.* **40**, 763–764.

Stupina, V. and Simon, A.E. (1997). Analysis *in vivo* of turnip crinkle virus satellite RNA C variants with mutations in the 3′ terminal minus-strand primer. *Virology* **238**, 470–477.

Stussi-Garaud, C., Garaud, J.-C., Berna, A. and Godefroy-Colburn, T. (1987). *In situ* location of an alfalfa mosaic virus non-structural protein in plant cell walls: correlation with virus transport. *J. Gen. Virol.* **68**, 1779–1784.

Su, L., Chen, L., Egli, M., Berger, J.M. and Rich, A. (1999). Minor groove RNA triplex in the crystal structure of a ribosomal frameshifting viral pseudoknot. *Nature Struct. Biol.* **6**, 285–292.

Sugimura, Y. and Matthews, R.E.F. (1981). Timing of the synthesis of empty shells and minor nucleoproteins in relation to turnip yellow mosaic virus synthesis in *Brassica* protoplasts. *Virology* **112**, 70–80.

Sugiyama, Y., Hamamoto, H., Takemoto, S., Watanabe, Y. and Okada, Y. (1995). Systemic production of foreign peptides on the particle surface of tobacco mosaic virus. *FEBS Lett.* **359**, 247–250.

Sullivan, M.L. and Ahlquist, P. (1997). *cis*-acting signals in bromovirus RNA replication and gene expression: networking with viral proteins and host factors. *Sem. Virol.* **8**, 221–230.

Sullivan, M.L. and Ahlquist, P. (1999). A brome mosaic virus intergenic RNA3 replication signal functions with viral replication protein 1a to dramatically stabilize RNA *in vivo*. *J. Virol.* **73**, 2622–2632.

Sulzinski, M.A. and Zaitlin, M. (1982). Tobacco mosaic virus replication in resistant and susceptible plants: In some resistant species virus is confined to a small number of initially infected cells. *Virology* **121**, 12–19.

Sulzinski, M.A., Gabard, K.A., Palukaitis, P. and Zaitlin, M. (1985). Replication of tobacco mosaic: VIII. Characterisation of a third subgenomic TMV RNA. *Virology* **145**, 132–140.

Sumner, J.B. (1926). The isolation and crystallisation of the enzyme urease. *J. Biol. Chem.* **69**, 435–441.

Sun, J.-H. and Kao, C.C. (1997a). Characterization of RNA products associated with or aborted by a virus RNA-dependent RNA polymerase. *Virology* **236**, 348–353.

Sun, J.-H. and Kao, C.C. (1997b). RNA synthesis by brome mosaic virus RNA-dependent RNA polymerase transition from initiation to elongation. *Virology* **233**, 63–73.

Sundaraman, V.P., Strömvik, M.V. and Vodkin, L.O. (2000). A putative defective interfering RNA from bean pod mottle virus. *Plant Dis.* **84**, 1309–1313.

Sunter, G. and Bisaro, D.M. (1989). Transcription map of the B genome component of tomato golden mosaic virus and comparison with A component transcripts. *Virology* **173**, 647–655.

Sunter, G. and Bisaro, D.M. (1991). Transactivation in a geminivirus: AL2 gene product is needed for coat protein expression. *Virology* **180**, 416–419.

Sunter, G. and Bisaro, D.M. (1997). Regulation of a geminivirus coat protein promoter by AL2 protein (TrAP): evidence for activation and derepression mechanisms. *Virology* **232**, 269–280.

Sunter, G., Hartitz, M.D. and Bisaro, D.M. (1993). Tomato golden mosaic virus leftward gene expression: autoregulation of gemini-virus replication protein. *Virology* **195**, 275–280.

Sunter, G., Sunter, J.L. and Bisaro, D.M. (2001). Plants expressing tomato golden mosaic virus AL2 or beet curly top virus L2 transgenes show enhanced susceptibility to infection by DNA and RNA viruses. *Virology* **285**, 59–70.

Susi, P. (1999). Replication in the phloem is not necessary for efficient vascular transport of tobacco mosaic tobamovirus. *FEBS Lett.* **447**, 121–123.

Susi, P. (2000). Dye-coupling in tobacco mesophyll cells surrounding growing tobacco mosaic tobamovirus-induced local lesions. *J. Phytopathol.* **148**, 379–382.

Suzuki, M., Kuwata, S., Masuta, C. and Takanami, Y. (1995). Point mutations in the coat protein of cucumber mosaic virus affect symptom expression and virion accumulation in tobacco. *J. Gen. Virol.* **76**, 1791–1799.

Suzuki, N., Kudo, T., Shirako, Y., Ehara, Y. and Tachibana, T. (1989). Distribution of cylindrical inclusion, amorphous inclusion, and capsid proteins of watermelon mosaic virus 2 in systemically infected pumpkin leaves. *J. Gen. Virol.* **70**, 1085–1091.

Swaans, H. and van Kammen, A. (1973). Reconsideration of the distinction between the severe and yellow strains of cowpea mosaic virus. *Neth. J. Plant Pathol.* **79**, 257–265.

Swanson, M.M., Brown, J.K., Poulos, B.T. and Harrison, B.D. (1992). Genome affinities and epitope profiles of whitefly-transmitted geminiviruses from the Americas. *Ann. Appl. Biol.* **121**, 285–296.

Sweet, J.B. (1975). Soil-borne viruses occurring in nursery soils and infecting some ornamental species of Rosaceae. *Ann. Appl. Biol.* **79**, 49–54.

Sweet, J.B. (1980). Hedgerow hawthorn (*Crataegus* spp.) and blackthorn (*Prunus spinosa*) as hosts of fruit tree viruses in Britain. *Ann. Appl. Biol.* **94**, 83–90.

Swinkels, P.P.H. and Bol, J.F. (1980). Limited sequence variation in the leader sequence of RNA4 from several strains of alfalfa mosaic virus. *Virology* **106**, 145–147.

Syller, J., Marczewski, W. and Pawlowicz, J. (1997). Transmission by aphids of potato spindle tuber viroid encapsidated by potato leafroll luteovirus particles. *Eur. J. Plant Pathol.* **103**, 285–289.

Sylvester, E.S. (1969). Evidence for transovarial passage of sow thistle yellow vein virus in the aphid *Hyperomyzus lactucae*. *Virology* **38**, 440–446.

Sylvester, E.S. (1973). Reduction of excretion, reproduction and survival in *Hyperomyzus lactucae* fed on plants infected with isolates of sowthistle yellow vein virus. *Virology* **56**, 632–635.

Sylvester, E.S. and Richardson, J. (1970). Infection of *Hyperomyzus lactucae* by sowthistle yellow vein virus. *Virology* **42**, 1023–1042.

Sylvester, E.S. and Richardson, J. (1971). Decreased survival of *Hyperomyzus lactucae* inoculated with serially passed sowthistle yellow vein virus. *Virology* **46**, 310–317.

Sylvester, E.S. and Richardson, J. (1981). Inoculation of the aphids *Hyperomyzus lactucae* and *Chaelosiphon jacobi* with isolates of sowthistle yellow vein virus and strawberry crinkle virus. *Phytopathology* **71**, 598–602.

Sylvester, E.S., Richardson, J. and Frazier, N.W. (1974). Serial passage of strawberry crinkle virus in the aphid *Chaetosiphon jacobi*. *Virology* **59**, 301–306.

Symons, R.H. (1978). The two-step purification of ribosomal RNA and plant viral RNA by polyacrylamide slab gel electrophoresis. *Aust. J. Biol. Sci.* **31**, 25–37.

Symons, R.H. (1997). Plant pathogenic RNAs and RNA catalysis. *Nucl. Acids Res.* **25**, 2683–2689.

Symons, R.H. and Randles, J.W. (1999). Encapsidated circular viroid-like satellite RNAs (virusoids) of plants. In: P.K. Vogt and A.O. Jackson (eds) *Satellites and Defective RNAs*, pp. 81–105. Springer, Berlin.

Symons, R.H., Haseloff, J., Visvader, J.E. *et al.* (1985). On the mechanism of replication of viroids, virusoids and satellite RNAs. In: K. Maramorosch and J.J. McKelvey, Jr. (eds) *Subviral Pathogens of Plants and Animals: Viroids and Prions*, pp. 235–263. Academic Press, Orlando, FL.

Symons, R.H., Hutchins, C.J., Forster, A.C., Rathjen, P.D., Keese, P. and Visvader, J.E. (1987). Self cleavage of RNA in the replication of viroids and virusoids. *J. Cell Sci. Suppl.* **7**, 303–318.

Syvänen, A.-C., Laaksonen, M.. and Soderlund, H. (1986). Fast quantitation of nucleic acid hybrids by affinity-based hybrid collection. *Nucl. Acids Res.* **14**, 5037–5048.

Szatmari-Goodman, G. and Nault, L.R. (1983). Tests of oil sprays for suppression of aphid-borne maize dwarf mosaic virus in Ohio sweet corn. *J. Econ. Entomol.* **76**, 144–149.

Tabara, H., Sarkissian, M., Kelly, W.G. *et al.* (1999). The *rde-1* gene, RNA interference and transposon silencing in *C. elegans. Cell* **99**, 123–132.

Tabler, M., Tsagris, M. and Hammond, J. (1998). Antisense and ribozyme mediated resistance to plant viruses. In: A. Hadidi, R.K. Khetarpal and H. Koganezawa (eds) *Plant Virus Disease Control*, pp. 79–93. APS Press, St. Paul, MN.

Tacke, E. Prüfer, D., Salamini, F. and Rohde, W. (1990). Characterization of a potato leafroll luteovirus subgenomic RNA: differential expression by internal translation initiation and UAG suppression. *J. Gen. Virol.* **71**, 2265–2272.

Tacke, E., Salamini, F. and Rohde, W. (1996). Genetic engineering of potato for broad-spectrum protection against virus infection. *Nature Biotechnol.* **14**, 1597–1601

Takahashi, H. and Ehara, Y. (1993). Severe chlorotic spot symptoms in cucumber mosaic virus strain Y-infected tobaccos are induced by a combination of the virus coat protein gene and two host recessive genes. *Mol. Plant Microbe Interact.* **6**, 182–189.

Takahashi, T. (1971). Studies on viral pathogenesis in plant hosts: I. Relation between host leaf age and the formation of systemic symptoms induced by tobacco mosaic virus. *Phytopathol. Z.* **71**, 275–284.

Takahashi, T. (1972a). Studies on viral pathogenesis in plant hosts. II. Changes in developmental morphology of tobacco plants infected systemically with tobacco mosaic virus. *Phytopathol. Z.* **74**, 37–47.

Takahashi, T. (1972b). Studies on viral pathogenesis in plant hosts. III. Leaf age-dependent susceptibility to tobacco mosaic virus infection in 'Samsun NN' and 'Samsun' tobacco plants. *Phytopathol. Z.* **75**, 140–155.

Takahashi, T. and Diener, T.O. (1975). Potato spindle tuber viroid. XIV. Replication in nuclei isolated from infected leaves. *Virology* **64**, 106–114.

Takahashi, W.N. (1956). Increasing the sensitivity of the local-lesion method of virus assay. *Phytopathology* **46**, 654–656.

Takahashi, W.N. and Rawlins, T.E. (1932). Method for determining shape of colloidal particles: application in study of tobacco mosaic virus. *Proc. Soc. Exp. Biol. Med.* **30**, 155–157.

Takamatsu, N., Ohno, T., Meshi, T. and Okada, Y. (1983). Molecular cloning and nucleotide sequence of the 30K and the coat protein cistron of TMV (tomato strain) genome. *Nucleic Acids Res.* **11**, 3767–3778.

Takamatsu, N., Ishikawa, M., Meshi, T. and Okada, Y. (1987). Expression of bacterial chloramphenicol acetyltransferase gene in tobacco plants mediated by TMV-RNA. *EMBO J.* **6**, 307–311.

Takamatsu, N., Watanabe, Y., Meshi, T. and Okada, Y. (1990a). Mutational analysis of the pseudoknot region in the 3 noncoding region of TMV RNA J. *Virology* **64**, 3686–3693.

Takamatsu, N., Watanabe, Y., Yanagi, H., Meshi, T., Shiba, T. and Okada, Y. (1990b). Production of enkephalin in tobacco protoplasts using tobacco mosaic virus RNA vector. *FEBS Lett.* **269**, 73–76.

Takamatsu, N., Watanabe, Y., Iwasaki, T., Shiba, T., Meshi, T. and Okada, Y. (1991). Deletion analysis of the 5′ untranslated leader sequence of tobacco mosaic virus RNA. *J. Virol.* **65**, 1619–1622.

Takanami, Y. (1981). A striking change in symptoms on cucumber mosaic virus-infected tobacco plants induced by a satellite RNA. *Virology* **109**, 120–126.

Takanami, Y., Kubo, S. and Imaizumi, S. (1977). Synthesis of single- and double-stranded cucumber mosaic virus RNAs in tobacco mesophyll protoplasts. *Virology* **80**, 376–389.

Takanami, Y., Nitta, N. and Kubo, S. (1989). A marked improvement of efficiency in infection of tobacco mesophyll protoplasts with plant viruses and virus RNAs by using polyethyleneimine as a polycation. *Ann. Phytopathol. Soc. Jpn* **55**, 324–329.

Takatsuji, H., Hirochika, H., Fukushi, T. and Ikeda, J.-E. (1986). Expression of cauliflower mosaic virus reverse transcriptase in yeast. *Nature (London)* **319**, 240–243.

Takatsuji, H., Yamauchi, H., Watanabe, S., Kato, H. and Ikeda, J.-E. (1992). Cauliflower mosaic virus reverse transcriptase: activation by proteolytic processing and functional alteration by terminal deletion. *J. Biol. Chem.* **267**, 11579–11585.

Takebe, I. (1977). Protoplasts in the study of plant virus replication. *Compr. Virol.* **11**, 237–283.

Takebe, I. and Otsuki, Y. (1969). Infection of tobacco mesophyll protoplasts by tobacco mosaic virus. *Proc. Natl Acad. Sci. USA* **64**, 843–848.

Takebe, I., Otsuki, Y. and Aoki, S. (1968). Isolation of tobacco mesophyll cells in intact and active state. *Plant Cell Physiol.* **9**, 115–124.

Takebe, I., Otsuki, Y., Honda, Y. and Matsui, C. (1975). Penetration of plant viruses into isolated tobacco leaf protoplasts. In: T. Hasegawa (ed.) *Proceedings of the First Intersectional Congress of the International Association of Microbiological Societies*, Vol. 3, pp. 55–64. Science Council of Japan, Tokyo.

Takemoto, Y. and Hibi, T. (2001). Genes Ia, II, III, IV and V of soybean chlorotic mottle virus are essential but the gene Ib is nonessential for systemic infection. *J. Gen. Virol.* **82**, 1481–1489.

Taliansky, M. and Barker, H. (1999). Movement of luteoviruses in infected plants. In: H.G. Smith and H. Barker (eds) *The Luteoviridae*, pp. 69–81. CAB International, Wallingford, UK.

Taliansky, M.E. and Robinson, D.J. (1997). Trans-acting untranslated elements of groundnut rosette virus satellite RNA are involved in symptom production. *J. Gen. Virol.* **78**, 1277–1285.

Taliansky, M.E., Atabekova, T.I. and Atabekov, J.G. (1977). The formation of phenotypically mixed particles upon mixed assembly of some tobacco mosaic virus (TMV) strains. *Virology* **76**, 701–708.

Taliansky, M.E., Atabekova, T.I., Kaplan, I.B., Morozov, S. Yu., Malyshenko, S.I. and Atabekov, J.G. (1982a). A study of TMV *ts* mutant Ni2519: I. Complementation experiments. *Virology* **118**, 301–308.

Taliansky, M.E., Malyshenko, S.I., Pshennikova, E.S., Kaplan, I.B., Ulanova, E.F. and Atabekov, J.G. (1982b). Plant virus-specific transport function: I. Virus genetic control required for systemic spread. *Virology* **122**, 318–326.

Taliansky, M.E., Malyshenko, S.I., Pshennikova, E.S. and Atabekov, J.G. (1982c). Plant virus-specific transport function. II. A factor controlling virus host range. Virology **122**, 327–331.

Taliansky, M.E., Robinson, D.J. and Murant, A.F. (1996). Complete nucleotide sequence and organization of the RNA genome of groundnut rosette umbravirus. *J. Gen. Virol.* **77**, 2335–2345.

Taliansky, M.E., Ryabov, E.V. and Robinson, D.J. (1998). Two distinct mechanisms of transgenic resistance mediated by groundnut rosette virus satellite RNA sequences. *Mol. Plant Microb. Interact.* **11**, 367–374.

Talley, J., Warren, F.H.J.B., Torrance, L. and Jones, R.A.C. (1980). A simple kit for detection of plant viruses by the latex serological test. *Plant Pathol.* **29**, 77–79.

Tamada, T. (1999). Benyviruses. In: A. Granoff and R.G. Webster (eds) *Encyclopedia of Virology*, 2nd edn, pp. 154–160. Academic Press, San Diego.

Tamada, T., Harrison, B.D. and Roberts, I.M. (1984). Variation among British isolates of potato leafroll virus. *Ann. Appl. Biol.* **104**, 107–116.

Tamada, T., Schmitt, C., Saito, M., Guilley, H., Richards, K. and Jonard, G. (1996). High resolution analysis of the readthrough domain of beet necrotic yellow vein virus readthrough protein: a KTER motif is important for efficient transmission of the virus by *Polymyxa betae. J. Gen. Virol.* **77**, 1359–1367.

Tamburro, A.M., Guantieri, V., Piazzolla, P. and Gallitelli, D. (1978). Conformational studies on particles of turnip yellow mosaic virus. *J. Gen. Virol.* **40**, 337–344.

Tamm, T. and Truve, E. (2000). Sobemoviruses. *J. Virol.* **74**, 6231–6241.

Tanguay, R.L. and Gallie, D.R. (1996). Isolation and characterization of the 102-kilodalton RNA-binding protein that binds to the 5′ and 3′ translational enhancers of tobacco mosaic virus RNA. *J. Biol. Chem.* **271**, 14316–14322.

Tanzi, M., Betti, L., de Jager, C.P. and Canova, A. (1986). Isolation of an attenuated virus mutant obtained from a TMV pepper strain after treatment with nitrous acid. *Phytopathol. Mediterr.* **25**, 119–124.

Taraporewala, Z.F. and Culver, J.N. (1996). Identification of an elicitor active site within the three-dimensional structure of the tobacco mosaic tobamovirus coat protein. *Plant Cell* **8**, 169–178.

Taraporewala, Z.F. and Culver, J.N. (1997). Structural and functional conservation of the tobamovirus coat protein elicitor active site. *Mol. Plant Microbe Interact.* **10**, 597–604.

Tatchell, G.M., Plumb, R.T. and Carter, N. (1988). Migration of alate morphs of the bird cherry aphis (*Rhopalosiphum padi*) and implications for the epidemiology of barley yellow dwarf virus. *Ann. Appl. Biol.* **112**, 1–11.

Tavladoraki, P., Benvenuto, E., Trinca, S., De Martinis, D., Cattaneo, A. and Galeffi, P. (1993). Transgenic plants expressing functional single-chain Fv antibody are specifically protected from virus attack. *Nature* **366**, 469–472.

Taylor, C.E. and Murant, A.F. (1968). Chemical control of raspberry ringspot and tomato black ring viruses in strawberry. *Plant Pathol.* **17**, 171–178.

Taylor, C.E. and Thomas, P.R. (1968). The association of *Xiphinema diversicaudatum* (Micoletsky) with strawberry latent ringspot and arabis mosaic viruses in a raspberry plantation. *Ann. Appl. Biol.* **62**, 147–157.

Taylor, D.N. and Carr, J.P. (2000). The GCD10 subunit of yeast eIF-3 binds the methyltransferase-like domain of the 126- and 183-kDa replicase proteins of tobacco mosaic virus in the yeast two-hybrid system. *J. Gen. Virol.* **81**, 1587–1591.

Taylor, G.A., Lewis, G.D. and Rubatzky, V.E. (1969). The influence of time of tobacco mosaic virus inoculation and stage of fruit maturity upon the incidence of tomato internal browning. *Phytopathology* **59**, 732–736.

Taylor, K.M., Spall, V.E., Butler, P.J.G. and Lomonossoff, G.P. (1999). The cleavable carboxyl-terminus of the small coat protein of cowpea mosaic virus is involved in RNA encapsidation. *Virology* **255**, 129–137.

Taylor, L.R. (1986). The distribution of virus disease and the migrant vector aphid. In: G.D. McLean, R.G. Garrett and W.G. Ruesink (eds) *Plant Virus Epidemics*, pp. 35–57. Academic Press, Orlando, FL.

Teakle, D.S. (1986). Abiotic transmission of southern bean mosaic virus in soil. *Aust. J. Biol. Sci.* **39**, 353–359.

Técsi, L.I., Maule, A.J., Smith, A.M. and Leegood, R.C. (1994a). Complex, localized changes in CO_2 assimilation and starch content associated with the susceptible interaction between cucumber mosaic virus and a cucurbit host. *Plant J.* **5**, 837–847.

Técsi, L.I., Maule, A.J., Smith, A.M. and Leegood, R.C. (1994b). Metabolic alterations in cotyledons of *Cucurbita pepo* infected by cucumber mosaic virus. *J. Exp. Bot.* **45**, 1541–1551.

Temmink, J.H.M. and Campbell, R.N. (1969). The ultrastructure of *Olpidium brassicae*. III. Infection of host roots. *Can. J. Bot.* **47**, 421–424.

Teng, P.S. (ed.) (1987). *Crop Loss Assessment and Pest Management*. American Phytopathology Society Press, St Paul, MN.

Tenllado, F. and Bol, J.F. (2000). Genetic dissection of the multiple functions of alfalfa mosaic virus coat protein in viral RNA replication, encapsidation and movement. *Virology* **268**, 29–40.

Tepfer, M. (1993). Viral genes and transgenic plants. *Biotechnology* **11**, 1125–1132.

Terryn, N. and Rouzé, P. (2000). The sense of naturally transcribed antisense RNAs in plants. *Trends Plant Sci.* **5**, 394–396.

Teycheney, P.-Y. and Tepfer, M. (2001). Virus-specific spatial differences in the interface with silencing of the *chs-A* gene in non-transgenic petunia. *J. Gen. Virol.* **82**, 1239–1243.

Teycheney, P.-Y., Aaziz, R., Dinant, S. *et al.* (2000). Synthesis of (–)-strand RNA from the 3′ untranslated region of plant viral genomes expressed in transgenic plants upon infection with related viruses. *J. Gen. Virol.* **81**, 1121–1126.

Thole, V. and Hull, R. (1996). Rice tungro spherical virus: nucleotide sequence of the 3′ genomic half and studies on the two small 3′ open reading frames. *Virus Genes* **13**, 239–246.

Thole, V. and Hull, R. (1998). Rice tungro spherical virus polyprotein processing: identification of a virus-coded protease and mutational analysis of putative cleavage sites. *Virology* **247**, 106–114.

Thole, V., Miglino, R. and Bol, J.F. (1998). Amino acids of alfalfa mosaic virus coat protein that direct formation of unusually long virus particles. *J. Gen. Virol.* **79**, 3139–3143.

Thomas, A.A.M., ter Haar, E., Wellink, J. and Voorma, H.O. (1991). Cowpea mosaic virus middle component RNA contains a sequence that allows internal binding of ribosomes and that requires eukaryotic initiation factor 4F for optimal translation. *J. Virol.* **65**, 2953–2959.

Thomas, C.L. and Maule, A.J. (1995a). Identification of the cauliflower mosaic virus movement protein RNA-binding domain. *Virology* **206**, 1145–1149.

Thomas, C.L. and Maule, A.J. (1995b). Identification of structural domains within cauliflower mosaic virus movement protein by scanning deletion mutagenesis and epitope tagging. *Plant Cell* **7**, 561–572.

Thomas, C.L. and Maule, A.J. (1999). Identification of inhibitory mutants of cauliflower mosaic virus movement protein function after expression in insect cells. *J. Virol.* **73**, 7886–7890.

Thomas, C.L., Perbal, C. and Maule, A.J. (1993). A mutation of cauliflower mosaic virus gene 1 interferes with virus movement but not virus replication. *Virology* **192**, 415–421.

Thomas, C.M., Hull, R., Bryant, J.A. and Maule, A.J. (1985). Isolation of a fraction from cauliflower mosaic virus-infected protoplasts which is active in the synthesis of (+) and (–) strand viral DNA and reverse transcription of primed RNA templates. *Nucl. Acids Res.* **13**, 4557–4576.

Thomas, J.E., Smith, M.K., Kessling, A.F. and Hamill, S.D. (1995). Inconsistent transmission of banana bunchy top virus in micropropagated bananas and its implications for germplasm screening. *Aust. J. Agric. Res.* **46**, 663–671.

Thomas, P.E. (1970). Isolation and differentiation of five strains of curly top virus. *Phytopathology* **60**, 844–848.

Thomas, P.E., Hassan, S. and Mink, G.I. (1988). Influence of light quality on translocation of tomato yellow top virus and potato leafroll virus in *Lycopersicon peruvianum* and some of its tomato hybrids. *Phytopathology* **78**, 1160–1164.

Thomas, P.E., Kaniewski, W.K. and Lawson, E.C. (1997). Reduced field spread of potato leafroll virus in potatoes transformed with potato leafroll virus coat protein gene. *Plant Dis.* **8**, 1447–1453.

Thomas, W. (1973). Control of *Olpidium brassicae*, the vector of cucumber systemic necrosis and bean stipple streak virus diseases. *N. Z. J. Exp. Agric.* **1**, 92–96.

Thompson, A.D. (1961). Effect of tobacco mosaic virus and potato virus Y on infection by potato virus X. *Virology* **13**, 262–264.

Thompson, D.L. and Hebert, T.T. (1970). Development of maize dwarf mosaic symptoms in eight phytotron environments. *Phytopathology* **60**, 1761–1764.

Thompson, G.A. and Schulz, A. (1999). Macromolecular trafficking in the phloem. *Trends Plant Sci.* **4**, 354–360.

Thompson, J.R. and García-Arenal, F. (1998). The bundle sheath-phloem interface of *Cucumis sativus* is a boundary to systemic infection by tomato aspermy virus. *Mol. Plant–Microb. Interact.* **11**, 109–114.

Thompson, S., Fraser, R.S.S. and Barnden, K.L. (1988). A beneficial effect of trypsin on the purification of turnip mosaic virus (TuMV) and other potyviruses. *J. Virol. Meth.* **20**, 57–64.

Thompson, W.R., Meinwald, J., Aneshansley, D. and Eisner, T. (1972). Flavonols: pigments responsible for ultraviolet absorption in nectar guide of flower. *Science* **177**, 528–530.

Thompson, W.W., Weier, T.E. and Drever, H. (1964). Electron-microscopic studies on chloroplasts from phosphorus deficient plants. *Am. J. Bot.* **51**, 933–938.

Thomson, A.D. (1961). Effect of tobacco mosaic virus and potato virus Y on infection by potato virus X. *Virology* **13**, 262–264.

Thomson, A.D. (1976). Virus diseases of *Solanum laciniatum* Ait. in New Zealand. *N. Z. J. Agric. Res.* **19**, 521–527.

Thomson, A.D. and Ferguson, J.D. (1976). Effect of varying the nutrient supply on response of pea plants to pea leaf roll. *N. Z. J. Agric. Res.* **19**, 529–533.

Thongmeearkom, P., Ford, R.E. and Jedlinski, H. (1976). Aphid transmission of maize dwarf mosaic virus strains. *Phytopathology* **66**, 332–335.

Thornbury, D.W. and Pirone, T.P. (1983). Helper components of two potyviruses are serologically distinct. *Virology* **125**, 487–490.

Thornbury, D.W., Hellmann, G.M., Rhoads, R.E. and Pirone, T.P. (1985). Purification and characterisation of potyvirus helper component. *Virology* **144**, 260–267.

Thottappilly, G., Dahal, G. and Lockhart, B.E.L. (1998). Studies on a Nigerian isolate of banana streak badnavirus. 1. Purification and enzyme-linked immunosorbent assay. *Ann. Appl. Biol.* **132**, 253–261.

Thresh, J.M. (1958). The spread of virus diseases in Cacao. *West Afr. Cacao Res. Inst. Tech. Bull.* **5**, 1–36.

Thresh, J.M. (1966). Field experiments on the spread of blackcurrant reversion virus and its gall mite vector (*Phytoptus ribis* Nal). *Ann. Appl. Biol.* **58**, 219–230.

Thresh, J.M. (1980). The origins and epidemiology of some important plant viruses diseases. In: T.H. Coaker (ed.) *Applied Virology*, Vol. VII, pp. 1–65. Academic Press, London.

Thresh, J.M. (1982). Cropping practices and virus spread. *Annu. Rev. Phytopathol.* **20**, 193–218.

Thresh, J.M. (1983). Progress curves of plant virus disease. *Adv. Appl. Biol.* **8**, 1–85.

Thresh, J.M. (1988a). Rice viruses and 'the green revolution'. *Aspects Appl. Biol.* **17**, 187–194.

Thresh, J.M. (1988b). Eradication as a virus disease control measure. In: B.C. Clifford and E. Lester (eds) *Control of Plant Diseases: Costs and Benefits*, pp. 155–194. Blackwell, Oxford.

Thresh, J.M. (1989a). Insect-borne viruses of rice and the green revolution. *Trop. Pest Manag.* **35**, 264–272.

Thresh, J.M. (1989b). Plant virus epidemiology: the battle of the genes. *NATO Adv. Res. Workshop: Recognition Response Plant Virus Interact., 1989*.

Thresh, J.M. and Otim-Nape, G.W. (1994). Strategies for controlling African cassava mosaic geminivirus. *Adv. Vector Dis. Res.* **10**, 215–236.

Tian, T., Rubio, L., Yeh, H.-H., Crawford, B. and Falk, B.W. (1999). Lettuce infectious yellows virus: *in vitro* acquisition analysis using partially purified virions and the whitefly *Bemisia tabaci*. *J. Gen. Virol.* **80**, 1111–1117.

Tien, P., and Wu, G. (1991). Satellite RNA for the biocontrol of plant disease. *Adv. Virus Res.* **39**, 321–339.

Tien, P., Zhang, X., Qiu, B., Qin, B. and Wu, G. (1987). Satellite RNA for the control of plant diseases caused by cucumber mosaic virus. *Ann. Appl. Biol.* **111**, 143–152.

Timchenko, T., de Kouchkovsky, F., Katul, L., David, C., Vetten, H.J. and Gronenborn, B. (1999). A single Rep protein initiates replication of multiple genome components of faba bean necrotic yellows virus, a single-stranded DNA virus of plants. *J. Virol.* **73**, 10173–10182.

Timchenko, T., Katul, L., Sano, Y., de Kouchkovsky, F., Vetten, H.J. and Gronenborn, B. (2000). The master Rep concept in nanovirus replication: identification of missing genome components and potential for natural genetic reassortment. *Virology* **274**, 189–195.

Timian, R.G. (1974). The range of symbiosis of barley and barley stripe mosaic virus. *Phytopathology* **64**, 342–345.

Timian, R.G. (1975). Barley stripe mosaic virus and the world collection of barleys. *Plant Dis. Rep.* **59**, 984-988.

Timmermans, M.C.P., Das, O.P. and Messing, J. (1992). Trans replication and high copy numbers of wheat dwarf virus vectors in maize cells. *Nucl. Acids Res.* **40**, 4047–4054.

Timmermans, M.C.P., Das, O.P. and Messing, J. (1994). Geminiviruses and their uses as extrachromosomal replicons. *Annu. Rev. Plant Physiol. Mol. Biol.* **45**, 79–112.

Tingey, W.M. and Laubengayer, J.E. (1981). Defense against the green peach aphid and potato leafhopper by glandular trichomes of *Solanum berthaultii*. *J. Econ. Entomol.* **74**, 721–725.

Tinsley, T.W. (1953). The effects of varying the water supply of plants on their susceptibility to infection with viruses. *Ann. Appl. Biol.* **40**, 750–760.

Tobiás, I., Rast, A.Th.B. and Maat, D.Z. (1982). Tobamoviruses of pepper, eggplant and tobacco: comparative host reactions and serological relationships. *Neth. J. Plant Pathol.* **88**, 257–268.

Todd, J.M. (1958). Spread of potato virus X over a distance. *Proc. Conf. Potato Virus Dis., 3rd, 1957*, pp. 132–143.

Todd, J.M. (1961). The incidence and control of aphid-borne potato virus diseases in Scotland. *Eur. Potato J.* **4**, 316–329.

Toedt, J.M., Braswell, E.H., Schuster, T.M., Yphantis, D.A., Taraporewala, Z.F. and Culver, J.N. (1999). Biophysical characterization of a designed TMV coat protein mutant, R46G, that elicits a moderate hypersensitive response in *Nicotiana sylvestris*. *Protein Sci.* **8**, 261–270.

Toh, H., Hayashida, H. and Miyata, T. (1983). Sequence homology between retroviral reverse transcriptase and putative polymerases of hepatitis B virus and cauliflower mosaic virus. *Nature (London)* **305**, 827–829.

Tollin, P. and Wilson, H.R. (1971). Some observations on the structure of the Campinas strain of tobacco rattle virus. *J. Gen. Virol.* **13**, 433–440.

Tollin, P. and Wilson, H.R. (1988). Particle structure. In: R.G. Milne (ed.) *The Plant Viruses. Vol. 4: The Filamentous Plant Viruses*, pp. 51–83. Plenum Press, New York.

Tollin, P., Wilson, H.R., Roberts, I.M. and Murant, A.F. (1992). Diffraction studies of the particles of two closteroviruses: heracleum latent virus and heracleum virus 6. *J. Gen. Virol.* **73**, 3045–3048.

Tomassoli, L., Ilardi, V., Barba, M. and Kaniewski, W. (1999). Resistance of transgenic tomato to cucumber mosaic cucumovirus under field conditions. *Molec. Breed.* **5**, 121–130.

Tomenius, K., Clapham, D. and Meshi, T. (1987). Localization by immunogold cytochemistry of the virus-coded 30K protein in plasmodesmata of leaves infected with tobacco mosaic virus. *Virology* **160**, 363–371.

Tomita, K. and Rich, A. (1964). X-ray diffraction investigations of complementary RNA. *Nature (London)* **201**, 1160–1164.

Tomlinson, J.A. (1962). Control of lettuce mosaic by the use of healthy seed. *Plant Pathol.* **11**, 61–64.

Tomlinson, J.A. (1981). Chemotherapy of plant viruses and virus diseases. In: K.F. Harris and K. Maramorosch (eds) *Pathogens, Vectors and Plant Diseases: Approaches to Control*, pp. 23–44. Academic Press, New York.

Tomlinson, J.A. and Faithfull, E.M. (1979). Effects of fungicides and surfactants on the zoospores of *Olpidium brassicae*. *Ann. Appl. Biol.* **93**, 13–19.

Tomlinson, J.A. and Faithfull, E.M. (1984). Studies on the occurrence of tomato bushy stunt virus in English rivers. *Ann. Appl. Biol.* **104**, 485–495.

Tomlinson, J.A. and Walker, V.M. (1973). Further studies on seed transmission in the ecology of some aphid-transmitted viruses. *Ann. Appl. Biol.* **73**, 293–298.

Tomlinson, J.A. and Ward, C. M. (1982). Selection for immunity in swede (*Briassica napus*) to infection by turnip mosaic virus. *Ann. Appl. Biol.* **101**, 43–50.

Tomlinson, J.A., Carter, A.L., Dale, W.T. and Simpson, C.J. (1970). Weed plants as sources of cucumber mosaic virus. *Ann. Appl. Biol.* **66**, 11–16.

Tomlinson, J.A., Faithfull, E.M. and Ward, C.M. (1976). Chemical suppression of the symptoms of two virus diseases. *Ann. Appl. Biol.* **84**, 31–41.

Tomlinson, J.A., Faithfull, E., Flemett, T.H. and Beards, G. (1982). Isolation of infective tomato barley stunt virus after passage through the human alimentary tract. *Nature (Lond.)* **300**, 637–638.

Tommiska, T.J., Hamalainen, J.H., Watanabe, K.N. and Valkonen, J.P.T. (1998). Mapping of the gene *Nx(phu)* that controls hypersensitivity resistance to potato virus X in *Solanum phureja* lvP35. *Theor. Appl. Genet.* **96**, 840–843.

Torbet, J. (1983). Internal structural anisotropy of spherical viruses studied with magnetic birefringence. *EMBO J.* **2**, 63–66.

Torbet, J., Timmins, P.A. and Lvov, Y. (1986). Packaging of DNA in cauliflower mosaic virus and bacteriophage Sd studied with magnetic birefringence. *Virology* **155**, 721–725.

Toriyama, S. (1986a). An RNA-dependent RNA polymerase associated with the filamentous nucleoproteins of rice stripe virus. *J. Gen. Virol.* **67**, 1247–1255.

Toriyama, S. (1986b). Rice stripe virus: prototype of a new group of viruses that replicate in plants and insects. *Microbiol. Sci.* **3**, 347–351.

Toriyama, S. and Peters, D. (1981). Differentiation between broccoli necrotic yellows virus and lettuce necrotic yellows virus by their transcriptase activities. *J. Gen. Virol.* **56**, 59–66.

Torrance, L. (1980). Use of protein A to improve sensitisation of latex particles with antibodies to plant viruses. *Ann. Appl. Biol.* **96**, 45–50.

Torrance, L. (1987). Use of enzyme amplification in an ELISA to increase sensitivity of detection of barley yellow dwarf virus in oats and in individual vector aphids. *J. Virol. Methods* **15**, 131–138.

Torrance, L. (1999). Pomoviruses. In: A. Granoff and R.G. Webster (eds) *Encyclopedia of Virology*, 2nd edn, pp. 1361–1364. Academic Press, San Diego.

Torrance, L. (2000). Genus *Furovirus*. In: van Regenmortel *et al.* (eds) *Seventh Report of the International Committee on the Taxonomy of Viruses*, pp. 904–908. Academic Press, San Diego.

Torrance, L. and Dolby, C.A. (1984). Sampling conditions for reliable routine detection by enzymelinked immunosorbent assay of three ilarviruses in fruit trees. *Ann. Appl. Biol.* **104**, 267–276.

Torrance, L. and Jones, R.A.C. (1982). Increased sensitivity of detection of plant viruses obtained by using a fluorogenic substrate in enzyme-linked immunosorbent assay. *Ann. Appl. Biol.* **101**, 501–509.

Torrance, L., Larkins, A.P. and Butcher, G.W. (1986a). Characterisation of monoclonal antibodies against potato virus X and comparison of serotypes with resistance groups. *J. Gen. Virol.* **67**, 57–67.

Torrance, L., Pead, M.T., Larkins, A.P. and Butcher, G.W. (1986b). Characterisation of monoclonal antibodies to a UK isolate of barley yellow dwarf virus. *J. Gen. Virol.* **67**, 549–556.

Torres-Pacheco, I., Garzón-Tiznado, J.A., Herrera-Estrella, L. and Rivera-Bustamante, R.F. (1993). Complete nucleotide sequence of pepper huasteco virus: analysis and comparison with bipartite geminiviruses. *J. Gen. Virol.* **74**, 2225–2231.

Torrigiani, P., Rabiti, A.L., Betti, L. *et al.* (1995). Improved method for polyamine determination in TMV, a rod-shaped virus. *J. Virol. Meth.* **53**, 157–163.

Toruella, M., Gordon, K. and Hohn, T. (1989). Cauliflower mosaic virus produces an aspartic proteinase to cleave its polyproteins. *EMBO J.* **8**, 2819–2825.

Toryama, S., Kimishima, T., Takahashi, M., Shumizu, T., Minaki, N. and Akutsu, K. (1998). The complete nucleotide sequence of rice grassy stunt virus genome and genomic comparisons with viruses of the genus Tenuivirus. *J. Gen. Virol.* **79**, 2051–2058.

Toth, R.L., Harper, K., Mayo, M.A. and Torrance, L. (1999). Fusion proteins of single-chain variable fragments derived from phage display libraries are effective reagents for routine diagnosis of potato leafroll virus infection in potato. *Phytopathology* **89**, 1015–1021.

Toussaint, A., Dekegel, D. and Vanheule, G. (1984). Distribution of *Odontoglossum* ringspot virus in apical meristems of infected *Cymbidium* cultivars. *Physiol. Plant Pathol.* **25**, 297–305.

Toussaint, A., Kummert, J., Maroquin, C., Lebrun, A. and Roggemans, J. (1993). Use of virazole to eradicate odontoglossu, ringspot virus from *in vitro* cultures of Cymbidium Sw. *Plant Cell Tiss. Org. Cult.* **32**, 303–309.

Toyoda, H., Oishi, Y., Matsuda, Y., Chatani, K. and Hirai, T. (1985). Resistance mechanism of cultured plant cells to tobacco mosaic virus. IV. Changes in tobacco mosaic virus concentrations in somaclonal tobacco callus tissues and production of virus-free plantlets. *Phytopathol. Z.* **114**, 126–133.

Traynor, P. and Ahlquist, P. (1990). Use of bromovirus RNA2 hybrids to map *cis*- and *trans*-acting functions in a conserved RNA replication gene. *J. Virol.* **64**, 69–77.

Traynor, P., Young, B.M. and Ahlquist, P. (1991). Deletion analysis of brome mosaic virus 2a protein: effects on RNA replication and systemic spread. *J. Virol.* **65**, 2807–2815.

Tremaine, J.H., Ronald, W.P. and Valcic, A. (1976). Aggregation properties of carnation ringspot virus. *Phytopathology* **66**, 34–39.

Trevathan, L.E., Tolin, S.A., Moore, L.D. and Orcutt, D.M. (1982). Total lipid, free sterol, free fatty acid, and triacylglycerol fatty acid content of tobacco mosaic virus-infected tobacco. *Can. J. Plant Sci.* **62**, 771–776.

Trudgill, D.L., Brown, D.J.F and McNamara, D.G. (1983). Methods and criteria for assessing the transmission of plant viruses by longidorid nematodes. *Rev. Nematol.* **6**, 133–141.

Truve, E., Aaspållu, A., Honkanen, J. *et al.* (1993). Transgenic potato plants expressing mammalian 2′-5′ oligoadenylate synthetase are protected from potato virus X infection under field conditions. *Biotechnology* **11**, 1048–1052.

Tsagris, M., Tabler, M. and Sänger, H.L. (1987a). Oligomeric potato spindle tuber viroid (PSTV) RNA does not process autocatalytically under conditions where other RNAs do. *Virology* **157**, 227–231.

Tsagris, M., Tabler, M., Mühlbach, H.-P. and Sänger, H.L. (1987b). Linear oligomeric potato spindle tuber viroid (PSTV) RNAs are accurately processed *in vitro* to the monomeric circular viroid proper when incubated with a nuclear extract from healthy potato cells. *EMBO J.* **6**, 2173–2183.

Tsai, M.-S., Hsu, Y.-H. and Lin, N.-S. (1999). Bamboo mosaic potexvirus satellite RNA (satBaMV RNA)-encoded P20 protein preferentially binds to satBaMV RNA. *J. Virol.* **73**, 3032–3039.

Tsugita, A. and Fraenkel-Conrat, H. (1960). The amino acid composition and C-terminal sequence of a chemically evoked mutant of tobacco mosaic virus. *Proc. Natl Acad. Sci. USA* **46**, 636–642.

Tsugita, A., Gish, D.T., Young, J., Fraenkel-Conrat, H., Knight, C.A. and Stanley, W.M. (1960). The complete amino acid sequence of the protein of tobacco mosaic virus. *Proc. Natl Acad. Sci. USA* **46**, 1463–1469.

Tu, J.C. (1973). Electron microscopy of soybean root nodules infected with soybean mosaic virus. *Phytopathology* **63**, 1011–1017.

Tu, J.C. (1977). Effects of soybean mosaic virus infection on ultrastucture of bacteroidal cells in soybean root nodules. *Phytopathology* **67**, 199–205.

Tu, J.C. (1997). Effect on necrotic and non-necrotic strains of bean common mosaic and bean yellow mosaic viruses on nodulation and root tip necrosis of bean (*Phaseolus vulgaris*). *Can. J. Phytopathol.* **19**, 156–160.

Tu, J.C. and Buzzell, R.I. (1987). Stem-tip necrosis: a hypersensitive, temperature dependent dominant gene reaction of soybean to infection by soybean mosaic virus. *Can. J. Plant Sci.* **67**, 661–665.

Tu, J.C., Ford, R.E. and Quiniones, S.S. (1970). Effects of soybean mosaic virus and/or bean pod mottle virus infection on soybean modulation. *Phytopathology* **60**, 518–523.

Tumer, N.E., Hwang, D.-J. and Bonness, M. (1997). C-terminal deletion mutant of pokeweed antiviral protein inhibits viral infection but does not depurinate host ribosomes. *Proc. Natl Acad. Sci. USA* **94**, 3866–3871.

Tumer, N.E., Hudak, K., Di, R., Coetzer, C., Wang, P. and Zoubenko, O. (1999). Pokeweed antiviral protein and its applications. *Curr. Top. Microbiol.* **240**, 139–158.

Turgeon, R., Beebe, D.U. and Gowan, E. (1993). The intermediary cell: minor-vein anatomy and raffinose oligosaccharide synthesis in the *Scrophulariaceae*. *Planta* **191**, 446–456.

Turina, M., Maruoka, M., Monis, J., Jackson, A.O. and Scholthof, K.B.G. (1998). Nucleotide sequence and infectivity of full-length cDNA clone of Panicum mosaic virus. *Virology* **241**, 141–155.

Turnbull-Ross, A.D., Mayo, M.A., Reavy, B. and Murant, A.F. (1993). Sequence analysis of the parsnip yellow fleck polyprotein: evidence for affinities with picornaviruses. *J. Gen. Virol.* **74**, 555–561.

Turner, D.R. and Butler, P.J.G. (1986). Essential features of the assembly origin of tobacco mosaic virus RNA as studied by directed mutagenesis. *Nucl. Acids Res.* **14**, 9229–9242.

Turner, D.R., Mondragon, A., Fairall, L., Bloomer, A.C., Finch, J.T., van Boom, J.H. and Butler, P.J.G. (1986). Oligonucleotide binding to the coat protein disk of tobacco mosaic virus: possible steps in the assembly mechanism. *Eur. J. Biochem.* **157**, 269–274.

Turner, D.R., Joyce, L.E. and Butler, P.J.G. (1988). The tobacco mosaic virus assembly origin RNA: functional characteristics defined by directed mutagenesis. *J. Mol. Biol.* **203**, 531–547.

Turner, D.R., McGuigan, C.J. and Butler, P.J.G. (1989). Assembly of hybrid RNAs with tobacco mosaic virus coat protein: evidence for incorporation of disks in 5′ elongation along the major RNA tail. *J. Mol. Biol.* **209**, 407–422.

Turner, D.S. and Covey, S.N. (1984). A putative primer for the replication of cauliflower mosaic virus by reverse transcription is virion-associated. *FEBS Lett.* **165**, 285–289.

Turner, D.S. and Covey, S.N. (1988). Discontinuous hairpin DNAs synthesized *in vivo* following specific and non-specific priming of cauliflower mosaic virus DNA (+) strands. *Virus Res.* **9**, 49–62.

Turner, N.E., O'Connell, K.M., Nelson, R.S. *et al.* (1987a). Expression of alfalfa mosaic virus coat protein gene confers cross-protection in transgenic tobacco and tomato plants. *EMBO J.* **6**, 1181–1188.

Turner, Ph.C., Watkins, P.A.C., Zaitlin, M. and Wilson, T.M.A. (1987b). Tobacco mosaic virus particles uncoat and express their RNA in *Xenopus laevis* oocytes: implications for early interactions between plant cells and viruses. *Virology* **160**, 515–517.

Turpen, T.H., Turpen, A.M., Weinzettl, N., Kumagai, M.H. and Dawson, W.O. (1993). Transfection of whole plants from would inoculated with *Agrobacterium tumefaciens* containing cDNA of tobacco mosaic virus. *J. Virol. Methods* **42**, 227–240.

Turpen, T.H., Reini, S.J., Charoenvit, Y., Hoffman, S.L., Fallarme, V. and Grill, L.J. (1995). Malarial epitopes expressed on the surface of recombinant tobacco mosaic virus. *Bio/Technology* **13**, 53–57.

Twomey, T.A. and Krawetz, S.A. (1990). Parameters affecting hybridization of nucleic acids blotted onto nylon or nitrocellulose membranes. *BioTechniques* **8**, 478–482.

Tzafrir, I., Torbert, K.A., Lockhart, B.E.M., Somers, D.A. and Olszewski, N.E. (1998). The sugarcane bacilliform badnavirus promoter is active in both monocots and dicots. *Plant Mol. Biol.* **38**, 347–356.

Uchino, H., Kanzawa, K. and Tamada, T. (1993). Effect of fluazinam on infections of sugar beets by *Polymyxa betae,* vector of beet necrotic yellow vein virus. In: C. Hiruki (ed.) *Proceedings of the Second Symposium of the International Working Group on Plant Viruses with Fungal Vectors*, pp. 153–156. American Society of Sugar Beet Technologists, Denver, CO.

Uegaki, R., Kobe, S. and Fujimori, T. (1988). Stress compounds in the leaves of *Nicotiana undulata* induced by TMV inoculation. *Phytochemistry* **27**, 365–368.

Ugaki, M., Ueda, T., Timmermans, M.C.P., Vieira, J., Elliston, K.O. and Messing, J. (1991). Replication of a geminivirus derived shuttle vector in maize endosperm cells. *Nucl. Acids Res.* **19**, 371–377.

Uhde, K., Kerschbaumer, R.J., Koenig, R. *et al.* (2000). Improved detection of beet necrotic yellow vein virus in DAS ELISA by means of antibody single chain fragments (scFv) which were selected to protease-stable epitopes from phage display libraries. *Arch. Virol.* **145**, 179–185.

Ukai, Y. and Yamashita, A. (1984). Induced mutation for resistance to barley yellow mosaic virus. *Jpn Agric. Res. Quart.* **17**, 255–259.

Ullman, D.E., Qualset, C.O. and McLean, D.L. (1988). Feeding responses of *Rhopalosiphum padi* (Homoptera: Aphidae) to barley yellow dwarf virus resistant and susceptible barley varieties. *Environ. Entomol.* **17**, 988–991.

Ullman, D.E., Westcot, D.M., Hunter, W.B. and Mau, R.F.L. (1989). Internal anatomy and morphology of *Frankliniella iccidentalis* (Pergande) (Thysanoptera: Thripidae) with special reference to interactions between thrips and tomato spotted wilt virus. *Int. J. Insect Morphol. Embryol.* **18**, 289–310.

Ullman, D.E., Sherwood, J.L. and German, T.L (1997). Thrips as vectors of plant pathogens. In: T. Lewis (ed.) *Thrips as Crop Pests*, pp. 539–565. CAB International, Wallingford, UK.

Umaharan, P., Padidam, M., Phelps, R.H., Beachy, R.N. and Fauquet, C.M. (1998). Distribution and diversity of geminiviruses in Trinidad and Tobago. *Phytopathology* **88**, 1262–1268.

Unge, T., Montelius, I., Liljås, L. and Ofverstedt, L.-G. (1986). The EDTA-treated expanded satellite tobacco necrosis virus: biochemical properties and crystallisation. *Virology* **152**, 207–218.

Unseld, S., Ringel, M., Konrad, A., Lauster, S. and Frischmuth, T. (2000). Virus-specific adaptations for the production of a pseudorecombinants virus formed by two distinct bipartite geminiviruses from Central America. *Virology* **274**, 179–188.

Uppal, B.N. (1934). The movement of tobacco mosaic virus in leaves of *Nicotiana sylvestris*. *Indian J. Agric. Sci.* **4**, 865–873.

Urban, C. and Beier, H. (1995). Cysteine tRNAs of plant origin as novel UGA suppressors. *Nucl. Acids Res.* **23**, 4591–4597.

Urban, L.A., Ramsdell, D.C., Klomparens, K.L., Lynch, T. and Hancock, J.F. (1989). Detection of blueberry shoestring virus in xylem and phloem tissues of highbush blueberry. *Phytopathology* **79**, 488–493.

Urban, L.A., Sherwood, J.L., Rezende, J.A.M. and Melcher, U. (1990). Examination of mechanisms of cross protection in nontransgenic plants. In: R.S.S. Fraser (ed.) *Recognition and Response in Plant–Virus Interactions*, pp. 415–426. Springer, Berlin.

Urcuqui-Inchima, S., Maia, I.G., Arrunda, P., Haenni, A.-L. and Bernardi, F. (2000). Deletion mapping of the potyviral helper component-proteinase reveals two regions involved in RNA binding. *Virology* **268**, 104–111.

Ushiyama, R. and Matthews, R.E.F. (1970). The significance of chloroplast abnormalities associated with infection by turnip yellow mosaic virus. *Virology* **42**, 293–303.

Uyeda, I., Lee, S.Y., Yoshimoto, H. and Shikata, E. (1987). RNA polymerase activity of rice ragged stunt and rice black-streaked dwarf viruses. *Ann. Phytopathol. Soc. Jpn* **53**, 60–62.

Uyeda, I., Kudo, H., Takahashi, T., Sano, T., Oshima, K., Matsumura, T. and Shikata, E. (1989). Nucleotide sequence of rice dwarf virus genome segment 9. *J. Gen. Virol.* **70**, 1297–1300.

Uyeda, I., Kimura, I. and Shikata, E. (1995). Characterization of genome structure and establishment of vector cell lines for plant reoviruses. *Adv. Virus Res.* **45**, 249–279.

Uyemoto, J.K. and Gilmer, R.M. (1972). Properties of tobacco necrosis virus strains isolated from apple. *Phytopathology* **62**, 478–481.

Uyemoto, J.K., Grogan, R.G. and Wakeman, J.R. (1968). Selective activation of satellite virus strains by strains of tobacco necrosis virus. *Virology* **34**, 410–418.

Vaden, V.R. and Melcher, U. (1990). Recombination sites in cauliflower mosaic virus DNAs: implication for mechanisms of recombination. *Virology* **177**, 717–726.

Vaeck, M., Reynaerts, A., Höfte, H. *et al.* (1987). Transgenic plants protected from insect attack. *Nature (Lond.)* **328**, 33–37.

Vaewhongs, A.A. and Lommel, S.A. (1995). Virion formation is required for the long-distance movement of red clover necrotic mosaic virus in movement protein transgenic plants. *Virology* **212**, 607-613.

Valdez, R.B. (1972). A micro-container technique for studying virus transmission by nematodes. *Plant Pathol.* **21**, 114–117.

Valle, R.P.C., Drugeon, G., Devignes-Morch, M.D., Legocki, A.B. and Haenni, A.-L. (1992). Codon context effects in virus translational readthrough: a study *in vitro* of the determinants of TMV and MoMuLV amber suppression. *FEBS Lett.* **306**, 133–139.

Valleau, W.D. (1946). Breeding tobacco varieties resistant to mosaic. *Phytopathology* **36**, 412.

Valleau, W.D. (1952). Breeding tobacco for disease resistance. *Econ. Bot.* **6**, 69–102.

Valverde, R.A. and Dodds, J.A. (1986). Evidence for a satellite RNA associated naturally with the U5 strain and experimentally with the U1 strain of tobacco mosaic virus. *J. Gen. Virol.* **67**, 1875–1884.

Valverde, R.A. and Dodds, J.A. (1987). Some properties of isometric virus particles which contain the satellite RNA of tobacco mosaic virus. *J. Gen. Virol.* **68**, 965–972.

Valverde, R.A., Dodds, J.A. and Heick, J.A. (1986). Double-stranded fibonucleic acid from plants infected with viruses having elongated particles and undivided genomes. *Phytopathology* **76**, 459–465.

van Beek, N.A.M., Lohuis, D., Dijkstra, J. and Peters, D. (1985). Morphogenesis of Festuca leaf streak virus in cowpea protoplasts. *J. Gen. Virol.* **66**, 2485–2489.

van Bel, A.J.E and Kempers, R. (1997). The pore/plasmadesma unit: key element in the interplay between sieve element and companion cell. *Prog. Bot.* **58**, 278–291.

van Belkum, A., Abrahams, J.P., Pleij, C.W.A. and Bosch, L. (1985). Five pseudoknots at the 204 nucleotide long 3' noncoding region of tobacco mosaic virus RNA. *Nucl. Acids Res.* **13**, 7673–7686.

van Belkum, A., Cornelissen, B., Linthorst, H., Bol, J., Pleij, C. and Bosch, L. (1987). tRNA-like properties of tobacco rattle virus RNA. *Nucleic Acids Res.* **15**, 2837–2850.

van Beynum, G.M.A., de Graaf, J.M., Castel, A., Kraal, B. and Bosch, L. (1977). Structural studies on the coat protein of alfalfa mosaic virus. *Eur. J. Biochem.* **72**, 63–78.

van Blokland, R., van der Geest, N., Mol, J.N.M. and Kooter, J.M. (1994). Transgene-mediated suppression of chalcone synthase expression in *Petunia hybrida* results from an increase in RNA turnover. *Plant J.* **6**, 861–877.

van Bokhoven, H., van Lent, J.W.M., Custers, R., Vlak, J.M., Wellink, J. and van Kammen, A. (1992). Synthesis of the complete 200K polyprotein encoded by cowpea mosaic virus B-RNA in insect cells. *J. Gen. Virol.* **73**, 2775–2784.

van Bokhoven, H., le Gall, O., Kasteel, D., Verver, J., Wellink, J. and van Kammen, A. (1993a). *Cis-* and *trans-*acting elements in cowpea mosaic virus RNA replication. *Virology* **195**, 377–386.

van Bokhoven, H., Verver, J., Wellink, J. and van Kammen, A. (1993b). Protoplasts transiently expressing the 200K coding sequence of cowpea mosaic virus B-RNA support replication of M-RNA. *J. Gen. Virol.* **74**, 2233–2241.

van Boxtel, J., Singh, B.B., Thottappilly, G. and Maule, A.J. (2000). Resistance of cowpea (*Vigna unguiculata* (L.) Walp.) breeding lines to blackeye cowpea mosaic and cowpea aphid-borne potyvirus isolates under experimental conditions. *J. Plant Dis. Protect.* **107**, 197–204.

van de Wetering, F., Goldbach, R. and Peters, D. (1996). Tomato spotted wilt tospovirus ingestion by first instar larvae of *Frankliniella occidentalis* is a prerequisite for transmission. *Phytopathology* **86**, 900–905.

van de Wetering, F., van der Hoek, M., Goldbach, R. and Peters, D. (1999). Differences in tomato spotted wilt virus vector competency between males and females of *Frankliniella occidentalis*. *Ent. Exp. Appl.* **93**, 105–112.

van den Heuvel, J.F.J.M. (1999). Fate of a luteovirus in the haemolymph of an aphid. In: H.G. Smith and H. Barker (eds) *The Luteoviridae*, pp. 112–119. CAB International, Wallingford, UK.

van den Heuvel, J.F.J.M., Verbeek, M. and van der Wilk, F. (1994). Endosymbiotic bacteria associated with circulative transmission of potato leafroll virus by *Myzus persicae*. *J. Gen. Virol.* **75**, 2559–2565.

van den Heuvel, J.F.J.M., Bruyere, A., Hogenhaut, S.A. *et al.* (1997). The N-terminal region of the luteovirus readthrough domain determines virus binding to *Buchneria* GroEL and is essential for virus persistence in the aphid. *J. Virol.* **71**, 7258–7265.

van den Heuvel, J.F.J.M., Hogenhout, S.A. and van der Wilk, F. (1999). Recognition and receptors in virus transmission by arthropods. *Trends Microbiol.* **7**, 71–76.

van der Krol, A.R., Lenting, P.E., Veenstra, J. *et al.* (1988). An antisense chalcone synthase gene in transgenic plants inhibits flower pigmentation. *Nature (Lond.)* **333**, 866–869.

van der Krol, A.R., Mur, L.A., Beld, M., Mol, J.N.M. and Stuitje, A.R. (1990). Flavenoid genes in petunia: addition of a limited number of gene copies may lead to a suppression of gene expression. *Plant Cell* **2**, 291–299.

van der Kuyl, A.C., Langereis, K., Houwing, C.J., Jaspars, E.M.J. and Bol, J.F. (1990). *Cis-*acting elements involved in replication of alfalfa mosaic virus RNAs *in vitro*. *Virology* **176**, 346–354.

van der Lubbe, J.L.M., Hatta, T. and Francki, R.I.B. (1979). Structure of the antigen from Fiji disease virus particles eliciting antibodies specific to double-stranded polyribonucleotides. *Virology* **95**, 405–414.

van der Meer, J., Dorssers, L., van Kammen, A. and Zabel, P. (1984). The RNA-dependent RNA polymerase of cowpea is not involved in cowpea mosaic virus RNA replication: Immunological evidence. *Virology* **132**, 413–425.

van der Plank, J.E. (1946). A method for estimating the number of random groups of adjacent diseased plants in a homogeneous field. *Trans. R. Soc. S. Afr.* **31**, 269–278.

van der Plank, J.E. (1948). The relation between the size of fields and the spread of plant disease into them. Part I. Crowd diseases. *Emp. J. Exp. Agric.* **16**, 134–142.

van der Vlugt, R., Allefs, S., de Haan, P. and Goldbach, R. (1989). Nucleotide sequence of the 3'-terminal region of potato virus YN RNA. *J. Gen. Virol.* **70**, 229–233.

van der Vlugt, R.A.A., Ruiter, R.K. and Goldbach, R. (1992). Evidence for sense RNA-mediated resistance to PVT^N in tobacco plants transformed with the viral coat protein cistron. *Plant Mol. Biol.* **20**, 631–639.

van der Vossen, E.A.G., Notenboom, T. and Bol, J.F. (1995). Characterization of sequences controlling the synthesis of alfalfa mosaic virus subgenomic RNA *in vivo*. *Virology* **212**, 663–672.

van der Want, J.P.H., Boerjan, M.L. and Peters, D. (1975). Variability of some plant species from different origins and their suitability for virus work. *Neth.J. Plant Pathol.* **81**, 205–216.

van der Wel, N.N. (2000). Interaction between the alfalfa mosaic virus movement protein and plasmodesmata. PhD thesis, University of Wageningen, Netherlands.

van der Wel, N.N., Goldbach, R.W. and van Lent, J.W.M. (1998). The movement protein and coat protein of alfalfa mosaic virus accumulates in structurally modified plasmodesmata. *Virology* **244**, 322–329.

van der Wilk, F., Korsman, M. and Zoon, F. (1994). Detection of tobacco rattle virus in nematodes by reverse transcription and polymerase chain reaction. *Eur. J. Plant Pathol.* **100**, 109–122.

van der Wilk, F., Verbeek, M., Dullemans, A.M. and van den Heuvel, J.F.J.M. (1997). Genome-linked protein of potato leafroll virus is located downstream of the putative protease domain of the ORF1 product. *Virology* **234**, 300–303.

van der Wilk, F., Verbeek, M., Dullemans, A. and van den Heuvel, J. (1998). The genome-linked protein (VPg) of southern bean mosaic virus is encoded by the ORF2. *Virus Genes* **17**, 21–24.

van Dijk, P., van der Meer, F.A. and Piron, P.G.M. (1987). Accessions of Australian *Nicotiniana* species suitable as indicator hosts in the diagnosis of palnt virus diseases. *Neth. J. Plant Pathol.* **93**, 73–85.

van Dun, C.M.P., Bol, J.F. and van Vloten-Doting, L. (1987). Expression of alfalfa mosaic virus and tobacco rattle virus coat protein genes in transgenic tobacco plants. *Virology* **159**, 299–305.

van Dun, C.M.P., van Vloten-Doting, L. and Bol, J.F. (1988a). Expression of alfalfa mosaic virus cDNAs 1 and 2 in transgenic tobacco plants. *Virology* **163**, 572–578.

van Dun, C.M.P., Overduin, B., van Vloten-Doting, L. and Bol, J.F. (1988b). Transgenic tobacco expressing tobacco streak virus or mutated alfalfa mosaic virus coat protein does not cross-protect against alfalfa mosaic virus infection. *Virology* **164**, 383–389.

van Emmelo, J., Ameloot, P., Plaetinck, G. and Fiers, W. (1984). Controlled synthesis of the coat protein of satellite tobacco necrosis virus in *Escherichia coli*. *Virology* **136**, 32–40.

van Etten, J.L. (1999). Algal viruses (*Phycodnaviridae*). In: A. Granoff and R.G. Webster (eds) *Encyclopedia of Virology*, 2nd edn, pp. 44–50. Academic Press, San Diego.

van Etten, J.L. and Meints, R.H. (1999). Giant viruses infecting algae. *Annu. Rev. Microbiol.* **53**, 447–494.

van Etten, J.L., Burbank, D.E., Kuczmarski, D. and Meints, R.H. (1983). Virus infection of culturable *Chlorella*-like algae and development of a plaque assay. *Science* **219**, 994–996.

van Etten, J.L., Schuster, A.M. and Meints, R.H. (1988). Viruses of eukaryotic *Chlorella*-like alga. In: Y. Koltin and M.J. Leibowitz (eds) *Viruses of Fungi and Simple Eukaryotes*, pp. 411–428. Dekker, New York.

van Griensven, L.J.L.D., van Kammen, A. and Rezelman, G. (1973). Characterization of the double-stranded RNA isolated from cowpea mosaic virus-infected *Vigna* leaves. *J. Gen. Virol.* **18**, 359–367.

van Hoof, H.A. (1970). Some observations on retention of tobacco rattle virus in nematodes. *Neth. J. Plant Pathol.* **76**, 329–330.

van Hoof, H.A. (1976). The bait leaf method for determining soil infestation with tobacco rattle virus-transmitting trichodorids. *Neth. J. Plant Pathol.* **82**, 181–185.

van Hoof, H.A. (1977). Determination of the infection pressure of potato virus YN. *Neth. J. Plant Pathol.* **83**, 123–127.

van Hoof, H.A. and Silver, C.N. (1976). Natural elimination of tobacco rattle virus in tulip 'Apeldoorn'. *Neth. J. Plant Pathol.* **82**, 255–256.

van Kammen, A. (1968). The relationship between the components of cowpea mosaic virus: I. Two ribonucleoprotein particles necessary for the infectivity of CPMV. *Virology* **34**, 312–318.

van Kammen, A. (1999). Beijerinck's contribution to the virus concept—an introduction. In: C.H. Calisher and M.C. Horzinek (eds) *100 Years of Virology. Arch. Virol.* **15** (Suppl.), 1–8.

van Kammen, A. and Eggen, H.I.L. (1986). The replication of cowpea mosaic virus. *BioEssays* **5**, 261–266.

van Kan, J.A.L., Cornelissen, B.J.C. and Bol, J.F. (1988). A virus-inducible tobacco gene encoding a glycine-rich protein shares putative regulatory elements with the ribulose bisphosphate carboxylase small subunit gene. *Mol. Plant Microbe Interact.* **1**, 107–112.

van Kooten, O., Meurs, C. and van Loon, L.C. (1990). Photosynthetic electron transport in tobacco leaves infected with tobacco mosaic virus. *Physiol. Plant.* **80**, 446–452.

van Lent, J., Storms, M., van der Meer, F., Wellink, J. and Goldbach, R. (1991). Tubular structures involved in the movement of cowpea mosaic virus are also formed in infected cowpea protoplasts. *J. Gen. Virol.* **72**, 2615–2623.

van Lent, J.W.M. and Verduin, B.J.M. (1985). Specific gold-labelling of antibodies bound to plant viruses in mixed suspensions. *Neth. J. Plant Pathol.* **91**, 205–213.

van Lent, J.W.M. and Verduin, B.J.M. (1986). Detection of viral protein and particles in thin sections of infected plant tissue using immunogold labelling. In: R.A.C. Jones and L. Torrance (eds) *Developments and Applications in Virus Testing*, Dev. Appl. Biol. 1, pp. 193–211. Association of Applied Biologists, Wellesbourne.

van Lent, J.W.M. and Verduin, B.J.M. (1987). Detection of viral antigen in semi-thin sections of plant tissue by immunogold–silver staining and light microscopy. *Neth. J. Plant Pathol.* **93**, 261–272.

van Loon, L.C. (1977). Induction by 2-chloroethylphosphonic acid of viral-like lesions, associated proteins and systemic resistance in tobacco. *Virology* **88**, 417–420.

van Loon, L.C. (1989). Stress proteins in infected plants. In: T. Kosuge and E.W. Nester (eds) *Plant–Microbe Interactions*, pp. 198–237. McGraw-Hill, New York.

van Loon, L.C. and Dijkstra, J. (1976). Virus-specific expression of systemic acquired resistance in tobacco mosaic virus- and tobacco necrosis virus-infected 'Samsun NN' and 'Samsun' tobacco. *Neth. J. Plant Pathol.* **82**, 231–237.

van Loon, L.C. and van Kammen, A. (1970). Polyacrylamide disc electrophoresis of the soluble leaf proteins from *Nicotiana tabacum* var. 'Samsun' and 'Samsun NN'. II. Changes in protein constitution after infection with tobacco mosaic virus. *Virology* **40**, 199–211.

van Loon, L.C. and van Strien, E.A. (1999). The families of pathogenesis-related proteins, their activities, and comparative analysis of PR-1 type proteins. *Physiol. Mol. Plant Pathol.* **55**, 85–97.

van Oosten, H.J., Meijneke, C.A.R. and Peerbooms, H. (1983). Growth, yield and fruit quality of virus-infected and virus-free Golden Delicious apple trees 1968–1982. *Acta Hortic.* **130**, 213–217.

van Pelt-Heerschap, H., Verbeek, H., Slot, J.W. and van Vloten-Doting, L. (1987a). The location of coat protein and viral RNAs of alfalfa mosaic virus in infected tobacco leaves and protoplasts. *Virology* **160**, 297–300.

van Pelt-Heerschap, H., Verbeek, H., Huisman, M.J., Loesch-Fries, L.S. and van Vloten-Doting, L. (1987b). Non-structural proteins and RNAs of alfalfa mosaic virus synthesized in tobacco and cowpea protoplasts. *Virology* **161**, 190–197.

van Poelwijk, F., Kolkman, J. and Goldbach, R. (1996). Sequence analysis of the 5' ends of tomato spotted wilt virus N mRNAs. *Arch. Virol.* **141**, 177–184.

van Regenmortel, M.H.V. (1975). Antigenic relationships between strains of tobacco mosaic virus. *Virology* **64**, 415–420.

van Regenmortel, M.H.V. (1982). *Serology and Immunochemistry of Plant Viruses*. Academic Press, New York.

van Regenmortel, M.H.V. (1984a). Recent advances in immunodiagnosis of viral diseases of crops. In: E. Kursak (ed.) *Applied Virology*, pp. 463–477. Academic Press, New York.

van Regenmortel, M.H.V. (1984b). Molecular dissection of antigens by monoclonal antibodies. In: N.J. Stern and H.R. Gamble (eds) *Hybridoma Technology in Agricultural and Veterinary Research*, pp. 43–82. Rowman and Allanheld, Totowa, NJ.

van Regenmortel, M.H.V. (1984c). Monoclonal antibodies in plant virology. *Microbiol. Sci.* **1**, 73–78.

van Regenmortel, M.H.V. (1986). The potential for using monoclonal antibodies in the detection of plant viruses. In: R.A.C. Jones and L. Torrance (eds) *Developments and Applications in Virus Testing*, Dev. Appl. Biol. 1, pp. 89–101. Association of Applied Biologists, Wellsbourne.

van Regenmortel, M.H.V. (1989a). Applying the species concept to plant viruses. *Arch. Virol.* **104**, 1–17.

van Regenmortel, M.H.V. (1989b). Structural and functional approaches to the study of protein antigenicity. *Immunol. Today* **10**, 266–272.

van Regenmortel, M.H.V. (1990). Virus species, a much overlooked but essential concept in virus classification. *Intervirology* **31**, 241–254.

van Regenmortel, M.H.V. (1992). The conformational specificity of viral epitopes. *FEMS Microbiol. Lett.* **100**, 483–488.

van Regenmortel, M.H.V. (1999). How to write the names of virus species. *Arch. Virol.* **144**, 1041–1042.

van Regenmortel, M.H.V. (2000). On the relative merits of italics, Latin and binomial nomenclature in virus taxonomy. *Arch. Virol.* **145**, 433–441.

van Regenmortel, M.H.V. and Dubs, M.-C. (1993). Serological procedures. In: R.E.F. Matthews (ed.) *Diagnosis of Plant Virus Diseases*, pp. 159–214. CRC Press, Boca Raton, FL.

van Regenmortel, M.H.V. and von Wechmar, M.B. (1970). A reexamination of the serological relationship between tobacco mosaic virus and cucumber virus 4. *Virology* **41**, 330–338.

van Regenmortel, M.H.V., Bishop, D.H.L., Fauquet, C.M., Mayo, M.A., Maniloff, J. and Calisher, C.H. (1997). Guidelines to the demarcation of virus species. *Arch Virol.* **142**, 1505–1518.

van Regenmortel, M.H.V., Fauquet, C.M., Bishop, D.H.L. *et al.* (eds) (2000). *Virus Taxonomy: Classification and Nomenclature of Viruses. Seventh Report of the International Committee on Taxonomy of Viruses*. Academic Press, San Diego.

van Rossum, C.M.A., Brederode, F.Th., Neeleman, L. and Bol, J.F. (1997a). Functional equivalence of common and unique sequences in the 3' terminal untranslated regions of alfalfa mosaic virus RNAs 1, 2 and 3. *J. Virol.* **71**, 3811–3816.

van Rossum, C.M.A., Reusken, C.B.E.M., Brederode, F.Th. and Bol, J.F. (1997b). The 3' untranslated region of alfalfa mosaic virus RNA 3 contains a core promoter for minus-strand synthesis and an enhancer element. *J. Gen. Virol.* **78**, 3045–3049.

van Tol, R.G.L. and van Vloten-Doting, L. (1979). Translation of alfalfa-mosaic-virus RNA1 in the mRNA-dependent translation system from rabbit reticulocyte lysates. *Eur. J. Biochem.* **93**, 461–468.

van Tol, R.G.L. and van Vloten-Doting, L. (1981). Lack of serological relationship between the 35K non-structural protein of alfalfa mosaic virus and the corresponding proteins of three other plant viruses with tripartite genomes. *Virology* **109**, 444–447.

van Tol, R.G.L., van Gemeren, R. and van Vloten-Doting, L. (1980). Two leaky termination codons in AMV RNA1. *FEBS Lett.* **118**, 67–71.

van Vloten-Doting, L. (1968). Verdeling van de genetische inforrnatie over de natuurlijke componenten van een plantvirus. PhD Thesis, University of Leiden.

van Vloten-Doting, L. (1975). Coat protein is required for infectivity of tobacco streak virus: biological equivalence of the coat proteins of tobacco streak and alfalfa mosaic viruses. *Virology* **65**, 215–225.

van Vloten-Doting, L. (1976). Similarities and differences between viruses with a tripartite genome. *Ann. Microbiol. (Paris)* **127**, 119–129.

van Vloten-Doting, L. and Jaspars, E.M.J. (1972). The uncoating of alfalfa mosaic virus by its own RNA. *Virology* **48**, 699–708.

van Vloten-Doting, L. and Jaspars, E.M.J. (1977). Plant covirus systems: three-component systems. *Compr. Virol.* **11**, 1–53.

van Vloten-Doting, L., Kruseman, J. and Jaspars, E.M.J. (1968). The biological function and mutual dependence of bottom component and top component of alfalfa mosaic virus. *Virology* **34**, 728–737.

van Vloten-Doting, L., Hasrat, J.A., Oosterwijk, E., van't Sant, P., Schoen, M.A. and Roosien, J. (1980). Description and complementation analysis of 13 temperature-sensitive mutants of alfalfa mosaic virus. *J. Gen. Virol.* **46**, 415–426.

van Vloten-Doting, L., Bol, J.F. and Cornelissen, B. (1985). Plant virus-based vectors for gene transfer will be of limited use because of the high error frequency during viral RNA synthesis. *Plant Mol. Biol.* **4**, 323–326.

van Wezenbeek, P., Verver, J., Harmsen, J., Vos, P. and van Kammen, A. (1983). Primary structure and gene organisation of the middle component RNA of cowpea mosaic virus. *EMBO J.* **2**, 941–946.

Vance, V.B. (1991). Replication of potato virus X RNA is altered in coinfections with potato virus Y. *Virology* **182**, 486–494.

Vance, V.B. and Beachy, R.N. (1984). Detection of genomic-length soybean mosaic virus RNA on polyribosomes of infected soybean leaves. *Virology* **138**, 26–36.

Vance, V.B., Berger, P.H., Carrington, J.C., Hunt, A.G. and Shi, X.M. (1995). 5' proximal potyviral sequences mediate potato virus X/potyviral synergistic disease in transgenic tobacco. *Virology* **206**, 583–590.

Vanderveken, J.J. (1977). Oils and other inhibitors of nonpersistent virus transmission. In: K.F. Harris and K. Maramorosch (eds) *Aphids as Virus Vectors*, pp. 435–454. Academic Press, New York.

Varrelmann, M. and Maiss, E. (2000). Mutations in the coat protein gene of plum pox virus suppress particle assembly, heterologous encapsidation and complementation in transgenic plants of *Nicotiana benthamiana*. *J. Gen. Virol.* **81**, 567–576.

Varveri, C. (2000). Potato Y potyvirus detection by immunological and molecular techniques in plants and aphids. *Phytoparasitica* **28**, 141–148.

Varveri, C., Candresse, T., Cugusi, M., Ravelonandro, M. and Dunez, J. (1988). Use of ³²P-labelled transcribed RNA probe for dot hybridization detection of plum pox virus. *Phytopathology* **78**, 1280–1283.

Vassilakos, N., MacFarlane, S.A., Welscher, B. and Brown, D.J.F. (1997). Exclusivity and complementarity in the association between nepo- and tobraviruses and their respective vector nematodes. *Mededl. Facult. Landbouwwetenschappen Univ. Gent* **62**, 713–720.

Vaucheret, H., Béclin, C., Elmayan, T. *et al.* (1998). Transgene-induced gene silencing in plants. *Plant J.* **16**, 651–659.

Vawter, L. and Brown, W.M. (1986). Nuclear and mitochondrial DNA comparisons reveal extreme rate variation in the molecular clock. *Science* **234**, 194–196.

Vela C., Cambra, M., Cortes, E. *et al.* (1986). Production and characterisation of monoclonal antibodies specific for citrus tristeza virus and their use for diagnosis. *J. Gen. Virol.* **67**, 91–96.

Veldee, S. and Fraenkel-Conrat, H. (1962). The characterisation of tobacco mosaic virus strains by their productivity. *Virology* **18**, 56–63.

Venekamp, J.H. and Beemster, A.B.R. (1980). Mature plant resistance of potato against some virus diseases. I. Concurrence of development of mature plant resistance against potato virus X and decrease of ribosome and RNA content. *Neth. J. Plant Pathol.* **86**, 1–10.

Vera, A., Daròs, J.-A., Flores, R. and Hernández, C. (2000). The DNA of a plant retroviroid-like element is fused to different sites in the genome of a plant pararetrovirus and shows multiple forms with sequence deletions. *J. Virol.* **74**, 10390–10400.

Verchot, J., Angell, S.M. and Baulcombe, D. (1998). *In vivo* translation of the triple gene block of potato virus X requires two subgenomic RNAs. *J. Virol.* **72**, 8316–8320.

Vercruysse, P., Gibbs, M., Tirry, L. and Hofte, M. (2000). RT-PCR using redundant primers to detect three viruses associated with carrot motley dwarf disease. *J. Virol. Methods* **88**, 153–161.

Verdaguer, B., de Kochko, A., Beachy, R.N. and Fauquet, C. (1996). Isolation and expression in transgenic tobacco and rice plants, of the cassava vein mosaic virus (CVMV) promoter. *Plant Mol. Biol.* **31**, 1129–1139.

Verdaguer, B., de Kocho, A., Fux, C.I., Beachy, R.N. and Fauquet, C. (1998). Functional organization of the cassava vein mosaic virus (CsVMV) promoter. *Plant Mol. Biol.* **37**, 1055–1067.

Verduin, B.J.M., Prescott, B. and Thomas, G.J. (1984). RNA-protein interactions and secondary structures of cowpea chlorotic mottle virus for *in vitro* assembly. *Biochemistry* **23**, 4301–4308.

Verhagen, W., van Boxsel, J.A.M., Bol, J.F., van Vloten-Doting, L. and Jaspars, E.M.J. (1976). RNA-protein interactions in alfalfa mosaic virus. *Ann. Microbiol. (Paris)* **127A**, 165–172.

Verma, H.N., Baranwal, V.K. and Srivastava, S. (1998). Breeding for resistance to plant viruses. In: A. Hadidi, R.K. Khetarpal and H. Koganezawa (eds) *Plant Virus Disease Control*, pp. 154–162. APS Press, St. Paul, MN.

Vértesy, J. and Nyéki, J. (1974). Effect of different ringspot viruses on the flowering period and fruit set of Montmorency and Pándy sour cherries: I. *Acta Phytopathol. Acad. Sci. Hung.* **9**, 17–22.

Verver, J., Goldbach, R., García, J.A. and Vos, P. (1987). *In vitro* expression of a full-length DNA copy of cowpea mosaic virus B-RNA: identification of the B-RNA encoded 24-kD protein as a viral protease. *EMBO J.* **6**, 549–554.

Verver, J., Le Gall, O., van Kammen, A. and Wellink, J. (1991). The sequence between nucleotides 161 and 512 of cowpea mosaic virus M RNA is able to support internal initiation of translation *in vitro*. *J. Gen. Virol.* **72**, 2339–2345.

Verver, J., Wellink, J., van Lent, J., Gopinath, K. and van Kammen, A. (1998). Studies on the movement of cowpea mosaic virus using the jellyfish green fluorescent protein. *Virology* **242**, 22–27.

Vetten, H.J. and Allen, D.J. (1983). Effects of environment and host on vector biology and incidence of two whitefly-spread diseases of legumes in Nigeria. *Ann. Appl. Biol.* **102**, 219–227.

Vetten, H.J., Lesemann, D.-E. and Dalchow, J. (1987). Electron microscopical and serological detection of virus-like particles associated with lettuce big vein disease. *J. Phytopathol.* **120**, 53–59.

Villegas, L.C., Druka, A., Bajet, N.B. and Hull, R. (1997). Genetic variation of rice tungro bacilliform virus in the Philippines. *Virus Genes* **15**, 195–210.

Virudachalum, R., Harrington, M., Johnson, J.E. and Markley, J.L. (1985). ¹H, ¹³C and ³¹P nuclear magnetic resonance studies of cowpea mosaic virus: detection and exchange of polyamines and dynamics of the RNA. *Virology* **141**, 43–50.

Visser, P.B. (2000). *Role of RNA 2-Encoded Proteins in Nematode-Transmission of Tobacco Rattle Virus*. PhD Thesis, University of Leiden, Netherlands.

Visser, P.B. and Bol, J.F. (1999). Nonstructural proteins of tobacco rattle virus which have a role in nematode transmission: expression pattern and interaction with viral coat protein. *J. Gen. Virol.* **80**, 3273–3280.

Visser, P.B., Brown, D.J.F., Brederode, F.T. and Bol, J.F. (1999a). Nematode transmission of tobacco rattle virus serves as a bottleneck to clear the population from defective interfering RNAs. *Virology* **263**, 155–165.

Visser, P.B., Mathis, A. and Linthorst, H.J.M. (1999b). Tobraviruses. In: A. Granoff and R.G. Webster (eds) *Encyclopedia of Virology*, 2nd edn, pp. 1784–1789. Academic Press, San Diego.

Visvader, J.E. and Symons, R.H. (1985). Eleven new sequence variants of citrus exocortis viroid and the correlation of sequence with pathogenicity. *Nucl. Acids Res.* **13**, 2907–2920.

Visvader, J.E. and Symons, R.H. (1986). Replication of *in vitro* constructed viroid mutants: location of the pathogenicity-modulating domain of citrus exocortis viroid. *EMBO J.* **5**, 2051–2055.

Vogel, R.H. and Provencher, S.W. (1988). Three-dimensional reconstruction from electron micrographs of disordered specimens: II. Implementation and results. *Ultramicroscopy* **25**, 223–240.

Voinnet, O., Vain, P., Angell, S. and Baulcombe, D.C. (1998). Systemic spread of sequence-specific transgene RNA degradation is initiated by localized introduction of ectopic promoterless DNA. *Cell* **95**, 177–187.

Voinnet, O., Pinto, Y.M. and Baulcombe, D.C. (1999). Suppression of gene silencing: a general strategy used by diverse DNA and RNA viruses of plants. *Proc. Natl Acad. Sci. USA* **96**, 14147–14152.

Vos, P., Jaegle, M., Wellink, J. *et al.* (1988a). Infectious RNA transcripts derived from full-length DNA copies of the genomic RNAs of cowpea mosaic virus. *Virology* **164**, 33–41.

Vos, P., Verver, J., Jaegle, M., Wellink, J., van Kammen, A. and Goldbach, R. (1988b). Two viral proteins involved in the proteolytic processing of the cowpea mosaic virus polyproteins. *Nucl. Acids Res.* **16**, 1967–1985.

Vriend, G., Verduin, B.J.M. and Hemminga, M.A. (1986). Role of the N-terminal part of the coat protein in the assembly of cowpea chlorotic mottle virus: a 500 MHz proton nuclear magnetic resonance study and structural calculations. *J. Mol. Biol.* **191**, 453–460.

Wagih, E.E. and Coutts, R.H.A. (1982). Peroxidase, polyphenoloxidase and ribonuclease in tobacco necrosis virus infected or mannitol osmotically-stressed cowpea and cucumber tissue. I. Quantitative alterations. *Phytopathol. Z.* **104**, 1–12.

Wagner, G.W. and Bancroft, J.B. (1968). The self-assembly of spherical viruses with mixed coat proteins. *Virology* **34**, 748–756.

Wagner, J.D.O. and Jackson, A.O. (1997). Characterization of the components and activity of sonchus yellow net rhabdovirus polymerase. *J. Virol.* **71**, 2371–2382.

Wagner, J.D.O., Chol, T.-J. and Jackson, A.O. (1996). Extraction of nuclei from sonchus yellow net rhabdovirus-infected plants yields a polymerase that synthesizes viral mRNAs and polyadenylated plus-strand leader RNA. *J. Virol.* **70**, 468–477.

Wagner, W.H. (1984). A comparison of taxonomic methods in biosystematics. In: W.F. Grant (ed.) *Plant Biosystematics*, pp. 643–654. Academic Press, Orlando, FL.

Wah, Y.F.W.C. and Symons, R.H. (1999). Transmission of viroids via grape seeds. *J. Phytopathol.* **147**, 285–291.

Wahyuni, W.S. and Randles, J.W. (1993). Inoculation with root nodulating bacteria reduces the susceptibility of *Medicago truncatula* and *Lupinus angustifolius* to cucumber mosaic virus (CMV) and addition of nitrate partially reverses the effect. *Austr. J. Agric. Res.* **44**, 1917–1929.

Waigmann, E and Zambryski, P. (1995). Tobacco mosaic virus movement protein-mediated protein transport between trichome cells. *Plant Cell* **7**, 2069–2079.

Waigmann, E., Lucas, W.J., Citovsky, V. and Zambryski, P. (1994). Direct functional assay for tobacco mosaic virus cell-to-cell movement protein and identification of a domain involved in increasing plasmodesmatal permeability. *Proc. Natl Acad. Sci. USA* **91**, 1433–1437.

Waigmann, E., Turner, A., Peart, J., Roberts, K. and Zambryski, P. (1997). Ultrastructural analysis of leaf trichome plasmodesmata reveals major difference from mesophyll plasmodesmata. *Planta* **203**, 75–84.

Waigmann, E., Chen, M.-H., Bachmaier, R., Ghoshroy, S. and Citovsky, V. (2000). Regulation of plasmodesmatal transport by phosphorylation of tobacco mosaic virus cell-to-cell movement protein. *EMBO J.* **19**, 4875–4884.

Wakarchuk, D.A. and Hamilton, R.I. (1985). Cellular double-stranded RNA in *Phaseolus vulgaris*. *Plant Med. Biol.* **5**, 55–63.

Wakarchuk, D.A. and Hamilton, R.I. (1990). Partial nucleotide sequence from enigmatic dsRNAs in *Phaseolus vulgaris*. *Plant Mol. Biol.* **14**, 637–639.

Walbot, V. (2000). Saturation mutagenesis using maize transposons. *Curr. Opin. Plant Biol.* **3**, 103–107.

Walden, R. and Howell, S.H. (1982). Intergenomic recombination events among pairs of defective cauliflower mosaic virus genomes in plants. *J. Mol. Appl. Genet.* **1**, 447–456.

Walker, H.L. and Pirone, T.P. (1972a). Particle numbers associated with mechanical and aphid transmission of some plant viruses. *Phytopathology* **62**, 1283–1288.

Walker, H.L. and Pirone, T.P. (1972b). Number of TMV particles required to infect locally or systemically susceptible tobacco cultivars. *J. Gen. Virol.* **17**, 241–243.

Walkey, D.G.A. (1968). The production of virus-free rhubarb by apical tip-culture. *J. Hortic. Sci.* **43**, 283–287.

Walkey, D.G.A. (1985). *Applied Plant Virology*. Heinemann, London.

Walkey, D.G.A. (1991). *Applied Plant Virology*, 2nd edn. Chapman & Hall, London.

Walkey, D.G.A. and Antill, D.N. (1989). Agronomic evaluation of virus-free and virus-infected garlic (*Allium sativum* L.). *J. Hortic. Sci.* **64**, 53–60.

Walkey, D.G.A. and Cooper, J. (1976). Heat inactivation of cucumber mosaic virus in cultured tissues of *Stellaria media*. *Ann. Appl. Biol.* **84**, 425–428.

Walkey, D.G.A. and Dance, M.C. (1979). The effect of oil sprays on aphid transmission of turnip mosaic, beet yellows, bean common mosaic and bean yellow mosaic viruses. *Plant Dis. Rep.* **63**, 877–881.

Walkey, D.G.A. and Freeman, G.H. (1977). Inactivation of cucumber mosaic virus in cultured tissues of *Nicotiana rustica* by diurnal alternating periods of high and low temperature. *Ann. Appl. Biol.* **87**, 375–382.

Walkey, D.G.A. and Webb, M.J.W. (1968). Virus in plant apical meristems. *J. Gen. Virol.* **3**, 311–313.

Walkey, D.G.A. and Whittingham-Jones, S.G. (1970). Seed transmission of strawberry latent ringspot virus in celery (*Apium graveolens* var. Dulce). *Plant Dis. Rep.* **54**, 802–803.

Walkey, D.G.A., Fitzpatrick, J. and Woolfitt, J.M.G. (1969). The inactivation of virus in cultured shoot tips of *Nicotiana rustica* L. *J. Gen. Virol.* **5**, 237–241.

Walkey, D.G.A., Brocklehurst, P.A. and Parker, J.E. (1985). Some physiological effects of two seed-transmitted viruses on flowering, seed production, and seed vigour in *Nicotiana* and *Chenopodium* plants. *New Phytol.* **99**, 117–128.

Walkey, D.G.A., Webb, M.J.W., Bolland, C.J. and Miller, A. (1987). Production of virus-free garlic (*Allium sativum* L.) and shallot (*Allium ascalonicum* L.) by meristem tip culture. *J. Hortic. Sci.* **62**, 211–220.

Walkey, D.G.A., Lecoq, H., Collier, R. and Dobson, S. (1992). Studies on the control of zucchini yellow mosaic virus in courgettes by mild strain protection. *Plant Pathol.* **41**, 762–771.

Wallin, J.R. and Loonan, D.V. (1971). Low-level jet winds, aphid vectors, local weather and barley yellow dwarf virus outbreaks. *Phytopathology* **61**, 1068–1070.

Walsh, I.A. (1986). Virus diseases of oilseed rape and their control. *Br. Crop Prot. Conf. Pest Dis.* **7A-3**, 737–743.

Walsh, J.A. (1992). Resistant watercress. *Grower* **118**, 18–21.

Walsh, J.A. (1998). Chemical control of fungal vectors of plant viruses. In: A. Hadidi, R.K. Khetarpal and H. Koganezawa (eds) *Plant Virus Disease Control*, pp. 196–207. APS Press, St. Paul, MN.

Walsh, J.A. and Tomlinson, J.A. (1985). Viruses infecting winter oilseed rape (*Brassica napus* ssp. *oleifera*). *Ann. Appl. Biol.* **107**, 485–495.

Wang, A., Carrier, K., Chisholm, J., Wieczorek, A., Huguenot, C. and Sanfaçon, H. (1999a). Proteolytic processing of tomato ringspot nepovirus 3C-like protease precursors: definition of the domains for the VPg, protease and putative RNA-dependent RNA polymerase. *J. Gen. Virol.* **80**, 799–809.

Wang, D. and Maule, A.J. (1992). Early embryo invasion as a determinant in pea of the seed transmission of pea seed-borne mosaic virus. *J. Gen. Virol.* **73**, 1615–1620.

Wang, D. and Maule, A.J. (1994). A model for seed transmission of a plant virus: genetic and structural analysis of pea embryo invasion by pea seed-borne mosaic virus. *Plant Cell* **6**, 777–787.

Wang, D. and Maule, A.J. (1995). Inhibition of host gene expression associated with plant virus replication. Science **267**, 229–231.

Wang, D. and Maule, A.J. (1997). Contrasting patterns in the spread of two seed-borne viruses in pea embryos. *Plant J.* **11**, 1333–1340.

Wang, D., MacFarlane, S.A. and Maule, A.J. (1997a). Viral determinants of pea early browning seed transmission in pea. *Virology* **234**, 112–117.

Wang, H. and Simon, A.E. (2000). 3'-end stem-loops of the subviral RNAs associated with turnip crinkle virus are involved in symptom modulation and coat protein binding. *J. Virol.* **74**, 6528–6537.

Wang, H. and Stubbs, G. (1993). Molecular dynamics in refinement against fiber diffraction data. *Acta Crystallog. Sect. A.* **49**, 504–513.

Wang, H., Planchart, A. and Stubbs, G. (1998a). Caspar carboxylates: the structural basis of tobamovirus disassembly. *Biophys. J.* **74**, 633–638.

Wang, H.-L., Wang, Y., Giesman-Cookmeyer, D., Lommel, S.A. and Lucas, W.J. (1998b). Mutations in viral movement protein alter systemic infection and identify an intercellular barrier to entry into the phloem long-distance transport system. *Virology* **245**, 75–89.

Wang, J. and Simon, A.E. (1999). Symptom attenuation by a satellite RNA *in vivo* is dependent on reduced levels of virus coat protein. *Virology* **259**, 234–245.

Wang, J., Carpenter, C.D. and Simon, A.E. (1999b). Minimal sequence and structural requirements for a subgenomic RNA promoter for turnip crinkle virus. *Virology* **253**, 327–336.

Wang, J.L. and Simon, A.E. (1997). Analysis of the two subgenomic RNA promoters for turnip crinkle virus *in vivo* and *in vitro*. *Virology* **232**, 174–186.

Wang, J.L. and Simon, A.E. (1998). Analysis of the two subgenomic RNA promoters for turnip crinkle virus *in vivo* and *in vitro*. *Virology* **232**, 174–186.

Wang, M.-B. and Waterhouse, P.M. (2000). High-efficiency silencing of β-glucuronidase gene in rice is correlated with repetitive transgene structure but is independent of DNA methylation. *Plant Mol. Biol.* **43**, 67–82.

Wang, M.-B., Abbott, D.C. and Waterhouse, P.M. (2001a). A single copy of a virus-derived transgene encoding hairpin RNA gives immunity to barley yellow dwarf virus. *Mol. Plant Pathol.* (in press).

Wang, M.-B., Abbott, D.C., Upadhyaya, N.M., Jacobsen, J.V. and Waterhouse, P.M. (2001b). *Agrobacterium tumefaciens* mediated transformation of an elite barley cultivar with virus resistance and reporter genes. *Aust. J. Plant Physiol.* (in press).

Wang, M.-B., Wesley, V., Finnegan, J., Smith, N. and Waterhouse, P.M. (2001c). Replicating satellite RNA induces sequence-specific DNA methylation and truncated transcripts in plants. *RNA* **7**, 16–28.

Wang, M.C., Lin, J.J., Duran-Vila, N. and Semancik, J.S. (1986). Alteration in cell wall composition and structure in viroid-infected cells. *Physiol. Plant Pathol.* **28**, 107–124.

Wang, P. and Tumer, N.E. (2000). Virus resistance mediated by ribosome inactivating proteins. *Adv. Virus Res.* **55**, 325–355.

Wang, P., Zoubenko, O. and Tumer, N.E. (1998c). Reduced toxicity and broad spectrum resistance to viral and fungal infection in transgenic plants expressing pokeweed antiviral protein II. *Plant Mol. Biol.* **38**, 957–964.

Wang, R.Y and Pirone, T.P. (1996). Mineral oil interferes with retention of tobacco etch potyvirus in the stylets of *Myzus persicae*. *Phytopathology* **86**, 820–823.

Wang, R.Y., Ammar, E.D., Thornbury, D.W., Lopez-Moya, J.J. and Pirone, T.P. (1996). Loss of potyvirus transmissibility and helper component activity correlate with non-retention of virions in aphid stylets. *J. Gen. Virol.* **77**, 861–867.

Wang, R.Y., Gergerich, R.C. and Kim, K.S. (1992). Noncirculative transmission of plant viruses by leaf-feeding beetles. *Phytopathology* **82**, 946–950.

Wang, R.Y., Gergerich, R.C. and Kim, K.S. (1994). Entry of ingested plant viruses into the hemocoel of the beetle vector *Diabrotica undecimpunctata howardi*. *Phytopathology* **84**, 147–153.

Wang, S and Miller, W.A. (1995). A sequence located 4.5 to 5 kilobases from the 5' end of the barley yellow dwarf virus (PAV) genome strongly stimulates translation of uncapped mRNA. *J. Biol. Chem.* **270**, 13446–13452.

Wang, S., Guo, L., Allen, E. and Miller, W.A. (1999c). A potential mechanism for selective control of cap-independent translation by a viral RNA sequence in *cis* and in *trans*. *RNA* **5**, 726–738.

Wang, S.P., Browning, K.S. and Miller, W.A. (1997b). A viral sequence in the 3' untranslated region mimics a 5' cap in facilitating translation of uncapped mRNA. *EMBO J.* **18**, 4107–4116.

Wang, Y., Gaba, V., Wolf, D., Xia, X.-D., Zelcher, A. and Gal-on, A. (2000). Identification of a novel plant virus promoter using a potyvirus infectious clone. *Virus Genes* **20**, 11–17.

Wanitchakorn, R., Hafner, G.J., Harding, R.M. and Dale, J.L. (2000). Functional analysis of proteins encoded by banana bunchy top virus DNA-4 to -6. *J. Gen. Virol.* **81**, 299–306.

Ward, A., Etessami, P. and Stanley, J. (1988). Expression of a bacterial gene in plants mediated by infectious geminivirus DNA. *EMBO J.* **7**, 1583–1587.

Ward, C.M., Walkey, D.G.A. and Phelps, K. (1987). Storage of samples infected with lettuce or cucumber mosaic viruses prior to testing with ELISA. *Ann. Appl. Biol.* **110**, 89–95.

Ward, C.W., Weiller, G.F., Shukla, D.D. and Gibbs, A. (1995). Molecular systematics of the *Potyviridae*, the largest plant virus family. In: A.J. Gibbs, C.H. Calisher and F. García-Arenal (eds) *Molecular Basis for Virus Evolution*, pp. 477–500. Cambridge University Press, Cambridge.

Warmke, H.E. and Edwardson, J.R. (1966). Electron microscopy of crystalline inclusions of tobacco mosaic virus in leaf tissue. *Virology* **30**, 45–57.

Wassenegger, M., Heimes, S. and Sänger, H.L. (1994). An infectious viroid RNA replicon evolved from an *in vitro*-generated non-infectious viroid deletion mutant via a complementary deletion *in vivo*. *EMBO J.* **13**, 6172–6177.

Watanabe, T., Honda, A., Iwata, A., Ueda, S., Hibi, T. and Ishihama, A. (1999). Isolation from tobacco mosaic virus-infected tobacco of a solubilized template-specific RNA-dependent RNA polymerase containing a 126K/183K protein heterodimer. *J. Virol.* **73**, 2633–2640.

Watanabe, Y. and Okada, Y. (1986). *In vitro* viral RNA synthesis by a subcellular fraction of TMV-inoculated tobacco protoplasts. *Virology* **149**, 64–73.

Watanabe, Y., Ohno, T. and Okada, Y. (1982). Virus multiplication in tobacco protoplasts inoculated with tobacco mosaic virus RNA encapsulated in large unilamellar vesicle liposomes. *Virology* **120**, 478–480.

Watanabe, Y., Meshi, T. and Okada, Y. (1984a). The initiation site for transcription of the TMV 30-kDa protein messenger RNA. *FEBS Lett.* **173**, 247–250.

Watanabe, Y., Emori, Y., Ooshika, I., Meshi, T., Ohno, T. and Okada, Y. (1984b). Synthesis of TMV-specific RNAs and proteins at the early stage of infection in tobacco protoplasts: Transient expression of the 30k protein and its mRNA. *Virology* **133**, 18–24.

Watanabe, Y., Meshi, T. and Okada, Y. (1987a). Infection of tobacco protoplasts with *in vitro* transcribed tobacco mosaic virus RNA using an improved electroporation method. *FEBS Lett.* **219**, 65–69.

Watanabe, Y., Morita, N., Nishiguchi, M. and Okada, Y. (1987b). Attenuated strains of tobacco mosaic virus. Reduced synthesis of a viral protein with a cell-to-cell movement function. J. Mol. Biol. **194**, 699–704.

Watanabe, Y., Kishibayashi, N., Motoyoshi, F. and Akada, Y. (1987c). Characterization of *Tm-1* gene action on replication of common isolates and a resistance-breaking isolate of TMV. *Virology* **161**, 527–532.

Watanabe, Y., Ogawa, T. and Okada, Y. (1992). *In vivo* phosphorylation of the 30-kDa protein of tobacco mosaic virus. *FEBS Lett.* **313**, 181–184.

Watanabe, Y., Ogawa, T., Takahashi, H. *et al.* (1995). Resistance against multiple plant viruses in plants mediated by a double stranded-RNA specific ribonucleases. *FEBS Lett.* **372**, 165–168.

Waterhouse, P.M. and Helms, K. (1985). *Metopalophium dirhodum* (Walker): a newly arrived vector of barley yellow dwarf virus in Australia. *Australas. Plant Pathol.* **14**, 64–66.

Waterhouse, P.M. and Upadhyaya, N.M. (1999). Genetic engineering of virus resistance. In: K. Shimamoto (ed.) *Molecular Biology of Rice*, pp.257–281. Springer, Tokyo.

Waterhouse, P.M., Gildow, F.E. and Johnstone, G.R. (1988). Luteovirus group. Commonwealth Mycological Institute/Association of Applied Biologists, *Descriptions of Plant Viruses*, No. 339.

Waterhouse, P.M., Graham, M.W. and Wang, M.-B. (1998). Virus resistance and gene silencing in plants can be induced by simultaneous expression of sense and antisense RNA. *Proc. Natl Acad. Sci. USA* **95**, 13959–13964.

Waterhouse, P.M., Smith, N.A. and Wang, M.-B. (1999). Virus resistance and gene silencing: killing the messenger. *Trends Plant Sci.* **4**, 452–457.

Waterhouse, P.M., Wang, M.-B. and Finnegan, E.J. (2001). Role of short RNAs in gene silencing. *Trends Plant Sci.* **6**, 297–301.

Waterman, M.S. (ed.) (1988). *Mathematical Methods for DNA Sequences*. CRC Press, Boca Raton, FL.

Waterworth, H.E. (1998). Certification for plant viruses: an overview. Breeding for resistance to plant viruses. In: A. Hadidi, R.K. Khetarpal and H. Koganezawa (eds) *Plant Virus Disease Control*, pp. 325–331. APS Press, St. Paul, MN.

Waterworth, H.E. and Hadidi, A. (1998). Economic losses due to plant viruses. In: A. Hadidi, R.K. Khetarpal and H. Koganezawa (eds) *Plant Virus Disease Control*, pp. 1–13. APS Press, St Paul, MN.

Waterworth, H.E., Tousignant, M.E. and Kaper, J.M. (1978). A lethal disease of tomato experimentally induced by RNA5 associated with cucumber mosaic virus isolated from *Commelina* from El Salvador. *Phytopathology* **68**, 561–566.

Waterworth, H.E., Kaper, J.M. and Tousignant, M.E. (1979). CARNA5, the small cucumber mosaic virus-dependent replicating RNA, regulates disease expression. *Science* **204**, 845–847.

Watson, L. and Gibbs, A.J. (1974). Taxonomic patterns in the host ranges of viruses among grasses, and suggestions on generic sampling for host-range studies. *Ann. Appl. Biol.* **77**, 23–32.

Watson, M.A. and Heathcote, G.D. (1965). The use of sticky traps and the relation of their catches of aphids to the spread of viruses in crops. *Rep. Rothamsted Exp. Stn.* pp. 292–300.

Watson, M.A. and Roberts, F.M. (1939). A comparative study of the transmission of *Hyocyamus* virus 3, potato virus Y and cucumber mosaic virus by the vector *Myzus persicae* (Sulz), *M. circumflexus* (Buckton) and *Macrosiphum gei* (Koch). *Proc. R. Soc. London* B **127**, 543–576.

Watson, M.A., Heathcote, G.D., Lauckner, F.B. and Sowray, P.A. (1975). The use of weather data and counts of aphids in the field to predict the incidence of yellowing viruses of sugar-beet crops in England in relation to the use of insecticides. *Ann. Appl. Biol.* **81**, 181–198.

Watts, J.W., King, J.M. and Stacey, N.J. (1987). Inoculation of protoplasts with viruses by electroporation. *Virology* **157**, 40–46.

Watts, L.E. (1975). The response of various breeding lines of lettuce to beet western yellows virus. *Ann. Appl. Biol.* **81**, 393–397.

Way, R.D. and Gilmer, R.M. (1963). Reductions in fruit sets on cherry trees pollinated with pollen from trees with sour cherry yellows. *Phytopathology* **53**, 399–401.

Wayadande, A.C. and Nault, L.R. (1992). Leafhopper probing behavior associated with maize chlorotic dwarf virus transmission in maize. *Phytopathology* **83**, 522–526.

Webb, D.R., Bonfiglioli, R.G., Carraro, L., Osler, R. and Symons, R.H. (1999). Oligonucleotides as hybridisation probes to localize phytoplasmas in host plants and insect vectors. *Phytopathology* **89**, 894–901.

Webb, J.E., Wallwork, J.A. and Elgood, J.H. (1978). *Guide to Invertebrate Animals*. Macmillan, London.

Weber, H. and Pfitzner, A.J.P. (1998). *Tm-2²* resistance in tomato requires recognition of the carboxy terminus of the movement protein of tomato mosaic virus. *Mol. Plant Microbe Interact.* **11**, 498–503.

Weber, H., Schultze, S. and Pfitzner, A.J.P. (1993). Amino acid substitutions in the tomato mosaic virus 30-kilodalton movement protein confer the ability to overcome the *Tm-2²* resistance gene in tomato. *J. Virol.* **67**, 6432–6438.

Webley, D.P. and Stone, L.E.W. (1972). Field experiments on potato aphids and virus spread in South Wales 1966/9. *Ann. Appl. Biol.* **72**, 197–203.

Weidemann, H.L. and Maiss, E. (1996). Detection of the potato tuber necrotic ringspot strain of potato virus Y (PVY^NTN) by reverse transcription and immunocapture polymerase chain reaction. *J. Plant Dis. Protect.* **103**, 337–345.

Weidemann, H.L., Lesemann, D., Paul, H.L. and Koenig, R. (1975). Das Broad Bean Wilt-Virus als Ursache für eine neue Vergilbungskrankheit des Spinats in Deutschland. *Phytopathol. Z.* **84**, 215–221.

Weiland, J.J. and Dreher, T.W. (1989). Infectious TYMV RNA from cloned cDNA: effects *in vitro* and *in vivo* of point substitutions in the initiation codons of two extensively overlapping ORFs. *Nucl. Acids Res.* **17**, 4675–4687.

Weiland, J.J. and Edwards, M.C. (1994). Evidence that the –a gene of barley stripe mosaic virus encodes determinants of pathogenicity to oat (*Avena sativa*). *Virology* **201**, 116–126.

Weiland, J.J. and Edwards, M.C. (1996). A single nucleotide substitution in the a gene confers oat pathogenicity to barley stripe mosaic virus strain ND18. *Mol. Plant-Microbe Interact.* **9**, 62–67.

Weiller, G.F., McClure, M.A. and Gibbs, A.J. (1995). Molecular phylogenetic analysis. In: A.J. Gibbs, C.H. Calisher and F. García-Arenal (eds) *Molecular Basis for Virus Evolution*, pp. 553–585. Cambridge University Press, Cambridge.

Weiner, H.L., Drayna, D., Averill, D.R., Jr. and Fields, B.N. (1977). Molecular basis for reovirus virulence: role of the S1 gene. *Proc. Natl Acad. Sci. USA* **74**, 5744–5748.

Weintraub, M. and Ragetli, H.W.J. (1970). Electron microscopy of the bean and cowpea strains of southern bean mosaic virus within leaf cells. *J. Ultrastruct. Res.* **32**, 167–189.

Welkie, G.W. and Pound, G.S. (1958). Temperature influence on the rate of passage of cucumber mosaic virus through the epidermis of cowpea leaves. *Virology* **5**, 362–371.

Welkie, G.W., Young, S.F. and Miller, G.W. (1967). Metabolite changes induced by cucumber mosaic virus in resistant and susceptible strains of cowpea. *Phytopathology* **57**, 472–475.

Wellink, J. and van Kammen, A. (1988). Proteases involved in the processing of viral polyproteins. *Arch. Virol.* **98**, 1–26.

Wellink, J. and van Kammen, A. (1989). Cell-to-cell transport of cowpea mosaic virus requires both the 58K/48K proteins and the capsid proteins. *J. Gen. Virol.* **70**, 2279–2286.

Wellink, J., Rezelman, G., Goldbach, R. and Beyreuther, K. (1986). Determination of the proteolytic processing sites in the polyprotein encoded by the bottom component RNA of cowpea mosaic virus. *J. Virol.* **59**, 50–58.

Wellink, J., Jaegle, M. and Goldbach, R. (1987a). Detection of a novel protein encoded by the bottom-component RNA of cowpea mosaic virus, using antibodies raised against a synthetic peptide. *J. Virol.* **61**, 236–238.

Wellink, J., Jaegle, M., Prinz, H., van Kammen, A. and Goldbach, R. (1987b). Expression of the middle component RNA of cowpea mosaic virus *in vivo*. *J. Gen. Virol.* **68**, 2577–2585.

Wellink, J., van Lent, J. and Goldbach, R. (1988). Detection of viral proteins in cytopathic structures in cowpea protoplasts infected with cowpea mosaic virus. *J. Gen. Virol.* **69**, 751–755.

Wellink, J., Verver, J., van Lent, J. and van Kammen, A. (1996). Capsid proteins of cowpea mosaic virus transiently expressed in protoplasts form virus-like particles. *Virology* **224**, 352–355.

Wellink, J., Le Gall, O., Sanfacon, H., Ikegami, M. and Jones, A.T. (2000). Family *Comoviridae*. In: M.H.V. van Regenmortel *et al.* (eds) *Virus Taxonomy: Seventh Report of the International Committee on Taxonomy of Viruses*, pp. 691–701. Academic Press, New York.

Wells, S.E., Hillner, P.E., Vale, R.D. and Sachs, A.B. (1998). Circularization of mRNA by eukaryotic translation initiation factors. *Mol. Cell* **2**, 135–140.

Wen, F. and Lister, R.M. (1991). Heterologous encapsidation in mixed infections among four isolates of barley yellow dwarf virus. *J. Gen. Virol.* **72**, 2217–2223.

Werker, A.R., Dewar, A.M. and Harrington, R. (1998). Modelling the incidence of virus yellows in sugar beet in the UK in relation to the numbers of migrating *Myzus persicae*. *J. Appl. Ecol.* **35**, 811–818.

Wetter, C. (1975). Tabakmosaikvirus und Para-Tabakmosaikvirus in Zigaretten. *Naturwissenshaften* **62**, 533.

Wetter, C. and Bernard, M. (1977). Identifizierung, Reinigung und serologischer Nachweis von Tabakmosaikvirus und Par-Tabakmosaikvirus aus Zigaretten. *Phytopathol. Z.* **90**, 257–267.

Wetzel, T., Dietzgen, R.G. and Dale, J.L. (1994). Genomic organization of lettuce necrotic yellows rhabdovirus. *Virology* **200**, 401–412.

Whalen, M.C., Stall, R.E. and Staskawicz, B.J. (1988). Characterisation of a gene from a tomato pathogen determining hypersensitive resistance in non-host species and genetic analysis of this resistance in bean. *Proc. Natl Acad. Sci. USA* **85**, 6743–6747.

Whenham, R.J. (1989). Effect of systemic tobacco mosaic virus infection on endogenous cytokinin concentration in tobacco (*Nicotiana tabacum* L.) leaves: consequences for the control of resistance and symptom development. *Physiol. Mol. Plant Pathol.* **35**, 85–95.

Whenham, R.J. and Fraser, R.S.S. (1981). Effect of systemic and local-lesion-forming strains of tobacco mosaic virus on abscisic acid concentration in tobacco leaves: consequences for the control of leaf growth. *Physiol. Plant Pathol.* **18**, 267–278.

Whenham, R.J. and Fraser, R.S.S. (1982). Does tobacco mosaic virus RNA contain cytokinins? *Virology* **118**, 263–266.

Whenham, R.J., Fraser, R.S.S. and Snow, A. (1985). Tobacco mosaic virus-induced increase in abscisic acid concentration in tobacco leaves: intracellular location and relationship to symptom severity and to extent of virus multiplication. *Physiol. Plant Pathol.* **26**, 379–387.

Whitcomb, R.F. and Black, L.M. (1961). Synthesis and assay of wound-tumor soluble antigen in an insect vector. *Virology* **15**, 136–145.

Whitcomb, R.F. and Davies, E.E. (1970). Mycoplasma and phytoarboviruses as plant pathogens persistently transmitted by insects. *Annu. Rev. Entomol.* **15**, 405–464.

White, J.L. (1982). Regeneration of virus-free plants from yellow-green areas and TMV-induced enations of *Nicotiana tomentosa*. *Phytopathology* **72**, 866–867.

White, J.L. and Brakke, M.K. (1983). Protein changes in wheat infected with wheat streak mosaic virus and in barley infected with barley stripe mosaic virus. *Physiol. Plant Pathol.* **22**, 87–100.

White, J.L. and Kaper, J.M. (1987). Absence of lethal stem necrosis in select *Lycopersicon* spp. infected by cucumber mosaic virus strain D and its necrogenic satellite CARNA5. *Phytopathology* **77**, 808–811.

White, J.L. and Kaper, J.M. (1989). A simple method for detection of viral satellite RNAs in small plant tissue samples. *J. Virol. Methods* **23**, 83–94.

White, J.L., Wu, F.-S. and Murakishi, H.H. (1977). The effect of low-temperature pre-incubation treatment of tobacco and soybean callus cultures on rates of tobacco- and southern bean mosaic virus synthesis. *Phytopathology* **67**, 60–63.

White, K.A. (1996). Formation and evolution of Tombusvirus defective interfering RNAs. *Sem. Virol.* **7**, 409–416.

White, K.A. and Morris, T.J. (1999). Defective and defective interfering RNAs of monopartite plus-strand RNA viruses. *Curr. Top. Microbiol.* **239**, 1–17.

White, K.A., Bancroft, J.B. and Mackie, G.A. (1992). Coding capacity determines *in vivo* accumulation of a defective RNA of clover yellow mosaic virus. *J. Virol.* **66**, 3069–3076.

White, P.S., Morales, F. and Roossinck, M.J. (1995). Interspecific reassortment of genomic segments in the evolution of cucumoviruses. *Virology* **207**, 334–337.

White, R.F. (1979). Acetylsalicylic acid (aspirin) induces resistance to tobacco mosaic virus in tobacco. *Virology* **99**, 410–412.

White, R.F. and Antoniw, J.F. (1983). Direct control of virus diseases. *Crop Prot.* **2**, 259–271.

White, R.F. and Sugars, J.M. (1996). The systemic infection by tobacco mosaic virus of tobacco plants containing the *N* gene at temperatures below 28 degrees C. *J. Phytopathol.* **144**, 139–142.

White, R.G., Badelt, K., Overall, R.L. and Vesk, M. (1994). Actin associated with plasmodesmata. *Protoplasma* **180**, 169–184.

Whitham, S., Dinesh-Kumar, S.P., Choi, D., Hehl, R., Corr, C. and Baker, B. (1994). The product of the tobacco mosaic virus resistance gene *N*: similarity to Toll and the interleukin-1 receptor. *Cell* **78**, 1101–1115.

Whitham, S., McCormick, S. and Baker, B. (1996). The *N* gene of tobacco confers resistance to tobacco mosaic virus in transgenic tomato. *Proc. Natl Acad. Sci. USA* **93**, 8776–8781.

Whitham, S.A., Yamamoto, M.L. and Carrington, J.C. (1999). Selectable viruses and altered susceptibility mutants in *Arabidopsis thaliana*. *Proc. Natl Acad. Sci. USA* **96**, 772–777.

Whitmoyer, R.E., Nault, L.R. and Bradfute, O.E. (1972). Fine structure of *Aceria tulipae* (Acarina: Eriophyidae). *Ann. Entomol. Soc. Am.* **65**, 201–215.

Whittaker, G.R. and Helenius, A. (1998). Nuclear import and export of viruses and viral genomes. *Virology* **246**, 1–23.

Wijkamp, I. and Peters, D. (1993). Determination of the median latent period of two tospoviruses in *Frankliniella occidentalis*, using a novel leaf disk assay. *Phytopathology* **83**, 986–991.

Wijkamp, I., van Lent, J., Kormelink, R., Goldbach, R. and Peters, D. (1993). Multiplication of tomato spotted wilt virus in its insect vector. *J. Gen. Virol.* **74**, 341–349.

Wijkamp, I., Almarza, N., Goldbach, R. and Peters, D. (1995). Distinct levels of specificity in thrips transmission of tospoviruses. *Phytopathology* **85**, 1069–1074.

Wijkamp, I., van de Wetering, F., Goldbach, R. and Peters, D. (1996). Transmission of tomato spotted wilt virus by *Frankliniella occidentalis*: median acquisition and inoculation access period. *Ann. Appl. Biol.* **129**, 303–313.

Wikoff, W.R., Tsai, C.J., Wang, G., Baker, T.S. and Johnson, J.E. (1997). The structure of cucumber mosaic virus: cryoelectron microscopy, X-ray crystallography, and sequence analysis. *Virology* **232**, 91–97.

Wilcoxson, R.D., Johnson, L.E.B. and Frosheiser, F.I. (1975). Variation in the aggregation forms of alfalfa mosaic virus strains in different alfalfa organs. *Phytopathology* **65**, 1249–1254.

Wildy, P. (1971). Classification and Nomenclature of Viruses. First Report of the International Committee on Taxonomy of Viruses. *Monogr. Virol.* **5**, 1–65.

Williams, R.C. and Wycoff, R.G.W. (1944). The thickness of electron microscopic objects. *J. Appl. Phys.* **15**, 712–716.

Williamson, K.I. and Taylor, W.B. (1958). The analysis of particle counts by the spray-drop method. *Br. J. Appl. Physiol.* **9**, 264–267.

Willison, J.H.M. (1976). The hexagonal lattice spacing on intracellular crystalline tobacco mosaic virus. *J. Ultrastr. Res.* **54**, 176–182.

Wilson, H.R., Tollin, P., Sawyer, L., Robinson, D.J., Price, N.C. and Kelly, S.M. (1991). Secondary structure of narcissus mosaic virus coat protein. *J. Gen. Virol.* **72**, 1479–1480.

Wilson, T.M.A. (1984a). Cotranslational disassembly of tobacco mosaic virus *in vitro*. *Virology* **137**, 255–265.

Wilson, T.M.A. (1984b). Cotranslational disassembly increases the efficiency of expression of TMV RNA in wheat germ cell-free extracts. *Virology* **138**, 353–356.

Wilson, T.M.A. (1985). Nucleocapsid disassembly and early gene expression by positive-strand RNA viruses. *J. Gen. Virol.* **66**, 1201–1207.

Wilson, T.M.A. (1986). Expression of the large 5′ proximal cistron of tobacco mosaic virus by 70S ribosomes during cotranslational disassembly in a prokaryotic cell-free system. *Virology* **152**, 277–279.

Wilson, T.M.A. (1993). Strategies to protect crop plants against viruses: pathogen-derived resistance blossoms. *Proc. Natl Acad. Sci. USA* **90**, 34–41.

Wilson, T.M.A. and Davies, J.W. (1994). New roads to crop protection against viruses. *Outlook Agric.* **23**, 33–39.

Wilson, T.M.A. and Glover, J.F. (1983). The origin of multiple polypeptides of molecular weight below 110,000 encoded by tobacco mosaic virus RNA in the messenger-dependent rabbit reticulocyte lysate. *Biochim. Biophys. Acta* **739**, 35–41.

Wilson, T.M.A. and McNicol, J.W. (1995). A conserved, precise RNA encapsidation pattern in Tobamovirus particles. *Arch. Virol.* **140**, 1677–1685.

Wilson, T.M.A. and Perham, R.N. (1985). Modification of the coat protein charge and its effect on the stability of the U1 strain of tobacco mosaic virus at alkaline pH. *Virology* **140**, 21–27.

Wilson, T.M.A. and Watkins, P.A.C. (1985). Cotransiational disassembly of a cowpea strain (Cc) of TMV: evidence that viral RNA-protein interactions at the assembly origin block ribosome translocation *in vitro*. *Virology* **145**, 346–349.

Wilson, T.M.A., Perham, R.N., Finch, J.T. and Butler, P.J.G. (1976). Polarity of the RNA in the tobacco mosaic virus particle and the direction of protein stripping in sodium dodecyl sulphate. *FEBS Lett.* **64**, 285–289.

Wimmer, E. Hellen, C.U. and Cao, X. (1993). Genetics of poliovirus. *Annu. Rev. Genet.* **27**, 353–436.

Windsor, I.M. and Black, L.M. (1973). Evidence that clover club leaf is caused by a rickettsia-like organism. *Phytopathology* **63**, 1139–1148.

Wingard, S.A. (1928). Host range and symptoms of ringspot, a virus disease of plants *J. Agric. Res.* **37**, 127–153.

Winter, E., Yamamoto, F., Almoguera, C. and Perucho, M. (1985). A method to detect and characterise point mutations in transcribed genes: amplification and overexpression of the mutant C-K1-*ras* allele in human tumor cells. *Proc. Natl Acad. Sci. USA* **82**, 7575–7579.

Wintermantel, W.M. and Schoelz, J.E. (1996). Isolation of recombinant viruses between cauliflower mosaic virus and a viral gene in transgenic plants under conditions of moderate selection pressure. *Virology* **223**, 156–164.

Wintermantel, W.M and Zaitlin, M. (2000). Transgene translatability increases effectiveness of replicase-mediated resistance to cucumber mosaic virus. *J. Gen. Virol.* **81**, 587–595.

Wintermantel, W.M., Anderson, E.J. and Schoelz, J.E. (1993). Identification of domains within gene VI of cauliflower mosaic virus that influence systemic infection of *Nicotiana bigelovii* in a light-dependent manner. *Virology* **196**, 789–798.

Wisler, G.C., Duffus, J.E., Liu, H.-Y. and Li, R.H. (1998). Ecology and epidemiology of whitefly-transmitted closteroviruses. *Plant Dis.* **82**, 270–280.

Wittmann, H.G. and Wittmann-Liebold, B. (1966). Protein chemical studies of two RNA viruses and their mutants. *Cold Spring Harbor Symp. Quant. Biol.* **31**, 163–172.

Wittmann, H.G., Wittmann-Liebold, B. and Jauregui-Adell, J.Z. (1965). Die primäre Proteinstruktur temperatur-sensitiver Mutanten de Tabakmosaikvirus. *Z. Naturforsch.* **20b**, 1224–1234.

Wittmann, S., Chatel, H., Fortin, M.G. and Laliberté, J.-F. (1997). Interaction of the viral protein genome linked to turnip mosaic potyvirus with the translational eukaryotic initiation factor (iso)4E of *Arabidopsis thaliana* using the yeast two-hybrid system. *Virology* **234**, 84–92.

Wobbe, K.K., Akgoz, M., Dempsey, D'M.A. and Klessig, D.F. (1998). A single amino acid change in turnip crinkle virus movement protein p8 affects RNA binding and virulence on *Arabidopsis thaliana*. *J. Virol.* **72**, 6247–6250.

Wobus, C.E., Skaf, J.S., Schultz, M.H. and de Zoeten, G.A. (1998). Sequencing, genomic localization and initial characterization of the VPg of pea enation mosaic enamovirus. *J. Gen. Virol.* **79**, 2023–2025.

Wolf, S., Deom, C.M., Beachy, R.N. and Lucas, W.J. (1989). Movement protein of tobacco mosaic virus modifies plasmodesmatal size exclusion limit. *Science* **246**, 377–379.

Wolfe, K.H., Li, W.-H. and Sharp, P.M. (1987). Rates of nucleotide substitution vary greatly among plant mitochondrial, chloroplast and nuclear DNAs. *Proc. Natl Acad. Sci. USA* **84**, 9054–9058.

Wolfe, K.H., Gouy, M., Yang, Y.-W., Sharp, P.M. and Li, W.-H. (1989). Date of the monocot–dicot divergence estimated from chloroplast DNA sequence data. *Proc. Natl Acad. Sci. USA* **86**, 6201–6205.

Wolstenholme, G.E.W. and O'Connor, M. (eds) (1971). *Strategy of the Viral Genome*. Churchill–Livingstone, Edinburgh.

Woo, Y.-M., Itaya, A., Owens, R.A. *et al.* (1999). Characterization of nuclear import of potato spindle tuber viroid RNA in permeabilized protoplasts. *Plant J.* **17**, 627–635.

Wood, G.A. (1983). Problems associated with the introduction of virus- and mycoplasma-free apple trees to New Zealand orchards. *Acta Hortic.* **130**, 257–262.

Wood, H.A. (1973). Viruses with double-stranded RNA genomes. *J. Gen. Virol.* (Suppl.) **20**, 61–85.

Woodford, J.A.T. and Gordon, S.C. (1978). Virus-like symptoms in red raspberry leaves caused by fenitrothion. *Plant Pathol.* **27**, 77–81.

Woodford, J.A.T., Gordon, S.C. and Foster, G.N. (1988). Sideband application of systemic granular pesticides for the control of aphids and potato leaf roll virus. *Crop Prot.* **7**, 96–105.

Woolston, C.J., Covey, S.N., Penswick, J.R. and Davies, J.W. (1983). Aphid transmission and a polypeptide are specified by a defined region of the cauliflower mosaic virus genome. *Gene* **23**, 15–23.

Woolston, C.J., Czapiewski, L.G., Markham, P.G., Goad, A.S., Hull, R. and Davies, J.W. (1987). Location and sequence of a region of cauliflower mosaic virus gene 2 responsible for aphid transmissibility. *Virology* **160**, 246–251.

Woolston, C.J., Reynolds, H.V., Stacey, N.J. and Mullineaux, P.M. (1989). Replication of wheat dwarf virus DNA in protoplasts and analysis of coat protein mutants in protoplasts and plants. *Nucl. Acids Res.* **17**, 6029–6041.

Worley, J.F. (1965). Translocation of southern bean mosaic virus in phloem fibres. *Phytopathology* **55**, 1299–1302.

Wright, E.A., Heckel, T., Groenendijk, J., Davies, J.W. and Boulton, M.I. (1997). Splicing features in maize streak virus virion, and complementary-sense gene expression. *Plant J.* **12**, 1285–1297.

Wrischer, M. (1973). The effect of ethionine on the fine structure of bean chloroplasts. *Cytobiologie* **7**, 211–214.

Wroth, J.M., Dilworth, M.J and Jones, R.A.C. (1993). Impaired nodule function in *Medicago polymorpha* L. infected with alfalfa mosaic virus. *New Phytol.* **124**, 243–250.

Wu, B. and White, K.A. (1999). A primary determinant of cap-independent translation is located in the 3'-proximal region of the tomato bushy stunt virus genome. *J. Virol.* **73**, 8982–8988.

Wu, B., Hammar, L., Xing, L. *et al.* (2000). Phytoreovirus *T* = 1 core plays critical roles in organizing the outer capsid of T = 13 quasiequivalence. *Virology* **271**, 18–25.

Wu, F.S. and Murakishi, H.H. (1979). Synthesis of virus and virus-induced RNA in southern bean mosaic virus-infected soybean cell cultures. *J. Gen. Virol.* **45**, 149–160.

Wu, G., Kang, L. and Tien, P. (1989). The effect of satellite RNA on cross-protection among cucumber mosaic virus strains. *Ann. Appl. Biol.* **114**, 489–496.

Wu, G.-J. and Bruening, G. (1971). Two proteins from cowpea mosaic virus. *Virology* **46**, 596–612.

Wu, J.-H. (1973). Wound-healing as a factor in limiting the size of lesions of *Nicotiana glutinosa* leaves infected by the very mild strain of tobacco mosaic virus (TMV-VM). *Virology* **51**, 474–484.

Wu, R.Y. and Su, H.J. (1991). Regeneration of healthy banana plantlets from banana bunchy top-infected tissues cultures at high temperature. *Plant Pathol.* **40**, 4–7.

Wu, R.Y., You, L.R. and Soong, T.S. (1994b). Nucleotide sequences of two single-stranded DNAs associated with banana bunchy top virus. *Phytopathology* **84**, 952–958.

Wu, S., Rinehart, C.A. and Kaesberg, P. (1987). Sequence and organisation of Southern bean mosaic virus genomic RNA. *Virology* **161**, 73–80.

Wu, X. and Shaw, J.G. (1996). Bidirectional uncoating of the genomic RNA of a helical virus. *Proc. Natl Acad. Sci. USA* **93**, 2981–2984.

Wu, X. and Shaw, J.G. (1997). Evidence that a viral replicase protein is involved in the disassembly of tobacco mosaic virus *in vivo*. *Virology* **239**, 426–434.

Wu, X. and Shaw, J.G. (1998). Evidence that assembly of a potyvirus begins near the 5' terminus of the viral RNA. *J. Gen. Virol.* **79**, 1525–1529.

Wu, X., Beachy, R.N., Wilson, T.M.A. and Shaw, J.G. (1990). Inhibition of uncoating of tobacco mosaic virus particles in protoplasts from transgenic tobacco plants that express the viral coat protein gene. *Virology* **179**, 893–895.

Wu, X., Xu, Z. and Shaw, J.G. (1994). Uncoating of tobacco mosaic virus RNA in protoplasts. *Virology* **200**, 256–262.

Wu-Scharf, D., Jeong, B.-R., Zhang, C. and Cerutti, H. (2000). Transgene and transposon silencing in *Chlamydomonas reinhardtii* by a DEAH-box RNA helicase. *Science* **290**, 1159–1162.

Wutscher, H.K. and Shull, A.V. (1975). Machine-hedging of citrus trees and transmission of exocortis and xyloporosis viruses. *Plant Dis. Rep.* **59**, 368–369.

Wyatt, S.D. and Kuhn, C.W. (1977). Highly infectious RNA isolated from cowpea chlorotic mottle virus with low specific infectivity. *J. Gen. Virol.* **35**, 175–180.

Wyatt, S.D. and Kuhn, C.W. (1980). Derivation of a new strain of cowpea chlorotic mottle virus from cowpeas. *J. Gen. Virol.* **49**, 289–296.

Wyatt, S.D. and Shaw, J.G. (1975). Retention and dissociation of tobacco mosaic virus by tobacco protoplasts. *Virology* **63**, 459–465.

Wyatt, S.D., Seybert, L.J. and Mink, G. (1988). Status of the barley yellow dwarf problem of winter wheat in eastern Washington. *Plant Dis.* **72**, 110–113.

Xie, Q., Sanz-Burgos, A.P., Guo, H., García, J.A. and Gutiérrez, C. (1999). GRAB proteins, novel members of the NAC domain family, isolated by their interaction with a geminivirus protein. *Plant Mol. Biol.* **39**, 647–656.

Xie, Z., Fan, B., Chen, C. and Chen, Z. (2001). An important role of an inducible RNA-dependent RNA polymerase in plant antiviral defense. *Proc. Natl Acad. Sci. USA* **98**, 6516–6521.

Xin, H.-W., Ji, L.-H., Scott, S.W., Symons, R.H. and Ding, S.-W. (1998). Ilarviruses encode a cucumovirus-like 2b gene that is absent in other genera within the *Bromoviridae*. *J. Virol.* **72**, 6956–6959.

Xiong, C., Muller, S., Lebeurier, G. and Hirth, L. (1982). Identification by immunoprecipitation of cauliflower mosaic virus *in vitro* major translation product with a specific antiserum against viroplasm protein. *EMBO J.* **1**, 971–976.

Xiong, C., Lebeurier, G. and Hirth, L. (1984). Detection *in vivo* of a new gene product (gene III) of cauliflower mosaic virus. *Proc. Natl Acad. Sci. USA* **81**, 6608–6612.

Xiong, Y. and Eichbush, T.H. (1990). Origins and evolution of retroelements based on their reverse transcriptase sequences. *EMBO J.* **9**, 3353–3362.

Xiong, Z. and Lommel, S.A. (1989). The complete nucleotide sequence and genome organization of red clover necrotic mosaic virus RNA1. *Virology* **171**, 543–554.

Xiong, Z., Kim, K.H., Giesman-Cookmeyer, D. and Lommel, S.A. (1993). The role of red clover necrotic mottle virus capsid and cell-to-cell movement proteins in systemic infection. *Virology* **192**, 27–32.

Xoconostle-Cazares, B., Xiang, Y., Ruiz-Medrano, R. *et al.* (1999). Plant paralog to viral movement protein that potentiates transport of mRNA into the phloem. *Science* **283**, 94–98.

Xu, Z., Anzola, J.V., Nalin, C.M. and Nuss, D.L. (1989b). The 3′-terminal sequence of a wound turnout virus transcript can influence conformational and functional properties associated with the 5′ terminus. *Virology* **170**, 511–522.

Yamaguchi, A. and Hirai, T. (1967). Symptom expression and virus multiplication in tulip petals. *Phytopathology* **57**, 91–92.

Yamamoto, K. and Yoshikura, H. (1986). Relation between genomic and capsid structures in RNA viruses. *Nucl. Acids Res.* **14**, 389–396.

Yamanaka, T., Ohta, T., Takahashi, M. *et al.* (2000). TOM1, an *Arabidopsis* gene required for efficient multiplication of a tobamovirus, encodes a putative transmembrane protein. *Proc. Natl Acad. Sci. USA* **97**, 10107–10112.

Yamaoka, N., Furusawa, I. and Yamamoto, M. (1982a). Infection of turnip protoplasts with cauliflower mosaic virus DNA. *Virology* **122**, 503–505.

Yamaoka, N., Morita, T., Furusawa, I. and Yamamoto, M. (1982b). Effect of temperature on the multiplication of cauliflower mosaic virus. *J. Gen. Virol.* **61**, 283–287.

Yamaya, J., Yoshioka, M., Meshi, T., Okada, Y. and Ohno, T. (1988). Expression of tobacco mosaic virus RNA in transgenic plants. *Mol. Gen. Genet.* **211**, 520–525.

Yamazaki, H. and Kaesberg, P. (1963a). Isolation and characterisation of a protein subunit of broadbean mottle virus. *J. Mol. Biol.* **6**, 465–473.

Yamazaki, H. and Kaesberg, P. (1963b). Degradation of bromegrass mosaic virus with calcium chloride and the isolation of its protein and nucleic acid. *J. Mol. Biol.* **7**, 760–762.

Yang, A.F. and Hamilton, R.I. (1974). The mechanism of seed transmission of tobacco ringspot virus in soybean. *Virology* **62**, 26–37.

Yang, C.-C., Liu, J.-S., Lin, C.-P. and Lin, N.-S. (1997a). Nucleotide sequence and phylogenetic analysis of a bamboo mosaic potexvirus isolate from common bamboo (*Bambusa vulgaris* McClure). *Bot. Bull. Acad. Sin.* **38**, 77–84.

Yang, T., Ding, B., Baulcombe, D.C. and Verchot, J. (2000). Cell-to-cell movement of the 25K protein of potato virus X is regulated by three other viral proteins. *Mol. Plant–Microb. Interact.* **13**, 599–605.

Yang, X., Hadidi, A. and Garnsey, S.M. (1992). Enzymatic cDNA amplification of citrus exocortis and cachexia viroids from infected citrus hosts. *Phytopathology* **82**, 279–285.

Yang, Y. and Klessig, D.F. (1996). Isolation and characterization of a tobacco mosaic virus-inducible myb oncogene homolog from tobacco. *Proc. Natl Acad. Sci. USA* **93**, 14972–14977.

Yang, Y., Shah, J. and Klessig, D.F. (1997b). Signal perception and transduction in plant defense responses. *Genes Dev.* **11**, 1621–1639.

Yarwood, C.E. (1951). Associations of rust and virus infections. *Science* **114**, 127–128.

Yarwood, C.E. (1952). The phosphate effect in plant virus inoculations. *Phytopathology* **42**, 137–143.

Yarwood, C.E. (1955). Deleterious effects of water in plant virus inoculations. *Virology* **1**, 268–285.

Yarwood, C.E. (1968). Sequence of supplements in virus inoculations. *Phytopathology* **58**, 132–136.

Yarwood, C.E. (1973). Quick drying versus washing in virus inoculations. *Phytopathology* **63**, 72–76.

Yassi, M.N. and Dodds, J.A.Q. (1998). Specific sequence changes in the 5′ terminal region of the genome of satellite tobacco mosaic virus are required for adaptation to tobacco mosaic virus. *J. Gen. Virol.* **79**, 905–913.

Yassi, M.N., Ritzenhaler, C., Brugidou, C., Fauquet, C. and Beachy, R.N. (1994). Nucleotide sequence and genome characterization of rice yellow mottle virus RNA. *J. Gen. Virol.* **75**, 249–257.

Yassin, A.M. (1968). Transmission of viruses by *Longidorus elongatus*. *Nematologica* **14**, 419–428.

Yassin, A.M. (1983). A review of factors influencing control strategies against tomato leafcurl virus disease in The Sudan. *Trop. Pest Manag.* **29**, 253–256.

Ye, X.S., Pan, S.Q. and Ku, J. (1989). Pathogenesis-related proteins and systemic resistance to blue mould and tobacco mosaic virus induced by tobacco mosaic virus, *Peronospora tabacina* and aspirin. *Physiol. Mol. Plant Pathol.* **35**, 161–175.

Yeh, S.-D. and Gonsalves, D. (1984). Purification and immunological analyses of cylindrical-inclusion protein induced by papaya ringspot virus and watermelon mosaic virus 1. *Phytopathology* **74**, 1273–1278.

Yeh, S.-D., Gonsalves, D., Wang, H.-L., Namba, R. and Chiu, R.-J. (1988). Control of papaya ringspot virus by cross protection. *Plant Dis.* **72**, 375–380.

Yeh, T.Y., Lin, B.Y., Chang, Y.C., Hsu, Y.H. and Lin, N.S (1999). A defective RNA associated with bamboo mosaic virus and the possible common mechanisms for RNA recombination in potexviruses. *Virus Genes* **18**, 121–128.

Yencho, G.C., Cohen, M.B. and Byrne, P.F. (2000). Applications of tagging and mapping insect resistance loci in plants. *Annu. Rev. Entomol.* **45**, 393–422.

Yepes, L.M., Fuchs, M., Slightom, J.L. and Gonsalves, D. (1996). Sense and antisense coat protein constructs confer high levels of resistance to tomato ringspot nepovirus in transgenic *Nicotiana* species. *Phytopathology* **86**, 417–424.

Yi, S.Y., Yu, S,H. and Choi, D. (1999). Molecular cloning of a catalase DNA from *Nicotiana glutinosa* L, and its repression by tobacco mosaic virus infection. *Mol. Cells* **9**, 320–325.

Yie, Y., Wu, Z.X., Wang, S.Y. *et al.* (1995). Rapid production and field testing of homozygous transgenic tobacco lines with resistance confered by expression of satellite RNA and coat protein of cucumber mosaic virus. *Transgen. Res.* **4**, 256–263.

Yi-Li, R., Wen-Li, C. and Rui-Fen, L. (1981). Studies on the rice virus vector small brown planthopper *Laodelphax striatella* Fallen. *Acta Entomol. Sin.* **24**, 290.

Yin, Y. and Beachy, R.N. (1995). The regulatory regions of the rice tungro bacilliform virus promoter and interacting nuclear factors in rice (*Oryza sativa* L.). *Plant J.* **7**, 969–980.

Yin, Y., Zhu, Q., Da, S., Lamb, C. and Beachy, R.N. (1997a). RF2a, a bZIP transcriptional activator of the phloem-specific rice tungro bacilliform virus promoter, functions in vascular development. *EMBO J.* **16**, 5247–5259.

Yin, Y., Vhen, L. and Beachy, R. (1997b). Promoter elements required for phloem-specific gene expression from the RTBV promoter in rice. *Plant J.* **12**, 1179–1188.

Yoon, H.Y., Choi, K.Y. and Song, B.D. (2000). Fluorometric assay of turnip mosaic virus NIa protease. *Anal. Biochem.* **277**, 228–231.

Yoshida, K., Goto, T. and Iizuka, N. (1985). Attenuated isolates of cucumber mosaic virus produced by satellite RNA and cross protection between attenuated isolates and virulent ones. *Ann. Phytopathol. Soc. Jpn* **51**, 238–242.

Yoshikawa, N., Poolpol, P. and Inouye, I. (1986). Use of a dot immunobinding assay for rapid detection of strawberry pseudo mild yellow edge virus. *Ann. Phytopathol. Soc. Jpn* **52**, 728–731.

Yoshikawa, N., Sasaki, E., Kato, M. and Takahashi, T. (1992). The nucleotide sequence of apple stem grooving capillovirus genome. *Virology* **191**, 98–105.

Young, B.D. and Anderson, M.L.M. (1985). Quantitative analysis of solution hybridization. In: B.D. Hames and S.J. Higgins (eds) *Nucleic Acid Hybridization: A Practical Approach*, pp. 47–71. IRL Press, Oxford.

Young, M.J., Daubert, S.D. and Shepherd, R.J. (1987). Gene I products of cauliflower mosaic virus detected in extracts of infected tissue. *Virology* **158**, 444–446.

Young, N.D. and Zaitlin, M. (1986). Analysis of tobacco mosaic virus replicative structures synthesized *in vitro*. *Plant Mol. Biol.* **6**, 455–465.

Ysebaert, M., van Emmelo, J. and Fiers, W. (1980). Total nucleoside sequence of a nearly full-size DNA copy of satellite tobacco necrosis virus RNA. *J. Mol. Biol.* **143**, 273–287.

Yung, K.-H. and Northcote, D.H. (1975). Some enzymes present in the walls of mesophyll cells of tobacco leaves. *Biochem. J.* **151**, 141–144.

Yusibov, V. and Loesch-Fries, L.S. (1998). Functional significance of three basic N-terminal amino acids of alfalfa mosaic virus coat protein. *Virology* **242**, 1–5.

Zabalgogeazcoa, I.A. and Gildow, F.E. (1992). Double-stranded ribonucleic acid in 'Barsoy' barley. *Plant Sci.* **91**, 45–53.

Zabalgogeazcoa, I.A, Cox-Foster, D.C. and Gildow, F.E. (1993). Pedigree analysis of the transmission of a double-stranded RNA in barley cultivars. *Plant Sci.* **83**, 187–194.

Zabel, P., Moerman, M., van Straaten, F., Goldbach, R. and van Kammen, A. (1982). Antibodies against the genome-linked protein VPg of cowpea mosaic virus recognize a 60,000 dalton precursor polypeptide. *J. Virol.* **41**, 1083–1088.

Zaccomer, B., Haenni, A.-L. and Macaya, G. (1995). The remarkable variety of plant RNA virus genomes. *J. Gen. Virol.* **76**, 231–247.

Zagorski, W., Morch, M.-D. and Haenni, A.-L. (1983). Comparison of three different cell-free systems for turnip yellow mosaic virus RNA translation. *Biochimie* **65**, 127–133.

Zahm, P., Seong-Iyul, R. and Klaus, G. (1989). Promoter activity and expression of sequences from Ti-plasmid stably maintained in mammalian cells. *Mol. Cell. Biochem.* **90**, 9–18.

Zaitlin, M. (1962). Graft transmissibility of a systemic virus infection to a hypersensitive host—an interpretation. *Phytopathology* **52**, 1222–1223.

Zaitlin, M. (1998). The discovery of the causal agent of the tobacco mosaic disease. In: S.-D. Kung and S.-F. Yang (eds) *Discoveries in Plant Biology*, pp. 106–110. World Scientific Publishing, Hong Kong.

Zaitlin, M. and Ferris, W.R. (1964). Unusual aggregation of a nonfunctional tobacco mosaic virus protein. *Science* **143**, 1451–1452.

Zaitlin, M. and Israel, H.W. (1975). Tobacco mosaic virus (type strain). *CMI/AAB Descriptions Plant Viruses* No. 151, pp. 1–5.

Zaitlin, M. and Palukaitis, P. (2000). Advances in understanding plant viruses and virus diseases. *Annu Rev. Phytopathol.* **38**, 117–143.

Zaitlin, M., Duda, C.T. and Petti, M.A. (1973). Replication of tobacco mosaic virus V: properties of the bound and solubilized replicase. *Virology* **53**, 300–311.

Zaitlin, M., Beachy, R.N. and Bruening, G. (1977). Lack of molecular hybridization between RNAs of two strains of TMV. A reconsideration of the criteria for strain relationships. *Virology* **82**, 237–241.

Zaitlin, M., Niblett, C.L., Dickson, E. and Goldberg, R.B. (1980). Tomato DNA contains no detectable regions complementary to potato spindle tuber viroid as assayed by solution and filter hybridization. *Virology* **104**, 1–9.

Zalloua, P.A., Buzayan, J.M. and Bruening, G. (1996). Chemical cleavage of 5'-linked protein from tobacco ringspot virus genomic RNAs and characterization of the protein-RNA linkage. *Virology* **219**, 1–8.

Zamore, P.D., Tuschi, T., Sharp, P.A. and Bartel, D.P. (2000). RNAi: double-stranded RNA directs the ATP-dependent cleavage of mRNA at 21 and 23 nucleotide intervals. *Cell* **101**, 25–33.

Zanotti, F.M.de A., Gibbs, M.J., Gould, E.A. and Holmes, E.C. (1996). A re-evaluation of the higher taxonomy of viruses based on RNA polymerases. *J. Virol.* **70**, 6083–6096.

Zanotto, P.M. de A., Gibbs, M.J., Gould, E.A. and Holmes, E.C. (1996). A reevaluation of the higher taxonomy of viruses based on RNA polymerases. *J. Virol.* **70**, 6083–6096.

Zaumeyer, W.J. and Meiners, J.P. (1975). Disease resistance in beans. *Annu. Rev. Phytopathol.* **13**, 313–334.

Zavriev, S.K. (1999). Carlaviruses. In: A. Granoff and R.G. Webster (eds) *Encyclopedia of Virology*, 2nd edn, pp. 238–242. Academic Press, San Diego.

Zavriev, S.K., Hickey, C.M. and Lommel, S.A. (1996). Mapping of the red clover necrotic mosaic virus subgenomic RNA. *Virology* **216**, 407–410.

Zech, H. (1952). Untersuchungen über den Infektionsvorgang und die Wanderung des Tabakmosaikvirus im Pflanzenkörper. *Planta* **40**, 461–514.

Zelcer, A., Zaitlin, M., Robertson, H.D. and Dickson, E. (1982). Potato spindle tuber viroid-infected tissues contain RNA complementary to the entire viroid. *J. Gen. Virol.* **59**, 139–148.

Zerfass, K and Beier, H. (1992). The leaky UGA termination codon of tobacco rattle virus RNA is suppressed by tobacco chloroplast and cytoplasmic tRNAs^{trp} with CmCA anticodon. *EMBO J.* **11**, 4167–4173.

Zettler, F.W. (1967). A comparison of species of Aphididae with species of three other aphid families regarding virus transmission and probe behaviour. *Phytopathology* **57**, 398–400.

Zettler, F.W. and Hartman, R.D. (1987). Dasheen mosaic virus as a pathogen of cultivated aroids and control of the virus by tissue culture. *Plant Dis.* **71**, 958–963.

Zeyen, R.J. and Banttari, E.E. (1972). Histology and ultrastructure of oat blue dwarf virus infected oats. *Can. J. Bot.* **50**, 2511–2519.

Zeyen, R.J., Stromberg, E.L. and Kuehnast, E.L. (1987). Long range aphid transport hypothesis for maize dwarf mosaic virus: history and distribution in Minnesota, USA. *Ann. Appl. Biol.* **111**, 325–336.

Zhan, X., Richardson, K.A., Haley, A. and Morris, B.A.M. (1993). The activity of the coat protein promoter of chloris striate mosaic virus is enhanced by its own and C1–C2 gene products. *Virology* **193**, 498–502.

Zhang, G. and Simon, A.E. (2001). Analysis of an RNA replication enhancer of a subviral RNA associated with turnip crinkle virus. Abstract W16–2, 20th Annual Meeting of the American Society for Virology.

Zhang, G., Slowinski, V. and White, A. (1999). Subgenomic mRNA regulation by distal RNA element in a (+)-strand RNA virus. *RNA* **5**, 550–561.

Zhang, G.C., Leung, C., Mardin, L., Rovinski, B. and White, K.A. (2000). *In planta* expression of HIV-1 p24 protein using an RNA virus-based expression vector. *Molec. Biotech.* **14**, 99–107.

Zhang, L., Zitter, T.A. and Palukaitis, P. (1991). Helper virus-dependent replication, nucleotide sequence and genome organization of the satellite virus of maize white line mosaic virus. *Virology* **180**, 467–473.

Zhang, S. and Klessig, D.F. (2000). Pathogen-induced MAP kinases in tobacco. In: *Results and Problems in Cell Differentiation*, Vol. 27, H. Hirt (ed.) *MAP Kinases in Plant Signal Transduction*, pp. 65–84. Springer, Berlin.

Zhang, S., Jones, M.J., Barker, P., Davies, J.W. and Hull, R. (1993). Molecular cloning and sequencing of coat protein-encoding cDNA of rice tungro spherical virus: a plant picornavirus. *Virus Genes* **7**, 121–132.

Zhang, W., Olson, N.H., Baker, T.S. *et al.* (2001). Structure of maize streak virus geminate particle. *Virology* **279**, 471–477.

Zhang, W.C., Yan, M.N. and Lou, C.H. (1990). Intercellular movement of protoplasm *in vivo* in developing endosperm of wheat caryopses. *Protoplasma* **153**, 193–203.

Zhang, Z.H. and Wu, L.P. (1998). Research and development of expressing antibodies in plants. *Prog. Biochem. Biophys.* **25**, 136–139.

Zhao, Y., Owens, R.A. and Hammond, R.W. (2001). Use of a vector based on potato virus X in a whole plant assay to demonstrate nuclear targeting of potato spindle tuber viroid. *J. Gen. Virol.* **82**, 1491–1497.

Zheng, H., Wang, G. and Zhang, L. (1997). Alfalfa mosaic virus movement protein induces tubules in plant protoplasts. *Mol. Plant–Microb. Interact.* **10**, 1010–1014.

Zhou, G. and Rochow, W.F. (1984). Differences among five stages of *Schizaphis graminum* in transmission of a barley yellow dwarf luteovirus. *Phytopathology* **74**, 1450–1453.

Zhou, G., Lu, X., Lu, H., Lei, J., Chen, S. and Gong, Z. (1999). Rice ragged stunt oryzavirus: role of the viral spike protein in transmission by the insect vector. *Ann. Appl. Biol.* **135**, 573–578.

Zhou, H. and Jackson, A.O. (1996). Analysis of *cis*-acting elements required for replication of barley stripe mosaic virus RNAs. *Virology* **219**, 150–160.

Zhou, X., Liu, Y., Calvert, L. *et al.* (1997). Evidence that DNA-A of a geminivirus associated with severe cassava mosaic virus disease in Uganda has arisen by interspecific recombination. *J. Gen. Virol.* **78**, 2101–2111.

Zhou, X., Liu, Y., Robinson, D.J. and Harrison, B.D. (1998). Four DNA-A variants among Pakistani isolates of cotton leaf curl virus and their affinities to DNA-A of geminivirus isolates from okra. *J. Gen. Virol.* **79**, 915–923.

Zhu, Y., Green, L., Woo, Y.-M., Owens, R. and Ding, B. (2001). Cellular basis of potato spindle tuber viroid systemic movement. *Virology* **279**, 69–77.

Zhuravlev, Yu.N., Reifman, V.G., Shumilova, L.A., Yudakova, Z.S. and Pisetskaya, N.F. (1975). Absorption of ^{32}P-labelled tobacco mosaic virus by isolated tobacco protoplasts and deproteinization of the virus in them. *Sov. Plant Physiol. (Engl. translation)* **22**, 941–943.

Ziemiecki, A. and Peters, D. (1976). The proteins of sowthistle yellow vein virus: characterization and location. *J. Gen. Virol.* **32**, 369–381.

Ziemiecki, A. and Wood, K.R. (1976). Proteins synthesized by cucumber cotyledons infected with two strains of cucumber mosaic virus. *J. Gen. Virol.* **31**, 373–381.

Zimmermann, D. and van Regenmortel, M.H.V. (1989). Spurious cross-reactions between plant viruses and monoclonal antibodies can be overcome by saturating ELISA plates with milk proteins. *Arch. Virol.* **106**, 15–22.

Zimmern, D. (1977). The nucleotide sequence at the origin for assembly on tobacco mosaic virus RNA. *Cell* **11**, 463–482.

Zimmern, D. (1982). Do viroids and RNA viruses derive from a system that exchanges genetic information between eukaryotic cells? *Trends Biochem. Sci.* **7**, 205–207.

Zimmern, D. (1983). An extended secondary structure model for the TMV assembly origin, and its correlation with protection studies and an assembly defective mutant. *EMBO J.* **2**, 1901–1907.

Zimmern, D. and Wilson T.M.A. (1976). Location of the origin for viral reassembly on tobacco mosaic virus RNA and its relation to stable fragment. *FEBS Lett.* **71**, 294–298.

Zink, F.W. and Duffus, J.E. (1972). Association of beet western yellows and lettuce mosaic viruses with internal rib necrosis of lettuce. *Phytopathology* **62**, 1141–1144.

Zink, F.W., Grogan, R.G. and Welch, J.E. (1956). The effect of percentage of seed transmission upon subsequent spread of lettuce mosaic virus. *Phytopathology* **46**, 622–624.

Zinovkin, R.A. and Agranovsky, A.A. (1998). Detection of the nonstructural p21 protein of beet yellows closterovirus *in vivo* with polyclonal antibodies to bacterially expressed protein. *Mol. Biol.* **32**, 928–931.

Zinovkin, R.A., Jelkmann, W. and Agranovsky, A.A. (1999). The minor coat protein of beet yellows closterovirus encapsidates the 5′ terminus of RNA in virions. *J. Gen. Virol.* **80**, 269–272.

Zlotnick, A., Aldrich, R., Johnson, J.M. *et al.* (2000). Mechanism of capsid assembly for an icosahedral plant virus. *Virology* **277**, 450–456.

Zrein, M., Burckard, J. and van Regenmortel, M.H.V. (1986). Use of the biotin–avidin system for detecting a broad range of serologically related plant viruses by ELISA. *J. Virol. Methods* **13**, 121–128.

Zuidema, D. and Jaspars, E.M.J. (1985). Specificity of RNA and coat protein interaction in alfalfa mosaic virus and related viruses. *Virology* **140**, 342–350.

Zuidema, D., Bierhuizen, M.F.A. and Jaspars, E.M.J. (1983). Removal of the N-terminal part of alfalfa mosaic virus coat protein interferes with the specific binding to RNA1 and genome activation. *Virology* **129**, 255–260.

Zuidema, D., Heaton, L.A. and Jackson, A.O. (1987). Structure of the nucleocapsid protein gene of sonchus yellow net virus. *Virology* **159**, 373–380.

Zulauf, M. (1977). Swelling of brome mosaic virus as studied by intensity fluctuation spectroscopy. *J. Mol. Biol.* **114**, 259–266.

Index